37th Annual USSD Conference 2017

It's a Small World: Managing Our Water Resources

Anaheim, California, USA
3 - 7 April 2017

Volume 1 of 3

ISBN: 978-1-5108-6027-8

Printed from e-media with permission by:

Curran Associates, Inc.
57 Morehouse Lane
Red Hook, NY 12571

Some format issues inherent in the e-media version may also appear in this print version.

Printed by Curran Associates, Inc. (2018)

United States Society on Dams (USSD)
9975 Wadsworth Pkwy.
Ste. K-2, 145
Westminster, CO 80021
USA

Phone: 303-792-8753
Fax: 303-792-8782

info@ussdams.org

Additional copies of this publication are available from:

Curran Associates, Inc.
57 Morehouse Lane
Red Hook, NY 12571 USA
Phone: 845-758-0400
Fax: 845-758-2633
Email: curran@proceedings.com
Web: www.proceedings.com

TABLE OF CONTENTS

VOLUME 1

SESSION 1A: CONFERENCE THEME

Reservoir Sedimentation and Sustainability 1
Timothy Randle, Sean Kimbrel, Kent L. Collins

Dam Safety and Resilient Water Management: A Case for Dynamic Reservoir Operations 9
Michael McMahon

The Nebraska Department of Natural Resources Coordinated Efforts in Flood Flows 10
Kristen Obermueller, Bob Beduhn, John Engel

SESSION 2A: ENVIRONMENT

Environmental Considerations in Dam Decommissioning 11
Kelly R. Schaeffer

Hydraulic Dredging of the Strontia Springs Reservoir — Lessons Learned 12
Douglas Raitt, Brett Cochran

Dam Decommissioning — Unique Approach by Conversion to a Levee 45
Zachary Whitten

The Prospects for Climate Scenarios in Hydropower Operations 56
Yulia Zakrevskaya, David Huard

SESSION 1B: PHYSICAL MODELING

Linear Weir Size Scale Effects: A Look at Flat Top Weirs 72
Blake Tullis, Kedric Curtis

Physical Model of Spillway and Reservoir Debris Interaction 73
Kent Walker, Jaron Hasenbalg, Kelly Brom

Boundary Dam Spillway No. 1 TDG Modifications 74
John Werner, James Rutherford, Andrew Bearlin, Peter Friz, Jeff Johnson, Daniel Kirschbaum, Joe Groeneveld, Justin Arnold

SESSION 2B: HYDROLOGIC MODELING

A Comprehensive Approach to Probabilistic Hydrologic Hazard Assessment and its Use to Support Risk-Informed Decisions 90
Shaun Carney, Keil Neff, Mel Schaefer, Bruce Baker

Meteorlogical Data Visualization and Analysis Using HEC-MetVue 91
Fauwaz Hanbali

Big Dam in a Small World: Reassessment of the Spillway Design Flood of Lake Sakakewea 92
Joshua Melliger

Application of Regional Hydrologic Criteria in the Rocky Mountains to Estimation of the PMF in Alberta 93
John Menninger, Seifu Guangul, Dan Hoffman, George Sabol, Syed Abbas

SESSION 1C: RISK

Approaches to Estimating Individual Risk for Dam Safety Tolerable Risk Evaluation 104
David S. Bowles, Jason Needham

Minimizing Bias in Geologic and Geotechnical Aspects of Risk Evaluation for Dams and Levees 139
Kerry D. Cato, Derek S. Morley

The Common Cause Adjustment in Dam Safety Risk Analysis: Scope, Purpose, and Applicability 158
Dom Galic

SESSION 2C: RISK

Potential Failure Mode Analysis – Insights and Lessons Learned for the Foundation of a Successful Dam Safety Management Program 174
Frank L. Blackett, Douglas D. Boyer

Epistemic vs Natural Uncertainty in Dam Safety Decision Making: Is it Fair Play? 194
Ignacio Escuder-Bueno, Adrián Morales-Torres, Jessica Castillo-Rodriguez

Consequence Estimation for a Large System of Dams: Our Experiences from TVA Projects 208
Michael M. Crouch, Anurag Srivastava, Morgan D. Ruark

Application of Risk Assessment to an Emergency Levee Design in an Urban Environment 221
Petros Armenis, Malcolm Barker, Peter Christensen, Graham Harrington

SESSION 1D: CONSTRUCTION I

Vibration Free Resonant Sheet Pile Driving in Dykes and Levees 245
Dane Reelie, Matthew James
Construction of Borinquen Dam 1E for the Panama Canal Expansion 246
Lelio Mejia, James Toose, Jorge Fernandez
Tsaile Dam — A Complete Approach Unearths a Safer Dam 267
Stephen Brown

SESSION 2D: CONSTRUCTION I

Replacement of Clear Lake Dam: Overcoming Challenges Constructing an RCC Dam on a Soil Foundation 268
Steve Jamieson, Bruce Cotie, Don Lopez, John Treacy
Dewatering — How to Address this Potential Achilles Heel for Construction 286
John W. France, G. Richard Bird, Gregg Batchelder Adams, Jessie Drayton Unterberg
Draft Tube Slab Analysis 304
Jesse Kropelnicki, Kareem Bynoe, David Schreck, Peter Bouchie
Core from Rock, Challenges of Core Construction for a Large High Hazard Dam 313
Michael Miller, Efrain Rondinel-Oviedo

SESSION 1E: RISK INFORMED H&H

Estimating the Reservoir Stage-Frequency Curve with Uncertainty Bounds for Cherry Creek Dam Using the Reservoir Frequency Analysis Software (RMC-RFA) 329
Haden Smith, John F. England Jr.
Recent Paleoflood Analyses for USACE Risk-Informed Dam Safety Evaluations 345
Keith I. Kelson, Christina M. Leonard, Brian M. Hall, John F. England Jr.
Risk Informed Freeboard Evaluation 357
Gregory S. Karlovits, David A. Margo

SESSION 2E: GATES AND VALVES

Fail Safe Design for Gated Structures Construction and Rehabilitation 368
Peter Davis, Mathew McGuire
Dams Inlet/Outlet Works Rehabilitation Program 377
Jack Husted, Sterling Klippel, Keith Lilley
Best Practices in a Tainter Gate Rehabilitation Program 378
Matt Moses, Layne Bukahir
Emergency Shutoff Valves 392
Kishen Prathivadi

SESSION 1F: EMBANKMENT DAMS I

A State Perspective on Tailings Dam Regulation and a Role for USSD and ASDSO 404
Charles Cobb
A Centrifuge Study of Co-Mixed Mine Tailings and Waste Rock Deposits 405
Nonika Antonaki, Tarek Abdoun, Inthuorn Sasanakul
Freeboard Assessment for Long Embankment Dams 420
Mike Gerlach

SESSION 2F: EMBANKMENT DAMS I

Offstream Reservoir Earthen Dam Seepage Systems — Design and Construction Challenges 421
Randall Bushey
Latest Compilation of Internal Erosion Incidents at Bureau of Reclamation Dams 422
William O. Engemoen
Modeling of Internal Erosion in Earthen Embankment Dam 440
Al Preston, Keil Neff, Chunling Li, Lucas de Melo
Turbid Discharge Incident at Cannonsville Dam 452
John H. Vickers, William H. Hover

POSTER SESSION

CONCRETE

Freezing and Thawing Durability of Laboratory and In-Situ Roller-Compacted Concrete Test Specimens 469
Timothy Dolen, Veronica Madera
Parametric Layout of Double-Curvature Arch Dams Using DIANA Finite Element Software 470
E. Goulas, G. Schreppers, M. Partovi, R.M. Gunn
Study of Uplift Pressure Effect in Roller Compacted Concrete Gravity Dam 487
Hichem Mazighi, Mustapha Kamel Mihoubi, Khaled Ghaedi, Zainah Ibrahim
Soil-Structure Interaction Dynamic Analysis for an Existing Old Earth-Rock Fill Dam with Concrete Core 505
Mahmood Seid-Karbasi, Upul Atukorala, Herb Hawson, Bruce Downing
RCC Post Construction Core Testing and Data Analysis For San Vicente Dam Raise Project 522
James Stiady, Andrew Oleksyn, Michael Rogers, Glenn Tarbox, Thomas Reynoldson, Gerard E. Reed III, Wade Griffis, Jim Zhou

CONSTRUCTION

Reconstruction of the Matala Spillway: Hydrotechnical and Safety Aspects 539
Ahmed Bouayad
Decommissioning Outdated Fish Bypass Systems and Preparation for Concrete Retrofit on Lower Granite Lock & Dam 550
Dirk deGroot
Sand Secant Pile Chimney Filter Installation at Pine Creek Dam 551
Bob Faulhaber, Michael Arnold
Execution of a Cut-Off Wall at the Dead Sea – Unique Conditions Require Unique Solution 562
Franz-WernerGerressen, Gai Lehrer, Israel Christian Scholz
Catawba Dam: The Long Road and Construction Update 577
Dennis Hogan
Water-Filled Bladders, an Innovative Bearing Interface for Arch Dam Bracing 578
Alexandre Lochu
Eductors — The Essential Alternative in Complex Dewatering Applications 586
Adam Richards, Harry Bagherzadeh
Application of Intelligent Compaction to Embankment Dam Construction 587
Robert V. Rinehart
Update No. 3 of the Calaveras Dam Replacement Project 602
John Roadifer, Michael Forrest, Erik Newman, Daniel Wade, Susan Hou, Tedman Lee, Carman Ng

VOLUME 2

Validated Integrated Compaction Monitoring at TVA's KIF Coal Combustion Product Stacking Facility 613
David J. White, Pavana Vennapusa, Brendan FitzPatrick, Jason F. Hill, Nicholas McClung, Eric Hageman
To Air is Human; To Succeed, Divine — Evolutionary Research on Grout Enriched RCC Air Entrainment 627
Jeremy Young, Eric Musselman

EMBANKMENT DAMS

Stability Analysis of a Concrete-Faced Rockfill Dam Using Parametric Evaluations 628
Christopher Conkle
Estimating the Peak Friction Angle of Sandy Soils In Situ with State-Based Overburden Normalized SPT Blow Counts 629
Robert A. Jaeger, Ian P. Maki
Comparison Analysis of the Behavior of Rock-Fill Dams with Clay Core at the Variation of Water Level in the Reservoir 645
Ljupcho Petkovski, Stevcho Mitovski
Using Non-Intrusive Geophysical Techniques in Embankment Dam and Levee Risk Assessment 658
Justin Rittgers, Cassandra Wagner, Bryan K. Simpson
Impact of Glacial Till Deposits — Structural Analysis of Green Mountain Dam 659
Hillery Venturini

HYDROLOGY AND HYDRAULICS

Using HEC-RAS 2 Dimensional Capability for Simplified Rapid Dam Break Analysis 660
Wesley Crosby
Hydraulic Modeling of a Complex Dam and Reservoir Project: A Case Study of Two Rivers Dam 661
Stephanie Doyal
Rainfall-Runoff Modeling for the Garrison Dam Drainage with HEC-HMS .... N/A
Jeff Harris, Kevin Denn, Melissa Larsen

An Innovative Temporary Bulkhead for the Ruskin Dam and Powerhouse Upgrade Project 662
Jorge Marin, Saman Vazinkhoo, Rafael Ibáñez-de-Aldecoa, Mark Mallet
Risk and Uncertainty Features in HEC-ResSim 674
George Modini, Fauwaz Hanbali
Methodology Applied for Mapping Flooded Areas of the Lajeado HPP 675
Pedro Pupim, Mauricio Santini, Milton Dall'Aglio
Leveraging Multi-Sensor Hydrologic Data for Model Validation 676
Kevin J. Ruswick, Gregory J. Daviero, Jonathon Ruff

LEVEES

Design of Levee Modifications to Accommodate Long-Term River Scour 696
Justin Dominguez, James Nickerson
Detection of Levee Voids at Aqueduct Crossings Using Geophysical Imaging Techniques 697
Zhihua Li, Atta Yiadom
Risk Analysis in a Levee Safety Context: The Upper Wood River Levee Case History 709
Chris J. Redell, Jose R. Lopez
Effects of Compaction on Erodibility Characteristics of Levees Due to Overtopping 720
Ali Asghari Tabrizi, M. Hanif Chaudhry, Jasim Imran

MONITORING

Preventing Catastrophic Failure of Underwater Infrastructure: Condition Assessments by ROV 734
Bill Sherwood, Bob Clarke
Automated Data Acquisition System Basics for Dam Safety Instrumentation Monitoring 735
Peter Zimmerman, Paul Booth, Greg Dutson

RISK

Failure of the Acude da Nacao Earth Dam: A Probabilistic Perspective 749
Danielli de Melo Moura, Rodrigo Borela, Gabriela Alvarado, Wilson Espinoza, Philippe L. Bourdeau

SESSION 3A: MONITORING

Operation and Maintenance of an Instrumentation Program 764
Amanda Sutter
FERC Best Practices for Dam Safety Performance Monitoring 765
Will Brown, Eric Kennedy, Justin Nettle, Emeruwa Anyanwu, Eric Gross, Jim McHenry, David Olson
Calculation of Threshold Limits for Dam Safety Instrumentation 787
Jack Husted, Travis Wylde, Sterling Klippel
Somerset Dam – Innovative Remote Monitoring for Time-Sensitive Breach Detection 800
Jerry Cross, Yogi Sookhu, Jud Donaghy

SESSION 4A: MONITORING

Geospatial Tools for Routine Dam Safety Monitoring and Response Applications During Emergencies 808
Yogesh Prashar, Atta B. Yiadom, Elizabeth Bialek, Rachel Wong
Special Considerations When Installing Automated Instrumentation at a Site with Access Limitations 809
William Walker
Instrumentation and Information Management at Boone Dam 810
Rozh Ameen, Raphael Siebenmann, Jamey Rosen

SESSION 3B: HYDRAULIC MODELING

Modeling Dam Breaches and Flood Inundation Using HEC-RAS 5.0's New 2D Capabilities 825
D.L. Silvertooth, M. Meshkat
Estimating the Probable Maximum Flood for Whittier Narrows Dam Using a 2D Hydrodynamic Modeling Approach 839
Brian M. Hall, Haden Smith, Reuben Sasaki
When an Adjacent Watershed Inundates Your Offline Reservoir: Ukumehame Reservoir Dam Removal Challenges and Design 855
Andrew J. Lynch, Stewart S. Vaghti, Dean B. Durkee
Atoka Dam and Spillway Modifications 867
Patrick Miles, Brad Kirksey, John Rutledge, Sam Samandi, John L. Hare, Andrew Mishler

SESSION 4B: SITE SPECIFIC PMP

Recent Extreme Storms and Their Relation to PMP 886
Bill Kappel, Doug Hultstrand, Geoff Muhlestein

Needs to Update Probable Maximum Precipitation for Determining Hydrologic Loading on Dams 906
Chandra Pathak

Site Specific Probable Maximum Precipitation Development 907
Jeff Harris, Ben Rohrbach

Advances In Watershed Modeling Inputs and Methods for Dam Safety Risk Analyses 924
Jason Caldwell, Nicole Novembre, Katie Laro, Tye Parzybok, Bruce Barker, Mel Schaefer, George Taylor, Wayne Gibson

SESSION 3C: EARTHQUAKES I

Use of the Conditional Mean for Improved Prediction of Ground Motion Intensity Measures for Embankment Dams 939
Richard J. Armstrong

Seismic Deformation Analysis of Blue Ridge Dam 954
Glenn J. Rix, Lucas de Melo, Christopher L. Saucier, Michael A. Morrison

A Case History Evaluation of State of Practice for Seismic Deformation Modeling of Earthen Dams 969
Khaled Chowdhury, Vlad Perlea, Raymond B. Seed, Michael Beaty, Ethan Dawson, George Hu

Comparison of Two Constitutive Models for Simulating the Effects of Liquefaction on Embankment Dams 1014
Jack Montgomery, Sam Abbaszadeh

SESSION 4C: EARTHQUAKES I

Seismic Evaluation of Lookout Point Embankment Dam in Oregon – Cascadia Subduction Zone Implications 1028
Andrew D. Kost, Kenji Yamasaki, David H. Scofield

Performance of Fena Dam During the 1993 Guam Earthquake 1043
Lelio H. Mejia

Seismic Evaluation of the Remediated Isabella Auxiliary Dam 1058
George Hu, David C. Serafini, Henri V. Mulder, Vlad Perlea

SESSION 3D: CONSTRUCTION II

Seismic Remediation of Perris Dam Using CDSM 1071
Wen Yan Chen, Michael Driller, Steven Friesen

Controlled Blasting for the Red Rock Hydroelectric Project 1090
Rachael V. Bisnett, James Valentyn, Nicholas Johnson

Upriver Dam Spillway Rehabilitation 1102
Ray Trudgeon, James H. Rutherford, Steve Burns, Val Kovalishyn, Ccee Chan, Nikou Jalayeri

Lakeside Ranch Stormwater Treatment Area – Lessons Learned During Design and Construction 1119
Danielle Neamtu, Stephen Whiteside, Stephen Blair

SESSION 4D: CONSTRUCTION II

Design and Repair of a Power Intake Gate Sealing Surface Under Challenging Constraints ..... N/A
Mike Carney, Don Thompson, Roland Lwiindi, Eli Lundberg, Joshua Phillips

Post-Tensioning in an Underwater Environment: Wanapum Dam + Monolith 4 Spillway Repair 1135
Paul Krumm

Balancing Water Resources and Power Generation for the Metlakatla Indian Community 1143
Jesse Hutton

SESSION 3E: CONCRETE I

Investigating the Structural Safety of Cracked Concrete Dams 1153
Glenn S. Tarbox, Robin Charlwood, Chris Hayes

A New Approach for the Design of the Köroglu RCC Arch/Gravity Dam 1167
Quentin Shaw

Upgrade and Spillway Modifications at Santa Anita Dam 1178
Sterling E. Klippel, Vik Iso-Ahola, Glenn S.Tarbox, Martina Gelo, George Manole

Understanding the Performance of Aging Infrastructure: A 99-Year-Old Multiple Arch Dam 1189
Vik Iso-Ahola, Glenn S.Tarbox, Mohammadreza Mostafa, Sean Ellenson, Munit Bector, Ali Rasekh

SESSION 4E: CONCRETE I

A Performance-Based Evaluation of Post-Tensioned Anchors Embedded Within a Concrete Gravity Dam 1206
Tae Ha Park, Samantha Hoang, Jonathan Lum, Ziyad Duron

On the Use of Discrete Lumped Element Modeling for Insight into Concrete Dam Behavior 1218
Jonathan Lum, Samantha Hoang, Tae Ha Park, Ziyad Duron, Guy Lund

VOLUME 3

Shear Strength of Concrete Lift Joints from Extensive Lab Testing Compared to Theoretical Results 1233
Dan D. Curtis, Husein Hasan, Gurinderbir S. Sooch

SESSION 3F: EMBANKMENT DAMS II

Seepage Distress and Canal Repairs on the California Aqueduct 1250
Leslie F. Harder, Jeanne Kuttel, Joseph Royer, Timothy Wehling, Joseph Burke, Robert Black, Michael Driller, Christina Kashiwada, Rob Barry, Raphael A. Torres

Geomembranes to Waterproof Embankment Dams 1317
John Wilkes, Alberto Scuero, Gabriella Vaschetti

Geotechnical Investigation of South Carolina Flood-Damaged Dams after the 2015 October Rainfall Event 1333
Inthuorn Sasanakul, Sarah L. Gassman, Charles E. Pierce, William Ovalle, Ryan Starcher, Emad Gheibi, Mostaqur Rahman

Rough River Dam: Phase 1B Exploratory Drilling and Grouting Program 1334
Conrad Ginther, Dmitri Ivanov, Doug Horvath, Rolando Benitez

SESSION 4F: EMBANKMENT DAMS II

Integrating a Multidisciplinary Field Investigation to Improve Assessment of Internal Erosion 1344
Cassandra M. Wagner, Justin B. Rittgers, Bryan K. Simpson

Laboratory Investigation of Backward Erosion Piping into Constricted Seepage Exits 1359
John Rice, Ibrahim A. Ibrahim

Evaluation of Internal Erosion Scenarios in an Embankment Dam 1360
R.J. Fannin, M. Li, C. Anderlini

SESSION 5A: MONITORING

Monitoring San Vicente Dam Raise During Initial Fill 1375
Andrew Oleksyn, Nathan Grima, Anna Kolakowski, Larry Rodrigues, Jim Zhou

First Filling Instrumentation Monitoring of the Folsom Dam Auxiliary Spillway Control Structure 1388
Kenneth R. Pattermann, George Hu

Effect of Antecedent Groundwater Level on Piezometric Response During Flood Events 1403
Chun-Yi Kuo, Richard B. Hockett, Troy S. O'Neal

SESSION 6A: MONITORING

Leak Detection In Earth Dams and Levees Using Optical Fiber Distributed Temperature Sensors 1417
Maxime Tatin, Vincent Lamour, Stéphane Bonelli, Aurélie Garandet

Elephant Butte Dam UAV Data Collection and Photogrammetry 1424
Matthew Klein

Proposed Guidelines for Conducting a Performance-Based Evaluation of a Concrete Dam 1425
Samantha Hoang, Tae Ha Park, Jonathan Lum, Ziyad Duron, Guy S. Lund

Reality Capture for Dam and Hydroelectric Facilities 1440
Adam Serock

SESSION 5B: RESERVOIR SEDIMENTATION

Numerical Sediment Transport Modeling of Operations at Paonia Reservoir, Colorado 1441
Sean Kimbrel

Impact of Climate Change on Sediment Accumulation in the Reservoirs of Embankment Dams: Case Study, Baldhill Dam, North Dakota 1442
Bahareh Shoghli, Yeo Howe Lim

Sediment Transport Through Lake Clarke and Lake Aldred 1443
Jonathan M. G. Viducich, Martin J. Teal

SESSION 6B: RESERVOIR OPERATIONS

To Release or Not to Release? Dam Operation Modeling Using Gridded Short-Term Forecast Rainfall 1460
Maged Aboelata

USACE CWMS: Case Studies of Water Management Decision Support Tools 1461
Ben Zoeller, Nathan Lefkovitz

Systems-Based Risk Analysis of Lake Mendocino Forecast-Informed Reservoir Operations with HEC-WAT 1462
Matthew McPherson, Leila Ostadrahimi

HEC-ResSim 3.3 — New Features to Support Complex Studies 1463
Joan Klipsch

SESSION 5C: EARTHQUAKES II

A Direct Finite Element Method for Nonlinear Analysis of Semi-Unbounded Dam-Water-Foundation Rock Systems 1464
Anil K. Chopra, Arnkjell Lokke

Numerical Assessment of Hydrodynamic Loads Induced During Seismic Interaction Between Reservoir and Concrete Dam 1465
Jerzy Salamon, Jonna Manie

Seismic Back Analysis of Monticello Arch Dam – Blind Prediction Workshop and Additional Analyses 1480
Emmanuel Robbe

SESSION 6C: EARTHQUAKES II

Non-Linear Time Domain Analysis of Ruskin Dam in LS-DYNA Including Reservoir-Foundation-Structure Interaction 1498
O. Penner, B. Bergman, S. Razavi-Darbar, D. Queen, P. Léger, Martin Leclerc, Yannick Boivin

Boca Dam Spillway Nonlinear Analysis and Validation Study 1513
Jennifer E. Huggins, Roman Koltuniuk, Miguel Rocha

Nonlinear Seismic Evaluation of Perris Dam Outlet Tower 1528
Yusof Ghanaat, Zachary Harper

Nonlinear Seismic Analysis of an Arch Dam Subjected to Very Intensive Earthquake Loading 1546
Gurinderbir S. Sooch, Dan D. Curtis

SESSION 5D: LEVEES

U.S. Levee Safety Coalition and International Community of Practice, Recent Activities and Updates 1564
Elena Sossenkina, Michael Sharp, Noah Vroman

USACE Levee Portfolio Report 1565
Noah Vroman

Levee Safety Guidance Comprehensive Document Review 1566
Michael Sharp

Inventory and Review of the Nation's Levees 1567
Jamie McVicker, Cathi Sanders

Levee Safety Practices in the Keystone State 1568
Andrew J. Orlovsky

SESSION 6D: LEVEES

Seismic Vulnerability Analysis of Slag Cement-Cement-Bentonite Seepage Cutoff Walls for Levees 1578
Vlad Perlea, Khaled Chowdhury, Mary Perlea, Michael Kynett

Advances in Design Procedures for Relief Well Lines in Variable Urban Settings 1604
Robert A. Jaeger, Richard A. Keizer Jr. Graham C. Bradner, Joseph Weber, Mark Stanley

Evaluation and Design of Irregularly-Positioned Relief Wells 1624
Joseph Weber, Mark Stanley, Robert Jaeger, Graham Bradner

Explaining Pore Pressure Changes Due to Piping Through Changes in Hydraulic Conductivity 1643
Carolyne Bocovich, Willem Kanning, Minal Parekh, Michael Mooney

Design and Construction of Feather River West Levee Improvements 1657
Christopher Krivanec, Les Harder, Michael Bessette

SESSION 5E: CONCRETE II

Simplified Nonlinear Seismic Analyses of Gravity Dams 1658
Elisa Messio, Wenbo Duan, Najib Bouaanani, Benjamin Miquel

Strontia Springs Dam: Methods to Better Understand Seismic Behavior 1673
Aimee L. Corn, Guy S. Lund, Darren J. Brinker, Erin Gleason

Cracking in Thin Arch Concrete Dam – Nonlinear Dynamic Structural Analysis 1685
Sherry Hamedian, Roman Koltuniuk

SESSION 6E: CONCRETE II

Comparison of Nonlinear Finite Element Analysis Methods Used to Simulate Post-Tensioned Anchorage in Concrete 1709
Vojislav Cvijanovic, Mark Schultz, Blake Dolve
Structural Inflection Points, and How They Enhance the Understanding of Structural Safety 1724
Guy S. Lund, Kevin Marshall, Todd Barber
Pseudo-Nonlinear Finite Element Analysis of Concrete Gravity Dams 1735
J. Taylor Perkins, Husein A. Hasan, Daniel A. Gilbert
Importance of RCC Mixture Lab Studies for the Design of Large Dams 1747
A. Garduno, M. Montero, F. Ballhausen

SESSION 5F: SAFETY & SECURITY

Managing Threats Through Effective Security Programs 1762
J. A. Gloschat, A. C. Baehr
Unmanned Aerial Vehicles – A Public Agency Approach to Deployment and Protection 1775
Katharine Killeen, John Rizzardo
The Economics of Dam Security Post 9/11 and the Threat to Consumer Confidence 1776
Alexander B. Shannon

SESSION 6F: SAFETY & SECURITY

Flood Risks from Spillways: Modeling and Communicating Non-Dam-Failure Impacts to Native American Tribes 1796
Rinda Tisdale, Matthew Young, Chris Clopper
City of Dubuque, Iowa: Defining Emergency Actions with a Levee Breach Evaluation 1803
Andrew McCoy, Deron Muehring, Rusty Jones
Public Safety and Unauthorized Extreme Activities at Dams 1827
Steven Barfuss, Brian M. Crookston, Kade Beck, Ryan Weller
Public Safety and Social Media 1828
Sharon Nicole Roach
Author Index

United States Society on Dams

USSD

It's a Small World: Managing Our Water Resources

37th Annual USSD Conference

Anaheim, California, April 3-7, 2017

On the Cover

Owned and operated by the California Department of Water Resources, Perris Dam is a 130-foot high earthen dam constructed in the early 1970's. The dam is located in Perris, California, near the San Jacinto fault zone and constructed on potentially liquefiable alluvial materials. Stability analyses found that the dam could suffer significant deformations during a large seismic event. Construction for a three- year seismic retrofit project began in late 2014.

U.S. Society on Dams

Vision

A world class organization dedicated to advancing the role of dam and levee systems and building the community of practice.

Mission

USSD, as the United States member of the International Commission on Large Dams, is dedicated to:

ADVOCATE:	Champion the role of dam and levee systems in society
EDUCATE:	Be the premier source for technical information about dam and levee systems
COLLABORATE:	Build networks and relationships to strengthen the community of practice
CULTIVATE:	Nurture the growth of the community of practice

Foreword

Included in these Proceedings are papers and/or abstracts of presentations during the **37th USSD Annual Meeting and Conference**, held April 3-7, 2017, in Anaheim, California.

The theme of the 37th Conference was *It's a Small World: Managing Our Water Resources*. The theme plays into the Conference venue in Anaheim, just blocks away from Disneyland, but more importantly, the theme points to how the world is connected to a finite amount of water resources, especially fresh water. Anaheim, like most cities in Southern California, has been affected by both extreme droughts and flooding that much of the western and southern United States has experienced during the last five years. Local, regional and statewide leaders, along with industry professionals, have been working together with a holistic approach to managing the limited water resources in many different forms: floods, groundwater and surface water. Thus, Anaheim was an ideal setting for sharing and learning from those who have effectively managed dwindling freshwater resources.

Growing populations combined with recurring extreme events as the "new normal" are putting unprecedented demands on our existing infrastructure, requiring creative, multifaceted solutions. With so many interconnected competing interests for our limited water resources (fish and wildlife, agriculture, energy, municipal water, recreation), collaboration from a diverse population is a necessity — just as the "It's a Small World" attraction in Disneyland showcases cultural diversity and international unity.

Levees and dams play a vital infrastructure role in water resources management, be it protecting the public from the impacts of excessive rainfall, or providing water storage that can also be used for recreation and other purposes. Global climate change is placing increased demands on these important structures, and thoughtful and integrated management is necessary to help ensure that future generations will have the necessary resources to enjoy this small world we all share.

Technical Sessions during the Conference focused on best practices and new technologies related to concrete dams; embankment dams and levees; risk; public safety and security; construction and rehabilitation; hydrology and hydraulics; monitoring; earthquakes; and environmental issues.

The presentations were selected from abstracts submitted in response to a Call for Papers, and include both oral and poster presentations. Authors are specialists with broad experience from government agencies, utilities, academia, water districts, consulting firms and private industry.

The Conference Planning Committee extends thanks and appreciation to the California Department of Water Resources, Division of Safety of Dams, host of the 37th Annual Meeting and Conference.

Special thanks are also extended to the Committee Members who selected the abstracts and reviewed the technical papers, and to the authors who prepared the papers included in the Proceedings.

RESERVOIR SEDIMENTATION AND SUSTAINABILITY

Timothy J. Randle, Ph.D., P.E., D.WRE[1]
Sean Kimbrel, M.S., P.E.[2]
Kent L. Collins, P.E.[3]

ABSTRACT

The continual or periodic management of reservoir sedimentation (clay, silt, sand, gravel, and cobble) for sustainable operations will help to ensure long-term viability of project benefits and avoid difficult and costly sediment problems for future generations. Reservoirs that become full of sediment may no longer have project benefits and have to be removed. Project benefits can be impaired by the burial of dam outlets, water intakes, boat ramps, and marinas long before a reservoir has completely filled with sediment.

Sustainable reservoir sediment management strategies can include the reduction of sediment loads reaching the reservoir (watershed management practices), prevention of sediment deposition within the reservoir (sediment bypassing or sluicing), removal of sediments already deposited in the reservoir (drawdown flushing, dredging, or excavation), or a combination of these strategies.

The Federal Advisory Committee on Water Information, Subcommittee on Sedimentation has passed a resolution encouraging Federal agencies to develop long-term reservoir sediment-management plans for the reservoirs that they own or manage. In addition, the Subcommittee has formed the National Reservoir Sedimentation and Sustainability Team to provide helpful information on these important topics. This team is composed of volunteer specialists from Federal agencies, universities, and consultants.

INTRODUCTION

The benefits of the nation's numerous reservoirs include such things as water storage for irrigation, municipal and industrial use, flood risk reduction, recreation, and hydroelectric power. However, all rivers transport sediment (e.g., clay, silt, sand, gravel, and cobble) in widely varying amounts, and sediment tends to deposit and accumulate in reservoirs over time because of the reduced flow velocity, turbulence, and shear stress through reservoir pools. The long-term process of sediment accumulation is known as reservoir sedimentation (Figure 1).

Typically, the outlet of a dam was planned to be above the rising sediment level over the project's sediment-design life, which was typically 50 or 100 years (Bureau of Reclamation, 2016). However, there were no plans to manage the reservoir sediment after this time period. The nation's dams are aging, but will continue to be relied upon for water supply and hydropower well into the future. If not managed, sediment and

[1] Bureau of Reclamation, P.O. Box 25007, Mail Code 86-68240, Denver, CO 80225, trandle@usbr.gov.
[2] Bureau of Reclamation, P.O. Box 25007, Mail Code 86-68240, Denver, CO 80225, skimbrel@usbr.gov.
[3] Bureau of Reclamation, P.O. Box 25007, Mail Code 86-68240, Denver, CO 80225, kcollins@usbr.gov.

woody debris accumulation will continue to decrease water storage capacity and the reliability of the reservoir water supply. Reservoir sedimentation is a worldwide problem, reducing storage capacity over time while the world population continues to increase.

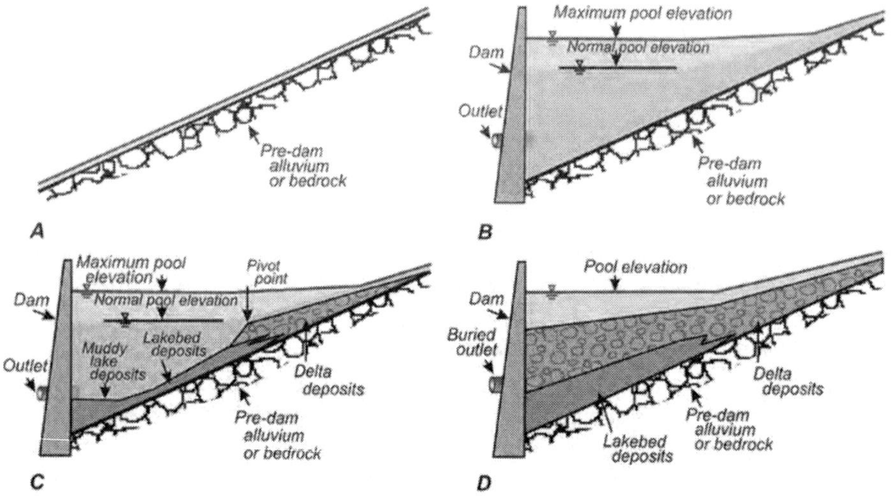

Figure 1. Process of reservoir sedimentation: A) original riverbed, B) new dam construction, C) sedimentation that is near the sediment-design life, and D) dam outlet works burial (Randle and Bountry, 2016).

If not managed, continued sedimentation threatens the project benefits of the Nation's reservoirs over time (Figure 2). Annandale (2013) estimates that net world-wide reservoir storage has been decreasing since the year 2000 due to sedimentation and, on a per capita basis, reservoir storage has been decreasing since 1980.

In addition to the reduction in water storage capacity, reservoir sedimentation will eventually bury dam outlets (Figure 3), water intakes, boat ramps, and marinas, and reduce the surface area for recreation (Figure 4). Reservoir sedimentation can also impair boat navigation. Sand or gravel can be very abrasive to dam outlets, turbines, and spillways (Figure 5). Sediment deltas at the upstream end of reservoirs tend to propagate upstream from the reservoir pool over time and increase ground water elevations and flood stage, which may affect upstream infrastructure and property. Reservoir sedimentation also prevents sediment from reaching the downstream river, which can lead to erosion of the downstream channel, the impairment of habitat for fish and wildlife, and a reduction of sediment delivered to coastal deltas.

Reservoirs cannot physically trap sediment indefinitely. As the reservoir becomes filled with sediment, inflowing sediment will be transported through the reservoir to the downstream channel, in an uncontrolled manner.

Figure 2. Sedimentation has almost completely filled the reservoir behind Matilija Dam near Ventura, CA.

Figure 3. Reservoir sedimentation has impaired the outlet at Paonia Dam near Paonia, CO (Bureau of Reclamation, 2016).

Figure 4. The reservoir delta has reduced the surface area available for recreation at Lake Powell near Hite, UT.

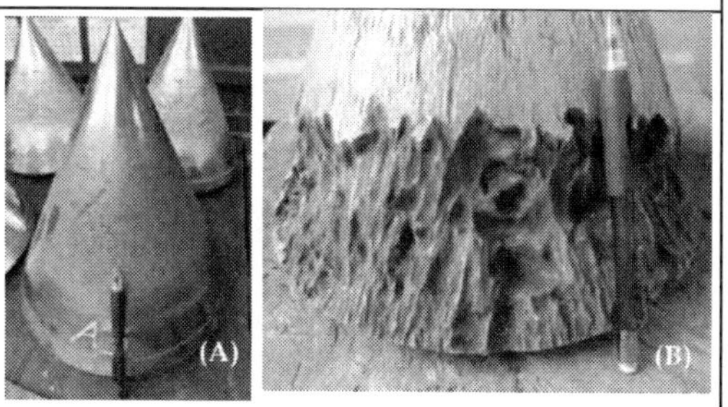

Figure 5. Sand abrasion can damage turbines in a matter of hours. A) Turbine after 10,000 hours of normal operations. B) Turbine after 24 hours of sand abrasion during extreme reservoir drawdown and flood inflow. Photographs Courtesy of Greg Morris.

The good news is that reservoir sediment management methods exist to sustainable manage reservoirs over the long term and include three basic categories (Kondolf, et al., 2014):

- Reduction of sediment loads reaching the reservoir (watershed management practices).
- Prevention of sediment deposition within the reservoir (sediment bypassing or sluicing).
- Removal of sediments already deposited in the reservoir (drawdown flushing, dredging, or excavation), or a combination of these strategies.

Each of these reservoir sediment management categories require a commitment and funding to implement. The option of doing nothing may appear attractive, but the cost of reservoir sediment management needs to be compared with the cost of lost reservoir benefits and the cost of reservoir retirement. Once a reservoir resource is lost to sedimentation, the cost to replace the lost benefits will include the cost of planning, design, land purchase, permitting, and construction to develop another dam and reservoir with fewer sites available than when the first dam and reservoir were developed. The cost of reservoir retirement often includes the cost of dam removal and downstream mitigation.

RESERVOIR SEDIMENTATION AWARENESS

For many people, reservoir sediment is out of sight and out of mind. Coarser sediments (sand and gravel) that deposit as deltas at the upstream end of the reservoir typically can't be seen from the dam and, when visited, tend to look like the river channel rather than the former reservoir (Figure 6 and Figure 7). Finer sediments (clay and silt) deposit along the reservoir bottom, and in the dead pool near the dam, and typically can't been visually seen through the reservoir pool.

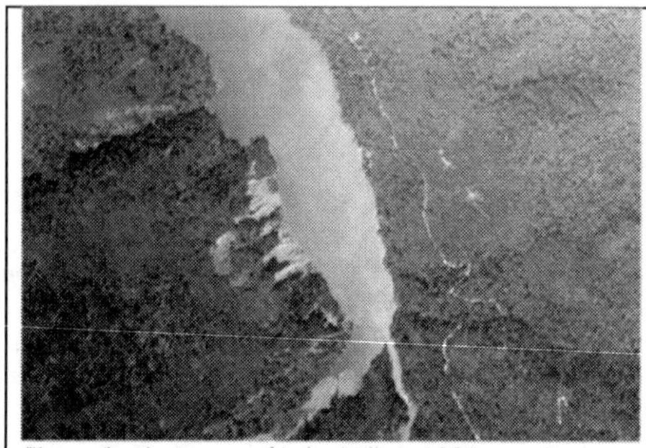

Figure 6. Upstream of Lake Mills in 1939 near Port Angeles, WA.

Figure 7. Upstream of Lake Mills and sediment delta in 2009 near Port Angeles, WA.

The first obvious signs of a reservoir sediment problem may be the plugging of a dam outlet or reservoir water intake with wood and sediment. This is analogous to human heart disease where the first obvious symptoms may be a stroke or heart attack. Bathymetric reservoir surveys are periodically needed to monitor the reservoir sediment accumulation over time and to estimate when the various reservoir facilities may be impaired by sediment. Without periodic monitoring, crisis management will eventually be necessary to continue dam and reservoir operations.

Fortunately, the cost of monitoring (bathymetric surveys) has been decreasing over time while the quantity and quality of data has increased. Older bathymetric surveys were more labor intensive and only measured data along widely spaced cross-sections of the reservoir identified by permanent monuments. Modern methods employ just a few people utilizing GPS survey-grade instruments for position and multi-beam depth sounders to measure the bathymetry in much more detail.

The interagency Subcommittee on Sedimentation has formed the National Reservoir Sedimentation and Sustainability Team to increase the awareness of important issues and focus on the following activities:

- Provide training courses and publically available web-based resources.
- Encourage reservoir sediment surveys and the storage of this data in national reservoir sedimentation database.

- Formulate a white paper on reservoir sedimentation and sustainability.

This team of volunteers is composed of specialists from Federal agencies, universities, and consultants.

RESERVOIR SEDIMENT MANAGEMENT SOLUTIONS

Planning studies for dams and reservoirs in the United States typically did not include plans and costs associated with sediment management to maintain the remaining reservoir storage beyond the sediment-design life. Most of the nation's dams are in the second half of their sediment-design life and many are reaching the end of their sediment-design life. Future planning studies will need to address the engineering, economic, and financial feasibility of potential sediment management actions, which will be necessary to maintain existing project benefits.

Sustainable reservoir sediment management strategies can include the reduction of sediment loads reaching the reservoir (watershed management practices), prevention of sediment deposition within the reservoir (sediment bypassing or sluicing), removal of sediments already deposited in the reservoir (drawdown flushing, dredging, or excavation), or a combination of these strategies:

- Watershed sediment management practices can reduce the sediment yield entering a reservoir. A wide range of methods can be employed that reduce landslides (Figure 8), soil erosion, stream-bank erosion, and the downstream transport of eroded sediments.
- Sediment bypass around the reservoir would keep sediments from entering the reservoir. For example, an upstream diversion weir could be constructed to divert river flows with high sediment concentrations into a tunnel or pipe that conveys the sediment around the reservoir and past the dam (Figure 9).

Figure 8. Landslide stabilization in Japan reduces the watershed sediment yield.

Figure 9. A tunnel was constructed along the Miwa Reservoir in Japan to bypass river flows with high sediment concentration.

- Passing inflowing sediments through the reservoir also would limit sediment deposition. The venting of turbidity currents through a low-level outlet in the dam (Figure 10) or drawing the reservoir down during periods of high sediment inflow would pass sediment through the reservoir.
- Sluicing of sediment during partial reservoir drawdown would help evacuate sediments near the sluice gates.
- Flushing of sediments previously deposited in the reservoir would help recover storage capacity. Emptying the reservoir is necessary to increase the flow velocity and sediment transport capacity through the reservoir (Figure 11). Downstream sediment concentrations can be very high during flushing.

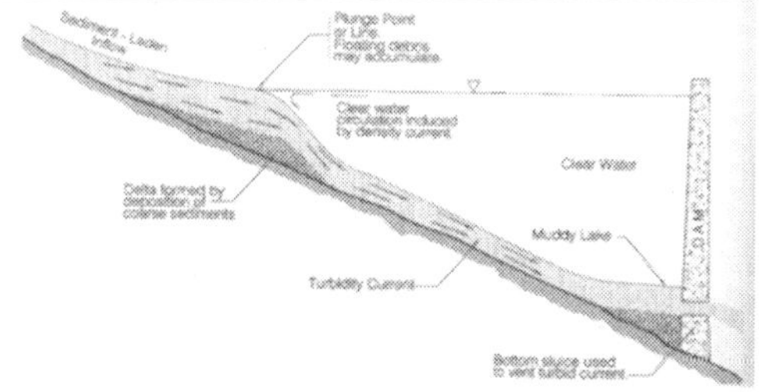

Figure 10. Sediment laden inflows can form a turbidity current along the reservoir bottom and it may be possible to vent this high sediment concentration through a low-level outlet in the dam (Morris and Fan, 1998).

Figure 11. Lowering the reservoir can result in the erosion and downstream flushing of sediments.

- Mechanical or hydraulic sediment removal also would help recover storage capacity. Sediment can be removed by hydraulic dredging (Figure 12), mechanical dredging (Figure 13), and dry excavation. Sediments can be conveyed from the reservoir by sediment slurry pipeline, truck transport, or conveyor belt. Removed sediments can be discharged into the downstream river channel or delivered to a disposal site.

Sediments passed through a reservoir to the downstream channel may slow, stop, or reverse channel erosion and degradation trends. Downstream channel aggradation may be acceptable so long as the aggradation does not cause unmitigated harm to people, property, or native species. The passing of annual sediment loads through or around the reservoir should be timed to benefit and avoid harm to native species. However, the downstream release of reservoir sediment may harm introduced sport fisheries that depend on the clear water releases from the dam. There may be feasible options (e.g. sediment slurry pipeline) for delivering sediment past a highly valuable sport fishery. For reservoirs that divert a significant portion of the stored water away from the downstream channel, the amount of sediment passed to the immediate downstream channel may have to be reduced or delivered farther downstream to avoid excessive channel aggradation.

Figure 12. Hydraulic dredging of reservoir sediments for transport through a sediment slurry pipeline.

Figure 13. Mechanical dredging of reservoir sediments and transport by truck.

CONCLUSIONS

The importance of monitoring reservoir sedimentation increases as most reservoirs are in the second half of their sediment-design life. A decade or more may be needed to plan and implement sustainable sediment management plans for a given reservoir. Failure to plan and implement a sustainable sediment management plan will lead to the eventual retirement of the dam and reservoir.

Sustainable reservoir sediment management may harm introduced sport fisheries, but those fisheries cannot be sustained if the dam outlet becomes plugged with wood and sediment or after the reservoir has substantially filled with sediment. Releasing sediments downstream of valuable fisheries may be a method to avoid impacts.

The cost and impacts of reservoir sediment management need to be compared against the eventual costs of retiring the reservoir, including the lost project benefits and the cost of constructing additional water storage elsewhere. Future generations should be considered when choosing a reservoir sediment management plan.

ACKNOWLEDGEMENTS

The ideas presented in this paper were developed through several workshops of the National Reservoir Sedimentation and Sustainability Team.

REFERENCES

Annandale, G. (2013). *Quenching the Thirst, Sustainable Water Supply and Climate Change*, CreateSpace Independent Publishing Platform, North Charleston, SC, 231 pages.

Bureau of Reclamation (2016). *The Knowledge Stream Reservoir Sedimentation and Sustainability Issue*, Spring 2016 Research and Development Office Newsletter, 2016-02, Denver, CO, 44 pages.

Kondolf, G.M.; Gao, Y; Annandale, G.W.; Morris, G.L.; Jiang, E.; Zhang, J.; Cao, Y.; Carling, P.; Fu, K.; Guo, Q.; Hotchkiss, R.; Peteuil, C.; Sumi, T.; Wang, H.W.; Wang, Z.; Wei, Z.; Wu, B.; Wu, C.; and Yang, C.T. (2014). "Sustainable sediment management in reservoirs and regulated rivers: Experiences from five continents." Earth's Future, 2, doi:10.1002/2013EF000184.

Randle, T.J., and Bountry, J. (2016). *Dam Removal Analysis Guidelines for Sediment - Version 1, Prepared for Subcommittee on Sedimentation, Federal Advisory Committee on Water Information*, Bureau of Reclamation, Sedimentation and River Hydraulics Group, Technical Report SRH-2016-38, September 2016, Denver, CO.

DAM SAFETY AND RESILIENT WATER MANAGEMENT: A CASE FOR DYNAMIC RESERVOIR OPERATIONS

Michael McMahon[1]

A holistic approach to water supply management can simultaneously provide for improved near-term and long-term guidance for water operations and efficiencies. This methodology, called Dynamic Reservoir Operations (DRO), will provide a means-to-an-end for adapting to changes in runoff timing and volume that are consequences of climate variability. The very nature of this methodology is designed to be proactive to a changing environment within the next several days, an entire water year, or even on a decadal basis. It takes the knowledge of what the global climate is doing, adds an advanced look at the current hydro-meteorology of a given basin through remote sensing, produces an accurate forecast from which operational decisions can be made, and then applies those decisions at the appropriate time. It is designed to adapt to what the hydro-meteorological conditions are telling us, both now and in the future.

The use of DRO could realistically improve water efficiencies by as much as 25% without a reduction in flood safety. Even a 5% increase in water management efficiency would mean thousands of acre-feet in increased system capacity. DRO will be discussed in the context of real-life water scenarios for the current water year and prior water years. This presentation will be designed to weigh the pros and cons of DRO methodologies in order to elicit an open discourse regarding its efficacy.

[1] HDR, michael.mcmahon@hdrinc.com

THE NEBRASKA DEPARTMENT OF NATURAL RESOURCES COORDINATED EFFORTS IN FLOOD FLOWS

Kristen Obermueller[1]
Bob Beduhn[2]
John Engel[3]

In September 2013, week-long rains in Colorado caused the Nebraska Department of Natural Resources (NDNR) to anticipate waters would inundate the South Platte River in Nebraska. To ensure surrounding communities were safe from flooding, NDNR successfully collaborated with the Colorado Division of Water Resources, Colorado Highway Patrol, Irrigation Districts, fire departments, governments, sheriffs, NWS, Natural Resources Districts, NDEQ, Nebraska Department of Roads, Nebraska Emergency Management Agency, Nebraska State Patrol, USACE, and the United States Geological Survey. Extensive coordination, education and communication was needed in previous years to accommodate the potential for a flooding scenario that would require immediate response. In 2008, NRDs and NDNR cultivated Integrated Management Plans (IMPs) for current and future water management that stipulated coordination to take advantage of flood flows due to previously missed opportunities. To assess the effectiveness of the IMPs in 2011, NDNR, NRDs, and irrigation districts tested the ability to divert flows for the benefit of ground water recharge. NDNR addressed administration issues and satisfied needs for all involved, resulting in resounding support. With groundwork laid for an emergency, NDNR was able to successfully recharge September 2013's flood waters in less than two days by using existing infrastructure to redirect water to reuse pits. The results of the recharged flood flows culminated into long-lasting benefits, including employing feedback loops with emergency responders, accurately assessed river levels, attenuating peak flows to prevent flooding, contributing to steady water supply in the river for years, and supporting sustainability of endangered wildlife.

This presentation will articulate how NDNR addressed the need for IMPs and the challenges of implementation for the September 2013 flood and present the subsequent communication tools that were developed to serve as future educational resources.

[1] HDR, Kristen.obermueller@hdrinc.com
[2] HDR
[3] HDR

ENVIRONMENTAL CONSIDERATIONS IN DAM DECOMMISSIONING

Kelly R. Schaeffer[1]

Dam decommissioning with dam removal is defined by the USSD's Guidelines for Dam Decommissioning Projects as "the full or partial removal of an existing dam and its associated facilities such that the statutory definition of a dam is no longer met or the structure no longer presents a downstream hazard"[2].
As part of our mission to determine the environmental sustainability of dams and levees, this presentation will highlight a few environmental issues that may be associated with the decision to decommission a dam or levee. This presentation will highlight water quality and vegetation/re-vegetation, fisheries habitat, and recreation and social community issues in dam decommissioning.

[1] Principal, Kleinschmidt Associates, Kelly.Schaeffer@KleinschmidtGroup.com
[2] 2015 U.S. Society on Dams, Guidelines for Dam Decommissioning Projects.

HYDRAULIC DREDGING OF THE STRONTIA SPRINGS RESERVOIR - LESSONS LEARNED

Douglas Raitt, PE, PMP[1]
Brett Cochran, EI[2]

ABSTRACT

The Strontia Springs Dam and Reservoir are located on the South Platte River in Jefferson County, Colorado. They are owned and operated by Denver Water. The 7,863 acre-foot reservoir is the forebay for the raw water intakes for Denver Water's Foothills Treatment Plant and the City of Aurora's raw water supply. Over two thirds of the raw water for the 1.4 million customers served by Denver Water flows through the Strontia Springs Reservoir.

Beginning on May 18, 1996, the Buffalo Creek Fire burned 12,000 acres of forest immediately upstream of the reservoir. On July 12, 1996, a severe storm dropped 2 1/2-inches of rain over a period of two hours on the fire damaged watershed. Due to the lack of vegetated cover, runoff was unusually intense and conveyed a significant amount of sediment, burned vegetation, and ash into the reservoir.

Following the Buffalo Creek Fire a series of annual bathymetric surveys of the reservoir were initiated to assess the quantity of sediment inflows. Alternatives to address the inflows were identified and hydraulic dredging was selected as a remedial measure in 2009 for the removal of 625,000 cubic yards of sediment. Hydraulic dredging of the reservoir commenced in May 2011 and was terminated in November 2011 after 228,000 cubic yards of material was removed.

This paper describes the investigation, the evaluated alternatives, the selection of hydraulic dredging and the factors that contributed to the ultimate shortfall in planned production. A number of "lessons learned" are also identified. Key technical design and construction challenges are noted as well as future challenges related to sediment management.

INTRODUCTION

A series of large and intense forest fires plagued the upstream watershed of the Strontia Springs Reservoir between 1996 and 2002. As a result, the upland forest and riparian vegetation that had stabilized much of the 833 square mile watershed immediately upstream of the reservoir was eliminated, and the area became much more susceptible to increased runoff due to heavy precipitation. Localized storms over the burn area resulted in several high intensity runoff events that conveyed significant amounts of sediment,

[1] Engineering Manager – Construction, Engineering Department, Denver Water, 1600 West 12th Ave., Denver , CO 80204, (303) 628-6426, douglas.raitt@denverwater.org
[2] Engineer II, Engineering Department, Denver Water, 1600 West 12th Ave., Denver , CO 80204, (303) 628-6070, brett.cochran@denverwater.org

burned vegetation, and ash into the reservoir causing a disruption to the primary raw water supply for metropolitan Denver on more than one occasion.

Therefore, Denver Water undertook a multi-year study of the vulnerability of the reservoir and developed alternatives to remediate the reservoir and protect the upstream watershed. This paper describes the investigation, the evaluated alternatives, the selection of hydraulic dredging and the outcome of that effort.

BACKGROUND

Prior to the construction of the Strontia Springs Dam, Denver Water operated the Platte Canyon Diversion Dam on the South Platte River at a location 3 miles downstream of the current dam site (Figure 1). This facility operated from its construction in 1911 until it was replaced as part of the Foothills Project in 1983.

During the operation of the original diversion dam, sediment accumulated behind the structure in the 125 acre-ft (AF) forebay. During a normal year from April 1 to October 1, 30,000 to 40,000 cubic yards (18 - 25 AF) of sediment accumulated behind the structure. Each spring the sediment from behind the diversion dam was sluiced over a period of 3 to 11 days at flow rates of 100 to 450 cubic feet per second for a total usage of 1,400 to 3,000 AF of water.[3] Four sluice gates were utilized to pass the sediment during peak runoff in the early spring of each year.

Figure 1. Platte Canyon Diversion Dam

The Strontia Springs Dam was originally conceived as the downstream diversion dam on the main stem of the South Platte River below the Two Forks Dam which ws proposed to be located several miles upstream of the Strontia Springs Dam site. A 3-mile long tunnel connects the Strontia Springs Reservoir to the 280 million gallon per day Foothills Treatment Plant (Figure 2). The Strontia Springs Dam and the Foothills Treatment Plant were both commissioned in 1983 completing a much needed upgrade of the water supply system for metropolitan Denver.

[3] W.M. Borland, Foothills Project – Strontia Springs Dam, Sedimentation Analysis, March 1978

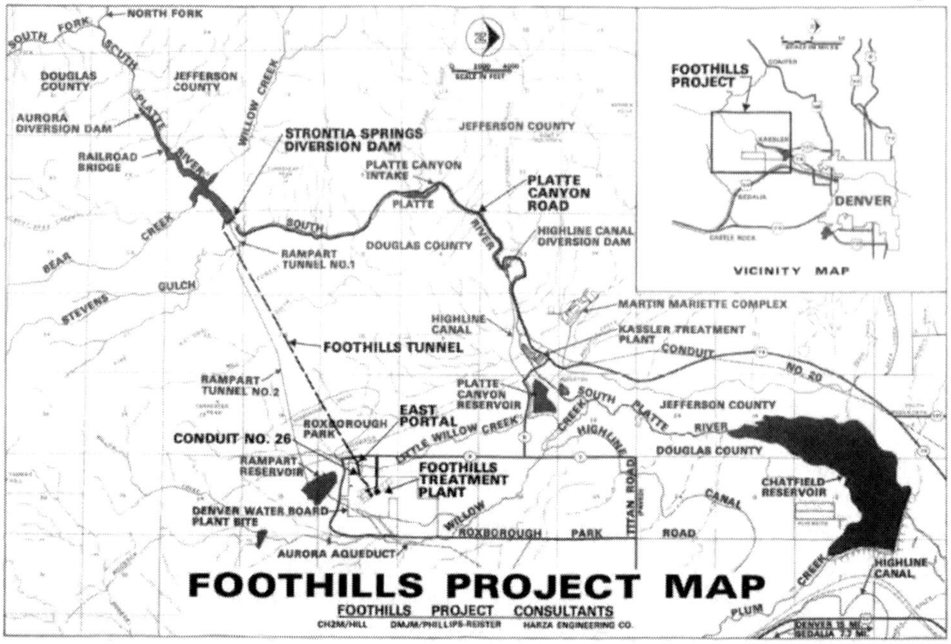

Figure 2. Foothills Project System Map

After many years of studies the proposed upstream 1,100,000 acre-foot Two Forks Dam and Reservoir was denied approval by the United States Environmental Protection Agency in 1990 and was never constructed[4]. Consequently, all sediment flows in the South Platte River have been deposited behind the Strontia Springs Dam in the reservoir pool.

STRONTIA SPRINGS DAM

Physical attributes of the Strontia Springs Dam:

General	
Location	Kassler, Colorado, USA
River	South Platte
Purpose	Diversion Storage
Completion Date	1982
(Figure 3 and 4 for Images)	
Dam	
Type	Double Curvature Arch
Maximum Structural Height	299 feet
Crest Length	650 feet

[4] Recommended Determination to Prohibit Construction of Two Forks Dam and Reservoir Pursuant to Section 404 (c) of the Clean Water Act, U.S. Environmental Protection Agency, Region VII, March, 1990

Reservoir

Normal Pool Elevation	6002 feet
Capacity at Elevation 6002	7,700 acre-feet
Surface Area at Elevation 6002	98 acres
Pool Length at Elevation 6002	1.7 miles
Contributing Drainage Basin	2,590 square miles

Figure 3 and 4. Strontia Springs Dam Construction 1979 through 1982

ORIGINAL SEDIMENT INFLOW ASSUMPTIONS

A study of sediment transport in the South Platte River was performed by Mr. Whitney. M. Borland in 1978. Interviews were conducted with Denver Water operating staff and an inspection of the upstream watershed was performed as part of the study. Characteristics of the sediment transport mechanisms in the river were evaluated based on the historical record of sediment management at the existing South Platte Diversion Dam. Borland described the sediment load as follows:

> "The South Platte River in the mountains is a comparatively clear stream with an average suspended sediment load of about 150 mg/liter. The only reason that sediment is a serious problem in this river is that it carries an unusually large bed load composed of 1.0 to 16.0 mm (l/32-inch to 5/8-inch) diameter bed material. This material is derived from the Pikes Peak Granite formation which outcrops and forms the soil mantle of over half of the directly contributing drainage area of 833 sq. mi. above the Strontia Springs Diversion Dam site...
>
> At the proposed Strontia Springs diversion site the slope of the river bed is 0.019 or about 100 feet per mile. Based upon estimates of bed material sluiced at the existing South Platte Intake Dam, photos, and sluice water required to clean the forebay, the bed load was calculated to be 81% of the suspended load, or 60,000 ton/year...
>
> The average bed load of 30,000 to 40,000 cubic yards annually will settle out in the forebay of any new diversion structure when velocities are less than 3 1/2 ft/sec., reducing the carrying capacity of conveyance channels and plugging conduits unless sufficient sediment storage capacity is provided. Because the Pikes Peak Granite is composed of about 25% quartz grain with a size of 1/8-inch

to 1/2-inch and a hardness of 7, it will readily erode concrete and steel when moving with a velocity in excess of 4 ft/sec. In order to divert water from a river having a comparatively large load of this type of material, the problem is whether to store the material in the reservoir or attempt to design a structure through which the material can be transported. If the latter choice is made, there is always the risk of failure of the structure and the consequent loss of water diversion capability which, in the case of a municipal water supply system, can be disastrous…

The depth of sediment at the dam after various periods of operation is dependent on the texture of the sediment, inflow-outflow relations, shape of reservoir and reservoir operation procedure. A log-log plot of the depth/capacity shows the reservoir to be a type II (Hill)[5]."

Borland also estimated the relationship between time and sediment accumulation in the reservoir. Using reservoir sediment trap efficiency by Gunnar Brune[6], sediment capture rates were estimated by Borland in Table 1:

Table 1. Estimated Sediment Capture by Borland

Elapsed Time (years)	Sediment Inflow – (acre-feet)	Trap Efficiency (%)	Trapped Sediment- (acre-feet)	Capacity at End of Period (acre-feet)	Depth of Sediment at Dam (feet)	Elevation of Sediment at Dam (feet)
0	0	0%	0			
10	520	77%	400	7,300	25	5810
20	1,040	76%	750	6,910	36	5821
40	2,090	75%	1,560	6,140	53	5836
60	3,120	75%	2,340	5,860	70	5855
80	4,160	74%	3,078	4,622	87	5872
100	5,200	73%	3,800	3,900	106	5891

In this sediment accumulation model the large outlet release system elevation of 5845 is covered by sediment during the 40-year to 60-year lifespan interval of the reservoir.

Watershed Impacts

[5] Distribution of Sediment in Large Reservoirs by W. M. Borland, C. R. Miller ASCE Vol. 84 No. HY 2 Proc. Paper 1587 April 1958 pp. 1587-1 to 1587-18.

[6] Trap Efficiencies by Gunnar Brune, Trans. A.G.U. Vo. 34 No. 3, June 1953

The watershed area above the Strontia Springs Dam and downstream of the next upstream dam in the South Platte River basin is 833 square miles. Land cover is distributed in Table 2 as follows[7]:

Table 2. Strontia Springs Reservoir Upstream Land Cover

Land Cover Type	Area (acres)	%
Barren	104,278	20%
Forest – Evergreen	395,615	75%
Ranches – Grassland	28,170	5%
Bermuda grass	1,236	<1%

Beginning in 1996, a series of wildfires in the watershed upstream of the Strontia Springs Reservoir burned a significant amount of the forested areas (Table 3).

Table 3. Fires in Strontia Springs Reservoir Watershed

Year	Fire	Acreage Burned - Acres
1996	Buffalo Creek	11,900
2000	High Meadow	10,800
2001	Polhemus	8,000
2002	Snaking	2,590
2002	Schoonover	3,860
2002	Hayman	137,760
2012	Lower North Fork	4,500
2012	Springer	1,145
2012	Trout Creek	25

Less than two months after the 1996 Buffalo Creek Fire was contained an unusually heavy precipitation event dropped 2 1/2 inches of rain (a 100-year event) over the burn area in 2 hours' time. Due to the high intensity of the burn, the steep terrain of the burn area, and the high erodibility of exposed ground, the resulting runoff had extremely high

[7] Basin Prioritization - Upstream Management Practices at Strontia Springs Reservoir, Golder Associates Inc., September, 2013

sediment loads which were conveyed to the reservoir (Figure 5 and 6). Significant quantities of burned vegetation mixed with the sediment laden runoff and created an untreatable condition in the water supply and a heavy sediment inflow to the reservoir.

As a result Denver Water initiated a study to quantify the upstream impacts to the watershed and investigate the sediment inflows to the reservoir. Subsequent fires in the watershed reinforced the need to develop watershed management techniques and alternatives to manage high sediment inflows.

Figures 5 and 6. Images of Strontia Springs Reservoir and Upstream Confluence July, 1996

SEDIMENTATION MITIGATION ALTERNATIVES

Between 1997 and 2006 14 alternatives to remove sediment from the reservoir were investigated. A few alternatives had different configurations, therefore, the total number of schemes studied was 18 (Table 4).

Table 4. Sediment Removal Alternatives

Alternative	Process	Disposition
1,2 & 3	Dredge and deliver sediment by a slurry pipeline, a conveyor, or trucks to a permanent upstream containment area.	Unfavorable – No suitable upstream disposal area.
4	Dredge and deliver sediment through a slurry pipeline to a permanent downstream containment area at Denver Water's Kassler facility.	Favorable – Property owned by Denver Water 6 1/2 miles downstream of reservoir.
5	Dredge and deliver sediment using trucks to a permanent downstream	Favorable – Property owned by Denver Water 6 1/2 miles

	storage area at Denver Water's Kassler facility.	downstream of reservoir. Slurry pumping required across reservoir.
6	Dredge and deliver sediment directly to river downstream of Strontia Springs Dam.	Unfavorable – USACOE Dredge and Fill Permit required. Water quality impacts to river anticipated. Possible disruption to municipal water supply.
7	Dredge and deliver sediment through a slurry pipeline to a temporary upstream containment area, haul sediment to Pine Junction.	Unfavorable – No suitable upstream disposal area.
8	Partial drawdown of reservoir. Excavate sediment and haul to permanent upstream containment area.	Unfavorable – No suitable upstream disposal area. Possible disruption to municipal water supply.
9 & 10	Dredge and deliver sediment through a slurry pipeline to a permanent containment area at Denver Water's Foothills Treatment Plant or Aurora's Rampart Reservoir.	Unfavorable – Possible disruption to municipal water supply.
11, 12	Complete or partial drawdown flushing of reservoir.	Unfavorable – Potentially complex regulatory review. Possible disruption to municipal water supply. Water quality impacts to river anticipated. Larger flushing outlet required.
13 & 14	Hydro-suction of sediment with delivery directly to river downstream of Strontia Springs Dam or delivery to Denver Water's Foothills Treatment Plant.	Unfavorable - Complex regulatory review. Possible disruption to municipal water supply. Water quality impacts to river anticipated. Technical feasibility concerns.

Denver Water also investigated alternatives to prevent sediment from entering the reservoir. These were not or; have not been pursued beyond the study phase due to the regulatory hurdles required to construct facilities in the main stem of the South Platte

River. (An Environmental Assessment or Environmental Impact Statement would likely be required to construct facilities in the river Table 5).

Table 5. Alternatives to Prevent Sediment Inflows to the Reservoir

Alternative	Process	Disposition
1	Construct a sediment pond at the confluence. Periodically clean and haul sediment to a permanent upstream containment area.	Available as a long term alternative if other options are not feasible.
2	Construct a sediment pond at the confluence. Periodically clean and haul sediment to Pine Junction.	
3	Construct sediment ponds at Spring Creek and Buffalo Creek tributaries. Periodically clean and haul sediment to a permanent upstream containment area.	
4	Construct sediment ponds along the South Platte at Nighthawk, Oxyoke and Trumbull. Periodically clean and haul sediment to a permanent upstream containment area.	

TRIBUTARY SEDIMENT CONTROL

Sediment control measures were employed following the devastating 2002 Hayman fire that bisected both the Strontia Springs Reservoir watershed and the upstream Cheesman Reservoir watershed (also Denver Water owned). This was the largest fire in Colorado's recorded history; it covered almost 138,000 acres. In an effort to intercept ash laden sediment before it entered the reservoir, two significant sediment control structures were constructed along main tributaries to the Cheesman Reservoir. The following details were developed for the Goose Creek tributary (Figures 7, 8 and 9).

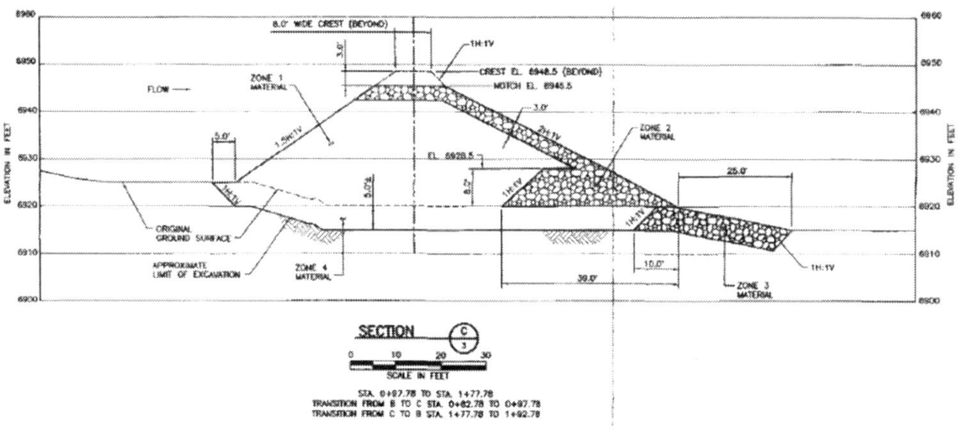

Figures 7. Section View of Goose Creek Sediment Control Dam

Figures 8 and 9. Pictures of Goose Creek Sediment Trap in Operation and During Maintenance

Denver Water self-performed construction of the Turkey Creek sediment trap and solicited a Design/Build contract for the Goose Creek sediment trap. A contract for the Goose Creek work was awarded in 2002; construction was completed in 2003. Following construction, sediment traps have been cleared periodically as runoff filled the basins.

Disposal of sediment was accomplished on Denver Water owned property in the immediate vicinity of the sediment trap. The cost history for construction and maintenance of the two tributary sediment traps is shown in Table 6:

Table 6. Cost History of Tributary Sediment Traps

Year	Turkey Creek Sediment Removed – (cubic yards)	Goose Creek Sediment Removed (cubic yards)	Total Sediment Removed (cubic yards)	Annual Cost	Unit Cost – ($/cubic yard)
2003	0	0	0	$871,562	

2004	0	0	0	$16,819	
2005	25,100	8,020	33,120	$334,028	$10.09
2006	15,420	14,160	29,580	$262,395	$8.87
2007	63,620	0	63,620	$288,129	$4.53
2008	34,250	18,875	53,125	$197,412	$3.72
2009	15,720	0	15,720	$277,931	$17.68
2010	0	0	0	$9,798	
Total	154,110	41,055	195,165	$2,258,074	$11.57

RESERVOIR SEDIMENT INVESTIGATION

Denver Water initiated studies of the Strontia Springs Reservoir sediment and upstream watershed immediately following the 1996 fire and extreme runoff events. Studies were also conducted with the USGS to evaluate the impacts of fires on the water quality and on the sediment properties of the runoff from the burned areas. The water supply disruption lasted only a few days; however, higher persistent levels of particulate and dissolved manganese were observed in the water supply. This led Denver Water to install a potassium permanganate injection system near the reservoir intake and upstream of the Foothills Treatment Plant to accelerate oxidation of the higher manganese levels in the raw water supply.

Hydrographic Surveys

After the floods of 1996, hydrographic surveys of the reservoir were initiated to develop an understanding of the amount and location of sediment accumulation. Over the years, Denver Water has utilized several approaches to measure and analyze data collected from hydrographic surveys of the reservoir. The latest equipment employed in this effort is listed below:

- 22-foot long Aluminum Survey Boat (Figure 10)
- Odom ES3 multi beam echo sounder
- Transducer Frequency (200 kHz)
- Hypack survey software Ver. 15.0.1.1
- SMC108 Motion sensor
- Trimble 852/552 GPS/Gleans System
- Post-Processing data in Hypack Office and volumes in AutoCAD Civil 3D 2014

Figure 10. Hydrographic Survey Boat

Using the survey boat, a two person crew conducts the surveys over five days. In-office post-processing of the data takes another five days.

Starting in 1996 a series of annual surveys were conducted to quantify sediment accumulation. The results of these early surveys are summarized in Figure 11:

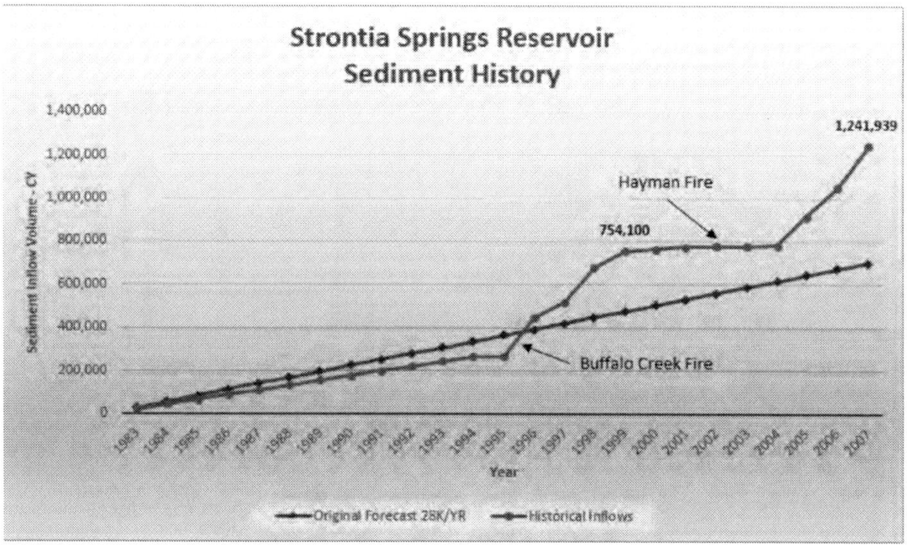

Figure 11. Sediment Inflows to Strontia Springs Reservoir Prior to 2008

By the mid-2000s, the rate of sediment accumulation was continuing to surpass the originally estimated inflows by substantial amounts. As a result, action was necessary.

SEDIMENT DEPOSIT INVESTIGATION

A geotechnical investigation was commissioned in 2007. A $90,000 contract was subsequently issued for a preliminary sediment removal design. This effort provided concepts for both upstream and downstream disposal options for sediment hydraulically dredged from the reservoir. The next step was the authorization of a $290,000 sediment investigation contract; added to the original scope of work in late 2007. In October, 2007, equipment was mobilized and sampling of the sediment in the proposed dredging area was completed. The equipment utilized for this operation included the following:

- A drill rig and support equipment mounted on a 40 foot by 40 foot Flexi-float barge system.
- Two each 40-foot long spud pipes for positioning. Positioning by two support boats.
- Location determined by GPS.
- Drilling was performed using a track mounted Mini-Sonic drill rig, equipped with a 6-inch diameter outer casing and a 4-inch o.d. (3.25-inch i.d.) inner sampling barrel

The sediment deposit was divided into 500-foot-long reaches; samples were taken along the center of the reservoir. Additional samples were acquired where the deposit widened out. Mechanical gradations, gradation/hydrometer tests, double hydrometer tests, Atterberg limit determinations, settlement column tests and organic content tests were performed on the samples. The cores were retained and samples were taken at representative locations for laboratory analysis. A detailed Sediment Investigation Report was published in 2008 (Figure 12).

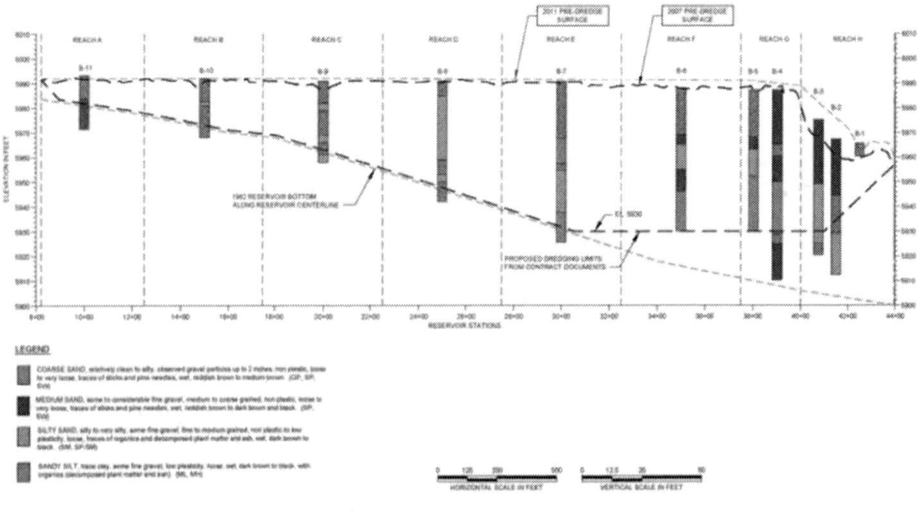

Figure 12. Sediment Sampling Profile from 2008 Sediment Investigation Report

The results of the sediment sample laboratory analysis and classification determination were summarized in a table that provided the estimated quantity of each type of sediment within each reach sampled (Figures 13, 14 and 15 and Tables 7 and 8).

Figure 13. Boxes with Sediment Cores

Reach	Borings within Reach	Volume of Sediment Type within Reach			
		Coarse Sand (cy)	Medium Sand (cy)	Silty Sand (cy)	Sandy Silt (cy)
A	B-11	3,385	4,872	0	0
B	B-10	15,697	0	1,452	0
C	B-9	32,842	0	14,755	0
D	B-8	40,480	0	58,251	0
E	B-7	50,081	0	32,553	42,569
F	B-6	41,210	23,727	21,229	38,713
G	B-4,5	12,545	20,699	5,645	23,836
H	B-1,2,3	29,482	76,385	28,142	0
Volume Totals (cy)		226,723	125,683	162,028	105,118
Percent of Total Volume (%)		36.6	20.2	26.2	17.0

Total Estimated Sediment Volume to be Dredged (cy) 619,552

Table 7. Sediment Quantities by Reach

Table 8. Sediment Particle Size Ranges

Material Type	Range of d_{50} values	Range of d_{85} values
Coarse Sand	1.5 mm – 4.2 mm	4.1 mm – 11 mm
Medium Sand	0.33 mm – 0.6 mm	0.67 mm – 2.8 mm
Silty Sand	0.08 mm – 0.23 mm	0.17 mm – 0.46 mm
Sandy Silt	0.04 mm – 0.06 mm	0.08 mm – 0.17 mm

Figure 14 and 15. Sedimentation Investigation, October 2007

Analytical tests were also performed on five samples to determine the dissolved metals and total dissolved solids concentrations. Dissolved iron and manganese were found in varying concentrations in the sediment pore water.

Lessons Learned:

- Core sediment classifications should be performed at intervals that do not exceed 10 feet of vertical core length.
- Bulk specific gravity tests should be performed to determine the average density of in-situ sediment so volumetric production rates can be correlated to slurry process densities.
- Sufficient chemical testing of sediment samples should be performed to accurately assess the characteristics of deposits that will be removed during dredging. Even non-hazardous but naturally occurring compounds will become concentrated during the slurry dewatering process.
- The Udden-Wentworth (Wentworth) soil classification system should be used for test results so descriptions are consistent with dredging industry and USACOE standard practices.

HYDRAULIC DREDGING CONTRACT PROCUREMENT

Following the completion of the sediment investigation phase, a hydraulic dredging contract was initiated to dispose of the sediment downstream from the dam and reservoir. This option was selected due to the availability of a 15 acre utility owned parcel at the lower end of the canyon below the reservoir and the lack of suitable disposal areas upstream.

After an industry wide search for qualified contractor teams, a group of five firms were pre-qualified based on experience, interest, and financial strength. Requests for Proposal (RFP) were then issued to the pre-qualified bidders in late 2009. The contract was structured such that the contractor was responsible for designing and operating the sediment removal system based on the sediment investigation study included with the RFP. The planned quantity for removal was 625,000 cubic yards. One year (2010) was allowed for equipment procurement and fabrication and one season (May 2011 through November 2011) was allowed for dredging. At an elevation of 6000 feet, ice on the reservoir limited the dredging operating season to seven months.

For various reasons, four of the pre-qualified teams dropped out and only one proposal was received in February of 2010. After a review of the proposal and incorporation of Value Engineering ideas, a contract was awarded in April 2010 for $30,046,500.

Lessons Learned:

- The RFP specified the use of electrically powered booster pumps on the land side portion of the transport system and identified a limited electrical power supply in the area. Diesel powered booster pumps with hospital grade exhaust systems were ultimately used due to the high cost and limited availability of electrically powered booster pumps. The diesel pumps were sufficiently quiet for the environmentally sensitive location of the project. The use of diesel pumps addressed power distribution system capacity limitations along the pipeline alignment.

- The pre-bid conference was held when the reservoir was iced over so bidders were unable to traverse the entire site. An earlier pre-proposal conference when the site was completely accessible would have been advantageous.

HYDRAULIC DREDGING PROCESS CONFIGURATION

The selected contractor's planned process for dredging is summarized as follows:

- Sediment excavation by diesel powered hydraulic cutterhead dredge to a depth of 60 feet on the reservoir
- Sediment transport across the reservoir via a floating 18-inch (nominal O.D.) HDPE pipeline
- Three each, floating diesel powered booster pumps located on the reservoir to maintain pipeline pressure
- Operation of a separation plant immediately downstream of the reservoir to remove sediment particles exceeding 1/4-inch from the sediment slurry
- Five each, land side diesel powered booster pumps located on the 6 1/2-mile long access road to maintain pipeline pressure
- Operation of a slurry separation plant at the terminus of the pipeline to remove remaining sediment particles from the sediment slurry
- Use of a settling pond to allow a separation of remaining fine particulates from the process effluent. A dosage of 10 mg/l of polymer based flocculant was originally anticipated to achieve the desired quality of the finished process water.
- Return of the finished process water to the raw water supply being conveyed to the downstream water treatment plant was the final step in the planned system.

The contractor's proposed system for hydraulic dredging consisted of the following elements (Figures 16, 17, and 18):

- 16-Inch Hydraulic Cutterhead Dredge
  - Dredge Supply 7650D Diesel Marlin - 104-feet long
  - 950 HP (Nominal) Prime Mover, 130 HP Cutter and Auxiliary
  - 40-Inch Impeller
  - 300+ Cubic Yards /Hour
- 14-Inch Hydraulic Cutterhead Backup Dredge
- Booster Pumps
  - Three Floating Diesel Pumps
  - Five Land Side Diesel Pumps,
  - 950 HP Each (Nominal), 7,650 GPM
- Slurry Pipeline
  - 18-Inch Nominal Diameter HDPE Pipeline, SDR 11 by WL Plastics, I.D. 14.531"

- 9,500 LF of Floating Pipeline
    - 36,000 LF of Land Side Pipeline
- Intermediate Scalping Plan
    - Remove 1/4-Inch+ Material from the Slurry immediately after reservoir conveyance
- Del Total Clean Desanding Plant at Filter Beds
- Water Treatment in Settling Beds, Polymer Chemical for Flocculation, and Bag Filters for Particulate Removal
- Discharge into Treatment Plant Raw Water Supply Conduit

Figure 16. Planned System Layout on the Reservoir

Figures 17 and 18. Land Side Booster Pump and Sand Slurry Separation Plant Layout

SLURRY TRANSPORT SYSTEM DESIGN

The requirements for the project allowed for one season to complete the removal of 625,000 cubic yards of sediment. The production rate for the sediment excavation and transport system was bound by the seven month window of opportunity on the reservoir for ice free operations. The system capacity configured for the project was as follows:

- Planned Operating Days in Season – 117 Days
- Hours Operated per Day – 20 Hours/Day
- Average Production Rate Required/Operating Hour – 267 Cubic Yards/Hour

The design criteria developed for the slurry conveyance system is provided in Table 9:

- SG – Specific Gravity
- WGCS – Well Graded Coarse Sand
- WGMS – Well Graded Medium Sand

Table 9. Original System Design Parameters from Contractor

Pre-Screen Design

Pipeline	Friction	GPM	SG	CV	Cubic Yards	Material
18 SDR11	0.06429	7250	1.219	0.1325	445	WGCS

Pre-Screen Expected

Pipeline	Friction	GPM	SG	CV	Production	Material
18 SDR11	0.0571	7250	1.175	0.106	346	WGCS

Station	Elevation	Friction	Elevation	Gain	Head	Pressure
0	6000	0.0	0.0	0	0.0	0
0	6000	0.0	0.0	180	180.0	78
500	6000	-28.6	0.0	0	151.5	66

Post-Screen Design

Pipeline	Friction	GPM	SG	CV	Cubic Yards	Material
18 SDR11	0.04933	7000	1.173	0.105	340	WGMS

Station	Elevation	Friction	Elevation	Gain	Head	Pressure
14500	5825	0.0	0.0	0	0.0	0
14500	5825	0.0	0.0	230	230.0	100
15000	5800	-24.7	29.3	0	234.7	102

The system design documents provided by the contractor showed sufficient capacity to meet the required production rate for the available calendar window. Booster pump calculations showed pumps were sufficiently sized to achieve desired system pressures and overcome anticipated line losses.

Lessons Learned:

- Hindsight tells us the D_{50} particle size assumed for the project system design was based on the Wentworth soil classification index, not the slightly different USCS classification index that was used for the sediment laboratory reports. A careful examination of the particle size distribution upon which friction line losses are calculated should be made early in the system design effort.
- The distribution of the various bands of coarse and finer grained sediment in the reservoir deposits should be considered when developing planned dredging production rates. Deposits near the surface in the upstream reaches are likely to be dredged initially, contain larger average particle size deposits and require much lower slurry pipeline specific gravities to maintain system pressures. System designs should anticipate the largest particle size deposits as controlling the system operation.

DREDGING EQUIPMENT MOBILIZATION

The original schedule for the project provided for one year for fabrication and delivery of the new booster pumps, the desanding and dewatering equipment, and the new 16-inch hydraulic dredge. In July 2010 crews mobilized to the site and began assembling the slurry transport pipeline on the access road to the dam. Fusing of the HDPE pipeline continued as materials were delivered and staged on the 6 1/2-mile access road towards the dam. Pipe was also assembled into 500-foot-long prefabricated lengths for use on the water the following season.

The normal weather pattern for the area results in a freeze of the reservoir surface starting in November and an ice off date of late March to early April. While the slurry pipeline was being assembled, land side booster stations were also installed. A 2-inch HDPE conduit was installed adjacent to the land side slurry pipeline for fiber optic cable which provided the telecommunications backbone for the process control system operated on the dredge (Figure 19).

Booster pumps, the main dredge, desanding, and slurry dewatering equipment were also fabricated at various manufacturers during this one year mobilization period prior to the ice off date in 2011. A backup dredge in the contractor's fleet was assembled on the water during the 2010 season and was moored at the reservoir over the 2010 to 2011 winter (Figure 20).

Figure 19 and 20. Fiber-optic HDPE Duct Installation and Assembled 14" Backup Dredge

Lessons Learned:

- One year is a reasonable timeframe for the procurement of new booster pump and dredge equipment.
- A suitable crane pad should be within reach of the equipment assembly area.
- Bubblers or agitators are required for winter storage of floating equipment on reservoirs that ice over.
- Aquatic nuisance inspections should be planned for floating equipment coming from areas where invasive species are prevalent.

WATER TREATMENT PROCESS DEVELOPMENT

The originally specified parameters for discharge water requirements from the slurry solids separation process were selected to minimize the impact on the water treatment plant facilities because the process return water was to be reused in the raw water supply. The parameters were as follows:

- Turbidity, < 50 NTU
- Manganese , < 0.05 mg/l
- < Drinking Water Maximum Contaminant Levels per U.S. EPA

In April 2010 following the ice retreat, the contractor initiated a sampling program of the sediment and pore water for the development of a treatability study. Thirteen 5-gallon buckets of sediment and six 5-gallon buckets of site water were used. The purpose of the study was to evaluate the estimated dosages of chemicals utilized in settling ponds located at the end of the process. The settled water would be reintroduced into the raw water supply conduits adjacent to the slurry separation plant site. This "closed system" would keep water used for dredging within the raw water supply system and minimize losses to the overall system water supply.

Sediment samples were processed through a #200 U.S. Standard Sieve with site water to remove particles greater than 75 micrometers. This process provided a diluted desanded sample closely representative of the overflow obtained from dredged sediment after being processed through a hydrocyclone. Sediment was diluted with site water to the specified percent feed solids; to simulate what is obtained through dredging operations. Sediment was diluted from 1% to 5% solids by weight slurry, with the majority of testing on 1% slurry material.

Polymer screening tests were performed using 100-mL samples of <#200 slurry. Various polymers of different type, molecular weight, and charge were added to the samples incrementally to determine the approximate dose range for a particular polymer, and to determine which polymers appeared to better enhance the flocculation and settling of suspended solid particles

In June of 2010, results of the treatability study were received. A sample of <#200 slurry was tested for total and dissolved Fe and Mn. The total Fe was 722 mg/L, dissolved Fe was 0.262 mg/L. This indicated that most of the iron was not in the dissolved form and could be easily removed through settling and filtration. The total Mn of the <#200 slurry was 59.1 mg/L, with 5.95 mg/L of dissolved. A sample of <#200 was treated with a cationic solution polymer, allowed to settle and sampled for total and dissolved Fe and Mn. The total Fe was reduced to 1.11 mg/L, with <0.083 mg/L of dissolved Fe. This further supported the analysis that the iron could be removed without chemical treatment. The total Mn was reduced to 6.95 mg/L through polymer treatment. However there was 6.08 mg/L of dissolved Mn remaining, indicating that chemical treatment was necessary to remove the dissolved Mn.

Further testing was performed using Potassium Permanganate (KMnO4) as an oxidizing agent in order to determine the optimum process for removing dissolved Mn. In one run, the Mn concentration was lowered to 0.033mg/L, with dissolved Mn at 0.019mg/L. This sample was treated with 20 ppm of polymer 7212 and allowed to settle. The supernatant was then decanted and treated with 10 ppm KMnO4. A rust colored precipitant was formed once the sample was treated with KMnO4. The sample was filtered through a 1μm filter to remove particulates and then bottled for analysis.

As a result of the treatability study, a further evaluation of the disposition of the process water and solid precipitate was made by both Denver Water and the contractor (Figure 21). New information was also discovered that indicated regulatory requirements for the disposal of precipitate would result in an expensive and unanticipated disposal cost of solids created in the water finishing process. As a result, the following actions were taken:

- Denver Water elected to eliminate treatment of the slurry dewatering process effluent and redirected the point of discharge from the planned treatment plant supply conduit to a point of discharge in an adjacent irrigation canal. No chemicals were added to the process.
- Therefore, the risk of excess dissolved manganese entering the raw water supply to the treatment plant was eliminated and the higher chemical treatment cost

avoided. In addition, industrial use permits would have been necessary had the addition of chemicals in the process stream occurred.

- Remnant suspended solids in the slurry process discharge were settled in the irrigation canal immediately downstream of the point of discharge and were later removed by machinery to the sediment stockpile areas.

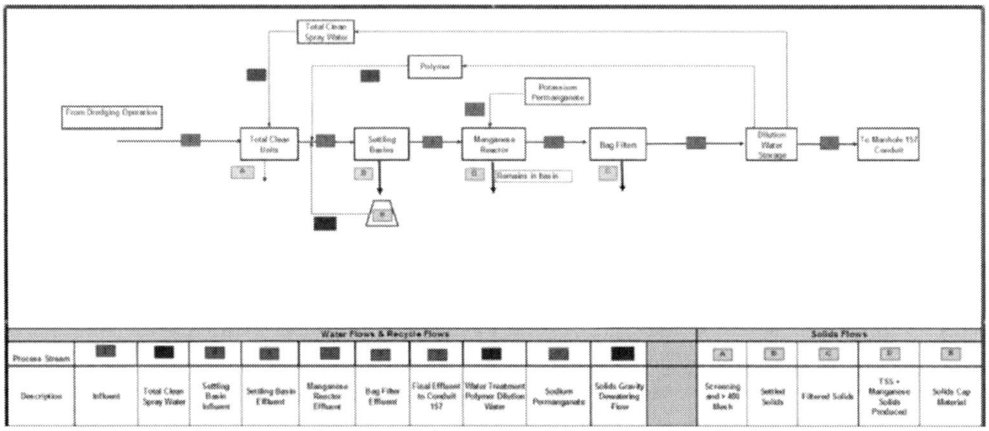

Figure 21. Treatment Process Schematic with KMnO4 – Not Used

Lessons Learned:

- The relatively low specific gravity of the slurry requires a large volume of water for dilution. Sufficient water needs to be allocated to the project in advance to support the planned volume of sediment removal.
- Water treatment may require substantial quantities of chemicals to treat process effluent. The amounts of these chemicals should be anticipated in the budgeting for the project. Disposal of precipitates may require costly landfill disposal.
- It is necessary to anticipate what process byproducts will be developed in the solids separation process that may fall under the jurisdiction of regulatory agencies. In addition, regulatory oversight and approval of the process anticipated for slurry dewatering should be incorporated into the project timeline and the project budget.

DREDGING STARTUP

Assembly of the equipment necessary to start dredging on the reservoir started in late March, 2011 following the seasonal retreat of ice. The main dredge, floating booster pumps and floating pipeline required transport, assembly and testing before production dredging could commence (Figures 22 – 27). This process took approximately five weeks with the first pumping of water taking place the first week of May. The complex system took approximately ten days to troubleshoot before solids were introduced into the system.

Figure 22. Main Dredge Assembly, April 2011

Figure 23. Marlin Dredge Operating June 2011

Figure 24. Floating Booster Pump

Figure 25. 1/4-Inch Scalping Plant and Booster Pumps

Figure 26. Land Side Booster Pump

Figure 27. Slurry Dewatering Plant – Original Configuration

Lessons Learned:

- Five weeks is a reasonable timeframe to mobilize a dredge; and booster pumps and assemble a pipeline on a reservoir over a 1 1/2 mile reach.
- Public access should not be allowed in the vicinity of the area being utilized for system assembly and operation.

- If sediment stockpiles are located in the floodplain, a Letter of Map Revision process for local agencies and FEMA should be anticipated.

DREDGING PROCESS SUMMARY

From the beginning of dredging operations, a shortfall in the production rate of sediment transport was observed (Table 10). Not only was the sustainable concentration of sediment in the slurry below the target specific gravity, the lost time due to system breakdown and plugging was significant.

Table 10. Dredging Process Summary

Period (2011)	Monthly Planned Dredged Quantity (cubic yards)	Cumulative Planned Dredged Quantity - CY(cubic yards)	Monthly Actual Dredged Quantity - CY(cubic yards)	Cumulative Actual Dredged Quantity - CY(cubic yards)
April	113,147	113,147		
May	107,759	220,905	5,431	5,431
June	118,534	339,440	22,199	27,630
July	107,759	447,198	42,492	70,122
August	123,922	571,121	48,672	118,794
September	53,879	625,000	37,741	156,534
October			45,931	202,466
November			25,534	228,000

- May 12 through June 23, System Startup. Numerous difficulties were encountered with system controls and balancing the individual legs of the system, clearing plugged lines, clearing accumulated sand at the scalper, weather delays due to lightning, and equipment breakdowns which caused numerous disruptions to dredging operations. 26 days of operation, 104 hours operated, uptime 23%, average production 105 cubic yards/hour.
- June 24 through Sept. 9, Clearing plugged lines, clearing accumulated sand at the scalper, weather delays due to lightning, and equipment breakdowns caused numerous disruptions to dredging operations. 643 hours operated, uptime 51%, average production 189 cubic yards/hour.
- Sept. 10 through Nov. 11, Clearing plugged lines, clearing accumulated sand at the scalper, equipment breakdowns, and burst pipeline incidents caused numerous

disruptions to dredging operations. 564 hours operated, uptime 47%, average production 148 cubic yards/hour.

In November 2011, the season ended when ice began forming at the inlet of the reservoir. Overnight freezing temperatures caused problems with seal water supplies. Denver Water chose not to renew dredging efforts in 2012 due to a number of factors including jurisdictional coordination, equipment wear and a limitation on access to the slurry pipeline right-of-way. Demobilization was initiated and the equipment was removed from the site by June 2012.

HYDRAULIC DREDGING PROCESS EVALUATION

Early in the dredging process, observations were presented by the contractor that sediment being removed from the reservoir appeared more coarse than originally anticipated. This coincided with difficulties passing the desired slurry concentration in the pipeline and plugging problems experienced at numerous locations in both the pre-screened and post-screened segments of the slurry pipeline.

An extensive investigation followed by both Denver Water and the contractor to ascertain the root cause of the issues that developed with the slurry transport system. The following information consists of the highlights of a comprehensive evaluation of the original assumptions that went into the system design and the shortcomings that were encountered.

Sediment Deposit Properties

The design of the sediment dredging and transport system assumed an overall average D_{50} particle size of 0.5 mm. This represented a composite of all layers of the targeted 60-foot layer at the top of the sediment deposit, combined in a single gradation. This did not consider the stratified character of the sediment body as deposited. The more coarse deposits are typically found in the upper layers of the sediment delta as the larger particle grains fall out in the slowing upstream pool first. The more fine grained particles that travel farther downstream into deeper water drop from suspension later in deeper water and are subsequently covered by coarser layers as the delta advances.

Initial dredging operations removed material from a 20-foot face at the top of the deposit in order to maximize production with a shallow dredge ladder angle. As a consequence, the D_{50} particle size of sediment removed from the top of the deposit was determined to be between 1.6 mm and 1.9 mm. This had a significant detrimental effect on the system which was designed to move much smaller average sized particles.

Sediment Classification

The laboratory tests conducted on sediment samples consisted of Atterberg limits· (ASTM 4318), gradation analysis (ASTM 422), and natural moisture content determinations (ASTM 02216). They were conducted to classify the soils according to the Unified Soil Classification System (USCS) (ASTM D 2487).

It became apparent during the investigation that the nomenclature used by the dredge system design team was consistent with the Udden-Wentworth (Wentworth) soil classification system, not the Unified Soil Classification System (USCS) that was used in the Sediment Investigation Report. The difference between the two systems is significant, particularly when designing slurry transport systems (Figure 28).

- Wentworth - Fines are differentiated from sands at 63 μm (ASTM No. 230 sieve); sands are differentiated from gravels and coarser materials at 2 mm (ASTM No. 10 sieve).
- USCS - Fines are differentiated from sands at 75 μm (ASTM No. 200 sieve); sands are differentiated from gravels and coarser materials at 4.75 mm (ASTM no. 4 sieve).

inches	U.S. Standard Sieve No.	mm	Unified Soil Classification System	Udden-Wentworth
		4026	boulders	boulders
		2048		
		1024		
		512		
12		300	cobbles	
10		256		cobbles
6		128		
3		75	coarse gravel	
		64		pebble gravel
		32		
1		25.4		
0.75		19	fine gravel	
		16		
0.5		12.7		
0.375		9.5		
		8		
0.25		6.35		
	4	4.76	coarse sand	
		4		granule
	10	2	medium sand	coarse sand
		1.6		
		1		
		0.5		medium sand
	40	0.425	fine sand	
		0.25		fine sand
		0.125		very fine sand
	200	0.074		
		0.0625		

[8]

Figure 28. USCS and Udden-Wentworth Systems Compared

The U. S. Army Corps of Engineers utilizes the Wentworth soil classification system as a standard for sediment evaluations[9].

[8] Chapter 3, Engineering Classification of Materials, National Engineering Handbook, Part 631, United States Department of Agriculture, January 2012

[9] Dredged Material Evaluation and Disposal Practices, Dredged Material Management Office, U.S. Army Corps of Engineers, Seattle District, November 2015

The most noticeable consequence of this nomenclature disparity was the incorporation of implied lower D_{50} particle size values in the piping line loss calculations used for the system booster pump spacing (Table 11).

Table 11. Sample Pipeline Calculation Using Wentworth Classification

Pre-Screen Expected						
Pipeline	Friction	GPM	SG	CV	Production	Material
18 SDR11	0.0571	7250	1.175	0.106	346	WGCS

Station	Elevation	Friction	Elevation	Gain	Head	Pressure
0	6000	0.0	0.0	0	0.0	0
0	6000	0.0	0.0	180	180.0	78
500	6000	-28.6	0.0	0	151.5	66

A coarse sand in Wentworth can have a D_{50} particle size of 0.50 mm to 2.00 mm; a coarse sand according to USCS can have a D_{50} particle size of 2.0 mm to 4.75 mm.

Slurry Transport Sensitivity to Particle Size

The subject of slurry flow in pipelines is quite complex and covered in detail in many publications by experts in the field. This paper does not delve into the theory but takes a broader approach to the subject and examines the factors that affect system friction losses and how to approach the subject with a basic understanding of the principals involved.

The ANSI standard for slurry pumps[10] provides an overview of rotodynamic (centrifugal) slurry pumps and the recommended criteria for their application in slurry transport systems. One of the exhibits in this publication provides the service class associated with slurry specific gravity at various D_{50} particle sizes (Figure 29).

[10] Rotodynamic Centrifugal Slurry Pumps (ANSI/HI 12.1-12.6-2016)

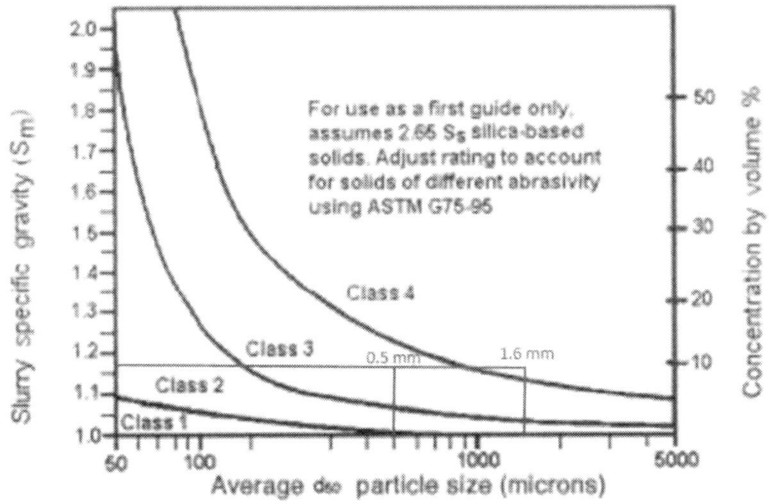

Figure 29. ANSI Service Classes for Slurry Pumping

This chart provides a rating for wear severity for centrifugal slurry pumps. The standard states that pumps operated in the Class 4 region can experience operating costs wherein the service wear parts cost approaches half of the total operating cost (capital + energy + wear) for the pump alone. The chart is annotated to show the average particle size likely encountered during the sediment removal process at Strontia as well as the design point – both within high wear zones of the standard.

Because the system design for the on-water pipeline used a lower average D_{50} particle size of 0.50 mm than the measured D_{50} of 1.6 mm to 1.9 mm, the friction losses in the pipeline were much higher than anticipated. This caused the pipeline pressures to drop faster, the average slurry velocity to slow, which then caused the larger particles to settle to the bottom of the pipeline. This subsequently led to reduced throughput and plugging. The following chart (Figure 30) illustrates the significant increase in friction losses that occurs with relatively small increases in the average slurry particle size.

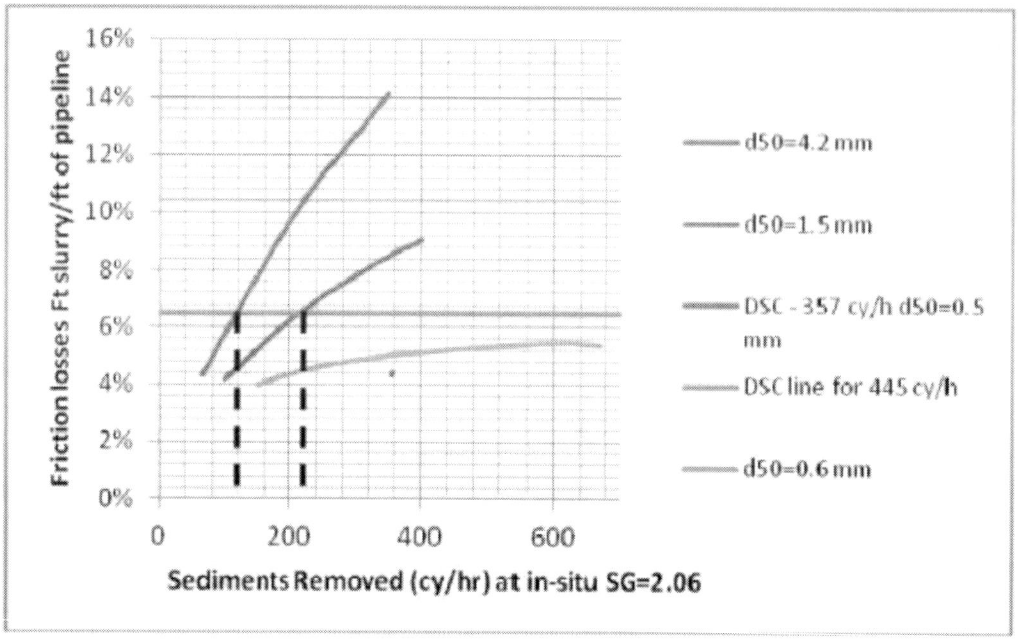

Friction Losses per unit pipe length for materials with various d_{50} at critical deposition velocity

Figure 30. Friction Losses for Various Average Particle Sizes in Slurries[11]

The same friction loss increase affected the land side pipeline. A smaller D_{50} particle size of 0.25 mm had been assumed for this segment of the system; however, screens at the scalper allowed a larger sized particle fraction though the intake. Since pipeline lengths between booster pumps were originally developed for lower line losses, the resulting reduction in system carrying capacity was significant. Again, when a slurry was introduced into the pipeline at the dredge and subsequently at the scalper with a specific gravity in excess of the capacity of the pipeline reach, the pipeline throughput was reduced and settling of slurry solids occurred. Specific failures related to the insufficient capacity included the following:

- Cavitation at booster pump intakes due to low suction side pipeline pressure.
- Plugging of the scalper due to uneven supply suction of the first land side booster pump.
- Plugging of pipelines due to the settlement of heavier particles falling below the critical deposition velocity.

Other System Issues

[11] Review and Failure Analysis of Dredging and Slurry Pipelines for the Strontia Springs Reservoir Sediment Removal Project, Baha Abulnaga, P.E., May 1, 2013

- Originally, a shaft seal water system was devised utilizing a filtered supply from the slurry pipeline at each booster pump. The filters proved incapable of providing a sufficiently clean source of shaft seal water which resulted in packing gland failures on numerous occasions. Ultimately separate cleaner reservoir and river sourced water supplies were put in place for each booster pump.
- There was insufficient pipe support on the water side floating pipeline. Heavier particles started to settle out in the floating pipeline when the specific gravity in the pipeline exceeded the capacity. Sags developed that exacerbated the sediment buildup and resulted in on-water pipeline plugs. These proved very difficult to find and clear, especially at night.
- The quartzite component of the sediment deposit was extremely abrasive. The HDPE pipeline developed numerous perforations after 100,000 cubic yards of throughput. The on-water pipeline proved most susceptible to wear as it carried the higher average sized D_{50} fraction. Abrasivity of the sediment particles is a parameter identified in the ANSI/HI publication that can affect the equipment selection. Steel pipe offers a more durable option for a high abrasivity slurry.

Lessons Learned:

- The system designer should verify that the soil classification system being used to select equipment and analyze system friction losses matches the particle size parameters that are the basis of system performance.
- The target specific gravity for the system operation should match the worst case D_{50} of the slurry likely to be encountered.
- The target specific gravity of the slurry should reflect the anticipated ANSI/HI wear class expected for the duration of system operation. A higher wear class system, though higher in production, will have a higher operating cost and more downtime for maintenance and repair.
- The allotted time frame for the project should be considered so that implied production rates match seasonal constraints, the target average specific gravity of the slurry, and a reasonable assumption about system downtime.
- The sediment body should be evaluated for the abrasivity of the planned slurry so that appropriate pipe materials are selected.
- Sufficient floats should be procured for floating pipelines to prevent sagging.
- A clean supply of water should be provided to each booster pump for the shaft seal water supply.
- The dredge operates most efficiently when the ladder can be traversed side to side at the maximum swing angle and with a 15-foot to 20-foot excavation face. Steeper ladder angles diminish production at deeper levels. If the reservoir can be lowered to optimize the ladder angle then higher production rates can be maintained. Anticipated production rates should be reduced if steep ladder angles will be required to complete dredging requirements.
- Any mid-point scalping process should be considered carefully as the introduction of an air gap in the system introduces a difficult balancing point on the supply side of the downstream booster pumps.

Cost Summary:

Table 11 provides a summary of the cost of the project as awarded for removal of 625,000 cubic yards of sediment. The project was structured with a fixed unit price basis with reimbursement for sediment removal based on the volume of sediment removed as measured by bathymetric survey. An escalation clause was included that allowed for reimbursement of diesel fuel expenses above an agreed to baseline fuel price.

Table 11. Original Contract Price Summary

Contract Element	Value
General Conditions	$2,445,015
Procurement (excluding Main Dredge)	$3,550,000
Mobilization and Equipment Setup	$5,860,929
System Operation	$16,260,556
Demobilization of Plant and Equipment	$1,930,000
Total	$30,046,500 ($48.07/CY)

The final volume of sediment removed in 2011 was 228,000 cubic yards. The final earned value for the project by the contractor was $18,571,361 ($81.45/CY). This includes a deduction agreed to for the elimination of water treatment, added pipeline construction for the alternate process effluent routing, and $702,000 for fuel escalation. Some project issues were addressed outside the original pay items but were not material to the final value of the project.

FUTURE SEDIMENTATION MITIGATION PLANS

Denver Water is investigating solutions to the continuing sediment issue within the reservoir. The options being studied include flushing the sediment through an abandoned bypass tunnel beneath the left abutment that was used during the construction of the dam, and mechanical excavation of the sediment from within the delta.

The flushing option would involve placing slide gates on both the upstream and downstream ends of the bypass tunnel and the removal of a 60-foot-long concrete plug that is located on the axis of the dam. Once the sediment has been flushed from the reservoir and into the river below, it will make its way downstream to a diversion structure located three miles below the dam. Mechanical excavation would then likely have to occur to remove the sediment from the river. The sediment would be dewatered, placed on trucks and transferred to a storage facility or sold for commercial use.

The mechanical option would involve the placement of excavators, dozers, cranes with clam-shell buckets, or a combination of such equipment into the delta area of the reservoir once the water level is lowered to expose the sediment. The sediment would be dewatered and placed on trucks or placed on a conveyer belt and transported away from the reservoir to a site more accessible to large trucks. It would then be transferred to storage sites, private or public, or a combination of private and public for permanent storage, or as with the flushing option, it could be sold for commercial use. One of the many issues with this option is that the upstream canyon leading into the reservoir is very narrow and major improvements need to be made to allow equipment and material transport into and out of the reservoir.

Regardless of the chosen method, Denver Water faces an ongoing challenge in the need to remove the sediment. Each option has many technical hurdles and many environmental and permitting issues that will need to be addressed, on local, county, state and federal levels.

CONCLUSIONS

While the dredging effort on the Strontia Springs Reservoir did not achieve the desired amount of sediment removal, a significant body of sediment was removed that extended its service life by many years. The Strontia Springs reservoir is relatively small and there are no realistic alternatives to its primary role in the water supply system for a major metropolitan area.

This project yielded useful data in the search for economic solutions to the universal issue of sediment accumulation that affects dams throughout the world. In the future, Denver Water will utilize the results of the 2011 dredging project to develop a sustainable approach to maintaining this critical piece of infrastructure.

REFERENCES

Abulnaga, Baha, P.E., Review and Failure Analysis of Dredging and Slurry Pipelines for the Strontia Springs Reservoir Sediment Removal Project, May 1, 2013

ANSI/HI, Rotodynamic Centrifugal Slurry Pumps (12.1-12.6-2016)

Borland, W. M. Foothills Project – Strontia Springs Dam, Sedimentation Analysis, March 1978

Borland, W. M and Miller, C. R., Distribution of Sediment in Large Reservoirs, ASCE Vol. 84 No. HY 2 Proc. Paper 1587 April 1958 pp. 1587-1 to 1587-18

Brune, Gunnar, Trap Efficiencies, Trans. A.G.U. Vo. 34 No. 3, June 1953

Golder Associates Inc., Basin Prioritization - Upstream Management Practices at Strontia Springs Reservoir, September, 2013

U.S. Army Corps of Engineers, Dredged Material Evaluation and Disposal Practices, Dredged Material Management Office, Seattle District, November 2015

United States Department of Agriculture, Chapter 3, Engineering Classification of Materials, National Engineering Handbook, Part 631, January 2012

DAM DECOMMISSIONING – UNIQUE APPROACH BY CONVERSION TO A LEVEE

Zachary Whitten, P.E.[1]

ABSTRACT

The Fredonia Flood Retarding Structure (FRS), a 1.8-mile long, 28-foot high earth embankment dam located in Fredonia, Arizona, is nearing the end of its 50-year design life. The Fredonia FRS is in need of rehabilitation due to severe embankment cracking, embankment erosion, and the potential for overtopping during the inflow design flood. The Arizona Department of Water Resources has declared the dam to be "Unsafe, Non-Emergency, Elevated Risk" in November 2006. The dam provides 100-year flood protection benefits to the Town of Fredonia.

Pre-design alternatives for the dam include: (1) Dam rehabilitation, (2) No Action – Sponsor's breach of the dam, and (3) Decommissioning by converting the dam to a levee. Critical to selection of a preferred alternative is the question of reusing the existing embankment and foundation for a full dam rehabilitation alternative or decommissioning to a levee alternative. Conversion of the dam to a levee is a unique and challenging decommissioning alternative. This paper will present a discussion of the dam alternatives and focus on the dam conversion to a levee given the economic, institutional, technical, and regulatory aspects of a dam versus a levee structure.

INTRODUCTION

The Fredonia Flood Retarding Structure (FRS) is an earthen flood control embankment dam constructed by the Natural Resources Conservation Service (NRCS). The structure was originally designed in 1966. Currently, the Town of Fredonia operates and maintains the dam under an operating agreement with the NRCS. The dam is located in Coconino County, Arizona, within the Town of Fredonia, Arizona (see Figure 1).

[1] Water Resources Engineer, Stantec Consulting Services, Inc. Phoenix, Arizona, zach.whitten@stantec.com

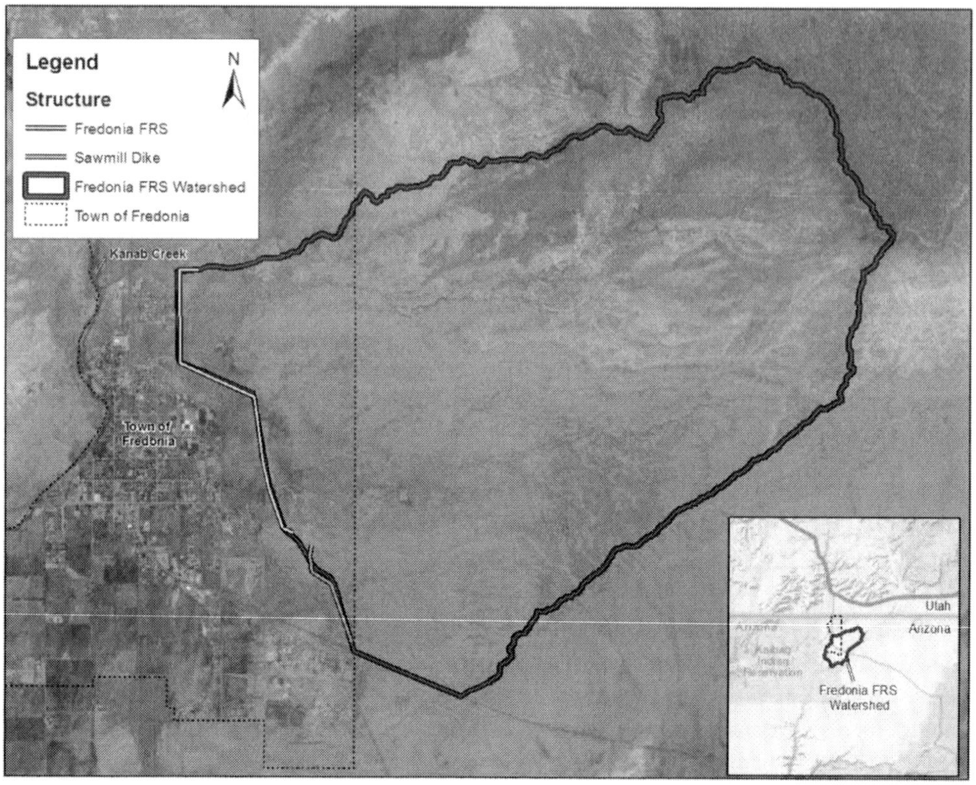

Figure 1. Project Location Map.

The structure was designed for a 50-year life span and was constructed with a 36-inch diameter principal spillway and a 450-foot wide earthen open-channel auxiliary spillway at the southern end of the dam embankment. The contributing watershed to the FRS is 7.62 square miles. The NRCS classifies the Fredonia FRS as a high hazard class structure, which means failure may result in the loss of life and serious damages to home, utilities, commercial buildings and transportation infrastructure. The Arizona Department of Water Resources (ADWR), the state's dam safety jurisdictional agency, classifies the dam as intermediate size and high hazard potential. Due to severe embankment cracking, embankment erosion and the potential for overtopping during the inflow design flood (IDF), ADWR assigned an "unsafe, non-emergency, elevated risk" condition to the Fredonia FRS.

For the reasons stated above, the Town of Fredonia and the NRCS recognized a need for the FRS to be rehabilitated or decommissioned. A pre-design "Planning Study" was conducted to determine the most practical and cost-efficient options for either rehabilitating or decommissioning the dam. The study concluded that the three more viable options were: (1) rehabilitate the dam to meet current NRCS and ADWR standards, (2) decommission the dam through a Sponsor's breach (breaching of the dam at the maximum section(s) rendering the dam a non-flood protecting structure) or (3) decommission the dam by converting the dam to a levee. After the Planning Study, the

Town of Fredonia and the NRCS contracted Stantec to conduct a more extensive study into the three alternatives. Figure 2 illustrates details of Fredonia FRS.

Figure 2. Fredonia FRS Overview Map.

EXISTING CONDITIONS

Hydrologic Conditions

Using current design procedures, the existing conditions of the dam were established to determine the level of effort needed to implement the three options. The Arizona Revised Statutes (A.R.S.) §45-1201 assigns the responsibility for supervision of the safety of dams to the Arizona Department of Water Resources (ADWR). ADWR's dam safety procedures are promulgated in the Arizona Administrative Code – Title 12, Chapter 15, Article 12. Article 12 of the Code defines the IDF as "the reservoir flood inflow magnitude selected on the basis of size and hazard potential classification for emergency spillway design requirements of a dam." The IDF requirements for the design of a high hazard potential dam for determining the spillway minimum capacity (for all dam size classifications) is defined as "for a high hazard potential dam, the applicant shall design the dam to withstand an inflow design flood that varies from 0.5 of the probable maximum flood (PMF) to the full PMF, with size increasing based on the persons at risk and potential for downstream damage." The Town of Fredonia and the NRCS selected the full PMF as the IDF of the Fredonia FRS.

ADWR's statewide Probable Maximum Precipitation Study for Arizona (PMP Study) was utilized to determine the Probable Maximum Precipitation (PMP) rainfall depth estimates and temporal distributions. The PMP Study published in 2013 developed a GIS-based tool that uses a GIS shapefile defining the drainage area of the study to compute the 6-hour Local, 72-hour Tropical and 72-hour Winter PMP rainfall depth and temporal distribution estimates. For the Fredonia FRS Watershed, the PMP rainfall depth estimates and temporal distributions are illustrated in Table 1 and Figure 3, respectively, below.

Table 1. Rainfall Depth Estimates.

Storm	Rainfall Estimate (in)
6-hour Local PMP	8.30
72-hour Tropical PMP	5.42
72-hour Winter PMP	5.42
100-yr, 24-hour*	2.91

*Estimate from NOAA Atlas 14 Point Precipitation Estimate

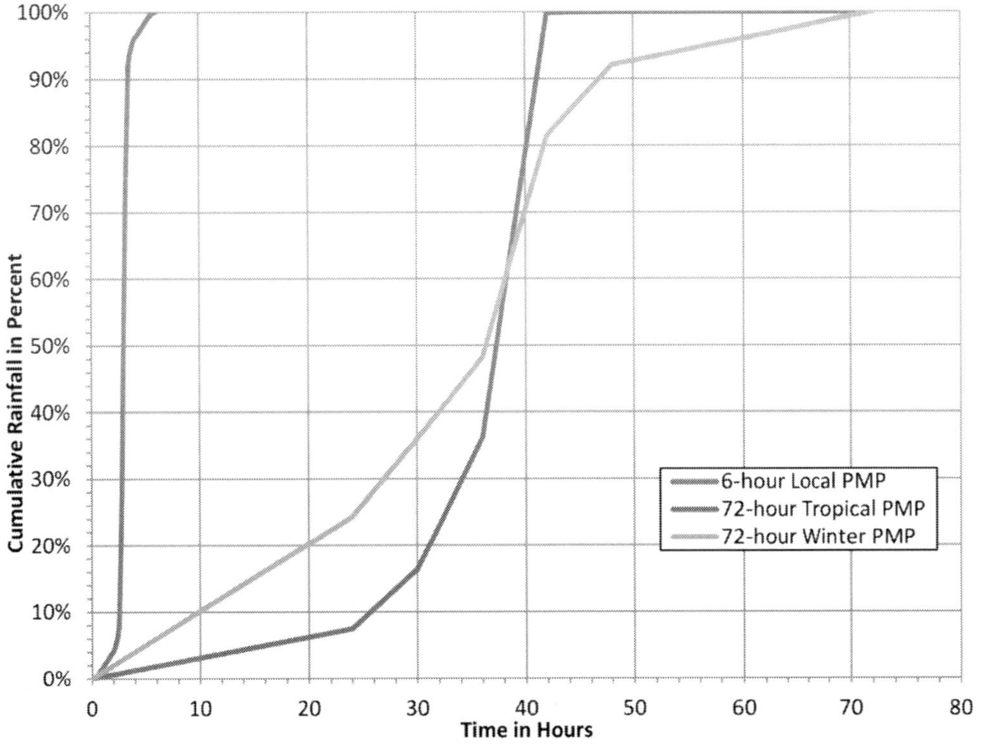

Figure 3. Rainfall Temporal Distribution Estimates.

The hydrology of the Fredonia FRS was conducted using NRCS methodology within NRCS's SITES computer program. The hydrology for the Fredonia FRS watershed is unique because there is not one major river or wash system draining into the impoundment area which is common for most dams. The Fredonia FRS watershed is defined by two separate washes: Landing Strip Wash (3.1-square mile watershed) at the northern portion of the dam and Cemetery Wash (2.7-square mile watershed) at the southern portion (washes are dry creek beds that experience temporary flows during flooding events). Due to the unique nature of two washes, the FRS watershed was split into multiple subbasins for the hydrologic analysis. Splitting the watershed into multiple subbasins will allow for a better understanding of the expected peak discharges and volume that will be entering the impoundment area at multiple locations across the 1.8 mile embankment. Figure 4 below illustrates the subbasin delineation for the Fredonia FRS hydrology. Table 2 below illustrates the peak discharge and volume generated by each of the three PMP storms as well as the 100-year, 24-hour rainfall event.

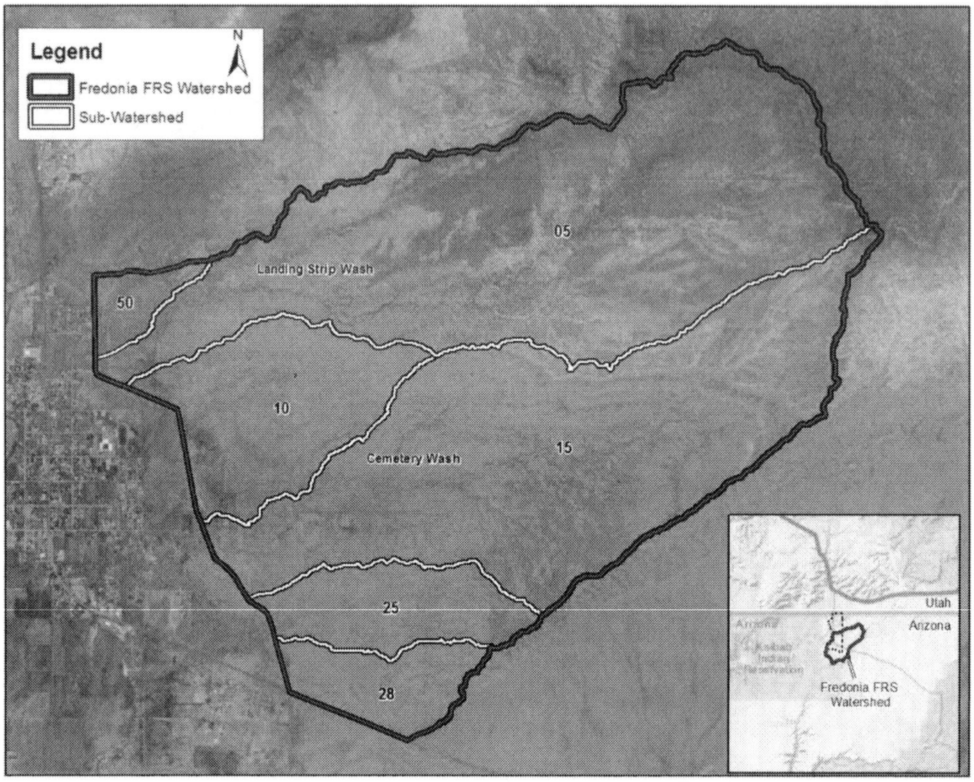

Figure 4. Subbasin Delineation Map.

Table 2. Peak Discharges for Storm Events.

Subbasin	Storm Event Peak Discharge in Cubic Feet per Second				Storm Event Volume in Acre-Feet			
	100-yr, 24-hr	6-hr Local PMP	72-hr Tropical PMP	72-hr Winter PMP	100-yr, 24-hr	6-hr Local PMP	72-hr Tropical PMP	72-hr Winter PMP
50	80	520	70	25	8	50	45	28
05	1,900	11,500	1,600	550	182	1,105	1,018	645
10	780	3860	430	150	52	302	279	178
15	1,770	10,270	1,380	480	166	967	892	569
25	330	1860	250	90	31	176	162	104
28	310	1550	180	60	21	122	113	72

Hydraulic Conditions

Also due to the unique nature of the watershed having multiple contributing washes into the flood pool, it was determined that level pool routing using HEC-1 or HEC-RAS computer programs would not be used for the hydraulic routing of the inflows through the impoundment. Instead, a two-dimensional flow analysis using the hydrodynamic flood routing program FLO-2D was utilized. FLO-2D allows the user to input flow hydrographs into the impoundment from multiple locations. This feature allows the model to more accurately simulate the dynamic situation occurring within the Fredonia FRS flood pool. Figure 5 below illustrates the results of the multiple PMP's and 100-year routing through the Fredonia FRS flood pool using FLO-2D. The figure is a profile view of the maximum water surface elevation at any time during the simulation along the FRS for each storm. The north side of the dam is on the left side of the profile and the south side is on the right.

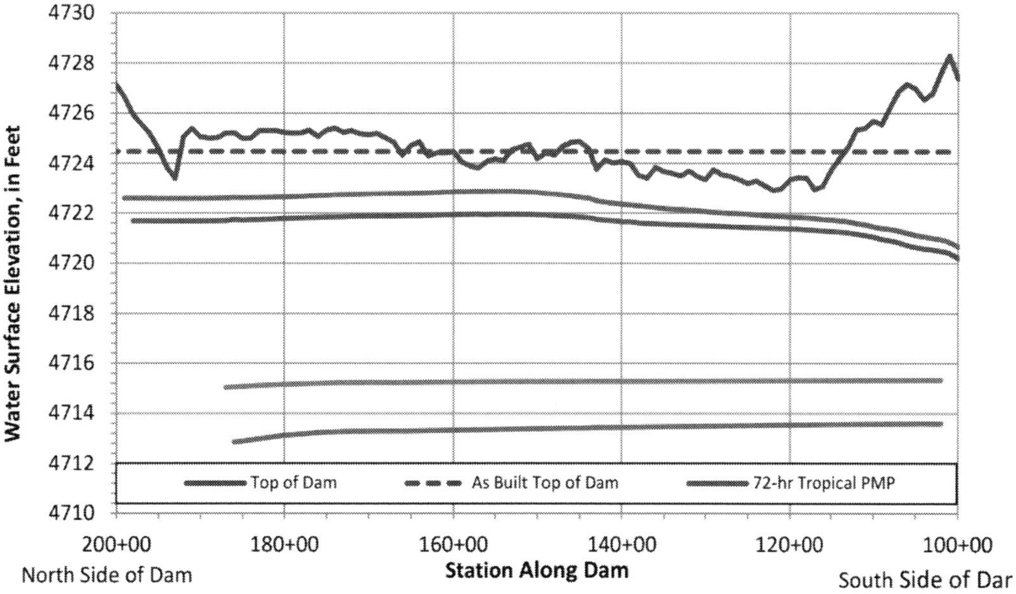

Figure 5. Existing Conditions Profile.

As Figure 5 indicates, the 6-hour PMP is the controlling storm for the Fredonia FRS and therefore, the IDF. Figure 5 also shows that the 6-hour PMP would require a dam raise to provide adequate freeboard required by ADWR. Therefore, rehabilitation of the FRS and maintaining the FRS as a dam, would require a geotechnical solution for the cracking and erosion issues, and a dam raise to safely pass the PMF with the required freeboard. It should also be noted that Figure 5 indicates that the FRS embankment has experienced non-uniform settling. From station 145+00 to station 115+00, the FRS has settled an of average 1 foot more than the northern end of the embankment. Figure 5 also show that the embankment at the end of the dam was raised. However, there is no documentation indicating the reasons why the dam was embankment was raised in this area.

ALTERNATIVE ANALYSIS

From the Planning Study, three alternatives were indicated as the most viable: (1) rehabilitate the dam to meet current NRCS and ADWR standards, (2) decommission the dam through a Sponsor's breach or (3) decommission the dam by converting the dam to a levee.

Dam Rehabilitation (1)

As described above, rehabilitation of the FRS would not only include both geotechnical solutions and a dam raise. However, due to multiple factors such as the estimated sediment volume being lowered by 52% compared to the 1966 estimate, it was determined that the crest of the auxiliary spillway could be without adversely affecting the 100-year flood protection provided by the structure or the Principal Spillway Hydrograph routing (unique storm event that the NRCS requires a dam route through the principal spillway without causing the auxiliary spillway to be inundated). According to the analysis, the auxiliary spillway could be lowered by more than 2 feet (from 4717.2 to 4715.0 ft NAVD 88). To estimate the effect of lowering the auxiliary spillway crest, the FLO-2D model was adjusted and the 6-hour PMP storm was reanalyzed. The results of the analysis can be seen in Figure 6 below. The solid line indicates the maximum water surface profile before the spillway was lowered and the dotted line illustrates the maximum water surface profile with the lowered spillway.

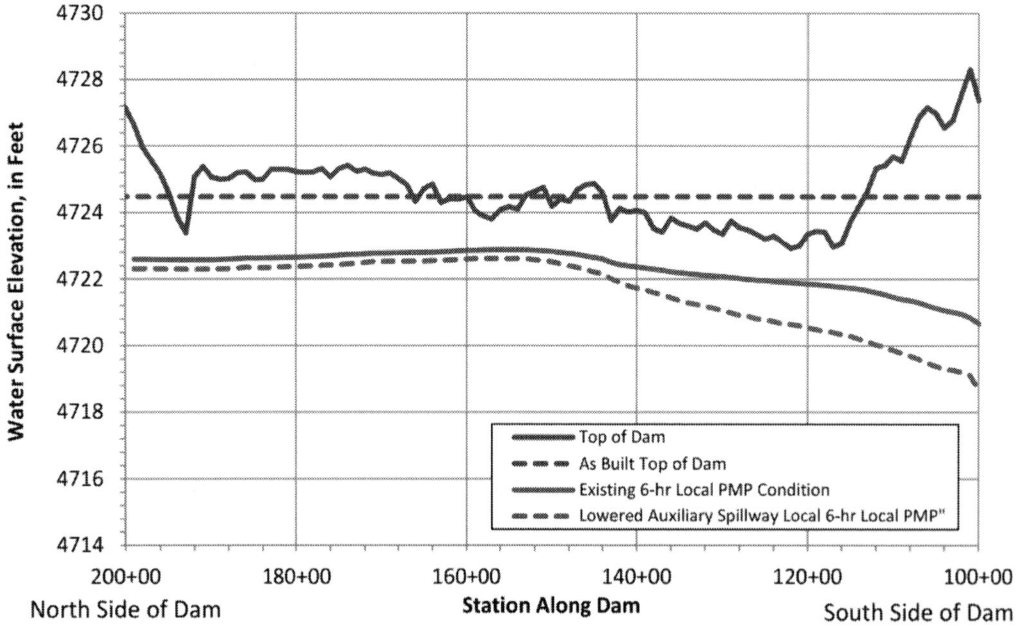

Figure 6. Lowered Auxiliary Spillway Results Comparison.

As Figure 6 indicates, the lowering of the auxiliary spillway crest has a significant impact to the southern portion of the dam, but a minimal impact at the northern portion. The variable maximum water surface elevation is caused by of the unique nature of the watershed having multiple washes affecting the flood pool and the lack of gradient in either direction within the flood pool. The flows from Cemetery Wash at the southern end of the dam are impacted by the lower spillway, but the flows from Landing Strip Wash at the northern portion are not affected. Due to the flat gradient within the flood pool, there is little to no head pushing flow from the north to the south along the 1.8-mile embankment. Therefore, the volume generated from Landing Strip Wash flows into the impoundment and builds up behind the embankment and "sits" rather than flow south toward the auxiliary spillway. The result is that the dam crest would still need to be raised at the northern and central portion of the dam. The estimated cost for the Option 1 Dam Rehabilitation is $18,140,601.

Sponsors Breach (2)

Option 2 was to breach the dam at the maximum section(s) to decommission the dam. The breaches will result in the dam no longer providing any flood protection for the Town of Fredonia. Due to the unique nature of the FRS watershed, it was determined that multiple breaches would need to be constructed to decommission the dam: one immediately downstream of Landing Strip Wash and another immediately downstream of Cemetery Wash. FLO-2D analysis indicated that a 30-ft bottom width, 6H: 1V side slope breaches would meet ADWR requirements for a sponsor's breach (less than 5ft of flow depth through the breach during a 100-year flooding event). It was estimated that the cost to construct Option 2 would be $452,328.

Conversion to a Levee (3)

The third alternative for the Fredonia FRS was to decommission the dam and convert it to a levee. This option would be completed by either 1) creating a breach at the northern end of the dam near the principal spillway and allowing water to flow to into Lost Springs Wash then eventually into Kanab Creek downstream, or 2) breaching the dam to the south near the auxiliary spillway and constructing a channel to convey flood water from the FRS to Kanab Creek. Since Kanab Creek is located approximately two miles from the southern end of the dam, thus requiring a long channel to convey flood water to Kanab Creek, breaching the dam at the southern portion of the dam was not considered for the detailed analysis. For the breach at the northern end of the dam, multiple breach widths were analyzed to determine what dimensions would pass the 100-year storm while conforming to ADWR requirements for a breach. The results indicated that the size of the breach did not have a large impact on the hydraulics of the impoundment area. The depth of water behind the embankment was similar when a 50-ft wide breach was in place compared to a 100-ft wide breach. Therefore, a 50-ft wide bottom width, 6H:1V side slope breach was selected. Figure 7 illustrates of the results of the levee alternative modeling. The solid line indicates the maximum water surface elevation of the 100-year, 24-hour storm with the existing conditions in place. The dotted line illustrates the maximum water surface elevation with a 50-ft wide breach near the principal spillway.

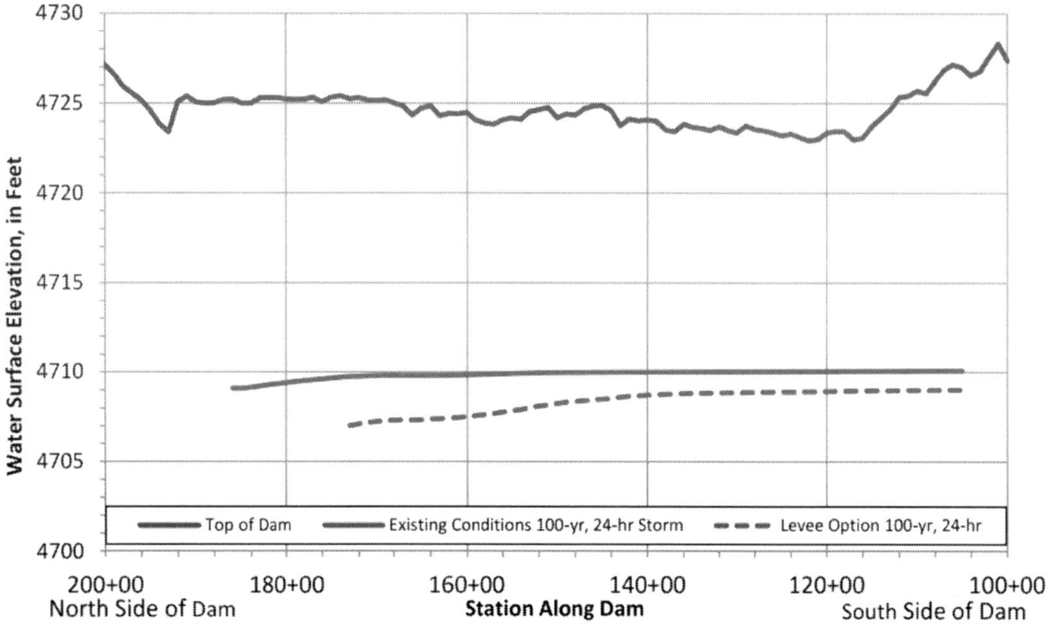

Figure 7. Levee Alternative Results Comparison.

The results of the breaching the dam near the principal spillway, are very similar to the results of lowering of the auxiliary spillway observed in the dam rehabilitation analysis, only flipped. The northern portion of the dam experiences a significant decrease in water surface elevation, while the southern portion experiences a minimal impact. The result is caused by the same condition: the lack of gradient along the flood pool resulting in a lack of head pushing flow from Cemetery Wash north to the breach. During the 100-yr, 24-hr simulation, the southern portion of the embankment was still inundated approximately 4 days after the storm has stopped producing runoff. As indicated in Figure 7 above, the maximum water surface elevation at the southern portion of is only 1 foot lower when Fredonia is a levee compared to when Fredonia is dam.

Part of the dam to levee conversion would include lowering the embankment from its current height to a height that would meet ADWR and FEMA criteria for levee freeboard, but not higher. As discussed in the existing conditions, the current embankment has experienced erosion and cracking that may cause the embankment to fail in the event of a large flooding event. Therefore, to reduce the risk of failure during the PMP and causing significant damage downstream, the current embankment will be removed down to the required elevation and the remaining embankment will be rehabilitated to reduce failure risk. The total cost for Option 3, conversion of the dam to a levee is $8,465,348.

CONCLUSION / PREFERRED ALTERNATIVE

The analysis indicated that the cost of Option 2, Sponsor Breach, was the lowest option that met ADWR and NRCS criteria. However, the Town of Fredonia would lose the flood protection that is currently provided by the FRS and would become subject to flooding during storm events and potentially be required to obtain flood insurance. Therefore, the Town determined that Option 2 was not a viable option for their priorities. Option 1, though the most robust option and provided the most flood protection for the Town of Fredonia, was at a cost that would be difficult for the Town to provide sufficient funds to complete construction then maintain the structure. Thus the most viable option that met the priorities and budget for the Town of Fredonia and the NRCS was Option 3, conversion of the dam to levee. This option allows the Town to continue to be protected against flooding during the 100-year event, but at a lower cost compared to Option 1.

At the time of this paper, the final design for the conversion from dam to levee had not been completed and the funding sources had not been finalized. It is expected that the construction for Option 3 will be completed by 2019.

THE PROSPECTS FOR CLIMATE SCENARIOS IN HYDROPOWER OPERATIONS

Yulia Zakrevskaya, P.Eng, M.Eng.[1]
David Huard, Ph.D.[2]

ABSTRACT

While short-term weather forecasts are commonly used by hydropower producers and dam operators, there is less experience using climate information at longer time scales. Paleo-climate studies using tree ring chronologies have recently been used to assess the risk of decadal droughts, and century scale projections from global and regional climate models are now being applied to assess future flood risks. Long range climate information is especially relevant to dam owners and manager due to their long life expectancy. One challenge with those applications is that the original data produced by climate scientists is often not directly usable by dam operators, civil engineers or resource planners. The recent field of climate services, which employs climate scientists, engineers and geographers, strives to translate the raw climate information and convert it into usable, relevant information. One of the main tasks of climate services is to assign confidence levels to climate products, identifying actionable information from research-grade hypotheses. This is especially challenging given the large uncertainties affecting climate projections. This paper presents examples of how long-range raw climate information can be used to inform planning and decision-making in the hydropower sector. In particular, the interactions between changes in flow regime and load seasonality are addressed to understand how they might affect dam storage requirements throughout the year. Results for a fictive power station in northern Québec suggest that considering a reduced demand for winter heating and an earlier spring flood, firm energy should increase and storage requirements should decrease to account for the expected climate conditions. This type of information can be useful to plan adaptation measures to cope with evolving conditions and ensure the safety and economic sustainability of infrastructures.

INTRODUCTION

Climate change is a planetary energy budget imbalance (von Schuckmann et al. 2016). The Earth continually intercepts solar radiation; a fraction is reflected back to space and the remainder is absorbed which heats the surface and the atmosphere. The Earth also emits blackbody radiation in the form of infrared. If more energy is intercepted than emitted, the earth's temperature increases until the outgoing radiation is balanced with the incoming solar radiation. If the Earth's emits more energy into space than it intercepts, the earth will cool until a new equilibrium is established. While this is an

[1]SNC-Lavalin Inc, 1801 McGill College 12th floor, Montréal, Qc, H3A 2N4, 514-393-8000, yulia.zakrevskaya@snclavalin.com
[2]Ouranos Inc, 550 Sherbrooke W., Montréal, Qc, H3A 1B9, 514-282-6464, huard.david@ouranos.ca

overly simplified version of the real system, it illustrates the fundamentals of climate change.

The current "global warming" phenomenon can be understood as the consequence of modifications to the radiative properties of the upper atmosphere. The so-called greenhouse gases (GHG) are opaque to some of the outgoing infrared frequencies emitted by the Earth while being largely transparent to incoming solar frequencies (Feldman et al. 2015). This asymmetry reduces the Earth's "cooling efficiency", in that some of the outgoing radiation that originally escaped to space is now absorbed and re-emitted back to Earth. The energy budget becomes positive and temperatures will increase until a new radiative equilibrium is reached.

Although this radiative equilibrium model was described by Svante Arrhenius in 1896, the influence of CO_2 as a control knob on the climate was largely dismissed until the fifties (Weart 2011). The arguments made against the climate influence of CO_2 would be refuted with the advent of quantum mechanics, high-resolution atmospheric absorption spectra and post-war research on the properties of the upper atmosphere (Plass 1956).

From there, the lure of weather control as a weapon in the cold war drove substantial funding into atmospheric physics, weather prediction models and climate models. These climate models became increasingly sophisticated. Starting from grid resolutions of 700 km and neglecting oceans and topography, climate models gradually improved in their realism. Current models now have resolutions of 100-200 km and describe ocean thermodynamics, clouds, sea ice, snow and glaciers, vegetation, carbon cycle, atmospheric chemistry, and many other processes thought to have an influence on climate.

Despite considering these additional physical processes, the basic conclusion that adding GHG to the atmosphere results in increased temperature still holds. The precise value of this climate sensitivity is up to debate, with estimates from models and paleoclimate records ranging from 1.5°C to about 6°C following a doubling of CO_2. Despite this uncertainty, there is no evidence in over 100 years of research that the atmosphere could cool as a result of increased GHG concentrations (Maslin & Austin 2012). Barring cataclysmic events such as major volcanic eruptions or a nuclear war, which would shoot aerosols in the upper atmosphere and reduce solar insolation, the Earth will warm with rising GHG concentrations.

Rising temperatures bring an array of related effects, hence the more general term climate change. Most regions of the world are already experiencing some of these changes, including higher mean temperatures (Ji et al. 2014), increased intense precipitations (Villarini et al. 2013), changes in long-term water availability (Rice et al. 2015) and rising sea levels (Sallenger et al. 2012). Because the changes in average conditions occur relatively slowly and gradually, they are often hard to perceive; more noticeable are the rare extreme events now occurring more frequently and that preoccupy civil security, insurers and reinsurers (Ward & Ranger 2010).

For water resource managers, however, these long-term changes in average water availability have important consequences. Systematic under- or over-allocation of water resources can have important economic and environmental impacts, hence the need to assess both the historical, current and future conditions of water systems. However, the translation of climate science into usable information is challenging. Climate models as well as observational records have inherent biases and uncertainties. Therefore, making sense of the various sources of sometimes conflicting information, passing judgement on which results are mature and which are "research-grade", and synthesizing and communicating a nuanced interpretation of the climate state-of-the-art, requires considerable training and experience.

To effectively bring rigorous climate information into the hands of managers and decision-makers is the goal of the emerging field of climate services. Climate service providers are usually regional organizations with a mandate to communicate with potential users of climate information such as resource managers and planners (Huard et al. 2014). Their job description is to understand the context in which those managers operate, document their weather and climate needs and work to either find existing climate products that meet those needs, or collaborate with academic, private and governmental scientists to improve or develop products and translate them into a form and language that speaks directly to clients.

Hydropower producers and dam managers have shown an early interest in climate services and often contributed to develop some of these services. Reservoir levels integrate the water balance equation over large regions, and since their historical levels are closely monitored, the small, gradual changes implied by climate change can be perceived early. For hydropower producers, even minute changes in inflows have direct economic consequences, and it's thus important to understand whether those changes are simply random fluctuations or long-term trends. In this paper, we present examples of work performed in collaboration with hydropower and dam managers to include future climate information into water resources management decisions.

BASIS FOR CLIMATE IMPACT STUDIES

One of the most popular types of climate services requested by water resource managers are impact studies. Typically, such studies target one specific watershed and describe the impacts climate change would have on variables of interest: liquid and solid precipitation, evapotranspiration, temperature, streamflow, aquifer storage, etc. These analyses are usually based on climate scenarios, defined as hypothetical yet plausible descriptions of future climate conditions. Climate scenarios are often constructed by combining historical records with future climate projections from climate models. One challenging aspect of analyzing impacts based on climate scenarios lies in the interpretation of the various sources of uncertainties that come with climate model projections. Nonetheless, the results can help water managers evaluate the plausible impacts of climate change on their infrastructure and operations and to identify options to maintain acceptable levels of service out of water infrastructures.

The climate scenarios used in impact analysis involve three broad categories of uncertainties : GHG emissions, structural model errors and natural variability (Hawkins 2010). For example, assume that an investor is interested in buying a water infrastructure asset, and wants to know the annual precipitation over a watershed in 2040. The first hypothesis a climate scientists would make is that this value will be distributed symmetrically around a climatological mean. The challenge is, of course, to estimate the shape of this distribution, as no existing tool, now and in the foreseeable future, can predict the exact value for 2040. The spread around the climatological mean is how we define the uncertainty due to natural variability.

A second hypothesis would be that GHG emissions up to 2040 would have an influence on the location and shape of the distribution, that is, climate change could decrease or increase the mean, or modify the spread of the distribution with respect to the historical climate. Here two sources of uncertainties are combined: the amount of GHG that will be emitted up to 2040, and the climate sensitivity to GHGs. The question of emission trajectories is squarely out of the realm of climate science and has more to do with policy, technology and economics. In fact, for climate modelers, emissions are a boundary condition – it's an input in the carbon cycle component of the model, which transforms emissions into concentrations, which then impact the radiative properties of the atmosphere. To understand the effect of GHG on the climate system, different hypothetical "emission scenarios" are thus evaluated: sudden release of GHGs, 1% increase per year, and a range of more realistic emission pathways. For example, modeling experiments from the latest round of the Coordinated Model Intercomparison Project (CMIP5) use four different scenarios of Radiative Concentration Pathways (RCP) RCP2.6, RCP4.5, RCP6.0 and RCP8.5, bracketing rapidly decreasing emissions (RCP2.6) to accelerating emissions (RCP8.5). Despite their differences, these four scenarios display similar climate impacts until 2040; only later in the century do meaningful differences appear.

The remaining source of uncertainty is model uncertainty. Climate models are simplified representations of the climate system; some of these simplifications are attributed to our incomplete understanding of the physical processes, and others are due to practical computing constraints, such as limited spatial resolution. Due to these limitations, some phenomena cannot be explicitly modeled (for example, cloud droplet formation) and are instead parameterized in bulk. Different modeling groups describe the same phenomena using different algorithms, some models include more processes than others, and in the end, each model provides a unique, but not fully independent (Knutti et al. 2013) representation of the climate system. Due to these model differences, future scenarios can be significantly different, especially at the regional scale. While each model can be described by its "climate sensitivity" - its global temperature response to a doubling of CO_2 concentrations - it is worthwhile to mention that there is no "climate sensitivity" parameter in models. Rather, the climate response to rising GHG concentrations is an emergent property of the model and the processes it describes.

To account for model differences, emission pathway hypotheses, and natural climate fluctuations, climate scenarios are usually ensembles built from multiple realizations from dozens of different models driven by different emission scenarios. This is done to make sure that the spectrum of possible climate future is, at least, partially covered, and to avoid the false confidence that could arise from analyzing a smaller subset. Climate analyses also usually present results in the form of long-term averages to reduce the influence of natural variability on the results.

EXAMPLES OF IMPACT STUDIES FOR HYDROPOWER DAM MANAGEMENT AND PLANNING

The signal to noise ratio of climate change varies considerably from region to region and depends as well on the variable considered. For example, the signal is much stronger for temperatures than precipitation, and among temperatures, the signal is stronger in the Arctic. The Icelandic power utility Landsvirkjun has thus been confronted with the impacts of increasing temperatures relatively early on. Indeed, its hydropower facilities are glacier fed, thus streamflow is directly tied to temperature instead of precipitation. Despite some initial skepticism in the organization, the utility started to take into account historical temperature changes in their operations. For instance, instead of using the entire historical records (50+ years) to evaluate production over the next years, they started by only using the 20 more recent years to better reflect changing conditions, knowing that the older records would be unlikely cold by today's standards. Building on experience, they later corrected their entire series to preserve the value of their historical observations but take into account the long-term changes to mean climate conditions over time. In a sense, they are using their record as the most reliable description of the natural variability, and the long term trend to describe the evolution of mean conditions. These corrected inflow series were used in the design of planned new assets. Indeed, turbine capacity for the Bùrfell and Hvammur plants were adjusted to take advantage of increased flows expected in 15 to 20 years (Ouranos 2016a).

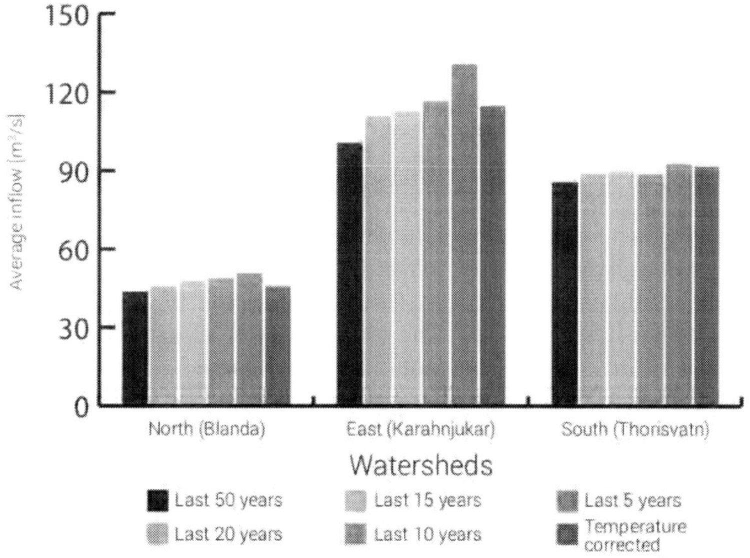

Figure 1: Average inflow in three Icelandic watersheds computed over different periods. Figure from Ouranos (2016).

Closer to home, the provincial power utility New-Brunswick Power conducted an economic and environmental assessment of its Mactaquac dam. Located just upstream from St-John, the power station has been operated since 1968 and requires major infrastructure work due to concrete deterioration resulting from Alkali-Aggregate Reaction. The utility is facing three options: repowering the dam by constructing a new powerhouse and spillway, retaining the headpond by building a new spillway while decommissioning the powerhouse, and restoring the river to its natural pre-dam state (Énergie NB Power 2016). To enlighten this decision, the utility commissioned climate scenarios to evaluate the impact of increased temperature on power demand for HVAC, future streamflow, environmental impacts on aquatic habitats and the future Probable Maximum Flood (PMF). Although the final decision is still pending, this is one of few integrated studies where climate change is considered both for generation, demand and environmental impacts.

Future PMF studies have also been conducted for dams in Manitoba, Québec and Ontario by other hydropower utilities (Ouranos 2015a). The results are, however, somewhat underwhelming due to model uncertainty and natural variability resulting in a large variability. The engineering approach used to estimate the PMF is based on running a hydrological model with the Probable Maximum Precipitation (PMP). This PMP is itself estimated by "replaying" the largest historical storms while assuming that the atmosphere at that time was almost saturated in water vapor. Adapting this approach to climate models yields results that are extremely variable from one simulation to the next due to the natural variability in the simulations' extremes. In some cases, future PMPs varied from -50% to + 50% compared to the reference value, leaving managers with little guidance on how to select an appropriate PMF for each dam site. This is not only an issue for dam owners, but also for regulatory agencies that will eventually require dams to be reasonably future proof. This type of results raise questions whether regulation should be based on system's risk analysis, better able to cope with uncertainties, or traditional standards-based design criteria.

On the production side, multiple studies have been conducted across Canada to evaluate the impacts of climate change on inflows (Boucher & Leconte 2013). While results vary with watersheds, some general conclusions can be made. One is higher winter flows. Precipitation remains liquid for longer due to the delayed onset of freezing temperature. The spring freshet also occurs earlier in the year, which often means that flows reach lower levels at the end of summer. Some results suggest that precipitation events would become more intense during the fall, with stronger flows, but the evidence is conflicting. In general, most Canadian watersheds should expect higher precipitation. Whether this will translate to higher flows overall is not always clear, due to stronger evapotranspiration. In northern areas such as James Bay, the streamflow projections consistently suggest increased annual streamflow of the order of 10-15% annually around 2050 (Ouranos 2015b).

In a jurisdiction like Québec where 95% of electricity is generated from hydropower, one major risk to consider is the occurrence of multiple years with below average inflows. Although plurennial storage is available, a few consecutive years of lower than normal flows would seriously hamper Hydro-Québec's capacity to meet demand, especially during the winter heating peak. To ensure power reliability, the provincial regulator requires Hydro-Québec to maintain reservoirs at levels that would allow the company to provide power in the event of the two- and four-year worst drought in the utility's recorded history. One question is whether that recorded history is still a reliable guide to evaluate current and future risks. The issue is that climate models still have considerable difficulty reproducing the amplitude and autocorrelation of the naturally occurring variability. To work around this, Ault et al. (2014) have proposed combining the long term trends on average rainfall that are generated by models with statistical models of natural variability fitted from long paleoclimate records. The hope is that by analyzing millennial time series, the natural variability reflects enough different conditions to encompass future conditions. Such a joint paleoclimate & model projection project has been launched for the region at the head of Québec's main hydropower watersheds.

On the demand side, again the picture is pretty clear. Increased temperatures imply lower heating demand in winter and higher summer demand for cooling. In winter peaking provinces such as Québec and Manitoba, everything else being equal, per capita demand decreases overall as a result of climate change, at least over this century. In fact, years ago Hydro-Québec started to account for climate change in its annual demand forecasts. Indeed, using the historical record, the utility consistently overestimated demand due to the rising temperature trend. By factoring in temperature trends, they were able to generate more accurate demand forecasts (Ouranos 2016b).

The studies described above about future production and demand are usually conducted independently. One issue we wanted to explore was whether or not the interactions between both issues had some non-intuitive features that could prove interesting to operators, and what they meant for storage.

INTERACTIONS BETWEEN CLIMATE CHANGE IMPACTS ON POWER DEMAND AND GENERATION

The evaluation of the impact of climate change on hydropower production is carried out using a fictional, but typical hydro system. This system is comprised of a single dam with a reservoir that was developed based on a real hydropower plant located on a river in the north of Quebec (Canada) between the 50th and 55th parallels. The hydro system has an installed capacity of 380 MW, a design discharge of 770 m^3/s, a live storage capacity of 4000 hm^3 and a maximum drawdown of 15 m.

For the purpose of this study, it is assumed that the power plant would be used for hydropower generation only and firm energy output would be guaranteed 100% of the time. Secondary energy will be produced only when the reservoir is full. Furthermore, a single hydro system was considered, i.e. there is no flow regulation upstream or downstream. From the demand standpoint, the simulated hydropower plant must meet the power demand by itself in accordance with the expected generation pattern. Finally, hydropower production is evaluated on a daily basis using a daily average power demand which allows calculating the energy produced every day.

The hydrometric data used for the study consists of a historic natural observed flow series from 1971 to 2000 in the watershed and 82 future climate runoff projections for 2041 to 2070 and 2071 to 2100, allowing to cover a wide range of possible future climates. The scenarios were derived from the historic data series using nine climate models combined with four RCP emission scenarios (RCP2.6, RCP4.5, RCP6.0 and RCP8.5). Figure 2 shows daily runoff hydrographs for a typical year from 1971 to 2000 and for emission scenarios prepared for the period 2041 to 2070. Each line represents a different scenario for the chosen year. The dark bold line shows the hydrograph from 1988 from the baseline scenario while the lighter lines show the hydrographs of the 82 climate projections for 2058. The graph illustrates the important variability of inflows during the spring flood season, especially in April during the recovery period. The flood starts approximately two weeks earlier for climate projections compared to the baseline scenario (a robust feature of most projections) while the peak discharge is higher (there's less confidence that this is a robust trend). Looking at the fall and summer of multiple years, daily runoff from climate projections is generally higher than that of the baseline scenario. The average annual inflows of climate projections for the period of 2041 to 2070 varies from -4% to +24% compared to the baseline scenario with an average of 728 m^3/s. For the period of 2071 to 2100, the average increases by about 12% compared to the previous period. These tendencies are consistent with the results of studies conducted across northern Canada.

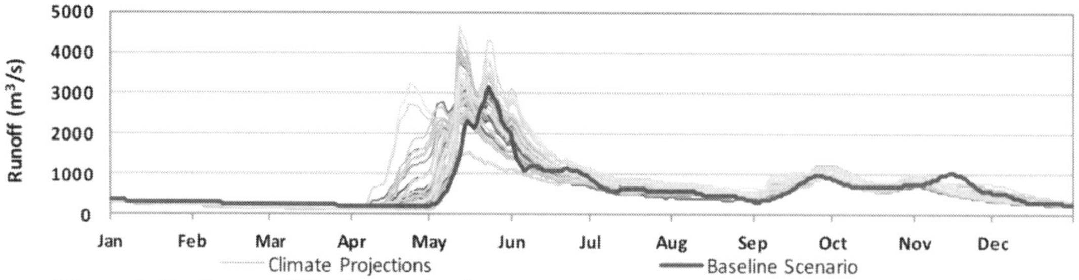

Figure 2: Daily runoff hydrographs for year 2058 according to multiple climate simulations, compared to the 1998 time series.

Two sets of analyzes are performed to examine the interactions between climate change impacts on power demand and generation. The first analysis evaluates the impact of climate change on hydropower generation. Using the typical hydro system, a simulation is run for the baseline scenario of natural inflows (1971 to 2000) representative of the current runoff conditions. Then, hydropower production is evaluated for climate scenarios representative of future runoff conditions disrupted by climate change for 2041 to 2070 and 2071 to 2100 periods. As illustrated in Figure 3, the results show that firm and total energy will increase in the future for both horizons with the increase of GHG emissions represented by RCP emission scenarios.

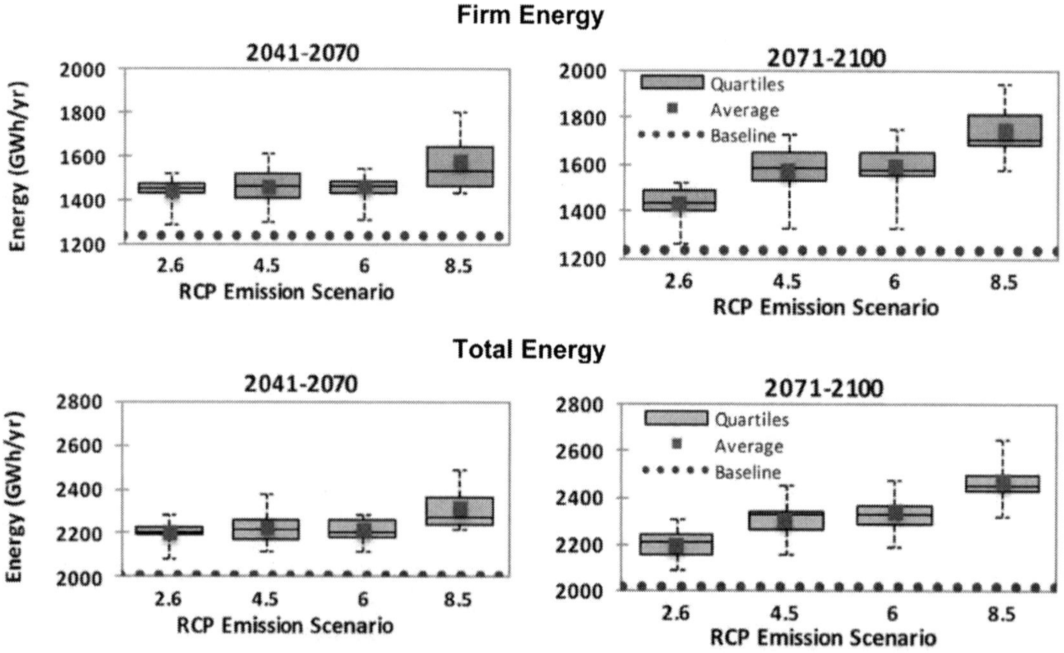

Figure 3: Firm and total energy for different RCP emission scenarios for the 2041-2070 and 2071-2100 periods.

The analysis of future inflows shows that average inflow runoff will increase in the future though it does not fully explain the increase in hydropower generation (particularly for the firm energy).

As shown in Figure 4, the duration of the critical period decreases with increasing emissions. The critical period, which typically spans from November to May, is defined as the lapse of time between the moment the reservoir is full and the moment the reservoir is empty. As the filling of the reservoir is governed by the starting date of the spring flood, an early flood would shorten the duration of the critical period. This period defines the maximum firm energy that can be generated by the system for a year and depending of the reservoir live storage. A roughly constant volume of water (live storage plus inflows which are generally negligible during winter time) processed over a shorter period of time will yield higher energy generation during this period and by extension the rest of the year. In other words, the decrease of the duration of the critical period will have little impact on the available volume of water during this period but a larger impact on the duration of the period over which this volume of water is processed.

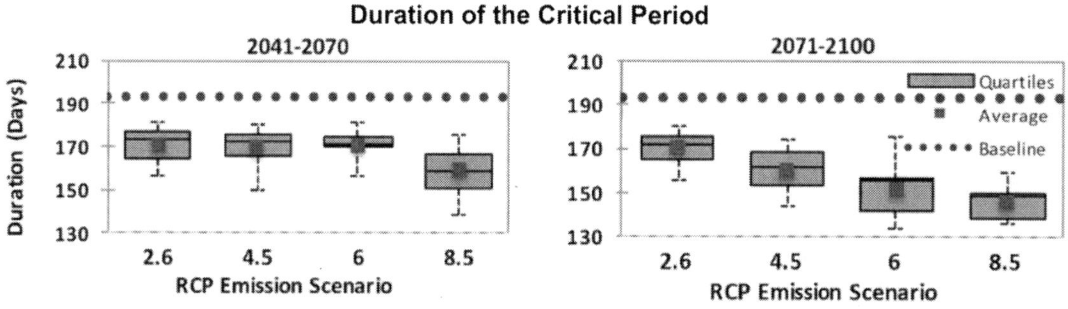

Figure 4: Duration of the critical period for RCP emission scenarios for the 2041–2070 and 2071–2100 periods.

A second analysis was performed to evaluate the impact of the variation of the power demand on power generation considering climate change. To do so, the previous simulations performed with the historical demand were repeated using a future demand pattern. The historical power demand pattern is typical of southern Québec, while the future pattern is typical of New England's climate. The historical demand is characterized by a strong demand in the winter mainly for heating while the future demand presents a decrease of consumption in the winter for heating and an increase of consumption in the summer for air conditioning all due to rising temperatures. The percent power demand on the y-axis represents the percentage of annual firm energy to be produced every month.

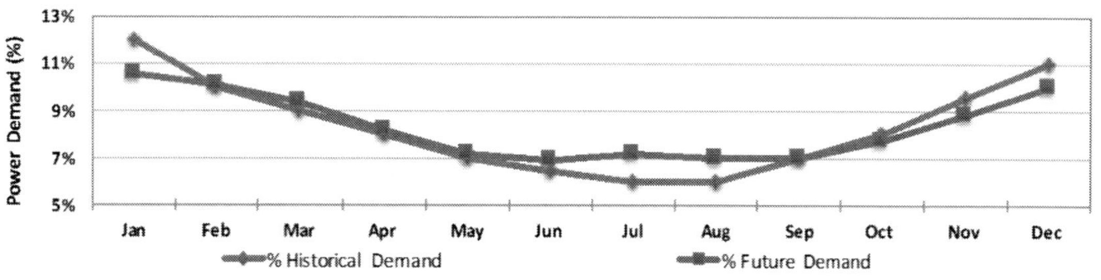

Figure 5: Monthly percent of historical and future power demand patterns in Québec.

As illustrated on the next figure, the simulation results show that faced with a lower demand in the winter (critical period) and a higher demand in the summer, the firm energy generated by the hydro system increases. Additionally, the generation pattern flattens to follow the shape of the future power demand pattern, as it is assumed firm energy is guaranteed 100% of the time.

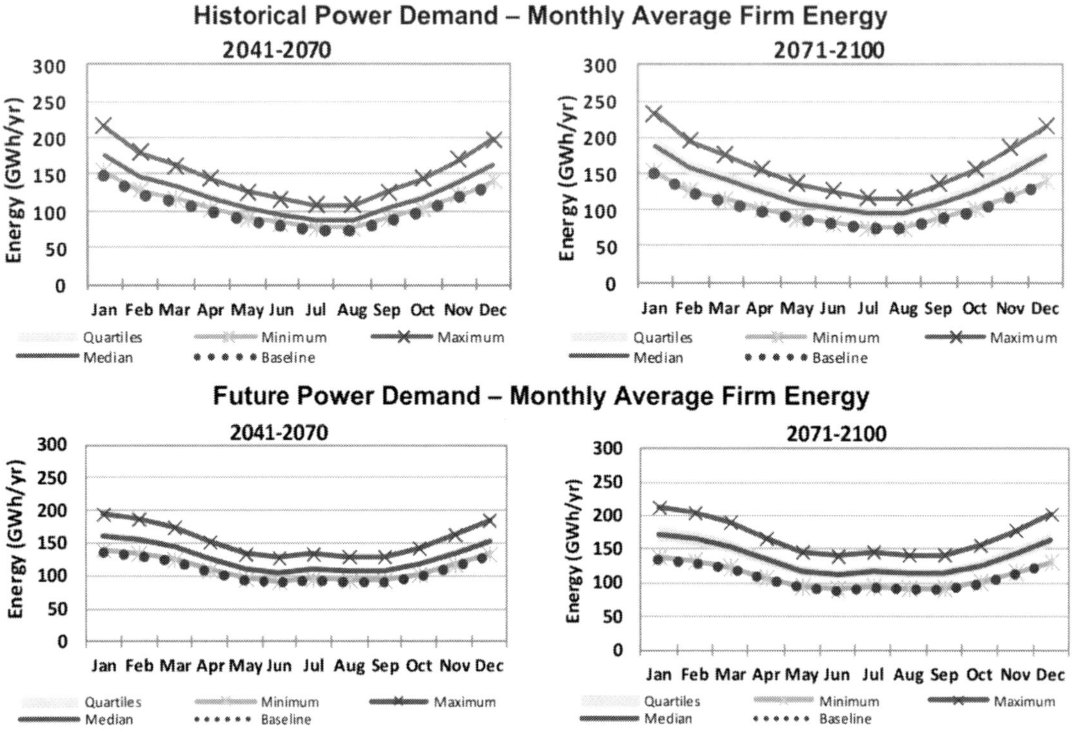

Figure 6: Historical and future power demand patterns - monthly average firm energy over the 2041–2070 and 2071–2100 periods.

Looking at the mean annual results, it is interesting to note that the firm energy of the baseline scenario and climate projections for the future power demand increases by a constant 4% compared to the historical demand. This constant variation of firm energy from one inflow series to the other when the power demand pattern is modified is due to the available volume of water during the critical period from November to May (volume stored in the reservoir plus inflows during the period). Assuming the critical period for an inflow series has the same duration for both simulations run with two different power demand patterns, the total energy generated during the critical period will be the same but the demand will be less restrictive during this period under the future demand pattern than the historical demand. For the rest of the year, generation is not limited by the availability of water so additional power can be generated. Overall, the shift of demand from winter to summer allows for higher generation. The 4% annual increase in firm energy is actually a result of 4% reduction in power demand between the historical and future power demand patterns during the critical period where water availability is limited.

LIVE STORAGE SENSITIVIY ANALYSIS

To gain further understanding of the interaction between future production, future demand and storage in the context of climate change, previous simulations were repeated with a live storage varying from 25% to 200% of the initial 4000 hm^3 reservoir capacity. The minimum and maximum operation levels remained the same. The inflow series have been evaluated for both historical and future power demand patterns.

Figure 7 presents the simulation results for all 83 scenarios of historical and future demand as well as both simulation periods (2041 to 2070 or 2071 to 2100). This figure shows that firm, secondary and total energy follow the same trend, i.e. as the live storage increases or decreases, firm and total energy vary almost linearly for both historical and future demands. During the critical period, when water availability is limited, there's a larger volume of water stored available for power generation which in turn translates into higher levels of energy output throughout the year. In parallel, secondary energy decreases with the increase of storage capacity improving inflows regulation and secondary energy is transformed into firm energy.

Energy generated under future demand is greater than energy generated using historical demand because, as explained earlier, for a critical period of equal duration, the future demand represents a smaller percentage of the total annual demand to be satisfied making it less restrictive. Also, as discussed earlier, the difference in firm energy between both demands is actually once again equal to 4%.

A larger storage implies a better regularization of inflows since water can be stored instead of being spilled. Eventually, as shown in the present case study for smaller reservoirs, energy generation is limited by the installed capacity. For reservoirs with larger storage, energy generation grows until inflows are regulated at 100%, i.e. when all the outflow is used to generate energy. The critical period lengthens when live storage is increased and shortens when it is decreased since the volume of water available during this period increases or decreases, respectively.

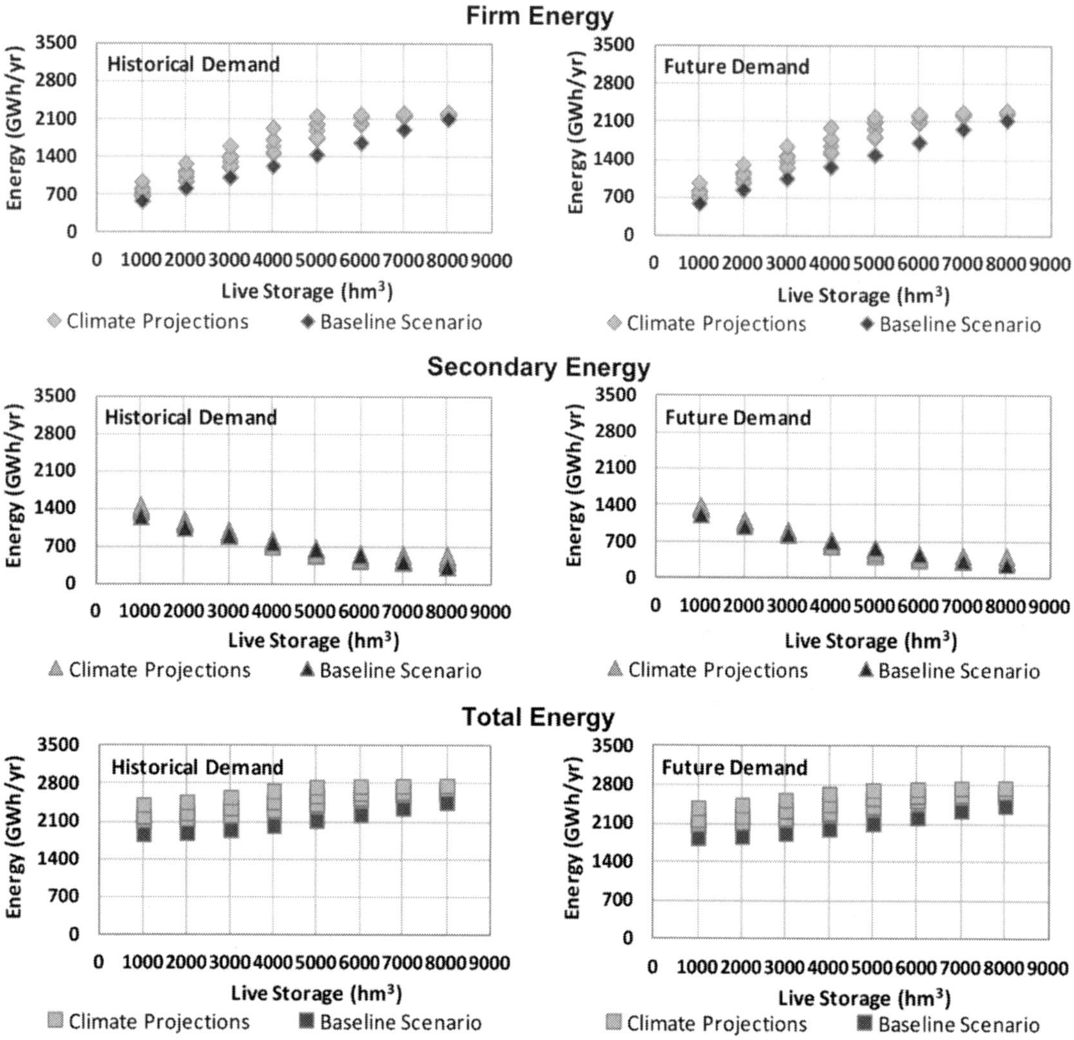

Figure 7: Historical demand and future demand - Firm, secondary and total energy for a varying live storage independent of the period.

CONCLUSION

The operation of dams and reservoirs are influenced by climate change impacts due to their long life expectancy. Whether these are changes to inflow and demand patterns, storage requirements, extreme events or natural variability, climate science can provide some information about those impacts based on theoretical arguments, empirical evidence or modeling experiments. In Northern Quebec, climate change will lead to an increase of the firm and total energy production due to the increase of natural inflows and the expected change in the demand pattern. In other areas, it will lead to a decrease of the energy production mainly due to a reduction of the natural inflows.

From a scientific standpoint, some climate change impacts are fairly certain, such as higher temperatures, higher evapotranspiration rates and more frequent intense rainfall events. On the other hand, other climatic variables that are pivotal to the operation and planning of dams, are still subject to considerable uncertainties and scientific consensus has not been reached. The question for water resource managers is how to take stock of the various sources of information, along with their varying levels of confidence to make good, risk-informed, decisions.

Many companies whose operations are either more climate-sensitive, hit by unexpected events or had reached high level of confidence in the projections for their watershed, have already taken action and put in place adaptive measures. Adaptation measures are highly diverse, some targeting structural modifications, others operational practices, and others have sought changes to regulatory regimes to help water managers cope with new or evolving conditions. Owners and investors also have concerns about the economic sustainability of water infrastructure. Large institutional funds are increasingly requiring climate change scoping in their due diligence studies. In a society that is more and more risk averse, and in which dams play such a critical role in terms of stability and economic development, the integration of climate information into dam and water resource management can only be expected to become mainstream.

ACKNOWLEDGEMENTS

The authors would like to express their thanks for the technical support of Nathalie Thiémonge, Catherine Guay and Michel Tremblay.

REFERENCES

Ault, T.R. et al., 2014. Assessing the risk of persistent drought using climate model simulations and paleoclimate data. *Journal of Climate*. Available at: http://dx.doi.org/10.1175/JCLI-D-12-00282.1.

Boucher, M.-A. & Leconte, R., 2013. Changements climatiques et production hydroélectrique canadienne: où en sommes-nous? *Canadian Water Resources Journal*, 38(3), pp.196–209. Available at: http://www.tandfonline.com/doi/abs/10.1080/07011784.2013.818297 [Accessed January

21, 2014].

Énergie NB Power, 2016. *Considering the future of Mactaquac*, Available at: http://search.proquest.com/docview/228498214?accountid=14723%5Cnhttp://cf5pm8sz21.search.serialssolutions.com/?&genre=article&sid=ProQ:&atitle=Considering+the+future+of+the+profession&title=Civil+Engineering&issn=08857024&date=2002-11-01&volume=72&issue=1.

Feldman, D.R. et al., 2015. Observational determination of surface radiative forcing by CO2 from 2000 to 2010. *Nature*, 519(7543), pp.339–43. Available at: http://www.ncbi.nlm.nih.gov/pubmed/25731165.

Hawkins, E., 2010. Our evolving climate : communicating the effects of climate variability. *Weather*.

Huard, D. et al., 2014. A Decade of Climate Scenarios – The Ouranos Consortium Modus Operandi. *Bulletin of the American Meteorological Society*, (August), p.140116121450005. Available at: http://journals.ametsoc.org/doi/abs/10.1175/BAMS-D-12-00163.1.

Ji, F. et al., 2014. Evolution of land surface air temperature trend. *Nature Climate Change*, 4(6), pp.462–466. Available at: http://www.nature.com/doifinder/10.1038/nclimate2223 [Accessed October 14, 2016].

Knutti, R., Masson, D. & Gettelman, A., 2013. Climate model genealogy: Generation CMIP5 and how we got there. *Geophysical Research Letters*, 40(6), pp.1194–1199. Available at: http://doi.wiley.com/10.1002/grl.50256 [Accessed October 14, 2016].

Maslin, M. & Austin, P., 2012. Uncertainty: Climate models at their limit? *Nature*, 486(7402), pp.183–184.

Ouranos, 2016a. Fine-Tuning Observations to Better Manage and Design Hydroelectricity Assets. In *Adaptation Case Studies in the Energy Sector – Overcoming Barriers to Adaptation*.

Ouranos, 2016b. New Normals ofr Energy Demand Forecasting. In *Adaptation Case Studies in the Energy Sector – Overcoming Barriers to Adaptation*.

Ouranos, 2015a. *Probable maximum floods and dam safety in the 21st century climate.*,

Ouranos, 2015b. *Vers l'adaptation Synthèse des connaissances sur les changements climatiques au Québec* 2015th ed., Montréal, Québec.

Plass, G.N., 1956. The influence of the 15μ carbon-dioxide band on the atmospheric infra-red cooling rate. *Quarterly Journal of the Royal Meteorological Society*, 82(353), pp.310–324. Available at: http://doi.wiley.com/10.1002/qj.49708235307.

Rice, J.S. et al., 2015. Continental U.S. streamflow trends from 1940 to 2009 and their relationships with watershed spatial characteristics. *Water Resources Research*, 51(8), pp.6262–6275. Available at: http://doi.wiley.com/10.1002/2014WR016367.

Sallenger, A.H., Doran, K.S. & Howd, P.A., 2012. Hotspot of accelerated sea-level rise on the Atlantic coast of North America. *Nature Climate Change*, 2(12), pp.884–888. Available at: http://www.nature.com/doifinder/10.1038/nclimate1597 [Accessed October 14, 2016].

von Schuckmann, K. et al., 2016. An imperative to monitor Earth's energy imbalance. *Nature Climate Change*, 6(2), pp.138–144. Available at: http://www.nature.com/doifinder/10.1038/nclimate2876 [Accessed October 14, 2016].

Villarini, G., Smith, J.A. & Vecchi, G.A., 2013. Changing Frequency of Heavy Rainfall over the Central United States. *Journal of Climate*, 26(1), pp.351–357. Available at: http://journals.ametsoc.org/doi/abs/10.1175/JCLI-D-12-00043.1.

Ward, R. & Ranger, N., 2010. Centre for Climate Change Economics and Policy TRENDS IN ECONOMIC AND INSURED LOSSES FROM WEATHER-RELATED EVENTS: A new analysis.

Weart, S., 2011. The Development of the Concept of Dangerous Anthropogenic Climate Change. *Oxford Handbook of Climate Change and Society*, (May), pp.67–81.

LINEAR WEIR SIZE SCALE EFFECTS: A LOOK AT FLAT TOP WEIRS

Blake Tullis[1]
Kedrick Curtis[2]

Weirs are commonly used as reservoir spillway discharge control, flow diversion, channel grade control, and flow measurement in open channels. Due to accurate head-discharge data acquisition limitations at the field scale, prototype weir performance data typically come from laboratory-scale model studies. Froude similitude laws are used to scale the model results to the prototype. In open channel flow, the equality of the inertial and gravitational force ratio at the model and prototype scales is the basis for scaling data between weir sizes. When other forces, such as surface tension and viscosity, become relevant in open channel flow applications (e.g., low-head weir discharges), Froude scaling alone does not account for this and some variation between the model and prototype performance at common Froude numbers would be expected. Hydraulic performance data size scale dependency, as influenced by the relative importance of surface tension and viscous forces, is referred to as size scale effects. At larger discharges (i.e., design flow levels), size scale effects are typically nonexistent. The objective of this study was to identify limiting low-head discharge conditions above which size scale effects become negligible for four sizes of geometrically similar flat top weirs, with weir heights equal to 610, 305, 152, and 76 mm, respectively. The test results were validated through good agreement with head-discharge data from the literature. Comparing the smaller weir performances to the 610 mm tall weir (prototype) showed that size scale effects indeed exist for flat top weirs and that the minimum upstream head, above which scale effects are negligible vary with the size of the model weir. The results from this study provide and indication of the level of uncertainty to be expected with scaling low-head model head-discharge data to prototype scale. The study also noted difference in the nappe aeration behavior as a function of size scale.

[1] Utah State University, blake.tullis@usu.edu
[2] Jones and Demille Engineering

PHYSICAL MODEL OF SPILLWAY AND RESERVOIR DEBRIS INTERACTION

Kent Walker[1]
Jaron Hasenbalg[2]
Kelly Brom[3]

Following a literature review of reservoir spillway debris by Kelly Brom in 2014 and an inventory / perceived susceptibility assessment of US Bureau of Reclamation projects by Jaron Hasenbalg the Hydraulic Investigations and Laboratory Services Group has constructed a 1:18 Froude scale physical model to test debris interactions on dam spillway structures in the Denver laboratory. The model will be used to determine the relative risk of plugging of two configurations each of morning glory (drop inlet) and gated ogee crest spillway types including pier configurations and bridge deck heights as appropriate. Debris will be introduced as single logs, clusters of 3-5 logs and as extensively connected log mats. Test metrics will include accumulated debris volume, location of clogging and changes to reservoir WSE due to debris clogging. The oral presentation will focus on Reclamation's inventory, case histories and literature review of past physical model studies of debris interaction and will also present preliminary results of the modelling.

[1] Bureau of Reclamation, kgwalker@usbr.gov
[2] Bureau of Reclamation
[3] Bureau of Reclamation

BOUNDARY DAM SPILLWAY NO. 1 TDG MODIFICATIONS

John Werner[1]
James Rutherford[2]
Andrew Bearlin[3]
Peter Friz[4]
Jeff Johnson[5]
Daniel Kirschbaum[6]
Joe Groeneveld[7]
Justin Arnold[8]

ABSTRACT

For the Boundary Total Dissolved Gas (TDG) Project, Hatch Associates Consultants Incorporated (Hatch) and Alden Research Laboratory (Alden) planned, designed and provided engineering support during construction to Seattle City Light (SCL) for modifications to Spillway No. 1 (SW1) on the left abutment of Boundary Dam. The Boundary Hydroelectric Project, owned and operated by SCL, is located in northeast Washington State. The modifications involved the installation of four post-tensioned (PT) baffle blocks with an upstream ramp structure. The project construction was completed in December 2016 with commissioning planned for early 2017.

This paper presents challenges faced in the design of SW1 modifications. The hydraulic design of the baffle blocks involved both physical and numerical modeling. The physical and numerical models, coupled with field studies, allowed TDG performance to be investigated for a wide range of baffle block and ramp arrangements.

Geotechnical conditions on the left abutment are challenging. A large rock wedge, formed by dominant fault structures, is present on the left abutment under SW1. As a result, the design team recognized that modifications could possibly affect the left abutment stability, and consequently the stability of SW1. To address this challenge, a detailed 3D model was developed to include the rock wedge, the thirteen existing anchors installed for slope stability in the early 1990s, new and existing drain holes, and the anchors for the new baffle blocks. Comprehensive rock stability analysis that accounted for existing conditions and the proposed TDG modifications confirmed that the modifications would not negatively affect foundation stability.

INTRODUCTION

The Boundary Hydroelectric Project (Project) is located on the Pend Oreille River in Northeastern Washington State. Boundary Dam is the third of five dams on the Pend Oreille River (Albeni Falls, Box Canyon, Boundary, Seven Mile, Waneta) upstream of

---

[1] Civil/Structural Engineer, Hatch, 6 Nickerson Street, Suite 101, Seattle, WA 98109, john.werner@hatch.com
[2] Project Manager, Hatch, 6 Nickerson Street, Suite 101, Seattle, WA 98109, james.rutherford@hatch.com
[3] Project Manager, Seattle City Light, Seattle, WA, andrew.bearlin@seattle.gov
[4] Engineering Geologist, Hatch, Vancouver, BC, peter.friz@hatch.com
[5] Geotechnical Engineer, Hatch, Vancouver, BC, jeff.johnson@hatch.com
[6] Mechanical Engineer, Seattle City Light, Seattle, WA, daniel.kirschbaum@seattle.gov
[7] Hydraulics Group Lead, Hatch, Calgary, AB, joe.groeneveld@hatch.com
[8] Physical Modeling Lead, Alden Hydraulic Laboratories, Redmond, WA, jarnold@aldenlab.com

---

the confluence with the Columbia River. The Project operates in a load-following mode, generating power during peak-load hours and curtailing generation during off-peak hours.

Figure 1 shows the key project features, consisting of an arch dam, reservoir, and underground power plant. Boundary Dam, constructed in the 1960s, is a variable-radius concrete arch dam with a structural height of 340 feet, a crest length of 508 feet, and a total length of 740 feet. The arch dam varies in thickness from eight feet at the crest, which is at elevation 2,004 feet (NAVD88) to 32 feet at the base at elevation 1,648 feet. The normal tailrace water surface elevation below the dam is 1,733 feet with approximately 261 feet of gross head for power purposes for a normal average upstream pool level.

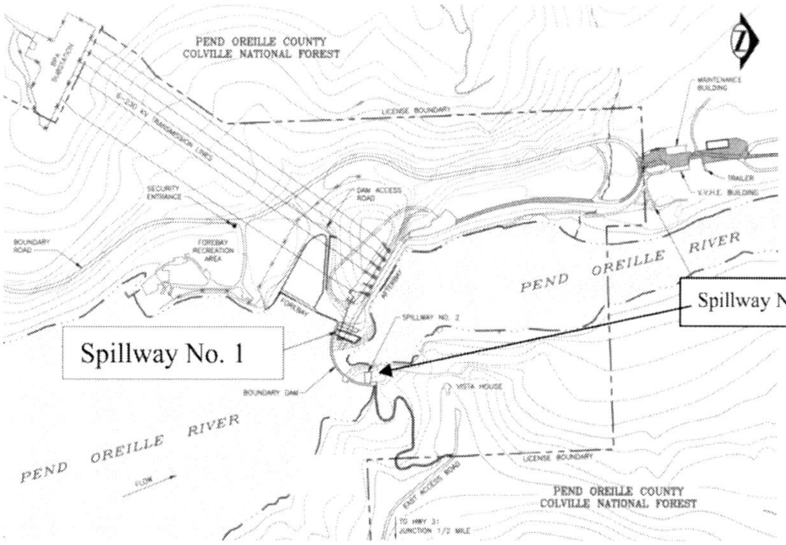

Figure 1. Boundary Dam Site Plan

Boundary Reservoir extends 17.5 miles upstream to Box Canyon Dam providing a total usable storage of approximately 41,000 acre-feet. The reservoir is operated from the normal maximum water surface elevation of 1,994 feet to 1,954 feet with a 40-foot operating range authorized by the current license. Figure 2 shows a downstream view of the dam looking upstream.

Since 2007 SCL has been developing measures to reduce total dissolved gas (TDG) levels in the tailrace of the Boundary Hydroelectric Project (Project) in order to comply with the Pend Oreille River total maximum daily load (TMDL) for TDG. The TMDL for TDG in the tailrace was established by Project Certification, Order No. 8872, issued by the Washington Department of Ecology in November 2011 and approved as part of new license requirements issued by the Federal Energy Regulatory Commission (FERC) in March 2013.

Figure 2. Downstream View of Boundary Dam Looking Upstream

The TDG Attainment Plan that is part of the FERC license water quality certification describes the proposed process for complying with the 110 percent TDG saturation standard (Ecology). City Light is required to identify all "reasonable and feasible" improvements that could be used to meet the state of Washington's 110 percent TDG saturation standard by evaluating operational and/or structural modification alternatives at the Project.

PLANNING OF TDG UPGRADES

The SCL TDG team has convened a series of design workshops to brainstorm and develop alternatives, review Project progress, discuss technical aspects of the work, evaluate alternatives and provide direction for ongoing development of operational and structural measures to reduce TDG downstream of the Project. The initial development of alternatives occurred during a two-day workshop in 2007, attended by SCL staff and an expert panel. During this initial workshop a large number of potential mitigation measures were proposed. SCL and the expert panel selected six alternatives that they considered to be feasible and likely to be effective, and presented these in relicensing study plan reports (SCL 2007 and SCL 2009); these are the alternatives on which the SCL TDG team has focused their subsequent design development efforts. Since 2009, the TDG mitigation design team holds an annual design workshop, and provides input to or attends periodic agency meetings.

Annual team meetings are a forum to review the year's progress and significant design advances, to brainstorm ideas and discuss challenges, and to establish the goals and prioritize the studies and design development for the following year.

The TDG abatement project alternatives identified during the design workshops held in 2009 were investigated from the perspective of hydraulic performance and engineering feasibility. The three short-listed alternatives selected during this process (SCL 2009) are as follows:

Throttle Sluice Gates:
- Flow capacity 12,000–18,000 cfs
- Deflector within water passage to protect gate slots

Roughen Sluice Flow:
- Flow capacity 16,000–20,000 cfs
- Deflectors at exit of jet from sluiceway to flatten jet trajectory.

Install Flow Splitters/Aerators on Spillways:
- Flow capacity 20,000–30,000 cfs
- Spillway deflectors, roughness elements, and/or dentated spillway

Based on both physical and numerical hydraulic analysis, as well as engineering considerations, modifications to Spillway No. 1 (SW1) and Spillway No. 2 (SW2) were selected for early implementation. SW2 modifications were scheduled first because SW2 had favorable foundation conditions that would require a simpler analysis than that of SW1. Additionally, SW2 was shorter, giving better access from the dam crest for construction of modifications.

Modifications to both spillways have consisted of adding Roughness Elements (REs) similar to baffle blocks to the spillway to induce turbulence, to accelerate break-up of the falling jet and to reduce jet penetration into the tailwater. The spillway REs were selected as the preferred first alternative to be implemented.

Modifications to SW2 were completed in December of 2014. A design/construction cycle for SW1 was completed following a process similar to the one followed with SW2, one year was required for design and another year was required for the construction process and commissioning.

HYDRAULIC ANALYSIS AND DESIGN

Resolution of many of the hydraulic design issues for this concept relied heavily on the results of both physical and numerical hydraulic models. Both were used in complementary roles to maximize their particular strengths. Each is briefly described below.

Physical Model Studies

A 1:25 scale physical hydraulic model of the Boundary Dam spill release facilities was constructed at Alden in late 2008 and commissioned in early 2009. While the physical model cannot replicate the prototype gas transfer process and cannot, therefore, directly predict TDG performance, it can be utilized to gain an understanding of the hydrodynamics of spill releases into the tailwater that influence the gas exchange process. Direct visualization of the flow patterns generated in the tailrace and the relative depth to which entrained air bubbles are driven by the flow provide insight into how project operations and structural modifications might be developed to mitigate TDG production. The physical model has advantages over the CFD and TDG predictive modeling tools described later in this report in that, once the model has been constructed, it can be used in a relatively quick manner (hours) to compare the relative effects of operations and geometric changes to facilities, as compared to the computation time (days or weeks) required for numerical solutions to be developed.

Since commissioning, the physical hydraulic model has been used to achieve the following specific items that contributed to the design of both SW1 and SW2 modifications:

- Develop rating curves for all flow outlets;
- Make velocity measurements in the project tailrace (for calibration of the CFD model);
- Characterize the plunge depth performance of the existing dam (Baseline Tests);
- Investigate spillway modifications including flip bucket treatments, spillway wall alterations, and roughness element placement, and;
- Optimize the configuration of roughness elements on the spillways to minimize plunge depth.

The physical model used in the design of SW1 confirmed the effectiveness of including a ramp upstream of the roughness elements to generally suppress vortex formations caused by flow separating from the roughness elements and to direct the tails of incidental vortex formations upwards into the flowfield, thereby protecting the spillway surface from cavitation damage. Figure 3 shows vortex formation in the model with an interim RE and ramp design. Figure 4 shows a similar location after the refinement of the RE and ramp design with no evidence of vortex formation.

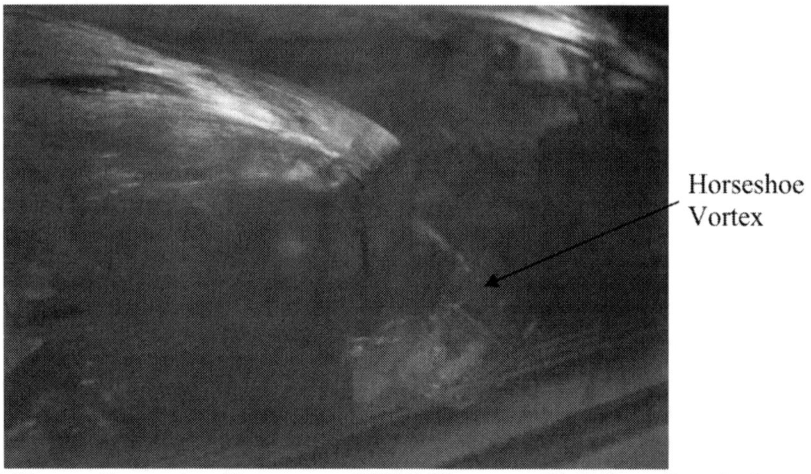

Figure 3. Vortex Formation Upstream of SW1 Elements (Full Gate Opening)

Figure 4. Ramp Installed Upstream of SW1 Elements (Full Gate Opening)

Computational Fluid Dynamics (CFD) Analysis

Full three-dimensional CFD models of SW1, SW2, a sluiceway, and the downstream plunge pool and river channel were developed in 2009. These models were utilized as part of the suite of design tools used to develop an understanding of the governing hydraulic and hydrodynamic processes driving gas exchange in the tailrace of the existing project under spill conditions, to develop the designs of structural TDG mitigation alternatives, and in combination with the TDG predictive model described in subsequent sections of this report, to predict the TDG performance of proposed TDG mitigation alternatives. The CFD models were used in conjunction with the physical model utilizing the relative strengths of the two analysis methods. The physical model was used to its greatest utility in iteratively developing physical geometry changes of the mitigation alternatives, while the CFD models were used to provide quantitative information in terms of hydrodynamic loads for design purposes and flow fields for use in TDG prediction.

Results from the CFD models are verified and validated against both physical model and field data (see Figure 5). CFD models were used extensively in the design of the SW1 modifications. Contributions have included:

- Analysis of numerous ramp and baffle block arrangements achieve reductions in TDG levels during variable spill flow;
- Estimation of hydraulic forces and pressures acting on the roughness elements;
- Estimation of pressures due to flow impingement on the left abutment under various operating scenarios;
- Investigation of the effect on tailrace flow fields associated with the addition of ramps on the spillway floor upstream of the roughness elements to suppress potential cavitation damage;
- Generation of hydrodynamic input to the TDG predictive tool for existing field data cases to validate the prediction tool; and

- Analysis to provide hydrodynamic input to the TDG predictive tool for baseline and modified operations.

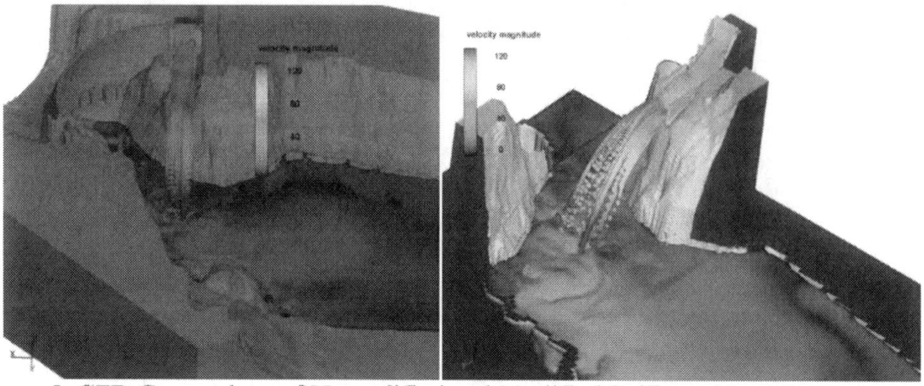

Figure 5. CFD Comparison of Unmodified and Modified Spillway No. 1 Jet Trajectories

TDG Prediction

To help evaluate the impact of the proposed SW1 modifications would have on downstream gas production, it was necessary to estimate expected gas transfer within the project plunge area, both for the existing or baseline condition, and with various mitigative works in place. The prediction of TDG performance was performed using the CFD model by employing a particle tracking method. The particle tracking methodology first involves application of a CFD model to simulate the movement of entrained air bubbles with time in the tailwater of the project when the spillway is operating. The hydrodynamic histories (velocity and pressure) of these bubbles are then exported to and used in a spreadsheet tool, which uses this data, in combination with gas transfer equations, to predict the equilibrium gas content of the tailrace receiving water.

The methodology of acquiring and using the required bubble history information from the CFD model was initially developed and tested for the Boundary tailrace situation in 2010 and 2011. Subsequently the predictive model was applied to replicate actual spill releases at the project, and the predicted TDG levels were compared with field measurements and found to be in reasonable agreement. In subsequent years the model has been further calibrated and successfully used to predict TDG performance of mitigation alternatives. The TDG model has made it possible to estimate the effect of specific roughness element arrangements and dimensions on TDG production in the tailrace. As an example, Figure 6 illustrates how the time dependent average bubble plunge depth is expected to change with the addition of the ramps and roughness elements to SW1 for a discharge of 20,000 cfs.

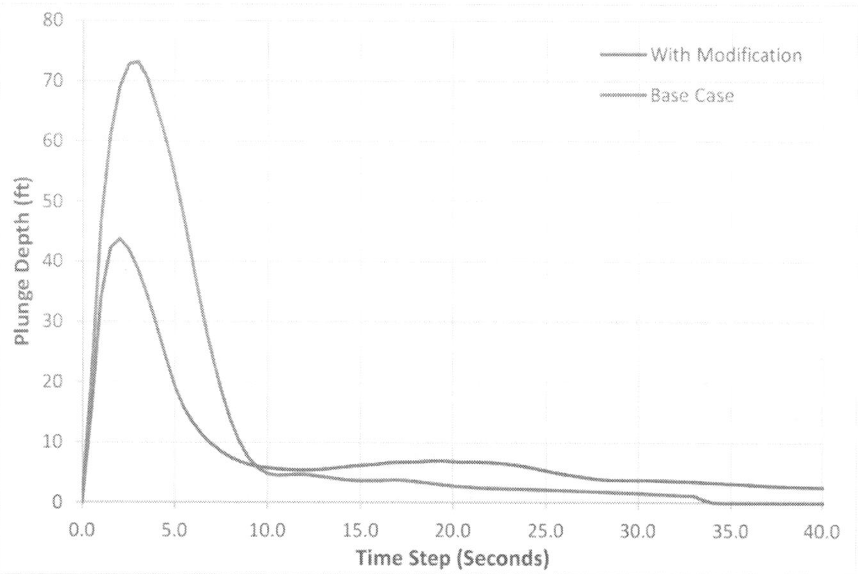

Figure 6. Time dependent air bubble plunge depth – Discharge of 20,000 cfs

STRUCTURAL ANALYSIS AND DESIGN

The key structural design objectives were to:

- Ensure that the modifications do not negatively affect the stability of the left abutment rock wedge under SW1.
- Ensure that the modifications are robust
- Maintain the original design intent of the spillway (most importantly discharge capacity)

Special attention was given to the existing drains and anchors embedded into the left abutment. LiDAR survey scans were performed at the left abutment and in the Level 6 Tunnel and Exploratory Adit. The LIDAR scan data was utilized to create a detailed surface mesh of the geological features and existing project features using AutoCAD 3D.

The 3D CAD model includes the left abutment rock surface, SW1 (including the spillway surface, training walls, gate and upstream entry geometry), Level 6 Tunnel, Exploratory Adit, existing 13 post-tensioned (PT) anchors, and drain holes. The new multi-strand PT anchors were designed to stabilize the roughness elements while at the same time fitting within the network of existing anchors and meet key structural design objectives.

Figure 7 shows a schematic representation of the roughness element loading diagram for normal hydraulic loading. The roughness elements were designed for full stagnation pressure on the upstream face of the element. Design of the roughness elements' anchorage considered normal operating flow acting on the installed features, maximum flow due to full gate discharge conditions, fatigue due to variable hydrodynamic loads during very turbulent flow conditions, and impact resistance due to log strikes.

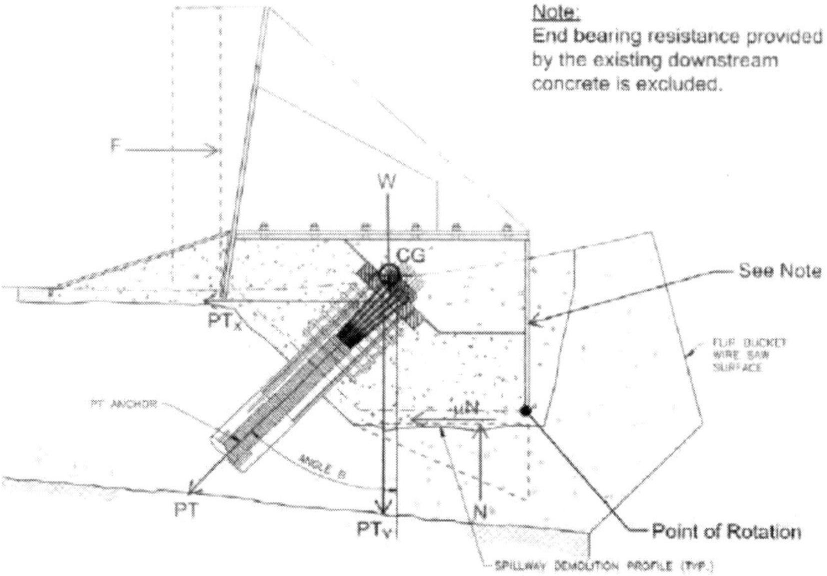

Figure 7. Roughness Element Loading Diagram

The SW1 modifications design addressed a full range of loads and site specific geotechnical parameters. Hydraulic loadings included hydrodynamic loading under different flow conditions and debris (impact) loading. Wind, snow and earthquake loadings were not significant for the design of the ramps and roughness elements that were considered in the design of the SW1 modifications.

The roughness elements were designed with a "break away" upper section so that under an extreme loading condition the lower portion of the structures and PT anchor would be protected. The design basis addressed this potential failure mode and required that the roughness element anchorage system be designed such that an impact load could not result in damage to the spillway surface that, in turn, could lead to significant erosion of the spillway structure and possibly degrade the foundation under sustained flow releases.

If the upper portion of a roughness element is sheared off or damaged due to impact, it is possible that, after a detailed inspection, the undamaged base could be used to attach a new roughness element top section using new fasteners. The fasteners at the interface between the top and bottom section of the roughness element were specially designed based on tests performed at the University of Washington (UW).

The PT anchor and fastener design also considered the possibility of fatigue during sustained or frequent flow releases. Spill events expose the roughness elements to chaotic, violent conditions that have rapidly varying pressures. Constant cycles of load can lead to fatigue failure. The anchor system was designed with an operating load large enough that the base always remains in compression even at the most extreme spill conditions (full stagnation pressure of a PMF event). Only an extreme impact event may decompress the base prior to fastener failure and breakaway of the upper element section.

Cavitation prevention was accomplished by the selection of robust construction materials-concrete-filled stainless steel fabrications. The cladding on the block has adequate surface durability to cavitation, and tunes the shape of the roughness element to reduce the potential for cavitation damage on the surfaces, and doubles as formwork for the placement of the concrete. A ramp was added upstream of each of the roughness elements to suppress vortex formation and to lift the vortices that do form away from the structure surfaces.

Design of Anchorage

The left abutment of the Project contains eight observe faults, with Faults 1 and 6X forming a large rock wedge that SW1 is partially founded on. Fault 6X is inclined at 49 degrees and forms the failure plane for the rock wedge. Faults 1 and 6X are shown on the site photo of SW1 in Figure 8. The roughness elements had their anchorage designed so that it relied on the rock below Fault 6X because of the potential for destabilizing the rock wedge. Multiple stability analyses were undertaken to ensure that any modifications did not adversely affect the stability of the rock wedge.

Figure 8. Site Photo of Left Abutment

A 3D CAD model of the left abutment was created to correlate information on the orientation of Faults 1 and 6X. It also included the position of existing drains and anchors within the left abutment, to ensure that any proposed modifications would not compromise those features. The 3D CAD model proved to be an invaluable tool to visualize existing anchors and optimize the orientation of the anchors for the modifications to SW1. In addition to anchors, four new drain holes were designed as part of this project. They were included in the event that grouting of the PT anchors communicates with existing drains and reduces their efficiency. Figure 9 is taken from the 3D CAD model and notes Faults 1 and 6X, defining the rock wedge as a separate color from the rest of the left abutment.

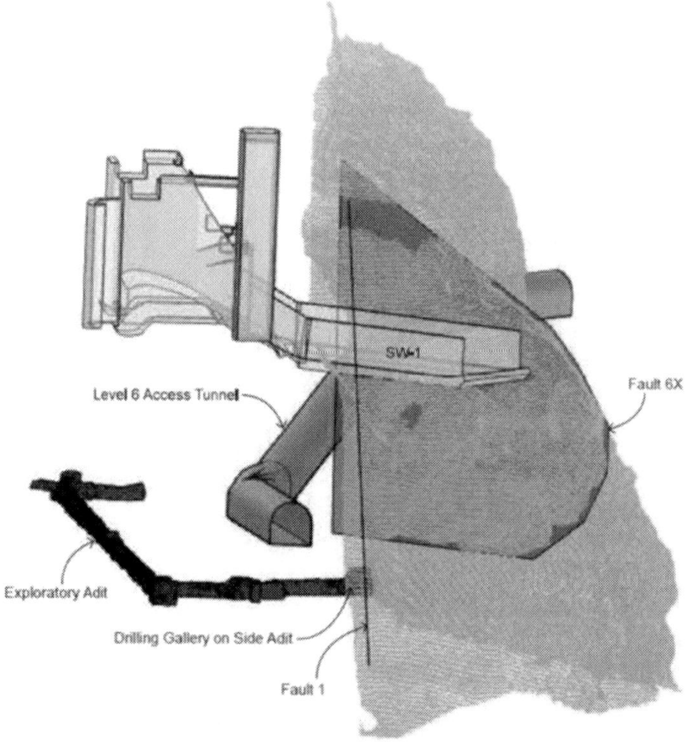

Figure 9. 3D Model Overview of Left Abutment Wedge with Faults

The analysis approach to check the rock block stability was based on a three-dimensional static limit-equilibrium analysis performed in a spreadsheet developed for the project. External loads and inertial forces acting on the block were collected and resolved into normal and sliding forces acting on Fault 6X, from which a Sliding Safety Factor (SSF) was determined. Factors that were included in the analysis include existing and proposed anchor PT forces and orientations, weight of Wedge 6X, weight of water on the spillway, hydrodynamic loading acting on the spillway and roughness elements, and inertial effects from a seismic loading.

In addition to this spreadsheet-based analysis, a set of analyses were completed using the Rocscience programs, RocPlane and Swedge. For all analyses, a friction angle of 54 degrees was used, estimated using Barton (1981) after determining that the joint's large-scale waviness controlled shear resistance rather than small-scale asperities. The load cases considered for this analysis were normal operating conditions as well as a seismic case with the peak ground acceleration applied as an external horizontal force with an azimuth towards the dip direction of the failure plane.

Based on the static limit-equilibrium analysis, it was concluded that no change in stability was to be expected with the addition of the four new proposed roughness element anchors. These results were not unexpected when considering the overall mass of the block and the existing anchor loads are compared to the contribution of the roughness element anchors. The Wedge 6X block resting on Fault 6X weighs approximately 108,000 kips while the existing anchors contribute a working load of 9,500 kips, only 9% of the total block rock mass. The proposed roughness element anchors would contribute even less anchorage force, at 2,700 kip, or 2.5% of the total rock mass.

The post-load analysis that included roughness element hydrodynamic loads, PT anchor loads, and water load displayed a small increase in the Wedge 6X stability factor of safety from 1.36 to 1.39 for static conditions and from 1.06 to 1.09 for seismic conditions. The static limit equilibrium analysis was checked using RocPlane and Swedge software analyses and the negligible effects of the proposed SW1 modifications were confirmed.

After determining that anchor installation would only marginally improve the stability of the rock mass resting on Fault 6X, design of the roughness elements could proceed. The roughness elements were designed with a "break away" upper section that required a fuse element to fail under extreme loading conditions. The problem with this is that while the fasteners have a minimum specified strength, their maximum strength can potentially much higher than expected, resulting in damage to the anchorage. The design methodology assumed that the bottom section of the roughness element would be undamaged and reusable, therefore a fastener with a known minimum and maximum strength was required.

Fastener Design and Testing

The Project decided to produce custom fasteners that met the requirements for the fuse within each roughness element. A lot of 9 12-foot-long 1-inch diameter ASTM A304L stainless steel round stock was acquired and several iterations of shear and tension tests were performed on it to categorize its material properties. This same lot of round stock also provided a pool that the finalized fasteners would be fabricated from, to ensure that the material properties would be constant between the material tested and the material that was installed.

Material and prototype fastener testing was performed at the UW's Structural Engineering Laboratory, and overseen by Professor John Stanton. The shear and tension tests were carried out on machined round stock threaded at the ends, with one or more reduced diameter sections in the interior, approximately 1/8" long. This reduced section could be adjusted based on the material strength, to ensure that the fastener was strong enough to resist normal loading of the RE, but would fail at a specific load to ensure the

PT anchors would remain undamaged. The specimens tested resembled the proposed fastener design so that geometry of the reduced section would be the same between the tested samples and the finalized fastener.

The fasteners were loaded in a combination of tension and shear, with the fastener location determining whether it experiences more tension or more shear. This required an understanding of the tensile and shear strength of the fasteners, so each was tested. Figures 10 and 11 show examples of tensile and shear specimens after they had been tested to failure.

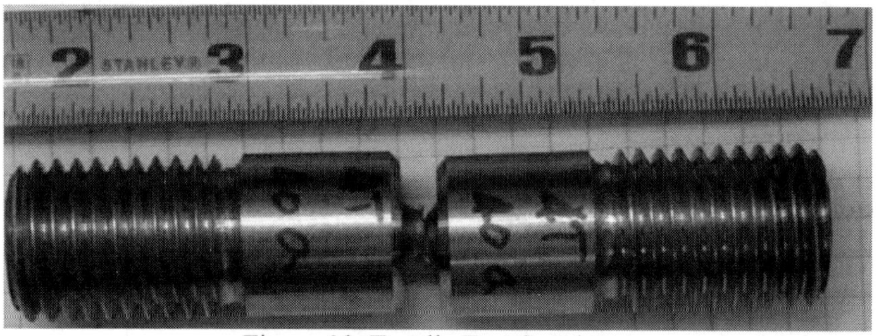

Figure 10. Tensile Test Specimen

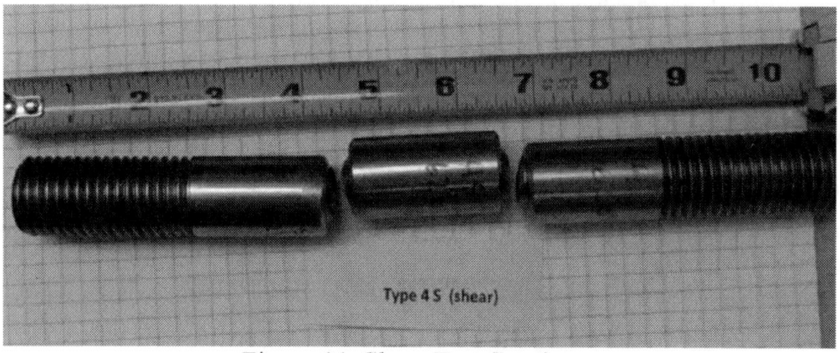

Figure 11. Shear Test Specimen

Tensile test specimens could be pulled in direct tension until failure, but the shear specimens required a more specialized approach. A Baldwin universal test machine was used in conjunction with a custom-designed test rig. The rig consisted of a shear box and a shear blade, as shown in Figure 12. Each shear specimen had two reduced sections and was placed in the rig so these sections fell in the shear planes. The Baldwin universal test machine applied compression to the rig until the shear specimen failed.

Figure 12. Shear Test Rig

Several rounds of testing were performed for different cross-sectional area, so that a curve could be constructed to relate different cross-sectional areas to shear and tensile strength. From this curve, a final fastener size was selected to be used in the design of the roughness element fuse mechanism. These fasteners could be installed with confidence that they would have adequate strength to resist the usual and unusual loadings the roughness elements would experience, but would fail before they damaged the PT anchorage.

CONSTRUCTION

Detailed design of SW1 modifications were finalized in early 2016. Through efficient contract administration by SCL, all major spillway modifications were completed by the end of 2016. The construction contract was awarded to the low bidder, Clearwater Construction & Management (Clearwater) of Spokane, WA. Clearwater was the general contractor responsible for the SW2 TDG modifications two years earlier.

Several challenges were experience during construction. Access and fall protection for work done at the very end of the spillway was difficult and in addition, later in the construction period discharges through SW2 had to be made by the City at some frequency. SW2 discharge periods frequently caused adverse conditions coating surfaces on SW1 with ice as the temperature dropped below freezing. Figure 13 shows RE post-tensioning procedure performed during a SW2 spill event. Figure 14 shows the completed roughness elements at the end of SW1, after 5 feet of the of the flip bucket had been removed with a wire-saw.

Figure 13. Post-Tensioning of Anchors for Roughness Elements

Figure 14. Completed Roughness Elements at end of SW1

CONCLUSION

CFD model and physical testing were instrumental in this design and gave the design team confidence that tailrace TDG would be reduced during discharges through SW1. Modifications to SW1 followed the same design process for modifications done to SW2 several years before. The performance of SW2 met expectations and has allowed further validation of computer and physical models. Once SW1 is commissioned, it will allow the TDG models to be further calibrated so they can be used to evaluate possible additional modifications to reduce TDG. The successful installation and operation of retrofit spillway roughness elements at Boundary Dam validated their use both to reduce total dissolved gas production and dissipate energy that could cause scour.

ACKNOWLEDGEMENTS

The authors would like to acknowledge the contribution of all of the SCL TDG team with special thanks to hydraulics experts Chick Sweeney of Alden Labs and Dr. Hank Falvey, an independent consultant.

REFERENCES

Barton, N.R. 1981. Shear Strength Investigations for Surface Mining 3rd Int. Conf. on Stability in Surface Mining. Vancouver, BC. Ch. 7: 171-192, AIME. 1981.

Ecology (Department of Ecology, State of Washington). 2012. Water Quality Program Policy Integrated Report. Olympia, WA. 2012.

SCL (Seattle City Light). 2007. Revised Study Plan for the Boundary Hydroelectric Project (FERC No. 2144). Seattle, WA. February 2007.

SCL (Seattle City Light). 2009. Updated Study Report. Boundary Hydroelectric Project (FERC No. 2144). Seattle, WA. March 2009

A COMPREHENSIVE APPROACH TO PROBABILISTIC HYDROLOGIC HAZARD ASSESSMENT AND ITS USE TO SUPPORT RISK-INFORMED DECISIONS

Shaun Carney[1]
Keil Neff[2]
Mel Schaefer[3]
Bruce Baker[4]

Two years ago the Tennessee Valley Authority (TVA) began a project to develop a framework for probabilistic hydrologic hazard assessment in support of the TVA Dam Safety Program. A regional point precipitation frequency analysis was performed for four different storm types to estimate the probabilities of extreme precipitation amounts. This analysis is now being extended to provide areal precipitation frequency estimates for specific watersheds for the various storm types. The precipitation frequency analysis will be used with a database of historical extreme storms to stochastically generate storms over the Tennessee valley for use in hydrologic modeling of floods.

TVA has developed a suite of hydrologic, hydraulic, and reservoir operation models that represent the TVA system. This includes a rule-based RiverWare operations model that represents the defined operating policy for over 30 reservoirs, including joint operations between multiple projects. The coupled system of models allows simulation of both natural flows and the operational response to historical or synthetically generated storms.

The Stochastic Event Flood Model (SEFM) will be coupled with this suite of models to estimate flood exceedance probabilities for projects throughout the TVA system. SEFM will be used to stochastically generate thousands of storms, sample initial hydrologic conditions from a 1000-year database created using the suite of models, and provide these data to the TVA's system model. Output will include hydrologic hazard curves for any point in the TVA system.

Plans are now being made to couple consequences analysis with the probabilistic hydrologic hazard analysis to provide a quantitative assessment of risk. The quantitative risk analysis framework will allow factors such as dam failure modes and gate operability to be incorporated into the stochastic simulation. The approach will provide TVA a basis for risk-informed decision-making in order to identify, prioritize, and justify structural and non-structural dam safety risk-reduction measures.

[1] Riverside Technology, Inc., shaun.carney@riverside.com
[2] Tennessee Valley Authority
[3] MGS Engineering
[4] MGS Engineering

METEORLOGICAL DATA VISUALIZATION AND ANALYSIS USING HEC-METVUE

Fauwaz Hanbali[1]

HEC-MetVue (Meteorological Visualization Utility Engine) is a new computer program being developed and managed by the U.S. Army Corps of Engineers (USACE) Hydrologic Engineering Center (HEC). HEC-MetVue can support visualizing and performing edits and analyses on a variety of meteorlogical data types (in particular Precipitation, Temperature, and Snow) and formats (including HEC-DSS, ASCII_GRID, NEXRAD, XMRG, NetCDF, GRIB, and PRISM). HEC-MetVue allows for temporal and spatial image aggregation, and provides a variety of image editing tools including: spatial extents trimming, data value editing and corrections, and storm translation and rotation.

In the USACE, HEC-MetVue applications straddle dam safety studies and real-time water management decision-support. HEC-MetVue currently includes tools for the development of Probable Maximum Precipitation (PMP) estimates as prescribed in the National Weather Service's Hydrometeorlogical Reports (HMRs) numbers 52 and 55a, where HMR52 covers the majority of areas in the United States east of the 105th Meridian and HMR55A covers areas in the U.S. between the Continental Divide and the 103rd Meridian. Future HEC-MetVue versions will feature implementations of HMRs tools to support PMP estimate development in the Northwestern and Southwestern regions of the U.S. Used to process observed and forecast meteorlogical data sets, HEC-MetVue also serves as a one of the primary components of the USACE Corps Water Management System (CWMS), a system that integrates real-time data and suite of modeling tools for real-time hydrologic and hydraulic simulation and scenario-based decision-support.

This presentation will provide an overview of HEC-MetVue and its capabilities.

[1] U.S. Army Corps of Engineers, Fauwaz.Hanbali@usace.army.mil

BIG DAM IN A SMALL WORLD: REASSESSMENT OF THE SPILLWAY DESIGN FLOOD OF LAKE SAKAKEWEA

Joshua Melliger[1]

Lake Sakakewea in North Dakota, formed following the closing of Garrison Dam in 1953, is the largest reservoir operated by the U.S. Army Corps of Engineers (USACE) with a total storage of about 24 million acre feet and contributing drainage area of 183,000 square miles. It is the second of six mainstem reservoirs located on the Missouri River operated by USACE. Seventy years have passed since the spillway design flood of the project was developed and documented within the 1946 Basis of Design Definite Project Report. Consequences of dam failure would be severe. As part of an evaluation of dam safety, USACE is leading an effort to reassess the spillway design flood. This presentation will focus on various technical components of the process and describe challenges overcome during the study. The development of seasonal probable maximum precipitation and flood events will be presented, including an assessment on historic snow water equivalent data to aide in determination of appropriate seasonal antecedent snowpack associated with expected precipitation. Meteorologic design assumptions made back in the 1940's will be compared to 2016 assumptions. Hydrologic modeling challenges for such a diverse and vast watershed will be presented, and flood results will be compared to the original spillway design flood.

[1] U.S. Army Corps of Engineers, joshua.j.melliger@usace.army.mil

APPLICATION OF REGIONAL HYDROLOGIC CRITERIA IN THE ROCKY MOUNTAINS TO ESTIMATION OF THE PMF IN ALBERTA

[1]John Menninger, PE
[2]Seifu Guangul, PhD, P.Eng.
[3]Dan Hoffman, PE
[4]George V. Sabol, PhD, PE
[5]Syed Abbas, P. Eng.

INTRODUCTION

Following the severe flooding in Alberta in June 2013, the Alberta government proceeded with the design of the Springbank off-stream reservoir (SR1) project to reduce the effects of future extreme floods on infrastructure, watercourses, and people in the City of Calgary and downstream communities. This project is designed as a dry pond (meaning there would be no permanent pool of water in the reservoir). The concept considers diverting extreme flood flow from the Elbow River into an off-stream storage reservoir where it would be temporarily contained and later released back into the Elbow River after the flood peak has passed. Project components include a diversion structure constructed across the Elbow River, and a diversion channel excavated through the nearby uplands to transport floodwater into an off-stream storage reservoir. The storage site includes an earthfill dam and a low level outlet structure incorporated into the dam to later release the stored water back into the Elbow River.

A key component of the SRI project is the estimation of the PMF for the design of the SR1 diversion dam on the Elbow River and the SR1 off-stream flood storage dam. For that purpose, an HEC-HMS model was developed and calibrated to represent the 2005 and 2013 flood conditions so that it can be used for PMF analysis. The model development included a comprehensive evaluation of appropriate methodologies and all relevant recorded data pertaining to the meteorological, hydrometric, and physical characteristics of the Elbow River Basin. The calibration process included adjusting model parameters so as to match the 2005 and 2013 floods. The calibrated model was applied to simulate the PMF by using PMP data that was developed for three locations: the 863 km^2 watershed upstream of the proposed SR1 diversion site, the 31 km^2 watershed upstream of the proposed SR1 dam site, and the 1,212 km^2 watershed upstream of Glenmore Reservoir. PMP data was developed and provided for use in estimating the PMF for two storm types: general storm (48-hour) and local storm (6-hour).

[1] John Menninger, PE Principal, Stantec Consulting Services Inc, 11687 Lebanon Road, Cincinnati, OH 45241-2012, john.menninger@stantec.com
[2] Dr. Seifu Guangul, Stantec Consulting Ltd., 500–311 Portage Avenue, Winnipeg, MB R3B 2B9, Canada, seifu.guangul@stantec.com
[3] Dan Hoffman, PE, Senior Project Engineer, Stantec Consulting Services Inc, 11687 Lebanon Road, Cincinnati, OH 45241-2012, Daniel.hoffman@stantec.com
[4] George V. Sabol, PhD, PE, Principal, Stantec Consulting Services Inc., 8211 S. 48th St, Phoenix, AZ 85044 george.sabol@stantec.com
[5] Syed Abbas, M.Sc., MBA, P.Eng., Director, Major Capital Projects Branch, Alberta Transportation, 4999 98 Ave, Edmonton, AB T6B 2X3, Canada, Syed.Abbas@gov.ab.ca

DESCRIPTION OF WATERSHED AND DATA

The Elbow River is located at the foothills of Alberta that flows through the Canadian Rockies towards the City of Calgary where it flows into Bow River. The Elbow River passes through a broad range of landscapes, from the alpine and sub-alpine high mountains to rolling boreal foothills to aspen parkland. In its relatively short flow length, the river drops with an average slope of one percent.
Annual floods in Elbow River are typically snowmelt. However, the major floods such as the 2005 and 2013 are the result of snowmelt augmented by intense rain storms.

Watershed characteristics for the study area are derived from a 1:50,000 (approximately 20 m x 20 m grid cells) digital elevation model (DEM) that covers the entire Elbow River Basin. The DEM was downloaded from Natural Resources Canada (GeoGratis 2015). The slope for the sub basin watercourses in the Elbow River Basin were derived from a topography map. The surficial geology of the Elbow River Basin was obtained from Alberta Geological Survey's digital data for the surficial geology of Alberta un-generalized digital mosaic. This GIS dataset is an organization of existing surficial map information for Alberta tiled into one layer (AGS 2013). Streamflow data were collected from Water Survey of Canada (WSC) and City of Calgary. Applied Weather Associates provided the PMP analysis and data (Applied Weather Associates 2015).

MODEL DEVELOPMENT AND CALIBRATION

A major factor in the development of the PMF was the hydrologic procedure to convert the applied rainfall (PMP) to rainfall excess and to the time distribution of runoff. Various methodologies for hydrologic watershed simulation were investigated. Of particular interest was to select the methodology that best represents the current state-of-the-practice for western Rocky Mountain watersheds. The method selected is the State of Colorado, Hydrologic Basin Response Parameter Estimation Guidelines (Sabol, 2008). Using those guidelines resulting in the use of the initial loss and uniform loss rate for the conversion of the PMP to rainfall excess, and dimensionless S-graphs for the unit hydrograph development. The Colorado guidelines provide data and procedures for the use and estimation of the necessary parameter values.

HEC-HMS was selected for the development of the Elbow River Basin hydrologic model. The model is available in the public domain and is widely applied to different hydrological studies in Canada and the United States. In this project, HEC-HMS was used to build and calibrate an event-based hydrological model for PMF estimation of the Elbow River Basin.

For the purpose of this study the Elbow River Basin was partitioned into eleven subbasins based primarily on topographic characteristics. However, vegetation, surficial geology, and land use were also considered to discretize the basin into meaningful hydrological units. Several hydrologic parameters were derived for each subbasin including length and slope of watercourses, area, elevation at centroid of the subbasins, and upstream and downstream elevations. Individual subbasins ranged in size from 3,120 ha to 35,300 ha. Figure 1 is a map of the delineated subbasins and the boundary of the Elbow River Basin.

Figure 1. Elbow River Basin: subbasin names as used in HEC-HMS model

A unit hydrograph approach was adopted to transform precipitation excess for each subbasin into surface runoff hydrographs. As such, unit hydrographs were developed for each subbasin using the US Bureau of Reclamation (USBR) synthetic unit hydrograph for the Rocky Mountains as defined in the State of Colorado guidelines (Sabol 2008). The coordinates of each unit hydrograph are derived from the basin lag (L_g) parameter, which is directly derived from topographic characteristics of each subbasin. A lumped parameter representing resistance to overland flow, K_n, was estimated for each subbasin, and the length of the longest watercourse (L), basin slope (S), and distance to the subbasin centroid (L_{ca}) were estimated from watershed data in the calculation of basin lag.

Once all the necessary parameters of the model were set, it was calibrated for two flood events: June 2005 and June 2013 flood that occurred in the Elbow River watershed. The 2005 flood was from June 1 at 8:00 am to June 9 at 7:00 am, whereas, the 2013 flood was from June 19 at 8:00 am to June 22 at 7:00 am. The model was calibrated using recorded runoff hydrographs for the Elbow River at Bragg Creek (WSC station ID 05BJ004) and Elbow River at Sarcee Bridge (WSC station 05BJ010) for both the 2005 and 2013 floods. The Bragg Creek station is located upstream of the proposed SR1 diversion site, while the Sarcee Bridge station is situated downstream of the diversion site, upstream of Glenmore Reservoir.

The contributing drainage area to the Bragg Creek Station is about 791 km^2 and includes the mountainous portions of the basin where both the 2005 and 2013 rainfalls were the heaviest. The contributing drainage area to the Sarcee Bridge Station is 1190 km^2 and represents nearly the full study area.

The model was calibrated by adjusting different model parameters pertaining to soil moisture, excess precipitation and channel. A comparison of the simulated and observed hydrographs at Bragg Creek for the 2005 flood is presented in Figure 2. Table 1 summarizes the accuracy of the match in terms of hydrograph peak, timing, and flood volume at Bragg Creek.

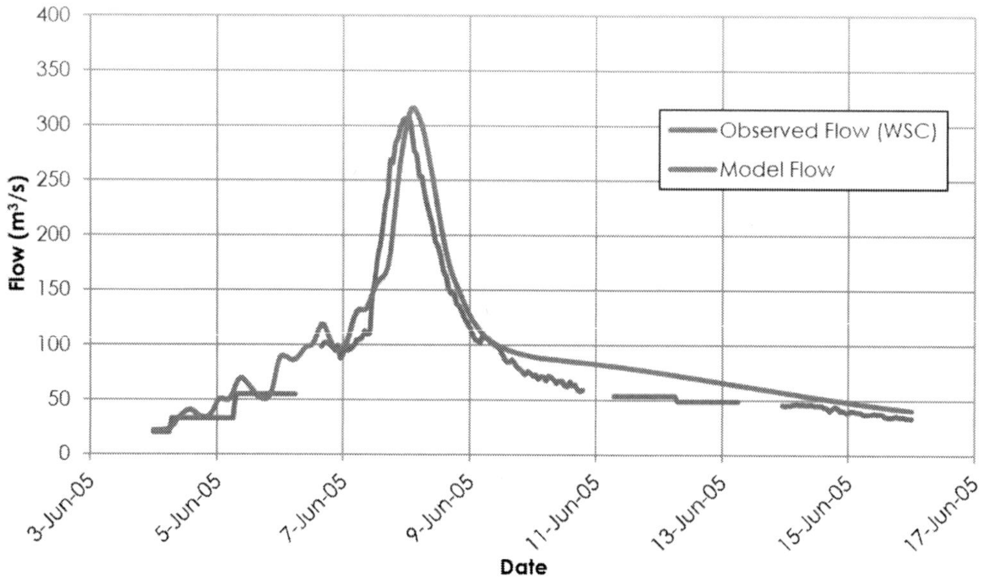

Figure 2. Observed and Calibrated Hydrographs at Bragg Creek for the 2005 Flood

Table 1: Calibration Accuracy for the 2005 Flood at Bragg Creek

Name	Peak Discharge (m^3/s)	Time of Peak	Volume (dam^3)[1]
Observed (WSC)	308.0	June 8, 2005 at 1:00	79,905
Calibrated Model	316.3	June 8, 2005 at 3:00	93,070
Percent Difference	+2.7%	-	+16.5%

[1] - Volume was calculated for the duration of simulation (June 4, 2005 at 00:00 to June 16, 2005 at 00:00)

Furthermore, comparison of the modeled and estimated hydrographs at Sarcee Bridge for the 2013 flood is presented in Figure 3. Table 2 summarizes the accuracy of the match in terms of hydrograph peak, timing, and flood volume at Sarcee Bridge.

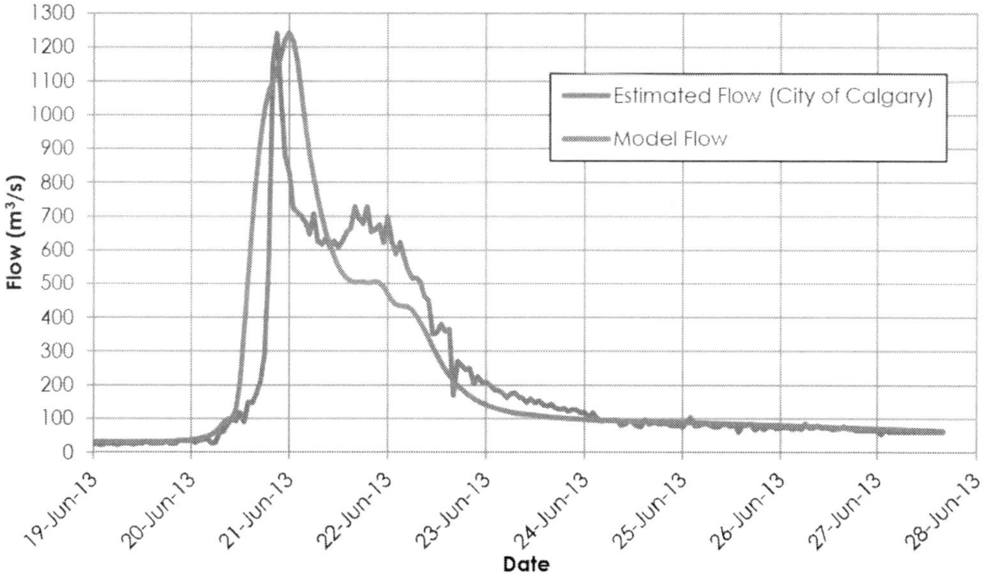

Figure 3: Estimated and Calibrated Hydrographs at Sarcee Bridge for the 2013 Flood

Table 2: Calibration Accuracy for the 2013 Flood Event at Sarcee Bridge

Name	Peak Discharge (m^3/s)	Time of Peak (UTC)	Volume (dam^3)[1]
Estimated (City of Calgary)	1240.4	June 21, 2013 at 5:00	157,308
Calibrated Model	1241.3	June 21, 2013 at 8:00	164,896
Percent Difference	+0.1%	-	+4.8%

[1] - Volume was calculated for the duration of simulation (June 19, 2013 at 08:00 to June 28, 2013 at 00:00)

Calibration of the HEC-HMS model was challenging, which was due to the uncertainty of the hydrometric data at the Bragg Creek and Sarcee Bridge gauging stations. The partial areal coverage and non-uniformity of rainfall used in calibration also played a role in the calibration process. Calibration was successful in adequately establishing the subbasin rainfall loss parameters, in refining the channel routing parameters, and in developing reasonable baseflow simulation methodology. However, actual rainfall for the 2005 and 2013 storms were highly variable in spatial distribution resulting in some subbasins receiving little rainfall and other subbasins receiving highly non-uniform rainfall. The consequences are that calibration of the unit hydrograph for the subbasins was tenuous since the basic unit hydrograph requirement of uniform rainfall over the subbasins is not achieved. Therefore, the model was recalibrated during the PMF simulation. That calibration was performed by adjusting the unit hydrograph parameters so that the simulated 100-year peak discharge and runoff volume for the input of the 100-year rainfall represented the calculated 100-year frequency flood peak and 7-day flood volume. This process is described further in the following sections.

PROBABLE MAXIMUM FLOOD (PMF) ESTIMATION

The PMF is defined as theoretically the largest flood resulting from a combination of the most severe meteorological and hydrologic conditions that could reasonably be expected to occur in a given area. The PMF is generally viewed as the flood resulting from a PMP, plus snowmelt where appropriate, applied to reasonable antecedent watershed conditions. Accordingly, the calibrated hydrologic model was applied to estimate the PMF for several viable PMP scenarios. A 100-year frequency rainfall as an antecedent condition and, in some cases, snowmelt were applied in the PMF simulations.

Gridded local storm PMP values were calculated for 6-hour durations, while general storm PMP values were calculated for 48-hour durations. The local storms were assessed for the area upstream of the SR1 diversion (863 km^2) and for the drainage area of the SR1 off-channel dam (31 km^2). The general storms were assessed for the entire watershed upstream of Glenmore Dam (1,212 km^2), as well as the area upstream of the SR1 diversion (863 km^2).

In regards to spatial distribution, the local storm PMP for the SR1 off-channel dam was centered over the drainage area of the SR1 off-channel dam. The PMP for the local storm upstream of the proposed SR1 diversion was spatially distributed using a representative severe local storm from the PMP database. The general storm PMP spatial pattern is based on orographic and moisture transposition factors of controlling storms (hereafter referred to as the orographic distribution). Therefore, a total of four different PMP scenarios were developed by AWA (see Table 3).

Table 3: Summary of PMP Scenarios

Scenario	Description
1	General storm PMP (48 hour) with orographic pattern over watershed upstream of proposed SR1 diversion (863 km^2)
2	General storm PMP (48 hour) with orographic pattern over watershed upstream of Glenmore Dam (1,212 km^2)
3	Local storm PMP (6 hour) with June 1966, Glen Ullin, North Dakota, maximum 1 hour spatial distribution centered over the watershed upstream of the proposed SR1 diversion (863 km^2)
4	Local storm PMP (6 hour) centered over subbasin W600 upstream of proposed SR1 Dam (31 km^2)

The procedures for selecting antecedent basin conditions vary among different agencies and hydrologists. The common practice in British Columbia (BC) and Alberta is *"... to precede the PMP with a 100-year 24-hour rainfall leaving a period of three days between the storms"* (Alberta Transportation 2004). While the shortest observed time interval between two severe rainfall events in the mountain and foothill areas of Alberta is on the order of 5-7 days, studies suggest that a time interval as short as three days is possible (Gerhard 2000). Based on the aforementioned, a decision was made to establish the basin antecedent conditions for the Elbow River prior to the PMP by introducing an antecedent storm, having a 100-year 24-hour rainfall, three days prior to the start of PMP.

Short duration (up to 24-hour) "point" (single station) rainfall amounts for various return periods are computed and published by Environment Canada, Meteorological Services Canada (MSC) for most airports and key meteorological sites across Canada. Currently there are no estimates of the 100-year, 24-hour rainfall amounts for larger area sizes. As such, it was decided that the estimation of the 100-year, 24-hour rainfall for the Elbow River Basin would be carried out by applying an area reduction factor (ARF) to the 1:100 year point rainfall values. For this project the ARF was based on the ratio of the 1,000 km^2 (approximately the drainage area of the Elbow River Basin) rainfall to 10 km^2 rainfall observed for major storms in Alberta. Point rainfalls are generally considered as representative of rainfall for a 10 km^2 area. It was further decided that the 100-year, short-duration point rainfall amounts to be used would be based on the rainfall amounts for Pincher Creek Airport. This Environment Canada meteorological station is the closest in proximity and physiographic characteristics to the Elbow River basin. It also has a relatively long period of record.

Alberta Transportation has analyzed depth-area-duration (DAD) curves of large storms in Alberta and has computed the mean DAD curve for the top 10, 20, and 50 storms. The ARF applied to adjust the previously computed antecedent point rainfall to a 1:100-year, 24-hour rainfall was estimated at 0.85 based on the ratio of the 1,000 km^2 to 10 km^2 rainfall for the top 20 large storms. This antecedent storm was applied three days prior to the local and general PMP for the full basin and area upstream of the SR1 diversion scenarios.

The moisture input from snowmelt during PMP is governed primarily by two factors: the snow-covered-area and the rate of melt. The snowmelt contribution to PMF then becomes simply the product of the snowmelt volume times the runoff coefficient. Snowmelt was applied to the general storms, not the local storms since severe convective storms cannot develop over large snowpack areas.

The maximum snowmelt during the 100-year, 24-hour antecedent storm and PMP was estimated by calculating the daily change in SWE (snow accumulation or depletion), during the four largest rainfall events. Investigation into the 2013 data shows that the largest observed snow depletion or melt was 69 mm and occurred during the four days surrounding the June 19 – 21. The largest single day melt was 32 mm on 20 June 2013.

The snowline elevation and the associated snow covered area for each subbasin of the Elbow River was calculated by plotting the SWE's prior to each large rainfall event against the snow pillow and snowcourse elevations so as to determine the lowest snowline elevation prior to each of the four large rainfall events.

The initial model run for PMF analysis was carried out using the hydrologic model calibrated to the 2005 and 2013 floods. That model run produced a peak flow of 1,215 m^3/s at the SR1 diversion site for the 100-year, 24-hour antecedent rainfall. However, the flood frequency analysis that was performed as part of the SR1 hydrologic study showed the estimated peak flow for a 100-year flood at the proposed SR1 diversion point was about 760 m^3/s. Therefore, the model calibrated for the 2005 and 2013 floods overestimated the 100-year flood event by approximately 60%.

Therefore, and given the reasons described previously, the model was recalibrated to match the peak flow using the 100-year, 24-hour antecedent rainfall with the one derived for the 100-year flood frequency value. The recalibration was performed by adjusting the K_n value within the recommended parameter range between values of 0.15 to 0.3. The processes of adjusting the K_n value showed that as the K_n was increased, the peak flow decreased; resulting in peak flow values closer to the 100-year flood frequency flow. A maximum K_n value of 0.3 resulted in a peak flow of 813 m^3/s. As 0.3 is the highest K_n value available within the recommended parameter range it was used for unit hydrograph development.

PMF SIMULATION RESULTS

PMF simulations were run for all four scenarios described in the previous sections using the calibrated model. The four scenarios differed primarily based on the PMP data but also on the antecedent rainfall, snowmelt, and unit hydrograph used in the model. Table 4 summarizes the detailed outline of each PMF simulation.

Table 4: Summary of Input Data for PMF Simulations

Scenario	Antecedent Rainfall	PMP	Unit Hydrograph	Snowmelt
1	100-year, 24-hour precipitation with ARF	General storm PMP with orographic pattern over watershed upstream of proposed SR1 diversion (863 km^2)	Rocky Mountain (applied to subbasins upstream of Bragg Creek) and Great Plains (applied to subbasins downstream of Bragg Creek)	Snowmelt contribution applied at Bragg Creek
2	100-year, 24-hour precipitation with ARF	General storm PMP with orographic pattern over watershed upstream of Glenmore Dam (1,212 km^2)	Rocky Mountain (applied to subbasins upstream of Bragg Creek) and Great Plains (applied to subbasins downstream of Bragg Creek)	Snowmelt contribution applied at Bragg Creek
3	100-year, 24-hour precipitation with ARF	Local storm PMP with a representative severe local storm spatial distribution centered over watershed upstream of proposed SR1 diversion (863 km^2)	Rocky Mountain (applied to subbasins upstream of Bragg Creek) and Great Plains (applied to subbasins downstream of Bragg Creek)	N/A
4	100-year, 24-hour precipitation	Local storm PMP centered over subbasin upstream of proposed SR1 dam (W600) (31 km^2)	Great Plains	N/A

For Scenarios 1, 2, and 3 peak flood and 7-day volume PMF results were reported at the proposed SR1 diversion site as well as at the Glenmore Dam. For Scenario 4, PMF results are reported at the proposed SR1 dam.

Hydrographs representing the PMF for scenarios 1 and 2 were generated at the proposed SR1 diversion site and Glenmore Dam (see Figure 4 and Figure 5). A detailed summary of the peak flow and 7-day volume for the PMF scenarios is given in Table 5.

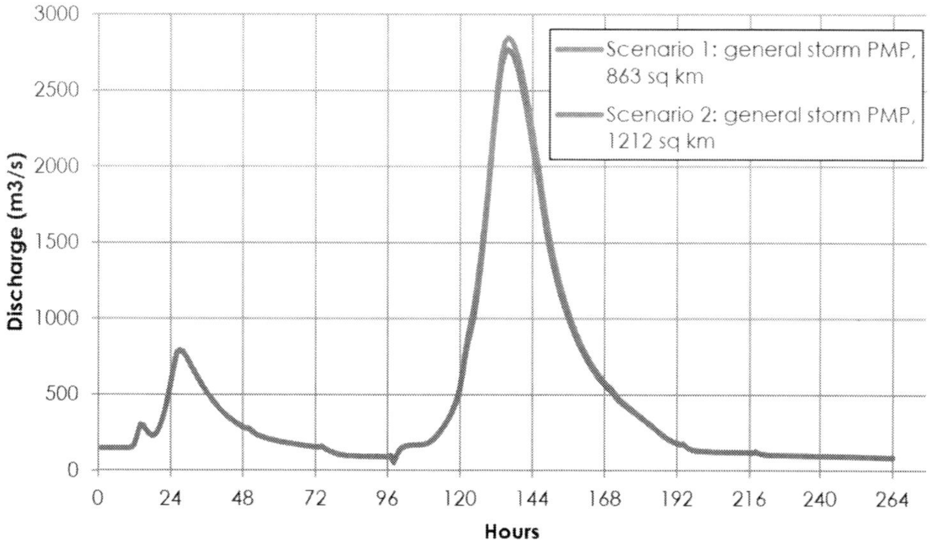

Figure 4: PMF Simulation Hydrographs for Scenarios 1 and 2 at the Proposed SR1 Diversion Site

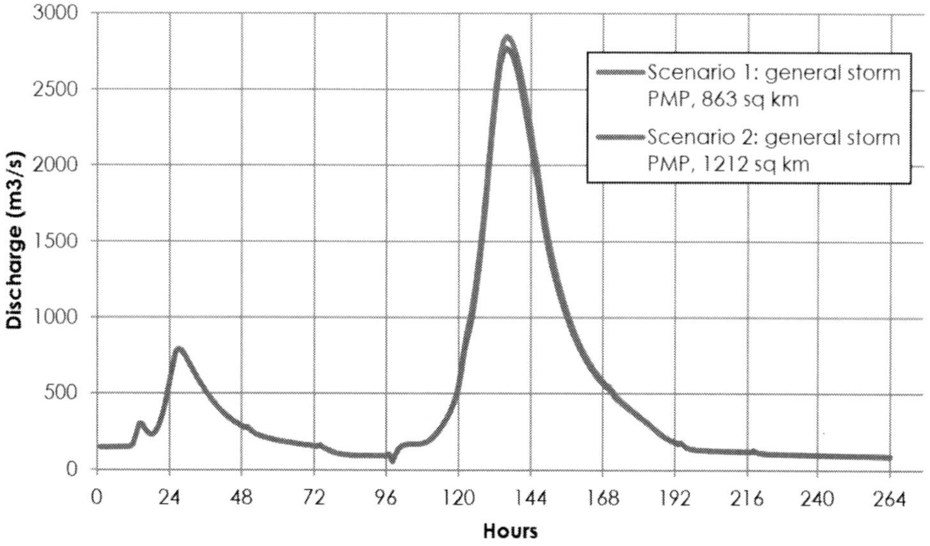

Figure 5: PMF Simulation Hydrographs for Scenarios 1 and 2 at Glenmore Dam

Table 5: PMF Results for Scenarios 1, 2, 3 and 4

Scenario	Peak Flow (m^3/s)			7-Day Volume (dam^3)		
	SR1 Diversion	Glenmore Dam	SR1 Off-Channel Flood Storage Dam	SR1 Diversion	Glenmore Dam	SR1 Off-Channel Flood Storage Dam
1	2,850	2,790	-	352,000	354,000	-
2	2,770	2,820	-	340,000	428,000	-
3	2,760	2,540	-	202,000	205,000	-
4	-	-	577	-	-	9,000

SUMMARY

A hydrological modeling approach for estimation of PMF was presented in this paper. The model parameters were primarily derived from physiographic characteristics and literature values. The HEC-HMS model was initially calibrated to the June 2005 and the June 2013 floods. For that purpose, AWA analyzed those storms and provided digital data for each subbasin that is representative of the actual temporal and spatial distributions of each of those storms. Due to limitations of the aerial extent of those storms and uncertainties in the streamflow data, the model calibration yielded preliminary conclusions. That calibration process was successful in developing appropriate rainfall loss parameter values, and in the development of appropriate watershed channel routing and baseflow methodologies. The calibration of the unit hydrograph methodology and parameter estimation was uncertain. This is attributed to the fact that historic rainfalls did not fully cover all of the model subbasins and rainfall intensities were not sufficiently uniform over the watershed and subbasins to meet the requirements of unit hydrograph theory. The final calibrated HEC-HMS model with PMP input for each of the four scenarios resulted in design PMF estimates at the SR1 diversion dam on the Elbow River and the SR1 off-channel flood storage dam. Although not a design requirement for the SR1 project, the PMF for Glenmore Dam was estimated as well.

Use of the State of Colorado guidelines resulted in reasonable estimates of the PMF for the Elbow River watershed in Alberta Canada. Those results were successfully compared to available data for sever regional floods. It is concluded that the Colorado guidelines can be successfully used for other hydrologically similar watersheds in the Rocky Mountains of the United States and Canada.

REFERENCES

Alberta Geological Survey (AGS). 2013. "Surficial Geology of Alberta: Ungeneralized Digital Mosaic (GIS data, polygon features)." Digital Data 2013-001.
Alberta Transportation. 2004. "Guidelines on Extreme Flood Analysis." Alberta Transportation: Transportation and Civil Engineering Division, Civil Projects Branch.
Applied Weather Associates, 2015. "Site-Specific Probable Maximum Precipitation Study for the Elbow River Basin-Springbank Off-Stream Storage Project." Pg230.
Elbow River Watershed Partnership. http://erwp.org/ [Accessed June 2016]

GeoGratis. 2015. http://geogratis.gc.ca/api/en/nrcan-rncan/ess-sst/3A537B2D-7058-FCED-8D0B-76452EC9D01F.html [Accessed February 2015]
Gerhard, Reuter. 2000. "Effects of Available Moisture on the Minimum Recurrence Interval between Two Major Rainstorms." Alberta Environmental Protection.
Sabol, G.V., Tierra Grande International, Inc. 2008. "State of Colorado Office of the State Engineer Dam Safety Branch Hydrologic Basin Response Parameter Estimation Guidelines." Denver, Colorado.

APPROACHES TO ESTIMATING INDIVIDUAL RISK FOR DAM SAFETY TOLERABLE RISK EVALUATION

David S. Bowles[1]
Jason Needham[2]

ABSTRACT

Dam safety tolerable risk guidelines have been defined for both individual and societal risk. Individual risk *is the chance that a particular individual … will be harmed* whereas societal risk characterizes *the scale of an accident in terms of numbers killed or harmed* (HSE 2012). However, societal risk (and annual probability of failure) has been the focus of tolerable risk evaluations for dams in the US. In contrast, individual risk is the sole focus in many other fields in many countries. Currently, USACE HEC-FIA and LifeSim software for estimating life loss are more suited for supporting decision-making based on the societal risk perspective.

This paper discusses the value of evaluating both individual and societal risk, and reviews some alternative ways of estimating individual risk, including considering actual persons vs. hypothetical persons. Two approaches are recommended for estimating individual risk using HEC-FIA or LifeSim in conjunction with DAMRAE risk analysis software. The first approach relies on information that is currently used for societal life-loss estimation. It is proposed for use in screening and issue evaluation study (IES) risk assessments. The second approach requires the construction of hypothetical persons. It is proposed for dam safety modification study (DSMS) risk assessments for informing decisions on selection of risk reduction measures and for judging the tolerability of residual incremental risk. A variation of the first approach would provide mapping of iso-lines of estimated individual risk, which could be useful for as-low-as-reasonably-practicable (ALARP) evaluations for DSMS risk assessments.

INTRODUCTION

Background and Importance of Individual Risk

The purpose of this paper is to review approaches to estimating individual incremental risk and to describe an approach that is suitable for use in the USACE dam and levee safety programs using the HEC-FIA and LifeSim software.

The USACE Engineering Regulation ER 1110-2-1156 for *Safety of Dams - Policy and Procedures* (USACE 2014) tolerable risk guidelines considers individual and societal (population) risks. As stated in HSE (2012) these are defined as follows:

[1]Managing Principal, RAC Engineers and Economists, LLC, Perry UT 84302 and Professor Emeritus of Civil and Environmental Engineering, Utah State University, Logan, UT 84322. David_S_Bowles@hotmail.com
[2] Consequences Specialist, Risk Management Center, US Army Corps of Engineers, Hydrologic Engineering Center, Davis, CA. Jason.T.Needham@usace.army.mil

... risks to people, rather than things such as property or the environment. Risks to people can be represented in two ways. Both are a combination of the likelihood of an event happening (e.g. an accident at a major hazard installation), and the possible consequences - in terms of harm to people:

- *'Individual risk' is the chance that a particular individual at a particular location will be harmed. It is usually described in numerical terms such as "a 1 in 18,000,000 chance of being killed by lightning". But assessment of individual risk does not take account of the total number of people at risk from a particular event.*
- *'Societal risk' is a way to estimate the chances of numbers of people being harmed from an incident. The likelihood of the primary event (an accident at a major hazard plant) is still a factor, but the consequences are assessed in terms of level of harm and numbers affected, to provide an idea of the scale of an accident in terms of numbers killed or harmed.*

The ER 1110-2-1156 Glossary defines individual risk as follows:

The increment of risk imposed on a particular individual by the existence of a hazardous facility. This increment of risk is an addition to the background risk to life, which the person would live with on a daily basis if the facility did not exist. (ANCOLD 2003)

On the topic of deciding whether it is appropriate to consider individual or societal risk, or both, HSE (2003) states that:

... it is important to consider the number of people that could be affected at the same time, from realisation of the hazards considered. It is always necessary to consider individual risk to determine whether the risk to the individuals of interest is acceptable, or not. If the hazards being considered have the potential to affect only individuals, or a few people at the same time it would be appropriate to only consider individual risk. However, if the hazards being considered have the potential to affect a large number of people at the same time, it would also be necessary to consider societal risk, even if the individual risk was estimated to be low.

To date, societal risk and annual probability of failure (APF) have been the focus of tolerable risk evaluation in the US. Therefore, life-loss estimation has focused on the societal risk perspective, which has been characterized and evaluated using the probability distribution of the total number of potential fatalities (i.e. F-N chart) and mean of that distribution, referred to as average annual life loss (AALL) or annualized life loss (ALL).

The evaluation of both individual and societal risks has value in different ways. Equity considerations that no person or group is disproportionately impacted by the risk of a dam or levee system breach can be examined through considering individual incremental risk. By considering societal risk the scale of potential life loss can be identified and taken into consideration in justifying the urgency, priority and magnitude of investment in risk reduction. Indeed, societal risk rather than individual incremental risk is used in estimating cost per statistical life saved and in cost benefit and incremental cost-effectiveness analyses. However, the understanding of the factors, which drive both individual and societal risk, can be valuable in formulating both structural and non-structural risk-reduction measures.

Since the numerical values of the USACE tolerable risk limit guidelines for APF and individual incremental risk are the same (i.e. 1 in 10,000 per year, see Figure 1), if the APF limit guideline is met, then the individual incremental risk limit guideline will also be met because individual incremental risk cannot be greater than APF. This is sometimes put forth as a reason not to estimate individual incremental risk. However, some of the reasons to estimate and evaluate individual incremental risk, include the following:

- The benefit of an improved ALARP evaluation that comes from the understanding that can result from estimating the spatial variation of individual incremental risk in the flood inundation area. This may help to identify cases for which individual incremental risk is high, even though it is not controlling the individual incremental risk evaluation; including the identification of inequities in the distribution of risk amongst certain types of individuals, such as the elderly, minorities who are poor or not English speaking, or those with handicaps, which affect their ability to receive and respond to evacuation warnings. In these situations, specific efforts can be made to reduce the risk as low as reasonably practicable (ALARP) for these individuals and groups, such as targeted warning and evacuation procedures.
- Another potential benefit of estimating individual incremental risk stems from the fact that individual incremental risk is less than or equal to the total annual probability of failure. This benefit applies to reservoirs with multiple dams (e.g. a main dam and several saddle dams such as Folsom Reservoir near Sacramento, California), long dams (e.g. Herbert Hoover Dike in Florida) and levee systems, which are long structures and often comprised of multiple structures. For such projects, it may be problematic to meet the APF guideline of 1 in 10,000 per year

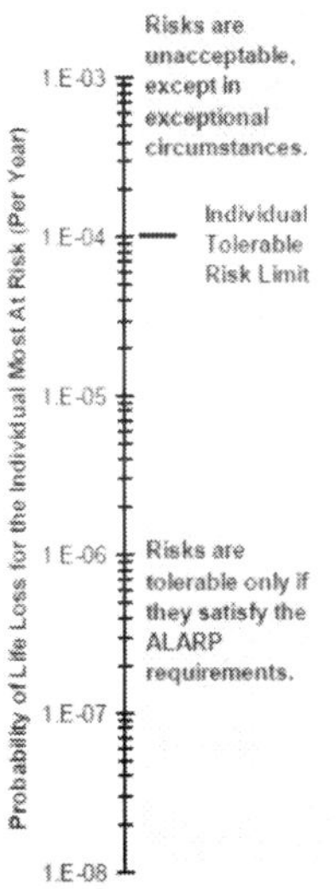

Figure 1. USACE Guideline for Incremental Risk (USACE 2014)

since the total of APF is a combination of the probabilities of failure for all dams on the reservoir or for the entire levee system. Even if the probabilities of failure for individual dams or for shorter levee reaches are below the APF limit guideline, they may accumulate to a probability of failure that significantly exceeds the APF limit guideline with no practical way to reduce the total to meet the APF guideline. For these types of structures individuals are likely not exposed to failures at all locations, but rather only to failures over levee reaches close to their location or failures at only one saddle dam or the main dam, which is located upstream of their location and not others for which a breach would flow into another valley. Therefore, in these cases it is only a part of the total APF that contributes to the individual incremental risk for each individual. Thus, an argument can be made for such long structures, that by focusing on satisfying the individual risk (and societal risk) tolerable risk limit guidelines and ALARP considerations, life safety has been given paramount consideration, even if the total APF for the entire project does not meet the tolerable risk limit guideline.

- Insights gained from estimating and evaluating individual incremental risk for long structures can also be valuable in prioritizing which dam on a reservoir or which part of a levee system (e.g. levee reach or closure structure) should be remediated first.

The examples given above have implications for how individual (incremental) risk (IR) is estimated and evaluated. Specifically, the first example would require that IR is estimated in sufficient detail to identify if disadvantaged groups are located within the flood inundation and to make sure that they are separately evaluated. In the second and third examples, IR would need to be estimated and evaluated at multiple structures on a reservoir or throughout a levee system and not just for a single "*person or group by location that is most at risk of loss of life due to dam breach*" for the entire reservoir or levee system. These issues and others affecting the details of how IR is estimated are discussed later in this paper.

USACE uses two simulation models for estimating life loss in its dam and levee system safety programs: HEC-FIA and LifeSim. Currently HEC-FIA is not well suited for supporting estimates of individual incremental risk; although an interim approach to using results from HEC-FIA to estimate individual risk has been developed. LifeSim is better suited to provide the necessary information for an individual incremental risk computation because it tracks individual people throughout a flood event. In addition, the life-loss computations currently performed in the Levee Safety Tool (LST), which is used for developing screening-level risk estimates for levee systems, yield only societal life-risk estimates. Therefore, LifeSim, which is just now becoming standard practice within USACE, is the life-loss estimation tool best suited for supporting individual risk estimates.

Section 5.3.5.3 of ER 1110-2-1156 provides important details on how the risk metric should be defined for estimating individual risk for use in evaluating the tolerable risk limit guideline for individual incremental life-safety flood risk for dams. There appears to be no requirement in ER 1110-2-1156 to evaluate individual risk for non-breach life safety flood risk. Also, for the purposes of this paper, tolerable risk guidelines for levee systems will be considered to be the same as those for dams. Separate guidance on tolerable risk is not provided for USACE staff, levee district staff or contractor personnel.

Extent of Use of Individual Risk

Consideration of societal risk has predominated in the US practice of dam and levee system safety risk assessments to date. In a review of US health and safety regulatory programs, Adler (2004) found that "*individual fatality risk plays a major role in our current system of health and safety regulation,*" although "*federal programs concerned with safety rather than health hazards generally seem to focus on "population risk"* (i.e. societal risk) *rather than "individual risk," and even health threats such as toxins, radiation, and pathogens are sometimes regulated with reference to "population risk."*

The Netherlands makes use of both individual and societal risk guidelines in its safety programs, such as those for the dikes (CUR TAW 1990 and ADD RECENT REFERENCE). In Australia, the ANCOLD (2003) guidelines on risk assessment for dams contain both individual and societal risk guidelines; although it is only recently that detailed guidance has been proposed for estimating individual risk for dams in Australia (Maslin et al 2014). The proposed approach is based on a Graham-type approach (Graham 1999) to estimating life loss and not a simulation approach as used by USACE. Reclamation (2011) views their annual probability of failure (APF) public protection guidelines as a conservative upper bound value of individual risk, and therefore, it does not estimate individual risk in its dam safety risk assessments.

However, the Health and Safety Executive (HSE) in the UK, which has been a world leader in the development of tolerable risk guidelines, focuses on individual risk, with minimal guidance on the evaluation of societal (population) risk. This is still the situation some 15 years after publication of its landmark document, *Reducing Risk, Protecting People* (HSE 2001), and despite extensive studies and consultation on the topic of societal risk. The HSE is not alone in its emphasis on individual risk: in fact, in many countries, there are many more examples of long-standing individual risk guidelines than societal risk guidelines.

The next major section contains a discussion of the estimation of individual incremental risk in the light of available USACE guidance and the approaches and guidance used by others to estimate individual risk for evaluation of life-safety tolerable risk. The third major section identifies and evaluates options for estimating individual incremental risk and describes suggested approaches for use in HEC-FIA and LifeSim for a range of flood scenarios including uncertainty. The final section contains a summary and conclusions.

A DISCUSSION OF INDIVIDUAL RISK ESTIMATION APPROACHES AND GUIDANCE

Overview

This section opens with a generalized concept for estimating individual incremental risk and identifies factors that affect individual risk. The USACE definition for individual risk, which is presented in the Introduction of this paper, and some additional USACE guidance on individual risk estimation and evaluation are analyzed and discussed in the second subsection to provide the basis for screening options for estimating individual risk and for developing the details for the proposed approaches to IR estimation. The issue of considering actual persons vs. hypothetical persons, including exposure considerations, is discussed in the following subsection and provides some additional basis for screening options and for developing the proposed approaches.

Some alternative constructs and risk metrics for individual incremental risk from other dam safety organizations and from health and safety evaluation and regulation in other fields are discussed in the final subsection. This is a focused discussion keeping in mind the purpose of this paper, which is to describe an approach for estimating individual incremental risk for the USACE dam and levee safety programs using HEC-FIA and LifeSim. It is not intended to be a general review of individual risk, which would be a major endeavor, and likely of limited relevance to dam and levee safety.

A Generalized Concept for Estimating Individual Incremental Risk

A generalized concept for estimating individual incremental risk, IR, for a person at risk of loss of life due to dam or levee system breach can be represented as follows:

$$IR = f(\text{Hazard, Performance, Exposure, Vulnerability}) \quad (1)$$

ER 1110-2-1156 defines the following five components of flood risk associated with dam or levee system breach that should be considered to estimate the flood risk as depicted in Figure 2:

1) **Hazard**: the external load on the dam or levee system (magnitude and likelihood of a hazard such as a flood or an earthquake) or an internal hazard, such as a flaw that can lead to the initiation of internal erosion;
2) **Performance**: the response of the dam or levee system to the hazard, which can lead to a breach or non-breach outcome;

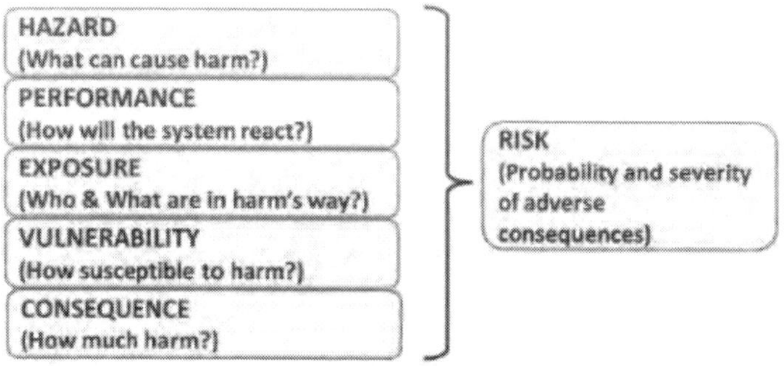

Figure 2. Components of Flood Risk for Dam and Levee System Breach (USACE 2014)

3) **Exposure**: the likelihood that people or objects (e.g. population at risk, property, infrastructure, etc.) may become subject to harm because of their location or some other relationship to the breach flood wave or other effects of a dam or levee system breach (this includes the likelihood of moving to another, perhaps safer, location);
4) **Vulnerability**: the susceptibility of an exposed person or object to harm; and
5) **Consequences**: a measure of the severity and scale of the harm caused (e.g. number of fatalities, dollar economic damages, environmental impacts, etc.).

For individual risk, the focus is on a type of consequence, that is, an individual fatality. The last three components listed above (exposure, vulnerability and consequences) are unique to the estimation of individual risk. The first two components (hazard and performance) are the basis for the estimating probability of failure of the dam or levee system, which also affects. the estimate of individual risk. Any interdependency between the first two components and the last three components should be considered when estimating individual risk.

Specifically, the likelihood and severity of exposure, vulnerability and consequences can each be dependent on the characteristics of the occurrence of a hazardous event or the response (performance) of the dam or levee system. For example, the magnitude of a flood that leads to a breach and the location, detectability, rate of development and mode of failure, can affect the likelihood and severity of exposure, vulnerability and consequences.

Table 1 lists factors that affect individual incremental risk grouped vertically by risk components and grouped horizontally by the following:

- Dam or levee system – brown font,
- Flood routing – blue font,
- Warning and evacuation systems – red font, and
- Individual, including shelter attributes based on their location – green font.

The factors in each column are entered in Table 1 using different color fonts for ease of tracking them later in the paper where alternative approaches for defining and estimating individual incremental risk are discussed for both breach and non-breach flood inundation cases. Exposure (e.g. time of the day, weekend/weekday, or season of the year) is not included in Table 1, although it can affect many of the factors that are included this table.

USACE Definition of Individual Incremental Risk

ER 1110-2-1156 defines individual incremental risk as follows:

> *The increment of risk imposed on a particular individual by the existence of a hazardous facility. This increment of risk is an addition to the background risk to life, which the person would live with on a daily basis if the facility did not exist. (ANCOLD 2003)*

Table 1. Factors Affecting Individual Incremental Risk

Risk Component and Definition	Factors affecting individual incremental risk			
	Dam or levee system	**Flood routing - Breach and Non-breach**	**Warning and evacuation systems**	**Individual including shelter attributes**
Hazard - *external load or internal condition that threatens the dam or levee system*	• Probabilistic flood hazard assessment • Probabilistic seismic hazard assessment • Initial reservoir level			
Performance - *response of the dam or levee system to the hazard, leading to a breach or non-breach outcome*	• Potential failure modes • Location of breach • Rate of breach development and size of breach			
Exposure - *likelihood that a person may become subject to harm at the time of arrival of a breach flood wave because of their location with respect to the dam or levee system*	• Detectability of failure mode • USACE notification decision time (Sorenson and Mileti 2015a)	• Extent and severity of flood inundation area (depth, velocity, etc.) • Travel time	• EMA warning issuance decision time (Sorenson and Mileti 2015a) • Warning messaging and channels • Evacuation system • Damage to infrastructure needed for warning or evacuation	• Presence (occupancy) in flooded area at time of warning issuance • Receipt of first alert (visual/ sensory impairments, etc.) (Sorenson and Mileti 2015b) • Time to initiate protective action (including non-response) (Sorenson and Mileti 2015c) • Personal characteristics (age, height, weight, physical condition, disability, etc.) • Mobility (including incarceration) • Successful evacuation out of flood inundation area • Successful vertical evacuation
Vulnerability - *susceptibility of an exposed person to harm*			1) Rescue if initially survive	• Fate of structure or vehicle • Highest accessible level in structure • Personal characteristics (age, height, weight, physical condition, disability, etc.)

In the context of this paper, a dam or a levee system is the hazardous facility referred to in the above definition. The risk that we are considering is associated with the breach of a dam or levee system due to all hazards and failure modes that are included in the risk analysis.

Section 5.3.5.3 of ER 1110-2-1156 (USACE 2014) provides the following important guidance on the specification of a risk metric for individual incremental risk for the USACE dam safety program:

1) Individual risk should be calculated as an **incremental risk** defined as follows: "*the increment of risk imposed on a particular individual by the existence of a hazardous facility* [i.e. dam or levee system]." "*This increment of risk is in addition to the background risk to life, which the person would live with on a daily basis if the facility did not exist.*" (USACE 2014, Glossary). However, this definition is at odds with (or perhaps qualified by) the definition of "*incremental risk*" presented in Section 2.4.5.4.1 of ER 1110-2-1156 as "the risk (likelihood and consequences) to the pool area and downstream floodplain occupants that can be attributed to the presence of the dam should the dam breach prior to or after overtopping, or undergo component malfunction or misoperation." For dams and levee systems it is conventional to define incremental risk as the difference in risk between the following two cases: i) the structure exists and consideration is given to the likelihood that it may breach or that it may malfunction and result in a non-breach unintended flooding condition (e.g. spillway gates for a dam do not open on demand or spillway gates experience a structural failure); and ii) the structure exists but it is assumed that it does not breach or malfunction. It is noted that the second case is not the condition that the facility does "*not exist,*" as stated in the USACE definition given above. The likely reason for the practice of using the non-breach case ii) for the dam or levee system in place to estimate incremental risk condition is that this is the situation that people have come to expect and that they have adjusted their pattern of life to. This suggests that the USACE definition of individual incremental risk given above should be modified by changing the last word in the definition from "*exist*" to "breach." This change is consistent with the definition of individual risk in Hartford and Baecher (2004).

   It is noted that all other "*background risk to life*" (e.g. the risk of death by cancer, a traffic accident or an act of violence) is associated with both cases i) and ii) described in the previous paragraph. Therefore, there is no need to calculate this background risk because it would be identical in both cases i) and ii), and it will net out in the calculation of incremental risk.

2) The "individual" that is considered should be an "**identifiable person or group by location.**" (USACE 2014, Section 5.3.5.3.1).

3) For the tolerable risk evaluation, the focus should be on the individual "that is **most at risk of loss of life** due to dam breach." (USACE 2014, Section 5.3.5.3.1). Clearly, individual incremental risk attributed to dam and levee systems varies

spatially throughout the flood inundation area and with factors such as life style and personal attributes, such as age and personal wellbeing or disabilities.

4) "**All exposure conditions**" for the individual should be considered. This seems to imply that the likelihood that persons would be present in the flood inundation area at the time that a first alert or warning to evacuate is issued should be considered as opposed to making an arbitrary assumption that the person is always present if that is not the case[3]. (USACE 2014, Section 5.3.5.3.1).

5) "**All failure modes** associated with **all loading or initiating events**, with due regard for non-mutually exclusive failure modes" should be considered. (USACE 2014, Section 5.3.5.3.1).

6) In the case of reservoirs with "**multiple structures**," and for tolerable risk evaluation, "*individual incremental risk should be checked below the main and each auxiliary structure (e.g., dike, levee, saddle dam)*". The logical extension of this statement to levee systems would seem to be that "*individual incremental risk should be checked*" at all locations protected by a levee, considering the effects of all breach locations or other points of ingress of water, such as a closure structure that remains open, in a levee system. (USACE 2014, Section 5.3.5.3.4).

7) "*The probability of life loss is based on*":

   a. "*the **probability of failure***",
   b. "*the **exposure factors** to characterize the day-night, seasonal, warning, or other exposure scenarios*",
   c. "*the **conditional probability of life loss given exposure** to the dam failure flood*" [vulnerability].

8) The level of detail that is appropriate for use in characterizing exposure factors should be "***decision driven***." (USACE 2014, Section 5.3.5.3.5).

Section 5.3.5.3 of ER 1110-2-1156 does not specifically mention how or if the following factors should be considered in estimating individual risk for USACE risk assessments:

1) **Individual attributes** such as age, disability, physical condition, height, access to transportation, access to social media, etc.,
2) **Actual persons vs. hypothetical persons**, and
3) **Uncertainty**.

Actual Persons vs. Hypothetical Persons

The term, "*identifiable person or group by location*," is used in ER 1110-2-1156 for describing the risk metric for individual incremental risk (see Introduction). The following quotation from HSE (2009) clarifies what is meant by an "*identifiable*"

[3] Note: Salami slicing is not addressed (see section on USACE Definition of Individual Incremental Risk).

individual or "*a particular person*" having "*some fixed relation to a hazard*" for the purpose of estimating individual risk. It also makes an important distinction between individual risk and the statistical risk to an individual averaged over an entire population.

> *The likelihood that a particular person in some fixed relation to a hazard (e.g. at a particular location, level of vulnerability, protection and escape) might sustain a specified level of harm.*
>
> *The frequency at which an individual may be expected to sustain a given level of harm from the realisation of specified hazards. For example, there may be an individual risk of 1 in a million that a particular person would be killed by an explosion at a major hazard near their home for every year that the person lives at that address.*
>
> *Note that this refers to a particular person or a typical member of a particular group e.g., 'Mr. Brown of Sycamore Villa' or 'an inhabitant of the Woodlands Estate'. There is also often a wider use of the term when the risk from a hazard is averaged over the whole population e.g., '5 000 road deaths per year implies an individual risk of 1 in 10 000 per year'*[4]*. This is a crude average and it conceals very wide variations between types of people (children, elderly, taxi-drivers, cyclists etc.). These variations should be borne in mind when comparing risks derived by QRA (*quantitative risk analysis*) with statistics of the risks from everyday life. Note that the level of individual risk is the same, whether there is actually just one person at risk or many people at the same risk.*

Under the HSE (2001) Tolerability of Risk (TOR) framework, individual risk is assessed in relation to hypothetical persons whose circumstances and exposure to the risk are representative of those of the main groups of people affected. This approach avoids various limitations associated with considering actual persons, in addition to the enormous effort that would be involved and the likely impracticability due to privacy issues. To ensure that all significant risks for a particular hazard are adequately covered, a number of hypothetical persons should be constructed.

HSE (2001) defines "*a hypothetical person as a hypothetical type of individual who is deliberately assumed to have some fixed relation to the hazard under consideration* e.g. *the person most exposed to it or a person living at some fixed point or with some assumed pattern of life.*"

HSE (2003) states that "*The aim in considering risks to hypothetical persons is to provide a 'full picture' of the risks generated by a hazard*" [i.e. dam or levee system breach] and that "*it is important that the hypothetical person is truly representative of the group of people they are designed to represent. ...for example, to estimate the risk over a typical working week.*"

[4] The implied calculation here is 5,000 road deaths divided by an approximate UK population of 50 million = 5,000/50,000 000 = 1 in 10,000 per year.

HSE (2001) states that "*by creating enough hypothetical persons to enable control measures to be put in place to protect all those exposed from the undesirable consequences of the hazard, taking account of the different populations exposed and the circumstances of their exposure. This technique has the merit of preventing risk being underestimated by making clear whether a generic assessment of the risks on its own is adequate, or whether it should be supplemented by other assessments pertaining to:*

- *particular groups of persons interacting with the hazard in a certain way or who are particularly vulnerable to it;*
- *a slice of time;*
- *particular locations.*

HSE (2001) provides the following discussion of how exposure ("*slice of time*" in the previous quote) can be addressed. It states that the concept of hypothetical persons "*deals elegantly with the phenomenon that exposure to many hazards is not uniform* [over time] *but comes in peaks and troughs. This, if present, must be factored in when determining the exposure of any exposed population by creating as necessary one or more hypothetical persons to take this into account. For example, the period of exposure of the hypothetical person could be time weighted and/or more than one hypothetical person could be constructed to deal with the various attributes of the exposure to the hazard.*"

Examples of the variation of exposure to dam and levee system breaches over time include the likelihood of floods being greater during certain times of the year, seasonal variations in recreational populations, and the greater difficulty associated with warning people at nighttime compared with daytime. The weights for "periods of exposure" referred to are typically termed "exposure factors" in dam and levee system risk assessment. An example would be for nighttime and daytime conditions to be weighted by 10/24 and 14/24, respectively, based on their fractions of a 24-hour day to which these conditions pertain.

Dr. Jean Le Guen, who is the principal author of the HSE's landmark publication, *Reducing Risk Protecting People* (HSE 2001), has written the following (Le Guen 2010) in regard to defining hypothetical persons for dams or levee systems:

> *"most of the hypothetical persons mentioned above would be protected by the one living permanently at the bottom of the levee, including the shopper – so not all persons will require special consideration."*
>
> 1. *A person who lives in the shadow of a dam or levee* [owner or operator organization], *who is there 24 hours of the day all year round, whose family and whose home are in close proximity to the dam.*
> 2. *An employee of the dam or levee, who does shift work, but whose work takes them onto the structures.*
> 3. *A member of the emergency services who might be involved in directing operations in the event of a crisis.*

4. *A person who lives further away from the dam or levee, but who might be affected if measures were taken to divert water to avoid overtopping.*
5. *A shopper in a development near the dam or levee who in the hypothesis is there throughout opening hours, though would be representing the reality of a series of shoppers visiting in sequence.*

In practice, measures to protect the most exposed of these hypothetical people in many cases will protect others, specifically, *most of the hypothetical persons mentioned above would be protected by the one living permanently at the bottom of the levee, including the shopper - so not all* hypothetical *persons will require special consideration*, or detailed analysis because a case can be made *a priori* that some are inferior to others in terms of their magnitude of individual risk.

Transient Individuals and "Salami Slicing"

However, for **transient individuals** HSE (2003) provides the following advice:

> *"Estimating the risk in cases of 'transient exposure' - avoiding 'salami slicing'. When considering the risk from a particular activity, situation or process to which individuals are exposed for only a short time, care is needed to ensure that an accurate picture of the risk is obtained. It is not appropriate to divide the time spent on the hazardous activity between several individuals and estimate the risk on this basis (the 'salami slicing' technique). For example, if any one person is only exposed to the hazard for a short time, but someone is always exposed, it would give a misleading picture of the risk to estimate the risk from this hazard by taking into account the exposure time of each individual. Instead, a truer picture of the risk would be obtained by constructing a hypothetical person who is exposed to the hazard 100% of the time (or for the proportion of time that any individual is exposed). In this case, it would be necessary to define precisely the hypothetical person's location/interaction with respect to the hazard."*

Some Alternative Constructs for Individual Risk

A range of approaches exist for estimating individual incremental risk. At one extreme, the individual incremental risk could be estimated and mapped throughout the flood inundation area to show how these estimates vary spatially. From that comprehensive picture of individual incremental risk, the "*person or group by location that is most at risk of loss of life due to dam breach*" could be identified. At the other extreme, one can attempt to identify the "*person or group by location that is most at risk of loss of life due to dam breach*" a priori, without the benefit of estimating how individual incremental risk varies spatially.

For the case of considering all persons in the flood inundation area, obtaining detailed information to fully characterize the exposure and vulnerability of each actual individual would be an enormous task, especially for large populations at risk. It is also unlikely that, even if available, the necessary information to support an actual-person approach could be obtained due to concerns about privacy. Therefore, if the goal is to estimate individual incremental risk throughout the flood inundation area some simplifications will be needed, especially for exposure and vulnerability factors.

Identifying the "*person or group by location that is most at risk of loss of life due to dam breach*" may be quite straightforward in some cases where there is clearly a small group that, because of its location and lifestyle, are most at risk. In other cases, it may require that several candidate individuals be identified, their individual incremental risk be estimated, and the one most at risk be identified from among the candidates. An approach to identifying these candidates is to use the principles of "hypothetical individuals" as proposed by the HSE (2001).

Zelinski (2014) identifies four alternative risk measures for individual risk, which he considers to be potentially applicable to dam safety. Each is summarized below and evaluated against USACE and other guidance on estimating IR to determine if it is a candidate for consideration as an approach for potential adoption by USACE:

- **Location-specific individual risk**: The risk to a hypothetical individual who is i) always present at a specific location, including when the dam or levee breach flood wave arrives at their location, and that ii) they are not protected from the effects of the flood wave. These two requirements would rule out consideration of this metric for use in the USACE dam and levee safety programs since exposure scenarios are not considered. As defined by these two requirements, this risk metric is largely a property of the locations for which it is estimated, and it can be represented on a map by contours of constant levels of IR based on considering many hypothetical individuals located throughout the flood inundation area with assumptions about their vulnerability.
- **Maximum location-specific individual risk**: The highest value of location-specific individual risk. Although the idea of obtaining a maximum value of IR is consistent with the USACE guidance for identifying the individual "*most at risk*" because this maximization is over the location-specific individual risk estimates, it is ruled out for use in the USACE dam and levee safety programs since exposure scenarios are not considered.
- **Individual risk averaged over the exposed population**: By estimating only an average value of IR, the variation in IR over the exposed population would be masked and hence, this is a significant limitation of this approach. Therefore, this risk metric is also ruled out for use in the USACE dam and levee safety programs since averaging over the exposed population does not meet the USACE requirement for identifying the individual "*most at risk.*" This risk metric is more suitable for estimating average annual life loss, which is used as a measure of societal risk, rather than IR, when it is defined for the individual "*most at risk.*" However, this approach may have some value if there is little variation in IR over the exposed population. What assumptions should be made for averaging? Same as for location specific IR?
- **Individual risk averaged over the duration of exposure**: Exposure is averaged over the year, thus if a person is present at a camp site for only one week in a year

their IR would be less, by a factor of 52, than for someone who is permanently present at the same location. This assumes no other differences between these two individuals. It also ignores the seasonality of floods, for example, such that if the person was only present in the non-flood season then it would be unreasonable to include the contribution to the probability of failure from floods in this time average. This approach considers exposure scenarios, which is a USACE requirement, but is subject to the concern about "salami slicing;" although this would not necessarily rule it out for use in the USACE dam and levee safety programs since the USACE guidance does take a position on "salami slicing." HEC-FIA and LifeSim software appear to avoid the "salami slicing" concern.

The first two examples of risk metrics for IR, listed above, would be expected to yield conservative estimates of IR due to the assumptions that the hypothetical individual is always present at a specific location and that they are not protected from the effects of the flood wave, thus giving no credit for shelter or successful evacuation. In contrast the last two examples could underestimate the "*the probability of life loss for the identifiable person or group by location that is most at risk of loss of life*" because of the effects of averaging over the exposed population or the duration of exposure.

In a review of "individual risk constructs" used by federal health and safety regulatory agencies Adler (2004) identified a range of approaches. Some of these are summarized below and evaluated against USACE and other IR guidance to determine if it is a candidate for consideration as an approach for potential adoption by USACE:

- **Reasonably maximally exposed individual**: "*EPA's rule for cleanups under the Superfund statute ... toxic waste dumps are to be remedied so that the lifetime fatality risk to the "maximally exposed individual" from carcinogens in the dump is within the range of 1 in 10,000 to 1 in 1 million. ... EPA focuses on risk given a "reasonable maximum exposure" rather than looking to the single maximally exposed individual.*"
- **Highly exposed individual**: "*EPA's decision to list a carcinogenic substance as a "hazardous waste," subject to stringent regulation under the Resource Conservation and Recovery Act, depends on the fatality risk that the substance would impose upon highly exposed individuals if discarded in unregulated landfills. If this risk exceeds 1 in 10,000, the substance is listed.*"
- **High-end consumer**: the FDA sets acceptable levels of microbial contaminants in foods with reference to the individual risk of illness of a high-end consumer.
- **Average individual**: In its ultimate safety goals governing the licensing and regulation of nuclear power plants, the Nuclear Regulatory Commission defines "*the individual risk of immediate death, resulting from an accidental release of radiation, as that incurred by the average person living near to a plant.*"

Apart from the last example, these individual risk constructs seem to be formulated to apply to the person "*most at risk,*" although they apply to a lifetime exposure rather than a per year exposure as considered in dam and levee safety tolerable risk evaluation.

Lifetime exposure is appropriate for the hazards to which the first three risk constructs apply since they require time for their harmful effects to develop.

The last example listed above is for exposure to "*an accidental release of radiation,*" which is a discrete event, similar to a dam or levee breach. However, it is also ruled out for use in the USACE dam and levee safety programs because it considers only the average person, presumably in terms of their vulnerability to the "*risk of immediate death,*" and does not meet the USACE requirement for identifying the individual "*most at risk.*" It is noted that considering the average person in this risk metric is not the same as estimating "*individual risk averaged over the exposed population,*" which is the third risk metric in the previous list.

Adler (2004) makes the case that individual risk can be considered to be a probability in the Bayesian sense rather than in the frequentist sense. He makes this case based on the important role of expert judgment (degree of belief) in developing a dose-response curve and then using it to predict individual exposures in a risk assessment for toxins. Hartford and Baecher (2004) point out that both the Bayesian and frequentist interpretations can be found in dam (and levee safety) risk assessment. They also make the case that it is valid to combine these two perspectives in dam safety risk assessment.

Sometimes the wording of a regulation or an individual risk guideline specifies how the effect of knowledge (epistemic) uncertainty should be addressed in implementation. Adler (2004) cites the example of the wording, "*reasonable certainty [of] no harm,*" in the EPA standard for possible carcinogens and pesticides that might cause other toxic effects [21 U.S.C. §346a(b)(2)(A)(ii)(2000)]. However, the qualifier, "reasonable" appears to be open to interpretation.

OPTIONS AND RECOMMENDED APPROACH FOR ESTIMATING INDIVIDUAL INCREMENTAL RISK FOR HEC-FIA AND LIFESIM

Overview

In the previous section, USACE and other guidance for defining an individual incremental risk metric were discussed and several risk metrics proposed for dams and other industries were presented. None of these met the USACE guidance, so it appears that a new risk metric will need to be proposed for use in the USACE dam and levee system risk management programs. However, since the context of this paper is to propose an individual incremental risk metric for use in the HEC-FIA and LifeSim software, we first review how these softwares, and related software (HEC-RAS and DAMRAE), currently handle factors that are related to those needed to estimate individual incremental risk. This review is presented in first subsection.

In the second subsection, we revisit the USACE and other guidance for defining a risk metric and identify some practical decisions or choices that must be made in formulating an approach for estimating individual incremental risk for HEC-FIA and LifeSim. These choices relate to the degree of effort that will be required to meet some of the USACE and non-USACE guidance for an individual incremental risk metric and the inevitable tradeoff between the degree of effort and the quality of the estimate. In this regard, it is important to keep in mind that USACE uses risk assessment for supporting different types of decisions in its dam and levee system risk management programs with several examples given in the next paragraph. The frame of reference for this discussion is provided by the following statement in ER 1110-2-1156 (USACE 2014, Section 5.3.5.3.5) that "*The level of detail ... should be "decision driven."*"

In the last subsection, we propose two approaches for estimating individual incremental risk, the first of which would rely mainly on information that is currently used for societal life-loss estimation and which is proposed for use in screening and IES risk assessments. The second approach would require the construction of hypothetical persons. It is proposed for use in a DSMS risk assessment for informing long-term decisions on the selection of risk reduction measures and for judging the tolerability of residual risk where a higher level of confidence is needed that individual incremental risk has been considered in an appropriate degree of detail. A variation of the first approach would provide IR mapping that could be useful for ALARP evaluation for DSMS risk assessments.

Current Role for USACE Software in Societal Risk Estimation

Since the context of this paper is to propose an individual incremental risk metric for use in the USCAE HEC-FIA and LifeSim software, we review here how these softwares currently handle factors that are related to those needed to estimate individual incremental risk. The following three types of USACE software, or similar software, have established roles in estimating and evaluating USACE tolerable risk limit guidelines for societal risk in the USACE dam and levee safety risk management program, which can be summarized as follows:

- **HEC-RAS** for flood routing for breach and non-breach cases,
- **HEC-FIA or LifeSim** for life-loss estimation, including warning and evacuation considerations, and
- **DAMRAE** for risk analysis calculations and evaluation of risk estimates against tolerable risk guidelines.

The following provides more detail on the current roles of these three types of software, which currently focus on societal risk:

- **HEC-RAS**: Provides estimates of flooding characteristics (i.e. depth and velocity vs. time) throughout the inundation area for flood scenarios defined by the risk analyst, such as follows:
  - Non-breach runs for a range of flood magnitudes.
  - Flood-related breach runs for a range of failure modes (and associated breach parameters including breach location), for a range of flood magnitudes.

- Non-flood related breach runs for a range of failure modes (and associated breach parameters including breach location), for a range of initial reservoir levels.

These flood scenarios may be run for the existing dam or levee system and for various structural and non-structural risk reduction alternatives. Currently there is no uncertainty mode in HEC-RAS and so uncertainty must be considered by running a range of sensitivity cases.

- **HEC-FIA or LifeSim**[5]: Provides estimates of life loss for individuals originating in structures throughout the inundation area, for combinations of the flood scenarios run in HEC-RAS and exposure scenarios defined by the risk analyst, such as follows:
  - Various times of the day at which a public warning is issued with their associated initial spatial distribution of the population at risk.
  - Various seasons of the year with their associated initial spatial distribution of the population at risk.

  The HEC-FIA and LifeSim models use the dynamic spatially-distributed flooding estimates from HEC-RAS and they also simulate other processes such as warning dissemination and response, protective action (mobilization), evacuation and the fate of structures. Estimates of uncertainties are provided when run in uncertainty mode.
- **DAMRAE**: Provides estimates of the probabilities of failure, average annual life loss, average annual economic loss, distributions of life loss and economic losses, and comparisons against USACE tolerable risk limit guidelines for APF and societal risk. Similar information, without the risk evaluations, is provided for the non-breach case. These results can be obtained for the existing dam or levee system and for various structural and non-structural risk reduction alternatives, for which estimates of risk reduction, benefit cost and cost effectiveness of risk reduction are also provided. The inputs to DAMRAE include the probabilities of external and internal (hazard) initiating events, system response (performance) probabilities and societal life loss and economic consequences for the various flood-exposure scenarios considered in HEC-RAS and HEC-FIA or LifeSim. These scenarios should be representative of a continuum of breach and non-breach flood magnitudes and a range of exposure conditions, such that DAMRAE can interpolate between the HEC-FIA or LifeSim scenario estimates to cover the full range of floods and exposure conditions. Estimates of uncertainties are provided when run in an uncertainty mode.

These three types of software also have a potential role in estimating individual incremental risk based on considering breach and non-breach flood inundation cases. Table 2 contains the factors affecting individual incremental risk, which were listed in Table 1, and assigns them to the three types of software. These factors can be tracked back to the groups of factors affecting individual incremental risk, which are shown in

[5] It is noted that there are some significant differences in the capabilities of HEC-FIA and LifeSim. These differences will be important to consider in the technical design for implementation. However, here the focus is on the role of these software in estimating and using life loss for societal tolerable risk evaluation, which is similar, and so the differences are not discussed here.

different columns in Table 1 by recognizing the different font colors that are assigned to each of the following four groups: dam and levee system (brown font); flood routing (blue font); warning and evacuation systems (red font); and individual including shelter characteristics (green). Exposure factors, which are not included in Table 1, are entered in several places in Table 2.

Table 3 summarizes the current ways in which factors affecting exposure and vulnerability aspects of individual incremental risk, from Table 2 (and Table 1), are addressed in HEC-FIA and LifeSim, keeping in mind the current focus on societal risk rather than individual risk. It is noted that several important relationships in HEC-FIA and LifeSim are based on characterizing population-wide variability rather than the characteristics of actual or hypothetical individuals. These include relationships for the protective action response and fatality probability distributions for flood lethality zones in HEC-FIA and LifeSim.

The precise role of each type of software for estimating individual incremental risk will depend on the specific approach that is selected for defining and estimating individual incremental risk. This can be expected to require additional types of inputs, model functionality and type of outputs for HEC-FIA and LifeSim.

Guidance and Options for Estimating Individual Incremental Risk

None of the alternative constructs for individual incremental risk reviewed in the previous section were found to satisfy the guidance in USACE policy and procedures and other guidance on estimating individual risk. In addition, the current HEC-FIA/LifeSim and DAMRAE software are designed for estimating life loss for societal tolerable risk evaluation rather than individual tolerable risk evaluation; and will therefore, require some modification to add the new capability for estimating and evaluating individual incremental risk.

The following discussion focuses on choices about whether, and to what degree in some cases, USACE and non-USACE guidance for an individual risk metric can or should be met in formulating a new individual incremental risk metric.

1) ***"Identifiable person or group by location"*** **- Actual vs. hypothetical persons:**
    a. **Actual persons** – to consider actual individuals would be impractical since the needed information is not available, and if it were there would be privacy issues about using it. Information available from Census and other databases provides information to characterize some relevant aspects of a population at risk statistically, and therefore, there is an "averaging" effect that reduces the variability below the individual to individual person variability. This would be expected to result in understating the extremes of unusual combinations of characteristics that are associated with a person who is the "individual most at risk." This would likely result in a

Table 2. Association of Factors Affecting Individual Incremental Risk and USACE Software and Associated Inputs

Risk Component and Definition	USACE Software		
	DAMRAE: **Dam and levee risk analysis**	**HEC-RAS:** **Flood routing - breach and non-breach**	**HEC-FIA or LifeSim:** **Life-loss estimation including warning and evacuation**
Hazard - *external load or internal condition that threatens the dam or levee system*	• Season of year • Probabilistic flood hazard assessment • Probabilistic seismic hazard assessment • Initial reservoir level		
Performance - *response of the dam or levee system to the hazard, leading to a breach or non-breach outcome*	• Potential failure modes • Location of breach	• Rate of breach development and size of breach	
Exposure - *likelihood that a person may become subject to harm at the time of arrival of a breach flood wave because of their location with respect to the dam or levee system*		• Season of year • Extent and severity of flood inundation area (depth, velocity, etc.) • Travel time	• Time of day, weekend/ weekday, season of year • Detectability of failure mode • USACE notification decision time • EMA warning issuance decision time • Warning messaging and channels • Evacuation system • Damage to infrastructure needed for warning or evacuation • Presence (occupancy) in flooded area at time of warning issuance • Receipt of first alert (visual/sensory impairments, etc.) • Time to initiate protective action (including non-response) • Personal characteristics (age, height, weight, physical condition, disability, etc.) • Mobility (including incarceration) • Successful evacuation out of flood inundation area • Successful vertical evacuation
Vulnerability - *susceptibility of an exposed person to harm*			• Rescue if initially survive • Fate of structure or vehicle • Highest accessible level in structure • Personal characteristics (age, height, weight, physical condition, disability, etc.)

Table 3. Current ways in which exposure and vulnerability factors of individual incremental risk are addressed in HEC-FIA and LifeSim

Risk Component and Definition	Factors affecting individual incremental risk	HEC-FIA	LifeSim
Exposure - *likelihood that a person may become subject to harm at the time of arrival of a breach flood wave because of their location with respect to the dam or levee system*	Time of day, weekend/ weekday, season of year	Exposure scenarios	
	Detectability of failure mode	Contributes to estimate of warning issuance time relative to time of initiation of breach	
	USACE notification decision time	Contributes to estimate of warning issuance time relative to time of initiation of breach	
	EMA warning issuance decision time	Contributes to estimate of warning issuance time relative to time of initiation of breach	
	Warning messaging and channels		
	Evacuation system	Simplified consideration as shortest perpendicular distance to inundation boundary and a constant travel speed; and wheatear the vehicle or person taking this route is overwhelmed by the flood.	Evacuation traffic modeling and wheatear the vehicle or person is overwhelmed by the flood.
	Damage to infrastructure needed for warning or evacuation	Not explicitly considered	
	Presence (occupancy) in flooded area at time of warning issuance	Numbers of people are assigned to structures based on structure occupancy type and generalized Census and other data bases and not for actual or hypothetical persons	
	Receipt of first alert (visual/sensory impairments, etc.)	Based on population-wide warning dissemination relationships as a function of warning system type and population-average activity profile. Being changed to an approach developed by Mileti and Sorenson (2015) and Sorenson and Mileti (2015d)	
	Time to initiate protective action (including non-response)	Based on population-wide mobilization relationships and a maximum mobilization percentage estimated or each community. Being changed to an approach developed by Mileti and Sorenson (2015) and Sorenson and Mileti (2015d)	
	Personal characteristics (age, height, weight, physical condition, disability, etc.)	Not considered apart from age – see Successful vertical evacuation under Exposure. These personal characteristics are embedded in population-wide relationships for the protective action response and fatality probability distributions for flood lethality zones in HEC-FIA/LifeSim	
	Mobility (including incarceration)	Not considered	
	Successful evacuation out of flood inundation area	See Evacuation system under Exposure	See Evacuation system under Exposure
	Successful vertical evacuation	Assumed that those under age 65 reach highest accessible level in structure and that those age 65 or older remain on the level in the structure to which they were originally assigned	
Vulnerability - *susceptibility of an exposed person to harm*	Rescue if initially survive	Not considered	
	Fate of structure or vehicle	Considered based on stability relationships for structures, people and vehicles and flood characteristics at the structure of vehicle location	
	Highest accessible level in structure	See Successful vertical evacuation under Exposure	
	Personal characteristics (age, height, weight, physical condition, disability, etc.)	See Personal characteristics under Exposure	

tendency to underestimate individual incremental risk when relying on these sources. However, such sources of information are readily available and they are currently used in HEC-FIA/LifeSim to address the societal risk perspective.

   b. **Hypothetical persons** – given the impracticality of considering actual individuals and the limitations of using Census and other databases, the hypothetical person concept developed by the HSE (2001) would likely yield the best quality estimates of IR when the additional effort to implement this approach is justified and when the approach is properly applied. This will require the development of guidance and the skills to construct hypothetical persons.

2) "***Individual most at risk***":
   a. This requirement rules out approaches that are based on identifying or considering only a person with average or median exposure since this would underestimate IR.
   b. It also rules out approaches that are deliberately overly conservative, such as those that assume that an individual is 100% exposed to the dam or levee system breach flood with no potential for evacuation, unless such a person actually exists. The most extreme example of a conservative approach is to assume that IR equals the APF. However, there may be some cases where overestimation is appropriate. An example would be where the individual tolerable risk limit guideline is met based on a conservative estimate of IR, but the extra effort to obtain a more realistic estimate is not justified. Even so, it would be questionable whether ALARP considerations have been met without some attempt to identify individuals who are most at risk, even if their risk is less than the IR tolerable risk limit guideline of 1 in 10,000 per year, because a reasonably cost-effective way may exist to further reduce their risk.
   c. A complete maximization process could be performed by estimating the IR for all actual members of the PAR and then finding the individual with the largest IR. However, such an approach would be problematic for the reasons related to considering actual person given in 1 above. A practical approach might be to use information from the Census and other databases, but as stated in 1 above, this would likely underestimate IR for the individual most at risk.
   d. Another approach to identifying the "*individual most at risk*" a priori, using judgment, or by selecting and estimating IR for a few actual individuals, who appear to be candidates for the "*individual most at risk,*" and then determining which has the largest estimated IR. Alternatively, the approach of considering just a few candidate individuals can be applied using hypothetical individuals.

3) **Spatial and population variability**:
   a. **Spatial variability** is important in estimating individual incremental risk because the severity of the flooding is generally greater near the dam or levee system structures but the warning time is generally less in these areas.
   b. **Population variability** is also important in estimating individual incremental risk. This includes individual attributes such as age, disabilities, physical

condition, height, access to transportation, access to social media, non-English speaking, etc., and differences in the way that an individual interacts with the world around them, such as reception to channels over which first alert messages are transmitted, the characteristics of the structure in which they are located, the characteristics of the evacuation route that an individual must take, etc.

c. **Exposure and vulnerability** are both affected by factors that are spatially variable and factors that vary over the population at risk. It is the fact that these types of variability exist that leads to IR being different across individuals, and therefore, the need to identify the "*individual most at risk*" in an IR tolerable risk evaluation, and the need to understand the effects of these variabilities on IR to evaluate ALARP opportunities for risk reduction.

d. **IR mapping** can be expected to provide insights as to how IR varies spatially, which might be affected by factors that vary over the population, such as areas where non-English speaking people live, or areas where evacuation options are poor perhaps due to routes that are subject to traffic congestion, or that pass through low lying areas that are subject to early flooding or traffic congestion.

4) ***"All exposure scenarios"***:

a. Exposure varies continuously as individuals move from one location to another and in and out of structures of different types and during an emergency as people evacuate. The initial exposure of people as represented in HEC-FIA and LifeSim is determined through their assignment to structures at the outset of a flood-exposure scenario simulation based on the time of day at which a public warning is issued. As the simulation proceeds, their locations, and therefore their exposure, may change if they take protective action, and this will therefore affect their vulnerability as a combination of their personal attributes and the attributes that determine flood lethality. As a practical matter, only a discrete number of exposure scenarios can be considered when using HEC-FIA and LifeSim to estimate life loss from the societal risk perspectives. These may be based on twelve 2-hour intervals in a typical 24-hour day or perhaps just day and night. They may also consider seasonal differences and weekend/weekday differences in exposure. However, these exposure scenarios that are reasonable from a societal or population risk perspective may not be representative of particular actual or hypothetical individuals and so this will need to be considered in the implementation of the new IR approach.

b. **Time-averaging of exposure** occurs when the variation of an individual's vulnerability over time is averaged by weighting their vulnerability in each time interval by the length of the entire time interval. The result is that their estimated IR will be less than for the time interval in which they have their highest vulnerability if that interval were considered separately. This consideration is discussed further under the proposed approaches.

c. ***"Salami slicing"*** occurs when, for example many people are exposed to the risk of a dam or levee system breach for short periods of time, but taken in the aggregate there is someone who is exposed over a longer period, which is the sum of the short periods for each individual exposure (e.g. transients and occupants of camp sites, a hotel, or a shopping center). HSE (2003) argue that "*it would give a misleading picture of the risk to estimate the risk from this hazard by taking into account the exposure time of each individual. Instead, a truer picture of the risk would be obtained by constructing a hypothetical person who is exposed to the hazard 100% of the time (or for the proportion of time that any individual is exposed).*" It is the opinion of the writers, that "*taking into account the exposure time of each individual*" is appropriate for estimating individual incremental risk and that the alternative approach of "*constructing a hypothetical person who is exposed to the hazard 100% of the time (or for the proportion of time that any individual is exposed)*" is more representative of a form of societal risk. This is because the particular individuals who are exposed as the result of a dam or levee system breach may be affected by the location in question and which individuals are exposed is random and unpredictable in nature. Interestingly, the current HEC-FIA and LifeSim calculations do not consider that different individuals are present at a shopping center, for example, in any time interval, and therefore it seems that these types of software use the alternative approach such that there is currently no concern with "*salami slicing*." However, it should be recognized that HEC-FIA and LifeSim currently take the societal risk perspective, and so it should not be surprising that they are using the alternative approach, which is consistent with the opinion of the writers.

5) **Uncertainty in IR estimates** arises from many sources. If Census and other databases are relied on as sources of information for estimating IR, then population variability (aleatory uncertainty) and epistemic of knowledge uncertainty would be combined for IR estimates. If separate vulnerability characteristics are developed for hypothetical persons, the uncertainty associated with these inputs would be epistemic or knowledge. However, in using uncertainty mode simulations for HEC-FIA or LifeSim to estimate IR for the hypothetical persons, additional sources of both variability and epistemic uncertainty would be introduced in estimates of the uncertainty associated with the resulting IR estimates.

6) **One or more proposed approaches for IR estimation**: Given the natural tension between level of effort for estimating individual incremental risk and for modifying existing software vs. the quality of the estimates (in terms of the discussion Tables 4 and 5), there appears to be merit to proposing more than one approach to IR estimation. This is further discussed in the next subsection.

<u>Proposed Approaches</u>

Following consideration above of the USACE and non-USACE guidance and various options for estimating individual incremental risk, the current role for USACE software in societal risk estimation, and the practical requirement that "*the level of detail ... should be "decision*

Table 4. Evaluation of the Proposed Approaches Against ER 1110-2-1156 (USACE 2014) Guidance on an IR Risk Metric

USACE Guidance on IR Risk Metric	Approaches 1a –Individual most at risk based on IR percentiles estimates and 1b - IR percentile mapping	Approach 2 - Individual most at risk based on hypothetical persons
1) Incremental risk	**Yes** - Readily achievable by running breach and non-breach scenarios in HEC-RAS, HEC-FIA/LifeSim, and DAMRAE. Risk models can be structured in DAMRAE to account for non-additivity of failure mode (e.g. probabilities due to dominance of one failure mode over another or common cause considerations).	
2) Identifiable person by location	**No** - Except by occupancy class	**Yes** - As representative hypothetical persons constructed by the risk analyst
3) Individual most at risk	**Unlikely** - Since this depends on the degree to which the subPar associated with structures include the individual most at risk in terms of exposure (see 4) and vulnerability (see 7c)	**Yes** - Providing the hypothetical persons constructed by the risk analyst include the individual most at risk
4) All exposure scenarios	**Unlikely to be representative** of the individual most at risk or other individuals by location since numbers of people assigned to structures are based on structure occupancy type and generalized census and other data bases in HEC-FIA/LifeSim and not for actual or hypothetical persons	**Yes** - Providing the hypothetical persons constructed by the risk analyst include representative exposure scenarios and HEC-FIA/LifeSim and DAMRAE are modified to consider these for hypothetical persons
5) All failure modes for all loading or initiating events	**Yes** - Readily achievable by running a range of breach scenarios representing all failure modes and all loading events (e.g. a range of flood magnitudes and initial reservoir levels) in HEC-RAS, HEC-FIA/LifeSim, and DAMRAE	
6) Multiple structures	**Yes** - Readily achievable by running breach scenarios representing all failure modes for all dam sections (or a representative range of levee system breach locations) in HEC-RAS, HEC-FIA/ LifeSim, and DAMRAE	
7a) Probability of failure	**Yes** – Readily achievable since this is estimated in DAMRAE	
7b) Exposure factors	See 4)	See 4)
7c) Vulnerability - Conditional probability of life loss given exposure to breach flood	**Unlikely to be representative** of individual most at risk or other individuals by location since consideration of vulnerability is embedded in population-wide relationships for the protective action response, fate of structures, and fatality probability distributions for flood lethality zones in HEC-FIA/ LifeSim and not for actual or hypothetical persons	**Yes** - Providing the hypothetical persons constructed by the risk analyst include representative relationships for protective action response, fate of structures, and fatality probability distributions for flood lethality zones and that HEC-FIA/ LifeSim and DAMRAE are modified to consider these relationships for hypothetical persons
8) The level of detail should be "decision driven"	**Yes** – Approach 1a is proposed as a reasonable compromise for screening and IES risk assessments; although it is to be expected that the percentile estimates will underestimate the individual incremental risk to the person most at risk. Approach 1b may provide IR map estimates that are useful for ALARP evaluations for DSMS risk assessments since the degree of underestimation will likely be less for IR mapping where the person most at risk is not the focus.	**Yes** – This approach is proposed as a reasonable degree of additional effort for DSMS risk assessments for evaluating or justifying long-term risk reduction measures or residual risk to provide a reasonable assurance that individual incremental risk has been adequately considered and especially for reservoirs with multiple dams, long dams and levee systems.

Table 5. Evaluation of the Proposed Approaches Against Non-USACE Guidance on an IR Risk Metric

Guidance	Approach 1a – Individual most at risk based on IR percentiles estimates	Approach 1b - IR percentile mapping	Approach 2 - Individual most at risk based on hypothetical persons
Hypothetical persons	**No** – based on statistical individuals assigned using Census or other databases.		**Yes** – hypothetical persons are considered.
Spatial variability	**Partially** – based on only population relationships not individual person relationships.	**Yes** – based on only population relationships not individual person relationships and displayed on IR maps but the extremes of IR may be underestimated due to the reliance on population data.	**Partially** - to the degree that hypothetical persons are defined across the inundation area.
Population variability	**Partially** – based on only population relationships not individual person relationships. Displayed in a cumulative distribution but the extremes of IR may be underestimated due to the reliance on population data.	**Partially** – based on only population relationships not individual person relationships and displayed on IR maps but the extremes of IR may be underestimated due to the reliance on population data.	**No** – except to the degree necessary to identify the person most at risk, assuming that hypothetical persons are defined to be representative of all candidate individuals for the person most at risk.
Time averaging of exposure	**No** – statistical individuals are assigned to a single structure that does not account for the possibility that they may be at one of several locations based on their pattern of life during the time associated with each exposure time scenario for which the HEC-FIA/LifeSim runs are made.		**Yes** – if multiple initial locations for each hypothetical person are combined into a single time-averaged IR estimate. If each of the multiple locations is evaluated separately, time averaging would not be involved.
"*Salami slicing*"	**No** – because people are assigned to structures and other locations without distinguishing that they are different individuals at different times		**No** – if a hypothetical person is defined to be present all the times that someone is present at a location such as a hotel, campsite or shopping center where "*salami slicing*" is an issue. **Yes** - if a hypothetical person is defined to be present only for the time that they personally would be present and not at any time that someone is present at a location such as a hotel, campsite or shopping center where "*salami slicing*" is an issue.
Uncertainty in IR estimates	**Yes** –based on population inputs rather than individual person inputs so the population variability contribution to the uncertainty would be expected to be underestimated. Also, uncertainties are included for loading and system response relationships in DAMRAE in its uncertainty mode.		**Yes** – uncertainty should be included in the exposure and vulnerability relationship for each hypothetical person. Also, uncertainties are included for loading and system response relationships in DAMRAE in its uncertainty mode.

driven" (USACE 2014), we propose the following approaches for estimating individual incremental risk for HEC-FIA and LifeSim:

a) Approach 1a – Individual most at risk based on IR percentile estimates,
b) Approach 1b – IR percentile mapping, and
c) Approach 2 - Individual most at risk based on hypothetical persons.

The first two approaches both rely mainly on the current information used for societal life-loss estimation in HEC-FIA and LifeSim. However, because of the statistical nature and other limitations associated with the Census and other data that is used, the first two approaches can be expected to underestimate individual incremental risk for the person most at risk, although this may be of less concern for Approach 1b than for Approach 1a, as discussed later. However, the second approach, which builds on the first, would produce maps that would provide information on the spatial variability of individual incremental risk. The third approach, when properly applied, should yield a better estimate of individual incremental risk for the person most at risk, but it will require additional effort to complete. Also, guidance will need to be developed on constructing hypothetical persons. All three approaches will require that some additional features be added to the HEC-FIA/LifeSim and DAMRAE software.

The three proposed approaches are evaluated against USACE and non-USACE guidance in Tables 4 and 5. Each of the three approaches is described in more detail in the following subsections. Technical design details for implementing each of the three approaches in HEC-FIA and LifeSim are presented in a report by Bowles (2016).

a) **Approach 1a – Individual most at risk based on IR percentile estimates:** In this approach, HEC-FIA or LifeSim would be run in uncertainty mode for a range of flood-exposure scenarios to provide estimates for the conditional probability of life loss for all statistical individuals, represented by subPar. These probability estimates are conditioned on the flood-exposure scenario for which HEC-FIA and LifeSim are run. The estimates should include the combined potential for life loss for statistical individuals in the structures to which they are originally assigned and during the evacuation process. The estimates would be in the form of the conditional probability of life loss for each statistical individual, calculated as the fraction of the total number of uncertainty runs for which each statistical individual is estimated to lose their life, based on the HEC-FIA or LifeSim inputs that are considered as uncertain.

   The conditional probability of life loss for all statistical individuals for all flood-exposure scenarios would be input to DAMRAE. For each statistical individual, a designation would be needed as an input to DAMRAE as to which dam on multi-dam reservoirs, or which components for a levee system, have the potential to flood the statistical individual as the result of each breach or non-breach flood scenario. This would include the possibility that more than one dam or more than one levee system component might affect the same statistical individual. DAMRAE would interpolate between the conditional probabilities of life loss for each flood-exposure case for each statistical individual, much in the same way that it currently does for societal risk estimates of life loss. These conditional probabilities would then be convoluted with the loading and system response probabilities to obtain a probability distribution of individual incremental risk as an annual probability of life loss for each statistical individual. For cases where a statistical individual can be affected by more than one dam on a reservoir or more than one component of a levee system, the product of the loading and system response probabilities would need to be appropriately combined accounting for correlation in the joint occurrence of breaches at multiple locations. These calculations should be performed using the uncertainty mode of DAMRAE. The details for this calculation are contained in Bowles (2016).

An uncertainty distribution for individual incremental risk as an annual probability for each statistical individual would be obtained from DAMRAE. The last step is to obtain the estimated individual incremental risk as an annual probability for each statistical individual for a selected IR percentile and then determine the maximum individual incremental risk across all statistical individuals. The selected percentile should be for a reasonably extreme value, such as the 99th percentile. A distribution of the estimated individual incremental risk for all statistical individuals could also be obtained for the selected IR percentile.

A stratified sampling of all statistical individuals for which the uncertainty distributions for the conditional probability of life loss are obtained in HEC-FIA or LifeSim could perhaps be developed if computational efficiency was a concern. The guidance for the stratified sampling would need to be developed but they should ensure that all candidates for the "*person most at risk*" are kept.

Approach 1a is proposed as a reasonable compromise in terms of effort required vs. quality of the estimate for screening and IES risk assessments that are not being used to evaluate or justify long-term risk reduction measures or residual risk; although it is to be expected that the percentile estimates will underestimate the individual incremental risk to the individual most at risk.

b) **Approach 1b – IR percentile mapping**: This approach is an extension of Approach 1a. The steps in the approach are identical to those described for Approach 1a until the last step when, instead of determining the maximum individual incremental risk over all statistical individuals, all estimates of individual incremental risk for each statistical individual for a selected percentile are output to a GIS file and associated with their position from one of the HEC-FIA/LifeSim runs so that a map of iso-IR contours can be developed and plotted. The selected percentile would not need to be an extreme value as should be used in Approach 1a, where the focus is estimating the individual most at risk. Instead the 50th percentile could be selected to map median estimates of IR. In addition, lower and higher percentiles (e.g. 25th and 75th percentiles) could be selected to provide an indication for the variability in the IR estimates for the inundation area. It would need to be determined how to map IR in areas where there are no structures or structures are sparse.

For Approach 1b, the stratified sampling option should not be used, since estimates of IR for all statistical individuals will be needed for mapping. As mentioned previously in this paper, maps as contours of constant levels of IR would help to identify areas of high IR. The causes of the high IR may take some investigation to identify and could be due to reasons such as areas where non-English speaking people live, which affect their ability to receive and respond to evacuation warnings, or areas where evacuation options are poor, perhaps due to routes that are subject to traffic congestion, or because they pass through low lying areas that are subject to early flooding. In these situations, specific measures could be identified to reduce the risk "as low as reasonably practicable" (ALARP) for these individuals and groups. Examples of such measures could include special targeting of non-English speaking groups with warning messages in their native tongue, establishing improved evacuation routes, or staging or starting evacuation of some areas earlier than others.

To assist in identifying the reasons for areas of high IR, it might be possible to develop some tools based on analyzing the association between variables in the HEC-FIA or LifeSim models and areas of high IR or high conditional probability of life loss.

Similar to Approach 1a, Approach 1b is expected to underestimate the individual incremental risk, although this may not be as serious an issue for Approach 1b since the focus of the mapping is explore the spatial variability of IR, rather to identify the individual most at risk. Approach 1b may, therefore, have value for DSMS studies for which ALARP is being evaluated and in situations where the local communities are interested in identifying non-structural risk reduction measures.

c) **Approach 2 – Individual most at risk based on hypothetical persons**: The steps for this approach are similar to those described for Approach 1a, except that instead of focusing on each subPar, representing statistical individuals, the focus is on a set of hypothetical persons. These hypothetical persons are constructed by the risk analyst, which will require that guidance be developed for the USACE dam and levee system risk management programs. The HSE (2003) concept of hypothetical persons is discussed earlier in this paper with some examples.

As a part of constructing each hypothetical person, the various inputs to HEC-FIA and LifeSim that characterize their exposure and vulnerability would need to be specifically defined for each hypothetical person, including knowledge uncertainties in these estimates. This contrasts with the approach currently taken in these models for societal risk (and for Approaches 1a and 1b), in which societal or population relationships are used rather than relationships for individuals. In addition, an initial location, defined as a structure in HEC-FIA or LifeSim, would need to be specified for each hypothetical person. The initial location should be multiple locations if there is a chance that, because of their pattern of life, the hypothetical person would be at more than one location during the time associated with each exposure time scenario for which the HEC-FIA/LifeSim runs are made. In this case, each location should be assigned a likelihood (initial location exposure factor) that the hypothetical person would be at that location, and these likelihoods should sum to 1.0 over all initial locations for a single hypothetical person. These likelihoods will be used to obtain a single estimate of IR considering all locations for each hypothetical person. For locations, such as hotels, camp sites or shopping centers, a separate hypothetical person could be defined who is present all the time that anyone is present to avoid the "*salami slicing*" concern.

The conditional probabilities of life loss for each hypothetical person would then be estimated for each flood-exposure scenario following the usual HEC-FIA/LifeSim procedures, except that the relationships that are developed for each hypothetical person would be used rather than the population relationships normally used in HEC-FIA/LifeSim. These probabilities would be conditioned on the flood-exposure scenarios for which HEC-FIA or LifeSim are run. However, the simulations should include all subPar (statistical individuals) normally considered so that the interactive effects, such as for traffic congestion, would be considered.

The conditional probabilities of life loss for all hypothetical persons for all flood-exposure scenarios would be input to DAMRAE and then the same procedure would be followed as described in Approach 1a. The calculations should be performed using the uncertainty modes of DAMRAE to obtain a probability distribution for individual incremental risk as an annual probability for each hypothetical person. The last step is to obtain the estimated individual incremental risk as an annual probability for each hypothetical person for a selected percentile and then determine the maximum individual incremental risk over all hypothetical persons. The selected percentile for Approach 2 need not be as extreme as for Approach 1a since the uncertainty in the estimates for the hypothetical person will not include population variability. Therefore, a 50th percentile of median estimate of IR may be appropriate, or if desired, a conservative estimate could be selected by using a higher percentile.

There is an additional step that will be needed for Approach 2 to complete the consideration of multiple initial locations for each hypothetical person. Since the different locations for a particular hypothetical person could be affected by breaches of different dams on a multi-dam reservoir, or by breaches of different levee reaches of a levee system, this additional step should take place in DAMRAE. The step involves combining the uncertainty distribution estimates of IR for each different initial location into a single IR uncertainty distribution estimate using the likelihoods for each initial location described above. This would represent an example for time-averaging of exposure; although, the IR estimates without time averaging could also be obtained from DAMRAE and evaluated separately.

It is noted that this additional step could take place in HEC-FIA or LifeSim if the hypothetical person were exposed to the breach of only a single dam or levee reach. However, since this would not always be the case, it is recommended that this additional step should be incorporated into DAMRAE.

Finally, it would be advisable to compare the estimated IR for the person most at risk based from Approach 1a with the estimate obtained from Approach 2 based on hypothetical persons. If the estimate from Approach 1a is higher than for Approach 2, the analyst should investigate the circumstances pertaining to the statistical individuals that control the estimate obtained from Approach 1a and assess if they are reasonable. However, the role for the selection of the IR percentile in Approach 1a and the difference between population relationship for statistical individuals in Approach 1a and hypothetical persons in Approach 2 must also be carefully considered.

Approach 2 is proposed as a reasonable degree of additional effort for DSMS risk assessments for evaluating or justifying long-term risk reduction measures or residual risk to provide a reasonable assurance that individual incremental risk has been adequately considered and especially for reservoirs with multiple dams, long dams and levee systems.

SUMMARY AND CONCLUSIONS

The overall purpose of this paper is to describe an approach for estimating individual incremental risk for the USACE dam and levee safety programs using the HEC-FIA and LifeSim simulation software. ER 1110-2-1156 (USACE 2014) defines individual incremental risk as follows:

> *The increment of risk imposed on a particular individual by the existence of a hazardous facility. This increment of risk is an addition to the background risk to life, which the person would live with on a daily basis if the facility did not exist.*

A background discussion on the importance and extent of use of individual risk is provided. Considerations for the estimation of individual incremental risk are discussed in the light of available USACE guidance and the approaches and guidance used by others. A minor but significant change in the USACE definition of individual risk is proposed by changing the last word in the definition given above from "*exist*" to "*breach.*" A generalized concept for estimating individual incremental risk is proposed and factors that affect individual risk are identified. Several alternative individual incremental risk constructs and risk metrics from dam safety and the health and safety regulatory field are discussed and screened in the light of USACE and other guidance. The issue of considering actual persons vs. hypothetical persons is discussed and provides some additional bases for screening options.

Various options for estimating individual incremental risk are evaluated. None of the available risk metrics reviewed were found to meet the USACE guidance. Therefore, it was determined that a new risk metric was needed for use in the USACE dam and levee system risk management programs. The HEC-FIA and LifeSim software and the related HEC-RAS and DAMRAE software were reviewed to identify how factors needed to estimate individual incremental risk are currently addressed.

The proposed approaches were developed with consideration to the inevitable tradeoff between the required degree of effort and the quality of the estimates. In this regard, it was important to keep in mind that USACE uses risk assessment for supporting different types of decisions in its dam and levee system risk management programs and that USACE (2014) directs that "*The level of detail ... should be "decision driven."*" Two approaches are proposed, the first of which would rely mainly on information that is currently used for societal life-loss estimation and which is proposed for use in screening and issue evaluation study (IES) risk assessments. The second approach would require the construction of hypothetical persons. It is proposed for use in cases where a higher level of confidence is needed in the individual risk estimates, including for dam safety modification study (DSMS) risk assessments for informing long-term decisions on the selection of risk reduction measures and for judging the tolerability of residual risk. A variation of the first approach would provide mapping of iso-lines of estimated individual risk, which could be useful for as-low-as-reasonably-practicable (ALARP) evaluations for DSMS risk assessments.

ACKNOWLEGEMENT

This paper is based on work conducted by RAC Engineers and Economists, LLC, with the USACE RMC under a subcontract with the HNTB+Kleinfelder JV issued under an Indefinite Delivery/Indefinite Quantity Contract with USACE.

REFERENCES

Adler, M.D. 2004. Against "Individual Risk": A Sympathetic Critique of Risk Assessment. Draft Public Law and Legal Theory Research Paper Series Research Paper No. 49 and Institute for Law and Economics Research Paper No. 04-01, University of Pennsylvania Law School. January.

ANCOLD (Australian National Committee on Large Dams). 2003. Guidelines on Risk Assessment. October. http://www.ancold.org.au/publications.asp.

CUR TAW. 1990. Probabilistic Design of Flood Defences. Centre for Civil Engineering Research and Codes, Technical Advisory Committee on Water Defences, Report 141.

Graham, W.J. 1999. A Procedure for Estimating Loss of Life Caused by Dam Failure. Report No. DSO-99-06, Dam Safety Office, US Bureau of Reclamation, Denver, CO.

Hartford, D.N.D. and G.B. Baecher. 2004. Risk and Uncertainty in Dam Safety. Thomas Telford.

HSE. 2001. Reducing Risks Protecting People - HSE's Decision Making Process. Risk Assessment Policy Unit. HSE Books, Her Majesty's Stationery Office, London, England. HSE 12/01. ISBN 0 7176 2151 0.

HSE. 2003. Good Practice and Pitfalls in Risk Assessment. Prepared by the Health & Safety Laboratory for the Health and Safety Executive. Research Report 151. ISBN 0 7176 2732 2.

HSE. 2009. Societal Risk: Initial Briefing to Societal Risk Technical Advisory Group.

HSE RR703 Research Report. Prepared jointly by the Health and Safety Laboratory and the Health and Safety Executive.

HSE. 2012. Proposals for Revised Policies to Address Societal Risk Around Onshore Non-Nuclear Major Hazard Installations. CD212 Consultative Document.

Le Guen, J. 2010. Legislation Applying to Reservoirs and Their Enforcement. Paper presented at the International Workshop on Exploration of Tolerable Risk Guidelines for Levee Systems sponsored by Headquarters, US Army Corps of Engineers, Alexandria, Virginia. March.

Maslin, K., M. Foster and L. McDonald. 2014. Estimating Individual Risk: A Pragmatic and Defensible Approach. ANCOLD.

Mileti, D., and J. Sorenson. 2015. Interview Schedule: Community Warning Issuance, Diffusion, and Protective Action Initiation Estimation.

Section 216. River and Harbor and Flood Control Act of 1970 (P.L. 91-611, HR 10877), 31 December 1970).

Sorenson, J. and D. Mileti. 2015a. First Alert and/or Warning Issuance Time Estimation for Dam Breaches, Controlled Dam Releases, and Levee Breaches or Overtopping.

Sorenson, J. and D. Mileti. 2015b. First Alert or Warning Diffusion Time Estimation for Dam Breaches, Controlled Dam Releases and Levee Breaches or Overtopping.
Sorenson, J. and D. Mileti. 2015c. Protective Action Initiation Time Estimation for Dam Breaches, Controlled Dam Releases, and Levee Breaches or Overtopping.

Sorenson, J. and D. Mileti. 2015d. Using Questionnaire Data to Select Curves for Local Community Issuance, Diffusion, and Protective Action Initiation for Dam Breaches, Controlled Dam Releases, and Levee Breaches or Overtopping.

Reclamation (Bureau of Reclamation). 2011. Interim Dam Safety Public Protection Guidelines: A Risk Framework to Support Dam Safety Decision-Making. Dam Safety Office, Bureau of Reclamation, Denver, Colorado. August.

USACE. 2014. Safety of Dams - Policy and Procedures. ER 1110-2-1156. March.

Zelinski, P.A. 2014. Setting Individual and Societal Risk Criteria for Safety of Dams. Proceedings of the Canadian Dam Association Annual Conference, Banff, Canada. October.

MINIMIZING BIAS IN GEOLOGIC AND GEOTECHNICAL ASPECTS OF RISK EVALUATION FOR DAMS AND LEVEES

Kerry D. Cato [1]
Derek S. Morley [2]

ABSTRACT

Many geological and geotechnical decisions in risk evaluation involve unknowns where predictions must be made. It is important to recognize the impact of unknowns and avoid letting bias influence the decision-making processes. Bias can impact decisions when adherence to existing models is given greater priority than updating models in light of new contrasting data or by oversimplifying complex conditions. For example, evaluation of foundation conditions and natural processes that affect dams and levees is often complicated by the difficulty and expense in accessing subsurface materials; as a result, geologic conditions often are interpolated between boreholes. This is just one of three types of predictions commonly made: (1) predicting a geological/geotechnical condition (where the answer may not be known now, but with additional information or during construction a "correct" answer can be determined); (2) predicting the future occurrence of an event or process, like a landslide, seepage, or earthquake (there is no way to identify the "correct" answer unless the event occurs); and (3) predicting a future action, such as likelihood of successful remediation construction for a dam. The latter is influenced not only by natural conditions and events but also by evolving policy and decisions by owners, agency administrators, contractors, and other human stakeholders, and in this case the "correct" answer is not predictable but is created by various human responses and interactions in conjunction with events. In situations in which there is often no clearly "correct" solution, a more balanced approach may result when technical professionals openly reveal areas of data gap, and where predictions are being made. To reduce the unintentional introduction of bias in the decision-making process, we can draw on lessons about bias from past traditional design experience and use those lessons as a precaution about bias going forward into the new paradigm of more widespread usage of risk-informed decision-making.

INTRODUCTION

This paper presents a concept with the implicit goal of generating questions, introspection, and discussion. We intend to promote consideration of how you – the risk-informed decision-making practitioner – evaluate data and the lack thereof to predict conditions, events, and outcomes during risk assessment. Further, we intend to promote consideration of where bias may enter risk-informed evaluations, and we hope to disseminate ideas for reducing such bias and its impacts. The profession and those who rely on our work would benefit from greater recognition of how bias enters the evaluation

[1] CEG, PG, PhD, Assistant Professor, Department of Geological Sciences, California State University San Bernardino, San Bernardino, CA, kerry.cato@csusb.edu.
[2] PE, Chief of Geosciences, USACE South Pacific Division, Dam Safety Production Center, Sacramento, CA, Derek.S.Morley@usace.army.mil

process in subtle ways and more importantly how bias can skew the results of conclusions.

Unknowns are the inherent 800-pound gorilla in our evaluations. Recognizing and dealing with how we make predictions about these unknowns has a huge effect on the risk analysis results. Instead of relegating these predictions to implicit background noise, this paper emphasizes the need to explicitly state the unknowns, the predictions we make about the unknowns, and the biases used in those predictions.

TYPES OF PREDICTIONS

In risk assessment we make predictions about the likelihood of what will or will not occur, and in risk-informed decision-making (RIDM) we use those predictions to guide our decisions (USACE, 2015). In this context a prediction is a foretelling that is based, to the extent feasible, upon scientific principles, data, and experience. Predictions are necessary due to unknowns; and inherently, because what we do involves unknowns, we must make predictions.

The unknowns, and thus the predictions, generally can be characterized as one of three types:

Type 1 - an already-existing physical condition that is not readily observable
Type 2 - a future physical event or occurrence
Type 3 - a future human action

Type 1 is an already-existing physical condition that is not readily observable. We may be able to see some portion of the feature, but not all of it. For the parts we cannot see, we must predict their condition – how they are composed and the condition of their constituents. Examples of Type 1 predictions are most soil and geologic conditions, some concrete conditions, some steel and conduit conditions, some mechanical and electrical conditions, and some vegetation conditions. All of these physical conditions actually exist, but we just don't know what they are or the extent to which they exist. Consequently, we "predict" their consistency, variability, integrity, and flaws. Ultimately there is a right answer (i.e., a condition that exists); we just cannot and do not know it at the beginning of the evaluation. We can invest in further investigation of the conditions (e.g., data mining, subsurface exploration, non-destructive structural testing, etc.) to improve our understanding of these conditions and reduce uncertainty to improve our predictions. But in most cases we still will not completely know the answer for certain and will need to make a prediction about the conditions that are present.

Type 2 is a future physical event or occurrence. We have some record or understanding of past events and occurrences, but we do not know exactly what will occur in the future. We must predict what events or occurrences will happen, how severe they will be, and how frequently they will occur. Examples of Type 2 predictions are seismic loading, hydrologic loading, debris loading, groundwater fluctuations, soil response to loading, structural response to loading, mechanical/electrical response to loading, vegetation changes, animal activity, and other occurrences that are future events. This type of prediction encompasses both loading and response – they are both future occurrences. The answer does not exist now and is not absolutely constrained to occur in a certain way, so we cannot know the answer now. We can only know the answer if the event occurs in the future. We can invest in further consideration of future occurrences (e.g., additional analyses and analytical modeling approaches, physical modeling, testing of site components, etc.) to improve our understanding of these types of occurrences and improve our predictions. But we still will not know the answer and will need to make a prediction about what future events will occur.

Type 3 is a future human action. We have some understanding of how most people tend to behave in various settings and circumstances, but we do not know exactly how any individual will act in any particular situation in the future, nor how any combination of such interactions will play out. However, in risk analyses, we must predict how individuals will act and how their collective actions will affect matters – and predict how beneficial or detrimental the actions will be with respect to what we are evaluating. Examples of Type 3 predictions are observation of distress, decision-making about responding to distress, distress intervention, communicating about distress, observation and communication about differing conditions, documentation of conditions and actions, response to schedule and budget pressures, response to political and public pressures, and other future actions that will be decided and taken by humans. We can invest in further consideration of future human actions (e.g., by the additional study of human response to various situations, interviews and observations of people involved, etc.) to improve our understanding of these types of human actions and improve our predictions. But we still will not know the answer and will need to make a prediction about how people will act.

Figure 1 illustrates these three types of predictions using conditions that may be encountered during a levee risk assessment and expressed on an event tree. In this example, we start with a high-water triggering event. The first prediction is a Type 1 prediction about geological/geotechnical conditions that may be present, specifically whether highly permeable sands underlie the levee. The second prediction is a Type 2 prediction about a future event or occurrence, specifically the response of whether sand boils form on the landside of the levee. The third prediction is a Type 3 prediction about a future human action, specifically about what detection and intervention action will be taken by humans in response to the preceding events.

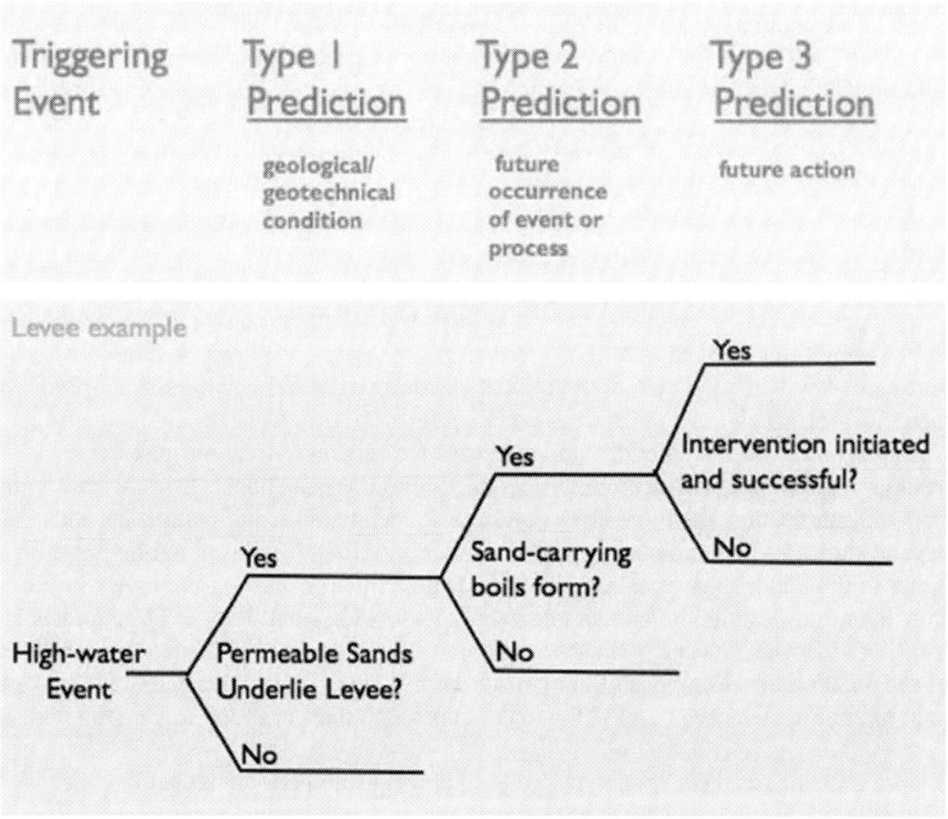

Figure 1. Simplified example event tree showing the three types of predictions.

THE WHATS OF BIAS

This paper focuses on how bias can be introduced and influence the process and results of risk analyses; thus, we need to define what bias is and describe what can lead to the introduction of bias in the risk evaluation process.

A general definition of bias is: *prejudice in favor of or against one thing, person, or group compared with another, usually in a way considered to be unfair or inaccurate.*

And in the scope of this paper, we shorten this definition to state: *prejudice in favor of or against one thing compared with another, usually in a way considered to be inaccurate.*

Bias can enter the risk evaluation process in a multitude of ways. It is often subtle and not intentional. It may occur when we oversimplify complex conditions. Bias may occur as a result of overuse of available data and reluctance to acquire new data (which is understandable because in the geotechnical world new data is often costly to acquire).

Bias may occur in how we model or how we think we understand a condition or process (non-existent data). Bias frequently occurs when adherence to existing predictive models is given greater priority as opposed to updating models in light of new contrasting data (e.g., preserving sacred cows or long held views).

In some cases, what may appear to be a bias can be well founded (i.e., prejudice in favor of or against one thing compared with another, but in a way that is accurate). For example, a group of practitioners will have common experiences that have shown certain conditions or processes are likely to occur and have developed an empirical knowledge base upon which to operate. Individuals have done this as well. However, the two knowledge bases may be somewhat dissimilar.

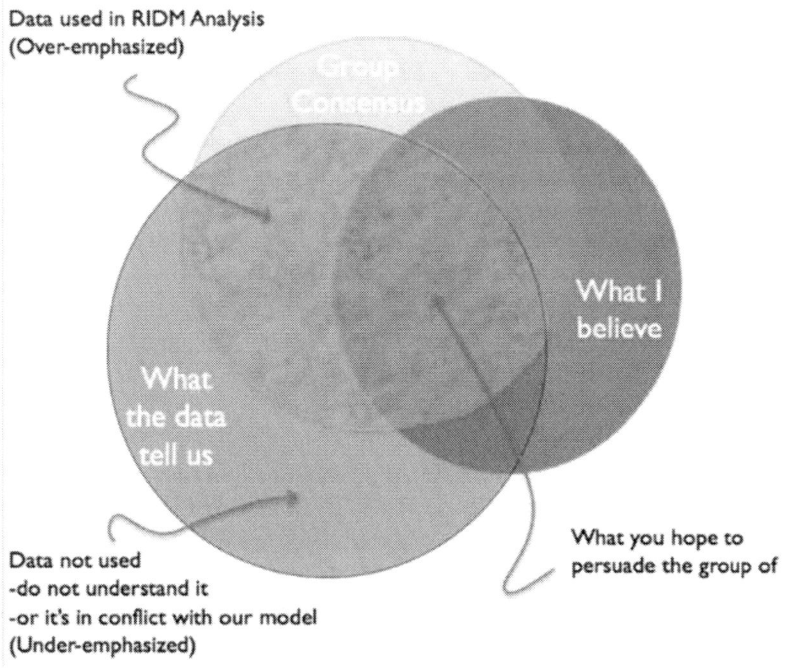

Figure 2. Visualization of how bias may exist in a risk-informed evaluation process and how this may be perceived differently by the group versus the individual.

Figure 2 is a visualization that shows how biases held by a group may differ from those held by a single member of the group. And we constantly encounter the situation whereby nature surprises us with a condition on process that we did not previously know existed. In Figure 2, this would be the data area outside of both the group and individual knowledge bases. In an ideal world, the basis for group decisions would include all the data. But often, some data are not known, are not understood, or do not fit our group or individual models or views, and therefore are not used. This has also been expressed in a different manner:

> ***"The eye sees only what the mind is prepared to comprehend."***
> Robertson Davies, Tempest-Tost

Regardless of how bias enters the process, its existence will prejudice the evaluation and the ultimate decision in one way or another. As humans, we generally understand that each individual brings unique observations, talents, skills, and, yes, personality to the table. As engineers and scientists most of us probably feel that individually, and as groups, we do a better-than-average job of letting logical processes dominate our decision-making processes. And with risk-informed decision-making we have created a process that works pretty well. However, this may not always be the case, and it is important to recognize where bias may exist and how it works against the logical model of data-driven processes.

In considering how bias can enter the risk-informed decision-making process, we discuss in this paper the following bias-affecting topics:

- **Use of Best Estimates**
- **Range of Data**
- **Bayesian Thinking**
- **Focus of Effort**
- **Worthy of Risk**
- **Fear of Reaction**

USE OF BEST ESTIMATES

We may introduce bias (typically unconservatively inaccurate bias) into our risk evaluation by using best-estimate characterizations of conditions or best-estimate values of parameters.

In risk assessment, we attempt to identify our best estimates of outcomes, as well as the associated uncertainty around the best estimates. Unlike design, in which we incorporate conservatism into our process to achieve a certain degree of reliability (at the expense of accuracy) in our product, in the risk assessment we attempt to achieve accuracy (at the expense of reliability). See Figure 3.

For example, when we consider risk for a footing foundation, we use our best-estimate value of the foundation soil strength (with no conservatism or factor of safety applied). This would represent our most accurate (or most likely) depiction of the conditions present. But when we design this footing foundation and we use the best-estimate value of the foundation soil strength (with no conservatism or factor of safety applied), then the foundation will have a 50% chance of failing. We avoid this, of course, because we do not want the foundation to fail – we want it to be reliable. So in design we sacrifice

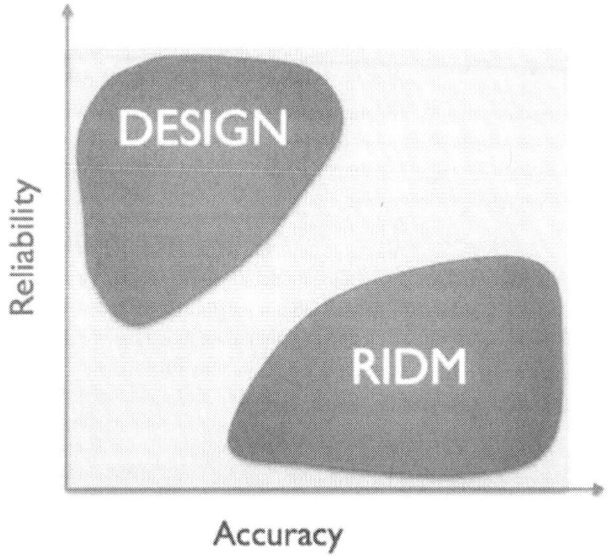

Figure 3. The difference in emphasis between design versus risk-informed decision making.

accuracy to achieve reliability. In risk assessment, we are most interested in accuracy, and so the practice involves developing best estimates and ranges associated with uncertainty.

However, we need to consider carefully the issue of *when* we should be using best-estimate values and *when* we should not. We are seeking best estimates of outcomes. Best estimates of outcomes often follow from best-estimate characterizations of conditions or best-estimate values of parameters; but this is not always the case. In many cases, best-estimate characterizations of conditions or best-estimate values of parameters provide an unconservatively inaccurate estimate of outcomes. This follows from the principal that, for the most part, things do not fail on the average; things fail at the weak point. And this is exacerbated when we look at weak-link structures, such as levees or floodwalls (i.e., if it fails at any location, the whole thing has failed to do its job).

For example, suppose we are considering a levee and its foundation susceptibility to underseepage, boiling, and backward erosion piping (BEP). Suppose our best-estimate value for any particular location is minor to unlikely susceptibility, and we believe the total range of values is from none to significant susceptibility along the length of the levee. The levee is highly unlikely to fail where the conditions are average; rather, it is most likely to fail at the weak spot. Our best-estimate value that we *should* use in our risk evaluation is the controlling value, i.e., the value at the weak spot. The best estimate of the *outcome* does not correspond to the best-estimate values of the *conditions*. Rather, the best estimate of the *outcome* corresponds to the best-estimate values of the *controlling conditions*.

Another example is seepage through geologic discontinuities in a dam abutment. We recognize that seepage doesn't flow in a straight-line from upstream to downstream; it follows a path of some tortuosity through the discontinuities. In characterizing that flow-path length, we consider the orientation of the discontinuities. We may plot our best-estimate values of the discontinuity sets on stereonets and use these best-estimate orientations to estimate the average path length from upstream to downstream. But water doesn't need to follow the average path length; in fact, it most likely will not – it will follow the shortest available path. Our best-estimate value for flow-path length that we should use in our risk evaluation is the controlling value, i.e., the shortest path available via discontinuities for the water to follow. Again, the best estimate of the *outcome* does not correspond to the best-estimate values of the *conditions*. Rather, the best estimate of the *outcome* corresponds to the best-estimate values of the *controlling conditions*.

RANGE OF DATA

We may introduce bias (typically unconservatively inaccurate bias) into our risk evaluation by predicting our best-estimate characterizations of outcomes based on limited data sets and not accounting for the impact of having limited data.

This bias occurs commonly in conjunction with the best-estimate parameters issue described above, even when we recognize that the best estimate of what will happen does not correspond to the best-estimate values of the conditions. When we are predicting an outcome that is based in part upon a parameter for which we are not using a best-estimate value of the conditions, but rather a best-estimate value of the controlling conditions, then the quality of our prediction is dependent on the amount of data upon which we are basing our prediction.

For example, suppose we evaluate the stability of a parapet wall on top of a dam. The wall has a footing foundation and its stability is dependent largely upon the strength of the foundation soil. Suppose we have only three soil strength measurements along the wall alignment, and they have friction angles of 33°, 35°, and 37°. Based on this data, one might conclude that the best-estimate friction value at any random location is 35°. However, this is another example of a weak-link structure, one that takes a weak area only a few tens of feet long to cause the wall to fail. "Aha!" we think, as we recognize that our best estimate of the outcome should not be based on the best-estimate value of conditions but instead on the best-estimate value of the *controlling* conditions. And so we conclude that the controlling conditions for this prediction are those soils that are the weakest foundation soils present under the wall (i.e., the weakest soil present across an area a few tens of feet long). With our limited data set of 33°, 35°, and 37°, we may be tempted to use 33° as our best-estimate value of the controlling conditions. But this soil strength parameter is likely to have a value distribution that is broader than represented by the data we have, as depicted in Figure 4. Maybe the value distribution is represented by a truncated normal distribution or

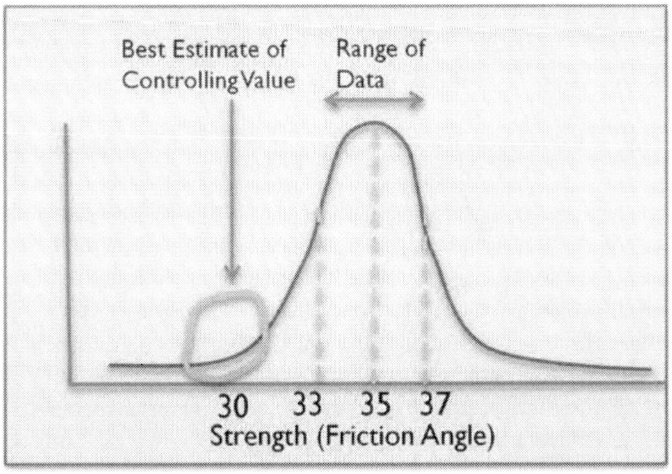

Figure 4. Estimating parameters beyond the range of data, or "thinking outside the data", is sometimes necessary when working with limited data sets.

something similar, and with only 3 data points, we'd expect that these values represent only the middle portion of that normal distribution. Based on these data and our expectation of the value distribution, we might expect that the weakest values present are below 30° and the near-weakest value (that is present across an area a few tens of feet) is about 30°. This *extrapolated* near-weakest value of 30° actually should be our best-estimate value of the controlling conditions. If instead we constrain our estimates to the data we have in hand, then we are overestimating the strength and producing unconservatively inaccurate predictions.

The obvious remedy to this problem is to collect more data, such that we have plenty of data to understand with confidence the range and distribution of parameters. But often this is not an option. We commonly are faced with parameters for which we don't have sufficient data for such understanding and do not have the opportunity to collect more data to create that degree of understanding and confidence. In those situations, it is important for us to consider not only the data we have, but to also consider the full, likely range and distribution of the parameters. In particular, we must consider how far the range likely extends *outside* of the range of data we have. And we need to consider that range and distribution with respect to the scale of the problem being considered, i.e., what is the physical (e.g., lateral) variability of the parameter and how does that compare to the physical characteristics controlling performance of the structure (e.g., tens of feet laterally for the stability of the wall). Often we will find that our best-estimate value of the controlling conditions should fall outside of the range of data we have in hand. In such cases, we need to extrapolate beyond the data we have to develop a best-estimate value of the controlling conditions. Although this may feel uncomfortable to some practitioners, we need to be able to "think outside the data" if we are to truly honor our best understanding of our practice and develop predictions of risk that minimize bias.

BAYESIAN THINKING

We may introduce bias (conservative or unconservative bias) into our risk evaluation by overtly adhering to predictive models and distributions rather than updating models as new information is acquired in subsequent investigations.

As our industry continues to develop its standards of practice and norms in risk assessment, we express our understandings of potential failure modes in event trees with increasingly familiar-looking component nodes and sometimes with increasingly familiar probability values for those nodes. These event trees and their values are based in part on empirical understanding of failure risks at sites that have been evaluated. However, as we evaluate each individual site, we need to be mindful of the unique conditions at that site. This includes the unique ways in which the various conditions interact and of the unique character of human involvement surrounding that site. We must be willing to adjust our models and value distributions based on what we observe from site-specific information and what we extrapolate from the data. If we don't, we bias our predictions away from the true nature of the site and force it to adhere to some norm or model that may not be applicable to our particular site.

And as we collect and incorporate new information to update our understanding at a site, we must consider the manner in which we incorporate the new information. Do we simply add the new information to the mix of information we already have, or do we use it to challenge ourselves and our understanding of the underlying nature of the information? In this developing arena of risk, and without comprehensive datasets upon which to base our assessments, we should think in Bayesian terms to leverage new information and maximize the advancement of our understanding.

For example, consider the likelihood of a significant landslide occurring on an embankment dam face due to (a) rain conditions, and (b) seismic shaking. We may estimate that the likelihood of (a) is 1 in 100, most likely near the left abutment, and of (b) is 1 in 10,000. Subsequently, suppose the site experiences numerous rain events, including some big ones, with one that triggers a slide in the embankment near the middle of the dam. We may interpret this to be generally consistent with our understanding of landslide risk, though expanding our area of concern from "near the left abutment" to a much greater portion of the dam. In effect we have added this new information and increased our knowledge base. But suppose the site experiences a moderate seismic event and that event triggers a slide in the dam face. We had estimated a 1 in 10,000 chance of a slide occurring, yet a slide did indeed occur when a (one) seismic event occurred. What are the odds that the 1-in-10,000 case just happened to occur? Not likely. Very slim actually. Much more likely is that our 1-in-10,000 estimate is grossly inaccurate, in error by orders of magnitude, and that the real probability is closer to 1 in 10, or 1 in 2. We should be updating our model and probability distribution to account for this new information. Because this site specific data is so significantly different from what we predicted, we should add some rigor to our update and apply Bayesian updating techniques to our numbers.

Further, what does this new information and updated probability at this dam indicate to us about our seismic estimates for other dams? Have we been applying the 1-in-10,000 philosophy, now demonstrated to be significantly and unconservatively inaccurate, to other dams in our portfolio of projects? How should an updated prediction at this dam impact our predictions for other dams? And, somewhat conversely, if conditions at those dams are similar enough to this dam to be affected by this prediction, then this prediction should be affected by what has happened at those dams. That is, if other dams have experienced seismic events and not experienced slides, then the probability is not quite as bad as would otherwise be indicated by the event at this dam. What we should be doing in every case is evaluating our experience versus what was predicted. Otherwise we risk not updating our models with actual data.

Within and across the industry, we are not especially well calibrated to probabilities of different aspects of performance. That is, while we have a sense of the number of dam failures per hour of service life of dams, we do not collect and evaluate system-wide data regarding various aspects of performance of dams. This might include boils versus high-water events, slides versus seismic events, etc. It is the norm to make forward predictions about likelihoods of occurrence of response given occurrence of loading. But most of this forward prediction is not informed by rigorous, collated past-performance data. We as an industry should compile data on frequency and extent of structure performance and failure events.

The discussion, thus far, has focused on loading and response of infrastructure and involves Type 1 and Type 2 predictions, the unknown physical conditions and future events, respectively. Imagine, though, our level of understanding and ability to calibrate for Type 3 predictions – those involving future actions by people. Our ability and efforts toward collating and calibrating probabilities associated with human involvement is even more rudimentary. Currently, we do a lot of speculating about human response and interaction, but we have relatively little compiled data upon which to base our estimates. We, as an industry, should make significant, coordinated efforts toward compiling data about our predictions, to update our thinking, our models, and our probability values.

FOCUS OF EFFORT

We may introduce bias (conservative or unconservative bias) into our risk evaluation by focusing on what we know a lot about, instead of focusing on what it is that drives the answer.

In any problem or evaluation, there are aspects that tend to be key and aspects that tend to be less important. And in approaching any problem or evaluation, we bring to the process our own experiences and expertise. When our own experience and expertise aligns well with the key aspects of the problem or evaluation at hand, we are in a good position. But often, our own experience and expertise doesn't align well with the key aspects of the problem or evaluation, and in these cases we are likely to introduce a discipline focus bias into the process, which creates its own inaccuracies in the results of our efforts.

Traditional evaluation and design efforts are prone to this focus bias, and so are risk assessment efforts.

For example, in the mid-2000's in the Central Valley region of California, the levee engineering community spent significant collective effort studying, exploring, testing, analyzing, evaluating, and debating what are appropriate soil strength values to use in levee stability evaluations and designs. The collective effort on soil strength seemed to dominate all other aspects of levee evaluation and design for some time. Not coincidentally, the levee engineering community in the Central Valley was populated largely by geotechnical engineers, many with significant experience and expertise in evaluating soil strength. Yet, in this region, the factors driving the performance of most of the levees is not strength and strength-related stability, but rather hydraulic conductivity issues, seepage-related problems, and erosion-related problems. It took a while for the collective professional community to channel the bulk of its collective energy on the real driver issues. A breakthrough in thinking occurred when geomorphologists, hydrogeologists, and other complementary disciplines began to play more prominent roles in the process. The levee engineering community, for all of its expertise and (acknowledged) valuable contributions, had spent a lot of time and energy focusing on what it knew a lot about instead of focusing on what it is that really drives the answer.

This discipline focus bias can be insidious and consequential. An example is the failure of Alexander Dam in Hawaii in 1930, which failed catastrophically during construction and killed 6 workers. Alexander Dam was being constructed as a hydraulic-fill dam, and it suffered a form of static liquefaction that led to massive sliding (Rodgers and Cato, 2016, 2017). At the time of failure, the embankment had been built to a height of approximately 95 feet, and about 275,000 yds^3 moved in the failed landslide mass (See Figure 5). After the failure, interviews with the workers indicated that drains in the area of the failure initiation had stopped flowing as early as 10 days prior to the failure and that some deformation of the embankment was observed. These indications of impending problems had indeed been observed, but were not an area of focus for any of the construction personnel. That these items were not an intended area of focus for the construction personnel is generally understandable. Still, if some of the construction personnel had contemplated the observed changes – even without understanding why the changes had

Figure 5. Post-failure section of Alexander Dam and photo of the failed embankment.

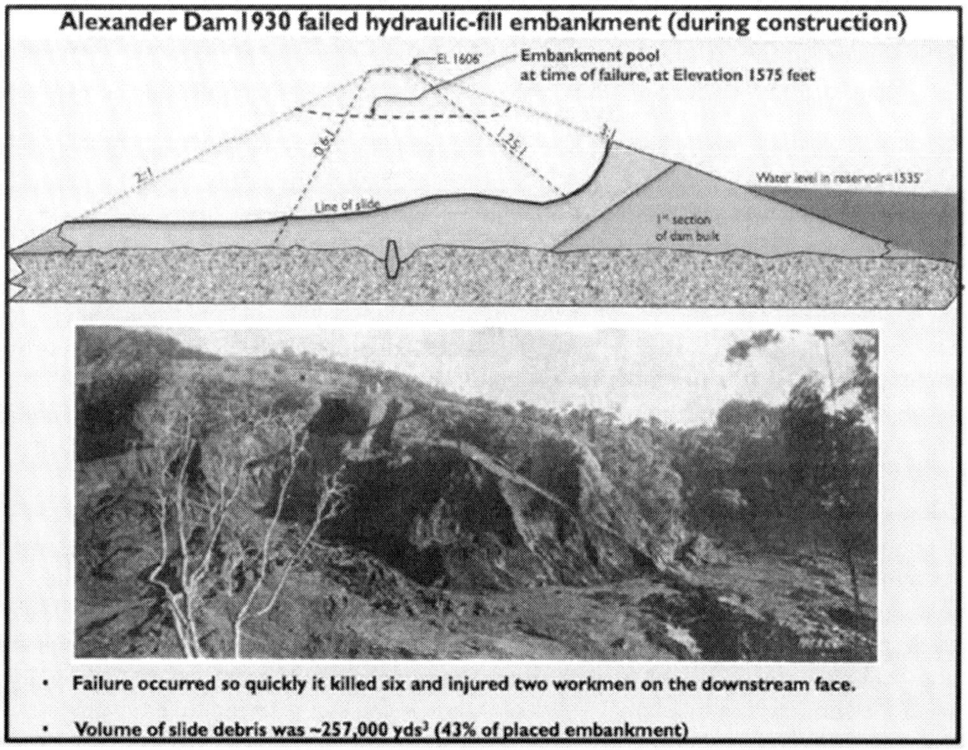

occurred – maybe they might have informed someone who would have understood the indications for what they were. In retrospect, one can only speculate that if this information had made it to the correct people, and that those people had taken action, it is conceivable that the deaths could have been prevented. But this example provides another view of how our individual and collective discipline focus can lead us away from that which is most important, even when it is right in front of us.

> ***"We all see only that which we are trained to see."***
> Robert Anton Wilson, Masks of the Illumniati

When we approach a risk assessment, we are prone to the same focus bias, and the resultant biases that tend to follow. If we focus on what we know a lot about instead of focusing on what it is that drives the answer, we spend our limited resources (project funds, project schedule time, intellectual processing time, debate time, etc.) where we receive little return. And more importantly, those limited resources are not invested in furthering our understanding of the key aspects that drive the answer. So our misdirected focus results in having less understanding, less data, and larger uncertainties.

For example, we may be performing risk assessment for an embankment dam that has a crack extending along and across its crest. If we have experience and expertise in evaluating flow through cracks and the potential for mobilization of embankment material, we may tend to focus our efforts on learning more about the variation in depth of the crack, its ability to remain open over its length, the composition of the material exposed along the crack, the tortuosity of potential flow through the crack, and so on. Yet, there is a good chance we are missing what it is that drives the answer, the truly key aspects. The crack exists; therefore something caused it to exist. What we *should* be focusing on is identifying the *mechanisms that caused the deformation* of the embankment that resulted in the presence of the crack. Is there a slide occurring within the embankment? Is there a slide occurring in the underlying abutment/foundation? Is the embankment suffering severe differential settlement along the abutment, with a portion hung up and a portion dropping? Is the embankment suffering severe differential settlement due to collapsing materials in a portion of the foundation? It is imperative that we ask these types of questions – about what the key issue(s) may be, not about what we have expertise in – to identify where our focus should be.

Of course, this is why we use multidisciplinary teams to evaluate and assess risk. We recognize that we need experts in a range of disciplines to identify and address the myriad issues we encounter at the wide variety of dams and levees we study. Still, we each must remain vigilant in regard to our individual and our collective team focus when approaching a site. We must be more than open-minded – we must actively "listen to what the site is trying to tell us". If we see a crack, the site is trying to tell us that dynamic stresses have resulted in observable strain and we need to understand why. Or, more generally, when we see a crack, the site is trying to tell us that something has happened or is continuing to happen that we need to understand if we wish to characterize risk at this site. The same can be said for settlement that is occurring where or when it was not expected, or periodically recurring seeps, or unusual instrumentation readings, or repeated malfunctions or repairs, or atypical as-built constructs, and so on. We must listen to what the site is trying to tell us and allow our focus to dwell upon those aspects, which will help focus our collective efforts on those aspects driving the answer.

WORTHY OF RISK

We may introduce bias (conservative or unconservative) into our risk evaluation by confidently and adroitly applying risk techniques to something we don't sufficiently understand.

Currently, we have a greater number of risk assessors who are well-versed in the philosophy, techniques, application, and tools of risk. As an industry, we are by far more broadly educated about risk and its application to our trade than we ever have been. Risk, as a lens to view our craft and subject matter, is a highly valuable tool, and leveraging the philosophy and techniques of risk assessment and risk management have taken our industry to a new level.

To that end, we must be "worthy of risk". Risk is a technique, a perspective, a tool, a lens, by which to view the thing (i.e., the subject matter); risk is not the thing itself. Understanding risk does not denote or connote understanding the thing. To apply risk in a meaningful manner to dams and levees, one must have a solid understanding of the thing as well as of risk. Its users must be experienced in evaluation, design, construction, maintenance, and repair of dams and levees. Only then can the technique be applied in a constructive and artful manner.

As risk becomes applied more frequently, to more and more projects and applications, the demand for risk assessors will continue to increase rapidly. The rate of increase in demand already appears to be exceeding the rate at which engineers can gain experience in evaluation, design, construction, maintenance, and repair of dams and levees. Consequently, the typical or representative amount of experience of a dam or levee assessor appears to be decreasing.

This trend exacerbates and increases occurrence of a bias, effected as confidently and adroitly applying risk techniques to something not sufficiently understood. This bias is not merely a lack of understanding, and thus opportunity for error (which may be conservative or unconservative). Less experienced risk practitioners are likely to make unconservatively inaccurate characterizations of conditions and occurrences. Their view of information is likely to be more academic or unknowingly idealistic. For example, they may believe that what is actually present in an as-constructed condition, and thus the basis for risk assessment, matches the design drawings and specifications. Those experienced in design and construction of dams and levees know that this is often not the case, and those truly experienced know where in a dam or levee the greatest deviations are likely to be found. Those less experienced may view things in a cross-sectional, 2-dimensional manner, and focus on the maximum section. Those more experienced know that 3-dimensional constraints and construction challenges result in worse conditions than those portrayed in cross-sections, and critical conditions are often found away from the maximum section (e.g., in/around key trench excavations extending up abutments, at/over oddly shaped or downstream-dipping abutment slopes, etc.). These differences tend to create an unconservatively inaccurate bias among those who are less experienced.

The preceding bias occurs when risk assessment is being performed for a dam or levee by personnel who do not have extensive traditional engineering understanding of dams or levees. This bias is exacerbated by the presence of special or niche conditions or concerns. For example, if seismic loading and response is an element of the evaluation, the quality of the risk assessment will depend in part on the experience and expertise of the risk assessors in seismic evaluation and engineering. If in this case the risk assessors do not have experience and expertise in seismic evaluation and engineering, then they are not "worthy of risk" for the seismic aspects of the project (i.e., if you don't understand seismic, you should not be applying risk to seismic). Similarly, if there is a landslide at a site, the quality of the risk assessment will depend in part on the experience and expertise of the risk assessors in landslide evaluation and engineering. If in this case the risk assessors do not have experience and expertise in landslide evaluation and engineering, then they are not "worthy of risk" for the stability aspects of the project (i.e., if you don't understand landslides, you should not be applying risk to landslides).

Somewhat obviously, the solution to this challenge is to bring a broad range of experts to the table. The range of represented expertise should cover all of the potential failure modes and all of the key, special, and niche topics. If seismic is at play, the process should involve people with seismic expertise. If landslides are potentially a factor, the process should involve people with landslide expertise. And so on. We must not be afraid to or hesitate to bring in experts from outside our own fields of expertise – even if they are related fields of expertise – to complement our risk assessment team.

FEAR OF REACTION

We may introduce bias into our risk evaluation by allowing ourselves to consider how others may react to our risk characterizations, which may alter how we characterize risk or present risk results.

In traditional design experience, a common quandary arises when a construction cost estimate exceeds the anticipated or acceptable value (for an example, see Neff, 2016). A previous lower estimate may have been used for budgeting purposes or as a basis for approval of the project, and the new higher estimate is likely to cause great difficulty for those managing the project. In such instances, it is tempting to reconsider the basis of the new estimate and revise it to be much closer to the previously disseminated estimate. Even without any deceitful intent, if estimators allow themselves to consider the adverse impacts of producing a higher estimate, bias is likely to creep into the estimating process. And for those who present and disseminate the new higher estimates, it is tempting to convey the results in a manner that downplays the changes or suggests that they are unlikely to occur.

Similarly, while characterizing risk or presenting risk results, practitioners may be concerned about reactions by managers, clients, infrastructure owners, policy-makers, other technical professionals, or the public. If risk practitioners allow themselves to consider reactions by other parties, bias may creep into the risk characterizations or the presentation of risk results.

A blunt type of example is when an owner or key stakeholder with a vested interest in a particular outcome is involved in the risk estimating process and with the risk assessment team. Of course, such participants may have individual bias in how they view conditions and contribute to value estimates. Additionally, even just mild pressure from such participants, or merely their presence, can introduce bias from others on the team. Even when not present, owners or key stakeholders with a vested interest in a particular outcome may result in bias in the process.

A more subtle example stems largely from the general lack of widespread understanding of risk. We commonly present the results of risk assessments not only to highly technical people but also to people who do not have an in-depth understanding of risk, probability, uncertainty, and related concepts. Truly risk-informed decision-making involves understanding not only our best estimate of risk values, but also of the uncertainty associated with our risk characterization and the ranges of estimates associated with key inputs. However, if we clearly and honestly present this information, it may appear to the untrained eye that the uncertainty and our range of estimates are absurdly large, which might lead such viewers to discount our work and our processes altogether. In the face of this possible response, many practitioners appear to have gone down the path of not presenting uncertainty information, or limiting what is presented to displaying only a range of best estimates. Doing so, however, may produce a bias in the risk-informed decision-making process – not in the development of risk values, but in the decision-making part of the process. By not accurately conveying the uncertainty information, we suggest a higher degree of certainty than actually exists. Ironically, we understate or completely fail to convey the risk associated with our estimate of risk. If we do not share this information, then the decision-makers we seek to inform are under-informed, and we may inadvertently mislead them into making decisions that are not aligned with their intent.

If our intent is to leverage risk to make more informed decisions, we should strive to avoid letting fear-of-reaction bias enter our process. To do so, all those involved in developing and presenting risk information must strive to honor the data and the science involved in the evaluation. Data gaps, contradictory information, uncertainties, and the likes should be openly discussed, documented, and rigorously carried through analyses and presented. We should accurately and comprehensively present risk and uncertainty information, and strive to educate recipients of the information regarding appropriate interpretation and use of the information. We should avoid alternative paths that involve gaming the system or otherwise altering the process or presentation to achieve an intended outcome, even if that intended outcome is a proxy for what we believe the "true" answer to be.

The moment we allow ourselves to be concerned about how others may react to our risk characterizations, we likely have compromised our ability to characterize risk and present risk results without bias. Even if we manage to personally avoid introducing bias into the process, there may remain the perception by others that bias has occurred, and thus in receiving and using our results the receiver will introduce a bias. It is therefore imperative that we not allow ourselves to be concerned about how others may react to our risk characterizations, but instead we remain disciplined about our process, our interpretation of data, and our application of scientific principles and empirical observations and experience.

And to further help ourselves avoid this bias, we should plan for and include peer review of our assessments. Knowing that our work will undergo scrutiny by our peers may help prevent this bias from entering our work, and the question-and-dialogue process involved in peer review may uncover this bias if it occurred. Peer review – specifically in the form of someone independent of the politics and pressures that might create this type of bias in the process – may help both the results themselves to be bias-free and also help prevent the impression of potential bias on the part of decision makers.

SUMMARY

In risk assessment, we make predictions about the likelihood of what will or will not occur, and in risk-informed decision-making (RIDM) we use those predictions to guide our decisions. In this context a prediction is a foretelling that is based, to the extent feasible, upon scientific principles, data, and experience. Predictions are necessary due to unknowns; and inherently, because what we do involves unknowns, we must make predictions.

The unknowns, and thus the predictions, generally can be characterized as one of three types:

- Type 1 - an already-existing physical condition that is not readily observable;
- Type 2 - a future physical event or occurrence;
- Type 3 - a future human action.

When making predictions in risk analysis, bias can skew or produce inaccurate results, bias being *prejudice in favor of or against one thing, and usually in a way considered to be inaccurate.* To reduce the unintentional introduction of bias in the decision-making process, we can draw on lessons about bias from past traditional design experience and use those lessons as a precaution about bias going forward into the new paradigm of more widespread usage of risk-informed decision-making.

We need to consider carefully as to when we should be using best-estimate values and when we should not. The best estimate of the *outcome* does not correspond to the best-estimate values of the *conditions*. Rather, the best estimate of the *outcome* corresponds to the best-estimate values of the *controlling conditions*.

Often we will find that our best-estimate value of the controlling conditions should fall outside of the range of data we have in hand. In such cases, we need to extrapolate beyond the data we have (to "think outside the data") to develop a best-estimate value of the controlling conditions.

We should think in Bayesian terms to leverage new information and maximize the advancement of our understanding. When site specific data is significantly different from what we predicted, we should add some rigor to our update and apply Bayesian updating techniques to our numbers. We as an industry should compile data on frequency and extent of structure performance and failure events, as well as for human involvement affecting risk.

If we focus on what we know a lot about instead of focusing on what it is that drives the answer, our limited resources are not invested in furthering our understanding of the key aspects that drive the answer. We must listen to what the site is trying to tell us and allow our focus to dwell upon those aspects, which will help focus our collective efforts on those aspects driving the answer.

Understanding risk does not denote or connote understanding the thing to which risk is being applied. Risk is a powerful tool, but to apply risk in a meaningful manner, one must have a solid understanding of the thing as well as of risk. To be "worthy of risk", its users must be experienced in evaluation, design, construction, maintenance, and repair of dams and levees. Additionally, we should bring in experts from outside our own fields of expertise – even if they are related fields of expertise – to complement our risk assessment teams to address specialty topics.

If risk practitioners allow themselves to consider reactions by other parties, bias may creep into the risk characterizations or the presentation of risk results. It is imperative that we not allow ourselves to be concerned about how others may react to our risk characterizations, but instead we remain disciplined about our process, our interpretation of data, and our application of scientific principles and empirical observations and experience.

CONCLUSION

Risk assessment and risk-informed decision-making is prone to various forms of bias. Bias can unknowingly result in inaccurate and unconservative risk estimates, undermining the validity and value of risk-informed decision-making. We, as individual risk practitioners and collectively as a profession, should be aware of our biases and work to mitigate, or at least communicate, these biases.

REFERENCES

Cato, K.D., and Rogers, J.D., 2017 (In Print), *Failure of the Alexander Dam Embankment on Kauai in 1930*: Environmental and Engineering Geoscience – Special Issue on Dams, scheduled publication August, 2017.

Neff, T., 2016, *"End Bias in Big Construction Project Predictions, Could game theory provide a better result than Bayes Theorem?":* Engineering News Record, July 18, 2016.

Rogers, J.D. and Cato, K.D., 2016, *Failure of the Alexander Dam Embankment on Kauai in 1930*: AEG NEWS Abstracts and Programs, p41.

US Army Corps of Engineers (USACE), 2015, *Best Practices in Dam and Levee Safety Risk Analysis*, US Society of Dams Workshop in Denver, CO, July 27-30, 2015, Notes dated 26 February 2015.

THE COMMON CAUSE ADJUSTMENT IN DAM SAFETY RISK ANALYSIS: SCOPE, PURPOSE, AND APPLICABILITY

Dom Galic, P.E.[1]

ABSTRACT

The risk analysis process is an integral part of many dam safety management programs. A key component of the process is the estimation of failure and life loss risk, which provides decision makers with information helpful in determining funding priorities within the inventory. When risks are independently estimated for different potential failure modes, the resulting total probability estimates often include a double-counted intersection probability. In some cases, the intersection probability is small enough that it can be ignored, while in others, the double counting can have a measurable impact on the interpretation of risk. This paper discusses the situations in which it may be beneficial to correct for the double counting of the intersection area, and provides an overview of the Common Cause Adjustment, the procedure used by the Bureau of Reclamation and others to correct for intersection. The mathematical basis for the adjustment process is explained, and the application of the process discussed in the context of both life loss and failure related risk estimation.

1. INTRODUCTION

The Bureau of Reclamation (Reclamation) is an agency of the United States Government that operates federal water projects in the 17 western states. Reclamation's inventory consists of hundreds of concrete and embankment dams, including such well-known facilities as Hoover Dam. While some of these structures are located away from cities and towns, many are in close proximity to populated areas, and balancing the safety of the public with the economic cost of potential dam safety improvements forms an essential component of Reclamation's risk management strategy. The details of the strategy can be found in the Dam Safety Public Protection Guidelines (Reclamation 2011a).

The risks associated with the routine operation of Reclamation's facilities are evaluated periodically. These risks can be plotted on a standardized chart, referred to for historical reasons as the fN chart (Figure 1). Two measures of risk are quantified for each facility, the Annualized Failure Probability (AFP) and the Annualized Life Loss (ALL). The AFP represents an estimate of the risk of failure, and is compared to the broad horizontal guideline in Figure 1; an AFP of greater than 1 in 10,000 is considered to indicate increasing justification to better understand or reduce risks. The ALL represents an estimate of life loss risk, and is compared to the diagonal guideline in Figure 1; an ALL of greater than 1 in 1000 is considered to indicate increasing justification to better understand or reduce risks. The rationale for the guidelines (as well as the meaning of the

[1] U.S. Department of the Interior, Bureau of Reclamation, Denver, Colorado

"box" shown in the lower right corner of Figure 1) is beyond the scope of this article. For more information, refer to Reclamation (2011b).

The example fN chart in Figure 1 includes three different plot markers. The large yellow marker represents the total risk associated with the facility; the superimposed "whiskers" measure uncertainty in the AFP and ALL directions (these bands are typically obtained from Monte Carlo simulation). The smaller markers represent individual potential failure modes (defined below). The spreadsheet used to generate the basic Reclamation fN chart is programmed to sum risks in both the AFP and ALL directions. In other words, the total AFP and total ALL are obtained as the simple sum of the estimated failure and life loss risks, respectively, of the individual potential failure modes. In most cases, this simplified approach provides credible estimates of risk. However, there are some scenarios in which not taking into account the nature of the relationships between different potential failure modes could alter the understanding of the total risk.

The term Common Cause Adjustment (CCA) can refer to any of the processes used by Reclamation and others to correct for the double-counting of intersection between individual PFMs. The term does not refer to a specific method, but each of the methods used is specific to a particular risk analysis scenario, and each is performed in accordance with the basic rules of probability theory. The CCA process is mentioned in the Best Practices in Risk Analysis training manual prepared jointly by Reclamation and the U.S. Army Corps of Engineers (Reclamation & USACE 2015), but the description is spread over several chapters of the document, and may be difficult to understand without context. The purpose of this paper is to provide a concise explanation of the CCA process, along with its mathematical basis, applicability, and scope. We begin with a brief review of two risk related concepts, the potential failure mode and statistical independence.

2. POTENTIAL FAILURE MODES AS EVENTS

A potential failure mode (PFM) represents a sequence of events that must occur for a particular failure mechanism to result in an uncontrolled release of the reservoir. In terms of the total risk of failure, it represents one of the ways in which the dam could plausibly fail. The PFM includes a trigger event (such as the occurrence of an earthquake, or the occurrence of an annual high pool), followed by a series of component events describing the response of the structure (e.g. foundation liquefaction is triggered, a slope failure occurs, deformations exceed the available freeboard) as well as a breach event. Because the "events" of the PFM are also events in the set theory sense, a probability can be assigned to each of them, one typically conditioned on the occurrence of the preceding events in the chain. The probability of the PFM occurrence event (the event equal to the intersection of the *n* events comprising the failure chain, including the breach event) is annualized by virtue of the load probability being annualized, and can be obtained using the basic intersection formula as:

$$P(PFM) = P(E_1E_2 \ldots E_n) = P(E_1)*P(E_2|E_1)*P(E_3|E_1E_2)* \ldots *P(E_n|E_1E_2 \ldots E_{n-1}) \quad \{1\}$$

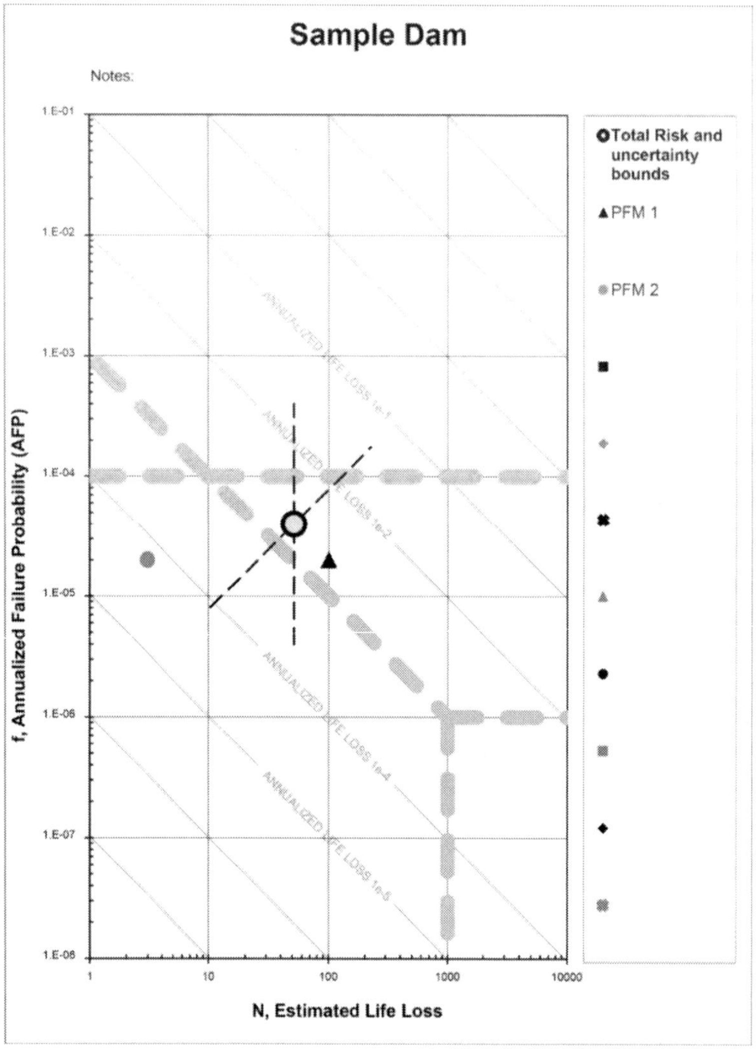

Figure 1. Standard Reclamation fN chart.

The probability of the PFM occurrence event is referred to as the PFM AFP (or AFP_{PFM}), as distinct from the AFP value associated with the facility as a whole (AFP_{Total}). In a Reclamation risk analysis, the conditional probabilities in Equation 1 would be estimated by a team of qualified subject matter experts. For a general discussion on the role of subjective probability estimation in engineering risk analysis, see Vick (2002).

For the purposes of this paper, we condense the event $E_1 E_2 \ldots E_n$ into the intersection of a trigger event T and a response event R. Equation 1 then reduces to:

$$P(PFM) = P(TR) = P(T)*P(R|T) \qquad \{2\}$$

where the P(R|T) term is referred to as the *conditional probability of failure*. It is important to understand that whereas R represents the intersection of events E_2 through E_n (to use the terminology of Equation 1), it is also a "complete" event in its own right. The same probability rules and equations that apply to events E_2 through E_n also apply to event R.

3. STATISTICAL INDEPENDENCE AND CONDITIONAL INDEPENDENCE

A pair of events E_j and E_k are statistically independent (SI) if the occurrence of one has no impact on the occurrence *probability* of the other. Mathematically, the events are SI if $P(E_j|E_k)$ is equal to $P(E_j)$. The concept, which can be generalized to any number of events, has several important implications, not the least of which is the simplification of the basic intersection formula. For *n* SI events,

$$P(E_1E_2 \ldots E_n) = P(E_1)*P(E_2)* \ldots *P(E_n) \qquad \{3\}$$

In practice, a set of static potential failure modes (i.e., PFMs not involving the occurrence of a comparatively remote triggering event) would be developed in a manner that ensured they remained "more or less" statistically independent of each other. Two closely related PFMs could be merged into a single PFM encompassing both failure mechanisms, or the focus could simply be shifted to the more critical failure mechanism. For a pair of SI PFMs with estimated AFPs close to guidelines (order of 1 in 10,000), the intersection probability given by Equation 3 would be quite small.

The fundamental rules of probability, including those related to statistical independence, apply not only at the level of the overall sample space, but also within the reconditioned sample space implied by the occurrence of a particular event (see e.g. Ang & Tang 1975). For the purposes of this paper, we consider a pair of potential failure modes PFM 1 and PFM 2 to be *conditionally independent* (CI) if $P(R_1|R_2T) = P(R_1|T)$, where R_1 and R_2 are the response events associated with PFMs 1 and 2, respectively, and T is the common triggering event. In other words, for CI PFMs, the conditional probabilities of events R_1 and R_2 are unaffected by the occurrence (or non-occurrence) of the other response event. For a set of *n* CI PFMs, Equation 3 can be written:

$$P(R_1R_2 \ldots R_n|T) = P(R_1|T)*P(R_2|T)* \ldots *P(R_n|T) \qquad \{4\}$$

As an example, consider a pair of response events A and B and an event Q representing the occurrence of a particular type of earthquake (Figure 2). Intuitively, the intersection events AQ and BQ would not be statistically independent, because the occurrence of one would imply that the earthquake common to both had occurred. However, at the conditional probability level (given the occurrence of Q), events A and B could be associated with completely unrelated failure mechanisms, or with relatively distant portions of the dam. In other words, the knowledge that a spillway structure built on rock had failed as the result of a 5000 year earthquake would not be likely to change a subjective probability estimate for foundation liquefaction under a 5000 year seismic loading.

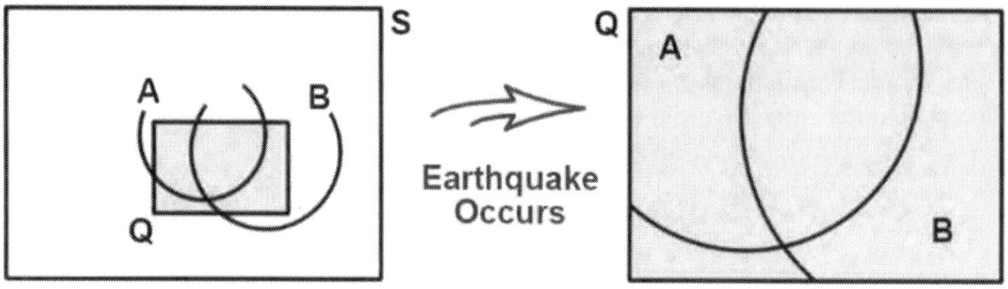

Figure 2. Conceptualized transformation of the sample space.

4. PURPOSE AND APPLICABILITY OF THE COMMON CAUSE ADJUSTMENT

Consider a set of individual PFM risk estimates (AFP_{PFM1} and ALL_{PFM1} through AFP_{PFMn} and ALL_{PFMn}) and a pair of total risk estimates (AFP_{Total} and ALL_{Total}). Since the direct estimation of risk occurs at the individual PFM level, it can be surmised that the risk associated with the facility as a whole must in some way be a function of the individual PFM risk estimates. As discussed in the introduction, the relationship that is typically assumed as a starting point by Reclamation (i.e. a simple additive one) may not be appropriate when the individual PFMs cannot realistically be considered mutually exclusive, to the extent that the interpretation of risk may be affected. At its most basic level, the Common Cause Adjustment amounts to the recognition that summing the individual risk estimates may not be appropriate for a given facility or risk estimate. In this case, the simple additive relationship must be modified to include a correction for intersection between the individual PFMs.

As discussed below, the AFP correction is quite straightforward when the PFMs are statistically independent or conditionally independent (the ALL correction is slightly more complex). In fact, the processes outlined in this paper are applicable to "reasonably" SI and CI potential failure modes only. If the controlling PFMs cannot be realistically viewed in this way, the processes outlined below cannot be used to correct for intersection. A set of PFMs that cannot "reasonably" be thought of as SI or CI should be approached in some other manner, such as by combining the PFMs into a single event tree (the sub-PFMs represented by parallel branches of an event tree are by definition mutually exclusive), if possible, or by focusing on the controlling PFMs only (when the estimated risks are unequal). If there are several important risk contributors and the nature of the dependence between the PFMs is understood, the AFP can also be bracketed using the unimodal bounds theorem (see e.g. Hill et al 2003).

The following two sections discuss the practical application of the CCA in the context of total AFP and ALL correction. However, there are many situations in which the extra work of performing the CCA is not justified by the resulting nominal increase in precision. The question of whether to even consider doing so must therefore be answered prior to delving into the mathematics. The list below provides broad guidance on when it may be appropriate to apply the CCA, assuming the conditions of the preceding paragraph are met:

- When the individual PFM AFP estimates are so high that their direct summation would result in a total AFP estimate close to 1.0 (this scenario would plausibly arise in exceptional circumstances only, such as in the context of an emergency dam modification).
- When either the total AFP or total ALL plots close to guidelines and the risk is controlled by two or more PFMs associated with the same load trigger (hence the "common cause") or triggers.
- When both the total AFP or total ALL plot below guidelines, the risk is controlled by two or more PFMs associated with the same load trigger, and the uncertainty crosshairs extend above the guidelines.

Since the probability of the intersection for SI events is greater than zero, the application of the CCA process can only result in an apparent reduction in the total estimated risk. However, due to the inherently low precision of subjective probability estimation, applying the CCA will not necessarily result in any change in the overall *interpretation* of risk. In considering the scenarios above, it should also be kept in mind that the objective of the risk analysis process is not to "end up below guidelines", but rather to gain a deeper understanding of the risk. By highlighting potential sources of risk inflation, the CCA can help estimators focus on the mechanisms actually driving the risk, which can be important in identifying the most effective corrective actions or the most optimal means of reducing uncertainty.

5. ADJUSTMENT OF THE ANNUALIZED FAILURE PROBABILITY

In the context of quantifying the risk of dam failure, the purpose of the CCA is to correct a total risk estimate (AFP_{Total}) that has been obtained through the direct summing of n individual PFM risk estimates (AFP_{PFMi}). The individual PFM risk estimates are assumed to have been estimated without any consideration of the remaining PFMs (beyond verification of statistical independence), or by an entirely separate risk team. Assuming they have been generated by a team of qualified experts, the individual PFM risk estimates (AFP_{PFMi}) are already "correct" and do not require any adjustment, though it must be understood that they will not sum directly to the corrected total AFP. To correct the total AFP for the double-counting of intersection, the adjustment is applied at the event space level (in the case of statistically independent PFMs) or at the response-event level (in the case of conditionally independent PFMs). The process is best illustrated through a pair of examples.

5.1 Example with two CI PFMs

Example Dam 1 is an embankment structure founded on soil (with a cutoff trench to rock), and has a gated side-channel spillway with a concrete inlet structure anchored to its rock foundation. To help generate power, the reservoir is kept high, with about 20 feet of water typically stored on the radial gates. Consider the following two PFMs associated with the occurrence of a 50,000-year plus earthquake:

- PFM 1: "A 50,000-year plus earthquake occurs, resulting in foundation liquefaction, sliding within the embankment, crest deformation in excess of the available freeboard, and an uncontrolled release of the reservoir."
- PFM 2: "A 50,000-year plus earthquake occurs, resulting in the failure of the steel anchors, an initial displacement of the inlet structure, the shearing of both sets of underdrains, the sliding failure of the inlet structure, and an uncontrolled release of the reservoir."

Each PFM can be decomposed into a triggering and a response event as follows:

- Triggering event (both PFMs): Q = "a 50,000-year plus earthquake occurs".
- Response event for PFM 1: A = "foundation liquefaction, sliding within the embankment, crest deformation in excess of the available freeboard occur, and an uncontrolled release of the reservoir occur".
- Response event for PFM 2: B = "failure of the steel anchors, an initial displacement of the inlet structure, the shearing of both sets of underdrains, the sliding failure of the inlet structure, and an uncontrolled release of the reservoir occur".

Response events A and B each represent the intersection of several component events. In practice, conditional probability estimates for the component events would already be available at the time of the CCA, which would ordinarily not be performed until a first-pass total AFP had been estimated. For the purposes of this example, assume that the conditional probability of response event A, written $P(A|Q)$, has been estimated by a team of embankment dam experts as 0.5, and that $P(B|Q)$ has been estimated by a team of concrete structure experts as 0.9. The probability of the triggering event follows directly from the definition of the exceedance probability, and $P(Q)$ is taken as 1/50,000.

Even without more detailed information, it is apparent that response events A and B would reasonably qualify as CI. The associated reservoir retention features are composed of and founded on different materials, and the breach mechanisms are unrelated. Although there may be subtle "dependency structures" present (Zielinsky 2014), their significance to the overall risk estimate will probably be limited if the essence of each failure process is captured by the PFM description. Also not relevant to the CCA is the question of whether the occurrence of response event A before B (or vice versa) will introduce dependence by lowering the reservoir elevation and thereby reducing the volume of the second release. Since there is no sense of time in a Venn diagram, the double counting of an intersection event can be addressed quite effectively without quantifying the specific timeline of a dual breach. In fact, the purpose of performing a risk assessment (as opposed to developing some kind of analytical model) is not to realistically capture the minute physical details of a failure process. It is simply to enumerate, in accordance with the rules of probability, the ways that a dam can realistically fail, so that the most effective means of reducing or better understanding the risk can be identified.

Once the response events have been identified as reasonably CI, the "size" of their intersection within the reconditioned sample space can be obtained from Equation 4. This in turn allows the total probability of the intersection, which must be subtracted from the simple sum of the AFP estimates to correct for double-counting, to be calculated. The five basic steps of the process are outlined below; in this case, the corrected AFP is approximately 30 percent lower than the uncorrected AFP:

1. $AFP_{PFM1} = P(Q)*P(A|Q) = 0.5/50{,}000$
2. $AFP_{PFM2} = P(Q)*P(B|Q) = 0.9/50{,}000$
3. $P(AB|Q) = P(A|Q)*P(B|Q) = 0.45/50{,}000$
4. Uncorrected total $AFP = P(Q)*P(A|Q) + P(Q)*P(B|Q) = 1.4/50{,}000$
5. Corrected total $AFP = P(Q)*P(A|Q) + P(Q)*P(B|Q) - P(Q)*P(AB|Q)$
   $= 0.95/50{,}000$

For simplicity, the above example uses a single critical load range, with earthquakes more frequent than the 50,000-year event not considered (or not considered to control the estimated risk). In practice, the load range is frequently broken up into multiple bins (e.g. Q_1 occurs, Q_2 occurs, etc.), with a conditional failure probability $P(R|Q_i)$ estimated for each. Assuming there is no overlap between bins, the associated intersection events (Q_1R, Q_2R, etc.) can be treated as mutually exclusive PFMs. In other words, the process outlined above can be separately applied within the reconditioned sample space of each load trigger Q_i, with no need to correct for intersection across the bins.

5.2 Example with three SI PFMs

For three events A, B, C, the probability of the union is given by the following equation:

$$P(A \cup B \cup C) = P(A) + P(B) + P(C) - P(AB) - P(AC) - P(BC) + P(ABC) \quad \{5\}$$

When the events in question are statistically independent, Equation 5 can be written as:

$$P(A \cup B \cup C) = P(A) + P(B) + P(C) - P(A)P(B) - P(A)P(C) - P(B)P(C) + P(A)P(B)P(C) \quad \{6\}$$

The first three terms in Equation 5 represent the uncorrected sum of the individual event probabilities, whereas the remaining terms provide a correction for the over-counting of the intersection. Note that the above equations can be generalized for any number of events.

In Dam Safety risk analysis, the total AFP is correctly represented by the probability of the union of the individual PFMs. However, for a given set of three PFMs, it would typically be interpreted as the sum of the first three terms of Equation 5. For PFMs that are statistically independent at the level of the sample space (i.e., SI rather than CI), this simplification is reasonable unless the individual PFM AFPs are very high. For example, let P(PFM 1) = 0.1, P(PFM 2) = 0.2. and P(PFM 3) = 0.3, where PFMs 1, 2, and 3 are SI; all three of these estimates would essentially imply a failure in progress. The resulting uncorrected total AFP would be 0.6 (60 percent chance of dam failure in a given year). The corrected total AFP, obtained using Equation 6, would be about 0.5, or about ten percent lower. This difference would likely have no impact on the conclusion that the dam was in dire need of corrective action.

6. ADJUSTMENT OF ANNUALIZED LIFE LOSS

Although it is considered an equally important measure of risk (Reclamation 2011a), the Annualized Life Loss represents a different kind of mathematical object than the AFP. For an individual PFM, the AFP is an intersection probability, whereas the total AFP associated with the facility is a union probability. In contrast, the ALL is calculated as "the product of the annualized failure probability and the life loss that is expected to result from failure" (Reclamation 2011a). In other words, for an individual PFM, $ALL_{PFM} = AFP_{PFM} * L_{PFM}$, where L_{PFM} is the life loss estimated for that PFM (see Reclamation 2015 for information on how the life loss is estimated).

The Reclamation fN chart shown in Figure 1 takes the ALL values entered for individual PFMs and sums them. For a set of *n* PFMs, each associated with an estimated life loss L_i, the ALL coordinate of the total marker is given by:

$$ALL_{Total} = P(PFM\ 1)*L_1 + P(PFM\ 2)*L_2 + \ldots + P(PFM\ n)*L_n \qquad \{7\}$$

The right side of Equation 7 is quite similar to the formula for the expectation of a discrete random variable (see e.g. Ang & Tang 1975). Provided the *n* PFMs are mutually exclusive, ALL_{Total} can be thought of as the expected value of life loss per year of operation (note that collective exhaustiveness is satisfied identically because the "balance" of the probability is not associated with any life loss).

What happens if the PFMs in Equation 7 are not mutually exclusive? The simple answer is nothing. Since the Annualized Life Loss is not *defined* as the expected value of life loss (Reclamation 2011a does not explicitly use this terminology), from the computational perspective it does not matter whether the analogy holds or not. Furthermore, since an expectation can be equal to any real number, there will be no striking violation of the axioms of probability, such as a probability estimate greater than 1.0, to warn of a flawed approach. In this sense, the applicability of the CCA process could quite easily be restricted to the correction of probability overlap within the total AFP.

There are two reasons why the extension of the CCA to Annualized Life Loss may be of interest. The first pertains to the meaning of the "societal" public protection guideline used by Reclamation and others (i.e., the diagonal line in Figure 1), which is intended to provide a sense of how much life loss risk society is willing to "accept" in exchange for societal benefits. Even without delving into the significance of the number selected for the threshold value, it can be argued that this quantity (0.001 lives per year) corresponds to an *expected* value. That is, in a society comprised of individuals having different levels of risk aversiveness, 0.001 would presumably correspond to roughly the average of what is considered "acceptable", rather than, say, the view of the least risk-aversive fifteen percent. In this sense, it would be desirable to make the ALL index correspond as nearly as possible to the expected value of life loss (rather than, for the analogy used, something closer to the 85 percentile). Second, in some cases the extension of the CCA to Annualized Life Loss may help increase a team's understanding of the factors driving the life loss risk. This could result in higher confidence even if the plotting position of the total risk marker did not significantly change.

As shown in Equation 7, the total ALL is a function of the individual PFM estimates, rather than of the total AFP. For this reason, knowledge of the adjusted total AFP will not necessarily be useful in adjusting the total ALL, and a second CCA will need to be performed at the individual-PFM level (with some exceptions). The three examples below are used to illustrate the CCA in the context of three basic types of life loss scenarios; other more complex scenarios are possible, but likely represent variations (or combinations) of these. As in the case of the individual PFM AFP estimates, the individual PFM ALL estimates remain correct, and do not require any kind of adjustment. It is the total ALL that may be affected by intersection between the PFMs, and that may need to be adjusted in order for its definition to coincide with that of the expected life loss.

In the examples below, the term Population at Risk (PAR) refers to the individuals inhabiting the projected inundation area of a given breach scenario prior to any warning being issued. In Reclamation practice, the size of the PAR (as distinct from the estimated life loss) is not explicitly adjusted to account for evacuation prior to or during the failure of the dam (Reclamation 2015). For simplicity, the examples also assume that for each of the facilities in question, there are only two controlling PFMs. Adjustment processes similar to the ones outlined below can be developed for cases where there are more than two controlling PFMs.

6.1 Scenario 1: Estimated life loss for the two PFMs is different, but the PAR associated with the smaller life loss estimate is a subset of the PAR associated with the larger life loss estimate

This life loss scenario arises when the critical PFMs involve significantly different breach sizes (but with discharge into the same downstream channel). In order to calculate the ALL using Equation 7, the two statistically independent (or conditionally independent) PFMs must be transformed into a set of mutually exclusive events. This can be accomplished by treating the intersection event as a separate PFM.

Recall that for Example Dam 1 (discussed in Section 5.1), the critical potential failure modes were considered to be CI. Assume that an estimated life loss of 100 individuals has been assigned to PFM 1, and an estimated life loss of 3 individuals to PFM 2. As before, the conditional probability of response event A (PFM 1) is 0.5, the conditional probability of response event B (PFM 2) is 0.9, and the estimated probability of the triggering event Q is 1/50,000. Within the reconditioned sample space (and using the syntax ¬E to denote the complement of E or "not E"),

- $P(AB|Q) = P(A|Q)*P(B|Q) = 0.45$
- $P(A\neg B|Q) = P(A|Q) - P(AB|Q) = 0.5 - 0.45 = 0.05$
- $P(B\neg A|Q) = P(B|Q) - P(AB|Q) = 0.9 - 0.45 = 0.45$
- $P(ABQ) = P(AB|Q)*P(Q) = 0.45/50{,}000$
- $P(A\neg BQ) = P(A\neg B|Q)*P(Q) = 0.05/50{,}000$
- $P(B\neg AQ) = P(B\neg A|Q)*P(Q) = 0.45/50{,}000$

The transformed PFMs defined by events ABQ, A¬BQ, and B¬AQ are mutually exclusive, sum directly to the corrected total AFP (0.95/50,000), and can be used directly with Equation 7. The only remaining question is what life loss value to associate with event ABQ (in which both the embankment and spillway fail). A reasonable answer is *the higher of the two life loss estimates*. Although it could be argued that the occurrence of both PFMs will result in a peak discharge greater than that of the higher-life loss PFM alone, this logic does not hold within the anachronistic event space of elementary set theory. The occurrence of two events in the Venn diagram sense is simply not the same thing as their simultaneous occurrence. In fact, it is quite easy to come up with examples for which the occurrence of two PFMs is mathematically possible (e.g. because they are SI) but for which a "simultaneous" failure could not possibly result in additive life loss (e.g. because they each must occur within the same limited physical space, such as a fuse plug). Since in applying the CCA we are correcting for a mathematical error, it is the mathematical meaning of the intersection that we are interested in, not the temporal. For this reason, the inundation area associated with the intersection PFM should be though of as the overlay, not the sum, of the individual inundation areas.

Using $L_{ABQ} = 100$, $L_{A\neg BQ} = 100$, and $L_{B\neg AQ} = 3$, the common cause adjusted ALL is obtained using Equation 7 as 45/50,000 + 5/50,000 + 1.35/50,000, or about 51/50,000. In contrast, the uncorrected ALL is about 53/50,000, in this case representing a relatively insignificant correction compared to the AFP correction (recall that in Section 5.1 the difference between the corrected and uncorrected total AFP was about 30 percent).

6.2 Scenario 2: Estimated life loss is equal for both PFMs

This is a special case in which it is possible to use the adjusted total AFP to calculate the adjusted total ALL. Example Dam 2 is a 600-foot long, 20-foot high embankment structure founded on soil (with a cutoff trench to rock). The embankment consists of an erosion-resistant plastic soil, but large differential settlements were observed shortly after construction ended. More recently, the results of an extensive field exploration program have suggested that the downstream shell of the dam is underlain by loose, sandy deposits. The critical PFMs involve foundation liquefaction, sliding-induced crest loss, and overtopping under seismic conditions (PFM 1), and internal erosion of the embankment under static conditions (PFM 2). Both PFMs would result in the total loss of the dam (down to the original ground surface), and it was assumed for both cases that no formal warning of an impending breach would be issued.

PFMs 1 and 2 involve different load triggers, different soils, and different performance loss mechanisms; as such, they can be identified as reasonably statistically independent (i.e. at the level of the overall sample space). Furthermore, since they both involve rapid breaches with the same physical dimensions, identical peak discharge inundation areas will apply to each PFM, and the life loss estimates should be quite similar. Based on the reasoning presented under Scenario 1, the life loss L associated with the intersection event PFM 1 ∩ PFM 2 would be the same as that associated with either of the PFMs individually. Transforming PFMs 1 and 2 into a trio of mutually exclusive events and applying Equation 7,

- $ALL_{Total} = P(PFM\ 1 \cap \neg PFM\ 2)*L + P(PFM\ 2 \cap \neg PFM\ 1)*L + P(PFM\ 1 \cap PFM\ 2)*L$
- $ALL_{Total} = [P(PFM\ 1 \cap \neg PFM2) + P(PFM\ 2 \cap \neg PFM\ 1) + P(PFM\ 1 \cap PFM\ 2)]*L$

Since the probability of the union of a set of mutually exclusive events is equal to the simple sum of their individual probabilities, the quantity in the brackets corresponds to the adjusted total AFP. In other words, when the estimated life loss for the controlling PFMs is similar, the adjusted total ALL can be obtained directly from the adjusted total AFP. What this means is that if application of the AFP CCA results in a certain percentage reduction in the total AFP, the same percentage reduction will also apply to the total ALL. In contrast, for the Scenario 1 example, the size of the ALL correction was negligible compared to that of the AFP correction. Based on this observation, in most situations, it may only make sense to apply an ALL CCA if life loss Scenario 2 applies.

6.3 Scenario 3: PAR associated with the two PFMs is additive

Example Dam 3 is comprised of a pair of small embankments located on opposite ends of a broad valley formed by ancient block faulting. Due to low precipitation in the summer, the reservoir is filled during the first major storm of the year, and kept high until the water is needed (the probability of the reservoir reaching the maximum elevation each year is close to 1.0). The north embankment is composed of clayey materials, and the south of sandy materials. The critical potential failure modes are PFM 1, "internal erosion by scour results in failure of the north embankment", and PFM 2, "internal erosion by backward erosion piping results in failure of the south embankment". The north embankment is located 3 km away from a town of 120 residents; 5 km further downstream is a large reservoir capable of retaining the breach flood. The area below the south embankment is relatively unpopulated, but there is a dam tender's residence and picnic area immediately downstream. The estimated AFPs are AFP_1 (for PFM 1) and AFP_2 (for PFM 2). The estimated life loss is L_1 for PFM 1 and L_2 for PFM 2.

Although each involves an internal erosion mechanism, PFM 1 and PFM 2 were considered by the risk team to be reasonably statistically independent. This conclusion was based on the following:

- The fact that the north embankment and south embankment are composed of different types of soils and susceptible to different types of internal erosion mechanisms.
- The fact that there are recreational users present at the downstream toe of the south embankment, but not at the north embankment (i.e. the knowledge that the failure in progress of the south embankment had been detected would not increase the probability of successful intervention for the other PFM).
- The fact that the reservoir is certain to reach the critical elevation every year (i.e. knowledge that one of the dams had failed would not imply a higher load probability for the other PFM).
- In general, when it was postulated that the north embankment was in the process of failing, none of the team members believed they could realistically modify any of the probability estimates developed for the south embankment PFM.

Once it was determined that the PFMs were reasonably SI, the following probabilities were calculated:

- $P(PFM\ 1 \cap PFM\ 2) = P(PFM\ 1)*P(PFM\ 2) = AFP_1*AFP_2$
- $P(PFM\ 1 \cap \neg PFM\ 2) = P(PFM\ 1) - P(PFM\ 1 \cap PFM\ 2) = AFP_1 - AFP_1*AFP_2$
- $P(PFM\ 2 \cap \neg PFM\ 1) = P(PFM\ 2) - P(PFM\ 1 \cap PFM\ 2) = AFP_2 - AFP_1*AFP_2$

Transforming the SI events PFM 1 and PFM 2 into three mutually exclusive events (PFM 1 ∩ PFM 2, PFM 1 ∩ ¬PFM 2, PFM 2 ∩ ¬PFM 1), noting that the total life loss associated with a joint failure would be additive (based on the "overlay" concept from Scenario 1), and inserting the relevant quantities into Equation 7 gives:

- $ALL_{Total} = P(PFM\ 1 \cap PFM\ 2)*(L_1+L_2) + P(PFM\ 1 \cap \neg PFM\ 2)*L_1$
  $+ P(PFM\ 2 \cap \neg PFM\ 1)*L_2$
- $ALL_{Total} = AFP_1*AFP_2*(L_1+L_2) + (AFP_1 - AFP_1*AFP_2)*L_1$
  $+ (AFP_2 - AFP_1*AFP_2)*L_2$
- $ALL_{Total} = AFP_1*AFP_2*L_1 - AFP_1*AFP_2*L_1 + AFP_1*AFP_2*L_2 - AFP_1*AFP_2*L_2$
  $+ AFP_1*L_1 + AFP_2*L_2$
- $ALL_{Total} = AFP_1*L_1 + AFP_2*L_2$

In other words, $ALL_{Total} = P(PFM\ 1)*L_1 + P(PFM\ 2)*L_2$, even though PFMs 1 and 2 are not mutually exclusive. When the potential failure modes are SI or CI and the population at risk is additive, the CCA amounts to the recognition that no PFM transformation is required for the total ALL from Equation 7 to equal the expected value of life loss for the facility.

7. RISK INTERPRETATION AND MATHEMATICAL EXACTITUDE

The total AFP is properly defined as the probability of the union of the individual PFMs. However, in practice, it is often approximated as the simple sum of the individual PFM probability estimates. Common cause adjustment provides a means of correcting the simple sum to compensate for the over-counting of the intersection area. The methods discussed in this article are applicable to statistically independent PFMs in general, and in particular to those whose response events are statistically independent given the occurrence of a common triggering event (conditionally independent PFMs). Although the total Annualized Life Loss represents a different kind of mathematical object than the total AFP, a similar set of methods can be developed to adjust it for intersection between the PFMs.

In the sense that they are specific to the situations under consideration and consistent with the rules of probability theory, the adjustment methods discussed in this article can be thought of as precise. Unfortunately, the same cannot be said of the quantities being adjusted. Neither the individual PFM annualized failure probabilities nor the individual PFM life loss estimates are exact quantities, nor can they ever become exact, no matter how much effort is expended during the modeling or field exploration stages. The concept of uncertainty is central to the topic of subjective probability estimation, with the ability to include bounds (a measure of what is not known) around a subjective probability best estimate providing an important means of identifying the most effective ways to improve the quality of the estimate. The significance of risk estimate uncertainty to the decision process as a whole is the reason that Reclamation typically plots uncertainty bands through the fN chart total risk marker (Figure 1). Although beyond the scope of this paper, the upper and lower limits of the uncertainty crosshairs can also be adjusted, if needed, using the methods described above.

The benefits of improving the nominal accuracy of a total risk estimate that remains inherently uncertain depend on the specifics of the situation. The example in Section 5.2, which involves an adjustment at the total probability level, results in a ten percent "improvement" in the accuracy of the (unrealistically high) total AFP. Taking into account the fact that the guidelines in Figure 1 are not considered hard boundaries (Reclamation 2011a), and given that the uncertainty crosshairs often span several log cycles, it is unlikely that an adjustment of this magnitude would have a significant impact on the overall interpretation of risk. Furthermore, because AFP estimates are typically much lower than those in the Section 5.2 example, this conclusion can be generalized to most situations involving potential failure modes that are SI over the entire sample space.

The situation is somewhat different when the PFMs in question are CI, as in the Section 5.1 example. In that example, the adjustment resulted in an approximately 30 percent reduction in the total AFP, possibly (but not necessarily) significant enough to result in an overall change in the interpretation of risk. However, if the conditional probabilities of response events A and B were 0.99 and 0.99 rather than 0.5 and 0.9, plausible if the load trigger is a strong earthquake, the adjusted total AFP (Figure 3) would be about 50 percent lower than the simple sum of the PFM AFPs (Figure 1). Similarly, if the conditional failure probability of a third PFM premised on the occurrence of the same load trigger was also 0.99, the adjusted total AFP would be about three times lower. By extension, for a set of n CI PFMs, the maximum apparent benefit of the CCA would be a factor of n risk reduction. With a high enough seismic hazard, the apparent reduction in the total AFP could thus potentially be significant in the case of a large reservoir with multiple retention structures.

There can thus arise situations where performing a common cause adjustment will help inform decision makers, or help improve a project team's understanding of what is actually driving the risk. However, it must be underscored that the precision of the numbers that are used as measures of risk is less important than the question of whether the overall interpretation makes sense. The desire to improve precision for the sake of improving mathematical exactitude is not alone a sufficient reason to perform a CCA. Whether "exact" or approximate, the numbers represent just one piece of the Dam Safety Case (Reclamation 2011a), and must be interpreted in the context of the other key considerations that apply, such as the overall condition of the dam, the potential for loadings and/or consequences to change, and the effectiveness of routine risk management activities.

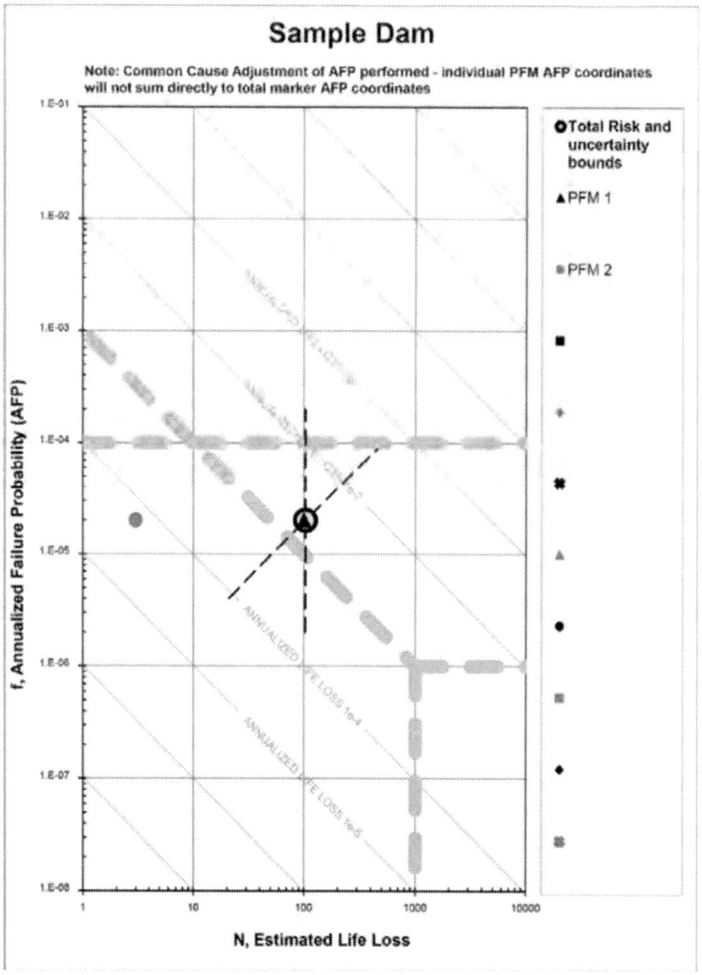

Figure 3. PFMs from Figure 1 with the total marker shown as common cause adjusted for AFP (assuming the PFMs are CI with respect to a 50,000-year earthquake and that each has a conditional failure probability of about 0.99). Note statement explaining why the fN chart does not appear to be adding "correctly". The reason that the total marker continues to plot at the same value of ALL as in Figure 1 is that no ALL CCA has been performed.

REFERENCES

Ang AHS & Tang WH (1975). Probability Concepts in Engineering Planning and Design: Volume 1, Basic Principles. John Wiley and Sons, New York, New York. 1975.

Bureau of Reclamation (2011a). Interim Dam Safety Public Protection Guidelines: A Risk Framework to Support Dam-Safety Decision Making. U.S. Department of the Interior, Bureau of Reclamation Dam Safety Office, Denver, Colorado. August 2011.

Bureau of Reclamation (2011b). Interim Rationale Used to Develop Reclamation's Dam Safety Public Protection Guidelines. U.S. Department of the Interior, Bureau of Reclamation Dam Safety Office, Denver, Colorado. August 2011.

Bureau of Reclamation (2015). RCEM – Reclamation Consequence Estimating Methodology. Interim Guidelines for Estimating Life Loss for Dam Safety Risk Analysis. U.S. Department of the Interior, Bureau of Reclamation. July 2015.

Bureau of Reclamation & USACE (2015). Best Practices in Dam and Levee Safety Risk Analysis. Training manual jointly prepared by the Bureau of Reclamation Technical Service Center and the U.S. Army Corps of Engineers Risk Management Center, Denver, Colorado. Version 4.0, July 2015.

Hill P, Bowles D, Jordan P, and Nathan R (2003). Estimating Overall Risk of Dam Failure: Practical Considerations in Combining Failure Probabilities. ANCOLD Bulletin, No. 127.

Vick SG (2002). Degrees of Belief. ASCE Press, Reston, Virginia. 2002.

Zielinsky PA (2014). Event Trees in the Assessment of Dam Safety Risks. Proceedings of the 2014 Annual Conference of the Australian National Committee on Large Dams (ANCOLD). Canberra, Australia. October 2014.

POTENTIAL FAILURE MODE ANALYSIS – INSIGHTS AND LESSONS LEARNED FOR THE FOUNDATION OF A SUCCESSFUL DAM SAFTY MANAGEMENT PROGRAM

Frank L. Blackett, P.E.[1]
Douglas D. Boyer, P.E., C.E.G.[2]

ABSTRACT

A Potential Failure Mode Analysis (PFMA) is a familiar and widely accepted process within the dam safety community. Although the concepts are generally well understood, its use and application is highly variable across different agencies, owners, and consultants. Those who routinely perform robust quantitative dam safety risk analyses clearly understand how to develop a detailed Potential Failure Mode (PFM). Others sometimes struggle with understanding all of the intricacies associated with developing a detailed PFM that will provide the best insight into the design, performance, and behavior of their dam, and how to apply PFMs to their dam safety program. Conducting a PFMA, developing PFMs, and placing PFMs into categories is not the culmination of the process; rather, it is the foundation for providing input to strategic dam safety management activities and actions for inspection programs, surveillance and monitoring activities, emergency action planning, site investigations, engineering studies, risk management, and others. Poorly developed, unclear, or confusing, and missing potential failure modes can result in focusing dam safety activities in lower priority areas and missing critical dam safety concerns altogether.

This paper is intended to provide some insights from lessons learned in conducting and participating in PFMA sessions and reviewing PFMA reports. This paper also addresses some commonly asked questions such as, "how many PFMs do I need for my project, why do I need to separate out all my PFM descriptions, and how do I manage all these PFMs?"

Disclaimer: The opinions expressed in this paper are those of the authors and do not represent official guidance or direction from the Federal Energy Regulatory Commission.

INTRODUCTION

The PFMA process has been a part of dam safety practice for nearly 20 years with improvements adopted into the process as more insight is gained from the study of dam safety incidents and failures. The primary purpose of the PFMA is to evaluate all possible ways a dam, and all its associated features, could fail that would result in some level of consequences to the project and/or the downstream reaches [1]. For the Federal Energy Regulatory Commission (FERC), a project failure is considered any uncontrolled

[1] Frank L. Blackett, P.E., Federal Energy Regulatory Commission, 100 First Street, Suite 2300, San Francisco, CA 94105 frank.blackett@ferc.gov
[2] Douglas D. Boyer, P.E., C.E.G, Federal Energy Regulatory Commission, 805 SW Broadway, Suite 550, Portland, OR 97205 douglas.boyer@ferc.gov

release of water, regardless of the potential for loss of life or economic impacts downstream of a dam. This is partially due to the fact that hydroelectric projects require water for the generation of power and therefore it is important to properly store and control the water.

Although used by Federal and state governments and a large portion of the general dam safety industry, there are highly variable opinions throughout the different users as to what is considered to be an appropriate PFMA. For agencies like the Bureau of Reclamation and the United States Army Corps of Engineers who have been using a risk-informed process, the importance of a well-developed PFM is very well understood [2]. Yet many others still develop incomplete PFMs, wrap multiple PFMs into one overarching PFM, exclude PFMs, or develop confusing PFM descriptions.

The key to a successful PFMA is the development of detailed and descriptive PFMs that provide the owner with an understanding of any potential weaknesses that may exist at the project that then allow them to take actions to reduce the possible threat to dam safety [3,4].

The development of PFMs and documentation of the PFMA process in a report is not intended to be the culmination of the process, instead it's intended to provide input and become the foundation of the management of dam safety activities and provide a thoughtful and transparent way to prioritize those activities [5].

POTENTIAL FAILURE MODES ANALYSIS PROCESS

Objective

Before starting the PFMA planning process, there are some important questions that must be asked to define the purpose and scope of the workshop. These questions include:

- What is the objective of the PFMA workshop? Is this the first PFMA of the project or a PFMA review? Is this a construction PFMA or a design PFMA?
- What expertise is required to conduct the PFMA? Operations staff? Outside consultants? Senior management? Subject matter experts?
- Is the project very large and complex that will require a large attendance or is it a smaller project requiring limited attendance?

It is very important to define the full objectives of a PFMA prior to starting any planning. A successful PFMA workshop requires proper planning and preparation by all participants. In preparation for a PFMA, all pertinent project records should be provided to the entire team to review. Chapter 14 of the FERC's Engineering Guidelines provides a good description of the PFMA process, including a list of project information and reports that should be provided to the team [1].

The FERC PFMA process, initiated in 2002, included a "reading" session where all participants would gather together in one location to read all of the project background information prior to participating in the PFMA workshop. This sometimes proved challenging when there are not multiple copies of the same documents and there are many participants. It also resulted in discussions that were best suited for the workshop setting rather than sidetrack the team members when they were trying to prepare for the workshop. With the ease of sharing electronic information in today's world, it is now considered more beneficial to provide all team members with access to all electronic files several weeks prior to the PFMA in order to allow them ample time to read the documents. The success of the PFMA is directly related to the proper preparation of the team members participating in the PFMA. Included below is more detailed discussion regarding specific details to ensure a successful workshop.

Logistics of Preparing for a PFMA

Very important in the preparation for the PFMA is the selection of the facilities to be used for the PFMA. A PFMA for a large project could last several days to a couple weeks, although a week is typical for most complex projects. Therefore, it is important to have a comfortable and ample-sized room with appropriate facilities to make the time spent together as comfortable as possible. The following items should be considered when planning for a successful PFMA:

- Additional Computer(s) and Projector(s)
- Full-sized drawings posted on walls
- Hard copy of all documentation
- Roomy and well-lit room
- Facilities for beverages and snacks
- Frequent breaks

It is often joked about having good snacks, lots of coffee, and frequent breaks at long meetings. A PFMA can become quite mentally exhausting, especially after several days of intense discussion, which make these items important to consider. Good working equipment and personnel will enable a more efficient PFMA by allowing the facilitator to focus on leading the group discussion rather than writing notes and sorting through documentation. A second operator with a separate projector and computer allows for the ease of bringing up drawings, photos, and searching for other information in the documentation to assist in answering questions raised during the development of a PFM. This second operator is also very useful for recording additional notes or items that either need to be discussed at another time or recorded in the final report.

PFMA Participants

PFMA teams typically vary in composition depending on the purpose, scope, complexity, technical issues, and other factors being evaluated. In general, PFMA teams typically include:

1. One or more facilitators (or co-facilitators)
2. One or more note-takers

3. Project personnel

Project personnel include members from the dam owner's organization, regulatory personnel, and others. From the dam owner's organization, it is important that operations and maintenance personnel and those responsible for performing monitoring functions of the dam be present. Other personnel may include personnel from emergency management agencies, current and former Part 12D independent consultants, engineering consultants with prior experience with the project, other regulators, and others that may possess relevant expertise or knowledge regarding the project.

If the PFMA team has been chosen properly, they will participate in the workshop with the appropriate expertise, an open mind, and a willingness to achieve the best possible results. This type of approach reaches a better conclusion than any of the individual members could have on their own and it is intended to reveal all potential dam safety concerns related to the project. However, sometimes the team may stumble along one or more of the following lines. The PFMA facilitator must recognize when this is occurring, and try to direct the group toward a more positive direction.

- A dominant individual may drive the team into their way of thinking by overwhelming them with filibustering or other techniques that eventually drive the entire team into thinking the way they do. It takes a fairly strong facilitator to deal with this, and usually requires emphasizing and bringing out the opposing point of view as well as drawing others into the conversation.

- People may not say what they actually believe or think for fear of appearing unknowledgeable, and will tend to go along with the rest of the group even though they have important input. This requires the facilitator to draw out their opinions by directing questions specifically at these individuals that are best suited to address the particular topic being discussed.

- A contrary individual may have valuable information even though their approach to communication may be difficult or challenging to the rest of the group. This information and these opinions should not be quickly dismissed without due consideration.

- The group gets tired due to the duration and rigors of the meeting and people agree just to get it done. The facilitator is not immune to this trap. If it is obvious that proper attention is not being paid to something, it is important to stop, take a break, and discuss ways to invest proper time for the evaluation. This may even require postponing the completion of the PFMA for a few days, like over a weekend.

<u>The Role of a Facilitator.</u> The facilitator's role in a PFMA is the most critical to the entire process and requires a significant amount of preparation. It is ideal for the facilitator to have read all of the documents regarding the project. This is particularly important if no one else has adequately prepared for the PFMA workshop. It is the role

of the facilitator to lead the group discussion during the finding of facts and discovery of all pertinent information when discussing and developing a PFM. Without a good understanding of the project, it makes it more challenging for the facilitator to draw the information out of the team members that is required to fully develop and understand a PFM. It could be argued that the facilitator should be able to develop each PFM by themselves, and while this could potentially be the case, there is no way they could know everything the team members know that is not included in the documentation, based upon their individual experiences and daily dealings with the project. It is also critical for the facilitator to remain neutral through the entire process.

The facilitator is responsible for the flow of the workshop to ensure that everyone stays focused and that the participants do not get sidetracked by discussions that are not relevant to the point at hand. It is often the case that there are several sidebar conversations going on during the group discussion. It is important that these are prevented and occur with the entire group so everyone has the same information. The facilitator is responsible for ensuring that all important facts are recorded for incorporation into the discussion during the workshop as well as the final PFMA report. As noted above, the facilitator is an important player in the entire PFMA process. Listed below are some of the qualifications a facilitator should possess.

Experience in the PFMA Process. A facilitator must have previous experience, or be a co-facilitator serving under the supervision and training of an experienced facilitator. They must fully understand the objective and requirements of the PFMA and what constitutes a fully developed and acceptable PFM.

Knowledge of Dam Safety and Precedents. Ideally, the facilitator(s) should have several years of dam design and inspection and/or dam safety experience in the type of project they are facilitating; e.g.... embankment dam or concrete dam. To a certain extent, the lack of expertise in one type of dam or another can be partially mitigated by having one or more team members that are experts in that field in order to assist in the technical aspects of the project. It is not necessary for them to be a design expert for the structures, but they must have a general understanding of how they operate and what case histories have taught the dam safety community with regards to design and operational weaknesses.

They should have an excellent understanding of dam incidents and failures and be able to draw on that knowledge during the PFMA process.

Independent/Objective. It is of utmost importance that the facilitator remains completely independent during the development of a PFM. Their role is not to act as a decision-maker, nor influence anyone's decision regarding the development of a PFM; it is to guide the team towards the different options available in the development of a PFM and the facts surrounding the different steps of a PFM. The facilitator should not make any comments or innuendos along the lines of how unbelievable or ridiculous a PFM is, or how "that would never happen" as discussions are taking place. It is totally appropriate for them to ask the typical, how, what, why, and when types of questions to get the team to think things through, as discussed below. They should help guide the team in a discussion that keeps them focused along the correct paths without directing them which path to take.

Too many times a facilitator has been heard making comments such as "That would never happen" or "You convince me that is a credible PFM and we can discuss it." Comments along these lines are not acceptable from the facilitator as they unduly influence the team members and discourage free and independent thinking. These types of comments are also not acceptable from the team members either. Only during a discussion of the likely and unlikely factors is discussion like this appropriate.

Interrogator. In conjunction with being independent, a good facilitator is able to lead the group discussion by asking questions that will help lead them down the most suitable path of a fully developed PFM. It is important that the facilitator remain independent when asking these questions. At each decision point in the process, a short discussion is often necessary to determine the different possibilities for the progression of the PFM. They can ask the group questions about which path is the most likely to progress and suggest the different options. This is why it is critical for the facilitator to have a good understanding of all the mechanisms that can present dam safety issues such as seepage and piping mechanisms, slope stability, overtopping erosion, and liquefaction related aspects of an embankment dam. If the project involves a concrete dam or concrete features, the facilitator must also have an understanding of sliding, overturning, structural strength, and general performance of concrete structures.

Motivational Speaker. This may seem like an odd characteristic for a facilitator. However, more often than not, there are several team members that are shy, reserved, or quiet during the discussions. This may be a result of their personality or their belief that they have nothing to offer to the process and are only there because their supervisor required their attendance. Quite often, these individuals are non-engineers or they are project operators that feel threatened by all the engineers and managers in the room and do not have the confidence to speak up. Those experienced in the PFMA process realize that these individuals often contribute some of the most critical information that comes to light during the workshop. These are the people who operate the facility day in and day out. They are the ones who know when the scripted message does not match the actual protocol used in the field. They could likely possess one bit of information that could prevent a dam safety emergency. Therefore, it is important for the facilitator to motivate these individuals to participate and play an active role in the discussion. This can often be done by asking them direct questions about their role in the project and emphasizing the importance of what they contribute to the overall understanding and safety of the project.

Referee. It is sometimes necessary for a facilitator to act as a referee during a discussion in order to successfully resolve differing opinions and uncertainties about certain aspects of the project. These differing opinions can result from a lack of understanding of the project or issue being presented and in some cases, could be someone attempting to corrupt the process. Depending upon the circumstances, it is possible that a team member is a major decision maker that does not want to spend money and will make every effort to convince the team that the PFM is not credible and a problem does not exist when it clearly does exist, or there is sufficient uncertainty to require additional investigation or research before making a final determination. In other cases, it may be the opposite, where a regional manager may need work for the staff and will try to convince the team that a PFM is credible or there is enough uncertainty to justify a recommended action, when the majority of the team and evidence suggests the PFM is not credible or so unlikely that no additional information is required. There could be any variation of these two extremes on the team. Therefore, it is critical for the facilitator to keep everyone on the same path of fact finding and sticking to the knowns. If unknowns are discovered during the discussion, additional studies or analyses may be in order to clearly identify whether the proposed PFM is credible.

PFMA WORKSHOP

Once prepared for the PFMA workshop, the workshop can be conducted in several different ways, although the end result should be the same. This paper presents what the authors consider to be the most efficient and methodical approach to the completion of both a new PFMA workshop and a review of an existing PFMA.

One question that is often asked in conjunction with Part 12D inspections at FERC-license facilities: Which should occur first, the PFMA workshop or the detailed dam safety inspection? The answer is - it depends. The dam safety inspections at the FERC are intended to be focused by the development of PFMs to help identify the critical parts of the project. However, if the project has not been inspected by all team members, it can be quite difficult at times for those team members to understand the physical layout and features of the dam in order to properly develop the PFMs. It is the author's opinion that the project should at least be visited prior to the PFMA. This does not necessarily require a detailed dam safety inspection, but it should at least involve an "arm-waving" inspection where the team members have had the opportunity to at least view the project from some vantage point or points where the operation and workings of the project can be described to those who have not previously been to the site. This helps provide a perspective on the magnitude of the project and physical layout that can influence the development of some of the PFMs.

Initial PFMA

As discussed above, it's critical to provide all the documentation to all team members prior to the workshop. It is typical that not all members of the team will read everything provided to them, although that is truly optimal. For this reason, it is important for the owner's representative to prepare a presentation for the team that summarizes all of the important information of the project. This presentation should be one of the first items on the PFMA agenda [1]. It should be appropriate for the complexity of the project and consider including the following items:

- Summary of original design

- Summary of construction with an emphasis on construction issues or other elements of construction that could influence the performance of the project
- Summary of significant modifications to the project since it was completed
- Summary of operational performance
- Summary and interpretation of instrumentation data and understanding of dam performance under all postulated loading conditions
- Findings of stability analyses
- Description of design flood and seismic studies
- Summary of the standard operation of the project

For complex projects, this presentation may require several hours. The length and depth of information should be scalable to the size of the project and the dam safety concerns associated with it. Once the presentation is completed, the PFMA workshop is ready to commence. As noted previously, there are different approaches to proceed with the PFMA, but the authors of this paper have found the following steps typically result in the best results.

Brainstorming PFMs. The first step in developing PFMs is to have a brainstorming session where the team attempts to determine every possible way that the dam could fail or an uncontrolled release of water to occur. This should be done for each loading condition. The owner may wish to include PFMs associated with operational failures of the mechanical and electrical portions of the project that do not result in an uncontrolled release of water or failure of the dam, but economically impact the project. This could include items such as failure of a turbine, pumps, gates, or similar that prevent them from generating power or providing irrigation or municipal water supplies.

It is recommended that the brainstorming session be methodical to prevent the possible oversight of a PFM. For example, one place to start with an embankment dam would be to brainstorm all possible PFMs associated with piping/internal erosion through the embankment. Once the team has completed that focused part of the project, they would proceed to the next focused part of the project, such as piping/internal erosion through the foundation of the embankment. This focused approach would be completed on each aspect of the project and under the different loading conditions. This helps prevent missing an important component of the project.

The following suggestions are provided to help more easily identify missing or incomplete PFMs:

- Separate identified potential failure modes by individual structure (e.g., main dam, auxiliary dike, spillway, outlet works, powerhouse, etc.) and loading condition (static, seismic, and hydrologic). Recognize that different loading conditions may activate the same failure mechanism and failure pathway. For example, if an internal erosion/piping, stability, or landslide potential failure mode exists as a potential failure mode under normal or static loading, it may be activated by unusual loading such as flooding or earthquake shaking. Thus, it may be a potential failure mode under all three loading conditions.

- Within the above categories, group similar-type potential failure modes together (e.g., piping/internal erosion PFMs, overtopping PFMs, gate PFMs, etc.). There may be different potential failure modes pathways that need to be considered depending on the site conditions. For example, piping through the embankment, piping from the embankment into the foundation, piping through the foundation; or overtopping of the main embankment leading to erosion of the embankment, overtopping (outflanking) of the embankment leading to erosion of the abutment materials; etc.

It is important to include, but also think beyond traditional analyses when identifying potential failure modes [2]. Dams are engineered systems and significant thought must be put into the details surrounding the interactions between the various features of a particular facility. Some of the greatest risks for uncontrolled reservoir release may be due to operational problems or potential failure modes that do not lend themselves to standard engineering calculations. Therefore, it is also important to have operational staff involved with the process.

An adequate job of identifying potential failure modes can be performed only after all relevant background information for a dam is diligently collected and thoroughly reviewed [1]. This includes information related to geology, design, analysis, construction, flood and seismic loading, operations, and performance monitoring. Photographs, particularly those taken during construction or unusual events, are often vital to identifying vulnerabilities.

Depending upon the size of the project, the brainstorming session could take a few hours to a day or more. It could also result in tens or more than 100 brainstormed PFMs, also dependent upon the size and complexity of the facility. A list of commonly considered mechanisms and pathways to consider when developing PFMs is included in Appendix 1.

Development of PFM Descriptions. In order for the team to have an adequate understanding of the potential failure modes, each PFM must be described in detail. As previously described, a PFM is a detailed description of a sequence of events, commencing with an initiating condition, progressing in a step-by-step manner, until a negative event occurs. Other papers and documents have discussed and presented examples of what should be included in a PFM description [1,2,4]. In summary, the potential failure mode must be described fully, from initiation through step-by-step progression to breach and uncontrolled release. There are three parts to the description [1,2]:

- **The initiator**. This is the loading or physical condition that leads to initiation of the potential failure mode. For example, this could include increases in reservoir due to flooding (perhaps exacerbated by a debris-plugged spillway), strong earthquake ground shaking, malfunction of a gate or equipment, deterioration, an increase in uplift, or a decrease in strength.

- **Failure progression**. This includes the step-by-step development of conditions that lead to the breach and uncontrolled release of the reservoir, or any uncontrolled release of water for FERC-licensed projects. The location where the failure is most likely to occur should also be highlighted. For example, this might include the path through which soils will be transported in an internal erosion situation, the location of overtopping in a flood, or anticipated failure surfaces in a sliding situation.
- **The resulting impacts**. The method and expected magnitude of the breach or uncontrolled release of the reservoir is also part of the description. This would include how rapid and how large the expected breach would be, and the breach mechanism. For example, the ultimate breach from an internal erosion failure mechanism adjacent to an outlet conduit might result from progressive sloughing and unraveling of the downstream slope as a result of flows undercutting and eroding the toe of the dam, until the reservoir is breached at which point rapid erosion of the embankment remnant ensues, cutting a breach to the base of the conduit.

During the development of a PFM description, it should be kept in mind that it is considered that there is a 100-percent chance of each step occurring as you move to the next step in the progression of events. **Only upon completion of the detailed description of the PFM is there a discussion regarding the likelihood of the PFM developing.** There is a tendency by many to make a determination of how likely the PFM is to occur when developing the step-by-step progress. This mindset must be avoided in order to adequately develop a PFM. However, there are always exceptions which prevent the entire process from becoming a prescriptive process and make it more of a rational thinking process.

As the team develops the step-by-step progression of a PFM, there are often points in the progression where there are numerous paths that could be taken to complete the development of the PFM. For example, seepage through an embankment can take many different paths, but a single path must be selected in order to fully develop the PFM. In cases like this, discussion is typically required for the team to decide which path is the most likely to occur. It is sometimes easy to determine which direction is the most likely, but there are instances where a significant amount of discussion is required to make a decision. It is important to document each option at each fork in the pathway, as it represents a totally different and separate PFM. Some facilitators will have the note taker create a separate list that captures the different paths the PFM could take in order to allow for discussion later. This list is sometimes referred to as a "parking lot" to keep the ideas until later in the workshop to ensure nothing important was missed.

It is commonly expressed by some that there is no need to separate all of the PFMs out individually and many of them can be "lumped" together. Some people claim that it has been easier to "lump" multiple failure mode pathways or multiple loading considerations into a single PFM rather than developing multiple individual PFMs. Unfortunately, this thinking has the potential to miss a critical element or elements in identifying potential vulnerabilities. The most important reason to individually identify and evaluate each

PFM is that when the team is trying to mentally track several different thoughts or branches of a PFM at one time, it is typical that some get confused about the exact path being discussed and discuss incorrect information about an element of the PFM. In some cases, this may not be significant, but it is not always possible to know what could be missing without having that unique discussion for each PFM. It's typical that once the discussion is held, many of the PFMs are very similar making it easy to document individually. It is also possible that once the discussion is held and it is found that different branches of a PFM result in identical consequences, it may be acceptable to lump some of them together. However, this decision should be made by a very experienced facilitator. Creating sketches and drawings of each PFM pathway greatly assists in keeping this clear to all team members.

As just noted, visual aids are also a very useful during the development of the PFM. This can consist of items as simple as a plan view of the project posted on the wall, photos of the project in the area of concern, or a sketch on a white board or flip chart. This is particularly useful when drawn by the individual who initially suggests the PFM being discussed. For example, it is often very beneficial to have visual aids to draw the actual path of internal erosion PFMs to assist the team in an understanding of the actual requirements for the PFM to fully develop. This concept is also applicable to most PFMs when attempting to determine which failure mechanism is being discussed. It is highly recommended that these sketches be included in the final PFMA report, which will be discussed later.

As potential failure modes are developed in a PFMA session, they should be:

- Well-described, for the reasons discussed above
- Comprehensive, in that ALL the ways that failure of the structure could occur are identified (there are no missing PFMs)
- Mutually exclusive. Separate, distinct potential failure modes are identified for each individual PFM pathway. Separate PFMs should be identified when any one of the following conditions are met:
    - Failure mode location/pathway is different
    - Loading condition/initiating event is different
    - Likelihood of failure is different
    - Monitoring for PFM is different
    - Breach location and geometry is different
    - Consequences are different
    - Risk reduction actions are different

At the conclusion of this step in the PFMA process, each potential failure mode should have a complete description of the proposed failure process from initiation to failure and should have a plan view and cross section depicting the failure mode pathway.

<u>Evaluating PFMs</u>. The potential failure mode identification process will generally result in a relatively long list of potential failure modes. This initial list of potential failure modes is called a candidate list of potential failure modes. Each potential failure mode

from this candidate list of potential failure modes must be evaluated against the following factors:

1. Is the potential failure mode physically possible?
2. Does the potential failure mode lead to failure (uncontrolled release of the reservoir)?

Physically impossible should be interpreted as the physical characteristics of the potential failure mode are such that development of the potential failure mode is physically impossible to happen and therefore could not lead to failure. This does not include PFMs the team decides are very remote, but could still physically occur.

If the answer to either of the above questions is no, then the candidate potential failure mode does not meet the definition of a credible potential failure mode. These candidate potential failure modes are ruled out and the justification for why they are no longer considered potential failure modes is documented in the report.

Example justifications for ruled-out potential failure modes include:

- Internal erosion/piping through auxiliary dike under normal loading. This candidate PFM was identified and discarded as a potential failure mode in that it is not physically possible to occur. The normal reservoir water surface is lower than the upstream toe of the dike. The dike embankment never sees the reservoir under normal loading.

- Failure of the Tainter gate under seismic loading. This candidate PFM was identified and discarded as a potential failure mode in that it does not lead to an uncontrolled release of the reservoir. The normal reservoir water surface is below the top of the concrete ogee section and water under normal loading is never on the upstream surface of the gate. Failure of the gate under seismic loading would not release the reservoir.

- Liquefaction of the foundation soils under seismic loading. This candidate PFM was identified and discarded as a potential failure mode after it was discovered from the construction records that all the foundation soil material was excavated to bedrock prior to construction of the embankment materials. No foundation soils exist below the embankment, therefore this potential failure mode is not physically possible.

The potential failure modes that remain from this evaluation process should be carried to the next steps in the PFMA process.

Subsequent steps in the PFMA process include [1]:

- Development of more likely-less likely factors
- Identification of PFM category (depending on regulatory requirements)
- Identification of potential short-term and long-term risk reduction measures

- Identification of surveillance and monitoring opportunities
- Identification of EAP opportunities
- Identification of dam safety inspection opportunities

The above steps are discussed in detail in Chapter 14 of FERC's Engineering Guidelines and will not be covered in this paper [1]. The last four items in the list above should form the basis of a dam safety management plan in which dam safety issues are monitored and tracked and prioritized [5].

Review of Existing PFMA

It is important to note that the PFMA process is dynamic. Of utmost importance in the PFMA workshop is maintaining an open and questioning mind by everyone involved with the project. For this reason, the Federal Energy Regulatory Commission, as part of the Part 12D process every five years, includes a detailed PFMA review. A PFMA review is intended to review any new information since the last PFMA review (or initial PFMA), any changes in the operation of the project, or any changes in the understanding of existing information since the last review that could impact the existing PFMs or result in the development of new PFMs.

Unlike an initial PFMA, a PFMA review is typically facilitated by the Independent Consultant (IC) selected to perform the Part 12D inspection and comprehensive review of the entire project. Of particular importance for the review is to focus on any new information, new analyses, instrumentation data, and anything else that may have changed in the understanding, operation, and/or performance of the overall project since the last PFMA review.

Although the primary responsibility of reviewing any new project information may seem to be focused on the IC, all participants in the PFMA review should review the same project information provided to the IC. For those intimately familiar with the project, it may not be as important to review historical information regarding the design and construction provided they have a good understanding of the project, but it is important to review the new information and performance of the project since the last PFMA review.

Similar to an initial PFMA, there are several approaches to performing a PFMA review. The authors present below what they consider to be the most methodical approach. Similar to an initial PFMA, it is important for the owner and IC to assess the extent of changes since the last PFMA and determine whether it would be beneficial to have a presentation to the team to explain the changes since the last review. This may not be necessary if there have been only minor changes, although this is always an option to consider to assist everyone in gaining an understanding of the project prior to discussing PFMs.

An approach used by some to prepare for a PFMA review is to provide a worksheet to all team members several weeks prior to the PFMA workshop to consider and review the existing PFMs and to brainstorm on their own for any new PFMs based upon changes they are aware of with the project. This will assist all team members in preparing for the workshop, but it should be pointed out that this is not intended to alleviate the IC of their responsibilities or be used as a substitute for team discussion when all are gathered together to discuss the PFMs. Any PFMA is intended to be a team effort and not an individual effort, with the group discussion taking place with sufficient detail to ensure that everyone on the team is satisfied with the review process.

The following process is recommended by the authors of this paper:

Review Existing PFMs. The first step of a PFMA review should be a review of each existing PFM that has been developed for the project, in detail, by asking questions similar to the following:

- Is the PFM adequately developed and detailed for all to fully understand?
- Does the PFM need to be revised and/or broken into multiple PFMs?
- Are all of the documented likely/unlikely (favorable/unfavorable) factors still relevant?
- Is there any new information to be added to the likely/unlikely (favorable/unfavorable) factors? Changed operation? Changed performance? Updated engineering analysis results?
- Does the Category of the PFM remain appropriate or should it be changed (for FERC PFMAs)? And if so, what should the new category be and why.
- Are there any additional risk reduction measures that should be identified?
- Are there additional surveillance and monitoring, dam safety inspection, or EAP activities that should be considered for adding to the surveillance and monitoring program?

This review should be performed for all categories of existing PFMs (Category I, II, III, and IV) as well as those not originally considered to be a "candidate PFM", or were classified as an "other consideration," or any other term used to describe brainstormed PFMs that were not fully developed. It is possible that information came to light since the last PFMA review that justifies fully developing one of the formerly undeveloped PFMs.

Once the existing PFMs have been fully discussed, an abbreviated brainstorming session should then be held, similar to that discussed above during an initial PFMA, to ensure that there are no additional PFMs that were previously missed. As indicated above, the PFMA process is a dynamic process and these PFMA reviews often result in the development of new PFMs not previously realized.

Just as in a new PFMA, all of the discussion should be documented and noted. The discussion should be incorporated into the report as discussed below. Without detailed notes incorporated into the PFMA report, it will be uncertain during subsequent PFMA reviews to determine what was discussed and what was not discussed. Failure to have everything documented will add time to the review in the future.

HOW MANY PFMS SHOULD I HAVE FOR MY DAM?

The answer to this question is quite simple: as many as you need. A more appropriate question would be: What PFMs are required to provide a full understanding of all the threats to the safety of my project. The answer to this question is a bit more difficult to answer, and yet identical to the previous question: as many as you need. Only upon reaching the end of a successful PFMA would you hopefully have your answer.

There are typically a lot more PFMs associated with an embankment dam than a concrete dam as a result of the uncertainties associated with soils and seepage related impacts to soils. Generally, concrete dams are constructed under more controlled and predictable measures than embankment dams. Therefore, it is not uncommon to have fewer PFMs for a concrete dam than an embankment dam. These are general thoughts and not definitive guidelines, as just like every dam, every PFMA is unique and only upon a detailed PFMA will the team have a sense of whether everything has been properly discussed and reviewed.

The primary objective of a PFMA review is the same as an initial PFMA, which is to fully identify and develop all PFMs that could pose a threat to the safety of the project.

PFMA REPORT

The PFMA report is the means to document the PFMs identified and described and the conclusions reached. The report serves as the foundation to manage the dam safety issues and concerns generated from each of the identified PFMs. The report can be organized in different ways and Chapter 14 of the FERC Engineering Guidelines contains more detailed information regarding the organizational structure of a PFMA report [1].

As mentioned earlier, the authors recommend organizing the potential failure modes by structure and by loading condition. In other words, all the PFMs developed under normal reservoir operating conditions for a particular structure (embankment, spillway, outlet works, dike, etc.) should be grouped and discussed separately in the report. Further grouping or organization of PFMs can be done based on similar characteristics. For example, internal erosion PFMs under static conditions for an embankment for pathways through the embankment, through the foundation, and from the embankment into the foundation could be discussed in separate subsections. Similar to the development of multiple similar PFMs where it is necessary to maintain a clear understanding of the details of one specific PFM at a time, the PFMA report is similar. A very methodical discussion of a project, loading condition by loading condition and mechanism by mechanism, keeps the readers focused on one general concept at the time rather than skipping around in a random or chaotic fashion that will confuse even the most focused reader.

There is flexibility in the preparation of the PFMA report. However, one item that is not an option and is considered to be one of the most critical components of the report is to provide a full and detailed documentation of all pertinent information and topics that were discussed during the workshop. Sketches or illustrations of each PFM location and pathway can provide clarity that is often difficult to convey in words as well and should be considered a necessity in PFMA reports. If something is not documented, important discussions will not be passed along to future PFMA review teams and could result in additional time and expense to research information that was previously discussed. The lack of documentation has also been known to result in the expenditure of large amounts of money that otherwise would not have been spent if the documentation was complete and accurate.

CONCLUSION

The PFMA process is considered to be the foundation of any dam safety program. In order to fully understand a dam, it is important to complete a detailed PFMA that identifies all the PFMs to fully analyze the project. Only by participating in one or more detailed PFMAs, facilitated by a well-experienced facilitator who understands the true value of the process, will some truly understand the value of the process. Others have learned the value by detecting a dam safety problem and resolving it before it became an actual failure. Some could argue that dam failures or incidents could have been prevented had a complete PFMA been completed. While difficult to confirm, this is not a risk any responsible dam owner should consider acceptable. Without a complete and comprehensive list of PFMs, one is not able to adequately demonstrate a comprehensive understanding of their dam, design a customized approach to dam safety management of the dam, thus reducing the potential risk of missing a critical dam safety element that could result in the loss of the project and the loss of lives downstream of their project.

REFERENCES

[1] Federal Energy Regulatory Commission, "Engineering Guidelines for Hydropower Projects, Chapter 14", January 2017.

[2] Bureau of Reclamation/US Army Corps of Engineers, "Best Practices in Dam and Levee Safety Risk Analysis, Chapter I-3, Potential Failure Mode Analysis", 2015.

[3] O'Leary, T.M., and G.A. Scott, "Semi-Quantitative Risk Characterization for Periodic Dam Safety Assessments," USSD Annual Meeting, 2013.

[4] Association of State Dam Safety Officials, "Potential Failure Mode Analysis as a Dam Safety Tool, Part I: Introduction to PFMA," Recorded Webinar, April 8, 2014.

[5] Association of State Dam Safety Officials, "Potential Failure Mode Analysis as a Dam Safety Tool, Part II: Application of PFMA in Dam Safety," Recorded Webinar, June 10, 2014.

APPENDIX 1

Common Mechanisms and Pathways to Consider when Developing Potential Failure Modes

NOTE: This list is not meant to represent a comprehensive list of all the ways a dam or appurtenant structure could fail. Each project is unique and requires a separate, detailed review of the project records and an understanding of the project operation and performance in order to identify appropriate potential failure modes for that particular project.

Embankment Dams

<u>Normal/Static Loading</u>

Backward erosion piping (BEP)
- Piping of embankment materials
- Piping of embankment materials into the foundation
- Piping, blowout and heave of foundation materials
- Piping of materials into drains/conduits

Concentrated leak erosion (CLE)
- Through cracks/defects in core
- Along contact with concrete structures
- Along conduits/penetrations
- Along bedrock contact

Contact erosion
- Erosion of fine particles from flow in an adjacent coarse layer within the foundation
- Erosion of fine particles from flow in an adjacent coarse layer between the embankment and the foundation

Suffusion (Internal instability)
- Erosion of fine matrix materials in a well graded or gap graded material in the embankment
- Erosion of fine matrix materials in a well graded or gap graded material in the foundation

Static slope instability of embankment
Static slope instability of foundation/abutments
Slope instability under rapid drawdown

Wave erosion of upstream slope
Runoff erosion/gullying of downstream slope

Landslide-induced wave leading to overtopping and erosion
Volcanic flow/air fall displacing reservoir contents leading to overtopping and erosion

Hydrologic/Flood Loading

Normal/Static Loading PFMs with added hydrologic loads

Internal erosion of embankment above core

Embankment overtopping and erosion
Embankment overtopping and erosion with wind and wave run up
Abutment outflanking and erosion
Structural failure/erosion of parapet wall foundation

Seismic/Earthquake Loading

Normal/Static Loading PFMs with added seismic loads

Liquefaction of embankment soils leading to overtopping and erosion
Liquefaction of foundation soils leading to overtopping and erosion
Deformation of embankment/foundation soils leading to overtopping and erosion
Deformation of embankment/foundation soils leading to cracking of the embankment and internal erosion

Fault offset in foundation leads to cracking and internal erosion
Seiche wave overtops embankment leading to erosion

Concrete-faced Rockfill Dam

Deformation and cracking of facing slab leads to internal erosion of embankment materials
Downstream slope failure resulting from piping of fine portion of "dirty" rockfill resulting in sinkhole development.
Failure of facing slab waterstops leads to internal erosion of embankment materials
Sliding instability of concrete plinth

(See Embankment Dam PFMs for additional PFMs)

Concrete Gravity and Arch Dams

Normal/Static and Hydrologic/Flood Loading

Concentrated leak erosion along foundation contact
Internal erosion through foundation materials

Static sliding instability along lift joints/base
Static sliding instability through foundation/abutments
Sliding instability due to abnormal loads (silt and ice)

AAR/ASR leads to cracking of dam and loss of strength

Overtopping leading to erosion of abutment/foundation materials
Overtopping of dam crest creates negative pressures and induces vibrations that lead to cracking and failure

Landslide-induced wave leading to overtopping and erosion

<u>Seismic/Earthquake Loading</u>

Normal/Static and Hydrologic/Flood Loading PFMs with added seismic loads

Fault offset in foundation leads to cracking of dam/erosion of foundation materials

Concrete Buttress Dams

Sliding instability of buttress at contact
Sliding instability of buttress through the foundation

Seismic instability and loss of lateral support of buttresses in cross canyon direction

Fault offset in foundation leads to cracking of dam/erosion of foundation materials

Spillway Structures

<u>Normal/Static and Hydrologic/Flood Loading</u>

Erosion/scour of soil/rock channels

Concentrated leak erosion along foundation contact
Internal erosion of spillway wall backfill
Internal erosion through foundation materials

Static sliding stability of crest structure along base
Static sliding stability of crest structure through foundation
Global sliding stability of spillway structure
Overturning of crest structure

Stagnation pressure causes jacking of spillway chute slab that leads to erosion
Uplift of spillway chute slabs/in adequate anchoring causes jacking of slab that leads to erosion
Stagnation pressure causes loss of material below chute slab, collapse of slab, and erosion
Overtopping of spillway chute/basin walls leads to erosion
Cavitation damage induced failure leads to erosion
Erosion and failure of stilling basin

Uplift and sliding of stilling basin

Debris plugging of approach channel/crest structure leading to overtopping and erosion

Seismic/Earthquake Loading

Seismic failure of pier(s)
Seismic failure of spillway walls
Seismic failure of spillway bridge
Foundation liquefaction/deformation of spillway

Outlet Works

Debris plugging/landslide at intake leads to inability to release flood waters and premature overtopping and erosion
Cavitation of outlet pipe leads to erosion and uncontrolled release of reservoir
Coating damage/corrosion of outlet pipe leads to erosion and uncontrolled release of reservoir
Rockfall/lining damage in gate chamber/downstream tunnel damages outlet pipe/control gates and leads to erosion and uncontrolled release of reservoir
Seismic failure of outlet tower leads to inability to close outlet gates and leads to uncontrolled release of the reservoir

Gates (spillway and outlet works), Stoplogs, and Flashboards

Structural failure of gate
  (Trunnion friction, corrosion, fatigue, excess demand)
Inability to operate gates due to:
  Loss of power (electrical supply, transmission)
  Inability to mechanically operate gates (failure of winches, hoists, hydraulics, chains, ropes, etc., ice restricts gate operation)
  Loss of access to dam/gates
AAR/ASR leads to cracking of dam and inability to operate spillway gates
Slope instability/rockfall damages gates/spillway structure
Operational/procedural errors, SCADA errors, and human error
Seismic failure of gates

Penstocks

Slope creep/landslide displaces penstock leading to rupture and uncontrolled release
Cavitation of penstock leads to erosion and uncontrolled release of reservoir
Coating damage/corrosion of penstock leads to erosion and uncontrolled release of reservoir
Rockfall damages penstock and leads to erosion and uncontrolled release of reservoir
Seismic failure of penstock supports leads to uncontrolled release of reservoir

EPISTEMIC VS NATURAL UNCERTAINTY IN DAM SAFETY DECISION MAKING: IS IT FAIR PLAY?

Ignacio Escuder-Bueno[1]
Adrián Morales-Torres[2]
Jessica Castillo-Rodríguez[3]

ABSTRACT

Nowadays there is a trend towards seeing dam safety as an active and ongoing management process rather than a static and deterministic statement. Tools such as risk analysis can be useful to answer some questions and assist owners to make risk-informed decisions. Risk assessment helps engineers to understand uncertainties in the project, and provides a logical process of identifying hazards, evaluating the seriousness of each hazard, and assessing the effectiveness of risk reduction measures.

In this framework, considering explicitly and independently natural and epistemic uncertainty in quantitative risk models allows one to understand the sources of uncertainty in risk results and to estimate the effect of actions, tests, and surveys to reduce epistemic uncertainty. In this work, different metrics are presented to analyze the effect that epistemic uncertainty has in the prioritization of investments based on risk results. These metrics allow consideration of the convenience of more uncertainty reduction actions, like site tests, surveys, or further analysis.

Finally, these metrics have been applied in the prioritization of risk reduction measures for four gravity dams in Spain. Results allow for a better understanding of how epistemic uncertainty about geotechnical resistant parameters is influencing the risk-informed decision-making process and they help to define the need for more tests and surveys in some of the dams.

INTRODUCTION

In recent years, risk assessment techniques have been developed and applied in the dam industry worldwide to inform safety governance (ICOLD 2005; ANCOLD 2003; SPANCOLD 2012; USACE 2014). These guidelines and recommendations have been developed within the tolerability of risk (TOR) framework, which was set out by UK's HSE (HSE 2001) for risk evaluation and management.

However, the contextual information provided above is considerably more complex than it may sound, resulting in theoretical and practical difficulties. Many of these difficulties are related to how uncertainties are explicitly considered today (in the context of risk

[1] Universitat Politècnica de València, Research Institute of Water and Environmental Engineering, Camino de Vera, s/n, 46022 Valencia, Spain. iescuder@hma.upv.es
[2] iPresas Risk Analysis. Avenida del Puerto, 180 1B 46023 Valencia. Spain. adrian.morales@ipresas.com
[3] iPresas Risk Analysis. Avenida del Puerto, 180 1B 46023 Valencia. Spain. jcastillo@ipresas.com

analysis) in opposition to the more traditional implicit treatment (in the context of state-of-the-art dam safety practice).

When risks associated with complex structures are analyzed, evaluation of uncertainty should play an important role in the analysis of the behavior of a constructed facility (Altarejos-García, Escuder-Bueno, and Morales-Torres 2015). In general, two sources of uncertainty are considered (Hoffman and Hammonds 1994; Ferson and Ginzburg 1996; Hartford and Baecher 2004) as shown in Figure 1:

- **Natural uncertainty or randomness:** Produced by the inherent variability in the natural processes. An example is the variability of the loads that the structure has to withstand, for instance, the variability in the potential floods magnitude that can occur. This type cannot be reduced, though it can be estimated.

- **Epistemic uncertainty:** Resulting from not having enough knowledge or information about the analyzed system. This lack of information can be produced by deficiency of data or because the structure's behavior is not correctly represented. The more knowledge that is available about a structure or system, the more this type of uncertainty can be reduced.

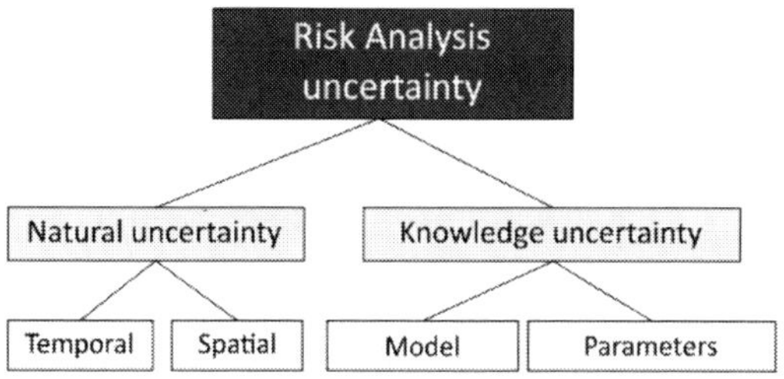

Figure 1: Taxonomy of uncertainty in risk analysis. Source: (Morales-Torres, Escuder-Bueno, et al. 2016)

In the dam safety field, both types of uncertainty are generally introduced in risk model inputs, without specifically distinguishing the effect of epistemic uncertainty and generally giving more importance to natural uncertainty. These results are very useful to prioritize risk reduction investments, but still two important questions for dam safety governance remain unanswered:

- How is epistemic uncertainty influencing the decisions made based on risk results?

- How potential uncertainty reduction measures (geotechnical tests, dam computational models, improvements in dam surveillance and monitoring…) can improve dam safety governance?

In order to solve these questions, this paper presents two numerical indicators named Indices of Coincidence that measure the effect of epistemic uncertainties in risk-informed decision making. These indices are computed comparing the effect of epistemic uncertainty in prioritization sequences of potential risk reduction measures. These sequences of measures are obtained with the procedure developed in (Morales-Torres, Serrano-Lombillo, et al. 2016).

UNCERTAINTY IN RISK MODELS FOR DAMS

Risk is the combination of three concepts: what can happen, how likely is it to happen, and what are its consequences (Kaplan 1997). Following this definition, in the dam safety field, risk is usually quantified with the following equation (Altarejos-García et al. 2012):

$$Risk = \int P(loads) \cdot P(response|loads) \cdot C(loads, response) \quad (1)$$

where the integral is defined over all the events under study, *P(loads)* is the probability of the different load events, *P(response|loads)* is the conditional probability of the structural response for each load event and *C(loads,response)* are the consequences of the system response for each load event. In the dam safety field, the system response analyzed is the dam failure. Consequences can be introduced in economic terms to obtain economic risk or in terms of potential loss of life, to obtain societal risk (Morales-Torres, Serrano-Lombillo, et al. 2016).

These terms of the equation are usually analyzed independently and they can be combined within a quantitative risk model to compute dam failure risk. In these risk models for dams, natural and epistemic uncertainties are not usually introduced separately. They are usually mixed in the probability input data introduced for the structural response with a mean conditional failure probability for each loading state (Morales-Torres, Escuder-Bueno, et al. 2016). This approach is called first-order probabilistic risk analysis (E. Paté-Cornell 2002) and it is the most common approach in risk-informed dam safety management (USBR and USACE 2015; SPANCOLD 2012; ANCOLD 2003).

First-order probabilistic risk analysis represents the Level 4 of complexity in the classification developed by (M. E. Paté-Cornell 1996). There is a higher level of complexity to fully represent both types of uncertainty (Level 5), called second-order probabilistic risk analysis. In this level, epistemic and aleatory uncertainties are introduced separately in the risk model, defining probability distributions for input data in the risk equation.

Hence, in a second-order probabilistic risk analysis, a risk probability distribution and a family of FN curves are obtained instead of a single value and curve, as explained in (E. Paté-Cornell 2002; Chauhan and Bowles 2001). The spread of risk probability distribution and the family of FN curves thus represents the degree of epistemic uncertainty in the risk assessment. In Figure 2, the type of risk results and risk representation are compared for first-order and second-order probabilistic risk analyses. Second-order probabilistic risk analysis is more common in other industries like the nuclear industry (EPRI 1994).

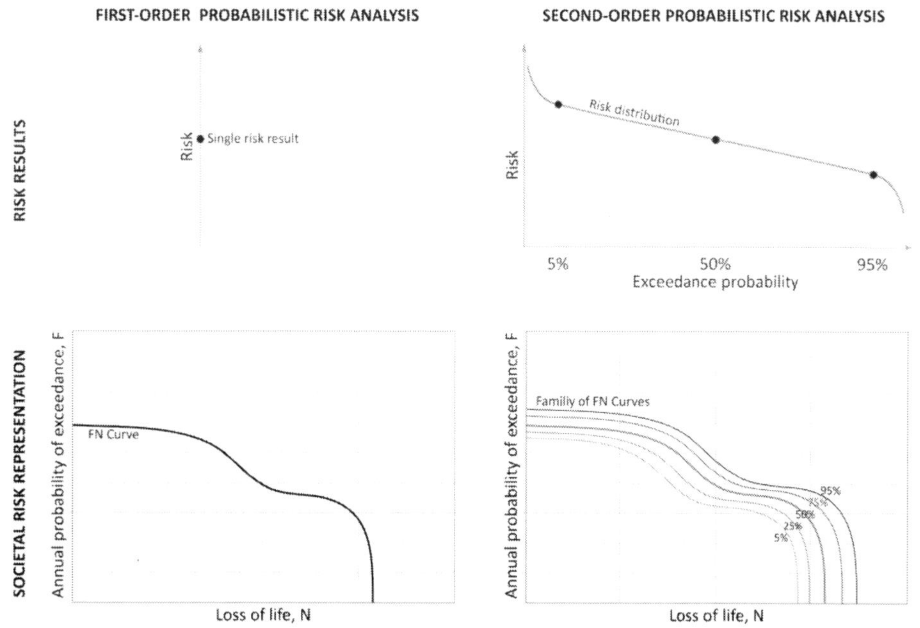

Figure 2: Comparison of risk results and risk representation between first-order and second-order probabilistic risk analysis.

RISK-INFORMED DECISION MAKING FOR DAM SAFETY MANAGEMENT

Once risk results are computed, they are used to inform dam safety management. In general, risk analysis to inform dam safety governance is contextualized within the Tolerability of Risk (TOR) framework developed by HSE (2001) for risk evaluation and management. This framework has been widely used worldwide to define risk-informed dam safety programs (ANCOLD 2003; USACE 2014; USBR 2011; SPANCOLD 2012). According to this framework, two basic principles are generally used to guide decision making (ICOLD 2005; HSE 2001):

- **Equity:** This principle is based on the premise that all individuals have unconditional rights to certain levels of protection. This principle is applied through the individual risk, which can be defined as the probability that at least one person dies as a result of the dam's failure (SPANCOLD 2012).

- **Efficiency or utility:** This principle arises from the fact that society possesses limited resources, which must be spent in the most efficient way. When considering several risk reduction measures, the one producing a higher risk reduction at a lower cost (the one that optimizes expenditure) should generally be chosen first.

When quantitative risk analysis is applied to inform safety management of portfolios of dams, a high number of results are obtained. In this context, risk reduction indicators have proved to be a useful tool to prioritize risk reduction measures (Bowles et al. 1999; Morales-Torres, Serrano-Lombillo, et al. 2016; Armando Serrano-Lombillo et al. 2016a).

These indicators are numeric values obtained for each potential risk reduction measure considered based on its costs and the quantitative risk reduction it provides. Risk reduction indicators are directly related with equity and/or efficiency principles (as shown in Figure 3) and they are computed based on risk results obtained for each considered measure with a first-order probabilistic risk analysis. In (Morales-Torres, Serrano-Lombillo, et al. 2016), a procedure to obtain prioritization sequences based on risk reduction indicators is introduced. In each step of the sequence, the measure with the lowest value of the indicator is chosen. It is demonstrated that risk indicators are a useful tool to prioritize a high number of investments in a portfolio of dams.

Equity Weighted Adjusted Cost per Statistical Life Saved (EWACSLS) combines equity and efficiency risk reduction principles. This indicator is presented in (Armando Serrano-Lombillo et al. 2016b) and it is computed with the following formulas:

$$EWACSLS = \frac{ACSLS}{\left(\frac{\max(r_i(base), IRL)}{\max(r_i(mea), IRL)}\right)^n} \quad (2)$$

$$ACSLS = \frac{C_a - (r_e(base) - r_e(mea))}{r_s(base) - r_s(mea)}$$

Where $r_s(base)$ is the risk expressed in loss of lives for the base case, $r_s(mea)$ is the risk in lives after the implementation of the measure, $r_e(base)$ is the economic risk of the infrastructure for the base case and $r_e\ (mea)$ is the economic risk after the implementation of the measure, C_a is the annualized cost of the measure including its annualized implementation costs, annual maintenance costs and potential changes in operation costs produced by the adoption of the measure, $r_i(base)$ is the individual risk for the base case expressed in years-1, $r_i(mea)$ is the individual risk in years^{-1} after the implementation of the measure, *IRL* stands for Individual Risk Tolerability Limit and *n* is a parameter that allows assigning a higher weight to either efficiency or equity in the prioritization process.

INDICES OF COINCIDENCE TO ANALYZE EPISTEMIC UNCERTAINTY IN DAM SAFETY DECISION MAKING

In dam safety management, two types of investments can be analyzed: risk reduction measures (higher outlets capacity, freeboard requirements…) and uncertainty reduction measures (geotechnical tests, dam computational models…). These two types of measures have a different impact on a risk probability distribution obtained by a second-order probability risk assessment. Risk reduction measures move the probability distribution downwards, while measures to reduce epistemic uncertainty produce a less steep risk distribution.

As explained in Section 3, current approaches to inform dam safety are focused on average risk results from first-order probabilistic risk analysis and natural uncertainty. For this reason, they are used to prioritize risk reduction measures but they do not analyze the effect of epistemic uncertainty. However, this type of uncertainty can influence decision making and prioritization sequences. For instance, in high epistemic uncertainty situations, the decisions made can change depending on the values considered within the epistemic uncertainty distributions.

As explained in Section 2, in a second-order probabilistic risk analysis a high number of risk results are obtained instead of a single risk value. The spread of these results indicates the existing epistemic uncertainty. Hence, a high number of risk results are obtained for the base case and for each risk reduction measure analyzed.

When these results are combined with the calculation of prioritization sequences explained in Section 3, a high number of sequences are obtained for each risk reduction indicator, instead of a single sequence for the average values. The differences between these high number of sequences indicate how epistemic uncertainty influences decision making. In a case where epistemic uncertainty is not influential, the order of the analyzed measures in all the sequences will be the same, while in a case with a high influence of epistemic uncertainty, there will be higher differences in the order of measures within the sequences. According to the authors, this is the key of dealing with epistemic uncertainty within dam safety management: analyzing how it can change the decisions made and when it is recommended to invest in reducing this type of uncertainty.

Based on this reasoning, two different indices have been developed and proposed in this paper to measure the effect of epistemic uncertainty in the calculation of prioritization sequences. These metrics are based on the difference in the order of measures between each sequence obtained with the results of a second-order probabilistic risk analysis and the reference sequence obtained with the average values from first-order risk analysis. The two indices developed are:

- **Index of Coincidence (IC):** It quantifies the difference in the order of measures between two sequences. For each step of the measure, it is computed with the division of the difference in the position of a sequence in the two itineraries and the maximum difference in the position that there could be.

- **Adjusted Index of Coincidence (AIC):** It is computed multiplying the Index of Coincidence in each step by a factor to preponderate the first measures of the sequence, since they are more important in the decision making process.

Thus, these indices of coincidence can be used to compare each implementation sequence obtained through a second-order probabilistic risk analysis with the reference implementation sequence obtained with a first-order probabilistic risk analysis. Hence, a high number of Indices of Coincidence are obtained, one for each sequence. The average Index of Coincidence of all these sequences is an indicator on how epistemic uncertainty is influencing decision making, since it indicates the differences in the order of measures that epistemic uncertainty could produce.

According to the authors' experience, Table 1 shows reference values of average Indices of Coincidence and what they could indicate when they are computed for a single source of uncertainty in the risk model.

Finally, Indices of Coincidence indicate the need for epistemic uncertainty reduction measures, so they are very useful for risk-informed dam safety management. In this sense, the effect of epistemic uncertainty reduction measures in the probability distributions introduced in the risk model can be estimated and Indices of Coincidence can be recomputed. Expected increments in average Indices of Coincidence of more than 5% indicate effective uncertainty reduction measures, especially when Indices of Coincidence are lower than 85%.

Table 1: Indicative meaning of average Index of Coincidence when computed for a single source of epistemic uncertainty.

Average Index of Coincidence value	Degree of influence of this source of epistemic uncertainty in measures prioritization
> 99%	Low
95% - 99%	Low-Medium
95% - 85%	Medium
85% - 75%	Medium-High
75% - 60%	High
< 60%	Efforts should be focused on reducing epistemic uncertainty before significant investments in risk reduction

CASE STUDY

In this section, the approach introduced in this paper is applied to inform safety management in four existing concrete gravity dams in Spain. The starting point for this case study is the risk models elaborated within a first-order probabilistic risk analysis performed on each dam. The analysis made is focused on the potential sliding of these dams and the epistemic uncertainty about the foundation resistant capacity. Hence, the analysis introduced in this case study is focused on this source of epistemic uncertainty within the risk model and how it can influence decision making.

Sliding safety management in these dams deals with the prioritization of potential risk reduction measures and/or investing in reducing uncertainty about the foundation. A total number of 20 potential risk reduction measures have been analyzed in the four analyzed dams. These structural and non-structural measures came from a list of actions already planned by the operators to improve dam safety, including improvements in operating rules, emergency procedures, outlet works, dam bodies and foundations.

First, in order to introduce epistemic uncertainty within the risk models for this failure mode, the procedure described in (Morales-Torres, Escuder-Bueno, et al. 2016) has been followed to obtain a family of fragility curves. In this case, since the uncertainty on the foundation resistance capacity is the main concern for the sliding failure modes in these dams, two independent random variables are considered within a Limit Equilibrium Model: friction angle and cohesion. Following the procedure detailed in (Morales-Torres, Escuder-Bueno, et al. 2016), a family of 1000 fragility curves has been obtained for each dam. For instance, Figure 5 shows the family of fragility curves obtained for dam A. The spread of this family is an indicator of the influence of the epistemic uncertainty in the results.

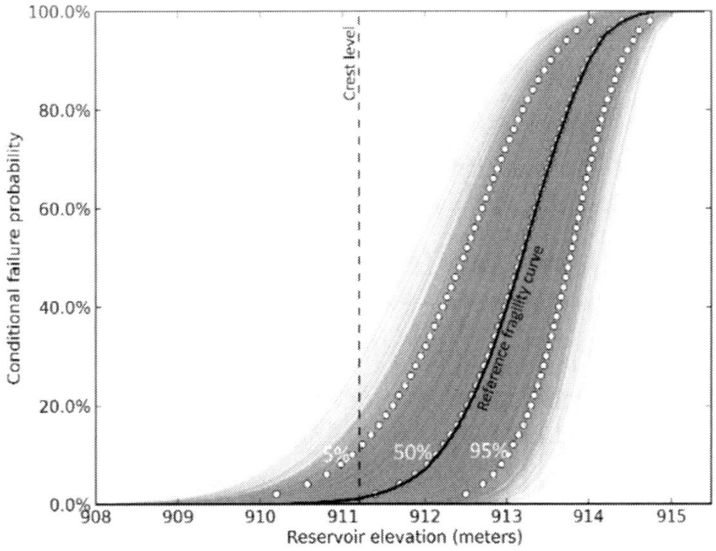

Figure 5: Family of fragility curves obtained for Dam A.

Second, the family of fragility curves has been introduced in the quantitative risk model elaborated for each dam in order to obtain a risk probability distribution for sliding failure. These risk models have been elaborated using iPresas Calc software (iPresas 2014), which is based on event trees to compute failure probability and risk. The risk model architecture of the four risk models is very similar. These risk models have been used to compute risk for the current situation and for the 20 risk reduction measures analyzed. Detailed procedures followed to derive consequences input data are explained in A Serrano-Lombillo, Morales-Torres, and García-Kabbabe (2012).

For each curve of the family of fragility curves of each dam, these risk models are used to compute failure probability, economic risk and societal risk. For each dam, societal risk results have been sorted to obtain the societal risk probability distribution shown in Figure 6. As can be observed in these graphs, societal risks are higher for Dam B and Dam C, while risk variations due to epistemic uncertainty are higher for Dam A and Dam D.

Third, once these risk distributions are obtained, Indices of Coincidence proposed and described in this paper have been computed to solve the key question of this paper: Is epistemic uncertainty influencing decision making? With this purpose, the 20 risk reduction measures analyzed have been prioritized following the procedure explained in (Morales-Torres, Serrano-Lombillo, et al. 2016), using the EWACSLS indicator (Armando Serrano-Lombillo et al. 2016b), combining equity and efficiency principles.

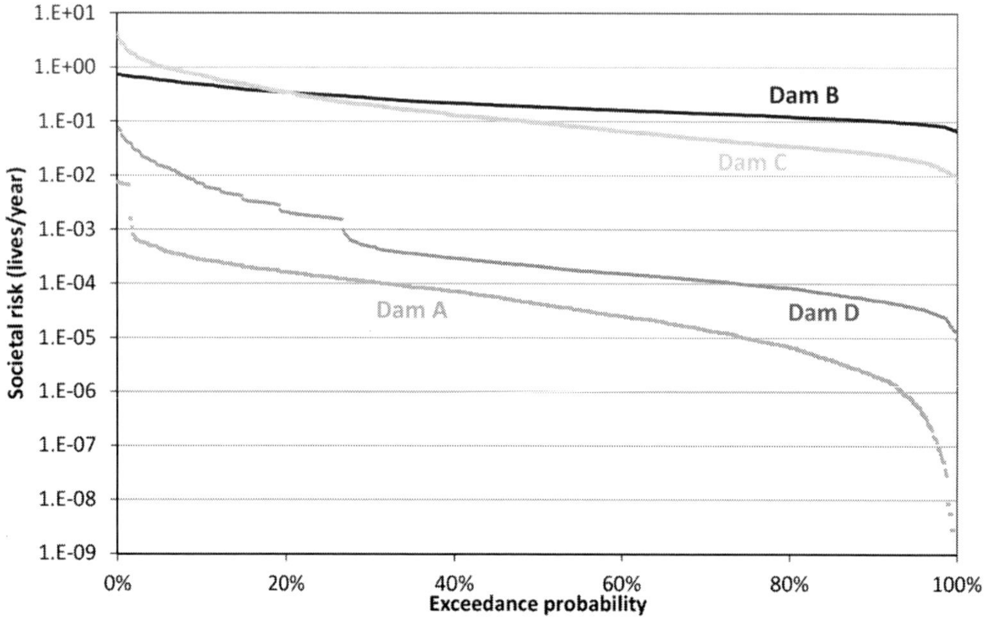

Figure 6: Societal risk probability distributions obtained for the four dams.

A reference implementation sequence of measures is obtained from the risk results, which in turn were obtained from the reference fragility curve in each dam. Next, 1000 implementation sequences were obtained combining the 1000 fragility curves with the risk results obtained for each dam. These 1000 sequences are compared with the reference sequence to obtain the average Indices of Coincidence shown in Table 2 (Base Case). As can be observed in this table, these indices have been computed after prioritizing measures for each dam independently and prioritizing the 20 measures together. Figure 7 shows the variation graphs of all sequences obtained for the prioritization of the 20 measures together.

As can be observed in Tables 2, Indices of Coincidence are lower for Dam B, which indicates that epistemic uncertainty has a higher influence on decision making, so uncertainty reduction actions are more recommended. In contrast, Indices of Coincidence for Dam A are close to 100%, which indicated that epistemic uncertainty has low influence on decision making. Indices of Coincidence of Dams C and D indicate a medium influence of epistemic uncertainty on results.

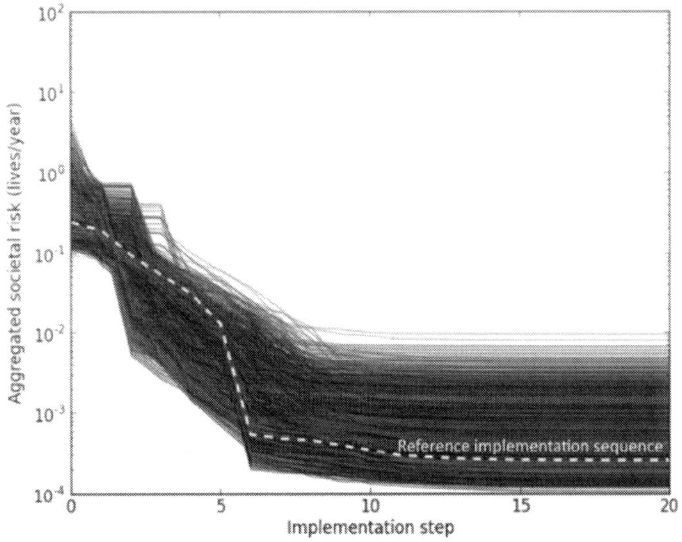

Figure 7: Variation graphs of the 1000 sequences obtained for the prioritization of the 20 measures together. Y axis represents aggregated societal risk of the four dams.

Finally, the potential effect of epistemic uncertainty reduction measures for the foundation resistance capacity, like geotechnical tests and detailed surveys, has been analyzed. With this purpose, the previous computations have been repeated but reducing by half the standard deviation of the epistemic uncertainty probabilistic. Thus, Indices of Coincidence have been recomputed for these cases as shown in Table 2. As can be observed the effect of reducing epistemic uncertainty in each dam has been independently analyzed in the individual sequences of each dam and in the sequences obtained combining the four dams.

Table 2: Indices of Coincidence: base case and after reducing epistemic uncertainty.

	Base case		Epistemic uncertainty reduction		Difference	
Epistemic uncertainty reduction	**Index of Coincidence**	**Adjusted Index of Coincidence**	**Index of Coincidence**	**Adjusted Index of Coincidence**	**Index of Coincidence**	**Adjusted Index of Coincidence**
Individual analysis						
Only in Dam A	99.35%	99.29%	99.94%	99.94%	0.60%	0.66%
Only in Dam B	79.86%	69.55%	83.19%	74.55%	3.34%	5.00%
Only in Dam C	87.42%	86.97%	89.38%	89.57%	1.96%	2.60%
Only in Dam D	94.11%	90.77%	96.74%	94.55%	2.63%	3.78%
Combined analysis						
Only in Dam A	86.95%	86.60%	87.76%	87.16%	0.80%	0.56%
Only in Dam B	86.95%	86.60%	87.23%	87.02%	0.27%	0.42%
Only in Dam C	86.95%	86.60%	88.35%	87.74%	1.40%	1.14%
Only in Dam D	86.95%	86.60%	88.22%	88.11%	1.26%	1.51%
All dams	86.95%	86.60%	91.28%	90.68%	4.33%	4.08%

Results show that reducing epistemic uncertainty in Dam C and Dam D would have a higher influence in the decision making process for the whole system of dams. In contrast, the effect of reducing epistemic uncertainty in Dam A is lower. Epistemic uncertainty reduction in Dam B has a high effect in the sequences obtained for this dam individually but its effect in the management of the four dams together is more limited. Hence, epistemic reduction actions are recommended when this dam is individually managed, but from the combined management point of view, these actions would be more recommended in Dams C and D.

If the results of reducing globally epistemic uncertainty for the four dams are analyzed, it can be concluded that these actions could be useful to support a better risk-informed decision making, since they provide an increment of Indices of Coincidence by 4%.

DISCUSSION AND CONCLUSIONS

Dams are located in natural and heterogeneous environment that cannot be characterized exactly. For this reason, unlikely to other industries, dam safety governance deals with higher natural and epistemic uncertainties since it is directly related with the nature behavior. For this reason, epistemic uncertainty can have a higher effect on decision making.

This paper introduces new metrics to analyze the influence of epistemic uncertainty in decision making for dam safety. This process is based on the results of a second-order probabilistic risk analysis, which requires separating natural and epistemic uncertainty within the risk model input data. Although this is not the most common approach in the dam safety field, the distinction between both types of uncertainty takes added importance for a proper dam safety management.

These metrics are computed by combining results of a second-order probabilistic risk analysis and prioritization of investments based on risk reduction indicators. The main identified discussion points about this procedure and the case study are:

- The case study introduced is focused on one source of uncertainty within the risk model: foundation resistant capacity. This approach of analyzing each source of epistemic uncertainty separately is more recommended since it provides a better understanding of what type of epistemic reduction actions are more effective. In any case, if epistemic uncertainty is included in all the nodes of the model, Indices of Coincidence can also be obtained although they will be lower, since the effect of different sources of epistemic uncertainty is combined.

- Hence, the introduced case study is focused on epistemic uncertainties in the second term of the risk equation: the system response, but Indices of Coincidence can also be used to analyze the effect of epistemic uncertainty in the other terms: loads probability and consequences.

- The effect on Indices of Coincidence on risk reduction measures that also help to reduce epistemic uncertainty could be analyzed. For example, improvements in the surveillance and monitoring system, since they help to detect the failure modes occurrence and increase the knowledge about the dam behavior.

- Although Indices of Coincidence have been developed within the dam safety management field, they could be also applied to analyze the effect of epistemic uncertainty in other fields.

In conclusion, the metrics proposed in this paper have significant advantages to inform dam safety governance, since they allow measuring the effect of epistemic uncertainty in decision making. Hence, they help to identify needs for reducing gaps in dam knowledge, giving value to measures that do not have a direct effect on average risk results.

REFERENCES

Altarejos-García, Luis, Ignacio Escuder-Bueno, and Adrián Morales-Torres. 2015. "Advances on the Failure Analysis of the Dam—Foundation Interface of Concrete Dams." *Materials* 8 (12): 8255–78. doi:10.3390/ma8125442.

Altarejos-García, Luis, Ignacio Escuder-Bueno, Armando Serrano-Lombillo, and Manuel Gómez de Membrillera-Ortuño. 2012. "Methodology for Estimating the Probability of Failure by Sliding in Concrete Gravity Dams in the Context of Risk Analysis." *Structural Safety* 36-37: 1–13. doi:10.1016/j.strusafe.2012.01.001.

ANCOLD. 2003. *Guidelines on Risk Assessment.* Hobart: Australian National Committee on Large Dams.

Bowles, D.S., A.M. Parsons, L.R. Anderson, and T.F. Glover. 1999. "Portfolio Risk Assessment of SA Water's Large Dams." Melbourne.

Chauhan, Ss, and Ds Bowles. 2001. "Incorporating Uncertainty into Dam Safety Risk Assessment." In *Proceedings of Risk Analysis in Dam Safety.* http://citeseerx.ist.psu.edu/viewdoc/download?doi=10.1.1.82.407&rep=rep1&type=pdf.

EPRI. 1994. "Methodology for Developing Seismic Fragilities. Electric Power Research Institute." Vol. 3.

Ferson, Scott, and Lev R. Ginzburg. 1996. "Different Methods Are Needed to Propagate Ignorance and Variability." *Reliability Engineering and System Safety* 54 (2-3): 133–44. doi:10.1016/S0951-8320(96)00071-3.

Hartford, Desmond N. D., and Gregory B. Baecher. 2004. *Risk and Uncertainty in Dam Safety.* Thomas Telford.

Hoffman, F. Owen, and Jana S. Hammonds. 1994. "Propagation of Uncertainty in Risk Assessments: The Need to Distinguish Between Uncertainty Due to Lack of Knowledge and Uncertainty Due to Variability." *Risk Analysis* 14 (5): 707–12. doi:10.1111/j.1539-6924.1994.tb00281.x.

HSE. 2001. "Reducing Risks, Protecting People - HSE's Decision-Making Process. Health and Safety Executive." Sudbury.

ICOLD. 2005. "Risk Assessment in Dam Safety Management. A Reconnaissance of Benefits, Methods and Current Applications. International Commission on Large Dams. Committee on Dam Safety. Bulletin 130." Paris: International Commission on Large Dams. Committee on dam safety. Bulletin 130.

iPresas. 2014. "iPresas Calc. User Guide." Valencia: iPresas Risk Analysis. http://www.ipresas.com.

Kaplan, S. 1997. "The Words of Risk Analysis." *Risk Analysis* 17 (4): 407–17.

Morales-Torres, Adrián, Ignacio Escuder-Bueno, Luis Altarejos-García, and Armando Serrano-Lombillo. 2016. "Building Fragility Curves of Sliding Failure of Concrete Gravity Dams Integrating Natural and Epistemic Uncertainties." *Engineering Structures* 125: 227–35. doi:10.1016/j.engstruct.2016.07.006.

Morales-Torres, Adrián, Armando Serrano-Lombillo, Ignacio Escuder-Bueno, and Luis Altarejos-García. 2016. "The Suitability of Risk Reduction Indicators to Inform Dam Safety Management." *Structure and Infrastructure Engineering* 12 (11): 1465–76. doi:10.1080/15732479.2015.1136830.

Paté-Cornell, Elisabeth. 2002. "Risk and Uncertainty Analysis in Government Safety Decisions." *Risk Analysis* 22 (3): 633–46. doi:10.1111/0272-4332.00043.

Paté-Cornell, M.Elisabeth. 1996. "Uncertainties in Risk Analysis: Six Levels of Treatment." *Reliability Engineering & System Safety* 54 (2-3): 95–111. doi:10.1016/S0951-8320(96)00067-1.

Serrano-Lombillo, A, A Morales-Torres, and L A García-Kabbabe. 2012. "Consequence Estimation in Risk Analysis." In *Risk Analysis, Dam Safety, Dam Security and Critical Infrastructure Management*, edited by Ignacio Escuder-Bueno, Enrique Matheu, Luis Altarejos-García, and Jesica T. Castillo-Rodríguez, 107–12. Leiden: CRC Press.

Serrano-Lombillo, Armando, Adrián Morales-Torres, Ignacio Escuder-Bueno, and Luis Altarejos-García. 2016. "A New Risk Reduction Indicator for Dam Safety Management Combining Efficiency and Equity Principles." *Structure and Infrastructure Engineering.*

SPANCOLD. 2012. *Risk Analysis as Applied to Dam Safety. Technical Guide on Operation of Dams and Reservoirs*. Vol. 1. Madrid: Professional Association of Civil Engineers. Spanish National Committe on Large Dams. http://www.spancold.es/Archivos/Monograph_Risk_Analysis.pdf.

USACE. 2014. "Safety of Dams - Policy and Procedures. Engineering and Design.

United States Army Corps of Engineers." Washington D.C.: United States Army Corps of Engineers.

USBR. 2011. *Dam Safety Public Protection Guidelines - A Risk Framework to Support Dam Safety Decision-Making*. United States Bureau of Reclamation. http://www.usbr.gov/ssle/damsafety/documents/PPG201108.pdf.

USBR, and USACE. 2015. "Best Practices in Dam And Levee Safety Risk Analysis. United States Bureau of Reclamation and United States Army Corps of Engineers." Denver: United States Bureau of Reclamation and United States Army Corps of Engineers.

CONSEQUENCE ESTIMATION FOR A LARGE SYSTEM OF DAMS: OUR EXPERIENCES FROM TVA PROJECTS

Michael M. Crouch, PE, CFM[1]
Anurag Srivastava, PhD[2]
Morgan D. Ruark, PE, PH, CFM[3]

ABSTRACT

The Tennessee Valley Authority (TVA) owns and operates more than 40 multipurpose dams in the Tennessee Valley. These dams are located in diverse hydro-meteorological settings and are upstream of both rural and high-population areas that differ in population density and economic status. TVA began to adopt a risk-informed dam safety management criteria in 2009 and completed the screening-level risk studies for most of its dams in 2014. As part of the continued updates to their dam risk assessments, TVA is incorporating more detail into the estimation of downstream life loss and economic consequence using the U.S. Army Corps of Engineers (USACE) Hydrologic Engineering Center's River Analysis System (HEC-RAS) and Flood Impact Analysis software (HEC-FIA). The consequence results can support both event-tree-based risk models and semi-quantitative risk assessment (SQRA) procedures in computing incremental risk. The consequence results also now consider a range of dam-loading conditions, potential failure modes, system response probabilities (SRPs), warning dissemination, and mobilization relationships. This paper highlights TVA's standardized consequence-modeling procedure as well as challenges associated with the modeling, including:

- Developing accurate and consistent downstream building, infrastructure, and Population at Risk (PAR) information for use in the modeling
- Modeling dam-loading conditions and downstream flow conditions that correspond to initiating events to obtain reasonable estimates of incremental consequence
- Developing realistic warning times, warning dissemination, and PAR mobilization information

INTRODUCTION

The Tennessee Valley Authority (TVA) owns and operates more than 40 multipurpose dams in the Tennessee Valley (Figure 1). These dams range from typical low-head power dams to very large multipurpose structures over 200 feet in height. These dams are upstream of both rural and high-population areas that differ in population density and economic status. The reservoir system has a variety of uses, including flood control, power production, and recreation. Operation of the system has been and is continually

1 Water resources project manager, RTI International, Fort Collins, CO 80528, mcrouch@rti.org
2 Senior water resources engineer, RTI International, Fort Collins, CO 80528, anuragsrivastav@rti.org
3 Hydrologist, River Management, Tennessee Valley Authority, Knoxville, TN, 37902, mdruark@tva.gov

optimized to accommodate power production, flood risk reduction, ecological requirements, and recreation.

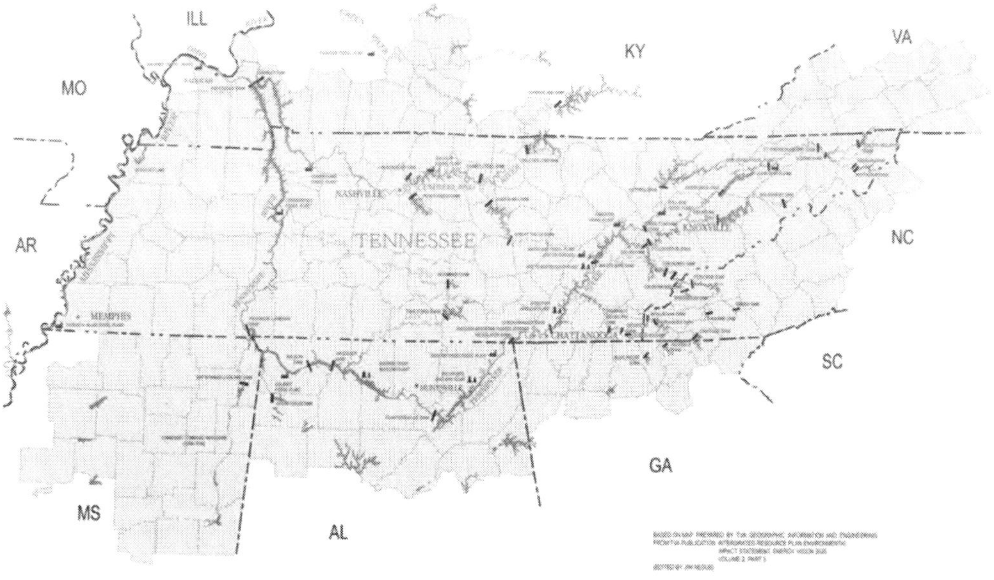

Figure 1: Tennessee Valley Map

Dams in the TVA system were originally designed using deterministic methods. However, TVA is moving to a risk-informed approach to prioritize dam safety modifications at its dam sites. The risk-informed approach requires reservoir loading and release probabilities as well as consequence information to determine risk for each dam. In 2014, TVA completed the screening-level risk studies for most of their dams. As part of the continued updates to their dam-risk assessments, RTI is working with TVA to develop more accurate consequence estimation models using HEC-RAS and HEC-FIA software to improve their event-tree-based risk models and Semi-Quantitative Risk Assessment (SQRA) studies. The new consequence models are tailored to accurately support the risk analysis and consider a range of dam loading conditions, potential failure modes, system response probabilities (SRPs), warning dissemination, and mobilization relationships.

SQRA uses consequence estimates for selected categories to develop the incremental risk matrix. SQRA provides an expedited procedure for risk analysis that is well suited for periodic assessments or for quickly identifying issues based on the maintenance priorities (O'Leary and Scott, 2013). An SQRA analysis may require less initial information about each given failure mode than a full quantitative risk assessment, with a greater degree of uncertainty associated with the failure probability. Full quantitative risk-analysis models, which are commonly based on event-tree analysis, also combine the consequence estimates with the probabilities of the initiating events (i.e., flood or seismic loading conditions), system response probabilities (SRPs), and exposure scenarios to produce the annualized estimates of consequences. Both types of risk assessments require a failure probability and level of consequence assigned to each identified failure mode. The probabilities are typically based on elicitation of subject matter experts and limited model developments.

This paper will discuss the background of why accurate consequence information is needed for risk assessment, and we will present some of the challenges in consequence estimation and the steps TVA is taking to overcome them. Through the course of estimating consequences for multiple dams, we have overcome several challenges to produce better results that support the risk assessment. These are: HEC-RAS scenario development to directly support the risk assessment; data considerations (structure information, population information, and exposure scenarios); population response information (warning time and mobilization); and HEC-FIA scenario setup.

CONSEQUENCE ANALYSIS OVERVIEW

HEC-FIA is used to estimate the consequences to support these assessments. The software is a standalone module that uses HEC-RAS output, along with Geographic Information System (GIS) layers, to compute potential economic and life loss that may occur during flood events. HEC-FIA provides a means to analyze both economic consequence and potential for loss of life due to various dam breach and flood scenarios. GIS data are used to represent the assets in the study area, especially homes and buildings. Other characteristics, such as the population's mobilization away from the flooded area, are determined by interviews with emergency managers for the downstream hazard area.

This methodology is highly detailed and provides the user with flexibility in adjusting inputs that impact the overall consequences associated with a flood event, such as changes in warning time, in-place warning systems, and population downstream of the project. Consequence estimation using this technique allows the user to evaluate risk-reduction measures that may reduce downstream consequences due to flooding or dam breach, with the possibility of reducing the risk associated with a particular dam.

Determining the frequency and probability of reservoir loading is challenging due to the number of dams in the system, TVA's operational possibilities, and the uncertainty in whether upstream failure will occur given a particular loading condition. Adding to the complexity is the determination of the downstream population at risk (PAR) and how they will react during a flood event, which can vary based on the time of year or time of day a breach occurs.

Currently each dam's risk is individually assessed, with downstream cascading failure considered when the studied dam's outflow creates a headwater condition downstream that will create a failure condition. For dams located downstream of headwater dams or other dams in the system, the upstream dams are assumed not to fail for the consequence analysis. The framework to develop risk as it relates to the entire system is currently under development, and consequence information created as part of the individual dam assessments will be leveraged for the system approach.

HYDRAULIC MODEL DEVELOPMENT FOR CONSEQUENCE ANALYSIS

Screening-level risk analyses for TVA dams were based on simplified relationships for life loss and economic consequence assessment (TVA, 2012). In some of the event trees that were developed to support these risk analyses, fixed consequence values were assigned to consequence branches that were connected with varying degrees of loading, subsequent events, and exposure scenarios. In other cases, linearly extrapolated relationships were synthesized based on a few modeled scenarios of peak discharge and non-breach consequences as well as peak-pool elevation and breach consequences. These fixed values and other simplified relationships were deemed appropriate by the subject matter experts and were also justified by the scope and efforts of screening-level risk analysis as prescribed by the USACE (USACE, 2014) and other federal agencies for dam safety risk analysis. The screening level risk assessment establishes the priorities for dam safety modifications, until these analyses are superseded by SQRA results. Now that TVA is updating these models for the next level of quantitative (event-tree based) and semi-quantitative risk analyses, more detailed consequence assessments considering the effects and interplay of reservoir loading conditions, mode of breach, exposure scenarios, warning time, transient population, and mobilization efficiency are needed.

Hydrologic Loading Consideration for Consequence Analysis

Reservoir loading and outflows are directly correlated to the hydrology of the surrounding region. Therefore, when developing scenarios to support risk assessment, we must first consider the hydrologic influences.

Downstream consequences are typically driven by dam outflow for non-breach cases and by reservoir volume for breach cases. During storm-induced breach events, the total consequence is a function of the release from the reservoir and the release from the breach. To accurately compute the incremental consequence (USACE, 2014), the reservoir outflows and downstream tributary inflows associated with the reservoir loading condition must be accurately defined. If downstream flows in the system (both non-breach and breach) are too low for the expected pool level, the computed incremental consequence may be too high.

Defining downstream flows associated with a dam's pool level in a multi-dam system such as TVA's becomes a challenge due to the variation in storms (spatial, intensity, and total depth) that may pass through the reservoir system. For current analyses, TVA assumes storms that produce outflow less than approximately a 500-year event are isolated to the watershed above the study dam. Outflows from the dam are based on its operational guide curve. For larger events, such as ½ the Probable Maximum Flood (PMF) or larger, a regional storm pattern is used that assumes other parts of the system that contribute to inflows downstream of the study dam are experiencing high inflows.

Determination of Hydraulic Model Scenarios to Support Risk Analysis

Each loading scenario that may occur at a dam will have a different breach and non-breach outflow, probability of occurrence, and probability of failure, which contributes to the total risk of the dam. Either a full quantitative risk analysis or SQRA approach can be used to define the total potential risk associated with a dam. An SQRA analysis uses incremental consequence information for a single scenario to define the risk associated with the dam, while a full quantitative risk assessment requires several relationships of consequence vs. peak outflow or reservoir pool elevation. For both procedures, consequence relationships may change based on failure mode and must be defined for all failure modes considered (e.g. flood, seismic, and normal operating conditions). It is common to develop a large number of model scenarios to support an event-tree risk analysis. The following paragraphs discuss how to develop hydraulic model scenarios to support an event-tree-based quantitative risk analysis.

The full range of flood events that may occur in any given year are used in a flood-event tree model so consequence information must be defined for the full range of flood events. Within an event-tree-based risk model, the range of flood events is expressed by the reservoir pool level. Each pool level has a corresponding Annual Exceedance Probability (AEP), which relates the pool level with the probability that it will be exceeded in any given year. Based on the evaluation of the structural performance of the dam, different branches representing the dam's potential failure modes and the environmental factors (e.g. storm event, earthquake, gate failure) leading to those failure modes are included in the event tree analysis. Each branch requires a consequence relationship, which is used to quantify its risk dependent on the exposure scenario (day/night, summer/winter) (Srivastava et.al., 2009).

Figure 2 presents a simple event-tree model, which considers a range of reservoir loading and various dam states that may occur during the loading: no-breach, overtopping failure, and piping failure. The range of flood loading (peak pool elevations) and AEP values for those loading conditions are defined in the Level 1 continuous branch of the event tree. The branches in Levels 2 and 3 calculate the depth of overtopping and peak discharge values for different peak-pool elevations, respectively. Level 4 includes a non-breach branch and two failure branches. The downstream population and their response to warning varies by the time of the day. Therefore, the event tree includes day and night exposure scenarios in Level 5 before the life loss and economic consequence branches in Levels 6 and 7, respectively.

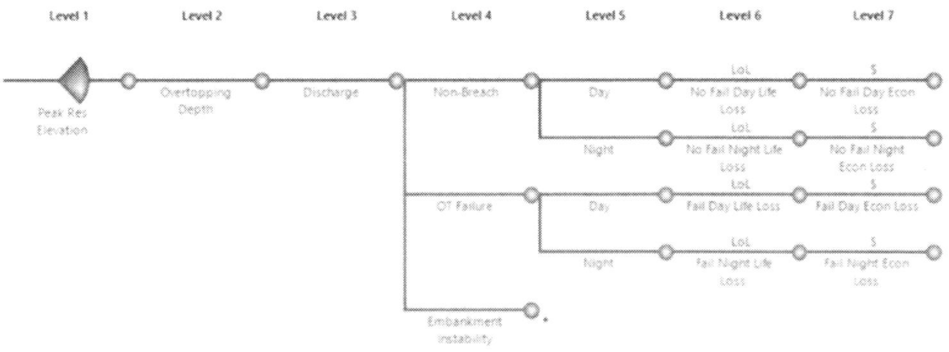

Figure 2: Example Flood Event Tree Model

For the consequence analysis, the unsteady HEC-RAS analysis provides the associated hydrograph outputs for various storm events at locations in the downstream hazard area. These outputs are obtained by routing the event inflow hydrograph through the reservoir in the HEC-RAS model and simulating the breach or no breach condition. The resultant hydrograph is then routed through the downstream valley. The HEC-RAS outputs are used in the HEC-FIA analysis to obtain the estimates of life loss and economic consequences for different scenarios, which vary with population at risk (PAR), warning time, mobilization rate, and evacuation efficiency. These complex hydrologic and sociological scenarios must be simplified and coded into the flood-event tree model as relationships of life loss and reservoir outflow or life loss and elevation at the time of failure. This requires multiple combinations of hydraulic and societal response scenarios.

For the example dam represented by the event tree model in Figure *2*, the range of pool levels or loading represented in the AEP relationship is from elevation 1677 to 1714.4 and is coded into Level 1 of the model. The flood-event tree model computes incremental model runs (through Level 7) for the entire range of pool levels. Corresponding outflows are computed using relationships provided in Levels 2 and 3. The relationship presented in Level 3 is taken from the operational guide curve presented Figure 3, which represents the expected outflow from the dam for normal loading conditions. For Level 4, SRPs are assigned for each failure mode. Exposure probabilities are assigned for each branch in Level 5. Each incremental model run has a corresponding annualized probability of economic damage and life loss based on the AEP, SRP, and exposure probabilities. The total annualized probability of consequence is determined by summing the incremental runs.

Within the flood-event tree model, consequence can be a function of reservoir outflow or reservoir elevation. In general, downstream consequences are expected to rise as the outflow from the dam increases or as volume in the reservoir increases. In the case of a non-breach event, life loss is primarily driven by the peak outflow, and therefore a life loss versus peak discharge relationship is typically assigned as an input to a non-breach consequence branch (Levels 6 and 7 of Figure *2*). In the case of a breach event, the failure mode and the volume in the reservoir, a function of reservoir elevation, are more indicative of consequence because they govern the outflow, therefore a relationship of reservoir elevation vs. consequence is applied for Levels 6 and 7 of the failure branches.

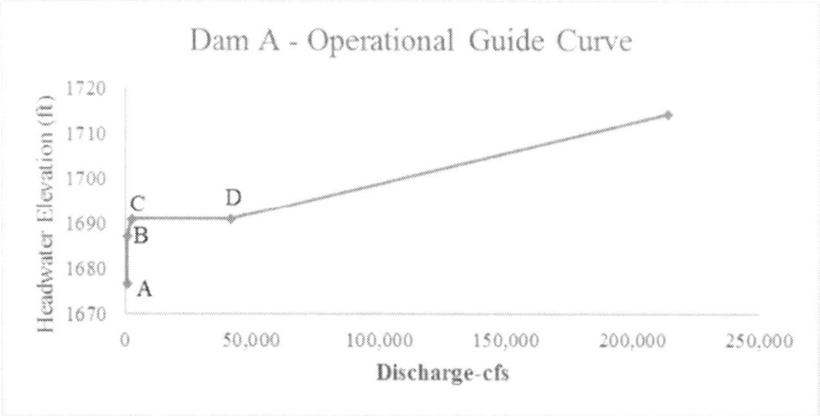

Figure 3: Operational Guide Curve for a Hypothetical Dam

Development of Non-Breach Consequence Relationships. The non-breach consequences represent the case where the flood event does not lead to the failure of the dam structure, but the outflow from the dam causes life loss in the downstream area. The first non-breach model scenario is developed to represent the threshold discharge, which is not expected to result in any consequences (Point A in Figure *3*). An intermediate outflow scenario is included between the low-outflow scenario and PMF outflow to detect any inflection point that may occur in the consequence estimates between these discharges. Lastly, a scenario at the PMF discharge is set to estimate the consequences associated with the PMF-level inflow. Within the event-tree model, the consequence from these scenarios are represented as a function of outflow. Table 1 lists the scenarios required to encompass the consequences for the non-breach branch of the event tree shown in Figure 2.

Development of Overtopping Consequence Relationships. To develop the relationship for the overtopping consequences, two model scenarios are required. The first scenario is set at the crest elevation of the dam to assess the lower limit of consequences in the case of overtopping failure. The second scenario is set to estimate the upper limit of the consequences caused by overtopping at the PMF pool elevation and corresponding discharge. Table *2* presents the required scenarios for the overtopping branch of the event tree. Within the event-tree model, the consequences from these scenarios will be represented as a function of pool elevation.

Development of Embankment Instability Consequence Relationships. The first scenario for the instability failure is set at the minimum pool elevation shown on the flood operating guide. The operating guide shows the minimum pool elevation at 1676.6 ft. with a minimum discharge of 900 cfs. Another scenario is defined at a higher pool elevation of 1687 ft., in which the discharge slightly increases to 1000 cfs, but significant increase in reservoir volume occurs. Additional scenarios are required to represent the consequence that will occur at pool elevation 1691 ft. At this pool level, reservoir elevation is not always indicative of the total outflow during a breach. Based on the operational guide curve shown in Figure *3*, a breach that occurs at elevation 1691 may have two peak outflows, dependent on the operational release that is occurring at the time of the breach. Therefore, two scenarios are required to depict the consequence that may occur at this elevation. Finally, a scenario is required to assess the consequence that may occur at the upper end of the AEP curve, the PMF pool elevation of 1714.4 ft. Table *3* lists the scenarios required for the instability failure mode.

Table 1. Scenarios for Non-Breach Life Loss Consequence

Peak Pool Elevation (ft)	Peak Discharge (cfs)
1676.6	900
1700	100,000
1714.4	214,220

Table 2. Scenarios for Overtopping Failure Mode

Peak Pool Elevation (ft)	Peak Discharge (cfs)
1713	200,000
1714.4	214,220

Table 3. Scenarios for Embankment Instability Failure

Peak Pool Elevation (ft)	Peak Discharge (cfs)
1676.6	900
1687	1,000
1691	2,700
1691	42,000
1714.41	214,220

CONSEQUENCE MODEL DEVELOPMENT

The HEC-FIA consequence models require HEC-RAS input based on the scenarios presented in the previous section, as well as numerous GIS datasets and response relationships to depict the flooding and behavior of the downstream PAR. Once the input data are compiled, the HEC-FIA model scenarios must be developed to represent the consequences consistently between the breach and non-breach cases. The following sections will detail important aspects of the data development and model setup for consequence assessment.

Data Considerations for Consequence Analysis

As with most engineering analyses, there are several inputs required for consequence analysis that are difficult to pinpoint due to their natural variability and limited data. The models we are developing for TVA consequence analysis attempt to refine the data inputs to produce realistic consequence estimates. Small errors with the assumptions in population can skew consequence estimates and reduce the validity of a risk assessment.

Downstream Structure Information. Initial consequence estimates performed by TVA utilized census-level information for structure locations and population, which proved to overestimate PAR for more frequent scenarios because structures were not accurately located.

TVA employs a team of GIS personnel to produce a detailed structure inventory for the entire Tennessee Valley. The team is using parcel information from municipalities in the hazard areas to estimate the value of buildings in a particular parcel to produce a GIS structure point dataset that has at the exact location of each structure based on aerial imagery. The foundation elevation for each structure is determined using LiDAR information and a visual stair count from Google Street View to the lowest finished floor of the structure. These dataset improvements allow for more accurate damage computations in the HEC-FIA model.

Population Estimates. In addition to the structure data point improvement, TVA is using a population generator that was developed by HEC to account for the transition of a hazard area's workforce from their residence to their workplace. The Population Utility for Disaggregated and Diurnally Distributed Inventories (PUfDaDDI) (HEC, 2015) appends workforce information available from U.S. Census Bureau's Quarterly Workforce Indicators (QWI) to the latest census information to improve population information in the hazard area.

TVA is also working to define the recreational population or transient population that may be in place below their dams based on information gathered from public agencies who own the parks and boat ramps in the downstream hazard area. Increased PAR may exist at downstream recreational sites during the summer season as compared to the winter season. In addition, variability of recreational population will occur from day to night. This recreational or "transient" population is included in scenarios corresponding to summer reservoir elevation with non-flood outflows. Scenarios modeled for flood events and winter pool levels use the base population plus a reduced transient population.

By careful refinement of the structure inventory and population used in the consequence model, TVA is improving the risk estimates for their dams.

Population Response Information. Mobilization: To compute life loss, HEC-FIA requires input representing the response of the downstream population during a flood event (Lehman and Needham, 2012). The Warning Issuance Time (WIT), Warning Issuance Delay (WID), First Alert/Warning Diffusion Time (FAT), and Protective Action Initiation Time (PAI) are used to define when the public is warned, how fast the warning spreads through the population, and how quickly and how much of the population responds to the warning. These variables are highly subjective and will vary based on what type of breach occurs and the time of day the breach occurs. Initial HEC-FIA model efforts utilized default settings within the model to make these estimates. TVA's current consequence modeling efforts use recently developed methods produced by USACE to account for each downstream counties' emergency notification procedure (Mileti and Sorenson, 2015).

TVA interviews downstream local emergency management officials to collect information on existing emergency plans, notification/warning procedures, communications resources, population demographics, etc. TVA conducts the interviews using the Interview Schedule developed by USACE (Mileti and Sorensen, 2015), and we enter the results into the USACE Elicitation of Experts (EOE) spreadsheet for computation of the WID, FAT, and PAI relationships for each county. These relationships are used as input to the HEC-FIA model for computing life loss for each impact area. The Interview Schedule is focused on local level emergency management and does not adequately represent the warning and response capabilities of major industries. Therefore, WID, FAT, and PAI relationships for major industries are selected from default curves provided by USACE.

Warning Time. The warning time is a critical factor in estimating the life-loss consequences. Depending upon the time delay between the breach detection at the dam site and the time when the flood wave arrives in the downstream area, the warning time can make a significant difference in the downstream life-loss estimates. For breach scenarios, warning issuance is typically dependent on when the breach is detected. For non-failure scenarios, warning issuance is dependent on when the TVA river forecast center is confident in the precipitation forecast and the corresponding release decisions.

TVA maintains a warning notification directory, which is used in the HEC-FIA modeling to determine the critical release at which a warning is issued. For the current level of analysis that TVA is undertaking, during storm event scenarios, warnings are typically assumed to be issued 24 hours before a high release from the dam and six hours before the breach reaches its catastrophic phase of formation. For sunny day failure or seismic failures where a breach is unlikely to be detected, a warning is assumed to be issued to the downstream PAR at the time the breach occurs.

To maintain consistency in warning between the non-breach and breach scenarios, a warning must be issued to the downstream PAR for the storm event. In the case of a breach event, another warning must be issued for the breach event. This is termed a double warning. For the breach event, the second warning can sometimes be of limited importance if the initial warning for the high-reservoir release is issued early enough. If the PAI curve applied to the downstream population approaches its maximum by the time the second warning is issued, only the PAR that remains in the flooded area can be warned the second time. The PAI curves developed for the TVA analysis often assume 85% to 90% of the downstream PAR mobilizes. In these cases, the downstream life loss can only be reduced by approximately 10% when issuing a second warning for the breach, which may be within the noise level of the modeling.

HEC-FIA Model Scenario Development

With refined model inputs, such as day/night population and event-based warning times, the consequence scenarios can be specifically tailored to the risk model using HEC-FIA. When setting up the HEC-FIA model, it is important to consider which assumptions are appropriate for each scenario. For example, one would expect a larger downstream population in a recreation area for a sunny day event than during a storm event. Warning time to the downstream population will likely vary based on the flood condition at the dam and breach type. For a breach scenario, it is more likely that an operator is at the dam and able to detect a breach occurring during a storm event than during a sunny day condition. During a seismic event, it may be unlikely that an operator will be at the dam and therefore unlikely that a warning will be issued until the breach occurs.

The matrix below (Table *4*) presents likely warning times and downstream population configurations used for the example scenarios developed as part of this consequence analysis. The consequence results from these scenarios are used to produce the relationships presented in Figure *4* a, b, and c and support the event tree based risk analysis.

Table 4: HEC-FIA Scenarios for Consequence Analysis

Scenario group	Peak Pool Elevation (ft)	Peak Discharge (cfs)	Population Assumption	Warning Time
Non-breach scenarios	1676.6	900	Base and reduced transient	0 hours
	1700	100,000	Base and reduced transient	24 hours
	1714.4	214,220	Base and reduced transient	24 hours
Overtopping breach scenarios	1713	200,000	Base and reduced transient	24 hours
	1714.4	214,220	Base and reduced transient	24 hours
Embankment instability breach scenarios	1676.6	900	Base and reduced transient	0 hours
	1687	1,000	Base and full transient	0 hours
	1691	2,700	Base and full transient	0 hours
	1691	42,000	Base and reduced transient	24 hours
	1714.41	214,220	Base and reduced transient	24 hours

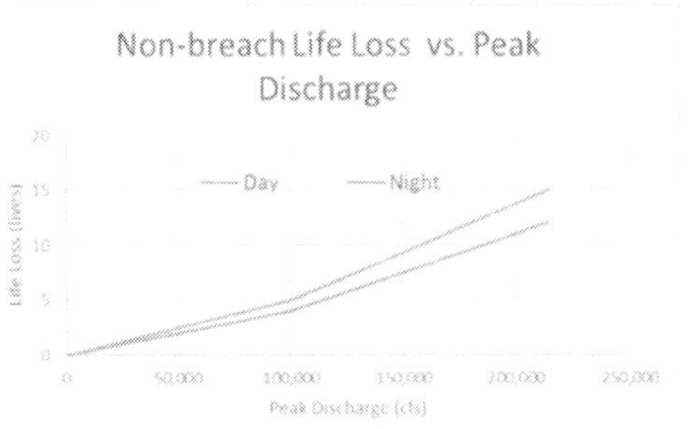

a. Non-breach life loss consequence

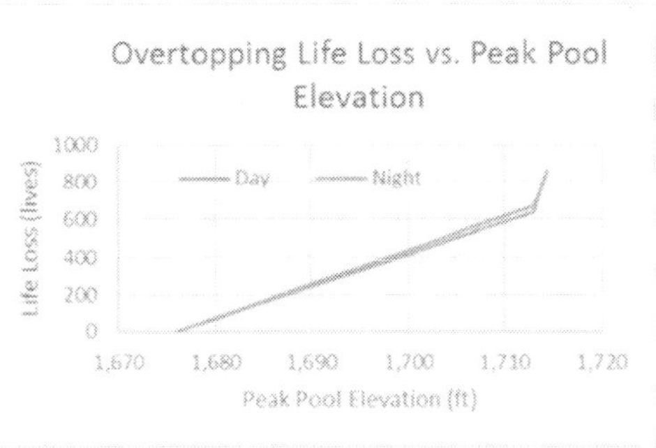

b. Overtopping life loss consequence

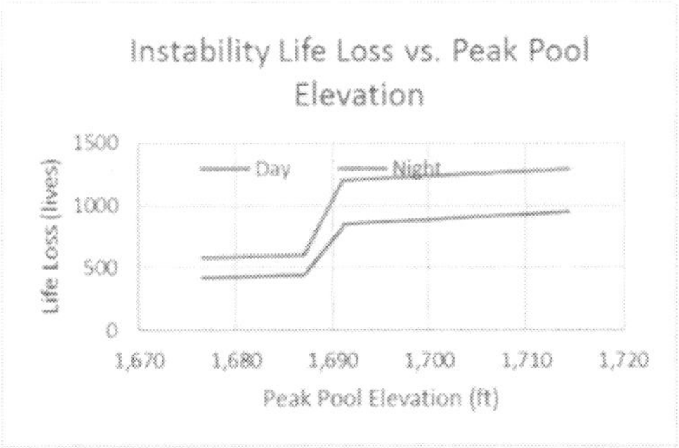

c. Instability life loss consequence

Figure 4: Life Loss Consequence Relationships for the Hypothetical Example

CONCLUSION

Recent developments in consequence assessment methods allow the user to tailor scenarios to the risk analysis. It is the responsibility of the agency to develop realistic model inputs for the scale of the study being performed. The consequence model, and ultimately the risk model, results hinge greatly on the input data and population response information. When assessing the risk of multiple dams in a system, developing a consistent approach that is applied to all dams is paramount to calculating results that are meaningful and can be used for decision making. The improved methods, while more expensive to implement upfront, can provide a valuable tool for making decisions about structural improvements and may create a cost savings in the long term.

REFERENCES

HEC (2015). Population Utility for Disaggregated and Diurnally Distributed Inventories (PUfDaDDI). December, 2015.

Lehman, W. and Needham, J. (2012). "Consequence estimation for dam failures," In Proceedings of 32nd Annual USSD Conference, New Orleans, LA. April 23-27, 2012.

Mileti, D. and Sorensen, J (2015). "Interview Schedule: Community Warning Issuance, Diffusion and Protective Action Initiation Estimation (DRAFT)," US Army Corps of Engineers, 2015.

O'Leary, T. M. and Scott, G. A. (2013). "Semi-quantitative risk characterization for periodic dam safety assessment," In Proceedings of the 33rd Annual USSD Conference, Phoenix, AZ. February 11- 15, 2013, pp 571-86.

Srivastava, A., Bowles, D.S., and Chauhan, S.S. (2009). Improvements to DAMRAE: A tool for dam safety risk analysis modeling. In Proceedings of the 29th ASDSO Conference on Dams (USA), Hollywood, FL. September, 2009, pp. 64-80.

TVA (2012). Initial screening report. River Operations Dam Safety Portfolio Risk Management Program, Tennessee Valley Authority. July 2012.

USACE (2014). "Safety of dams – policy and procedures. Engineering Regulations 1110-2-1156," USACE, Washington, DC 20314. October, 2014.

APPLICATION OF RISK ASSESSMENT TO AN EMERGENCY LEVEE DESIGN IN AN URBAN ENVIRONMENT

Petros Armenis[1]
Malcolm Barker[2]
Peter Christensen[3]
Graham Harrington[4]

ABSTRACT

The Canterbury Earthquake Sequence in September 2010 and February 2011 caused large areas of land to change by differing amounts throughout Christchurch, New Zealand. Land levels fell by more than 300 mm in some areas. This increased flood risk in the tidal reaches of the Avon River. Urgent repairs were completed with the objective to restore the tidal river defenses to a crest level equivalent to a 1% AEP tide level. This work needed to be completed prior to impeding spring tides.

The levees will be required for up to 20 years and then probably be rebuilt on a new alignment. To better understand the risks ass[1]ociated with the ongoing reliance of the levees for flood protection in the interim, a risk assessment was undertaken using conventional Australian National Committee on Large Dams (ANCOLD) practices and levee design procedures. Careful consideration was made to the performance of the existing levees under seismic, flood and tidal loading from which the societal and individual risk profiles were derived. The work included the following:

1. The identification of critical sections along both sides of the 11 km levee river alignment through consideration of the foundation and embankment construction
2. Combining seismic events with tides and flood events with tides using levee lifetimes of 1, 5, 10 and 20 years
3. Consideration of overtopping failure and piping through the levee embankment, foundation or tree roots and narrowed embankment sections owing to trees being blown over
4. The application of CIRIA Levee handbook and the USBR piping toolbox
5. Evaluation of risk reduction with upgrade options

This paper will present the levee design and the process applied for the analysis of the levee and the upgrade options selection.

[1] Dams Engineer, Dams, Tunnels and Geotechnics Service Group, GHD Pty Ltd, Brisbane, QLD, Australia, 4000, petros.armenis@ghd.com
[2] Principal Dams Engineer, Dams, Tunnels and Geotechnics Service Group, GHD Pty Ltd, Brisbane, QLD, Australia, 4000, malcolm.barker@ghd.com
[3] Technical Advisor, Stormwater and Land Drainage Rebuild, Christchurch City Council (CCC), Christchurch, New Zealand, 8013, peter.christensen@ccc.govt.nz
[4] Senior Surface Water Planner, Water and Waste Planning Team, Christchurch City Council (CCC), Christchurch, New Zealand, 8013, graham.harrington@ccc.govt.nz

INTRODUCTION

Significant damage was caused to the levees (also known as stopbanks) that line the Avon River by the September 2010 and February 2011 Christchurch earthquakes. Emergency repairs were required to repair the levees before the upcoming Spring Tides. A contractor was engaged by the Christchurch City Council (CCC) with the objective to restore the river defenses initially to a minimum crest level and then later to a raised crest level to approximately the original design level. Construction continued between March and June 2011 with the aim of utilizing a variety of levee forms.

A "generic levee design' was developed by the original designers utilizing a cut off drain, 1 in 4 slopes and an approximate crest width of 1 m. A "dirty' pit run was developed to construct the temporary levees. The dirty pit run was developed by blending 3 different materials, one of which had significant fines content. Limited space meant the standard design could not be used in all areas along the lower Avon levees. Sand bags were utilized in small areas where there was virtually no space without encroaching the adjacent road and infrastructure. Some areas had room for an aggregate stop bank but there was not enough space for heavy machinery to undertake the construction. As a result of this, some portions of the levee do not have cut off drains or engineered foundations and they have not been compacted using compaction equipment.

Following the original construction, minimal maintenance was undertaken prior to this study. Maintenance that had been undertaken comprised of periodic crest level surveys and subsequent re-grading of areas less than the original required minimum.

At the commencement of this study, it was clear that the existing levees would be needed for up to an additional 20 years at which time they would be probably relocated and rebuilt for a longer service life which would include allowance for sea level rise and stability in future earthquakes. GHD was engaged by CCC to investigate the risks, benefits and costs associated with the ongoing reliance on the Avon River existing stopbanks, for flood protection in the tidal reaches of the Avon River for a design life of up to 20 years. In order to achieve this, a risk assessment approach was used as described in this paper.

LEVEE DEFINITION & IDENTIFICATION OF LEVEE CRITICAL SECTIONS

Desktop Review of Existing Levee Embankment Configuration and Ground Conditions

During the original temporary levee construction, it was agreed with CCC that for ease and rapid rate of construction, the standard levee configuration would be constructed as follows:

- Minimum crest elevation of RL 10.8 m;
- Trapezoidal cross section, crest width of 2.5 m and side slopes of 1:4 (V:H);

- Cutoff trench typically of depth 0.3 m to 1.5 m and 2.0 m wide to be taken into the original levee or founding material;
- With material comprising silty gravel with maximum particle size 200 mm and containing approximately 15% fines. The material was reasonably well graded and was easily compacted. The gravel/cobble component comprised rounded or sub-rounded material;
- The material was sourced from a number of quarries and was blended at the contractor's yard. The material was placed and compacted to approximately 95% of maximum modified dry density; and
- The permeability of this material as measured in the laboratory and an in situ measurement showed that the permeability ranged from 10^{-9}m/s to 10^{-6}m/s

Typical gradings of the levee fill material are shown in Figure 1 below.

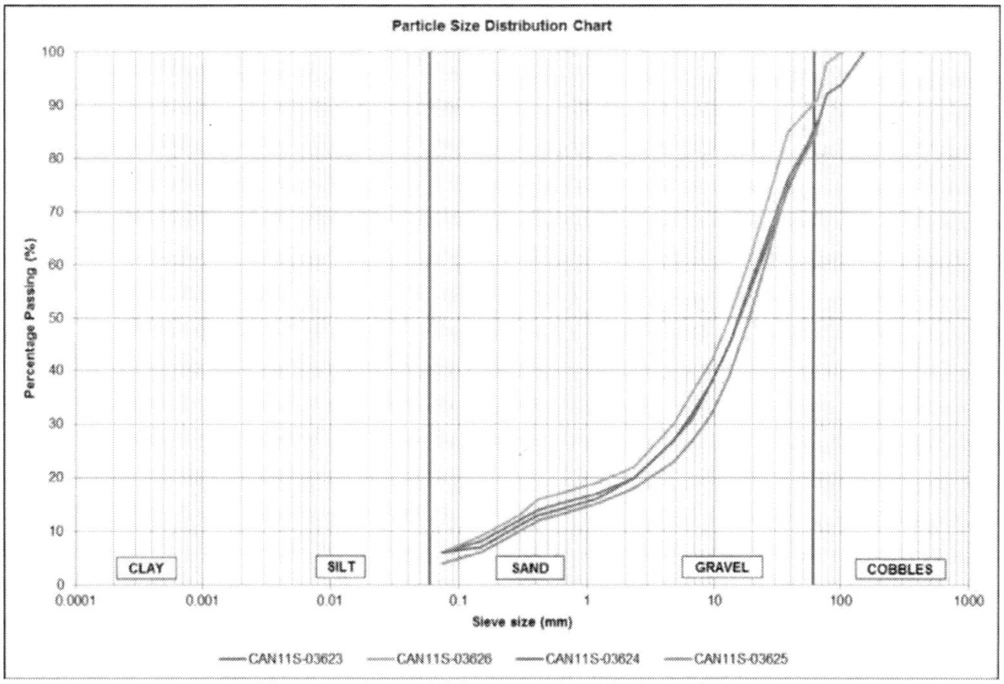

Figure 1. Levee fill material typical gradings

Photographs of the constructed Levee are shown in Figure 2 and Figure 3 below. Due to the working space constraints, certain sections of the levee were not constructed in accordance with the standard configuration where crest widths as little as 1.0 m were constructed. In some instances, side batters were as steep as approximately 1:1, or even near vertical if retained by Diamond Pro Block or portable segmented concrete barrier retaining walls. Compaction had also been compromised in some areas and in almost all locations compaction of the side slopes had not been performed. This results in superficial cracking of the slopes that may worsen through water ingress.

Geogrid, Triax TX160, and Bidim Geofabric had been used in some areas, particularly those with poor founding conditions. Sandbags were placed on the levee crest at several areas where the width of the levee was narrow owing to space constraints. In addition to this, the levees were constructed around and adjacent to trees which were protruding the crest and levee side slopes.

Figure 2. Photograph of Typical Levee Section 15

Figure 3. Photograph of levee topped up with eroded sand bags and protruding trees

Site Inspection and Risk Workshop

A walkover of the site was undertaken involving staff members from GHD and CCC. The purpose of the site inspection was to evaluate the existing condition of the levees and identify critical sections and appurtenant structures for the risk analysis.

The following key items were identified along portions of the levees:

- The absence of cut off drains
- Low points along the levee walls
- Trees growing on and adjacent to the levees
- Deteriorated sand bags on the crests of the levees
- Slumping and lateral spreading
- Longitudinal cracks on the downstream face of the levees
- Seepage through the foundation sands

A total of 19 discrete sections were identified for analysis in the risk assessment. These location are shown in plan in Figure 4 below:

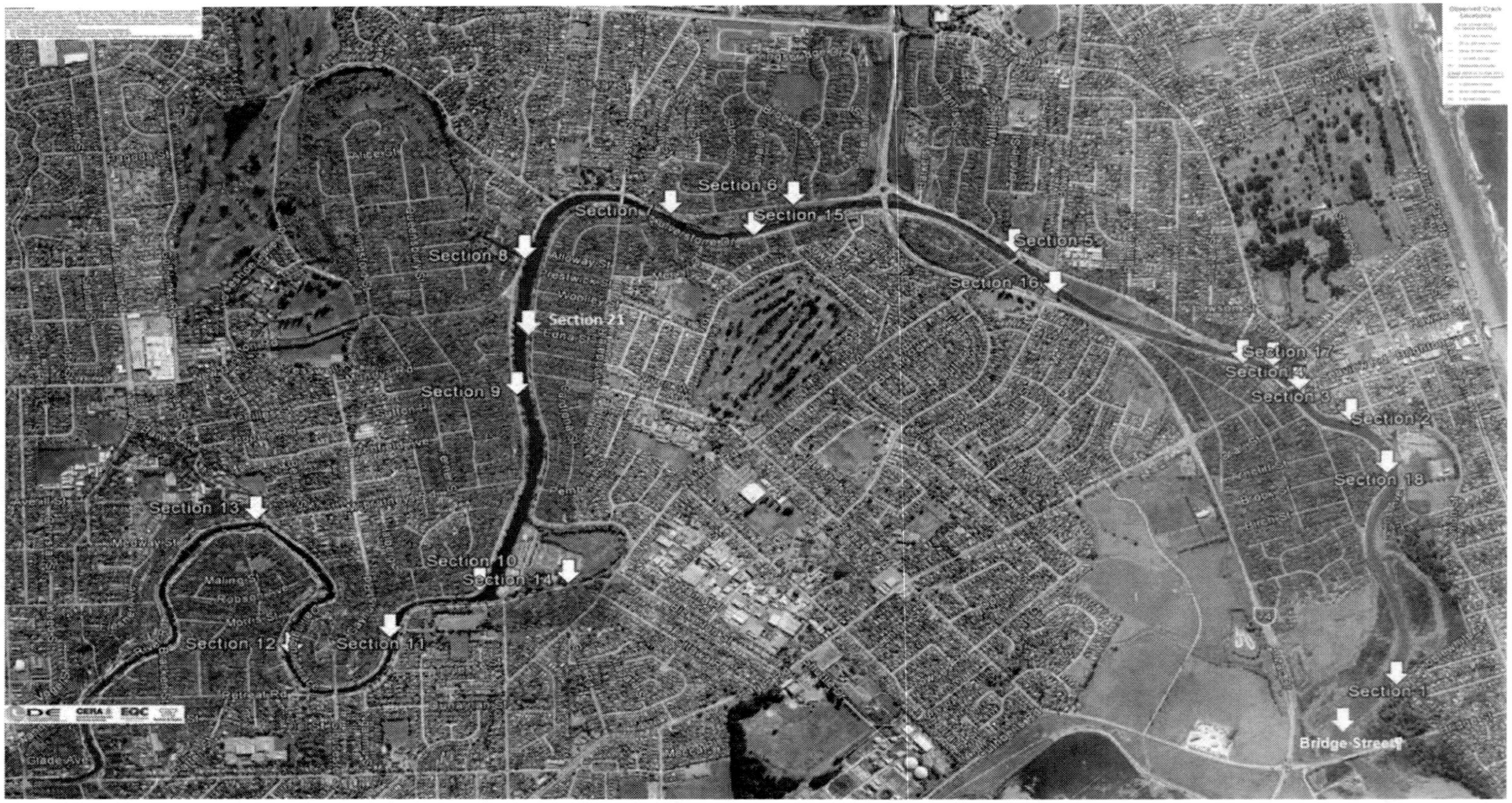

Figure 4. Aerial view of Avon River and locations of critical levee sections adopted for the study

Following the site inspection, a workshop was undertaken with GHD and CCC with the purpose of discussing the following:

- The risk assessment criteria and the acceptable risk levels with CCC
- The consequences of failure based on the available inundation mapping

Acceptable risk tolerance An exercise of defining the acceptable risk tolerance of the levees was undertaken It was agreed that a design life up to 20 years was reasonable to consider for the risk assessment of the medium term structure. **Table 1** illustrates the risk criteria for the levees from which it was determined that a 10% probability of occurrence was acceptable and tidal recurrence intervals above 1 in 200 were not required to be evaluated.

Table 1. CCC Risk Tolerance (probability of event occurring within design life of the structure)

Design Life	Tide Average Recurrence Interval (ARI) Years						
	2	5	10	20	50	100	200
1	50.0%	20.0%	10.0%	5.0%	2.0%	1.0%	0.5%
2	75.0%	36.0%	19.0%	9.8%	4.0%	2.0%	1.0%
5	96.9%	67.2%	41.0%	22.6%	9.6%	4.9%	2.5%
10	99.9%	89.3%	65.1%	40.1%	18.3%	9.6%	4.9%
20	100.0%	98.8%	87.8%	64.2%	33.2%	18.2%	9.5%

Inundation consequences An assessment of the loss of life caused by levee failure was carried out as part of the risk assessment. The Reclamation Consequence Estimating Methodology (RCEM 2014) was used to undertake the assessment. To carry out this assessment, flooding extents of three discrete water levels were assessed using a bathtub inundation flood model. An example of the inundation flood modelling is provided in Figure 5 below. It should be noted that green properties signify existing inhabited properties and red properties signify properties that were evacuated following the 2010 and 2011 earthquakes. The evacuated properties are no longer inhabited and therefore do not currently represent a Population at Risk (PAR).

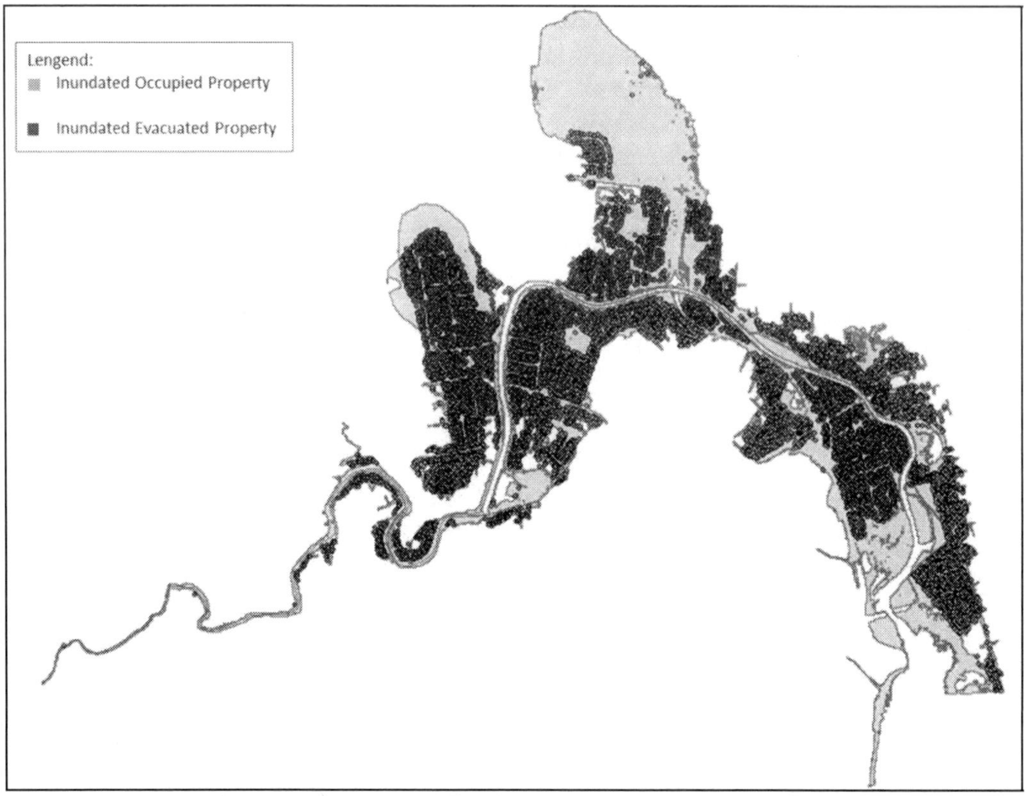

Figure 5. Example of flood inundation mapping output

HAZARD ANALYSIS

The Avon Levees are subject to seismic and hydrological flood loading conditions. In addition to this, the levees are also subject to the tidal influence of the Avon River.

Tidal Influence

Tide fluctuations along the Avon River vary significantly between the maximum and minimum water level on the levees. The peak tidal levels, however, do not vary significantly, as shown on Figure 6 . Tides up to the 1 in 200 AEP event were considered as required for the CCC risk tolerance.

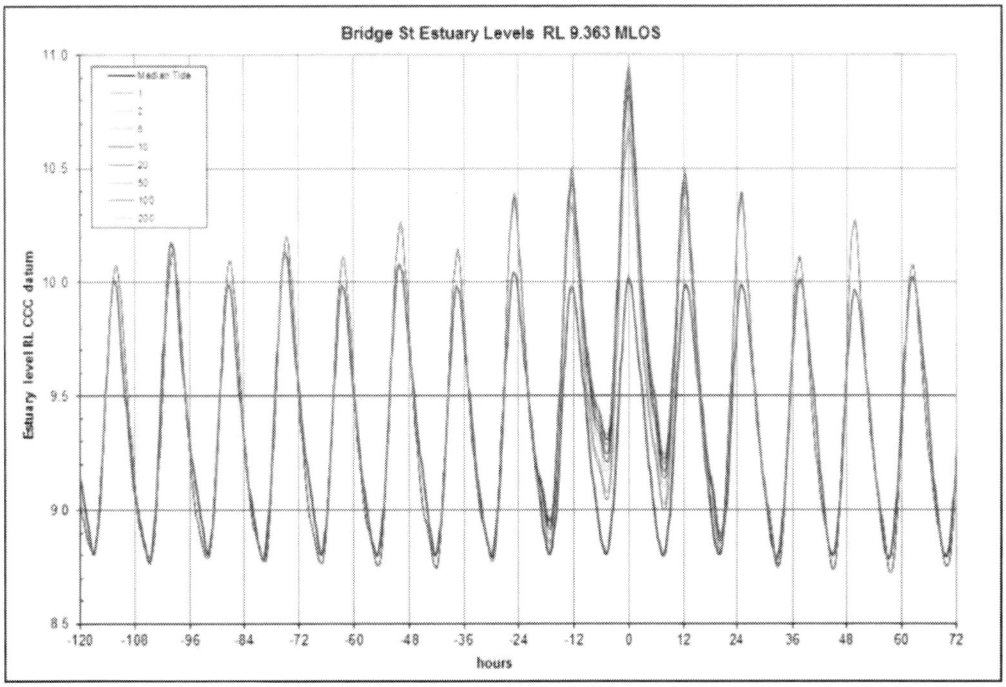

Figure 6. Tidal Level Data at Bridge Street (A Specific Location Along the Avon River Alignment)

Flood Loading

Hydrological models were provided by CCC to GHD for up to the 1 in 200 AEP event. The water levels associated with the floods at each section of interest is presented in **Figure 7** below.

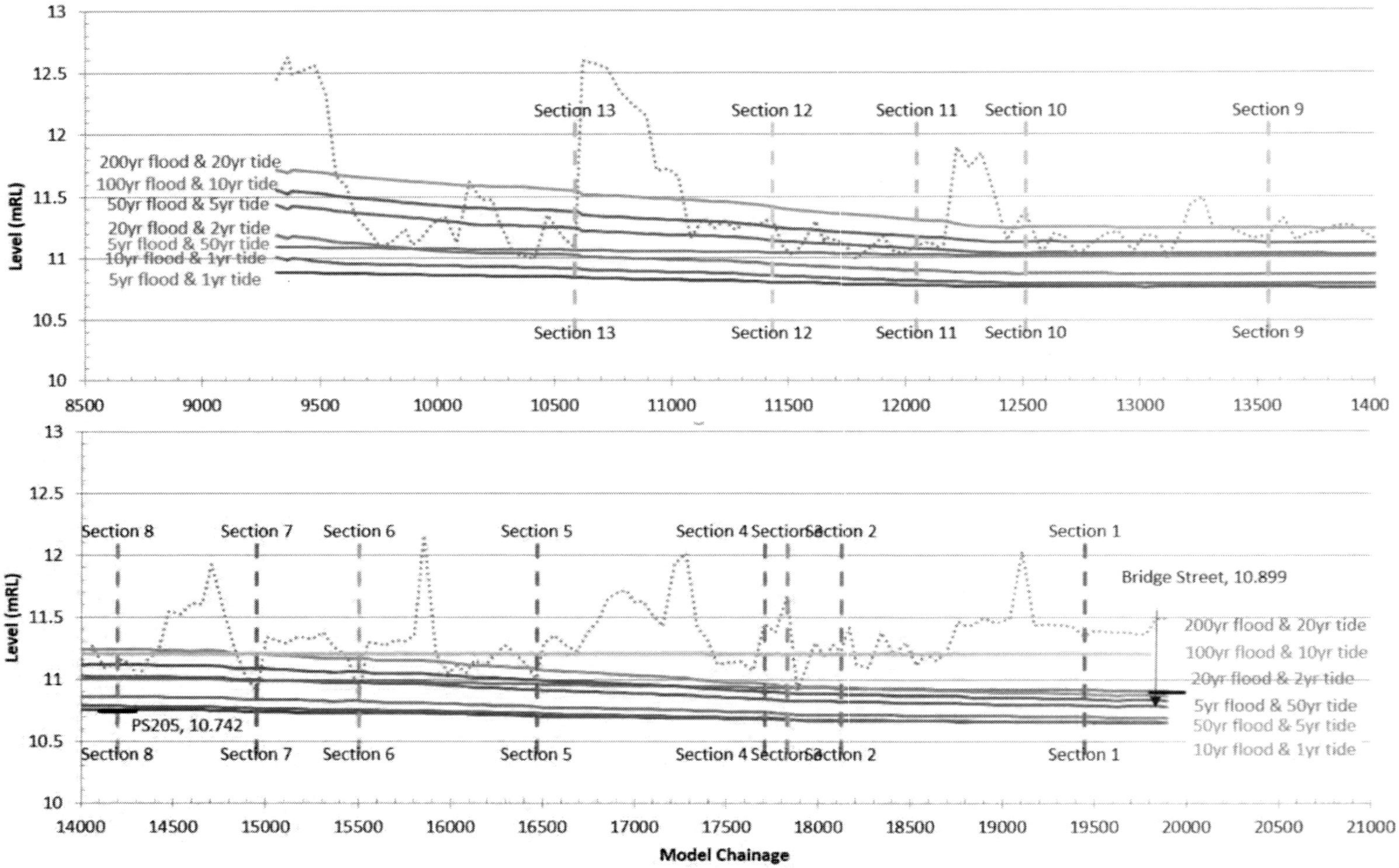

Figure 7. Flood and River Bed Levels Along the River

Flood and Tide Coupling

Larger tidal events have the potential to drown out smaller flood events near the mouth of the river. Between Bridge Street and Chainage 17,900, the 1 in 50 AEP Tide with the 1 in 5 AEP indicates an almost linear Hydraulic Grade Line (HGL). Considering all of the other flood events in the data set, the HGL was almost the same in this section, hence tidal water level influence was considered dominant over flood water level influence downstream from Chainage 17,900.

For all flood events provided in the data set, a difference in HGL slope could be seen when comparing Chainage 14,300 to 17,900 and Chainage 17,900 to Bridge Street (Chainage 19,700). Considering the largest available flood event (the 1 in 200 AEP flood with the 1 in 20 AEP Tide) and the 1 in 50 AEP tide coupled with the 1 in 5 AEP flood, between Chainage 9,300 to 17,900 it could be seen that the larger flood was creating higher water levels.

Conversely, smaller floods were creating lower water levels than the larger tides in this location. Hence, for the flood events coupled with different tidal events, this section was considered as changing point and flood levels were compared against the tidal levels with no flood influence and the greater water level was adopted for the event under consideration. An example of this process is shown in Figure 8 below.

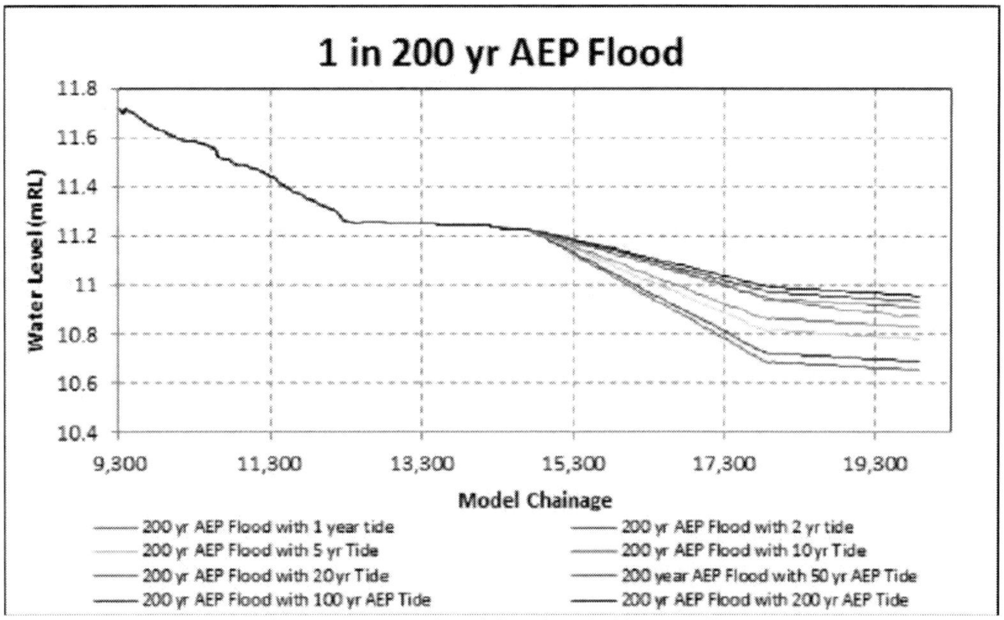

Figure 8.Example of flood and tide coupling

Seismic Loading

Seismic loading for the risk assessment was adopted from a literature source describing the seismic hazard of the Canterbury Region, New Zealand (Stirling et al. 2008). Spectral acceleration of 0 seconds was adopted for the seismic loading considered in the risk assessment. The Peak Ground Acceleration (PGA) for the seismic events under consideration are shown in Figure 9.

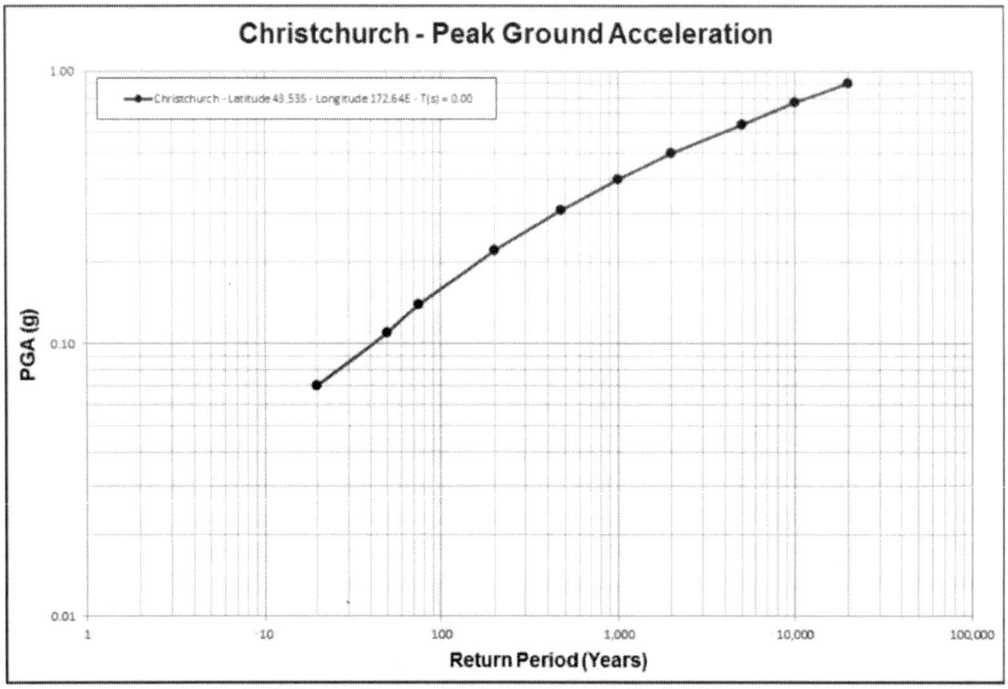

Figure 9. PGA vs Return Period Adopted for Risk Assessment

FAILURE MODES ANALYSIS

The following failure modes were evaluated in detail for the risk analysis.

- Overtopping
  - Seismic deformation loss of freeboard and overtopping
  - Floods or tides overtopping the gravel embankment
  - Floods or tides overtopping the sandbag sections

- Piping
  - Seismic cracking
  - Cracks in the embankment due to differential movement
  - Through the sand foundation
  - Through rotted tree roots
  - Through narrowed section caused by trees blowing over and damaging the embankment

Slope instability was evaluated and found to be of significantly lower likelihood than the above failure modes and was dismissed for further analysis.

Overtopping

Overtopping failures were assessed where the water level in the Avon River exceeded the crest height of the levee under consideration. Overtopping flow up to 500 mm flow depth was assessed as this was close to the maximum caused by the flood events under consideration in this risk assessment.

Sections which had existing sandbags were assessed taking the top of the sand bag as the reported levee crest level from the LIDAR data provided to GHD by CCC.

The potential for overtopping erosion failure was evaluated using data from "The International Levee Handbook", (CIRIA 2013) as follows.

The critical velocity that would likely cause erosion of the levee crest was evaluated for the levee fill material. Two scenarios were considered (see Figure 10 below)

- Erosion of the gravelly levee fill material
- Erosion of the sandbags that form the crest of the levee (note: portions of the levees topped up by sandbags including both cementitious and non-cementitious sandbag fill materials which were considered separately)

Figure 10. (a) Levee Gravelly levee fill material (b) Sandbags that Form the Crest of the Levee

The following critical erosion velocities were estimated for the levee materials:

Levee material zone	Estimated critical erosion velocity (m/s)
Gravelly fill	1.5
Cementitious sandbag	1.5
Regular sand bags	0.5

The overtopping failure probabilities were assessed for the applicable flood and tide loading conditions. The outcomes of this assessment are summarized in Figure 11.

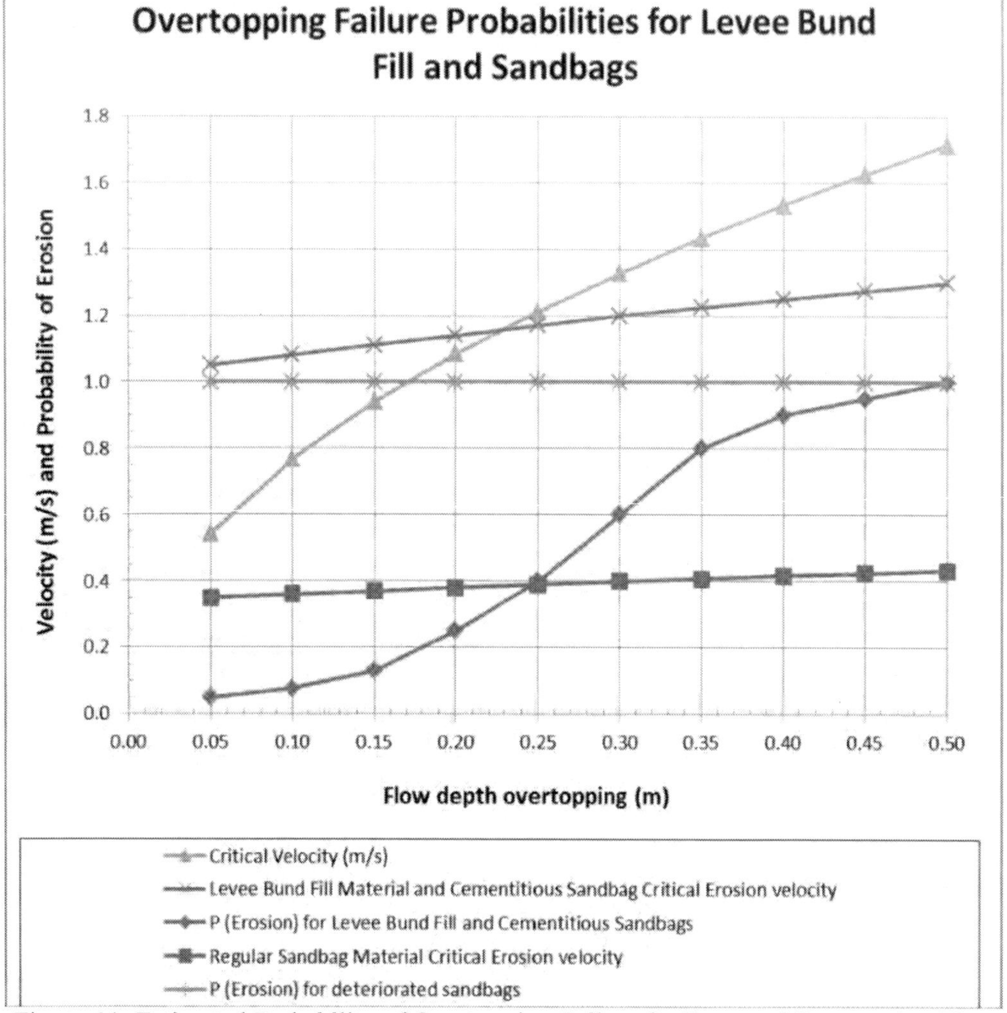

Figure 11. Estimated Probability of Overtopping Failure for Range of Overtopping Flow Depths

Piping

Piping through the embankment and the foundation was primarily assessed using the Piping Toolbox (USACE et al 2008) and Sellmeijer (2011). The evaluation of the piping failure modes were mostly based on the generic sequence of events presented in Figure 12.

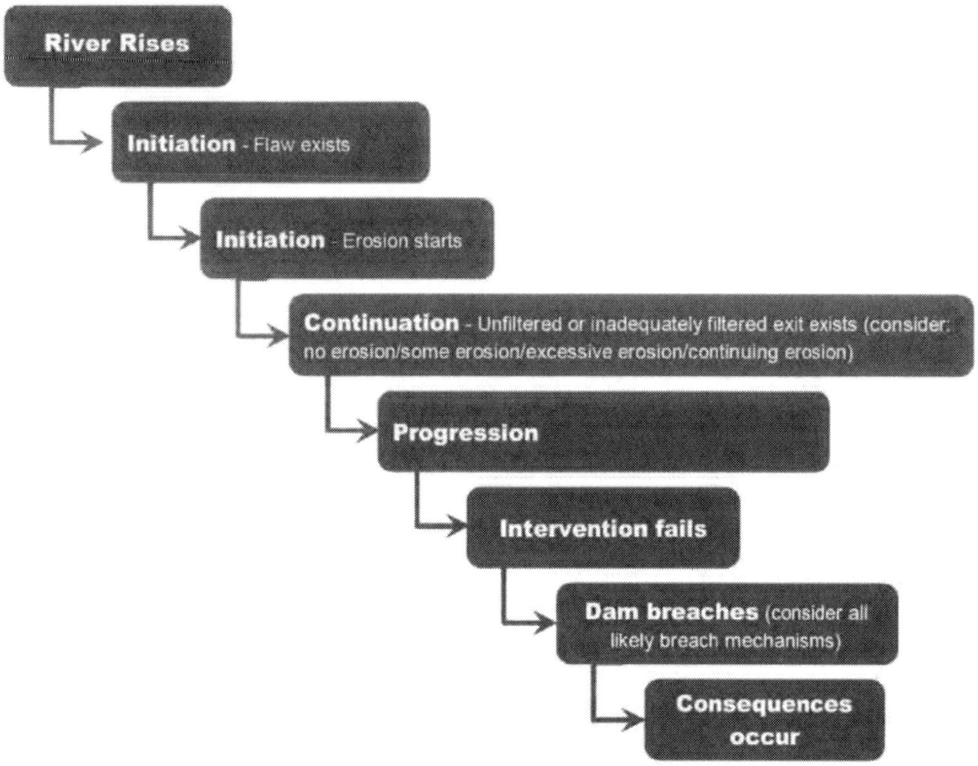

Figure 12. Generic sequence of events for piping failure modes analysis

Cracking The mechanisms for cracking that were assessed as applicable for the levees were:

- Cracking caused by differential foundation settlement
- Seismically induced cracking
- Cracking along defects adjacent to tree roots through the embankment and/or the foundation of the levees

The piping toolbox (USACE et al 2008) was used to estimate the size and depths of the cracks for the mechanisms listed above.

Hydraulic gradients The hydraulic gradients used to assess the likelihood of piping through the embankment were calculated for a range of partition levels. Following the seismic events, cracks were observed at various locations along the levee alignment on both the left and right banks. Transverse cracks were generally observed to be diagonal to the axis of the levee rather than perpendicular hence the seepage length was taken as three times the transverse width (perpendicular to the axis of the levee). The estimated piping initiation level was taken as the levee crest level after settlement (initiated by seismic loading) minus half of the original height of the levee. The estimation of hydraulic gradients for the piping assessment is shown schematically in Figure 13.

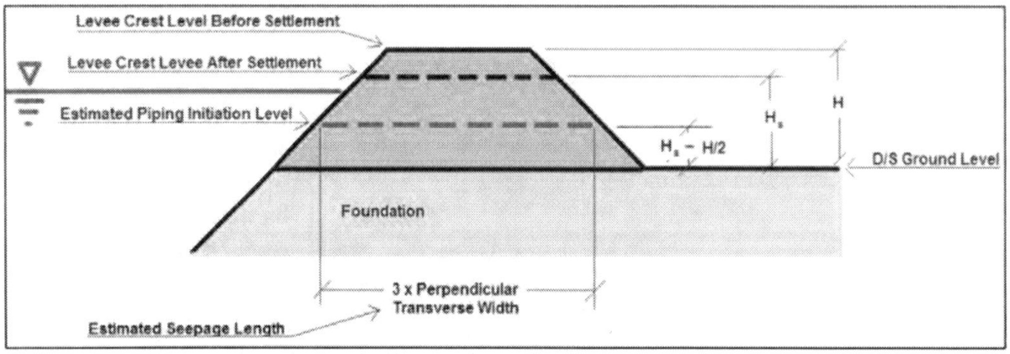

Figure 13. Schematic illustration of hydraulic gradient calculation

Piping initiation The probability of piping initiation was assessed using the Piping Toolbox (USACE et al 2008) and Sellmeijer (2011). For embankment piping mechanisms assessed using the piping toolbox, the probability of piping initiating in a crack through the embankment given an average hydraulic gradient was estimated for the cracks at various depths within the levees. Table 5.29 of the piping toolbox (USACE, 2008) was used to estimate the probability of erosion initiating for different seepage gradients under different sized crack widths:

Table 2. Estimation of initiation probabilities for ML or SM with <30% fines soil types (Adopted from Table 5.29 of the Piping Toolbox)

Estimated Crack Width (mm)	Probability of initiation of erosion for different seepage gradients							
	Average Hydraulic Gradient							
	0	0.01	0.1	0.25	0.5	1	2	5
0	0	0	0	0	0	0	0	0
0.5	0	0.00005	0.025	0.1	0.3	0.475	0.5	0.5
1	0	0.0001	0.05	0.2	0.6	0.95	1	1
2	0	0.001	0.1	0.6	0.9	1	1	1
5	0	0.005	0.6	1	1	1	1	1
10	0	0.01	0.9	1	1	1	1	1
25	0	0.1	1	1	1	1	1	1
50	0	0.15	1	1	1	1	1	1
75	0	0.2	1	1	1	1	1	1

The probability of piping through the foundation sands was assessed using Sellmeijer (2011) by estimating a critical hydraulic gradient for piping. The general levee geometry and water levels used to estimate the critical hydraulic gradient required to initiate piping is shown in Figure 14. A summary of the estimated piping initiation probabilities is also shown in Figure 15.

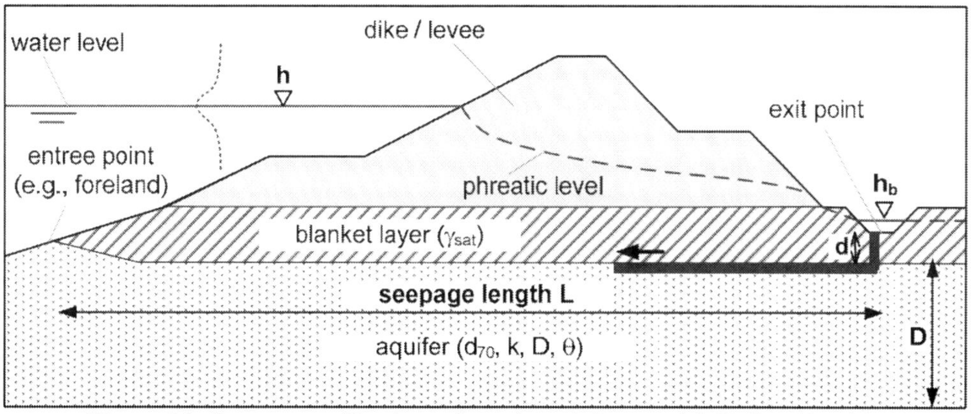

Figure 14. Geometry of backward erosion piping model (Sellmeijer, 2011)

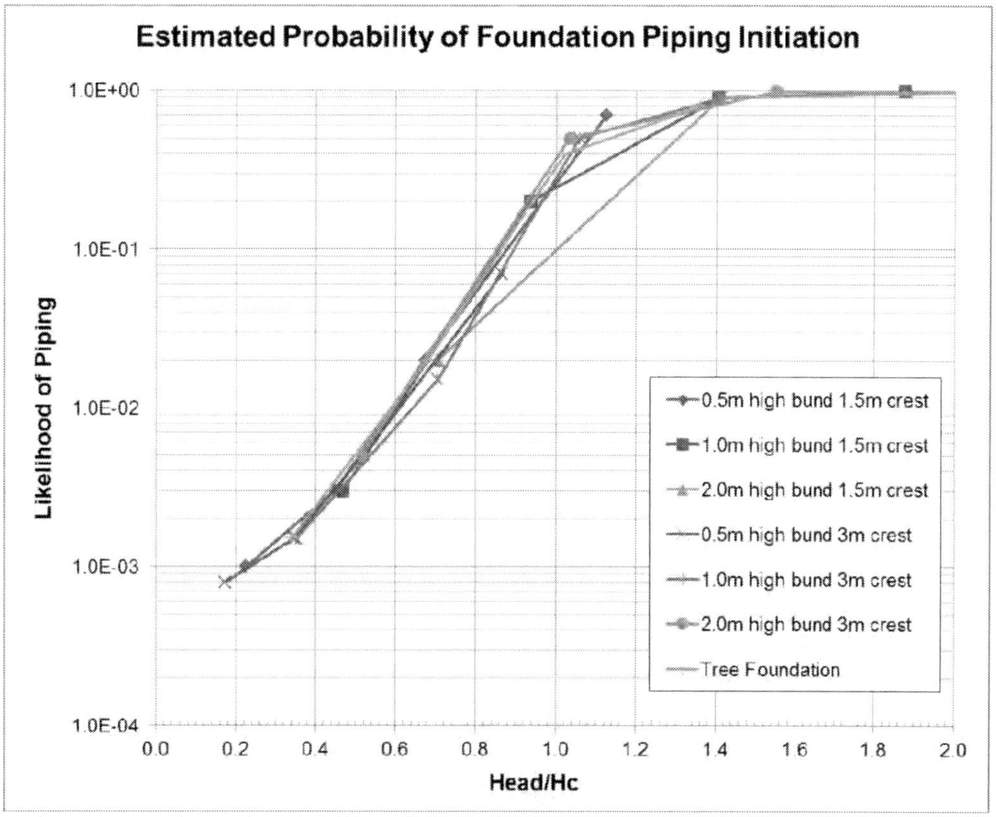

Figure 15. Estimated probability of piping initiating through the foundation sands for a range of levee geometries

<u>Piping continuation</u> Continuation is the phase where the relationship of the particle size distribution between the base (core or infill materials within the foundation) and the filter controls determines whether or not erosion will continue. No filter materials make up the

fill of the levee bunds and therefore, a probability of 1 was assigned to the occurrence of this event.

Piping progression Progression is the third phase of internal erosion, where hydraulic shear stresses within the eroding soil may or may not lead to the enlargement of the pipe. Increases of pore pressure and seepage occur in this phase. The main issues are whether the pipe will collapse and whether upstream zones may control the erosion process by flow limitation or crack filling. The likelihood of progression was evaluated using Table 11.1 of the Piping Toolbox copied below.

Table 11.1 – Probability of a soil being able to support a roof to an erosion pipe

Soil Classification	Percentage Fines	Plasticity of the Fines	Moisture Condition	Likelihood of Supporting a Roof
Clays, sandy clays (CL, CH, CL-CH)	> 50%	Plastic	Moist or saturated	1.0
ML or MH	>50%	Plastic or non-plastic	Moist or saturated	1.0
Sandy clays, Gravely clays, (SC, GC)	15% - 50%	Plastic	Moist or Saturated	1.0
Silty sands, Silty gravels, Silty sandy gravel (SM, GM)	> 15%	Non plastic	Moist Saturated	0.7 to 1.0 0.5 to 1.0
Granular soils with some cohesive fines (SC-SP, SC-SW, GC-GP, GC-GW)	5% to 15%	Plastic	Moist Saturated	0.5 to 1.0 0.2 to 0.5
Granular soils with some non plastic fines (SM-SP, SM-SW, GM-GP, GM-GW)	5% to 15%	Non plastic	Moist Saturated	0.05 to 0.1 0.02 to 0.05
Granular soils, (SP, SW, GP, GW)	< 5%	Non plastic Plastic	Moist and saturated Moist and saturated	0.0001 0.001 to 0.01

Notes: (1) Lower range of probabilities is for poorly compacted materials (i.e. not rolled), and upper bound for well compacted materials.
(2) Cemented materials give higher probabilities than indicated in the table. If soils are cemented, use the category that best describes the particular situation.

Piping Intervention fails Failure to intervene is the fourth phase of the failure pathway and this considers whether the internal erosion failure mechanism will be detected and whether intervention and repair will successfully stop the failure process. Given the rapid response to the previous seismic events, the likelihood of not intervening was taken to be 0.5 for the smaller seismic and flood events to 0.9 for the larger events.

Piping Related Breach Levee Breach is the final phase of internal erosion and the following four phenomena were considered:

- Gross enlargement of the pipe (which may include the development of a sinkhole from the pipe to the crest of the embankment).

- Slope instability of the downstream slope.

- Unravelling of the downstream face.

- Overtopping (e.g. due to settlement of the crest from suffusion and/or due to the formation of a sinkhole from a pipe in the embankment).

No differentiation was made with respect to the breach mechanism for the risk analysis, however, given the low height of the Levee and construction material, the most likely breach mechanism is expected to be sloughing or unravelling for which the likelihood was evaluated using Table 13.12 of the Piping Toolbox. This indicated that the Probability could have been between 0.1 to 1, depending on the amount of seepage likely to pass through the embankment zone. The probability of breach was therefore, taken to be 0.5 for the low flood events to 0.9 for the largest flood event.

RISK ANALYSIS OUTCOMES & CONCLUSIONS

Risk Analysis Outcomes of the Existing Levee Sections

The annual probability of failure was assessed for the 19 sections under consideration. The analysis considered a design life of 1, 5, 10 and 20 years as required by CCC. Annual probabilities were estimated combining the likelihood of the hazard loading (ie. Floods, Tides and Earthquakes) with the likelihood of the adverse response (ie. Piping, Overtopping, etc) given the hazard loading that occurs. Annual failure probabilities for the existing configuration of the 19 sections under consideration are shown in Figure 16 and Figure 17. The contribution of each failure mode to the sections under consideration was also assessed and plotted in Figure 18 to identify the major contributor to the risk. The risk analysis results clearly show that the risk was dominated by flood overtopping with the sections having sandbags contributing the highest proportion of the risk.

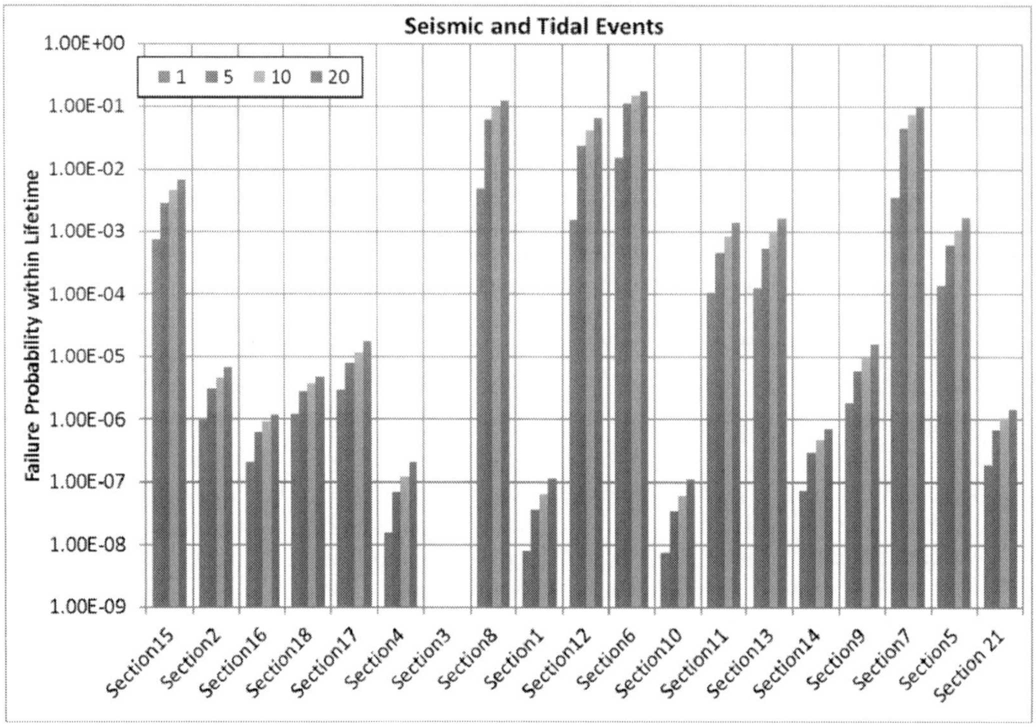

Figure 16. Avon levees Seismic and Tidal Events Total Probability of failure for sections within 1, 5, 10, 20 year lifetimes

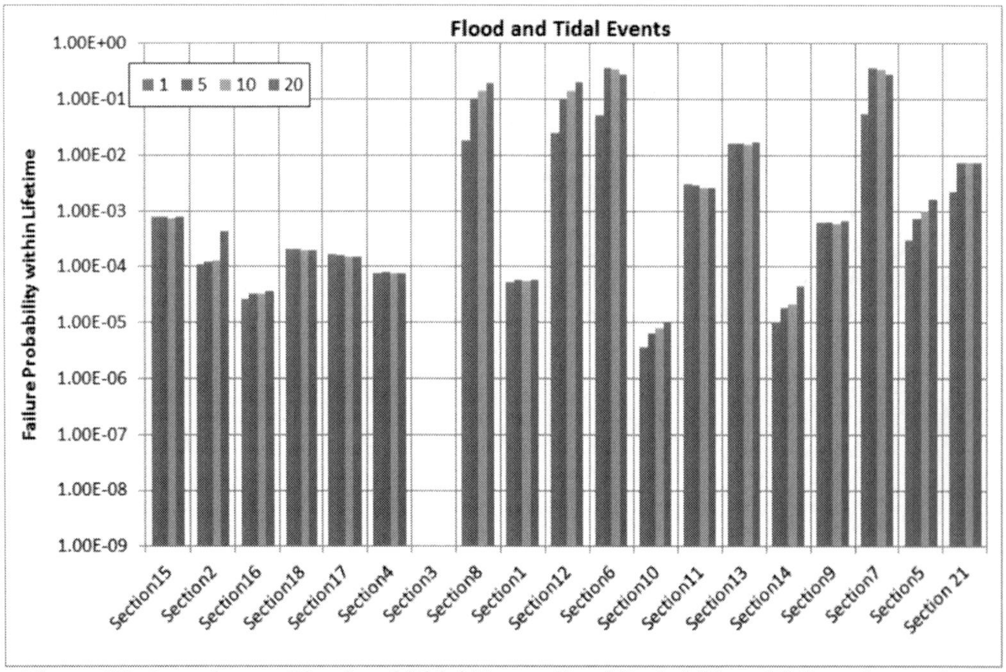

Figure 17. Avon levees Flood and Tidal Events Total Probability of failure for sections within 1, 5, 10, 20 year lifetimes

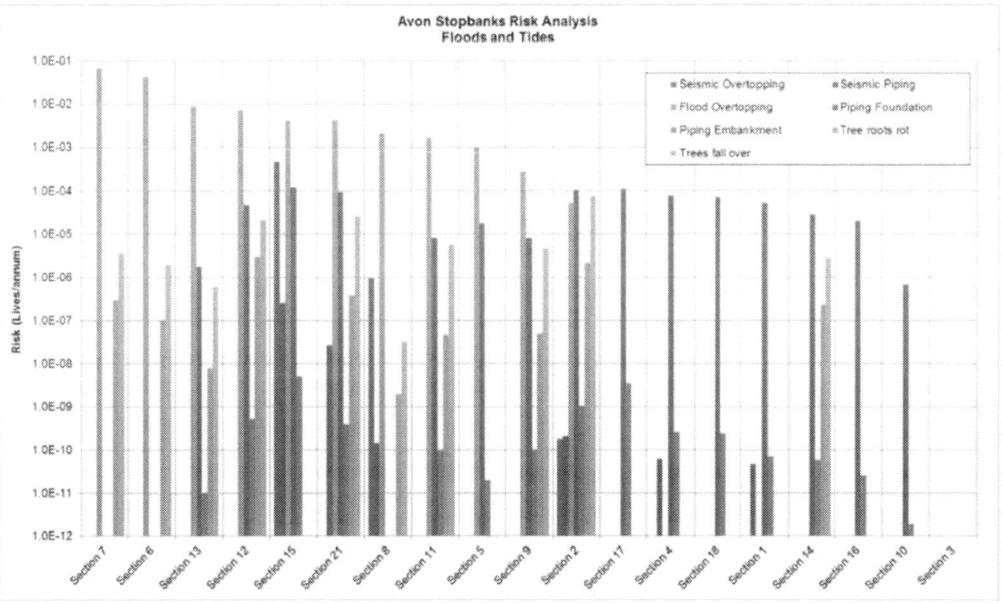

Figure 18. Avon levees Annual Risk (Lives/yr) for each failure mode and Section location for Floods and Seismic Events

Societal and Individual Risk Profile of Existing Levee Sections

The societal risk was calculated for the existing Stopbank and is presented in Figure 19. Seismic and flood events were not combined for the assessment of societal risk but were coupled with tidal loading. The results indicated that the risk for floods coupled with tides was above the tolerable limit for which upgrade works were required.

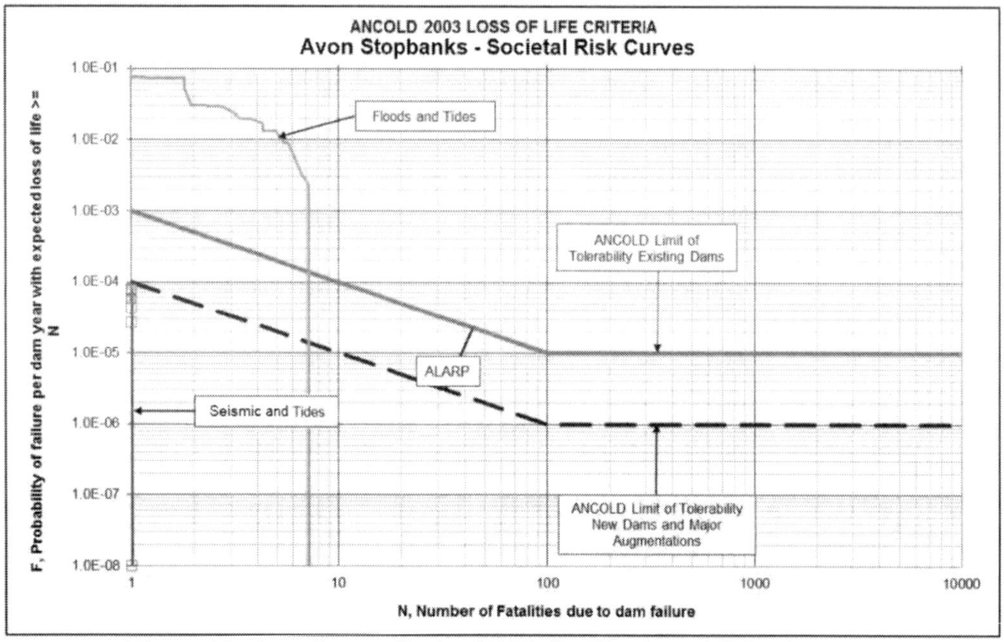

Figure 19. Avon levees Societal Risk for Floods and Seismic events

The Individual risk was calculated for each section under consideration, as shown in Figure 20, It was determined that Sections 6, 7, 8 and 12 were at or exceeded the ANCOLD limit of tolerability of 1E-4 lives/annum.

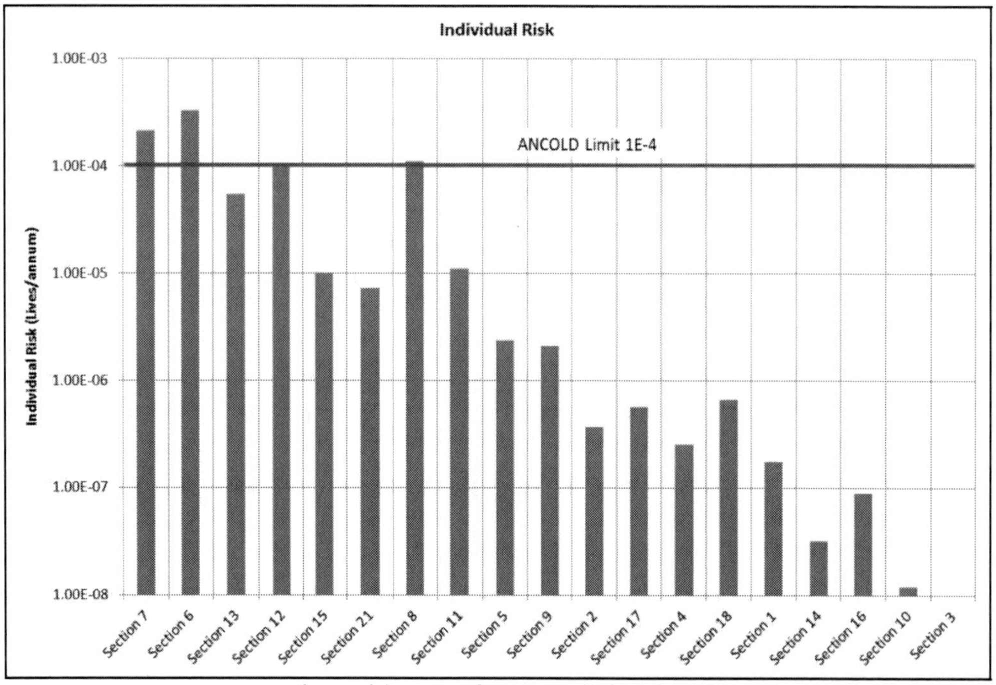

Figure 20. Avon levees Individual Risk

Levee Upgrade Options and Risk Reduction Measures

Given the highest risk was associated with floods overtopping the levees or tides overtopping the embankments following a seismic event, the most significant risk reduction could be achieved by raising the stopbanks floods up to the 1 in 200 AEP. The levee raise heights required to lower the risk to acceptable limits ranged from 40 mm to 460 mm. It was therefore recommended to raise the levees to at least these heights.

The resulting Societal risk after completion of the upgrade works is shown in Figure 21, which indicated that the risk was reduced to below the ANCOLD Tolerable limit and further upgrade works should be considered based on the ALARP principle. The Individual risk is also presented in Figure 22 showing that the risks were acceptable after upgrade.

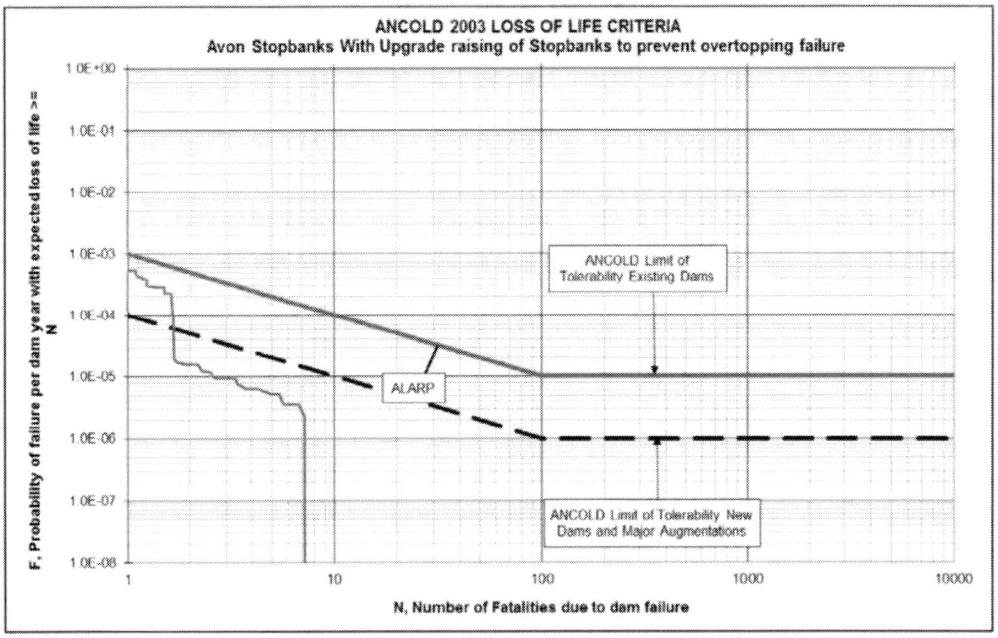

Figure 21. Avon levees Societal Risk after raising levees to prevent overtopping

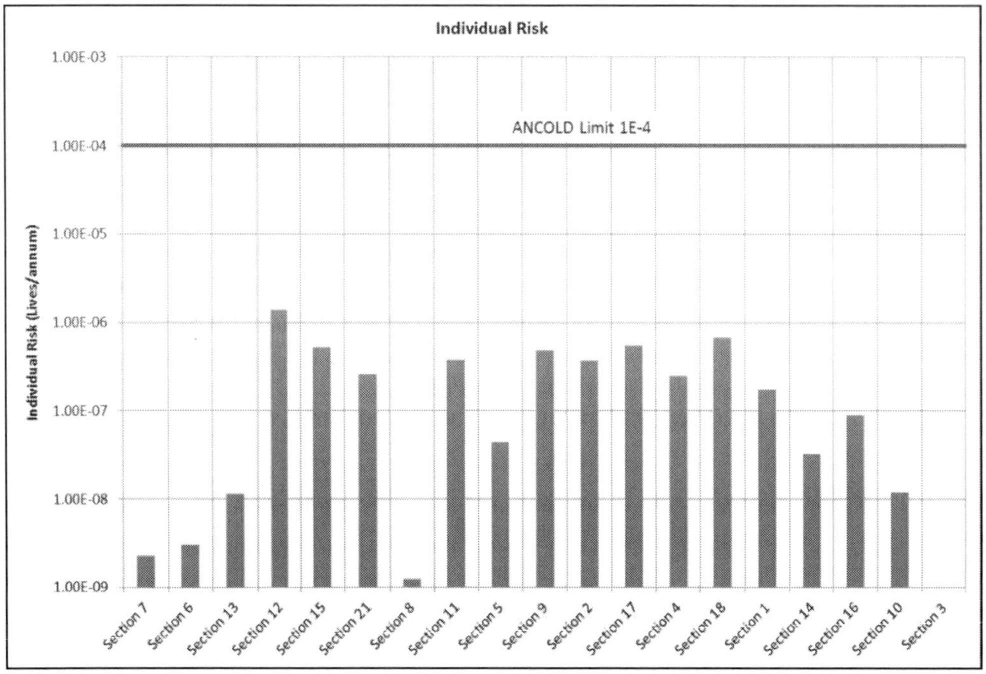

Figure 22. Avon levees Individual Risk for Floods, Seismic and Tidal events after raising levees to prevent overtopping

REFERENCES

Sellmeijer et al. (2011) – "Fine-tuning of the backward erosion piping model through small-scale, medium-scale and IJkdijk experiments", European Journal of Environmental and Civil Engineering, 15/8 2011: 1139-1154.

Stirling, M et al (2008), "Seismic Hazard of the Canterbury Region, New Zealand: New Earthquake Source Model and Methodology" – Bulletin of the New Zealand and Society for Earthquake Engineering, Vol. 41, No. 2

U.S. Department of the Interior Bureau of Reclamation (USBR), (2014), "RCEM – Reclamation Consequence Estimating Methodology – Interim – Guidelines for Estimating Life Loss for Dam Safety Risk Analysis"

USBR (August, 2008), "Risk Analysis for Dam Safety –A unified method for estimating probabilities of failure of embankment dams by internal erosion and piping (Piping Toolbox)

USBR (February, 2014), "RCEM – Reclamation Consequence Estimating Methodology"

VIBRATION FREE RESONANT SHEET PILE DRIVING IN DYKES AND LEVEES

Dane Reelie[1]
Matthew Janes[2]

Dams, dykes and levees are susceptible to ground vibrations and disturbance. Recent advances in high frequency or Resonant pile driving have proven that high production sheet piling using a vibrator is possible in sensitive dams and levees with imperceptible ground vibrations. The method uses tuned, high frequency vibration to resonate (treating the pile as an axial spring) and rapidly drive the pile into the ground. The method is proven to drive with zero ground vibration and zero ground disturbance, ensuring the integrity of the dam or levee is maintained. No flushing, predrilling or removal of soils is required. Case histories from sensitive dyke rehabilitation in Holland are presented where 18 m sheet piles are driven into sensitive, saturated dykes adjacent to fragile, historical masonry structures on shallow foundations. Measured ground vibrations are less than 1 mm per sec (<1/25 of an inch) as close as 2 m (6 ft). Spectral analysis shows all frequencies produce less than 0.2 mm/sec (1/100 of an inch) ground velocity. High productions result in reasonable costs and less disruption to levee occupants and the community.

[1] Resonance Technology International Inc., mj@rtii.ca
[2] Resonance Technology International Inc.

CONSTRUCTION OF BORINQUEN DAM 1E FOR THE PANAMA CANAL EXPANSION

Lelio Mejia[1]
James Toose[2]
Jorge Fernandez[3]

ABSTRACT

The recently completed Panama Canal Expansion Project required construction of a 6.7-km-long channel at the Pacific entrance to the Panama Canal, to provide navigation access from the new Post-Panamax locks to the Gaillard Cut section of the Canal. The new channel required construction of four dams, known as Borinquen Dams 1E, 2E, 1W, and 2W. The dams retain Gatun Lake, the main waterway of the Panama Canal, approximately 11 m above the level of Miraflores Lake and 27 m above the Pacific Ocean.

The largest of the dams, Dam 1E, is 2.4 km long and up to 32 m high. It abuts against Fabiana Hill at the dam's southern end, and against the original Pedro Miguel Locks at the northern end. This paper provides an overview of the construction of Dam 1E including the sequencing of the works, borrow of the embankment materials from required channel excavations and other sources, and key aspects of construction of the embankment, its foundation, and seepage cutoffs. The paper also describes the most salient design changes required during construction.

INTRODUCTION

The recently completed Panama Canal Expansion Project required construction of a 6.7-km-long channel at the Pacific entrance to the Panama Canal, to provide navigation access from the new Post-Panamax locks to the Gaillard Cut section of the Canal. The new channel required construction of four dams, known as Borinquen Dams 1E, 2E, 1W, and 2W. The dams retain Gatun Lake, the main waterway of the Panama Canal, approximately 11 m above the level of Miraflores Lake and 27 m above the Pacific Ocean. A view of the 5-km-long, northern portion of the new channel and of Dam 1E is shown in Figure 1.

The dams were designed as rockfill embankments with central impervious cores of residual and alluvial soils flanked by filter and drain zones of processed sands and gravels sourced from crushed rock (Mejia et al, 2011). Design and construction of the dams posed multiple challenges, including: 1) variable foundation conditions with occasional unpredictable weak features, 2) use of residual soils derived from rock weathering as core materials, 3) a wet tropical climate with a 4-month-long dry season, and 4) geologic faults across the dam foundations.

[1] Geosyntec, Oakland, CA, USA, LMejia@Geosyntec.com
[2] AECOM, Melbourne, Australia, James.Toose@aecom.com
[3] Autoridad del Canal de Panamá, Balboa, Panamá, JFernandez@pancanal.com

The largest of the dams, Dam 1E, was built as part of the fourth contract for construction of the Pacific Access Channel (PAC), which included excavation of 3.8 km of the new channel. The contract was awarded in January 2010 to a consortium of ICA of Mexico, FCC of Spain, and MECO of Costa Rica (CIFM). URS (now AECOM) was engaged by the Panama Canal Authority (Autoridad del Canal de Panamá, ACP) to assist ACP with inspection and design engineering services during construction of Dam 1E.

Figure 1. View of Pacific Access Channel (PAC) and Dam 1E, looking North

Construction of Dam 1E included the following main project elements: 1) erection of a 1.7-km-long, 19-m-high, cellular sheetpile cofferdam, 2) installation of a 30-m-long, 18-m-deep, triple-row, jet-grout cutoff wall, 3) construction of a 460-m-long, 18-m-deep, cement-bentonite slurry cutoff wall, 4) dewatering and excavation of the dam foundation, 5) treatment and geologic mapping of the foundation, 6) injection of a 2.4-km-long, double-row grout curtain, 7) placement of a 2.4-km-long, 5.3-million-cubic-meter, zoned rockfill embankment, 8) installation of performance monitoring instrumentation, and 9) construction of a 97-m-long, 26-m-deep, secant-pile wall to provide closure against the structure of the Pedro Miguel Locks.

This paper provides an overview of the construction of Dam 1E including the sequencing of the works, borrow of the embankment materials from the required channel excavations and other sources, and the key aspects of construction of the above project elements. The paper will also describe the most important design changes required during construction of the dam. A full description of the dam construction is provided by URS (2015a).

DAM LAYOUT

Borinquen Dam 1E was designed as a central earth core and rockfill embankment. To adequately control potential seepage and provide internal drainage, zones of chimney filters and drains flank the core. In addition, filter and drain blankets extend along the entire foundation of the outboard (downstream) rockfill shell (Mejia et al., 2011).

The dam is 2,420 m long and has a crest 30 m wide at elevation 32.30 m (project datum is close to mean sea level, MSL). The southern end of the dam abuts against Fabiana

Hill, and the north end is connected to the northern monolith of the Pedro Miguel Locks by a 97-m-long secant pile cutoff wall, known as the North Tie-In wall. Figure 2 shows the general arrangement of Dam 1E and the PAC. A typical cross section of the dam is shown in Figure 3. The as-built embankment volume is approximately 5,290,000 m^3.

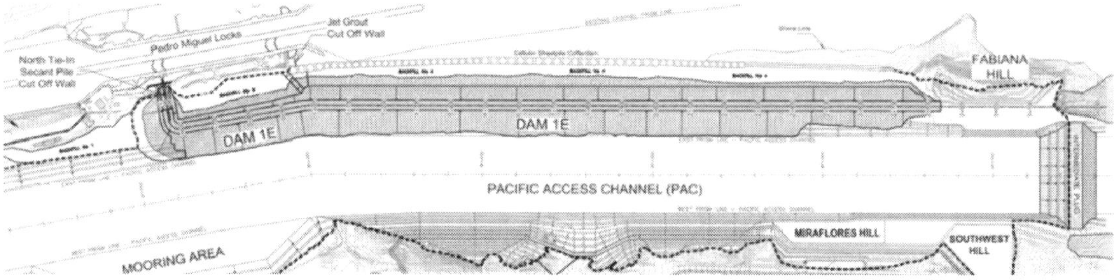

Figure 2. Layout of Borinquen Dam 1E and Pacific Access Channel (PAC)

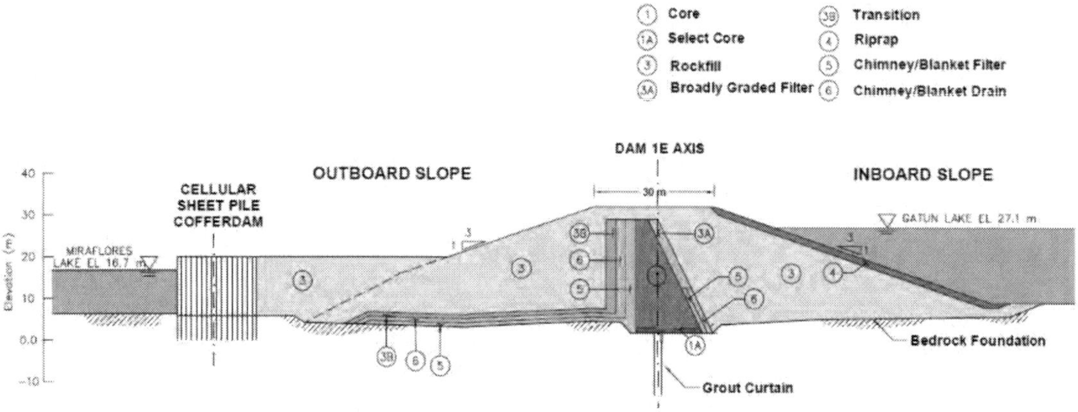

Figure 3. Typical Cross Section of Borinquen Dam 1E

CELLULAR COFFERDAM AND JET-GROUT WALL

A cellular sheetpile cofferdam and a jet-grout wall were constructed to isolate the construction area from Miraflores Lake, and allow dewatering of the PAC and dam foundation excavations. In addition, a cement-bentonite cutoff wall was installed to reinforce the existing Pedro Miguel Saddle dam, which retained Gatun Lake.

Cellular Sheet-Pile Cofferdam

The cofferdam extends along the outboard toe of Dam 1E from the southern wingwall of the Pedro Miguel Locks to Fabiana Hill (Figures 2 and 3). It consists of a 1,300-m-long segment of sheetpile cells and a 460-m-long segment of interlocked Z-section sheets. The latter segment closes through fill that extends to the foot of Fabiana Hill. The cells have diameters between 17.9 m and 21.8 m, and a height of 19 m. Circular cells are connected by intermediate arc cells of compatible diameters.

Construction of the cofferdam began in August 2010 with dredging of the foundation to remove lake-bottom sediments over a 1,100-m length of the alignment. The sheetpiles were aligned using a template, and were inserted progressively by self-weight, followed by driving with a vibratory hammer, in turn followed by a diesel impact hammer. The sheets were driven to their design tip elevation of -1.0 m or to refusal in rock. Unsuitable materials were removed from within the cells, and these were backfilled with free-draining gravel (Figure 4). After backfilling, the west face of the cofferdam was buttressed with a rock berm placed through water. The work was completed in May 2011. Further description of the cofferdam construction is presented in URS (2012).

Figure 4. Backfilling of cellular sheetpile segment of cofferdam

Jet-Grout Cutoff Wall

At the north end of the cofferdam (Figure 2), a jet-grout wall was installed between the cofferdam and the southern wing-wall of the Pedro Miguel Locks, to provide a cutoff to seepage from the lake. The cutoff consisted of a 30-m-long, 18-m-deep, triple-row, jet-grout wall. The inboard (upstream) row was constructed first, followed by the outboard row, in turn followed by the middle row. Three passes were made for each row to install primary columns followed by secondary columns. Grout was injected at a rate of 350 kg/m (controlled through the rotation and rise rate of the drill rod) at a pressure of approximately 350 bars.

Cement-Bentonite Cutoff Wall

The Pedro Miguel Dam closed the former gap between Paraiso Hill and the Pedro Miguel Locks to retain Gatun Lake. Because this dam was not designed for the water head difference expected during construction, a cutoff wall was required to control seepage and prevent piping through the dam, or its soil foundation, and subsequent loss of Gatun Lake. This cutoff wall is approximately 458 m long and 18 m deep.

The wall was excavated and constructed in panels using a clamshell. A cement-bentonite slurry was continuously pumped into each panel during excavation until the final depth was reached. The wall penetrated through the dam fill and underlying alluvial soils to terminate at least 1 m into La Boca Formation rock.

CHANNEL AND DAM FOUNDATION EXCAVATIONS

Excavation of the channel inboard of the dam began in June 2010 and blasting of rock started the following August. Following completion of the cellular cofferdam, the dam footprint was dewatered in mid-June 2011, and removal of muck from the former lake bed commenced shortly afterwards. Bulk excavation of the dam foundation started near the south end and progressed north. The foundation excavation was nearly complete by late 2012. Between 1 and 2 m of material were left above the foundation objective level for removal during final excavation. Figure 5 shows the dam foundation after bulk excavation and initial placement of rockfill on the inboard foundation. The volume of channel and foundation excavation totaled 26,989,000 m^3.

Figure 5. Dam 1E foundation with cellular cofferdam in foreground, PAC excavation in background, and rockfill shell on the inboard dam foundation

DAM EMBANKMENT MATERIALS

The dam embankment materials were sourced from the channel excavation and reserve borrow areas of residual soil. Additional volumes of residual soils, rockfill, and sound basalt for aggregate production were obtained from external sources. An additional 3,910,000 m^3 was obtained from sources outside the channel limits, as modifications to the contract.

Rockfill

Rockfill was sourced from the channel excavations, Miraflores Hill (Figure 2), and disposal stockpiles from previous PAC excavation contracts. The Basalt and Pedro Miguel Formations encountered in the channel yielded suitable rockfill (Zone 3 in Figure 3). The basalt was a medium hard to very hard, strong to very strong, massive or columnar jointed rock. The welded tuff and agglomerate facies of the Pedro Miguel Formation were also strong and sufficiently durable for use as rockfill. The fine-grained tuffaceous facies of that formation were found to be weak and subject to slaking under repeated cycles of wetting and drying, and were excluded from use in dam construction.

Approximately, 1,571,000 m^3 of rockfill was obtained from disposal stockpiles left by earlier PAC excavation contracts. Those stockpiles contained a mix of basalt and welded agglomerate rock. Rockfill from the stockpiles had higher fines content than rockfill from the channel blasting and excavation operations. The disposal stockpiles were also used for backfill between the dam embankment and the cofferdam (Figure 3).

Core Material

The core material (Zone 1) was sourced from alluvial and residual soils derived from the weathering of local bedrock. Alluvial sources included deposits encountered in the dam foundation excavation and in a reserve borrow area adjacent to the Cocoli River. The alluvial soils classified as clays, sandy clays, silts, and sandy silts of high plasticity (CH/MH). The natural water content of the alluvial soils was generally high, up to 12% above the standard optimum.

Residual soils of Pedro Miguel agglomerate and of basalt were sourced from the channel excavations and from the Miraflores Hill 2 borrow site, respectively. Residual soils from siltstones and sandstones of the La Boca Formation and from diorite were obtained from other borrow sites. The weathering profile in Miraflores Hill 2 was relatively shallow with a high proportion of cobbles, requiring selective excavation in areas of sufficient thickness and lateral extent. A source of La Boca Formation residual soil several meters thick was developed on the east side of the Canal, approximately 5.0 km north of the site.

The core material often classified as a sandy silt of high plasticity (MH), plotting just below the A line on the Casagrande plasticity chart. Liquid limits of the materials were in the range of 48 to 87, and plasticity indices were in the range of 14 to 49. The fines content (minus 75-micron particles) ranged from 33 to 92% and averaged about 71%.

The core soils were mined with excavators, hauled to stockpiles in 40-ton articulated dump trucks, and spread with bulldozers in 300-mm lifts. The alluvial borrow sources provided relatively homogeneous materials that required little mixing. The residual soil borrow sources were more variable and required mixing and removal of oversize rock. As the stockpiles were raised, they were sealed with a smooth drum roller, and slope-graded to minimize rainwater infiltration. The alluvial soils were typically at the upper limit of the water content specification and workability. The in-situ water content of the residual soils was often drier than the minimum specified compaction water content, requiring addition of water in the stockpile.

Ten test fills were completed on the core materials, one for each borrow source, to demonstrate compliance with the specifications. They consisted of a minimum of three lifts for compaction and testing. The water content of the lifts was progressively increased to find the upper limit at which the minimum specified undrained shear strength of the compacted material could be achieved.

Filter and Drain Material

Filter and drain materials were produced by onsite crushing and screening of basalt excavated from the channel and were also imported from offsite sources. Up to seven crushing plants were mobilized for the on-site production of fine filter (Zone 5), drain (Zone 6), and transition (Zone 3A/B) materials. To meet the embankment demand, Zone 5 and Zone 6 materials were also obtained from offsite quarries.

Considerable difficulties were encountered in configuring and calibrating the onsite plants, and in selecting suitable feed materials to achieve specification compliance. Compliant production of Zone 5 materials began in January 2012, nearly a year after equipment mobilization. To facilitate production, the Contractor requested a few changes in the specified gradations, which were approved by the Engineer. The main changes consisted of increasing the maximum fines content from 3 to 5% in Zone 5, from 2 to 3% in Zone 6, and from 3 to 5% in Zone 3A.

To verify compliance with the specification, quality control testing was performed at the plant, at the stockpile, and within the placed lifts in the embankment. Final approval of material was based on gradations taken from material placed in the embankment. Aggregate production was completed in mid-June 2015.

FOUNDATION DEWATERING

Dewatering of the foundation excavation was accomplished by pumping from 43, 150-mm-diameter deep wells with tip elevations of -10 m. The wells were successful in drawing down the groundwater to at least 1 m below the base of the core and allowing preparation of the foundation. Groundwater levels were monitored with 50 standpipe piezometers distributed across the foundation. All wells were decommissioned by removal of the screens and tremie or pressure grouting of the holes.

FOUNDATION PREPARATION

Excavation within the core foundation commenced in May 2012 and continued until September 2014. The final excavation and cleaning of the foundation for the outboard Zone 5 started in August 2013 and continued until June 2015. During excavation of the trench for the core foundation, several sections of the outboard cut slope slumped. These failures required regrading the slope locally to inclinations between 2H:1V and 3H:1V to remove the failed masses and stabilize the slope. No failures occurred on the inboard slope. The slumps developed on adversely oriented features such as bedding shears or joints, which caused wedge failure of the trench walls (Figure 6).

Geologic Mapping

Dam 1E is underlain predominantly by Miocene-age sedimentary and volcanic rocks, including sandstone, siltstone, clay shale, basalt, agglomerate, and tuff (consolidated volcanic ash). The sedimentary rock is low-strength La Boca Formation consisting of interbedded, sandstone, siltstone, clay shale and conglomerate units and underlies the northern three-quarters of the dam foundation. Near the southern end, the La Boca Formation is overlain by Pedro Miguel Formation consisting of interbedded volcaniclastic sandstone with minor siltstone units (URS, 2015b).

Figure 6. Example failure of outboard core trench slope

The foundation geology was mapped to confirm compliance with the specification and to determine treatment requirements (Figure 7). After treatment and final cleaning, the foundation surface was inspected and approved for placement of embankment fill. Geologic mapping confirmed that the minimum rock strength and weathering objectives were achieved beneath the core and shoulders. Where geologic units at foundation grade did not meet the specified requirements (e.g. some weak lignite beds and sheared clay shale beds) foundation treatment was specified.

Figure 7. Geologic mapping of the dam core foundation

Design Changes

Excavation widening at fault crossings. The design of Dam 1E included provisions to widen the chimney filter and drain zones within 50 m of fault traces deemed capable of the design fault displacement. Based on the geologic investigations carried out for design, two branches (East and West) of the Pedro Miguel Fault capable of such displacement were mapped to cross the dam foundation during design (Mejia et al., 2011).

Geologic mapping during construction revealed multiple minor faults throughout the foundation (URS, 2015b). A prominent fault trace was encountered crossing the dam axis about 200 m north of where the West Branch of the Pedro Miguel Fault was mapped before construction (Schug et al., 2016). Two fault traces associated with the Pedro Miguel fault were encountered near where the East Branch ('Stewart' Strand) of the fault had been mapped at the foot of Fabiana Hill. Those faults were judged capable of the design fault displacement and the filter and drain widening provisions of the design were implemented in their vicinity. No other significant faults were encountered in the foundation that required widening of the core trench excavation.

PAC slope instability. Immediately after initial excavation, instability developed over a 100-m length of the PAC cut slope along the inboard toe of Dam 1E, about one-third of the dam length north of Fabiana Hill (between approximate Stations 2+080 and 2+180, Figure 2). The cut slope failed towards the floor of the channel leaving back-scarps extending approximately 5 m back from the top of the cut, close to the design location of the dam toe.

Subsequent field investigations showed that the slope was cut in weak, fissured rock with adverse dips associated with faulting. Investigation trenches indicated that the failures were translational slumps with movement on basal bedding planes with thin clay gouge, dipping up to about 30 degrees to the west (out of slope).

These geologic conditions were deemed to pose a significant instability hazard to the dam. Thus, the foundation excavation design was revised over a 550-m length (between Stations 1+900 and 2+450) by removing the sheared rock and excavating a wide bench 2 m below the level of the PAC floor along the inboard dam toe. The excavation was backfilled with compacted rockfill to become part of the inboard dam shoulder. The required bench width varied between 20 and 35 m, based on stability analyses using selectively fully-softened and residual strengths for various foundation rock features.

FOUNDATION GROUTING

The grout curtain extended the full length of the dam to a nominal depth of 15 m. The curtain consists of two rows of holes, 1.5 m inboard (Row B) and 1.5 m outboard (Row A), astride of the dam axis. The Row-A holes dip 70 degrees to the north whereas the Row-B holes dip to the south. A 3-m-deep concrete cutoff wall was installed along Row B. In addition, multiple rows of "stitch grout" holes were installed in shear zones across the dam axis (URS, 2015c).

Super primary holes were drilled at 24-m intervals along the curtain, to a 25-m depth below the foundation. Primary holes, spaced at 6 m on centers, and secondary holes (at the mid points) were mandatory, resulting in a maximum 3-m spacing between holes along both rows. The design provided for "split-spaced" holes to be triggered if the grout take in any of the mandatory holes exceeded a predetermined quantity.

The grout was specified to consist of Type III cement, API 13A grade bentonite, potable water, superplasticizer, and silica fume. Holes were grouted using an ascending stage split-space method. Grouting pressures of 28.3 kPa/m were used for depths less than 18 m, and of 45.2 kPa/m, for greater depths. As foundation conditions became better understood and survey data became available, grouting pressures were adjusted to manage ground movement. Reduced grouting pressures were used at the south abutment.

A closure criterion of 25 kg of cement injected per meter of grout stage was established initially for grout take. Based on verification testing, which indicated no discernible decrease in the effectiveness of the curtain, that criterion was revised to 35 kg/m, to improve production. Criteria such as communication between holes, surface leakage, and hole collapse were used to add split-spaced holes.

Grouting work started in late May 2012 with a 30-m-long test section. The curtain was completed in December 2014. Approximately 5950 holes were drilled and grouted, including stitch grouting, and 2,842,000 kg of cement were injected into the ground.

FOUNDATION TREATMENT AND CLEANING

Treatment of the core foundation was applied in two stages. To prepare fault and shear zones for stitch grouting, initial rock shaping and dental concrete were completed before grout curtain installation. The final treatment, performed after the grout curtain was completed, consisted of minor dental concrete, shotcrete, and slush grouting.

Cleaning of the foundation surface to remove loose and encrusted materials was achieved with compressed air and water. The fine grained, non-welded clay shale units within the La Boca formation deteriorated rapidly after exposure. This required trimming and cleaning of the foundation immediately before placement of embankment materials.

Concrete Cutoff Wall

Following initial cleaning and mapping of the core foundation, the concrete cutoff wall was installed along Row B of the grout curtain. The wall is 3 m deep and 0.6 m wide and was to extend initially along the length of the foundation in La Boca Formation. During foundation excavation, the rock mass of the Pedro Miguel Formation was observed to be extensively fractured and the wall was extended to the full length of the foundation.

The wall trench was excavated with a trenching machine with a 0.65-m-wide blade. Compressed air, water, and suction were used to remove rock fragments, water, and debris at the base of the trench prior to backfilling with concrete (Figure 8). Adversely oriented discontinuities (bedding, joints, faults) in the bedrock caused multiple blocks and wedges up to 2 m wide to collapse from the outboard wall at the top of the trench. These collapses were typically repaired with dental concrete.

Figure 8. Cleaning of concrete cutoff wall trench

Shear Zones

The core foundation contained numerous shear zones and faults ranging from less-than-50-mm-wide shear zones to sheared rock and gouge zones up to 1.5 m wide. The gouge materials typically consisted of low plasticity silt and sandy soil, and included up to 5-cm-thick seams of high plasticity clays and silts, in which polished surfaces could be separated by hand. The shear zones and faults were treated by over-excavating them to a depth three times their width, up to a maximum depth of 1.5 m. The features were cleaned with compressed air or water and backfilled with concrete.

Weak Beds

Beds of highly fractured, weak rock were encountered in the La Boca Formation, generally in the finer grained units (clay shales). The beds were often sheared through their full thickness with polished surfaces. Several beds of dark gray to black, weak, carbonaceous clay shale and siltstone, interbedded with lignite, were also encountered (Figure 9). Generally, the rock quality of these beds did not improve with depth.

To provide a suitable surface for placing compacted core, these weak and sheared beds were covered with a minimum thickness of 0.2 m of backfill concrete. Shotcrete was used on inclined fractured and weak rock surfaces, where regular concrete could not be vibrated in place.

Shoulder Foundation Treatment and Cleaning

Once suitable rock was exposed, the outboard shell foundation was cleaned by trimming with an excavator fitted with a smooth-edged bucket. Significant shear zones and faults were treated by over-excavating the sheared materials to a depth three times the width of the feature and backfilling the excavation with compacted filter material. The excavation depth was generally limited to 1.5 m for large features. Within the treatment zone, cleaning with compressed air was used to ensure a good contact between the filter material and potentially erodible materials within the shear zone.

Figure 9. Cleaning of lignite bed for placement of backfill concrete

EMBANKMENT CONSTRUCTION

Construction Sequence

Because placement of the core could not be started until the nearby grout curtain was completed, the inboard shell was placed up to a height of 17 m in advance of the core, over a significant length of the dam. Once grouting was completed, core placement commenced in July 2013 at the northern end of the dam, and progressed south. A second core placement front was started in December 2013 near the south end (Figure 10). The outboard filter blanket was started in August 2013 in two simultaneous fronts, north and south. Embankment construction was completed in July 2015.

The chimney filter zones were generally placed ahead of the adjacent core and rockfill zones. After placing a lift of filter material, a lift of core was placed in between the chimney filter zones. Upon placement of the core lift, the adjacent rockfill zones were placed and compacted. However, often the rockfill zones were advanced one lift above the adjacent filters and pushed against them, once the filter placement caught up.

Embankment Core

A zone of select core material (Zone 1A) up to 1 m thick was placed at the base of the core to provide a good contact with the foundation. Alluvial soil without rock fragments was typically used for this zone. A 300-mm-thick lift was placed first to avoid damaging the foundation. A 200-mm loose thickness was used for subsequent lifts within Zone 1A. The materials were compacted with 8 passes of a weighted front-end loader (Figure 11). In confined areas inaccessible to the loader, gasoline-powered hand tampers were used with a reduced lift thickness of 150 mm.

Figure 10. Dam embankment construction, looking north from Fabiana Hill

Figure 11. Compaction of core Zone 1A with front-end loader

The main body of the core (Zone 1) was placed in 225-mm-thick loose lifts. The earthfill was hauled to the embankment from the stockpiles in 40-ton articulated dump trucks, and was dumped and leveled to the required lift thickness by dozers. Each lift was disked (Figure 12) and compacted with a self-propelled, quadruple-drum, pad-foot compactor or with a rubber-tired, single-drum, pad-foot roller.

The specified upper limit on compaction water content was +12% above the standard optimum. However, the earthfill became effectively unworkable at a water content of approximately +8%, and required removal of lifts exhibiting deep rutting (close to or deeper than the thickness of the lift). Furthermore, at water contents higher than +8%, the minimum specified undrained shear strength of 75 kPa could not be achieved reliably.

Figure 12. Disking of Zone 1 core material

A large portion of Zone 1 had to be placed during the wet season, making control of water content critical. Compaction and sealing of lifts with a cross fall, prior to approaching rains, proved effective in limiting ponding of water and infiltration of rainfall on the core working surface. Despite these measures, more than 40,000 m^3 of core material had to be removed from the fill and returned to the stockpile.

Compaction control was achieved by quality control and assurance field testing and inspection. Compaction was indexed by measuring the undrained shear strength of each lift using a field shear vane in accordance with standard ASTM D2573 (Figure 13). This proved to be a rapid and reliable quality control test method. Periodically, field density tests were also performed as part of the quality assurance program, using the sand replacement method (ASTM D1556) at the same locations tested with the shear vane.

The in-place average water content of Zone 1 was +5.7% above the standard optimum. The average vane shear strength was 114 kPa. The relative compaction of the material averaged 93.2% (ranging between 85 and 99%) of the standard maximum dry density (ASTM D698).

<u>Filter and Drain Zones</u>

The chimney filter, drain, and transition zones flanking the core were widened from their design width of 2.5 m to 4.5 m within 50 m of the main fault crossings described above. Depending on access constraints, the materials were placed by one of the following methods: a) spreading with an excavator, b) end-dumping from haul trucks, and c) delivery by telescoping belt and tremmie hose (Figure 14). Once spread, the materials were leveled with excavators, bulldozers, or handheld screeds, and compacted by 4 passes of a 12-ton, smooth-drum, vibratory, self-propelled roller. Ample water was added to the materials during placement and spreading.

Figure 13. Vane shear testing of compacted core material

Protective mats were placed to allow haul trucks to cross over the chimney filters and drains. These methods proved effective at minimizing contamination and breakdown of the materials. Traffic over the blankets was avoided to the extent possible.

Figure 14. Placing filter material with telescoping belt and tremmie hose

Rockfill

Shoulder rockfill was placed by end-dumping from large trucks on an advancing front and by spreading out and leveling the material in 0.9-m lifts with a bulldozer. Water was added with water cannons or water trucks during spreading (Figure 15). The rockfill was compacted by 6 passes of a 12-ton, smooth-drum, vibratory, self-propelled roller. Eleven in-situ density and gradation tests were completed in the rockfill. An average dry density of 2,310 kg/m^3 was measured using standard ASTM D5030.

Figure 15. End-dumping, spreading, and watering of shoulder rockfill

Riprap

Riprap (Zone 4) was placed as protection against erosion on the inboard face of the embankment. Riprap was sourced by selectively stockpiling columnar basalt excavated from the PAC, Miraflores Hill, and Southwest Hill. The material was placed using a 30-ton excavator with a general-purpose bucket (Figure 16). The rockfill was trimmed to the design slope to allow accurate placement of the riprap over the trimmed face.

Figure 16. Riprap placement on trimmed face of rockfill

NORTH TIE-IN CUTOFF WALL

The North Tie-In Cutoff Wall completes the Dam 1E water barrier between the dam embankment and the Pedro Miguel Locks. The wall is a plastic concrete secant-pile structure, 97 m long and up to 26 m deep. It penetrates at least 1 m into sound rock of the La Boca Formation, and overlaps vertically with the grout curtain to form a seepage cutoff through the backfill of the locks structure and the underlying bedrock.

The design required that the secant-pile layout provide a minimum wall thickness of 0.6 m. The Contractor selected a pile diameter of 1.4 m with a center-to-center spacing of 0.86 m, resulting in an overlap thickness of 1.1 m. This layout provided ample margin to meet the specified minimum wall thickness, while allowing for potential deviation of the piles from vertical.

Plastic concrete with an unconfined compressive strength between 750 and 1400 kPa, and a maximum permeability of $5x10^{-7}$ cm/sec at 28 days, was specified as the wall material. The Contractor carried out a laboratory test program to select mix proportions that would meet the specified strength, permeability, and slump requirements. The selected mix consisted of 437 kg of water, 29 kg of bentonite, 140 kg of cement, and 1340 kg of aggregate per cubic meter. This mix yielded a 28-day compressive strength between 1080 and 1100 kPa, and a permeability between $3.6x10^{-7}$ and $5.1x10^{-7}$ cm/sec.

A drill rig equipped with a cutter-tooth auger bit and a hydraulic oscillator was used to drill and case the pile holes. The oscillator successfully drove temporary steel casing to support the holes in the cobble and boulder backfill of the Pedro Miguel Locks and the auger excavated the materials out (Figure 17). The shaft bottom was carefully cleaned with a trap-door bucket auger.

Figure 17. Excavation of secant piles

A mobile concrete plant was used to batch the plastic concrete and dispatch the materials into concrete agitator trucks that finished mixing the batch components before tremie-placement in the piles. During pile production, the batch plant was not able to maintain calibration, creating variability in the strength and permeability of the concrete. An increased frequency of quality control testing was introduced and piles that failed to meet the minimum specified strength of 750 kPa and other requirements were replaced. Fourteen piles out of 112 were replaced for strength, permeability or other deficiencies.

Closure of the secant pile wall against the concrete monolith of the locks was made by two closure shafts drilled against the monolith. One of the shafts was re-drilled to provide a broad contact with the monolith (Figure 18). The gap between the closure shafts and the monolith wall was intended to be closed with three 850-mm-OD connection shafts. The connection shafts were cored to intersect each other and to overlap the monolith wall and the closure shafts. Camera inspection showed that the north closure shaft had deviated from the monolith wall, and a fourth connection shaft was drilled to ensure a complete seal. The closure and connection shafts were backfilled with lean concrete.

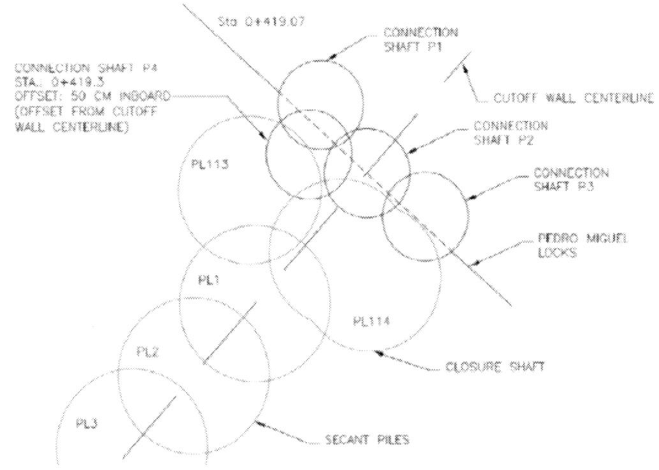

Figure 18. Layout of closure and connection shafts of secant-pile wall

Upon completion of the wall, the concrete guides were removed, the top of the wall was trimmed to the design elevation, and the embankment was completed over the wall alignment. The joints between piles were delineated at the surface by a thin smear of clay/bentonite around the perimeter of the secondary piles (Figure 19). The contact between piles and the integrity of the piles was verified at depth with drillholes. Core recovery and camera inspection of the holes showed a good contact between piles, and between the foundation rock and the base of the piles.

INSTRUMENTATION

The design called for installing instrumentation to monitor the performance of the dam as the embankment was completed, during filling of the PAC, and during long-term operation of the dam. The following instrumentation was installed as embankment construction progressed:

- Surface survey monuments at 50 m intervals along the inboard crest edge and the embankment downstream toe, to measure vertical and horizontal displacements.
- Inclinometers through the dam shoulders at three locations where significant bedding plane shears were observed in the foundation.
- Settlement sensors at two locations along the dam core, to measure settlement at three levels within the core.
- Vibrating wire piezometers at regular intervals along the dam, to measure water pressures in the foundation beneath the outboard shell and toe.
- Standpipe piezometers within the fill at the outboard toe, to measure water levels within the outboard backfill.
- Standpipe piezometers to measure groundwater levels within Fabiana hill near the location of the Pedro Miguel Fault treated with stitch grouting.
- Accelerographs on the dam crest, at the Pedro Miguel Locks monolith, and at the Fabiana hill abutment, to record future earthquake accelerations.

Figure 19. Trimming the top of the cutoff secant-pile wall

The settlement sensors and vibrating wire piezometers were routed to terminals on the embankment crest. An Automated Data Acquisition System (ADAS) was installed to allow automated reading of the instruments and wireless data transfer to ACP's offices.

CHANNEL FILLING

The north tie-in secant pile wall was the final element of the dam to be completed. After the last piles in the wall were built and allowed to cure sufficiently, filling of the PAC commenced. Filling lasted 101 days including a holding period of 28 days with the water level at elevation 22.0 m. The reservoir water level reached the design elevation of 27.10 m on 16 November 2015, without incident. The instruments were read and the data were reviewed daily during filling. The measured response of the dam to reservoir filling was well within design expectations.

CONCLUSIONS

Borinquen Dam 1E is a critical component of the Panama Canal and is vital to the Canal's operation. This paper presents an overview of the construction of the dam, including the chronology and sequencing of the works, borrow of embankment materials from required excavations and other sources, and key aspects of construction of the embankment, its foundation, and seepage cutoffs. The most salient design changes required during construction were also described.

Construction of the dam was a complex undertaking involving multiple and diverse project elements. Several challenges were encountered during the work, such as: weak foundation conditions including cross-cutting geologic faults, core materials of variable residual and alluvial soils, and a wet tropical climate with a short dry season. These challenges were met and the dam was constructed successfully through: a) close collaboration between the Contractor (CIFM), the Owner and Construction Manager (ACP), and the Design Engineer (URS), b) diligent inspection and quality assurance testing of the works, c) flexibility in the design concept, d) diligent investigation and solution of unanticipated issues, and e) timely decision-making.

The reservoir was filled without incident and the dam has been performing satisfactorily since it was completed in September 2015. The response of the dam to reservoir filling and its performance under the permanent embankment and reservoir loads is consistent with the design assumptions and has been well within expectations.

ACKNOWLEDGEMENTS

The Authors thank the Autoridad del Canal de Panamá (ACP) for permission to publish this article. They would also like to acknowledge the large team of professionals from ACP and URS (the first author's former employer) who participated in the construction of Borinquen Dam 1E and who worked diligently to ensure its successful completion.

REFERENCES

Mejia, L., Roadifer, J., Forrest, M., Abrego, A., DePuy, M., 2011, Design of the dams of the Panama Canal Expansion, Proceedings of U.S. Society on Dams (USSD) Annual Conference, San Diego, California, April.

Schug, D., et al., 2016, Fault investigations during Borinquen Dam 1E Construction, Panama Canal Expansion, Proceedings of U.S. Society on Dams (USSD) Annual Conference, Denver, Colorado, April.

URS, 2012, Review of Cofferdam QA/QC Records and As-Built Records, Report Prepared for Autoridad del Canal de Panamá (ACP), February.

URS, 2015a, Panama Canal Expansion Program, PAC-4 Borinquen Dam 1E, Final Construction Report, Report Prepared for ACP, December.

URS, 2015b, Panama Canal Expansion Program, PAC-4 Borinquen Dam 1E, Final Geology Report, Report Prepared for ACP, November.

URS, 2015c, Panama Canal Expansion Program, PAC-4 Borinquen Dam 1E, Final Report - Foundation Drilling and Grouting, Report Prepared for ACP, December.

TSAILE DAM - A COMPLETE APPROACH UNEARTHS A SAFER DAM

Stephen Brown[1]

Tsaile Dam is located in northeastern Arizona on the Navajo Nation, near Chinle, Arizona. Numerous previous evaluations and studies had been performed to develop the design modifications for Tsaile Dam, a 60-foot high zoned embankment dam, to address the following significant dam safety deficiencies for this high hazard dam:

- Internal erosion caused by a deteriorated CMP outlet works that included multiple holes along the conduit.
- Excessive seepage and potential liquefaction of alluvial foundation soils.
- Potential poorly compacted embankment at steep abutment rock contacts.
- Inadequate dam access during spillway operation.

Instead of a conventional rehabilitation that could include enlargement of the downstream stability berm, slip-lining of the CMP outlet conduit with an HDPE pipe, and installation of a 4-foot wide pedestrian bridge across the spillway to provide access to the dam, a comprehensive remedy was developed to completely eliminate the potential failure modes, as follows:

- Complete removal of the dam embankment down to bedrock, and replacement with a zoned earthfill section to address seepage and seismic issues;
- Complete removal of the CMP outlet conduit, intake and outlet structure, and replacement with a new outlet concrete-encased ductile iron conduit, concrete intake tower, and energy dissipator that are all founded on rock; and
- Installation of concrete box culverts under BIA Road 8077 where it crosses the spillway channel to provide safer dam access by equipment and safety personnel during flood events, and the subsequent elimination of the proposed spillway pedestrian bridge.

During construction, 20-foot high overhanging rock cliffs were revealed in the foundation, large erosion holes were found adjacent to the CMP outlet pipe, and extensive alluvium was found on the original stream bed. These features could only have been identified and fully addressed by the complete reconstruction of the dam, which was completed in 2015.

[1] Kleinfelder, sbrown@kleinfelder.com

REPLACEMENT OF CLEAR LAKE DAM: OVERCOMING CHALLENGES CONSTRUCTING AN RCC DAM ON A SOIL FOUNDATION

Steve Jamieson, P.E.[1]
Bruce Cotie[2]
Don Lopez, P.E.[3]
John Treacy, P.E. [4]

ABSTRACT

The replacement of Clear Lake Dam involved unique design and construction challenges. The original 110-year-old dam was a timber crib and embankment dam with a history of seepage and inadequate spillway issues. The dam is located above the Town of Georgetown, Colorado in a narrow, glaciated, mountain valley. The dam is a high hazard-potential dam regulated by both the Federal Energy Regulatory Commission and the Colorado Dam Safety Branch. The water impounded by Clear Lake Dam provides important water supply, hydroelectric generation and recreation benefits.

In 2011, sinkholes were observed on the upstream slope and significant leakage was discovered in the outlet works conduit. The dam owner voluntarily restricted the reservoir and developed interim operating guidelines to maintain the reservoir restriction while the design team developed a range of dam rehabilitation alternatives. The owner ultimately determined that the best alternative was to replace the dam with a new Roller Compacted Concrete (RCC) dam.

This paper summarizes the risk analysis process used during design and some of the key technical design features including:

- Use of a three-dimensional (3D), AutoCAD model to help visualize the construction of key dam features within the narrow dam site;
- Design of a robust filter to control seepage in the soil foundation;
- Unique foundation construction procedures to minimize settlement, cracking, and RCC dam contraction joint leakage; and
- Use of a 3D, Computational Fluid Dynamics (CFD) model to evaluate converging flows on the downstream slope during dam overtopping.

[1] President, W. W. Wheeler & Associates, Inc., 3700 South Inca Street, Englewood, CO 80110, steve.jamieson@wwwheeler.com.
[2] Plant Manager, HydroWest, Xcel Energy, 18201 West 10th Avenue, Golden, CO 80401, bruece.e.cotie@xcelenergy.com .
[3] Project Engineer, W. W. Wheeler & Associates, Inc., 3700 South Inca Street, Englewood, CO 80110, don.lopez@wwwheeler.com.
[4] Associate Engineer, W. W. Wheeler & Associates, Inc., 3700 South Inca Street, Englewood, CO 80110, john.treacy@wwwheeler.com.

INTRODUCTION

Clear Lake Dam is part of the Georgetown Hydroelectric Project located approximately 50 miles west of Denver, Colorado. The project is owned and operated by Public Service Company of Colorado (PSCo) doing business as Xcel Energy. The dam is regulated by the Colorado Division of Water Resources, State Engineer's Office, Dam Safety Branch (SEO) and the Federal Energy Regulatory Commission (FERC). The location of the project is shown on Figure No. 1.

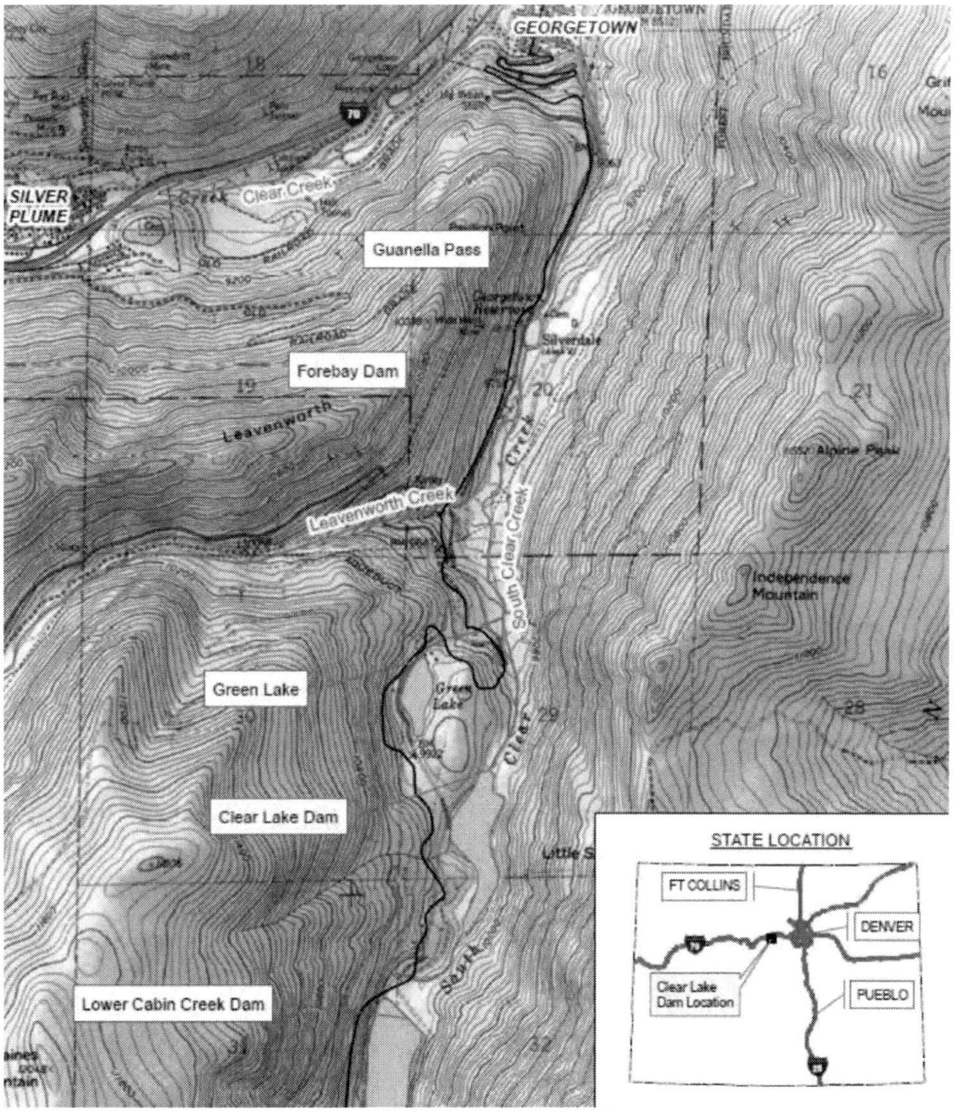

Figure 1. Project Location Map.

BACKGROUND INFORMATION

Project History

The original dam was a small, high hazard-potential embankment dam with a total active storage capacity of approximately 700 acre-feet. The dam was constructed in a narrow valley immediately downstream of a natural lake with a normal high water line at Elevation of 9875[5]. The original dam was constructed in 1902 as a timber crib dam and the dam has been enlarged and modified several times to address issues associated with foundation seepage, outlet works leakage, and inadequate spillway capacity. The dam was constructed on a foundation consisting of unstratified landslide debris, rockfall, and glacial till deposits composed of silts, sands, gravels, cobbles, and boulders. A photo of the old dam prior to demolition is provided on Figure No. 2.

Figure 2. Old Clear Lake Dam in 1999

Based upon a 1905 filing map, the original dam appears to have been about nine feet high. The outlet works was reportedly constructed at a depth of about 20 feet below natural ground (Hammer, 2002). There is very little documentation of most of the dam modifications, particularly the early modifications. Between 1903 and 1916, a concrete cutoff wall was constructed along the upstream face of the dam to reduce seepage. In 1936, a sheet pile facing was constructed along Figure 2. Old Clear Lake Dam in 1999

[5] All elevations are reported in feet above the 1929 vertical datum.

the upstream crest of the dam. In 1956, leakage through the dam had become so severe that the upstream slope of the dam was excavated and a new upstream sheet pile and concrete cutoff wall was constructed. In 1967, a new spillway was constructed and riprap facing was placed on the downstream slope of the dam. In 1996, PSCo constructed a reinforced concrete service spillway in the downstream slope, made additional outlet works modification, and constructed a grouted riprap overtopping protection system over the entire downstream slope and crest of the dam to act as the emergency spillway during the Probable Maximum Flood (PMF). The outlet works modifications included moving the control gates from a timber shored gate shaft in the dam to a new a new reinforced concrete gate tower constructed 80 feet upstream of the dam. During the reservoir filling after the 1996 modifications, cloudy discharge was observed flowing out of the left side of the downstream dam toe and vortices were observed in the reservoir, indicating potential leakage in the outlet conduit and voids under the grouted riprap. In 1997, PSCo lowered the reservoir and lined the upstream 50 feet and downstream 180 feet of the outlet works conduit with an insituform liner. A grouted soldier pile wall was also constructed in the left abutment to minimize seepage and the potential for sinkhole development in the left abutment.

In 2009, several piezometers were installed in the old dam. The subsequent piezometer data indicated that the 1997 soldier pile was not effective in controlling seepage. The piezometer data also indicated that the phreatic surface in the dam was generally about 10 feet higher in the left side of the dam than the right side of the dam. This higher left abutment phreatic surface was postulated to be the result of seepage flowing from Green Lake into the left abutment of Clear Lake Dam. As shown on Figure No. 1, Green Lake is located about 1,200 feet left of the left abutment of Clear Lake Dam and the water surface in Green Lake is about 100 feet higher in elevation than the foundation of Clear Lake Dam.

2011 Sinkholes

In early October of 2011, seven sinkholes were observed along the upstream slope of the dam and adjacent reservoir bottom after an early drawdown of Clear Lake. Additional investigations completed in October of 2011 confirmed that each sinkhole could convey significant seepage flows that ranged from 35 to more than 85 gallons per minute (gpm) (Wheeler, 2011). The sinkholes were all located within about 50 feet of the outlet works conduit. When dye was poured into the sinkholes, it was observed exiting in the downstream section of the outlet works conduit and downstream of the dam in less than an hour. This indicated a very direct connection from the sinkholes to exit points in the outlet conduit and immediately downstream of the dam. Internal video camera inspections also documented a significant leak into the outlet conduit occurring approximately 50 feet downstream of the gate tower. As a result of the identified sinkholes and outlet works leakage, PSCo voluntarily restricted the storage in Clear Lake Dam to Elevation 9840, which is the lowest elevation that water can be controlled by gravity in Clear Lake.

Existing Dam Potential Failure Modes

As part of the FERC Part 12D inspection process six potential failure modes were identified in the old dam and four of these potential failure modes were related to the effectiveness of the left abutment foundation cutoff (HDR, 2012). The six Potential Failure Modes identified at Clear Lake Dam are summarized in Table No. 1. Potential Failure Mode (PFM) Nos. 4 and 5 were particularly concerning to PSCo because both of these failure modes were categorized as Importance Category I in FERC's qualitative risk analysis system. PFM No. 4 was associated with foundation piping and PFM No. 5 was associated with piping along the outlet conduit. Both of these potential failure modes had high downstream consequences associated with the Population-at-Risk in the Town of Georgetown, located about three miles downstream of the dam.

Table 1 - Summary of 2012 FERC PFMs for the Old Clear Lake Dam

Potential Failure Mode No.	Consequence of Failure	Potential Failure Mode Description	Importance Category
PFM No. 1	High	Major flooding initiating erosion of unprotected areas of the upstream slope or abutments.	II
PFM No. 2	High	Major flooding initiating scour of areas downstream of the grouted riprap.	II
PFM No. 3	Moderate to High	Major flooding initiating excessive uplift and failure of sections of the grouted riprap	II
PFM No. 4	Highest	Foundation piping resulting in uncontrolled release of water or dam failure	I
PFM No. 5	Highest	Piping failure initiated along the outside of the outlet conduit	I
PFM No. 6	Low	Landslide activation that overtops the dam	IV

Alternatives Evaluation

Several Clear Lake Dam rehabilitation alternatives were considered by PSCo in 2012 to address the key PFMs summarized in Table No. 1 (Wheeler, 2012). Alternatives associated with other potential dam sites were eliminated from consideration based on permitting and licensing issues, as well as the need to maintain PSCo's existing senior water rights in Clear Lake. In accordance with Colorado Water Law, constructing the new dam more than 200 feet from the old dam could put PSCo's Clear Lake water rights in jeopardy. These water rights were valued at more than $25,000 per acre-foot. The Clear Lake Dam alternatives considered ranged from No Action to various levels of repair to address the dam seepage and outlet works leakage issues to complete dam removal and replacement.

PSCo selected the more expensive and more reliable dam replacement alternative because the water impounded by Clear Lake Dam has a significant senior water right value to PSCo beyond power generation for the Georgetown Hydroelectric Project and recreation. Water in Colorado is administered under the prior appropriation doctrine and an explanation of Colorado Water Law is outside of the scope of this paper. However, it is important to note that PSCo's dam replacement decision was based on the value of the Clear Lake water rights because water in Clear Lake can be used in conjunction with other PSCo water rights to provide a reliable alternative water supply for cooling water in PSCo's Denver Metro power plants, during drought periods. With increasing development and demand for water in the Clear Creek and South Platte River Basins, which flow through the Denver Metro area, the value of Clear Lake water storage is expected to appreciate and become more valuable in future years.

The selected dam replacement alternative was to remove the old dam and replace Clear Lake Dam with a new RCC dam supported on a layer of highly compacted select granular fill placed above the prepared foundation subgrade. The new RCC dam was designed to bear directly on the uniform layer of Select Fill placed to a minimum depth of six feet under the entire dam foundation. The Select Fill was mainly processed from on-site soils as a well graded, 3-inch-minus material to match the gradation of the existing foundation soils. Placement of the Select Fill layer between the dam and the foundation soils also provides more uniform support that limits differential settlement and cracking in the new dam. The selection of an RCC dam also allowed potential failure modes associated with leakage into or out of the outlet works to be eliminated because the new outlet works could be constructed entirely within the RCC section of the new dam. An RCC dam also allowed for safe dam overtopping of depths that approached 8 feet over the narrow, 214-foot-long, dam crest during the PMF inflow.

DESIGN OF THE NEW RCC DAM

The new RCC dam was designed to be constructed on essentially the same foundation footprint as the old dam and was designed to store water at the same Normal High Water line, Elevation 9875, as the old dam. Pertinent data for the new RCC dam are provided in Table No. 2. A three-dimensional drawing of the new dam is provided on Figure No. 3.

Table 2 - New Clear Lake Dam Pertinent Data

Dam Crest Elevation	9875.0
Top of Parapet Elevation	9877.0
Dam Crest Length (feet)	214
Service Spillway Width (feet)	80
Dam Crest Width (feet)	15
Downstream RCC Face	0.8H:1V
Outlet Works Inlet Elevation	9840.0
Stilling Basin Elevation	9850.0
Low Level Outlet Discharge Elevation	9835.0

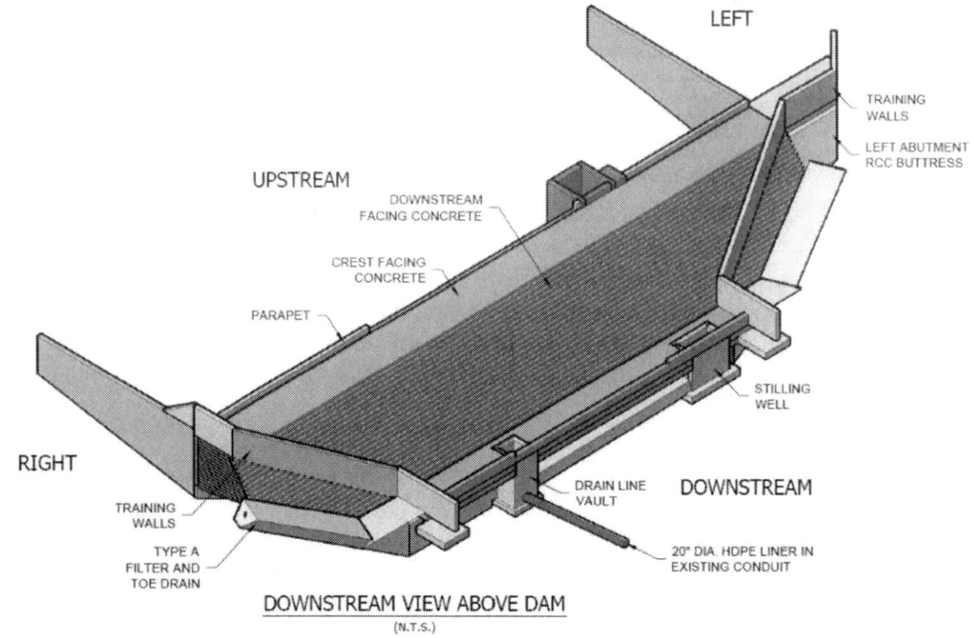

Figure 3. Three-dimensional Project Features Drawing

The three-dimensional drawing shown on Figure No. 3 was a very valuable tool that helped the design team, regulators and construction contractors understand how the dam would be constructed in this narrow dam site.

Spillway Design

The service spillway section is located in the center of the dam and includes an 80-foot-wide section between two-foot-high parapet walls constructed on the upstream edge of the dam crest. The parapet walls serve as the emergency spillway control section for dam overtopping during very extreme flood events. The dam crest is 15 feet wide with an ogee shape on the downstream edge of the crest. A 10-foot-long stilling basin with a three-foot-high end sill wall was provided at the downstream toe of the RCC dam to contain the hydraulic jump associated with the 100-year flood event. For flood events larger than the 100-year flood discharge, the tailwater begins to submerge the downstream face of the dam due to the narrow channel downstream of the dam. As a result, the energy from larger spillway discharges is dissipated in the high tailwater pool that will drown out the hydraulic jump. The spillways were designed to be overtopped by the local storm PMF which has a peak inflow of approximately 11,500 cubic feet per second (cfs), and a design discharge at the dam of approximately 10,500 cfs. This storm event would overtop the narrow dam length by nearly 8 feet with a unit discharge of about 50 cfs/foot. Nine-foot-high training walls were constructed at the abutment contacts to eliminate the potential for high velocity overtopping flows to erode the soil abutments of the dam. The training walls converge down from 214 feet apart at the downstream edge of the dam crest to 124 feet apart at the stilling basin at the downstream toe of the dam.

The Wheeler design team consulted with Dr. Hank Falvey and Alden Laboratories, Inc. (Alden) to perform a three-dimensional (3D) computational fluid dynamics (CFD) model of the overtopping flows. The 3D model was used to assess the effects of the converging flows that would overtop the dam during extreme precipitation events. The 3D CFD model showed that during the PMF, the abutment training walls on the downstream face of the dam are overtopped by up to 2.5 feet with localized velocities outside of the training walls at this discharge approaching 30 feet per second. To mitigate the potential failure mode associated erosion of the soil outside of the training walls during this extreme flood event, RCC buttresses were constructed outside of training walls to allow overtopping during extreme flood events without erosion occurring in the soil abutments. The total time that the abutment training walls is overtopped during the PMF is less than 2.5 hours. Although some surficial riprap downstream of the training walls may be moved if the training walls are ever overtopped, the RCC buttress backfill will stop any scour outside of the training walls.

Outlet Works and Gate Tower

The outlet works is located left of the service spillway section and the upstream gate tower is integral to the upstream face of the dam to be accessible by light maintenance vehicles from the dam crest.

The outlet works includes parallel 36-inch-diameter and 12-inch-diameter steel outlet conduits encased in reinforced structural concrete that convey flows through the dam to a stilling well, located in the stilling basin at the toe of the RCC dam. The inlet invert of both conduits is set at Elevation 9840. Flows lower than about 20 cfs are conveyed under the stilling basin from the stilling well to a drain vault through an 18-inch-diameter steel drain line. From the drain vault these routine outlet discharges are conveyed through an existing 24-inch-diameter outlet conduit to an existing meter vault located approximately 150-feet downstream of the dam. The existing outlet pipe was lined with an 18-inch-diameter HDPE pipe rather than replacing it because of the construction challenges associated with excavating and replacing this pipe with cuts of over 20 feet deep in the narrow valley downstream of the dam. Outlet works discharges in excess of about 20 cfs will rise upwards in the stilling well and discharge into the stilling basin. The 36-inch-diameter outlet conduit can be used to meet the Colorado SEO's drawdown criteria that requires lowering the top five feet of the reservoir in five days. The 12-inch-diameter outlet conduit is used to release minimum stream flows and most of the routine operational releases from Clear Lake.

The gate tower is equipped with a 48-inch-square, cast-iron, sluice gate that functions as a guard gate. The guard gate will be operated in either the fully open or fully closed position. A 36-inch-square, cast-iron control sluice gate is mounted on the upstream end of the 36-inch-diameter outlet conduit and a 12-inch-diameter knife gate is used to control discharges into the 12-inch-diameter conduit. The control gates are mounted on the downstream wall of the gate tower and the guard gate is mounted on the inside of the right wall of the gate tower to allow better access for maintenance.

RCC Dam Design

The minimum 90-day compressive strength of the RCC was required to be 2,500 psi. The RCC was completely encased on the upstream face, crest, and downstream face with structural concrete to provide long-term freeze-thaw protection and enhance durability during overtopping flows. A minimum 28-day compressive strength of 4,500 psi was required for the structural concrete.

An ultimate bearing capacity of 55,100 pounds per square foot (psf) was calculated for the RCC dam foundation, resulting in an allowable bearing capacity of 18,300 psf with a factor of safety of 3. A lower maximum allowable bearing pressure of 5,000 psf was used for design of the dam to minimize settlement. A typical section of the new RCC dam is provided on Figure No. 4. The stability and stress analysis indicated that the maximum stress on the soil foundation was below 3,200 psf after evaluating usual, unusual (flood), and extreme (seismic) loadings on the dam.

A simple thermal analysis was performed due to the relatively low ambient temperatures at the site and the relatively small size of the dam to determine the number of transverse contraction joints in the dam. The thermal analysis indicates that the conservative maximum differential temperature the dam RCC mass will experience will be approximately 65 degree Fahrenheit. Based on this net change in temperature the maximum calculated total crack width for the crest length of the dam was approximately 0.35 inches. Assuming a 0.15-inch average crack occurs at each formed joints, the number of joints required for the main section was three with a maximum spacing between joints of 70 feet. Water stop and joint sealant was provided in the upstream facing at each of the contraction joints and a unique HDPE liner detail was provided in the foundation of the dam under each contraction joint to prevent seepage associated with a crack up to 0.5 inches thick in the contraction joints. A detail of the HDPE liner is provided on Figure No. 5. This detail, coupled with the robust toe drain system described below, is intended to prevent soil from the foundation of the dam from moving through the RCC contraction joints.

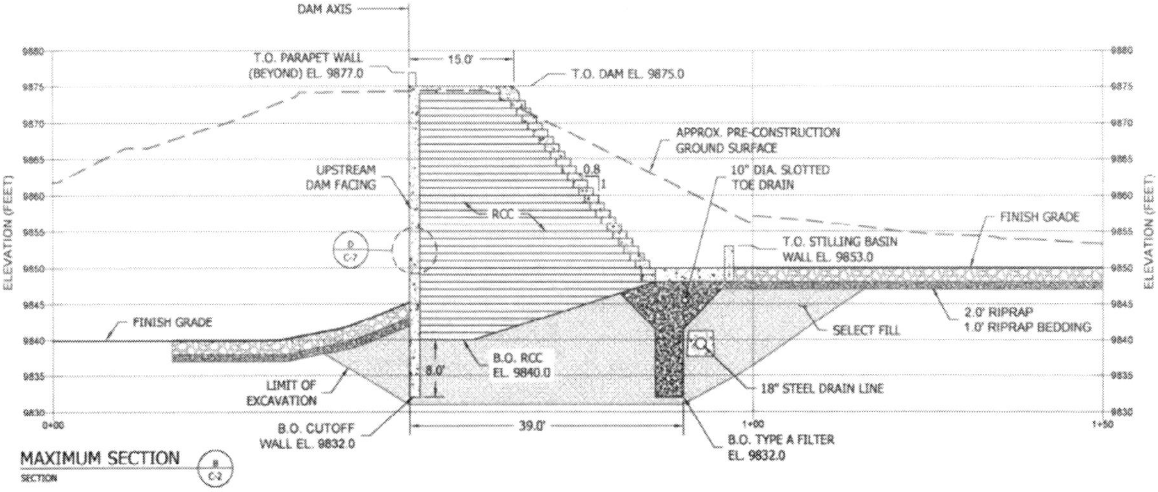

Figure 4. Typical Section of the New Clear Lake Dam

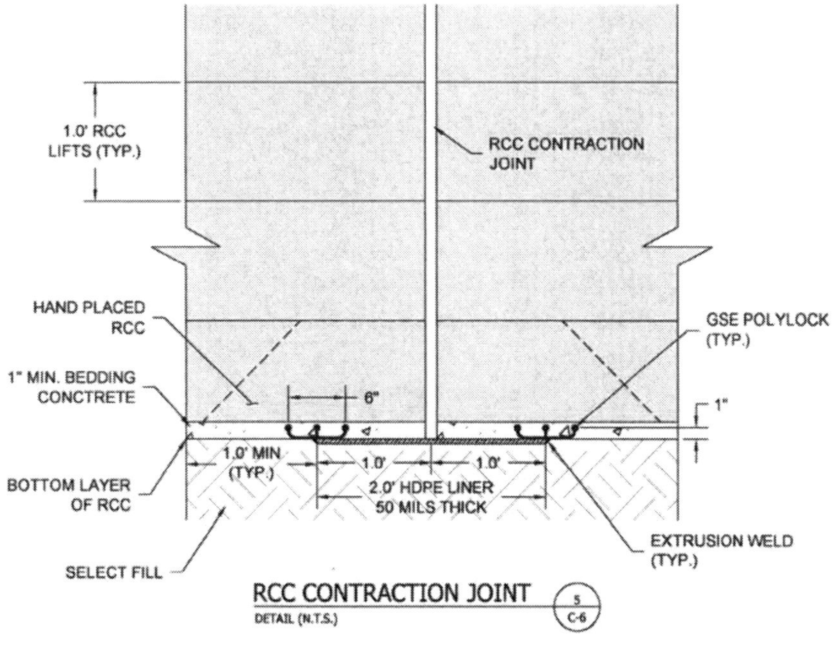

Figure 5. Foundation Seepage Prevention Detail at RCC Contraction Joints

A robust toe drain system was constructed along the downstream toe and in the abutments of the new dam. The toe drain system was designed to collect and filter seepage from the foundation under the dam as well as seepage in the abutments. The toe drain system consists of a commercially produced filter with a ten-inch-diameter slotted PVC collection pipe embedded within the filter material. The filter gradation was designed to be filter compatible with the Select Fill material that was used for the new dam foundation. Flow from the toe drain system discharges into two weir boxes located in the drain vault where the flow can be measured through two V-notch weirs that were constructed inside the drain vault. The toe drain is shown on Figure Nos. 3 and 4.

DESIGN AND CONSTRUCTION PFMA SUMMARY

As required by FERC, a Design and Construction PFMA workshop was held on June 18, 2013 for the Clear Lake Dam Replacement Project. The workshop used the 60 percent complete design documents. A key conclusion of the workshop was that the new RCC dam would eliminate concerns associated with piping and seepage through the old dam section and leakage through the outlet works, thus eliminating PFM Nos. 4 and 5 associated with the old dam. Some additional design details were identified during the 2013 PFMA workshop to address potential failure modes developed during the workshop for the new RCC dam. These additional details included:

- Constructing RCC buttress backfill outside of the training walls to prevent erosion associated with overtopping the training walls during the PMF.

- Including the detail provided on Figure No. 5 to prevent movement of foundation soils into RCC contraction joints.

- Confirmation that the 8-feet-deep upstream cut-off wall and 60-foot-long abutment cut-off walls add value to reduce potential failure modes associated with foundation and abutment seepage.

- Confirmation that the robust toe drain system constructed throughout the foundation and abutments of the dam will be effective in controlling seepage and minimizing the risk of potential failure modes associated with foundation piping.

Six key PFMs were identified for the new dam during the 2013 Design and Construction PFMA workshop. A summary of these PFMs and their associated importance categories is provided in Table No. 3. Key design issues identified in the workshop were addressed in the final design of the dam replacement. Five of the six identified PFMs were classified by Importance Category IV. A robust foundation filter and drainage system was included in the final design to address the one PFM associated with the potential for foundation piping. This PFM was classified as Category II/III in the 2013 Design and Construction PFMA report.

Table 3 - Summary of 2013 Design and Construction PFMA Workshop PFMs

Failure Mode No.	Consequence of Failure	Potential Failure Mode Description	Importance Category
PFM No. 1	Moderate to High	Seepage in the new dam foundation or abutments that removes fines in the soil matrix creating seepage channels that grow large enough to cause internal erosion and piping that causes a sliding failure due to destabilization of the dam section or a release of the active reservoir storage.	II/III
PFM No. 2	Low to High	A landslide occurs in the reservoir rim that is large enough to create a seiche wave that overtops the dam and empties the reservoir or erodes the new dam abutment support leading to a sliding failure of the dam section.	IV
PFM No. 3	Low to High	A joint in the new RCC section fails or there is a settlement crack in the dam section that is large enough to lead to piping in the dam foundation or abutments that exits above the dam's filter system. Undetected piping progresses in an undetected manner to the point that a sliding failure of the dam section occurs due to a destabilization of the dam section or a release of the active reservoir storage.	IV
PFM No. 4	Low to Moderate	A failure of one of the outlet works control gates occurs resulting in an uncontrolled release of the active storage of the reservoir through the outlet works of up to about 200 cubic feet per second (cfs). This uncontrolled release either drains the reservoir or causes property damage in downstream communities.	IV
PFM No. 5	Low to Moderate	Scour from large spillway discharges well in excess of 200 cfs resulting in head cutting in the discharge channel downstream of the dam that undermines the dam foundation leading to a sliding failure of the dam section.	IV
PFM No. 6	Low to Moderate	Significant movement of the abutment cut-off walls due to an extreme seismic event, landslide, or loss of soil that leads to piping in the abutments that opens a discharge tunnel that exits around or above the dam's filter system that empties the active storage in the reservoir.	IV

PROJECT CONSTRUCTION

The removal of the old dam and construction of the new dam occurred over three construction seasons in 2014, 2015 and 2016.

2014 Construction Season

During 2014, the contractor installed temporary bypass pumps and piping around the dam site and constructed a cofferdam upstream of the old dam. The contractor also started Select Fill and RCC Aggregate production and demolished the old dam to Elevation 9850.

Figure 6. Demolition of the Old Dam in 2014

2015 Construction Season

During the 2015 construction season, the dam was excavated to Elevation 9832 and foundation preparation was completed. The outlet works and gate tower was constructed and the bottom of the RCC and the foundation drain system was constructed to Elevation 9851 before cold weather shut-down the project on January 31, 2016.

Figure 7. 2015 Construction of the Bottom of the RCC Dam

2016 Construction Season

The RCC and structural concrete placements in the new dam were completed in 2016. The outlet gates, instrumentation, and other appurtenances were also completed in 2016 and the dam was considered to be substantially complete at the end of September. The State Dam Safety Branch provided a conditional first fill approval in October, 2016.

Figure 8. 2016 Completion of the RCC Dam

CONSTRUCTION CHALLENGES

Some of the construction challenges associated with removing the existing dam and constructing the replacement dam in a narrow high mountain valley are summarized below.

Access Challenges

Access to the project was from Guanella Pass Road. Access from the north is through the Town of Georgetown, and was limited to 40-foot trailer lengths through town followed by several tight switchbacks up the valley. Larger construction equipment, such as cranes, crushing equipment and batch plants had to be mobilized from the south, over Guanella Pass, which is closed from Thanksgiving through Memorial Day.

The construction site at the dam was very small with only about one acre located immediately upstream of the dam. This created several logistical challenges for the contractor, forcing the construction sequencing to follow more of a linear process than parallel processes. This also limited the amount of heavy construction equipment that could safely operate at the dam site.

Weather Challenges

Clear Lake Dam was constructed just below 10,000 feet in the Rocky Mountains. The dam in located on in a valley that runs from north to south with high mountains on the east and west sides. The typical construction season is considered to be June through October in any given year. Due to the orientation of the project, direct sunlight hours are limited at the dam.

Heavy snowfall in April and May of 2015 and 2016 led to higher and later spring runoffs than historical averages. The contractor was required to install and maintain more reservoir bypass pumping capacity for longer than was originally anticipated during the bidding process.

Geological Challenges

The dam is located on a soil foundation that has been described as "internally unstable." This means that the fine grained portion of the soil matrix is not filter compatible with the coarse portion of that same matrix. The fine portion tends to wash out and leaves nested clusters of rock behind.

During the 2015 construction season it was discovered that the downstream tail-water resulting from the reservoir bypass flows were flowing back into the upstream foundation excavation through the soils downstream of the dam. Several weeks were spent relocating reservoir bypass discharges further downstream before the foundation excavation could resume. In addition, a significant filter was added to the excavation slopes downstream of the dam to help mitigate potential washing of the fine portion of the foundation soils downstream during normal operations.

INITIAL DAM PERFORMANCE

The contractor substantially completed the project on September 29, 2016 and the State Dam Safety Branch provided a conditional fill right to Elevation 9870, 5 feet below the service spillway crest on October 4, 2016. As of the writing of the draft of this paper, the storage rights are out of priority, meaning that PSCo has not yet begun to store water according to the initial filling plan. Reservoir inflows are flowing through the outlet works. The water surface in the reservoir has been generally around Elevation 9843, or three feet above the inverts of the control gates in the gate tower. The downstream tailwater has been flowing at about Elevation 9842 at the discharge end of the low level outlet works. This appears to generally confirm the calculated one-foot of head loss through the outlet works that was calculated during the design of the new dam.

Piezometers

The reservoir has been drawn down to Elevation 9850 or lower for the last five years. There are 10 piezometers installed in the dam and abutments. The initial piezometer data indicates that there is elevated ground water in both abutments. The right abutment has a water table of about 9854, or 11-feet higher than the current reservoir Elevation of 9843. It has long been speculated that seepage from Green Lake, which is located just behind the hillside that forms the left abutment of Clear Lake Dam, is a significant contributor to the observed elevated ground water in the left abutment. This theory is confirmed by the piezometers in the left abutment that indicates that the ground water is at about Elevation 9850, or 7 feet higher than the current reservoir.

Two piezometers completed within the RCC dam indicate that there is water at about Elevation 9852. This is consistent with the elevations of minor seeps that have been observed exiting the downstream steps and the upstream face of the dam. The source of these seeps appears to be the ground water from both abutments.

Toe Drains

Two toe drains were constructed at the downstream toe of the RCC dam section and they both discharge into drain vault at Elevation 9845.3. During construction the left toe drain was initially observed to be flowing at about 60 gallons per minute (gpm). At that time Green Lake was at or near it's highest elevation as measured on it's staff gauge. At the completion of the Clear Lake Dam construction, discharges in the left toe drain had reduced to about 6.5 gpm as measured with a bucket and stopwatch. Green Lake's staff gauge was indicating that Green Lake was about five feet lower at that time.

Cracking

Five crack gages have been installed over the contraction joints in the RCC dam. Three gages are installed across the dam crest, one at each of the three main RCC dam contraction joints (CTJ). Only the center CTJ has been observed to crack through the dam. One gage is installed on the downstream steps on the center CTJ and the final gage is installed where the center CTJ crosses the downstream stilling basin. Additional gages will be installed if additional cracking is observed. The crack gage data indicates minimal cracking in the dam to date, with the only significant cracking observed at the center contraction joint. The maximum opening of this joint during the coldest part of the winter of 2016/2017 was measured at 9 mm (0.35 inches). The minimal cracking observed to date may be a result of constructing the RCC dam during the colder periods of the year, which limits thermal cracking. In addition, because the dam was constructed over two construction seasons the foundation settlement was allowed to occur in smaller increments over a longer period of time, which may have limited settlement cracking.

Constructing the RCC dam on a well compacted, uniform soil foundation may also contribute to reduced differential settlement and cracking in the dam because the soil foundation can be more easily shaped to reduce geometric stress concentrations in the dam foundation. Most RCC dams are constructed on rock foundations and essentially "adhered" to the rock with abutment and foundation bedding mortar. As the dams on rigid rock foundations expand and contract stress concentrations can develop in these foundation contact areas. It was postulated by some members of the design team and the Board of Consultants that this stress concentration due to expansion and contraction may be minimized on this soil foundation.

CONCLUSIONS

Key conclusions and lessons learned from the successful completion of this project are provided as follows:

1. It is possible to construct smaller RCC dams on soil foundations, but seepage control in the dam foundation and differential settlement control need to be carefully considered in the design.

2. It is common to design contraction joints in RCC dams to control the extent and location of cracking that will occur in any concrete dam. It is important to provide design features to minimize the potential for foundation piping into the contraction joints of the RCC dam. This issue was controlled at Clear Lake Dam with a unique limited HDPE liner located under the RCC contraction joints in combination with a robust filter and toe drain system constructed in the dam foundation and abutments.

3. The 3-D AutoCAD model of the dam and its key appurtenances was a very valuable tool to help explain how the dam was to be constructed to regulators and construction contractors. The CAD model also proved to be an important tool that expedited the 3D CFD modeling

4. The Construction Potential Failure Modes workshop that was held with FERC and Colorado SEO staff, the design team, and the owner was very valuable and resulted in the inclusion of several design details that reduced the risks associated with the new dam.

ACKNOWLEDGEMENTS

The authors gratefully acknowledge the contributions of some of the key project participants including:

1. Jim Talbot, Chairman, Board of Consultants
2. Ken Hansen, Board of Consultants
3. Dan Hertel, Board of Consultants
4. Larry Nuss, Board of Consultants
5. Jeremy Franz, Colorado Dam Safety Branch
6. Art Reis, Federal Energy Regulatory Commission
7. Sonny Buczek, Project Manager, Xcel Energy
8. Pat Martinez, Project Superintendent, Xcel Energy
9. Greg Monley, Project Geotechnical Engineer, Kumar & Associates, Inc.
10. Chris Jones, Project Geologist, Kumar & Associates, Inc.
11. David Blanchette, Project Structural Engineer, SM&RC Structural Engineers, Inc.
12. Hank Falvey, Hydraulics Consultant
13. Dan Gessler, 3D CFD modeling, Alden Research Laboratory, Inc.
14. Dustin Bennetts, Project Sponsor, ASI Constructors, Inc.
15. Mike Atwood, Project Manager, ASI

REFERENCES

1. Colorado Division of Water Resources (CO DWR, 2007) *Rules and Regulations for Dam Safety and Dam Construction*, Denver, CO, January 2007.

2. Federal Energy Regulatory Commission (FERC, 2010) *Engineering Guidelines for the Evaluation of Hydropower Projects*, 2010.

3. Greg Hammer (Hammer, 2002) *If it Ain't Broke, Don't Fix It*, Paper presented at Dam Safety 2002.

4. HDR Engineering, Inc. (HDR, 2012), *Georgetown Hydroelectric Project, Clear Lake Dam, FERC Project No. 2187-CO, Third Independent Consultant's Safety Inspection Report,* May 2012.

5. W. W. Wheeler and Associates, Inc. (Wheeler, 2011) *Clear Lake Dam Supplemental Evaluation of Left Abutment Grouted Soldier Pile Wall*, October 14, 2011.

6. W. W. Wheeler and Associates, Inc. (Wheeler, 2012) *Final Draft Clear Lake Dam Decision Report,* October 2012.

DEWATERING – HOW TO ADDRESS THIS POTENTIAL ACHILLES HEEL FOR CONSTRUCTION

John W. France P.E., D.GE, D.WRE [1]
G. Richard Bird, P.E.[2]
Gregg Batchelder Adams, P.E.[3]
Jessie Drayton Unterberg, P.E.[4]

ABSTRACT

Construction dewatering can be a critical aspect of dam construction or rehabilitation, both with regard to schedule and safety. A number of potential dewatering methods are available: sumps, well points, deep wells, eductor wells, etc. Which methods are applicable for which conditions? What are the limitations of the various methods? How do you design a construction dewatering system? This paper will address those questions. Another important question is should the engineer design the dewatering system or leave the system to be designed by the contractor? In the authors' opinion there is no universally correct answer to that question. This paper will review the benefits and drawbacks for each of these approaches to dewatering design and provide guidance on when to use each. The paper closes with brief discussions of example projects illustrating successful implementation of dewatering systems.

INTRODUCTION

For new dam construction, dewatering is often required for excavations into alluvial river bed soils. For dam rehabilitation, dewatering is sometimes required for excavations to construct seepage collection systems or appurtenant structure modifications, such as outlet works extensions. In both cases, successful dewatering is critical to construction schedules and construction quality. In the case of dam rehabilitation, dewatering can also be critical to dam safety if a full or significant reservoir pool is maintained during construction. In numerous cases, ineffective or inadequate dewatering has led to serious construction delays and increased cost.

Construction specifications often place the responsibility for dewatering on the construction contractor. This practice can be effective, but the authors suggest that the designer needs to take responsibility for guiding the contractor in the correct direction as well as for specifying performance requirements, so that dewatering is effectively addressed. Further, the authors argue that there are instances when it is reasonable for the designer, together with the owner, to take on the responsibility of actually designing the

[1] Vice President and Hydropower, Dams, and Levees Technical Practice Network National Leader, AECOM, 6200 South Quebec Street, Greenwood Village, CO, 80111, john.france@aecom.com
[2] Principal Geotechnical Engineer, AECOM, 1001 Highlands Plaza Drive West, St. Louis, MO, richard.bird@aecom.com
[3] Supervisory Civil Engineer, U.S. Army Corps of Engineers, Risk Management Center, 12596 West Bayaud Avenue, Lakewood, CO, gregg.batchelderadams@usace.army.mil
[4] Civil Engineer, Geotechnical Engineering, AECOM, 4225 Executive Square, Suite 1400, La Jolla, CA, 92037, jessie.drayton@aecom.com

dewatering system(s) and specifying installation methods. Depending on the approach to risk sharing among the parties to the construction, there may be significant cost implications (either positive or negative) when the contractor is involved in the design of the dewatering system.

This paper includes discussions of the following topics: 1) purposes of dewatering, 2) dewatering methods, 3) considerations for design of dewatering systems, and 4) considerations related to owner design versus contractor design of dewatering systems. The paper concludes with brief discussions of examples of successful dewatering systems for dam rehabilitation.

PURPOSES OF DEWATERING

The purposes of installing dewatering systems are to facilitate construction of subsurface structures and features below ambient groundwater levels by: 1) increasing excavation stability, 2) preventing heave, uplift, and blowout, and 3) preventing internal erosion of soils. Construction below ambient groundwater levels for new dams or dam rehabilitation should not be attempted without adequate control of groundwater and subsurface hydrostatic pressures.

A properly designed and implemented dewatering system will:

- Intercept seepage that would otherwise emerge from the slopes or at the bottoms of excavations.
- Increase stability of excavated slopes by lowering pore water pressures.
- Improve excavation and backfill characteristics of sandy soils.
- Reduce lateral loads on excavation support systems, if they are used.

Design of dewatering is not a precise science; rather, it is a combination of "art" and science. Trial and error and adaptation of the system in the field are often required. The originally designed and installed dewatering system is not always sufficient, and adjustments must be made based on observed performance. Success of a dewatering system is not dependent only on analysis, but also on practical experience and adaptation to actual conditions. Dewatering efforts should be directed by an appropriately experienced person. This person may not necessarily be an engineer. Superintendents and technicians employed by dewatering companies often have excellent practical experience that may be immensely valuable in implementing and adapting dewatering systems. It is recommended that during preparation of construction specifications the designer consult individuals with practical experience in dewatering.

DEWATERING METHODS

Methods available for dewatering include:

- Sumps and ditches.
- Wellpoints and high vacuum wellpoints.
- Eductor wells and wellpoints.

- Deep wells, sometimes sealed for application of vacuum.
- Drainage ditches with perforated collector pipes.
- Vertical drains.
- Electro-osmosis.
- Barrier walls and bottom seals.

Each of these methods is discussed briefly below.

Sumps and Ditches

Sumps and ditches are generally effective for limited dewatering in soil and rock strata that are not easily erodible, such as clean gravels and clean sandy gravels. Sumps and ditches will be most effective if there is limited or no recharge. This method is generally inadvisable for pervious and semi-pervious (silty sand, silt) soils with recharge where the groundwater must be lowered by more than a few feet. However, there have been some exceptions to the limitation of a few feet of drawdown for sumps and ditches. With highly engineered partially penetrating sheet pile cofferdams or other barriers, adequate interior berms to extend the seepage paths from the source of seepage to the sump / ditch system, maintenance of uniform excavation slopes, and placement of filters at and below exiting seepage on the excavation slopes, sumps and ditches have been used to lower groundwater tens of feet in relatively clean sand on some major projects (White and Prentis, 1950).

Sumps can range from relatively simple installations to more complex installations, as shown in Figure 1. Sumps must be designed and constructed to prevent internal erosion of surrounding soils. This requires properly sizing openings in the sump pit structures and may also require installation of properly graded filter and drainage aggregates (clean sands and gravels) around the sumps.

Limitations of sump and ditch installations include:

- Drainage can be relatively slow in sand and gravel soils with high fines contents.
- Wet conditions can remain during excavation and backfilling.
- Sumps and ditches require space within the excavation and can increase the excavation footprint.

Workers skilled in construction and operation of sumps and pumps are required to provide enough water control and prevent internal erosion.

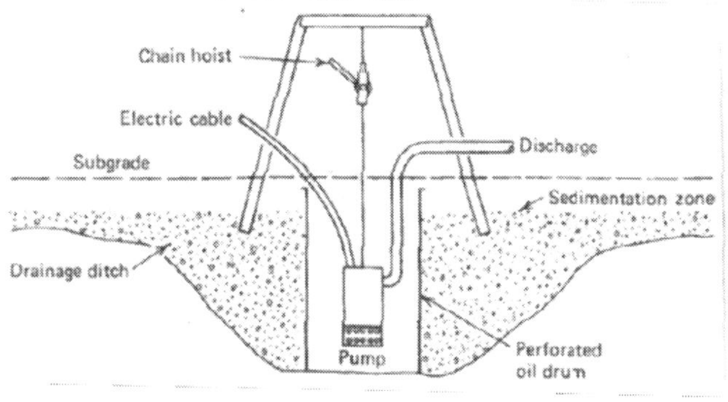

a) Simple sump

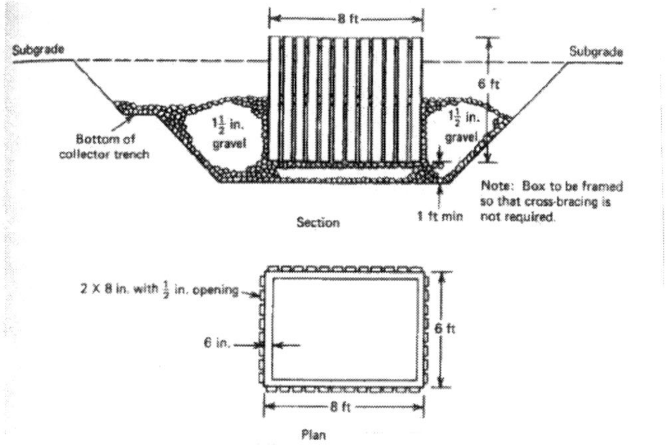

b) More complex sump

Figure 1. Example Sump Installations (Powers, 2007).

Wellpoints and High Vacuum Wellpoints

Wellpoints are well suited to pervious soils and consist of individual wellpoints in the ground connected to a pumping system on the ground surface. By definition, a dewatering wellpoint includes a drawdown pipe so that water entering the wellpoint screen above the bottom of the drawdown pipe is forced to flow downward in the annulus between the drawdown pipe and the wellpoint screen. Wellpoints are typically installed in rings or lines, with center to center spacing varying from 3 feet to 10 feet. Screens, meshes, and or geotextiles are included in each wellpoint to prevent infiltration of surrounding soils during pumping, and a select sand filter is placed in the annulus between the wellpoint and the drilled or jetted hole in which it is installed. Wellpoints are typically 6 inches or less in diameter. There are two types of wellpoints typically used for construction dewatering: vacuum wellpoints and eductor wellpoints.

Vacuum wellpoints operate by applying vacuum to the wellpoint from a pumping system at the ground surface. In an ideal wellpoint system, the vacuum is regulated such that the wellpoints either will not break suction or every wellpoint is installed at a depth (30 feet or more) below the pump suction such that breaking suction is not possible. High vacuum wellpoints, used only in very fine silty soils that do not readily drain by gravity, are designed to break suction in order to develop a partial vacuum in the sand filter around the wellpoint screen and riser pipe. To minimize air entry, the holes for the wellpoints are sealed around the riser pipes, preferably against a continuous clay stratum. Because high vacuum wellpoints are intended to break suction, providing adequate air handling capacity in the pumping system is crucial. If seals are ineffective, the most practical way to overcome excessive air entry is generally to increase the air handling capacity of the pumping system. An example of a high vacuum wellpoint installation is shown in Figure 2.

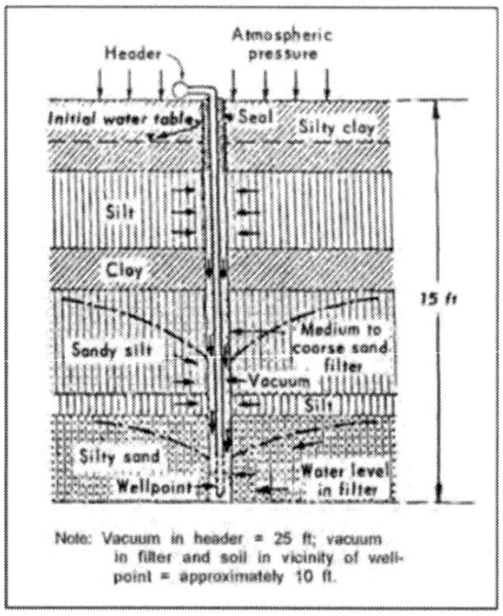

Figure 2. High Vacuum Wellpoint Installation in Silty Soils (Leonards, 1962).

The amount of drawdown of the groundwater level achievable by a vacuum wellpoint system is limited. The theoretical limit is atmospheric pressure minus the absolute lowest pressure the vacuum system can deliver. At sea level, atmospheric pressure is about 30 inches of mercury, and it decreases with increasing elevation. Typical vacuum pumps have an absolute pressure limit of about 5 inches of mercury, although some special pumps may achieve as low as 3 inches of mercury. Hence for a typical system, the theoretical limit of groundwater drawdown is about 25 inches of mercury or about 29.5 feet of water head. This absolute limit decreases by about 1 foot for each 1,000 feet of elevation. So, at 5,000 feet elevation the theoretical limit of drawdown is about 23.4 feet, and at 10,000 feet elevation it is about 18.5 feet. In practice, maximum drawdown depths are less, because of less than perfect operating systems. At sea level, groundwater drawdown with vacuum wellpoints is typically limited to between 20 and 25 feet.

These drawdown limits discussed in the previous paragraph are for drawdown at the wellpoint location. Because water flow creates a cone of depression around the wellpoint, drawdown between wellpoints and within the excavation will be less. Typically, the maximum achievable drawdown within an excavation for a vacuum wellpoint system is

15 feet or less. If drawdown greater than 15 feet is required, a multistage wellpoint system could be used, as illustrated in Figure 3. In multistage wellpoint systems, the upper stages typically are not pumped after lower stages are installed and pumped.

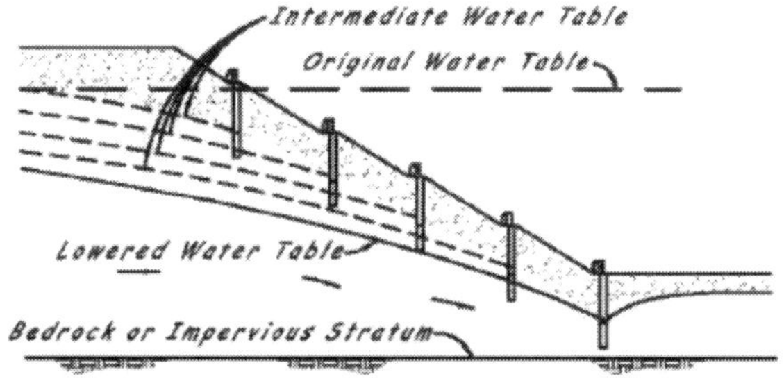

Figure 3. A Multistage Vacuum Wellpoint Installation (Terzaghi, Peck, and Mesri, 1996).

Eductor Wells and Wellpoints

Another alternative for groundwater drawdown greater than 15 feet is an eductor well, as shown in Figure 4, or an eductor wellpoint. An eductor wellpoint consists of a conventional dewatering wellpoint that is attached to the suction of a concentric pipe (as opposed to a twin pipe) eductor. With an eductor well or wellpoint, vacuum can be developed for the full length of the filter above the educator, if the eductor breaks suction *and* if the air flow to the filter is sufficiently low. Eductor systems can lower the groundwater level as much as 100 feet below the top of the excavation. Because eductor pumps are inefficient (typically 30 to 35 percent), these systems work best in situations where the volume of pumped water per wellpoint is less than about 2 gallons per minute. An additional drawback of an eductor system is that the air and water flow capacity is limited to the total capacity of the eductor. If too much air enters the filter or well screen, it will not be possible to maintain a reasonably high vacuum in the filter. An alternative to eductors when water flows are higher than 2 gallons per minute or when air flow is too high to maintain a vacuum in the filter is a sealed deep well system (discussed below), with or without applied vacuum.

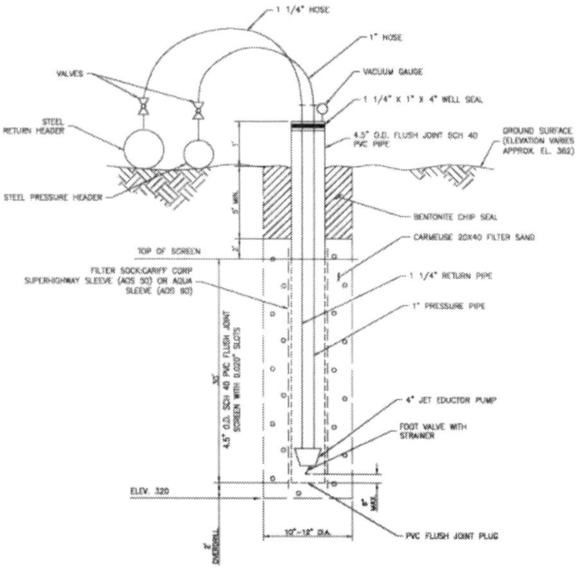

Figure 4. An eductor well (AECOM Technical Services)

Deep Wells

Deep wells consist of well casings extending below foundation bottom excavation with submersible or lineshaft turbine pumps installed inside the well casings, as shown in Figure 5. The deep wells need to be large enough in diameter to accommodate pumps that can provide sufficient capacity, which will be a function of the specific subsurface conditions at the site.

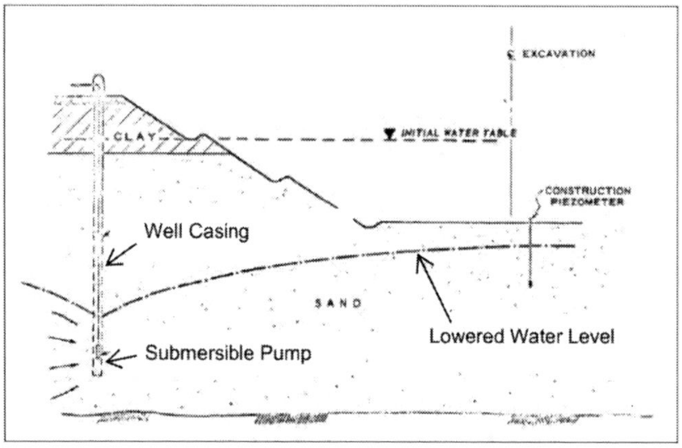

Figure 5. A Deep Well Installation (Adapted from Terzaghi, Peck, and Mesri, 1996).

Deep wells are usually most suitable in moderate to high permeability soils, where the cone of depression is relatively flat and the effect of the individual deep well is felt at significance distance from the well location. Large flows can be accommodated using high capacity pumps. In the right conditions, deep wells can lower the groundwater hundreds of feet. Deep wells installed around the periphery of an excavation can provide pre-drainage for the full depth of the excavation. However, there are circumstances, as discussed below, where the deep well system alone may not be sufficient to provide full depth pre-drainage.

Similar to wellpoints, a cone of depression develops around the pumped deep well. If a relatively low permeability boundary exists near the bottom of the excavation, as illustrated in Figure 6, the collection capacity of the deep wells is limited by the submergence of the well screen when the system is pumped. If the submergence is marginal or inadequate, supplemental dewatering within the excavation may be required to achieve the required drawdown, also as shown in Figure 6.

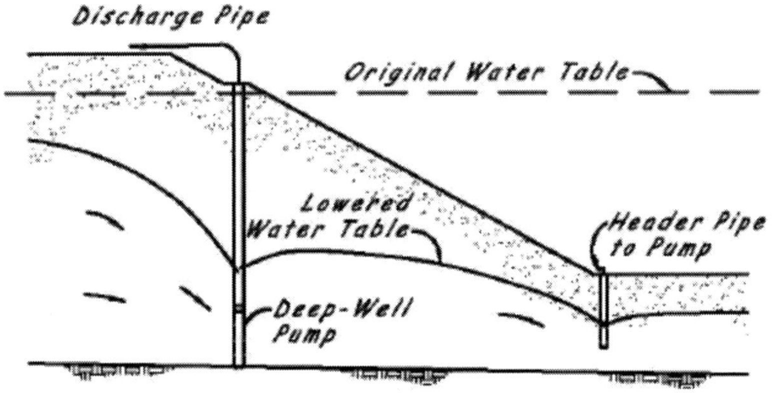

Figure 6. Supplemental Dewatering Required If Low Permeability Boundary Exists Near Bottom Of Excavation (Terzaghi, Peck, and Mesri, 1996).

Deep wells can also be adapted to dewatering fine-grained soils, similar to those dewatered using high vacuum wellpoints, eductor wells, and eductor wellpoints, by sealing the annulus of each deep well above the filter pack and applying a vacuum to the well casing. The well is pumped using a low capacity submersible pump, the casing is sealed against the discharge pipe and pump cable using a compression-type sanitary well seal. The smallest commercially available submersible pump has a capacity of about 5 to 6 gallons per minute. Since the steady flow to the well in fine soils is likely to be below the pump capacity, automatic level controls are used to cause the pump to operate cyclically. As long as the number of starts per day is less than the motor manufacturer's criteria, the pump and motor will be covered by the manufacturer's warranty. Vacuum is applied to the wells using electrically driven vacuum pumps installed at the ground surface. The only limit to the lift that can be achieved is the head capacity of the submersible pump that is selected. In many cases, deep wells with applied vacuum will be more economical to install and operate than a comparable eductor well system, and the air handling capacity can be increased, if necessary, either by changing the vacuum pump capacity or increasing the number of vacuum pumps connected to the well system.

Drainage Trenches with Perforated Collector Pipes

Drainage trenches with perforated collector pipes have been used to dewater excavations for constructing pipelines and other linear structures. The pipes are typically wrapped in a geotextile and installed with specially equipped trenchers, as shown in Figure 7. The pipes installed through this method are connected to a wellpoint pump at the ground surface or a submersible pump in a vertical or inclined riser connected to the horizontal perforated collector pipe.

Figure 7. Trench and collector pipe installation (Courtesy of DeWind Trenching).

Vertical Drains

Vertical drains can be effective in dewatering perched groundwater in semi-pervious (silty sand or silt) or pervious strata overlying hydraulically isolated pervious strata with adequate submergence for well or wellpoint screens. The vertical drains intercept seepage in the upper stratum and conduct it to the lower dewatered stratum. Historically, vertical drains consisted of sand columns, called sand drains, however, in the past few decades, manmade wick drains or earthquake drains, see Figure 8, have become the more common technology for vertical drains. The potential effectiveness of vertical drains is illustrated in Figure 9.

Figure 8. Installation of Earthquake Drain by Hollow Mandrell Advanced with Vibratory Hammer (Hayward-Baker Inc.).

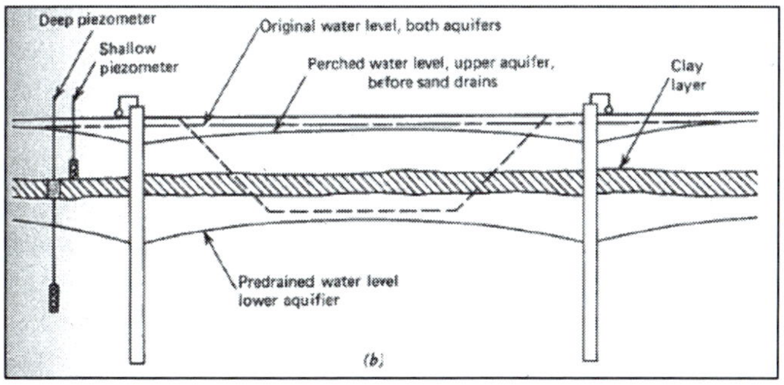

a) Without vertical drains.

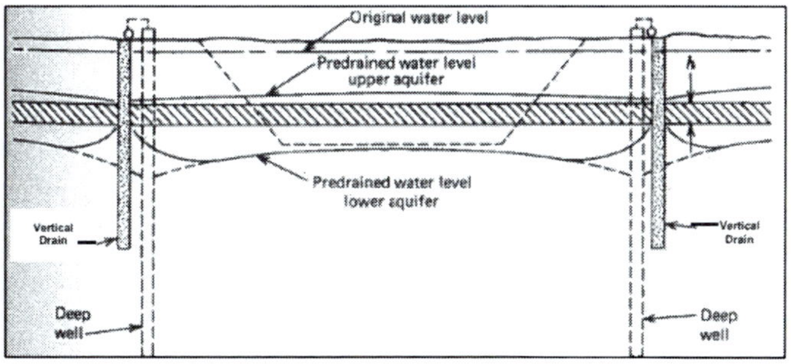

b) With vertical drains.

Figure 9. Potential Effectiveness of Vertical Drains (Powers, 2007).

Electro-osmosis

In low permeability clays and silts, electro-osmosis can be used for dewatering. An electrical current is established in the soil using a system of electrodes. Water migrates from the positive electrodes (anodes) to the negative electrodes (cathodes). The cathodes serve as well points where the migrating water can be extracted. An electro-osmosis system is illustrated in Figure 10.

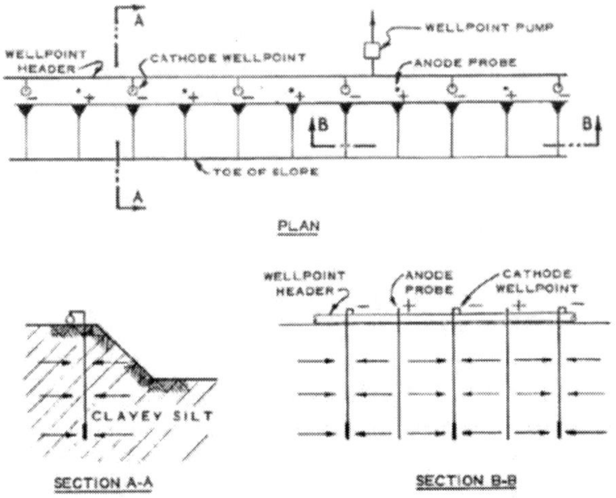

Figure 10. An Electro-Osmosis Installation (U.S. Army, 1985).

Barrier Walls and Bottom Seals

In certain circumstances, barrier walls around an excavation can be used to reduce the demands on a dewatering system. Barrier walls can reduce the seepage into the excavation and simplify dewatering. Barrier walls are most effective when there is a very low permeability stratum below a more pervious stratum. The barrier wall can be extended into the very low permeability stratum, cutting off all or most of the seepage into the excavation. Collection and removal of underground water will still be required to remove water stored in the more pervious strata and to remove seepage or leakage through or beneath the wall.

Possible technologies for barrier walls include:

- Sheetpiles.
- Continuous walls constructed with slurry support methods.
- Concrete panel or secant pile walls.
- Soil mix walls.
- Grouting.
- Ground freezing.

Discussion of barrier wall technologies is beyond the scope of this paper, but has been discussed in two other publications by one of the authors and his colleagues (France, 2014 and Kula et al, 2016).

In some cases, bottom seals can be used together with barrier walls. The bottom seal resists uplift pressures and prevents upward seepage into the bottom of the excavation. A bottom seal installation is shown in Figure 11. Bottom seals are not practical for large open excavations, but they can be effectively used for smaller excavations for structures such as stilling basins or pump / valve houses.

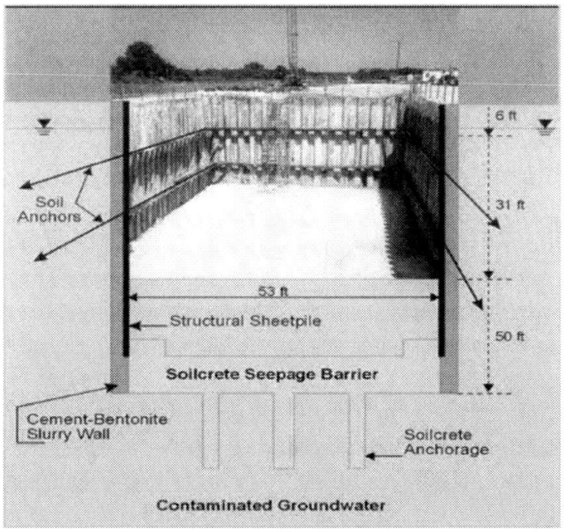

Figure 11. A Jet Grout Bottom Seal Combined with a Barrier Wall (Hayward-Baker Inc.).

CONSIDERATIONS FOR DESIGN AND TESTING OF DEWATERING SYSTEMS

Design of dewatering systems requires:

- Investigations and data collection.
- Analysis.
- Selection of dewatering system and equipment.
- Evaluation of power requirements.
- Specification of dewatering requirements.

Investigations and data collection should be sufficient to understand the geology of the area around and beneath the excavation. Of specific interest is an understanding of the extents, thicknesses, stratification, and permeabilities of the various soil and rock strata. Conventional subsurface investigation techniques, such as borings and penetrometer soundings, piezometers and observation wells, laboratory tests, and geophysical investigations will provide much of the needed information. In addition, in-situ borehole permeability tests can be very helpful in identifying hydraulic properties of subsurface strata. However, caution should be exercised in placing too much reliance on borehole permeability test results. In critical dewatering situations, field pumping tests on wells with an adequate number of monitoring piezometers should be considered. These tests, if properly conducted, provide the best indication of how a production dewatering system will perform. The advice of individuals with practical dewatering experience can be invaluable in deciding whether pump testing is appropriate.

Groundwater tables and piezometric conditions need to be carefully evaluated, including consideration of seasonal or operational variations of groundwater conditions. The chemical and biological characteristics of the groundwater also need to be understood, as these characteristics could result in corrosion or fouling of well or wellpoint screens, sand filters, pumps, and piping.

It is also important to understand the likely sources of recharge to the dewatering system and their probable fluctuations during the dewatering period. The design and configuration of the dewatering system may be significantly different if the water to be intercepted is primarily from the downstream channel, rather than from the reservoir, which can be the case if the seepage barrier within the dam and its foundation is very effective. In some cases, dewatering flow requirements can be reduced by using pipes to discharge reservoir diversion flows farther downstream, hence, drying up the stream near the dam and reducing recharge to the dewatering system.

Analysjs of dewatering systems can be completed using numerical analyses, flow nets, or simplified equations and charts. The degree of sophistication of the analyses can be tailored to match the complexity of the geology and the criticality of the dewatering system. If a pumping test is performed on a confined aquifer with measurements of piezometric levels, the resulting data can be used in the calibration of a numerical model. A report should be prepared to document the test data as well as engineering analyses of the test data. With the wide usage of personal computers and computer programs, the engineer's first thought is often to use a numerical analysis program for design of a dewatering system. While this may be appropriate in some cases, there are many situations where simplified equations and charts provide an equal or superior solution, along with a more fundamental understanding of how the dewatering system will work. Designers are encouraged to match the analysis method to the needs of the particular application. Again, the advice of individuals with practical dewatering experience can be invaluable in selecting the appropriate analysis methods.

Based on the data and analysis, the components of the dewatering system (e.g. types of wells, numbers and depths of wells, well spacing, etc.) are selected and configured. In designing the system, redundancy should be considered in terms of the types and numbers of backup and reserve components to be available on site. Selection of the type of dewatering system depends on the average collection capacity of the dewatering device selected for the deepest excavation that will be made, considering the positions of less permeable interfaces and the hydraulic interference between all of the dewatering devices in the system. Laterally extensive impermeable strata close to the excavation bottom can be used in conjunction with a cofferdam constructed with interlocking sheetpiles (or other types of water barriers and support systems) to minimize seepage into the excavation and temporarily retain the soils while the new construction is completed. If there is sufficient submergence for deep wells, this type of system can be used with or without sheetpile cofferdams. If required excavations are narrow with respect to the depth of the pervious formation, partially penetrating wells or wellpoints can sometimes be used to achieve adequate drawdown, while at the same time minimizing the total flow required for drawdown within a shored excavation. Trench boxes often enable the safe construction of sumps in situations where interface seepage is a problem at the required subgrade elevations.

An important aspect of the system is power. The selection and specification of the power supply for the dewatering system should be based on availability and reliability of power sources and other considerations, such as possible noise limitations. Typical sources of power are engine power, line electrical power and generator power. Engine and generator power can be noisy, and may not be acceptable in some settings. For critical dewatering systems, on site, automatically switched backup power should be provided, so that the risk of a dewatering system outage is reduced. For highly critical systems, it may also be appropriate to require on site, full time, human monitoring of the system.

After system components and requirements are identified, specifications need to be prepared for either a contractor-designed or engineer / owner-designed dewatering system. The specifications need to include all pertinent performance and installation requirements. Full-scale system tests with flow and drawdown measurements and analysis of such tests, to evaluate the effectiveness of the system in achieving the specified performance criteria are always advisable. Provisions should be made in project specifications and in budgeting for engineering during construction. An engineering memorandum report should be prepared to document the test data as well as the engineering analyses of the test data.

ENGINEER / OWNER DESIGN VERSUS CONTRACTOR DESIGN

Both contractor-designed and engineer / owner-designed dewatering systems can be successful. Typically, the designer has studied the site for a significant period of time and likely understands the site conditions and their relationship to dewatering better than the contractor can be expected to understand them in the short period allowed for preparation of a bid. Therefore, it is incumbent on the engineer to impart his or her knowledge to the greatest degree practical.

In the authors' opinion, there are some circumstances when it is reasonable and prudent for the owner / engineer to take on the responsibility of designing the dewatering system. Factors and possible scenarios that could lead an owner / engineer to take on the responsibility to design the dewatering system include:

- The dewatering system performance is critical for providing for public safety.
- The dewatering system performance is schedule critical – there is no time for extensive trial and error in implementation of the dewatering system.
- Subsurface conditions are complex and contractors may be tempted at bid time to omit dewatering a deeper pervious layer that requires pressure relief for bottom stability of the excavation.
- The successful prime contractor bids the project using a dewatering subcontractor proposal that is inadequate and is reluctant to increase the scope of dewatering beyond that included in the subcontractor's bid, or the contractor

and dewatering contractor select an overly expensive system to reduce risk.

- The successful prime contractor gets no bids from competent, responsible dewatering subcontractors at bid time and estimates the cost of dewatering using his own opinion of the required dewatering scope.
- An engineer/owner designed system provides a basis to bid for contractors and, therefore, removes most of the uncertainty from this bid item, which is typically highly variable, thus potentially reducing claims.

If it is ultimately decided that the dewatering system will be contractor-designed, the specifications should be as explicit as possible regarding dewatering requirements. If wells, wellpoints, or other special measures are believed to be necessary, then this should be clearly stated in the specifications. In all cases, performance requirements should be clearly identified – e.g. piezometric heads must be lowered to at least 2 feet below the bottom of the excavation, as demonstrated with piezometers.

EXAMPLES

The authors cite the following four dewatering projects as examples of successful dewatering for dam rehabilitation:

- Rye Patch Dam, Nevada.
- Deer Flat Dams, Idaho.
- Confidential project, southeastern United States.
- Little White River Dam, South Dakota

Rye Patch Dam, Nevada

Rye Patch Dam was modified in the early 1990s to address concerns related to seismic instability. The modifications involved an excavation at the downstream toe to replace liquefiable foundation material and install filters and drains. The design was developed by Woodward-Clyde Consultants, an AECOM legacy firm, under contract to the U.S. Department of the Interior, Bureau of Reclamation.

Because the available construction period was short, it was decided that the dewatering system would be owner / engineer-designed. The system consisted of deep wells, supplemented by sumps and ditches to address residual seepage inflows. The design required 1) backup wells to be available on site for installation, if required, and 2) a backup diesel generator power supply system with an automatic switchover system.

The dewatering system was very successful, and the project was completed on schedule. The backup wells were not required.

Deer Flats Dams, Idaho

The Deer Flat Dams were modified in the 2000s to address internal erosion concerns related to the two outlet works at the site. Because of concerns regarding one of the outlet works structures, an emergency choke filter berm was placed upstream of the structure. It was subsequently decided to completely reconstruct that outlet works structure. The design was prepared by Reclamation.

It was decided that dewatering would be an owner / engineer design, because of concerns with schedule and public safety, as the construction was completed with a substantially full reservoir. The system consisted of deep wells installed inside of a sheetpile barrier wall, as shown in Figure 12. The system performed extremely well, and the excavation was very dry. Construction was completed on schedule.

Figure 12. Deer Flat Dams Modification Project (Bureau of Reclamation).

Confidential Project, Southeastern United States

This project consisted of construction of a filtered toe drain buttress at the downstream toe of a coal combustion residuals impoundment. The design for this project was completed by AECOM as a minimum required design. Dewatering was required to allow for excavation into the downstream toe of the embankment. The final system was contractor-designed. The specifications identified a minimum system consisting of a line of eductor wells and required lowering the groundwater to 2 feet below the bottom of the excavation. The contractor adopted the minimum required system, which successfully dewatered most of the site. Supplemental sumps and pumps and well point dewatering were required to achieve the required dewatering at one location. The dewatering adjustments were made expeditiously and the project was completed on schedule.

Little White River Dam, South Dakota

Little White River Dam was modified in 2007 to remediate internal erosion adjacent to the outlet works/service spillway structure and severe headcutting near the emergency spillway. The design required excavation into fine sandy silt and silt foundation material to construct these appurtenant structures. The design was prepared by GEI Consultants, Inc. and construction oversite was performed by URS Corporation, a legacy AECOM firm, both under contract with the U.S. Department of Interior, U.S. Fish and Wildlife Service.

A dewatering system was required in order to construct the new appurtenant structures on a stable foundation. Since the reservoir was drained prior to construction, the dewatering system was contractor-designed and submitted for approval by the engineer / owner. The specifications required 1) a system consisting of sumps, wellpoints, deep wells or a combination of these systems to construct foundations on a dry, stable foundation, 2) the groundwater be lowered a minimum of 2 feet below the bottom of the foundation, and 3) a backup power supply system with an automatic switchover system. The contractor elected to initially use sumps, which did not adequately dewater the silty foundation material and resulted in a very unstable (water mattress) foundation. The contractor then elected to use 20-foot-long wellpoints spaced at about 3 foot on-center along the entire left side (upstream to downstream) of the excavation (see Figure 13). A few wellpoints were initially installed without a sand filter around the wellpoints, which upon testing, completely clogged with silt. Ultimately a sand filter was installed around the wellpoints. This wellpoint system created a stable foundation and the project was successfully completed on schedule.

Figure 13. Dewatering at Little White River Dam, South Dakota.

CLOSURE

Dewatering requirements for dam construction can be complicated, but they also can be critical to project success. Engaging people with the appropriate experience in selection, design, installation, and operation of the system is vital to successful dewatering.

REFERENCES

Cedergren, H.R., 1989, *Seepage, Drainage, and Flow Nets*, Third Edition, John Wiley & Sons, Inc., New York.

France, J.W., 2014, "Embankment Dam Seepage Modification – Choices and Considerations," Proceedings of the 2014 Annual Conference of the United States Society on Dams, San Francisco, CA.

Kula, J.R., France, J.W., Bird, R. and Pinciotti, R.D., 2016, "A Comparison of Methods for Managing Seepage through Embankment Dams," Proceedings of Dam Safety 2016, the Annual Conference of the Association of State Dam Safety Officials, Philadelphia, PA.

Leonards, G.A., 1962, *Foundation Engineering*, McGraw-Hill Book Company, Inc., New York.

Powers, J.P, 2007, *Construction Dewatering and Groundwater Control – New Methods and Applications*, Third Edition, John Wiley & Sons, Inc., New York, NY.

Terzaghi, K., Peck, R.B., and Mesri, G.,1996, *Soil Mechanics in Engineering Practice*, Third Edition, John Wiley & Sons, Inc., New York, NY.

White, L. and Prentis, E.A., 1950, *Cofferdams*, Second Edition, Columbia University Press, New York and London.

U.S. Army, 1985, TM5-818-5 (Change 1), *Dewatering and Groundwater Control,* Headquarters, Departments of the Army, Air Force, and Navy, Washington, DC,

DRAFT TUBE SLAB ANALYSIS

Jesse Kropelnicki[1]
David Schreck[2]
Kareem Bynoe[3]
Peter Bouchie[4]

ABSTRACT

This paper summarizes the results of a multi phased draft tube slab investigation conducted in 2014-2015 by WSP | Parsons Brinckerhoff (PB) for a 423 MW plant. Together with the Owner, PB determined that an investigation was needed to decide if it was safe to work in a completely dewatered draft tube, with a thin, under-reinforced floor slab during required maintenance, or if remediation measures were warranted.

Based on discussions between the Owner and PB, a multi-phased approach was developed to address the structural adequacy of the draft tube slab. The initial phase determined factors of safety for the draft tube slab using a simple ACI based one-way slab approach. Results of this analysis indicated very low capacity; limited by the reinforcement and factors of safety which were not acceptable. The second phase utilized a failure modes based approach to determine factors of safety, but this approach also determined that results were not adequate for the Owner's safety standards. The third phase focused on a geotechnical evaluation of the foundation underneath the draft tube slabs to better understand rock quality and the potential for uplift pressure. This phase included laboratory testing of the original 1930's rock cores from the construction of the powerhouse. The last phases focused on a finite element model (FEM) failure analysis of the concrete draft tube slab and included steel reinforcing bars not included in previous analyses of the draft tube slab.

The final results of the multi-phased analysis indicated that the factor of safety for the governing condition, with the existing slab foundation drains assumed not to relieve uplift pressure, is more than adequate and the consequence of postulated failure is not as significant as initially thought.

[1]Jesse Kropelnicki, P.E., Principal Engineer, WSP | Parsons Brinckerhoff, 75 Arlington St., Boston, MA 02116, (617) 960-4975, kropelnicki@pbworld.com
[2]David Schreck, P.E., Regional Dam Safety Engineer, Brookfield Renewable Energy Group, 1 Powerhouse Road, Conestoga, PA 17516, david.schreck@brookfieldrenewable.com
[3]Kareem Bynoe, Supervising Engineer, WSP | Parsons Brinckerhoff, 851 SW Sixth Avenue, Portland OR 97204, (503) 417-1363, bynoe@pbworld.com
[4]Peter Bouchie, P.E., Lead Engineer, Parsons Brinckerhoff, 75 Arlington St. Boston, MA 02116, (617) 960-4810, bouchie@pbworld.com

SAFE HARBOR DAM

The Safe Harbor Hydroelectric Project (Project) is located in the village of Safe Harbor, Pennsylvania on the Susquehanna River approximately 32 miles above its mouth at Chesapeake Bay and approximately eight miles upstream of the Holtwood Dam. The Project is owned and operated by Brookfield Renewable Energy Group (BREG) and consists of a concrete gravity dam and powerhouse structure that extends across the Susquehanna River. The left side of the dam is in Lancaster County, and the right side is in York County, Pennsylvania respectively. Beginning at the east shore, the project is composed of the non-overflow East Shore Bulkhead, the Powerhouse (which includes a non-overflow powerhouse bulkhead, an intake section, and a non-overflow connecting bulkhead), a fish passage system, the East Spillway, the non-overflow Island Bulkhead, the West Spillway, and the non-overflow West Shore Bulkhead. Lake Clarke serves as the Project reservoir with an area of 11.5 square miles and an approximate length of ten miles.

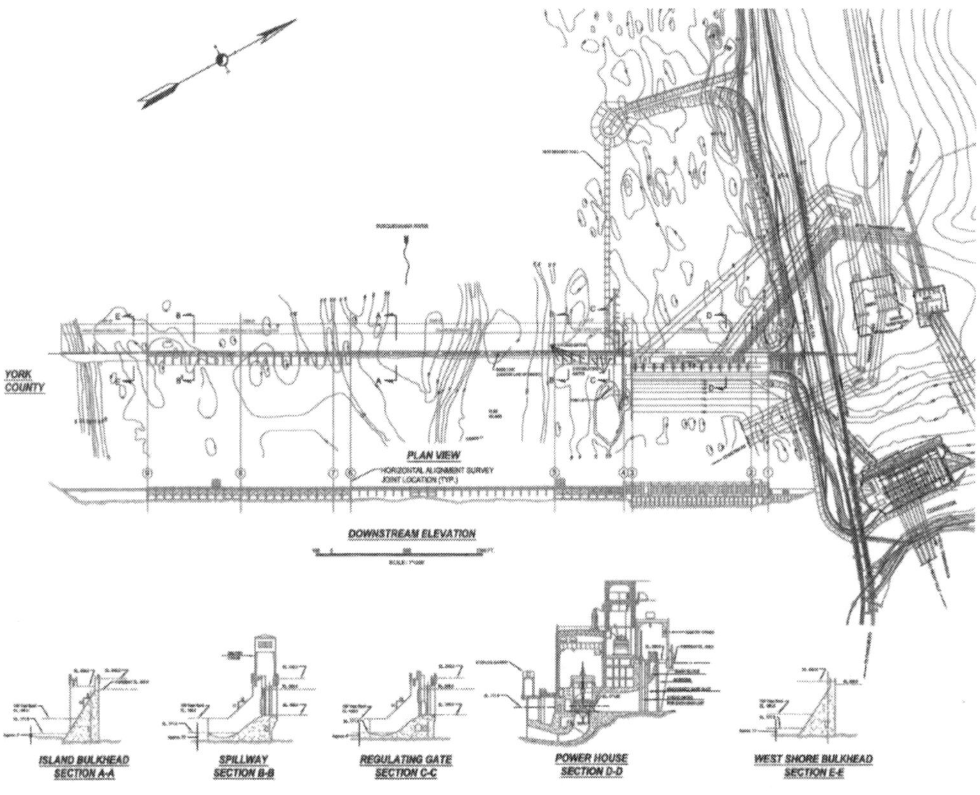

Figure 1. Safe Harbor Dam, plan, elevation, and sections.

The Safe Harbor station is used for the production of hydroelectric power and to provide regulated storage for users downstream.

The two spillways, east and west, have a total length of 454 feet and 1356 feet, respectively. The East Spillway is equipped with vertical gates in all of its seven bays. The West spillway is comprised of 24 spillway bays, each equipped with vertical single-leaf flood gates. The gates of both spillways are 48-feet-tall by 35-feet-wide. The gates are operated by three gantry cranes, available for both the east and west spillways.

Figure 2. Safe Harbor Dam, viewed from the left abutment.

There are six non-overflow retaining structures including a 977-foot-long powerhouse and five non-overflow bulkheads. The five non-overflow bulkheads include a 238-foot-long East Shore Bulkhead, a 113-foot-long powerhouse bulkhead (which is included in the powerhouse), a short 58-foot-long connecting bulkhead (which is also included in the Powerhouse), a 1,356-foot-long Island Bulkhead, and a 487-foot-long West Shore Bulkhead. The powerhouse is comprised of the 113-foot-long powerhouse bulkhead, fourteen 62 foot turbine-generator bays, and the 58-foot-long connecting bulkhead.

The powerhouse has 14 generating units. There are three water passages at each main unit, with clear distances between piers of about 16 feet. There are two service units referred to as Units 41 and 42. The two service units each have a single water passage. Each water passage is equipped with a head gate, and trash rack. Six main generating units were initially installed during the period of 1931 to 1934. The seventh unit was installed in 1940. These seven units are identified as the Old Units. Between 1982 and 1986, the Powerhouse was expanded in bays provided during initial construction with the addition of five new generating units (New Units). The addition of these units increased the total station generating capacity to 417.5 MW with a maximum powerhouse discharge of approximately 110,000 cfs.

The intake section of the unit block substructure extends approximately 116.5 feet upstream of the centerline of the units. The height of the structure extends from approximately El. 160 at bedrock to El. 232 at the operating deck for a total of 72 feet.

DRAFT TUBE SLAB

The draft tube concrete floor slabs were built utilizing drain holes to relieve uplift pressure during a dewatered condition. The draft tube slabs rest on uneven rock and therefore vary in thickness. In 2014, an attempt to clean the drain holes was made. However, since it was not known how successful this was in making the drains effective, the drain holes were assumed ineffective for the purposes of stability and slab safety. The assumption that the drains are ineffective has the largest impact on the adequacy of the floor slabs during a dewatered condition when maintenance may be required. With this in mind and based on discussions with the owner, a multi-phased approach was developed to address the structural adequacy of the draft tube slabs for potential dewatered conditions during unit maintenance or upgrades. Each phase was approached with an increasing complexity of analysis.

Phase I Analysis

There were three primary tasks associated with the New and Old units for Phase I. Based on the fact that the draft tubes were de-watered numerous times in the past for maintenance without failure, it was assumed that the slabs were sufficiently capable of resisting the possible governing pressures under the slabs with a factor of safety of 1.0. The first task was to back-calculate the existing capacity of the floor slabs for a factor of safety equal to 1.0 and determine the additional amount of resistance required to ensure acceptable results. The applied load was assumed based on the FERC method of calculating uplift since actual uplift conditions were unknown. It was also assumed that the slab behaves as a one-way reinforced beam, based on ACI-318 definitions for required reinforcement. The second task was to confirm that the draft tube slabs and the powerhouse act independently from each other such that a slab failure would not affect the powerhouse stability. The third task was to determine various factors of safety by evaluating the sensitivity of the input parameters including tailwater elevations, measured intake drain efficiencies, uplift pressure, additional resistance, slab depths, concrete strengths, failure modes for the New and Old units, and critical locations of failure within the draft tubes.

Although the Phase I analysis did not yield acceptable safety factors for the concrete slab analysis when reasonably conservative assumptions were made for the input parameters and results evaluated against customary ACI-318 criterion, it was determined that the draft tube slab acts independently of the powerhouse for the purposes of stability. It was also determined that unknown additional resistance exists (cohesion or higher concrete strengths) or the magnitude of uplift pressure applied to the draft tube slab is less than the assumed per the FERC approach and further investigation was recommended. Lastly, it was determined that the slab for the Old units governed since the slab for the New units is secured to the rock with 10-foot-long vertical rebar dowels.

Phase II Analysis

Since favorable results were not achieved in Phase I, Phase II was implemented to further refine the analysis of the draft tube slab. Due to the complexity of this evaluation and the inability for conservative input assumptions to provide results which pass the acceptance criteria, each subsequent phase used a more detailed approach and refined input parameters to obtain more accurate results.

Phase II was based primarily on three tasks. The first task involved the evaluation of the slabs with and without the rock acting together as one mass for bending, beam shear and punching shear failure mechanisms. The second task included developing more detailed slab failure modes based on a geotechnical evaluation which identified better than expected quality rock. This evaluation involved performing a geotechnical study of the original construction core logs, field pressure tests, detailed review of available construction photographs, and a review of the 1980's expansion program staff's observations and photographs to assess the degree to which good quality, intact rock exists just below the slabs. This included determining rock joint spacing, the minimum and maximum rock prism geometry (based on the joint spacing), estimating the rock joint cohesion and tensile strength based on construction photographs, existing boring logs and published literature. The third task determined factors of safety for various partially dewatered draft tube elevations and the fully dewatered condition. Based on the refined failure modes and a more refined reinforced concrete analysis the safety factors were determined.

The analysis concluded that the draft tube slab is stable under normal loading conditions assuming an additional force of at least 21 psi exists between the slab and rock base. The analysis also concluded that under a partially dewatered draft tube, it is stable (with no cohesion). Neither one of these conclusions was considered acceptable based on the owner's and PB's criterion. Therefore, a phase III was implemented.

Phase III Analysis

Since favorable results were not achieved in Phase II, Phase III was performed and based primarily on developing a better understanding of the rock quality and the applied uplift pressures as more refined input to the stability analysis

In order to make an assessment of the intact rock that existed before being covered by the draft tube slabs, PB evaluated in greater detail: original 1929 construction boring logs, photos of the rock cores, original construction photographs of the New and Old draft tubes, photographs of PB's field observations of the exposed rock surface downstream of the west spillway, and observations by project personnel of the rock surface at the New units before the concrete slab was cast on the rock for the 1980's construction. The 1929 boring logs identified that the rock at the powerhouse is about 90% Gneiss and 10%

Schist, with Schist only at the outside edge of the powerhouse near Unit 1. Therefore, the geotechnical evaluation was performed assuming the rock below the draft tube slab is Gneiss throughout.

As discussed above, due to the existence of dowels anchoring together a 10-foot-deep zone of rock below the new units, the concrete slab above the rock would resist uplift. The old units do not have dowels into the rock below the concrete slab and do not have the uplift pressure resistance compared the new units. Therefore, the old units governed for the geotechnical evaluation.

An abbreviated list of the conclusions determined based on the initial geotechnical evaluation includes:

- Existing project information did not provide definitive rock capacity values that could be used to assess risk with completely dewatered draft tubes.
- The project information provided a general understanding of the rock joint configuration but not joint strength.
- Based on rock mechanics it was expected that the joint capacity may decrease over time due to dewatering the units multiple times.
- Although the actual draft tube slab concrete strength was not known the total resisting strength is not highly sensitive to the concrete strength.
- Based on the previous analysis, and conservative assumptions, the calculated factor of safety is large enough to allow the draft tube to be dewatered, but only partially.
- Based on the previous analysis, and conservative assumptions, the calculated factors of safety are not large enough to allow the draft tubes to be fully dewatered.

As part of the geotechnical evaluation, the 1930's rock cores from the original construction were inspected, samples were chosen, a test program was developed, and the samples were tested in a laboratory. The laboratory tests performed were unconfined compression, tensile, and permeability, all in general accordance with ASTM standards. The average of the successfully completed compression tests was 11,500 psi, the average of the successfully completed tensile tests was 584 psi, and the average of the successfully completed permeability tests was 3.7 E-07 cm/sec. Overall these values indicated good quality, intact rock thereby decreasing the likelihood of significant uplift via rock joints. PB felt that this approach to utilizing rock cores from construction was a better alternative to more invasive, and expensive, present day coring.

The laboratory testing program quantified the qualitative strength observations from draft tube construction photos and spillway outcrops. It also confirmed why during the 1980's expansion program, staff members did not see any significant amount of water seeping through the rock into the draft tubes prior to placing the concrete slab. This test program was important since records for rock testing during construction were not available, and there is now a valuable baseline of important rock properties to use for future stability evaluations or construction projects.

PHASE IV ANALYSIS

Since favorable and definitive results were not achieved in the previous phases, Phase IV was conducted and based primarily on a ductile reinforced concrete failure mechanism for a two-way slab versus a factor of safety stress analysis using a one-way reinforced concrete beam. This phase included four evaluations.

The first evaluation was to estimate the permeability of the rock, rock joints, concrete, and drains based on assumed values, to better quantify the magnitude of potential water inflow. This resulted in a much lower rock permeability than expected. The much lower rock permeability indicated that even if the slab was not bonded to the rock the potential inflow would not be significant during a failure scenario.

The second evaluation involved analyzing the slab for two conditions; one with drains relieving pressure (11 foot maximum span between drains) and one with drains not relieving pressure (25 foot maximum span between piers) since only 5 of the 12 units have had the drains cleaned recently.

The third evaluation included analyzing the assumed unreinforced and reinforced concrete slabs, since the slab reinforcement present, based on the construction drawings and photos, is below the minimum required reinforcement per ACI-318. Specifically, for the old units 1-7 the reinforcement was 1 inch diameter bars at 3 foot on center spacing in the direction parallel to the pier and ¾ inch diameter at 3 foot on center spacing in the direction perpendicular to the pier. For the New units 8-12 the reinforcement was #9 bars at 12 inches on center each way at the top face of the slab.

The fourth evaluation included incorporating a shear friction end condition for the slab to ensure a ductile bending based failure mode. This was done since the construction photos showed that even though the slab reinforcement is doweled into the piers, shear keys in the piers were not provided. Therefore, the end condition is retrained by the reinforcing dowels as well as shear friction produced during bending and end rotation of the slab.

All of the above evaluations were completed using increasingly refined assumptions and hand calculation methods consistent with ACI's reinforced and unreinforced methods, but did not yield acceptable results based on the owner's and PB's criterion.

Finite Element Model (FEM)

Since the refined hand evaluations did not yield favorable results for bending and shear for both reinforced and unreinforced concrete slabs, without cleaned drains, a nonlinear FEM was developed to confirm that additional resistance, and adequate safety factors existed. The FEM had several advantages, which are described in greater detail below.

The slab FEM had restraints against the piers that allow compression but not tension. This more accurately matched the condition of the pier/slab interface while conservatively neglecting the amplitude of the rock projections. The rock projections would reduce the ability of the bottom of the slab to pull away from the piers as the center was pushed upwards.

A conservative allowable concrete compressive strength value along with a conservative allowable tensile strength value were used for the governing slab without cleaned drains, thus allowing the slab to crack and redistribute stress, which is more realistic. The model was also created to allow concrete elements that exceed the concrete compressive strength to crush.

Shear friction mechanism was used at the slab ends to resist the uplift pressure in absence of a shear key, based on the ACI-318 approach. The FEM showed that the vertical force is initially resisted by the reinforcement in shear, then as the slab begins to rotate, a large compressive force is developed against the pier which creates the shear friction required to maintain a stable end condition.

Figure 3. FEM Model Stress Distribution & Interface Elements at Edges

Acceptance Criteria

Condition	Criteria
Concrete - Compression	2,550 psi based on ACI 318 compared against minimum principal stress
Concrete - Tension	273 psi based on ACI 318 input into FEM for cracking evaluation
Reinforcement Axial Stress	40 ksi based on ACI 318 against maximum axial stress

Using the conservative FEM, adequate factors of safety were achieved for the governing concrete slab with ineffective drains assumed, and the draft tube fully drained. These factors of safety were evaluated against the stress based reinforcing acceptance criteria above as well as the ability of the model to create a stable end condition via friction. The FEM was considered a conservative simulation of the actual conditions since geometric nonlinearities were not included in the analysis and a conservative constant slab thickness was used vs. an undulating bottom surface, which tends to lock in the slab with the rock foundation and provides additional resistance.

SUMMARY AND CONCLUSIONS

The FEM structural analysis results indicated that the factor of safety for the FERC's method of calculating uplift pressure, for the governing condition of fully drained draft tube, and with the drains assumed ineffective for relieving uplift pressure, is equal to or greater than 1.7. The uplift pressure was based on the normal headwater at elevation 228 feet, with tailwater at the upper end of the normal range at elevation 171.5, and the field measured minimum intake drain efficiency of 15% (note, the intake drains are independent of the draft tube drains). Other elevations may be acceptable but were not included in the evaluation.

These factors of safety indicated that entry into both the old and new fully dewatered draft tubes is safe. Included with the acceptable determination was the requirement that upon entry, the drain holes in all units not recently cleaned, be thoroughly cleaned, and the drain holes in units that were recently cleaned, be inspected and cleaned as needed. For subsequent days of work, it was also recommended that staff working in the dewater draft tubes make an overall observation at the beginning of each shift, especially of the drain holes, to assess if there were any changed flow conditions.

CORE FROM ROCK, CHALLENGES OF CORE CONSTRUCTION FOR A LARGE HIGH HAZARD DAM

Michael J. Miller, P.E.[1]
Dr. Efrain Rondinel-Oviedo[2]

ABSTRACT

The construction of Linga Dam required approximately 1.6 million cubic meters of core material. Locating an economical source of clay would prove extremely difficult during the exploration and final design phase of the dam. The Cerro Verde Mine had two active pits from which they produced copper ore, the Cerro Verde and Santa Rosa pits. From past work on the mine site, engineers knew that isolated areas of hydrothermally altered leach cap rock, granodiorite, located over the ore body had the ability to generate, when broken down and wetted, clayey materials. The challenges faced by the design and construction team would be to identify the locations of raw materials with suitable clay content, fragment, excavate and process the raw materials such that it was suitable for use as core material, and then transport place and compact the core material into the body of the dam.

PROJECT BACKGROUND

As part of its production expansion, the Cerro Verde Mine (SMCV), located outside of Arequipa, Peru, required the construction of a 200-meter high dam, Linga Dam, to store start up and operations water and to serve as the starter dam for its tailing storage facility. Feasible options for the Linga Dam construction included rockfill dams constructed with central clay or asphalt cores; or rockfill dams faced with clay, asphalt, concrete or a geomembrane. The most economical and reliable dam for this site would prove to be a central clay core rock fill dam.

During the design significant challenges arose in the identification of a suitable source of material to use as the Zone 3 - Core and Zone 7 - Select Core materials for dam construction. In this paper we will detail the processes and procedures developed and implemented to produce suitable core materials and the construction of the core for Linga Dam.

This paper will also detail how these challenges were surmounted by the team and will discuss test fill programs, QA/QC testing of core materials with large diameter aggregates and the seepage observations during the reservoir filling.

DAM DESIGN AND CONSTRUCTION

Linga dam is a 200-meter high central core rockfill dam located in a highly seismic area and designed to resist a magnitude 9.0 interface earthquake, and a magnitude 8

[1] Principal Engineer, MWH – Denver Colorado, michael.miller@us.mwhglobal.com
[2] Supervising Geotechnical Engineer, MWH – Lima, Peru, Efrain.A.Rondinel-Oviedo@mwhglobal.com

intraslab earthquake. The dam is used to store a fresh water reservoir that provides start-up and operations water for a new concentrator plant and serves as the starter dam for the tailing storage facility. Fill placement began in January of 2014 and continued until the crest of the dam reached the design elevation of 2570 masl on August 8, 2015.

The slope of the upstream rockfill face of the dam is 1.5 horizontal to 1.0 vertical (1.5H: 1.0V), and the slope of the downstream rockfill face varies between 2H: 1V and 3.5H: 1V. The upper portion of the fill has the steeper slope of 2H: 1V and the slope change occurs at elevation 2460 masl. The crest of the dam is 15 meters wide, the crest length is about 660 meters and the structural height of the dam is 200 meters, measured from the downstream toe. Additionally, there is a 15 meter wide bench at elevation 2560 masl on the downstream slope to support the underflow distribution piping. A typical cross section of Linga Dam is shown in Figure 1.

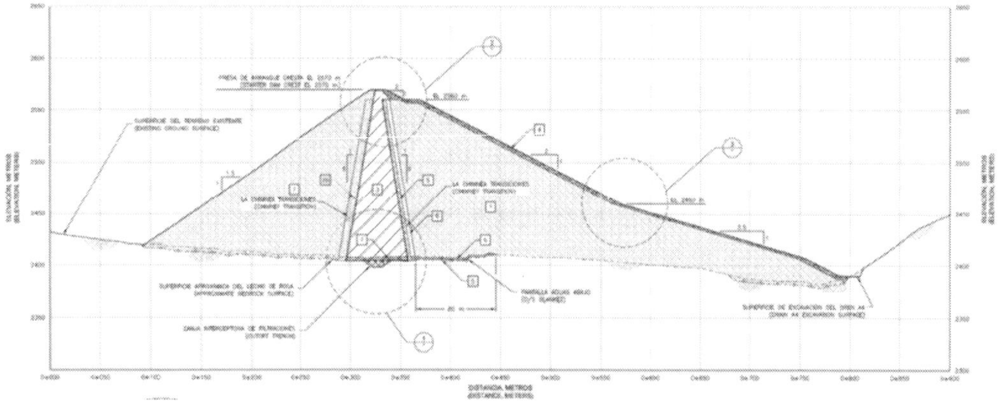

Figure 1 – Typical Linga Dam Cross Section showing the Different Zones

Linga Dam is designed as a zoned rockfill embankment dam consisting of seven different material zones. The seven material zones that constitute the dam are:

- **Zone 1: Rockfill.** This material forms the main body of the dam. The material was predominantly obtained from rock quarries located upstream and downstream of the embankment.
- **Zone 2B: Upstream Transition Material.** The purpose of Zone 2B is to reduce the risk of particle migration from the central core material (Zone 3) to the upstream shell rockfill (Zone 1) and to serve as a "crack stopper" if a crack were to develop in the core of the dam. Zone 2B material was produced by screening, crushing and washing alluvium excavated from alluvial borrow areas upstream of the dam.
- **Zone 3: Core.** The purpose of the low-permeability central core zone is to limit potential seepage from the impoundment through the dam. The Zone 3 material was sourced from localized areas of cohesive materials removed from the west side of the Santa Rosa Pit at the Cerro Verde Mine.

The source material (unprocessed material) consists of material blasted from the leach cap in the west portion of the Santa Rosa pit, above the ore body. The blasted material was hauled and stockpiled for subsequent screening. Power screens were used to produce 6-inch minus cohesive materials.

- **Zone 7: Select Core.** The Zone 7 material was used to backfill the cutoff trench in the valley bottom and was also placed in a 4-meter wide zone in contact with and adjacent to bedrock at the abutments and at the perimeter of the cutoff trench. The Zone 7 material was produced from the same source as Zone 3, but was screened to have a maximum size of 1½ inches. The 1½ inch maximum size was specified to enhance the bond at the cutoff trench and abutment contacts.
- **Zone 4: Downstream Face Veneer.** The purpose of Zone 4 is to reduce the potential for migration of the underflow tailing sands into the downstream rockfill (Zone 1) shell. Zone 4 material was produced by screening alluvium primarily excavated from an alluvial borrow area located downstream of the dam.
- **Zone 5: Downstream Transition / Filter.** Zone 5 is part of a two-phase filter designed to control the migration of fine particles from the Zone 3 Core into the Zone 6 Downstream Transition / Drain; and in an unlikely case of a continuous crack through the core, to limit the risk of migration of impounded tailing to the downstream rockfill shell. Zone 5 material was produced by screening, crushing and washing alluvium excavated from borrow areas located upstream of the dam.
- **Zone 6: Downstream Transition / Drain.** Zone 6 is a part of a two-phase filter designed to meet filter criteria between the Zone 5 Downstream Transition Filter and the downstream Zone 1 Rockfill shell. Zone 6 material was produced by crushing and screening alluvium excavated in borrow areas located upstream of the dam.

The quantities of the (compacted) material placed during the construction of Linga Dam are:

• Zone 1: Rockfill	10,686,000 m^3
• Zone 2B: Upstream Transition Material	323,508 m^3
• Zone 3: Core	1,498,800 m^3
• Zone 4: Downstream Face Veneer	613,920 m^3
• Zone 5: Downstream Transition / Filter	220,293 m^3
• Zone 6: Downstream Transition / Drain	215,929 m^3
• Zone 7: Select Core	84,595 m^3
• Blanket Drain (Zones 5 and 6)	87,058 m^3
Total Approx. Volume of Linga Dam Fill:	13,730,103 m^3

ZONE 3: CORE AND ZONE 7: SELECT CORE MATERIAL REQUIREMENTS

The Zone 3 Core material is designed as a low permeability material, clay (CL), clayey sand (SC) or clayey gravel (GC), located above the Zone 7 Select Core, which was placed adjacent to the rock abutments and in the cutoff trench. Zone 3 is located in the central part of the dam and is designed to provide seepage control through the dam during operation. The core was designed to have a maximum permeability of $1x10^{-5}$ cm/sec with the quantity of fines being important to the core in fulfilling this function. The required percentage passing the No. 200 sieve must be greater than 15%, with a minimum 10% plasticity index (PI), a minimum 20% liquid limit (LL) and a maximum 15 cm size. The placement moisture content was to be between optimum moisture content and 2% above optimum and the material was to be compacted to not less than 98% of the laboratory maximum dry density as defined by standard proctor testing (ASTM D-698).

The Zone 7 Select Core material is a low permeability material used as backfill for the cutoff trench in the valley bottom and along the abutments, and was also placed between the prepared foundation rock and Zone 3 along the fringes of the core at the dam abutments. Zone 7 is designed to limit seepage passing through the core and at the foundation interface during operations. The Zone 7 Select Core material had the same material property requirements as that of the Zone 3 Core material above with the exception that the Zone 7 material had a maximum 38 mm particle size. Also, the placement moisture content for the Zone 7 material was to be between +2% optimum moisture content and +4% optimum moisture content for the first meter (horizontally) from the abutment, and then between optimum and +2 % of optimum.

CHALLENGE

Locating an economical source of clay proved extremely difficult during the exploration and final design phases of the project. No source of clay material was available near the dam site. Alternatives examined to supply an acceptable source of core material included the following:

- Spent heap leach material
- Off-site source of clay material hauled to the dam site
- Bentonite soil amendment
- Onsite source of clay

<u>Spent Heap Leach Material</u>: Not materially viable for use as core and was rejected as a potential core source. This material had insufficient fines content and very low plasticity indices. Also, the average gradation of the material suggested that this material was internally unstable under hydraulic gradients if placed as a low permeable fill. A test fill was constructed with the Spent Heap Leach Material that demonstrated this material to be extremely difficult to work with due to its narrow band of optimum to +2% of optimum moisture. For these reasons this alternative was abandoned.

Off Site Clay Source: The only known source of suitable clay materials, in sufficient quantities, for core material was located approximately 48 km to the east of the dam site off of the SMCV property. The cost of purchasing and hauling this material to the dam site for use as clay core made this option economically unviable.

Bentonite Soil Amendment: Considered as an alternative for manufacturing core material. This option would have included purchasing and hauling bentonite from a supplier in the city of Lima and then mixing it with onsite soils to manufacture core material. This alternative proved less expensive than hauling off site materials to the dam site but would prove far more costly than identifying an onsite source of core material. This alternative was kept as a fall back measure if sufficient quantities of onsite core materials was not identified.

Onsite Source of Clay: SMCV has two active pits from which they source copper ore, the Cerro Verde pit and the Santa Rosa pit, located approximately 9 km north of the dam site. From past work at the mine, Engineers knew that isolated areas of hydrothermally altered argillaceous leach cap rock, granodiorite, located over the ore body had the ability to generate clayey materials when broken down and wetted. The leach cap was a waste product of the mining process and was deposited in waste dumps near the open pits. It was decided to further investigate this alternative as a means of obtaining material suitable for use as core. The following are the means and methods of the process of constructing the core for Linga dam:

PROCESS TO IDENTIFY GENERAL LOCATIONS OF SUITABLE RAW MATERIALS

An initial Geologic reconnaissance, performed in November and December of 2011, indicated that the areas with the highest potential for producing suitable raw feed material for the production of Zone 3 Core and Zone 7 Select Core were the leach cap rock in areas along and in proximity to the primary fault zones located at the west and northwest side of the Santa Rosa pit. In general these areas consisted of highly fractured and altered/weathered leach cap bedrock, Yarabamba Granodiorite, in proximity to the main traces of fault zones that transect the SMCV mine. The rock in these areas tended to have higher degrees of fracturing and mineral alteration (i.e. clayey mineralogy and in-situ degradation), and when blasted had the highest potential for generating PI>10 material that could be stockpiled and utilized as raw feed material from which Zone 3 Core and Zone 7 Select Core could be manufactured.

Further investigations conducted in the spring of 2012 were performed based on the findings from initial geologic reconnaissance. These investigations included:

- Field reconnaissance and mapping of the western Santa Rosa pit walls
- Evaluation of available spatial geological, geomechanical, and mineralogical data from SMCV
- Review and evaluation of SMCV exploration drill hole data
- Collection of field samples for laboratory testing.

The results of these further investigations indicated that suitable zones of raw feed materials for the manufacture of core materials within the Santa Rosa pit generally had higher fines content, with fines that had a higher plasticity value. In these zones the rock also tended to be weaker and exhibited hydrothermal alterations. Unsuitable zones were also observed that were composed of waste rock that tended to be highly sandy and/or "boney" coarse materials.

Laboratory testing was performed and focused on properties that gave an indication of the geotechnical characteristics of the raw feed and waste materials. Index testing was performed on samples, including Atterberg limits and grain size distribution. Testing on raw feed material for core included permeability testing and proctor testing on remolded samples to assess the potential hydraulic conductivity and optimum moisture content and density to be used during construction of a future test fill.

TEST FILLS TO ASSESS SUITABILITY OF RAW FEED MATERIALS FROM SANTA ROSA PIT

During the August / September 2012 period two test fills using Zone 3 Core and Zone 7 Select Core materials were constructed using raw feed materials from the Santa Rosa pit. The Zone 3 and Zone 7 test fills were performed using materials harvested from the northwest corner of the Santa Rosa pit. Material in the pit that was identified as having plastic fines was excavated and hauled to the test fill area for processing. The feed material for the Zone 3 Core was screened on a 6-inch static grizzly until approximately 5000 m^3 was obtained. In a similar manner the feed material for the Zone 7 Select Core was screened on a 1½-inch static grizzly until approximately 5000 m^3 was obtained. Laboratory testing of gradation, Atterberg limits and moisture density relations were done on each 300 to 500 m^3 of material produced.

After screening, the materials were moisture conditioned in stockpiles and then the test fills were constructed for both Zones. The Zone 3 test fill was constructed with five 30 cm loose lifts and the Zone 7 test fill was constructed with six 30cm loose lifts. Each lift surface was, at a minimum, 45 m by 15 m giving sufficient area to operate the large compaction (CAT 825 Compactor) and spreading (CAT 14M Motor Grader) equipment that would be needed to construct the test fills. In general the following were evaluated for each test fill:

- Material Placement and Compaction
    - In-situ Density and Moisture
    - Particle Breakdown / Gradation
    - Relative Compaction vs. Lift Thickness
- Compactive Effort vs. Density
- In-situ Permeability
- Observation Trenches
    - Particle segregation (nesting) within and between lifts
    - Particle interference during compaction

Indication of excessive cracking
Uniformity of moisture conditioning
Particle breakdown of large size aggregates
QA/QC Field Testing Techniques and Methods

The results and observations from these two test fills indicated that select raw feed material located in the north and west rims of the Santa Rosa Pit were suitable for use in the processing/manufacturing of Zone 3 Core and Zone 7 Select Core materials.

PROCESS OF GATHERING SUFFICIENT QUANTITIES OF MATERIAL

During the process of the development of an open pit mine, mine plans are formulated detailing the short-term and the long-term operations of the mine. These plans divide the mine into projects, or areas where excavation of ore and waste rock will occur. The purpose of this plan is to both provide a blueprint for mining operations and to allow for the projection of the economic output of the mining process. One of the constraints was that the construction team had to harvest raw core materials within the schedule of the short term mine plan.

Once general areas of suitable raw core materials were identified in the Santa Rosa pit the construction team implemented a system by which the suitability of raw material within each blast project would be assessed. To delineate zones of suitable raw core material in each blast project the drill cuttings from each blast hole were examined and the plasticity of the cuttings were evaluated both visually and to verify the visual assessment by laboratory testing. On average one Atterberg limits test was run per 10 to 15 blast holes visually assessed. As shown on Figure 2, cuttings that exhibited a plasticity index of greater than 12 were assigned a green designation; cuttings that exhibited a plasticity index between 8 and 12 were assigned a blue designation; cuttings that exhibited a plasticity index less than 8 were assigned a red designation.

The typical depth of a blast project was approximately 15 m over an area of about 400 to 700 m^2. Each blast was designed such that the 80% size (P80) of the material was less than 6 in. with a powder factor typically of 0.35 kg/mt. The blast holes were 8-inches in diameter, drilled vertical 15 m deep, on 2 m spacing's.

Occasionally, if the areas of the blast project was considered suitable, the powder factor was increased to improve the fragmentation of the raw feed materials. Engineering judgment was used to divide the project up in a manner such that the areas delineated by red and blue blast holes, unsuitable material, was hauled to the waste dump for disposal while the area delineated by the green and blue blast holes, suitable material, was hauled to the raw feed processing area for manufacturing Zone 3 Core or Zone 7 Select Core materials.

The mine designated the core material as a specific material classification, entered that classification into their Dispatch and then assign that material type for haulage to the core material stockpile / processing area. In this way the mine made the mining of raw core material part of their regular job and it was easy for them to produce and track performance. Given favorable geologic conditions, most mining operations can make core material in this manner and use their control systems such that production is routine and easy.

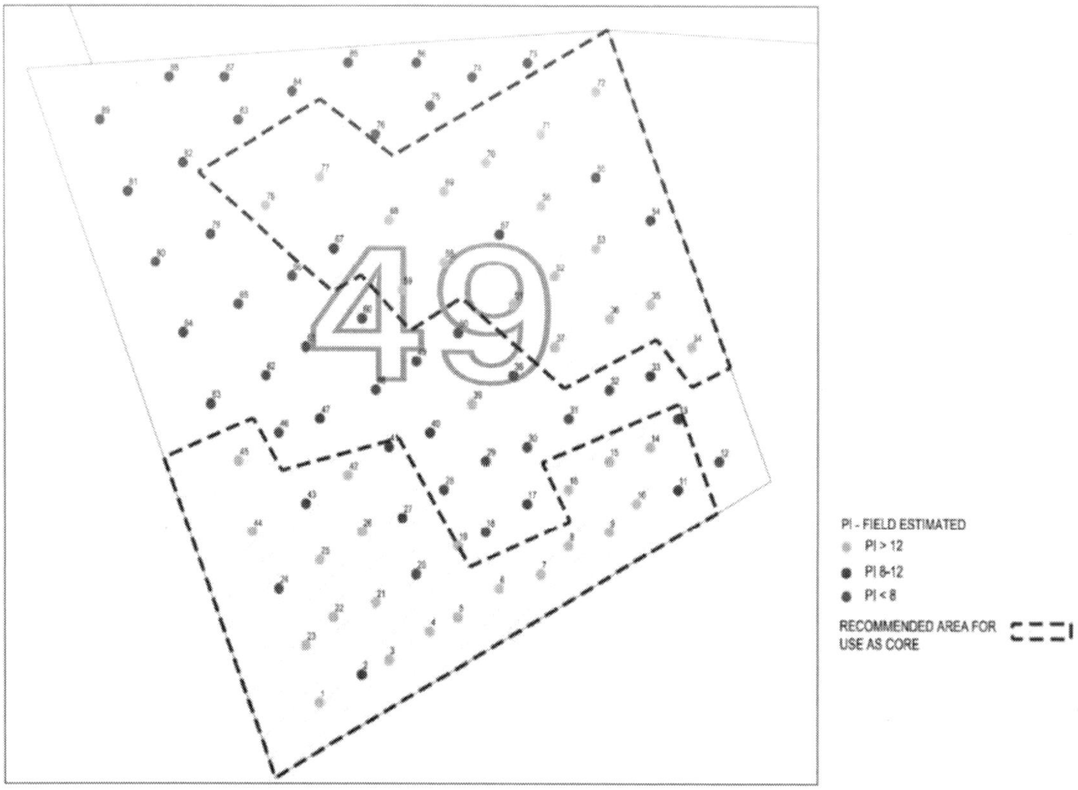

Figure 2 – Plasticity Assessment of a Typical Blast Project

PROCESSING OF RAW CORE MATERIALS

SMCV contracted SuperMix of Arequipa Peru to process the Zone 3 Core and Zone 7 Select Core materials. The raw feed material harvested from the west edge of the Santa Rosa pit was fed into Mestos Locotrack 620 and Terex Finley 595 mobile screening plants to produce the minus 15cm Zone 3 Core Materials and into a Mestos Locotrack ST 3.5, a Mestos Locotrack 620 and a Terex Finley 390 Hydratrak mobile screening plants to produce the minus 3.8cm Zone 7 Select Core Material. By visual inspection areas of segregated coarse materials were wasted prior to processing. The waste factors observed during processing of the Zone 3 and Zone 7 materials averaged 16% and 57% respectively. The Zone 3 and Zone 7 materials were then hauled to a temporary staging area by SuperMix to await haulage of the processed Core and Select Core materials to the Moisture Conditioning Area by SMCV using CAT 793 haul trucks.

Moisture conditioning of the core and select core was done by SMCV in the Ling Valley approximately 3 kilometers upstream of the dam. The CAT 793 haul trucks off loaded the Core and Select Core materials. SMCV then used CAT 992 loaders to stockpile the materials in loose piles/windrows 4 to 5 m in height by about 16 m in width. SMCV then installed sprinkler irrigation systems to distribute water across the top of the stockpiles, thus allowing the water to penetrate to the bottom of the stockpiles. This process extended over a period of 4 to 6 weeks and allowed the material to cure and further break down the argillaceous materials present in the Zone 3 material thus increasing the fines content of the core. At any one time there was about 100,000 m^3 of Zone 3 material moisture curing in the Linga Valley. The Construction Teams intent was that all water added to the core and select core materials would be done in the stockpile area and no water would be added for compaction purposes on the fill, with the exception that water was added to moisten lift surfaces that had dried out due to exposure to the sun and wind.

Larger size particles that segregated along the base and outside edges of the stockpiles were removed during the moisture conditioning phase. It was found that if this segregated rock was removed the Zone 3 Core material generally met the gradation requirement of 15% of the material passing the #200 sieve. Prior to hauling the core and select core to the dam each material was mixed with a CAT 992 loader to evenly distribute the moisture within the material and eliminate any segregated pockets which might be present.

TEST FILL PRIOR TO INITIAL CORE AND SELECT CORE CONSTRUCTION

During the September / October 2013 a test fill using Zone 3 Core material was constructed. The purpose of this test fill was to 1) verify that the core material was internally stable; and 2) assess the break down process by evaluating the effect of compaction and discing of the core materials.

The Zone 3 test fill was constructed with four 30 cm loose lifts Each lift surface was, at a minimum, of 45 meters by 15 meters giving sufficient area to operate the large compaction (CAT 825 Compactor) and spreading (CAT D8 Dozer and CAT 14M Motor Grader) equipment that would be needed to construct the test fills. In general the following were evaluated for the test fill:

- Material Placement and Compaction
  - In-situ Density and Moisture
  - Particle Breakdown / Gradation
  - Relative Compaction vs. Lift Thickness
  - Compactive Effort vs. Density
  - Internal stability utilizing the Burenkova method by Wan and Fell
  - Effectiveness of use of discing equipment
- In-situ Permeability
- Observation Trenches
  - Particle segregation (nesting) within and between lifts

Particle interference during compaction
Indication of excessive cracking
Uniformity of moisture conditioning
Particle breakdown of large size aggregates
QA/QC Field Testing Techniques and Methods

The results and observations from the test fill indicated that the Zone 3 material was internally stable and suitable for use as core material. Other conclusions resulting from the test fill included:

- After compaction, typically:
  - The percentage of fines increased 3.3% to 5.7%
  - The percentage of sands increased up to 5.3%
  - The percentage of gravels generally decreased
  - The maximum size of the gravels decreased
  - Gravels were surrounded by a matrix of sands and clayey fines
- The disc was recommended to moisture condition and mix Zone 3 material to increase sand and fines generation during compaction.
- Analyses using the Modified Burenkova Method (Wan and Fell, 2008) demonstrated that the Zone 3 material with a maximum size of 6 in was not susceptible to internal instability after moisture conditioning, discing, and compaction to the specified density.
- The procedures used to process the Zone 3 - Core produced materials that met the intent of the design.
- Pre-moisture conditioning was necessary to break down the sand and gravel sized argilic material that were present in both the Zone 3 and Zone 7 materials.

ZONE 3 AND ZONE 7 FOUNDATION PREPARATION

Zone 3 – Core Material – Foundation Preparation: The Zone 3 Core always had Zone 7 Select Core Material between it and the contacts with the bedrock foundation and abutments. Zone 3 material was never placed directly on or against the bedrock foundation or abutments. The lift preparation of the Zone 3 material required that the previous lift of Zone 3 or Zone 7 material was roughened with the pad feet of the CAT 825H compactor and moistened prior to the placement of the next lift.

Zone 7 – Select Core Material – Foundation Preparation: Prior to the placement of Zone 7 – Select Core Material in the cutoff trench and on the foundation bedrock upstream and downstream of the cutoff trench, the bedrock surface was cleaned and treated. Cleaning was performed by manually removing loose rock using bars, picks, shovels and demolition hammers. Rock and soil was removed by sweeping and using compressed air and pressurized water to remove the loose material and clean the rock surface to prepare it to receive dental concrete and slush grout.

After cleaning, dental concrete and/or slush grout was placed in small fractures and cracks in the rock where the Zone 7 - Select Core Material could not be placed and adequately compacted. Overhanging and steep slopes were flattened during excavation using a hydraulic hammers and/or excavators in order to obtain an appropriately shaped foundation surface such that Zone 7 material could be placed and compacted at the abutment contacts and meet the requirements of the design intent.

Dewatering of the cutoff trench at the valley bottom was accomplished using a series of diversion channels and sumps. Submersible pumps were placed in the sumps to evacuate the seepage water that accumulated. In areas of the upstream face and bottom of the right abutment contact of the cutoff trench where seepage was observed, contact grouting was performed using a thick Type HE cement grout mix (0.6:1 mix) to reduce the seepage inflow into the cutoff trench excavation. During the initial phases of placement of the Zone 7 material in the cutoff trench, the sumps were abandoned by backfilling with structural concrete.

As part of the dam foundation treatment, foundation grouting was performed starting from the bottom of the cutoff trench in the valley bottom and working upslope along both abutments.

HAULAGE, PLACEMENT AND COMPACTION

Zone 3 Core material was generally hauled to the Starter Dam using 20-m^3 dump trucks and on occasion CAT 789 Haul trucks. The Zone 3 material was spread in 30-cm loose lifts using a CAT D8 and/or Komatsu D155 bulldozers. The Zone 3 material was compacted using CAT 825H soil compactors. Typically the compactor made a minimum of 12 passes to reach 98% compaction as defined by a Standard Proctor (ASTM D698). Occasionally, when one of the two CAT 825H compactors was unavailable due to maintenance, a Bomag BW 219 DH Padfoot compactor was used. The lift preparation of the Zone 3 material required that the lower lift was always scarified or roughened and moistened prior to the placement of the next lift. When the lift surface started to dry as a result of exposure to the sun and wind prior to the placement of the subsequent lift, the lift surface was moistened using 5000-gallon water trucks. The fines content of the Zone 3 material generally increased by 2 to 5% during compaction.

The following equipment was generally used for hauling, distribution, placement and compaction of Zone 3 Core Material:

- 02 – CAT 825H Soil Compactor (32 tons)
- 01 – Bomag BW 219 DH Padfoot (19 ton) compactor
- 01 – Komatsu Bulldozer (D155 AX)
- 01 – Caterpillar Bulldozer (CAT D8 or D6)
- 01 – Volvo (FMX 440) and 01 Mercedes Benz (Actros 4144) 5,000-gallon water trucks
- 10 – Volvo (FMX 440) or Mercedes Benz (Actros 4144) 20 m^3 dump trucks
- 04 – CAT 789/793 haul trucks

- 01 – CAT 992 wheel loader

Zone 7 – Select Core Material: Zone 7 material was hauled to the Starter Dam using 20 m^3 dump trucks. The Zone 7 material was spread in 30-cm loose lifts using CAT D8 and Komatsu D155 bulldozers. The Zone 7 material was compacted using CAT 825H soil compactors. Typically the compactor made a minimum of 12 passes to reach 98% compaction as defined by the standard Proctor (ASTM D698). Adjacent to the abutments the Zone 7 material was spread in a 15 cm lift, and compacted with Wacker type compactors and a 2.5 ton smooth drum roller. The moisture content of the material within one meter of the abutment was increased such that it was +2% to +4% above optimum moisture content as defined by ASTM D-698 to provide a material that was more ductile at the abutment contacts. The lift preparation of the Zone 7 material required that the previous lift was always scarified and moistened prior to the placement of the subsequent lift. When the lift surface started to dry as a result of exposure to the sun and wind prior to the placement of the subsequent lift, the lift surface was moistened using 5000-gallon water trucks.

The equipment used for haulage, placement, moistening conditioning, and compaction of Zone 7 Select Core material was as follows:

- 02 – CAT 825H Soil Compactor (32 tons), same equipment used for Zone 3
- 01 – Bomag BW 219 DH Padfoot (19 ton) compactor
- 01 – Komatsu D155 AX bulldozer
- 02 – Volvo (FMX 440) or Mercedes Benz (Actros 4144) 5000-gallon water trucks
- 10 – Volvo (FMX 440) or Mercedes Benz (Actros 4144) 20 m^3 dump trucks
- 01 – JCB 3CPlus backhoe (92 HP)
- 01 – Bomag BW120 AD-4b roller (2.5 ton)
- 02 – Wacker Neuson vibratory rammers (2.4 Hp)
- 01 – Vibratory plate (140 kg)

HAUL TRUCK TRAFFIC ON CORE SURFACE

In general large haul trucks, CAT 789's and CAT 793's, were not permitted to cross the constructed Core Zone of the dam. On several occasions it was necessary to allow unloaded haul trucks to cross the core to access the upstream works. When these rare crossing occurred the area of the crossing was excavated to depths of 0.75 to 1 m. The Engineer would inspect the exposed surface and subsurface to verify that any distress, such as cracks and fissures in the core due to traffic were removed and replaced with compacted moisture conditioned material. Light haulage vehicles were allowed to cross the core and any contamination or damage of the core caused by haulage activities was removed and replaced with suitable materials.

On May 1, 2015, due to scheduling conflicts and work area constraints SMCV was forced to close the downstream rock quarry that had been supplying the Zone 1 rockfill to construct the downstream shell. This closure forced SMCV to utilize the upstream rock quarry to provide the Zone 1 Rockfill material needed to finish construction of the downstream shell. This closure would require SMCV to cross the core with fully loaded CAT 789 haul trucks with payloads of about 177 metric tons.

The engineer evaluated the conditions and required that SMCV construct an earth bridge over the core to limit distress that may leave cracks and fissures in the core if left untreated. The earth bridge was conservatively sized to limit the allowable bearing load of the fully loaded CAT 789 haul trucks to less than 6000 psf. The bridge dimensions were generally 2 m high by 12 m wide at its top and its length extended the full width of the core and filter zones. The bridge was constructed with Zone 3 core material. The bridge would allow about 5 to 6 m of fill to be placed before it had to be relocated and the area backfilled with the appropriate zoned materials. The subgrade beneath the bridge was removed a minimum depth of 0.5 m as directed by the Engineer, and inspected for cracks. Contaminated Zone 3 bridge material was removed and wasted while uncontaminated material was allowed to be used to backfill the gap left by the operation. The bridge was never reconstructed in the same location during construction.

QA/QC TESTING

Zone 3 – Core Material – QA/QC Results: The Zone 3 Core material was a low permeability clayey, sandy gravel (GC) located beneath the crest of the Starter Dam. The purpose of the low permeability central core zone was to limit potential seepage from the impoundment through the Starter Dam. The specifications indicated that percentage passing sieve No. 200 must be higher than 15%, and that the material have a minimum 10% plasticity index (PI), a minimum 20% liquid limit (LL) and a maximum particle size of 150 mm size. The placement moisture content was to be between optimum moisture content and 2% above optimum and the material was to be compacted to not less than 98% of the laboratory maximum dry density (ASTM D-698). The following is a summary of the QA/QC testing done on the Zone 3 Core.

Zone 3 – Core Material – Gradation: Zone 3 test results showed that the placed material had a maximum 15-cm particle size, with a fines content (percent passing the No. 200 sieve) in the range between 9.6% and 25.2% with an average of 17.1 %. At the start of Zone 3 placement, the fines content of the core material was observed to occasionally be below 15%. This situation was remediated by eliminating gravel and cobble size material that segregated from the stockpiles and with additional mixing in the stockpiles prior to loading the material into haul trucks. Above elevation 2445 masl the fines content was consistently above 15%. For the compacted material, the plasticity index (PI) ranged between 12 and 19 with an average of 15, and the liquid limit (LL) between 28 and 35 with an average of 31. Given the large sample size (400 kilos) and the time required to perform the gradation tests, smaller samples of 100 kg were used. Frequent testing was performed to compare the variation between the gradation tests for 400 kg samples and for 100 kg samples. It was found that the variation in the fines was minimal (less than 1%) and it was considered acceptable to use 100 kg samples.

Permeability tests were performed using the Guelph Permeameter on the compacted layers. The hydraulic conductivity values obtained had a geometric mean of 2.6e-6 cm/s. Three of the permeability tests were performed at elevations ranging from 2412.5 to 2419.5 masl where the fines contents were below the specified 15% values and those tests showed permeability results ranging from 2.1e-6 cm/s to 3.5e-6 cm/s indicating sufficiently low permeability values were being obtained.

Zone 3 – Core Material – Moisture Content: The moisture content of the Zone 3 material was measured and included the whole sample (minus 15 cm). Moisture content of the placed and accepted material ranged between 5.8% and 11.0%, with an average of 9.0%. The optimum moisture content of the Zone 3 material, as defined by ASTM D 698, ranged between 6.0% and 8.4% with a 7.1% average. Moisture values lower than optimum were corrected in the field by ripping the lift, adding additional water and mixing the material with a motor grader to uniformly increase the moisture content prior to re-compaction of the lift.

Zone 3 – Core Material – Density Tests: The density of the Zone 3 Core placed material was calculated correcting for the plus 3/4–inch material (ASTM D 4718 – Standard Practice for Correction of Unit Weight and Water Content for Soils Containing Oversize Particles). The corrected values of maximum dry density and optimum moisture of the selected Proctor curve, as defined by ASTM D698, were used. A total of 54 Proctor tests were taken of the core material. The Proctor curve used to calculate the percent compaction and the optimum moisture content, was selected from a family of Proctor curves based on One Point-Proctor Tests performed on each density test sample in the field. The test frequency to measure field density was a minimum of one test for every 1,500 m^3 of Zone 3 placed. Because the maximum particle size was 15 cm, the water replacement method ASTM D 5030 (Standard Test Methods for Density of Soil and Rock in Place by the Water Replacement Method in a Test Pit) was used to measure the in-place density. The in-place dry density values observed in the field ranged from 2,100 kg/m^3 to 2,320 kg/m^3 with an average dry density of 2,200 kg/m^3. When a density test failed, the lift under evaluation was re-compacted, or the moisture content modified, and reworked until a compaction percentage of 98% or higher was achieved. The maximum percentage of maximum proctor compaction calculated was 105%, the minimum was 98% and the average was 99.4%.

Zone 7 – Select Core Material – QA/QC Results: The following is a summary of the QA/QC testing done on the Zone 7 Select Core

Zone 7 – Gradation: Zone 7 test results show that the placed material had a maximum 38 mm particle size, with a fines content passing the No. 200 sieve in the range between 19.1 and 33.1% with an average of 23.2%. When values of less than 20% of fines were observed in the field the Zone 7 material was remixed to eliminate zones of segregation and the area in question retested. For the compacted material, the plasticity index (PI) ranged between 12 and 18 with an average of 15 and the liquid limit (LL) ranged between 26 and 35 with an average of 31. This material generally classified as clayey gravel (GC) as per ASTM D 2487. Permeability tests were performed using the Guelph Permeameter on the compacted layers. The hydraulic conductivity values obtained had a geometric mean of 2.2e-6 cm/s.

Zone 7 – Moisture Content: The moisture content of the Zone 7 material was measured and included the whole sample. Moisture content of the placed material ranged between 7.0% and 11.9%, with an average of 9.3%. The optimum moisture content of the Zone 7 material, as defined by ASTM D 698, ranged between 7.3% and 10.6% with an 8.4% average. Moisture values lower than optimum values were corrected in the field by ripping the lift of material, adding water and mixing the material with a motor grader to uniformly increase the moisture content, and then re-compacting the lift. The placed material had a moisture content ranging between optimum moisture content and +2% of optimum with an average of 0.9% above optimum moisture content. As discussed above, the material placed within 0.50 m of the abutment had moisture contents +2% to +4% of optimum moisture.

Zone 7 – Density: The Density of the Zone 7 Select Core material was calculated correcting for the plus 3/4–inch material for the in-place material. The corrected values of maximum dry density and optimum moisture of the selected Proctor curve, as defined by ASTM D698, were used. A total of 61 Proctor tests were taken of the Select Core material. The Proctor curve, to evaluate the percent compaction and the optimum moisture content, was selected from a family of Proctor curves based on One-Point Proctor Tests performed on a material sample from the excavated select core material for each density test performed in the field. The specified testing frequency to monitor field density was a minimum of one test for every 1500 m^3 of Zone 7 placed, but in general one density test was done at each abutment for every lift of Zone 7 material resulting in a more frequent testing frequency. In-situ field densities were measured utilizing sand cone methods as defined by ASTM D-1556. The in-place dry density values observed in the field ranged from 2,039 kg/m^3 to 2,221 kg/m^3 with an average dry density of 2,109 kg/m^3. When a density test indicated a failing result, the lift under evaluation was re-compacted, and/or the moisture content modified, and reworked until a compaction percentage of 98% or higher was achieved. Based on the density test results, the maximum percentage of compaction was 103.5%, the minimum was 97.9% and the average was 99.5%.

RESERVOIR FILLING

In late January of 2015 the filling of the startup reservoir was begun. At this time the dam construction had been completed to approximate elevation 2490 with 80 meters left to construct. Due to drought conditions and water demand from mine operations the availability of water was sparse and often erratic at times. Because the varying quantities of water available during reservoir filling some flexibility in the fill rates was built into the process. The designer required an average daily rise in the in the water elevation of 0.3 m with a maximum daily rise of 0.5 m. The maximum allowable elevation gain per week was held at 2.1 m per any consecutive period of 7 days. After a period of 8 months of filling and just prior to the start of tailing deposition on the downstream toe of the dam, the observed seepage rate was relatively low with a maximum measured seepage of 180 liters per minute corresponding to a head of approximately 67 m of water on the base of the core.

REFERENCES

Wan, Chi F and Robin Fell. 2008, "Assessing the Potential of Internal stability and Suffusion in Embankment Dams and Their Foundations." *Journal of Geotechnical and Geoenvironmental Engineering*. Vol. 134 No. 3. March 2008, pp. 401-407

ESTIMATING THE RESERVOIR STAGE-FREQUENCY CURVE WITH UNCERTAINTY BOUNDS FOR CHERRY CREEK DAM USING THE RESERVOIR FREQUENCY ANALYSIS SOFTWARE (RMC-RFA)

Haden Smith, P.E.[1]
John F. England, Jr., Ph.D., P.E., P.H., D.WRE[2]

ABSTRACT

The U.S. Army Corps of Engineers (USACE) Risk Management Center (RMC) developed the Reservoir Frequency Analysis software (RMC-RFA) to facilitate hydrologic hazard assessments within the USACE Dam Safety Program. RMC-RFA produces a reservoir stage-frequency curve with uncertainty bounds by utilizing a deterministic flood routing model while treating the seasonal occurrence of the flood event, the antecedent reservoir stage, inflow volume, and the inflow flood hydrograph shape as uncertain variables rather than fixed values. In order to quantify both the natural variability and knowledge uncertainty in reservoir stage-frequency estimates, RMC-RFA employs a two looped, nested Monte Carlo methodology. The natural variability of the reservoir stage is simulated in the inner loop defined as a realization, which comprises many thousands of events, while the knowledge uncertainty in the inflow volume-frequency distribution is simulated in the outer loop, which comprises many realizations.

Cherry Creek Dam is a USACE dam located in Denver, CO, and is considered one of the highest priority dams in the USACE portfolio. Cherry Creek Dam is currently being evaluated in a USACE Dam Safety Modification Study (DSMS). For this study effort, a comparative stage-frequency curve analysis was performed using RMC-RFA to evaluate the potential risk reduction benefits of four spillway design alternatives. Each reservoir stage-frequency simulation in RMC-RFA took less than five minutes of run time, permitting the rapid evaluation of the change in overtopping frequency for each design alternative. This application of RMC-RFA illustrates the efficiency and efficacy that can be gained from performing Monte Carlo simulations that explore a more extensive variety of possible flood scenarios at the reservoir as compared to traditional deterministic approaches.

INTRODUCTION

There are two primary components of randomness in reservoir stage exceedance probabilities: natural variability and knowledge uncertainty. Natural variability is best described as the effect of randomness and is a function of the system (Vose 2008). It is not reducible through either study or further measurement. For example, a flow-frequency curve describes the natural variability in flow.

[1]Hydrologic Engineer, US Army Corps of Engineers, Risk Management Center, 12596 West Bayaud Ave, Suite 400, Lakewood, CO 80228, Cole.H.Smith@usace.army.mil.
[2]Lead Hydrologic Engineer, US Army Corps of Engineers, Risk Management Center, 12596 West Bayaud Ave, Suite 400, Lakewood, CO 80228, John.F.England@usace.army.mil.

Knowledge uncertainty is the lack of knowledge about parameters that characterize the system being modeled. Knowledge uncertainty can be reduced through further measurement or study. For example, the confidence intervals, or uncertainty bounds, around a flow-frequency curve describe the knowledge uncertainty in the statistical flow-frequency curve parameters.

The Reservoir Frequency Analysis software (RMC-RFA) produces a reservoir stage-frequency curve with uncertainty bounds by utilizing a deterministic flood routing model while treating the inflow volume, the inflow flood hydrograph shape, the seasonal occurrence of the flood event, and the antecedent reservoir stage as uncertain variables rather than fixed values. In order to quantify both the natural variability and knowledge uncertainty in reservoir stage-frequency estimates, RMC-RFA employs a two looped, nested Monte Carlo methodology. The natural variability of the reservoir stage is simulated in the inner loop defined as a realization, which comprises many thousands of simulated flood events, while the knowledge uncertainty in the inflow volume frequency distribution is simulated in the outer loop, which comprises many realizations. The basic construct of the simulation procedure employed by RMC-RFA is illustrated in Figure 1.

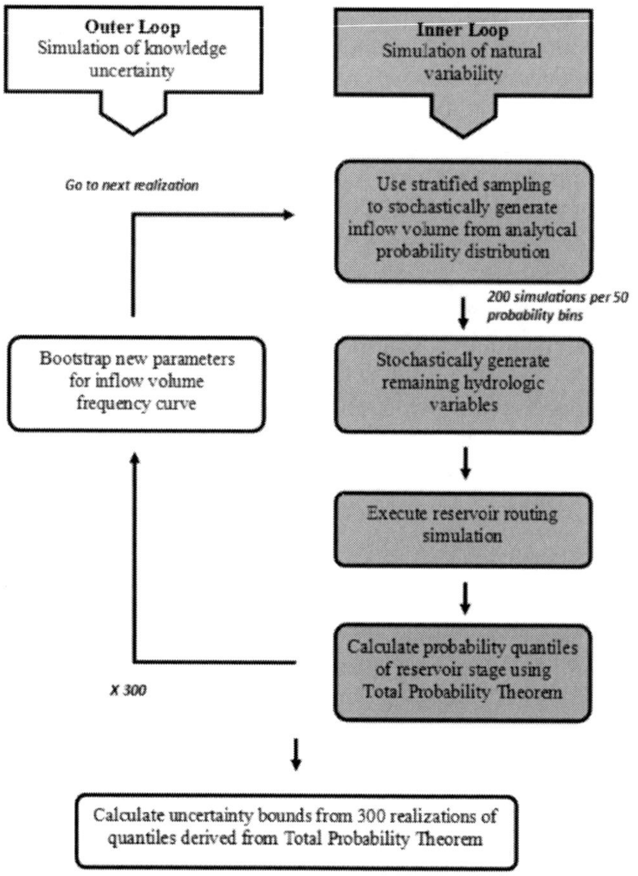

Figure 1. Flowchart of the steps involved in the RMC-RFA simulation.

STRATIFIED SAMPLING APPROACH

Standard Monte Carlo sampling procedures are computationally inefficient. The bulk of the computation burden is expended on sampling flood events in the range of exceedance probabilities that are not typically important in risk assessments. For example, in a Monte Carlo simulation with 10,000 events, there are only 1,000 events that provide any information about flood events with an annual exceedance probability less than 1 in 10.

Consequently, to ensure computational effort is focused on extreme flood events, RMC-RFA uses a *stratified sampling* approach based on procedures outlined in (Nathan and Weinmann, 2013) and (Nathan, Jordan, et al. 2016). The procedure involve first dividing the inflow volume-frequency curve into 50 bins uniformly spaced over the Extreme Value Type I (EVI) probability domain. Within each bin, 200 inflow volumes are stochastically sampled. For each sampled inflow volume, a reservoir routing simulation is executed using a set of model parameters (as described in Table 1) that are sampled using standard Monte Carlo procedures. This process is performed for each of the 200 sampled inflow volumes within each of the 50 bins, for a total of 10,000 reservoir routing simulations. The routing results from the 10,000 simulations are combined using the Total Probability Theorem to produce a reservoir stage-frequency curve.

With only 10,000 simulations, accurate reservoir stage-frequency results can be attained within the exceedance probability range of 0.99 to $\sim 10^{-8}$ using the stratified sampling approach. Conversely, with only 10,000 simulations, standard Monte Carlo sampling procedures will only offer accurate results within the range of 0.99 to $\sim 10^{-4}$.

Table 1. Model parameters treated as random variables.

Input Parameter	Dependency	Statistical Distribution	Sampling Approach
Inflow Volume	Independent	Analytical	Stratified
Inflow Hydrograph Shape	Independent	Empirical	Monte Carlo
Flood Season	Independent	Empirical	Monte Carlo
Reservoir Starting Stage	Flood Season	Empirical	Monte Carlo

THE PARAMETRIC BOOTSTRAP

A significant portion of the knowledge uncertainty in the reservoir stage-frequency curve is due to sampling error, which is caused by observing the sample instead of the population. The sampling error is the difference between the sample statistic used to estimate the population parameter, such as the mean or variance, and the actual but unknown value of the parameter. Sampling error is a function of the record length of observed reservoir inflow, stage, and discharge data. As the record length increases, the sampling error decreases; thus, the knowledge uncertainty decreases.

The uncertainty due to sampling error can be estimated using the *bootstrap*, which was originally introduced in 1979 as a computer-based method for estimating standard errors by Dr. Bradley Efron in his paper "Bootstrap methods: Another look at the jackknife" (Efron 1979). Typical applications of the bootstrap method use the empirical cumulative distribution function as the approximating distribution for the observed data. However,

RMC-RFA uses the *parametric bootstrap*, which is a variation of the general bootstrap, in which the approximating distribution for the observed dataset is an analytical probability distribution. The bootstrap procedure in RMC-RFA involves the following steps:

1. Randomly sample *N* inflow volumes from a user-defined volume-frequency curve (*N* is equal to the effective record length of the annual maxima inflow volume data).
2. Estimate the statistical parameters of interest (such as mean, standard deviation, and skew) from the bootstrapped sample of size *N*.
3. Using the new statistical parameters, fit a new inflow volume-frequency curve.
4. Repeat steps 1 through 3 for a sufficiently large number of realizations in order to derive uncertainty bounds (300 realizations were used for this study).

An illustration of steps 1 through 3 is show below in Figure 2. Histogram representing the uncertainty in skew (of log) developed from 300 bootstrap realizations is shown in Figure 3. For greater details on the parametric bootstrap, see Efron's "An Introduction to the Bootstrap" (Efron and Tibshirani, 1998).

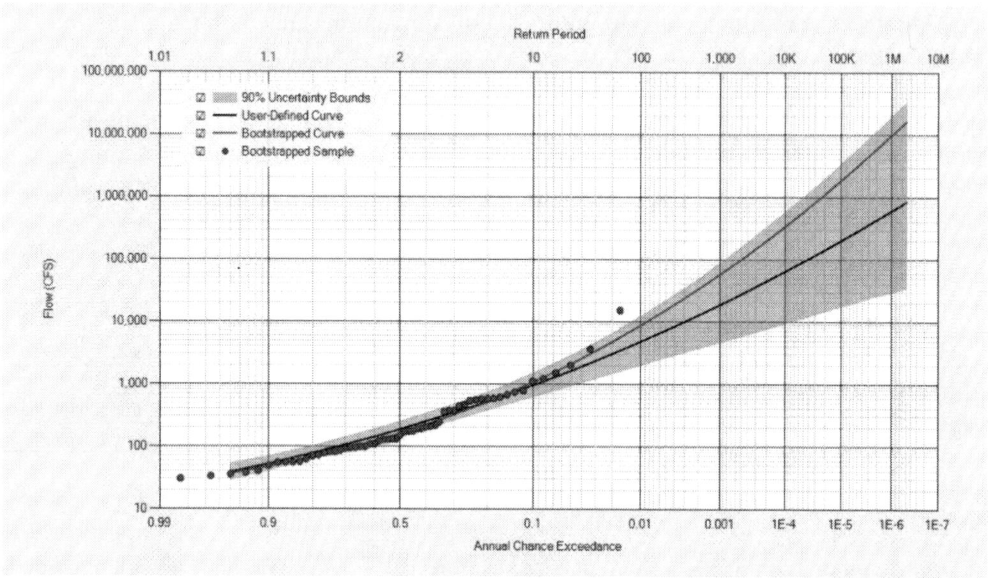

Figure 2. One bootstrapped sample derived from a user-defined Log Pearson Type III distribution.

In RMC-RFA, inflow volume-frequency relationships can be defined using the Log Normal, Log Gumbel, Log Pearson Type III, or Log Generalized Extreme Value distribution, allowing the user to also assess the range of uncertainty in reservoir stage-frequency due to statistical model selection. An example of the uncertainty due to statistical model selection is provided in Figure 4.

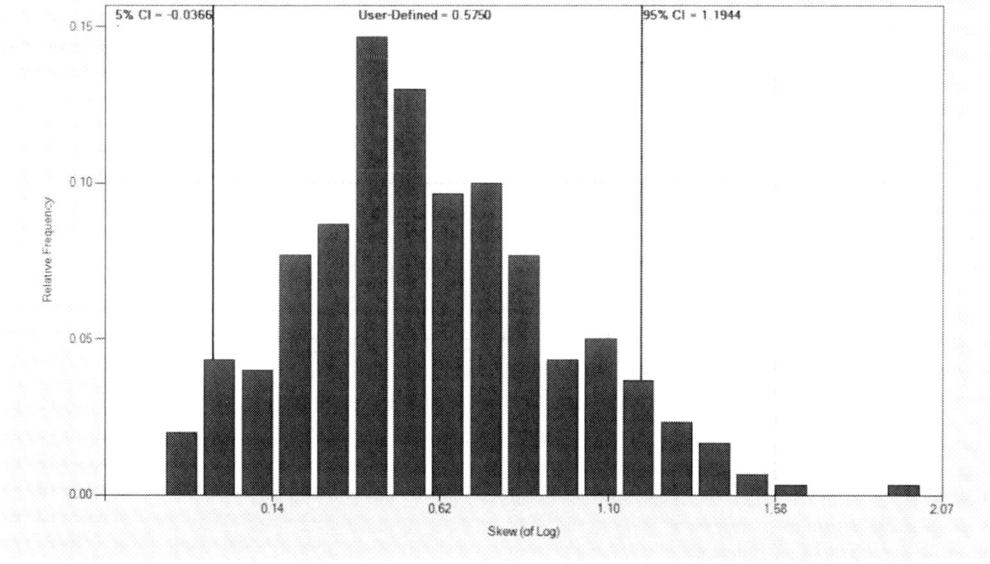

Figure 3. Histogram of 300 bootstrap realizations of skew (of log). The solid lines represent the 5% and 95% percentiles of the histogram.

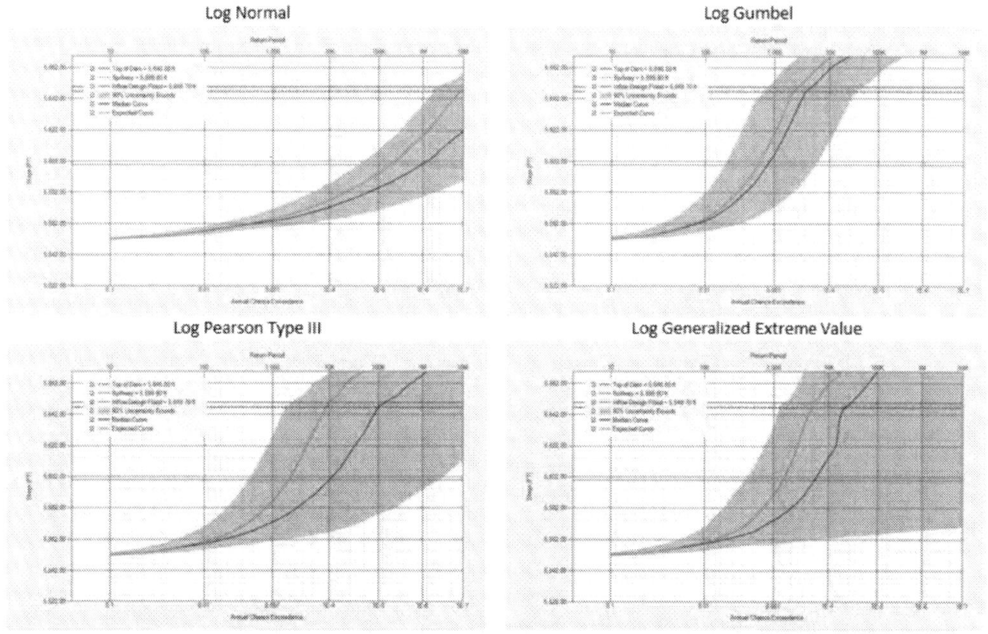

Figure 4. RMC-RFA allows four different analytical probability distributions for inflow volume, enabling the user to assess the range of uncertainty in reservoir stage-frequency due to statistical model selection.

CHERRY CREEK DAM

Cherry Creek dam is located on Cherry Creek in Arapahoe County, Colorado, southeast of the city of Denver. The dam site is 11.4 miles above the confluence of Cherry creek with South Platte River. The elongated leaf-shaped Cherry Creek drainage area has an average width of about 11 miles and a length of approximately 57 miles, with the main channel located approximately through the center of the watershed as seen in Figure 5. Its confluence with the South Platte River is located in the heart of the business district of Denver; and the lower 5.5 miles of the creek courses through the residential and industrial districts of the city. Below the confluence with Cherry Creek, the South Platte River flood plain contains the heaviest concentration of economic development within the Denver area. The reach in Denver proper contains railroad facilities, dense concentrations of industrial and commercial properties and occasional older residential areas. Downstream of Denver proper, the flood plain is principally rural with agriculture being the major land use.

Figure 5. Map of the Cherry Creek Watershed.

Dam Safety Modification Alternatives

The four alternatives being considered for the dam safety modification are summarized in Table 2. A summary of relevant hydrologic features for each alternative is provided in Table 3. The stage-storage relationship as well as the stage-discharge rating curves for each alternative are shown in Figure 6 and Figure 7, respectively. It can be seen that the variation in discharge capacity between each alternative is limited. The spillway configuration is the same for Alternative 1, Alternative 2 and Alternative 4. Whereas, Alternative 3 has a higher spillway crest than the others. Downstream channel capacity limits the release through the primary outlet works, thus limiting the degree to which the Water Control Plan (WCP) can be altered.

Table 2. Descriptions of structural alternatives evaluated with RMC-RFA.[3]

Alternative 1	There is no changes to the WCP. However, there is a structural change to the spillway. The original spillway design had a crest elevation of 5,599.8. The current crest elevation is 5,610.5 due to sloughing. The spillway will be returned to its original crest elevation. The Inflow Design Flood (IDF) will still overtop the dam by 2.9 feet.
Alternative 2	This is a WCP change to increase the allowable releases through the primary outlet works from 5,000 cfs to 7,000 cfs. The spillway will be returned to its original crest elevation. The Inflow Design Flood (IDF) will still overtop the dam by 2.9 feet.
Alternative 3	This is a WCP change to increase the allowable releases through the primary outlet works from 5,000 cfs to 7,000 cfs. The embankment will also be raised 7.1 feet to contain the PMF accounting for risk based freeboard requirement of 1.5 feet. However, the current spillway crest of 5,610.5 will be maintained.
Alternative 4	This is a WCP change to increase the allowable releases through the primary outlet works from 5,000 cfs to 7,000 cfs. The embankment will also be raised 4.2 feet to contain the PMF accounting for risk based freeboard requirement of 1.5 feet. The spillway will be returned to its original crest elevation.

Table 3. Summary data for alternatives.

Alternative	Dam Raise (ft)	Dam Crest (NAVD88)	Spillway Crest (NAVD88)	IDF Elevation (Peak Stage)	Max Spillway Flowrate (CFS)
1	0	5,646.80	5,599.80	5,649.70	51,525
2	0	5,646.80	5,599.80	5,649.70	51,525
3	7.1	5,653.90	5,610.50	5,652.40	46,920
4	4.2	5,651.00	5,599.80	5,649.00	51,525

[3] These alternatives are examples of ones currently being considered and subject to change.

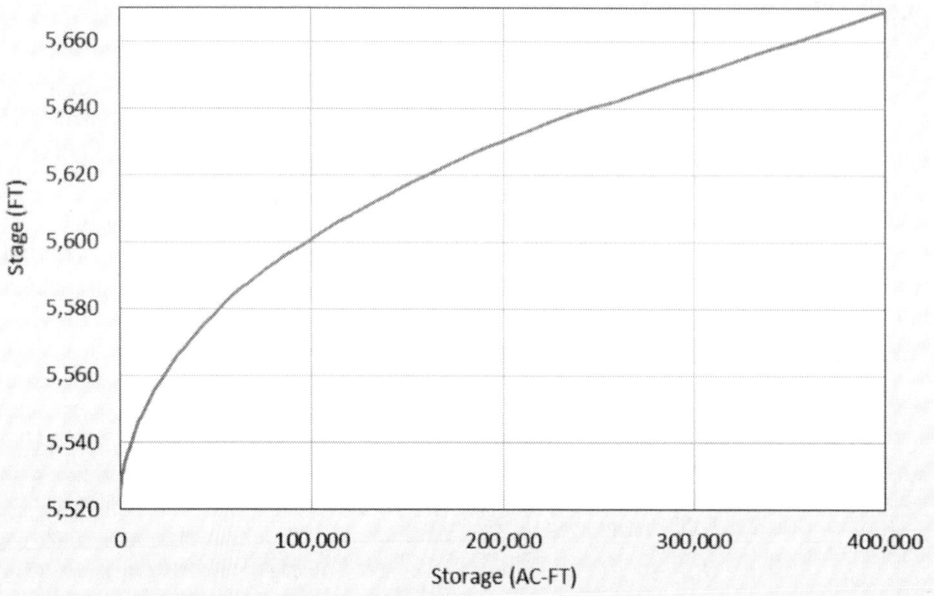

Figure 6. Stage-storage relationship for all alternatives.

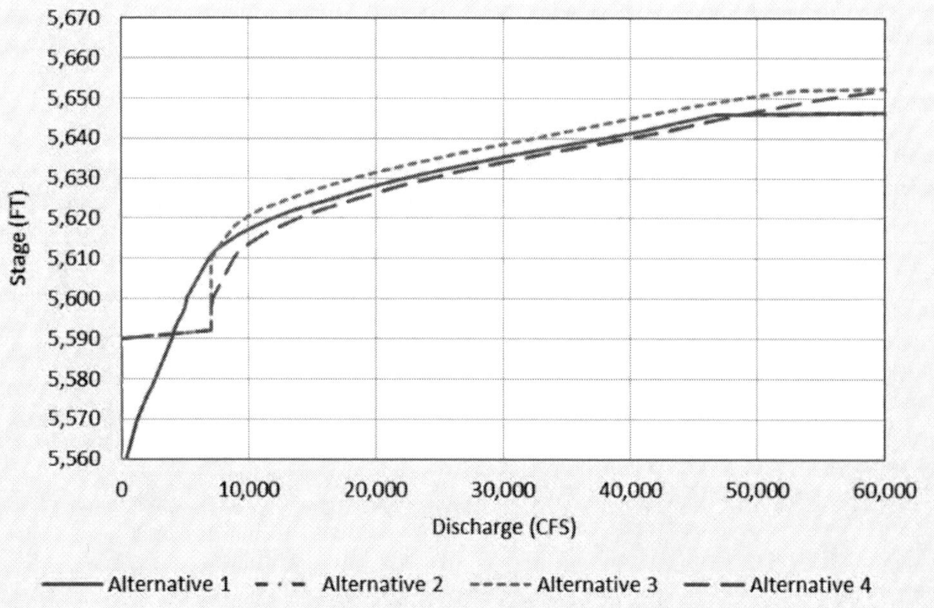

Figure 7. Stage-discharge relationships for all alternatives.

Volume-Frequency Curve

The critical inflow duration was determined to be one day. The volume-duration-frequency curve was derived using daily average inflow data at the dam site for water years 1957-2015, with an effective record length of 59. The volume-frequency was simulated using a Log Pearson Type III distribution with a mean (of log) of 2.32, a standard deviation (of log) of 0.499, and a skew (of log) of 0.575.

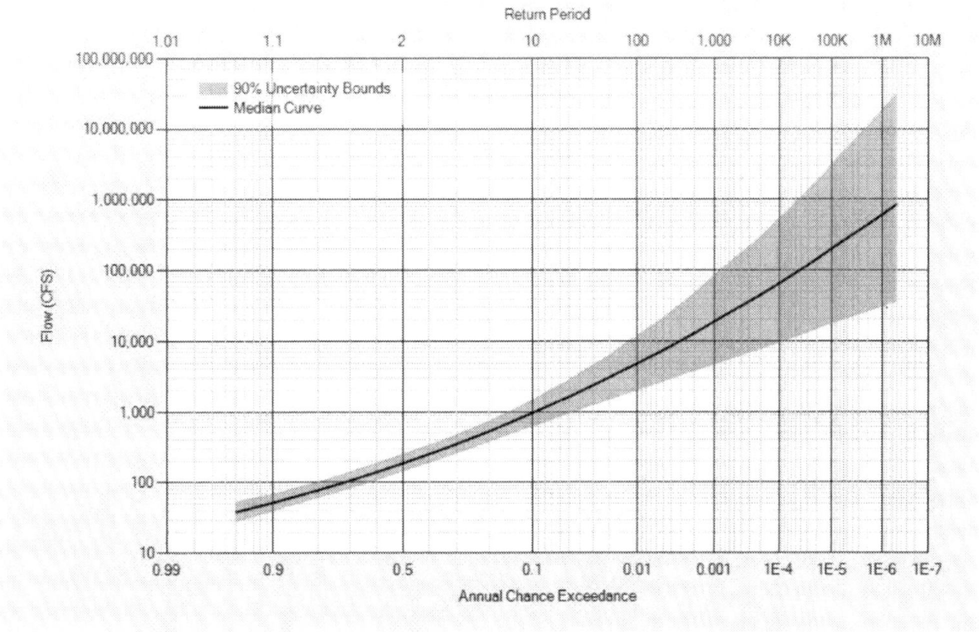

Figure 8. Volume-frequency curve as plotted in RMC-RFA.

Inflow Hydrograph Shapes

Inflow hydrographs from the June 1965 and May 1973 events, as depicted in Figure 6, were used for rescaling the sampled inflow flood events. The 1965 event had an hourly-average peak inflow of 58,000 cfs, and the 1973 event had an hourly-average peak inflow of 4,700 cfs. The 1965 event had a very short critical duration but a much larger peak inflow. Both hydrographs suggest the critical inflow duration is 24 hours or less. The two hydrograph shapes were given equal probability of occurrence since there is limited hydrograph data available for large events in the watershed to suggest a different weighting scheme. Results indicate that the stage-frequency curve is not sensitive to the hydrograph shapes.

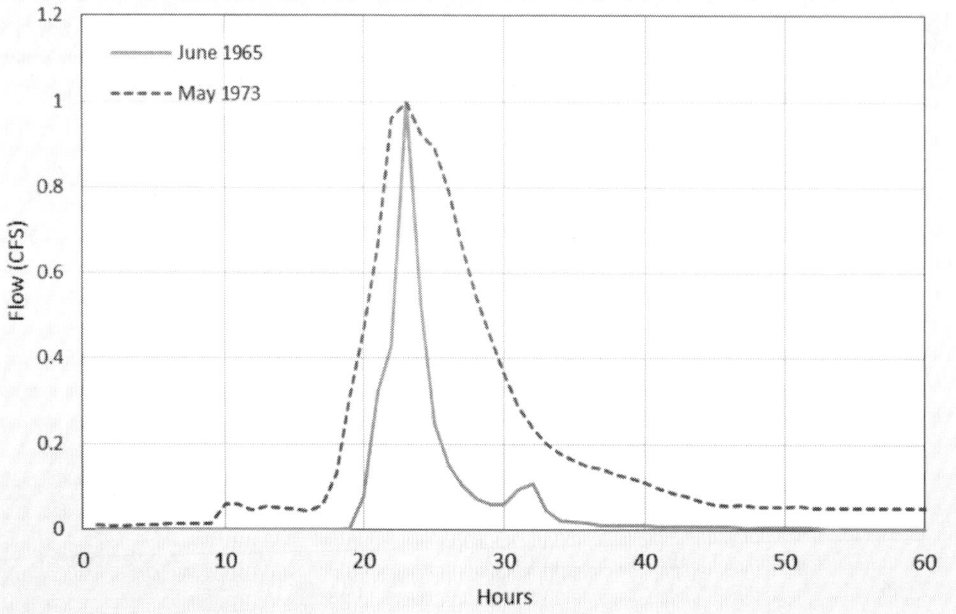

Figure 9. The June 1965 and May 1973 inflow hydrographs normalized by peak flow.

Flood Seasonality

The term *flood seasonality* is intended to describe the frequency of occurrence of floods on a seasonal basis. The flood seasonality is used to sample a starting reservoir stage consistent with the time of year when floods are likely to occur, which can be very important for reservoirs that operate with seasonal fluctuation in pool for purposes such as flood control and water supply. For the purposes of RMC-RFA, a rare flood is defined as any event where the inflow volume exceeds some user specified threshold. The flood threshold is usually chosen as a flow that corresponds to a specified frequency level (i.e. the 5-year or 10-year return period) for a specified critical duration to provide a metric for counting the number of flood occurrences in the historic record. The flood seasonality histogram developed for this analysis can be seen below in Figure 10. This is consistent with the regional climate. March, April and May bring snowmelt driven inflows. Starting in early summer, increases in tropical moisture result in short late-afternoon thunderstorms. The tropical moisture decreases in the late fall as artic air combines with moisture from the Pacific Northwest bringing snowfall to the region.

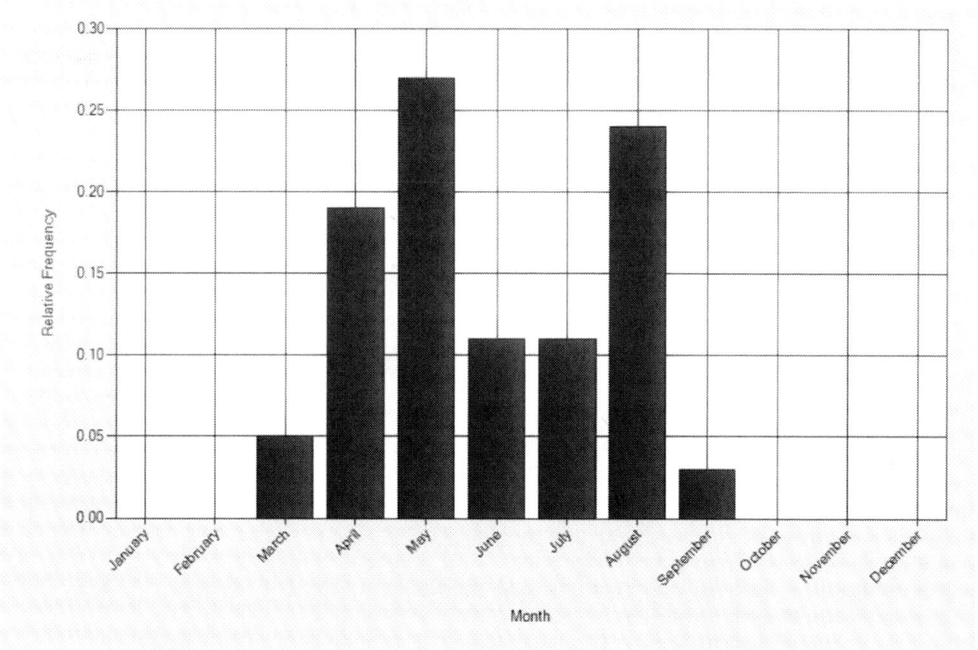

Figure 10. Flood seasonality histogram for Cherry Creek Dam.

Reservoir Starting Stage Duration

Stage duration curves represent the percent of time during which specified reservoir stages are exceeded at a given location. Ordinarily, daily variations in stages are inconsequential, so duration curves are typically developed using observed daily average reservoirs pools. *Reservoir starting stage* duration curves represent the percent of time during which antecedent reservoir stages are exceeded. Starting stage duration curves are developed by first filtering observed daily average stages so that they only represent typical starting stages based on a pool change threshold. Then, the filtered data set is sorted by month or season. This is done to avoid a certain degree of "double counting" flood events.

For this analysis, historical daily stage data was deleted from the record set if the pool was rising more than a 0.2 feet per day. The pool data was also filtered to remove the rising limb, peak, and receding limb of any stage hydrograph that is 30 days in duration or less. The starting stage elevation data and monthly stage duration curves used for this analysis can be seen in Figure 11 and Figure 12, respectively.

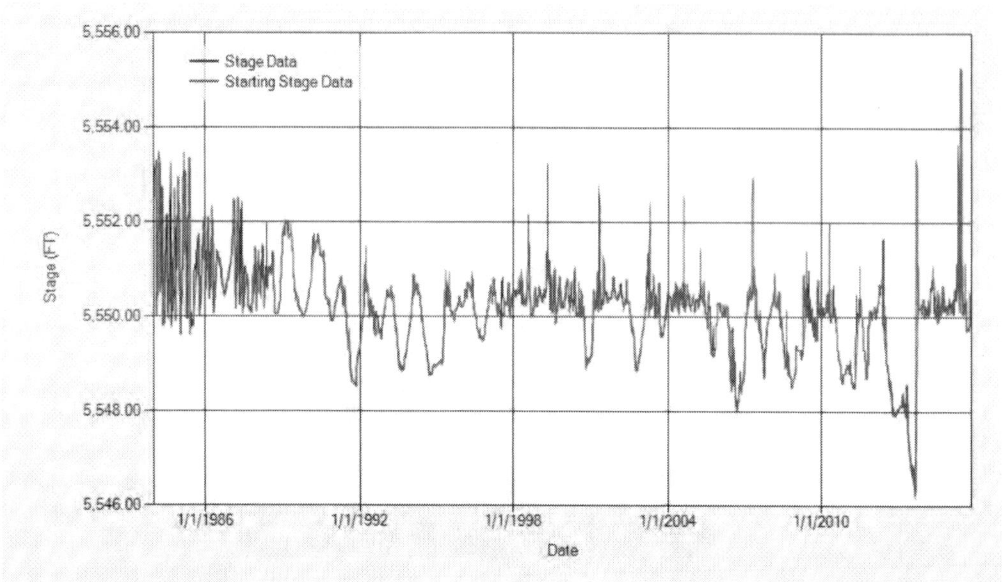

Figure 11. Comparison of historical stage elevations and starting stage data.

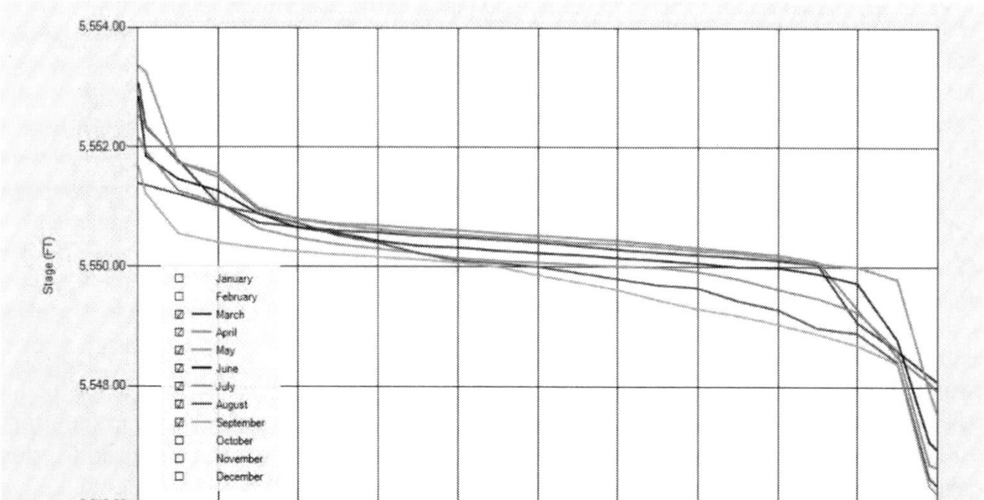

Figure 12. Monthly starting reservoir stage duration curves for Cherry Creek.

STAGE-FREQUENCY RESULTS

Figure 13 shows the reservoir stage frequency results for Alternative 1. The *median curve* represents the uncertainty in stage-frequency due to natural variability, the *uncertainty bounds* represent the uncertainty in stage-frequency due to knowledge uncertainty, whereas the *expected curve* represents the combined uncertainty due to both natural variability and knowledge uncertainty.

A high degree of knowledge uncertainty due to a short record length results in an asymmetrical distribution for both very high and very low reservoir stages. As a result, the median curve does not adequately represent the long tail of the probability distribution. Therefore, instead of using the median to represent the "best-estimate" probability of exceedance, the mean is used for this analysis. The expected curve is considered the "best-estimate" because it reflects the relative likelihood of all probabilities of a stage exceeding a certain predefined stage, rather than the point where 50% of the exceedance probabilities lie either above or below the median. The expected curve implies that on average the estimated exceedance probability for a given reservoir stage is correct.

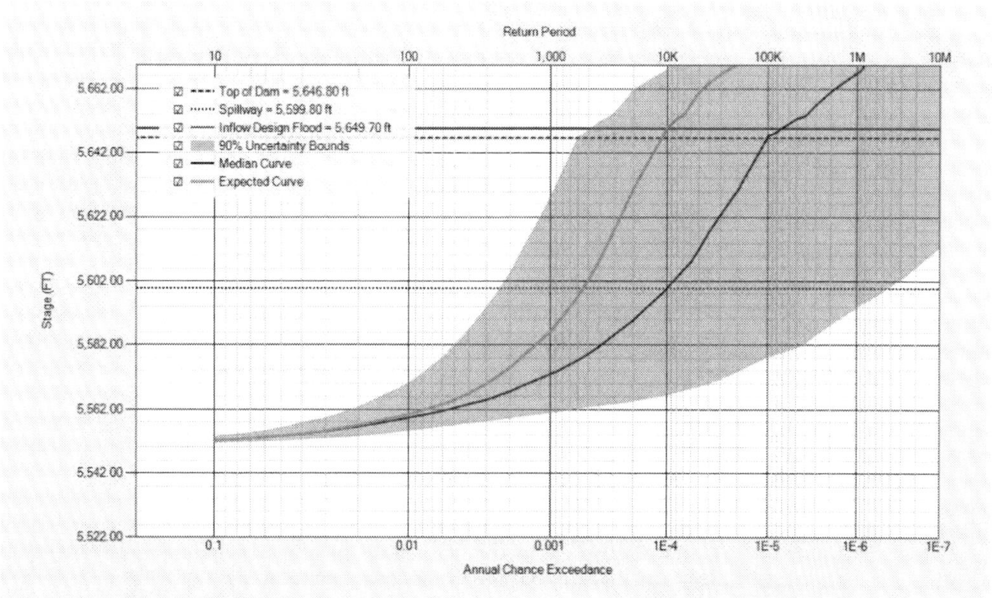

Figure 13. Reservoir stage-frequency curve results for Alternative 1.[4]

[4] The inflow volume-frequency curve in Figure 8 and the stage-frequency curves shown in Figure 13 and Figure 15 are examples to illustrate the RMC-RFA software. These frequency curves were not used in final dam safety modification decisions for Cherry Creek Dam.

Figure 14 shows an example of an overtopping flood event that occurred while simulating Alternative 1. The event occurred in a realization that had the following characteristics:

- The inflow volume-frequency curve had a mean (of log) of 2.21, a standard deviation (of log) of 0.393, and a skew (of log) of 0.799.
- The sampled 1-day inflow volume was 330,378 cfs, which equates to a ~3 million year event.
- The 1-day inflow volume was scaled using the 1973 hydrograph shape.
- The flood event occurred in June.
- The reservoir starting stage was 5,553.3 ft, which is exceeded less than 1% of the time.
- The resultant peak stage was 5,653.33 ft.

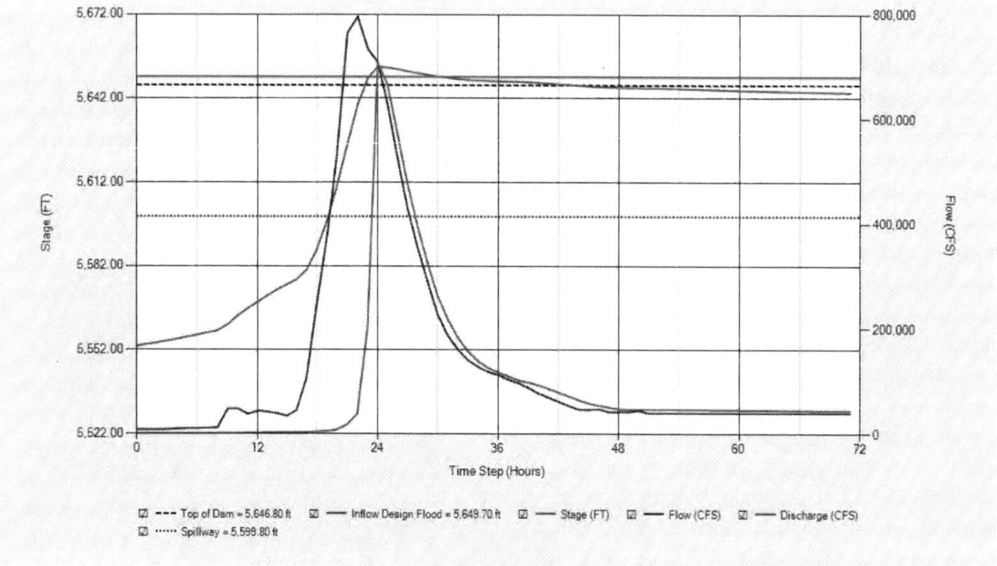

Figure 14. Example of a simulated overtopping flood event.

This event illustrates the ability of the RMC-RFA simulation to create plausible extreme flood scenarios that will cause overtopping.

The expected reservoir stage-frequency curve results for all alternatives are provided in Figure 15. Each simulation only required five minutes of runtime.[5]

[5] Runtimes are based on the following computer specifications:
Intel Core i7-4900MQ CPU @ 2.80 GHz (4 cores – 8 threads) and 32 GB of RAM

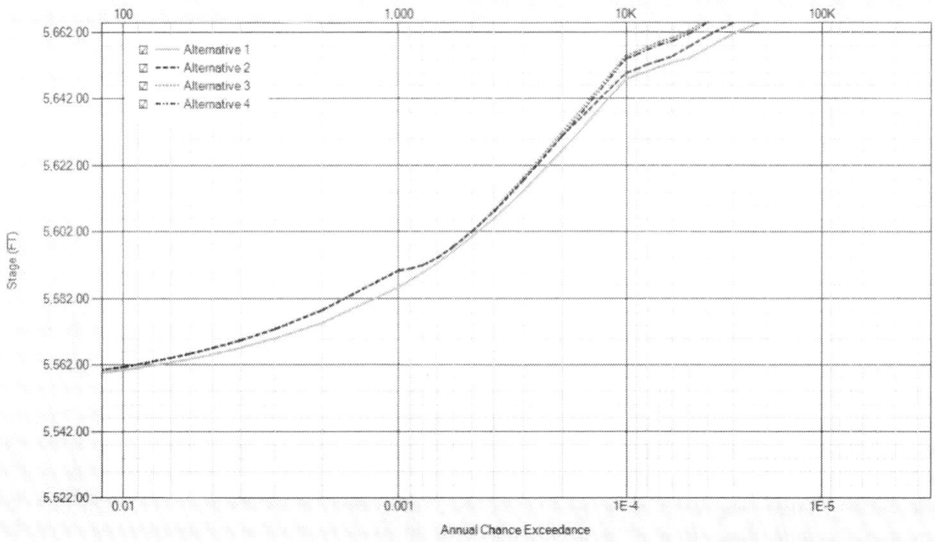

Figure 15. The *expected* reservoir stage-frequency curve results for all alternatives.

CONCLUSION

The results from the Cherry Creek Dam analysis indicate that there is very little incremental risk reduction benefits between each alternative. The difference in overtopping flood hazard between each of the design alternatives is insignificant considering the large uncertainty in the reservoir stage-frequency estimates.

Each alternative was simulated with 300 realizations, each comprising 10,000 stochastic flood events (for a total of 3 million flood events), requiring only five minutes of runtime for each simulated alternative. This application of RMC-RFA illustrates the efficiency and efficacy that can be gained from performing Monte Carlo simulations and the added value of exploring a more extensive variety of possible flood scenarios at the reservoir as compared to more traditional deterministic approaches.

REFERENCES

Efron, Bradley. 1979. "Bootstrap methods: another look at the jackknife." *The Annals of Statistics* 7 (1): 1-26.

Efron, Bradley, and Robert J Tibshirani. 1998. *An Introduction to the Bootstrap.* Boca Raton: CRC Press LLC.

Nathan, Rory, and Erwin Weinmann. 2013. *Autralian Rainfall And Runoff Discussion Paper: Monte Carlo Simulation Techniques.* Engineers Australia: Water Engineering.

Nathan, Rory, Phillip Jordan, Matthew Scorah, George Kuczera, Melvan Schaefer, and Erwin Weinmann. 2016. "Estimating the exceedance probability of extreme rainfalls up to the probable maximum precipitation." *Journal of Hydrology* (Journal of Hydrology).

U.S. Army Corps of Engineers (USACE) (2013) Hydrologic Risk Assessment and PMP Coincident Flows. Cherry Creek Dam Safety Modification Study, Hydrologic Engineering Branch, Engineering Division, Omaha District, February 2013, 64 p.

Vose, David. 2008. *Risk Analysis: A Quantitative Guide.* West Sussex, England: John Wiley & Sons, Ltd. .

RECENT PALEOFLOOD ANALYSES FOR USACE RISK-INFORMED DAM SAFETY EVALUATIONS

Keith I. Kelson, C.E.G.[1]
Christina M. Leonard, M.S.[2]
Brian M. Hall, M.Eng.[3]
John F. England, Ph.D.[4]

ABSTRACT

This paper summarizes recent USACE paleoflood analyses conducted for dam safety evaluations, and comments on lessons learned from working in a wide range of geomorphic and hydrologic conditions across the country. Our goal was to estimate the magnitude and timing of rare floods in order to reduce uncertainties in flow frequency analyses. We completed a screening of 52 dams to identify sites with potential for yielding paleoflood data, using (a) hydrologic criteria based on estimated overtopping hazard, (b) geomorphic criteria based on the production and preservation of flood sediments, and (c) programmatic criteria based on consequence estimates and known uncertainties in existing dam evaluations. Our screening identified three viable sites: the Missouri River near Garrison Dam (North Dakota), the West River near Ball Mountain Dam (Vermont), and the Middle Fork Willamette River near Lookout Point Dam (Oregon). Subsequent field work included characterizing and dating fluvial terraces that represent past flood stages or non-inundation since deposition. We used 2D hydraulic modeling to estimate discharges required to form or inundate these features, and then developed perception thresholds for flow frequency analyses. Not surprisingly, we found that the USACE dam portfolio contains a mix of both good and less-than-perfect sites for paleoflood analysis, and that adaptive, customized investigative approaches work well for developing paleoflood data. We also found that uncertainties in paleodischarges are strongly influenced by unknown floodwater depths and the occurrence of ice-affected river stages. Overall, we conclude that the magnitudes and frequencies of rare floods can be estimated for dam safety evaluations, even at less-than-ideal sites, by integrating geomorphic, hydrologic, hydraulic, and statistical methods in a collaborative effort.

INTRODUCTION

The U.S. Army Corps of Engineers (USACE) conducts dam safety evaluations for more than 700 dams across the United States, using a risk-informed decision framework. This ongoing effort uses the frequency and magnitude of historic hydrologic events, as well as hydrologic and meteorological modeling, to estimate rare large floods not captured in the historic record. This paper summarizes screening and feasibility-level analyses of selected river reaches to obtain evidence for or against rare large floods. A primary goal

[1] US Army Corps of Engineers, Sacramento Dam Safety Production Center, *keith.i.kelson@usace.army.mil*
[2] US Army Corps of Engineers, Sacramento District, *christina.m.leonard@usace.army.mil*
[3] US Army Corps of Engineers, Risk Management Center, *brian.m.hall@usace.army.mil*
[4] US Army Corps of Engineers, Risk Management Center, *john.f.england@usace.army.mil*

of these analyses is to provide data on magnitudes and frequencies of rare floods, to develop longer, more robust hydrologic records for reservoir inflow frequency analyses and dam safety evaluations.

Paleofloods are past floods that have no direct hydrological measurements (i.e., gage measurement) or observations. A record of paleofloods can be preserved in geologic and geomorphic features on the landscape, as direct or indirect evidence of large prehistoric discharges. The science of paleoflood hydrology uses evidence of ancient floods, combined with physical hydraulic principles, to estimate amounts and distributions of paleodischarges (Costa, 1978; Baker, 2008), and requires collaboration among the disciplines of geology, geomorphology, hydrology, hydraulics, and statistics. Several paleoflood analyses in the western U.S. have benefitted dam safety assessments (Ostenaa et al., 1996, House et al., 2002; England et al., 2010; Dworak et al., 2014). This paper summarizes the application of these past works to USACE dam safety evaluations.

Paleoflood analyses utilize two general techniques for identifying and characterizing evidence of past large flood stages: paleostage indicators and paleohydrologic bounds (Figure 1; House et al., 2002). Paleostage indicators (PSI) are landforms or deposits directly related to high discharges, such as fluvial bedforms, gravel bars, or slackwater deposits (SWD) (House et al., 2002; Harden et al., 2011; Greenbaum et al., 2014). PSI provide minimum paleostage elevations, which are used to estimate paleodischarges through physics-based modeling of water depth and velocity. Numerical and relative dating of PSI yields data on paleoflood timing for flow frequency analyses.

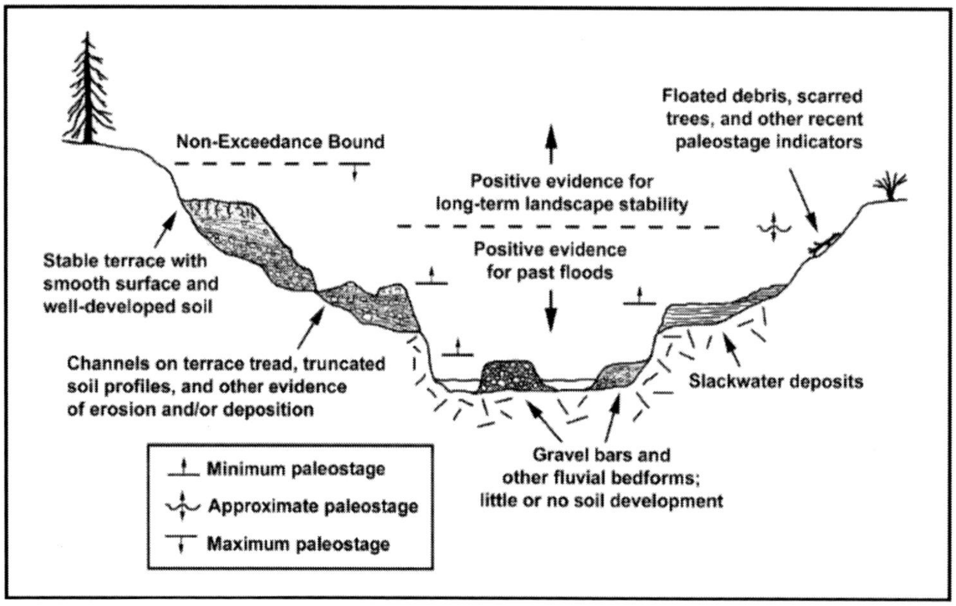

Figure 1. Schematic geomorphic cross-section of key features used in paleoflood analyses to estimate or constrain maximum flood stages (England et al., 2010).

Another technique for constraining maximum paleoflood discharge is to identify "Non-Exceedance Bounds" (NEB), which are deposits, landforms or other features that provide direct evidence of long term landscape stability, or direct evidence of an absence of flood erosion or deposition (Figure 1). Unlike slackwater deposits or other PSI, NEB may be preserved in non-ideal depositional conditions across a wide range of fluvial systems, and thus are useful along reaches where PSI are not fortuitously preserved. Characterizing NEB commonly involves identifying stable geomorphic surfaces (e.g., abandoned floodplains, tributary alluvial fans), estimating ages of the features, and using hydraulic models to estimate the discharge needed to inundate the surface and cause perceptible erosion or deposition (Figure 1). Numerical and relative age dating techniques provide estimates of the elapsed time since formation of the NEB, thus providing a time interval during which the paleostage (and associated paleodischarge estimate) has not occurred.

INITIAL PALEOFLOOD SCREENING

We completed an initial screening of 52 USACE facilities undergoing current safety assessments, with a goal of identifying sites that (a) could benefit from additional information on the frequencies and magnitudes of rare floods, and (b) have a high potential for yielding paleoflood data. Our intent was to use a rapid, initial compilation of site specific characteristics to rank the "paleoflood potential" of the sites. Hydrologic criteria included qualitative or quantitative estimates of the potential for peak reservoir elevations to approach or overtop dam crest elevations. Geomorphic criteria included qualitative measures of watershed sediment production potential, depositional site geology, and local valley confinement (O'Connor et al., 2014). Estimating watershed sediment production involved a GIS-based analysis of watershed bedrock materials that could provide sandy and silty alluvium that might serve as a record of large floods. For example, watersheds underlain by poorly consolidated sandstone or sandy glacial outwash tend to produce sandy and silty sediment, and were ranked highly for screening. Depositional site geology was estimated via the presence of previously mapped alluvial terraces or tributary alluvial fans, or other qualitative assessments of local geomorphic conditions directly adjacent to each dam. For example, site conditions that favor deposition and preservation of sandy materials over long time periods were favored over conditions dominated by landscape instability. Qualitative assessment of relative valley confinement involved review of topographic data and photographic imagery to identify the presence of narrow, bedrock confined reaches that would likely harbor evidence of paleoflood deposits (PSI) or limits (NEB) (Figure 1).

The sites were grouped into three qualitative classes for yielding paleoflood information: likely, possible, or unlikely (Figure 2). The "likely" sites were subjected to programmatic criteria, such as the possibility that paleoflood data could be developed in time to be useful in near-future risk assessments. Three sites across the country (Figure 2) were selected for feasibility paleoflood analyses, representing a large variability in hydrologic and geomorphic settings (Kelson et al., 2016). The selected sites included a site in the western US where, presumably, traditional paleoflood techniques (e.g., House et al., 2002; Harden et al., 2011) could be used. It was acknowledged that the sites in the central and eastern US could require adjustment and adaptation of established techniques.

Overall, the breadth of hydrologic and geomorphic settings for the three sites (Table 1) provides advantages as well as challenges for obtaining well-constrained paleoflood data.

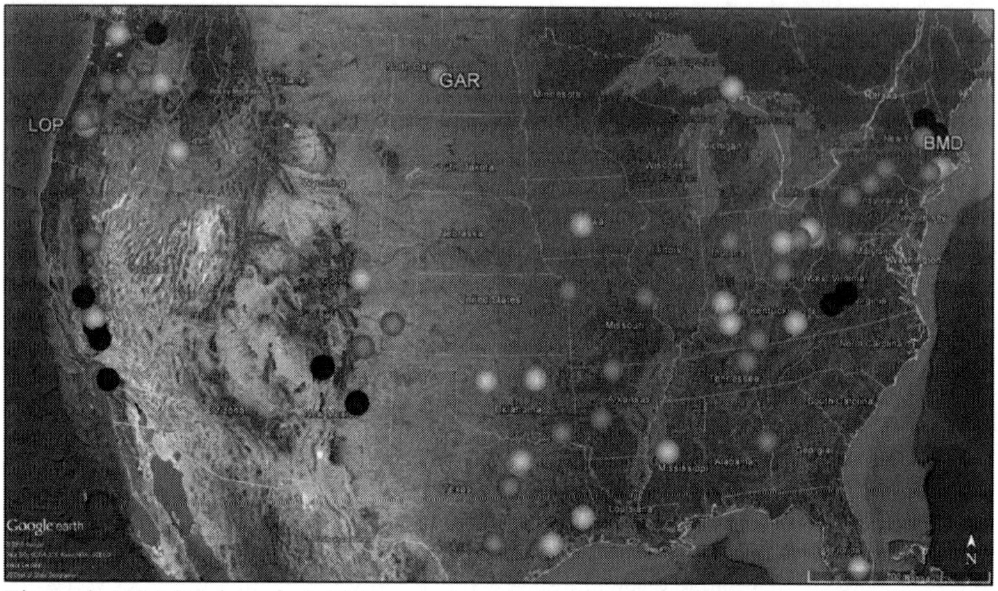

Figure 2. USACE dams in initial screening; sites classified as likely (dark blue), possible (medium blue), or unlikely (light blue) to yield paleoflood data. Sites selected for further analyses: LOP, Lookout Point Dam; GAR, Garrison Dam; BMD, Ball Mountain Dam.

Table 1. Summary of geomorphic and hydrologic conditions at three USACE locations selected for paleoflood analyses.

Watershed / site characteristics	Garrison Dam Missouri River Bismarck ND	Ball Mountain Dam West River Jamaica VT	Lookout Point Dam Middle Fk Willamette River Oakridge OR
Watershed sediment production	sandy glacial deposits in watershed; local sandstone produces sand and silt	sandy glacial deposits in watershed; local volcanic bedrock produces cobble gravel	sandy glacial deposits in watershed; local gneiss bedrock produces cobble gravel
Site depositional characteristics	possible SWD in tributary mouths; fluvial terraces	possible SWD in alcoves; fluvial terraces	possible SWD at constrictions; fluvial terraces
Site preservational characteristics	continuous terraces; no alcoves, no caves; broad valley	discontinuous terraces; alcoves, no caves; narrow resistant bedrock valley	discontinuous terraces; alcoves, no caves; narrow resistant bedrock valley; landslides
Datable geoloigc materials	sandy alluvium (OSL); organic material (C14); relative soil development	sandy alluvium (OSL); organic material (C14); relative soil development	sandy alluvium (OSL); organic material (C14); relative soil development
Unregulated flood hydrograph	slow rise/fall; long flood duration	moderate flood duration	moderate flood duration
Channel stationarity	non-stationary channel; historic migration and incision	stable channel; resistant bedrock valley	stable channel; resistant bedrock valley
Altered river stages	historic ice jams; historic flood during construction	unaltered	unaltered

FEASIBILITY LEVEL PALEOFLOOD ANALYSES

Feasibility paleoflood analyses at the three sites used similar approaches, customized based on site-specific conditions. Overall, the analyses included delineating and dating geologic and geomorphic PSI and NEB, estimating paleodischarges based on these features, and using the ranges of discharges and ages for flow frequency analyses. As part of preparing for fieldwork at the selected sites, we used available imagery (e.g., GoogleEarth) to identify locations of SWD, fluvial terraces, or other possible PSI, and developed initial 2D hydraulic models to estimate water surface elevations for flood-of-record discharges and channel shear stresses. These data guided the field reconnaissance by identifying locations where preservation of PSI would be most likely. During field visits, we searched for SWD, mapped fluvial terraces, and identified other relevant geomorphic features (e.g., alluvial fans), usually using topography derived from airborne LiDAR data (NAVD88; with resolution and accuracy both less than 3 ft) (Figure 3). Flood stratigraphic characteristics were documented via natural or excavated exposures, and dated via relative soil-profile development and multiple radiometric (carbon-14) and/or Optically Stimulated Luminescence (OSL) analyses (Figure 4).

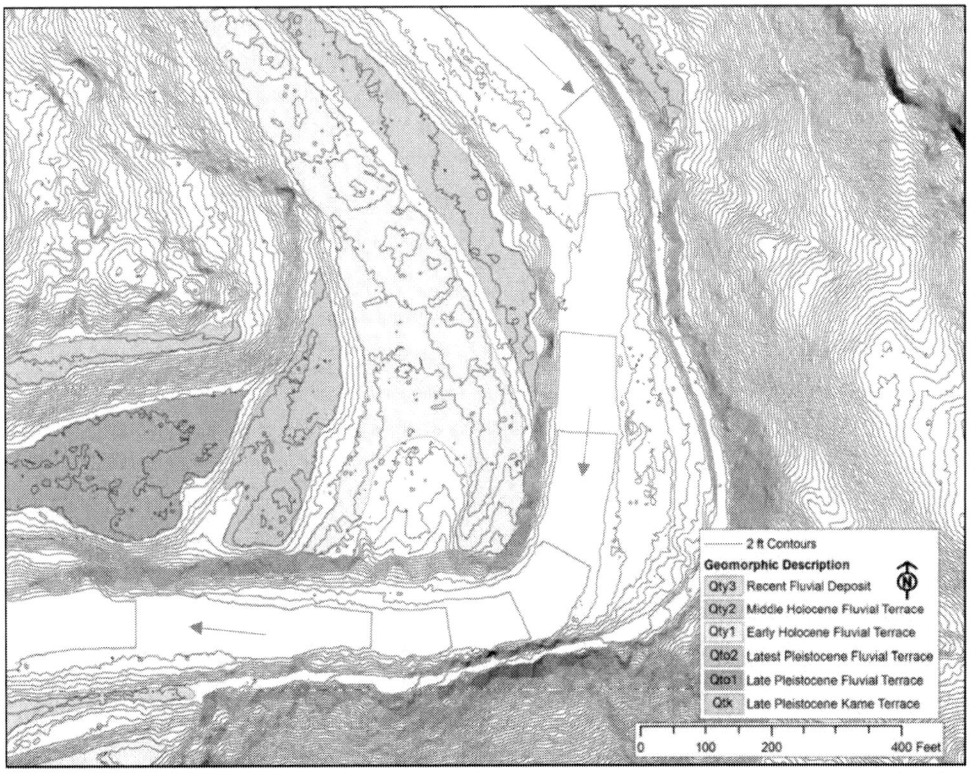

Figure 3. LiDAR-derived topographic map of West River valley downstream of Ball Mountain Dam (VT), showing suite of nested fluvial terraces on western side of a river bend. The terrace deposits were interpreted as PSI to estimate paleostages, or as NEB to limit paleostages.

Figure 4. Photographs of (left) terrace along western bank of Missouri River near Garrison Dam, looking north toward higher, older terrace (C. Leonard for scale); and (right) exposure of interbedded sand and silty flood deposits underlying lower terrace.

With field-based geomorphic data on flood-related deposits (PSI) and or stable surfaces (NEB), we used HEC-RAS v5.0 two-dimensional (2D) hydraulic modeling software to develop a range of paleodischarges that would inundate the features. Model boundary conditions were set using inflow hydrographs and energy slopes typical for each study reach, and were calibrated to historic water-surface elevations and a range in Manning's *n* values. Down-valley profiles from the 2D model were compared with PSI or NEB elevations to estimate paleodischarges associated with the dated features (Figure 5).

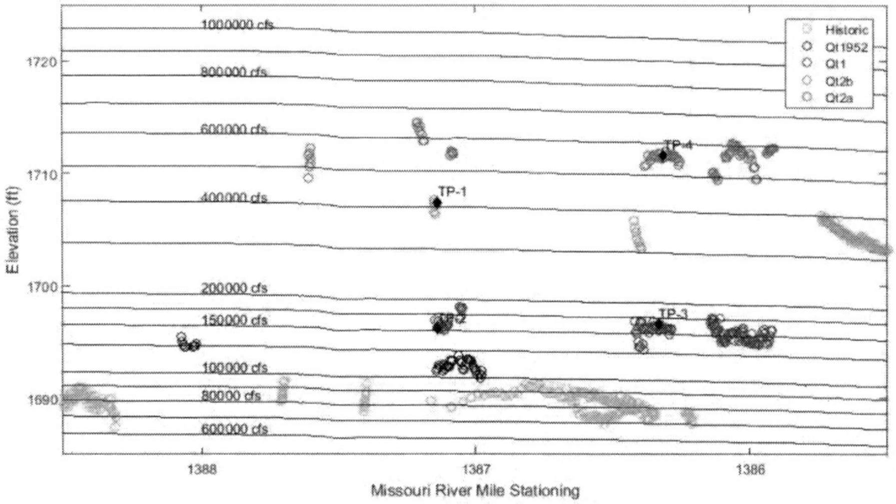

Figure 5. Profiles of modeled water surface elevations for flood discharges (solid black lines are best estimates; model uncertainties not shown), point elevations on terraces (correlated by same-color dots), and stratigraphic exposures (TP-1 to TP-4).

Paleoflood discharges from hydraulic modeling were combined with PSI or NEB ages to develop perception thresholds for prehistoric time intervals. Perception thresholds were merged with the historical streamflow record using the Expected Moments Algorithm (EMA, Cohn, 2012) in the statistical software program HEC-SSP v2.1 (USACE, 2016). EMA provides the statistical framework for extending flow frequency analyses over longer geologic time scales (e.g., Harden et al., 2011), in order to assess hydrologic loading at annual exceedance probabilities relevant to dam safety risk evaluation.

SOME LESSONS LEARNED

The Screening Process Works Well

The process of portfolio screening was useful for identifying and ranking reaches in assessing rare hydrologic loading for USACE dam safety evaluations. Site ranking helped focus limited resources on sites where paleoflood data could reduce uncertainties in hydrologic loading. We are encouraged that the screening process identified several sites that have moderate or high potential for yielding paleoflood data, and that the three sites yielded reasonable geomorphic evidence of PSI and NEB. The USACE portfolio includes dams in a range of geomorphic and hydrologic settings, many of which are not ideal for "proof-of-concept" paleoflood research. Because the site selection process combined the likelihood to yield fruitful paleoflood data with the need to characterize hydrologic hazard, each analysis involved both challenges and opportunities specific to the particular reach. The portfolio includes several locations that appear very good for paleoflood research, but which were not selected because paleoflood data would not have affected dam safety risk evaluations. For the selected sites, development of reasonable data on extent and age of paleofloods suggests that the screening process was effective.

Adaptive Investigation Approach

Because the variability of geomorphic and hydrologic conditions at USACE dams across the country presents unique challenges and opportunities for developing paleoflood information, each site required a customized technical approach and set of analytical tools. Our approach involved adapting the level of investigation according to the available time and budget, the perceived benefits to hydrologic loading estimates, and an acceptable degree of uncertainty. In cases where time and/or resources are limited, feasibility-level analyses may be justified for obtaining initial information that place rough bounds on rare paleodischarges. Limited analyses may also be appropriate for identifying key sources of uncertainties and focusing subsequent analyses on the primary contributors to hazard. All components of paleoflood analyses, from identifying and delineating PSI and NEB, to quantitative discharge estimation, to flow frequency analysis, to incorporating results into risk assessments, are all scalable according to time and resources. While a "quick and dirty" reconnaissance may not produce adequate data at some sites, a years-long research-level investigation likewise may be inappropriate. The approach and level of effort for a paleoflood analysis at any reach can (and should) be customized, to balance cost, time, and anticipated reductions in uncertainties.

For example, our initial approach for field data collection was influenced by recent successes in identifying and dating paleoflood slackwater deposits (SWD) in riverside caves (Harden et al., 2011) and steep, arid canyons (Greenbaum et al., 2014). We developed initial hydraulic models to identify locations where bedrock alcoves, overhanging ledges, or mouths of small tributaries would favor preservation of SWD. Subsequent field reconnaissance encountered a few historic SWD at some of these locations, but no unequivocal prehistoric SWD that could be used for paleoflood analysis. At all three sites, we deemed that searching for SWD would be less likely to yield useful data than identifying fluvial terraces as PSI and/or NEB. We attribute the absence of SWD in all three feasibility sites to poor preservation of unconsolidated silty deposits along steep valley walls. As noted by House et al. (2002), SWD preservation requires ideal site conditions (e.g., riverside caves) and landscape stability over hundreds or thousands of years. Although such sites may exist in some watersheds, we concluded that characterizing terrace remnants as PSI or NEB had a higher likelihood for providing paleoflood data than exhaustive searches for SWD. An absence of SWD at high-likelihood locations helped shift our fieldwork focus toward delineating fluvial terrace deposits as PSI and/or NEB. We view that using an adaptive approach was key for developing paleoflood data in these limited feasibility analyses.

Detailed Topographic Data (LiDAR)--An Essential Tool

During the three analyses, we found that use of detailed topographic data was critical for identifying and delineating PSI and NEB. Because many geomorphic features were often limited in extent and covered by dense vegetation, defining the elevations and distributions of PSI and NEB requires high-resolution topographic data. For example, initial field geomorphic mapping along the West River near Ball Mountain Dam originally used standard USGS topographic maps (with 20 ft contours) and identified several discontinuous terraces for estimating paleoflood stages. Subsequent acquisition of high-resolution LiDAR data (accuracy less than 3ft) helped identify several additional terrace remnants, refine elevations of PSI in exposures, and correlate terrace remnants for defining paleostages along the entire river reach. These data greatly improved the elevation profiles of PSI and NEB, provided higher resolution 2D hydraulic modeling of paleostages, and ultimately improved paleodischarge estimates.

Floodwater Depths Affect Paleodischarge Estimates

A key assumption in interpreting paleodischarges from PSI and NEB is that their elevations closely approximate paleostages of floodwaters. Geomorphic mapping of flood deposits on fluvial terraces, coupled with high-resolution topographic data, provides a minimum paleostage and thus a minimum paleodischarge. The actual paleodischarge can be estimated using other PSI, such as floodwater scars preserved on tree trunks or other high-water marks, or by estimating floodwater depths needed to transport sediment of a specific grain size (for PSI) or erode a stable surface (for NEB). Because of uncertainty in floodwater depths, PSI profiles provide minimum discharge estimates, whereas NEB provide maximum discharge estimates. Developing defensible information on depths of paleodischarges proved difficult at the feasibility level.

Assessing the range in uncertainty in paleoflood discharges is important for flow frequency analysis and risk assessment. Field-based mapping and hydraulic modeling both have sources of uncertainty related to limited (epistemic) knowledge and inherent (aleatory) variability in river systems. Because of inherent variability in terrace elevation profiles (i.e., terraces are not perfectly planar), we developed down-valley profiles for each set of correlative terrace remnants. The variability in elevations of terrace remnants provides a qualitative view of the degree of uncertainty in water surface elevations, when compared to smooth profiles produced by 2D hydraulic modeling (Figure 5). To address epistemic uncertainty in the hydraulic modeling, sensitivity simulations were performed using variations in Manning's n value to estimate discharge uncertainties along the reach profiles. By considering both the inherent water surface variability (e.g., by using terrace profiles) and our incomplete knowledge (e.g., by sensitivity analyses of the hydraulic modeling), we attempted to capture a reasonable level of total uncertainty in discharges by selecting a best-fit line through a terrace profile to represent a preferred discharge value. We estimated a range in discharge values by considering the range in elevations of the terrace remnants and an uncertainty band around each calculated discharge profile.

Beware of Ice Jams

Another source of uncertainty in calculating paleodischarges from PSI and NEB is the potential effect of ice jams on river stages. Accumulation of ice at natural (or artificial) constrictions causes riverine backwater and elevated river stages while the ice is in place. Sudden or intermittent release of ponded water and sediment can produce flood deposits or highwater marks controlled by the location, thickness, and duration of ice blockage, rather than discharge. As an ice jam releases, thick accumulations of ice on the water surface may violate the basic assumption of open-channel flow used in most hydraulic models. In addition, historic peak flood discharges are often estimated via extrapolation of stage-discharge rating curves developed from measurements made at low or moderate flow during ice-free conditions. Applying ice-affected stages to rating curves developed from ice-free conditions may yield artificially high estimates of peak discharge during ice-affected conditions. Similarly, discharges may be overestimated from PSI elevations if the possibility of ice-affected stages is not acknowledged. Hence, sites affected by localized ice accumulations present special challenges to paleoflood analyses.

The Missouri River between Garrison Dam and Bismarck has a history of flood stages influenced by ice jams, presenting challenges to interpretations of PSI, NEB, and stage-discharge rating curves at nearby streamflow gages. Peak stages during the 1952 flood of record were affected by an ice jam located approximately 40 mi downstream of Garrison Dam, and field discharge measurements were hampered during high stages. Because of the possibility of prehistoric ice jams along the Missouri River near Garrison Dam, our analysis wrestled with capturing the level of uncertainty in paleodischarges as well as historic, pre-dam peak discharges known to have been affected by ice jams. A paucity of historic discharge measurements at or near peak flows (especially during ice-jam events) raises the possibility of large uncertainties in historical peak discharge estimates as well as paleodischarge estimates. We addressed these uncertainties by treating calculated paleodischarges as maxima that are constrained by elevations of non-inundated features (e.g., alluvial fans, eolian deposits).

CONCLUSIONS

Paleoflood analyses recently completed by USACE in North Dakota, Vermont, and Oregon help improve hydrologic loading estimates for three dams, and help validate the use of paleoflood data in dam safety evaluations. The three dams lie in watersheds that collectively have a broad range of hydrologic and geomorphic characteristics. We collected site-specific geomorphic evidence for large past floods (i.e., PSI) and for non-inundation over time scales of hundreds to thousands of years (i.e., NEB), and calculated discharges associated with the elevations of these features. Collectively, these analyses show that the USACE dam portfolio spans a large range in site conditions and contains a mix of both good and less-than-perfect sites for assessing rare floods. Overall, a key lesson learned is that at sites where paleoflood information would be most helpful for USACE dam safety evaluations, site depositional and preservation conditions often are not ideal, and many sites have inherent uncertainties that would generally discourage "proof-of-concept" research. Because paleoflood data can greatly help constrain hydrologic loading estimates, there is inherent value in developing these data if uncertainties are adequately captured. Thus, feasibility paleoflood analyses can contribute significantly to hydrologic loading estimates and dam safety risk evaluations.

However, limited field investigations must be adaptive and opportunistic to take advantage of site-specific conditions. Technical challenges encountered during the feasibility studies made assessments difficult but not insurmountable. A lack of detailed topographic data at the start of field work was overcome by adapting field techniques and updating interpretations when better topographic data become available. Other technical challenges, such as uncertainties in floodwater depths, or the presence of historic ice jams, are inherent to many sites, and may be addressed by acknowledging and capturing uncertainties in the analysis. The presence of these and other sources of uncertainty should not discourage risk assessments away from paleoflood analyses, but instead should be treated as opportunities to improve the understanding of hydrologic and hydraulic conditions within each watershed. By integrating geomorphic, hydrologic, hydraulic, and statistical methods in a collaborative effort, these feasibility studies show that the magnitudes and frequencies of rare floods can be estimated, even at less-than-ideal sites, and used in risk-informed dam safety evaluations.

ACKNOWLEDGEMENTS

Support for these paleoflood efforts has been via the USACE Dam Safety Program, and the USACE Hydrology, Hydraulics, and Coastal Community of Practice. We thank Robert Simrall, John Remus, David Margo, and Ed Stowasser (USACE) for guidance, and Haden Smith, Nicholas Koutsunis, Chris Nygaard, Chris Humphrey, and Stephen Potts (USACE) for hydraulic modeling or field assistance. Angela Duren (USACE) is contributing to ongoing inflow frequency analyses for the sites. Drs. Tess Harden and Shannon Mahan (USGS) provided excellent contributions in the field and in laboratory analyses. Review comments by David Margo, Russell Graham, and Justin Pearce improved the paper and are greatly appreciated.

REFERENCES

Baker, V., 2008, Paleoflood hydrology: Origin, progress, prospects: *Geomorphology*, v. 101, p. 1-13.

Cohn, T., 2012, User Manual for Program PeakfqSA, Flood frequency analysis with the Expected Moments Algorithm.

Costa, J.E., 1978, Holocene stratigraphy in flood frequency analysis; *Water Resour. Res.* 14, 626–632. doi:10.1029/WR014i004p00626

Dworak, F., England, J., Elsner, M., and Klinger, R., 2014, Modeling extreme flood hydrology for Grand Coulee Dam: https://acwi.gov/sos/pubs/3rdJFIC/Contents/11F-Dworak.pdf

England, J.F. Jr., Godaire, J.E., Klinger, R.E., and Bauer, T.R., 2010, Paleohydrologic bounds and extreme flood frequency of the Arkansas River Basin, Colorado, USA. *Geomorphology*, 124, doi:10.1016/j.geomorph.2010.07.021, pp. 1-16.

Greenbaum, N., T. M. Harden, V. R. Baker, J. Weisheit, M. L. Cline, N. Porat, R. Halevi, and J. Dohrenwend, 2014, A 2000 year natural record of magnitudes and frequencies for the largest Upper Colorado River floods near Moab, Utah; *Water Resour. Res.*, 50, 5249–5269, doi:10.1002/ 2013WR014835.

Harden, T.M., O'Connor, J.E., Driscoll, D.G., and Stamm, J.F., 2011, Flood-frequency analyses from paleoflood investigations for Spring, Rapic, Boxelder, and Elk Creeks, Black Hills, western South Dakota: *U.S. Geol. Surv. Sci. Invest. Rpt.* 2011-5131, 136 p.

House, P. K.; Webb, R.H.; Baker, V.R.; Levish, D.R. (eds.), 2002, Ancient floods, modern hazards: Principles and applications of paleoflood hydrology: *Water Science and Application*, v. 5, American Geophysical Union, 385 p.

Kelson, K., and 10 others, 2016, Initial USACE paleoflood analyses in a risk-informed decision-making framework [abstr.], *in* Harden, T., and O'Connor, J., Proceedings, 5th International Paleoflood Conference, Rapid City, South Dakota, September, 2016.

O'Connor, J.E., Atwater, B.F., Hohn, T.A., Cronin, T.M., Keith, M.K., Smith, C.G., Mason, R.R., 2014, Assessing Inundation Hazards to Nuclear Powerplant Sites Using Geologically Extended Histories of Riverine Floods, Tsunamis, and Storm Surges: *U.S. Geol. Surv. Sci. Invest. Rpt.* 2014-5207, 65 p.

Ostenaa, D., Levish, D., O'Connell, D., 1996, Paleoflood study for Bradbury Dam, Cachuma Project, California. U.S. Bureau of Reclamation Seismotectonic Report 96-3.

U.S. Geological Survey, 2016, Guidelines for Determining Flood Flow Frequency, Bulletin 17C: Techniques and Methods 4-Bxx [August, 2016 review draft]

U.S. Army Corps of Engineers, 2016, HEC-SSP Statistical Software Package v2.1, User's Manual, [*http://www.hec.usace.army.mil/software/hec-ssp/documentation/HEC-SSP_20_Users_Manual.pdf*]

RISK INFORMED FREEBOARD EVALUATION

Gregory S. Karlovits, P.E, P.H, CFM[1]
David A. Margo, P.E.[2]

ABSTRACT

Freeboard provides a margin of safety against overtopping failure of dams. Evaluation of freeboard requirements at a particular site must consider many factors including the probability of high winds from a critical direction combined with high reservoir levels. Current deterministic methods and guidance can result in significant over or under prediction of freeboard requirements and overtopping risks. This leads to either over or under investment in dam safety remedial actions. Common issues include the use of all direction wind statistics, the use of prescribed frequency wind events, ignoring the joint probability of high wind and high reservoir level, and neglecting flood seasonality. The solution to these issues is presented in a scalable risk informed framework using three case studies. The first case study demonstrates a screening approach using a directionally dependent wind speed frequency analysis. The intermediate approach used for the second case study combines a wind hazard model with a stochastic reservoir stage model. The third case study applies a detailed approach that combines a tropical storm model, a hydrodynamic wave generation model, and a continuous simulation hydrologic model. Results of the case studies will be presented to demonstrate the varying levels of effort needed to support credible estimates of freeboard and overtopping risk.

BACKGROUND

Freeboard is an important design concept used to inform the selection and accommodation of inflow design floods for dams. It is generally defined as the vertical distance between a specified stillwater level and the top of the dam. Freeboard allowances are typically added to the static design flood water level to address various sources of uncertainty such as wind generated overwash, misoperation or malfunction of spillway gates, debris blockage of spillways, settlement, climate change, watershed response characteristics, and design flood characteristics. This paper focuses on the estimation of freeboard allowances for wind generated overwash.

From a design perspective, several factors need to be considered when assessing wind generated overwash. First, an appropriate design wind condition must be selected to coincide with the design flood event. Second, an acceptable amount of overwash must be estimated to maintain the integrity of the dam, project operation, and personnel safety. Current deterministic design criteria does not do an adequate job of addressing these factors. The designer is often left to choose between using conservative criteria (e.g. maximum wind with zero overwash) (US Army Corps of Engineers, 1997), arbitrary

[1] Statistical Hydrologist, Corps of Engineers Risk Management Center, 12596 West Bayaud Avenue, Lakewood, CO 80228, gregory.s.karlovits@usace.army.mil, 303-963-4531.
[2] Lead Civil Engineer, Corps of Engineers Risk Management Center, 1000 Liberty Avenue, Pittsburgh, PA 15222, david.a.margo@usace.army.mil, 412-667-6629

criteria (e.g. 10 percent hourly exceedance wind speed) (US Bureau of Reclamation, 2013), or no specific criteria at all (e.g. designer selects an "appropriate" wind speed). These approaches can lead to significant over or under investment in measures to achieve dam safety objectives. A risk informed approach can resolve this problem by addressing the issue directly and providing a clear path to selecting a freeboard allowance. The remainder of this paper will demonstrate the use of a risk informed approach by presenting three case histories.

CASE HISTORY 1

Dam A is a zoned earth fill embankment with a crest length of 3500 feet, a height of 140 feet, and a drainage area of 390 sq. mi. The reservoir serves multiple purposes including flood damage reduction, irrigation, recreation, and hydropower generation. An update of the inflow design flood resulted in a design water level less than 1 foot below the dam crest. Previous estimates of the freeboard allowance used the return intervals and ASCE-7 (American Society of Civil Engineers, 2013) structural design wind speeds shown in Table 1. The results of these analyses suggest that a freeboard allowance on the order of 6 to 8 feet was required to prevent overwash. The dam does not have this amount of freeboard. The likelihood of an extreme flood combined with high winds and the potential life loss resulted in overwash being identified as a risk driving potential failure mode.

Table 1. ASCE-7 Wind Speed

Return Interval	1 Hour Wind Speed
10	48
25	52
50	56
100	60

It was recognized that the freeboard estimate used to assess overwash risks combined with the assumption that adverse winds would occur coincident with an extreme reservoir level might be unreasonably conservative. This could drive unnecessary investment in remedial action to prevent overwash. A screening level analysis was undertaken to further explore the overwash potential failure mode. A frequency analysis of annual maximum winds was performed using data from a nearby observation station. The analysis considered both the direction and seasonality of the wind. A Galton distribution was fit to the observed data for annual maximum 1 hour wind speed conditional on the flood season (November through March) and originating from three north-northeast wind directions (20°, 30° and 40°). The wind directions were selected based on the fetch direction and length required to generate significant wave action at the dam. An example fit for annual maximum winds is shown in Figure 1 for the 30° direction. Notice that the wind speeds conditioned on direction and season are about 70% less than the all direction wind speeds from ASCE-7 in Table 1. Figure 2 shows a boxplot for all observed 1 hour wind speeds conditional on direction. Notice that high wind speeds are not typically observed during the flood season in a direction that is adverse to the dam, generally from

15° to 45°. This result is consistent with the current understanding of the causative meteorology for floods in the region. The predominant inflow wind direction for moisture transport is from the southwest direction (225°).

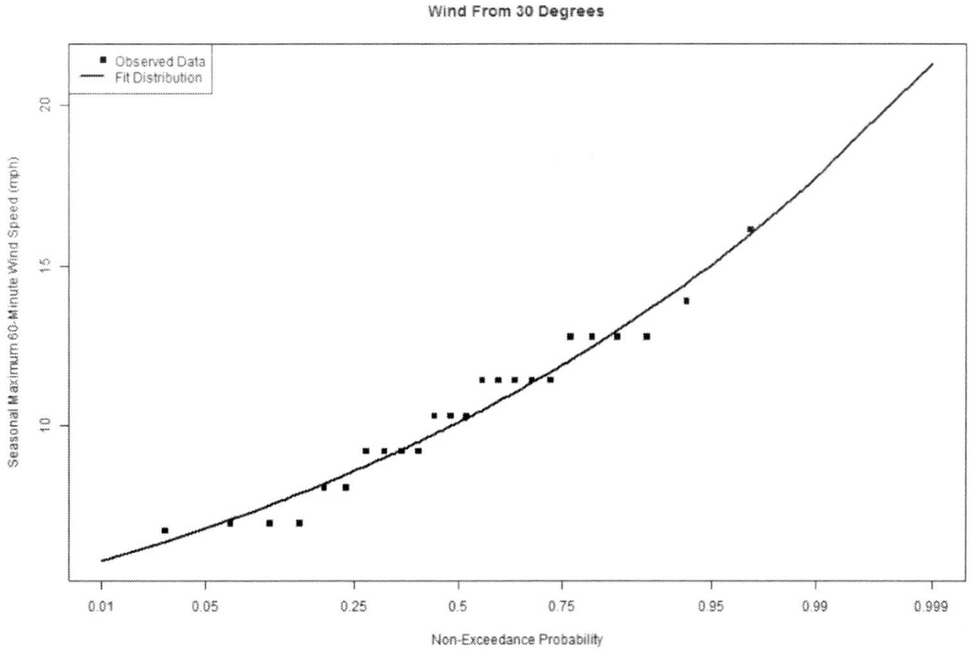

Figure 1. Frequency of Flood Season Maximum 1 Hour Wind Speed from 30° During the Flood Season

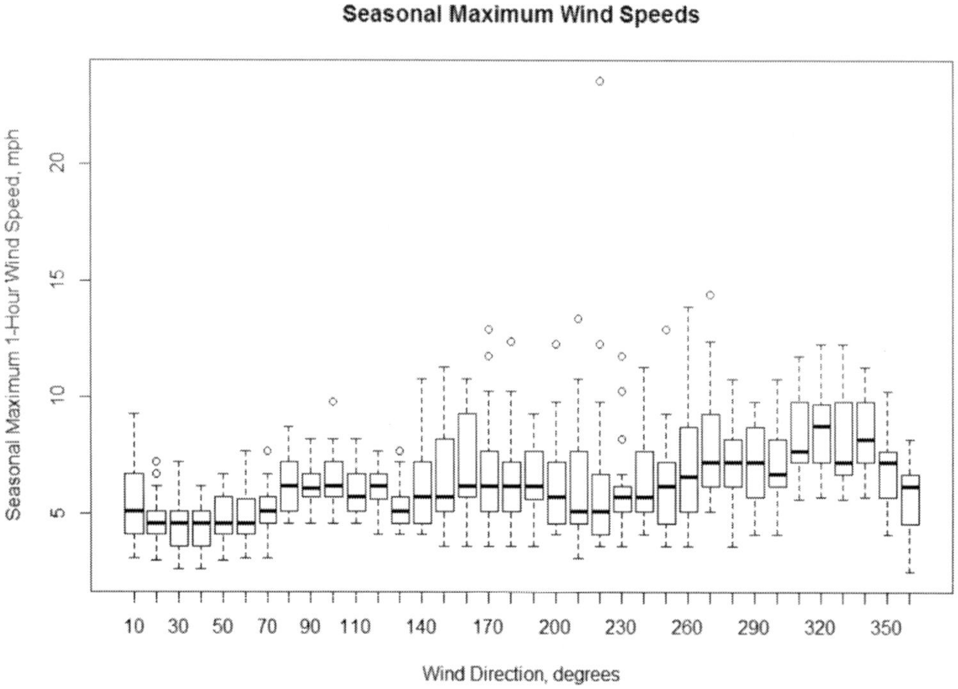

Figure 2. Boxplot of 1 Hour Wind Speed Conditional on Wind Direction

The revised design wind speeds are listed in Table 2. Using these values results in a freeboard allowance of less than 2 feet with a negligible amount of overwash. Based on the results of this screening analysis, it was concluded that previous estimates of freeboard allowance were significantly overestimated and overwash could be ruled out as a risk driving potential failure mode. The wind speeds required to generate significant overwash are estimated to have a return interval significantly greater than 100 years. The joint probability of adverse winds coincident with a stage approaching the top of the dam is even more remote with a return interval greater than 10 million years. It is concluded that the overwash potential failure mode is not a risk driver because it poses a risk that is negligible relative to other potential failure modes. This conclusion is supported by current understanding of the causative meteorology for extreme floods in the region which suggests that winds from a north-northeast direction do not provide a sufficient source of moisture to produce significant winds coincident with an extreme flood.

Table 2. Site Specific Wind Speed

Return Interval	1 Hour Wind Speed
10	14
25	15
50	17
100	18

CASE HISTORY 2

Dam B is a rolled earth fill embankment with a crest length of 14,300 feet, a height of 140 feet, and a drainage area of 385 sq. mi. The reservoir serves multiple purposes including flood damage reduction, recreation, and fish and wildlife enhancement. An update of the inflow design flood resulted in a design water level approximately 3 feet higher than the dam crest. Overtopping was identified as a risk driving potential failure mode and a dam raise was formulated as a possible remedial action. A preliminary estimate suggested a minimum freeboard allowance for wind generated waves on the order of 3 to 5 feet.

An intermediate level of analysis was undertaken to refine the freeboard allowance estimate. A stochastic wind model was coupled with a stochastic flood model to estimate the joint probability of extreme flood stages and wind generated waves. The flood model uses inflow volume frequency relationships derived from period of record, historic, and paleoflood information combined with observed flood hydrograph shapes to generate annual maximum flood events. These events are then routed through the reservoir using a stochastically generated initial reservoir level and month of occurrence to obtain stage hydrographs. An annual peak stage frequency relationship can be derived from these hydrographs. For each flood event generated, a sequence of hourly wind speed and direction is generated by the wind model. This wind data is then coupled with the hourly reservoir stages obtained from the flood model so that hourly wind setup and run-up can be estimated. The resulting wind influenced stage hydrographs can then be used to derive an annual flood and wind influenced peak stage frequency relationship. The frequency relationships with and without wind influence can then be compared to inform selection of a freeboard allowance.

The wind model generates a sequence of winds that could occur at any given time rather than considering only the annual maximum wind. The joint probability of winds and extreme reservoir stages is established by a priori assuming a flood is occurring and then generating a sequence of hourly wind data consistent with historic observations of speed and direction. An autoregressive model framework is used to generate hourly wind speeds and directions that are dependent on the preceding hour. A sequence of hourly wind direction is generated by inverting a generalized linear model with a Gaussian margin with parameters that are dependent on the wind speed and direction from the prior step in the simulation. Wind direction for the current hour is obtained from the wind direction for the previous hour modified by a change in direction defined by a random sample from the generalized linear model. A sequence of hourly wind speed is generated from a three parameter gamma distribution with a logarithmic transform of wind speed (i.e. Log Pearson Type III). Distribution parameters are dependent on the month of the flood event and the direction and speed from the prior step in the simulation. A lag minus 1 autocorrelation coefficient is used to provide a smooth transition between successive hours of generated wind speed. A comparison between observed and simulated wind direction for the month of March is shown in Figure 3. Wind directions adverse to the dam generally fall in the range of 110° to 160°. The model tends to under predict the frequency of winds from the north and south and over predict the frequency of winds from the east and west. The model generates winds from a hazardous direction more frequently than the observed record suggesting that the model is conservative (i.e. it will over predict the freeboard allowance). Similar comparisons for wind speed suggest that the model does a good job of replicating the observed record.

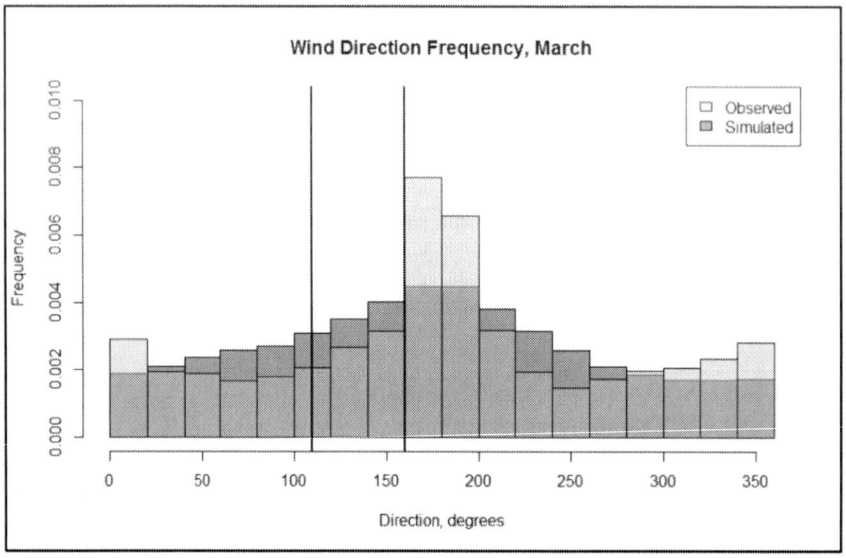

Figure 3. Simulated Histogram of 1 Hour Wind Speed Conditional on Wind Direction

A total of five million flood events were simulated using the flood and wind models. The model computed hourly wind setup and wave run up for events where the wind originated from an adverse direction (110° to 160°). The maximum hourly stage was retained for each event in the simulation for both stillwater and wind induced stages. In more than 90% of the simulated events, the maximum stillwater stage was equal to the maximum wind induced stage because the maximum wind did not occur coincident with the maximum stillwater stage. This illustrates the rarity of coincident peak stages and peak winds. The average increase in peak stage due to wind for all five million simulations was about 0.2 feet and the maximum was about 4 feet. The resulting peak stillwater and wind induced stage frequency relationships shown in Figure 4 are indistinguishable. The results suggest that any increase in flood hazard and risk due to wind generated waves and overwash is negligible. A risk informed case could be made to provide zero freeboard allowance for wind induced waves. This would reduce the height of a remedial dam raise by 3 to 5 feet saving significant cost. Future improvements to the modeling framework might include a longer memory autoregressive process (greater than 1 hour), use of regional wind observations, and quantification of parameter uncertainty for wind generation.

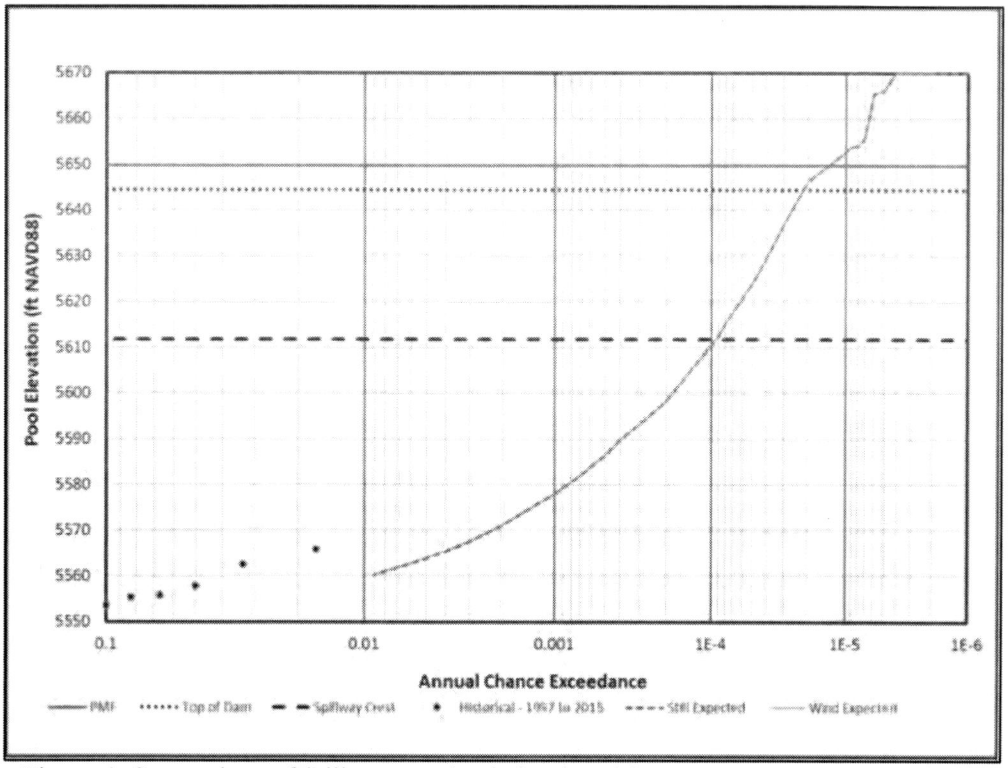

Figure 4. Comparison of Stillwater and Wind Induced Stage Frequency Relationships.

CASE HISTORY 3

Dam C is an earthen embankment with a crest length of 143 miles, a height of 30 feet, and a drainage area of 5200 sq. mi. The reservoir serves multiple purposes including flood damage reduction, navigation, water supply, preservation of fish and wildlife, and water quality. The flood hazard is significantly influenced by peak stages generated from a combination of both flood and wind events. A deterministic analysis for the original design evaluated three combinations of reservoir stage and wind to estimate the freeboard allowance. The controlling case was a standard project hurricane following a 100 year flood with a stillwater stage of 21 feet, an hourly peak wind speed of 85 mph, and a freeboard allowance of 14 feet.

A risk informed study was conducted to identify risk driving potential failure modes and formulate remedial measures. The number of combinations is large in part because adverse winds can originate from any direction, set up and run up varies with location along the structure, and peak stages are influenced by long duration sequences and combinations of events. An integrated modeling framework was developed to combine a stochastic weather generator to generate precipitation and wind events, a hydrologic model to transform precipitation to inflow and route through the reservoir, and a hydrodynamic model to generate set up and run up from generated winds using observed and synthetic wind field patterns.

The weather generator simulates a mixed population of events by explicitly simulating both ordinary precipitation events and tropical storm events at a daily time step. Because the tropical storm events are modeled explicitly, the effect of wind associated on stages derived from a long sequence of rainfall due to both ordinary and tropical meteorology can be evaluated. Generated rainfall sequences are applied to the hydrologic model to obtain stage hydrographs and peak stillwater stages. Key parameter uncertainties in both the weather and flood models were explicitly quantified. The stillwater stage frequency relationship is shown in Figure 5.

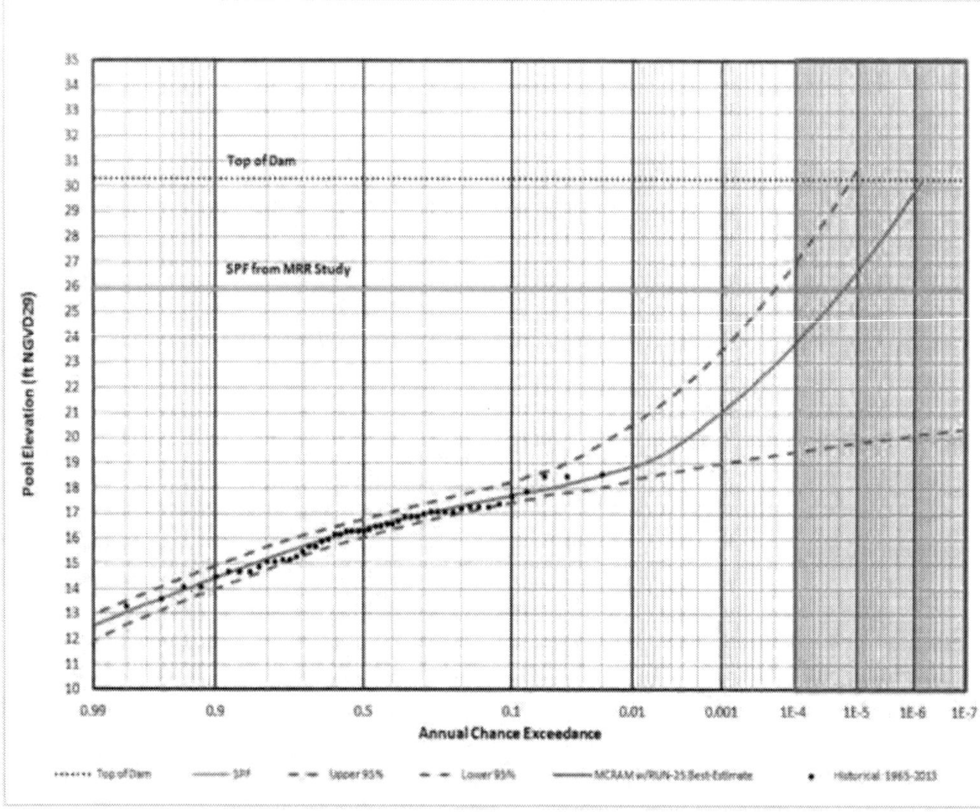

Figure 5. Stillwater Stage Frequency Relationship

The joint probability of flood and wind induced stages was explicitly modeled by sampling tropical storm winds from a set of about 7000 synthetic tropical storm events. Look up tables of set up and run up were developed using the hydrodynamic model based on a range of values for wind speed and direction. Computational time made the explicit modeling of set up and run up for each generated tropical event impractical. The generated wind events were combined with the stillwater stage hydrographs and the hydrodynamic look up tables to obtain set up stage hydrographs. From these hydrograph, annual maximums could be obtained to develop stage frequency relationships for wind influenced stages at key locations along the dam. The wind influenced stage frequency curve is shown in Figure 6 with the stillwater curve also shown for reference. Wave properties were then used to estimate the cumulative overwash volume at key locations so that overwash potential failure modes could be evaluated.

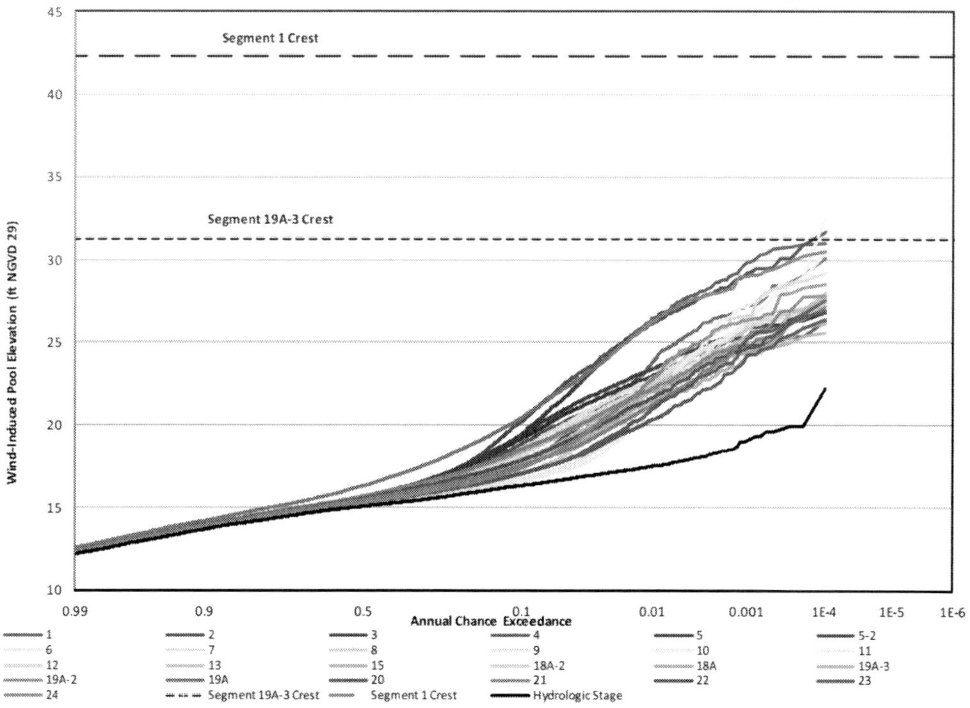

Figure 6. Wind Influenced Stage Frequency Relationship

Segment 24, shown in Figure 7, has the highest overwash hazard potential despite not having the highest set up or the lowest crest elevation. This is partly due to the steepness of the embankment contributing to more wave run up and the orientation of the segment being adverse to a more likely direction of high winds.

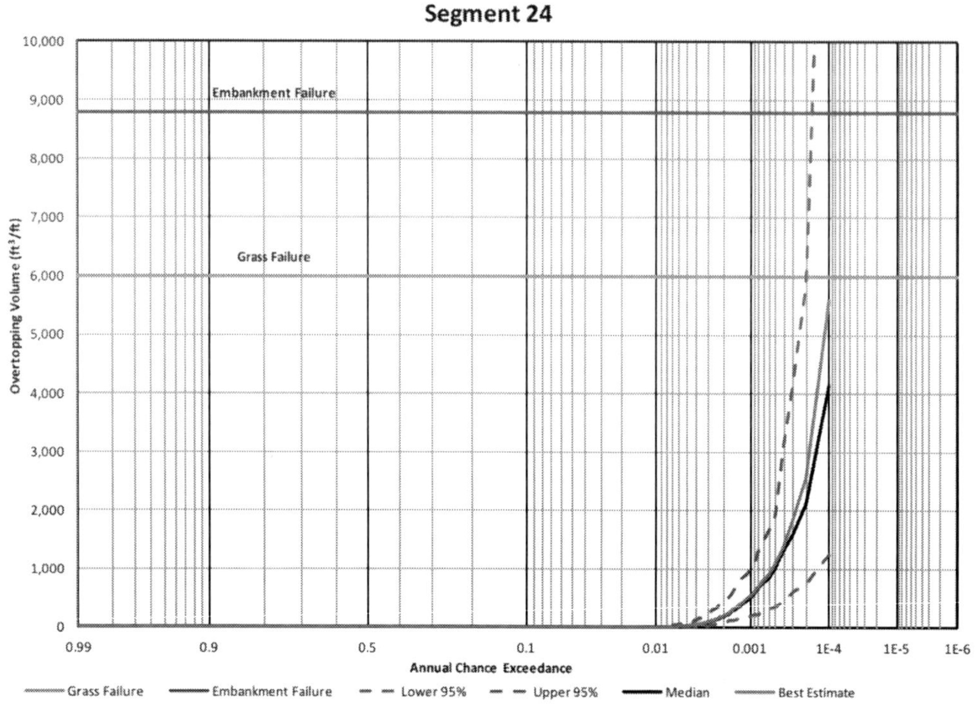

Figure 7. Overwash Frequency Relationship for Segment 24

The results in Figures 5, 6, and 7 were used in the risk estimate to evaluate overtopping as a possible risk driving potential failure mode. Results of the risk estimate suggest that overtopping risks are relatively low based on the low likelihood of significant overwash and low potential for life loss. Outcomes of the risk assessment included a decision to not make significant investment to mitigate overtopping risks.

PATH FORWARD

The three case histories presented in this paper provide a scalable framework for a risk informed evaluation of freeboard allowance and associated overwash potential failure modes. Results for the first two case studies suggest that freeboard allowances for wind generated waves may be negligible in cases where high winds in an adverse direction are not likely to occur during flood events. This may be a common situation for inland reservoirs not aligned with the direction of flood generating weather patterns and having relatively short flood durations on the order of hours to days. In general, it may be reasonable to screen out overwash potential failure modes for these types of projects. The third case study suggests that wind may be a critical factor to consider in cases where reservoirs are aligned with flood generating weather patterns, are influenced by tropical weather, and have relatively long flood durations on the order of weeks to months. Because of the many confounding variables that might affect peak stages, a more comprehensive evaluation of wind affected peak stages may be appropriate to adequately characterize the flood hazard and associated risks.

REFERENCES

American Society of Civil Engineers. (2013). Standards ASCE/SEI 7-10. *Minimum Design Loads for Buildings and Other Structures.*

US Army Corps of Engineers. (1997). Engineer Manual 1110-2-1420. *Hydrologic Engineering Requirements for Reservoirs.* Washington, DC.

US Bureau of Reclamation. (2013). Design Standards No. 14. *Appurtenant Structures for Dams (Spillways and Outlet Works).*

FAIL SAFE DESIGN FOR GATED STRUCTURES CONSTRUCTION AND REHABILITATION

Peter Davis, PE[1]
Mathew McGuire, PE[2]

ABSTRACT

There are approximately 16,000 gated structures in the United States from small diversion facilities to large Hydro-Electric Dams. The gates associated with these structures have been designed based upon evolving engineering manuals as well as design engineer preference. Over the past several decades there have been gate failures including both structural and mechanical components. The purpose of this paper is to discuss design approach and recommendations to move toward fail safe designs.

One case study will be used to identify the pertinent design issues. Recommendations for design approach will be included as well as a checklist to identify risk of minor to catastrophic failures.

Pete Davis, PE has over 40 years experience in the inspection, design and construction of heavy movable structures including lock and dam gates as well as movable bridges. Matt McGuire, PE has 16 years of inspection and design experience for hydraulic structures.

INTRODUCTION

There are approximately 16,000 gated structures in the United States from small diversion facilities to large Hydro-Electric Dams. The gates associated with these structures have been designed based upon evolving engineering manuals as well as design engineer preference. Many of these structures are over 50 years old and are just now undergoing major rehabilitation. Over the past several years there have been gate failures including both structural and mechanical components. These failures can be attributed to various root causes. The purpose of this paper is to discuss design approach and recommendations to move toward fail safe designs.

What is Fail Safe?

"Capable of compensating automatically and safely for a failure" – American heritage College Dictionary, 3rd Edition

The owners of gated structures need to determine what is the appropriate "fail safe condition". This condition is highly site specific. As an example, if gate hoist machinery fails, and the gate drops, is that the fail safe condition or not. Climate Change along with

[1] Vice President, Heavy Movable Infrastructure Program Leader, HDR Inc., 1037 Raymond Blvd., Suite 1400, Newark, NJ 07102, Pete.Davis@hdrinc.com
[2] Western Region Heavy Movable Structure Leader, HDR Inc., 1001 SW 5th Ave., Portland, OR 97204, Matthew.McGiure@hdrinc.com

increased sensitivity to liability issues is increasing the significance of gate failure(s). The goal of this paper is to improve awareness of potential failure modes and specific steps that can be taken to mitigate risk. The focus of the paper is on mechanical systems; however the electrical control systems are of equal importance regarding fail safe design.

The first question to be asked is "what are the risks of gate failure", and what constitutes a failure. Is the inability to move the gate a failure, or is dropping the gate a failure or is structural collapse considered a failure. When most gate hoist systems were designed a comprehensive analysis of potential failure modes and their consequences were evaluated under different cost/risk considerations than are acceptable today. While a truly fail safe design may not be practical, safer designs can be achieved in an economical manner.

One of the design standards for the United States Army Corp of Engineers (USACE) that governs the design of gated structures is the Engineering Manual (EM)1110-2-2610 "Mechanical and Electrical Design for Lock and Dam Operating Equipment" (June 2013). The update to this manual, in 2013, included component selection criteria as well as recognizing additional load cases. A supplementary standard which has been referenced by EM1110-2-2610 is the AASHTO LRFD Moveable Highway Bridge design Specifications. Specifically, the EM makes reference to ASSHTO for the design of engineered chain, but AASHTO also provides additional insight and formulas to calculate fatigue life, component allowable stresses, and recommendations on gear geometry. All of this additional AASHTO content could be applied to more than just engineered chain design for gate and gate hoists.

It is likely that over the next several years, the number of gate failures may rise due to age, and environmental events. Recent failures include rope connections, coupling failures due to misalignment, structural failures and damage from environmental events. Owners have performed repairs and in some cases upgrades. When performing a rehabilitation project, the following issues should be considered:

Photo 1. An Example of a Spillway Gate Structure

1) Have the operating parameters of the structure changed?
2) How often do the gates need to be operated?
3) Is the equipment operated on a regular interval?
4) Does the hoisting system have redundancy – are there single points of failure?

5) What are the potential failure modes?
6) What are the system risks of one or more gate failures?

In order to investigate these questions, three case studies will be utilized.

This facility is a dry dam and has six vertical service gates with dedicated hoist machinery and six emergency gates which are serviced by a gantry crane. The emergency gates are used to allow maintenance of the service gate or in emergency situations where the service gate is not functional. The service gates are raised and lowered in a slot by the hoist machinery. Three gates are used to control flow through the two channels. The purpose of this facility is for flood control during high water events.

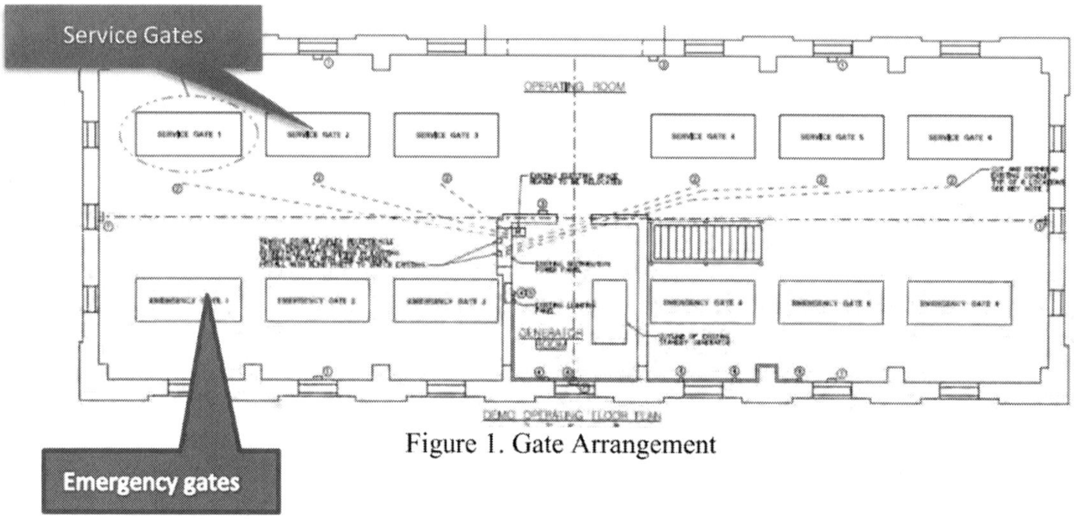

Figure 1. Gate Arrangement

The vertical service gates are raised and lowered by wire ropes attached to the gate and wound around a hoist drum. The drum and sheaves are rotated by a gearbox and electric drive motor (see figure 2). The drive motor has a drum brake for control.

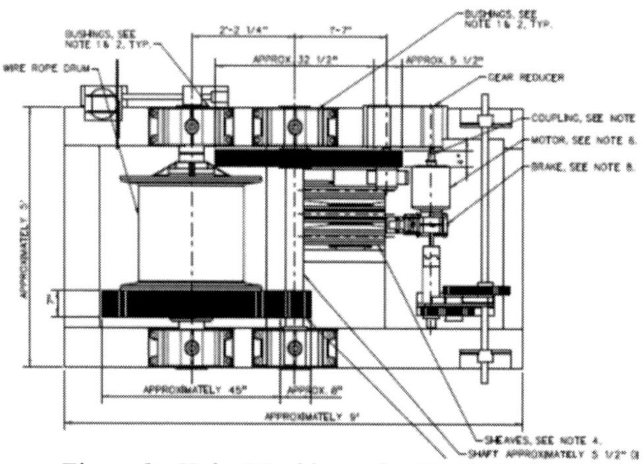

Figure 2. Hoist Machinery for Vertical Gate

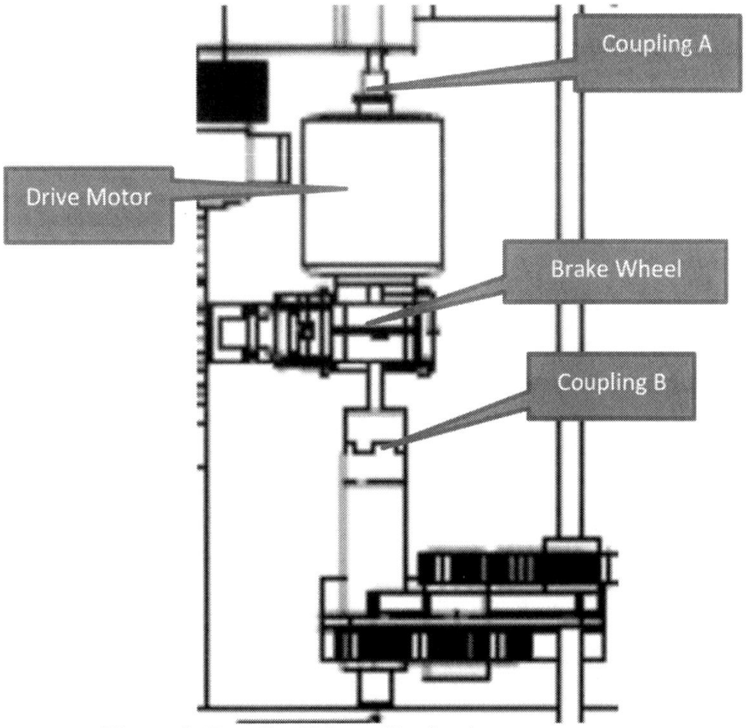

Figure 3. Detail of Motor/Brake Arrangement

The hoist machinery for this gate does not have inherent redundancy. If coupling "A" fails, then the gate is not restrained. The brake will be decoupled from the gate, potentially resulting in the gate dropping. Does this gate dropping constitute a catastrophic failure or merely an inconvenience? The answer to this question can only be answered by a system wide risk analysis. One potential solution to reduce the risk of a single component failure would be to relocate the brake adjacent to the hoist drum. The most recent revision of EM 1110-2-2610 makes reference to the brake being located on the input or output shaft of the first reducer. The challenge for the engineer is to determine the cost/risk benefit of modifying the hoist design to locate the brake closer to the load. For this application, the brake could have been mounted to the drum shaft or the intermediate shaft.

The other feature of this design is the brake wheel is mounted on the motor shaft extension. As the brake is applied, the shaft will bend. The most recent EM identifies fatigue as a design consideration where the original document did not. As older hoist systems are rehabilitated, fatigue details should be mitigated wherever possible. The simple remedy for this fatigue prone detail would have been to install a bearing as close to the wheel as possible, thus reducing the unbraced shaft length.

It is recommended that when existing hoist machinery is being reused or rehabilitated in kind, the design details should be reviewed to determine if they meet current design standards. If the details do not meet the most recent code, then a risk analysis should be performed to determine if the cost to modify the design is appropriate.

Photo 2. Typical Tainter Gate

In the Western US there have been gate structural and mechanical failures. When a gate fails structurally, the natural inclination is to strengthen the original design. While this approach is reasonable, the mechanical and electrical systems require a capacity review and modification where appropriate. These M&E reviews should be performed at the component level (see figure 4). Figure 4 denotes the output from an evaluation. Three load cases were evaluated (normal operation, one side lifting and motor overload). The point of the evaluation is that while some components may be within design guidelines (safety factor greater than 1) others may not. The rating used as pass/fail denotes whether the components meets the allowable stress, fatigue, or deflection criteria .Notice that speed reducer 2 does not meet the defined criteria for alternative T2a and T4 but does meet the criteria for T3.

Table 1. M&E Component Analysis by Load Case

HDR

Project: HOWARD HANSON DAM REHAB Computed: MGW Date: 11/30/2015
Subject: MACHINERY ANALYSIS Checked: SAM Date: 01/04/2016
Task: 1 SUMMARY Page: 3 of: 4
Job #: 269602 No:

1.7 Summary Table

Machinery Component	T_{2a}			T_3			T_4		
	Value	Rating	F.S.	Value	Rating	F.S.	Value	Rating	F.S.
3 COUPLINGS									
Coupling 17	107 ft-lbs	OK	3.92	370 ft-lbs	OK	1.14	340 ft-lbs	OK	1.24
Coupling 18	108 ft-lbs	OK	11.6	373 ft-lbs	OK	3.37	340 ft-lbs	OK	3.70
Coupling 19	16 ft-lbs	OK	32.4	28 ft-lbs	OK	18.8	45 ft-lbs	OK	11.6
4 REDUCERS									
Speed Reducer 2	16 ft-lbs	NG	0.89	28 ft-lbs	OK	1.03	45 ft-lbs	NG	0.63
Speed Reducer 3	107 ft-lbs	OK	1.02	370 ft-lbs	NG	0.59	340 ft-lbs	NG	0.64
5 LINE SHAFT									
Shaft Stress	214.4 psi	OK	10.4	739.2 psi	OK	7.0	673.8 psi	OK	7.7
Angular Deflection	6.16 deg	NG	0.53	21.24 deg	NG	0.15	19.36 deg	NG	0.17
Fatigue Check	0.02	OK	INF	-	-	-	-	-	-
6 BRAKE CHECK									
Brake Torque	24 ft-lbs	OK	13.4	-	-	-	-	-	-
8 GEAR BENDING									
Pinion AGMA Bending Stress	66,584 psi	NG	0.97	-	-	-	-	-	-
Pinion Lewis Bending Stress	22,334 psi	NG	0.99	-	-	-	-	-	-
Gear AGMA Bending Stress	57,072 psi	OK	1.02	-	-	-	-	-	-
Gear Lewis Bending Stress	16,059 psi	OK	1.12	-	-	-	-	-	-
9 GEAR PITTING									
Pinion Pitting Contact Stress	205,015 psi	NG	0.979	-	-	-	-	-	-
Gear Pitting Contact Stress	205,015 psi	NG	0.878	-	-	-	-	-	-
10 GEAR OVERLOAD									
Pinion Overload	-	-	-	-	-	-	84,702 psi	NG	0.75
Gear Overload	-	-	-	-	-	-	60,901 psi	NG	0.74
11 DRUM SHAFT STRESS									
Bending - Fillet 1	6,177 psi	OK	1.43	21,293 psi	OK	1.06	26,621 psi	NG	0.85
Bending - Keyway	4,066 psi	OK	1.54	14,017 psi	OK	1.14	17,524 psi	NG	0.91
Bending - Max. Moment	7,395 psi	OK	2.04	25,491 psi	OK	1.51	31,868 psi	OK	1.21
Bending - Fillet 2	332 psi	OK	26.49	1,146 psi	OK	19.63	1,433 psi	OK	15.70
Shear - Fillet 1	3,089 psi	OK	1.76	10,647 psi	OK	1.30	13,310 psi	OK	1.04
Shear - Keyway	2,033 psi	OK	1.33	7,009 psi	NG	0.99	8,762 psi	NG	0.79
Shear - Max. Moment	3,697 psi	OK	2.04	12,745 psi	OK	1.51	15,934 psi	OK	1.21
Shear - Fillet 2	166 psi	OK	32.64	573 psi	OK	24.19	716 psi	OK	19.35
Shear - Max. Shear	1,097 psi	OK	6.86	3,783 psi	OK	5.09	6,438 psi	OK	2.99
12 DRUM SHAFT FATIGUE	Keyway governs and is shown below								
AASHTO Shaft/Trunnion Size	0.71	OK	INF	-	-	-	-	-	-
AASHTO Comb. Shear/Bending	0.71	OK	INF	-	-	-	-	-	-
Soderberg	0.71	OK	INF	-	-	-	-	-	-
Modified Goodman	0.71	OK	INF	-	-	-	-	-	-
ASME-Elliptic	0.51	OK	INF	-	-	-	-	-	-
13 DRUM SHAFT BRG.									
Bearing Pressure	635 psi	OK	6.30	2,189 psi	OK	6.17	2,736 psi	OK	4.93
Bearing PV	141 psi-fpm	OK	212	487 psi-fpm	OK	62	-	-	-
14 PINION SHAFT BRG.									
Bearing Pressure	1,362 psi	OK	2.94	4,695 psi	OK	2.88	4,804 psi	OK	2.81
Bearing PV	910 psi-fpm	OK	32.98	3,136 psi	OK	9.57	-	-	-

Speed Reducer Example

The second case to discuss is multiple gates operated by shared hoist machinery. Figure 5 is an example of this machinery arrangement. Note that there are no brakes on this side of the gate at gate A (if the brake is located on the adjacent pier, then risk is reduced). The motor is shared between the gates. The risk of this design is that if one of the couplings on the transverse shaft fail, then gate A or B can become uncontrolled. In addition there are multiple single failure points within this type of system. As discussed previously, the engineer should determine the risk associated with hoist failure and develop a cost risk based decision on how to proceed with hoist repairs and /or replacement. For this example, installing a brakes closer to the lifted load and in redundancy would significantly reduce risk of uncontrolled gate movement.

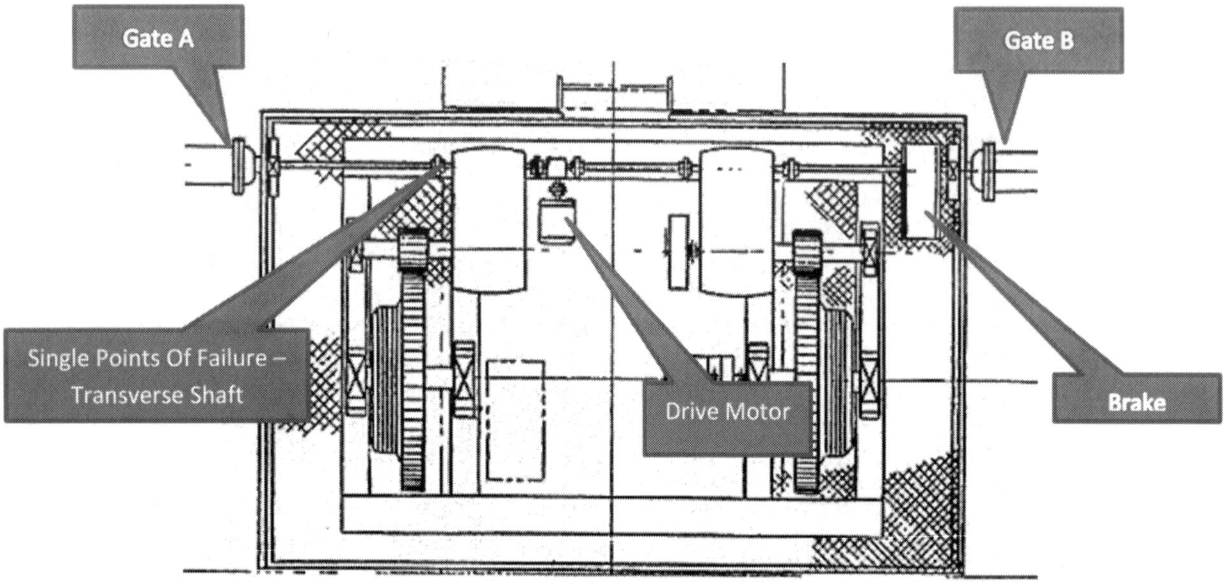

Figure 4. Typical Multiple Gate Hoist Machinery

Figure 5. Updated Hoist Design Reducing

Figure 5 is a new design hoist system. While this design layout is fundamentally similar to previous designs, the two stages of worm reducers (difficult to back feed) have a dedicated motor brake. In this design, the factors of safety for components where failure would result in an uncontrolled gate were increased. It is typical for the hoist rope factor of safety to be 5. The individual components between the brake and the gate should also have an increased safety factor. During the initial design risk assessment, the design factors of safety should be established.

At some facilities, the hoist machinery is a movable hoist (know in the industry to some as a 'mule') which is used to operate a series of individual gates with the single hoist. For a dam with multiple gates, the mule system presents significant risk for fast occurring high water events. The mule may not have enough time to repeat the task of raising a gate and traverse to the next gate. Mules may not be sized for operating gates under flow, which may also be caused by quickly rising waters. Another risk of mules is that the act of connecting to a gate, operating a gate, and disconnecting from a gate present multiple points at which a mule or a gate may become jammed, affecting the operation of subsequent gates. The Erie Canal experienced a situation during tropical storm "Irene" in 2011 where the mule system on multiple movable dams did not have sufficient capacity to raise a single gate under a high water event and could not be deployed rapidly enough to open multiple gates before the high water arrived at a dam.

An analysis to improve movable hoist reliability should consider several aspects. The number of hoists to gates and the access of the hoists to gates (if a hoist should fail) should be considered. Connections of hoists to gates should be designed such that a gate could be 'dogged' (or held in place by the structure) such that a hoist can be disengaged from a gate should it become jammed or inoperable. Movable hoists should also be sized to handle adverse loads such as loads due to high water levels and uneven, one sided lifting of gates.

Photo 3. Erie Canal Movable Dam Gate Hoist

CONCLUSION:

As hoist machinery reaches the end of its useful life, the facility owners have an opportunity to reduce risk by moving designs toward a failsafe condition. The definition of fail safe should be defined in the current and future anticipated context of the facilities. A fail safe condition should be defined as whether the gate does not move when a mechanical or electrical control failure occurs or does the gate 'fail' to a lowered or raised position. Once this determination is made, then the risk associated with the current or proposed design can be evaluated. It is likely that a 100% failsafe design cannot be economically achieved; however potential failure modes can be identified. These failure modes can be mitigated by design, inspection, periodic maintenance and component replacement. The following list of questions should be considered while deciding to rehabilitate or replace a dam gate hoist system.

1) What is the risk of a gate operational failure
    a. Has a risk assessment been performed?
    b. What is the fail safe condition today and in the future?
    c. Is this facility subject to increased or decreased flows in the future?
2) Can this risk be mitigated by:
    a. Adding brakes at an appropriate location.
    b. Increasing the design safety factor of specific components.
    c. Designing the hoist system for potential failure modes (jammed gate on one side, jammed gate where drive motor goes to 300% design torque, control system failure resulting in gate going to full open or closed position, etc.).
3) Can the risk of existing equipment be mitigated by inspection and or maintenance practices?

DAMS INLET/OUTLET WORKS REHABILITATION PROGRAM

Jack Husted[1]
Sterling Klippel[2]
Keith Lilley[3]

The Los Angeles County Department of Public Works (LACDPW) owns and operates 14 major dams equipped with a variety of outlet works. Flood control and storm water capture operations require significant variations in reservoir level and frequent releases through the outlet valves. Reliability is critical. Most of the dams are over 80 years old and much of the outlet works had surpassed their useful lives. LACDPW performed extensive investigations to determine the best plan for modifying and rehabilitating the outlet works at each dam, and developed long-term program to upgrade all deficient gates, valves, actuators, penstocks, protective coatings, intake structures, electrical, and control systems at the dams. Each dam is different and requires unique solutions. In some cases, existing vintage valves and gates are of impressive quality and can be restored or rebuilt with minimal modifications. In other cases, equipment is simply due for replacement. Upgrades also look beyond in-kind replacement, and seek out innovative solutions, sustainable practices, and new technologies. For example, new valves and updated operating protocols have improved sediment management and increased the capacity for storm water capture, and low flow outlets can be designed to support downstream habitat. Working with field staff, the program also identifies ways to improve operations, maintenance, and safety. This paper discusses the development and implementation of this program, involving civil, mechanical, and electrical disciplines, and exemplifies the challenges, lessons learned, and case studies for these rehabilitations.

[1] Los Angeles County Department of Public Works, jhusted@dpw.lacounty.gov
[2] Los Angeles County Department of Public Works
[3] Los Angeles County Department of Public Works

BEST PRACTICES IN A TAINTER GATE REHABILITATION PROGRAM

Matt Moses, P.E.[1]
Layne Bukhair, P.E.[2]

ABSTRACT

Radial arm tainter gates are among the most common type of spillway gates for dams in the United States. Since the heyday of large reservoir construction in the 1950's through 1970's, many tainter gate spillways are now near or beyond the end of their original design life. The tainter gate failure at Folsom Dam in 1995, resulting in the loss of nearly 40 percent of Folsom Lake, highlighted the importance of maintaining and rehabilitating these aging structures. Several owners of large, high hazard dams in Texas have recently undertaken major spillway gate rehabilitation programs, including the following projects: Buchanan Dam – 37 gates (2006 – present); Toledo Bend Dam – eleven gates (2010 – present); Lake Fork Dam – five gates (2012 – 2014); Lake Conroe Dam – five gates (2012 – 2014); Palmetto Bend Dam – twelve gates (2013 – present); and Choke Canyon Dam – seven gates (2014 – present). This paper outlines the primary components of a successful tainter gate rehabilitation program using examples from the projects mentioned above to illustrate various approaches. These include program planning, detailed initial inspections, structural analysis, alternative evaluation, developing a rehabilitation design to address the observed conditions, and overseeing a construction program which helps define the need for additional repairs. The lessons learned and best practices developed over the course of these projects will also be summarized.

INTRODUCTION

Project Descriptions

When completed in 1938, Buchanan Dam was the largest multiple-arch dam in the world. The dam impounds Lake Buchanan, which stores approximately 875,000 acre-ft at a normal pool level. The dam is over two miles long and consists of multiple concrete arches, a powerhouse, three gated spillways, an uncontrolled spillway, and an embankment. The three gated spillways have a total of 37 tainter gates of various sizes. The spillway located closest to the powerhouse has seven large tainter gates (40 feet wide and 25 feet high). The next spillway to the north has fourteen small tainter gates (33 feet wide and 15 feet high) and the third spillway has sixteen small tainter gates (33 feet wide and 15 feet high). The dam is operated by the Lower Colorado River Authority (LCRA).

Construction of Toledo Bend Dam and reservoir was completed in 1967. The reservoir impounds approximately 4.5 million acre-feet of surface water along the Texas-Louisiana border. The reservoir is the largest man-made body of water in the South. The facility functions as a water supply source across the respective state's lines, serves as a recreational reservoir, and is used to produce hydroelectric power. The main features of

[1] Structural Engineer, Freese and Nichols, Inc., Austin, TX mm@freese.com
[2] Structural Engineer, Freese and Nichols, Inc., Austin, TX

the dam include the primarly eathern embankment, three dikes, a powerhouse, and gated spillway. The spillway is controlled by twelve radial tainter gates and each radial gate is 40 feet wide by 29 feet high. The dam is operated by the Toledo Bend Project Joint Operation (TBPJO) which is a conjunction of the Sabine River Authority of Texas (SRA-TX) and the Sabine River Authority of the State of Louisiana (SRA-LA).

The construction of Lake Fork Dam was completed in 1980. The reservoir impounds approximatley 675,000 acre-feet of water. The dam functions as a water supply source and as a recreational reservoir. The main features of the dam include an eathern embankment and a gated spillway. The spillway is controlled by five radial tainter gates and each radial gate is 40 feet wide by 20 feet high. The dam is operated the Sabine River Authority of Texas (SRA-TX).

The construction of Lake Conroe Dam was completed in 1973. The reservoir impounds approximatley 430,000 acre-feet of water. The dam functions as a water supply source and serves as a recreational reservoir. The main features of the dam include an eathern embankment, reinforced concrete gated spillway, and a service outlet. The spillway is controlled by five radial tainter gates and each radial gate is 40 feet wide by 30 feet high. The dam is operated the San Jacinto River Authority of Texas (SJRA).

Palmetto Bend Dam was constructed in 1980 under the banner of the Palmetto Bend Water Resource Project; a joint effort accomplished by the Texas Water Development Board (TWDB), the Bureau of Reclamation, and the Lavaca-Navidad River Authority (LNRA). The reservoir impounds approximately 170,000 acre-feet of water, providing both water supply and recreation. The dam is comprised of a rolled earthen embankment, a reinforced concrete gated spillway, outlet works on opposite sides of the spillway, and a separate river outlet works. The spillway is controlled by twelve radial tainter gates with each gate, and each radial gate is 35 feet wide by 23 feet high. The dam is now owned and operated by LNRA.

Figure 1 below depicts a typical tainter gate, along with nomenclature for the typical primary components that make up a tainter gate.

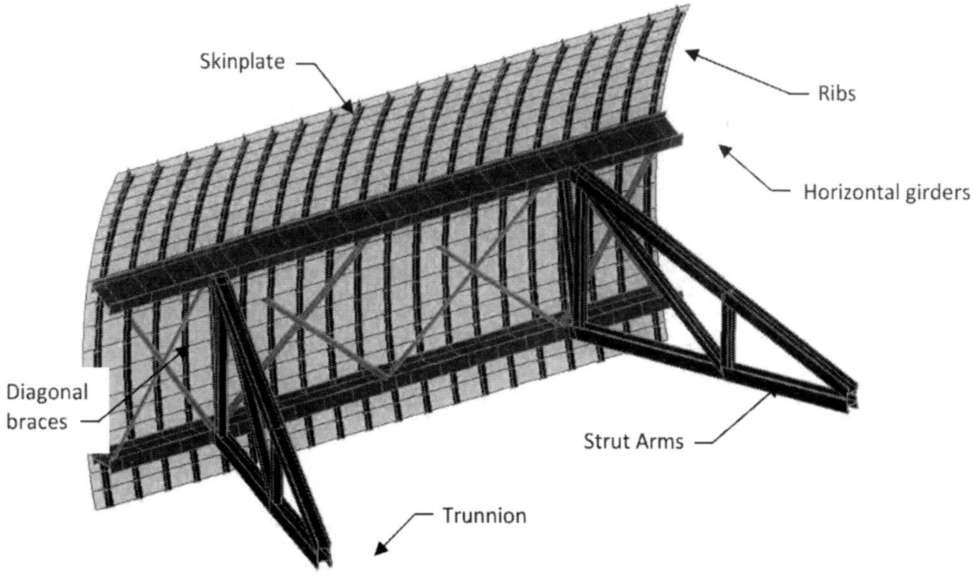

Figure 1: Typical Tainter Gate Components

CONDITION ASSESSMENT

Purpose

The first step in the path towards spillway gate rehabilitation is to perform an overall condition assessment. Condition assessments generally include:

- A detailed inspection and documentation of each radial tainter gate, including an evaluation of the condition of the existing coating system, structural members, connections, trunnions, lifting assemblies, and seals. This involves utilizing engineers trained in high angle rope access skills and equipment (Figure 2);

- An evaluation of the gate operating system for the spillway gates, which involves visual observations and operational testing of the mechanical and electrical components of the gate hoists.

- A structural evaluation and analysis of the gates under various loading conditions, both from an "as-built" condition as well as a current condition (evaluating member sections with cross sectional losses);

Figure 2: Detailed Climbing Inspection

Climbing Inspections and Evaluation of Gate Operating Systems

Prior to beginning a gate inspection, stop logs are installed upstream of the first gate to be inspected and tested. During a typical first day of inspection, a safety training session takes place first, followed by a briefing of anticipated detailed activities for the full inspection. Sequencing of stoplog installation and movement is planned to facilitate efficient inspection and testing of multiple gates. Typical gate work involves inspection of the upstream side of the gate and operational testing first, so that the stoplogs can be moved to the next gate while inspection of the downstream side occurs.

Focus on the upstream side of the gate is primarily on the condition of the skinplate coating system, seals, and lifting assemblies, including wire ropes and lifting brackets. Focus on the downstream side of the gate is also on the coating system, as well as the structural condition of strut arms, girders, ribs, braces, seals, guide shoes or rollers, and trunnion assemblies.

Attention is also directed to whether there is uniform contact made by the side seals against the pier faces, whether there is uniform clearance between the guide shoes or rollers and pier faces, and the levelness of the bottom sill relative to the bottom of the gate. These observations are made as an early indicator of even rope tension distribution.

Mechanical and electrical hoist inspections focus on operational performance. Operational testing involves full travel open and close. Attention is given to irregular noises induced by the equipment's bearings, motors, and brakes. Visual observations conducted on hoisting equipment components are aimed at detecting items such as: signs of damage at the motors, brakes, drums and pillow block bearings, leaks within the gearbox housing and through the couplings, loose or damaged bolts, and wearing/damage to the wire rope assemblies. In addition to looking for each of the aforementioned issues, a visual inspection is performed to check for appropriate lubrication and contact patterns within each of the gate's open gear boxes. The tainter gate electrical motors and control panels are typically evaluated both visually and with an ammeter and a volt meter at initial opening. The visual evaluation of the electrical motors and control panels consists of visually inspecting each of the motors and control panels for any excessive heating, burn marks and wire insulation abrasions.

Estimations of potential ranges of trunnion pin friction are made with the use of a laser mounted on the trunnion and a target mounted on the backside of the skinplate. Measurements of movement of the laser against the target are taken during initial gate opening. These deflections measurements are used to estimate pin friction for evaluation in the computer modeling of the gates.

During climbing inspections, the upstream face of the skinplate is typically found to be in fair condition with heavy organic build up. The majority of corrosion is usually found at and below the water line. Rust blisters are often visible across the face of the skin plate. Once the blisters are removed, the skinplate displays a noticeable concave shape about the same diameter of the blister (Figure 3). The rust blisters can make the skinplate brittle and can create weak points. In some extreme cases, evidence of rust blisters observed on the upstream face can be seen on the downstream side of the gates. This is observed when the downstream coating swells in line with the upstream blister, forming a bubble. In some instances, these bubbles can be punctured all the way through to the upstream side of the skinplate.

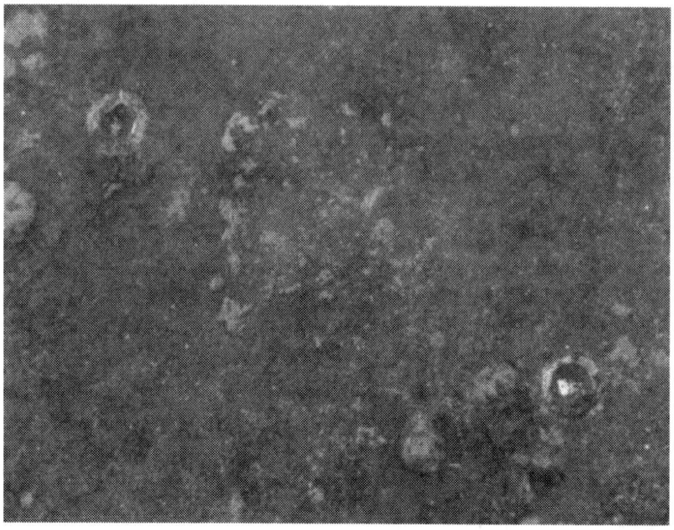

Figure 3: Skinplate Rust Blisters

Strut arms are typically found to be in fair to good condition, with localized areas of corrosion and isolated pitting, mostly near connections. The primary contributing factor to corrosion appears to be blockage of drain holes which restricts effective drainage of water, resulting in water accumulation and pooling within the strut arms.

In typical gate geometries, horizontal girders span the width of each gate and support the vertical tees of the skinplate. As visible from the pier, these girders may appear to be in good condition. However, moderate to extensive damage is regularly observed during the close-up inspection. An example of this is shown in Figure 4, along the upstream flange of the girder nearest to the skinplate. This damage seems to occur in primarily in a region along the girders spanning between the strut arms and the piers, and appears to be a result of water leakage along the side seals. Typical of most gates, the most extensive damage occurs along the bottom girders. Continual wetting and drying of this area, in combination with coating failure and no further protection from corrosion, lead to extensive section loss of the upstream flanges.

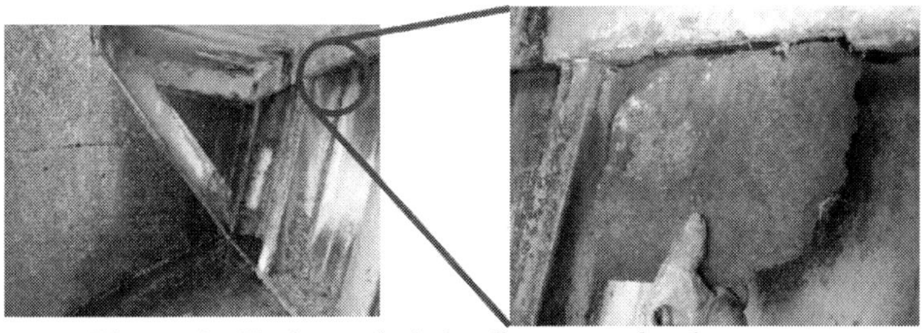

Figure 4: Horizontal girder flange section loss

Interior vertical ribs are typically found to be in fair to good condition as a whole. Exterior vertical ribs closest to the pier faces, however, are typically found to be in fair to poor condition, with holes of varying diameter often observed. This is usually the result of spraying water from leaking side seals which leads to ponding water, specifically in areas of coating failure. Vertical ribs near the bottom of the gate are also often found to be in poor condition, with severe section loss typically within the webs (Figure 5). This is often the result of leakage through drain holes in the lower girders into regions of the gate where coating failure has occurred.

Figure 5: Holes in Rib

The wire rope attachments often contain a heavy buildup of organic material which can make inspection difficult. It has been found that this is one area that typically requires further inspection and evaluation once blasted during the initial stages of the rehabilitation effort.

Overall, detailed climbing inspections are considered to be very useful, however, not all deficiencies in the gates can be thoroughly observed. This is primarily due to areas of standing water, spraying water, organic buildup and the presence of the existing coating system; all of which inhibit detailed inspection and measurement. From experience, it has been found that additional localized deficiencies are often uncovered once the gates are blasted.

Structural Analyses

As part of the condition assessment, a computer model of a typical gate is developed, typically using design drawings and then confirming member sizes and dimensions from the climbing inspection. The gate model is created as a representation of all of the radial gates present within the spillway. Steel material properties are based on data listed on the design drawings and an understanding of the type of steel used in radial gate construction at the time the radial gates were fabricated. The entire gate geometry is modeled to account for unsymmetrical loading. Modeling the entire gate also provides a check on the structure against the loads applied since supports and the reactions at the trunnions should be equal for symmetrical loading conditions. See Figure 6 for isometric views of a typical finite element computer gate model.

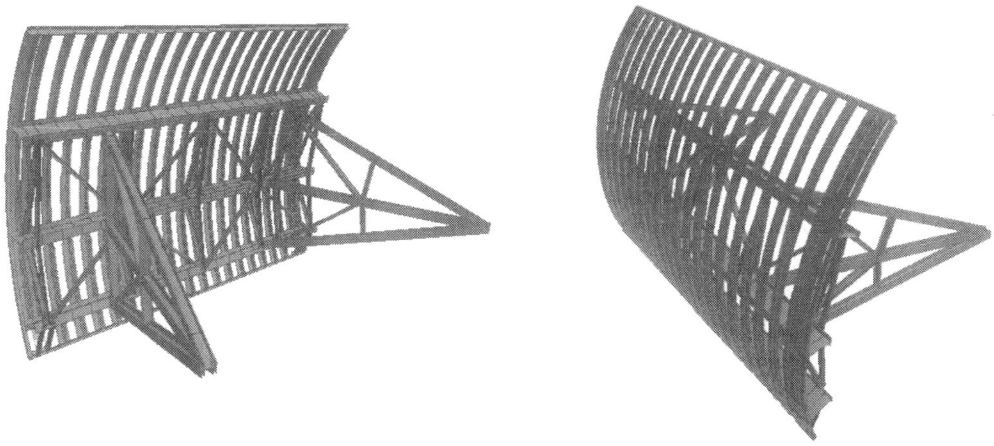

Figure 6: Typical finite element model isometric views

Stresses in the radial gate members are established for a number of loading conditions as defined by the United States Army Corps of Engineers (USACE). Up to seven load cases are typically evaluated, made up of defined combinations of individual loads. These include gravity, hydrostatic pressure based on varying water surface levels, wave, wind and operation (trunnion pin friction, side seal friction, and wire rope pressure). The gates are evaluated fully closed, fully open, and in operation. Results of the detailed inspection are incorporated by creating a separate reduced section finite element gate model. The results of the analyses for both the as-built and reduced section models are compared to allowable and yield stresses as well as American institute of Steel Construction (AISC) code strength requirements.

Report of Findings

From the above activities, a typical report of findings is developed and presented to the owner. The report includes a summary of field observations, measurements, and data collected, gate structural analysis design criteria, results, and conclusions, and recommendations for necessary rehabilitation with associated opinions of probable construction cost for repairs or strengthening.

TYPICAL REHABILITATION RECOMMENDATIONS

From our experiences, following physical inspections and structural analyses, the existing tainter gates we have evaluated have been found to be in generally satisfactory condition overall. Often though, specific members and components are found to require repair by either strengthening or replacement, resulting from observed deficiencies and/or inadequate strength capacity.

Typical gate and hoisting system maintenance and modifications have consisted of the following:

- Removal of existing coatings and recoating.

- Rebuilding corroded areas of horizontal girders due to corrosion and subsequent section loss (Figure 7 depicts horizontal girder segment replacement at end of girder).

- Cut out and replacement of portions of vertical/horizontal ribs observed to have holes through the web or that have experienced severe section loss in the flanges and/or webs (Figure 8 depicts vertical rib segment replacement at the bottom of the gate).

- Strengthening of strut arms at connections to the lower horizontal girder, resulting from section loss or inadequate strength capacity (Figure 9 depicts installation of cover plates).

- Replacement of web stability braces, due to corrosion and subsequent section loss along the members and at the connections.

- Replacement of lower knee braces on each side of each gate, due to corrosion and subsequent section loss along the members and at the connections.

- Installation of additional members and/or member strengthening to account for updated loads.

- Increasing the size of drain holes in the webs of the horizontal girders, adjacent to the piers, and in the webs of the strut arms at connections. Ponding of water is typically observed at these locations.

- Repairing corroded welds at connections throughout the gates.

- Replacement of gusset plates, due to corrosion and subsequent section loss.

- Replacement of side and bottom seals, seal clamp bars, and associated fasteners as necessary.

- Replacement of lifting assembly components, including hitch plates, equalizer bars, and pins.

- Wire rope cleaning, and in some cases complete rope and socket replacement either as a result of age or condition.

- Maintenance to mechanical equipment, including drum gear cleaning and lubrication, oil replacement, seal replacement, coupling replacement, motor bearing replacement, motor and worm gear re-alignment, brake shoe replacement, and provision of guidelines for regular monitoring and maintenance of all equipment.

Figure 7: Horizontal Girder Repairs (view from below)

Figure 8: Typical Rib Repairs

Figure 9: Typical Strengthening of Strut Arm

PILOT PROGRAM APPROACH

A great amount of information on the gates can be gathered from a climbing inspection and structural analysis. However, given areas of standing water, spraying water, coating failures and localized organic buildup that are often present during initial inspections, not all areas can be inspected in detail to the level required to define the full extent of damage and all required repairs. In addition, it is not possible to efficiently and accurately measure gate member section losses, or remaining metal thicknesses, due to combinations of member corrosion and scaling, and in the presence of the existing coating system. Accurate measurement can only be done by removing loose material and the existing coating system by abrasive blasting to a "near white" surface condition.

To gain a more thorough understanding of the condition of the gates, to better define the complete rehabilitation requirements, and to better estimate overall construction costs and schedules, it is often advantageous to develop contract documents for the rehabilitation of a single gate, to serve as the pilot program for all gates within a given spillway or project. Some of the benefits of this approach are discussed in the following paragraphs.

If required repairs to all gates within a spillway are developed under a single rehabilitation contract based on the climbing inspection results and structural analysis alone, any unidentified repairs determined after abrasive blasting of the first gate would be considered additional work at an additional cost to the owner. These unidentified repairs clearly have cost implications that are magnified by the total number of gates to be rehabilitated.

Often, funds for large scale rehabilitation of all gates within a spillway or project are not immediately available. Funding by fiscal year is often deemed more reasonable and manageable by the owner. Following completion of a pilot project, it is often much more financially feasible to schedule rehabilitation of the remaining gates within a project in gate sets over the course of several consecutive fiscal years. The final cost realized at the completion of the pilot program can assist with budgeting and scheduling of how many gates can be completed in subsequent fiscal years.

PILOT PROGRAM EXECUTION

In the majority of our inspections performed to date, the condition of individual gates across the spillway have been found to not vary greatly. The gate generally considered to be in the worst condition is selected to be the first gate thoroughly evaluated; as this gate is considered the pilot gate. This is done to effectively 'capture' the widest range of anticipated or potential rehabilitation measures required on the project.

Following mobilization, with the stoplogs in place, the contractor sets up scaffolding and a containment system is installed around the pilot gate. After side and bottom seal removal, the first major effort is abrasive blasting of the gate to a "near white" (SSPC SP10) surface condition (Figure 10). Once blasted and cleaned, a prime coat is applied to protect the steel from spot rusting. At this point in time, a detailed inspection of the blasted gate is performed by the engineer.

Figure 10: Girder in Blasted Condition

The focus of the inspection is primarily on specific locations identified during the condition assessment along primary gate members, member connections, wire rope connections at the skinplate, the skinplate itself, and other hard to access locations. Often, as noted previously, much of the damage is focused along the sides of the gates adjacent to the piers, primarily as a result of spraying water from leaking side seals and failing coating systems. Member thicknesses are re-measured at critical areas. For example, along horizontal girders, particularly towards the ends of the gate where large section losses are sometimes identified during the climbing inspection, remaining member thicknesses are measured and the extent of required further repairs are defined. Additional section losses are also often identified at connections and in hard to access locations; particularly in areas that were originally observed to have standing water, most often due to plugged drain holes.

During the climbing inspection, heavy organic buildup is often observed on the wire rope lifting assemblies, and specifically around the hitch pin plates that attach the ropes to the skinplate. Once blasted, some hitch pin plates have been observed to be heavily pitted with substantial section loss (Figure 11). In some instances, it is necessary to remove and replace the plates in kind. Pitting on the upstream face of the skinplate is often found to be more extensive than anticipated. Previous pitting that has been filled is removed during abrasive blasting, leaving all pitted areas exposed.

Figure 11: Wire rope hitch pin plate – before and after blasting

Following all structural repairs, the gates are recoated, new side and bottom seals, clamp bars and fasteners are installed, and each gate is operationally tested in dry and wet conditions and placed back in service. Rehabilitation work on the remaining gates is then scheduled all at once or over the course of several fiscal years, depending on the availability of funds.

The duration of a pilot program varies but typically can be completed within three to four months. Rehabilitation of subsequent gates can typically be done in two to three months per gate on average. Rehabilitation costs per gate vary, but the cost per gate has averaged around $300,000 to $350,000 per gate. This number can of course fluctuate depending on the extent of damage and amount of repairs, access, weather, and containment requirements.

LESSONS LEARNED FROM PILOT PROGRAMS

Following numerous gate rehabilitation projects, we have found that in general the final pilot gate repairs do not significantly deviate from what is defined in the original pilot program contract documents. However, identified changes during the pilot program do compound as the number of gates within a spillway increase, and thus there is the potential for substantial cost increases to the overall project. The ability to inspect one gate after blasting has proved to be invaluable. It allows the engineer to further refine the required repairs, observe in detail areas that were difficult to access as a result of standing and spraying water and heavy organic buildup, and much better understand the condition of specific areas in question. These findings can then be applied to the rehabilitation of the remaining gate sets; thus, minimizing unknowns, defining the work as thoroughly as possible, and better controlling costs. Again, any number of additional repairs identified on one gate can result in considerably higher costs when multiplied by a large number of gates within a given spillway, so identifying these additional repairs early is highly beneficial to all parties.

Contractors have been able to gain a better understanding of the nature of the rehabilitation work to be performed; the pilot program has served somewhat as a learning experience for working on a spillway gate project. Often, from our experience, the type of contractor that pursues this type of rehabilitation has experience in recoating projects and structural repairs, primarily on water storage tanks, but not often specifically on gated structures on a dam. The benefit of the learning process occurring on a pilot gate is that it equips a contractor with more knowledge and confidence in pursuing the remaining rehabilitation gate sets. The experience gained leads to more efficient processes during subsequent gate set rehabilitation work for containment and scaffolding placement, coating removal and recoating, bracing techniques for structural repairs, and seal installation and adjustments. And finally, a contractor's comfort with the process can keep unit prices down, and overall costs down, in subsequent gate sets.

The pilot program also gives the owner confidence in anticipating and planning of subsequent gate set rehabilitation work, specifically related to costs and schedules. It should be noted that performing this work over several fiscal years under separate gate sets does result in a recurring mobilization cost, however, the mobilization effort and cost for this type of rehabilitation project is relatively small and the benefits of a pilot program far outweighed this small additional cost. Another benefit to the owner is high contractor interest in a pilot program, as it potentially leads to the rehabilitation of many additional gates. Bid prices have been found to be quite competitive as contractors want to get in early in the process.

EMERGENCY SHUTOFF VALVES

Kishen Prathivadi, P.E., PMP[1]

ABSTRACT

Butterfly valves are widely used for flow control applications; however, in the penstock valve systems used in Hydropower stations, the butterfly valve is essentially used to protect the powerhouse in case of rupture. These are designed to withstand large forces encountered during closing under excessively high flow velocity and pressure. The entire structural design of the valve is based on rupture velocity. Under this condition, the valve closure develops large reaction forces caused by hydraulic torque generated by large flow velocities.

Typically the valve system would have ancillary support systems like the bypass valves, air release valves, anti-vacuum valves, trip mechanism and over velocity sensing devices. The most critical component in the control interlock and actuation element of the penstock system is the over velocity sensing device and trip mechanism. The primary function is to monitor the flow velocity through the penstock valve and generate an electrical signal with corresponding mechanical action in the event the flow exceeds the pre-set value. The required action is to convert the signal into a mechanical force large enough to be able to release the latch mechanism, so as to initiate valve closure.

Modern trends are to provide a local Programmable Logic Controller (PLC) system for the emergency shutoff equipment which would communicate with the master PLC system within the main control panel and provide all operational information and receive control commands. The PLC system controls and monitors the shutoff valve on the basis of a stored digital program of instructions.

INTRODUCTION

Emergency Shutoff Butterfly Valves find applications in hydroelectric and irrigation plants and water works to limit the damages caused by excess velocities and/or pressure in pipelines supplying a hydroelectric turbine or more frequently in case of a pipe rupture due to earthquakes, excessive stresses etc.

Typically, the penstock valve system consists of a collection of different valve types with the butterfly valve being the primary valve for safety, maintenance, shut-off and flow control. Butterfly valves are essentially emergency closing devices to protect the powerhouse in case of penstock rupture. As a result of this duty, they have to be designed to withstand very large forces encountered during closing under excessively high flow velocity and pressure.(Ramos et al.,) The penstock valve is also used to maintain the turbine inlet valve apart from the turbine itself. In addition, by closing the penstock valve, a large amount of dewatering time can be saved for the headrace tunnel above the valve when the

[1] Engineering Manager, Sewer Authority Mid-Coastside, Half Moon Bay, California, kishen@samcleanswater.org

units below are shut down for inspection and servicing.

Arrangement of Hydro Power Plant

In a hydropower plant, the shallow water intake is equipped with a coarse trash rack which prevents trees, branches, debris and stones from entering the conduit system to the turbine.

An intake gate is installed to shutoff the water delivery when the conduit system has to be emptied. In addition, a small gate may be installed for draining leakage through the main gate. In high-head power plants there is typically a head race tunnel between the water intake and the pressure shaft. A sand trap is located at the end of the head race tunnel. At the downstream end of the headrace tunnel there is a surge chamber system. The function of the surge chamber is rapidly to reduce water hammer pressure variations and keep the mass oscillations, caused by load changes, within acceptable limits and decrease the oscillations to stable operation as soon as possible. Before the water enters the pressure shaft it passes through a fine trash rack. This is the last protection of the valve and the turbine against floating debris or smaller stones if the sand trap is full.

The pressure shaft may be either lined or unlined. Where the rock is of sufficiently high quality the shafts are normally unlined. A steel penstock connects the shaft with the valve and at the upstream end an automatic isolating valve is normally installed; which closes automatically if a pipe rupture occurs.

A typical arrangement of a hydropower plant is shown in Fig 1.

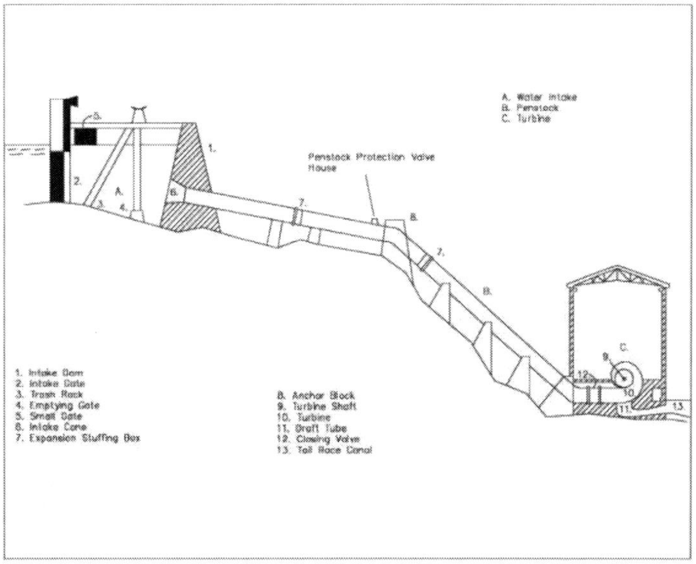

Figure 1. Arrangement of Hydro Power Plant (Jain, 2013)

EMERGENCY SHUTOFF VALVE

Features

The Emergency shutoff valve consists of a flanged butterfly valve with a rubber seated disc, adjustable and replaceable against wear and damage. The valve is fixed to the inlet pipe vertically and supported on a concrete pedestal, free to move axially due to thermal expansion and axial hydraulic forces. Current practice is to provide a double eccentric design to accomplish high degree of tightness on closure and to reduce wear and damage to the rubber seal on the blade periphery. The emergency valve closing is by counterweight actuation and opening is by oil pressure hydraulic cylinder actuation. The hydraulic cylinder is connected to the counterweight lever assembly.

The first step in the penstock valve design is to determine the rupture velocity. The velocity is depended on many factors such as length of penstock downstream, pipeline friction coefficient of the penstock, upstream pipe configuration including the Head Race Tunnel and the maximum head in the reservoir. Rupture can take place anywhere along the penstock length. The magnitude of rupture is also a critical consideration that needs to be predicted and the type of rupture will decide the flow coefficient and the corresponding rupture velocity (Majmudar, 1989). Interestingly, rupture velocity depends on many factors and so do many other design parameters on rupture velocity.

Rupture velocity, therefore is a very important parameter in sizing and selection of servo cylinder apart from structural design of the body and trunnion and the connecting linkage. The uprooting foundation force is also the function of rupture velocity. The entire structural design of the valve is based on rupture velocity. Under this condition, the valve closure develops large reaction forces caused by hydraulic torque generated by large flow velocities and pressures. When the valve closes against such heavy hydraulic torque, the oil in the cylinder is forced out and the resultant pressure rise in the servo mechanical cylinder is many times higher than the oil pressure required to operate the valve. For example, a cylinder meant to operate 600-700 pounds per square inch (psi) oil pressure from the pumping unit, may encounter internal pressure of 3000-3500 psi under rupture velocity closure.

The parameters required for estimating the rupture velocity are data of penstock length and configuration downstream of the valve, thickness of penstock at various sections and turbine runaway flow. Primarily, the cost of the valve is very sensitive to the rupture velocity. If the rupture velocity is not adequately estimated it would defeat the purpose of having a penstock protection valve, even though other design criteria may be met.

Ancillary support systems of the Emergency Shutoff Valve

Typically, the construction of the major parts of the Penstock Valve would be as follows:

Valve Body. The valve body is made of cast steel or stress relieved welded steel and have corrosion resistant sealing seats to receive the disc rubber seal. Shoe pads need to be provided on the valve body for transmitting the vertical load to the sliding pads and concrete foundation. The body is so arranged that no weight or vertical thrust from the valve or contained water is applied to the penstock. The support shoe pads shall permit adequate movement in the upstream or downstream direction upon valve closure or due to temperature variations. For large diameters the valve body is generally made in two sections for bolted connections, split horizontally to facilitate assembly, site disassembly, handling and shipping.

Valve Disc. The valve disc is made of annealed cast steel or stress relieved welded steel and is designed to provide a low friction profile. The disc profiles include lattice construction or through flow blade. Flat slab blades are generally adequate for sizes within 72 inches depending on the head condition but for larger sizes the through flow blade is best recommended. The disc is of streamline section to minimize local eddy currents and sudden velocity changes. All surfaces of the disc in contact with water shall be smooth and free from hollows, depressions, cracks or projections that might cause pitting due to cavitation.

The disc and trunnion shafts maybe integrally cast and designed to adequately resist the forces acting directly on the disc when in closed position with utmost rigidity to insure minimum deformation. If it is not integrally cast, the disc shall be bored and key seated to receive the trunnion shaft, which shall be held in position by keys and pins.

Valve Shaft. The valve shafts not integrally cast with the disc shall be forged steel. The valve shaft is key seated at both ends to lever of the operating mechanism. The shaft safely transmits the hydraulic load on the disc to the valve body and the forces from the operating mechanism to the disc.

Valve Bearings. The main shaft bearing housing is of cast steel or welded steel plate. If the bearing housing is cast steel typically it would be cast integral with the valve housing. The bearing is arranged to permit horizontal adjustment to the valve disc. The bearing is generally made of bronze, grease lubricated type or stainless steel backed with PTFE.

Operating Mechanism. The cylinder is secured to a weighted operating lever to serve as a dashpot to regulate the closing speed of the valve. The cylinder needs to have a cushion at the end of the stroke. The weight operator would have sufficient weight of cast iron or steel plates at the end of the valve lever together with hydraulic unbalance to assure closure of the valve under the most adverse conditions. Locking devices that engage the valve lever is provided to hold the valve in the open position or closed position against the full force of the operating cylinder or the hydrostatic head.

The valve is opened and kept in open position by means of hydraulic servomotor actuated by oil pressure. The Electric motor operates the pump only when the valve is to be opened. Closure of the valve is by counterweight.

Bypass Valves. The bypass line is required for filling the penstock and for equalizing the pressure on both sides of the valve disc before opening the valve. For this purpose, a bypass valve typically consists of a needle valve and an isolating gate valve between the makeup pipe work on either side of the penstock valve. The needle valve ensures cavitation free flow and the guard valve is required for shutting off penstock pressure from the bypass line for maintenance purposes.

Sizing of the bypass line is a function of penstock filling rate and the length of the penstock. Usually one inch of bypass per foot of penstock diameter is a rule of thumb and provides economical proportions as large needle valves tend to be more expensive.

The needle valve is usually motor operated but at times hydraulic operation of the needle valve is considered as a small amount of pressurized oil is always available without any extra margin on the hydraulic power pack capacity.

<u>Air release valve.</u> Adequate capacity of the Air release valve is required to be provided to discharge the air displaced by penstock filling. Sizing is extremely critical since undersized valve can produce enormous amount of noise and vibrations.

<u>Anti-vacuum valve.</u> This valve works very infrequently but its reliability in the event of sudden load rejection by turbine and/or penstock burst is vital. This is the most important and critical sub system for the protection of penstock from atmospheric pressure when the flow separation takes place in the event of valve closure.

Normally anti-vacuum valve is sized based on turbine runaway speed and standby is provided in the event one valve fails to respond. Sizing is also based on penstock vacuum strength, detailed layout of the penstock and its structural integrity.

A combination of Air/ Vacuum Release Valves could also be provided where the advantages lie in low initial cost, maintenance of only one equipment and larger volumes of air can be handled.

<u>Over Velocity Sensing Device and Trip Mechanism.</u> This is the most critical control interlock and actuation element of the Penstock system for the over-velocity sensing function.

The primary function is to monitor the flow velocity through the penstock valve and generate an electrical signal with corresponding mechanical action in the event the flow exceeds the pre-set value.

The required action is to convert the signal into a mechanical force large enough to be able to release the latch mechanism, so as to initiate valve closure.

In early days, simple and classical methods were used for over velocity tripping device. Most popular was the paddle trip which consisted of a suspended paddle in the flow path which would be actuated by excessive velocity to trip the latch mechanism. Due to the presence of weed and pebbles in water, the trip mechanism would de-latch prematurely causing the turbine to trip without any reason. This would also require periodic maintenance of mechanical joints and bearings.

An improvement was the differential pressure switch type tripping device. This device has pressure taps upstream and downstream of the valve. In the event of excess velocity, the differential pressure switch would close the contacts to energize a battery-operated solenoid valve, which would allow the penstock water to operate the trip cylinder on the penstock valve latch mechanism. The differential pressure set point is field adjustable and the emergency closure circuit incorporates a timer which is adjustable. In the event that the pressure transmitter's signal sustains itself for a predetermined time the shutoff valve closes and the 'Piping High Velocity' alarm would display. The battery remains constantly energized from the main power source through a trickle charger in the control panel. In the event the battery runs down to an unacceptable level, an alarm is provided in the control room. The pressure switch is housed in a weather proof enclosure to provide years of inactiveness but it is possible to check operability of pressure switch by periodically reducing the differential pressure setting to normal velocity when the penstock valve can be checked for closure through tripping mechanism as a routine plant maintenance schedule.

It is customary for setting the differential pressure switch corresponding to turbine runaway flow velocity so that in the event of sudden load rejection, penstock valve will also provide instantaneous closure.

This device cannot be used for buried penstocks and requires straight length of exposure for penstock upstream and downstream of several meters , typically 2.5D and 5.0D respectively.

The Ultrasonic flow detection and measuring device is the most popular sensing device and is based on the acoustic method of discharge measurement in which the propagation of velocity of an acoustic wave and flow velocity are summed in a vector form. The flow meter utilizes the multiple parallel path transit time flow measurement technique.

The flow meter is connected by signal cables to multiple pairs of transducers mounted in a pipe at specific elevations. The device consists of a set of transducers and a set of receivers. The transducers send sound waves and the difference in the sound velocity in the two directions is recorded.

The unit is designed to allow removal of entire transducer for repair, replacement or cleaning without dewatering the pipe. There is virtually no maintenance required and on site recalibration is possible. This type of device is slightly more expensive than the differential pressure device and also requires skilled assistance for installation and commissioning.

Operating Principle

The velocity at each elevation is determined using the differential travel time method in which an acoustic pulse travels downstream faster than a pulse travels upstream. A pulse of sound traveling diagonally across the flow in a downstream direction will be accelerated with the velocity component of the water and, conversely, a pulse traveling diagonally upstream will be decelerated by the water velocity.

The acoustic path layout is shown in Fig 2.

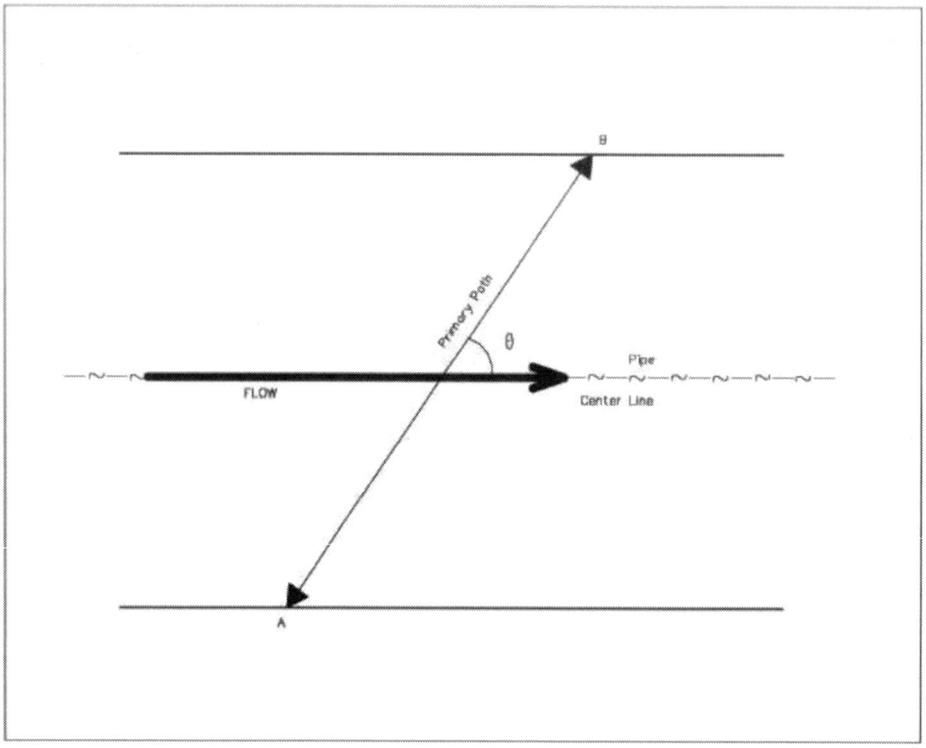

Figure 2. Acoustic Path Layout (Accusonic)

This method of measurement is described as follows:

$$T1 = L / C\text{-}V\cos \mathbf{f}, \quad T2 = L/C + V \cos \mathbf{f} \qquad (1)$$

where T1 is the travel time of the acoustic pulse between transducer B and transducer A in seconds (sec).

T2 is the travel time of the acoustic pulse between transducer A and transducer B in sec

C is the speed of sound in water in meters per second (m/sec)

V is the velocity of water in m/sec

f is the angle between the acoustic path and the direction of water flow in degrees.

The above equations are solved for V, independent of C, yielding:

$$V = (T1\text{-}T2)\, L / (\, T1T2 \times 2 \cos \mathbf{f}) \qquad (2)$$

Therefore, the velocity of water at the acoustic path can be calculated knowing the path length L and path angle **f** and measuring the time for the acoustic pulse to travel between the transducers in the upstream and downstream directions (IEC 41).

Typically four pairs of transducers are spaced in the pipe to give four parallel acoustic paths.

Trip Mechanism. The conventional trip mechanism is to provide penstock water operated cylinder. This is a simple, economical, reliable and proven method. No external power is required and it is ideal for high penstock pressures. The only disadvantage is that water from the penstock needs to be suitably filtered to prevent ingress of silt in the cylinder.

A typical Trip cylinder latch mechanism is shown as Fig 3.

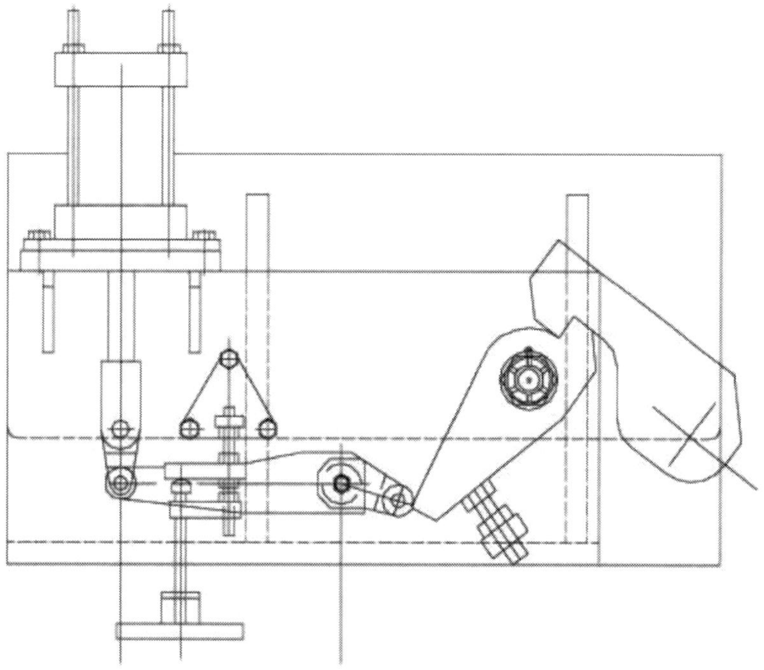

Figure 3. Trip Cylinder / Latch Mechanism

In the absence of the latch mechanism, it is customary to provide a drift restoring mechanism for the emergency shutoff valve. The drift control circuit automatically restores the valve to its full open position if there is a leakage in the hydraulic system. The valve drift sensing signal will come from a proximity switch integral with the hydraulic cylinder. The restoring circuit generally has a timer which is field adjustable. Failure to restore the valve to its full open position within a pre-determined time would illuminate the 'Drift Trouble' light on the control panel.

OPERATION OF EMERGENCY SHUTOFF VALVE

Normal Opening

When an open command from the local pushbutton or remote panel is given, the bypass valve opens and pressure on the downstream side equalizes. The differential pressure switch gives signal to open the valve. This starts the main pump in the hydraulic power pack, energizes a solenoid and the valve opens fully actuating a limit switch. The main pump stops and the solenoid is de energized with the crank arm of the valve being mechanically latched.

An auxiliary event occurs of the air release valve opening automatically when the filling of the penstock starts and closes when the penstock is full.

Normal Closing

Closing requires energizing the solenoid valve on the trip cylinder. The de-latching is activated and the valve begins to close due to the counter weight.
The normal closing takes place in dual speed- first 90% travel in about 20% of the closing time and the balance 10% travel in about 80% of the time. This division is site adjustable and is based on system pressure rise behavior.

Sequence of Events due to Emergency Closing

This is initiated by one of the following events, either a burst penstock or the turbine rejects load and main inlet valve fails to close.

In either case, the water velocity through fully open valve will exceed a preset value. This increase will be sensed by over velocity trip mechanism which in turn will de-latch the valve held in fully open position. The closing time comes down by up to 50% due to higher hydrodynamic torque.

Sequence of events in penstock system leading to complete emergency closure of Penstock Protection Valve (PPV):

- Whenever there is occurrence of an emergency condition like a burst penstock or turbine load rejection and the main inlet valve fails to close beyond a set point the velocity of water through the penstock rises instantly and the static pressure starts falling.

- This velocity increase passes through a pre- set value and the over velocity sensing device is activated. Annunciation now starts and the solenoid valve on the trip mechanism gets energized. The trip mechanism now operates and de-latches the valve blade. The butterfly valve starts closing rapidly aided by hydrodynamic torque.

- The oil in the main servo cylinder is forced out to the tank in the hydraulic power pack.

- Oil pressure in the servomotor rises rapidly, well above the normal delivery pressure of the hydraulic power unit. When flow separation starts in downstream line, anti-vacuum valve opens to admit air. At about 20% travel from fully closed position, self-closing hydrodynamic torque reaches the peak and oil pressure in servo motor rises up to about 200% of normal delivery pressure of hydraulic power unit while the oil is still being forced out. Counter weight and blade eccentricity add to the value of the shaft torque and consequent pressure rise.

- Simultaneously, reaction force on the valve foundation causes large upward thrust trying to lift the valve out of the foundation. The reaction forces on the servo cylinder causes downward thrust.

- At about 15% travel from fully closed position, the output signal on valve position transmitter passes through a preset value and energizes another solenoid valve in return flow oil path, with more restricted passage and the valve closure slows down.

- At about 10% travel from fully closed position, the internal cushion in the servo cylinder comes into effect and the closing slows down further.

- Virtual flow isolation takes place at 95% of fully closed position.

- Valve closes fully and the close limit switch is activated stopping the annunciation.

INTERLOCKS IN AN EMERGENCY SHUTTOFF VALVE

The following interlocks are common in any penstock protection system:

- Valve opening is precluded unless pressures on Upstream and Downstream are balanced.

- Locking pin engaged will prevent the valve from opening or closing.

- Over velocity automatically closes the valve.

- Low oil level in the reservoir prevents hydraulic power unit motor start.

- Standby pump starts automatically on failure of the main pump.

- High oil pressure will stop hydraulic pump motor.

SAFETY AND RELIABILITY

The penstock protection valve is primarily meant to ensure safety of the pipe and the power house, but it remains inoperative for most of its operating life.

High safety factors are required in the structural/ mechanical design and the elements performing safety related functions should be of high reliability and duplicated to ensure action against possible malfunction.

The popular safety features being adopted include:

- Duplication of Over Velocity Trip Devices B either two sets of pressure switches or two sets of ultrasonic flow detection devices at different angles.
- Extra capacity for the Anti Vacuum Valve.
- Velocity Fuse in the Cylinder line.
- Manual override for Trip Cylinder.
- Separate solenoid valves for trip cylinder for normal closing and emergency closing.

MODERN TRENDS

Modern trends are to provide a Programmable Logic Controller (PLC) system which would communicate with the main PLC system within the main control panel and provide all operational information and receive control commands.

The PLC system and the local control panel would control, monitor and process the collected data of the hydraulic valve and other equipment like gates , fish screens, etc. It would primarily communicate with the master PLC system within the main control panel and provide all operational information and receive control commands. Generally, the systems operational logic would be within the local PLC and with no dependence on the main PLC. All commands are typically duplicated within the local and main control panels. Data transfer is effected at all times and under any operational mode of operation.

The PLC system controls and monitors the emergency shutoff valve on the basis of a stored digital program of instructions and various input process variables and discrete status contacts and output analog or discrete control signals and externally selectable manual switches.

The most common programming method is the ladder logic instructions and all programming, monitoring, searching and editing are accomplished using Windows programming software.

The communication to the main control panel is generally performed through a fiber optic redundant data highway system.

All commands generated from the PLC system is confirmed from the relevant feedback signals from the commanded equipment within a predefined period (for example the opening of the valve is confirmed by its open limit switch). The failure of such a confirmation is properly alarmed and the system is reset to a safe position which is the closed position of the valves and the off position of the equipment. Any change of the process status caused by an action of an equipment is timed for its occurrence. The unconfirmed status change is properly alarmed and the system is reset to a safe position.

REFERENCES

Accusonic. Multiple path transit time flowmeters. http://www.adsenv.com/sites/adsenv.com/files/datasheets/Accusonic-Principle-of-Operation.pdf.

IEC 41. Abstract of Appendix J. Acoustic Method of Discharge Measurement.

Jain, Dikshitkumar, 2013. http://eeerockss.blogspot.com/2013/05/arrangement-of-hydropower-plants.html.

Majumdar A.A, 1989. Penstock protection valve for hydro electric generating unit. India Journal of Power and River Valley Development.

Ramos, Helena, A. Betamio De Almeida, Didia Covas. Design Criteria Applied to Pressurised Gravity Systems.

A STATE PERSPECTIVE ON TAILINGS DAM REGULATION AND A ROLE FOR USSD AND ASDSO

Charles Cobb[1]

Tailings dam design and operation are universal subjects addressed both globally and regionally within a relatively small community of specialists. Organizations, businesses and individuals host conferences and workshops; publish guidelines, books and papers; and design tailings storage facilities to address a variety of technical situations and site conditions at faraway places around the world. South Africa, Canada and Australia are recognized leaders in these discussions. However, the regulation of tailings dams involves those same complex topics from the perspectives of a morass of regulatory agencies within the jurisdiction of the mine. The United States focused the dialogue on the safety of tailings dams after the Buffalo Creek Mine disaster killed 125 people in 1972. That event was the catalyst for the National Dam Safety Act of 1972, which led to state dam safety programs across the nation, many which regulate tailings dams. Since then, the International Commission on Large Dams and other groups formed and contributed to the conversation, including government agencies such as the USEPA, and organizations such as the Canadian Dam Association, the USSD, and ASDSO. Although tailings dams represent a small fraction of the population of dams in the world, the value of mining to local, national and global economies, the technical advancements in tailings dam design, and the operational and long-term performance requirements, demand a continuous discussion to ensure the safety of these dams that now include some of the largest man-made structures on the planet. A brief review of the history of tailings dam regulation in the US and the contributions of many organizations to this important subject will highlight a state regulatory perspective of current issues. USSD and ASDSO are uniquely positioned to facilitate the necessary and on-going dialogue between the many stakeholders to accomplish the common objective of safe tailings dams.

[1] Alaska Department of Natural Resources, charles.cobb@alaska.gov

A CENTRIFUGE STUDY OF CO-MIXED MINE TAILINGS AND WASTE ROCK DEPOSITS

Nonika Antonaki[1]
Tarek Abdoun[2]
Inthuorn Sasanakul[3]

ABSTRACT

Centrifuge tests were performed on mine tailings alone and then mixed with waste rock at the Center for Earthquake Engineering Simulation (RPI). In the field, the tailings are thickened (dewatered) and hydraulically deposited into the containment structure in layers. The first objective of the centrifuge testing was to evaluate the consolidation behavior of a medium-height tailings stack deposited at the pumping water content (59% in this project). Tailings were prepared in layers and consolidated in the centrifuge, thus allowing instrumentation of each layer before consolidation of the complete impoundment. Shaking Table tests were conducted on the centrifuge in order to evaluate the liquefaction potential of mildly sloped (4%) consolidated mine tailings. Earthquake-induced liquefaction is one of the most common causes of tailings dam failures. The need for improvement of both the consolidation parameters and the shear strength of the deposit arose. One of the alternative mine waste management methods that improves physical stability and conserves space is co-disposal of mine tailings and waste rock. In this study, co-mixing of the materials at a specified ratio of dry mass (waste rock to tailings) prior to disposal was examined. The behavior of the homogeneous mixtures was found to improve compared to that of mine tailings alone with respect to consolidation rate, settlement accumulation, as well as slope stability and response to dynamic loading. This paper presents two centrifuge tests performed on sloped mixtures of tailings and waste rock at different mixture ratios.

INTRODUCTION

Mining activity produces large quantities of both tailings with high water content and generally dry waste rock. Tailings are commonly discharged as a slurry with slow consolidation properties and low shear strength, often causing failures in tailings impoundments due to physical instability. In contrast, the waste rock is characterized by high shear strength and is commonly disposed of in large dumps. The unsaturated conditions allow weathering of the waste rock, which may cause long-term acid drainage and metal leaching. Blending the two materials could produce self-sealing deposits with high shear strength, low compressibility and density higher than either material on its own, thereby reducing the total volume and surface area requirements for impoundment design and final reclamation (Bussiere 2007, Wilson et al. 2009).

---

[1] Geotechnical Engineer, PhD, WSP|Parsons Brinckerhoff, New York, NY 10119, antonakin@pbworld.com, nonika.a@live.com, 518-526-0215.
[2] Professor, Associate Dean for Research and Graduate Programs, CEE, Rensselaer Polytechnic Institute, Troy NY, United States. E-mail: abdout@rpi.edu
[3] Assistant Professor, Cee, University Of South Carolina, SC, United States.E-Mail: Sasanaku@Cec.Sc.Edu

To date, most experimental research on mine tailings has consisted of small scale laboratory tests, such as simple or direct shear tests, triaxial tests, etc. (Aubertin et al. 1996, Dimitrova et al. 2011, Garga et al. 1984, Highter and Tobin 1980, Highter and Vallee 1980, Proskin et al. 2010, Qiu et al. 2001, Sanin et al. 2012, Wijewickreme et al. 2005, etc). Laboratory investigations have also been conducted on mixtures by a few researchers (Leduc et al. 2003, Wickland et al. 2005, Wickland et al. 2006, Wickland et al. 2010). In Australia, the co-disposal of coarse and fine coal wastes by combined pumping was pioneered in 1990 at the Jeebropilly colliery in southeastern Queensland (Morris et al. 1997). An extensive literature review revealed very few tests of mine tailings using a geotechnical centrifuge.

In this study, fine-grained tailings from the pilot plant for a planned copper-gold mine in an earthquake-prone zone of South America were studied by means of centrifuge tests. The main objectives of the research were to model the deposition of tailings in layers on the centrifuge, examine the consolidation behavior in terms of time, pore pressure dissipation rate and settlement and, finally, to evaluate the liquefaction potential and slope stability under dynamic loading. The importance of whether a sloped surface can be maintained lies in the potential increase in stored volume within a specific impoundment area. Due to long consolidation time, large settlement and clear signs of liquefaction after a few cycles of dynamic loading, the need for improvement became evident after the first few tests. The same testing scenario was repeated using a mixture at a selected ratio of waste rock to tailings by dry mass for improvement and comparison purposes. The focus of this paper is on the comparison of two tests, one using a mixture at a 2.4:1 mixture ratio (waste rock to tailings by dry mass), and one using a mixture at 3:1 ratio, in terms of response to two harmonic base motions (different amplitude).

OBJECTIVES AND TEST DESCRIPTION

In previous centrifuge tests, tailings and mixed deposits were constructed by uniform layer disposal and time was allotted for consolidation of each layer as well as of the completed deposit (Antonaki et al. 2013, 2014a, b, 2015, 2016a, b). About 60% of the tailings material passed the #200 mesh. Waste rock particles with dimensions in the range of 20 - 40 cm (prototype) were considered adequately representative of field conditions. More details on the materials tested can be found in the aforementioned publications. Dynamic response was examined under nearly flat-ground conditions for two levels of harmonic base excitation (referred to as Motion 1 and Motion 2) consisting of 50 cycles and applied in flight by a shaking table attached to the centrifuge basket. The last two tests of this study aimed to test mixtures in a more critical scenario, where the materials were mixed uniformly and then dumped in a steeply sloped pile. The possibility of eliminating the retaining structure, or even reducing its height, could be highly advantageous to the mining industry and was hence investigated. Contrary to tests presented in previous publications, the impoundments were not levelled and there was no waiting period between layers. The models were built, instrumented and soon after subjected to dynamic earthquake loading while at 80 g centrifuge acceleration. Mixture ratios of 3:1 and 2.4:1 were selected for these tests, respectively, as these mixtures performed in a satisfactory manner when previously tested under more favorable conditions. It was nevertheless deemed reasonable that the "just-filled" case (2.4:1) and a slightly more conservative case (3:1) be further studied.

Model preparation is illustrated in Figure 1. Mine tailings were mixed at 59% water content and then co-mixed with waste rock at the specified ratio of waste rock to tailings by dry mass. The container base was lined with high-friction sandpaper to avoid any sliding between the mine waste dump and the aluminum during dynamic loading. Adhesive measuring tape was used to keep track of the height and slope of the models. The material was then deposited using scoops as shown. Pore pressure sensors were embedded in the material and accelerometers were both embedded and glued on the container.

Figure 1. Stages of model preparation: mixing the mine tailings, co-mixing with waste rock, material deposition on the centrifuge, sensor installation and target placement for video tracking.

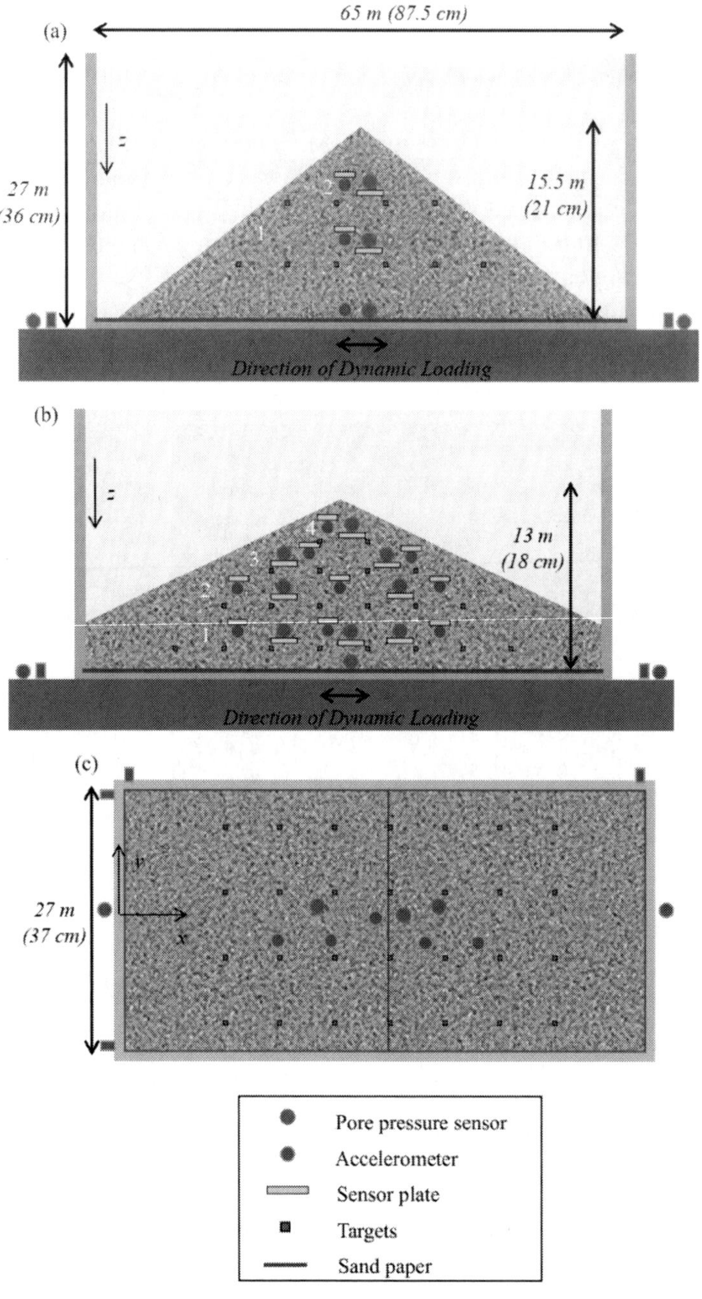

Figure 2. Model set-up (a) for free-standing 3:1 mixture, (b) for 2.4:1 mixture and (c) plan view of 2.4:1 mixture (not to scale).

Bright plastic targets were placed perpendicularly to the Plexiglas wall and vertically embedded at the surface, as exhibited in Figure 1 and schematically in Figure 2. Lipstick cameras were used to monitor the side Plexiglas window and the high-speed camera was used to capture the surface view. Videos were then used in conjunction with video tracking software and surface displacements in the x and y directions were obtained.

Both tests were performed in a long rigid container with dimensions as shown in Figure 2 in prototype and model scale. The scale factor being used is 74, as the model is slightly lifted by the height of the shaking table. Figure 2 (a) presents the configuration applied to the 3:1 mixture, which was able to form a steep slope with no contact to the "retaining" walls. The tests were considered as two-dimensional and dynamic loading was applied in the depicted direction. Due to the narrow width and low water content of the 3:1 mixture, only three layers of sensors were installed along the centerline, including the base sensors. The model was then spun up to try and avoid excessive drying and segregation. Starting maximum height was roughly 15.5 m (prototype) and average inclination was 34° or 70%. The slope was steeper near the toe. Figure 2 (b) presents the same for the 2.4:1 mixture. A less steep slope and wider footprint had to be adopted for this material and more sensors were installed and spread out. Initial maximum height and average inclination were about 13 m and 16° or 30%, respectively. In this case complete elimination of the retaining structure was not be feasible, but a shorter wall or dam was simulated by allowing the material to be partially supported by the container walls, as seen in Figure 2 (b). Embedded sensors were all attached to floating plates, similarly to previous tests, for stability. Sensors attached to the box included x, y and z accelerometers as shown.

TEST RESULTS

Both mixtures were subjected to harmonic 50-cycle motions at prototype frequency of 0.9 Hz. Maximum acceleration was in the region of 0.06 – 0.07 g and 0.13 g for Motions 1 and 2 correspondingly. A few smaller motions were applied prior to the main events and are not included herein. Figure 3 exhibits comparable views of the slopes before and after the tests. The inclination of the 3:1 mixture did not alter dramatically, as can be seen in parts (a) and (c) of the figure. The final slope was measured at 30° or 58% and final height was approximately 15 m. The slope was more uniformly inclined after the test and the footprint was found slightly enlarged. As observed from Figure 3 (b) and (d), the 2.4:1 mixture experienced more settlement and change in inclination. Maximum height went down to less than 12 m and the inclination was reduced to 4° or 7%. Some slope failure occurred during centrifuge acceleration to 80 g, as will be shown and discussed.

Figure 3. (a) 3:1 and (b) 2.4:1 mixture before test; (c) 3:1 and (d) 2.4:1 mixture after test.

Based on the condition of the 3:1 dump after the test a few observations can be made. Large areas of the surface were completely dry, while a considerable amount of water was gathered at the toe. Drainage or collection ponds would have to be provided for the water in the field, if no retaining structure was constructed. The dry area extended almost to the toe. Smaller dry areas were visible at the top of the 2.4:1 dump after the test, while deeper pools of water collected at the toe. There was some segregation between tailings and rock particles in the 3:1, but that became far more noticeable in the 2.4:1 mixture. The surface at the toe was covered by a layer of fines and the water level was about four times higher in this case.

Figure 4 is a compilation of side views of the 2.4:1 model. From top to bottom, Figure 4 (a) shows the slope before the test, (b) was taken after spin-up to 80 g and (c) was captured after the test. Most of the settlement and movement towards the toes occurred prior to dynamic loading due to spin-up. The slope self-stabilized at an inclination of 14% or 8° and the applied motions caused it to further be levelled to 7% or 4°. It became evident that this material would require to be retained.

Figure 4. View of slope (a) before spin-up, (b) before dynamic loading and (c) at the end of the test for the 2.4:1 mixture. The bottom of the box is not visible and, combined with the overlap between pictures, leads to the exaggeration of the model length.

The 3:1 mixed deposit settled during centrifuge acceleration and then almost remained undeformed when the motions were applied. This conclusion is supported by the high speed camera snapshots of the top obtained before and after each motion and exhibited in Figure 5. Parts (a), (b) and (c) of the figure are practically identical. The only notable difference is the expansion of the dry areas. It can be observed that some of the targets came out of the slope due to deformation while spinning up. Figure 6 presents the corresponding images for the 2.4:1 mixed deposit. The images are undistinguishable before and after Motion 1. When the stronger Motion 2 was applied, the model clearly settled, as is apparent in Figure 6 (c) from the shiny surface caused by the water level rise. Some of the targets moved downslope and are no longer visible in the last image. The cable of the accelerometer placed at the surface also came out due to surface settlement. Video tracking of the high speed videos gave residual displacements in the range of 15 – 140 cm in the x direction and 15 – 70 cm in the y direction, corresponding to the middle region of the high speed camera view or the area around the head of the slope. The targets further from the top became invisible and could not be tracked as they were covered by water.

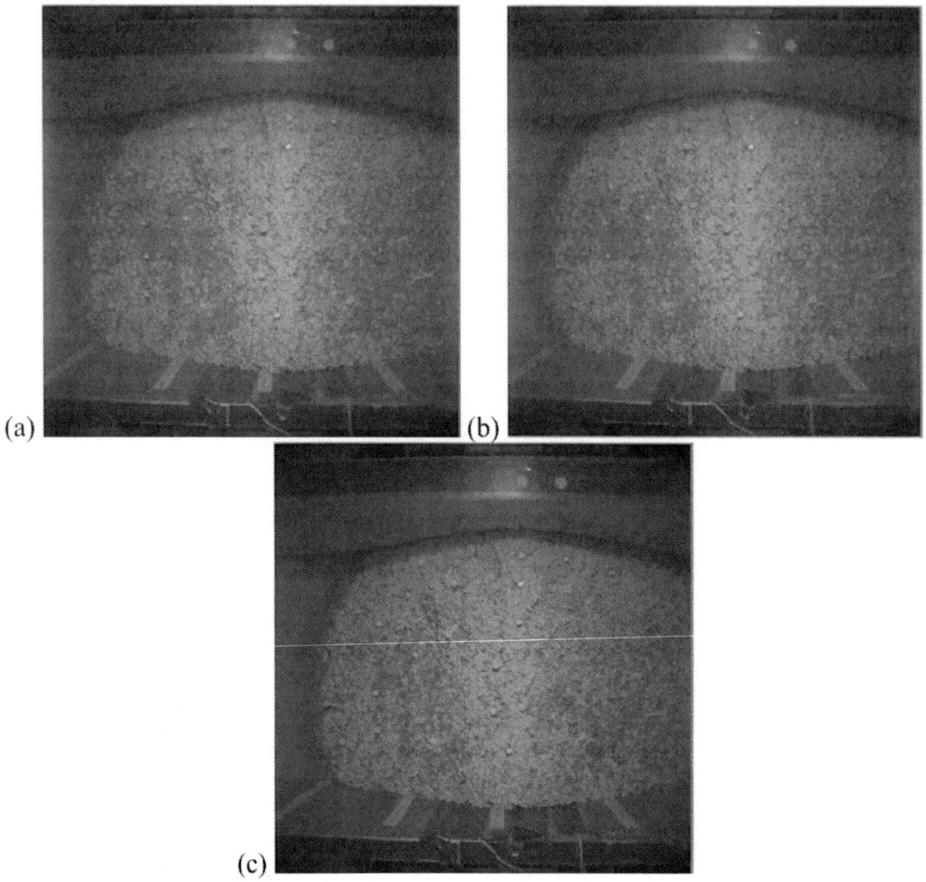

Figure 5. High speed camera view of slope (a) before Motion 1, (b) before Motion 2 and (c) at the end of the test for the 3:1 mixture.

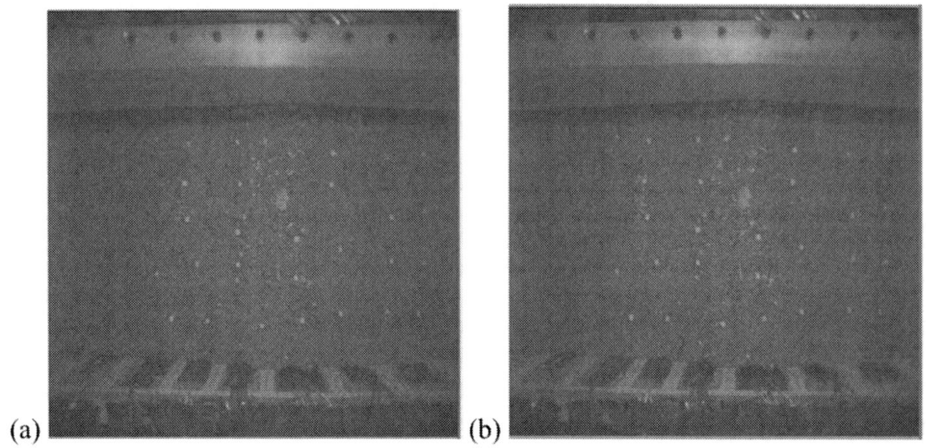

(c) 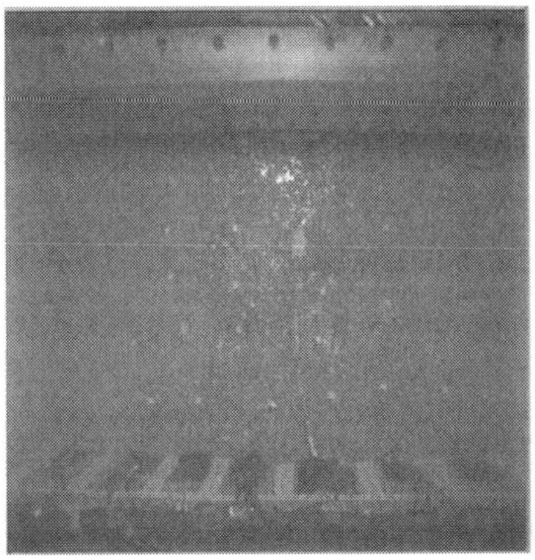

Figure 6. High speed camera view of slope (a) before Motion 1, (b) before Motion 2 and (c) at the end of the test for the 2.4:1 mixture.

Figures 7 – 10 present acceleration and pore pressure build-up ratio ($r_u = \Delta u/\sigma'_v$) time histories within the two deposits. Effective stress was calculated using average unit weight and the measured water level and was used to calculate the plotted r_u. Note that r_u values in the vicinity of 1 indicate liquefaction. Figures 7 and 8 show the response of the 3:1 and 2.4:1 mixtures to Motion 1, respectively. Figures 9 and 10 compare the response of the two mixtures to Motion 2. Base excitation is included at the bottom of all figures. Motions reached their maximum amplitude during the first ten cycles and dissipated during the last ten. Motion 1 refers to acceleration amplitude of 0.06 – 0.07 g and Motion 2 to 0.13 g or so. These values correspond to filtered acceleration data. Input motions were not identical for the two tests but were pondered comparable. Sensors were selected at analogous depths for direct comparison purposes and final sensor depth is incorporated in the upper left corner of each graph. Soil acceleration plots are located at the left side of the graphs and r_u plots at the right.

It is evident from Figure 7 that liquefaction did not occur in the 3:1 deposit. Acceleration was somewhat amplified through the mine waste, with slightly higher amplification occurring at 9.6 m from the surface than at 2.2 m. The dump sustained Motion 1 and pore pressure build-up was practically zero. This outcome was expected, as this mixture was not fully saturated. At 3:1 mixture ratio, the layered deposit had not exhibited any signs of liquefaction either, but some build-up was measured due to higher degree of saturation under almost level-ground conditions. Figure 8 is only mildly different than Figure 7. The 2.4:1 deposit also amplified the base motion, with amplification rising towards the surface in this case. Maximum pore pressure build-up ratios of 0.15 and 0.17 were measured at 9.6 and 2.6 m from the surface, respectively. Similarly to the 3:1 dump, no lateral slope displacement was measured during Motion 1.

Figure 9 exhibits similar behavior of the 3:1 mixture when subjected to stronger Motion 2. Base excitation was amplified through the deposit, more so at 9.6 m than at 2.2 m from the surface. Pore pressure essentially did not build-up for this motion either. Significant amplification is also noticeable in Figure 10, with marginally higher acceleration measured closer to the top of the dump. No signs of liquefaction are present in the acceleration, but considerable pore pressure built up in the deposit. Maximum r_u of 0.55 was measured at 9.6 m and 1 at 2.6 m from the surface. Based on sensors located between the two depths, maximum r_u was in the range of 0.7. Below 9.6 m the maximum dropped to about 0.5. It was concluded that a zone along the sloped surfaces of the dump, with a thickness of 2 – 4 m, liquefied and moved downslope. The top 2 m at the center of the dump were unsaturated as estimated from the shallowest pore pressure sensor and confirmed by the observed water level. That was noticeable in Figure 6 (c), where the central top region of the deposit remained above the water level after Motion 2 was applied.

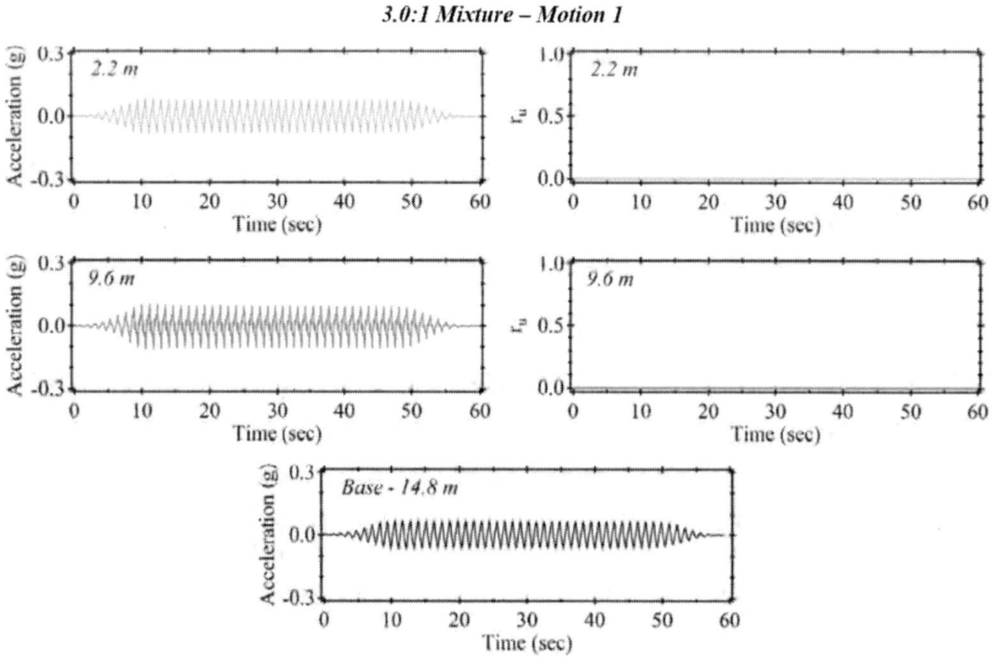

Figure 7. Base excitation and acceleration and pore pressure build-up ratio (r_u) time histories at two depths during Motion 1 for the 3:1 mixture.

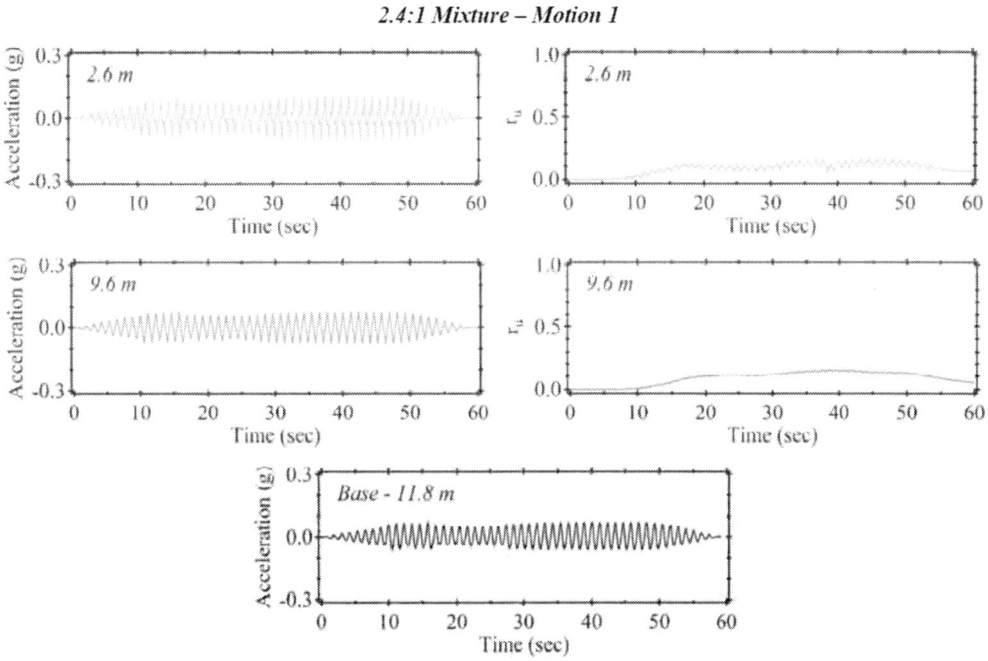

Figure 8. Base excitation and acceleration and pore pressure build-up ratio (r_u) time histories at two depths during Motion 1 for the 2.4:1 mixture.

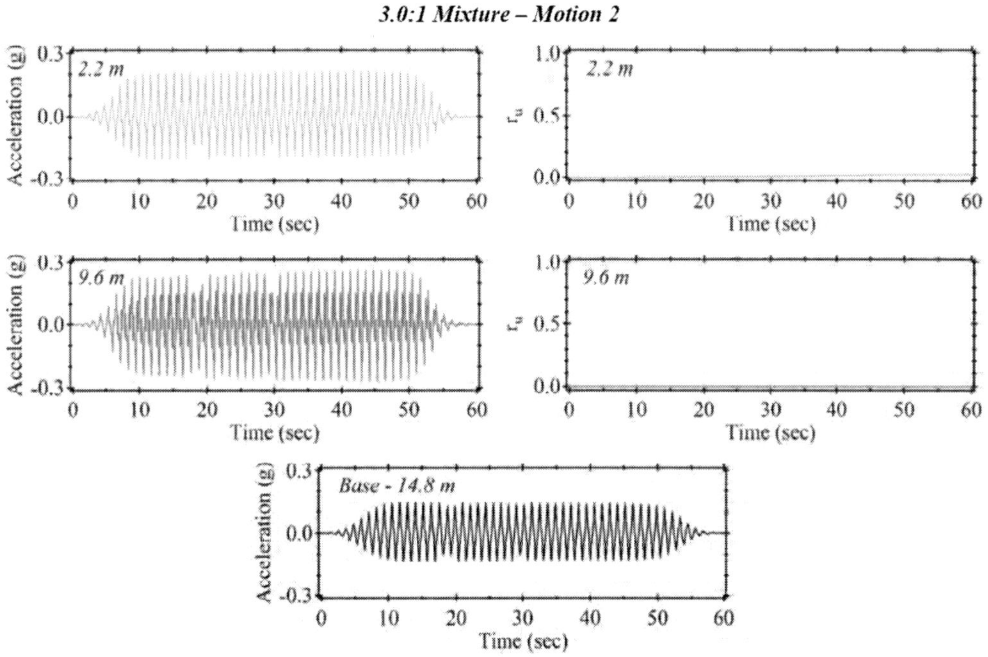

Figure 9. Base excitation and acceleration and pore pressure build-up ratio (r_u) time histories at two depths during Motion 2 for the 3:1 mixture.

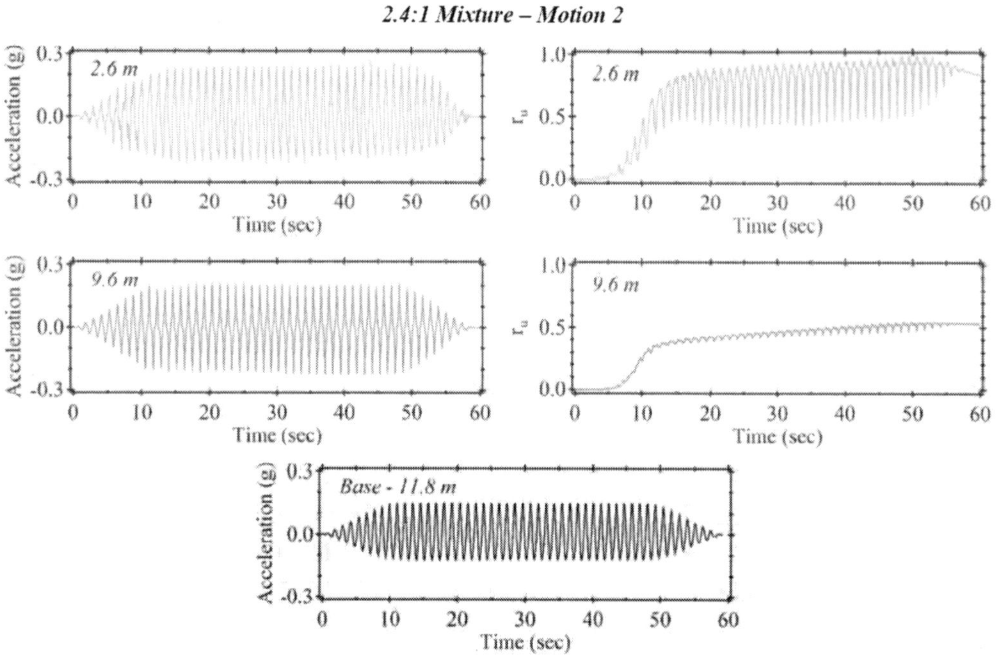

Figure 10. Base excitation and acceleration and pore pressure build-up ratio (r_u) time histories at two depths during Motion 2 for the 2.4:1 mixture.

The liquefied zone is more clearly visualized in Figure 11. Red stars correspond to liquefied depth calculated from maximum pore pressure build up and are placed at the final location of each sensor. The red curve roughly delineates the border between liquefied and non-liquefied soil. One sensor was deemed inconsistent and is not included. One of the included points does not seem to quite follow the general trend and is hence included but not taken into account when drawing the curved line. The liquefied region was somewhat thicker near the top and near the toes.

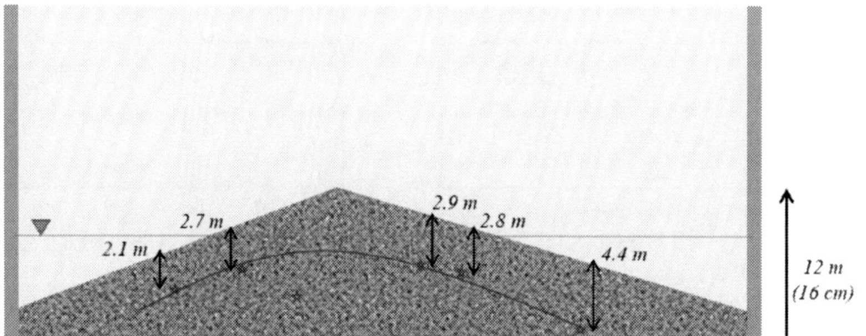

Figure 11. Estimated depth of liquefaction based on pore pressure readings during Motion 2 for the 2.4:1 mixture.

SUMMARY – CONCLUSIONS

Two mixed deposits at mixture ratios of 3:1 and 2.4:1 (uniform waste rock to fine tailings by dry mass) were built on the centrifuge at their natural angle of repose. Initial tailings water content was 59% and a ratio around 2.3:1 was estimated to lead to a "just-filled" waste rock skeleton once some of the water drained out. The 3:1 deposit was able to support itself and form slopes at average inclination of 34°, with actual values higher near the toe and lower near the top. The 2.4:1 deposit required some support from a retaining structure, simulated by the walls of the container, and formed slopes at 16°, with the same tendency to flatten near the top after deposition. Initial heights of the dumps were 15.5 (3:1) and 13 m (2.4:1), respectively. Both deposits were instrumented with accelerometers and pore pressure transducers and were accelerated to 80 g to be subjected to two levels of harmonic seismic loading. Maximum base acceleration was 0.06 – 0.07 g and 0.13 g for Motions 1 and 2, respectively, and was maintained for about 30 out of 50 cycles at 0.9 Hz. Previous shaking tests on layered consolidated deposits with mildly sloped surfaces (tailings and mixtures) proved satisfactory behavior of these two mixtures compared to separate tailings and waste rock disposal. The presented testing sequence was performed to further evaluate the two mixtures.

Interesting observations regarding segregation of the materials were made in these last two tests. Water drained out at the toe and fines created a thin layer at the surface of the toe of the 2.4:1 deposit, indicating that tailings washed out through the rock voids. That was not observed in previous tests as the materials were levelled after co-mixing and deposition. Perhaps lower tailings water content could reduce segregation. A very small amount of fines came out at the toe of the 3:1 deposit. Surface dryness was very noticeable at the surface after the test. Some dryness was observed at the top of the 2.4:1 deposit as well. Both deposits settled to a more stable configuration while the centrifuge was accelerating. The 3:1 deposit stabilized at 30° and the 2.4:1 deposit at 8°.

The 3:1 mixture practically remained undeformed during dynamic loading and its final height was just under 15 m. Acceleration was amplified through the material but pore pressure practically did not build up. No noticeable deformation or liquefaction signs were observed in the 2.4:1 deposit for Motion 1 (0.06 – 0.07 g), even though some pore pressure built up. However, the deposit settled to a final inclination of 4° and final height just under 12 m when subjected to Motion 2 (0.13 g). A 2 – 4 m thick layer liquefied along the slopes of the 2.4:1. One of the objectives of these tests was to assess the feasibility of free-standing mixed deposits. That proved to be feasible for both earthquake intensities in the case of the 3:1 mixture at 30°, as it led to a partially saturated rock skeleton and employed the shear strength of the waste rock. The 2.4:1 mixture would not economically allow the elimination of the retaining structure in a seismically active area, as it was stable at a low angle of 8° for Motion 1 and it settled to 4° after stronger Motion 2.

ACKNOWLEDGEMENTS

The authors appreciate the support of Golder Associates Ltd..

REFERENCES

Antonaki, N., Abdoun, T. & Sasanakul, I. (submitted 2016a). Centrifuge Tests on Co-mixing of Mine Tailings and Waste Rock. *Journal of Geotechnical and Geoenvironmental Engineering.*

Antonaki, N., Abdoun, T. & Sasanakul, I. (submitted 2016b). Consolidation and Dynamic Response of a Layered Mine Tailings Deposit in Centrifuge Tests. *Geotechnical Testing Journal.*

Antonaki, N., Abdoun, T., Sasanakul, I., Sanin, M.V., & Puebla, H. (2015). Centrifuge Modeling of Consolidation and Dynamic Loading of Fine-grained Mine Tailings. Proceedings from *XV Panamerican Conference on Soil Mechanics and Geotechnical Engineering*. Buenos Aires, Argentina. IOS Press.

Antonaki, N., Sasanakul, I., Abdoun, T., Sanin, M.V., Puebla, H., & Ubilla, J. (2014a). Centrifuge Modeling of Deposition, Consolidation and Seismic Loading of Fine-grained Mine Tailings. Proceedings from *Tailings and Mine Waste 2014.* Keystone, CO. Taylor & Francis Group.

Antonaki, N., Sasanakul, I., Abdoun, T., Sanin, M. V., Puebla, H., & Ubilla, J. (2014b). Centrifuge Modeling of Deposition and Consolidation of Fine-grained Mine Tailings. Proceedings from *Geo-Congress: Geo-Characterization and Modeling for Sustainability.* Atlanta, GA: ASCE.

Antonaki, N., Sasanakul, I., Abdoun, T., Sanin, M.V., Puebla, H., & Ubilla, J. (2013). Centrifuge Modeling of The Consolidation Evolution of Fine-Grained Mine Tailings. Proceedings from: *Tailings 2013: First International Seminar on Tailings Management.* Santiago, Chile: Gecamin, Conferences for Mining.

Aubertin M., Bussiere B., Chapuis R.P. (1996). Hydraulic Conductivity of Homogenized Tailings from Hard Rock Mines. Canadian Geotechnical Journal, 33: 470-482.

Bussiere B. (2007). Colloquium 2004: Hydrogeotechnical properties of hard rock tailings from metal mines and emerging geoenvironmental disposal approaches. Canadian Geotechnical Journal, 44: 1019–1052.

Dimitrova R.S., Yanful E.K. (2011). Undrained Strength of Deposited Mine Tailings Beds: Effect of Water Content, Effective Stress and Time of Consolidation. Geotechnical and Geological Engineering, 29: 935-951.

Garga V., McKay L.D. (1984). Cyclic Triaxial Strength of Mine Tailings. Journal of Geotechnical Engineering, 110 (8): 1091-1105.

Highter W.H, Tobin R.F. (1980). Flow Slides and the Undrained Brittleness Index of Some Mine Tailings. Engineering Geology, 16: 71-82.

Highter W.H, Vallee R.P. (1980). The Liquefaction of Different Mine Tailings under Stress-controlled Loading. Engineering Geology, 16: 147-150.

Leduc M., Smith M.E. Tailings Co -Disposal™ Innovations for Cost Savings and Liability Reduction. The Latin America Mining Record, July/Aug. 2003.

Morris P.H., Williams D.J. (1997). Co-disposal of washery wastes at Jeebropilly colliery, Queensland, Australia. Trans. Instn. Min. Metall. (Sect. A: Min. Industry), 106.

Proskin S., Sego D., Alostaz M. (2010). Freeze–thaw and consolidation tests on Suncor mature fine tailings (MFT). Cold Regions Science and Technology, 63: 110-120.

Qiu Y., Sego D.C. (2001). Laboratory properties of mine tailings. Canadian Geotechnical Journal, Vol. 38: 183-190.

Sanin M.V., Puebla H., Eldridge T. Cyclic behavior of thickened tailings. Proceedings of Tailings and Mine Waste 2012, Keyston, Colorado.

Stone K.J.L., Randolph M.F., Toh S., Sales A.A. (1994). Evaluation of Consolidation Behavior of Mine Tailings. Journal of Geotechnical Engineering, ASCE, 120: 473-490.

Wickland B.E., Wilson G.W. (2005). Self-weight consolidation of mixtures of mine waste rock and tailings. Canadian Geotechnical Journal, 42: 327–339.

Wickland B.E., Wilson G.W., Wijewickreme D., Klein B. (2006). Design and evaluation of mixtures of mine waste rock and tailings. Canadian Geotechnical Journal, 43: 928–945.

Wickland B.E., Wilson G.W., Wijewickreme D. (2010). Hydraulic conductivity and consolidation response of mixtures of mine waste rock and tailings. Canadian Geotechnical Journal, 47: 472–485.

Wijewickreme D., Sanin M.V., Greenaway G. (2005). Cyclic Shear Response of Fine-grained Mine Tailings. Canadian Geotechnical Journal, 42: 1408-1421.

Wilson G.W., Wickland B.E., Palkovits F., Longo S. The implementation of paste rock systems for preventing acid rock drainage from waste rock and tailings impoundments. Submitted for oral presentation, 8th ICARD, 2009, Skelleftea, Sweden.

FREEBOARD ASSESSMENT FOR LONG EMBANKMENT DAMS

Mike Gerlach[1]

Long embankment dams are relatively common in the southwestern U.S. These types of structures can generally be considered as low head dams but with large flood storage volumes. Inflow into the reservoir area tends to be non-uniformly distributed along the length of dam which can extend for several miles. The outlet works are typically located at the ends of the dam which can result in complex hydraulic routing of the inflow through the reservoir and to the outlet works. Assessing the adequacy of the freeboard for these types of structures requires sophisticated hydraulic modeling. There are several numerical models that can be used to simulate the hydraulic routing for these types of conditions. Each model has unique limitations, mathematical algorithms and require certain assumptions that may make it more applicable for a given situation. So, how does the engineer choose the appropriate model? The Vineyard Road Flood Retarding Structure is a 4.5 mile long earthen embankment dam located near Phoenix, Arizona. Currently, the dam has insufficient spillway capacity to pass the inflow design flood. The Vineyard Road Flood Retarding Structure is used as a case study to investigate different two-dimensional hydraulic routing computer software and the sensitivity to input parameters in the computation of the maximum water surface elevation and the corresponding freeboard on the structure.

[1] Stantec, mike.gerlach@stantec.com

OFFSTREAM RESERVOIR EARTHEN DAM SEEPAGE – SYSTEMS DESIGN AND CONSTRUCTION CHALLENGES

Randall Bushey[1]

Offstream reservoir earthen embankment seepage systems are integral to the stability and safety of a dam, protection of downstream persons, infrastructure, and the environment, and safe operations and maintenance of the reservoir. The seepage system is typically composed of an earthen embankment, low permeability foundation or clay barrier, geomembrane, cutoff wall, connection between the geomembrane and the cutoff wall, drainage collection network, and monitoring devices. Design of seepage systems is undertaken with limited site knowledge compared to the knowledge gained during the actual construction of the dam. The reservoir embankment plans, specifications, and other contract documents should, therefore, account for the varied experience of contractors, necessary continued investigations, and unknown site characteristics that may be discovered during earth moving that were not evident during the geotechnical pre-design and design investigations. The use of Potential Failure Modes Analysis (PFMA) during the 60% design phase is recommended as a tool to address these factors.

Three offstream reservoir projects in Florida will be used as examples of different seepage systems, geomembranes, cutoff wall designs, seepage/drainage collection materials, and constructability approaches used to meet changing site conditions not fully evident during the design investigations. The connections between the seepage system components are critical to the effectiveness of the system. The performance of the geomembrane/cutoff wall connection is specifically dependent on the site materials, construction approach, and the contractor's quality control program.

Long term project success depends upon: constructability of the design; planned continued site investigations during construction; specified thoroughness of the contractor's quality control program and personnel; vetting of the construction materials employed by the contractor; comprehensive review of field decisions; and the planned ability to monitor the effectiveness of the seepage barrier systems performance.

[1] CH2M, rbushey@ch2m.com

LATEST COMPILATION OF INTERNAL EROSION INCIDENTS AT BUREAU OF RECLAMATION DAMS

William O. Engemoen, P.E.[1]

ABSTRACT

The Bureau of Reclamation (Reclamation) has an inventory of several hundred dams, including more than 230 embankment dams plus many more dikes. The ages of the dam embankments range from recently constructed to more than 100 years old. During the 100-plus year history of Reclamation, there have been numerous instances of the initiation of internal erosion in Reclamation embankment dams or their foundations. An updated study of incidents has been completed, building on past efforts to compile and study the causes of the incidents and better understand the category (location) and specific mechanism involved. This paper will tabulate the incidents in terms of the categories and mechanisms considered, and develop the historical rates of initiation of internal erosion at Reclamation dams, both in terms of category and mechanism. The insights from this study of incidents can be used when evaluating risks of internal erosion failure of embankment dams.

INTRODUCTION

Quantitative risk analysis is becoming more widespread as an important tool for dam safety practitioners. With this tool, evaluators consider all potential failure modes at a dam, and then estimate the likelihood that these failure modes could lead to dam failure. All load ranges are considered, including failure modes under flood loading, earthquake loading, and normal reservoir operations. A key failure mode for any embankment dam is internal erosion, which is considered to have been the causative mechanism in nearly half of all historical failures of embankment dams (Foster et al. 1998).

Evaluating the probability of initiation of internal erosion within an embankment dam or its foundation is very difficult. At this time, deterministic approaches are limited and not particularly robust given all the factors that must be considered but cannot be captured with engineering analyses or laboratory testing. Laboratory testing of small specimens to develop erosion properties and similar data may not be representative of the weak link or true condition in the spatially vast embankment/foundation system. It is unlikely that instrumentation will be located in the exact, critical flow path to provide the necessary data on gradients and velocities. Similarly, seepage models may not be representative of the key hydraulic conditions that would drive the development of an internal erosion potential failure mode along a long and potentially tortuous seepage path through variable materials and/or potential defects that likely vary in aperture and orientation as well. Instrumentation, laboratory testing, and analyses can provide insights into the potential for internal erosion to develop at a dam, and application of these efforts can be a prudent part of a dam safety evaluation. However, practitioners should be aware of the limitations

[1]Geotechnical Engineer – Geotechnical Services Division, Bureau of Reclamation, P.O. Box 25007, Denver, Colorado, wengemoen@usbr.gov

of these methods in terms of ability to estimate the likelihood of initiation of internal erosion.

Reclamation believes the study of internal erosion case histories can provide some degree of "ground truth" or empiricism/precedence to the evaluation. Estimated historic rates of internal erosion initiation can provide risk teams with a relative range or average value for various types of internal erosion (i.e., a "starting" or "anchoring" point). Reclamation in the past has typically based the likelihood of initiation of internal erosion on the documented historical rate of internal erosion failures and incidents, specifically work by the University of New South Wales (UNSW) (e.g., Table 12 from Fell and Wan 2005), and adjusted upward or downward based on site specific factors. Past reviews of Reclamation internal erosion incidents have been made (Engemoen and Redlinger, 2009; Engemoen, 2011). A more current review was conducted in 2015-2016 resulting in a description of each incident and a recalculation of base rate frequencies given the review and additional occurrences of internal erosion in the past few years. This review and recalculation has been peer reviewed by experienced geotechnical staff to incorporate a variety of perspectives and ensure reasonable judgments.

This study reviewed performance at 233 Reclamation embankment dams and dikes which comprise the vast majority of the major embankments at Reclamation facilities. These structures range in age from 6 to 110 years, with an average age of 63 years and a median age of 62 years. The total number of dam-years of operation (through year 2016) at these 233 embankments is 14,611, so it represents a significant data set.

The following are general observations from this most recent study:

- Reviews of Reclamation internal erosion incidents indicate there have been a total of 97 known incidents including one failure (Teton Dam). Internal erosion incidents have occurred throughout the history of Reclamation embankments, and sometimes multiple times at the same dam. The total number of dams that have experienced incidents is 61, or about 1 in every 4 Reclamation embankment dams. Of these 61 dams, 26 experienced more than one incident.
- **These incidents are not limited to first filling but can occur at any time in a dam's life**. About 30 percent of Reclamation incidents have occurred during the first five years of reservoir operation, and 70 percent of all incidents have occurred after more than five years of successful operation. At least two incidents occurred after more than 90 years of successful operation.
- The incidents have also not been limited to older or deteriorated dams; newer, modern dams have also had incidents. For those dams that experienced incidents after first filling, approximately half of the incidents occurred in dams that were more than 40 years old, and the other half in dams that were less than 40 years old.
- For this study, defining a seepage event as an "incident" considered the severity of the seepage and distress, as well as Reclamation's response to the event. Typically, the incidents reported in this study involved significant seepage (and often manifestation of particle transport as discussed in the following bullet), and

in most cases led to some type of remediation effort (such as grouting, installation of filters, etc.).

- For each incident, the evidence for developing internal erosion was categorized as either "excessive seepage" or "particle transport." The use of particle transport was limited to those cases where clear evidence of internal erosion was noted; such as the presence of sinkholes, voids, sand boils that were moving soil particles, or turbid seepage water. Of the total 97 incidents/failures at Reclamation embankments, there have been a total of 66 cases where particle transport was observed.

INTERNAL EROSION BY CATEGORY

For the purposes of categorizing internal erosion incidents within Reclamation, each incident was first classified into one of five categories (or locations of the developing internal erosion):

1) Internal erosion solely through the embankment
2) Internal erosion solely through the foundation
3) Internal erosion of embankment into the foundation (or along the contact)
4) Internal erosion into or along a conduit
5) Internal erosion into a drain

It is important to note that the category of internal erosion of embankment into foundation has two subcategories. The first condition is when the primary seepage path is through the embankment, and embankment soils are transported into unfiltered portions of the foundation (soil or rock). The second condition is where the primary seepage path is through a pervious foundation (soil or rock), and the seepage has sufficient velocity to begin eroding and transporting embankment soils at the contact with the foundation.

The following table portrays the number of incidents at Reclamation embankments by category (location).

Table 1. Category of Internal Erosion Incidents at Reclamation Embankments

Category of internal erosion	Incidents/Failures with definitive particle transport	Incidents/Failures with excessive seepage and perhaps sand boils	All incidents and failures
Embankment only	5	4	9
Foundation only	28	16	44
Embankment into foundation	7	11	18
Into or along conduit	8	0	8
Into drain	18	0	18
TOTAL	**66**	**31**	**97**

The following observations from Table 1 are of note.

- Approximately 45 percent of the tabulated internal erosion incidents have involved internal erosion through the foundation, likely due to the significant number of Reclamation dams without a fully penetrating cutoff over their entire length, the pervious nature of the foundation materials leading to significant seepage, and the presence of erodible soils in the foundation. See additional discussion under the paragraph heading "Type of Foundation Cutoff."
- Of the 44 foundation incidents, 10 involved glacial soils, and 5 were attributed to bedrock seepage. The remainder, or majority, of the incidents were in soil foundations comprised of alluvium, colluvium, eolian, or landslide deposits.
- When considering both internal erosion through the foundation and internal erosion from embankment into foundation, the foundation has played a role in at least two-thirds of all Reclamation incidents.
- The relatively low frequency of internal erosion through the embankment incidents might be explained by Reclamation's use of wide cores (long seepage path) often flanked by shells of sands/gravels/cobbles (providing some filtering capability). In addition, nearly 60 percent of Reclamation embankment cores are comprised of plastic soils ($PI \geq 7$) and another 7 percent feature an impermeable reservoir liner or cutoff wall.
- The frequency of internal erosion into/along conduits is much lower than reported by Foster et al. 1998. This may be explained by the conscientious effort to specially compact against Reclamation conduits and the reliance on formed concrete conduits (as opposed to precast concrete, CMP, etc.). In addition, the number of dams with a conduit plays a role, and is discussed further in "Presence of Conduits in Embankments."
- The relatively high number of internal erosion into drain incidents may be due to decades of relatively poor design details for drains (open jointed pipe, brittle pipe materials, coarse gravel envelopes, and thin filters).

INTERNAL EROSION BY MECHANISM

In addition, each incident of internal erosion was also categorized by the type of failure mode believed to be represented by the incident. Although the literature discusses many different types of internal erosion and varying definitions, Reclamation uses the following four descriptions to describe the main mechanisms of internal erosion.

Piping - Classic "backward erosion piping" occurs when soil erosion begins at a seepage exit point, and erodes backwards through the dam or foundation, with surrounding soil providing a support (roof) to keep the developing "pipe" open. Four conditions are needed for development of piping: (1) a concentrated leak/source of water (of sufficient quantity and velocity to erode material), (2) an unprotected seepage exit point, (3) erodible material in the flow path, and (4) material capable of supporting a pipe or a roof. If these conditions all develop, uncontrolled erosion begins to create a "pipe" within the embankment or foundation.

Internal Migration – "Internal migration," also referred to as stoping in the literature, can occur when the soil is not capable of sustaining a roof or a pipe, or there is an insufficient horizontal gradient to induce upstream/downstream erosion. Soil particles are eroded and a temporary void grows until the overlying soils can no longer provide support, at which time the roof collapses. This mechanism can be repeated progressively until the void shortens the seepage path and leads to uncontrolled erosion and ultimately to breach of the dam. Alternately, the progressive stoping could take out the dam crest and lead to an overtopping failure. With this mechanism, sinkholes are frequently observed at or near the point where erosion initiates.

Scour – "Scour," or contact erosion, occurs when tractive seepage forces along a surface (e.g. a crack within the soil, adjacent to a wall or conduit, along the dam/foundation contact, or through a pervious layer or bad construction lift) are sufficient to move soil particles into an unprotected area. Once this begins, failure from a process similar to piping or simply progressive erosion could occur.

Suffusion/suffosion – Both suffusion and suffosion are potential mechanisms in internally unstable soils (i.e. certain broadly graded or gap graded coarse-grained soils). Within Reclamation, "suffusion" occurs when the finer particles of a soil are eroded through a matrix of coarser particles, leaving behind a coarsened and more permeable soil skeleton. For suffusion to occur the soils typically need to be primarily coarse-grained containing at least 70 to 80 percent gravel or large size particles; these coarse particles provide the point-to-point contact creating a network or skeleton through which the finer particles will move through. The resulting coarser and more pervious soil structure remaining may no longer be filter compatible with adjacent upstream soils and thus provide an unfiltered exit at which point internal erosion may initiate. A closely related mechanism, "suffosion," occurs in soils where the coarser particles are floating in a matrix of finer particles. It takes a higher gradient with this mechanism (as opposed to suffusion) to wash out the fine soils due to the higher stresses existing between the fine particles; thus it is judged to be a less likely mechanism. In addition, while volume change may not occur in suffusion, volume change is highly likely to occur in suffosion.

It is important to note that internal erosion due to any of these mechanisms is not always rapid, and can be a gradual process, taking decades instead of hours or days. In some cases, the internal gradients are only at critical levels for short periods of time, and erosion is thus intermittent or episodic.

Admittedly, the assignment of an internal erosion mechanism to a past incident requires considerable judgment – in many cases a definitive understanding of just what type of process or mechanism occurred is not clear. Furthermore, some incidents may well involve a combination of mechanisms.

The following table portrays the number of incidents at Reclamation embankments by type of mechanism.

Table 2. Postulated Internal Erosion Failure Mechanisms Involved in Incident

Internal erosion mechanism	Incidents/Failures with definitive particle transport	Incidents/Failures with excessive seepage and perhaps sand boils	All incidents and failures
Backward erosion piping	16	6	22
Internal migration	24	0	24
Scour	17	20	37
Suffusion/Suffosion	9	5	14
TOTAL	**66**	**31**	**97**

The following observations from Table 2 are of note.

- Approximately 40 percent of all incidents are suspected to have involved scour, with piping and internal migration mechanisms each accounting for about one fourth of the total.
- Suffusion/suffosion is the least common mechanism, believed to have accounted for about 14 percent of the total incidents. This low total probably reflects a lesser number of embankments/foundations with internally unstable soils.
- 11 of the 14 suspected suffusion/suffosion incidents occurred in the foundation. 7 of the 14 suspected suffusion/suffosion incidents involved glacial soils, typically broadly graded soils that are internally unstable.
- 16 of the 37 suspected scour incidents involved bedrock seepage through discontinuities.
- 15 of the 22 piping incidents occurred in dam foundations where only a partial penetrating cutoff was constructed.

INFLUENCE OF RESERVOIR LEVEL AT TIME OF INCIDENT

There is a widely held belief that most internal erosion incidents in embankments or foundations initiate at threshold reservoir levels that create sufficient hydraulic gradients or velocities to begin eroding susceptible soil particles. Thus, practitioners are likely to be more concerned about internal erosion at high reservoir levels. In fact, based on Reclamation incidents, internal erosion can manifest at varying reservoir levels.

A special condition involving reservoir levels involves the initial filling of the reservoir. Initial filling refers to the first time, shortly after completion of construction, that the reservoir is filled to its normal operational level (frequently the top of active conservation capacity at Reclamation reservoirs). Other studies have pointed out the frequency at which embankments experience internal erosion incidents during first filling. At Reclamation, 23 of the 97 incidents, or about 24 percent, occurred during initial filling of the pool. Obviously this points out the need for careful surveillance during initial reservoir filling (or re-filling after a modification).

However, at Reclamation, 74 incidents occurred during operations after first filling. The following table portrays the level of the reservoir at the time of these 74 observed incidents at Reclamation embankments.

Table 3. Reservoir Level at Time of Incident (for embankments after first filling)

Reservoir Level	Number of Incidents	Percentage of Incidents
Above normal operational level	11	15 %
At normal operational level *	52	70 %
Below normal operational level	11	15 %
TOTAL	**74**	

*Note: At normal operational level typically means within 2 feet of the normal annual high pool

The following observations from Table 3 are of note.

- At Reclamation embankments that successfully completed initial reservoir filling and were in normal operational status, **approximately 85 percent of internal incidents were observed at reservoir levels at or below normal levels.**
- Only 15 percent of internal erosion incidents after first filling were observed during higher than *normal* reservoir levels. At most Reclamation facilities, *"normal"* levels refer to the top of active conservation or joint use that the pool routinely reaches in most years. At a few facilities where the top of active is rarely reached, normal was assumed to be the typical upper levels reached in the past. Part of the reason for a low occurrence at higher reservoir levels can be explained by the observation that Reclamation reservoirs generally do not experience unusually high pool increases, due in part to the use of uncontrolled spillways and careful reservoir operations in advance of forecasted precipitation events. In fact, both the average and median historical maximum reservoir level experienced to date at the 233 Reclamation embankments in this study was computed to be only 1.5 feet above the normal operational level.
- Some of the incidents in this study clearly suggest that internal erosion had been slowly progressing for years or even decades. Reclamation believes that many internal erosion processes may take years to progress to the point of showing indications of distress or obvious particle transport. Thus, the reservoir level at the time of the incident may not reflect the key hydraulic head responsible for the initiation of internal erosion. **It is possible that internal erosion may occur during the several-week period of typical high reservoir levels at Reclamation facilities, and then stop for the season. In other words, internal erosion can be an episodic event that progresses to a limited extent only at normal high pools, year after year.**

INFLUENCE OF AGE OF EMBANKMENTS

The list of incidents at Reclamation embankments clearly shows that internal erosion can occur at any time in the life of an embankment. For those dams that completed a successful first filling, it was found that the average and median age of the embankments that experienced an internal erosion incident was approximately 40 years.

Figure 1 is a histogram portraying the number of incidents by age of dam at the time of incident. Tables 4 and 5 portray the age of the dam (or age after significant modifications to a dam) at the time of each incident (Table 4 by category and Table 5 by mechanism).

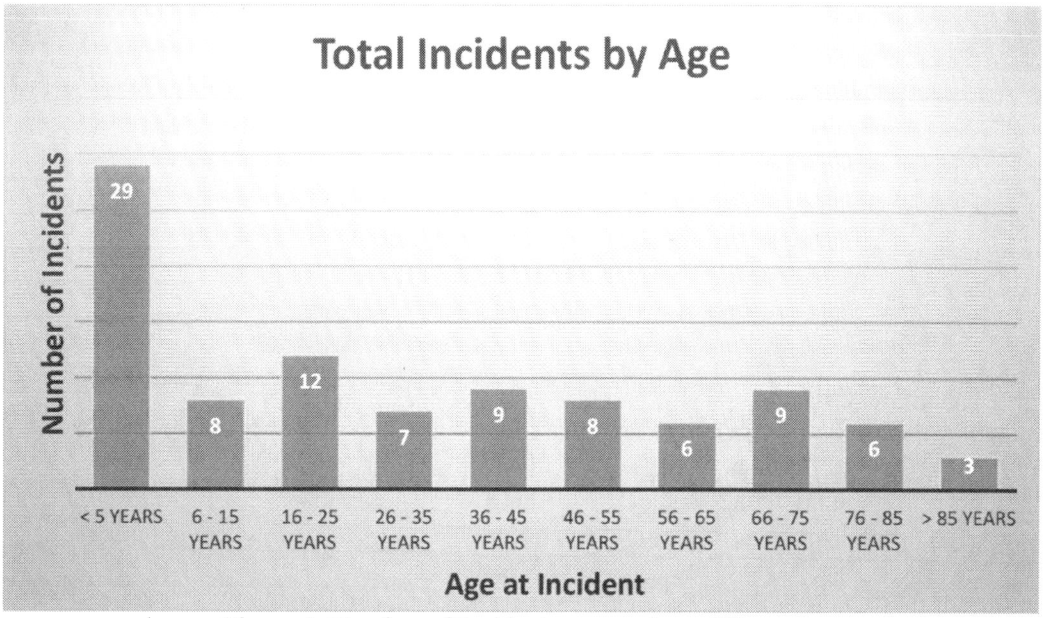

Figure 1. Number of Incidents versus Age of Dam

Table 4. Age of Dam at Incident and Category of Internal Erosion

Dam Age at Incident	Number of Embankment Incidents	Number of Foundation Incidents	Number of Emb/Fnd Incidents	Number of Conduit Incidents	Number of Drain Incidents	Total Number of Incidents
≤ 5 years	1	16	9	1	2	**29**
6-15 years	0	4	3	0	1	**8**
16–25 years	0	5	1	0	6	**12**
26-35 years	1	3	2	0	1	**7**
36-45 years	2	4	1	0	2	**9**
46-55 years	1	2	1	1	3	**8**
56-65 years	0	4	1	1	0	**6**
66-75 years	3	3	0	1	2	**9**
76-85 years	1	2	0	2	1	**6**
>85 years	0	1	0	2	0	**3**
Totals	**9**	**44**	**18**	**8**	**18**	**97**

Table 5. Age of Dam at Incident and Mechanism Type

Dam Age at Incident	Number of Piping Incidents	Number of Internal Migration Incidents	Number of Scour Incidents	Number of Suffusion Incidents	Total Number of Incidents
≤ 5 years	6	5	14	4	**29**
6-15 years	1	2	5	0	**8**
16–25 years	5	5	2	0	**12**
26-35 years	2	1	3	1	**7**
36-45 years	3	2	3	1	**9**
46-55 years	0	3	2	3	**8**
56-65 years	2	1	3	0	**6**
66-75 years	0	3	4	2	**9**
76-85 years	2	2	0	2	**6**
>85 years	1	0	1	1	**3**
Totals	**22**	**24**	**37**	**14**	**97**

The following observations from Figure 1 and Tables 4 and 5 are of note.

- Incidents are much more likely to occur in the first five years of reservoir operation (which includes first filling of the reservoir).
- However, incidents continue to occur beyond 5 years, with little significant decline in rate of incidents after 5 years.
- About half of the incidents involving scour occurred in the first 15 years of operational history.
- More than half of the embankment/foundation incidents occurred during the first 15 years. This is similar to the preceding bullet since all Reclamation embankment/foundation incidents were judged to result from scour.
- The observation that conduit incidents occur later in the life of a dam is likely due to gradual deterioration of the conduit and the fact that most Reclamation conduit incidents were a very gradual process (limited by the constriction of the defect).
- However, all mechanisms tend to occur throughout the operational history; i.e. they are as likely to occur late as early.

INFLUENCE OF SOIL PLASTICITY

For this incident study, "plastic" soils were considered to have a Plasticity Index (PI) greater than or equal to 7 percent. Conversely, cohesionless or low plasticity soils were considered to have a PI less than 7. It is recognized that selecting a particular PI value to differentiate between plastic and non-plastic behavior in internal erosion is subject to interpretation. A value of 7 corresponds to the Idriss-Boulanger value in liquefaction analyses to differentiate between "clay-like" and "sand-like" behavior.

The following table shows the plasticity characteristics of Reclamation embankment dams. All 233 embankments were evaluated for core plasticity, while foundation plasticity was only considered for those embankments that did not feature a fully penetrating cutoff of foundation soils (as described later under "Type of Foundation Cutoff" paragraph).

Table 6. Nature of Embankment and Foundation Soils at Reclamation Embankments

Location of soils	**Plastic soils (PI ≥ 7) or Core Wall or Liner**		**Cohesionless or Low Plasticity soils (PI < 7)**		**Variable soils (contains both high and low PI)**	
	# of Dams	**Percent**	**# of Dams**	**Percent**	**# of Dams**	**Percent**
Embankment	153	66 %	80	34 %	-	-
Foundation	13	12 %	49	45 %	46	43 %

The following observations from Table 6 are of note.

- Approximately two thirds of all Reclamation embankment cores would be considered relatively plastic, with only one third considered relatively cohesionless or of no to low plasticity.
- Foundation conditions are quite different from core conditions, with only 12 percent of embankment foundations consisting solely of plastic soils. A total of 88 percent of Reclamation foundations (those that do not feature a complete cutoff) either consist solely or partially of cohesionless or low plasticity soils.

When considering the 97 instances of reported internal erosion incidents in this study, a significant majority (approximately 80 percent) of incidents involved cohesionless or low plasticity soils (PI < 7). When looking only at the 66 incidents that displayed definitive evidence of particle transport, 83 percent of these cases involved cohesionless or low plasticity soils. Of the eleven incidents out of 66 cases with definitive particle transport involving plastic soils (PI ≥ 7), there were no incidents involving piping or suffusion. Nine of the incidents involving plastic soils were judged to be scour and two judged to be internal migration. This is in line with other research findings that indicate it takes a high gradient (higher than exists in most embankment dams) to initiate piping in plastic soils, and that suffusion is generally limited to cohesionless soils.

Based on Reclamation experience, internal erosion is significantly less likely to occur if the embankment or foundation soils being considered have a PI ≥ 7.

PRESENCE OF CONDUITS IN EMBANKMENTS

A very sizable portion of historical internal erosion incidents and failures in embankment dams have resulted from the presence of a conduit or similar penetrating feature. To accurately determine the frequency of incidents involving conduits, the database of 233 embankments was analyzed to determine the number of conduits and associated dam-years of operation at these Reclamation features. From this study, it was found that:

- A total of 124 of the 233 embankments featured a conduit constructed within the embankment. This amounts to 53 percent of the embankments, and results in 7,858 dam-years of operation.
- The remaining embankments featured tunnel outlet works, outlets in concrete portions of the dam, or no outlet at all.
- Interestingly, 7 of the 8 internal erosion incidents involving a conduit actually featured a "conduit." One of the incidents involved internal erosion into a tunnel outlet works (Willow Creek Dam in Montana). However, the tunnel was in soft rock with less than 20 feet of cover, so in a sense was similar to a conduit.

TYPE OF FOUNDATION CUTOFF

Because of the large number of incidents involving internal erosion of foundation materials, a closer look was given to the types of cutoffs constructed at the 233 Reclamation embankments in this study. It was found that:

- A total of 125 embankments, or 54 percent, featured a full cutoff in the foundation (which means a cutoff to bedrock for the entire dam length, fully cutting off seepage through overburden). Included in this number were four cases of lined reservoirs.
- A total of 108 embankments, or 46 percent, featured a partial cutoff in the foundation. Included in this number were 3 embankments with no cutoff. Often times, the valley section featured a cutoff trench fully extended to bedrock, while the abutments featured a cutoff that only partially penetrated soil deposits. The number of dam-years of operation for embankments with partial cutoffs totaled 7,191.
- Of the 44 Reclamation incidents involving internal erosion of the foundation, 41 of them (93 percent) involved an embankment with a partially penetrating foundation cutoff. The other three involved karstic or highly pervious bedrock foundations (i.e., non-typical foundation conditions for a Reclamation dam).

DEVELOPMENT OF BASE RATE FREQUENCIES OF INITIATION OF INTERNAL EROSION

An estimate of the rate of initiation of internal erosion at a typical Reclamation embankment dam can be obtained by dividing the number of incidents and failures by the number of dam-years of operation. Since our current inventory of embankment dams contains all dams that have experienced a first filling (the youngest dam is 6 years old as of 2016), of specific interest are incidents at dams that have survived first filling. (For a dam safety evaluation of a new dam, these base rate frequencies reflecting the likelihood of initiation would not be appropriate and could be much higher. Furthermore, when considering the likelihood of initiation of internal erosion at flood levels, these base rate frequencies should be adjusted upward based on considerations such as potential for flaws in upper portion of the dam, amount of increased gradient expected, etc.)

When considering the potential for incidents after first filling, there seems to be a significant break at an age of 5 years as shown on Figure 1. Thus, the number of incidents after 5 years of operation will be considered in developing base rate frequencies for dams that have survived first filling. Furthermore, since Reclamation is particularly concerned with the risk of internal erosion at normal operation levels, another filter will be to consider only those incidents that occurred at or below normal reservoir levels. From this Reclamation incident database, there are 58 incidents that occurred at or below normal pool levels at dams that had at least 5 years of reservoir operation. These applicable incidents are summarized in Table 7.

Table 7. – Number of Incidents of Internal Erosion at Reclamation Embankment Dams (after 5 years of operation and at or below normal pool level)

	Backward Piping	**Internal Migration**	**Scour**	**Suffusion/ Suffosion**	**TOTALS**
Embankment	0	1	5	1	**7**
Foundation	5	5	7	6	**23**
Emb into Fnd	0	0	5	0	**5**
Into/along Conduit	2	4	1	0	**7**
Into Drain	5	9	0	2	**16**
TOTALS	**12**	**19**	**18**	**9**	**58**

The total number of operational dam-years was obtained by identifying each of the 233 Reclamation embankments considered in this study, determining their present age (to year 2016), and summing all ages. In this manner, the number of dam-years of operation at Reclamation facilities was calculated to be 14,611. For all categories not involving the foundation only or into/along a conduit, this number of dam-years is considered to represent the operational history of the Reclamation embankments in the study.

However, adjustments are needed when considering internal erosion in the foundation and into/along conduits. As the incident database reflects, internal erosion through the foundation only is likely limited to embankments with partially penetrating cutoffs (with only 3 exceptions involving unusual bedrock/foundation conditions). Therefore, the appropriate operational history to use would be the number of dam-years associated with those Reclamation embankments featuring partial cutoffs – this value was computed to be 7,191 dam-years. Similarly, internal erosion into/along the conduits is only a viable scenario (with very rare exceptions) for those embankments that feature a penetrating conduit within the embankment. The number of dam-years associated with these specific embankments total 7,858.

Using these values of dam-years of operation and the number of incidents (under normal operations and after 5 years of operation) identified by the study, the following tables present the estimated historical rate at which erosion has initiated (and continued to progress to at least some degree). Table 8 is by category and Table 9 is by mechanism.

Table 8. Historical Rate of Initiation of Internal Erosion at Reclamation Embankments Based on Category

Internal erosion category	Estimated historical rate of erosion initiation*	
	Incidents/Failures with definitive particle transport	**All incidents/failures**
Embankment only	$3.4x10^{-4}$	$4.8x10^{-4}$
Foundation only	$2.2x10^{-3}$	$3.2x10^{-3}$
Embankment into foundation	$1.4x10^{-4}$	$3.4x10^{-4}$
Into/Along conduit	$8.9x10^{-4}$	$8.9x10^{-4}$
Into drain	$1.1x10^{-3}$	$1.1x10^{-3}$
TOTAL	$\mathbf{4.7x10^{-3}}$	$\mathbf{6.0x10^{-3}}$

***Note: See later discussion of whether these data include more than just "initiation"**

Table 9. Historical Rate of Initiation of Internal Erosion at Reclamation Embankments Based on Mechanism

Internal erosion mechanism	Estimated historical rate of erosion initiation*	
	Incidents/Failures with definitive particle transport	**All incidents/failures**
Backward erosion piping	$1.0x10^{-3}$	$1.3x10^{-3}$
Internal migration	$1.9x10^{-3}$	$1.9x10^{-3}$
Scour	$1.2x10^{-3}$	$1.8x10^{-3}$
Suffusion/suffosion	$6.2x10^{-4}$	$1.0x10^{-3}$
TOTAL	$\mathbf{4.7x10^{-3}}$	$\mathbf{6.0x10^{-3}}$

***Note: See later discussion of whether these data include more than just "initiation"**

It is easy to question whether this review of past internal erosion incidents, admittedly not an intensive research effort (consisting of years), is a reasonable portrayal of performance at Reclamation embankments. Another way to consider this frequency question might be to note that it is not unusual, on average, to see 1 or maybe even 2 new "incidents" of unusual seepage or piezometric behavior, new sand boils, or new sinkholes each year within the Reclamation inventory of embankment dams. Assuming 1.5 incidents per year with 233 embankments equates to an annual frequency of $6.4x10^{-3}$. This alternate approach to estimating a base rate frequency of the initiation of internal erosion happens to be similar to the summary value obtained from the incident study – although far from definitive, this does support a measure of confidence in the reasonableness of these frequency data.

Rather than directly use the values reflected in these tables, it is recognized that additional adjustments may better reflect the ranges of potential "best estimate" values given the variables and uncertainties involved with categorizing internal erosion events. One key uncertainty deals with whether the historical base rate of incidents portrayed above reflects more than just the "initiation" phase of the internal erosion process. In other words, initiation may have occurred in more Reclamation embankments than these catalogued because the process never "progressed" far enough to manifest symptoms like those detected or observed in the 97 cases.

It is difficult to estimate an additional number of dams where internal erosion may have initiated but did not continue or progress, and thus remained undetected. The original UNSW study of world-wide dams (Foster et al. 1998) assumed the number of "unreported" incidents of initiation was probably in the range of 2 to 10 times the number of reported incidents. Given Reclamation's reporting and documentation capabilities, it would seem likely that the lower portion of this range would be more applicable. In addition, there are some cases where the true source of the seepage and any signs of particle transport could not be conclusively tied to the reservoir. This is especially true in the cases of internal erosion through foundation and into drains, as other seepage sources (hillside, tailwater, etc.) may have provided the driving hydraulic head.

Furthermore, rather than specifying a single value, it appears to make more sense to suggest a range of **best estimate** values. The term "best estimate" is used since the true range of initiation of internal erosion probably spans several orders of magnitude. The lower end of the best estimate range is based on applying a multiplier of 1.25 to the observed 46 cases of definite particle transport, which assumes 25 percent more cases of definitive particle transport may have occurred and not noticed or reported. It seems rather unlikely that definitive evidence would not have been observed; thus a relatively small multiplier. (However, this multiplier was not applied to "foundation only" and "into drain" categories due to the belief that some incidents may have been attributed to seepage sources other than the reservoir.) The upper end of the best estimate range is based on a doubling of all 58 reported incidents, including the 12 that did not manifest any particle transport. Thus, the upper range values assume that only about 40 percent of all cases of definitive initiation of internal erosion have actually been observed and documented within Reclamation.

These adjusted values shown in the following tables are proposed as "starting points" or an empirical reference point for considering the probability of the initiation of internal erosion for Reclamation dams (or in an inventory of dams similar to Reclamation's). It should be noted that this inventory includes approximately 200 dams constructed prior to the failure of Teton Dam without well designed filters. These tables are considered "summary" tables in that they provide a broad view of overall base rate frequencies.

Table 10. Proposed Best Estimate Values of Annual Probabilities of Initiation of Internal Erosion by Category

Type of internal erosion	Range of initiation probability
Embankment only	$4x10^{-4}$ to $1x10^{-3}$
Foundation only*	$2x10^{-3}$ to $6x10^{-3}$
Embankment into foundation	$2x10^{-4}$ to $7x10^{-4}$
Into/Along conduit**	$1x10^{-3}$ to $2x10^{-3}$
Into drain*	$1x10^{-3}$ to $2x10^{-3}$

***Note: "Foundation only" and "Into drain" lower values were not adjusted by the 1.25 multiplier given the belief that some incidents may have been attributed to seepage sources other than the reservoir**

Table 11. Proposed Best Estimate Values of Annual Probabilities of Initiation of Internal Erosion by Mechanism

Type of internal erosion	Range of initiation probability
Backward erosion piping	$1x10^{-3}$ to $3x10^{-3}$
Internal migration	$2x10^{-3}$ to $4x10^{-3}$
Scour	$2x10^{-3}$ to $4x10^{-3}$
Suffusion/suffosion	$8x10^{-4}$ to $2x10^{-3}$

The incident database can be summarized further by examining the types of mechanisms within the general categories. In other words, for example, how frequently have we observed piping in a foundation, or scour in the embankment? Table 7 presented earlier (but shown again below for continuity) portrays the number of incidents for a specific mechanism/category.

Table 7. Number of Incidents of Internal Erosion at Reclamation Embankment Dams (after 5 years of operation and at or below normal pool level)

	Backward Piping	Internal Migration	Scour	Suffusion/ Suffosion	TOTALS
Embankment	0	1	5	1	**7**
Foundation	5	5	7	6	**23**
Emb into Fnd	0	0	5	0	**5**
Into/along Conduit	2	4	1	0	**7**
Into Drain	5	9	0	2	**16**
TOTALS	**12**	**19**	**18**	**9**	**58**

This table can provide insight for risk teams as to what type of internal mechanism has been suspected in the various categories of internal erosion incidents observed at Reclamation dams. Observations from this table include:

- Scour is suspected to be the main cause of internal erosion of the embankment at Reclamation dams. This scour is typically thought to occur in cracks or similar defects in an embankment.

- All four mechanisms are suspected in Reclamation incidents of internal erosion in the foundation, with no one mechanism significantly more likely.
- Scour is suspected to have been the likely mechanism involved in all embankment/foundation internal erosion incidents at Reclamation dams. This is generally attributed to erosion of the embankment soils by seepage flowing along a jointed bedrock surface or in a coarse/pervious foundation layer.
- Internal migration is the most likely mechanism suspected in Reclamation cases of both internal erosion into/along a conduit and internal erosion into a drain. Typically, these incidents have resulted in sinkholes stoping upward from the internal erosion initiation location.
- Although not a particularly likely mechanism, suffusion/suffosion has been observed predominantly in foundation soils.

CONSIDERATIONS FOR USAGE OF THE BASE RATE FREQUENCY TABLES

1. These ranges shown in the tables are considered to be "best estimates" – not the reasonable low and reasonable high. The true range of reasonable low to reasonable high estimates likely contains more uncertainty.

2. These best estimates ranges shown in the tables are considered to represent a typical Reclamation embankment dam (wide central core, granular shells, no engineered filters, limited foundation treatment). Higher or lower estimates of initiation probability will often be appropriate if conditions at the specific dam being evaluated are judged to be better or worse than the "average" condition at a Reclamation dam. For example, dams with very low hydraulic gradients and minimal seepage or with modern design features may lead to lower estimates of initiation, while dams with appreciable seepage or a history of concerns may warrant higher estimates.

3. The incidents used to develop these values were limited to only Reclamation embankment dams, so use on embankments designed/constructed by others should consider how well those dams compare to Reclamation practices and the "typical" Reclamation embankment dam, as well as similarity of reservoir operations.

4. Tables 10 and 11 provide a reasonable range of best estimates which can be used as an initial starting point or anchor for estimating the probability of initiation of internal erosion. Risk teams should consider both the category range and the mechanism range in establishing their initial estimate. Risk teams should study Table 7 to see what types of mechanisms for the categories of internal erosion are suspected in Reclamation's past incidents.

5. Approximately 80 percent of the Reclamation incidents featured soils with no or low plasticity (PI<7). If postulated failure modes involve low plasticity soils, there is no need to consider higher estimates since these soils are well represented by the incidents. However, with more plastic soils, lower initiation rates may be considered. Note also that no Reclamation incidents of piping or suffusion/suffosion have occurred in plastic soils.

6. The portrayed base rate frequencies have been developed for dams with more than 5 years of operational history and operating at normal reservoir levels. Thus, when evaluating new dams, consideration should be given to using somewhat higher values of initiation probability. Similarly, when estimating hydrologic risks, higher values of initiation probability would be expected when a dam is exposed to future higher reservoir levels.

7. Simply referring to the tabulated best estimates from Reclamation's history of incidents is not sufficient in evaluating the probability for erosion to initiate. Instead, site conditions must be considered in order to determine whether there are features, conditions, or behaviors present at a given site that will influence the potential for erosion to initiate.

CONCLUSIONS

Internal erosion has been responsible for nearly half of the historic failures of embankment dams. A study of 233 Reclamation embankment dams indicates that about 1 in 4 dams in the Reclamation inventory have experienced an internal erosion incident. These episodes can occur at any time in a dam's life, even after decades of successful operation. Furthermore, most Reclamation incidents have not occurred at unusually high reservoir levels. This temporal aspect to the probability of initiation of internal erosion is difficult to incorporate into typical engineering or risk analyses.

For dam safety professionals, the risks associated with internal erosion can be very difficult to predict. This is in part due to the lack of any deterministic processes to accurately calculate the factor of safety against internal erosion. Further, any evaluation also cannot rely on past performance, as case histories include numerous incidents or failures of embankment dams that had been in service for tens of years without any alarming signs of distress.

One of the most difficult aspects of estimating the risk of internal erosion involves the probability that internal erosion will "initiate." Researchers have developed approaches that include measuring the critical shear stress and hydraulic gradient necessary to initiate internal erosion in particle soils, usually involving a small sample of representative embankment or foundation soil. However, for the dam safety practitioner, the difficult leap of faith is to assume that laboratory tests represent the critical "weak link" in an embankment-foundation system where internal erosion is most likely to occur.

Reclamation considers that base rate frequencies obtained from the operation of Reclamation dams for over 100 years can provide valuable insights for how similar embankments may perform in the future, particularly when many embankment dams may perform well for decades before experiencing an incident. The development of tables of suggested best estimate rates of initiation based on empirical evidence serve as an important starting point for risk teams constructing an event tree to estimate the probability of failure.

REFERENCES

Engemoen, W. & Redlinger, C. 2009. Internal erosion incidents at Bureau of Reclamation dams, USSD Annual Conference, Nashville, Tennessee.

Engemoen, W. 2011. Bureau of Reclamation experiences with internal erosion incidents, Meeting of the European Working Group on Internal Erosion, Brno, Czech Republic.

Fell, Robin and Wan, Chi Fai. 2005. Methods for estimating the probability of failure of embankment dams by internal erosion and piping in the foundation and from embankment to foundation. UNICIV Report No. R-436, University of New South Wales, Sydney, Australia.

Foster, M.A., Fell, R. and Spannagle, M. 1998. Analysis of embankment dam incidents, UNICIV Report No. R-374, University of New South Wales, Sydney, Australia.

MODELING OF INTERNAL EROSION IN EARTHEN EMBANKMENT DAM

Al Preston, Ph.D, P.E[1]
Keil Neff, Ph.D., P.E.[2]
Chunling Li, Ph.D., P.E.[3]
Lucas de Melo, Ph.D., P.E.[4]

ABSTRACT

In 2014, a high hazard embankment dam in South East Appalachia developed a sinkhole at the toe and additional inspections revealed muddy water seeping from the riverbank into the tailrace, indicating potential internal erosion. Consequences of dam failure included probable loss of human life and significant economic and environmental consequences. Numerical modeling of the geotechnical and hydrotechnical components of a potential internal erosion failure was conducted to assess downstream risk due to potential dam breach and to examine potential mitigation strategies.

The Windows™ Dam Analysis Modules – Version C (WinDAM C) numerical model was used to evaluate the potential internal erosion failure and provide hydrograph input for downstream flow-routing modeling. A bootstrapping technique was used to estimate realistic ranges of soil erodibility based on standard soil test results and using relationships from the literature. The WinDAM C model provided predictions of the peak downstream flow rate and time of occurrence, downstream warning time, and the final size of the breach in the embankment. The results demonstrated that maintaining the reservoir at a low pool elevation was the most effective strategy for minimizing the risk associated with internal erosional dam failure. Other mitigation strategies, such as opening the spillways at the onset of failure, were also assessed. The results of the modeling provided critical information to better assess the dam safety risk and facilitate decisions on mitigation strategy.

INTRODUCTION

In 2014 a high hazard earthen embankment dam in South East Appalachia exhibited signs of internal erosion. A sinkhole had developed at the toe and additional inspections revealed muddy water seeping from the riverbank into the tailrace. Consequences of dam failure included probable loss of human life and significant economic and environmental consequences. The dam is 160 ft high and in addition to the earthen embankment section, it also consists of non-overflow concrete sections, a concrete spillway section, and a powerhouse. The dam is constructed on top of a karst foundation which is susceptible to sinkhole development. However, this paper focuses on the failure associated with internal erosion of the foundation soil/embankment soil. Potential failure

[1] Geosyntec Consultants, 448 S. Hill St., Ste 1008, Los Angeles, CA 90013, apreston@geosyntec.com
[2] Bureau of Reclamation, Denver Federal Building, Bldg 67, 5th Floor, PO Box 25007, Denver, CO 80225, kneff@usbr.gov
[3] Geosyntec Consultants, 10211 Wincopin Circle, Floor 4, Columbia, MD 21044, cli@geosyntec.com
[4] Geosyntec Consultants, 10211 Wincopin Circle, Floor 4, Columbia, MD 21044, ldemelo@geosyntec.com

associated with sinkhole development in the bedrock is outside the scope of this paper. Due to concerns of potential failure associated with internal erosion, the dam owners and operators commissioned the authors to embark on an internal erosion numerical modeling study to quantify downstream effects and assess mitigation strategies. The focus of the modeling was to assess the current risk and evaluate temporary interim mitigation strategies. The assessment of permanent remedial actions is outside the scope of this paper.

APPROACH

The Windows™ Dam Analysis Modules – Version C (WinDAM C) numerical model, jointly developed by the United States Department of Agriculture, the U.S. Army Corps of Engineers, and Kansas State University (Visser et al. 2015), was used to conduct the internal erosion and hydraulic modeling. The internal erosion modeling module of WinDAM C assumes that a horizontal and rectangular internal erosion conduit has formed within the dam or its foundations. The user sets the elevation and initial size of the conduit. The user also specifies the inflow hydrograph to the reservoir, reservoir stage-storage curve, tailwater rating curve, and principal and auxiliary spillway rating curves to simulate the hydraulic routing through the dam and over the spillways, and to calculate the dynamic upstream headwater and tailwater elevations. The difference between the upstream head and tail water elevations drives the flow through the conduit as estimated from hydraulic calculations based on conduit length, conduit size, and resistance. The WinDAM C program estimates the soil erosion due to the flow and calculates the growth of the conduit as the embankment material is eroded. The enlarged conduit then feeds back into the hydraulic calculation in a fully-coupled simulation. The erosion of the conduit may eventually lead to collapse of the pipe roof and complete breaching of the embankment dam, when the conduit has enlarged to a point at which the shear stress on the pipe roof exceeds the defined undrained shear strength.

To conduct the analysis, results of conventional geotechnical testing were first used to develop a range of estimates for the soil erodibility that has fundamental importance on the simulation results. These were then used to perform a sensitivity analysis. A range of reservoir operating conditions, including various pool elevations, inflows, and spill releases, were evaluated in potential dam failure scenarios. The WinDAM C model provided prediction of the downstream hydrograph, from which summary metrics were calculated, including peak downstream flow rate and time of occurrence and downstream warning time. In addition, the final size of the breach in the embankment was estimated. Based on these results, the effects of various operating conditions were assessed.

ESTIMATION OF ERODIBILITY

The erodibility of the soil is an important parameter used in the WinDAM C model and affects the model results significantly. The soil erodibility, which is also referred to as the surface detachment coefficient, is used to compute the rate of erosion as the average shear stress induced by the flow exceeds a critical shear stress using the equation as follows:

$$E_r = k_d(\tau_e - \tau_c) \quad (1)$$

where: E_r =erosion rate (ft/hr);
k_d = soil erodibility (ft/hr/psf);
τ_e = hydraulically applied shear stress (psf); and
τ_c = critical shear stress (psf).

The embankment and foundation soils at the site of this study were mostly identified as CH (high-plasticity clay) or CL (lean clay), and occasionally MH (high-plasticity silt).

Ideally, the soil erodibility is obtained from in-situ jet erosion tests (Clark and Wynn 2007). When jet erosion test results are not available, there are empirical correlations for estimating the soil erodibility from typical geotechnical properties. For this evaluation, the equation provided by the *National Engineering Handbook* by USDA (2011) was used:

$$k_d = \frac{5.66\gamma_w}{\gamma_d} exp\left[-0.121{p_c}^{0.406}\left(\frac{\gamma_d}{\gamma_w}\right)^{3.1}\right] \quad (2)$$

where: k_d = soil erodibility (ft/hr/psf);
γ_d = dry unit weight;
γ_w = unit weight of water (in same units as dry unit weight), and
p_c = percent of clay (particles finer than 0.002 mm).

The critical shear stress (τ_c) is the flow-induced shear stress above which soil erosion will occur. Empirical correlations proposed by Smerdon and Beasley (1961) were used:

$$\tau_c = 0.16{I_w}^{0.84} \quad (3)$$

$$\tau_c = 0.493\times10^{0.0182p_c} \quad (4)$$

where: τ_c = critical shear stress in Pascal;
I_w = plasticity index (%);
D_{50} = mean particle size (m); and
p_c = percent of clay.

Table 1 shows a summary of relevant geotechnical parameters for embankment and foundation soils. The derived erodibility and critical shear stress using the above equations are shown in Table 2. In a preliminary sensitivity analysis using WinDAM C, the value of k_d in equation (1) was found to have the most significant effect on the modeling results, while the value of τ_c had only minimal effects (since, except for the initiating stages of failure, the hydraulically applied shear stress, τ_e, is much larger than τ_c). Thus a simple estimated value (τ_c = 0.09 psf) calculated as the average of the estimates from equations (3) and (4) was used for τ_c, while additional analyses were used to characterize k_d as described below.

WinDAM C uses a single value of k_d to represent the entire dam, and therefore the value of k_d used should be a realistic estimate of an average, rather than extreme high or low values that may exist in isolated samples. As such, a Monte Carlo bootstrapping method was developed to estimate a realistic range for the *average* erodibility parameter based on repeated combinations and re-sampling of over 100 geotechnical test data points, which inherently takes into account the non-linearity of equation (2). Three values were determined for k_d, based on the simulated distribution for the average k_d. A mid-range value of 0.57 ft/hr/psf (mean of the average distribution), a high-range value of 0.72 ft/hr/psf (mean plus three standard deviations of the average distribution), and a low-range value of 0.42 ft/hr/psf (mean minus three standard deviations of the average distribution) were selected and used for sensitivity analysis. Hanson et al. (2010) presented typical range of k_d for engineered fills based on field test results, and k_d was found to be dependent on the percentage of clay, compaction effort, and water content at compaction. The selected values of k_d in this analysis are in line with the reported values by Hanson et al. (2010) for standard to low compaction effort and dry of optimum water content at compaction, and are considered conservative.

Table 1. Summary of Geotechnical Parameters

Parameter	Range	Average
total unit weight, γ(pcf)	112.1 – 119.3	117.5
dry unit weight γ_d (pcf)	82.5 – 106.2	92.7
plasticity index (%)	15 – 59	36
percent clay (%)	32.4 – 71.2	48

Table 2. Summary of Erodibility Parameters

Parameter	Range	Average
erodibility (Equation 2), k_d (ft/hr/psf)	0.10 – 1.39	0.49
critical shear stress (Equation 3), τ_c (psf)	0.033 – 0.103	0.068
critical shear stress (Equation 4), τ_c (psf)	0.034 – 0.204	0.091

MODEL SET-UP AND SCENARIOS

The idealized dam cross section, as set-up within the constraints of the WinDAM C model, as well as the headwater and tailwater elevations, are shown in Figure 1. The analysis assumed a conduit formed at the base of the dam, which is the worst case in terms of maximizing hydraulic head through the conduit. The WinDAM C model was used to evaluate the time available to take action and the consequence of a potential dam breach. In the analysis, the initial conduit width and height were both assumed to be 0.1 ft at the beginning of the simulation. Preliminary sensitivity analysis showed this initial conduit size only affects the time until noticeable seepage occurs (i.e. 10 cfs, as estimated by the dam operators), but does not affect the erosion rate and downstream hydrographs after this time.

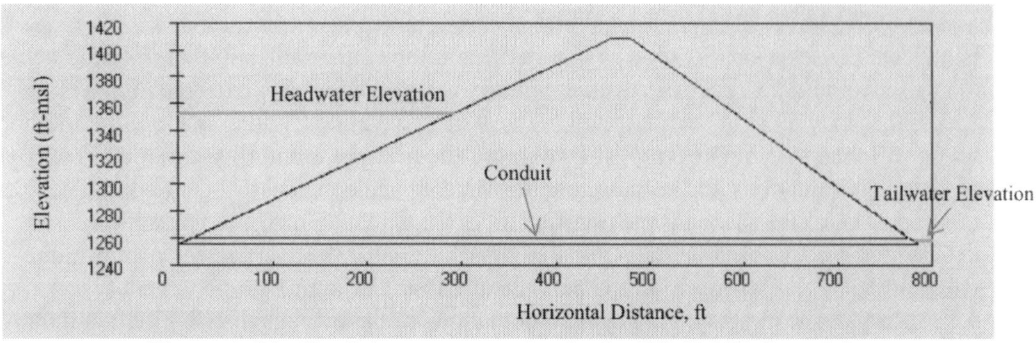

Figure 1. Idealized dam cross section in the WinDAM C model

The analyses considered various parameters and operation scenarios, summarized as:

- Low, mid, and high-range values of the soil erodibility.
- Reservoir inflows;
  - at a constant rate of 1,800 cfs (representing a normal operating condition),
  - with a flood hydrograph with peak flow of 28,700 cfs (largest storm since 1974),
- Reservoir at four different initial headwater elevations (to assess drawdown strategy);
  - 1340, 1352.5, 1362, and 1382 ft-msl
- Three different dam discharge operation strategies;
  - operating two of the turbines at maximum sustained load (MSL),
  - partially-opened spillway gate limiting maximum discharge to 5,000 or 20,000 cfs, with turbines operated at MSL, and
  - unlimited spillway discharge, with turbines operated at MSL.

RESULTS AND FINDINGS

The evaluation focused on the downstream effects, including downstream peak flow rates, downstream water elevations, and the timing to when peak or damaging flow occurs (i.e., warning time). The results and major findings are summarized below.

Sensitivity to Soil Erodibility k_d

For purposes of sensitivity analysis, three WinDAM C simulations were conducted where the k_d values were varied as follows: 0.42 (low-range), 0.57 (mid-range) and 0.72 (high-range) ft/hr/psf, while the headwater elevation was held constant at 1352.5 ft-msl.

Figure 2 plots the simulated downstream total flow rate for the three scenarios. Due to faster growth of the conduit, higher values of k_d resulted in higher peak flow rates and decreased time to reach peak flow. The effect of variability in k_d from the mid-range value is expected to result in changes in peak flow rate of up to ±15%, and changes in time to peak flow of up to ±2 hours.

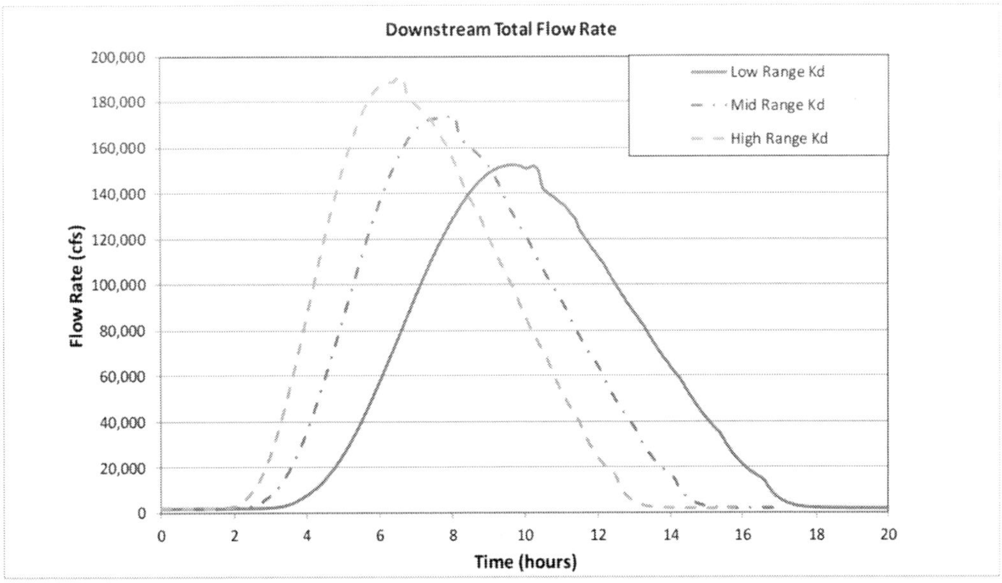

Figure 2. Downstream total flow rate vs. time for varying k_d
(inflow = 1,800 cfs, turbine = 1,800 cfs, headwater elevation = 1,352.5 ft-msl)

Effect of Headwater Elevations

Figure 3 shows the downstream total outflow hydrograph for four different initial headwater elevations and the mid-range value for k_d (i.e., 0.57 ft/hr/psf). The peak flow rate decreases three-fold from approximately 360,000 cfs to 120,000 cfs as the initial headwater elevation is lowered from 1,382 ft-msl to 1,340 ft-msl. Similar three-fold reductions in peak flow rate are obtained at other values of k_d, as summarized in Figure 4 that plots the peak downstream flow rate verses initial headwater elevation for the full range of k_d. These three-fold reductions due to lowering headwater elevations indicate that operating the reservoir at lower pool elevations is an effective strategy to mitigate downstream risks.

The warning time, defined as the time between when significant, noticeable internal erosion flow (assumed to be 10 cfs) is observed and the time when total downstream outflow reaches 100,000 cfs (a threshold chosen to represent significant downstream consequences), can be calculated from the downstream hydrographs. At the time of the study, the dam was inspected every 4 to 6 hours. The threshold for noticeable internal erosion flow (i.e., 10 cfs) was selected based on discussion with the dam owners and operators and hydrogeologic modeling conducted by other consultants. Based on the hydrographs presented in Figure 3, the warning time was estimated to range from 2.6

hours (when the initial headwater was 1,382 ft-msl) to 4.9 hours (when the initial headwater was lowered to 1,340 ft-msl). Thus, operating the reservoir at lower headwater elevation allows longer warning time for taking action.

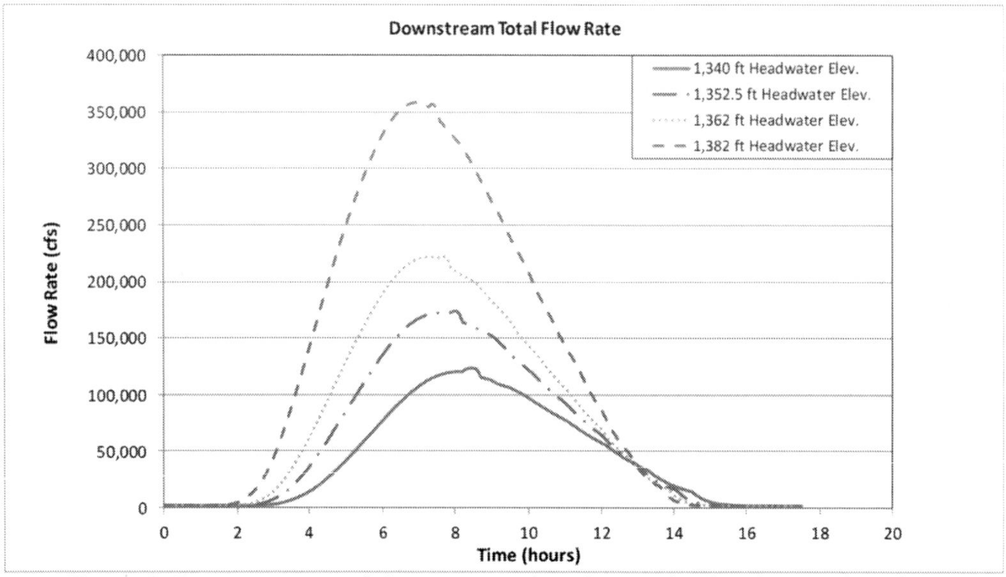

Figure 3. Downstream total flow rate vs. time for varying headwater elevations (inflow = 1,800 cfs, turbine = 1,800 cfs, k_d =0.57 ft/hr/psf)

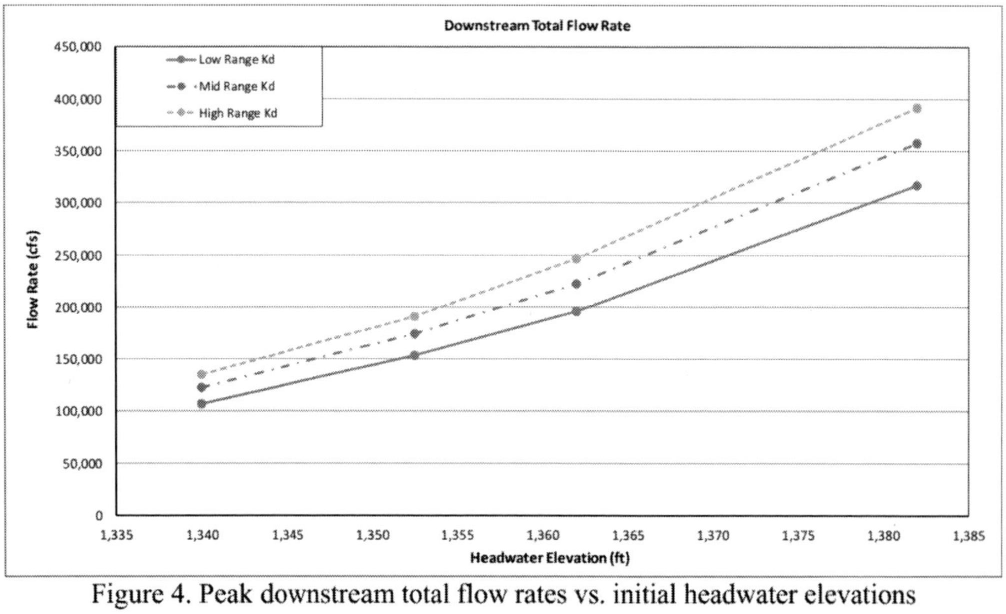

Figure 4. Peak downstream total flow rates vs. initial headwater elevations (inflow = 1,800 cfs, turbine = 1,800 cfs, k_d = 0.42, 0.57, and 0.72 ft/hr/psf)

Finally, the damage to the embankment dam can be evaluated by examining the final conduit dimensions as predicted by WinDAM C. Figures 5 and 6 show the final conduit height and width for various initial headwater elevations and k_d values. It is noted that WinDAM C is able to model full collapse of the conduit roof, but in these simulations the full collapse was not predicted to occur. It was found that the final conduit size can be reduced by approximately 30% by lowering the initial headwater elevation to 1,340 ft-msl, compared to keeping the initial headwater elevation at 1,382 ft-msl. This shows that maintaining a low headwater elevation also helps to minimize the damage to the embankment.

In summary, operating the reservoir at lower headwater elevation is an effective mitigation strategy that can significantly decrease peak downstream flow rates, increase warning time, and reduce ultimate damage to the embankment.

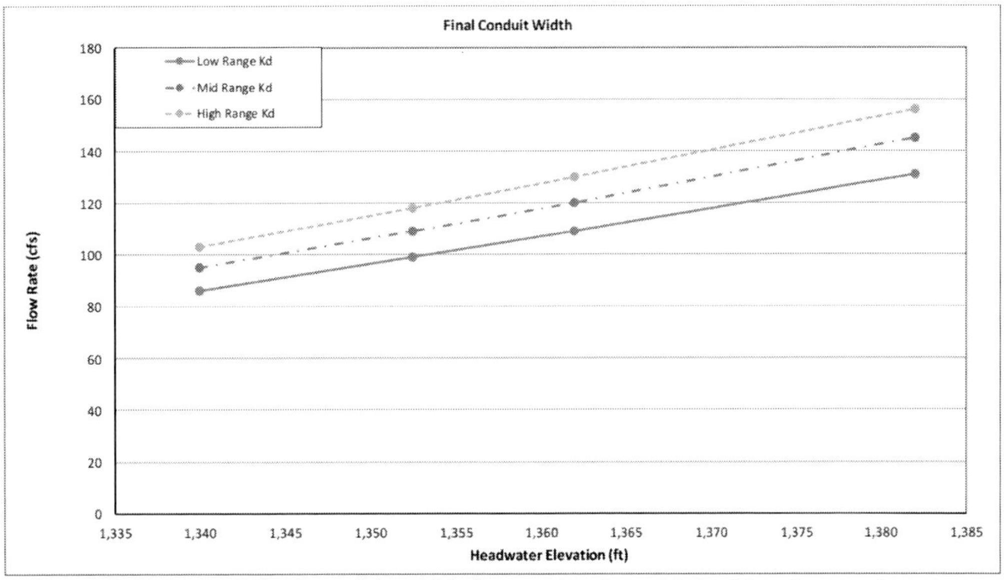

Figure 5. Estimated final conduit width for varying k_d and initial headwater elevations (inflow = 1,800 cfs, turbine = 1,800 cfs, k_d = 0.42, 0.57, and 0.72 ft/hr/psf)

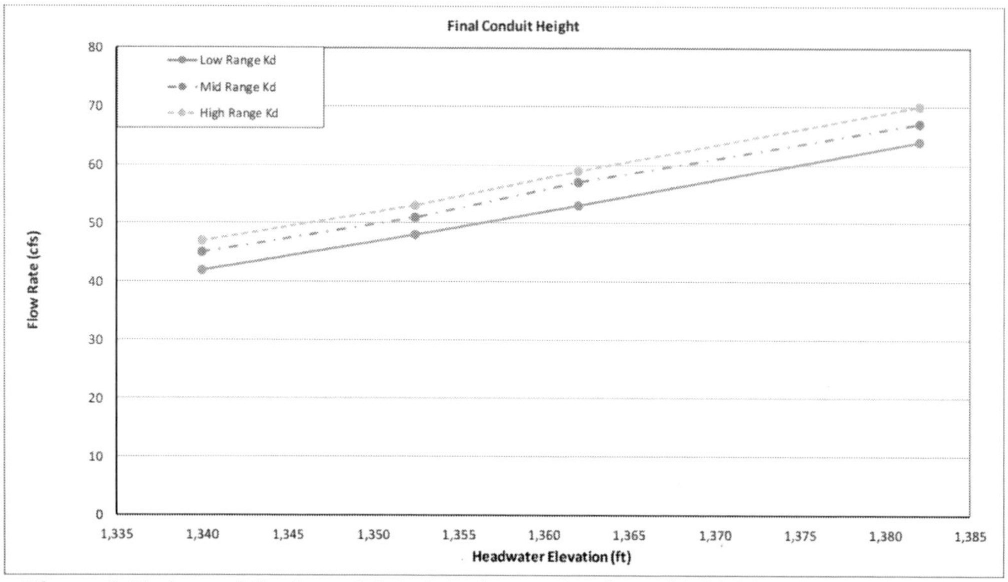

Figure 6. Estimated final conduit height for varying k_d and initial headwater elevations (inflow = 1,800 cfs, turbine = 1,800 cfs, k_d = 0.42, 0.57, and 0.72 ft/hr/psf)

Effect of Spillway Operations

Three different dam discharge operation strategies (including operating the turbine at maximum sustained load (MSL), operating at MSL in addition to partially-opening the spillway gate while limiting spillway discharge to 5,000 or 20,000 cfs, and operating at MSL with unlimited spillway discharge) were evaluated. The modeling results shown in Figures 7 and 8 (high and low-range k_d, respectively) indicate that when the initial headwater elevation is at 1,382 ft-msl, opening the spillway up to a capacity of 5,000 or 20,000 cfs does not substantially alter the magnitude and timing of the peak total downstream flow. Fully opening the spillway resulted in approximately a 20% reduction in peak flow rate, but decreased the warning time to zero, since the downstream flow exceeds 100,000 cfs as soon as the spillway is fully opened. These results indicate that while the decision to fully open the spillway could be made during an emergency condition, relying on this strategy is not as effective as operating the reservoir at lower headwater elevation.

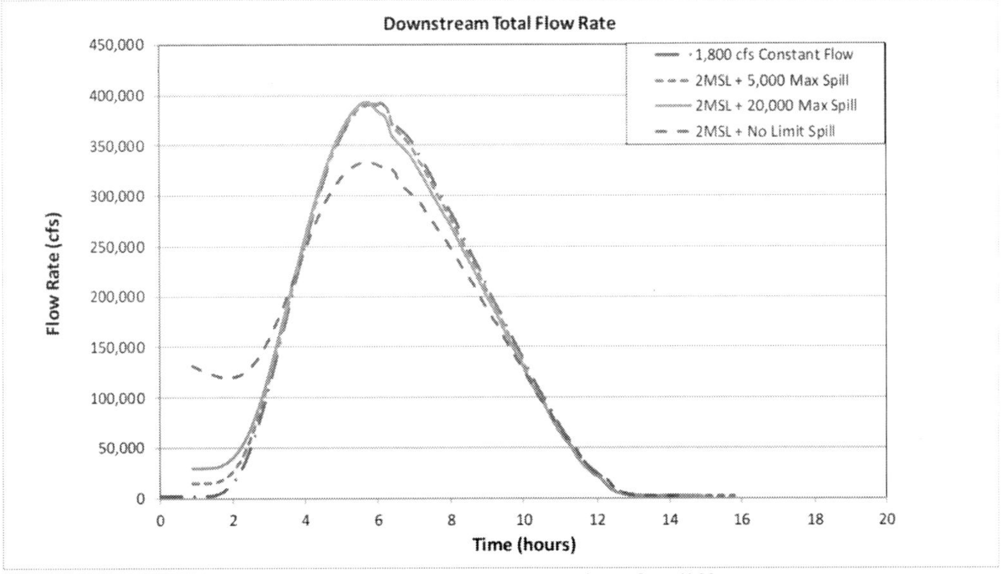

Figure 7. Downstream total flow rate vs. time for different responses (inflow = 1,800 cfs, k_d = 0.72 ft/hr/psf, headwater elevation = 1,382 ft-msl)

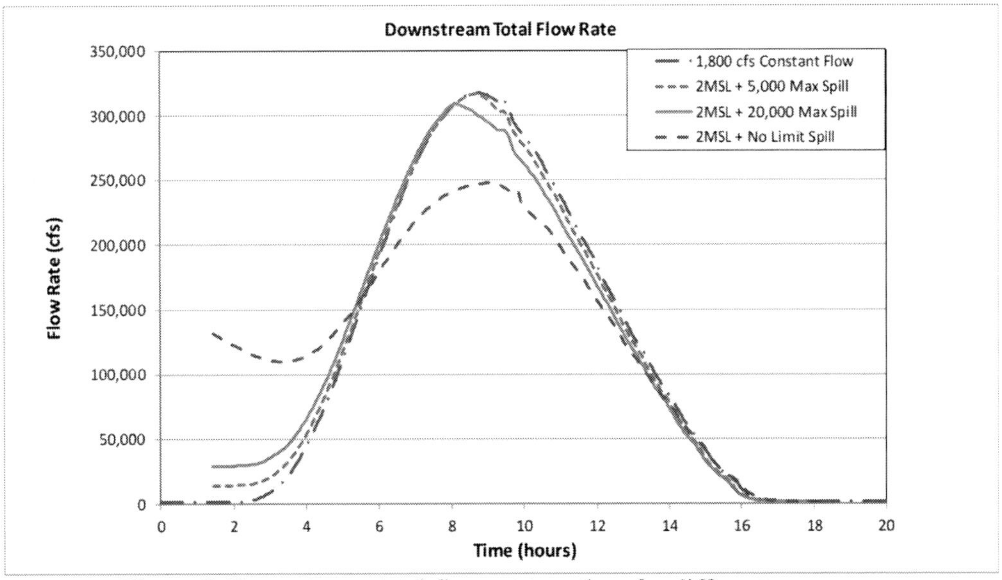

Figure 8. Downstream total flow rate vs. time for different responses (inflow = 1,800 cfs, k_d = 0.42 ft/hr/psf, headwater elevation = 1,382 ft-msl)

Effect of Storm Inflow

WinDAM C was also used to evaluate the effect of a large upstream storm inflow by dynamically applying an inflow hydrograph (peak inflow of 28,700 cfs) while the initial headwater elevation was at 1352.5 ft-msl. The effect of the inflow was to raise the headwater elevation to approximately 1360 ft-msl, after which failure proceeded and resulted in similar peak downstream flow rates as the static simulation with initial headwater elevation of 1362 ft-msl. That is, the inflow hydrograph did not have any direct dynamic effect on the dam failure and results, other than to raise the initial headwater elevation. These results may not hold for higher initial headwater elevations and larger storm inflows, particularly if the storm leads to dam overtopping.

SUMMARY

The WinDAM C model provides a useful tool for assessing the downstream effects of a potential dam breach due to internal erosion. The results of the WinDAM C modeling for this dam in South East Appalachia indicate that the most effective strategy for reducing the risk associated with internal erosion is to operate at low pool level. The modeling indicated that increasing the controlled outflow through the turbine or spillway once the internal erosion reaches a noticeable level (assumed as when the internal erosion flow reaches 10 cfs) was not as effective as operating the reservoir at lower headwater elevation. The results of the modeling allowed the dam owners and operators to assess the risk and evaluate temporary interim mitigation strategies (e.g., lowering of the headwater elevation), while they evaluated and planned for permanent remedial actions (i.e., structural measures, such as grouting, installing barrier wall or filters and external rip-rap).

REFERENCES

Clark, L. A., Wynn, T. M. (2007). Methods for Determining Streambank Critical Shear Stress and Erodibility: Implications for Erosion Rate Predictions. *Transactions of the ASABE*. Vol. 50(1):95-106.

Hanson, G. J., Tejral, R. D., Hunt, S. L., & Temple, D. M. (2010). Internal erosion and impact of erosion resistance. In *Proceedings of the 30th US Society on dams annual meeting and conference* (pp. 773-784).

Smerdon, E. T., and R. P. Beasley. (1961). Critical tractive forces in cohesive soils. Agric. Eng. 42(1): 26-29.

USDA (2011) National Engineering Handbook, Appendix 52 D: Erodibility Parameter Selection for Soil Material Horizons (Surface Detachment Coefficient and Headcut Erodibility Index), 210-VI-NEH, draft, October 2011, U. S. Department of Agriculture.

Visser, K., R. D. Tejral, and M. L. Neilson (2015). WinDAM C Earthen Embankment Internal Erosion Analysis Software. In *Proceedings of the 3rd Joint Federal Interagency Conference on Sedimentation and Hydrologic Modeling*, April 19-23, 2015, Reno, Nevada, USA (pp. 1831-1841). https://acwi.gov/sos/pubs/3rdJFIC/index.html (accessed 2/1/2017).

TURBID DISCHARGE INCIDENT AT CANNONSVILLE DAM

John H. Vickers, P.E. [1]
William H. Hover, P.E. [2]

ABSTRACT

On July 8, 2015, the New York City Dept. of Environmental Protection (DEP) discovered a turbid discharge at the toe of Cannonsville Dam, a 2,800 ft. long, 175 ft. max. height, earthen dam constructed from 1955 – 1965. The Dam impounds the W. Branch Delaware River, has a storage capacity of 96.7 billion gallons and is an important component of the NYC water supply system which supplies 1.1 billion gallons daily to 9 million people (Figures 1 and 2). Reservoir was at full pool level.

The turbid discharge developed during test borings for foundation design of a proposed hydro- electric plant at the toe of the dam under a license issued by the Federal Energy Regulatory Commission (FERC). Turbid discharge resulted from open bore holes penetrating a confining geologic layer into an artesian formation in the dam foundation. DEP notified FERC and NY State Dam Safety officials and responded immediately to the potential dam safety incident. Actions included cessation of drilling at the dam, around the clock monitoring on site including parameters such as potential settlement, contingency planning in case the situation worsened, reservoir drawdown from full pool, public notification and emergency contracting for engineering consultant and construction contractor support. DEP assembled a team of engineering consultants and additional recognized experts and received active assistance from FERC. The team used a collaborative process to develop, decide upon and implement a rapid mitigation strategy consisting of pressure relief wells and a repair strategy to permanently stop the turbid discharge. Seven pressure relief wells with deep well pumps were installed by the end of July to eliminate or neutralize upward flow from the artesian zone. The turbid discharge stopped by August 1, 2015, at which time DEP ceased active reservoir drawdown (up to 15 feet). Permanent closure of the open boreholes were successfully completed using compaction grouting methods, and the relief well pumps were turned off early in the morning of August 28, 2015. As of May 16, 2016 the reservoir returned to normal pool and the repairs continue to hold with no further turbid discharge.

This paper reviews the dam owner's response; monitoring and early warning mechanisms; efforts to warn and inform the public; the collaborative approach to analyze the situation and design mitigation and repairs; repair and test methods; the monitoring plan; and lessons learned. The following questions are explored: How can dam owners assure that prudent research and contractor preparations are taken before subsurface exploration on or around dams? What actions can dam owners take before an incident

[1] Chief, Water Operations Division, NYC Environmental Protection, Bureau of Water Supply, Source Water Operations, Grahamsville, NY 12740 jvickers@dep.nyc.gov
2 Sr. Principal, GZA GeoEnvironmental, Inc., 249 Vanderbilt Ave., Norwood, MA 02062 whover@gza.com

occurs that will assist them to respond and recover from an incident should one occur? What actions can dam owners take during an incident response and recovery that will help them achieve an acceptable solution?

BACKGROUND

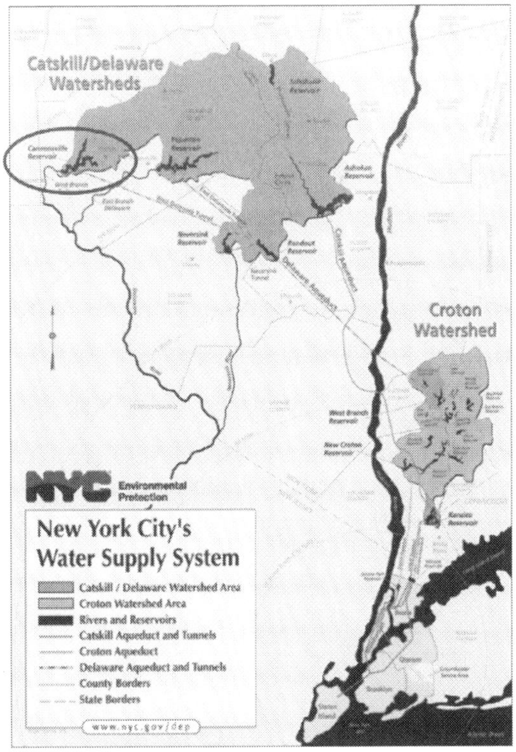

Figure 1 – New York City Water Supply System

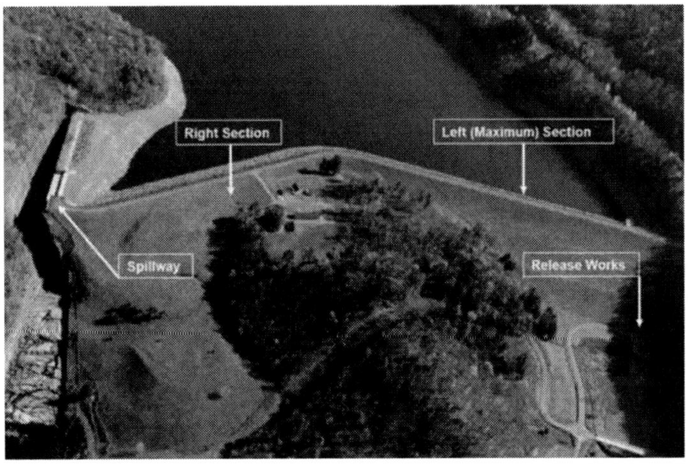

Figure 2 – Aerial View of Cannonsville Dam – Focus Area At Lower Right

Figure 3 shows stone rip-rap armoring the upstream slope with stone fill below topsoil/turf on the downstream slopes. Benches with catch basins pass surface drainage through the near-surface rock fill. The rock fill toe passes surface drainage and dam seepage to the tailwater..

.

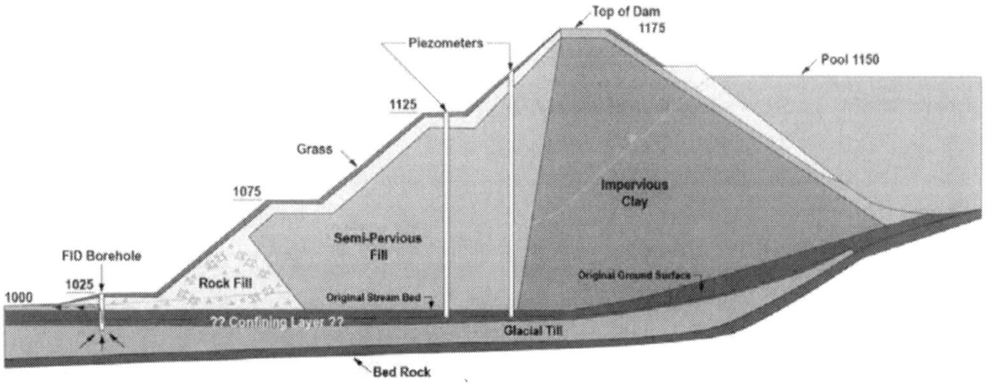

Figure 3. Inferred Dam Section During the Incident

DEP filed for and FERC granted to NYC a license on May 13, 2014 to construct a 14.1 MW hydroelectric plant, with powerhouse located next to the release chamber (Figure 4). A hydropower consultant developed a plan to collect subsurface information and install piezometers to estimate seepage flows and measure pore water pressures in the powerhouse foundation (Figure 5). Eight borings were proposed and three were partially completed.

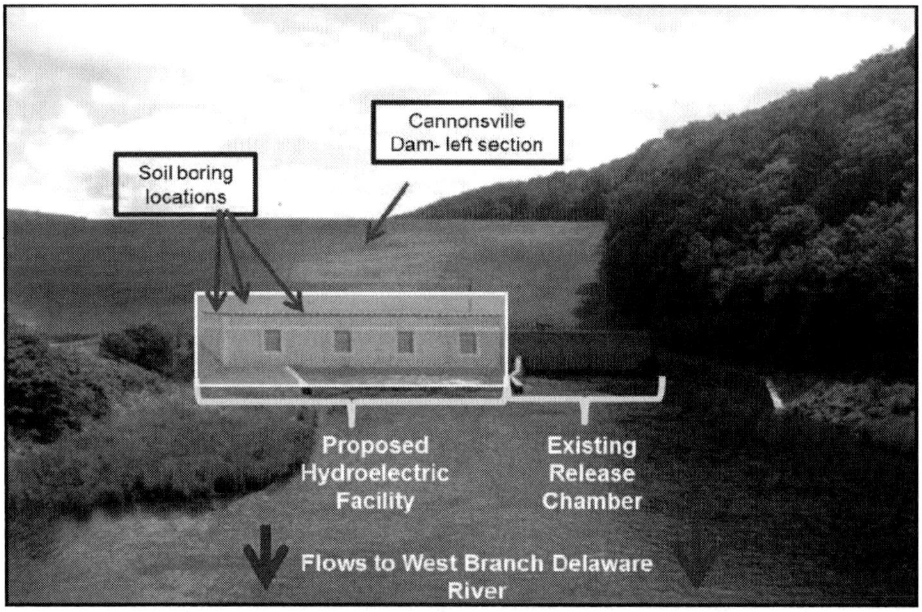

Figure 4. Conceptual Rendering of Proposed Powerhouse

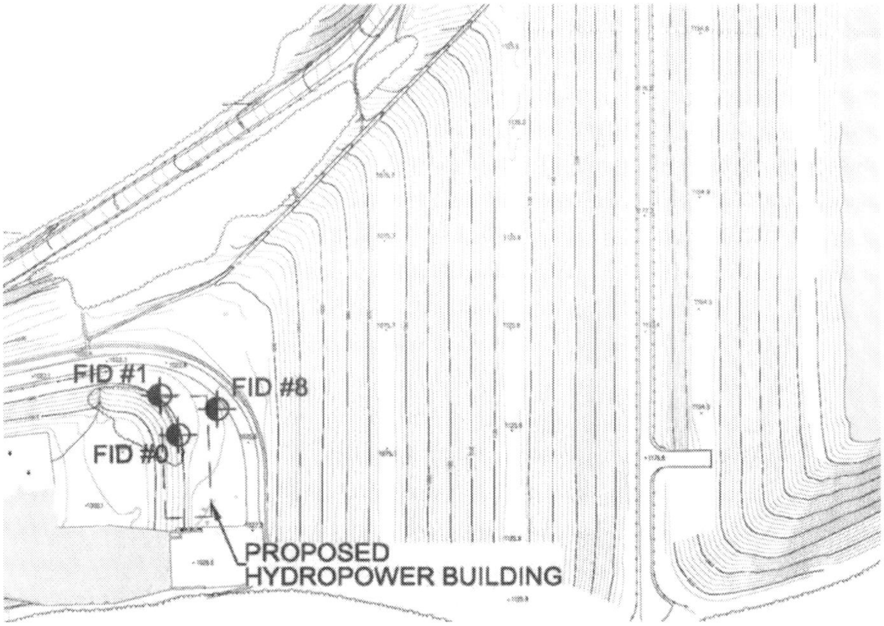

Figure 5. Locations of Three Borings for Proposed Hydroelectric Plant

INCIDENT

On July 8, 2015, turbid discharge was emanating from the toe into tailwater during the 3rd of 8 boreholes (FID-0) (Figures 5 and 6). FID-8 began on July 1, resuming on July 5 and was drilled through the rock toe via hollow stem augers (HSA) to 52 feet and uncased to about 62 feet by mud rotary methods. The driller had difficulty backfilling FID-8 on July 6 by pouring 9 bags (450 lbs) of bentonite chips, 1 cy of 1 in. stone, 3 cy of sand and 2 bags of bentonite chips into the borehole from ground surface after auger removal. FID-1 was drilled on July 6 and 7 with HSA to 29 feet and uncased to 61 feet by mud rotary methods. FID-1 was backfilled using bentonite mud and cuttings. After HSA removal the hole collapsed to 6 feet below grade and was backfilled with bentonite chips. FID-0 was drilled on July 7 and 8 in the same manner as FID-1. The borehole was filled with 60 gallons of cement grout bentonite grout prior to HSA removal, followed by borehole collapse to 6 feet and backfill with bentonite chips.The belief on July 8 was that drill cuttings, "recirculated mud" and backfill from previous holes were mixing with drainage and seepage causing turbid discharge. DEP stopped the borings to evaluate. After turbidity continued for 2 more days, with significant drops in deep piezometer levels beneath the dam, DEP notified FERC and NY State DEC Dam Safety and mobilized staff to the normally unoccupied site for 24-7 monitoring. DEP notified state, county and town officials and emergency managers. DEP contacted consultants who performed a comprehensive dam safety assessment in the past, to respond. In consult with FERC, DEP initiated reservoir drawdown on July 13 at a rate of 1,000 MGD (max. sustainable release); and increased max.supply diversion to 450 MGD.

RESPONSE

DEP took quick steps to mitigate risk and enable successful resolution. DEP mobilized staff, equipment and materials and, using the Emergency Action Plan as a guide, notified emergency managers, elected officials, regulating agencies and other stakeholders. DEP stockpiled sand, stone, rip-rap, geotextiles and sandbags; and brought in staff, tents, potable water (no amenities available at this remote location), lighting, communications trailers and other means to provide 24-7 on site monitoring. DEP surveyors established control points to ensure the dam profile had not changed and erected staff gages in tailwater and on the toe area of turbid discharge to monitor changes in discharge water surface and tailwater elevations with automated pressure transducer (Figure 7). DEP began daily inspections

Figure 6. Turbid Discharge at Downstream Toe – July 13, 2015

Figure 7. Staff Gages (Tailwater and Turbid Discharge)

and photo documentation. The Automated Data Acquisition System reading interval of the piezometer array installed in 2000 was changed from daily to hourly, with twice daily analyses. DEP established deviation alarms and SOPs to notify engineers of rapid change of any piezometer. DEP built sediment collection boxes to examine grain size distribution of sediment (Figure 8). DEP installed a constant monitoring turbidimeter and provided data to the Bureau of Water Supply (BWS) control room. BWS requested communications assets from NY State as there is no cellular service at the dam. Verizon and AT&T established satellite voice and data communications until local telephone company upgraded to allow data and voice over internet phones. DEP erected generator powered lights and closed circuit television with camera feed to BWS and deployed satellite telephones. The division chief assimilated daily inspection and meter reading data and issued written situation reports to the project team to document site conditions and any changes.

Figure 8. Sediment Collection Box (36 in. x 12 in.)

Following is a timeline of selected events from the incident.

July 8, 2015	Turbid discharge observed
July 10, "	DEP began 24-7 monitoring on - site
July 13, "	Initiated reservoir drawdown; notified public officials
July 18, "	Selected specialty contractor; First public notification meeting
July 20, "	Submitted relief well design; DEP began 24/7 remote CCTV monitoring. NYSOEM communication trailer arrived - satellite phone, internet service
July 22, "	Verizon communication trailer - 6 phones, internet, improved cell service
July 24 "	Begin drilling Relief Well No. 4; Board of Consultants meeting no. 1
July 28 '	Second larger rig begins drilling Relief Well No. 1
August 2, 2015	Confirmed visually that Relief well pumping eliminated turbidity
August 4-5 "	Board of Consultants meeting no. 2
August 25, "	Compaction grouting of FID Boreholes completed
August 28, "	Termination of relief well pumping

PUBLIC OUTREACH

The dam is key to the safety of 80,000 people in the dam failure inundation area. DEP initiated communications to timely disseminate information to the public. DEP included daily opportunities to ask questions and receive answers about public safety, emergency prepared- ness, the state of the dam and progress of repairs, preventing rumors and building credibility with downstream neighbors. DEP hosted 12 public meetings, in communities spanning 130 miles of the Delaware River. DEP compiled an email list of officials, business owners, and residents along the Delaware. Updates were sent daily during the height of the incident. Updates and weekly photographs were posted to DEP's Facebook page. To increase the audience for Facebook posts. DEP "tagged" key media outlets, nonprofits and other groups. Updates were designated for community news. Facebook post sharing allowed updates to be seen by tens of thousands of people and provided DEP an opportunity to answer questions. Traditional media shared information about the Dam and repair work. DEP compiled a list of newspapers, radio and TV stations and news websites that covered riverside communities in NY, PA and NJ. Many outlets and towns did not usually receive information from DEP. DEP asked the Associated Press to ensure its story was posted on the wire for all three states. DEP created a central page on its water supply website to post updates, including information on downstream releases, reservoir storage, inundation maps for downstream areas, and other useful information. A cell phone number of DEP's information officer was shared in all updates and at all public meetings. Site tours were provided for elected leaders and emergency response officials, bolstering confidence in the information DEP provided.

ANALYZING THE PROBLEM

While DEP mobilized, agency senior leaders used emergency procurement means to hire William Hover, P.E. of GZA GeoEnvironmental Inc., the consultant who had performed a thorough engineering assessment of the dam and was very familiar with it, to assist with the problem. DEP hired 3 additional expert consultants (Joseph Ehasz of AECOM, Frederick Rhyner of Mueser Rutledge and Stephen Whiteside of CDM/ Smith) with the advice and approval of FERC, as the independent Board of Consultants (BOC) to help analyze the situation and provide feedback on DEP plans to mitigate and resolve.

DEP engineers reviewed the piezometer data and discovered that deep piezometers in the dam foundation had 13 to 14 foot drops in head coincident with drilling of boreholes FID#1 on July 7 and FID#0 the next day and then remained relatively steady thereafter (Figure 9). An artesian well located about 200 yards downstream that has historically discharged freely through a 1 inch orifice and approximately 30 inches upward, had dropped to just a few inches above the orifice (Figure 10). The artesian well and the new boreholes were hydraulically connected in some manner.

DEP, its consultants, FERC and DEC performed an abbreviated Potential Failure Mode Analysis on site on July 14, 2015. Of concern was the potential for a piping condition to form and if not halted, leading to dam failure. Larger soil particles could potentially be transported and deposited into the rock fill toe without detection. The rock fill toe, which covers a large area and which propagates into the riverbed, made construction of a measuring weir and filter blanket or other surficial mitigating measures impractical. DEP engineers believed that the 3 boreholes pierced a confining layer under the rock toe into an artesian zone, causing water to flow upward and erode soil from within the boring, creating turbid discharge through the rock toe. (Figure 11 - Sketch by Jeffrey Helmuth, P.E.). The consultant agreed that this was the most plausible theory, while BOC and FERC agreed that this was a plausible theory. To help confirm this theory, DEP collected samples from the borings and the turbid discharge and sent them to a laboratory, the RJ Lee Group, for analysis. RJ Lee's microscopy studies suggested that the material in the turbid discharge was coming from the confining layer of silts and clays located below the rock fill toe and above the artesian zone. Neither bentonite nor Portland cement was observed, confirming that the boring backfill was not the source of turbidity. DEP searched its archives to learn more about dam and foundation construction. Historical photographs helped the project team to understand where the relic riverbed was located; the dense nature of glacial till foundation materials; that higher rates of seepage might occur through seams in the bedrock along either side of the old valley but that overall seepage volumes would be small; and that the downstream artesian well had been used during construction. It was learned that Dr. Arthur Casagrande was a project consultant.

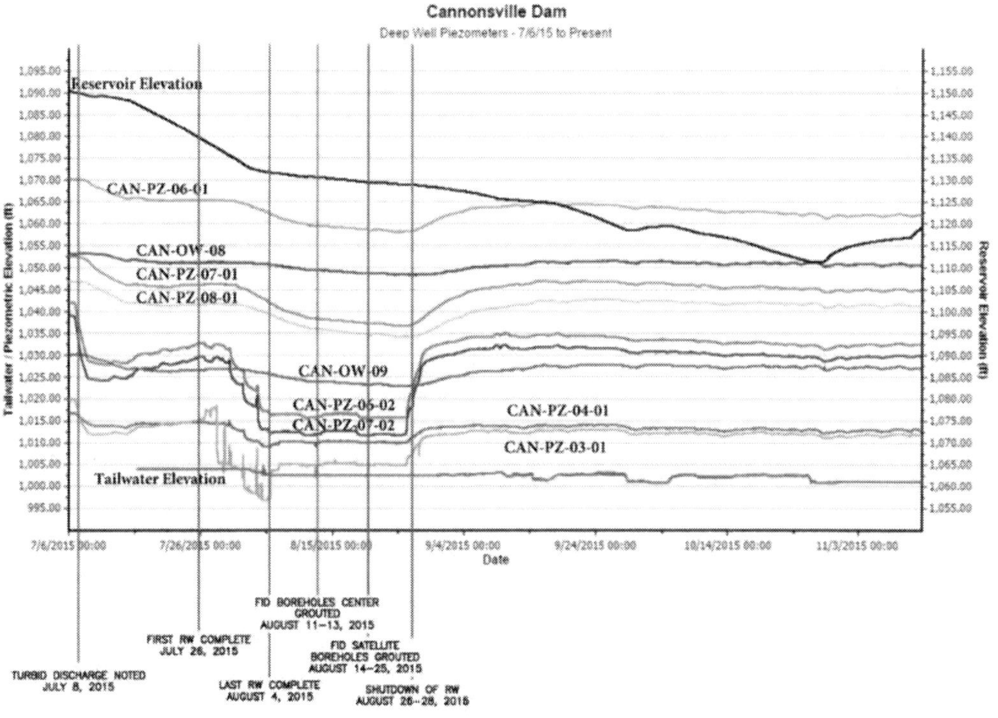

Figure 9. Decreased Piezometer Levels Concurrent With FID Borings

Figure 10. Artesian Well Before (left) and After (Right) New Borings

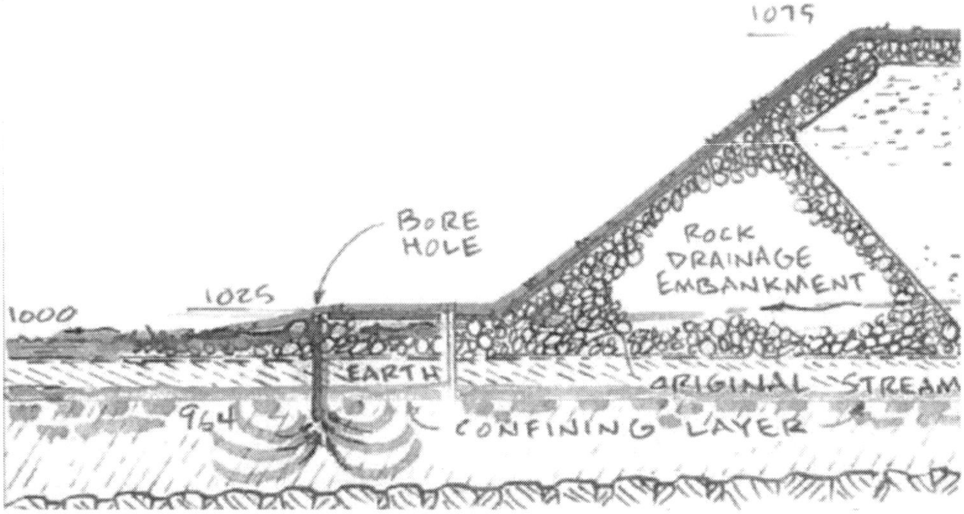

Figure 11 – Sketch by Jeffrey Helmuth, P.E. Illustrating Mechanism of Artesian Flow

MITIGATION

The project team brainstormed alternatives and all agreed that installation of pressure relief wells would be followed by plugging the boreholes. The relief wells would depressurize the aquifer to halt the turbid discharge and permit the boreholes to be plugged. DEP also tapped the knowledge of four reputable specialty geotechnical contractors. On July 17, 2015 DEP and project team members met with 4 contractors individually and listened to their approach to relief well installation and borehole plugging. Potential remediation measures were discussed to provide a basis for rational selection of a highly qualified contractor with the resources to respond rapidly. Potential solutions discussed included large diameter over-boring the original boreholes, high mobility pressure grouting and low mobility compaction grouting. Three of the four contractors independently felt compaction grouting to be the best solution to plugging the boreholes. DEP entered into an emergency contract with Moretrench to install relief wells and plug boreholes.

Using the experience of the GZA consultant, DEP archives, contractor's recommendations, and a collaborative effort with state and federal regulators and Board of Consultants, the project team settled on 10 temporary relief wells positioned in an arc up-gradient from the boreholes (See Figure 12). The relief wells were designed in a drainage curtain configuration upstream of the proposed powerhouse and along the right abutment at 40 ft. spacing for the 1st stage and 20 ft. spacing for a 2nd stage. Pumps would increase pressure relief, and the wells could be maintained for future use if appropriate. Relief well holes were drilled through permanent steel casings to control artesian pressure. The contractor recommended sonic drilling methods to rapidly advance the larger diameter casing through the rock toe with little or no water or air flushing which could exacerbate the condition The wells were constructed with slotted PVC screen and solid PVC riser. The slotted section was backfilled with filter sand on top of which a bentonite plug was installed. The annular space above the screened zone was backfilled with cement grout. Well screens were set below the silt-clay confining

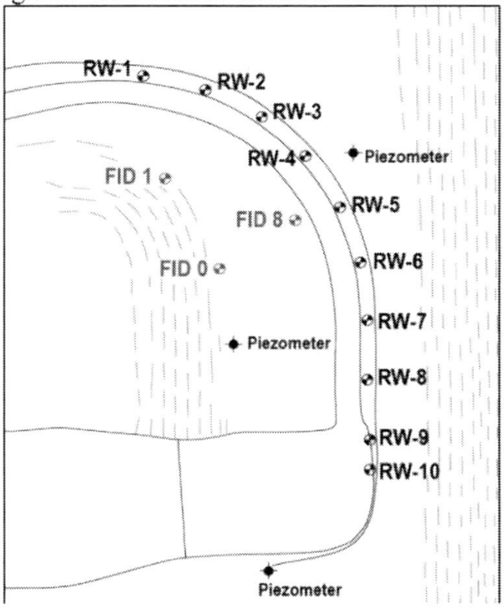

Figure 12 – Relief Well Location Plan

layer to relieve the artesian stratum (Figure 13). Each relief well accommodated an electric submersible pump with a PVC discharge pipe fitted with a sampling port and flow meter and then connected to a collection header discharging into a sedimentation tank and then into an existing catch basin. Ready availability of sonic drill rigs and permanent casing for borehole installation limited screened diameters. Two different sizes of relief wells, with 3-inch and 4-inch diameter finished well screens, were installed. Ten (10) inch sonic drill casing allowed a permanent 8-inch steel casing to be grouted into the confining soil strata below the rockfill before advancing the hole into soils under artesian head. A careful evaluation was essential to provide a casing with adequate seal into the confining layer, before encountering the artesian stratum. During initial drilling, the extent and depth of the artesian zone was unknown.

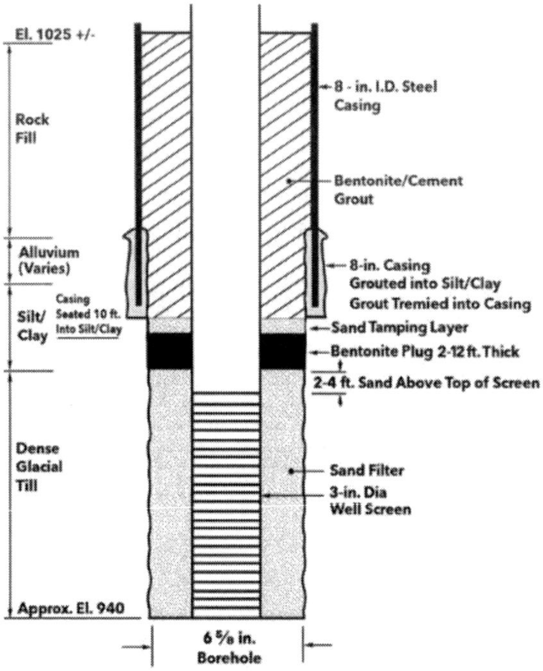

Figure 13. Relief Well Detail

Sonic drilling allowed continuous soil sampling and custom design of screened zones for each well. RW-1 was screened in clean sands and gravels associated with an ancient buried river bed under artesian pressure, encountered beneath the dam crest and upper downstream berm, deep within the glacial till foundation during 2001 explorations. The old artesian water supply well located downstream of the dam, RW-2 and RW-3 were screened in the same stratum.RW-4 through RW-7 did not encounter the sand-gravel stratum and were screened in the till or till – like material just above bedrock.The first well drilled was RW-4; completed on July 28, 2015. This well was selected based on the proximity of a nearby piezometer and availability of the first sonic drilling rig which was a smaller machine. RW-1, closest to FID-01 (borehole showing 1st piezometer reaction), was designed to intercept the ancient riverbed. RW-1, 2 and 3 awaited the larger sonic rig to allow larger well screens and pumps. RW-7 was the 2nd well established before arrival of the larger drill rig. RW-1 showed 14 gpm artesian flow, RW-7 did not. When RW-4 started pumping at 4 gpm, turbid discharge slightly declined. RW-7 initially yielded 12 gpm but quickly ran dry. On August 1, RW-1, 2 and 3 were pumping 50, 25 and 30 gpm. The toe was depressurized and turbidity was no longer observed (Figures 14, 15). The relief wells were declared a success the next morning. Pumping only from the deep artesian stratum using RW-1, 2 and 3 revealed that it was the source of water causing turbid discharge. Further confirmation of the source of the erosive flow was provided when pumping was briefly stopped for electrical re-wiring and turbid seepage re-emerged. The construction photo (Figure 16) indicates the relic river channel corresponding to the stratified sands, silts, gravels and clays above of the silt-clay confining layer and underlying pervious artesian zone, and the location of the higher yielding wells and the piezometers that were highly responsive to pumping.

Figure 14. Staff Gage and Sediment Box Flow Rate, Turbidity Diminished July 31, 2015 – RW-1 at 50 gpm

Figure 15. Tailwater – No Turbid Discharge August 2, 2015

As a result of the successful relief well pumping, authorization to resume normal reservoir operations was granted by FERC. DEP stopped deliberate reservoir drawdown (15 feet in 18 days) by returning both the release and water supply diversion to normal seasonal rates of flow. The reservoir continued to decline more gradually due to normal reservoir operation.

Figure 16. Dam Construction - Note Former Shallow River Channel

REMEDIATION

The FID borehole remediation plan was developed concurrent with relief well installation. The team selected compaction grouting to restore pre-breach hydraulic conditions by sealing the boreholes through the confining layer, replacing eroded ground washed out by artesian flow. The precise location and verticality of the FID boreholes was unknown. The team decided that the grouting should modify the ground within a target area rather than at the precise borehole location. Determination of the outcome of the grouting program would require indicators such as piezometric levels and tailwater turbidity.

Low mobility grouting (LMG) was selected for borehole closure. LMG employs low slump mortar-like grout injected under pressure through an open-ended steel casing to displace and densify the surrounding soil. The grout forms a bulbous mass. Strategically placed grout bulb "columns" within the target zone displace and densify the surrounding soil, as well as fills larger voids. The technique inherently results in ground displacement and movements.

The Phase 1 LMG grouting plan included an attempt to re-trace each exploratory hole and also three holes surrounding each FID borehole location on a triangular pattern (Figure 17). The re-traced holes were to clean out and backfill the FID boreholes and the surrounding three holes were to squeeze the FID boreholes closed if they could not be found and re-traced. The grouting plan included stand by time for dissipation of locally elevated pore-pressures. The LMG casing was advanced by internal flush duplex drilling to the target depth. The casing was then retracted in 2-ft (0.6-m) lifts as the grout was injected. For each lift, injection continued until pressure refusal, a pre-determined target grout volume was reached, or surface heave was observed. The team agreed that borehole re-tracing, cleaning out and grouting would be the most reliable closure expected. Special measures were taken to give this approach the greatest success. Relatively flexible, specially selected 2.5 inch drill casing with a lost point tri-cone 3.5-inch diameter bit was advanced first with positive water flush so as to follow the existing FID borehole as closely as possible. The approach worked very well at each FID borehole location. Minimal resistance to the advancement of the drill string and an abrupt increase in density at termination depth of each FID holes was a conclusive indicator that the boreholes were re-traced to full depth.

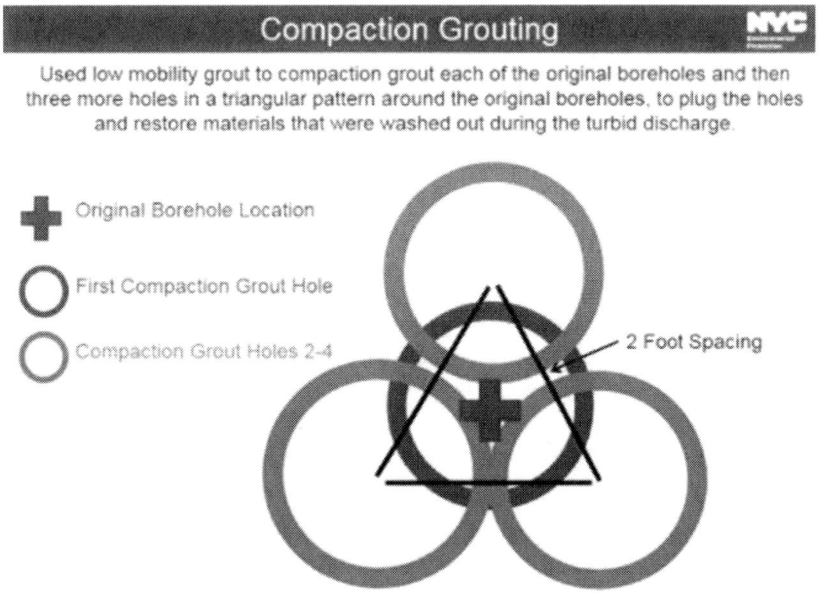

Figure 17. LMG Borehole Closure Pattern

The lost-point bit was dropped at the bottom of each of the 3 boreholes and the hole flushed with water and pressure grouted with a stiff mortar-like material. The high pressure grouting began 3 -5 feet below the bottom of FID boreholes. This assisted in densifying soils which may have been lifted or loosened as a result of upflow. Higher grout pressures, up to 500 psi, were utilized beneath the confining stratum of hard silt-clay and underlying glacial till to replace eroded ground and compact loosened soils. Lower pressure grouting, at 50 to 300 psi, was performed with an LMG grout of greater slump [4 to 6 inches] in the "erodible zone" of soils comprised of the stratified sands, silts, gravels, and clays above the silt-clay confining layer and below the rockfill and the foundation interface at the bottom of the rockfill. Pressures were lowered to reduce the potential for damaging the erodible and confining layers due to the high forces that accompany LMG grouting.

Potential hydro-fracturing of the confining layer required careful selection of the grout mix and the grouting pressure. The grout mix was relatively dry to keep the grout from acting too "liquid." The grouting pressure needed to be high enough to close the borehole, but not cause hydraulic fracturing, nor lift the strata grouted, potentially causing new groundwater flow paths. A value of 500 psi was several times lower than the undrained shear strength of the silt-clay confining layer, and lower than the shear strength of the underlying soils and was selected for trial use based on these considerations and engineering judgment. Based on successful trial use, including vertical surveys which demonstrated that lifting was not occurring, the 500 psi value was adopted.

Following grouting of the 3 FID boreholes, grouting was performed on a triangular pattern around each FID borehole, based on the volumes of grout injected and concern about radial voids or loosened zones beneath the confining layer. Figure 18 shows the grout volumes injected vs depth at each FID borehole. The highest grout volume was observed at FID-1, which was suspected by the project team to be the most problematic based on its hydraulic connection to the artesian layer (Figure 18).

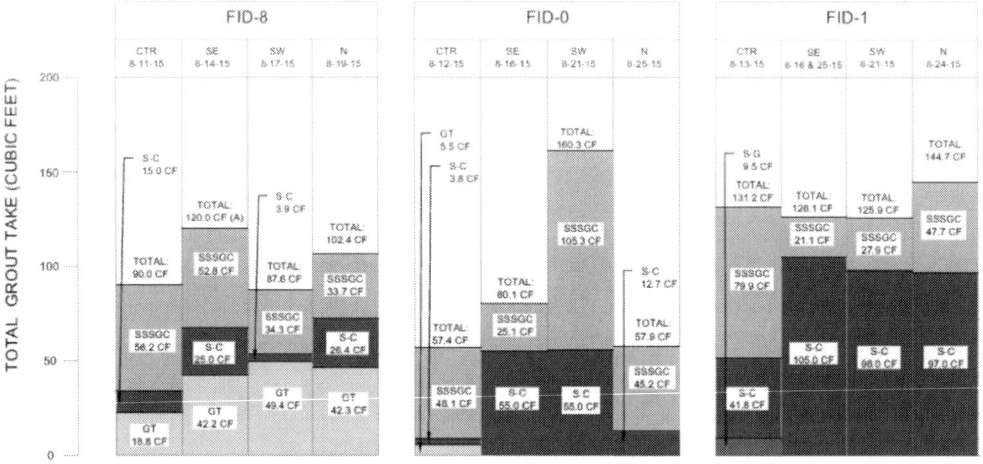

SSSGC = Stratified Sand, Silt, Gravel, and Clay; S-C = Silt-Clay;
GT = Glacial Till; S-G = Sand and Gravel

Figure 18. Grout Volumes Vs. Depth at Each FID Borehole Repair Location

Observations during drilling of the FID boreholes provided confidence that the boreholes had been retraced. Observations during grouting indicated that the ground had been tightened. After 24 hours grout cure time, the relief wells were deactivated one at a time. Immediate piezometer response (Figure 9) suggested that pre-breach conditions had been restored. On August 30, 2015, after all relief wells were turned off, the project team determined that the grouting had been successful, but long-term verification of recharge of the piezometer levels as the pool returned to historic levels would be more conclusive.

RECOVERY

DEP developed a phased monitoring program to check the success of the borehole remediation. The first phase continued the same level of 24/7 on site monitoring and lasted two weeks after the borehole repairs were completed, ending on September 11, 2015, at which time DEP partially demobilized from the site. The second phase lasted approximately 10 weeks and included daily inspections, remote visual monitoring, turbidimeter and enhanced piezometer reading frequency and analysis (which had been changed to 5 minute intervals to support relief well installation and LMG processes). DEP continued weekly Situation Reports, without issues. Following submission of the Forensic Report on November 25, 2015, monitoring entered the third phase. DEP decreased inspections to twice weekly and issued reports monthly. Throughout the third phase, piezometer readings rose towards pre-incident levels as the reservoir level returned to full pool. On May 16, 2016, the project team met on site to reassess the incident and plan for disposition of the relief wells. The team agreed that DEP could return to pre-incident levels of monitoring and to maintain the relief wells in a stand-by status pending decisions on the future hydro-electric project.

LESSONS LEARNED

The root cause of the incident was the improper preparation for and execution of the FID boreholes, using uncased drilling techniques which led to unsuccessful borehole closure. Contributing factors included lack of understanding of subsurface conditions at the toe of the dam, selection of an inexperienced lead field geotechnical engineer for the boring operation, and lack of contingency planning and preparations in case of artesian conditions. Successes included prior engineering assessments and installation of piezometers, periodic data collection and addition of the ADAS which enabled quick determination of areas of the dam reacting to the turbid discharge. Routinely conducted dam safety training programs prepared DEP to quickly react with confidence and mobilize staff, materials and equipment. Successes included the existence of a well laid out Emergency Action Plan which helped to guide public notifications and contingency planning. The multi-point public information and awareness push informed local elected leaders, businesses and residents, while preventing anxiety and instilling confidence with downstream stakeholders. The collaborative decision making process with regulators, consultants and the contractor led to good decisions and resulted in mutual trust and confidence that supported timely field changes.

The hydro-electric project is on hold pending further feasibility analysis in light of the artesian conditions below the proposed power house. DEP has taken steps to prevent recurrence of a similar incident at Cannonsville Dam and other locations. DEP has updated its Drilling and Boring checklist to ensure open hole methods are not used in areas subject to artesian conditions. This includes the base of all dams, dikes and aqueducts.

CONCLUSIONS

The turbid discharge incident at Cannonsville Dam was a hard fought skirmish which required quick, informed and collaborative decision making. The success of the project team prevented the risk from developing into something much worse. DEP was quick to respond and, with the help and support of FERC and NYS Dam Safety and other local and state officials, gathered a team of experts that used their diverse knowledge, skills and abilities to respond, analyze the problem, mitigate and remediate the problem, and develop and implement plans for long term success. It is our hope that the lessons learned will assist other dam owners and dam safety professionals.

CONTRIBUTING AUTHORS

NYCDEP - Thomas DeJohn, BWS Dam Safety Engineer, Paul Costa, P.E. BEDC Portfolio Manager

Anthony Garagliano BWS Instrumentation Engineer

GZANY – Chad Cox, P.E. Associate Principal, Laurie Gibeau, P.E. Assistant Project Manager, James Guarente, P.E. Sr. Project Manager

FREEZING AND THAWING DURABILITY OF LABORATORY AND IN-SITU ROLLER-COMPACTED CONCRETE TEST SPECIMENS

Timothy Dolen[1]
Veronica Madera[2]

Roller-compacted concrete (RCC) is a no-slump concrete with a dry consistency that makes it difficult to obtain a well distributed, entrained air-void system for freezing and thawing (FT) durability. However, in spite of the difficulty of obtaining FT durability, RCC is still being used in FT environments for new construction, concrete dam modifications, and embankment dam overtopping protection. Often, the RCC is protected with a layer of conventional air-entrained concrete, or the section is increased to allow for "sacrificial" concrete; both of which increase the cost. The Bureau of Reclamation conducted several RCC mix design investigations with air-entrained RCC and was able to obtain sufficient FT durability in some applications. In a more extensive research program, Reclamation investigated the FT durability and air-void parameters of fresh RCC (1) immediately after mixing, (2) after depositing and spreading, and (3) using "push-tube sampling," immediately after compaction. Additionally, Reclamation drilled cores from several RCC dams for strength and FT durability testing.

The authors will summarize Reclamation history of air-entrained RCC, factors that promote successfully obtaining suitable air-void parameters for FT durability, and new developments in RCC technology that can lead to better FT durability.

[1] Dolen and Associates, tpdolen@yahoo.com
[2] Bureau of Reclamation

PARAMETRIC LAYOUT OF DOUBLE-CURVATURE ARCH DAMS USING DIANA FINITE ELEMENT SOFTWARE

E. Goulas[1]
G. Schreppers[2]
M. Partovi[3]
R. M. Gunn[4]

ABSTRACT

In this paper, an optimized and automated process to generate a finite element model of double-curvature arch dams including a portion of the foundation within any topographical profile that is suitable for such dams, by means of the Python scripting language, is presented. In the presented approach a limited number of parameters control the entire shape of the model which can be efficiently varied in the design process. The dam body and footprint excavation volumes, stresses and deformations are automatically calculated for a wide range of loading conditions. Different analyses, such as phased construction, local and global stability, eigen-value and non-linear time-history can be rapidly and efficiently performed with the model. This approach makes it possible to create realistic three-dimensional computer models already at the preliminary design stage of arch dams, whereas until now such techniques are typically only employed at the detailed design stage due to the time and cost implications related to such studies. A parabolic shaped double-curvature arch dam has been implemented into the script, however other dam types and geometric shapes could also be readily introduced into the code based on the same underlying principles described herein. The processes described herein are mainly targeted towards new dam builds, however existing dam models can also be readily generated as demonstrated in the validation of the script using an existing structure. Finally, a dam site location, "Vikos Gorge" from the Pindus Mountains of northern Greece, is chosen to illustrate the developed procedure allowing the potential optimum location of a double-curvature arch dam to be selected.

INTRODUCTION

Arch dams are solid impermeable barriers made of unreinforced mass concrete which can be shaped in both the horizontal and vertical planes based on specific sets of geometric functions. In terms of concrete volume, arch dams offer an economic solution compared to other dam types provided the valley profile and geological conditions are suitable for such a structure. Implicitly, reduced concrete volume also means less formwork. To design an arch dam, an initial dam layout is used, usually obtained from empirical methods, and it is reshaped until the essential boundary constraints are satisfied within the design objectives. The application of finite element analysis techniques during this iterative process is arduous and often postponed to the final detailed design stage whereby the basic parameters such as location, height and orientation of the dam have

[1] DIANA FEA BV, Delft, the Netherlands, +31-883426200, E.Goulas@dianafea.com.
[2] DIANA FEA BV, Delft, the Netherlands, +31-883426200, G.Schreppers@dianafea.com.
[3] DIANA FEA BV, Delft, the Netherlands, +31-883426200, M.Partovi@dianafea.com.
[4] Swiss Federal Office of Energy, Bern, Switzerland, from February 1st 2017, russell.gunn@bg-21.com.

already been predefined. Hence, there are important cost implications in terms of mass concrete and rock excavation volumes using such an empirical approach.

In addition, the ability to rapidly modify the dam geometry in a semi-automated manner and perform further design verifications during the construction process following the discovery of new geological and/or geomechanical conditions has been necessary for several arch dam projects [4], [5] thus demonstrating the importance and the underlying motivation behind developing such a design tool as the one presented herein. Indeed, during the construction of the Luzzone Dam [5], a rock slide occurred on the left bank and the excavations in this region had to be deepened and the dam geometry adjusted in such a way as to not disturb the already constructed parts of the dam. The dam geometry was fortunately defined by main and side parabola geometric shapes (the latter referred to as side fillets used for local section thickening in both the horizontal and vertical planes) which greatly facilitated the reshaping of the structure. This experience and many more dams constructed throughout the world in a similar manner, greatly justify the use of the dam geometric definition developed in the software application, although the underlying principles described herein can be readily applied to other dam types and geometries.

In this paper an automated computational procedure is proposed to design double-curvature arch dams using Python scripting language for the definition of the dam geometry and DIANA Finite Element Software for computational analyses. This method follows the general guidelines of the USACE' publication [6] but it is modified so that a more realistic model can be generated from the beginning of the design process. In addition, the number of parameters needed to define the geometry of the arch dam has been simplified allowing a rapid manual adjustment of the dam layout that goes a long way towards the optimization of both concrete and excavation volumes.

The proposed method was developed in the scope of a master project in Delft University of Technology and the main goal was to use Finite Element Modeling, FEM, from the beginning of the design stage. Due to time constraints, the script was created to design thin arch dams in narrow valleys and not to evaluate existing structures. Furthermore, only parabolic shaped arch dams are generated. The use of other arch dam shapes such as circular or elliptical dam definitions can be integrated by enhancing the existing script. This method should be considered as the basis for a new process which aims to automate the design of double-curvature arch dams considering the essential aspects, like dam thickness, valley profile, dam shape etc.

MODIFIED PARABOLIC EQUATION

To describe the necessary minimum set of geometric functions for the definition of the arch dam, the modified parabolic equation (1) is introduced. Using this equation, the shape of the dam can be adjusted both at the design and especially during the build phase of the project whereby unexpected geological conditions can be handled by deepening the foundation depth and hence extending the arches more into the foundation without compromising the entire shape of the structure. Hence, it becomes difficult for arch dam shapes other than parabolic to respect the well-known USBR 30° rule [6] in the case that foundation deepening is required. However, parabolic shaped arch dams may not always be the best solution as they can result in a more expensive design if they are medium-thick or thick and placed in wide shaped valleys (developed crest length to dam height ratio equal to or greater than 5). In this paper, the modified parabolic equation will be used in the following sections.

$$f(x)=\frac{\left(x-x_{sym}\right)^2}{2p}+y_{sym}=\frac{1}{2p}x^2-\frac{x_{sym}}{p}x+\left(\frac{x_{sym}^2}{2p}+y_{sym}\right)=ax^2+bx+c \tag{1}$$

Where, $a=\frac{1}{2p}, b=-\frac{x_{sym}}{p}, c=\frac{x_{sym}^2}{2p}+y_{sym}$ are the constants of the parabola,

$\left(x_{sym}, y_{sym}\right)$ are the coordinates of the symmetry point of the parabola,

$\left(x_{sym}, y_{foc}\right)$ are the coordinates of the focus point of the parabola and

p is the distance between the symmetry and the focus point [Figure 1].

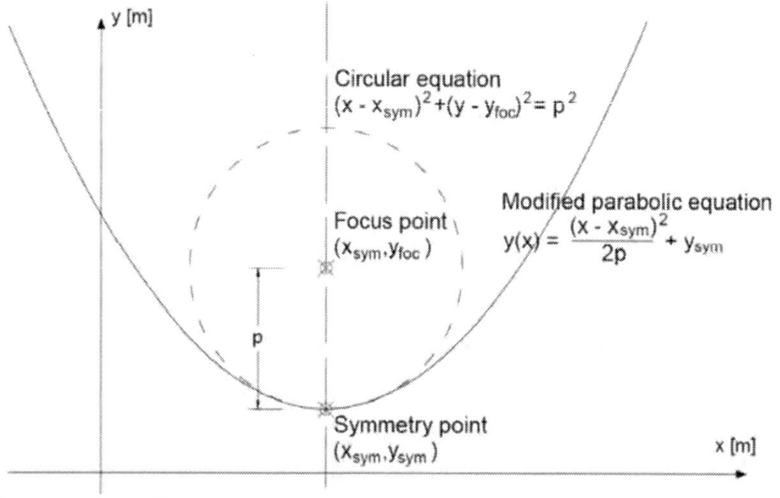

Figure 1. Modified parabola with symmetry and focus point in plan plane

MAIN PARTS OF A DOUBLE-CURVATURE ARCH DAM

The geometry of a double-curvature arch dam can be described by four main parts; the crown cantilever section, the main parabolas (horizontal arches), the side parabolas (fillets) and the dam-rock interface. By distinguishing these parts the dam-geometry can be fully described and reshaped if necessary [2].

Crown Cantilever Section

The crown cantilever section is the maximum height vertical cross-section of the dam and it is usually located in the bed of the river stream. The upstream and downstream lines contain the symmetry points of the parabolas which describe the main upstream and downstream faces respectively [Figure 2]. The initial crown cantilever section is selected based on experience. However, the key elements in the selection of the shape are the upstream heel stresses and crest downstream overhang such that self-weight tensile stresses are limited to less than 1 MPa throughout the crown section based on a 2D analysis. In general, the initial crown cantilever section provided by the USACE' publication [6] satisfies this constraint.

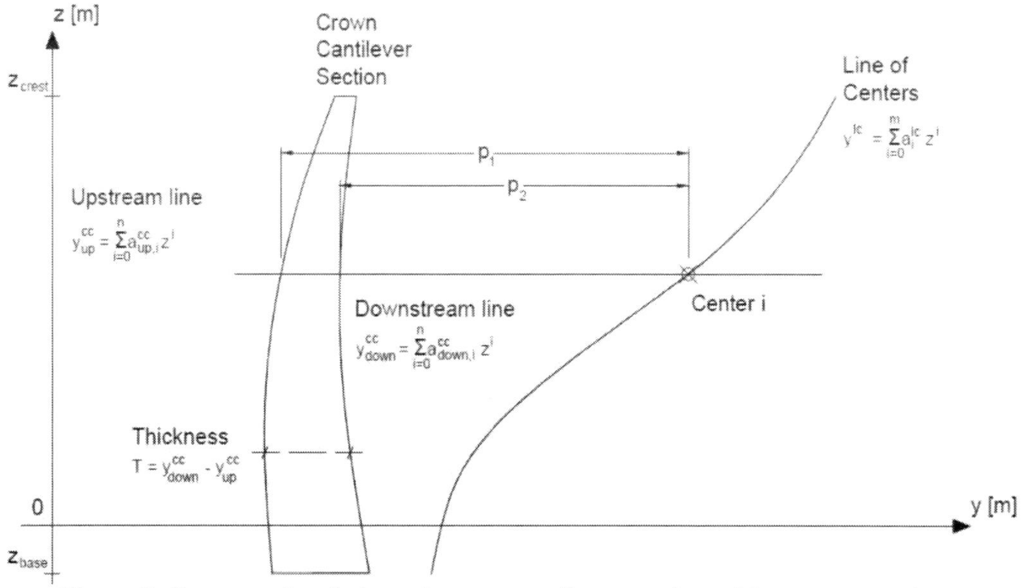

Figure 2. Cross-sectional view of crown-cantilever section with upstream and downstream faces, line of centers and focus distances p_1 and p_2 to upstream and downstream line respectively [2]

Main Parabolas (Horizontal Arches)

The main parabolas are the two sets of parabolas which define the upstream and downstream face of the dam. The two faces are described when the equations from the crown cantilever section and the line of centers are known. Both faces use the same line of centers so the thickness along the x-y section remains uniform for every elevation. Parabolic shaped arch dams have been used for both narrow and wide valley canyons, however, it is true to say that this shape may not be suitable to very wide canyons.

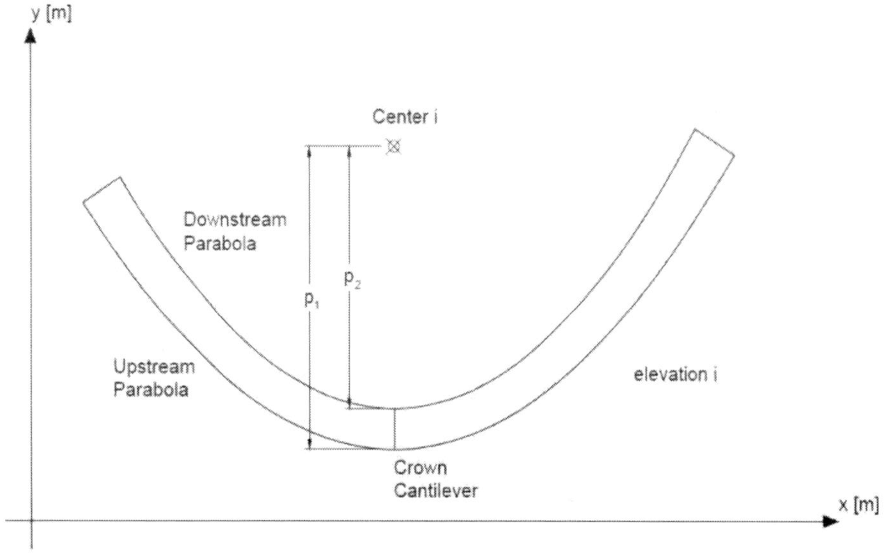

Figure 3. Upstream and downstream main parabolas in plan plane [2]

Side Parabolas (Fillets)

In the case where abutment thickening is required, two more sets of parabolas, one for each side of the banks, are used on the downstream face and are named side parabolas. Side parabolas are dependent on the main downstream parabolas and they add the needed abutment thickening at every elevation. A special technique is followed to create them such that a uniform increase in abutment thickening over the entire dam is achieved avoiding possible local geometric anomalies which may result in high stress concentrations in certain locations of the dam.

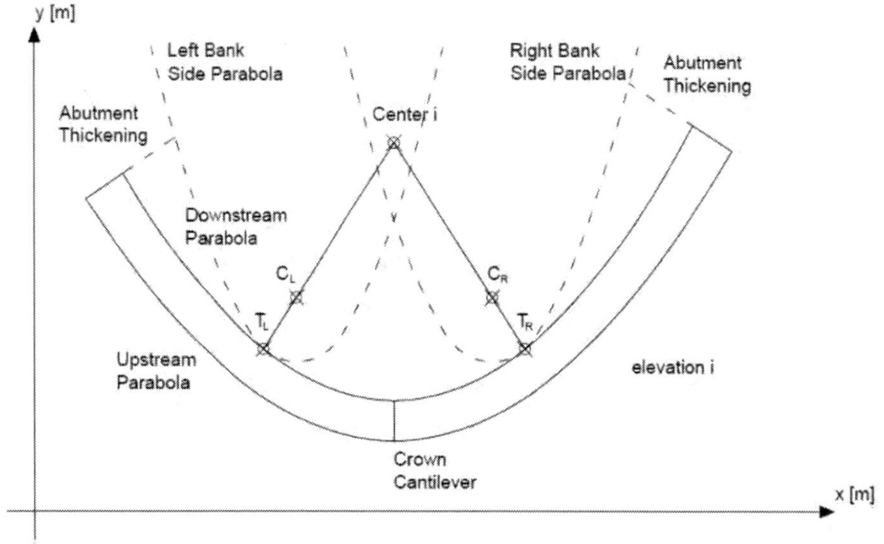

Figure 4. Plan view with outline of parabolic thickening at sides [2]

Dam-rock interface

The dam body is embedded in the rock. The foundation area is the dam-rock interface surface and it is defined by the end points of the parabolas of the upstream and downstream faces. This part is very important for the correct definition of the dam-geometry as there must be no gaps between the dam and the valley profile. Moreover, the embedment of the dam in the rock also defines the stability of the dam during the construction phase (non-grouted joints and hence, independent blocks) and the final grouted joints resulting in the monolith dam structure.

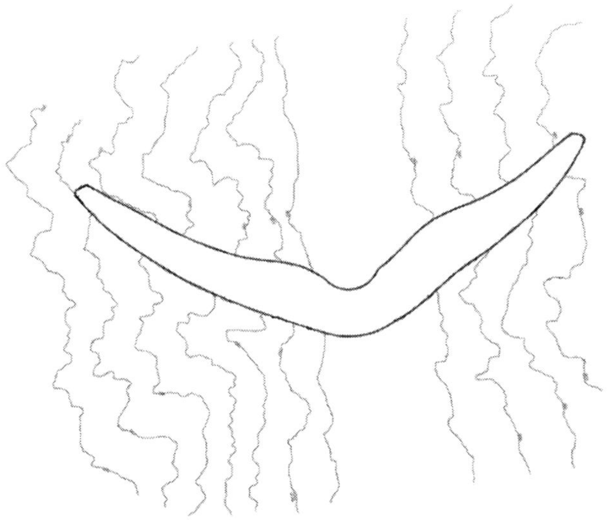

Figure 5. Plan view of the dam-rock interface [2]

GENERAL WORKFLOW

In the general design work-flow of a double-curvature arch dam two different stages are distinguished in the Engineer Manual [6] as described below.

Preliminary design stage

The first stage is called Preliminary Design Stage. During this stage the best fitting dam along the chosen valley must be found. Hence, different dam sites must be checked where the number of the sites is defined by the designer. The sites should be selected based on two important aspects; the canyon profile and the foundation characteristics. Then, a dam design is performed for each site whereby the best fitting dam must be generated in terms of two predominant factors; the dam concrete and excavation volume. A preliminary analysis can be performed as well to evaluate the maximum deformations and stresses, stability of the dam cantilevers, etc. The best fitting dams of the selected sites are compared to each other so the best alternative is found in terms of cost and performance, which is used for the next stage.

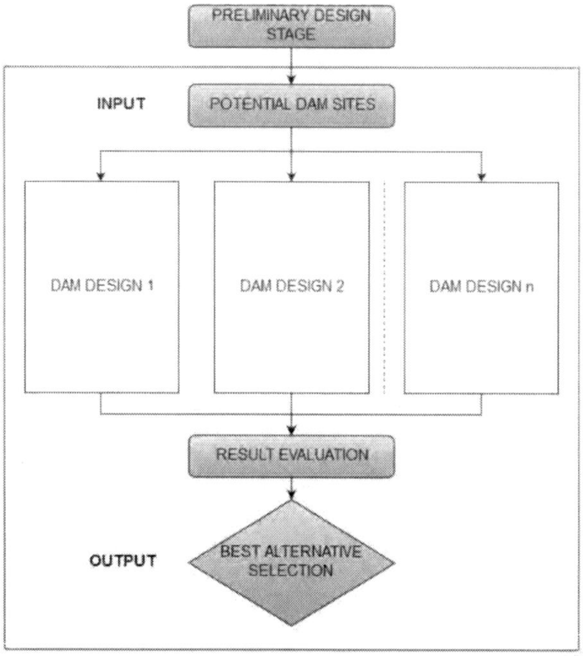

Figure 6. Work-flow for preliminary design stage [2]

Final design stage

The second stage is called Final Design Stage. The preliminary best alternative is used during this stage and it is reshaped until the allowable stress levels are reached. The main goal is to add the least concrete and excavation volumes to reduce the principal tensile stresses to a safe level. Several variables are used to change the preliminary shape of the best alternative and they are called final variables. Structural interface elements can be inserted into the model to simulate possible sliding between the dam and the rock and no-tension behavior. Finally, a non-linear static analysis may be performed and the critical stress areas are located, taking out the stress singularities, and evaluated based on the design constraints. The nonlinearities are located only in the interface elements with no-tension bedding while the properties of the concrete and the rock remain elastic [3], [7]. When the analysis results are not satisfactory, the variables must be redefined and the aforementioned steps repeated. The variable values are changed directly by the designer so he/she has full control of the design process. The optimal dam shape is highly dependent on the designer's judgment and experience.

In the design of new concrete dams where the material properties are unknown and need to be defined on a trial basis, the expected stresses must be checked against compressive and tensile strengths. As the allowable tensile strength is implicitly related to compressive strength and loading conditions, the design process for new dams is iterative. In this paper, the tensile strength is assumed to be less than 1 MPa as only static loading is considered. The allowable compressive stress should not exceed the compressive strength of the a priori selected concrete considering the factors of safety that are relevant to the design load cases combinations (normal, exceptional, extreme seismic, etc.). In the same procedure, additional constraints such as maximum displacements can also be considered.

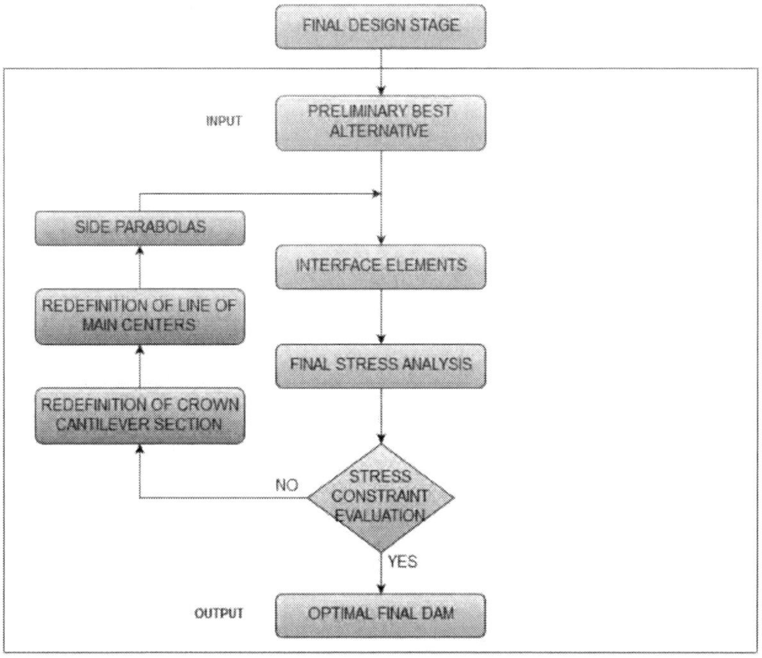

Figure 7. Work-flow for final design stage [2]

Preliminary variables

A number of variables are defined which control the geometry of the model. The variables that are used during the Preliminary Design Stage are [2].

$y_{start}\left[meters\right]$	Left Hand Side (LHS) y-coordinate of the highest upstream parabola and it controls the location of the dam. Its value must be inside the borders of the topographic surface.
$H\left[meters\right]$	Height of the dam (from the base elevation to the crest elevation), defined from the hydrologic data.
$z_{base}\left[meters\right]$	Elevation of the dam base. If there is no information for the base elevation, it should be about 8 meters below the riverbed.
$\theta_{axis}\left[\text{degrees}\right]$	Angle of the dam axis and represents the orientation of the dam. Using this variable, the entire dam is rotated respective to the riverbed.
$L_{excavation}\left[meters\right]$	Assumed initial uniform excavation length and it shows how deep is the dam inside the rock, a positive value is required.
$L_{excavation}^{\min}\left[meters\right]$	Minimum excavation length that is allowed for the dam, a positive value is required.

Final variables

The variables that are used during the Final Design Stage are categorized based on the main parts [2].

For efficiency reasons, it is assumed that the reshaping of the crown cantilever section is defined by three correction parameters which reshape the downstream line while the upstream line is fixed [Figure 8]. In this way, possible anomalies in the main faces are avoided and the crown cantilever is always in the same location.

$cc\_down\_tol\_crest[meters]$	Tolerance that is added in the y-coordinate of the downstream crown cantilever point at the crest elevation.
$cc\_down\_tol\_045H[meters]$	Tolerance that is added in the y-coordinate of the downstream crown cantilever point at $0.45H$ above the base elevation.
$cc\_down\_tol\_base[meters]$	Tolerance that is added in the y-coordinate of the downstream crown cantilever point at the base elevation.

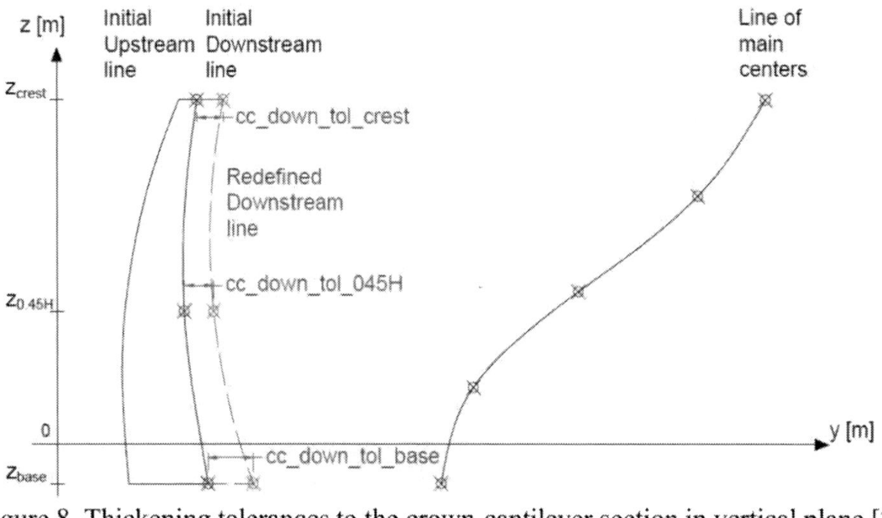

Figure 8. Thickening tolerances to the crown-cantilever section in vertical plane [2]

The line of main centers is redefined from the following variables which influence the shape of the main parabolas and consequently the main faces [Figure 9]. The distance between the crown cantilever section and the line of main centers, at any elevation, determine how steep the respective parabola will be.

$center\_tol\_z\_crest[meters]$	Tolerance that is added in the y-coordinate of the initial main center point at the crest elevation.
$center\_tol\_z\_075H[meters]$	Tolerance that is added in the y-coordinate of the initial main center point at $0.75H$ above the base elevation.

$center\_tol\_z\_050H[meters]$	Tolerance that is added in the y-coordinate of the initial main center point at $0.50H$ above the base elevation.
$center\_tol\_z\_025H[meters]$	Tolerance that is added in the y-coordinate of the initial main center point at $0.25H$ above the base elevation.
$center\_tol\_z\_base[meters]$	Tolerance that is added in the y-coordinate of the initial main center point at the base elevation.

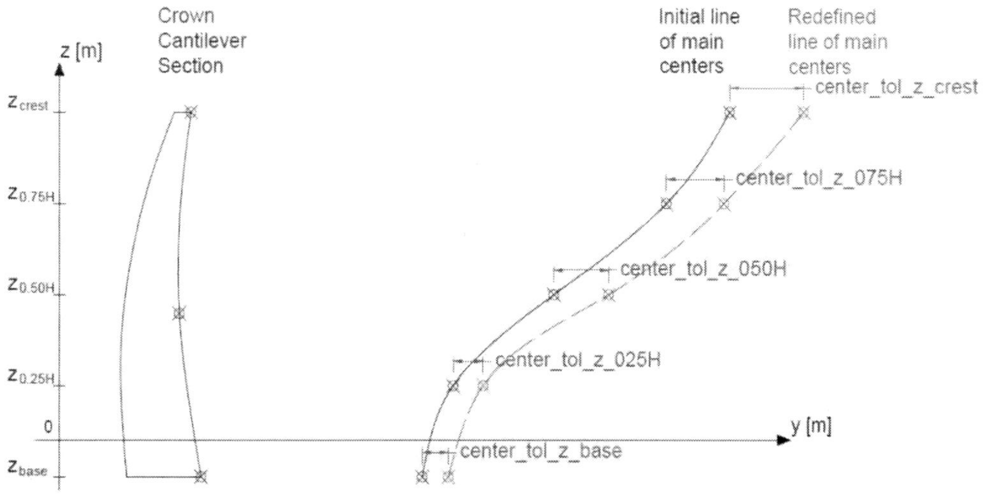

Figure 9. Offset-tolerances to the line of main-centers in vertical plane [2]

The side parabolas are introduced from the following variables which add the required concrete in the downstream face of the dam. For each side, the line of tangent points and the line of side centers must be defined [Figure 10]. Having these two lines the set of side parabolas is computed avoiding possible local anomalies around the dam foundation.

$control\_angle[-]$	Proportion which defines the side angles. The variable is multiplied with the main angles along the dam height and its range is from 0 to 1.
$control\_crest[-]$	Proportion which defines the side center point at the crest elevation, its range is from 0 to 1.
$control\_050H[-]$	Proportion which defines the side center point at elevation 050H above the base, its range is from 0 to 1.
$control\_base[-]$	Proportion which defines the side center at the base elevation, its range is from 0 to 1.

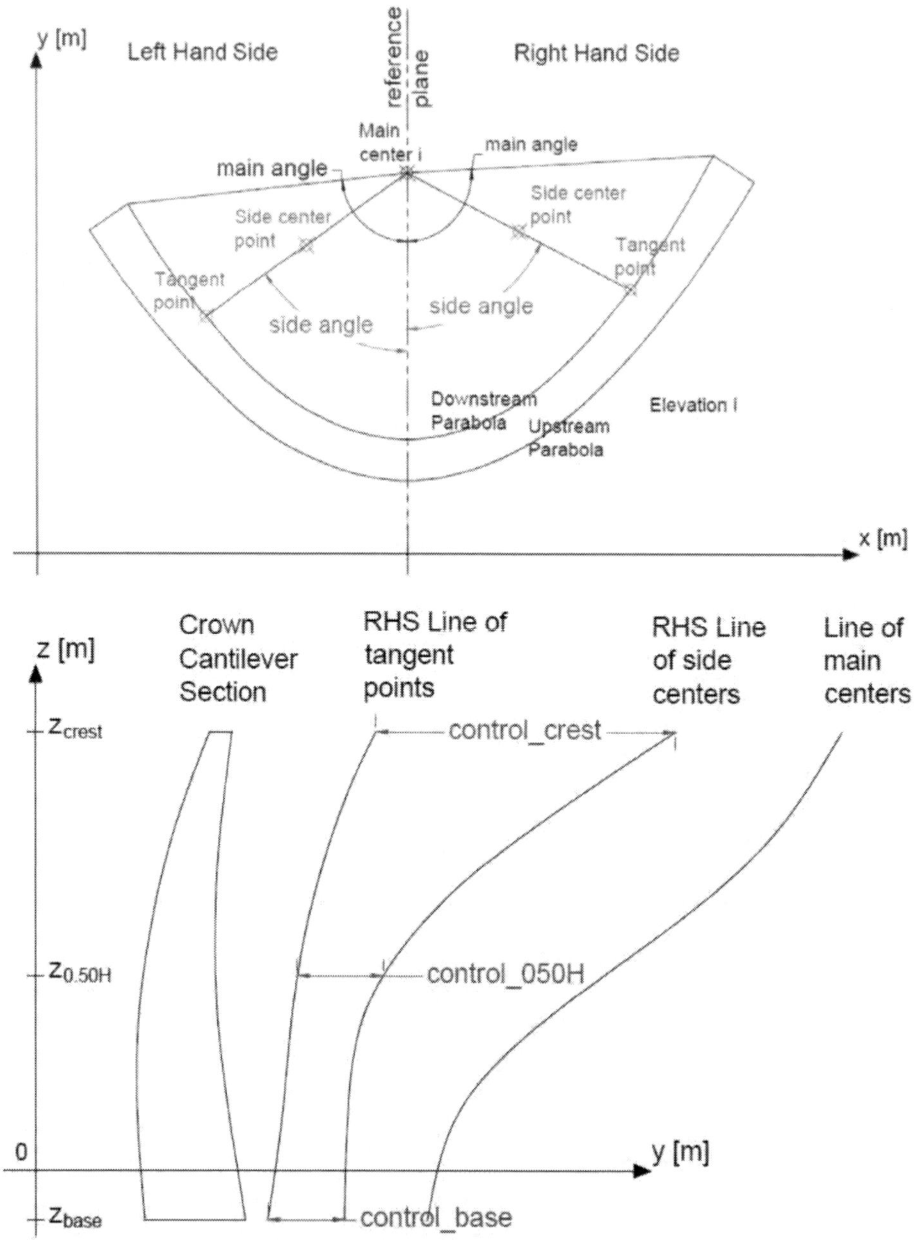

Figure 10. Illustration of meaning of side-parabola areas and thickening functions [2]

EXAMPLE: VIKOS GORGE

To demonstrate the aforementioned process, an example is given. Firstly, a topographic surface must be chosen that is long and wide enough in plan, so that a sufficient number of potential dam sites can be checked. The "Vikos Gorge" from the Pindus Mountains of northern Greece was selected. The Vikos Gorge has a length of about 20 km, an elevation ranging from 450 to 1600 m above sea level and a width ranging from 400 m to a few meters at its narrowest part of the gorge. Only a part of the topographical map is used with a length of 2 km hence satisfying the criteria to design a double-curvature arch dam.

Preliminary design stage

Starting from the Preliminary Design Stage, more than 100 different dam site locations were checked aiming to find three locations based on the general shape of the dam so all the potential dam shapes can be compared. Location A generates an unsymmetrical dam where the Right Hand Side (RHS) is longer than the Left Hand Side (LHS), Location B a symmetrical dam and Location C, an unsymmetrical dam where the Left Hand Side (LHS) is longer than the Right Hand Side (RHS) as shown below.

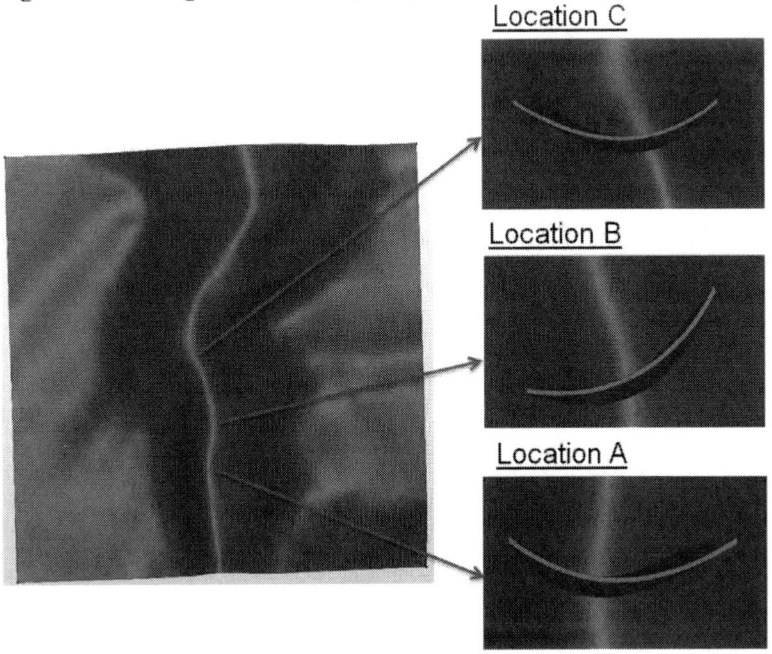

Figure 11. Plan view of the topografic surface of Vikos Gorge with 3 potential dam locations [2]

In order to find the most economical dam the preliminary parameters were defined. At the same time the different dam options can be compared with each other. For this reason, the dam height, the uniform excavation length and the minimum excavation length have the same values for all the locations. The value of the axis angle must be checked for each location so that the most economical model is generated, the valid range is found and then the final value is selected. The elevation of the dam base is chosen to be 8 m below the riverbed for each location noting that sediment deposits in narrow gorges can be important, but for the time being are not predominant to demonstrate the use of the design tool.

Table 1. Parameters for 3 alternative dams in preliminary stage process

	Location A	Location B	Location C
y_{start} (m)	-400	-240	60
H (m)	250	250	250
z_{base} (m)	42	38	38
θ_{axis} (°)	-25 °..6 °→-10 °	-14 °..40 °→24 °	-6 °..20 °→8 °
$L_{excavation}$ (m)	30	30	30
$L_{excavation,min}$ (m)	10	10	10

Figure 12 shows the volumes of the dam-body and excavation for Location B for different axis angles. There are three different colors for the lines of the dam and excavation volumes. The colors indicate if the solution is well positioned in the rock. Hence, only the values with the green color are used which give the valid range of the axis angle. From this range, the value which gives the most economical model is chosen and it is the best fitting dam for Location B.

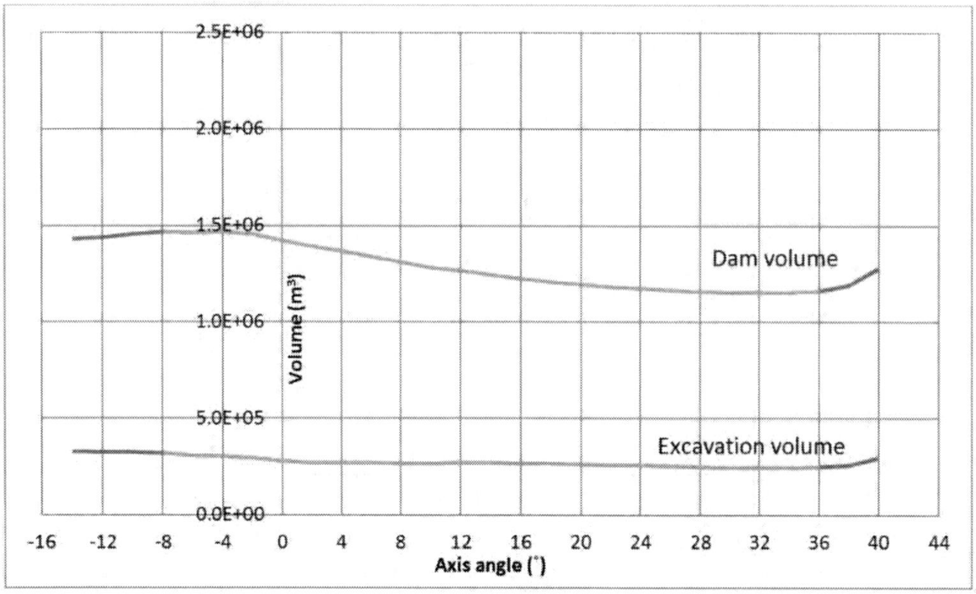

Figure 12. Volumes of dam-body and excavation for various axis-angle of the dam for Location B [2]

At this stage, a finite element model can be generated for each location and it can be used to generate the preliminary results. The model is recommended to be as simple as possible because the preliminary results may not be so realistic due to stress singularities occurring around the common surfaces of the dam with the rock. For this reason, linear material properties are used in the model, while their values are chosen by the designer [Table 2]. An automatic meshing procedure was applied for both foundation and dam, which generates pre-dominantly eight-node isoparametric brick elements, but also six-node wedge-elements, five-node pyramid elements and four-node tetrahedron elements. The meshing procedure optimizes the shape of the element such that the differences of length of the edges of one element are minimal. Furthermore, a structural linear static analysis is performed where the water pressure on the dam and the self-weight are taken into account.

Table 2. Preliminary material properties for Finite Element Model

	E (GPa)	v	ρ (kg/m^3)
Rock	18.0	0.27	0
Concrete	25.5	0.20	2400

To find the best alternative the best fitting dam for each location is compared, the predominant factors are the concrete and excavation volumes. Hence, Location B apparently is the best alternative as it can be seen in Figure 13.

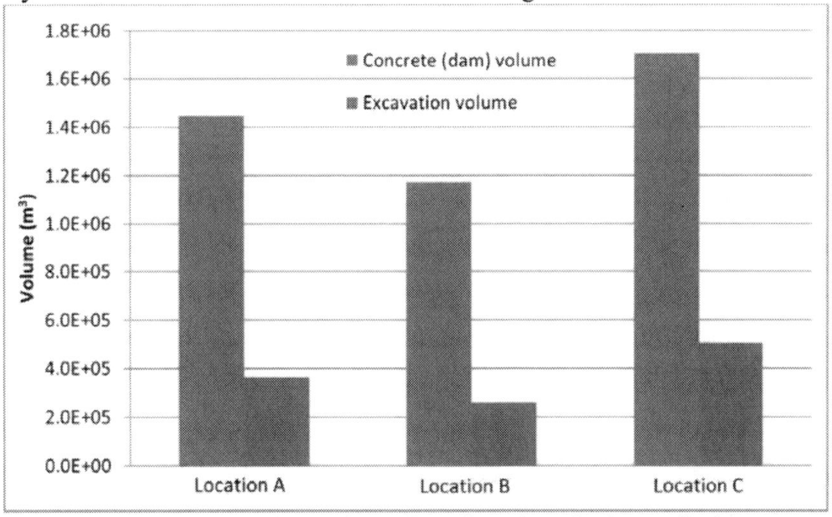

Figure 13. Dam and excavation volume for 3 locations in the preliminary stage [2]

Final design stage

Having selected the best alternative along the valley, the Final Design Stage starts. The main goal is to reshape Location B until the optimal dam in terms of stress levels is obtained. An iterative process starts, where the final variables are changed progressively adding a small amount of concrete in the critical areas each time. There are three critical areas for static loading; the downstream central area and the upstream side areas which are located close to the dam-foundation line for each bank. This aims to reduce the critical in-plane principal stresses until the design constraints are satisfied. The optimization is manual and the final shape of the dam is dependent on the designer's judgement and experience.

More than 60 different models were tested to obtain the optimal final dam shape for Location B. First, the tensile principal stresses must be reduced from the downstream central area of the dam. To achieve this, the crown cantilever section is thickened where more concrete volume is needed around the base elevation. However, changing only this part is not enough, so the line of main centers must be changed simultaneously creating steeper parabolas where is required to resist even more against the water pressure. Then, the side parabolas are used to reduce the tensile principal stresses in the upstream side areas by adding concrete around the dam-foundation. Regarding the compressive stresses, there was no need to take them into account during the analysis of the arch dam, as they were always below the allowable stress levels.

The material properties of rock, concrete and interfaces were defined by the designer. The analyses show that when more flexibility, or no-tension, is assumed in the interface-elements between dam-body and rock, the peak-stresses in the dam-body reduce significantly [Figure 14]. For the moment, the entire width from upstream to downstream of the dam-rock interface has been modified. However, it is logical to select interface-elements just for the upstream third of the section which are directly in tension.

Table 3. Final material properties for Finite Element model

	E (GPa)	f_c (MPa)	v	ρ (kg/m^3)
Rock	18.0	-	0.27	0
Concrete (C 30/37)	32.0	30	0.20	2400
	K_n (GPa)	K_t (GPa)		
Interface	3200	320		

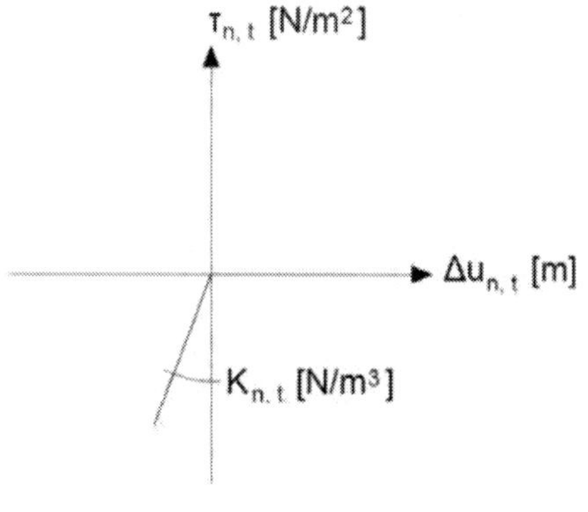

Figure 14. Constitutive model for interfaces in normal and tangential direction [2]

From the 60 models, only 4 characteristic models are listed in the table below. Iteration 0 shows the results of the initial model, Iterations 1 and 2 display two intermediate models and Iteration 3 shows the final model which is concluded to be the optimal solution.

Table 4. Final parameters for 4 characteristic iterations in final design stage

Location B				
	Iteration 0	Iteration 1	Iteration 2	Iteration 3
Side parabolas				
control_angle (-)	-	0.667	0.667	0.667
control_crest (-)	-	0.875	0.875	0.750
control_050H (-)	-	0.750	0.625	0.500
control_base (-)	-	0.625	0.500	0.500
Crown cantilever section				
cc_down_tol_crest (m)	-	0	3	4
cc_down_tol_045H (m)	-	3	5	6
cc_down_tol_base (m)	-	10	20	30
Line of main centers				
center_tol_z_crest (m)	-	0	0	-5
center_tol_z_025H (m)	-	0	-5	-15
center_tol_z_050H (m)	-	-5	-5	-25
center_tol_z_075H (m)	-	-5	-15	-35
center_tol_z_base (m)	-	-10	-25	-45
Volumes				
Concrete (m^3)	1.173E+06	1.365E+06	1.564E+06	1.806E+06
Excavation (m^3)	2.590E+05	3.430E+05	4.191E+05	5.892E+05
Tensile stresses	(Allowable tensile stress: 1.000E+06)			
Upstream max (N/m^2)	1.861E+06	1.541E+06	1.107E+06	9.227E+05
Downstream max (N/m^2)	3.517E+06	2.447E+06	1.695E+06	8.778E+05
Compressive stresses	(Allowable compressive stress: concrete strength class)			
Upstream-Downstream min (N/m^2)	-23.710E+06	-21.040E+06	-19.870E06	-14.320E+06

CONCLUSION

Using a scripting language such as Python greatly facilitates the Computed-Aided Design process allows the semi-automated generation of a Finite Element model for double-curvature arch dams considering the essential design objectives and constraints.

For narrow valleys with strong rock, any realistic topography, at any location, can be processed by the script to define the layout of a double-curvature arch dam in an efficient way.

The modelling effort is drastically reduced, because only topographic data and a limited number of dam geometric design variables are required.

The developed process can generate many double-curvature arch dam profiles by simply modifying a few essential design parameters in a short time allowing accurate quantitative checks of dam and excavation volumes as well as stresses and deformations for different load case combinations.

Even with a fine mesh, some thousands of elements, the model generation can be done in less than 1 hour performing a strength analysis. In this way, more models can be analyzed and evaluated in less time.

DISCLAIMER

The statements and opinions herein are those of the authors and do not necessarily reflect a stance position of the Swiss Federal Office of Energy, Bern, Switzerland.

REFERENCES

[1] DIANA User-Manual release 10.1. *DIANA FEA | Analysis and Design*. [ONLINE] Available at: http://dianafea.com/. 2016. [Accessed 17 October 2016].

[2] E. Goulas. "Design of Double-curvature Arch Dams in Terms of Geometric and Stress Constraints by using Script-Based Finite Element Modelling", M.Sc. Thesis Delft University of Technology. The Netherlands. Delft. 2016.

[3] ICOLD. "Bulletin B155, Guidelines for Use of Numerical Models in Dam Engineering". France. Paris. 2013.

[4] R. M. Gunn, C. Garrido, A. and Fankhauser. "The heightening of an arch-gravity dam in Switzerland", International Congress on Conservation and Rehabilitation of Dams. Spain. Madrid. 2002.

[5] R. M. Gunn. "Stability of a dam abutment including seismic loading", 10th International Benchmark on the Numerical Analysis of Dams, Theme C Formulator. France. Paris. 2009.

[6] U.S. Army Corps of Engineers. "Engineer Manual 1110-2-2201, Arch Dam Design". USA. Washington. 1994.

[7] U.S. Army Corps of Engineers. "Engineer Manual 1110-2-6051, Time-History Dynamic Analysis of Concrete Hydraulic Structures". USA. Washington. 2003.

STUDY OF UPLIFT PRESSURE EFFECT IN ROLLER COMPACTED CONCRETE GRAVITY DAM.

Hichem Mazighi[1]
Mustapha Kamel Mihoubi [2]
Khaled Ghaedi[3]
Zainah Ibrahim[4]

ABSTRACT

In recent past, researches have done to properly investigate the uplift pressure effect at the base level of concrete dams, in particular for a cracked-base situation under static conditions. This paper attempts to implement two and three-dimensional Finite Element (FE) Models for nonlinear static analysis. To aid the aim, Kinta Roller Compacted Concrete (RCC) gravity dam is selected as a case study and FE ABAQUS software is used to investigate the responses of the dam under ful reservoir hydrostatic pressure and uplift effect as well as the effect of different densities of the dam for the stability purpose. In addition, Concrete Damaged Plasticity (CDP) model is conducted to monitor the crack propagation in the dam. The results show that, the uplift pressure affects considerably the relative displacement more than the hydrostatic pressure. The relative displacements increase when the concrete density of the dam goes up unlike the maximum and minimum principal stresses and crack propagation.

INTRODUCTION

Dam stability analysis is an important factor before and after construction. The hydrostatic and uplift pressures can generate tensile stresses near the upstream and the foundation faces successively, possibly exceeding the strength of the material and causing a horizontal crack or opening of construction joints. These huge structures normally have cracks in practical service caused by previous earthquakes, construction conditions, or temperature effects (Barani et al., 2016). For a concrete dam under its probable maximum flood, the hydrostatic pressure acting inside the crack reduces the resistance and increases the penetration of water which exerts a uplift pressure (Zhu and Pekau 2007).

The water pressure inside the cracks greatly reduces the structural strength of the concrete gravity dams (Bhattacharjee and Leger 1995).

Many researchers have analyzed the effect of the crack propagation in concrete dams (Bruhwiller and Saouma 1995, Plizzari 1997, Slowik and Saouma, 2000; Javanmardi et al., 2005; Pekau and Zhu, 2008; Shi et al., 2013; Pandey et al., 2014; Ghaedi et al., 2015).

[1] Phd student, Laboratory Water Resource Mobilization and Enhancement (LMVR), National High School for Hydraulic (ENSH), BP 31 Blida, ALGERIA, h.mazighi@ensh.dz
[2] Prof., Laboratory Water Resource Mobilization and Enhancement (LMVR), National High School for Hydraulic (ENSH), BP 31 Blida, ALGERIA, mihkam@ensh.dz
[3] Research Assistant, Dept. of Civil Engineering, University of Malaya, 50603, Malaysia; Hormoz Beton Firm, Bandar Abbas City, Iran, khaledqhaedi@yahoo.com)
[4]Senior Lecturer, Dept. of Civil Engineering, University of Malaya, 50603, Malaysia, zainah@um.edu.my

The FE analysis technique is the most commonly used method of analysis for different hydraulic structures, including dams.

Skrikerud and Bachmann (1986), have studied the crack propagation of the Koyna gravity dam by using single crack model. The results showed that there was a relationship between aggregate forces and surface of openings. Ayari (1990) investigated the fracture mechanics based model and discrete crack closure model for Koyna dam under dynamic loading in transient condition.

Guanglun et al., (2000) proposed a mathematical model based on the nonlinear crack band theory to investigate the dynamic fracture behavior of gravity dams in two-dimensional FEM.
Also, they presented the finite element remesh for the front cracks via shifting the element edge couples of cracks in direction of the tensile stresses. The smeared crack model was used to inspect the nonlinear dynamic response of dams considering reservoir water effect under earthquake excitations (Mazloumi et al., 2012 ; Ayari, 1990).

Zhu and Pekau (2007) employed and adopted the Incremental Displacement Constraint Equations (IDCE) model along the crack to consider the behavior of dynamic contact states in the cracks.

The damping model of IDCE was validated in dynamic contact conditions for flexible and rigid bodies. The obtained results revealed very attractive occurrences such as peak rocking direction, jumping and large damping effect of multi cracks on the peak residual sliding. Researchers also discussed about seismic behavior of dams by implementation of two-dimensional finite element modeling, (Calayir and Karaton, 2005; Akköse and Simsek 2010, Jiang and Du, 2012; Mazloumi et al., 2012, Zhang et al., 2013, Paggi et al., 2013; and others. However, the effects of base crack opening on dams have been neglected in the above studies.

Ghaedi et al., 2016 analyzed the shape and size of the effects of the gallery on RCC dams considering opening inside the dam body under earthquake excitation. Meanwhile, the interaction of the reservoir dam with the stationary foundation is implemented. For this purpose, two dimensional (2D) Finite Element Model (FEM) is used for nonlinear dynamic analysis by means of finite element software, ABAQUS. In addition, Concrete Damaged Plasticity (CDP) model is also implemented to inspect the tensile damage of the dam during earthquake excitation. Kinta RCC dam of Malaysia is considered as a case study in analysis.

Based on the above literature, many researchers have considered various aspects of dams analyses. However, it seems that the simultaneous effect of uplift pressure and concrete density of dam body has not been thoroughly considered. Therefore, present paper tries to investigate the concurrent effect of uplift pressure, hydrostatic loading and concrete density of dam body in 2D and 3D FE modelling on dams.

THEORETICAL MODELING

Concrete Damaged Plasticity (CDP) Theory

The linear assumption may not be suitable for seismic analysis of the RCC dams (Zhang and Wang, 2013). In order to explain the complicated mechanical response of the concrete materials under seismic excitations, many constitutive approaches have been proposed including damage model, anisotropic damage and isotropic damage model. A fundamental constitutive model was proposed by Lubliner et al., (1989) and modified by van Mier and Fenves, (1998).

The method explaining the nonlinear behavior of each combinatorial material in a multiphase composite material is generally used in the cracking analysis for concrete dams. This model factorizes the uniaxial strength functions into two divisions to stand for the permanent degradation of stiffness and deformation. The model assumes two major failure mechanisms for concrete materials, the first for cracking and the second one for crushing in tension and compression, respectively.

In the incremental theory of plasticity the strain tensor (ε) is divided into two parts including the elastic strain (ε^e) and plastic strain (ε^p) in which for the linear elasticity can be written as:

$$\varepsilon = \varepsilon^e + \varepsilon^p \tag{1}$$

The variables (ε^e, ε^p, k) are assumed to be identified at time (t). With mentioned information, for stress tensor the following can be obtained:

$$\sigma = (1-d)\bar{\sigma} = (1-d)E_0(\varepsilon - \varepsilon^p) \quad and \quad d = d(k) \tag{2}$$

In which d is the scalar stiffness degradation variable which can be in the range of 0 (undamaged) to 1 (fully damaged); E_0 is the undamaged elastic stiffness for concrete material. The failure mechanism of the material associated with the damage, thus, reduction of the elastic stiffness that supposed a function of the internal variable k including of the compressive and tensile variables, namely $k=(k_t, k_c)$. Function of damages consisting tension (d_t) and compression (d_c) which are nonlinear functions computed by uniaxial response in compression with practical data. Hence, the effective stress is determined as:

$$\bar{\sigma} = \frac{\sigma}{1-d} = E_0(\varepsilon - \varepsilon^p) \tag{3}$$

Modeling Cracking in a Gravity Dam

The Sg. Kinta Dam project is part of the stage II development of the Ipoh Water Supply on mainland Malaysia that is located approximately 12 km from the city of Ipoh, which is turn located about 200 km north of Kuala Lumpur. It is the first Roller Compacted Concrete (RCC) dam which is located in Malaysia (Allan et al., 1999, Ghaedi K. et al., 2015). A finite element representation of the Kinta RCC dam including the foundation is presented in Figure 1.

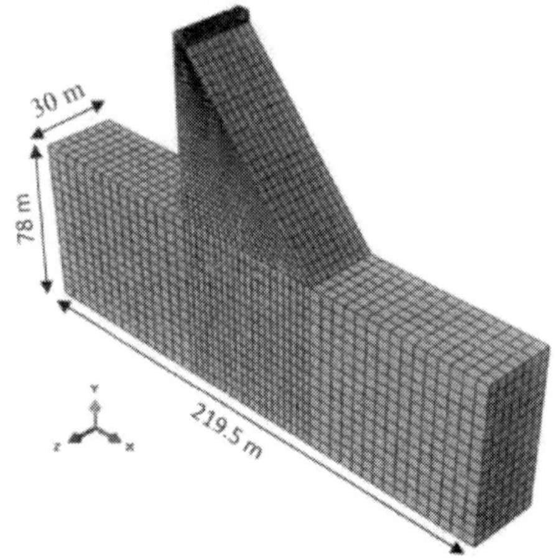

Figure 1. Dam body –foundation system in 3D FEM

In this case study, the uplift and hydrostatic pressures illustrated in Figure 2 that are considered to evaluate the static analysis of the dam.

The drainage system is neglected to study the effect of uplift pressure on the stability of the dam in the most devastated case

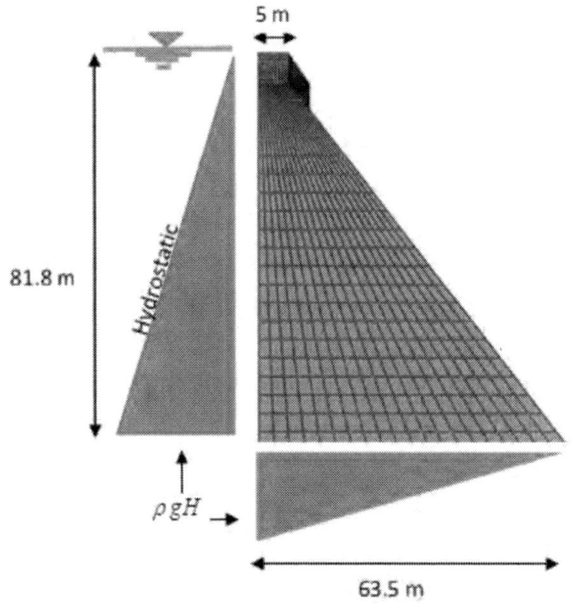

Figure 2. Principal loading applied to the dam

The modeling is developed using the finite element software, ABAQUS (version 6.14). This software is used for different nonlinear static and dynamic analysis such as water wave and seismic loadings. The boundary conditions of the dam in 3D are shown in Figure 3

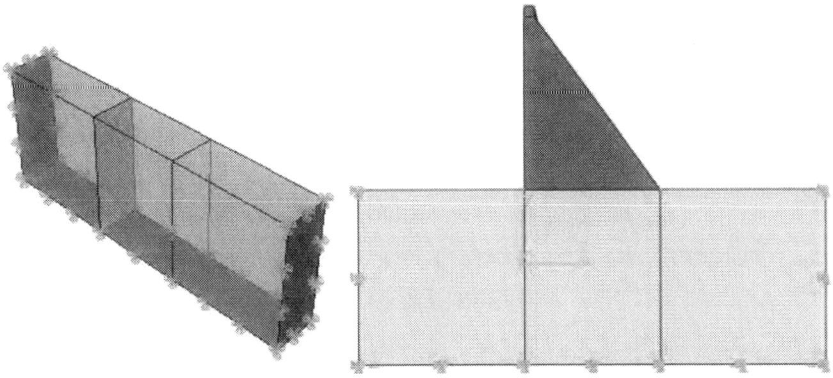

Figure. 3. Boundary conditions of the model

The model's 3D isoparametric elements are implemented to discretization of the Kinta dam. The boundary conditions of the dam in 3D is shown in Figure. 3 and the details of discretization can be seen in Table 1 (GHD, 2002).

Table 1 : Material Properties of the Kinta RCC Dam

Characteristics of the Discretization		
Section	Dam	Foundation
Number of node	8547	6528
Number of element	7200	5250
Element Type	C3D8R	C3D8R
Material properties		
Material properties	Foundation	RCC Body
Poisson's ratio υ	0.2	0.2
Young modulus E (N/m^2)	0.30E+11	0.23E+11
Mass density γ (Kg/m^3)	2650	2386
Tensile strength DTS f_t(MPa)	2.25	2.5
Tensile strength DTS f_t(MPa)	2.25	2.5

The different densities of concrete of RCC dam body defined in Table 2, are considered to evaluate the static analysis of the dam.

Table 2 : Different case study of concrete in the Kinta RCC dam depending on the binder content for the corresponding density

Cases	1	2	3
Mass density γ (Kg/m^3)	1800	2000	2300

RESULTS AND DISCUSSIONS

Relative Displacement Responses

The relative displacement in both horizontal and vertical direction due to hydrostatic and uplift pressures respectively of the dam heel confirm that the maximum value in X direction is 0.14 cm and 0.46 cm for Y direction. As the values indicated, the difference of 69.5% occurred for relative displacement for Y direction in compared to X direction. The state of the relative displacement of the dam heel is indicated in Figure 4.

Fig. 5 shows the displacement as a function of time from 0 pressure to the peak at the dam heel due to the hydrostatic pressure and the pressure respectively.

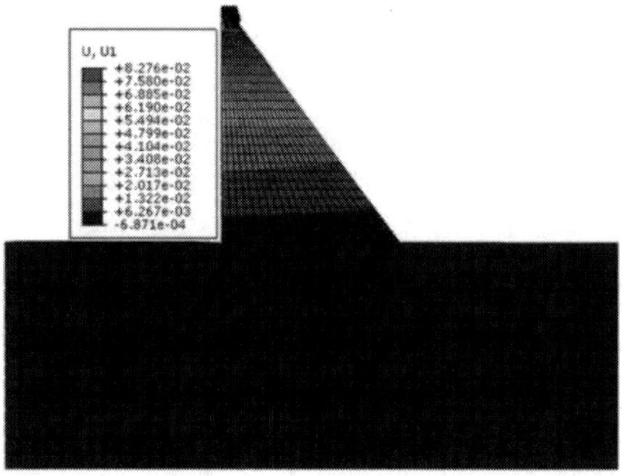

a) Displacement due to hydrostatic pressure

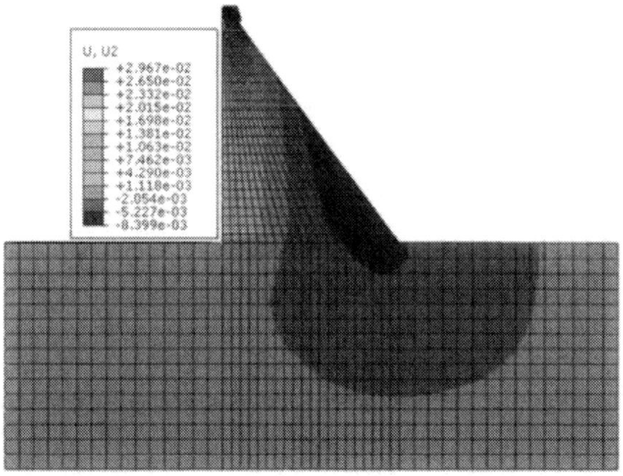

b) Displacement due to uplift pressure

Figure 4. Relative displacement of the dam heel

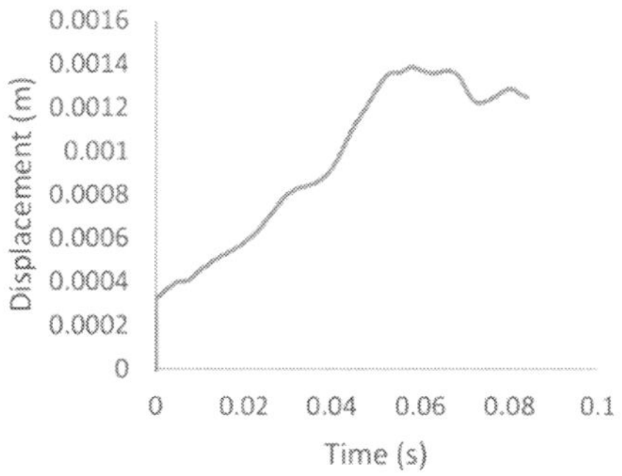

a)

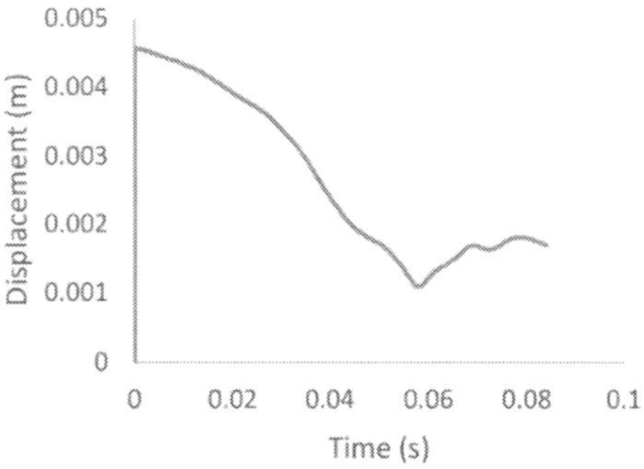

b)

Figure 5: Displacement time history analysis of the dam heel
a) Hydrostatic effect b) Uplift pressure effect

Stress

The maximum and minimum principal stresses for Kinta RCC dam in 2D and 3D model are illustrated through the figures 6 and 7. As indicated in these figures, the maximum stress for is 2.430 MPa and the minimum stress is 3.45 MPa.

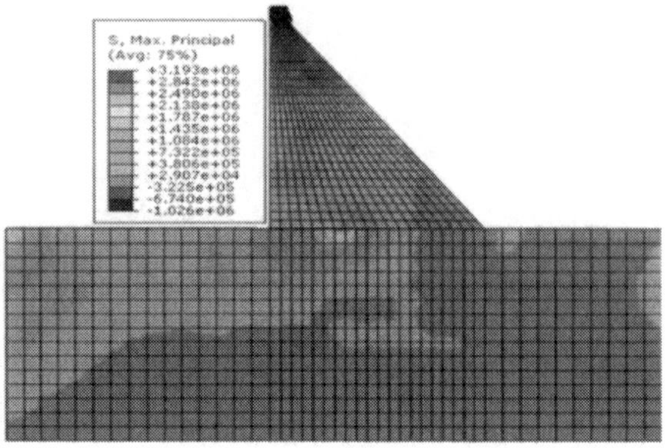

a) Max. Principal Stress

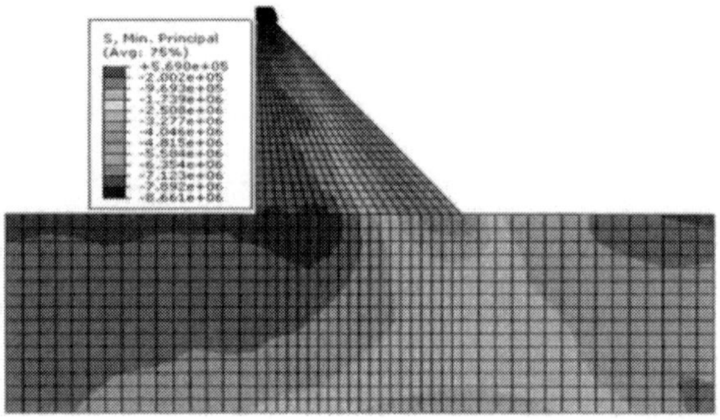

b) Min. Principal Stress

Figure 6. Stress on dam in 2D

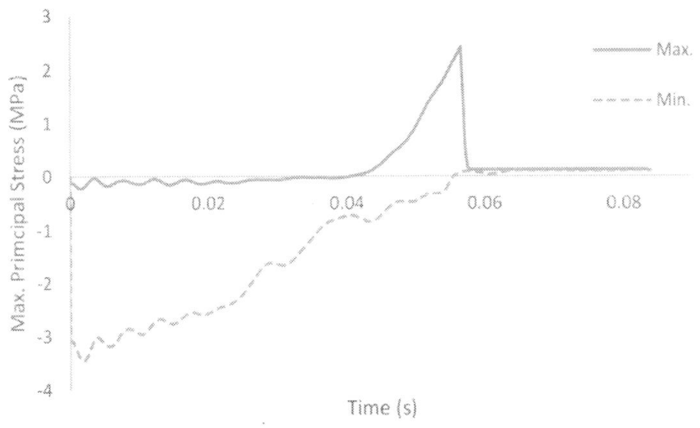

Figure 7. Time history analysis of maximum and minimum principal stresses

Tensile Damage and Crack Pattern in RCC Dam Body

The crack propagation of the dam body is observed at 3 levels of the dam body and formed from the upstream face which propagates toward downstream side. It is because of the pressure applied in the crack due to hydrostatic pressure. The most critical cracks occurs in the heel elements owing to the concurrent effect of hydrostatic and uplift pressures in those elements as shown in Figure 8.

Figure 8. Tensile cracks in the dam body due to Mutual effect of hydrostatic and uplift pressure

The relative displacement of the dam heel due to hydrostatic and uplift pressures for the three cases are investigated and illustrated in figures 9 and 10. It shows that when the concrete density of dam body increases the relative displacement increases too for both directions.

Table 3 indicates the peak relative horizontal and vertical displacements of the Kinta RCC dam for the 3 cases.

Table 3: Peak relative horizontal and vertical displacements of the Kinta RCC dam for the 3 cases.

Cases / Displacements (mm)	1	2	3
Horizontal	0.17	0.17	0.20
Vertical	- 0.42	- 0.42	- 0.45

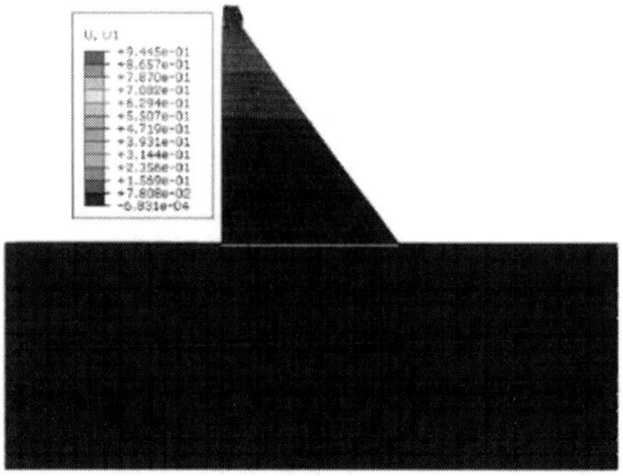

a) γ= 1800 kg/m^3

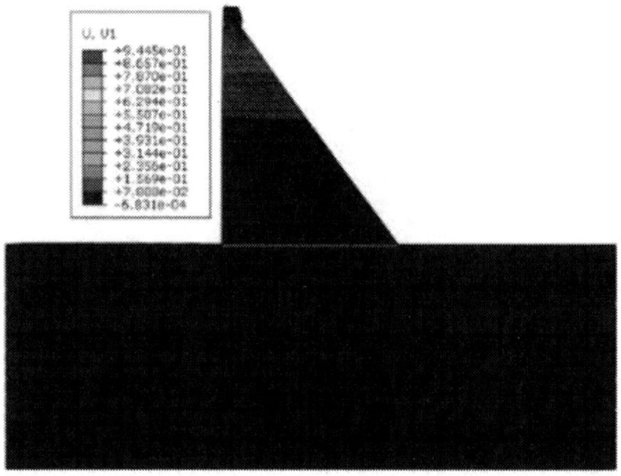

b) γ= 2000 kg/m^3

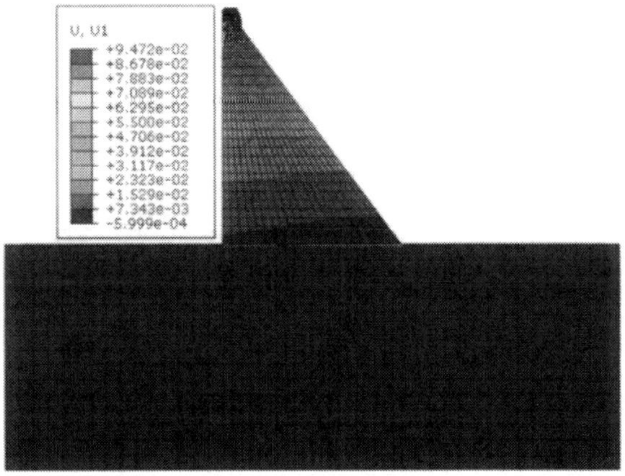

c) γ= 2300 kg/m^3

Figure 9. Relative displacement of the dam heel due to hydrostatic pressure

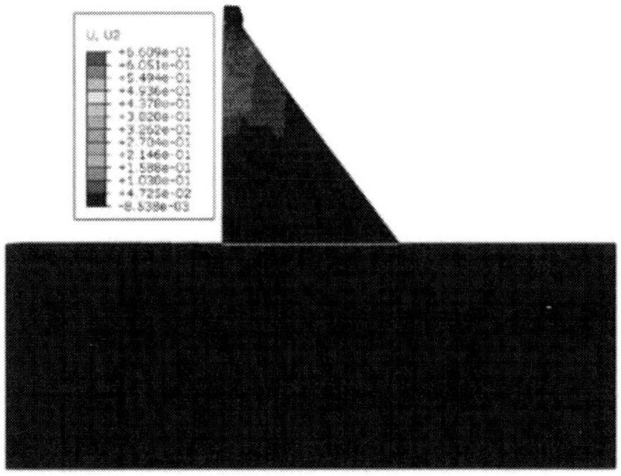

a) γ= 1800 kg/m^3

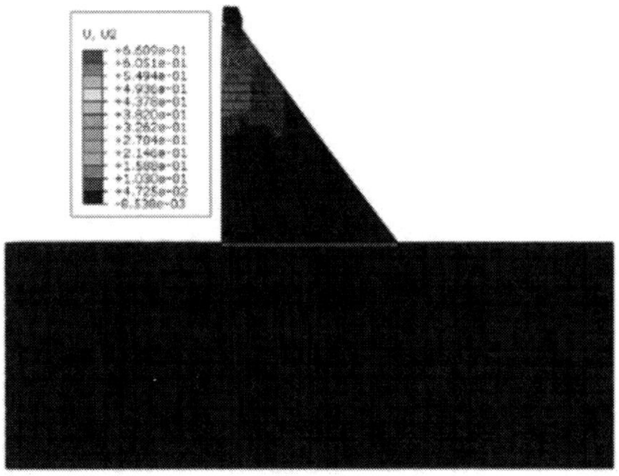

b) γ= 2000 kg/m^3

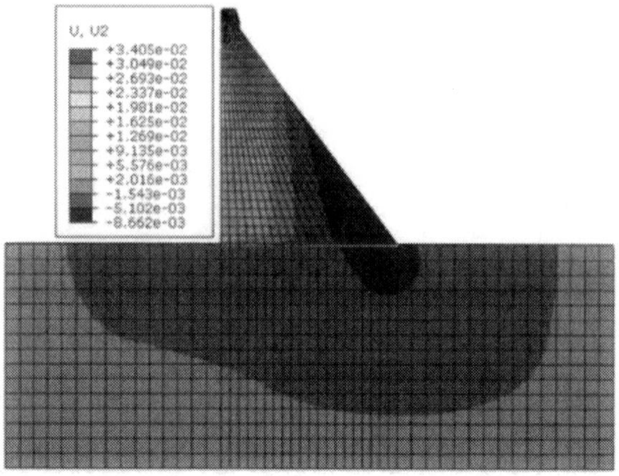

c) γ= 2300 kg/m^3

Figure 10. Relative displacement of the dam heel due to uplift pressure

The stresses in the RCC dam body for the 3 cases are also investigated and illustrated in Figures 11 and 12. It is shown that the maximum and minimum principal stresses are inversely proportional to the concrete density of dam body, when the density increases the maximum principal stress on the foundation decreases and the same for minimum principal stress.

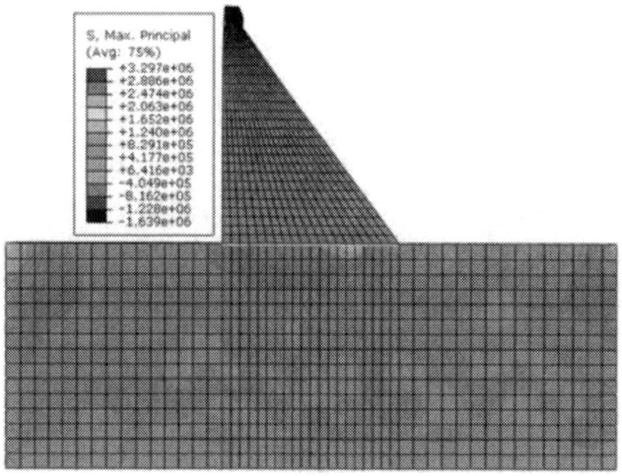

a) γ= 1800 kg/m^3

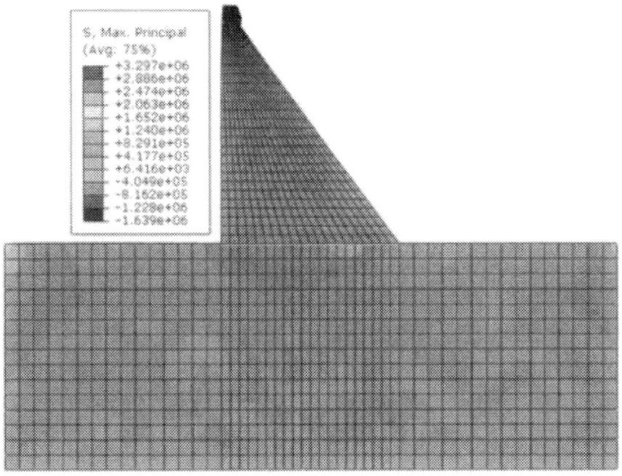

b) γ= 2000 kg/m^3

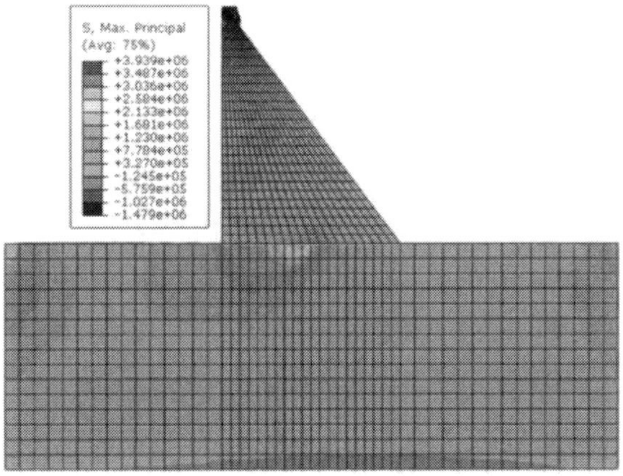

c) γ= 2300 kg/m^3

Figure 11. Maximum principal Stress on dam in 2D

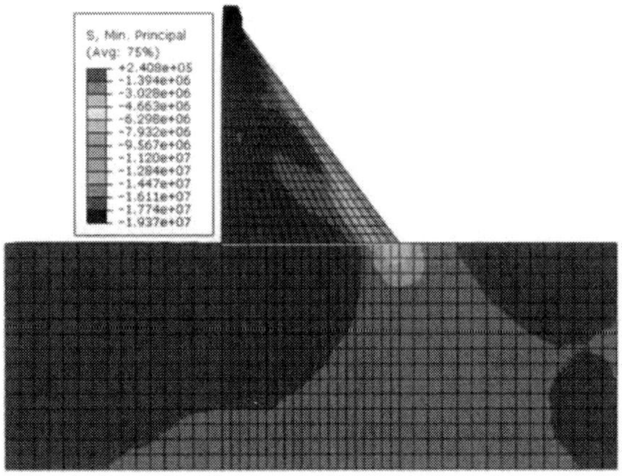

a) γ= 1800 kg/m^3

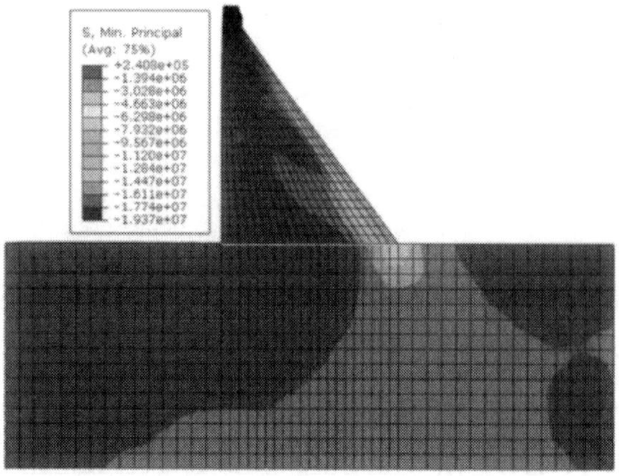

b) γ= 2000 kg/m^3

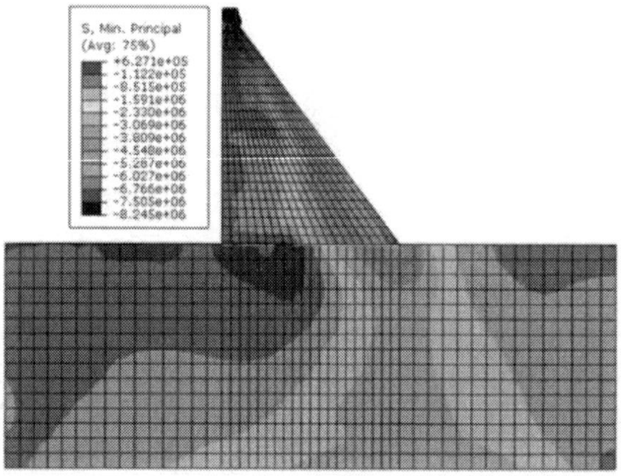

c) γ= 2300 kg/m^3

Figure 12. Minimum principal Stress on dam in 2D

The damage propagation of the Kinta RCC dam is inspected for different conditions, the starting point of the cracking is formed upstream face and it propagates toward downstream side, it is observed when the concrete density of dam body increases over the dam becomes more stable as Figure 13 shows.

a) γ= 1800 kg/m^3

b) γ= 2000 kg/m^3

c) γ= 2300 kg/m^3

Figure 13. Tensile cracks generated in body dam under the simultaneous hydrostatic pressure and uplift pressure effect without drains.

CONCLUSION

In this study an attempt has been made to inspect the concurrent effect of hydrostatic and uplift pressure as well as density changes of the dam body using 2D and 3D models. For this purpose, Kinta RCC dam located in Malaysia has been chosen and nonlinear time history analysis has been conducted. The obtained results have been presented in terms of bottom nodal and relative displacement, tensile crack pattern, maximum and minimum principal stresses. Based on the analysis of 3 different cases of density changes of dam body, the following conclusions can be written:

- Nodal displacement of the heel in vertical displacement due to uplift pressure is 69.5% greater than the horizontal displacement due to the hydrostatic pressure. This shows the importance of effect of uplift pressure in analysis of concrete gravity dam.
- Different densities do not significantly changes the maximum and minimum principal stresses of the dam.
- The critical cracks happen at the heel zone when the hydrostatic and uplift pressures are the most highest.
- As density of the dam body increases, the displacements in both directions increase.
- As density of the dam body increases, the tensile damage decreases.
- A techno-economic study must be done to choice the best variant to reduce the occurrence of crack. In this regard, there are many solutions among them such as grout injection which increases the concrete density of dams.

In general, from above discussions it can be concluded that the uplift pressure affects considerably the analysis of concrete gravity dams and must be taken into consideration mostly in relative displacement and crack propagation, also the concrete density of dam body must be selected carefully.

ACKNOWLEDGMENTS

We gratefully acknowledge those responsible for the civil engineering department of the University of Malaya for the cooperation and support for the implementation of Research Internships and fruitful exchange of information.

REFERENCES

Abaqus Inc. (2010). Abaqus theory manual, Version 6.10. Abaqus Inc. (n.d.).

Akköse, M. and im k, E. (2010). "Non-linear seismic response of concrete gravity dams to near-fault ground motions including damwater-sediment-foundation interaction." Applied Mathematical Modelling, Vol. 34, No. 11, pp. 3685-3700.

Mazloumi, A., Ghaemian, M. and Noorzad, A. (2012). Nonlinear seismic analysis of rcc dam considering orthotropic behavior of layers, In International Symposium On Dams For a Changing World.

Ayari, M. L. (1990). "A fracture mechanics based seismic analysis of concrete gravity dams using discrete cracks." Engineering Fracture Mechanics, Vol. 35, Nos. 1/2/3, pp. 587-598

Barani O. R., Majidaie S., and Mosallanejad M.,: "Numerical Modeling of Water Pressure in Propagating Concrete Cracks" *Journal of Engineering Mechanics,* 0733-10.1061/(ASCE)EM.1943-7889.0001048.

Bhattacharjee, S. S., and Léger, P. (1995). "Fracture response of gravity dams due to rise of reservoir elevation." J. Struct. Eng., 10.1061/(ASCE)0733-9445(1995)121:9(1298), 1298–1305.

Brühwiler, E., and Saoma, V. E. (1995a). "Water fracture interaction in concrete. Part I: Fracture properties." ACI Mater. J., 92(3), 296–303.

Brühwiler, E., and Saoma, V. E. (1995b). "Water fracture interaction in concrete. Part II: Hydrostatic pressure in cracks." ACI Mater. J., 92(4), 383–390.

Calayir, Y. and Karaton, M. (2005). "Seismic fracture analysis of concrete gravity dams including dam–reservoir interaction." Computers & Structures, Vol. 83, Nos. 19-20, pp. 1595-1606

GHD (2002). Study of restrictions on RCC temperature, Stage 2 development of Ipoh water supply.

Guanglun, W., Pekau, O. A., Chuhan, Z., and Shaomin, W. (2000). Seismic fracture analysis of concrete gravity dams based on nonlinear fracture mechanics, Engineering Fracture Mechanics, 65.

Javanmardi, F., Léger, P., and Tinawi, R. (2005). "Seismic water pressure in cracked concrete gravity dams: Experimental study and theoretical modeling." J. Struct. Eng., 10.1061/(ASCE)0733-9445(2005)131:1(139), 139–150.

J. G. M. van Mier (1997). Fracture process of concrete: Assessment of material parameters for fracture models, Boca Raton, Florida: CRC Press, Inc.

Jiang, S. and Du, C. (2012). Seismic stability analysis of concrete gravity dams with penetrated cracks, Vol. 5, No. 2007, pp. 105-119.

Ghaedi K., Jameel M., Zainah I., and Khanzaei P. : "Seismic Analysis of Roller Compacted Concrete (RCC) Dams Considering Effect of Sizes and Shapes of Galleries", 2016, 10.1007/s12205-015-0538-2.

Ghaedi K., Javanmardi A., Gordan M., Khatibi H., Abdollah Mohammadi: "Application of 2D and 3D finite element modelling of gravity dams under seismic loading", The 3rd National Graduate Conference (NatGrad2015), Universiti Tenaga Nasional, Putrajaya Campus, 8-9 April 2015.

Lee, J. and Fenves, G. (1998) : "Plastic-damage model for cyclic loading of concrete structures." Journal of Engineering Mechanics, Vol. 124, No. 8, pp. 892-900.

Lubliner, J., Oliver, J., Oller, S., and Oñate, E. (1989) : "A plastic-damage model for concrete." International Journal of Solids and Structures, Vol. 25, No. 3, pp. 299-326.

Paggi, M., Ferro, G., and Braga, F. (2013). "A multiscale approach for the seismic analysis of concrete gravity dams." Computers & Structures, Vol. 122, pp. 230-238

Pandey K.K., Kumar N., Kumar V., "Formulation of transient uplift pressure in cracks of concrete gravity dams", *i-manager's Journal o Civil Vol. No. 1,* 16-25, 2015

Pekau, O. and Zhu, X. (2008). "Effect of seismic uplift pressure on the behavior of concrete gravity dams with a penetrated crack." J. Eng.Mech., 10.1061/(ASCE)0733-9399(2008)134:11(991), 991–999.

Plizzari, G. (1997). "LEFM applications to concrete gravity dams." J. Eng. Mech., 10.1061/(ASCE)0733-399(1997)123:8(808), 808–815.

Skrikerud, P. and Bachmann, H. (1986). "Discrete crack modeling for dynamically loaded, unreinforced concrete structures." Earthquake Engineering & Structural Dynamics, Vol. 14, No. 2, pp. 297-315.17.

Slowik, V., and Saouma, V. E. (2000). "Water pressure in propagating concrete cracks." J. Struct. Eng., 10.1061/(ASCE)0733-9445(2000)126:2(235), 235–242.

Zhang, S., Wang, G., and Yu, X. (2013). "Seismic cracking analysis of concrete gravity dams with initial cracks using the extended finite element method." Engineering Structures, Vol. 56, pp. 528-543.

Zhu, X., and Pekau, O. A. (2007). "Seismic behavior of concrete gravity dams with penetrated cracks and equivalent impact damping." *Engrg. Struct.*, 29(3), 336–345

SOIL-STRUCTURE INTERACTION DYNAMIC ANALYSIS FOR AN EXISTING OLD EARTH-ROCK FILL DAM WITH CONCRETE CORE

Mahmood Seid-Karbasi, Senior Engineering Analyst (PhD, P.Eng.)[1]
Upul Atukorala, Principal, Senior Geotechnical Engineer (PhD, P.Eng.)[2]
Herb Hawson, Principal, Senior Geotechnical Engineer (MSc, P.Eng.)[2]
Bruce Downing, Principal, Senior Geotechnical Engineer (MSc, DIC, P.Eng.)[2]

ABSTRACT

An existing cascading dams complex, located in a park just upstream from currently residential/commercial area within a city in western Canada comprises two earthfill dams namely: Upper and Lower Dams and have a high seismic hazard exposure. The dams were constructed in the early 1900s to supply water to a nearby area. The Lower Dam has been assigned "Extreme" consequence category whereas the Upper Dam has been assigned a "Very High" consequence category based on Canadian Dam Association Dam Safety Guidelines 2007 (2013 Edition).

This paper outlines the project components and analyses carried out to assess the seismic stability and performance of the Lower Dam. The Lower Dam is a rockfill dam with a 1.2 m thick, vertical concrete core wall. Mine/process waste material was placed on the downstream face of the dam sometime after construction with an additional toe berm/filter layer placed over the lower half of the downstream face in 1980. The dam is approximately 24 m high with a crest length of about 77 m. The width of the dam crest varies from about 5m at the right abutment to more than 20 m at the left abutment. The upstream and downstream embankments slope at 2.2H:1V and 1.8H:1V, respectively.

A rigorous dynamic soil structure interaction (SSI) analysis was carried out using FLAC^{2D} computer program for the lower dam to assess the seismic behavior of the dam when subjected to shaking levels corresponding to 10,000-year (equivalent to the Maximum Credible Earthquake, MCE) and 2,475-year return periods. The analyses showed that under "as-is" conditions the dam would undergo large deformations in the upstream and downstream shells when subjected to the 10,000-yr and 2,475-yr ground motions. The downstream shell undergoes excessive displacements exhibiting shallow failure for the MCE ground motions. Driven by the unbalanced inertial forces, the concrete wall tends to rotate towards the reservoir, but its vertical displacements are not large enough to cause loss of freeboard in the reservoir. Therefore catastrophic overtopping failure is unlikely.

INTRODUCTION

A cascading dams system comprising two earthfill dams i.e. the Upper and Lower Dams are located in a river valley within the city park in western Canada. The Lower Dam is

[1] Golder Associates Ltd., Suite 200 - 2920 Virtual Way, Vancouver, BC, V5M 0C4, Canada, mseidkarbasi@golder.com.
[2] Golder Associates Ltd., Suite 200 - 2920 Virtual Way, Vancouver, BC, V5M 0C4, Canada.

located downstream of the Upper Dam, and was constructed in 1910 to supply water for the nearby operating colliery neighborhood when it was in full production. At present, the classification has been assigned "Extreme" for the Lower Dam and "Very High" for the Upper Dam consequence categories as per CDA Guidelines (2013). Figure 1 shows the aerial view of the dams and the two reservoirs.

Figure 1. Aerial View of the Cascading Dams Complex.

The river valley is narrow with steep slopes within the park boundaries with exposed conglomerate bedrock and till on the abutments. Some thin colluvium has been noted at the base of the valley slopes. Soft/loose unconsolidated channel deposits are present at the base of the valley between the two dams and downstream of the Lower Dam; they are of limited lateral extent due to the narrow and steep sided nature of the river valley. They appear to only be present when the valley bottom flattens below steeper reaches of the channel where bedrock is exposed. Some drilling investigations were completed at both dams in 1978. Limited supplementary investigations were carried out as part of a study in 2014 to collect additional data on the concrete core and downstream shell materials. The results of these investigations are used in this assessment.

The main focus of this investigation is the Lower dam seismic performance. The study results of the complex remediation and associated risks with each remedial measure can be found in Hawson et al. (2017) and Roberds et al. (2017), respectively. This paper presents results of SSI dynamic analysis carried out to assess the seismic stability of the Lower dam and estimate the anticipated deformations of the concrete core under strong seismic shaking that corresponds to a 10,000-yr return period using a two-dimensional dynamic analysis approach. Additional analysis were also carried out for strong shaking that corresponds to a 2,475-yr return period to assess the dam performance under lower shaking level. Time-domain analyses were carried out using the commercially available computer program, $FLAC^{2D}$ (*Itasca*, 2008).

THE LOWER DAM AND ITS FOUNDATION

The Lower Dam is a rockfill dam with a 1.2 m thick, vertical, concrete core wall. Mine/process waste material was placed on the downstream face of the dam sometime after construction with an additional toe berm/filter layer placed over the lower half of the downstream face in 1980. The lower dam is approximately 24 m high with a crest length of about 77 m. The width of the dam crest varies from about 5m at the right abutment to more than 20 m at the left abutment. The upstream and downstream embankments slope at 2.2H:1V and 1.8H:1V, respectively.

The Lower Dam is mainly comprised of rock/random fill in U/S and D/S shells that are separated by the concrete core wall as seepage cut-off for the dam. The rockfill placed in the lower portion of the D/S shell is overlain by fills comprising cinders and slag, which change to a mixture of sand and gravel at the D/S toe. The dam is founded on strong conglomerate bedrock sloping at about 4 to 18 degrees with the horizontal direction. Figure 2 shows typical dam cross-section inferred from original design and available documentations from subsequent modifications over the dam long history.

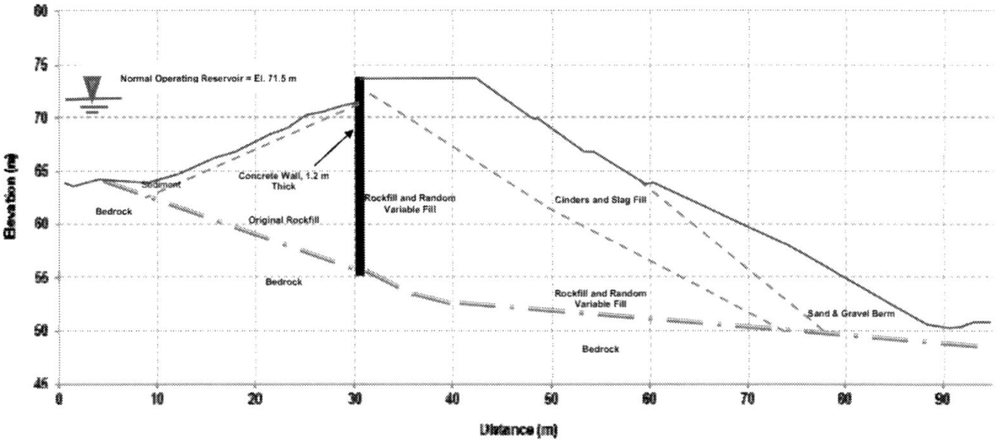

Figure 2. Inferred Maximum Cross-Section of the Lower Dam.

The geotechnical conditions at the Lower Dam site were established using data collected from the different geotechnical investigations completed at the site. The 2014 investigations also included downhole shear wave velocity measurements, non-invasive Ground Penetrating Radar (GPR) profiling of the concrete core to investigate the possible presence and spacing of rebar, and recovering concrete core samples for detailed inspection and strength testing.

TYPES OF MATERIALS IN THE DAM BODY ZONES

The dam body is inferred to consist of five (5) distinct material zones:

Zone 1: Coarse-grained rockfill and random variable fill (downstream) – This zone in the Lower Dam is comprised of coarse-grained rockfill and random variable fill materials placed during original construction of the dam in 1911. Very limited data is available on the quality and in-situ relative density of rockfill from the investigations completed to date. Some Standard Penetration Tests (SPTs) were carried out during the 1978 investigation. The measurements indicate erratic blow counts influenced by the material size. Therefore, the N values are not considered to be representative of the in-situ materials. The coarse-grained rock fill materials are not susceptible to excess pore water pressure generation and/or liquefaction during earthquake loading. These fills have been encountered in the downstream shell of the dam placed to a maximum elevation of about +72 m forming a slope of about 1.9H: 1V.

Zone 2: Coarse-grained rockfill (upstream) – No information is available on the condition and in-situ relative density of the rockfill upstream of the core wall. For purposes of the dynamic analysis of the dam, these coarse-grained rockfill materials were assigned parameters similar to Zone 1. As Zone 1 fills, these coarse-grained rock fill materials are not susceptible to excess pore water pressure generation and/or liquefaction during earthquake loading.

Zone 3: Granular fills consisting of cinders and slag on top of the downstream rockfill shell (Zone 1), cinders, ash, slags and sands from mining works have been placed on top of Zone 1 rockfill and random variable fills by end-dumping to flatten the downstream shell slope. This zone is about 5m thick at the dam crest and extends to the lower part of the downstream shell. This zone consists of very loose sandy soils with N-values varying from 1 to 4 blows/0.3 m. This zone was characterized as a material exhibiting "sand-like" behavior with an equivalent $(N_1)_{60}$ of about 3 blows/0.3m. These fills are inferred to be in a moist to dry state, and as such, they are not expected to undergo liquefaction and generation of excess pore water pressures in a manner similar to water saturated loose granular fills as a result of earthquake loading.

Zone 4: Sand and gravel berm at the downstream toe, the berm has been constructed in 1980 to buttress the slope and provide lateral support. The fills consist of mine process waste materials. The as-built drawings indicate that a toe berm/filter layer has been constructed at the interface between Zones 2 and 3. This zone has been inferred to consist of medium dense granular soils with an equivalent SPT $(N_1)_{60}$ of 20 blows/0.3m.

It is envisaged that the Zone 4 soils are not prone to generation of excess pore water pressure and liquefaction during earthquake shaking.

The 1.2 m wide concrete core wall located in the center of the dam serves as a seepage cut-off wall. Concrete Mass Density (average of dry and wet) was considered as 2300 kg/m^3. A value of 30 MPa for compressive strength of concrete was used for this assessment. The rockfill embankments both upstream and downstream of the concrete wall provide lateral confinement and support for the core wall. The upstream fills are about 2.0 m lower than the downstream fills. The wall is uncovered about 0.6 m at the dam crest.

FOUNDATION AND WATER CONDITIONS

The upstream and downstream shells and the concrete seepage cut off wall comprising the dam are inferred to be directly founded on bedrock that slopes at about 3H:1V towards downstream. The recent investigations indicate that the foundation consists of strong conglomerate bedrock with an average shear wave velocity (V_s) greater than 760 m/s. The foundation is inferred to be intact and no seepage issues have been observed or noted during the dam safety reviews.

The water level of at Normal Operating Reservoir (NOR) Level is at elevation of +71.5m as seen in Figure 2. In the downstream shell, the inferred water level was taken at the top of the bedrock surface, with no water flowing through the downstream shell.

SEISMICITY AND GROUND MOTIONS

The seismicity at the site results from the offshore subduction of the Juan de Fuca Plate beneath the North America Plate. The ground motion parameters applicable for the dynamic analysis of the Lower Dam were established based on studies already completed for several nearby dams. The project schedule did not permit the completion of a comprehensive site-specific seismic hazard assessment.

"Firm-ground" peak horizontal acceleration applicable for a return period of 10,000-yrs was estimated by combining all available data on peak ground acceleration, PGA as a function of the annual probability of exceedance. Based on the available data, the PGA that corresponds to a return period of 10,000-yrs was established as approximately equal to 0.8 g. Figure 3a shows peak ground acceleration vs. return period. The Uniform Hazard Response Spectrum (UHRS) established for the Lower Dam is shown on Figure 3b. This spectrum was established by uniformly scaling the spectrum to a PGA = 0.8 g. The spectra are generally similar in shape when plotted in the Log_{10} (T) - Log_{10} (Sa) space. As seen, the developed spectrum for this dam site shows a good match after T=0.1 (sec.) with the spectrum established for Dam No. 3 from a comprehensive site-specific hazard assessment. In the absence of such assessment for this site, the input acceleration time-histories already developed (by others during a previous phase of the project) for the 3,000-yr were uniformly scaled relative to the PGA to establish the input motions for the 2,475-yr and 10,000-yr return period ground motions.

Figure 3c shows the modified acceleration time-history record of CC0 earthquake for "firm-ground" condition (i.e. Class "C") used in these analyses.

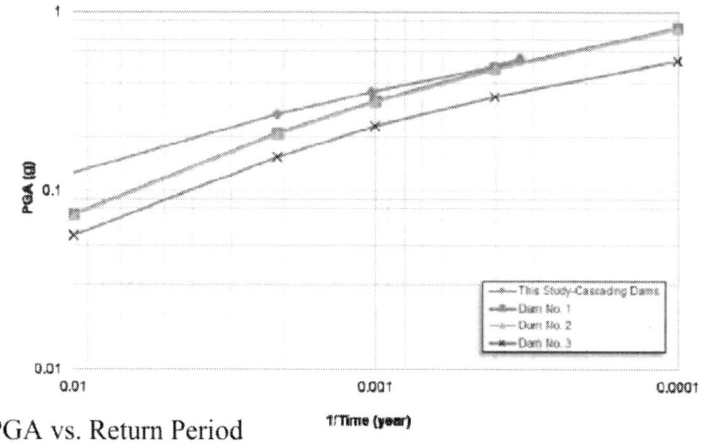

a) PGA vs. Return Period

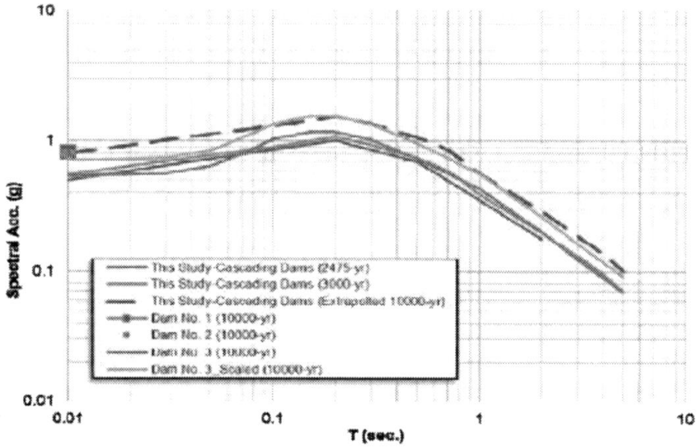

b) Spectral Acceleration (D = 5%)

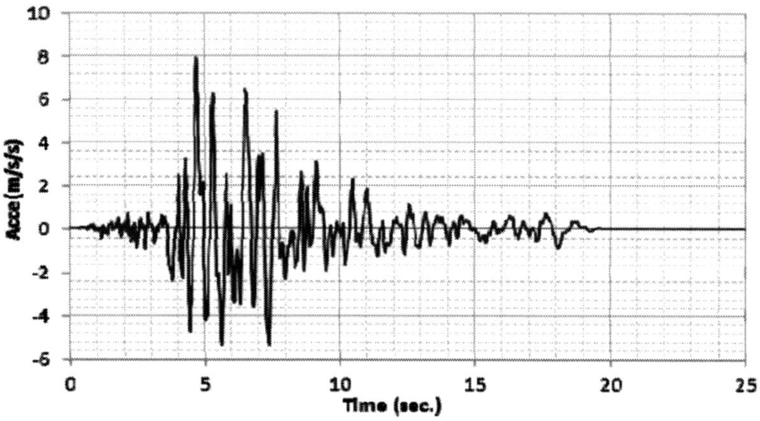

c) Acceleration Time-History (CC0)

Figure 3. PGA vs. Return Period, Spectral Acceleration and Earthquake Record

DYNAMIC ANALYSIS OF THE DAM-FOUNDATION SYSTEM

The response of the dam-foundation system to earthquake shaking was evaluated in two stages. In the first stage, the dam body and its foundation were analyzed under gravity loads (i.e. in the static mode) with the steady state water level that corresponds to the pool at normal operating reservoir level in order to establish the pre-earthquake stress state within the material zones comprising the dam body.

Thereafter, the analysis was switched to the dynamic mode. Considering the coarse-grained nature of the materials comprising the dam, generation of excess pore water pressures and liquefaction are not anticipated in these materials. Hence, the seismic response analyses were carried out using a total stress-based approach.

The dynamic analysis of the dam was conducted using the commercially available finite difference computer code FLAC^{2D} (V6, *Itasca,* 2008). FLAC has several built-in widely used constitutive models (e.g. elastic and Mohr-Coulomb) and also the possibility to implement user-defined constitutive models. The program can be used for dynamic soil-structure analysis carried out in the time domain.

The dynamic response of the dam was assessed for a 2-D cross section of the dam, which corresponds to the deepest section of the valley. Plane strain conditions were assumed and 3-D effects resulting from the valley geometry were ignored.

The earthquake ground motions were applied as time-histories of shear stresses at the base of the model with a compliant (i.e. "quiet") boundary specified at the base. The dam was extended some 50 m and 90 m upstream and downstream of the core wall, respectively, and free-field boundaries were imposed along the left and right hand boundaries. The time-histories of shear stresses were established from the "outcropping" earthquake acceleration time-history records (i.e. without any overburden effects).

FLAC MODEL AND MATERIAL PARAMETERS

The 24m-high Lower Dam with its foundation (bedrock) was discretized using a finite difference grid comprising 30 × 95 zones in the vertical and horizontal directions, respectively. The nominal thickness of the material zones was 1 m. The model extended about 100m and 30m from the upstream and downstream toe areas of the dam, respectively over a total length of 140 m. Figure 4 shows the FLAC finite difference model along with the different material zones considered in the analyses.

Geo-Materials Parameters

The analysis was initiated with an elastic model as a first step to define the stress-level dependent parameters, and a Mohr-Coulomb model in the second step. For the elastic analysis, the small-strain shear modulus, G_0, of the materials was estimated using the correlation developed by Seed et al. (1986). The corresponding static shear modulus, G_{st} was established as $G_0/9$. The Poisson's ratio μ was taken as 0.33 that corresponds to an at rest earth pressure coefficient, $K_0 = 0.5$.

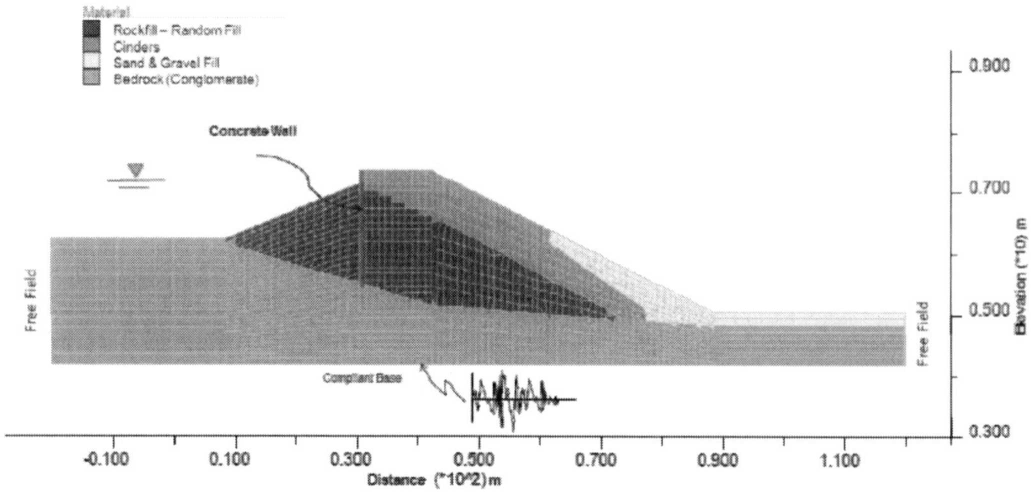

Figure 4. FLAC Model of the Dam-Foundation System.

The strength parameters of the rockfill were estimated using the data base found in Leps (1971) and others (e.g. Barton & Kjaersnly, 1981 and Saboya & Byrne, 1993) for rock-fill dams. Stress level-dependent frictional angles (φ) were assigned for the different rockfill zones.

The small-strain shear modulus, G_0 for rockfill (Zones 1 & 2) was estimated based on a stiffness number, i.e. $K_{2\text{-max}} = 100$ which is in agreement with the values for similar materials suggested by Uddin & Gazetas and Seed et al. (1986). For other granular materials in the dam body (i.e. in Zone 3; cinders and slag and Zone 4; sand and gravel) the shear stiffness was computed based on an estimated $(N_1)_{60}$ value using the following equations:

$$K_{2\text{-max}} = 20 \cdot [(N_1)_{60}]^{0.33} \quad (1)$$

$$G_0 = 21.7 \cdot Pa \cdot K_{2\text{-max}} \cdot [\sigma_{c'} / Pa]^{0.5} \quad (2)$$

$$\varphi = \varphi_1 - \Delta\varphi \cdot \log_{10} (\sigma_{c'}/Pa) \quad (3)$$

where, $K_{2\text{-max}}$ = stiffness parameter, $\sigma_{c'}$ = effective confining stress, Pa = atmospheric stress (=~100 kPa), φ_1 = reference friction angle at 100 kPa, and $\Delta\varphi$ = friction angle increment for every log cycle of stress level change.

The coarse-grained rockfill, variable random fill, sand and gravel fill, and cinder and slag fills were modeled using the built-in Mohr-Coulomb Constitutive model in FLAC with stress-level dependent stiffness parameters. Bedrock was modelled as an elastic material and its stiffness G_0, was computed based on an inferred shear wave velocity of 1,200 (m/s) that represents "Class B" site conditions as per NBCC 2012.

In the dynamic analysis, in order to take into account the reduction in shear modulus as a function of shear strain, Sig3 model, the hysteretic model (built-in the FLAC program) was used. This three-parameter hysteretic model for the granular sandy soils was assigned for material Zones 3 and 4 to simulate the non-linear behavior of these soils. The hysteretic model parameters for the rockfill in material Zones 1 and 2 were calibrated with the data found in the literature for similar materials, e.g. (Yasuda & Matsumuto (1993) and Yasuda et al. (2003). Figure 5 compares the results from the FLAC simulations with the laboratory data from Yasuda & Matsumuto (1993).

Concrete Core Wall

The 1.2 m thick concrete core wall was modeled using beam elements that interacted with the rockfill zones via interface elements. The seismic response of the concrete core wall was evaluated for two scenarios:

1. Considering elastic behavior (i.e. unlimited yield moment capacity); and
2. Imposing a limit on the yield moment capacity (M_P) = 600 kNm to account for tensile cracking threshold of the concrete wall.

The 1.2 m thick concrete core wall was treated as beam structural element fully embedded in the rockfill materials in both sides. A summary of the input parameters established for the different zones is provided in Table 1 below.

. Table 1. Material Properties Used in FLAC Analyses

Material Zone	Unit Weight (kN/m³)	$(N_1)_{60}$ [1] (blows/0.3m)	$K_{2\text{-max}}$	φ' (Deg.)	$\Delta\varphi_1$ (Deg.)	G_0 [2] (kPa)	B [3] (kPa)	E (kPa)
1. Rockfill/Random Fill (Zones 1, 2, and 5)	20.0	----	100[4]	40	5.0	Var.	2.67·G	----
2. Cinders & Slag Fills (Zone 3)	18.5	3	29	33	----	Var.	2.67·G	----
3. Sand and Gravel (Zone 4)	19.0	20	54	35	----	Var.	2.67·G	----
4. Conglomerate Bedrock	26.0	----	----	----	----	3.8e6	----	----
5. Concrete Wall[5]	23.0	----	----	----	----	----	----	1.13e7
6. Cement-Stabilized Region (Zone 6)[6]	20.0	----	----	----	----	4.2e5	2.67·G	500·UCS

Notes:
1) Representative $(N_1)_{60}$ values inferred from limited previous data.
2) Small strain shear modulus, G_0, of dam body zones were estimated based on $G_0 = 21.7\ K_{2\text{-max}}.P_a.(\sigma'_m/Pa)^{0.5}$
3) Bulk modulus, B was estimated for condition of K_0=0.5
4) Rock fill stiffness parameters were estimated based on assumed $K_{2\text{-max}}$ = 100
5) A plastic moment of 600 kNm was assigned in some of the analyses carried out
6) Soil-cement strength and stiffness properties were estimated based on UCS=2000kPa and E=500 · UCS, respectively

In addition to the "as-is" conditions, another set of analysis was conducted for the upgraded dam where a 5 m thick surface zone of the downstream shell of the dam was improved with soil-cement to strengthen the shell so that the dam could withstand overtopping due to flooding. Table 1 also shows the properties of the improved zone.

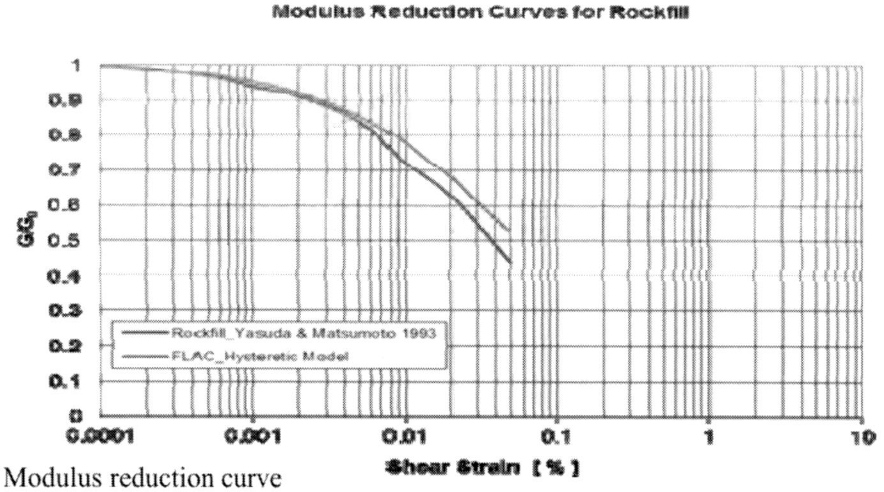

b) Modulus reduction curve

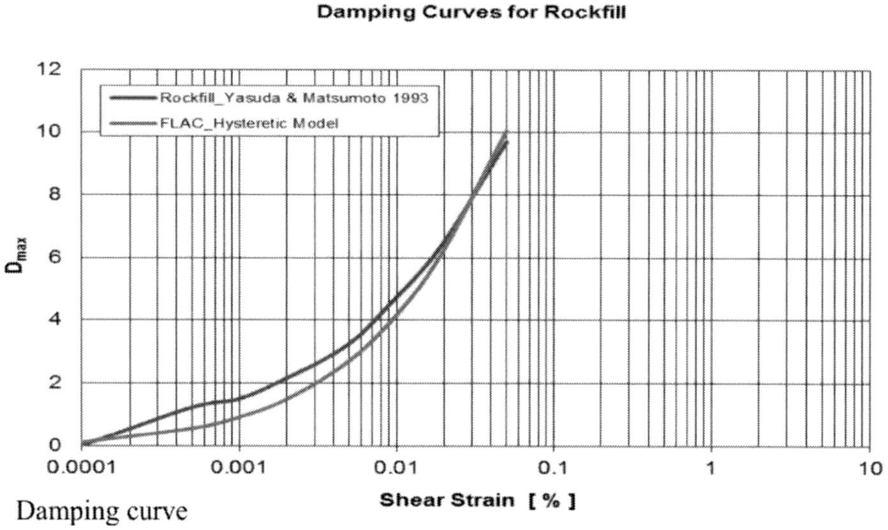

a) Damping curve

Figure 5. Comparison of FLAC-Sig3 Hysteretic Model with Laboratory Data.

RESULTS AND DISCUSSIONS

As mentioned earlier, the analyses were first conducted for the "as-is" condition and then for the improved case where the performance was examined under seismic loading conditions. The results of the "as-is" case are presented and discussed in the following sections.

"As-Is" Case

After establishing the pre-earthquake in-situ stress state for gravity loading of the dam, the performance of the Lower Dam was assessed for ground motions corresponding to 10,000-yr and 2,475-yr return periods. Figure 6 shows the computed seismic response of the dam in terms of lateral displacements at the end of shaking for an earthquake ground motions with a 10,000-yr return period. The results indicate that large displacements in the order of several meters occur in both the upstream and downstream shells of the dam. The computed permanent displacements are larger in the downstream shell than those in the upstream shell.

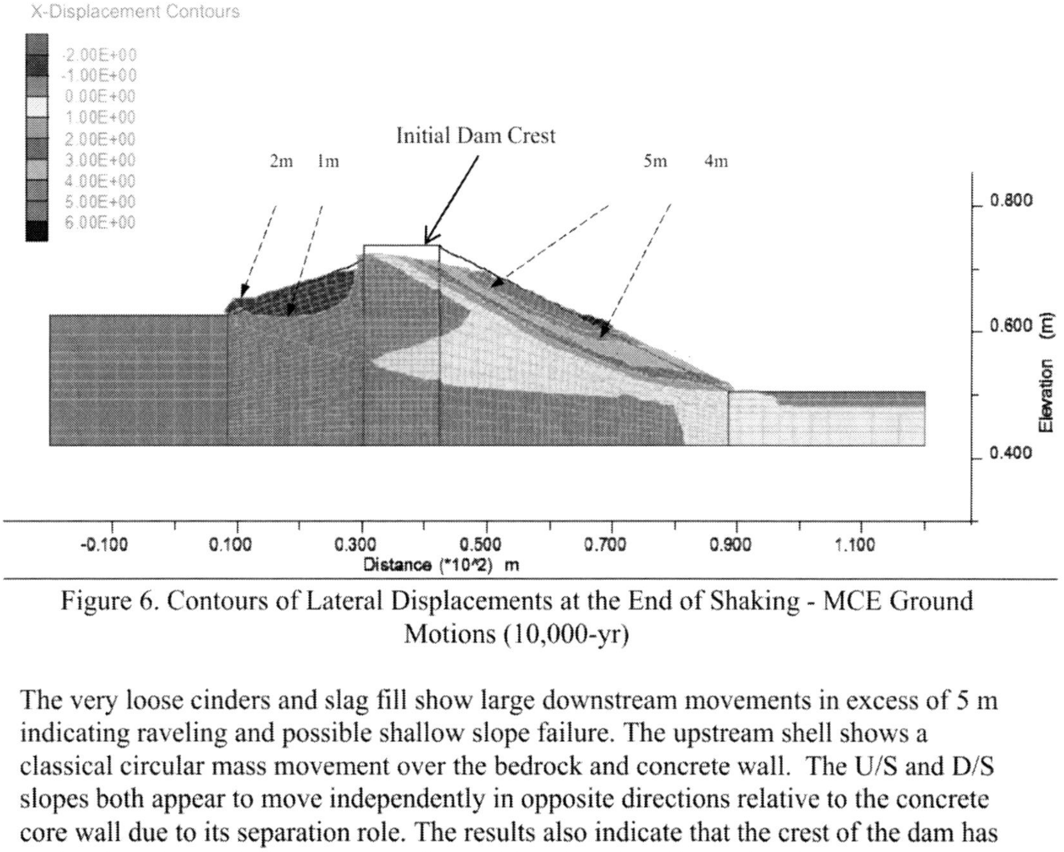

Figure 6. Contours of Lateral Displacements at the End of Shaking - MCE Ground Motions (10,000-yr)

The very loose cinders and slag fill show large downstream movements in excess of 5 m indicating raveling and possible shallow slope failure. The upstream shell shows a classical circular mass movement over the bedrock and concrete wall. The U/S and D/S slopes both appear to move independently in opposite directions relative to the concrete core wall due to its separation role. The results also indicate that the crest of the dam has settled from its original pre-earthquake position (i.e. elev. 73.5m) by about 2 m. Figure 7 shows the residual bending moments and displacements in the wall at the end of shaking. As may be seen, the wall moves towards the reservoir rotating almost uniformly about the heel. As a result, high bending moments are induced in the lower part of the wall. The wall also undergoes a vertical displacement of about 150mm; however, it is inferred that this vertical displacement or loss of free board may not result in an over-topping failure when the reservoir level is not above NOR.

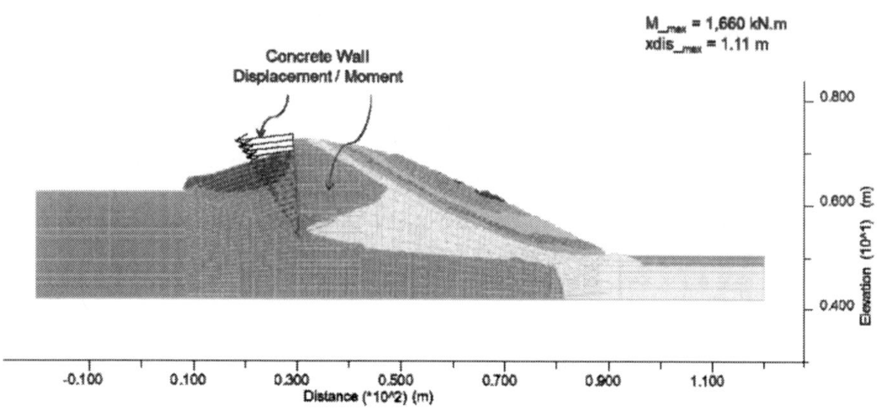

Figure 7. Residual Concrete Wall Displacements and Moment, MCE Motions (10,000-yr)

The transient displacements and bending moments induced in the wall during strong shaking are higher than the residual values as the inertia effects of the earthquake no longer exist at the end of shaking. Figure 8 shows the time-history of lateral displacement of a structural node at the top of the wall along with that at the base of the model (input motion). It shows that the maximum transient displacements in the wall can be as high as 1.2 m.

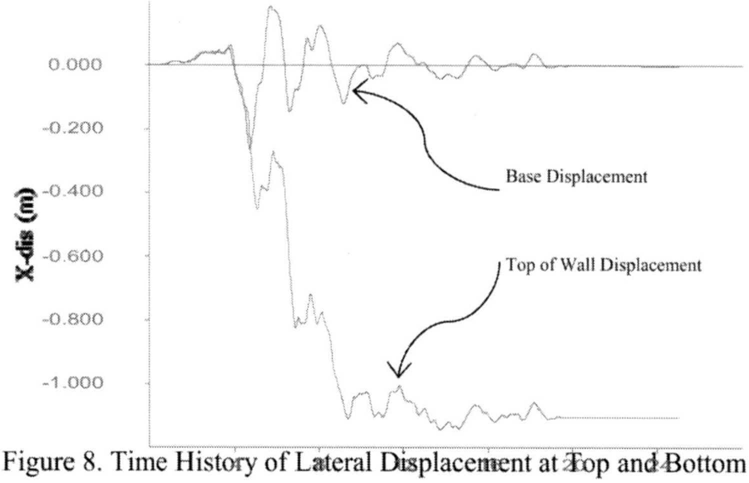

Figure 8. Time History of Lateral Displacement at Top and Bottom of the Wall

A second set of analyses for this condition were conducted with a plastic moment of 600kNm for the concrete wall to account for the threshold cracking tensile stress due to its flexural behavior. Limiting the plastic moment capacity of the wall resulted in larger lateral displacements of the wall. The maximum bending moment was limited to the plastic moment capacity of the wall. Figure 9a shows the contours of lateral displacements of the dam along with the distribution of the induced displacements and bending moments in the concrete core wall. The results indicate that large displacements of the dam shell relative to the wall have resulted in some slippage at the interface between the wall and rockfill. Figure 9b shows the distorted mesh superimposed on the initial dam boundary in the vicinity of the wall. As can be seen, some misalignment of the geo-material elements on both sides of the wall is evident in the deformed mesh. This indicates that slippage has occurred along the wall-dam shell contact surfaces during earthquake shaking.

A third set of analyses were carried out for shaking level corresponding to the 2,475-year return period that corresponds to a PGA = 0.5g. The deformation pattern of the dam and concrete wall were similar.

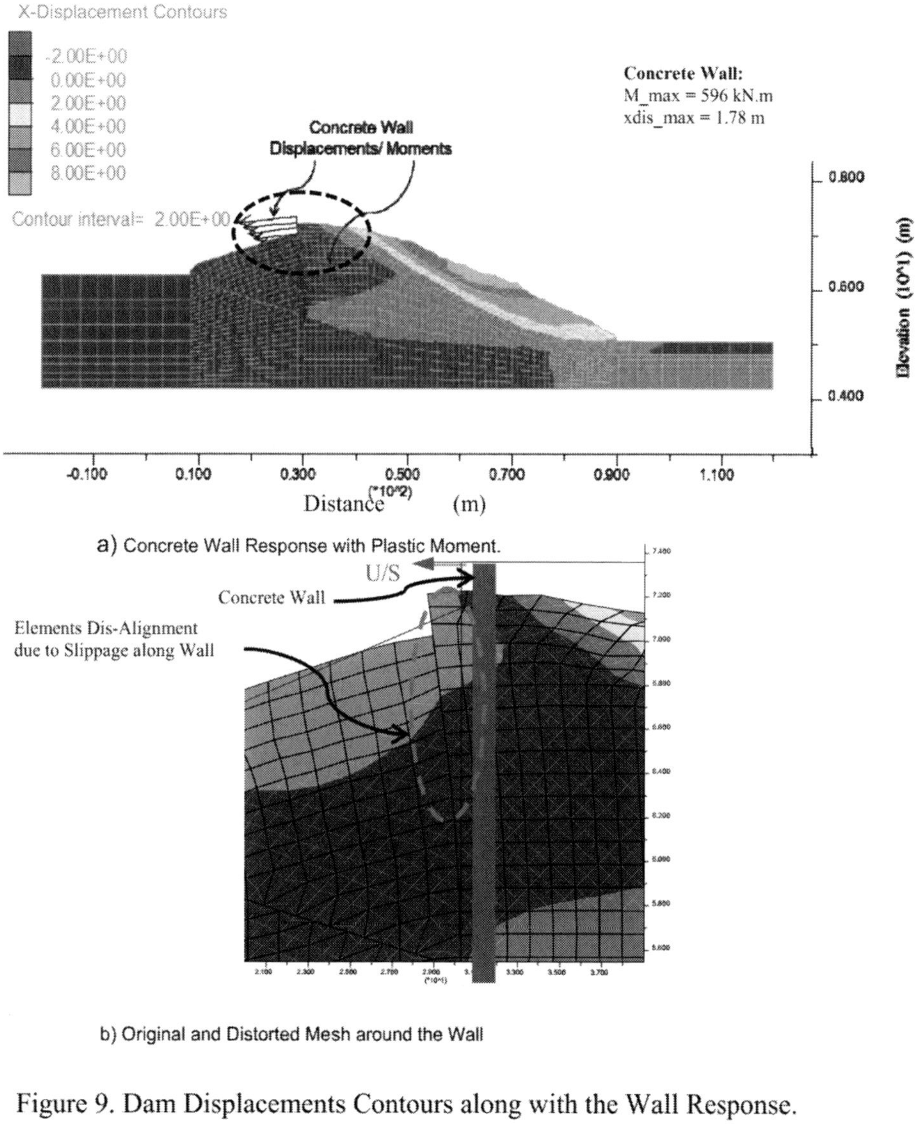

Figure 9. Dam Displacements Contours along with the Wall Response.

Table 2 summarizes the results of the different seismic analyses carried out for the Lower Dam in terms of the concrete core wall displacements computed.

Table 2. Summary of Results of Dynamic SSI Analyses, As-Is Condition

Shaking Level	Concrete Wall Plastic Moment Capacity (kNm)	Computed Horizontal Displacement of the Concrete Wall (mm) At the End of Shaking	
		Horizontal	Vertical
10,000-yr Ground Motion	-----	~1,100	~140
10,000-yr Ground Motion	600	~1,785	~185
2,475-yr Ground Motion	-----	~440	~60
2,475-yr Ground Motion	600	~820	~70

Improved Case

One of the options considered for remediation of the Lower Dam was to allow the dam to be over-topped during the PMF. In order to improve the hydraulic stability of the dam during over-topping, it was considered stiffening and strengthening the upper 2 to 3 m of the downstream shell currently consisting of cinder and slag fills. This can be achieved in-situ using the soil mixing technique. The thickness of the soil-cement zone increases to about 5 m at the downstream side of the dam crest and extends about 6 m downstream of the toe of the dam. The concrete core wall and the cement stabilized zone will be separated/isolated by an approximately 1.0 m wide zone of compressible clay-type material (e.g. vermiculite) to minimize the effects of pounding of the two units together to reduce the risk of damage to the core wall during strong shaking. Therefore the analyses were extended to investigate the dam performance when improved.

The soil-cement zone was assigned a uniaxial compression strength of 2 MPa in the analyses and was modelled using the Mohr-Coulomb Constitutive model with a cohesion of 1,000 kPa (1/2 of UCS). The stiffness parameters of this zone were estimated based on FHWA guidelines proposed for such materials (Bruce, 2000). Published modulus reduction and damping curves for soil-cement stabilized zones were used in the analyses (e.g. Saxena et al. 1988 and Acar & El-Tahir, 1986 and Pestana et al., 2006).

The soft isolator zone was treated as an elastic material with much lower stiffness compared to that of the adjacent geo-material zones in order to minimize their interaction with the concrete wall (i.e. minimal strength for wall-isolator interface).

Figure 10a shows the contours of the computed lateral displacements for the improved case along with the induced displacement and bending moment distributions in the concrete wall when subjected to the 10,000-yr ground motions. As can be seen, the displacements of the downstream shell are reduced significantly due to the high strength of the soil-cement material. The soil-cement stabilized zone also acts as a shear key at the toe of the dam which assists the stability. The upstream slope displacements, however, remained unchanged from the previous analyses. The isolator zone behind the concrete wall facilitates independent movement of the wall and improved the stability of the downstream shell by allowing the formation of a gap between the wall and the geo-material as depicted in Figure 10b.

Analyses were carried out imposing a plastic moment of 600 kNm for the concrete wall. The results indicated that the deformation pattern of the wall is essentially the same and over-topping at NOR condition is not predicted.

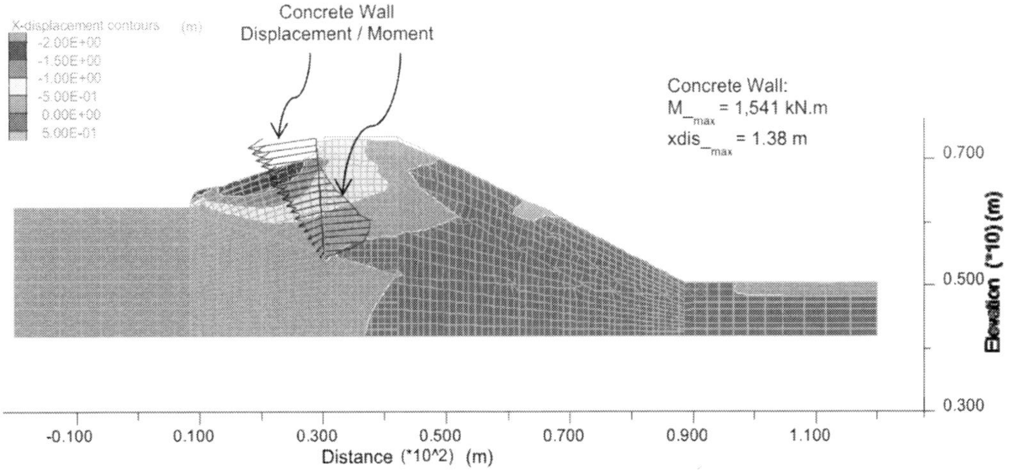

a) Lateral Displacement Contours and Wall Response, Improved Case (MDE event)

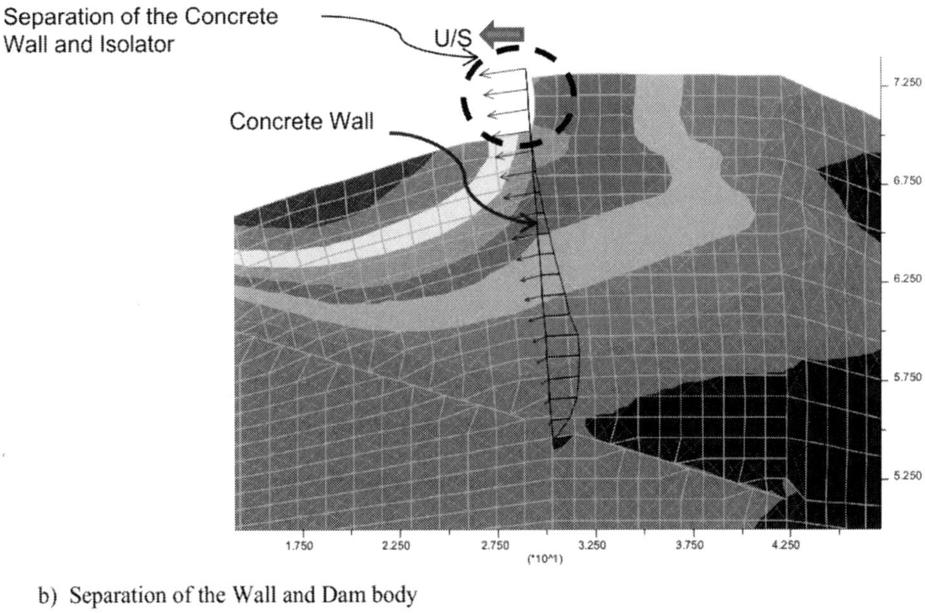

b) Separation of the Wall and Dam body

Figure 10. Response of Dam and Concrete Wall for "Improved Case" (MDE Motions (2,475-yr)

CONCLUSIONS

The results of a series of 2-D plain-strain dynamic soil-structure interaction analyses of the Lower Dam of a cascading dams system are presented in this paper. The analyses indicate that the dam would undergo large deformations in the upstream and downstream shells when subjected to the 10,000-yr and 2,475-yr ground motions without any remedial measures in place or under "as-is" conditions.

The downstream shell undergoes displacements in excess of 4 m in the very loose cinder and slag fills exhibiting shallow failure when subjected to the 10,000-yr ground motions. Driven by the unbalanced inertial forces, the concrete wall tends to rotate towards the reservoir, but its vertical settlements are not large enough to cause a loss of freeboard in the reservoir.

When a plastic moment capacity was imposed on the concrete wall, the computed displacements were larger. This result is consistent with the expected performance. The plastic moment capacity is controlled by the tensile cracking stress.

The seismic performance of the remediated dam (Hardened Option - the construction of a soil-cement zone over the entire downstream shell to some 2 to 3 m depth), improved with limited distortions occurring in the downstream shell and did not change the seismic shaking performance of the concrete core wall.

The results of dynamic analyses reported herein correspond to one cross section of the Lower Dam taken at the deepest section of the valley and ignoring the 3-D effects of the valley across which the dam is built.

REFERENCES

Acar, Y. B., and El-Tahir, E. A. 1986. "Low strain dynamic properties of artificially cemented sand". *ASCE Journal of Geotechnical Engineering,* 112, pp.1001 - 1015.

Barton, N., and Kjaernsli, B. 1981. "Shear Strength of Rockfill", *ASCE Journal of the Geotechnical Engineering Division,* V. 107(GT7): 873-891.

Hardin, B. O., and Drenvich, V. P. 1972. "Shear Modulus and Damping in Soils: Design Equations and Curves", *ASCE Journal of the Soil Mechanics and Foundation Engg. Division,* V. 98(SM7), pp. 667-692.

Hawson H, Downing B, & Simzer J. 2017."Development of Remediation Options for an Ageing Dam" Under review for the 85th Annual Meeting of International Commission on Large Dams (ICOLD) held in Prague, Czech Republic, 3-7 July 2017.

ITASCA, 2008. "Fast Lagrangian Analysis of Continua (FLAC)", Version 6, *User's Guide.* Itasca Consulting Group, Inc.

Leps, T.M. 1970. "Review of Shear Strength of Rockfill", *ASCE Journal of the Soil Mechanics and Foundation Engg. Division,* V. 96(SM4), pp. 1159-1170.

Pestana, J.M., and Salvati, L. A. 2006. "Small-Strain Behavior of Granular Soils. I: Model for Cemented and Uncemented Sands and Gravels". *ASCE Journal of Geotechnical and Geoenvironmental Engineering,132, pp. 1071-1081.*

Roberds, W., Downing, B. and Hawson, H. 2017. "A Risk-Based Approach to optimize Cascading Dams Remediation", Under review for the 85th Annual Meeting of International Commission on Large Dams (ICOLD) held in Prague, Czech Republic, 3-7 July 2017.

Saboya J.F. and Byrne. PM. 1993 "Parameters for Stress and Deformation Analysis of Rockfilll Dams", *Canadian Geotechnical Journal,* V. 30, pp.690-701.

Seed H. B., and Idriss, I. M. 1970. "Soil Moduli and Damping Factors for Dynamic Response Analyses", *Report No. EERC 70-10,* Earthquake Engineering Research Center, University of California, Berkeley.

Seed, H.B., Seed, R.B., Lai, S.S. and Khamenehpour, B. 1985. "Seismic Design of Concrete-Face Rockfill Dams", *In Proceedings*, Symposium on Concrete-Face Rockfill Dams- Design, Construction and Performance, ASCE, New York, N.Y. pp. 459-478.

Seed, H. B., Wong, R. T., Idriss, I. M., Tokimatsu, K. 1986. "Moduli and Damping Factors for Dynamic Analyses of Cohesionless Soils", *ASCE Journal of the Geotechnical Engg.,* V. 112(GT11), pp. 1016-1032.

Uddin N. and Gazetas, G. 1997. "Dynamic Response of Concrete-Faced Rockfill Dams to Strong Excitation", *ASCE Journal of the Geotechnical Engineering Division,* V. 1210(GT2): 185-197.

Yasuda, N. and Matsumuto, N. 1993. "Dynamic Deformation Characteristics of Sands and Rockfill Materials", *Canadian Geotechnical Journal,* V. 30, pp.747-757.

Yasuda, N. Yamaguchi, Y. and Matsumuto, N. 2003. "Dynamic Deformation and Strength Properties of Rockfill Materials", *In Proceedings*, International Congress on Large Dams, ICOLD, Montreal, Q83-R34 V. 3, pp. 577-594.

RCC POST CONSTRUCTION CORE TESTING AND DATA ANALYSIS FOR SAN VICENTE DAM RAISE PROJECT

James Stiady, PhD, PE, GE[1]
Michael Rogers, PE[3]
Thomas Reynoldson[5]
Wade Griffis, PE[7]
Andrew Oleksyn, PE[2]
Glenn Tarbox, PE, PE[4]
Gerard E. Reed III, PE[6]
Jim Zhou, PE[8]

ABSTRACT

The San Diego County Water Authority (SDCWA) recently completed the San Vicente Dam Raise Project (SVDR) to increase local water storage capacity in San Diego County, California as the final phase of the $1.5 billion Emergency & Carryover Storage Project (E&CSP). The E&CSP provides an increase in water supply reliability in case of catastrophic failure to the delivery system, including major earthquakes as well as a more flexible conveyance system. The SVDR project raised the original 220-foot high concrete gravity dam (90,063 AF of storage) by 117 feet, increasing reservoir storage capacity by 152,000 AF. The dam raise utilized more than 600,000 cubic yards of roller-compacted concrete (RCC), making it the largest dam raise in the world using RCC. To confirm the quality of the as-constructed RCC materials, SDCWA developed a post-construction Coring and Testing Program to recover over 1,500 linear feet of 6-inch diameter RCC cores (3 times MSA of 2 inches) and performed more than 1,000 direct tensile and compressive strength tests. This paper presents an in-depth data analyses of the test results and discusses the limitations of the coring process. The interpretation of the test results to assist in evaluating whether the in-situ RCC material meets the design intent of the raised structure is also presented.

INTRODUCTION

The Emergency & Carryover Storage Project (E&CSP) developed by SDCWA is one of the most innovative infrastructure achievements in the normally arid southern California region. The final phase of the E&CSP was the San Vicente Dam Raise (SVDR) Project that is a critical achievement for the dam industry as it provides many lessons learned in the design, construction and operation of a raised concrete dam, especially through the use of roller-compacted concrete (RCC). SVDR construction was started in 2010 with RCC placement for the dam raise completed during the period of 2011-2012. The overall project completion was in 2014. The paper by Rogers, *et al* (2016) presented the description of the coring program and summary of the test results of concrete coring

[1] Kleinfelder, 5761 Copley Drive # 100, San Diego, CA 92111, jstiady@kleinfelder.com.
[2] San Diego County Water Authority (SDCWA), 4677 Overland Ave., San Diego, CA 92123, AOleksyn@sdcwa.org.
[3] MWH/Stantec, 9665 Granite Ridge Dr #220, San Diego, CA 92123, Michael.F.Rogers@mwhglobal.com.
[4] MWH/Stantec, 2353 130th Ave NE Suite 200, Bellevue, WA 98005, Glenn.S.Tarbox@mwhglobal.com.
[5] Kleinfelder, 1801 California St. #1100, Denver, CO 80202, treynoldson@kleinfelder.com
[6] SDCWA, 4677 Overland Ave., San Diego, CA 92123, JReed@sdcwa.org.
[7] SDCWA, 4677 Overland Ave., San Diego, CA 92123, wgriffis@sdcwa.org.
[8] SDCWA, 4677 Overland Ave., San Diego, CA 92123, JZhou@sdcwa.org.

performed to extract over 1,500 lineal-feet of 6-inch diameter cores with a total of 803 direct tension and 265 unconfined compression tests performed. This paper presents more detailed analyses and discussions of the test data that supplement those in the previous paper (Rogers, 2016).

RCC DESIGN STRENGTHS FOR DAM RAISE

The raised San Vicente Dam was designed as a concrete gravity dam. The original dam was constructed using about 312,000 CY of conventionally placed mass concrete. The main dam was raised using about 600,000 CY of RCC. In addition, a saddle dam was constructed in a topographic low along the reservoir ridge using 12,500 CY of RCC.

The design team analysis indicated that the maximum vertical tensile stress resulting from seismic loading provided the RCC strength requirements for the raised section of the dam. The distribution of the tensile stress is shown in Figure 1. A more detail analysis was discussed in the previously published papers by Rogers (2016), Zhou (2009), Tarbox (2009), and Zhou (2007). The analysis indicated that a dynamic tensile strength of 210 psi should be used as the criteria for RCC strength requirement across the RCC lift joint.

For the RCC mix design, a target average compressive strength of 4,800 psi was selected to provide compatibility with the original mass concrete section having average compressive strength of 4,000 psi and a minimum static tensile strength of 140 psi.

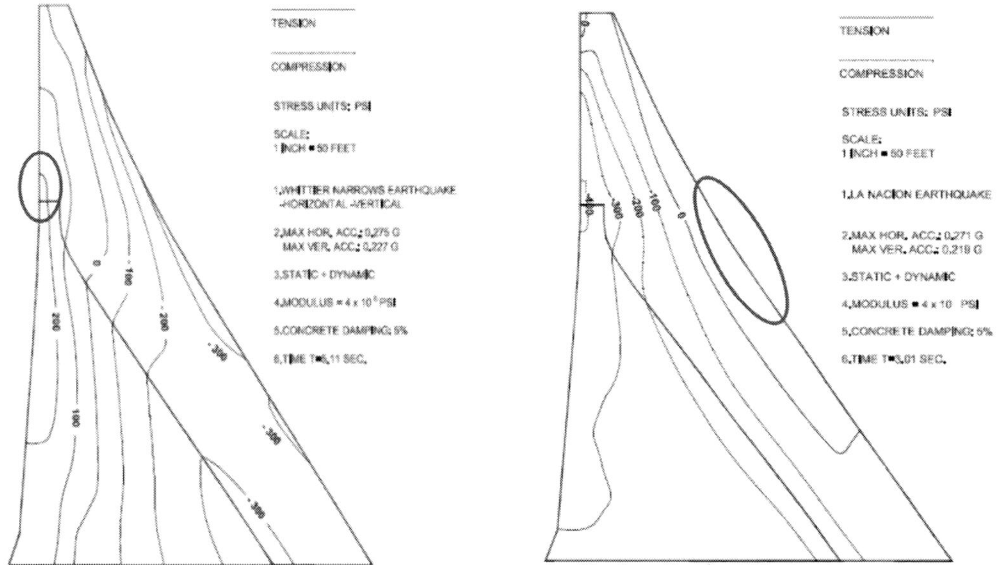

Whittier Narrows Eq. f_t = 210 psi on top of old dam

La Nacion Eq. f_t = 90 psi at downstream face of the new RCC

Figure 1. Vertical Stress Contour under Design Dynamic Loading

CORING PROGRAM DEVELOPMENT

The SVDR coring program was developed to confirm that the as-built RCC raised section would provide long-term performance of the structure satisfactory to the RCC design strength requirements. The structural analysis indicated that the performance of the dam is critically dependent on the tensile strength capacity at the following areas of the completed dam:

- the sloping interface joint of RCC placed against the original dam
- the sloping interface joint of RCC placed against abutments
- the horizontal lift joint of RCC placed atop previously-placed RCC

Grout-enriched-vibratable RCC (GEVR) was used at the all sloping interfaces. The presence of the additional grout in the GEVR would enhance the bond between the new RCC material and the original dam concrete or the in-situ bedrock, which was confirmed by several test specimens recovered from the GEVR interfaces and the test results. The remaining critical areas for this coring program, therefore, were the horizontal lift surfaces with specific testing for the tensile strength of these joints.

Following completion of the dam raise construction, a coring program was developed to primarily test the lift joints, including some areas of the parent RCC material. The program was developed to test the RCC after it had reached the age of at least 2 years, which corresponds to the target design strength gain of the RCC materials.
The final coring program consisted of a total of 3 Test Holes and 12 Core Production Holes. Discussion on the test holes was presented by Rogers (2016). The extracted cores from the test holes were not intended for tensile testing but were kept in a curing room as spares. They were ultimately used for compression testing discussed later in this paper.

The core production program was divided into 3 phases. Preliminary results were reviewed at the end of each phase, which guided changes to subsequent phase coring and testing. Core locations for phases 1 and 2 were mostly as planned. However, phase 3 locations were modified and compression testing was added for a portion of the recovered cores. The coring locations of the twelve production holes are summarized in Table 1 and are shown graphically on Figure 2.

ANALYSIS OF RCC CORING RECOVERY

Despite the best efforts to optimize the drilling parameters using the test holes, unintentional breaks occurred within core runs during recovery. Unintentional breaks are a common occurrence during rock, conventional concrete, and RCC coring. Photographs showing a typical unintentional break at the joint surface and in parent material are provided in Figure 3(a) and 3(b), respectively.

The distribution of the unintentional breaks among the joint types is summarized in Table 2. As anticipated, 78 percent (or 164 of 209) of the unintentional break occurred at the joint surface because the joint surface is expected to be weaker than the parent materials.

Table 1. Summary of Core Production Holes

Core ID	Location	Phase	Approximate Station	Hole Top Elev. (ft)	Orientation of Hole	Approx. Depth (ft)
C1	Top of Dam	3	4+43	776	Vertical	120
C2	Top of Crane Pad at Tower	1	3+30	776	Vertical	76
C3	Top of Dam	1	4+20	776	Vertical	120
C4	Top of Dam	1	5+20	776	Vertical	120
C5	Top of Crane Pad at Right Side of Spillway	1	5+55	776	Vertical	166
C6	Top of Dam	2	10+25	776	Vertical	120
C7	Top of Dam	3	12+21	776	Vertical	120
C8	Downstream Steps	2	Below Spillway Crane Pad-Right Side 5+85	650 +/-	Inclined	70
C9	Top of Dam	3	Below Spillway Crane Pad-Right Side 5+84	776	Vertical	146
C10	Top of Crane Pad at Left Side of Spillway	2	9+25	776	Vertical	166
C11	Downstream Steps	2	Below Spillway Crane Pad-Right Side 5+85	650	Vertical	70
C12	Downstream Steps	3	Below Spillway Crane Pad-Right Side 5+87	650	Vertical	70
Phase 1: 482 Linear Feet			Phase 2: 426 Linear Feet		Phase 3: 456 Linear Feet	
					Grand Total: 1,364 Linear Feet*	
Note: * The test holes presented by Rogers (2016) were not included in this table.						

The occurrence of 22 percent (or 45 of 209) of the unintentional breaks on the parent material may have indicated the excessive stress generated during coring operations. A review of Table 2 indicated that about 65 percent (or 754 of 1,167) of joints encountered during coring were warm joints and 33 percent (or 389 of 1,167) were hot joints. Of the 164 unintentional joint breaks, 126 were at warm joints. This equates 77 percent which is reasonably consistent with the 65 percent warm joints encountered. The overall joint break rate for all vertical core extracted was 14 percent (or 164 of 1,167), which is reasonable for RCC. Various joint types occurred during construction due to the exposure duration to ambient temperature resulting into differences in surface preparation prior to receiving the subsequent RCC lift. A more detail discussion about joint types was presented in the previously published papers by Rogers (2016) and Jubran (2012).

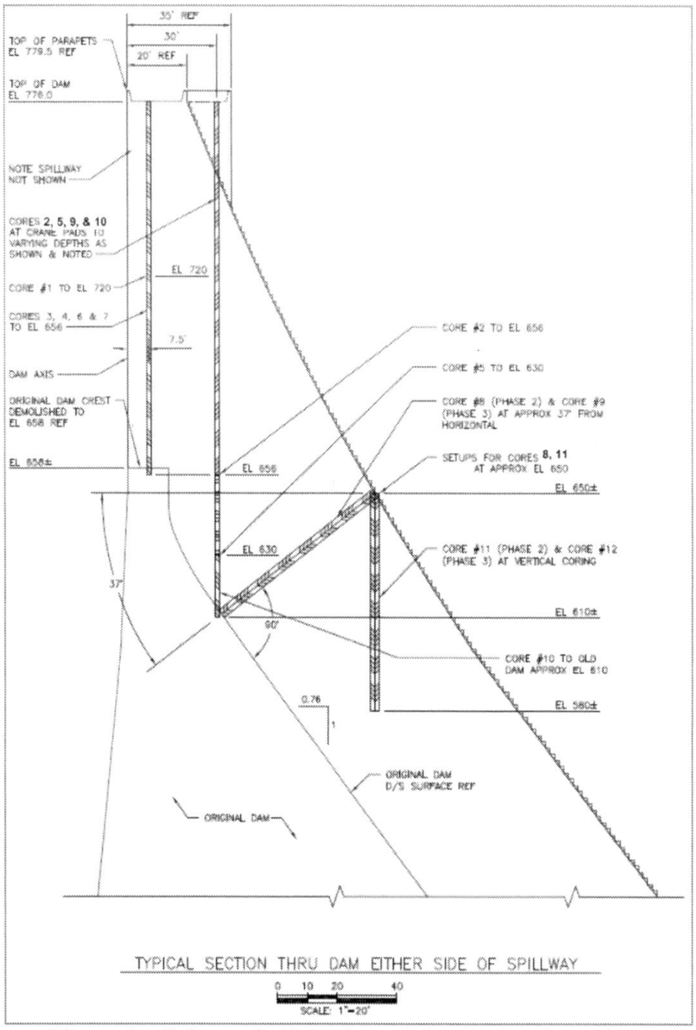

Figure 2. Schematic Drawing of the Coring Locations

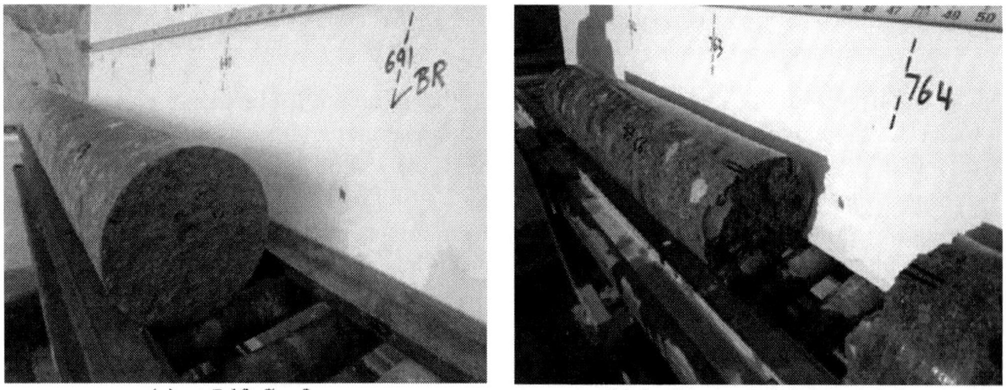

(a) at Lift Surface (b) in Parent RCC Material

Figure 3. Unintentional Break in the RCC Cores

Table 2. Distribution of Unintentional Breaks with Joint Type

RCC Lift Joint Type	Total Joints Encountered During Coring	Number of Unintentional Breaks		Percentage of Unintentional Break from Total Encountered	
		at the Joint	at Parent Materials	At the Joint (%)	At Parent Materials (%)
Hot	389	36	11	9	3
Warm	754	126	32	17	4
Cold	1	0	0	0	0
Super Cold	23	2	2	9	9
Total	1,167	164	45	14	4

Observation of the coring process indicated that unintentional breaks may have occurred for various reasons, including the drilling (drill bit rotation, downward pressure, core advancement, and mast wobbling), core run termination (lifting, spring grip, separation, and vibration), core handling (lifting and changing the orientation from vertical to horizontal), and core ejection process (sliding and possibly bending); all of which could have induced stresses into the RCC cores. The other reason would have been the RCC material strength variation. Detailed discussions were presented by Rogers (2016).

DIRECT TENSILE TEST AND RESULTS

Direct tensile tests were performed on core specimens in accordance with the U.S. Army Corps of Engineers test procedure CRD-C 164 by the testing laboratory. The specimen was 6 inches in diameter and about 12 inches in height. The specimen was cut from the 6-inch diameter core. The machine used to pull the specimen in tension was a Riehle FH-400 universal testing machine. The machine setup included a steel plate and rod for reaction support, a set of shackles, and a set of eye bolts for metal cap connection. The other side of the metal cap had a ½-inch threaded hole that was used to attach the eye bolts to the cap. The eye bolt was connected to a series of shackles that were looped together to form a chain at each end of the specimen. To minimize the eccentricity of the system, during setup, the alignment of the very top and the very bottom eye bolts was adjusted and verified using a plumb-bob. The system setup using a series of shackles and top-and-bottom eye bolt alignment verification would minimize the effects of torsion and eccentric loading to the test results. The specimens were pulled apart at a controlled loading rate of 35±5 psi per second until the sample pulled apart into two pieces. For each specimen, the peak load was recorded and the break location was visually identified and designated as a joint surface or a RCC parent material.

Direct Tensile Testing on Horizontal RCC Lift Joint

The tensile strength of the horizontal lift joint of RCC is critical during the earthquake loading. For the purpose of evaluating the tensile strength of the lift joint, the grout line indicating the joint surface was placed in the middle of the specimen when marking the core for cutting. If the grout line was not visually identifiable, the elevation mark was assumed as the location of the joint surface. The intention was that the actual joint would be evaluated and the effects of end caps were minimized during tensile testing.

When each direct tensile test result is considered to represent any location and any lift joint elevation, the direct tensile test results performed in this coring program can be summarized in Table 3. The direct tensile tests were performed on a total of 803 specimens. The average age of the tested specimens was 927 days, with the oldest being 1,134 days and the youngest being 782 days. The average of the direct tensile strength across the joint was 181 psi. As expected, the specimens broken at the parent materials during testing showed higher average direct tensile strength (197 psi) than those broken at the identifiable joint surface (156 psi). The table also indicates that approximately 61 percent (485 of 799) of the joint surfaces from testable samples were stronger than the parent materials. As each tested specimen contained joint surface and parent materials, the weaker part broke first under tension. Thus a break in the parent materials indicated that the joint surface was stronger than the parent materials within the specimen.

A graph showing the statistical distribution of the direct tensile strength test results is provided in Figure 4. A similar graph showing the statistical distribution of the direct tensile strength test results for specimen broken at the joint surface and those broken at the parent materials is provided in Figure 5. The graph in Figure 5 also indicates the majority of break plane of the specimens with less than 50 psi direct tensile strength occurred at the joint surface, as anticipated.

The highest demand for tensile strength capacity would occur in the raised RCC section above the original dam, as shown in Figure 1. The required static tensile strength would be about two thirds (2/3) of the dynamic tensile stress occurred during the simulated earthquake event. A comparison between the direct tensile test results and the required static tensile strength in raised RCC section is plotted against the elevation of the RCC section and shown in Figure 6. A review of Figure 6 indicated that 96 percent of the average direct tensile test results were higher than the required static tensile strength. The average and maximum range of the direct tensile results for each elevation was about 171 psi and 360 psi, respectively. The range indicated the high variation of direct tensile test results for each elevation. The range does not seem to vary with the elevation; nor does the low test results.

Table 3. Summary of Direct Tensile Test Results

Parameter	No. of Test	Average Tensile Strength (psi)	Standard Deviation (psi)
All direct tensile test results excluding the contact with the original dam conventional concrete	799	181	78
Broken at identifiable joint surface only	314	156	75
Broken at parent materials only	485	197	75
Contact between GEVR and the original dam conventional concrete	4	233	4

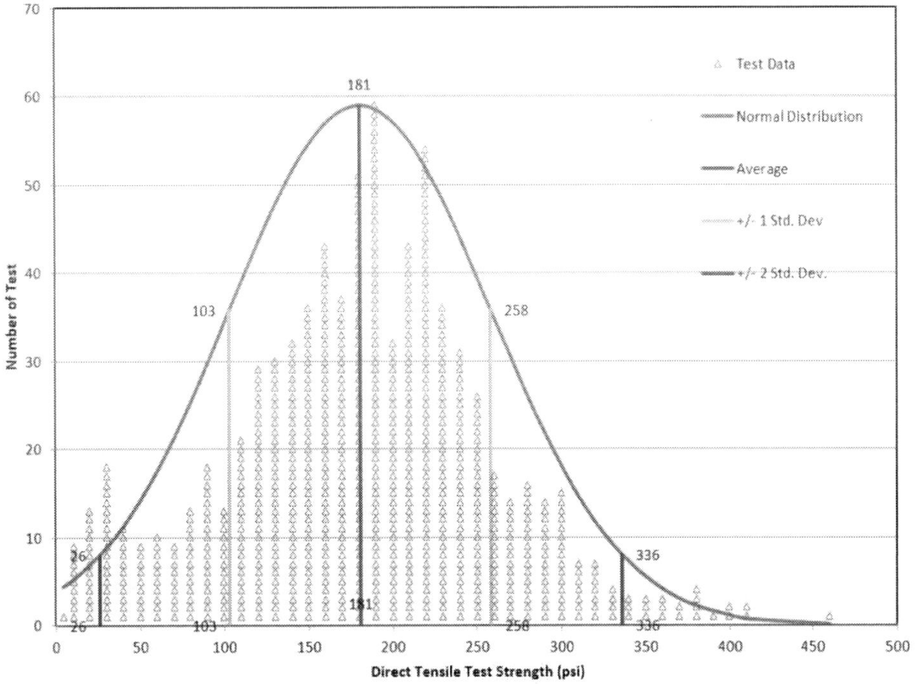

Figure 4. Statistical Distribution of All Direct Tensile Test Results

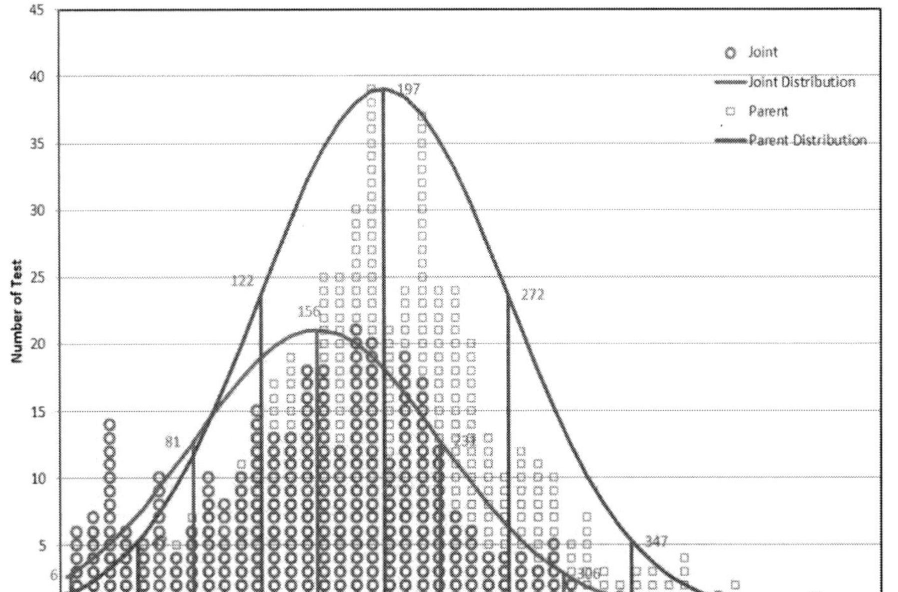

Figure 5. Statistical Distribution of Direct Tensile Test Results Based on Break Surface

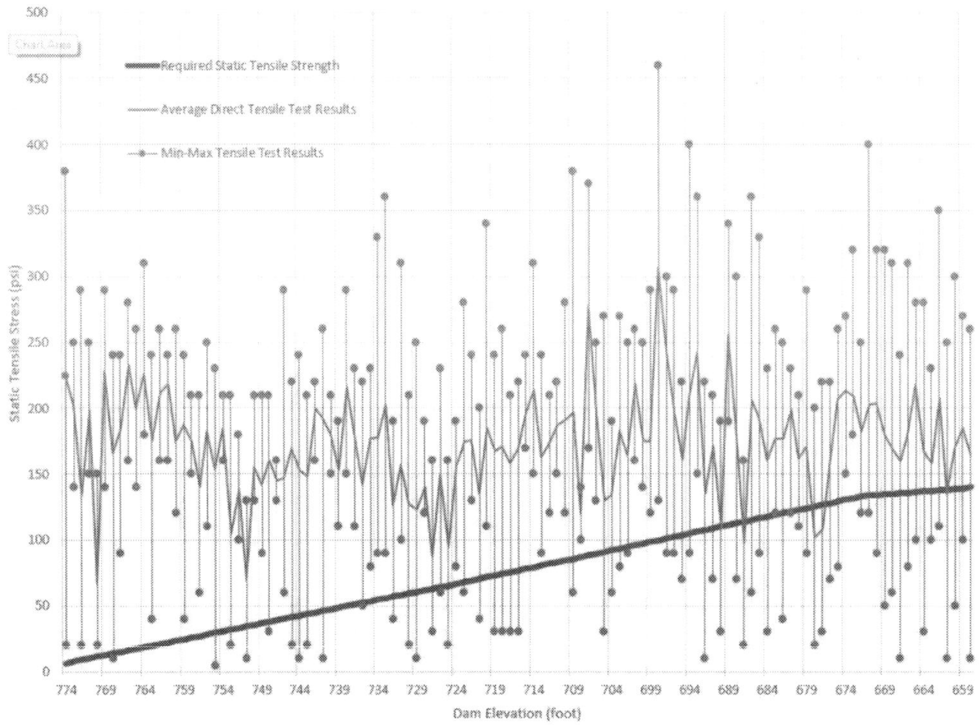

Figure 6. Static Tensile Strength in the Raised RCC Section

Direct Tensile Testing on Contact Surface with the Original Dam

During construction, the contact between the new RCC and the original dam was prepared by hydro blasting the original conventional concrete surface. Then RCC was placed against that surface with GEVR. In the coring program, the contact was planned to be evaluated at 6 coring locations: C3, C4, C6, C7, C10, and C8. At locations C3, C4, C6, and C7, a horizontal contact plane with the original dam crest was obtained. At location C10, an angled contact plane with the original dam downstream surface was obtained. At location C8, the drilling was conducted at about 37 degree from horizontal and a perpendicular–to-the-core run contact plane was retrieved.

Contact specimens from locations C3, C4, C6, and C7 were retrieved successfully and were evaluated in direct tensile testing. Test results indicated that the average direct tensile strength of the contact was 233 psi as shown in Table 3. A photograph showing a typical break plane of the contact is provided in Figure 7(a). Most of tested interface samples broke along the interface material. Contact between RCC and original conventional concrete at location C10 was retrieved successfully. As planned, the contact was in a steep angle at about 7 inches long as shown in Figure 7(b). Due to the steepness of the contact, the sample was stored and not tested. The contact at location C8 was retrieved successfully. However, due to the presence of a 5-inch diameter rock in the original dam concrete near the contact surface that fractured the core, a testable specimen could not be produced. Nevertheless the visual appearance of the rest of the interface contact seemed as good as those shown in other cores at the GEVR interface, which was consistent with the results obtained during the trial placement.

(a) Typical Break Surface

(b) Steep Contact

Figure 7. Contact between RCC and Original Dam Concrete

Comparison with Previously Performed Direct Tensile Testing

Before this post-construction coring program, SDCWA performed direct tensile testing on specimens collected from the RCC trial placement and the spillway flip bucket area which was the lower portion of the dam at elevation 495 feet. Comparison with the previous results are summarized in Table 4. Review of the table indicates the following:

- The coefficient of variation (COV) of the 2-½ year test results was lower than that of the one year. This may indicate that the drilling, preparation, and testing process during the post construction coring program had less impact to the test results than the process during the flip bucket coring program.
- The average of direct tensile strength at 2-½ years appeared to be lower than that at 1 year. However, our analyses indicated that the difference between the average of direct tensile strength at 2-½ years and that at 1 year was not statistically significant. This observation indicated that there was no significant gain in the tensile strength from 1 to 2-½ years.
- There is an indication that the direct tensile strength reached its ultimate strength at about one year regardless the joint types.

Table 4. Comparison of Previously Performed Direct Tensile Test Results

Item	Post Construction (2.5 years)	Flip Bucket (1 year)	Trial Placement (90 days)
Average direct tensile strength (psi)	181	214	139
Standard deviation (psi)	78	118	48
Coefficient of variation, COV (%)	43	55	35
Number of test	799	30	26
Average specimen age (days)	927	365	98
Main dam elevation (foot)	580 to 774	445 to 495	n/a[a]

Note: a. Trial Placement was conducted at separate location, not within the main dam footprint.

Direct Tensile Strength and Maturity Factor for RCC Lift Joint

The RCC lift joint surface was prepared according to the joint classification that was developed based on its maturity factor measured in degree Fahrenheit – Hour units. The direct tensile strength of the (visually identifiable) lift joints in relation with the maturity factor is summarized in Table 5. The distribution of the joint direct tensile strength is shown in Figure 8. Review of the table and the graph indicates the following:

- The hot joint was typically stronger than other joint types because the subsequent lift was placed before the joint surface reached the initial set. The hot joint was often not well identified in the specimen based on the appearance; and sometimes it might have been classified as the parent material for direct tensile test break.
- The super cold joint exhibited the lowest direct tensile strength across the joint surface, but still met the minimum strength requirement.
- The hot and warm joints exhibited significant direct tensile strength gain, approximately 40 percent, after about 90 days of age.
- The cold and supercold joints exhibited approximately 10 to 15 percent of the gain in the direct tensile strength after 90 days. Those joint surfaces were treated with grout before placing the next RCC lift. Grout, which is a mixture of cement and water, typically reaches ultimate strength at early age (earlier than 28 days).

Table 5. Summary of Direct Tensile Test Results in Relation with Maturity Factor

Joint Type	Maturity Factor (°F-hr)	Post Construction (2.5 years)		Flip Bucket (1 year)		Trial Placement (90 days)	
		Average (psi)	No. of Test	Average (psi)	No. of Test	Average (psi)	No. of Test
Hot	< 1,000	207	280	245	13	144	14
Warm	1,000 to 2,500	168	497	202	11	120	3
Cold	2,500 to 3,500	170	7	187	3	153	3
Super Cold	>3,500	137	15	153	3	130	6

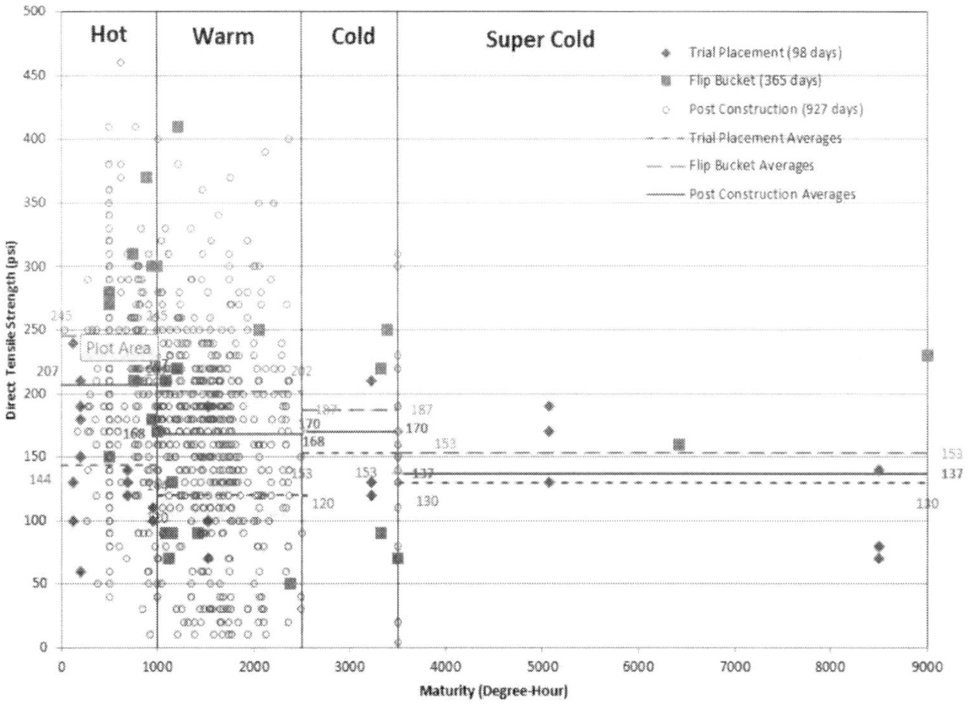

Figure 8. Direct Tensile Test Results Distribution with Joint Maturity Factor

COMPRESSIVE STRENGTH TEST AND RESULTS

Lift Joint Surface Impact on Compressive Strength

The compressive strength test was performed on a total of 265 core specimens. The average age of the test specimens was 1,028 days; the oldest was 1,129 days; and the youngest was 896 days. The test results are summarized in Table 6. The average compressive strength was 3,315 psi. Compressive test specimens were typically prepared without lift joint surface, and the core was cut at about the lift joint surfaces. Test specimens specifically with a lift joint surface identified were also prepared and tested for comparison. The average compressive strength of the specimen with lift joint and that without lift joint were 3,431 and 3,300 psi, respectively. As anticipated, the effect of lift joint surface on the compressive strength was statistically negligible as long as the surface was perpendicular to the specimen wall.

Core and Cylinder Compressive Strength

The relationship between core and cylinder compressive strength test results is shown in Figure 9. This relationship was established by pairing a core compressive strength test result with a cylinder compressive strength test result from specimens prepared using the fresh RCC mix samples collected from the same lift elevation and/or on the same placement day. Review of Figure 9 indicated the following:

- The ratio of core-to-cylinder compressive strength from post construction, flip bucket, and trial placement specimens were 0.58, 0.56, and 0.70, respectively. The compressive strength ratio from specimens obtained from the main dam (0.58, 0.56) appeared to be consistent although the specimens were tested at different age.
- The relationships of RCC compressive strength test ratio of core-to-cylinders (56 to 70 percent) are below the general range (80 percent minimum to 130 percent maximum) expected for the project. The expected range was developed using the data from another SDCWA's project (Olivenhain Dam) constructed in 2003.
- As the average compressive strength tested in the original San Vicente dam was about 4000 psi, the design intent set a target to match that strength. A target cylinder strength of 4800 psi at 2 to 3 years or a core-to-cylinder ratio of 83 percent was expected for the SVDR.
- The average of post-construction cylinders compressive strength tested at 2 years was 5,628 psi or about 17 percent higher than the target RCC mix compressive strength.
- The average of post-construction cores tested at 2 years was 3,351 psi or about 17 percent lower than expected core compressive strength of 4,000 psi. The ratio of averaged core-to-cylinder compressive strength was about 58 percent.

We have not been able to explain why the core-to-cylinder ratio is below the expected range for this project. Review of the construction documents revealed no anomaly in the cement and fly ash quantities, the in-place density of the compacted RCC materials, and the density of the cores and cylinders. Nevertheless the compressive strength meets the design intent of 1,200 psi for the static strength requirement of the RCC portion of the raised dam. That being said, the RCC is expected to continue gaining long-term compressive strength based on our test data from Olivenhain Dam where the strength gained approximately 5 to 8 percent between 3 and 10 years of age.

Table 6. Comparison of Previously Performed Compressive Strength Test Results

Item		Post Construction	Flip Bucket	Trial Placement
Average Compressive Strength (psi)	Core	3,315	3,193	2,647
	Companion Cylinder	5,628	5,680	3,700
Standard Deviation (psi)	Core	759	416	751
	Companion Cylinder	571	393	476
Coefficient of variation, COV (%)	Core	22.9	13.0	28.3
	Companion Cylinder	10.2	6.9	12.9
Average age (days)	Core	1028	365	131
	Companion Cylinder	730	365	91
Number of specimens		265	6	53
Main dam elevation (foot)		582 to 773	466 to 491	n/a[a]

Note: a. Trial Placement was conducted at separate location, not within the main dam or saddle dam footprint

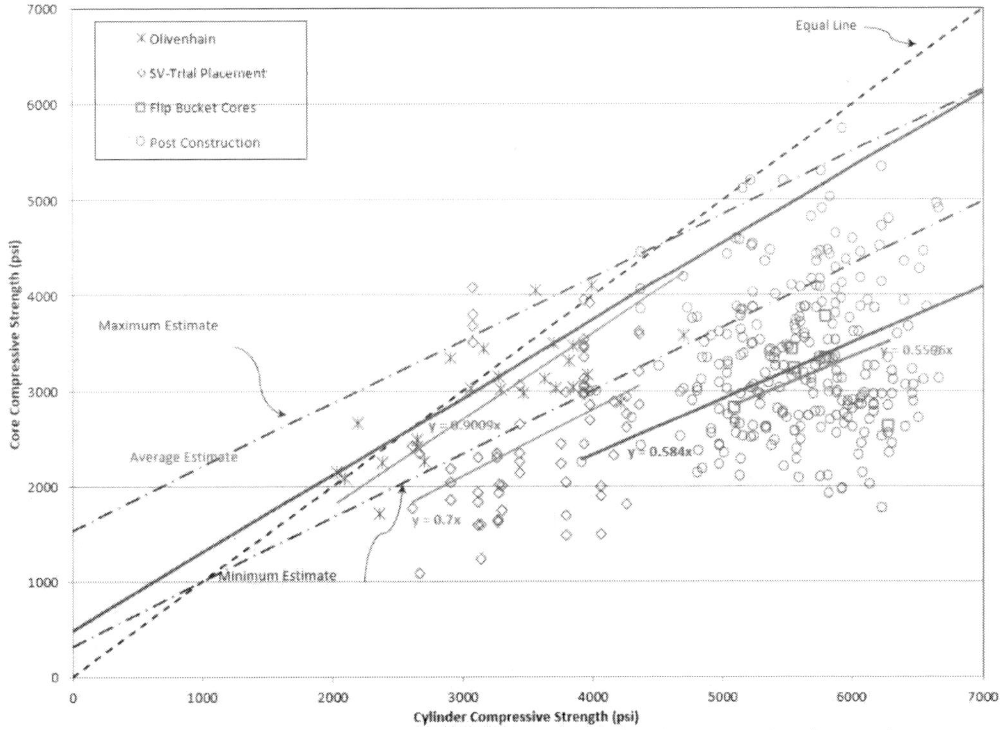

Figure 9. Relationship between Core and Cylinder Compressive Strengths (Shown with Estimated Range developed by Dunstan)

CORE TENSILE STRENGTH AND CORE COMPRESSIVE STRENGTH

The relationship between core tensile and core compressive strength test results is shown in Figure 10. The relationship was established by pairing direct tensile strength and compressive strength test results performed on the same vertical lift elevation, but not necessarily at the same horizontal location due to the fact that only one core specimen could be produced for tensile or compressive strength testing at any given location. Review of the graph indicated that the ratio of core tensile to core compressive strength from main dam and trial placement specimens were 0.057 and 0.081, respectively. The difference in the ratios may be due to data size and age of the tested specimens. The core specimens from flip bucket were not included in Figure 10 because each direct tensile and each compressive strength test were performed on different lift elevation.

The results of core tensile and core compressive strength at the San Vicente dam is comparable to those reported by Dolen (2011) for other dams. For example, the ratio for Upper Stillwater RCC dam was 5.1 percent. The RCC mix design for Upper Stillwater was a high paste mix similar to San Vicente. Dolen (2011) reported that direct tensile strength of parent mass concrete tested by USBR averaged about 5.2 percent of the compressive strength and ranged from about 2.9 to 7 percent. Dolen (2011) also presented the results of several hundred concrete direct tensile tests at USBR concrete dams.

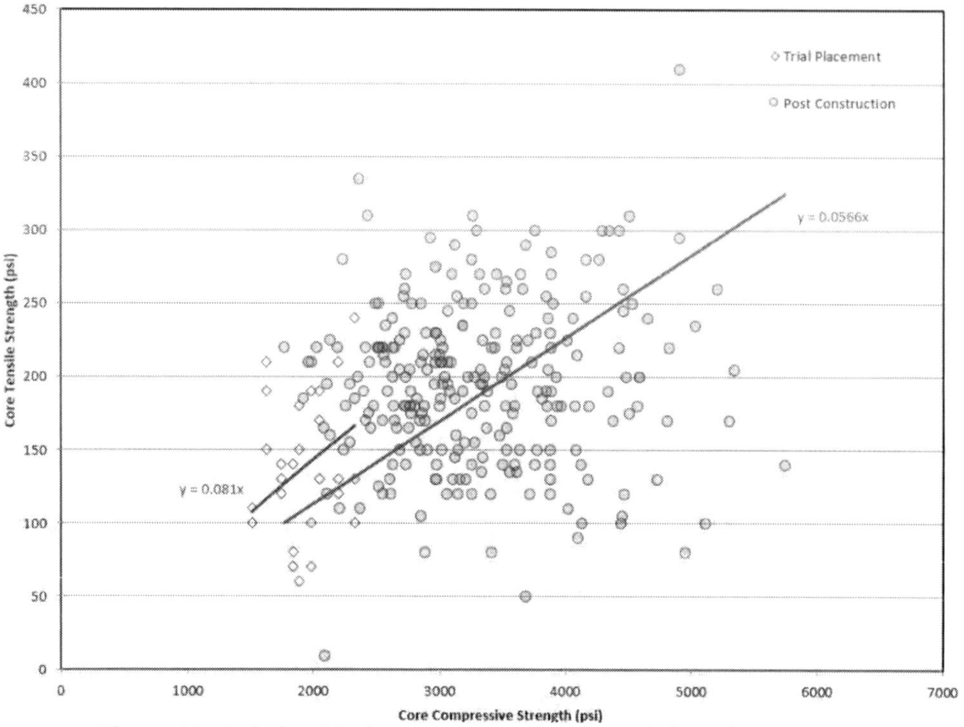

Figure 10. Relationship between Core Tensile and Core Compressive Strengths at San Vicente Dam Raise Project

CONCLUSIONS

The post-construction RCC Coring and Testing Program represents the overall condition of the in-situ RCC materials and interface joints. More than 1,500 linear feet of 6-inch diameter core were collected and more than 1,000 direct tensile tests were performed.

Based on a review of the testing results in this post-construction RCC Coring and Testing Program, we concluded the following:

- Based on the direct tensile strength tested in recovered cores, the average dynamic tensile capacity for RCC lift lines, parent materials and interface joints of the raised San Vicente Dam range between 234 and 350 psi, which are greater than the computed maximum dynamic tensile stress of 210 psi in the raised portion of the dam under design seismic conditions.
- All compression strength test results were greater than the computed compressive stresses required by the design criteria.
- Therefore, the in-situ RCC materials in the raised San Vicente Dam meet the tensile and compressive strength requirements based on the results of this investigation.

In addition, we made the following specific observations about the RCC tensile and compressive strength based on the test results and analyses:

- Direct tensile strength of the RCC parent materials and lift surfaces seemed to reach its ultimate strength at 1 year of age.
- Direct tensile strength of the RCC parent materials and at the hot and warm lift surfaces reached approximately 70 and 60 percent of their full strength, respectively, at 90 days of age.
- Direct tensile strength at the cold and supercold lift surfaces reached approximately 85 to 90 percent of their full strength at 90 days.
- Horizontal lift joint has little effect on the compressive strength across the lift.
- Core-to-cylinder compressive strength ratio was 0.58, lower than the expected value of 0.83 for the project, without any identifiable explanation.
- Although the core-to-cylinder ratio was lower than the expected, the ratio of 0.58 was comparable to the ratio of 0.51 from Upper Stillwater project as reported by Dolen (2011).
- The ratio of direct tensile strength to compressive strength of the RCC cores was 0.057, in general agreement with the previous range published by USBR.

ACKNOWLEDGEMENTS

The authors gratefully acknowledge the contributions of the State of California, Department of Water Resources, Division of Safety of Dams (DSOD) – David Gutierrez, Chief; Mark Schultz, Design Branch; Anna E Kolakowski, Field Engineering Branch – for their support and partnership on this project.

REFERENCES

1. M. Rogers, A. Oleksyn, G. Tarbox, T. Reynoldson, J. Reed, W. Griffis, J. Zhou, J. Stiady, "Post-Construction Coring Evaluation of RCC at the San Vicente Dam Raise Project", Proc. of ASDSO Annual Conference, Philadelphia, PA, September 2016.

2. Tarbox, G.S.; Reed, G.E.; Griffis, J.W. "Reaching New Heights at San Vicente," The International Journal on Hydropower & Dam Construction, V.65, Issue 7, 2013

3. Jubran, M.; Zhou, J.; Grant, R.; Stiady, J.; and Oleksyn, A., "San Vicente Dam Raise Project – Tensile Strength Testing on Existing Dam and Trial Placement", Proc. of 32nd USSD Annual Meeting and Conference, New Orleans, LA, April 2012.

4. Dolen, T., "Selecting Strength Input Parameters for Structural Analysis of Aging Concrete Dams", Proc. of 31st USSD Annual Meeting, San Diego, CA, April 2011.

5. Zhou, J.; Rodgers, M.F.; Keller, T.O.; Meda, J. "Roller Compacted Concrete For Raising San Vicente Dam," Int'l Commission On Large Dams, 23rd ICOLD Congress, June 2009

6. Tarbox, G.S.; Ahola, V.I.; Rogers, M.F.; Zhou, J., "Structural Design Optimization with Raising San Vicente Dam." 23rd ICOLD Congress, Q.90, R.3, Brazil, June 2009.

7. Zhou, J.; Shoaf, J.A.; Rogers, M.F.; Tarbox, G.S.; Dunstan, M.R.H. "Designing the Tallest RCC Dam Raise for San Vicente Dam in California." Proc. of Fifth International Symposium on RCC Dams, Guiyang, China, Nov. 2007.

8. US Army Corps of Engineers, 1995, "Gravity Dam Design", EM 1110-2-2200.

RECONSTRUCTION OF THE MATALA SPILLWAY: HYDROTECHNICAL AND SAFETY ASPECTS

Ahmed Bouayad, P.Eng.[1]

ABSTRACT

The Matala hydroelectric development is located near the city of Matala in Angola. It was built in the 1950s. The power station has an installed capacity of 42 MW. The dam is a 1,030 m (3,380 ft) long and 14 m high concrete gravity structure built across the Cunene River. The original spillway consisted of 29, 17.5 m (57 ft) wide bays equipped with flap gates.

The concrete structures of the dam and spillway are affected by Alkali Aggregate Reaction (AAR) phenomenon, which has caused partial or complete blockage of the flap gates. This has resulted in a highly unsafe situation for the dam and for the population living downstream. A new gated spillway was designed and built using current dam safety practice.

The new spillway is equipped with eight radial gates with a different hydraulic operation as compared to the original flap gates. Opening of the radial gates may add significant flow to the river downstream, a situation the downstream population is not familiar with.

This paper describes the hydraulic and safety issues of the project focusing on the following:

- Spillway design flood;
- Hydraulic design and analysis of downstream flow conditions;
- Flow diversion during construction and associated safety concerns;
- Safety issues during construction;
- Emergency management during construction, and;
- Emergency preparedness plan for post-construction conditions.

Despite the difficulties encountered, particularly in the local context in which the safety aspects are not a high priority and due to the lack of regulation; challenges related to safety were successfully met during construction and an alert system has been implemented and tested for the operational phase.

[1] SNC-Lavalin, Montreal, e-mail: ahmed.bouayad@snclavalin.com

GENERAL

Introduction

The Matala hydroelectric development is located approximately 1,000 km to the southeast of Angola's capital city, Luanda. The Matala Dam was commissioned in 1954 on the Cunene River, on a series of rapids close to the town of Matala. The objectives assigned to the Matala hydroelectric facility were:

- To generate hydroelectric power for the towns of Matala and Lubango; and
- To provide storage for the primary water supply of local developments and irrigation of adjacent farmlands.

The hydroelectric facility spans the entire Cunene River and has a total length of approximately 1,035 m as shown in Figure 1. The original layout included, from the left to the right bank (looking downstream), the following structures:

- An overflow weir section comprised of six bays separated by piers (piers 1 to 7) totaling 120 m (394 ft) in length with an average height of approximately 8.5 m (28 ft);
- A spillway section equipped with 29 steel flap gates separated by piers (piers 7 to 36), with a total length of 580 m (1,903 ft) and an average height of approximately 11 m (36 ft);
- A 15 m (49 ft) high concrete gravity section that is approximately 108 m long, which incorporates the intake for the water supply and irrigation system;
- An integrated intake-powerhouse structure approximately 65 m long equipped with trashracks, stoplogs and closure gates (radial type); and
- A short concrete gravity section approximately 22 m long with a maximum height of 11 m (36 ft).

The powerhouse, connected to the intake structure with a short penstock, is currently equipped with three 13.6 MW generating units for a total installed capacity of 40.8 MW.

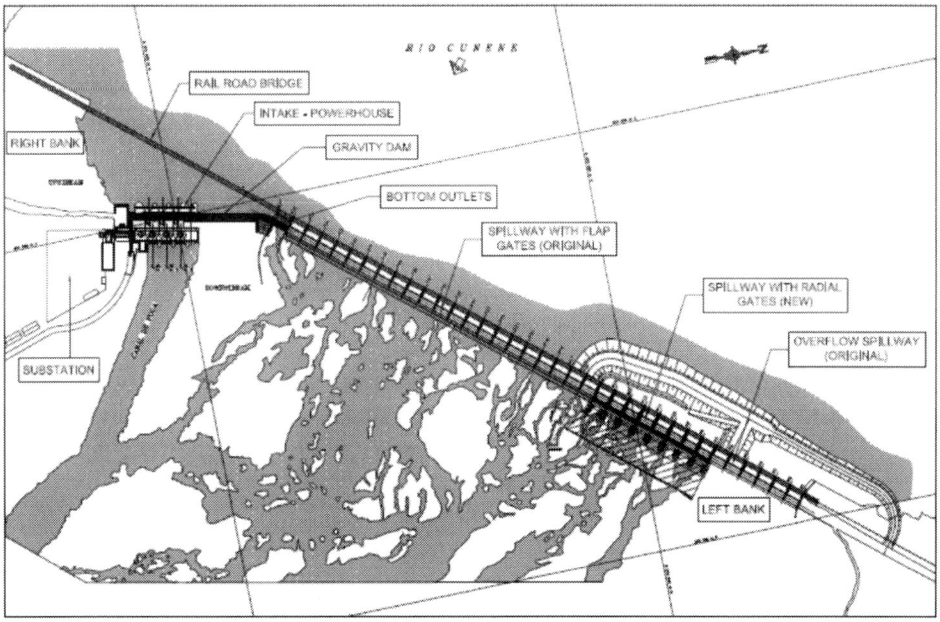

Figure 1. General Layout of the Matala Hydroelectric Facility

Rehabilitation and Modernization Project

Due to the presence of the AAR phenomenon in the original concrete structures of the Matala Dam (i.e., dam and spillway), the spillway piers and gravity section have deformed over the years. This ensuing deformation has limited the range of operation of the original steel flap gates and reduced the discharge capacity of the spillway.

In 2007, Empresa Nacional de Electricidade (ENE) awarded to SNC-Lavalin an Engineering, Procurement and Construction (EPC) contract for rehabilitation works that include the construction of a new gated spillway.

The main aspect of the rehabilitation works is the construction of a new spillway equipped with eight radial gates (width: 17.5 m (57 ft), height: 12 m or (39 ft)). Other works included in the project are:

- Replacement of the remaining flap gates with concrete overflow sections;
- Rehabilitation of the concrete structures affected by AAR;
- Evaluation (capacity and structural integrity) of the concrete bridge structure;
- Reinforcement of the bridge structure, as required, subsequent to the evaluation;
- Installation of dam monitoring equipment; and
- Grouting works.

The rehabilitation project started in 2011 and the excavation and concrete works were initiated in 2012. The radial gates were installed and commissioned in 2014.

Hydrotechnical Scope

The hydrotechnical activities were provided to the project during the EPC process to assist with the design and safety aspects of the project. The hydraulic team has assisted with the following aspects:

- Determine the new spillway design flood and hydraulic capacity,
- Prepare a flow diversion strategy during the construction works,
- Undertake hydrotechnical analyses to clearly identify the safety issues for both the construction phase and the operation phase of the dam,
- Prepare flood inundation maps and emergency preparedness; and
- Mitigation measures.

SPILLWAY DESIGN FLOOD

Original Spillway Hydraulic Capacity

Little information from the original design of the spillway is available. However, a hydrological study completed in 1966 mentioned that the spillway at Matala was designed to handle a discharge of 5,700 m^3/s (201,000 cfs).

Another study undertaken afterwards established the design flow of the spillway at 5,350 m^3/s (189,000 cfs). The recurrence of this flood event was not mentioned; however, the design rainfall was qualified as having a recurrence interval greater than 1:100-years.

Dam Safety Regulations

Spillways are safety devices for dams. Their design flow is often determined by dam safety regulations based on the potential damage caused by failure of the dam. In Angola, however, there is no legislation on dam safety. Nonetheless SNC-Lavalin, for the safety of the downstream population, has opted to use an approach based on several existing international regulations (Ref. to 9) to validate the design flow.

Design Flow of the New Spillway

No consistent criteria exist internationally for the selection of the Inflow Design Flood (IDF) for a river control development. The most common trend followed by a large number of countries consists in adopting a dam failure consequence based classification and to determine the IDF according to this classification. . For a given flood, the consequences of the failure of a dam are evaluated as the incremental consequences that would be observed between two cases:

- Damages observed during the passage of the flood, with no dam failure; and
- Damages observed during the passage of the flood, with dam failure occurring at the most critical time of flood conditions.

No incremental consequences are generally considered for a given flood if the incremental rise in the water level caused by the dam failure is small. The corresponding value specified in Quebec legislation, which is 30 cm (1ft), has been adopted in the analyses for the Matala spillway.

For dams where failure would result in a certain number of fatalities, the IDF is either the Probable Maximum Flood (PMF), as is the case of regulations adopted in the USA and the UK; or a flood with a magnitude equal to some proportion between the 1:1,000-year flood and the PMF, as is the case of regulations adopted in most European countries, South Africa, Canada, Australia and New Zealand (Ref. 1 to 9).

Based on the results of dam break analyses and damage assessment, a compilation of IDFs that would be selected for the conditions present at Matala Dam for countries for which dam safety criteria could be obtained was prepared. This compilation is presented in Table 1.

As indicated by the compilation presented in Table 1, the majority of legislations and guidelines considered in this study specify an IDF with a recurrence interval close, equal or lower than the 1:10,000-yr flood. The 1:10,000-year flood with peak inflow of 3,465 m^3/s (122,000 cfs) is lower than the proposed design flood flow of 5,350 m^3/s (189,000 cfs).

Table 1. IDF for conditions equivalent to Matala Dam

Country	IDF definition	IDF peak outflow (m^3/s)
Canada (CDA) (Ref. 1)	1/3 between 1000-yr and	4,313 (152,000 cfs)
USA (FERC) (Ref. 2)	PMF	7,500 (265,000 cfs)
Australia (Ref. 3)	10,000-yr	3,465 (122,000 cfs)
New Zealand (Ref. 4)	10,000-yr	3,465 (122,000 cfs)
South Africa (Ref. 5)	1,200-yr	2,910 (103,000 cfs)
Austria (Ref. 7 and 8)	5,000-yr	3,400 (120,000 cfs)
Finland (Ref. 6)	10,000-yr	3,465 (122,000 cfs)
France (Ref. 7 and 8)	1,000-yr	2,720 (96,000)
Germany (Ref. 7 and 8)	10,000-yr	3,465 (122,000 cfs)
Italy (Ref. 7 and 8)	1,000-yr	2,720 (96,000 cfs)
Norway (Ref. 7 and 8)	1,000-yr	2,720 (96,000 cfs)
Portugal (Ref. 7 and 8)	1,000-yr	2,720 (96,000 cfs)
Spain (Ref. 7 and 8)	1,000-yr	2,720 (96,000 cfs)
Switzerland (Ref. 7 and 8)	1.5 times 1,000-yr	4,080 (144,000 cfs)

HYDRAULIC DESIGN OF THE NEW SPILLWAY

As mentioned previously in this paper, the original main spillway is equipped with 29 flap gates. Each flap gate consisted of a steel leaf hinged at bearings at the lower edge (refer to Fig. 2). A counterweight attached to the upper portion of the gate through a lever arm kept the leaf in the closed position when the upstream water level was low. At high water levels, the gates opened due to water pressure acting on the gates which exceeded the counterweight force.

For the new spillway, eight of the 29 flap gate bays were replaced by radial gates (refer to Fig. 4) and the remaining 21 flap gate bays were removed and converted to overflow weirs (refer to Fig. 3).

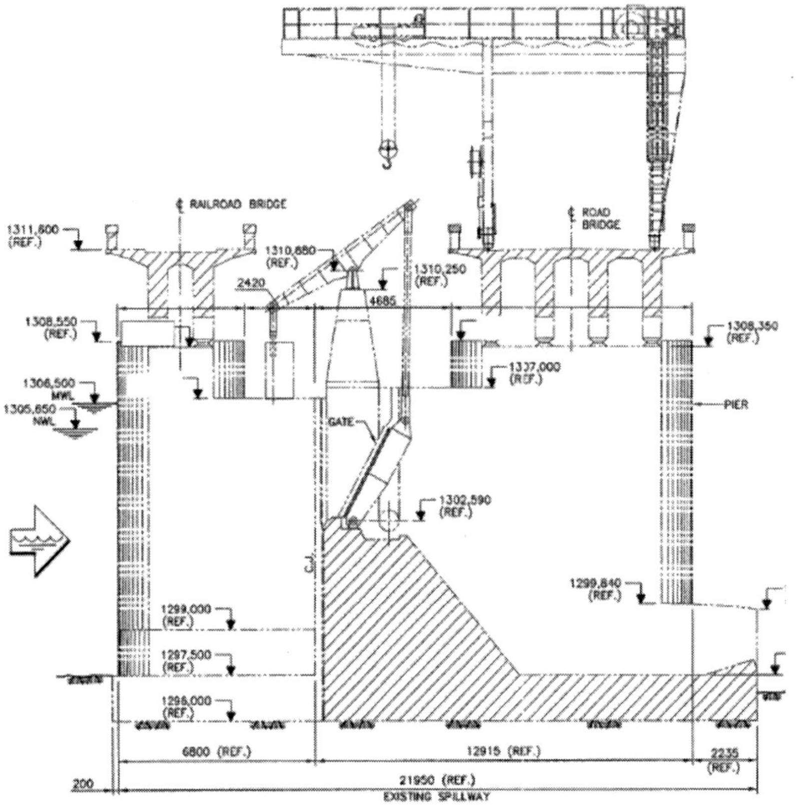

Figure 2. Section view of original spillway with flap gate and counterweight.

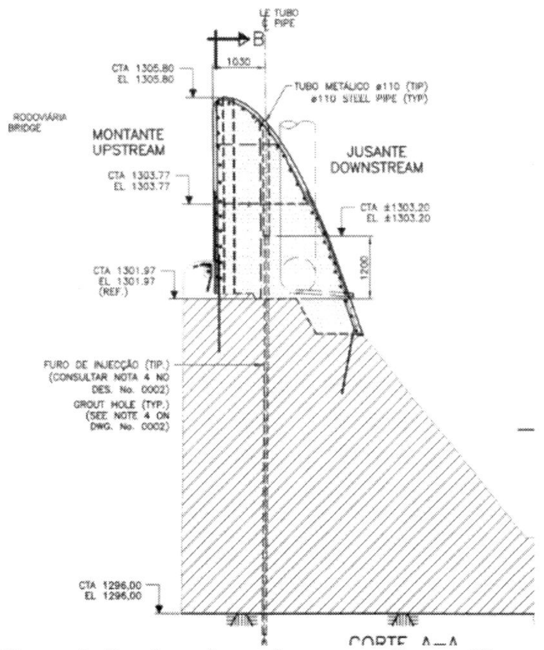

Figure 3. Section view of new overflow spillway.

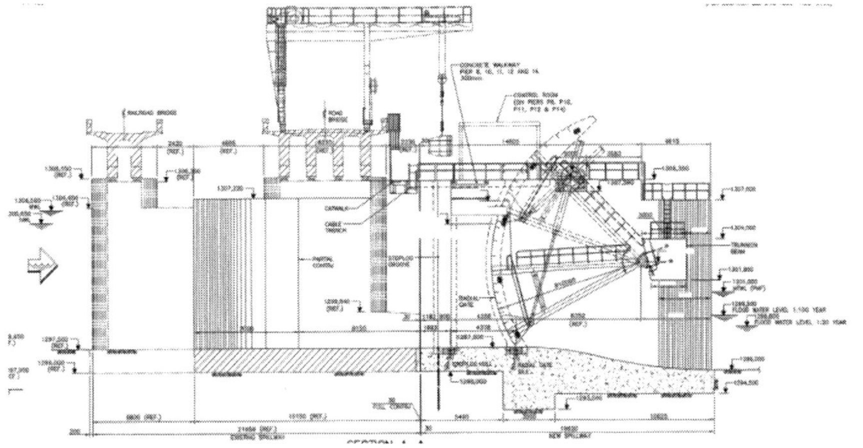

Figure 4. Section view of new radial gates spillway.

The hydraulic design of the new spillway has the objective to ensure that the new spillway has the required hydraulic capacity and that the flow conditions near the structures are acceptable. Both aspects were analyzed through the extensive use of the Computer Fluid Dynamic (CFD) modeling tool Flow 3D developed by Flow Science (Ref. 10). CFD analyses were used for determining optimal geometry of the concrete weirs and rock excavations downstream of the spillway (Fig. 5).

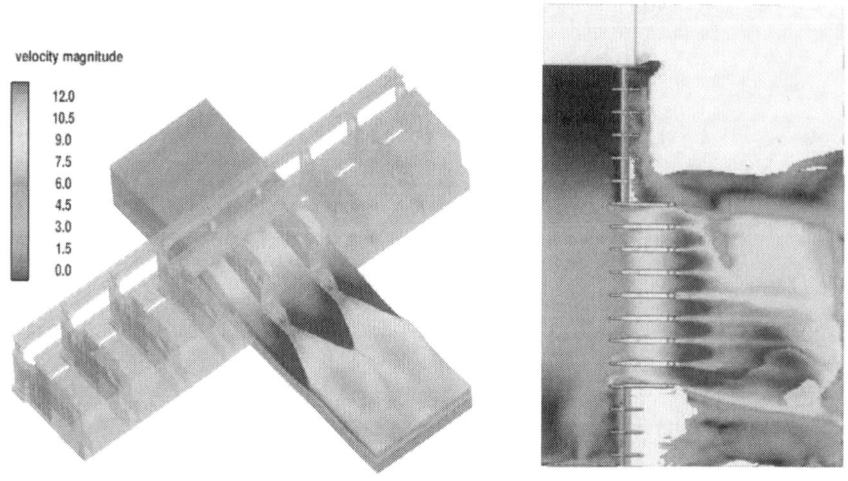

Figure 5. Radial gate spillway flow behavior and velocity magnitude pattern. Oblique view (left) and plan view (right)

FLOW DIVERSION DURING CONSTRUCTION

During construction of the new spillway equipped with radial gates, a 10 m high cofferdam was built upstream of the work area, as shown in Figure 1. The river flow was diverted through the powerhouse and through the remaining flap gates. Ten flap gates were temporarily repaired to be fully operational. At the end of the rehabilitation process, all flap gates were decommissioned and replaced with overflow weirs behind stoplogs (Fig. 3).

Construction of the cofferdam caused significant reduction of the spillway discharge capacity at the site during the work period which extended over two flood seasons. Therefore, appropriate measures had to be taken to protect the cofferdam against overtopping and to alert the population downstream in case of an unexpectedly large flood event.

SAFETY ASPECTS

Socio-Economic Activities Downstream of the Dam

The Matala dam plays an important role in the socio-economic plans for the population living downstream of the dam. It allows irrigation of a 3,500 ha cultivated area through a 42 km long irrigation canal which conveys water from the reservoir. Villagers also draw much of their water needs from the regulated Cunene River.

The most densely populated area near the dam is the town of Matala, with approximately 33,000 inhabitants, located on the left bank of the reservoir. There are also several villages downstream of the dam mainly on the left bank of the river. There is a low-lying municipality of Kapelongo at risk of being flooded, 20 km south of the dam. On the right bank, there are also a few clusters of dwellings.

On both banks, however, there are many accesses to the river that are used by the locals to collect water for drinking and laundry purposes. Fishing and recreational activities are also common at several locations along the river. Since there are no bridges downstream of the dam, the locals use boat ferries to cross the river at a few locations.

Figure 6. Photos taken at the beach area downstream of the Matala dam.

Review of Potential Hazards to Areas Downstream of the Dam

Three flood risk situations were identified for evaluation:

- Inundation risk during construction;
- Inundation risk during testing of radial gates;
- Inundation risk during the operation of the dam and spillway after construction.

During the construction works, the main risk to the population downstream of the dam is the failure of the cofferdam by flood overtopping. To mitigate this risk, a study of the cofferdam failure was conducted and flood maps were prepared as well as an internal emergency preparedness plan. The analyses indicated that in the case of a breach at the cofferdam, the water level may suddenly raise by 4 m (13 ft) in about 30 min at the beach area located about 2 km (71,000 ft) downstream of the dam.

Dam break analysis and flood wave propagation in the valley downstream on the dam were carried out using HEC-RAS software (Ref. 11). The section of the river modeled is about 30 km (19 miles) long. The Figure 7 below shows the simulated water level variation with the time at 3 locations downstream of the dam. The results of those simulations were used to prepare flood inundation maps for the analyzed scenarios.

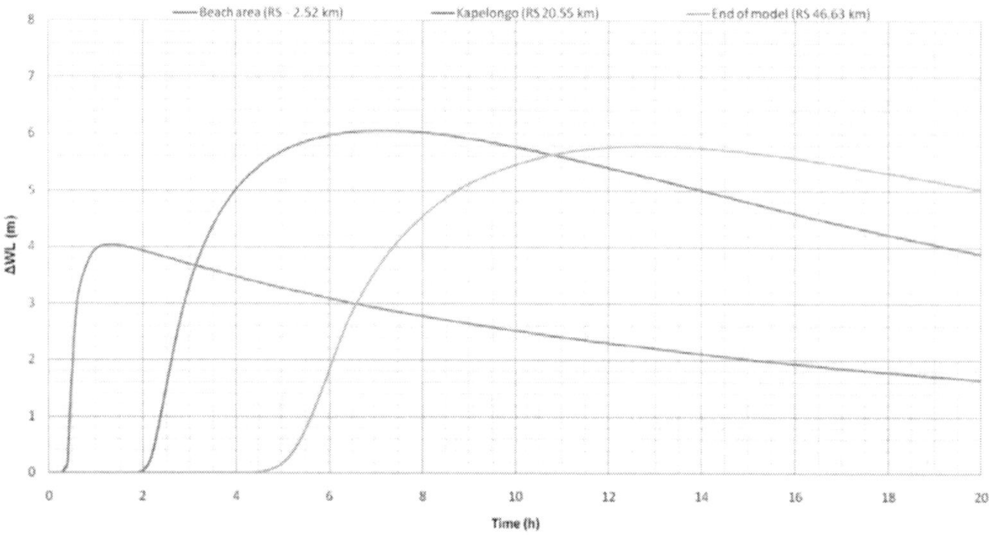

Figure 7. Cofferdam failure under normal operating conditions. Stage hydrographs at locations downstream of the dam.

Similar analyses were undertaken to determine the effects of the opening of radial gates, for testing after the completion of the installation, to downstream areas. The increase in water surface level at the most impacted area (beach area located at about 2 km downstream of the dam) was found to be 1 m in about 30 min. To reduce the impact of this sudden raising of water levels in an area where there are usually children swimming in the river, the following mitigation measures were adopted: installation of a siren warning system; and dispatch of a guard to be physically present at the site and connected to the dam operators via radio.

For the operation phase of the dam and the spillway, the identified risks are related to the opening of the radial gates, especially when more than 4 gates are opened. To help the owner to manage these situations, flood inundation maps were developed for each scenario and an emergency preparedness plan (EPP) was prepared for several scenarios. The EPP included public awareness, communication evacuation procedures.

CONCLUSION

There are still many countries and states where there are no regulations on dam safety. This is generally the case in developing countries where security issues are not a high priority.

During the design and the constructions of dams, engineers are required to choose design and safety standards for their projects. In the case of the project for the reconstruction of the spillway of Matala dam, SNC-Lavalin has opted for high standards for the spillway design flood and for emergency management during and after construction.

Dam break analysis were carried out to support the hazard classification of the Matala dam and to determine its spillway design flood. Results of the dam break analysis were also used to prepare flood inundation maps for the emergency preparedness plan.

REFERENCES

1. Canadian Dam Association, "Dam Safety Guidelines", *2007.*

2. Federal Energy Regulatory Commission (FERC), "Engineering Guidelines for the Evaluation of Hydropower Projects" Chapter 2 "Selecting and Accommodating Inflow Design Floods for dams", 1993.

3. Australian committee on Large Dams (ANCOLD), "Guidelines on the consequence Categories for Dams", 2012.

4. New Zealand Society on Large Dams (NZSOLD)," New Zealand Dam Safety Guidelines", 2000.

5. Government of South of Africa, "Water and Dam Safety Legislation", 2001.

6. Centre for Economic Development, Transport and the Environment of Finland, "dam safety Guide", 2012.

7. Unknown, "Classification of Dams and Inflow Design Flood Requirement in Europe"

8. ICOLD, "Flood Evaluation and Dam Safety", 2015.

9. ICOLD European Club, "Dam Legislation", 2007.

10. Flow Science, "Flow 3D software", flow3d.com.

11. Hydrologic Engineering Center, "HEC-RAS software release 4".

DECOMMISSIONING OUTDATED FISH BYPASS SYSTEMS AND PREPARATION FOR CONCRETE RETROFIT ON LOWER GRANITE LOCK & DAM

Dirk deGroot[1]

Fishery science has changed considerably since the construction of the Lower Granite Lock and Dam in 1965. Located on the Snake River, the most upstream dam to have a fish ladder, Lower Granite Dam has undergone several renovations to better protect fish from the power generating units; the most recent being completed in February, 2016. Construction of the Lower Granite Juvenile Fish Facility Phase 1a required a tunnel boring machine to be used to widen the collection channel, running internally through the length of the dam, to such a degree that the dam would be structurally compromised. Structural stability was able to be maintained with the placement of approximately 8,000 CY of reinforced concrete, below the water surface, in the bulkhead slots housing the outdated fish bypass systems.

Global Diving & Salvage, Inc. (GDS), working as a subcontractor performed the decommissioning and removal of the existing, outdated, fish bypass system, installation of 54 EA 6,000# concrete plugs, 18 EA 85,000# steel/concrete bulkhead slot plugs, and approximately 630 CY of tremie concrete. A single 9 man dive team utilizing surface decompression tables performed "penetration dives" into the penstock to anchor and grout the plugs in place. ROV operators worked in tandem to provide QA/QC to meet tolerances for the placement of these structures. All work was completed from the deck of the dam, immediately adjacent to an active road way.

After GDS teams and equipment were out of the way, Garco Construction, Inc. dewatered the 18 bulkhead slots, installed the remaining concrete, and tunneled through the dam.

[1] Global Diving & Salvage, Inc., ddegroot@gdiving.com

SAND SECANT PILE CHIMNEY FILTER INSTALLATION AT PINE CREEK DAM

Bob Faulhaber (PE)[1]
Dr. Michael Arnold (PhD)[2]

ABSTRACT

The Pine Creek Dam Modification project is located just outside of Valliant, OK. The prime contract holder was Heeter Construction, Inc. with Bauer Foundation Corp. constructing all of the deep foundation works. The United States Army Corps of Engineers (USACE) determined the 13 foot diameter concrete outlet conduit was leaking causing erosion issues inside the dam embankment; therefore, the purpose of the project was to address four major work tasks: 1. Install a concrete cut-off wall around the 13 foot diameter outlet conduit 2. Re-install the existing vertical sand filter drain along with filling potential voids in the embankment 3. Excavate and install a horizontal downstream invert filter 4. Install a 12 foot diameter steel liner inside the outlet conduit. Bauer Foundation Corp. assisted Heeter Construction, Inc. with the first two major work tasks.

This paper describes task two, the installation of a sand secant pile chimney filter to depths approaching 120 feet deep. Bauer's design utilized 5 foot diameter fully cased excavations similar to the excavations executed for the concrete cut-off wall and used the tremie method to convey the sand to the bottom of the excavation. It was a requirement of the USACE that the drilling through the embankment into bedrock was completed essentially dry – i.e. no drilling fluid was allowed and the embankment was locally dewatered via a complex series of dewatering wells.

INTRODUCTION

This report describes the construction of the 21 chimney filter columns completed and the 9 void filler piles, as shown in Figure 1. The continuous chimney filter is approximately 50 feet long and is parallel to the cutoff at an offset of 28.5'. The void filler piles were non-continuous and were constructed in two rows centered over the conduit with different offsets downstream of the continuous chimney filter. The exception to this is VFA-6 which was located to the right of the chimney filter. Straight, non-scalloped guide walls were installed prior to the chimney filter works. No guide wall was used for the void filler piles.

[1] Bob Faulhaber, bob.faulhaber@bauerfoundations.com
[2] Dr. Michael Arnold, Michael.arnold@bauerfoundations.com

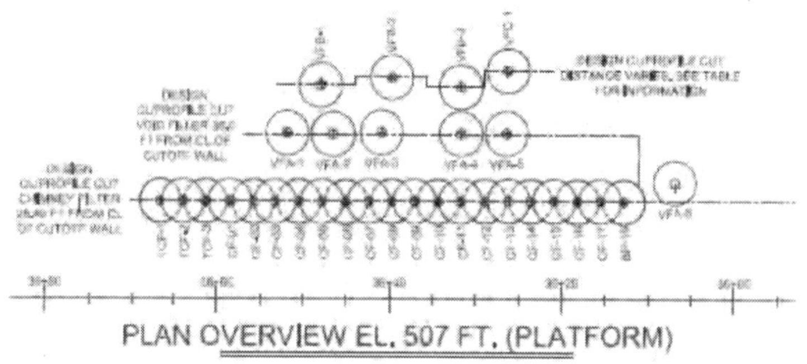

Figure 1: Column and Pile Layout

GROUND CONDITIONS

Pine Creek Dam was designed as a zoned earth fill dam with random fill and an impervious fill core, based on USACE (2014), as shown in Figure 1. The impervious fill core consists of predominantly clayey materials with about 59% fines, 40% sand and minor portions of gravel. It can be classified as sandy lean clay (CL). The random fill has in average 50% sand, 47% fines and 3% of gravel. Its classification ranges from SM, SC, CL, CL-ML, SC-SM to ML. A three-foot wide chimney filter is located just downstream of the impervious fill core and concrete plug. A five-foot horizontal filter is located around the conduit, which was placed in a trench cut into the bedrock. The bedrock at the conduit is predominantly a medium to hard fine grained, very hard quartzitic sandstone. There are numerous thin shale seams interbedded with the quartzitic sandstone beds.

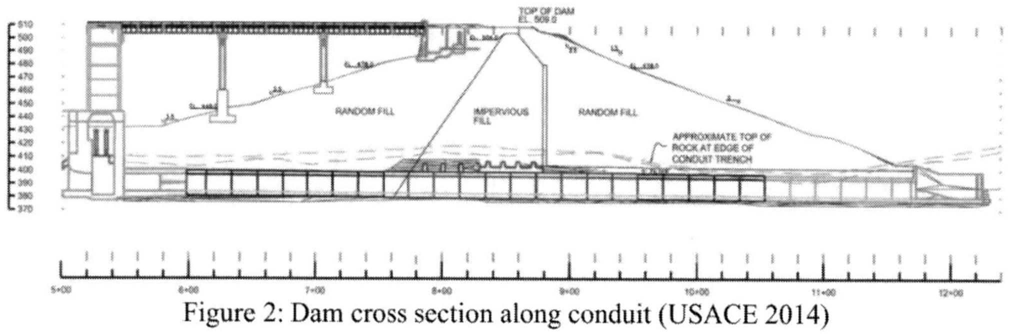

Figure 2: Dam cross section along conduit (USACE 2014)

CONSTRUCTION PROCESS

Excavation

Installation of the chimney filter piles involved drilling approximately 60-inch-diameter columns at a center to center spacing of about 30 inches. The average achieved wall thickness was 49 inches and the minimum wall thickness was 42 inches. BAUER used a BG-40 for the excavation of the columns in the chimney filter and void filler piles, drilling through soil, sand and rock. The excavation was fully encased using steel casing with the casing advanced by the drill. The excavations were completed in the dry, i.e. they did not use any drilling fluid.

To begin the excavation, the surveyor marked the proposed center point of each column on the ground using conventional surveying techniques. Chimney filter elements used a template bolted to a guide wall to assist the BG-40 with starting the excavation at the design location. No guide wall was necessary for void filler elements therefore the template was not used. The surveyor would mark the center location of the pile, and two offsets were measured perpendicular to each other from the center location. The BG-40 would move the starter casing visually close to the center, then the distance from the casing's outside diameter to the offsets was measured to determine if the operator needed to make any adjustments.

The BG-40 drilling rig with Kelly tools was used for excavation of the columns. The photos below show the BG-40 with some of the tools that were used during excavation. A 60-inch-diameter starter casing was installed and advanced. As drilling commenced, and the depth of the pile increased, more casing was added to keep the excavation fully encased. As the casing was advanced, the excavation was cleaned out with the appropriate drilling tools. The on-board instruments provided the operator, CQC, and QA staff with drilling data. Offsets from the design location were used to measure the location of the casing with respect to the design location. CQC staff logged the drilling of each column on the Column Excavation Log and filled out a Quality Control checklist.

Figure 3: BG-40 with Auger

Figure 4: BG-40 and Oscillator being used to Advance the Casing

Figure 5: BG-40 with Roller Barrel

As-Built Measurements

When excavation was completed, the verticality was measured by both the DIS (Drilling Inclination System) and the RIS (Rope Inclination System). The RIS was used as a primary method of measurement, and the DIS was used to confirm the data from the RIS.

The verticality of the columns at depth were measured relative to the actual location at the ground surface. Survey data show that the deviation of the excavation was within acceptable tolerances for wall width on all piles. The overlap requirement of 0.3333 times pile diameter which equals 20 inches was met for all piles with the exception of CF-9. CF-9 deviated such that the bottom few feet of the pile had an overlap that was as much as ½ inch less than the requirement of the specifications. Ultimately, the pile was accepted as constructed and no remediation was necessary.

As-built measurements took place after casing refusal but before cleaning or rock excavation, if necessary, to prevent the RIS sledge from getting trapped at bottom of the excavation. The as-built measurements obtained for each column include center point at ground surface and interpolated results every ten feet. The as-built center of the column was determined by land survey of prisms located on the RIS frame when placed on the casing. Then the sledge angle was used to determine the center point at ground surface as recorded by the RIS. The as-built verticality of each column was measured by the following techniques:

1. Rope Inclination System (RIS) – A centralizing carriage was lowered into the casing. The carriage was suspended from an inclination measuring unit mounted on top of the casing. When the carriage was at predetermined depths in the casing, the inclination of the Kevlar rope used to suspend the carriage was measured using a bi-axial inclinometer. The horizontal displacement of the rope at the top of the carriage was calculated using the measured inclination. Figure 5 shows the RIS set up on the top of the casing being used to measure a pile.

Figure 6: Performing RIS Survey

2. Drilling Inclination System (DIS) – A cylinder with an outer diameter that is similar to the inner diameter of the casing was lowered into the casing. The cylinder was suspended from a cable connected to an auxiliary winch on the

BG-40 mast. When the cylinder was at predetermined depths in the casing, the inclination of the auxiliary winch cable was measured using the total station. The horizontal displacement of the cable at the top of the cylinder was calculated using the measured inclination. Figure 6 below shows a DIS measurement being taken. The weight has been lowered into the excavation using a cable attached to an auxiliary winch on BG-40. The measurement is taken by surveying the cable attached to the weight.

Figure 7: Performing DIS Survey

The RIS was intended as the primary system, with the DIS providing support. BAUER used the as-built column alignment to evaluate whether adjacent columns needed to be adjusted to provide the minimum overlap dimensions. BAUER CQC staff confirmed the RIS and DIS were performed and reviewed and compared the data generated by the RIS with the DIS data for each column.

Bottom Cleaning

Bottom cleaning was performed after the element geometry surveys were performed and the excavation was complete. If the excavation was to the top of the conduit cleaning took place immediately after the element geometry surveys were complete. However, if the excavation was to be drilled into rock, further excavating took place after the geometry surveys, and bottom cleaning was done after the excavation was complete. The general procedure for cleaning the bottom of the excavations was to first use the baler to clean the bulk of the material out that may be in the bottom of the excavation, then to use the wire brush to clean the bottom of the excavation as much as possible. The brush was sent down until only clean sand was observed in the bristles. Each time material was brought out of the hole either by the baler or the brush, the material was inspected by both the BAUER QDI and USACE staff as shown in Figure 7 below. Information regarding the material was not logged. The purpose was to look for anything unusual and to check whether the sand was wet or not. It was not possible to get the bottom of the excavation completely clean, i.e., the brush always brought up some sand. On a few elements the cleaning was suspended due to sand coming into the bottom of the excavation or the presence of water in the bottom of the excavation. In these cases the USACE directed normal backfilling procedures to take place immediately. When the cleaning was approved by the USACE, BAUER staff then sounded the hole in three locations to get the final depth.

Figure 8: Using Baler to Clean the Bottom of an Excavation. BAUER and USACE Staff Inspecting Material

Evaluation

Evaluation of the pile construction was not conducted within the chimney filter. Instead, a series of verification borings were directed to be placed downstream of the chimney filter. A split spoon sampling method was used to determine if voids were present in the location of the borings. The USACE evaluated the initial verification borings to determine the location of six void filler piles (VFA-1 through VFA-6). After the installation of these void filler piles, further verification borings were then directed to determine if further void filler piles would be required. The USACE then directed three additional void filler piles (VFB-1 through VFB-3). The USACE directed the installation of one more void filler pile (VFC-1) after water was introduced into a dewatering well in an attempt to remove a blockage and get the pump back pumping.

Backfill Material and Placement

Fine filter sand was used to backfill the chimney filter and void filler piles. The sand was received from Martin Marietta, and was required to conform to the specifications listed below in Table 5. Before sand was used for backfill, it was tested by American Engineers, a third party agency contracted by the USACE. Heeter also supplied an outside testing agency, Terracon, to test the sand.

Sieve Size	Percent Passing by Weight
3/8"	100
#4	95-100
#8	80-100
#16	50-85
#30	25-60
#50	5-30
#100	0-10
#200 (After Compaction)	0-2

Table 5 - Fine Filter Sand Specs (Modified C 33/C33M Fine Aggregate). This table was taken from the Technical Specifications Section 31 23 00.00 20 paragraph 2.1.4

The placement of sand normally took place after bottom cleaning had been approved by the USACE and the QC reports were signed by all parties (BAUER, Heeter, and the USACE). Sand was placed in the piles using the tremie method. BAUER field staff assembled the tremie string (see Figure 9) so that the bottom of the string was no more than five feet from the bottom of excavation. Sand was scooped into hoppers using a mini-excavator and weighed (See Figure 10), then picked up by an extendable forklift, the hopper was secured to the forklift and then dumped into the tremie hopper. After a few loads were placed, the tremie string was lifted by the crane and the hole was sounded to determine whether or not to shorten the tremie string. The tremie string was shortened five feet at a time keeping the bottom of the tremie string within five feet of the sand. BAUER QC staff recorded the number of loads placed as well as the size of the hopper used for placing each load, sounded depths, and tremie and casing pulls. When the casing became embedded 30 feet, the tremie string was removed from the excavation by either the crane or BG-40 and a section of casing was pulled using the oscillator and in some cases the oscillator and the BG-40 were needed to pull casing. However, due to the difficulty in pulling casing on some holes, the shorter sections of casing were pulled before achieving thirty feet of embedment while still maintaining the minimum ten feet of embedment. The depth to sand was sounded before and after a section of casing was pulled to document the amount that the sand dropped. The sand typically dropped approximately 3.5' to 4.0' after a 16.4 ft. length of casing was pulled. After completing the casing pull, water was added to the hole through the tremie pipe to compact the sand. The hole was also sounded before and after adding water to determine whether or not the sand was compacted. In most cases compaction was confirmed after 50 – 80 gallons were added to the excavation. This process of adding sand and shortening tremie and casing at the appropriate intervals was continued until the sand level in the hole reached the top of the oscillator platform (i.e., approximately three feet above the top of the guidewall/ground surface). This provided enough sand so that when the final section of casing was pulled the sand level was even with the ground surface.

Figure 9: Placing Sand in Excavation

Figure 10: Assembling Tremie String

Figure 11: Loading and Weighing Sand

Figure 12: Cutting Tremie during Sand Pour

CONCLUSION

This project was the first time the USACE attempted to complete sand secant piles to depths of over 110 feet. Bauer successfully completed all 21 of the pile installations for the chimney filter wall, exceeding the specified minimum thickness of 3.0 feet by averaging 4.1 feet thick. In addition to the chimney filter secant piles, a series of exploratory borings and void filler piles were required to locate and fill with sand any voids downstream of the chimney filter. The void filler piles specifications were essentially the same as the chimney filter with the exception of the requirement to create a continuous wall and were similarly met by Bauer.

REFERENCES

United States Army Corps of Engineers (USACE 2014): Pine Creek Lake Dam Geotechnical Baseline Report.

EXECUTION OF A CUT-OFF WALL AT THE DEAD SEA –UNIQUE CONDITIONS REQUIRE UNIQUE SOLUTION

Franz-Werner Gerressen[1]
Gai Lehrer[2]
Christian Scholz [3],

ABSTRACT

The Dead Sea Works Ltd., Beer Sheva, Israel operates several large-scale salt pans at the "lowest place on earth". The salt content of the Dead Sea water (Brine) is up to 33%. The mineral composition differs significantly from seawater. In order to ensure operation, dike heights have to be adjusted accordingly. The northernmost salt pan (PAN 5) covers an area of approximately 75 km^2 with a capacity of about 150-200 million m^3. The dike, which runs parallel to the Jordanian border, has an approximate length of 18 km. Soil conditions show an alternating sequence of salt and clay layers. The salt layers are very porous, which causes sinkholes due to revers erosion.

Therefore, besides heightening, a rehabilitation of the dike was also required. This was performed in two steps. In the first step, the cracks and voids in the dike body were closed by dynamic compaction. The second step was the installation of a cut-off-wall to a depth of up to 34 m. The unique conditions required a unique approach. Therefore, a new two phase cut-off with an internal sheet pile wall was developed and introduced. Due to the high salt content of the brine, in addition to the aggressive desert climate, new rules had to be established in terms of cut-off materials and mechanical engineering. Upon completion, 550,000 m^2 of cut-off-wall have been installed

INTRODUCTION

From Lebanon to the Red Sea there is a 1,000-kilometer-long rift valley. The poorly vegetated strip of land extends from north to south along the border between Israel and Jordan. In the widening of the rift valley lakes such as the Sea of Galilee and the Dead Sea have formed.

A large salt lake of approximately 800 km^2 known as the "Dead Sea" is the lowest sea on earth – with an altitude of 400 m below sea level. The Dead Sea is separated into a northern and a southern part. The northern part currently has a maximum depth of 370 m. It is fed by the Jordan River, but due to the desert climate it is subject to significant evaporation. In the last three decades, the surface of the sea has shrunk by about a third. Consequently, there is no balance between inflow and evaporation/ withdrawals.

[1] Dipl. Ing., BAUER Maschinen GmbH, Germany , Franz-Werner.Gerressen@bauer.de
[2] B.Sc., Y.LEHRER Engineering Ltd., Israel, gai.lehrer@gmail.com
[3]Dr. Ing., Wayss & Freytag Spezialtiefbau GmbH, Germany, Christian.Scholz@wf-ib.de

The southern basin has a maximum depth of 6 m. The northernmost salt pan of the southern basin (PAN 5) has an area of approximately 75 km^2 and contains a volume of about 150-200 million m^3. It is bordered by dikes 3/5, 4/8, 5/8 and slopes naturally to the west. Dikes 4/8 and 5/8 have a total length of 18.2 km (see Figure 1).
The first stage of the dike was constructed in the 1960s and since then has been raised in three stages. The dike was built as a gravel dike with an internal Lime Carbonate core.

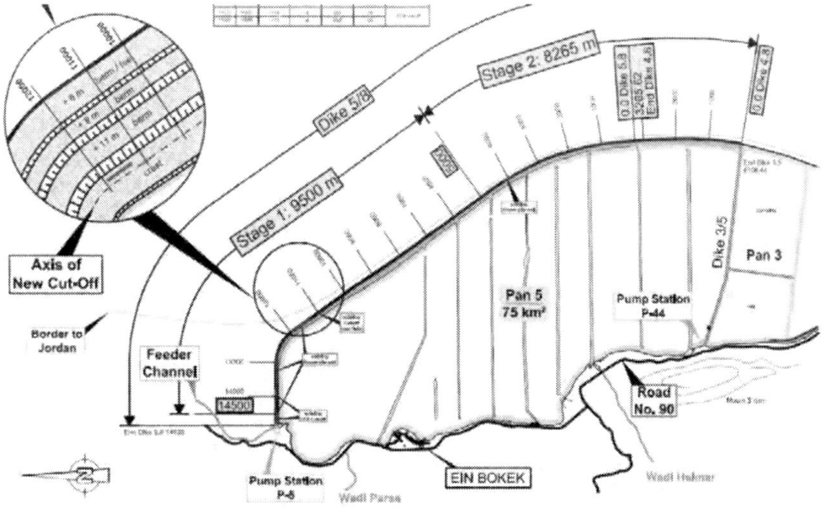

Figure 1. Information about PAN 5 and containment dike

REHABILITATION OF THE DIKES

Over the past years severe damage has occurred along dike 5/8. Cracking along the upstream and downstream shoulder of the crest and along the centerline of the crest coupled with sinkholes in the areas of the cracks has been observed. Seepage at the downstream toe was noticed in several sections.

The original dike design was based on a low permeability base and internal core. The internal core was planned to be connected to a low permeability lime carbonate layer between the upper and lower salt layers. The alternating salt layers, which were formed by the crystallization of brine, are described as very porous. Due to the large volume pore spaces connecting together, the core of the dike was subject to reverse erosion. As a result of the erosion and the related reduction in the sealing effect, seepage losses over the past years had been rising. Seepage was observed over a large area on the downstream side of the dike.

The current rehabilitation of the dike was carried out according to a concept of Wittke Consulting Engineers Aachen (WBI) as a two-step solution.
In a first step, the cracks and voids in the dike body were closed by means of Dynamic compaction.

In a second step of rehabilitation, an up to 34.2 m deep and 0.8 m wide diaphragm wall along the entire length of the dike 4/8 and 5/8 was to execute. The diaphragm wall was designed as a cut-off wall with internal sheet piles and covers both the critical load cases (earthquake), as well as sealing against seepage.

SPECIFIC FEATURES OF THE PROJECT

General

Following a one year tender phase, in April 2011, the contract to install the 18 km long cut-off wall was awarded to the Lehrer-Brückner-Partnership. The mobilization of equipment was done from May to August 2011, followed by a one month field test. During the project, for the first time in Israel, trench cutters were mobilized. Accordingly, it was necessary to have very intensive and open discussions with all project participants before and during all phases of the project. All trades required detailed planning and quality management to meet the very high demands on the construction with respect to making the project a success.

Environmental Conditions / Engineering Standards

The site is located in the Negev desert. The project conditions at the Dead Sea are characterized by very special and extreme environmental conditions:
- Extremely salty water (density $\rho \approx 1.25 - 1.30$ g/cm^3), in which the works are to be done,
- Very high ambient and brine temperatures of up to $T = 50$°C in summertime,
- Very strong storms and rains in winter, which go hand in hand with temperatures near the freezing point (at night).

Due to the specific ground composition and materials, new methods and procedures had to be developed, which significantly differ from known standards. Furthermore, it should be emphasized that the general standard used EN1538, does not describe plastic concrete wall with internal sheet piles and the related concreting procedure of concreting on both sides of the sheet pile wall. Therefore new procedures needed to be created.

CONSTRUCTION OF THE CUT-OFF WALL

General

Because of the specifics described previously, an intensive testing period was commissioned before the actual start of construction work. It was carried out over a period of several weeks, reviewing and optimizing the intended equipment and procedures.

Ground Condition

Figure 2 shows an example of a section of the dike body and all relevant soil layers. In general, it should be noted, that both the dike body and the ground underneath are very heterogeneous. Even within the shortest distances soil layer thickness, width and or soil properties can change significantly.

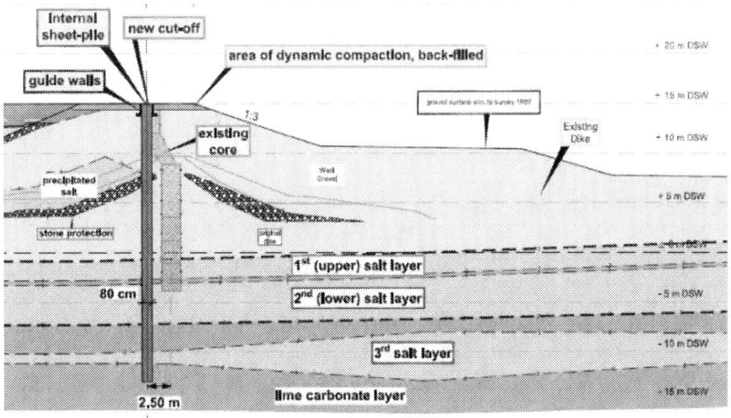

Figure 2: Cross-section of the dike and ground composition

Body of the dike: The dike body consists mainly of "Wadi Gravel". Wadi Gravel is gravelly sand with fines, which was sourced from the adjacent dried out river valleys. The material was typically not screened before placing and therefore contains diverse grain fractions up to $d = 70$ cm. The stone consists of limestone and flint, as well as dolomite. The material is classified as "very abrasive". The dike core consists of "Lime Carbonate", which can be characterized from a soil mechanics point of view as clay/silt. The clay core was constructed through the 1st salt layer, which is located below the dike footprint and is connected to the 2nd Lime Carbonate layer.

Ground composition:

The ground underneath the dike body consists of an alternating sequence of Lime Carbonates and precipitated salt. Immediately below the dike body there are two very strong, partially permeable layers of salt (1st salt and 2nd salt layer). These can be up to several meters thick and are only interrupted by relatively thin layers of Lime Carbonate. The Lime Carbonate layers are classified as slightly permeable. The salt deposits are hard, porous and locally fractured and characterized as highly permeable. The uniaxial compressive strength (taken on core samples) is up to $q_u = 13$ MN/m^2 [2].

Groundwater (brine):

The Pan level is approximately 1-1.5 m below the dike crest. The brine density changes according to the progress in the precipitation process. Along the construction stage, the density increases from $\rho = 1.25$ to 1.33 g/cm^3. The salinity is in a range of 25% - 33% (in comparison, to the Mediterranean Sea where the salinity is on average approximately 4%).

Construction Materials

General:

Due to the lack of available potable water in the Negev desert, the use of brine water was considered a must. In order to distinguish these materials from standard materials the terms "Phase I-Material" for slurry and "Phase II-Material" for concrete were selected.

Phase I-Material

The slurry must fulfil the function to act as temporary support to the trench walls and also must be able to be pumped considerable distances. The material basically consists of the Brine from the Dead Sea and Lime Carbonate. The density of the Phase I-Material was specified by the client to be $\rho = 1.4 - 1.5$ g/cm^3. Those values are much higher than that the density of fresh slurry ($\rho < 1.10$ g/cm^3) typically known from the standards. The handling of the Phase I-material proved to be one of the main aspects of the project. In general, three stages of development/improvement can be defined:

- Initial Phase I-Material production,
- Attempts to extend the lifetime of the material, and
- Everyday processing of material.

Initial Phase I-Material production:

At the beginning of the test phase, extensive mixing tests were carried out in order to develop a method for mixing Lime Carbonate with brine. The brine was taken directly from Pan 5. The Lime Carbonate was mined near the site in large scale quarry pits. Initially it was examined, and a determination made whether the natural material could be placed directly into the mixing plant. These tests led to unsatisfactory results, since the cohesive lumps of clay could not be dispersed in the slurry. In the second step, the Lime Carbonate in the borrow areas was extensively stripped and dried. Afterwards the dried material was crushed and fed directly to the mixing plant (see Figure 3).

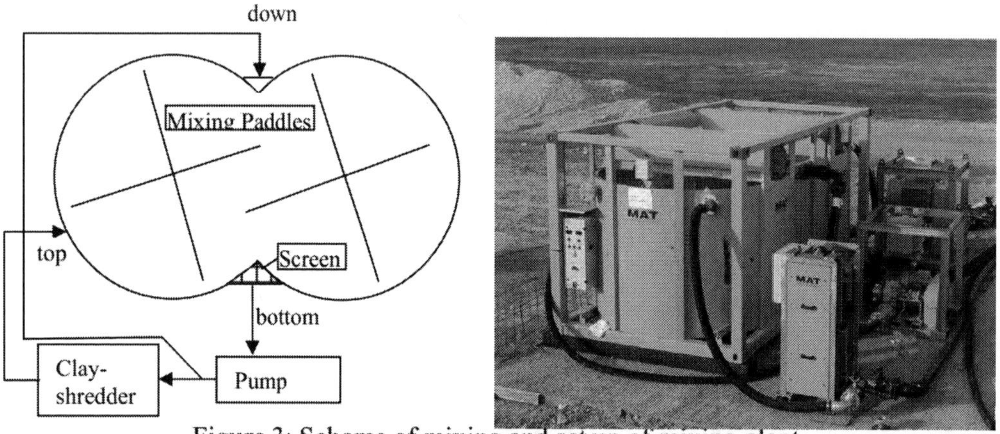

Figure 3: Scheme of mixing and set up of mixing plant

However, some significant contamination of Lime Carbonate with various stones proved to be problematic, since it led to considerable wear and tear. As a result of extensive testing, it was possible to prepare stable and homogeneous slurries with similar physical properties to traditional Bentonite slurries (see Table 1).

Density [g/cm^3]	Marsh-Value [sec]	Liquid limit [N/mm^2]	Sedimentation after 24 hrs [%]	Sand content [%]
1,40	34	9	3	2,5

Table 1: Physical Properties of Phase I-Material

The Phase I-Material is charged during the mixing process with soil, pumped to a central de-sanding unit where sand sized particles are removed. Figure 4 shows the range of the particle size distribution of Clay/Lime Carbonate. Regular de-sanding plants have a cut point at 60 µm. Thus, the existing fine silt could not be separated, the density of the Phase I-Material increased rather fast and workability decreased. As soon as the contractual density limit of $\rho = 1.50$ g/cm^3 was reached, the material must be exchanged and therefore used material was disposed of. Due to the amount of used slurry being disposed of, and considering the short lifetime of fresh material, the cost and effort to produce Phase I-Material was considered very high and uneconomical.

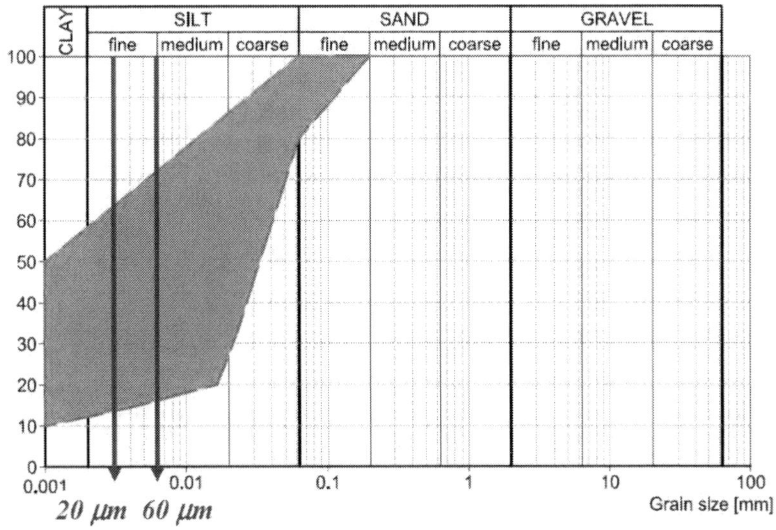

Figure 4: Range of grain size distribution of Lime Carbonate

<u>Attempts to extend the lifetime of the material:</u>
It was determined that in order to extend the lifetime of the Phase I-Material more of the fine silt should be extracted. A secondary cleaning step – using smaller cyclones - was added to the system (see Figure 5). The cut point was lowered to 20 µm

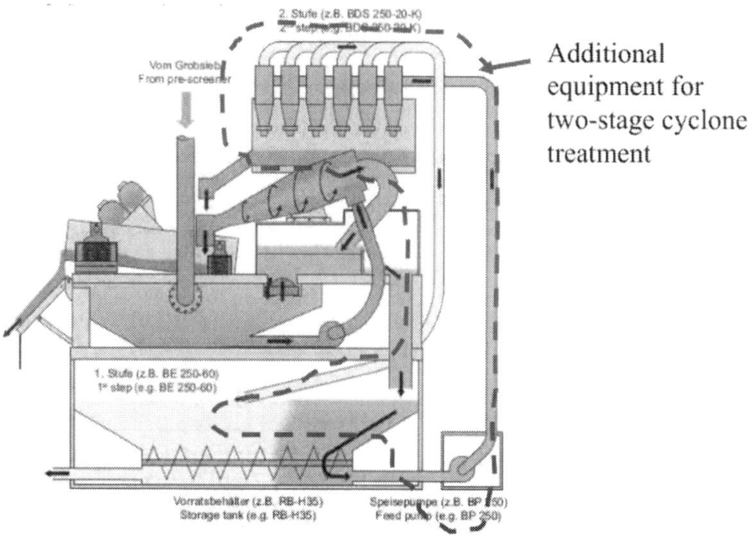

Figure 5: Scheme of enhanced fine de-sanding

Now, the slurry could be treated in such a way, that a longer life time was achieved and workability was maintained.

Everyday processing of material:
Since even the extended recycling of the slurry was not efficient, the method of treatment was reassessed. The general idea was to keep the fines from the Lime Carbonate, which were charged during the cutting process within the slurry, and to re-mix it. The approach is described below. The physical characteristics are listed in Table 2. For the processed slurry, the characteristics are defined as having a density of ρ= 1.50 g/cm$^{3.}$ These characteristics were determined when double cycloning was not utilized.

	Density [g/cm^3]	Marsh-Time [sec]	Liquid Limit [N/mm^2]	Sedimentation after 24 hrs [%]	Sand Content [%]
Fresh Slurry	1,40	34	9	3,0	2,5
Process Slurry (without double cycloning)	1,50	45	35	1,5	1,5

Table 2: Physical properties of Phase I-Material

Comparing the characteristics of fresh and processed slurry even at the upper limit, in terms of sedimentation and sand content, the processed slurry showed better values than the fresh slurry. In order to reduce the density, and consequently to improve the pumpabilty of the material, brine was added to the slurry during the de-sanding process and the slurry was re-mixed. Instead of disposing of slurry, which became too thick, the slurry was recycled.

The volume of brine was determined by calculation and confirmed by field testing. Subsequently, salt water pumps were installed in Pan 5 and lines routed directly to the de-sanding plants, taking care that the density would never become less than ρ= 1.40 g/cm^3 thereby ensuring trench stability. The pumps and lines were sized so that even accidental sudden addition of brine could not force the density of the slurry below ρ = 1.40 g/cm^3, which could cause stability problems in the open trench.

The advantages of the described procedure were:

- Mixing of the slurry was completely eliminated (therefore the efforts for the time critical collection, processing and transportation of clay could be eliminated).
- Slurry was recycled and kept within the defined boundaries. Originally it was planned to replace the slurry before concreting with clean slurry, but this step could be completely eliminated, which provided an enormous saving in both time and equipment.
- Double cycloning for treatment of slurry could also be eliminated.

Phase II-Material:

Phase II-Material has similar properties to Plastic Concrete. In the final construction stage the purpose of the material was to protect the sheet piles against corrosion. The Phase II-Material consisted essentially, of special cement, brine from the Dead Sea and Wadi Gravel.

In lab- and field tests a material was developed which:

- Had sufficient workability in a fresh state,
- Achieved strength of sufficient erosion stability
- Provided a sufficient difference in density to Phase I-Material to ensure the displacement of Phase I-Material.

The criteria for the flow diameter was defined as 55-62 cm. The uniaxial strength after 28 days reached q_u = 5-10 MN/m^2. The fresh concrete density was approximately ρ = 2.25 g/cm^3. The material could displace the Phase I-Material sufficiently by using tremie placement. The relatively low liquid limit of the Phase I-Material had a positive impact.

The brine content in the mix caused a delay in the hydration of the cement. Therefore, the processing time was extended when compared to known concrete mixes. As a welcome side effect, it was found that adding cooling agents could be avoided, even at high outside temperatures such as $T > 45°C$. However, at low temperatures the effect led to delays in the construction process.

Sheet piles and stop end elements:

The sealing concept of the dike rehabilitation was to use a thin sheet pile profile, which was centrally installed in the trench. The cut-off wall was constructed by means of 9.3 m long panels. The waterproof and force-locked connection between adjacent panels was achieved via a tubular stop end element, whereby sheet pile interlocks were welded on both sides.

Execution of the Cut-Off

Logistics on site:
The site was designed to allow construction to move along a line. The dike crest width varied between 12 to 20m. The berms were also utilized in the construction process.

The construction site was operated on a six day working week, 24-hours a day basis. When considering utilizing up to five trench cutter units, one challenge was to identify and source the significant materials so that possible sources of conflict and confusion could be minimized. Two concepts were established when dealing with logistical issues, central and decentralized units

Central units:
Mixing of Phase II-Material was done in a central mixing unit, consisting of four concrete plants (see Figure 6) and a significant number of truck mixers.

Figure 6: Concrete plants on site

To meet the average daily demand for Wadi gravel (2,500 tons) and cement (400 tons), a considerable logistics operation and fleet of trucks was required.

Decentralized units:
The cut-off works were performed locally in self-contained units. Machinery and personnel were assigned to the individual units. Each individual unit was assigned a discrete section of the dike. A complete unit consisted of:

- Backhoe excavator for pre-excavation,
- Trench cutter and de-sanding plant,
- Plant for processing and storing of Phase I-Material,
- Service crane and truck mounted crane as well as various auxiliary and earthmoving equipment.

In view of limited space, the dike crest was only used for cut-off excavation and installation of materials. The berms were used by concrete trucks and for all kinds of deliveries and discharges.

Whilst the trench cutter and service units were employed on the dike crest, the de-sanding unit and the plant for processing and storing of Phase I-Material was set up on the first berm (see Figure 7).

Figure 7: Set up of Trench cutter, de-sanding plant and slurry pools

<u>Logistics within the cutter units:</u>
Each cutter unit was assigned a discrete section (see Figure 8). The cut-off wall was constructed by means of the starter-consecutive-method. This involved installing starter panels and working continuously from alternating sides.

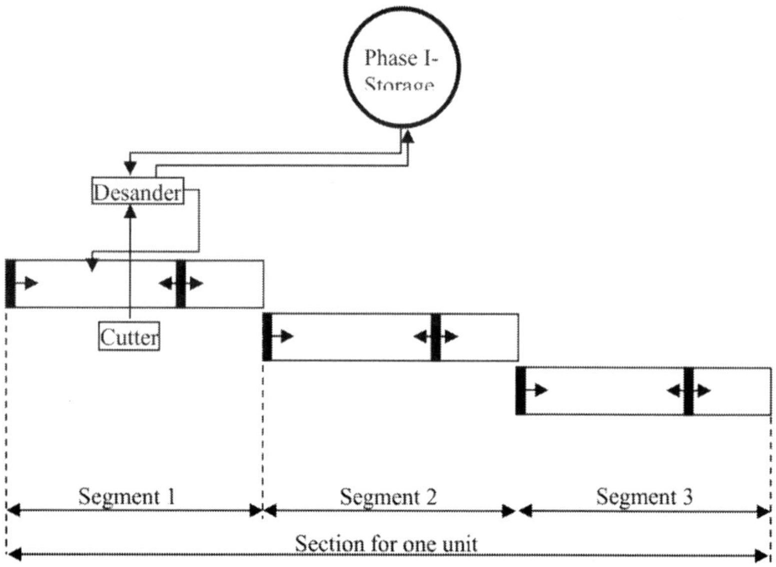

Figure 8: Logistic within a section

The length of each segment was adapted to suit the pumpability of the Phase I- Material. The maximum distance was dependent on the pumpable distance from the trench cutter to the de-sanding plant. The average total length of a segment was about 450 m. After completion of a segment the de-sanding plant was moved in one piece, utilizing a special heavy duty forklift (see Figure 9). As the total length of a working section for a single trench cutter was up to 1,350 m, upon completion of each section, the entire unit, including the large slurry storage pool, was moved to a new section (see Fig. 12).

Figure 9: Transportation of entire de-sanding plant and 750m^3 steel pool

The following steps were applied for the cut-off wall production:

- Guide wall installation,
- Pre-excavation by backhoe and excavation of trench by cutter until the final depth,
- Installation of the sheet pile elements, stop end elements and protective profile,
- Concreting by Tremie method and
- Construction of clay cap.

Excavation method:

A major challenge was to establish an excavation technology that allowed excavating both within the Wadi Gravel of the dike body, which contained large boulders along with harsh salt layers and some very soft layers of clay. Basically, excavation by means of using grabs (clam shells) in the Wadi Gravel was possible and had been partially performed previously. In a test field in 2008, tests were carried out to intersect the salt layers by grabs but this proved almost impossible due the strength of the salt layers. Due to the failure of the grabs, the client explicitly requested the use of trench cutters to excavate the salt layers. Finally, it was decided to execute the complete excavation using trench cutters only. Initially a 4 m deep pre-excavation was done by hydraulic excavators. This represented a significant simplification of equipment and logistics. In tests, the use of full face cutter wheels proved to be most economical with respect to performance and wear (see Fig. 10).

Figure 10: Utilized trench cutter Bauer BC40 on base carrier Liebherr HS 885

The high abrasiveness of the Wadi Gravels in conjunction with the omnipresent brine was a particular problem for cutting. Figure 11 shows the processed material at the coarse screen on the de-sanding unit. In particular, the limestone and flint inclusions caused increased wear on the cutting tools and de-sanding units.

Figure 11: Cuttings from Coarse screen of de-sanding plant

During summer, Phase I-Material and trench excavation equipment was exposed to temperatures of up to T ≈ 50 - 55°C. This resulted in an increased cooling demand on the hydraulic motors for the cutter gear boxes and mud pumps. In addition, all parts of the equipment which came into contact with the brine, had to be cleaned with potable water continuously.

Another challenge was the existence of large fractures within the salt layers, which led to significant slurry losses. When large losses were noticed the area was immediately backfilled. These sections ere reworked at a later stage.

Installation of sheet piles and stop end elements:

After panel excavation and corresponding quality control tests of the slurry, sheet piles, stop end elements and protection profiles were installed. The handling of thin sheet pile wall elements and protection profiles required special cable traverses and equipment to minimize bending moments during lifting. Figure 12 shows the erection of a "connection profile". Such connection profiles were used in closing panels for compensating of tolerances.

Figure 12: Erection of a special sheet pile connection profile

Concreting of panels / Construction of clay cap:

In the final step, the Phase II-Material was placed using the Tremie-Method. Pairs of tremie pipes were installed on both sides of the sheet pile wall. Casting was carried out in a way that the level of Phase II-Material was equal on both sides of the sheet piles to avoid bending. Special hoppers were designed and used to split the volume of one concrete mixer to two tremmie pipes equally (see Figure 13).

Figure 13: Casting of Phase II-Material using special hoppers

After hardening of the Phase II-Material the guide walls were removed and a clay cap was built (see Figure 14).

Figure 14: Cut-off Wall Head and construction of the clay cap

CONCLUSION

Project experience showed that the unique conditions required a unique approach to construction. The development of a new system of two phase cut-off with an internal sheet pile wall was successfully introduced and was proven for an area of 550,000 m² of Cut-off Wall. Due to the high salt content of the brine, in addition to the aggressive desert climate, new ways of dealing with supporting slurries, cut-off materials and equipment had to be found.

The cutter technology showed its capability to deal with the difficult conditions as well as slurry handling and extreme ground conditions.

REFERENCES

Deiche und Schutz der Hotelanlagen im Bereich eines 75 km2 großen Verdunstungsbeckens im Süden des Toten Meeres, Israel, Prof. Wittke et. Al., Tagungsband Baugrundtagung München, 2010

Dikes 5/8 and 4/8, Specifications and scope of work, Construction of Cut-off, Dead Sea Works LtD, 2011

DIN EN 1538, Execution of special geotechnical work –Diaphragm walls, DIN Deutsches Institut für Normung e. V., Berlin. Beuth Verlag GmbH, 2010

CATAWBA DAM: THE LONG ROAD AND CONSTRUCTION UPDATE

Dennis Hogan[1]

In response to the water scarcity issues that are evident throughout the Southeastern United States, a joint venture by Union County, NC and Lancaster County Water and Sewer District (LCWSD), SC undertook the design of a 100 foot high, 1,500 foot long zoned earth embankment dam to impound a new 100 million gallon pumped storage reservoir. The pumped storage scheme upgrades the existing pump station, piping, and smaller existing reservoir.

The project included an extensive geotechnical investigation in the challenging Piedmont Geology. The focus of the investigation was to maximize the use of on-site materials for the embankment construction and minimize external impacts. A cutoff wall will be installed from the existing valley surface, through saprolite, to the top of bedrock to provide for seepage protection in the foundation. The embankment core will be built onto the embedded cutoff wall. Elastic silts and clays from on site will be processed and utilized as the core of the earth embankment dam, and on-site alluvial deposits will be utilized as best possible for processing filter and shell materials.

This project was brought to 90% design in 2011 before being caught up in permitting delays. The project recently came back to life and design was completed. It was then bid, re-designed, re-bid, and will now go to construction in September 2016. We will examine the permitting and procurement processes, and update the cutoff wall and early embankment construction in the Spring of 2017.

[1] Black & Veatch, hogandj@bv.com

WATER-FILLED BLADDERS, AN INNOVATIVE BEARING INTERFACE FOR ARCH DAM BRACING

Alexandre Lochu[1]

ABSTRACT

Some existing arch dams can require the construction of a supporting downstream structure to improve their behavior. The bracing structure is generally rigidly connected to the dam, but in some cases the designer can seek a flexible connection that can adapt to the deformation mode of each structure and avoid hard points emergence and stress concentration. This has already been done at Kölnbrein dam in Austria, by the use of rubber pads.

The paper presents a new solution to realize a flexible connection, allowing an excellent control of the forces transferred from the dam to its bracing structure. This patented solution also affords the distinct advantage of not requiring unloading the dam in order to load the bearings. This solution, based on the use of water-filled bladders, is under development at detailed design stage for a 50 m high arch dam. Some tests will be conducted in 2017 on full scale prototypes in order to qualify bladders' suppliers and check the installation process of the bladders.

INTRODUCTION

Because of their original design – for example arch dams in large valleys – dams are sometimes affected by non typical behavior, such as unstabilised irreversible displacements. In some cases, these troubles can necessitate bracing of the dam. The regulatory evolution, in terms of flood control or seismic protection, can also lead to a need for reinforcement.

A typical solution for bracing an existing concrete dam consists of building a downstream concrete structure to help support the dam, which will reduce its displacements as well as the stress applied to its foundation surface. This solution has already been used on dams such as Kölnbrein dam in Austria or Les Toules dam in Switzerland.

In this kind of project, the control of the load that is transferred from the existing dam to the downstream bracing structure is critical. Some reinforcement designs, such as Les Toules, require a fully rigid connection between the dam and its reinforcement, in order to make both structures behave as a monolith. In some other cases, such as for Kölnbrein dam where the massive reinforcement structure buttresses the lower section of the existing dam, the connection must be flexible in order to adapt to the deformation mode of each structure, namely the existing dam and the downstream bracing, avoiding hard points emergence and stress concentration. Consequently, the designer of the Kölnbrein bracing used more than 600 neoprene pads in order to bear the dam on its reinforcement

[1] Civil Engineer, Civilworks Department, EDF-CIH (Electricité de France - Hydro Engineering Center), Savoie Technolac – 73373 Le Bourget du Lac – France, alexandre.lochu@edf.fr

structure. The bearings developed for Kölnbrein have proved to be effective, but some disadvantages can be highlighted:

- the loads transferred by the neoprene pads depend on their wedging, whose determining requires very accurate modelling. Moreover, the transferrable load is quite limited in the lower part of the dam because of the limited displacements at its toe;
- the loading of the reinforcement, and every modification of the wedging of the pads, necessitate the unloading of the dam by emptying it;
- the monitoring of the actual transferred loads is arduous.

EDF has patented a new concept invented by the author of this paper, which avoids these disadvantages and offers some distinct advantages. The solution allows:

- an excellent control of the loads applied between both structures, whatever their relative displacements are;
- an automatic and passive adaptation of the transferred loads to the level of the reservoir;
- the loading of the reinforcement structure irrespective of the thermo-mechanical stress state of the dam during its commissioning, which allows the scheme to continue operating without being dewatered and avoids all the associated costs and environmental impact related thereto.

THE CONCEPT

The main idea is to use big inflatable bladders filled with water as interface elements between the downstream face of the dam and the upstream face of the reinforcement structure. The bladders act as jacks, applying opposite forces which only depend on the surfaces and pressures, but not (at first order) on the distance between the bearing surfaces. With this solution, the action of the reinforcement upon a dam is directly controlled in terms of load, and not in terms of displacement, unlike with elastic bearings. This aspect considerably facilitates the numerical modelling of the behaviour of both structures, which can be decoupled (for static loads).

The second point of the concept is that by hydraulically connecting the inflatable bladders to the upstream reservoir, the pressure profiles on the upstream face of the dam and inside the bearing devices are the same (according to the communicating vessels principle, and whatever the fitting level). The load transfer is thus proportional to the level of the reservoir, and adapts totally passively to the hydraulic upstream load on the dam.

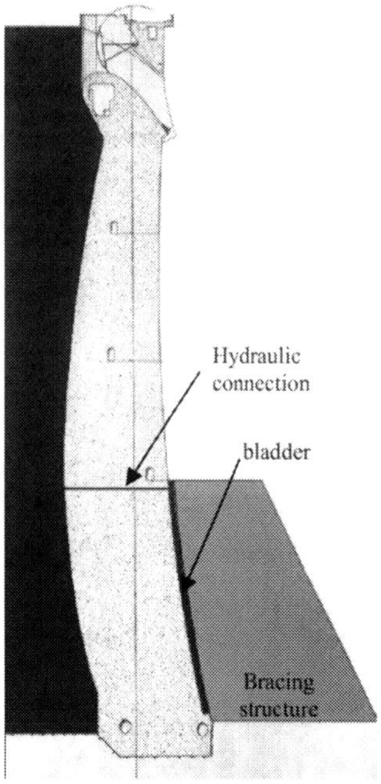

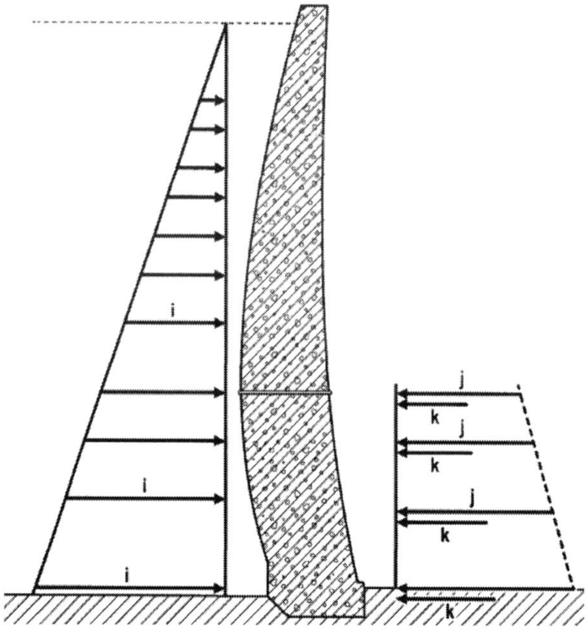

Figure 1 (left). Cut-away drawing on a bladder connected to the reservoir.
Figure 2 (right). Pressure profiles applied on the upstream face (i) and on the downstream face of the dam (j = pressure in the bladders, k = mean pressure on the face, depending on the bladder's surface).

Moreover, if the hydraulic connection is permanent, then the system operates fully automatically. This also allows for compensation of the potential leaks of the bladders in order to avoid (or limit) pressure drops.

FROM CONCEPT TO DETAILED SOLUTION

Based on this concept, EDF's Hydro Engineering Centre is currently developing a detailed design to reinforce a 50 m high arch dam with a 22 m high downstream structure. The name of the dam is not reported in the paper for confidentiality reasons, the owner awaiting the administration's authorisation to carry out the project before communicating further.

The major questions to be solved by the design studies are discussed in the following sections.

General layout and sizing of the bladders

The bladders could theoretically be made-up with any shape, and be laid out with any geometry, but regarding manufacturing rationalization and installation aspects, the best solution appears to be a layout with vertical strips.

The actual sizes of the bladders are also limited on the one hand by manufacturing and handling issues, and on the other hand by strength reasons.

The most critical aspect of the bladders is their thickness, which is actually the distance between the bearing surfaces of the dam and of its reinforcement structure. This distance corresponds to the diameter (Ø) of the free edges of the bladders, and the tension (T) in the membrane is directly proportional to it and to the internal water pressure (p), according to the well-known formula:

$$T = p\text{Ø}/2 \qquad (1)$$

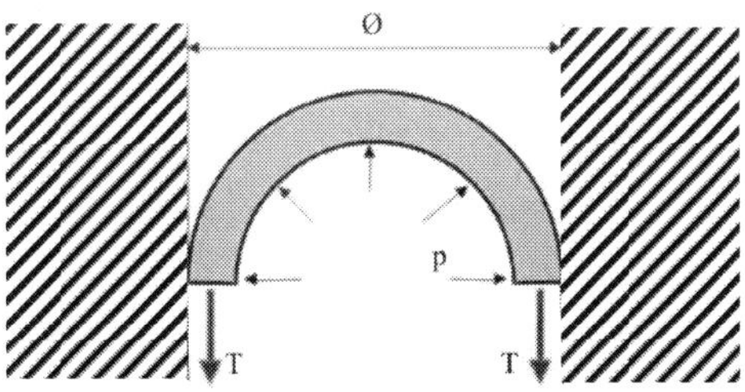

Figure 3. Typical cross section (not to scale) illustrating forces applying on free edges of the bladders.

An 80 cm thickness seems to us the minimum acceptable size in order to allow access to the inner space between the bearing surfaces, for exceptional maintenance purposes. Under a 5 bars working pressure (corresponding to the 50 m water height in the reservoir), the tension in the membrane is hence 200 daN/cm.

From our point of view, the most appropriate technology to fabricate the membrane is one using rubber coated fabrics. This technology is already used for rubber dams, which are in some aspects quite similar to the bearing bladders. The usual global safety factor for the sizing of rubber dams' membranes is between 8 and 9. This factor takes into account concentration factor, creep, weathering and the chemical aggressiveness of the environment. The adaptation of the partial safety factors to the bladders' actual environment is aiming at this stage at a 5-6 global safety factor in order to maintain a safety factor of 3 at the end of the 20 years design lifetime. Therefore the initial strength of the coated fabric should be 1 000 -1 200 daN/cm for our project. Despite the fact that design tensions in rubber dams are usually lower than this, some references such as Ramspol barrier in the Netherlands prove that this strength is attainable. Indeed the initial strength of Ramspol coated polyamid fabric was ca. 2 000 daN/m.

It is also to be noted that the thicker the bladders are, the more space is wasted between them because of their free edges. A given mean pressure and a given thickness involve thus a minimum width of the bladders. In our case the aimed mean pressure goal is 70 % of the upstream pressure, which means the bladders must cover 70 % of the downstream face of the dam (along the height of the reinforcement structure).

In the light of the previous considerations, the following layout has been adopted for the project: 3.3 m wide bladders arranged along a vertical axis, the total height being divided into 2 rows in order to limit the length of the bladders to 11 m. This layout, illustrated by Fig. 4, allows the transference onto the downstream reinforcement structure of approximately 50 % of the total hydraulic load applied on the arch.

Fig. 4 also shows that the bladders will bear on the downstream face of the dam through concrete beams (drawn in green). These beams will be built in order to verticalise and offset the bearing surfaces. Indeed the double curvature of the dam, which makes the crest overhang the face, and which is different for each radial cross section, would make the construction of the reinforcement structure and the installation of the bladders much more complicated without these concrete beams. The same provision is also retained for the bearing surfaces onto the reinforcement structure, in order to create vertical access wells between each couple of bladders.

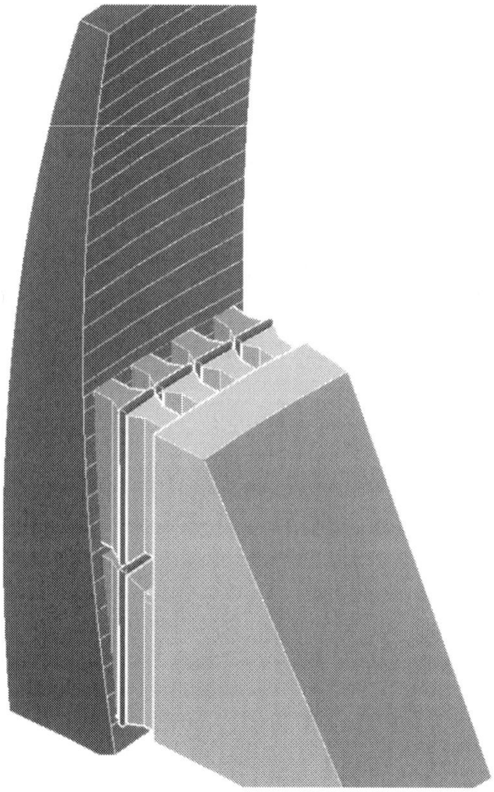

Figure 4. 3D view of the layout of the bladders (drawn in magenta) for one block.

Water supply network

Pressurizing of each bladder will be ensured by a fitting placed on top of it. Pressurizing only requires that the bladder be connected to the reservoir through one small hose, as demonstrated by Pascal's barrel experiment. The design of the network is in fact based on reliability considerations. The first retained principle is to dispatch the bladders between several independent networks, and to connect adjacent bladders to different networks in order to avoid a total loss of pressure on several blocks in case of failure of one network, as illustrated in Fig. 5.

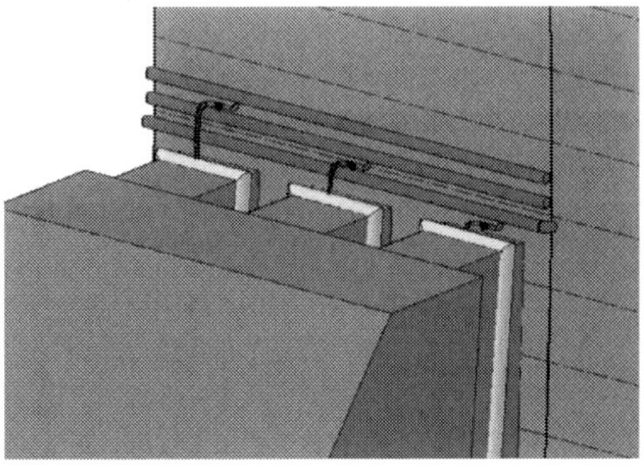

Figure 5. Illustration of the principle of separate networks.

Another fitting will be placed at the bottom of each bladder, for initial water filling.

The diameters of the pipes, hoses and fittings, and especially the ratio between them will determine the velocity in the different sections, and so the pressure drops along the hydraulic circuit. The diameters are thus defined by the acceptable losses of transferred loads for a given size of the failure point.

Our analysis leads to a diameter between 100 and 150 mm for the main pipes, 50 mm for the top pressurizing hoses, and 25 mm for the bottom filling fittings.

Depending on the specific environment of the dam, another function of the water supply network must also be to provide water treatment to avoid proliferation of bacteria and/or heating to prevent freezing. To that end, a closed loop water circulation can be achieved inside the bladders by using the bottom fittings.

Installation

The empty bladders will weigh about 1 tonne each. Depending on the road accessibility to both the dam's crest and the crest of the reinforcement structure, the handling of the bladders can require the installation of permanent dedicated lifting tools. For instance it can be a grappler on the dam's crest and a gantry crane on top of the reinforcement structure. In that case it is important to find a solution to limit the height of the bladders when packed for handling. One solution could be to roll them onto a reel. A low pressure air pre-filling will provide for correct shaping before water filling. Unless a bottom supporting cradle is installed, which seems unnecessary under working conditions, the handling fittings will have to support the weight of water required in order to achieve self-support by friction.

Vandalism and ageing

One of the factors holding back the development of rubber dams in France is due to its (supposed) sensitivity to vandalism. In the case of inflatable bearing bladders, it is worth noticing that the bladders won't be reachable by passers-by, and can be easily fully protected by building a protection roof. Such a roof can also slow down the ageing of the rubber coating of the bladders by protecting them from UV, thermal and rain exposure.

Qualification tests

Several coated fabric product manufacturers have been contacted and have proposed solutions matching our strength and size requirements.
The next step now is to perform tests on full scale prototypes. Beyond the questions of strength and water tightness of the bladders, the goal of these full scale tests will be to check some aspects difficult to accomplish on a small scale model. So the testing will require building a vertical dedicated bench test in order to correctly simulate the behavior of the bladders and to check/adapt/define in particular:

- the handling and installation procedure;
- the filling procedure (how to master the final position of the bladder, avoid membrane folds and exhaust the air?);
- the actual final sizes, which depends on creep, friction/sliding;
- the need for a bottom supporting cradle (will the bladder progressively slide down?);
- the dynamic behavior under seismic load (on a smallest scale bladder);
- the actual initial safety factor (on a smallest scale bladder).

The qualification of the bladders will be completed by ageing laboratory tests.

PERSPECTIVES

The ongoing detailed studies for the first application of the concept so far confirm its feasibility.

This inflatable bearing solution seems to be a promising answer for some dam reinforcement projects, by affording in particular:

- an excellent control of the loads applied between both structures;
- an automatic and passive adaptation of the transferred loads to the level of the reservoir;
- the avoidance of dewatering during commissioning, with the associated economical and environmental benefits.

The reinforcement of a ca. 100 m high dam seems achievable with inflatable bladders made of rubber coated fabrics. Other technologies should probably be explored for higher dams.

ACKNOWLEDGEMENTS

The author sincerely thanks Thomas Pinchard (EDF-CIH) for his confidence in this new concept and for the contribution he has made to its development under his bracing project.

REFERENCES

Bärenthaler, G. et al., "Remedial project for Kölnbrein arch dam – Design and construction", Ostereichische Draukraftwerke, Carinthia, Austria, 1991.

Van Breukelen, M., "Improvement and scale enlargement of the inflatable rubber barrier concept", MSc. Thesis, TU Delft, Netherlands, 2013.

EDUCTORS - THE ESSENTIAL ALTERNATIVE IN COMPLEX DEWATERING APPLICATIONS

Adam Richards[1]
Harry Bagherzadeh[2]

As the nation's dam infrastructure continues to age and the number of high hazard dams significantly increases, eductor dewatering systems will play a vital role stabilizing construction risks and costs for dam rehabilitation projects. Eductor systems allow for the upgrading of dams and levees to current design standards by controlling groundwater and providing safe and stable working environments where geotechnical and hydraulic conditions are commonly complex. Questions provided below provide a basic understanding of eductors and how they provide an effective dewatering solution for the most challenging construction projects:

What is an eductor dewatering system?
What site and project conditions warrant an eductor system?
Why are eductor systems a better alternative than other dewatering systems for dam projects?
How will eductors lower construction costs and risks?
What are the system costs?

Multiple case studies involving dam rehabilitation projects will be presented to answer the questions above.

[1] Griffin Dewatering, adam.richards@griffindewatering.com
[2] Griffin Dewatering

APPLICATION OF INTELLIGENT COMPACTION TO EMBANKMENT DAM CONSTRUCTION

Robert V. Rinehart, Ph.D., P.E.[1]

ABSTRACT

Intelligent Compaction, also termed Continuous Compaction Control, is an emerging technology in the US and holds significant potential benefit for the dam construction community. Utilizing GPS and vibration monitoring equipment installed on the roller compactor, along with an on-board computer, Intelligent Compaction (IC) tracks compaction operations and progress in real-time. IC can be implemented in a variety of approaches, including verifying adherence to method specifications through documenting the number of passes, determining the over- or under-compacted zones for spot testing purposes, monitoring the compaction curve to establish when a material is fully compacted, or correlating the roller measured values to traditional spot test values such as density. Intelligent Compaction technology is commercially available and can be implemented on a wide variety of compaction equipment including vibratory rollers, pad foot and sheepsfoot rollers, double drum asphalt rollers, and rubber tire pneumatic rollers. This paper provides an introduction to the technology and presents detailed descriptions of how it can be employed in the various aspects of dam-related earthwork construction. It is recommended that the embankment dam engineering community start to implement IC on a proof-of-concept basis on upcoming filter placement projects and that simpler approaches be implemented first.

INTRODUCTION

The current state of practice for earthwork construction control on embankment and roller compacted concrete (RCC) dam projects could benefit from existing technology known as Intelligent Compaction (IC). Current earthwork specifications usually take one of two forms. For materials that are suitable for in-situ density testing, specifications typically mandate that a single density measurement (spot test) be carried out per a certain volume of fill placed – typically on the order of hundreds or thousands of cubic yards depending on the nature of material and the placement. Traditionally, nuclear density gage and sand cone density tests have been performed. This spot testing approach, while potentially providing an accurate measurement of density at the specific spot tested, leaves in excess of 99% of the fill untested. In other cases, where in-situ density testing is impossible or very costly (e.g., rockfill, pea gravel), specifications may prescribe a certain number of passes of the compaction equipment be completed. The requirements for developing these method specifications and for showing adherence to them can be challenging. As will be described in the following sections, IC offers solutions to these problematic areas of earthwork – under IC, quality control/quality assurance (QC/QA), including verifying adherence to method specifications, can be carried out with essentially 100% coverage. Roller-integrated measurement of soil conditions during compaction has been used as a QC/QA tool in European earthwork practice since the late 1970s (Forsblad, 1980;

[1] Civil Engineer, Bureau of Reclamation, PO Box 25007, Denver, CO, 80225; rrinehart@usbr.gov

Thurner and Sandstrom, 1980). Termed Continuous Compaction Control (CCC) in Europe and Roller-Integrated Compaction Measurement (RICM) in some US literature, the technology consists of: (1) monitoring roller drum vibrations and/or machine operating parameters and orientation, (2) estimating soil properties through mathematical modeling, and (3) documenting the spatial distribution of the roller-measured soil properties using onboard computing and GPS – essentially, the compactor itself also serves as the QC/QA testing device as depicted in Figure 1. Feedback control technology to automatically optimize drum vibration parameters (i.e., vibration amplitude and frequency) based on the detected soil condition has been available since the late 1990s, but is currently not recommended for use in practice, as detailed later in this paper. This automatic control represents an opportunity for true "intelligence," but is in its infancy. Therefore, the CCC or RICM nomenclature may be more appropriate given the state of the art, but to avoid confusion with the majority of US literature, the term IC is used here.

This paper gives a basic description of the current state-of-the-art of IC technology, describes the applicability of IC to various embankment dam construction activities, and presents the various methods to implement IC into earthwork construction control.

Figure 1. Summary of typical IC equipment, highlighting (a) instrumented roller with vibration sensors and GPS, (b) onboard computer, and (c) roller-measured data map.

INTELLIGENT COMPACTION SYSTEMS

Roller Measurement Values

Roller measurement values (MVs) form the basis of IC. Through monitoring of roller drum vibrations or other machine operating parameters, and mathematical or numerical modeling of the roller-soil system, information about the soil condition is determined in real time onboard the roller compactor. Several different MVs are available, with differing levels of sophistication and physical basis. While MVs often have significant physical meaning and some are even presented with engineering units, in the current state-of-the-art, it is more common to treat them as relative compaction indices. Note that the development and interpretation of MVs is an area of ongoing research.

The primary manufacturers of IC rollers include Ammann of Switzerland (offered under the Case brand in the US and referred to hereafter as Ammann/Case), Bomag and Hamm of Germany, Dynapac and Volvo (formerly Ingersoll Rand) of Sweden, Sakai of Japan, and Caterpillar of the US. As summarized in Table 1, the six roller MVs available for use include: (1) compaction meter value (CMV), developed by Geodynamik and used by Dynapac, Caterpillar, and Volvo; (2) compaction control value (CCV), a refinement of CMV developed and used by Sakai; (3) soil modulus E_{vib}, developed and used by Bomag; (4) soil stiffness k_s, developed and used by Ammann/Case; and (5) machine drive power (MDP), developed and used by Caterpillar. CMV, CCV, E_{vib}, and k_s are based on drum vibration measurements and, thus, are only applicable on vibratory rollers. MDP is not based on vibration and can be employed on both vibratory and static rollers (including padfoot and sheepsfoot rollers). In addition, private companies are starting to offer retrofit equipment, enabling traditional rollers to be upgraded to IC rollers. For more detail regarding the derivation of MVs refer to Mooney et al., 2010a.

It is important to note that the highway construction industry (both in the United States and abroad) has been the major driver behind the development of IC technology. In recent years, there has been a movement to make pavement design mechanistic-empirical (ME) rather than purely empirical. ME pavement design relies heavily on the stiffness or modulus of the pavement components, including subgrade and base course materials, and as a result, there has been an emphasis on developing QC/QA technologies that can

Table 1. Commercially available roller MV systems.

Roller MV	Manufacturers	Parameters used to determine roller MV
Compaction Meter Value, CMV	Caterpillar, Dynapac, Hamm, Volvo	In the frequency domain, ratio of vertical drum acceleration amplitudes at operating vibration frequency and its first harmonic
Compaction Control Value, CCV	Sakai	In the frequency domain, algebraic relationship of multiple vertical drum vibration amplitudes, including operating frequency, and multiple harmonics and subharmonics
Stiffness, k_s	Ammann/Case	Vertical drum displacement, drum-soil contact force
Vibration Modulus, E_{vib}	Bomag	Vertical drum displacement, drum-soil contact force, employs cylinder resting on elastic half space model to convert stiffness to modulus
Machine Drive Power, MDP	Caterpillar	Hydraulic pressure, engine operating condition, ground slope measurements, and machine velocity used to determine rolling resistance

directly assess stiffness/modulus. Accordingly, many MVs have been developed as measures of stiffness/modulus. This is in contrast to basing QC/QA on measurements of density – which is a surrogate to the stiffness/modulus properties integral to ME pavement design. Density is of prime importance for many aspects of earth dam construction, and the challenge of relating stiffness-based MVs to density is discussed in more detail later.

Onboard Computing and Documentation

Another key feature of IC is the documentation system, which provides a spatial-temporal record of MVs and roller location. Using onboard GPS and computing, the roller MVs for each pass, number of passes, time the roller operated over the location, roller speed, and vibration amplitude, etc., are all immediately available for any point in any lift of the compacted structure. Even if the MV portion of the measurement system is not employed, the documentation system can still be used to confirm adherence to a method specification by tracking the number of roller passes.

IC rollers have onboard computers and can display the MV data map in real time as shown in Figures 1 and 2. These onboard maps provide instant feedback to operators and inspectors regarding the condition of the current and/or previous lifts. Further, equipment manufacturers provide office software for performing further analysis and documentation of compaction activities as shown in Figure 3.

Automatic Feedback Control

Another component of IC is automatic feedback control of drum vibration amplitude and frequency. Modern roller drum technology coupled with the IC measurement systems allows for the implementation of feedback control to continuously adjust the drum vibration in real time in response to the measured soil condition. The impetus for this is to optimize the compaction process and automatically avoid overcompaction, and while this is a good approach in theory, it is not yet fully functional (Mooney et al., 2010a).

Benefits

The embankment dam community could realize many benefits from IC. Several benefits fall under the umbrella of higher quality. Another important benefit is the ease and completeness with which the as-built compacted structure can be digitally documented.

IC enables QA with 100% coverage of the compacted area. Roller MV data or pass counts, either of which can form the basis for acceptance, exist wherever the roller has operated. This is a substantial improvement over traditional spot testing which covers much less than 1% of the compacted area, and allows inspectors to hone in on problem areas quickly and effectively. It also results in a final product with lower uncertainty and risk because 100% of it has been inspected.

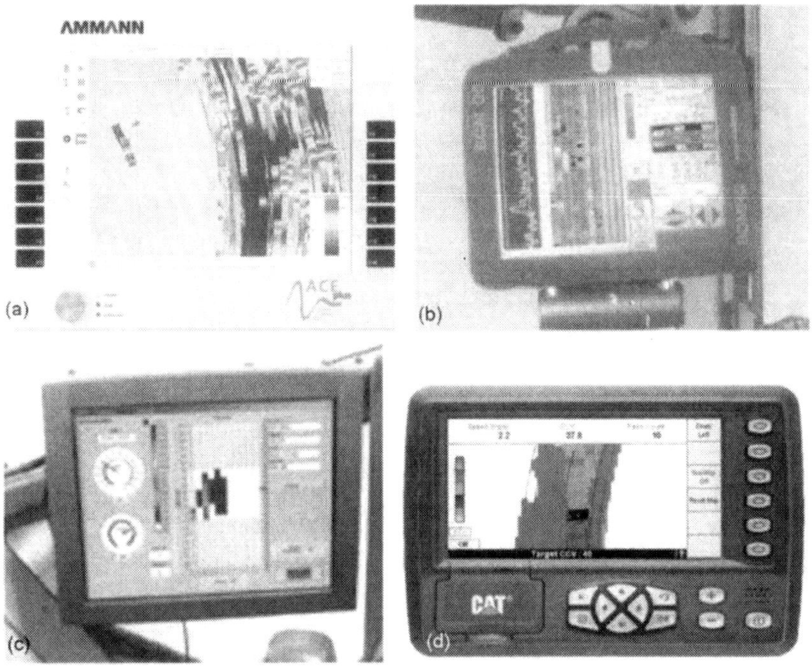

Figure 2. Examples of onboard personal computers with interactive displays from (a) Amman/Case, (b) Bomag, (c) Dynapac, and (d) Caterpillar

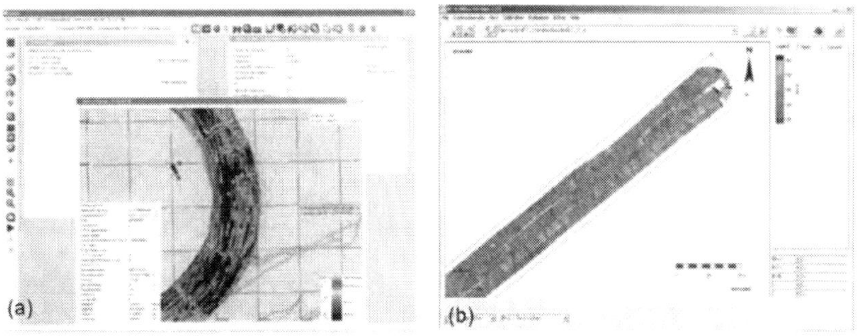

Figure 3. Example screenshots of office software from (a) Amman/Case and (b) Bomag.

Benefits include:

1) Implementing IC decreases dependence on density testing (sand cone, nuclear density gage), as the roller becomes the primary QA tool. This is desirable for several reasons, including avoiding the costs associated with the extra personnel required to perform spot tests, the costs associated with purchasing, maintaining, and storing nuclear gages, as well as the training and licensing of operators. There is increasing risk associated with storing nuclear material, and many agencies are

looking for ways to reduce the use of nuclear density gages for this reason. Further, the many safety hazards associated with performing sand cone density tests in an active construction area would be mitigated by performing IC-based QA.

2) IC technology provides GPS position-indexed and time-stamped QA data for the entire constructed area. This comprehensive electronic data can be stored long-term and easily referenced in the event issues arise requiring investigation into the as-built facility.

3) It is expensive and difficult to reliably perform density testing on materials such as pea gravel rockfill, and RCC. These materials are often used during embankment construction and have traditionally been controlled by method specifications rather than rigorous density testing. However, IC is applicable to these materials. If it is still desired to use a method rather than performance specification, IC can be used during the development of the method specification, and to rigorously verify adherence to it.

4) IC also has the potential to help avoid overcompacting and damaging granular filter materials, as roller operators receive immediate feedback about the state of compaction. The operator of the roller compactor is able to view the roller MV data in real time on a monitor onboard the roller. These data are typically color coded (i.e., green = good, red = bad; see Figures 1-3) to give the operator immediate feedback regarding where more compactive effort is needed and where compaction is likely finished, allowing for more efficient and more uniform compaction operations.

5) IC, with its 100% coverage, allows for the assessment of uniformity of compaction. Uniformity is not typically specified or controlled by earthwork specifications mainly due to the limited number of spot test data points available. IC would enable the direct assessment of uniformity.

Fundamental Challenges of Roller Measurement Values

There are several fundamental challenges with the current roller measurement systems that must be understood (and taken into account if possible) to ensure successful implementation of IC. These issues represent areas of active research, and significant advances are expected in the coming years.

1) It well know that any given roller MV is likely representative to a depth exceeding the current lift of material being compacted. Vibration-based MVs are a composite measurement representing soil properties from the surface down to a depth termed the measurement depth. Research by Rinehart and Mooney (2009) showed that for thick, homogenous earthfills (e.g., 2+ meters of compacted material of the same composition), the measurement depth of a roller MV varies between 0.8–1.2 meters from low to high drum vibration amplitude.

These findings were based on field studies and involved in-situ stress and strain instrumentation. Other research (Musimbi et al., 2011) shows that for layered situations, the measurement depth of roller MVs depends on vibration amplitude, modulus ratio of the materials involved, and the thickness of the top layer. In all practical cases, MVs are influenced by both the current layer (i.e., material of interest) and the previously compacted underlying material(s). This implies that a soft (undercompacted) zone in the previous lift could be manifest as a low MV measured on the current lift. Vennapusa and White (2016) show an example of this phenomena using the MDP MV. Measurement depth is a key issue that needs to be considered during any implementation of IC.

2) Another key is that roller MVs are typically dependent on drum vibration amplitude. For example, it is well documented that higher vibration amplitudes lead to higher CMV and CCV values (Adam and Kopf, 2004; Mooney et al., 2010a). Amplitude dependence for a layered soil structure is more complicated and depends on the nature of the soils in question (see Rinehart and Mooney, 2009). The amplitude dependence of MVs is one of the major issues facing truly intelligent compaction because of the possibility of the amplitude continually varying to optimize compaction leading to the MV varying for otherwise constant soil conditions. Therefore, current best practice is to mandate that a single, consistent amplitude and frequency be used for QC/QA documentation. It is recommended that the embankment dam construction community adopt such an approach – initial compaction passes could be carried out at whatever amplitude the contractor deems best (even including variable amplitude IC), but later passes (i.e., when the soil is near its final state and the roller data will be used for QC/QA purposes) should all be performed at a single, consistent, low vibration amplitude (e.g., 0.5–0.7 mm).

3) A third issue pertinent to the adoption of IC technology is that all current MVs are to some degree reflective of the involved soil's stiffness. Empirically, MVs can be correlated to soil density or percent compaction – but these relationships are nothing more than empirical correlations and are often challenging to identify. The key is the relationship between a soil's moisture-density and moisture-stiffness curves, and these relationships are soil specific and not well understood. The literature (e.g., Mooney et al., 2010a, Vennapusa and White, 2016) has shown that correlations between MVs and density are often of poor quality. In the transportation community (the fundamental driver for the development of IC), this is less of an issue, as current construction methodology is trending toward stiffness/modulus-based QC/QA and away from density-based QC/QA. Regarding the construction of embankment dams, density is often of prime importance, and it is therefore unlikely that practice will deviate from density-based QC/QA. Ongoing research and development of improved MVs has potential to alleviate this issue.

EXISTING IC PRACTICES

European Practice

Specifications for QA of earthwork compaction using IC were first introduced in Austria (1990), Germany (1994), and Sweden (1994). The International Society of Soil Mechanics and Geotechnical Engineering (ISSMGE) has endorsed the Austrian specifications for IC (ISSMGE, 2005) based on work by Adam (2007). The Austrian/ISSMGE and German specifications each permit multiple options for using IC in earthwork compaction QA. The most common and simplest approach uses IC to identify weak areas for evaluation via spot testing (e.g., static plate load test, light weight deflectometer, or density via nuclear gage or sand cone). Acceptance is based on these weak areas meeting contract-defined requirements. If roller-identified weak areas meet acceptance, it is inferred that the rest of the area meets acceptance too. This basic approach is the only IC option approved for use in Sweden and is the most commonly used approach in Germany and Austria.

The more complicated CCC-based QA specifications involve correlating roller MVs to spot tests in a defined calibration area. If a suitable correlation is found, a target roller MV is determined from the MV versus spot test regression equation. Acceptance is based on comparison of roller MV data collected in a production area to the target roller MV. Based on a survey of European practice, the calibration approach is challenging to implement and requires a high level of onsite knowledge, and is therefore not commonly used. A complete description of the European specifications is provided in Mooney et al., 2010a and a brief summary is provided by White et al., 2011.

U.S. Practice

In the United States, federally funded research has investigated how IC might be implemented in practice (Mooney et al., 2010a; Chang et al., 2011), and State Departments of Transportation (DOTs) are actively developing guide specifications for the use of IC on highway construction projects (e.g., TxDOT, 2008; MnDOT 2007, 2010, White et al., 2011). The approaches recommended to date mainly follow established European practice, relying on roller data maps to identify weak zones for spot testing, establishing correlations between MVs and spot test results in test strips, or using the GPS portion of the IC technology to confirm adherence to method specifications.

Applicable Materials

As summarized in Table 2, IC is applicable to a wide range of materials. The selection of the type of MV employed (i.e., vibration based or rolling resistance based) depends on the material type and is important, as not all MVs work for all types of soils or are available on all types of rollers. Note that while vibratory padfoot rollers are available, some research has shown that current roller MVs may be unreliable for vibratory padfoot rollers for normal U.S. earthwork practices (Mooney et al., 2010a). Other research has shown that with very careful placement and moisture conditioning of the compaction lift,

Table 2. Summary of the applicability of IC to various soil types.

Material	Vibratory-based MVs applicable?	Rolling resistance-based MVs applicable?	Comments
Sand and gravel	Yes	Potentially	Many methods exist for implementing IC in materials that are best compacted with smooth drum vibratory rollers.
Rockfill	Yes	Potentially	
Silty or clayey sands and gravels	Potentially	Yes	Vibratory MV-based IC is applicable if smooth drum vibratory compaction is the preferred method. Alternatively, vibratory MV-based IC could be used for QA purposes in a "proof-roll" type approach. Rolling resistance-based IC with static padfoot or sheepsfoot rollers can be used. Material variability may pose problems.
Silts, clays	Potentially	Yes	Vibratory compaction is not appropriate for these soils, but vibratory MV-based IC could be used for QA purposes in a "proof-roll" type approach. Rolling resistance-based IC with static padfoot or sheepsfoot rollers can be used. Moisture conditioning is critical to achieve adequate and uniform compaction with these materials, and moisture variability will adversely affect IC measurements.

good understanding of the sublift material, and by employing multivariate regression that IC can be performed with vibratory-based MVs from padfoot rollers (Chang et al., 2011).

APPLICABILITY OF INTELLIGENT COMPACTION TO EMBANKMENT DAM CONSTRUCTION

Previous Application to Dams

There are limited discussions about the application of IC to earth dams in the literature. IC was employed during construction of the Shuibuya Dam in China (Kloubert, 2008). Shuibuya Dam is a concrete-faced rockfill dam (CFRD) constructed from 2002–2008 and is 764 feet tall and contains approximately 20.4 million cubic yards of fill materials. The fill material contained boulders up to 24 inches in diameter and was placed in lifts ranging from 32 to 40 inches thick. The construction specification required a compacted dry density of 134.2 pounds mass per cubic foot. Conducting density tests in such a material (e.g., via test pit with water or sand replacement) is extremely labor intensive and time consuming. Through constructing test fills, engineers were able to verify that the roller MV (in this case, Bomag's E_{vib}) correlated to dry density, and a method specification was subsequently developed. IC resulted in a higher quality and more uniform product, as QC data were available for 100% of the compacted volume rather than much less than 1% of the volume as would be the case with density spot testing. In addition, by eliminating long delays in compaction work associated with density spot testing, the contractor was able to meet a demanding production schedule of over 19,000 cubic yards per day using three IC rollers (Kloubert, 2008).

Intelligent Compaction Implementation Methods

IC could be employed in several different fashions. The following presents recommendations in order from most simple to most complex. The simpler methods are preferred over the more complex ones particularly during the transitional phases of adoption when contractors and other construction personnel are becoming familiar with IC technology.

As will be discussed, several of the IC implementation strategies rely on test strips or fills. The purpose of these fills is to help establish the relationship between MVs and spot test results (e.g., density) before production compaction begins, or to define the number of passes required by a method specification for the project-specific roller and soils. It is preferable that test fills consist of the placement and compaction of at least three lifts atop a representative subgrade. Alternatively, with agreement of all parties, the first few lifts of production compaction can constitute the test fill.

Using Intelligent Compaction Rollers to Verify Adherence to Method Specifications. Regardless of whether or not the MV portion of the IC technology is used, the GPS documentation system can still prove useful. The number of roller passes and roller operating parameters (e.g., vibration amplitude, frequency, and roller speed) can all be recorded throughout the compaction process. The ability to ensure robust adherence to a method specification would result in an increase in quality of compaction activities where method specifications are commonly employed (e.g., pea gravel, rockfill, RCC). This is the simplest way to implement IC.

Selecting Spot Test Locations from Roller Measurement Value Data Maps. Another simple way to implement IC (either vibratory or rolling resistance based) is to use the MV map to discern spot testing locations. In cases where only a minimum required density has been specified, spot testing can be focused in the least compacted areas according to the MV map. In cases where upper and lower limits have been established (e.g., filter sand), the MV map can be used to locate the least and most compacted zones according to the roller MV. In either case, if the tested areas pass spot testing requirements, the remaining areas pass by association. This method should prove more stringent than current practice, as spot test locations are not picked randomly, but rather selections are informed by the MV data collected over the entire production area. This method of IC is applicable to many materials, including materials that are traditionally very difficult or expensive to spot test (e.g., rockfill). This approach has been used successfully in Europe and is the most commonly used method there. This method could also be employed using a proof roll methodology, and therefore not every roller on the site would need to be IC equipped. When the contractor feels compaction is nearly complete, the IC roller could perform a single over the area to collect MV data. The MV map from the proof roll serves as the basis for selecting spot test locations, and as a means for documentation.

When using this method of IC, it is critical to establish that a positive regression relationship exists between the MV and density – that is, high MVs correspond to high densities and vice versa. This is not always the case (typically because of soft or variable subsurface and moisture conditions) and needs to be verified in the field by performing a relatively small test fill prior to production compaction operations or by increased frequency of spot testing in the early stages of a project (i.e., by using the first few lifts of the fill as the test section). The relationship should be verified or re-established periodically or whenever conditions change (e.g., fill material, underlying layers, etc.) by performing a regression analysis between MVs and spot test results.

Compaction Control via Relative Change in Measurement Values. It is possible to use the percent change (increase or decrease) in MVs between subsequent passes to judge the degree of compaction. The MVs over a large area can be averaged and the change in the average from pass to pass can be examined, or a spatial analysis can be performed to examine more local change in MVs. The latter is more complicated and error prone. If the MVs do not appreciably change between subsequent passes, it can be inferred that the soil is as compacted as possible with the given roller and roller operating conditions, and compaction operations can cease. Precedence for this type of approach exists in the form of using the nuclear density gage to test select embedment for pipelines (Howard, 2011).

A test fill should be compacted prior to production compaction to show that the roller and roller operating conditions being used are capable of achieving suitable compaction. This method of IC is applicable to many materials, including materials that are traditionally very difficult or expensive to spot test (e.g., rockfill).

An alternative and simpler approach to using the percent change in roller MVs is to chart the MV-based compaction curve (i.e., average MV versus pass number) during the compaction of a test strip or during the first few lifts. The number of passes required to reach full compaction (i.e., no change in MV from pass to pass) can be determined from this initial testing and can be applied via a method specification in the production compaction area. As before, it should be verified that the density (or other) requirements are met in the test fill. This approach has been used successfully on a large dam in the past (Kloubert, 2008).

Calibration of Roller Measurement Values to Soil Properties of Interest. There are several discussions in the literature regarding attempting to perform a rigorous regression analysis between roller MVs and soil spot test results (summaries of available literature are presented in White et al., 2007; Mooney et al., 2010a; and Change et al., 2011). Once this regression is established, a "target-MV" is selected to coincide with the desired degree of compaction, and this target-MV forms the basis of QC/QA. Any location with an MV greater than the target-MV (which coincides with the specified minimum density) meets acceptance, and locations with MVs below the target-MV would fail. While seemingly more elegant, this approach is complicated and problematic and is not recommended for implementation at this time. This approach has been used successfully in Europe, but is not common. It requires uniform materials as well as contractors and site engineers with many years of detailed experience.

It is worth noting that one area where the calibration approach holds promise is in relating MVs to the *E'* composite modulus commonly used in pipeline embedment (Howard, 2009). Given that vibratory MVs are measures of material stiffness, it is reasonable to expect that good correlations exist between MVs and *E'*, or that with more sophisticated machine-soil interaction models, that *E'* could be measured directly. While the majority of commercially available IC equipment is in the form of heavy rollers, some work has been performed to install IC on plate vibrators and walk-behind rollers (Anderegg et al., 2006), and Ammann now offers an IC-enabled vibratory plate compactor.

CONCLUSIONS

IC is an emerging earthwork compaction technology in the United States. The soil condition is continuously monitored via instrumented rollers and spatially recorded via onboard GPS and personal computers. While there are several challenges with the measurement systems themselves (all of which are the topics of ongoing research and many of which can be addressed), the technology holds tremendous benefits for embankment dam and RCC construction, including:

1) Avoidance of over- and under-compaction
2) More uniform compaction and the ability to assess uniformity
3) Reduction in the number of density spot tests required, with the possibility of eliminating density testing during production compaction
4) More robust adherence to method specifications, including for RCC
5) QA data with 100% coverage, and ease of documentation
6) Ability to perform QA on materials that are very difficult or expensive to spot test such as pea gravel and rockfill

While there are many ways in which IC can be implemented, it is recommended that the industry start with the more simple approaches. Using the GPS documentation features of IC, owners can more strictly and more easily enforce method specifications. Further, using the roller MV data map for a given lift, inspectors can focus spot testing on roller-identified undercompacted or overcompacted zones – rather than choosing locations at random. This will provide a higher quality product. Finally, the roller MV data can be used to assess the uniformity of any given compaction area.

As a next step, it is recommended that IC be implemented alongside traditional QC/QA for upcoming embankment and RCC projects. This work should include the construction of representative test fills. This approach will give construction personnel and contractors experience with the technology while ensuring that a consistent level of quality is achieved via the traditional QC/QA methods. It will also provide opportunities to fine tune best practices regarding IC on embankment dam projects. All major equipment manufacturers have IC equipment available on heavy single drum rollers, and equipment is also available to retrofit existing rollers; therefore, requiring this equipment on jobs should not represent an unusual burden. For the time being, no single MV is preferred over another, and the specific measurement system (i.e., manufacturer) does not need to be specified.

REFERENCES

Adam, D. (2007). "Roller Integrated Continuous Compaction Control (CCC) Technical Contractual Provisions & Recommendations." Design and Construction of Pavements and Rail Tracks: Geotechnical Aspects and Processed Materials, A.G. Correia, Y. Momoya, and F. Tatsuoka, eds., Taylor & Francis Group, London, UK, pp. 111–138.

Adam, D. and F. Kopf. (2004). "Operational devices for compaction optimization and quality control (Continuous Compaction Control & Light Falling Weight Device)." Proceedings of the International Seminar on Geotechnics in Pavement and Railway Design and Construction, December 2004, Athens, Greece, pp. 97–106. (Invited paper).

Anderegg, R., D.A. von Felten, and K. Kaufmann. (2006). "Compaction Monitoring Using Intelligent Soil Compactors." Proceedings: GeoCongress 2006, ASCE, Atlanta, February 2006.

Bräu, G., K. Hartman, and G. Pelz. (2004). "Flächendeckende Prufung der Verdichtung (FDVK)-baupraktische Umsetzung und verfahrens-bezogene Verdichtungsanforderungen. (CCC testing of compaction - implementation in construction practice and procedure-related compaction specifications)." Lehrstuhl und Prüfamt für Grundbau, Bodenmechanik und Felsmechanik der Technischen Universitat München. Heft 897. München,Germany. (In German).

Chang, G., X. Qinwu, J. Rutledge, B. Horan, L. Michael, D.J. White, and K.R.P. Vennapusa. (2011). "Accelerated Implementation of Intelligent Compaction Technology for Embankment Subgrade Materials, Aggregate Base, and Asphalt Pavement Materials." Final Report: Federal Highway Administration Pooled Fund Study DTFH61-07-C-0032, 263 p.

Forssblad, L. (1980). "Compaction meter on vibrating rollers for improved compaction control." Proceedings of the International Conference on Compaction, Vol. II, Paris, pp. 541–546.

Howard, A. K. (2009). "Composite E' (Modulus of Soil Reaction)." Proceedings of Pipelines 2009, ASCE San Diego, CA, August 2009.

Howard,A. K. (2011). "Target Values of Compacted Gravel for Pipelines." Proceedings of Pipelines 2011, ASCE, Seattle, WA, July 2011.

International Society for Soil Mechanics and Geotechnical Engineering (ISSMGE) (2005). "Roller-Integrated Continuous Compaction Control (CCC), Technical Contractual Provisions – Recommendations." International Society for Soil Mechanics and Geotechnical Engineering: Geotechnics for pavements in transportation infrastructure.

Kloubert, H.J. (2008). "Shuibuya Rockfill Dam – Bomag Compaction Technology in use on World's Highest Rockfill Dam." White Paper: Bomag GmbH, 6 p.

Minnesota Department of Transportation (Mn/DOT) Specification 2106, (2007). "Excavation and Embankment – Quality Compaction by IC, LWD, & Test Rolling (Pilot Specification for Embankment Grading Materials)." Minnesota Department of Transportation, 11 p.

Minnesota Department of Transportation (Mn/DOT) Specification 2105/2106, (2010). "2010 Pilot Continuous Compaction Control (CCC) Method." Minnesota Department of Transportation, 4 p.

Mooney, M.A.. R.V. Rinehart, N.F. Facas, O. Musimbi, D.J. White, and P.K.R. Vennapusa. (2010a). "Intelligent Soil Compaction Systems." NCHRP Report 676, Transportation Research Board, Washington, D.C., 178 p.

Musimibi, O., J. Berger, M.A. Mooney, and R.V. Rinehart. (2011). "Analysis of Drum Vibration on Layered Media Using an Iterative Contact Width, Contact Force and Stiffness Algorithm." International Journal of Numerical and Analytical Methods in Geomechanics, submitted for publication April 27, 2011.

Rinehart, R.V. and M.A. Mooney. (2009). "Measurement Depth of Vibratory Roller-Measured Soil Stiffness." Geotechnique, Vol. 59, No. 7, pp. 609–619.

Scherocman, J., S. Rakowski, and K. Uchiyama. (2007). "Intelligent Compaction, Does it Exist?" Proceedings of the Annual Conference-Canadian Technical Asphalt Association, Issue 52, pp. 373–398.

Thurner, H. and A. Sandström. (1980). "A new device for instant compaction control." Proceedings of International Conference on Compaction, Vol. II, Paris.

Texas Department of Transportation (TxDOT). (2008). "Quality Compaction Using Intelligent Compaction Rollers." Special Specification CSJ: 0172-09-031, Texas Department of Transportation, 4 p.

Vennapusa, P. K. R., and White, D. J. (2016). "Machine Drive Power Based Roller-Integrated Compaction Measurements for Cohesive Embenkment Construction," in proceedings: Geo-Chicago 2016 Geotechnical Special Publication 272, American Society of Civil Engineers, Chicago, August 2016.

White, D., M. Thompson, and P. Vennapusa. (2007). Field Validation of Intelligent Compaction Monitoring Technology for Unbound Materials. Final report submitted to Minnesota Department of Transportation, Iowa State University, Iowa. April.

White, D. J., Vennapus, P. R. K., and Gieselmen, H. H. (2011). "Field Assessment and Specification Review for Roller-Integrated Compaction Monitoring Technologies," Advances in Civil Engineering, Hindawi Publishing Corporation, Volume 2011, Article ID 783836.

UPDATE NO. 3 OF THE CALAVERAS DAM REPLACEMENT PROJECT

John Roadifer[1]
Michael Forrest[2]
Erik Newman[3]
Daniel Wade[4]
Susan Hou[5]
Tedman Lee[6]
Carman Ng[7]

ABSTRACT

Calaveras Dam is a major component of the San Francisco Public Utilities Commission (SFPUC) Hetch Hetchy Regional Water System. Since 2001, in response to seismic stability concerns from the California Division of Safety of Dams about this 90-year-old hydraulic fill dam, the SFPUC has lowered Calaveras Reservoir to about 39 percent of its 96,850-acre-foot capacity. To restore reservoir capacity, a replacement dam and new appurtenant works were designed to remain functional after the design earthquake, which is a magnitude 7¼ maximum credible earthquake (MCE) on the Calaveras Fault located 0.3 mile from the dam. The MCE peak ground acceleration would be 1.1 g. The dam site, which is located in the active and geologically complex Coast Range, contains multiple secondary faults (classified as conditionally active or inactive), Tertiary sedimentary bedrock, a diverse suite of Franciscan assemblage rocks, and numerous active, dormant, and inactive landslides.

Construction of the replacement dam, which started in August 2011, is now in its fifth year; the project is approximately 75 percent complete. During the past year, key features of the project have been completed including the spillway, intake shaft and tower, and outlet works. Excavation for the dam foundation was also completed and currently placement of embankment fill is underway.

This paper provides an update of the project status as well as some of the recent engineering challenges that have arisen at this geologically complex site. The challenges included completion of foundation grouting, dam foundation cleanup and treatment of

[1] Sr. Project Manager, AECOM, 300 Lakeside Drive, Suite 400, Oakland, CA 94612, (510) 874-1732, john.roadifer@aecom.com
[2] Vice President, AECOM, 300 Lakeside Drive, Suite 400, Oakland, CA 94612, (510) 874-3012, michael.forrest@aecom.com
[3] Project Engineer, AECOM, 300 Lakeside Drive, Suite 400, Oakland, CA 94612, (510) 874-1732, erik.newman@aecom.com
[4] Director, Water System Improvement Program, San Francisco Public Utilities Commission, 525 Golden Gate Ave., San Francisco, CA 94102, (415) 554-1853, dwade@sfwater.org
[5] Project Manager, San Francisco Public Utilities Commission, 525 Golden Gate Ave., San Francisco, CA 94102, (415) 551-4666, shou@sfwater.org
[6] Project Engineer, San Francisco Public Utilities Commission, 525 Golden Gate Ave., San Francisco, CA 94102, (415) 551-4866, tclee@sfwater.org
[7] Assistant Project Engineer, San Francisco Public Utilities Commission, 525 Golden Gate Ave., San Francisco, CA 94102, (415) 551-4866, cng@sfwater.org

many existing exploratory adits dating from 1911 that underlie the downstream toe of the dam, development of a complex quarry to produce hard rockfill for the dam embankment while separating out unusable soft shale, and placement of embankment fill on an irregular foundation surface.

INTRODUCTION

Since construction began in 2011, modifications of the project due to the complex geology of the site as well as other engineering challenges through the end of 2015 have been discussed in Forrest, et al. (2014), Forrest, et al. (2015), and Roadifer, et al. (2016). Forrest, et al. (2014) discussed characterization of a large slide complex on the left abutment and mitigations that included flattening the slope above the spillway crest and designing new disposal sites in the reservoir, mitigation of an unanticipated thick zone of weak foundation material below a rockfill disposal site dike with deep soil mixing columns, and stabilization of a wide fault zone exposed in a portion of the spillway cut. Forrest, et al. (2015) discussed continuing characterizing of the large left abutment slide complex within the dam and spillway foundations and mitigations that included building the spillway foundation back up to grade using backfill concrete and construction of a 50-foot-high tieback wall to support a portion of the slide that needed to remain in place to provide construction access for spillway construction. Forrest, et al. (2015) also discussed stability analyses and reinforcement of the 400-foot-high stilling basin excavation in the Franciscan Complex and protection of concrete structures against sulfate and acid attack from the geologic formations. Roadifer, et al. (2016) included additional discussion of analyses and reinforcement of the 400-foot-high stilling basin excavation and modifications that were made to the design of the 78-inch outlet pipe under the dam to minimize damage due to sympathetic movement of a series of bedrock faults found along its alignment during excavation.

During the past year, the spillway, intake shaft and tower, and outlet works have been completed. Excavation for the dam foundation has also been completed and placement of embankment fill is currently underway. The replacement dam is an earth- and rockfill dam that is founded directly on bedrock. The bedrock foundation at the left abutment is Temblor Sandstone and Franciscan Complex at the valley bottom and right abutment. The upper 40 feet of the left abutment of the dam is a concrete gravity wall.

The replacement dam section (Figure 1) is based on materials available at the site and on results of stability and seismic deformation analyses discussed below. The crest width of the 220-foot-high dam will be 80 feet and is controlled by the core width required for the potential future enlargement up to 386,000 acre-feet (NMWS elevation 890 feet). The dam section consists of a rockfill upstream shell (Zone 5) and an earthfill downstream shell (Zone 4) supporting a central clay core (Zone 1). A 15-foot-thick chimney filter (Zone 2) and 10-foot-thick drain (Zone 3) separate the core from the downstream shell. The core is separated from the upstream shell by an 8-foot-thick upstream filter (Zone 2A) and an 8-foot-thick coarse filter (Zone 5A). A blanket drain system underlies the downstream shell. The blanket drain is a filter/drain/filter (Zone 2/3/2) on the abutments and a filter/drain/rockfill/filter (Zone 2/3/5/2) in the valley bottom. The upstream slope of the dam is protected against erosion by a layer of riprap (Zone 6).

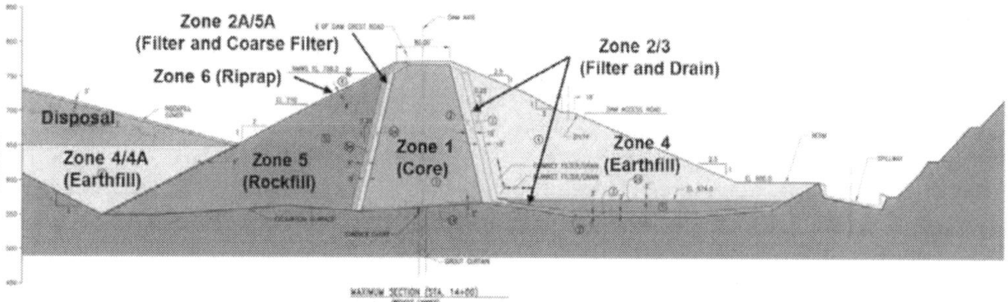

Figure 1. Section of Calaveras Replacement Dam.

Core, earthfill, rockfill, and riprap materials are being sourced from on-site excavations and borrow areas. The filter, coarse filter, upstream filter, and drain materials are being imported from commercial sources.

FOUNDATION GROUTING

Final design of foundation grouting generally consisted of a 2-row grout curtain (see Figure 1) with a short 4-row reach across the unconformable contact of the Temblor Sandstone and the Franciscan Formation. During final design there were indications in the exploratory borings that the contact might contain highly pervious in areas where the contact might not be well bonded. As construction progressed, these areas turned out to be the base of the Slide B (Forrest, et al. 2015). The contact was in fact observed to be well-bonded everywhere it was exposed and the 4-row reach was reduced to 2 rows.

Installation of the grout curtain along the dam axis was completed during 5 different periods of time between 2012 through 2016 (see Figure 2 for progress as of November 2016.). Typically, grouting starts in the valley bottom and progresses up the valley slopes. Due to schedule issues, transverse grout curtains were installed at three locations to allow grouting to begin midway up the left abutment and right abutment slopes before the valley bottom was excavated. The transverse grout walls were 48 foot in length and included 3 primary grout holes and 2 secondary grout holes. The purpose of the transverse grout curtains was to provide a grouted area of rock against which grouting of production holes could act as they proceeded from the transverse curtain up the abutment.

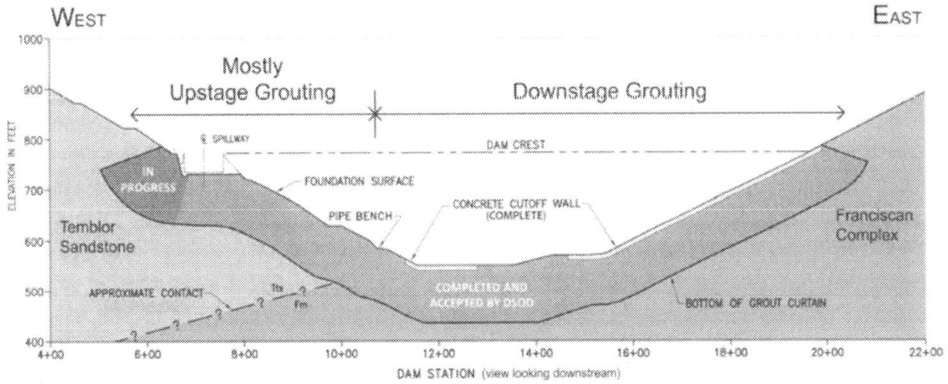

Figure 2. Profile of grout curtain as of November 2016.

In the valley bottom, the grouting was accomplished in two phases with Phase 1 being installed through approximately 40 feet of alluvium and fill, and Phase 2 occurring after excavation was complete in the early summer of 2016. During the Phase 2 drilling and grouting program, surface leakage and water takes exceeding the split spacing criteria for grouting was observed to be generally following a sheared contact between a large block of Franciscan shale and a block of silicious schist. Closure of the curtain in this area required an additional row of holes along the dam axis. Given the location of the contact at the valley bottom two additional production lines of grout holes (Rows C and D) were added upstream of Rows A and B to provide a wider grouted zone in this area.

The grouting approach and results in the Franciscan Formation differed from the Temblor Sandstone. The Franciscan Formation, being a chaotic mixture of harder blocks of greywacke, blueschist, greenstone, and serpentinite within a sheared shale matrix, is a generally tight formation. Grouting was typically performed using a down-staging approach due to the weak rock. Grout takes were typically low, less than 5 lbs/lf (see Figure 3). The Temblor Sandstone, being more brittle than the Franciscan Formation, contained more fractures that were able to transmit water and grout. Grouting was typically performed using an upstaging approach. Grout takes were typically greater than for the Franciscan Formation (see Figure 3).

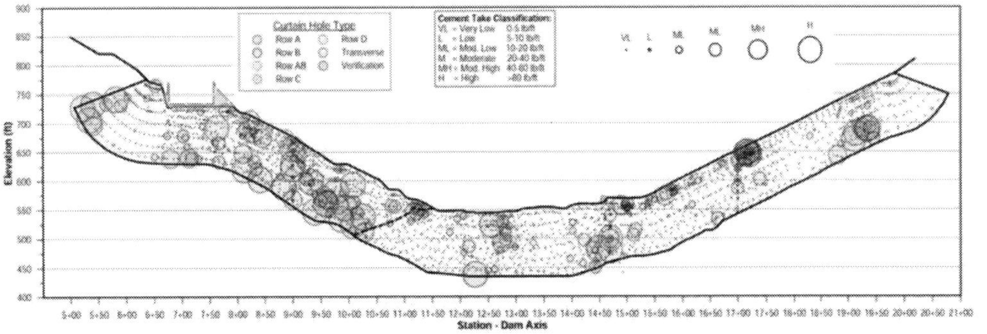

Figure 3. Plot of grout takes along Rows A and B.

In addition to the grout curtain, several features (primarily shear zones) were identified in the core foundation that required stitch grouting across the features. A total of three features identified in the left abutment required 232 stitch holes and five features identified in the right abutment required 78 stitch holes.

At completion, the grouting curtain involved drilling and grouting a much larger number of holes that had significantly lower grout take than what was estimated during design. In general, the discontinuities in the rock mass were small enough that grout travel was limited requiring a greater number of split holes and at some locations additional lines between the two designed lines. The number of holes in the grout curtain was 686 holes compared with an estimated 422 holes. The length of drilling grout holes (including redrilling) for the program was about 88,450 lineal feet compared with an estimated 48,000 lineal feet. Grout takes in the Temblor sandstone and Franciscan averaged 8.0 pounds/foot and 3.4 pounds/foot compared with an estimated 46 pounds/foot and 29 pounds/foot, respectively.

TREATMENT OF EXPLORATORY ADITS

Originally, a potential dam site located approximately 1,500 feet downstream of the existing dam was explored for an arched concrete gravity dam between 1905 and 1911. Explorations were primarily conducted by excavation of a large number of shafts and adits through the alluvium in the valley bottom and colluvium on the valley slopes into the underlying Franciscan bedrock. Following excavation, approximately 30 of the shafts and adits were found to be located within or near the toe of the replacement dam (see Figure 4). The lengths of remnant adits following excavation ranged from 15 to 180 feet.

Treatment of the adits varied depending on their location. Adits that were located on the valley walls within the limits of the internal drainage system of the dam were plugged with concrete at the entrance and allowed to drain into the drain blanket through a perforated drainage pipe installed through plug. Adits located on the valley walls outside of the internal drainage system, but within the dam foundation, were plugged with concrete at the entrance and backfilled through the plug with controlled low strength material (CLSM). A couple of these adits that were located above the internal drainage system but within the footprint of the Zone 4 facilities berm at the toe of the dam were piped to allow drainage of seepage to the downstream toe of the dam. Shafts located in the valley bottom were excavated and backfilled with Zone 5A/Zone 2A or with concrete. Adits in the valley bottom were generally excavated to break through the top of the adit and then backfilled with Zone 5/Zone 2A or with concrete. In one case, a portion of an adit was too deep to break through and was backfilled with CLSM through 6-inch diameter holes drilled through the top of the adit.

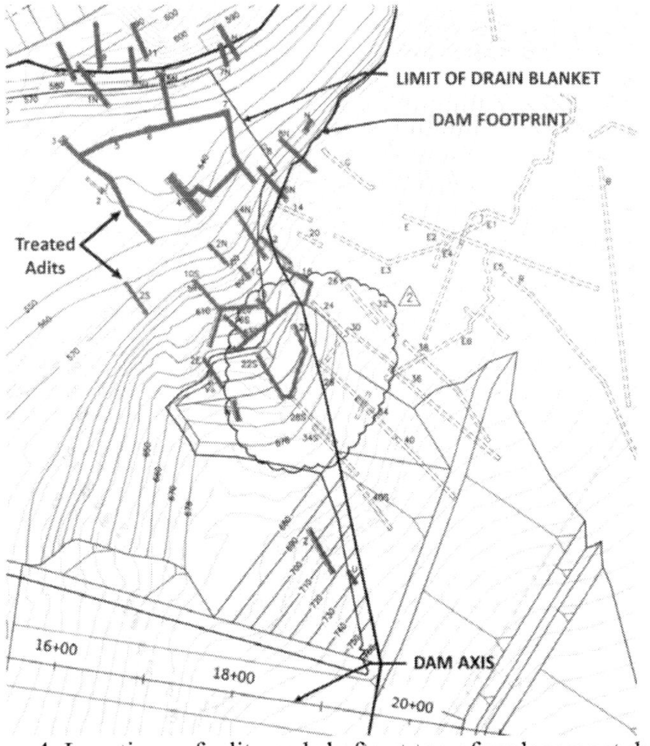

Figure 4. Locations of adits and shafts at toe of replacement dam.

The adits were carefully mapped during the final excavation process to confirm that all were found and treated as described above prior to placement of embankment fill. All of the adits and shafts that were shown in the records from 1905 through 1911 were accounted for, although some discrepancies in their locations were noted during their treatment.

DAM FOUNDATION TREATMNENT AND INITIAL CORE PLACEMENT

The foundtion in the valley bottom consists of the Franciscan formation hard rock (blueschist and greenstone) and soft rock (melange shale and serpentinite). Due to this variable rock hardness, the foundation has an irregular surface, with peaks and valleys.

In the Zone 1 core foundation area, the valleys are being filled with low-medium plastic clay, built up lift by lift. The first three feet of core materials are placed at optimum to 3 percent wet of optimum water content with the initial lift being a 12-inch thick loose lift and the succeeding lifts being 6-inch thick loose lifts. This material is being compacted into the foundation irregularities by the wheels of a Cat 966 front-end loader (with full bucket).

Foundation surface treatment consisted of dental excavation to clean out shears and joints wider than 2 inches and backfilling them with dental concrete. Features less than 2 inches wide were treated by filling them with slush grout. Depths of treatment were taken typically down to three times the widths of the foundation features being treated.

The core foundation was initially cleaned for inspection and geologic mapping for review and approval by the Designer's engineering geologist and the California Division of Safety of Dams (DSOD). This was done in an area-by-area basis from top down as the foundation was excavated. Prior to Zone 1 core placement, the foundation was given a final clean-up (Figure 5). The Zone 1 core foundation has to be completely cleaned to provide for a bond between the core and rock foundation. As such, the foundation is cleaned to remove all particles of soil, loose rock and grout spill until a bare rock foundation is produced. Just prior to core material placement, the foundation is wetted to provide for a moistened surface. This final clean-up is also performed on an area-by-area basis from the bottom up. In areas that are prone to slaking (e.g., mélange shale), the foundation was covered by core materials within 24 hours, but in no case was the foundation allowed to remain exposed for more than 72 hours prior to placing core materials.

Upstream and downstream of the core foundation, under the shell zones of the dam, the foundation was cleaned to remove windrows and areas of loose materials, but the foundation was not required to be clean the level of the Zone 1 core foundation.

Figure 5. Foundation clean-up.

QUARRY DEVELOPMENT

Borrow Area B quarry, located about 1,000 feet downstream of the replacement dam, is the source of rockfill materials for the Zone 5 upstream shell of the replacement dam. Investigations of Borrow Area B included 6 borings during final design, 14 borings drilled by contractor in 2011, and an additional 16 core borings drilled in late 2015. The borings drilled in 2011 were borings required by the contract for the purpose of providing additional data to aid the contractor in planning development of the borrow area. The borings drilled in 2015 were performed to investigate a shear zone that could present a possible slope stability issue, and evaluate other discontinuities that might impact the stability of the final excavation as well as to investigate for the presence of large masses of material unsuitable for rockfill.

The Franciscan Complex in Borrow Area B generally consists of a major portion of the rock mass being composed of blueschist (BS) and/or greenstone (GS). Variable amounts of mélange matrix (Fm) consisting of inter-mixed shale, serpentinite, greywacke, and siliceous schist typically envelop blueschist/greenstone blocks, which range in size from feet to hundreds of feet. Two larger zones, which primarily consist of mélange matrix, have been identified within the Borrow Area B rock volume. The first area (Mass A) is a mass of mélange matrix and siliceous schist in the east-central area of the borrow area that was identified during final design and further defined during the 2011 drilling. Data from the 2015 borings did not change the approximate geometry and extent of Mass A but delineated a second area (Mass B) high in the northwestern portion of the borrow area (Figure 6).

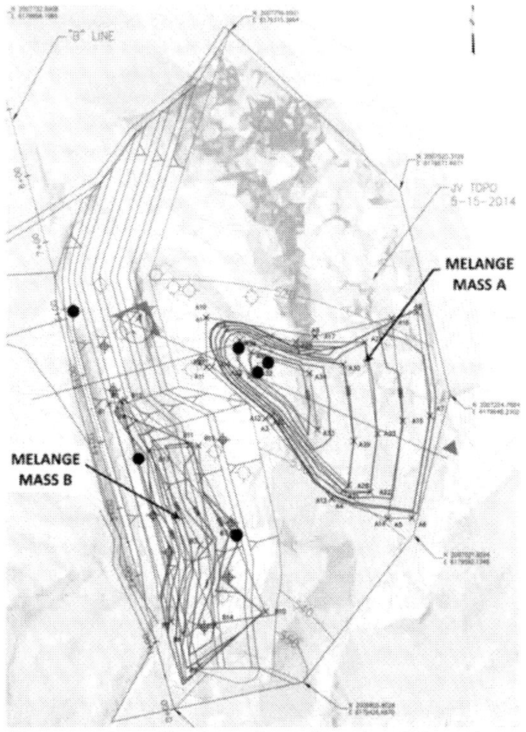

Figure 6. Borrow Area B plan view showing Mass A and Mass B.

In addition to the delineation of the large masses of mélange matrix material (Masses A and B), the 2015 exploration program provided data for the design of slope stabilization measures in the borrow area. Initial excavation into the upper portions of the Franciscan material in the southern half of the quarry had identified some unstable blocks formed by curvilinear discontinuities within the harder blueschist and greenstone materials. The discontinuities defining these blocks were not visible during mapping, and only became apparent after the block detached during or shortly after blasting. The 2015 exploration program confirmed the potential for the curvilinear discontinuities in the southern portion of the borrow area and indicated rock that was less fractured and discontinuities that were generally not adverse to the slope in the northern half of the borrow area. There is no long-term stability criteria specified for the quarry slope because the area, which is outside of any operations for the reservoir, will be abandoned after the end of construction and the area is not accessible to the public. However, the potential for the curvilinear block failures in the southern half of the borrow area during construction created an unacceptable safety risk.

To mitigate this safety risk, stabilization measures in the form of pattern bolting were designed to preemptively support these curvilinear blocks because they could not be identified during mapping. The design for the bolting used a joint strength enveloped based on the Barton (1973) approach. The joint compressive strength (JCS) was measured using a Schmidt rebound hammer on a number of exposed faces of the curvilinear discontinuities, and the joint roughness coefficient (JRC) was estimated by measuring asperity amplitudes on these same discontinuities. Based on the lowest measure JCS of 11.5 ksi and the lowest measured JRC of 9, a nonlinear strength envelope was developed which had secant friction angles on the order of 50-60 degrees. This strength envelope was used to design the pattern bolting necessary to support a hypothetical block with a height of about 50 feet (the vertical bench spacing) at a minimum factor of safety of about 1.3. This factor of safety was judged to be adequate for safety during construction. The block failures thus far were on the order of 10-15 feet and factors of safety including the slope reinforcement for blocks of this size were greater than 2.0. The area where this preemptive bolting would be required is shown on red on Figure 7.

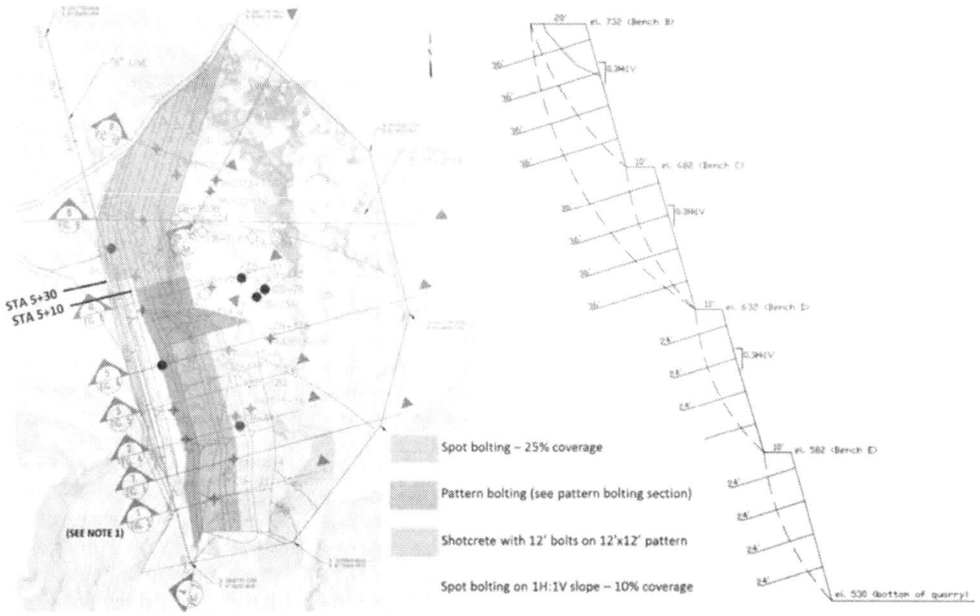

Figure 7. Borrow Area B stabilization.

Also shown on Figure 7 is the portion of the excavation slope where fracture spacing decreased in the northern portion of the quarry slope (in green) that is anticipated to require spot bolting. Where the permanent slope is anticipated to pass through Mass B (in blue on Figure 7) the slope was flattened from 0.5H:1V to 1H:1V and stabilization provided by placing shotcrete on the weaker mélange matrix to reduce the potential for exposure and weathering over the duration of construction to create rockfall debris which could be a safety hazard. Below Mass B, the permanent slope will continue to the bottom of the quarry at a 1H:1V slope through blueschist and greenstone that is likely to contain curvilinear discontinuities. Spot bolting was determined to provide sufficient support for potential blocks on the flattened slope.

CONCLUSION

Excavation for the dam foundation is complete. The spillway, intake shaft and tower, and outlet works have also been completed. Control of the reservoir through the outlet works is scheduled to begin on December 15, 2016. Embankment construction has begun at Calaveras Dam Replacement Project, and the dam is scheduled to be topped out in the spring of 2018.

ACKNOWLEDGEMENTS

The authors of this paper acknowledge the substantial efforts and contributions made to the formulation, approval, and implementation of the engineering solutions presented in this paper by the staffs of the Designer (AECOM and SFPUC), the San Francisco Planning Department, the Construction Manager (Black and Veatch), the Contractor

(Dragados, Flatiron, and Sukut Joint Venture), the Calaveras Technical Advisory Panel, and the California Division of Safety of Dams.

REFERENCES

Barton, N. (1973). "Review of a New Shear-Strength Criterion for Rock Joints," *Engineering Geology*, 7: 287-332.

Forrest, M., Roadifer, J., Wade, D., Hou, S., and Tang, G. (2014). Update on the Calaveras Dam Replacement Project, Alameda County, California, Proceedings of the 34th USSD Conference, San Francisco, California, April 2014.

Forrest, M., Roadifer, J., Wade, D., Hou, S., Lee, T., and Ng, C. (2015). Update No. 2 of the Calaveras Dam Replacement Project, Alameda County, California, Proceedings of the 35th USSD Conference, Louisville, Kentucky, April 2015.

Roadifer, J., Forrest, M., Wade, D., Hou, S., Lee, T., and Ng, C. (2016). Calaveras Dam Replacement Project, Providing a Seismically Reliable Water Source, Alameda County, California, Proceedings of the 36th USSD Conference, Denver, Colorado, April 2016.

AUTHOR INDEX

Abbas, Syed 93
Abbaszadeh, Sam 1014
Abdoun, Tarek 405
Aboelata, Maged 1460
Adams, Gregg Batchelder 286
Alvarado, Gabriela 749
Ameen, Rozh 810
Anderlini, C. 1360
Antonaki, Nonika 405
Anyanwu, Emeruwa 765
Armenis, Petros 221
Armstrong, Richard J. 939
Arnold, Justin 74
Arnold, Michael 551
Atukorala, Upul 505
Baehr, A. C. 1762
Bagherzadeh, Harry 586
Baker, Bruce 90
Ballhausen, F. 1747
Barber, Todd 1724
Barfuss, Steven 1827
Barker, Bruce 924
Barker, Malcolm 221
Barry, Rob 1250
Bearlin, Andrew 74
Beaty, Michael 969
Beck, Kade 1827
Bector, Munit 1189
Beduhn, Bob 10
Benitez, Rolando 1334
Bergman, B. 1498
Bessette, Michael 1657
Bialek, Elizabeth 808
Bird, G. Richard 286
Bisnett, Rachael V. 1090
Black, Robert 1250
Blackett, Frank L. 174
Blair, Stephen 1119
Bocovich, Carolyne 1643
Boivin, Yannick 1498
Bonelli, Stéphane 1417
Booth, Paul 735
Borela, Rodrigo 749
Bouaanani, Najib 1658
Bouayad, Ahmed 539
Bouchie, Peter 304
Bourdeau, Philippe L. 749
Bowles, David S. 104
Boyer, Douglas D. 174
Bradner, Graham 1624
Bradner, Graham C. 1604
Brinker, Darren J. 1673
Brom, Kelly 73
Brown, Stephen 267
Brown, Will 765
Bukahir, Layne 378
Burke, Joseph 1250
Burns, Steve 1102
Bushey, Randall 421
Bynoe, Kareem 304
Caldwell, Jason 924
Carney, Shaun 90
Castillo-Rodríguez, Jessica 194
Cato, Kerry D. 139
Chan, Ccee 1102
Charlwood, Robin 1153
Chaudhry, M. Hanif 720
Chen, Wen Yan 1071
Chopra, Anil K. 1464
Chowdhury, Khaled 969, 1578
Christensen, Peter 221
Clarke, Bob 734
Clopper, Chris 1796
Cobb, Charles 404
Cochran, Brett 12
Collins, Kent L. 1
Conkle, Christopher 628
Corn, Aimee L. 1673
Cotie, Bruce 268
Crookston, Brian M. 1827
Crosby, Wesley 660
Cross, Jerry 800
Crouch, Michael M. 208
Curtis, Dan D. 1233, 1546
Curtis, Kedric 72
Cvijanovic, Vojislav 1709
Dall'Aglio, Milton 675
Daviero, Gregory J. 676
Davis, Peter 368
Dawson, Ethan 969
De Melo, Lucas 440, 954
deGroot, Dirk 550
Dolen, Timothy 469
Dolve, Blake 1709
Dominguez, Justin 696
Donaghy, Jud 800
Downing, Bruce 505
Doyal, Stephanie 661
Driller, Michael 1071, 1250
Duan, Wenbo 1658
Durkee, Dean B. 855
Duron, Ziyad 1206, 1218, 1425
Dutson, Greg 735
Ellenson, Sean 1189
Engel, John 10
Engemoen, William O. 422
England Jr., John F. 329, 345
Escuder-Bueno, Ignacio 194
Espinoza, Wilson 749
Fannin, R.J. 1360

AUTHOR INDEX

Faulhaber, Bob 551
Fernandez, Jorge 246
FitzPatrick, Brendan 613
Forrest, Michael 602
France, John W. 286
Friesen, Steven 1071
Friz, Peter 74
Galic, Dom 158
Garandet, Aurélie 1417
Garduno, A. 1747
Gassman, Sarah L. 1333
Gelo, Martina 1178
Gerlach, Mike 420
Gerressen, Franz-Werner 562
Ghaedi, Khaled 487
Ghanaat, Yusof 1528
Gheibi, Emad 1333
Gibson, Wayne 924
Gilbert, Daniel A. 1735
Ginther, Conrad 1334
Gleason, Erin 1673
Gloschat, J. A. 1762
Goulas, E. 470
Griffis, Wade 522
Grima, Nathan 1375
Groeneveld, Joe 74
Gross, Eric 765
Guangul, Seifu 93
Gunn, R.M. 470
Hageman, Eric 613
Hall, Brian M. 345, 839
Hamedian, Sherry 1685
Hanbali, Fauwaz 91, 674
Harder, Les 1657
Harder, Leslie F. 1250
Hare, John L. 867
Harper, Zachary 1528
Harrington, Graham 221
Harris, Jeff 907
Hasan, Husein 1233
Hasan, Husein A. 1735
Hasenbalg, Jaron 73
Hawson, Herb 505
Hayes, Chris 1153
Hill, Jason F. 613
Hoang, Samantha 1206, 1218, 1425
Hockett, Richard B. 1403
Hoffman, Dan 93
Hogan, Dennis 577
Horvath, Doug 1334
Hou, Susan 602
Hover, William H. 452
Hu, George 969, 1058, 1388
Huard, David 56
Huggins, Jennifer E. 1513
Hultstrand, Doug 886
Husted, Jack 377, 787
Hutton, Jesse 1143
Ibáñez-de-Aldecoa, Rafael 662
Ibrahim, Ibrahim A. 1359
Ibrahim, Zainah 487
Imran, Jasim 720
Iso-Ahola, Vik 1178, 1189
Ivanov, Dmitri 1334
Jaeger, Robert 1624
Jaeger, Robert A. 629, 1604
Jalayeri, Nikou 1102
Jamieson, Steve 268
Janes, Matthew 245
Johnson, Jeff 74
Johnson, Nicholas 1090
Jones, Rusty 1803
Kanning, Willem 1643
Kappel, Bill 886
Karlovits, Gregory S. 357
Kashiwada, Christina 1250
Keizer Jr., Richard A. 1604
Kelson, Keith I. 345
Kennedy, Eric 765
Killeen, Katharine 1775
Kimbrel, Sean 1, 1441
Kirksey, Brad 867
Kirschbaum, Daniel 74
Klein, Matthew 1424
Klippel, Sterling 377, 787
Klippel, Sterling E. 1178
Klipsch, Joan 1463
Kolakowski, Anna 1375
Koltuniuk, Roman 1513, 1685
Kost, Andrew D. 1028
Kovalishyn, Val 1102
Krivanec, Christopher 1657
Kropelnicki, Jesse 304
Krumm, Paul 1135
Kuo, Chun-Yi 1403
Kuttel, Jeanne 1250
Kynett, Michael 1578
Lamour, Vincent 1417
Laro, Katie 924
Leclerc, Martin 1498
Lee, Tedman 602
Lefkovitz, Nathan 1461
Léger, P. 1498
Lehrer, Gai 562
Leonard, Christina M. 345
Li, Chunling 440
Li, M. 1360
Li, Zhihua 697
Lilley, Keith 377
Lim, Yeo Howe 1442

AUTHOR INDEX

Lochu, Alexandre 578
Lokke, Arnkjell 1464
Lopez, Don 268
Lopez, Jose R. 709
Lum, Jonathan 1206, 1218, 1425
Lund, Guy 1218
Lund, Guy S. 1425, 1673, 1724
Lynch, Andrew J. 855
Madera, Veronica 469
Maki, Ian P. 629
Mallet, Mark 662
Manie, Jonna 1465
Manole, George 1178
Margo, David A. 357
Marín, Jorge 662
Marshall, Kevin 1724
Mazighi, Hichem 487
McClung, Nicholas 613
McCoy, Andrew 1803
McGuire, Mathew 368
McHenry, Jim 765
McMahon, Michael 9
McPherson, Matthew 1462
McVicker, Jamie 1567
Mejia, Lelio 246
Mejia, Lelio H. 1043
Melliger, Joshua 92
Menninger, John 93
Meshkat, M. 825
Messio, Elisa 1658
Mihoubi, Mustapha Kamel 487
Miles, Patrick 867
Miller, Michael 313
Miquel, Benjamin 1658
Mishler, Andrew 867
Mitovski, Stevcho 645
Modini, George 674
Montero, M. 1747
Montgomery, Jack 1014
Mooney, Michael 1643
Morales-Torres, Adrián 194
Morley, Derek S. 139
Morrison, Michael A. 954
Moses, Matt 378
Mostafa, Mohammadreza 1189
Moura, Danielli De Melo 749
Muehring, Deron 1803
Muhlestein, Geoff 886
Mulder, Henri V 1058
Musselman, Eric 627
Neamtu, Danielle 1119
Needham, Jason 104
Neff, Keil 90, 440
Nettle, Justin 765
Newman, Erik 602
Ng, Carman 602
Nickerson, James 696
Novembre, Nicole 924
O'Neal, Troy S. 1403
Obermueller, Kristen 10
Oleksyn, Andrew 522, 1375
Olson, David 765
Orlovsky, Andrew J. 1568
Ostadrahimi, Leila 1462
Ovalle, William 1333
Parekh, Minal 1643
Park, Tae Ha 1206, 1218, 1425
Partovi, M. 470
Parzybok, Tye 924
Pathak, Chandra 906
Pattermann, Kenneth R. 1388
Penner, O. 1498
Perkins, J. Taylor 1735
Perlea, Mary 1578
Perlea, Vlad 969, 1058, 1578
Petkovski, Ljupcho 645
Pierce, Charles E. 1333
Prashar, Yogesh 808
Prathivadi, Kishen 392
Preston, Al 440
Pupim, Pedro 675
Queen, D. 1498
Rahman, Mostaqur 1333
Raitt, Douglas 12
Randle, Timothy 1
Rasekh, Ali 1189
Razavi-Darbar, S. 1498
Redell, Chris J. 709
Reed III, Gerard E. 522
Reelie, Dane 245
Reynoldson, Thomas 522
Rice, John 1359
Richards, Adam 586
Rinehart, Robert V 587
Rittgers, Justin 658
Rittgers, Justin B. 1344
Rix, Glenn J. 954
Rizzardo, John 1775
Roach, Sharon Nicole 1828
Roadifer, John 602
Robbe, Emmanuel 1480
Rocha, Miguel 1513
Rodrigues, Larry 1375
Rogers, Michael 522
Rohrbach, Ben 907
Rondinel-Oviedo, Efrain 313
Rosen, Jamey 810
Royer, Joseph 1250
Ruark, Morgan D. 208
Ruff, Jonathon 676

AUTHOR INDEX

Ruswick, Kevin J. 676
Rutherford, James 74
Rutherford, James H. 1102
Rutledge, John 867
S.Tarbox, Glenn 1178, 1189
Sabol, George 93
Salamon, Jerzy 1465
Samandi, Sam 867
Sanders, Cathi 1567
Santini, Mauricio 675
Sasaki, Reuben 839
Sasanakul, Inthuorn 405, 1333
Saucier, Christopher L. 954
Schaefer, Mel 90, 924
Schaeffer, Kelly R. 11
Scholz, Israel Christian 562
Schreck, David 304
Schreppers, G. 470
Schultz, Mark 1709
Scofield, David H. 1028
Scuero, Alberto 1317
Seed, Raymond B. 969
Seid-Karbasi, Mahmood 505
Serafini, David C. 1058
Serock, Adam 1440
Shannon, Alexander B. 1776
Sharp, Michael 1564, 1566
Shaw, Quentin 1167
Sherwood, Bill 734
Shoghli, Bahareh 1442
Siebenmann, Raphael 810
Silvertooth, D.L. 825
Simpson, Bryan K. 658, 1344
Smith, Haden 329, 839
Sooch, Gurinderbir S. 1233, 1546
Sookhu, Yogi 800
Sossenkina, Elena 1564
Srivastava, Anurag 208
Stanley, Mark 1604, 1624
Starcher, Ryan 1333
Stiady, James 522
Sutter, Amanda 764
Tabrizi, Ali Asghari 720
Tarbox, Glenn 522
Tarbox, Glenn S. 1153
Tatin, Maxime 1417
Taylor, George 924
Teal, Martin J. 1443
Tisdale, Rinda 1796
Toose, James 246
Torres, Raphael A. 1250
Treacy, John 268
Trudgeon, Ray 1102
Tullis, Blake 72
Unterberg, Jessie Drayton 286
Vaghti, Stewart S. 855
Valentyn, James 1090
Vaschetti, Gabriella 1317
Vazinkhoo, Saman 662
Vennapusa, Pavana 613
Venturini, Hillery 659
Vickers, John H. 452
Viducich, Jonathan M. G. 1443
Vroman, Noah 1564, 1565
Wade, Daniel 602
Wagner, Cassandra 658
Wagner, Cassandra M. 1344
Walker, Kent 73
Walker, William 809
Weber, Joseph 1604, 1624
Wehling, Timothy 1250
Weller, Ryan 1827
Werner, John 74
White, David J. 613
Whiteside, Stephen 1119
Whitten, Zachary 45
Wilkes, John 1317
Wong, Rachel 808
Wylde, Travis 787
Yamasaki, Kenji 1028
Yiadom, Atta 697
Yiadom, Atta B. 808
Young, Jeremy 627
Young, Matthew 1796
Zakrevskaya, Yulia 56
Zhou, Jim 522, 1375
Zimmerman, Peter 735
Zoeller, Ben 1461

United States Society on Dams (USSD)
9975 Wadsworth Pkwy.
Ste. K-2, 145
Westminster, CO 80021
USA

ISBN 978-1-5108-6027-8

9 781510 860278

CURRAN ASSOCIATES INC.
proceedings.com

37th Annual USSD Conference 2017

It's a Small World: Managing Our Water Resources

Anaheim, California, USA
3 - 7 April 2017

Volume 2 of 3

37th Annual USSD Conference 2017

It's a Small World: Managing Our Water Resources

Anaheim, California, USA
3 - 7 April 2017

Volume 2 of 3

ISBN: 978-1-5108-6027-8

Printed from e-media with permission by:

Curran Associates, Inc.
57 Morehouse Lane
Red Hook, NY 12571

proceedings.com

Some format issues inherent in the e-media version may also appear in this print version.

Printed by Curran Associates, Inc. (2018)

For permission requests, please contact United States Society on Dams (USSD)
at the address below.

United States Society on Dams (USSD)
9975 Wadsworth Pkwy.
Ste. K-2, 145
Westminster, CO 80021
USA

Phone: 303-792-8753
Fax: 303-792-8782

info@ussdams.org

Additional copies of this publication are available from:

Curran Associates, Inc.
57 Morehouse Lane
Red Hook, NY 12571 USA
Phone: 845-758-0400
Fax: 845-758-2633
Email: curran@proceedings.com
Web: www.proceedings.com

TABLE OF CONTENTS

VOLUME 1

SESSION 1A: CONFERENCE THEME

Reservoir Sedimentation and Sustainability 1
Timothy Randle, Sean Kimbrel, Kent L. Collins

Dam Safety and Resilient Water Management: A Case for Dynamic Reservoir Operations 9
Michael McMahon

The Nebraska Department of Natural Resources Coordinated Efforts in Flood Flows 10
Kristen Obermueller, Bob Beduhn, John Engel

SESSION 2A: ENVIRONMENT

Environmental Considerations in Dam Decommissioning 11
Kelly R. Schaeffer

Hydraulic Dredging of the Strontia Springs Reservoir — Lessons Learned 12
Douglas Raitt, Brett Cochran

Dam Decommissioning — Unique Approach by Conversion to a Levee 45
Zachary Whitten

The Prospects for Climate Scenarios in Hydropower Operations 56
Yulia Zakrevskaya, David Huard

SESSION 1B: PHYSICAL MODELING

Linear Weir Size Scale Effects: A Look at Flat Top Weirs 72
Blake Tullis, Kedric Curtis

Physical Model of Spillway and Reservoir Debris Interaction 73
Kent Walker, Jaron Hasenbalg, Kelly Brom

Boundary Dam Spillway No. 1 TDG Modifications 74
John Werner, James Rutherford, Andrew Bearlin, Peter Friz, Jeff Johnson, Daniel Kirschbaum, Joe Groeneveld, Justin Arnold

SESSION 2B: HYDROLOGIC MODELING

A Comprehensive Approach to Probabilistic Hydrologic Hazard Assessment and its Use to Support Risk-Informed Decisions 90
Shaun Carney, Keil Neff, Mel Schaefer, Bruce Baker

Meteorlogical Data Visualization and Analysis Using HEC-MetVue 91
Fauwaz Hanbali

Big Dam in a Small World: Reassessment of the Spillway Design Flood of Lake Sakakewea 92
Joshua Melliger

Application of Regional Hydrologic Criteria in the Rocky Mountains to Estimation of the PMF in Alberta 93
John Menninger, Seifu Guangul, Dan Hoffman, George Sabol, Syed Abbas

SESSION 1C: RISK

Approaches to Estimating Individual Risk for Dam Safety Tolerable Risk Evaluation 104
David S. Bowles, Jason Needham

Minimizing Bias in Geologic and Geotechnical Aspects of Risk Evaluation for Dams and Levees 139
Kerry D. Cato, Derek S. Morley

The Common Cause Adjustment in Dam Safety Risk Analysis: Scope, Purpose, and Applicability 158
Dom Galic

SESSION 2C: RISK

Potential Failure Mode Analysis – Insights and Lessons Learned for the Foundation of a Successful Dam Safety Management Program 174
Frank L. Blackett, Douglas D. Boyer

Epistemic vs Natural Uncertainty in Dam Safety Decision Making: Is it Fair Play? 194
Ignacio Escuder-Bueno, Adrián Morales-Torres, Jessica Castillo-Rodriguez

Consequence Estimation for a Large System of Dams: Our Experiences from TVA Projects 208
Michael M. Crouch, Anurag Srivastava, Morgan D. Ruark

Application of Risk Assessment to an Emergency Levee Design in an Urban Environment 221
Petros Armenis, Malcolm Barker, Peter Christensen, Graham Harrington

SESSION 1D: CONSTRUCTION I

Vibration Free Resonant Sheet Pile Driving in Dykes and Levees 245
Dane Reelie, Matthew James
Construction of Borinquen Dam 1E for the Panama Canal Expansion 246
Lelio Mejia, James Toose, Jorge Fernandez
Tsaile Dam — A Complete Approach Unearths a Safer Dam 267
Stephen Brown

SESSION 2D: CONSTRUCTION I

Replacement of Clear Lake Dam: Overcoming Challenges Constructing an RCC Dam on a Soil Foundation 268
Steve Jamieson, Bruce Cotie, Don Lopez, John Treacy
Dewatering — How to Address this Potential Achilles Heel for Construction 286
John W. France, G. Richard Bird, Gregg Batchelder Adams, Jessie Drayton Unterberg
Draft Tube Slab Analysis 304
Jesse Kropelnicki, Kareem Bynoe, David Schreck, Peter Bouchie
Core from Rock, Challenges of Core Construction for a Large High Hazard Dam 313
Michael Miller, Efrain Rondinel-Oviedo

SESSION 1E: RISK INFORMED H&H

Estimating the Reservoir Stage-Frequency Curve with Uncertainty Bounds for Cherry Creek Dam Using the Reservoir Frequency Analysis Software (RMC-RFA) 329
Haden Smith, John F. England Jr.
Recent Paleoflood Analyses for USACE Risk-Informed Dam Safety Evaluations 345
Keith I. Kelson, Christina M. Leonard, Brian M. Hall, John F. England Jr.
Risk Informed Freeboard Evaluation 357
Gregory S. Karlovits, David A. Margo

SESSION 2E: GATES AND VALVES

Fail Safe Design for Gated Structures Construction and Rehabilitation 368
Peter Davis, Mathew McGuire
Dams Inlet/Outlet Works Rehabilitation Program 377
Jack Husted, Sterling Klippel, Keith Lilley
Best Practices in a Tainter Gate Rehabilitation Program 378
Matt Moses, Layne Bukahir
Emergency Shutoff Valves 392
Kishen Prathivadi

SESSION 1F: EMBANKMENT DAMS I

A State Perspective on Tailings Dam Regulation and a Role for USSD and ASDSO 404
Charles Cobb
A Centrifuge Study of Co-Mixed Mine Tailings and Waste Rock Deposits 405
Nonika Antonaki, Tarek Abdoun, Inthuorn Sasanakul
Freeboard Assessment for Long Embankment Dams 420
Mike Gerlach

SESSION 2F: EMBANKMENT DAMS I

Offstream Reservoir Earthen Dam Seepage Systems — Design and Construction Challenges 421
Randall Bushey
Latest Compilation of Internal Erosion Incidents at Bureau of Reclamation Dams 422
William O. Engemoen
Modeling of Internal Erosion in Earthen Embankment Dam 440
Al Preston, Keil Neff, Chunling Li, Lucas de Melo
Turbid Discharge Incident at Cannonsville Dam 452
John H. Vickers, William H. Hover

POSTER SESSION

CONCRETE

Freezing and Thawing Durability of Laboratory and In-Situ Roller-Compacted Concrete Test Specimens 469
Timothy Dolen, Veronica Madera

Parametric Layout of Double-Curvature Arch Dams Using DIANA Finite Element Software 470
E. Goulas, G. Schreppers, M. Partovi, R.M. Gunn

Study of Uplift Pressure Effect in Roller Compacted Concrete Gravity Dam 487
Hichem Mazighi, Mustapha Kamel Mihoubi, Khaled Ghaedi, Zainah Ibrahim

Soil-Structure Interaction Dynamic Analysis for an Existing Old Earth-Rock Fill Dam with Concrete Core 505
Mahmood Seid-Karbasi, Upul Atukorala, Herb Hawson, Bruce Downing

RCC Post Construction Core Testing and Data Analysis For San Vicente Dam Raise Project 522
James Stiady, Andrew Oleksyn, Michael Rogers, Glenn Tarbox, Thomas Reynoldson, Gerard E. Reed III, Wade Griffis, Jim Zhou

CONSTRUCTION

Reconstruction of the Matala Spillway: Hydrotechnical and Safety Aspects 539
Ahmed Bouayad

Decommissioning Outdated Fish Bypass Systems and Preparation for Concrete Retrofit on Lower Granite Lock & Dam 550
Dirk deGroot

Sand Secant Pile Chimney Filter Installation at Pine Creek Dam 551
Bob Faulhaber, Michael Arnold

Execution of a Cut-Off Wall at the Dead Sea – Unique Conditions Require Unique Solution 562
Franz-WernerGerressen, Gai Lehrer, Israel Christian Scholz

Catawba Dam: The Long Road and Construction Update 577
Dennis Hogan

Water-Filled Bladders, an Innovative Bearing Interface for Arch Dam Bracing 578
Alexandre Lochu

Eductors — The Essential Alternative in Complex Dewatering Applications 586
Adam Richards, Harry Bagherzadeh

Application of Intelligent Compaction to Embankment Dam Construction 587
Robert V. Rinehart

Update No. 3 of the Calaveras Dam Replacement Project 602
John Roadifer, Michael Forrest, Erik Newman, Daniel Wade, Susan Hou, Tedman Lee, Carman Ng

VOLUME 2

Validated Integrated Compaction Monitoring at TVA's KIF Coal Combustion Product Stacking Facility 613
David J. White, Pavana Vennapusa, Brendan FitzPatrick, Jason F. Hill, Nicholas McClung, Eric Hageman

To Air is Human; To Succeed, Divine — Evolutionary Research on Grout Enriched RCC Air Entrainment 627
Jeremy Young, Eric Musselman

EMBANKMENT DAMS

Stability Analysis of a Concrete-Faced Rockfill Dam Using Parametric Evaluations 628
Christopher Conkle

Estimating the Peak Friction Angle of Sandy Soils In Situ with State-Based Overburden Normalized SPT Blow Counts 629
Robert A. Jaeger, Ian P. Maki

Comparison Analysis of the Behavior of Rock-Fill Dams with Clay Core at the Variation of Water Level in the Reservoir 645
Ljupcho Petkovski, Stevcho Mitovski

Using Non-Intrusive Geophysical Techniques in Embankment Dam and Levee Risk Assessment 658
Justin Rittgers, Cassandra Wagner, Bryan K. Simpson

Impact of Glacial Till Deposits — Structural Analysis of Green Mountain Dam 659
Hillery Venturini

HYDROLOGY AND HYDRAULICS

Using HEC-RAS 2 Dimensional Capability for Simplified Rapid Dam Break Analysis 660
Wesley Crosby

Hydraulic Modeling of a Complex Dam and Reservoir Project: A Case Study of Two Rivers Dam 661
Stephanie Doyal

Rainfall-Runoff Modeling for the Garrison Dam Drainage with HEC-HMS .... N/A
Jeff Harris, Kevin Denn, Melissa Larsen

An Innovative Temporary Bulkhead for the Ruskin Dam and Powerhouse Upgrade Project 662
Jorge Marin, Saman Vazinkhoo, Rafael Ibáñez-de-Aldecoa, Mark Mallet
Risk and Uncertainty Features in HEC-ResSim 674
George Modini, Fauwaz Hanbali
Methodology Applied for Mapping Flooded Areas of the Lajeado HPP 675
Pedro Pupim, Mauricio Santini, Milton Dall'Aglio
Leveraging Multi-Sensor Hydrologic Data for Model Validation 676
Kevin J. Ruswick, Gregory J. Daviero, Jonathon Ruff

LEVEES

Design of Levee Modifications to Accommodate Long-Term River Scour 696
Justin Dominguez, James Nickerson
Detection of Levee Voids at Aqueduct Crossings Using Geophysical Imaging Techniques 697
Zhihua Li, Atta Yiadom
Risk Analysis in a Levee Safety Context: The Upper Wood River Levee Case History 709
Chris J. Redell, Jose R. Lopez
Effects of Compaction on Erodibility Characteristics of Levees Due to Overtopping 720
Ali Asghari Tabrizi, M. Hanif Chaudhry, Jasim Imran

MONITORING

Preventing Catastrophic Failure of Underwater Infrastructure: Condition Assessments by ROV 734
Bill Sherwood, Bob Clarke
Automated Data Acquisition System Basics for Dam Safety Instrumentation Monitoring 735
Peter Zimmerman, Paul Booth, Greg Dutson

RISK

Failure of the Acude da Nacao Earth Dam: A Probabilistic Perspective 749
Danielli de Melo Moura, Rodrigo Borela, Gabriela Alvarado, Wilson Espinoza, Philippe L. Bourdeau

SESSION 3A: MONITORING

Operation and Maintenance of an Instrumentation Program 764
Amanda Sutter
FERC Best Practices for Dam Safety Performance Monitoring 765
Will Brown, Eric Kennedy, Justin Nettle, Emeruwa Anyanwu, Eric Gross, Jim McHenry, David Olson
Calculation of Threshold Limits for Dam Safety Instrumentation 787
Jack Husted, Travis Wylde, Sterling Klippel
Somerset Dam – Innovative Remote Monitoring for Time-Sensitive Breach Detection 800
Jerry Cross, Yogi Sookhu, Jud Donaghy

SESSION 4A: MONITORING

Geospatial Tools for Routine Dam Safety Monitoring and Response Applications During Emergencies 808
Yogesh Prashar, Atta B. Yiadom, Elizabeth Bialek, Rachel Wong
Special Considerations When Installing Automated Instrumentation at a Site with Access Limitations 809
William Walker
Instrumentation and Information Management at Boone Dam 810
Rozh Ameen, Raphael Siebenmann, Jamey Rosen

SESSION 3B: HYDRAULIC MODELING

Modeling Dam Breaches and Flood Inundation Using HEC-RAS 5.0's New 2D Capabilities 825
D.L. Silvertooth, M. Meshkat
Estimating the Probable Maximum Flood for Whittier Narrows Dam Using a 2D Hydrodynamic Modeling Approach 839
Brian M. Hall, Haden Smith, Reuben Sasaki
When an Adjacent Watershed Inundates Your Offline Reservoir: Ukumehame Reservoir Dam Removal Challenges and Design 855
Andrew J. Lynch, Stewart S. Vaghti, Dean B. Durkee
Atoka Dam and Spillway Modifications 867
Patrick Miles, Brad Kirksey, John Rutledge, Sam Samandi, John L. Hare, Andrew Mishler

SESSION 4B: SITE SPECIFIC PMP

Recent Extreme Storms and Their Relation to PMP 886
Bill Kappel, Doug Hultstrand, Geoff Muhlestein
Needs to Update Probable Maximum Precipitation for Determining Hydrologic Loading on Dams 906
Chandra Pathak
Site Specific Probable Maximum Precipitation Development 907
Jeff Harris, Ben Rohrbach
Advances in Watershed Modeling Inputs and Methods for Dam Safety Risk Analyses 924
Jason Caldwell, Nicole Novembre, Katie Laro, Tye Parzybok, Bruce Barker, Mel Schaefer, George Taylor, Wayne Gibson

SESSION 3C: EARTHQUAKES I

Use of the Conditional Mean for Improved Prediction of Ground Motion Intensity Measures for Embankment Dams 939
Richard J. Armstrong
Seismic Deformation Analysis of Blue Ridge Dam 954
Glenn J. Rix, Lucas de Melo, Christopher L. Saucier, Michael A. Morrison
A Case History Evaluation of State of Practice for Seismic Deformation Modeling of Earthen Dams 969
Khaled Chowdhury, Vlad Perlea, Raymond B. Seed, Michael Beaty, Ethan Dawson, George Hu
Comparison of Two Constitutive Models for Simulating the Effects of Liquefaction on Embankment Dams 1014
Jack Montgomery, Sam Abbaszadeh

SESSION 4C: EARTHQUAKES I

Seismic Evaluation of Lookout Point Embankment Dam in Oregon – Cascadia Subduction Zone Implications 1028
Andrew D. Kost, Kenji Yamasaki, David H. Scofield
Performance of Fena Dam During the 1993 Guam Earthquake 1043
Lelio H. Mejia
Seismic Evaluation of the Remediated Isabella Auxiliary Dam 1058
George Hu, David C. Serafini, Henri V. Mulder, Vlad Perlea

SESSION 3D: CONSTRUCTION II

Seismic Remediation of Perris Dam Using CDSM 1071
Wen Yan Chen, Michael Driller, Steven Friesen
Controlled Blasting for the Red Rock Hydroelectric Project 1090
Rachael V. Bisnett, James Valentyn, Nicholas Johnson
Upriver Dam Spillway Rehabilitation 1102
Ray Trudgeon, James H. Rutherford, Steve Burns, Val Kovalishyn, Ccee Chan, Nikou Jalayeri
Lakeside Ranch Stormwater Treatment Area – Lessons Learned During Design and Construction 1119
Danielle Neamtu, Stephen Whiteside, Stephen Blair

SESSION 4D: CONSTRUCTION II

Design and Repair of a Power Intake Gate Sealing Surface Under Challenging Constraints .... N/A
Mike Carney, Don Thompson, Roland Lwiindi, Eli Lundberg, Joshua Phillips
Post-Tensioning in an Underwater Environment: Wanapum Dam + Monolith 4 Spillway Repair 1135
Paul Krumm
Balancing Water Resources and Power Generation for the Metlakatla Indian Community 1143
Jesse Hutton

SESSION 3E: CONCRETE I

Investigating the Structural Safety of Cracked Concrete Dams 1153
Glenn S. Tarbox, Robin Charlwood, Chris Hayes
A New Approach for the Design of the Köroglu RCC Arch/Gravity Dam 1167
Quentin Shaw
Upgrade and Spillway Modifications at Santa Anita Dam 1178
Sterling E. Klippel, Vik Iso-Ahola, Glenn S.Tarbox, Martina Gelo, George Manole
Understanding the Performance of Aging Infrastructure: A 99-Year-Old Multiple Arch Dam 1189
Vik Iso-Ahola, Glenn S.Tarbox, Mohammadreza Mostafa, Sean Ellenson, Munit Bector, Ali Rasekh

SESSION 4E: CONCRETE I

A Performance-Based Evaluation of Post-Tensioned Anchors Embedded Within a Concrete Gravity Dam 1206
Tae Ha Park, Samantha Hoang, Jonathan Lum, Ziyad Duron

On the Use of Discrete Lumped Element Modeling for Insight into Concrete Dam Behavior 1218
Jonathan Lum, Samantha Hoang, Tae Ha Park, Ziyad Duron, Guy Lund

VOLUME 3

Shear Strength of Concrete Lift Joints from Extensive Lab Testing Compared to Theoretical Results 1233
Dan D. Curtis, Husein Hasan, Gurinderbir S. Sooch

SESSION 3F: EMBANKMENT DAMS II

Seepage Distress and Canal Repairs on the California Aqueduct 1250
Leslie F. Harder, Jeanne Kuttel, Joseph Royer, Timothy Wehling, Joseph Burke, Robert Black, Michael Driller, Christina Kashiwada, Rob Barry, Raphael A. Torres

Geomembranes to Waterproof Embankment Dams 1317
John Wilkes, Alberto Scuero, Gabriella Vaschetti

Geotechnical Investigation of South Carolina Flood-Damaged Dams after the 2015 October Rainfall Event 1333
Inthuorn Sasanakul, Sarah L. Gassman, Charles E. Pierce, William Ovalle, Ryan Starcher, Emad Gheibi, Mostaqur Rahman

Rough River Dam: Phase 1B Exploratory Drilling and Grouting Program 1334
Conrad Ginther, Dmitri Ivanov, Doug Horvath, Rolando Benitez

SESSION 4F: EMBANKMENT DAMS II

Integrating a Multidisciplinary Field Investigation to Improve Assessment of Internal Erosion 1344
Cassandra M. Wagner, Justin B. Rittgers, Bryan K. Simpson

Laboratory Investigation of Backward Erosion Piping into Constricted Seepage Exits 1359
John Rice, Ibrahim A. Ibrahim

Evaluation of Internal Erosion Scenarios in an Embankment Dam 1360
R.J. Fannin, M. Li, C. Anderlini

SESSION 5A: MONITORING

Monitoring San Vicente Dam Raise During Initial Fill 1375
Andrew Oleksyn, Nathan Grima, Anna Kolakowski, Larry Rodrigues, Jim Zhou

First Filling Instrumentation Monitoring of the Folsom Dam Auxiliary Spillway Control Structure 1388
Kenneth R. Pattermann, George Hu

Effect of Antecedent Groundwater Level on Piezometric Response During Flood Events 1403
Chun-Yi Kuo, Richard B. Hockett, Troy S. O'Neal

SESSION 6A: MONITORING

Leak Detection In Earth Dams and Levees Using Optical Fiber Distributed Temperature Sensors 1417
Maxime Tatin, Vincent Lamour, Stéphane Bonelli, Aurélie Garandet

Elephant Butte Dam UAV Data Collection and Photogrammetry 1424
Matthew Klein

Proposed Guidelines for Conducting a Performance-Based Evaluation of a Concrete Dam 1425
Samantha Hoang, Tae Ha Park, Jonathan Lum, Ziyad Duron, Guy S. Lund

Reality Capture for Dam and Hydroelectric Facilities 1440
Adam Serock

SESSION 5B: RESERVOIR SEDIMENTATION

Numerical Sediment Transport Modeling of Operations at Paonia Reservoir, Colorado 1441
Sean Kimbrel

Impact of Climate Change on Sediment Accumulation in the Reservoirs of Embankment Dams: Case Study, Baldhill Dam, North Dakota 1442
Bahareh Shoghli, Yeo Howe Lim

Sediment Transport Through Lake Clarke and Lake Aldred 1443
Jonathan M. G. Viducich, Martin J. Teal

SESSION 6B: RESERVOIR OPERATIONS

To Release or Not to Release? Dam Operation Modeling Using Gridded Short-Term Forecast Rainfall 1460
Maged Aboelata
USACE CWMS: Case Studies of Water Management Decision Support Tools 1461
Ben Zoeller, Nathan Lefkovitz
Systems-Based Risk Analysis of Lake Mendocino Forecast-Informed Reservoir Operations with HEC-WAT 1462
Matthew McPherson, Leila Ostadrahimi
HEC-ResSim 3.3 — New Features to Support Complex Studies 1463
Joan Klipsch

SESSION 5C: EARTHQUAKES II

A Direct Finite Element Method for Nonlinear Analysis of Semi-Unbounded Dam-Water-Foundation Rock Systems 1464
Anil K. Chopra, Arnkjell Lokke
Numerical Assessment of Hydrodynamic Loads Induced During Seismic Interaction Between Reservoir and Concrete Dam 1465
Jerzy Salamon, Jonna Manie
Seismic Back Analysis of Monticello Arch Dam – Blind Prediction Workshop and Additional Analyses 1480
Emmanuel Robbe

SESSION 6C: EARTHQUAKES II

Non-Linear Time Domain Analysis of Ruskin Dam in LS-DYNA Including Reservoir-Foundation-Structure Interaction 1498
O. Penner, B. Bergman, S. Razavi-Darbar, D. Queen, P. Léger, Martin Leclerc, Yannick Boivin
Boca Dam Spillway Nonlinear Analysis and Validation Study 1513
Jennifer E. Huggins, Roman Koltuniuk, Miguel Rocha
Nonlinear Seismic Evaluation of Perris Dam Outlet Tower 1528
Yusof Ghanaat, Zachary Harper
Nonlinear Seismic Analysis of an Arch Dam Subjected to Very Intensive Earthquake Loading 1546
Gurinderbir S. Sooch, Dan D. Curtis

SESSION 5D: LEVEES

U.S. Levee Safety Coalition and International Community of Practice, Recent Activities and Updates 1564
Elena Sossenkina, Michael Sharp, Noah Vroman
USACE Levee Portfolio Report 1565
Noah Vroman
Levee Safety Guidance Comprehensive Document Review 1566
Michael Sharp
Inventory and Review of the Nation's Levees 1567
Jamie McVicker, Cathi Sanders
Levee Safety Practices in the Keystone State 1568
Andrew J. Orlovsky

SESSION 6D: LEVEES

Seismic Vulnerability Analysis of Slag Cement-Cement-Bentonite Seepage Cutoff Walls for Levees 1578
Vlad Perlea, Khaled Chowdhury, Mary Perlea, Michael Kynett
Advances in Design Procedures for Relief Well Lines in Variable Urban Settings 1604
Robert A. Jaeger, Richard A. Keizer Jr. Graham C. Bradner, Joseph Weber, Mark Stanley
Evaluation and Design of Irregularly-Positioned Relief Wells 1624
Joseph Weber, Mark Stanley, Robert Jaeger, Graham Bradner
Explaining Pore Pressure Changes Due to Piping Through Changes in Hydraulic Conductivity 1643
Carolyne Bocovich, Willem Kanning, Minal Parekh, Michael Mooney
Design and Construction of Feather River West Levee Improvements 1657
Christopher Krivanec, Les Harder, Michael Bessette

SESSION 5E: CONCRETE II

Simplified Nonlinear Seismic Analyses of Gravity Dams 1658
Elisa Messio, Wenbo Duan, Najib Bouaanani, Benjamin Miquel
Strontia Springs Dam: Methods to Better Understand Seismic Behavior 1673
Aimee L. Corn, Guy S. Lund, Darren J. Brinker, Erin Gleason

Cracking in Thin Arch Concrete Dam – Nonlinear Dynamic Structural Analysis 1685
Sherry Hamedian, Roman Koltuniuk

SESSION 6E: CONCRETE II

Comparison of Nonlinear Finite Element Analysis Methods Used to Simulate Post-Tensioned Anchorage in Concrete 1709
Vojislav Cvijanovic, Mark Schultz, Blake Dolve
Structural Inflection Points, and How They Enhance the Understanding of Structural Safety 1724
Guy S. Lund, Kevin Marshall, Todd Barber
Pseudo-Nonlinear Finite Element Analysis of Concrete Gravity Dams 1735
J. Taylor Perkins, Husein A. Hasan, Daniel A. Gilbert
Importance of RCC Mixture Lab Studies for the Design of Large Dams 1747
A. Garduno, M. Montero, F. Ballhausen

SESSION 5F: SAFETY & SECURITY

Managing Threats Through Effective Security Programs 1762
J. A. Gloschat, A. C. Baehr
Unmanned Aerial Vehicles – A Public Agency Approach to Deployment and Protection 1775
Katharine Killeen, John Rizzardo
The Economics of Dam Security Post 9/11 and the Threat to Consumer Confidence 1776
Alexander B. Shannon

SESSION 6F: SAFETY & SECURITY

Flood Risks from Spillways: Modeling and Communicating Non-Dam-Failure Impacts to Native American Tribes 1796
Rinda Tisdale, Matthew Young, Chris Clopper
City of Dubuque, Iowa: Defining Emergency Actions with a Levee Breach Evaluation 1803
Andrew McCoy, Deron Muehring, Rusty Jones
Public Safety and Unauthorized Extreme Activities at Dams 1827
Steven Barfuss, Brian M. Crookston, Kade Beck, Ryan Weller
Public Safety and Social Media 1828
Sharon Nicole Roach
Author Index

VALIDATED INTEGRATED COMPACTION MONITORING AT TVA'S KIF COAL COMBUSTION PRODUCT STACKING FACILITY

David J. White, Ph.D., P.E.[1]
Pavana Vennapusa, Ph.D., P.E.[2]
Jason F. Hill[3]
Eric Hageman[4]
Nicholas McClung, P.E.[5]
Brendan FitzPatrick P.E.[6]

ABSTRACT

The Tennessee Valley Authority (TVA) recently implemented state-of-the-art validated integrated compaction monitoring (VICM) at selected coal combustion product (CCP) impoundment and stacking facilities to provide real-time compaction quality assurance monitoring. This paper describes project results at the TVA Kingston Fossil Plant where the VICM program was designed to improve field process control, reduce the risk of placing poorly compacted CCP materials, and improve cost to rate payers by improving construction efficiency. Results are presented for placement of approximately 500,000 cubic yards of coal fly ash in a new landfill over a period of 9 months (January to October 2016). The landfill material originated from a previously staged landfill constructed from the clean-up of the 2008 Kingston dredge cell failure. Compaction results are presented as color-coded geospatial maps of VICM measurements and control charts of in situ spot test measurements over time with reference to the project target requirements. Rapid in situ tests for measurement of moisture, density, elastic modulus, and shear strength, and a field geotechnical mobile lab for onsite testing and evaluation of test results were setup for this project. A key finding form this project is that the calibrated VICM mapping allowed real-time identification of "soft" areas which aided in timely decision making to improve compaction control and rework. The overview of technologies and analysis described in this paper are of interest for other construction operations requiring definitive and real-time compaction control such as landfills, embankments, levees, and dams.

INTRODUCTION

This paper summarizes the results of a compaction monitoring program carried out TVA's Kingston Fossil (KIF) Facility in Kingston, Tennessee. The purpose of the project

[1] President, Ingios Geotechnics, Inc., P.O. Box 101, Northfield, MN, 55057, 515-509-7587, david.white@ingios.com
[2] Lead Engineer, Ingios Geotechnics, Inc., P.O. Box 101, Northfield, MN, 55057, 515-509-7587, pavana.vennapusa@ingios.com
[3] Program Manager, R&QA, Tennessee Valley Authority, 1101 Market St LP 5E-C, Chattanooga, TN 37402, 423-802-7771, jfhill0@tva.gov
[4] Project Engineer, HDR Engineering, Inc., Chattanooga, TN, eric.hageman@hdrinc.com
[5] Manager, Risk and Quality Assurance, Tennessee Valley Authority, namcclung@tva.gov
[6] Director of Operations, Ingios Geotechnics, Inc., P.O. Box 101, Northfield, MN, 55057, brendan.fitzpatrick@ingios.com

was to implement an innovative compaction monitoring program to control the Coal Combustion Product (CCP) compaction operations, document and verify CCP compaction results, and provide independent quality assurance testing. Prior work at TVA's dry stacking operations demonstrated compaction control and monitoring improvements with evaluations of selected compaction monitoring technologies (Christopher et al., 2013; Christopher et al., 2015; White et al., 2012a; White et al, 2012b; White et al., 2013; White et al., 2015a, White et al., 2015b).

A TVA-supplied roller (Caterpillar CS56B) used for ash compaction by TVA's ash handling contractor was instrumented with an Ingios Validated Integrated Compaction Monitoring (VICM) monitoring system and used for real-time monitoring and mapping during placement of both Ballfield ash and production ash. Figure 1 shows the roller instrumented with an in-cab computer monitor. The work completed under the project scope included multiple calibrations of the VICM monitoring equipment on the CCPs to refine the calibration accuracy for on-site material placement.

One of the key outcomes of the project was to provide "guidance to the roller operators" on pass/fail conditions so the changes to the compaction effort and/or moisture content of the CCPs could be adjusted. The pass/fail was based on meeting a target value that is established from field calibration to meet the permit criteria (95% relative compaction). Figure 2 provides an example of the resulting on-board VICM mapping results.

Figure 1. TVA Roller Instrumented with VICM system

Figure 3 shows an example compaction report prepared after mapping a given layer. Typically, mapping occurred after several loads (20 to 50+) of material were hauled, spread, leveled with a box blade, moisture conditioned. Normally, a minimum of four roller passes was required to compact the specified maximum 12 in. loose lift of CCP material. In some cases, as many as 14 passes demonstrated compaction improvement. The roller operators can view the compaction results and progress (i.e., color changes) during compaction.

Figure 2. Operator display in VICM instrumented roller showing calibrated mapping results in real-time (blue indicates XMV ≥35, red indicated XMV ≤ 10).

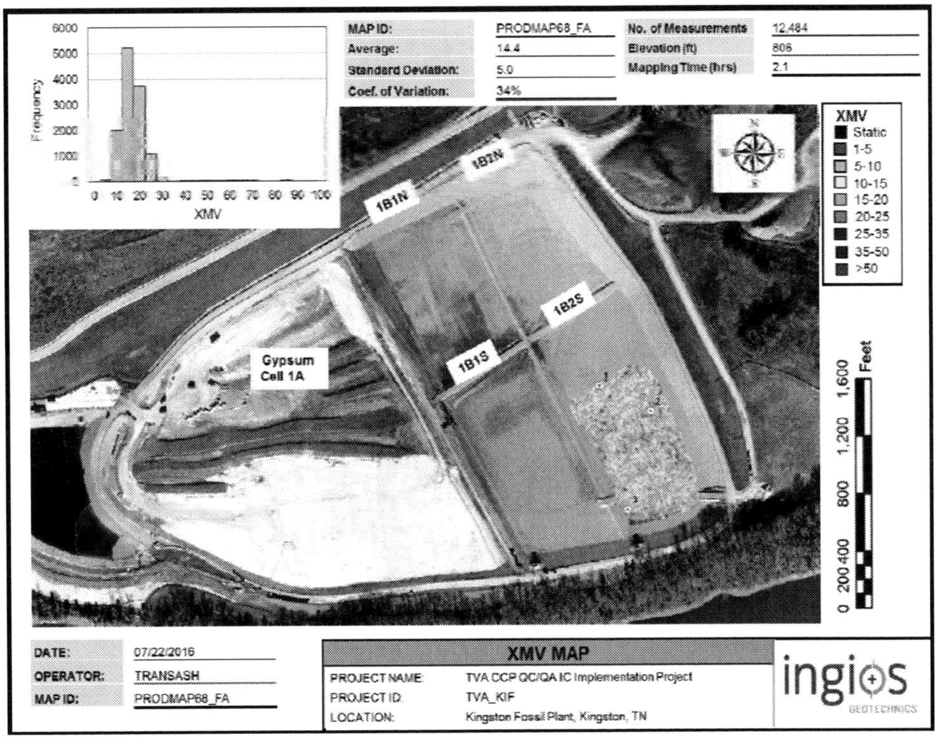

Figure 3. Example VICM compaction report Ballfield fly ash - July 22, 2016 (XMV = compaction index value)

CONSTRUCTION MONITORING AND FILL RECORD

During placement of approximately 500,000 cubic yards of CCPs in KIF cell 1B, periodic global navigation satellite system (GNSS) topographic surveys were collected to document fill height. The surveys were accomplished using a real time kinematic (RTK) global positioning system (GPS) rover (~ 2 cm accuracy) connected to TVA location base station and traversing the project site in a four-wheeler using nominal 10 ft spacing between lanes. The topographic survey data was used to generate a time series of surface renderings. Average elevation data from each survey was used to estimate the overall fill thickness in KIF cells 1B1S and 1B2S (see Figure 4). The topographic surveys typically took 10 to 30 minutes per cell. Having a spatial record of the surface elevation provides independent documentation of the amount of CCP material placed and compacted and the overall slope of the surface for each survey.

After completing each topographic survey, a webcam image was associated by time-stamp to the survey record. A dedicated webcam was programmed to collect time-lapse images every 2 hours for the duration of this project. Figure 5 shows selected time-lapse images of the landfill material and construction activities between March and August 2016. For future projects, additional options for generating topographic records are being considered including use the VICM machine elevation map based on RTK corrected GNSS coordinates.

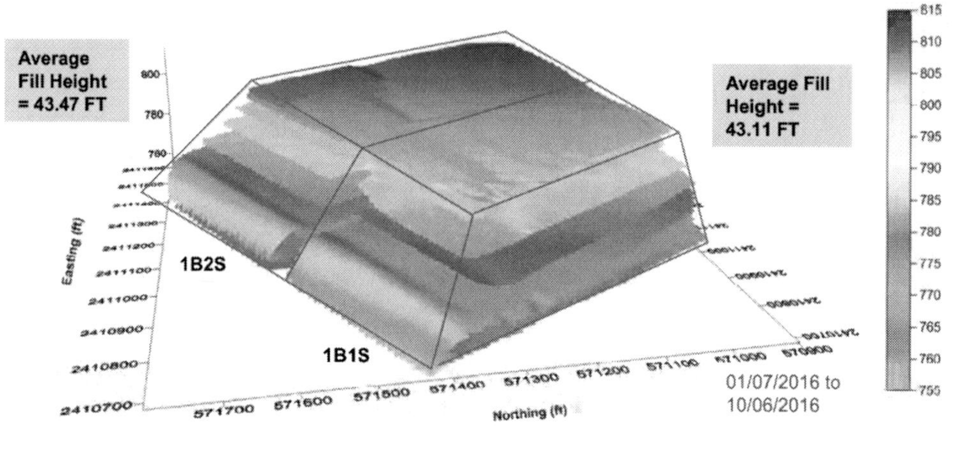

Figure 4. Topographic surface renderings for 300 days of landfilling at KIF involving placement and compaction of approximately 500,000 cubic yards of CCPs

Figure 5. Time-lapse of CCP placement in landfill (top) March 07, 2016 with 2 ft fill, (middle) June 06, 2016 with 20 ft fill, and (bottom) August 12, 2016 with 40 ft fill

VICM CALIBRATION

Developing site and material specific calibrations for VICM results verifies that the compaction conditions for CCPs are within compaction target limits and that the roller is outputting certified results. Establishing the calibration results from in situ testing is critical to implementation of VICM operations. At KIF, VICM mapping was initiated January 19, 2016 and continue to October 31, 2016. Calibration results establish the relationships between the VICM sensor measurements and independent in situ test measures.

Figure 6 shows the calibration test results for KIF Ballfield ash and gypsum in comparison to the elastic modulus measured with plate load test results (see Figure 7). Results show that the R^2 value is 0.9, indicating a quality fit to the data, and a standard error (SE) of about 14,000 psi. The statistical summary of this model demonstrates that the VICM results provide a statistically significant and high quality prediction of elastic modulus. Based on the calibration results, the VICM calibration reports are color coded (e.g., Figure 3) to the calibration test results (elastic modulus, shear strength, density, moisture content, shear wave velocity, and penetration resistance). By color coding the results, the roller operator can quickly determine if the compaction target limits have been met and if changes in the compaction process controls need to be changed.

In situ testing including determination of penetration resistance, compaction density, moisture content, and shear wave velocity are used to establish site and material specific calibration results to provide the roller operator with a menu of preselected mapping results to assess compaction quality.

Based on the experience at KIF, a standard operating procedure (SOP) was developed to establish calibration results and when re-calibration is needed (new material type or change in conditions that is not accounted for in the subsequent calibrations). With added calibration test results, the ability to estimate various engineering parameters values of the compacted materials improves because the data set becomes more robust. This is a significant advantage of VICM operations, where the calibration settings in the system are improved and updated with time. The current calibration program requires initial calibration every 3 months and after one year, calibration is added at least once every 4000 hrs or if the conditions produce results that are outside the existing calibrations.

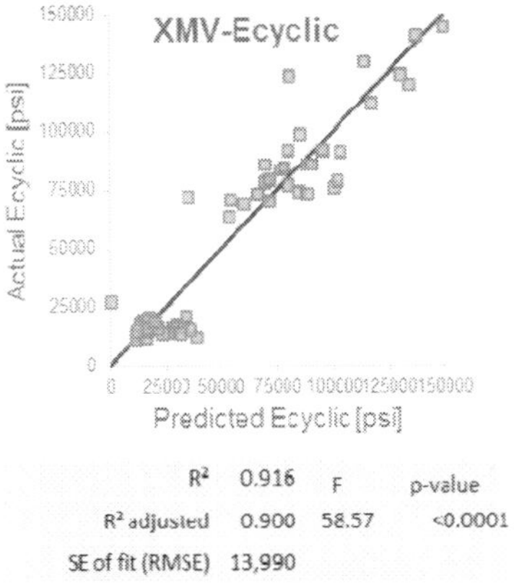

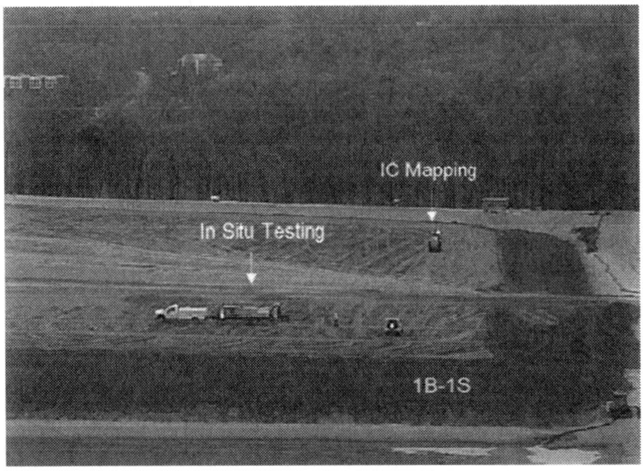

Figure 6. Example VICM calibration testing using independent testing methods (E_{cyclic} represents the elastic modulus determined from cyclic plate load tests)

IN SITU TESTING

In situ compaction testing at KIF was selected to provide key engineering parameter values to ensure that the CCPs were placed in a high-density condition and provided adequate strength with low potential for post-construction settlement. Test methods were selected that would provide measurements that were efficient to perform and provided independent measurements of important CCP landfill compaction conditions.

The test methods selected included well-established methods including the drive cylinder (DC) method to measure density and moisture content and the dynamic cone penetrometer (DCP) to determine penetration resistance profiles, along with advanced plate load testing (PLT) (see Figure 7) to determine the elastic modulus and deformation properties, the Borehole Shear test (BST) to directly measure effective stress shear strength parameter values, and the shallow Shear Wave test (SWT) to determine the shear wave velocity and shear modulus.

Standard Operating Procedures were developed for selected test methods for the KIF VICM program including interpretation of results, test frequency, and safe operating guidelines. Table 1 was developed as a comparison between the different test methods based on experience at this project.

Figure 7. Example VICM calibration testing using independent testing methods

Table 1. Example VICM calibration testing using independent testing methods

Test Device	Parameter Measured	Testing Time (Min.)	Testing Methods
Plate Load Test (PLT)	Elastic Modulus, E Permanent Settlement, δ_p	~30	ASTM D1195, ASTM D1196
Dynamic Cone Penetrometer (DCP)	California bearing Ratio (CBR), Bearing Capacity (q_u), and Lift Thickness	~8	ASTM D6951
Borehole Shear Test (BST)	Effective Stress Friction Angle (ϕ') and Cohesion (c')	~15	Handy Geotechnical Instruments Manual
Shear Wave Velocity (SWT)	Shear Wave Velocity (V_s) and Shear Modulus (G)	~2	Ingios Geotechnics Manual
Drive Cylinder (DC) Method	Density (γ_d) and Moisture Content (w %)	~8	ASTM D 2937

DATA SUMMARIES USING CONTROL CHARTS

All of the in situ test results were incorporated into control charts for monitoring engineering values with time as a means to quickly assess quality and identify if trends in the compaction quality required different quality control requirements. Figure 8 shows the in situ moisture content and dry unit weight for the duration of the project. A trend line (polynomial shown here) demonstrates that the moisture content was highest in the March to April timeframe and decreased until later summer. The dry unit weight trend was the opposite of the moisture content.

Figure 9 shows three key control charts: XMV values (key measurement values derived from VICM system), the in situ moisture content minus the standard Proctor optimum moisture content, and the percent relative compaction. For each of these plots target limits (dashed lines) show the position of the field test results compared to the desired targets. By keeping continuous records of these parameters, control measures can be implemented in the field to improve the field results. Additional control charts not shown here included the BST, DCP, and weather records.

DAILY OPERATIONS

The following bullet points outline primary testing operations for a typical day/week as part of the TVA intelligent compaction (IC) program. This work was completed collaboratively by field engineers/technicians. Standard operating procedures (SOPs) developed for each of the daily work items were implemented for safe execution of the tasks.

In brief, daily/weekly operations involved one or more of the following tasks:

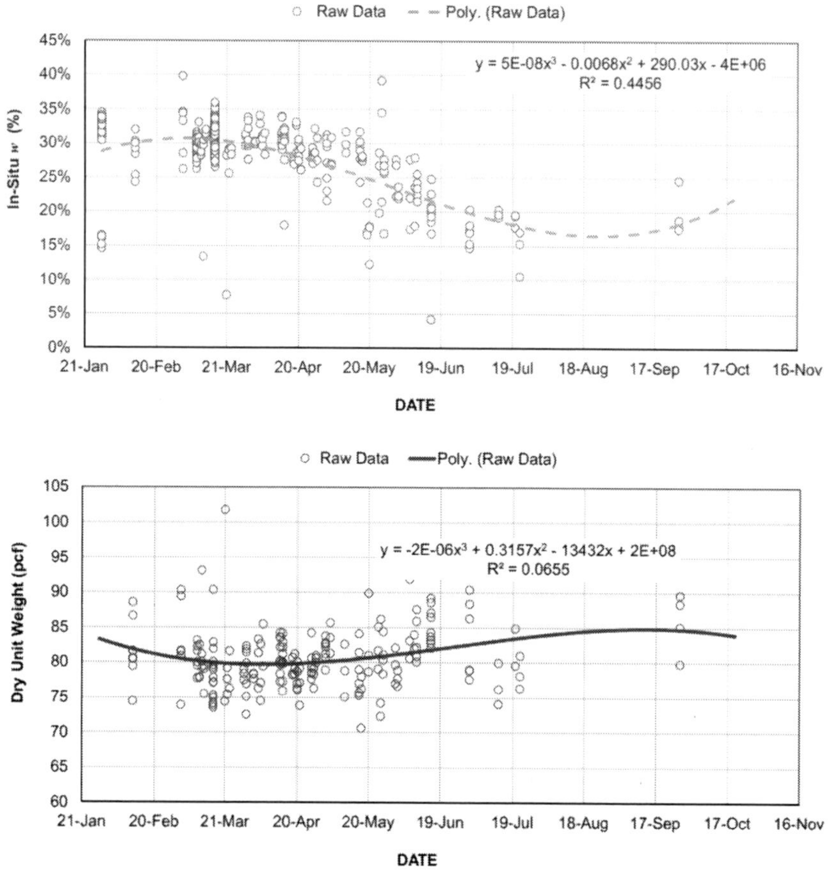

Figure 8. Example VICM calibration testing using independent testing methods

- At the beginning of each work day, conduct a daily safety briefing with the ash handling contractor with special focus on status of ash placement and field testing schedule and activities. Fill out 2-minute safety cards as needed for field testing activities. Complete any additional safety related exercises and complete corresponding documentation to be placed in safety binder.
- Place GPS units in the trucks (at the start of each work week). This activity involves placing a small magnetic mount GPS unit in the cab of the contractor's trucks prior to hauling operations. The GPS units are retrieved from the trucks at the end of the work week for data download.
- Collect moisture content samples from CCP materials placed in the stacking area(s) and measure the in-place compaction density using a drive core cylinder, in situ shear strength properties using the bore hole shear test, and penetration resistance profiles using the dynamic cone penetrometer. (# of daily test/samples as needed per project team guidance)

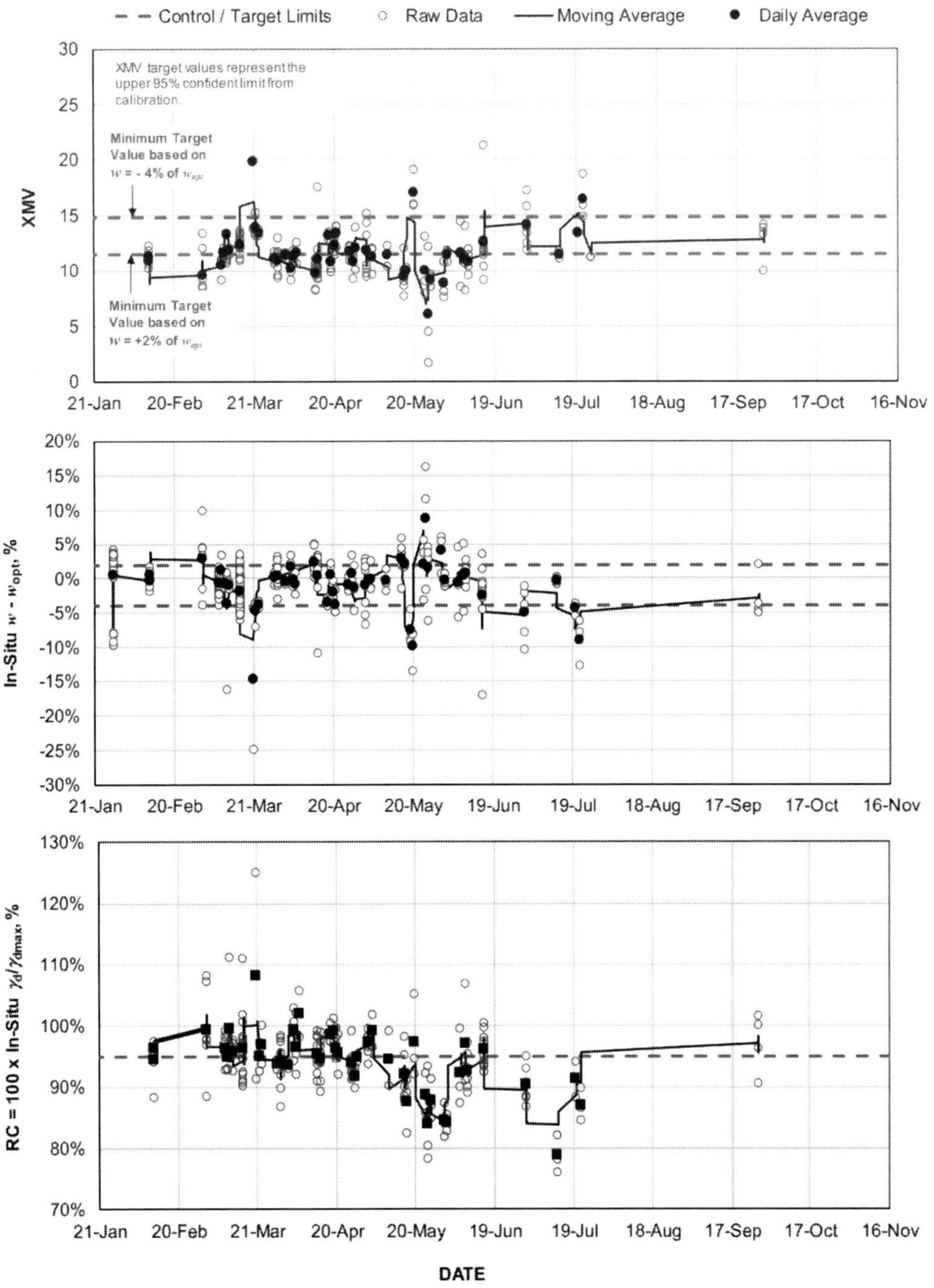

Figure 9. Example VICM calibration testing using independent testing methods

- Collect two to three 5-gallon buckets of CCP material for lab testing. (# of daily test/samples as needed per project team guidance)
- Determine the compacted lift thickness of the placed CCP layer periodically using hand-held rover GPS units (as needed per project team guidance).
- Perform the following tests in the on-site lab test trailer (temperature controlled self-contained lab trailer): moisture content, specific gravity, proctor moisture-density relationships, consolidation, and unconfined compressive strength.
- Monitor the contractor VICM compaction operations and review "maps" of VICM records produced daily/weekly. Periodically inspect the VICM system kits and verify calibrations.
- Collect photos of the construction operations and in situ testing operations. Monitor the work site remote cameras and periodically inspect to ensure on-going operation/functionality.
- On a quarterly basis, perform on-site calibration of the VICM compactor using in situ plate load testing and other test measurements as needed.

In addition to these daily/weekly tasks, at the beginning of the project we installed and setup the VICM equipment kits, remote cameras, and on-site lab trailer.

SUMMARY OF LESSON LEARNED

Following are the key lessons learned from implementing VICM on the KIF project:

- The use of VICM systems provides the operator with a greater understanding of real-time material behavior during placement. This real-time visualization (on-board computer screen) and feedback allows the operator to efficiently adjust the number of passes, focus on specific problem areas, add water for moisture conditioning, or other measures to enhance compaction.
- Communication with the VICM roller operator is imperative to understanding the complete picture of the compaction situation in the field. For instance, the operator often provides useful feedback about working conditions and field operations that help explain the results obtained during the IC mapping.
- Having a designated roller operator trained on IC operations adds to the quality of the results and benefits the overall project by having personal investment into quality compaction.
- Roller operator could quickly learn using the Ingios VICM system. With experience during the project, the operator was able to make independent decisions on areas that needed attention based on the color-coded onboard display.
- VICM mapping when used consistently provides much higher frequency data collection (100% spatial coverage) and significantly greater quality control measurements throughout the field placement. (IC data generates ~1,000,000 the measurement data compared to traditional spot test measurement therefore greatly reducing risk of building in defects to the ash stack).

- Moisture content remains a critical aspect regarding compaction consistency. The proactive work at the silo level being done at KIF appears to have significant impacts on compaction.
- VICM monitoring, combined with proper field calibration and verification testing (plate load testing, shear wave velocity testing, density/moisture, dynamic cone penetration, and shear strength testing), effectively detects changes in moisture contents that directly impact the degree of compaction being achieved.
- It was found that sometimes partial lifts did not receive complete compaction during night shift operations and sometimes due to miscommunication between night/day shift operators. These was readily identified via VIMC mapping. Additional compaction effort with heavy pneumatic rolling was required to improve the areas.
- Use of control charts proved effective in monitoring in situ VICM and point test results to detect changes in material conditions.
- VICM calibration results demonstrate high R^2 values greater than 0.9 for calibration testing, which leads the industry in this capacity.
- Compaction productivity can be improved using IC. Data collected during production compaction operations indicated compaction productivities at 35% to 50% due to "idle" times as the material was being placed and spread.
- Future implementation of VICM for CCP compaction operations will require improvements to the real-time data sharing, such as a web-based "dash-board", new protocols to qualify acceptance based on the calibrated IC results, and continued training for the ash handling contractors to improve their understanding and use of the IC maps to improve efficiency and quality of compaction.

ACKNOWLEDGEMENTS

The results of this paper were developed from the compaction monitoring program carried out at the TVA's Kingston Fossil (KIF) Facility in Kingston, Tennessee under Contract No. 9676.

REFERENCES

Christopher, B. R., White, D. J., Sanchez, R. (2013). "Performance based QC/QA specifications for TVA's CCP stacking facilities: Part I. Evaluations of performance requirements," 2013 World of Coal Ash (WOCA) Conference, April 22-25, Lexington, KY.

Christopher, B. R., White, D. J., Christopher, D. R., Sanchez, R., McClung, N. (2015). "Large scale box tests for calibration of fly ash and gypsum performance properties," 2015 World of Coal Ash (WOCA) Conference, May 5-7, Nashville, TN.

White, D. J., Vennapusa, P., Hageman, E., Christopher, B., McClung, N., Sanchez, R. (2015a). "Assessment of new QC/QA compaction monitoring program at TVA's coal combustion product stacking facilities: A case study," 2015 World of Coal Ash (WOCA) Conference, May 5-7, Nashville, TN.

White, D. J., Vennapusa, P., Hageman, E., Christopher, B., McClung, N., (2015b). "Effects of micromorphology and chemical composition on densification of CCPs," 2015 World of Coal Ash (WOCA) Conference, May 5-7, Nashville, TN.

White, D. J., Christopher, B. R., Sanchez, R. (2013). "Performance based QC/QA specifications for TVA's CCP stacking facilities: Part II. Proposed specifications," 2013 World of Coal Ash (WOCA) Conference, April 22-25, Lexington, KY.

White, D.J., Vennapusa, P., Gieselman, H., Miller, K., Harland, J. (2012a). "Quality Compaction Field Research: TVA Widows Creek Fossil Plant," Final Report Submitted to TVA, Center for Earthworks Engineering Research, Iowa State University, Ames, IA.

White, D.J., Vennapusa, P., Gieselman, H., Miller, K., Harland, J. (2012b). "Quality Compaction Field Research: TVA Cumberland Creek Fossil Plant," Final Report Submitted to TVA, Center for Earthworks Engineering Research, Iowa State University, Ames, IA.

TO AIR IS HUMAN; TO SUCCEED, DIVINE - EVOLUTIONARY RESEARCH ON GROUT ENRICHED RCC AIR ENTRAINMENT

Jeremy Young[1]
Eric Musselman[2]

Early experience on roller compacted concrete (RCC) dams in the 1980s showed a tendency for seepage to develop along lift lines. Designers responded with upstream facing systems as a watertight barrier, adding cost and complexity to RCC dams. One economical facing material, first tested at Yantan cofferdam in 1987, is Grout Enriched RCC (GERCC). Use of GERCC has been limited in the U.S., primarily due to concern over its low freeze-thaw resistance. There has been limited success integrating air entraining admixtures into GERCC to improve freeze-thaw resistance.

Schnabel Engineering provided a grant, supplied specialty RCC sampling and testing equipment, and organized a team of three RCC experts from Schnabel who collaborated with Villanova University to evaluate grout formulations combining air entraining and supportive admixtures, and supportive construction techniques to improve freeze-thaw resistance. The primary goal was to advance GERCC air entrainment to be technically viable and cost effective for RCC construction in freeze-thaw climates. The research team performed initial studies between 2013 and 2015, focusing on laboratory evaluation of grout formulations to stabilize the GERCC air void system. A field trial was conducted using air entrained GERCC mixes and prototype placement techniques. Results of the initial study, presented in several papers throughout 2016, indicated that producing a homogeneous mixture of air entrained GERCC is feasible with commercially available admixtures and construction techniques. Additional research was recommended to achieve thorough mixing of the RCC and air entrained grout.

In the Fall of 2016, the research team will initiate Phase 2 of the research program to improve vibration energy, mechanically mix grout and RCC, and improve RCC-grout penetration. Phase 2 research may trigger modification to the Phase 1 GERCC mix design. This paper will provide an update on status of the next phase of research and present preliminary findings and recommendations.

[1] Schnabel Engineering, jyoung@schnabel-eng.com
[2] Villanova University

STABILITY ANALYSIS OF A CONCRETE-FACED ROCKFILL DAM USING PARAMETRIC EVALUATIONS

Christopher Conkle[1]

The authors were retained by the owner of a rockfill embankment dam as part of an ongoing Dam Safety and Risk Assessment Program (DSRAP) to better understand the risks associated this dam. The dam, an approximately 500 foot long concrete-faced rockfill embankment dam, as part of a larger complex hydroelectrc system was built the early 1900's.

Due to limited information about materials within the dam, static stability analysis included an in-depth desk study to establish typical, lower, and uppper bound rockfill static and dynamic material proerties based on literature data, anecdotal information from the Dam operators, historical information, and photographic evidence. A parametric static stability evaluation was performed to identify critical slip surfaces and material properties. Results of the parametric static stability analysis were used to evaluate stability in required scenarios and identify critical slip surfaces for seismic stability analysis.

Seismic stability analysis involved selecting a suite of near-field and far-field strong ground motions scaled to a site-specifc conditional mean spectrum (CMS). The selected ground motions were then applied to a QUAD4MU model to estimate accelerations throughtout the dam cross section. The QUAD4MU model material properties were parametrically varied to identify the critical set of material properties. The appropriate Dam response time-history was selected based on the identified cross section in the static stability analysis and used to evaluate potential crest displacement with a Newmark sliding block approach.

The authors provided typical, lower, and upper bound estimates of expected seismic displacement of the Dam crest based on combinations of material properties and ground motions identified in the parametric analysis. The analysis allowed the dam's owner to identify which unknown material properties had the largest effect on displacements and therefore warranted additional investigation.

[1] Geosyntec Consultants, cconkle@geosyntec.com

ESTIMATING THE PEAK FRICTION ANGLE OF SANDY SOILS IN SITU WITH STATE-BASED OVERBURDEN NORMALIZED SPT BLOW COUNTS

Robert A. Jaeger, PE, PhD[1]
Ian P. Maki, PE[2]

ABSTRACT

Relationships for estimating the peak friction angle of sandy soils from standard penetration test (SPT) blow count often involve correcting the blow count to a reference stress of one atmosphere with an overburden normalization (or correction) factor. Existing relationships between SPT blow count and peak friction angle have typically been derived from databases of calibration chamber tests, in situ tests, and accompanying laboratory tests. Published in situ datasets of SPT blow count and peak friction angle are not well-constrained due to limitations with respect to sand characteristics and stress range. Development of relationships between SPT blow count and peak friction angle based on these datasets thus often requires extrapolation. This paper presents a design relationship between state-based overburden-normalized SPT blow count and peak friction angle, which is a companion to a similar relationship developed for the cone penetration test. A state-based overburden normalization approach is adopted because it provides both a rational basis for interpreting the peak friction angle in sandy soils from blow count, and a critical state-based method for extrapolating these relationships to conditions outside the range currently covered by available peak friction angle data. The design relationship between state-based overburden-normalized blow count and peak friction angle is developed based on published results from in situ and laboratory tests on undisturbed sandy soil samples, calibration chamber standard penetration tests, and a semi-empirical relationship between SPT blow count and peak friction angle.

BACKGROUND

Relationships between standard penetration test (SPT) blow count (N) and sand properties have been developed over the years based largely on empirical studies (e.g., Terzaghi and Peck 1948, Skempton 1986, Kulhawy and Mayne 1990, Hatanaka and Uchida 1996, Mayne 2006). These studies have shown the SPT blow count in sand primarily depends on the sand's strength (typically characterized in terms of peak friction angle, ϕ'_{peak}), stiffness, and dilatancy, which in turn depend on density, stresses, fundamental soil characteristics (i.e., mineralogy, grain size distribution, angularity), and environmental factors (i.e., aging, cementation).

Been and Jefferies (1985) and Bolton (1986) both demonstrated a single parameter could account for the combined influence of confining stress and void ratio on the strength and dilatancy of sand, albeit different expressions were used in each publication. The single parameter used by Been and Jefferies (1985) was the initial state parameter (ξ_0), which is

[1] Senior Engineer, GEI Consultants, Rancho Cordova, California, CA 95670, rjaeger@geiconsultants.com.
[2] Design Engineer, Department of Water Resources Division of Safety of Dams, California, CA 95818, Ian.Maki@water.ca.gov.

the difference between the initial void ratio (e_0) and the void ratio on the critical state line (e_{cs}) at the same initial mean effective stress, p'_0 (Figure 1). Initial conditions falling below the critical state line (i.e., negative ξ_0 values) are dense-of-critical and exhibit generally dilative behaviors when sheared, whereas initial conditions falling above the critical state line (i.e., positive ξ_0 values) are loose-of-critical and exhibit contractive behaviors when sheared. Boulanger (2003) later proposed an initial relative state parameter index (ξ_{R0}), defined as ξ_0 divided by the difference between the maximum void ratio (e_{max}) and the minimum void ratio (e_{min}). Thus ξ_{R0}, like ξ_0, would also be expected to correlate directly to the peak friction angle and dilatancy of sands.

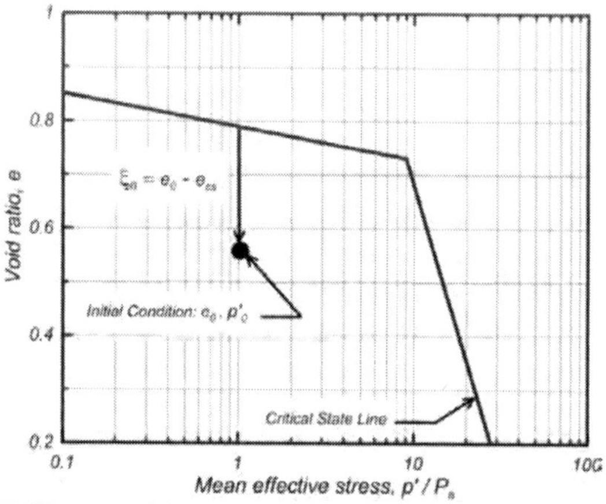

Figure 1. Example Bilinear Critical State Line of Sand and Definition of Initial State Parameter

Boulanger (2003) derived an expression for the ξ_{R0} parameter by applying the state parameter approach from Been and Jefferies (1985) to a critical state line defined from the Bolton (1986) stress-strength-dilatancy framework. The expression derived by Boulanger (2003) is

$$\xi_{RO} = \frac{\mathrm{R}}{Q - \ln\left(100 \frac{(1+2K_0)\sigma'_{vo}}{3}\frac{}{Pa}\right)} - D_{RO} \tag{1}$$

where Q and R are empirical factors found to be about 10 and 1, respectively, for quartz and feldspar sands (Bolton 1986), σ'_{vo} is the initial effective vertical overburden, K_0 is the ratio of the horizontal to the vertical effective stresses, P_a is the atmospheric pressure (taken as 100 kPa), and D_{R0} is the initial relative density.

This paper presents a relationship between ϕ'_{peak} and a state parameter-based (or state-based) overburden normalized and energy corrected SPT blow count. First, background on a framework for state-normalization of SPT blow count is presented. Second, two commonly used semi-empirical relationships between ϕ'_{peak}, stress, and density are then reviewed. Third, these relationships are then used to develop a state-based relationship between SPT blow count, stress, and ϕ'_{peak} based on (1) a database of experimental in situ SPT data with laboratory testing on undisturbed sand samples (Hatanaka and Uchida

1996, Mayne 2006) and (2) calibration chamber SPT for three clean sands (Bieganousky and Marcuson 1976, 1977).

FRAMEWORK

Relationships between SPT Blow Count, Soil Density, and Stress

The variation of SPT blow count (N) with σ'_{v0} can be isolated from the variation of N with density by projecting (or normalizing) N to a standard reference stress while all other factors are held constant. In this paper, SPT blow counts are also corrected to an energy-ratio of 60% and are henceforth referred to as N_{60}. The current study adopts the normalizations developed by Boulanger (2003), Boulanger and Idriss (2004), and Idriss and Boulanger (2008) to either ξ_{R0} or D_{R0}. The stress normalization approach in terms of ξ_{R0} can be expressed as:

$$(N_{1\xi})_{60} = N_{60} C_{N\xi} \tag{2}$$

$$C_{N\xi} = \left(\frac{P_a}{\sigma'_{v0}}\right)^{m_\xi} \tag{3}$$

where $(N_{1\xi})_{60}$ is the N_{60} that a soil at a given ξ_{R0} would have at a reference stress of $\sigma'_{v0} = P_a$, $C_{N\xi}$ is the mapping function that accounts for the effects of stress on N_{60} for the same soil at the same ξ_{R0}, and m_ξ is a fitting exponent. The choice of reference stress is conceptually arbitrary, but $\sigma'_{v0} = P_a$ is typically in the stress range of practical interest and is commonly adopted for interpretation of in situ test results.

The adopted state-based stress normalization (Boulanger 2003) uses stress normalizations in terms of D_{R0} or e_0 as an intermediate step. In terms of D_{R0} or e_0, the stress normalization of N_{60} can be expressed as:

$$(N_{1e})_{60} = N_{60} C_{Ne} \tag{4}$$

$$C_{Ne} = \left(\frac{P_a}{\sigma'_{v0}}\right)^{m_e} \tag{5}$$

where $(N_{1e})_{60}$ is the N_{60} that a soil at a given D_{R0} or e_0 would have at a reference stress of $\sigma'_{v0} = P_a$, C_{Ne} is the mapping function that accounts for the effects of stress on $(N_{1e})_{60}$ for the same soil at the same D_{R0} or e_0, and m_e is a fitting exponent. In much of the literature on foundation and liquefaction engineering (i.e. Liao and Whitman 1986, Youd et al. 2001, Idriss and Boulanger 2008), $(N_{1e})_{60}$ is commonly written as $(N_1)_{60}$ and C_{Ne} is commonly written as C_N; the addition of the subscript e is intended to identify the underlying assumption of stress normalization at a respective e_0 (or equivalently D_{R0}).

Boulanger (2003) developed the following regression for the fitting exponent m_e based on numerical analyses of cone penetration testing by Salgado et al. (1997):

$$m_e = 0.784 - 0.521 D_{R0} \tag{6}$$

Boulanger (2003), and later Idriss and Boulanger (2008), adopted the following regression between D_{R0} and $(N_{1e})_{60}$:

$$D_{RO} = \sqrt{\frac{(N_{1e})_{60}}{C_d}} \tag{7}$$

where C_d is a fitting parameter taken as 46 by Idriss and Boulanger (2008), but has been shown to vary between about 38 and 91 for clean sands (Boulanger 2003).

Boulanger and Idriss (2004) developed an expression for the mapping function $C_{N\xi}$ based on the Bolton (1986) stress-strength-dilatancy relationship. The Boulanger and Idriss (2004) $C_{N\xi}$ factor can be written as:

$$C_{N\xi} = C_{\xi} C_{Ne} \tag{8}$$

where,

$$C_{\xi} = \left(\frac{D_{RO} - \Delta D_{R,CS}}{D_{RO}}\right)^2 \tag{9}$$

$$\Delta D_{R,cs} = \frac{1}{Q - ln\left(100 \frac{1+2K_o}{3} \frac{\sigma'_{vo}}{P_a}\right)} - \frac{1}{Q - ln\left(100 \frac{1+2K_o}{3}\right)} \tag{10}$$

where K_0 is the ratio of horizontal to vertical initial effective stresses. As described by Boulanger (2003), the parameter $\Delta D_{R,cs}$ defines the difference in relative density on the critical state line from the in situ stress to the reference stress (one atmosphere).

Examples of the relationships between N_{60}, σ'_{v0}, and either ξ_{R0} or D_{R0} using the form in Equations 2-10 as proposed by Boulanger (2003) and Boulanger and Idriss (2004) are shown in Figures 2a and 2b. The relationships in Figure 2a show N_{60} versus σ'_{v0} for a sand at seven different values of ξ_{R0} (-0.6, -0.5, -0.4, -0.3, -0.2, -0.1, and 0.0) with $C_d = 46$, $Q = 10$, $R = 1$, and $K_0 = 0.5$. These N_{60} versus σ'_{v0} graphs are nearly linear in a log-log scale and illustrate how the stress exponent m_ξ (i.e., the slope of the lines) decreases with increasing density (i.e., more negative ξ_{R0}). The corresponding dependence of $C_{N\xi}$ on ξ_{R0} is shown in Figure 3 for D_{R0} of 35%, 55%, and 75% at σ'_{v0} of 1 atmosphere.

The relationships in Figure 2b show N_{60} versus σ'_{v0} for the same properties used in Figure 2a, but for six different D_{R0} values: 10%, 20%, 40%, 60%, 80%, and 100%. The slopes of these N_{60} versus σ'_{v0} graphs, in a log-log scale, correspond to the stress exponent m_e. Similar to the trends for constant ξ_{R0}, the stress exponent m_e decreases with increasing D_{R0} (i.e., steeper lines in Figure 2b). The corresponding dependence of C_{Ne} on D_{R0} is shown in Figure 3 for D_{R0} of 35%, 55%, and 75%.

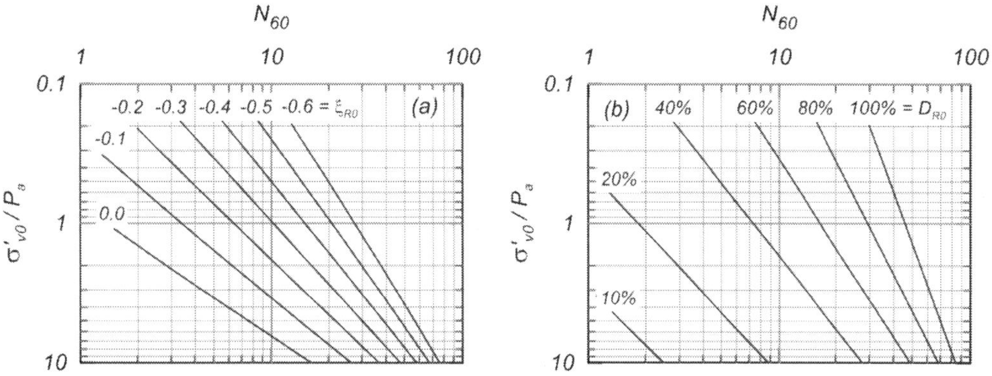

Figure 2. Relationship between N_{60}, Initial Vertical Effective Stress (σ'_{v0}), and (a) Initial Relative State Parameter (ξ_{R0}), and (b) Initial Relative Density (D_{R0})

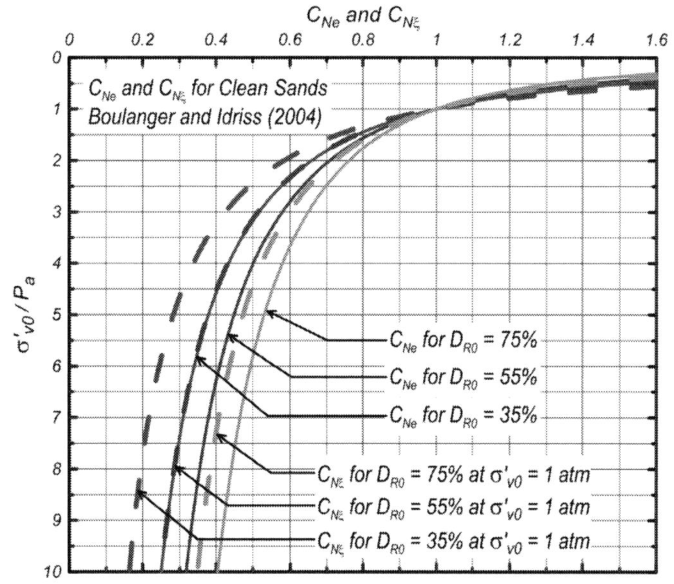

Figure 3. C_{Ne} and $C_{N\xi}$ versus Initial Vertical Effective Stress (σ'_{v0})

Relationships between Peak Friction Angle, Density, and Stress in Sands

Several frameworks have been proposed for estimating the drained shear strength of coarse grained soils (e.g., Been and Jefferies 1985, Bolton 1986, Collins et al. 1992). Generally these relationships compute the peak friction angle (ϕ'_{peak}) as a function of the critical state friction angle (ϕ'_{cs}), e_0 (or D_{R0}), p'_0, and material-specific parameters. As previously discussed, p'_0 and e_0 can also be described in a single term, ξ_0 or ξ_{R0}. In terms of ϕ'_{cs} and ξ_0, Collins et al. (1992) suggested the following fit to the data presented by Been and Jefferies (1985):

$$\phi'_{peak} = \phi'_{cs} + A\left(e^{-\xi_0} - 1\right) \geq \phi'_{cs} \quad (11)$$

where A is a soil type parameter in the range of 0.6 to 0.95 (in radians). Figure 4a illustrates the relationship between peak friction angle and σ'_{v0} as a function of ξ_0 for $K_0 = 0.5$, $A = 0.8$, and $\phi'_{cs} = 32$ degrees, which is representative of the sands evaluated by Collins et al. (1992).

Similar to Been and Jefferies (1985) and Collins et al. (1992), Bolton (1986) identified the peak friction angle of sands is dependent on both void ratio and stress. The relationship presented by Bolton (1986) for triaxial compression is given in terms of critical state friction angle, mean effective stress at failure, and D_{R0}:

$$\phi'_{peak} = \phi'_{cs} + 3\left[D_{R0}\left(Q - \ln(p'_f)\right) - R\right] \geq \phi'_{cs} \tag{12}$$

where Q and R are empirical factors found to be about 10 and 1, respectively, for quartz and feldspar sands (Bolton 1986), and p'_f is the mean effective stress at failure in kPa (herein assumed to be twice the initial vertical effective stress, per Kulhawy and Mayne 1990). Bolton (1987) later recommended limiting the use of Equation 12 to p'_f greater than 150 kPa when laboratory shear strength data at lower p'_f values for a material of interest is unavailable. Based on the recommendation by Bolton (1987), Equation 12 can be rewritten as:

$$\phi'_{peak} = \phi'_{cs} + 3\left[D_{R0}\left(Q - \ln\left(\max(p'_f, 150\ \text{kPa})\right)\right) - R\right] \geq \phi'_{cs} \tag{13}$$

Figure 4b illustrates the Bolton (1986, 1987) relationship (Equation 13) between ϕ'_{peak}, σ'_{v0}, and D_{R0} for Q = 10, R = 1, and $\phi'_{cs} = 32$ degrees. Note that for a constant D_{R0} (or equivalently a constant $(N_{1e})_{60}$), ϕ'_{peak} is generally not constant and decreases with increasing stress. Conversely for a constant ξ_{R0} (or equivalently a constant $(N_{1\xi})_{60}$), ϕ'_{peak} is constant and does not change as stress changes.

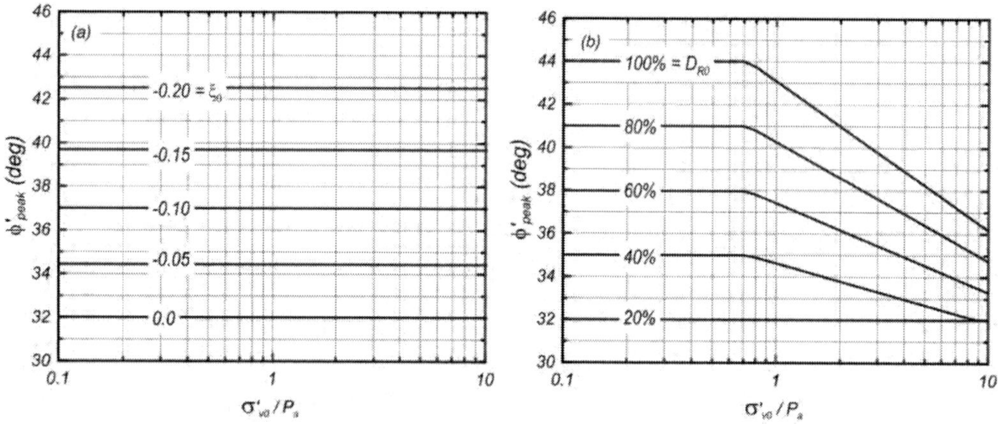

Figure 4. Variation of Peak Friction Angle (ϕ'_{peak}) with Initial Vertical Effective Stress (σ'_{v0}) based on the Relationships by (a) Collins et al. (1992) after Been and Jefferies (1985), and (b) Bolton (1986, 1987)

DATA SOURCES

The relationship between SPT blow count and peak friction angle is examined with data from two sources: (1) databases of experimental in situ SPT data and laboratory tests on undisturbed sand samples, and (2) calibration chamber SPT data. The experimental in situ SPT data and laboratory tests on undisturbed sand samples were taken from databases compiled by Hatanaka and Uchida (1996) and Mayne (2006). The calibration chamber SPT data was taken from Bieganousky and Marcuson (1976, 1977).

In Situ Tests

The databases of experimental in situ SPT data and laboratory tests on undisturbed sand samples compiled by Hatanaka and Uchida (1996) and Mayne (2006) were used for the present study. The Hatanaka and Uchida (1996) database included 12 sands from six sites in Japan, where eleven sands were natural deposits and one sand was from a fill (Table 1). All samples were collected using an in situ frozen sampling technique, as described in Hatanaka and Uchida (1996). The sands had σ'_{v0} values between 0.4 and 1.4 atm, with an average of about 0.8 atm. Five of the 12 sands were clean sands (less than about 5% fines), four sands had between about 5 and 12% fines, and three samples had fines content of about 12% or higher.

Table 1. Characteristics of Undisturbed Sand Samples (adapted from Hatanaka and Uchida 1996)

ID	Sand	Description	Fines Content	D_{R0}	σ'_{v0} (atm)	N_{60}	ϕ'
Clean Sands (5 of 12 Sands)							
1	IK1	Fill	1.3-2.4%	34%	0.68	12	36.0°
2	IK2	Holocene	2.1-4.6%	57%	0.97	22	38.2°
3	KY2	Volcanic	1.8-2.9%	59%	0.68	7	31.0°
4	KY3	Volcanic	1.4-2.3%	59%	0.78	8	35.0°
5	NA	Pleistocene	4.1-5.7%	81%	1.26	13	32.7°
Sand with ≥ 5% Fines (7 of 12 Sands)							
6	KG	Volcanic	3.6-9.6%	70%	0.49	14	40.4°
7	NG1	Pleistocene	4.3-9.4%	74%	0.39	23	43.4°
8	NG2	Pleistocene	7.0-10.4%	81%	0.68	20	37.8°
9	KY1	Volcanic	3.6-11.0%	72%	0.58	7	39.0°
10	NG5	Pleistocene	13.0-16.2%	79%	1.36	13	35.1°
11	KA1	Holocene Volcanic	13.5-15.5%	81%	0.78	6	28.0°
12	KA2	Holocene Volcanic	9.1-30.4%	78%	0.87	5	30.0°

Mayne's database included tests on 15 sands from China, Japan, Canada, Norway, and Italy (Table 2). As shown in Table 2, 10 sands were natural deposits, 3 sands were hydraulic fills, and 2 were tailings. Fourteen of the 15 sands were obtained with frozen sampling techniques; the remaining sample was obtained with a Mazier tube. Fourteen of the 15 in situ tests had σ'_{v0} values between 0.42 atm and 1.8 atm, with the remaining test at σ'_{v0} = 5.16 atm for Mildred Lake sand (10% fines). The average σ'_{v0} value was about 1.3 atm. Eleven of the 15 sands are clean sands (less than 5% fines) and the other four sands have fines contents between 8 and 15%. The particle shapes varied from subrounded and subangular to angular (Mayne 2006).

Table 2. Characteristics of Undisturbed Sand Samples (adapted from Mayne 2006)

ID	Sand (Location)	Description	Sampling Method	Fines Content	D_{R0}	σ'_{v0} (atm)	N_{60}	ϕ'
Clean Sands (11 of 15 Sands)								
13	Kowloon (China)	Hydraulic Fill	Mazier Tube	1%	46.5%	1.80	28	38.1°
14	Yodo (Japan)	Natural Alluvial	Frozen	1.9%	60.2%	1.02	27	42.4°
15	Yodo (Japan)	Natural Alluvial	Frozen	0.3%	51.9%	1.23	39	38.4°
16	Yodo (Japan)	Natural Alluvial	Frozen	2.1%	37%	1.43	37	39.1°
17	Natori (Japan)	Natural Alluvial	Frozen	0.2%	77.2%	0.87	47	40.9°
18	Tone (Japan)	Natural Alluvial	Frozen	3.8%	69.1%	0.84	29	41.7°
19	Edo (Japan)	Natural Alluvial	Frozen	0.4%	44.3%	0.51	16	39.7°
20	Massey (Canada)	Natural Alluvial	Frozen	<5%	32.5%	1.2	11	36.6° [a]
21	Kidd (Canada)	Natural Alluvial	Frozen	<5%	29.8%	1.6	16	37.1° [a]
22	Holmen (Norway)	Natural Alluvial	Tube/frozen	2%	30.5%	1.1	1	33.2°
23	Gioia Tauro (Italy)	Natural Coarse Sand	Frozen	0.7%	42%	0.42	19	41.5°
Sand with ≥ 5% Fines (4 of 15 Sands)								
24	LL Dam (Canada)	Tailings	Frozen	8%	40.3%	1.0	5	39.4° [a]
25	Mildred Lake (Canada)	Hydraulic Fill	Frozen	10%	43.6%	5.16	41	39.0° [a]
26	Highmont Dam (Canada)	Tailings	Frozen	10%	37.4%	1.38	6	41.5°
27	J-pit (Canada)	Hydraulic Fill	Frozen	15%	42.7%	0.55	2	34.7° [a]

[a]Value reported by Mayne (2006) revised based on original reference.

Jaeger and Maki (2016) reevaluated the laboratory test data for several sands in the Mayne (2006) database and revised peak friction angles for the Massey, Kidd, LL Dam, Mildred Lake, and J-pit sands. As discussed by Jaeger and Maki (2016), the revised values were selected based on review of the available laboratory data for the sands and the reported friction angles in the original references (Wride and Robertson 1997a,b,c). The peak friction angles for the Massey, Kidd, and Mildred Lake sands decreased by 0.1°, 0.2°, and 0.6°, respectively, relative to values listed in Mayne (2006). The peak friction angles of the LL Dam and J-pit sands were increased by 0.3° and 2.0°, respectively, relative to values listed in Mayne (2006). The revised peak friction angles were adopted for the present study.

The combined datasets from Hatanaka and Uchida (1996) and Mayne (2006) includes 27 sands. All but one of the sands had σ'_{v0} values between 0.4 and 2.0 atm; the remaining sand had a σ'_{v0} of about 5.2 atm. Thus, the combined dataset is relatively constrained for initial vertical effective stresses between 0.5 to 2.0 atmospheres, but is not well-constrained for initial vertical effective stresses greater than about two atmospheres.

Calibration Chamber Tests

Calibration chamber SPT data for Reid Bedford, Platte River, and Standard Concrete sands was obtained from Bieganousky and Marcuson (1976, 1977). The calibration chamber tests were performed under initial vertical effective stresses of 0.7, 2.7, and 5.4 atm by use of hydraulic rams applied at the top of the sample. Reid Bedford sand was tested at overconsolidation ratios (OCRs) of 1 and 3, whereas the other sands were only tested for OCR of 1. The calibration chamber had a diameter of 4 feet and a height of 6 feet. The walls of the chambers were stacked steel rings with rubber spacings contained within the rings in order to reduce soil arching in the chamber.

It is likely that the calibration chamber SPT data is influenced by the artificial boundary conditions imposed by the chamber. Calibration chamber corrections (CCCs) have been proposed to convert calibration chamber test cone penetration test (CPT) tip resistance values to free-field conditions (e.g. Houlsby and Yu 1990, Schnaid and Houlsby 1991, Parkin and Lunne 1982, Been et al. 1986, Salgado et al. 1998), but similar corrections have not been developed for the SPT. Based on the proposed CCCs for CPT data with a rigid lateral boundary, rigid bottom boundary, and constant stress imposed on the top boundary, the SPT blow counts measured by Bieganousky and Marcuson (1976, 1977) are expected to be largely influenced by the boundaries. The influence of the chamber is expected to increase as the relative state becomes smaller (i.e., p'_0 decreases or D_{R0} increases). However, these effects could not be quantified.

Relevant soil properties of the evaluated calibration chamber sands are provided in Table 3 below. The three sands are similar in that they are clean, uniformly graded sands with median grain sizes ranging from 0.37 to 0.53 mm, coefficients of uniformity ranging from 1.5 to 2.0, minimum void ratios of 0.49 to 0.60, and maximum void ratios of 0.79 to

0.91. Critical state parameters were available for Reid Bedford sand (Jefferies and Been 2006), but not the other two sands.

Table 3. Relevant Properties of Calibration Chamber Test Sands

Property	Reid Bedford	Platte River	Standard Concrete
Median grain size, D_{50}[a]: mm	0.25	2.0	0.5
Maximum void ratio, e_{max}[a]	0.86	0.61	0.6
Minimum void ratio, e_{min}[a]	0.54	0.35	0.38
Critical state friction angle, ϕ'_{cs}[b]	32°	N/A	N/A
Coefficient of Uniformity[a]	1.6	5.3	2.1
Angularity[a]	Subangular to Subrounded	Subrounded	Subrounded to Well Rounded

[a]Data from Bieganousky and Marcuson (1976, 1977)
[b]Data from Jefferies and Been (2006)

RESULTS

Relationship between $(N_{1\xi})_{60}$ and Initial State Parameter

The relationship between $(N_{1\xi})_{60}$ and ξ_{R0} is shown in Figure 5 for the in situ and calibration chamber data. The $(N_{1\xi})_{60}$ values were calculated using Equations 1-10. The ξ_{R0} values for the six Canadian sands in Table 1 were calculated as the initial state parameters reported by Wride et al. (2000) divided by (e_{max} - e_{min}) and are shown as blue points in Figure 5 below. For the remaining in situ data (red points) and all calibration chamber data (green points), the ξ_{R0} values were estimated using Equation 1 for a Q of 10, R of 1, and K_0 of 0.5.

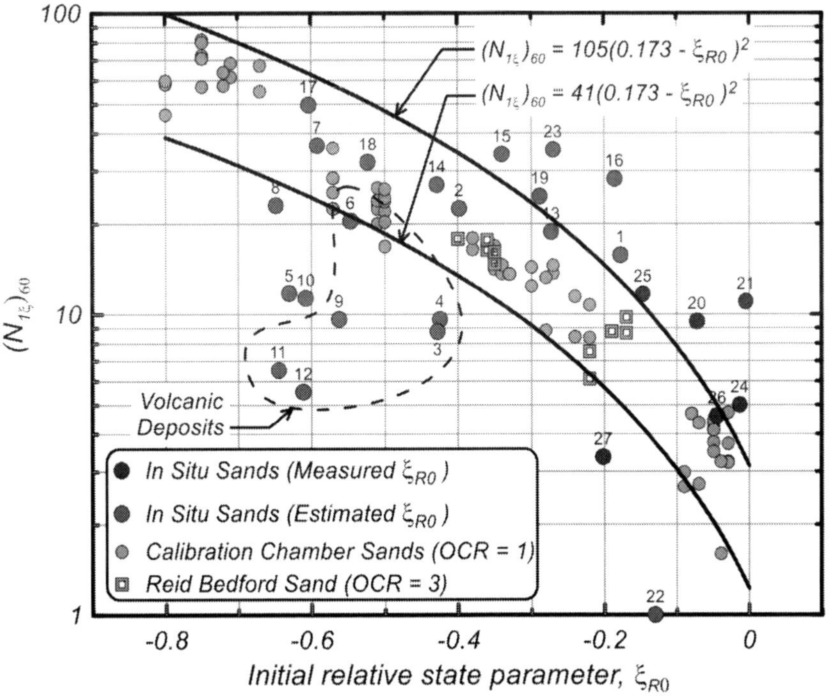

Figure 5. Relationships between $(N_{1\xi})_{60}$ and Initial Relative State Parameter (ξ_{R0}) for In Situ and Calibration Chamber Data

As shown in Figure 5, the $(N_{1\xi})_{60}$ values generally increased as the ξ_{R0} values decreased (i.e., as the soils became more dense-of-critical), as expected. The calibration chamber data plotted as a band approximately bounded by the two trendlines shown in the figure above. The trendlines correspond to C_d values of 41 and 105 input into Equations 1 and 7 at a vertical effective stress of 1 atmosphere, where $(N_{1\xi})_{60}$ is equal to $(N_{1e})_{60}$. The lower-bound C_d value of 41 is consistent with the recommendation by Meyerhof (1957) for clean sands and is slightly below the C_d value of 46 recommended by Idriss and Boulanger (2008). The upper bound C_d value of 105 is above the expected range of 38 to 91 for clean sands, as previously published in the literature (i.e., Cubrinovski and Ishihara 1999, Boulanger 2003).

The in situ data generally followed a trend similar to the calibration chamber data, but with a number of points plotting either significantly above or below. The differences between the in situ and calibration chamber results is likely due to a combination of factors, including calibration chamber boundary effects and differences in the mineralogy, gradation, and particle characteristics of the sands.

For ξ_{R0} values greater than -0.35, the in situ data generally plotted above the calibration chamber data, but with two exceptions: J-pit sand (point 27) and Holmen sand (point 22). J-pit sand is a hydraulically filled tailings material with a relatively high fines content (15% fines). The J-pit sand point is believed to plot lower than the others because of its measured low shear stiffness (or high compressibility) relative to the other sands in the

Mayne (2006) dataset (shear stiffness was not available for the Hatanaka and Uchida data points). Holmen sand is a very loose, clean sand that was sampled using thin-walled tube samplers (Mayne 2006), which may have densified the material. This potential densification of the material during sampling would have led to a lower calculated ξ_{R0} than exists in situ, thus contributing to the apparent bias in Figure 5.

At ξ_{R0} values less than -0.35, half (seven) of the in situ data points were consistent with the calibration chamber data, but the other half of the in situ points had $(N_{1\xi})_{60}$ values about 5 to 20 less than the calibration chamber data at the same ξ_{R0}. Five of the seven points below the calibration chamber data were from volcanic deposits, which suggests differences in the mineralogy, gradation, and particle characteristics of the sands. However, this could not be confirmed from the source publications. Additionally, all but two of the volcanic sands (points 3 and 4) had fines contents as high as 10-30%, which would be expected to reduce the $(N_{1\xi})_{60}$ value compared to cleaner sands because sands with fines typically have higher compressibilities than clean sands.

Relationship between $(N_{1\xi})_{60}$ and Peak Friction Angle

The relationship between $(N_{1\xi})_{60}$ and ϕ'_{peak} was plotted for the in situ data and the calibration chamber data for Reid Bedford sand at OCRs of 1 and 3, as presented in Bieganousky and Marcuson (1976). The ϕ'_{peak} values for the in situ sands were measured from laboratory tests on undisturbed samples. The ϕ'_{peak} values for Reid Bedford sand were estimated from the Bolton (1986, 1987) equation (Equation 13) based on the relative density and confining stress of each test combined with the reported ϕ'_{cs} of 32°, and typical Bolton (1986) Q and R values of 10 and 1, respectively. For comparison, two semi-empirical lines representing the combination of the Idriss and Boulanger (2008) approach for estimating N_{60} and the Bolton (1986) approach for estimating ϕ'_{peak} were also added using $C_d = 46$, $Q = 10$, and $R = 1$ for ϕ'_{cs} of 28° and 35°.

For both the calibration chamber and in situ data, the ϕ'_{peak} values tended to increase as $(N_{1\xi})_{60}$ values increased, as expected. The calibration chamber data and most of the in situ data plotted within the semi-empirical trendlines for the combined Idriss and Boulanger (2008) and Bolton (1986) approach. The in situ sand data points generally plotted within the upper portion of the calibration chamber data or above, with relatively few exceptions. As previously discussed, differences between the in situ and calibration chamber data can be attributed to differences in the mineralogy, gradation, particle characteristics of the sands, and the lack of calibration chamber corrections of the calibration chamber data.

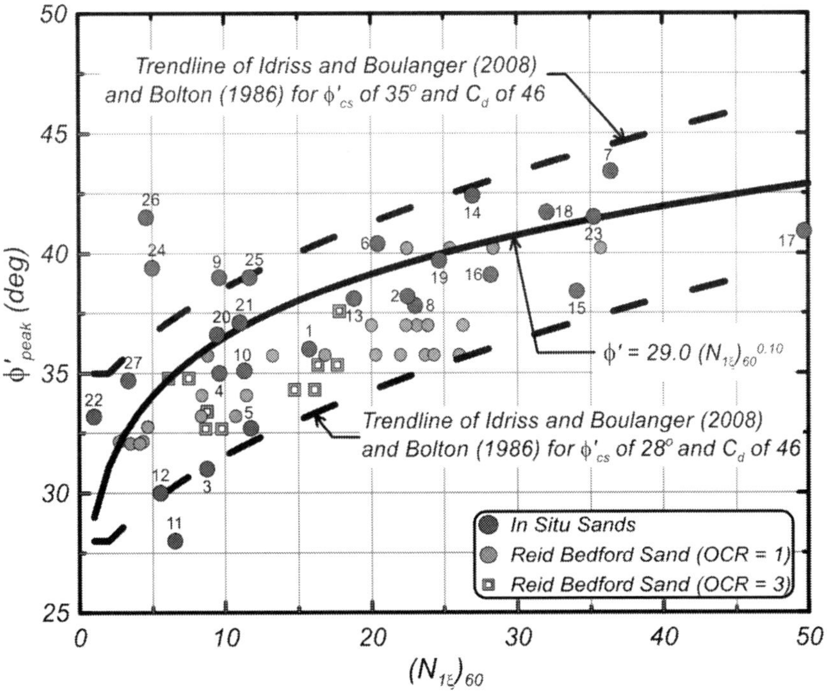

Figure 6. Relationship between Peak Friction Angle (ϕ'_{peak}) and $(N_{1\xi})_{60}$

Two of the in situ sands plotted significantly above the general trend of the calibration chamber and in situ data: LL Dam (point 24) and Highmont Dam (point 26). These two sands were hydraulically placed fills with angular particles and appreciable amounts of mica, feldspar, and illite (Wride and Robertson 1997c). The angular particles would likely lead to increased friction angles, whereas the elevated amounts of mica, feldspar, and illite would lead to reduced $(N_{1\xi})_{60}$ values as compared to the other less compressible sands evaluated in this study (i.e., Mayne 2014).

A representative fit was developed based on the approximate median of the in situ data, but with consideration of the calibration chamber data and a semi-empirical trend combining the Idriss and Boulanger (2008) approach for estimating N_{60} with $C_d = 46$ with the Bolton (1986) approach for estimating ϕ'_{peak} with Q = 10, R = 1, and $\phi'_{cs} = 33°$. The selected fit is defined by the following regression:

$$\phi'_{peak} = 29.0\left(N_{1\xi}\right)_{60}^{0.10} \tag{14}$$

CONCLUSIONS

A relationship between peak friction angle and state-based overburden-normalized SPT blow count was developed based on in situ test results, calibration chamber standard penetration test results, and a semi-empirical model of standard penetration test blow count and peak friction angle. A state-based overburden normalization approach was adopted because it provides both a rational basis for interpreting the peak friction angle in sandy soils from N_{60}, and a critical state-based method for extrapolating these relationships to conditions outside the range currently covered by available peak friction angle data. The results indicated state-based overburden-normalized SPT tip resistance is strongly correlated to to the initial state parameter (or relative state parameter) and peak friction angle, as expected based on critical state soil mechanics and published trends from laboratory testing of sands.

ACKNOWLEDGEMENTS

The authors are grateful for the data made available by Hatanaka and Uchida (1996) and Mayne (2006). The opinions expressed in this paper are solely those of the authors, and do not necessarily represent the views of GEI Consultants, Inc. or the California Department of Water Resources, Division of Safety of Dams.

REFERENCES

Been, K., and Jefferies, M. G. (1985). "A state parameter for sands." *Géotechnique,* 35(2): 99–112.

Been, K., Crooks, J. H. A., Becker, D. E., and Jefferies, M. G. (1986). "The cone penetration test in sands: Part I, state parameter interpretation." *Géotechnique, 36*(2): 239-249.

Bieganousky, W. A., and Marcuson, W. F. (1976). "Laboratory standard penetration tests on Reid Bedford Model and Ottawa Sands." US Army Engineer Waterways Experiment Station Soils and Pavement Laboratory Research Report 8-76-2.

Bieganousky, W. A., and Marcuson, W. F. (1977). "Laboratory standard penetration tests on Platte River Sand and Standard Concrete Sand." US Army Engineer Waterways Experiment Station Soils and Pavement Laboratory Research Report 8-76-2.

Bolton, M. D. (1986). "The strength and dilatancy of sands." *Géotechnique*, 36(1): 65-78.

Bolton, M. D. (1987). "Discussion on The strength and dilatancy of sands." *Géotechnique*, 37(2): 219-226.

Boulanger, R. W. (2003). "High overburden stress effects in liquefaction analyses." *J. Geotechnical and Geoenvironmental Eng.,* ASCE 129(12): 1071–082.

Boulanger, R. W. and Idriss, I. M. (2004). "State normalization of penetration resistance and the effect of overburden stress on liquefaction resistance." In *Proc. 11th SDEE and 3rd ICEGE*, Berkeley, CA.: 484-491.

Collins, I. F., Pender, M. J., and Yan, W. (1992). "Cavity expansion in sands under drained loading conditions." *Int. J. Numer. Anal. Meth. Geomech.*, 16: 3–23.

Cubrinovski, M., and Ishihara, K. (1999). Empirical correlation between SPT N-value and relative density for sandy soils. Soils and Foundations, 39(5): 61-71.

Hatanaka, M. and Uchida, A. (1996). "Empirical correlation between cone resistance and internal friction angle of sandy soils", *Soils and Foundations*, Vol. 36(4): 1-10.

Houlsby, G. T. and Yu, H. S. (1990). "Finite element analysis of the cone pressuremeter test." In *Pressuremeters*, London: Thomas Telford, pp. 221-229.

Idriss, I. M. and Boulanger, R. W. (2008). "Soil liquefaction during earthquakes." Monograph MNO-12, Earthquake Engineering Research Institute, Oakland, CA, 261 pp.

Jefferies, M. G. and Been, K. (2006). "Soil Liquefaction – A Critical State Approach", Taylor and Francis Group.

Jaeger, R. A. and Maki, I. P. (2016). "Estimating the Peak Friction Angle of Sandy Soils In Situ with State-Based Overburden Normalized CPT Tip Resistance." In *Proc. 36th Annual USSD Conference: Celebrating the Value of Dams and Levees - Yesterday, Today and Tomorrow*, pp. 865-882.

Kulhawy, F. H. and Mayne, P. W. (1990). "Manual on estimating soil properties for foundation design." Report EL-6800. Electric Power Research Institute, Palo Alto, CA.

Liao, S. C. and Whitman, R. V. (1986). "Overburden correction factors for SPT in sand." *J. Geotech. Engrg.*, 112(3): 373–377.

Mayne, P. W. (2006). The Second James K. Mitchell Lecture Undisturbed sand strength from seismic cone tests. *Geomechanics and Geoengineering: An International Journal, 1*(4): 239-257.

Meyerhof, G. G. (1957). Discussion on research on determining the density of sands by spoon penetration testing. In *Proc. 4th Int. Conf. on SMFE*, pp. 110.

Parkin, A. and Lunne, T. (1982). "Boundary effects in the laboratory calibration of a cone penetrometer in sand." *Proc. 2nd European Symposium on Penetration Testing*, Amsterdam, Vol. 2: 761–768.

Salgado, R., Boulanger, R. W., and Mitchell, J. K. (1997). "Lateral stress effects on CPT liquefaction resistance correlations." *J. Geotech. Geoenviron. Eng.*, 123(8): 726–735.

Salgado, R., Mitchell, J. K., and Jamiolkowski, M. (1998). "Calibration chamber size effects on penetration resistance in sand." *J. Geotech. Geoenviron. Eng.*, 124(9): 878–888.

Schnaid, F. and Houlsby, G. T. (1991). "An assessment of chamber size effects in the calibration of in-situ tests in sand." *Géotechnique,* 41(3): 437-445.

Skempton, A. W. (1986). "Standard penetration test procedures and the effects in sands of overburden pressure, relative density, particle size, ageing and overconsolidation." *Géotechnique,* 36(3): 425-447.

Terzaghi, K. & Peck, R.B. 1948. Soil Mechanics in Engineering Practice, 1st Edition, John Wiley and Sons, New York

Wride, C. E. and Robertson, P. K. (1997a). "CANLEX: Phases I and III data review report (Mildred Lake and J-Pit Sites)."

Wride, C. E. and Robertson, P. K. (1997b). "CANLEX: Phase II data review report (Massey and Kidd Sites, Fraser River Delta)."

Wride, C. E. and Robertson, P. K. (1997c). "CANLEX: Phase IV data review report (LL Dam and Highmont Dam Sites, Highland Valley Copper Mine)."

Wride, C.E., Robertson, P.K., Biggar, K.W., Campanella, R.G., Hofmann, B.A., Hughes, J.M.O., Kupper, A., Woeller, D.J. (2000). "Interpretation of in situ test results from the CANLEX sites." *Canadian Geotechnical Journal,* **37**: 505-529.

Youd, T. L., Idriss, I. M., Andrus, R. D., Arango, I., Castro, G., Christian, J. T., Dobry, R., Finn, W. D. L., Harder, L. F., Hynes, M. E., Ishihara, K., Koester, J. P., Liao, S. S. C., Marcuson, W. F., Martin, G. R., Mitchell, J. K., Moriwaki, Y., Power, M. S., Robertson, P. K., Seed, R. B., and Stokoe, K. H. (2001). "Liquefaction resistance of soils: summary report from the 1996 NCEER and 1998 NCEER/NSF workshops on evaluation of liquefaction resistance of soils." *J. Geotechnical and Geoenvironmental Eng.*, 127(10): 817–33.

COMPARISON ANALYSIS OF THE BEHAVIOR OF ROCK-FILL DAMS WITH CLAY CORE AT THE VARIATION OF WATER LEVEL IN THE RESERVOIR

Ljupcho Petkovski, PhD[1]
Stevcho Mitovski, PhD[2]

ABSTRACT

By application of advanced numerical methods, the behavior of embankment dams is successfully analyzed, almost for all typical states of static loading. The occurrences, not fully clarified during the static analysis of some types of embankment dams, are caused by the water effect. Such occurrences include: (a) hydraulic fracturing phenomenon of the coherent material (earth dams and earth-rock dams) during full reservoir with established steady seepage and (b) appearance of softening and weakening (or the latest term collapse settlement) of the rock material at the upstream dam shell in case of rock-fill dams with core, during the rapid filling of the reservoir. The topic of the research, whose results are included in the paper, is clarification of the stress-strain state for rock-fill dams with central waterproof element at variation of the water level in the reservoir. This issue, stated in the dam engineering in 70's of the last century, was analyzed in several occasions, but the general conclusion is that the effect of water saturation of the rock material is not fully explained. In this paper are presented results from the comparison analysis of the variation of the water level of the reservoir, formed by construction of Kozjak Dam, on the river Treska, a tributary of the river Vardar. It is a rock-fill dam with slightly inclined clay core, with structural height of 130.0 m, the highest dam in the Republic of Macedonia. The first filling of the reservoir took place in 2003-2004 and the dam behavior was monitored by surveying methods and by installed instruments in the dam body.

INTRODUCTION

By applying contemporary numerical methods, based on Finite Element Method (FEM), the mathematical models for successful research of the behavior of fill dams are developed, almost for any typical state of static loading. The only occurrences that are yet to be fully clarified at static analysis of fill dams are those caused by the water effect. These include the possible occurrence of hydraulic fracturing phenomenon in coherent materials (within earth and earth rock dams) at the stage of full reservoir through established steady seepage flow, and the possible softening and weakening (or collapse settlement) of the rock material in the upstream dam shell in case of rockfill dams with core during the stage of reservoir rapid filling.

[1] Professor, University "Sts Cyril and Methodius", Civil Engineering Faculty, Blvd. Partizanski odredi 24, 1000, Skopje, Republic of Macedonia, petkovski@gf.ukim.edu.mk
[2] Assistant Professor, University "Sts Cyril and Methodius", Civil Engineering Faculty, Blvd. Partizanski odredi 24, 1000, Skopje, Republic of Macedonia, s.mitovski@gf.ukim.edu.mk

The objective of this research, is to clarify the stress-deformation state of rockfill dams with central impermeable element during variation of the water level in the reservoir. This issue, raised in dam engineering since the 1970s, was investigated on several occasions, but all attempts ended by stating that the modeling was not properly done and according to the latest knowledge in this field. The general conclusion is that the saturation effect of the rock material is still present. The aim of this paper is to contribute to the selection of most favorable numerical models for simulation of the response of embankment dams with shells of rock material and core of coherent material under the effect of variation of the water level in the reservoir. To accomplish the specified goal, a comparison of few models has been done, differing by the constitutional laws for the stress-strain dependence. The contribution of this research refers to the recommendations of more advanced numerical model for study of the stress-deformations state in case of rockfill dams with core during the stage of rapid filling of the reservoir.

THE WATER EFFECT IN CASE OF STATIC RESPONSE OF EARTH ROCK DAMS WITH CORE

The research on water effects on earth-rock dams subjected to static loading was limited until the 1970s, due to the application of the classical methods, based on Theory of elasticity and Theory of plasticity. Therefore, these methods were gradually replaced by advanced methods – Finite element method (FEM), developed in the last three decades of the twentieth century, thanks to the pioneering work in this field by Zienkiewicz and Clough (ICOLD 1978). The improvement of the calculation technique (hardware and software) in the beginning of the 21th century has enabled the advanced numerical methods, based on the FEM, to fully replace the classical methods in the engineering practice. The focus of the current research is to clarify the response of rockfill dams with earth core under the effect of variation of water level in the reservoir. The variation of water level causes various effects on the stress-strain state of the fill dams. In the following paragraphs, a short overview of water effect for typical loading states of the structure during construction and in service period is given.

During rockfill dam construction, water effects are relevant for the zones of coherent material, and can be modelled using FEM. The coherent materials are applied in layers with optimal humidity, meaning that the pores are fully filled with water. By increasing the load (through placing of the upper layers) at the first moment, the full load is accepted by the pore water and pore pressure occurs. During time, a pore pressure dissipation occurs by leaching of the water. Such slow hydrodynamic process of pore pressure release, followed by an increase of the effective stresses and material settlement is called consolidation. The FE simulation of the development (increase and dissipation) of the pore pressure is performed according to two concepts: analysis of total stresses and analysis of effective stresses. According to the first (simplified) method, a consolidation pore pressure, caused by the change of the total stresses, can be generated by application of total stresses or in non-drained conditions in case of low permeable coherent materials. The stress-deformation state, as well as the development (generation and dissipation) of the pore pressure, are most realistically determined by analysis of the effective stresses, meaning that the coherent material is treated in drained conditions. In the response of the structure according to the second approach, three components are included: (1) mechanical and elastic properties, (2) hydraulic properties (seepage coefficient and volume content of the water) and (3) time factor, or dynamic of dam construction.

For the state of reservoir filling, the water effect is most simply manifested in case of rockfill dams with facing made of artificial material. This is a case where the water is outside of the dam body and acts as external pressure. Far more complex is the response of the rockfill dams with diaphragm and earth rock dams with core of coherent material, due to the water effect. At stage of rapid filling of the reservoir, the water rapidly fills the pores of the rock material in the upstream dam shell and causes the following effects: (1) softening of the submerged non-coherent material, causing additional settlements, (2) alleviation of the permeable material (increase of the total stresses and pore pressure and decrease of the effective stresses) and (3) action of force due to the hydrostatic pressure along the upstream face of the core/diaphragm.

Similar water effects, but with opposite action, occur in the state of rapid drawdown of the water level in the reservoir. The initial state for the stage of rapid drawdown of the reservoir is the state of steady seepage, for which pore pressure generated seepage in the earth material. Lowering of the water level in the reservoir causes change of the pore pressure in the earth material. It should be noted that the lowering of the reservoir water level during a period of few days to few weeks is rapid, compared to the time consuming hydrodynamic process of slow leaching of the water from the saturated low permeable earth material. Similar to the stage after dam construction (before first filling of the reservoir), at the stage of rapid drawdown of the reservoir excess pore pressure occurs, thus initiating consolidation process (manifested by decrease of the pore pressure, raise of the effective stresses and settlements). The difference in the consolidation processes for both stages consists in the factor that causes the excess pore pressure. During construction, the excess pore pressure is caused by the loading of the upper layers, while at the stage of reservoir rapid drawdown, the variation of the boundary hydraulic conditions generates the excess pore pressure.

For the state of full reservoir, with established steady seepage through the earth material, the stresses distribution depends on the following factors: dam composition and geometry, material parameters and dam type. In case of fill dams, there is regular occurrence of stress transfer, thus enabling unloading of some elements and overloading of others. If in a particular zone, the value of the pore pressure exceeds the value of the total normal stresses, a zone with negative effective pressure occurs. It can lead to hydraulic fracturing, manifested by appearance of fissures in the coherent material. Most common reason for occurrence of hydraulic fracturing is the non-uniform settlement of zones by materials of different stiffness properties, where the softer material "hangs" on the stiffer material and thus transferring part of its stresses. The next potential reason is distinct unevenness and different inclinations in the foundation of the coherent material, causing uneven settlements and zone of shattered and unloaded material. According to Penman researches from 1975, the occurrence of the hydraulic fracturing at earth materials is possible if the pore pressure in within the interval of maximal and minimal total normal stress. According to other authors, for occurrence of the fissuring of coherent material, beside the negative effective normal pressure, an appropriate non-homogeneity of the material is required, thus referring to the finding that such phenomena is not fully clarified yet.

From the above specified short overview of the water effect at the response of dams for typical loading states in static conditions, it can be concluded that for research of the fill dams behavior (during construction and in service period) standard numerical models are distinguished. Such models are successfully applied in case of various dam types and almost for all loading states [Petkovski L., Tančev L., Mitovski S., 2007]. For most of the dams (fill and concrete) for the reservoir filling state are obtained displacements in downstream direction, which in fact is intuitively expected by the researchers. It was one of the reasons why in the past were adopted minor curving of the dam crest in layout, placing the convex face towards the reservoir, even in the case of fill dams, even though they are gravitational structures. Such solutions for earth-rock dams with central or slightly inclined core in cross section and with curved shape in layout with convex face towards the reservoir were very popular in the middle of the 20th century. Most likely, the designer's prediction was that the downstream horizontal displacement, under influence of the reservoir filling would cause additional compaction of the local material that would provide improved safety of the fill dams.

By more precise measurements of the horizontal displacements in the interior part of the dam, for the state of reservoir first filling, in case of rockfill dams with core/diaphragm is registered unusual occurrence in the cross section axis. Such occurrence, registered by inclinometer, was noted by Marshal and Ramirez in 1967, and referred on the first filling of the reservoir at El-Infiernillo Dam, Mexico [ICOLD, 1986, Bulletin 53]. Such occurrence is manifested by upstream horizontal displacement of the dam during first filling of the reservoir up to 50% of the height. The further raising of the water level in the reservoir causes downstream displacement in the lower part up to 30% of the dam height, while the crest is still displaced upstream. By reaching the normal water level in the reservoir, the dam axis is displaced in downstream direction, with maximal intensity at 50% of the dam height. Such unusual bidirectional horizontal displacement at the crest in dependence of the water level in the reservoir results from the complex water effect in the upstream shell of the rockfill dams with core. The first attempt for analysis of the displacement phenomena at rockfill dams with core during first filling was by Nobari and Duncan in 1972, where following components were superposed: water load, softening and weakening of the fill material. The softening of the local material by the saturation process is confirmed with triaxial testing for dams Oroville (USA) and Beliche, (Portugal), and in the numerical multistage experiment should be modelled by variation of the stiffness and strength parameters.

The further research (modelling and measuring) of the non-usual bidirectional horizontal displacements of rockfill dams with core, for the stage of first filling, systemized in the publications of the most authoritative institution in dam engineering worldwide [ICOLD, 1988, Bulletin 94], point out that such phenomena is not yet clarified in full. Such issue, raised in the dam engineering in the 70-ties of the 20th century was analyzed in several occasions [ICOLD, 1994, Numerical analysis of dams, Volume III] and it was stated that it is not modelled properly [ICOLD, 2001, Bulletin 122]. In the latest publications, the terms "softening and weakening" are most often replaced by "collapse settlement" [ICOLD, 2013, Bulletin 155], but the effect of the saturation of the rock material is still present.

ANALYSIS OF THE STATE OF RAPID FILLING OF EARTH-ROCK DAMS WITH CORE

The state of stress in the dam body for stage of first filling of the reservoir can be analyzed by numerical models, differing by: (a) constitutive law for the dependence stress-strain and (b) the boundary condition on where to apply the hydrostatic pressure – the upstream slope of the dam or upstream face of the waterproof element. If the boundary condition is upstream slope of the dam, a far more realistic picture is obtained for the distribution of the maximal normal stresses (total and effective), that is the base for all further structural (static and dynamic) analyses of the fill dam. Therefore, the purpose of the current research is to contribute at the choice of most favorable numerical model for simulation of the stress-deformation state in case of earth-rock dams with shells of rock material and central waterproof element, under influence of the reservoir filling. In order to achieve the specified aim a comparison of the specified particular models is done.

The comparison of the different numerical models for research of the stress state in case of fill dams during the stage of reservoir first filling is illustrated by the results from static analysis of dam Kozjak, on river Treska, right tributary of the river Vardar, Republic of Macedonia (fig. 1, 2 and 3). It is a rockfill dam with slightly inclined clay core, with structural height of 130 m, the highest dam in Macedonia. The dam was constructed in 2000, the first filling of the reservoir took place in 2003-2004 and the dam behavior was monitored by surveying methods and by installed instruments in the dam body.

Figure 1. Map of the Republic of Macedonia.

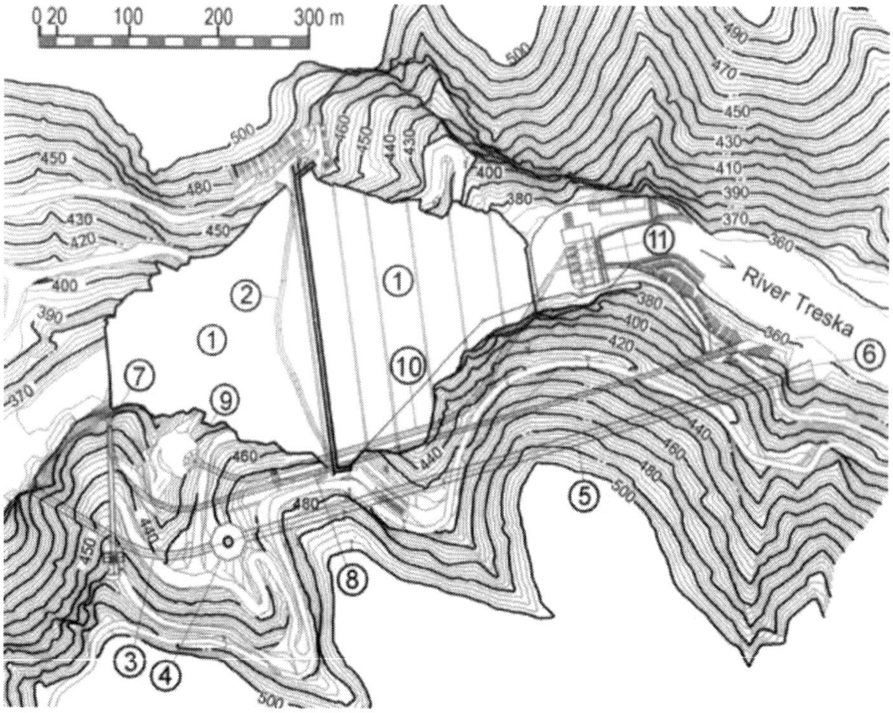

Figure 2. Layout of the hydraulic scheme with Kozjak Dam. (1) Dam body; (2) grouting gallery; (3) diversion tunnel; (4) morning glory (shaft) spillway; (5) spillway tunnel; (6) flip bucket; (7) intake structure of the bottom outlet; (8) bottom outlet tunnel; (9) intake structure of the power plant; (10) water supply tunnel for the power plant; (11) tail raise channel.

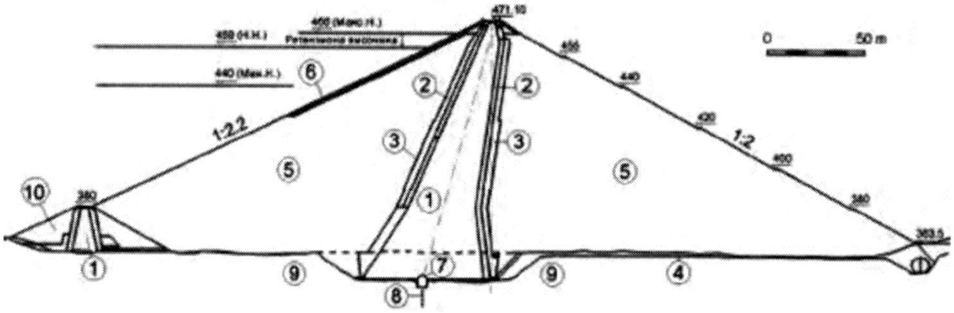

Figure 3. Main cross-section of Kozjak Dam, section no. 17. (1) Clay core; (2) first transition zone; (3) second transition zone; (4) river deposit; (5) rockfill shells (limestone); (6) arranged slope protective stones; (7) grouting gallery; (8) grout curtain; (9) rock foundation (limestone); (10) gravel in the upstream cofferdam.

NUMERICAL MODEL CALIBRATION

For calibration of the numerical model for analysis of the state of first filling are used data from technical monitoring of the dam for construction state and consolidation process of the dam. The respective strength parameters are adopted by using numerous data from control laboratorial testing of the placed local materials. The elastic parameters, by variable elasticity modulus with change of the effective stresses, are set by meeting the criteria on minimization of the difference between the measured and simulated values. The following key measured values are registered within the dam monitoring, directly after dam construction: maximal settlement of 1.3 m, maximal pore pressure at dam core foundation (400-700) kPa and crest settlement, caused by consolidation, shortly before the first filling, of 0.1 m. The typical cross section no. 17, representative for plane static analysis, is discretized with grid of 947 finite elements, connected in 947 nodes (fig. 4).

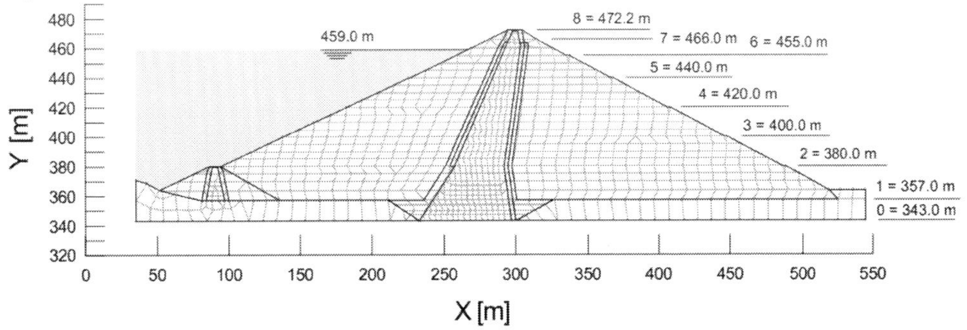

Figure 4. Numerical model for cross section no. 17, discretized by finite elements

For obtaining of the initial state before reservoir filling, a model with effective stresses is applied, by coupling mechanical and hydraulic response of the structure in real domain. The state after dam construction is simulated in 28 loading increments for period of 920 days (fig. 5). The consolidation state is simulated in one loading increment, by 14 calculation exponential steps, for period of 1,043 days. From the distribution of: settlements after construction (fig. 6), pore pressure after construction (fig. 7) and incremental consolidation settlements (fig. 8), it can be concluded that the model calibration is properly done.

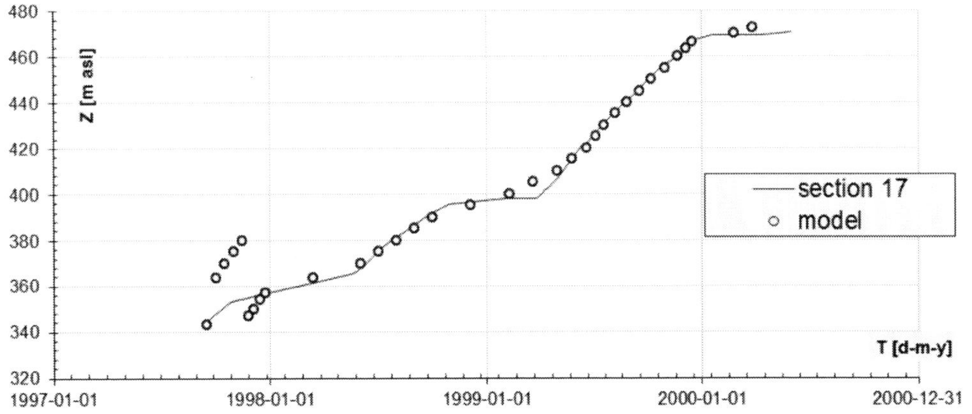

Figure 5. Dam construction simulation.

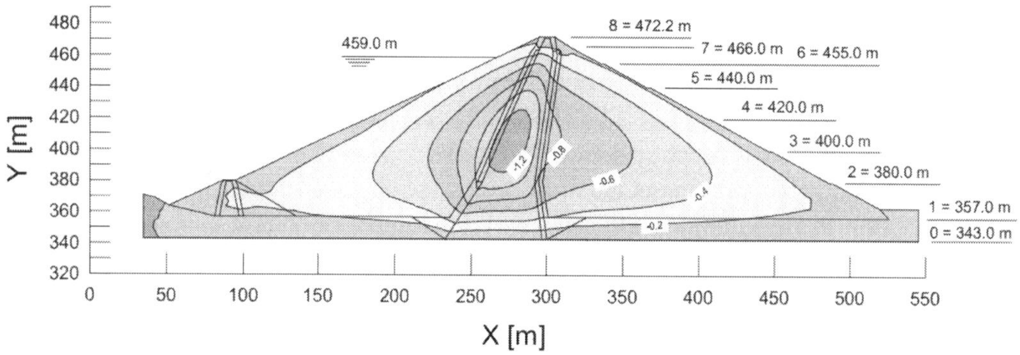

Figure 6. State after dam construction, Ydisplacement (-1.338) - (+0.012) m

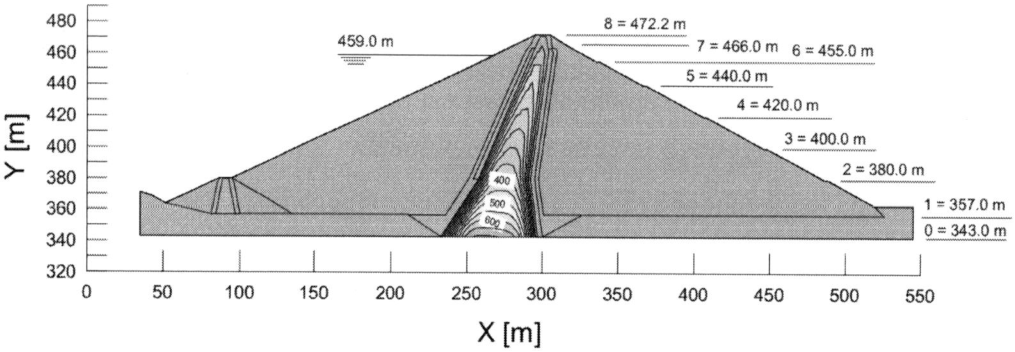

Figure 7. State after dam construction, Pore water pressure, (-0.0) - (+710.5) kPa

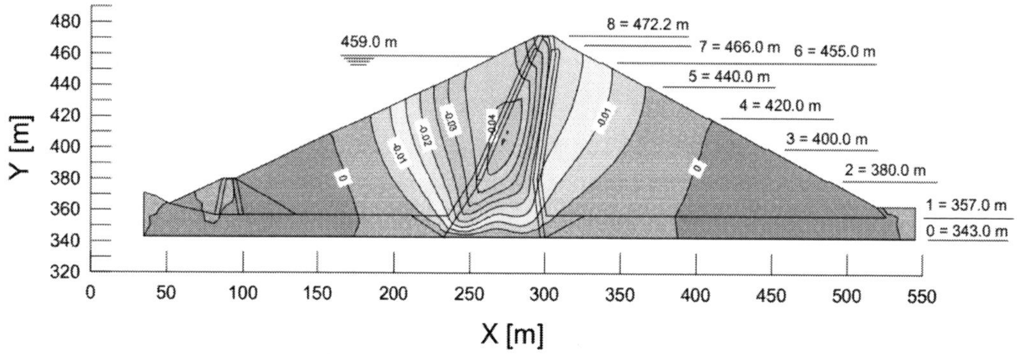

Figure 8. State upon consolidation, dY displacement (-0.045) - (+0.001) m

ANALYSIS OF THE STATE OF FIRST FILLING OF THE RESERVOIR

By using the calibrated model, the state of first filling is simulated in period of 516 days, by linear increase of 15 calculation steps of the water level from 343.0 m asl to 459.0 m asl (fig. 9). The generation and dissipation of the pore pressure in the three typical stages (fig. 10), by total lasting of 2,479 days, mostly matches with the measured values, that is contributing the conclusion regarding the regularity of the numerical experiment.

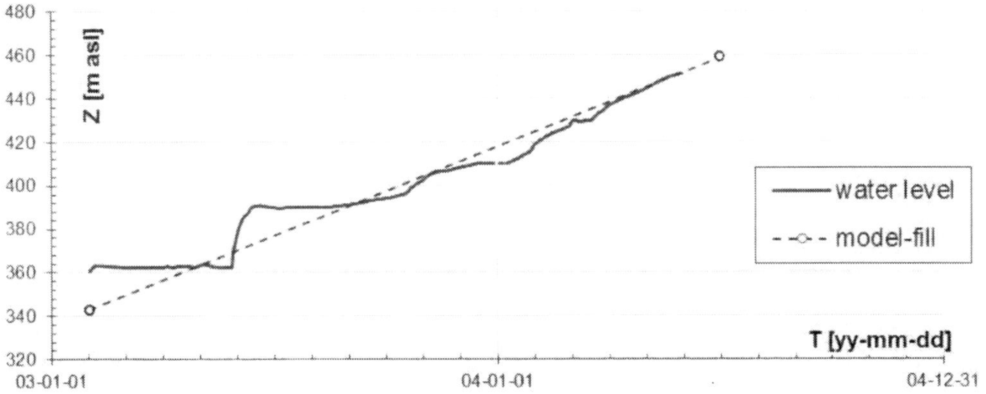

Figure 9. Simulation of the reservoir first filling

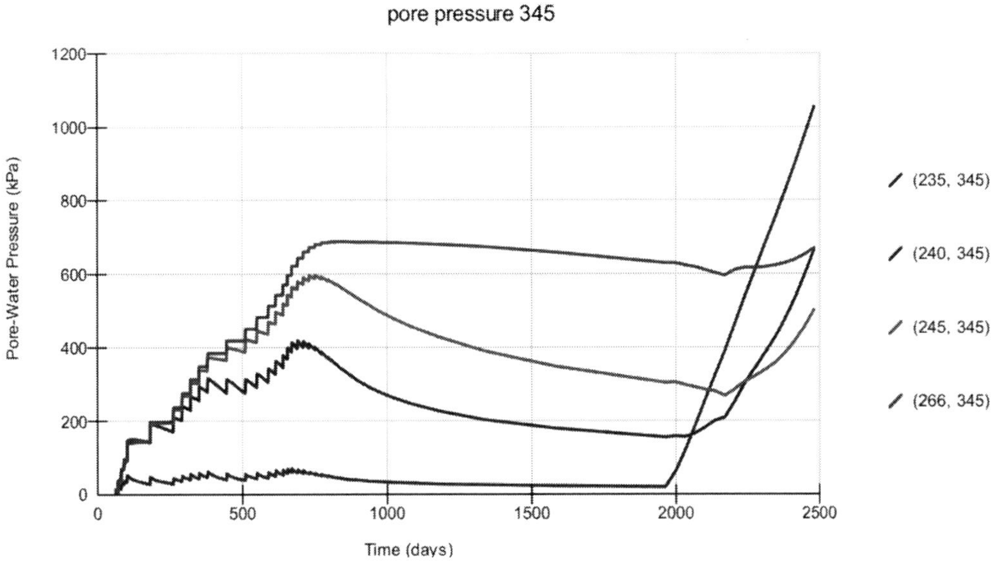

Figure 10. Generation and dissipation of the pore pressure at core foundation, at elevation 345.0 m asl, with coordinates X = {235, 240, 245, 266} m

The unusual bidirectional displacements of the central waterproof element, in dependence of the water level, are displayed for the upstream and downstream face of the core, fig. 11 and fig. 12. By the pattern and distribution of the horizontal displacements, it can be concluded that the response of the earth rock dam during first filling is properly simulated, and that the state of the effective stresses upon reservoir filling (fig. 13) can be successfully applied for analyses of the following static and dynamic loadings of the structure.

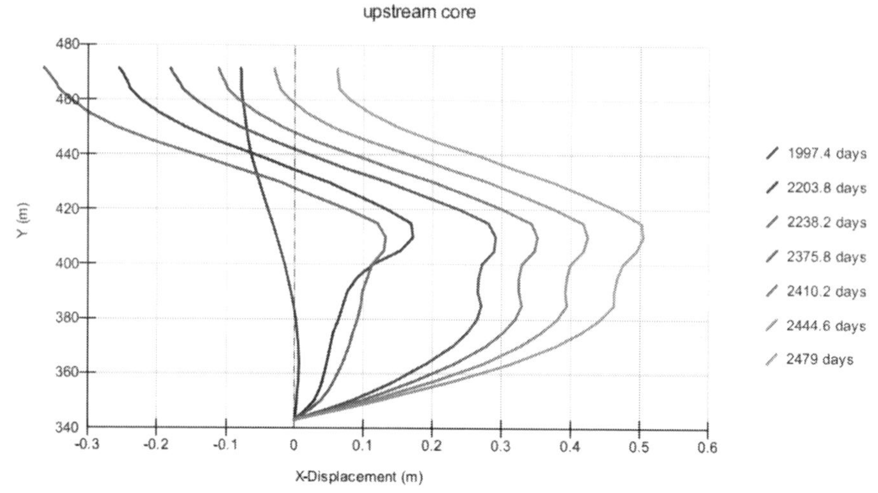

Figure 11. Development of horizontal displacements along upstream face of the core during first filling of the reservoir

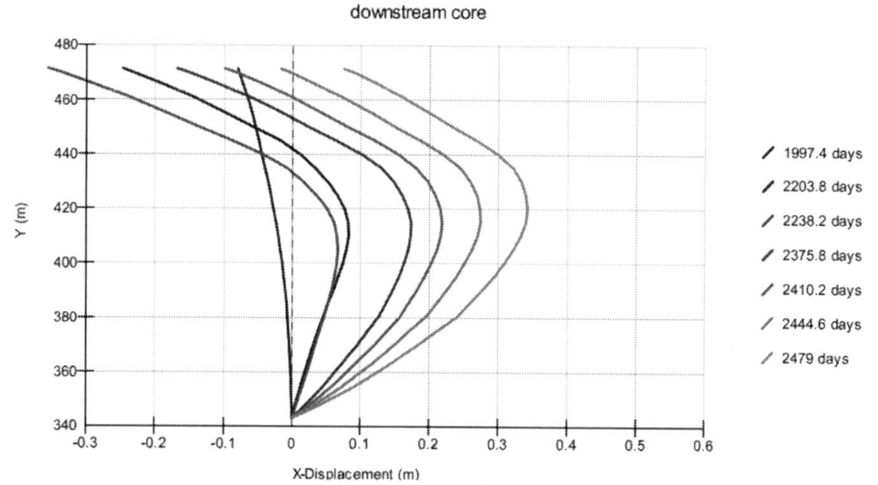

Figure 12. Development of horizontal displacements along downstream face of the core during first filling of the reservoir

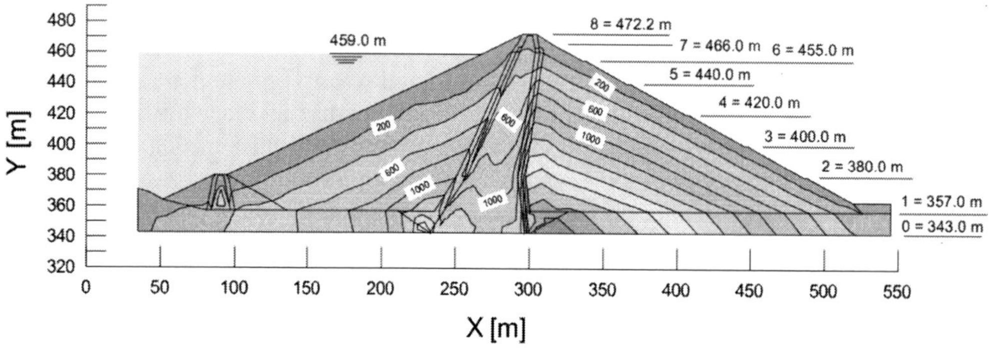

Figure 13. Distribution of effective vertical stresses, (10.14) - (+2,590) kPa after reservoir filling

The change of the total stresses results from the additional load from the increase of the volume weight and external hydrostatic pressure along the upstream slope, while the lowering of the stiffness of the rock material, due to the lowering of the effective stresses, is affecting the displacements. The effective stresses are difference of the total stresses and neutral water pressure (according to the laws of hydrostatic) in the pores of submerged materials upstream of the core. The softening of the material is actually caused by the lowering of the elasticity modulus, due to lowering of the effective stresses. The reason why in reality, by water saturation of the upstream shell (or by lowering of the effectives stresses) raising does not occur (elastic response) is the superposition of at least three effects: (a) increased stiffness at unloading, (b) downstream displacement caused by the basic load – hydrostatic pressure and (c) occurrence of "collapse settlement". The third effect is manifested by settlements of the coarse material after submerging in water, due to the decrease of its strength parameters, crushing of the grains edges - phenomena intensively researched in the last two decades [Alonso et al., 2005; Oldecop and Alonso, 2007; Roosta and Alizadeh, 2012]. The results for the partial horizontal displacements (fig. 10 and 11), where in the axis of the inclined core, the maximal horizontal displacement is approximately 40 cm in the intermediate part of the dam, and at crest approximately 10 cm, according to the pattern are similar with obtained values by monitoring of earth-rock dams with central core and are appropriate to the numerical analysis of rockfill dam Konsko (Gevgelija, Republic Macedonia) with asphalt core [Petkovski L., Tančev L., Mitovski S., 2013].

CONCLUSION

The response of fill dams to the action of static loads is a complex issue that in most cases cannot be solved by physical law, but is assessed by numerical models. The inclusion of models instead of laws means that for analysis of a dam, the models (based on different approximations) are not mutually excludible but in contrary, they contribute to a better understanding of the prototype behavior. By comparison of the results from the considered models (elastic and non-elastic, with constant and variable elasticity modulus) for the behavior of the earth rock dams with slight inclined core, during the stage of reservoir first filling, the following two observations can be outlined. First, by applying elastoplastic model with variable elasticity modulus in dependence of the effective normal stresses are obtained patterns of bidirectional displacements in the axis of the waterproof element, verified by monitoring of real structures. Second, by applying of boundary hydraulic condition along the upstream slope of the dam, a realistic distribution of the maximal main normal stresses (total and effective) is obtained, which is the base for all further structural (static and dynamic) analyses of the fill dam. Namely, this concept for displacements is in correlation with the calculated stresses, meaning they result from the change in the effective stresses, as difference between the total stresses and neutral water pressure.

ACKNOWLEDGEMENTS

The notes cover the results of an investigation carried out by the authors involved in the scientific-research project, titled "Comparative analysis of stress-deformation state of rockfill dams with core/diaphragm at variation of the water level in reservoir", financed by Sts Cyril and Methodius University, Skopje, for 2015/16. The authors wish to express their gratitude to the members of the research team from the Civil Engineering Faculty in Skopje within Sts Cyril and Methodius University and HPP Crn Drim, Struga for the cooperation. The authors would also like to express their appreciation to AD ELEM, Skopje for their permission for using the data from the technical monitoring of dam Kozjak, on river Treska, Republic of Macedonia.

REFERENCES

Alonso, E.E., S. Olivella and N. M. Pinyol. 2005. A review of Beliche Dam, Géotechnique 55, No. 4, p.p.267–285

Geo-Slope SIGMA/W, 2012. "Stress/deformation analysis", GEO-SLOPE International Ltd., Calgary, Alberta, Canada

ICOLD, 1978, Bulletin 30, Finite elements methods in analysis and design of dams, Paris, France

ICOLD, 1986, Bulletin 53, Static analysis of embankment dams, Paris, France

ICOLD, 1988, Bulletin 94, Computer software for dams - Validation, Paris, France

ICOLD, 1994, Numerical analysis of dams, Volume III, Theme B1 "Evaluation of pore pressure and settlements of an embankment dam under static loadings", Paris, France

ICOLD, 2001, Bulletin 122, Computational procedures for dam engineering - Reliability and applicability, Paris, France

ICOLD, 2013, Bulletin 155, Guidelines for use of numerical models in dam engineering, Paris, France

Oldecop, L.A. and Alonso, E.E. 2007. Theoretical investigation of the time-dependent behaviour of rockfill. Géotechnique, 57 (3), pp.289-301

Petkovski L., Tančev L., Mitovski S., 2007. 06 "A Contribution to the standardization of the modern approach to assessment of structural safety of embankment dams", 75th ICOLD Annual Meeting, International Symposium "Dam Safety Management, Role of State, Private Companies and Public in Designing, Constructing and Operation of Large Dams", 24-29 June 2007, St.Petersbourg, Russia, Abstracts Proceedings p.66, CD-ROM

Petkovski L., Tančev L., Mitovski S., 2013.10 "Comparison of numerical models on research of state at first impounding of rockfill dams with an asphalt core", International symposium, Dam engineering in Southeast and Middle Europe - Recent experience and future outlooks, SLOCOLD, 16-17.10.2013 Ljubljana, R.Slovenia, ISBN 978-961-90207-9-1, Proceedings, 106-115

Roosta M. R., and Alizadeh A. 2012. Simulation of collapse settlement in rockfill material due to saturation. International Journal of Civil Engineering, Vol. 10, No. 2, June 2012, pp.93-99

USING NON-INTRUSIVE GEOPHYSICAL TECHNIQUES IN EMBANKMENT DAM AND LEVEE RISK ASSESSMENT

Justin Rittgers[1]
Cassandra Wagner[2]
Bryan K. Simpson[3]

Water management and dam safety professionals are constantly faced with a myriad of engineering challenges and uncertainties in risk analysis that are only exacerbated from a lack of relevant subsurface information. From chemical degradation and voids behind concrete structures, to concentrated seepage and internal erosion of massive earthen embankments, many issues commonly faced by this community involve scales-lengths that can span from pore-scale to basin-scale. Thus, it can be very difficult and expensive to obtain adequate and useful information about a subsurface feature or process of concern from drilling alone. For example, the chance of finding a localized defect within a large structure by blind drill-hole placement is very small, and there could be great risk associated with a defect remaining undiscovered or being improperly characterized due to insufficient data coverage. Furthermore, drilling can pose added risk of inducing damage or failure of a structure (e.g., drilling near cutoff walls, hydraulic fracturing, compaction of embankment materials, contaminating filters, blowouts, etc).

Consequently, appropriate selection and implementation from a variety of geophysical techniques can provide invaluable information and guidance during subsurface investigations and risk analysis efforts. Examples include geophysical imaging to help identify and characterize defects or other structural features, to monitor defects for progression, to help guide drilling efforts to intercept or avoid specific features, to help interpolate material properties/features between wells, and to help appropriately parameterize commonly used equations (e.g., the Sellmeijer or blanket equations) or subsequent modeling efforts (e.g., FLAC or hydrogeologic models). This talk will present an overview of many geophysical techniques applicable to the water management and dam safety communities, and will provide general capabilities and limitations for engineers and management to consider during future risk analysis and exploration efforts. Examples will be drawn from a series of recent research efforts and surveys conducted at Reclamation structures.

[1] Bureau of Reclamation, jrittgers@usbr.gov
[2] Bureau of Reclamation
[3] Bureau of Reclamation

IMPACT OF GLACIAL TILL DEPOSITS - STRUCTURAL ANALYSIS OF GREEN MOUNTAIN DAM

Hillery Venturini[1]

Green Mountain Dam, completed in 1943 as a major storage feature of the Colorado-Big Thompson Project, is a zoned earth-fill embankment with a concrete gate structure and spillway located on the left abutment. The spillway consists of an unlined approach channel, a concrete-lined inlet channel, gate structure controlled by three radial gates, and a downstream concrete lined chute. The reservoir is used to regulate replacement irrigation water for the Colorado River and water for power generation and recreational use.

Drill hole cores samples along the spillway floor, were used to verify filter compatibility of the embankment and foundation materials adjacent to and beneath the spillway. Gradations confirmed the presence of zone 1 and 2, impervious and semi-pervious material, respectively, with data limited to drill locations. Lower strength Glacial Till material removed and replaced according to construction documentation, was identified to bedrock depth based on a single drill hole near the gate structure. Therefore the presence of Glacial Till material may be located beneath the upstream edge of the spillway gate structure, daylighting downstream of the chute. Under seismic loading, stability of the Glacial Till material is unknown. Pre-existing settlement of the gate structure in a rotational manner during construction, may have been due to this underlying material.

An LS-DYNA finite element structural analysis was performed on the spillway and dam, to analyze the stability of the spillway gate structure considering soil structure interaction under dynamic conditions. The model included the concrete gate structure, steel radial gates, soil material underlying the gate structure, foundation, and earth zoned embankment. Concrete was represented by two non-linear material models, which incorporated reinforcement to access the global behavior of the structure. Soil adjacent of the spillway gate structure was represented by non-linear material properties, while the embankment dam was represented by linear elastic material properties.

[1] Bureau of Reclamation, hventurini@usbr.gov

USING HEC-RAS 2 DIMENSIONAL CAPABILITY FOR SIMPLIFIED RAPID DAM BREAK ANALYSIS

Wesley Crosby[1]

The USACE Modeling Mapping and Consequences Production Center (MMC) provides hydraulic modeling, mapping and consequence analysis for USACE dams in support of the USACE Dam Safety and Critical Infrastructure Protection and Resilience (CIPR) Programs. The MMC has developed processes, tools and standards for creating dam breach hydraulic models for use in emergency action plans (EAP), during real-time flood events, and in support of the Corps Dam Safety and Security programs. The MMC-developed standards have been used to provide dam failure modeling for over 400 USACE dams and multiple flood events, involving over 1000's of stream miles throughout the continental U.S. and Alaska.

This presentation will provide examples of how to use the new two dimensional (2D) capabilities within the Hydrologic Engineering Center's River Analysis System (HEC-RAS) to perform simplified rapid dam break analysis. It will illustrate case studies on how the simplified rapid development of dam break models can be used for screening purposes to enable a more risk informed decision on how to prioritize which dams would need additional analysis. In addition, it will demonstrate how to produce rapid 2D inundations for dam breaks and flooding during a real-time flood event. In October of 2015 the MMC was asked to demonstrate the USACE's capability in producing rapid dam break inundations for the flooding in South Carolina. This case study will be presented to show how HEC-RAS 2D helped to accomplish this task in relatively short time.

[1] U.S. Army Corps of Engineers, wesley.a.crosby@usace.army.mil

HYDRAULIC MODELING OF A COMPLEX DAM AND RESERVOIR PROJECT: A CASE STUDY OF TWO RIVERS DAM

Stephanie Doyal[1]

The USACE Modeling, Mapping, and Consequences Production Center (MMC) provides hydraulic modeling, mapping, and consequence analysis for US Army Corps of Engineers (USACE) dams and levees in support of USACE Dam and Levee Safety Programs and Critical Infrastructure Protection and Resilience Programs (CIPR). In 2015, the MMC was tasked with creating a dam break model of the Two Rivers Dam, located near Roswell, NM, in support of USACE Albuquerque District's Periodic Assessment of the dam. Multiple challenges were encountered during the modeling process including varying physical characteristics of the reservoir, limited terrain data, and significant out-of-channel flows. This presentation will provide an overview of the project and the successes and failures that were encountered during the modeling effort.

The Two Rivers Dam and Reservoir Project consists of two earthen dams and a reservoir that acts as one pool during high stages and as two separate pools during low stages due to a dike constructed within the reservoir. This presentation will demonstrate the different methods available to model this situation and why the final method was chosen. In addition, significant out-of-channel flows occur downstream of the dam, which are difficult to accurately depict in a typical 1-dimensional model. This presentation will demonstrate how the 2-dimensional capabilities of HEC-RAS were utilized to model these flows and the effect of input variables on model run time and output. This presentation will also outline how the terrain data required by this 2-dimensional model was sourced, and the processes used to merge the available limited terrain data sets into one usable digital elevation model (DEM). Other challenges presented by the karst environment, limitations of HEC-RAS modeling software, and arid climate will also be discussed.

[1] U.S. Army Corps of Engineers, stephanie.a.doyal@usace.army.mil

AN INNOVATIVE TEMPORARY BULKHEAD FOR THE RUSKIN DAM AND POWERHOUSE UPGRADE PROJECT

Jorge Marín[1]
Saman Vazinkhoo[2]
Rafael Ibáñez-de-Aldecoa[3]
Mark Mallet[4]

ABSTRACT

Ruskin Dam and Powerhouse, located near Mission in British Columbia, were originally constructed in 1930 and BC Hydro is currently in the midst of a major seismic upgrade of the entire facility while maintaining normal operations. A major aspect of the work includes replacing the existing 7-bay spillway with a new 5-bay spillway while maintaining the same or greater flood routing capacity. To complete this task, and meet BC Hydro's construction flood passage and earthquake design requirements, the replacement is being completed in three phases behind a metal bulkhead. The bulkhead bears against the existing piers and dam sections to create a watertight barrier in front of the workspace.

The design concepts provided in the bid documents required intermediate supports to transfer the large bulkhead forces (especially under the design earthquake) to the dam. This would have resulted in an extended construction schedule in order to systematically remove and replace the piers and gates. For this challenging task, an innovative design was developed using a truss and hanger system that eliminated the need for intermediate supports and shortened the construction schedule substantially.

INTRODUCTION

Ruskin Dam is one of the three BC Hydro facilities in the Alouette-Stave-Ruskin Hydroelectric System. The dam, a concrete gravity structure founded mainly on bedrock, and the associated powerhouse were constructed in 1929 and 1930. BC Hydro is currently upgrading the entire 80 years old facility under different contracts. One major component of the main civil contract (Contract "A") is the refurbishment of the spillway of the concrete dam, which includes demolition of existing eight piers and seven Tainter gates (10.06 m x 8.52 m), and installation of six new piers for five wider new gates (13.50 m x 8.52 m) (see Fig. 1 and 2), with their corresponding operating equipment, as well as construction of new spillway control facilities on the new piers and on the left abutment [BC Hydro, 2011].

To carry out the refurbishment while maintaining the facility in operation, which implies keeping the reservoir pool several meters above the spillway ogee, and maintaining the

[1] Dragados Canada Inc., 10400 Hayward Street, Mission, BC V4S 1H8, jjmarin@dragados-canada.com
[2] British Columbia Hydro, 6911 Southpoint Dr., Burnaby, BC V3N 4X8, saman.vazinkhoo@bchydro.com
[3] Dragados USA Inc., 701 5th Avenue, Suite 6550, Seattle, WA 98104, ribanez@dragados-usa.com
[4] Flatiron Constructors Canada Ltd., 10400 Hayward Street, Mission, BC, mmallet@flatironcorp.com

ability to pass the 1:2000 year construction design flood, it is necessary to install, in three consecutive phases, a temporary, watertight, metal construction bulkhead. This bulkhead acts as a shield allowing working conditions in-the-dry to replace the piers and gates in each of the phases.

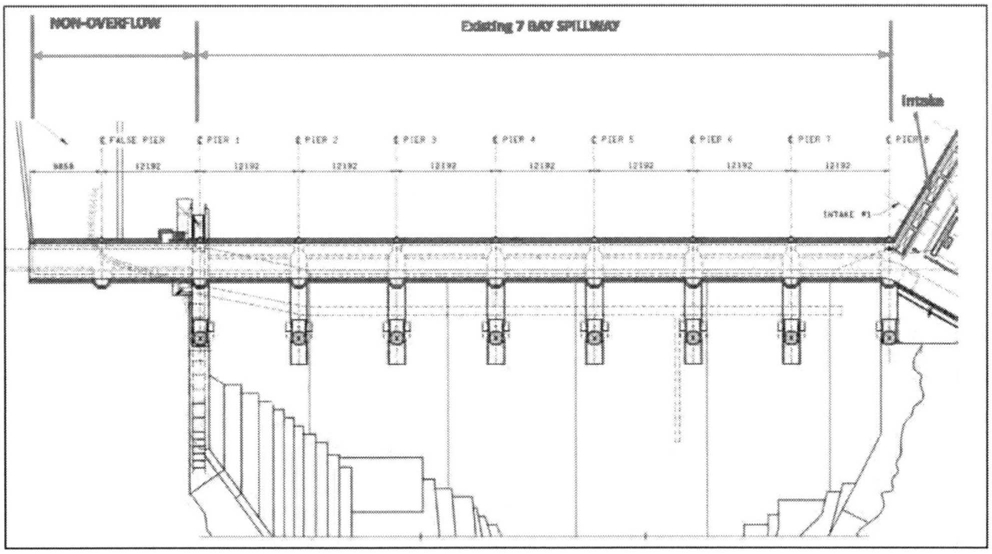

Figure 1. Existing spillway configuration

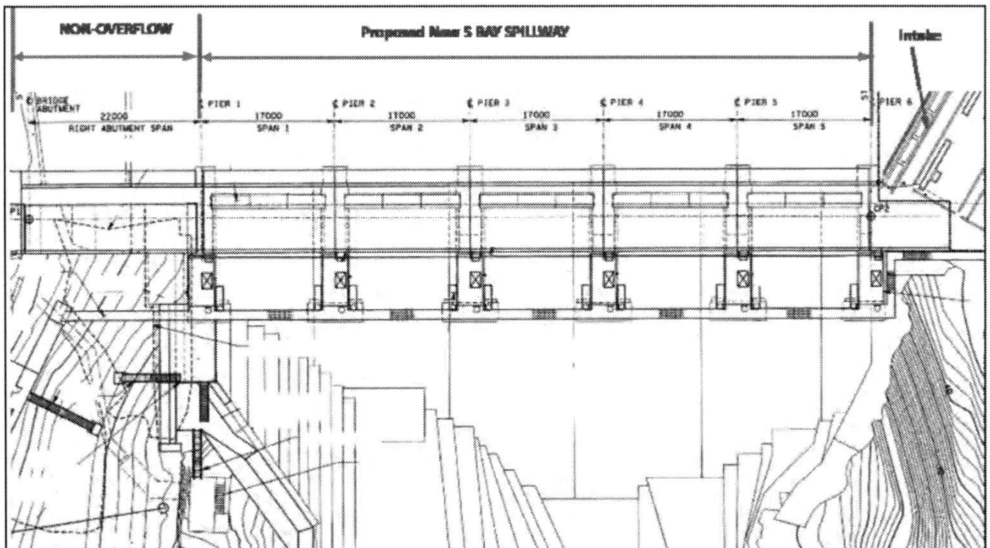

Figure 2. Proposed spillway configuration

Once work on the spillway gates is completed, the bridge on top of the spillway, formerly a single lane, will be replaced with a two-lane deck and pedestrian walkway.

This paper first reviews the carefully studied design of the bulkhead, that transmits large loads to different parts of the existing dam and spillway structure to which it is connected. A detailed sequence of the assembly of the bulkhead for one of the phases is subsequently described, followed by an explanation of sequencing of the three different phases needed to substitute all the piers and radial gates. Finally, this paper presents a comprehensive list of the main works executed in all phases once sheltered by the bulkhead.

REQUIREMENTS AND KEY DETAILS OF THE DESIGN OF THE BULKHEAD

In order to maintain the spillway gates and powerhouse in operation the bulkhead must provide no less than the equivalent spillway capacity provided by 2 new gates and 1 existing gate. Gates providing this flow capacity must be available for full and unrestricted operation. In order to achieve this requirement the work was divided in 3 phases, with a minimum of 2 new gates and 1 existing gate, or 4 existing gates available at any time.

The bulkhead is formed by two vertical trusses, one horizontal truss, and a sheet pile wall hanging from the downstream chord on the horizontal truss. The whole system is suspended by two hanger rods on each side and has full bearing contact with the dam face in the lower part of the bulkhead through the total bulkhead length, to maintain watertightness (see Fig. 3). The total dimensions of the bulkhead (for phases 1 and 2) are 39.6 m wide by 12.0 m high. It is designed to meet BC Hydro's construction flood (1:2,000 year event) and earthquake design (1:475 year event) requirements [Bittner-Shen, 2013].

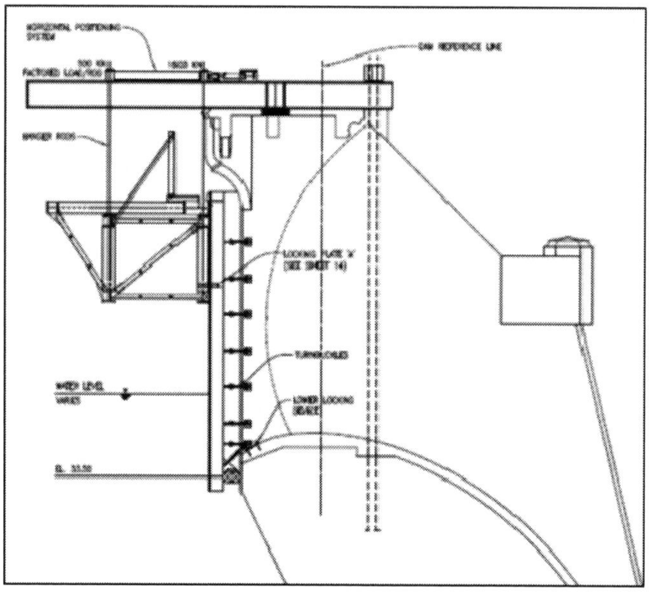

Figure 3. Bulkhead vertical section at pier

One of the Contractor's innovations in this design was the use of the trusses to eliminate mid-point support requirements on existing or new piers, which allowed the project schedule to be shortened significantly as compared to the base design provided in the bid documents.

The watertightness at the bulkhead bottom is achieved using 12 x 12 timbers with neoprene between timbers; at the sides against the piers it is achieved with a grouted inflatable fabric fire hose (see Fig. 4 and 5). At the downstream side of the bulkhead 80 locking devices (turnbuckles) are installed to guarantee watertightness.

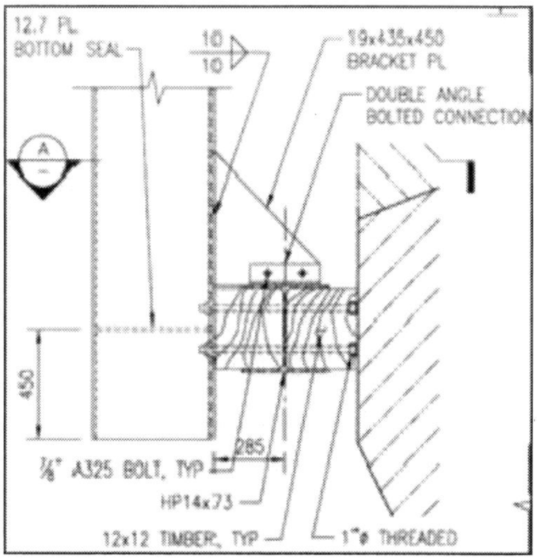

Figure 4. Bottom seal

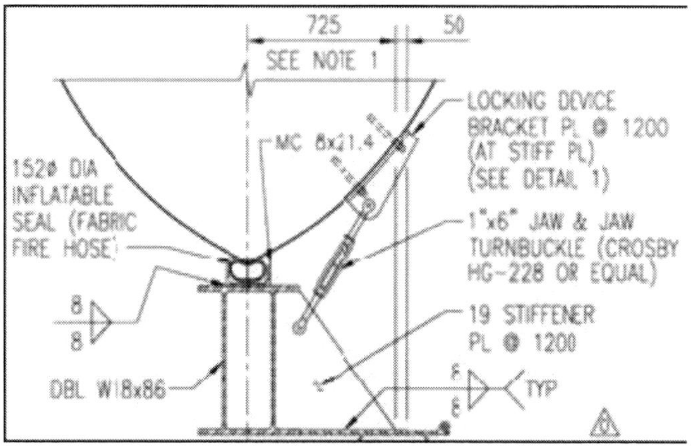

Figure 5. Side seal

An analysis of the loads transmitted by the bulkhead to the dam was performed and, as a result, additional anchoring works for the existing piers and the right abutment non-overflow dam section was needed to counteract the overturning moment introduced by the hanging bulkhead. A total of 6, 3-66 mm-bar bundles, post-tensioned and fully grouted were installed in the right abutment non-overflow dam section, while 23, 63 mm, fully grouted passive anchors were installed in the load bearing piers.

BC Hydro performed concrete tests in 1996 on samples taken from the dam gallery [Li, 1996]. Additional concrete testing was conducted as part of recent construction activities on samples recovered from cores drilled in the right abutment non-overflow section, for the installation of the post-tensioning anchors mentioned above, as well as on large diameter samples (457 and 305 mm Ø) taken during demolition of the existing piers. The latter testing program was intended to obtain material properties of existing mass concrete for use in a 3D Finite Element Model (FEM) seismic analysis [Razavi Darbar et al., 2016]. In all testing programs the compressive strength obtained was significantly higher (double to triple) than the original design strength of 14 MPa at 28 days. These test results provided confidence in the selected values for the compressive and tensile strengths for the bulkhead design.

INSTALLATION OF THE BULKHEAD FOR ONE OF THE PHASES

For the bulkhead installation, a detailed Step-by-Step Procedure and Erection Manual (with more than 100 steps) was prepared. The bulkhead was assembled on a 40 m x 6 m flexifloat with the aid of a 230 t crawler crane on a different flexifloat (see Fig. 6, 7 and 8). The general assemblage sequence commenced with the assembly of the vertical trusses, followed by the assembly of the horizontal truss, and finally the sheet piles wall. After completion of the assemblage, the bulkhead was erected to its final position at the bearing points, and then, the seals were fitted. Divers were required for placement of the bottom seal.

Figure 6. Bulkhead end sheet pile and seal pre-assembly

Figure 7. Bulkhead sheet piles wall assembly

Figure 8. Bulkhead erection

PHASES NEEDED TO REPLACE ALL PIERS AND RADIAL GATES

The work is divided into three different phases. In Phases 1 and 2 three of the existing gates (EGs) are removed and two new gates (NGs) installed, while in the third phase only one EG is removed and one NG is installed, as shown in Fig. 9 and 10 (EP refers to existing piers while NP refers to new piers).

At the moment of writing this paper, the works are in an advanced stage of Phase 2.

Figure 9. Bulkhead relocation to Phase 2

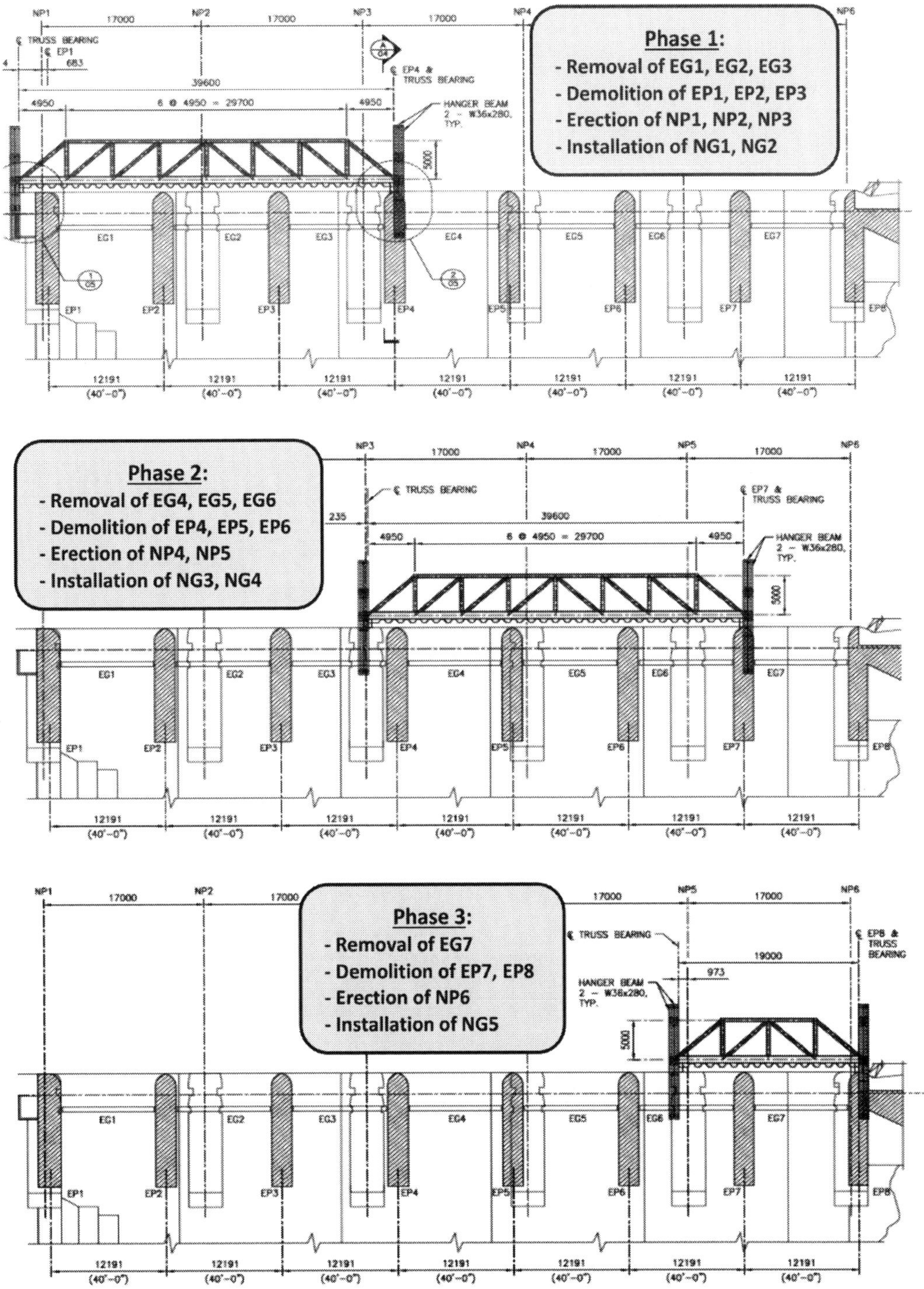

Figure 10. Sequence of Phases 1, 2 and 3

DESCRIPTION OF THE MAIN WORKS EXECUTED BEHIND THE BULKHEAD

A concise list of the main works to be executed in all phases, once sheltered by the bulkhead (see Fig. 11), includes:

- Demolition of the existing spillway bridge
- Removal of the existing gates
- Demolition of the existing piers
- Cut-out for footings for the new piers and new sill beams
- Drilling and installation of threadbar dowels for new piers
- Construction of new piers
- Installation of new radial gates and stoplogs
- Construction of new control buildings for each gate on top of the piers
- Spillway face rehabilitation (remove old shotcrete and replace with new)
- Construction of the new two-lane spillway bridge with a pedestrian walkway

Figure 11. Aerial view of the works behind the bulkhead during Phase 1

In addition to the bulkhead, it has been necessary to employ several engineered platforms, some of them anchored to the very steep chute of the existing spillway, in order to access and perform the above work (see Figs. 11, 12, 13 and 14).

A particularly challenging task is to avoid contamination of the Stave River downstream of the dam. All slurries and contaminated waters originating as a consequence of the demolition of existing concrete piers (mostly cutting with diamond wire saw), removal of old shotcrete, and construction and curing of new concrete structures, have to be collected at the toe of the spillway in isolation ponds and pumped to a treatment facility.

Construction efforts during all three phases require close coordination between the Contractor and BC Hydro's plant staff, given that the existing hydropower plant must remain operational during construction.

Figure 12. Installation of New Gates NG1 and NG2 during Phase 1

Figure 13. Aerial view of the works behind the bulkhead during Phase 2

Figure 14. Installation of New Gates NG3 and NG4 during Phase 2

CONCLUSION

From the early stages of design, BC Hydro recognized that one of the biggest challenges of this project would be the construction of a bulkhead that would provide efficient construction of the new piers and gates. BC Hydro retained two experienced consulting firms and developed conceptual designs for the bid documents to address some of the challenges presented by this problem. These two conceptual designs, as well as the designs presented by other bidders, required intermediate supports while the design presented by the winning bidder, Flatiron-Dragados joint venture, employed the concepts described herein and created significant cost and schedule savings for the project. The elimination of the intermediate supports was achieved largely through the use of the innovative hanger beam concept.

REFERENCES

Bittner-Shen Consulting Engineers Inc. "Ruskin Dam Temporary Bulkhead Design". April 2013.

British Columbia Hydro and Power Authority (BC Hydro). "Ruskin Generating Station Upgrade, Upper Dam and Powerhouse Structure, RFP #943". July 2011.

Li, J. Testing Concrete Cores from Ruskin Dam. Powertech Labs Inc. July 1996.

Razavi Darbar, S., Queen, D., Hatton, C, Dolen, T. and Bartojay, K. "Static and Dinamic Mass Concrete Material Properties of a Concrete Gravity Dam". 36th Annual USSD Conference, Denver, CO, April 2016.

Figure 15. Aerial view of the project

RISK AND UNCERTAINTY FEATURES IN HEC-RESSIM

George Modini[1]
Fauwaz Hanbali[2]

HEC-ResSim is the reservoir system simulation software developed by the U.S. Army Corps of Engineers (USACE) Hydrologic Engineering Center (HEC), and is one of the primary tools used in USACE for water management decision support, watershed planning, and risk and uncertainty (R&U) analysis. Dam safety studies for risk assessment on USACE dams prompted the development of Monte Carlo features in HEC-ResSim since 2010, as well as the integration of HEC-ResSim R&U analysis in the HEC-WAT/FRA (Watershed Analysis Tool Flood Risk Analysis) software suite. The HEC-ResSim Monte Carlos features allow for the definition of random variable types that include: Input Time Series, Time Series Multipliers, Reservoir Rule Parameters, Reservoir Initial Conditions, and Rating Curves at river control points. As part of the random variable definition, a variety of probability distributions are available for use in the Monte Carlo sampling. Correlated random variable sampling is also supported in HEC-ResSim. The application of HEC-ResSim R&U analysis can vary in the required level of detail and level-of-effort for execution depending on the required level of risk assessment for a dam safety project.

This paper will introduce the Monte Carlo capabilities in HEC-ResSim and various types of applications.

[1] U.S. Army Corps of Engineers, george.c.modini@usace.army.mil
[2] U.S. Army Corps of Engineers

METHODOLOGY APPLIED FOR MAPPING FLOODED AREAS OF THE LAJEADO HPP

Pedro H F Pupim[1]
Mauricio Santini[2]
Milton Dall'Aglio[3]

Changing the natural state of the rivers, through the construction of dams in watercourses, generates the need for complex studies for this size management structures due to the high risk involved in the storage and control of large volumes of water, because the potential damage associated with the rupture of dams may result economic, social and environmental impacts, even loss of human life.

Through hydrodynamic studies, associated with hypothetical dam failure, it is possible to carry out simulations of wave propagation, which allows to obtain flood areas maps downstream reservoirs, collaborates in adopting mitigation measures of risk and planning of emergency actions.

With an installed capacity of 902 MW, Lajeado HEPP is located on the Miracema do Tocantins city, in Tocantins State, northern of Brazil. The plant is located on the Tocantins River, has a reservoir with a storage capacity of 5.2 billion cubic meters and 185 square kilometers drainage area. Discharge devices have the flow capacity of 49,870 m^3/s, corresponding to ten-thousand-year flow rate, average flow rate of 2,532 m^3/s and maximum flow recorded of 28,558 m^3/s.

This paper aims to compare the methodologies MCT and Hec-Has for the propagation of flood wave generated by hypothetical break flow rate of the Lajeado Dam.

[1] Civil Engineer at EDP and Master Student at Universidade Estadual Paulista, Ilha Solteira, SP, Brazil, pedro.pupim@edpbr.com.br
[2] Civil Engineer and Master Student at Unesp, Ilha Solteira, SP, Brazil, mauriciosantinijr@gmail.com
[3] Doctor and Associate Professor at Unesp, Ilha Solteira, SP, Brazil, milton@dec.feis.unesp.br

LEVERAGING MULTI-SENSOR HYDROLOGIC DATA FOR MODEL VALIDATION

Kevin J. Ruswick, PE, CFM[1]
Gregory J. Daviero, PhD, PE[2]
Jonathon Ruff, PE[3]

ABSTRACT

The City of Plattsburgh, NY (City) manages an aging water supply infrastructure including Mead Reservoir Dam. An Engineering Assessment was performed in 2010 that included hydrologic modeling predicting the dam would overtop by two feet during the Spillway Design Flood (SDF). A spillway rehabilitation project including replacement of the existing 40-ft wide ogee spillway with a 2-cycle labyrinth and enlarged chute/stilling basin was subsequently designed and let to bid. All bids received greatly exceeded the allocated $4M budget and the project was put on hold.

The City retained Schnabel Engineering to re-evaluate the hydrology and need for spillway capacity improvements. The previous modeling utilized the SCS Unit Hydrograph Methodology which is known to be conservative with its prediction of peak flows. Schnabel's modeling was performed utilizing the Snyder methodology and focused on model validation for significant historic floods, including Hurricane Irene. Hourly precipitation data were available from land-based gaging stations but were inconsistent due to significant sustained wind and gust speeds. To assess the storm's spatial and temporal variations, National Weather Service (NWS) multi-sensor hourly gridded rainfall data were leveraged. This data set uses both radar and land-based gage data to develop rainfall depth estimates on a 4km x 4km grid. Using the multi-sensor gridded rainfall data set, model results using the Snyder methodology correlated much better with observed reservoir levels than the SCS methodology. Model predictions using gridded hourly rainfall data and the Snyder methodology resulted in an SDF that is less than half of the previously utilized SCS based value. This will result in the design and construction of a new spillway at a significantly reduced cost and demonstrates both the importance of hydrologic model calibration/validation and the need to leverage the technological advances in hydrologic precipitation data.

INTRODUCTION AND BACKGROUND

Mead Reservoir Dam (NID: NY00237) is a water supply dam owned by the City of Plattsburgh, New York. The dam was reportedly constructed in 1922-23 and is comprised of an earthen embankment and a concrete gravity spillway section. The dam

[1]Senior Associate, Schnabel Engineering, 28 Corporate Drive, Clifton Park, NY 12065
(518) 348-8575, kruswick@schnabel-eng.com
[2]Principal, Schnabel Engineering, 28 Corporate Drive, Clifton Park, NY 12065
(518) 348-8575, gdaviero@schnabel-eng.com
[3] Environmental Manager, City of Plattsburgh, 41 City Hall Place, Plattsburgh, NY 12901
(518) 536-7519, Ruffj@cityofplattsburgh-ny.gov

has a maximum height of approximately 60 feet (ft) and is classified by the New York State Department of Environmental Conservation (NYS DEC) as a Class C – High Hazard dam. The dam is approximately 615 ft in length including a 40-ft wide concrete gravity ogee spillway section. There is an embankment section approximately 360-ft long to the right (SW) of the spillway section, and an embankment section approximately 215 ft in length to the left (NE) of the spillway section. The embankment reportedly includes a concrete core wall that is 8 inches thick at the top and 18 inches thick at the base. Recent survey of the dam crest indicates that the crest is un-level with crest elevations ranging from 542.7 to 545.2. The principal spillway is a 20-ft high, 40-ft wide concrete overflow weir with an ogee-shaped crest at an elevation of approximately 536.1. The ogee spillway discharges to an upper plunge pool that is approximately 50-ft wide and 20-ft long. The spillway chute is approximately 50-ft wide, 215-ft long, and discharges to a lower plunge pool/stilling basin. The lower plunge pool is approximately 50-ft long and widens to approximately 70 ft at the 4-ft tall end sill. The original design drawings and the 1981 Phase I inspection report also indicate that there is an "emergency relief weir" located in the shorter embankment section to the left (NE) of the principal spillway. Details on the specific design and capacity of the "emergency relief weir" are not available, but detailed site survey data does indicate a depressed portion of the embankment crest between the spillway and left abutment.

An Engineering Assessment (EA) was conducted by an engineering consultant in April 2010. As part of this EA, hydrologic and hydraulic analyses were conducted to assess whether the existing spillway was in compliance with current regulatory requirements in regards to spillway capacity. As an existing High Hazard dam, Mead Reservoir Dam is required by the New York State Department of Environmental Conservation (NYSDEC) to have a spillway capable of safely conveying the Spillway Design Flood (SDF) which is 50 percent of the Probable Maximum Flood (PMF). The results of the 2010 EA indicated that the existing spillway capacity was inadequate and that the embankment would overtop by an average of approximately 1.6 ft during the SDF. Subsequent to the 2010 EA, the original consultant was retained to design remedial repairs to the dam including spillway capacity improvements, replacement of the spillway chute/energy dissipater, and new intake piping and valves. The proposed spillway capacity improvements included a new 2-cycle labyrinth weir and leveling of the dam crest to a minimum elevation of 545.0. Based on the hydrologic and hydraulic modeling, the proposed spillway would be able to convey the peak flow from the SDF with approximately one foot of freeboard.

Detailed design drawings and specifications were developed for the proposed remedial measures and the project was advertised to bid. Bids between $5.9M and $8.5M were received, all exceeding the allocated $4M budget; subsequently all bids were rejected and the project was put on hold. Schnabel Engineering was retained by the City of Plattsburgh in October 2013 to perform an independent evaluation of the hydrology and hydraulics of Mead Reservoir Dam.

PREVIOUS HYDROLOGIC AND HYDRAULIC ANALYSES

Mead Reservoir Dam – Phase I Inspection Report (1981)

As part of the National Dam Safety Program, a Phase I Inspection Report for Mead Reservoir Dam was completed in August 1981 by the New York District of the Army Corps of Engineers. The Phase I Inspection included hydrologic and hydraulic modeling to evaluate compliance with spillway capacity requirements. Pertinent data from the Phase I report in regards to the dam is summarized in Table 1.

Table 1. Summary of Dam Data from Phase I Inspection

Drainage Area	6.39 square miles
Top of Dam	541.5
Crest of Emergency Relief Weir	540.5
Primary Spillway Crest	533.0
Normal Pool Storage	1,228 acre-feet
Maximum Storage (EL 540.5)	1,895 acre-feet
Maximum Spillway Discharge (EL 540.5)	2,827 cfs

The USGS datum is reported as the Plan Datum (Design Drawings) + 16.5 ft. It is presumed that the adjustment is based on the normal pool elevation of 533.0 which is indicated on USGS 7.5-Minute topographic mapping. The hydrologic modeling was conducted using the USACE's HEC-1 computer program. The model utilized the Initial (1.0 inches) and Constant Loss Rate (0.1 inches/hour) to represent abstraction and infiltration. The Snyder Unit Hydrograph Methodology was used to transform excess rainfall into direct runoff. The Snyder Method utilizes two hydrologic parameters to define the unit hydrograph for a basin: C_t, which represents a basin topography coefficient that typically ranges from 1.8 to 2.2; and C_p, which is a peaking coefficient that typically ranges from 0.4 to 0.8. The values used in the model for Mead Reservoir Dam were 2.0 for C_t and 0.625 for C_p. These values generally fall in the middle of the recommended range. Rainfall data used in the analysis was based on NOAA's HMR-33, which preceded the current HMR-51/HMR-52. A rating curve was used to represent the spillway capacity at various reservoir elevations. Weir discharge coefficients used in the rating curve ranged from 3.2 at low head (0.5 ft) to 3.72 at heads greater than or equal to 3.0 ft. The model was executed for various percentages of the PMF ranging from 50 percent to 100 percent. The Phase I model results presented for the 50 percent PMF predicted a peak inflow of approximately 3,290 cfs, a peak outflow of approximately 2,606 cfs, and a peak reservoir elevation of approximately 540.1. This is approximately 0.4 ft below the crest of the emergency relief weir and 1.4 ft below the main embankment crest.

Engineering Assessment (2010) and Engineering Design Reports (2012)

The City hired another consultant who conducted hydrologic and hydraulic modeling for Mead Reservoir Dam as part of the 2010 Engineering Assessment (EA) and 2012

Engineering Design Report (EDR). Pertinent data from their analysis is provided below.

Table 2. Summary of Dam Data from 2010 Engineering Assessment

Drainage Area	6.65 square miles
Top of Dam	544.3
Crest of Emergency Relief Weir	542.7
Primary Spillway Crest	536.1
Normal Pool Storage	1,461 acre-feet
Maximum Storage (EL 544.3)	2,292 acre-feet
Maximum Spillway Discharge (EL 544.3)	2,292 acre-feet

Note that the spillway and dam crest elevations reported in the 2010 EA differ by approximately 3.1 ft with the Phase I report. The EA values are based on a field survey performed in 2009 and referenced to the NGVD 1929 datum which should be consistent with the USGS Datum. The 2010 EA speculates that the 533.0 elevation listed on USGS mapping as the normal pool for Mead Reservoir may be correct, but the normal pool is commonly below the spillway crest. In the absence of field survey data for the 1981 Phase I report, it is likely that the 533.0 elevation was subsequently interpreted as representing the spillway crest elevation in deriving the datum adjustment from the design plans to the USGS datum (NGVD 1929).

The analysis reported in the EA utilized the Bentley PondPack hydrologic model to develop and route the SDF through Mead Reservoir. The watershed was represented in the model as a single basin. The model utilized the SCS Curve number methodology to represent hydrologic losses, and used the SCS Unit Hydrograph methodology to transform excess rainfall into direct runoff. The SCS curve number was computed based upon analysis of hydrologic soils groups and land use within the watershed. The computed curve number used in the analysis was 56. The SCS unit hydrograph method utilizes a single hydrologic parameter, the Time of Concentration (T_c). The T_c represents the time it takes for rain falling in the most hydrologically distant point in the watershed to reach the basin outlet. The T_c was computed to be approximately 3.5 hours. Rainfall data used in the analysis was the 72-hour Probable Maximum Precipitation (PMP) based on NOAA's HMR-51 and HMR-52 methodologies. The 72-Hour PMP used in the analysis had a total rainfall depth of 29.8 inches. The primary spillway was represented as an overflow weir in the model with a constant discharge coefficient of 3.42. The dam crest was represented by two weirs representing the embankments to the left and right of the primary spillway. The field survey-based topographic mapping was used to represent embankment crest elevations.

The model was executed with the 72-hour PMP rainfall data, and the resulting runoff hydrograph was then multiplied by 0.5 to develop the 50 percent PMF inflow design flood hydrograph. The inflow hydrograph was then routed through Mead Reservoir using the computed stage versus discharge relationship for the existing spillway, and the computed stage versus storage relationship for the reservoir. The modeling assumed the starting reservoir pool elevation was 536.1, equivalent with the primary spillway crest. The model results for the 50 percent PMF storm event predicted a peak inflow of approximately 8,826 cfs, a peak outflow of approximately 7,840 cfs, and a peak reservoir elevation of approximately 545.9. Based on the model results, the 50 percent PMF storm event would overtop the existing dam embankment by an average of approximately 1.6 ft, and the existing spillway does not have adequate capacity to safely convey the regulatory SDF. Subsequently, two alternatives were analyzed to increase spillway capacity resulting in the development of design drawings for a new 2-cycle labyrinth weir.

Schnabel Hydrologic and Hydraulic Modeling (2014)

Schnabel was retained by the City to review previous analyses and perform an independent hydrologic and hydraulic analysis to assess spillway capacity compliance for Mead Reservoir Dam. The revised Schnabel modeling was performed using the USACE's HEC-HMS (Version 4.1) hydrologic model software.

Table 3. Summary of Dam Data used in 2014 Schnabel H&H Analyses

Drainage Area	6.55 square miles
Top of Dam	544.3
Crest of Emergency Relief Weir	542.7
Primary Spillway Crest	536.1
Normal Pool Storage	1,461 acre-feet
Maximum Storage (EL 544.3)	2,292 acre-feet
Maximum Spillway Discharge (EL 544.3)	3,658 cfs

USGS StreamStats was used to generate a watershed delineation of Mead Reservoir Dam and was found to be about 6.55 square miles. This is similar to the watershed size from the Phase I study (6.39 sq. mi.) and the 2010 EA (6.65 sq. mi.). Figure 1 presents the watershed boundary overlain on USGS topographic mapping. Infiltration and hydrologic losses in the model were represented using the SCS Curve Number (CN) methodology. A composite CN was computed for the watershed using GIS-based datasets for Hydrologic Soils Group mapping (Figure 2). The watershed is generally comprised of Type A and B soils with lesser amounts of Type C soils. Landcover in the watershed is dominated by deciduous and mixed forest with minor residential and agriculture areas. The composite CN was computed to be 54.

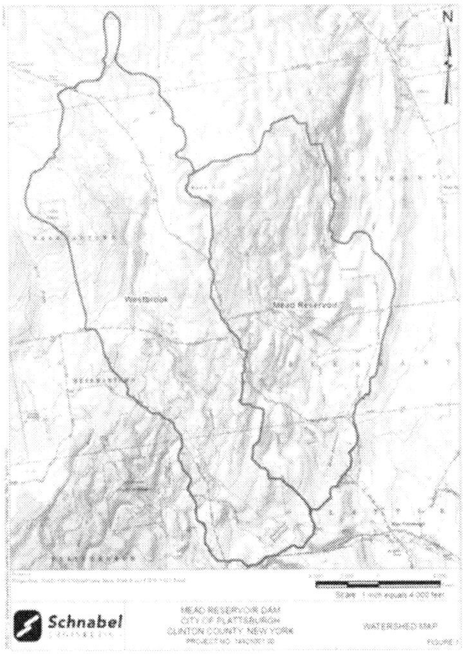

Figure 1. Watershed Map

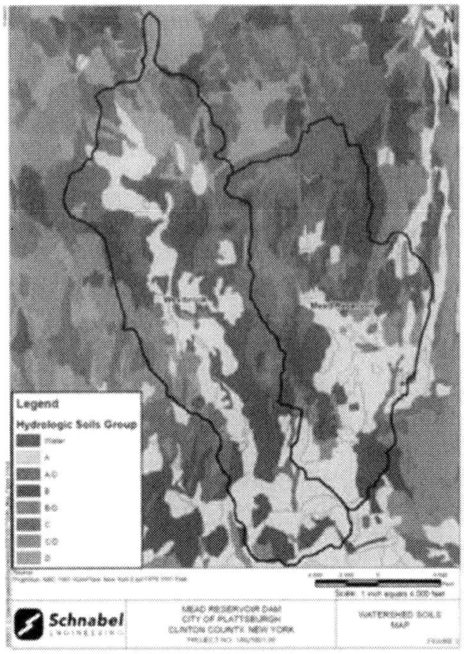

Figure 2. Soils Map

For the transformation of excess rainfall into runoff, the refined analyses utilized the Snyder Unit Hydrograph Methodology. The SCS unit hydrograph methodology utilized in the 2010 and 2012 analyses has often been found to be overly conservative in terms of peak flow predictions. From the authors' perspective, the Snyder Unit Hydrograph Method and/or the Clark Unit Hydrograph method have been found to better represent the rainfall-runoff process for extreme storm events; however these methodologies typically require calibration to develop hydrologic parameters. As the Snyder Unit Hydrograph method had already been applied to this watershed, that methodology was selected for the current analysis with the intent to validate previous hydrologic parameters or refine the parameters through calibration. As noted previously, the Snyder Unit Hydrograph Method generally uses two hydrologic parameters to define the unit hydrograph for a basin: C_t, which represents a basin coefficient that typically ranges from 1.8 to 2.2; and C_p, which is a peaking coefficient that typically ranges from 0.4 to 0.8. We developed an estimate for C_t based on the Taylor and Schwartz methodology *(Dodson & Associates, Hands-on HEC-1, June 1992).* This method was developed in 1952 based on analysis of 20 watersheds in the North and Middle Atlantic States.

$$C_t = \frac{0.6}{\sqrt{S}} \tag{1}$$

Where:

S = watershed slope (feet per foot)

An estimate of the watershed slope was computed using the USGS StreamStats application and found to be approximately 0.073 feet/feet. This slope gives a value for C_t

of 2.2 which is slightly greater than the 2.0 used in the USACE's Phase I study, but still within the recommended range. Using this value, a watershed lag time (T_L) of 5.1 hours was computed using the Snyder watershed lag equation:

$$T_L = C_t(L \times L_{ca})^{0.3} \qquad (2)$$

Where:

C_t = Coefficient representing variations in watershed topography (2.2)
L = Watershed length in miles (6.5 miles)
L_{ca} = Length along main stream to watershed centroid (2.6 miles)

The values for L and L_{ca} were obtained from USGS topographic mapping. A value of 0.625 was used for the peak coefficient (C_p), consistent with the 1981 Phase I report and within the recommended range of 0.4 to 0.8.

<u>Reservoir Routing</u>

The existing primary spillway at Mead Reservoir Dam is an uncontrolled overflow weir with a width of 40 ft and a crest elevation of 536.1 (NGVD 1929). Based on the original design drawings, the spillway has a 2:1 (H:V) sloping upstream face and a rounded downstream face with a 3.0'-radius (See Figure 3).

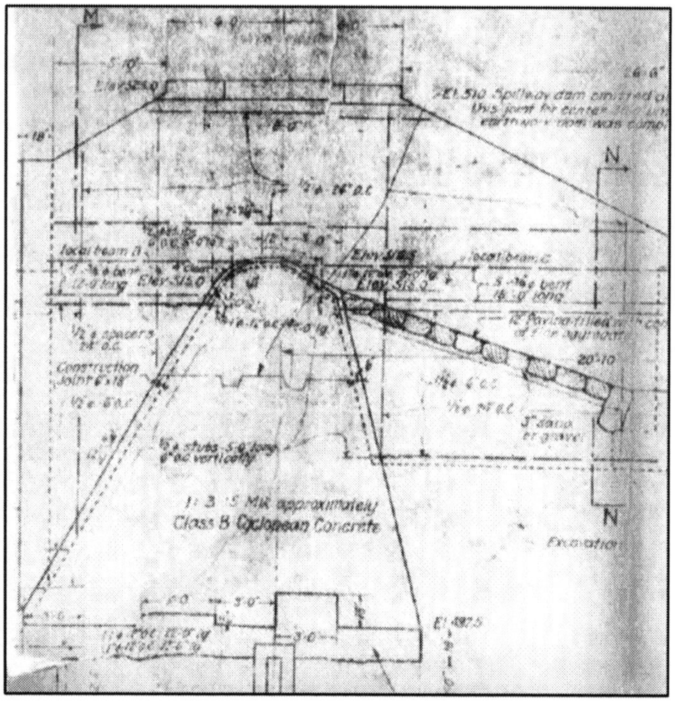

Figure 3. Original Design Drawing Spillway Section

Weir coefficients for the existing spillway were assigned based on review of Brater and King's *Handbook of Hydraulics* (1976) for Weirs of Irregular Section. The book notes that these irregular weir sections were evaluated at the hydraulics laboratory of Cornell University in 1903 to determine the coefficients of discharge for weirs modeled after various types of dams. An excerpt from the Handbook of Hydraulics depicting the various weir crest shapes included in the research study is provided in Figure 4.

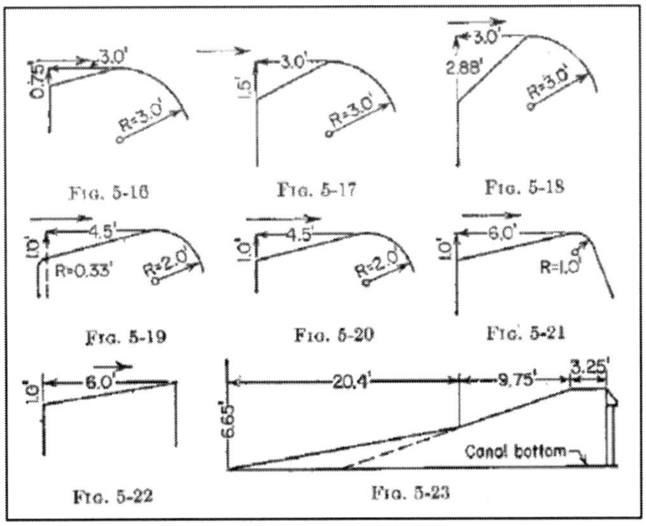

Figure 4. Weir Sections from Experiments at Cornell University

Note that the weir section labeled *Fig 5-17* matches the design drawings for the Mead Reservoir spillway: a sloping upstream approach with a length of 3.0', and a vertical drop of 1.5'. The radius of curvature of the downstream face is also a match at 3.0'. Considering that the plans for Mead Reservoir were completed in 1923, it is likely that the research performed at the Cornell University hydraulics laboratory (1903) was used as the basis of design. Weir coefficients associated with this weir crest cross section were summarized in *Table 5-13* from the *Handbook of Hydraulics* (Figure 5).

Table 5-13. Values of C in the Formula $Q = CLH^{3/2}$ from Experiments at Cornell University

Experiments were made on models resembling existing dams (except that the last two experiments were made on actual dams).

No. of figure	Length of model in feet	Head in feet, H									
		0.5	1.0	1.5	2.0	2.5	3.0	3.5	4.0	4.5	5.0
5-16	7.94	....	3.30	3.32	3.36	3.40	3.43	3.48	3.53	3.62	3.72
5-16	15.97	3.32	3.44	3.46	3.42	3.41	3.46	3.50			
5-17	7.98	....	3.38	3.46	3.51	3.55	3.58	3.62	3.68	3.74	3.83
5-17	15.97	3.22	3.48	3.61	3.67	3.70	3.72				
5-18	15.97	3.15	3.45	3.64	3.75	3.82	3.87	3.88			
5-19	15.97	3.23	3.34	3.43	3.52	3.59	3.64				
5-20	15.97	3.18	3.30	3.37	3.42	3.46	3.49	3.52	3.54		
5-21	15.97	3.28	3.50	3.54	3.52	3.36	3.31	3.30			
5-22	15.97	3.53	3.54	3.55	3.50	3.35	3.27	3.25	3.25		
5-23	15.93	3.13	3.14	3.10	3.14	3.20	3.26	3.31	3.37		
5-24	....	3.09	3.11	3.33							
5-25	....	....	3.30								

5-44

Figure 5. Discharge Coefficients from Weir Experiments at Cornell University

The values associated with *Figure 5-17* from the table were used to assign weir coefficients at various head values for the Mead Reservoir spillway. Values at head depths greater than five feet were extrapolated. The computed weir discharge values at various reservoir elevations are summarized in Table 4. Note that while the original design drawing shows a spillway crest elevation of 516.5, that elevation was based on a local datum, and the recent survey during the 2010 EA indicates a spillway crest elevation of 536.1.

Table 4. Mead Reservoir Spillway Discharge Rating Table

Elevation (NGVD 1929)	Head (ft)	C	Q (cfs)
536.1	0	3.30	0
537.1	1	3.38	135
538.1	2	3.51	397
539.1	3	3.58	744
540.1	4	3.68	1,178
541.1	5	3.83	1,713
542.1	6	3.85	2,263
543.1	7	3.87	2,867
544.1	8	3.89	3,521
545.1	9	3.91	4,223
546.1	10	3.93	4,971

Emergency Relief Weir

As depicted on the original design drawings, the Mead Reservoir Dam was constructed with an Emergency Relief Weir located in the left embankment, approximately 60 ft to the northeast of the primary spillway (Figure 6.). The design drawings indicate that the Emergency Relief Weir has a length of approximately 163 ft and a crest elevation of EL 524.0, which is 1.5 ft lower than the dam crest of EL 525.5. Adjusting to the NGVD 1929 datum provides an Emergency Relief Weir crest elevation of approximately EL 543.6 and dam crest elevation of approximately EL 545.1. The 2010 EA does not note the presence of the emergency relief weir; however, the PondPack hydrologic model does include a dam crest to the left (northeast) of the spillway that is lower than the main embankment. A plot of the crest profiles for the main embankment and left embankment based on the 2009 field survey is provided in Figure 7. While not a level overflow 1.5 ft below the main embankment crest, the left embankment is consistently lower than the main embankment and has similar horizontal dimensions as the design drawings. The Emergency Relief Weir was included in the HEC-HMS hydrologic model as a dam top based on the surveyed profile points and a weir coefficient of 2.6.

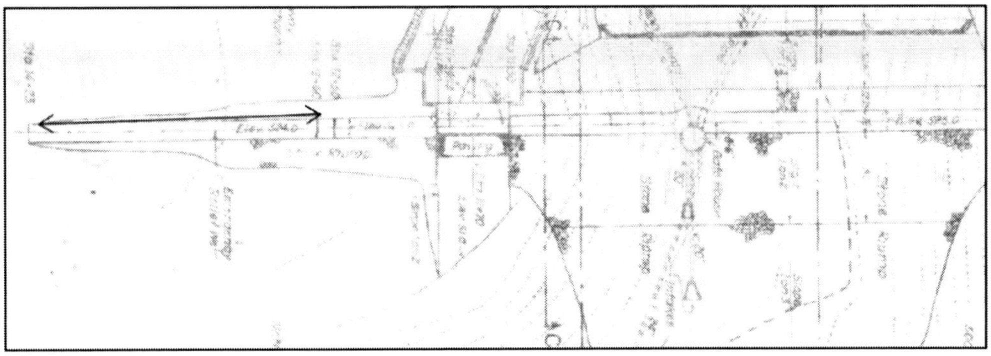

Figure 6. Excerpt from Original Design Drawings – Plan View of Emergency Relief Weir

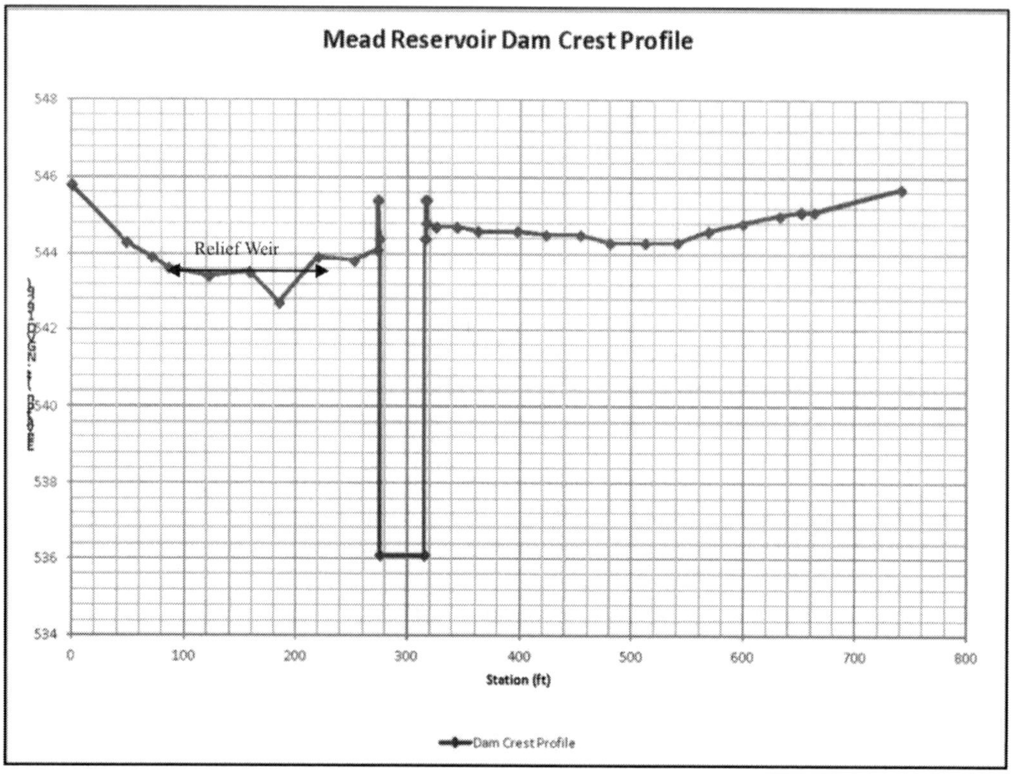

Figure 7. Dam Crest Profile from 2009 Survey Data

Reservoir Storage

Table 5 summarizes the storage rating curve used in the HEC-HMS model to route the inflow design flood through Mead Reservoir.

Table 5. Mead Reservoir Stage vs. Storage Rating Table

Elevation (NGVD 1929)	Storage (ac-ft)
489.7	0.0
491.5	2.01
506.5	142.7
521.5	572.8
533.0	1,206.9
540.0	1,781.6
550.0	2,967.6

Spillway Design Flood Evaluation

The Schnabel HEC-HMS model was executed for the 72-hour PMF using the 29.8 inches of total rainfall consistent with the 2010 EA. The model results for the 50 percent PMF storm event predicted a peak inflow of approximately 3,951 cfs, a peak outflow of approximately 3,235 cfs, and a peak reservoir elevation of approximately 543.5. Based on the model results, the dam can pass the 0.5 PMF storm event with approximately 0.8 ft of freeboard for the main embankment; however, there would be flow over the Emergency Relief Weir with a maximum depth of up to 0.8 ft and maximum flow rate of approximately 50 cfs.

Table 6 presents the results of the Schnabel hydrologic model for the 0.5 PMF storm event in comparison with the results from the previous analyses. The USACE's Phase I study and the Schnabel model both utilized the Snyder Unit Hydrograph methodology, while the 2010 EA used the SCS Unit Hydrograph methodology. Both the USACE's Phase I analysis and the Schnabel analysis concluded that the existing spillway could pass the 0.5 PMF storm event, while the 2010 EA concluded that the dam would overtop during that storm event. Note that the USACE's Phase I study utilized HMR-33 rainfall data, while the EA and Schnabel analyses used HMR-51/HMR-52 rainfall data.

Table 6. Mead Reservoir – Comparison of Studies for 0.5 PMF Storm Event

Study	Inflow (cfs)	Outflow (cfs)	Peak Reservoir Elevation (ft)
USACE's Phase I (1981)	3,290	2,606	543.2
EA (2010)	8,826	7,840	545.9
Schnabel (2014)	3,951	3,235	543.5

Model Validation

The results of the revised hydrologic modeling predict peak flows and associated reservoir elevations significantly less than the modeling performed for the 2010 EA. Based on the authors' past experience, this difference is common when comparing modeling of the same watershed using SCS Unit Hydrograph methodology versus alternate unit hydrograph methodologies including the Snyder Unit Hydrograph methodology. In an effort to validate and/or calibrate the Snyder Unit Hydrograph methodology and the hydrologic parameters used in the analysis better represent the rainfall-runoff response of the watershed during extreme storm events than the SCS unit hydrograph, a review of nearby stream gages and precipitation gages was conducted to identify historic storm events to analyze with the hydrologic models. Large rainfall-induced local floods were identified for November 9, 1996, June 26· 1998, and August 28, 2011 (Tropical Storm Irene). The City of Plattsburgh did have reservoir elevation measurements (Mead Reservoir and Westbrook Reservoirs) for each of these three storm events, however detailed rainfall data was only available for the 2011 Tropical Storm Irene. Subsequently in an effort to maximize the available reservoir elevation and rainfall data, the hydrologic modeling was expanded to include the adjacent Westbrook Reservoir No. 2, which is another water supply reservoir owned and operated by the City.

Westbrook No. 2 Watershed

The watershed for Westbrook Reservoir No. 2 is adjacent to the west side of the Mead

Reservoir watershed. USGS StreamStats was used to generate a watershed delineation for Westbrook Reservoir No. 2 and was found to have an area of approximately 8.26 square miles. Soils mapping and land use mapping were used to compute a composite runoff curve number for the watershed and was found to be approximately 61. The Snyder Unit Hydrograph methodology was also used to simulate the rainfall-runoff response of the Westbrook No. 2 watershed. As described in Section 3.1, the Taylor-Schwartz equation was used to compute a value of the Snyder parameter, C_t (2.4) based on average watershed slope from USGS Stream Stats (330 ft/mile). The value of C_t was then used in the Snyder watershed equation to compute a watershed lag time (T_L) of 6.78 hours. A value of 0.625 was used for the Snyder peaking coefficient (C_p), consistent with the value used for Mead Reservoir Dam. Stage versus storage and stage versus discharge relationships for Westbrook Reservoir No. 2 were taken from the Engineering Assessment performed for that facility.

August 28, 2011 (Hurricane Irene)

Hurricane Irene made landfall in North Carolina on August 27, 2011, and subsequently tracked northward up the east coast. By the time it reached New Jersey, the hurricane was downgraded to a Tropical Storm. While the top wind speeds dropped below hurricane force, the storm produced intense rainfall and resulted in extreme flooding in portions of eastern New York. The remnants of Hurricane Irene passed through the Plattsburgh area in the afternoon/evening of August 28, 2011. Figure 8 presents total rainfall depths along the East Coast prepared by the NOAA Hydrometeorological Prediction Center.

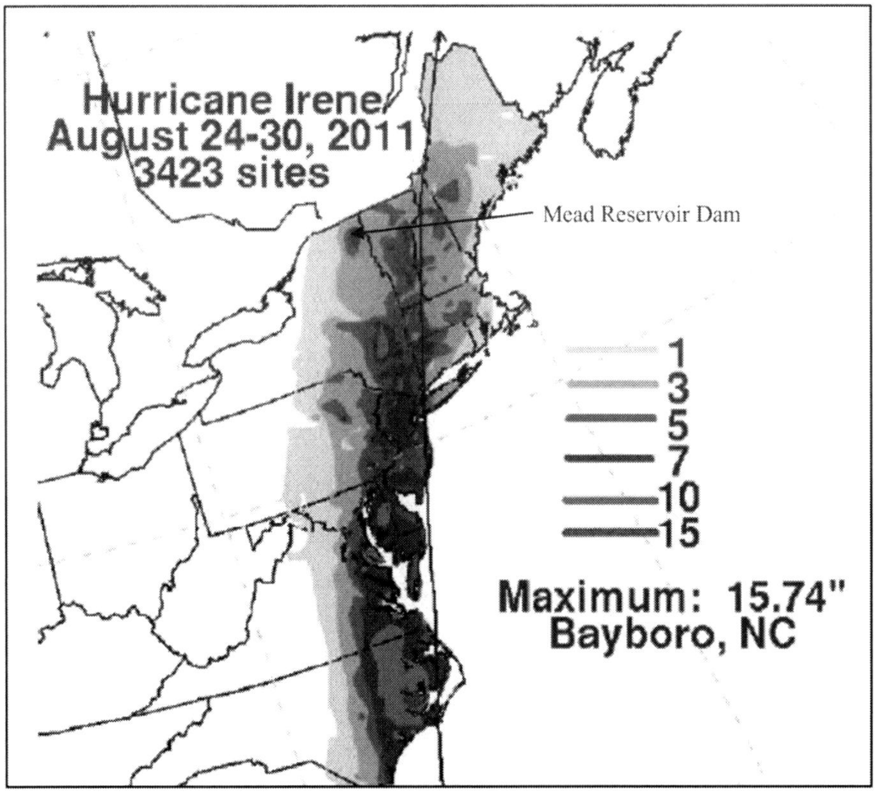

Figure 8. Hurricane Irene Rainfall Totals

Note the area of northeastern NY near Plattsburgh that has a small zone where total rainfall depths were approximately 7 inches. For application in the hydrologic model, hourly rainfall would be required to develop the inflow hydrograph and route it through Mead Reservoir and Westbrook Reservoir No. 2. Hourly rainfall data was first obtained from the National Climatic Data Center for the Plattsburgh International Airport precipitation gage. This gage is located approximately 8 miles southeast of Mead Reservoir and measured a total of 3.77 inches of rainfall. However, as can be seen in Figure 8, there is a good deal of variability in potential rainfall depths around Plattsburgh, with greater potential rainfall depths over the Mead Reservoir and Westbrook Reservoir No. 2 watersheds. In an effort to better understand the potential spatial variability in rainfall depths over the subject watersheds, gridded hourly rainfall data for the storm was obtained from the Earth Observing Laboratory of the National Center for Atmospheric Research. The data is classified as Stage IV multi-sensor data projected onto 4km grids and accumulated for hourly, 6-hourly, and 24-hourly mosaics. The data is collected from regional Doppler radar gages and has undergone manual quality control by the National Centers for Environmental Protection Environmental Monitoring Center to remove bias and validate with ground measurements. Figure 9 presents the 24-hourly rainfall depths for August 11, 2011, at the centroid of each grid cell overlain on the Mead Reservoir and Westbrook No. 2 watershed boundaries.

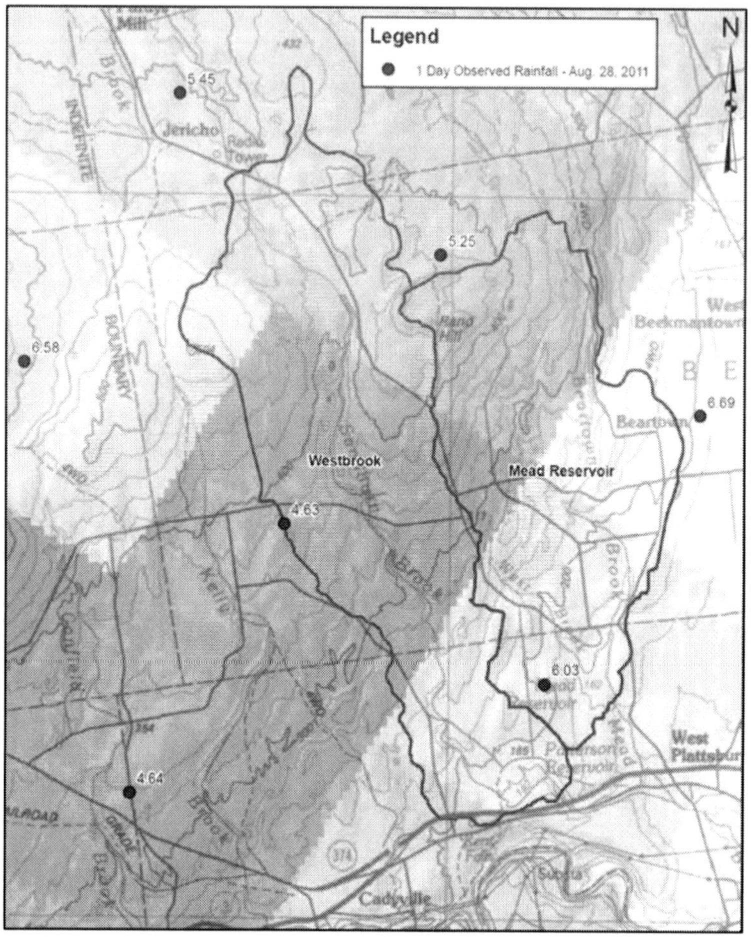

Figure 9: Gridded 24-Hourly Rainfall Totals for August 28, 2011

As can be seen in Figure 9, 24-hour rainfall depths do vary in the watersheds with a maximum of approximately 6.69 inches and a minimum of approximately 4.63 inches. Based on our understanding of the precipitation gage technology and limitations during extreme storm events, it is the authors' opinion that the Stage IV multi-sensor gridded data provides a better representation of the spatial and temporal distribution of rainfall that fell in the watersheds when compared with hourly point data measured during Tropical Storm Irene at the Plattsburgh International Airport located almost 10 miles to the southeast of the watersheds. Subsequently, hourly grids were downloaded and an example of the gridded rainfall data is presented in Figure 10. Note that rainfall depths are presented in millimeters and the values are for 7 PM on August 28, 2011. This was one of the peak hourly rainfall depths reported for the storm event.

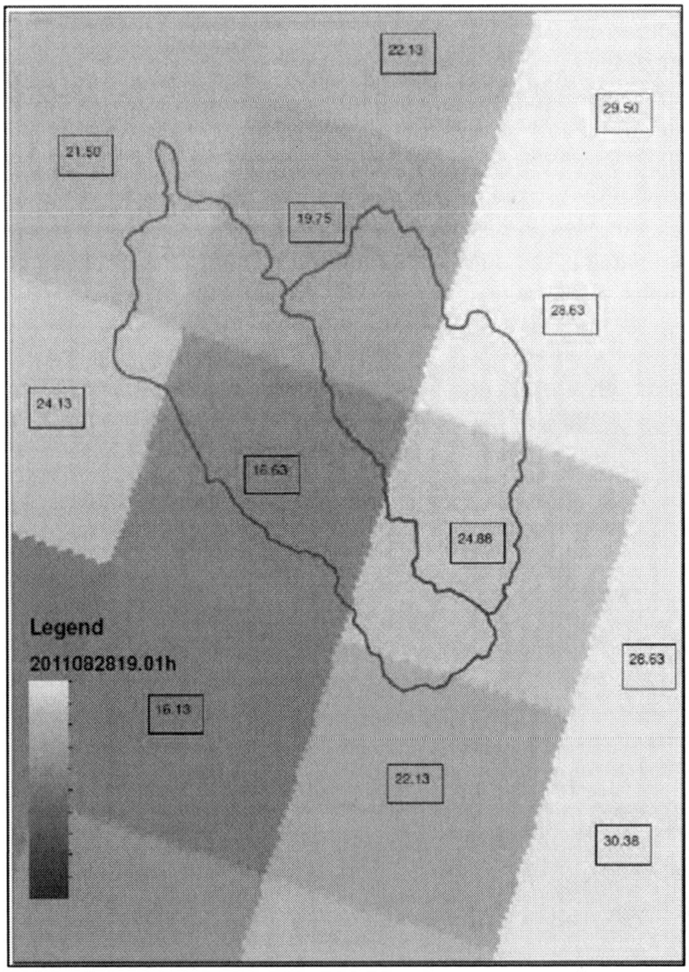

Figure 10. Gridded Rainfall Totals (mm) for 7 PM on August 28, 2011

The rainfall depths from each cell were used to develop a composite hourly rainfall depth based upon the area weighted average of the cells within each of the two watersheds. Based on the gridded rainfall data, the Mead Reservoir watershed had a composite 24-hour rainfall depth of 5.82 inches and the Westbrook No. 2 watershed had a composite 24-hour rainfall depth of 5.23 inches, with the vast majority of the rain falling between 8 AM and 8 PM. To put the magnitude of the event into context, the 24-hour 100-year design rainfall depth from NOAA Atlas 14 is approximately 5.12 inches and the 24-hour 500-year design rainfall depth is approximately 7.34 inches. Thus, the Tropical Storm Irene event for the Mead Reservoir and Westbrook No. 2 watersheds was well in excess of a 100-year recurrence interval storm event and is suitable for validation of the watershed response to extreme storm events.

Tropical Storm Irene Model Results

The gridded composite rainfall data for Tropical Storm Irene was input to the HEC-HMS hydrologic model for Mead Reservoir and Westbrook Reservoir No. 2 and was executed using both the Schnabel (Snyder Unit Hydrograph) and the 2010 EA (SCS Unit Hydrograph) basin models. Table 7 presents a summary of the results of the two basin models as well as observed values for peak reservoir elevation and outflow. Reservoir stage hydrograph plots are presented in Figures 11 and 12 for Mead Reservoir and Westbrook Reservoir No. 2, respectively. Note that the peak observed elevations (red data point) were estimated based on the slopes of the rising limb and receding limb of the stage hydrographs. This was required as the last field measurements of reservoir elevations were recorded at 8:30 PM on August 28, 2011, and the next reading wasn't recorded until 7:45 AM on August 29, 2011. While the rain had generally stopped by 8:30 PM on August 28, 2011, the reservoir was still rising and is estimated to have reached peak elevation at around midnight on August 29, 2011.

Table 7. Tropical Storm Irene Model Results

Parameter	Mead Reservoir			Westbrook Reservoir No. 2		
	Observed	SCS	Snyder	Observed	SCS	Snyder
Peak Elevation (ft)	537.5	539.1	538.1	599.7	601.0	599.8
Peak Inflow (cfs)	N/A	1,145	579	N/A	1,280	667
Peak Outflow (cfs)	229	699	381	546	994	565

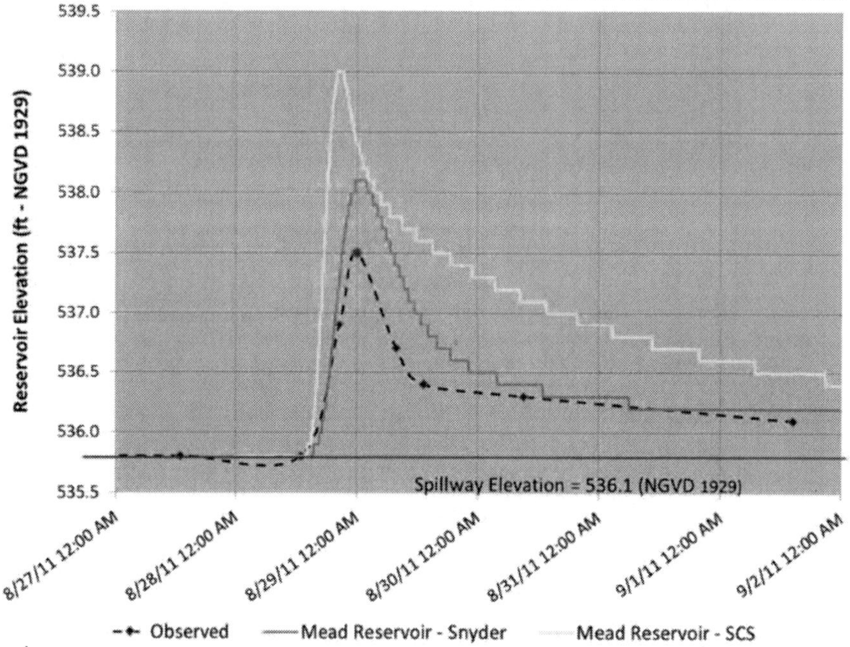

Figure 11. Stage Hydrographs for Tropical Storm Irene – Mead Reservoir

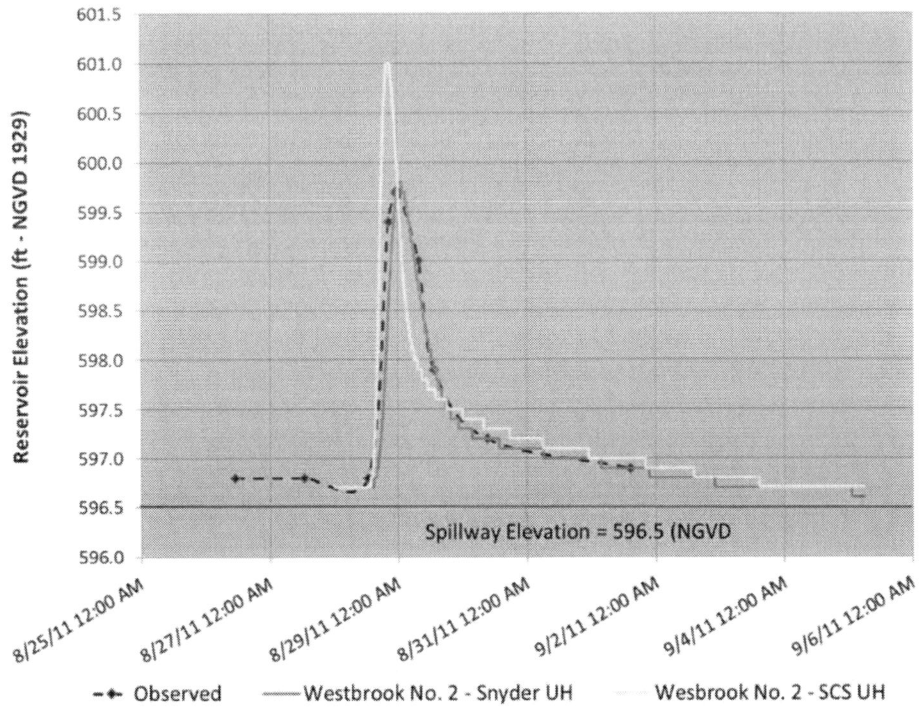

Figure 12. Stage Hydrographs for Tropical Storm Irene – Westbrook Reservoir No. 2

The model results for Mead Reservoir indicate that both the Snyder Unit Hydrograph and SCS Unit Hydrograph predicted peak reservoir elevations greater than the estimated observed peak. However, the Snyder Unit Hydrograph approach does replicate the timing of the observed hydrograph better, and the peak elevation is much closer to observed values when compared with the SCS Unit Hydrograph results.

The model results for Westbrook Reservoir No. 2 show excellent correlation for the Snyder Unit Hydrograph methodology. The results from the SCS methodology show a peak elevation much higher than the estimated observed peak and the shape of the hydrograph is much narrower, indicating an apparent "flashier" watershed response than was observed.

In general, the Snyder Unit Hydrograph methodology better replicated observed reservoir elevations for both Mead Reservoir and Westbrook Reservoir No. 2 during Tropical Storm Irene. Note that no adjustments to hydrologic parameters were made to better fit with observed data. A reduction in the CN and/or C_p could have allowed better correlation for Mead Reservoir. However, the values used for the Snyder Unit Hydrograph were computed using the same GIS-based methodology for Mead Reservoir and Reservoir No. 2. Subsequently the geographic location of the adjacent watersheds and similar topographic relief suggest that an adjustment to parameters at Mead Reservoir should be accompanied by a corresponding reduction for Westbrook Reservoir No. 2. As the model results for Westbrook No. 2 showed nearly perfect correlation, it was concluded that retaining the existing values for hydrologic parameters offers the best combination of adequately replicating the rainfall-runoff response of the watershed while retaining a level of conservatism.

Design Storm Comparison

As Tropical Storm Irene was the only historic storm where both detailed rainfall data and reservoir elevations were available, additional comparisons of model output for standard design storms with USGS StreamStats values were conducted to help support the application of the proposed Snyder Unit Hydrograph methodology and parameters for Mead Reservoir Dam. Rainfall data (depth and distribution) for the 100-year and 500-year storm events were obtained from the Northeast Regional Climate Center. Table 8 presents a summary of peak flows for both Mead Reservoir and Westbrook Reservoir No. 2 for the 100-year and 500-year recurrence interval events.

Table 8. Design Storm Model Results Comparison

Peak Flow (cfs)	Mead Reservoir			Westbrook Reservoir No. 2		
	StreamStats	SCS	Snyder	StreamStats	SCS	Snyder
100-Year	560	986	387	578	1484	610
500-Year	736	2423	960	752	3287	1305

As can be seen in Table 8, the model results with the SCS Unit Hydrograph methodology predicts peak flows typically two to four times higher than the USGS StreamStats estimates. The peak flows from the models using the Snyder Unit Hydrograph also predict higher flows than StreamStats for all but the Mead Reservoir 100-year storm; however, the magnitudes of the peaks are much closer to the StreamStats values than the SCS Unit Hydrograph models. While this analysis doesn't represent an actual calibration as the USGS peak flows are estimates from regional regression equations; however, it does help support that the model results using the SCS Unit Hydrograph typically over predict peak flows.

CONCLUSIONS

Hydrologic analysis in the support of dam spillway design may best be described as an inexact science. In evaluating spillway capacity of a dam, it is unlikely that two independent water resources engineers would develop the exact same spillway design flood for the dam. Hydrology is founded in science but the practical application often relies heavily on the engineer's experience, training, as well as project budget. Unit Hydrograph methodologies were developed to assist a hydrologic engineer in evaluating the rainfall-runoff response of a watershed. Early methodologies include the Snyder and Clark Unit Hydrograph Methodologies. These methods offer the user flexibility in the prediction of peak flows as well as the timing of peak flows and shape of the hydrograph; however typically require calibration/validation for parameter values. The SCS Unit Hydrograph was later developed as a user friendly methodology that requires a single hydrologic parameter in an effort to increase the consistency in design of numerous watershed flood control structures throughout the country. While the SCS unit hydrograph methodology is easy to use and has been applied in many dam safety designs and evaluations, many users have found that the method often is conservative in over-predicting peak flows when compared with observed values. Nevertheless, many engineers continue to apply the SCS unit hydrograph methodology due to the ease of use and the unfamiliarity with older unit hydrograph methodologies as well as lack of budget for model calibration/validation. These calibration/validation studies rely heavily on precipitation data and observed stream flows at gaging stations or reservoir elevation data.

The analyses presented in this paper attempt to demonstrate the value of hydrologic model calibration/validation to historic storm events and in particular the need to leverage the latest technologies from the national meteorological agencies. Traditional precipitation and stream gaging stations rarely are located at a fine enough resolution to apply to a small watershed such as that of Mead Reservoir Dam. Leveraging emerging hydrologic precipitation data such as gridded multi-sensor data can be a very cost-effective solution in terms of application to hydrologic modeling and ultimately to the owner of a dam facing a spillway rehabilitation project. The NWS multi-sensor gridded rainfall data is used on a daily basis for the River Forecast Centers located throughout the United States. This data is available free to the public and while it may represent a different approach to rainfall data of many dam safety engineers, this type of emerging technology can be a critical tool in cost-effectively implementing safety improvements to the nation's aging dams and appurtenant infrastructure.

REFERENCES

Brater, E.F. and King, H.W. (1976). *Handbook of Hydraulics (Sixth Edition).* McGraw-Hill, Inc., New York, NY.

Dodson & Associates. (1992). "Hands-On HEC-1." Houston, TX.

Everleth, J.A., Adnams, K.E. (2012). "Engineering Design Report – Mead Reservoir Dam." CHA, Albany, NY.

Phase I Inspection Report, Mead Reservoir Dam. (1981). New York District Corps of Engineers, New York, NY.

DESIGN OF LEVEE MODIFICATIONS TO ACCOMMODATE LONG-TERM RIVER SCOUR

Justin Dominguez[1]
James Nickerson[2]

The Green River in is a meandering watercourse that passes through the City of Kent, Washington. Levees have been constructed along the banks of the Green River since the late 1800's, and many of these are at or very close to the tops of the riverbanks. The levees were originally constructed to protect agricultural lands, but in recent years the City of Kent and surrounding area has become an industrial center and one of the largest transportation hubs for goods in the Western US.

The Green River is very dynamic, and frequent floods undercut and saturate the riverbank slopes and deepen the river channel causing instability and sloughing of the riverbanks. The levees have historically performed well during flood events, but numerous, costly repairs to the riverbanks have been required to maintain the integrity of the levees.

In 2010, the City of Kent began engaging consultants to evaluate the levee systems protecting the City for accreditation by FEMA in the National Flood Insurance Program. The evaluations considered ongoing lateral and vertical movement of the river channel and subsequent riverbank instability due to high velocity flows. A number of levee segments were identified that were only marginally stable and that would likely soon become unstable due to riverbank scour and sloughing. This paper will outline the methodology developed to incorporate future river scour into the riverbank stability evaluation and mitigation. This paper will also discuss several repair alternatives that were developed and implemented to improve stability and meet the design criteria requirements for accreditation of the levee system.

[1] GEI Consultants, Inc., jdominguez@geiconsultants.com
[2] GEI Consultants, Inc.

DETECTION OF LEVEE VOIDS AT AQUEDUCT CROSSINGS USING GEOPHYSICAL IMAGING TECHNIQUES

Zhihua Li[1], PhD, PE
Atta Yiadom[2], PE

ABSTRACT

This paper presents the results of a recent geophysical imaging investigation to detect voids in levees where crossed by the Mokelumne Aqueducts in the Sacramento–San Joaquin Delta (Delta) region. Recent surficial evidence of voids in the Delta levees due to rodent activities and settlement at the aqueduct crossings have been observed at a number of locations. Accurately identifying the spatial distribution of potential voids is necessary to develop remedial measures to prevent piping damage of the levees. To avoid using invasive techniques such as trenching the levees, aqueduct owner East Bay Municipal Utility District (EBMUD) decided to use non-invasive geophysical imaging techniques to image levee interiors to detect small-scale voids at several Delta levee aqueduct crossing locations. Electrical resistivity tomography and induced polarization techniques were employed to detect potential soil voids and to determine the extent of void distribution. The results indicate that voids may exist in the Delta levees at aqueduct crossings. Further action of employing thin-probing detection techniques will be taken to confirm the existence of voids, followed by filling the confirmed voids with low-pressure grouting. The effect will be examined with further geophysical imaging.

INTRODUCTION

The Mokelumne Aqueducts are water conveying pipelines of East Bay Municipal Utility District (EBMUD), a publicly-owned utility district that supplies water to about 1.4 million people in 35 municipalities on the eastern side of the San Francisco Bay in Northern California. EBMUD's primary water source is Pardee Reservoir on the Mokelumne River, about 90 miles east of the service area. Water from Pardee Reservoir is conveyed by Mokelumne Aqueducts that cross levees protecting the flood plains in the Sacramento–San Joaquin Delta (Delta) region (Figure 1).

The Delta soils are composed of mineral sediments delivered by the Sacramento and San Joaquin Rivers and of peat material derived from decaying marsh vegetation (Galloway et al., 1999). Decomposition of organic carbon in the peat soils has caused land subsidence. Figure 2 shows the mechanism of land subsidence due to decomposition of mineral-rich peat soils.

[1] Associate Civil Engineer, Geotechnical Engineering Section, East Bay Municipal Utility District, 375 11th Street, Oakland, CA 94607, zhihua.li@ebmud.com
[2] Senior Civil Engineer, Geotechnical Engineering Section, East Bay Municipal Utility District, 375 11th Street, Oakland, CA 94607, atta.yiadom@ebmud.com

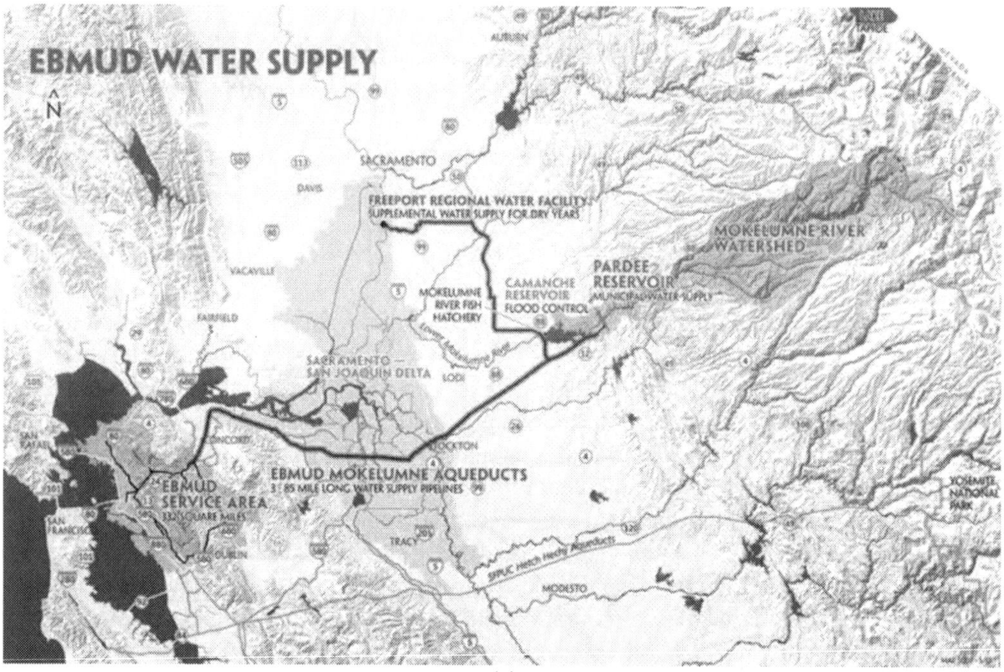

(a)

(b)

Figure 1. (a) EBMUD Water Conveyance System; (b) A Typical Aqueduct-Crossing-Levee Location in the Sacramento-San Joaquin Delta.

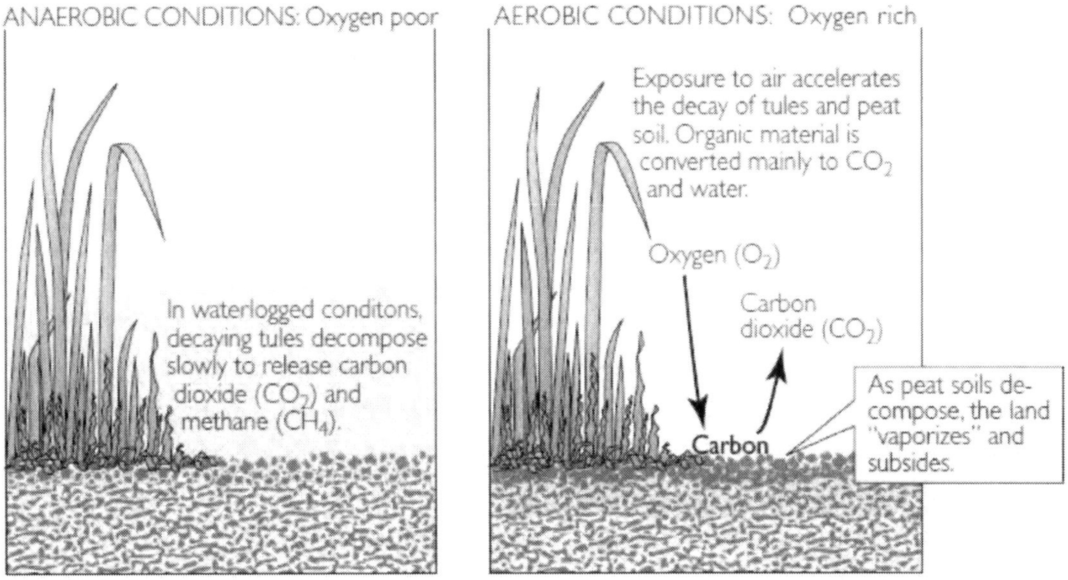

Figure 2. Mechanism of Delta Land Subsidence Due to Decomposition of Peat Soil (Galloway et al., 1999).

The subsidence of land creates problems for structures and levees built over them. The levees, built in the 1800s using hand tools and largely completed by 1930 for protecting reclaimed farmlands from flooding, are highly vulnerable to natural hazards. Most of the levees are also used as public roads and are frequently loaded, especially during the busy farming seasons. In addition, the water levels are year-round high to the levees built on top of the subsiding lands. In the past, there have been levee failures that posed potential threats to the aqueducts and the reclaimed lands in the area.

At the aqueduct crossing locations, the aqueducts are supported on piles. With gradual soil subsidence occurring under the embedded pile caps, cavities could be created. As the Delta is a lush habitat for animals, their activities could also introduce holes in the levees. Recently, surficial evidences of voids in the Delta levees have been observed at a number of aqueduct crossing locations due to rodent activities and settlements. If continuous voids or tunnels exist in the levees underneath the water surface, they could become seepage paths which may lead to piping and could damage to the levees. The challenge is to develop an approach to accurately identify the distribution of the voids, if any, at these crossings and for necessary remedial measures to prevent piping damage of the levees.

APPROACH

One way to identify voids at the crossing locations is by trenching, just as used in geology for investigating an active fault with surface rupture. However, as the levees in the Delta are vulnerable and sensitive, trenching could pose potential risks to the existing levees and EBMUD's crucial aqueducts. Instead, we decided to use geophysical imaging techniques to non-invasively detect the potential small-scale soil voids at the Delta levee aqueduct crossing locations.

Geophysical imaging is a technique based on applied physics that investigates the subsurface As established tools for the oil and gas industry, many geophysical imaging techniques involve solving inverse problems; i.e., making inferences about the subsurface based on the data collected at or near the ground surface. In almost all cases these measured data are only indirectly related to the properties of the subsurface that are of interest, so an inverse problem must be solved in order to obtain estimates of the physical properties underneath the ground surface (Barhen et al., 2000). Figure 3 shows the general idea of an electrical forward and inverse problem.

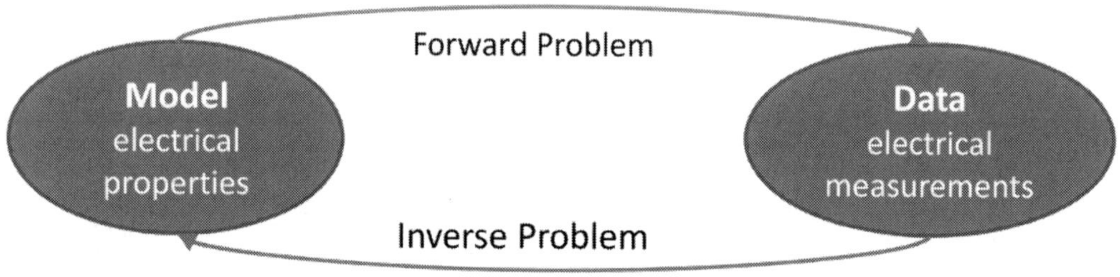

Figure 3. Electrical Forward and Inverse Problem.

We first evaluated several imaging techniques including Ground Penetrating Radar (GPR). Becasue GPR uses high-frequency sources and the penetrating depth is limited for the purpose of this investigation, we decided to utilize low frequency geophysical imaging techniques, such as Electrical Resistivity Tomography (ERT), that achieve a deeper penetration depth. We hired Tremaine & Associates, Inc., to perform the requested survey. As the purpose of this investigation is to detect potential soil voids in the areas adjacent to the embedded metal aqueduct pipelines, the presence of the conductive metal pipelines obviously complicates the “host medium.” As a consequence, two non-invasive geophysical imaging techniques were evaluated and chosen given the subsurface heterogeneities at different site locations: ERT and Induced Polarization (IP) methods (Elwaseif and Miele, 2016).

ERT and IP methods determine the spatial distribution of the low frequency resistivity and capacity characteristics of the medium, respectively. ERT is a well-established non-invasive technique to detect embedded “anomalies” which show electrical resistivity characteristics in contrast to its hosting medium (Telford et al., 1990; Daily et al., 2004; Rings et al., 2008; Kemna, 2000; Binley and Kemna, 2005). Since its early development, this technique has been widely used in many industrial applications including site characterization, hydrogeology, geophysical prospecting, environmental and structural inspections. The surveys are relatively easy to carry out, instrumentation is cost-effective, data processing tools are widely available, and the relationships between resistivity and material properties, such as porosity and moisture content, are well established. The implementation of the method involves injecting current using two current electrodes and measuring potential variations between two voltage electrodes (Figure 4), with the goal to reconstruct the resistivity profile as a function of depth from the potential measurements.

Basically, from the current I and potential V values, an apparent resistivity value is calculated.

$$\rho = k\frac{V}{I} \tag{3}$$

where k is a geometric constant that depends on the dimensions and arrangement of the four electrodes. The spacing of the electrodes, the electrode array coverage, and the scanning sequence will determine the ''effective depth'' to which the current will penetrate, and the resolution of the image.

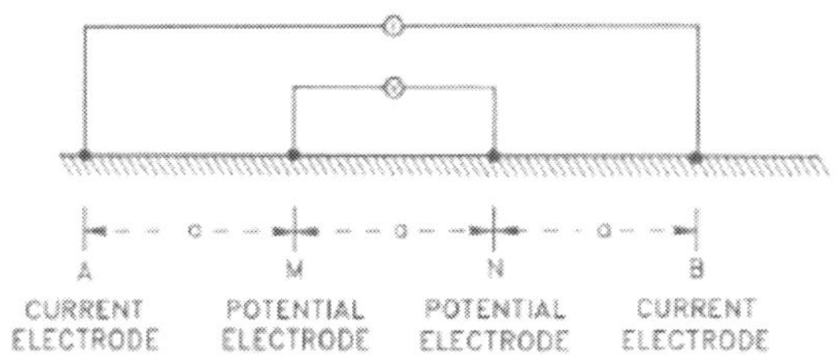

Figure 4. General Principle of ERT Measurements (Example: Wenner Arrangement, Dobrin et al., 1988).

The IP method, also called the Complex Conductivity (CC) method, is given by the ratio of the recorded voltage to the injected current amplitude (resistance), and the voltage decay over a period of time (chargeability), once the current is switched off (Binley and Kemna, 2005; Flores Orozco et al., 2015) (Figure 5). IP method is deployed in a similar manner as ERT, and also adopts the four-electrode measurement approach with two current electrodes and two voltage electrodes. Whereas the resistivity of the soil is principally controlled by electrical conduction within the pore fluid, IP is strongly affected by processes at the fluid-grain interface. As a result, IP method permits additional information about the spatial variability in lithology and grain-surface chemistry. Therefore, IP methods can be effective used in detecting lithology variations. In this investigation, IP method was employed at the same time with ERT survey, with the goal of differentiating the embedded conductive pipeline.

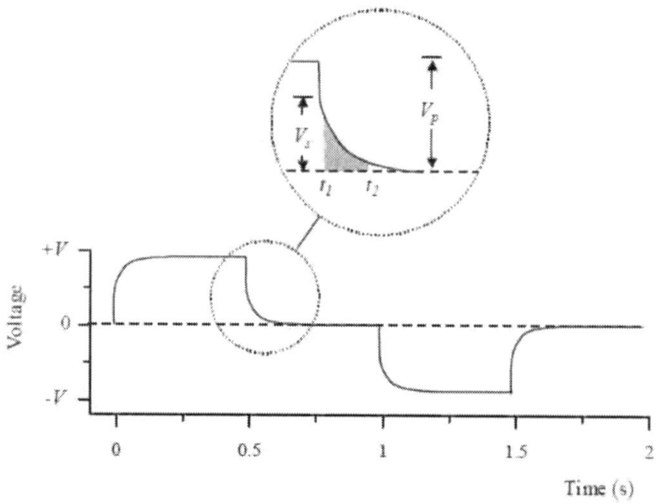

Figure 5. Time Domain Induced Polarization Measurements (Binley and Kemna, 2005)

DATA ACQUISITION AND PROCESSING

A technical report documenting data acquisition, data processing and findings was submitted to EBMUD (Elwaseif and Miele, 2016). This section was based on both the discussion between EBMUD and our consultants through all stages of the survey (personal communications) and the EBMUD report (Elwaseif and Miele, 2016).

In the survey, an IRIS Instruments Syscal Pro 10-channel imaging system equipped with 96 electrodes was employed for both ERT and IP data acquisitions. The survey locations are the Delta levees at the Mokelumne Aqueduct crossings in Contra Costa County and San Joaquin County. A typical survey area at a crossing location is roughly 45 meters long and 6 meters wide. In order to acquire high resolutions to detect potential voids, transmitting and receiving electrodes were placed with small intervals, which varied from 0.2 to 0.5 meters. The depth of investigation reached about 10 meters, which, in most cases, was beyond the landside or waterside levee toe. This investigation depth covered the critical portions of the levee, which were the lower part of the levee underneath the embedded aqueduct pipes. This portion is below the water level so connected voids could lead to water penetrating through the levee, eroding the levee embankment materials, and compromising the integrity of the levees and aqueducts. Figure 6 shows an IRIS Instruments Syscal Pro 10-channel imaging system and a typical setup for data acquisition.

(a) (b)

Figure 6. (a) An IRIS Instruments Syscal Pro 10-channel Imaging System; and (b) A Typical Setup for Data Acquisition.

Eighteen-inch-long customized stainless steel electrodes were employed in the survey and inserted into holes drilled through asphalt or compacted gravel levee surface. Normally about four lines oriented longitudinally to the levees were surveyed. One or two short IP and ERT profiles were collected at each study location using small electrode spacing (about 0.25 meters to obtain higher resolution information regarding the depth and vertical extent of the aqueducts. Seven-meter long wires were fabricated to connect the electrodes to the main instrument cable so that it could remain off the levee without obstructing traffic (Figure 6).

As the ultimate goal of the investigation is to identify voids by comparing the distribution of electrical properties of the subsurface, the objective of the data processing (inversion, or solving inverse problem) was to resolve a model containing a structure consistent with expectations regarding the likely distribution of a physical property within the subsurface in terms of electrical resistivity, while providing a set of theoretical measurements (forward problem) that fit the measured data to some acceptance standards. In order to make the geophysical surveys sensitive to small-scale voids up to the investigated depth, and to prevent the electrical current from being channeled within the aqueducts, Elwaseif and Miele (2016) developed a measurement sequence file that consists of multiple current and potential array configurations.

The IP data was collected simultaneously along with the ERT data. The main goal of IP survey is to provide information about the vertical extent of the aqueducts as high chargeability anomalies with well-defined geometries. The vertical extent of aqueducts is useful in guiding the inversion to find the correct subsurface electrical resistivity distribution model by giving it some known information about the subsurface. In this case the inversion was given a tip that at the upper and lower boundaries of each aqueduct pipeline there are sharp changes in terms of resistivity (Figure 7, email communication by Elwaseif, 2016).

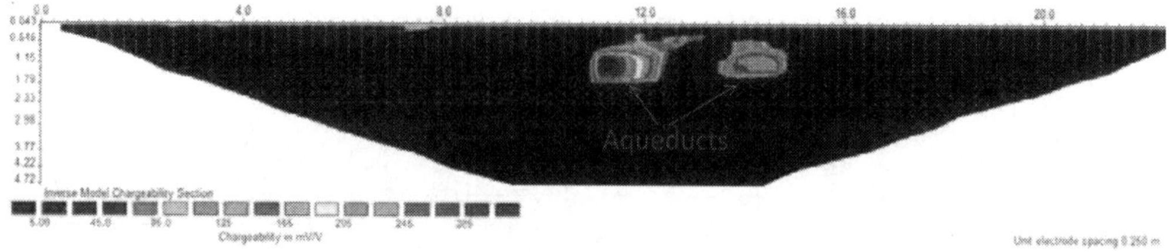

Figure 7. Inverted Induced Polarization Model (email communication by Elwaseif, 2016)

Data processing was based on an iterative routine involving determination of a two-dimensional (2D) simulated model of the subsurface, which was then compared to the observed data and revised. Convergence between theoretical and observed data was achieved by non-linear least squares optimization. How well the observed and calculated theoretical models agree indicates the degree of validity of the true resistivity model by the final root-mean-squared (RMS) error.

RESULTS

Geophysical imaging results illustrating potential voids in levees at aqueduct crossings were obtained. The target of interest, i.e., the voids, is expected to show electrical resistivity contrast compared to the surrounding levee materials. Figure 8 shows a typical survey profile and Figure 9 shows the ERT results at one of the aqueduct crossing locations, manifesting the resistivity reconstructions along the vertical cross sections under five survey profiles P1 through P5 (from west to east).

(a) (b)

Figure 8. Lower Jones Road Aqueduct Crossing: (a) Survey Profiles (Google Earth, Elwaseif and Miele, 2016); (b) Photo (Facing East)

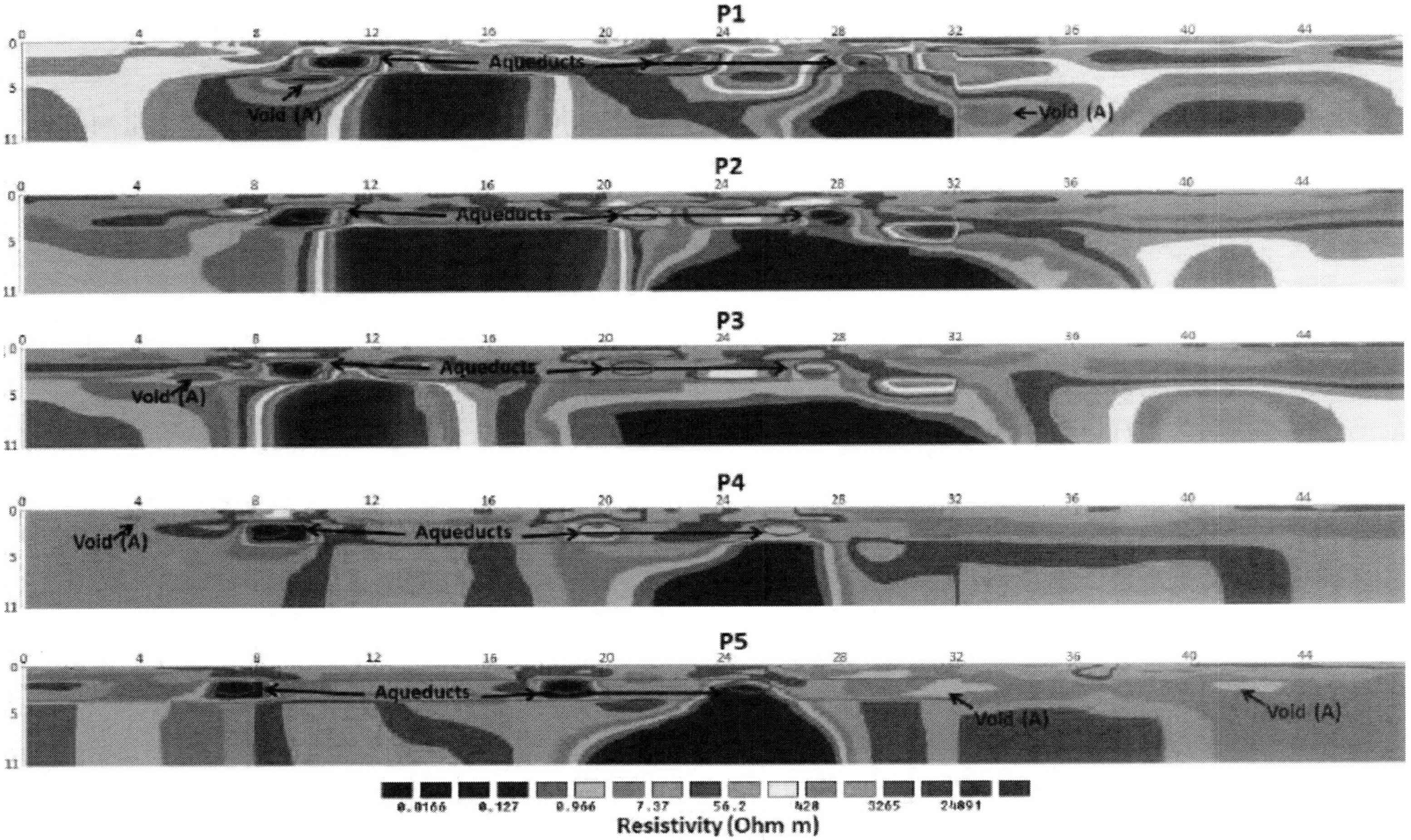

Figure 9. Resistivity Profiles for Lower Jones Road Aqueduct Crossing – Facing East (Elwaseif and Miele, 2016).

In general, the subsurface electrical resistivity distribution reveals that the levee material at the aqueduct crossing is heterogeneous. This could be related to different levee materials (e.g., sand, clay, and peat) when it was constructed, how dense the material was compacted to, or what is the moisture content in the materials, etc. For example, the high resistivity material shown on P1 to P3 (in orange to purple) could indicate well compacted levee material with very low moisture content, while on P4 and P5 replaced with less well compacted materials with increased moisture content (in dark green colors).

From Figure 9, six discrete voids were detected, among which four voids (shown in P3, P4 and P5) are relatively shallow; i.e., depth from surface is less than 4 meters. There are no potential voids shown on P2. On Profile P1, one void is shown to the north of Aqueduct # 3 with the bottom of the void at a depth of approximately 5 meters and another void in the deeper portion of the levee at a depth of approximately 10 meters. This deep void is not seen on other profiles. That is to say, it is not likely continuous. The three red ovals enclosing low resistivity denote the reconstructed cross sections of conductive Mokelumne Aqueducts # 3, #1 and #2 (in sequence from north to south) at a depth of approximately 2 meters beneath the levee surface. The low resistivity under Mokelumne Aqueduct # 3 (in blue) could be due to ground saturation with water leakage from the aqueduct, which also occurs not too far to the same aqueduct to the east.

At other crossing locations, potential voids were also detected using ERT with IP methods. Detailed images and descriptions are not illustrated here.

To confirm the detected potential voids, we plan to use a 1-inch diameter probe which will be inserted gradually into the ground while at the same time injecting water with low pressure from its tips. If sudden loss of water occurs, it is assumed that a void exists at that depth. Then the same 1-inch diameter probe will be employed to inject grout with low pressure to fill the identified/confirmed voids.

CONCLUSION

Geophysical imaging techniques were employed to detect small-scale soil voids in levees at Mokelumne Aqueduct crossing locations in the Sacramento-San Joaquin Delta region in Northern California. As the levees in the region are sensitive, the nondestructive geophysical imaging methods provided an efficient approach for the initial evaluation of potential voids. Approaches were carefully examined for high resolution detection specific to the unique environment and condition of the Delta levees at the aqueduct crossings. Electrical Resistivity Tomography and Induced Polarization were employed with the inversion algorithm developed suitable to the described complexities. Results reveal that potential voids may exist in the aqueduct-crossing levees. Confirmation of the detected potential voids followed by grout filling with low pressure is planned to minimize the possibility of piping damage to the levee and the aqueducts.

ACKNOWLEDGEMENTS

Tremaine & Associates conducted the geophysical investigation.

REFERENCES

Barhen, J., Berryman, J.G., Borcea, L., Dennis, J., de Groot-Hedlin, C., Gilbert, F., Gill, P., Heinkenschloss, M., Johnson, L., McEvilly, T., Moré, J., Newman, G., Oldenburg, D., Parker, P., Porto, B., Sen, M., Torczon, V., Vasco, D., Woodward, N.B., (2000). "Optimization and Geophysical Inverse Problems," *Technical Report.* Ernest Orlando Lawrence Berkeley National Laboratory, San Jose, California.

Binley A. and Kemna A. (2005). "DC resistivity and induced polarization methods." In: *Hydrogeophysics* (eds Y. Rubin and S.S. Hubbard), pp. 129–156, Springer.

Bhattacharya, B., and Ghosh, R. (1988). "A method of inversion and its application to VES measurements." In B.B. Bhattacharya, Ed., *Advances in Geophysics*, Oxford and IBH Publishing Co., 43-64.

Daily W, Ramirez A, Binley A, LaBrecque D. (2004). "Electrical resistance tomography." *The Leading Edge* 23: 438–442.

Dobrin, M., and Savit, C. H. (1988). *Introduction to Geophysical Prospecting (Fourth Edition)*, McGraw-Hill Book Company.

Elwaseif, M. and Miele, M., (2016), *Letter Report for Mokelumne Aqueduct Levee Crossings Study*, submitted to East Bay Mud Utility District.

Elwaseif, M. (2016), Email communication. October 21, 2016.

Flores Orozco, A., Velimirovic, M., Tosco, T., Kemna, A., Sapion, H., Klaas, N., Sethi, R.,Bastiaens, L., (2015). "Monitoring the Injection of Microscale Zerovalent Iron Particles for Groundwater Remediation by Means of Complex Electrical Conductivity Imaging." *Environ. Sci. Technol.* 49, 5593–5600. doi:10.1021/acs.est.5b00208.

Galloway, D., Jones, D.R., and Ingebritsen, S.E. (1999) Eds., *U.S.Geol. Surv. Circ.* 1182.

Google Inc. *Google Earth* (Version 7.1.7.2606) [Software].

Kearey, P., Brooks, M., (1991). *An Introduction to Geophysical Exploration, Second Edition*, Blackwell Scientific Publications, Oxford, pp. 216.

Kemna, A., (2000). *Tomographic Inversion of Complex Resistivity – Theory and Application.* Ruhr-Universität Bochum, Bochum, pp. 196.

Rings, J., Scheuermann, A., Preko, K., Hauck, C., (2008). "Soil water content monitoring on a dike model using electrical resistivity tomography." *Near Surface Geophysics* 6 (2), 123–132.

Telford, W.M., Geldart, L.P., and Sheriff, R.E. (1990). *Applied Geophysics, 2nd ed.,* Cambridge Univ. Press.

RISK ANALYSIS IN A LEVEE SAFETY CONTEXT: THE UPPER WOOD RIVER LEVEE CASE HISTORY

Chris J. Redell, PE[1]
Jose R. Lopez, PE[2]

ABSTRACT

The science of Risk Analysis utilizes a framework that thoroughly evaluates the level of risk and determines what actions, if any, need to be taken to manage that risk. The framework explicitly and implicitly recognizes all the contributing elements of the risk, to include hazard, performance, and consequences. Risk Assessment, Management, and Communications are essential components of the risk framework that work in conjunction to yield more effective decisions about infrastructure investments. All three pieces were implemented within the Wood River-Mel Price Levee Deficiency project in order to better understand and take action to manage the risk that exists. Through a systematic evidence based approach, the nature, likelihood, and magnitude of risk was determined with due regards for uncertainty. Based on these findings, decisions were made to select, implement, and monitor a solution that will reduce the risk to tolerable levels. This framework not only allowed for a more cost effective solution but also allowed for decision making with an outcome-based approach as opposed to an action based approach.

This case history will present how the risk analysis framework was successfully implemented for the Wood River-Mel Price Levee Remediation Project. It will cover how the framework was applied to a project intended to reduce the risk of failure due to underseepage performance issues discovered in 2009. This paper will discuss the history and background of the project, present the potential failure modes which required action, and the solution that was developed in collaboration with the sponsor, utilizing the tools of risk assessment, risk management, and risk communication. It will show how the application of this risk framework within an engineering and decision making context demonstrates improved effectiveness with regards to life safety and investment decisions.

PROJECT BACKGROUND

The Melvin Price Lock and Dam is located 21 miles north of St. Louis, Missouri, and two miles below Alton, Illinois, between the mouths of the Missouri River and the Illinois River. This study is focused on the reach of Upper Wood River levee adjacent to the lock and dam from project station 52+50 to 126+00 as shown in Figure 1. The spatial relationship between the Melvin Price Locks and Dam, the Wood River Levee, and the areas of underseepage concern is presented within Figure 2.

[1] Chris J. Redell, Geotechnical Design Engineer, St. Louis District, USACE, 1222 Spruce Street, St. Louis, MO., chris.j.redell@usace.army.mil.
[2] Jose R. Lopez, Chief Geotechnical Design Section, St. Louis District, USACE, 1222 Spruce Street, St. Louis, MO., jose.r.lopez@usace.army.mil.

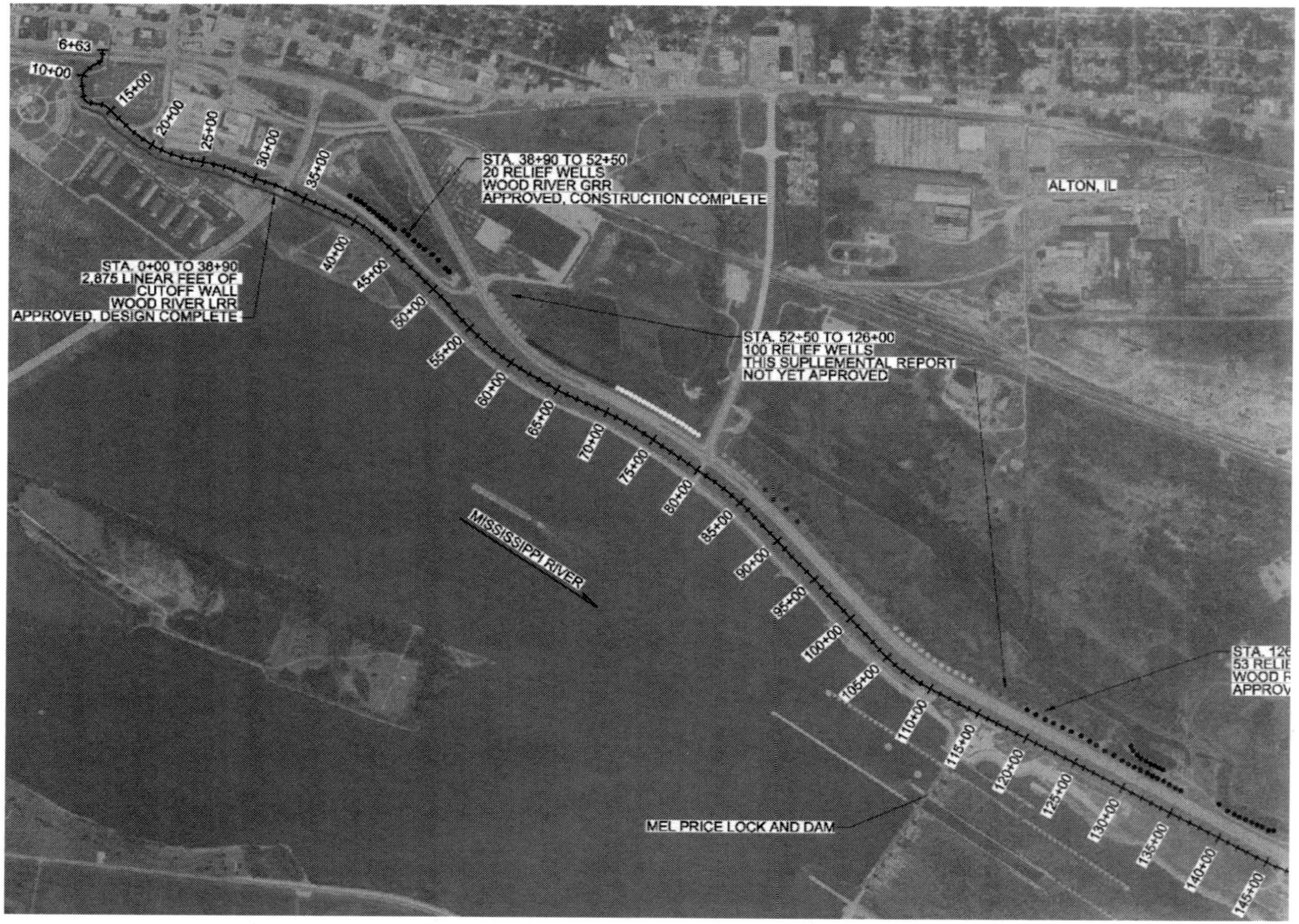

Figure 1: Map of Completed Underseepage Features in Upper Wood River

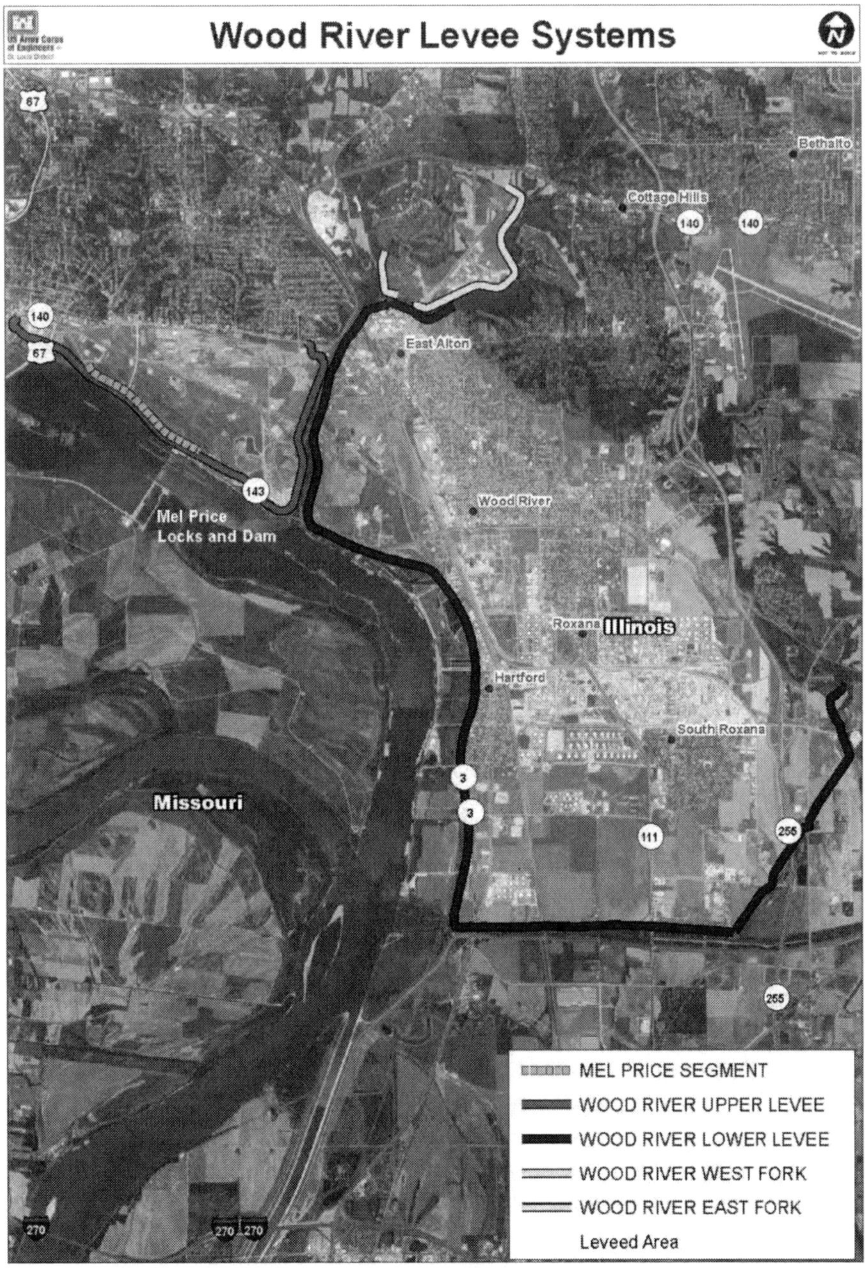

Figure 2. Wood River Levee System with Mel Price Segment

From its construction until the year 1989, the upstream most 2.2-miles, of the Upper Wood River levee between Upper Mississippi River miles 203 to 200.8 (project station 0+00 to 115+00) was located within the tailwater of Lock and Dam #26. Lock and Dam #26 was constructed in 1936 at river mile 203. With the completion of the Melvin Price Locks and Dam at river mile 200.8 (2.2 miles downstream of the original LD #26) and the subsequent raising of the navigation pool in 1989, the 2.2-mile length of the Wood River levee (between project station 0+00 to 115+00) was now located within the permanent navigation pool of the new lock and dam. Because the pool was extended two miles downstream, the river elevation along this stretch of levee increased by 12-15 feet from elevation 404 to 407 to elevation 419 ft.

The Wood River Levee project was originally authorized by the 1938 Flood Control Act and between 1949 and 1960, the St. Louis District issued numerous construction contracts to build/improve levees in the Wood River Drainage and Levee District. As originally authorized, the project includes approximately 21-miles of mainline levee, 170 relief wells 70 of which are in the Mel Price Reach, 26 closure structures, 41 gravity drains, and 7 pump stations. The levee is authorized to provide protection to flood height equal to a level of 52-feet on the St. Louis gage plus 2-feet of freeboard, this corresponds to a flood elevation ranging from 443 to 443.5 ft.

The original underseepage analysis of the Wood River Levee system was completed in October 1956 and is presented in the Corps Technical Manual 3-430 "*Investigation of Underseepage Alton to Gale, Illinois*". This analysis looked solely at the underseepage regime created by the maximum flood height that corresponded to the urban flood elevation (52-feet on the St. Louis gage plus 2-feet). The 1956 analyses predicted the need for positive seepage controls for the design flood height. These positive seepage controls were to be installed at various locations throughout the project. The analyses recommended installation of relief wells at various spacing throughout the levee system resulting in the installation of 170 relief wells in the early 1960s.

In 2009, excessive underseepage with unfiltered exits and sand boils were discovered while survey crews were in this area gathering additional points for a recent LiDAR survey that was conducted along this reach of Upper Wood River levee. At the time of discovery, the navigation pool was under normal flow conditions (the river was not in a flood condition and was being regulated by the dam). A study to investigate potential solutions to the underseepage problem was initiated in 2010. A Limited Reevaluation Report (LRR) was prepared and approved on August 27, 2012 by the Commander, MVD. The LRR recommended a 4,700-foot slurry trench cutoff wall and 55 new relief wells to control underseepage up to the authorized design flood (54 feet on the St. Louis gage). The cost of the recommended plan at the time was $31.8 million. Design of this plan was initiated in 2013 and the first phase involved performing subsurface exploration and a limited pump test to better characterize the subsurface conditions. These efforts revealed that the aquifer permeability values estimated in the LRR were higher and that relief wells alone would not be sufficient in order to meet the USACE criteria in EM 1110-2-1913. Additionally, once the initial exploration was under way MVS realized that construction durations for a slurry trench would escalate due to the discovery of a dense cobble and boulder zone that was prevalent in the area. These findings inherently increased the original cost estimate by over 5 times. The St. Louis District evaluated this new information and determined that a risk informed process should be used in order to reformulate the original tentatively selected plan (TSP). Thus, a baseline condition risk assessment (BCRA) of the levee system was conducted starting in October 2014 to evaluate the incremental risks and provide a basis for better decisions regarding rehabilitation considering the levee safety risk framework as shown in Figure 3 for the Upper Wood River-Melvin Price levee reach.

Figure 3. Levee Safety Risk Framework

RISK ASSESSMENT

To characterize the existing risk and primary risk drivers for the Upper Wood River Levee, a panel of 7 engineers and 1 geologist from the St. Louis District (MVS) and the MVD Levee Safety Production Center, with support from the Risk Management Center (RMC) performed a Baseline Condition Risk Assessment (BCRA) between August 2014 and May 2015. The BCRA report was prepared in general accordance with the methods and policies described in ER 1110-2-1156, Safety of Dams – Policy and Procedures (USACE, 2014) and Best Practices in Dam and Levee Safety Risk Analysis (USBR/USACE, 2015). As part of the BCRA, an assessment of the existing risk condition was performed to determine the significant and credible failure modes for the levee system. Detailed probability estimates to account for loading probability (i.e. hazard), performance and consequences were limited to these significant and credible failure modes, which include internal erosion and overtopping failure modes in the reach of Upper Wood River Levee from Sta. 52+50 to Sta. 126+00. The intent of the risk assessment was to identify the risk, estimate the risk and characterize the uncertainty of the systems performance under flooding conditions. This information would be used to help communicate with stakeholders and risk managers in order to facilitate development of better informed risk management options (RMOs) with due regards for uncertainty and to yield a more effective decision with regards to infrastructure investments. All three pieces of the levee safety risk framework were implemented within the Wood River-Mel Price Levee Deficiency project in order to better understand and manage the risk that exists.

INCREMENTAL RISKS

Tolerable risk guidelines are used in risk management to help inform the process of characterizing and judging the significance of estimated risks obtained during the risk assessment process. Tolerable risks are those that society is willing to live with to achieve or obtain certain benefits. Intolerable risks are those that need to be actively managed or controlled. Within the USACE framework, risks that are above these tolerable limits are determined to warrant some form of management action. USACE ER 1110-2-1156 (Safety of Dams) establishes a guideline which displays the societal tolerable risk limit for average annualized life loss (AALL). Additionally, the ER provides a chart that relates annual probability of failure (APF, assigned the letter "f") against the average incremental life loss (assigned the letter N). This is commonly referred to as the f-N chart and is displayed in Figure 5 for this risk assessment of equal importance. This chart is used to help identify what probable failure modes (PFMs) provide the greatest contribution to the total risk. It allows assessors and managers to target and tailor specific actions for risk management. The line on the chart which defines the societal tolerable risk limit is referred to in this report as the tolerable risk reference line (TRRL). Risks that are plotted on the chart above this line are considered intolerable.

The risk assessment team considered all significant and credible failure modes within the Upper Wood River Levee system. Backward Erosion Piping or BEP, referred in the report as PFM 1, and Overtopping with Breach, referred in the report as PFM 15, were determined to be the primary risk drivers for Upper Wood River. PFM 1 was determined to be slightly above the TRRL with and without intervention. PFM 15 was determined to be below the TRRL. The annualized probability of failure (APF), average annualized life loss (AALL), and average incremental life loss (N) for PFM 1 and PFM 15 are shown in Figure 5. Due to the improvements within the levee system made by the local sponsor in the other parts of Upper Wood River, PFM 1 is the primary risk driver and is primarily contained within the levee reach of concern. Therefore, the results of the risk assessment indicated that the levee had a greater chance of failing prior to overtopping (due to underseepage) than it did of failing due to overtopping flows eroding the levee cross-section.

CONFIDENCE AND MAJOR UNCERTAINTIES

A key piece of Risk Analysis is the explicit acknowledgement and consideration for uncertainty. Uncertainty can be commonly divided into knowledge uncertainty (KU) and natural variability (NV). For the most part knowledge uncertainty can be reduced with additional data gathering, analyses or other analytical tools but natural variability can only be understood and not readily reduced.

For this risk assessment, the uncertainties were only evaluated for the levee system's primary risk driver. Uncertainty was considered for the three components of the conceptual risk equation: loading, system response (i.e., performance), and consequences. The risk assessment results helped the team determine that the vast majority of uncertainty ascribed to the underseepage failure mode was related to initiation of internal erosion as shown in Figure 4. The risk assessment team recognized the uncertainty in the state of the practice with regards to BEP analysis and utilized current best practices to inform subjective probability estimates. For this specific part of the failure mode event tree the risk assessment team elicited the lowest reasonable, the most likely, and the highest reasonable values. These elicitations provided a range of uncertainty for the system response.

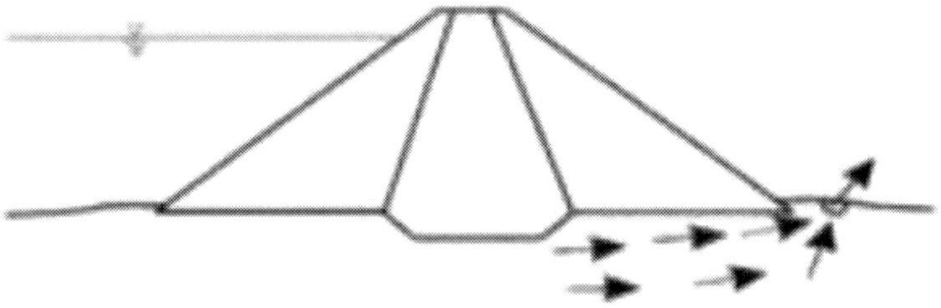

Figure 4: Event Tree - Initiation

RISK CHARACTERIZATION

Information on the elicited annual probability of failure and the expected annual life loss is combined to produce a probability distribution of estimated life loss encompassing all failure modes and all population exposure scenarios. This information is used to graphically depict the current risk with respect to the tolerable risk reference line and is typically plotted on an F-N chart (not shown). Based on the incremental risk f-N chart (see Figure 5), a Levee Safety Action Classification (LSAC) of 3 for breach prior to overtopping was assigned by the risk managers. The LSAC 3 (Moderate Urgency and Risk) rating was approved by the LSOG and the recommendation was forwarded to HQUSACE in September 2015.

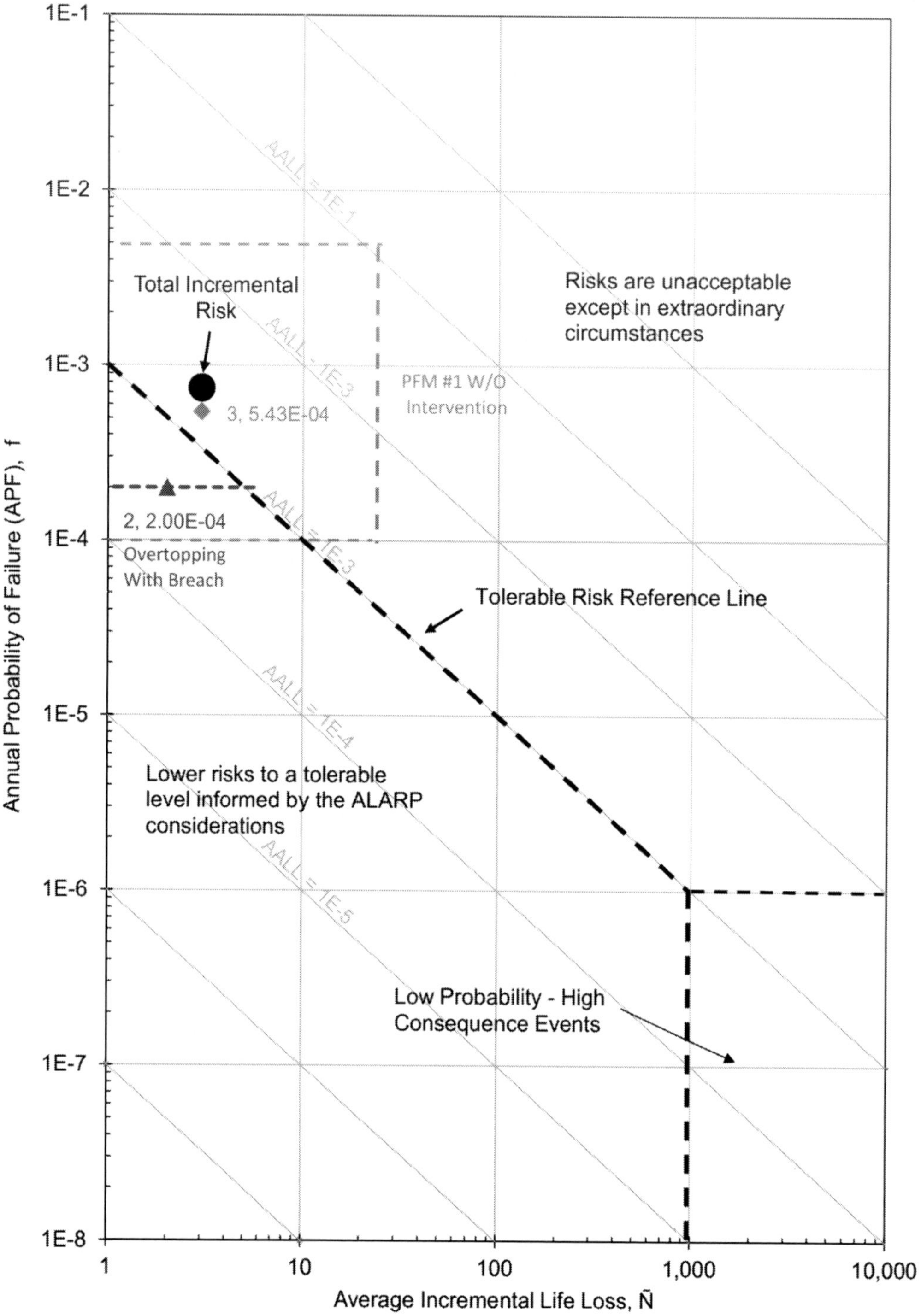

Figure 5: f-N Chart for Upper Wood River Levee

RISK MANAGEMENT

After the risk assessment was completed and the known risks were properly characterized the results were discussed with MVS, the MVD Levee Safety Production Center, Risk Management Center and senior management at the HQ Levee Safety Program. The recommendations for risk management activities developed after consideration of risk assessment results and uncertainties are summarized below:

1. Moderate level of monitoring and surveillance along the levee reach for seepage should continue to confirm modeling assumptions and performance during high-water events.

2. Interim operation plan for the Upper Wood River levee system should be revised based the understanding of risks and failure modes described in the risk assessment report.

3. Complete final designs, plans, and specifications to address the underseepage issues along the levee reach. The final design should include proper abandonment and replacement of the original relief wells. The following should be considered regarding replacement and/or improving the original relief wells:
    a. The relief wells have exceeded their design life. The current relief well efficiency is below 80% and cannot be improved due to the delicate condition of the wood stave screens.
    b. The current relief wells are not effective during low flow regulated river conditions and do not provide the designed or intended pressure relief. Lowering the flow lines on the majority of the wells will allow excess pressures to be relieved.
    c. Point gradient analyses shows the potential for active functioning relief wells to minimize localized horizontal gradients at the well line even though the blanket has been comprised and will limit horizontal progression of BEP.
    d. A properly designed relief well system will provide the necessary seepage control for all anticipated load cases. If needed, the system's capacity can be temporarily increased by pumping using airlifting or mechanical techniques.
    e. Replacing and maintaining a relief well centric solution will lower the overall risk of the Mel Price Reach below tolerable guidelines.

4. The results of this risk assessment should be a supporting document for National Flood Insurance Program (NFIP) accreditation activities of the levee system.

5. Continue to communicate the levee system risks to local emergency management officials and the local sponsor. The emergency preparedness plan may be updated based on the results on the risk assessment.

FUTURE WITHOUT PROJECT CONDITION

If no Federal action is taken to correct the underseepage design deficiency, this reach of Upper Wood River levee will continue to experience excessive seepage that results in a risk to the system that is above a tolerable level. While the limited BCRA established the current risk to the system, the risk assessment team did not quantify a change in risk for the future without project condition. Because the underseepage problem already indicates an intolerable risk, it was determined that this risk is actionable and should be addressed with a permanent solution. The solution should lower the risk to a point below the tolerable risk guideline at a cost commensurate with life safety risks.

While quantitative estimates of future levee performance or life loss were not developed, general conclusions about the future without project can be drawn from the current conditions. In general, there is limited developable land in the inundation area. There is a proposed Marriott hotel development which would increase day and nighttime population at risk and increase economic consequences as a result of inundation. Additionally, the Alton Business Park's site plan has 75 acres of land cleared for development inside the floodplain. The Wood River Coal Power Facility (owned by Dynegy) is planned to be closed in 2016, which would result in reduced potential for life safety and economic damages. Given both proposed new construction and closure of existing facilities, growth in the population and structures at risk is expected to be minimal.

This reach of levee contains 70 wood-stave relief wells that were installed in the 1950s. In addition to inadequate design for the existing seepage condition, these existing wells have also lost efficiency and have started to fail. It is reasonable to expect the levee performance related to underseepage to continue to degrade in the future if no project is implemented.

FUTURE WITH PROJECT CONDITION

The objective of the chosen seepage remedial solution for a future with project condition is to address and reduce the risk to the system associated with the Underseepage failure mode. The recommendations discussed within the risk management team indicate that a relief well centric solution is the "least cost alternative" that will meet the tolerability of life safety. The project's design was selected using a risk informed-decision making process in general accordance with ER 1110-2-1156 and USACE Best Practices for Dam and Levee Safety amongst other USACE technical manuals. The objective of the selected plan, was to reduce the level of risk below USACE Tolerable Risk Guidelines. The selected plan was designed and informed by a methodology that takes into account, not only the performance and potential failure modes that cause the increased risk to the system, but also accounts for the consequences of analyzed failure modes and hydraulic loadings.

The selected RMO consists of 100 new 10-inch diameter stainless steel relief wells that discharge at elevation 409 feet. Through finite element modeling it was estimated that lowering the relief well discharge elevations results in more pressure relief and decreases the likelihood of progression. In addition, an access road for installation and maintenance will be built as well as associated modifications to the drainage ditch to allow for the excess water generated to be directed into the ponding area. The 100 new relief wells extend from station 53+00 to 118+00 and were analyzed using GeoStudio's Seep/W with differing water elevations for vertical factor of safety throughout this 6500 feet in 8 representative geological cross sections. Each factor of safety was reported and the relief wells were shifted accordingly so that their location obtained the highest factor of safety and decreased the likelihood of backward erosion piping.

CONCLUSION

The risk analysis framework utilizes the best science, techniques and understanding of infrastructure systems in order to make effective and efficient risk management decisions. The framework and analytical process explicitly and implicitly recognizes all the contributing aspects of the risk, including the consequences of the levee not performing as expected/intended and facilitates identification, characterization and communication of the risks. The Wood River-Mel Price Levee Risk Assessment was an application of the framework within the Levee Safety Program of the USACE. Reliance on standards based and deterministic plan formulation had led to inefficient risk management measures with excessive costs. The results of the risk assessment allowed the St. Louis District and USACE as a whole to really understand the failure mode and the associated uncertainty with regards to levee performance and combined with consequence assessments helped paint a more complete picture of the risk. With this information USACE was able to more effectively and efficiently select risk management alternatives with estimated cost-avoidances of approximately $100 million.

Additionally, USACE was able to more effectively and accurately communicate the level of risk associated with the infrastructure system to the public and the non-federal entity in charge of O&M. The risk management option selected was sufficient to reduce the levels of risk within tolerable limits with due consideration for uncertainty, in particular, accounting for the low life loss potential associated with this area as well as with the documented and observed distress due to seepage issues. The application of the risk framework processes used within this engineering project demonstrates improved effectiveness with regards to life safety and investment decisions.

REFERENCES

USBR/USACE, 2015, Best Practices in Dam and Levee Safety Risk Analysis, July 2015.

USACE, YEAR, Corps Technical Manual 3-430 "Investigation of Underseepage Alton to Gale, Illinois"

USACE, ER 1110-2-1156-2-1556, Safety of Dams – Policy and Procedures, 31 March 2014.

EFFECTS OF COMPACTION ON ERODIBILITY CHARACTERISTICS OF LEVEES DUE TO OVERTOPPING

Ali Asghari Tabrizi[1]
M. Hanif Chaudhry[2]
Jasim Imran[3]

ABSTRACT

Erodibility is a function of geotechnical properties of the embankment material (e.g., compaction level, water content, cohesion) and is a key parameter in studying and modeling of earthen embankment failure. In this study, four laboratory tests are conducted on overtopping of a homogeneous, non-cohesive embankment with variations of the compaction level under controlled hydraulic conditions. The breach shape evolution is monitored quantitatively using a sliding rods technique and image processing. Breach characteristics (i.e. time history of the breach top width and breach depth along the centerline of the crest, breach outflow, and breached volume) are estimated and compared for all four tests. From the results of the experiment, non-dimensional relationships are developed in horizontal and vertical directions which express the widening and deepening rate of the breach as a function of the excess Shields number for different compaction levels. Finally, two non-dimensional relationships are developed using regression analysis which express the deepening and widening rate as a function of compaction energy.

SIGNIFICANCE AND BACKGROUND

Embankments are primarily constructed to provide protection from flood, although they may be used for water storage, hydropower generation, and recreation, and they can be categorized into dams and levees depending on their orientation with respect to the general flow direction. Most of the levees are formed by embankments composed of natural erodible materials and they may fail due to various reasons especially under extreme conditions (ASCE/EWRI Task Committee on Dam/Levee Breaching 2011). The frequency of the extreme hydrological events (e.g., rainstorms and hurricanes) has increased in recent years (Kakinuma and Shimizu 2014), thereby increasing the risk of earthen embankment failures and flooding which may cause human fatalities and extensive economic and social stress. For example, the levee failures in New Orleans during Hurricane Katrina in 2005 and more recently, the dam and levee failures in the Midlands, South Carolina during Hurricane Joaquin in October 2015 resulted in considerable damages. Overtopping is the main cause of the earthen embankment failures and therefore this type of failure mechanism is investigated in this study.

[1] Senior Staff Engineer, Schnabel Engineering, Greensboro, NC 27407. Email: ali.a.tabrizi@gmail.com
[2] Mr. and Mrs. Irwin B. Kahn Professor and Associate Dean (International Program and Continuing Education), Department of Civil and Environmental Engineering, University of South Carolina, Columbia, SC 29208. Email: chaudhry@sc.edu
[3] Professor, Department of Civil and Environmental Engineering, University of South Carolina, Columbia, SC 29208. Email: imran@email.sc.edu

Erodibility is a key parameter in modeling and predicting the failure of earthen embankments which is crucial for preparing emergency action plans and risk management. The erodibility of an embankment depends on the properties of the embankment material (e.g., compaction level, compaction water content, soil texture, and cohesion). Erodibility plays a major role in earthen embankment failures. However, there are only few experimental studies investigating the effect of this element and parameters such as compaction on the earthen levee failure. Most of the embankment breach studies in the past were on dam failure which is significantly different from the levee breach scenario (Wahl 2007; Kakinuma and Shimizu 2014). Also, most of these studies focused only on the widening process and ignored the initial deepening stage of the failure. For example, Coleman et al. (2002), Hunt et al. (2005), Visser (1998), Pugh (1985), and Tinney and Hsu (1961) have studied dam embankment failures due to overtopping with focusing only on the widening process. Kakinuma and Shimizu (2014) conducted a series of field-scale levee breach overtopping experiments, but the scope of their study was on the breach widening stage. However, the initial deepening stage is of great importance to have an accurate prediction of the failure process. Tabrizi et al. (2015) conducted a levee breach experiment for a non-cohesive, non-compacted embankment and estimated the erodibility coefficients in both vertical and horizontal directions.

The aim of this research is to investigate the effects of the compaction level on the erodibility of a levee due to overtopping. This study fills the gap in the existing literature by analyzing the deepening and the widening stages separately. Four experiments of levee breach failure are conducted in the Hydraulics Laboratory, University of South Carolina, by varying the compaction level of embankment material.

PROJECT DESCRIPTION

The plan and schematic views of the experimental setup are shown in Figure 1. The test setup was explained in detail in Tabrizi et al. (2015). Four Baumer ultrasonic probes are used in this experiment to precisely monitor the water surface elevation at different locations, one on top of the crest, one in the main channel upstream of the levee, one in the main channel downstream of the levee, and one in the main channel at downstream end.

Prior to the embankment overtopping tests of compaction effect, Standard Proctor Compaction tests (ASTM D698) are carried out to determine the optimum water content and maximum dry unit weight of the embankment material. The optimum water content is 5.2% and the corresponding maximum dry unit weight, $\gamma_{d,max}$, is 15.44 kN/m^3. The trapezoidal earthen embankments of uniform sand with a mean diameter of 0.55 mm and a water content of 5.2% are placed in four loose layers approximately 10 cm thick. Each layer is then compacted by a 4.54 kg rammer with the release height of 5 cm with different number of blows per layer, N_b, for each case - 0 (no compaction), 2, 4, and 10. After the compaction, the levee is trimmed to the final shape. To initiate the overtopping, a pilot channel, 0.05 m deep and 0.10 m wide, is carved at one third of the length from the upstream edge of the earthen levee.

The inflow discharge is kept constant at 0.05 m^3/s for all the tests. The breach shape evolution is monitored using a front camera and a sliding rods technique developed by the authors (Figure 2) along with image processing.

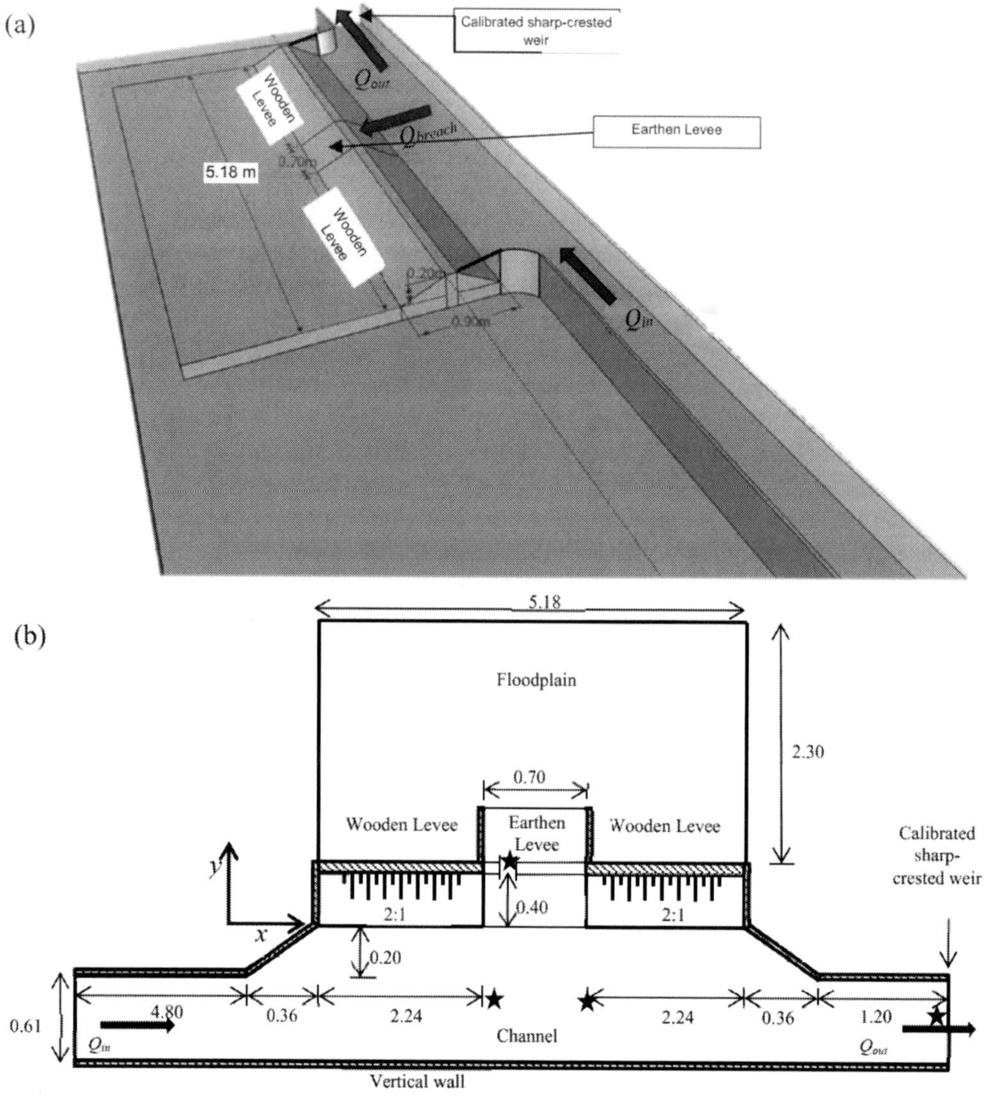

Figure 1. Plan view of the experimental setup: (a) schematic view; and (b) plan view. Location of the Baumer ultrasonic devices are indicated by stars (dimensions are in meters)

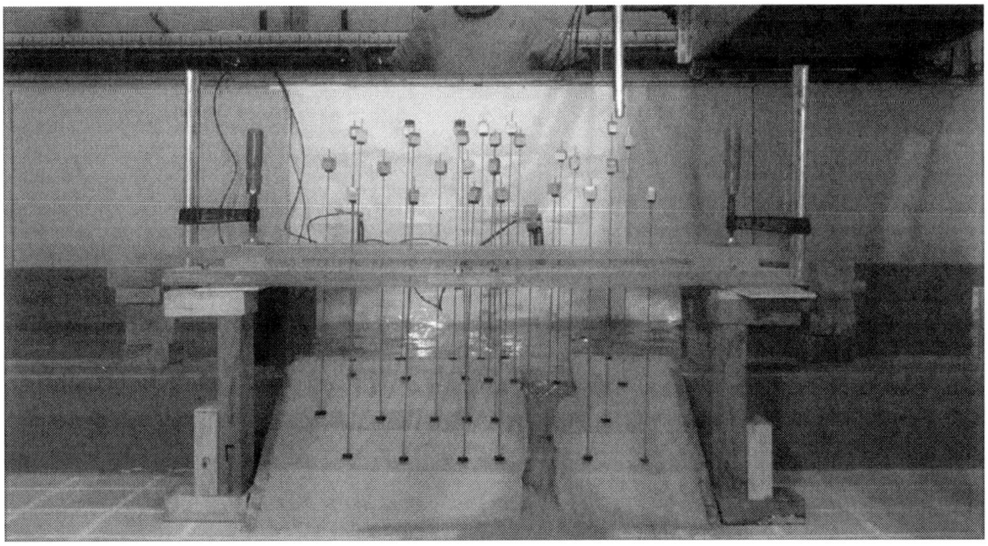

Figure 2. Sliding rods technique to monitor the breach shape evolution (Test case with N_b = 4 at time = 22 s)

RESULTS

Breach shape evolution

Time history of the breach top width, *W*, and the maximum breach depth, *D*, along the centerline of the levee crest are presented in Figures 3 and 4 for all the tests, respectively. The breach top width remains relatively constant at the initial stage of the failure process and then starts to accelerate, while the changes in maximum breach depth include acceleration and deceleration phases. The breach width, maximum breach depth, widening rate, and deepening rate decrease with the compaction level for a given time step. The three-dimensional breach shape evolution is estimated from the sliding rods technique and is compared for all the four tests.

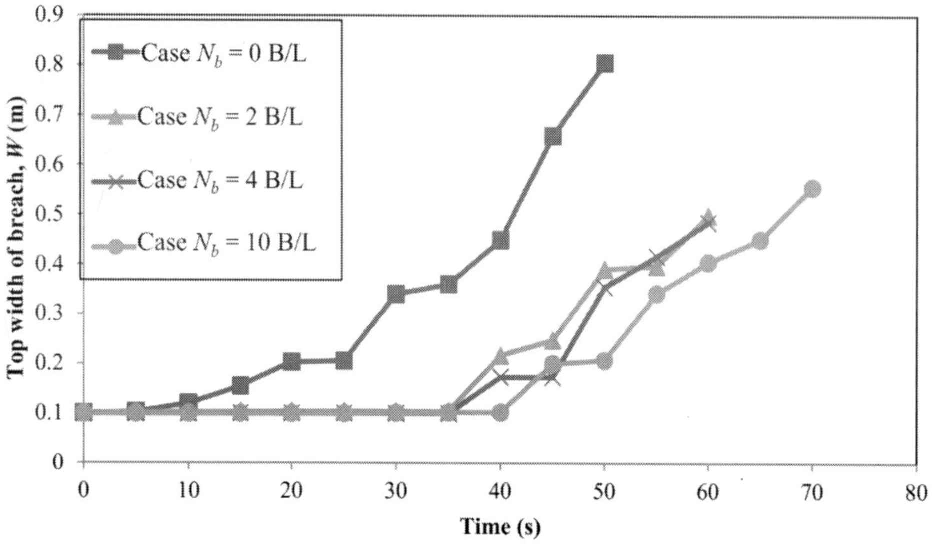

Figure 3. Time-history of the breach top width along the centerline of the levee crest for four levels of compaction

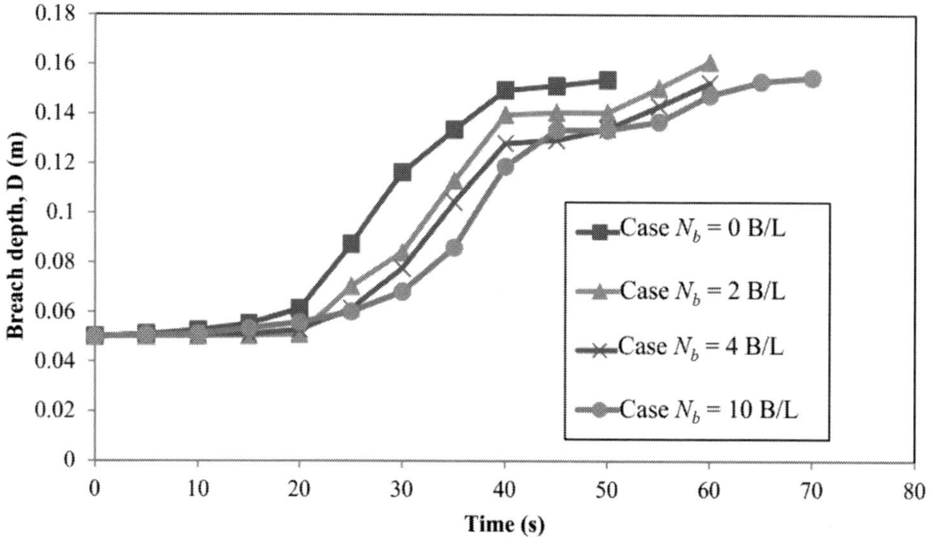

Figure 4. Time-history of the breach depth along the centerline of the levee crest for four levels of compaction

Figure 5 shows the temporal evolution of the breach for the case with $N_b = 10$. The erosion starts on the downstream face of the levee, it retrogrades from the downstream edge to the upstream edge of the levee crest, and it continues with the degradation of the crest. Then the breach starts to widen and shortly the widening accelerates. The observed failure stages (i.e., levee breach initiation, onset of widening, and widening acceleration) are consistent with those from Kakinuma and Shimizu (2014) large-scale experiments. However, the last stage of widening deceleration is not observed in the current study due to the limited length of the earthen section.

Duration of the failure increases with compaction as expected. It is observed from the figures that as the compaction increases, the irregularity of the breach shape increases. This can be due to the fact that as the compaction increases, the head-cut erosion dominates as compared to the surface erosion, which is the controlling erosion mechanism in the test with no-compaction.

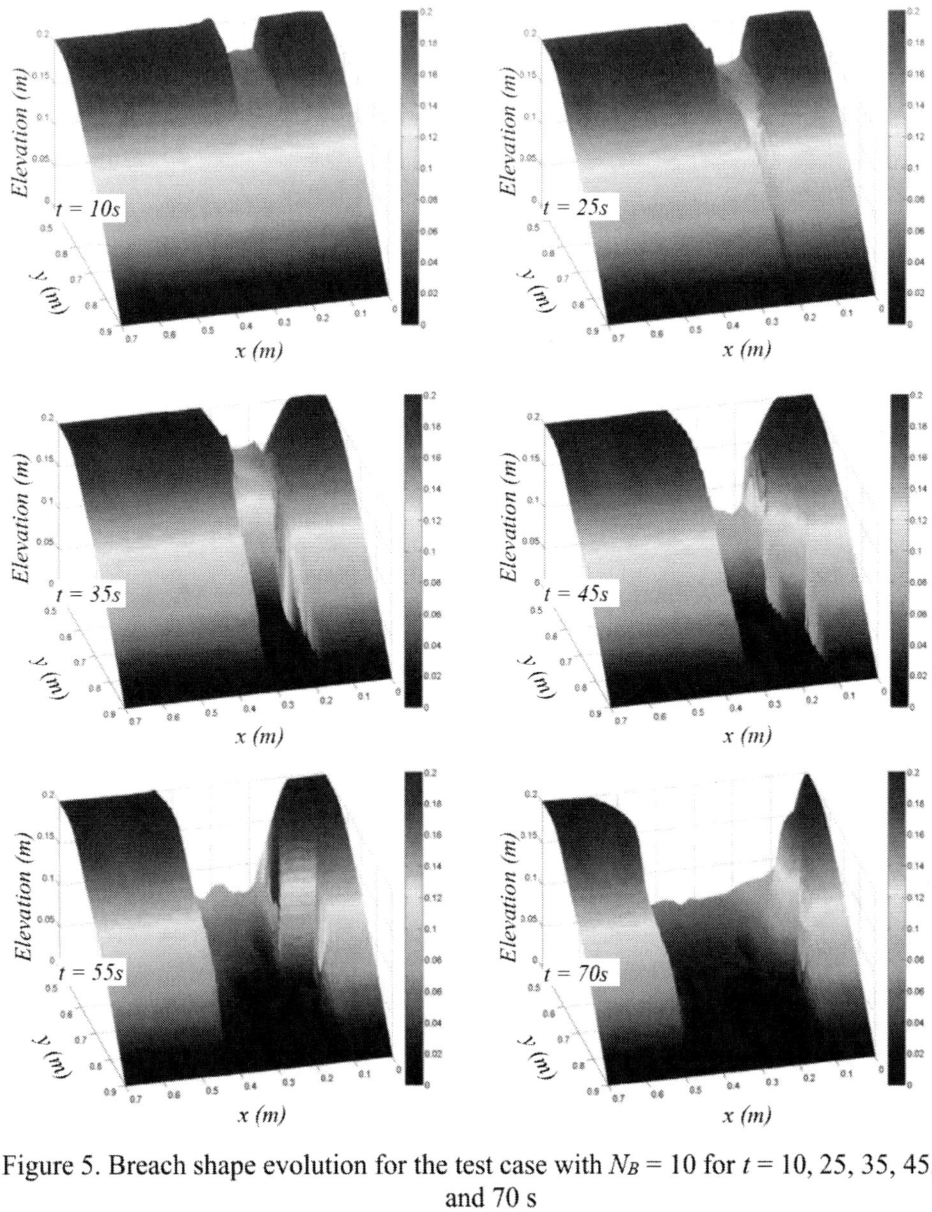

Figure 5. Breach shape evolution for the test case with $N_B = 10$ for $t = 10, 25, 35, 45, 55$, and 70 s

The time-history of the accumulated breach eroded volume for the downstream half of the breach is shown in Figure 6 for all the tests as measured by the sliding rods technique. The time rate of change of the breached volume is almost the same for the compacted tests, while it is higher for the non-compacted embankment.

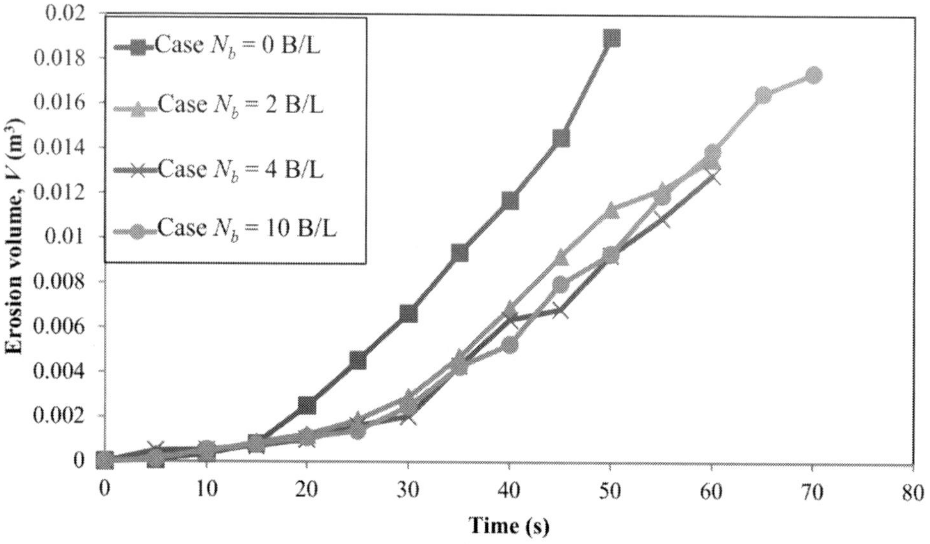

Figure 6. Cumulative breach eroded volume with time for four levels of compaction

The changes of the submerged area along the centerline of the levee crest versus time are calculated using the recorded water surface elevations on top of the pilot channel and are presented in Figure 7 for all the tests. A similar trend is observed for the compacted and non-compacted tests; the submerged area is negligible until about t = 20 s and then it starts to accelerate with almost a constant slop. However, the submerged area is larger at any given time for the non-compacted test.

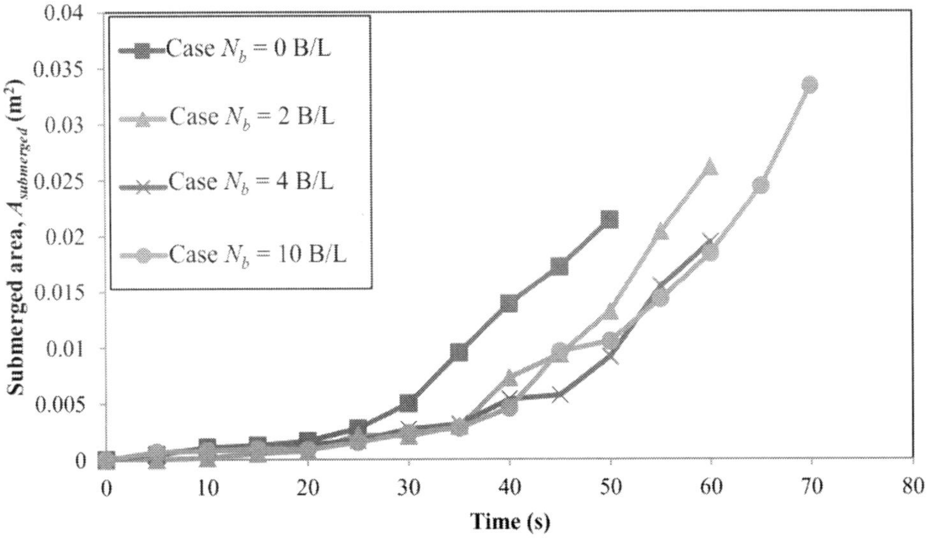

Figure 7. Time-series of the submerged area along the centerline of the levee crest for four levels of compaction

Breach outflow discharge

The flow discharge over the sharp-crested calibrated weir at the downstream end of the flume is monitored, and the breach outflow hydrograph is computed by balancing the weir overflow and the constant inflow discharge for all the tests (Figure 8). The breach outflow includes acceleration and deceleration phases. It remains relatively small in each test at the initial stage of the failure and then starts to accelerate and increase with almost a constant rate, then it decelerates and reaches roughly a constant value until the end of the failure process (0.035 m^3/s for all the tests). However, the time rate of breach overflow changes is higher for the non-compacted test as compared to the compacted tests. These changes correspond to the deepening phase of the failure as well. Besides, at a certain point (i.e., at about $t = 50$ s in the non-compacted test and at about $t = 70$ s in the compacted tests) the breach outflow exceeds the weir flow in all the tests.

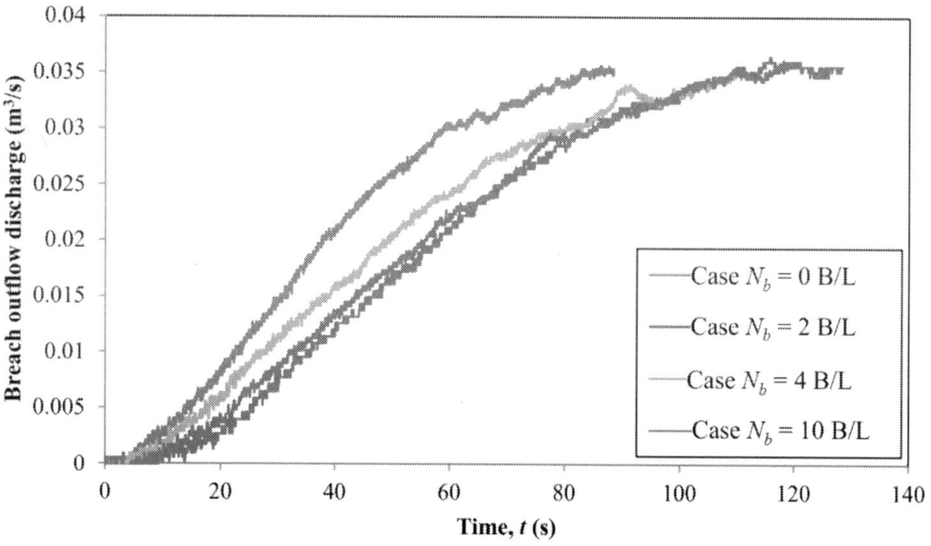

Figure 8. Breach outflow discharge for four levels of compaction

Formula for breach morphology as widening and deepening rates

The non-dimensional relationship between the deepening rate and widening rate of the breach cross-section, along the crest centerline, with the excess shear stress can be expressed in a form similar to the Meyer-Peter and Muller bedload formula as:

$$\epsilon_d^* = \alpha_d(\tau_{ed}^* - \tau_c^*)^{\beta_d} \tag{1}$$

$$\epsilon_w^* = 2\alpha_w(\tau_{ew}^* - \tau_c^*)^{\beta_w} \tag{2}$$

where ϵ_d^* and ϵ_w^* are the non-dimensional erosion rates for deepening and widening stages of the failure, respectively (* denotes the normalized form, and *d* and *w* denote deepening and widening, respectively); α_d, β_d and α_w, β_w are the correspondent erodibility coefficients; τ_{ed}^* and τ_{ew}^* are the non-dimensional applied shear stresses on bed and side walls of the breach opening, respectively; and τ_c^* is the non-dimensional critical shear stress and is calculated using the relation proposed by Parker et al. (2003). The normalized erosion rates, ϵ_d^* and ϵ_w^*, are calculated as

$$\epsilon_d^* = (\frac{\Delta D}{\Delta t})^* = \frac{\Delta D}{\Delta t} \times \frac{1}{\sqrt{RgD_g}} \tag{3}$$

$$\epsilon_w^* = (\frac{\Delta W}{\Delta t})^* = \frac{\Delta W}{\Delta t} \times \frac{1}{\sqrt{RgD_g}} \tag{4}$$

where $\Delta D/\Delta t$ and $\Delta W/\Delta t$ are deepening and widening rates, respectively; t is time; R is sediment submerged specific gravity; g is gravitational acceleration; and D_g is the mean diameter of the soil. The non-dimensional deepening and widening rates are calculated every 5 s from the experimental observations for each test.

Similar to the procedure expressed by Hunt et al. (2005), a rectangular cross-section and a critical flow depth are assumed along the centerline of the levee crest. These assumptions are used to estimate the applied shear stress on the bed, τ_{ed}, and on the side walls of the breach, τ_{ew}, as

$$\tau_{ed} = \gamma_w d S_f \tag{5}$$

$$S_f = \frac{gn^2}{(d_c)^{\frac{1}{3}}} \tag{6}$$

$$\tau_{ew} = 0.7\tau_{ed} \tag{7}$$

where γ_w is the specific weight of water; d is the average water depth across the breach cross section which is assumed to be the critical depth, d_c; S_f is the energy slope which is calculated from the Manning equation; and n is the Manning coefficient. Eq. (7) assumes a rectangular cross section and approximates the shear stress on the wall as a proportion of the shear stress on the bed (Chow 1959).

Incorporating the observed data of the observations of the breach widening and deepening processes into the aforementioned relations, the erodibility relations can be obtained in vertical and horizontal direction for each compacted embankment test. Figures 9 and 10 show the test results of the correlation between the deepening and widening rate with the excess shear stress, respectively for four levels of compaction. The correlations are expressed by their corresponding fitted equations in the plots for different compaction levels. The results of the coefficients of these equations are summarized in Table 1.

Table 1. Erodibility coefficients for breach widening and deepening

Test	**N_b (B/L)**	**γ_{dry} (kN/m^3)**	**C_e^***	**α_d**	**β_d**	**α_w**	**β_w**
1	0	12.01	0	0.10	0.87	0.53	2.67
2	2	14.29	2.88	0.07	0.75	0.17	2.77
3	4	15.29	5.77	0.06	0.71	0.28	4.06
4	10	15.49	14.42	0.02	1.10	0.12	5.87

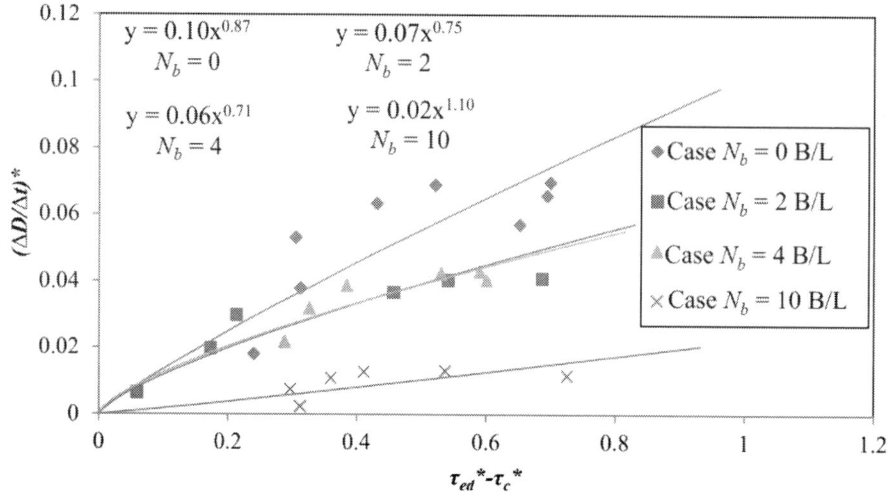

Figure 9. Correlation of the breach deepening rate along the crest centerline and the excess shear stress on the bed

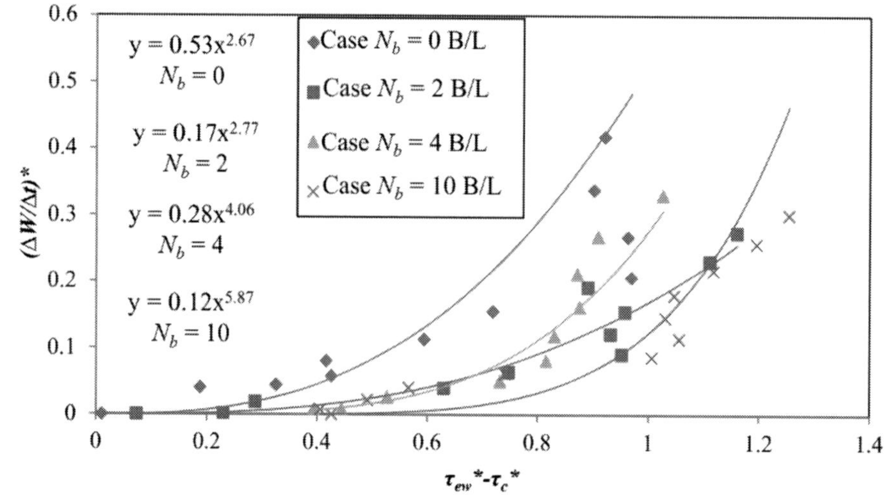

Figure 10. Correlation of the breach widening rate along the crest centerline and the excess shear stress on the sidewalls

To have a more general variable representing compaction, normalized compaction energy, $C_e^* = C_e/(\rho_{dry,max} \times h)$, is used herein, corresponding to each N_b, where

$\rho_{dry,max}$ is the maximum dry density from the proctor compaction test; h is the height of the levee; and C_e is compaction energy which is calculated as

$$C_e = \frac{N_b \times Number\ of\ layers \times Hammer\ weight \times Release\ height}{Soil\ volume\ under\ hammer} \tag{8}$$

Using regression analysis, the non-dimensional relationships of the breach deepening and widening rate as a function of the compaction energy were obtained as:

$$\epsilon_d^* = 0.068e^{-0.158C_e^*}(\tau_{ed}^* - \tau_c^*)^{0.462} \tag{9}$$

$$\epsilon_w^* = 0.544e^{-0.043C_e^*}(\tau_{ew}^* - \tau_c^*)^{2.67} \tag{10}$$

To verify the validity of the proposed relations, i.e. Eqs. (9) and (10), predicted and observed results of the breach deepening and widening rates are compared for different tests in Figures 11 and 12, respectively. A relatively better prediction is observed for the breach widening rate as compared to the deepening rate.

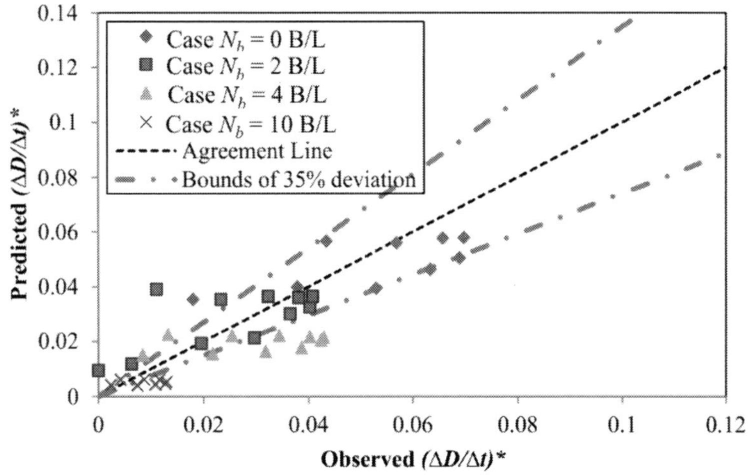

Figure 11. Comparison of predicted and observed breach widening rate along the crest centerline

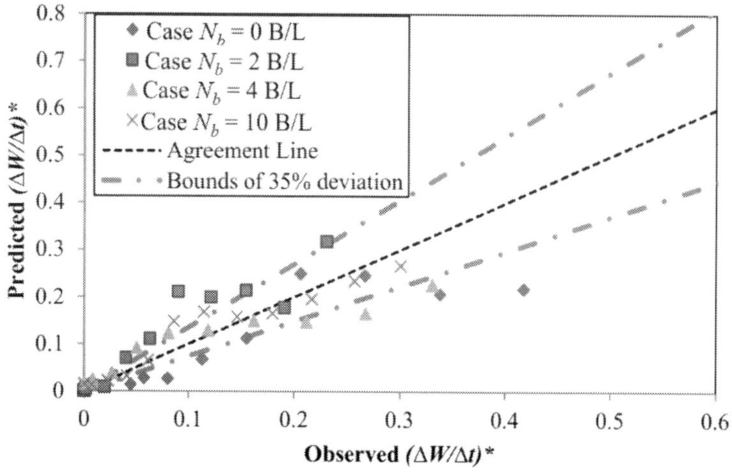

Figure 12. Comparison of predicted and observed breach widening rate along the crest centerline

CONCLUSION

The effects of compaction on the overtopping failure of a non-cohesive levee are investigated by conducting a series of laboratory tests. Comparing the non-compacted test with compacted tests, the compaction energy of the embankment material is found to be a controlling factor in the failure process. However, the compaction effect diminishes as the compaction effort increases since the compaction level is near optimum.

The observed first three levee failure stages in this study (i.e., levee breach initiation, onset of widening, and widening acceleration) are consistent with those observed in the large-scale levee breach experiments of Kakinuma and Shimizu (2014), except the last stage of widening deceleration which is not captured in the current study due to the limited length of the earthen section. A non-symmetric breach development is observed in all the experiments as the downstream bank of the breach is struck by higher shear stresses compared to the upstream side which results in a faster erosion on the downstream side-wall of the breach.

To quantify the breach morphology, two non-dimensional equations are proposed which express the deepening and widening rates of a breach along the crest centerline as a function of the excess shear stress and compaction level of the embankment material. The proposed models are then compared against the measurements from the laboratory tests presented in the current study. A relatively better prediction is observed for the breach widening rate as compared to the deepening rate. To achieve more reliable predictive models, full-scale experiments are required with more variable soil compositions and compactions.

ACKNOWLEDGMENTS

The author would like to thank the National Science Foundation (NSF), the student Scholarship awarded by the United States Society on Dams (USSD), and the SPARC Graduate Fellowship sponsored by the Office of the Vice President for Research, University of South Carolina for the financial support to this study.

REFERENCES

ASCE/EWRI Task Committee on Dam/Levee Breaching (2011). "Earthen Embankment Breaching." J. Hydraul. Eng., 137(12), 1549–1564.

Chow, V. T. (1959). Open-Channel Hydraulics. New York, N.Y.: McGraw-Hill.

Coleman, S. E., D. P. Andrews, and M. G. Webby. 2002. Overtopping breaching of noncohesive homogeneous embankments. J. Hydraulic Eng. 128(9): 829-838.

Hunt, S. L., Hanson, G. J., Cook, K. R., and Kadavy, K. C. (2005). "Breach widening observations from earthen embankment tests." Trans. ASAE, 48(3), 1115–1120.

Kakinuma, T. and Shimizu, Y. (2014). "Large-Scale Experiment and Numerical Modeling of a Riverine Levee Breach." J. Hydraul. Eng., 140(9), 04014039.

Parker, G., Toro-Escobar, C. M., Ramey, M. and S. Beck, (2003). "The effect of floodwater extraction on the morphology of mountain streams." J. Hydraul. Eng., 129(11), 885-895.

Pugh, C. A. (1985). "Hydraulic model studies of fuse plug embankments." REC-ERC-85-7. Denver, Colo.: U.S. Bureau of Reclamation.

Tabrizi, A. A., Elalfy, E., LaRocque, L.A., Chaudhry, M.H., Imran, J. (2015). "Experimental Modeling of Levee Failure Process due to Overtopping." Proc. Dam Safety 2015 Conference, New Orleans, LA.

Tinney, R. E., and H. Y. Hsu. (1961). "Mechanics of washout of an erodible fuse plug." J. Hydraulics Division of ASCE 87(3): 1-30.

Visser, P. J. (1998). Breach growth on sand-dikes. In Communications on Hydraulic and Geotechnical Engineering. Report No. 98.1. Delft, The Netherlands: Delft University of Technology.

Wahl, T. L. (2007). "Laboratory investigations of embankment dam erosion and breach processes." Rep. T032700-0207A, EA Technologies (CEATI), Montréal, Canada.

PREVENTING CATASTROPHIC FAILURE OF UNDERWATER INFRASTRUCTURE: CONDITION ASSESSMENTS BY ROV

Bill Sherwood [1]
Bob Clarke[2]

For major infrastructure, it is generally accepted that you can't have risk assessment without condition assessment. This is no less true for underwater structures. Unfortunately, however, there is still a serious gap in our general thinking concerning condition assessments for things that can't normally be seen - until something fails.

This presentation will address the topic of the importance of accurate and comprehensive condition assessment of underwater infrastructure by Remotely Operated Vehicle (ROV), with particular attention to recent advances in underwater sensor technology. The paper will begin with an overview of specific case histories of recent underwater infrastructure failure, the various responses considered and attempted, and the end results. High resolution video, 2D and 3D sonar, laser dimensioning, and advanced photogrammetry will be discussed. Rendered examples of each data type used in real world underwater survey scenarios will be included, with discussion about which sensor is considered most applicable in various situations and conditions. Finally, the paper will address the next wave of underwater technology development: advances in remote robotic repair.

[1] ASI Group – ASI Marine, bsherwood@asi-group.com
[2] ASI Group – ASI Marine

AUTOMATED DATA ACQUISITION SYSTEM BASICS FOR DAM SAFETY INSTRUMENTATION MONITORING

Peter Zimmerman, PE, PG[1]
Paul Booth, PE[2]
Greg Dutson[3]

ABSTRACT

Dam safety surveillance and monitoring commonly includes instruments for measuring deformation, porewater and uplift pressure, seepage, seismic acceleration, and other parameters. Common methods for recording instrumentation data range from manual readings to fully automated data acquisition systems (ADAS). Automated systems are sometimes dismissed as a monitoring option because they are thought to be overly complex with an internal function that is not well understood. The general performance of the ADAS is straightforward with sensors linked to data acquisition modules through site telemetry, and communication that allows remote monitoring from the user's desktop or mobile device. ADAS doesn't make sense for every dam, but the system can bring advantages over traditional manual instrument monitoring when used appropriately. Automation allows adjustable recording intervals that can be easily adjusted for special events and for different instruments. Alarm notifications, commonly sent via text and email, may be programmed to immediately alert the dam owner of instrument threshold exceedance. The ADAS does not require personnel to be on site for routine measurements, which is ideal for remote dams and will lead to cost reductions for field personnel and reporting. Automation allows consistency of readings so that data variations due to changes in personnel and measurement techniques are eliminated. ADAS allows near real time access to data and charts, and the project data can be viewed remotely from anywhere. Successful ADAS deployment includes considering and addressing challenges inherent in the technology. The ADAS introduces some complexity to the monitoring program and the system must include protection from environmental factors such as lightning. Computer network security must be maintained, and the owner's Information Technology team will want to be involved in ADAS planning and deployment. The initial ADAS deployment could include significant cost and this must be considered with any potential long-term savings and risk reduction.

INTRODUCTION

Dam safety surveillance and monitoring commonly includes instruments for measuring parameters including but not limited to deformation, pore water and uplift pressure, seepage, and seismic acceleration. Instrumentation data at a given dam may be recorded manually, data recording may be completed using a fully automated data acquisition systems (ADAS), or some combination of manual readings and ADAS may be used. Automated systems are sometimes dismissed as a monitoring option because they are

[1] Canary Systems, Inc. 7151 Lee Hwy, Ste 600, Chattanooga, TN 37421, peter@canarysystems.com
[2] Canary Systems, Inc. 7151 Lee Hwy, Ste 600, Chattanooga, TN 37421, paul.booth@canarysystems.com
[3] Canary Systems, Inc. 5 Gould Road, NH 03257, greg@canarysystems.com

thought to be overly complex with an internal function that is not well understood. The actual performance of an ADAS is straightforward with sensors linked to data acquisition modules through site telemetry, and communication that allows remote monitoring from the user's desktop or mobile device.

Dam safety instrumentation should be thought of and designed as a system that is tailored to monitor the potential failure modes (PFMs) identified at the dam. The Army Corps of Engineers (USACE) engineering and design manual for Instrumentation of Embankment Dams and Levees correctly points out that "every instrument system is unique and a significant amount of engineering judgment must be applied to each" (USACE, 1995). Advantages and disadvantages of various data acquisition methods should be considered during the instrument system design phase and as monitoring progresses. ADAS might not have been considered when a dam's monitoring system was established, but advances in technology, heightened monitoring requirements, increased sensitivity to or awareness of risk, and/or changes in personnel demands might make ADAS an attractive monitoring alternative at an existing dam. ADAS doesn't make sense for every dam, but the system can bring advantages over traditional manual instrument monitoring when used appropriately.

DATA ACQUISITION BASICS

Engineers and geologists commonly identify dam performance parameters that will be monitored, select the type of instruments to be installed, and identify the instrument locations. A sometimes-neglected aspect of instrumentation system design is the method for data acquisition. As part of the instrumentation system design planning, the next questions that should be answered include the following:

1) How do you want to access the instrument data?
2) What are the site, organizational, and computer network constraints for site access and communication?

The data acquisition system is designed to meet the monitoring objectives given the constraints identified during the planning phase of the design. Accessing the instrument data may be accomplished by the following general methods:

- Manual data acquisition – physically read an instrument and record the data
- Automated data acquisition – collect instrumentation data using a datalogger and transmit the data to a database for local and remote monitoring
- Hybrid acquisition system – a hybrid system incorporates both manual and automated data collection techniques

Manual Data Acquisition

Instruments that require a person to take a measurement or reading and record the data may be referred to as having manual data acquisition. Examples of manually reading instruments include measuring water level in an open standpipe piezometer using a water level meter, recording pore water pressure from a vibrating wire piezometer using a handheld readout device, measuring water level behind a seepage collection weir using a staff gage, and using an inclinometer probe to measure displacement of grooved inclinometer casing. The thing that these examples have in common is that a qualified person is required to travel to the site and physically take the reading. The instrument reading typically is then entered in a spreadsheet or database where it can be incorporated into charts and reports for monitoring and analysis.

Manually read instruments are monitored on a schedule that is assigned in the dam's surveillance and monitoring plan. While critical instruments, instruments that have exceeded a threshold value, or instruments during unusual loading conditions may be read more often, it is common to record manual data monthly or quarterly. Alignment and settlement surveys are likely to be performed bi-annually, annually, or at some greater interval.

Automated Data Acquisition

Instruments with signal output may be used in a data collection system that is programmed to collect readings, store the readings in a datalogger, and transmit the data to an onsite or remote location for near-real time monitoring. The data collection and transmission in a fully automated system is accomplished without active human interaction.

The pieces of an automated monitoring system may be generally categorized into the following items that are also shown in the Figure 1 schematic:

1) Instrument data collection point(s)
2) On-site communication between the instrument(s) and the datalogger
3) Datalogger
4) Connection or telemetry from the on-site datalogger to a computer server and database located either on or off-site
5) Instrumentation database and monitoring software. The database is accessed by local and remote users for display, monitoring, and reporting activities.

Instrument Data Collection Point. Automated instruments may include vibrating wire (VW) products, load cells, transducers, micro-electro-mechanical (MEMs) sensors, accelerometers, flowmeters, thermocouples, thermistors, optical encoders, weather sensors, total stations, global positioning satellite (GPS) modules, and many others. When describing a general data acquisition organization, the type of instrument is not

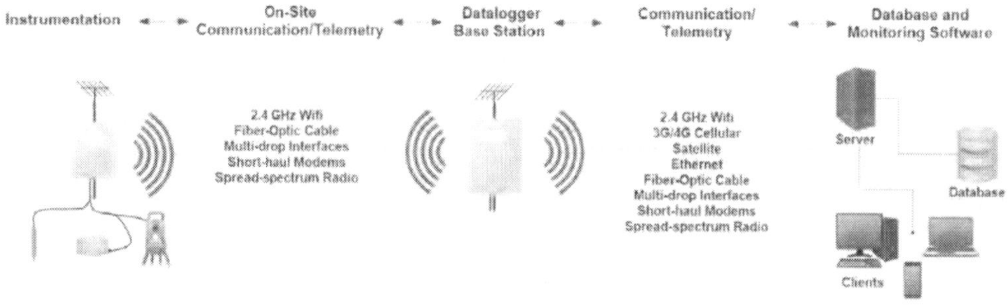

Figure 1. Generalized ADAS functional schematic.

critical to understanding the system concept. Specific instruments are discussed in later sections.

Signals from the instrument may be transmitted directly to a datalogger or may first pass through associated hardware such as a vibrating wire analyzer module. The instrument may be hardwired to the datalogger at more compact sites or it can be connected by telemetry at larger projects. Regardless of how the instrument is connected into the system, the next step in the automated data collection process is to send the instrument data to a datalogger.

On-Site Communication. The datalogger communicates with the instrument data collection points to control the frequency of readings, collect data, and store the data. On-site communication may be through cable, fiber optics, radio, Wi-Fi, or cellular network.

Measurement and Control Datalogger. The datalogger is an electronic device that measures sensors at a specific scan rate, processes the data, and records data over time. The datalogger transmits the instrumentation data for remote monitoring, analysis, and reporting.

Data Telemetry. Data is sent from the site to the user's network for remote monitoring. The user may use this same connection to remotely program the datalogger for different recording intervals or other changes. Connection is made through an internet network connection on site when available, but data may be transmitted by satellite, telephone landline, or cellular transceiver.

Instrumentation Database. The data from the datalogger is collected in the database on the user's network. Specialized computer software is used to organize instrument data with hierarchical relationships that make data storage and viewing efficient. The user may then configure access to the database based on monitoring needs and network access restrictions. The most advanced software allows viewing of the entire project database in a web browser via intranet (as well as the Internet).

Hybrid Monitoring Systems

It is common for instrumentation monitoring systems to incorporate elements of manual and automated data collection. There are several common examples of instrumentation systems that incorporate hybrid data acquisition. The dam may have some instruments that require manual readings while others are collected automatically. Another example of a hybrid collection system is a dam with automatic collection of instrumentation data using a datalogger, but the data is held within the datalogger for intermittent local download. Some instruments have an integral datalogger so that the device may be programmed for a sampling schedule, deployed, and then collected later for data download to a computer or mobile device.

Monitoring systems that include both automated and manually collected data offer some challenges for data organization and reporting. The automated and manually collected data must be synchronized within the database, and some owners use a mobile application (i.e. an application running on an iOS or Android mobile device) to bridge the gap between data collection in the field and the hosted project database (Zimmerman, 2016). One software solution has the field technician visit each instrument location, scan the instruments Quick Response (QR) code to identify it, then manually take a reading which is entered and saved to the mobile device. The survey is finished by automatically synchronizing the data saved on the mobile device with the project database.

The example of a dam with automatic collection of instrumentation data using a datalogger that stores the data for later download would be expected to have data collection functions like a fully automated system. The main difference between this hybrid system and the fully automated system is in the method of transmitting the data. Where the fully automated system will transmit the data from the datalogger to the instrumentation database, the hybrid system requires the user to physically access the datalogger for the data download.

Water level loggers are an example of a common instrument with integrated dataloggers. The loggers are typically deployed for short to medium durations when short-term fluctuations in water level might otherwise be missed by an intermittent manual sampling schedule. Water level loggers can be very useful for deployment during relatively short-term events that require high data density such as pumping tests, dewatering, or reservoir drawdown. Users may download the data from the device at the end of the deployment or intermittently as needed.

DAM SAFETY INSTRUMENTATION AND DATA RECORDING

Dam safety instrumentation systems are commonly used to monitor seepage, movement, piezometric and hydraulic pressure, and seismic acceleration. The Federal Energy Regulatory Commission (FERC) Engineering Guidelines for the Evaluation of Hydropower Projects offers a general definition of instrumentation as consisting of electrical and mechanical instruments or systems used to measure pressure, flow, movement, stress, strain, and temperature (FERC, 2015). The USACE Engineering

Manual for Instrumentation of Embankment Dams and Levees groups instruments into those used for measuring piezometric pressure, deformation, total stress, temperature, seismic events, and seepage. The USACE Engineering Manual on Instrumentation for Concrete Structures identifies similar categories of measurement while detailing measurements such as strain, stress, and pressure that are more common to concrete structures (USACE, 1987). Various instruments are used within these broad categories to monitor and evaluate dam performance and safety.

Many manually read instruments can be incorporated into an ADAS or a similar instrument may be substituted. For instance, a VW piezometer may be read with a hand-held readout display or its data may be recorded by an automated datalogger. In some cases, the only difference between an automated or manually accessed seismic recorder is whether the time history data is stored in the device or stored and then transmitted out via telemetry. Instruments below are grouped into general categories of devices that are used to monitor piezometric and hydraulic pressure, seepage and leakage, and movement. Common manually read instruments are listed in the first column of each table while automated instruments of the same type or that have similar general use are listed in corresponding rows in the second column. The tables are not intended to be exhaustive, but to instead show commonly used options for automating data collection.

Piezometric and Hydraulic Pressure

Piezometric pressure is commonly monitored in dam embankments and foundations using piezometers and observation wells. Water pressure in concrete dams and their foundations may be monitored using closed standpipe piezometers which are referred to as uplift cells or pore pressure cells. It is relatively straightforward to upgrade a manually read piezometer or uplift cell by suspending a VW piezometer or transducer in the open standpipe or within the uplift cell. Table 1 compares manually read and automated instruments.

Table 1. Manually read instruments for monitoring piezometric and uplift pressure compared to automated instrument alternatives.

Instrument - Manually Read	Instrument - Common Automated Alternative
Piezometer – Porewater Pressure	
Open standpipe piezometer - water level indicator	VW Piezometer (mounted in standpipe or grouted in place)
VW Piezometer – hand-held readout	VW Piezometer – automated
Pneumatic piezometer	Pneumatic piezometer – automation is possible but not common due to complexity and expense
Uplift Pressure	
Uplift Cell or Closed Standpipe Piezometer – Pressure gage	VW Piezometer or Pressure Transducer (in place of pressure gage)

Seepage and Leakage

Seepage or leakage at a dam may be collected and channeled through weirs or flumes for calculation of water flow quantity or rate. The water level behind the weir or flume is measured for use in a hydraulic equation for flow calculation. Manual measurement of water level is typically accomplished using a staff gage mounted behind the weir or with a tape measure. Automated systems may replace the staff gage with a pressure transducer, VW piezometer, VW weir monitor, or other instrument for measurement of water level. Table 2 compares manually read and automated instruments for seepage and leakage monitoring.

Table 2. Manually read instruments for monitoring seepage and leakage compared to automated instrument alternatives.

Instrument - Manually Read	**Instrument - Common Automated Alternative**
Flow Volume/Rate	
Weir or Parshall Flume – Upstream water level measured using staff gage	VW Piezometer or VW Weir Monitor (in place of staff gage)

Movement

Monitoring for dam movement and deformation may include everything from measurement of very small movements using strain gages to monitoring larger movements using survey techniques. A multitude of both general and specialized instruments for deformation and movement monitoring are available. While a complete description of movement monitoring devices is beyond the scope of this paper, it is safe to say that there is an automation-capable instrument available for most needs. Common applications such as crack monitoring and determination of displacement using a slope inclinometer are generally straightforward candidates for automation. Even complex tasks such as alignment and settlement surveying may be automated using a robotic total station or fixed global positioning system (GPS) technology. Recent advances in GPS equipment allow cost-effective use of fixed GPS modules on all types of dams. Table 3 compares manually read and automated instruments for movement and deformation.

Table 3. Manually read instruments for monitoring movement and deformation compared to automated instrument alternatives.

Instrument - Manually Read	**Instrument - Common Automated Alternative**
Surface Movement and Deformation	
Survey Methods (Alignment and Leveling/Settlement)	- Automated Total Station - Fixed GPS Modules
Settlement Plates	VW Settlement Sensor
Extensometers (single or multiple point) measured using dial indicator or micrometer	Extensometers (single or multiple point) with VW Displacement Transducers

Instrument - Manually Read	Instrument - Common Automated Alternative
Probe Inclinometer (in grooved casing)	- In-Place Inclinometer (MEMS or VW) - Shape Accelerometer Array (SAA) - Time Domain Reflectometry (TDR) cable
Plumb Line, Inverted Pendulum	- Plumb Line, Inverted Pendulum (with electronic pendulum readout) - Tiltmeter (VW or MEMS)
Strain gage/VW strain gage	Strain gage/VW strain gage
Crackmeter/Jointmeter	Crackmeter/Jointmeter (VW)
Strandmeters (VW)	Strandmeters (VW)

ADVANTAGES OF ADAS

Instrumentation and monitoring is a key part of dam safety and performance evaluation. FERC engineering guidance on instrumentation and monitoring emphasizes that timely collection and evaluation of instrument data is important. ADAS allows collection, processing, and display of data within minutes. Long term cost reduction and lower staffing requirements are typical with an ADAS in place. The ADAS promotes greater risk reduction due to the dam engineer and owner having access to more information in a more timely and actionable manner than what a non-automated system provides.

More (and More Accurate) Information

ADAS allows an adjustable recording interval for all instruments. The data acquisition rate may be increased during special events such as floods or construction before returning to a normal recording interval at other times. Even if the ADAS is programmed to record data at a rate of just one reading per day, the number of data points will be far greater than a non-automated site where instruments are read monthly or quarterly.

The data collected with the ADAS is frequently of higher quality than manually collected data due to the elimination of human error in measurement and recording. Transposed numbers, incorrect data entries, changes in measurement technique due to personnel changes, and other human factors are eliminated. Data is collected reliably and consistently. Issues with difficult or expensive personnel access to remote dams will not limit the ADAS function.

The accuracy and consistency of manual data that is collected on the same site as an ADAS may be improved by using the ADAS tools available with at least one commercially available system. This ADAS software system includes mobile applications to bridge the gap between data collection in the field and the project database. The mobile application (app) allows seamless collection and storage of manually read instrumentation data, and has the additional advantage of allowing error identification in the field instead of in the office. The app will display the previous readings and will notify the user when the current reading is outside of the normal range of values. The field technician can then immediately take another reading instead of remobilizing to the instrument later for data verification. The data collected using the app is immediately synchronized with the project database and is available for review and reporting (Zimmerman, 2016).

ADAS allows integration of different types of instrumentation data. A piecemeal approach to automated data acquisition might result in piezometer data being acquired by one acquisition system while seismic data is acquired and processed by a proprietary system sold by the seismic recorder manufacturer. Instead of relying on multiple data acquisition platforms, ADAS software is available that will easily integrate the seismic recorders at the hardware level all the way through to auto-delivery of processed seismic event data. This type of comprehensive platform for all instrumentation data allows the user to quickly compare data from various instruments without the complexity of running multiple software and database platforms.

More Timely

Manual data collection requires the field technician to travel to the site, record the data, return from the field, and then transfer the data to a spreadsheet or database format. Compare this process to the ADAS which allows the data to be immediately available, stores the data in one database, and allows automatically generated charts and reports to simultaneously be viewed from multiple locations.

An important feature of ADAS software is the ability to set multiple levels of alarms or thresholds for each instrument. For example, piezometers might be assigned alarms that correspond to piezometric elevations within an embankment correlating to slope stability factors of safety less than 1.5. Typical alarms are assigned based on high or low values or on rates of change. Notifications may be configured to automatically alert assigned personnel by email or text when the alarm values are met or exceeded. The notification email can even be configured to include a chart or data report from the instrument. The user can then log into the system at the site or remotely to check dam instrumentation performance. The system may be configured to send an automated notification email with a predetermined output when new data is added to the database or on some other programmed schedule. This provides a prompt to the engineer to check the ongoing performance of critical instruments.

More Actionable

ADAS provides more data at a faster rate than a non-automated system. Thus, the engineer or owner can more quickly determine that a problem is developing and move to address the issue. Data can be accessed whenever or wherever needed. Even issues with the system components themselves may often be diagnosed and fixed from a remote location.

Project Example: Slope Stability Monitoring During Construction. Slope stability monitoring during construction is an example of receiving actionable instrumentation data. Whether a non-automated slope inclinometer is measured daily, weekly, or monthly, the data density is much less than an inclinometer with in-place sensors and a datalogger programmed to acquire and transmit data at an interval of minutes. The ADAS has the additional advantage of having defined alarm levels and automated notification if those levels are exceeded. Action may be taken immediately to address the issue that caused the alarm.

In-place inclinometers were used to monitor slope movement in potentially unstable zones that were at risk to be affected by nearby excavation in the following project example. Six Geokon MEMS in-place inclinometer arrays were installed and data acquisition was controlled using a Campbell Scientific datalogger. Data was transmitted from the site using a cellular modem and Canary Systems MLWeb software was used to monitor the data in near-real time. Alarm levels based on displacement and rates of change were assigned to each instrument prior to commencement of construction activities. The software display, shown in Figure 2, was configured with visual status indicators to show the alarm status of the instruments. Instruments with cumulative displacement below alarm levels had their data displayed in green data fields while instrument data above an alarm level displayed in a red data field.

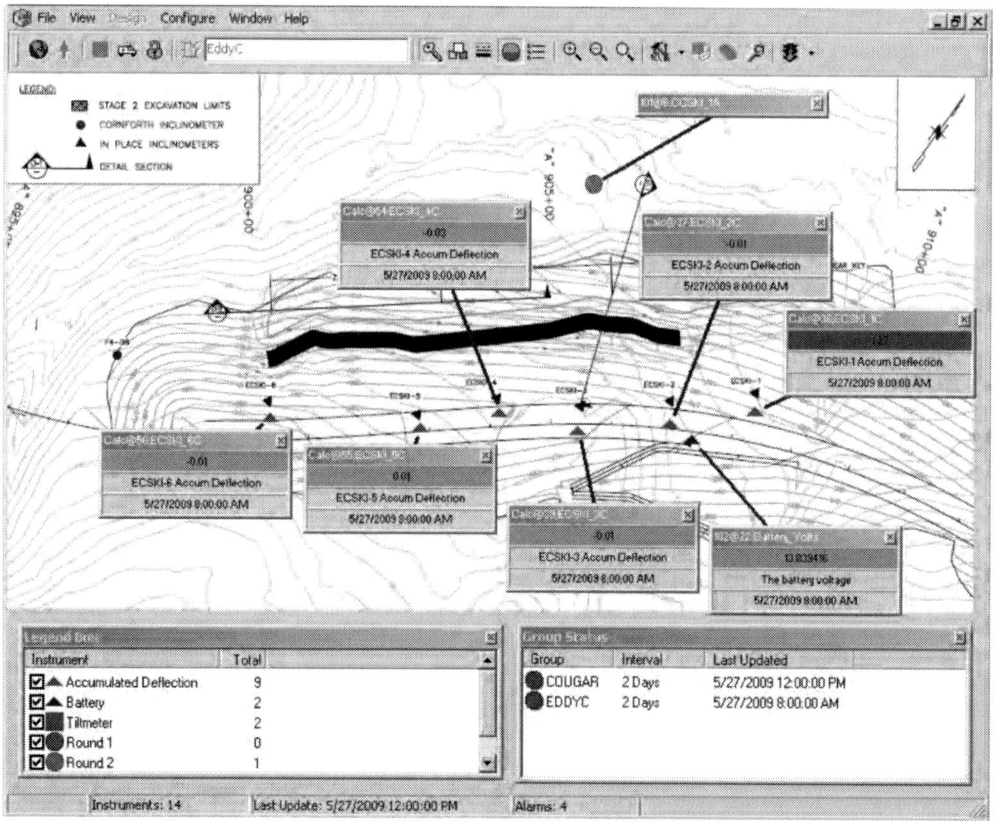

Figure 2. Plan view of in-place software display for inclinometer monitoring of slope movement adjacent to construction.

Figure 3 shows actionable slope displacement due to construction activity. The deflections in the three boxes were recorded during construction activities. Immediate notifications when deflections exceeded action levels allowed the engineer to suspend construction activities and reconsider the rate of excavation.

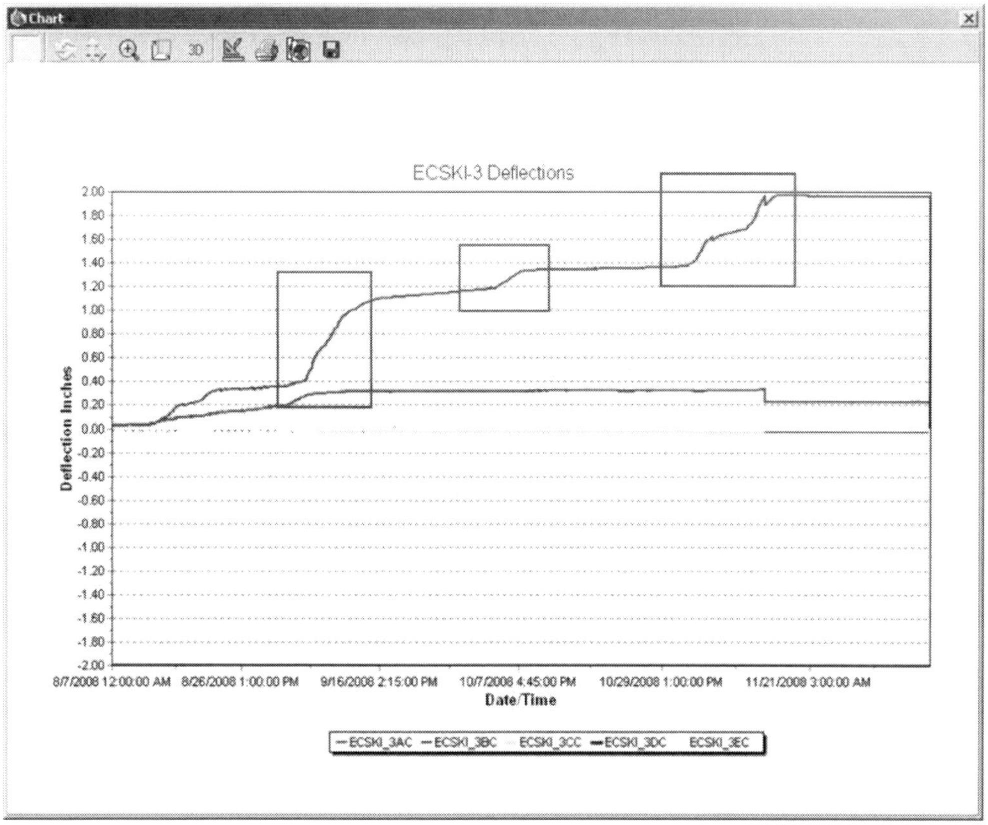

Figure 3. Slope displacement due to construction activity.

Economical and Efficient

Implementation of an ADAS has the immediate effect of reducing field personnel and reporting costs. The cost savings with time can be especially dramatic for remote sites and complex sites with many instruments. The USACE identifies trends of decreasing dam safety manpower and resources within their organization and lists an advantage of ADAS as replacing lost manpower (USACE, 1995). Similar trends may be observed in other government organizations as well as within the industry. Instrumentation reporting can be time consuming when data is collected in multiple spreadsheets and multiple formats. ADAS database and reporting software offers a more efficient and effective method for reporting (Zimmerman, 2016).

ADAS CHALLENGES

ADAS use within the dam safety environment includes challenges that must be recognized and addressed during consideration and planning. Many of the technological limitations of past decades are addressed in modern automated systems. Data storage, communication, and computer processing speed are listed as limitations in instrumentation manuals published 10 to 20 years ago, but these limitations have largely been overcome by the tremendous advances in the communication and computer fields. ADAS challenges today include initial investment cost, maintenance, and the reduced presence of personnel on site for routine observations due to fewer opportunities or need for data collection site visits.

Initial Investment Costs

ADAS installation requires more up-front investment than a non-automated solution. Project planning to ensure an appropriate system design and implementation is critical. Equipment, installation, and database software configuration costs are typical for ADAS deployment. Initial costs should be compared to the long-term cost reductions due to reduced personnel costs as well as the risk reduction advantages offered by an automated system.

Maintenance Costs

Long-term funding for maintenance and repair of ADAS components should be included in budgets and in cost comparisons. Modern instrumentation, datalogger, and communication gear is generally robust, but the ADAS introduces a higher level of complexity to monitoring and associated maintenance should be expected. Instruments installed in wet environments such as concrete dam drainage galleries will experience corrosion, batteries will eventually require replacement, and other issues may arise with time. The system should include built in protections from potential disruption by environmental issues such as lightning and wind, but occasional damage may occur even with safeguards in place.

Reduced Personnel Time on Site

Visual observation of the dam occurs on each instrumentation data collection visit. FERC (FERC, 2015) correctly observes that qualitative visual observations complement the quantitative information gathered from the instruments. A disadvantage of an ADAS, therefore, is that the opportunities for visual observation of the dam are reduced as personnel time on site is reduced. It is important that the instrumentation program with ADAS be combined with a robust inspection program.

CONCLUSION

Automated instrumentation monitoring systems are sometimes dismissed as an option because they are thought to be overly complex with an internal function that is not well understood. However, the general performance of an ADAS for dam safety instrumentation monitoring is straightforward with sensors linked to data acquisition modules through site telemetry, and communication that allows remote monitoring from the user's desktop or mobile device. Instruments capable of automation are readily available and many equivalent or superior options typically exist for replacement of manually read instruments.

The advantages and limitations of various data acquisition methods should be considered during the instrument system design phase and as monitoring progresses. ADAS can provide more data than a non-automated instrumentation system and the data can be monitored in near-real time from anywhere in the world. ADAS instrumentation data is often of better quality than a non-automated alternative due to elimination of human error in collection and reporting tasks. With increased data quantity, quality, and speed, the ADAS-provided data is more actionable and risk may be reduced. An instrumentation program with ADAS should be combined with an appropriately robust inspection program. Fewer personnel visits for routine manual data collection means that fewer "eyes" are on site for visual observations. Regardless of the data acquisition system that is used, limitations can be reduced with appropriate planning and implementation.

Automation requires more up-front investment than a non-automated system, but it may provide significantly better risk reduction, lower long-term costs, and lower staffing requirements. Reporting software that is part of an ADAS should offer the ability to create report templates which can be re-used each time a report is needed. With the click of a button, the user can generate a comprehensive multi-page report that includes text, data charts, and pictures. Engineers can then concentrate on identifying data trends and verifying dam performance rather than on the mechanics of collating data and creating charts.

Manual collection of data is well understood and is often the default option in both new and existing projects. However, advances in ADAS technology make automated data collection a relevant option at many dams. ADAS might not have been considered when the original dam monitoring system was established, but advances in technology, heightened monitoring requirements, increased sensitivity to or awareness of risk management, and/or changes in personnel demands might make ADAS an attractive monitoring alternative at an existing dam. ADAS doesn't make sense for every dam, but the system can bring advantages over traditional manual instrument monitoring when used appropriately.

REFERENCES

FERC (2015) *Engineering Guidelines for the Evaluation of Hydropower Projects, Chapter 9, Instrumentation and Monitoring*

USACE (1995) *EM 1110-2-1908 Engineering and Design, Instrumentation of Embankment Dams and Levees*

USACE (1987) *EM 1110-2-4300 Engineering and Design, Instrumentation for Concrete Structures*

Zimmerman, P., Jordan, D., and Newell, V. (2016) Dam Safety Instrumentation Data Acquisition, Monitoring, and Data Visualization Improvements at the Rocky Mountain Hydroelectric Plant. *ASDSO Dam Safety 2016 Conference Proceedings.*

Zimmerman, P., Jordan, D., and Newell, V. (2016) Technology Solutions for Dam Safety Instrumentation Data Collection and Reporting. *The Journal of Dam Safety*, 14(2), 41-44.

FAILURE OF THE ACUDE DA NACAO EARTH DAM:
A PROBABILISTIC PERSPECTIVE

Danielli de Melo Moura[1]
Rodrigo Borela[2]
Gabriela Alvarado[3]
Wilson Espinoza[4]
Philippe L. Bourdeau[5]

ABSTRACT

In the Northeastern part of Brazil, where water resources are scarce, a number of earth dams have been constructed to create small multipurpose reservoirs that support local economies. Recent changes in the regional hydrologic cycle have caused a series of intense rainstorm events that, by increasing repeatedly the hydraulic load, affect the safety of these structures. During one of these flooding episodes in June 2010, the Acude da Nacao Dam, a 12m (39ft) tall and 240m (787ft) long compacted earth fill structure, breached and as a result the Bom Conselho city was partially submerged, with its main water supply interrupted. Rehabilitation efforts concentrated on a geotechnical site survey for the purpose of reconstruction, without a detailed investigation of the mechanism of failure. In this paper, the Acude da Nacao dam failure is revisited using available data from the retrofit project in a probabilistic approach. Slope stability and internal erosion are analyzed for cross-sections, representative of the breached area, and probabilities of failure are obtained for the respective mechanisms. Two techniques were used in the probabilistic computations: the first order-second moment approximation, based on Taylor series development and Rosenblueth's point estimate method. The results indicate that piping, resulting from internal erosion, was the most likely cause of failure. The reliability of the retrofitted dam is also evaluated using the same methodology. The analysis shows improved reliability with the new design, should the same flooding event occur again. However, the risk of hydraulic failure remains a concern, given the uncertainty of hydrological nature.

INTRODUCTION

Earth dams of moderate height, whose main purpose is to store water for irrigation and provide drinking water to local communities, are very common structures in Brazil. Construction of such dams is fairly affordable since the embankment fill material can often be borrowed near the site, and their design is considered relatively simple.

---

[1] Lyles School of Civil Engineering, Purdue University, 550 Stadium Mall Drive, West Lafayette, IN 47906, +1 765 838-9546, ddemelom@purdue.edu.
[2] School of Civil and Environmental Engineering, GaTech, 790 Atantic Dr. NW, Atlanta GA 30332, formerly at Lyles School of Civil Engineering, Purdue University, +1 765 476-6945, rborelav@gatech.edu.
[3] Lyles School of Civil Engineering, Purdue University, 765 772-8041, alvarad9@purdue.edu.
[4] School of Civil and Environmental Engineering, GaTech, 790 Atantic Dr. NW, Atlanta GA 30332, formerly at Lyles School of Civil Engineering, Purdue University, 919 207-7176, wespinoza6@gatech.edu.
[5] Lyles School of Civil Engineering, Purdue University, +1 765 494-5031, bourdeau@ecn.purdue.edu.

In comparison with concrete dams, earth dams are more ductile structures but they are vulnerable to failure due to overtopping, internal erosion and slope instability.

Typical overtopping and related failures in cohesive material happen as turbulent flow erodes the toe of the dam on the downstream side, creating a front that propagates upstream. This mechanism is similar to scouring observed in streams. In the case of an earth dam, large portions of the material would detach from the downstream slope, eventually causing a breach. Alternatively, several cases have been reported in the literature, in which dams sustained overtopping events without loss of stability and only limited damage (Powledge, et al., 1989). This indicates that overtopping does not necessarily imply structural failure of a dam (Wahl, 2004).

Failure by slope instability for in-service earth dams is generally related to the effect of pore pressure increase within the structure. At flood stage, rising of reservoir level and the related phreatic surface within the body of the dam generate higher pore water pressure, lessens the resistance to sliding and may lead to a potential failure surface to be activated.

Failure may also originate from internal erosion and piping. Higher reservoir and phreatic surface levels would also result in higher hydraulic gradients within the dam. Seepage forces along channels of preferential flow may become sufficient to initiate local instability at the downstream face. The induced loss of material may then evolve, through a retrogressive erosion process, into a stage of piping of such extent that finally the dam breaches.

The exact mechanism of failure of the Acude da Nacao dam had not been clearly identified. Though the hypothesis of scouring could be discarded, as will be discussed in the next section, reports did not elucidate the respective roles played by the other two possibilities, slope instability and internal erosion. The question being still open while the dam was reconstructed, concern may be raised on the effectiveness of a rehabilitation project conducted without full knowledge of the sequence of events that had led to failure of the original structure. The present study is an attempt to address this question, a posteriori, through the means of analysis and numerical modeling. Both the original, failed dam and the newly reconstructed structure are analyzed for comparative purpose. Because of the considerable level of uncertainty on material parameters, and also in order to achieve better objectivity in making comparative assessment, a probabilistic approach was adopted.

ORIGINAL ACUDE DA NACAO DAM

Description

The Acude da Nacao dam was built in the 1980s across the Papacacinha River, located in Bom Conselho, in the state of Pernambuco, Brazil. The structure was originally composed of a homogenous compacted earth fill, 10m (33ft) high in its tallest section,

240m (787ft) long and had a 1(V):2.5(H) slope on its downstream face, and resting on bedrock.
There was also a 30m (98ft) long spillway. According to residents of the area, the height of the dam including its masonry spillway was increased by 2m (6ft) in 1998 but there is no record of a design study in view of this modification. At that time, four rectangular openings with approximate dimensions of 1.0m x 1.5m (3ft x 5ft) were included just below the crest, to serve as uncontrolled spillway gates. Information available on the embankment fill material is very limited and comes from post-failure testing performed as part of the retrofitting project, as the same borrow pit that had been the source of construction material for the original dam was going to be used again. Data on particle size, consistency and shear strength from those three samples are shown in Table 1. The data base is too sparse for any significant statistics being extracted, but the important scattering in the shear strength indicates significant uncertainty is attached to these properties.

Table 1. Characteristics of soils used as embankment fill in both the original dam and the retrofitting project

Gravel (%)	Coarse sand (%)	Medium sand (%)	Fine sand (%)	Silt (%)	Clay (%)	LL (%)	PL (%)	PI (%)	c' (kPa)	ϕ'
9	13	23	18	7	30	32	18	14	32	32°
9	4	18	18	8	43	32	21	11	21	36°
1	5	21	34	7	32	32	18	14	33	33°

In 2010, intense rain in the area led to rapid filling of the reservoir. The spillway gates were not capable of discharging the excess water, and overtopping took place, ensued by breaching of the dam. After the initiation of failure, the whole structure was overthrown by the flood with an estimated peak flow of 170m^3/s (6,000cfs). The breached section is pictured in Figure 1.

Figure 1. Breached section of the Acude Da Nacao dam, with lowered reservoir on the right side (Source: Projetec / SEINFRA, 2011)

In evaluating the potential mechanisms of failure, scouring, as a direct consequence of the overtopping event, was the first mode under consideration. Methods to assess the susceptibility to scouring cover a wide spectrum ranging from empirical approaches (MacDonald and Langridge-Monopolis, 1984 and Wahl, 2004), soil erosion hydraulic models (Powledge et al., 1989), combinations of the aforementioned methods (Xu and Zhang, 2009) and numerical models. One such model is implemented in the software WinDAM B, offered by the National Resources Conservation Services (NRCS). However, given the known characteristics of this dam, scouring as the direct cause of failure was ruled out for the following reasons: (1) the crest was protected by a pavement layer; (2) the downstream slope was protected by dense vegetation; (3) laboratory test results indicated the embankment fill material was a non-dispersive soil.

Two potential failure mechanisms remained under consideration, instability of the downstream slope and hydraulic failure due to internal erosion leading to piping; these are the focus of the present analysis.

Slope Instability

During a prolonged flooding period the reservoir water level reaches its peak and the phreatic surface in the body of the dam rises. This generates larger gradients and seepage forces, as well as larger pore pressure in the embankment earth fill. For stability of the downstream slope of the dam, these conditions constitute a critical situation. The analysis was conducted using commercially available software GeoStudio (Geo-Slope 2014) modules Seep/W (for seepage) and Slope/W (for slope stability). The following simplifications were made:

- Both seepage and stability computations were two-dimensional analyses of the cross-section that actually breached during the flooding event. Dimensions and boundary conditions are shown in Figures 2 and 3, respectively for seepage and stability.
- Seepage was analyzed under steady-state flow conditions.
- All material properties were defined as homogeneous and isotropic. This means short-range spatial variability or local heterogeneity is not captured in the model. Particular implications in a probabilistic context are discussed herein in the concluding section.
- The most critical of the potential sliding surfaces was searched using mean values of the mechanical parameters. Then, for this surface only, selected parameters were considered as random variables and the related probability of failure was computed.

The analysis was twofold: at first, seepage at flood stage through the dam was computed for determining the patterns of flow including pore pressure distribution, and then slope stability computations were performed using Bishop's method of slices. The outcome from seepage analysis, relevant to slope stability, is the distribution of pore pressure shown in Figure 2. It is noted that for the conditions considered here, steady-state flow with homogeneity and isotropy, pore water pressure and hydraulic gradients are independent of hydraulic conductivity, thus only mechanical properties were modeled as random variables in order to assess the probability of failure. The safety factor, FS, related to the most critical sliding surface was defined as performance function of three random variables, saturated unit weight (γ), cohesion (c') and angle of internal friction (ϕ'). The First Order Second Moment (FOSM) technique, based on Taylor series approximation of the performance function around the mean values of the variables (Hahn and Shapiro, 1967), was used to compute the expected value and variance of the safety factor. The function was then modeled as log-normally distributed, and the probability of failure was computed as:

$$P_f = P[FS(\gamma, c', \phi') < 1] \quad (1)$$

In its analytical form, the FOSM provides expected value and variance of a performance function, y, of N random variables, x_i $(i = 1 \text{ to } N)$, in Equations (2) and (3), respectively, where μ, σ and ρ denote mean values, standard deviations and coefficients of correlation of the variables. In Equation (3) partial derivatives of the function are to be calculated at the mean values of the variables.

$$E[y] = y(\mu_{x_i}, \ldots, \mu_{x_N}) \quad (2)$$

$$V[y] = \sum_{i=1}^{N}\left[\left(\left.\frac{\partial y}{\partial x_i}\right|_{\mu_{x_i}}\right)^2 \sigma_{x_i}^2\right] + 2\sum_{i \neq j}^{N-1,N}\left[\left.\frac{\partial y}{\partial x_i}\right|_{\mu_{x_i}} \left.\frac{\partial y}{\partial x_j}\right|_{\mu_{x_j}} \rho_{x_i,x_j}\sigma_{x_i}\sigma_{x_j}\right] \quad (3)$$

In the Bishop method of slices the safety factor is solution of an implicit equation which must be resolved by iterations. Furthermore the potential sliding surface may not have a simple geometric shape. Thus it is not practical to formulate Equation (3) in closed form. A finite difference form of the variance proposed by Wolff (1995) and later popularized by Duncan (2000), in which finite increments on each side of the variables mean values are chosen equal to their standard deviations, was used to approximate Equation (2). The first two moments of the safety factor are thus obtained as:

$$E[FS] = FS(\mu_\gamma, \mu_{c'}, \mu_{\phi'}) \quad (4)$$

$$V[FS] = \left[\left(\frac{FS_{\phi'+} - FS_{\phi'-}}{2\sigma_{\phi'}}\right)^2 \sigma_{\phi'}^2 + \left(\frac{FS_{c'+} - FS_{c'-}}{2\sigma_{c'}}\right)^2 \sigma_{c'}^2 + \left(\frac{FS_{\gamma+} - FS_{\gamma-}}{2\sigma_\gamma}\right)^2 \sigma_\gamma^2\right] +$$

$$\left[2\left(\frac{FS_{\phi'+} - FS_{\phi'-}}{2\sigma_{\phi'}}\right)\left(\frac{FS_{c'+} - FS_{c'-}}{2\sigma_{c'}}\right)\rho_{\phi'c'}\sigma_{\phi'}\sigma_{c'} + 2\left(\frac{FS_{\phi'+} - FS_{\phi'-}}{2\sigma_{\phi'}}\right)\left(\frac{FS_{\gamma+} - FS_{\gamma-}}{2\sigma_\gamma}\right)\rho_{\phi'\gamma}\sigma_{\phi'}\sigma_\gamma +\right.$$

$$\left.2\left(\frac{FS_{\gamma+} - FS_{\gamma-}}{2\sigma_\gamma}\right)\left(\frac{FS_{c'+} - FS_{c'-}}{2\sigma_{c'}}\right)\rho_{c'\gamma}\sigma_\gamma\sigma_{c'}\right] \quad (5)$$

In Equation (5), subscripts (+) and (-) indicate FS is computed for values of the corresponding variable equal, respectively, to its mean plus one standard deviation ("higher" values in Table 3) and its mean minus one standard deviation ("lower" values in Table 3).

Input data for the three random variables are given in Table 2. These estimates were based on in-situ and laboratory tests that preceded the retrofitting project, together with coefficients of variation and coefficients of correlation previously reported in the literature for similar parameters (e.g. Harr, 1987, Amundaray, 1994). They represent fairly conservative estimates in consideration that the small number of tests did not allow true statistics be extracted. Detail of computation for evaluating Equations (4) and (5) is presented in Table 3.

Table 2. Random variable input data for original dam stability analysis

Material property	**Mean value (μ)**	**Standard deviation (σ)**
Unit weight, γ	22 kN/m^3	3 kN/m^3
Cohesion, c'	32 kPa	10 kPa
Friction angle, ϕ'	32°	6°
Material properties	**Correlation (ρ)**	
γ and ϕ'	+ 0.5	
c' and ϕ'	- 0.5	
γ and c'	0	

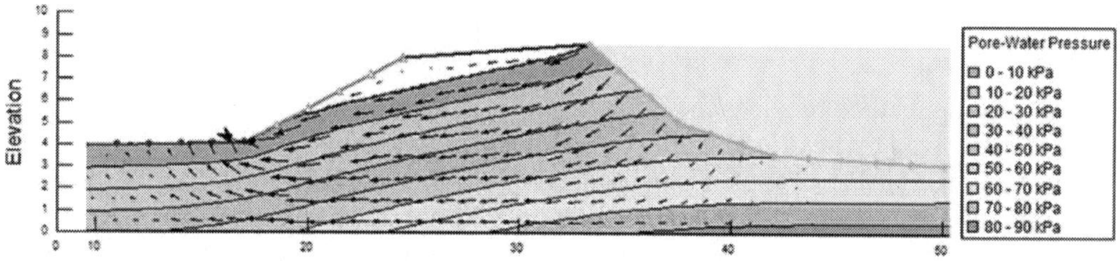

Figure 2. Seepage analysis for the original dam breached section (Arrows represent gradient vectors)

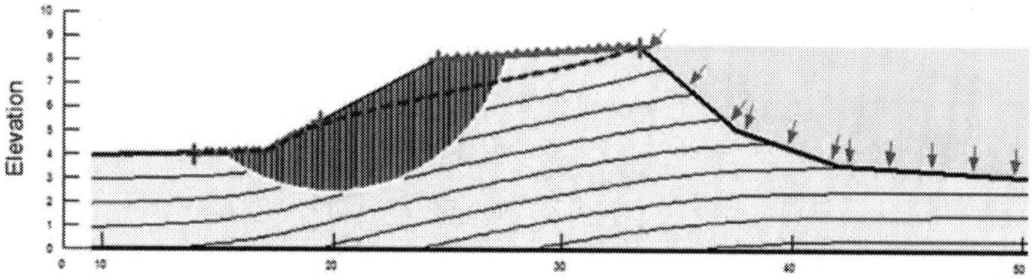

Figure 3. Slope stability analysis for the original dam breached section (Sub-horizontal lines relate to pore pressure distribution; see Figure 2)

Table 3. Performance function computations for slope stability of original dam

Analysis	ϕ'	c'(kPa)	γ(kN/m³)	Factor of safety	
Mean values	32°	32	22	FS	3.80
Lower friction	26°	32	22	$FS_{\phi'-}$	3.48
Higher friction	38°	32	22	$FS_{\phi'+}$	4.16
Lower cohesion	32°	22	22	$FS_{c'-}$	3.07
Higher cohesion	32°	42	22	$FS_{c'+}$	4.53
Lower unit weight	32°	32	19	$FS_{\gamma-}$	4.03
Higher unit weight	32°	32	25	$FS_{\gamma+}$	3.62

The resulting expected value of the slope safety factor was $E[FS] = 3.80$ and its variance was $V[FS] = 2.51$ (i.e. a standard deviation of 1.58 and a coefficient of variation of 42%). This corresponds to a probability of failure (Equation 1) P_f, of the order of 10^{-3}, a small value consistent with reports that slope instability is not a frequent cause of failure of small earth dams (Foster et al., 2000).

Internal erosion

At flood stage, seepage forces may become large enough to mobilize embankment fill material near the downstream face and initiate a backward internal erosion process which, eventually would lead to complete piping and breaching. It is generally agreed that, in unstructured earth dams (i.e. constructed using a single type of fill material), once the process has started at the downstream face it is extremely likely to evolve into complete piping. Accordingly, initiation of internal erosion at the open downstream face was considered in the present case to be the criterion for hydraulic failure. This would translate in the maximal gradient (I_{max}) exceeding the critical gradient (I_{cr}), and the probability of failure being defined as:

$$P_f = P[I_{max} > I_{cr}] \tag{6}$$

The maximal gradient was found in the seepage analysis to be located near the toe and acting horizontally with magnitude $I_{max} = 0.493$. For a homogeneous fill dam with sloped downstream face and where the highest hydraulic gradient is nearly horizontal at the toe, the critical gradient can be estimated according to Kovács (1981), as

$$I_{cr} = \left(\frac{\gamma_{sat}-\gamma_w}{\gamma_w}\right)\left(\frac{\tan(\phi')-\tan(\beta)}{1+\tan(\phi')\tan(\beta)}\right) \tag{7}$$

where, γ_{sat} is the saturated unit weight of the dam fill material, γ_w the unit weight of water and β is the slope angle of the downstream face.

The first two moments of the critical gradient, a function of two correlated random variables (γ_{sat} and ϕ') were computed using the Point Estimate Method (PEM) proposed by Rosenblueth (1975), see also Harr (1987). Given $y = y(x_i, \ldots . x_N)$ a function of N

random variables, the PEM does not approximate the function, in contrast with the FOSM method, but it substitutes each continuous variable with a discrete variable which takes two values (the point estimates, Equations 10 and 11) and associated probabilities (Equations 8 and 9). Equivalence between the continuous and discrete variable is based on equality of their first three moments. Then, point estimates of the function are calculated for each permutation of the variable point estimates. For a function of N variables discretized at two points each, there are 2^N such permutations (Equation 12) and associated probabilities (Equation 13). The function expected value and variance are given in Equations (14 an (15), respectively. The method is considered superior to the FOSM method when the function is strongly non-linear, although it is computer-intensive for problems involving a large number of variables.

$$P_{x_{i+}} = \frac{1}{2}\left[1 + \frac{\beta_1}{|\beta_1|}\sqrt{1 - \frac{1}{1+\left(\frac{\beta_1}{2}\right)^2}}\right] \tag{8}$$

$$P_{x_{i-}} = 1 - P_{x_{i+}} \tag{9}$$

$$x_{i+} = \mu_{x_i} + \sigma_{x_i}\sqrt{\frac{P_{x_{i-}}}{P_{x_{i+}}}} \tag{10}$$

$$x_{i-} = \mu_{x_i} - \sigma_{x_i}\sqrt{\frac{P_{x_{i+}}}{P_{x_{i-}}}} \tag{11}$$

$$y_{ijk..N} = y(x_{i\mp}, \dots . x_{N\mp}) \tag{12}$$

$$P_{y_{ijk..N}} = \left(\prod_{i=1}^{N} P_{x_{i\mp}}\right)\left(1 + \sum_{i,j=1,2}^{N-1,N} \mp\rho_{ij}\right) \tag{13}$$

$$\boldsymbol{E[y] = \sum_{1}^{2^N} (y_{ijk..N}\, P_{y_{ijk..N}})} \tag{14}$$

$$V[y] = E[y^2] - (E[y])^2 \tag{15}$$

Input data for γ and ϕ' are in Table 2. In absence of specific information the coefficients of skewness in Equation (8) were assumed to be zero. Detail of calculation is shown in Table 4. Computed expected value of the critical gradient was $E[I_{cr}]$= 0.24 and its variance $V[I_{cr}]$= 0.03 (i.e. a standard deviation of 0.17 and coefficient of variation of 71%).

The resulting probability of hydraulic failure (Equation 6), with a lognormal distribution for the critical gradient, was P_f= 0.92 (or 92%). Comparison of this result with the probability of failure by slope instability provides a strong indication that internal erosion was the leading mechanism to the dam failure.

Table 4. PEM computations for original dam internal erosion analysis

Probability concentrations and point estimates	$P_{\gamma\pm}$	$P_{\varphi\pm}$	$\gamma_{sat\pm}$ **(kN/m³)**	$\phi'_{\pm}$	$P_{Icr\pm\pm}$	$I_{cr\pm\pm}$
Permutation #1	0.50	0.50	19.0	26.0°	0.38	0.068794
Permutation #2	0.50	0.50	25.0	26.0°	0.13	0.113709
Permutation #3	0.50	0.50	19.0	38.0°	0.13	0.272165
Permutation #4	0.50	0.50	25.0	38.0°	0.38	0.449858

RETROFITTED DAM

Because loss of the embankment had been limited to the breached section of the dam, and the remaining of the structure was still in acceptable condition, retrofitting the dam was decided instead of a complete reconstruction. In addition to rebuilding the failed section, the surface of the slopes and crest were scrapped overall, and additional material was placed in order to raise the dam by approximately 2m (6ft) over the pre-failure datum.

The upstream slope was made at an inclination of 2.5(H):1(V) and was protected by a layer of riprap. The downstream was compacted to 2(H):1(V) but with the addition of a granular filter and drainage layer, 1.5m (5ft) thick, covering the bottom half of the slope. A new reinforced concrete spillway was also constructed, 9m (29ft) high and 70m (229ft) long. The total length of the dam is now approximately 280m (918ft) and its crest is 10m (30ft) wide. A picture of the downstream face is shown in Figure 4 and the new cross-section is provided in Figure 6. New embankment material was borrowed from the same local pits that had already been used for the original embankment construction (see Table 1).

Figure 4. View of the retrofitted dam: downstream face with filter layer. New spillway is seen in the background (Source: first author's personal collection)

Analyses were performed for assessing probabilities of failure of the retrofitted dam subject to the same reservoir level than during the flooding event that had caused the past failure. Slope instability and internal erosion were considered again, and using the same methods of analysis than for the original dam.

Slope Instability

As shown in Figure 5, seepage analysis was performed in order to determine the pore pressure distribution that was needed for stability analysis, without accounting for the filter layer. As it is made of open-graded highly permeable material, this layer was considered to have no significant influence on the flow patterns. Accordingly, computation was performed with the downstream boundary being located along the interface between the embankment fill and the filter.

However, the mechanical effect of the filter layer was not disregarded while performing stability analyses, as it acts as a stabilizing berm at the toe of the dam (Figure 6). Additional parameters to characterize the filter granular material are given in Table 5. Parameters relating to the dam fill material are the same than those already used for the original dam. Detail of the FOSM computation is given in Table 6.

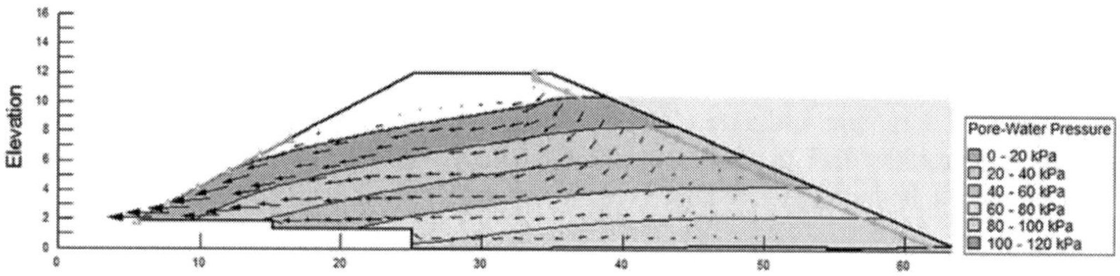

Figure 5. Seepage analysis of retrofitted dam (Arrows represent gradient vectors)

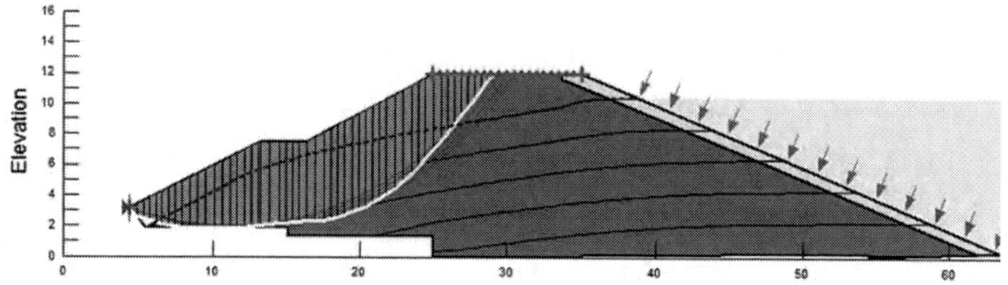

Figure 6. Slope stability analysis of retrofitted dam (Sub-horizontal lines relate to pore pressure distribution; see Figure 5)

Table 5. Random variables input data for analysis of retrofitted dam

Soil property	Mean value (μ)	Standard deviation (σ)
Embankment unit weight, γ_D	22 kN/m³	3 kN/m³
Embankment cohesion, c'_D	32 kPa	10 kPa
Embankment angle of internal friction ϕ'_D	32°	6°
Filter unit weight, γ_F	19 kN/m³	3 kN/m³
Filter angle of internal friction ϕ'_F	34°	6°
Material properties	**Coefficient of Correlation (ρ)**	
γ_D and ϕ'_D	+ 0.5	
c_D' and ϕ_D'	- 0.5	
γ_F and ϕ'_F	+ 0.5	

Table 6. Performance function computations for slope stability of the retrofitted dam

Analysis	ϕ'_D	c'_D (kPa)	γ_D (kN/m³)	ϕ'_F	γ_F (kN/m³)	Factor of safety	
Mean values	32°	32	22	34°	19	FS	2.70
Lower emb. friction	26°	32	22	34°	19	$FS_{\phi'D-}$	2.38
Higher emb. friction	38°	32	22	34°	19	$FS_{\phi'D+}$	3.07
Lower cohesion	32°	22	22	34°	19	$FS_{c'D-}$	2.33
Higher cohesion	32°	42	22	34°	19	$FS_{c'D+}$	3.07
Lower emb. unit weight	32°	32	19	34°	19	$FS_{\gamma D-}$	2.86
Higher emb. unit weight	32°	32	25	34°	19	$FS_{\gamma D+}$	2.58
Lower drain friction	32°	32	22	28°	19	$FS_{\phi'F-}$	2.69
Higher drain friction	32°	32	22	40°	19	$FS_{\phi'F+}$	2.72
Lower drain unit weight	32°	32	22	34°	16	$FS_{\gamma F-}$	2.64
Higher drain unit weight	32°	32	22	34°	22	$FS_{\gamma F+}$	2.76

Resulting moments for the safety factor were: expected value $E[FS]$=2.7, variance $V[FS]$=0.279 (i.e. a standard deviation of 0.528 and coefficient of variation 19.6%). With a lognormal distribution of FS, the probability of failure is $P_f = 2.5\ 10^{-7}$, an extremely low value.

Internal Erosion

In this case the maximal gradient at the embankment –filter interface was compared to the critical gradient. In other terms, the criterion for hydraulic failure is the initiation of internal erosion, independently of the filtration. However the presence of the filter layer has the beneficial effect of increasing the overburden stress in the region where the gradient is maximal, near the interface. This, in turn, contributes to increasing the critical gradient. The solution presented by Kovács (1981) for this particular configuration was used in a slightly modified form,

$$I_{cr} = \left[\frac{(\gamma_{sat\,D}-\gamma_w)+\left(\frac{\gamma_F \times h_F}{L}\right)}{\gamma_w}\right]\left(\frac{\tan(\phi'_D)-\tan(\beta)}{1+\tan(\phi'_D)\tan(\beta)}\right) \tag{16}$$

where (h_F) is the overburden height of filter material above the maximal gradient location, equal to 1.5m (5ft), and (L) is the horizontal distance over which the gradient is maximal. Parameter (L) is difficult to determine with accuracy because it is affected by the discretization mesh in the seepage finite element solution. For this reason, (L) was treated as an additional random variable, in addition to mechanical parameters. Its mean value was estimated to be 1.5m (5ft) on the basis of seepage analysis output, and a standard deviation of 0.2m (8in) was considered to represent the related uncertainty.

The critical gradient was analyzed as a function of four random variables (γ_D, ϕ'_D, γ_F, L) and its expected value and variance were computed using the PEM. Input data relating to material properties are given in Table 5. Detail of the computation is given in Table 7.

Table 7. PEM computations for retrofitted dam internal erosion analysis

Probability concentrations and point estimates	$P_{\gamma D}$	$P_{\varphi'}$	$P_{\gamma F}$	P_L	γ_D (kN/m³)	φ_D'	γ_F (kN/m³)	L (m)	P_{Icr}	I_{cr}
Permutation #1	0.5	0.5	0.5	0.5	19.00	26°	16	1.3	0.0937	0
Permutation #2	0.5	0.5	0.5	0.5	25.00	26°	16	1.3	0.0312	0
Permutation #3	0.5	0.5	0.5	0.5	19.00	38°	16	1.3	0.0312	0.573
Permutation #4	0.5	0.5	0.5	0.5	25.00	38°	16	1.3	0.0937	0.697
Permutation #5	0.5	0.5	0.5	0.5	19.00	26°	22	1.3	0.0937	0
Permutation #6	0.5	0.5	0.5	0.5	25.00	26°	22	1.3	0.0312	0
Permutation #7	0.5	0.5	0.5	0.5	19.00	38°	22	1.3	0.0312	0.717
Permutation #8	0.5	0.5	0.5	0.5	25.00	38°	22	1.3	0.0937	0.841
Permutation #9	0.5	0.5	0.5	0.5	19.00	26°	16	1.7	0.0937	0
Permutation #10	0.5	0.5	0.5	0.5	25.00	26°	16	1.7	0.0312	0
Permutation #11	0.5	0.5	0.5	0.5	19.00	38°	16	1.7	0.0312	0.483
Permutation #12	0.5	0.5	0.5	0.5	25.00	38°	16	1.7	0.0937	0.607
Permutation #13	0.5	0.5	0.5	0.5	19.00	26°	22	1.7	0.0937	0
Permutation #14	0.5	0.5	0.5	0.5	25.00	26°	22	1.7	0.0312	0
Permutation #15	0.5	0.5	0.5	0.5	19.00	38°	22	1.7	0.0312	0.593
Permutation #16	0.5	0.5	0.5	0.5	25.00	38°	22	1.7	0.0937	0.717

Values of zero in the last column of Table 7 occurred for permutations where the point estimate of the angle of internal friction (ϕ'_D) was smaller than the interface slope angle

(β), and would have produced, through Equation (16), negative values for the critical gradient point estimates. In such instances, the point estimates distribution of (I_{cr}) was truncated to zero.

Resulting moments are, for the critical gradient, a mean value $E[I_{cr}]$ = 0.34 and variance $V[I_{cr}]$ = 0.122 (i.e. standard deviation of 0.35 and coefficient of variation of 102%. The probability of internal erosion initiation, for the computed maximal gradient I_{max} = 0.516, is P_f = 0.818 (or 82%), which is still a very high value. However, it should be noted that this probability of hydraulic failure was computed independently of the filtration action of the granular layer on the downstream face, only its self-weight plays a role in the analysis. Therefore this result should be interpreted as a conditional probability of failure, in the event that the filter performance is unsatisfactory.

CONCLUSION

A new assessment of the Acude da Nacao Dam failure strongly supports the hypothesis of internal erosion being the controlling mechanism that led to catastrophic breaching of the structure in 2010. This conclusion is based on probabilistic analyses of the two most likely scenario, (1) downstream slope instability and (2) internal erosion initiated by excessive hydraulic gradient at the open face, both of these mechanisms being exacerbated by the peak of hydraulic load during an extreme flooding episode.

Retrofitting of the damaged earth dam, including the construction of a new spillway and the addition of a granular filter layer on the downstream face, produced a more reliable structure. However probabilistic analyses conducted using the same methodology than for the original dam, and for the same two potential failure mechanisms, indicate internal erosion is still a concern. Computed probability of initiation of piping at the interface between embankment and filter (82%) is only marginally smaller than the probability of hydraulic failure of the original dam (92%). Of course, these two values are not truly comparable because the retrofitted dam is equipped with a downstream filter layer which was absent from the original dam. Thus the probability of internal erosion of 82% must be understood as a conditional probability of failure, should the granular filter fail to retain mobilized particles of the embankment fill. The concern is that so much of the whole system's safety relies on a single component performance, as compared to how much more reliable a redundantly engineered system could be.

A number of simplifications were made in order to perform the analysis in a practical way. The two most important, because of their potential influence on the results, were (1) to model both seepage flow and slope stability as two-dimensional problems, and (2) representing uncertainty in material properties using bulk random variables, and in doing so ignoring the spatial variability of these materials at smaller scale including local heterogeneity. It should be noted that these two issues are linked, because two-dimensional modeling of a geotechnical problem is incompatible with a correct random field description of spatial variability. Considering a two-dimensional cross-section as representative of a geotechnical structure implicitly assumes all other parallel cross-sections are identical and therefore, material properties are perfectly autocorrelated in the transverse direction (See discussion by Auvinet, 2005, on this topic). In the present case study, if data were available on the spatial variability of material properties and their autocorrelation structure, three-dimensional modeling including these features would improve the accuracy of the analysis, but it is doubtful the conclusions with respect to the relative weight of the two modes of failure would be significantly altered.

ACKNOWLEDGEMENTS

Danielli de Melo Moura and Rodrigo Borela were supported by the Science without Borders program (SwB/CAPES) with funding provided by the Brazilian Federal Government. Data were provided by the Secretaria de Infraestrutura de Pernambuco (SEINFRA). This study was undertaken by the first four authors as a project assignment, in the *CE 583 Earth Retaining Structures and Slopes* course taught at Purdue University by Philippe Bourdeau.

REFERENCES

Amundaray, J.I. (1994). *Modeling Geotechnical Uncertainty by Bootstrap Resampling*, Ph.D. Thesis, School of Civil Engineering, Purdue University.

Auvinet G. (2005). Probabilistic approach in geotechnical engineering practice. *Proceedings, 4eme conference nationale sur la fiabilite des materiaux et structures (JNF'05), Clermont-Ferrand, Universite Blaise Pascal, France, Keynote Lecture; 9p.* [in French].

Duncan, J.M. (2000). Factors of Safety and Reliability in Geotechnical Engineering. *Journal of Geotechnical and Geoenvironmental Engineering, ASCE, 126 (4)*, 307-316.

Foster, M., Fell, R., & Spannagle, M. (2000). The statitstics of embankment dam failures and accidents. *Canadian Geotechnique Journal, 37*, 1000-1024.

Geo-slope (2014). GeoStudio Modeling Software, GEO-SLOPE International Ltd, http://www.geo-slope.com/

Hahn G.J. and Shapiro, S.S. (1967*). Statistical Models in Engineering*. Wiley, N.Y.

Harr, M.E. (1987). *Reliability-based Design in Civil Engineering*, Mc-Graw Hill, N.Y.

Kovács, G. (1981*). Seepage Hydraulics*, Elsevier Scientific, Amsterdam, 349-379

MacDonald, T. C., & Langridge-Monopolis, J. (1984). Breaching Characteristics of Dam Failures. *Journal of Hydraulic Engineering, ASCE, 110 (5)* , 567-586.

Rosenblueth, E. (1975). Point Estimates for Probability Moments. *Proceedings, National Academy of Science, USA, 72 (10)*

Powledge , G. R., Ralston, D. C., Miller, P., Chen, Y. H., Clopper, P. E., & Temple, D. M. (1989). Mechanics of Overflow Erosion on Embankments. II: Hydraulic and Design Considerations. *Journal of Hydraulic Engineering, ASCE, 115 (8)*, 1056-1075.

Wahl, T. L. (2004). Uncertainty of Predictions of Embankment Dam Breach Parameters. *Journal of Hydraulic Engineering, 130 (5)*, 389-397.

Wolff, T.F. (1995). *Introduction to Probability and Reliability Methods for Use in Geotechnical Engineering, Appendix B.* Department of the Army, U.S Army Corps of Engineers, Washington D.C. Technical Letter ETL 1110-2-547.

Xu, Y., & Zhang, L. (2009). Breaching Parameters for Earth adn Rockfill Dams. *Journal of Geotechnical and Geoenviromental Engineering, ASCE, 135 (12)*, 1957-1970.

OPERATION AND MAINTENANCE OF AN INSTRUMENTATION PROGRAM

Amanda Sutter[1]

This presentation will cover many of the topics that will be included in the upcoming White Paper by the USSD Committee on Monitoring of Dams and Their Foundations, *Operation and Maintenance of an Instrumentation Program*.

Some of the topics that will be featured in the White Paper are the responsibilities and qualifications of personnel, documentation required, results that should be expected, and how to handle anomalous readings. The presentation will not be specific to any instrument types or reading systems. It will cover the general process of operating and maintaining a monitoring program.

[1] U.S. Army Corps of Engineers, Amanda.J.Sutter@usace.army.mil

FERC BEST PRACTICES FOR DAM SAFETY PERFORMANCE MONITORING

Will Brown, P.E.[1]
Eric Kennedy, P.E.[2]
Justin Nettle, P.E.[3]
Emeruwa Anyanwu, D. Engr.[4]
Eric Gross, P.E.[5]
Jim McHenry, P.E.[6]
David Olson, P.E.[7]

ABSTRACT

The primary purpose of dam safety performance monitoring is to identify initiation or progression of failure in time to allow intervention and prevent disaster. Recently, dam safety performance monitoring has triggered successful intervention at several Federal Energy Regulatory Commission (FERC) regulated hydropower dams. Evaluation of the performance monitoring data following these dam safety incidents indicates that there may have been opportunities for earlier intervention. Intervening earlier can both lower the likelihood of failure and reduce remediation costs. Understanding performance monitoring data after dam safety incidents is relatively easy compared to the challenge of correctly identifying and understanding the problem in advance. Meeting this challenge requires creative thinking, critical evaluation and continued pursuit of the understanding of performance data, history and design of the structure. Success is more likely if we critically evaluate past incidents and apply the lessons learned.

INTRODUCTION

Performance monitoring of dams can be a highly challenging task, especially when unexpected instrumentation data does not have a corresponding Potential Failure Mode (PFM) or visible changes that explain the reading. How do you determine if the cause of an unexpected reading is poor quality control, lack of instrument maintenance, a benign change in conditions, or initiation of an unidentified PFM? This is a very important question, because action is needed if the dam is failing, but we do not want to expend valuable time and resources needlessly.

Looking at case studies of dam failures and major safety incidents can provide valuable insight into how we look at unexpected instrumentation data. This point is demonstrated by an incident that recently occurred at a FERC regulated gravity dam. The operator

[1] Branch Chief, FERC, Division of Dam Safety and Inspections, Duluth, GA, william.brown2@ferc.gov
[2] Civil Engineer, FERC, Division of Dam Safety and Inspections, Portland, OR, eric.kennedy@ferc.gov
[3] Civil Engineer, FERC, Division of Dam Safety and Inspections, Portland, OR, justin.nettle@ferc.gov
[4] Civil Engineer, FERC, Division of Dam Safety and Inspections, New York, NY, pe.anyanwu@ferc.gov
[5] Civil Engineer, FERC, Division of Dam Safety and Inspections, Chicago, IL, eric.gross@ferc.gov
[6] Civil Engineer, FERC, Division of Dam Safety and Inspections, Washington, DC, james.mchenry@ferc.gov
[7] Civil Engineer, FERC, Division of Dam Safety and Inspections, Chicago, IL, david.olson@ferc.gov

noticed misalignment in the crest. This triggered a crest survey and dive inspection that resulted in lowering of the reservoir until the dam could be remediated. The incident can be seen as a success of the monitoring program from the standpoint that the problem was detected by visual surveillance and timely intervention prevented failure of the dam.

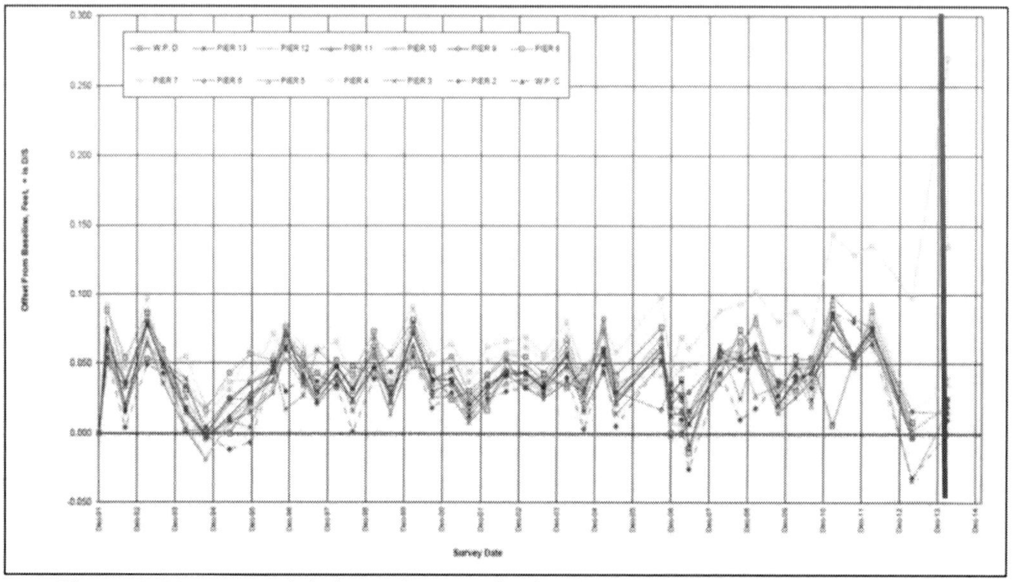

Figure 1. Unexpected deformation monitoring data (red line indicates time when movement was identified through visual observation)

However, anyone looking at the results of the deformation surveys of this dam (shown in Figure 1) can see that there was a trend going on for years. It is easy to see now that this problem could have been detected earlier, reducing the likelihood of failure and the cost of the repair. The data in Figure 1 was reviewed by the owner, multiple consultants, and FERC inspectors prior to the large differential movement that was identified visually by the operator, yet no one identified it as a problem. No action was taken to determine if the trends and differential movement shown in Figure 1 was expected behavior or an indication of poor performance. To learn from this incident, we must critically evaluate how we look at performance monitoring data.

Looking across FERC's portfolio, there are numerous dams with unexpected readings and trends. Frequently these readings and trends are ignored or dismissed. In most cases, years go by and there is no adverse response. This is because many of these readings are the result of poor quality control, lack of instrument maintenance, or a benign change in conditions. However, some of these instruments may be telling us that the dam is in a state of distress, that a potential failure mode has initiated and become a failure in progress, and we are not recognizing it. Our challenge is to identify these dams and take action before we have additional major dam safety incidents or failures. We will meet this challenge only by learning from the past by and critically evaluating all performance data that is outside the expected range.

BACKGROUND

In 2005, FERC adopted the Potential Failure Modes Analysis (PFMA) and established the Dam Safety Performance Monitoring Program (DSPMP) for regulated hydropower dams, requiring many dams to prepare Dam Safety Surveillance and Monitoring Plans (DSSMPs) and Reports (DSSMRs). In the ten-year history of the formal program, many lessons have been learned through success and failure of the importance of a robust, targeted surveillance and monitoring program based on a dam's unique potential failure modes. To share the lessons learned with FERC stakeholders, FERC's Division of Dam Safety and Inspections (D2SI) staff organized monitoring workshops in August 2015 and March 2016. The purpose of the workshops was to review FERC's guidelines on DSPMP, including recent changes, and discuss monitoring best practices. The workshops included practical exercises and case studies presented by the licensees or dam owners.

While preparing and participating in these monitoring workshops, FERC staff compiled and presented best practices related to various aspects of performance monitoring. The best practices are demonstrated through a series of short case studies with discussion. To focus attention on best practices, the dams in the case studies are not named. In some cases, the plots or figures may be taken out of context in order to make a point. Many of the concepts presented below are not original; they have been previously published or practiced within the dam safety community including at FERC regulated dams. However, these practices have not been implemented at many other dams. Because of this, we have an opportunity to improve the safety of many dams with more widespread use of these best practices.

WELL-DEFINED POTENTIAL FAILURE MODES

The main purpose of surveillance and monitoring within a dam safety program is to provide a means for the early detection of adverse behavior, typically as defined by the identified potential failure modes. For this reason, one of the most important aspects of successful dam safety surveillance and monitoring is the initial identification and full characterization of the potential failure modes associated with the dam and other water retaining features at a project. FEMA defines a potential failure mode as:

"A physically plausible process for dam failure resulting from an existing inadequacy or defect related to a natural foundation condition, the dam or appurtenant structures design, the construction, the materials incorporated, the operations and maintenance, or aging process, which can lead to an uncontrolled release of the reservoir." [8]

Potential failure modes are not limited to those that result in a catastrophic dam failure; they should also consider an uncontrolled release of water without complete failure. To properly monitor potential failure modes they first have to be identified via a PFMA and then fully developed.

[8] FEMA (2004)

A credible potential failure mode must identify an initial condition and include one or more changes that lead to failure. The initial condition is any combination of material properties and loads that could allow a failure to develop including, but not limited to, a design flaw, construction defect, operating circumstances, or unforeseen loads. The changes may be a material failure, an increase in loads, or an operational activity and may be either sudden or gradual.

For example, with a typical piping failure of an earth embankment dam, the initial vulnerable condition could be the existence of an unfiltered exit from the dam's core. A change such as increased reservoir head could increase seepage at the exit point. The increased seepage flow could begin to erode and move soil particles, form a soil pipe into the dam, and erode a void in the dam's core. Collapse of the void and subsequent erosion could cause loss of freeboard, full breach of the embankment, and an uncontrolled reservoir release.

When evaluating potential failure modes for a dam, it is important to see the dam not as an isolated structure, but as part of an overall system. This "system" includes the dam as well as the foundation and abutments; the pipelines, penstocks and spillways that pass flow through, over, or around the dam; the mechanical features such as gates, valves, and turbines that control flow through a dam; the electrical systems that provide power to those mechanical features; the control and data acquisition systems that monitor and control the facilities operation; and the humans who maintain and operate the dam. Any part of the system has the potential to introduce a condition that could lead to failure or could allow a failure to progress.

The development of a DSSMP to monitor a potential failure mode has to take into consideration several factors including, but not limited to, the following:

- At what point in the potential failure mode process can anomalous performance or failure mode initiation be detected?
- Can the potential failure mode be detected with instrumentation before an actual dam failure occurs?
- Can the potential failure process be detected visually before an actual dam failure occurs?
- Are trained personnel available to review collected data and information, in a timely manner, in order to detect indications of initiation/development of the potential failure mode?
- How quickly will the potential failure mode develop once detected?
- How frequently should the monitoring be performed?
- Will there be time to take action once the potential failure mode is detected?
- What action items will follow if a potential failure mode is indicated in surveillance and monitoring?

It is important to understand the complete sequence of each identified potential failure mode to incorporate the right combination of visual monitoring and instrumentation into the surveillance and monitoring program. For example, if an increase in uplift pressure on a concrete gravity dam can result in a sliding failure, the surveillance and monitoring program should include some combination of: visual monitoring for signs of movement or displacement such as offset concrete joints or cracking; regular surveying of the dam; joint meters at vertical joints; foundation drain flow metering; and water pressure measurements beneath the dam. With this program, changes in the foundation drain flow would likely be the first indication of a condition that could initiate the identified potential failure mode, as a change might indicate opening of the dam-foundation contact, caused by an increase in uplift pressure that may also be picked up by pressure measurement devices. The joint meters would provide the first indications of actual displacement taking place followed by visual observations and the survey. If the visual observations and instrumentation were all showing similar indications, the confidence that the potential failure mode has initiated would be high.

Even with an extensive instrumentation program and appropriately defined threshold and action levels, the data needs to be evaluated in a manner consistent with the potential failure mechanism. Consider **Error! Reference source not found.**, which shows the foundation drain efficiency, which is a measure of the uplift reduction created by the drain system calculated using the geometry of the dam, headwater, tailwater, and the foundation pore water pressure. A high drain efficiency means the uplift on the dam is low, resulting in greater stability and higher factors of safety. The drain efficiency values in Figure 2 are for a gated ogee spillway section, as measured twice per year using packers and pressure gages. In this plot, every data point being evaluated is significantly greater than the threshold and action levels (75 and 50 percent, respectively). If evaluating the performance of the spillway with this plot, relative to a sliding stability failure on the dam-foundation interface due to excessive uplift pressure, one would most likely conclude that the performance is acceptable.

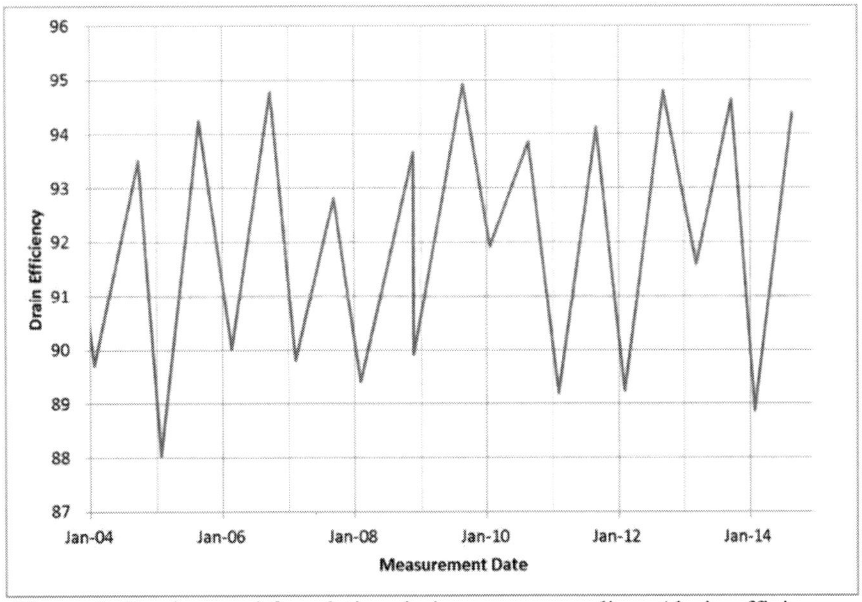

Figure 2. Semi-annual foundation drain pressure readings (drain efficiency, spillway-average).

However, to produce this plot, the measured drain efficiency was averaged across the entire spillway, which comprises 22 separate monoliths with 115 drain holes. Figure 3 shows the calculated drain efficiency for each hole; this plot was presented in the DSSMR for reference but was not thoroughly considered with regards to the PFM. In order to activate the failure mechanism, only one monolith has to experience uplift pressures high enough to become destabilized and release the reservoir. The uplift pressure applied to monolith 19 is irrelevant when evaluating the stability of monolith 5, and it should be clear that the two monoliths are experiencing different uplift conditions. The PFM itself was written in such a manner that multiple loading conditions were combined; the concept that monoliths can behave independently was lost by the time the monitoring data was processed. Figure 4 shows the data when examined by monolith, averaging only the drain pressure measurements within a given monolith.

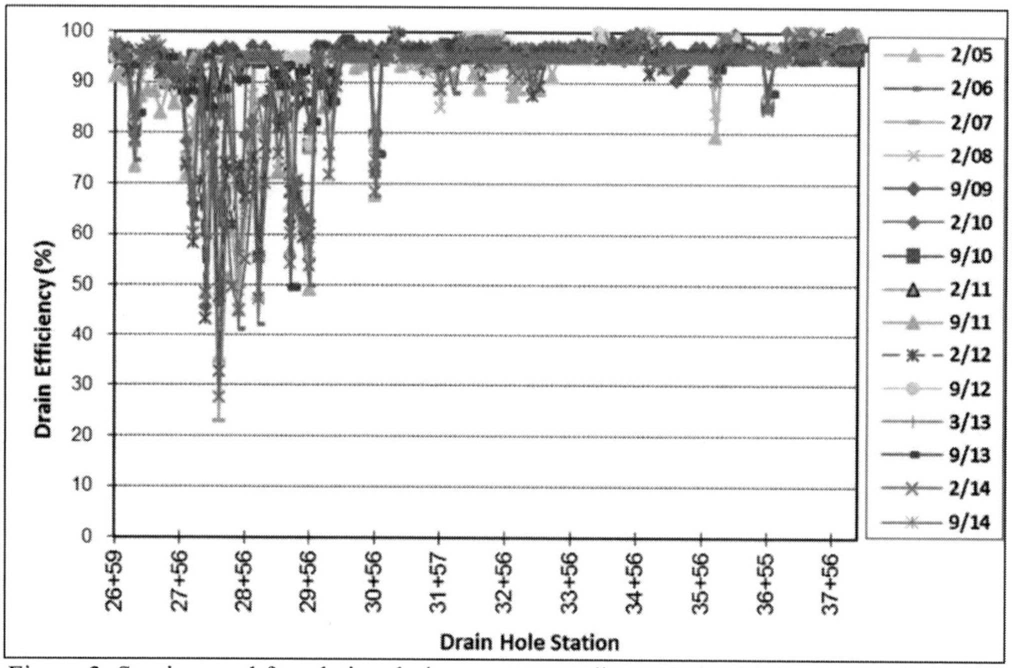

Figure 3. Semi-annual foundation drain pressure readings (drain efficiency in each hole).

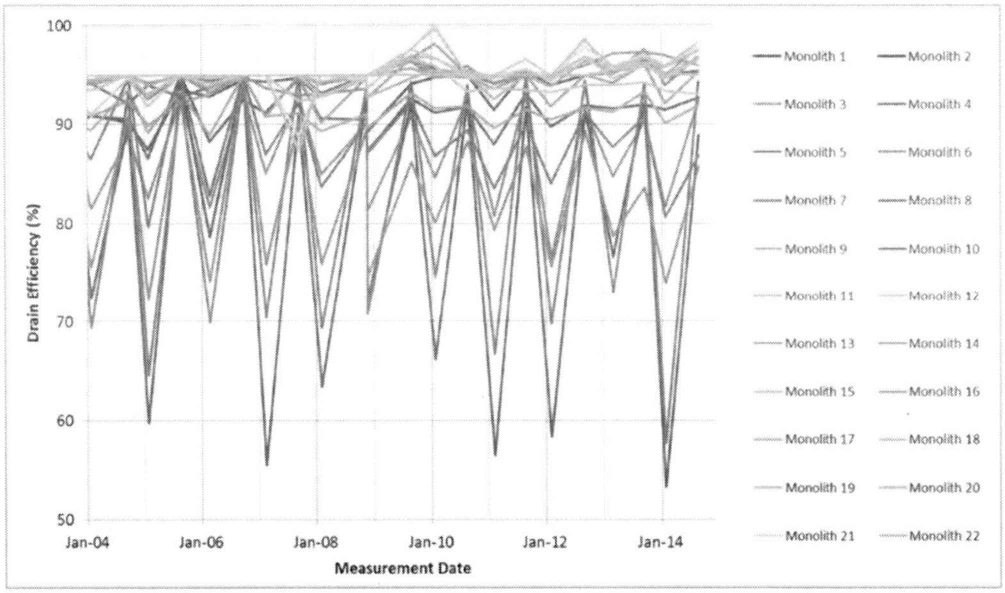

Figure 4. Semi-annual foundation drain pressure readings (drain efficiency, monolith-average).

Two important details should be readily apparent to the reader:

- First, the seasonal effect is more significant than previously thought. While the spillway-average plot shows a maximum seasonal variation of 7 percent, the monolith-average plot shows a variation of approximately 40 percent.
- Second, and more importantly, the drain efficiency in several monoliths approaches or drops below the threshold level (75 percent) and in some cases approaches the action level (50 percent).

By averaging the data across the entire dam, these differences in performance were obfuscated and the DSSMR presented the PFM as being far more improbable than it is. This may be clear to everyone in hindsight, but it is not unusual for large amounts of similar data to be made more manageable by using an average for evaluation. The original PFM was written to consider the "spillway section," and while the alignment survey necessarily considers monoliths individually, the PFM was not defined enough for anyone to previously consider a worst-case drain efficiency for evaluation.

Subsequently, this licensee has enhanced the monitoring program for these several monoliths: drain pressures have been measured more frequently during the winter, plans have been developed to install vibrating wire piezometers upstream and downstream of the drains in each monolith, and the DSSMR contains several more plots that consider monoliths individually, evaluate seasonal variation, and take into account design differences between the left and right half of the spillway.

GENERAL HEALTH MONITORING

While monitoring for identified potential failure modes is the best method for preventing dam failure, doing so requires identification of the conditions that could potentially introduce a failure mechanism at a dam. Identification of these conditions is based on what we know or what we can infer about the structure and its foundation from design and construction records, visual observations, loading conditions, and historical performance. Unfortunately, the reality is that even under the absolute best conditions the process for identifying and developing potential failure modes is constrained by the information that is available for that structure and the knowledge and ability of those completing the PFMA. Conditions that could develop into a potential failure mode can be missed if there is no possible way to know that they exist with the information at hand or if the potential failure mode analysis process was somehow deficient. These unknown conditions can be the most dangerous issues relative to dam safety because they can develop into potential failure modes that cannot be anticipated and therefore can't be directly monitored.

For example, reviewing design documents, construction records and the performance history of a concrete gravity structure may show the structure was well designed and constructed and has no history of movement. If factors of safety against sliding are very high and no adverse conditions are indicated or can be inferred, a sliding type failure mode may not be considered possible for the dam. However, there still remains the possibility that something might have been missed or simply could not have been identified with the information available. Perhaps there is a weak clay seam in the foundation that was not identified by the geologic exploratory program and is not in any way visible from the surface. If such a condition existed it could introduce a potential failure mechanism that was not accounted for during design. Without knowing or being able to infer that the condition was present, a potential failure mode would not be developed and could not be monitored directly.

The purpose of general health monitoring is to attempt to account for these unknown conditions that could develop into a potential failure mode that has not been previously identified for a structure. General health monitoring should capture any performance criterion that could change over time (e.g., seepage, settlement, movement, etc.) and, depending on the situation, could be as simple as regular visual inspection. The level of general health monitoring required is site-specific and depends on the type of dam, foundation conditions, past behavior of the structure, the level of comfort with the understanding of conditions at the dam, and perhaps even the consequences of failure. Since the goal of general health monitoring is to identify changes in performance that may not be tied directly to an identified PFM, a minimum level of visual surveillance and basic instrumentation should be maintained at all dams.

A well-known (and therefore named) example of general health monitoring showing a developing failure mode is the fracture at Wanapum Dam. The PFMA had identified sliding stability failure modes at the base and within the foundation, but not within the body of the dam. The instrumentation included several pneumatic piezometers, flow and pressure readings on the foundation drains, and a semi-annual alignment survey. Because the activated failure plane was located above the drainage gallery, the piezometers and foundation drains provided no indication of it occurring; however, the alignment survey provided clear indication for at least a few years prior to visual observation of movement at the crest. Even before that day, operators noticed a significant increase in leakage from opposite corners of adjacent spillway gates – the gates had moved along with the shared pier.

The important lesson to take away from this example is not that the surveillance and monitoring program failed; rather, that the general health monitoring (alignment survey and visual observations) did succeed in identifying the issue, albeit at a very late stage in the progression. It is important to have general health monitoring in place but we also have to listen to what it tells us and investigate anything unusual.

USE OF THRESHOLD AND ACTION LEVELS

During the recent FERC monitoring workshops, there was a great deal of debate regarding the definitions of threshold and action levels. During the discussion, FERC staff pointed out that the definitions in the Engineering Guidelines are general to allow dam owners to apply judgment in defining instrument levels that require action. These terms and how they are used in a monitoring program to improve dam safety should be defined in the DSSMP. Some participants advocated the use of the CEATI "Performance Action States" presented in Figure 5 for development of thresholds and action levels. FERC staff pointed out that this approach to classifying readings outside the expected range is consistent with the guidance in Chapter 14 of the Engineering Guidelines[9]. However, this system may not be appropriate for all dams and judgment should be used to determine if this system would improve dam safety for individual projects.

The current best practice for establishing threshold and action levels is to use a combination of historic readings and design basis values to determine levels triggering action. Individual instrument readings, trends identified from multiple readings, or visual observations can trigger action that may include:

- Reading verification
- Special inspection
- Evaluation by an engineer
- Increase in monitoring
- Formal investigation
- Implementation of risk reduction measures

Each action should include a specified time period for implementation and notification procedures. When possible, each instrument should have high and low level bands on the expected range.

[9] FERC (2017)

ACTION STATE	DESCRIPTION	RESPONSE
NORMAL – PERFORMANCE AS EXPECTED		
NORMAL – GREEN	Observations and measurement indicate expected and acceptable values.	Continue inspection, monitoring, and maintenance program.
THRESHOLD LEVEL		
CAUTION – YELLOW	One or more indicators of performance are above expected values.	Review the data for reliability. Meet with Evaluation Team to decide what to do. Inform all involved parties of the current condition and the recommended plan of action. Take steps to reduce chance that reading will exceed the Limit Level.
LIMIT LEVEL		
ALERT – RED	One or more indicators are above the Limit Levels established for each instrument.	Inform all parties to stop any work in affected area. Implement contingency plan. Develop safe steps to proceed.

Figure 5. CEATI Performance Action States[10] is one approach to identifying readings outside the expected range, classifying the level of concern and determining the action that is triggered

Threshold and action levels should identify readings that do not reflect expected performance. Figure 6 presents piezometer data where exceedance of a horizontal threshold is expected behavior when the reservoir rises. Depending on the project, a horizontal threshold may be frequently exceeded when the reservoir rises. Frequent exceedance of threshold levels that does not trigger alarm may desensitize dam safety personnel to the importance of threshold exceedance. Additionally, this system may not effectively identify readings outside the expected range. This approach to developing threshold and action levels may not be successful at dams where the reservoir levels do not fluctuate significantly.

When threshold or action levels are linked to PFMs, special plots or figures may be appropriate for evaluating the levels. One example of this is presented in Figure 7.

[10] CEATI (2012)

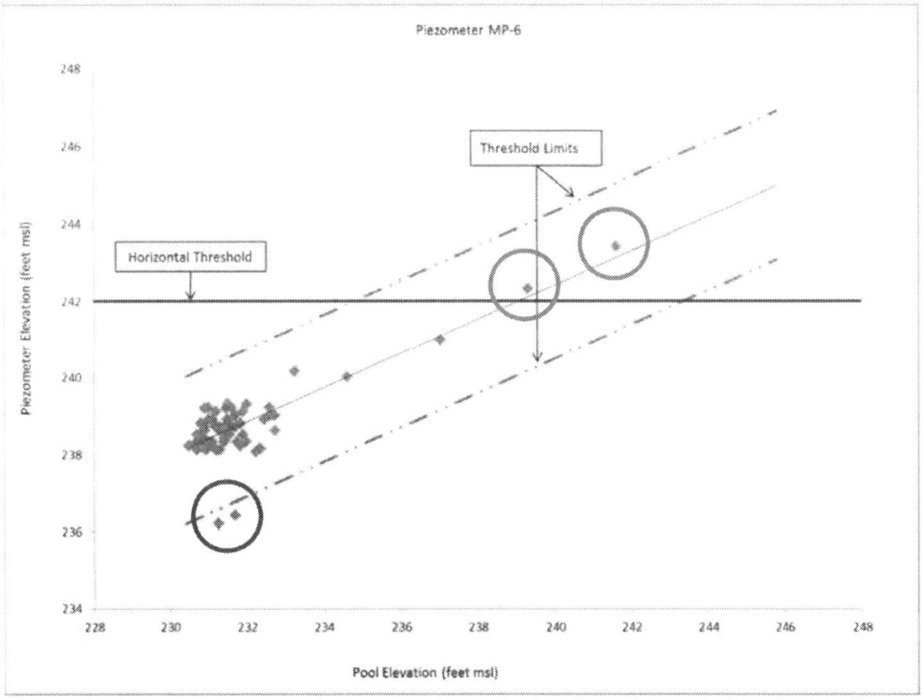

Figure 6. Comparison of horizontal and correlation-based thresholds. In this figure, expected readings (in green circles) exceed the horizontal threshold, but the unexpected low readings (in red circle) only exceed the correlation based thresholds.

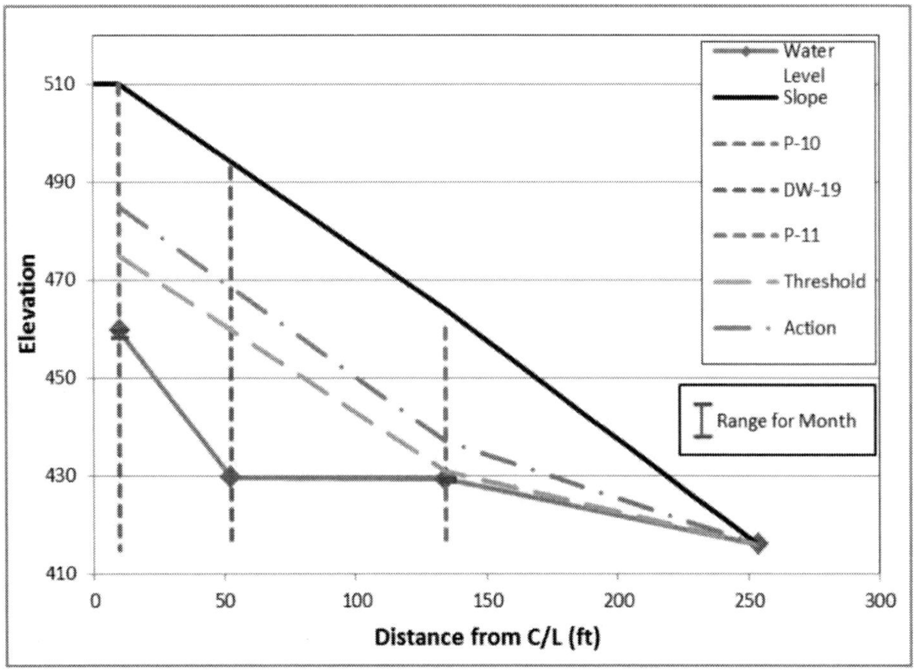

Figure 7. Threshold and actions levels developed for a stability related PFM

MAINTENANCE OF INSTRUMENTS

While an "Equipment Maintenance Program" for instrumentation is recommended in *FERC Engineering Guidelines for the Evaluation of Hydropower Projects*, review of recent DSSMRs demonstrates that the instrumentation maintenance programs for many FERC regulated dams are inadequate.[11] A good maintenance program is essential for ensuring reliable and accurate instrument readings. Poor instrument maintenance can result in trends and anomalous readings that do not reflect the actual performance of the dam. These readings can result in misunderstanding, and a waste of time and resources investigating the cause of the unexpected performance data. Examples of poor maintenance include:

- Deformation plots that show movement and investigation later shows the monument was disturbed by maintenance equipment;
- Piezometer plots that show gradual increases for years and readings return to normal after the instrument is flushed;
- Weir flow plots with anomalies where investigation later shows that the unexpected readings were the result of debris accumulation in the weir box;
- Extensometer elongation seeming to indicate movement of blocks within the foundation when the heads may have been impacted by falling rock, ice, or been damaged by ice jacking; and
- Survey data showing the dam crest rising steadily for several years with no visible distress, when the issue may be an unstable benchmark.

A preventive maintenance program will greatly reduce or eliminate the unexpected performance data described above. This will increase confidence in and reliability of the instrument readings and prevent unnecessary evaluation and investigation of performance monitoring data.

QUALITY CONTROL PROCEDURES AND TRAINING

A good performance monitoring quality control program is integral with an Owner's Dam Safety Program and begins with training of all involved personnel. If project staff understands what the data is used for and the implications of inaccurate or unexpected readings, they are much more likely to take greater care in taking the readings and respond appropriately when there are readings outside the expected range.

Poor quality control at the time of the reading can cause problems later. If the Chief Dam Safety Engineer reviews data months after it's collected, there is no way to tell if an anomalous reading is the result of measurement error or indication of poor performance; the chance to perform another reading as validation has long since passed. As discussed in other sections of this paper, anomalous readings must be evaluated immediately to determine if there is development of an identified or unidentified PFM. Anomalous

[11] FERC (2014).

readings that are the result of poor quality control can result in wasted time and resources by licensee and FERC staff performing unnecessary evaluations and investigations.

For this reason, verification of readings as they are taken is essential. This means that the technician must have an understanding of the expected range for the instrument, whether that is based on the threshold and action levels or historical performance. Technicians should be trained on the appropriate action to take when readings fall outside the expected range, or exceed thresholds or action levels. This approach is straight forward when instruments have constant maximum or minimum thresholds but it may be more complex for variable thresholds. For example, action levels and thresholds related to drain efficiency are dependent on both the headwater and tailwater elevations at the time of reading. When this is the case, DSSMPs can incorporate tables or graphs that show the threshold or action levels for various conditions, but this makes real-time evaluation by the technician more complex. This information could be conveniently included on the back of the data collection worksheet. Another solution would be to input headwater and tailwater levels in a spreadsheet to calculate the threshold or action levels for the day of the measurement. For projects with data acquisition and analysis systems, the system can be programed to calculate the drain efficiencies from piezometer readings and alert dam safety personnel when action levels or thresholds are exceeded.

Several examples of poor quality control were presented by FERC staff during recent monitoring workshops. Figure 8 is a piezometer plot that shows a varying frequency of measurements; additionally, there are unexpected drops in the piezometer in 2012 and 2013. Discussion with the technician revealed that they were often overwhelmed by other tasks and did not have time to complete the monthly readings. The technician did not understand the implications of the unexpected readings and did not verify the measurements as they took them. The licensee immediately implemented improved quality control procedures and investigated the unexpected readings. A new PFM was identified and the licensee increased the frequency of readings, prepared special reports, and conducted periodic meetings with FERC staff to discuss results. After three years of readings in the expected range, it was concluded that the unexpected drops are most likely due to poor quality control as opposed to indications of a developing failure mode, such as the collapse of an internal erosion pipe.

Figure 9 is another example of poor quality control. FERC staff noticed a perfect linear relationship when they prepared a correlation plot for this piezometer. After further investigation, the licensee found that staff had fabricated the data from December 2009 to March 2012. The licensee immediately revised their quality control procedures and conducted training on the importance of dam safety surveillance and monitoring. This apparent fabrication of performance data by the licensee's staff put the safety of the dam at risk. If there were adverse conditions during this time period, the monitoring program may not have detected the problem. This example demonstrates the need for training so that staff understand the importance of the readings. It also demonstrates that there is no substitute for spot-checking, with management or engineers periodically participating in the monitoring activities. This emphasizes the importance of monitoring and provides an opportunity to review field procedures for taking and verifying measurements.

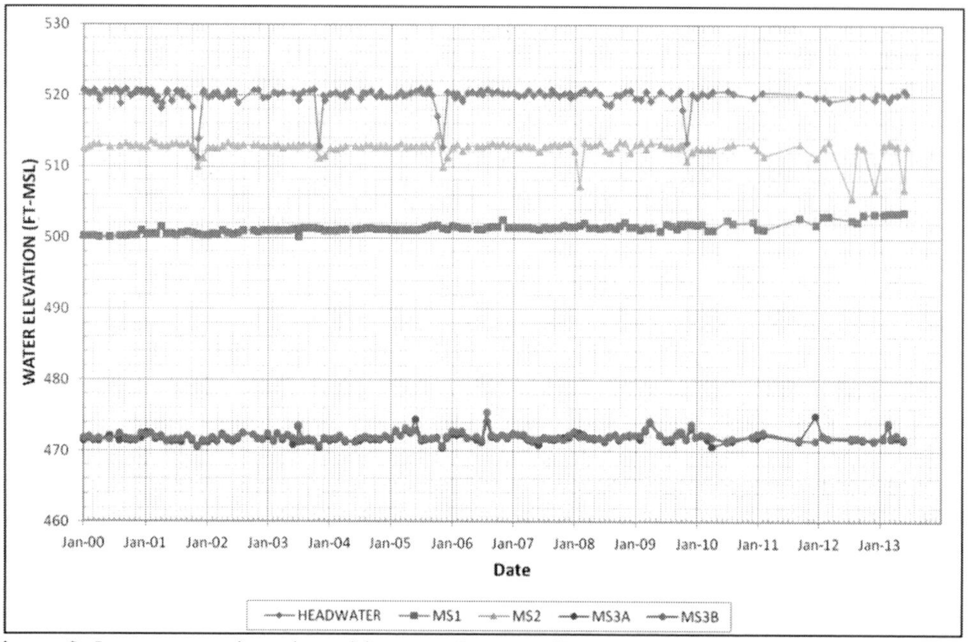

Figure 8. Instrumentation plot with anomalous readings that may have resulted from poor quality control.

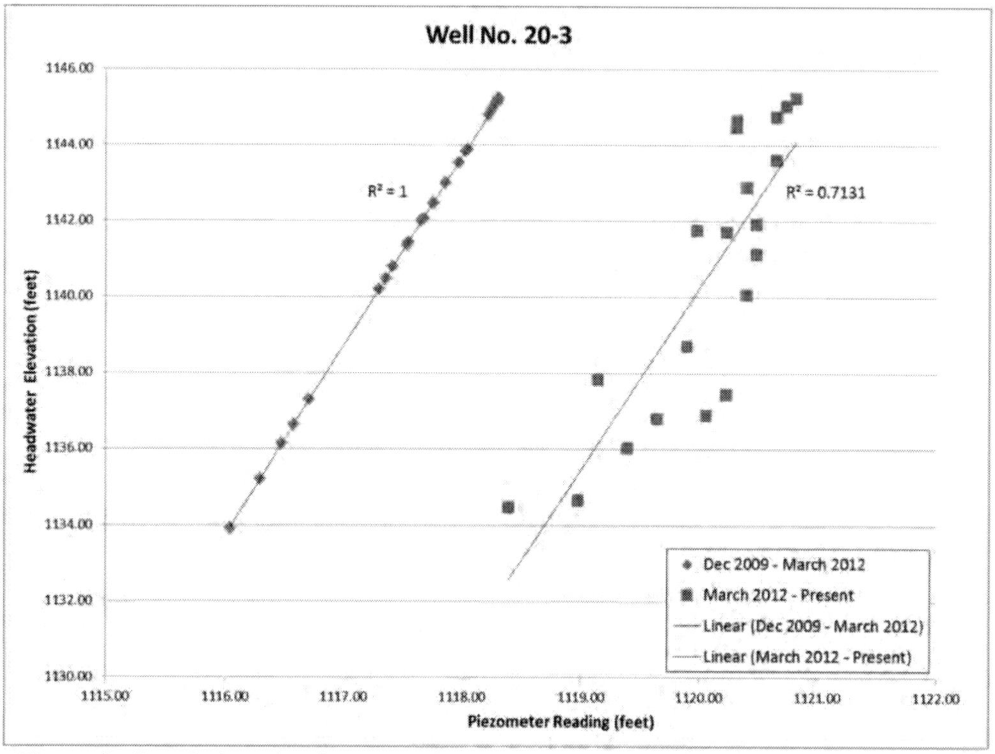

Figure 9. Correlation plot demonstrating poor quality control and training. Technician apparently fabricated data from December 2009 to March 2012.

TIMELY EVALUATION OF DATA

As discussed above, the purpose of a DSPMP is to allow timely intervention to prevent dam failure. Yet, many existing DSSMPs do not include time requirements for responding to readings that exceed established threshold and action levels. Review of many DSSMRs indicate that there are cases when trends and threshold exceedance are only evaluated annually during preparation of the DSSMR. Best practices for timely evaluation of data were discussed at the recent FERC monitoring workshop and include identifying the critical monitoring activities that are time sensitive in the DSSMP. The DSSMP should include a table or list with the time sensitive activities that describe the appropriate action and the timeline in which it should be completed. While many times exceedance of a threshold or action level may not be directly connected to initiation of an identified PFM, evaluation of the data should be given a high priority to ensure that it does not indicate initiation of an unidentified PFM. The DSSMP should establish standards for timely review to ensure that delays do not jeopardize dam safety.

In some cases, instrument readings that exceed a threshold or action level may indicate that a failure is in progress. These instrument readings should be identified during the PFMA process and will typically be associated with Category I, Highlighted PFMs. Depending on the time available, the instrument readings may be automated. In a case study presented during the workshop, a licensee presented their monitoring system that they developed for a high hazard canal. The PFM of concern was a landslide into the canal that could trigger overtopping failure of the canal embankment. In some sections of the canal there were residences immediately downstream of the embankment. The licensee installed redundant real-time monitoring of the forebay levels. When action levels are exceeded, the system issues text message notifications to the operators and triggers automatic closure of the headgates to prevent overtopping of the canal embankment.

GRAPHICAL PRESENTATION OF DATA

When interpreting monitoring data, rarely will trends or anomalies become clear to the reviewer by staring at tabulated values; presenting the information graphically is critical to our ability to evaluate data. It is important to consider the identified PFMs when developing plots; however, we must not have a myopic view of the surveillance and monitoring data as applying only to PFMs. As discussed previously, general health monitoring is important as it can reveal adverse behavior that has not been considered in the PFMA. For instrumentation to serve the purpose of PFM-specific and general health monitoring, the data should be looked at in multiple ways; one of the benefits of regulating many types of dams and agencies across the country is that the FERC gets to see many different methods for presenting data. Some best practices identified by FERC staff and discussed during the recent workshops include:

- **Correctly identify all instruments that can monitor identified PFMs**. For example, many owners correctly identified visual surveillance and seepage flow measurement for monitoring piping, but left out deformation surveys and piezometer readings, which have been used to identify piping problems at other FERC regulated dams. Sharing of case studies can complement ongoing PFMA reviews to make sure that each DSPMP considers all related instrumentation for each PFM.

- **Include threshold values and action levels in all plots.** This allows the reviewer to identify readings outside the expected range and take the appropriate action.

- **Annotate plots to explain anomalous data.** Once the cause of readings outside the expected range is determined, it is important to include this information in the plots to aid in future review. This will improve efficiency of reviews and direct attention away from data that exceeds a threshold or action level but is expected performance. An example of a piezometer time history plot with thresholds and annotation of data is presented in Figure 10.

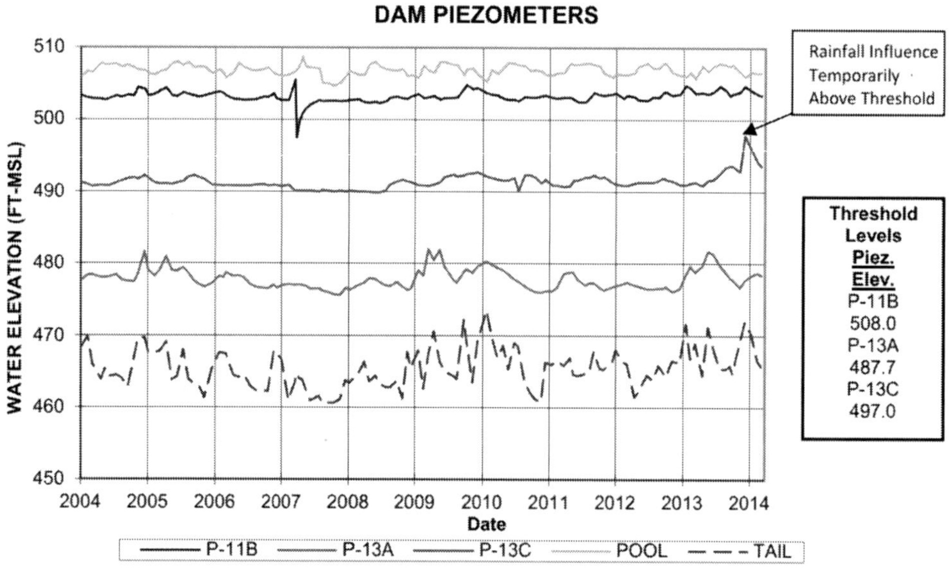

Figure 10. Time history plot that includes annotation of anomalous reading and threshold levels.

- **Managing mountains of data**. Some FERC regulated dams have hundreds of piezometers, deformation monuments and other instruments, producing enormous amounts of data. This makes is difficult to identify unexpected performance in the readings when it occurs. Condensing data from nearby instruments or piezometer readings from the same foundation layer may help review. Some owners include a location map on each data plot so that the reviewer recognizes the location of the instrument. Other owners include the expected range as a shaded band on each plot so that anomalous readings are apparent. When the data becomes unmanageable, automated data collection and analysis systems may improve both efficiency and the overall DSPMP.

- **Time history plots alone are often not enough**. Piezometer readings for instruments that monitor stability, piping or cutoff performance should be plotted in a cross section in addition to the time history plot. The time history plot is useful for identifying trends or anomalous readings. However, reviewers are unable to understand the context of the reading without seeing it in a cross-sections. For example, high artesian pressures during elevated reservoir conditions may not be apparent from the time history plot. Figure 11 shows a cross section plot with the critical failure surfaces. This plot is helpful for evaluating the piezometer readings related to stability, since the foundation and embankment have different phreatic levels and critical slip surfaces. For dams with varying headwater or tailwater, correlation plots can be a valuable tool for identifying unexpected performance or confirming design assumptions.

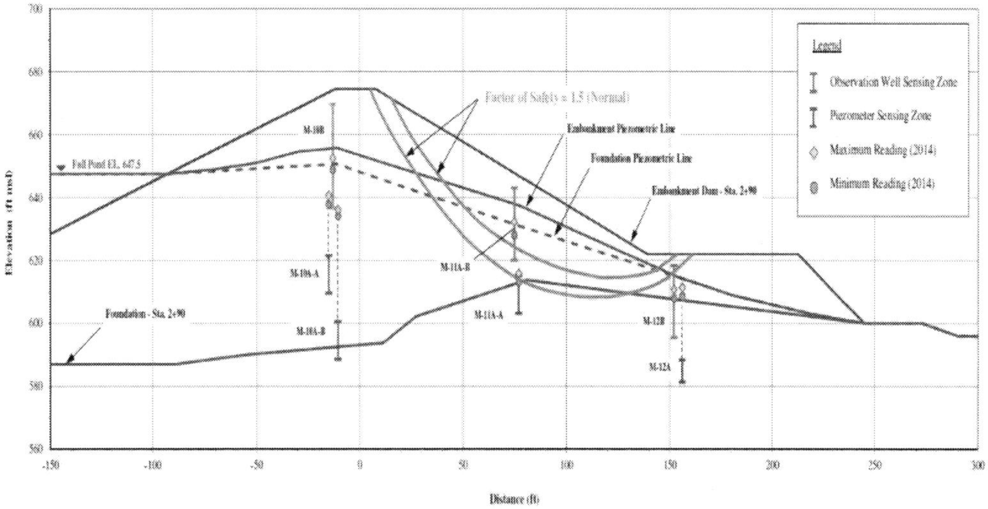

Figure 11. Cross section that shows piezometer readings relative to the phreatic surfaces for two different critical slip surfaces.

- **Create special plots to evaluate trends or readings outside the expected range.** Widely available software includes powerful data analysis tools. For example, a statistical analysis combined with a correlation plot of adjacent monolith differential movement for the gravity dam discussed in the introduction of this paper produced the plots found in Figure 12 and Figure 13, below. A plot like this could have been used to highlight the problem in advance to justify an investigation prior to the large differential movement that resulted in a major dam safety incident.

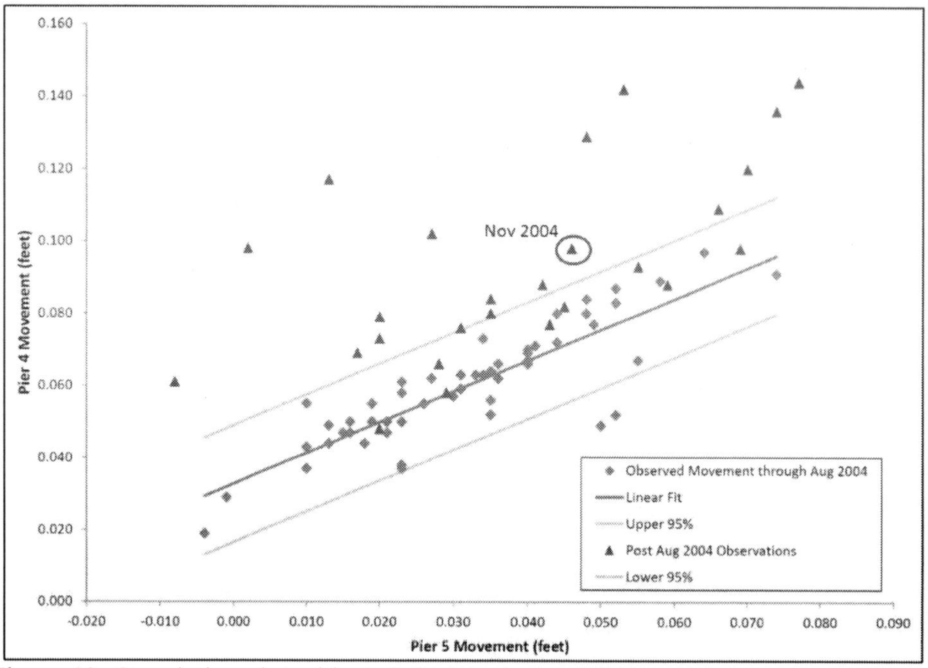

Figure 12. Correlation plot with statistical analysis of differential movement between monoliths 4 and 5.

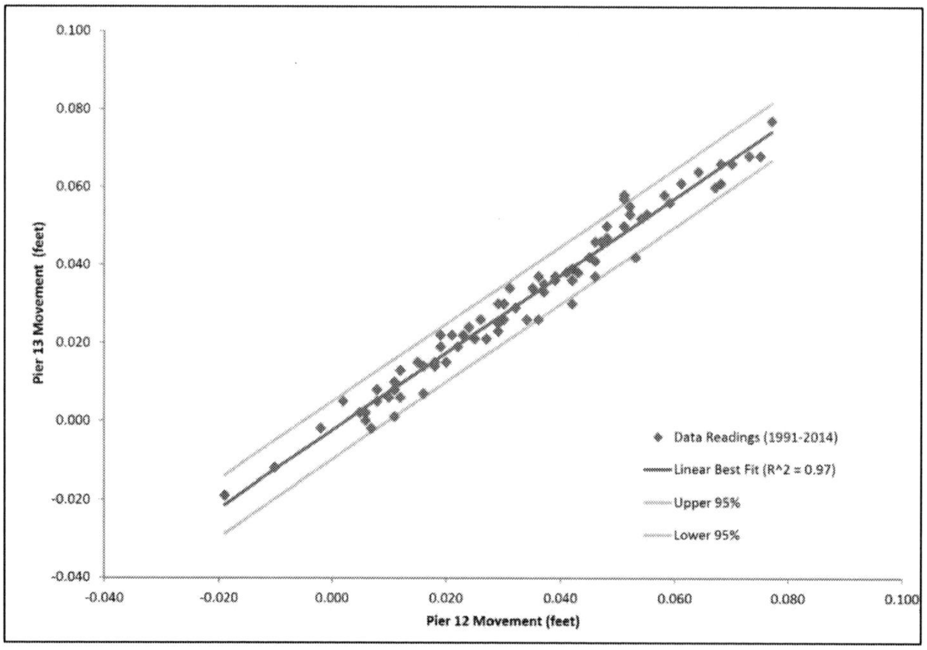

Figure 13. Correlation plot with statistical analysis of differential movement between monoliths 12 and 13.

As an example of best practices, consider a zoned earthen embankment with several piezometers installed in the core and downstream shell. Perhaps the phreatic surface in the downstream shell is very responsive to the reservoir elevation and stability analyses have shown there is a critical level at which the slope could become unstable. The types of plots that could be developed for this case include:

- Correlation plots to show how closely the headwater elevation and measured piezometric head are related;
- Time-series plots that can reveal long-term trends or even how quickly the downstream piezometers respond after a change in reservoir elevation; and
- Cross-section plots showing the piezometer installation details (location in embankment, tip elevation, screened range), the measurement, and the interpreted phreatic surface compared with the critical surface from the analysis.

Each of the above plots provides certain benefits to the reviewer. If the correlation plot shows no link between headwater elevation and measured piezometric head, fluctuations may be related to tailwater, precipitation, or another phenomenon. Time-series plots give a general idea of whether the system is behaving the same as it has in the past; whether that is acceptable or not can be seen on the cross-section plots showing the critical phreatic surface. This accomplishes the goal of evaluating a PFM (slope instability due to a high phreatic surface) as well as general health monitoring, just by looking at the data in different ways.

PREPARATION FOR EXTREME EVENTS

Most existing DSSMPs have a general statement about increased monitoring during floods or following seismic events. These DSSMPs often do not specify which instrument readings should be taken or what should be inspected at various flood levels and seismic events. Floods and significant seismic events are a busy time for operations staff and "general plans" can lead to missed opportunities. Detailed site-specific plans should be developed and incorporated into the DSSMP to monitor identified PFMs and confirm design or record analysis assumptions. Table 1 is an example of a well-structured plan for monitoring during a flood; project management should recognize that this would likely require additional staff on duty during the flood, in order to visually monitor and collect data while performing other critical tasks such as operating spillway gates.

Table 1. Example Surveillance and Monitoring Plan for Flood Events

Reservoir Level (ft)	Visual Inspection Frequency	Piezometer Readings Frequency	Drain Flow Measurement Frequency
Up to 500	Weekly	Monthly	Monthly
500-505	Daily	Weekly	Weekly
505-515	Hourly	Daily	Daily
> 515	Continuous	Daily	Daily

Note: Scour surveys competed following flow events greater than 100K cfs.

One case study presented during the workshop highlighted the importance of developing detailed monitoring plans for extreme events. During the PFMA review for this dam, the review team evaluated the possibility that high tailwater pressures were transmitted to the embankment at an unfiltered drain outlet in the tailrace training wall. The review team postulated that as the tailwater receded, the high phreatic levels in the embankment would trigger piping at the unfiltered drain outlet. Prior to this PFMA review, the licensee's DSSMP included increased visual monitoring during flood events, but operations staff did not observe or document inspection of the drain during the lowering of the tailwater.

The team recommended increased visual monitoring of the drain, which was subsequently conducted during the next high flow event with emphasis on observing the drain as the tailwater receded. As a result, the licensee observed active piping from the drain as suspected. This confirmed understanding of the defect and triggered remediation plans for the drain.

EVALUATING TRENDS AND ANOMALIES

During the recent FERC monitoring workshops, participants discussed the importance of evaluating trends and anomalies in the performance data. Many participants expressed concern that trends and anomalous data that are not understood could be symptoms of unidentified PFMs, as discussed in the General Heath section above. Many DSSMRs submitted to FERC continue to identify trends or unexpected readings, but provide no evaluation of the readings in relation to identified or unidentified PFMs. To rule out the possibility that a PFM has initiated, it is important that we go beyond identifying the unexpected readings to gain an understanding of what the readings are trying to tell us about the performance of the structure.

An important first step is to look for visual clues to changes in performance. When there are unexpected readings for a particular structure, the personnel conducting visual inspections should be aware of this condition to focus attention to potential visual changes that could be related to the instrument readings. The findings of these inspections should be documented in the DSSMR as part of the evaluation of the unexpected readings.

As discussed above, evaluation of quality control procedures, testing or performing maintenance of instruments may result in questions regarding the validity of the readings. Creating various plots and data analysis may also help gain an understanding of the readings. When the cause of the readings remains unexplained, review of construction photographs, drawings, project history and past Part 12 reports may help provide an explanation.

As an example, during a recent evaluation of piezometer data at a project, FERC engineers noted that the foundation piezometers were much more responsive to the reservoir elevation than the embankment piezometers. Review of the construction photographs revealed that there was an alluvial sand foundation layer exposed upstream that helped explain the behavior (Photograph 1). We are increasingly reliant on the historical record – as-built drawings, test data and reports, correspondence, written accounts, and photographs – to provide valuable insight and background information. This new understanding not only resolved questions about the performance data, but also provided important information for the next PFMA review.

At another dam, review of project drawings resulted in identification of a new PFM during evaluation of unexpected piezometer drops. In another case, a twenty-year old Part 12 inspection report helped bring understanding and perspective to unexplained embankment settlement.

Photograph 1. Construction photograph revealed exposed alluvial sand foundation layer

After extensive evaluation of the data and project history, there are cases where the cause of the unexpected readings remains unexplained. When this is the case, the owner must decide if increased monitoring frequency, formal investigation, installation of additional instruments, or other action is appropriate.

CONCLUSION

Performance monitoring of dams can be a highly challenging task, especially when unexpected instrumentation data does not have a corresponding Potential Failure Mode (PFM) or visible changes that explain the reading. Best practices discussed at recent FERC monitoring workshops can be implemented to help meet this challenge. The first step is a thorough PFMA with detailed PFMs. The DSPMP should be designed and implemented to monitor for initiation of these PFMs, considering the failure mechanisms involved and how they might present as measureable data, as well as the general health of the dam. To increase confidence in data and reduce or eliminate unexpected readings that do not reflect the performance of the dam, the DSPMP should include routine maintenance and quality control procedures. Threshold and action levels should be established to help identify unexpected readings that could represent PFM initiation, instrument failure, or other poor performance and should trigger specific action by the dam owner. Readings outside of the expected range that are not understood should be investigated. What we have learned from recent dam safety incidents justifies diligent pursuit of understanding to determine if the readings are the result of a benign change of conditions or initiation of an unidentified PFM. Ignoring or delaying investigation of unexpected readings may result in costly consequences.

ACKNOWLEDGEMENTS

The authors would like to thank all participants from the 2015 and 2016 FERC monitoring workshops who contributed to the concepts discussed in this paper.

REFERENCES

FEMA (2004), *Federal Guidelines for Dam Safety, Glossary of Terms*, Washington, D.C.

FERC (2017), *Engineering Guidelines for the Evaluation of Hydropower Projects*, Appendix J of Chapter 14, "Dam Safety and Surveillance and Monitoring Plan Outline" Washington, D.C.

CEATI (2012), *Dam Safety Performance Monitoring and Data Management – Best Practices*. Quebec, Canada.

CALCULATION OF THRESHOLD LIMITS FOR DAM SAFETY INSTRUMENTATION

Jack Husted, PE[1]
Travis Wylde, PE[2]
Sterling Klippel, PE[3]

ABSTRACT

A proper dam safety monitoring program involves numerous measurements and extensive data collection for instruments such as piezometers, contraction joints, crack meters, and survey monuments. It is necessary to establish an expected range for these measurements and a basis for comparison. This range can be set as instrument threshold limits, which provide an indicator for dam safety concerns and a quick validation of measurements following major storms or seismic events. This paper discusses a simple statistical methodology for calculating threshold limits based on a confidence interval. Piezometer levels and leakage rates tend to vary based on reservoir level, so the threshold limits must reflect this relationship in the equation. The calculated threshold limits use a conservative range, and if any analysis has determined a smaller allowable range based on uplift pressure, allowable saturation, or other parameters, these values will supersede the statistically calculated limits.

The Los Angeles County Department of Public Works (LACDPW) owns and operates 14 major dams located above highly populated communities in a seismic zone. LACDPW implements a proactive dam safety program and emphasizes emergency response protocols. Recently, LACDPW developed this methodology and established updated threshold limits for all instruments with adequate data. Additionally, the LACDPW set a standardized format for graphical plots, equations, and a system for automated dam safety alarms. The calculated threshold limits provide a consistent and technically defensible approach for dam safety instrument monitoring.

INTRODUCTION

Dam safety instruments monitor the hydraulic conditions and movement in the dam and can provide an indicator for dam safety issues. Measurements fluctuate with changes in reservoir level and temperature, but should remain within an expected range. Dam safety personnel must monitor the data closely for sudden changes or adverse trends. Instrument threshold limits designate values for the expected range and provide a basis for comparison of measurements. Threshold limits can provide a quick validation of measurements following major storms or seismic events. They can identify erroneous

[1] Civil Engineer, Los Angeles County Department of Public Works, 900 S Fremont Ave, Alhambra, CA 91803, jhusted@dpw.lacounty.gov
[2] Associate Civil Engineer, Los Angeles County Department of Public Works, 900 S Fremont Ave, Alhambra, CA 91803, twylde@dpw.lacounty.gov
[3] Senior Civil Engineer, Los Angeles County Department of Public Works, 900 S Fremont Ave, Alhambra, CA 91803, sklippel@dpw.lacounty.gov

data due to a malfunctioning instrument or human error. They can also help to identify long term trends that may not be apparent when observing only recent data.

The Los Angeles County Department of Public Works (LACDPW) owns and operates 14 major dams located above highly populated communities in a seismic zone. These dams include various types of concrete and embankment dams with a variety of instrumentation. LACDPW implements a proactive dam safety program and emphasizes emergency response protocols. Recently, LACDPW developed a methodology and established updated instrument threshold limits for all dam safety instruments with adequate data.

LACDPW collects instrumentation data both by routine manual measurements and by automated data acquisition systems (ADAS). Dam operators manually measure the instruments on a regular basis. Most measurements are taken weekly or monthly, but may vary based on reservoir elevation. The dam operators submit the data to the dam safety engineers using an online form, and the engineers will process and review the data. The ADAS transmits data automatically via internet or radio connections. The ADAS is also monitored by the dam safety engineers, and the measurements are regularly compared with the manual measurements to ensure proper calibration of the instruments.

SETTING THRESHOLD LIMITS

There are multiple approaches that could be considered for setting threshold limits. First and foremost, one must consider any applicable dam safety studies that are based on engineering analysis for the stability of the dam. This could include calculated maximum uplift pressure, embankment saturation, allowable settlement, and other parameters. One may then look to historical data to determine the actual expected range for an instrument. Thresholds limits based on historical data should always be more conservative than the allowable values derived from an engineering analysis.

Selection of the Dataset

Setting threshold limits based on historic data requires several years of data to characterize annual trends and seasonal variations. LACDPW has generally used the most recent ten years of data. However, it is important to compare all historic values to determine if there have been any major changes in instrument readings. If so, these changes should be analyzed to determine if there are any concerns or need for instrument maintenance. In other cases, changes may be a result of a physical change such as modification of the instrument, a rehabilitation project, or site changes due to a seismic or geologic event. The dataset used to calculate the threshold limits should begin after any notable changes had occurred.

The data should also cover the full range of operation for the reservoir. Some flood control facilities are almost always dry and may not have data available with reservoir levels. In this case, it may not be feasible to calculate threshold limits from historic data. Similarly, some water storage facilities are held at a steady level with limited data for varying reservoir levels. It should also be noted that instruments may require time to stabilize in conditions of rapidly varying reservoir levels. Finally, it is important to review the data and eliminate any potential outliers. The goal is to set threshold limits that represent normal conditions, so it is most conservative to remove any questionable data.

The data is plotted over a span of time to observe the annual trends and historical maximum and minimum and to identify potential factors that influence the measurements, such as reservoir level and seasonal changes. For instruments that monitor movement, it is usually observed that the fluctuations are seasonal due to thermal expansion and contraction. For instruments that monitor hydraulic conditions, the measurements usually depend most on reservoir level. The thresholds need to reflect these relationships.

Figure 1 shows a sample ten-year data plot for a piezometer. It can be observed that the ten years of data characterizes seasonal variations in reservoir levels and piezometer response. The data should then be plotted as a scatter plot of instrument versus reservoir level, as shown in Figure 2. In this plot, it can be observed that there is a direct relationship between reservoir level and piezometer level.

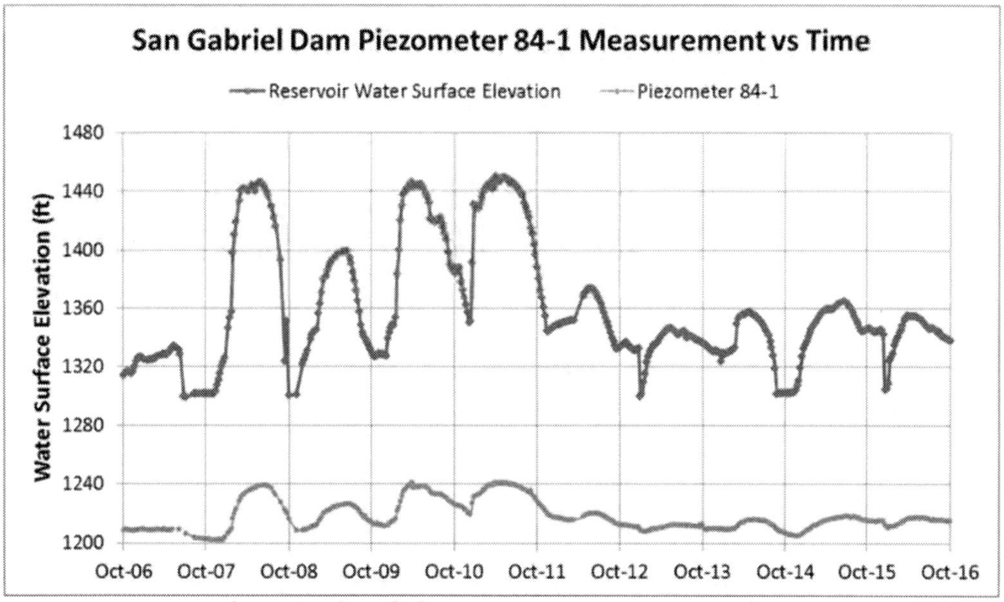

Figure 1. Plot of Piezometer Measurement vs. Time

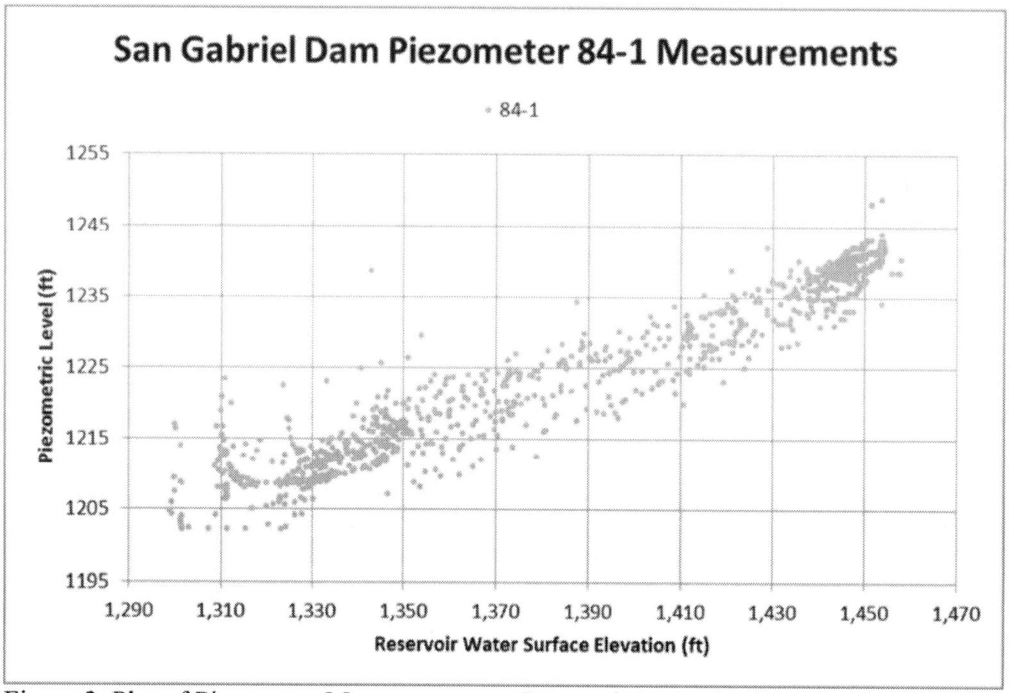

Figure 2. Plot of Piezometer Measurement vs. Reservoir Water Surface Elevation

Alternative Methodologies

Historic Maximum/Minimum

The upper and lower threshold limits can be set to match the historic maximum or minimum. This method assumes that the dam performed well at these values, and therefore they may be considered as acceptable levels. However, any extreme outliers must be excluded to account for potential errors in the data. This is a simple approach, but does not address the relationship to reservoir level. Additionally, there could be an overall change to the dam or instrument since the historic minimum or maximums occurred, in which case the threshold should be based on more current data.

Graphical Approach

Another simple method to set threshold limits is a graphical approach. The data is plotted as a scatter plot such as joint measurement versus time or piezometer level versus reservoir level. The upper and lower threshold lines can then be drawn in manually using engineering judgment for the expected range. For a conservative threshold, there should likely be some outliers in the plot. This manual approach is simple and can be easily adapted to a variety of data. However, this method relies entirely on the engineer's judgment, and could be very inconsistent. The plots also do not provide discrete values for the thresholds, so it is not ideal for inputting to automated systems. Figure 3 shows an example of a manually prepared piezometer response chart.

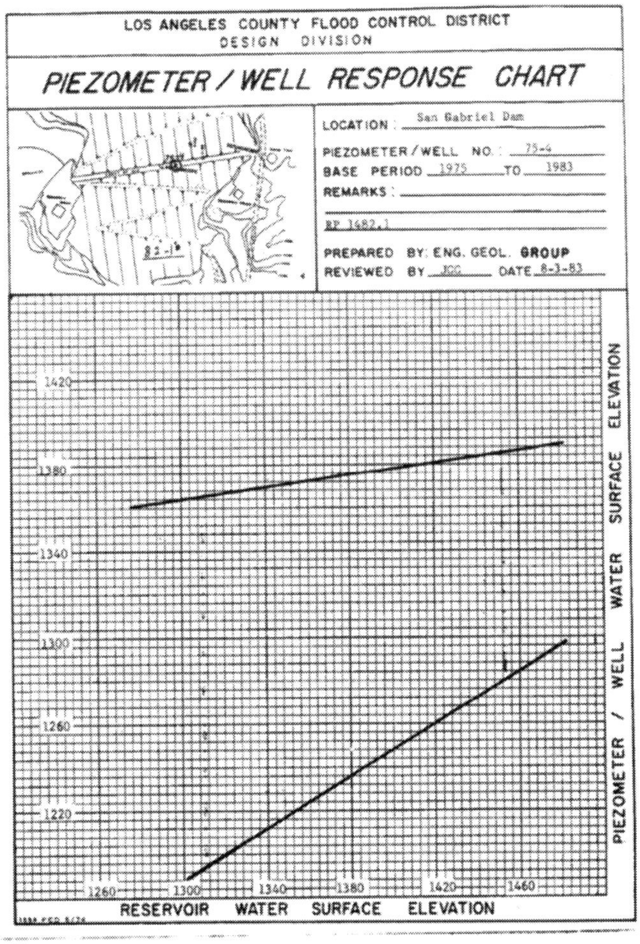

Figure 3. Graphical Piezometer Response Chart (Los Angeles County – 1983)

Statistical Approach

The preferred approach is to conduct a basic statistical analysis to determine a confidence interval for threshold limits. This provides consistency and repeatability, and produces an equation which can be input to automated systems for alarms and reporting. The statistical approach can either consider deviations from the average or an equation that describes the relationship of the instrument to the reservoir.

LACDPW THRESHOLD METHODOLOGY

LACDPW developed a statistical methodology for setting the threshold limits for instruments that are mostly independent of reservoir and for those that are dependent on reservoir level. In both cases, it was found that instrument data fits fairly well for a normal distribution curve since a majority of data points are grouped near the average or expected value. The 99 percent normal distribution confidence interval was selected to set the threshold range. The most recent 10 years of data was used to conduct this analysis. When analyzing the data, spurious values were removed from the calculations to more accurately determine the trend and confidence intervals. Less than one percent of the data was removed in most cases. If potential outliers are not removed, the standard deviation could be excessive, and the resulting threshold limits could allow for erroneous or hazardous conditions.

Instruments Independent of Reservoir Level

It was observed that joints and crack measurements generally follow a seasonal trend due to thermal expansion and contraction, but do not show a strong relationship to reservoir level. Therefore, the threshold limits are based on the average measurement and did not consider an equation to describe the normal fluctuations. The limits are calculated as the 99% Confidence Interval, plus and minus 0.05 inches, and rounded to the hundredth of an inch.

$$TL_{upper} = \bar{x} + 2.575\sigma + 0.05$$

$$TL_{lower} = \bar{x} - 2.575\sigma - 0.05$$

TL = Threshold Limit

$\bar{x}$ = Arithmetic Mean

σ = Standard Deviation

$Z\sigma = 2.575\sigma$ for 99% confidence interval

Data for joint measurements fits fairly well to the normal distribution with a majority of data points grouped near the mean, as shown by the histogram in Figure 4. Therefore, the 99% confidence interval for normal distribution can be used to describe the regular range of seasonal fluctuations. This range is considered acceptable movement for a particular joint. An additional range of 0.05 inches outside the confidence interval accounts for minor outliers. This was necessary to provide a reasonable range for data with very small standard deviations. Any movement greater than 0.05 inches outside the regular range is considered significantly measurable movement for threshold limit criteria. Figure 5 shows a graphical representation of the threshold limits for a typical joint measurement.

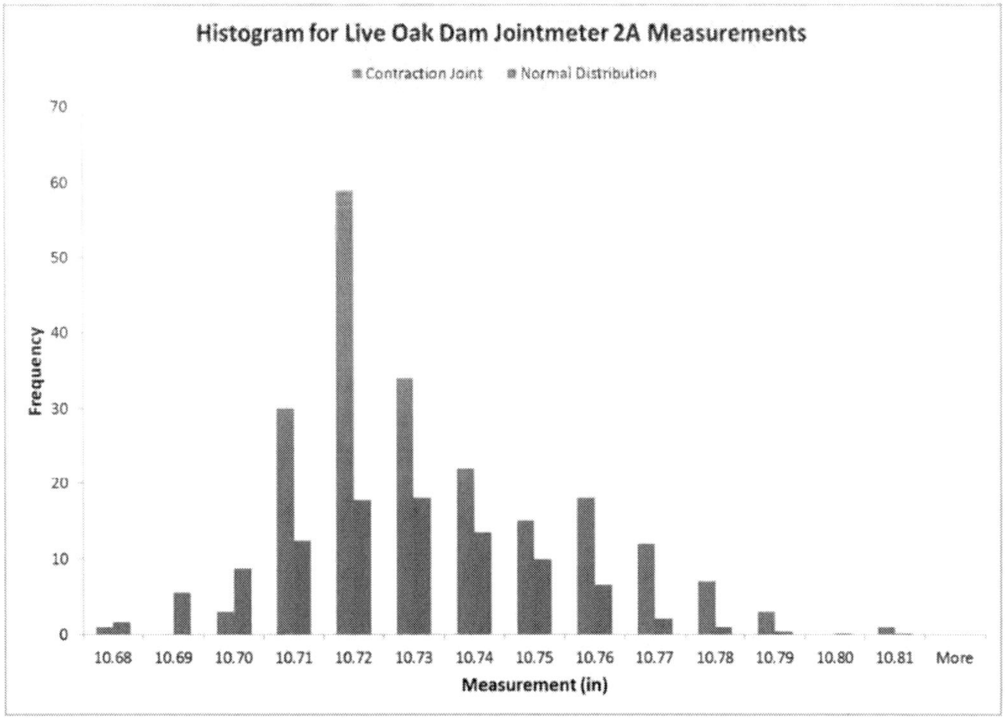

Figure 4. Histogram of joint measurements

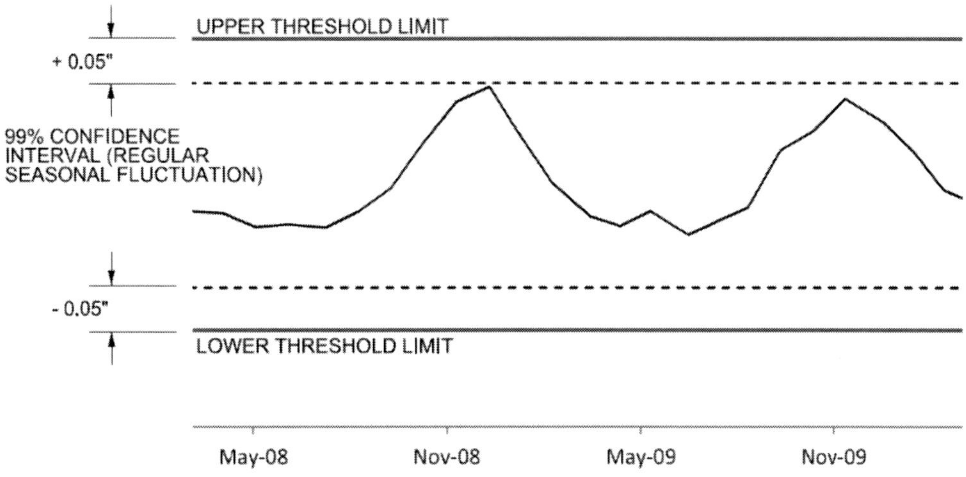

Figure 5. Graphical Representation of a Typical Joint Measurement

Reservoir Dependent Instruments

For piezometers or leakage measurements that relate directly to the reservoir level, the historical data is plotted versus reservoir level, and a second order polynomial best fit line is calculated for the data. The best fit equation could be developed by various mathematical approaches or software. Microsoft Excel was selected as a simple, user-friendly system to develop trend lines and equations. The equation for the best fit line is then used to calculate expected values for piezometer readings. The standard deviation of the actual piezometer readings from the expected value is then calculated. The range for each piezometer is calculated as the 99 percent confidence interval based on the standard deviation. The range value is then added to the best fit line equation to create the upper and lower threshold limits.

The following guidelines are used to establish threshold limits for special conditions:

1. In cases where the calculated threshold exceeds the reservoir level for certain piezometers, the threshold will be set to equal the reservoir level or according to the engineer's judgment.
2. In cases where the calculated value falls below the bottom of the piezometer, the threshold will be set to equal the bottom of piezometer.
3. In cases where the data exhibits a distinct trend or excessively small standard deviation, the thresholds will be set according to the engineer's judgment.
4. If a dam safety analysis has been performed to determine a smaller allowable range based on uplift pressure, allowable saturation, or other parameters, these values will supersede the statistically calculated threshold limits.

For example, for a piezometer at San Gabriel Dam (reservoir bottom El. 1280 and spillway El. 1453), let x = measured reservoir water surface elevation (RWSE) and y = piezometric elevation for datum 1 through n. Calculate the best fit second order polynomial line equation for the data set. Then calculate the expected piezometric elevation for each reservoir water surface elevation.

Best fit equation: $y_{ei} = Ax_{mi}{}^2 + Bx_{mi} + C$ (1)

Where:
y_{ei} = expected piezometric value from best fit equation for a given RWSE
x_{mi} = measured RWSE for a set of piezometer readings
A, B, C = constants from best fit equation (constants should not be significantly rounded)

Then, to find the standard deviation:
For $i = 1{:}n$

$$T_i = (y_i - y_{ei})^2 \quad (2)$$

$$\sigma = \sqrt{\frac{\Sigma T_i}{n}} \quad (3)$$

Where:
Threshold Range = Best fit line ± range for 99 percent confidence
= Best fit line ± $Z\sigma$
= Best fit line ± 2.575σ

Z = Tolerance interval for normal distribution, which is equal to 2.575 for 99 percent confidence

The piezometer data fits fairly well to the normal distribution with a majority of data points grouped near the expected values, as shown by the example histogram in Figure 6. Figure 7 shows an example scatter plot of the piezometer data with calculated threshold limits.

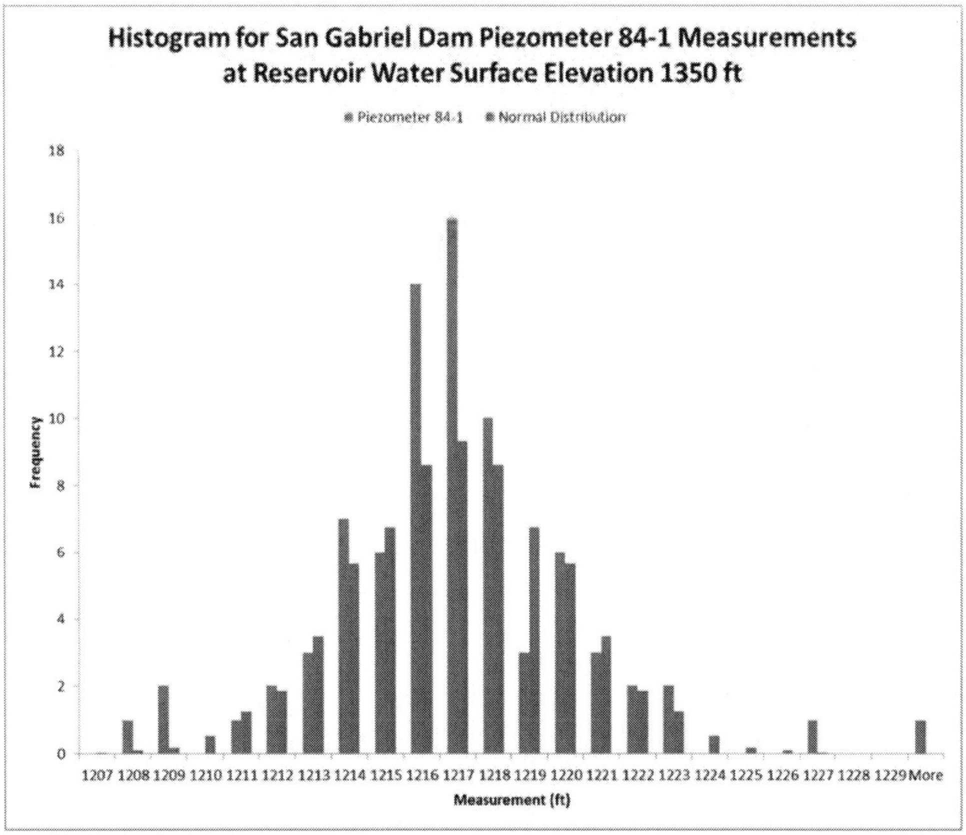

Figure 6. Histogram of piezometer data vs. expected values

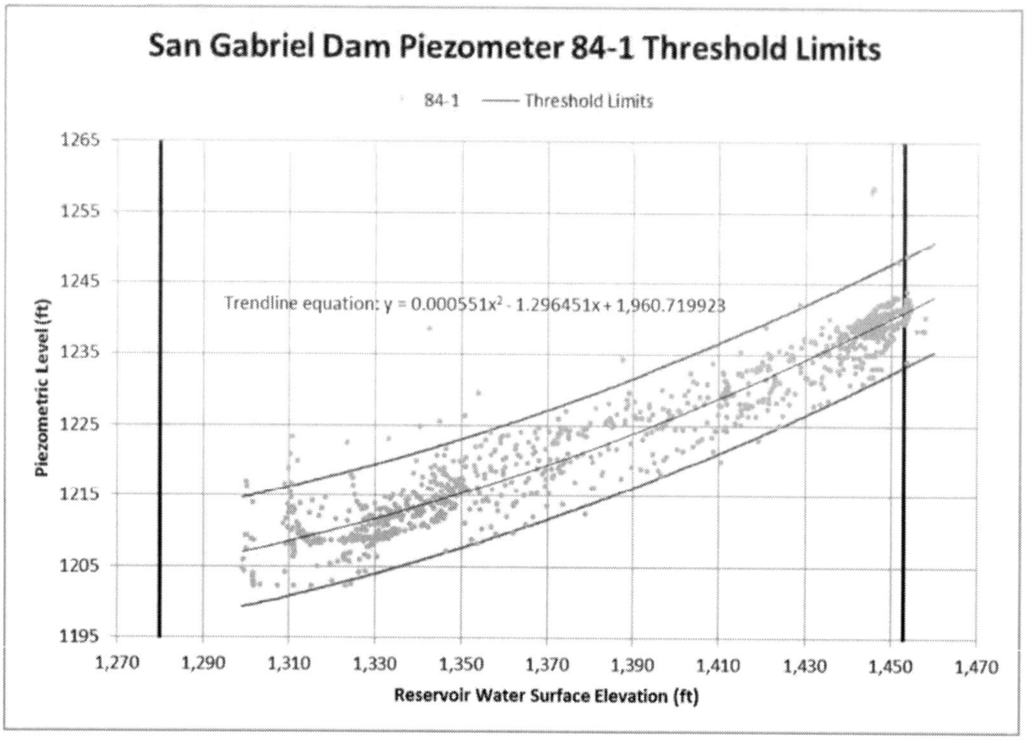

Figure 7. Typical Piezometer Threshold Graph

IMPLEMENTATION

LACDPW began progressively implementing the threshold limits for each of the County's fourteen major dams. All of these dams have existed for many years and in most cases extensive data was available. However, LACDPW dams operate primarily for flood control and stormwater capture. Flood control facilities are often dry through much of the year and fill with water for short periods of time during storm events. For some of these facilities there is not sufficient data for the full range of reservoir levels. In these cases, instrument data could not effectively be analyzed to produce statistical threshold limits.

After calculating and reviewing the thresholds, plots were created in a standardized format, shown in Figure 8. The sheets include a general location map, photograph of the instrument, installation dates, dimensions, and material data. The threshold limits were presented to the dam operators who perform the measurements with a basic training on how to reference and check the new threshold limit sheets.

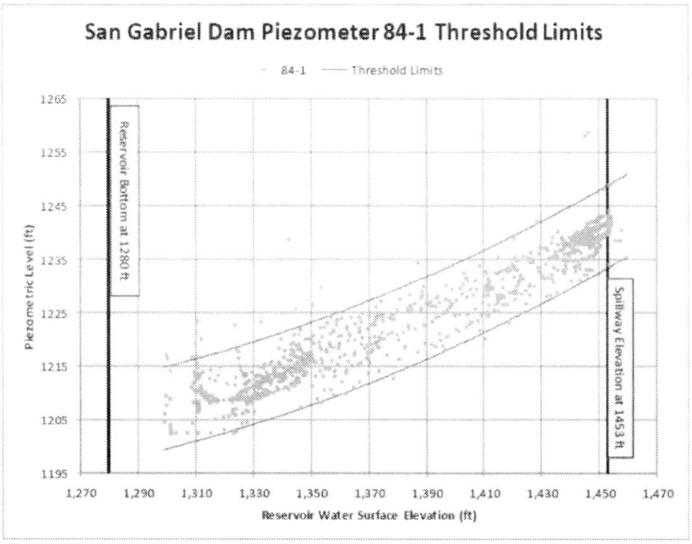

LOCATION

PHOTOGRAPH

Piezometer	Location	Attitude	R.P. (ft)	Depth (ft)	Measures	Perforations (ft)	Measures	DADAS
84-1	Crest of the dam to the left of midway	Vertical	1482.9	406	Uplift Pressure	391-406	Uplift pressure in the foundation	Yes

R.P. = Reference Point

Figure 8. Final Instrument Threshold Chart

Automation of Threshold Limits

Upon completion of the new threshold limits, the next step was to incorporate the limits into automated alarms for both the manual measurements and automated data acquisition systems. For the manual measurements, LACDPW utilizes an online form for dam operators to submit measurements to the dam safety engineers. The online forms were programmed to check the measurements before being submitted. If a measurement is beyond the threshold limit it will be highlighted in red. The dam operator will immediately see the warning and recheck the measurement. A sample form is shown in Figure 9 with highlighted exceedance. Often the threshold exceedance is found to be erroneous and this validation allows for correction of these errors. If the measurement is verified to be out of range, the dam operator can submit an incident report and immediately notify the dam safety engineers to investigate.

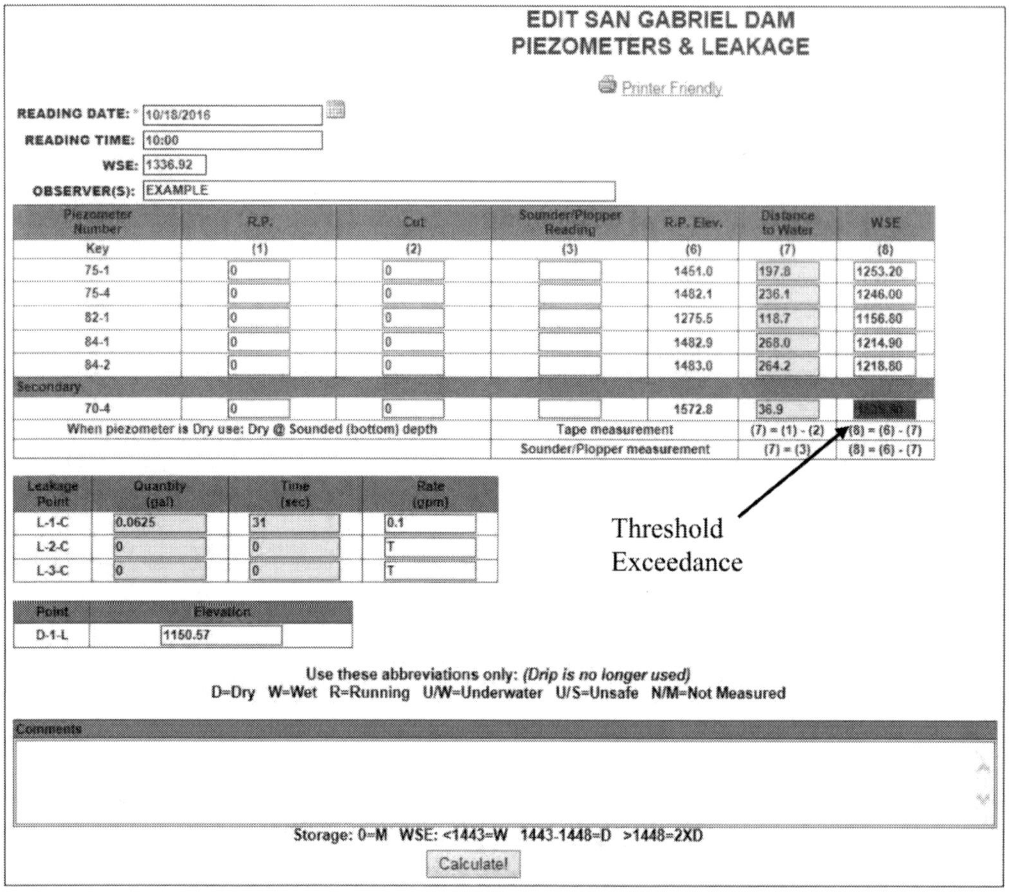

EDIT SAN GABRIEL DAM
PIEZOMETERS & LEAKAGE

Printer Friendly

READING DATE: * 10/18/2016
READING TIME: 10:00
WSE: 1336.92
OBSERVER(S): EXAMPLE

Piezometer Number	R.P.	Cut	Sounder/Plopper Reading	R.P. Elev.	Distance to Water	WSE
Key	(1)	(2)	(3)	(6)	(7)	(8)
75-1	0	0		1451.0	197.8	1253.20
75-4	0	0		1482.1	236.1	1246.00
82-1	0	0		1275.5	118.7	1156.80
84-1	0	0		1482.9	268.0	1214.90
84-2	0	0		1483.0	264.2	1218.80
Secondary						
70-4	0	0		1572.8	36.9	[illegible]
When piezometer is Dry use: Dry @ Sounded (bottom) depth			Tape measurement		(7) = (1) - (2)	(8) = (6) - (7)
			Sounder/Plopper measurement		(7) = (3)	(8) = (6) - (7)

Leakage Point	Quantity (gal)	Time (sec)	Rate (gpm)
L-1-C	0.0625	31	0.1
L-2-C	0	0	T
L-3-C	0	0	T

Point	Elevation
D-1-L	1150.57

Use these abbreviations only: *(Drip is no longer used)*
D=Dry W=Wet R=Running U/W=Underwater U/S=Unsafe N/M=Not Measured

Comments

Storage: 0=M WSE: <1443=W 1443-1448=D >1448=2XD

Calculate!

Figure 9. Online Form Highlighting Threshold Exceedance

LACDPW is currently updating the automated data acquisition system and will incorporate automatic alarms for any threshold exceedance. The threshold equations and limits will be programmed into the software. If an exceedance is measured, the software will automatically generate an email to LACDPW dam safety engineers. In this event, the dam operators would be requested to measure the instrument manually and verify the issue. The dam safety engineers would then follow up with an investigation. This provides a continuous watch for the condition of the dam including nights, weekends, storm events, and earthquakes.

CONCLUSION

Threshold limits are a useful tool to monitor and validate dam safety instrumentation measurements. The threshold limits provide a basic guide for the acceptable range of measurements, but it is important to recognize that any sudden changes or unusual trends may still be cause for concern, even if the measurement is within the threshold limits. A statistical approach to generate threshold limits provides a straightforward process by which new thresholds can be established and existing thresholds can be evaluated and updated. With variations in the methodology, the approach can be applied to both reservoir dependent measurements, such as piezometers, and reservoir independent measurements, such as contraction joints. Calculation of threshold limits requires a sufficient amount of data covering the full range of operating levels and seasonal conditions for the dam. The calculated limits must also be evaluated with engineering judgment to confirm the values are reasonable and conservative relative to the safety of the dam. LACDPW is successfully implementing a monitoring program to include thresholds limits with advances in data recording and automated monitoring, which could allow for early detection and quick response to abnormal conditions and possible dam safety issues.

SOMERSET DAM – INNOVATIVE REMOTE MONITORING FOR TIME-SENSITIVE BREACH DETECTION

Jerry Cross[1]
Yogi Sookhu[2]
Jud Donaghy[3]

ABSTRACT

Tropical Storm Irene wreaked havoc on the Deerfield River in Southern Vermont. Ten inches of rain fell in 12 hours producing floods approaching the half-PMF. Roads were washed out. Power and telecommunication lines were knocked down. All land and cell based communication to TransCanada's hydro stations was lost. Operators had to be dispatched to each power plant and dam. Other than voice communication by radio, control room operators were blind to the conditions at the hydro sites, with one exception. Communication with the satellite based system installed at Somerset Dam functioned normally relaying water level information throughout the storm.

This paper describes innovative technology installed at a remote location that continually monitors the dam for seepage and sudden changes in water levels to provide instantaneous detection, verification and reliable satellite communication with TransCanada's control center.

PROJECT DESCRIPTION

TransCanada's Deerfield River Project consists of eight developments located on the Deerfield River from Somerset Reservoir in Windham County, Vermont, to just below Gardners Falls in Franklin County, Massachusetts. Three of the developments are located in Vermont (Somerset, Searsburg and Harriman) and five in Massachusetts (Sherman, Deerfield No.5, Deerfield No.4, Deerfield No 3 and Deerfield No.2).

Somerset Dam is located on the East Branch of the Deerfield River, and is the furthest upstream dam on the river. The project consists of a storage reservoir, dam, outlet works and spillway. There are no power generating facilities at this development. There are also no power or telecommunication lines within nine miles of the site.

Somerset Reservoir is about 5½ miles long and one mile across at its widest point, with a surface area of 1,500 acres and gross storage of 57,000 acre-feet. The normal operating range is from elevation 2113 feet to 2128 feet.

Somerset Dam, with a crest elevation of 2147 feet, is a semi-hydraulic earth-fill dam that is approximately 2100 feet long with a maximum height of about 110 feet.

[1] Chief Civil Engineer, Gotham Analytics LLC (formerly Regional Engineer at FERC New York Regional Office), Englewood, NJ 07631, jerry@gothamanalytics.net
[2] CEO, Gotham Analytics LLC, Englewood, NJ 07631, yogi@gothamanalytics.net
[3] Chief Dam Safety Engineer, TransCanada, Concord, NH 03301, jud_donaghy@transcanada.com

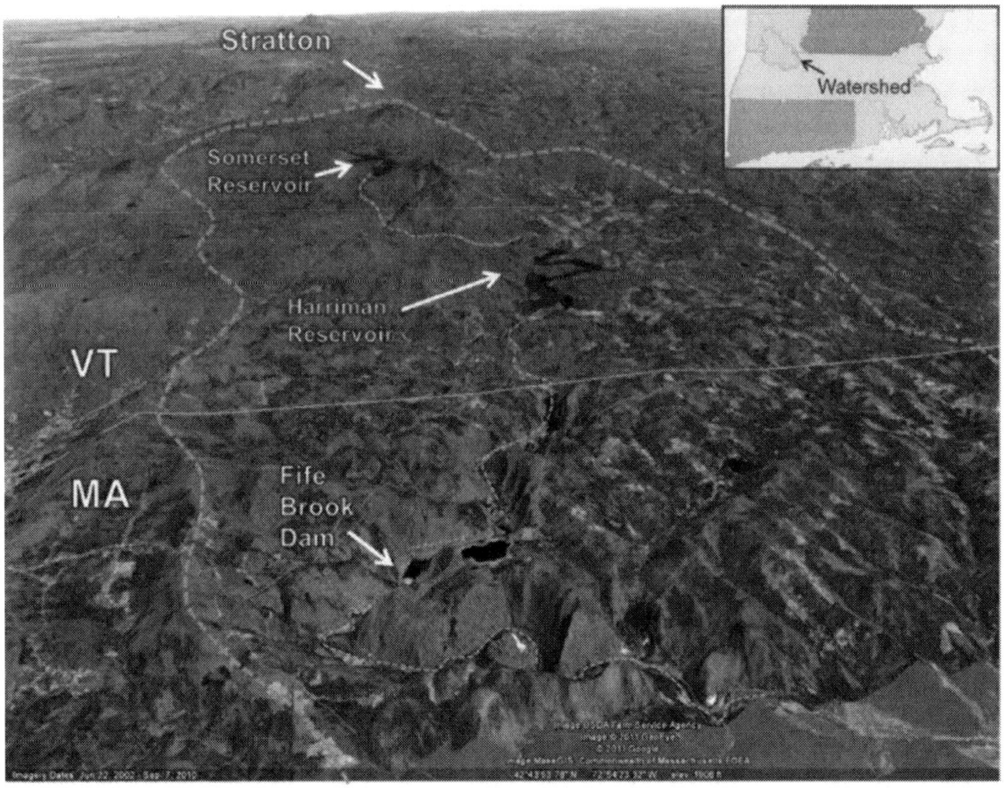

Figure 1. Project Location Map

The main outlet works, located in the gatehouse at the southeastern (left) end of the dam has two gated 48-inch diameter pipes used to control reservoir discharge and minimum flow. These valves are operated remotely via radio or by on-site personnel and powered by an on-site emergency generator.

The emergency spillway is a side channel spillway with 3-foot high flashboards located on the northwestern (right) end of the dam. It has a permanent crest elevation of 2133.6 feet, and is about 192 feet long and seven feet high. The spillway discharges into a manmade channel that is about 700 feet long and empties back into the river downstream of the dam.

Tropical Storm Irene (August 2011)

In August 2011, Tropical Storm Irene (See Figure 2) swept through the Deerfield River watershed, and produced nearly a foot of rain in 12 to 16 hours. Runoff from the heavy rainfall produced record floods approaching the ½ PMF (Probable Maximum Flood) event. Destruction from flooding was rampant washing out roads and bridges, damaging homes and businesses, and eroding dam and canal embankments.

Figure 2. Tropical Storm Irene

Figure 3. Sherman Dam Hydro Plant Access Road

Figure 4. Sherman Dam Toe Erosion

Figure 5. River Road along Deerfield 5 Canal

Two powerhouses were flooded so badly that it took several years to clean the soil; remove, design and replace ruined electrical equipment; and restore the powerhouses to service.

Figure 6. Interior of Deerfield 3 Powerhouse

Power and telecommunication lines were also knocked down including all land and cell based communication. All of TransCanada's hydroelectric projects were damaged, communication was knocked out, and all of the SCADA systems were incapacitated. Other than voice communication by radio, control room operators were blind to the conditions at the hydro sites. Due to TransCanada's foresight, operators were dispatched to each of their hydro stations prior to the peak of the storm. Had they delayed, all of the projects would have been inaccessible because of the washed out roads.

The road leading to Somerset dam was already washed out when operators attempted to travel to the dam. Fortunately, all of the remote monitoring equipment continued to send valuable information to TransCanada's control center and to their Chief Dam Safety Engineer's smart phone. Throughout the storm, water level data was displayed on the dashboard (See Figure 7) verifying that Somerset Dam was safe

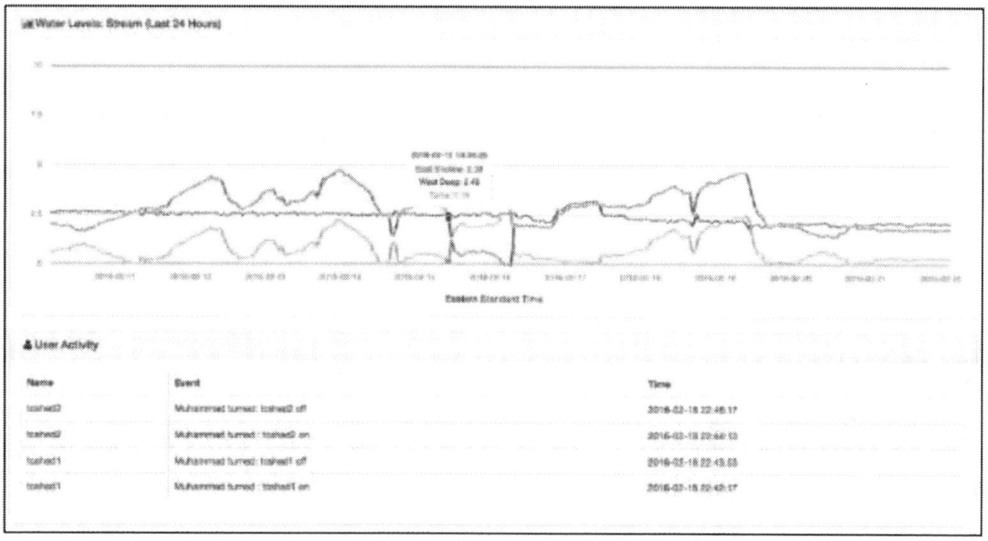

Figure 7. Typical Plot of Headwater and Tailwater at Somerset Dam

EMERGENCY ACTION PLANNING

Downstream Hazard Potential

Dam breach analyses determined that the Inflow Design Flood (IDF) is the Probable Maximum Flood (PMF). Inundation maps for a dam breach under both sunny day and PMF conditions are included in the Emergency Action Plan (EAP). A significant number of residential structures will be impacted by a dam failure under both conditions.

The current hazard potential classification for the Somerset Development is High Hazard. Dam failure during the PMF will result in an outflow of approximately 600,000 cfs, and a rise in river stage of 36 feet immediately downstream of the dam. Failure of Somerset

Dam during the PMF causes the overtopping and potential failure of all downstream dams from Searsburg to Deerfield No. 2 due to more than two feet of overtopping.

Time Sensitiveness of the Emergency Action Plan

A piping failure of Somerset Dam under normal conditions will also cause the overtopping failure of Searsburg Dam and impact to residential areas in the Town of Searsburg in less than 1½ hours. Due to Somerset Dam's remote location, a dam failure could go without detection until after the residential areas are impacted.

The Federal Energy Regulatory Commission (FERC) requires that Detection, Verification, Notification, and Evacuation be completed at least 15 minutes prior to impacts from the breach flood wave. Without some form of remote monitoring and rapid verification of a breach, this requirement would be unachievable.

Detection, Verification and Notification

TransCanada was in need of equipment that would function in a remote location without an external power source and that could provide nearly instantaneous detection and verification of a dam failure. Gotham Analytics installed their AquaEdge™ system that provides a solar powered control system that receives sensor information, processes that information, and transmits the processed data to a dashboard that can be viewed in the control room, or on a smart phone. Prior to Tropical Storm Irene, water level sensors were installed at Somerset Dam. One was placed in the reservoir at the spillway and one at a point downstream in the river where any potential breach water would pass. After Tropical Storm Irene, seven thermal imaging cameras were installed to provide a comprehensive view of the dam.

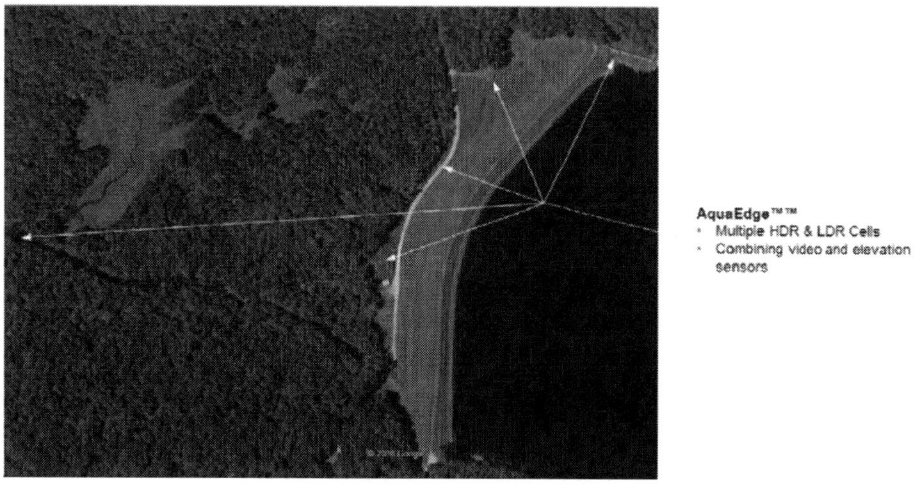

Figure 8. Somerset Dam

Figure 9. LDR Cell at Tailwater Gage

Remote Detection. Water level pressure sensors were installed in the reservoir and in the stream about ½ mile downstream. Using this instrumentation a dam failure would be detected immediately by a sudden rise in tailwater. The drop in headwater would also be detected, but to much less magnitude due to the size of the reservoir.

After Tropical Storm Irene, seven thermal imaging cameras were installed to supplement the water level sensors. A dam failure caused by piping would be detected by a change in temperature on the downstream face caused by water emanating from the embankment. However, the thermal cameras will pick up an increase in seepage long before a piping failure develops. Thermal imaging cameras have the advantage of detecting changes in temperature such as cold water emanating from a warm embankment, even during a moonless night. They have also been able to penetrate more than 12-inches beneath the snow on the embankment to detect warmer water on a cold background. Thermal cameras have the added benefit to detect unauthorized visitors, day or night.

Figure 10. Thermal Image of access road and embankment toe

Remote Verification. Redundancy of the instrumentation allows the equipment to complement each other. If the water level sensors detect a failure, the thermal imaging cameras will provide the verification. If a thermal imaging camera detects a dam failure, the other cameras and the water level sensors will provide the verification. In addition, a video camera is was also included in the package to provide verification if the failure occurs during daylight hours.

Notification. The installation at Somerset Dam transmits the data and images to a dashboard via satellite communications. If the water level sensors detect a level that exceeds the threshold level, and alarm will sound in the control room at TransCanada's control center, 60 highway miles away. Operators in the control center will switch on the cameras to verify the failure, and immediately start making the calls in the notification flow chart included in the Emergency Action Plan.

The AquaEdge™ system also has the capability to initiate a reverse 911 call down, or to send an audio message to a series of telephone numbers.

System Resilience and Reliability

As previously discussed, all of the remote monitoring equipment continued to send valuable information to TransCanada's control center and to their Chief Dam Safety Engineer's smart phone throughout Tropical Storm Irene. The water level data was displayed on the dashboard verifying that Somerset Dam was safe. After several years of operation, data and imagery continue to successfully monitor Somerset Dam.

CONCLUSIONS

Hurricane Irene caused $Billions in Damages. All of the SCADA and communications for the TransCanada projects failed, except for the remote monitoring equipment installed at Somerset Dam, which continued to send data throughout the storm.

Following Hurricane Irene and to comply with the FERC requirements for time sensitive response times, TransCanada expanded the on-site equipment to include Thermal and Optical Cameras. Those efforts have reduced the time for detection, verification, and notification to 2-3 minutes.

Future enhancements will include:

- Pressure Sensors in Weir Boxes,
- Temperature Probes in Reservoir &
- Optical Camera at Tailwater Gage Location

The years since installation of the remote monitoring equipment have proven the equipment to be robust and invaluable in monitoring a remote dam site. It has also provided a level of confidence in the safe operation of the dam.

GEOSPATIAL TOOLS FOR ROUTINE DAM SAFETY MONITORING AND RESPONSE APPLICATIONS DURING EMERGENCIES

Yogesh Prashar, PE, GE[1]
Atta B. Yiadom, PE[2]
Elizabeth Bialek, PE[3]
Rachel Wong[4]

This paper discusses a proposed approach using Geospatial tools (ArcGIS) to evaluate the fragility of embankment dams in the East Bay Municipal Utility District (EBMUD) Service Area. Dam Safety monitoring and emergency response is tightly interrelated. The current state of practice for monitoring, response and recovery of Dams as it is performed by many agencies and dam owners is has not widely caught up with the available roboust geospatial tools. In California many if not most dams are located in in moderate to high seismic areas. As part of its Dam Safety Program, EBMUD had already completed the seismic stability evaluation of most of the 26 dams. Many of the dams in this study are under the jurisdiction of the California Department of Water Resources, Division of Safety of Dams (DSOD). EBMUD has developed Marconi, open-source emergency management software that integrates seismic model results with emergency response that and import ShakeCast data maintained by USGS (Prashar et. al., USSD 2012). These data can be exported and visualized into ArcGIS database.

The approach of using ArcGIS to improve dam safety monitoring was developed in-house as a pilot project with 4 dams of varying size and with different levels of instrumentation. The instruments include piezometers (vibrating wire and traditional), survey monuments, toe drain and underdrain flow measurement stations, rain gauges, strong motion accelerograms, and other instruments and outworks. The point data were geo-located using high precision satellite based surveying methods. All of the instruments and critical appurtenances were then mapped onto the GIS database and several layers were created including aerial photographs, topographic contours, and all instruments at the dam sites. The instruments have been read and EBMUD has a history of 25 or more years of monitoring for most of the instruments. The presentation will show how large volumes of numerical data can be rapidly visualized geospatially to assess and evaluate the condition of a Dam both during normal operations and after natural disaster such as an earthquake. The approach can easily be applied and implemented by water agencies, utilities or entities with critical infrastructure in areas of high seismicity.

[1] Associate Civil Engineer, East Bay Municipal Utility District, Engineering and Construction Department, Engineering Services Division, 375 11th Street, MS #610, Oakland, CA 94607.
[2] Senior Civil Engineer, East Bay Municipal Utility District, Engineering and Construction Department, Engineering Services Division, 375 11th Street, MS #610, Oakland, CA 94607.
[3] Engineering Services Division Manager, East Bay Municipal Utility District, Engineering and Construction Department, , Engineering Services Division, 375 11th Street, MS #503, Oakland, CA 94607.
[4] Senior GIS Drafter, East Bay Municipal Utility District, Engineering and Construction Department, Engineering Services Division, 375 11th Street, MS #805, Oakland, CA 94607.

SPECIAL CONSIDERATIONS WHEN INSTALLING AUTOMATED INSTRUMENTATION AT A SITE WITH ACCESS LIMITATIONS

William Walker[1]

Automated data acquisition systems (ADAS) are a proven tool for mitigating some risk associated with large, high risk dams. The near real time access that ADAS provides to performance data is critical when evaluating potential signs of distress related to known potential failure modes. This level of access to data is not warranted at all dams and should be limited to structures during new construction, major rehabilitation projects, or other structural modifications. Major dam safety modifications in the United States Army Corps of Engineers (USACE) inventory such as Wolf Creek Dam, Bluestone Dam, and East Branch Dam have all successfully implemented ADAS technology to enhance monitoring during invasive foundation treatment. Effective design and installation of an ADAS requires detailed planning and coordination to ensure that appropriate risk drivers are targeted. This process can be challenging when working in extremely remote areas or locations where site access is limited. Issues can be further compounded when as-built documents for existing instrument installations are not available. In 2015 USACE was tasked by the Department of State (DoS) to assist in the planning and installation of an ADAS at a high risk dam in the Middle East. Successful installation of the ADAS at this site required unique techniques to overcome site access and logistical limitations.

[1] U.S. Army Corps of Engineers, william.g.walker@usace.army.mil

INSTRUMENTATION AND INFORMATION MANAGEMENT AT BOONE DAM

Rozh Ameen, P.E., PMP[1]
Raphael Siebenmann, P.E.[2]
Jamey Rosen, M.Sc., P.G.[3]

ABSTRACT

The appearance of a large sinkhole on October 20, 2014 at Boone Dam, followed by turbid seepage observed in the tailrace a few days later, triggered emergency response actions by Tennessee Valley Authority (TVA), which included: immediate lowering of the headwater level, installation of a filter at the tailrace of the dam consisting of sand, TDOT No. 57 Stone, TDOT No. 2 Stone, and armoring with grout bags, continuous surveillance and monitoring of the project, a thorough review of historical records, an extensive subsurface exploration and instrumentation program, and exploratory grouting. These programs have and will continue to generate a significant amount of data that, along with the archives of historical data, need to be reviewed and accessed, often in real time, by a broad range of stakeholders.

This paper will provide an overview of the development and implementation of the state of the art Information Management System (IMS) and Instrumentation Data Acquisition System (IDAS) to support the monitoring and construction activities. These systems include the use of web technology to consume and visualize high-frequency instrumentation data, and the use of Geographical Information Systems (GIS) technology to allow access to disparate data streams in a common geospatial context. The paper will describe how TVA is currently using these tools in the development of a hydrogeological model, in the monitoring of the embankment during exploratory grouting, and finally to ensure the most important objective is met, which is to cause no harm during investigation and construction activities.

INTRODUCTION

Background

Boone Dam is a hydroelectric project located on the South Fork Holston River near Johnson City, Tennessee, and is owned and operated by TVA. Constructed between 1951 and 1953, it provides hydroelectric power, floodwater control, water quality, recreation, aquatic ecology and water supply to local communities.

[1] Rozh Ameen, PMP, P.E., Tennessee Valley Authority, 1101 Market Street Chattanooga, TN 37402, 423-751-7319, ramohamadameen@tva.gov
[2] Raphael Siebenmann, P.E., Geosyntec Consultants, 1255 Roberts Boulevard Kennesaw, GA 30144, 678-202-9555, RSiebenmann@Geosyntec.com
[3] Jamey Rosen, M.Sc., P.G., Geosyntec Consultants, 130 Stone Road West Guelph, ON N1G 3Z2, 519-515-0871, JRosen@Geosyntec.com

The dam is constructed on limestone foundations and is comprised of two major structural components, shown on Figure 1a: a concrete gravity dam across the original river channel and an earthfill embankment on the right abutment and right rim. Figure 1b shows plan view and downstream elevations of Boone Dam. When originally constructed, the embankment section was approximately 750 feet long at crest elevation (El.) 1400 ft msl, with a crest width of 20 feet. The original configuration included a homogeneous rolled earth fill embankment with an upstream slope of 3 feet horizontal to every 1 foot vertical (3H:1V) and downstream slopes of 2H:1V to a berm at El. 1365 and 2.5H:1V below the berm. The embankment is in line with the concrete dam for a distance of 60 feet, then curves downstream to its terminus in the right abutment.

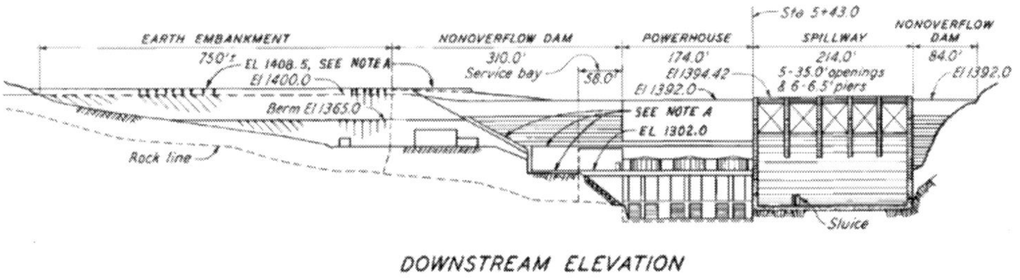

Figure 1a. Elevation Section of Boone Dam

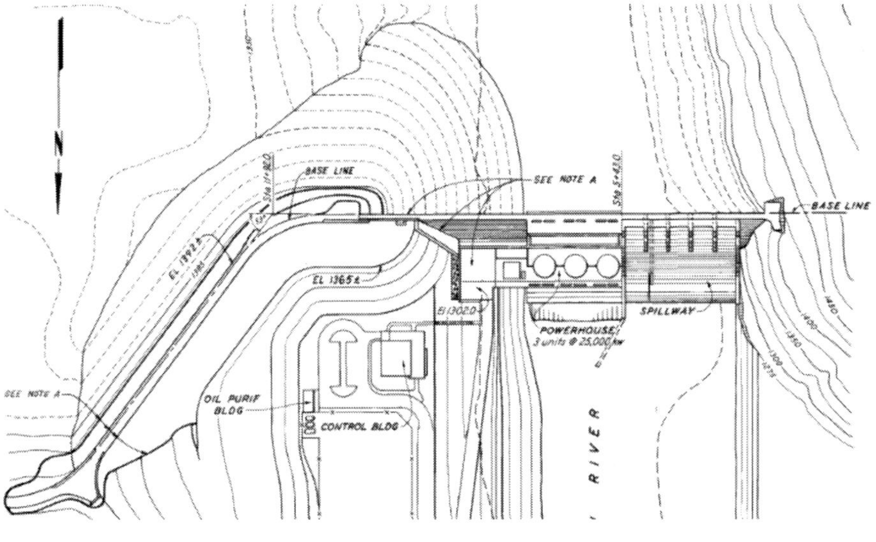

Figure 1b. Plan view of Boone Dam

Modifications to the embankment to allow the dam to safely pass the Probable Maximum Flood (PMF) were completed in 1984. The modification included raising the embankment crest with an impervious core and rockfill slopes 8.5 feet to El. 1408.5, with a modified crest width of 16 feet. Figure 2 shows a typical embankment section with the PMF modifications.

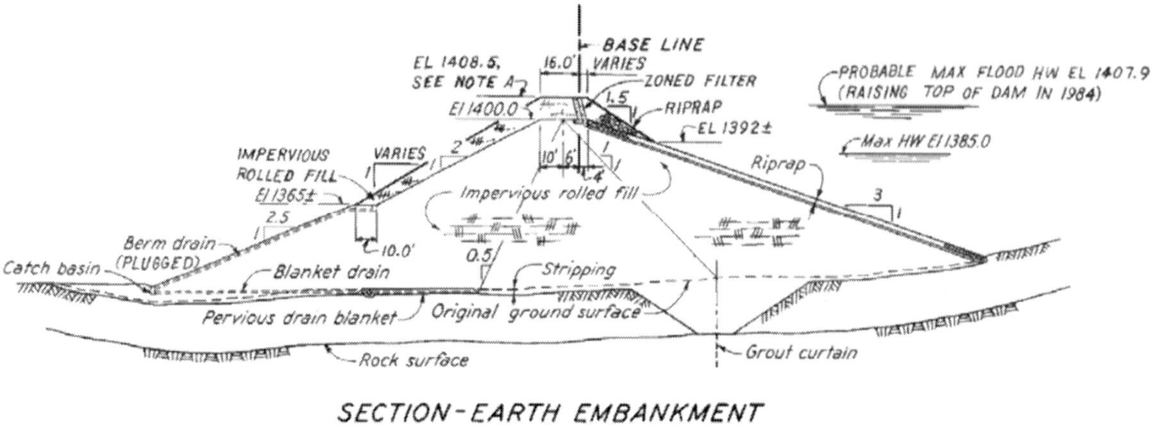

Figure 2. A typical Earth Embankment Section post PMF modification

The dam was raised to the crest elevation of 1,408.5 feet for the total height of about 146 feet above the tail water level at 1,262 feet. The average operating water level at the headwater pool is about 1,385 feet for the total head difference of 123 feet.

<u>Sinkhole and need for remediation</u>

A few years after construction, the earthfill dam started to experience some seepage issues (Figure 3). First signs of seepage emerged at the downstream toe of the dam in 1955 followed by other seepage or wet spots emerging from 1960s though 1990s. These events were also accompanied by a number of minor settlements in various areas of the dam embankment.

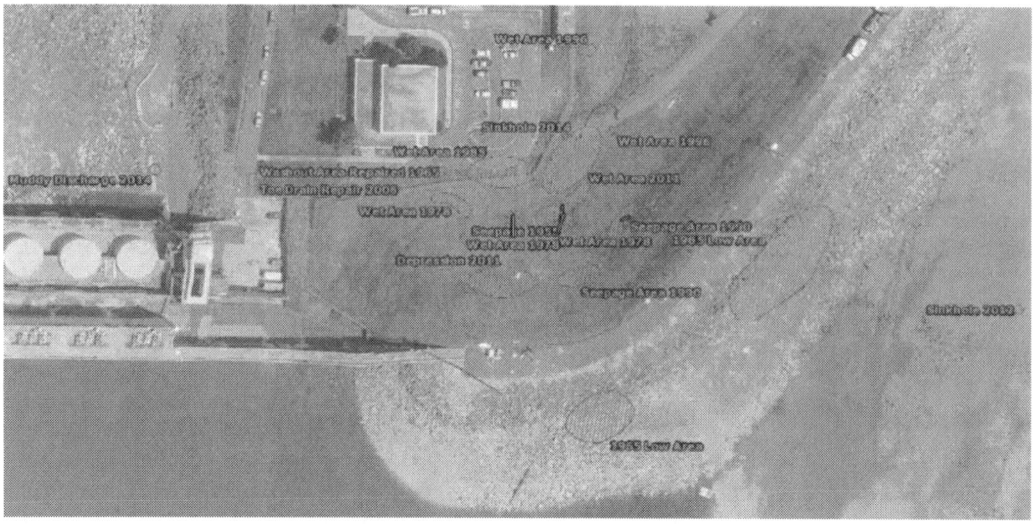

Figure 3. Historic Inspection Observations

More recently, in 2012 a sinkhole appeared upstream of the dam in the beach area (Figure 4). On October 20, 2014 a large sinkhole appeared near the downstream toe area (Figure 5) followed by a turbid seepage in the tailrace area on October 26, 2014 (Figure 6).

Figure 4. 2012 Sinkhole on upstream Beach

Figure 5. 2014 Sinkhole in Control Building Parking Lot as Discovered

Figure 6. Muddy Plumes in Tailrace

Remedial activities

The recent 2014 events triggered the emergency response action by TVA, which included: immediate lowering of the headpool water level, 24/7 dam surveillance, instrumentation and monitoring, physical assessment of the seepage issues, immediate construction of a filter berm at the tailrace of the dam, and exploratory grouting made up of Low Mobility Grouting (LMG) and High Mobility Grouting (HMG) (Figure 7)

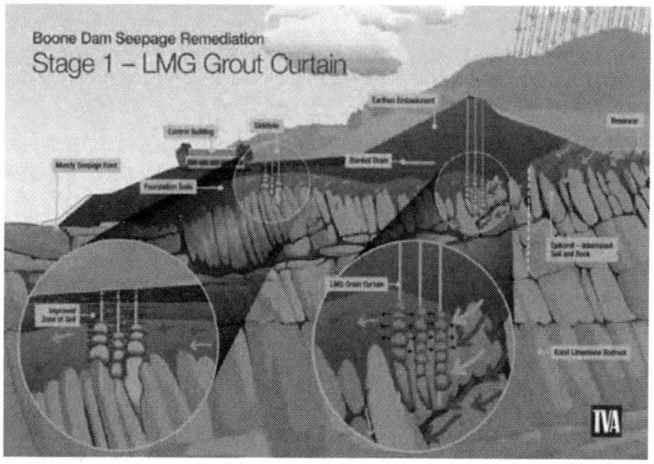

Figure 7. Anticipated illustration of subsurface groundwater movement in response to Exploratory Grouting

The need for an IMS and IDAS

As part of the site characterization of the subsurface conditions and seepage mechanism at Boone Dam, TVA identified the need for a tool to aid in the development of conceptual mitigation alternatives. This tool collectively includes the IMS and IDAS.

TVA developed the IMS to compile, manage and present (i.e., visualize and make accessible) historic data, as well as data collected and generated as part of the exploratory and construction activities for the mitigation efforts. The IMS hosting and reporting capabilities are utilized as part of Construction Quality Assurance (CQA) during construction to aid and facilitate construction documentation reviews between the owner, engineer, and contractor.

TVA developed the IDAS to manage, annotate, and report real time collected instrumentation data to aid in monitoring during exploratory and construction phase of the Boone Dam Project, This tool is utilized extensively during drilling and grouting operations to provide engineering team and grouting professionals the ability to make real time decisions.

BOONE DAM INFORMATION MANAGEMENT SYSTEM

The Boone Dam Information Management System (IMS) is a collection of technologies and tools that facilitates the transfer, compilation, storage, processing, analysis, and reporting of data associated with the reconstruction efforts at Boone Dam. The IMS consists of an Enterprise database (the database), a collection of Geographic Information System (GIS) projects and tools, a website, a File Transfer Protocol (FTP) site, a custom mobile data entry application built in Microsoft Access, and several other reporting and data collection tools.

Enterprise Database

The database is built in Microsoft SQL Server version 2008R2. The database schema was designed based on the intended geospatial analysis of data, and for consistency with the schema of incoming data streams. This allows the raw files generated in the field (e.g., grout monitoring data generated in the GroutIT system or borehole log data generated using gINT software) to be imported into the database in an automated fashion with minimal remapping or recalculating of database fields.

Data in the database are accessed using live database "views" which normalize, transform and calculate the data into a usable format. For instance, total grout takes by injection stage are aggregated and calculated from raw GroutIT inputs and presented to the user in web-based tables and graphical representations. These views allow data collected in the field to be made available in the database in near-real time.

Data are appended into the database using several methods. Generally, raw data files generated in the field are uploaded by field personnel to the FTP site, from where they are automatically imported to the database using SQL Server Integration Services (SSIS). Database tables requiring manual input are linked to the website, allowing users to enter data directly into the database via web-based forms.

GIS

Data in the database with a spatial component are analyzed, visualized and accessed using the GIS. The GIS is made up of several tools, including multiple web-based interactive maps and a 3D model. The webmaps access data in the database and in non-database GIS files (e.g., raster surfaces and base maps) developed using ESRI ArcGIS software tools. The 3D model is accessible on desktop computers using ESRI's ArcScene Viewer, and a simplified 3D model is available as a web tool. Collectively, the components of the GIS empower users to interactively view and query data within the database in a spatial and temporal context.

- Users interact with the web-based GIS viewers by:toggling on and off or adjusting then transparent of data layers;
- interactively view the tabular data associated with a given feature by clicking on it (e.g., the grout take at a specific grout hole;
- searching for features by name; and
- selecting features to view by applying custom queries or filters.

2D GIS View Options

The GIS is accessible via three different view schemes. The first, Planimetric View (Plan View), is displayed in the project coordinate system (NAD 1983 StatePlane Tennessee FIPS 4100 Feet) (Figures 8 and 10). This top-down view is best utilized for general Site orientation, and viewing surface features and data. Hyperlinks in selected parts of this view allow for selected subsurface data, such as a graphical representation of grout takes, to be accessed directly. Static documents, such as drilling logs in PDF format are also accessible via hyperlinks from this view.

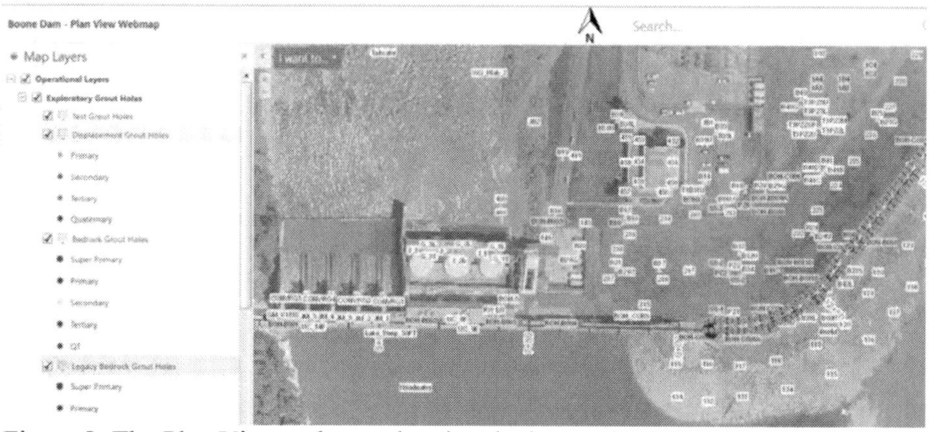

Figure 8: The Plan View webmap showing the interactive table of contents used to toggle layers on or off

The Profile View utilizes the Dam Stationing as the X-Axis, and Elevations as the Y-Axis (Figures 9 and 10). It allows for the 2-dimensional viewing of subsurface data along the dam alignment. Selected data are projected onto this centerline for visualizing within this view. Geologic contacts along the alignment have been interpolated from contact depths to visualize lithology in this view.

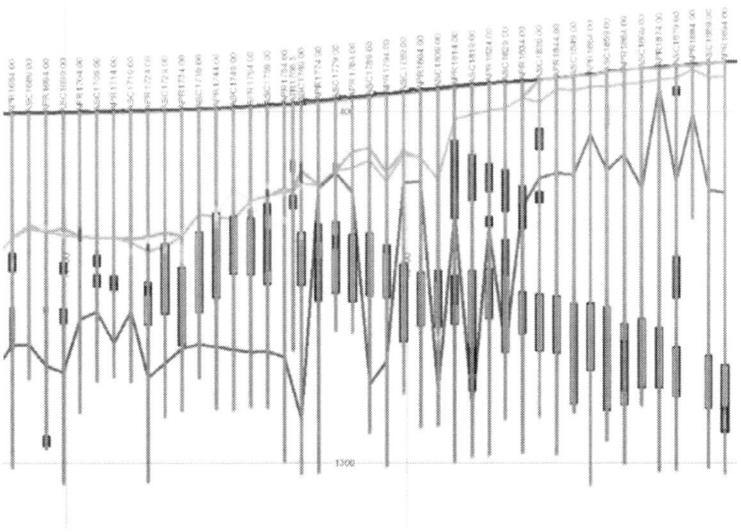

Figure 9. The Profile View webmap, showing grout holes symbolized by their grout take, interpreted geological layers, and geologic observations

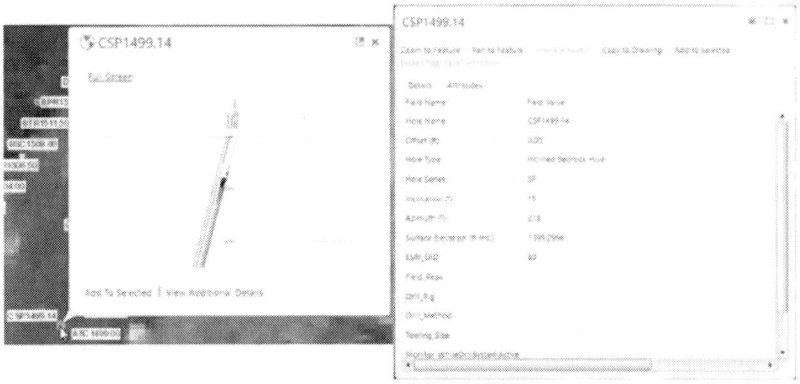

Figure 10. Viewing attributes from the GIS

A series of Transverse Profiles (i.e., perpendicular to the alignment) have also been selected for inclusion in the IMS. These profiles can be visualized from the Transverse Profile webmap, and use the Transverse Stationing as the X axis and Elevation as the Y-Axis.

<u>Automated Static GIS Outputs</u>. Static outputs in PDF format of the Plan and Profile view GIS are automatically generated on a daily frequency, and are available for download

from a link on the IMS website (Figure 11). These drawings are used by the project engineers for offline analysis, which include lithology, borehole information, and grout takes.

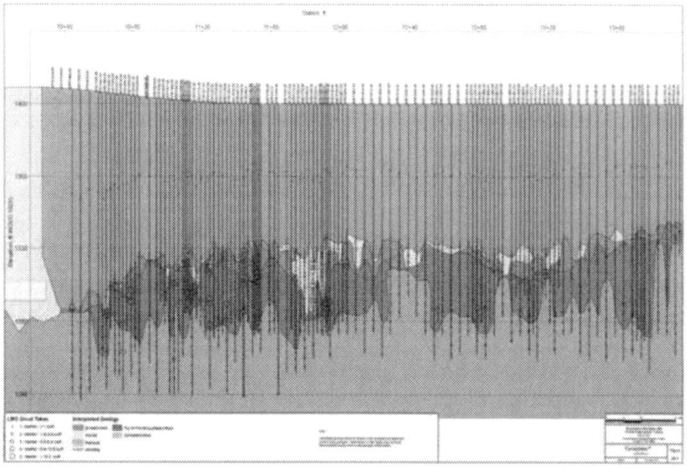

Figure 11. Example static PDF output from the profile view GIS (A-Line)

<u>3D GIS Model</u>. The 3D model includes data collected from historic geologic data sources, grout hole logging, boreholes, in-situ permeability tests, and instrumentation readings intergrated with geological mapping of the dam site, including geological discontinuities, bedding, shear zones, joint sets, into a 3D hydrogeological model of the embankment foundation materials, including the right rim and downstream area.

<u>Temporal Data</u>. Instrumentation (i.e., piezometer) data can be viewed in the Plan View webmap on a "time-aware" GIS layer (Figure 12). When this layer is made visible in the webmap, a time slider appears which allows the user to adjust and watch a simple visualization of piezometers that may have triggered action alarms alongside any drilling or grouting activities on site. This visualization helps highlight action alarms that may have been caused by on-site activity. By combining the temporal selection with other query tools, project team members can assess the instrument response associated with drilling, grouting, or other construction activities.

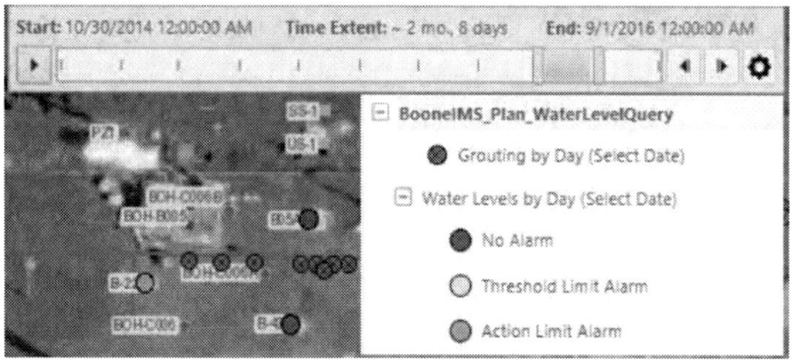

Figure 12. Viewing a time-aware layer in the Plan View webmap

Website

The project website serves as a "portal" into the IMS for most users (Figure 13). The website contains links to all web-accessible content in the GIS, tools for online document management, linked tables to the database, online reports, and access to files hosted on the FTP site.

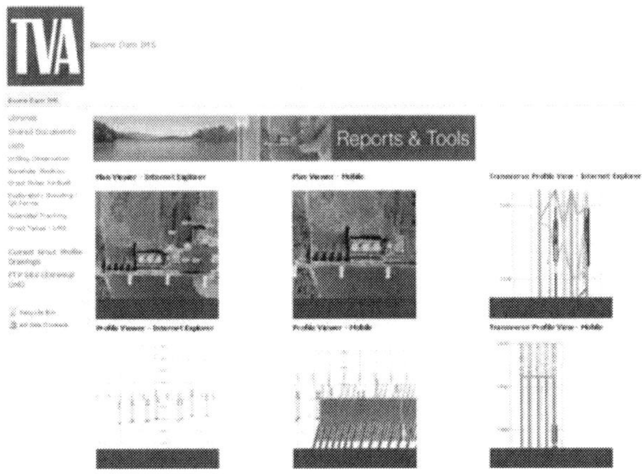

Figure 13. The IMS website landing page

GIS Tools.
With the exception of the detailed 3D GIS model, all GIS content is available from the website. This includes links to the plan, profile, and transverse profile view webmaps. Similarly, the simplified web accessible 3D GIS model is available from the website.

Tabular Grout Data. The website includes reports and spreadsheet-style tables that allow interactive access to data in the database (Figure 14). These tables, for example, return calculated data from the database (i.e., tabular calculated grout takes). These tables can be sorted, filtered, or exported to desktop file formats (e.g., Microsoft Excel) easily for offline analysis.

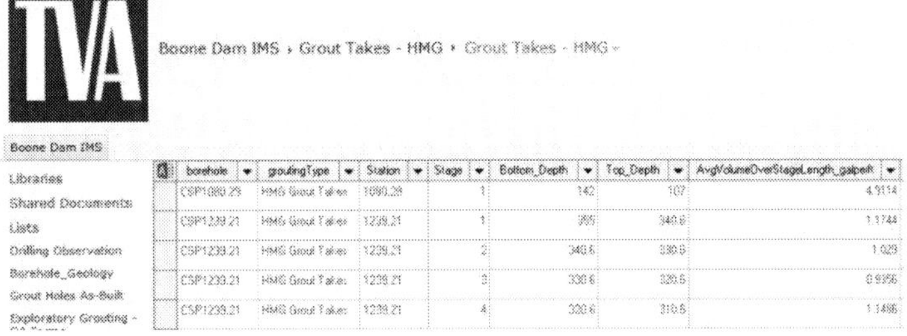

borehole	groutingType	Station	Stage	Bottom_Depth	Top_Depth	AvgVolumeOverStageLength_galperft
CSP1090.29	HMG Grout Takes	1090.29	1	142	107	4.9114
CSP1239.21	HMG Grout Takes	1239.21	1	355	340.6	1.1744
CSP1239.21	HMG Grout Takes	1239.21	2	340.6	330.6	1.029
CSP1239.21	HMG Grout Takes	1239.21	3	330.6	320.6	0.9356
CSP1239.21	HMG Grout Takes	1239.21	4	320.6	310.6	1.1486

Figure 14. Tabular grout takes data as viewed from the website

Graphical Grout Trend Data. Time series graphs of data collected from the grout monitoring system (GroutIT) are available from the website (Figure 15). These graphs show the volume, flowrate, gauge pressure, effective pressure and the maximum target pressure of grout injected for any given stage as a function of the measurement frequency. These reports are available from the GIS and tabular grout data via hyperlinks, or can be generated "on-demand".

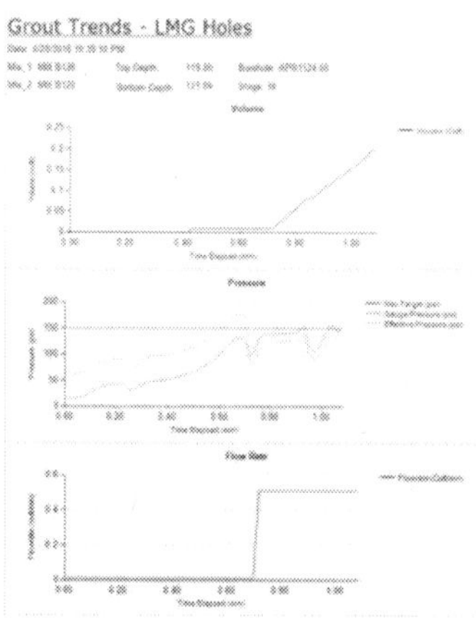

Figure 15. The grout trends report as it appears on the website

Data Entry

Several tools are deployed to allow field personnel to enter data into the database. These include web-based tools which allow direct data entry into database tables (Figure 16) and field tools which compile data into text files which are uploaded to the database via the FTP site (Figure 17).

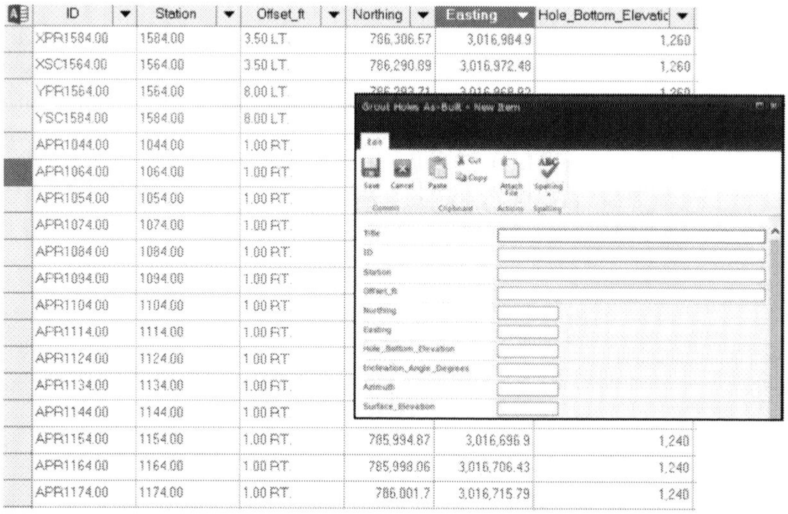

Figure 16. Web-based Spreadsheet with form view for grout hole data entry

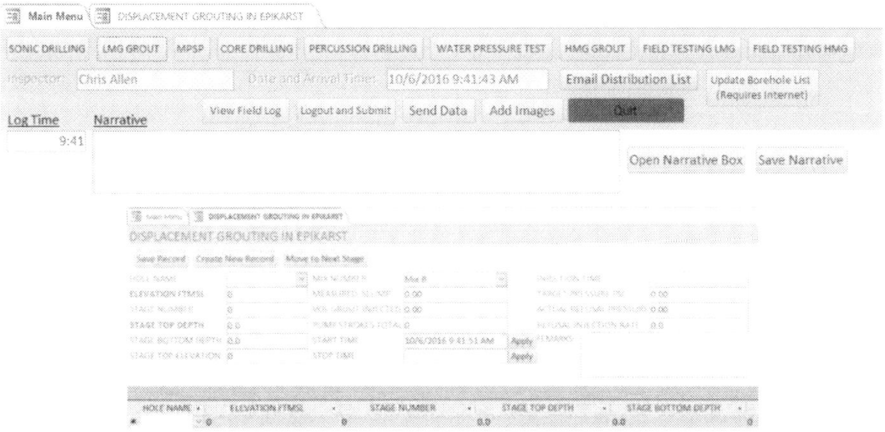

Figure 17. Field Tool for logging of Quality Assurance Data

BOONE DAM INSTRUMENTATION DATA ACQUISITION SYSTEM

Introduction

The IDAS was developed to manage, analyze, annotate, and report real-time geotechnical instrumentation data using commercially off the shelf software (COTS) that was configured and customized for project needs. The primary COTS software used include Eagle.IO, Microsoft SQL Server, and a variety of ESRI software, including ArcGIS Server, ArcGIS GeoEvent Processor, and ArcGIS Online. The system allows users to view these data in intuitive interactive reporting tools with a variety of display options, and that include temporal, spatial, and tabular views. The system also generates alarms through email and text messages based on defined alarm thresholds. Email notifications for threshold exceedances can be set up on a per user and per data stream basis or a data source basis.

The dam safety instrumentation network at Boone Dam includes a wide variety of instruments, including:

Instrument Type	Quantity	Location	Reading Frequency
Automated Total Stations (ATS)	2 ATS and 86 prisms	Upstream and downstream earthen embankments	30 minutes (min)
Piezometers	172	Various depths along earthen embankment	1 min and 5 min, depending on location
Inclinometers	7	Along earthen embankment	3 daily and 4 weekly
Flow meters	4	Gallery drain and embankment drain	60 min and 1 min respectively
Crack meters	5	Turbine pits and generator floor	60 min
Growth meters	6	Left abutment of spillway	2 min
Uplift Cells	17	Below powerhouse and spillway	60 min
Convergence meters	5	Spillway bay	2 min
Joint meters	6	Upstream spillway deck	2 min
Water Quality	3	Upstream and downstream of dam	5 min

Data Transfer

Data from this network is transmitted electronically to both an FTP site for archiving and processing, as well as Eagle.IO, an online data management platform designed to rapidly deliver real-time data. The automated instrumentation data consists primarily of *.dat files, from the dam's Campbell Scientific based instrumentation hardware. The IDAS also incorporated historical data provided by TVA, which is an export of Boone Instrumentation data from TVA's Canary System.

Functionality

The key functionality and differentiator provided by the IDAS is the rapid processing and visualization of water levels from hundreds of sensors during grouting operations using customized dashboards. These dashboards can load very rapidly, on the order of seconds, allowing users to quickly make decisions regarding site operations during investigation and remediation.

Dashboards were prepared for each set of instruments, but are also routinely updated and modified to meet ongoing project operations with an easy to use, drag and drop user interface that can be modified "on the fly" without significant programming experience on the part of the users. Importantly, because these dashboards are accessible via the web, they do not require specialized software installations and these dashboards can be accessed using mobile devices, such as smart phones and tablets.

A variety of dashboards are provided below (Figures 18 – 21):

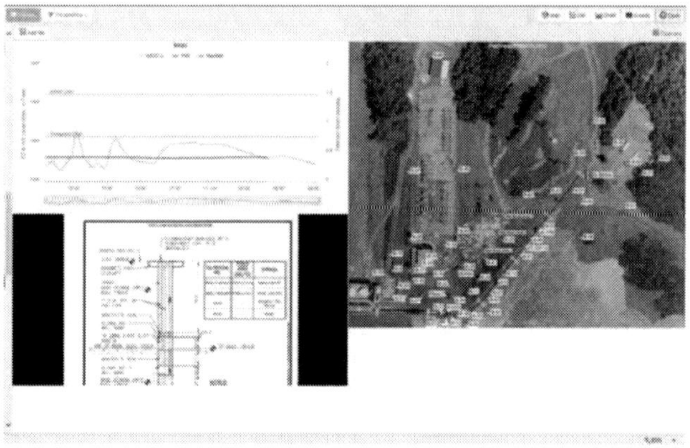

Figure 18. Piezometers, which includes plan view map and instrument detail

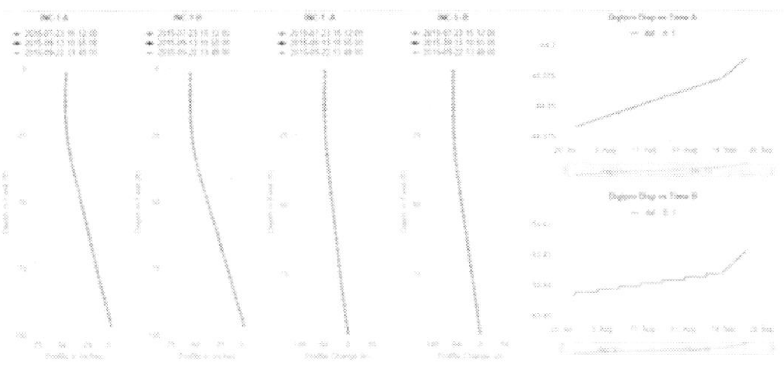

Figure 19. Inclinometer Charting

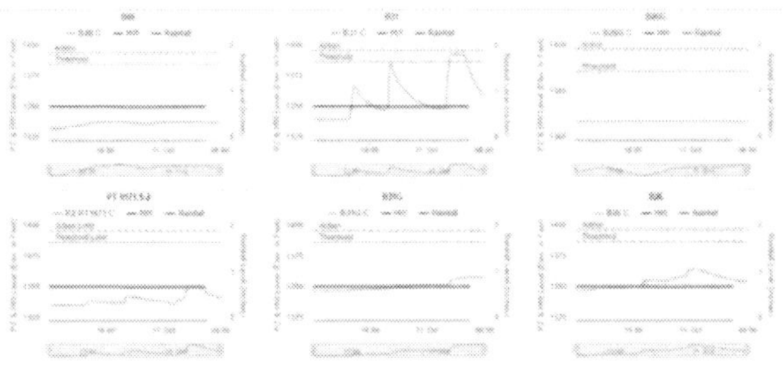

Figure 20. Piezometer dashboards that allows the monitoring of multiple piezometers near a grout injection

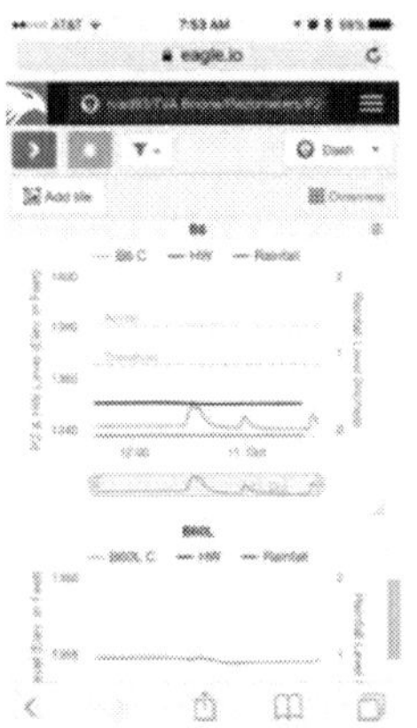

Figure 21. Mobile Access

In addition to the display of real-time data using charts, instrumentation data is also made accessible using an interactive web application, the ArcGIS™ Online IDAS Web App. The web app, a secure online portal to the data developed using ESRI®'s Web AppBuilder, consists of a variety of tools that are typical of an online GIS application. It includes a layer list control, which allows users to turn map layers on and off, such as historical boring layouts, contours, and sinkhole investigation layers. The web app is continually updated as new data are received from the site. Electronic data is posted to a FTP site, where it is automatically downloaded on a continuous basis.

The IDAS web app provides access to the underlying real-time instrumentation data. Clicking on an instrument location will generate a popup window that provides the current data available for instruments at that location, as well as a hyperlink to a time trend graph of recent data.

Data Use and Re-Use

There is a significant benefit to storing instrumentation in an organized and accessible way using industry standard data management software. Because the instrumentation data is stored in a MS SQL Server database, the data is available for a wide variety of downstream analysis and reporting tasks, including integration with IMS (real-time layers in Viewer). The IMS is able to re-use data stored and managed by the IDAS without further data management burdens.

Challenges and Advances

One challenge in the development of this system was in the selection of COTS components and site-specific integration of those components in the minimum allowable time. Considerable time and expense can be spent on software development and customization rather than using and configuring existing available software. Overcoming this challenge on this site required careful planning, testing, and frequent communication of both short term and long term strategic goals by the project team.

TVA's existing Boone Dam instrumentation monitoring program, as is the case for many dams, was meant for long-term monitoring in support of stability analysis and not necessarily for monitoring of construction activities. The use of this IMS and IDAS at Boone Dam brings two advancements or, at a minimum, reinforces the following technological trends in the industry including:

- integration and management of construction and instrumentation in a single framework (i.e., integration between the IDAS and IMS), and
- rapid delivery of real-time instrumentation data in seconds as opposed to minutes.

The combination of the IMS and IDAS is viewed as a cost-effective solution that enhances the existing long-term monitoring solution used at the site.

CONCLUSIONS

The remedial activities at Boone Dam generate a significant amount and variety of data and thus require sophisticated systems to manage and assess these data. The IMS and IDAS allow these data to be rapidly organized, analyzed, visualized, and made accessible to TVA and its contractors and are being used to guide and manage site operations.

MODELING DAM BREACHES AND FLOOD INUNDATION USING HEC-RAS 5.0'S NEW 2D CAPABILITIES

D.L. Silvertooth, P.E., CFM [1]
M. Meshkat, Ph.D., P.E., CFM [2]

ABSTRACT

The U.S. Army Corps of Engineers owns and operates over 700 dams. Risk assessments are conducted periodically to understand the potential downstream impacts and to inform management strategies, if dam failure were to occur. Accurately simulating breaches, inundation depths, flooding extents, and arrival times are critical, especially in heavily urbanized and flat alluvial floodplains where risk is high and flow paths may vary. Multiple one-dimensional (1D) and two-dimensional (2D) models have been used for evaluating breach scenarios and flood inundation, but are typically either standalone 1D models or 2D models that evaluate the breach scenarios external to the simulation. The Hydrologic Engineering Center's River Analysis System (HEC-RAS) version 5.0, which was publically released in March 2016, is capable of simulating breaches and both 1D and 2D unsteady flow calculations, including combinations of 1D and 2D. The purpose of this paper is to compare the varying computational modeling techniques with HEC-RAS to simulate dam breaches and the downstream impacts of potential dam failure scenarios for two dams of similar characteristics.

INTRODUCTION

The purpose of this study is to compare modeling techniques to simulate dam breaches and determine the downstream impacts of potential dam failure scenarios for two dams in highly urbanized areas and flat alluvial floodplains.

The U.S. Army Corps of Engineers (Corps) owns and operates over 700 dams and there are countless additional structures throughout the country that are either privately owned or within other jurisdictions. As these structures age and development increases downstream, the need to understand structural vulnerabilities and potential risk becomes increasingly important, which requires better tools and techniques to evaluate the consequences. The Corps and other entities such as the Bureau of Reclamation and the Federal Energy Regulatory Commission have developed risk-informed approaches to assessing dams and are currently evaluating the structures within each of their respective portfolios (Boyer, 2016; Scott, 2011; USACE, 2014).

The Hydrologic Engineering Center's River Analysis System (HEC-RAS) is a widely used hydraulic modeling software that is frequently utilized in risk assessments and has been able to perform dam and levee breaching analysis since the release of HEC-RAS

[1] Senior Hydraulic Engineer, Los Angeles District, U.S. Army Corps of Engineers, 915 Wilshire Blvd., Suite 930; Los Angeles, CA 90017; *213.452.3569*; david.l.silvertooth@usace.army.mil
[2] Senior Engineer, Amec Foster Wheeler Environment & Infrastructure, Inc., 3800 Ezell Road, Suite 100, Nashville, TN 37211; *615.333.0630*; masoud.meshkat@amecfw.com

version 3.1 in November 2002 (Brunner, 2003). HEC-RAS version 3.1 utilized 1D unsteady flow computations available in previous releases of HEC-RAS (Brunner et al., 2000).

The highly urbanized and flat alluvial floodplains downstream of many of the Nation's structures, including the dams evaluated in this study, add complexity to understanding and quantifying the risk of flooding. Therefore, tools such as multidimensional numeric models are critical to determine flow paths, inundation extents, and the risk to life and property. Until recently, many dam breach studies that required multidimensional analysis conducted the breaches using 1D HEC-RAS then utilized the resulting breach hydrographs to conduct 2D simulations with other available software (Ackerman et al., 2008).

The HEC-RAS version 5.0 was publically released in March 2016 and has the ability to compute 2D unsteady flow calculations. The program uses an Implicit Finite Volume algorithm and is able to solve either the full 2D Saint Venant equations or the 2D Diffusion Wave equations. This version of HEC-RAS maintains all previous features and is able to have combined 1D and 2D models, affording the potential to have various reaches or portions of the study domain analyzed either using 1D or 2D calculations based upon the expected flow conditions (Brunner et al., 2015).

The number of dams undergoing risk assessments require efficient and effective means of analyzing failure scenarios. This often includes utilizing existing models and previously conducted analyses in an effort to streamline the evaluation process, prevent duplication of efforts, and avoid unnecessary use of limited funding resources. The capability of HEC-RAS version 5.0 to combine 1D and 2D model elements allows for this efficiency, while also providing the capability to model complex hydraulic conditions that require multidimensional computations. This study provides a demonstration of the two modeling approaches, a 1D and 2D combined model versus a 2D standalone model, and the comparisons of their applications that can be used as a case study for future risk assessments.

ALTERNATIVE MODELING APPROACH COMPARISON

Evaluation Tool

The HEC-RAS version 5.0 was used to route flood hydrographs and simulate failure scenarios for both dams in this study. Prior to simulating the failure and downstream inundation, four breach methods were evaluated externally from the modeling effort to determine the breach parameters. The dam failure analyses were simulated under three hydrologic loading conditions: 1) a Security Scenario Pool, 2) Top of Active Storage Pool, and 3) the Maximum High Pool. The Security Scenario Pool condition reflects a reservoir pool elevation with a 1% Annual Chance Exceedance (ACE) frequency. The Top of Active Storage Pool condition reflects a reservoir pool elevation equal to the spillway crest elevation. The Maximum High Pool condition reflects the reservoir design pool elevation, where the potential breach represents a worst case dam failure scenario.

Study Comparison

One dam, which from hereon will be referred to as Dam 1, was modeled using a combined 1D and 2D model. This model utilized an existing 1D model of the channel downstream of Dam 1, which was incorporated into the combined 1D and 2D model. The combined 1D and 2D unsteady flow model employs the 1D model elements for the dam embankment and downstream prismatic channel system, while modeling the overbank floodplains with 2D computational gridded mesh. The banks of the downstream channel are modeled as lateral structures, which serve as the connections between the 1D and 2D components of the model. The second dam, which from hereon will be referred to as Dam 2, was modeled using an entirely 2D computational gridded mesh. The 2D mesh for Dam 2 included the downstream prismatic channels and the overbank floodplains. The dams varied in size and storage capacity, but both are located upstream of flat alluvial floodplains and are within heavily urbanized environments with potentially significant consequences downstream of the structures.

Dam 1

Dam 1 was built in the mid-twentieth century and is an approximately 1,300-foot long, 50-foot high, earth-fill embankment with an impervious earth core. The crest width is approximately 20 feet. The upstream and downstream face slopes are both 1V on 2H from the toe to the crest and are covered with a 3-foot layer of loose rock to protect against erosion. The structure includes an uncontrolled, 110-foot wide broad-crested reinforced concrete spillway at the left abutment of the dam and a 60-inch diameter reinforced concrete gated outlet, which discharges into the spillway channel downstream of the dam. The spillway extends approximately 90 feet upstream and about 470 feet downstream from the axis of the dam. The spillway varies in width, converging from a width of 110 feet at the axis of the dam to a width of 30 feet at the downstream extent, where it transitions to join the downstream channel. The reservoir design storage capacity, expected during a Maximum High Pool scenario, is approximately 1,200 acre-feet (AF). The downstream channel is a rectangular concrete lined open channel and was constructed after the construction of the dam. The channel discharge capacities and configurations vary throughout the channel system, but remain in the order of magnitude of 10,000 cubic feet per second (cfs) within the consequence zone.

Dam 2

Dam 2 was built in the mid-twentieth century and is an approximately 10,500-foot long, 100-foot high, earth-fill embankment with an impervious earth core. The axis of the dam flows a gentle curve in order to connect the abutments of the dam with a prominent rock outcrop located near the center of the dam. The crest width is approximately 30 feet. The upstream face of the dam has a slope of 1V on 3H and is covered with a 2 feet 6-inch layer of riprap over a 6-inch spall blanket. The downstream face has a varying slope consisting of 1V on 6H for the bottom third of the embankment from a rock toe, a slope of 1V on 5H for the middle third, and a slope of 1V on 3H for the upper third to the dam crest. The downstream face is not protected with stone. An approximately 300-foot wide ogee spillway structure, with six 3-foot wide equally spaced piers, is located near the center of the dam on a prominent rock outcrop just west of the downstream channel. The outlet conduits, eight gated and two ungated, penetrate the overflow spillway section, located symmetrically with respect to the spillway centerline and aligned to discharge into the downstream channel. The gated conduits are located in the center of the outlet section in two groups of four. The reservoir design storage capacity, expected during a Maximum High Pool scenario, is approximately 53,000 AF. The downstream channel is a rectangular concrete lined open channel and was constructed after the construction of the dam. The channel discharge capacities and configurations vary throughout the channel system, ranging from 10,000 to 100,000 cfs within the consequence zone.

BREACH SCENARIOS

Breach conditions were evaluated prior to beginning the modeling effort and were conducted external to HEC-RAS. The calculated breach parameters are a function of geometric characteristics, geotechnical properties, expected hydrologic loading conditions, and the operations of the dams. Breaches resulting from piping failure through the main embankment were assumed to occur in conditions where the water level was below the top of dam elevation. Breaches resulting from overtopping failure through the main embankment were assumed to occur in conditions where the water level was above the top of dam elevation. Pertinent data for the breach scenarios were compiled from as-builts, design memorandums, and operations manuals.

Breach Equations

Four breach parameter methods were evaluated for both dams to determine the most appropriate failure progression. MacDonald and Langridge-Monopolis (Equations 1 through 3), Von Thun & Gillette (Equations 4 and 5), and two variations of Froehlich from 1995 (Equations 6 and 7) and 2008 (Equations 8 and 9) were evaluated externally to HEC-RAS. All regressions were derived using data sets from multiple earthen dams with cores similar to the study's dams (Froehlich, 1995; Froehlich, 2008; MacDonald and Langridge-Monopolis, 1984; Von Thun and Gillette, 1990). Previous studies have compared these breach parameter methods and applications to dam safety studies using HEC-RAS (Gee, 2009).

$$V_{eroded} = 0.0261\ (V_{out} * h_w)^{0.769} \tag{1}$$

$$t_f = 0.0179\ (V_{eroded})^{0.364} \tag{2}$$

$$W_b = \frac{V_{eroded} - h_b^2(CZ_b + h_b Z_b Z_3/3)}{h_b(C + h_b Z_b/2)} \tag{3}$$

$$B_{ave} = 2.5\ \mathrm{h}_w + C_b \tag{4}$$

$$t_f = 0.015\ h_w \tag{5}$$

$$B_{ave} = 0.1803\ (K_o\ V_w^{0.32}\ h_b^{0.19}) \tag{6}$$

$$t_f = 0.00254\ (V_w^{0.53}\ h_b^{-0.90}) \tag{7}$$

$$B_{ave} = 0.27\ (K_o\ V_w^{0.32}\ h_b^{0.04}) \tag{8}$$

$$t_f = 63.2\left(\frac{V_w}{h_b^2}\right)^{0.5} \tag{9}$$

Where;

B_{ave}	=	Average Breach Width
C	=	Crest width of the top of the dam
C_B	=	Von Thun & Gillette Coefficient, which is a function of the reservoir size
g	=	Gravitational acceleration
h_b	=	Height from the top of the dam to the bottom of the breach
h_w	=	Depth of water above the bottom of the breach
K_o	=	Constant (1.4 for overtopping failure, 1.0 for piping)
t_f	=	Breach formation time
V_{eroded}	=	Volume of material eroded from the dam embankment
V_{out}	=	Volume of water that passes through the breach
V_w	=	Reservoir volume at time of the failure
W_b	=	Bottom width of the breach
Z_1	=	Average slope of upstream face of dam
Z_2	=	Average slope of upstream face of dam
Z_3	=	$Z_1 + Z_2$
Z_b	=	Side slopes of breach, 0.5 for the McDonald method

All regression equations provided breach parameters that yielded similar max water surface elevation when comparing the Maximum High Pool scenario. However, both dams ultimately utilized the 2008 Froehlich regression equation to estimate average breach size and development time. The 2008 Froehlich regression equations yielded breach width and formation times within the range of the minimum and maximum values of the respective beach parameters also derived from the MacDonald-and-Langridge-Monopolis method and Von-Thun-and-Gillette method. The Froehlich (2008) regression equation also resulted in a slightly higher water surface elevation directly downstream from the dams and was considered to be the most conservative breach regression equation.

Hydrologic Loading Conditions

Both dams are located in a Mediterranean climate region with warm summers and mild winters. Normal annual precipitation ranges from about 12 inches in the lowlands to more than 40 inches in the upper portions of the watersheds. Both dams operate as dry reservoirs.

Starting reservoir pool elevations, inflow conditions into the reservoir, and tributary flow into the modeled reaches downstream of the dams, can significantly impact consequence results. These inputs were determined using the best available hydrologic data, which included historic stream flow, reservoir pool elevation records, the inflow design floods reflecting the calculated probable maximum flood (PMF), and facility operation information.

Exceedance duration elevations were developed and calculated for the three loading conditions using the Hydrologic Engineering Center's Data Storage System (HEC-DSS). Hydrologic data from the past 40 years were used in the analysis. Inflow and outflow conditions for both dams are constant and represent a typical discharge from the dams for each loading condition.

The dam failure analyses were simulated under three hydrologic loading conditions: 1) a Security Scenario Pool (1% ACE pool elevation), 2) Top of Active Storage Pool (spillway crest elevation), and 3) the Maximum High Pool (design pool elevation).

MODEL DEVELOPMENT

Unique HEC-RAS models were developed for each dam and downstream conditions. The best available data were used to develop the geometries and flow files associated with each of the models.

Both dams utilized existing 1D in-channel only models that reflected current conditions downstream of the dams. Existing models were utilized either in the final breach model, in the development of the terrain, and/or in the development of the 2D computational mesh.

Dam 1 is significantly smaller than Dam 2 in both embankment size and storage capacity. Due to the smaller size of the embankment, the expected breach locations, and the resulting breach discharges, Dam 1 incorporated the existing 1D cross-sections into the breach model. The 1D cross-sections were also used to create a unique terrain for the model within RAS Mapper. The Dam 1 breach model added 2D overbank flow areas to simulate breakout flows on either side of the downstream prismatic channel, creating a combined 1D and 2D model. The banks of the downstream channel are modeled as lateral structures, which serve as the connections between the 1D and 2D components of the model. Volumetric flow rates resulting from the breach scenarios for Dam 1 were within the same order of magnitude as the channel conveyance capacity downstream and the configuration of the engineered channel was expected to distribute breach flows uniformly over the channel banks, which was accurately reflected by the lateral structures in the model.

Dam 2 is larger than Dam 1 in both embankment geometry and storage capacity; therefore, the resulting breach flow volumes were greater. The breach discharges would significantly overwhelm the downstream channel capacity and conveyance of flow to the overbanks via lateral structures was determined inappropriate; therefore, a 2D flow area was developed for the entire model domain downstream of the dam. The existing 1D model of the channel downstream of the dam was utilized in the development of the terrain.

Terrain

Dams 1 and 2

HEC-RAS has the ability to support the use of a digital terrain model (DTM) for representing a surface of a model domain. HEC-RAS also has the ability to generate unique terrains from multiple surfaces or existing 1D cross-sections.

Both dam models utilized existing 1D in-channel cross-section models that reflected as-built geometry conditions of the respective downstream channel systems. These 1D cross-sections were merged with a 3-meter Interferometric Synthetic Aperture Radar (IFSAR) Digital Elevation Model (DEM) to generate unique terrains for both model domains.

Storage Areas

Dams 1 and 2

Storage areas were used to model the reservoir pool for both Dam 1 and Dam 2. Elevation-storage relationships were assigned based upon design memoranda and best available reservoir data. The storage areas were connected to either the Inline Structure or 2D Area Connection feature representing the dam in each of the models.

Inline Structures and Storage Area/2D Area Connections

The Inline Structure and Storage Area/2D Area Connection tools within HEC-RAS were used interchangeably to reflect the dam embankments and routing operations. These tools were also used to simulate the breach progression and flow routing through the breach. All structure characteristics, including the geometry and flow operations were obtained from as-built and current operational data. All breach parameters were developed and are outlined in the Breach Scenarios discussion.

Dam 1

The Inline Structure used in the Dam 1 model reflects the dam embankment and was connected directly to the 1D in-channel cross-sections immediately downstream of the dam. The Inline Structure reflected as-built drawings data for the dam geometry, spillway and gate structures. This connection simulated the breach, then calculated and communicated the breach flows to the downstream 1D cross-sections.

Dam 2

The 2D Area Connection used in the Dam 2 model reflected the dam embankment and was connected directly to the 2D flow area immediately downstream of the dam. This connection simulated the breach, then calculated and communicated breach flows to the downstream 2D flow area.

1D Cross-Sections

Dams 1 and 2

1D cross-sections were only used in the Dam 1 model simulation. 1D cross-sections were utilized in the development of the Dam 2 model terrain, but were not used in the final simulation model.

The Dam 1 cross-sections were imported from an existing calibrated model of the channel system and reflected as-built conditions. Cross-sections were located throughout the system to ensure geometric characteristics of the channel were reflected, including changes in geometry such as expansions and contractions. As noted within the Bridge Section, bridges within the Dam 1 model were also modeled using 1D cross-sections. Instabilities associated with unsteady-state analysis, especially dam breach modeling, can frequently be attributed to the simulation time-step and inappropriately spaced cross-sections. The Courant condition (Equation 10) was used to ensure spacing intervals, as well as the computational time-step, were suited to the routing of the flood wave (Brunner, 2003).

$$C_r = V_w\left(\frac{\Delta t}{\Delta x}\right) \leq 1.0 \tag{10}$$

Where;

C_r	=	Courant Number
V_w	=	Wave Speed
Δt	=	Computational time-step
Δx	=	Distance between cross-sections

2D Computational Gridded Meshes

Two-dimensional hydraulic modeling with HEC-RAS is based on a computational grid (mesh) that preprocesses hydraulic property tables from the underlying terrain. This retains the detail of the terrain instead of averaging the terrain into one data point per cell.

Selecting the grid cell size for a 2D flow area is an iterative process that depends on the underlying terrain, velocity of flow, and the spatial extents of the model domain.

Dam 1

The model developed for Dam 1 utilized in-channel only cross-sections reflecting the as-built geometry of the downstream channel and consists of two 2D flow areas representing the left and right overbank floodplains on either side of the downstream channel. The 2D computational meshes reflected the IFSAR terrain data. For the Dam 1 model 50-foot x 50-foot, 100-foot x 100-foot, 200-foot x 200-foot, and 500-foot x 500-foot rectangular computational grids were tested. The 100-foot and larger grids were determined to be inappropriate due to the unstable and oscillating velocities when run with an appropriate time-step. The 50-foot grid was selected over the larger grids due to locations of steeper slopes of the terrain and rapidly changing water surface elevations. In addition, the smaller grid cell size more accurately captures high ground, reducing the number of break lines in the model.

The break lines tool within HEC-RAS was used to refine the 2D flow areas. Break lines are added to computational meshes in order to align the cell edges with high ground and to represent significant features in the floodplain such as berms, roadways, and channel features. Without break lines flows may cross a high ground barrier or exit the channel prematurely. ArcGIS was used to develop a shapefile of the break lines which was then added to the HEC-RAS model.

While the grid cell size was set to 50-foot x 50-foot (structured square grids), some grid cells are unstructured (not square) and have less than 50-foot x 50-foot grid cell size. There are approximately one hundred thousand grid cells in the Dam 1 computational mesh, encompassing approximately 13 square miles. A computational time-step of 5 seconds was used to meet the Courant condition.

Dam 2

The Dam 2 model consists of one 2D flow area encompassing the entire downstream domain. The upstream boundary is just downstream of Dam 2 and the downstream boundary is at the Pacific Ocean in order to include all areas that could be inundated. The 2D flow area also includes coincident tributary flows from 2 tributaries. For the Dam 2 model 50-foot x 50-foot, 100-foot x 100-foot, 200-foot x 200-foot, and 500-foot x 500-foot rectangular computational grids were initially tested. The 200-foot and larger grids were ruled out because the 200-foot grid produced unstable, oscillating velocities when run with an appropriate time-step. The 100-foot grid was selected over the 50-foot grid due to the domain size, which was significantly larger than that of Dam 1, while also accommodating any steep slopes within the terrain and instabilities due to rapidly changing water surface elevations. In addition, the smaller grid cell size more accurately captures high ground, reducing the number of break lines in the model. While the grid cell size was set to 100-foot x 100-foot (structured square grids), some grid cells are unstructured (not square) and have less than 100-foot x 100-ft grid cell size. Break lines were used extensively in the upper reaches of the 2D domain to ensure proper flow paths immediately downstream of the structure. There are approximately five hundred thousand grid cells in the computational mesh, encompassing approximately 370 square miles. A computational time-step of 30 seconds was used to meet the Courant condition.

Roughness Coefficients

Dams 1 and 2

Manning's roughness coefficients (n-values) were used to estimate friction losses in 1D and 2D hydraulic calculations. The HEC-RAS software allows the modeler to spatially vary n-values across the width of each 1D cross-section and throughout the 2D flow areas.

The Manning's n-values for the channel downstream of both dams were estimated from aerial photography and as-built information. The channels immediately downstream are both rectangular concrete lined open channels so the n-values reflected a relatively smooth surface of 0.014. Based upon the uniform channel material one Manning's n-value was estimated to reflect the entire channel system. The overbank Manning's n-values within the 2D flow areas were assigned with the aid of the National Land Cover Dataset (NLCD) and were verified using landuse information obtained from aerial photography. The NLCD provides nationwide data on land cover at the 30-meter resolution. The NLCD assigns all land cover into one of 16 categories; therefore, it is less precise than digitizing terrains manually, but accounts for sufficient variation in the terrain. The overbank areas for both dams consisted of heavily developed urban conditions; therefore, given the size of the study areas, and the relative consistency of the terrain throughout the watershed, the NLCD data was used for determining the spatial extents of the n-values in both models' 2D flow areas. Manning's n-values were increased immediately downstream of the dam to account for accumulation of debris and turbulence created during the breach.

Bridges

Multiple bridges exist downstream of both dams; however, bridges were only included if they were determined to have a significant impact on flow characteristics resulting from the breach scenarios. Embankments, roadways, and bridges that were within the floodplain were reflected in the terrain of the 2D computational meshes.

Dam 1

The model developed for Dam 1 utilized in-channel only cross-sections reflecting the as-built geometry of the downstream channel. The in-channel geometry included over 30 bridges, which were modeled using blocked obstructions and lidded cross-sections. The bridges were modeled using blocked obstructions and lidded cross-sections due to the model instabilities that resulted from the supercritical flow regime and high velocity flows. The blocked obstructions and lidded cross-sections were included to ensure that in-channel hydraulics accurately captures effects of the structures.

Dam 2

The Dam 2 model did not include bridges, culverts, or in-channel hydraulic structures. The location of the breaches and the large volumetric flow rates associated with the failure scenarios were unlikely to be significantly impacted by the in-channel structures; therefore, the structures were not included.

Lateral Structure

Lateral structures define the location and extent of natural channel banks and man-made structures where water can flow over either into overbank areas or back into the channel. Lateral structures have been available tools within HEC-RAS for several versions and are not unique to HEC-RAS version 5.0 (Jensen et al, 2000). However, within HEC-RAS version 5.0, the lateral structures can now also serve as the connection between 1D cross-section and an adjacent 2D flow area. Flow exceeding the elevation of the lateral structure (channel bank height) is calculated when exiting the channel system and entering the 2D flow area with the computed headwater in the 1D channel and the tailwater in the 2D flow area. Dependent upon the flow conditions in the 1D channel and 2D overbank areas, these calculations can also be made to reflect flows reentering the channel from the overbanks areas.

Dams 1 and 2

Lateral structures were used in the Dam 1 model to reflect the channel banks and connect the 1D portion of the model to the 2D flow areas. Station-elevation data, which defines the lateral structure's geometric profile, was extracted from the known as-built bank elevations. Access and maintenance roads that are approximately 20 feet wide parallel the channel downstream of Dam 1 so a weir width of 20 feet was applied to the lateral structures. Levees do not exist along the downstream channel; however, in some locations the banks do slope toward the adjacent floodplain. Weir coefficients for the lateral structures varies from 0.5 to 2.0, depending upon the geometric configurations of the top of channel, banks, adjacent terrain, and the flow characteristics entering and exiting the system over the lateral structures. Multiple lateral structures were developed to reflect the right and left banks. The lateral structure lengths were adjusted to ensure appropriate distribution of flows and stability of the model. The flat alluvial floodplain and resulting relatively flat lateral structure profile slope caused instabilities when using structures that reflected the entire channel length. As a result, lateral structures were limited to approximately 1,000 feet in length to prevent large volumes of flows from rapidly entering or exiting the system and computationally "shocking" the simulation. Levee breach analysis was not incorporated into this study and all elevated banks were assumed to be rigid and remain in-place in the event of overtopping.

Lateral structures were not used in the Dam 2 model because it was an entirely 2D model.

Tributaries

Dams 1 and 2

Tributaries which experience greater than 10 miles of backwater flooding are typically modeled as separate reaches in dam breach studies. Both Dam 1 and Dam 2 discharge into engineered channels designed for supercritical flow conditions; therefore, backwater is not expected. Nonetheless, tributary flows were added to both models to reflect approximately 50% channel capacity flows, in an effort to simulate storm conditions without influencing the volume of damaging breakout flows.

Downstream Boundary Conditions

Dams 1 and 2

The downstream boundary conditions used for both models was normal depth that was derived from preliminary non-breach scenario simulations of the model and is a function of channel geometry and flow characteristics. Both dams are located in a coastal basin; however, due to the proximity to the ocean, the influence of the tidal elevation is not expected to significantly impact flood inundation extents or depths and was determined to be unnecessary for the evaluation.

RESULTS

Both models and approaches yielded reliable results and valuable information for on-going risk assessments, while demonstrating the ability of HEC-RAS version 5.0 to appropriately utilize existing 1D models for breach and inundation analyses based upon the structural, floodplain, and flow conditions.

Due to the sensitivity of these analyses and the consequences associated with the safety of life and property, specific breach and inundation results are not available for public release at this time.

The results of this study and similar investigations estimate consequences and are used as prioritization tools that support decisions regarding additional analyses and detailed studies. In the case of the over 700 Corps owned and operated dams, those assets identified as high-consequence facilities through initial consequence assessments can be assigned higher priority for conducting detailed flood inundation studies or detailed risk assessments. The consequence analyses can effectively inform decision-makers about those facilities within the area that should receive particular attention from the emergency management community because of their potential for significant impacts at the local and regional levels.

CONCLUSION

As the U.S. dam infrastructure continues to age, the Corps and other entities, public and private alike, need tools to properly assess the risk of structures and accurately evaluate downstream consequences. The HEC-RAS version 5.0 is a widely used hydraulic modeling software that is capable of 1D and 2D computations and dam breach simulations that are needed for such risk assessments. This study exemplifies the versatility of the software based upon the structural and floodplain conditions, downstream channel configurations, and expected breach scenarios. Moreover, the study demonstrates the ability to utilize and combine existing 1D models with 2D flow areas to effectively and efficiently simulate breaches and downstream inundation required to properly identify consequences in highly urbanized and flat alluvial floodplains.

ACKNOWLEDGEMENTS

A number of people guided, contributed to, and reviewed the study. The authors would like to offer their appreciation to all of them for their valuable support and insights during this effort, especially Victor J. Andreas, P.E., Brian Dela-Barre, Ph.D., P.E, and Simon C. Evans, P.E., CFM.

REFERENCES

Ackerman, C., Fleming, M., and Brunner, G. (2008). Hydrologic and Hydraulic Models for Performing Dam Break Studies. *World Environmental and Water Resources Congress 2008*: pp. 1-11.

Boyer, D., and Allerton, W. (2016). Implementing a Risk-Informed Approach to Dam Safety in a Regulatory Environment. *United States Society of Dams 2016.*

Brunner, G., Jensen, M., and Piper, S. (2000). Unsteady Flow Routing with the Hydrologic Engineering Center River Analysis System (HEC-RAS). *Building Partnerships*: pp. 1-7.

Brunner, G. (2003). Dam and Levee Breaching With HEC-RAS. *World Environmental and Water Resources Congress 2003.*

Brunner, G., Piper, S., Jensen, M., and *Chacon, B. (2015). Combined 1D and 2D Hydraulic Modeling within HEC-RAS. World Environmental and Water Resources Congress 2015*: pp. 1432-1443.

Froehlich, D. C. (1995). Embankment Dam Breach Parameters Revisited. *Water Resources Engineering, Proc. 1995 ASCE Conf. on Water Resources Engineering, New York*: pp. 887-891.

Froehlich, D. (2008). Embankment Dam Breach Parameters and Their Uncertainties. *J. Hydraul. Eng.*: pp. 1708-1721.

Gee, D. (2009). Comparison of Dam Breach Parameter Estimators. *World Environmental and Water Resources Congress 2009*: pp. 1-10.

Jensen, M., Piper, S., and Brunner, G. (2000) Lateral Weir and Split Flow Optimization with the Hydrologic Engineering Center's River Analysis System (HEC-RAS). *Building Partnerships*: pp. 1-7.

MacDonald, T. and Langridge-Monopolis, J. (1984). Breaching Characteristics of Dam Failures. *J. Hydraul. Eng.*, pp. 567-586.

Scott, G. (2011). The Practical Application of Risk Assessment to Dam Safety. *GeoRisk 2011*: pp. 129-168.

U.S. Army Corps of Engineers, "Safety of Dams – Policy and Procedures", ER 1110-2-1156, Washington, DC, March 2014. Available at: http://www.publications.usace.army.mil/Portals/76/Publications/EngineerRegulations/ER_1110-2-1156.pdf

Von Thun, J. L., and Gillette, D. R. (1990). "Guidance on Breach Parameters", Internal Memorandum, U.S. Dept. of the Interior, Bureau of Reclamation, Denver, CO.

ESTIMATING THE PROBABLE MAXIMUM FLOOD FOR WHITTIER NARROWS DAM USING A 2D HYDRODYNAMIC MODELING APPROACH

Brian M. Hall, P.E.[1]
Haden Smith, P.E.[2]
Reuben Sasaki, P.E.[3]

ABSTRACT

Whittier Narrows Dam is a U.S. Army Corps of Engineers (USACE) dam located at the southern limit of the San Gabriel Valley, approximately 10 miles east of downtown Los Angeles, CA. The Rio Hondo and San Gabriel River flood flows are impounded by the dam. The Whittier Narrows Dam watershed is approximately 566 square miles and contains extremely varied regions ranging from remote mountains to highly urban. Whitter Narrows Dam is a high hazard dam, central to the Los Angeles County flood control system, and is currently being evaluated in a USACE Dam Safety Modification Study (DSMS). The Probable Maximum Flood (PMF) for Whittier Narrows Dam was estimated utilizing a lumped rainfall-runoff model in conjunction with a two-dimensional (2D) hydrodynamic model. HEC-RAS 5.0.3 software, which allowed for 2D unsteady flow surface water hydraulics calculations, was used to estimate the PMF by routing the excess precipitation and the baseflow derived from the HEC-HMS rainfall-runoff model for each respective subbasin. The HEC-RAS 2D model domain spanned the entire watershed and modeled the complex urban drainage, overbank depression storage regions, the operations for 14 dams located within the watershed, and the backwater effects at the confluences. Modeling the rainfall-runoff response and surface water hydraulics with HEC-RAS 2D mitigated the modeling limitations associated with the linear runoff response of unit hydrographs. Estimating the PMF for Whittier Narrows Dam using a 2D hydrodynamic modeling approach greatly improved the understanding of extreme flood hydrology and hydraulics within the watershed as compared to traditional lumped hydrologic models.

INTRODUCTION

Whittier Narrows Dam is located at the southern limit of the San Gabriel Valley, approximately 10 miles east of downtown Los Angeles (Figure 1). The reservoir is located at a natural gap in the hills that form the southern boundary of the San Gabriel Valley, in Los Angeles County, California, approximately 7.5 miles downstream from Santa Fe Dam. The Rio Hondo and the San Gabriel River flow through this gap and flood flows are constrained by the dam. There are several Los Angeles County Department of Public Works (LACDPW) reservoirs upstream from Whittier Narrows Dam. Outlet

[1] Hydraulic Engineer, U.S. Army Corps of Engineers, Dam Safety Modification Mandatory Center of Expertise, Lakewood, CO 80228, Brian.M.Hall@usace.army.mil
[2] Hydrologic Engineer, U.S. Army Corps of Engineers, Institute for Water Resources Risk Management Center, Lakewood, CO 80228, Cole.H.Smith@usace.army.mil
[3] Hydraulic Engineer, U.S. Army Corps of Engineers, Los Angeles District, Los Angeles, CA 90017, Reuben.A.Sasak@usace.army.mil

releases from Santa Fe Dam (USACE dam) flow into Whittier Narrows Dam. The Whittier Narrows Dam watershed is approximately 566 square miles and contains extremely varied regions ranging from remote mountains to highly urban.

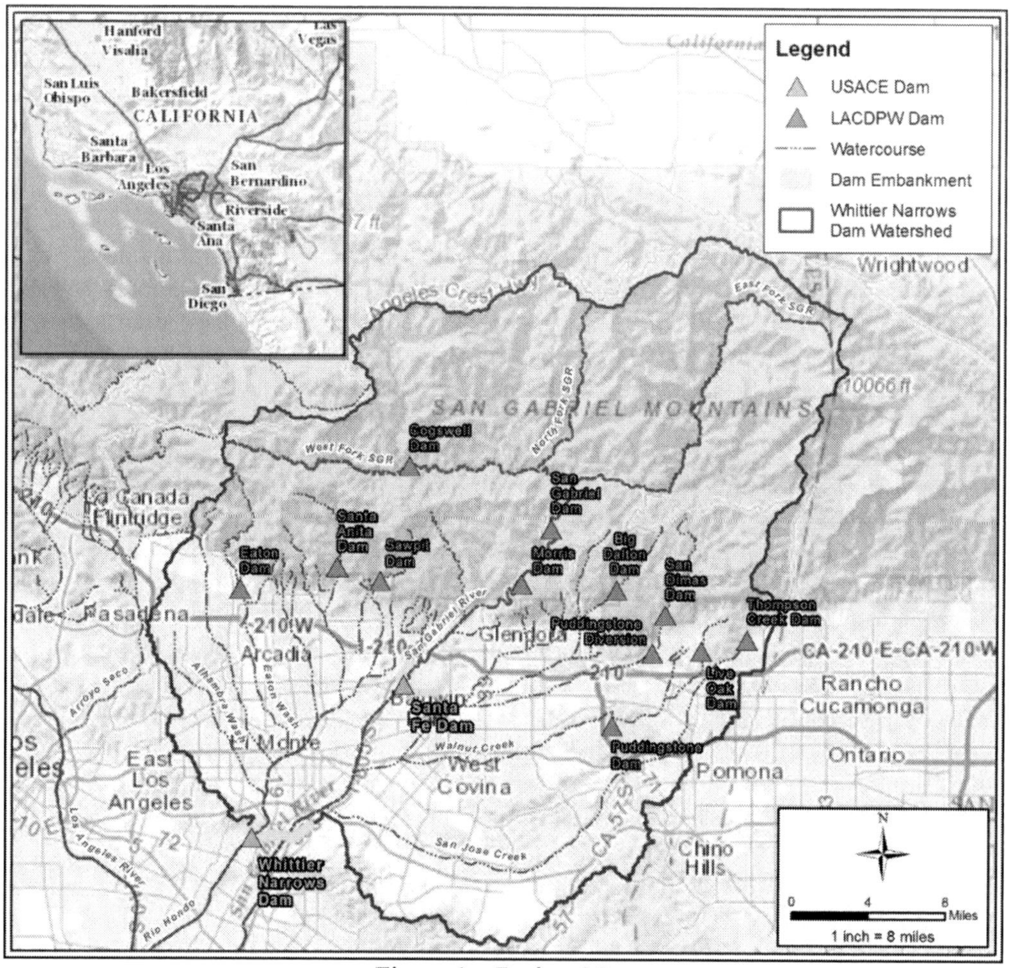

Figure 1. Project Map

The construction of Whittier Narrows Dam was authorized by the Flood Control Act of 1941 for the primary purpose of flood control. Recreation is a secondary purpose, as authorized by the Flood Control Act of 1944. A third purpose of the dam was set forth by the Chief of Engineers in 1956 for water conservation. The U.S. Army Corps of Engineers (USACE) owns, operates, and maintains the dam and all associated flood control facilities. Construction began in March 1950 on the first major contract for the project, and the final major contract was completed in March 1957. The dam consists of an approximately 16,960 foot-long earthen embankment and two gated outlet structures. The outlet works discharges into the Rio Hondo, and the spillway discharges into the San Gabriel River (Figure 2). The reservoir is normally empty and a "crossover weir" within the reservoir diverts lower flows from the San Gabriel River to the Rio Hondo.

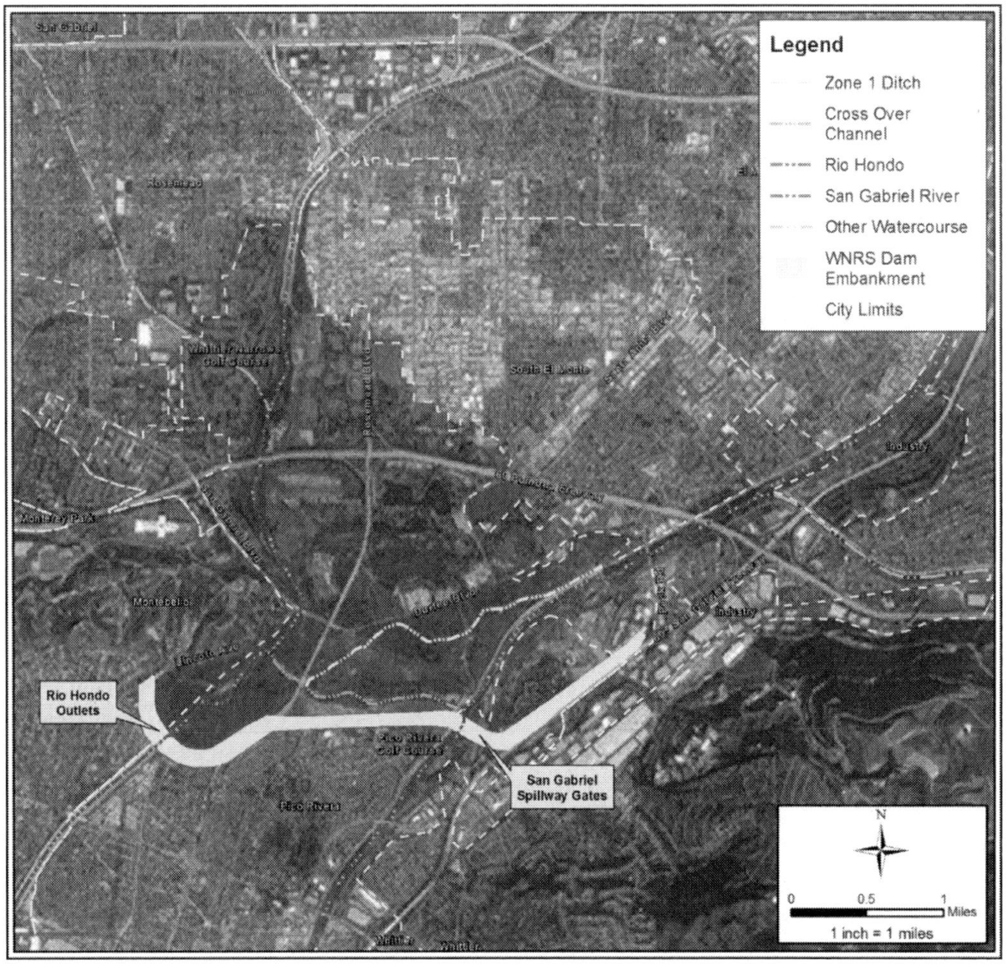

Figure 2. Whittier Narrows Dam

Whitter Narrows Dam is a high hazard dam, central to the Los Angeles County flood control system, and is currently being evaluated in a USACE Dam Safety Modification Study (DSMS). The Probable Maximum Flood (PMF) was re-evaluated as part of the DSMS.

Engineering Regulation (ER) 1110-8-2 *Inflow Design Floods for Dams and Reservoirs* requires that reservoir unit hydrographs for inflow design floods and PMFs be peaked 25 to 50 percent to account for the fact that unit hydrographs are usually derived from smaller floods (U.S. Army Corps of Engineers, 1991). Unit hydrographs are based on an assumption of linearity, meaning that if two units of input fall in the same time step, there will be twice the output response. However, in nature the rainfall-runoff response behaves nonlinearly. Thus, the peaking factors are required to account for the nonlinearity that is expected to occur during extreme rainfall events.

According to Beven (Beven, 2012), nonlinear rainfall-runoff responses are primarily due to two causes: 1) rainfall and runoff behaves nonlinearly because the wetter the antecedent soil conditions prior to a unit of rainfall, the greater the runoff volume that will be generated; and 2) nonlinearity results from the change of flow velocities and celerities with discharge. Average flow velocities and celerities will increase with flow nonlinearly. As shown by Minshall (Minshall, 1960), higher velocities mean that the runoff will get to a measurement point faster, and faster celerities mean the shape of the unit hydrograph will change as runoff increases.

Unfortunately, ER 1110-8-2 does not provide any guidance on selecting a peaking factor that most appropriately accounts for the mechanisms that cause nonlinearity. Arbitrarily choosing a peaking factor that is too low or too high could result in a PMF that is under or overly conservative. Therefore, in order to appropriately account for nonlinearity in the rainfall-runoff response, the PMF for Whittier Narrows Dam was estimated utilizing a lumped rainfall-runoff model in conjunction with a two-dimensional (2D) hydrodynamic model.

Hydrologic Engineering Center River Analysis System (HEC-RAS) 5.0.3 software, which allowed for 2D unsteady flow surface water hydraulics calculations, was used to estimate the PMF by routing the excess precipitation and the baseflow derived from the Hydrologic Engineering Center Hydrologic Modeling System (HEC-HMS) rainfall-runoff model for each respective subbasin. Figure 3 depicts the PMF modeling workflow process.

The nonlinearity in runoff due to antecedent soil conditions was accounted for by using HEC-HMS, and the nonlinearity due to changes in flow velocities and celerities was accounted for by using the HEC-RAS 2D unsteady flow surface water calculations. This modeling approach minimizes the limitations associated with unit hydrograph peaking factors.

The HEC-RAS 2D model domain spanned the entire watershed and modeled the complex urban drainage, overbank depression storage regions, the operations for 14 dams located within the watershed, and the backwater effects at the confluences. Modeling the rainfall-runoff response and surface water hydraulics with HEC-RAS 2D mitigated the modeling limitations associated with the linear runoff response of unit hydrographs. Note that results are preliminary and review is ongoing.

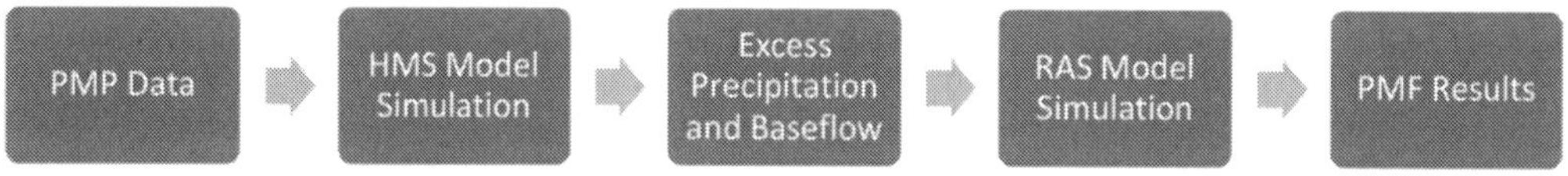

Figure 3. PMF Workflow

MODEL FORMULATION

HEC-HMS was utilized to determine precipitation loss and baseflow for each respective subbasin within the Whittier Narrows watershed. Initial HMS parameters were obtained from the 1991 Los Angeles County Drainage Area (LACDA) Hydrology Technical

Report (U.S. Army of Corps of Engineers Los Angeles District, 1991). Parameters and model geometry were modified to include the most current data available which included re-delineation of subbasins, updated reservoir storage curves, and updated dam discharge rating curves. Figure 4 depicts the HMS model geometry.

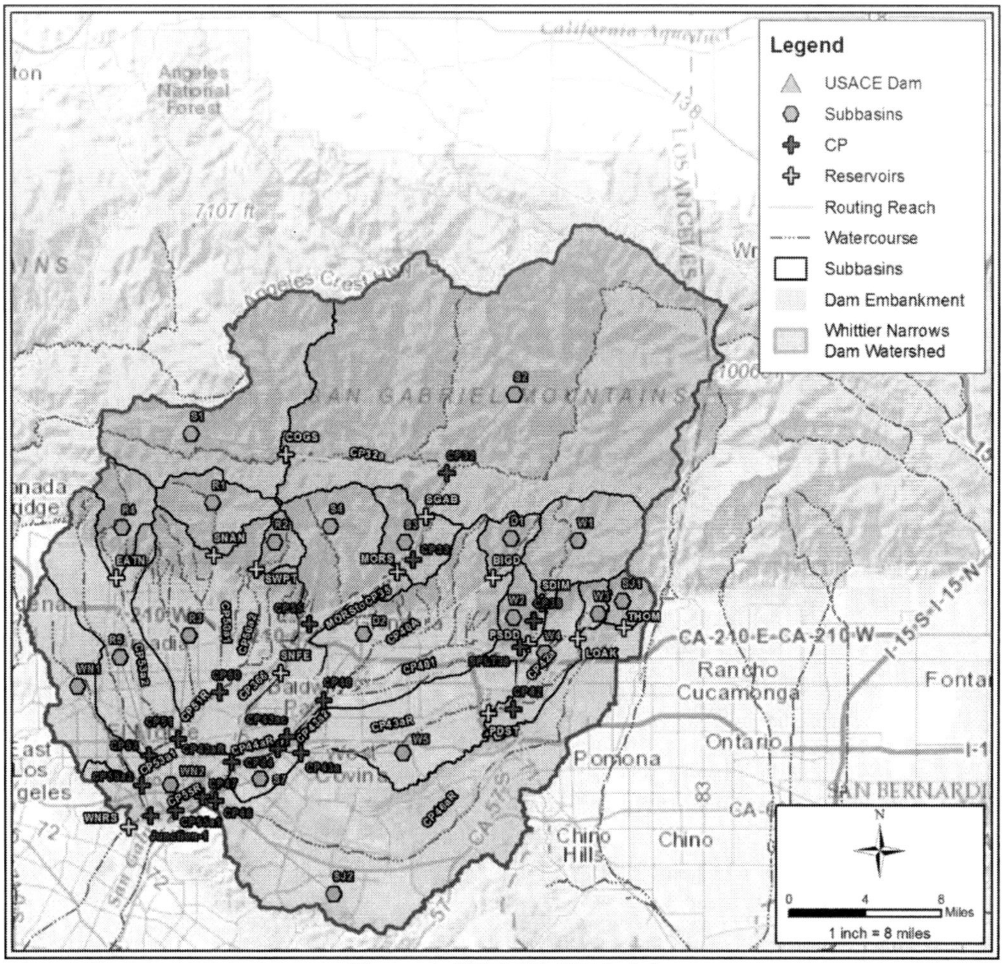

Figure 4. HEC-HMS Model

HEC-RAS 5.0.3 software was used to route the excess precipitation and the baseflow derived from the HEC-HMS rainfall-runoff model for each respective subbasin using the 2D Diffusion Wave equation set. The excess precipitation computed from HMS was uniformly applied to each respective 2D subbasin. The baseflow was introduced at the outlet of each respective 2D subbasin. Figure 5depicts the RAS model geometry.

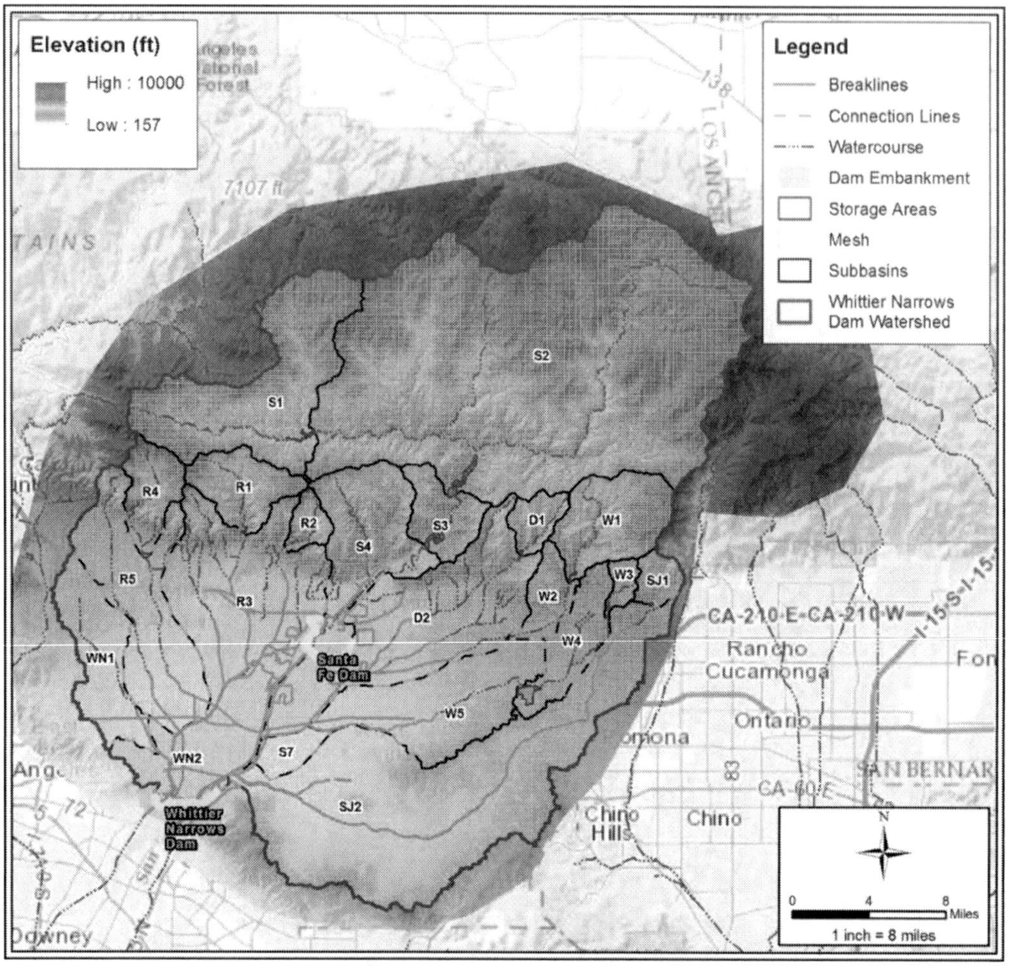

Figure 5. HEC-RAS Model

The 2D model domain encompassed the entire watershed which included the operations of Whittier Narrows Dam (USACE), Santa Fe Dam (USACE), and twelve dams owned by Los Angeles County Department of Public Works including Cogswell Dam, San Gabriel Dam, and Morris Dam.

Manning's roughness coefficients for the model were assigned utilizing a composite land cover dataset. The base land cover dataset was obtained from the National Land Cover Database (NLCD 2011). For the urban region spanning from Santa Fe Dam to Whittier Narrows Dam a higher resolution land cover dataset was developed that superseded the NLCD dataset. A composite terrain dataset was developed for the model domain. The base digital elevation model (DEM) was obtained from the Los Angeles Regional Imagery Acquisition Consortium (LARIAC 2006) and had a resolution of 10 ft. High resolution survey data obtained from the USACE Los Angeles District was utilized for Whittier Narrows Dam (2011, 1 ft resolution) and Santa Fe Dam (2010, 3 ft resolution) regions. Bathymetry data sets for the gravel pits near Santa Fe Dam were obtained from the City of Irwindale, CA. Mesh breaklines were extensively utilized to capture key features such as road embankments, bridges, culverts, and overbank depression storage (specifically gravel pits).

CALIBRATION AND VALIDATION

The HEC-HMS model and HEC-RAS 2D model were calibrated to two historical events and validated to one historical event. The three historical events were the largest events in which sub-daily gage data exists across the watershed. Table 1 summarizes the events utilized for calibration and validation.

Table 1. Calibration and Validation Summary of Events

Event	Event Start	Event End	Calibration/Validation	Basin Average Precipitation (in)
January 2005	1/6/05	1/12/05	Calibration	16.19
February 2005	2/17/05	2/26/05	Calibration	12.74
December 2010	12/18/10	12/23/10	Validation	12.83

Seven key target locations across the watershed were evaluated for the calibration and validation events. A "weight-of-evidence" approach, where both statistical tests and graphical comparisons were employed, was used for evaluating model performance. Calibration and validation results from the two most important target locations are shown below in Figure 6 and Figure 7. Figure 6 depicts the San Gabriel Inflow (observed, HMS model, and RAS model) at Whittier Narrows Dam for the two calibration events and one validation event. Figure 7 depicts the Rio Hondo Inflow (observed, HMS model, and RAS model) at Whittier Narrows Dam for the two calibration events and one validation event.

From the results shown in Figure 6 and Figure 7, it is evident that both the HEC-HMS and HEC-RAS models adequately simulate reservoir inflow behavior at Whittier Narrows. However, the following limitations were the primary sources of modeling error encountered during the calibration and validation effort:

1) Upstream reservoir operations: Limited historical gate operation data for the upstream reservoirs made it challenging to accurately model the outlet works of those control structures. However, the model performs well at simulating the upstream dam spillway flows that occur during the PMF.
2) Stormwater management networks: The complex urban stormwater management networks that exists in the greater Los Angeles area is challenging to model accurately during smaller rainfall events. The 2D model geometry cannot capture all of the efficient stormwater conveyance systems that currently exist, resulting in modeled peak flows that are often lower than observed flows.
3) Uniformly distributed precipitation data: The current version of HEC-RAS does not permit the use of gridded excess precipitation. Therefore, lumped excess precipitation derived in HEC-HMS must be applied uniformly across each 2D region. This limits the ability of the model to accurately account for the spatial and temporal interactions of the complex urban drainage networks and channels. Consequently, the model does not do well at modeling the timing of peaks at the

confluence of these drainage networks, resulting in modeled peak flows that are lower than observed peak flows.

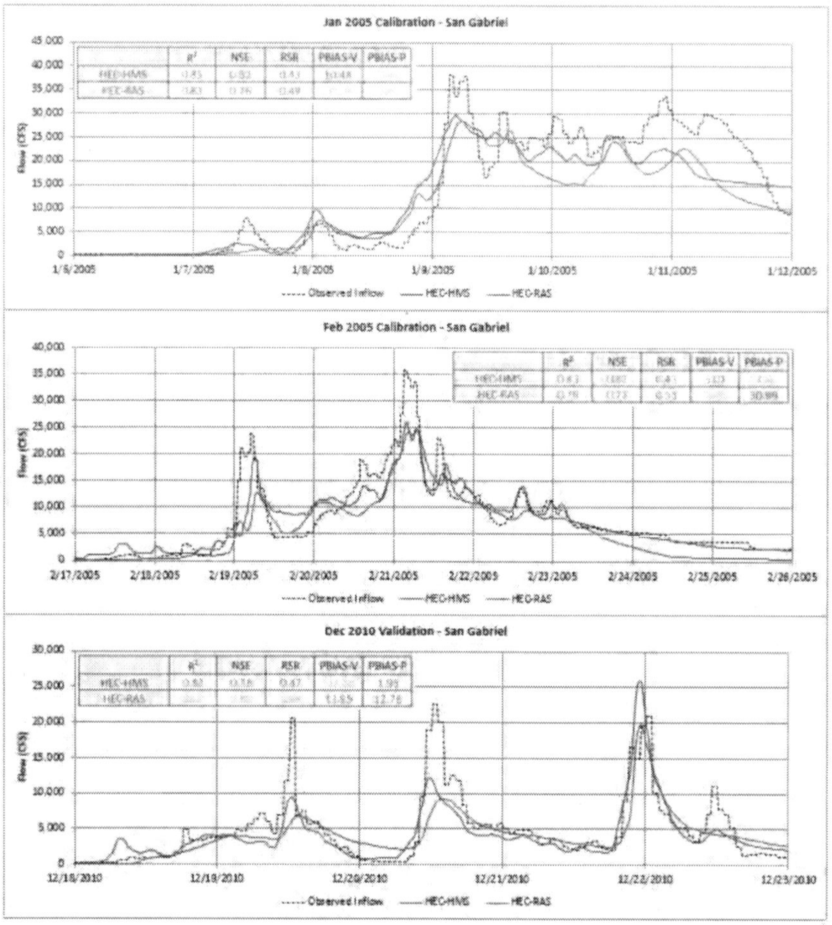

Figure 6. Whittier Narrows Dam San Gabriel Inflow

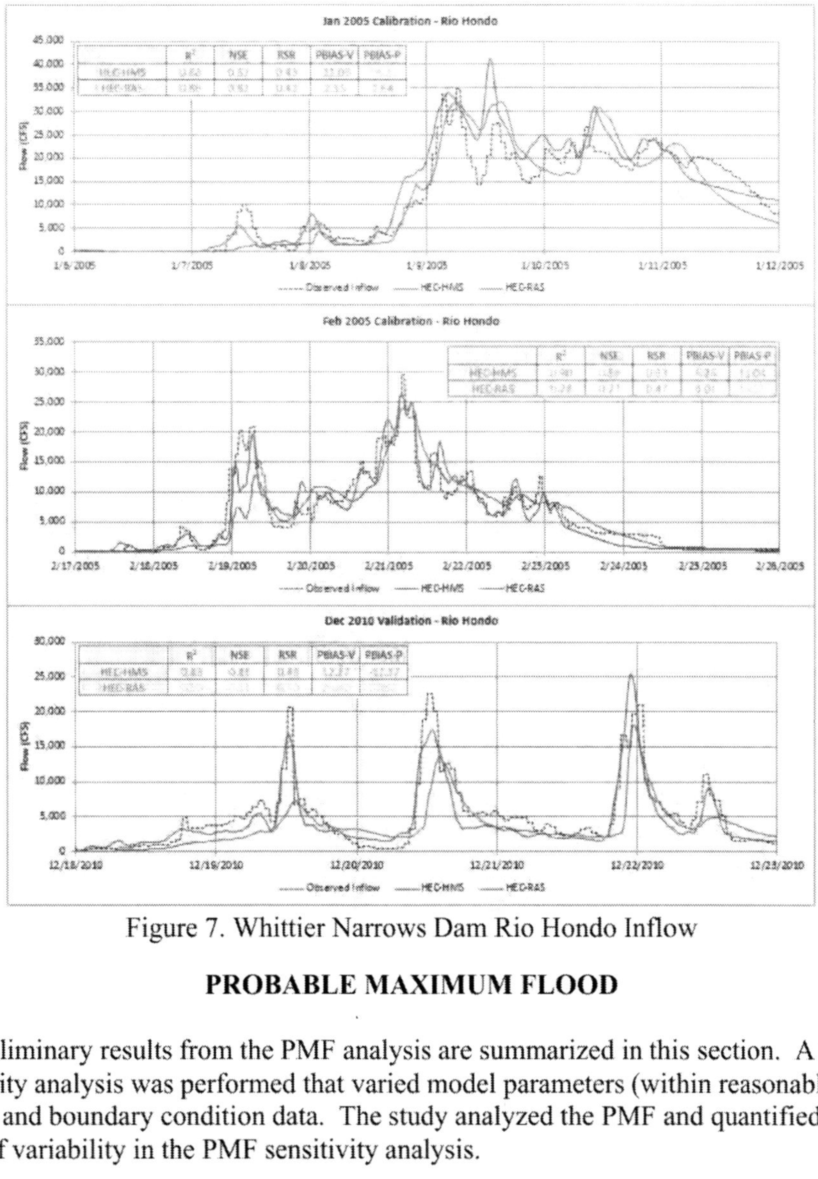

Figure 7. Whittier Narrows Dam Rio Hondo Inflow

PROBABLE MAXIMUM FLOOD

The preliminary results from the PMF analysis are summarized in this section. A PMF sensitivity analysis was performed that varied model parameters (within reasonable ranges) and boundary condition data. The study analyzed the PMF and quantified the range of variability in the PMF sensitivity analysis.

Table 2 summarizes the thirteen PMF plans that varied model parameters and boundary conditions. Four different reservoir and groundwater starting pools; 99.9%, 90%, 50%, and 10% non-exceedance pools were evaluated. Three variations in precipitation, 1) HMR59, 2) 90% HMR59, 3) and 110% HMR59 were utilized as boundary conditions. The existing condition (EC) and future condition (FC) Whittier Narrows Dam crest profiles were evaluated. Manning's roughness coefficients (N-values) were simulated for the following 5 conditions: 1) calibrated parameter set, 2) entire domain reasonably low, 3) entire domain reasonably high, 4) mountain domain reasonably low and valley domain reasonably high, and 5) mountain domain reasonably high and valley domain reasonably low. Constant loss rates were simulated utilizing the best estimate PMF calibrated parameter set, low values, and high values. From the *Antecedent Flood Study* (U.S. Army Corps of Engineers Los Angeles District, 2014) the lag time between the antecedent event and the main flood event was determined to be 1 day. A sensitivity test was performed utilizing a 3 day lag time. Plan ID P02 represents the recommended PMF simulation plan. P02 model parameters were based on data obtained from the analysis of rainfall-runoff relations from historical floods and corresponds to those considered reasonably likely to occur during storms of PMF magnitude. P02 antecedent starting pools correspond to a reasonably high flood control storage pool (greater than 50%) based on the historical flood data.

Table 2. PMF Plans

Plan ID	Reservoir & Groundwater Starting Pools	PMP	WNRS Crest Profile	N-Values	Constant Loss Rate	Lag between Antecedent and Main Flood Event
P01	90.0%	HMR59	EC	Calibrated	Calibrated	1 Day
P02 Recommended PMF	**90.0%**	**HMR59**	**FC**	**Calibrated**	**Calibrated**	**1 Day**
P03	99.9%	HMR59	FC	Low	Low	1 Day
P04	50.0%	HMR59	FC	Calibrated	Calibrated	1 Day
P05	10.0%	HMR59	FC	High	High	1 Day
P06	90.0%	90% of HMR59	FC	Calibrated	Calibrated	1 Day
P07	90.0%	90% of HMR59	FC	High	High	1 Day
P08	90.0%	110% of HMR59	FC	Calibrated	Calibrated	1 Day
P09	90.0%	110% of HMR59	FC	Low	Low	1 Day
P10	90.0%	HMR59	FC	Calibrated	Calibrated	3 Day
P11	90.0% Additional gravel pit storage based on LARIAC volume	HMR59	FC	Calibrated	Calibrated	1 Day
P12	90.0%	HMR59	FC	Low-Mtn, High-Valley	Calibrated	1 Day
P13	90.0%	HMR59	FC	High-Mtn, Low-Valley	Calibrated	1 Day

The Probable Maximum Precipitation (PMP) was developed based on the procedures outlined in HMR58 and HMR59. This followed the previous 2013 PMP study. The PMP precipitation temporal pattern was developed by breaking the pattern in 6-hr periods. The PMP precipitation was spatially distributed over the watershed using the 100-year, 72-hour NOAA Atlas 14 frequency grid. The basin average PMP for the Whittier Narrows watershed is 37.6 inches. Figure 8 depicts the spatial pattern and temporal pattern using HMR58/59.

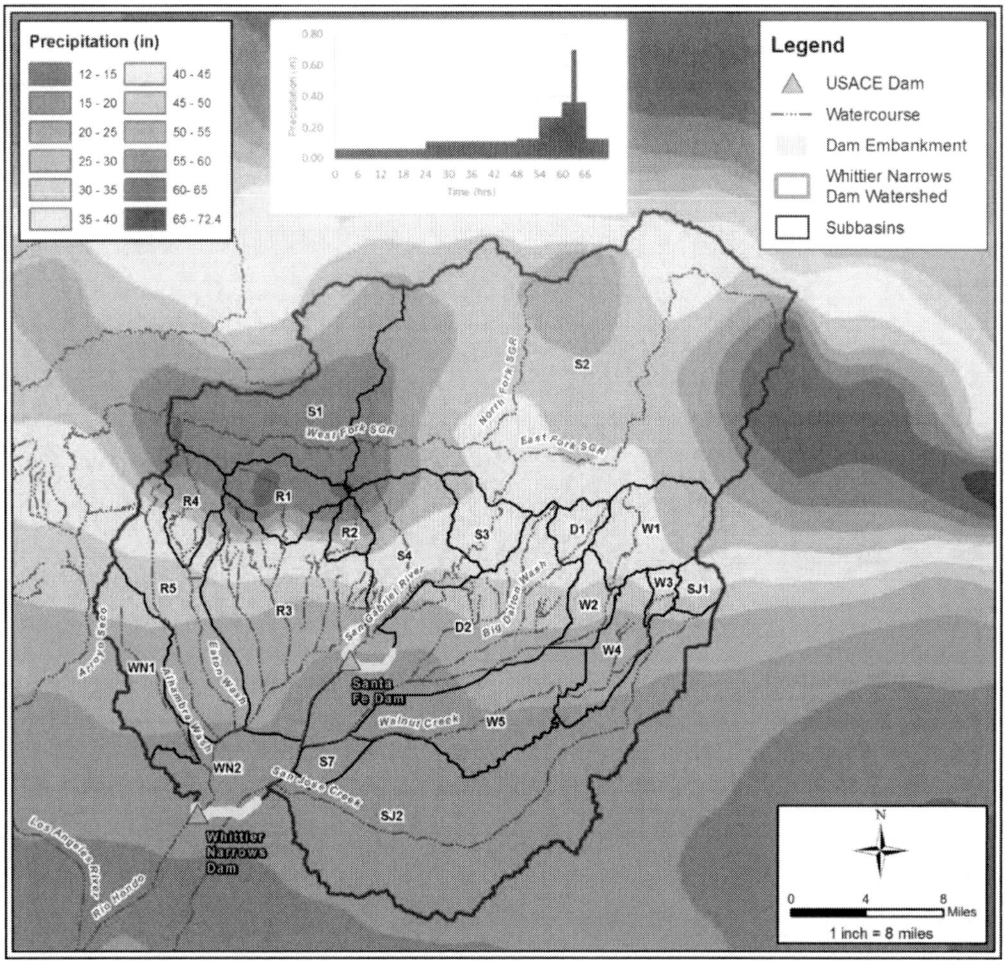

Figure 8. PMP Spatial & Temporal Pattern

Figure 9 depicts the maximum velocity from the recommended PMF plan (P02) output for the region downstream of Santa Fe Dam to Whittier Narrows Dam. Note the significant overland flow, activation of the Santa Fe Dam Spillway, overbank depression storage, gravel pit storage, and backwater influence at the confluence regions.

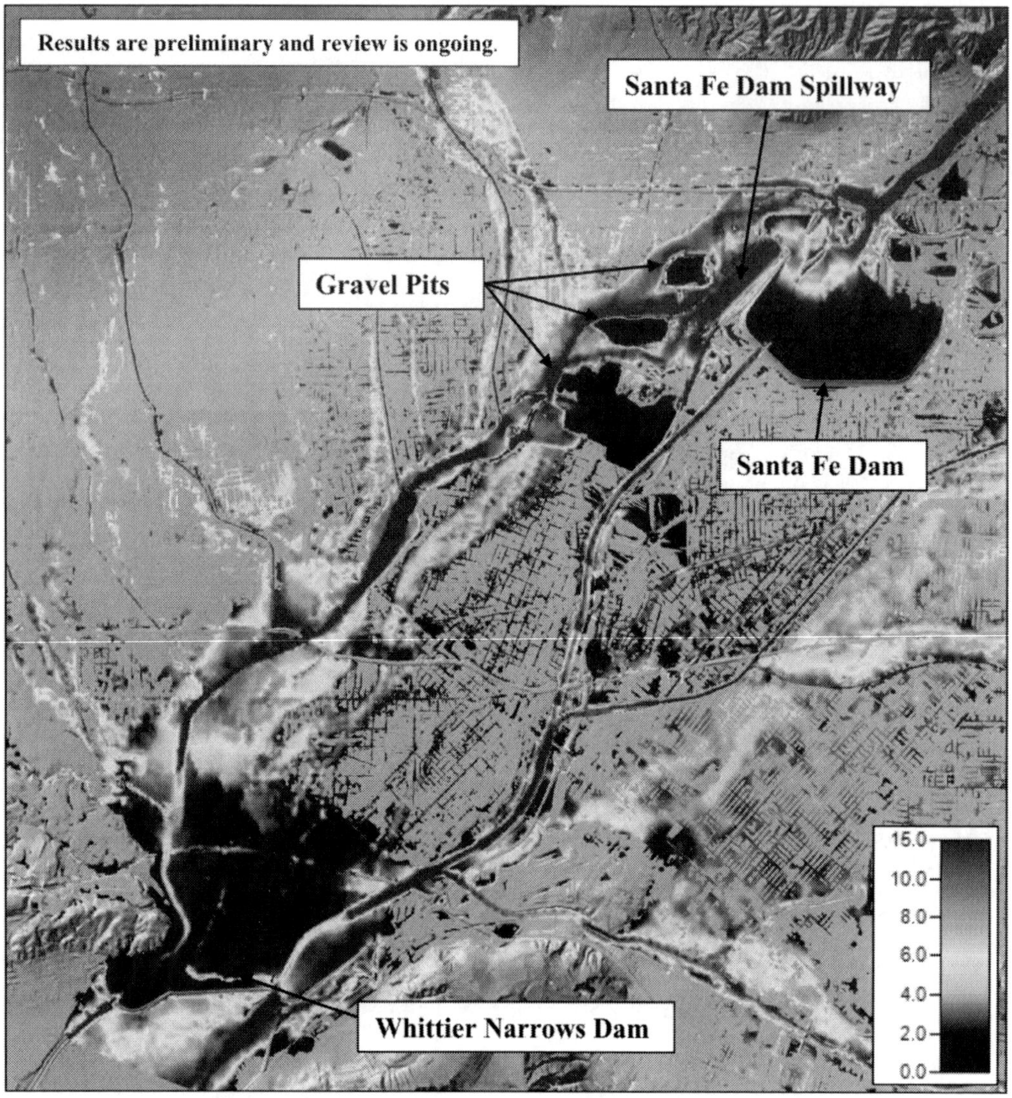

Figure 9. PMF – Whitter Narrows Maximum Velocity

The PMF results are summarized in Table 3. All of the PMF plans resulted in overtopping at Whittier Narrows Dam. The recommended PMF plan (P02) estimated a peak headwater elevation of 244.2 ft NAVD88 (2.9 ft of overtopping). PMF plan P07 estimated the lowest peak headwater elevation of 242.9 ft NAVD88 (1.7 ft of overtopping) and plan P09 estimated the highest peak headwater elevation of 245.0 ft NAVD88 (3.8 ft of overtopping). Figure 10 depicts the PMF peak inflow range based on the sensitivity analysis at Whittier Narrows Dam. Figure 11 depicts the PMF peak headwater elevation sensitivity analysis and FC crest elevation at Whittier Narrows Dam. Note that results are preliminary and review is ongoing.

Table 3. PMF Results

Plan ID	RAS Peak Elevation (NAVD88)	RAS Peak Inflow (cfs)	Overtopping (ft)	Overtopping Duration (hr)
P01	243.3	556,000	4.8[1]	7.25[1]
P02 Recommended PMF	**244.2**	**556,000**	**2.9**	**5.75**
P03	244.5	593,000	3.2	6.00
P04	244.2	556,000	2.9	5.75
P05	243.7	503,000	2.4	5.50
P06	243.5	494,000	2.3	4.75
P07	242.9	454,000	1.7	4.25
P08	244.7	622,000	3.4	6.75
P09	245.0	670,000	3.8	7.50
P10	244.2	556,000	2.9	5.75
P11	244.1	547,000	2.8	5.50
P12	244.0	536,000	2.7	6.00
P13	244.1	554,000	2.9	5.75

[1] Based on EC Low Point Crest Invert 238.5 ft NAVD88
Results are preliminary and review is ongoing.

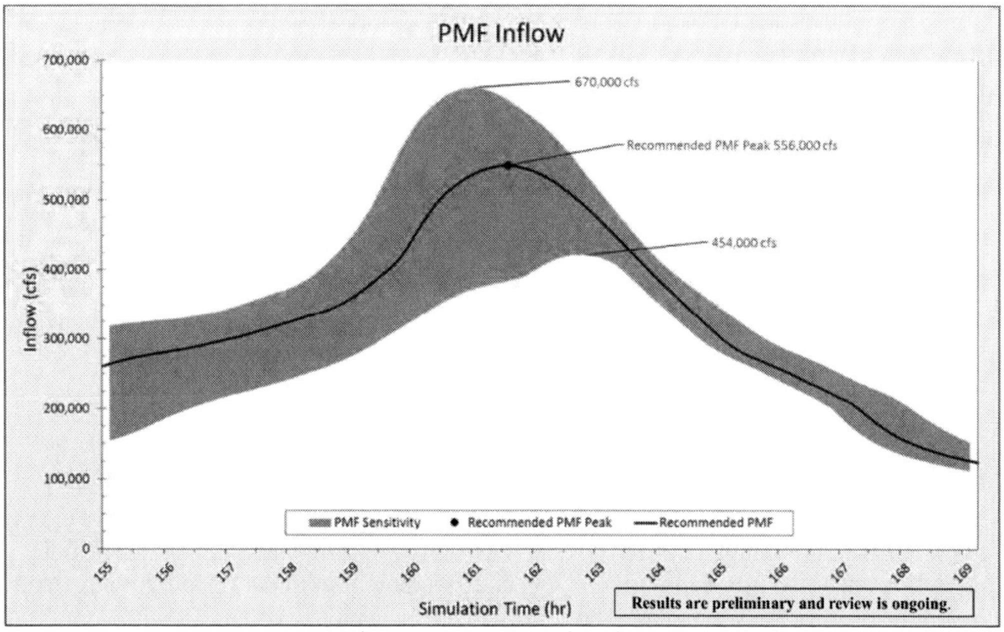

Figure 10. PMF Peak Inflow

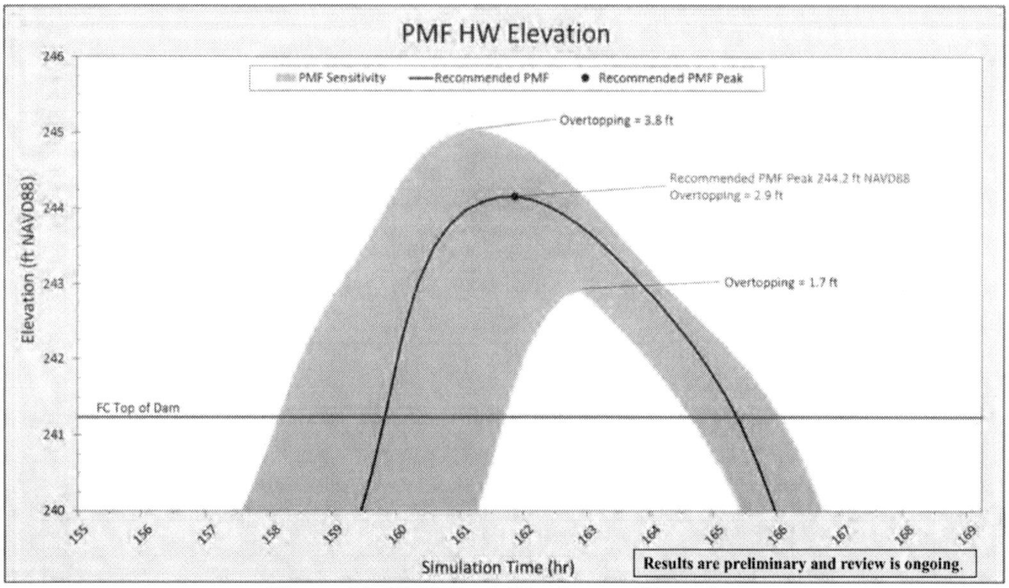

Figure 11. PMF Peak Headwater Elevation

CONCLUSION

Table 4 compares the HMS model output to the RAS model output for PMF simulations P01-P09. For the recommended PMF plan (P02), the RAS model simulated a 1.4 ft higher peak stage and 30.3% higher peak inflow as compared to the HMS model. For PMF studies where peak headwater elevation is sensitive to peak inflow, the design team should consider utilizing a hydrodynamic routing method to better estimate the peak inflow and headwater elevation at the reservoir. The 2D equation set utilized in RAS provided a physics based modeling approach for routing extreme large magnitude events that traditionally required arbitrary peaking factors (25-50% either at the dam or individual subbasins) to the unit hydrograph routing methods. Modeling the rainfall-runoff response and surface water hydraulics with HEC-RAS 2D mitigated the modeling limitations associated with the linear runoff response of unit hydrographs and also allowed for backwater interaction at the dams, confluences, and subbasin outlets.

Additionally, the 2D hydrodynamic model provided detailed output including: a PMF hydrograph, peak inflow to the dam, peak water surface elevation, outlet discharge, spillway discharge, overtopping discharge, embankment flanking discharge, and velocity magnitudes. Estimating the PMF for Whittier Narrows Dam with a 2D hydrodynamic modeling approach greatly improved the understanding of extreme flood hydrology and hydraulics within the watershed. The results from this analysis will assist with dam safety modification alternative evaluation, leading to more optimal risk reduction measures. Note that results are preliminary and review is ongoing.

Table 4. RAS vs. HMS Comparison

Simulation ID	RAS Peak Elevation (NAVD88)	HMS Peak Elevation (NAVD88)	RAS Peak Inflow (cfs)	HMS Peak Inflow (cfs)	Peak Elevation Difference (ft) RAS - HMS	Peak Flow Difference (%) RAS - HMS
P01	243.3	242.4	556,000	427,000	0.9	30.2
P02	244.2	242.8	556,000	427,000	1.4	30.3
P03	244.5	243.1	593,000	452,000	1.4	31.1
P04	244.2	242.8	556,000	426,000	1.4	30.4
P05	243.7	242.3	503,000	403,000	1.4	24.7
P06	243.5	240.8	494,000	371,000	2.7	33.2
P07	242.9	239.2	454,000	351,000	3.7	29.5
P08	244.7	243.5	622,000	482,000	1.2	28.9
P09	245.0	243.7	670,000	504,000	1.3	33.0

Note: Elevations are rounded to the nearest tenth and flows are rounded to the nearest thousand. Results are preliminary and review is ongoing.

ACKNOWLEDGEMENTS

The authors gratefully acknowledge the contributions of G.S. Karlovits and J.F. England (both with U.S. Army Corps of Engineers) for their work on this study.

REFERENCES

Beven, K. (2012). *Rainfall-Runoff Modeling: The Primer.* West Sussex, UK: John Wiley & Sons.

Minshall, N. (1960). Predicting storm runoff on small experimental watersheds. *ASCE Journal of the Hydraulics Division*, 17-38.

National Oceanic and Atmospheric Administration. (1998). *Hydrometeorlogical Report No. 58 Probable Maximum Precipitation for California.*

National Oceanic and Atmospheric Administration. (1999). *Hydrometeorlogical Report No. 59 Probable Maximum Precipiation for California.*

U.S. Army Corps of Engineers. (1991). *Engineer Regulation 1110-8-2 Inflow Design Floods for Dams and Reservoirs.*

U.S. Army Corps of Engineers Los Angeles District. (1957). *Reservoir Regulation Manual for Whittier Narrows Flood Control Reservoir.*

U.S. Army Corps of Engineers Los Angeles District. (2013). *Draft Whittier Narrows Dam Probable Maximum Flood and Baseline Hydrology Update in Support of Dam Safety Modification Study.*

U.S. Army Corps of Engineers Los Angeles District. (2014). *Antecedent Flood Study for Whittier Narrows Dam.*

U.S. Army Corps of Engineers Los Angeles District. (2014). *Site-Specific Probable Maximum Flood Study for Whittier Narrows Dam.*

U.S. Army of Corps of Engineers Los Angeles District. (1991). *Los Angeles County Drainage Area Final Feasibility Interim Report: Part I Hydrology Technical Report Base Conditions.*

WHEN AN ADJACENT WATERSHED INUNDATES YOUR OFFLINE RESERVOIR: UKUMEHAME RESERVOIR DAM REMOVAL CHALLENGES AND DESIGN

Andrew J. Lynch, PE, CFM[1]
Stewart S. Vaghti, PE, CFM[2]
Dean B. Durkee, PhD, PE[3]

ABSTRACT

Ukumehame Reservoirs Nos. 2 & 3 are a pair of offline reservoirs impounded by earthen embankments located on the island of Maui, Hawaii. The reservoirs were constructed during Hawaii's sugar plantation era as a part of a statewide network of reservoirs and irrigation ditches to support local agriculture. In recent decades with the decline of sugar production, these and similar reservoirs are often neglected and have fallen into disrepair.

The reservoirs are classified as high-hazard due to a residential structure immediately downstream of the No. 3 embankment. In June 2008, a significant seep was discovered at the toe of the No. 3 embankment. Following this discovery, the reservoirs were drained and the dam owners decided to remove the reservoirs by breaching the embankments.

Hydrologic and hydraulic analyses led to the discovery that the probable maximum flood (PMF) within the adjacent stream overtops the embankments and inundates the reservoirs. We were challenged with designing breach sections to accommodate flows from the watersheds directly contributing to the reservoirs as well as flows from the overbank of the adjacent stream considering what we had learned about the PMF.

To better understand this condition, a 1D unsteady HEC-RAS model was created to simulate pre- and post-breach conditions. This analysis assisted in presenting the design of the reservoir removal to the owner and the State Dam Safety Division. A 2D HEC-RAS model was subsequently prepared for comparison with the 1D HEC-RAS model and the 2D model agreed with the conclusions drawn from the 1D model.

INTRODUCTION

History

Beginning approximately 180 years ago, sugar plantations and farming became a significant and integral part of the economy of the Hawaiian Islands. The first commercially successful sugarcane plantation was founded in Koloa on the island of Kaua'i in 1835 (U.S. National Park Service). Commercial production of sugar quickly grew to outpace the manpower of the local islanders, and beginning in 1850, foreign laborers from China and Japan were imported to meet the demand.

---

[1] Project Engineer, Gannett Fleming, Phoenix, AZ 85012, alynch@gfnet.com
[2] Senior Project Manager, Gannett Fleming, Phoenix, AZ 85012, svaghti@gfnet.com
[3] Vice President, Gannett Fleming, Phoenix, AZ 85012, ddurkee@gfnet.com

As the production of sugarcane continued to increase, the Hawaiian Islands underwent significant transformation in order to meet the demands created by the new industry. Island-wide networks of ditches, tunnels, canals and reservoirs were constructed in order to intercept, store and deliver rainfall runoff to the sugarcane fields. The irrigation network was so successful that it enabled cultivation on the leeward side of the islands, which are much drier than the opposite windward side.

The transformation of the local drainage patterns came at significant cost to native Hawaiians, who were powerless to stop the loss of a resource that they depended on for their own agricultural uses. Historically, Hawaiians were self-sufficient, relying for centuries on the cultivation of, among other things, the kalo (Hawaiian for taro) plant. Kalo, similar to sugarcane, requires a large amount of water for cultivation and the production of taro reduced drastically due to the rise of sugarcane farming.

Sugar production in Hawaii has been in decline since the 1970s. Despite having once been the primary economic driver for the state of Hawaii, the final remaining commercial-scale sugar farm ceased operations on December 12, 2016 (Associated Press).

Project Background

Ukumehame Reservoirs Nos. 2 & 3 are situated on the southwestern side of Maui, approximately 8 miles southeast of the city of Lahaina (see Figure 1). The reservoirs and the adjacent Ukumehame Ditch were constructed in 1905 (Wadsworth) by the Honolua Ranch & Pioneer Mill Company.

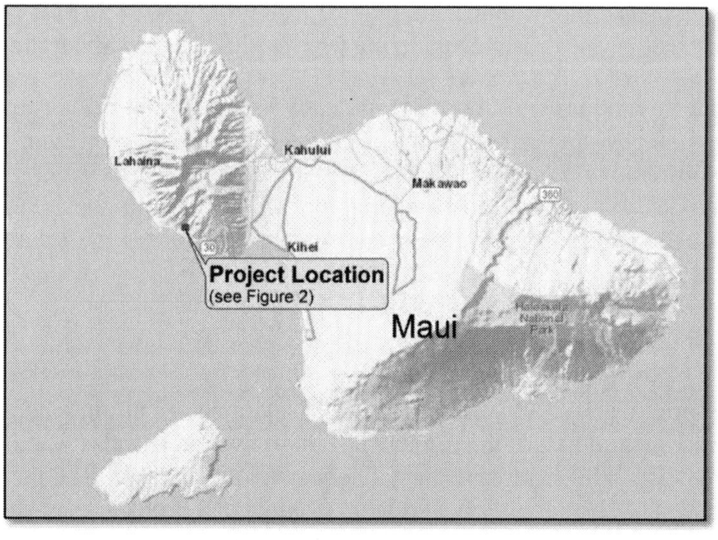

Figure 1 Island of Maui, HI

The reservoirs are side-by-side, offline reservoirs which are situated within Ukumehame Gulch and immediately adjacent to the main stream channel centerline (see Figure 2), which runs along the southeastern side of the reservoirs and discharges into the Pacific

Ocean approximately 0.8 miles away. The reservoirs were historically filled by directing water from the manmade Auwai Ditch (which runs along the right slope of the Ukumehame Gulch, looking downstream) into the reservoirs by way of a diversion structure.

Due to the existence of a residence immediately downstream of the Reservoir No. 3 embankment and active residential construction between the dams and the ocean outfall from the stream, the reservoirs are classified as high-hazard, meaning that the failure of one or both of the embankments could result in probable loss of human life and significant property damage.

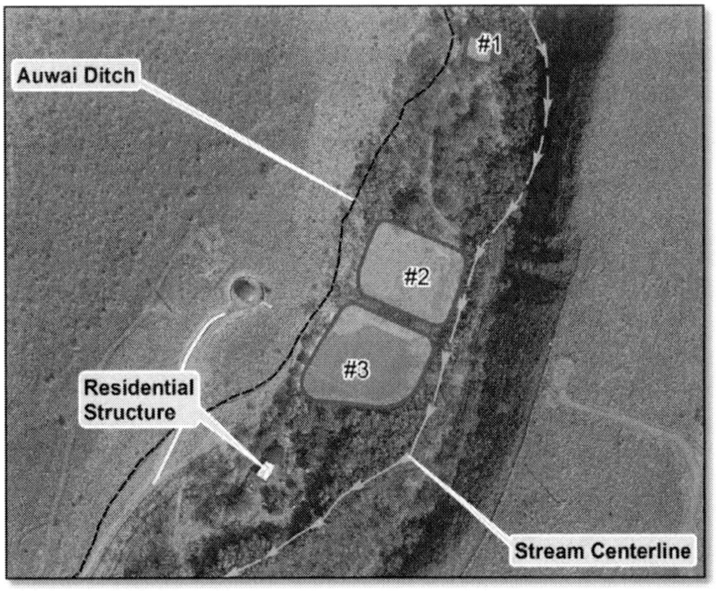

Figure 2 Project Location Map (imagery source: Google Earth)

The original areas of active sugar cultivation have been abandoned and sold off for residential construction. It is unclear how long the reservoirs were actively in use providing irrigation water to the sugar plantations; however, in recent decades, and with the decline of sugar production on the island, the reservoirs have fallen into disrepair with significant vegetation growing on the embankment slopes and within the reservoir bottoms (see Figure 3).

Figure 3 Bottom and Embankments of Reservoir #2 Showing Significant Vegetation Growth

During an inspection in June 2008, a significant seep was discovered within the Reservoir #3 embankment (see Figure 4). The flow rate was estimated to be 30-50 gpm at the time of measurement (Matsuda) and was clear, indicating embankment material was not being lost at a high rate. As a result of these findings, the State of Hawaii Department of Land and Natural Resources (DLNR) Dam Safety Division required that the level in the reservoirs be immediately lowered and steps be taken to evaluate the severity of the situation out of concern that the seep could develop into a piping failure, and ultimately into a catastrophic loss of the reservoir.

Following the discovery of the seep, Gannett Fleming prepared a Condition Assessment Report in 2009 (Gannett Fleming, Inc.) which included three options for addressing the deficiencies; rehabilitating the structures, converting to a flow-through system as a temporary measure, and breaching the reservoirs. The dam owners chose to pursue a full breach of both of the reservoirs. The embankment between Reservoirs No. 2 and No. 3 would be breached and the embankment between No. 3 and the stream channel would be breached as shown in Figure 5.

Figure 4 Seep Discharging from the Toe of the Reservoir #3 Embankment

Figure 5 Proposed Breach Design

HAWAII DAM SAFETY REGULATIONS

The basis for our analysis and design is Hawaii Administrative Rules Chapter 13-190.1. These rules provide design and analysis requirements based on size and hazard classification. Because the Condition Assessment Report included both a rehabilitation and a breach alternative, we evaluated both the PMF and the 100-year floods for this study. The rules state that all high-hazard dams, regardless of height or capacity, shall be able to pass the full PMF without overtopping the embankments (Hawaii Department of Land and Natural Resources §13-190.1-4(c)). In the event that the reservoirs would have been rehabilitated, this rule would apply. The rules also state that for removal an embankment shall be breached to allow passage of the 100-year flood with a maximum depth of five feet anywhere within the breached channel section (§13-190.1-21(b)(2)); this rule provided the basis for the breach design.

HYDRAULIC ANALYSIS

Model Setup

We had determined during the hydrologic and hydraulic analyses for the Condition Assessment Report that the PMF within Ukumehame Gulch overtops the embankments and inundates the reservoirs while the 100-year flow remains within the main stream channel. Even though we were proposing to breach the embankments and were therefore only required to evaluate the passage of the 100-year storm event, DLNR Dam Safety requested that we analyze the PMF condition passing through the breaches and determine the maximum depth of flow within the reservoirs. The concern was expressed that the remaining embankments could detain some volume of water, however briefly, which may still pose a risk to the downstream residence.

We determined that an unsteady HEC-RAS model would be necessary to model the various interrelated parts, such as differential timing of the incoming hydrographs, effects due to the storage capacity of the reservoirs, etc. The following model configuration was developed:

- The Ukumehame Gulch is modeled by 34 cross sections cut from a recent survey. All HEC-RAS geometry was cut and exported from Civil 3D 2014.
- The north and east sides of Reservoir 2 are modeled with a lateral weir structure that can exchange flows back and forth with the stream channel.
- The south and east sides of Reservoir 3 are modeled with a separate lateral weir structure that can exchange flows back and forth with the stream channel. This lateral weir reflects the geometry of the proposed breach.
- Reservoirs 2 and 3 are each modeled as storage areas with independent stage-storage relationships.
- Reservoirs 2 and 3 each have directly contributing inflow hydrographs from the previous hydrologic analysis.
- Reservoirs 2 and 3 are connected to each other by way of a storage area connection. The cross section of the storage area connection reflects the proposed breach between Reservoirs 2 and 3.

A schematic of the model and a graphic of the HEC-RAS geometry file are shown in Figure 6 and Figure 7, respectively. It is important to note that the lateral weirs and the storage area connection allow both positive and negative flow. Also, because HEC-RAS calculates the storage areas using level pool routing, there is no time delay between flows entering and leaving a storage area. This is not a significant issue due to the small size of these reservoirs; however, under different circumstances the travel time within a reservoir may be important to capture.

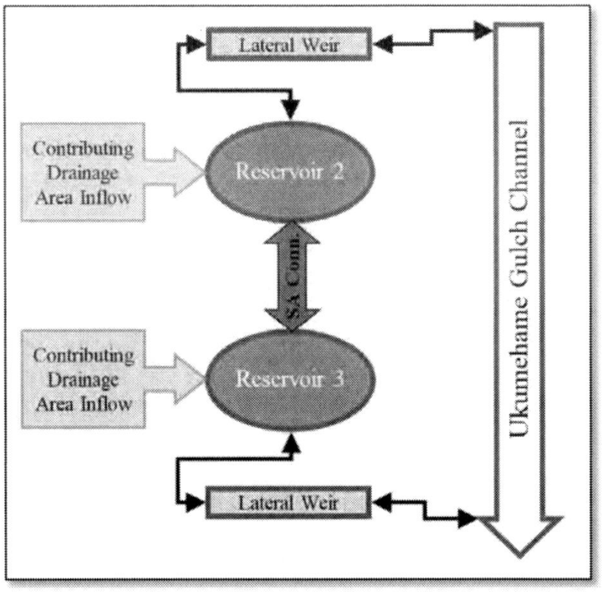

Figure 6 Unsteady HEC-RAS Connectivity Schematic

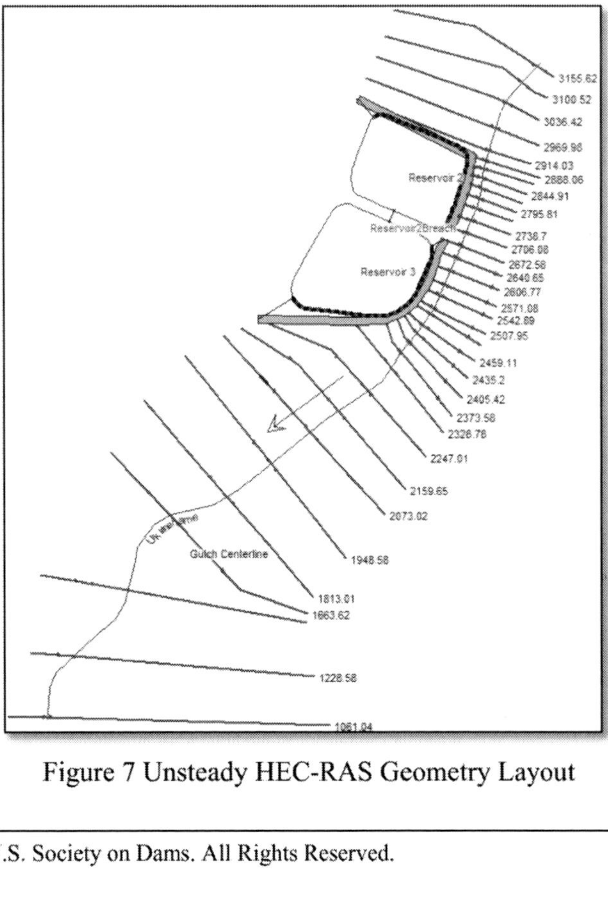

Figure 7 Unsteady HEC-RAS Geometry Layout

Analysis of HEC-RAS Output

HEC-RAS provides many good analysis and visualization tools which facilitate quick review and understanding of the model results; these tools can be particularly useful when evaluating an unsteady model. For our purposes with this model, however, we needed to be able to visualize the output in ways HEC-RAS can't natively do. For example, we wanted to display the water surface elevations at a HEC-RAS cross section within the stream channel and the water surface elevation (WSEL) in the adjacent reservoir at each timestep within the model so that we could evaluate the differential loading on the embankment. To accomplish this, we used Excel to create graphs to visualize custom views of the HEC-RAS model results.

We accomplished this by importing select cross section data and HEC-RAS stage hydrograph data into Excel. By importing the data from each timestep, we were able to create an interactive spreadsheet where we could step through the model. Figure 8 and Figure 9, below, are examples of the graphs we produced. Each graph includes a scroll bar to interactively step through the model results, which are dynamically updated from the HEC-RAS output.

These figures show the WSEL within the stream channel alongside the WSEL within the storage area(s) during the PMF event.

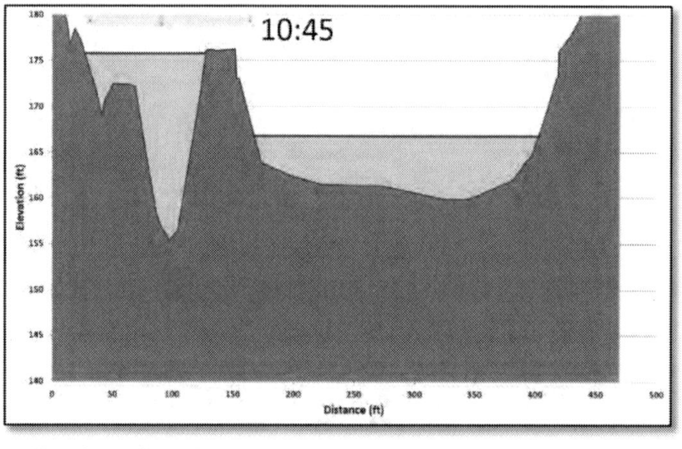

Figure 8 Visualization of HEC-RAS Cross Section WSEL (left) and Adjacent Storage Area WSEL (right)

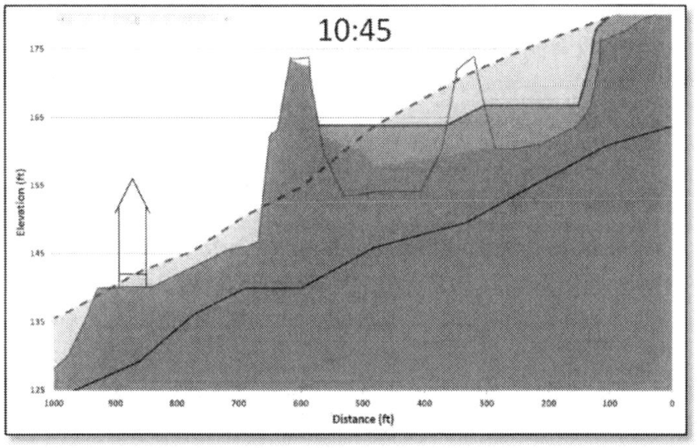

Figure 9 Visualization of HEC-RAS Stream Profile WSEL (dashed line / light blue) and Adjacent Storage Areas WSEL (solid line / dark blue)

Results of the Analysis

As stated above, the concern expressed by Hawaii DLNR was the potential for the remnant embankment to fail during the PMF, and the sudden release of any temporarily-impounded water. The graphical representations shown above indicate that for the majority of the flow event, the WSEL within the stream channel is higher than the WSEL within the reservoirs. This implies that if any failure were to occur, it would more likely be the embankment failing *into* the reservoirs and not the other way around.

We presented these results to Hawaii DLNR over web conference. We were able to demonstrate that even during the PMF, the depth and volume within the reservoirs does not appear to create a condition where the remnant embankments could fail and release a surge of water into the downstream reach. With the concurrence of Hawaii DLNR, we proceeded to complete the breach design based on the 100-year inflows from the upstream watersheds only.

Comparison to 2D HEC-RAS Model

Following the completion of the design and the then-recent release of HEC-RAS 5.0 with 2D modeling, we created a comparison model. The 2D grid was created from data exported from our Civil 3D design surface, and the inflow hydrographs used in the 1D model are identical to those used within the 2D model. The maximum WSEL within the reservoirs matched within a reasonable level, and the results predicted by the 1D model are reflected within the 2D model. Figure 10 shows a snapshot of the 2D modeling results. This graphic depicts the greater depths within the channel section (darker blue) compared to the depths within the adjacent reservoir.

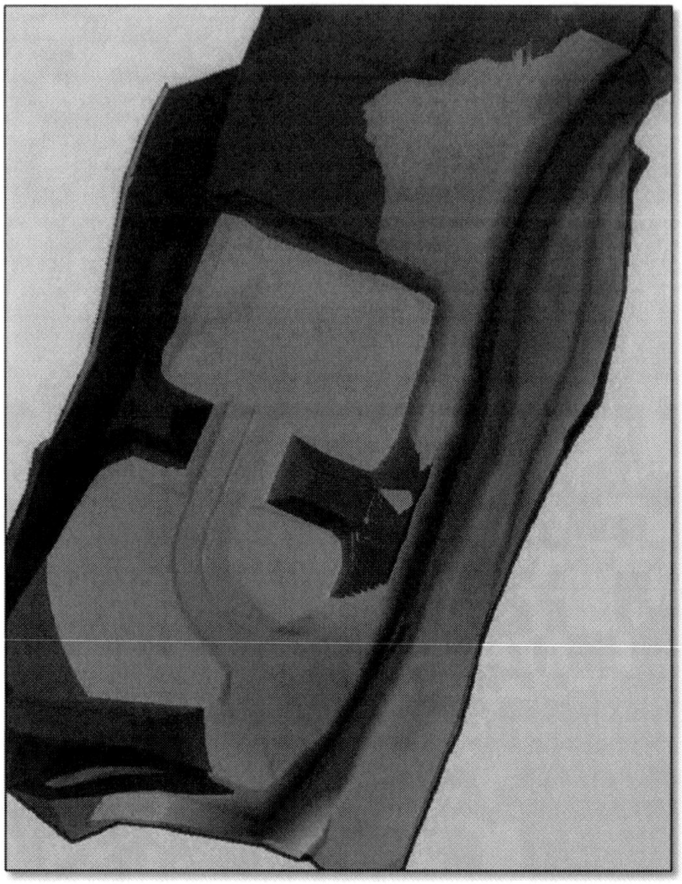

Figure 10 Graphical Output of 2D HEC-RAS Model results

Recommendation for Additional Investigation

The 2D HEC-RAS model has not been advanced beyond the initial comparative analyses described above. In the future, this model can be updated with additional detail such as spatially variable roughness and refinement of the two dimensional grid. While it is not expected for the results to match precisely between the 1D and 2D models, close agreement will lend credibility to the initial one dimensional analysis performed for this project.

CULTURAL SENSITIVITY

Immediately to the north of Reservoir No. 2 sits a lo'i, or a kalo patch, which is still being cultivated to this day. This lo'i is accessed by crossing the Reservoir Nos. 2 and 3 embankments and is owned by the family of Ekolu Lindsey, who continues the centuries-old tradition of cultivating kalo.

Figure 11 Lo'i (foreground) next to Reservoir #2 (background, prior to draining) (Maui Cultural Lands - used with permission)

The existence and maintenance of this lo'i is very important to preserve a part of the Hawaiian heritage. In recognizing this importance, we took additional steps in the preparation of the design and construction documents in order to prevent disturbance of this valuable piece of land, to maintain the water source from the nearby ditch, and to maintain access for the local residents who care for it. These steps may also include informing local community members at a public forum of the planned construction activities and revegetating the disturbed areas with a seed mixture of only native grasses and plants. We are conscious of what can be a delicate situation and want to strike an appropriate balance between the need to remedy the deficient dam and public safety hazard, and the need to preserve Hawaii's natural resources for those who cherish them.

CONCLUSION

Following the discovery of a dam safety issue at Ukumehame Reservoir Nos. 2 & 3, we were presented with a unique opportunity and challenge. This project required some understanding of the local culture and history, and sensitivity to the needs of DLNR, our client and the local residents.

The analysis of the Ukumehame Reservoirs was a bit atypical; offline reservoirs which nonetheless receive a substantial inflow volume from an adjacent stream. A concern was expressed that, even though the embankments were to be breached, could the capacity of the breached section restrict flows as they pass through the reservoir? And if so, could a failure of the remnant embankments during a flood event release a surge of water and still pose a threat to people or property downstream? In order to understand this situation, an unsteady HEC-RAS model was created which allowed us to understand the how flows within the main stream and reservoirs relate. This allowed us to suggest that a failure of the remnant embankments would likely not release a surge of water downstream because the reservoir WSEL is lower than the adjacent stream WSEL for the majority of the flow duration.

Additionally, this project required sensitivity to other, non-technical issues such as the cultural and historical significance of the project area, and how the proposed modifications to the reservoirs will impact the local residents. For this project to be a success once construction is complete, it will require more than a design which simply addresses the need to breach the reservoirs; it will require reaching out to local residents, educating them about the project, learning about their concerns and preparing into a design which does its best to satisfy the needs of those involved.

ACKNOWLEDGEMENTS

Thank you to Mark Lake for contributing to the project background by coming through on records requests and thank you to Ekolu Lindsey for sharing insights into the cultural importance of the Ukumehame Reservoirs and your personal experiences with the adjacent lo'i.

REFERENCES

Alexander & Baldwin, Inc. *Alexander & Baldwin Announces Transition Of Hawaiian Commercial & Sugar Company To A Diversified Farm Model*. 6 January 2016. 03 October 2016. <http://phx.corporate-ir.net/phoenix.zhtml?c=85663&p=irol-newsArticle&ID=2127059>.

Associated Press. "End of an era: Hawaii's last sugar mill wraps up final harvest." Honolulu: Honolulu Star-Advertiser, 12 December 2016. <http://www.staradvertiser.com/2016/12/12/business/business-breaking/end-of-an-era-hawaiis-last-sugar-mill-wraps-up-final-harvest/>.

Gannett Fleming, Inc. "Condition Assessment Report - Ukumehame Reservoir Nos. 2 & 3." April 2009.

Hawaii Department of Land and Natural Resources. "Hawaii Administrative Rules chapter 13-190.1." Honolulu, 14 March 2012.

Matsuda, Edwin, et al. "State of Hawaii - DLNR Dam Safety Inspection Sheet for Ukumehame Reservoirs." Dam Inspection Report. June 16, 2008.

Maui Cultural Lands. *Ukumehame*. 2015. <http://mauiculturallands.org/projects/ukumehame>.

U.S. National Park Service. *Old Sugar Mill of Koloa*. n.d. 03 October 2016. <https://www.nps.gov/nr/travel/Asian_American_and_Pacific_Islander_Heritage/Old-Sugar-Mill-of-Koloa.htm>.

Wadsworth, Harold Anderson. "Irrigation Census of Hawaii with some Comparisons with Continental United States." *Hawaiian Sugar Planters' Association*. Honolulu: Advertiser Publishing Co., Ltd., 1935.

ATOKA DAM AND SPILLWAY MODIFICATIONS

Patrick Miles, P.E.[1]
Brad Kirksey, P.E.[2]
John Rutledge, P.E.[3]
Sam Samandi, P.E.[4]
John L. Hare, P.E.[5]
Andrew Mishler, E.I.T.[6]

ABSTRACT

Atoka Dam is currently owned and operated by the Oklahoma City Water Utilities Trust and is a critical component of the Trust's water supply system. Completed in 1959, the dam is a zoned earthen embankment approximately 80 feet tall and 800 feet long, located 100 miles southeast of Oklahoma City. The spillway is a 324-foot-long side channel spillway that spills into a stilling basin, which has suffered severe damage as various components of the concrete chute have broken off and moved during multiple storm events. The competent rock in the area has allowed the spillway structure to survive multiple floods even with a heavily damaged stilling basin. The dam in its current condition passes about 35% of the PMF, well short of the 75% state requirement.

Freese and Nichols, Inc. performed an alternatives analysis on the dam and spillway to review modification alternatives to meet Oklahoma dam safety regulations and to rehabilitate the damaged spillway. A wide range and combinations of conceptual modification alternatives were considered to repair or replace the spillway, and two-dimensional HEC-RAS was used to study the capacity and flow patterns of the existing spillway and each alternative. The steep, rocky terrain greatly limited available space for expansion and complicated potential construction sequencing. The alternatives reviewed required raising the embankment between 2.5 and 11.5 feet, which was complicated by the existing water supply pump station at the downstream toe. In all, the alternatives considered resulted in a matrix of 15 potential modification alternatives.

The paper presents a case study of the unique situation faced at Atoka Dam and the variety of intricate and innovative alternatives reviewed to develop a solution.

[1] Freese and Nichols, Inc., 4055 International Plaza, Suite 200, Fort Worth, Texas 76109, 817-735-7236, patrick.miles@freese.com.
[2] Freese and Nichols, Inc., 4055 International Plaza, Suite 200, Fort Worth, Texas 76109, 817-735-7535, brad.kirksey@freese.com.
[3] Freese and Nichols, Inc., 4055 International Plaza, Suite 200, Fort Worth, Texas 76109, 817-735-7284, john.rutledge@freese.com.
[4] Oklahoma City Water Utilities Trust, 420 West Main Street, Suite 500, Oklahoma City, Oklahoma 73102, 405-297-3877, sam.samandi@okc.gov.
[5] Oklahoma City Water Utilities Trust, 420 West Main Street, Suite 500, Oklahoma City, Oklahoma 73102, 405-297-3681, larry.hare@okc.gov.
[6] Oklahoma City Water Utilities Trust, 420 West Main Street, Suite 500, Oklahoma City, Oklahoma 73102, 405-297-2094, andrew.mishler@okc.gov.

INTRODUCTION

Atoka Dam, which forms the Atoka Reservoir, is currently owned by the City of Oklahoma City (City) and is operated and maintained by the Oklahoma City Water Utilities Trust (Trust). It is a critical component of the City's water supply system. Atoka Dam, completed in 1959, is located slightly more than 100 miles southeast of Oklahoma City on North Boggy Creek five miles northeast of Atoka, Oklahoma. The zoned, earthen embankment dam is approximately 80 feet tall and 800 feet long with a dam crest elevation of 600.0 feet and a masonry wave wall up to elevation 602.5 feet. The dam embankment has a 3H:1V slope on the upstream face, protected by rock riprap and a 2H:1V slope on the downstream face with grass erosion protection. Despite the steep slope on the downstream face, the dam has no known history of slides or signs of slope failures or instability.

The spillway is a 324-foot-long side channel spillway that spills into a varying width chute, with a maximum width of about 53 feet. The spillway crest is shown in Figure 1. The stilling basin of the spillway has suffered damage over several years as various components of the concrete chute have broken off and moved during multiple storm events. Figure 2 shows a view of the remaining portions of the chute looking downstream. Though the competent rock in the area has allowed the spillway structure to survive multiple floods even with a heavily damaged stilling basin, the structure is currently in a condition that cannot provide for any confidence in its long-term future use and needs to be replaced.

Figure 1. Side-Channel Spillway Crest

Figure 2. Damaged Spillway Chute

The first step of this study developed updated hydrologic and hydraulic computer models in order to provide a current estimate of the dam and spillway's compliance with state spillway capacity rules. The models were then used to review various modification alternatives for the dam and/or spillway that would enable the project to become compliant as well as repair the heavily damaged spillway stilling basin. A field inspection and limited geotechnical exploration were also performed to help identify the current condition of the structures involved.

HYDROLOGIC AND HYDRAULIC ANALYSIS

The hydrologic and hydraulic model for Atoka Dam utilizes a combined approach with HEC-HMS and HEC-RAS for routing flood events through the contributing drainage area to the Atoka Reservoir. The overall basin delineated for the original USGS model is further subdivided into eight drainage basins representing approximately 171.6 square miles, or 109,800 acres. A separate model for developing spillway rating curves was created utilizing two-dimensional HEC-RAS.

The Atoka spillway has a crest elevation of 590 feet with a total length of 324 feet. The spillway discharges through a concrete-lined chute along the downstream toe of the dam into a stilling basin and then down North Boggy Creek. The existing conditions discharge rating curve was calculated using a two-dimensional hydraulic model in HEC-RAS, which provided two primary findings: 1) In existing conditions, velocities through the side-channel spillway are significantly slowed toward the left one-third to one-half of the crest length, resulting in ineffective flow and reduced spillway capacity; and 2) The discharge capacity of the spillway chute accentuates the ineffective flow problem by choking down flows through the chute and increasing tailwater immediately downstream from the spillway crest.

Figure 3 provides a screen capture from the two-dimensional model demonstrating the reduced velocities (blue to green color) near the left (north) end of the spillway crest, compared with the normally expected high velocities (yellow to red color) near the right (south) end of the spillway.

Figure 3. Two-Dimensional Hydraulic Model Results (Existing Conditions)

A range of frequency flood events, from the 100-year, or 1% annual chance, to the 1,000-year flood, was analyzed for existing conditions and each of the Atoka Dam spillway modification alternatives to demonstrate the functionality of the existing spillway and proposed modifications under a range of extreme flood events. Flood model results are shown in Table 1 for the 100-year, 500-year and 1,000-year flood events under existing conditions.

Table 1. Frequency Flood Event Results (Existing Conditions)

Event	Peak Lake Elev (feet)	Peak Discharge (cfs)
100-yr	596.2	15,345
500-yr	598.5	19,799
1,000-yr	599.6	21,832

The Probable Maximum Flood (PMF) was modeled for existing conditions and each modification alternative with the reservoir at the normal pool elevation. According to Oklahoma Water Resources Board (OWRB) regulations, Atoka Dam is classified as a large-size, high-hazard potential structure and, as such, is required to safely pass 75% of the PMF to be in compliance with the state regulations. The routing results show a peak lake level of 610.6 feet for existing conditions with a peak discharge of 43,900 cfs during the 75% PMF event. For this scenario, it is assumed that the top of dam would be theoretically raised such that the flood event would be contained and all discharges are released through the existing spillway; therefore, no overtopping flows were accounted for in these computations. A similar run of the model showed that the dam in its current condition passes 35% of the PMF with the top of the dam at elevation 600.0 feet, with a peak discharge of 22,600 cfs. As noted in the previous section, this is roughly equivalent to the 1,000-year flood. If the top of the wave wall is considered the top of the dam, the structure passes 44% of the PMF at elevation 602.5 feet, with a peak discharge of 27,400 cfs. Both are well short of the 75% requirement of the OWRB. A summary of the PMF results is shown in Table 2.

Table 2. Probable Maximum Flood Results (Existing Conditions)

Event	Peak Lake Elev (feet)	Peak Discharge (cfs)
35% PMF	600.0	22,600
44% PMF	602.5	27,400
75% PMF	609.6	43,900

FIELD INVESTIGATION FINDINGS

A detailed field investigation of the dam resulted in the following key observations and conclusions:

1. The structure has suffered considerable scour and damage to the downstream end of the spillway chute, as shown in Figure 4.

2. The current end of the chute is at an expansion joint where the foundation keyway still maintains fair contact with the rock, as shown in Figure 5. The key is displaced but is not at an immediate risk of being undermined. The drain pipes are also exposed, but are still functional.

3. Overall, the majority of the concrete in the chute appears to be in fair to good condition. There was consistent damage at the joints, with numerous failures of the water stops and spalling at the joints. At the upstream end of the chute across from the crest, more voids and spalling were noted. Several locations have exposed rebar with noticeable corrosion, though no signs of leakage were noticed.

Figure 4. Scour Hole Below Spillway Chute, Looking Upstream

Figure 5. Downstream End of Damage Chute

4. The vertical wall immediately downstream from the crest shows more deterioration than the rest of the chute, with several locations showing leakage through voids in the concrete. Rebar is also exposed in some of these voids. Much of the leakage is high up on the wall and on the crest, demonstrating full lake head on the downstream side of the crest. The algae growth demonstrates that this is a normal condition for long periods of time, suggesting that corrosion of the steel in

portions of the wall is likely to be extensive. At the north end of the crest, cracking and bulging of the crest slab also exists in the downstream direction. This demonstrates that the hydraulic pressure from the lake has caused this relatively thin slab to fail. In addition, there are multiple vertical displacements of the crest slabs, up to a maximum of about four inches, in this location. This suggests that portions of the crest slab have settled and/or some have shifted upwards due to shifting of the material on which it is founded. Either way, it indicates overload and failures of portions of the crest slabs.

5. The embankment berm on the south end of the reservoir appears to be in good condition, in spite of excessive vegetation growth on the slopes.

As a result of this visual inspection, a review of the drawings, and subsequent design calculations, the following conclusions were reached:

1. The majority of the chute appears to be in fair to good condition. Repairs should be feasible, provided the reinforcing steel is in good condition. The chute will not meet today's industry standards for structural design under design loads from the original design, let alone for the additional loading from the higher discharges considered with the current design storm. This can likely be handled with extensive anchoring to the competent rock below with a concrete overlay that engages the anchor heads. Repairs of the joints and spalling would be affected as part of the overlay.

2. The crest wall is in the worst condition of the various components of the spillway. Due to the extensive leakage, voids, and numerous isolated failure locations, it is unlikely that much of the crest structure could reasonably be salvaged. It is clearly overloaded and likely has relatively extensive corrosion of the reinforcing steel. It is assumed that the entire crest wall will need to be replaced in reviewing the various alternatives.

SLOPE STABILITY ANALYSES

Slope stability analyses were performed using the Spencer method for the existing embankment dam slope to assess the impact of increasing the top of dam and the potential maximum reservoir storage to as high as Elevation 611.0 feet. The analyses were performed for an embankment cross section approximating the original creek channel or the tallest portion of the embankment, using the computer model SLOPE/W developed by GEO-SLOPE International Ltd. The subsurface profiles for the embankment cross section are inferred from the results of the historic and most recent geotechnical borings and take into consideration prior embankment modifications and the observed site geology. The results of the analysis are described in Table 3, along with the recommended minimum factors of safety provided in Corps of Engineers manual EM 1110-2-1902.

Table 3. Summary of Slope Stability Analyses

Loading	Slope Analyzed	Calculated Factor of Safety	Recommended Factor of Safety
Steady State Seepage	Downstream – Shallow Failure	2.1	1.5
	Downstream – Deep Failure	1.9	1.5
Rapid Drawdown	Upstream	2.0	1.3

MODIFICATION ALTERNATIVES

Alternatives were developed to simultaneously resolve two key issues for the Atoka Dam. These are to repair the damaged spillway and to modify the dam and spillway as needed to meet OWRB criteria for flood passage. All the alternatives require a combination of spillway improvement and an appropriate raising of the dam embankment that would combine to allow the structure to pass the required flood. No practicable alternatives were found that allowed for the required design flood to be passed with the current top of dam. Multiple alternatives were reviewed for each component of the project. Two alternatives were reviewed for replacing the stilling basin, both of which involve protecting the exposed foundation material. Two general alternatives were also reviewed for the chute: 1) Repairing the chute as needed, and 2) Completely replacing the chute to pass the higher flows. Because the inspection found that the spillway crest is likely in a structural state that would not be effective to repair and will need to be replaced regardless, two alternatives were reviewed for a new crest. One alternative replaces the crest in essentially its current location and the other alternative replaces the crest 50-feet upstream to improve the hydraulic performance of the existing chute. Each of these is assumed to be coupled with the option to repair the existing chute. The crest reconstruction options do not significantly increase the capacity of the spillway and would therefore need to be combined with a significantly higher increase in the dam embankment height compared to the alternatives that completely replaced the spillway.

Three alternatives were reviewed as a complete replacement of the spillway crest with a new configuration, a gated spillway and two versions of a labyrinth spillway. These are intended to dramatically increase the spillway capacity and allow for a smaller dam embankment raise. These are each combined with the alternative for replacing the spillway chute.

In addition, the higher lake levels created by the larger design storm, regardless of the spillway alternative, will create an overflow of the berm located in the south end of the reservoir. Flow through this area could potentially reduce the flow needed through the service spillway, but would produce added flood risks on the adjacent Muddy Boggy Creek in the town of Atoka. Alternatives for handling the south end, either to allow or block potential flood flows were also reviewed in conjunction with the spillway alternatives.

Stilling Basin Repair Alternatives

The stilling basin for the Atoka spillway has been destroyed by a series of flood events that displaced the downstream portions of the chute over a period of years. The foundation rock is quite strong and erosion resistant but is subject to weathering over time. Occasional removal of large rock pieces continues to occur with further weathering and significant spillway discharges. The high-quality rock prevented the damage from being worse and will serve as a good foundation for the repairs. However, the rock cannot provide a long term suitable surface and needs to be protected. Three repair alternatives were considered. The first would be to reconstruct a hydraulic jump stilling basin. The other two would be to protect the existing exposed foundation material to prevent further undermining and erosion with either Roller Compacted Concrete (RCC) or reinforced concrete combined with frequently spaced anchors intended to tie the concrete to the rock and engage the mass of the foundation.

The purpose of the stilling basin is to dissipate the energy of the spillway discharges before they are released into the downstream channel. Though the original design calculations were not available or reviewed, it is clear from the damage that there is insufficient tailwater for a properly performing hydraulic jump. This may have been due to erosion of much of the area downstream from the end of the chute. As this erosion process continued, it eventually undermined the various downstream components of the chute and displaced them. Reconstruction of the hydraulic jump was reviewed and calculations confirmed that, in its current condition, tailwater is insufficient to produce a hydraulic jump within the basin. It would be feasible to build a weir downstream from the new stilling basin to force the higher tailwater, but it would need to be approximately 20 feet high, which would create a new erosion concern downstream from the new weir as well as possible increased flooding risk for the pump station. For these reasons, a new hydraulic jump basin is not considered viable. For the various chute alternatives, an option to widen the chute by 40 feet was considered, for which a hydraulic jump basin would be viable with existing tailwater. However, it is estimated to cost an additional $3 million and is not considered financially viable. Based on these results, the hydraulic jump basin was not considered for further evaluation. The two protection options are considered equivalent and a choice between them will be made during final design.

Table 4. Summary of Stilling Basin Repair Alternatives

	Pros	Cons	Action
Reconstruct Hydraulic Jump Basin	Fully dissipate high energy from spillway flows	Very high initial cost estimate	Discard due to cost and flood risk
Protect Exposed Foundation Material with RCC	Protect exposed rock from further weathering/erosion	High energy flows dissipate in plunge basin	Carry forward to Final Design
Protect Exposed Foundation Material with Reinforced Concrete	Protect exposed rock from further weathering/erosion	High energy flows dissipate in plunge basin	Carry forward to Final Design

Spillway Chute Alternatives

The spillway chute is a rectangular concrete channel that turns sharply from the spillway crest toward the natural channel downstream. The chute crosses under a bridge for the publicly accessible Lake Road and varies in width from about 45 feet wide at the downstream end of the spillway crest to about 53 feet under the bridge to about 43 feet wide at the current downstream end of the chute. The original end of the chute was about 33 feet wide. Much of the chute is in fair to good condition, though extensive work would be needed to repair the joints and numerous spalls and cracks. In addition, the chute would be structurally undersized by today's structural codes and standards for flow in a full channel. However, because the current regulatory design flood of 75% PMF is much higher the design flood used in the original design, the peak water depths in the chute will be much deeper than the existing chute walls but would be contained by the excavated rock slopes above the walls. In order to address these issues, two alternatives, repairing and replacing the chute were considered. For the repair alternative, it is assumed that the walls on both sides will be anchored to the rock behind it. The anchors will have heads exposed in the chute and a new, relatively thin slab of reinforced concrete will be placed over the anchor heads to fully engage the rock mass and to strengthen the walls. The same treatment would be applied to the floor slab as well due to joint issues and concerns about the long-term viability of the drains under the slab. The spillway appears suitable for the repairs as indicated, but there are unknowns that would need to be confirmed during the final design phase, such as the integrity of the reinforcing steel. If significant corrosion exists, then some of the existing concrete would need to be replaced rather than repaired in place, likely increasing the costs for the alternative significantly. Additional slope protection or wall extensions would be needed to contain and protect the area from the higher flows at the new design storm level of 75% of the PMF.

For the spillway alternatives that involved a full spillway replacement, a new chute was assumed. Based on the available plans and topography, it is estimated that the chute can only be widened a maximum of 40 feet, which will be 93 feet at its widest. Hydraulic analysis showed that this is sufficient for the capacity needed to pass the additional flows associated with the new spillway options with minimal submergence of the various crest alternatives reviewed. Each of the alternatives that replace the chute will likely require a full replacement of the existing bridge, so this was assumed in the associated cost estimates.

Table 5. Summary of Spillway Chute Alternatives

	Pros	**Cons**	**Action**
Repair Chute in Existing Configuration	Reduced construct cost/care of water sequencing	Reduced hydraulic capacity	Combine with crest reconstruction alternatives (1 and 2)
Widen Chute by Approx. 40ft	Additional hydraulic capacity	Additional construction cost	Combine with full replacement alternatives (3, 4, and 5)

Spillway Crest Alternatives

A full range of alternatives was reviewed to look at possible ways to provide the needed spillway capacity for the dam to pass the design flood 75% of the PMF. All alternatives include replacing the crest which was deemed to be unsafe to repair in its current condition. Alternative 1 is a replacement of the crest in place, using the existing wall as a form with the new extension anchored to the rock behind the wall. This also assumes a thin slab overlay with anchors drilled into the rock below or behind the slab and wall, as appropriate, similar to that described for the chute repairs. The intention of this alternative is to leave the existing spillway intact as much as possible. Because this does not increase spillway capacity, it requires the largest subsequent increase in the dam height, from the current elevation 600.0 feet to as high as 611.5 feet.

The second spillway crest option, Alternative 2, is similar to the first, except that the control weir is moved 50 feet out into the lake. A hydraulic model of the spillway was used to review the hydraulic flow characteristics. Moving the weir further away from the chute significantly reduces the energy losses associated with the flow being redirected down the chute, increasing the efficiency of the weir by about 20%. However, the chute capacity is still constrained and the hydraulic improvement for the design flood limited, reducing the needed increase in the top of the dam to a maximum of 609.5 feet.

Three other alternatives were reviewed that completely replace the spillway and the chute in order to maximize the spillway discharge and minimize the subsequent needed increase in the top of dam.

The first of these, Alternative 3, is to replace the current spillway with a new gated spillway. Since the approach channel to the spillway crest is at a consistent elevation of 585.0 feet, the gated structure is assumed to have a crest of 585.0 feet with seven 40-foot wide leaf gates that will open by lowering down and allowing the water to pass across the top. The structure will need piers in between the gates which will make the total width similar to the existing crest and an operating bridge over the top for access and operation. The option is combined with the proposed 40-foot chute widening and provides sufficient discharge capacity to reduce the new required top of dam down to 605.5 feet. However, a gated spillway is considerably more expensive than other comparable alternatives and requires much more operations and maintenance costs, as well as potential flood liability risk due to the operational issues. Therefore, the gated spillway alternative is not considered viable.

The second of the full replacement options, Alternative 4, is a labyrinth spillway. Employing a curved 14 cycle configuration for the labyrinth would improve the hydraulics associated with the flow entering the chute as well as reduce the construction costs over a straight alignment. The selected full length of 248 feet matches the capacity of the fully widened chute option. The additional discharge from a longer labyrinth weir would be lost due to submergence from the chute reaching its capacity. This configuration requires a top of dam elevation of 606.5 feet, one foot higher than the gated option and 6.5 feet above the existing crest. However, it provides considerable benefits over the gated option as it is much more economical and eliminates the risks and costs associated with the operation of the gates. One drawback to the labyrinth is related to its increased efficiency. Because the labyrinth has relatively high discharges at lower levels, it reduces the 100-year flood level on the lake by approximately 1.7 feet but also increases the 100-year discharges downstream of the dam by about 60%. While the curved labyrinth spillway provides very efficient discharge capacity, increasing the downstream flood risk is undesirable.

The final option, Alternative 5, attempts to address this issue, by using a compound labyrinth configuration, as shown in Figures 6 and 7. The lower level will be a 4-cycle labyrinth at elevation 590.0 feet, the normal water level. This is sized to approximate the current 100-year flood level on both the lake and in the downstream channel. The longer section will be a 12 ½-cycle labyrinth at elevation 596.0 feet that will only begin to engage at the 100-year flood level. This upper weir will act as an emergency spillway, engaging only during extreme flood events. The greater total length makes up for the lower capacity relative to the initial labyrinth alternative, resulting in a similar required top of dam elevation of 606.5 feet. It also has the added advantage of requiring less excavation and a shorter weir height for the long section due to the limited total overtopping depth.

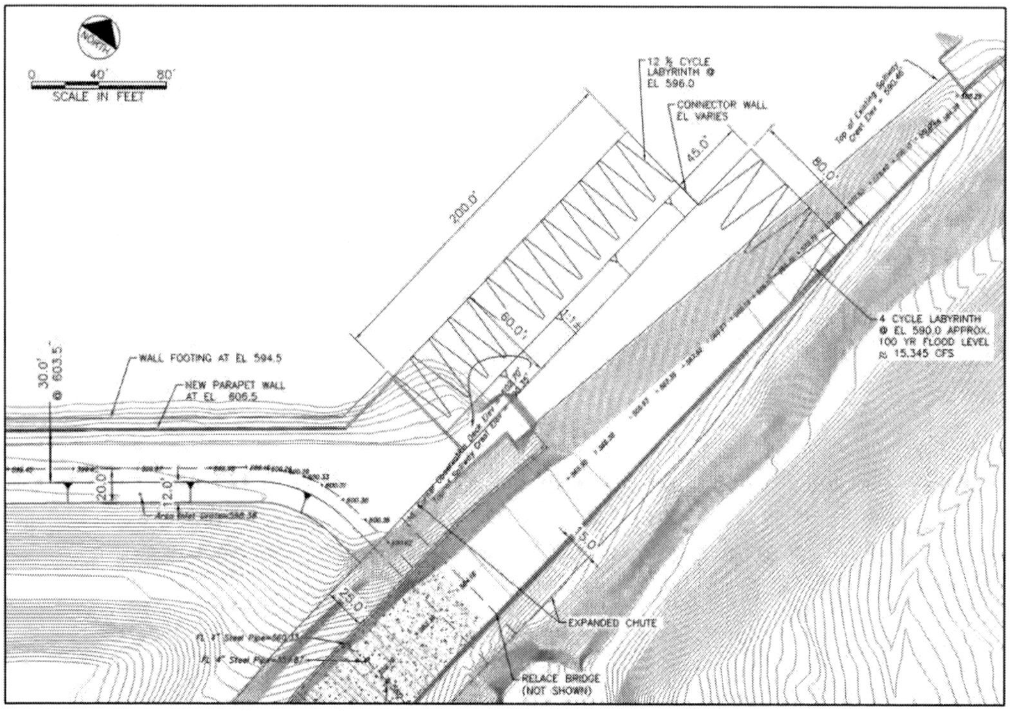

Figure 6. Compound Labyrinth Weir (Plan View)

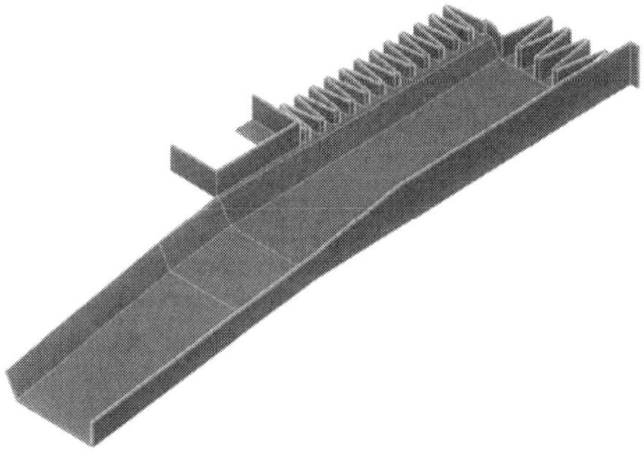

Figure 7. Compound Labyrinth Weir (Isometric View)

Table 6. Summary of South Reservoir End Alternatives

	Pros	Cons	Action
Alt 1 – Repair Spillway Crest in Current Location	Least impact/ change from existing conditions	No increased capacity; highest dam raise	Carry through conceptual cost estimate
Alt 2 – Replace Spillway Crest 50ft Into Reservoir	Approximate 20% increased capacity over Alternative 1	Low capacity; high dam raise	Carry through conceptual cost estimate
Alt 3 – Replace with Gated Spillway	Sufficient capacity for smallest dam raise	High construction, O&M costs; operations liability	Carry through conceptual cost estimate
Alt 4 – Replace with Curved Labyrinth	Sufficient capacity for small dam raise	Increase discharge by approximately 60% for 100-year	Carry through conceptual cost estimate
Alt 5 – Replace with Two-Stage/ Compound Labyrinth	Sufficient capacity for small dam raise; constrain 100-year discharge to match existing	Some dam raise still required; higher cost than replacement alternatives	Carry through conceptual cost estimate; Recommended alternative

South Reservoir End Alternatives

A low point exists at the south end of the reservoir. This was noted in the original design and a berm was built up to elevation 602.0 feet to prevent any floodwaters from being released in this direction. Any higher lake levels will release flow out of the south end and this discharge will flow across State Highway 69 creating flooding concerns there and on down to Muddy Boggy Creek, aggravating potential flood levels on the creek in the town Atoka as well.

All of the spillway and chute options being considered require raising the existing top of the dam to a level above 602.0 feet. Therefore, the higher flood water elevations on the south end and their effect on the existing berm were considered. The option to allow flood flows to be discharged from the south end is technically viable as it acts only in very rare events and will reduce the needed size and cost of the modified spillway, chute, and the raised dam elevation.

Three options were considered for the treatment of the south end of the reservoir in place of the existing berm. One is to raise the berm elevation sufficiently to contain the increased design flood level of the 75% PMF, eliminating any discharges and flood liability risks from the area. The second is to leave the berm "as is" and allow for higher lake levels to be released through the approximately 1,000-foot-wide opening, serving as an emergency spillway. The third option is to replace the existing berm with a fuse plug, which is an embankment designed to wash out when overtopped, thus opening up significantly more conveyance and therefore discharge capacity once the lake levels reach a designated level. These planned wash outs can be staged with concrete separator walls so only a portion of the fuse plug erodes at one time. The top of the fuse plug will be set so that it does not operate until a flood greater than a 1,000-year flood has occurred. Potential discharges through this area may impact a proposed pump station site. Careful coordination of the two designs will be needed if either of the two emergency spillway options are selected.

These last two options for an emergency spillway on the south end of the reservoir increase the total capacity of the combined spillway system at the dam and reduce the needed additional height to the existing dam. However, allowing flow through the south end will add significantly to the costs if flood easements are acquired for the area impacted, which is likely between 1,500 and 2,500 acres. Because an emergency spillway on the south end would not operate unless a flood of magnitude greater than the 1,000-year flood occurs, the likelihood of ever flooding these properties is low. However, the potential liability is very high in such an occurrence. Whether the costs of such potential liability is measured in the cost of actually acquiring the flood easements or estimated as some fraction of the property values, the wide extent of potential land impacted is high enough that it was decided that alternatives allowing flow through the south end would not be adopted.

Table 7. Summary of South Reservoir End Alternatives

	Pros	**Cons**	**Action**
Raise Berm to Contain Design Flood	No additional downstream impacts	Less discharge capacity, higher dam raise	Carry forward to Final Design
Allow Flows Over Existing Berm as Emergency Spillway	Additional spillway discharge capacity	Land acquisition/flood easement; downstream impacts	Discard due to cost/downstream impacts
Allow Flows through Fuse-Plug Emergency Spillway	Additional spillway discharge capacity	Land acquisition/ flood easement; downstream impacts	Discard due to cost/downstream impacts

Embankment Alternatives

As described above, all proposed spillway modifications require a significant increase in the reservoir storage to pass the new design flood and meet state regulations for the 75% PMF design flood. This can only be achieved by raising the existing crest of the dam. The maximum needed increase in the top of dam elevation was estimated to be up to elevation 611.5 feet, an 11.5-foot raise of the current dam crest. Because of the existence of the pump station at the downstream toe of the dam, raising the dam crest solely with compacted fill is not considered feasible. Through discussions with the Owner, raising the top of the dam through the sole use of a tall parapet wall was also discarded as it would block all views of the lake from the road and would be a difficult obstruction for maintenance of the upstream slope and for access to the intake tower.

Therefore, a combination of compacted fill and a parapet is needed and a general configuration was developed that includes a retaining wall on the upstream slope combined with compacted fill behind it that raises the top of the embankment. The retaining wall would have a three-foot extension serving as a road barrier as well as additional flood retention for the dam. The retaining wall was assumed to have a six-foot-wide footing on the lake side that would serve as an access walkway for the front face of the dam as well as to the walkway for the intake tower. The 611.5 top of dam elevation, the tallest of the configurations considered, would actually have this walkway at elevation 590.0 feet, the normal water level, which would mean it would be frequently submerged and would face difficult design constraints given its height and location at the normal water level. For these reasons, this was considered the maximum possible dam raise for this configuration and shorter raises would be strongly preferable for similar cost ranges.

Figure 8 shows a typical configuration, applied to a top of dam of 606.5, which corresponds to the recommended alternative moving forward to final design.

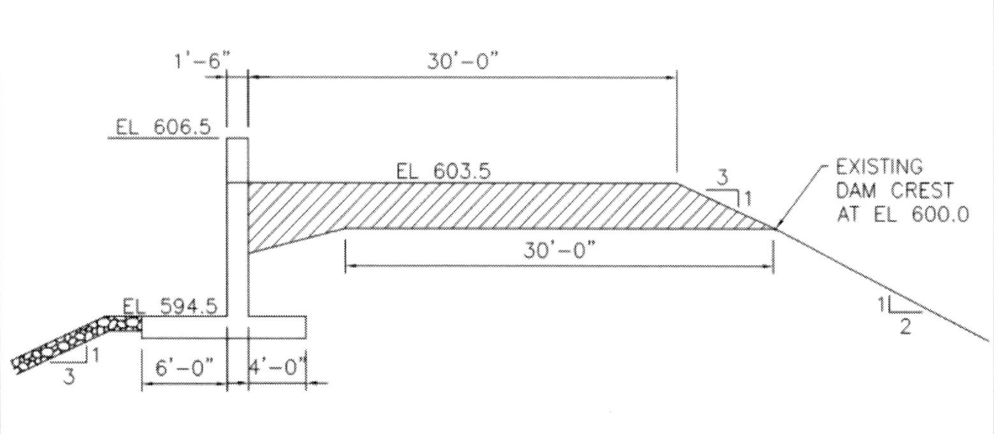

Figure 8. Dam Raised to Elevation 606.5 feet

Table 8. Summary of Embankment Alternatives

	Pros	Cons	Action
Raise Top of Dam with Parapet Wall	Avoids conflict with pump station	Obstructed view of lake from public access road	Discard due to obstructed view
Raise Top of Dam with Compacted Fill	Lower preliminary cost estimates	Extended slope conflicts with existing pump station at toe	Discard due to pump station conflict
Raise Top of Dam with Combined Wall and Fill	Balances concerns from both wall and fill alternatives	---	Carry forward to Final Design

HYDRAULIC REVIEW OF ALTERNATIVES

A wide range of conceptual modification alternatives was considered during the study. They generally can be defined based on the spillway crest alternative. These are:

1. Repaired Existing Spillway with existing chute dimensions
2. Repaired Chute with Spillway Crest offset 50 feet upstream (into the lake)
3. Gated Spillway with widened chute
4. Curved Labyrinth Spillway with widened chute
5. Compound Labyrinth Spillway with widened chute

Each of the spillway crest alternatives was evaluated utilizing a two-dimensional hydraulic model. In some cases, such as the labyrinth spillway alternatives, the model was only used to determine a tailwater rating curve within the spillway chute. Standard numerical calculations were used for the spillway crest itself, and additional detailed hydraulic modeling, including a physical model study, will be a significant component of the final design process.

However, the two-dimensional model does provide a visual comparison of the velocity distribution and relative hydraulic efficiency the proposed spillway configurations. The hydraulic model results shown in Figure 9 compare the existing conditions spillway with the Compound Labyrinth Spillway proposed as Alternative 5. As before, blue to green colors represent low velocities, while yellow to red represent high velocities.

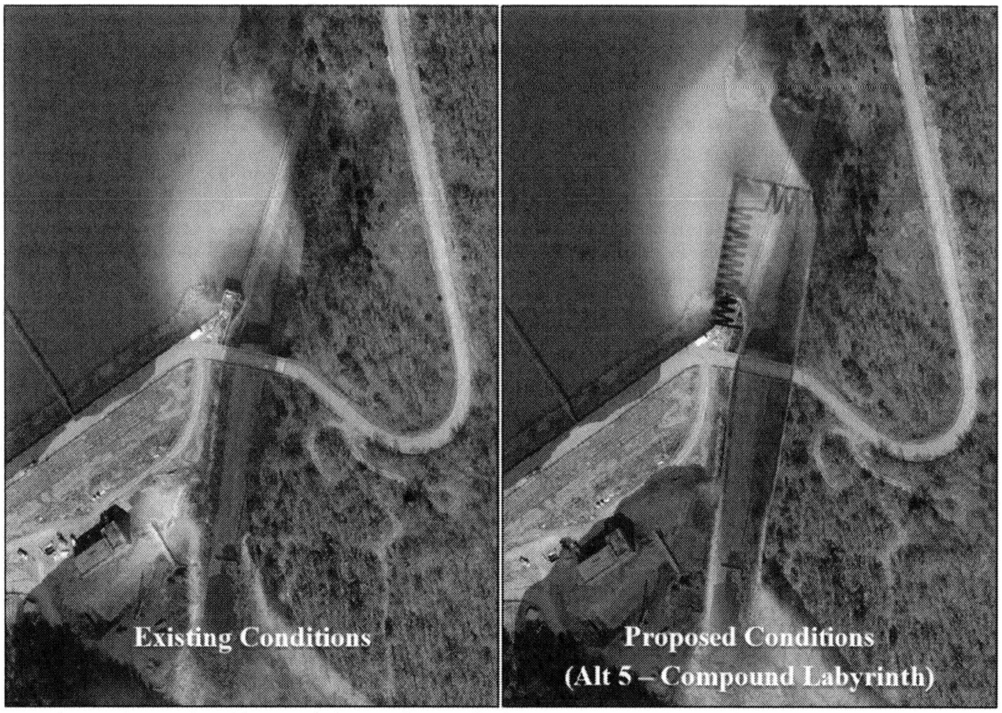

Figure 9. Hydraulic Model Results (Existing & Proposed Conditions at 40,000 cfs)

Flood routing model development focused on flows ranging from the 100-year flood to the 75% PMF Design Flood. The models were used to route these floods through the spillway for various alternatives and combinations. The results from the 100-year storm event runs for existing conditions and each alternative are shown in Table 9. These results assume no flows through the south end of the reservoir because the 100-year flood event does not reach the emergency spillway elevation.

Table 9. 100-Year Flood Model Results

Principal Spillway Configuration	Peak Elevation (feet)	Spillway Discharge (cfs)
Existing Conditions	596.2	15,345
Alt 1 – Repaired Spillway	596.6	14,788
Alt 2 – Repaired Spillway with 50' Offset	596.3	17,389
Alt 3 – Gated Spillway	N/A	N/A
Alt 4 – Curved Labyrinth Weir	594.5	24,735
Alt 5 – Compound Labyrinth	596.5	16,339

Table 10 contains the results of the 75% PMF model runs for the five spillway alternatives. For all scenarios, it is assumed that the top of dam will be theoretically raised such that the flood event will be contained; therefore, no overtopping flows are accounted for in these results.

Table 10. Probable Maximum Flood Model Results (75% PMF)

Principal Spillway Configuration	Peak Elevation (feet)	Spillway Discharge (cfs)
Existing Conditions	610.6	43,949
Alt 1 – Repaired Spillway	611.2	41,630
Alt 2 – Repaired Spillway with 50' Offset	609.0	48,912
Alt 3 – Gated Spillway	605.2	65,360
Alt 4 – Curved Labyrinth Weir	606.2	74,500
Alt 5 – Compound Labyrinth	606.4	75,803

ALTERNATIVES COST COMPARISON AND RECOMMENDATION

Table 11 summarizes the estimated total construction costs associated with the different alternatives. No engineering or construction management fees are included.

Table 11. Estimated Costs for Atoka Dam Rehabilitation

Principal Spillway Configuration	Estimated Costs ($ Millions)
Alt 1 – Repaired Spillway	21.9
Alt 2 – Repaired Spillway with 50' Offset	23.1
Alt 3 – Gated Spillway	32.4
Alt 4 – Curved Labyrinth Weir	26.8
Alt 5 – Compound Labyrinth	25.6

Though the spillway appears suitable for the repairs as indicated, there are unknowns that would need to be confirmed such as the integrity of the reinforcing steel. If significant corrosion exists, then some of the existing concrete would need to be replaced rather than repaired in place, likely increasing the costs for the alternative significantly. In addition, the repair alternatives included a very large increase in the embankment height with up to a 21.5-foot tall retaining wall at the lake edge. The long-term performance of this wall on a saturated clay embankment is questionable. Because of these concerns and the relatively slight increase in costs to obtain a reconstructed spillway with a moderate embankment increase, the compound labyrinth appears to provide the best balance of risk reduction and long term durability. In addition, when compared to the single level labyrinth, it eliminates the potential negative impact to 100-year flood levels downstream.

CONCLUSION

The paper describes the numerous alternatives and combinations of alternatives developed to address the primary concerns at the Atoka Dam. These are to modify the dam to meet Oklahoma dam safety regulations, as administered by the OWRB, and to rehabilitate the spillway which has suffered significant damage, including the total loss of the stilling basin. The recommended alternative is the Compound Labyrinth Spillway with widened spillway chute and erosion protection for the exposed stilling basin. Final design of these modifications is underway with additional detailed hydraulic modeling, including a physical model study, and further review of construction sequencing and estimated costs to be completed later this year.

REFERENCES

Geo-Slope International Ltd., *Seepage Modeling with SEEP/W 2007*, Fourth Edition, 2009.

Geo-Slope International Ltd., *Stability Modeling with SLOPE/W 2007*, Fourth Edition, 2009.

Oklahoma Water Resources Board, Dam Safety Program: *Hydrologic and Hydraulic Guidelines for Dams in Oklahoma*, August 2011.

U.S. Army Corps of Engineers, Engineering Manual EM 1110-2-1902, *Slope Stability*, 2003.

RECENT EXTREME STORMS AND THEIR RELATION TO PMP

Bill Kappel[1]
Doug Hultstrand[2]
Geoff Muhlestein[3]

ABSTRACT

Extreme rainfall during June 2013 along the foothills of the Rocky Mountains west of Calgary caused extensive flooding in the region, leading to the costliest natural disaster in Canadian history at the time. A few months later, a similar extreme storm produced record rainfall and devastating floods along the Front Range of the Rockies from southern Wyoming, through Colorado and extending into southern New Mexico. More recently, during October 2015, a perfect combination of meteorological conditions resulted in record rainfall and hundreds of dam failures in South Carolina. So, has something changed, are PMP-type events becoming more common? Applied Weather Associates has analyzed each of these events using our Storm Precipitation Analysis System (SPAS), providing rainfall data at high spatial and temporal resolutions for use in Probable Maximum Precipitation (PMP) development. PMP is utilized by hydrologists to derive the Probable Maximum Flood and the observed storm data for hydrologic model calibration. This presentation will provide comparisons of each of these extreme events to historic storms, which control PMP and discuss how they relate to PMP values in the region where they are transpositionable.

INTRODUCTION

Probable Maximum Precipitation (PMP) has been used as a design basis for many state- and federally-regulated dams since the 1940s. Combined with a rainfall-runoff model for converting precipitation to a streamflow hydrograph, PMP is used to predict the timing, volume, and peak flow associated with extreme flood events at a dam. Designers obtain PMP values from a patchwork of hydrometeorological reports (HMRs) produced by the U.S. Weather Bureau or its successor, the National Weather Service (NWS), or from site-specific studies performed by private contractors.

Although the PMP continues to be widely applied as a design standard, there is no corresponding national-scale effort to update decades-old HMRs. New PMP studies since the 1990s are limited to regional or site-specific studies. Many recent site-specific studies have produced PMP values significantly different from the HMR values. Reasons for the differences include analysis of storms that have occurred since the publications of the HMRs, improved storm adjustment procedures, and technology advances such as weather

[1] Chief Meteorologist, Applied Weather Associates, PO BOX 175, Monument, Colorado, 80132, billkappel@appliedweatherassociates.com
[2] Senior Hydrometeorologist, Applied Weather Associates, PO BOX 175, Monument, Colorado, 80132, dhultstrand@appliedweatherassociates.com
[3] Senior GIS Analyst, Applied Weather Associates, PO BOX 175, Monument, Colorado, 80132, gmuhlestein@appliedweatherassociates.com

radar and geographic information system (GIS) software. Applied Weather Associates (AWA) has completed more than 100 PMP studies over the last 20 years (Figure 1). These studies have continued to advance the science and understanding of extreme rainfall and application of PMP values. Data sets have been continuously updated and current understanding of meteorological science are incorporated into the PMP derivation process.

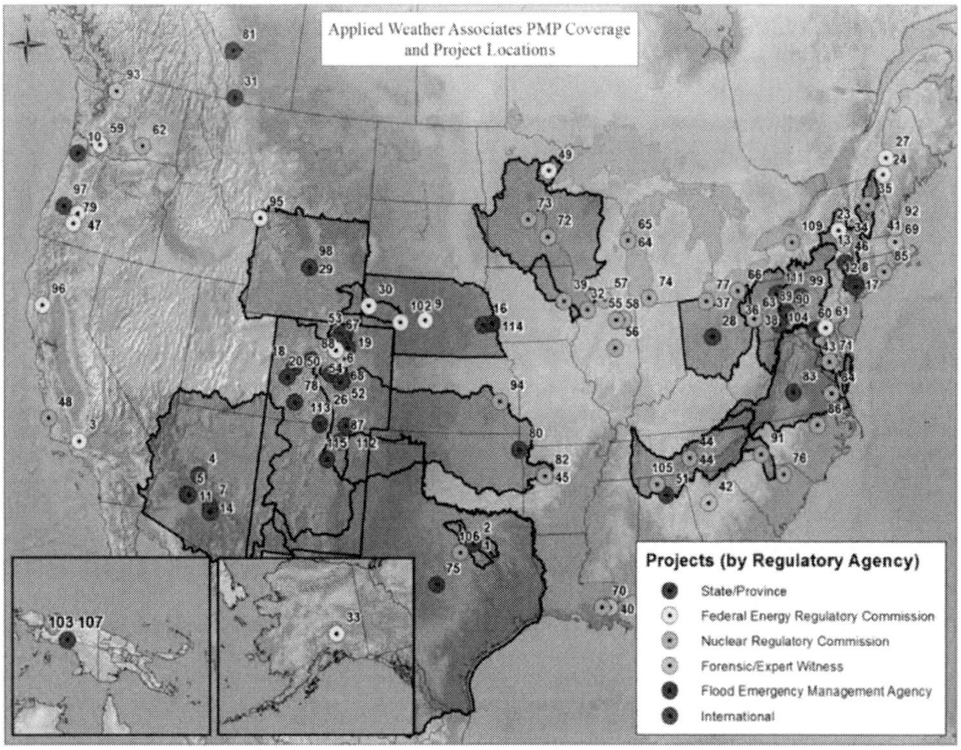

Figure 1. Location of PMP studies completed by AWA

PMP Background

The practice of using extreme rainfall values for computing design floods began in the 1930s. The upper limit to precipitation potential first was called the maximum possible precipitation (MPP), and the U.S. Weather Bureau's Hydrometeorological Reports from 1940s to the 1950s used this designation. In 1956, federal agencies replaced the MPP with the PMP, reasoning that MPP carried too strong a connotation of a physical upper limit to precipitation and too little acknowledgement of the uncertainty of such estimates.

The NWS's Hydrometeorological Report 59, published in 1999, gives the current definition of PMP: *...theoretically, the greatest depth of precipitation for a given* *duration that is physically possible over a given storm area at a particular geographic location at a certain time of year.*

The science and technology of meteorology have not progressed to a level that permits direct computation of the theoretical greatest depth of precipitation. Instead, a procedure based on historic storms is the standard practice (Myers 1967). Precipitation totals from the most intense storms that have occurred over a region represent the lowest potential levels of PMP. A basic assumption in the procedure is that the record of extreme storms is sufficiently large to represent the most efficient storm mechanisms but not the optimum available moisture that would accompany a PMP event (Myers 1967) (WMO, 1986). Storm efficiency is the ability of a storm to convert atmospheric moisture to precipitation through complex cloud physics mechanisms.

History of the Development of the HMRs

The current set of HMRs was developed by the Weather Bureau and NWS over a period of more than 20 years and represents a variety of techniques and methods. The oldest HMR that is still considered current is HMR 49, published in 1977. Other current HMRs were published between 1978 and 1999. The series covers 50 states and Puerto Rico (Figure 2).

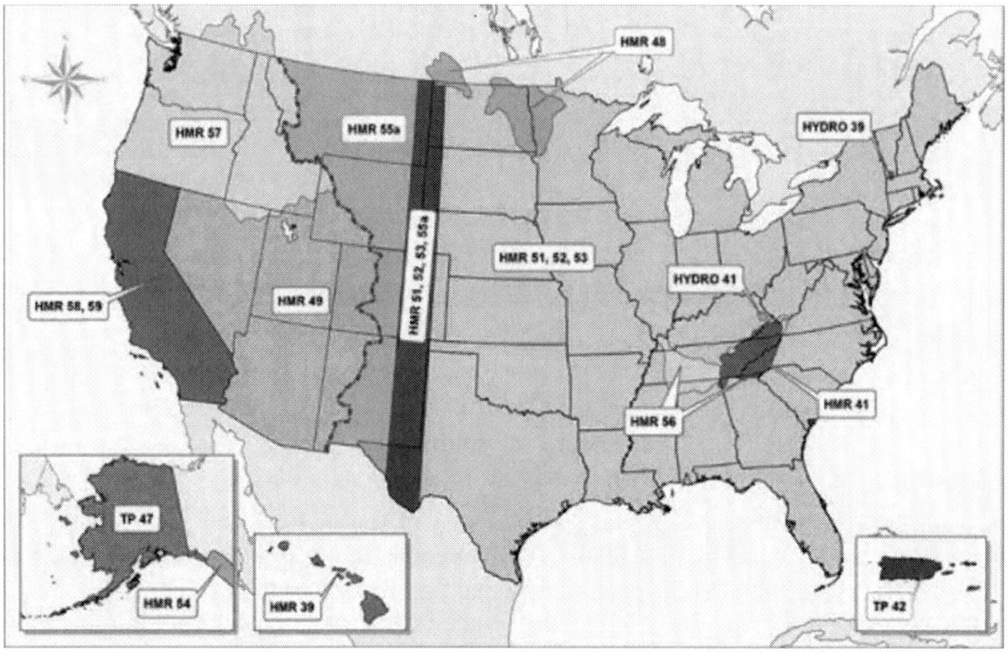

Figure 2. Coverage of current National Weather Service Hydrometeorological Reports

A comprehensive database of historic storms is critical in developing PMP estimates for a region. The Weather Bureau and U.S. Army Corps of Engineers completed numerous analyses of extreme rainfall events through the 1950s. These efforts included collecting, quality controlling, and archiving many official and unofficial rainfall observations for each storm. One of the results of a storm analysis is a depth-area-duration table, showing the maximum depth of rainfall over various areas and for various durations during the storm. Since the mid-1950s, with the exception of site-specific studies and western U.S. storms analyzed for HMRs 57 and 59 (Corrigan et al. 1999), only a few storms have been analyzed, the most recent being Hurricane Agnes in 1972. The most recent storm used in any of the HMR series occurred in 1995, and there remains a large backlog of unstudied storms.

The procedures used to develop generalized PMPs vary considerably among the reports. For example, HMR 49 specifies a ratio method to determine spatial and temporal characteristics of storm rainfall and has an additive procedure for orographic (mountain range) effects. In this method, a set of standard additive factors are applied across all storm areas and durations. Later HMRs use analyses of individual storm depth-area-duration rainfall patterns to evaluate spatial and temporal rainfall distributions and employ a multiplicative factor to address orographic effects (Hansen et al. 1994). HMR 51, which covers the eastern two-thirds of the U.S., does not include any orographic adjustments but identifies two "stippled regions" in which the authors acknowledge that terrain effects were not evaluated and suggest that future NWS studies examine these regions more closely (Schreiner and Riedel, 1978). Unfortunately, no such studies have been completed.

An important parameter used in the development of PMP is dew point temperature, a measure of atmospheric moisture. The HMRs used persisting dew point temperatures, mostly for the 12-hour duration. This is the highest dew point temperature that persists for a 12-hour period. (Another way to say this is the lowest dew point temperature that occurs at any time during the 12-hour period). However, a 1993 study of PMP in Michigan and Wisconsin (Tomlinson 1993), sponsored by the Electric Power Research Institute and a group of hydro project owners, challenged this assumption. Authors of this study noted that the 12-hour persisting dew point was inappropriate for large thunderstorm systems and consistently produced rainfall values that were about 20 percent too large. Instead, the authors used average dew point temperatures for periods that were similar to the precipitation duration. The merit of this approach was acknowledged by the Weather Service in HMR 57 in the form of a recommendation for future study (Hansen et al, 1994).

Storm Precipitation Analysis System, A New Way To Analyze Storms

Over the past two decades, various advances have been made in the measurement and analysis of extreme rainfall events. For example, NWS NEXRAD (NEXt generation RADar) provides new opportunities to analyze temporal and spatial rainfall distributions, and GIS technology is being used to develop storm depth-area-duration tables. There also have been significant improvements in the meteorological understanding of rainfall production mechanisms. HYSPLIT model data that determine atmospheric moisture trajectories provide reliable moisture inflow vectors for individual storms, especially useful for moisture sources over oceans (Draxler and Rolph, 2013). Some of these advances were used in the 1990s in HMRs 57 and 59.

HMRs 57 and 59 use storm depth-area-duration tables produced by a "mini-storm" analysis procedure developed at the Bureau of Reclamation. This procedure automated and standardized many tasks that previously were performed subjectively (Hansen et al, 1994). Improving and expanding on this procedure, a program called the Storm Precipitation Analysis System (SPAS) has been developed jointly by Applied Weather Associates LLC and Metstat Inc. to take advantage of NEXRAD data and GIS technology (Parzybok and Tomlinson, 2006). Advanced quality control procedures and interpolation schemes are also incorporated.

SPAS results have been compared with previous NWS storm analyses and with an analytically constructed storm pattern where the rainfall distributions were known exactly (Parzybok and Tomlinson, 2006). For the analytical storm, SPAS produced the exact solution. For the NWS-analyzed storms, most values in the depth-area-duration tables agreed within 5 percent. The SPAS program has been applied in site-specific PMP studies and extreme storm analyses throughout the U.S. It uses NEXRAD data when available but also can be used to analyze conventional rain gauge data. The program generates depth-area-duration tables, rainfall mass curves, and maps of rainfall (isohyetal maps) for historic storms. Historically, storm analyses have been primarily dependent on subjective procedures. Although SPAS is highly automated, meteorologists interact to provide quality control and adjustments to the rainfall analysis procedures.

Storm analyses using NEXRAD are producing maximum rainfall amounts that were not observed using rain gauges. NEXRAD records rainfall continuously across an entire storm area, whereas in the past the spatial distribution of rain had to be inferred from a network of rain gauge points. For example, in the Gorham, Maine, storm of October 1996, the largest rain gauge observation was just over 19 inches, but NEXRAD data showed a storm total of 23 inches several miles offshore. One of the justifications for making conservative assumptions in developing the HMRs is the belief that rain gauges often did not capture the most intense storm cells. SPAS using NEXRAD data now captures and analyzes these intense storm cells that can occur between rain gauges.

Are HMRs Too Conservative?

The HMR series is the standard source for PMP values across the U.S. Many regional and site-specific studies have also been completed. Most of these studies have been accepted by the Federal Energy Regulatory Commission (FERC) and state regulators for computing the PMF at hydropower and water storage facilities. In addition, several PMP studies that have been developed as part of flood hazard re-evaluations for nuclear sites are currently under review by the Nuclear Regulatory Commission (NRC). The first study completed by a private consultant and accepted by FERC was the 1993 Michigan-Wisconsin regional PMP study (Tomlinson 1993). This resulted in significant reductions in PMP. The greatest reduction relative to HMR 51, around 20 percent, applied to storm durations up to 12 hours and area sizes up to 500 square miles. However, for larger area sizes and longer durations the PMP values were larger by about 5% relative to HMR 51.

Since then, AWA studies have resulted in reduction from the appropriate HMR of over 50% (e.g. Kappel et al, 2014) and in rare occasions increases compared to the appropriate HMR (e.g. Kappel et al, 2015). In almost all cases, updated studies produce reductions compared to the appropriate HMR. This is a result of refined, site-specific evaluations where explicit transposition limits are applied to storms, updated maximizations and storm adjustments are applied, and inappropriate combinations of conservatisms, such as combining storm types into one PMP design storm, are corrected.

One of the most important aspects of the studies completed by AWA over the years is the continuous updating of the storm data base used to compute the values. Because the PMP development process is a storm-based process, it is critically important to ensure the storm data are representative of all PMP-type storms that could potentially affect the given values. As part of this process, AWA has analyzed more than 600 extreme rainfall events since 2002 using the SPAS program (Figure 3). These storm results have been used in all PMP studies and extensively reviewed and evaluated.

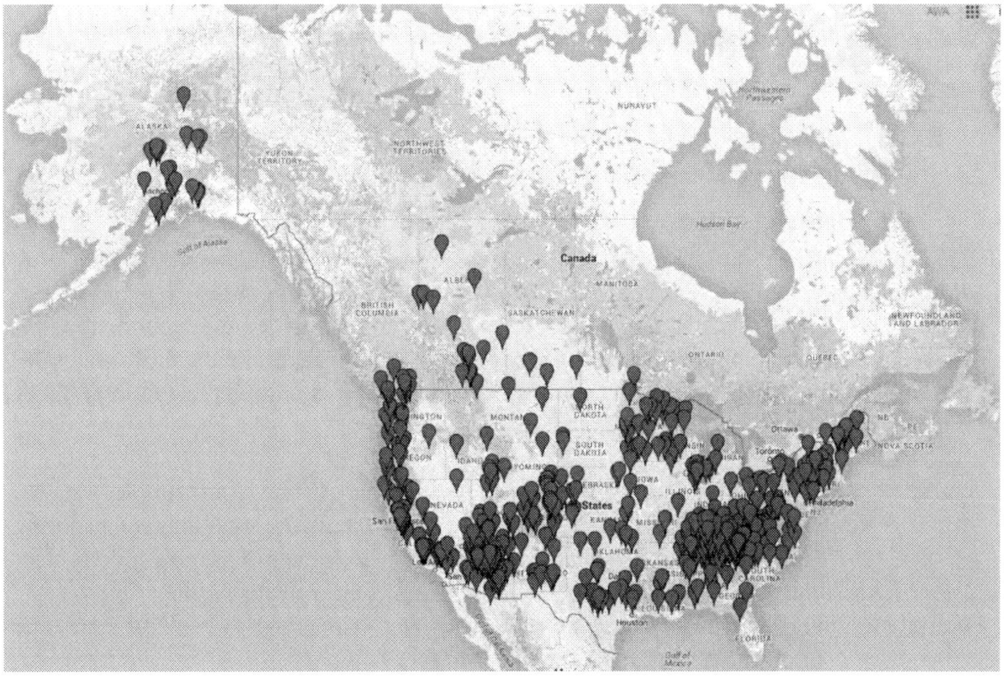

Figure 3. SPAS storm analysis locations

Considering these results, should we draw the conclusion that the PMP values in the HMR series are consistently too large? Not necessarily. The HMRs provide generalized rainfall values that are not basin-specific and tend to represent the largest PMP values across broad regions. Site-specific PMP studies incorporate basin characteristics that are specific to the topography and local climate of the watershed being studied.

When to Consider a New Study

There is no objective way to assess the accuracy of PMP estimates in the HMR series. When considering whether a new study is warranted, meteorologists apply professional judgment and experience to consider a number of factors, including:

- The difference between PMP values and maximum observed rainfall over the surrounding meteorologically homogeneous region;
- The number and severity of recorded storms in the region, if a long record is available;
- Limitations on storm transposition in the region;
- The number, character, and interrelationship of maximizing steps;
- The reliability of any model used to relate rainfall to other meteorological variables; and
- The probability of exceeding the individual meteorological variables used in the models, taking care to avoid excessive compounding of rare events (Hansen et al. 1977).

Many of these issues are better addressed in some HMRs than in others. The oldest report, HMR 49, used methods that have been replaced in all subsequent publications. The second oldest, HMR 51, covers a large area — two-thirds of the lower 48 states — but provides no adjustments for barrier moisture depletion and limited elevation effects on moisture.

Arbitrary geographic boundaries also affect the use of the HMRs. For example, a dam owner in Nevada uses the 30-year-old HMR 49, while across the state line in California the applicable HMR is 20 years newer. At the moment, however, it seems unlikely that further HMR updates are forthcoming from the NWS. The website for the Hydrometeorological Design Studies Center notes:

NOAA's National Weather Service has provided PMP guidance and studies since the late 1940s at the request of various federal agencies and with funding provided by those agencies. In recent years that funding has diminished and gradually ceased. As a result we are unable to continue our PMP activities.

Therefore, dam owners, regulators, and designers who require PMP estimates need to carefully consider issues such as the currency of the storm database and the scientific and technological advances that were, or were not, applied in a given HMR. In some situations, it will be evident that the HMR does not represent local meteorology well. In such situations, experience has shown that a site-specific or detailed regional study has the potential to produce significantly different estimates than the prevailing HMR.

COMPARISON OF RECENT EVENTS AGAINST PMP

Several recent extreme rainfall events have produced devastating flooding, property damage, and lost of life across various parts of North America. However, are these storms unusual when compared to past similar events and do they exceed PMP? Three storms in particular are discussed in this section and compared to PMP in locations where they were used and against previous controlling storms. The first storm is the June 2013 rainfall and flood event near Calgary, Alberta. This produced at the time what was the costliest natural disaster in Canada. The second event was along the Colorado Front Range a few months later, in September 2013. This event produced widespread flooding and destruction extending from New Mexico through Southern Wyoming, with the focus of devastation between Fort Collins and Denver, CO. Just over two year later, extreme rainfall occurred in South Carolina as a hurricane offshore pumped large amounts of moisture into the region that combined with a frontal system in the area and produced record rainfall over a several day period. The resulting floods devastated the region. All three events overtopped and destroyed numerous dams in the affected regions. In most cases, the dams were not constructed to pass the Probable Maximum Flood. So, was PMP exceeded or were these just extreme events that can be expected to happen over time?

Calgary, Alberta June 2013 Storm Event

The Calgary, Alberta event in June 2013 was a classic set up for heavy rainfall in the region. An area of low pressure moved slowly through the region, tapped into moisture that originated from the Gulf of Mexico and combined with the topography of the region to enhance and focus the rainfall. In this region, these storms are most common from late May through late June. However, in most cases, the storms move through quickly, and lack efficient storm dynamics and/or access to significant moisture. This extreme event produced 13.78 inches of rainfall at its storm center over a 3-day period (Figure 4). The rainfall was highly influenced by terrain and the initial interaction of the first major ridges rising from the plains (Figure 5). Although this was an extreme rainfall event and resulted in disastrous flooding and damage, it did not control PMP values in the basins that were immediately affected. For example, as part of the site-specific PMP that AWA completed as a sub consultant to Stantec for the Elbow River Basin and Springbank Off Stream Storage Project (Kappel et al. 2015), the controlling storms were the infamous Gibson Dam, MT June 1964 and Veteran, Alberta storm of June 1973. The 1964 event resulted in over 19 inches of rain in two days (Figure 6), and the 1973 storm resulted in 9.56 inches of rain in just over 24 hours. Therefore, the June 2013 storm was not unprecedented in its rainfall accumulation, but there was a significantly larger amount of infrastructure that was vulnerable to flooding. In addition, the storm's rainfall accumulation occurred in an ideal location to maximize runoff in the basin affecting Calgary. Another way to visualize this event is to instead consider what would have happened if the Gibson Dam, MT June 1964 storm had occurred in the same location as the June 2013 event. All else being equal, it is likely that significantly higher amounts of flood runoff and damage would have occurred.

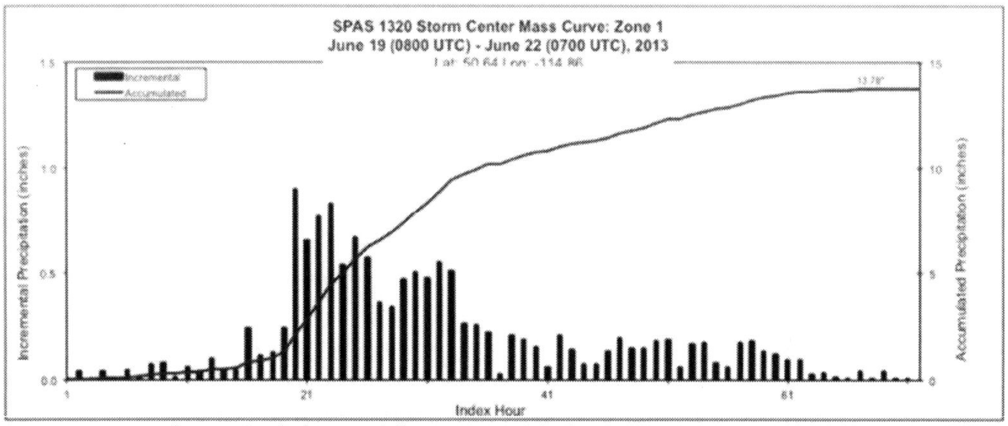

Figure 4. Calgary, Alberta June 2013 storm center mass curve

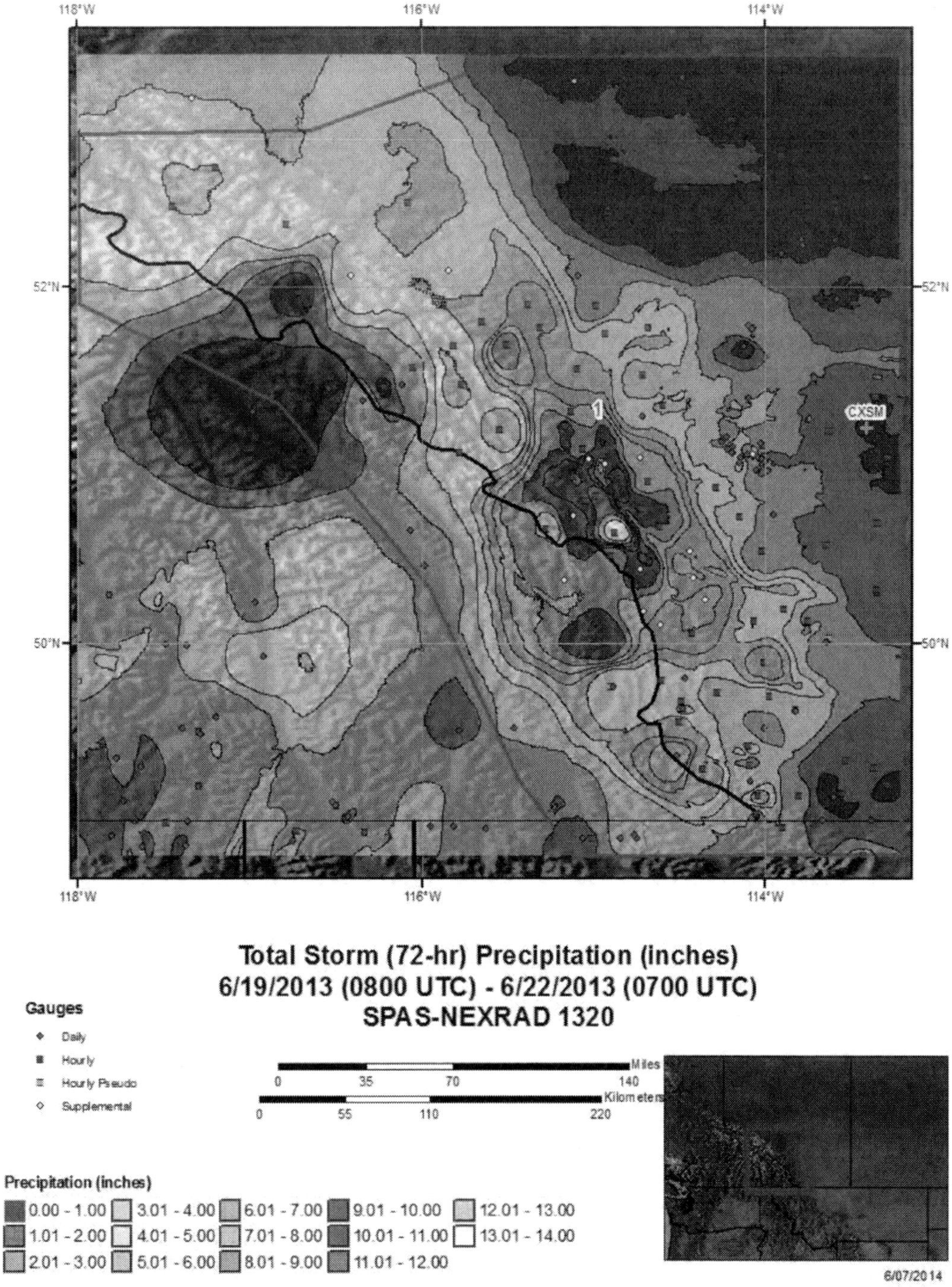

Figure 5. Calgary, Alberta June 2013 total storm isohyetal pattern

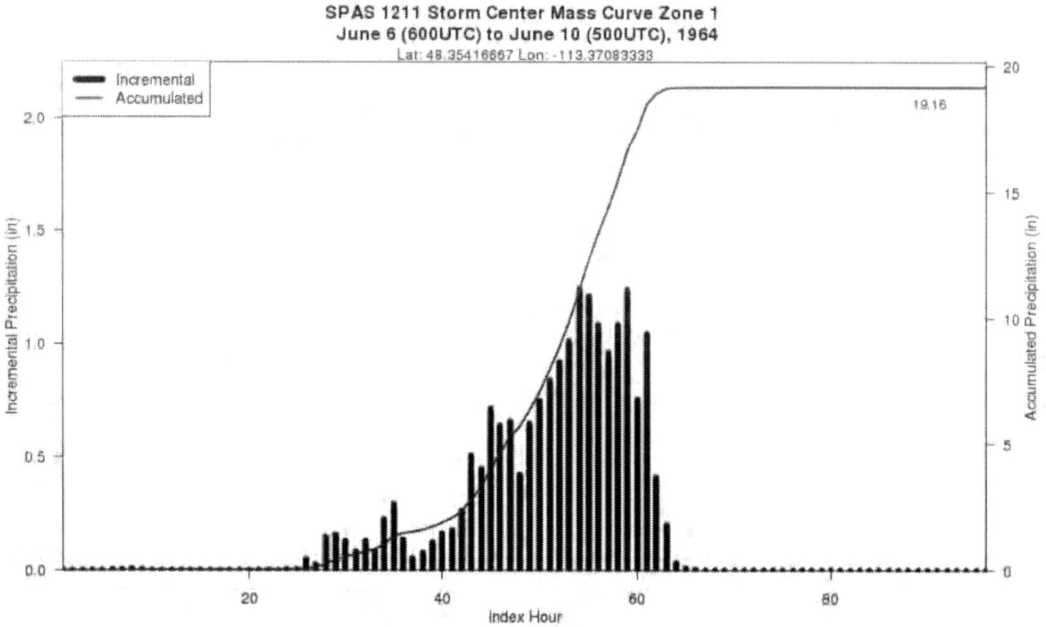

Figure 6. Gibson Dam, MT June 1964 storm center mass curve

September 2013 Front Range Storm Event

Only three months after the Calgary, Alberta event of June 2013, a similar storm brought widespread rainfall and flooding across a vast region extending from the Front Range and Eastern Plains of southern New Mexico through southern Wyoming (Figures 7 and 8). The meteorology of the storm was almost very similar to the June 2013 event. Again, an area of low pressure moved slowly through the region, tapped into moisture that originated from the Gulf of Mexico and combined with the topography of the region to enhance and focus the rainfall. In this region, these storms are most common from May through June and again in September. Areas along the Front Range of Colorado from the Colorado Springs region through Denver and Boulder, and higher elevations in the foothills, were hardest hit (Gochis, 2015: NOAA, 2013). The event was unusual in that there was nearly continuous rainfall for seven days (Figure 9). In addition, there was very little convection associated with the event. Instead anomalously high levels of atmospheric moisture released by a combination of atmospheric dynamics and terrain where responsible for the record rainfall. Like the June 2013 event, these factors were maximized for the region and time of the year. In fact, the storm resulted in a new 24-hour rainfall accumulation record for the state of Colorado (11.85" accumulated at the Fort Carson Army Post near Cheyenne Mountain, CO). The duration of rainfall was unusual for the region and didn't fit the standard duration (72-hours or less) of PMP values. Annual Exceedance Probabilities of the rainfall accumulation show that this storm was most rare at the longer durations (Figure 10). Because of this, the storm generally did not control when compared to standard HMR 51 and 55A PMP values for durations up to 72-hours. In part, this is because there have been several other extreme events in similar regions which produced large rainfalls over durations of 1-, 2-, and 3-

days. Examples of extreme storms in the Colorado Front Range region include May-June 1894, June 1921, May 1935, May 1955, and May 1969. However, if the runoff hydrology of a large basin in the region indicates durations longer than 72-hours could control the PMF, this would be an important storm to consider, and potentially controlling of large basin PMP.

Total Storm Precipitation (inches)
September 8 (800 UTC) - September 18 (700 UTC), 2013

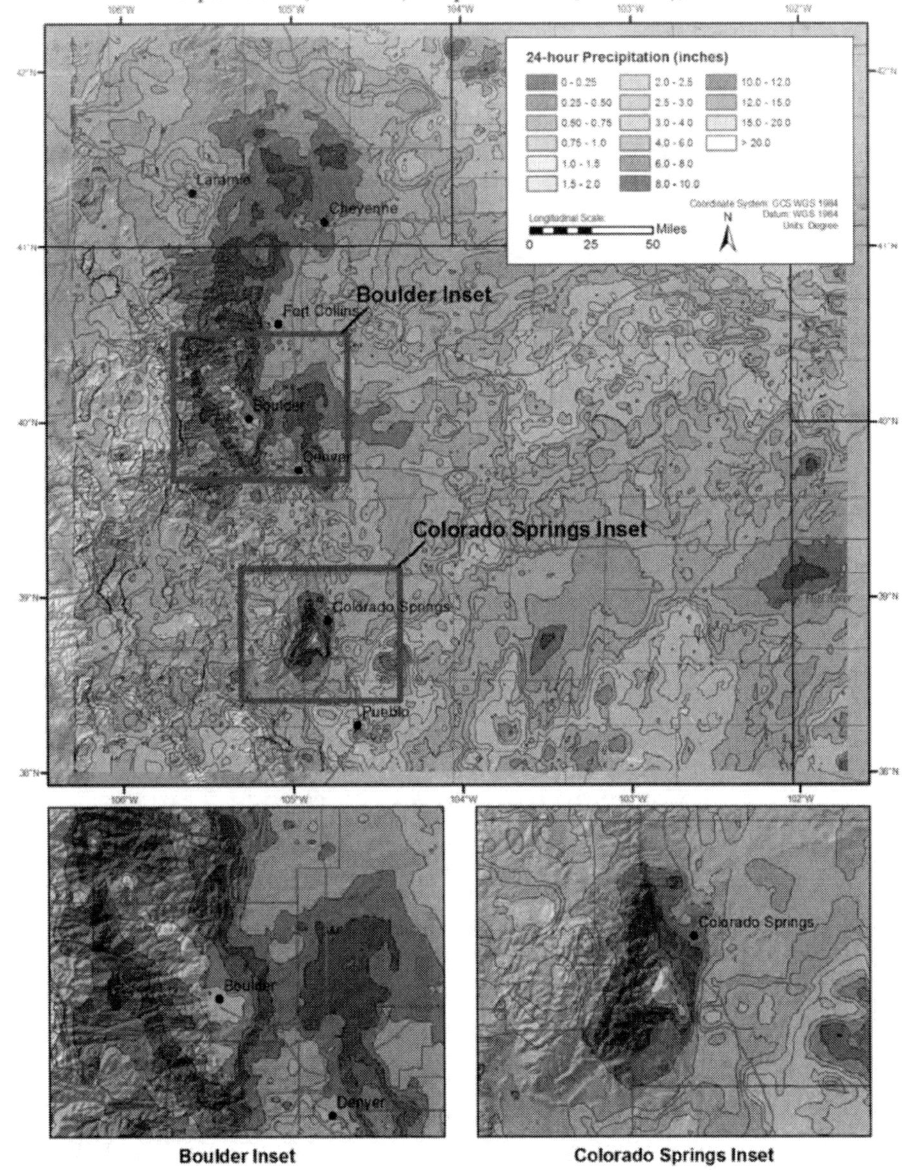

Figure 7. Total storm isohyetal for the Colorado and Wyoming rainfall centers

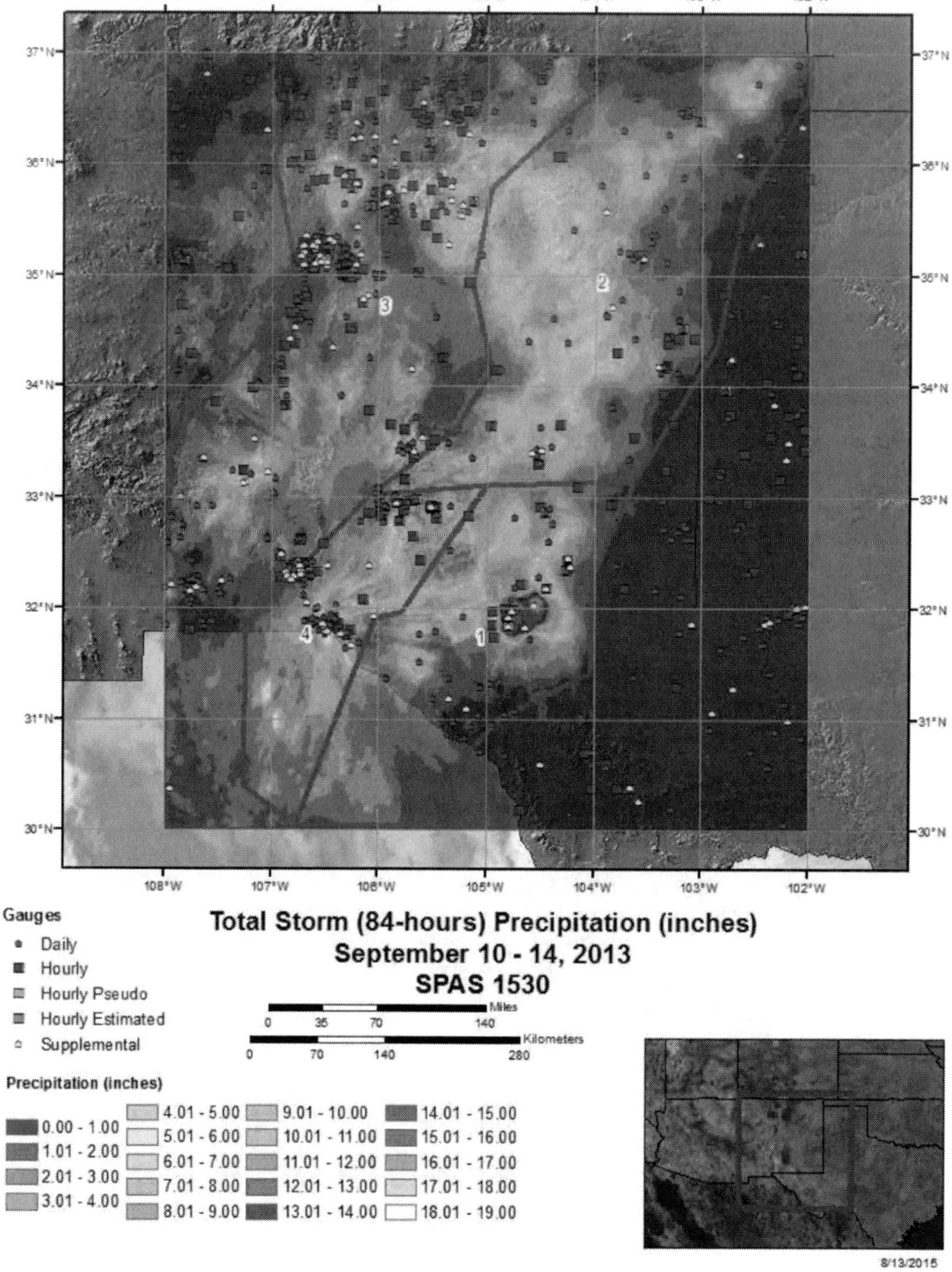

Figure 8. Total storm isohyetal for the New Mexico rainfall centers

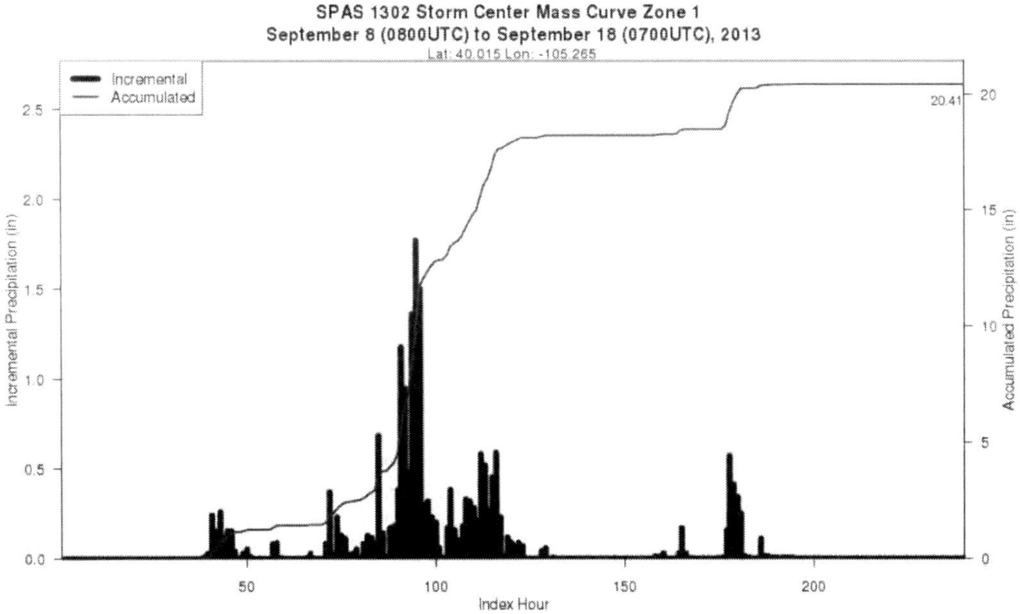

Figure 9. September 2013 storm center mass curve near the Boulder, CO center

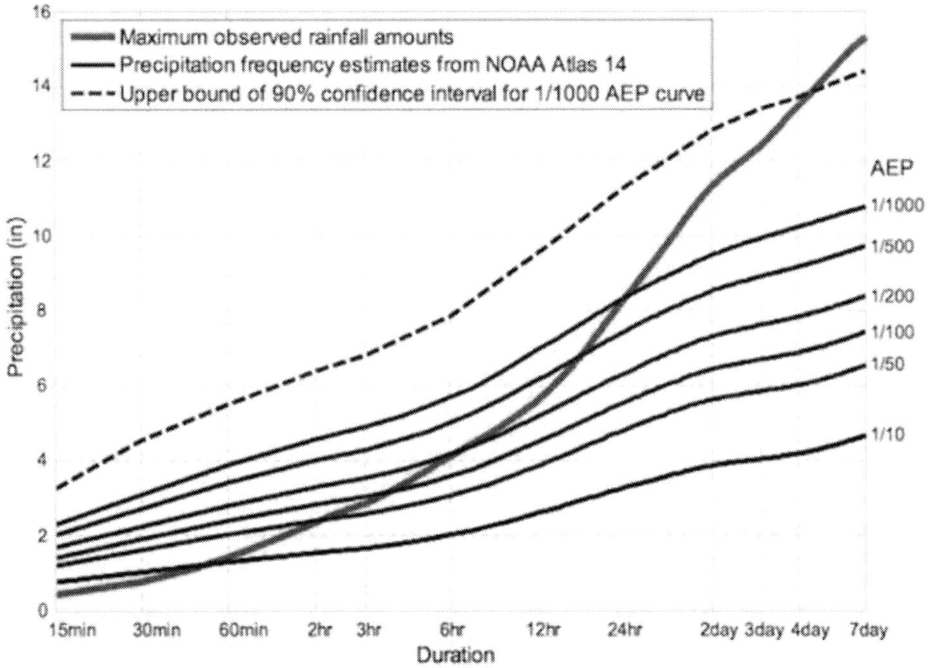

Figure 10. Maximum observed rainfall amounts in relationship to the corresponding precipitation frequency estimates from NOAA Atlas 14 at the Justice Center gauge, Boulder, CO. Image from http://www.nws.noaa.gov/oh/hdsc/aep_storm_analysis/8_Colorado_2013.pdf

South Carolina October 2015 Storm Event

This storm produced record flooding and was associated with a constant feed of tropical moisture emanating from Hurricane Joaquin, which was located over the northern Bahamas at the time (Figure 11). High levels of moisture from tropical systems are a common occurrence in this region. What made this event so unique and devastating was the fact that this moisture also interacted with a stalled front in the region. This provided an extra focusing mechanism and enhanced the rainfall. In addition, the front and general weather pattern remained over the same area for several days, allowing the record amounts of rainfall to accumulate.

Extensive and widespread flooding occurred across Colleton, Dorchester, Berkeley, and Charleston counties in South Carolina. The most significant flooding occurred in areas along and near smaller creeks and streams, especially those that were tributaries to larger rivers such as the Edisto, Ashley, Cooper, and Santee. In addition to impacting thousands of homes and businesses, floodwaters also damaged many of the area roads and bridges, causing significant travel disruptions that lasted for multiple days.

This extreme event produced nearly 28 inches of rainfall at its storm center over a 4-day period (Figure 12). The rainfall was concentrated over the immediate coastal regions of South Carolina around Charleston and Mt Pleasant, with decreasing amounts going inland. This confirms that coastal convergence and frontal process were most important for rainfall production (Figure 13). In these locations the AEP was greater than a 1000-years across a wide region (Figure 14). Comparisons of this storm's maximized rainfall against PMP depths that were derived for a river basin in North and South Carolina just prior to the occurrence of this storm event showed that at area sizes greater than 1000-square miles and durations longer than 48-hours, the 2015 storm event would control PMP values. Other important historic storms that control PMP in the region include Vade Mecum, NC August 1908 and Alta Pass, NC July 1916, and Yankeetown, FL September 1950.

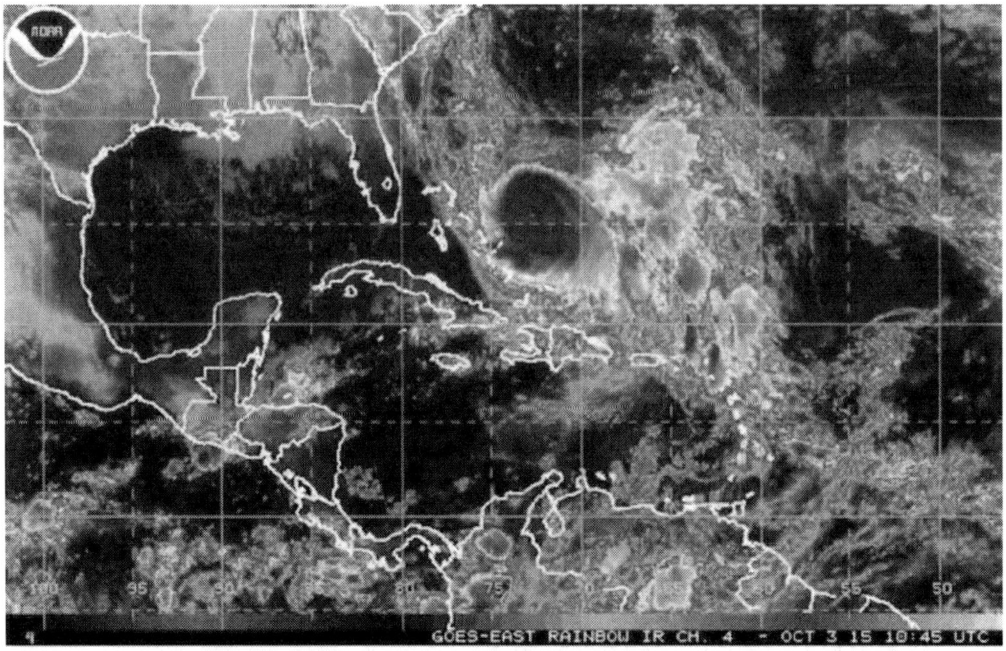

Figure 11. NOAA infrared satellite image from 1045UTC, October 3, 2015 showing the connection between Hurricane Joaquin and the rainfall in South Carolina

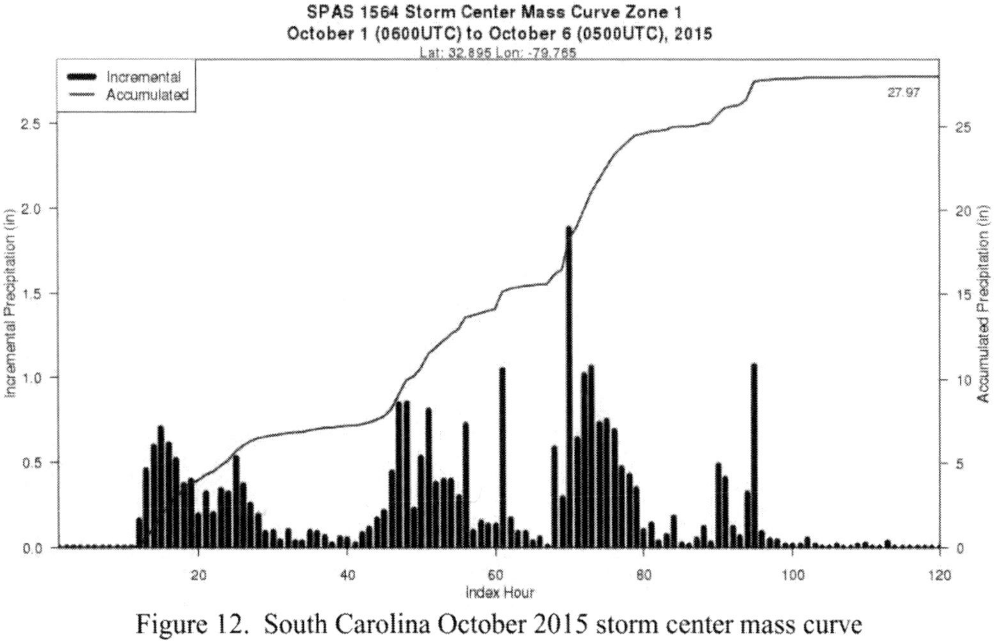

Figure 12. South Carolina October 2015 storm center mass curve

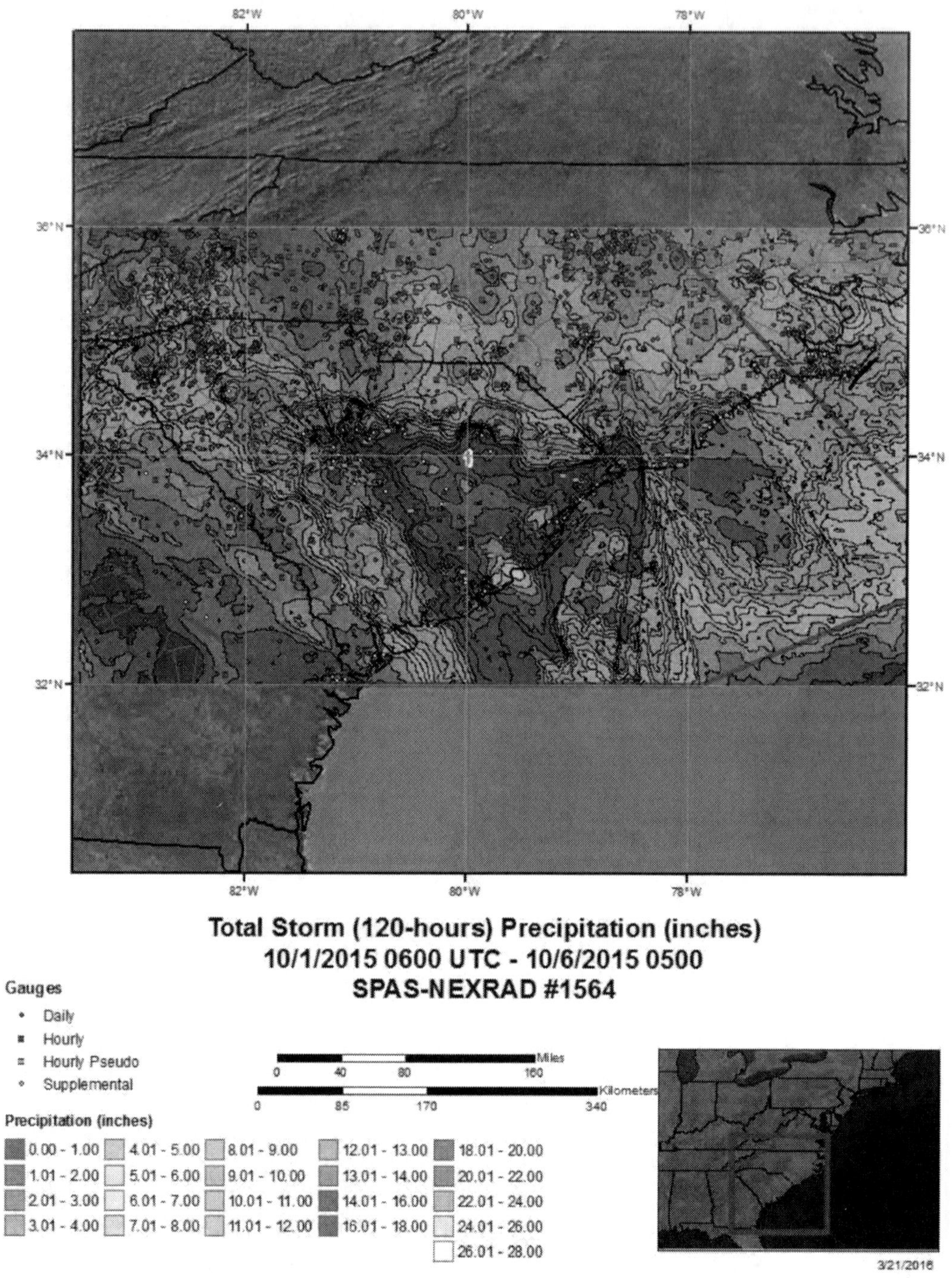

Figure 13. South Carolina October 2015 total storm isohyetal pattern

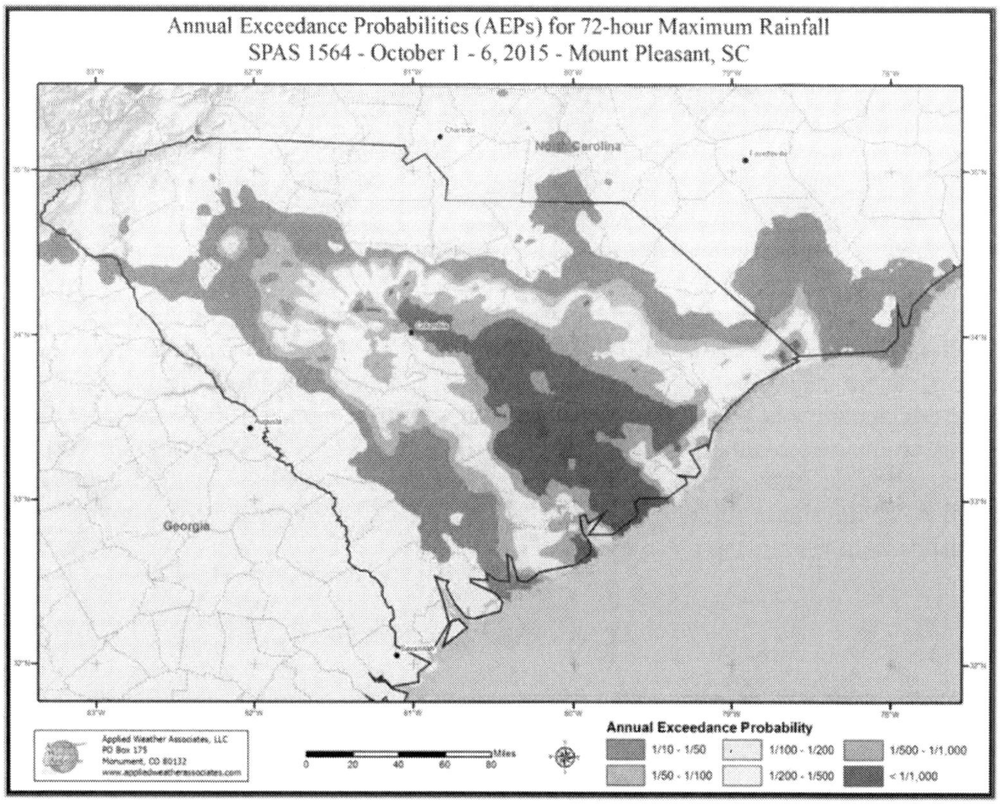

Figure 14. October 2015 AEP for the 72-hour rainfall across South Carolina. Red colors are >1000-year AEP.

SUMMARY

Recent extreme rainfall events have resulted in devastating flooding, damage, and numerous fatalities across many locations in the United States. In some cases, these new storms have controlled PMP values in updated sites-specific PMP investigations, while in many other cases, previously-controlling storms still control today. In this paper, three recent extreme storms were investigated and discussed in relation to PMP. These included Calgary, Alberta June 2013; Colorado Front Range September 2013; and South Carolina October 2015. Each of these storms was important for developing PMP in the regions where they occurred, understanding flood runoff characteristics of extreme storms, and evaluating performance of dams. Other recent controlling storms in the Central and Eastern United States (which AWA has analyzed and used in updated PMP development), include storms such as Madison County, VA June 1995, Hurricane Floyd September 1999, Nashville, TN May 2010, Hurricane Irene August 2011, Duluth, MN June 2012, Islip, NY August 2014, Baton Rouge, LA August 2016. In the Western United States, recent storms controlling of PMP include Opal, WY August 1990, December 1996-January 1997 in Northern California, June Lake, WA February 1996, Index WA November 2006, Pe Ell, WA December 2007, Southern New Mexico and

West Texas September 2014, Phoenix, AZ September 2014, South Central Texas October 2015.

However, none of these storms was considerably different than other storms in each region that previously controlled PMP values. In fact, many of the large historic storms events from the late 1800's through the 1960's still control PMP values today. These include storms such as Hearne, TX September 1899, Thrall, TX September 1921, Quinault Ranger Station, WA January 1935; Smethport, PA July 1942; Hoegee's Camp, CA January 1943; Warner, Ok May 1943; Gibson Dam, MT June 1964.

Therefore, when updated PMP studies combine recent events having an abundance of high quality storm data with past PMP-controlling events, the resulting values can be utilized with the highest confidence in evaluating and designing infrastructure. The mixed population of controlling storms extending back to the late 1800's also addresses non-stationarity, given that the climate cycles of the previous 150 years are inherently captured in the data sets. Therefore, changes in the climate over the useful lifetime of current PMP values (30-50 years) have already been captured and analyzed in the existing data sets.

ACKNOWLEDGMENTS

This work was conducted in collaboration with the support of personnel from Applied Weather Associates including Doug Hultstrand, Geoff Muhlestein, Jake Rodel, Dana McGlone, and Kristi Steinhilber.

REFERENCES

Corrigan, P., Fenn, D.D., Kluck, D.R., and J.L. Vogel, 1999: Probable Maximum Precipitation Estimates for California. *Hydrometeorological Report No. 59*, U.S. National Weather Service,
National Oceanic and Atmospheric Administration, U.S. Department of Commerce, Silver Spring, MD, 392 pp.

Draxler, R.R. and Rolph, G.D., 2013. HYSPLIT (HYbrid Single-Particle Lagrangian Integrated Trajectory) Model access via NOAA ARL READY Website (http://ready.arl.noaa.gov/HYSPLIT.php). NOAA Air Resources Laboratory, Silver Spring, MD.

Gochis, D., R. Schumacher, K. Friedrich, N. Doesken, M. Kelsch, J. Sun, K. Ikeda, D. Lindsey, A. Wood, B. Dolan, S. Matrosov, A. Newman, K. Mahoney, S. Rutledge, R. Johnson, P. Kucera, P. Kennedy, D. Sempere-Torres, M. Steiner, R. Roberts, J. Wilson, W. Yu, V. Chandrasekar, R. Rasmussen, A. Anderson, and B. Brown, 2015: The Great Colorado Flood of September 2013. Bull. Amer. Meteor. Soc., 96, 1461–1487.

Hansen, E.M., F.K. Schwarz, and J.T Reidel (1977). Probable Maximum Precipitation Estimates, Colorado River and Great Basin Drainages, Hydrometeorological Report No. 49, U.S. Department of Commerce, Silver Spring, Md.

Hansen, E.M., D.D. Fenn, P. Corrigan, and J.L Vogel (1994). Probable Maximum Precipitation, Northwest States, Hydrometeorological Report No. 57, U.S. Department of Commerce, Silver Spring, Md.

Kappel, W.D., Hultstrand, D.M., Muhlestein, G.A., Steinhilber, K., McGlone, D., Parzybok, T.W, and E.M. Tomlinson, December 2014: Statewide Probable Maximum Precipitation (PMP) Study for Wyoming.

Kappel, W.D., Hultstrand, D.M., Muhlestein, G.A., Steinhilber, K., McGlone, D., Lovisone, S., and B. Lawrence, June 2015: Site-specific Probable Maximum Precipitation for the Elbow River Basin, Alberta, Canada.

Kappel, W.D., Hultstrand, D.M., Muhlestein, G.A., Steinhilber, K., McGlone, D., Parzybok, T.W., and B. Lawrence, November 2015: Statewide Probable Maximum Precipitation for Virginia.

Myers, V.A. (1967). "The Estimation of Extreme Precipitation as the Basis for Design Floods: Resume of Practice in the United States," *Proceedings of the Leningrad Symposium on Floods and their Computation*, Volume 1, International Association of Scientific Hydrology, Gentbrugge, Belgium, pages 84-10.

Manual for Estimation of Probable Maximum Precipitation (1986). Operational Hydrology Report, Report No.1, World Meteorological Organization, Geneva, Switzerland.

National Oceanic and Atmospheric Administration (NOAA), September 2013: Exceedance Probability Analysis for the Colorado Flood Event, 9-16, September 2013, http://www.nws.noaa.gov/oh/hdsc/aep_storm_analysis/8_Colorado_2013.pdf.

Schreiner, L.C., and J.T. Riedel (1978). Probable Maximum Precipitation Estimates, United States East of the 105th Meridian, Hydrometeorological Report No. 51, U.S. Department of Commerce, Silver Spring, Md.

Parzybok, Tye W., and Edward M. Tomlinson, "A New System for Analyzing Precipitation from Storms," Hydro Review, Volume 25, No. 3, May 2006, pages 58-65.

Tomlinson, E.M. (1993). Probable Maximum Precipitation Study for Michigan and Wisconsin, Volume 1, EPRI TR-101554, North American Weather Consultants, Salt Lake City, Utah.

NEEDS TO UPDATE PROBABLE MAXIMUM PRECIPITATION FOR DETERMINING HYDROLOGIC LOADING ON DAMS

Chandra Pathak[1]

Probable Maximum Precipitation (PMP) is theoretically the greatest depth of precipitation for a given duration that is physically possible over a given size storm area at a particular geographical location at a certain time of the year. It is used to develop inflow flood hydrographs, known as Probable Maximum Flood (PMF), as design standard for dams and determining hydrologic loadings on existing dams. PMP estimation methodology was developed in the 1930s and 40s when many dams were constructed in the US. The procedures to estimate PMP were later standardized by the World Meteorological Organization (WMO) in 1973 and revised in 1986.

In the US, PMP estimates were published in a series of Hydrometeorological Reports (e.g., HMR55A, HMR57, and HMR58/59) by the National Weather Service since 1950s. In these reports, storm data up to 1980s were used to establish the current PMP estimates. Since that time, we have acquired additional meteorological data for 30 to 40 years, including newly available radar and satellite based precipitation data. These data sets are expected to have improved data quality and availability in both time and space. In addition, significant numbers of extreme storms have occurred and selected numbers of these events were even close to or exceeding the current PMP estimates, in some cases.

In the last 50 years, climate science has progressed and scientists have better and improved understanding of atmospheric physics of extreme storms. However, applied research in estimation of PMP has been lagging behind. Alternative methods, such as atmospheric numerical modeling, should be investigated for estimating PMP and associated uncertainties. It would be highly desirable if regional atmospheric numerical models could be utilized in the estimation of PMP and their uncertainties, in addition to methods used to originally develop PMP index maps in the existing hydrometeorological reports.

[1] U.S. Army Corps of Engineers, chandra.s.pathak@usace.army.mil

SITE SPECIFIC PROBABLE MAXIMUM PRECIPITATION DEVELOPMENT

Jeff Harris, P.H.[1]
Ben Rohrbach, P.E.[2]

ABSTRACT

The Corps of Engineers uses the Probable Maximum Precipitation (PMP) to compute the Probable Maximum Flood (PMF). The PMF is used for spillway adequacy studies among other items. Currently, several Hydrometerological Reports (HMR) exist that include instructions, and data for computing the PMP. However, many of these are outdated and are thirty, or more, years old. The National Weather Service does not currently have funding or personnel to update these HMR documents. This necessitates the need for others to perform site specific analyses on an as-needed basis. The Nashville District, Corps of Engineers has such a need for Center Hill Dam. This paper will provide an overview of the steps necessary to generate a site-specific PMP for the Center Hill Dam. It will include steps on accessing historic rainfall, generating Depth-Area curves, applying transpositioning and maximizing criteria from applicable HMR documents and providing results.

INTRODUCTION

The goal of this study is to review and update the Probable Maximum Precipitation (PMP) for the Caney Fork River basin (also referred to as the Center Hill basin) upstream from Center Hill Dam, TN. Center Hill Dam is located in DeKalb County, TN on the Caney Fork River, a tributary of the Cumberland River. The project is located approximately 26.6 miles upstream of the confluence with the Cumberland River. The Caney Fork drainage is located on the western side of the Appalachian Mountains. This study was performed by WEST Consultants, Inc. under contract with U.S. Army Corps of Engineers, Nashville District (LRN). LRN specifies Locks and Rivers Division, Nashville District.

Purpose

The results of this study are intended to be used by LRN for calculation of the Probable Maximum Flood (PMF) runoff from the Caney Fork River basin as inflow to Center Hill Dam. This study provides information for determining if the PMP developed in 1989 spillway adequacy study using criteria in Hydrometerological Report No. 51 (HMR51) is still applicable to the Caney Fork River basin. This determination is made by analyzing historic rainfall events which have occurred in the general climatologic region of Center Hill Dam, since the development of HMR51, and are transposable to the Caney Fork

---

[1] Senior Hydrologist, WEST Consultants, Inc., 101 Parkshore Drive, Folsom, CA, 95630. jharris@westconsultants.com

[2]Chief, Hydrology and Hydraulics Branch, US Army Corps of Engineers, Nashville District, 110 9th Street, Nashville, TN, 37203. Ben.Rorhbach@usace.army.mil

basin. Additionally, storm events used in the development of HMR51 are also included to determine their possible influence. Storm events will be analyzed to see if their inclusion in the PMP computations could have an impact on PMP developed in 1989. The location of Center Hill dam and the drainage basin is shown in Figure 1.

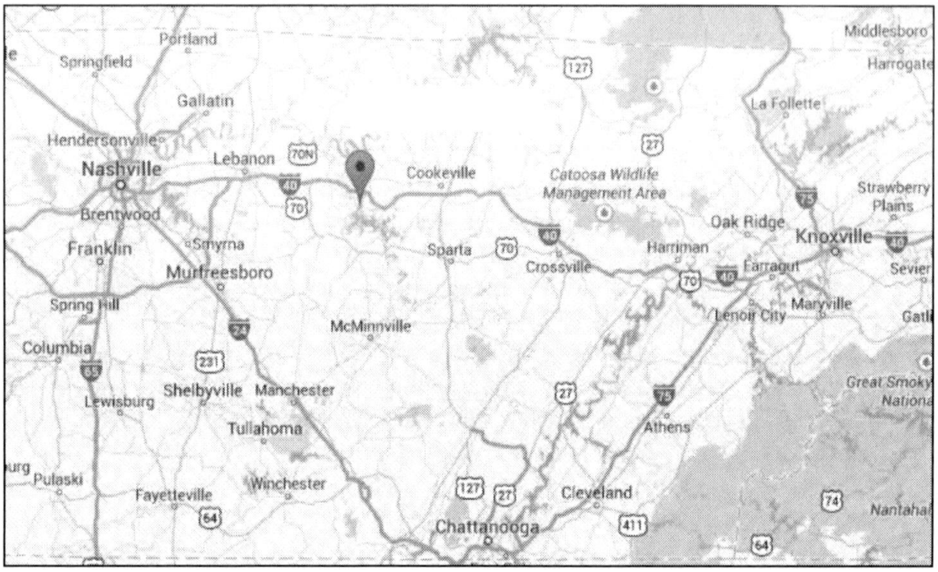

Figure 1 - Center Hill Dam Location and Caney Fork Drainage Basin (C/O Google Maps)

LRN has developed a Hydrologic Modeling System (HMS) model for the Caney Fork basin that will incorporate the PMP and compute the PMF inflow into Center Hill. This study provides the subbasin specific PMP rainfall amounts for use in the HMS model.

Background

In December 2008, Center Hill Dam received a Dam Safety Action Classification (DSAC) rating of 1 during the Screening Portfolio Risk Assessment. A DSAC 1 rating implies that the dam is Unsafe and Critically near failure or poses Extreme high risk. The causes for DSAC-1 range from geotechnical to hydraulic. In the case of Center Hill, the main cause for the rating was foundation deficiencies. The PMF for Center Hill Dam is based on a 1989 spillway adequacy study (USACE 1987, USACE 1989) completed by the Nashville District utilizing National Weather Service Hydrometeorological Report (HMR) No. 51 "Probable Maximum Precipitation Estimates, United States East of the 105th Meridian" (1978) and temporal and spatial distribution guidance from HMR No. 52 "Application of Probable Maximum Precipitation Estimates, United States East of the 105th Meridian" (1982). Since the Probable Maximum Precipitation (PMP) for Center Hill Dam was developed 25 years ago and significant investment is being made for dam safety modifications, it was recommended the PMP and PMF be reviewed and updated for spillway analysis.

Center Hill Dam is a multi-purpose project of the Cumberland River Basin System which provides flood control, hydropower production, commercial navigation, recreation, fish and wildlife enhancement, water quality, and water supply benefits for the region. Dam construction was started in 1942, halted in 1943 due to World War II, resumed in 1946 and then closed in 1948. The reservoir was impounded in 1949 and the last of three hydropower turbines was installed in 1951. Center Hill Dam controls a drainage area of 2,174 square miles. Total reservoir capacity is 2,092,000 acre-feet at maximum pool elevation of 685 NGVD29. The seasonal summer pool is maintained at elevation 648 with a storage capacity of 1,330,000 acre-feet. Seasonal winter pool is slightly lower at elevation 632.

Selection of Key Storms

The first step in developing a site specific PMP is to select key storms that impact the Center Hill drainage area. Essentially, events are selected which have occurred since HMR51 was published. The November 2010 After Action Report (AAR) (USACE 2010), prepared by the Corps of Engineers, on the May 1-3, 2010 flood event in the Cumberland River basin, included a discussion of historical floods in the Cumberland basin. A section of the AAR is included, below.

".... The track and direction of large flood-producing storms in the Cumberland Watershed generally parallel the southwest to northeast orientation for the major part of the basin. Almost all floods on the mainstem of the Cumberland River occur in the period from late November to mid-May, mainly because precipitation amounts are greatest during that time of year and hydrologic conditions are conducive to excessive runoff......"

Based on the information provided in the AAR and the fact that HMR51 indicates no contribution from storms on the eastern side of the Appalachians should be included, all events analyzed for the Caney Fork/Center Hill Basin were based on events streaming from the Gulf, including hurricane and tropical events, or storms emanating in the Ohio Valley region. Events were selected by downloading NCDC precipitation data and plotting the data to determine time periods of large rainfall.

After reviewing tropical, hurricane and synoptic storm data, including an analysis of storm tracks, 15 events were selected for analysis.

Development of Depth Area Curves

Following the selection of the key storms, a depth area analysis was performed for each storm. After accumulating the precipitation data for each storm, grids were created with ArcGIS using precipitation totals. Utilizing the gridded rainfall data for each event and an ellipsis pattern encompassing 10 to 6,500 square miles, depth-area data values were extracted using ArcGIS. The ellipsis shapefile was intersected with the rainfall grid to develop the depth-area values. The ellipsis was moved to various locations for each event in order to maximize the rainfall for the event. The events analyzed are listed in Table 1.

Review of the storm tracks for the events that were analyzed supports the statement in the May 2010 Post Flood Report that the general storm track is southwest to northeast.

Table 1 - Selected Storm Events

Event	Event Total Rainfall (in.)	Station	Name
September 2011	11.7	Berry 3 NW, AL	Tropical Storm Lee
April 2011	13.9	near Poplar Bluff, MO	
May 2011	10.4	near Jonesboro, AR	
May 2010	18.6	Near Centerville, TN	Nashville Flood
August 2005	7.1	Woodbury, KY	Hurricane Katrina
September 2004	9.1	Jasper, IN	Hurricane Ivan
May 2004	7.2	Calhoun City, MS	
May 2003	9.9	Monterey, TN	
September 2002	12.3	Booneville, MS	Tropical Storm Isidore
March 1997	13.6	Greenville, TN	
October 1995	10.1	Birmingham Airport, AL	Hurricane Opal
May 1994	7.9	Knoxville McGhee Tyson Airport, TN	
March 1989	4.4	Rend Lake Dam, IL	
September 1982	5.6	Grenada Dam, MS	Tropical Storm Chris
March 1975	10.2	Springfield Experimental Sta. TN	

Storm Transposition and Development

The three steps involved in transposing the key storms to the Caney Fork basin include in-place maximization (IPM), horizontal transposition (HT) and vertical transposition factors (VT). The steps outlined in HMR 51, HMR 55A, and the WMO Manual on PMP were all adhered to during the storm transpositioning process. When studying heavy rainfall events, the IPM is designed to factor in "how much larger" or "how much more moisture" could an extreme storm have contained under additional extreme climatological conditions. This step requires an analysis of the weather conditions of the event itself as well as a review of climatological studies.

In order to determine the IPM factor, an investigation of the weather patterns associated with each storm was conducted. Archived weather maps were accessed from the NOAA website. In addition, HYSPLIT (Hybrid Single Particle Lagrangian Integrated Trajectory Model) was utilized to determine the air inflow trajectory to the storm.

Once the moisture inflow trajectory was identified, weather observation data from along that trajectory were gathered from the NCDC archives in order to determine each storm's representative 12-hour persisting dew point temperature. The *Climatic Atlas of the United States* (1968) was used as a reference for the maximum 12-hour persisting dew point temperature. The IPM factor is determined from a ratio of the precipitable water available at the 12-hour persisting dew point temperature and the precipitable water available at maximum 12-hour persisting dew point temperature. The 12-hour persisting dewpoint is the standard measure from HMR55A. It is not related to the storm duration. Shorter, local events may only use one representative station.

The most recent update from the NCDC to the Climatic Atlas was in 2002, but only includes (1) Mean Dew Point Temperatures, (2) Mean Maximum Dew Point Temperature, and (3) Mean Minimum Dew Point Temperature. Therefore, the Internet archive of the Climactic Atlas, Figure 2, was used to obtain the 12-hr maximum persisting 1000 MB dewpoint.

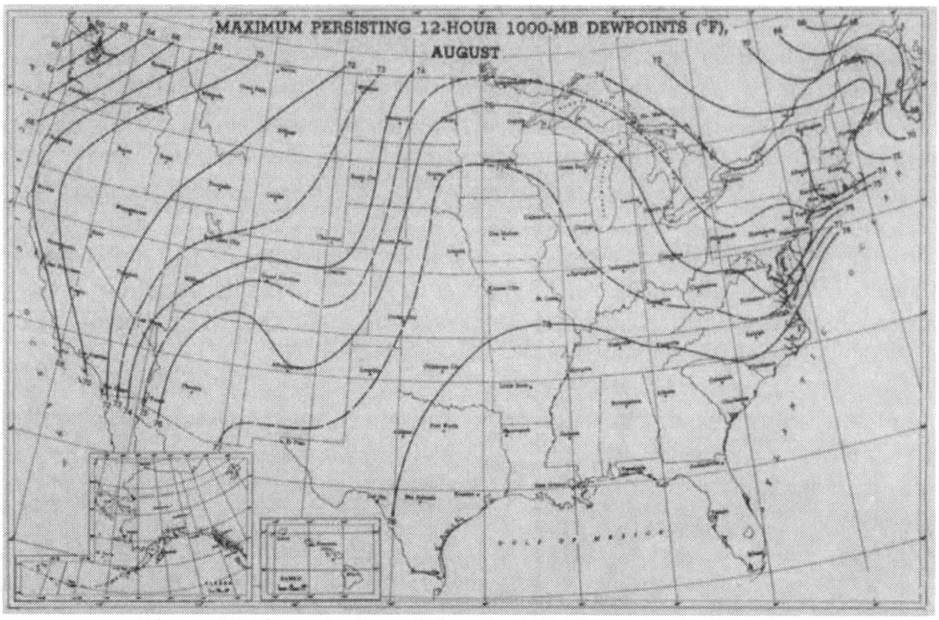

Figure 2 – Sample from Climatic Atlas Internet Archive Page

Using the May 2010 Cumberland River event as an example, the maximum rainfall occurred in west central Tennessee near the town of Centerville. Using weather maps, Figure 3, and HYSPLIT, the trajectory was observed to be south to north through Alabama. Accordingly, dewpoint information was acquired at (1) Tuscaloosa Municipal Airport; (2) Huntsville International/C.T. Jones Field; (3) Northwest Alabama Regional Airport; (4) Columbus AFB; and, (5) Nashville International Airport.

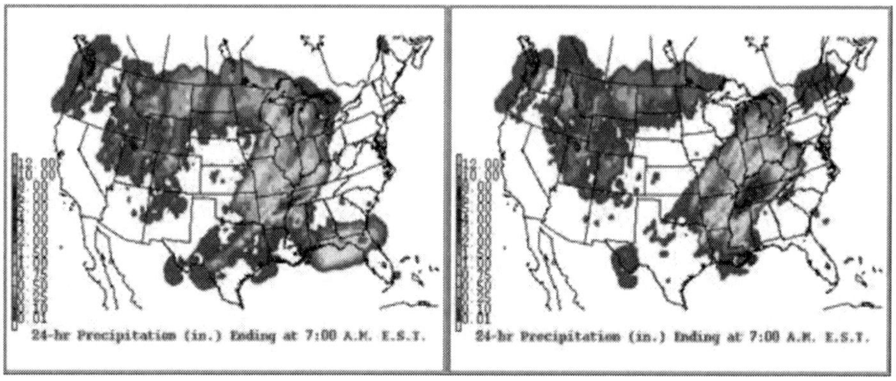

Figure 3 - May 1-2, 2010 Rainfall (May 1 on the left)

Computing a running 12-hour average from these 5 sites, adjusted to 1000mb, the 12-hour persisting dewpoint (T_d) was determined to be 69°F for May 2nd. The maximum 12-hr T_d value at the moisture source was read from the online 1968 Climatic Atlas.

Adjusting the May 2nd value by the 15 day seasonal adjustment results in a 12-hr maximum persisting 1000mb dewpoint of 76°F. The May and June maximum persisting T_d values of 74.5 and 77.2, respectively, represent the value from the Climatic Atlas at approximately the center of the 5 stations used. Table 2 represents the values used to compute the 12-hr maximum T_d at the moisture source.

Table 2 – 12-Hr Maximum Persisting Dewpoint at Moisture Source

June Maximum Persisting T_d	77.2
May Maximum Persisting T_d	74.5
Date of Event	2-May
Adjust 15 days	17-May
Days in May	31
June T_d - May T_d	2.7
Degrees per Day	0.09
Days between 17 May and 1 June	15
Degree Change	1.31
Max Persisting T_d	**76**

The In-Place Maximization (IPM) is computed using the above computed Max persisting T_d value and the Precipitable Water (PW) Tables found in Appendix C of HMR55A. The elevation at the storm center (700ft) was used with the PW tables in HMR55A, to produce PW values of 2.86 and 2.02 for 76 and 69°F, respectively. In order to remain consistent with HMR51 and 55A, an upper limit of 1.5 is placed on the IPM.

$$IPM = PW\ (76^{\circ}F \text{ at } 700') \ /\ PW\ (69^{\circ}F \text{ at } 700') = 2.86''/2.02'' = 1.42$$

The Horizontal Transposition (HT) factor is an adjustment to compensate for a change in available moisture due to the horizontal transpositioning of the storm centroid to the Center Hill centroid. From HMR55A, the HT is limited to a factor of 1.2. The elevation at the centroid is 900 ft. The basin centroid for Center Hill was calculated to have a 12-hr maximum persisting T_d of 75°F for 17 May as shown in Table 3. Table 3 computations follow the same steps as those for Table 2.

Table 3 – 12-Hr Maximum Persisting Dewpoint at Center Hill Centroid

June Maximum Persisting T_d	76.6
May Maximum Persisting T_d	73.6
Date of Event	2-May
Adjust 15 days	17-May
Days in May	31
June T_d - May T_d	3
Degrees per Day	0.10
Days between 17 May and 1 June	15
Degree Change	1.45
Max Persisting T_d	75

The horizontal transposition (HT) equation from HMR55A is:

HT=PW(75°F @ Center Hill centroid @ 700') / PW(76°F @ storm centroid @ 700')

HT=2.72 / 2.86

HT= 0.95

The values used are shown in Table 4.

Table 4 - Horizontal Transposition Values

Elevation to use (ft)	700
12-Hr Max Persisting T_d at Center Hill Centroid	75
PW @ Elevation to use (from HMR55A)	0.18
PW @ Ceiling (from HMR55A)	2.9
PW	2.72
Horizontal Transposition (HT)	0.95

The final transpositioning, vertical (VT), takes into account the change in height from the storm centroid to the Center Hill centroid. This study incorporates the same method outlined in HMR55A and does not take into account the first 1000 feet in elevation difference between the storm center and the basin centroid. Additionally, as discussed in HMR51 and HMR55A, the elevation of any barrier between the storm centroid and basin centroid, which is higher than the basin centroid, is used as the elevation of the basin centroid. There is no barrier so there is no change in elevation. From HMR55A, the HT is limited to a factor of 1.2. The vertical transposition (VT) equation from HMR55A is:

VT=0.5+0.5*(PW 75°F @ Center Hill centroid @ 700') / PW(75°F @ Center Hill centroid @ 700')

VT=2.72 / 2.72

VT = 1.0

The values used are shown in Table 5.

Table 5 - Vertical Transposition Values

Elevation to use (ft)	700
12-Hr Max Persisting T_d at Center Hill Centroid	75
PW @ Elevation to use (from HMR55A)	0.18
PW @ Ceiling (from HMR55A)	2.9
PW @ 700'	2.72
PW @ 700' (From Table 4)	2.72
Vertical Transposition (VT)	1.0

The final Transposition Factor (TF) equation from HMR55A is:

TF=IPM * HT * VT = 1.42 * 0.95 * 1.00 = 1.35

The depth-area values for each event were multiplied by their respective TF in order to produce results that could be compared to the Center Hill PMP values. The maximized DA values for each of the events, including March 1975, are shown on Figure 4.

Some of the HMR events used in the generation of HMR51 are also shown, for comparison, on Figure 4. It is obvious from Figure 4 that HMR storm #47 was a very significant event when transposed to the Center Hill basin (See discussion on Storm #47 in Comparison to Previous Studies). The 72-hr rainfall actually exceeds the 72-Hr Zonal PMP at drainage area sizes greater than Center Hill. The 48-hr May 2010 event and the

60-hr HMR Storm 8 both exceed the 6-Hr PMP value in the 500-600 sq. mile range and the 12-hr PMP in the range of the Center Hill basin size. The 72-Hr September 2011 event exceeds the 6-hr PMP value at about 1,500 sq miles. Other events exceed PMP values at drainage sizes larger than Center Hill.

All the events which exceed the 6- and 12-hr PMP values were for durations much greater than the PMP duration which they exceeded. For example, the May 2010 event was a 48-hour duration event. It only exceeded the 6- and 12-hr PMP values. The HMR51 Storm 8 represents 72-Hour values. September 2011 was a 48-hour event.

It is important to note that the selected storms were all of durations greater than 12 hours. However, it is important to determine if any of these storms had durations that would exceed the PMP values. Therefore, depth-duration analyses were performed. Figure 5 portrays the depth duration results for the Center Hill basin size (2,174 sq mi). The red GIS Zonal PMP curve represents the latest PMP developed by Nashville District using GIS and digital versions of the HMR51 nomographs and is the latest PMP for Center Hill drainage.

Results and Comparison with Previous Studies

Based on the depth-area curves shown on Figure 4, six events were selected (September 2011, May 2011, April 2011, May 2010, March 1997 and March 1975) for additional analysis. Depth-duration data were developed for these events. These events were selected since they were judged to have a possible impact on the shape of the envelope curve. Results depicted in Figure 5 represent depth-duration curves at 2,174 sq. mi, which is the size of the Caney Fork drainage above Center Hill dam. These curves indicate that none of the selected events exceed the most recent GIS Zonal PMP values at this basin size. The depth-duration analysis included 6 storms used in the development of HMR51. The storm that comes closest to the current PMP is Storm #47 (March 1929) from HMR 51.

Storm #47 achieved 100% of PMP at the 72-hr duration. The percentage of PMP at each duration continually rose until reaching 100%. HMR storm #8 (April 1900) also exhibited large percentages at shorter durations but reduced as the duration increased. The most recent event to show large percentages of the PMP was the May 2010 event. This was essentially a 48-hr event. At 48-hours it had reached 75% of PMP. Based on percentages of these events versus the existing PMP, it appears that a possible reduction in the existing PMP may be warranted. This will be explored later in this report.

HMR Storm #47 occurred March 11-16, 1929 with the centroid of rainfall near Elba, AL. This event was used in the development of rainfall for the southeast in HMR51. Even though HMR51 mapping indicates that this event was not transposed farther north, the use of the storm influenced (increased) the data north of this area. In effect, this event was "transposed" out of the area of influence and impacted the smoothing of the curves in this area. The result of this would be inflated rainfall values in the Center Hill area resulting in inflated PMP amounts. After reviewing this information, it was decided to remove Storm #47 from further consideration in the development of the site specific PMP for Center Hill. The removal of this event from site specific analysis will eliminate the artificially inflated rainfall values and result in lower site specific PMP values when compared to the current PMP values.

For these same reasons, HMR51 Storms #2 and #54 were dropped from consideration.

Storms 1, 8 and 80 were reviewed and determined they are valid events that should be considered in the site specific analysis.

The overall impacts of HMR51 Storms 2, 47 and 54 were to artificially increase the existing GIS Zonal PMP storm amount for Center Hill Basin. Taking this into consideration, a logical conclusion could be that the site specific PMP may be lower than the existing HMR51 GIS Zonal PMP.

Orographic Impacts

Orographic impacts, hence increased rainfall, are generated as an air mass is lifted as it moves over elevating terrain such as a mountain. As the air is lifted, it cools. As the air cools its ability to retain moisture is inhibited. Therefore, the moisture falls to the ground.

The 1989 PMP evaluation for Center Hill included a 4.6% factor, in order to be consistent with studies in adjacent basins and account for orographic impacts. This factor was applied to the rainfall over the entire basin in 1989. In lieu of applying a factor to the entire basin area, a suggested method would be to use the results from the NOAA Atlas 14 analysis. HMR51 includes information on areas orographically influenced by the Appalachian Mountains. Orographic impacts could be computed by taking a ratio of subbasin to total basin Atlas 14 rainfall and applying the ration to the PMP. However, the Center Hill basin is nearly completely outside this area of Appalachian lifting. Therefore being outside of this area, orographic influences were not applied to this PMP analysis.

Conclusions and Recommendations

The primary focus of this effort was to make a determination if the existing PMP for Center Hill Basin needs to be updated. The recommended update could be to raise, lower or keep the same. For this effort, 18 events were analyzed. These included tropical and extra-tropical events. After evaluation of the 18 events, three were dropped based on criteria in HMR51. HMR51 Storm No. 47 matched the 72-hour PMP value but was subsequently discounted.

Based on the discussion on the HMR51 events earlier in the Results and Comparison with Previous Studies and how storms 2, 47 and 54 have inflated the HMR51 values, it is recommended that the existing PMP for the Caney Fork basin be reduced.

Figure 6 shows Depth Duration data for the 2,174 sq mi Center Hill basin. Included are the existing PMP (GIS Zonal) and a 10% and 15% reduction of the GIS Zonal amounts. Based on the discussion earlier in this report about the influence of the storms used in the development of HMR51 it is recommended that the existing PMP be reduced by 10% to 15%. When the Zonal PMP values are input to HMR52, rainfall values are reduced due to centroid of rainfall and orientation of the rainfall ellipse. For comparison purposes, Figure 6 also includes the HMR52 rainfall resulting from a 10% reduction applied to the Zonal PMP. This curve is nearly identical to the 15% reduction HMR51 PMP values. Figure 6 also includes historic events. The HMR52 curve resulting from the 10% reduction is above these historic events. It could be argued that the values could be reduced more. However, a 15% reduction would result in the HMR52 values just above the moisture maximized HMR51 Storm #8. In essence, this says that the basin would have already experienced a moisture maximized event which is very close to the PMP.

Therefore, this report recommends a 10% reduction to the PMP. This will keep the HMR51 and resulting HMR52 values above the maximized historical events and at the same time remove much of the high bias that was introduced into HMR51 by using the three events that were subsequently removed from this analysis. The PMP rainfall will be computed with HMR52 with the 10% reduction accounted for on the ST record in the HMR52 input file.

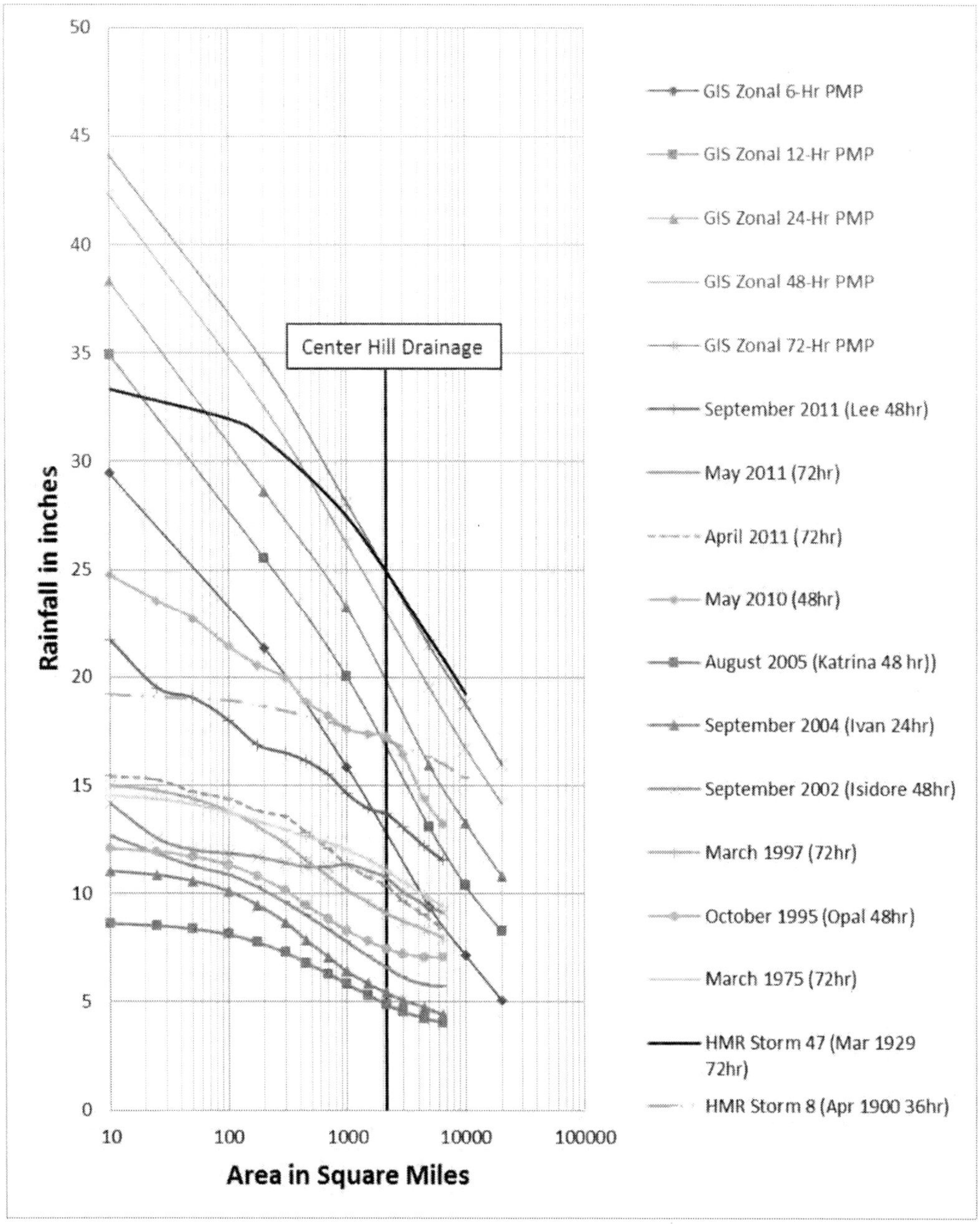

Figure 4 – Maximized and Transposed Depth-Area Curves for Selected Storms.

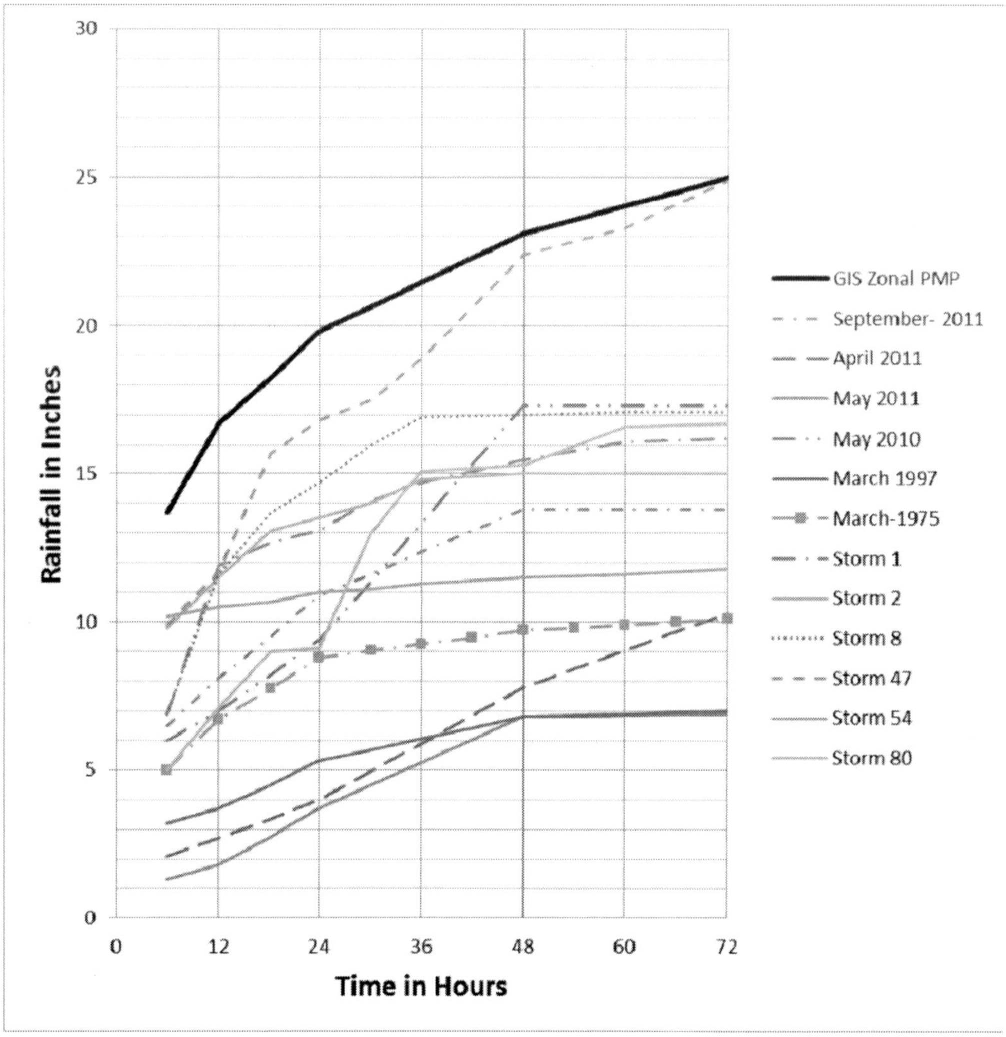

Figure 5 - Maximized and transposed Depth-Duration Curves at 2,174 sq. mi.

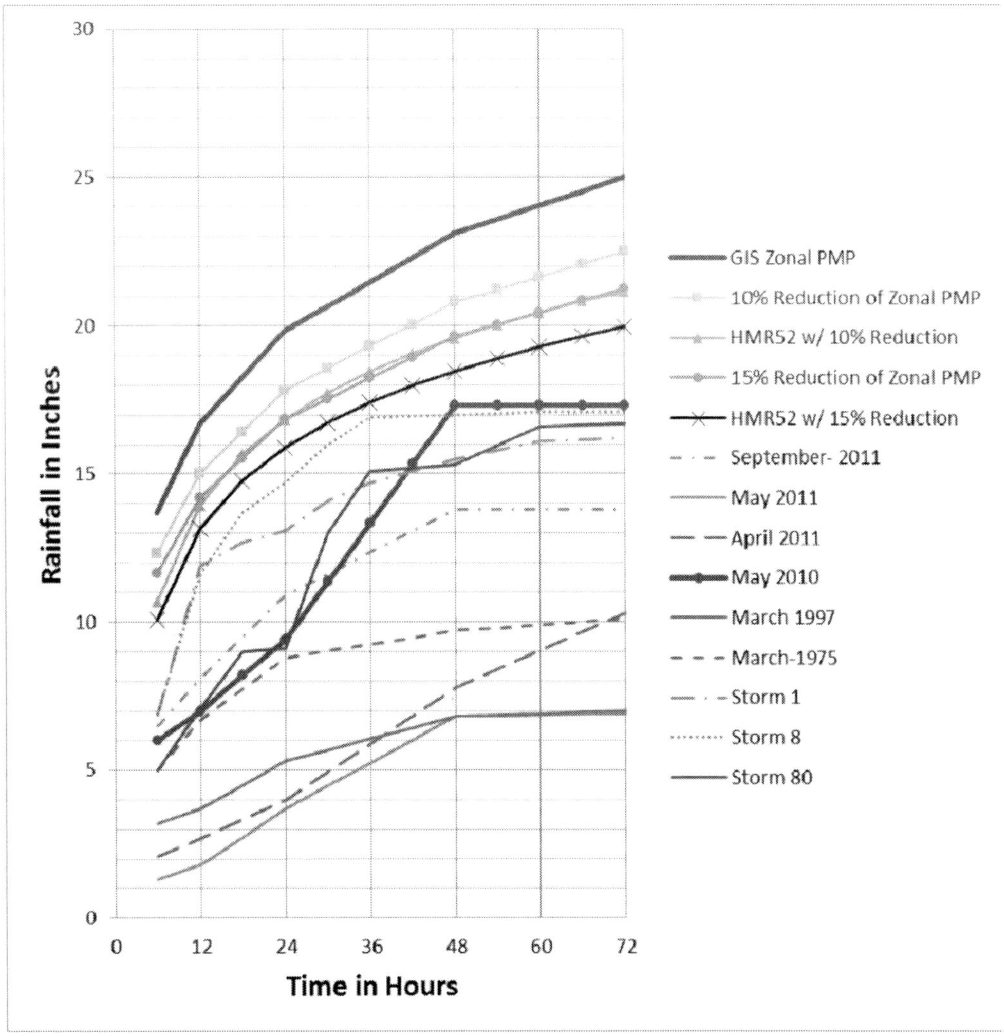

Figure 6 - Depth-Duration Curves for 2,174 Sq. Mi. Center Hill Basin

REFERENCES

HMR 41: June 1965, "*Probable Maximum and TVA Precipitation over the Tennessee River Basin above Chattanooga*"

HMR 47: May 1973, "*Meteorological Criteria for Extreme Floods For Four Basins in the Tennessee and Cumberland Watersheds*"

HMR 51: June 1978, "*Probable Maximum Precipitation Estimates, United States East of the 105th Meridian*"

HMR 52: August 1982, "*Application of Probable Maximum precipitation Estimates – United States East of the 105th Meridian*"

HMT 55A: June 1978, "*Probable maximum Precipitation Estimates – United States Between the Continental Divide and the 103rd Meridian*"

HMR 56: *October 1986, "Probable Maximum and TVA Precipitation Estimates with Areal Distribution for Tennessee River Drainages Less than 3,000 mi^2 in Area*"

HYDRO 41: August 1985, "*Probable Maximum Precipitation Estimates for the Drainage above Dewey Dam, Johns Creek, Kentucky*"

Petterssen, S., 1956, Weather Analysis and Forecasting, Volume II, Weather and Weather Systems, Second Edition, 266 pp.

Schwarz, F.K., 1973: Addendum to Hydrometeorological Report No. 45, National Weather Service, National Oceanic and Atmospheric Administration, U.S. Department of Commerce, Silver Spring, MD, 22 pp.

USACE 1984: "*HMR52 Probable Maximum Storm (Eastern United States) User's Manual*"

USACE 1986, "Request for Review of Hydrometeorological Branch, National Weather Service, of PMP Estimates for Wolf Creek Dam, Kentucky".

USACE 1987, "PMP Estimate for Center Hill Spillway Adequacy Study, 19 August 1987".

USACE 1989, "Evaluation of the 1989 Probable Maximum Flood Analysis for Center Hill"

USACE 2011: *"Chena River Basin, Alaska: Probable Maximum Precipitation Study"*

USACE 2010: "*After Action Report, May 2010 Flood Event Cumberland River Basin, 1-3 May 2010, Great Lakes and Ohio River Division"*

USACE 2014: "*Bluestone Basin PMP Review Final Report*

U.S Climatic Atlas, U.S. Department of Commerce, June 1968

ADVANCES IN WATERSHED MODELING INPUTS AND METHODS FOR DAM SAFETY RISK ANALYSES

Jason Caldwell[1], Nicole Novembre[2]
Katie Laro[3], Tye Parzybok[4]
Bruce Barker[5], Mel Schaefer[6]
George Taylor[7], Wayne Gibson[8]

ABSTRACT

While FERC and other regulators/operators continue to utilize traditional deterministic (PMP/PMF) analyses, there is interest and motivation to apply probabilistic analyses. Probabilistic analyses provide information for performing Risk-Informed Decision-Making (RIDM) for dam remediation, reservoir operations, monitoring and prioritization of dam safety activities. Evaluation of hydrologic risks to dams and infrastructure requires examination of extreme precipitation events for application in hydrologic models. These models require detailed information on precipitation, temperature, and hydrologic parameters to characterize the dynamics of hydrometeorological processes (e.g., snowmelt, soil moisture, runoff). Quantifying and understanding the characteristics of extreme precipitation events is critical in identifying the key drivers of significant floods. Invaluable research from the meteorological and engineering communities is steadily enriching the literature with new data and methods for RIDM applications; however, practical application lags state-of-the-science approaches. Here, we discuss several state-of-the-science/engineering applications of new data and methods: (i) the necessity of storm typing for meteorological and hydrologic analyses; (ii) advances in stochastic hydrologic modeling; and (iii) meteorological data sets that support both deterministic efforts and probabilistic analyses within the RIDM framework. These advances are an important consideration for dam safety personnel when performing site-specific, regional, statewide, or national products in support of RIDM for dam safety.

---

[1] Chief Hydrometeorologist, MetStat, Inc., 2950 East Harmony Road, Suite 392, Fort Collins, CO 80528, jcaldwell@metstat.com.
[2] Chief Hydrologic Engineer, MetStat, Inc., 2950 East Harmony Road, Suite 392, Fort Collins, CO 80528, nnovembre@metstat.com.
[3] Senior Meteorologist, MetStat, Inc., 2950 East Harmony Road, Suite 392, Fort Collins, CO 80528, klaro@metstat.com
[4] President/Chief Meteorologist, MetStat, Inc., 2950 East Harmony Road, Suite 392, Fort Collins, CO 80528, tyep@metstat.com
[5] Principal, MGS Engineering Consultants, Inc., 7326 Boston Harbor Road NE, Olympia, WA 98506, bbarker@mgsengr.com
[6] President, MGS Engineering Consultants, Inc., 7326 Boston Harbor Road NE, Olympia, WA 98506, mgschaefer@mgsengr.com
[7] President, Applied Climate Services, 5478 SW Philomath Blvd, Corvallis, OR 97333, taylorgh@comcast.net
[8] Scientist/Programmer, Applied Climate Services, 5478 SW Philomath Blvd, Corvallis, OR 97333, gibsonwp@comcast.net

---

INTRODUCTION

Dam safety officials are increasingly moving to include risk-informed decision-making (RIDM) methods, creating a need for probabilistic flood estimates. A key piece of any hydrologic hazard assessment (HHA) for dam safety is a hydrologic model that simulates, at the most basic level, the generation of streamflow, changes in storage, and dam operations within a specific watershed. Screening-level HHAs typically involve a single deterministic model simulation to generate the probable maximum flood (PMF) from a derived probable maximum precipitation (PMP) using Hydrometeorological Reports (HMRs) or a site-specific PMP study. Screening-level studies are used to prioritize the need for a higher-level HHA, which typically involve stochastic hydrologic modeling with detailed meteorological and hydrological analyses, such as:

- identification and reconstruction of historically significant precipitation and flood events;
- watershed-specific precipitation-frequency estimates;
- time-series and climatologies of meteorological and hydrologic parameters (e.g., snow cover/water equivalent, evapotranspiration, antecedent precipitation, freezing level heights, reservoir elevation, soil moisture deficit, etc.);
- development, calibration, and validation of a hydrologic model; and,
- thousands of simulations to generate hydrologic frequency estimates.

Stochastic hydrologic modeling allows combination of multiple hazard curves based on storm type, leading to the development of magnitude-frequency relationships for quantities of interest, such as the maximum one-hour reservoir elevation (Figure 1) developed using the Stochastic Event Flood Model (SEFM; Schaefer and Barker, 2015).

The complexity of dam safety studies varies depending on the objective, ranging from screening-level studies that characterize risk based on deterministic data to stochastic modeling efforts to evaluate probabilities of specific failure modes. Intermediate analyses exist, such as the Australian Rainfall-Runoff method (Nathan and Weinmann, 2016) and other extrapolation and scaling techniques. Members of the dam safety community, therefore, require solutions that are cost-effective, time-efficient, high-quality, and scientifically-sound, that scale with the complexity of the study. In this paper, we describe advances in the realms of meteorology and hydrology critical to meeting this need (Parzybok et al., 2016).

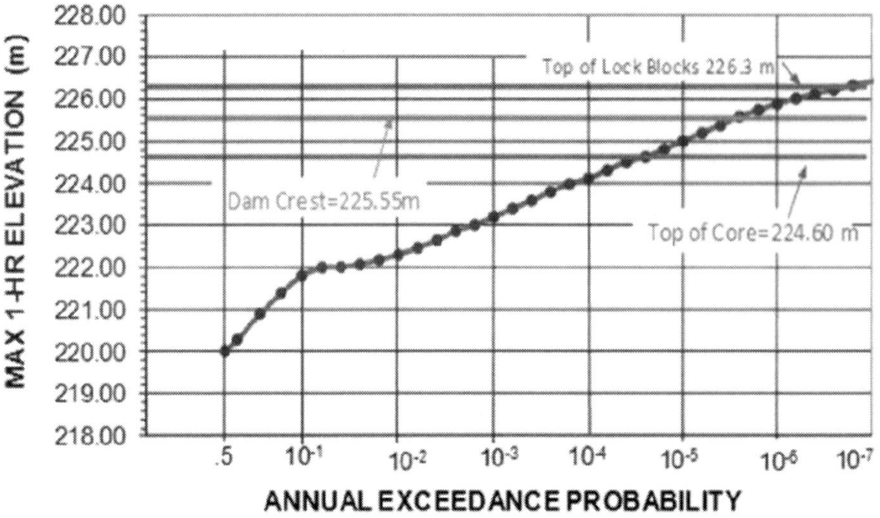

Figure 1. Hydrologic hazard curves from the Stochastic Event Flood Model (SEFM) for a dam safety evaluation, indicating the annual exceedance probability related to key infrastructure elevations at the site. Risk-informed decision making (RIDM) requires this type of complex analysis to provide annual exceedance probabilities for evaluating total risk using decision trees.

ADVANCES IN STOCHASTIC HYDROLOGIC MODELING

Storm Typing

Historically, precipitation-frequency relationships applied in hydrologic models utilize the all-season estimates from the Hydrometeorological Design Studies Center (HDSC) in the multiple volumes of NOAA Atlas 14. Mixed distributions of storm types contribute to the magnitude of precipitation and shape of the frequency curve due to differences in the mean, variance, and skew for a specific storm type. For example, in the southeastern United States (MGS et al., 2014), meso- to synoptic-scale storms could be estimated by a Generalized Extreme Value distribution, while the means varied significantly. Tropical storms were relatively infrequent in the region, resulting in different distributional characteristics. Combination of the multiple storm type-specific hydrologic hazard curves provides improved evaluation of risk. Additional details are provided in the 'Advances in Regional Precipitation-Frequency Analysis' section, later in this paper.

Performing a storm typing analysis involves discretizing each day in the historical record as a specific type (e.g., local storm (LS), mesoscale storm (MEC), mid-latitude cyclone (MLC), or tropical storm (TSR)). Each storm type has different spatial and temporal storm characteristics and generally results in differing flood responses with regards to flood peak discharge, runoff volume and hydrograph shape for watersheds of various sizes. Realistic assessment of hydrologic loadings and development of Hydrologic Hazard Curves for risk assessment requires the proper preservation of the physics represented by the watershed precipitation-frequency relationship; the spatial and temporal storm characteristics; and hydrologic modeling approach for a given storm type. This results in a more informative assessment of hydrologic loadings and reduces uncertainties in the assessment of reservoir and spillway performance for assessing hydrologic risks.

Stochastic Event Flood Model (SEFM)

The SEFM simulates thousands of years of flood annual maxima by utilizing a calibrated deterministic flood model and treating the input parameters as variables. The variables are chosen from ranges of values specified by the user based on historical data to model realistic watershed behavior. The goal is to create synthetic simulations of reality, not to create a maximum or worst-case scenario condition. Figure 2 shows the process of developing flood-frequency relationships while accounting for uncertainty.

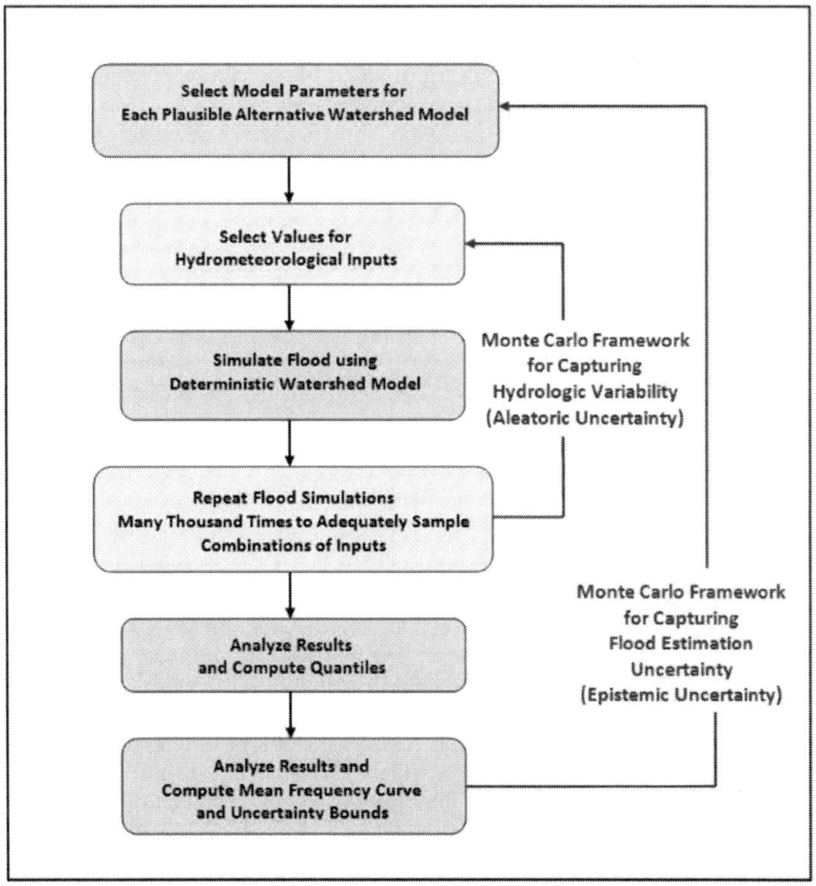

Figure 2. Flowchart for the Monte Carlo framework for stochastic flood and uncertainty analysis.

Each SEFM simulation represents an annual peak event. Precipitation volume is sampled from a probability distribution developed using an L-moments regional statistical method (described later in this paper). The precipitation volume is distributed spatially and temporally using storm type templates developed from storm analyses of past events. Initial states for each simulation (i.e. antecedent precipitation, antecedent snowpack, antecedent snowpack, soil moisture deficit) are resampled from long series of simulated values produced by continuous simulation.

SEFM utilizes a sampling procedure based on the Total Probability Theorem (Nathan et al., 2004) that reduces the number of simulations required to estimate rare exceedance probabilities. For example, the approach allows for exceedance probabilities to 10^{-6} to be estimated with only 10,000 stochastic flood simulations because without this procedure an exceedance probability of only 10^{-4} is achievable.

The use of the stochastic approach allows for the development of separate hydrologic hazard curves for flood peak flow, maximum reservoir level, flood runoff volume and any other flood characteristic that can be obtained from the outputs of a watershed model. For example, frequency information about maximum reservoir levels is important for use in hydrologic risk assessments for dams because it accounts for flood peak flow, runoff volume, hydrograph shape, initial reservoir level, and reservoir operations (Figure 3).

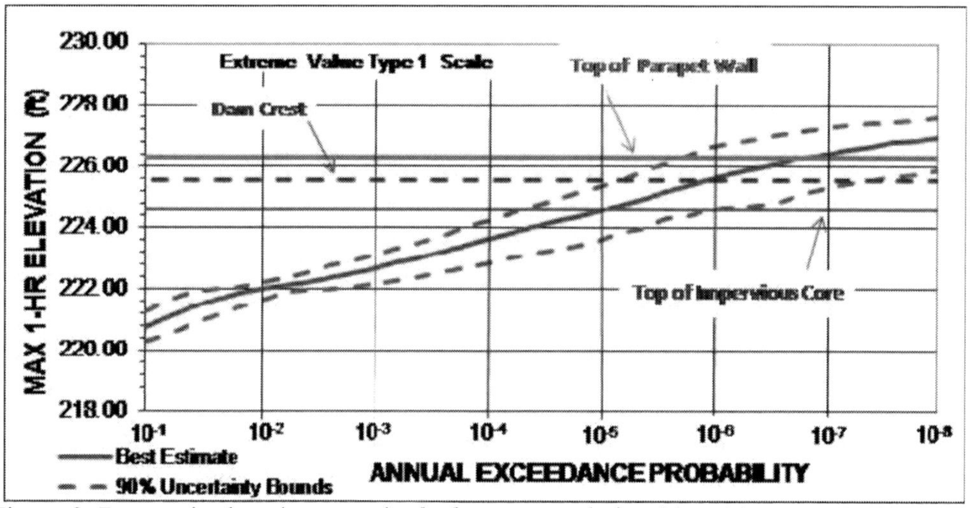

Figure 3. Reservoir elevation magnitude-frequency relationship with uncertainty bounds.

The stochastic flood simulations can also be used to represent dam failure modes such as those related to flood capacity (e.g., reservoir elevation or flood runoff volume), spillway gate operability (e.g., modifications to gate operations), and other failure modes related to flood capacity, such as piping failures. In each case, the failure can be represented by a combination of chance elements, as well as, functional relationships to critical flood characteristics, such as the depth and duration of overtopping.

ADVANCES IN METEOROLOGY

High-Resolution Radar-Rainfall Estimation

Meteorological analyses of large precipitation events are useful in a variety of applications related to dam safety, from determining depth-area-duration (DAD) relationships to inform PMP calculations, to producing spatial and temporal patterns of precipitation for application in hydrologic model calibration, verification, and simulation. Historically, these analyses would have involved collection of gauge data and manual contouring of isohyets and disaggregation of daily to sub-daily time steps using nearby gauges to derive DAD relationships. Since the mid-1990s, however, National Weather Service radars have offered the capability to convert reflectivity to precipitation rate in a quantitative precipitation estimate (QPE) product.

In recent years, the National Weather Service installed dual-polarization (dual-pol) radar at each office across the United States. The value of dual-pol is the ability to estimate the size, shape, and phase of precipitation (e.g., rain drops vs. snowflakes vs. hail, etc.; Parzybok et al., 2015). Hence, calculated precipitation rates are more accurate. In addition, radar and satellite products are higher spatial and temporal resolution, allowing sophisticated algorithms (e.g., MetStorm®; Hendricks et al., 2015; Laro et al., 2015; Laro et al., 2016) to produce sub-hourly, 250-meter gridded precipitation. The algorithms integrate satellite, radar, and precipitation gauges to produce multisensor QPE grids. Satellite data aid in the spatial representation of precipitation in complex terrain, where gauges are typically sparse and radar is unavailable due to beam blockage. Applications of these data include historical storm analysis for hydrologic modeling (Figure 4), real-time monitoring for operational decision making, and assessment of basin average precipitation-frequency relationships (described later in this paper). Other outputs from the algorithms include: DAD plots and tables; average recurrence interval maps; gauge data catalogs; validation plots; error statistics; mass curve plots and tables; and, a storm report – all of which are archived in a national storm database.

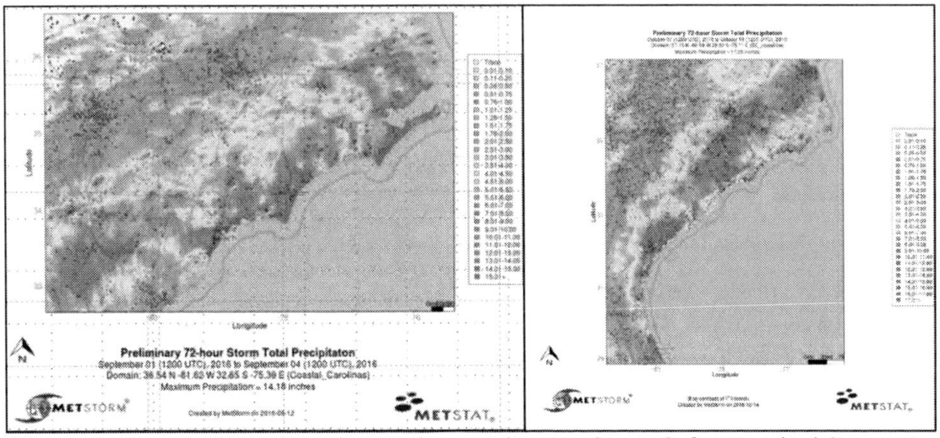

Figure 4. Historical precipitation analyses using MetStorm® for Tropical Storm Hermine (left) and Hurricane Matthew (right) that resulted in record flooding across eastern portions of North and South Carolina in September and October 2016

Applications of Numerical Weather Prediction Models

Numerical weather prediction (NWP) models are multi-dimensional computer models of the atmosphere and ocean/land surface that provide the ability to reconstruct historical weather and climate conditions from observed data (i.e., reanalysis), as well as, to predict future conditions (i.e., forecast). Computer power and speed continue to increase rapidly, allowing complex applications like NWP models to provide continuous time series of key meteorological variables at spatial and temporal resolutions suitable for use in hydrologic modeling. Some examples of reanalysis datasets of interest to hydrologic hazard assessments include:

- Snow Data Assimilation System (SNODAS; NOHRSC, 2004)
- NOAA 20th Century Reanalysis (Compo et al., 2011)
- NCAR/NCEP Reanalysis (Kalnay et al., 1996)
- North American Regional Reanalysis (Mesinger et al, 2004)
- Climate Forecast System Reanalysis (Saha et al., 2010)

These reanalysis datasets offer a wide array of potential applications in dam safety studies. For deterministic modeling efforts, scientists at NOAA and the University of Colorado (Alexander et al., 2015; Hughes et al., 2014) evaluated the predominant moisture pathways into the western United States and the reduction in moisture availability relative to distance from the coast and terrain-blocking (Figure 5).These products provide objective criteria to support subjective decisions on the definition of transposition regions, moisture adjustment factors, and orographic adjustment factors - key variables in PMP development. Additionally, higher-resolution products can also be used to generate precipitable water, dewpoint temperature, and sea surface temperature climatologies for in-place maximization factors in a PMP study (Caldwell et al., 2011).

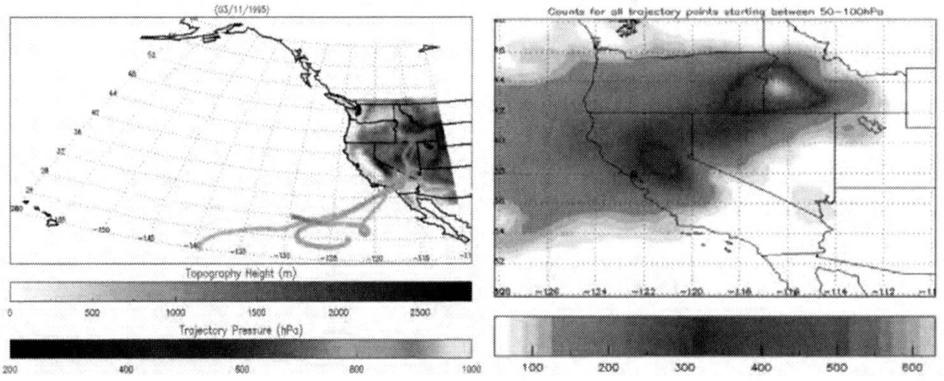

Figure 5. Back-trajectory analysis (left) for March 1995 precipitation event in southern ID and back-trajectory counts for large precipitation events in southern ID. Underlying precipitable water data is available at 6-hour increments for each back-trajectory. Source: Alexander et al. (2015).

For stochastic modeling efforts used in RIDM efforts, a suite of hydrometeorological variables are required as described previously. Statistical distributions of these variables can be easily calculated and provide spatially and temporally continuous meteorological information (e.g., Figure 6), making the reanalysis a valuable tool in locations with sparse observational data.

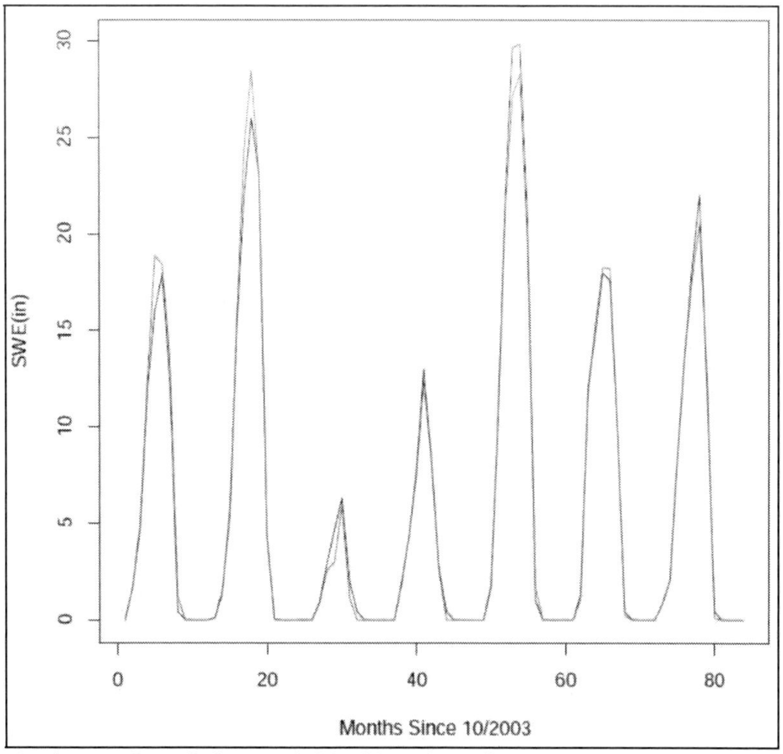

Figure 6. Time series of observed (black) and predicted (red) snow water equivalent generated using SNODAS data.

In addition, the reanalysis products offer historical time series of freezing level height, 1000mb air temperature, solar radiation, winds, and humidity - variables important for modeling the dynamics of snowmelt, evapotranspiration, and other complex processes in hydrologic models (Sankovich et al., 2013; Figure 7). The 1000mb air temperature and freezing level height are required to generate storm spatial and temporal templates for use in stochastic models like SEFM. In addition, distributional parameters of each of these variables serve as the basis for resampling techniques to provide variability in the Monte Carlo procedures.

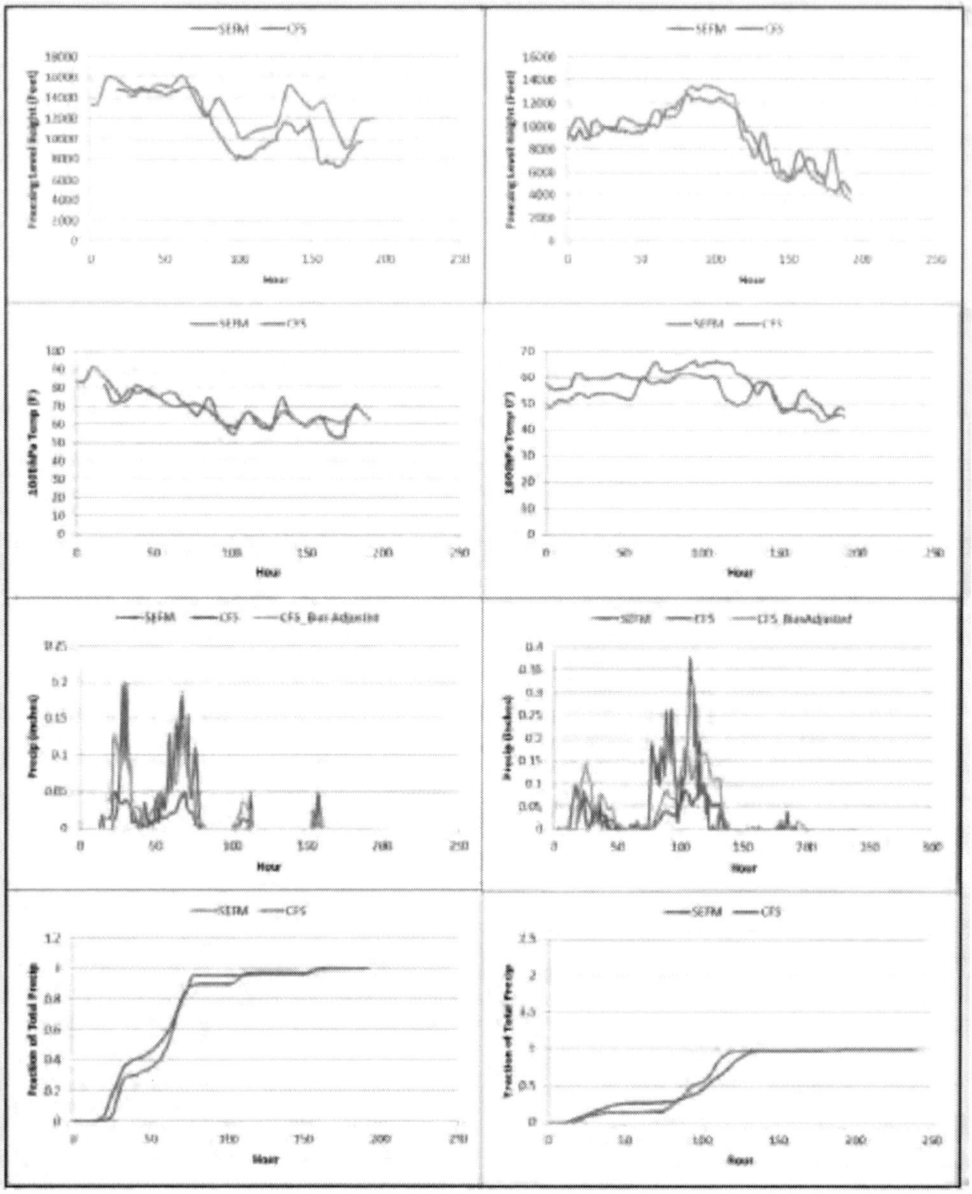

Figure 7. Comparison of manual (SEFM) vs reanalysis (CFS) of freezing level, 1000mb temperature, and precipitation at Friant Dam, California, during storm events in 1982 (left) and 1996/1997 (right).

ADVANCES IN REGIONAL PRECIPITATION-FREQUENCY ANALYSIS

NOAA Atlas 14 provides regional precipitation-frequency analyses (Figure 8) for much of the United States at multiple durations and return periods, though these analyses are all-season estimates and, therefore, include a mixed population of storm types (e.g., local convective storms (LS), mesoscale systems (MEC), mid-latitude cyclones (MLC), and tropical storms (TSR).

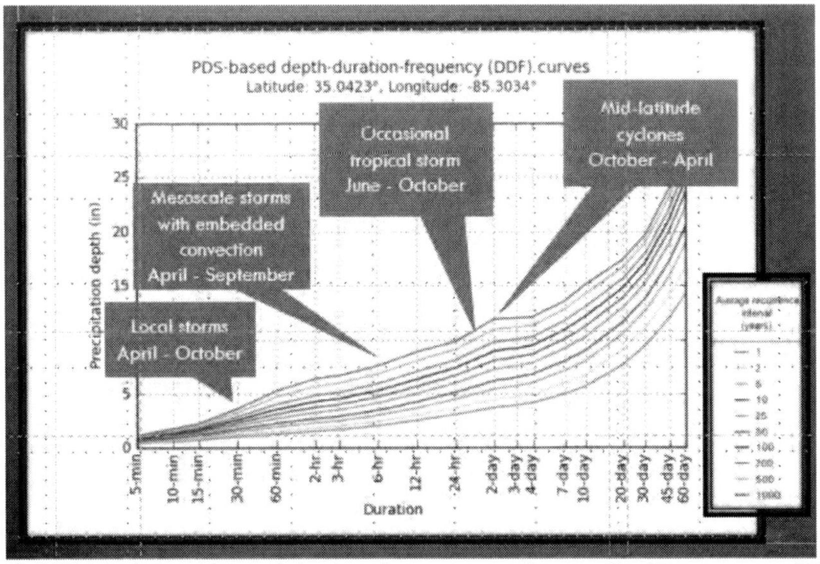

Figure 8. NOAA Atlas 14 precipitation-frequency estimates for Chattanooga, TN with examples of mixed distributions for each storm type.

Spatial and temporal characteristics of storms from each type vary significantly and the hydrologic risk for a specific dam is typically driven by a subset of these, depending on the watershed characteristics and dam design. For example, a 40,000-square mile basin is unlikely to have a hazard related to a local storm; however, a 10-square mile basin is more likely focused on the rapid influx of volume over a short time span associated with a thunderstorm event.

In recent years, precipitation frequency analyses based on specific storm types has become integral in accurately quantifying the temporal and spatial precipitation for different stochastic scenarios. To conduct such an analysis, a database of daily storm types is required. Reanalysis data, as described earlier, provides a long-term record of atmospheric conditions that enable definition of daily storm type conditions based on meteorological parameters such as precipitable water, convective available potential energy (CAPE), upper-level wind patterns, and surface pressures (Figure 9). Additionally, the "footprint" size and magnitude of precipitation on the ground, as inferred from gauge data, is a critical element in the storm typing algorithm.

This time series of storm types can then be used to isolate annual maxima from each storm type for use in developing multiple precipitation frequency relationships and, hence, storm/flood-type specific hydrologic hazard curves. Combination of these mixed population events is a necessity for areas with mixed populations, particularly for extreme events.

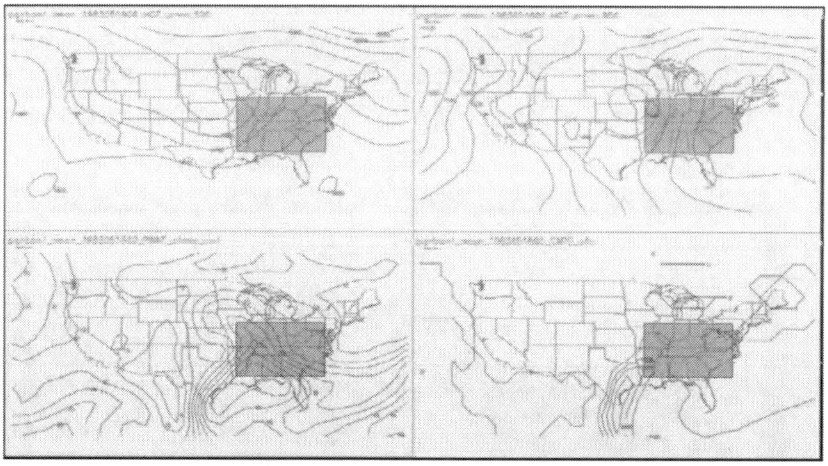

Figure 9. Four panel plot of reanalysis data on May 19, 1983 used for manual storm typing process prior to automated storm typing for each day.

In addition to the point precipitation-frequency estimates, a watershed precipitation-frequency relationship must be defined for each storm type and basin size of interest due to the variability in the spatial coverage of precipitation associated with the LS, MEC, MLC, and TSR events. Storm analysis software (e.g., MetStorm®) provides the opportunity to generate spatial and temporal patterns of precipitation from historical events of each type, such that relationships between point precipitation amounts at key stations and basin-average precipitation can be developed for specific durations. In a recent study in the Tennessee Valley (MGS et al., 2014), daily storm typing was coupled with MetStorm® software to generate point-to-area relationships for various watershed sizes, ranging from tens of square miles to over 10,000 square miles (Figure 10).

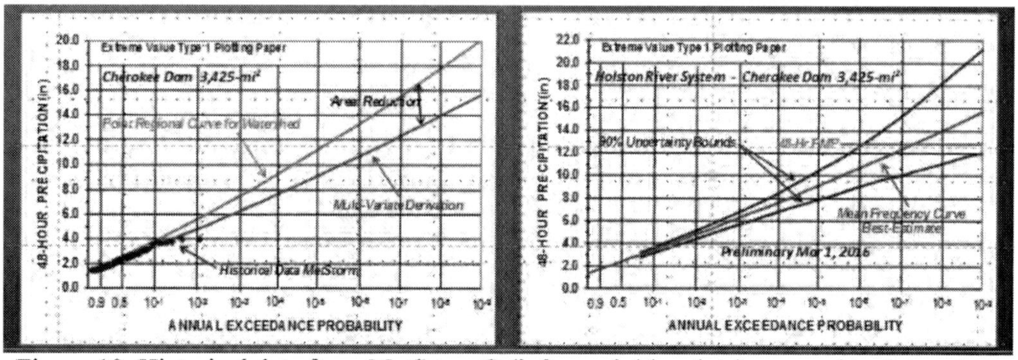

Figure 10. Historical data from MetStorm® (left panel, blue dots) are used to determine the areal reduction relationship with larger reductions at the rarer annual exceedance probabilities (AEPs). Uncertainty analysis (right) is developed relative to the watershed precipitation-frequency relationship.

The uncertainty analysis for point precipitation is computed for all parameters which are used in quantile estimates for precipitation at selected AEPs. This includes uncertainty in the at-site mean, regional L-Cv, regional L-Skewness and the regional probability distribution. The uncertainty analysis is conducted in manner that allows computation of the mean frequency curve (best-estimate) and uncertainty bounds, along with identification of the uncertainty variance for all contributing components (Figure 11).

Uncertainty bounds for extreme AEPs are relatively narrow compared to uncertainty bounds for other natural phenomena. This is attributable to:

- the use of storm typing to assemble highly homogeneous datasets for a given storm type;
- the reduction of sampling variability for the regional L-moment statistics by using homogeneous sub-regions; and,
- further reduction in sampling variability by using large regional datasets and regional solutions for the at-site mean, L-Cv and L-Skewness, resulting in improved spatial coherence of the L-moment statistics.

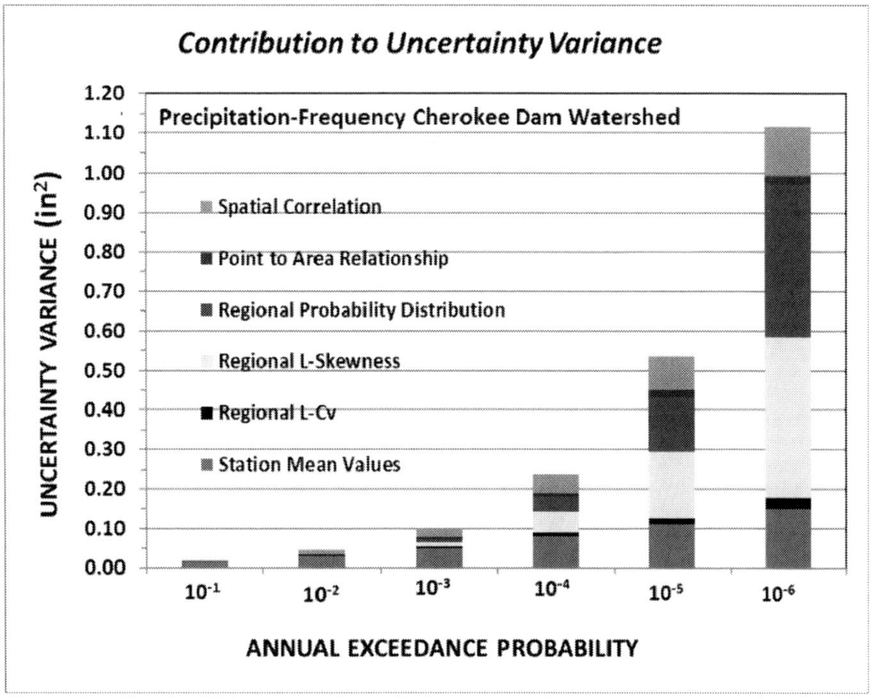

Figure 11. Stacked histogram for relative uncertainty variance associated with precipitation-frequency estimates for the Cherokee Dam watershed.

SUMMARY

Probabilistic flood-frequency estimates including uncertainty bounds for hydrologic parameters are essential for risk-based analyses to make safety, policy, operation, and budgetary decisions. Because the record of observation of natural events is very short compared to the required return intervals, stochastic modeling methods are necessary to create many thousands of years of synthetic events that represent the true behavior of the watershed and dam of interest. Additionally, the use of stochastic flood modeling outputs to perform failure mode analyses is extremely valuable. Stochastic modeling of rare events with very low annual exceedance probabilities is becoming more computationally efficient thanks to the use of the Total Probability Theorem, and can therefore be more widely used for hydrologic studies.

Additionally, modern software and hydrometeorological datasets are available to the dam safety community that offer expeditious compilation of the multiple inputs required for stochastic modeling. Most importantly, perhaps, is the availability of a rich database of precipitation gauges coupled with reanalysis data to determine storm type for application in precipitation-frequency analysis, selection of spatial and temporal patterns to represent the full range of storm types for a given region, and to perform historical analyses for watershed-specific precipitation-frequency relationships.

Increases in computational efficiency have led to the advancements in stochastic modeling methods, the evolution of storm-typing as a key element for RIDM studies, and the development of storm analysis software and reanalysis products. Together, these items collectively serve to provide a wide range of hydrologic scenarios with reduced costs and enhanced results from HHAs to use within the RIDM process.

REFERENCES

Alexander, M.A., J.D. Scott, D. Swales, M. Hughes, K. Mahoney, and C.A. Smith, 2015: Moisture pathways into the U.S. Intermountain West associated with heavy winter precipitation events, J. Hydromet., 16, 1184-1206.

Caldwell, R.J., Sankovich, V.L. and England, J.F. Jr. (2011) Synthesis of Extreme Storm Rainfall and Probable Maximum Precipitation in the Southeastern U.S. Pilot Region, for the Nuclear Regulatory Commission, Office of Nuclear Regulatory Research. Bureau of Reclamation, Denver, CO, December, 39 p.

Compo, G.P. et al., 2011: The Twentieth Century Reanalysis Project. Quarterly J. Roy. Meteorol. Soc., 137, 1-28. DOI: 10.1002/qj.776

Hendricks, A.D., T.W. Parzybok, K. Laro, and W. Mokry, 2015. MetStorm®: Near Real-Time Storm Precipitation Analytics. 2015 GIS in the Rockies, Denver, CO, September 23-24, 2015.

Hughes, M., K. Mahoney, P. Neiman, B. Moore, M. Alexander, and F.M. Ralph, 2014: The landfall and inland penetration of a flood-producing atmospheric river in Arizona, Part II: Sensitivity of modeled precipitation to terrain height and atmospheric river orientation.

Kalnay et al.,1996. The NCEP/NCAR 40-year reanalysis project, Bull. Amer. Meteor. Soc., 77, 437-470.

Laro, K.L., T.W. Parzybok, A.D. Hendricks, 2016. A MetStorm® Analysis on the Lower Mississippi River Valley Storm of March 2016. National Hydrologic Warning Council Transmission. April 2016.

Laro, K.L., T. W. Parzybok, A. Hendricks, and W. Mokry, 2015. MetStorm®: Near Real-Time Storm Precipitation Analytics. 11th National Hydrologic Warning Council, Indianapolis, IN, June 15-18, 2015.

Mesigner et al., 2004. North American Regional Reanalysis (NARR): A long-term, consistent, high-resolution climate dataset for the North American domain, as a major improvement upon the earlier global reanalysis datasets in both resolution and accuracy, submitted to Bull. Amer. Met. Society (BAMS).

MGS Engineering Consultants, MetStat, Applied Climate Services and Riverside Technology, Regional Precipitation-Frequency Analyses for Mid-Latitude Cyclones, Mesoscale Storms with Embedded Convection, Local Storms and Tropical Storm Remnant Storm Types in the Tennessee Valley Watershed, prepared for Tennessee Valley Authority, Dec 2014.

Nathan, R.J., P.E. Weinmann, Sinclair Knight Merz, Hydrology: Science & Practice for the 21st Century, Vol 1. Proc. British Hydrol. Soc., London, 186-193, 2004.

Nathan, R.J. and P.E. Weinmann, E, 2016, Estimation of Very Rare to Extreme Floods, Book 8 in Australian Rainfall and Runoff - A Guide to Flood Estimation, Commonwealth of Australia.

National Operational Hydrologic Remote Sensing Center. 2004. *Snow Data Assimilation System (SNODAS) Data Products at NSIDC*, [list the dates of the data used]. Boulder, Colorado USA: NSIDC, http://dx.doi.org/10.7265/N5TB14TC.

Parzybok, T.W., J. Caldwell, K.L. Laro, M. Schaefer, G. Taylor, E. Tomlinson, 2016. Advances in Extreme Precipitation Methodologies. 2016 ASDSO Dam Safety National Conference, Philadelphia, Pennsylvania, September 11-15, 2016.

Parzybok, T.W., N.A. Lock, C.W. Porter, K. Laro, A.D. Hendricks, and W. Mokry, 2015. Near Real-time Storm Precipitation Analytics Using Dual-pol Radar Data. American Meteorological Society 37th Conference on Radar Meteorology, Norman, OK, September 13-18, 2015.

Saha et al., 2010. The NCEP Climate Forecast System Reanalysis, Bull. Amer. Met. Society (BAMS), 1015-1057.

Sankovich, V.L., R.J. Caldwell, J.F. England, 2013: Investigation of Storm Temporal Patterns using Gridded Meteorological Datasets for Hydrologic Modeling, Annual Meeting of the American Meteorological Society, Austin, TX, Poster 48.

Schaefer, M.G., B.L. Barker, 2015. SEFM, Stochastic Event Flood Model User's Manual, MGS Software LLC Olympia WA. 2001, 2009, 2015.

USE OF THE CONDITIONAL MEAN FOR IMPROVED PREDICTION OF GROUND MOTION INTENSITY MEASURES FOR EMBANKMENT DAMS

Richard J. Armstrong, PE, PhD[1]

ABSTRACT

Earthquake-induced embankment-dam deformations are affected by various ground motion characteristics. These ground motion characteristics can be described by various ground motion intensity measures, such as spectral acceleration, peak ground velocity, and arias intensity. A design target level for one or more of these intensity measures may be predicted as part of a seismic evaluation of a dam. The conventional approach for setting the design target levels in dam-engineering practice is to select the target value of each intensity measure to represent uniformly a specified percentile level. An alternative to this conventional approach is to select a single intensity measure, called the conditioning intensity measure, that relates well to embankment-dam response and then to apply this percentile level to that particular intensity measure only. The average values of the other intensity measure targets are then selected, given (or "conditioned on") the value of the conditioning intensity measure. This new design target is termed the conditional mean. In this paper, the steps for computing the conditional mean are described. It is shown that unlike the conventional approach, the conditional mean is probabilistically correct and represents the observed intensity measures from recorded ground motions.

INTRODUCTION

The seismic response of an embankment dam depends on various characteristics of the earthquake ground motion. These ground motion characteristics can be described by various ground motion intensity measures, such as spectral acceleration, peak ground velocity, and arias intensity. When a seismic hazard analysis is conducted, the intensity measures deemed important to the dam are identified, after which the potential distribution of each intensity measure is predicted using ground motion prediction equations. The design target level of each intensity measure is then determined according to deterministic or probabilistic seismic hazard analyses. For dams, deterministic seismic hazard analyses are still common, and the intensity measure levels are set to be a certain number of standard deviations away from the mean prediction based on the percentile level chosen. For example, for high-hazard dams, the California Division of Safety of Dams uses an 84th-percentile target level that corresponds to one standard deviation above the mean prediction (Fraser and Howard 2002). Once this percentile level is selected, the decision often made in dam-engineering practice is to apply this percentile level uniformly when setting the target level for all intensity measures. Thus, for an 84th-percentile target loading, all spectral ordinates—as well as other intensity measures, such as peak ground velocity and arias intensity—all represent 84th-percentile predictions.

[1] Assistant Professor, California State University, Sacramento, 6000 J Street, Sacramento, CA 95819-6029, richard.armstrong@csus.edu

An alternative approach to a uniform percentile level for all intensity measures is to select a single intensity measure that relates well to embankment-dam response and to apply the selected percentile level only to this intensity measure. The average values of the other intensity measure targets are selected according to the percentile level of this single intensity measure. The average value of each intensity measure is termed the conditional mean, and the single intensity measure in which the percentile level is selected is termed the conditioning intensity measure.

The conditional mean of multiple intensity measures has been used as an alternative design target for seismic evaluations. For spectral acceleration only, the conditional mean is commonly known as the conditional mean spectrum (CMS) and has been shown by Baker (2011) to be probabilistically consistent and able to produce design levels consistent with recorded ground motions. In addition, the conditional mean was used by Bradley (2010) as a key target in the generalized conditional intensity measure approach.

The purpose of this paper is to (1) describe the steps involved in computing the conditional mean of multiple intensity measures, and (2) compare the design target using the conditional mean to the design target using the conventional approach. A seismic hazard analysis of an example embankment dam will be used throughout this paper to aid in both the computations and the comparisons. In addition, example calculations of the conditional mean are included at the end of this paper to help those readers interested in replicating some of the results contained herein. For the sake of brevity, certain probabilistic fundamental concepts and proofs are not included in this paper. Readers interested in these topics are encouraged to see Baker and Cornell (2005), Bradley (2010), and Baker (2011).

COMPUTATION OF INTENSITY MEASURES FOR EXAMPLE DAM

The example embankment dam consists of a zoned embankment with both shells underlain by loose alluvium (Fig. 1). The dam is located 5 km from an active strike-slip fault that is capable of a magnitude 7.5 earthquake. In the bedrock, the shear wave velocity in the upper 30 m (V_{s30}) is 600 m/s. Also, the embankment dam has a predominant period of approximately 0.4 sec. In the design seismic event, it is expected that the alluvium will liquefy and could result in significant embankment deformations (e.g., vertical crest deformations).

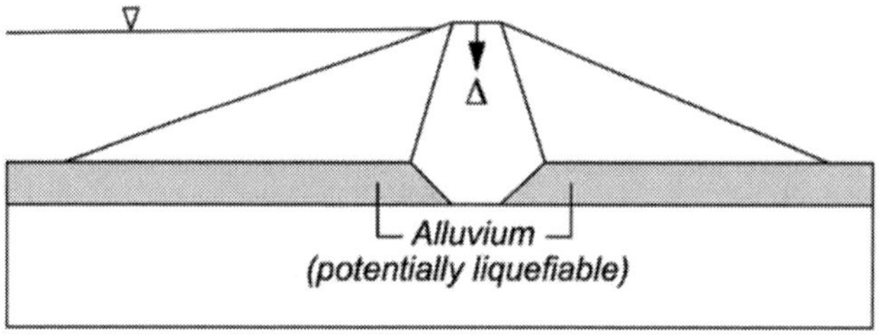

Figure 1. Ground motion development for an embankment dam underlain with potentially liquefiable alluvium.

For embankment dams—or, more generally, slopes—the ground motion intensity measures that have been found by others (Bray and Travasarou 2007, Saygili and Rathje 2007, Beaty and Perlea 2012, and Wang and Du 2012) to relate to deformations are: spectral acceleration, peak ground velocity, arias intensity, cumulative absolute velocity, and significant duration between 5% and 95% of arias intensity. Note that arias intensity and cumulative absolute velocity are similar time-integrated intensity measures (square versus absolute value in integral) with a strong statistical relationship between each (i.e., correlation coefficient of 0.923—Campbell and Bozorgnia 2013). In California, the California Division of Safety of Dams has historically emphasized the use of arias intensity as an important intensity measure for ground motion development to analyze embankment dams (Howard et al. 2008). Therefore, this paper focuses on the use of arias intensity instead of cumulative absolute velocity. This leads to the use of spectral acceleration, peak ground velocity, arias intensity, and significant duration as intensity measures that relate to embankment deformations.

Ground motion prediction equations, developed by researchers based on previous earthquake data, can be used to predict distributions of the aforementioned intensity measures. These ground motion prediction equations are a function of the earthquake's moment magnitude, rupture distance, style of faulting, and other parameters. The majority of ground motion prediction equations predict an intensity measure that has a lognormal distribution and can be expressed as:

$$\ln IM_i \sim \mathcal{N}\left(\mu_{\ln IM_i}, \sigma_{\ln IM_i}\right) \quad (1)$$

where $\ln X \sim \mathcal{N}(\mu_{\ln X}, \sigma_{\ln X})$ is the shorthand notation for X having a lognormal distribution with mean $\mu_{\ln X}$ and variance $\sigma_{\ln X}$. With the notation used in Equation (1), IM_i represents the ith intensity measure, a lognormally distributed random variable with mean $\mu_{\ln IM_i}$ (median, $\widetilde{IM}_i$ equals $e^{\mu_{\ln IM_i}}$) and standard deviation $\sigma_{\ln IM_i}$. Note that IM_i has a lognormal distribution, and $\ln IM_i$ has a normal distribution. The fact that IM_i has a lognormal distribution is stressed in the "$\ln IM_i$" term in the subscript of the mean and standard deviation. Using the notation for IM_i, random variables representing spectral acceleration, peak ground velocity, arias intensity, and significant duration are SA, PGV, AI, and $D595$, respectively. In contrast to IM_i, im_i represents a specific value of IM_i. Using this notation, specific values for spectral acceleration, peak ground velocity, arias intensity, and significant duration are sa, pgv, ai, and $d595$, respectively.

For the example embankment dam, the distributions of SA, PGV, AI, and $D595$ for the earthquake scenario are shown in Fig. 2 in terms of the mean ($\mu_{\ln IM_i}$) and mean plus and minus one standard deviation ($\mu_{\ln IM_i} \pm \sigma_{\ln IM_i}$). Predictions for SA and PGV are from Campbell and Bozorgnia (2008), AI is from Campbell and Bozorgnia (2013), and $D595$ is from Bommer et al. (2009). In total, 22 intensity measure distributions are predicted: 19 spectral ordinates between 0.01 and 5.0 seconds, plus peak ground acceleration, arias intensity, and significant duration. The 19 spectral periods considered are listed in the example calculations.

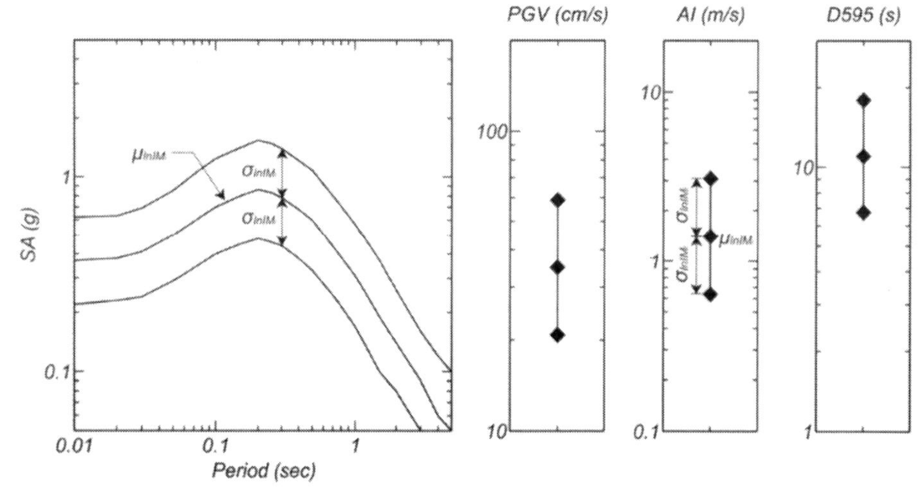

Figure 2. Results for $\mu_{\ln IM_i}$ and $\mu_{\ln IM_i} \pm \sigma_{\ln IM_i}$ for all intensity measures considered.

COMPUTATION OF THE CONDITIONAL MEAN

The conditional mean—or, more generally, the conditional distribution—is determined according to conditional probability theory. It states that given two events, A and B, the probability of B given that A has occurred, $P[B|A]$, is equal to:

$$P[B|A] = \frac{P[B \cap A]}{P[A]} \tag{2}$$

where $P[B \cap A]$ is the probability that events A and B have occurred and $P[A]$ is the probability that event A has occurred. Equation (2) can be extended to random variables X and Y according to:

$$f_{X|Y=y} = \frac{f_{XY}}{f_Y} \tag{3}$$

where $f_{XY}$is the joint probability density function of two random variables X and Y, $f_Y(y)$ is the marginal distribution of Y, and $f_{X|Y=y}$ is the conditional distribution of X given $Y = y$. In terms of the lognormally distributed random variable IM_i, the conditional distribution of $\ln IM_i$, given that an observation of another intensity measure IM^* equals im^* has occurred, is defined as $\ln IM_i|IM^* = im^*$ and can be expressed as:

$$\ln IM_i|IM^* = im^* \sim \mathcal{N}\left(\mu_{\ln IM_i|IM^*=im^*}, \sigma_{\ln IM_i|IM^*=im^*}\right) \tag{4}$$

where IM^* is the conditioning intensity measure and im^* is the specific value of IM^*. As implied in Equation (5), $\ln IM_i|IM^* = im^*$ is a lognormally distributed random variable with conditional mean $\mu_{\ln IM_i|IM^*=im^*}$ and conditional standard deviation $\sigma_{\ln IM_i|IM^*=im^*}$.

Note that the distribution of $\ln IM_i | IM^* = im^*$ changes with the value of $IM^* = im^*$; thus, $\ln IM_i | IM^*$ is a function of im^*, or $\ln IM_i | IM^* = im^*$. Also note that IM^* is a lognormally distributed random variable and equals a single IM_i with mean $\mu_{\ln IM^*}$ and standard deviation $\sigma_{\ln IM^*}$. The computation of $\mu_{\ln IM_i | IM^* = im^*}$ and $\sigma_{\ln IM_i | IM^* = im^*}$ follows from probability theory as:

$$\mu_{\ln IM_i | IM^* = im^*} = \mu_{\ln IM_i} + \sigma_{\ln IM_i} \rho_{\ln IM_i, \ln IM^*} \left(\frac{\ln im^* - \mu_{\ln IM^*}}{\sigma_{\ln IM^*}} \right) \tag{5}$$

and

$$\sigma_{\ln IM_i | IM^* = im^*} = \sigma_{\ln IM_i} \sqrt{1 - \rho_{\ln IM_i, \ln IM^*}{}^2} \tag{6}$$

where $\rho_{\ln IM_i, \ln IM^*}$ is the correlation coefficient between $\ln IM_i$ and $\ln IM^*$. The correlation coefficient has been determined between various intensity measures based on comparison of observations from recorded ground motions. For example, the correlation coefficient has been calculated between spectral acceleration and arias intensity (i.e., $\rho_{\ln SA, \ln AI}$) and between peak ground velocity and arias intensity (i.e., $\rho_{\ln PGV, \ln AI}$) by Campbell and Bozorgnia (2013). It has also been calculated between significant duration and arias intensity (i.e., $\rho_{\ln D595, \ln AI}$) by Bradley (2015). The values of these correlation coefficients are shown in Fig. 3 (note that the correlation coefficient between arias intensity and itself is 1.0).

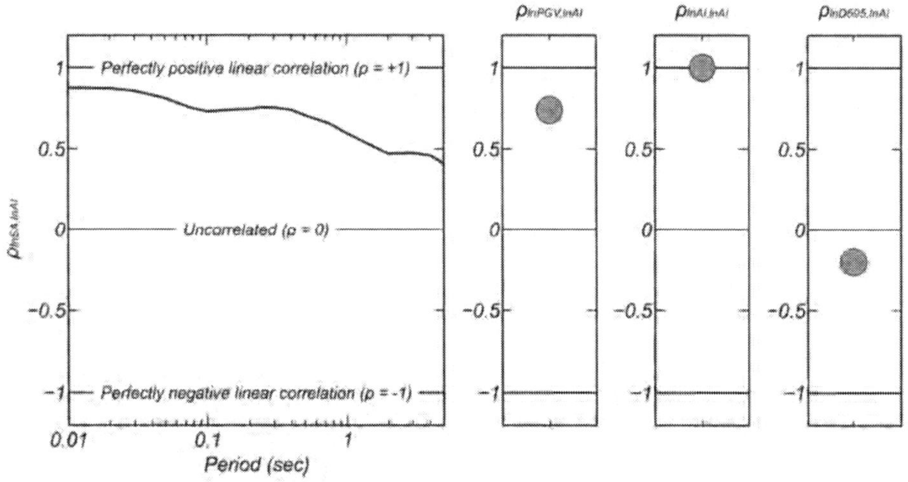

Figure 3. Correlation coefficient between various intensity measures.

The conditional mean is computed in Fig. 4 for the example embankment dam. When computing the conditional mean, arias intensity was set as the conditioning intensity measure (i.e., $IM^* = AI$) because arias intensity is an example of an intensity measure that relates well to the liquefaction-induced embankment deformations such as expected for this dam (Beaty and Perlea 2012). The conditional mean was computed given $IM^* = AI = ai$ for ai equal to 1.4, 3.1, and 6.8 m/s (corresponds to 0, 1, and 2 standard deviations above the mean prediction of arias intensity, or a 50th, 84th, and 98th percentile, respectively). Example calculations of computing the conditional mean given $ai = 3.1$ m/s are included for reference at the end of this paper.

The assemblage of the response spectrum, peak ground velocity, arias intensity, and significant duration characterizes the conditional mean design target. Three of these design targets are shown in Fig. 4. They represent a family of intensity measure targets conditioned on (or given) ai equal to 1.4, 3.1, and 6.8 m/s. Each conditional mean target represents the expected mean of the logarithm of each intensity measure for recorded ground motions ai equal to 1.4, 3.1, or 6.8 m/s. The expected mean is represented by the $\mu_{\ln IM_i|IM^*=im^*}$ term in Equation (5) with $IM^* = AI$ and $im^* = ai$. The reason that the mean is representative of recorded ground motions with $IM^* = AI = ai$ is because the correlation coefficient $\rho_{\ln IM_i,\ln IM^*}$ in Equation (5)—which controls $\mu_{\ln IM_i|IM^*=im^*}$—was computed based on the observed correlation between IM_i and IM^* from recorded ground motions. Therefore, the conditional mean given $ai = 3.1$ m/s represents the expected mean value of each intensity measure observed from recorded ground motions with $ai = 3.1$ m/s. Similarly, the conditional mean given $ai = 6.8$ m/s represents the expected mean value of intensity measures observed from recorded ground motions with $ai = 6.8$ m/s.

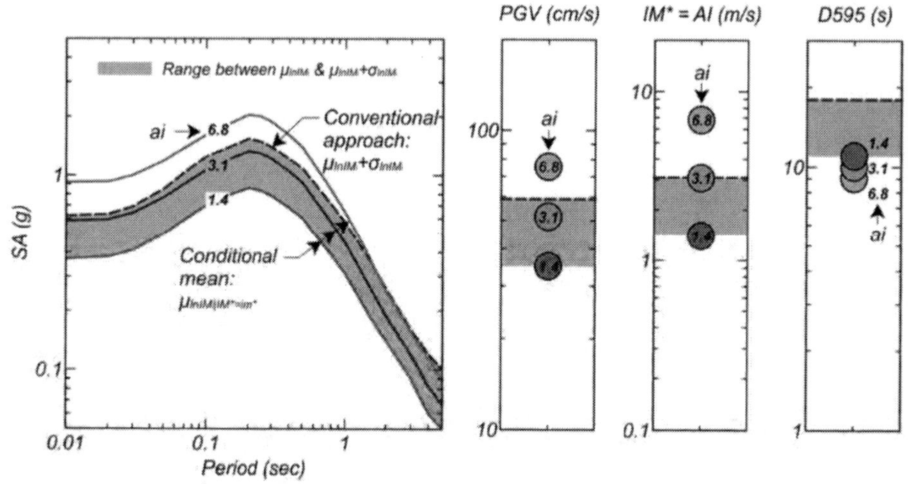

Figure 4. Design targets of multiple intensity measures based on the conditional mean and conventional approaches.

In the case of $ai = 3.1$ m/s, which corresponds to a prediction one standard deviation above the mean prediction of arias intensity, it is seen in Fig. 4 that the conditional mean ($\mu_{\ln IM_i|IM^*=im^*}$) of each intensity measure is between the unconditional mean ($\mu_{\ln IM_i}$) and the unconditional mean plus one unconditional standard deviation ($\mu_{\ln IM_i} + \sigma_{\ln IM_i}$). The variation of $\mu_{\ln IM_i|IM^*=im^*}$ between $\mu_{\ln IM_i}$ and $\mu_{\ln IM_i} + \sigma_{\ln IM_i}$ is because no intensity measure is perfectly positive linear correlated ($\rho_{\ln IM_i,\ln IM^*} = 1$) or uncorrelated ($\rho_{\ln IM_i,\ln IM^*} = 0$) with the conditioning intensity measure (see Fig. 3). By comparing Fig. 3 to Fig. 4, it is seen that as the correlation coefficient between each intensity measure and IM^* approaches 1.0, the conditional mean approaches the unconditional mean plus one unconditional standard deviation. When the correlation coefficient between each intensity measure and IM^* approaches 0, the conditional mean approaches the unconditional mean. Finally, as the correlation coefficient between each intensity measure and IM^* is less than 0 (e.g., $\rho_{\ln D595,\ln IM^*} = -0.20$), the conditional mean approaches the unconditional mean minus one unconditional standard deviation. These same trends would occur for $ai = 6.8$ m/s, which corresponds to a prediction two standard deviations above the mean prediction of arias intensity. However, in this case, the one unconditional standard deviation referenced previously would be replaced by two unconditional standard deviations.

Noting that $\mu_{\ln IM_i} + \sigma_{\ln IM_i}$ represents the design target from the conventional approach for an 84th-percentile prediction, it can been seen in Fig. 4 that the conditional mean given an 84th-percentile prediction of arias intensity (i.e., $ai = 3.1$ m/s) is below the conventional approach design target at all intensity measures except arias intensity itself. For the conditional mean to meet well the design targets from the conventional approach for spectral acceleration and peak ground velocity, it was found that the required percentile level of arias intensity needed to be increased from 84th to 93rd percentile (or 3.1 m/s to 4.4 m/s).

Deformation analysis results with ground motions based on both targets are needed to assess the practical significance of the difference seen in Fig. 4 between the design targets from the conventional approach and the conditional mean. It would be expected that the effect of the difference between the two targets for intensity measures that correlate strongly to deformation will have a significant effect on the deformations, whereas intensity measures that do not correlate strongly to deformations, regardless of magnitude in the difference between the two design targets, would have a minimal effect on the deformations. This type of analysis work is planned by the author in the future.

Although the practical significance of these differences in design targets will not be addressed in this paper, comments will be made on critical aspects of the targets themselves. In particular, the probabilistic correctness of each target and the representativeness of recorded ground motions will be addressed in the next two sections.

PROBABILISTIC CORRECTNESS OF DESIGN TARGETS

Consider again an 84th-percentile prediction level. By using an 84th-percentile level for a single intensity measure, there should only be a probability of 16% (or 0.159 rounded to three significant figures) of observing an earthquake with an intensity measure that exceeds this level. Thus, for a single intensity measure, it is probabilistically correct that the probability of exceedance of this single intensity measure for an 84th-percentile prediction is 0.159. However, when multiple intensity measures are constrained to this

84th-percentile prediction level, the probability of observing these multiple intensity measures that simultaneously meet 84th-percentile level targets is actually less than 0.159. The probability of exceeding multiple intensity measures simultaneously can be determined by computing the joint probability. Computing the joint probability requires knowledge of the mean, standard deviation of each intensity measure, and covariance between all intensity measures. The joint probability of exceedance is computed in Fig. 5a for one through six intensity measures simultaneously exceeding an individual probability of exceedance of 0.159 (i.e., 84th-percentile level or $\mu_{\ln IM_i} + \sigma_{\ln IM_i}$). In this computational exercise, six intensity measures were considered: AI, $SA(0.4)$, PGV, PGA, $SA(0.1)$, and $SA(0.5)$. The probability of exceedances shown in Fig. 5a would be different with other intensity measures or other combinations of these intensity measures; however, the trends seen would be similar. In addition to the six intensity measures considered, the joint probability of exceedance of all 22 intensity measures considered for the example embankment dam (i.e., 19 spectral ordinates, plus peak ground velocity, arias intensity, and significant duration) simultaneously exceeding $\mu_{\ln IM_i} + \sigma_{\ln IM_i}$ is also computed and shown in Fig. 5a.

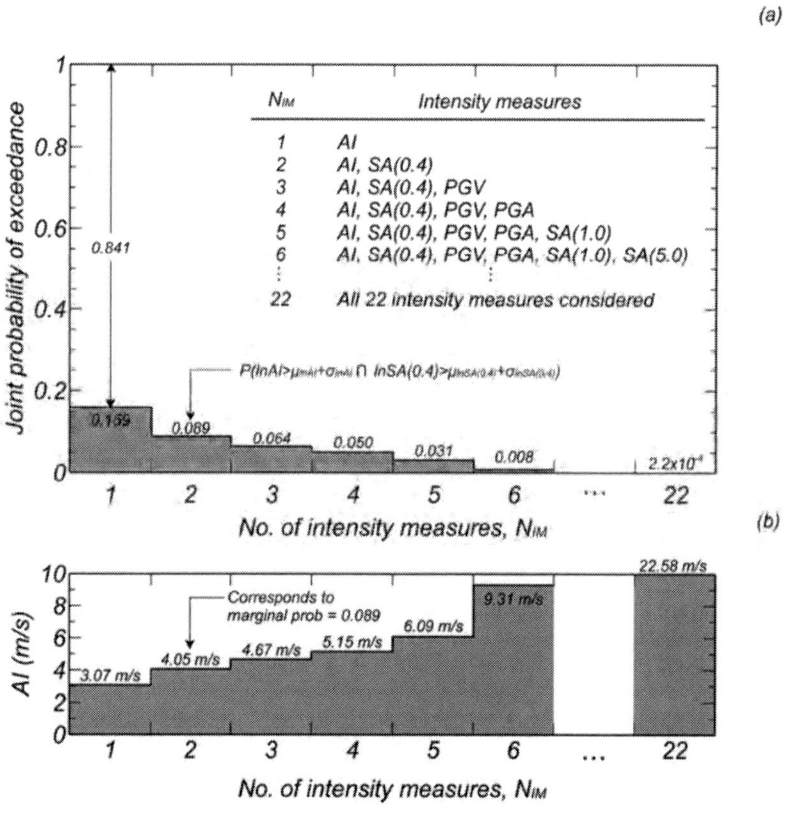

Figure 5. Computation of the joint probability of exceedance for $\ln IM_i > \mu_{\ln IM_i} + \sigma_{\ln IM_i}$ of multiple correlated intensity measures for the example embankment dam (a) and the corresponding equivalent $AI = ai$ (b).

As seen in Fig. 5a, the joint probability of simultaneously exceeding multiple intensity measures is less than the probability of exceeding any individual intensity. For example, the joint probability of AI and $SA(0.4)$ simultaneously exceeding $\mu_{\ln IM_i} + \sigma_{\ln IM_i}$ is 0.089, or a 91.1 percentile level, whereas the probability of exceedance of $\mu_{\ln IM_i} + \sigma_{\ln IM_i}$ for only AI or $SA(0.4)$ is 0.159. The probability of exceedance continues to decrease with each additional intensity measure. Finally, the joint probability of all 22 intensity measures simultaneously exceeding $\mu_{\ln IM_i} + \sigma_{\ln IM_i}$ is 2.2×10^{-4}, or a 99.98 percentile level. If more than 22 intensity measures were considered, the joint probability of simultaneously exceeding $\mu_{\ln IM_i} + \sigma_{\ln IM_i}$ would continue to decrease.

The design target developed by the conventional approach represents, therefore, a scenario that will essentially never happen (i.e., a joint probability of exceedance near 0 or a percentile level near 100). However, other scenarios with the same probability of exceedance could be much more damaging to the structure considered. For instance, for the example embankment dam in which arias intensity relates well with embankment deformations, using only arias intensity as the design target with the same probability of exceedance as multiple intensity measures simultaneously could potentially produce significantly more damage to the embankment dam. As an example, the equivalent value of arias intensity corresponding to the joint probability of one through six intensity measures is shown in Fig. 5b. These large levels of arias intensity will likely produce more damage to the example embankment dam than having multiple intensity measures with the same probability of exceedance represented by the conventional approach.

In contrast to the conventional approach, the conditional mean represents the expected mean intensity measure given a specific value of the conditioning intensity measure. Thus, if the conditioning intensity measure is chosen to correspond to a probability of exceedance of 0.159 (84th percentile), then the design target of the conditional mean correctly represents the expected mean intensity measures levels given an 84th-percentile conditioning intensity measure.

REPRESENTATIVENESS OF RECORDED GROUND MOTIONS WITH DESIGN TARGETS

The likelihood of observing intensity measures from recorded ground motions near the target will directly impact the selection and modification of recorded ground motions using either the design target from the conventional approach or the conditional mean. No recorded ground motion will directly meet the design target at all intensity measures. Instead, recorded ground motions will be above or below the design target at one or more intensity measures. Recorded ground motions are then selected from similar previous earthquake events. These ground motions are subsequently modified using amplitude or spectrum-matching techniques to meet the target. It is desirable to minimize the impacts of these modifications, so the necessary scaling factor (SF) used to modify linearly the acceleration time histories of the recorded ground motions to the design targets should be within an acceptable range. For example, Watson-Lamprey and Abrahamson (2006) suggested a SF limit between a factor of 2 to 4.

As an example, consider a single recorded ground motion from an earthquake event similar to that from the seismic hazard analysis of the example embankment dam. The intensity measures observed from this recorded ground motion are compared in Fig. 6a to the uniform $\mu_{\ln IM_i} + \sigma_{\ln IM_i}$ target from the conventional approach corresponding to the 84th percentile at all intensity measure targets. Comparing the intensity measures observed from the recorded ground motion to the 84th-percentile prediction target, it is seen that the intensity measures observed from the recorded ground motion are near the 84th-percentile prediction for spectral acceleration at periods less than 0.6 secs and arias intensity but significantly below the 84th-percentile prediction for the other intensity measures. In contrast, the intensity measure observed from the same ground motion are compared in Fig. 6b to the conditional mean target, and it is seen that the conditional mean more closely resembles the intensity measure values observed from the recorded ground motion.

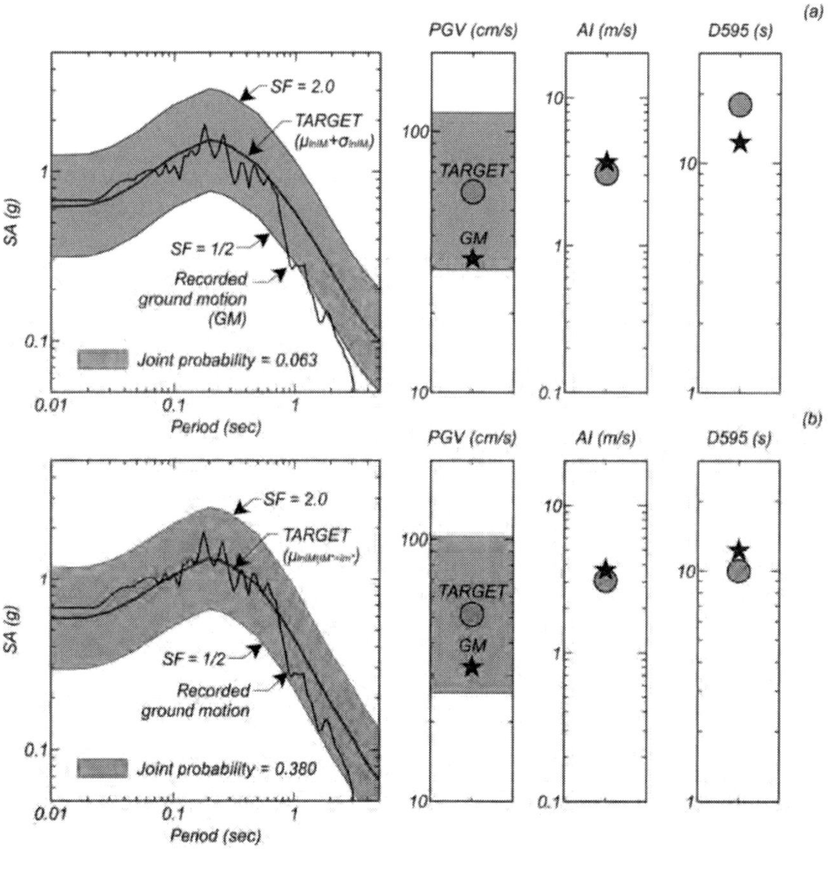

Figure 6. 84th-percentile design targets from the conventional approach (a) and using the conditional mean (b) compared to those intensity measures observed from a recorded ground motion.

The probability of finding a recorded ground motion similar to the one shown in Fig. 6 within a specified range above and below the design target can be computed. As an example, consider the range defined by a SF = ½ and 2 times the design target—a level of practical significance in terms of selection of appropriate candidate-recorded ground motions. With this range in design targets, probabilistic calculations were performed to compute the probability of observing intensity measures between the SF = ½ and 2 times the design target value of the 19 spectral ordinates as well as the peak ground velocity (representing the gray region shaded in Fig. 6). Similar to the computation of probabilities in the previous section, the joint probability must be calculated; and in the case of the conditional mean, the conditional joint probability is actually calculated.

As seen in Fig. 6, the probability of finding an actual recorded ground motion with observed intensity measures that are between a SF of ½ and 2 is much more likely with the conditional mean than with the conventional approach. With the conventional approach, the joint probability is 6.3%; for conditional mean, the joint probability 38.0%. As such, it is expected that with the conditional mean, significant more recorded ground motions will be found that are acceptably close to the design target and that will require less scaling or modification than with the conventional approach.

COMMENTS ON THE CONDITIONING INTENSITY MEASURE

With the conditional mean as the design target, the hazard level (e.g., percentile level) relates solely to the conditioning intensity measure, IM^*. Therefore, the selection of the IM_i used for IM^* is vital for a meaningful seismic evaluation. IM^* must relate to the engineering demand parameter of interest—for example, vertical crest deformation—because the expectation is that as $IM^* = im^*$ increases, so should the engineering demand parameter. More specifically, as stated by Kramer (2008), IM^* should be unbiased, consistent, robust, efficient, and sufficient. For stiff embankment dams in which significant strength-loss is not expected, IM^* should relate well to the SA at the first-mode period of the structure. However, for embankment dams founded on liquefiable alluvium, other non-SA intensity measures such as AI, CAV, and $\sqrt{AI \cdot D595}$ have been found to relate best to response (Beaty and Perlea 2012), and one of these intensity measures should be considered equal to IM^*. The selection of the most appropriate IM^* for embankment dams is still an area of ongoing research.

For the preceding example embankment dam, $IM^* = AI$ because of the expected effect of liquefaction on embankment deformations. Note that if another IM^* had been chosen, the conditional mean of the intensity measures would change. Consider, for example, if IM^* equals the SA at the period of the embankment dam, which is 0.4 seconds. The new conditional mean could be computed given a specific value of $IM^* = SA(0.4)$ and would result in a different target than when $IM^* = AI$. Any difference between the two conditional means would be attributed to the different correlation coefficients between IM^* equal to AI and SA at T = 0.4 seconds and the other intensity measures. Deformation analyses of the embankment would need to be conducted to assess the impacts of the two differing sets of targets on embankment deformations.

CONCLUSIONS

The earthquake-induced deformations of embankment dams are related to various earthquake ground motion characteristics. In this paper, spectral acceleration, peak ground velocity, arias intensity, and significant duration were used to characterize different components of earthquake shaking that relate to embankment dam deformations. Design targets were developed for an example embankment dam based on a conditional mean given a conditioning intensity measure and using a conventional approach.

For the conventional approach, it was shown that the hazard level assigned was misrepresentative of the actual hazard level of the target. In addition, the actual hazard level was not unique; it varied with the number of intensity measures. It would also be expected to vary with the type of intensity measures as well as the earthquake magnitude, distance, etc. used to predict these intensity measures. As a result, the use of such a target is problematic from a design standard or regulatory point of view; it cannot be enforced consistently from project to project because each project would represent a different hazard level. In contrast, the design targets developed using the conditional mean represented the assigned hazard level and defined a unique probabilistic level of hazard. Although it is true that the conditional mean is dependent on which conditioning intensity measure is selected, the selection of the conditional intensity measures and its effects on the conditional mean are well defined. Finally, for ground motion selection and modification considerations, the target developed from the conditional mean was more representative of recorded ground motions and would therefore minimize the negative effects of scaling to these targets.

Given the aforementioned benefits of the conditional mean, this design target could be considered as an alternative design standard or regulatory policy target to the conventional approach. As an example in a deterministic analysis, the design target for embankment dams with liquefaction concerns could be defined as the conditional mean given a set percentile arias intensity (or another preferred conditioning intensity measure); and for embankment dams without liquefaction concerns, the design target could be defined as the conditional mean given a set percentile level at the period of the embankment dam. For probabilistic analysis, a similar standard or policy could be defined, instead specifying a desired return period.

ACKNOWLEDGEMENTS

The intensity measure distributions for the earthquake scenario considered were developed previously by Mr. Marvin Woods. In the preparation of this paper, the comments and suggestions by Dr. Erik Malvick, Mr. Ian Maki, and Mr. Vojislav Cvijanovic from the California Division of Safety of Dams, as well as by Dr. Tadahiro Kishida from PEER, were particularly valuable and are greatly appreciated. In addition, discussions with Professor Ghazan Khan from the California State University, Sacramento on the probabilistic analysis contained herein was beneficial and is also greatly appreciated.

EXAMPLE CALCULATIONS

The example calculations below are based on $IM^* = AI = ai = 3.1$ m/s. All calculations were completed with two significant figures.

IM_i	$\widetilde{IM}_i = e^{\mu_{\ln IM_i}}$	$\mu_{\ln IM_i}$	$\sigma_{\ln IM_i}$	$\rho_{\ln IM_i, \ln IM_j}$	$\widetilde{IM}_i \| IM^* = im^* = e^{\mu_{\ln IM_i \| IM^*=im^*}}$
$SA(0.01)$	0.37	-0.99	0.52	0.88	0.59
$SA(0.02)$	0.38	-0.97	0.51	0.87	0.59
$SA(0.03)$	0.41	-0.89	0.52	0.86	0.64
$SA(0.05)$	0.50	-0.69	0.53	0.81	0.77
$SA(0.075)$	0.61	-0.49	0.56	0.76	0.94
$SA(0.1)$	0.70	-0.36	0.57	0.73	1.06
$SA(0.15)$	0.80	-0.22	0.57	0.74	1.22
$SA(0.2)$	0.86	-0.15	0.58	0.75	1.34
$SA(0.25)$	0.83	-0.19	0.58	0.76	1.28
$SA(0.3)$	0.78	-0.25	0.58	0.75	1.21
$SA(0.4)$	0.68	-0.39	0.58	0.74	1.04
$SA(0.5)$	0.60	-0.51	0.59	0.71	0.91
$SA(0.75)$	0.41	-0.89	0.60	0.66	0.61
$SA(1.0)$	0.31	-1.17	0.61	0.60	0.45
$SA(1.5)$	0.19	-1.66	0.67	0.52	0.27
$SA(2.0)$	0.14	-1.97	0.62	0.47	0.19
$SA(3.0)$	0.09	-2.41	0.58	0.48	0.12
$SA(4.0)$	0.06	-2.81	0.69	0.46	0.08
$SA(5.0)$	0.05	-3.00	0.69	0.41	0.07
PGV	35.0	3.56	0.52	0.74	51.42
AI	1.4	0.34	0.79	1.00	3.10
$D595$	11.0	2.40	0.49	-0.20	9.97

REFERENCES

Baker, J.W. (2011). The conditional mean spectrum: a tool for ground motion selection. *Journal of Structural Engineering*, 137(3), 322–31.

Baker, J.W., and Cornell, C.A. (2005). A vector-valued ground motion intensity measure consisting of spectral acceleration and epsilon. *Earthquake Engineering and Structural Dynamics*. 34,1193–1217.

Beaty M.H., and Perlea V.G. (2012). Effect of ground motion characteristics on liquefaction modeling of dams. *ASCE GeoCongress*.

Bommer J.J., Stafford .P.J, Alarcón J.E. (2009). Empirical equation for the prediction of significant, bracketed, and uniform duration of earthquake ground motion. *Bulletin of the Seismological Society of America*. 99(6), 3217–33.

Bradley B.A. (2010). A generalized conditional intensity measure approach and holistic ground motion selection. *Earthquake Engineering and Structural Dynamics*, 39(12), 1321–42.

Bradley B.A. (2015). Correlation of arias intensity with amplitude, duration and cumulative intensity measures. *Soil Dynamics and Earthquake Engineering*, 78, 89–98.

Bray J.D., and Travasarou T. (2007). Simplified procedure for estimating earthquake-induced deviator slope displacement. *Journal of Geotechnical and Geoenvironmental Engineering*, 133(4), 381–92.

Campbell K.W. and Bozorgnia Y. (2008). NGA ground motion model for geometric mean horizontal component of PGA, PGV, PGD and 5% damped linear elastic response spectra for periods ranging from 0.01 to 10 s. *Earthquake Spectra,* 24(1), 139–171.

Campbell K.W. and Bozorgnia Y. (2013). A comparison of ground motion prediction equations for arias intensity and cumulative absolute velocity developed using a consistent database and functional form. *Earthquake Spectra*, 28(3), 931–41.

Fraser W.A. and Howard J.K. (2002). Guidelines for use of consequence-hazard matrix and selection of ground motion parameters. *California Division of Safety of Dams*.

Howard J.K., Fraser W.A., and Schultz M.G. (2008). Probabilistic use of arias intensity in geotechnical earthquake engineering. *Geotechnical Earthquake Engineering and Soil Dynamics IV*, Geotechnical Special Publications, (181).

Kramer S.L. (2008). Performance-based earthquake engineering: opportunities and implications for geotechnical engineering practice. *Geotechnical Earthquake Engineering and Soil Dynamics IV*, D. Zeng, M. Manzari, and D. Hiltunen, eds., Geotechnical Special Publication No. 181, ASCE, NY.

Saygili G and Rathje E.M. (2007). Empirical predictive models for earthquake-induced sliding displacements of slopes. *Journal of Geotechnical and Geoenvironmental Engineering*, 134(6), 790–803.

Wang G and Du W. (2012). Estimating earthquake-induced slope displacements using vector ground motion intensity measures. *Proceedings of 15th World Conference on Earthquake Engineering*, Lisbon, Portugal; 10pg.

Watson-Lamprey, J. and Abrahamson, N. (2006). Selection of ground motion time series and limits on scaling. *Soil Dynamics and Earthquake Engineering*, 26, 477–82.

SEISMIC DEFORMATION ANALYSIS OF BLUE RIDGE DAM

Glenn J. Rix, Ph.D., P.E.[1]
Lucas de Melo, Ph.D., P.E.[2]
Christopher L. Saucier, Ph.D., P.E.[3]
Michael A. Morrison, P.E., PMP[4]

ABSTRACT

Blue Ridge Dam (BRD) was constructed in the 1920s on the Toccoa River near Blue Ridge, Georgia, and was subsequently purchased by the Tennessee Valley Authority (TVA) in 1939. BRD is approximately 985-ft long and 172-ft high and retains a useful storage volume of approximately 183,900 acre-feet. The dam is a semi-hydraulic fill earthen embankment constructed by placing uncompacted "dumped" fill from local, residual soil sources on the upstream and downstream faces. The dumped fill was subsequently "washed" using water cannons to mobilize the fines, creating a "sluiced fill" zone and a central "puddled core" of fine-grained material. The resulting embankment materials are transitional soils that straddle the boundary between non-plastic, coarse-grained and moderate plasticity, fine-grained soils. The dam is located in a low-to-moderate seismic hazard area near the Eastern Tennessee Seismic Zone. Because of the nature of the construction and location of BRD, TVA decided to evaluate seismic deformations of BRD. An extensive site investigation program was used to characterize the embankment soils, including the use of normalized (i.e., SHANSEP) monotonic and cyclic strength parameters and stress history profiles. PM4SAND, an advanced constitutive model available for FLAC® 7.0 (Fast Lagrangian Analysis of Continua) was calibrated using the monotonic and cyclic laboratory results and used for numerical analyses. A suite of earthquake time histories representing high- and low-frequency ground motions and two return periods (3,000 and 10,000 years) was employed in the analyses.

INTRODUCTION

Blue Ridge Dam (BRD) was constructed between 1925 and 1930 by the Toccoa Electric Power Company on the Toccoa River near Blue Ridge, Georgia and was subsequently purchased by TVA in 1939. BRD is approximately 985-feet (ft) long and 172-ft high and retains a useful storage volume of roughly 183,900 acre-feet within the Blue Ridge Lake reservoir. The elevation of the crest of the dam is 1,710 ft above mean sea level (ft-msl). The seasonal water surface elevation of Blue Ridge Lake typically ranges from approximately 1,665 ft to 1,688 ft-msl, with a maximum pool level of elevation 1,691 ft-

[1] Geosyntec Consultants, Inc., 1255 Roberts Boulevard NW, Kennesaw, GA 30144, 678-202-9546, grix@geosyntec.com.
[2] Geosyntec Consultants, Inc., 10211 Wincopin Circle, Floor 4, Columbia, MD 21044, 410-910-7634, ldemelo@geosyntec.com.
[3] Tennessee Valley Authority, 1101 Market Street, LP 1F-C, Chattanooga, TN 37402, 423-751-7999, clsaucier@tva.gov.
[4] Tennessee Valley Authority, 1101 Market Street, LP 1F-C, Chattanooga, TN 37402, 423-751-3631, mamorrison0@tva.gov.

msl. The original powerhouse is located at the toe of the embankment and is supplied by a 12-ft-diameter slip-lined steel penstock under the embankment. A second small generating unit was added in 1995 and is located east of the original powerhouse. This unit is connected to the main penstock via a 4-ft-diameter, pile-supported, above-ground steel penstock. Figure 1 shows the site plan depicting the main features of BRD.

Figure 1. Blue Ridge Dam.

The BRD site is located in the western portion of the Blue Ridge Physiographic Province. According to Sowers and Richardson [1983], soils in the Blue Ridge are products of chemical weathering of igneous and metamorphic rocks. This chemical weathering produces so-called residual soils containing clay minerals, hydrous micas, iron oxides, and carbonates. BRD is located within a gradual river bend of the Toccoa River. Flood plain deposits consist primarily of micaceous sandy silt overbank deposits with varying fractions of clay and sand. These soils overlie a thin stratum of residual soil deposits that extend downward to approximately elevation 1,537 ft-msl where weathered rock is encountered.

Original Construction

BRD was constructed using a semi-hydraulic fill method and residual soils from local borrow areas. The dam was constructed by placing uncompacted "dumped fill" on the upstream and downstream sides of the embankment. This fill was then washed with water cannons to convey the soils, creating a "sluiced fill" or "washed fill" zone and a central "puddled core" of fine-grained material. The dam was constructed above mostly weathered rock with approximately one to five ft of overburden soils removed, except under the puddled core, where a 15-ft deep trench was excavated into competent bedrock. Additional, recent drilling has confirmed that a portion of the dumped fill near the downstream toe and mid-slope was placed over an approximately 10-ft thick layer of alluvium and residual soil deposits.

Recent Modifications

In 1995, the crest was raised from elevation 1,705 ft-msl to elevation 1,710 ft-msl; a riprap blanket was placed along the raised crest; and a concrete parapet wall was constructed with a top-of-wall elevation at 1,713 ft-msl. The small generating unit and associated penstock described above were also constructed at this time.

In 2010, the Seismic Remediation Project (SRP) started with the goal of providing a configuration of upstream and downstream rockfill buttresses deemed necessary to ensure the dam's safe performance during the design seismic events. The downstream buttress consisted of a 20- to 25-ft thick zoned filter and rockfill to improve the seismic stability. During SRP construction, the amount of rockfill was reduced because the observed deformations caused by placement of the rockfill were larger than expected. Figure 2 shows a typical cross section representing the current geometry of the embankment near the maximum section. Geosyntec Consultants, Inc. (Geosyntec) was retained by TVA in late 2013 to evaluate the seismic stability and deformation of BRD for the current configuration of the embankment under the design seismic events.

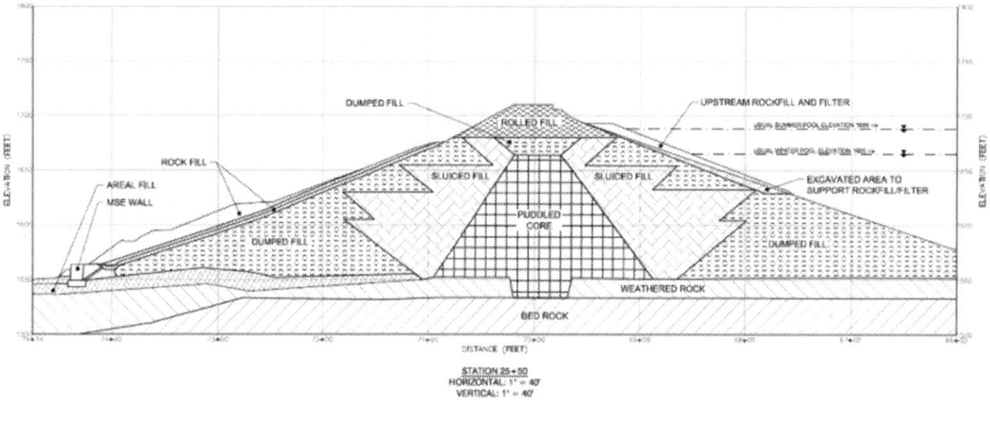

Figure 2. Typical Cross Section.

FIELD AND LABORATORY INVESTIGATIONS

To supplement historical data, a geotechnical investigation was conducted by Geosyntec in 2014 at five locations on the downstream face of the embankment and targeted the dumped fill, sluiced fill, puddled core, and alluvium. Additionally, one boring was located near the downstream toe of the dam for the purpose of characterizing the in situ seismic wave velocity profile of the foundation rock and performing acoustic televiewer measurements. The investigation on the downstream face of the dam consisted of seismic cone penetration test (CPT) soundings to obtain the shear wave velocity profiles in the embankment soils and geotechnical borings advanced using the mud-rotary technique for the collection of disturbed and undisturbed soil samples. Undisturbed soil samples were collected utilizing an Osterberg piston sampler and Shelby tubes modified to flatten the tip cutting angle from approximately 45 degrees to 5 degrees to minimize the amount of soil disturbance during sampling (DeGroot and Ladd, 2012).

The purpose of the ensuing laboratory test program in 2014 was to determine index and engineering properties of selected soil specimens from dumped fill, sluiced fill, puddled core, and alluvium materials. The laboratory testing program specifically targeted:

- index and physical properties;
- stress history and compressibility characteristics;
- drained and undrained monotonic shear strength characteristics; and
- cyclic strength characteristics.

The latter tests were a critical aspect of the effort to understand the seismic performance of the embankment and foundation soils and included stress-controlled cyclic direct simple shear (DSS) tests with and without initial bias shear stress and with post-cyclic monotonic testing.

A supplemental field investigation was performed in 2015 to collect additional high-quality undisturbed samples of the dumped fill at two boring locations on the downstream face. Additional laboratory tests were also performed to evaluate the peak and residual undrained shear strength and cyclic strength of the dumped fill, and to assess if the monotonic and cyclic strength of the dumped fill exhibit normalized behavior consistent with the Stress History and Normalized Soil Engineering Properties (SHANSEP) method developed by Ladd and his coworkers (Ladd and Foott, 1974; Ladd et al. 1977; and Ladd, 1991). The supplemental laboratory testing program included the following:

- constant-rate-of-strain consolidation (CRSC) tests to evaluate the stress history of the dumped fill and assist in the selection of the consolidation stress levels to use in the shear strength tests;
- monotonic DSS tests to evaluate the peak undrained shear strength; and
- additional cyclic DSS tests to evaluate the cyclic strength as a function of OCR and the residual (post-cyclic) undrained shear strength.

MATERIAL PROPERTIES

Index Properties

As noted above, the materials used to construct the embankment were obtained from local borrow areas consisting of residual soils. The dumped fill and sluiced fill are heterogeneous mixtures of materials with Unified Soil Classification System (USCS) categories including micaceous silty sand (SM) with gravel, sandy silt (ML), and low-plasticity clay (CL). The fines content ranges from 10 to 90 percent. Additionally, the Atterberg Limits indicate approximately 50 percent of the soils are non-plastic; the plasticity index (PI) of the remaining 50 percent is between 1 and 25. Overall, the index properties of the sluiced fill are similar to those of the dumped fill, and the washing of the sluiced fill does not appear to have resulted in a distinct material. The puddled core is classified as silty sand (SM) and sandy silt (ML to MH) with a fines content between 25 and 95 percent. Only approximately 4 percent of puddled core specimens were classified as non-plastic. The remaining specimens present values of LL between 30 and 60 and values of PI between 1 and 20.

The index properties presented above suggest that the embankment soils straddle the boundary between non-plastic or low-plasticity, coarse-grained soils and moderate-plasticity, fine-grained soils. Furthermore, the results of X-ray diffraction analyses indicated the total amount of muscovite and non-muscovite mica ranges approximately from 10 to 30 percent by weight. Both minerals are present in large enough quantities that they are likely to exert a strong influence on the mechanical behavior of the soils.

Geosyntec also used the CPT material index, I_c (Robertson, 2004 and 2009), to evaluate the embankment soils. The I_c values generally exceed 2.6 and indicate that the dumped fill had a mechanical behavior that is generally comparable to the behavior of silts and clays. Similar results were obtained for the puddled core and sluiced fill. The estimated material types from the CPTs and the soil classification from the borehole samples do not always agree. In some instances, the CPT results indicate silts and clays where the results from sample classifications indicate silty sands. It appears that for those soil layers that classify as silty sands in the boreholes, fines were present in large enough quantities to exert a strong influence on the mechanical behavior of the sands and to cause these materials to behave more like silts or clays.

Undrained Shear Strength

The undrained shear strength, s_u, design profiles of the embankment materials were evaluated from a CPT-based empirical correlation (Mayne, 2014) to estimate the preconsolidation stresses, σ'_p, of the soils, and using the SHANSEP method for evaluating the strength characteristics of soils. An example is shown in Figure 3. The two left panels of Figure 3 show the σ'_p profiles as estimated from the CPTs performed in the dumped fill using the following expression (Mayne, 2014):

$$\sigma'_{p,CPT} = 0.33(q_t - \sigma_{vo})^{m'} \left(\frac{p_{atm}}{100}\right)^{1-m'} \tag{1}$$

where, $\sigma'_{p,CPT}$ is the preconsolidation stress, q_t is the total tip resistance, σ_{vo} is the total overburden stress, p_{atm} is the atmospheric pressure in the same units as q_t and σ_{vo}, and m' is the yield stress exponent calculated using the CPT material index, I_c:

$$m' = 1 - \frac{0.28}{1+(I_c/2.65)^{25}} \tag{2}$$

The σ'_p measurements from incremental and CRSC consolidation tests performed on samples from boreholes in the vicinity of the CPTs are also included in the left panel of Figure 3. The laboratory σ'_p measurements are slightly larger than the $\sigma'_{p,CPT}$ values. Both measured and estimated σ'_p values are larger than the in situ vertical effective stresses, σ'_{vo}, resulting in OCR values for dumped fill that generally range between 1.5 and 4.0.

The right panels of Figure 3 show the s_u profiles of dumped fill calculated using the SHANSEP relationship and the estimated $\sigma'_{p,CPT}$ values as follows:

$$s_u = S \times \sigma'_{vo} \times \left(\sigma'_{p,CPT}/\sigma'_{vo}\right)^m \tag{3}$$

where s_u is the undrained shear strength of the soil, S (= 0.25) is the normalized undrained shear strength ratio for normally consolidated dumped fill, $\sigma'_{p,CPT}$ is the preconsolidation stress estimated from the CPT data, σ'_{vo} is the in situ effective vertical stress, and m (= 0.86) is the empirical coefficient for dumped fill. Site-specific values of S and m were obtained from the monotonic DSS tests performed on dumped fill.

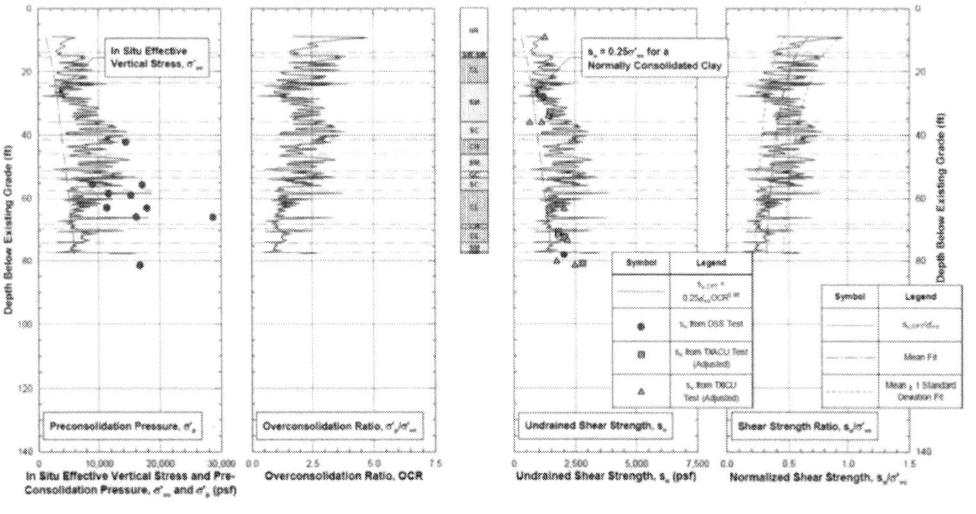

Figure 3. Preconsolidation Stress, OCR, and Undrained Strength Profiles

The s_u measurements from the DSS tests, and the CK_0UC and CIUC performed on samples from boreholes in the vicinity of the CPTs are also included in the third panel of Figure 3. The strength tests were performed using samples reconsolidated to the estimated in situ effective vertical stress prior to shearing. The s_u measurements from the CK_0UC and CIUC tests were adjusted to approximate a mode of shearing for DSS conditions using the recommendations by Mayne and Kulhawy (1990) to allow for comparison with the results from the DSS tests and CPTs. The s_u measurements from the strength tests appear to be slightly less than the mean SHANSEP-estimated s_u values from Equation 3, which may be attributable to sample disturbance. The results indicate normalized shear strength values, s_u/σ'_{vo}, that generally range between 0.4 and 0.6. The fourth panel of Figure 3 shows the mean and $\pm$ one standard deviation lines fit to the s_u/σ'_{vo} estimates to facilitate their implementation in subsequent stability and deformation analyses. The same procedure was used to calculate the undrained shear strength profiles of the sluiced fill and puddled core.

NUMERICAL DEFORMATION ANALYSES

Earthquake Ground Motions

Geosyntec performed seismic deformation analyses for two levels of ground motion specified by TVA, corresponding to uniform hazard response spectra (UHRS) with return periods of 3,000 and 10,000 years (yrs). For each return period, low- and high-frequency ground motions were developed for spectral frequencies of 1 Hz and 10 Hz, respectively. At each of the two spectral frequencies, three horizontal time histories from McGuire et al. (2001) were spectrally matched to the target spectrum. The UHRS and low- and high-frequency response spectra are shown in Figure 4. The significant durations of the low-frequency motions were greater because they represent large-magnitude earthquakes. Thus, a total of 12 ground motions were used in the seismic deformation analyses, i.e., 2 return periods × 2 spectral frequencies × 3 time histories. The corresponding vertical acceleration time histories were also used in the seismic deformation analyses.

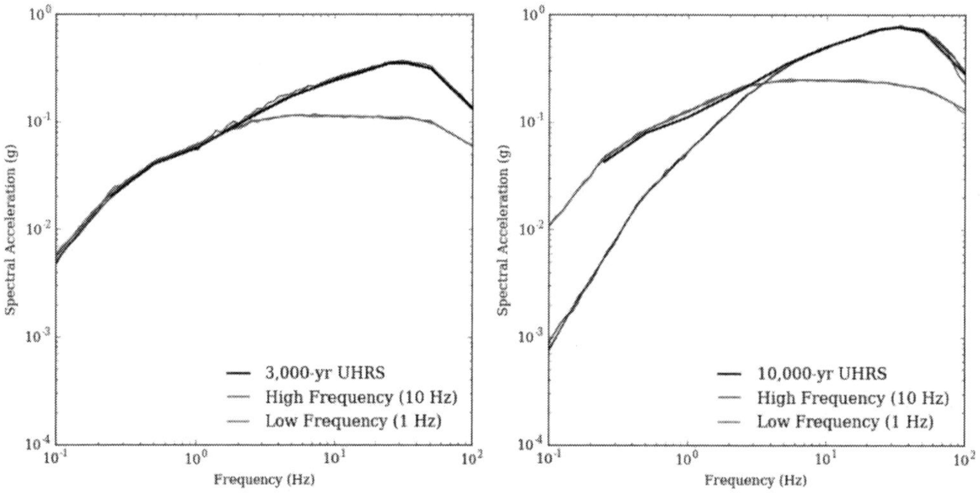

Figure 4. Ground Motions Used for Seismic Deformation Analyses

Constitutive Model Calibration

Geosyntec employed the PM4SAND v. 3.0 model (Boulanger and Ziotopoulou, 2015) for the seismic deformation analysis of BRD. PM4SAND is a modified version of a plasticity model originally proposed by Dafalias and Manzari (2004) to better predict the response of sandy soils to laboratory testing and to fit published design relationships and correlations. The soil response is compatible with critical state soil mechanics. To calibrate the model, Geosyntec used the cyclic stress-strain and vertical effective stress vs. shear stress path obtained from the cyclic DSS tests performed on each of the BRD materials. The calibration focused on matching the cyclic strength curves at various OCR values.

Figure 5 shows one example of the comparisons of the CSR vs. cyclic shear strain and CSR vs. normalized vertical effective stress from the cyclic DSS tests and the PM4SAND model predictions for tests at a variety of OCRs and CSRs. The agreement between the measured and predicted response is considered to be very good. Figure 6 shows a comparison between the experimentally measured cyclic strength values for various OCRs and the curves predicted using PM4SAND. The agreement between the experimental and predicted cyclic strength curves is considered to be very good.

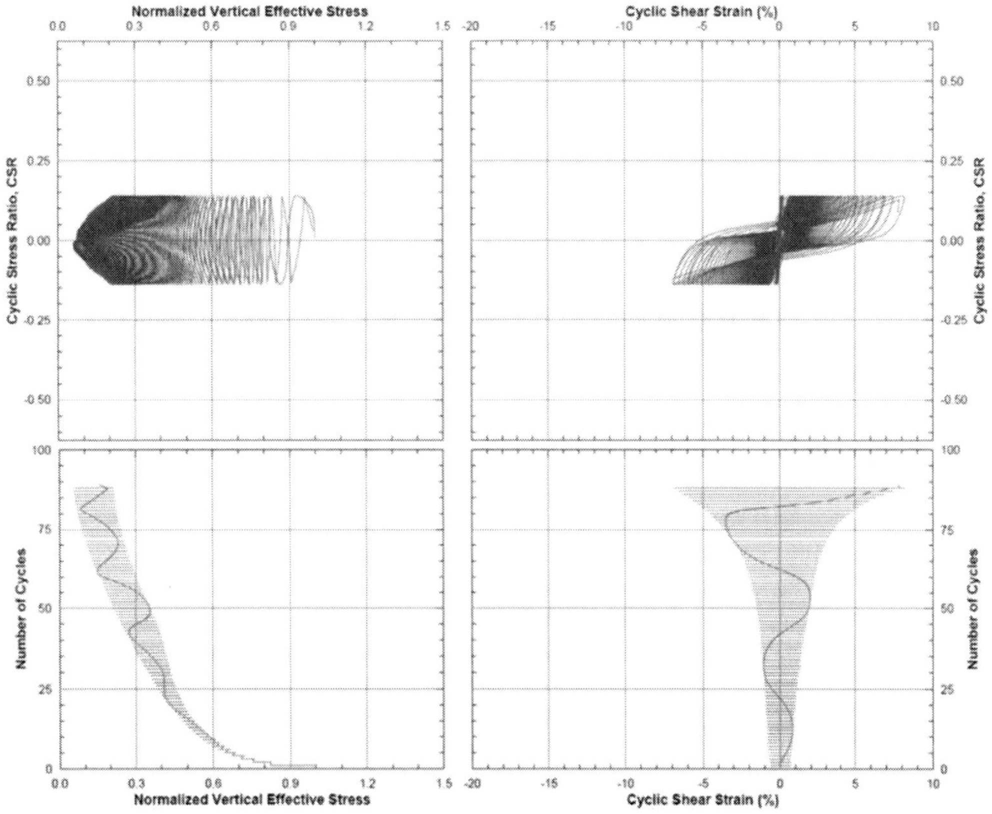

Figure 5. Comparison between PM4SAND Simulations (red) and Experimental Cyclic DSS Test Results (black)

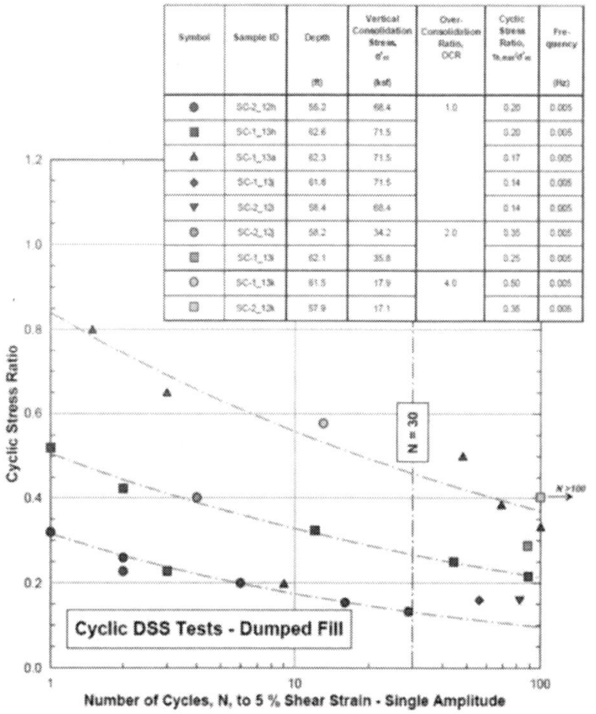

Figure 6. Comparison between Simulated and Experimental Cyclic Strength Curves for OCR = 1.0, 2.0, and 4.0

PM4SAND was also used to simulate cyclic DSS tests performed using an initial static shear stress such as is encountered in the embankment. Values of α = 0.1 and 0.2 were simulated where $\alpha = \tau_{xy}/\sigma'_{vc}$. As expected, increasing the initial static shear stress decreases the cyclic strength in a manner consistent with the results reported by Idriss and Boulanger (2008) for fine-grained soils. As a further check of the PM4SAND model predictions, the monotonic undrained shear stress vs. shear strain predicted by PM4SAND for OCRs of 1.0, 2.0, and 4.0 were compared with the experimental data and the shear modulus reduction curves (G/G_{max}) for the PM4SAND simulations were compared with the empirical relationship developed by Darendeli (2001) using the appropriate mean effective stress, plasticity index, and OCR. The simulated secant shear stiffness agrees very well with the expected behavior based on Darendeli (2001).

Based on comparisons between the monotonic and cyclic behavior from experiments on dumped fill and PM4SAND model predictions, Geosyntec considers the PM4SAND model to accurately represent the behavior of the embankment and foundation soils. The results of the calibration demonstrate the flexibility of adapting PM4SAND, a constitutive model developed to predict the response of sandy soils, to test results obtained for transitional soils.

Model Initialization

The analyses were performed using FLAC® 7.0, a 2D, explicit finite difference program capable of simulating the behavior of geomaterials (Itasca, 2011). The FLAC® model was generated by shaping the grid to model boundaries. Only quadrilateral zones were used in the model to avoid numerical problems associated with triangular zones during plastic flow. As the model involves plastic deformations, the aspect ratio of zones was limited to 1:3 throughout the model with even better quality zones in the regions where plastic deformations are expected. The maximum length of zones in any direction was set as 2.5 ft. Using a criterion of at least six grid points per wavelength during seismic loading, this zone dimension allows accurate modeling of frequencies up to 25 Hz given the minimum shear wave velocity of 400 ft/s. The model was configured for effective stress analyses by enabling the groundwater option. Seepage analyses were performed first to determine the phreatic surface and hydrostatic pore pressure distribution in the embankment. The headwater elevation was set as 1,687 ft-msl and the tail water at elevation 1,542 ft-msl approximately 100 ft from the downstream toe. The upstream slope above the headwater elevation and the entire downstream slope were assigned zero pore pressure.

A simplified procedure was used to establish the initial state of stress within the model. First, the phreatic surface was determined by using fluid flow only calculations as described above. Next, vertical stresses were installed in the model using an initial estimate based on depth below ground surface. Horizontal stresses were calculated assuming a lateral stress ratio (K_0) equal to 0.5. Gravity was turned on and the model was solved to equilibrium assuming it as an elastic material. The horizontal stresses are updated again assuming $K_0 = 0.5$ to account for changes in vertical stresses during the initial solving step. The constitutive model was changed to an elastic-perfectly plastic, Mohr-Coulomb model, and the model was solved to equilibrium again. Finally, the constitutive model for dumped fill, sluiced fill, puddled core and alluvium was switched to PM4SAND and the model was cycled to equilibrium. The model was fixed in the x-direction at the sides and in both the x- and y-directions at the base. The bulk modulus of water was set to zero during this phase to avoid any change in pore pressures due to volumetric strains incurred while cycling the model for equilibrium. The displacements were reset to zero at the end of this stage and water bulk modulus was restored to 4.18×10^7 psf.

Dynamic Analyses

After the static stresses were initialized, the displacement fixities at the sides and base of the model were removed. Quiet boundaries were applied at the base of the model where the model is supported by means of reaction forces calculated from static equilibrium and dashpots are added in both x- and y-directions to absorb outgoing waves. For the sides, free-field boundaries were used. Free-field boundaries are essentially quiet boundaries with additional one-dimensional site response run in parallel for each side. The difference in motion between the model boundary and 1D column is converted to forces and applied back to the model boundary to avoid leakoff of wave energy. Groundwater flow was turned off during dynamic shaking to simulate undrained response.

A stiffness-proportional Rayleigh damping of 0.1 percent at 5 Hz was used to reduce high-frequency numerical noise. Stiffness-only proportional damping was chosen to avoid over-damping lower frequencies. Furthermore, 0.1 percent at 5 Hz was chosen to avoid over-damping frequencies above 10 Hz. As a result, some high-frequency noise (> 25 Hz) was observed in acceleration time histories near the crest. Shear stress and shear strain histories are not significantly affected because they are controlled by the velocity time history, which is less sensitive to high-frequency noise.

The ground motions described previously contained frequencies up to 50 Hz. Modeling such high frequencies places stringent requirements on zone size in the model to accurately model wave propagation. This increases both the number of zones and reduces the dynamic time step, resulting in drastic increase in simulation times. However, geotechnical structures such as embankments typically do not respond to frequencies greater than 20 Hz. To keep run times reasonable, the acceleration time histories were low-pass filtered using a fourth-order Butterworth filter with a cut-off frequency of 25 Hz.

The input ground motion was applied as a shear stress time history for the horizontal component and a normal stress history for the vertical component of each ground motion. The outcrop acceleration time histories were integrated to obtain velocity time histories. The outcrop histories need to be divided by a factor of two to convert them to incoming motion, but also need to be multiplied by two to account for half the energy being absorbed by dashpots at the quiet boundary. These two factors cancel each other out and the stress histories were obtained by multiplying the outcrop velocity time histories by a factor of ρV_S for shear stress and ρV_P for normal stress where ρ is the total mass density and V_S and V_P are the shear and compression-wave velocities, respectively.

Results

Figure 7 shows representative results for the permanent deformations of the embankment at the end of shaking for return periods of 3,000 and 10,000 yrs. For the ground motions with a return period of 3,000 yrs, the maximum horizontal deformations are 1.0 ft on the upstream face of the dam and 0.75 ft on the downstream face. For the ground motions with a return period of 10,000 yrs, the maximum horizontal deformations are 4.0 ft on both the upstream and downstream faces, except for a small area near the crest where the deformations exceed 5.0 ft. The significance of these deformations near the crest is of lower concern because they occur above the elevation of the phreatic surface. For both return periods, larger total deformations were observed for the low-frequency ground motions, likely because of the longer duration of these motions.

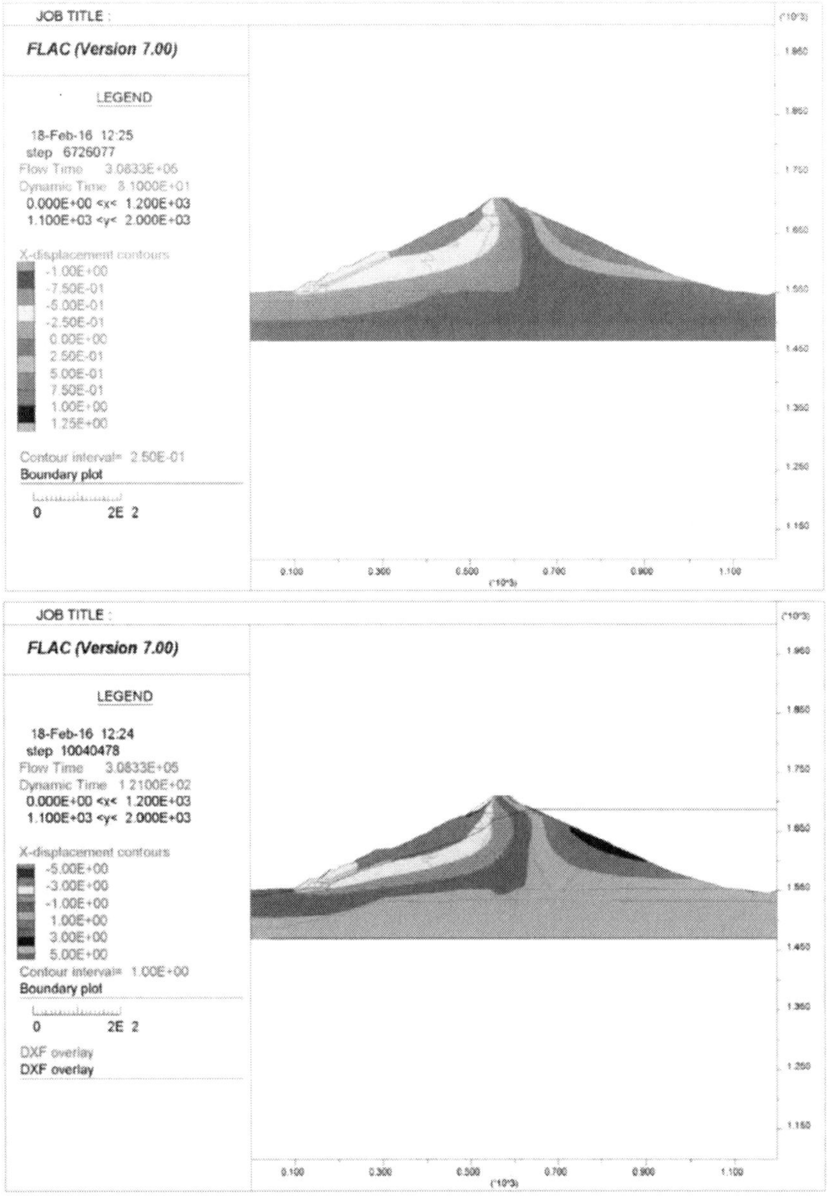

Figure 7. Permanent Horizontal Deformation for Low-Frequency Ground Motions with a Return Period of (a) 3,000 and (b) 10,000 Yrs

Values of excess pore pressure ratios (r_u) for the ground motions with a return period of 3,000 yrs are less than 0.25 for most of the embankment. There is a zone of negative excess pore pressure near the crest on the upstream face, likely due to the high values of OCR near the surface. There are also zones of higher r_u (up to 0.5) near the downstream and upstream toes of the dam. The patterns of r_u for the ground motions with a return period of 10,000 yrs are generally similar. For the 3,000-yr ground motions, the shear strains are less than 1 percent for most of the embankment. There are isolated zones with higher shear strains (up to 2 percent) near the crest, along the upstream face, and near the downstream toe. Larger calculated strains are observed for the 10,000-yr ground motions. Within much of the embankment, maximum shear strains of up to approximately 3 percent are present, with calculated shear strains exceeding 5 percent near the crest, near the downstream toe, and along the upstream face.

Comparison with Empirical Data on Seismic Performance of Dams

Figure 8 compares the calculated crest settlement for BRD (expressed as the normalized crest settlement, which is the crest settlement divided by the height of the embankment) with results from other embankment dams compiled by Swaisgood (2014). The normalized crest settlements are shown as a function of the free-field peak ground acceleration (PGA) at the location of the dam. The calculated normalized crest settlements for the 3,000-yr and 10,000-yr high-frequency ground motions at BRD are comparable with the observed performance of other embankment dams, including other hydraulic fill embankments. The calculated crest settlements for the 3,000-yr and 10,000-yr low-frequency ground motions are greater than the observed performance of other dams, but this is likely attributable to the long duration of the low-frequency motions at BRD.

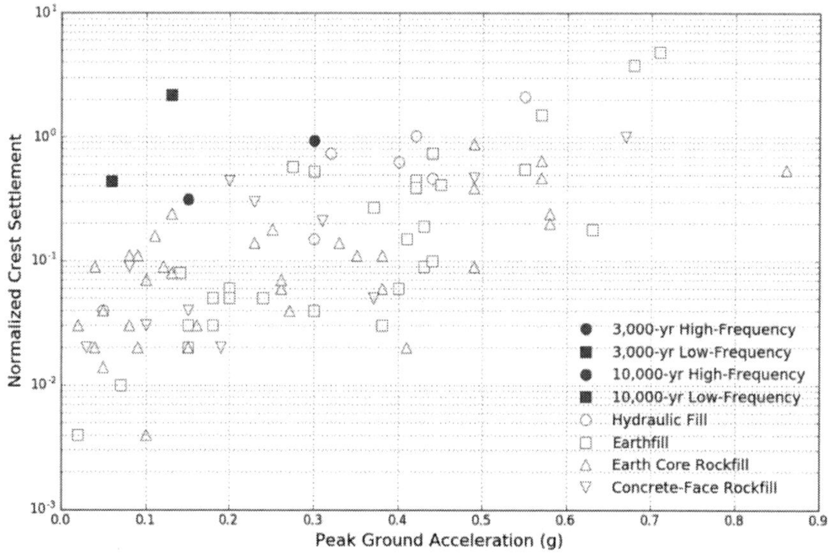

Figure 8. Comparison of Calculated Crest Settlements for BRD with Empirical Performance Data from Swaisgood (2014).

CONCLUSIONS

The following are the key findings from the seismic deformation analysis of Blue Ridge Dam:

- Based on the review of historical geotechnical data and results obtained as part of the field and laboratory investigations conducted by Geosyntec, the embankment soils straddle the boundary between non-plastic or low-plasticity, coarse-grained soils and moderate-plasticity, fine-grained soils.

- Monotonic and cyclic laboratory tests performed on high-quality specimens indicate that the undrained shear strength of the dumped fill follows normalized behavior consistent with SHANSEP.

- The calibrated PM4SAND constitutive model predicts the monotonic and cyclic stress-strain response of BRD soils with reasonable accuracy, including the effects of overconsolidation and initial static shear stress.

- Numerical deformation analyses suggest that the maximum permanent displacements caused by seismic loading would be approximately 1.5 ft or less for ground motions with a return period of 3,000 yrs. For ground motions with a return period of 10,000 yrs, the maximum calculated displacements are approximately 4.5 ft.

- The calculated permanent deformations caused by high-frequency ground motions agree well with the expected performance based on empirical data. Calculated deformations for low-frequency ground motions are larger, likely because of the long duration of the motions.

ACKNOWLEDGEMENTS

The first two authors are grateful for the support provided by TVA during the execution of the project. The authors are also grateful for the guidance provided by Professor Ross W. Boulanger and Dr. William F. Marcuson, III throughout the project.

REFERENCES

Boulanger, R.W. and Ziotopoulou, K. (2015). "PM4SAND (Version 3.0): A Sand Plasticity Model for Earthquake Engineering Applications," Report No. UCD/CGM-15/01, Center for Geotechnical Modeling, University of California at Davis, March, 108 pp.

Dafalias, Y.F. and Manzari, M.T. (2004). *Simple plasticity sand model accounting for fabric change effects*, Journal of Engineering Mechanics, ASCE, 130(6), 622-634.

Darendeli, M.B. (2001). *Development of a new family of normalized modulus reduction and material damping curves*. Ph.D. thesis, The University of Texas, Austin.

DeGroot, D.J. and Ladd, C.C. (2012). "Site characterization for cohesive soil deposits using combined in situ and laboratory testing." *Geotechnical Engineering State of the Art and Practice: Keynote Lectures from GeoCongress 2012*, Geotechnical Special Publication No. 226, K Rollins and D. Zekkos, Eds., ASCE GeoInstitute, pp. 565-608.

Idriss, I. M., and Boulanger, R. W. (2008). *Soil Liquefaction During Earthquakes,* Monograph No. MNO-12, Earthquake Engineering Research Institute, 237 p.

Itasca Consulting Group (2011). FLAC® 7.0 (Fast Lagrangian Analysis of Continua) User's Guide, Minneapolis.

Ladd, C.C. (1991). "Stability Evaluation During Staged Construction (22nd Terzaghi Lecture)," *Journal of Geotechnical Engineering*, 117(4), 540-615.

Ladd, C.C. and Foott, R. (1974). "New Design Procedure for Stability of Soft Clays," *Journal of the Geotechnical Engineering Division*, 100(GT7), 763-786.

Ladd, C.C., Foott, R., Ishihara, K., Schlosser, F., and Poulos, H.G. (1977). "Stress-deformation and strength characteristics: SOA report." Proc., 9th Int. Conf. on Soil Mechanics and Foundation Eng., Tokyo, 2, 421-494.

Mayne, P.W. (2014). "Generalized CPT Method for Evaluating Yield Stress in Soils," *GeoCharacterization and Modeling for Sustainability*, GSP 234, American Society of Civil Engineers, Reston, VA.

Mayne, P.W., and Kulhawy, F.H. (1990). Manual on Estimating Soil Properties for Foundation Design, Report No. EL-6800, Electric Power Research Institute.

McGuire, R.K., Silva, W.J., and Costantino, C.J. (2001). Technical Basis for Revision of Regulatory Guidance on Design Ground Motions: Hazard- and Risk-consistent Ground Motion Spectra Guidelines (NUREG/CR-6728), U.S. Nuclear Regulatory Commission, Washington, DC.

Robertson, P.K. (2004). Evaluating soil liquefaction and post-earthquake deformations using the CPT. *Geotechnical and Geophysical Site Characterization*, Vol. 1 (Proc. ISC-2, Porto), Millpress, Rotterdam: 233-252.

Robertson, P.K. (2009). "Interpretation of Cone Penetration Tests: a Unified Approach," *Canadian Geotechnical Journal*, 46(11), 1335-1355.

Sowers, G.F., and Richardson, T. L., (1983). Residual soils of Piedmont and Blue Ridge, *Transportation Research Record 919*, TRB Washington DC, pp 10-16.

Swaisgood, J. R. (2014). "Behavior of embankment dams during earthquakes," *Journal of Dam Safety*, 12(2), 35-44.

A CASE HISTORY EVALUATION OF STATE OF PRACTICE FOR SEISMIC DEFORMATION MODELING OF EARTHEN DAMS

Khaled Chowdhury, PE, GE[1]
Vlad Perlea, PhD, PE[2]
Raymond B. Seed, PhD[3]
Michael Beaty, PhD, PE, GE[4]
Ethan Dawson, PhD[5]
George Hu, PE[6]

ABSTRACT

Liquefaction-induced seismic deformation analyses address one of the major potential failure modes for earthen dams. The State of Practice has evolved over the last several decades with regard to soil liquefaction triggering relationships, post-liquefaction strength evaluation, constitutive models, and analysis tools. FLAC (Fast Lagrangian Analysis of Continua) is a widely used numerical modeling tool for dynamic analyses of dams using the finite difference method. However, based on selection of soil modeling parameters, different triggering curves, post-liquefaction strength relationships, and constitutive models, the resulting estimated deformations of dams can vary significantly. As the estimated deformations are used to perform risk assessment of dams to (1) evaluate current conditions, (2) inform mitigation design, and (3) implement interim risk reduction measures, the impacts of these differences can be significant. This paper presents analyses of a field performance case history employing a number of different sets of analytical models and relationships. The Upper San Fernando Dam (USFD) experienced well-documented moderate deformations during the 1971 San Fernando earthquake. This well-documented case history is a good candidate for assessing impacts of numerical modeling approaches for dynamic analyses of dams. Several constitutive models (UBCSAND, Roth/URS, and Mohr-Coulomb) for liquefiable and non-liquefiable soils were utilized to evaluate effects of (1) model selection, (2) parameter selection details, (3) triggering relationships, and (4) post-liquefaction strength evaluation approaches. Previous numerical modeling efforts for the USFD case history have been performed mainly to evaluate or validate individual constitutive models. This study employs multiple models and relationships, and it compares analytical results with observed field performance for a suite of selected approaches to develop insights regarding differences between the various analysis approaches employed, and the usefulness and reliability of these types of complex engineering assessments.

[1] PhD Candidate, University of California, Berkeley, California, and Senior Geotechnical Engineer, United States Army Corps of Engineers, South Pacific Division Dam Safety Production Center, 1325 J Street, 12th Floor, Sacramento, CA 95814-2922, Khaled.Chowdhury@usace.army.mil
[2] Senior Geotechnical Practitioner, AECOM, 2020 L St, Sacramento, CA 95811, vlad.perlea@aecom.com
[3] Professor, Civil and Environmental Engineering Department, University of California, Berkeley, California, rmseed6@aol.com
[4] Principal Engineer, Beaty Engineering, LLC, Beaverton, OR. mhb@beatyeng.com
[5] Senior Geotechnical Engineer, AECOM, Oakland, CA 95612. ethan.dawson@aecom.com
[6] Civil-Geotechnical Engineer, United States Army Corps of Engineers, South Pacific Division Dam Safety Production Center at Sacramento District, 1325 J Street, 12th Floor, Sacramento, CA 95814-2922, George.Hu@usace.army.mil

INTRODUCTION

Liquefaction-induced seismic deformation is one of the major potential distress and failure modes examined in seismic evaluation of dams. Liquefaction-induced deformations can range from limited or no deformation cases to very large deformations (such as flow failures). Excessive deformations in earthen dams can pose risk with regard to both life loss and economic consequences. The State of Practice for liquefaction-induced deformation analyses has evolved over the last several decades with regard to liquefaction triggering relationships, post-liquefaction strength evaluation, constitutive models, and analysis tools. It is important to apply these tools to back-analyses of field performance case histories in order to evaluate their effectiveness and reliability across different ranges of conditions.

This paper presents a suite of analyses of one significant performance case history, employing different combinations of analysis models and relationships. This is one of three case histories that are being back-analyzed as part of this overall effort. The three case histories under study have been selected for (1) availability of suitable data, (2) availability of well-documented seismic performance, (3) differing ranges of complexity, (4) differing levels of seismic excitation, and (5) differing ranges of observed or projected seismic performance.

The Upper San Fernando Dam (USFD) experienced well-documented small to moderate deformations during the 1971 San Fernando earthquake. This range of performance (small to moderate deformations and displacements) is important for seismic evaluation of existing dams as well as for design and implementation of mitigation of dams considered to be potentially at risk. It is also important for mitigation design, as mitigation often results in expected small to moderate deformations and it is important that these do not pose unacceptable levels of risk. Estimated seismic deformations are also often used to develop interim reservoir restrictions, until overall mitigation can be implemented, and thus they can have consequences with regard to interim reservoir operations and water supply and/or power generation.

Post-earthquake studies of the USFD have produced significant amounts of laboratory and field data. As a result, this field performance case history is a good candidate for assessing impacts of numerical modeling approaches and details on the results of dynamic analyses of dams. Several constitutive models (UBCSAND, Roth/URS, and Mohr-Coulomb) for liquefiable and non-liquefiable soils are being utilized in these current studies to evaluate effects of (1) model selection, (2) parameter selection details, (3) liquefaction triggering relationships, and (4) post-liquefaction strength evaluation approaches on these types of deformation analyses. This work is ongoing, and this paper presents a suite of analyses based on various approaches, but does not yet address the full range of models being studied.

Figure 1. Upper San Fernando Dam after the 1971 San Fernando Earthquake (Steinbrugge Collection, NICEE library, U.C. Berkeley)

DEFORMATION MEASUREMENTS AT USFD

The Upper San Fernando Dam (USFD) suffered liquefaction-induced damage, and displacements, during the 1971 San Fernando Dam on February 9, 1971. Figure 1 shows a photograph of the USFD after the earthquake. Serff (1976) summarized horizontal and vertical displacements measured at six locations (including five survey monuments) on or near to the maximum height cross-section of the dam. These locations were: (1) on the crest parapet wall, (2) at the mid-point of the downstream slope of the upper rolled embankment fill, (3,4) at the upstream and downstream ends of the crest of the downstream berm, (5) at the mid-point of the downstream berm slope, and (6) at the downstream toe. Figure 2 shows locations shows locations of these displacement measuring points, and the measured displacements (both vertical and horizontal) at these locations. The dam crest displaced 4.9 feet toward the downstream side, and the maximum measured displacement was a downstream translation of 7.2 feet at the hinge point at the top of the downstream berm. The maximum settlement measured was 2.5 feet at the crest parapet wall.

The California Department of Water Resources (DWR, 1989) also presented additional settlement measurements along crest at four more locations, and on the downstream berm and slope at four additional locations. Figure 3 shows the locations and values of these additional settlement measurements.

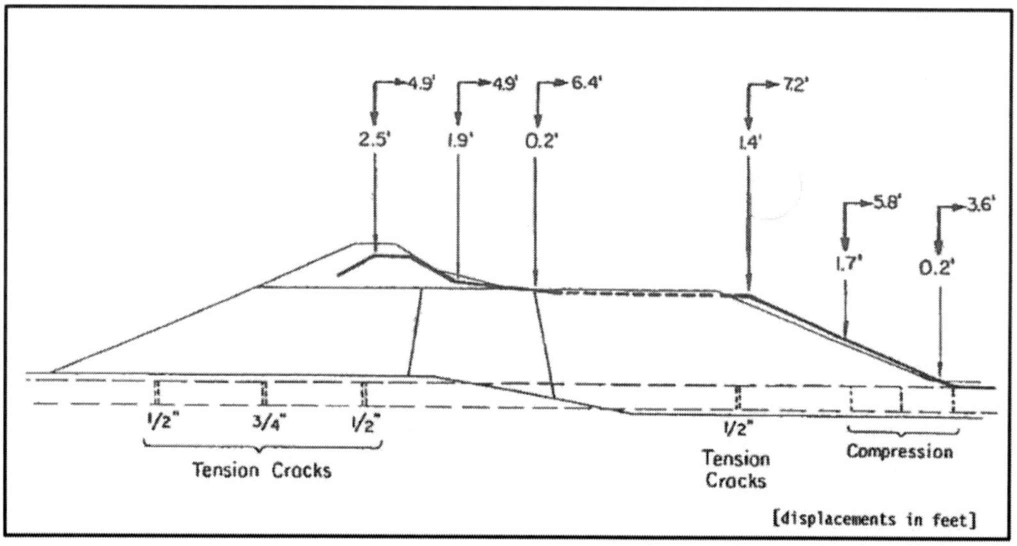

Figure 2. Displacements Measured after the 1971 San Fernando Earthquake (Serff, 1976)

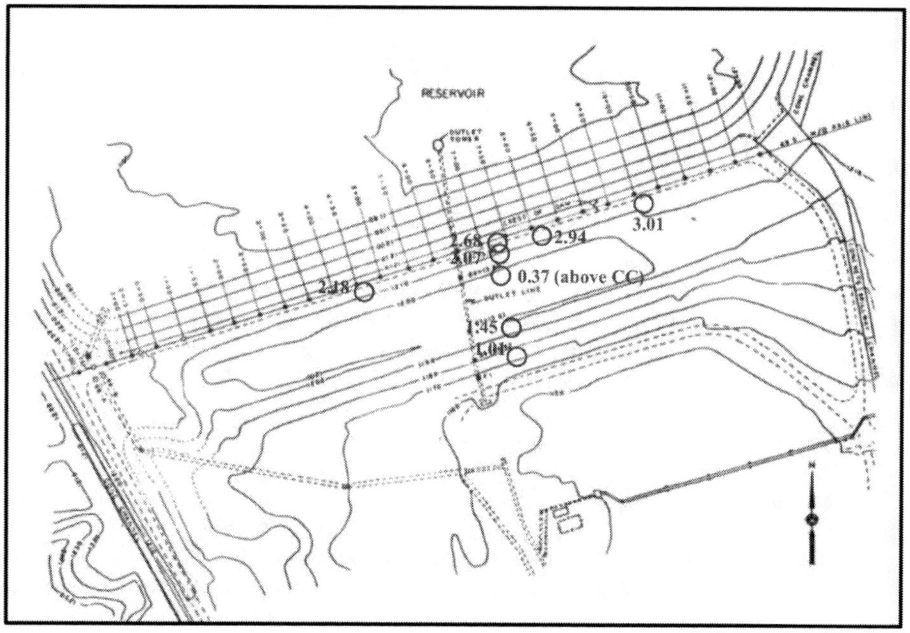

Figure 3. Plan view of USFD with settlement measurement locations and settlement magnitudes measured in feet (DWR, 1989)

ANALYSIS SECTION AND DEFORMATION ANALYSES PLAN

Seed et al. (1973) prepared an idealized cross-section (Figure 4) of the USFD that was utilized, sometimes with variations as to details, by a number of researchers for subsequent numerical modeling and analyses of the USFD field performance case history. This cross-section does not occur at any location, but was judged to be generally representative of conditions at and near the central portion of the dam (near the maximum height cross-section).

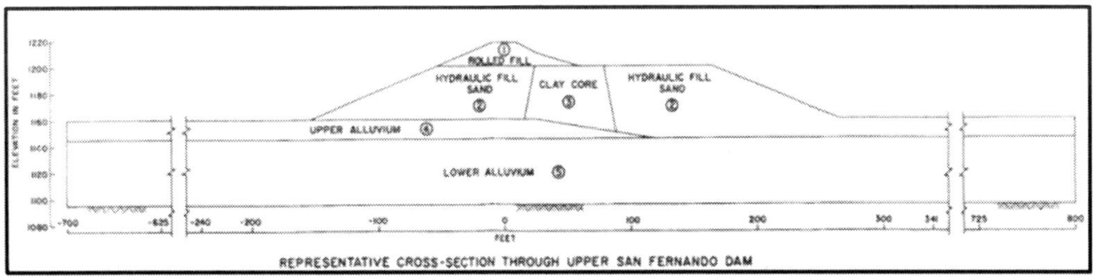

Figure 4. Representative Cross-Section of USFD (Seed et al., 1973)

Seed et al. (1973) also presented three detailed actual transverse cross sections across the dam that include pre-earthquake and post-earthquake conditions. Figure 5 shows the locations of these three transverse cross-sections in plan view. One of these actual cross-sections, Section B-B', was selected for use in analyses for this study as it is located near center of the dam, nearly at the maximum height cross-section, and it aligns closely with the surveyed lateral and vertical displacements presented previously in Figure 2. Figure 6 shows both the pre-earthquake (dashed lines) and post-earthquake configurations (solid lines) of this Cross-Section B-B′, and it is this central section that will be studied in the back-analyses presented in this paper.

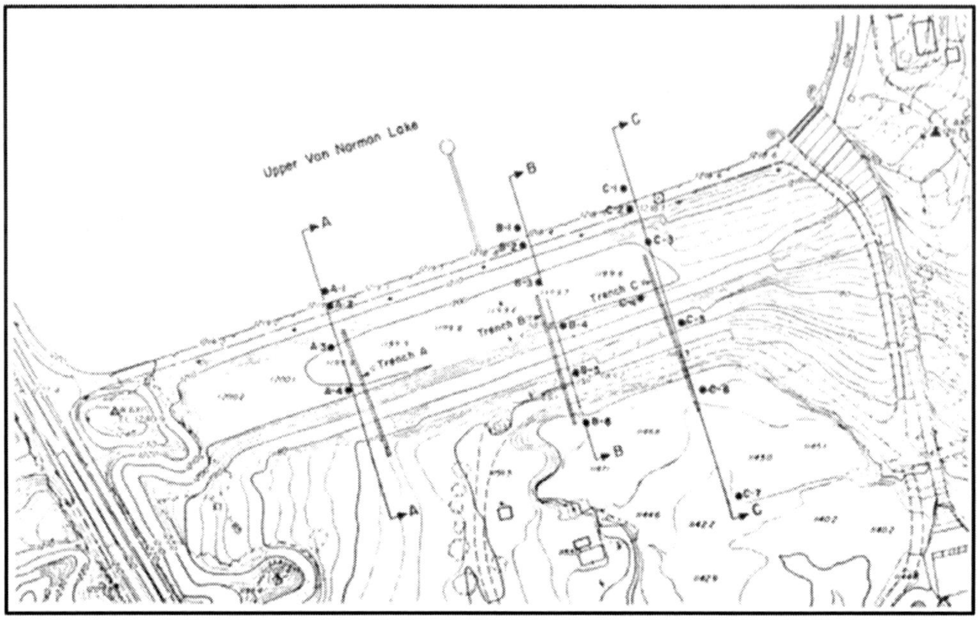

Figure 5. Plan view of USFD with cross-sections, borings, and trench locations from the post-earthquake study. (Seed et al., 1973)

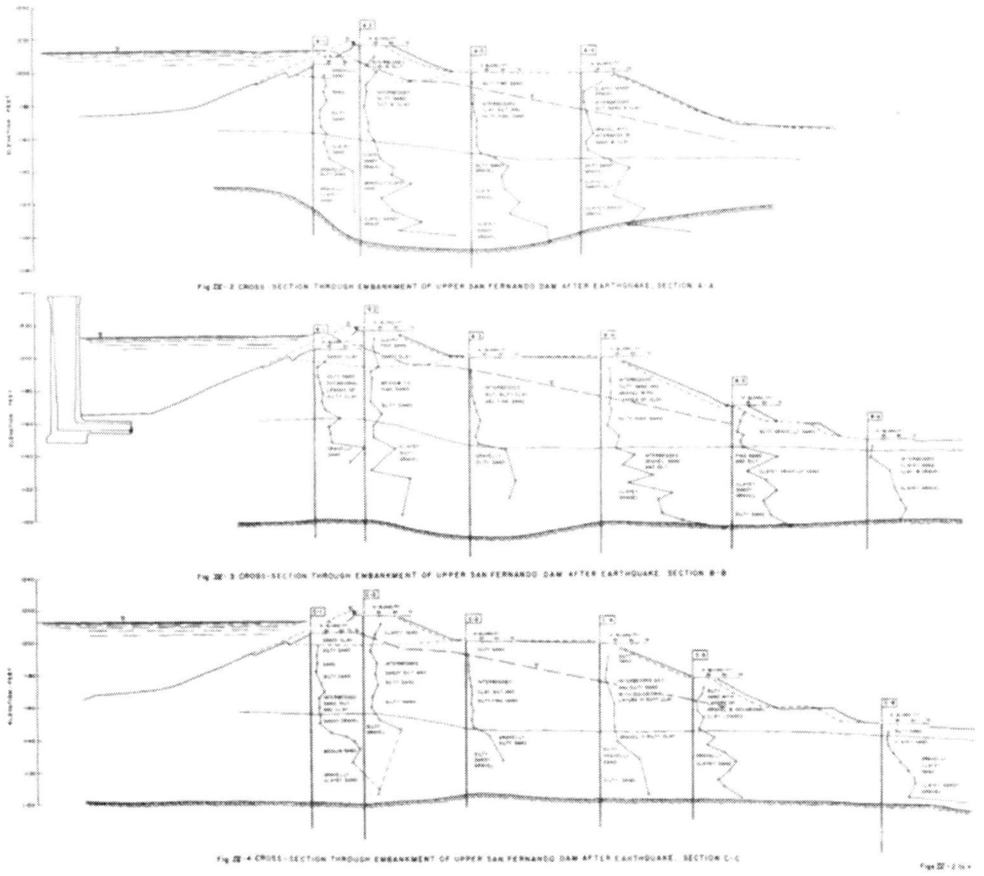

Figure 6. Observed deformed shapes of the Upper San Fernando Dam at Sections A-A', B-B′, and C-C′ (Seed et al, 1973)

Over the past several decades a number of constitutive and/or behavioral models have been developed for nonlinear dynamic analyses of seismic soil response including cyclic pore pressure generation effects, and these have been increasingly applied to seismic deformation analyses of earthen embankments such as conventional earth and rockfill dams, levees, and tailings dams. These analytical tools have been applied within both finite element analysis and finite difference analysis frameworks.

These current studies employ a limited set of selected analytical models, and all of them are applied within a finite difference analysis framework using the code FLAC (Fast Lagrangian Analysis of Continua); (Itasca, 2011). This paper presents the results of a suite of six sets of analyses. The material behavior models employed in these analyses are listed in Table 1, and they will be explained further. Each model must be "parametrized" to model (1) soil liquefaction "triggering" behavior (cyclic pore pressure generation) and (2) post-liquefaction residual strengths.

There are a number of empirical relationships available for parametrization of the selected material behavior models for each of these two important elements of overall behavior, and the relationships employed in the suite of analyses presented herein are also listed in Table 1.

Table 1. Summary analytical modeling schemes for the USFD analyses presented.

Analysis ID	Description
Analysis 1: -UBCSAND	Constitutive model for liquefiable soils – UBCSAND Constitutive model for non-liquefiable soils – Mohr-Coulomb Liquefaction Triggering - Youd et al. (2001) K_σ – Youd et al. (2001) Post Liquefaction Strength, S_r – Seed and Harder (1990) [Median Sr applied by post-shaking FISH function]
Analysis 2: -UBCSAND	Similar to Analysis 1-UBCSAND, except Post Liquefaction Strength, Sr – Weber et al. (2015) [50th percentile Sr applied by post-shaking FISH function]
Analysis 3: -UBCSAND	Similar to Analysis 1-UBCSAND, except Post Liquefaction Strength, Sr – Idriss and Boulanger (2015) [Sr with significant void redistribution curve applied by post-shaking FISH function]
Analysis 4: -Roth/URS	Constitutive model for liquefiable soils – Roth/URS Constitutive model for non-liquefiable soils – Mohr-Coulomb Liquefaction Triggering – Youd et al. (2001) K_σ –Youd et. al. (2001) Post Liquefaction Strength, Sr – Seed and Harder (1990) [Median Sr applied during-shaking]
Analysis 5: -Roth/URS	Constitutive model for liquefiable soils – Roth/URS Constitutive model for non-liquefiable soils – Mohr-Coulomb Liquefaction Triggering – Cetin et al. (2016) K_σ –Youd et al. (2001) Post Liquefaction Strength, Sr – Weber et al. (2015) [50th percentile Sr applied during-shaking]
Analysis 6: -Roth/URS	Constitutive model for liquefiable soils – Roth/URS Constitutive model for non-liquefiable soils – Mohr-Coulomb Liquefaction Triggering –Boulanger and Idriss (2014) K_σ – Idriss and Boulanger (2008) Post Liquefaction Strength, Sr – Idriss and Boulanger (2008/2015) [Sr with significant void redistribution curve applied during-shaking]

These studies are ongoing, and additional analyses will be performed using a broader range of combinations of both triggering relationships and post-liquefaction residual strength relationships, as well as at least one additional constitutive model, and that will lead to further combinations of analytical models, triggering and post-liquefaction strength characterizations, and other analytical details.

INPUT PARAMETER DEVELOPMENT

After the 1971 San Fernando earthquake, the Division of Safety of Dams (DSOD) of the State of California Department of Water Resources (DWR) and the Los Angeles Department of Water and Power (LADWP) sponsored and performed a post-earthquake investigation program for both the Upper and Lower San Fernando Dams. For the Upper San Fernando Dam, this consisted of 17 SPT borings with 231 SPT measurements, several trenches, cross-hole geophysical studies, and laboratory testing. The National Science Foundation (NSF) sponsored a grant to the University of California at Berkeley (UCB) and to the University of California at Los Angeles (UCLA) for the analytical studies as well as portions of the laboratory testing program. The investigation program and analytical studies were directed by the late Professors H.B. Seed of U.C. Berkeley and K.L. Lee of UCLA, and they are summarized in Seed et al. (1973, EERC 73-2). The parameters for this study were developed based on a re-evaluation of the data presented in EERC 73-2, subsequent research studies performed, and current State of Practice procedures for parameter evaluation and modeling employing the analytical tools employed in the analyses presented.

Input Ground Motions

Input ground motion time history is one of the most important variables for seismic analyses of liquefaction-induced deformations. Reliable site-specific strong ground motion recordings for the USFD during the 1971 San Fernando Earthquake are not available. The unavailability of a site-specific ground motion record will continue to be one of the major sources of uncertainties in seismic back-analyses of the performance of the USFD.

In the absence of any recorded strong motion at the USFD site, ground motion recordings obtained at an abutment instrument station located at the nearby Pacoima Dam were modified by Seed et al. (1973), after evaluation of ground motion records obtained at different sites in the surrounding region during the San Fernando earthquake. The "abutment" strong ground motion recording at Pacoima Dam was obtained at a site located about 55 feet above the left abutment of the 365 foot high dam. The two horizontal components of ground motion at this station both showed maximum accelerations of approximately 1.25g and the vertical component showed a maximum acceleration of 0.72g. Figure 7 shows a plot of maximum accelerations recorded at rock sites at different distances from the epicentral region in the San Fernando earthquake. Based on this figure, Seed et al. (1973) suggested that a maximum horizontal acceleration of the order of 0.55g to 0.6g in the vicinity of the Upper and Lower San Fernando Dams would be reasonable. Seed et al. (1973) first modified the Pacoima abutment records by reducing the maximum pulse of the Pacoima Dam abutment record to a maximum value of 0.9g, and then further scaled down the entire modified record to a PGA of 0.6g. This modified and scaled record has since been widely utilized as an "input" ground motion by different researchers in performing seismic back-analyses of both USFD and LSFD [e.g.

Seed and Harder (1990), Inel et al. (1993), Moriwaki et al. (1998), Beaty (2001), etc.]. Figure 8 shows acceleration and velocity components of this modified Pacoima record used as an input motion for USFD analyses.

This modified input ground motion was developed at a time when near-field directivity effects (e.g. pulse and fling step) were not yet well understood, and it should be recognized that this modified input record might have been missing (or under-representing) some potentially important site-specific ground motion characteristics. This will be discussed further. The use of this ground motion, however, is nonetheless useful both for purposes of evaluating effects of different combinations of different models and parameters, as well as for comparing analytical results and findings of the current study with previous studies.

SPT Blow Counts

A total of 17 SPT borings were performed across three transverse (upstream to down-stream) cross-sections, and these provided a total of 231 SPT blow counts at different locations and in different strata. Figure 7 shows a plan view of the field boring locations.

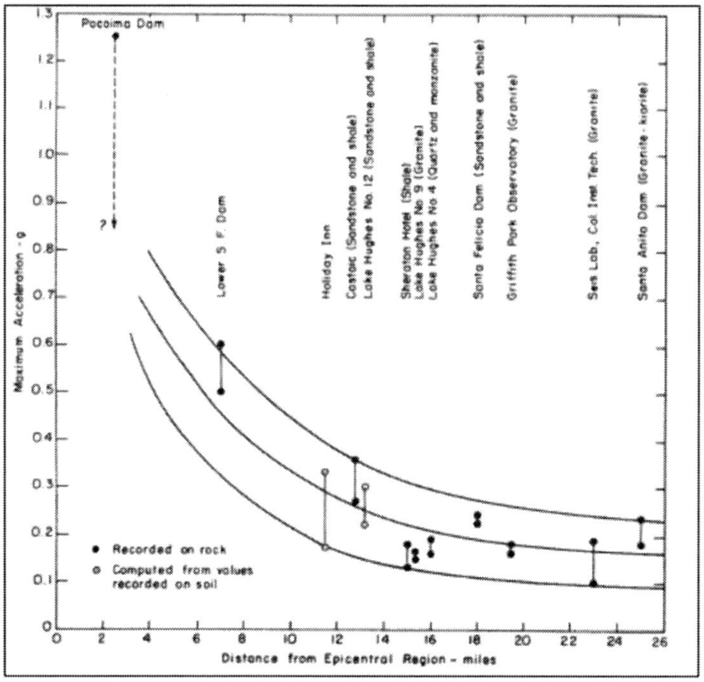

Figure 7. Maximum accelerations recorded on rock during the San Fernando Earthquake (Seed et al. 1973)

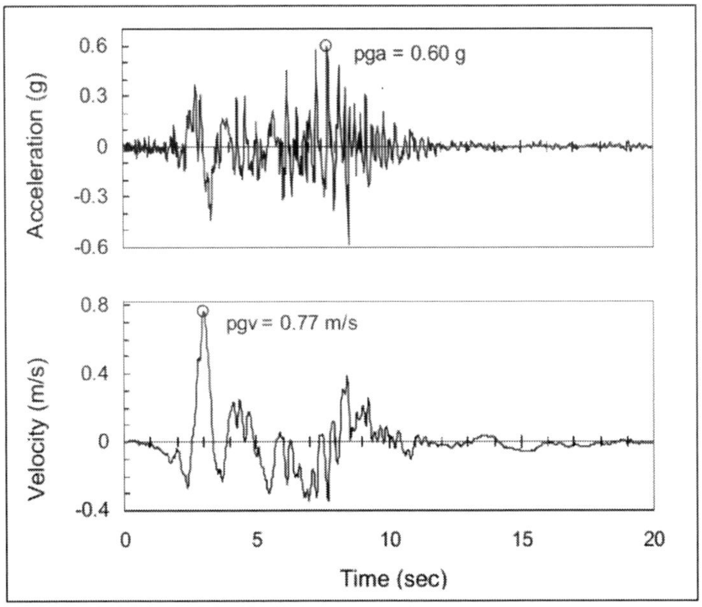

Figure 8. Pacoima Dam record from the 1971 San Fernando Earthquake as modified by Seed (1973) for analyses of the Upper San Fernando Dam; (Beaty, 2001)

In most previous studies, data from 11 borings from (a) upstream, (b) crest, and (c) downstream locations were utilized to develop average $(N_1)_{60}$ and $N_{1,60,CS}$ values for USFD seismic back-analyses, and properties were considered to be similar in the potentially liquefiable material zones on both the upstream and downstream zones at each elevation range. In this current study, the upstream and downstream sections will be separated and characterized on a more localized basis.

Figure 9 shows a cross-section of the USFD, and the principal soil zones modeled in these current studies. The soil strata in these current studies (and in Figure 9) are identified using the following nomenclature:

Hydraulic Fill Upstream - HFU (HFU-1 = upper, and HFU-2 = lower)

Hydraulic Fill Downstream - HFD (HFD-1 = upper, HFD-2 = middle, and HFD-3 lower)

Upper Alluvium - UA

Lower Alluvium - LA

Rolled Fill - RF

Clayey Core - CC

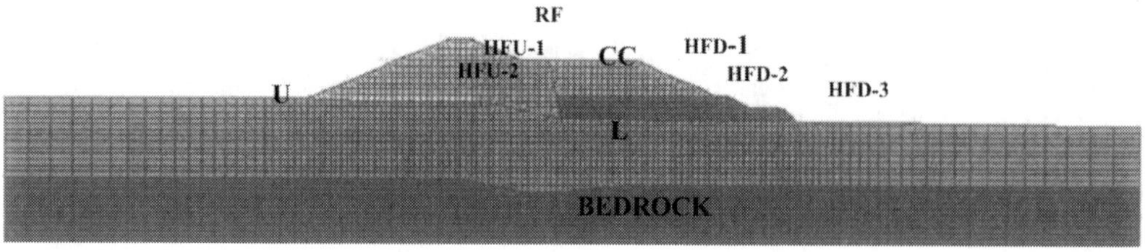

Figure 9. Geotechnical Model for Seismic Deformation Analyses in This Current Study

California's DWR performed an additional evaluation of the USFD data as part of the post-1975 Oroville earthquake investigation for Oroville Dam (DWR, 1989). In this evaluation, DWR used settlement values measured at the crest and the top deck of the downstream shell to estimate post-earthquake consolidation (densification), and reduced average blow counts by 4.5 blows/foot to obtain better estimates of pre-earthquake conditions. Beaty (2001) correctly noted that that the settlements measured would include both (1) vertical settlement from kinematic movement of the slide mass, and (2) volumetric post-earthquake consolidation (densification), and suggested that a reduction of between 2 and 4 blows/foot would be more appropriate. Beaty reduced average blow counts by 3 to obtain best estimates of average "pre-earthquake" blow counts for his back-analyses. Beaty and Byrne (2011) re-processed the data set, producing nearly the same results, and then made a fines adjustment ($\Delta N_{1,60}$) as a function of their estimated average 25% fines content (and based on the fines correction relationship recommended by Seed et al. (1984) and Youd et al. (2001) to develop fines-adjusted $N_{1,60,CS}$ values. Table 2 presents a summary of the average blow counts developed by DWR (1989), Beaty (2001) and Beaty and Byrne (2011).

Table 2. SPT $(N_1)_{60}$ and $N_{1,60,CS}$ values from the DWR (1989), Beaty (2001), and Beaty and Byrne (2011) studies

Soil Layer	DWR Study (1989)		Beaty (2001)	Beaty and Byrne (2011)	
	Average Post-Earthquake $(N_1)_{60}$	Average Pre-Earthquake $(N_1)_{60}$	Median Pre-Earthquake $(N_1)_{60}$	Median Pre-Earthquake $(N_1)_{60}$	Median Pre-Earthquake $N_{1,60,CS}$
HFU-1 (EL.1200-EL. 1170)	13.5	9	9	10	15
HFU-2 (El. 1170-1160)	17.5	13	15	14.5	19.5
HFD-1 (EL.1200-EL. 1170)	13.5	9	9	10	15
HFD-2 (EL.1170-EL. 1160)	17.5	13	15	14.5	19.5
HFD-3 (EL.1160-EL. 1145)	17.5	13	15	13	18

In these current studies, crest and upstream borings were treated separately from downstream borings, and SPT data were separately processed to develop characterizations for the upstream and downstream zones. Blowcount data within each of the zones were fully re-processed, and then a blow count reduction of 2 blows/foot was applied to post-earthquake $(N_1)_{60}$ values to obtain best estimates of pre-earthquake $(N_1)_{60}$ values for both upstream and downstream layers. This reduction factor is similar to that employed in a recent study by Weber et al. (2015), and differs by only 1 blow/foot from that employed by Beaty and Byrne (2011). This slight difference between the adjustment made by Beatty and Byrne (2011) is based on the observation that deviatoric deformations may have contributed slightly more strongly to observed settlements, so that volumetric densification would have been slightly smaller. SPT blow counts were screened to include SPT from only non-plastic layers, and an average (or representative) fines content of 25 percent (similar to Beaty and Byrne, 2011) was used for evaluation of clean-sand correction factors to develop $N_{1,60,CS}$ values.

In these current studies, the upper Alluvium layer on the upstream side was separated from the downstream side lower hydraulic fill layer. These two layers are located at approximately the same elevations, but overburden-corrected blow counts from the upstream side unit are higher than those of the downstream side unit.

The corrected and normalized $N_{1,60}$ blow counts for these current studies were developed by applying the equipment, energy, procedural and effective overburden corrections of each of the three liquefaction triggering relationships of (1) Youd et al. (2001), (2) Idriss and Boulanger (2014), and (3) Cetin et al. (2016), and the resulting $N_{1,60}$ values are presented in Table 3. Further adjustments for fines contents to develop $N_{1,60,CS}$ values were then applied based on each of these three methods as well, and the resulting $N_{1,60,CS}$ values are presented in Table 4.

Table 3. SPT Blow Counts [$(N_1)_{60}$] for This Study, Based on Equipment, Energy, Procedural and Effective Overburden Corrections as per Each of the Liquefaction Triggering Approaches Listed

Soil Layer	Youd et al. (2001)		Idriss and Boulanger (2014)		Cetin et al. (2016-personal communication)	
	Average Post-Earthquake $(N_1)_{60}$	Average Pre-Earthquake $(N_1)_{60}$	Average Post-Earthquake $(N_1)_{60}$	Average Pre-Earthquake $(N_1)_{60}$	Average Post-Earthquake $(N_1)_{60}$	Average Pre-Earthquake $(N_1)_{60}$
HFU-1 (EL.1200-1170	11.6	9.6	11.6	9.6	11.6	9.6
HFU-2 (EL. 1170-1160)	14.7	12.7	14.8	12.8	14.7	12.7
UA (EL.1160-1145)	16.1	14.1	16.6	14.6	16.1	14.1
HFD-1 (EL.1200-1170)	12.3	10.3	12.1	10.1	11.9	9.9
HFD-2 (EL.1170-1160)	13.7	11.7	13.5	11.5	13.6	11.6
HFD-3 (EL.1160-1145)	14.6	12.6	14.7	12.7	14.6	12.6

Table 4. Fines Corrected $N_{1,60,CS}$ SPT Blow Counts for This Study

Soil Layer	Youd et al. (2001)		Idriss and Boulanger (2014)		Cetin et al. (2016)
	Average Pre-Earthquake $N_{1,60,CS}$ for Triggering	Average Pre-Earthquake $N_{1,60,CS}$ for Sr	Average Pre-Earthquake $N_{1,60,CS}$ for Triggering	Average Pre-Earthquake $N_{1,60,CS}$ for Sr	Average Pre-Earthquake $N_{1,60,CS}$ for Triggering and Sr
HFU-1 (EL.1200-1170)	13.8	11.6	13.4	11.6	11.8
HFU-2 (EL. 1170-1160)	18.6	14.7	17.9	14.8	15.5
UA (EL.1160-1145)	19.7	16.1	19.3	16.6	16.8
HFD-1 (EL.1200-1170)	16.0	12.3	15.2	12.1	12.7
HFD-2 (EL.1170-1160)	17.5	13.8	16.6	13.5	14.3
HFD-3 (EL.1160-1145)	18.6	14.6	17.8	14.7	15.4

Finally, as an illustrative example, Figure 10 shows the processed $N_{1,60,CS}$ values developed based on the equipment, energy, procedural and effective overburden corrections of Cetin et al. (2016) for six of the principal soil units analyzed in these studies, and this figure serves to illustrate the variability of the individual values in each unit. Similar figures can be presented for each of the other two approaches (Youd et al., 2001, and Boulanger and Idriss, 2014), and variability within each unit would be similar.

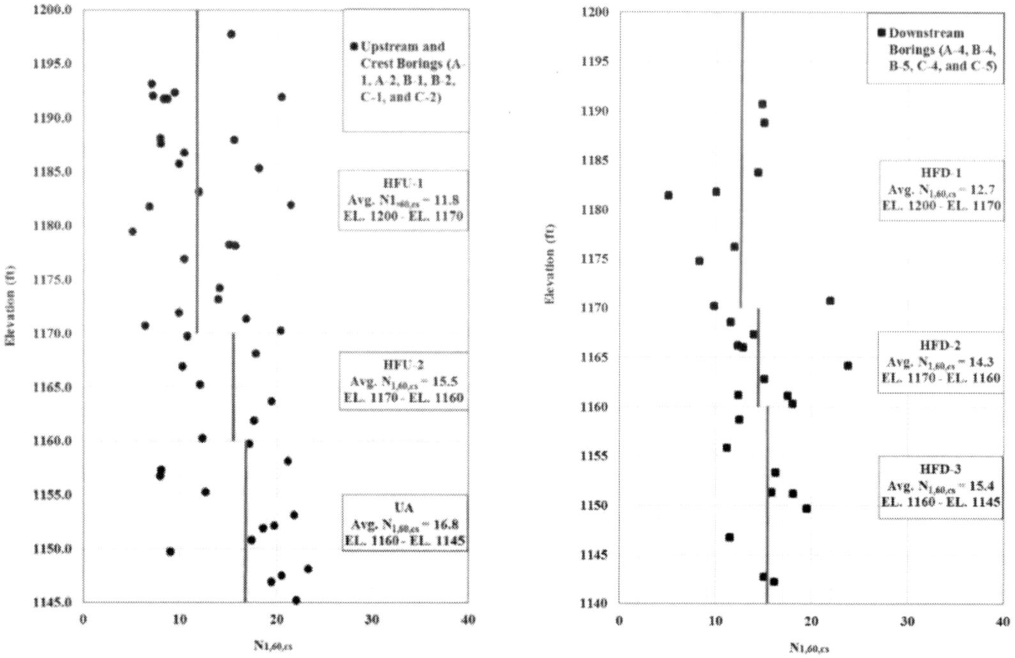

Figure 10. SPT Blow Counts (Post-Earthquake) using Cetin et al. (2016) for energy, equipment, procedural and effective overburden stress corrections, and adjustment for earthquake-induced densification.

Relative Density

Relative density is one of the most important parameters for seismic evaluations. However, there are inherent difficulties in estimating the relative density of coarse-grained soils based on in-situ penetration tests.

It is convenient to estimate relative density (D_R) from $(N_1)_{60}$ values for coarse-grained soils. Studies of Skempton (1986) and Cubrinovski and Ishihara (1999) were evaluated as part of this current study. Skempton (1986) evaluated Meyerhof's equation

$$(N_1)_{60} = (a + b)\ D_R^2 \qquad \text{[Equation 1]}$$

where the parameters a and b are constants for a particular sand within a range of $0.35 < D_R < 0.85$ and $0.5\ \text{atm} < \sigma'_v < 2.5\ \text{atm}$. Skempton (1986) has shown that the parameters a and b (combined as $a + b = C_d$) tend to increase with increasing grain size, with increasing age of deposits, and with increasing overconsolidation ratio. Skempton (1986) suggested that for normally consolidated natural sands, an approximate value of $C_d = (N_1)_{60} / D_R^2 \simeq 60$ would produce similar results as the relative density relationship of Terzaghi and Peck (1967). He recommended use of $C_d \approx 55$ for fine sand and 65 for coarse sand. Skempton also considered ageing effects. Table 5 presents C_d values proposed by Skempton (1986) considering effects of aging.

Table 5. Effect of Aging on C_d Values (Skempton, 1986)

	Age: Years	$C_d = (N_1)_{60}/D_R^2$
Laboratory Tests	10^{-2}	35
Recent Fills	10	40
Natural Deposits	$>10^2$	55

Cubrinovski and Ishihara (1999) compiled results of 61 relative density measurements performed on samples of sands and gravels with SPT data of known $N_{1,78}$ blowcounts (based on an average Japanese SPT energy ratio of 78 percent). The results were summarized to develop relative density relationships based on D_{50}, $e_{max} - e_{min}$, and $N_{1,78}$ values. The following equations are based on the relationship of Cubrinovski and Ishihara (1999) for 78 percent hammer energy ratio, but with the Cubrinovski and Ishihara (1999) equation for C_d adjusted here for the 60 percent SPT hammer energy ratio (ER) in Equation 4.

$$e_{max} - e_{min} = 0.23 + 0.06/D_{50} \quad \text{[Equation 2]}$$

$$C_d = 9/(e_{max} - e_{min})^{1.7} \text{ for ER=78\%} \quad \text{[Equation 3]}$$

$$C_d = 11.7/(e_{max} - e_{min})^{1.7} \text{ for ER=60\%} \quad \text{[Equation 4]}$$

Cubrinovski and Ishihara (1999) suggest that C_d values can vary between 10 and 100. In this current study, the database of Cubrinovski and Ishihara (1999) was re-evaluated for a range of interest more suited for seismic evaluation of liquefaction-related behavior of soils with $e_{max} - e_{min}$ of between 0.3 and 0.7 and C_d values of between 20 and 70. $(N_1)_{78}$ values for soils conforming to this range in their larger data set were converted to $(N_1)_{60}$ values corresponding to 60 percent hammer energy ratio, and only D_R values of less than 100 percent were considered. A total of 28 values within this range remained, and these were used in developing a sub-relationship between C_d and $e_{max} - e_{min}$. Figure 11 shows plots of (1) $e_{max} - e_{min}$ vs. D_{50} and (2) C_d vs. $e_{max} - e_{min}$ from Cubrinovski and Ishihara (1999). Figure 12 shows C_d vs. $e_{max} - e_{min}$ plot based on re-evaluation of Cubrinovski and Ishihara (1999) data within the constrained ranges described above. The following equation was developed in this study to relate void ratio range with C_d.

$$C_d = 23.334(e_{max} - e_{min})^{-0.793} \quad \text{[Equation 5]}$$

It should be noted that use of D_{50}-based $e_{max} - e_{min}$ may be an uncertain way to characterize or represent a soil layer, due to uncertainty in developing a representative D_{50} and also considering that $e_{max} - e_{min}$ may be dependent on both D_{60} and D_{10} values. It may be more appropriate to estimate $e_{max} - e_{min}$ from commonly used values for soil types, such as presented in Cubrivoski and Ishihara (1999) and Mitchell and Soga (2012).

As numerical modeling approaches are increasingly being simplified, and parameter selection is increasingly being correlated with $(N_1)_{60}$-based relationships, a sensitivity study based on a range of possible D_R values for the same $(N_1)_{60}$ value and different C_d values may be appropriate for critical projects.

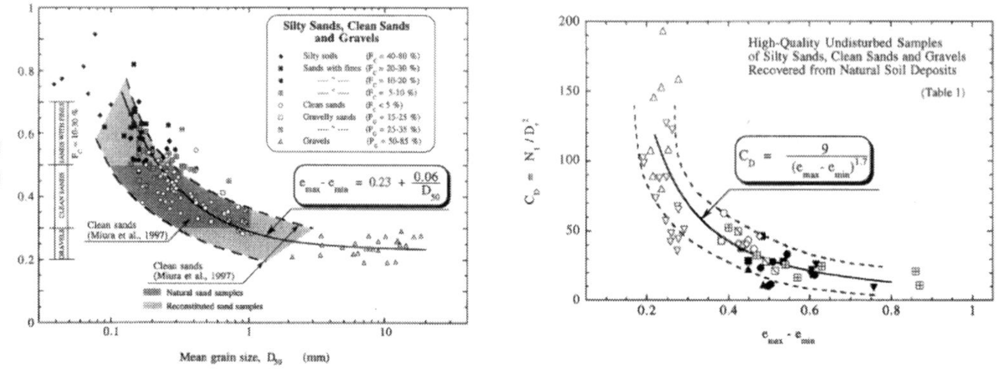

Figure 11. Plots of (a) e_{max} - e_{min} vs. D_{50} and (b) C_d vs. e_{max} - e_{min} from Cubrinovski and Ishihara (1999)

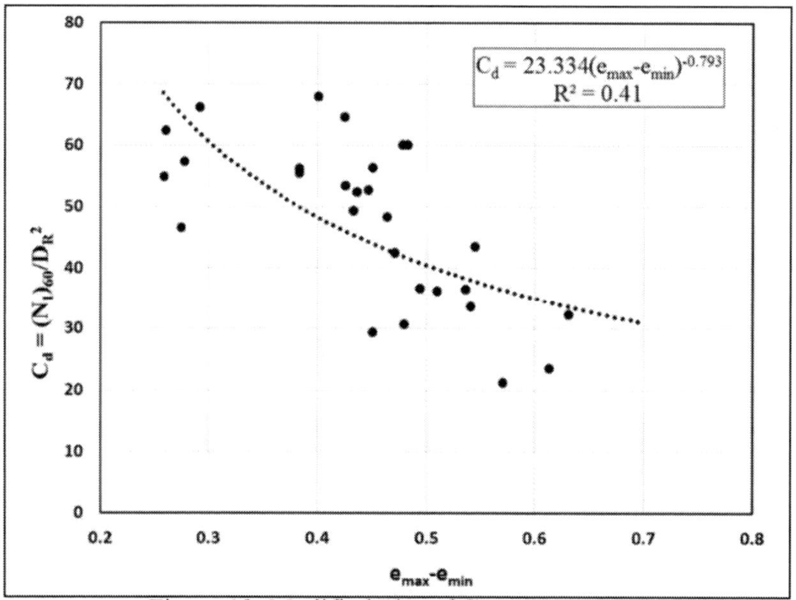

Figure 12. Modified Plot of C_d vs. $e_{max} - e_{min}$

Table 6 shows a comparison of C_d values for different ranges of void ratio: (1) as recommended by Cubrinovski and Ishihara (1999) based on 78% SPT energy ratio (ER), (2) modified C_d values based on 60% ER, and (3) the further modified relationship developed in this study for a more constrained range of soils. The resulting further modified C_d values show a better correlation with Skempton's (1986) recommendations, which may be due in part to the more constrained range of soil material types to which it is likely to apply.

Table 6. Max to Min Void Ratio Ranges and C_d Values

Void Ratio Range from Cubrinovski and Ishihara (1999)				C_d from Cubrinovski and Ishihara (1999) with ER = 78%	C_d from Cubrinovski and Ishihara (1999) with ER = 60%	C_d from modified data sub-set from Cubrinovski and Ishihara (1999) (current study)	C_d values from Skempton (1986)
Soil Type	Fines Content (%)	Gravel (%)	e_{max} - e_{min}				
Silty Soils	40-80	< 5	> 0.7	< 17	< 21	< 31	55 for Silty Sand 65 for Coarse Sand
Silty Sand	20-30	< 5	0.60 - 0.70	21 to 17	28 to 21	35 to 31	
Silty Sand	10-20	< 5	0.50 - 0.60	29 to 21	38 to 28	40 to 35	40 for Recent Fill 55 for Natural Soil
Silty Sand	5-10	< 5	0.45 - 0.55	35 to 25	45 to 32	44 to 37	
Clean Sand	< 5	< 5	0.30 - 0.50	70 to 29	91 to 38	61 to 40	
Gravelly Sand	< 5	15-35	0.30 - 0.40	70 to 43	91 to 56	61 to 48	

Table 7. Relative Density for USFD from Post-Earthquake Measurements

Layer	$e_{max} - e_{min}$ for coarse sand samples (Seed et al. 1976, Figure IV-7)	C_d using Equation 3 (this study)	Post-Earthquake $(N_1)_{60}$ from Cetin et al. (2016)	Relative Density, D_R (%) using C_d (Modified Cubrinovski and Ishihara, this study)	Measured and Estimated Relative Density, D_R (%) from Seed et al. (1973)
HFU-1	0.41 to 0.59	47.3 to 35.5	11.6	49.5 to 57.2 (average 53.4)	Direct Measurement D_R = 50 to 54 Percent
HFU-2			14.7	55.7 to 64.4 (average 60.1)	
UA			16.1	58.3 to 67.4 (average 62.9)	Estimated Values D_R = 45 to 65 Percent
HFD-1			11.9	50.1 to 57.9 (average 54.0)	
HFD-2			13.6	53.6 to 61.9 (average 57.8)	
HFD-3			14.6	55.5 to 64.2 (average 59.9)	

Relative density data were developed for soils at the USFD from post-earthquake laboratory and field studies (Seed et al. 1973). A comparison between directly measured D_R values with estimated SPT-based values from the modified relationship developed for this study (based on Cubrinovski and Ishihara (1999), but with a reduced dataset and a more limited range of applicability) is shown in Table 7. Based on post-earthquake in-situ testing and engineering evaluations, Seed et al. (1973) suggested that the relative density in the hydraulic fill typically ranges between 45 percent and 65 percent, which is comparable to the direct measurement values (50 to 54 percent). Average relative density values estimated based on the modified Equation 4 as described above vary between 53.4 percent and 62.9 percent, and these match well with the Seed et al. (1973) estimates and also with their actual measurements.

Hydraulic Conductivity and Pre-Earthquake Phreatic Surface

Hydraulic conductivity values and seepage analyses are important to establish the pre-earthquake steady-state seepage conditions, which are in turn important to characterize the baseline (pre-earthquake) static effective stress conditions. In this study, a range of hydraulic conductivity values were initially developed using the equation of Chapuis (2004), which is based on d_{10} and density values. These hydraulic conductivity values were then varied in a series of seepage analyses in FLAC to develop parameters that produce a good match with the pre-earthquake phreatic water surface elevations measured at three piezometers across a transverse cross-section. These piezometers were located approximately 20 feet, 45 feet, and 175 feet downstream of the crest centerline, and the piezometer levels were recorded as El. 1198, El. 1196, and El. 1194. Table 8 shows a summary of hydraulic conductivity values for USFD analyses in this study. There may be different combinations of hydraulic conductivity values that would result in phreatic surfaces that also match reasonably well with recorded data, however for this study the Table 8 values were considered appropriate considering that these provide a very good match with pre-earthquake piezometer readings, and they require relatively little modification from the initial values estimated based on the Chapuis (2004) relationship. Hydraulic conductivity values used by Beaty and Byrne (2011) are also presented for comparison.

Table 8. Hydraulic Conductivity Values for USFD Analyses

Layer	Beaty and Byrne (2011)				This Study			
	K_h cm/sec	K_v cm/sec	K_h/K_v	Nominal Porosity	K_h cm/sec	K_v cm/sec	K_h/K_v	Nominal Porosity
RF	1.0E-5	1.0E-5	1	0.5	4.0E-5	1.0E-5	4	0.5
CC	1.0E-5	1.0E-5	1	0.5	4.0E-5	1.0E-5	4	0.5
HFU (1 and 2)	1.0E-3	1.0E-4	10	0.5	1.0E-3	1.0E-4	10	0.5
HFD (1 and 2)	1.0E-3	1.0E-4	10	0.5	1.0E-3	1.0E-4	10	0.5
HFD-3	1.0E-3	1.0E-4	10	0.5	1.0E-3	1.0E-4	10	0.5
UA	5.0E-2	5.0E-3	10	0.5	4.0E-5	1.0E-5	4	0.5
LA	5.0E-2	5.0E-3	10	0.5	4.0E-5	1.0E-5	4	0.5

Strength and Stiffness Parameters

Strength and stiffness parameters for USFD were developed by Seed et al. (1976) based primarily on laboratory testing of samples collected after earthquake. These values have also been used by a number of different researchers in subsequent studies. In this current study, strength and stiffness parameters were re-evaluated. As laboratory testing in coarse-grained liquefiable soils is not routinely performed, strength and stiffness parameters were developed using commonly used relationships that are mainly either

$(N_1)_{60}$ - based or D_R - based. The following relationships were used to develop stiffness parameters.

Shear Modulus, $G_{max} = 21.7\ K_{2max}\ P_a\ (\sigma_m'/P_a)^{1/2}$ (Seed et al., 1986) [Equaation 6]

where $K_{2max} = 20\ N_{1,60,CS}^{0.33}$ (Seed et al. 1986) [Equation 7]

or $K_{2max} = 0.6\ D_R + 15$ (Byrne et al. 1987) [Equation 8]

or $K_{2max} = 625\frac{OCR^k}{0.3+0.7e^2}$ where k is a function of PI and stress history

(Hardin, 1978) [Equation 9]

Equation 7 was originally developed for clean sand conditions. In this study, $N_{1,60,CS}$ values were used, as this better incorporates effects of fines on blow counts and resulting shear modulus. Table 9 presents a summary of drained shear strength parameters and Table 10 presents dynamic stiffness parameters developed for each of the principal geological units of Cross-Section B-B′ for use in seismic deformation analyses performed using FLAC.

Table 9. Drained Shear Strength Parameters for USFD Analysis

Layer	Dry Unit Weight γ_d (pcf)	Moist (γ_m) and Saturated (γ_{sat}) Unit Weight (pcf)	Cohesion, c' (psf)	Friction Angle, ϕ' (degrees)
RF	110	128/134	200	36
CC	96	114/120	Su=0.25 σ'_{v0}	
HFU-1	102	120/126	0	37
HFU-2	102	120/126	0	37
HFD-1	102	120/126	0	37
HFD-2	102	120/126	0	37
HFD-3	102	120/126	0	37
UA	103	122/128	200	36
LA	108	124/130	100	37
ROCK	-	134/140	-	-

Table 10. Dynamic Stiffness Parameters or Indicative Indices for USFD Analysis

Layer	K_{2max} for Cetin et al. (2016) $N_{1,60,cs}$	Shear Modulus Number (K_{ge}) using $N_{1,60,cs}$ based equation (Seed and Idriss, 1990)	D_R (Modified Cubrinovski and Ishihara, 1999 equation) as used in this study)	Shear Modulus Number (K_{ge}) using D_R-based equation (Byrne et al., 1987)	e and OCR	Shear Modulus Number (K_{ge}) using e and OCR-based equation (Hardin, 1978)	Selected K_{ge} for Analysis
RF					e = 0.53 OCR = 1	1254	1254
CC					e = 0.76 OCR = 1 PI = 20	893	893
HFU-1	45.2	980	48.5	957			980
HFU-2	49.2	1072	55.8	1052			1072
UA	50.7	1101	58.8	1091			1101
HFD-1	46.3	1004	50.3	967			1004
HFD-2	48.1	1044	53.6	1020			1044
HFD-3	49.3	1070	55.6	1049			1070
LA			80.0	1367			1367
Note: A rounding of calculated values may be reasonable.							

Soil Liquefaction Triggering Relationships

In constitutive models of liquefied soils, each analytical model needs to be calibrated (or parametrized) to agree with a pre-selected soil liquefaction triggering relationship. These calibrations are generally performed using a single element direct simple shear (DSS) model with uniform stress or uniform strain conditions.

Three different relationships were employed in these current studies to estimate cyclic resistance ratio (CRR). These were all empirical (field performance-based) triggering relationships based on back-analyses of earthquake field performance liquefaction case histories from level ground sites. The three relationships employed were those of (1) Youd et al. (2001), (2) Boulanger and Idriss (2014) and (3) Cetin et al. (2016). Youd et al. (2001) adopted the Seed et al. (1985) triggering relationship, with an adjustment at $N_{1,60,CS}$ values of less than about 5 blows/foot. Figure 13 presents all three of these relationships in terms of liquefaction triggering curves normalized to field conditions corresponding to $\sigma_v' = 1$ atmosphere and initial static shear stress on a horizontal plane = zero. The back-analyzed field data upon which each of these relationships are based are shown by solid "dots" for cases in which soil liquefaction occurred (was triggered) and by open circles for sites at which soil liquefaction did not trigger. All three relationships, and their underlying data, have been re-plotted to the same vertical and horizontal scales for clarity of cross-comparisons. In this study, the P_L=50 % line, corresponding to a factor of safety against liquefaction of about 1.0 (Cetin et al., 2016) is used when the Cetin et al. (2016) relationship was employed. An equation approximately corresponding to the P_L = 20 percent boundary curve was used for the Idriss and Boulanger (2014) triggering relationship.

Each of these three triggering relationships require several secondary sub-relationships for application to field conditions. The first of these is an adjustment of CSR required to trigger liquefaction as a function of magnitude-correlated duration of shaking (or number of equivalent uniform loading cycles). This is applied as a magnitude-correlated scaling of liquefaction resistance using a scaling factor K_M. Fortunately, all three relationships employ similar sub-relationships here, and all three produce largely similar values of K_M for this case history (with M_w = 6.6) so that this is not a significant source of differences between the three triggering relationships for this case. The impact of these magnitude relationships in the time-domain FLAC analyses performed is incorporated through model calibration for triggering behavior (including cyclic pore pressure generation) and their interactions with ground motion effects.

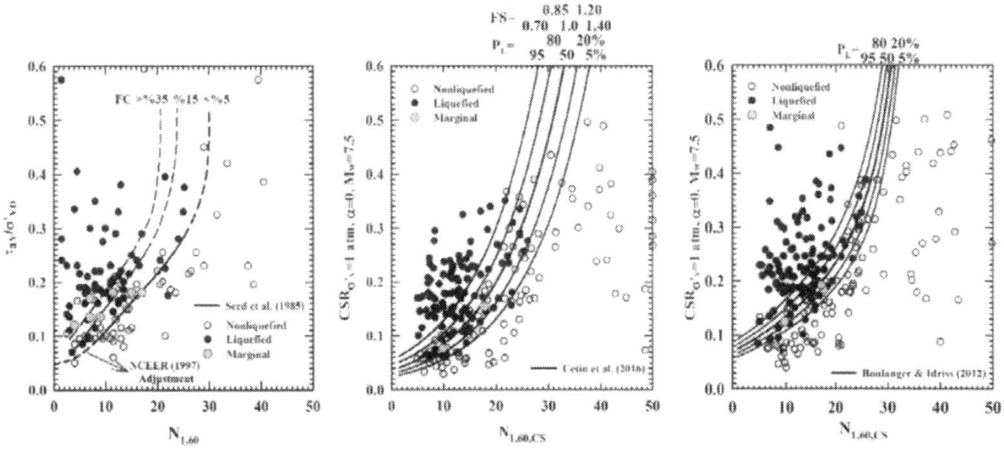

Figure 13. Liquefaction triggering relationships as proposed by (a) Seed et al. (1985) as modified slightly by Youd et al.(2001), (b) Cetin et al. (2016), and (c) Boulanger and Idriss (2012).

Each also requires an adjustment of liquefaction triggering resistance for effects of initial effective overburden stress. This is implemented with an adjustment factor (K_σ). Figure 14 (from Cetin et al. 2016) shows the available laboratory test data (based on undrained cyclic triaxial and cyclic simple shear laboratory testing) for development of K_σ. It also shows the proposed K_σ relationships of (1) Youd et al. (dark gray solid and dashed lines for $D_r \approx 60\%$ and 40%, respectively) and (2) Idriss and Boulanger (blue solid and dashed lines for $D_r \approx 60\%$ and 40%, respectively). Cetin et al. (2016) performed regressions of the large field case history database of liquefaction triggering performance field case histories and developed K_σ as represented by the red lines in Figure 14. These are valid only over the range of the field performance case histories, which were confined to relatively shallow depths due to uncertainties as to whether or not deeper soil strata liquefied. Cetin et al. (2016) concluded that the K_σ relationship of Youd et al. (2001) best matches the back-analyzed (and regressed) field performance data. The field performance case history data produce results that differ somewhat from the laboratory data, and it is field performance that engineers seek to analyze and model. Cetin et al. recommend the use of the K_σ relationship of Youd et al. up to initial vertical effective overburden stresses of approximately 2 atmospheres, which is the limit of the accuracy of the field case history-based regressed relationship. At higher effective stresses, they note that engineers should be able to select K_σ relationship behaviors based on their own judgment, but that these K_σ relationships should not be strongly refuted by the field case history-based regressed relationship for $\sigma'_v \leq 2$ atmospheres. Depth effects (K_σ) are an area where the various different triggering relationships currently available continue to differ, and with potentially significant consequences for large dams where effective overburden stresses can be high. Analyses performed in these current studies employ the K_σ of Youd et al. (2001) for both the Youd et al. (2001) and the Cetin (2016) triggering relationships, and they employ the K_σ relationship of Idriss and Boulanger (2010) for the Boulanger and Idriss (2012, 2014) triggering relationship.

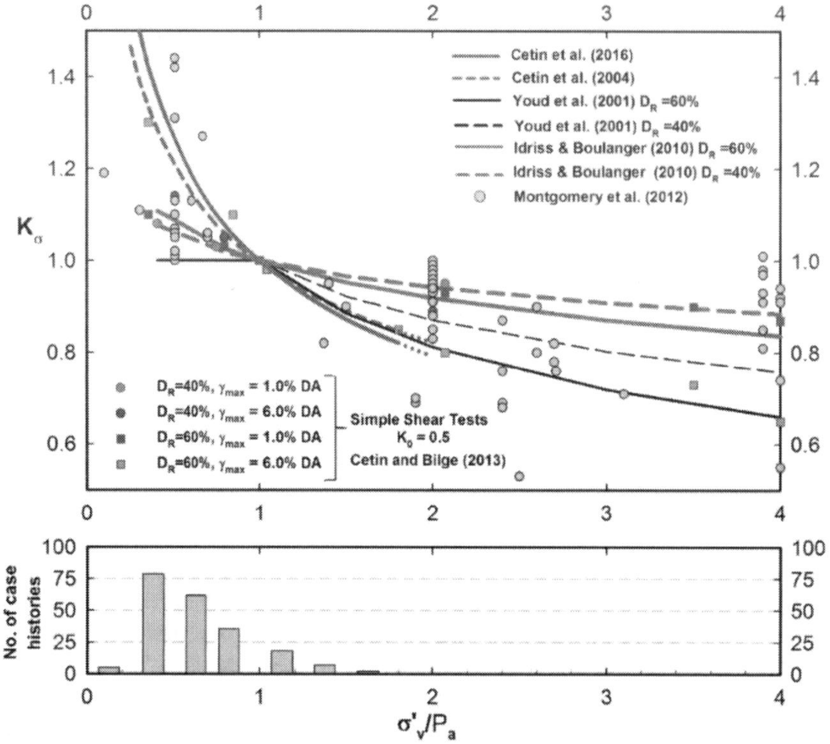

Figure 14. The recommended K_σ relationships of (1) Youd et al. (2001) as appended to the triggering relationship of Seed et al. (1985), (2) Boulanger and Idriss (2012), and (3) Cetin et al. (2016).

Post-Liquefaction Residual Strength

Post-liquefaction residual strength (S_r) is implemented differently by different analysts, and the choices here can also depend on the constitutive models and analysis approaches employed. In UBCSAND, a post-liquefaction residual strength is typically utilized at the end of strong ground motion shaking. Since the estimated residual strength in any element may be less than the strength mobilized at the end of shaking by the UBCSAND constitutive model, imposing the residual strength can lead to additional post-shaking deformations related to stress redistribution and/or instability. The estimated additional deformations are often small unless the embankment is marginally stable or fully unstable after the residual strengths are imposed.

The Roth model utilizes post-liquefaction residual strength at the onset of liquefaction in each individual element, and thus essentially "caps" the minimum strength (and the maximum strength) that can be mobilized with a value of post-liquefaction strength based on any one of the several S_r relationships selected for these current studies.

The post-liquefaction residual strength relationships of Seed and Harder (1990), Idriss and Boulanger (2008), and Weber et al. (2015) were employed in these current analyses, as shown in Table 1. These are all empirically based on back-analyses of liquefaction-induced slope failure field case histories, and these S_r relationships are illustrated in Figures 15, 16, and 17. The post-liquefaction residual strength relationship of Seed and Harder (1990) predicts S_r as a function of $N_{1,60,CS}$ and has no modification factors as a function of initial effective overburden stress. An approximately 50th percentile value within their recommended range (see Figure 14) was employed. The post-liquefaction strength relationships of Idriss and Boulanger (2008) and Weber et al. (2015) use a S_r/σ'_v relationships, and thus depend on effective stress (σ'_v) as well as on $N_{1,60,CS}$ values. In the relationship of Idriss and Boulanger (2008), for a given $N_{1,60,CS}$ value the ratio of S_r/P is fixed and does not vary as a function of initial effective stress. The lower relationship in Figure 16, which addresses soils with likely void redistribution effects, was employed in view of the layering or stratification of both the hydraulic fill units and the upper alluvium. The S_r relationship of Weber et al. (2015) is also a function of both $N_{1,60\text{-}CS}$ and also initial effective vertical stress (σ'_v), but the ratio of S_r/P varies systematically as a function of initial effective vertical stress. A pore pressure ratio of $r_u \geq 0.7$ was used for application of S_r values for Analyses 1, 2, and 3.

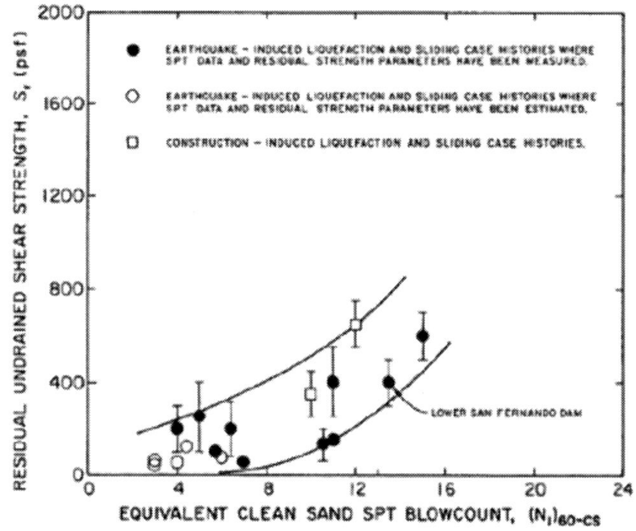

Figure 15. Variation of post-liquefaction residual strength, S_r as a function of fines adjusted SPT penetration resistance, $N_{1,60\text{-}cs}$. (Seed and Harder, 1990)

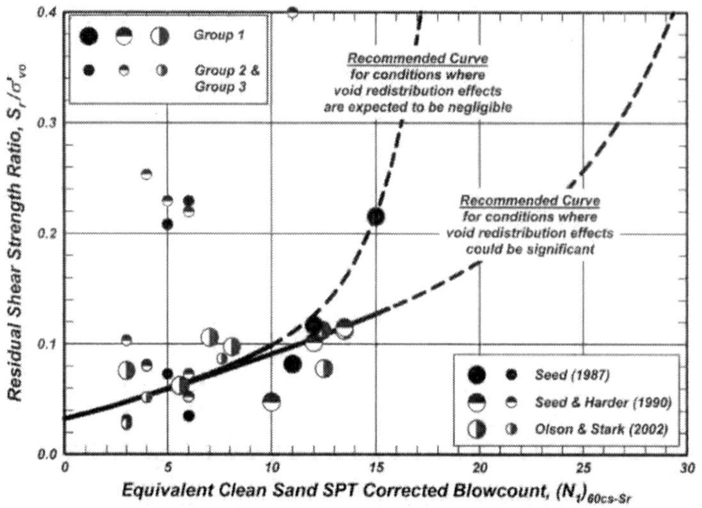

Figure 16. Variation of post-liquefaction residual strength, S_r as a function of fines adjusted SPT penetration resistance, $N_{1,60,CS}$ (Idriss and Boulanger, 2015)

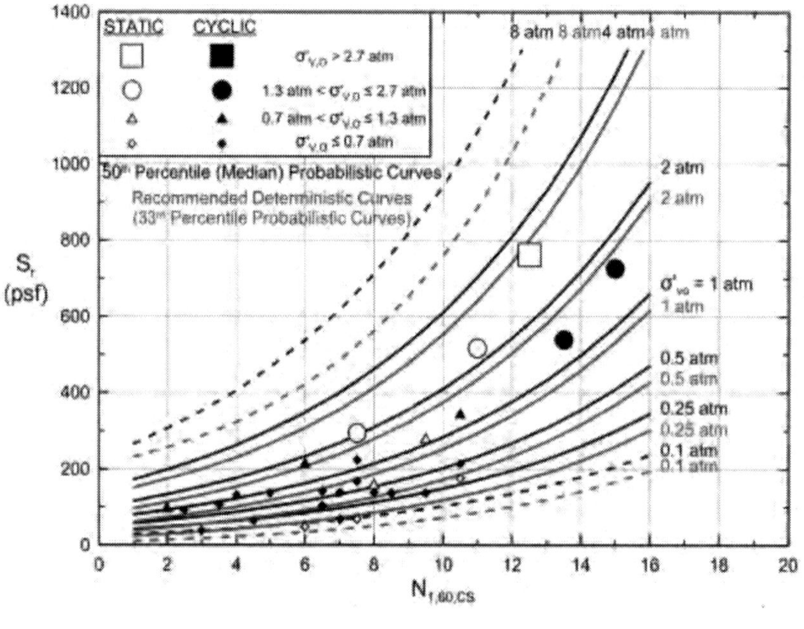

Figure 17. Graphical representation of 33rd percentile values of S_r (red lines) and 50th percentile values of S_r (black lines) from probabilistic relationships (Weber et al., 2015) [Sr/P was used]

Numerical instability can give rise to large surficial displacements somewhat analogous to surface sloughing in these types of dynamic analyses. These are fictitious movements. A common approach to mitigate this is to assign a minor amount of "minimum" shear strength that applies to surface and near-surface elements. This is often applied as a minimum S_r. The minimum value of S_r applied here should be selected (1) such that it is barely large enough to prevent surficial (and near surface) elements from experiencing excessive deformations during shaking, but (2) small enough that it does not provide potentially significant strength such that the surface and near surface elements form a strong skin or buttress that can suppress potential deformations otherwise driven by the main mass of the embankment. This can be checked by varying the minimum S_r value employed and checking for these effects. A minimum S_r value of 75 lbs/ft^2 was selected for this study considering blow counts and different S_r relationships, and checks were made to ensure that this did not adversely impact the engineering results of the analyses performed.

UBCSAND MODEL APPROACH AND INPUT PARAMETERS

UBCSAND (Beaty, 2001, Beaty and Byrne, 2011) is a widely used soil constitutive model for dynamic analyses of liquefiable soils. The most recent version of UBCSAND is 904aR, which was developed by modifying the original version 904a during work on Success Dam for the USACE in California (Beaty and Byrne, 2011). It is this most recent version that was used in these current studies.

UBCSAND is a two-dimensional effective stress plasticity model for use in analyzing geotechnical structures such as earthen dams. The model predicts the shear stress-strain behavior of the soil using an assumed hyperbolic relationship, and estimates the associated volumetric response (and pore pressure response) of the soil skeleton using a flow rule that is a function of the current stress ratio, η. The model can be used in a fully coupled fashion where the mechanical and groundwater flow calculations can be performed simultaneously.

In UBCSAND 904aR documentation (Beaty and Byrne, 2011), a set of input parameters have been developed to represent the response of a hypothetical generic sand. These parameters provide reasonable estimates of stiffness and capture the liquefaction response in terms of the cyclic resistance ratio (CRR) as per the liquefaction triggering relationship of Youd et al. (2001), the effect of overburden stress on liquefaction as captured by the K_σ factor of Youd et al. (2001), and interpretation of the effects of initial static shear stress bias (K_α) as per Idriss and Boulanger (2003).

Analysis Case 1 was performed using recommended relationships and parameters as per UBCSAND904aR. Analysis Case 2 is similar to Analysis Case 1, except that the post-liquefaction strength relationship of Weber et al. (2015) was implemented. Analysis Case 3 was again similar, except that the post-liquefaction strength relationship of Idriss and Boulanger (2008/2015) was implemented. Project-specific stiffness parameters, as shown on Table 10, were used for Analysis Cases 1, 2 and 3.

ROTH/DAMES & MOORE/URS MODEL APPROACH AND INPUT PARAMETERS

The Roth model (also known as the Dames and Moore/URS model) is a practice-oriented model initially developed by Wolfgang Roth at Dames and Moore and later modified by Ethan Dawson and Lelio Mejia (1999) of URS. The Roth model is a Mohr-Coulomb (linearly elastic and perfectly plastic) soil model coupled with an empirical pore pressure generation scheme. Analyses using this simplified procedure also incorporated K_σ and K_α factors based on Youd et al. (2001).

Figure 18 schematically presents the cyclic pore-pressure generation scheme. In this scheme, the shear stress time history of each element is monitored and shear stress cycles are counted. The model monitors the shear stress in each element on horizontal planes σ_{xy} rather than a shear stress invariant. As soon as a stress cycle is detected the excess pore pressure is incremented by an amount dependent on the cyclic stress ratio amplitude of that cycle. The generated excess pore pressure u_e is described in terms of the pore pressure ratio:

$$r_u = u_e/\sigma'_v \qquad \text{[Equation 10]}$$

Half cycles are detected from shear stress reversals. The cyclic stress τ_{cy} amplitude of each half cycle is half the difference between the preceding peak and valley. The increment of pore pressure ratio Δr_u^i for a half cycle is calculated the number of uniform cycles that are required for complete liquefaction triggering ($r_u = 1.0$) at a cyclic stress ratio, CSR_i, using the following relationship

$$\Delta r_u^i = 0.5/N_L^i \qquad \text{[Equation 11]}$$

and the increment in pore pressure u_e is then

$$\Delta u_e^i = \Delta r_u^i \, \sigma'_v \qquad \text{[Equation 12]}$$

The increasing pore pressure is accompanied by decrease in effective stress and a resulting decrease in shear modulus using the following relationship

$$G = G_{initial}\,(1 - r_u)^{0.5} \qquad \text{[Equation 13]}$$

The model incorporates the appropriate post-liquefaction residual strength (S_r) as the minimum strength value, as shown in Figure 18, in a two-segment failure envelope.

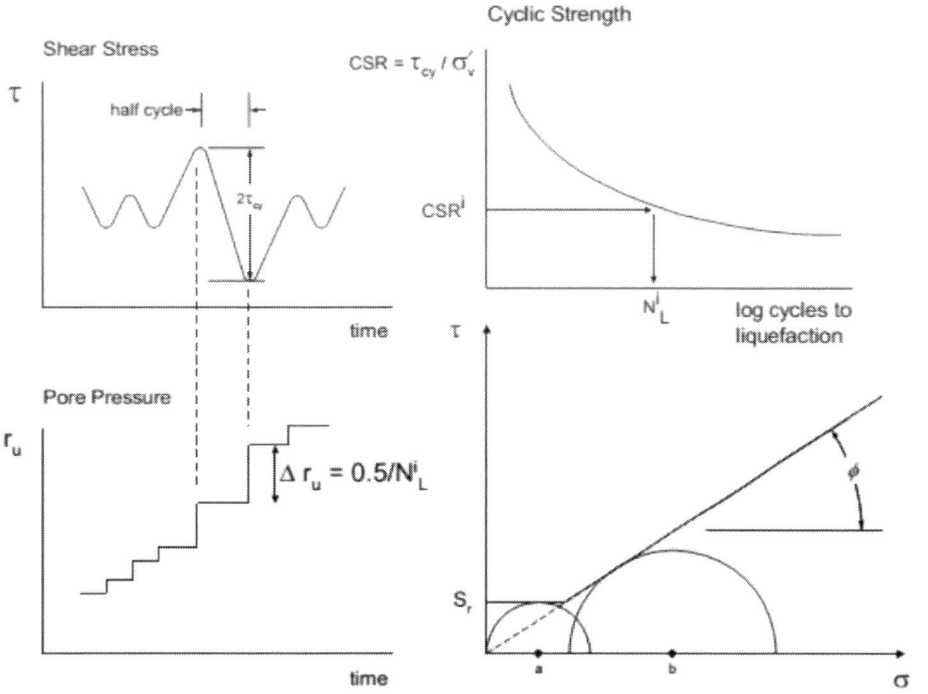

Figure 18. Pore pressure generation procedure, cyclic strength curve, and shear strength for the Roth/Dames and Moore/URS Model

The main input parameters for Roth/Dames and Moore/URS model include (1) CRR_{15}, the cyclic stress ratio required for liquefaction in 15 cycles for a causative earthquake magnitude of M_w=7.5 and at an initial effective vertical overburden stress of 1 atmosphere, and (2) a B factor, derived from the Magnitude Scaling Factor (MSF) or Correction Factor for Magnitude Duration (K_{MW}). The CRR_{15} is used in the program using the following relationship:

$$CSR = CRR_{15} (N_L/15)^{-1/B} \quad \text{[Equation 14]}$$

Table 11 shows the CRR_{15} factors employed in these analyses, and the B factor values for Analysis Cases 3, 4, and 5 are shown below:

B = 2.97 for Analysis Case 4 [Youd et al. (2001)] and 6 [Idriss and Boulanger (2014)], based on Idriss (1999)

B = 2.324 from Cetin et al. (2016) for Analysis Case 5

Table 11. CRR_{15} for Roth/URS Model

Soil Layer	Youd et al. (2001) [used in Analysis Case 4]		Boulanger and Idriss (2012) [used in Analysis Case 6]		Cetin et al. (2016) [used in Analysis Case 5]	
	Average Pre-Earthquake $N_{1,60,CS}$ for Triggering	CRR_{15}	Average Pre-Earthquake $N_{1,60,CS}$ for Triggering	CRR_{15}	Average Pre-Earthquake $N_{1,60,CS}$ for Triggering	CRR_{15}
HFU-1 (EL.1200-1170)	13.8	0.148	13.4	0.143	11.8	0.104
HFU-2 (EL. 1200-1170)	18.6	0.199	17.9	0.182	15.5	0.143
UA (EL.1160-1145)	19.7	0.212	19.3	0.198	16.8	0.159
HFD-1 (EL.1200-1170)	16.0	0.171	15.2	0.158	12.7	0.112
HFD-2 (EL.1170-1160)	17.5	0.187	16.6	0.170	14.3	0.129
HFD-3 (EL.1160-1145)	18.6	0.198	17.8	0.182	15.4	0.141

COMPARISON OF DEFORMATION ANALYSIS RESULTS WITH OBSERVED FIELD PERFORMANCE

Analyses 1, 2, and 3

The first three sets of analyses presented here are Analysis 1, Analysis 2, and Analysis 3. As shown previously in Table 1, these analyses were performed in FLAC using the UBCSAND model for potentially liquefiable layers and a Mohr-Coulomb model for non-liquefiable layers. The analytical model was parameterized to match the liquefaction triggering relationship of Youd et al. (2001) in these three analyses. In Analysis 1 the post-liquefaction strength (S_r) modeled was based on the median values of the Seed and

Harder (1990) S_r relationship, in Analysis 2 the 50th percentile S_r relationship of Weber et al. (2015) was modeled, and in Analysis 3 the recommended S_r/P curve of Idriss and Boulanger (2015) for significant void redistribution curve was used.

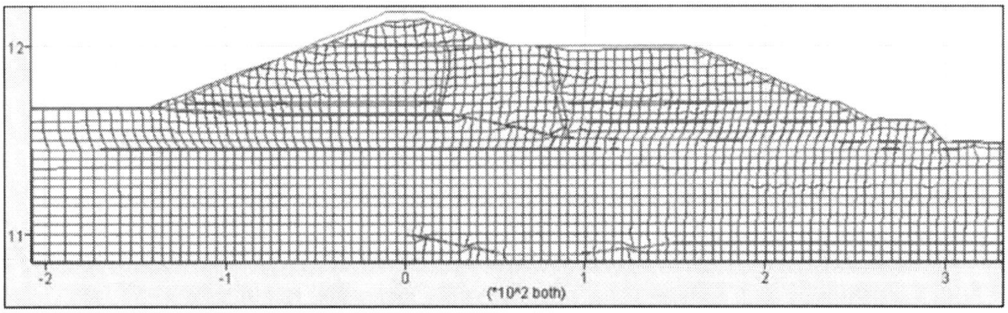

Figure 19. Final deformed shape from Analysis 1 [UBCSAND with Youd et al. (2001) triggering relationship and Seed and Harder (1990) S_r]

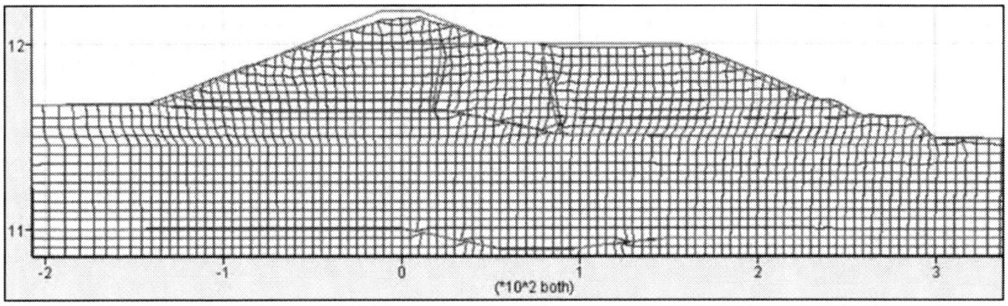

Figure 20. Final deformed shape from Analysis 2 [UBCSAND with Youd et al. (2001) triggering relationship and Weber et al. (2015) S_r]

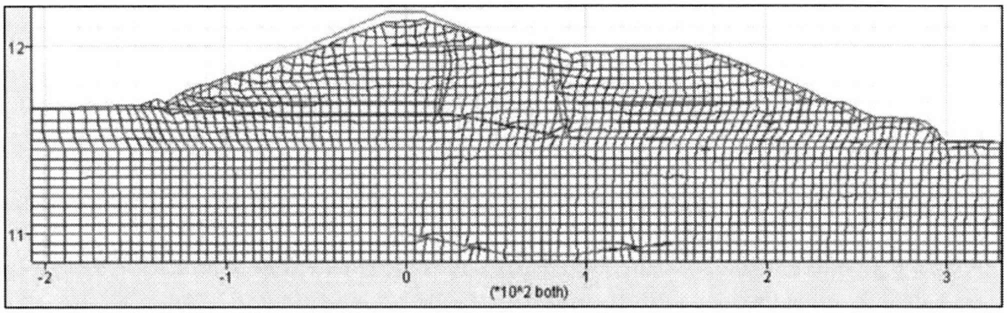

Figure 21. Final deformed shape from Analysis 3 [UBCSAND with Youd et al. (2001) triggering relationship and Idriss and Boulanger (2008/2015) S_r]

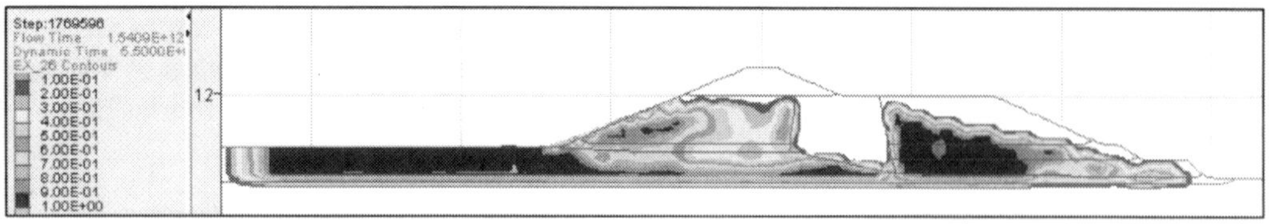

Figure 22. Pore pressure ratio (r_u) plot from Analyses 1, 2, and 3 at the end of shaking.

Figures 19, 20 and 21 show the final displaced meshes at the end of Analyses 1, 2, and 3, respectively. All three analyses showed moderate crest settlement and moderate lateral displacements towards the downstream side. Upon close examination, it can also be seen that moderate lateral displacements towards the upstream side also occurred near the upstream toe.

Figure 21 presents contours of cyclic pore pressure development during Analyses 1, 2 and 3. Because the same liquefaction triggering relationship was modeled, and because S_r does not affect cyclic pore pressure development in the UBCSAND analysis framework employed here, the pore pressure development contours are identical for Analyses 1, 2 and 3.

Figures 22 and 23 show summaries of calculated displacements of the slope faces along the full cross-section. Figure 22 summarizes vertical displacements, and Figure 23 summarizes horizontal displacements. In both figures, the calculated displacements are compared to the measured and observed displacements. The red squares in these figures show the measured displacements of the survey monuments shown previously in Figure 2, and the blue triangles show additional displacements scaled from Cross-Section B-B′ of Figure 6.

The displacements shown in Figures 22 and 23 are presented as both "end of shaking" displacements, and also "end of analysis" displacements. After strong shaking had been completed, zones with high pore pressure ratios ($r_u \geq 0.7$) were next assigned post-liquefaction shear strengths and the model was allowed to slump further. In this conservatively enveloped post-liquefaction strength modeling scenario, the embankment and foundation were statically stable on both the upstream and downstream sides for all three analysis cases.

The displacements and deformations of Analyses 1 and 2 as evidenced in Figures 19 and 20, and as summarized in Figures 22 and 23, are almost entirely the result of cyclic pore pressure generation and softening, coupled with the cyclic inertial loading of the earthquake. Most of these deformations occur during application of strong shaking, and the additional (post-shaking) deformations due to slumping that occurred when post-liquefaction strength S_r is applied are essentially negligible. Only very limited additional slumping occurred on the downstream side after cessation of strong shaking in both Analyses 1 and 2. This can be most clearly seen in Figures 22 and 23, where little change occurred from the "end of shaking" displacements (lavender symbols) to the final displaced condition.

These displacements agree reasonably well with the observed behavior of the embankment. Calculated (and observed) displacements were "moderate", and indicated successful performance of the dam with regard to safe retention of the reservoir. The magnitudes of both calculated vertical displacements and calculated horizontal displacements downstream of the crest are generally in good agreement with observed performance.

The magnitudes of calculated horizontal and vertical displacements upstream of the crest were not surveyed, nor were they well documented, so actual performance was poorly recorded or poorly characterized on the upstream side of the dam away from the crest region. All three analyses "predict" small to moderate lateral translation towards the upstream toe, accompanied by some associated vertical displacements (or settlement), except at the toe where minor (upwards) toe bulging is predicted (as with the downstream toe region).

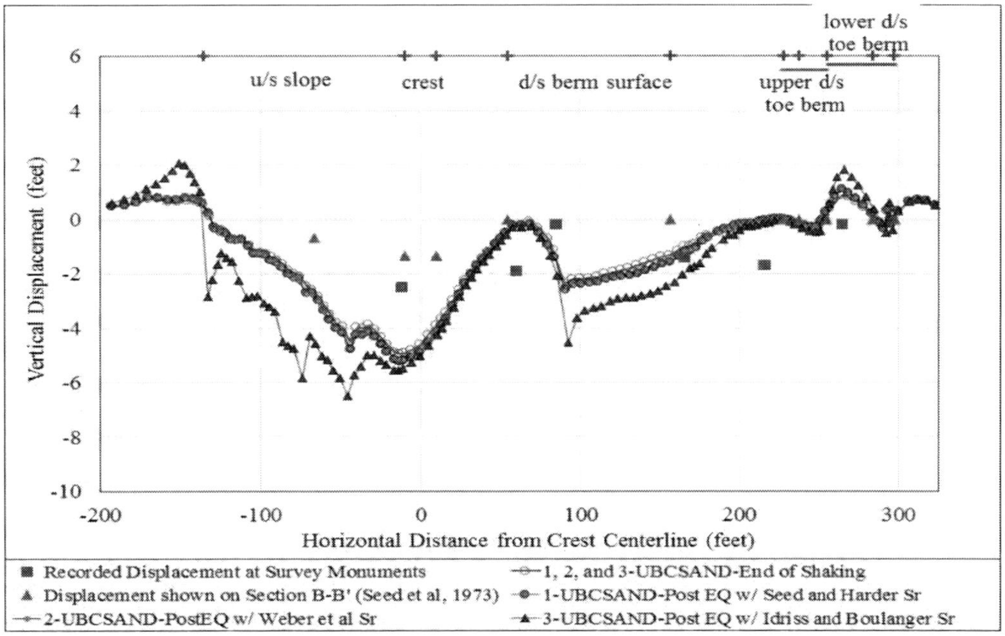

Figure 22. Comparison of vertical displacement from Analyses 1 and 2 with observed displacements

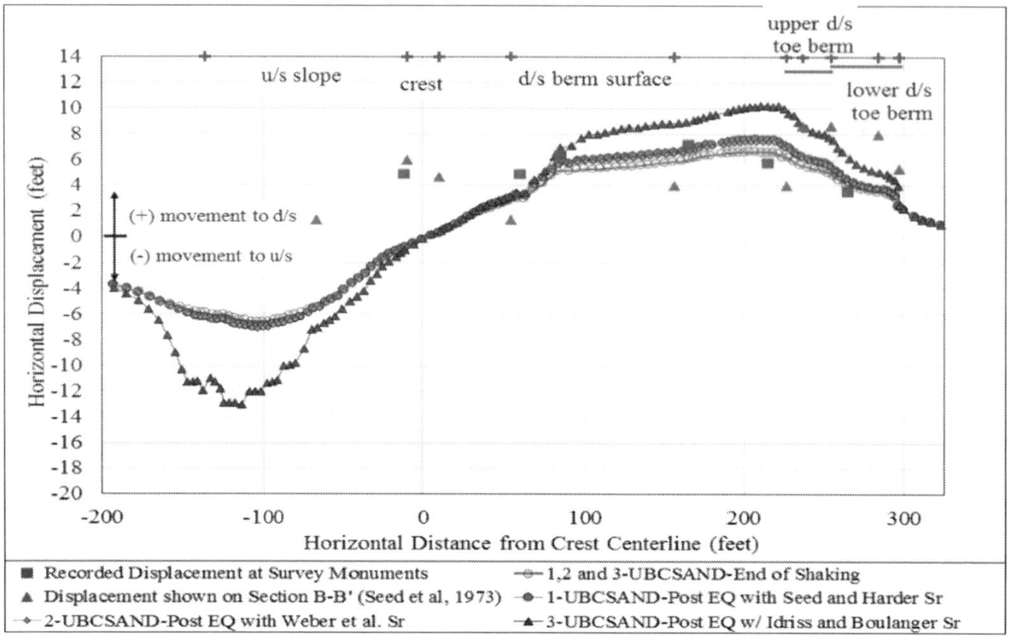

Figure 23. Comparison of horizontal displacement from Analyses 1 and 2 with observed displacements

On balance, all three analyses appear to provide useful engineering assessments of the expected seismic performance of the dam for this seismic loading.

Closer inspection of the results shows that all three analyses fail to predict the observed 4.5 to 6 feet of lateral translation of the crest towards the downstream direction, and this displacement was not well captured in these analyses. This may not be a short-coming of either this type of analysis, nor of the analytical models employed. Instead, this may represent an overall downstream "lurching" of the embankment, such as might be caused by one or more large velocity pulses. Evidence for this might include (1) the fact that the analyses performed here tended to largely capture the observed crest settlements, but without any uni-directional overall embankment translation, and (2) no previous nonlinear seismic deformation analyses have correctly captured this lateral crest displacement either. This is not conclusive, but it does call into question the scaling and appropriateness of the "modified" input motion employed in the analyses presented here; and this is one of the reasons for performing additional seismological modeling studies of the details of the actual fault rupture during the 1971 San Fernando, including detailed modeling of the near-field fault sources, to ascertain whether the input motion might warrant revisiting.

As discussed earlier in this paper, Seed (1973) had reduced the size of the large pulse in the ground motion recorded nearby at the abutment of Pacoima dam, before then further down-scaling the motion to be generally compatible with observed attenuation in this region during the 1971 San Fernando Earthquake. That was done at a time when near-field directivity effects were poorly understood. With our modern understanding of near-field directivity effects, the large "pulse" recorded appears to have been potentially reasonable, and it may be that the input record should be re-processed and re-scaled. These overall studies are ongoing, and the seismological re-evaluation of the input motion is a part of that ongoing effort.

Figures 22 and 23 also show a difference between the results of Analyses 1 and 2 vs. those of Analysis 3 with regard to the magnitudes of final deformation and displacement that is difficult to detect in Figures 19, 20 and 21. Analysis 3 produced larger final displacements than the other two analyses, and larger final displacements than were observed. These are still "moderate" displacements, and they would likely represent useful engineering predictions in forward analyses for evaluation or design purposes. As shown most clearly in Figures 22 and 23, these larger final displacements the result of "post-shaking" assignment of S_r, and resulting additional slumping, as all three analyses produced identical displacements "during shaking". Application of S_r in Analyses 1 and 2 produced essentially negligible additional post-shaking slumping displacements, but the S_r relationship employed in Analysis 3 produced somewhat larger post-shaking displacements.

Analyses 4, 5, and 6

Analyses 4, 5 and 6 were performed with the Roth/URS constitutive model for liquefiable soils. As shown in Table 1, Analysis 4 employed (1) the liquefaction triggering relationship of Youd et al. (2001), (2) the K_σ relationship recommended by Youd et al. (2001), and (3) the post-liquefaction residual strength (S_r) relationship of Seed and Harder (1990). Analysis 5 employed (1) the triggering relationship of Cetin et al. (2016), (2) the K_σ relationship of Youd et al. (2001) as recommended by Cetin et al. (2016), and (3) the post-liquefaction S_r relationship of Weber et al. (2015). Analysis 6 employed (1) the triggering relationship of Idriss and Boulanger a (2008), (2) the K_σ relationship of Idriss and Boulanger (2008) and (3) the post-liquefaction S_r relationship of Idriss and Boulanger (2008/2015). All these models used Idriss and Boulanger (2008) K_α relationship.

Figures 24 and 25 show the results of Analysis 4 in terms of (1) final deformed mesh and (2) pore pressure ratios generated by the seismic loading. Figures 26 and 27 present the same sets of results for Analysis 5, and Figures 28 and 29 present the same sets of results for Analysis 6. Figures 30 and 31 then present summary plots comparing (1) calculated vs. measured/observed vertical displacements for all three analyses, and (2) calculated vs. measured/observed horizontal displacements for all three analyses.

Figures 27, 29 and 31 show that all three analyses predicted generation of high cyclically-induced pore pressures near the bases of both the upstream and downstream hydraulic fill zones. Analysis 5 (employing the triggering relationship of Cetin et al. (2016)) produced the highest levels of cyclic pore pressure generation, Analysis 6 (employing the triggering relationship of Idriss and Boulanger (2008/2014)) produced the lowest levels of cyclic pore pressure generation, and Analysis 4 (employing the triggering relationship of Youd et al. (2001)) produced levels of cyclic pore pressure generation intermediate between those of the other two analyses. All three analyses produced extensive cyclic pore pressure generation, and the resulting softening and/or strength reductions (coupled with inertial shaking) produced moderate deformations and displacements in all three analyses.

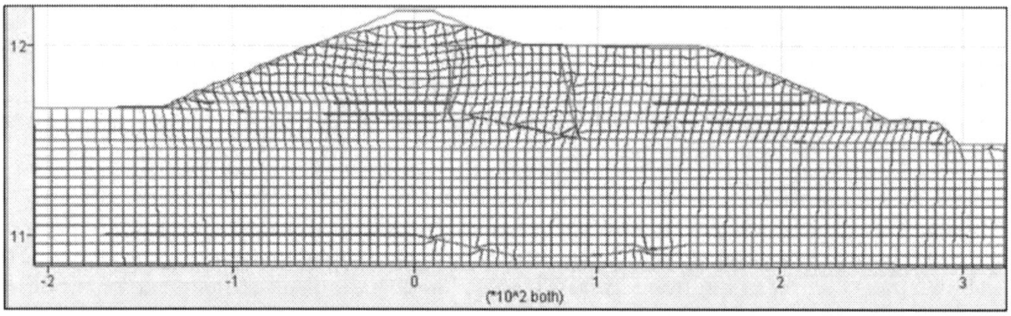

Figure 24. Deformed shape from Analysis 4 [Roth/URS with Youd et al. (2001) liquefaction triggering and Seed and Harder (1990) Sr]

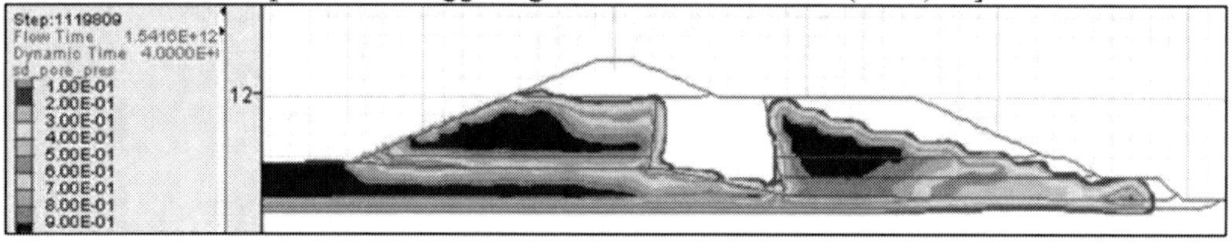

Figure 25. Pore pressure ratio (r_u) plot from Analysis 4 [Roth/URS with Youd et al. (2001) liquefaction triggering and Seed and Harder (1990) Sr]

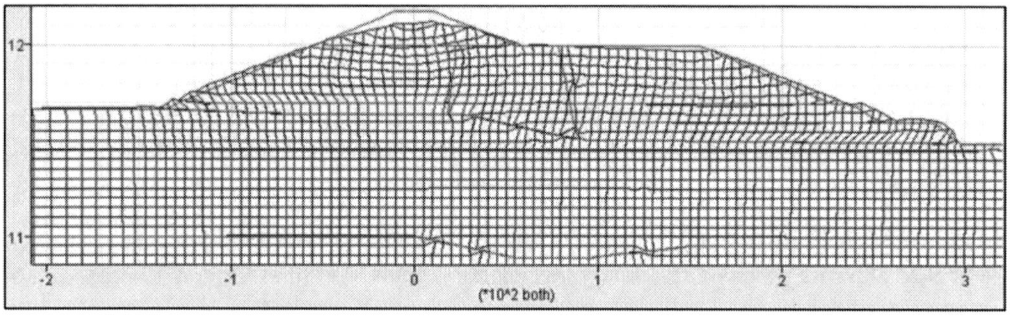

Figure 26. Deformed shape from Analysis 5 [Roth/URS with Cetin et al. (2016) liquefaction triggering and Weber et al. (2015) S_r]

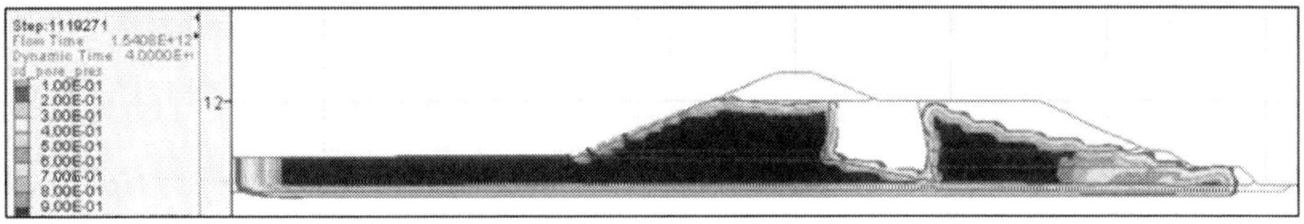

Figure 27. Pore pressure ratio (r_u) plot from Analysis 5 [Roth/URS with Cetin et al. (2016) liquefaction triggering and Weber et al. (2015) S_r]

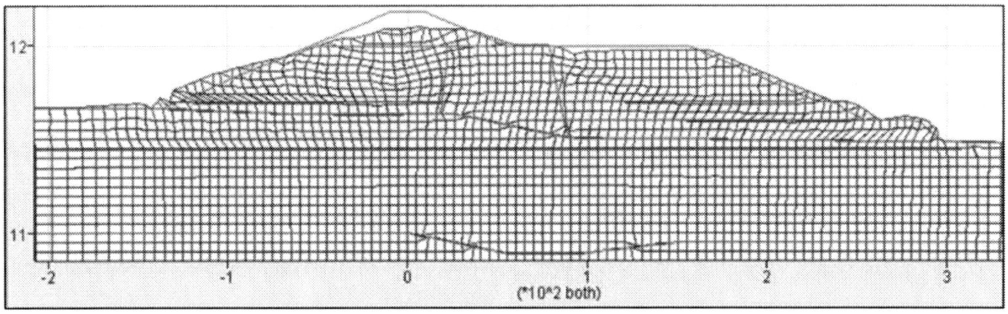

Figure 28. Deformed shape from Analysis 6 [Idriss and Boulanger et al. (2008) liquefaction triggering and S_r]

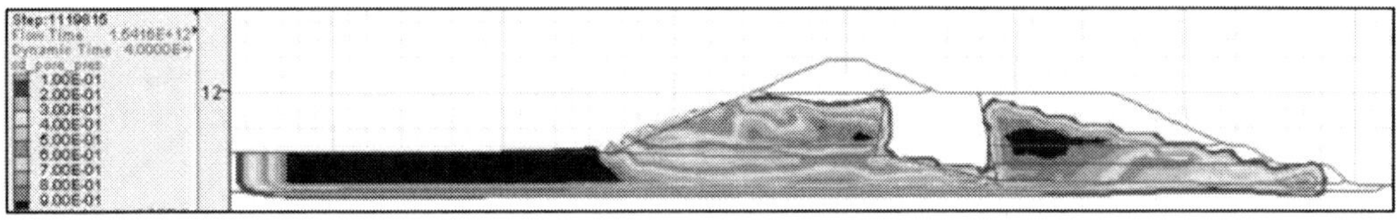

Figure 29. Pore pressure ratio (r_u) plot from Analysis 6 [Idriss and Boulanger et al. (2008) liquefaction triggering and S_r]

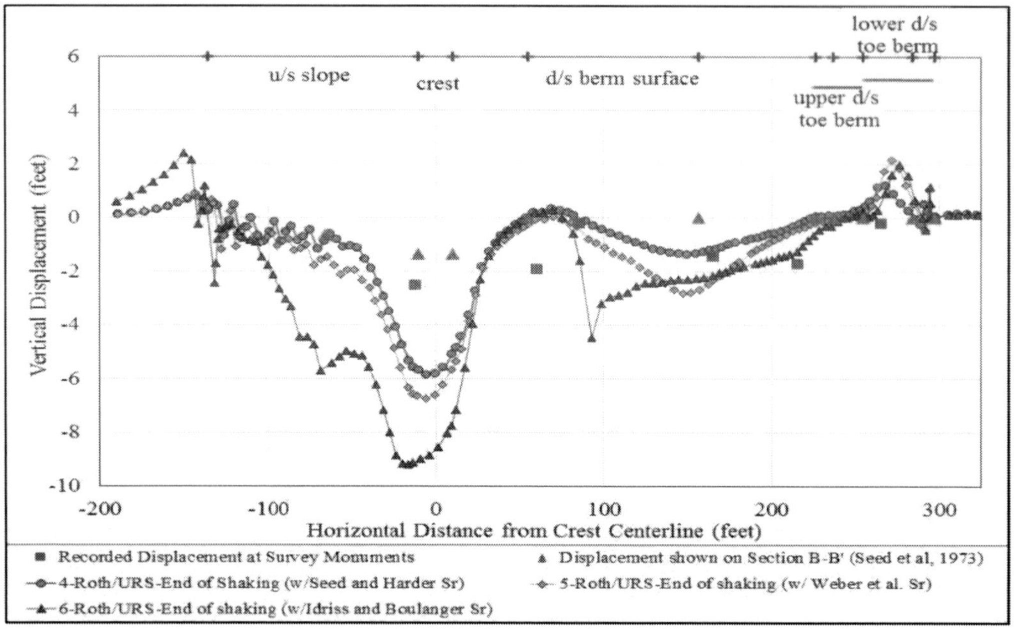

Figure 30. Comparison of vertical displacement from Analyses 3, 4, and 5 with observed displacements

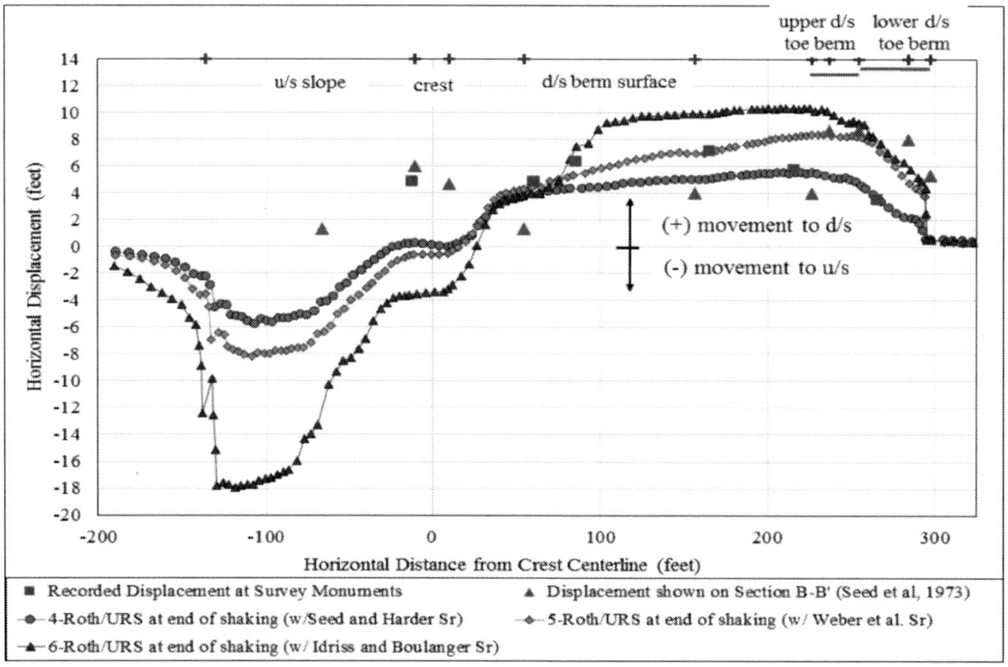

Figure 31. Comparison of horizontal displacement from Analyses 4, 5, and 6 with observed displacements

The magnitudes of the resulting calculated displacements were not necessarily correlated only with cyclic pore pressure generation. As shown in Figures 24, 26 and 28, and more clearly in the summary plots of Figures 30 and 31, Analysis 6 produced the largest predicted deformations and displacements despite having produced the lowest levels of cyclic pore pressure generation. The other two analyses predicted noticeably higher levels of cyclic pore pressure generation, especially Analysis 5, but both Analyses 4 and 5 produced lower levels of seismically-induced displacements.

The explanation for this appears to lie in the post-liquefaction strengths used. The post-liquefaction strength employed in Analysis 5 results in nearly marginal static stability on the upstream side when combined with the cyclically-generated pore pressure fields of Figure 29.

Examining Figures 30 and 31, it can be seen that calculated vertical displacements generally matched well with observed/measured displacements on the downstream side of the dam, but that all three of the analyses again over-predicted settlements at the crest, and all three again failed to capture the approximately 4.5 to 6 feet of lateral translation towards the downstream direction that occurred at the crest. Analyses 4 and 5 appear to represent useful engineering predictions of performance for purposes of either evaluation or design, and Analysis 6 provides a somewhat conservative prediction of observed performance for this case history.

SUMMARY

The performance of the Upper San Fernando Dam has long been an important case history with regard to evaluating the accuracy and reliability of modern seismic analysis methods when applied to liquefaction-related assessments of predicted performance for dams with likely moderate displacements and deformations. In performing back-analyses of such case histories, it is important to maintain a broad perspective and to examine multiple potential causes of observed phenomena; both with regard to observed field performance, and also with regard to "observed" analytical prediction results. The following is a summary of important observations that can be made based on the suite of analyses presented herein:

- It is important to correctly assess both liquefaction triggering (and cyclic pore pressure generation) as well as post-liquefaction strength (S_r) for evaluating expected seismic performance of a dam.
- The analyses have shown the importance of performing a post-liquefaction stability (and deformation) evaluation. In practice, this is sometimes omitted when deformations and displacements "during shaking" are only small to moderate.
- The USFD is a well-known case history that has been back-analyzed multiple times over the past four decades. Over that time span it has been generally assumed that the successful performance of this dam during the 1971 San Fernando Earthquake was primarily due to adequately minimal cyclic pore pressure generation. Now it appears at least possible that this successful performance might instead be due to adequate post-liquefaction strength (and stability) on the downstream side of the dam.

- None of the analyses presented here, and none of the analyses previously performed, correctly captured the approximately 4.5 to 6 feet of lateral translation of the dam crest toward the downstream direction. This suggests a need to re-evaluate the likely appropriateness of the "modified" input motion employed in these analyses performed to date.

Overall, it is noted that a majority of the suites of analyses presented herein would be judged to represent a useful basis for engineering assessment of the expected seismic performance of this dam in an earthquake of this intensity. Also that the details clearly matter. It is important to get the right answer for the correct reasons. It is also important not to make significant unconservative errors, or potentially financially costly over-conservative errors; and so developing as full an understanding as possible of the principle mechanisms expected to influence field performance and the abilities of any analytical framework (and models and relationships) to suitably capture those behaviors is important.

These studies are presented as work in progress, and they are still ongoing. Further efforts will include (1) re-evaluation of the suitability of the widely used input strong motion employed in the analyses presented in this paper, and (2) employment of at least one additional nonlinear constitutive model and analysis approach.

ACKNOWLEDGEMENTS

The data evaluation and analyses were performed by the primary author as part of his Ph.D. studies at the University of California Berkeley with valuable input the co-authors. The authors are grateful to a number of experts who have graciously been willing to discuss this ongoing work and to offer guidance and advice. These include a number of additional experts on seismic deformation analyses of dams, and also a number of the lead authors' colleagues at the U.S. Army Corps of Engineers.

REFERENCES

Beaty, M.H. (2001). A Synthesized Approach for Estimating Liquefaction-Induced Displacements of Geotechical Structures. A thesis submitted in partial fulfillment of the requirements for the degree of Doctor of Philosophy, The University of British Columbia (Vancouver), Canada.

Beaty, M.H. and Byrne, P.M. (2011). Documentation Report: UBCSAND Constitutive Model on Itasca UDM Website. February. http://www.itasc-udm.com/

Boulanger, R.W. and Idriss, I.M. (2014). CPT and SPT Based Liquefaction Triggering Procedures. Report No. UCD/CGM-14/01, Center for Geotechnical Modeling, Department of Civil and Environmental Engineering, College of Engineering, University of California, Davis.

Byrne, P.M., Cheung, H., and Yan, L. (1987). Soil Parameters for Deformation Analysis of Soil Masses. Canadian Geotechical Journal.

Byrne, P.M., Park, S.S., Beaty, M., Sharp, M.K., Gonzalez, L., and Abdoun, T. (2004) Numerical Modeling of Liquefaction and Comparison with Centrifuge Tests". Canadian Geotechical Journal, Vol. 41(2): 193-211.

Cetin, K. O., Seed, R. B., Kayen R. E., Moss, R. E. S., Bilge H. T., Ilgac M., and Chowdhury, K. (2016), "Summary of SPT-Based Field Case History Data of The Updated 2016 Database", Report No: METU / GTENG 08/16-01, Middle East Technical University, Soil Mechanics and Foundation Engineering Research Center.

Cetin, K.O., Seed, R.B., Kiureghian, A. D., Tokimatsu, K., Harder, L.F., Kayen, R. E., and Moss, R.E.S. (2004). Standard Penetration Test-Based Probabilistic and Deterministic Assessment of Seismic Soil Liquefaction Potential. Journal of Geotechnical and Geoenvironmental Engineering, ASCE.

Chapuis, R. (2004). Predicting the saturated hydraulic conductivity of sand and gravel using effective diameter and void ratio. Canadian Geotechnical Journal, 41:787-795 (2004). doi: 10.1139/T04-022.

Cubrinovski, M. and Ishihara, K. (1999). Empirical Correlation Between SPT-N Value and Relative Density for Sandy Soils. Soils and Foundations, Japanese Geotechical Society, Volume 39, No. 5, pp 61-71, October.

Cubrinovski, M. and Ishihara, K. (2001). Correlation between penetration resistance and relative density of sandy soils. Proceedings of International Society for Soil Mechanics and Geotechnical Engineering International Conference, Istanbul, Turkey.

Dawson, E.M. and Mejia, L.H. (2012). Updates to a Practice-Oriented Liquefaction Model. Geo-Congress 2012, Geo-Institute, ASCE, Oakland, California.

Department of Water Resources (1989). The August 1, 1975 Oroville Earthquake Investigation, Supplement to Bulletin 203-78.

Harder, L. F. Jr. and Boulanger, R.W. (1997). Application of $K\sigma$ and $K\alpha$ Correction Factors. Proceedings of the NCEER Workshop on Evaluation of Liquefaction Resistance of Soils, Edited by Youd, T.L. and Idriss, I.M. Technical Report NCEER-97-0022. National Center for Earthquake Engineering Research.

Hardin, B.O. (1978). The Nature of Stress-Strain Behavior of Soils. Volume 1 of Earthquake Engineering and Soil Dynamics – Proceedings of the ASCE Geotechical Engineerng Division Specialty Conference (pp 3-90), June 19-21, Pasadena, California; Sponsored by Geotechnical Engineering Division of ASCE.

Hatanaka, M. and Uchida, A. (1996). Empirical Correlation Between Penetration Resistance and Internal Friction Angle of Sandy Soils. Soils and Foundations, Japanese Geotechical Society, Volume 36, No. 4, pp 1-9.

Idriss, I. M., and Boulanger, R. W. (2015). "2nd Ishihara Lecture: SPT- and CPT-based relationships for the residual shear strength of liquefied soil." Soil Dynamics and Earthquake Engineering, 68, 57-68, 10.1016/j.soildyn.2014.09.010.

Idriss, I.M. and Boulanger, R.W. (2008). Soil Liquefaction During Earthquakes. EERI MNO-12.

Idriss, I.M. and Boulanger, R.W. (2010). SPT-Based Liquefaction Triggering Procedures. Report No. UCD/CGM-10/02, Center for Geotechnical Modeling, Civil and Environmental Engineering, University of California, Davis.

Inel, S., Roth, W.H., and De Rubertis (1993). Non-Linear Dynamic Effective Stress Analysis of Two Case Histories. Proceedings of the Third International Conference on Case Histories in Geotechnical Engineering. St. Louis, Missouri.

Itasca (2011). FLAC – Theory and Background

Mitchell, J. K. and Soga, K. (2005). Fundamentals of Soil Behavior. ISBN: 978-0-471-46302-3. Wiley.

Moriwaki, Y., Tan, P. and Ji., F. (1998). Seismic Deformation Analysis of the Upper San Fernando Dam under the 1971 San Fernando Earthquake. In P. Dakoulas, M. Yegian, and R.D. Holtz (Eds.), Geotechnical Earthquake Engineering and Soil Dynamics III, ASCE Geotechnical Special Publication No. 75 Vol. 2, Proceedings of a Specialty Conference (pp. 854-865). Seattle: ASCE.

Olson, S. M. and Stark, T. D. (2002). Liquefied Strength Ration from Liquefaction Flow Failure Case Histories, Canadian Geotechnial Journal, Volume 39, pp. 629-647.

Seed, H.B., Idriss, I.M., Makdisi, F., and Banarjee, N. (1975). Representation of Irregular Time Histories by Equivalent Uniform Stress Series in Liquefaction Analyses. Report No. EERC-75-29, Earthquake Engineering Research Center, College of Engineering, University of California, Berkeley.

Seed, H.B., Lee, K.L., Idriss, I.M., Makdisi, F. (1973). Analysis of the Slides in the San Fernando Dams During the Earthquake of Feb. 9, 1971. Report No. 73-2, June 1973, Earthquake Engineering Research Center, College of Engineering, University of California, Berkeley.

Seed, H.B., Tokimatsu, K., Harder, L.H., and Chung, R.M. (1984). The Influence of SPT Procedures in Soil Liquefaction Resistance Evaluations. Report No. UCB/EERC-84/15, Earthquake Engineering Research Center, College of Engineering, University of California, Berkeley.

Seed, H.B., Wong, R.T., Idriss, I.M., and Tokimatsu, K. (1996). Moduli and Damping Factors for Dynamic Analyses of Cohesionless Soils. Journal of Geotechical Engineering.

Seed, R.B. and Harder, L.F. (1990). SPT-Based Analysis of Cyclic Pore Pressure Generation and Undrained Residual Strength. in J.M Duncan ed., Proceedings, H. Bolton Seed Memorial Symposium, University of California, Berkeley, Vol. 2. pp. 351-376.

Serff, N, J. (1976). Earthquake-Induced Deformations of Earth Dams. A dissertation submitted in partial satisfaction of the requirements for the degree of Doctor of Philosophy in Engineering – Civil and Environmental Engineering in the Graduate Division of the University of California, Berkeley.

Skempton, A.W. (1986). Standard Penetration Test Procedures and the Effects in Sands of Overburden Pressure, Relative Density, Particle Size, Ageing, and Overconsolidation. Geotechnique 36, No. 3, pp 425-447.

Terzaghi, K. and Peck, R. B. (1967). Soil mechanics in engineering practice. Wiley.

Weber, J. P. (2015). Engineering Evaluation of Post-Liquefaction Strength. A dissertation submitted in partial satisfaction of the requirements for the degree of Doctor of Philosophy in Engineering – Civil and Environmental Engineering in the Graduate Division of the University of California, Berkeley.

Weber, J. P., Seed, R.B., Pestana, J. M., Moss, R. E. S., Nweke, C., Deger T.T., and Chowdhury, K. (2015). Engineering Evaluation of Post-Liquefaction Strength. Department of Civil and Environmental Engineering, University of California, Berkeley. Prepared for the U.S. Nuclear Regulatory Commission.

Youd, T.L., Idriss, I.M., Andrus, R.D., Arango, I., Castro, G., Christian, J.T., Dobry, R., Finn, W.D.L., Harder, L.F. Jr., Hynes, M.E., Ishihara, K., Koester, J.P., Liao, S. S. C., Marcuson III, W. F., Martin, J. R., Mitchell, J. K., Moriwaki, Y., Power, M.S., Robertson, P.K, Seed, R.B., Stokoe II, K. H. (2001). "Liquefaction resistance of soils: summary report from the 1996 NCEER and 1998 NCEER/NSF workshops on evaluation of liquefaction resistance of soils." Journal of Geotechnical and Geoevironmental Engineering, ASCE.

COMPARISON OF TWO CONSTITUTIVE MODELS FOR SIMULATING THE EFFECTS OF LIQUEFACTION ON EMBANKMENT DAMS

Jack Montgomery, Ph.D.[1]
Sam Abbaszadeh, Ph.D. P.E.[2]

ABSTRACT

Evaluating the potential for earthquake-induced liquefaction under and beneath embankment dams continues to be a significant challenge for geotechnical earthquake engineers. A variety of methods are available to estimate the potential deformations associated with liquefaction and range from relatively simple analytical models to fully coupled numerical analyses using advanced constitutive models. Multiple constitutive models are currently available which can simulate the cyclic response of sandy soils, but these models use different formulations and calibration procedures which can lead to very different response at the element level. These differences may be exacerbated when elements are consolidated under large confining and/or shear stresses such as those present under embankment dams. This paper seeks to examine how model differences at the element level influence the overall response of a hypothetical embankment dam. This paper utilizes nonlinear deformation analyses (NDAs) to examine the effects of the choice of constitutive model on the response of a hypothetical embankment dam founded on liquefiable alluvium. The paper briefly describes each constitutive model under consideration and their respective calibration procedures. Two-dimensional NDAs are utilized to examine how the different constitutive models affect the overall response of the dam. Effects of soil density, shaking magnitude and ground motion selection on differences in the model responses are examined using a limited sensitivity study. Implications of these findings for practice are discussed.

INTRODUCTION

Soil liquefaction during earthquakes continues to be one of the most complicated and controversial topics in geotechnical earthquake engineering. Methods to examine the effects of liquefaction on embankment dams range from relatively simple analyses, such as Newmark sliding block methods, to more complex nonlinear deformation analyses (NDAs) which rely on advanced constitutive models to capture the cyclic response of the soils (e.g., Perlea and Beaty 2010). Multiple constitutive models are currently available to model liquefaction behavior in sandy soils, but these models use different constitutive approaches which can lead to very different response at the element level especially under large shear stresses such as those found beneath embankment dams (e.g., Ziotopoulou et al. 2014). Comparisons of these models at the system level are less common, but without these comparisons it is not possible to determine how differences at the element level affect overall differences in the response of an overlying structure.

[1] Assistant Professor, Department of Civil Engineering, Auburn University, Auburn, AL, USA jmontgomery@auburn.edu
[2] Senior Geotechnical Engineer, MWH now part of Stantec, Tempe, AZ, USA Sam.Abbaszadeh@mwhglobal.com

This paper seeks to examine the effect of two different constitutive models on the response of a hypothetical embankment dam founded on liquefiable alluvium. Each of the constitutive models is briefly described along with a discussion of the calibration procedures. Two-dimensional NDAs are then utilized to examine how the different constitutive models affect the overall response of the dam. Effects of varying soil density, shaking intensity and ground motion variability on differences in the model responses are also examined. Implications of these findings for practice are discussed.

CONSTITUTIVE MODELS

One of the key components of a NDA is the choice of constitutive model. Several different models are available to evaluate the effects of liquefaction (e.g., Wang and Makdisi 1999, Dawson et al. 2001, Yang et al. 2003, Beaty and Byrne 2011, Boulanger and Ziotopoulou 2015) and each varies in the complexity of the formulation and the ease of calibration. This paper will focus on comparing two of these models: UBCSAND (Version 904aR, Beaty and Byrne 2011) and PM4Sand (Version 3, Boulanger and Ziotopoulou 2015). Both models are implemented into FLAC (Itasca 2011) as a user defined material in a dynamic link library (DLL) provided by the respective authors. Each model will be briefly discussed in the following sections.

UBCSAND

UBCSAND is an effective stress plasticity model which was developed to simulate the behavior of sandy soils under seismic loading (Byrne et al. 2004, Beaty and Byrne 2011). The model is stress-ratio controlled and adopts a Mohr-Coulomb yield surface along with a non-associated flow rule that predicts the shear stress-strain response of the soil using a hyperbolic relationship. The associated volumetric response is predicted based on the relative density of the soil, which is correlated to a clean sand Standard Penetration Test resistance [$(N_1)_{60cs}$] and the model can capture associated pore pressure generation during cyclic loading when used in coupled fluid mechanical simulations. The version of the model used in this study is 904aR which is fully documented by Beaty and Byrne (2011).

PM4SAND

The constitutive model PM4Sand is a stress-ratio controlled, critical state compatible, bounding surface plasticity model developed primarily for earthquake engineering applications (Boulanger and Ziotopoulou 2015, Ziotopoulou and Boulanger 2016). PM4Sand adopts the general framework of the model proposed by Dafalias and Manzari (2004), but has been modified to be easy to calibrate and to better capture the important aspects of the liquefaction response of sandy soils. The formulation is able to model the volumetric response of the soil to cyclic loading and to directly simulate the associated pore pressure generation when used in coupled fluid mechanical simulations. This study used Version 3 which is fully documented by Boulanger and Ziotopoulou (2015).

Calibration of Constitutive Models

The calibration procedure for constitutive models is a critical component of any numerical simulation. The two constitutive models selected for examination in this study both have default calibrations which are provided by the developers to approximate common design relationships important for liquefaction analyses. Comparisons of these relationships to the results from single element simulations are provided in the respective manual for each models. The default calibrations were utilized in this study because they are able to provide a reasonable approximation of the cyclic strength of sandy soils across a range of densities and its variation with overburden and static shear stresses. For many projects little to no information beyond penetration resistances of the critical layers is available and therefore the default calibrations provided with the models give a basis to capture the general trends that are important to modeling liquefaction under dams. In this respect, this paper provides a comparison of the models as they would likely be used on a project without site-specific laboratory data. One drawback to using the default calibrations is that the constitutive models will likely have different cyclic strengths and stiffnesses for the same $(N_1)_{60cs}$ value due to the different design relationships used for developing the calibrations for each model. This source of uncertainty is acknowledged and comparisons using a more detailed calibration may be explored in future studies. The process for using the default calibrations for each model is briefly described below.

UBCSAND has five primary input parameters which must be specified by the user, but the developers have provided a series of relationships which relate each of the parameters to an $(N_1)_{60cs}$ value. This greatly simplifies the calibration process for the user as only one input parameter must be specified in order to use the model. The default calibration is shown in the UBCSAND manual to produce reasonable approximations of the liquefaction triggering correlations recommended by Youd et al. (2001). All other parameters retained the default values recommended by Beaty and Byrne (2011).

PM4Sand has three primary input parameters (D_R, G_o, h_{po}) which must be set by the user. The relative density (D_R) and shear modulus coefficient (G_o) which were mapped to $(N_1)_{60cs}$ values following the procedure provided by Boulanger and Ziotopoulou (2015). The last parameter h_{po} does not have a default value and must be calibrated for each $(N_1)_{60cs}$ value. This was done using single-element direct simple shear simulations to match the CRR for M = 7.5 and σ'_{vc} = 1 atm (i.e., $CRR_{M7.5,\sigma'=1}$) based on the SPT-based liquefaction triggering correlation of Idriss and Boulanger (2008). The calibration was based on the $CRR_{M7.5,\sigma'=1}$ value being equivalent to 15 equivalent uniform loading cycles being required to trigger 3% peak shear strain (at which $r_u = \Delta u/\sigma'_{vc} \approx 1.0$) in an element consolidated to σ'_{vc} = 1 atm (101 kPa) with coefficient of lateral earth pressure at rest K_o = 0.5. The other model parameters retained the default values recommended in Boulanger and Ziotopoulou (2015).

NUMERICAL SIMULATIONS

The response of the two constitutive models discussed above will be evaluated through the analysis of a hypothetical embankment dam (Figure 1). The embankment is 45-m tall

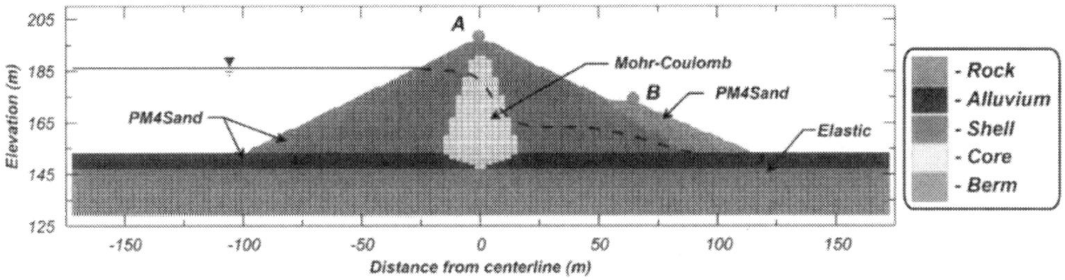

Figure 1. Dam geometry used for the numerical simulations. Points A and B indicate recording locations for deformations shown in Figures 6 - 9.

and consists of a clay core, sand shells, and a berm at the downstream toe. The average slopes of the embankment are 2H:1V while the downstream berm slope is 3H:1V. The embankment shells are underlain by a 6-m-thick layer of potentially liquefiable recent alluvium. The core is extended down to the underlying bedrock and the boundaries of the model are extended beyond the toes of the dam in order to minimize the effects of these boundaries on the response of the dam. The model geometry and input files are based on those used by Boulanger et al. (2015) and Boulanger and Montgomery (2015).

Soil Properties

Different zones within the dam are represented using different constitutive models and properties which are summarized in this section. The embankment shells, downstream berm and alluvium are modeled using either PM4Sand or UBCSAND. The alluvium was assigned a uniform $(N_1)_{60cs}$ of either 10, 15 or 20 depending on the simulation. The shells were modeled as uniform with properties corresponding to $(N_1)_{60cs}$ values of either 25 or 35. The downstream berm was assigned an $(N_1)_{60cs}$ of 35 in all simulations. The core was modeled as a Mohr-Coulomb material with an undrained shear strength (s_u) equal to 0.35 times the vertical effective consolidation stress (σ'_{vc}), and a shear modulus (G) which was proportional to the square root of mean effective stress (p') and equal to 43 MPa for p' = 101 kPa. The bedrock was modeled as linear elastic with G = 1,800 MPa. The analyses included Rayleigh damping of 0.5% for the shells, berm and alluvium, 1% for the bedrock, and 5% for the core, all at a frequency of 3 Hz. A Poisson's ratio of 0.35 was used for all zones. The permeability of the shells and alluvium was 5.0E-4 cm/s, the core was 5.0E-5 cm/s and the bedrock was 5.0E-6 cm/s. The berm remained dry in these simulations.

Boundary Conditions

Static stresses were established in the dam by constructing the model in layers with fixed boundaries along the sides and bottom of the model. Constant head boundary conditions are used to apply the reservoir (fixed at 10-m below the crest of the dam, as shown in Figure 1) and to fix the water table at the downstream edge of the model at 1-m below the ground surface. The ground motions (described in the next section) for the dynamic simulation were applied to the base of the model using a compliant base and free-field boundary conditions were used along the sides of the model. A detailed discussion of these boundary conditions is provided by Itasca (2011).

Ground Motions

Three ground motions were selected from among the suite of rock motions recommended by Baker et al. (2011) to approximate a strike-slip event with a moment magnitude (M_w) of 7 at distance of 10 km (Table 2). These motions were selected to represent a range of spectral shapes as show by their linear-elastic response spectra in Figure 2. The motions are input into the simulation as bedrock outcrop motions, from which the shear stress time series applied to the compliant base is computed. These motions were scaled linearly to match the desired outcrop PGA and some intensity measures are listed in Table 2.

Table 2. Properties of the selected ground motions.

NGA Record Sequence Number	Event Name	Station	M_w	Hypo. Distance (m)	Scaled PGA = 0.5g		
					Arias Intensity (m/s)	Cumulative Absolute Velocity (m/s)	Significant Duration (sec)
1619	'Duzce, Turkey'	'Mudurnu'	7.14	43.83	3.65	7.75	25.18
572	'Taiwan SMART1(45)'	'SMART1 E02'	7.3	72.91	5.14	17.0	26.95
265	'Victoria, Mexico'	'Cerro Prieto'	6.33	35.48	1.25	20.4	12.32

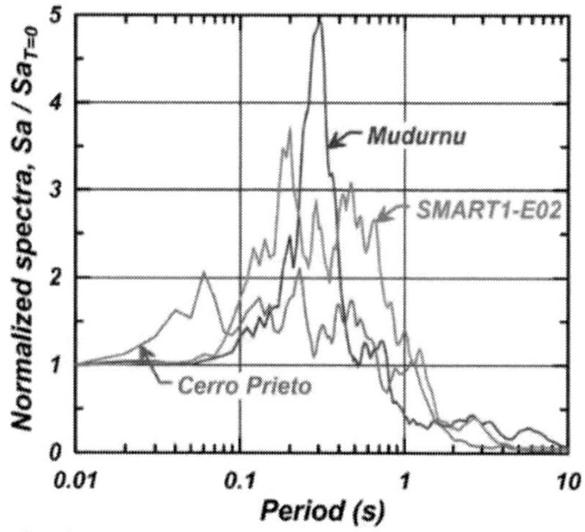

Figure 2. Elastic response spectra of the selected ground motions.

SIMULATION RESULTS

The primary focus of the numerical simulations was observing the response of the dam during shaking and to assess differences in the response from the two constitutive models. A post-shaking stability analysis was not performed as part of this study. The dynamic response at the crest of the dam for both models is shown in Figure 3 for simulations using the Mudurnu motion scaled to an outcrop PGA of 0.5g. The response of the two models is similar, although UBCSAND shows larger peak accelerations prior to 7 seconds when liquefaction begins to be triggered within the alluvium (Figure 4).

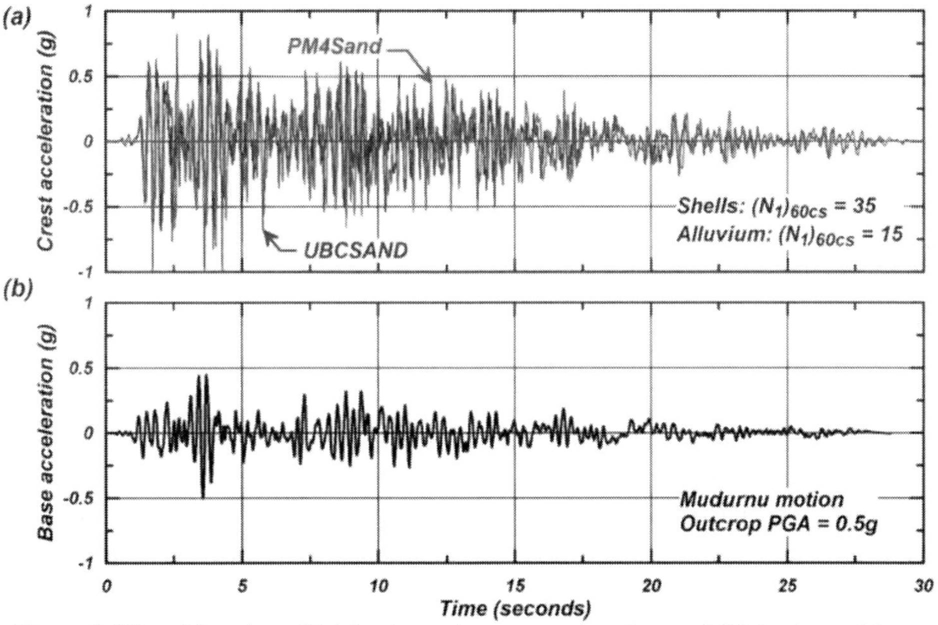

Figure 3. Time histories of (a) horizontal crest acceleration and (b) horizontal base acceleration.

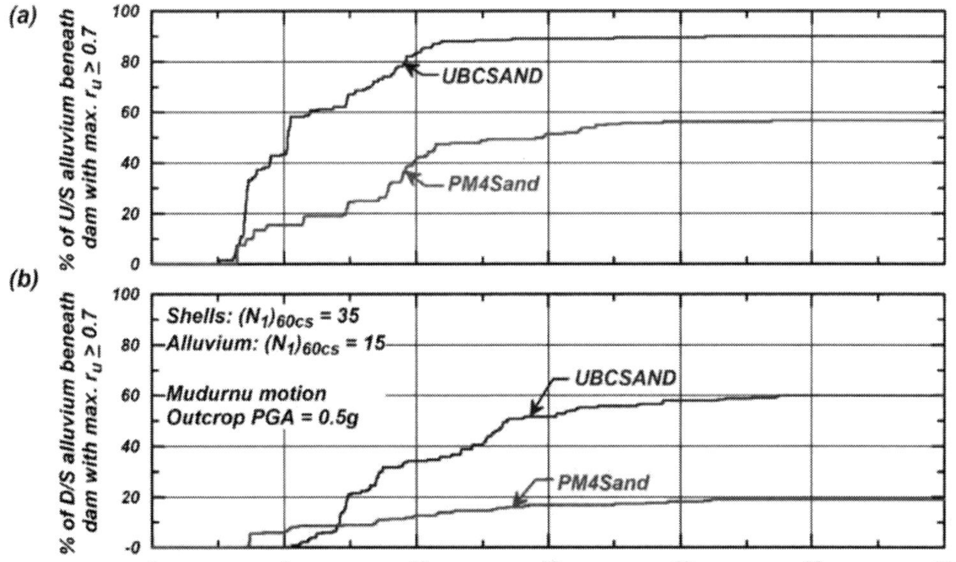

Figure 4. Time histories showing the percentage of elements in the alluvium under the (a) upstream (U/S) and (b) downstream (D/S) shells which have reached a maximum excess pore pressure ratio (r_u) of 0.70 or greater during the simulation.

The rate of excess pore pressure generation in the alluvium varies both between the upstream and downstream and between the two models. Figure 4a shows a time history of the percentage of the alluvium under the upstream shell of the dam which has reached a maximum excess pore pressure ration (r_u) of 0.70. This threshold can be used along with a shear strain criterion to judge liquefied elements for a post-shaking stability analysis (Boulanger et al. 2015). The two constitutive models show different rates of excess pore pressure generation with UBCSAND generating high r_u values in a larger percentage of the alluvium. Both models show less excess pore pressure generation in the downstream alluvium (Figure 4b). Differences in the responses of the two models and between the two sides of the dam can likely be explained by differences in the stresses on these elements and how the models account for the effects of confining and static shear stress on liquefaction triggering (i.e., K_σ and K_α).

Generation of excess pore pressure results in a loss of strength and stiffness within the soil which can lead to significant deformations of the overlying structure. The deformation patterns of the embankment dam in these simulations can be observed from the contours of shear strain at the end of shaking (Figure 5). In both simulations, the primary movement is upstream which is expected due to the buttressing effects of the downstream berm. The simulation using PM4Sand (Figure 5a) shows shear strains concentrated near the top of the alluvium as the bottom of the layer did not develop high excess pore pressures. The simulations with UBCSAND show more uniform pore pressure generation and therefore shear strains across the alluvium. The UBCSAND simulations also show slightly more shear strain under the downstream berm and slightly less shear strain under the upstream shell than the simulations using PM4Sand.

The deformations patterns discussed above can be further examined by looking at the vertical settlement of the crest and lateral displacement of the top of the downstream berm during shaking (Figure 6). The magnitude of the deformations was similar for both models with PM4Sand producing slightly greater crest settlements and UBCSAND producing slightly larger displacements of the downstream berm. This is consistent with the deformation patterns shown in Figure 5 and demonstrates that the differences in excess pore pressure generation between the two models had little effect on the overall deformations of the dam in this simulation.

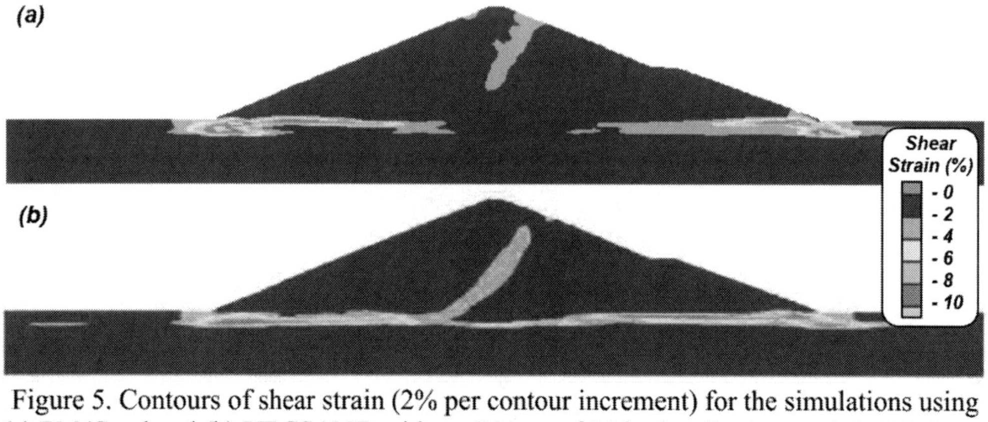

Figure 5. Contours of shear strain (2% per contour increment) for the simulations using (a) PM4Sand and (b) UBCSAND with an $(N_1)_{60cs}$ of 15 in the alluvium and the Mudurnu motion with an outcrop PGA of 0.5g.

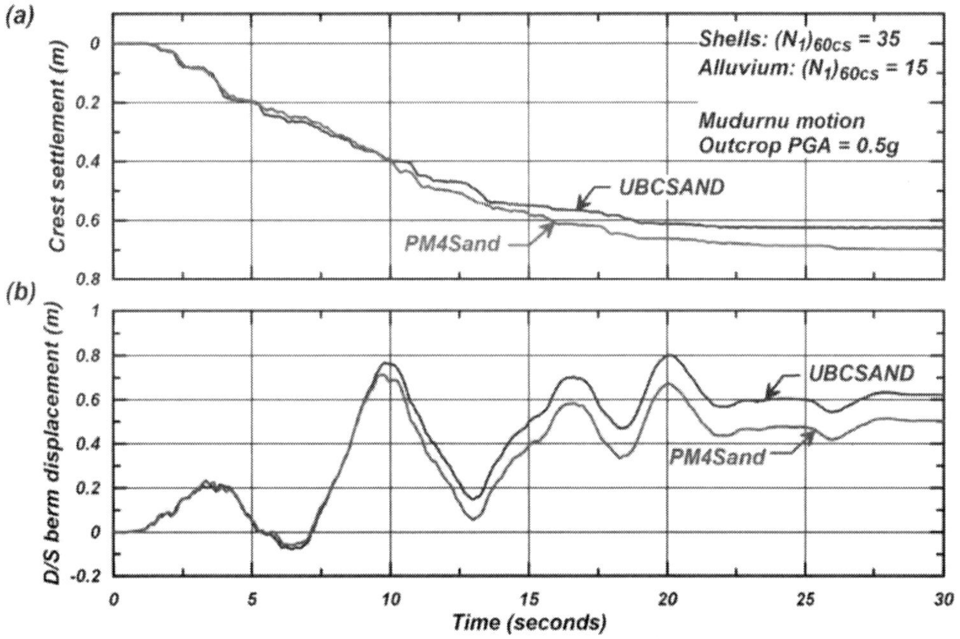

Figure 6. Time histories of (a) crest settlement (Point A in Figure 1) and (b) downstream (D/S) berm displacement (Point B in Figure 1).

The baseline analyses showed little difference in the overall deformation patterns produced by the two constitutive models, but the rate of excess pore pressure generation was different. This is likely due in part to differences between the default calibrations. The magnitude of this difference and its effect on the overall response are expected to depend on the cyclic strength of the soil and the magnitude of cyclic loading. This dependence will be explored in the following sections using the two additional ground motions described previously (Table 2 and Figure 2).

Effect of $(N_1)_{60cs}$ in the Alluvium

Additional simulations were performed using $(N_1)_{60cs}$ values of either 10, 15 and 20 for the alluvium. Results were examined in terms of crest settlement (Figures 7a, c, and e) and lateral displacement of the D/S berm (Figures 7b, d and f). Results for the least intense motion (Cerro Prieto) are presented at the top while results for the most intense ground motion (SMART1 E02) are presented at the bottom. This same pattern is used in subsequent figures as well. PM4Sand generally showed larger deformations for the denser alluvium (20-60% larger) while UBCSAND showed more deformation for the looser alluvium (0-60% larger). These differences could likely be reduced through a more detailed calibration of the two models to ensure consistent cyclic strengths for a given $(N_1)_{60cs}$ value. It is also important to note that the differences in deformations between the two models are small compared with the differences between the ground motions which are all scaled to the same PGA and intended to represent the same seismic source. Ground motion variability is often considered to be one of the largest sources of uncertainty in liquefaction analyses and these results are consistent with that observation.

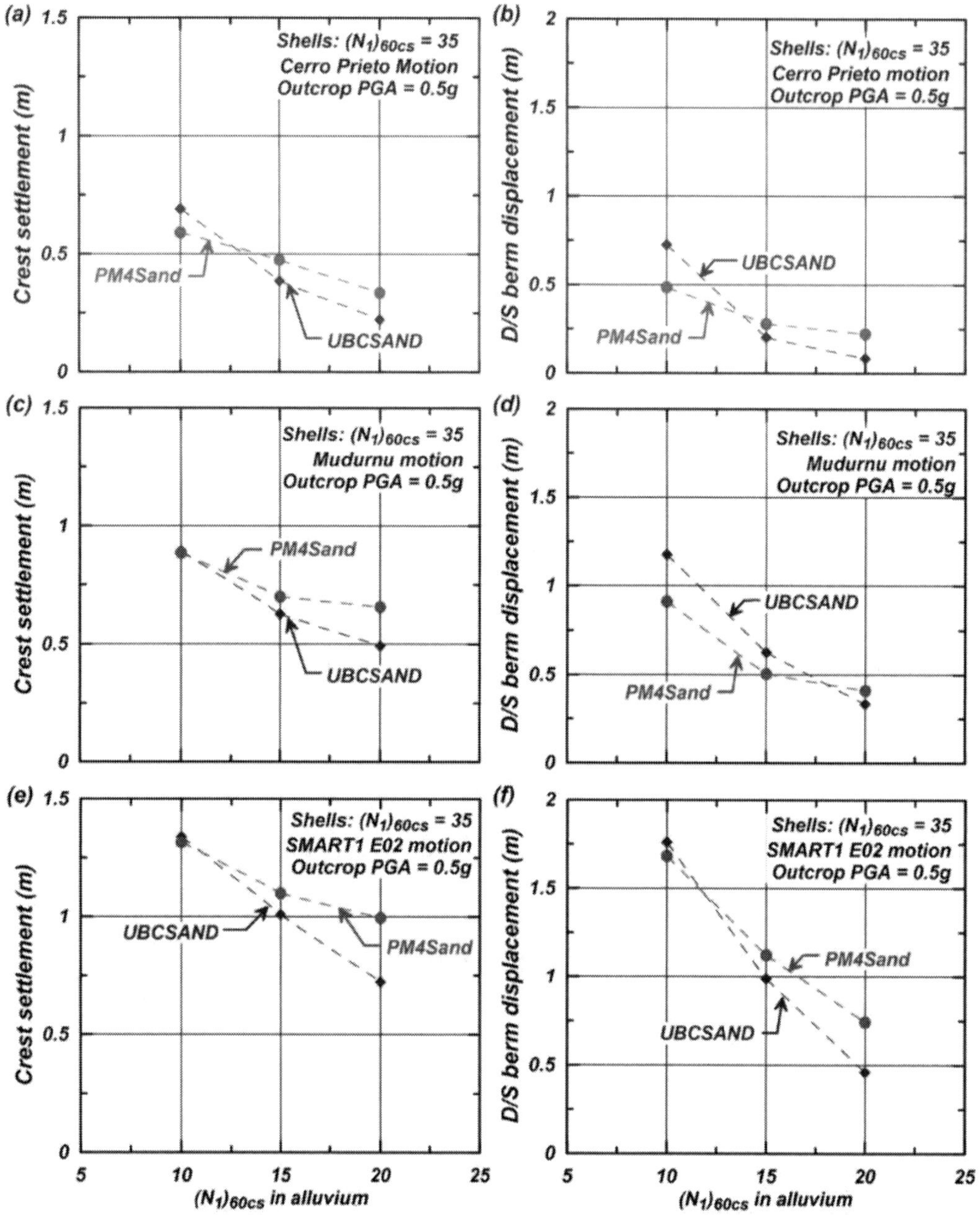

Figure 7. Comparison of crest settlement (Point A in Figure 1) and downstream (D/S) berm displacement (Point B in Figure 1) results using UBCSAND and PM4Sand for three input motions all scaled to a peak outcrop acceleration (PGA) of 0.5g.

Effect of $(N_1)_{60cs}$ in the Shells

The previous analyses considered a well-compacted shell which generated relatively small strains during shaking. To examine the impact of the shell density on the overall displacement, the $(N_1)_{60cs}$ value in the shells was lowered from 35 to 25. This led to additional excess pore pressure generation in the shells, but not enough to trigger significant liquefaction within the shell. Crest settlements (Figures 8a, c, and e) from both models increased between 15 and 45% with the largest increases coming for the simulations with the dense alluvium [$(N_1)_{60cs}$ = 20 where the additional strains in the shell have a larger impact on the total deformation. The displacements of the downstream berm (Figures 8b, d, and f) were largely similar to the results using the denser shell with

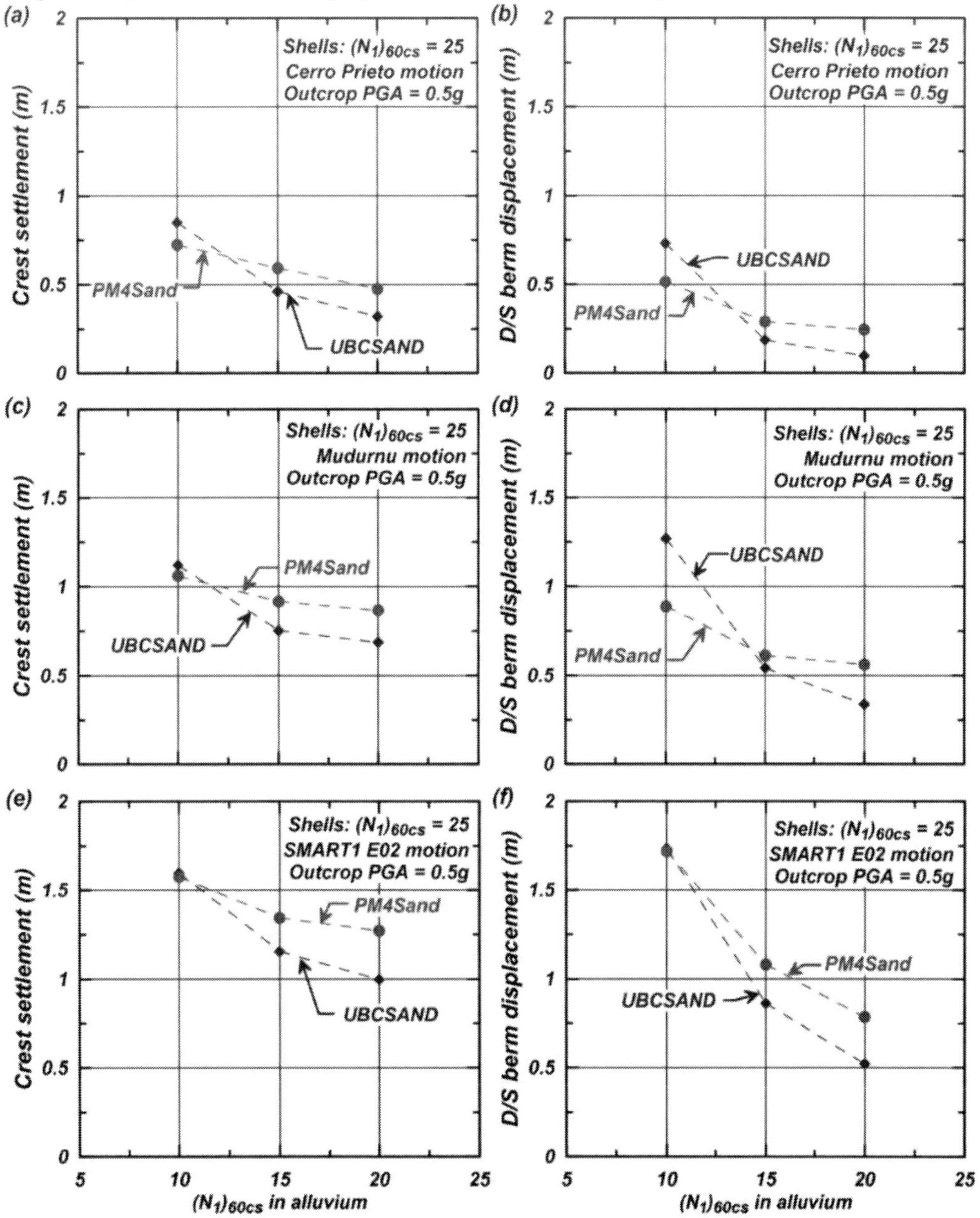

Figure 8. Comparison of crest settlement (Point A in Figure 1) and downstream (D/S) berm displacement (Point B in Figure 1) results using UBCSAND and PM4Sand for an outcrop PGA of 0.5g and a $(N_1)_{60cs}$ of 25 for the embankment shells.

differences of less than 15% between the two sets of simulations. This is expected as movement of the berm is driven primarily by the response of the alluvium. Overall, the pattern of results between the two models was not significantly affected by using the looser shell although the magnitude of differences increased for some of the simulations.

Effect of PGA

The effect of shaking amplitude was examined by increasing the outcrop PGA of the selected motions by 50% to 0.75g. Deformations result from these simulations, which used the denser shell [$(N_1)_{60cs}$ =35] are shown in Figure 9. The increase in PGA resulted in significantly larger settlements and displacements for all the simulations as expected. The effect of increasing the PGA on the results from the two models was mixed with some combinations showing larger differences (e.g., Figure 9d) and others showing the results moving closer together (Figure 9c). The largest difference between the two models can be seen for the crest settlement on the denser alluvium. The larger PGA begins to generate more strain within the dam shells which is more pronounced in the PM4Sand simulations than the UBCSAND simulations.

DISCUSSION

The results discussed previously show that while the individual deformation results from simulations using PM4Sand and UBCSAND may vary as much as 60% the overall pattern of results was similar. The differences between the constitutive models were on average significantly less than differences resulting from changing the $(N_1)_{60cs}$ values, the ground motion or the magnitude of loading. For example, the three input motions used in this study produced crest settlements which varied by as much as a 300% with all other input parameters held constant. This emphasizes the important role of examining the effects of ground motion variability when performing NDAs.

This study used the default calibration parameters for the two constitutive models which likely led to differences in cyclic strengths for a given $(N_1)_{60cs}$ value. These differences result from the developers of the models making different choices when selecting their default calibrations. A more detailed calibration could have removed some of these differences and would have likely reduced some of the differences in results between the two models. This study focused on the default calibrations as they are commonly used when information to perform a more detailed calibration is not available. Despite these differences in calibration, the overall pattern of results was similar between the models.

Many other parameters would also be expected to affect the results and therefore the comparison between the two models including additional shaking levels, alternative ground motions, thickness of the alluvium, height of the dam, strength of the downstream berm, and consideration of spatial variability in the material properties. It would also be informative to include additional constitutive models in these comparisons. The authors plan to explore some of these factors in future studies and it is hoped that this study provides a framework for others to perform similar comparisons as well.

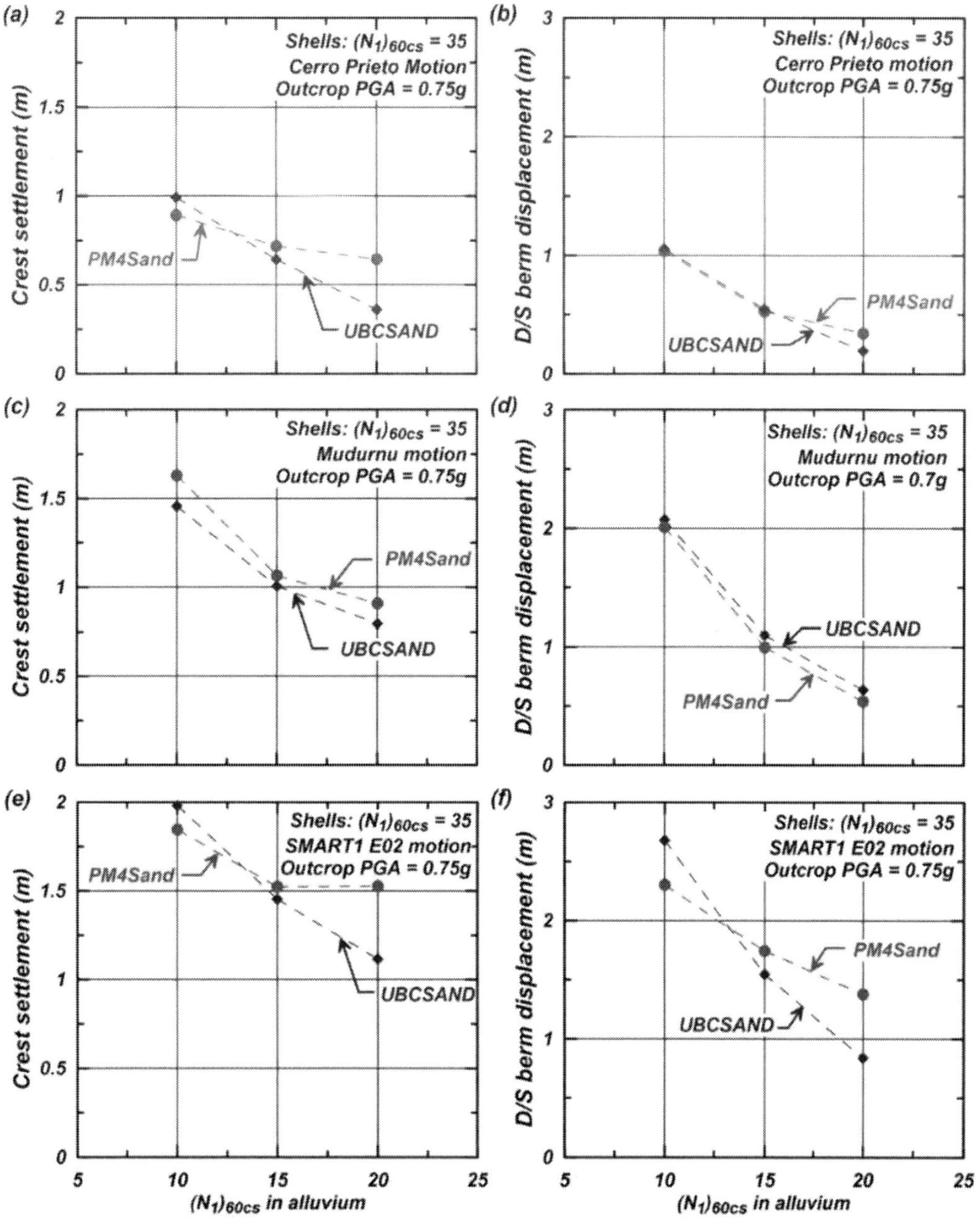

Figure 9. Comparison of crest settlement (Point A in Figure 1) and downstream (D/S) berm displacement (Point B in Figure 1) results using UBCSAND and PM4Sand for an outcrop PGA of 0.75g and a $(N_1)_{60cs}$ of 35 for the embankment shells.

CONCLUSIONS

This study examined the response of two constitutive models through examination of a hypothetical embankment dam on a liquefiable alluvial deposit. The two models, PM4Sand and UBCSAND, are both used in the dam engineering practice and have default calibrations which produce results consistent with commonly used design relationships for liquefaction problems. Deformations results from the two models were compared for a baseline case using a $(N_1)_{60cs}$ of 15 in the alluvium and 35 in the shells. This simulation showed that, although the constitutive models produced different patterns of excess pore pressure generation, the overall response of the dam was similar with PM4Sand showing slightly more crest settlement and UBCSAND showing slightly more displacement of the downstream berm.

A limited sensitivity analysis was performed to examine the effect of changing the $(N_1)_{60cs}$ of the alluvium and shells, different ground motions and a higher PGA. The results from this set of analyses showed that UBCSAND tended to estimate more deformations for the weaker alluvium while PM4Sand estimated higher deformations for the denser alluvium. The difference in crest settlement and berm displacement between the two models ranged from 0-60% depending on the simulation, but the overall response of the two models for the range of parameters examined was similar. The differences between the models were also relatively small compared with the differences between the three ground motions used in this study. These results show that despite differences in the formulation and calibration of these two models, the deformation results for the cases examined here were similar. This should give additional confidence in the use of these models for evaluating the effects of liquefaction on embankment dams.

ACKNOWLEDGEMENTS

The FLAC model of the hypothetical dam used in this study was based on models which were previously developed using input and files from Professors Ross Boulanger and Richard Armstrong. Their assistance and guidance during development of the original model is greatly appreciated.

REFERENCES

Baker, J. W., Lin, T., Shahi, S. K., and Jayaram, N. (2011). "New ground motion selection procedures and selected motions for the PEER transportation research program." Report PEER 2011/03, Pacific Earthquake Eng. Research Center, University of California, Berkeley.

Beaty, M. H. and Byrne, P. M. (2011). "UBCSAND constitutive model: Version 904aR." Documentation report on Itasca UDM Web Site, February 2011.

Boulanger, R. W., and Montgomery, J. (2015). "Nonlinear deformation analyses of an embankment dam on a spatially variable liquefiable deposit." Proc., 11th Int. Conf. on Earthquake Geotechnical Eng., Christchurch, New Zealand.

Boulanger R. W., Montgomery J. and Ziotopoulou K. (2015) "Nonlinear deformation analyses of liquefaction effects on embankment dams." Perspectives on Earthquake Geotech. Eng., A. Ansal and M. Sakr, eds., Geotech., Geological and Earthquake Eng., Springer; 37: 247-283.

Boulanger, R. W., and Ziotopoulou, K. (2015). "PM4Sand (Version 3): A sand plasticity model for earthquake engineering applications." Report No. UCD/CGM-15/01, Center for Geotech. Modeling, Department of Civil and Environ. Eng., University of California, Davis, CA, 112 pp.

Byrne, P., Park, S., Beaty, M., Sharp, M., Gonzalez, L., and Abdoun, T. (2004). "Numerical modeling of liquefaction and comparison with centrifuge tests." Canadian Geotech. J., 41(2), 193–211.

Dafalias, Y. F., and Manzari, M. (2004), "Simple plasticity sand model accounting for fabric change effects," J. of Engineering Mechanics, ASCE, 130(6), 622-634.

Dawson, E. M., Roth, W. H., Nesarajah, S., Bureau, G., and Davis, C. A. (2001). "A practice oriented pore pressure generation model."2nd FLAC Symposium on Numerical Modeling in Geomechanics. Lyon, France.

Idriss, I. M., and Boulanger, R. W. (2008). Soil liquefaction during earthquakes. Monograph MNO-12, Earthquake Eng. Research Institute, Oakland, CA, 261 pp.

Itasca (2011). "FLAC, Fast Lagrangian Analysis of Continua, User's Guide, Version 7.0." Itasca Consulting Group, Inc., Minneapolis, MN.

Perlea, V. G., and Beaty, M. H. (2010). "Corps of Engineers practice in the evaluation of seismic deformation of embankment dams." Fifth International Conf. on Recent Advances in Geotech. Earthquake Eng. and Soil Dynamics, San Diego, CA, paper SPL-6.

Wang, Z.L. and Makdisi, F. I. (1999). "Implementing a bounding surface hypoplasticity model for sand into the FLAC program", FLAC and Numerical Modeling in Geomechanics, A.A. Balkema, Netherlands, 483-490.

Yang, Z., A. Elgamal and E. Parra, (2003). "Computational model for cyclic mobility and associated shear deformation." J. Geotech. and Geoenviron. Eng., 129(12), 1119-1127.

Youd, T. L. et al. (2001). "Liquefaction resistance of soils: summary report from the 1996 NCEER and 1998 NCEER/NSF workshops on evaluation of liquefaction resistance of soils." J. Geotech. Geoenviron. Eng., 127(10), 817-833.

Ziotopoulou K and Boulanger, R. W. (2016). "Plasticity modeling of liquefaction effects under sloping ground and irregular cyclic loading conditions." J. Soil Dynamics and Earthquake Eng., 84:269-283, 10.1016/j.soildyn.2016.02.013.

Ziotopoulou, K., Maharjan, M., Boulanger, R. W., Beaty, M. H., Armstrong, R. J., and Takahashi, A. (2014). "Constitutive modeling of liquefaction effects in sloping ground." Tenth U.S. National Conf. on Earthquake Eng., Frontiers of Earthquake Eng., July 21-25, Anchorage Alaska.

SEISMIC EVALUATION OF LOOKOUT POINT EMBANKMENT DAM IN OREGON – CASCADIA SUBDUCTION ZONE IMPLICATIONS

Andrew D. Kost, E.I.[1]
Kenji Yamasaki, P.E.[2]
David H. Scofield, P.E., C.E.G.[3]

ABSTRACT

Awareness of the hazard posed by a Cascadia Subduction Zone (CSZ) event to critical infrastructure in the Pacific Northwest has led to a reassessment of the seismic resiliency of dams and other assets. This paper describes a study of a 250-foot-high zoned embankment and concrete gravity dam located in western Oregon. Predicted ground motions have increased significantly from those assumed in design over 60 years ago, and the population at-risk downstream has grown. The study was commissioned to evaluate the dam's performance during seismic events of varying return periods for use within a probabilistic risk assessment framework.

A two-dimensional finite element model was developed to perform an equivalent linear analysis of the dam using three acceleration time histories from subduction zone events. Seismic displacements were evaluated using Newmark's decoupled approach. The approach accounted for potential cyclic softening of the clay core of the dam; however, this effect generally proved to be small due to the confinement of the rockfill shell. Despite the increased magnitude of applied ground motions, calculated displacements were small (on the order of inches) for 1,000- and 2,500-year motions. For the much larger 10,000-year motions, however, displacements for shallow slip surfaces become large (on the order of feet), and the assumptions of the equivalent linear model and the decoupled method become less tenable.

INTRODUCTION

Western Oregon is located within the region that will be subjected to strong ground shaking during a Cascadia Subduction Zone (CSZ) megathrust earthquake. Although the seismic environment of the western United States has been studied for nearly a century, it is only in the last two to three decades that the existence and magnitude of subduction zone events have become well-established and better understood. Increased awareness of the hazard posed by a subduction zone event has led to a reassessment of many dams and other critical assets in the Pacific Northwest.

[1] Staff Engineer, Cornforth Consultants, Inc. | 10250 SW Greenburg Rd Suite 111, Portland, OR 97223 | akost@cornforthconsultants.com
[2] Senior Associate Engineer, Cornforth Consultants, Inc. | 10250 SW Greenburg Rd Suite 111, Portland, OR 97223 | kyamasaki@cornforthconsultants.com
[3] Civil Engineer, U.S. Army Corps of Engineers, Portland District, Geotechnical Design Section | P.O. Box 2946, Portland, OR 97209 | David.H.Scofield@usace.army.mil

Lookout Point Dam (Figure 1) is a 250-foot-high, zoned embankment and concrete gravity dam located within the populated Willamette Valley of western Oregon, about 22 miles upstream of Eugene-Springfield metropolitan area. The dam is owned and operated by the U.S. Army Corps of Engineers (USACE) and was constructed between 1951 and 1954. The primary purposes of the dam are power generation and flood risk reduction. The project is also used for secondary purposes including recreation, downstream water supply, and water quality management.

Figure 1. Oblique Aerial View of Lookout Point Dam.

As part of the USACE routine dam safety program, the dam undergoes periodic inspections and risk assessments, which are similar in scope to dams regulated by other federal agencies. In 2012, a Seismic Safety Review (SSR) for Lookout Point Dam identified significant changes to seismic design criteria and predicted ground motion intensity since the previous review was completed in 1981. At the same time, property development and population growth downstream of the project have increased the potential consequences of a dam failure. Consequently, a full Risk Assessment was recommended for the project. This paper describes key aspects, results, and conclusions of a study commissioned in advance of the Risk Assessment to evaluate specific seismic performance criteria for the embankment portion of Lookout Point Dam.

APPROACH

USACE assesses risk using a probabilistic approach. Following the Seismic Safety Review, a Semi-Quantitative Risk Assessment (SQRA) was used to identify several potential failure modes that appeared to be the primary risk drivers for the project. One significant geotechnical failure mode identified was the potential for cyclic softening of the impervious core under a long-duration CSZ event, leading to reduction of crest height through embankment deformations, resulting in overtopping of the dam. A related risk was that embankment deformations could compromise the performance of the filter zone, leading to internal erosion. An improved estimate of potential seismic displacements was required to fully assess the risk.

A study was undertaken to refine the estimate of permanent seismic deformations of the zoned embankment portion of the dam. Analysis results must be in probabilistic form to be compatible with the USACE risk assessment framework. To this end, analyses examined a wide range of possible input parameters including pool level, return period, ground motions, potential shear surfaces, and material strengths. A method was developed to evaluate the potential of the clay core to soften under cyclic earthquake loads, and the consequences of softening were also evaluated using a parametric approach.

This paper demonstrates the performance implications for an embankment dam subjected to ground motions representing a potential Cascadia Subduction Zone event. Analysis results are presented for the most likely scenario—maximum conservation pool with best estimates of the strength of the dam materials under both static and dynamic conditions—to highlight the key findings of the study. Dam performance is primarily expressed in terms of permanent seismic displacements. Because the embankment dam includes a downstream filter zone, one key criterion is that seismic deformations must be less than the width of the filter to preserve the integrity of the dam. The effect of permanent displacements is also evaluated with respect to reduction of crest height.

METHODS

A two-dimensional finite element model was developed to perform an equivalent linear analysis of dam response during a subduction zone event. Seismic displacements were subsequently evaluated using Newmark's decoupled approach in conjunction with the average acceleration determined from the finite element model. Model development and analysis methods are summarized in the following sections.

Dam Geometry and Composition

USACE has maintained detailed, as-built construction drawings of Lookout Point Dam (USACE, 1981). Cross-sections at regular intervals along the dam axis were used to develop a three-dimensional model of the embankment dam using Sketchup modeling software. An orthogonal view of the model with the location of the cross-sections is shown in Figure 2. The embankment dam is zoned, featuring a clay core, downstream filter, and widely graded rockfill shells. At maximum height, the crest is approximately 250 feet above the foundation and 24 feet wide. The upstream slope varies from 2.5H:1V along the lower slope to 2H:1V along the upper slope. The downstream slope grades from 2.5H:1V at lower elevations, to 2H:1V mid-height, to 1.75H:1V at the upper elevations. A 2-foot-thick dumped stone revetment armors the upstream slope.

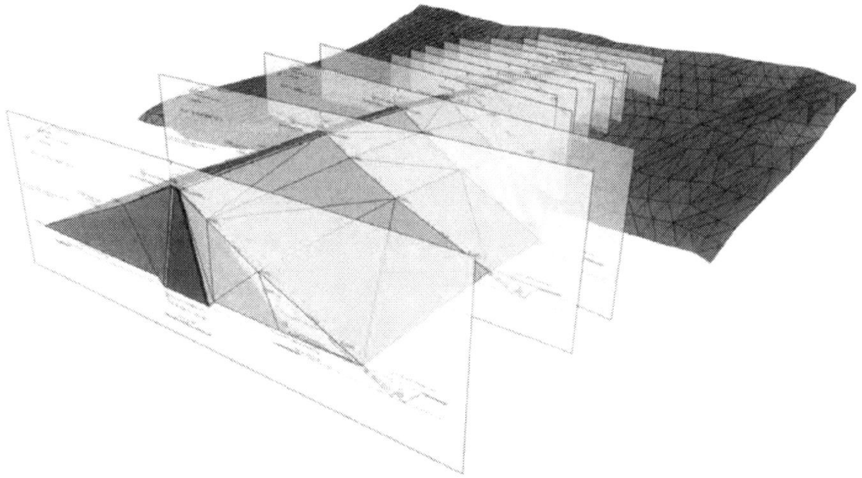

Figure 2. Sketchup Model of Lookout Point Dam.

Detailed post-construction reports also describe the construction materials and foundation materials for the dam (USACE, 1959). The impermeable core consists of a high plasticity clay obtained from a nearby borrow area and compacted approximately 1.5% wet of optimum. The downstream filter is 8 feet thick and was constructed using processed alluvial gravel from the reservoir area upstream of the dam. The steeply inclined core and filter are supported by rockfill shells comprised of poorly graded gravel and sandy gravel, with cobbles and small boulders up to 18 inches in diameter. Fines content was typically less than 5% by weight, and in no case greater than 12%.

The rockfill shells of the dam are founded on alluvial materials consisting primarily of gravel, cobbles, and boulders. A surficial silt deposit 5 to 10 feet thick was stripped from the surface during construction. The remaining gravel deposits left in-place are 5 to 40 feet thick, with an average thickness of approximately 25 feet, and overlie bedrock. Bedrock generally consists of andesite and tuff-breccia underlain by basalt, with interbedded flow breccia also present. Overburden deposits were excavated to found the clay core and filter on firm bedrock.

Finite Element Model

A two-dimensional finite element model was developed of the maximum-height section of the dam using the GeoStudio software suite, a product of Geo-Slope International of Calgary, Alberta. The model was first used to evaluate the steady-state seepage pore water pressure distribution in SEEP/W using the maximum conservation pool (16 feet below the crest). These pore pressures were incorporated into QUAKE/W to calculate the in-situ stress state under static conditions, perform the ground response analysis, and evaluate seismic displacements. Maximum element size is 10 feet, or approximately one fourth of a 25-Hz wavelength, to maintain computational accuracy in the ground response analyses. The finite element model, with boundary conditions corresponding to initial (static) stress state, is shown in Figure 3.

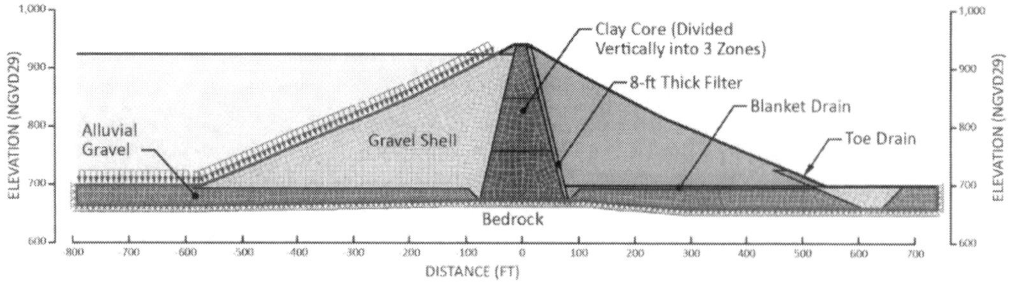

Figure 3. Finite Element Model of Lookout Point Dam at Maximum Section.

The ground response software, QUAKE/W, only permits the use of rigid boundaries. A rigid boundary will reflect downward-propagating waves upward and back into the model. Therefore, the model boundary for applying the acceleration time history was set at the top of bedrock, where the existing impedance contrast will tend to produce this type of wave reflection. Nodes along the base of the model were fixed in both the x- and y-axes. Nodes along lateral boundaries were fixed in the x-axis only to calculate the initial stress state and fixed in the y-axis only for the ground response analyses.

Material Properties

A thorough laboratory testing program was performed in advance of design and as quality assurance testing during construction. The results are summarized in reports by USACE (1954 and 1959). These data formed the basis of the material properties selected for the modeling effort. For instances when laboratory test data was not available, the property value was estimated from published literature or by correlation to a known property. No additional subsurface investigations, laboratory testing, or in-situ testing was performed for this study. Material properties are summarized in Table 1.

The undrained strength of the clay core must be determined in order to consider the potential effects of softening during a long-duration, CSZ event. In the absence of test data, the undrained strength at the beginning of shaking was estimated through correlation to the in-situ stress state, using the correlation developed by Jamiolkowski et al. (1985):

$$s_u = 0.23\sigma'_v \cdot OCR^{0.8} \tag{1}$$

In Equation 1, σ'_v represents the effective vertical stress, OCR is the overconsolidation ratio, and s_u is the undrained strength in stress units consistent with σ'_v. The equation shows that the undrained strength of a clay material is a function of the stress history and stress profile. Depending on depth, the consolidation stress profile can be controlled either by the field compaction process or by the weight of overlying materials and pore pressure. Near the crest of the dam, the vertical stress imparted by the compaction effort will exceed the geostatic stresses, although the compaction process should not be considered to be 100 percent efficient (that is, the apparent pre-consolidation pressure is typically less than the magnitude of vertical stress imparted by the compaction effort).

Table 1. Material Properties.

Material	Unit Weight (pcf)	Static Strength	Dynamic Strength	G_{max} (psf)
Foundation Gravel	148	c' = 0 psf ϕ' = 40.4°	c' = 0 psf ϕ' = 37.5°	$1{,}000K_2\sqrt{\sigma'_m}$ [c] $K_2 = 120$ [a]
Impervious Core	110 – 118 (varies w/ depth)	c' = 560 psf ϕ' = 25.2°	See text	$900s_u$ [b]
Filter	148	c' = 0 psf ϕ' = 40.4°	c' = 0 psf ϕ' = 40.4°	$1{,}000K_2\sqrt{\sigma'_m}$ [c] $K_2 = 120$ [a]
Blanket Drain	148	c' = 0 psf ϕ' = 38.0°	c' = 0 psf ϕ' = 35.1°	$1{,}000K_2\sqrt{\sigma'_m}$ [c] $K_2 = 120$ [a]
Shell	148	c' = 0 psf ϕ' = 40.4°	c' = 0 psf ϕ' = 37.5°	$1{,}000K_2\sqrt{\sigma'_m}$ [c] $K_2 = 120$ [a]

Notes: [a] See Kramer (1996).
[b] See Egan and Ebeling (1985) and Kramer (1996).
[c] Mean effective stress, $\sigma'_m = \frac{1}{3}(\sigma'_v + 2K_o\sigma'_v)$

In the lower two thirds of the clay core, geostatic effective stresses control. The magnitude of these stresses is dependent on the pore pressures, which in turn depend on the time assumed for the dam to reach steady-state seepage. Settlement monitoring of Lookout Point Dam during and following construction, together with records of pool level during the first filling, provided a basis to estimate the degree of consolidation occurring prior to reaching steady state seepage.

A best estimate of in-situ stress was taken as the larger of 80 percent of the compaction stress applied and the geostatic stresses calculated when assuming that 90 percent consolidation of the clay core occurred prior to reaching steady state seepage. The resulting effective stress profile was used with Equation 1 to determine a best estimate of undrained strength with depth. The core was subsequently divided into three zones with constant undrained shear strength. The best estimate of the monotonic undrained strength was 1,900 psf for the upper zone, 3,030 psf for the middle zone, and 4,860 psf for the lower zone. Each zone comprises approximately one third of the clay core.

The undrained strength of the core was reduced for dynamic analyses to account for potential consequences of soil softening under cyclic loads. The magnitude of this reduction is discussed under the "Cyclic Softening" subheading below.

Dynamic analyses also considered the potential of excess pore water pressure generation in saturated granular soils. Previous analyses have concluded that the foundation soils possess a low liquefaction potential. Granular materials placed for the filter, blanket, and shells were well-compacted and are also unlikely to generate significant excess pore water pressure during a seismic event. To account for some pore water pressure generation, the shear strength of saturated granular materials was reduced by 10 percent for dynamic analyses.

Time History Selection

The regional seismicity of the Willamette Valley was recently studied by AMEC Geomatrix and Quest Structures (2009). The study established deterministic and probabilistic response spectra for Lookout Point Dam, along with 12 other dams in the Willamette Valley. Uniform hazard response spectra (UHRS) were developed for various return periods. The study also recommended a suite of recorded subduction zone and crustal motions with associated scaling factors to match the UHRS for each return period considered. The results were similar to the USGS 2014 national seismic hazard mapping (Peterson et al., 2014).

For this study, three subduction zone time histories were selected from the recommended motions and scaled to match the 1,000-year, 2,500-year, and 10,000-year UHRS. Details of these recorded time histories are provided in Table 2. The target response spectra for the three return periods considered are shown in Figure 4, along with the spectral acceleration for the three ground motions scaled to the 2,500-year return period. The spectral acceleration for recorded motions at other return periods is omitted for clarity.

Table 2. Subduction Zone Ground Motions.

Event (Year)	Recording	Magnitude	Duration (sec)	Scaled PGA (g)		
				1,000-yr	2,500-yr	10,000-yr
Valparaiso (1985)	U.F.S.M. D.I.C., 70°	8.0	79.37	0.156	0.272	0.505
Michoacan (1985)	Caleta de Campo, 180°	8.0	50.64	0.140	0.244	0.453
Synthetic	Geomatrix (1993)	9.0	297.7	0.143	0.250	0.465

Note: PGA = Peak Ground Acceleration

Equivalent Linear Approach

The response of the embankment dam to the applied ground motion was evaluated using the equivalent linear approach in QUAKE/W. The equivalent linear method provides a means of representing the non-linear dynamic behavior of soil with a linear secant modulus. The shear strain magnitude must be assumed and then checked to implement this approach, making it iterative. The equivalent linear method relies on a pair of curves to describe the non-linear dynamic soil behavior. One curve describes the modulus degradation (G/G_{max}) as a function of shear strain. The second curve describes the damping ratio (representing energy dissipation) as a function of shear strain.

This study relied on published modulus degradation and damping curves. The behavior of the clay core was modeled using curves by Vucetic and Dobry (1991) for clay with PI = 30. All granular materials were modeled using the mean curves for sand by Seed and

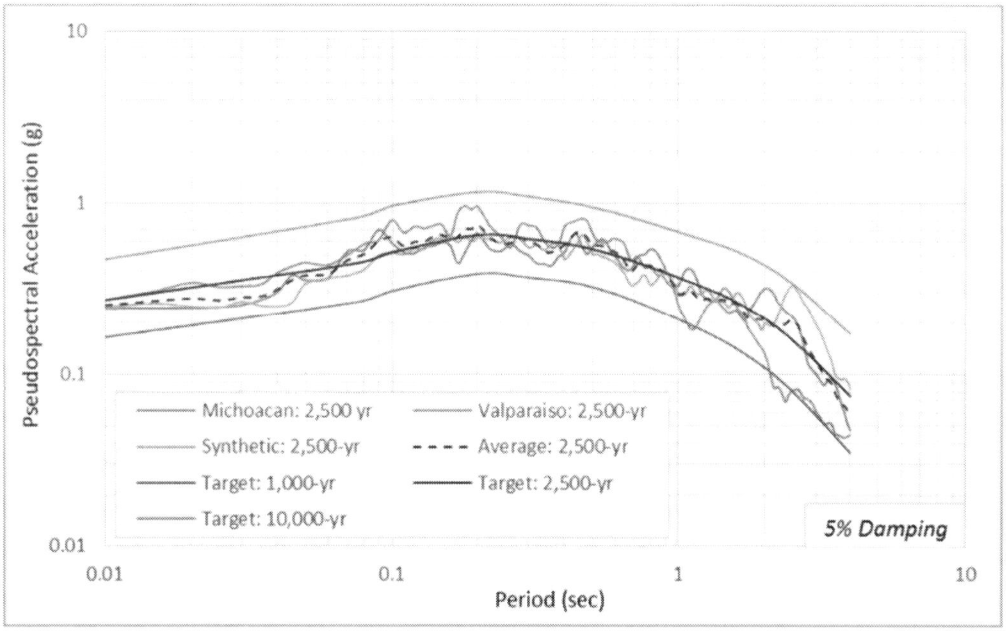

Figure 4. Target and Scaled Spectral Acceleration.

Idriss (1970). Values for the small-strain modulus of the soil, G_{max}, were determined using the correlations shown in Table 1.

One characteristic of the equivalent linear method is that all deformations are elastic. Consequently, the method cannot be used to calculate permanent seismic slope displacements. However, the method provides a reasonable approximation of the accelerations and shear stresses throughout the dam as a function of time. These results can be used to determine the average horizontal acceleration of a slide mass for use in a decoupled analysis with other methods, such as the Newmark sliding block method, to estimate permanent displacements.

Cyclic Softening Potential of Clay Core

Cyclic softening is defined as the onset of significant strains or strength loss in saturated clays and plastic silts during an earthquake (Idriss and Boulanger, 2008). An evaluation of cyclic softening includes two components: the potential of the soil to soften during cyclic shear, and the consequences for the material (i.e., strength loss or strain) if softening occurs.

A detailed evaluation of these two components generally requires specialized laboratory testing, which was beyond the scope of this study. Instead, the authors developed a simplified screening method to assess the likelihood of the core to soften during a seismic event. Several points within the clay core were selected to evaluate the potential for cyclic softening. For a given seismic event, the shear stress time history on a horizontal plane is determined at a given point from the finite element model, and the maximum shear stress at that point, τ_{max}, is noted. The irregular shear stress time history is then converted to an equivalent uniform stress time history at an amplitude of $0.65\tau_{max}$ using the procedure established by Seed et al. (1975). The number of uniform stress cycles required to cause cyclic softening is obtained from curves published by Idriss and Boulanger (2008) and compared with the number of equivalent uniform stress cycles from the previous step to determine if softening is likely to occur. A more detailed description of this method will be published in a forthcoming paper.

If the method described above indicates that softening is likely to occur, professional judgement was used to estimate the consequences of cyclic softening in the absence of testing. The undrained strength of the core under monotonic loading was reduced by 70 percent when softening was determined to be likely to occur. For scenarios where the clay core is determined to have a low likelihood of softening, a modest reduction of 20 percent was still applied to the monotonic strength to account for uncertainty in the screening method used. This approach is believed to be conservative, since the clay core was compacted near its optimum water content (resulting in a low liquidity index) using modern construction equipment.

Critical Shear Surfaces

A potential shear surface must be selected to calculate seismic displacements using the Newmark method. For this study, three potential shear surfaces were selected for both the upstream and downstream slopes of Lookout Point Dam. This paper will discuss only the upstream shear surfaces, which were calculated to have the largest seismic displacements.

The three potential shear surfaces begin near the crest of the dam and exit at points in the upper, intermediate, and lower zones of the dam, respectively. Entry and exit points were allowed to vary within these zones, and the critical surface was identified as the shear surface with the lowest yield acceleration (that is, the smallest average acceleration that initiates slope movement). The critical surface was determined using a pseudostatic approach, where the yield acceleration corresponds to the horizontal seismic coefficient required to reduce the factor of safety to 1.0. The resulting critical shallow, intermediate, and deep shear surfaces for the upstream slope are shown in Figure 5.

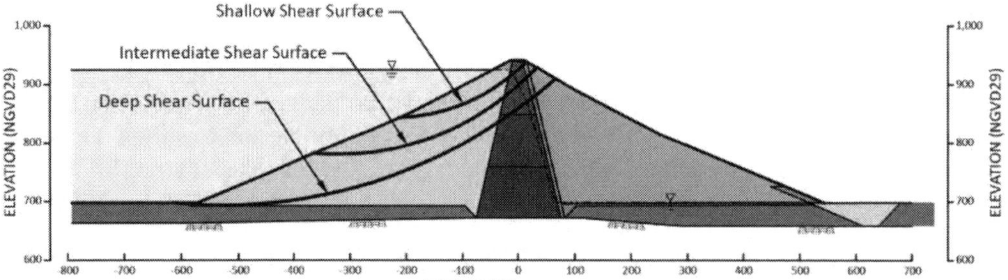

Figure 5. Critical Upstream Shear Surfaces.

Seismic Displacement Calculations

Permanent seismic displacements were calculated using the Newmark method, a decoupled approach in which the ground acceleration is calculated independently of the ground displacement. The GeoStudio software permits direct integration of the accelerations from QUAKE/W into the slope stability program SLOPE/W to perform this calculation. For each critical shear surface considered, the average acceleration time history is determined for the slope mass bounded by the shear surface and compared to the yield acceleration previously determined. Each time the average slope acceleration exceeds the yield acceleration, the area beneath the acceleration curve is double-integrated to determine the displacement. The sum of the incremental displacements gives the permanent seismic displacement along the shear surface.

RESULTS AND DISCUSSION

The following paragraphs present modeling results, validate these results through comparison with published values and alternative analysis methods, and discuss the implications of analysis results for dam performance.

Static Analyses

While the purpose of these analyses was not to evaluate the static stability of the dam, the static factor of safety does provide a useful index of the margin of safety that presently exists against additional destabilizing forces. With the pool at the maximum summer conservation level and assuming steady-state seepage, the static factor of safety against upstream slope failure varied from approximately 2.0 for shallow shear surfaces to 2.3 for deep shear surfaces.

Yield Acceleration

The yield accelerations for the critical upstream shear surfaces, which are shown in Figure 5 above, were calculated using both the softened and unsoftened cyclic strength of the clay core. Yield acceleration values are summarized in Table 3 on the following page. Reduction of the core strength produces the largest effect on the shallow shear surface due to the length of the shear surface passing through the core. The difference is smallest for the deep surface, which derives its resistance primarily from the rockfill shell.

Ground Motion Amplification

Ground motions applied at the base of the finite element model amplified as they propagated upward. An example of this amplification is illustrated in Figure 6 on the following page, which presents the peak horizontal acceleration contours for the 2,500-year Valparaiso motion. The applied motion had a PGA of 0.27 g, which amplified to 0.56 g at the crest of the dam. An examination of the contours shows that a modest damping effect occurs within the central portion of the dam, while most of the amplification occurs within approximately 40 feet of the ground surface.

Table 3. Yield Acceleration.

Shear Surface	Yield Acceleration (g)	
	Unsoftened $s_{u,cyclic}$	Softened $s_{u,cyclic}$
Shallow	0.21	0.15
Intermediate	0.21	0.18
Deep	0.22	0.20

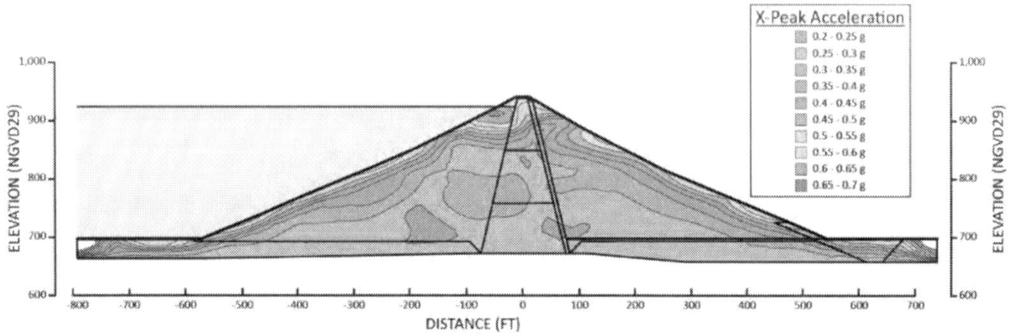

Figure 6. Peak Horizontal Acceleration Contours for 2,500-year Valparaiso Motion.

The amplified ground motions and associated amplification ratios are summarized in Table 4. These ratios compared favorably with observed amplification ratios reported by Mejia (2012).

Table 4. Ground Motion Amplification.

Event	1,000-year Motion			2,500-year Motion			10,000-year Motion		
	PGA Base	PGA Crest	Amp. Ratio	PGA Base	PGA Crest	Amp. Ratio	PGA Base	PGA Crest	Amp. Ratio
Valparaiso	0.16	0.44	2.8	0.27	0.56	2.1	0.51	0.75	1.5
Michoacan	0.14	0.40	2.9	0.24	0.57	2.4	0.45	0.73	1.6
Synthetic	0.14	0.28	2.0	0.25	0.41	1.6	0.46	0.62	1.3
Average	***0.15***	***0.37***	***2.5***	***0.25***	***0.51***	***2.0***	***0.47***	***0.70***	***1.5***

A simplified displacement analysis was performed using the Makdisi and Seed (1977) method to verify the reasonableness of the seismic displacements, as will be discussed in a later section. This method also estimates the peak horizontal acceleration at the crest and can be used as an additional means of verification. The Makdisi and Seed method calculated the PGA at the crest to be 0.59 for the 2,500-year Valparaiso motion, 0.57 for Michoacan, and 0.50 for Synthetic. These values also are in good agreement with the accelerations calculated in the finite element model.

Softening of Clay Core

The cyclic softening potential was evaluated at multiple points within the clay core. Using the method outlined in this paper, cyclic softening is considered likely to occur in the upper half of the core for the 2,500-year and 10,000-year ground motions. The upper half is considered unlikely to soften under the 1,000-year motion, and the lower half of the core is unlikely to soften under any of the motions for the return periods considered.

Seismic Displacements

Seismic displacements were calculated with the Newmark method using the yield acceleration which corresponds to the softened or unsoftened state of the clay core, as appropriate. The resulting permanent seismic displacements are summarized in Table 5.

Table 5. Calculated Seismic Displacement of Upstream Shear Surfaces.

Event	Permanent Seismic Displacement (inches)								
	1,000-year Motion			2,500-year Motion			10,000-year Motion		
	S	I	D	S	I	D	S	I	D
Valparaiso	0.3	0.0	0.0	11	0.1	0.0	74	9	0.4
Michoacan	2.3	< 0.1	0.0	22	1.7	0.0	79	16	3.1
Synthetic	1.1	0.0	0.0	20	0.4	0.0	110	8	0.0
Average	***1.2***	***0.0***	***0.0***	***18***	***0.7***	***0.0***	***88***	***11***	***1.2***

Note: S = Shallow shear surface, I = Intermediate shear surface, D = Deep shear surface

Displacements computed using the Makdisi-Seed simplified analysis method were compared with results from the Newmark method. The comparison was performed for the 2,500-year return period motions at an early stage of the analyses, prior to incorporation of the core softening potential. The upper bound of the Makdisi-Seed displacements was 3.4 to 12.4 inches, depending on the input motion. These values are in good agreement with the range of 3.0 to 9.5 inches calculated by the Newmark method for the shallow surface without consideration of the cyclic softening of the clay core.

The potential consequences of the calculated displacements can be evaluated by comparison with the available freeboard and width of the filter zone. Table 5 shows that computed displacements are small (less than 1 inch) for intermediate and deep shear surfaces for the 2,500-year and smaller motions. Computed displacements are larger for the shallow shear surface, but even the 18 inches of displacement calculated for the 2,500-year motion is unlikely to significantly affect the performance of the 8-foot-thick filter or to appreciably reduce the 16 feet of freeboard. In general, the conservative approach applied to the softened shear strength of the core did not result in excessive displacements. This result is due in large part to the support of the gravel shells on both sides, which are free-draining, well-compacted, and assumed to maintain most of their strength during the seismic event.

The computed displacements for the shallow surface under the 10,000-year motion are significantly larger. If the calculated displacements of 6 to 9 feet were to occur, it is likely that the performance of the filter would be compromised at that location. However, consider that the shallow shear surface intersects the filter at or above pool level near the crest of the dam (Figure 5). A large offset at this location may not be as critical to the overall performance of the dam as a large offset at a lower elevation, and the analyses show that displacements decrease rapidly with depth. A more important consideration may be the Newmark method's reliance on a decoupled approach in which slope acceleration and displacement are evaluated independently. The validity of this approach becomes less tenable at such large displacements. However, the results still provide a useful index, indicating that seismic displacements may be large.

CONCLUSIONS

A two-dimensional, finite element model was developed for a 250-foot high, zoned embankment dam using material descriptions, laboratory data, and as-built drawings from design and construction records. The equivalent linear procedure was used to analyze the embankment dam using three acceleration time histories scaled to match the response spectra of three return periods. The Newmark method was used to calculate permanent seismic displacements considering the cyclic softening potential of the clay core.

The analyses indicated that the upper half of the clay core is likely to soften under cyclic loading from a subduction zone event with a return period of 2,500 years or greater. Calculated seismic displacements are largest for the upstream slope, decrease with depth of the failure surface, and increase with intensity of the ground motions. However, even with conservative modeling of the softened strength, calculated displacements are not excessive (less than 22 inches), except for displacements along the shallow shear surface under the 10,000-year motion. This result is strongly influenced by the support of the gravel shells, which are assumed to maintain most of their strength during a seismic event.

The calculated seismic displacements were significantly larger (6 to 9 feet) for the shallow shear surface under the 10,000-year motion. At displacements of this magnitude, the equivalent linear ground response analysis and the Newmark method's decoupled approach of evaluating slope displacements becomes less tenable, and uncertainty in the computed displacements is greater. However, the results still provide a relative indication that seismic displacements may be large.

ACKNOWLEDGEMENTS

The authors wish to acknowledge the U.S. Army Corps of Engineers for furnishing historical documents pertaining to Lookout Point Dam and granting permission for this publication.

REFERENCES

AMEC Geomatrix and Quest Structures (2009). Regional Seismic Hazard Assessment: Willamette Valley in the Pacific Northwest Region. Report prepared for the U.S. Army Corps of Engineers, Washington, D.C., 24 February 2009.

Egan, J.A. and Ebeling, R.M. (1985). Variation of Small-Strain Shear Modulus with Undrained Shear Strengths of Clays. Presented at Second International Conf. on Soil Dynamics & Earthquake Engineering onboard the liner Queen Elizabeth 2, June 1985.

Idriss, I.M. and Boulanger, R.W. (2008). Soil Liquefaction During Earthquakes. *Report No. MNO-12*, Earthquake Engineering Research Institute, Oakland, CA.

Jamiolkowski, M., Ladd, C. C., Germaine, J. T., and Lancellotta, R. (1985). New developments in Field and Laboratory Testing of Soils. In *Proceedings of the 11th International Conference on Soil Mechanics and Foundation Engineering*, San Francisco, Vol. 1, pp. 57 – 153.

Kramer, S.L. (1996). *Geotechnical Earthquake Engineering*. Prentice Hall, Upper Saddle River, NJ.

Makdisi, F.I. and Seed, H.B. (1977). A Simplified Procedure for Estimating Earthquake-Induced Deformations in Dams and Embankments. *Report No. EERC 77-19*, Earthquake Engineering Research Center, University of California, Berkeley.

Mejia, L.H. (2012). "Analysis of Dynamic Response." Presented at short course entitled *Seismic Design and Analysis of Embankment Dams*, Seattle, WA, May 11, 2012.

Peterson, M.D., Zeng, Y., Haller, K.M., McCaffrey, R., Hammond, W.C., Bird, P., Moschetti, M., Shen, Z., Bormann, J., and Thatcher, W. (2014). Geodesy- and geology-based slip-rate models for the Western United States (excluding California) national seismic hazard maps. *U.S. Geological Survey Open-File Report 2013–1293*, 80 pp., *http://dx.doi.org/10.3133/ofr20131293*.

Seed, H.B. and Idriss, I.M. (1970). Soil Moduli and Damping Factors for Dynamic Response Analyses. *Report No. EERC 70-10*, Earthquake Engineering Research Center, University of California, Berkeley.

Seed, H.B., Idriss, I.M., Makdisi, F., and Banerjee, N. (1975). Representation of Irregular Stress Time Histories by Equivalent Uniform Stress Series in Liquefaction Analyses. *Report No. EERC 75-19*, Earthquake Engineering Research Center, University of California, Berkeley.

U.S. Army Corps of Engineers (1954). Lookout Point Dam Permanent Record Samples Summary. Prepared by Office of the District Engineer, Portland, Oregon, May 1954.

U.S. Army Corps of Engineers (1959). Review of Soils, Design, Construction, and Performance Observations, Lookout Point Dam, Oregon. *Technical Report No. 3-502.* Prepared by U.S. Army Engineer Waterways Experiment Station, Vicksburg, Mississippi, May 1959.

U.S. Army Corps of Engineers (1981). Lookout Point Lake, Report of Dam Safety Assurance Study. Prepared by Office of the District Engineer, Portland, Oregon, February 1981.

Vucetic, M. and Dobry, R. (1991). Effect of Soil Plasticity on Cyclic Response. *Journal of Geotechnical Engineering*, ASCE, 117(1), pp. 89-107.

PERFORMANCE OF FENA DAM DURING THE 1993 GUAM EARTHQUAKE

Lelio H. Mejia, PhD, PE, GE[1]

ABSTRACT

The potential for softening and strength loss of cohesionless and clay-like soils during strong, long-duration earthquake shaking is a key issue in the seismic safety assessment of embankment dams located in tectonic subduction regions of the world, and in regions where the hazard from large magnitude earthquakes is significant. Limited published experience is available regarding the performance of dams underlain by fine-grained soils subjected to large earthquakes. This paper presents a case history on the observed performance of Fena Dam during the magnitude 7.8, 1993 Guam Earthquake, which was generated by subduction of the Pacific Plate under the Philippine Plate. Located on the Island of Guam, the dam is an earth and rockfill embankment founded on deep, young alluvial and lacustrine deposits. The paper describes the observed performance of the dam during the earthquake, presents analyses of the dam's seismic stability, and discusses implications of the observed behavior for practice.

INTRODUCTION

Liquefaction of cohesionless soils and softening of clayey soils during strong, long-duration earthquake shaking are key considerations in the seismic safety evaluation of embankment dams located in seismic regions where the hazard from very large magnitude earthquakes is significant, such as the US Pacific Northwest tectonic subduction region. Limited experience has been published in the technical literature regarding the seismic performance of embankment dams underlain by fine-grained soils subjected to strong, long-duration earthquake motions. Thus, case histories are needed that can shed light on the behavior of those types of soils in dam foundations when subjected to such earthquake motions.

This paper presents a case history of the observed performance of Fena Dam during the magnitude M 7.8, August 8, 1993 Guam Earthquake, which was generated by subduction of the Pacific Plate under the Philippine Plate, along the Mariana Trench. Located on the Island of Guam, about 8 miles southwest of the City of Agana and 5 miles southeast of Apra Harbor, the dam is a 135-foot-high earth and rockfill embankment founded on deep, young alluvial and lacustrine deposits. Views of the dam's upstream and downstream slopes are shown in Figure 1. The dam is owned and operated by the U.S. Navy.

The dam's performance during the earthquake represents an excellent case history on the behavior of an embankment dam founded on silty and clayey soils shaken by a major subduction earthquake. This paper describes the observed performance of the dam and presents analyses of its seismic response, which provide insight into the dam's observed behavior. The paper also discusses implications of the observed behavior and of the analyses results for engineering practice.

[1] Senior Principal, Geosyntec, 1111 Broadway, Oakland, CA, 94607, LMejia@Geosyntec.com

Figure 1. Views of upstream (left) and downstream (right) slopes of Fena Dam

TECTONIC AND GEOLOGIC SETTING

Guam is the largest and southern-most of the Mariana Islands, a volcanic ridge above the Mariana Subduction Zone. This is an active tectonic plate boundary where the Pacific Plate dives under the Mariana Plate, a microplate along the eastern edge of the Philippine Plate, at a conversion rate of about 60 mm/year (Sella et al., 2002). Guam lies about 80 km northwest of the Mariana Trench, which is the surface expression of the subduction.

The dam and reservoir are located in the upper part of the Fena River valley in south-central Guam. The dam closes a natural throat in the valley where two spurs from the valley slopes form the dam abutments. The embankment is founded on deep alluvium and lacustrine deposits of Holocene age. Well-cemented tuffaceous sandstone and volcanic conglomerate bedrock underlies the alluvium and forms the abutments. Other bedrock units in the area include tuff, tuff breccia, and basalt flows (Tracey et al., 1964).

Small faults, shear zones, and prominent joints have been mapped in the dam and reservoir area. In addition, three prominent crustal faults crisscross the island (the Talofofo fault zone, the Adelup fault, and the Tamuning-Yigo Fault zone). However, none of the island faults show evidence of recent seismic activity (WCC, 1995).

THE 1993 GUAM EARTHQUAKE

With a moment magnitude of M 7.8, and a surface wave magnitude of M_s 8.1, the August 8, 1993 earthquake was the largest and most damaging to affect Guam in recent history. As shown in Figure 2, its source was an approximately 40-km-wide and 120-km-long rupture within the subduction zone. The fault plane was interpreted to strike at about N65E, approximately parallel to the strike of the Mariana Trench, and dip about 15 degrees to the northwest. The rupture was a shallow-dipping thrust toward the northwest with a small strike-slip component, which began about 40 km south of the dam, at a depth of about 50 km, and propagated to the northeast over about 32 seconds (Tanioka, et al., 1993; EERI, 1995). The main shock was followed by a relatively large number of aftershocks. Global Positioning System (GPS) measurements indicated that after the earthquake the island had displaced about 25 cm to the southeast and subsided about 10 cm (Beavan et al., 1993).

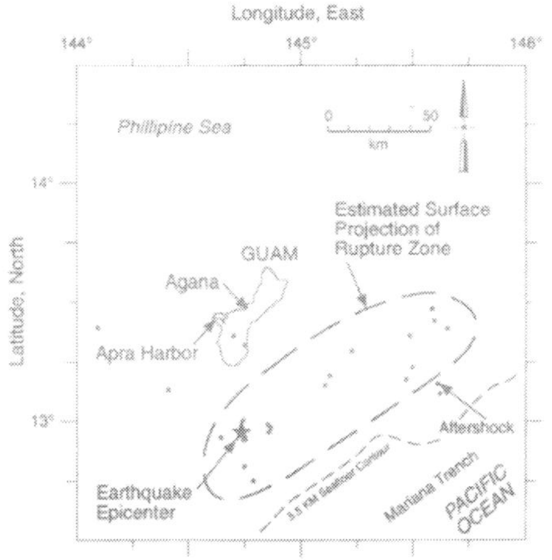

Figure 2. Location of M 7.8 August 8, 1993 Guam Earthquake

No strong ground motion records were obtained on the island during the earthquake due to malfunction of local instruments. For the dam site, the attenuation relationships of Crouse (1991) and Youngs et al. (1988) yielded estimates of rock ground motions characterized by a peak horizontal acceleration of about 0.2 g. Those estimates are consistent with the observed performance of structures and equipment throughout the island, such as damage at the base of single-column, concrete bus-stop shelters, which indicated that peak accelerations in a large portion of the island were on the order of 0.15 to 0.25 g (EERI, 1995). Newspaper and personal accounts indicate that the duration of strong ground shaking was about 60 seconds.

DAM DESCRIPTION

The maximum transverse section of the dam and a section along its longitudinal axis are shown in Figure 3. The dam is a zoned earth and rock embankment with a wide central impervious core, which impounds an 8,400 acre-feet reservoir. From the base of the core to the crest, the embankment has a structural height of 135 feet. It rises 85 feet above the original valley floor and has a total volume of 1,600,000 yd^3. The dam crest is 30 feet wide, 1,050 feet long, and is capped by a 20-foot-wide paved road. The crest elevation provides approximately 15 feet of freeboard above the spillway crest.

Near the center of the valley, the dam is founded on alluvium and lacustrine deposits up to 95 feet thick. On both abutments, the core was founded on rock while the shells were placed on residual soils with limited stripping. In the valley, the core was founded in a 50-foot-deep trench excavated into the alluvium. The core trench is 200 feet wide at the base (see Figure 3).

The core is flanked by filters and rockfill shells with outer upstream and downstream slopes of 3:1 and 2:1 (H:V). As shown in Figure 3, the filters wrap beneath the rockfill, and are founded on the alluvium. Filter material was also used to cap the core and build the crest camber. Materials from required excavations not used in the main embankment were placed in random fill berms upstream and downstream of the shells to a maximum thickness of about 30 feet. A single line grout curtain was installed along the longitudinal axis of the dam (WCC, 1995).

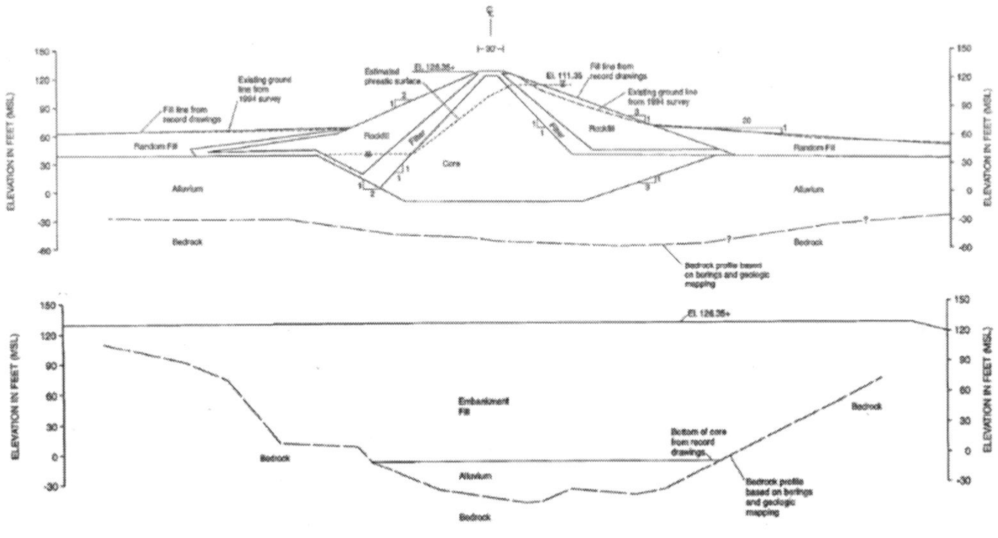

Figure 3. Transverse (top) and longitudinal (bottom) sections of Fena Dam

Seepage that reaches the downstream filter blanket is collected at a toe drain in a 10-inch-diameter concrete pipe, and discharges at the toe of the dam into a 36-inch pipe that leads to the river, approximately 800 feet downstream from the dam toe.

The spillway is a straight, rectangular, reinforced-concrete, open channel 900 feet long and 80 feet wide. The channel was cut over a topographic ridge of sound tuffaceous sandstone on the right abutment, and has a discharge capacity of 17,500 cfs. The structure ends in a 130-foot-long, 30-foot-deep stilling basin. The outlet works have a discharge capacity of 60 cfs and consist of a reinforced concrete intake structure, a reinforced-concrete tunnel conduit, and a terminal structure. The tunnel runs through the tuffaceous sandstone of the right abutment and is about 600 feet long.

EMBANKMENT CONSTRUCTION

Construction of the dam began in July of 1949 with excavation of the core trench. The embankment was topped out in June of 1951 and the reservoir was filled within one year after. The core materials were specified as select borrow clay and were placed in 6-inch lifts and compacted with sheepsfoot rollers or crawler tractors. The moisture content was kept between 42 and 50 percent, except in the upper 26 feet where the specification was relaxed to allow the maximum water content that would result in a minimum in-place dry density of 65 pcf.

The filter materials were specified as free-draining, graded quarry limestone with a maximum size of 10 inches, and were thoroughly wetted and compacted in 12-inch lifts by four passes of a 20-ton crawler tractor. The rockfill was quarry run limestone rock specified to be a free-draining mixture of rock fragments, cobbles, and boulders with a maximum size of 1 yd^3. Rock spalls were specified not to exceed the amount needed to fill-in the voids in the coarse material. The rockfill was dumped and leveled in 3-foot layers without compaction or sluicing. The random fill berms were placed in 12-inch layers and compacted with crawler tractors.

DAM PERFORMANCE BEFORE THE 1993 EARTHQUAKE

Overall, the dam's performance since construction has been satisfactory. During construction, and within a few years after, the dam experienced significant settlements due to consolidation of the alluvium under the weight of the embankment. Foundation settlements were measured with base plates installed at the foundation after excavation of the core trench. Surface settlements were measured with survey monuments installed as the fill was raised. The settlements showed an increasing trend with increasing thickness of the underlying alluvium and increasing weight of the overlying embankment.

Through construction and topping-off of the embankment, the maximum foundation settlements measured about 3.3 feet near the maximum section of the dam, over the deepest alluvium beneath the embankment toes. About 3 feet of foundation settlement was recorded beneath the crest near the dam maximum section over the same period. In the 3 years following embankment construction, an additional 0.45 feet of foundation settlement was registered beneath the dam crest. The surface settlements over the same 3-year period approached 1.2 feet and 0.8 feet on the crest upstream and downstream edges. These measurements indicate about 0.75 feet and 0.35 feet of compression settlement of the upstream and downstream rockfill shells over the same period. Surface settlements were continuing at a small rate as of the August 8, 1993 earthquake.

Continuing settlements at the crest have been associated with recurrent longitudinal cracking of the crest road pavement. The settlement data and pattern of observed cracking indicate that at least 40 to 50 percent of the post-construction settlement measured at the crest edges has resulted from compression of the rockfill shells. Vertical cracks a few inches wide, but no deeper than the limestone fill overlying the core, have been observed from 5 to 10 feet upstream and downstream of the crest centerline since construction. Historically, the cracks downstream of the centerline have been less severe than those upstream.

Pneumatic piezometers installed in the core and foundation during construction indicated that: 1) no hazardous excess pore pressures developed in the core or the foundation during construction, 2) excess pore pressures dissipated quickly, and 3) pore pressures reached equilibrium rapidly upon reservoir filling. The piezometers were abandoned shortly after 1955, but other instruments installed since indicate normal phreatic levels (see Figure 3).

Before August 8, 1993, no major earthquakes had shaken the dam strongly and its seismic performance had been satisfactory, particularly during notable regional events of magnitudes 7.0 and 6.1 in 1975 and 1983 (WCC, 1995).

MATERIAL CHARACTERISTICS

Foundation

As shown in Figure 3, the mid-section of the dam's core is founded on up to 45 feet of alluvium whereas the abutments are founded on sandstone bedrock. The upstream and downstream shells are founded on thick alluvium up to 90 feet thick.

Primarily lacustrine and flood plain deposits, the alluvium soils are predominantly medium-stiff to stiff clayey silts of high plasticity with interbedded layers of silty sands. Figure 4 shows the variation of fines content, Atterberg limits, and water content within the alluvium. The plasticity index of the materials ranges broadly between about 10 and 70, the water content varies between about 45 and 85 percent, and the liquidity index is between about 20 and 90 percent.

The normalized SPT blow count, $(N_1)_{60}$, of the materials in the upper 40 feet of the deposit is less than 10 and averages about 7. Below, the blow count ranges from about 5 to 20 and the average increases from about 7 to about 15 over a depth of 50 feet. The average dry unit weight of the materials is relatively uniform with depth and increases slightly from an average of about 60 pcf near the ground surface (approximately elevation 40 feet above MSL) to about 65 pcf at a depth of 90 feet. Consolidated-undrained triaxial compression (CIU) tests indicate that the ratio of undrained shear strength to effective consolidation pressure, S_u/p', is about 0.35. The corresponding ratio for simple shear loading is about 0.23.

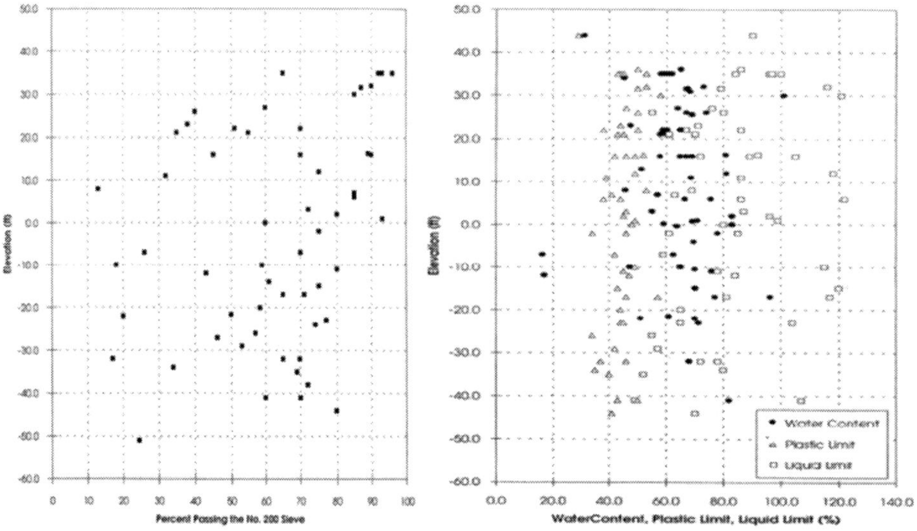

Figure 4. Variation of measured index properties within foundation alluvium

Sandy soils within the deposit classify mostly as SM materials. Thus, although the alluvium is predominantly plastic clayey silt soil, the stratum is interspersed with sandy materials that would be considered susceptible to liquefaction, if their fines were to be of low plasticity. Based on the observed increase in blow counts with depth, the stratum may be characterized as consisting of two units: an upper and a lower alluvium, defined by a contact at elevation -10 feet approximately. The lower alluvium was chosen as the base for the core in the design of the dam.

Embankment

Sampling of the embankment core recovered a very stiff silt of high plasticity (MH) with a plasticity index ranging between about 40 and 70, liquid limits between 80 and 120, and water contents between about 40 and 60, just above the plastic limit. The normalized SPT blow count, $(N_1)_{60}$, ranged between about 15 and 25 with a representative average of about 20, approximately constant with depth. Such a trend is also observed in the dry and total unit weights of the material, which are narrowly distributed and average about 70 pcf and 105 pcf, respectively.

The shell rockfill consists of limestone boulders up to 3 feet in dimension with a clayey sandy gravel matrix. The overall mass is free-draining and is characterized by rock-to-rock contact between large particles. The matrix consists of about 50 to 60 percent gravel, 20 to 30 percent sand, and 16 to 18 percent fines. The filter material consists of graded quarry limestone up to 10 inches in dimension and is dense and free draining.

DAM RESPONSE TO THE M 7.8 1993 EARTHQUAKE

An approximately 500-foot-long section of the crest road upstream shoulder slumped by up to 2.9 feet during the earthquake. As shown in Figure 5, the slumping created a 2-foot-wide crack with a 3-foot-deep scarp near the contact between the road shoulder and the pavement. This contact lies approximately above the upstream edge of the core. The earthquake also widened pre-existing longitudinal cracks and induced new longitudinal cracks along the crest (Figure 5). The reservoir level was 30 feet below the spillway crest at the time of the earthquake.

Figure 5. Upstream crest shoulder slump (left) and crest road cracks (right)

No major sliding of the upstream slope, or transverse cracking of the crest, was observed. The crest road shoulder slump did not affect the integrity of the core, and the overall stability of the rockfill shell and of the foundation was maintained. The crest road shoulder slump and crest cracks were repaired within a few days after the earthquake.

Post-earthquake topographic and bathymetric surveys of the embankment, extending well beyond the downstream and upstream toes of the dam, showed about a 5-foot depression of the upstream slope ground surface relative to the fill line from the design drawings, near the mid-section of the dam (Figure 3). Relative depressions of the upstream slope surface up to about 2 feet deep were also measured at cross sections cut near the crest quarter points, where the upstream shell is founded on bedrock. The observed upstream crest edge settlements suggest that the measured depressions may have been caused in large part by the earthquake shaking.

Settlements along the crest upstream edge exceeded 2 feet over a length of about 300 feet, whereas settlements of the downstream edge were less than about 0.3 feet. The settlements of the downstream edge were accompanied by horizontal displacements downstream of up to 0.25 feet. Measured horizontal displacements and settlements at the toe of the downstream rockfill shell and along the downstream berm were also small, typically less than 0.2 feet (WCC, 1995).

The survey measurements confirmed that sliding or significant spreading of the embankment did not occur, and suggest that the observed slumping of the crest road shoulder resulted from volumetric compression of the upstream rockfill shell during the earthquake shaking. Such mechanism of deformation is consistent with settlement patterns of the rockfill shells before the earthquake, the fact that the rockfill materials were not compacted, and observations at other dumped-rockfill dams during past earthquakes. The significant difference in settlement between the upstream and downstream shells is likely due to the fact that a larger volume of the upstream rockfill was saturated at the time of the earthquake.

Although dense vegetation precluded close inspection of the downstream slope and random fill berm, no evidence of significant movements or cracking or of excessive seepage in these areas could be detected immediately after the earthquake or after the areas were cleared a few months later. Similarly, no significant damage to the spillway or outlet works was reported.

No evidence of excessive seepage in downstream areas could be detected immediately after the earthquake or in the weeks thereafter. A 2- to 4-foot rise in water levels was measured 5 days after the earthquake in three piezometers at the dam crest, relative to measurements 4 days before the earthquake.

ANALYSIS OF DAM RESPONSE

To provide insight into the dam's observed performance during the August 8, 1993 earthquake, the embankment seismic response to the earthquake was analyzed using procedures representative of current engineering practice. This exercise also served to evaluate the ability of such analysis procedures to explain the dam's performance.

Analyses were performed for the maximum section of the dam using two-dimensional analysis techniques and the acceleration time history shown in Figure 6, which was selected to represent the rock motions estimated to have occurred at the dam site during the earthquake. Analyses were also performed for motions with an amplitude about 1.6 times larger than those shown in Figure 6, and for the motions recorded at station 'La Union' during the M_s 8.1 1985 Mexico Earthquake, scaled by a factor of 1.8.

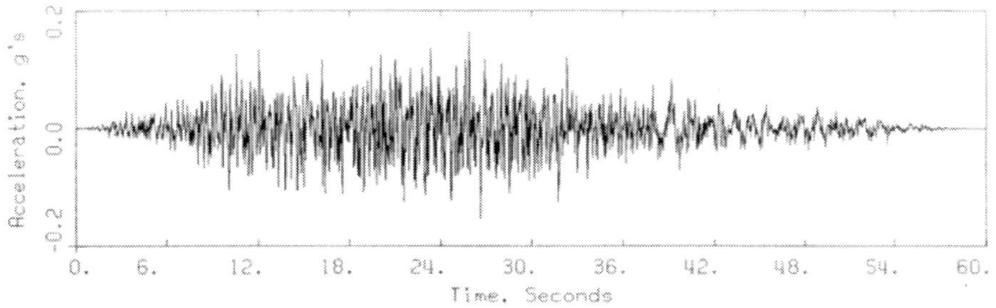

Figure 6. Accelerogram selected to represent rock motions during 1993 earthquake

If it were founded directly on rock, the embankment would generally be expected to perform well under earthquake shaking, based on experience regarding the seismic performance of dams. The core material and transition zones were well compacted and have a high seismic resistance. Although the shell rockfill is loose to medium dense, the material is free draining and unlikely to lose significant strength, despite being susceptible to densification under seismic shaking. Thus, the key issue regarding the dam's capacity to withstand earthquake loads is the behavior of the foundation alluvium.

Using the general Seed-Lee-Idriss approach (Seed, 1979), analyses were performed to evaluate the potential for liquefaction and cyclic softening of sandy and clayey soils in the foundation alluvium. Finite element procedures were used to calculate the stresses in the embankment and foundation due to normal, gravity and seepage loads and due to earthquake shaking. The earthquake-induced stresses were compared with the cyclic resistance of the foundation materials against development of excess pore pressures, liquefaction triggering, and cyclic softening. In addition, seismic deformations calculated from nonlinear dynamic finite element analyses were compared with the observed dam deformations.

Figure 7 shows the calculated effective cyclic stress ratio, τ_{eff}/σ'_v, throughout the embankment and foundation for the 1993 Guam Earthquake time history. The effective shear stress, τ_{eff}, corresponds to 0.65 times the peak horizontal shear stress, τ_{max}, calculated from equivalent-linear dynamic analyses. The vertical effective stress, σ'_v, was calculated from static stress analyses. As shown in Figure 7, the calculated cyclic stress ratio in the foundation beneath the core and downstream is between 0.1 and 0.2, whereas it is between about 0.2 and 0.3 upstream. The analyses for the motions equal to the 1993 earthquake motions scaled by a factor of 1.6 yielded cyclic stress ratios of about 0.2 beneath the core and downstream, and between about 0.3 and 0.5 upstream.

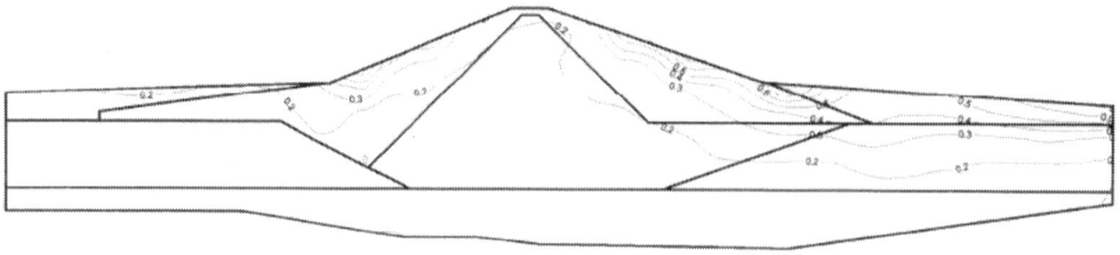

Figure 7. Effective cyclic stress ratio, τ_{eff}/σ'_v, for the 1993 Guam Earthquake

Liquefaction Triggering

Based on their construction placement and gradation and plasticity characteristics, it may readily be concluded that the embankment materials are not susceptible to liquefaction. That conclusion also applies to the clayey silts in the alluvium, as those materials have liquid limits ranging from 50 to 120 and plasticity indices ranging from about 15 to 70. It is less clear, however, why no evidence of liquefaction in the alluvium silty sands was observed after the earthquake.

The silty sands have over 20 percent fines with plasticity indices higher than 15, and an average normalized SPT blow count, $(N_1)_{60}$, of about 7 in the upper 40 feet of the alluvium (see Figure 4). Below that depth, they have an average $(N_1)_{60}$ of about 10 to 12. Figure 8 shows a comparison between the cyclic stress ratio induced by the earthquake and the cyclic resistance of the sands estimated using the correlations developed by Seed et al. (1985) for non-plastic sands.

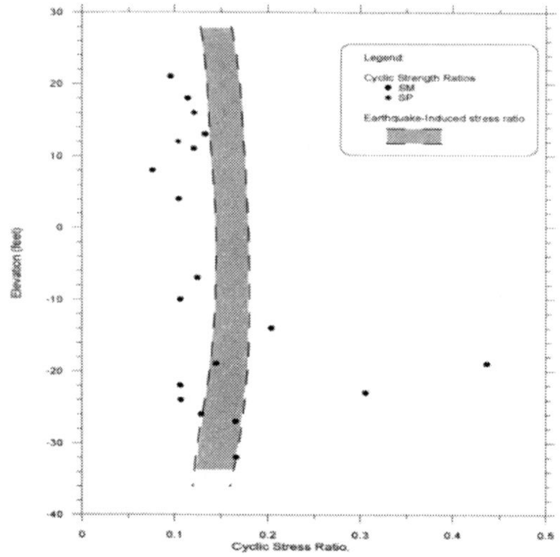

Figure 8. Earthquake-induced stress and cyclic resistance ratios for 1993 earthquake

Considering that the earthquake-induced stress ratio exceeds significantly the estimated cyclic resistance (Figure 8), it may be concluded that the alluvium sands would likely have liquefied early during the earthquake, if their fines were predominantly non-plastic. Because no evidence of overall foundation strength loss was observed after the earthquake, it may be concluded that either the silty sands are not susceptible to liquefaction, or they are not sufficiently continuous to affect the overall strength of the foundation, even if they were to liquefy.

Cyclic Softening

Cyclic triaxial compression tests were used to evaluate the behavior of the alluvium clayey silts under cyclic loading. The tests showed that the materials exhibit clay-like behavior and confirmed that they are not susceptible to liquefaction, although they can develop significant shear strains and pore pressures under strong cyclic loading.

Figure 9 shows the triaxial cyclic stress ratio required to cause various levels of double-amplitude axial strains for several loading cycles, measured in the cyclic triaxial tests. These results were used to estimate the cyclic resistance of the materials in terms of the shear cyclic stress ratio required to cause shear strains of 3.75% over a number of loading cycles for the conditions of effective overburden pressure and sustained shear stress in the dam foundation.

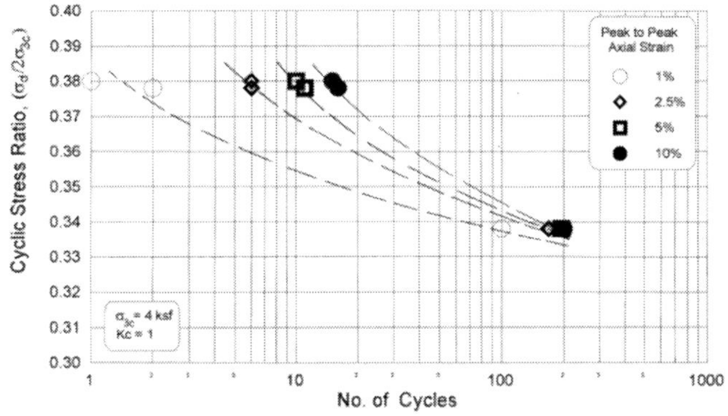

Figure 9. Results of cyclic triaxial tests on clayey silt alluvium

To estimate potential earthquake-induced shear strains and excess pore water pressures in the foundation, the cyclic resistance of the materials was compared to the calculated earthquake-induced stress ratio shown in Figure 7. In general, calculated factors of safety against development of shear strains of 3.75% were well above 1.0 throughout the alluvium, indicating a low potential for permanent foundation seismic strains and deformation (WCC, 1995).

The calculated shear strains in the foundation are shown in Figure 10. Integration of such strains with depth yields estimated lateral surface displacements for the downstream and upstream berms of about 0.5 and 1 feet, respectively. The associated settlement of the core at the crest is about 0.3 feet. These deformations are larger than those observed during the earthquake by a factor of 2 to 3.

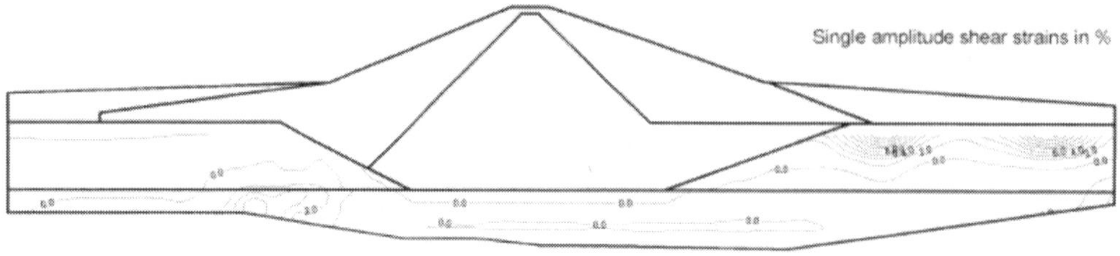

Figure 10. Calculated shear strains in foundation for 1993 Guam earthquake

Likewise, the calculations indicated a low potential for developing high excess pore pressures during the earthquake. Together with the results of post-cyclic compression strength tests, the calculated excess pore pressures were used to estimate the expected reduction in undrained shear strength in the alluvium after the earthquake. Analyses of post-earthquake stability using the reduced strengths indicated adequate factors of safety, in good agreement with the observed dam performance after the 1993 earthquake.

The analyses of expected shear strains for the motions scaled by a factor of 1.6 yielded significantly larger deformations with an estimated settlement of the core at the crest of about 2.5 feet. Nonetheless, the post-earthquake stability analyses with correspondingly reduced strengths, and with small-strain strengths, indicated adequate stability.

Nonlinear Analyses

Nonlinear dynamic analyses of the dam were used to obtain more refined estimates of the expected seismic deformations during the 1993 earthquake. The analyses were performed using a multiple-nested yield surface model (Salah-mars and Kavazanjian, 1992) to simulate the nonlinear stress-strain behavior of the embankment and foundation soils. This analysis model has been shown to yield calculated seismic deformations in good agreement with observed performance, when clayey materials control the response of a dam (Mejia et al., 1990).

Figure 11 shows the horizontal displacements at the downstream and upstream toes of the embankment, and the vertical displacement at the crest, calculated for the 1993 earthquake. Permanent horizontal displacements of about 4 and 6 inches are calculated at the downstream and upstream toes of the embankment. The calculated settlement of the crest is about 4 inches. The overall pattern of deformations corresponds to spreading of the embankment at its base due to shear deformation of the foundation alluvium.

The calculated displacements from the nonlinear analysis are higher by a factor of about 2 than the displacements observed after the earthquake. However, they are somewhat smaller (within a factor of 1.5) than those calculated by integration of shear strains calculated using the Seed-Lee-Idriss approach.

Nonlinear analyses for the motions equal to the 1993 earthquake motions scaled by a factor of 1.6, and for the scaled 'La Union' time history, yielded deformations about twice as large as those calculated for the 1993 earthquake. Parametric analyses for these higher motions with a 'lower bound' estimate of the foundation strengths yielded deformations about 1.5 times higher than those for the 'best estimate' strengths. In all cases the embankment remained stable after the earthquake shaking (WCC, 1995).

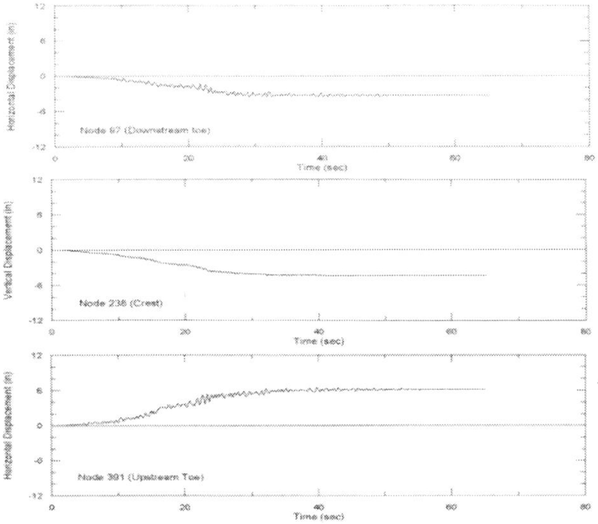

Figure 11. Calculated displacements from nonlinear analysis for 1993 Guam earthquake

DISCUSSION

Rockfill dams with central clay cores built on dense soil or rock foundations have performed well during large earthquakes. Although founded on deep, young fine-grained alluvium, Fena Dam withstood the M 7.8 August 8, 1993 Guam Earthquake with damage limited to a 3-foot slump of the crest road shoulder (due to settlement of the upstream rockfill shell) and minor longitudinal cracking at the crest. Settlements of the core and deformations of the downstream shell and berm were small, and only modest repairs to the dam crest were required after the earthquake.

Clearly, the overall strength of the dam foundation was maintained. This indicates that isolated zones of silty sands in the foundation alluvium did not liquefy, or that such zones are not sufficiently continuous to affect the overall strength of the foundation, even if they liquefied. The absence of any surface evidence of liquefaction suggests that the silty sands did not liquefy, which might be explained based on the high plasticity of their fines fraction. The observed performance also indicates that earthquake-induced strains in the alluvium clayey silt were small and that the earthquake-induced stresses were well below the seismic strength of the materials.

Considerable insight into the dam's observed performance was gained through analyses of seismic response, liquefaction triggering, and seismic deformations. The analyses indicate that the dam's acceptable performance can be ascribed to the predominantly clayey nature of the foundation alluvium, and the low susceptibility of the foundation materials to liquefaction. The analyses also suggest that, had the intensity of earthquake shaking been much higher than was estimated for the earthquake, significantly larger deformations may have developed, but the overall stability of the dam would have been maintained.

The simulation exercise also served to evaluate the ability of current methods of analysis to predict the observed dam performance. The analyses of liquefaction triggering using current methods for sands with fines of low plasticity suggest that liquefaction should have triggered in the foundation. This result does not appear to be consistent with the observed performance, and suggests that the alluvium silty sands may not be susceptible to liquefaction because of the high plasticity of their fines.

In general, the results of the analyses of earthquake-induced strains in the alluvium clayey silt using the Seed-Lee-Idriss approach, and the nonlinear analyses of deformations are consistent with the observed performance. This finding suggests that, if judiciously applied, such analysis methods can yield reasonable estimates of dam seismic deformations under similar situations.

CONCLUSIONS

The observed performance of Fena Dam during the M 7.8 August 8, 1993 Guam Earthquake represents an excellent case history on the behavior of dams underlain by deep, young clayey silt alluvium when subjected to long duration shaking from a subduction earthquake. The dam withstood the 1993 earthquake with limited damage that only required modest repairs after the earthquake. Other than significant compression of the loose upstream shell rockfill, overall deformations of the dam were small.

Insight into the dam's performance was gained through analyses using procedures representative of current practice. The analyses indicate that the dam's acceptable performance can be ascribed to the predominantly clayey nature of the foundation alluvium, and the low susceptibility of the foundation materials to liquefaction.

The simulations also served to evaluate the ability of current methods of analysis to predict the observed dam performance. In general, the results of the analyses of earthquake-induced strains using the Seed-Lee-Idriss approach, and the nonlinear analyses of deformations are consistent with the observed performance. This finding suggests that, if judiciously applied, such analysis methods can yield reasonable estimates of dam seismic deformations under similar conditions.

ACKNOWLEDGMENTS

This paper is dedicated to the memory of Dr. Said Salah-mars, a great engineer and friend, who contributed significantly to the development of nonlinear analysis models for soils and to the analyses of Fena Dam. The author also wishes to recognize the contributions of Mr. David Gross and Drs. Joseph Sun, John Rice, and Paul Somerville to the study, and the cooperation of the U.S. Navy. The support of AECOM (the author's former employer) during preparation of this paper is also gratefully acknowledged.

REFERENCES

Beavan, J., 1993, Surface deformation due to the M 7.8 Guam Earthquake of August 8, 1993, Abstracts of AGU Fall Meeting.

Crouse, C.B., 1991, Ground motion attenuation equations, for earthquakes on the Cascadia Subduction Zone, Earthquake Spectra, v. 7, pp 201-236.

Earthquake Engineering Research Institute (EERI), 1995, Guam earthquake of August 8, 1993 – Reconnaissance Report, Earthquake Spectra, Supplement to Volume 11, April.

Mejia, L.H., Sun, J.I., Salah-Mars, S., Moriwaki, Y., and Beikae, M., 1990, Nonlinear dynamic response analysis of Lexington dam, Proceedings of SMIP92, California Division of Mines and Geology, Sacramento, CA, May.

Salah-mars, S., and Kavazanjian, E., 1992, A virtual surface concept for nested yield surface plasticity, Int. Journal for Numerical and Analytical Methods in Geomechanics, Vol. 16, No. 11, 779-796.

Seed, H.B., 1979, Considerations in the earthquake-resistant design of earth and rockfill dams, Geotechnique, Vol. 29, No. 3, September, pp 215-263.

Seed, H.B., Tokimatsu, K., Harder, L.F., and Chung, R., 1985, Influence of SPT procedures in soil liquefaction resistance evaluations: Journal of Geotechnical Engineering Division, ASCE, v. 111, no. 12

Sella, G.F., Dixon, T.H., and Mao, A., 2002, REVEL—A model for recent plate velocities from space geodesy: Journal of Geophysical Research, v. 107, no. B4, p. ETG 11–1–ETG 11–30.

Tanioka, Y., Satake, K. and Ruff, L., 1993, Mechanism of the Guam earthquake, August 8, 1993, Abstracts of the Meeting of the Seismological Society of Japan, Tottori, August.

Tracey, J.I., Stark, S.O., Doan, J.T., and May, H.G., 1964, Geology and hydrology of Guam, U.S. Geological Survey, Professional paper 403-A, 104 p.

Woodward-Clyde Consultants (WCC), 1995, Fena Dam Seismic Study, Report prepared for Naval Facilities Engineering Command, U.S. Navy, Marianas (Guam), June.

Youngs, R.R., Day, S.M., and Stevens, J.L., 1988, Near-field ground motions for large subduction earthquakes, Proceedings specialty conference Earthquake Engineering and Soil Dynamics II, ASCE, pp 445-462.

SEISMIC EVALUATION OF THE REMEDIATED ISABELLA AUXILIARY DAM

George Hu, P.E.[1]
David C. Serafini, P.E. G.E.[2]
Henri V. Mulder, P.E.[3]
Vlad Perlea, P.E.[4]

ABSTRACT

The dynamic response of the proposed remediation of Isabella Lake Auxiliary Dam was evaluated using a finite difference software, Fast Lagrangian Analysis of Continua (FLAC2D version 7). Analyses were performed for the maximum design earthquake (MDE), a 10,000-year earthquake, and 84th-percentile maximum credible earthquake (MCE). Dynamic response of the potentially liquefiable soils was evaluated using a fully coupled effective stress constitutive model, UBCSAND. Four different post-triggering residual strength correlations (Seed and Harder (1990), Idriss and Boulanger (2008), Kramer and Wang (2015), and Weber (2015)) were used in this evaluation. The various residual strength correlations produced a large range in predicted post-earthquake crest settlement, ranging from 0 to 19 feet. This paper presents the numerical analysis of the dynamic response of the potentially liquefiable soils and the effects on the resulting deformations based on different residual strength correlations.

INTRODUCTION

Isabella Reservoir is located on the Kern River in the southern Sierra Nevada mountain range 34 miles upstream of the City of Bakersfield in Kern County, California. Isabella Dam was constructed from 1948 to 1953 and consists of two embankment dams: a Main Dam on the Kern River and an Auxiliary Dam crossing Hot Springs Valley to the east of the Main Dam. The Main Dam is 185 feet high and 1,695 feet long; the Auxiliary Dam is 100 feet high and 3,260 feet long. The town of Lake Isabella is just one mile downstream of the Auxiliary Dam. Isabella Dam has been designated as a Dam Safety Action Classification I project by the U.S. Army Corps of Engineers due to high risk from seismic, seepage, and hydrologic (overtopping) failure modes. To address the significant failure modes for the Auxiliary Dam, the dam will be modified to include a crest raise of 16 feet and a stabilizing buttress constructed against the downstream slope of the dam with portion of the downstream shallow liquefiable foundation alluvium removed. Figure 1 shows the existing conditions and the proposed remediated section at Station 58+00 as modeled in FLAC (Itasca, 2011).

[1] USACE SPD Dam Safety Production Center, 1325 J Street, Sacramento, California, George.Hu@usace.army.mil.
[2] USACE SPD Dam Safety Production Center, 1325 J Street, Sacramento, California, David.C.Serafini@usace.army.mil.
[3] USACE SPD Dam Safety Production Center, 1325 J Street, Sacramento, California, Henri.V.Mulder@usace.army.mil.
[4] Retired from USACE, perlea@msn.com.

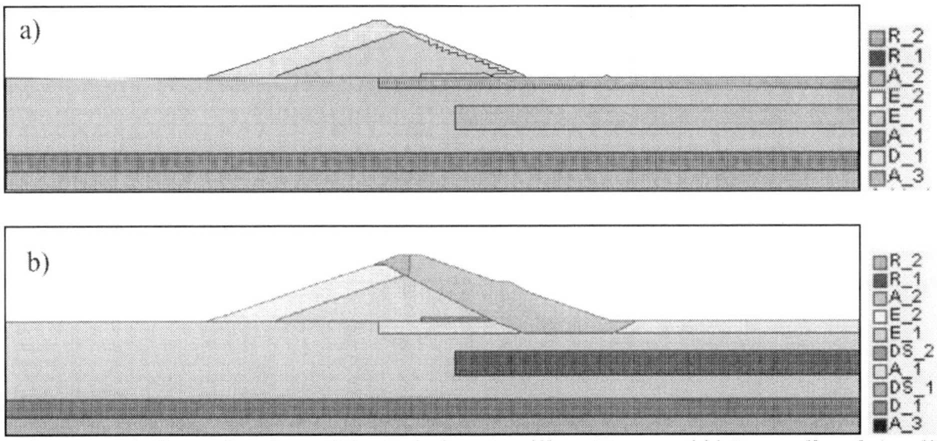

Figure 1. a) Cross-Section of the Existing Auxiliary Dam and b) Remediated Auxiliary Dam at Station 58+00 in FLAC

The majority of the Auxiliary Dam foundation is characterized by unconsolidated alluvial fan deposits derived from the eastern side of the Hot Spring Valley. Based on trenches near the downstream toe, these alluvial fan deposits primarily consist of silty sand and clayey sand. Site investigations and geotechnical analyses identified the alluvial fan deposits as being susceptible to liquefaction. The maximum section of the Auxiliary Dam embankment with the thickest alluvial section occurs between Station 56+00 and Station 58+00. The cross section at Station 58+00 was one of the sections evaluated for the remediated Auxiliary Dam and will be discussed in this paper.

Seismic hazard is high due to the active Kern Canyon Fault, which passes underneath the Auxiliary Dam near the right abutment and is capable of producing a maximum event of moment magnitude (Mw) 7.7 with a PGA of 0.68g (84th percentile maximum credible earthquake) based on probabilistic and deterministic seismic hazard study by URS (2016). The location of the dam and faults considered in the seismic hazard study, including Kern Canyon Fault, are shown in Figure 1. Kelson et. al. (2010) provides a detail discussion on the seismic hazard characterization of the Kern Canyon Fault.

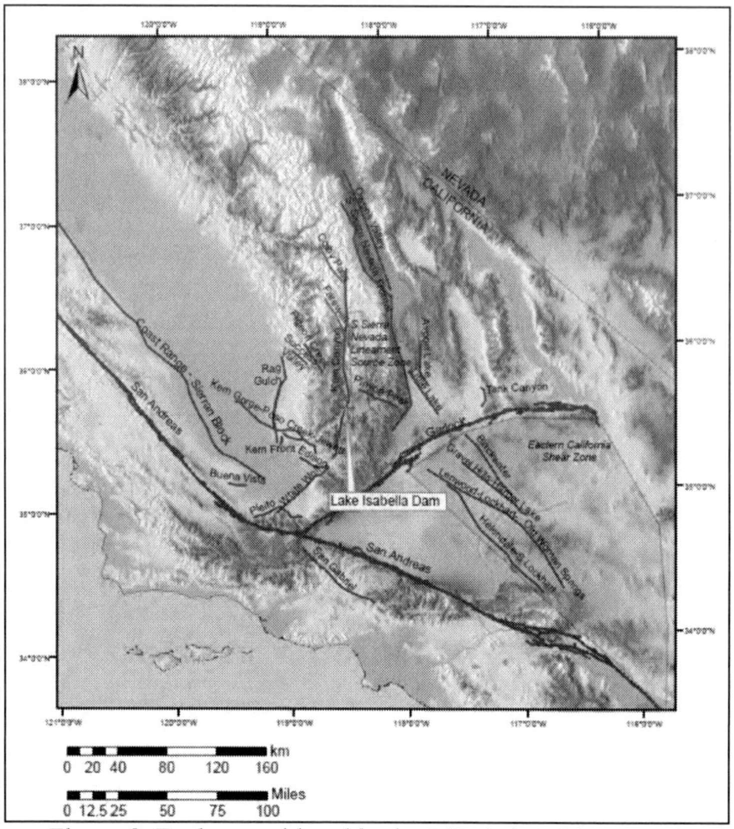

Figure 2. Faults considered in the PSHA (AECOM 2016)

GROUND MOTION INPUT

URS performed the probabilistic and deterministic seismic hazard analysis and developed the ground motions for Lake Isabella Main Dam and Auxiliary Dam. The seed ground motions used for the MCE (84th percentile), 10,000-year, and MDE events were NGA #143, 1978 Tabas, Iran, Tabas Station, NGA #1148, 1999 Kocaeli, Turkey, Arcelik Station, and NGA #1511, 1999 Chi-Chi, Taiwan, TCU76. All three-components (2 horizontal and 1 vertical) of the seed ground motions were spectrally matched to the target uniform hazard response spectrum (UHRS). Using UHRS as the target spectra is likely conservative, since it implies that large amplitude spectra will occur at all periods in one single ground motion. Properties of the spectrally matched time histories are presented in Table 1.

The ground motions were developed using Vs30 of 1,500 m/s, which is similar to the shear wave velocity of the intact rock being modeled (approximately 1,400 m/s). Since the acceleration time histories were outcrop bedrock motions developed with a similar Vs30 as the intact rock (Group R_2 in Figure 1) being modeled, these outcrop acceleration time histories were converted to traction time histories and applied at the base of the quiet boundary without deconvolution. The compressional wave velocity was updated to account for the effect of pore water.

Table 1. Characteristics of Spectrally-Matched Time Histories

Earthquake	Component	PGA	PGV	5-95% AI (m/s)	5-95% Duration (Sec)
MDE					
Tabas, Tabas	H-1	0.3	17.1	0.7	18.1
	H-2	0.3	18.8	0.8	18.7
	V	0.27	15.9	0.7	15.2
Kocaeli-Arcelik	H-1	0.3	21.3	0.5	11.3
	H-2	0.29	25.2	0.6	10.5
	V	0.25	12.6	0.8	13.9
Chi-Chi TCU076	H-1	0.29	22.5	1.3	30.6
	H-2	0.29	17.8	1.1	30.0
	V	0.26	16.2	1.0	29.7
10,000-year					
Tabas, Tabas	H-1	0.54	33.2	2.5	18.2
	H-2	0.53	36.3	2.6	18.8
	V	0.60	38.3	3.6	15.4
Kocaeli-Arcelik	H-1	0.55	38.3	1.8	11.1
	H-2	0.54	46.8	2.0	10.4
	V	0.46	23.4	2.6	13.9
Chi-Chi TCU076	H-1	0.54	40.5	4.5	30.6
	H-2	0.53	33.1	3.9	30.0
	V	0.48	27.4	3.3	29.7
84th Percentile MCE					
Tabas, Tabas	H-1	0.67	57.3	4.3	18.1
	H-2	0.65	55.9	4.3	18.6
	V	0.60	38.3	3.6	15.4
Kocaeli-Arcelik	H-1	0.69	55.4	2.9	11.0
	H-2	0.68	72.1	3.4	10.4
	V	0.59	33.6	4.2	13.8
Chi-Chi TCU076	H-1	0.70	70.7	7.7	29.9
	H-2	0.66	43.4	6.8	29.6
	V	0.60	36.1	5.3	29.9

MATERIAL CHARACTERIZATION

Embankment Material

The homogeneous Isabella Auxiliary Dam embankment is composed predominantly of silty sand (SM), clayey sand (SC), and silty, clayey sand (SC-SM). The embankment was constructed in phases. The first phase was built from the right abutment to Station 60+35 and from 64+25 to the left abutment. The Phase 1 section was built to crest elevation of 2620.76 ft. The second phase construction completed the embankment between Station 60+35 and 64+25 and raised the entire embankment section to its current elevation of 2637.26 ft.

Field investigation and laboratory data showed no significant difference in material composition and classification between the two phases embankment fill; however the standard penetration testing (SPT) blow counts are generally higher in Phase 2 section. The $N_{1,60}$ ranged between 5 and well in excess of 50. However less than 10% of the blow counts were below $N_{1,60}$ of 15. The 33rd percentile $N_{1,60}$ were 26 and 33 for Phase 1 and

Phase 2 material, respectively. Based on the review of construction compaction effort, strength tests, field and laboratory density data, and $N_{1,60}$ data, it was determined that the embankment material is dilative and would not liquefy and experience significant strength loss during shaking.

Foundation Alluvium

Two primary stratigraphic alluvium units were identified Qa1 and Qa2 at analyzed section of the Auxiliary Dam. Qa2 unit was further divided into upper and lower members. In general the foundation valley fill consists predominantly of sand with varying amount of low-plasticity fines (CL, ML, and CL-ML) and minor amounts of gravel. The foundation soils gradually become finer and more plastic as a section moves away from the deposit source on the left abutment towards the right end of the embankment.

Unit Qa1 is the oldest and most clay-rich deposit; it overlies the granitic bedrock along the dam site. It is primarily composed of clayey sand and silty, clayey sand with some areas of silty sand and small pockets of poorly graded sand with silt. The fines content ranged from 25% to 40%. It is the thickest deposit and covers the entire foundation area of the dam, except the right abutment. It reaches to a depth of 140 feet near the right side of the valley bottom.

The Qas sand member is considered to be Holocene in age and is composed of poorly graded sand with silt and silty sand. The fines content ranged from 5% to 15%.

The Qa2 lower member overlies the Qas member and is composed primarily of silty sand with fines content of 20% to 30%. The deposit is 5 to 15 feet thick and overlies most of the dam site except the right abutment. The Qa2 upper member overlies the Qa2 lower member and is similar to Qa2 upper member in lateral extent, clay content, and grain size. However Qa2 upper member appears to be looser than the Qa2 lower member.

Grain size distribution data, Atterberg limits data, and exploratory data such as standard penetration tests (SPT), cone penetrometer tests (CPT), and shear wave velocity data were obtained and reviewed. The review of the data suggest two layers at Station 58+00 may be potentially susceptible to liquefaction under strong shaking. Based on evaluation of SPT, CPT and shear wave velocity data, representative $N_{1,60\text{-}cs}$ of 23.0 and 23.5 were selected for the shallow and deeper liquefiable layers, respectively. Figure 3 shows the alluvium units and the $N_{1,60\text{-}cs}$ of the potentially liquefiable layers at Station 58+00. Other materials were not considered to be liquefiable.

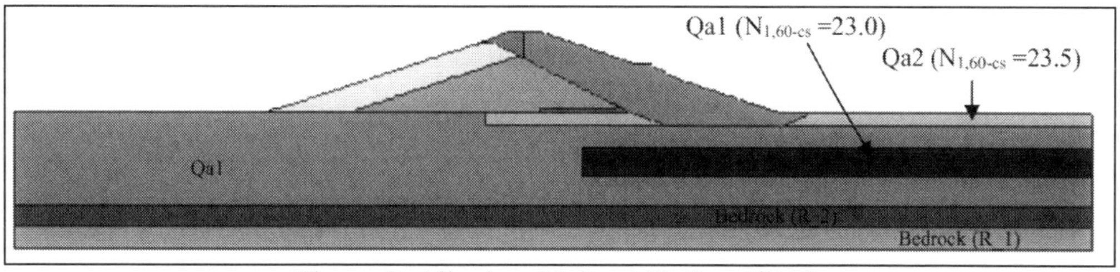

Figure 3. Alluvium Units at Station 58+00

NUMERICAL ANALYSIS OF DAM

To obtain insight to the dynamic performance of the proposed remediated Auxiliary Dam embankment, a two-dimensional numerical model of dam was created in FLAC2D version 7. In general the analysis procedure consisted of three stages: static, dynamic, and post-earthquake stage. A brief summary of the modeling procedure is presented here.

Static Stage

The embankment was modeled in FLAC with more than 12,000 zones. The zones were 5-foot wide and ranged from 2 feet in height for the thin liquefiable materials to about 10 feet in less critical layers. All the liquefiable zones were 5 feet or less in height. To obtain a reasonable realistic distribution of effective stresses at the start of the dynamic analysis, the excavation and construction of the existing embankment and remediation work, such as the excavation of the foundation, dewatering, and the construction of the new buttress, were modeled in FLAC. Mohr-Coulomb model with stress-dependent stiffness properties were used in this stage for all materials, except for the bedrock layers (R_1 and R_2).

The initial stresses in the foundation without the existing dam were determined using a hydrostatic groundwater condition with the groundwater elevation 2538 feet. Once equilibrium was reached, a single row of zones was added to the model and was brought to equilibrium before adding a new row, until the existing dam was constructed. The effects of reservoir loading and unloading were simulated by performing a decoupled mechanical-groundwater flow process. Although coupled flow process will capture the changes in effective and shear stress more properly, decoupled flow process should provide reasonable stress states given the significant gain in computational time and that the embankment is not liquefiable. Reservoir loading was increased in four increments up to elevation 2609 feet (gross pool). The reservoir was then lowered to elevation 2582 feet and finally 2568 feet, which is the proposed reservoir level for remediation construction. The model was brought to equilibrium until it reached stead-state seepage condition at each increment reservoir loading before increasing or decreasing the reservoir level. The following proposed modification activities were then modeled in FLAC:

- Dewatering at the downstream toe to lower the phreatic surface to elevations 20 feet below the bottom of the proposed excavation was simulated;
- Excavation of the shallow liquefiable alluvial foundation;

- Removal of portion of the existing downstream slope embankment; and
- Construction of the new buttress

When the modifications to the dam was complete, the reservoir was brought back to elevation 2609 feet in three incremental loadings. The new filters were not modeled in the analysis. This was considered to be reasonable since the embankment is not liquefiable.

Dynamic Stage

Before starting the dynamic stage, the potentially liquefiable material was changed from Mohr-Coulomb constitutive model to UBCSAND. The non-liquefiable materials were left unchanged. UBCSAND constitutive model is considered a fully coupled, effective stress model that considers the effects of shear-induced pore pressures during shaking. Since no laboratory cyclic stress-strain data were available for the liquefiable material, generic input parameters for the UBCSAND model were adopted. Select parameters were based on the estimated $N_{1,60\text{-}cs}$.

Post-Earthquake Stage

As the final stage of the analysis, post-earthquake response of the dam was evaluated to estimate the potential deformation that could result from loss of strength of the liquefied materials. At the end of shaking, the motion at the base of the model was stopped and solved for an additional five seconds of quiet time to allow for velocities and accelerations to decrease and displacements to stabilize. The liquefiable zones with excessive pore water pressure ratio exceeding 0.7 and strain greater than 5% were assigned with post-liquefaction strengths. These zones were changed from UBCSAND to Mohr-Coulomb model. Four different post-liquefaction strength estimates were selected for this study: Seed and Harder (1990), Idriss and Boulanger (2008), Kramer and Wang (2015), and Weber (2016). The Seed and Harder residual strength was the "one-third" value as recommended for deterministic analyses. Kramer and Wang residual strength was the 33rd percentile value as recommended by deterministic analyses. After assigning new reduced strengths to liquefied zones, analysis continued in dynamic mode until equilibrium is established and deformation, if any, was stabilized.

RESULTS AND DISCUSSIONS

A summary of key results from the FLAC analyses are presented here. Figure 4 and Figure 5 show the effective vertical stresses and pore water pressure (units are psf) at the start of the dynamic stage, respectively.

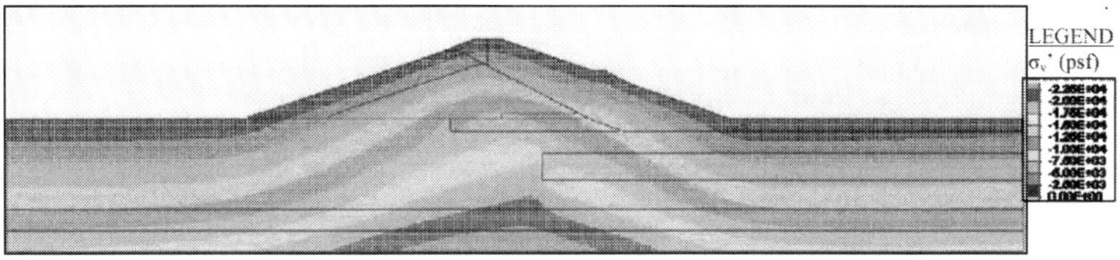

Figure 4. Effective Vertical Stress Contours (in psf) at the Beginning of Dynamic Stage

Figure 5. Pore Pressure Contours (in psf) at the Beginning of Dynamic Stage

The extent of liquefaction was evaluated in the two liquefiable foundation alluvium layers through excess pore pressure development. Figure 6 shows the excess pore pressure ratio for one of the time histories: 84th percentile MCE Chi-Chi H-2. The result shows that practically the entire liquefiable layers had developed high excess pore pressure midway into the shaking. Excess pore pressure increased significantly between 22.5 and 45 seconds, which coincides to a large velocity pulse during this period of the ground motion record. The excessive pore pressure remained generally elevated with little dissipation by the time it reaches the end of shaking. Chi-Chi H-1 shows similar development of excessive pore pressures in the liquefiable layers. On the other end of the spectrum Kocaeli time histories showed the much slower, late staged development of excess pore pressures and only in limited areas. The 10,000-year ground motions had similar pattern of excess pore pressure development, but with lesser extent. The MDE ground motions saw little to no excess pore pressure development. Predictably based on the observations of the excess pore pressure development, Chi-Chi time histories resulted in the highest deformations, while Kocaeli time histories the least for the same hazard level. The observed response is contributable to the fact that Chi-Chi ground motions had the longest duration (90 seconds) compared to Tabas-Tabas (33 seconds) and Kocaeli (30 seconds). Furthermore the Arias intensity is also the highest for Chi-Chi ground motions followed by Tabas and Kocaeli, in decreasing order.

Figure 7 shows that maximum shear strain is generally concentrated mainly in the deep liquefiable layer and the shallow liquefiable layer under the embankment. This further indicates that high pore pressures that had led to softening of the liquefiable alluvium are the driver for the seismic response of the embankment.

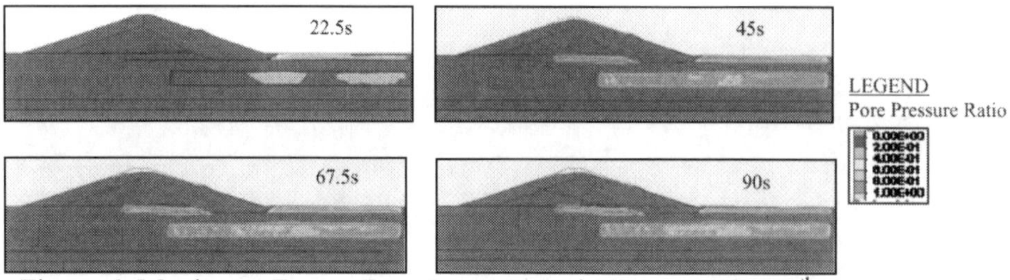

Figure 6. Maximum Excess Pore Pressure Ratio Contours for 84th Percentile MCE Chi-Chi H-2 Time History at 22.5 sec, 45 sec, 67.5 sec, and end of shaking (90 sec)

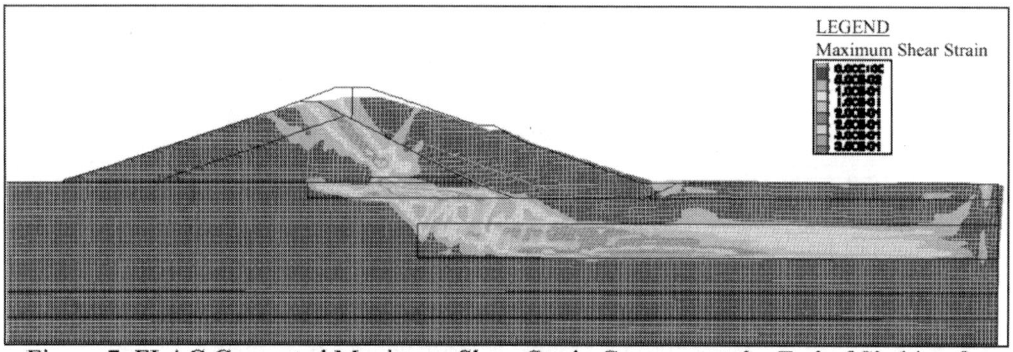

Figure 7. FLAC Computed Maximum Shear Strain Contours at the End of Shaking for 84th Percentile MCE Chi-Chi H-3 Time History

Figure 8 shows the amount of crest settlement at the end of shaking for all the ground motions. The amount of crest settlement ranged from 0.1 feet (MDE ground motions) to 11.2 feet (MCE ground motions). Amongst the MCE ground motions, the crest settlement ranged from 1 foot (Kocaeli H-1) to 11.2 feet (Chi-Chi H-2). Chi-Chi time histories led to the highest settlement for the same hazard level (i.e. MDE, 10,000year, and MCE). Although the ground motions were spectrally matched to the same UHRS, this highlights the importance of using multiple ground motions in a seismic evaluation.

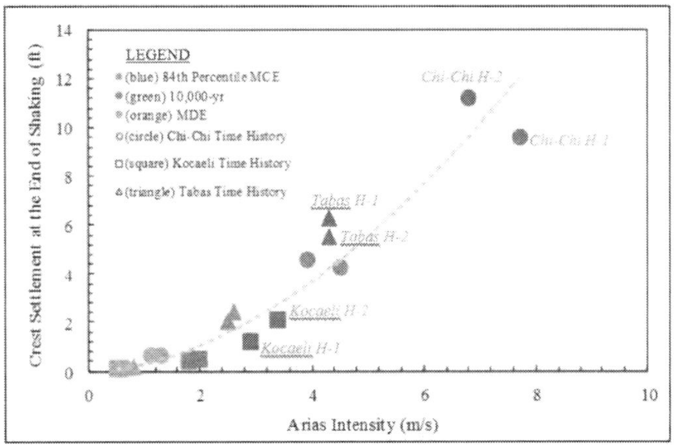

Figure 8. Crest Settlement at the End of Shaking versus Arias Intensity

Post-Earthquake Response

Computed crest settlement at the end of shaking and the final crest settlement at the end of the post-earthquake stage ranged from 0 to 29 feet (Figure 9) depending on the selected definition of residual strength and extent of liquefaction due to various simulated earthquakes. For reference, Figure 9 also shows the crest settlement at the end of shaking. In general ground motions that led to crest settlement less than about a foot of settlement at the end of shaking did not result in additional post-earthquake settlement. However, as the amount of end-of-shaking crest settlement increases, so does the post-earthquake crest settlement (Figure 10).

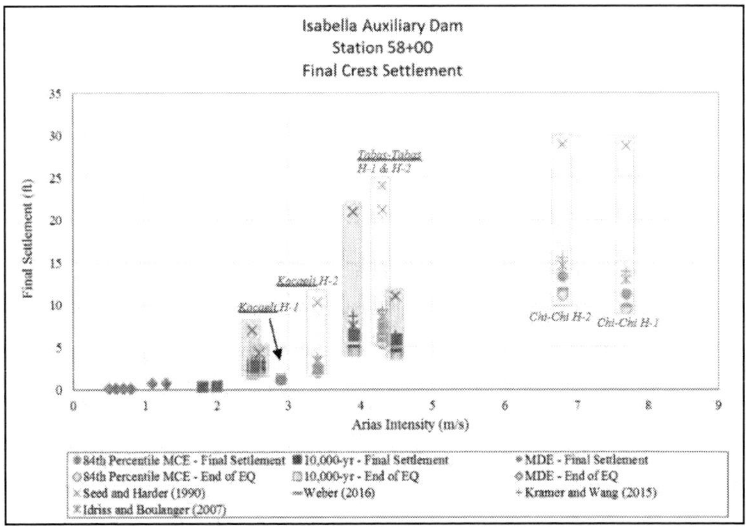

Figure 9. Final Crest Settlement Estimates versus Arias Intensity

Using Seed and Harder (1990) residual strength resulted in the highest additional crest settlement from the post-earthquake stage of up to 19 feet. On the other end of the spectrum, Weber (2015) resulted in up to about one foot of additional post-earthquake settlement. Kramer and Wang (2016) and Idriss and Boulanger (2008) resulted in similar post-earthquake crest settlement between 0 and 4 feet. Since Seed and Harder (1990) definition of residual strength does not vary with vertical effective stress, this suggests that liquefied materials with relatively high initial vertical stress is controlling the post-earthquake behavior in the FLAC model. Figure 11 shows the residual strength versus vertical confining stress for the deeper liquefiable alluvial layer, Qa1. The red shaded area is the approximate range of initial vertical effective stresses for this layer. It can be seen that within this range the residual strength esimate is highest using Weber (2015) definition while Seed and Harder (1990) the lowest.

Residual strength case histories used for the development of residual strength relationships are limited to blow counts ($N_{1,60\text{-}cs}$) up to 16 and high initial vertical effective stress up to about 4 atm. Residual strengths outside of these conditions are generally extrapolations, and thus the wide range of variability in residual strength prediction. The lack of case history with high blow count and vertical effective stress may suggests potentially low susceptibility of sustained excessive pore pressure development. Furthermore, large displacement (e.g. flow failure) may be necessary for

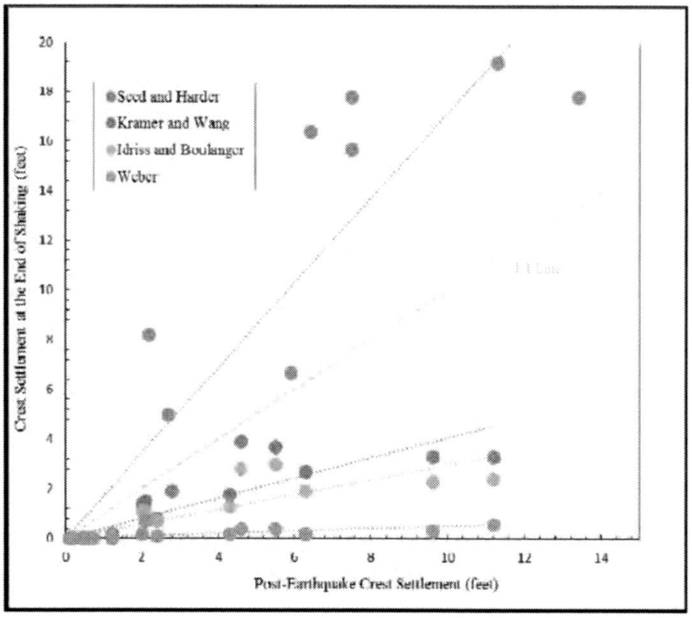

Figure 10. Settlement at the End of Shaking versus Additional Post-Earthquake Crest Settlement

the mobilization of residual strength. The lack of case history with large displacement or flow failure due to liquefaction of a relatively deep layer could then suggest that the likelihood of residual strength mobilization is low. In any case it appears prudent to evaluate post-liquefaction stability using multiple relationships to capture the range of variability and uncertainty. Lastly, the modeling of additional deformations after the earthquake in FLAC was considered an approximation since it is probably conservative to assume reduction of strength in all liquefied zones at the same time.

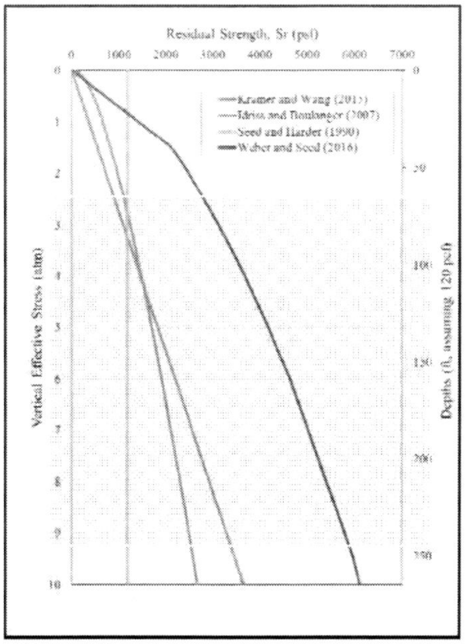

Figure 11. Residual Strength versus Initial Vertical Effective Stress for Deeper Alluvium Layer (Qa1)

CONCLUSIONS

The dynamic response of proposed remediated Isabelle Lake Auxiliary Dam was evaluated using FLAC2D. UBCSAND constitutive model was used to evaluate the dynamic behavior of the liquefiable material. Three spectrally matched ground motions were used for this evaluation. The crest settlement at the end of each time histories ranged from 1 to 11 feet for the same hazard level (i.e. uniform response spectra).

Four different post-liquefaction strength relationships were used for this evaluation. The crest settlement from the post-earthquake stage ranged from 0 to 19 feet, depending on the residual strength definition and time history of the same hazard level. The variability found in this evaluation suggests that using multiple ground motions and residual strength definitions is prudent in a seismic evaluation where seismic hazard is high and liquefiable material is present.

REFERENCES

URS (AECOM). (2016). Updated Probabilistic Seismic Hazard Analysis and Development of Time Histories, Lake Isabella, Kern County, California.

Idriss, I. M., & Boulanger, R. W. (2015). 2nd Ishihara Lecture: SPT- and CPT-based relationships for the Residual Shear Strength of Liquefied Soils. *Soil Dynamics and Earthquake Engineering*, 57-68.

Itasca. (2011). FLAC – Fast Lagrangian Analysis of Continua, Version 7.0. Itasca Consulting Group, Inc., Minneapolis, Minnesota.

Kelson, K. I., Simpson, D. T., Rose, R. S., & Serafini D.C. (2010). Seismic Hazard Characterization of the Kern Canyon Fault for Isabella Dam, California. US Society of Dams.

Kramer, S. L., & Wang, C. H. (2015). Empirical Model for Estimation of the Residual Strength of Liquefied Soil. *Journal of Geotechnical and Geoenvironmental Engineering, ASCE.*

Seed, R. B., & Harder , L. F. (1990). SPT-Based Analysis of Cyclic Pore Pressure Generation and Undrained Residual Strength. *Proc. H. Bolton Seed Memorial Symposium* (pp. Vol 2, 351-376). Bi-Tech Publishing Ltd.

Weber, J. P. (2015). *Engineering Evaluation of Post-Liquefaction Strength, PhD. Dissertation.* University of California, Berkeley.

SEISMIC REMEDIATION OF PERRIS DAM USING CDSM

Wen Yan (Grace) Chen[1]
Michael Driller[2]
Steven Friesen[3]

ABSTRACT

Lenses of potentially liquefiable soil have been identified within the Perris Dam foundation which may lose strength during a strong seismic event. The remediation was designed primarily to limit deformation of the dam. It includes the use of Cement Deep Soil Mixing (CDSM) to strengthen the foundation and to construct a downstream berm to stabilize the dam. The construction of the project started in Fall 2014 and is anticipated to take approximately 3 years. The remediation covers over 5,000 linear feet of the downstream toe of the dam. The CDSM production work has been performed by two CDSM rigs using two working shifts per day over six months. The installation of the CDSM was completed in April, 2016. Over 320,000 cubic yards of CDSM was placed in the ground. Predrilling was performed at all element locations prior to placing CDSM. The CDSM contractor chose to use the same rigs for both predrilling and mixing. All CDSM walls were installed successfully and were very well mixed. Some low 28-day strengths were observed in areas with high ground water level/ high seepage and in areas where high amounts of water were injected into the ground, but the elements eventually met the strength requirements after re-coring and retesting at later dates. This paper describes the design of the CDSM and focuses on the construction aspects of the work, including findings from the bench scale testing program, the field validation test section, and the production work. The CDSM work for Perris Dam is believed to be one of the largest CDSM projects (single contract) to date in the United States.

INTRODUCTION

Perris Dam is a zoned earth fill dam constructed between 1970 and 1974 with a length of 2.2 miles, a maximum height of 128 feet, and a capacity of 131,450 acre-feet of water. It is located in Riverside County, California at the terminus of the East Branch of the State Water Project. The dam has a sloping sandy clay core (Zone 1) with a 10-foot deep cutoff trench and a silty sand shell (Zone 2) with a downstream chimney and blanket drain (Zone 3 and Zone 4) as shown in Figure 1. The foundation at Perris Dam consists of up to approximately 290 feet of alluvium underlain by granitic bedrock. The Perris Dam Foundation Study Report, prepared by the Department of Water Resources (DWR, 2005), indicated the presence of thin sandy layers in the dam foundation that are potentially susceptible to liquefaction and severe loss of strength during a large

[1] Engineer, Water Resource, California Department of Water Resources, Sacramento, 1416 9th Street, CA 95814, WenYan.Chen@water.ca.gov
[2] Senior Engineer, Water Resource, California Department of Water Resources, Sacramento, 1416 9th Street, CA 95814, Mike.Driller@water.ca.gov
[3] Engineer, Water Resource, California Department of Water Resources, Sacramento, 1416 9th Street, CA 95814, Steven.Friesen@water.ca.gov

earthquake event. Many of the liquefiable soils are located in the upper 20 feet of the foundation with isolated deeper pockets. Liquefaction of the foundation soils could potentially result in slope failure of the embankment dam and uncontrolled reservoir release from Lake Perris. A reservoir water level restriction of 25 ft below the maximum water surface elevation has been in place since 2005 to maintain safe operation of Perris Dam and Lake until the final repair is implemented.

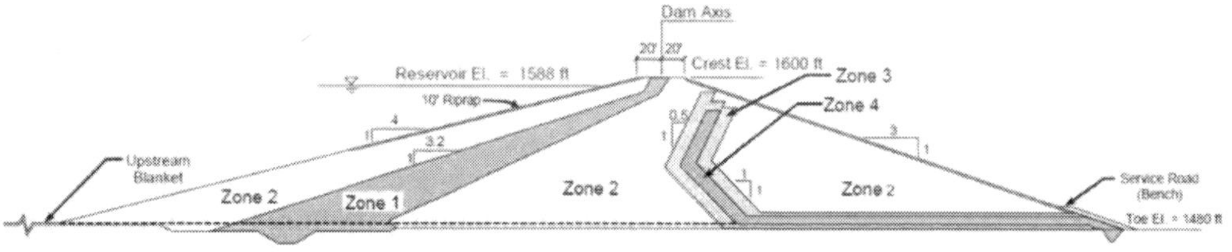

Figure 1. Perris Dam Typical Section

A remedial alternative study completed by DWR in 2007 recommended removal and replacement of up to 25 feet of the foundation along with construction of a berm over the dam toe. CDSM was also recommended to treat deeper portions of the foundation. In 2010, two test sections were constructed in order to evaluate CDSM and excavation construction methods. CDSM methods were largely successful, but the contractor had significant difficulties dewatering the foundation prior to performing the shallow excavation. A particular concern was that water levels rebounded very quickly after dewatering pumps were shut off. After consideration of the test section findings, it was decided that major excavation near the dam toe would be unsafe and that CDSM would be required to strengthen the shallow foundation rather than attempting to remove and replace these soils. The seismic performance of the existing and modified dam was evaluated using FLAC models. The design of the remediation was focused on optimizing the size of the berm and CDSM in order to reduce the deformation of the dam to acceptable levels. The alternatives were compared on a cost versus performance basis and the most cost-effective alternatives were identified. Based on the results, the final size of the berm and CDSM were selected. Figure 2 shows a typical cross section of the final remediation design completed in 2014.

A previous paper that detailed the design of the CDSM was published by Chen et. al., 2015. The current paper mainly focuses on the findings from the CDSM construction, which included a laboratory bench scale testing program, a full scale Field Validation Test Section, and six months of production work.

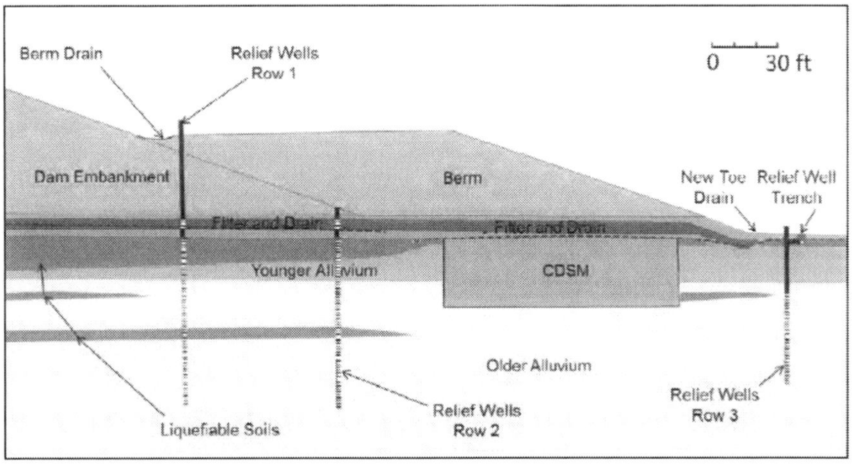

Figure 2. Typical Remediated Section

CDSM DESIGN, CONFIGURATION, AND DEPTH OF TREATMENT

CDSM Strength Requirement

The CDSM design strength was based on two main factors: 1) replacement ratio (RR); and 2) required shear strength. For this project, the replacement ratio was defined as the total area of the transverse and longitudinal walls within a cell divided by the total area of the cell. Typical replacement ratios are in the range of 30 percent to 60 percent for similar type projects in highly seismic areas. After consideration of previous projects and the balance between replacement ratio and strength, a replacement ratio of 42 percent was selected for the CDSM design. While lower treatment ratios with higher CDSM strength could achieve the same average strength at a potentially lower cost, the authors favor a higher replacement ratio to reduce the amount of untreated soil and provide a more reliable CDSM strength.

The seismic performance of the existing and modified dam was evaluated in 2012 using FLAC models (Friesen et.al. 2014). The CDSM was first modeled as a stiff elastic material (uncracked CDSM case). The elements representing the CDSM were monitored for the maximum induced shear stress during the earthquake. The design shear strength of the CDSM was then selected according to the maximum induced shear stresses, so that the majority of the CDSM would remain elastic. The models suggested that the required shear strength be 165 psi for the CDSM walls with a replacement ratio of 42 percent. To produce the required shear strength, the design CDSM average unconfined compressive strength (UCS) was determined to be 500 psi.

CDSM unconfined compressive strengths generally follow a normal statistical distribution. Data from previous projects indicate that the use of an absolute minimum strength requirement to control the lower end of the strength often tends to push the distribution too far toward the higher strengths. For the CDSM design, 10 percent of the test results were allowed to be lower than 300 psi (60 percent of the design strength).

According to other projects, this approach produced soil-cement with average values closer to the intended design average and reduced the waste of CDSM material.

Two strength requirements were specified for the CDSM production work based on every three consecutive full-depth continuous cores:

1. The average UCS shall be at least 500 psi after 28 days of curing.
2. No more than 10 percent shall exhibit an UCS of less than 300 psi.

DWR specified the design strength as the average 28-day strength, rather than a 56-day or 90-day strength, to reduce delays in terms of placing material over the completed CDSM walls. Because CDSM gains strength over time, the current in situ strength is likely well above the design strength.

CDSM Configuration and Depth of Treatment

CDSM cells, consisting of transverse walls (perpendicular to the dam axis) and longitudinal walls (parallel to the dam axis), were installed along the toe of the existing dam for approximately 4,100 feet between Stations 73+00 and 114+00 (Figure 4). A total of 72 cells and approximately 320,000 cubic yards of CDSM material were installed. There are nine cell types. Each type varies in length, in the transverse direction, to fit in the CDSM defined limits. In the transverse direction, the length of each cell ranges from 120.5 feet at the two ends of the treatment area to 134 feet in the most critical remediation area located in the middle section of the dam.

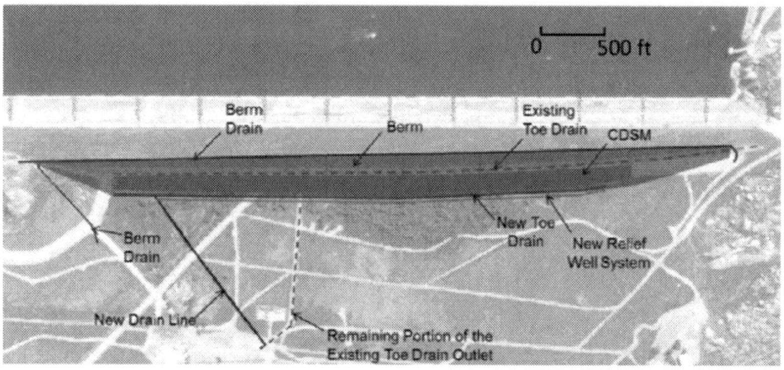

Figure 4. Layout of CDSM Treatment

CDSM was installed with an alternate pattern of shallow and deep cells as shown in Figure 5. All the walls from the shallow cells and the longitudinal walls from the deep cells extended no greater than 35 feet below the original ground surface. All transverse walls from the deep cells extended to a depth of 40 to 65 feet below the original ground surface. The shallow cells were designed to intercept the shallow potentially liquefiable layers. The deeper CDSM walls were designed to intercept some isolated deeper pockets. Figure 6 shows the typical shallow and deep cells. The shallow cells and the shorter longitudinal walls within the deep cells were designed to allow more open areas for seepage passing under the CDSM treatment area, and to reduce the potential for increased pressures developing under the dam.

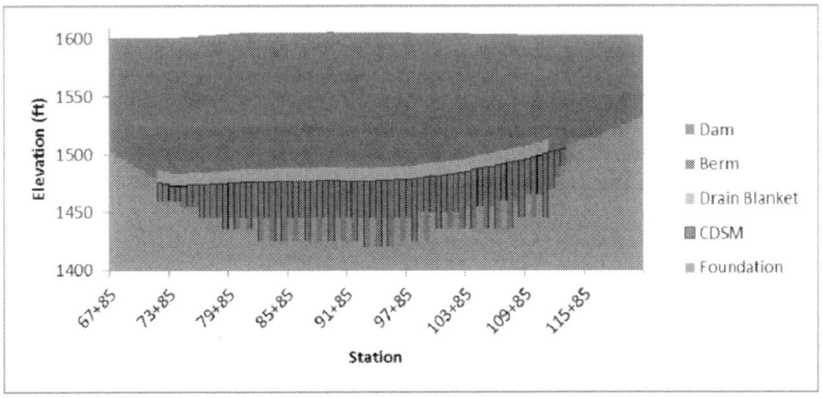

Figure 5. Profile of Dam Remediation

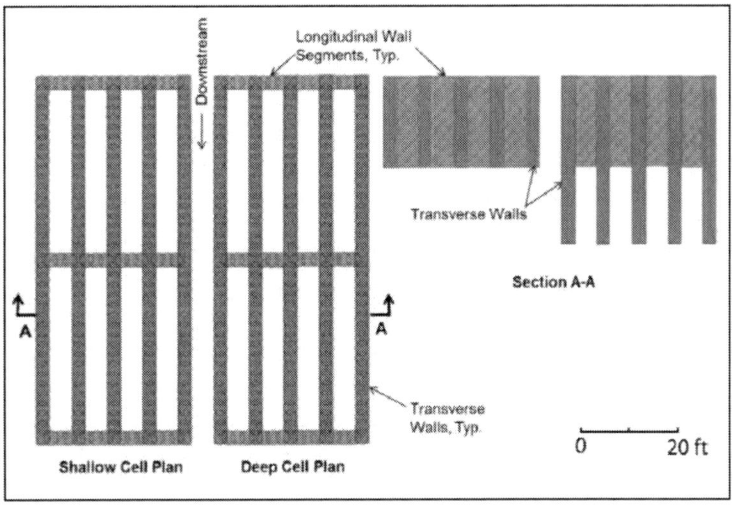

Figure 6. Typica Shallow and Deep Cell Layout

LABORATORY BENCH SCALE TESTING PROGRAM

A Construction Phase CDSM Testing Program was conducted prior to the production work of the CDSM installation. It included a laboratory bench-scale testing program and a field validation test. The bench-scale testing program was conducted to determine the design mixes to be used for the field validation test. Samples were collected by the CDSM Contractor from the dam foundation and three series of bench scale testing were conducted. Samples were tested in pairs at 7, 14, 28, 56, and 90 days of curing (Table 1). The testing results were unexpected. The influence of cement dosage, which ranged between 160 kg/m^3 and 300 kg/m^3, on strength was generally poor, and all batches, except for one, that were cured at a temperature of 72°F, produced strengths less than the specified average UCS of 500 psi at 28-days of curing. Only the batches prepared using an elevated curing temperature of 115°F produced strengths greater than the specified average. Ultimately, all samples continued to gain strength over time and exceeded 500 psi for tests performed at 90 days of curing.

The laboratory bench-scale testing program did not demonstrate that a particular mix design would reliably achieve the specified strength. However, based on previous experience, The CDSM Contractor believed that CDSM columns would be cured underground with a temperature greater than 115°F, and they expected the 28-day UCS from the field validation test section would meet the strength requirement.

Table 1. Bench Scale Testing Series

Series	Cement Dosage (kg/m3)	Water : Cement	Cement Source	Curing Temperature	Total Number of Samples Prepared
1	160, 230, 300	1, 1.2, 1.4, 1.6	Mojave	72 to 74°F	144
2	160, 230, 300	1.2, 1.4	Rillito	72 to 74°F	72
3	230, 300	1.2	Mojave	72 and 115°F	12

FIELD VALIDATION TEST SECTION

The field validation test section, which consisted of two test walls and one test cell, was installed to determine the final design mix that would produce the required UCS for the production walls. In addition, the test provided information to calibrate the equipment, to demonstrate the ability of the equipment to thoroughly mix the foundation soil with cement grout, and to demonstrate the verticality and proper overlapping of elements. The test walls and test cell were all 60 feet deep and were installed from 5 feet below the original ground surface. Locations are shown in Figure 6.

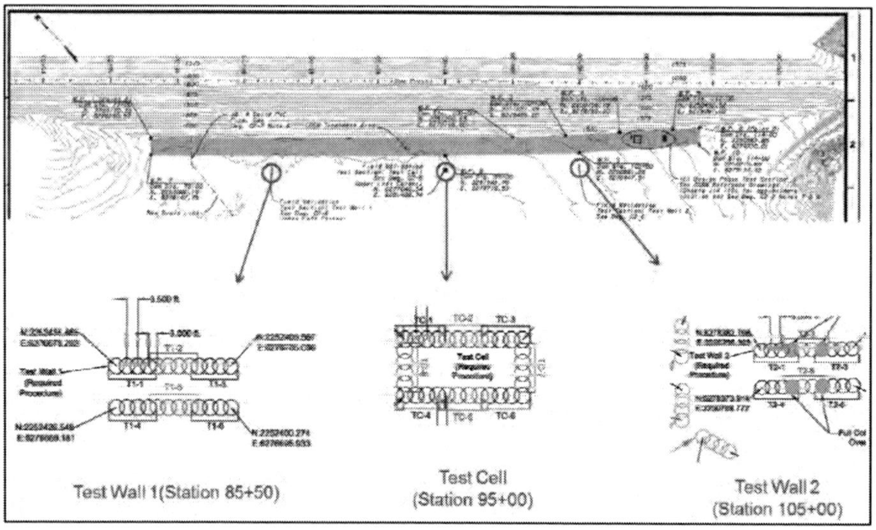

Figure 6. Field Validation Test Section Locations

A total of 20 test elements were installed, including six from each of the two test walls and 8 from the test cell. The test elements were installed with different combinations of cement dosages and final water to cement ratios (Table 2). The water to cement ratio (W:C) for the cement grout was maintained at 0.9, while the final in place water to cement ratio was estimated to vary between 1.1 and 1.25 depending on the cement dosage, which varied between 190 kg/m^3 and 320 kg/m^3.

A total of 187 cored samples (QC samples) were tested by the CDSM Contractor's independent laboratory for 28-day UCS and 167 cored samples (QA samples) were tested by DWR's onsite laboratory for confirmation. The 28-day UCS results in Figure 7 show a relatively large scatter in average strength. As expected, the strength generally decreased with increasing water to cement ratio and increased with increasing cement dosage. Figure 7 also shows that the average strength from all design mixes, regardless of the water to cement ratio and the cement dosage being used, exceeds the requirement of 500 psi based on the test results from the QC samples. The strength is much higher than that from the bench scale testing and suggests that high curing temperatures underground helped the soil-cement material gain strength sooner.

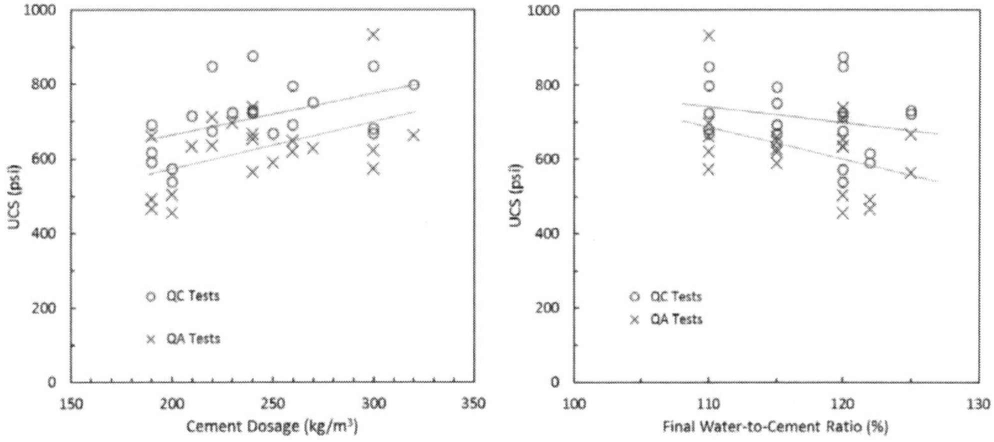

Figure 7. Average 28-Day UCS vs. Water to Cement Ratio and Cement Dosage

Figure 7 shows that the QA samples generally show lower strength than the QC samples. This is likely caused by sample preparation differences between DWR's onsite lab and the CDSM Contractor's independent lab. The CDSM Contractor's lab used both sulfur caps and spherical bearing plates for each cored sample to be tested. Initially, DWR tried different ways to prepare the ends – no end caps at all (just trimmed ends if necessary), end caps without trimming, end caps with trimming, trimming with spherical bearing plates, and end caps with spherical bearing plates (the same way as that prepared for the CDSM Contractor's samples). As shown in Figure 8, the strength from the QA samples was generally lower than the QC samples until DWR's lab applied both the sulfur caps and the spherical bearing plates to the sample, similar to the CDSM Contractor's laboratory. The strength is the lowest when the samples were tested without any preparation to the ends.

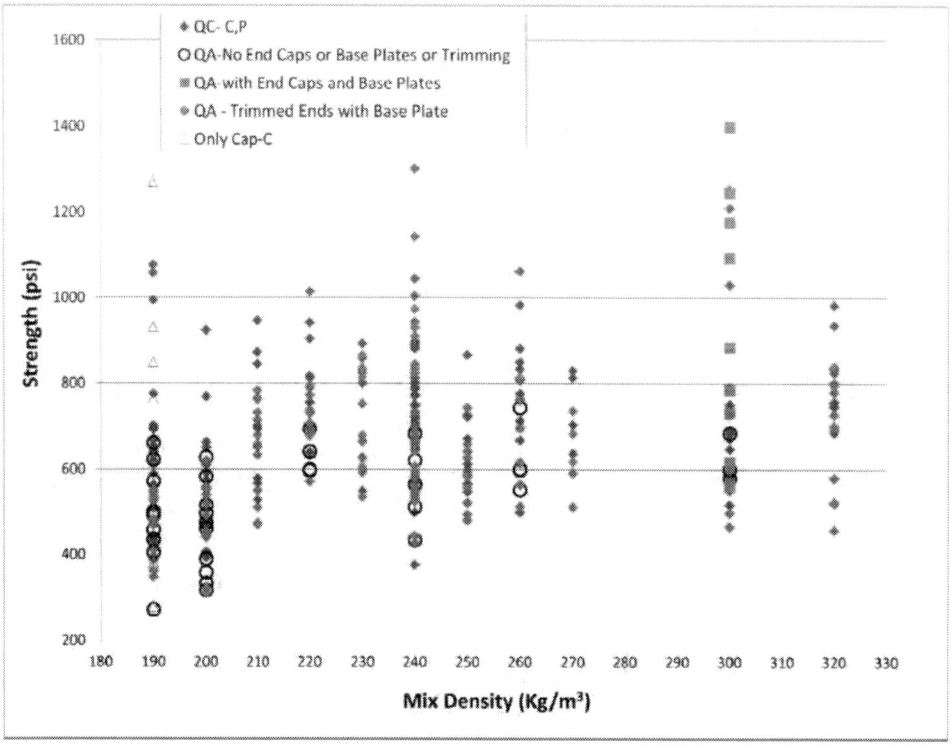

Figure 8. UCS from Individual Cored Samples from the Test Section

Long term UCS testing was performed on samples with 90 days of curing by DWR. Results are compared in Table 2 with the 28-day UCS. The average strength gain from 28 days to 90 days of curing varied from 6 percent to 49 percent with an average of 20 percent. Strength gain from Test Wall 2 was more than 10 percent higher than that from Test Wall 1 and the Test Cell. In addition, Test Wall 2 has the lowest average 28-day strength amount the three sets. The results also show that samples with higher cement dosage do not necessarily produce higher strengths and there does not seem to be a correlation between cement dosage and strength gain over time.

The project required all production CDSM cells to be excavated down to the design depths and filter and drain materials to be placed on top of the CDSM. Therefore, one of the objectives of the field validation test section was to ensure that an intact CDSM top could be obtained after excavating the hardened CDSM material to the desired depth. Approximately three weeks after the installation of the test elements, the two test walls and the test cell were dewatered and excavated to expose the top 5 feet of the CDSM elements. All exposed CDSM elements were well mixed, intact, and well aligned (Figure 9).

Table 2. Test Section Strength Gain

	Element	Dose (kg/m^3)	Final W:C	28-Day$_{Avg}$ (psi)	90-Day$_{Avg}$ (psi)	Strength Gain (28 to 90 days) (%)
Wall 2	T2-3	190	122	616	709	15
	T2-4	200	120	572	706	23
	T2-5	220	120	675	850	26
	T2-2	240	125	730	1092	49
	T2-6	260	115	691	908	31
	T2-1	300	110	681	844	24
		Average		661	853	28
Cell	TC-1	190	115	691	779	13
	TC-2	210	120	715	840	18
	TC-3	230	110	724	839	16
	TC-8	240	125	721	863	20
	TC-4	250	115	668	851	27
	TC-5	270	115	751	798	6
	TC-6	300	110	668	764	14
	TC-7	320	110	798	860	8
		Average		717	824	15
Wall 1	T1-1	190	122	590	739	25
	T1-4	200	120	538	621	15
	T1-5	220	120	849	905	7
	T1-2	240	120	876	973	11
	T1-6	260	115	794	996	25
	T1-3	300	110	849	1006	19
		Average		749	873	17

Figure 9. Exposed CDSM Test Wall 1

PRODUCTION WORK

Two CDSM rigs and two 10-hour working shifts were utilized during production work. The CDSM Contractor chose to use the CDSM rigs for both pre-drilling and mixing operations. Four 5-foot diameter augers were attached to each of the two rigs to perform the work. CDSM production work started at the end of October, 2015 and was completed in late April, 2016. Production work was completed approximately five months ahead of schedule.

Figure 10. CDSM Work at Perris Dam

Pre-drilling

Pre-drilling CDSM column locations was required for the CDSM work. The 2010 CDSM Design Phase Test Section suggested that the penetration rates during CDSM mixing could be highly uneven without predrilling. The Perris Dam foundation contains cemented materials in between potential liquefiable soils at different locations and depths. Pre-drilling helped to loosen up the cemented soils and assisted to maintain the uniformity of the soil-cement product during CDSM mixing. The CDSM Contractor used pressurized air and water to help the augers penetrate through the dense material during pre-drilling. The air pressure was regulated to remain at 120 psi or less and water was injected at a rate of 8 gal/ft or less. The use of air and water, however, created other problems. Air and water venting was observed during pre-drilling. Examples include water pulsing out of dewatering wells to a height several feet above the ground surface (Figure 11) and water bubbles being observed in nearby ponded water.

Figure 11. Water pulsing out of dewatering wells during pre-drilling

More frequent and more intensive venting happened in locations with higher groundwater and more seepage. Less to no venting was observed in locations with lower groundwater and less seepage. Concerns from DWR included the potential for hydrofracturing foundation soils, possible grout migration, and damage to newly installed CDSM columns. Based on field observations, after an "isolated" element (elements which are more than 12 feet from the nearest predrilled element) was predrilled, it would act as a vent to allow pressure to be released through the loosened soil. Consequently, the potential for venting from subsequent pre-drilling was diminished. A venting mitigation plan was developed by the CDSM Contractor to reduce the potential harm from air venting. The mitigation plan included limiting the amount of air that could be injected into the ground and increasing the distance between the dam toe and the isolated elements. The pre-drilling operation was closely monitored by both DWR and the CDSM Contractor for signs of distress caused by the pressurized air. The CDSM Contractor followed the air mitigation plan and most air/water venting happened in areas close to the pre-drilling operation. Cores were well intact and no cracks and fractures were observed due to air and water venting.

CDSM Mixing

Based on the results from the field validation test section, the CDSM Contractor started with a more conservative cement dosage of 240 kg/m^3, but reduced it to 210 kg/m^3 after the first month of production. The water to cement ratio of the grout was kept at 0.9 for the entire installation. Cement dosage was later raised up again to approximately 240 kg/m3 after quite a few failing tests occurred in the high groundwater/ high seepage areas. Figure 12 shows the results of the average UCS vs cement dose from the test section.

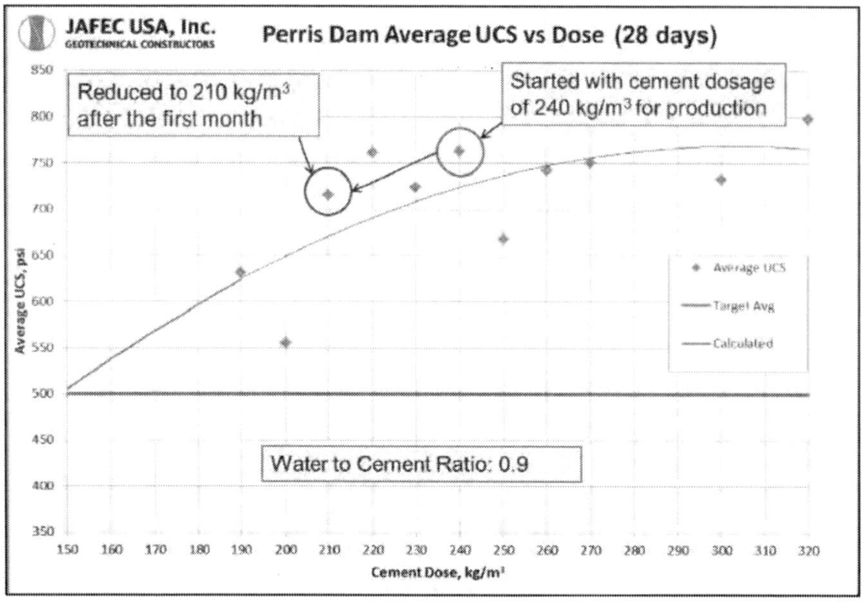

Figure 12. UCS vs Cement Dosages from the Field Validation Test Section (figure provided by JAFEC USA)

Sampling

Samples from both wet-grab and coring methods were performed at the beginning of the production. However, it was very difficult to obtain wet-grab samples from the high strength CDSM material. The CDSM Contractor tried a total of three different methods to obtain wet-grab samples, but none of the methods could effectively lower the sampler to the required depth and retrieve the wet-grab samples. This slowed down the production work greatly. Considering that wet-grab samples only provided information on early strength and were not part of the evaluation for the acceptance criteria, DWR and the CDSM Contractor jointly decided to replace wet-grab samples with additional core samples. This change came after installing the first 22 cells. In total, 123 wet-grabs were performed from the 22 cells and close to 600 samples were tested for 7, 14, and 28 day strengths.

Full depth continuous cores were obtained for compressive strength testing at 14, 28, and 90 days. The 14-day cores were the replacement of the wet-grab samples. One 14-day core was retrieved from each of the deep cells after the wet-grab sampling was terminated. The 28-day cores were obtained to evaluate the specified recovery, uniformity, and strength criteria. Typically, four cores were retrieved from each cell per specifications. In addition to the 28-day cores, the specifications also allowed DWR to request coring at 10 additional locations. DWR eventually requested seven cores that were cored at 90 days of curing. A total of 407 full depth continuous cores were collected. All cores were intact with recovery rates of nearly 100 percent. Small unmixed pockets were sometimes observed in the cores, but all cores met the uniformity criteria defined in the specifications.

Strength Testing

UCS vs. the Amount of Water and Mixing Energy Introduced to the System. During production, 2310 QC cored samples were tested by the CDSM Contractor's independent laboratory for 28-day UCS to evaluate if the CDSM material met the strength acceptance criteria. DWR's onsite laboratory also tested 1340 cored samples at 28 days of curing for quality assurance purposes. Figure 13 shows the 28-day UCS across the entire CDSM treatment zone. Most of the average UCS met the specification requirements, while some low strength elements were encountered. The low strength elements were mainly located in areas with high ground water level and high seepage. Other low strength elements occurred in locations where more water was injected than required during pre-drilling to help break up the dense to cemented material. Oftentimes, the Contractor increased the amount of cement grout in the areas where more water had been injected. However, the UCS was typically more sensitive to water than to the amount of grout being used. The 28-day strength was still lower than required in many of the elements even though more cement grout was injected. Re-coring was performed on many of the low strength elements. This typically occurred one to three weeks after the 28-days of curing. Eventually all retested UCS samples met the required average strength of 500 psi.

High strength CDSM occurred where shallow bedrock was encountered. The CDSM cells in those locations were installed at the beginning of the project with a more conservative cement dosage of 240 kg/m^3. Because bedrock was encountered at shallow depths, many of the columns were essentially mixed twice due to the requirement to mix the bottom 10 feet of each CDSM column twice (bottom mixing). Therefore, more grout and more mixing energy were introduced to the CDSM columns. In addition, these areas had lower ground water levels and less seepage. All these factors contributed to higher strength CDSM in locations where shallow bedrock was encountered.

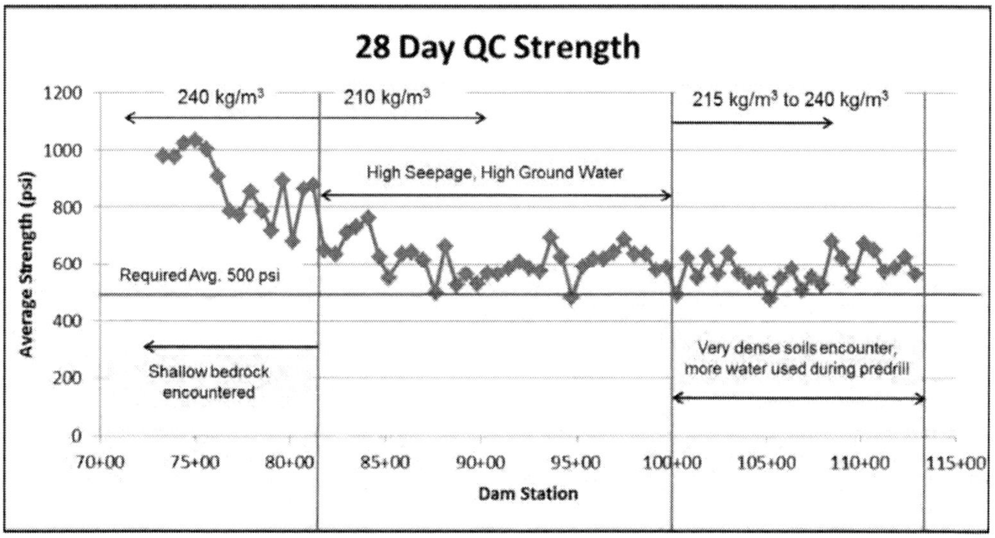

Figure 13. 28-day UCS across the CDSM Treatment Zone

UCS vs. Curing Conditions. Results show that laboratory curing temperature affected the strength from wet-grab samples much more than it affected the strength from cored samples. At the beginning, all wet-grab samples and all QC cored samples were stored in the CDSM Contractor's laboratory prior to testing at the specified curing temperature of 73 +/- 3°F. However, many wet-grab samples exhibited strengths lower than required after 28-days of curing. The CDSM Contractor was concerned that at that temperature, they would repeat the failing test results from the bench scale testing program and therefore, he requested to raise the curing temperature to 115°F to better reflect the in situ conditions. DWR eventually accepted the request, but maintained the specified curing (73°F) temperature for the QA cored samples that were stored in DWR's on-site laboratory. As shown in Figure 14, after elevating the curing temperature, the average wet grab sample strength increased over 40 percent at 7 and 14 days of curing and increased over 20 percent at 28 days of curing.

On the other hand, the 28 day UCS for QA and QC cored samples are comparable to each other despite the fact that one was cured at 73°F and the other one was cured at 115°F. Figure 15 shows the strength difference between QC and QA samples from each of the cells versus the number of days the samples were cured in the lab before testing. As shown, in many cases, the QC samples that were cured at higher temperatures have lower strengths than the QA samples that were cured at lower temperatures. In summary, a higher curing temperature in the laboratory has a marked effect on strength for wet-grab samples, but not for cored samples.

Strength results were also compared between cores cured underground and cores cured in the lab. DWR selected seven CDSM element locations to core and test after 90 days of curing. The strength results were compared to the 90-day strength from the samples that were cored at 28 days of curing, but were stored in the lab until they were tested at 90 days. As shown in Figure 16, the results show that the 90-day strength of the 90-day cores was almost 30 percent higher than the 90 day strength from the 28-day cores that were cured in the laboratory. For the 90-day cores, the CDSM material was cured underground with a high curing temperature without being interrupted until they were cored and tested at 90 days. The strength behavior indicates that once the CDSM material is harden and cores are retrieved from the ground, the curing temperature in the lab will have a small effect on strength gain. The samples cured in the ground without being interrupted had significantly more strength gain when compared to those being cured in laboratory.

Coefficient of Variation. The coefficient of variation (CV) of the strengths from the cored samples from all 72 cells was 0.30, which indicates that the strength variation between cells is considerably small and the mixing is consistent between cells. The overall CV can be further reduced if not considering the installation between Stations 73+00 and 81+00, where shallow bedrock was encountered and the CDSM strengths are substantially higher than other area because of the reasons stated earlier. The CV of the strength from different cores within individual cells are very small and are generally less than 0.1.

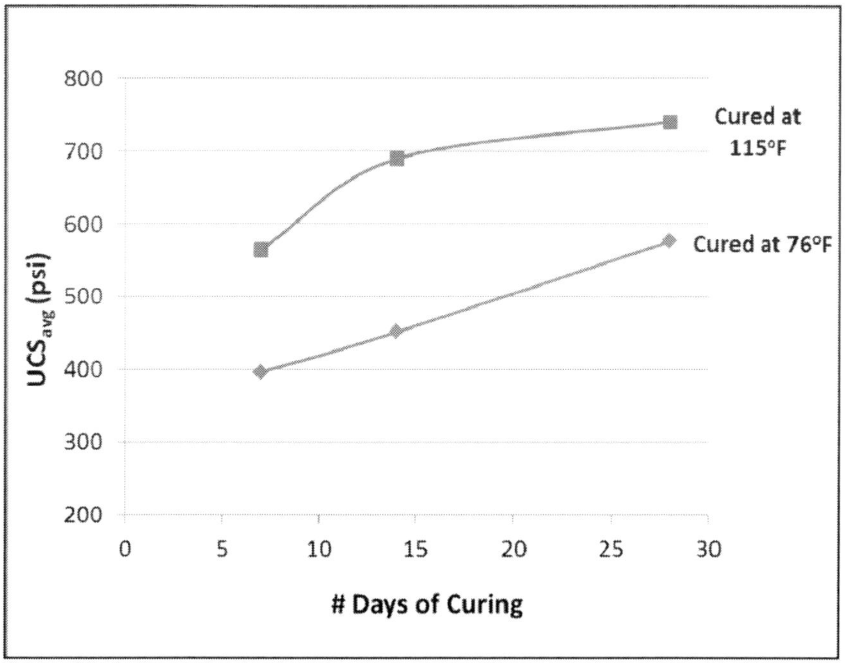

Figure 14. Wet-grab samples UCS vs. Curing Temperature

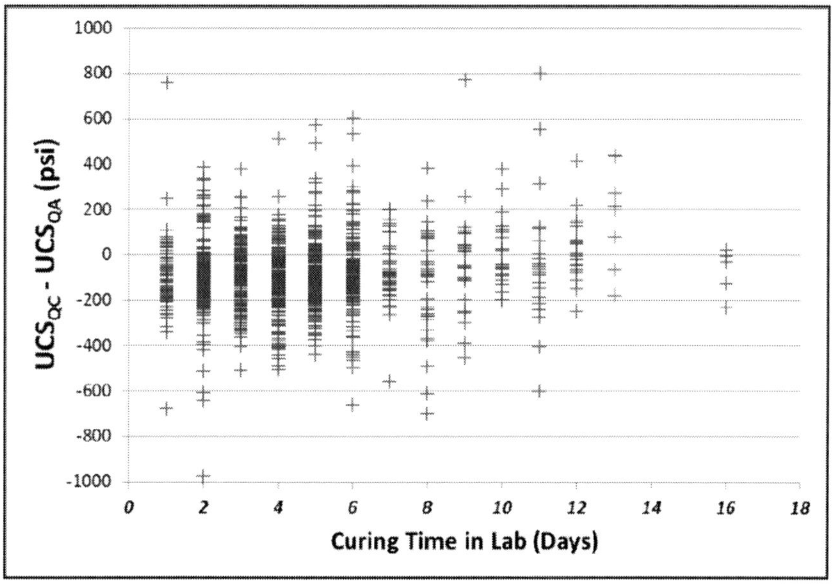

Figure 15. 28 days UCS$_{QC}$ – UCS$_{QA}$ vs. Curing Time in Lab (UCS testing on QA samples was performed by DWR and on QC samples was performed by the CDSM Contractor)

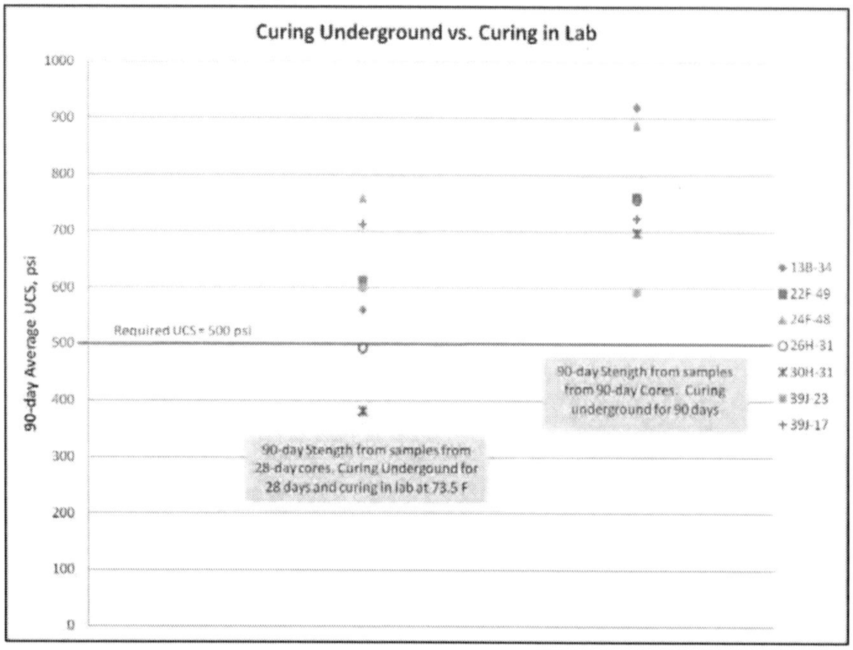

Figure 16. 90-day Strength Comparison

Wet-grab strength vs cored strength. The 28-day wet-grab sample strength was compared to the 28-day cored sample strength. As shown in Figure 17, the average 28-day strengths between the wet-grab and cored samples are comparable, with the cored samples approximately 14 percent higher. The finding is different than that from the 2010 Design Phase Test Section, in which the wet-grab samples yielded substantially higher results than the cored samples. DWR feels that the difference is mainly due to the preparation of the samples. In 2010, the wet-grab material was pushed (i.e. "packed") by hand into the plastic molds. It is likely that the prepared cylinders resulted in samples that were denser than the in-place material. On the other hand, impacts from coring, transporting, and preparing the cored samples would tend to disturb the quality of the samples and reduce their strengths. In this project, the CDSM Contractor prepared both the wet-grab and the cored samples very carefully. A very low coring speed was used and all cores retrieved were well mixed and intact. Wet-grab samples were not pushed and forced into the cylinder, but rather a careful tapping technique was used.

Strength Gain. In addition to quality assurance UCS testing at 28 days of curing, DWR also performed long term testing at 90, 180, and 360 days of curing for the cored samples. The testing at 90 days and 180 days of curing has been completed. Over 1300 samples were tested at 90 days and over 1200 samples were tested at 180 days. The testing at 360 days was started in November, 2016 and will be completed in mid-April, 2017. Figure 18 shows the strength gain of the samples from 14 days of curing to 180 days of curing. Strength increased more than 35 percent from 14 days to 28 days of curing, about 25 percent from 28 days to 90 days, and only 3 percent from 90 days to 180 days, indicating that the strength levels off.

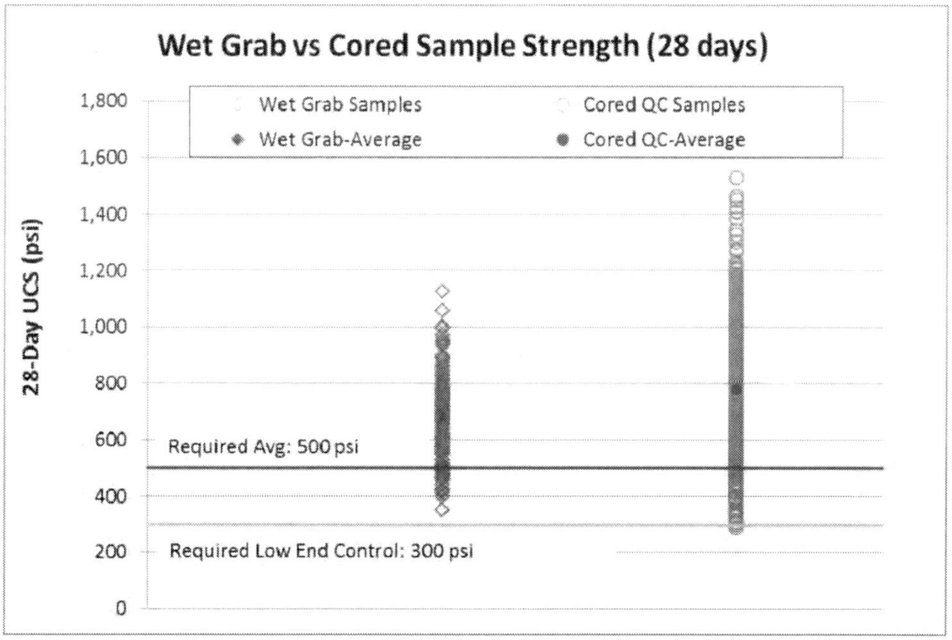

Figure 17. Wet-grab vs Cored Sample Strength at 28 days of Curing

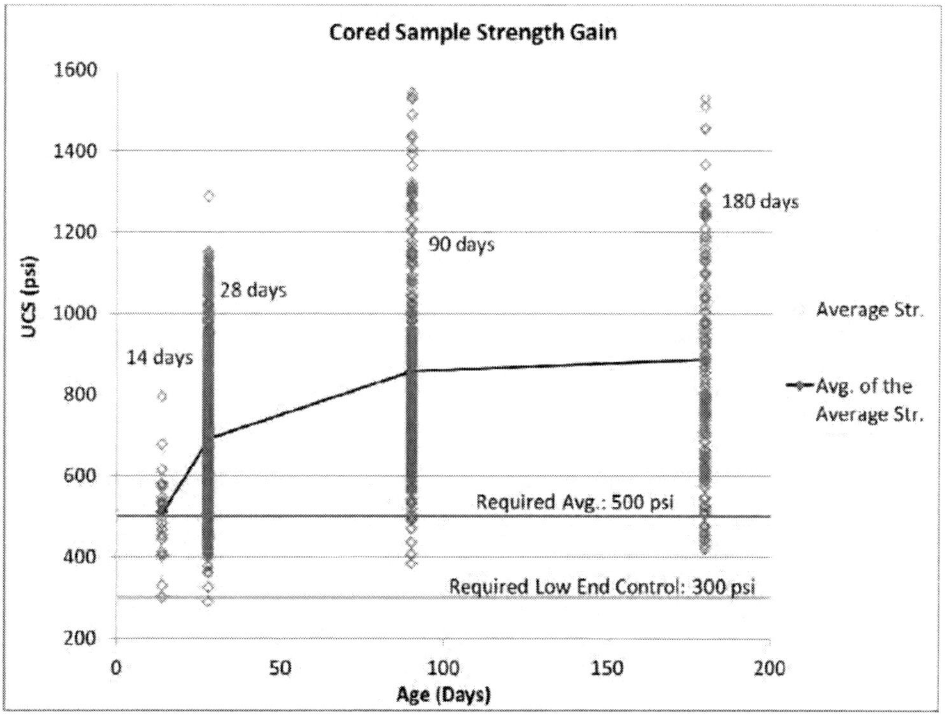

Figure 18. Cored Sample Strength Gain over Time

CONCLUSIONS

The CDSM work for the Perris Dam Remediation Project was successfully completed. Over 320,000 cubic yards of soil-cement was installed immediately downstream of the dam toe between Station 73+00 and Station 114+00 to strengthen the potentially liquefiable foundation soils. The work included a bench scale testing program, a field validation test section, and production work to install a total of 72 deep and shallow CDSM cells. The findings are summarized as follows:

1. The results of the bench-scale testing did not correlate well with the actual strength measured in the field. The bench-scale tests under predicted the field testing results because most of the samples were cured under lower temperatures in the lab than in the field.

2. The strength of the CDSM material was much more sensitive to the amount of water being applied to the system than the cement dosage. Low strengths were observed in areas where higher amounts of water were injected and in areas with high groundwater seepage.

3. The UCS testing results from the bench-scale testing program, the field validation test section, and the production work showed that curing temperature had varying influence on the strength of the sample depending on the curing environment. CDSM, cured in the ground, continued to gain strength over time. But curing temperature had little effect on strength gain after CDSM material was harden and cored from the ground and cured in the laboratory. Wet-grab samples, cured in the lab, produced higher strengths when cured at higher temperatures.

4. The testing results from the test section showed that end caps and plates can be applied to the cored samples prior to testing to ensure that the two ends of the samples are flat and level. Use of caps likely results in an increase in strength, but not using the caps could result in uneven ends and potential for a decrease in strength.

5. The strength test results showed that the CDSM material gained most of its strength in the first three months. The rate of strength gain was much less after that.

6. The coefficient of variation of the CDSM strength from all cells was small, which indicates that the strength variation throughout the treatment area was small.

7. Wet-grab sample strength and cored sample strength were comparable to each other for this project. However, that is not always the case. Good sampling methods and good sample preparation are very important factors in obtaining reliable wet-grab and cored sample test results. Compared to wet-grab samples, cores represent the final condition in the ground and were considered the more reliable source for strength testing. The experience at Perris Dam also indicated

that wet-grab samples were very difficult to obtain in the high strength CDSM materials, and the process was very time consuming.

8. Using pressurized air and water during pre-drilling effectively helped loosen up the dense material in the foundation and helped produce uniform CDSM material during mixing. However, the method of using pressurized air and water has the potential to damage nearby structures. The amount of air and water to be injected into the ground should be regulated and closely monitored during pre-drilling to minimize the potential damages.

9. The excavation of the CDSM test section and the production walls to expose the CDSM top showed even two months after the installation of the CDSM, the CDSM could be trimmed to an intact, level surface.

ACKNOWLEDGEMENTS

The authors wish to express appreciation to the CDSM Contractor, JAFEC USA, for a successful project and for being professional, honest, responsible, and hardworking during the CDSM installation.

The authors would like to thank Dr. David Yang, Dr. Donald Bruce, Dr. George Filz, DSOD, Project Geology from DWR, Levee Repair Project Headquarter from DWR, DWR's on-site Laboratory, JAFEC's Independent Laboratory – Leighton Group, AECOME, LCI (Lettisci), IT (Infra Terra), Ismael Echeverria, Daniel Cimini, Krishnan Athipotta-Variam, and the General Contractor, Pulice Construction Inc. for their contributions to the CDSM work at Perris Dam. This CDSM project was a showcase of good teamwork.

REFERENCES

Chen, W., Friesen, S., Driller, M. (2015). *Seismic Remediation of Perris Dam*. 2015 DFI Deep Mixing Conference, San Francisco.

DWR. (2005). *Perris Dam Foundation Study*. California Department of Water Resources.

Friesen, S., Balakrishnan, A., Driller, M., Beaty, M., Arulnathan, R., Newman, E., Murugaiah, S. (2014). *Lessons Learned from FLAC Analyses of Seismic Remediation of Perris Dam*. 34th USSD Annual Meeting and Conference, San Francisco.

CONTROLLED BLASTING FOR THE RED ROCK HYDROELECTRIC PROJECT

Rachael V. Bisnett, P.E.[1]
James Valentyn, P.E.[2]
Nicholas Johnson[3]

ABSTRACT

The powerhouse for the Red Rock Hydroelectric Project is being constructed at the toe of the existing U.S. Army Corps of Engineers' Red Rock Dam immediately adjacent to the existing stilling basin retaining wall. Drill and blast rock excavation for the powerhouse and tailrace channel was performed within 5 feet of the existing structures to a depth of 40 feet into bedrock. Approximately 26,000 cubic yards of rock was excavated.

Given the close proximity of the excavation to the existing USACE structures, cellular cofferdam, and construction excavation support structures, blasts were designed to achieve peak particle velocities less than 2 inches per second measured at multiple points around the excavation. For blasts within 25 feet of structures, additional control measures were utilized to minimize the potential for excessive ground motions and overbreak which could undermine or otherwise compromise the structures.

Development and approval of all blast plans required close coordination between the Contractor, their drilling and blasting subcontractor, the Engineer, and the USACE. The resulting drilling and blasting program achieved the intent of minimizing impacts to the structures.

INTRODUCTION

The Red Rock Hydroelectric Project (Project) is currently under construction at the existing US Army Corps of Engineers (USACE) Red Rock Dam on the Des Moines River near Pella, Iowa. At peak generation, the Project will generate up to 55 MW with an average annual energy output of 178 gigawatt-hours. The Project is being developed by Missouri River Energy Services (MRES) as agent for Western Minnesota Municipal Power Agency.

The powerhouse for the Project is being constructed at the toe of the existing dam immediately adjacent to the existing stilling basin retaining wall, as shown in Figure 1 and Figure 2.

[1] MWH now part of Stantec, 350 N. Orleans St., Suite 1301, Chicago, IL 60654, 312-831-3188, Rachael.V.Bisnett@mwhglobal.com
[2] Ames Construction, 37West 2100 South, West Valley City, UT 84120, 801-977-8012, JamesValentyn@amesco.com
[3] Hoover Construction Co., 302 S. Hoover Rd., Virginia, MN 55792, 218-741-3280, nick@hoovercon.com

Figure 1. Aerial of Powerhouse Location at Toe of Dam (MRES)

Figure 2. Powerhouse Location Prior to Blasting

Geologic units at the Project include the Warsaw Limestone and the St. Louis Limestone of Mississippian age, as shown in Figure 3. The oldest unit exposed is the Warsaw Limestone, which consists of dolomitic and argillaceous shales. At the site, the Warsaw is present below El. 638. It is overlain by the St. Louis Limestone, which is the upper bedrock unit at the site. The St. Louis Limestone consists of alternating beds of limestone, sandstone, dolomite, and discontinuous basal evaporite beds. It is exposed as the top of rock at the Project site, extending from approximately El. 674 to El. 638. The powerhouse and tailrace channel excavation is up to 40 feet deep into rock through the St. Louis Limestone and terminating in the Warsaw Limestone.

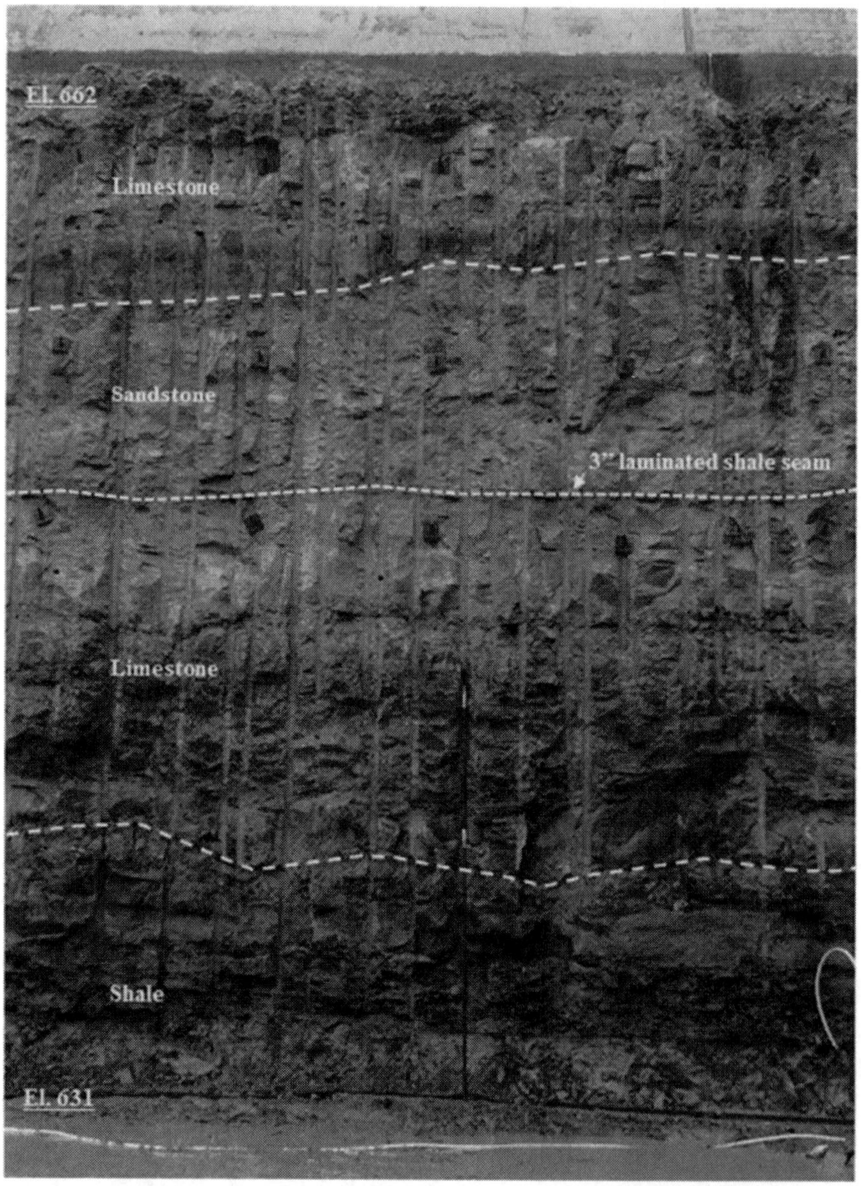

Figure 3. St. Louis and Warsaw Limestones below El. 662

ROCK EXCAVATION OVERVIEW

Ames Construction (Ames), the general construction contractor for the construction of the Project, self-performed the ripping of softer upper strata of rock with a D10 bulldozer equipped with a single shank ripper to an average elevation of El. 665.6 across the powerhouse and tailrace areas. Drill and blast rock excavation was performed by Hoover Construction Co. (Hoover) under subcontract to Ames.

Hoover Construction, Ames' drilling and blasting subcontractor, began blasting on October 28, 2015 and concluded on December 20, 2015; and rock excavation and foundation preparation continued through February 2016. Approximately 26,000 cubic yards of rock was excavated.

Drilling was performed using top hammer drills. Line drilling was initially performed around the perimeter of the rock excavation to the final foundation elevation with 2.5 inch diameter holes spaced at 18 inches center-to-center. Production blast holes were typically 3 inch diameter. Burden and spacing of the holes was typically 5 feet.

Blasts were designed to achieve peak particle velocities less than 2 inches per second measured at multiple points around the excavation. Emulex 900 Series, Blastex 2x16, 1 pound and ½ pound booster explosives and nonelectric detonators were used. Powder factors ranged from 0.61 to 1.34. Two decks were typically used for production blast holes. Blasts were initiated remotely. All blasts were covered with rubber tire mats to prevent flyrock.

Blasting and excavation occurred in three lifts, each around 11 feet in depth, to reach the foundation elevation of El. 631 across most of the powerhouse. A fourth lift was required to excavate one deeper corner of the powerhouse to El. 621.

BLAST PLANNING

Prior to commencement of any drilling and blasting operations, Ames was required to prepare and submit a Blasting Methods Plan to MWH, the Engineer for the Project, and the USACE for review and acceptance. The Blasting Methods Plan outlined Ames' planned approach to the blasting activities, including:

- Details of the required communication, coordination and monitoring efforts between Ames/Hoover, MWH, and USACE regarding possible impacts of the blasting upon day-to-day operations of the existing dam and spillway;
- Sequence and schedule of blasting, including the general method of developing the excavation, lift heights, etc.;
- Specifics of a typical blast to be implemented in each of the following areas:
    - At the area where blasting is intended to be started;
    - At the deepest rock cut area;
    - At the perimeter of the rock cut, where perimeter control procedures are required, including details of perimeter control procedures proposed for use

- Details of an audible advance signal system to be employed at the job site as a means of informing that a blast is about to occur; and
- Plan for monitoring of vibration and noise due to blasting.

After initial review of the Blasting Methods Plan, a pre-activity coordination meeting was held at the site between Ames, Hoover, MWH, MRES, and the USACE to discuss and finalize details such as the individual blast plan review process, data collection and review, and public notifications.

Individual blast plans were reviewed and accepted by on-site engineers from MWH and the USACE on a daily basis as blasting progressed. The review process was facilitated by frequent discussions between Ames, Hoover, MWH, and the USACE.

BLAST MONITORING

Vibrations and air-blast overpressure monitoring were required at the following locations, as shown in Figure 4:

- Secant pile wall supporting the powerhouse excavation;
- Toe of the existing stilling basin wall;
- Inboard side of the cellular cofferdam;
- County Highway T15 crossing the crest of the dam; and
- Campground adjacent to the site (air-blast overpressure only).

Figure 4. Blast Monitoring Locations (Braun Intertec, 2016)

A maximum 2 inch per second peak particle velocity was established at all locations. Air-blast overpressure was limited to 120 linear decibels.

The contract specifications also required a seismologist experienced in the control and monitoring of blasting vibration and noise to supervise the establishment of the blast monitoring program and provide regular review of the monitoring results.

A statistical approach to seismic limitations using standard statistical regression methods (Ayyub and McCuen, 2003) was applied. The model was built to match known and published results as summarized in the Blaster's Handbook (ISEE, 2011). The maximum peak particle velocity, PPV (in/sec), was initially estimated based on the following equation from well-known industry references (Oriard, 2005 and Edwards and Northwood, 1960):

$$PPV = 242\times\left[\frac{D}{\sqrt{W}}\right]^{-1.6} \tag{1}$$

where, D is the shortest distance from the blast to the point of interest (ft) and W is the maximum weight of explosive per delay of 8 milliseconds or more (lbs).

Equation 1 was used as the basis to build a model for the Project's local geological conditions. This model was updated after every blast as part of the blast monitoring plan. A regression was performed using R to find the rock attenuation parameters (Clark et al., 2000), and a script was created to generate a consistent report as the data set was updated.

The script analyzed input data from each recorded blast, particularly the PPV and scaled distance (SD). PPV and SD data was transformed into log form, then a normal linear regression was applied. The results were converted out of log form to give the best fit line for the data to be plotted on a log-log plot, as shown in Figure 5. With a coefficient of determination (r^2) value of 83% was considered reasonable to assume that SD predicts PPV based on our data (Ayyub and McCuen, 2003).

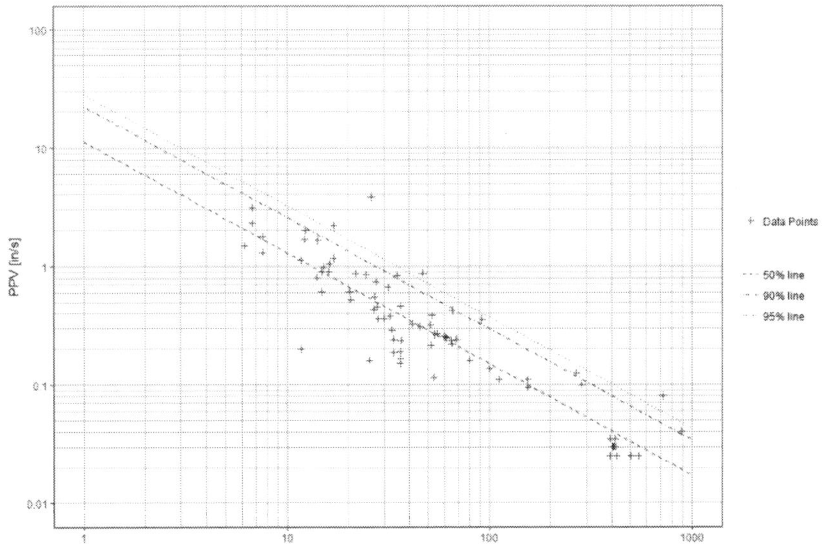

Figure 5. Project-Specific PPV vs. SD Regression

EXCLUSION ZONE BLASTING

The contract specifications initially precluded the use of blasting for rock excavation within 25 feet (exclusion zone) of the following structures, as shown on Figure 6:

- temporary secant pile wall for powerhouse excavation;
- existing stilling basin wall; and
- cellular cofferdam.

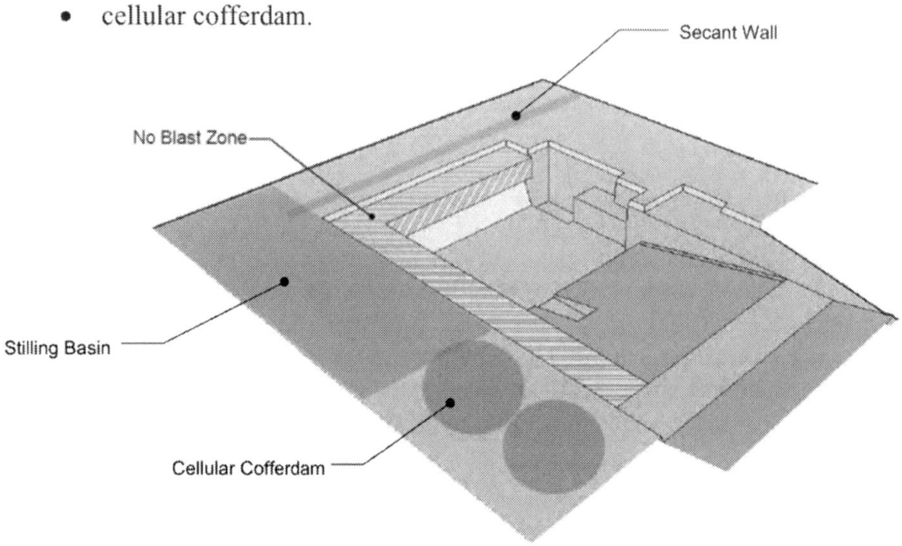

Figure 6. Blast Exclusion Zone (Hoover, 2015a)

Ames proposed a test blasting program to evaluate the use of blasting within the exclusion zone. This would permit approximately 5,400 cubic yards, or about 20%, of the rock to be removed via blasting instead of mechanical methods. Test blasts to evaluate burden, spacing, and charge for potential exclusion zone blasts were carried out as part of the production blasting program. As part of their proposal, Hoover developed a seismic regression to analyze the results and allow continuous revisions to the recommended maximum weight of explosives based on the results of previous blasts.

MWH and the USACE approved the use of blasting for rock excavation within the exclusion zone based on the test program with the following provisions:

- Additional line drilling was performed along a line offset 1-foot to the interior of the excavation in advance of blasting adjacent to the toe of the dam's existing stilling basin wall; and
- An additional monitoring point was added to the stilling basin wall to provide a more comprehensive evaluation of the behavior of the structure during blasting.

Prior to performing work in the exclusion zone, the Project-specific script was used to estimate how many pounds of explosives could safely be used per delay at specified offset distances to maintain PPV values less than 2 in/s. To this end, the standard equation with the regression parameters for log-log plotting was rearranged to solve for weight of explosives (Equation 2). This maximum estimated pounds per delay was updated throughout the Project.

$$W = \left(\frac{D}{\left(\frac{PPV}{K}\right)^{\frac{1}{m}}} \right)^2 \quad (2)$$

Blasts within the exclusion zone were designed for the maximum pounds per delay in each row to fall within the 95% confidence interval computed from the seismic monitoring data collected from prior blasts. A table (Table 1) was prepared to summarize the maximum allowable pounds per delay based on offset distance. This simplified the preparation of the blast plans by allowing for flexibility in the drill pattern and variable loading within the shot.

Table 1. Excerpt from Blast Planning Table (Hoover, 2015b)

distance	PPV	50%	90%	95%
15	2	5.58	1.31	**0.81**
19	2	8.96	2.09	**1.3**
23	2	13.13	3.07	**1.9**
27	2	18.09	4.23	**2.62**

The regression was updated by Hoover throughout blasting, as shown in Figure 7.

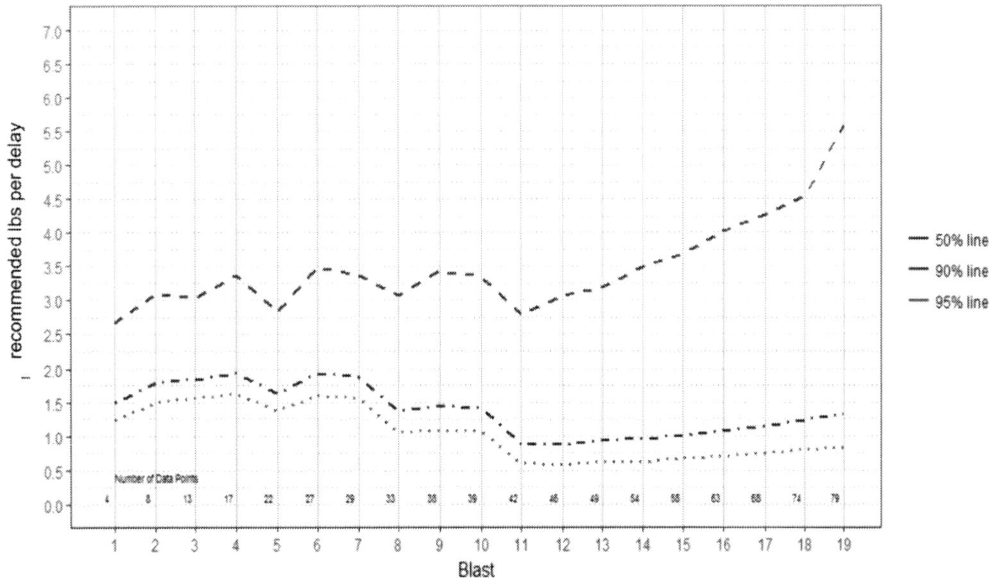

Figure 7. Regression Results per Blast (pounds per delay at 15 feet offset after each blast) (Hoover, 2015b)

The final equation calculated by the regression for PPV was:

$$PPV = 28.11\times\left[\frac{D}{\sqrt{W}}\right]^{-0.94} \tag{3}$$

Rearranged in terms of weight of explosives and based on the 2 in/s limit, Equation 3 can be written as:

$$W = \left(\frac{D}{\left(\frac{2}{28.11}\right)^{\frac{1}{-0.94}}}\right)^2 \gg \left(\frac{D}{16.683}\right)^2 \tag{4}$$

In addition to the site-specific seismic regression, control measures for blasting within the exclusion zone and near other final excavation surfaces included:

- Boosters were typically used because they provided more flexibility for the use of smaller explosives in the blast design.
- Burden and spacing were reduced to 4 feet in some areas.
- Additional decking (up to four decks) with delays between decks was used to reduce the weight per delay.
- Subdrilling was precluded on the final floor of the excavation.

RESULTS AND CONCLUSIONS

Blasting within 25 feet of critical dam safety structures was successfully performed by utilizing blast control measures such as added line drilling, boosters, reduced burden and spacing, and lower pounds per delay. Supplemental monitoring was also performed with additional instrumentation and visual inspections of the structures before and after blasting.

The excavated rock walls were generally observed to be undisturbed by the blasting methods. Half-cast holes were visible and backbreak was not noted on the walls, as shown in Figure 8 and Figure 9. Where shotcrete was placed to protect the rippable, weathered upper sandstone unit, it was similarly undisturbed by blasting. After excavation, the pattern excavation support (rock bolts) initially specified was able to be reduced due to the quality of the vertical faces and favorable joint spacing and orientation.

Out of a total 173 monitoring records, there were only six instances where the measured peak particle velocity exceeded the limit of 2 inches per second, and no instances of air-blast overpressure exceedance. The corresponding maximum dynamic displacements associated with the exceedance events were negligible, as summarized in Table 2.

Table 2. Blasting Results for Exceedance Events (Braun, 2016)

Blast No.	Location	Distance (ft)	PPV (in/sec)	Frequency (Hz)	Dynamic Displacement (in)
6	Secant Pile Wall	39	2.005	> 100	0.0032 [a]
8B	Secant Pile Wall	17	2.200	> 100	0.0035 [a]
11	Stilling Basin Wall	29	3.825	> 100	0.0061 [a]
12A	Stilling Basin Wall	13	3.095	> 100	0.0049 [a]
12B	Stilling Basin Wall	13	2.295	> 100	0.0036 [a]
23	Stilling Basin Wall	13	2.120	73	0.0046

Note: [a] Dynamic displacement calculated for frequency of 100 Hz.

Figure 8. Typical Wall Condition after Blasting below Secant Pile Wall

Figure 9. Typical Wall Condition after Blasting below Stilling Basin Wall

The final powerhouse excavation profile below the stilling basin and secant pile walls is shown in Figure 10.

Figure 10. Completed Powerhouse Excavation

ACKNOWLDEGEMENTS

The Authors would like to thank MRES for supporting the publication of this paper and acknowledge the contributions also made by the Federal Energy Regulatory Commission to the review of the Blasting Methods Plan, particularly related to the exclusion zone.

REFERENCES

Ayyub, B.M. and R.H. McCuen (2003). *Probability, Statistics, and Reliability for Engineers and Scientists*. Boca Raton, Chapman & Hall/CRC.

Braun Intertec (2016). Red Rock Hydroelectric Blast Measurement Report – Final Blasting Report. Prepared for Ames Construction.

Clark, D., B. Larsson, and G. Lande (2000). "Vibration: Its Effect and Measurement Techniques At or Near Dwellings." International Society of Explosives Engineers.

Edwards, A.T. and T.D. Northwood (1960). "Experimental Studies of Effects of Blasting on Structures." The Engineer No. 211, September 30.

Hoover Construction (2015a). Red Rock Hydroelectric Project Blasting Methods Plan. Prepared for Ames Construction.

Hoover Construction (2015b). Seismic Regression Calculation. Prepared for Ames Construction for the Red Rock Hydroelectric Project.

International Society of Explosives Engineers (ISEE) (2011). *Blaster's Handbook 18th Edition*. International Society of Explosives Engineers.

MWH (2016). Red Rock Hydroelectric Project Foundation Report. Prepared for Western Minnesota Municipal Power Agency.

MWH (2013 and 2015). Red Rock Hydroelectric Project Specifications – Excavation. Prepared for Western Minnesota Municipal Power Agency.

Oriard, L. (2005). *Explosives Engineering, Construction Vibrations and Geotechnology*. International Society of Explosives Engineers.

UPRIVER DAM SPILLWAY REHABILITATION

Ray Trudgeon, P.E. [1]
James H. Rutherford, P.E. [2]
Steve Burns, P.E. [3]
Val Kovalishyn, P.E. [4]
Ccee Chan, P.Eng. [5]
Nikou Jalayeri, P.Eng. [6]

ABSTRACT

Upriver Hydroelectric Project, operated by the City of Spokane, is located on the Spokane River. The dam, built in 1937 consists of eight spillway bays with eight radial gates supported by reinforced concrete piers. The overall width of the spillway is 229 feet with an upstream – downstream monolith length of 70 feet and height of 55 feet measured from the spillway bridge deck to the foundation contact.

Hatch was selected by the City of Spokane to provide comprehensive engineering services that have included alkali-aggregate reaction (AAR) investigations, dam stability analysis, trunnion friction investigations, hydraulic analysis of spillway and apron baffle blocks, design of extensive rehabilitation measures, and engineering support during rehabilitation construction. Computational Fluid Dynamics modeling was used to estimate hydraulic loading at the baffle blocks and investigate potential changes in hydraulic performance due to adding a topping slab on the downstream spillway apron.

Based on investigations by Hatch, it was concluded that pier degradation has, to a large extent, been due to differential ARR expansion above and below the construction joint at El. 1915.5 ft. Since the trunnion anchors are located in the pier concrete above the cracked construction joint at El. 1915.5 ft, the stability of the upper portion of the piers was of concern because they support the gate trunnions and much of the gate load.

This paper will present the challenges faced during the investigations, engineering and construction phases of the project. Engineering issues have included cracking and apparent differential expansion above and below pier at the pier construction joints, high trunnion friction measurements for some of the gates, eroded spillway apron surfaces, damaged spillway and apron baffle blocks, potentially high hydrodynamic loads on the spillway baffle blocks and elevated piezometric pressures at one of the abutments. Rehabilitation construction was completed in November 2016.

[1] Hatch Project Engineer, 6 Nickerson Street, Suite 101, Seattle, WA 98109, raymond.trudgeon@hatch.com
[2] Hatch Project Manager, 6 Nickerson Street, Suite 101, Seattle, WA 98109, james.rutherford@hatch.com
[3] City of Spokane Dam Safety Engineer, 2701 N Waterworks Street, Spokane, WA 99212, sburns@spokanecity.org
[4] Hatch Senior Mechanical Engineer, 1066 West Hastings, Suite 400, Vancouver, British Columbia, V6E 3X2, Canada, val.kovalishyn@hatch.com
[5] Hatch Structural Engineer, 1066 West Hastings, Suite 400, Vancouver, British Columbia, V6E 3X2, Canada, ccee.chan@hatch.com
[6] Hatch Hydraulics Engineer, 1066 West Hastings, Suite 400, Vancouver, British Columbia, V6E 3X2, Canada, nikou.jalayeri@hatch.com

INTRODUCTION AND BACKGROUND

Upriver Dam Hydroelectric Project is located in Spokane, WA on the Spokane River and is owned and operated by the City of Spokane (City). Construction of the existing Upriver Dam was completed in 1937. It is located immediately downstream of the original timber crib and rock fill dam that was constructed in 1895. The project consists of an earthen embankment with an emergency fuse plug spillway, concrete gravity spillway, power canal and two (2) powerhouses with a combined rated capacity of 17.7 MW.

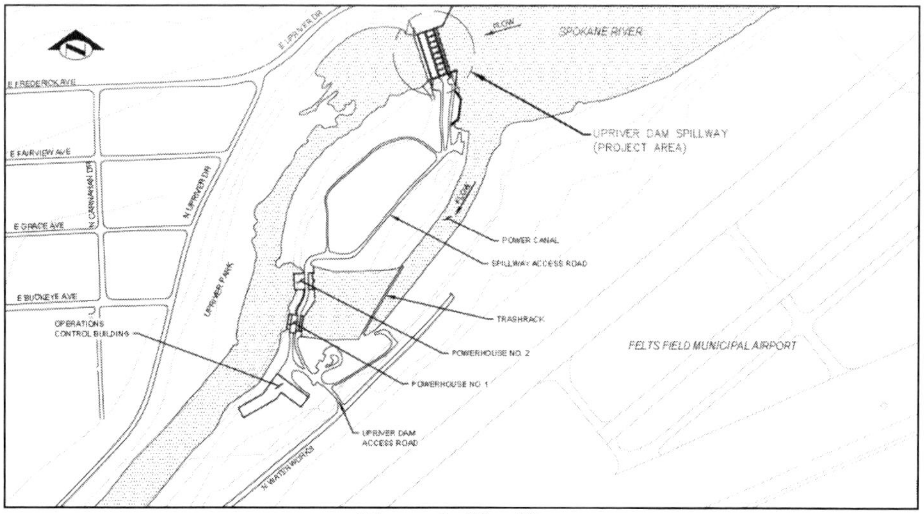

Figure 1. Site Plan

The concrete gravity spillway section is founded on alluvial soils and is 229 ft wide and 70 ft in length (upstream to downstream), and is 38 ft in height. There are eight bays and gates are numbered 1 through 8 (right to left looking downstream), with pier/abutments labeled A through I (also right to left looking downstream). Each bay is controlled by a 26 ft wide by 17 ft tall radial gate. The spillway crest elevation (El) is 1910 ft (all elevations are referenced to City Datum) and normal forebay pool is maintained at approximately 1927 ft.

The primary spillway apron at El 1895 ft has sixteen (16) total 8 ft x 8 ft x 8 ft spillway baffle blocks. These baffle blocks are located immediately downstream of the end of the spillway piers and are the primary means of energy dissipation at the spillway. A secondary apron at El 1892 ft extends 56 ft downstream of the primary apron and contains forty (40) total 3 ft x 3 ft x 4 ft apron baffle blocks. Pressure relief (PR) drains and weepholes are located at the primary and secondary aprons as measures to reduce uplift pressure. Additional drains were also previously installed in the ogee crest of Bays 7 and 8 as well as at the left abutment to reduce uplift. A general arrangement of the spillway is shown below in Figure 2.

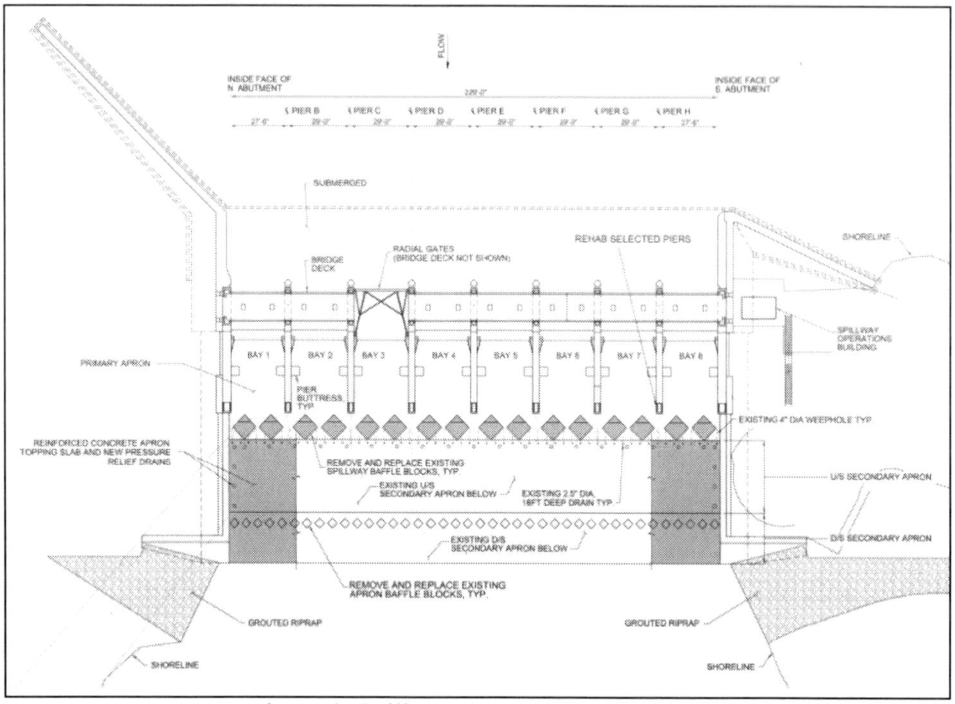

Figure 2. Spillway General Arrangement

Significant cracking and spalling has been observed at the reinforced concrete piers and baffle blocks. Repairs, especially to the spillway baffle blocks, have been performed frequently. In addition, sections of the secondary apron concrete surface have experienced erosion and reinforcement has become exposed. Although the project is relatively small in size, the foundation conditions, lack of complete as-built information, progressive erosion of the downstream apron, issues with alkali-aggregate reaction (AAR), variability in piezometric levels over time, and issues with hydraulic gate trunnion friction have created a complex engineering challenge requiring detailed investigations and analysis design of appropriate repairs and careful monitoring of construction.

PIER INSPECTIONS AND TESTING

Concrete deterioration of the spillway piers has been visible for many years. Repairs to the top of the piers were performed in the late 1980s. It has been observed that the more significant cracking is present at the top of the piers, above the construction joint (CJ) at El 1915.5 ft. As part of the field investigations phase of the project, core samples were obtained at the piers in order to perform AAR testing and petrographic analysis. The core samples were taken from above and below the CJ to assess AAR conditions separately in the two areas. During the site inspection it was also observed that some of the piers showed a horizontal downstream displacement of concrete above CJ at El 1915.5 ft. Photo 1 shows both the spillway piers and baffle blocks prior to rehabilitation.

Photo 1. Spillway Piers and Baffle Blocks Prior to Rehabilitation

The AAR testing of concrete cores included petrography of the concrete, Damage Rating Index (DRI), Soluble Alkalis Test and a Residual Expansion Test. The petrography examination revealed the presence of reactive fine and coarse aggregate. Products of alkali aggregate reaction were found in each of the samples and the DRI estimates show that, for both piers tested, the top core samples were found to have a higher damage rating than the lower core samples. All of the samples tested for soluble alkalis were above 2 kg/m3, which is the value considered a threshold for sustained AAR growth in large dams. The fine and coarse aggregate were found to potentially provide additional alkalis that could contribute to AAR. The residual expansion tests were not conclusive, but based on experience with similar projects and the available test results, we would expect only relatively low future expansion.

There were areas of rust staining along the CJ, indicated that the reinforcing steel might be losing cross-sectional area due to corrosion. When the core samples were obtained for AAR testing, the reinforcement at CJ EL 1915.5 ft was also exposed in several locations. In the four locations where the reinforcing was exposed for inspection it was found that the surface of the reinforcing had only minor discoloration due to corrosion. Calipers were used to measure the cross sectional area of the reinforcement and it was found that there was no appreciable loss of cross-sectional area.

Mapping of significant surface cracks on each pier was completed in 2015. In locations accessible from the spillway apron, the crack widths were measured using a pocket comparator and noted on the crack maps. This tool consists of a small magnifying glass with a reticle showing gradations to the nearest 0.01 mm. The location and width of significant cracks not accessible from the spillway apron were estimated. The 2015 crack maps will be used as a base line to investigate changes in pier cracking after construction is completed in 2016.

During inspections of the piers it was realized that there was the substantial cracking and evidence of movement along the horizontal construction joint at El. 1915.5. Figure 3 shows the three-dimensional Auto CAD used to determine the center of gravity and concrete weights used to check sliding stability along the cracked joint at El. 1915.5.

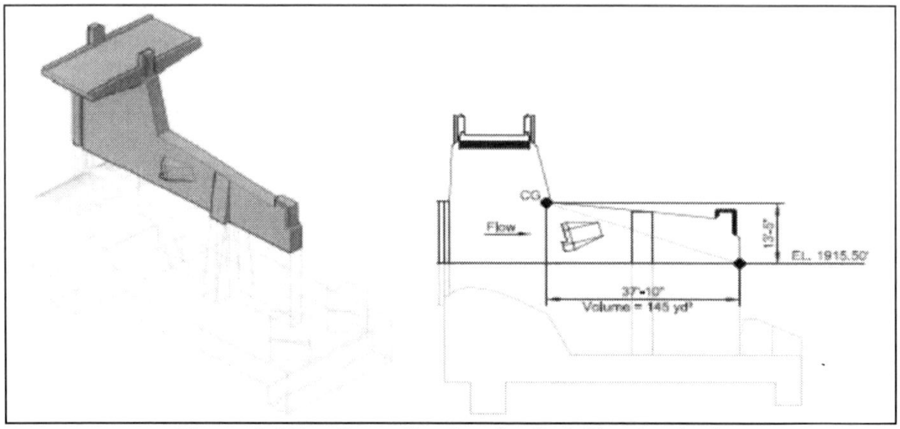

. Figure 3. Pier Cross Section for Stability Analysis

Stability analysis for the pier above the CJ at El. 1915.5 under normal, earthquake, and flood discharge conditions was performed. The analysis included the gate loadings that are resisted by trunnions located above the analysis plane. Analysis results showed that if the reinforcing steel crossing the failure plane is included, the SSF was significantly greater than the FERC guidelines limiting criteria of 1.5 for all load cases. The pier was also found to be stable and resists overturning.

SPILLWAY MONOLITH STABILITY ANALYSIS

Spillway Performance and Stability Analysis Setup

Due to the significant surface erosion at the secondary apron, it was determined that a new high-strength reinforced concrete apron topping slab (ATS) should be placed over the top of the existing slab to improve abrasion resistance and increase service lifetime. Placement of the ATS required that all existing pressure relief drains and weepholes be extended to the top of the new topping slab. Although the project was initially intended to only address reinforced concrete deterioration, it was recognized that if any new PR drains were to be installed they would need to be in place prior to the placement of the ATS.

The Upriver Dam spillway drain system originally consisted of only 4" diameter, 45-degree inclined weepholes installed during the original construction of the primary apron. During a FERC Part 12D Safety Inspection in August 1988, evidence of transport and discharge of fine sediment was observed downstream of one of the inclined primary apron weepholes and in the vicinity of a weephole in the left abutment wall. The mitigation measures to address the piping of fine material completed in 1989 included installation of screens in the existing 45-degree apron drains and installation of a row of vertical 2 ½" diameter polyvinyl chloride (PVC) drains extending 16 ft through the upstream side of the secondary apron slab (GeoEngineers, 2010).

Increased uplift pressures were observed at piezometers D7 and D11 near the left abutment wall in July 1989. In response to the observed uplift pressures, additional 2-1/2" diameter vertical PVC drains were installed through the upstream side of the secondary apron between the existing vertical drains installed a year earlier. Three horizontal drains were also installed into the left abutment wall extending approximately 8 ft into the left abutment soil (Geoengineers, 2010). Chemical grouting has also been performed at the apron slab and left abutment to fill possible voids and reduce seepage.

In addition to additional drainage, a seepage barrier was extended from the upstream face of the spillway structure to approximately 105 feet upstream on the floor of the forebay. The seepage barrier, installed in 1989, reduces uplift pressures by increasing the overall seepage path. In most locations, it consists of a polyethylene membrane covered with a 5" thick reinforced concrete slab (Geoengineers, 2010). Through the review of construction photographs it appears that the membrane extends along the stream face of spillway wingwalls to the edge of the fuse plug at the left abutment, while the reinforced concrete extends upstream of the spillway bays, see Photo 2 below. It appears that these masonry wing walls were repaired in locations with shotcrete at the time of the membrane installation.

Photo 2. 1989 Seepage Barrier Construction

Hatch reviewed the potential failure modes (PFMs) related to the spillway monolith sliding and pier bending and shear loading under cross valley loading. In addition, stability analyses were performed using the historic piezometric data recorded by the City to estimate uplift. The Upriver Dam has a total of 30 active piezometers, including 22 located in the immediate vicinity of the dam spillway. The operational loading condition assumed that one bay is dewatered and the adjacent bay is spilling.

Uplift and Seepage Analysis

Plots of piezometric data collected between 1999 and 2015 were used as a basis to estimate and evaluate uplift. It was found that Spillway Bays 1-2, 4-5, and 7-8 could be analyzed using data from specific piezometers:

Spillway Bays	Associated Piezometer Data
1-2	D2, D3, and D8
4-5	D4, D5, and D9
7-8	D6, D24 and D10

In the cross-valley direction, piezometers are aligned in upstream (D2, D4, and D6), middle (D3, D5, and D7) and downstream (D8, D9, and D10) rows. Review of piezometer data in the cross-valley direction allowed changes in performance on the left abutment to be better understood.

The seepage path considered for the migration of water from upstream to downstream of the spillway is illustrated in Figure 4. The rate of head loss between the piezometer locations was assumed to vary linearly along assumed straight line seepage path segments.

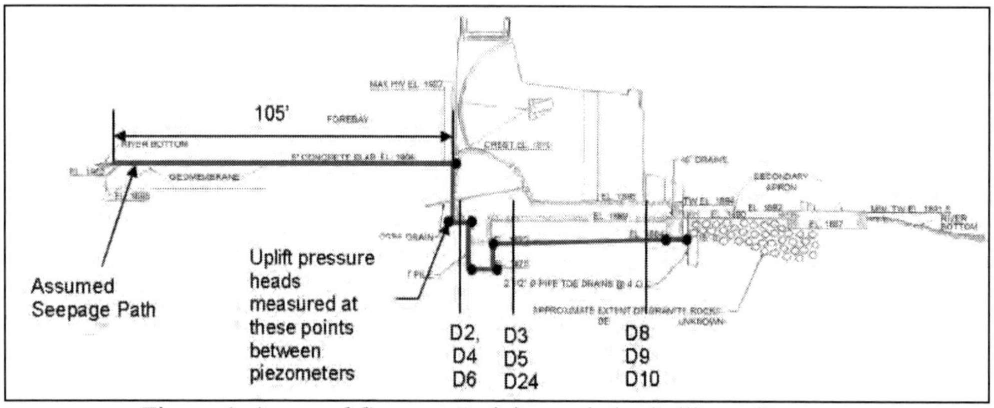

Figure 4. Assumed Seepage Path beneath the Spillway Structure

On this basis, uplift pressure profiles for each of the spillway monoliths were developed using the historical piezometric information. SEEPW software was used to construct a flow net beneath the spillway structure using the identified maximum piezometric levels. Piezometric data for Bays 4 and 5 (North Section), and 7 and 8 (South Section) were used to develop flow nets considering the effects of the upstream seepage barrier and the downstream 2-1/2" diameter vertical PVC drains installed through the upstream side of the secondary apron. Flow nets were developed to evaluate the following cases:

- Uplift Condition 1: Effective upstream seepage barrier, ineffective downstream drains;
- Uplift Condition 2: Effective upstream seepage barrier, effective downstream drains;
- Uplift Condition 3: Ineffective upstream seepage barrier, effective downstream drains; and
- Uplift Condition 4: Ineffective upstream seepage barrier, ineffective downstream drains

A representative flow net for the south section, Uplift Condition 3 is shown below in Figure 5.

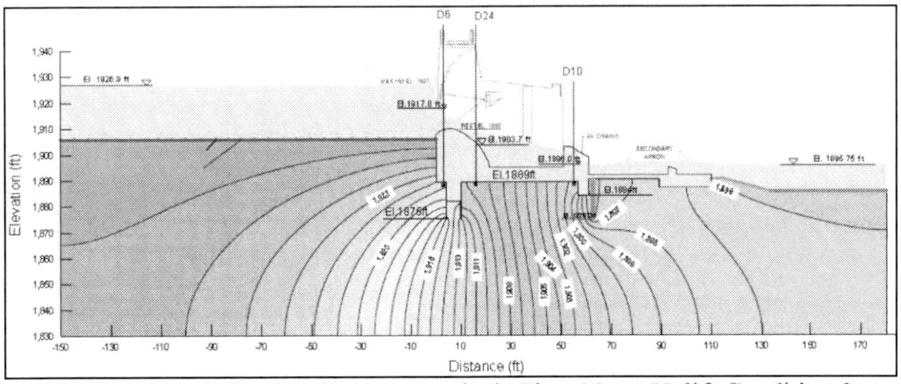

Figure 5. South Section Seepage Analysis Flow Net – Uplift Condition 3

The comparison of the seepage analysis flow net to historical piezometric information indicated that the existing conditions for the north section (Bays 4 and 5) the upstream seepage barrier and downstream drains are effective. The south side Bays (7 and 8) that are at the upstream seepage barrier were found to be less effective while the downstream drains are effective.

Spillway Stability Analysis

The spillway monolith was idealized for the stability analysis as a single spillway pier with half of a spillway bay on either side of the pier (equivalent to one spillway bay with a 29 ft width as measured between the centerline of two piers). Figure 6 shows an isometric view of the typical spillway cross-section used in this analysis. Three-dimensional Auto CAD was used to determine the center of gravity and concrete weights shown on this figure.

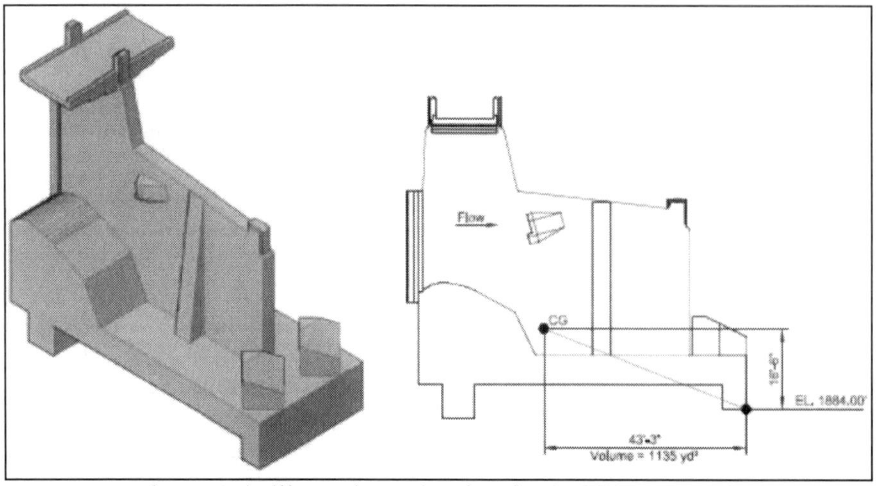

Figure 6. Spillway Cross Section for Stability Analysis

The stability analysis assumed that the spillway dam structure behaves as a two-dimensional rigid block subjected to hydrostatic and soil pressures. The analysis was performed for normal pool conditions and a range of tailwater levels based on maximum piezometric conditions. The uplift pressures determined in the uplift analysis were used. Figure 7 below shows the loading conditions at the spillway section.

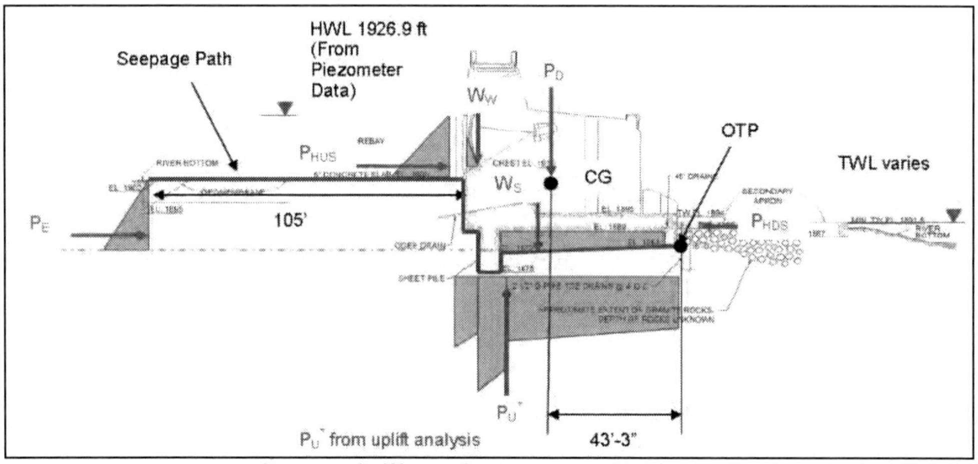

Figure 7. Spillway Structure Force Diagram

The estimated Sliding Safety Factor (SSF) for Bays1 and 2 and 4 and 5 was greater than 1.5 for normal load conditions and meets the limiting criteria given by the Federal Energy Regulatory Commission's (FERC) guidelines. However, the estimated SSF for Bay 7 and 8 was slightly less than 1.5 for normal loading due to higher uplift pressures and doesn't quite meet FERC guidelines. New and replacement pressure relief drains (27 total) were installed as part of the Upriver rehabilitation. Figure 8 shows the typical spacing of the new pressure relief drains.

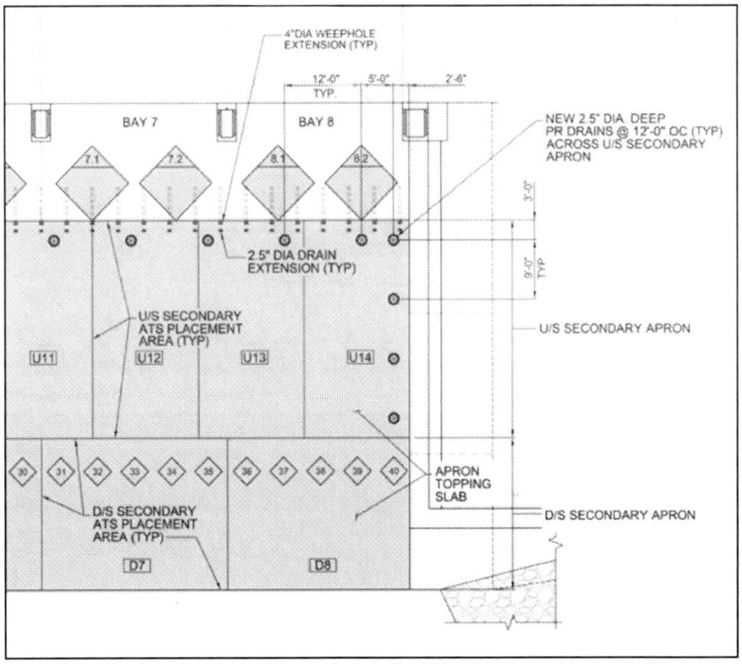

Figure 8. Typical Pressure Relief Drain Installation Plan

The new pressure relief drains are expected to only provide a small reduction in uplift pressure on the base of the spillway. Pressure relief is usually more effective further upstream. The City elected to replace the downstream PR drains since it could easily be accomplished during construction of the ATS. The City will continue to monitor the piezometric information to assess the changes in uplift pressures and overall stability. The monitoring program will focus on the left abutment with a follow-up investigation to look for high-permeability zones along the upstream membrane interface with the dam / dam embankment.

Pier Cross-Valley Stability Analysis

A pier analysis was carried out to evaluate the cross-valley capacity of the spillway piers under the condition where one bay is dewatered and the adjacent bay is spilling. The existing spillway pier wall was modeled with the STRAND7 structural analysis program. The pier wall was modeled using plate elements. Fixed connections were assumed at the intersection of the pier wall with the spillway slab. The ogee crest was also considered to rigidly support the pier wall, as it spans the entire width of the bay and the pier cannot experience bending at the location of the ogee. Figure 9 shows a view of the pier wall modeled in STRAND7.

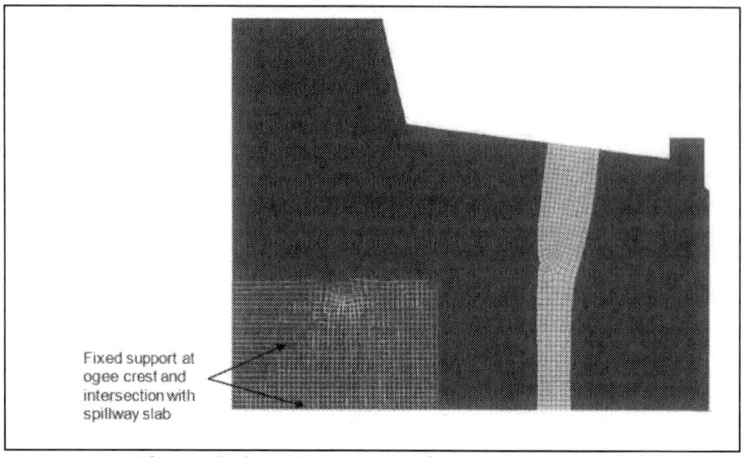

Figure 9. STRAND7 Pier Structural Model

The analysis considered an open spillway gate next to a closed spillway gate resulting in cantilever bending of the pier between the two gates. The water spilling over the pier was approximated as a linear distribution varying from headwater elevation at the upstream end to tailwater elevation at the downstream end of the pier. The pier was analyzed for normal conditions and normal earthquake conditions and results indicated that it is stable in both loading conditions.

HYDRAULIC ANALYSIS OF SPILLWAY STILLING BASIN

The configuration of stilling basin and baffle blocks is unconventional. Typical stilling basin energy dissipation and hydraulic jump equations would not be able to accurately describe the flow conditions and expected hydraulic jump characteristics. Therefore, FLOW-3D, a commercial computational fluid dynamics (CFD) model, developed by Flow Science Incorporated of Los Alamos, New Mexico, was selected to investigate the complex configuration and resulting hydraulic conditions. One of the major strengths of FLOW-3D is its ability to accurately model problems involving free surface flows. It is robust, handling transitions between sub-critical and super-critical flow within a single model set up. These capabilities made it well suited for simulating the varied and complex flow conditions that were anticipated in the stilling basin at Upriver.

The Upriver CFD model investigation concentrated on analyzing the hydraulics and energy dissipation characteristics in the stilling basin and in the immediate tailrace for the existing apron and the proposed case with an approximately 9" higher apron elevation. The CFD model was also used to estimate the hydrodynamic loading at the spillway, apron baffle blocks and secondary apron. These forces were used in the analysis and design of the replacement baffle block and apron topping slab anchorages and reinforcement.

The relatively long spillway and expected flow characteristics on the apron allowed for the use of a "sectional" CFD model configuration to reduce computational time; a single bay and two adjacent half bays on either side of the Upriver Dam spillway.

CFD Results

The results are qualitative in that proper validation of the model could not be performed without either field data or physical model data, which were not available. Therefore, the modeling was performed under the assumption that if the results for the proposed case showed no significant differences to the results for the existing case, then the hydraulic performance of the proposed secondary apron topping slab is adequate to dissipate the energy on the apron. Figure 10 shows an isometric view of the results of the existing apron configuration during expected PMF flow conditions. The figure represents a snapshot in time (not time-averaged), however averaged results showed that the post-modification case was very similar to the existing case. In both cases, the primary hydraulic jump is contained on the stilling basin and the downstream flow characteristics are similar with reasonably low flow velocities along the bottom, indicating similarly low tailrace erosion potential.

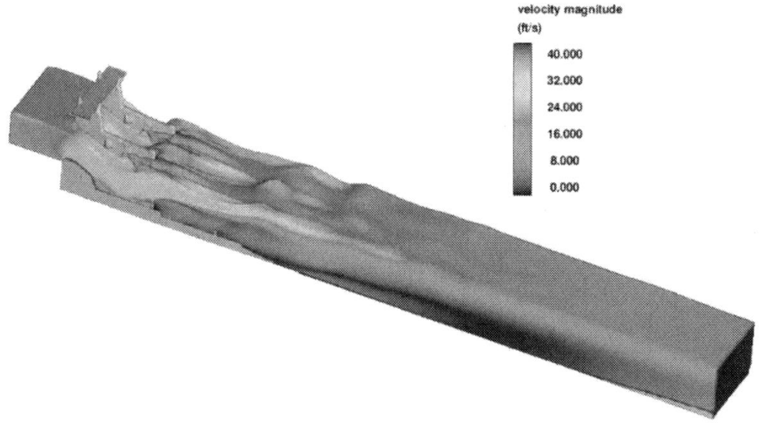

Figure 10. Isometric View of CFD Model Results – Existing Case

TRUNNION FRICTION MEASUREMENTS

To meet FERC Guidelines, dam owners are required to perform regular gate inspection and trunnion friction measurements at their facilities. All eight of the Upriver radial gates are of original construction. The City had only performed minor maintenance since original installation. Hatch performed inspections, analysis and trunnion friction testing of the gates in 2014/2015.

Based on the City's records it appears that the gates were manufactured from ASTM A7-1934 steel with a minimum yield point of 33 ksi. A trunnion friction testing program for the Upriver gates was implemented during the field investigations phase of the work. Trunnion friction coefficients were estimated based on structural modeling of the gates that related radial arm deflections to trunnion frictions, and input field measurements of deflection taken during gate operation.

Deflection measurements at the radial gate arms were obtained using electronic dial indicators (sensors) temporarily installed at the midpoint of each arm between the trunnion and the connection of the cross bracing. These sensors were fixed to one end of an aluminum angle (sensor arm) that served as a cantilever, with the other end clamped to the top flange of the radial arm near the trunnion, see Photo 3.

Photo 3. Deflection Measurement Equipment Attached to the Radial Gate Arm

The sensors transmitted the deflection readings to recording equipment located at the bridge deck. Measurements were plotted versus time to show the deflection of each arm during gate opening and closing as shown below in Figure 11. After each test, the equipment was easily transferred to the next gate, allowing for multiple gates to be tested in quick succession.

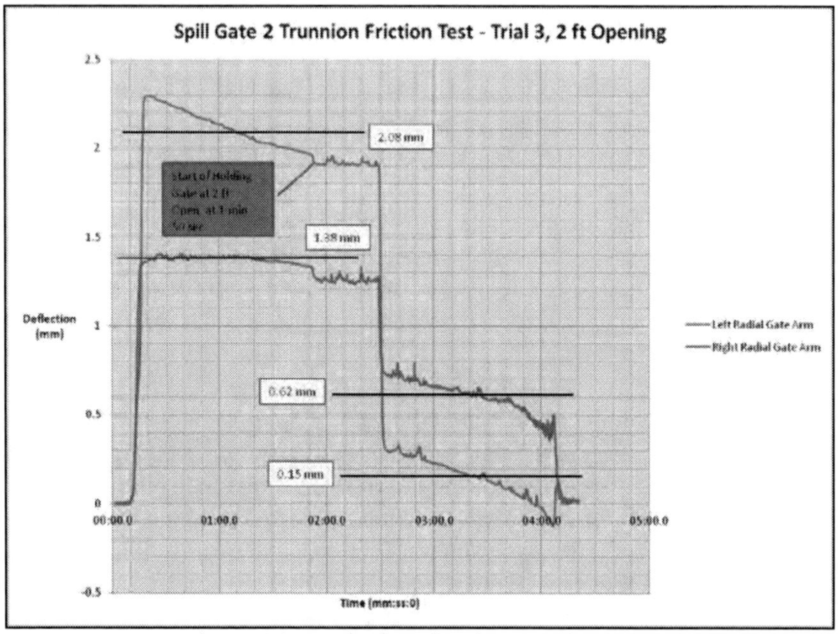

Figure 11. Typical Deflection vs Time Plot

A structural model (Figure 12) of the radial gate was developed in the RISA 3D structural analysis program, with member sizes based on original design drawings and as-built measurements. The cantilever sensor arm was included in the structural model of the gate to enable a direct determination of the theoretical deflection at the point of measurement. Gate weights and hydraulic loading were input to the model to obtain the trunnion pin loading and allow radial arm deflections to be related to trunnion friction. This testing method was used to identify radial gates with high and potentially problematic trunnion friction coefficients.

Load cases for normal conditions, earthquake conditions and flood discharge were included in the gate analysis. The gates were also analyzed for a range of trunnion coefficients under normal load conditions, and up to a trunnion coefficient where a member failure was possible. For the earthquake and flood discharge load cases, the gate was analyzed in the closed position. The highest load carrying demands for the radial arms was at the connection of the lower radial arm to the vertical girder, at a trunnion friction factor over 1.0 under normal load conditions. The gates were found to be stable under earthquake and flood conditions but some locally high stresses were found at the lower radial arm to vertical girder connection. Based on the analysis, an additional gusset plate was designed to reinforce the lower arm to vertical girder connection. The gusset installation is planned for a later phase of work.

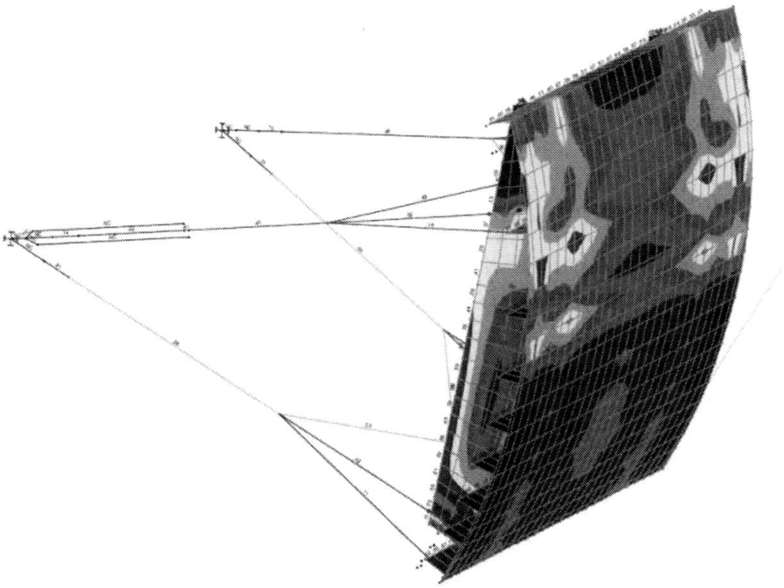

Figure 12. Structural Model of the Radial Gate

The use of radial arm deflection measurements to estimate trunnion friction allows for fast setup and installation of monitoring equipment, repeatable testing using high accuracy instruments, and direct measurement of member deflection as compared to strain measurement at isolated locations. Deflection limits were established based on gate analysis prior to testing.

The complete field component of the trunnion friction testing at Upriver Dam included testing all eight radial gates in adverse weather conditions in a combined time of approximately 15 hours. A few gates with high trunnion friction coefficients were identified, which allowed the City to prioritize gate maintenance. Since measurements were recorded for both radial arms, unbalanced cable lift tensions could be identified in case one side lags behind the other during operation. Based on gate inspections, analysis and trunnion friction measurements, minor modifications to the radial gates were recommended and repairs are planned to be implemented in 2017 after other aspects of the spillway rehabilitation project are completed.

REHABILITATION DESIGN

Hatch performed the field investigations, design, construction cost estimating, and bidding and construction management assistance for the Upriver Dam Spillway Rehabilitation Project. The primary objective of the project was to implement comprehensive repairs at the spillway to extend its longevity.

The spillway piers and spillway baffle blocks exhibit AAR-driven expansion cracking. During the design process, reinforced concrete repair measures for the baffle blocks were reviewed and it was concluded that baffle block replacement was economically justified. It was noted that the City had to frequently repair the top portion of the spillway baffle blocks. It appears that there is a plane of weakness due to the continued expansion below the repair so that impact loads frequently caused the top of the baffle blocks to fail. The replacement baffle blocks were designed with high strength, fiber-reinforced concrete to improve shatter resistance for debris impact. Epoxy injection grouting and bolting were performed to address the pier cracking.

To address surface erosion, exposed reinforcing, and accelerating deterioration, a reinforced concrete apron topping slab was designed to protect the secondary apron. The ATS concrete was designed with silica fume in the mix to improve hardness and resistance to abrasion during spill. Hooked reinforcing bars were installed at 3 feet on center to tie the old apron concrete to the new ATS. Pressure relief drains were designed to provide increased drain efficiency at the secondary apron and reduce uplift. The work also included the cleaning and extension of existing PR drains and weepholes to the top surface of the new ATS.

The City required that four fully functioning spillway bays be available with energy dissipation at all times during construction, if an unexpected flow release was necessary. This was accomplished by construction of first the north half of the spillway and then the south half.

CONSTRUCTION

The City awarded the contract to McMillen Jacobs Associates (Contractor) in January 2016. The City performed full-time site inspection for the project, with construction management and inspection assistance from Hatch. Environmental compliance monitoring and materials testing were performed by Spring Environmental Inc. and Budinger Associates, Inc. respectively.

The Contractor mobilized to site and began construction at the start of low flow conditions in early July 2016. The north side of the spillway was sequenced to be constructed prior to dewatering and the beginning of construction at the south side. Construction was completed within the scheduled window provided by the City. The main challenges included drilling and installing drains through and into an alluvial foundation with underlying timber crib materials, and controlling water on the existing apron so the ATS could be placed. Flow from existing and new drains and weepholes required collection and rerouting via piping around the work areas and leakage at the spillway apron concrete required patching to allow for effective dewatering. Following the installation of dewatering measures, construction of the reinforced concrete rehabilitation measures at the apron and piers proceeded. Substantial completion was achieved in late October 2016 and the project was completed on time and on budget.

Photo 2. Completed Spillway Rehabilitation (November 2016)

REFERENCES

City of Spokane. (2005). *Upriver Dam Hydroelectric Project - FERC No. 3074-WA - Supporting Technical Information.* City of Spokane.

City of Spokane. (2008). *Upriver Dam Hydroelectric Project FERC No. 3074-WA - Supporting Technical Information.*

Federal Energy Regulatory Commission (FERC). (2005). *Engineering Guidelines for the Evaluation of Hydropower Projects.* Washington, DC: Federal Energy Regulatory Commission (FERC).

Geoengineers. (2010). *Seepage Erosion Evaluation.* Redmond, Washington.

LAKESIDE RANCH STORMWATER TREATMENT AREA – LESSONS LEARNED DURING DESIGN AND CONSTRUCTION

Danielle Neamtu, P.E.[1]
Stephen Whiteside, P.E.[2]
Stephen Blair, P.E.[3]

ABSTRACT

The Lakeside Ranch Stormwater Treatment Area (LRSTA) is a key component of the Northern Everglades program to reduce the phosphorus loading to Lake Okeechobee. CDM Smith is providing design and construction oversight of the 2,000-acre emergent-vegetation STA. The project is currently in the second phase of construction after addressing numerous scheduling, physical, and regulatory constraints.

Previous STA projects constructed south of Lake Okeechobee were built on sites with minor topographic relief. However, the Lakeside Ranch property has a 13-foot elevation change across the site. The final design accommodated the topography, along with other environmental and economic constraints, by utilizing an intricate system of levees, canals, ditches, and hydraulic structures to distribute flows evenly across the STA. After eagles' nests and cultural resources were identified in a central area bisecting the site, the STA had to be designed for construction in two phases. The design allowed the first phase to operate as a standalone system and as a fully integrated component of the system following completion of the second phase. Uninterrupted operation during construction of the second phase was also required.

The design of the embankments, seepage collection ditches, and seepage control measures needed to address the local site geology and environmental constraints. The STA includes eight treatment cells, numerous control structures, a 5,000-foot-long discharge canal, 9,000 feet of canal improvements, more than 20 miles of earthen levees, and a 250-cfs intake pump station.

Construction of the first phase started in April 2009 and was completed in July 2012. The second phase is currently under construction and is scheduled to be completed in February 2018. This paper will discuss the innovative design concepts and analysis techniques and will describe the lessons learned during both construction phases especially with regard to seepage control.

---

[1] CDM Smith, 5400 Glenwood Ave, Suite 400, Raleigh, North Carolina, 27612, 919-325-3500, neamtudk@cdmsmith.com.
[2] CDM Smith, 5400 Glenwood Ave, Suite 400, Raleigh, North Carolina, 27612, 919-325-3500, whitesidesl@cdmsmith.com.
[3] CDM Smith, 621 N.W. 53rd Street, Suite 265, Boca Raton, Florida, 33487, 561-571-3800, blairsh@cdmsmith.com.

PROJECT BACKGROUND AND DESCRIPTION

Lake Okeechobee covers more than 730 square miles – a vast source for drinking water, home to aquatic life, and an active recreational landmark. Suffering from decades of phosphorous build up from non-point source pollution runoff, the state of Florida through the South Florida Water Management District (SFWMD) is working in cooperation with the U. S. Army Corps of Engineers (USACE) to implement a comprehensive program to reduce phosphorous loads on the lake as part of its efforts to improve the quality of the entire Everglades system.

CDM Smith developed the Basis of Design Report (BODR) for the Taylor Creek/Nubbin Slough system as part of the Lake Okeechobee Fast Track (LOFT) Project. The overall goal of the project is to reduce the mass and concentration of total phosphorus entering Lake Okeechobee. The project included an evaluation of the existing conceptual design to value engineer the greatest benefits for the available funding. One of the recommendations was the diversion of flows from agricultural lands to a series of wetland treatment systems working in coordination with the water management system of navigable locks, canals, flood control gates, and pump stations to capture and treat phosphorus. One key element of this evaluation was the Lakeside Ranch Stormwater Treatment Area (LRSTA) phosphorous reduction project, a component of the Northern Everglades and Estuaries Protection Program. The 4.5-square-mile LRSTA is intended to serve as a large regional facility that provides treatment of stormwater runoff and consists of three key project components that, as a system, will facilitate the removal of an estimated 19 metric tons of phosphorous annually from captured stormwater prior to discharge into Lake Okeechobee.

The project included modeling, design, development of comprehensive system operations manuals, preparation of construction bid packages, and construction oversight as well as implementation of an extensive water quality sampling and monitoring system. CDM Smith's BODR evaluated the capture and treatment of stormwater runoff. These flows along with other redirected stormwater flows will be conveyed via re-routing into the LRSTA, a 1,760-acre emergent wetland treatment area on 2,710 acres of land with 8 cells, numerous hydraulic control structures, a 5,000-foot discharge canal, 9,000 feet of canal improvements, and over 20 miles of earthen embankments. Stormwater flows are routed into the STA via a new 250 cfs intake pump station.
The layout of the STA is shown on Figure 1. Water is routed to the STA via a series of canals and pumped into a distribution channel that evenly distributes the flow into the STA. The water flows from cell to cell in a general southeasterly direction with a pair of gated weir structures at each cell divide that provides flexible control of the flow. The water ultimately flows into a collection channel and is then conveyed via a newly constructed triple box culvert under U.S. Highway 98 and where it discharges into the L-47 canal. The entire LRSTA site is enclosed with a 6-foot-high perimeter berm. The cells are divided by 4-foot-high interior berms. Each interior berm includes an emergency overflow spillway.

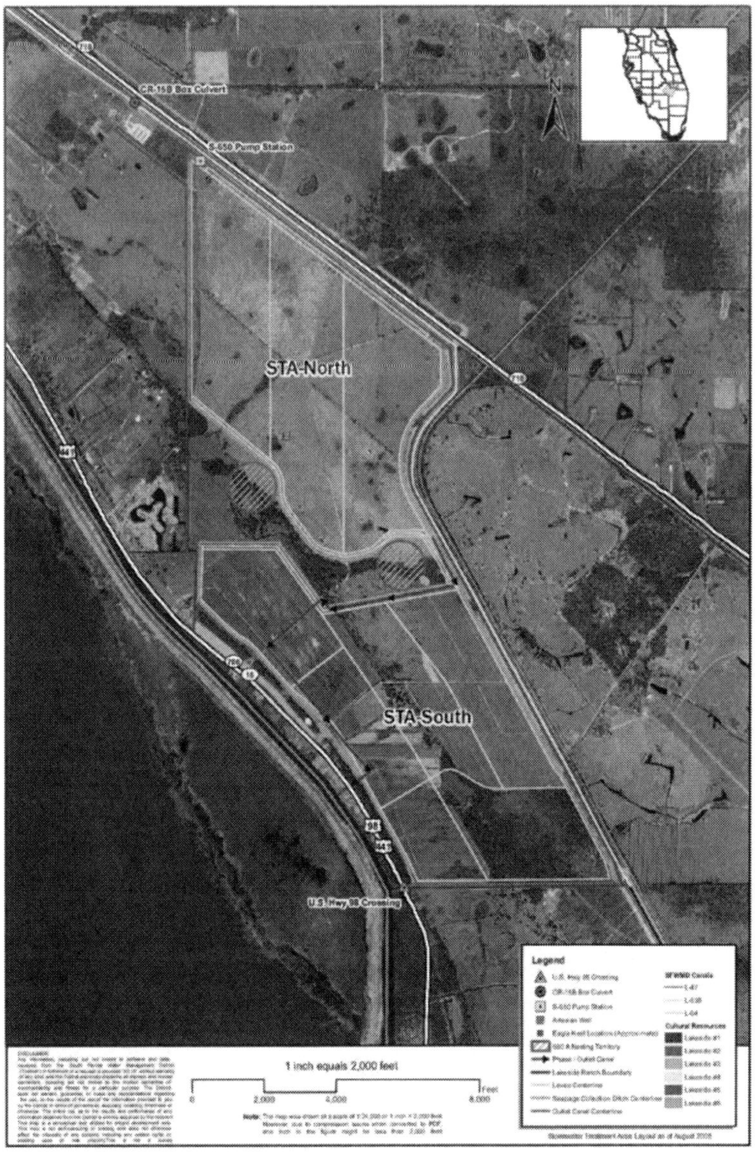

Figure 1. LRSTA Site Plan

Seepage collection canals are included along the perimeter of the LRSTA. The LRSTA design includes a provision for recirculation of lateral seepage and recirculation of STA discharge during dry conditions to account for hydration need of LRSTA. A flow of 50 to 100 cfs is required for the hydration need.

Project Design Challenges

The LRSTA project faced many design challenges that had to be addressed to ensure the project was both environmentally sound and cost-effective in terms of dollars per pound of phosphorous removed. The District had already purchased the property and the designers' task was to optimize the site and integrate the STA into the existing complex system of canals, pump stations, and various hydraulic control structures. Some of the major challenges the design team met included:

- Modeling the complex hydraulic and hydrologic systems and water quality throughout the watershed
- Developing a civil site design that balanced cut and fill over a 1,760-acre footprint that had a 13-foot elevation change. The total cut and fill volume is about 3 million cy.
- Designing an STA that provided the required hydraulic residence time to effectively remove the phosphorous while keeping embankment height under six feet to avoid its classification as a dam;
- Avoiding impacts to designated cultural resource areas, wetlands, eagle nesting sites, and contaminated areas shown on Figure 1. Designing a two-phase STA where each phase could operate independently or as a single system;
- Reversing the flow and improving the capacity of an existing canal; and
- Constructing impoundment embankments with soils containing fine sands that are susceptible to piping.

CDM Smith utilized multiple modeling tools to evaluate and solve each of the design issues identified (e.g., fast-track schedule, hydraulic constraints, seepage control for embankments and ditches, potential off-site mounding of groundwater, dry-condition hydration of the STAs, logistics and coordination of multiple project components, and budget limitations). These models included big-picture and detailed design tools at different resolutions and time scales to facilitate design, analyses, and quality checking.

A schematic of the models used and the interrelationships is included in Figure 2. The model tools included:

- United States Geological Survey (USGS) MODFLOW Model for groundwater hydrology and hydraulics for evaluation of seepage control options and groundwater mounding;
- Geo-Slope SEEP/W for seepage management system design;

- EPA Storm Water Management Model Version 5 (SWMM 5) for surface water hydrology and hydraulics;
- Soil and Water Environment Technology (SWET) Watershed Assessment Model View (WAMView) for long-term continuous simulation of daily flows and total phosphorus loads;
- USACE Hydrologic Engineering Center (HEC) River Analysis System (RAS) dam break module;
- SFWMD Dynamic Model Stormwater Treatment Area 2 (DMSTA2) for STA evaluation and design; and
- ISEE System's Systems Thinking Experimental Learning Laboratory with Animation (STELLA) for system and operations evaluations.

Due to large elevation changes across the site (13 feet of elevation difference across the site from north to south and more than 9 feet from east to west), a terraced grading plan was developed to minimize earthwork quantities. This terraced cell configuration required the development of an intricate system of canals, ditches, and hydraulic control structures to distribute flows evenly throughout the system. This design concept was unique to SFWMD STA projects because other SFWMD STA sites were located south of Lake Okeechobee in the Everglades flow-way where the existing topography was significantly flatter. These STAs were designed and constructed as a single compartment with much simpler hydraulics. They essentially have a single perimeter embankment surrounded by a seepage control ditch. In contrast, the LRTSA design required three parallel treatment "trains", each consisting of multiple treatment cells operating in series. Based on detailed modeling and discussions with both the SFWMD and USACE, the treatment system uses a series of adjustable hydraulic control structures (weirs) that will maintain the water depths in the cells over a range of operational depths that were determined to provide the best phosphorous removal benefit under the given conditions.

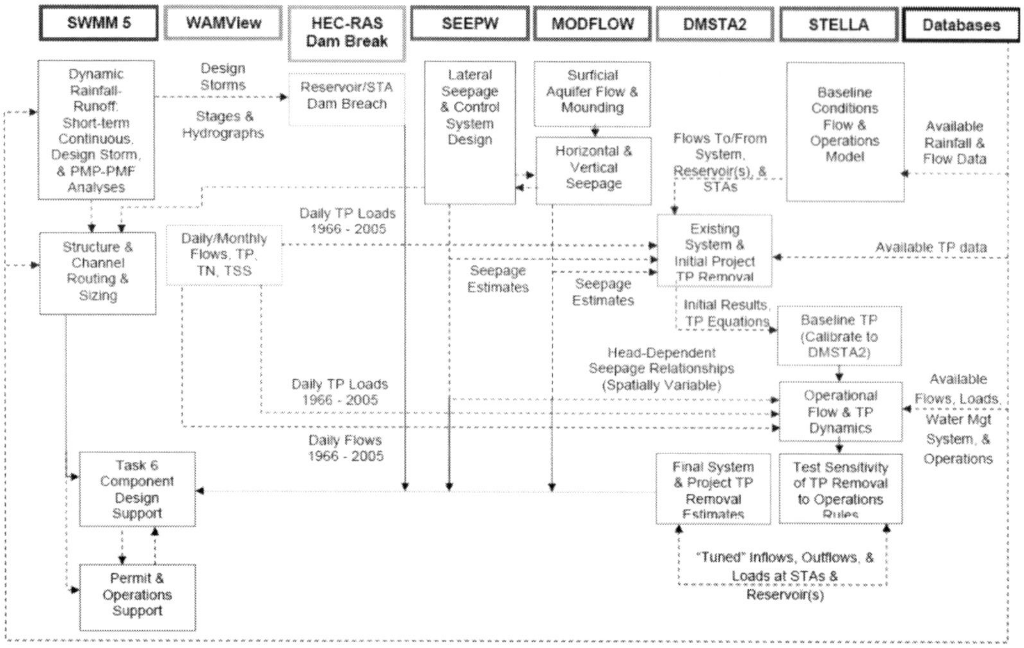

Figure 2. LOFT BODR Coordination Chart

Project Implementation

To address cultural resources, fish and wildlife, wetland, and other environmental concerns, the Project was divided in two phases. The final design included a 2,700-acre emergent wetland STA consisting of 9,000 feet of canal improvements, a 250-cfs intake pump station, a 450-cfs recirculation/ flood control pump station, seepage control channels, discharge, spreader, and collector ditches, water control structures, and 20 miles of earthen levees.

Phase 1 is the northern portion (LRSTA-North) and was completed in July 2012. Phase 1 (LRSTA-N) consisted of a three-cell, terraced configuration with an intricate system of canals, ditches, and structures to evenly distributes flows throughout the system to maintain water depths conducive for STA vegetation health and total phosphorus (TP) removal that are also safe for dam/ levee requirements and manageable for seepage control (Figure 3).

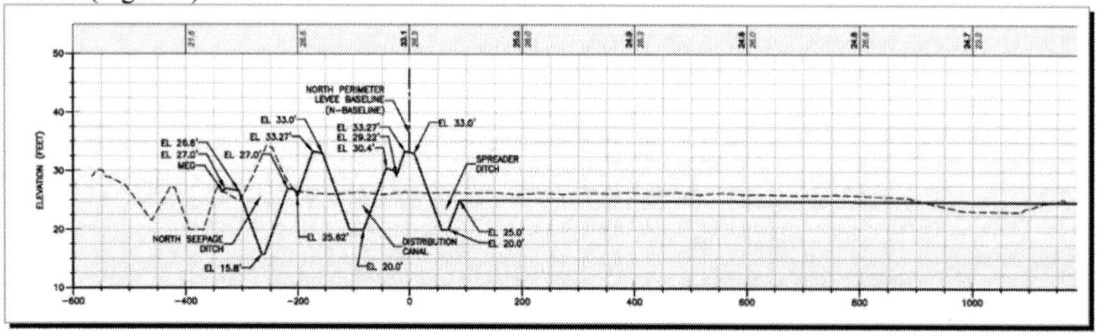

Figure 3. Terraced Cell Configuration from Final Design

Phase 2 (STA-South) is currently under construction and projected to be completed in February 2018. Phase 2 (STA-South) will consist of a five-cell, terraced configuration with canals, ditches, and structures similar to Phase 1. Phase 2 will also include the permanent outfall.

DESIGN

Design Overview

When construction is complete, the LRSTA will encompass approximately 2,700 acres (1,760 acres of stormwater treatment area (STA)) and include an eight-cell STA, distribution and outlet canals, and water control structures. The layout has been divided into two phases and the design is structured so that the first phase (STA-N) will be able to operate independently of the second phase (STA-S).

The existing site topography is the predominant factor influencing the layout of the LRSTA. The Lakeside Ranch property has up to 13 feet in elevation change from north to south and 9 feet from east to west. From a hydraulic performance perspective, it is desired that the STA cells have no slope perpendicular to the flow direction and only a slight slope parallel to the flow direction. To meet this objective, it was recognized that the LRSTA needed to be separated into multiple treatment trains and terraced in order to reduce earthwork volumes. After several iterations, a three-treatment-train system was established as the basis of the LRSTA layout. Over 20 miles of earthen embankments will be constructed to create the 3 treatment trains and 8 terraced treatment cells.

For STA-N, an above-ground distribution canal conveys water from the pump station to the cells of STA-N. During Phase 2, water will be distributed from STA-N to STA-S via an above-ground canal system. Due to topography differences, the system will consist of two parallel canals, one which services the combined flows from the west side of STA-N to STA-S and one which services the east side.

A seepage collection ditch will be provided around the perimeter of the LRSTA to collect seepage and control offsite groundwater mounding. Due to the site topography, the ditch will be terraced from north to south, and weirs will be placed in the ditch to maintain a normal operating water depth approximately equal to the surrounding groundwater surface and reduce off-site groundwater mounding. The weir locations and elevations were determined based on groundwater modeling performed to evaluate off-site impacts. It should be noted that, in addition to seepage control, the east seepage collection ditch will also reduce potential impacts to an adjacent natural gas line. The ditch will lower the downgradient piezometric pressures and, as a result, reduce the uplift forces acting on the gas line. This will in turn eliminate the need for anti-flotation anchors on the gas line.

The LRSTA design includes the use of variable-weir slide gates as part of the hydraulic controls for the STA cells. The depth of water is important in STA design because water elevations that are too high or too low over extended periods of time negatively influence the treatment performance. Therefore, managing the design water depth maximizes phosphorous removal. Control structures are required between the STA cells to assist in maintaining the water depths within the design depth envelope of 12 to 24 inches and ideally to 18 inches, the optimum depth for emergent aquatic vegetation design conditions. If the target depth for desired vegetation is changed, the weir elevations can be adjusted to achieve the required water depth.

On-site Borrow Materials

One of the first design considerations for the project was the use of the onsite fine sand soils for construction of the impoundment berms. These soils were spread throughout the site and were susceptible to piping. A detailed borrow source mining specification was developed based on a comprehensive geotechnical investigation. The specification, coupled with onsite monitoring and testing during construction, will promote a better blending of the excavated soils and provide a more-suitable soil gradation for the embankment construction.

Freeboard Evaluation

Wave set-up and wave run-up analyses were performed as part of the freeboard evaluation to confirm the proposed 6-foot height for the Lakeside Ranch STA-S embankments, the maximum allowable height to avoid classification as a dam. Both Phase 1 and Phase 2 design treatment flows of 100 and 250 cubic feet per second (cfs) were considered in the freeboard analysis. Freeboard is defined as the required vertical distance above maximum water surface level (MWSL) to contain wind set-up and wind-generated waves without exceeding acceptable embankment overwash rates. MWSL was considered as the routed 100-year, 24-hour event on top of normal operating pool levels (which range from a Normal Pool depth of 18 inches to the Maximum Operating depth of 28 inches for the STA cells).

Wave run-up is dependent on several factors including STA design water depth, design wind speed, fetch distance, embankment slope angle, and embankment slope surface roughness, which depends on the type of erosion protection.

Both short-term (no vegetation) and long-term (sustained emergent vegetation) conditions were considered in the evaluation.

The analyses of the 6-foot perimeter embankments indicated overwash rates up to 0.03cfs/ft for short-term conditions, which exceeded the acceptable rate of 0.01 cfs/ft that was established based on field tests performed on embankments of other SFWMD projects (the C-43 and CC-44 embankments as presented in 2007 ASDSO Dam Safety Proceedings in the paper "Over the Top – Results from Overwash Testing on Embankment Dams in South Florida). Additional computations showed that reducing the maximum operating depth from 28 inches to 24 inches decreased the maximum overwash rate to 0.01 cfs/ft. Therefore, the initial operating levels in the STA cells were recommended to be maintained at 24 inches prior the grow-in period for the sustained emergent vegetation.

The analyses of the 4-foot internal embankment indicated an overwash rate of 0.76 cfs/ft for short-term conditions. Unfortunately, lowering the operating depth to 24 inches did not result in acceptable overwash rates for internal berms. To mitigate the high overwash rates for the internal embankments, it was decided to protect the 12- to 14-foot-wide crests and the embankment slopes with turf reinforcement mats overlain by sod. Based on discussions with North American Green, for an overwash rate of 0.8 cfs/ft, typical turf reinforcement can provide an allowable shear stress factor of safety of 1.0 for unvegetated conditions. Therefore, the calculated overwash rates for the internal embankments in a short-term post-construction unvegetated condition are considered acceptable if turf reinforcement is utilized. Turf reinforcement is also used on the interior slope of the perimeter embankments.

Because the proposed STA cells will be operated to support sustained emergent vegetation, increased bottom roughness (and wave damping) due to the vegetation was considered for long-term conditions. The wave run-up calculations considering vegetation were performed using the VBET program developed by Lori Hadley of the USACE Jacksonville District (SAJ) in conjunction with STWAVE. Wave heights are reduced using Manning's coefficients to account for the bottom roughness of cells due to vegetation. STWAVE transforms the wave height from grid cell to grid cell applying the dissipation at each grid cell location. If the fetch is sufficiently long, the dissipation will exceed the wind energy input and effectively damp all the energy from the wave system resulting in a significant wave height of zero. From observations in shallow-water vegetated areas to date, this appears to be a good representation of field conditions, particularly for emergent vegetation. Based on the results of the analyses, the minimum required freeboard with the maximum operating depth of 28 inches is anticipated to be the default 3 feet. The fetch at each cell is sufficiently long that the dissipation due to vegetation will exceed the wind energy input for the shallow water depths of the STA cells resulting in significant wave heights and overwash rates of practically zero for both the perimeter and the interior embankments.

Slope Stability and Seepage Control

The design cross section has 3H:1V slopes with a 12- to 14-foot-wide crest for vehicular use. The crest of the perimeter embankments is 6 feet above the adjacent cell floor, and the crest of the interior embankments is 4 feet above the higher adjacent cell floor. For economy of construction the embankments have homogeneous cross sections of the insitu soils which are typically fine sands and slightly silty to occasionally silty sands (SP, SP-SM, and SM).

Factor of Safety against Piping: Based on literature review, including EM 1110-2-1901, EM 1110-2-1913, TR GL-97-2, and ETL 1110-2-569, recommended factors of safety against piping range from 2 to 5. Higher factors of safety are recommended for embankment heights greater than 6 feet and when subsurface information is limited. For embankments less than or equal to 6 feet in height and where subsurface information is ample, factors of safety of 2 to 3 may be acceptable.

Based on the USACE's experience on similar water treatment/storage projects in Florida, the USACE recognized that lower risk exists at the bottom of drainage canals, and they recommended an allowable vertical exit gradient of 0.5, corresponding to a factor of safety of 1.7 for a saturated soil unit weight of 115 pcf. Based on the guidance from the USACE, the abundance of data for this project, and engineering judgment, CDM recommended minimum factors of safety of 3 at the toe of perimeter embankments and 2 for all canals/ditches and interior embankments.

Factor of Safety for Slope Stability: The factors of safety required by the USACE (EM 1110-2-1902) were used to assess the embankments. The conditions that were analyzed include: end-of-construction, long-term or steady-state seepage at maximum pool, rapid drawdown, and earthquake.

The USACE requirements were modified for the seepage collection ditches. Rapid drawdown in the perimeter seepage collection ditches was not considered applicable because the water levels in the ditches are controlled and in some cases are lower than ambient groundwater. Also the minimum factor of safety for the long-term condition with steady-state seepage through the embankment was decreased from 1.5 to 1.3. This decision was based on the calculated exit gradients in the seepage collection ditches having factors of safety against a piping failure of at least 2.0, and the potential failure surfaces being very shallow and confined to the ditch slopes, so the integrity of the STA embankments was not affected.

Summary: Because the STA embankments are not classified as dams, reduced factors of safety are justifiable which results in significant savings in the construction costs. The final values of the safety factors were selected balancing construction costs against long-term system costs such as maintenance and reliability.

Method of Analysis: The analyses for the project were performed using the SEEP/W and SLOPE/W version 2007 software programs, distributed by GEO-SLOPE International, Ltd. Pore pressures from the SEEP/W analyses were imported into the SLOPE/W analyses. The exit gradients were computed by hand because gradients taken directly from boundary nodes in the SEEP/W model are very sensitive to mesh size and do not converge with increasing mesh density. After discussions with both the USACE and GEO-SLOPE, a manual method of computation was established using the equipotential and flow lines generated by SEEP/W because these are independent of the mesh size. To determine exit gradients by hand, the total head difference between the upstream and downstream water levels was divided into 20 equipotential drops, and the exit gradient was calculated using the last two equipotential lines at the downstream end. The hand-calculated exit gradient was divided into horizontal and vertical components based on the angle of flow exiting the ground surface at the node of maximum exit gradient generated by SEEP/W

Soil Parameters: A north-south cross section through the site is shown in Figure 3. The subsurface soil profiles and the corresponding soil properties were based on the geotechnical and hydrogeological investigations which were performed in four phases as the project design was developed. The investigations included Standard Penetration Test

(SPT) borings, Piezocone (CPTU) soundings, test pits, "muck" probes, an aquifer pump test, piezometers with slug tests, electrical resistivity surveys, and laboratory tests.

The friction angle for the upper 15 feet of sands was a critical parameter in the seepage and slope stability analyses. Statistical analysis of more than 863 SPT N-values in this stratum and published correlations between SPT N-values and friction angles resulted in values of 31 and 32 degrees. The friction angles obtained from the triaxial tests and correlations with CPTU data were considerably higher, 35 to 39 degrees. The lower values of the friction angles were selected for the seepage and stability analyses because they were more in line with standard practice and experience in the project area. Even these lower values were higher than those used by the SFWMD on similar projects. The unit weights for the various soils were based on laboratory tests and local experience. The hydraulic conductivities for the sand layers were developed during the extensive groundwater modeling studies. For the portion of the site underlain by discontinuous layers of caprock and/or shell, slug test data were used in conjunction with layer thickness information from the adjacent SPT borings to estimate the hydraulic conductivities of these strata.

<u>Cross sections:</u> Because the STA site includes 2,700 acres, analyses had to be performed for various subsurface profiles and project geometries. These included cross sections of the perimeter embankments with adjacent seepage collection ditches, distribution canals and auxiliary berms on the four sides of both the north and south STAs as well as cross sections of the interior embankments.

The analyses were performed initially using best estimates of the global soil parameters and subsurface conditions. Additional analyses were then run considering soil layers with lower strengths and hydraulic conductivities (as had been observed occasionally during the field explorations) to evaluate the sensitivity of the results.

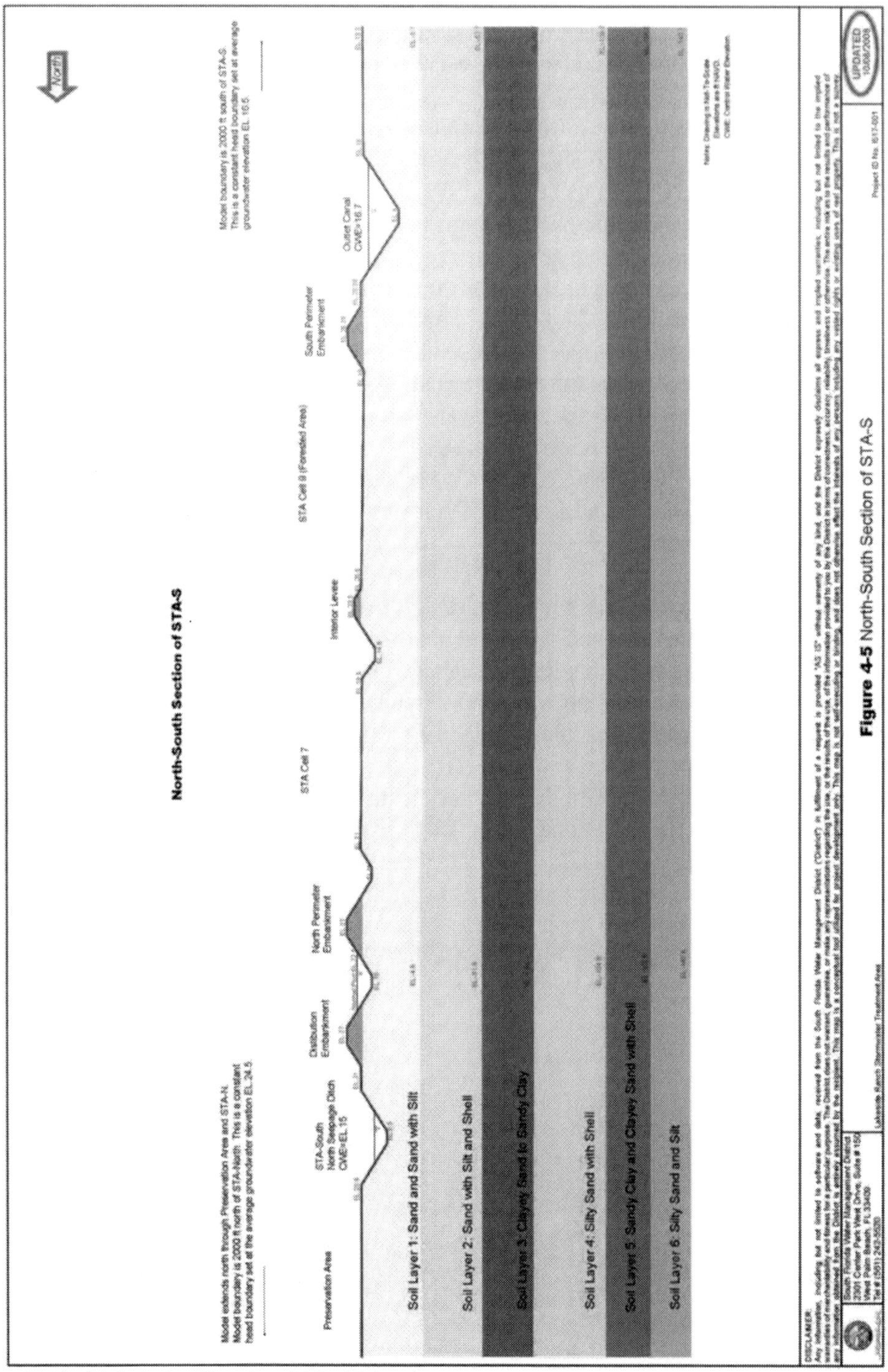

Figure 4. North – South Cross Section Through STA-S

Scenarios were also run considering the presence of caprock and shell layers individually and in combination. Discontinuous caprock was modeled as a 1-foot-thick layer of 10-foot-long rock segments spaced at 10-foot intervals and separated by soil. The location of the rock segments was varied by changing the rock segments to soil and the soil to rock to evaluate the impact of diverse soil conditions at the seepage collection ditches.

Results of Analyses with Best Estimate Input: The analyses indicated that seepage faces were confined to the seepage collection ditches; no seepage breakouts occurred on the downstream (outside) slopes of the embankments. However, the computed phreatic surface was within one to two feet of the embankment toe in several scenarios.
The slope stability analyses resulted in acceptable factors of safety for the perimeter embankments for all the analyzed conditions. One interior embankment was found to have a low factor of safety in the steady-state seepage condition. A representative SEEP/W analysis is shown in Figure 5.

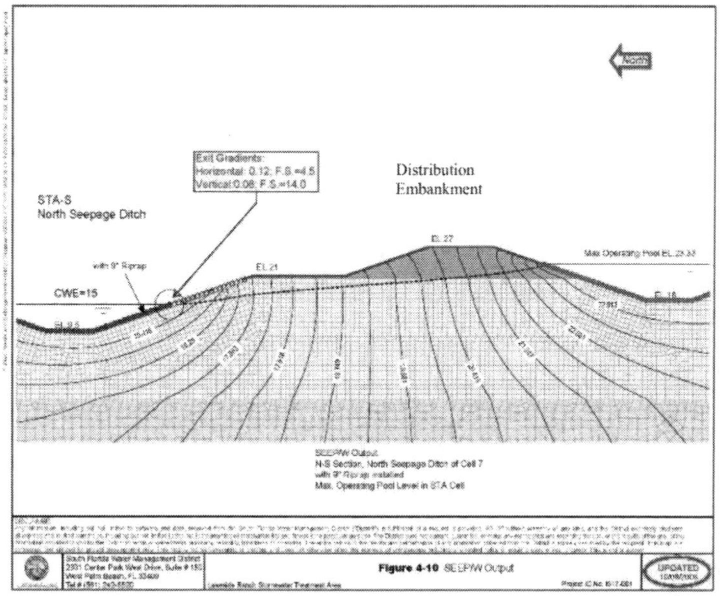

Figure 5. Representative SEEP/W Analysis

The seepage collection ditches were more problematic as the calculated factors of safety for both piping and slope stability failures were low for four perimeter cross sections. In all instances a cross section with a low factor of safety against piping also had a low factor of safety for slope stability. Due to the long length of the ditch, the economic impact of the low safety factors is significant.

Results of Additional Analyses: The analyses with the less favorable soil parameters generally had lower, but still acceptable, factors of safety indicating that the system had an adequate margin of safety even in localized zones with poorer subsurface conditions. In the case of a marginal seepage collection ditch, low factors of safety from the additional analyses influenced the decision to install riprap along an entire section of the ditch.

The caprock and shell layers showed some increase in seepage flows and decrease in the safety factors but nothing warranting a design change.

The sensitivity analyses provided a greater level of confidence that the design was adequate, neither overly conservative nor too risk-prone.

Design Modifications: Several design modifications were made to reduce gradients and improve stability in the locations with low safety factors. For the interior embankment, flattening the embankment slope adjacent to the lower cell to 3.5H:1V provided the required factors of safety against piping and for slope stability. However, flatter slopes were not an option for the seepage collection ditches due to geometric restraints, i.e. the ditches could not be widened

For the seepage collection ditches, the options of constructing a drainage trench just upstream of the seepage ditch or lining the STA side of the seepage collection ditch with riprap were evaluated. While both methods improved the situation, the riprap option was significantly more effective. The design was revised to include a 9-inch-thick layer of riprap on the STA side of three of the four perimeter seepage collection ditches with low safety factors.

For the fourth seepage collection ditch with low safety factors, a change of the ditch geometry was possible in order to avoid the extra cost of installing riprap. Raising both the ditch bottom and the water level in the ditch by one foot increased the factor of safety against piping and improved the slope stability.

The cross sections where the phreatic surface was too shallow at the toe of the embankment were improved by raising the grade of the exterior bench between the embankment and the seepage collection ditch.

The selected design modifications resulted in substantial increases in the safety factors for the seepage collection ditches at a relatively low cost.

CONSTRUCTION

Phase 1 Overview

Phase 1 (STA-N) began in April 2009 and was completed in July 2012. The project faced several challenges during construction including embankment/canal slope failures and erosion related to dewatering and seepage. After a heavy precipitation event, severe erosion and breaching of an internal embankment section was observed at a control structure. Although the failure was not observed during the precipitation event, it was believed to be an internal erosion (piping) failure along the structure sidewalls. Several side slope failures in newly constructed canals also occurred during construction when the dewatering system was abruptly turned-off before canals were flooded/filled to balance differential water levels.

Operational status began in August 2013 after wetland grow-in. The goal was to maintain relatively steady flow rates with average water depths of 18 inches to allow establishment and maintenance of wetland vegetation. Maximum operating depths are 28 inches and minimum operating depths are 6 inches during drought conditions. Figure 6 shows an aerial view of the LRSTA during the emergent vegetation grow-in period.

Figure 6. Aerial View of LRSTA with Emergent Vegetation during Grow-in.

Although grow-in and nutrient removal for Phase 1 has been successful thus far, a couple of items have been noted during operation and maintenance that needed to be addressed.

First, severe erosion and sloughing of riprap occurred at a major canal weir near the pump station. Both a storm drain discharge pipe location and insufficient embedment of the weir into the bank appeared to be contributing factors to this condition.

Second, the presence of a hardpan in the vicinity of seepage canals resulted in vegetation die-off (due to seepage) and minor erosion of sandy soils (in some locations). A hardpan layer was encountered during the subsurface investigation and subsequent construction, but was discontinuous across the site. However, where hardpan was encountered above the water levels in the canal excavations and extended back towards the levee, it trapped seepage flows above the layer, resulting in seepage daylighting in the canal side slopes. The daylighting of seepage above the canal water levels did not allow vegetation to properly establish and sod died off in several locations. In addition, minor erosion of the sandy soils above the hardpan was noted in a few seepage locations.

Design Changes Implemented for Phase 2

As a result of both construction issues and operations/maintenance observations during initial Phase 1 (STA-N) operations, several design changes were implemented for Phase 2 (STA-S) design.

A filter sand diaphragm and was added to all the inlet, outlet, and internal control structures for the downstream-third of the discharge culverts to reduce the potential for a piping failure along the culverts. The filter diaphragm consisted of a 2-foot-thick sand layer below and on the sides of the culverts wrapped in a geotextile fabric.

Dewatering was specified in further detail on both the drawings and in the specifications to highlight the importance of maintaining water levels below foundation subgrades during construction and gradually discontinuing the dewatering systems and flooding canals to allow water levels to stabilize in a controlled manner.

Weirs in all the canals were extended further into bank to reduce potential for piping/erosion at the interface between channel sides and weirs. For the weir location adjacent to the pump station, the storm drain outlet was redirected away from the weir and the weir and sheetpiling was extended further into the bank on both sides of the canal.

Additional seepage modeling was conducted to identify potential mitigation measures for locations where hardpan was encountered in the seepage canal excavation above the normal water levels for the canal. Riprap, filter trench drains, and hardpan removal were all considered. For locations where riprap was present on the canal side slopes due to other project design parameters (i.e. flow rates in the canals), riprap was deemed sufficient to control seepage flows above the hardpan layer. For all other locations, hardpan removal within 10 feet of the canal slope face was selected as the most economical solution.

CONCLUSION

The STA design was accomplished in the fast-track timeframe in order to implement the Northern Everglades program and to realize the environmental benefits as quickly as possible. Innovative design concepts and analysis techniques were used to develop a practical and cost-effective project that evolved to incorporate lessons from construction and operations/maintenance of Phase 1 (STA-N). Construction of Phase 2 (STA-S) is currently in progress, and information from the construction will be included in the presentation at the conference.

POST-TENSIONING IN AN UNDERWATER ENVIRONMENT: WANAPUM DAM + MONOLITH 4 SPILLWAY REPAIR

Paul Krumm[1]

ABSTRACT

Concerning post-tension underwater anchors required for remedial repairs on Monolith 4, two anchor systems were evaluated in part with the accelerated design process due to the emergency nature of the project. Pre-encapsulated solid bar anchors were selected due to complications envisioned with grouting and post-tensioning multiple strand anchors in an underwater environment. Sixty-nine bar anchors were installed through the ogee spillway of which most were installed below water and on the upstream side of the radial gates.

Extensive coordination and planning were required to adequately prepare, perform and to install the anchors. Of the techniques employed in performing the emergency repair work on the spillway, drilling and installation of the underwater anchors proved to be the most challenging. The process required innovative construction techniques and extensive pre-planning by the drillers, divers, engineers and project management team. Many rehearsals were conducted, in the dry, on a full scale mock-up of an underwater anchor installation along with the grouting, load testing, anchor lock-off and completion.

Drilling and installing anchors were performed from work platforms suspended 80 feet above the surface of the ogee spillway using specialized tooling for drilling the boreholes and recessed sockets 50 feet below water. Significant consideration was given to environmental concerns. To prevent grout fluid loss into the water specialized packers designed to seal around the bar anchor were used. Throughout the process a collaborative approach with the various project teams fostered creative solutions to overcome unique challenges in design and construction to successfully complete the underwater anchors.

INTRODUCTION

Wanapum Dam is a Federal Energy Regulatory Commission (FERC) licensed hydroelectric project and the most upstream development of the Priest Rapids Hydroelectric Project. The dam is owned and operated by the Grant County Public Utility District (District). The dam is located on the Columbia River, in central Washington, six miles south of the town Vantage. In the early 1960s when the Wanapum Dam was built, it featured spillway gates nearly seven-stories tall which at the time were the highest in the world pushing the state of engineering design practice. After more than 50 years in operation, in late February 2014, dam operations personnel noticed a misalignment in the bridge deck curbs and handrails at Pier 4 of the spillway. Surveys indicated the top of Pier 4 had moved more than 1.5 inches downstream. A 65-foot long by 2-inch wide open crack spanning the full width of Monolith 4 along a lift joint was discovered by divers inspecting the upstream face of the ogee spillway. The District

[1] Krumm, Paul, Operations Manager, Nicholson Construction Company, North Salt Lake, Utah 84054 paul.krumm@nicholsonconstruction.com

immediately initiated drawdown of the pool to relieve loading on the monolith and pressure within the crack.

The spillway is a reinforced concrete structure with 12 radial gates. A concrete stilling basin slab extends beyond the spillway. Each radial gate is 50 feet wide by 67 feet high. Each of Monoliths 2 through 12, are 65 feet wide, have a 15-foot wide center pier with one-half of a gate bay on each side, and are separated from their adjacent monoliths by un-keyed and un-grouted contraction joints. The typical ogee height is nominally 55 feet with a grouting and drainage gallery, but deepens to 75 feet on the right end of the spillway. Monoliths 1 and 13 comprise the right and left ends of the spillway, respectively, and contain only one-half of a gate bay.

An extensive root cause analysis determined that errors in the initial design calculations and late changes to the final spillway design led to a condition that increased the tensile stresses in the concrete at the upstream face, that ultimately resulted in the formation of a crack along a horizontal lift joint located at elevation 485, about 85 feet below the normal pool level. Remedial measures for overcoming the tensile forces would dictate not only the final design solution but also the sequence and method for how the remedial repair work would have to be constructed.

The District, its design consultant, MWH, and the FERC performed potential failure mode (PFM) analyses to understand the Monolith 4 failure mechanism as well as to develop PFMs that could occur during repair. Amongst several of the failure modes that were evaluated, the PFM analyses determined that tensile stress develops on the upstream face of Monolith 4 under normal pool hydrostatic load, cold weather temperature (winter contraction) increase tensile stress significantly, opening the crack cause the reinforcing steel on the upstream face of the ogee to yield and eventually fail, the weight of concrete above the crack is not sufficient to resist the increase in uplift, and installation of a maintenance bulkhead to dewater a bay causes eccentric, critical loading.

Key requirements for static stability for the Wanapum Spillway were that overturning and sliding equilibrium be maintained with adequate factors of safety per FERC guidelines. In accordance with such requirements, MWH developed a post-tensioned anchoring design program to satisfy overturning and sliding stability requirements for static and seismic load conditions for Monolith 4. The remedial design required installing three post-tensioned 61-strand rock anchors near the upstream end of the pier; eight 3-in. post-tensioned bar rock anchors at the upstream side of the ogee; twelve 2.5-in. and 3-in. post-tensioned bar concrete anchors upstream of the gate near the contraction joints; and twenty 3-in. post-tensioned bar concrete anchors perpendicular to the face of the downstream ogee. Bar anchors conformed to ASTM A722, Grade 150 ksi specifications. The 3-in. bars have a design load of 581 kips (60% GUTS). The computed bond length for the 3-in. bar rock anchors was 14-ft. with total lengths ranging from 79-ft. to 101-ft. All total, Monolith 4 had 40 bar anchors to help 'stitch' closed the crack. Anchor placement is shown in the section view of the Monolith 4 ogee anchor layout in Figure 1, below.

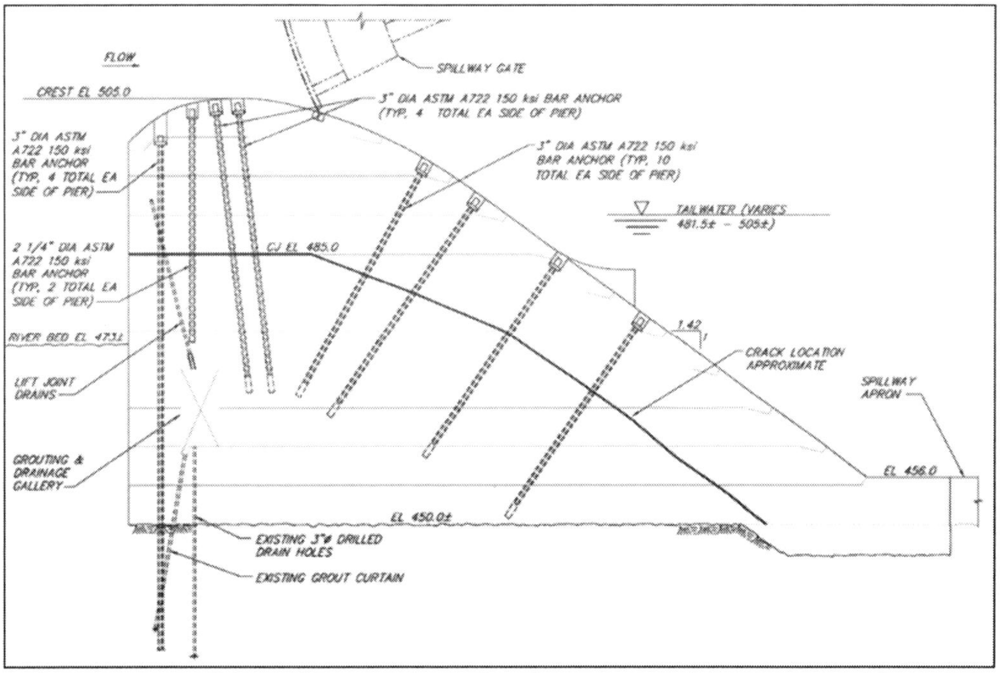

Figure 1 – Monolith 4 spillway anchors placed upstream and downstream of the radial gate.

PFM analysis also concluded certain failure modes could occur in other monoliths and therefore, as a preventative measure, post-tensioned anchors would be installed for all the monoliths. Although monoliths 3 and 5 thru 12 had no apparent damage, precautionary design philosophy was to preclude cracking from propagating downstream in order to prevent a failure mode similar to Monolith 4. The resulting design for these monoliths entailed three post-tensioned, multi-strand rock anchors at the upstream end of each pier of the undamaged monoliths and two post-tensioned bar rock anchors at the upstream side of the ogee to satisfy overturning and sliding stability requirements for operating load conditions.

Due to the urgent nature of implementing the remedial repairs as quick as possible, the post-tension anchor design for the ogee spillway was in progress as construction of the anchors for the piers was already underway. Early on in the project, it was yet to be determined whether or not anchors located upstream of the spillway gate would need to be installed prior to the interim pool raise. If installation was necessary prior to raising the pool elevation, these anchors would have to be installed 'in the wet,' otherwise; the spillway bay could be dewatered first to allow construction 'in the dry.' Ultimately, the PFM analysis dictated that the anchors would need to be installed prior to placing the maintenance bulkhead and dewatering the spillway bay.

Among the techniques used to install the post-tensioned anchor systems on the Wanapum Dam, the underwater bar anchors posed the most challenging aspect of the work. Most of the bar anchors were drilled, installed, grouted and post-tensioned underwater using specialty drilling equipment and working in close coordination with diving crews.

Working remotely in an underwater environment without the benefit of visual observation or 'hands-on' manual operating for control or adjustment, required innovative methods to install the anchors along with extensive planning and rehearsals of the work on the part of the project team that included engineers, drillers, divers, crane operators and numerous support personnel to ensure flawless execution of the work. The anchor installation, grouting and load testing process would have to be performed remotely with engineers observing the work via remote underwater cameras, while dive teams performed the underwater work like setting up the testing frame, hydraulic stressing jack and dial gages for monitoring bar elongation.

As part of the accelerated design process two options were considered for the underwater ogee anchors – tendon anchors and bar anchors. Bar anchors were selected in lieu of a strand tendons due to a more simplified installation and tensioning process, specifically concerning the stressing and lock-off processes of a bar anchor versus that of a tendon anchor in an underwater environment. A bar anchor is locked-off simply by a screwing down a hex nut after the bar is tensioned to design load, however; the lock-off mechanism for a tendon anchor is a bit more complex in that each strand of the tendon anchor utilizes a 3-part wedge where seating in the anchor head is critical to ensure a strand is properly locked-off. With multiple strand anchors, it would be nearly impossible for divers to peer into a 48-in. deep recessed pocket to verify the wedges of each strand was properly seated.

Work platforms, shown in Figure 2, below, spanned 65 feet between the existing nose piers of the spillway. Suspended over the water, the platforms allowed for staging the drilling equipment upstream of the spillway gates. From the deck of the platform to the collar of the borehole at the ogee surface measured 80 feet. For the most upstream anchors installed in the ogee spillway, the borehole was 7 feet from the upstream face of the spillway, or midway between the upstream face and the drainage gallery. The ceiling of the gallery was at an elevation 33 feet lower than the elevation of the anchor borehole collar.

Like the drilling program for the tendon anchors, a pilot hole was used to establish the trajectory of the 8.5 inch diameter borehole for the bar anchors. At 10 feet intervals the pilot hole was checked for azimuth and verticality using a gyroscope survey instrument. Because underwater drilling poses an environmental risk, controls were implemented to contain drill cuttings at the collar with a diverter designed to channel drill fluids to the surface. The diverter used large casing to act as a conduit to route cuttings to the surface and into containment bins, as shown in Figure 3 below.

The anchor requires that the bearing plate sits on an even surface to uniformly transfer the load to the anchored structure. Typically, this is accomplished by casting a bearing pad using high strength self-leveling grout but, because these pockets were underwater placement of the bedding grout was not possible. The site team needed devise another way to achieve a flush surface. To achieve a smooth bearing surface, an impregnated diamond milling bit was manufactured to fit into the recessed pocket that allowed grinding the concrete to a flat, polished bearing surface. This was done after first drilling the recess using a 20-in. diameter narrow kerf diamond core bit and then divers removing the concrete with chipping hammers.

Figure 2 – Overwater work platform upstream of the radial spillway gate.

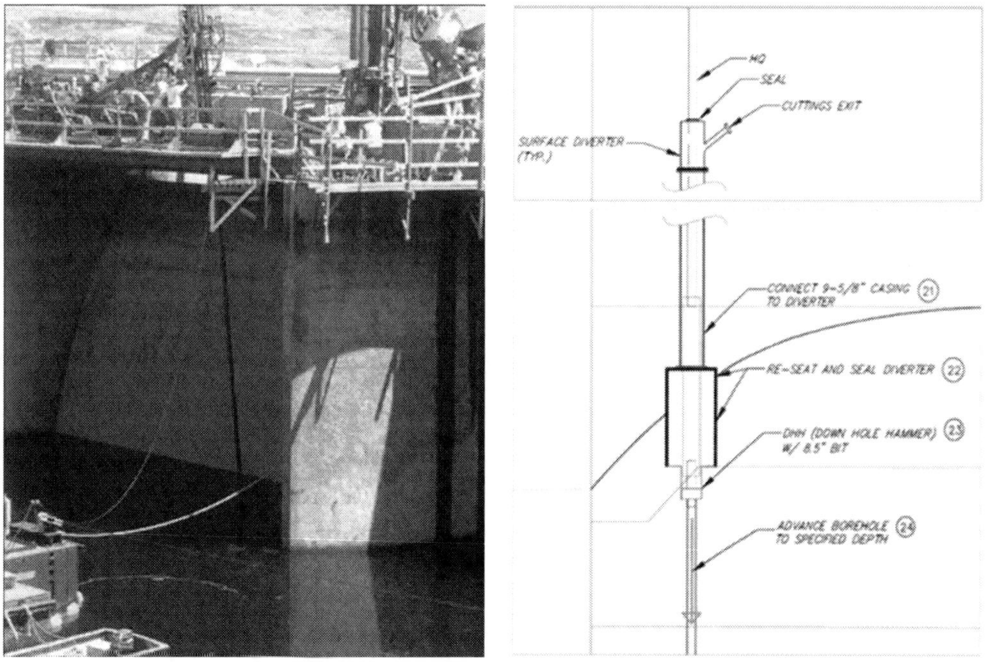

Figure 3 – Drilling borehole for bar anchor (right) and diagram of drilling diverter (left).

The next step in involved installation and grouting. A bar anchor was preassembled at the right abutment of the spillway where a special hanger bracket attached to the wall of the spillway permitted pre-assembling the bar anchor its full length. This method of pre-assembling was selected because it did not require tying up valuable barge crane time to assemble the anchor over the borehole. This freed up the barge crane to support other drilling and anchor activities. When each anchor was fully assembled, it was ferried to the spillway with the barge crane (Figure 4) where the anchor was lowered into the borehole. Dive teams assisted with guiding the anchors into the borehole and prepping the anchor for underwater grouting.

A custom grout packer was designed to isolate the borehole and bar anchor from the pool to prevent grout spillage into the water body. The grout packer also ensured quality placement of the primary grout bond zone whereas grouting the bond zone using conventional tremie line placement would have been susceptible to placement issues.

Figure 4 – Moving a bar anchor from the assembly area to the spillway.

The final step involved stressing the bar anchor, Figure 5. All the stressing equipment used for this crucial step was custom fabricated specifically for this project. The exception to this was the hydraulic jack. Due to the 4-ft. depth of the recessed anchor pocket, special tooling and lifting devises were fabricated to safely set the bearing plate and nut, as well as the stressing chair.

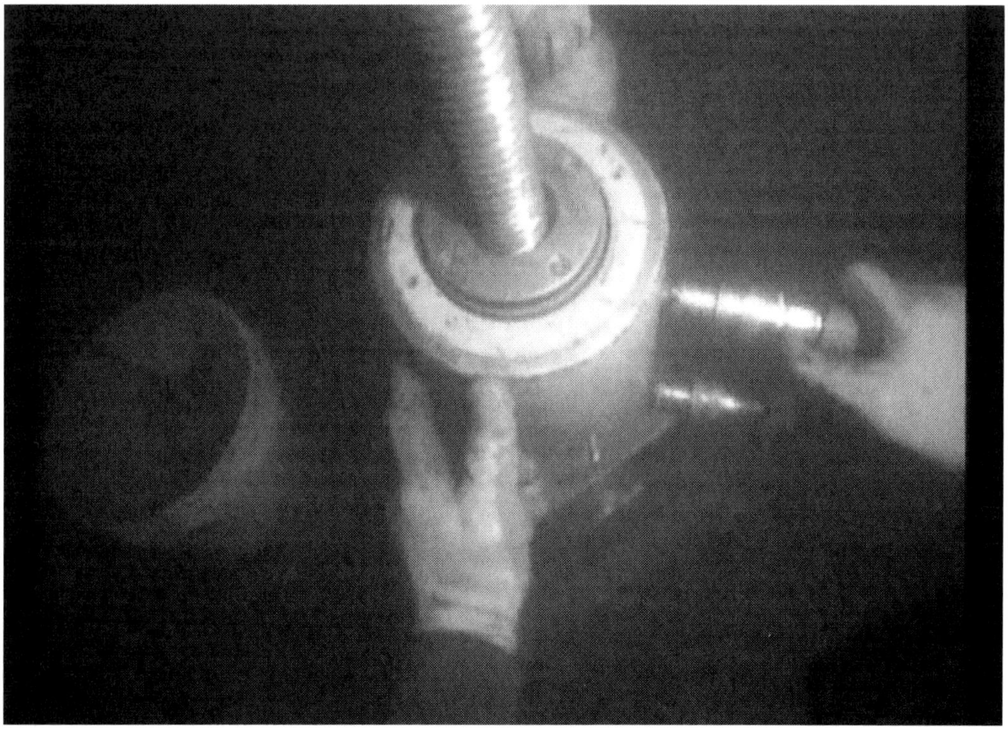

Figure 5 – Divers setting up a hydraulic jack to stress a bar anchor.

Once the stressing chair was set, the divers used a custom wrench to torque down the bar hex nut and lock-off the anchor. Lock-off was followed by a lift-off test to verify that the bar was properly locked off at the correct load. During stressing, divers read dial gauges directly to the engineers who recorded the test data from the deck of the barge. Not only did this effort take an extraordinary amount of coordination, but it took weeks of planning to prepare.

The installation, grouting and stressing of an underwater bar anchors was successful due to appropriate planning and coordination between all parties involved prior to starting the work. The project team performed mock-up drills of these processes on deck in order to ensure communication and coordination of every team member before the first bar was installed. The mock-ups and team coordination allowed the construction process to be performed quickly and safely without any learning issues. After the first bar was successfully installed, the construction process remained the same for the remaining anchors.

While anchors may have been installed in underwater environments on other projects, the urgency of implementing the remedial work on this project placed added emphasis on preplanning and mock rehearsals to proof the innovative approaches to accomplish the work successfully. The underwater anchor work required extraordinary coordination and planning to adequately prepare and successfully complete the anchors. The collaborative approach by the project team fostered creative solutions to overcome unique challenges in the design, planning, and construction phases of the underwater anchors.

REFERENCES

[1] *Stabilizing Wanapum – Lessons Learned from Discovering and Repairing a Substantial Crack.* Hunter I., Simpson R., Thompson D., Barrie R. Hydro Review Magazine – feature article, September 2016, ASDSO Dam Safety National Conference, 2016, Philadelphia, PA.

[2] *Wanapum Spillway Crack Development Root Cause Analysis.* Jones A., Snyder K., Charlwood R., Marshall K., Campbell D. Paper presented at the ASDSO Dam Safety National Conference, September 2015, New Orleans, LA.

[3] *The Wanapum Dam Spillway Incident: Response, Remediation and Lessons Learned.* Mishalanie, D., Rollins, G., Simpson, R., Hunter, I., Thompson, D. Paper presented at the ASDSO Dam Safety National Conference September 2015, New Orleans, LA.

[4] *Repairing the Wanapum Spillway Crack.* Krumm, P., Stein, A., Hepworth, B., Hughes, A. Paper presented at the ASDSO Dam Safety National Conference, September 2015, New Orleans, LA.

[5] Post-tensioning Institute "Recommendations for Prestressed Rock and Soil Anchors", 4th Edition, 2004. (5th Edition 2014).

[6] Post-tensioning Institute "Post-tensioning Manual", 6th Edition, 2006.

BALANCING WATER RESOURCES AND POWER GENERATION FOR THE METLAKATLA INDIAN COMMUNITY

Jesse Hutton[1]

ABSTRACT

In 2015 the remote Alaskan Metlakatla Indian Community suffered water shortage problems due to lack of seasonal precipitation and over consumption of water resources. Chester Lake is the sole source of municipal water for the tribe, and provides 25% of the island's hydropower. This combination allowed water levels to recede dangerously close to the penstock inlet and the tribe determined that an extension of the penstock deeper into the lake was the most cost effective and timely solution to ensure that the community would not run out of water.

This paper discusses the many challenging aspects of this project and the critical planning required due to the remote location and virtually no access by vehicles. All equipment needed to be airlifted via helicopter to the worksite to complete dam modifications, rock and concrete demolition, as well as float and sink pipeline installation. Ongoing communication with the Indian Community during all stages of the project, from constructability review to completion, was necessary to prevent disruption of this critical resource.

INTRODUCTION

The Metlakatla Indian Community (MIC) is located on Annette Island and is the only Indian Reserve in the State of Alaska. The MIC Annette Island Reserve exists by the authority of the Constitution and By-laws of the MIC as approved on August 23, 1944 by the Secretary of Interior and MIC as an Indian Tribe organized under provisions of the Indian Reorganization Act. Annette Island Reserve is held in trust by the United States for the benefit of MIC, the Secretary of Interior has delegated responsibility to MIC to prescribe rules and regulations governing the use of Annette Island Reserve.

Primary industries include land, hydroelectric power generation, commercial fishing, timber, and water resources. The livelihood of the residents of Annette Island rely strongly on the potable water and stored energy resources of Chester Lake and the Chester Lake Dam.

The concrete dam, built in the mid 1980's, is located a few miles northeast of Metlakatla and is very remote. The site can only be accessed via narrow hiking trail or helicopter. The crest elevation is 850 feet.

Location of Annette Island in relation to Ketchikan, AK is shown below in Figure 1.

[1] Ballard Marine Construction, Inc., 727 S 27th St., Washougal, WA 98671, Jesse.Hutton@BallardMC.com

Figure 1 Annette Island and image of penstock leading from dam to the Metlakatla Community

TECHNICAL ISSUE

In the summer of 2015 the Metlakatla Indian Community (MIC) suffered water shortage problems due to lack of seasonal precipitation and over consumption of water resources. Chester Lake is the sole source of municipal water, and provides 25% of the islands hydropower. The primary penstock is 20 feet below the crest of the dam and during summer months' lake draw-down receded dangerously close to the penstock inlet.

To answer to this urgent and unique technical issue the owner, and their engineers, brought together an experienced team to address the problem. A key initiative during the emergency response stage was early contractor involvement and after an expedited RFP process Ballard Marine Construction was selected as the technical/constructability resource for the project. This allowed the design to be vetted by a team with dam rehab experience in remote locations. Balancing operational requirements with community needs, safety, budget, human and equipment resources were all considered as part of the initial response.

This design/build team was comprised of:

- Metlakatla Indian Community
  - Hydro Operations
  - Water Operations
  - Natural Resources
- Bureau of Indian Affairs
- Americas Energy Services
- Kleinschmidt Engineering
- Ballard Marine Construction

SOLUTION

It was decided that extending the existing penstock deeper into the lake was the most cost effective and timely solution to ensure that the community did not run out of water.

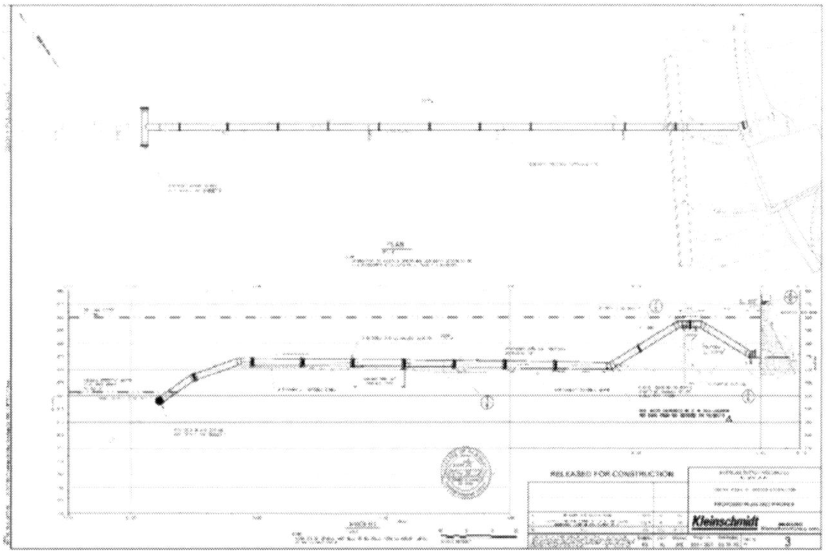

Figure 2 Penstock Extension Layout

To add to the complexity, Chester Lake is the primary source of potable water and stored energy. These important resources had to be balanced during operational activities requiring facility shut downs. The owner, design and construction teams worked together closely to mitigate this risk to the remote tribal community.

Through early contractor involvement and collaboration, alternate solutions were identified that mitigated potential water supply and power generation disruptions during construction.

As an example, the design called for a steel pipe spool to be inserted into the existing penstock and welded into place. Based on remote access and constructability issues the original concept was modified to eliminate all underwater welding. This mitigated a very labor intensive operation using divers that would require extended outages and also mitigated the risks associated with anchoring the penstock extension to an embedded structure in the dam of unknown condition. The revision included the design of a steel adapter plate with a flange spool installed on the face of the dam to accommodate attachment and support of the 250' penstock extension. An underwater picture of the adapter plate from the dive inspection is shown below in Figure 3.

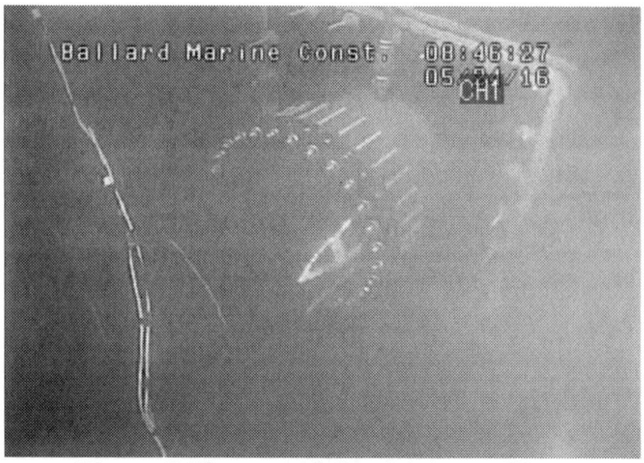

Figure 3 Flange and Steel Adapter Plate

Involving the construction contractor during the design phase allowed for additional design modifications to the pipe ballast system, concrete/rock anchor systems and other components to accommodate installation processes that considered reduced operational windows and a schedule driven by water resource needs while also maintaining design-life and dam safety.

Design, Constructability & Planning

The penstock extension design had to accommodate unique site conditions. The Chester Lake Dam is the third water control structure built in the intake channel. Two existing structures were in place upstream of the dam – the previous concrete dam and an old timber crib cofferdam. Both structures were obstacles to the extension of the penstock and created constructability issues that had to be overcome by the design and construction teams. It was decided that the penstock extension design would include a "speed bump" riser section in the new pipe that would span the crest of the previous concrete dam. The pipe also had to be laid through a previously installed gap in the timber crib shown in Figure 4, making pipeline alignment critical.

Figure 4 Timber Crib Structure

Figure 5 "Speed Bump" Penstock Extension Riser

Sourcing and allocating heavy construction resources in a remote location such as Metlakatla was challenging – the design and construction team worked in conjunction with the Tribe to identify local equipment resources on the island that could be used for construction to save on cost. Some of the standard construction equipment was rented locally in Ketchikan, AK and the specialty equipment was sourced from Ballard Marine Construction facilities in Seattle, WA. Specialty equipment included commercial dive gear, underwater video equipment, hoisting and lifting winches, engineered lift bags for float and sink operations, underwater ultrathermic steel burning equipment and life support/safety equipment. All the equipment was mobilized to Annette Island via air freight and barge and once it passed QC inspection was then airlifted to the remote dam site via helicopter.

An important planning task was to identify the quantity and qualifications of labor needed for the construction project. It was imperative to control costs and use local workers with experience and fuel the tribal economy. It was the tribe's desire to contribute as much labor as possible. Ballard met with multiple tribal departments including Water, Power, and Natural Resources to understand local resources and then formulated a labor plan for MIC. This included utilization of local labor for pre-construction fabrication, construction of pre-cast concrete units, pre-construction assembly of piping components, management of the helicopter landing zone, rigging, forklift and small equipment operations, etc. Ballard communicated expectations and orientation/training to local personnel through on-site training by construction personnel with the specialized experience required for this project. This helped to meet the Tribes goals and ensure safe and productive operations using local a local labor force. Ballard supplied construction management personnel, experienced topside construction technicians, commercial divers, and administrative personnel from its offices in Seattle, WA, and Portland, OR to support equipment and personnel coordination as well as health and safety initiatives.

Balancing Resources during Operations

Due to the nature of the work outages and lock-out of the existing penstock and water supply lines was required. As noted earlier, Chester Lake is both the primary potable water source and the hydro energy source for the tribal community, and despite the design changes and advanced construction techniques it was necessary to carefully plan and execute a water conservation outreach program for the community. The primary goal was to communicate the need for water conservation throughout the life of the project. MIC communicated this to the residents through mailings, tribal council meetings and community events. The community came together during the project to meet the conservation goals required for successful completion of the job.

Construction

Ballard construction managers and operational teams worked with the local design team, Bureau of Indian Affairs, and the Tribe to implement the resource management plan and construct the dam modifications using highly experienced and qualified construction crews to perform the civil marine and underwater installation activities. Local construction crews helped with topside activities and were used wherever possible to support the local economy.

Figure 6 Local labor was utilized wherever possible

Close communication with the tribe related to potable water and energy resources management throughout the life of the project was needed to protect the interests of the community. The previously mentioned community outreach plan was crucial because the water treatment plant operated by the tribe only had the capacity to store enough potable water for about 8 hours combined with the differential pressure safety issues created by the draw through the penstock, the lock out / tag out process had to be carefully planned. Each day the Ballard project manager communicated with the water treatment plant operator to plan the daily construction/lock-out schedule based on in-water construction activities and stored water levels.

Figure 7 Typical Site Condition at Dam Crest

The safety of crews, equipment and long lead materials were a primary focus of the construction team. The project and construction site was very challenging from a Health and Safety perspective. The typical site conditions at the dam pictured above in Figure 7 show the "laydown area" availability and the lack of improved structures and surfaces from which to work. During the planning phase, all stakeholders worked diligently to formulate an effective job hazard analysis, emergency management plan and risk matrix to ensure all risks were identified. Mitigation plans were formulated from the analysis and in place during all operations.

The entire operational crew hiked to the construction site from the base camp every morning and back down every night. The trail was unimproved and required personnel to climb over rocks, downed trees, and other obstacles. Close attention was paid to crew fitness, lighting, and trail conditions during daily "mobilization" to make sure that even this operation was performed with the highest level of safety awareness.

With divers working on the upstream side of the operational dam, differential pressure issues were assessed and taken very seriously. Two existing siphons, the penstock opening and the spillway were all identified as potential area of diver entrapment.

Working space at the dam was very restricted. Leveling pads for equipment had to be constructed on uneven rock slopes to accommodate positioning of construction equipment such as generators, hydraulic power units, air compressors, drilling equipment and diving systems.

Heavy lift helicopter operations were required to support the construction effort. While these operations are inherently risky when working with restricted space and uneven surfaces they become treacherous. All equipment and materials, including all penstock extension pipe and the intake structure were staged at the landing zone and pre-rigged ahead of helicopter operations to ensure expedited movement to the dam. All loads were labeled for proper flight sequencing. Both the landing zone team and the construction team at the dam were briefed and trained on helicopter long-line operations.

Figure 8 Pipe Section mid-transit to the work site.

Weather in the region is very dynamic and can change quickly. High winds and low visibility were the main concern and affected both helicopter and diving operations during the project.

One of the primary construction challenges was assembling the penstock extension piping, complete with ballast, in the reservoir without the support of cranes or traditional heavy equipment. Due to lack of construction equipment and material laydown area materials were "wet stored" on the lake bottom and staged for construction activities as shown below in figure 9. This included the penstock extension adapter plate, pipe mounts, pipe spools, pre-cast concrete ballast and the intake structure. Divers then retrieved the materials throughout the course of the job using engineered lift bags to move materials into position for installation. This approach required the work be performed at high water reservoir water levels.

Figure 9 "Wet Storage" of Materials

Divers demolished the existing trash rack system and installed the adapter plate on the face of the dam. The entire perimeter of the plate was sealed with a two-part epoxy grout to prevent air entrainment during periods of low water.

The penstock extension pipe was built one spool section at a time in the reservoir using a combination of winches mounted on the dam and lift bags to align the sections. Divers worked to flange the pipe sections and install the precast ballast blocks. Once the entire extension pipe was built and supported with lift bags it was floated into position on the pipe alignment which was previously marked using surface buoys. Due to the "speed bump" section, the extension pipe was floated into position in the flat orientation and then rotated once it spanned the old concrete dam. The extension pipe was then connected to the adapter plate and submerged in place on the pipeline alignment. To improve structural integrity of the system, divers installed pipe supports and sand bags to mitigate free span areas of the pipeline.

With the penstock extension in place, the new intake structure now draws water about 28 feet lower than the original penstock.

CONCLUSION

The penstock modification was installed in time to mitigate the risk of the summer dry cycle and performed as designed for the tribal community. The project was performed safely, under budget, met the original construction schedule and was executed while closely managing the natural resources of the Metlakatla Indian Community.

The needs of the tribal community to balance water resource while improving a critical water resource structure while in operation could only be accomplished in the manner described above by the early involvement of a technically competent and specifically experienced contractor. While not all dam rehabilitation projects are in such a remote location, with such stringent schedule requirements or with so few redundant resource options; they are all critical pieces of infrastructure throughout the world. Care must always be taken by the owners of these facilities to not only select the teams that will provide the experience and technical ability to safely execute the work but to involve them early in the contracting process to improve dam safety, reduce cost, and accelerate schedules.

Due to the extensive planning, the commitment, and diligence of the entire team to safe operations there were no injuries.

ACKNOWLEDGEMENTS

This technically challenging and critical project could not have been accomplished without the excellent efforts of all the involved personnel. As always we appreciate the diligence of our clients, our employees and our subcontractors in adhering to the safe work practices required by Ballard Marine Construction.

INVESTIGATING THE STRUCTURAL SAFETY OF CRACKED CONCRETE DAMS

Glenn S. Tarbox, P.E., P. Eng., S.E.C.B., F. ASCE[1]
Robin Charlwood, PhD, P.E.[2]
Chris Hayes, P. Eng.[3]

ABSTRACT

CEATI's Dam Safety Interest Group has recently completed (published) a research report titled, "Investigating the Structural Safety of Cracked Concrete Dams" [1]. The objective of the CEATI report was to provide dam owners, operating staff and engineers with a basic framework and practical tools for identifying and evaluating cracks in concrete dams during field inspections and making wise decisions consistent with sound international practices in the assessment of cracks in concrete. An international team of engineers experienced in the design of concrete dams was assembled to perform the analysis and prepare the report presenting a practical step by step process for the evaluation of cracking in concrete dams and the determination of appropriate steps to address any issues identified by that process and to establish and carry out a dam crack management program. Sound practice requires that engineers and operating staff of dams be able to identify if a particular type of crack can lead to a dam safety problem, whether it is merely one to be routinely observed, or is a maintenance problem that needs attention.

Presented in this paper, and based on the research report, is a crack management approach including crack evaluation and management logic, potential cracking failure modes, and risk management of a cracked dam. Seven steps of a crack management logic flowchart are presented and discussed briefly. The seven steps are (1) Discovery, (2) Crack Characterization, (3) Dam Type and Construction Methods, (4) Root Cause Analysis, (5) Case Histories, (6) Potential Failure Modes Analysis and (7) Managing the Cracking. Underlying the discussions in the paper is the identification of eight "root causes" of cracks, (a root cause of crack formation being the physical mechanism or chemical process that creates a tensile or shearing stress field that exceeds the tensile or shear strength of the concrete resulting in cracking of the concrete). Included among the root causes are structural behavior, foundation and abutment behavior, shrinkage, thermal effects, freeze-thaw effects, expansive chemical reactions, earthquakes, and corrosion of rebar and embedded parts.

INTRODUCTION

The objective of the CEATI [1] report was to provide dam owners, operating staff and engineers with a basic framework and practical tools for identifying and evaluating cracks in concrete dams during field inspections and making wise decisions for

[1] MWH Global, now part of Stantec, Bellevue, Washington, USA, glenn.s.tarbox@mwhglobal.com
[2] Robin Charlwood & Associates, PLLC, Freeland, Washington, USA, robincharlwood@gmail.com
[3] CEATI International, Montreal, Quebec, Canada, chris.hayes@ceati.com

appropriate and safe management of the structures consistent with sound international practices.

APPROACH

The approach was to assemble an international team of engineers experienced in the design of concrete dams to perform the analysis and prepare the report presenting a practical step by step process for the evaluation of cracking in concrete dams and the determination of appropriate steps to address any issues identified by that process and to establish and carry out a dam crack management program. Sound practice requires that dam engineers and operating staff of dam owners be able to identify if a particular type of crack can lead to a dam safety problem, whether it is merely one to be routinely observed, or is a maintenance problem that needs immediate attention.

Drawing upon the research report, the following are presented: a crack management approach including crack evaluation and management logic flow chart, types of cracks, root causes of cracks, potential cracking failure modes in dams, the use of cracked dam case histories process and critical actions for the management of cracks in dams.
To evaluate either a new or existing crack(s) in a dam, the logic diagram provided in Table 1 presents seven recommended steps for users to follow. The steps outlined require a knowledgeable, experienced and trained team to be fully effective.

The approach to investigations of cracking in dams is intended to apply to conventional vibrated concrete (CVC) and roller compacted concrete (RCC) dams including mass concrete in appurtenant structures such as spillways, integral to the structure of the dams.

Table 1. Crack Evaluation and Management Logic Flow Chart – Seven Steps

1. Identify cracking in dam concrete

↓

2. Document key crack characteristics:
   - Location
   - Shape
   - Shape parameters
   - Environment/Features

↓

3. Review dam type, age, design basis and construction methods to identify potential root causes of cracking at time of construction

↓

4. Seek to identify the "root cause" of the cracking

↓

5. Conduct Potential Failure Mode Assessment if an apparent safety issue and assign a Risk Category using FERC system

↓

6. Compare the cracking with available case histories

↓

7. Establish appropriate response to manage the cracked dam:
   - If Category I or II: Plan and implement monitoring and risk reduction measures.
   - If Category III: Plan investigations and/or analysis program and upon completion repeat Step 7 and implement appropriate measures.
   - If Category IV: No further action required.

CRACKING OF CONCRETE

Types of Crack

The process starts with a general characterization and documentation of the cracks by size (length, width and depth), open, closed or hairline, and individual or part of a pattern of cracks.

Open cracks that penetrate deeply into or through the body of a dam regardless of location are usually more critical than shallow, hairline cracks that can be seen on horizontal, vertical or sloping surfaces such as a parapet wall or roadway although surface cracks may be important indicators of internal mechanisms such as chemical expansion. ***Cracks that offset concrete*** on either side are frequently significant because they indicate differential movements within a structure regardless of location. ***Cracks that occur in the concrete of the dam either parallel or perpendicular to its foundation*** are significant because all types of dams rely on a sound bond with their foundations. Any ***cracks through which water is leaking*** are also significant because the passage of water is unintended and indicative of an abnormal condition requiring investigation to understand the cause of cracking and the source of the water.
Observation of cracks and/or deterioration in a dam should lead to a detailed crack inspection. Inspection should include categorized observations that can be consistently applied at subsequent inspections.

Table 2 is provided as a guide for characterizing and documenting cracking. It is suggested that these descriptors are used in conjunction with an inspection sheet similar to that provided in ACI 201.1R [2] (see example in Appendix A). Information gathered with this outline can then be carried forward to assist in the determination of the root causes of observed cracking. This table is intended to also provide the user with the necessary means to document the observed crack, and as reference descriptors for determining if any changes have occurred at future observations. It is recommended that notes be taken in this format as a standard reference.

Determination of crack opening as cited in Table 2 under *Local shape parameters* can be measured using a simple device called a crack meter as shown in Figure 1.

Figure 1. Example of simple crack meter (MWH, 2012).

Table 2. Guide for Characterizing and Documenting Observations at Cracks

Applicable to a crack visible along the face of a non-reinforced concrete mass

Localization	Code	Comments
isolated in the concrete mass	I	
along a construction **joint**	J	
ending on a block **side** / joint	S	
ending at a block **angle**	A	
General shape	Units	Comments
developed length	m	
straight / curved shape	I, C, S	straight, curve one side, S-shape
maximum opening	mm	
opening variation	mm/m	indicates a virtual rotation center
orthotropy	%	in % from 0 (straight) to 100% (stepped)
small scale shape, amplitude	mm	first sinusoidal shape approximation
small scale shape, period	mm	second sinusoidal shape approximation
smoothness		use Barton models *(at right)*
Local shape parameters at each point)	Units	Comments
opening	mm	
orientation / surface	°	not easily detected - 90° if perpendicular
orientation / vertical or reference	°	azimuth degrees
offset inside face	mm	provide orientation (clockwise / counter-CW)
offset perpendicular to face	mm	provide orientation (up out / up in)
depth	mm	possible only through interception borehole
Environment features	Units	Comments
oxydation		change in colour versus depth / distance
cuts or debonds aggregates		age of concrete when crack occurred
humidity / water flow		
infilling		calcite, mud, ...

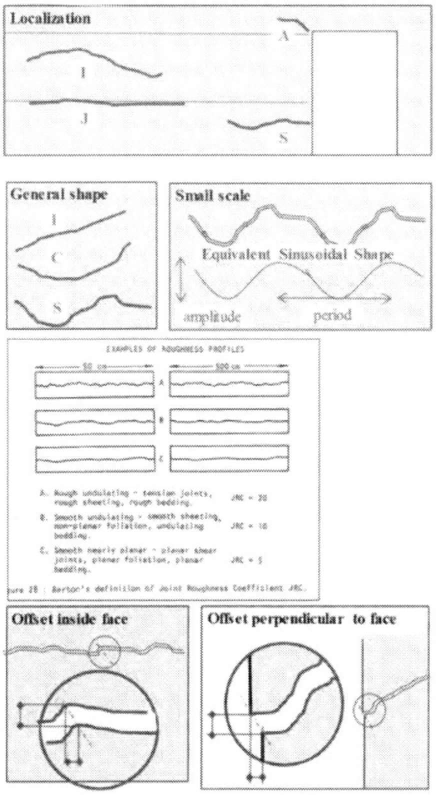

Root Causes of Cracks

A root cause of crack formation refers to the physical mechanism or chemical process that creates a tensile or shearing stress field that exceeds the tensile or shear strength of the concrete resulting in cracking of the concrete. These mechanisms can be external and/or internal forces. A crack in and of itself is a detectable (usually visible) manifestation of these phenomena indicating that the strength of the concrete has been exceeded either locally or more generally. Once the crack occurs, the causal stresses are relieved or diminished sufficiently to arrest the crack unless reinitiated.

Examples of external force mechanisms applied to a dam that may cause cracking include hydrostatic pressure from the reservoir or ground shaking from earthquake.

Internal forces that cause cracking in concrete by volume changes can be caused by a variety of mechanical, physical and chemical processes such as stresses from applied loads, moisture variations, temperature variations, cement hydration, carbonation and phenomena like sulfate attack or alkali-aggregate reaction. Some are characterized by a volume decrease (shrinkage) while others by a volume increase (swelling).

Separate sections are presented in the CEATI Report describing the types of cracking for each of the "root causes" along with general remarks about the potential impact on the dam structure behavior. Eight root causes of cracking are identified and discussed in the CEATI Report: structural behavior, foundation and abutment behavior, three types of shrinkage, thermal effects, freeze-thaw effects, chemical expansion, earthquakes and corrosion of rebar and embedded parts. Each root cause is described in terms intended to help inspectors make visual identification in the field. While initial cracking can be due to one or more of the root causes noted, once a crack has formed it becomes a weak point and other causes can act to amplify the damage.

An important assist to categorizing the root cause of cracks is to place the dam in the appropriate era of concrete construction development discussed by Tim Dolan [3]. Many older dams are in need of rehabilitation due to shortcomings in early concrete mixing, delivery and consolidation techniques. Any one or a combination of these shortcomings may contribute to poor quality concrete, e.g. concrete having been "sluiced" and/or "chuted" in some early cases as shown in Figure 2 and result in cracking. This step can be helpful in pointing an analyst towards likely causes or help eliminate spurious issues.

Figure 2. Concreting operations at Salmon Creek Dam, Alaska circa 1914. Photos courtesy AEL&P.

Case Histories of Cracking in Dams

Although there are a small number of catastrophic concrete dam failures, there are many cases where significant cracking has occurred without "failure" and has been reported. Many are reported in ICOLD congresses and symposia which are a primary source of information. The following Professional Society and Incident Databases were also consulted: ICOLD, USSD, FERC, Stanford NPDP, and ACCEL of Paris.

Typical PFMs that may develop as a result of cracking from the identified "root causes" are presented in the CEATI report. The descriptive headings used for each case history were: Project Description, Relevance of Case Study, Description of cracking, Root Cause, and PFM description. The following typical example for cracking initiated by earthquake loading is excerpted from the CEATI Report for Koyna Dam:

Koyna Dam || Koyna Nagar, Satara District, India

Dam Type: Gravity || *Height*: 103.2m || *Crest Length*: 807.2m || *Crest Width*: 14.8m || *Construction Date*: 1964

Project Description: Koyna Dam is a large gravity dam located in India. During construction, changes to the design had to be made, resulting in an atypical profile for this gravity dam. In December 1967, a magnitude 6.5 earthquake caused cracking at these irregular changes in section.

References: [4] [5] [6]

Relevance of Case Study: Cracking extending through the entire thickness of the dam near the base of the "chimney" section following a M6.5 earthquake attributed to Reservoir-induced Seismicity (RIS). This is one of the classic cases of this phenomenon.

Description of Cracking: Large structural cracks occurred at the changes in section on both faces of the dam. Fine cracks also occurred in similar vicinity. An approximately horizontal crack at the change in cross section occurred on the upstream and downstream side of the monolith. Fine cracks were noticed on either side of the main crack. Leakage from some of the cracks was also found, indicating cracking across the section. This allowed for potential for toppling of chimney section and loss of reservoir.

The Root Cause: The earthquake initiated cracking at the irregular changes in the cross section.

The PFM: Sliding of the upper portion of the dam, "chimney section," under full water level and eventual failure of this section.

Potential Cracking Failure Modes in Dams

A potential failure mode analysis (PFMA) is an effective procedure to assess the significance of cracking exhibited in dams and, if necessary, provide a rational basis to develop a dam safety management plan. The PFMA process is a state-of-the-art technique for dam safety assessment which was developed by USBR and others and has been used extensively by FERC since 2003 and has gained usage worldwide. FERC's PFMA Guidance Document [7] outlines the necessary steps to seek to ensure that all potential failure modes (PFMs) are identified and adequately addressed in a systematic procedure and leads to a risk categorization scheme. The risk categorization is used to determine the appropriate level of actions to manage the cracking. It is recommended that either the FERC process, or another equivalent process as may be prescribed outside of the United States, be used to assess the significance of cracking.

Managing Cracked Dams

Once it is determined that a significant crack(s) exists in a concrete dam, it is necessary to manage the extent and changes that can occur. In this section, each of the root causes of cracks, as introduced previously under Root Causes of Cracks are cited as shown in Column 1 of Table 3. Three critical actions are listed in the table for the management of cracks in dams: (1) options for analyzing the causes and effects of cracks, (2) means of monitoring the status of cracks and (3) options for controlling and/or reducing risks associated with the presence of the cracks in the dam.

Table 3. Managing Cracked Concrete Dams

Root * Cause	Description	Analysis Options	Monitoring **	Risk Reduction Options
1	Structural Behavior: structural distress for any reason regardless of root cause including flooding and overtopping loads.	Refer to FERC Engineering Guidelines, Chapter VIII	Plumblines, Inverted Plumblines, Extensometers (MPBX), Foundation Deformation Meters, Joint Meters, Tape Gauges, Collimation, EDM, Triangulation, GPS, Levels (survey monuments), LiDAR, Inclinometers, Tiltmeters, Crack Gauges, Strain Meters, Accelerometers, Piezometers, Weirs, Flumes, Flowmeters	Structural distress might call for reducing the reservoir water level thus reducing the stresses in the dam. Update risk assessment covering all structural design parameters. Pre-stressed tendons.
2	Foundation and Abutment Behavior	Two- and Three-Dimensional Stability Analysis, FEA	Foundation deformation meters, Extensometers (MPBX), Inclinometers, Tiltmeters, Piezometers, Mapping, Stereographic Projections, Joint Rosettes,	Grouting, Drainage, Pre-stressed Tendons, Shear Keys, Shotcrete, Dental Concrete, Reshaping (benching, flattening slopes), Toe Blocks, Berms
3	Drying Shrinkage	N/A	Physical inspection, crack gauges, chronological photos	Sealing, epoxy grouting
4	Thermal Effects	Thermal FEA	Physical inspection, Crack gauges, Chronological photos, Thermocouples, Thermistors, Thermometers	Cooling, Heating, Insulation

Root * Cause	Description	Analysis Options	Monitoring **	Risk Reduction Options
5	Freeze-Thaw Effects	Structural FEA	Visual Surveillance, LiDAR surveys, Coring, Chronological photos, Petrographic Analysis	Sealing: Upstream Membrane, Shotcrete, Restore original lines and grades via build back, Epoxy Grouting
6	Expansive Chemical Reactions	Structural FEA	Physical Inspection, Chronological photos, Crack mapping, Petrographic Analysis, Plumblines, Inverted Plumblines, Extensometers (MPBX), Collimation, EDM, Triangulation, GPS,	Slot Cuts; Pre-stressed Tendons, Water Proofing, Membranes, Grouting, Drainage
7	Earthquake	Structural FEA	Review ground motion data, Post-earthquake physical inspection and damage assessment	Site specific seismic hazard assessment, Pre-stressed Tendons, Grouting
	Corrosion of Rebar and Embedded Parts	N/A	Visual Surveillance and Physical Inspection, Chronological photos, Samples	Chip, saw-cut surrounding concrete to grit blast steel or remove damaged metal and replace steel (epoxy coated rebar) or embedded part and re-concrete. Corrosion inhibitors can also be included in the concrete mix such as, calcium nitrite.

* Numbered cause corresponds to those listed previously under Root Causes of Cracks.

** Refer to "ASCE Guidelines for Instrumentation and Measurements for Monitoring Dam Performance" and CEATI's "Dam Safety Performance Monitoring and Data Management – Best Practices" [8].

Analysis Options

The approach used in the CEATI Report is to provide guidance on when analysis is recommended and suggestions of useful tools without emphasis on the details of the tools. Detailed descriptions of the analytical methods are presented elsewhere in technical papers, academic research reports, user manuals of computer programs offered by commercial software vendors, etc. contained in the professional literature.

One of the main analytical approaches to evaluate cracking is the use of finite element analysis (FEA). FEA has the potential to assist in developing an understanding of cracking mechanisms and their potential impacts. While not a cure-all tool, FEA can simulate numerous phenomena that may impact a dam structure during its lifetime. Depending on the severity of the impacts and complexity of the phenomena under evaluation, either linear or nonlinear analyses may be appropriate with FEA computer programs. For instance, the inclusion of load redistribution due to cracking can be important or the use of appropriate material behavior laws to model the time and stress dependent effects of expansive chemical reactions can be essential.

Monitoring

Means and methods of monitoring various crack types are discussed briefly in the CEATI report and shown in Table 3. The list of instrument types presented in the Table for Root Cause 1- Structural Behavior, are used in general for monitoring concrete dams. A tailored list of instruments applicable to particular root causes are shown for each subsequent root cause.

Monitoring cracks is accomplished by visual inspection of the cracks combined with photographic records to document the condition of cracks, measurements of the behavior of the dam structure and its foundation due to cracking using instruments and instrumentation systems, and plotting the recorded data chronologically to display patterns and changes in measurements with time. The reservoir level and ambient air and water temperatures should also be recorded simultaneously, as the structural behavior data is gathered and plotted on the same charts.

The identification of a "Threshold Level" (recheck reading, investigate cause of reading beyond the threshold limit) and an "Action Level" (reading is beyond normal range, indicates a potential adverse situation and requires an action to rectify situation and if necessary, to preserve the safety of the water retaining structures) are effective monitoring tools and can be shown on the monitoring data plots as triggers beyond which certain actions are required by a dam owner's organization.

A detailed treatment of instrument types, their purpose, operating principle, and applicability to measuring particular material and structural behavior of concrete dams is covered in the American Society of Civil Engineers (ASCE) publication "Guidelines for Instrumentation and Measurement for Monitoring Dam Performance." The guideline is currently under revision and is expected to be reissued in 2017. Similar information is provided in the CEATI publication "Dam Safety Performance Monitoring and Data Management – Best Practices" [8].

Risk Reduction Measures

A summary of typical risk reduction options that have been taken by dam owners is also presented in Table 3. Risk reduction measures can be achieved by affecting: (1) the probability of an applied load through reservoir management practices (drawdowns and filling schedules) and spillway gate operation protocols, (2) the probability of a particular response of the dam by redesigning, repairing, strengthening, grouting, sealing, slot-cutting, insulating or re-configuring the dam and foundation, reducing uplift pressures by adding or improving drainage, and (3) the probability of an unacceptable downstream consequence by installing early warning systems, revising emergency response plans, restricting human use of public lands, flood plain set-asides, zoning restrictions, etc. Dam owners have used one or a combination of such risk reduction measures to lower the risk of a serious dam incident to within tolerable limits. No general guidelines for implementing emergency actions are provided in the CEATI Report as the need to take any such action must be determined on a case-by-case basis.

CONCLUSIONS

- Concrete dams crack for a variety of reasons, and it is important for owners to be able to assess the significance of observed cracks within a dam structure in a timely and effective manner. The identification of cracking or other anomalous behavior may be by the owner's operating staff or engineering resources. Their causes and effects are not always obvious. Some are superficial and have limited consequences while others may have structural, operational or maintenance consequences which need to be addressed. Cracks can occur during the initial hydration process of the fluid concrete and during the service life of a dam as structures age. The investigation process and tools must address such diverse scenarios.

- Once determined that a crack(s) exists in a concrete dam, best practice throughout the industry has shown that owners seek to manage the cracks and the mechanisms causing them. There are three principal activities that responsible owners follow to manage cracking in their dam(s): (1) analyze the causes and effects of the cracks, (2) establish means and methods of monitoring the status of cracks, and (3) decide which option(s) to select for remediation of the situation from among all available options for controlling and/or reducing risks associated with the presence of cracks in the dam.

- An effective approach to conducting an assessment of cracks in a dam has been proposed using a practical framework taking into consideration (1) the design concept of the dam, (2) the constituent materials used in the concrete mix, (3) the means, methods and sequence of constructing the dam, (4) the environment within which the dam was built and operates, and (5) the behavior of the dam under the loads to which it is subjected.

- Dam owning institutions, public agencies and private companies seek logical bases for planning required investigations and remedial actions to cure issues of cracking in concrete dams. To that end, it is critical to the success of such endeavors that dam owning entities understand that no two dams are the same and that, no singular, universal approach necessarily applies for assessing how and why cracking occurs in concrete dams. A tested process of diagnosis, analysis and planning that has evolved among leading dam owning organizations uses standardized observation, classification and recording systems for the cracks coupled with guides to recognize possible "root causes" of specific mechanisms that may be present in a cracked dam.

- If a critical situation exists where a crack(s) is discovered that threatens the safety of the dam, it must be immediately communicated to the dam safety engineer for response and action. Action may require a quick decision to immediately lower the reservoir level to reduce a calculated amount of pressure off the dam in order to reduce the stresses in the dam and stabilize the structure. This action is a first step to stabilize the situation until any additional risk reduction measures might be determined to be needed. In a parallel track, action must be taken to plan and undertake investigations and analyses of alternative measures to be studied as necessary major remedial action(s) as outlined in Table 2-1.

ACKNOWLEDGEMENTS

MWH, now part of Stantec prepared the research report for CEATI International under the leadership of Chris Hayes, Vice President CEATI International with contributions from an international team of concrete dam specialists drawn from ICOLD and USSD Committees. The study was led by Glenn Tarbox with Robin Charlwood and contributions by Alain Carrere from France and MWH, now part of Stantec staff from the USA, including Vik Iso-Ahola and Jennifer Fordney. Michael Manwaring of MWH, now part of Stantec, provided overall review.

The investigators are grateful to CEATI for the opportunity to have worked on this interesting issue, authored this technical paper and for contributions from the following individuals:

- Gus Tjoumas CEATI Technical Advisor
- Project Monitors
  - Eric Bourdarot of EDF,
  - Anders Sjodin of ELFORSK,
  - Dave Hart and Tony Deakin of Environment Agency, and
  - Jim Wagoner of OPG

REFERENCES

[1] "Investigating the Structural Safety of Cracked Concrete Dams," CEATI Report No. T122700-0226, April 2014.

[2] American Concrete Institute (ACI), "Guide for Conducting a Visual Inspection of Concrete in Service (ACI 201.1R)," 2008.

[3] T. P. Dolen, "Historical Development of Durable Concrete for the Bureau of Reclamation," United States Bureau of Reclamation, Denver, CO.

[4] International Commission on Large Dams (ICOLD), "Bulletin 107: Concrete Dams - Control and Treatment of Cracks," ICOLD Committee on Concrete for Dams, Paris, France, 1997.

[5] L. Nuss, N. Matsumoto and K. Hansen, "Shaken, but Not Stirred: Earthquake Performance of Concrete Dams," in *Proceedings of the 32nd U.S. Society on Dams Annual Conference, New Orleans*, 2012.

[6] A. Mansouri, M. A. Lashteh Neshaei and R. Aghajany, "Fracture analysis of gravity dam under earthquake induced loads," *Journal of Applied Sciences and Environmental Management, Volume 15,* pp. 317-325, June 2011.

[7] Federal Energy Regulatory Commission, "Engineering Guidelines for the Evaluation of Hydropower Projects," 28 June 2010. [Online]. Available: http://www.ferc.gov/industries/hydropower/safety/guidelines/eng-guide.asp. [Accessed 7 January 2013].

[8] CEATI International Dam Safety Interest Group (DSIG), "Dam Safety Performance Monitoring and Data Management – Best Practices (CEATI Report No. T082700-0210)," CEATI International, Montreal, Quebec, Canada, 2012.

A NEW APPROACH FOR THE DESIGN OF THE KÖROĞLU RCC ARCH/GRAVITY DAM

Quentin Shaw, (CEng, PrEng, PhD)[1]

ABSTRACT

In China, almost 30% of all RCC dams are arch or arch/gravity structures and consequently, there are more RCC arch type dams in that country than there are RCC dams in any other country. More than 25 years after the first RCC arch was completed, however, only a few dams of this type have so far been constructed outside China.

In general terms, most RCC arch dams have been designed on the basis of the same principles applied for conventional mass concrete (CVC) arch dams, with RCC providing a method for a simpler and possibly more rapid horizontal construction. By comparison with CVC, the inclusion of post-cooling pipe systems is much more complicated and inconvenient in horizontally-placed RCC and while such systems have been included in some RCC arches, in others alternative compensatory measures have been employed to avoid this requirement; such as the use of lightly burnt MgO powder to compensate for autogenous cementitious paste shrinkage and the inclusion of short and other joint systems to accommodate increased arch displacements.

Ascertaining that certain types of RCC exhibit particularly low levels of stress-relaxation creep has created opportunities to develop more efficient and perhaps simpler RCC arch dam types and a number of such arch dams have successfully been constructed in temperate climates over the past decade, without the requirement for joint grouting.

In this paper, the author demonstrates how low stress-relaxation creep RCC was used in combination with strategic and targeted post-cooling and joint grouting to develop an innovative RCC arch/gravity solution at the Köroğlu RCC arch/gravity dam in the relatively extreme climate of north-east Turkey.

This development demonstrates that low stress-relaxation creep RCC can be applied to gain time and cost savings for arch structures under a broad range of climatic conditions.

INTRODUCTION

The Köroğlu dam is situated on the Kura River in the north-east of Turkey, close to the town of Ardahan and a short distance before the river crosses the border and flows into Georgia. Construction at the dam started in mid 2015, with an initial objective to complete RCC placement by the end of 2016. Problems with aggregate supply, however,

[1] Director: ARQ Consulting Engineers, 6 Daventry Street, Lynnwood Manor, Pretoria, South Africa, quentin@arq.co.za.

resulted in a slower RCC placement than anticipated and with a construction season limited to 7 to 8 months by the extreme winter temperatures, completion is now anticipated in mid 2017. The final dam will be approximately 95 m in height, will indicate a crest length of approximately 370 m and will contain approximately 560 000 m^3 of RCC. The dam is being built by Ünal Construction for its sister company EBD Enerji and the adjacent Köroğlu and Kotanli HEPPs will be operated as a system to generate a maximum of 130 MW of hydropower.

Figure 1. Kotanli HEPP September 2015

As a consequence of relatively poor foundation conditions and less than ideal topography, an efficient, thin arch structure was never realistically possible for the Köroğlu site. Nevertheless, the structure under construction indicates a concrete saving of approximately 30%, compared with an equivalent gravity dam structure.

In this paper, the innovative approach and application of materials behaviour are illustrated, by means of which it was possible to adapt a technology only previously applied in temperate climates to more extreme climatic conditions.

BACKGROUND

Typical RCC Arch Dams

The majority of RCC arch dams constructed to date have been designed on the basis of similar principles to those applied for conventional mass concrete arch dams. The latter are constructed as a series of vertical blocks, through which thermal contraction is accommodated with adjacent blocks shrinking away from each and the joint in between subsequently being filled with grout. Where natural cooling is insufficient to bring the concrete temperature down sufficiently before joint grouting is required, post-cooling is applied through the internal circulation of chilled water.

To replicate similar conditions in horizontally placed RCC, the cross section of the structure at the location of a joint is fully de-bonded using a system including grouting facilities, which is placed in the RCC. In some cases, post-cooling systems are installed to reduce temperatures and to allow joint grouting before impoundment, while in other cases, secondary, or re-groutable systems are included to facilitate additional phases of joint grouting once the full hydration heat dissipation has been achieved. Depending on the design approach applied, the configuration of the arch and the site climatic conditions, induced joints that do not de-bond the full cross section and/or short joints in the upstream face are also included to accommodate the increased arch displacements associated with higher total temperature drop loads. While the installation of post-cooling pipes in conventional concrete (CVC), which is more workable and vibrated with immersion pokers, is relatively simple, it is substantially more difficult in less workable RCC, which is compacted by rollers in relatively thin horizontal layers and over which heavy equipment must pass. Consequently, while post-cooling is commonly applied in CVC arch dams, it is generally avoided unless essential in RCC arch dams.

Low Stress-Relaxation Creep RCC Arch Dams

Measured observation of higher-workability fly ash-rich RCCs has demonstrated exceptionally low stress relaxation creep during the hydration cycle. Whereas stress relaxation creep of between 150 and 250 microstrain might typically be expected in conventional dam mass concrete, an effective value of zero has been observed on two RCC arch/gravity dams [Shaw, 2012]. While direct stress-relaxation creep in fly ash-rich RCC is particularly low, other effects related to horizontal construction have been demonstrated to give rise to the realization of an effective zero stress-relaxation creep.

With substantially reduced relaxation creep during the hydration temperature rise (and associated thermal expansion), horizontal, cross-canyon compression stresses remain. In the case of a curved arch structure, arches become longer and thinner and accordingly progressively more flexible with height and consequently the retained compression stresses cause increasing expansion and the structure effectively leans progressively further upstream, as illustrated in Figure 2. The greater flexibility in the upper arches reduces constraint, which further contributes to a reduction in compressive stress and consequently stress-relaxation.

Figure 2. Indicative behavior of RCC arch with low SRC during construction

As the wider, lower part of the structure subsequently dissipates hydration heat, cools and contracts, the cross-canyon compression stresses are released and the structure consequently tends to tilt forward, which in turn progressively increases compression stresses in the upper arches. With the majority of arch action in a heavy arch occurring in the upper part of the structure, analyses have demonstrated a related benefit of this effect equivalent to approximately 40 to 50 microstrain. A similar condition has been created at certain RCC arch dams in China, whereby a degree of autogenous concrete expansion has been developed through the use of lightly burnt MgO powder.

The foregoing situation creates a significant opportunity for the application of arch dams without joint grouting, as the applicable structural temperature drop is effectively substantially reduced.

The Optimal RCC Arch Dam

The primary benefit associated with an arch dam is a reduction in concrete volume made possible through the more efficient structural utilization of concrete. RCC construction requires simplicity to facilitate speed and consequently, the benefit of reduced concrete volume for an RCC arch must be weighed against potentially increased construction complexity, and/or a loss in construction efficiency, and potential delays in impounding caused by the need to wait for cooling to allow joint grouting. By contrast, the most efficient arch dams are possible in steep sided valleys, with low crest length/height ratios and in such circumstances, access restrictions favor vertical construction, using cable cranes, rather than horizontal construction, where difficult plant access and materials delivery routes and systems must frequently be changed and/or realigned, often compromising achievable placement rates. Vertical construction and the easier accommodation of galleries, etc, implies consolidation grouting on steep abutments can be more simply accommodated in CVC arch dams than in RCC arches, with placement rates having consequently been compromised at certain RCC arches.

For RCC arch dams, where higher lift joint tensile strengths are often required, high cementitious content RCC has been universally used to date. Such RCCs will tend to indicate a total cementitious materials content of approximately 200 kg/m^3, which is higher than the cementitious content for the general conventional concrete in an arch dam of more than 200 m height. Although the RCC would undoubtedly contain more pozzolan as part of the cementitious material than would be the case for the CVC, the pozzolan generally applied will be fly ash and while fly ash was traditionally substantially lower cost than Portland cement, currently only a marginal cost benefit often exists.

Average monthly placement rates for RCC on large dams of upwards of 120 000 m^3 per month have been demonstrated on a few of the larger and simpler RCC dams in recent years. It is, however, important to remember that such rates were actually achieved in CVC at Hoover Dam in the 1930s. Furthermore, conventional mass concrete placement rates averaging approximately 100 000 m^3 per month have repeatedly been achieved at large conventional concrete, double-curvature arch dams in China in recent years using cable cranes and vertical construction.

As a consequence of the above, it is clear that RCC construction is unlikely to universally represent the optimal solution for all arch dams. While RCC construction has largely replaced CVC construction over the past 3 decades for concrete gravity dams, this situation is unlikely to occur, or to be applicable for arch dams. The best sites for arch dams will generally be more suited to vertical CVC construction than horizontal RCC construction, while more complex, structurally efficient arch geometries are more practically achievable applying CVC construction, rather than RCC construction.

With CVC arch dams more suited to the narrowest and most efficient dam sites and RCC gravity dams the optimal solution for wide valley sites, RCC arch dams are most likely to represent the optimal solution in a middle ground and consequently to most commonly indicate a structural section similarly between the two extremes.

To represent the optimal solution for a particular site, an RCC arch must offer a reduced concrete volume and an earlier construction completion, compared to the equivalent RCC gravity structure, while relatively rapid and simple construction must be possible to ensure that a possibly more structurally efficient CVC arch is not more competitive. Considering all of the above, RCC arch dams are most often likely to be "heavier" arch dams; thick arches and arch/gravity structures, as illustrated in Figure 3.

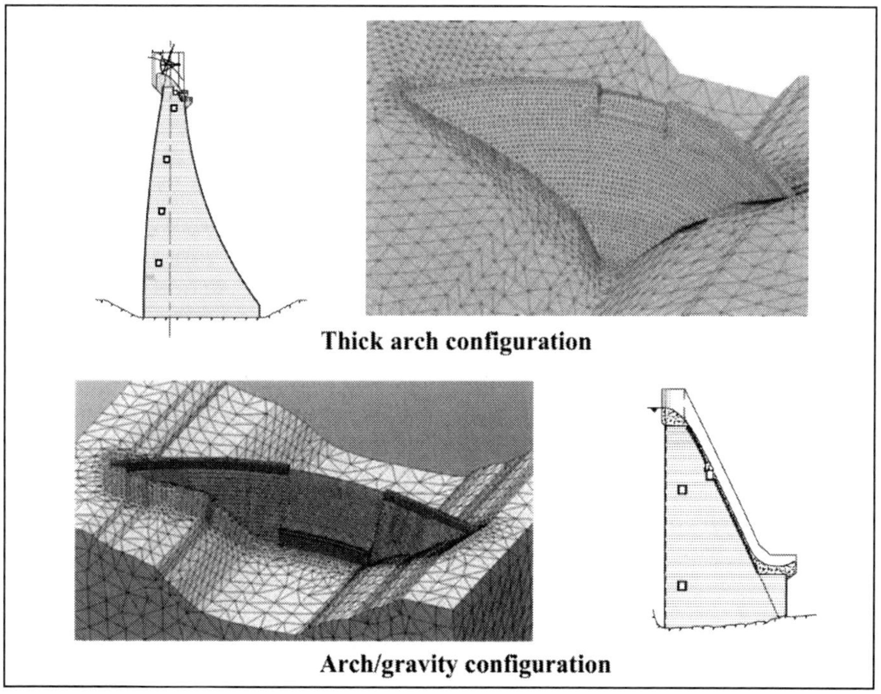

Figure 3. Typically more suitable RCC arch dam types

Whereas a thin, double-curvature arch dam acts as something of a shell, or dome, with maximum arch stresses usually at around 2/3 dam height, arch dams in wider valley sites indicate maximum arch action at the crest and blend cantilever and arch action. These structure types generally indicate broader sections, particularly towards the base, commensurate with the requirement for stiffer cantilevers, and the heavier sections and associated simpler geometries are often quite suited to simple, rapid RCC construction.

THERMAL EFFECTS & DESIGN

General

With RCC arch dams generally likely to be heavier, or thicker structures and arch action required before impoundment, a number of issues must be addressed; all related to the fact that the natural cooling to release the hydration heat will take a number of years. In a best-case scenario of low stress-relaxation creep and a temperate climate, arch stresses will progressively reduce as the structure cools, but the structure can be designed to remain fully serviceable after the full hydration heat has been dissipated. In a worst-case scenario of high stress-relaxation creep and an extreme climate, the process of shrinkage associated with cooling can result in a sufficient loss in arch action to the extent that the structure becomes unstable. Before such a situation is allowed to develop, the (induced) joints in the structure must be grouted to ensure appropriate and adequate structural continuity. In view of the fact that such grouting is most effective when the joints are as wide open as possible, it is typically either necessary to artificially cool the dam structure, with post-cooling, and to grout before impoundment, or to draw down the impounded water level and grout once sufficient natural cooling has occurred.

Specific critical loading

Typically in the case of an arch dam, the greater the cooling that occurs, the higher the impact on the dam structural action. Particularly when low stress-relaxation creep RCC is used, however, the long term cooling will not always represent the critical thermal loading condition in a heavy-section RCC thick arch, or arch/gravity structure. For such structures, with their stiff-cantilever configuration, the upper sections, where all arching occurs, will cool substantially more rapidly than the lower, more massive sections. During construction, as hydration increases the dam temperature, the crest is displaced upstream. Due to the differential cooling of the upper and lower sections of the dam, the upstream displacement remains while the upper section of the dam cools and the joints open, implying a greater downstream displacement of the crest is required to close joints and incur structural arch action. This effect can give rise to deleterious vertical tensions on the upstream face of the dam. However, it also creates opportunities for the strategic grouting of the arch structure, as arch stresses in the grouted upper structure will continue to increase as the temperature in the base continues to drop, further implying that any requirement for re-grouting, or subsequent grouting of joints will be eliminated.

THE DESIGN OF KÖROĞLU ARCH/GRAVITY DAM

Climatic conditions

As indicated in Table 1, the Köroğlu dam is located in an area with a particularly cold climate, with mean monthly temperatures never exceeding 20°C, but dropping below 0°C for four months of the year.

Table 1: Koröğlu Dam Site Estimated Monthly Temperatures

Month	Jan	Feb	Mar	Apr		May	Jun	Jul	Aug	Sep	Oct	Nov	Dec	Average
Mean Temp (°C)	-9.9	-8.3	-2.4	5.8		10.2	14.0	17.8	17.8	13.8	7.3	0.3	-6.4	5.0

Mean annual precipitation in the general area of the Köroğlu dam site is approximately 470 mm, which occurs throughout the year, with elevated figures between April and August, peaking in June.

Cementitious materials & RCC behavior

The cementitious materials used for the RCC of Köroğlu dam comprised a blend of ordinary Portland cement and a Trass natural pozzolan in the proportions 85 kg/m^3 + 130 kg/m^3, respectively. The same RCC mix was used in the construction of the smaller Kotanli Dam during 2014 and 2015, as illustrated in Figure 1. While the Kotanli dam structure only requires arch action under seismic loading, the strain behaviour of the RCC was monitored through instrumentation, primarily as a means to provide design information for Köroğlu Dam, for which arch action is required under normal loading.

The results of the instrumentation data at Kotanli Dam demonstrated early strain behaviour comparable to that of fly ash-rich RCC, suggesting minimal stress-relaxation creep during the hydration process and whilst quantitative evaluations will not be possible until greater RCC cooling has occurred, it could be stated with certainty that any stress-relaxation creep that might develop would not be of a similar order to that generally evident in CVC.

Dam structural design

Early structural analyses indicated that all arch action would essentially be eliminated through the degree of temperature drop to be experienced at Köroğlu Dam and consequently confirmed that joint grouting would be essential to ensure the ongoing structural integrity and safety of the dam. Long term structural temperature drop is typically defined by the difference between the zero stress temperature (or closure temperature, T3) and the final long term winter temperature, after all hydration heat has been dissipated (T4). In the case of Köroğlu Dam, this temperature drop will be of the order of 15 to 20°C. Correspondingly, the design challenge lay in identifying a realistic and cost-effective manner to achieve appropriate structural behavior, in the knowledge that the RCC for construction indicates minimal stress-relaxation creep.

A solution was conceived whereby a relatively small portion of the dam structure would be post-cooled and the joints in this zone would be grouted immediately prior to impoundment. While cooling pipes are not realistically compatible with RCC construction, water chillers indicate high capital and power running costs and consequently, it was considered advantageous to minimize the necessary installations within the RCC and to avoid, if possible, artificially chilled water. It was accordingly proposed that a realistically achievable solution would be to install HDPE cooling pipes in every 5th RCC layer and to use natural river water, which rarely exceeds 8°C in temperature, for cooling. To eliminate the possibility of compromising impermeability, it was considered appropriate not to install cooling pipes in a 6 m wide strip against the upstream face of the dam. The area of the dam over which post-cooling was applied is illustrated in Figure 4.

STRUCTURAL/THERMAL ANALYSIS

Post-cooling analyses

To demonstrate the feasibility of the proposed solution, it was first necessary to establish whether a realistically achievable post-cooling arrangement would develop the necessary cooling and whether strategic, rather than general, post-cooling would not develop deleterious stresses within the dam structure during construction. The first step in the thermal analysis involved evaluating the cooling effect of circulating 8°C water in pipe loops placed at 1.5 m vertical intervals and with a horizontal spacing of 1.5 m.

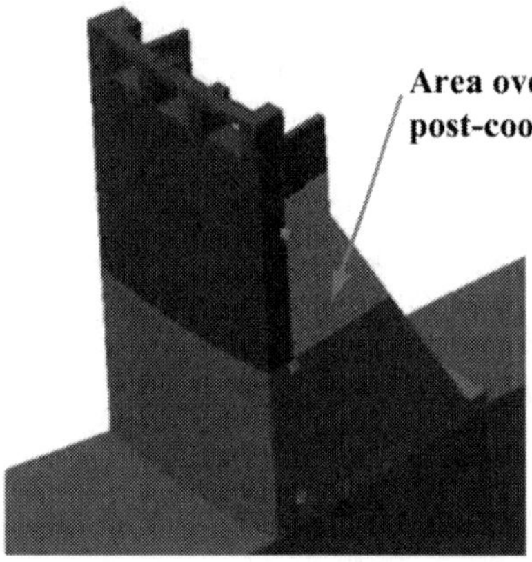

Figure 4. Zone post-cooled

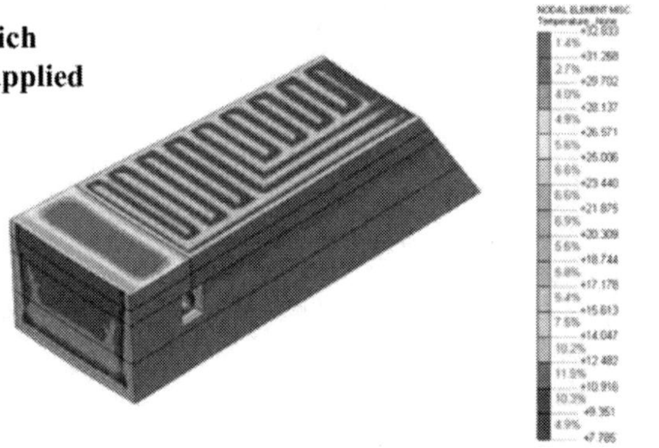

Figure 5. "Block" for evaluation of post-cooling

Simplifying the generation of hydration heat on the basis of a single 1.5 m deep layer and modelling a block depth limited to 13.5 m (see Figure 5) to reduce analysis time, post-cooling water circulation was simulated for 14, 28, 42 and 56 days and the respective temperature distributions were evaluated. The temperature data from the 3-dimensional block simulation was subsequently applied to a 2-dimensional sectional representation of the dam and the heating/cooling analysis was applied on the basis of the intended construction schedule and the associated external climatic seasonal temperature variations. Subsequently, temperature profiles for the dam at the end of construction were developed, as indicated in Figure 6.

On the basis of the derived temperature distributions, the structure was analyzed for consequential stresses, assuming a worst-case zero stress-relaxation creep scenario. Whilst the thermal study demonstrated little additional cooling benefit beyond 28 days water circulation, maximum associated tension stresses did not exceed 1.2 MPa, which is within the tensile strength of the parent RCC.

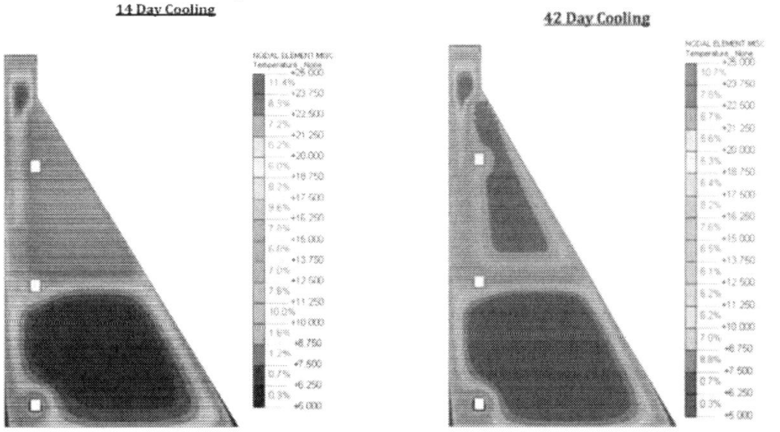

Figure 6. Temps. after 14 & 42 days post-cooling

Joint opening & grouting studies

The findings of the initial thermal analyses were transferred into a 3-dimensional analysis of the dam structure, applying interface elements at each of the induced contraction joints. These joints were assigned a limiting tensile strength of 250 kPa, allowing the joints to fail in tension and open for higher levels of stress. For the subsequent analyses, two levels of stress relaxation creep (SRC) were considered; 0 and 50 microstrain, with the latter modelled by raising the zero stress temperature (T3) by 5°C (equivalent to a thermal expansion coefficient of 10 x 10^{-5}) above the placement temperature (T1).

The temperature distribution conditions at placement and at end-construction were compared, applying the two SRC scenarios, and the respective joint openings were evaluated at the latter time step. Post-cooling applications for 14, 28 and 42 days were also compared. With a 0.5 mm joint opening assumed as sufficient to allow grouting, the analyses demonstrated that 14 days of cooling was not realistically sufficient to adequately open the induced joint. On the other hand, 28 days of post-cooling was demonstrated to be quite sufficient, even for a zero SRC scenario.

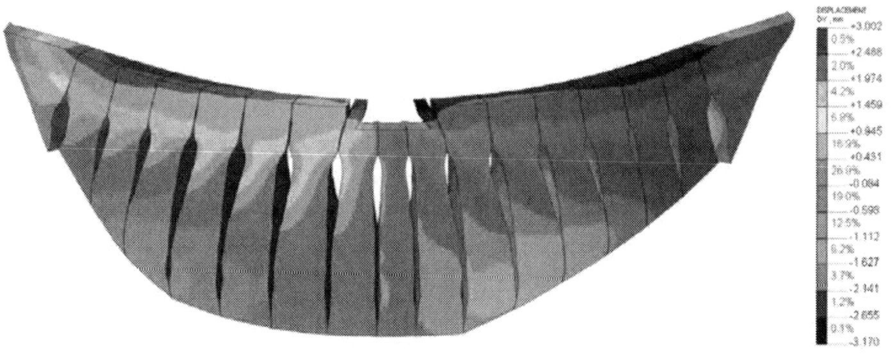

Figure 7. Joint opening at end of construction for 28 day post-cooling & 0 SRC

Long-term response

Considering all joint openings exceeding 0.5 mm to be grouted, applying the above-mentioned two SRC scenarios and 28 and 42 post-cooling, the full 3-dimensional dam structure was subsequently subjected to climate cycles and hydrostatic load for a simulated 20 year period and the crest displacement and stress conditions were evaluated at various time step increments. Grouting of the joints was simulated by revising the zero stress temperature (T3) at the point grouted to the actual applicable temperature at the end of construction, or time of grouting, while for the remainder of the dam cross section, T3 was equated to the placement temperature (T1) in the case of 0 SRC, and T1 + 5°C, in the case of 50 microstrain SRC.

ANALYSIS FINDINGS

Post-cooling analyses

While the analyses indicated very little difference between a 28 and 42 post-cooling, fully acceptable structural behavior was demonstrated whether 0, or 50 microstrain SRC was assumed. A higher degree of arch stress in the lower sections of the arch was indicated for the higher SRC, although similar crest displacements were apparent (25 - 27 mm).

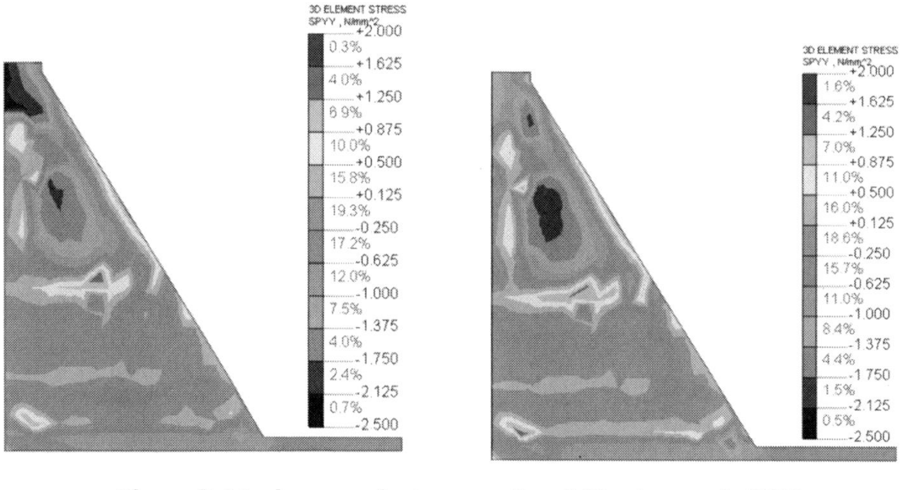

Figure 8. Maximum arch stresses – 0 and 50 microstrain SRC

The usual evaluation of arch performance on the basis of upstream and downstream face stress distributions was not possible in the proposed scenario, due to the fact that an internal arch strut is created. However, as demonstrated in Figure 9, the continuity of thrust movement laterally through the arch structure can clearly be seen on a plot on which every second block is removed. In reality, an arch gravity dam will always indicate an arch within an arch, with only part of the structure transferring load laterally, but, as can be seen, full dam stability is assured with maximum arch stresses of just 2.5 MPa, which is an order of magnitude lower than the actual RCC compressive strength.

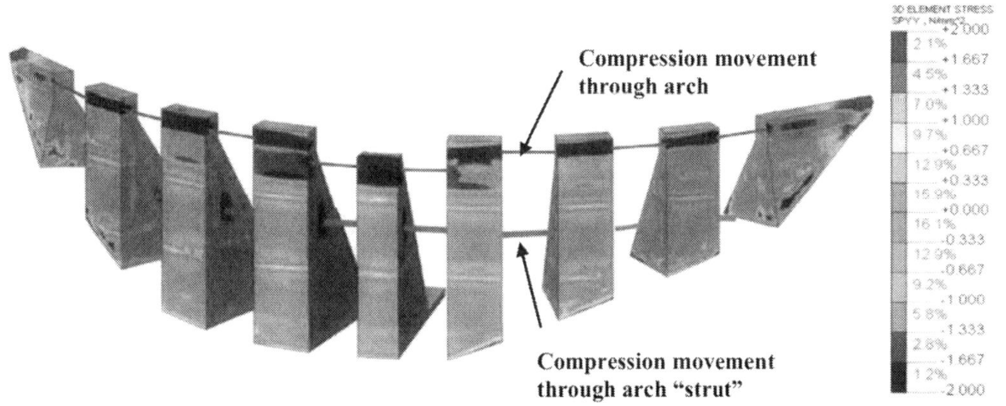

Figure 9. Illustration of arch compression movement through grouted structure

In the case of Köroğlu Dam, applying a cooling temperature of 8°C for analysis represents a conservative assumption, due to the water temperature being likely to be lower during the most important stages of cooling. Furthermore, the analyses were additionally conservative in not taking into account the upstream movement of the dam structure that will occur as a result of the lower sections of the dam retaining almost full hydration heat at the time that the joints of the upper section of the dam structure are grouted.

CONCLUSION

The analyses presented confirm the appropriateness of the solutions currently being implemented to ensure effective structural function of an RCC arch/gravity dam in a relatively extreme climate. Strategic, rather than general post-cooling and joint grouting, representing approximately 28% of the RCC volume, has been demonstrated to be fully appropriate, while the use of the benefits of low stress-relaxation creep to allow an increase in compression as the structure continues to cool further confirms the opportunities for the beneficial exploitation of this characteristic. A technology that was initially developed in temperate climates has accordingly been demonstrated to indicate a more general application for simple, heavy section RCC arch dams. Retaining simplicity and avoiding the time and cost complications that can be associated with RCC arch dams will allow increased economy and an increased competitiveness of this dam type.

REFERENCES

ICOLD. Bulletin 126 Update. Roller compacted concrete dams. State of the art. Pre-publication 2016.

SHAW, Q.H.W. The Influence of Low Stress Relaxation Creep on the Design of Large RCC Arch and Gravity Dams. 6th Symposium on Roller Compacted Concrete Dams. Zaragosa, Spain October 2012.

UPGRADE AND SPILLWAY MODIFICATIONS AT SANTA ANITA DAM

Sterling E. Klippel, P.E.[1]
Vik Iso-Ahola, P.E.[2]
Glenn S.Tarbox, P.E.[3]
Martina Gelo, P.E.[4]
George Manole, P.E.[5]

ABSTRACT

Santa Anita Dam is 225 feet high concrete arch dam completed in 1927 near Los Angeles, California, and is owned and operated by the Los Angeles County Department of Public Works (LACDPW) for flood management and stormwater capture. LACDPW's Dams Rehabilitation Program assessed the hydraulic and seismic safety of the dam, which identified deficiencies in the spillway capacity and the ability of the dam to resist seismic loads. As a result, the State of California Department of Water Resources, Division of Safety of Dams (DSOD) required an increase in the spillway capacity and either buttressing the dam or restricting the maximum normal pool reservoir elevation.

Safety improvements for the dam were developed to address the increased probable maximum flood (PMF) and maximum credible earthquake (MCE). The initial concept development phase considered several dam modification options, including five spillway augmentation concepts and two seismic rehabilitation concepts. The seismic rehabilitation concepts were validated using a three dimensional (3D) finite element model. However, due to excessive costs associated with buttressing the dam, LACDPW addressed immediate seismic concerns by constructing a riser connected to a free-flowing outlet to limit the maximum reservoir storage to a seismically safe elevation. The subsequent rehabilitation design focused on selection and development of a new spillway. 3D Computational Flow Dynamics (CFD) modeling of the selected notching (spillway) alternative was performed to validate hydraulic capacity and trajectory of the discharge nappe.

The dam rehabilitation includes cutting of a notch in the center of the dam, installation of an ogee crest with flip bucket, outlet works modifications, and upgrades to the valves and controls systems. This paper reviews the spillway rehabilitation design development and summarizes the 3D structural evaluation as well as the overall rehabilitation and upgrade of the 90 year old dam.

[1] LACDPW, Alhambra, California, USA, sklippel@dpw.lacounty.gov
[2] MWH Global, Walnut Creek, California, USA, vik.iso-ahola@mwhglobal.com
[3] MWH Global, Bellevue, Washington, USA, glenn.s.tarbox@mwhglobal.com
[4] MWH Global, Walnut Creek, California, USA, Martina.Gelo@mwhglobal.com
[5] LACDPW, Alhambra, California, USA, gmanole@dpw.lacounty.gov

INTRODUCTION

Santa Anita Dam is located on the Big Santa Anita Creek approximately four miles north of the City of Arcadia, Los Angeles County, California. The dam, constructed in 1924 through 1927, is owned and operated by the Los Angeles County Department of Public Works (LACDPW). The dam is a constant-angle variable-radius concrete arch constructed for the purpose of flood control and stormwater capture. The dam's crest height above the streambed at the downstream toe of the dam is 225 feet (dam crest El. 1,325), and the crest length of the arch (including the abutments) is 612 feet. The dam had an original storage capacity of 1376 acre feet (AF) with a tributary watershed of 10.8 square miles. Stormwater captured at this facility is later released and diverted into downstream spreading grounds for local groundwater recharge with an average annual recharge of 2,300 AF. Local communities rely on this local groundwater supply for a significant portion of their water supply. The dam is one component of a system of conjunctive use facilities in Santa Anita Canyon. A photo of the dam is shown in Figure 1.

Figure 1 – Santa Anita Dam

The hydraulic and seismic safety of the dam was initially evaluated by the International Engineering Company (IECO, 1977), Los Angeles County Department of Public Works (LACDPW, 1978) and the State of California Department of Water Resources, Division of Safety of Dams (DSOD, 1980). These evaluations played a role in identifying the need for safety improvements for two types of possible events:

(1) The probable maximum flood (PMF), and
(2) The maximum credible earthquake (MCE).

Subsequently, as part of the LACDPW Dams Rehabilitation Program, MWH performed a hydrological study to update the PMF and deterministic seismic hazard assessment to update the MCE. MWH's study was updated by DSOD, and the PMF was established at 26,100 cfs, or about 5 times greater than the existing spillway and combined outlet capacities. The MCE was estimated as a 7.25 magnitude earthquake at 0.5 km from the dam, producing a 1.1g peak ground acceleration (pga) at the dam site. Utilizing the updated MCE, MWH re-evaluated the seismic stability of the dam, performing 3D finite element method (FEM) dynamic analyses of the dam structure, which indicated that the dam would be severely overstressed during the MCE, subject to cracking and at risk of uncontrolled release of the reservoir. The concern for seismic stability was further exacerbated by an ongoing phenomena of Alkali-Aggregate Reaction (AAR) occurring in the concrete, which causes expansion and cracking of the concrete. As a result of the re-evaluations, DSOD required an increase in the spillway capacity and strengthening of the dam or restricting the maximum normal pool reservoir elevation.

INITIAL STUDIES

MWH performed an initial feasibility study in which several dam modification options were considered, including five spillway augmentation concepts and two seismic rehabilitation concepts, aimed at mitigating the PMF and MCE, respectively. The five spillway augmentation concepts included variations of a notch concept placed at the center of the dam arch. The notch alternatives included using an upstream parapet wall at either end of the dam crest to create a central notch, and variations of a two-level notch to concentrate low flows in the most heavily armored downstream area while spreading the most extreme floods for energy dissipation purposes. Consideration of a two-level notch was prompted by a previous hydraulic model study for Santa Anita Dam (Burke and Fuller, 1942) that advocated spreading out the PMF to achieve better energy dissipation. All concepts were sized to be capable of passing the 26,100 cfs PMF without overtopping. Although the parapet wall concept was considered most economical, further consideration of the parapet wall concept was eliminated due to concerns in attaching the wall to the AAR affected concrete.

The seismic rehabilitation concepts included two options: 1) determining a "safe" reservoir elevation for the existing dam that allowed for damage during the MCE but no risk of uncontrolled release of the reservoir, and 2) strengthening of the dam to resist seismic loading. Option 1 required FEM analysis to determine the safe reservoir elevation. Option 2 considered several conceptual alternatives, including vertical post-tensioned anchors, upstream, and downstream "buttressing" (i.e., thickening) of the dam. These options were discussed with LACDPW and DSOD prior to conducting FEM analysis. Due to cost and regulatory requirements that restrict the use of "active" systems to stabilize dams (i.e., post-tensioned anchors), the downstream buttress alternative was selected for preliminary evaluation. MWH developed two 3D FEM structural models to investigate these options as shown in Figures 2 and 3.

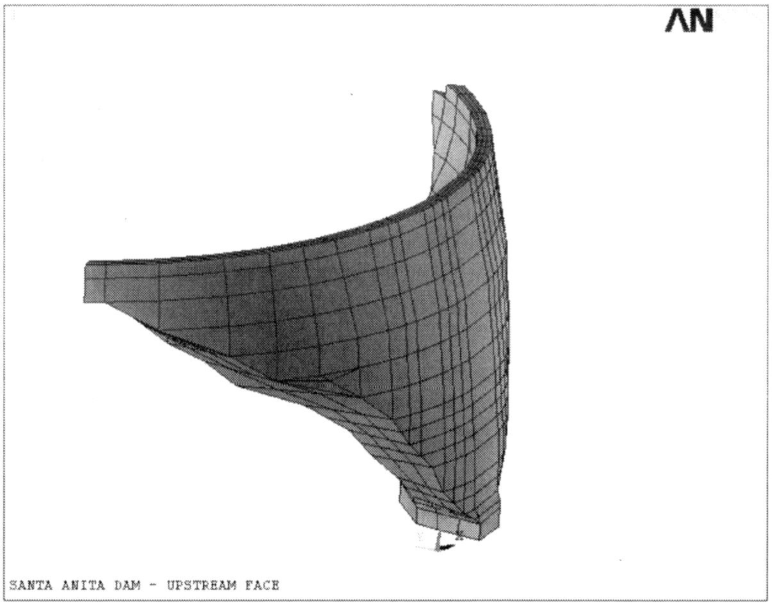

Figure 2 – Santa Anita Dam Finite Element Model

The analysis revealed that Option 1 would require a permanent reservoir restriction, lowering the maximum operating pool by 94 feet. The Option 2 analysis optimized the size of buttress based on constructability, cost, and the gained benefit of additional reservoir storage above the restricted elevation. As a result a buttress that was about 40 feet above the restricted elevation, but below the dam crest was proposed (shown in Figure 3).

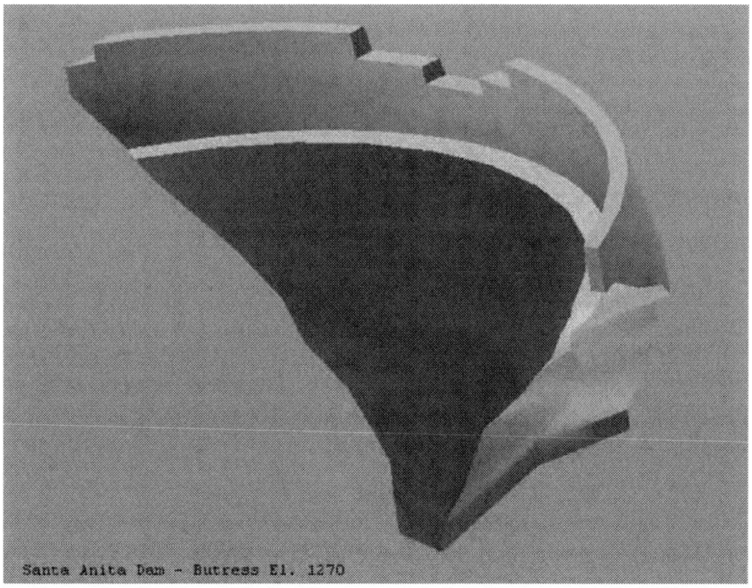

Figure 3 – Santa Anita Dam Finite Element Model, Buttress Rehabilitation Concept (no mesh shown for clarity)

CONCEPT VALIDATION

Upon consideration of the initial modification concepts, LACDPW requested MWH to further develop the two rehabilitation concepts to a 30% design level, which were advanced as a "Partial Rehabilitation" option and an "Intake Riser Modification" option. The "Partial Rehabilitation" (Figure 4) concept included modifications to the existing dam by thickening the dam with the buttress that was validated in the FEM analysis Option 2 above. Also, in order to add capacity to pass the PMF, the concept included a notch spillway centered on the downstream channel and a 6-foot diameter octagonal orifice spillway exiting through the buttress to permanently maintain the reservoir level at the top of buttress.

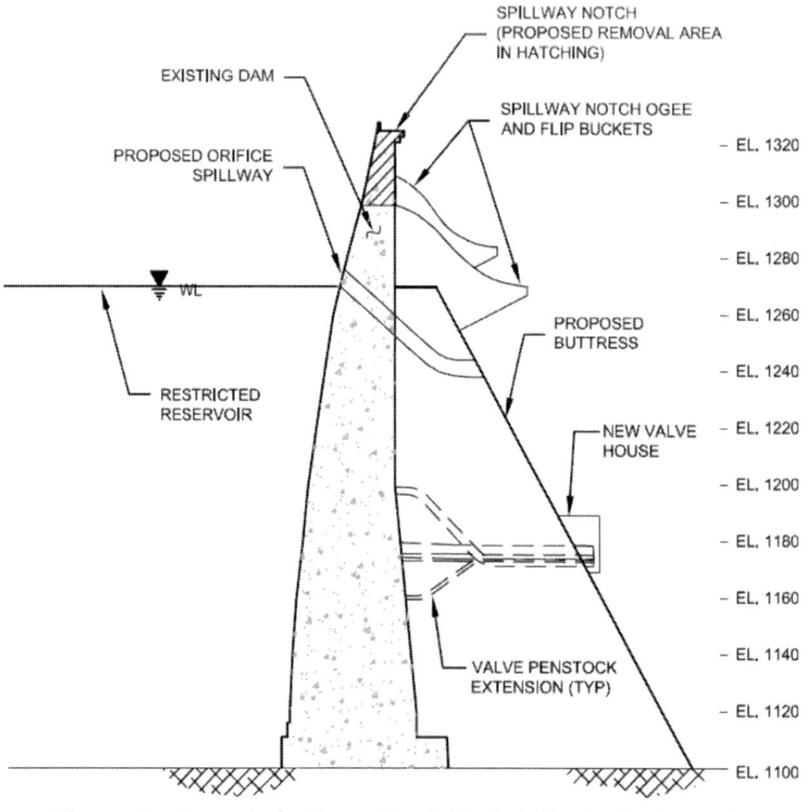

Figure 4 – Santa Anita Dam "Partial Rehabilitation" Alternative

The "Intake Riser Modification" (Figure 5) included construction of a notch spillway, as well as modifications to the existing sluiceway intake riser to maintain the aforementioned 94-foot permanent reservoir restriction to reduce risk of an uncontrolled release under seismic loading. The riser modifications included removal of the upstream gate that controls flow into the sluiceway, replacement of the gate onto the outside of the new sluiceway riser, and debris rack rehabilitations. Removal of the sluiceway gate assured that the facility would self-regulate its water surface elevation and maintain it at a safe level.

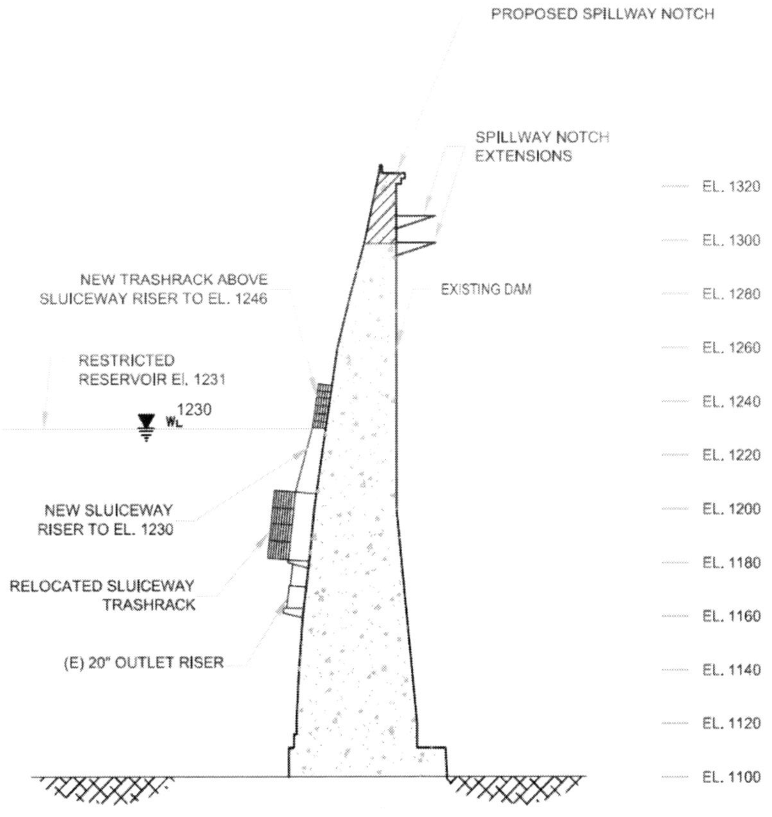

Figure 5 – Santa Anita Dam "Intake Riser Modification" Alternative

As part of the validation studies, a detailed FEM analysis of the "Intake Riser Modification" alternative with spillway notch and orifice was performed, including dynamic time-history analyses using the design "Maximum Credible Earthquake" (MCE) with peak ground accelerations of approximately 1.4g. Similar to the unmodified FEM analysis model, the stress output in the notched spillway showed zones of tensile stress that exceeded the dynamic tensile capacity of the concrete. It was observed that the stress distribution in the dam varied somewhat due to the presence of the notch (see Figure 6 for side-by-side comparison), but overall the pattern of peak stress was exhibited in the central portion of the arch near the crest of the dam and primarily limited to the horizontal arch direction. The zones of overstress above the established restricted reservoir elevation were investigated in a separate non-linear analysis. Results from the FEM analysis demonstrated overall dam stability but suggested that potential cracking in the dam during the MCE may occur in the upper elevations of the dam above the restricted reservoir level precluding a sudden or uncontrolled release of reservoir water. As a precautionary measure and in order to stabilize the portion of the arch above the notch, vertical tendon anchors were installed at a narrow vertical cantilever of concrete created on one side of the notch due to a contraction joint.

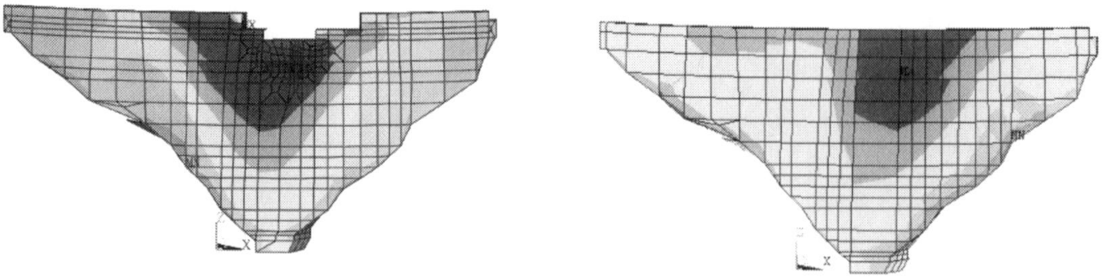

Figure 6 – Santa Anita Dam Finite Element Model – Comparison of Peak Tensile Arch Stress (red) during MCE for "Intake Riser Modification" (left) vs. Unmodified Dam (right)

REHABILITATION DESIGN

Following the 30% design of the intake riser modification concept, LACDPW completed the design of the "Intake Riser Modification" option (riser modification only, no notch) and hired a contractor to complete the modifications (Figure 7). By completing this portion of the modification concept, LACDPW satisfied regulatory criteria for seismic safety by establishing a permanent restricted reservoir at El. 1230. Construction of the new riser was conducted in parallel with reservoir sediment removal totaling 330,000 cubic yard (205 AF) to allow for lower reservoir operations and prevent plugging of outlets with sediment and debris.

Figure 7 - Riser Modification Construction

Subsequent to the riser modifications, rehabilitation design of the dam to address passage of the PMF was initiated. MWH's spillway notch sizing evaluations during the preliminary studies, which utilized standard hydraulic calculations and LACDPW's ResRoute reservoir routing software, provided the basis for notch dimensioning in the modification design, but did not sufficiently characterize the location and extent of the downstream impact area. Therefore, a 3D computational flow dynamics (CFD) model was developed for the dam, notch, and downstream areas to model the spillway discharge impact area. The 3D CFD model provided the ability to optimize the placement of the notch at a location where the spill trajectory was concentrated onto the most heavily armored locations, minimizing risk of excessive erosion of the abutments. In addition, the CFD model was used to validate and refine the notch dimensions (width and depth). A snapshot of the 3D CFD spillway discharge is shown on Figure 8.

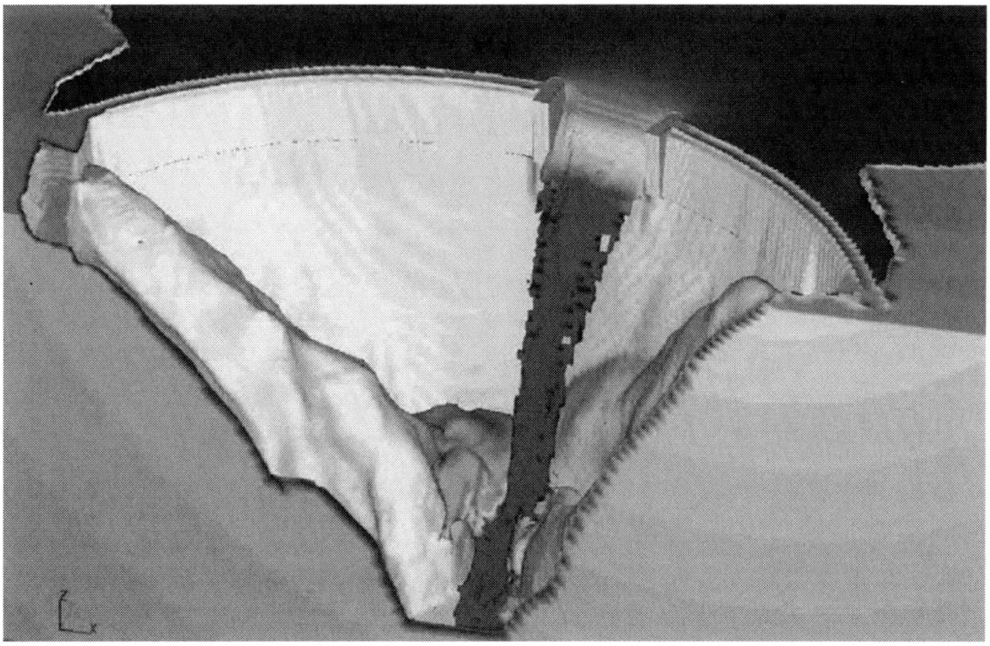

Figure 8 – Spillway Notch CFD Model Discharge

The final design of the notch included a 74-foot wide by 20-foot deep[6] notch cut near the center of arch, and installation of an ogee crest with flip bucket to eject spillway discharges away from the base of the dam (Figure 9). In order to achieve the new ogee and flipbucket shape, placement of about 20-feet of mass concrete into the notch is required. The ogee structure is attached to the dam with 10-foot dowels drilled at 1-foot on center and embedded 5-feet into the existing dam on both the upstream and downstream edges of the notch. Also, a pedestrian access bridge is installed across the notch to allow for maintenance access to both sides of the notch.

[6] The notch is cut vertically 40-feet to allow for construction of the 20-foot ogee and flipbucket structure.

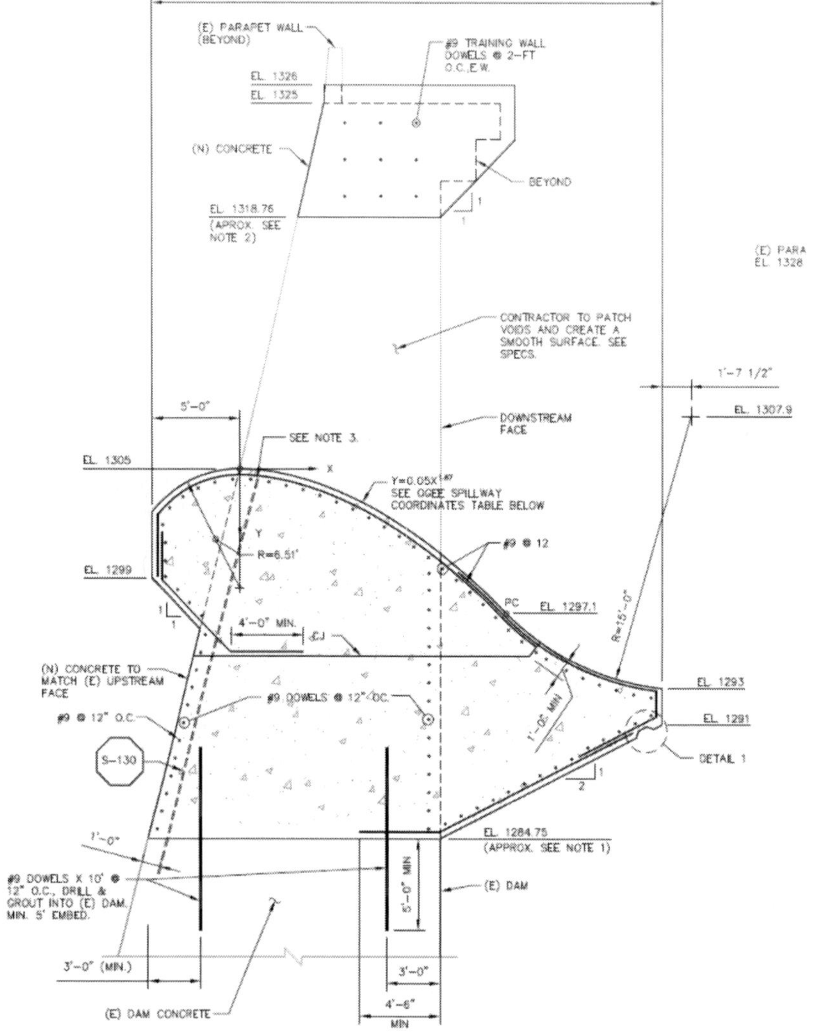

Figure 9 – Spillway Notch Section with Ogee and Flipbucket

While the primary goal of the project was to augment spillway capacity, LACDPW also included several additional design features into the rehabilitation scheme:

- Replacement of aging valves to restore flexibility to reservoir control.
- A new water system to replace the aging existing system.
- A new garage in the parking area for storage of maintenance equipment.
- Relocation of a hydraulic power unit from the dam crest to the control house for improved access.
- A new cableway hoist system at the dam crest with increased lifting capacity
- New generator and fuel supply tank to provide onsite emergency power.

- New electronic controls for improved, centralized control, including remote operation and monitoring.
- New lighting system onsite for improved visibility of site facilities.
- New catwalks and OSHA compliant stairs to improve safe access.
- New heliport for helicopter access to the site.
- Access road repaving and improvements.

SANTA ANITA CANYON SYSTEM OVERVIEW

To adjust for loss of water storage at Santa Anita Dam, LACDPW developed a suite of projects in Santa Anita Canyon to maintain flood control and maximize stormwater capture operations. In addition to Santa Anita Dam, the following other facilities are being modified:

- Santa Anita Headworks Division Structure
- Sierra Madre Spreading Grounds
- Santa Anita Spreading Grounds
- Santa Anita Debris Dam

As a result of this integrative approach, the State of California awarded a $20 million Proposition 1E grant to support projects at each of these facilities.

CONCLUSION

Final design of the Santa Anita Dam Spillway Modification Project (Project) was completed in the summer of 2016 and construction is expected to begin in early 2017. Completion of the design phase for this project concludes a multiple decade effort that started with initial evaluation and problem identification in the late 1970s, restarted with analyses and evaluations utilizing updated computer-based simulations in 1999 and the early 2000s, and progressed through a phased alternatives evaluation and design implementation through 2016.

The alternatives identified to address the updated MCE and PMF (if unaddressed, would respectively overstress and overtop the dam) were thoroughly evaluated and selected based on technical merit, constructability, and economic benefit. The Project final design included riser modifications to permanently restrict the reservoir elevation to address the MCE and installation of a spillway notch and ogee with flip bucket to pass the PMF. Upon completion of construction in 2017/2018, the rehabilitated dam will fully meet regulatory requirements for the MCE and PMF, as well as provide a modern facility with upgraded and remotely operated controls, new valves, backup power, improved emergency and maintenance access including a heliport, and energy efficient lighting.

A holistic approach was taken by Los Angeles County Public Works to upgrade and modify the system of flood control and water conservation facilities in Santa Anita Canyon. This included engaging stakeholders in the early stages of project development and evaluating various modifications to increase overall system resiliency and efficiency.

REFERENCES

IECO, 1977, International Engineering Company Inc. and Converse Davis Dixon Associates. Final Report On Investigation And Reanalysis Of Santa Anita Dam – Volumes 1 And 2, prepared for Los Angeles County Flood Control District, August, 1977.

Los Angeles County Flood Control District (1942), Report on Santa Anita Dam Model Studies, by M.F. Burke and E.S. Fuller, Sept. 28, 1942.

Los Angeles County Flood Control District (1978), Santa Anita Dam Spillway Modifications, June 1978.

State of California Department of Water Resources – Division of Safety of Dams (1980), Phase 1 Inspection Report for Big Santa Anita Dam, prepared for US Army Corps of Engineers – Sacramento District, National Dam Inspection Program, June 1980.

UNDERSTANDING THE PERFORMANCE OF AGING INFRASTRUCTURE: A 99-YEAR-OLD MULTIPLE ARCH DAM

Vik Iso-Ahola, P.E.[1]
Glenn S.Tarbox, P.E.[2]
Mohammadreza Mostafa, Ph.D., P.E.[3]
Sean Ellenson, E.I.T.[4]
Munit Bector, P.Eng.[5]
Ali Rasekh, Ph.D., P.E., P. Eng.[6]

ABSTRACT

Lake Eleanor Dam is a multiple arch reinforced concrete structure, built in 1918 within Yosemite National Park, California, owned and operated by the San Francisco Public Utility Commission. The maximum height of the dam is 61 feet above the river bed and its crest length is 1260 feet. Cracks, leakage, and concrete deterioration at the dam are evident and have driven the need for a comprehensive condition assessment. A detailed field inspection was performed in October 2015, which included concrete coring and testing, as well as innovative use of three-dimensional (3D) laser scanning that helped facilitate 3D modeling of the dam and established detailed 3D reference points for use in future monitoring and baseline comparison. Further, the site seismicity and Probable Maximum Flood (PMF) were updated using the latest methodologies, and the dam was analyzed using 3D Finite Element Analysis (FEA).

In order to provide updated earthquake ground motions for use in the seismic stability analysis of the dam, a site-specific seismic hazard assessment was performed to determine the Maximum Credible Earthquake. Given the complex yet repeating geometry of each arch bay, only a partial FEA model was developed of the three tallest arch bays in the dam. To achieve a realistic structural response of the partial model, the analysis required application of appropriate boundary conditions at the ends of the model in order to simulate the effects of the truncated longitudinal portions of the dam. A discussion of the applied boundary conditions is provided, including sensitivity analysis results and comparison to similar studies by others.

The dam inspection, performance history evaluation, concrete testing results, use of laser scanning, and updated PMF studies are also summarized in a discussion of the identified needs for extending the life of the historic structure.

[1] MWH Global, Walnut Creek, California, USA, vik.iso-ahola@mwhglobal.com
[2] MWH Global, Bellevue, Washington, USA, glenn.s.tarbox@mwhglobal.com
[3] MWH Global, Walnut Creek, California, USA, Mohammadreza.Mostafa@mwhglobal.com
[4] MWH Global, Walnut Creek, California, USA, Sean.Ellenson@mwhglobal.com
[5] MWH Canada, Vancouver, B.C., Canada, Munit.Bector@mwhglobal.com
[6] MWH Canada, Vancouver, B.C., Canada, Ali.Rasekh@mwhglobal.com

INTRODUCTION

Lake Eleanor Dam is located within Yosemite National Park, about 5 miles west of O'Shaughnessy Dam and 2 miles east of Cherry Valley Dam. It is located on Eleanor Creek at a point roughly 3 miles upstream from the confluence of Cherry Creek and 7.5 miles upstream from the point where the Cherry Creek meets the Tuolumne River. Lake Eleanor is hydraulically connected to Cherry Lake by the Eleanor-Cherry Tunnel (and the Cherry Pump Station) that is located north and upstream of the dam on the west side of Lake Eleanor. Lake Eleanor Dam was built in 1918 and is currently operated to provide water for power generation by transferring water to Cherry Valley Reservoir via the Eleanor-Cherry Tunnel that was constructed in 1952. The reservoir and dam, shown in Figure 1, are owned and operated by the San Francisco Public Utilities Commission (SFPUC).

Figure 1. Lake Eleanor and Lake Eleanor Dam.

The dam itself is a multiple arch reinforced concrete dam with a maximum height of 61 feet above the stream bed and a crest length of 1,260 feet. The central portion of the dam consists of 20 arches, each with a span of 40 feet, while the two ends of the dam consist of gravity sections and a concrete overflow spillway. Each arch is laterally braced with reinforced concrete struts between buttresses, and an 11-foot wide reinforced concrete bridge spans the length of the crest providing access across the dam and a platform for visual surveillance.

The upstream and downstream views of the dam are shown in Figure 2.

Figure 2. Upstream (Left) and Downstream (Right) views of Lake Eleanor Dam.

Over the years the dam's steel handrails, concrete rail posts, concrete bridge/roadway, and downstream horizontal struts have cracked and deteriorated. Also, there has been continual leakage through horizontal cold joints in the dam. Various repairs were made in 1996 to remediate these issues, including localized removal and patching of deteriorated concrete, sealing the upstream face of the dam with a 4-inch thick layer of shotcrete, applying shotcrete reinforced with anchor welded wire mesh on top of the arches, replacing the guard posts and handrails along the roadway, and repairing and resurfacing the roadway.

In late 2015, an investigation was undertaken to assess the current condition of the dam, including the following:

- Concrete coring program on the upstream and downstream faces to assess deterioration and update concrete strengths
- Gathering data and analyzing the Probable Maximum Precipitation (PMP) to update the Probable Maximum Flood (PMF)
- Assessment of the current surveillance and monitoring plan
- Laser scanning of the entire structure
- Seismic hazard assessment to determine the Maximum Credible Earthquake (MCE)
- Condition assessment of the dam for various loading scenarios (operational, seismic and PMF) using finite element analysis
- Development of a needs assessment report

FIELD INSPECTION

In order to develop an assessment of the changing conditions of the dam an extensive field inspection was performed, including visual inspection, limited sampling and testing of the concrete in the dam, 3D laser scanning, and a review of the current surveillance and monitoring plan. The information was obtained to provide inputs and refinements for the structural analysis and to provide recommendations for future capital improvement projects that would extend the life of the dam.

The field inspection, 3D scanning, and concrete coring activities were performed in October 2015. The field inspection consisted of a two-day site walk to investigate the dam crest, upstream and downstream faces of the arch barrels, buttresses, spillway, outlet works and valves, and foundation. Field notes, photographs, personnel interviews with the Watershed Keeper and Dam Safety Engineer, and cracking patterns visible by 3D laser scans were recorded, evaluated and used for this study.

Visual Inspection

A detailed visual inspection was performed in accordance with all requirements outlined in the FERC Part 12 guidelines. While the SFPUC reservoirs on the Tuolumne River do not fall under FERC's jurisdiction, all FERC inspection guidelines were followed as general good practice.

During the observation, all cracks in the foundation and abutments, spillway, struts, buttresses and the downstream face of the dam were mapped and compared to a crack mapping study performed in 1994. The upstream face of the dam has since been coated with a layer of shotcrete and thus the inspection could only gather new data on cracks and pitting found in the outer shotcrete layer.

Figure 3. Efflorescence and minor cracking found in downstream face of arch.

Since the reservoir was dewatered during the inspection, no data regarding the seepage through the shotcrete was collected during the October 2015 site visits. However, SFPUC inspection photos and videos from June 2015 were provided for review. These were reviewed for comparison against observations made during the field inspection and it was noted that seepage through lift joints is common when the reservoir is filled. While dam seepage exists, it was generally found to be consistent with previously documented observations (i.e., no significant changes observed). Future investigations and mitigation plans will be performed regarding the observed seepage of the dam.

The visual inspection observed varying degrees of deterioration of the dam. The inspection found delamination, minor crack growth and efflorescence throughout many of the concrete arches (Figure 3). The horizontal struts were found to be in generally poor condition, with rebar exposure, deterioration and crack development, and corrosion of the reinforcing steel within the members. The bridge deck contained many cracks, pitting and delamination of the bridge girders, along with corrosion and exposure of the reinforcing steel (Figure 4). The foundation was found to be sound and stable, with some scour and undermining of minor structures.

Figure 4. Deterioration of walkway and exposed reinforcement (left) Deterioration of spillway buttress (right)

Concrete Coring and Compressive Tests

In order to best approximate the condition and strength of the concrete for use in the structural analyses, concrete coring and sampling was performed at nine locations on the dam. Upon completion of coring activities, the cores were transported to a laboratory to perform concrete uniaxial compression testing and petrographic analysis on the concrete cores in accordance with ASTM C 42 & ASTM C 856, respectively. Nine representative core samples were selected based on depth and suitability for testing. The aggregate size of three core samples were greater than 40% the core diameter, thus were considered invalid per ASTM standards. The core testing results are shown below in Table 1.

Table 1. Concrete core testing data.

Core	Test	Comp. Strength (psi)
Upstream 1	Compression	3770
Upstream 2	Compression	4560
Upstream 3	Petrographic	N/A
Upstream 4	Compression	4220
Downstream 1	Petrographic	N/A
Downstream 2	Compression	4610
	Average	4290

Based on a prior conceptual engineering study performed in 1993, the average uniaxial compression strength for the concrete was reported as 3600 psi. The average value presented in Table 1 is about 20% higher, which is likely due to exclusion of several test samples in the average (the average with *all* samples is 3,457 psi). Given the wide range of the 1993 testing results (2010 psi to 6620 psi), it is suspected that core sample test results with aggregate sizes larger than 40% of the core diameter were inappropriately included.

The petrographic analysis test results found that no deleterious reactions were detected (i.e., no direct evidence of alkali-silica reactivity) and that the concrete was sound and of good quality.

3D Laser Scanning

3D laser scanning was performed in order to capture a high degree of detail on the upstream and downstream faces of the dam and provide reference information to supplement visual inspection. The 3D scanning consisted of a Leica Scan Station P30 Terrestrial Stationary LiDAR Scanner set up within each arch bay of the dam on the downstream, on the dam crest, and in the drained reservoir for scans of the upstream face. The 3D scan mapped the dam at ¼-inch accuracy. Color HD photos were integrated within each scan and were mapped onto the resulting laser scan "point cloud" of 3D scanned points on the dam surface.

The captured "point cloud" data was processed and drafted as a 3D model in AutoCAD, transferred and processed further in Inventor, and then imported into the 3D finite element structural model of the dam (Figure 5).

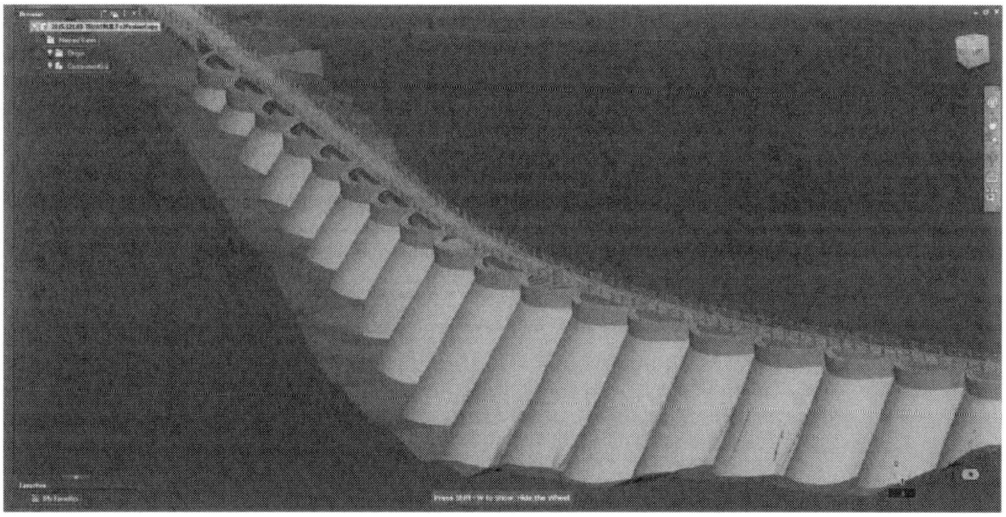

Figure 5. 3D AutoCAD Model of Lake Eleanor Dam
(developed from point cloud data).

Surveillance and Monitoring Assessment

The dam surveillance and monitoring program (SMP) was reviewed during the field inspection in order to assess the safety of the dam for continued operation. The dam surveillance program consists of routine inspections, annual inspections, five-year inspections, and special inspections after unusual events (e.g., earthquakes or floods). The documented procedures in the SMP were found to be clear and comprehensive for monitoring the behavior of the dam. The procedures consist of standardized forms and follow practices conducted consistent with other dam owners and recommendations provided by the United States Society on Dams (USSD).

PROBABLE MAXIMUM FLOOD

No PMF studies had been previously prepared for the Lake Eleanor Dam. The Probable Maximum Flood (PMF) for Lake Eleanor Dam was developed for the site in order to better understand the worst case loading that can be experienced by the dam and other appurtenant structures. The PMF results from the Probable Maximum Precipitation (PMP) and a complex set of other factors that affect runoff including snowpack, factors affecting snowmelt, unit hydrographs, loss rates, initial reservoir levels, and the spatial and temporal distribution of precipitation over the watershed. The PMF was determined on a monthly basis to find the maximum reservoir water level. The studies are generally based on the methodology provided in the most recent FERC guidelines available for PMF determination. The following are the highlights of the PMP/PMF study:

- The maximum PMP depth occurs during the November through March period.
- On Average, the 24-hour general storm PMP was 23.88 inches and the 72-hour general storm PMP was 38.82 inches.

- The local storm (thunderstorm) PMP was 5.61 inches for 6 hours, but it did not result in the critical PMF inflow condition.
- The estimated maximum PMF water level occurs in December.
- The peak PMF inflow at Lake Eleanor Dam is estimated at 70,659 cfs, which would result in a maximum reservoir level at El 4,668.4 feet.
- The peak outflow at Lake Eleanor Dam is estimated at 69,882 cfs.
- The peak water level during the PMF at Lake Eleanor Dam would be 6.4 feet above the crest of the dam.

Based on the results of the PMF update, it was determined that the spillway capacity is inadequate to pass the PMF by over an order of magnitude, therefore subjecting the dam to overtopping flows. Given that the dam is subject to overtopping events on an approximate 10 to 20 year frequency, the study warranted the need to address the potential for foundation erosion and risk to the dam foundation.

SEISMIC HAZARD ASSESSMENT

A site-specific seismic hazard assessment was performed to update the Maximum Credible Earthquake (MCE) and associated ground motions for Lake Eleanor Dam. There had been no previous seismic study performed at this site. Site-specific time histories were developed based on the updated ground motions, which were used in the dynamic finite element analyses performed as part of the stability evaluation of the dam.

The seismic hazard assessment included seismic source characterization and a deterministic seismic hazard analysis (DSHA). The MCE was selected based on the results of the DSHA and the guidelines set forth by the California Department of Water Resources, Division of Safety of Dams (DSOD). Time histories were selected by performing a search of the Pacific Earthquake Engineering Research (PEER) Center's Strong Ground Motion database and plotting to determine spectral shape. Three recorded motions were selected for spectral matching to the MCE ground motions and one representative motion was used for the finite element analysis of the dam.

Time histories were developed for the 84th percentile minimum earthquake based on the parameters defined in the DSHA. Representative earthquake ground motions were selected from the PEER database and then modified to match a magnitude 6.25 earthquake event located at a distance of 15 km with a duration of 14 seconds for an assumed VS30 of 569 m/s. A total of three sets of three component (fault normal, fault parallel, and vertical) spectrum-compatible time histories were developed for the minimum earthquake event. The site specific acceleration time histories are shown below in Figure 6.

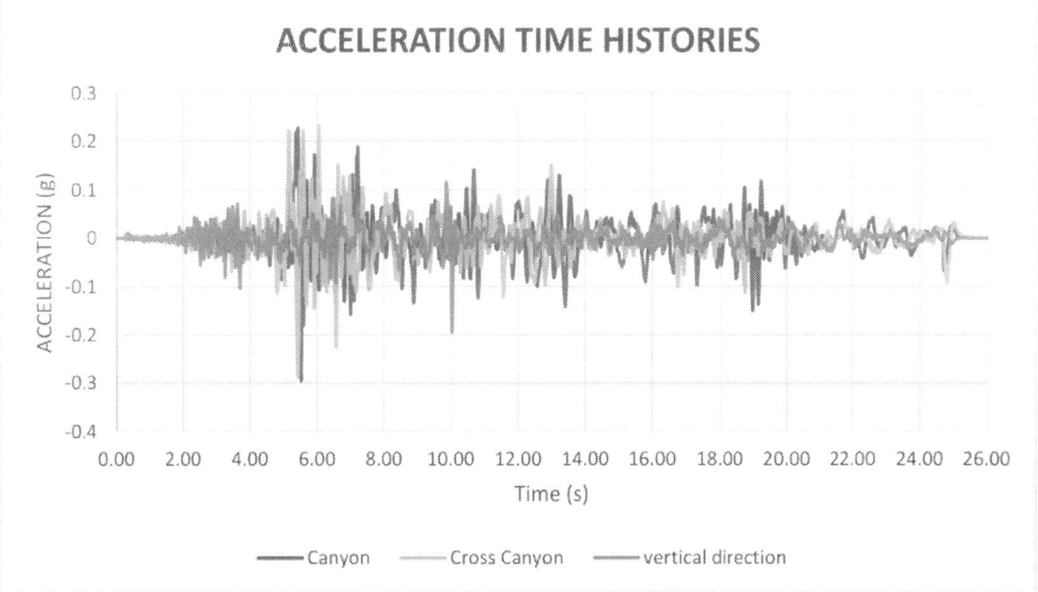

Figure 6. Acceleration Time Histories used for the SHA at Lake Eleanor Dam.

The time histories developed for the Lake Eleanor project were truncated to reduce run time on the seismic deformation analysis. The truncation involved removing approximately 2 to 3 seconds at the beginning of each time history and removing approximately 35 to 50 seconds at the end of the time histories (depending on the ground motion being considered). The truncation was done in a way such that the energy generated by the ground motion was preserved. The Arias intensity (energy of the ground motion) for each component of the truncated time histories was checked to verify that there was no significant difference from the originally developed time histories. The 5-95% significant duration (duration over which 5% to 95% of the Arias intensity occurs) for each component was also checked to verify that there was no significant difference from the originally developed time histories. The time histories were baseline corrected after truncation.

STATIC AND DYNAMIC ANALYSES OF DAM STRUCTURE

A 3D finite element model was developed to perform the static and dynamic analyses for usual, unusual, and extreme loading conditions for the dam. The analyses included the loading derived from the PMF update and earthquake from the seismic hazard assessment performed as part of this study. USBR and USACE guidelines were used as references for the stress evaluations of the concrete (USBR 1981, 2006; USACE 1981, 1994, 2003).

Model Setup

The three-dimensional static and linear dynamic finite element analyses of Lake Eleanor Dam were performed using the computer program ANSYS® Workbench™ V16.2.

ANSYS is a state-of-the-art commercially available general finite element method (FEM) program, which is widely used as a research and development tool in a number of civil/structural engineering applications, including static and dynamic stability evaluations of dams.

As previously mentioned, the geometry of the dam was extracted from the 3D scanning data gathered during the field inspection, which was processed to develop a 3D AutoCAD file. This 3D AutoCAD model was exported into Autodesk® Inventor™ 2015 3D modeling software for additional geometry processing and then into ANSYS, which was subsequently used as the basis for creating the FEM mesh of the dam.

Due to the repeating nature of each multiple arch bay, it was possible to model only a segment of the dam and produce accurate results. Based on work by USBR on the Bartlett Dam, it was discovered that a "sliced" model yielded acceptable stress results compared to a full model when the boundaries allowed some flexibility in the cross-canyon direction (Nuss, 1995). For our analysis, only the tallest three full arch bays and two half arches and the associated buttresses and struts were modeled (Figure 7). The bridge at the crest of the dam and the struts between buttresses on the downstream were also modeled to represent additional stiffness in the cross-valley direction of the dam. Sensitivity analyses were performed to measure the effects of constraining motion at the boundaries, which is discussed further in the following section.

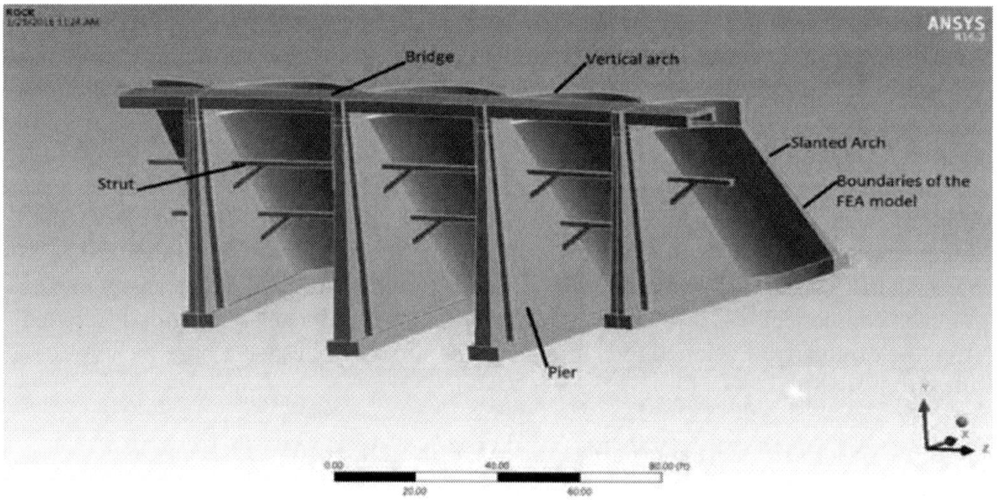

Figure 7. 3D View of the Dam and Rock, Downstream View.

To represent the dam foundation, a large massless body was modeled with dimensions per guidelines in reference 9 (USACE, 2003). In the absence of available test data for the rock properties of the foundation, the modulus of deformation of the rock (E_f) was approximated based on low-end values for the plutonic (mostly granitic) rock found in the area. The mechanical properties of the concrete (E_c) were based on the concrete core test data. With an Ef/Ec = 0.39 the foundation was assumed to be flexible, and modeled to extend at least two times the dam height in the upstream, downstream and cross canyon direction, shown in Figure 8 and Figure 9.

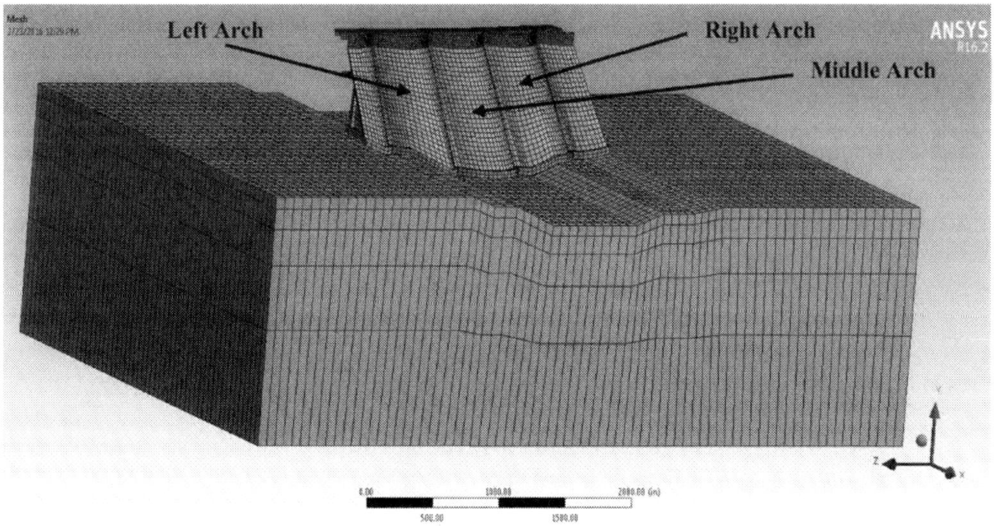

Figure 8. 3D View of the Dam and Rock, Upstream View

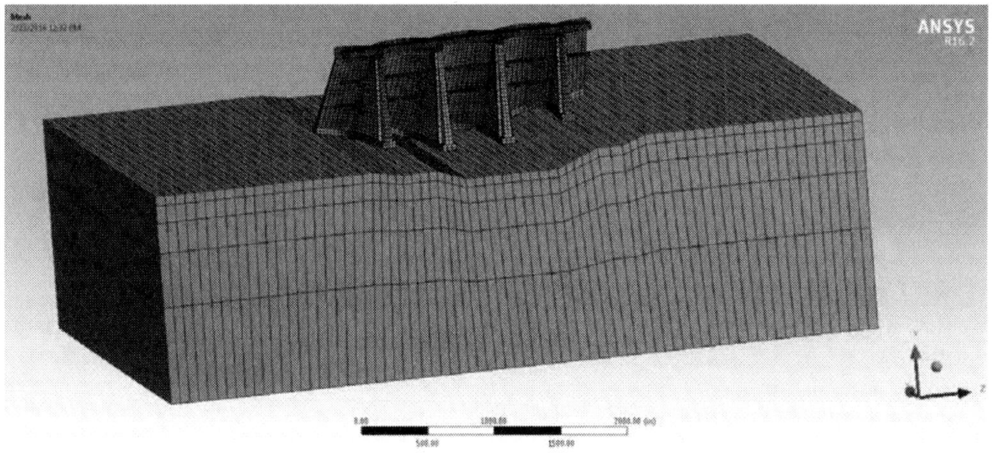

Figure 9. 3D View of the Dam and Rock, Downstream View.

The material properties for the concrete and rock were assumed to remain in the linear regime throughout all the analyses. Due to the cracks on the bridge and struts surfaces, their stiffness values were reduced down to 25% per ACI-318. Also, all the nodes of the dam and the rock were bonded together and no opening of joints at contact surfaces (e.g., dam-foundation contact, lift joints) were considered. This was based on the assumption that the stress level at the joints were smaller than the strength of the joint and nonlinear behavior would not occur. This assumption requires that for all load cases, the demand stresses remain below the capacity of the structure, as opposed to a nonlinear analysis which allows for opening of joints and cracking of concrete.

Model Boundary Conditions

Given that a partial model of the dam was developed, the response of the dam is sensitive to the boundary conditions in the finite element model. Based on the work for Bartlett Dam (Nuss, 1995), a set of two boundary conditions (run within two separate FEA models) were considered, where:

a. The half arch edges of the dam (edges shown in Figure 7, labeled as "Boundaries of FEA model") are free to move relative to the foundation in the cross valley direction. On the other hand, their rotation is constrained through use of a remote boundary. In this case all three components of seismic velocities are applied to only the boundaries of the foundation.
b. The half arch edges of the dam are completely constrained against relative movement (with respect to the foundation) in the cross valley direction and also rotation. This is achieved through application of the cross-valley seismic velocities to the sliced faces of the dam while at the same time all three components of the seismic velocities are applied to the boundaries of the foundation.

The more conservative boundary condition from above (boundary condition "a") was used and results are presented herein. Given that the boundary conditions only simulate the behavior of the entire dam, it is noted that the actual behavior of the dam is expected to be somewhere between the two mentioned boundary conditions. Also, given that the foundation rock is considered massless in the FEA model and includes a displacement boundary condition (i.e., no non-reflecting boundaries), the results are considered conservative, since this type of foundation model will reflect seismic waves which can increase stresses in the model 20 to 30-percent.

Load Cases

Following reference 9 (USACE, 2003), the analyses were performed for usual, unusual, and extreme loading conditions as detailed in Table 2.

Table 2. Finite Element Analysis Load Cases.

Load Case	Loading Type	Description
Usual	External Hydrostatic Pressure	Normal Maximum Reservoir W.S. El. (Head = 62 ft.)
	Gravity	Self-Weight of Concrete (32.2 ft/s^2)
	Added Mass	Shotcrete (150 lb/ft^3)
Unusual	External Hydrostatic Pressure	PMF Reservoir W.S. El. (Head = 67.4 ft.)
	Gravity	Self-Weight of Concrete (32.2 ft/s^2)
	Added Mass	Shotcrete (150 lb/ft^3)
Extreme	Seismic	MCE (M_w = 6.25)
	External Hydrostatic Pressure	PMF Reservoir W.S. El. (Head = 62 ft.)
	Gravity	Self-Weight of Concrete (32.2 ft/s^2)
	Added Mass	Shotcrete (150 lb/ft^3)

The weight of the foundation was not included in this analysis, as the foundation was assumed to be massless. In addition, silt and snow/ice loading was neglected in the model, as no significant silt was observed at the site.

Finite Element Analysis Results

The results for the static analyses (usual and unusual) indicate that the computed demand forces will not exceed the estimated tensile and compressive capacities of the concrete in the dam, which were calculated using the factors of safety recommended by FERC for arch dams (FERC 1999). The envelope of maximum principal stresses for the usual load case is shown below in Figure 10.

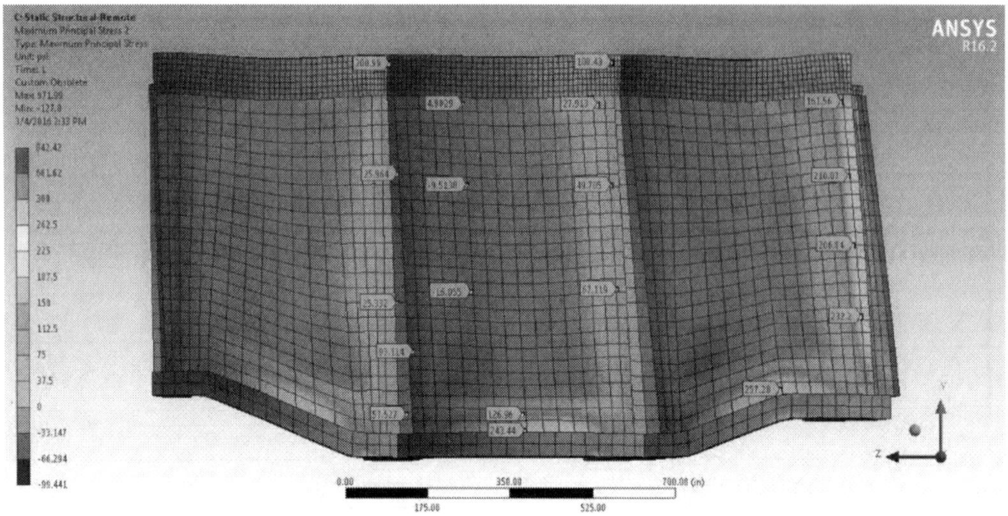

Figure 10. Usual Case. Principal Stresses on the Upstream Faces.

The analysis results for the bridge and struts considering the extreme (MCE) load case indicate that the internal forces (shear, moment-axial interaction) are estimated to remain within the capacity of the reinforced concrete. The analyses also indicate that the buttresses are estimated to be adequate against the forces induced by the MCE event (Figure 11).

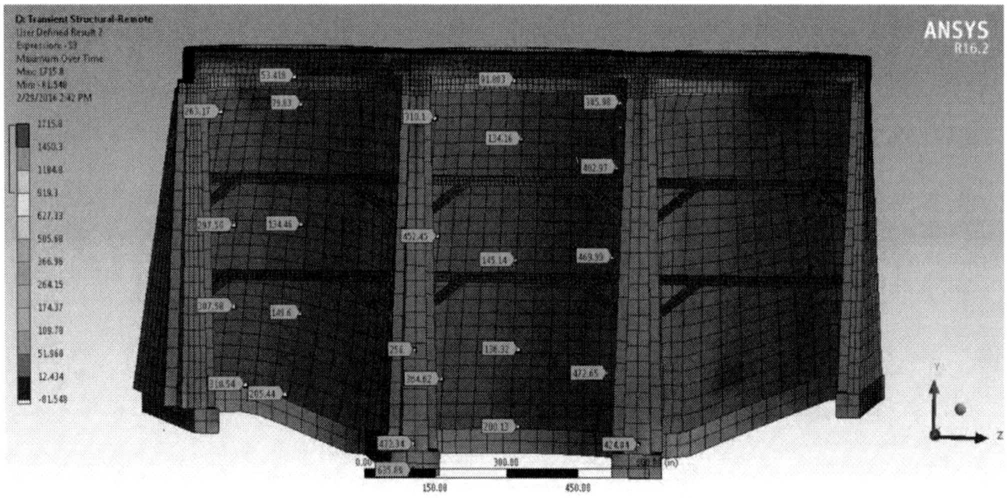

Figure 11. MCE. Envelope of compressive stress on the downstream arches and buttresses.

The results of the analyses for the extreme load case (Figure 12) indicate that the estimated tensile capacity of the concrete of the dam will not be exceeded (e.g., DCR < 1.5), considering the factors of safety set forth by reference 9 (USACE, 2003). Due to the effects from the boundary conditions applied to the model and the complicated geometry within the dam (intersections of arches, bridge, and buttresses near the crest), some stress concentrations occurred in the FEA model in localized areas. Since these areas are limited in extent and their potential overstress is very brief and superficial, and considering the fact that the dam is reinforced (which adds to the capacity of the concrete), it was concluded that no significant damage would occur to the dam under the MCE earthquake loading, or pose the threat of uncontrolled release of the reservoir during and after the MCE earthquake approximated for this study.

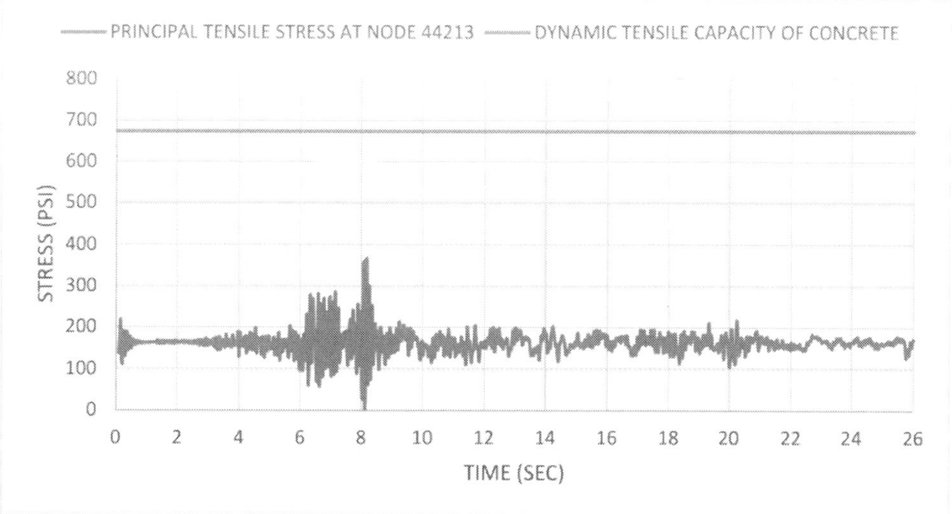

Figure 12. MCE - Maximum Principal Stress Time History of node in Dam Body.

The effect of the bridge and struts on the seismic performance of the dam was also investigated by varying their stiffness in the analyses. The analyses showed that the struts do not have significant impact on the cross-canyon response of the dam due to their size and location within the arches, while the bridge serves a crucial role in the cross-canyon stability of the dam during an MCE event.

Due to the criticality of the bridge in the seismic evaluation, a separate analysis was performed analyzing the bridge with AASHTO loading (AASHTO 2010). The analysis found that the bridge was not capable of carrying the vertical loads defined in AASHTO. Moreover, the surface cracks at the intersections of the deck with the buttresses provide an indication of distress on the bridge due to vertical loads, yet smaller than those prescribed by AASHTO. Future evaluation and retrofitting of the bridge was found to be necessary to ensure the longevity of the dam.

INSPECTION RESULTS AND IDENTIFIED NEEDS

Utilizing the results of the field inspection, laboratory testing, seismicity update, PMF development, and finite element analysis, a needs assessment evaluation and report was provided to identify key components of the dam that required upgrade and/or further evaluation. It was determined during the assessment that semi-regular flood overtopping events leading up to and including the PMF present the highest risk to the integrity of the dam. Also, elimination of leakage through cracks, bridge and strut repair, and valve upgrade were identified as needs to be addressed to extend the life of the aging facility.

In order to address the identified needs, the study provided several options to be considered for upgrades or refurbishment of the facility, including:

- Geomembrane lining of arch faces to eliminate leakage through the arch barrels
- Repair of the bridge on the dam crest to avoid future deterioration and overload from vehicular access, as well as to preserve the stability of dam during the MCE event (critical for seismic performance).
- Replacement/upgrade of existing outlet valves as well as the valve operating equipment, including provision for future remote operation.
- Increased spillway capacity to limit overtopping during flood events (capacity need to be determined after performance of a dam break study and determination of the Inflow Design Flood, which may be less than the PMF).
- Concrete lining of the downstream rock foundation adjacent to the dam to prevent scouring during overtopping events, which would undermine dam stability.

SUMMARY

Based on the intensive field inspection along with the static and dynamic analyses of Lake Eleanor Dam, it was concluded that the dam is in good condition, with some minor areas for improvement. The field observations, review of the surveillance and monitoring program, PMF development, seismic hazard update, and finite element analysis of the structure were used to develop a needs assessment that identified modifications required to extend the life of the 99-year old multiple arch dam.

The static and dynamic analyses of the structure found that the dam is stable and is able to withstand any usual, unusual or extreme seismic loading scenarios. The analysis utilized a truncated model, which was evaluated by performing sensitivity analyses to understand the change in structural response by varying bridge and strut stiffness, model geometry, and boundary conditions. The truncated model saved computational time during the study and allowed for faster iterations during the sensitivity evaluations.

SFPUC will be conducting in the near future the next phase of evaluation to address the needs at the dam in an alternatives analysis study. This study will be performed to select alternatives that best address the needs based on operational, economic, environmental, and constructability considerations.

ACKNOWLEDGEMENTS

The authors would like to thank the project owner, SFPUC, for its gracious support in the development of this study.

REFERENCES

1. American Association of State Highway and Transportation Officials (AASHTO), 2010, LRFD Bridge Design Specification, Fifth Edition, 2010.

2. American Concrete Institute (ACI), 2012, ACI 318-11 Building Code Requirements for Structural Concrete and Commentary, Michigan, 2012.

3. ASTM C 42- Standard Test Method for Obtaining and Testing Drilled Cores and Sawed Beams of Concrete.

4. ASTM C 856- Standard Practice for Petrographic Examination of Hardened Concrete.

5. Federal Energy Regulatory Commission (FERC), 1999, Engineering Guidelines for the Evaluation of Hydropower Projects, Chapter 11 - Arch Dams, 1999 .

6. Nuss, Larry, 1995, Comparison between complete and slice finite element models of a multiple arch and Buttress dam, ASCE, Water Power 95, San Francisco, 1995.

7. United States Army Corps of Engineers (USACE), 1994, ETL 1110-2-365 Nonlinear, Incremental Finite Element Analysis of Massive Concrete Structures, August 31, 1994.

8. United States Army Corps of Engineers (USACE), 1997, ETL 1110-2-542 Thermal Studies of Mass Concrete Structures, May, 30, 1997.

9. United States Army Corps of Engineers (USACE), 2003, EM 1110-2-6051 Time-History Dynamic Analysis of Concrete Hydraulic Structures, 2003.

10. United States Bureau of Reclamation (USBR), 1981, EM 34, Control of cracking in mass concrete structures, 1981.

11. United States Bureau of Reclamation (USBR), 2006, State-of-Practice for the Nonlinear Analysis of Concrete Dams at the Bureau of Reclamation, 2006

12. Zangar, C.N., 1952, Hydrodynamic Pressures on Dams Due to Horizontal Earthquake Effects, U.S. Department of the Interior, Bureau of Reclamation, May, 1952.

A PERFORMANCE-BASED EVALUATION OF POST-TENSIONED ANCHORS EMBEDDED WITHIN A CONCRETE GRAVITY DAM

Tae Ha Park[1]
Samantha Hoang[2]
Jonathan Lum[3]
Ziyad Duron, PhD[4]

ABSTRACT

Post-tensioned anchors are placed within the part of a dam where an operator considers a structural reinforcement to be necessary. Various methods have been developed to determine anchor functionality; however, they tend to be costly, potentially dangerous and destructive to the anchors. A non-destructive method is proposed to evaluate the status of post-tensioned anchors based on Performance-Based Testing (PBT). A Cold Gas Thruster (CGT) is placed on a concrete gravity dam, and the broad-band, transient response to high-amplitude, short-duration impulsive load is measured near the location of anchors. The same data is taken on non-anchored blocks as well in order to differentiate anchor-specific dynamic behavior from that of the monolith.

In order to characterize the status of the anchor, a high frequency range is targeted for the investigation. Spectral analysis is used to identify the different anchor modes excited, and a simple, continuous model of an anchor is constructed to confirm the observed behavior. A match between the data and the model is achieved within 5 percent. Analysis techniques are described for developing a preliminary inspection guideline and a performance indicator to monitor anchor status in a dam. Limitations to the current approach are discussed, and a better procedure for data acquisition is proposed to ensure more consistent data on the anchored blocks.

INTRODUCTION

The failure of a dam can be catastrophic to the safety of the population situated downstream due to the sheer size of a dam and the amount of water it holds. Therefore, it is imperative to monitor and prevent all dam failure mechanisms. Some potential failure modes include the overturning or downstream sliding of dam monoliths due to either a structural vulnerability to water pressure under normal operation or seismic disturbance (e.g., Brown 2015). In order to reinforce its structural weakness, post-tensioned anchors are placed within the part of a dam where failure is most likely to occur. Theoretically, tensioned anchors that connect the ground and the top of the dam are expected to better immobilize the monoliths and mitigate the vibrational amplitude due to the earthquake. However, there has been few field studies that demonstrate the effectiveness of anchors as a preventive device in the event of earthquake. Additionally, past field studies regarding anchor functionality have been limited to lift-off testing (e.g., Heslin et al.

[1] De Pietro Engineering Fellow, Harvey Mudd College, Claremont, CA, 91711, tpark@hmc.edu
2 De Pietro Engineering Fellow, Harvey Mudd College, Claremont, CA, 91711, shoang@hmc.edu
3 De Pietro Engineering Fellow, Harvey Mudd College, Claremont, CA, 91711, jlum@hmc.edu
4 Professor of Engineering, Harvey Mudd College, Claremont, CA, 91711, duron@hmc.edu

2009), in which a hydraulic pump is used to measure the load at which the wedge plate and the bearing plate are separated. Lift-off testing can measure the actual load of the anchors, but it can be dangerous and destructive, since the load capacity of corroded anchors are not known, and applying the force larger than the capacity load can potentially break them. The other method under study is to investigate the dispersive wave behavior (e.g., Poiroux, 2011) which involves tapping the exposed portion of an anchor to deduce the tension of the embedded trunnion anchor.

In this work, a non-destructive method is developed to evaluate the anchor functionality of Shaver Lake Dam from the Performance-Based Testing (PBT). PBT, explained in detail by Goldkamp et al. (2014), incorporates a Cold Gas Thruster (CGT) to induce high-amplitude, short-duration impulse-like load to the dam. The impulse response measured on both anchored and non-anchored blocks can be analyzed and compared to differentiate anchor-specific characteristics in frequency domain. It will be shown that the difference in responses between anchored and non-anchored blocks is detected in high frequency range beyond 100Hz. For such high frequency behavior represented by a number of steel anchors embedded within concrete monoliths, the finite element model can be extremely laborious to build and may present inaccurate results. Therefore, a simple continuous model composed of Euler-Bernoulli beam under tension and a tip mass is studied in order to validate the field studies. These dynamics responses are used to analyze two factors that can significantly undermine the performance of anchors in case of need: loss of tension and corrosion.

OVERVIEW OF SHAVER LAKE DAM PBT STUDY

The data set studied in this paper was obtained during the summer of 2015 on Shaver Lake Dam in Fresno County, California. Shaver Lake Dam is a concrete gravity-arch dam, and 45 post-tensioned anchors were installed in December 2003 at each end of the gravity sections, specifically in Blocks 1 through 5 and Blocks 27 through 35 (Figure 1). Blocks 1 to 5 are at the right side of the dam when looking downstream, and Blocks 27 to 35 are at the left side. Figure 1 also shows where the tri-axial accelerometers were placed. Specifically, two accelerometers are placed right across the gap between the adjacent monoliths, so there are total 6 locations across three gaps between blocks 25, 26, 27, and 28. In this paper, the location of the accelerometer will be noted as the following: if the accelerometer is located on Block 26, but it is placed right next to the gap facing Block 27, the location will be termed "Block 26 near Block 27."

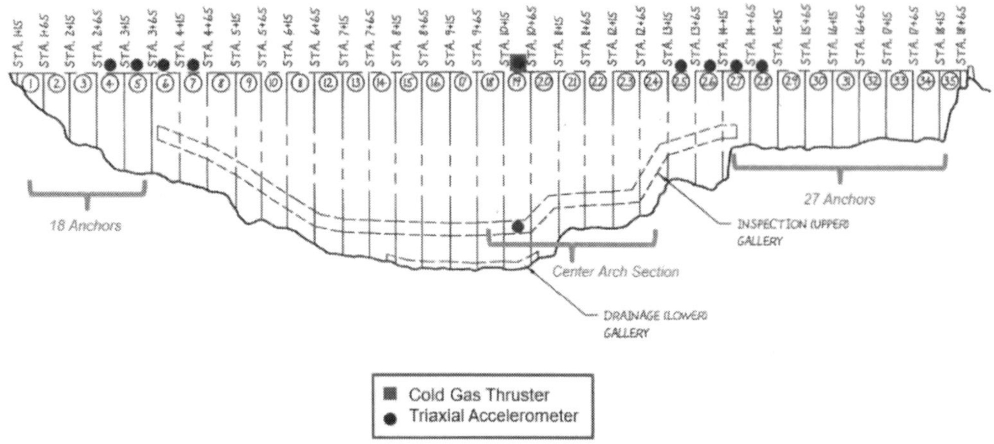

Figure 1: Side-view of Shaver Lake Dam and the location of CGT and accelerometers

In this data set, the responses of anchored blocks are measured *globally*. In the present anchor study, the term *global* refers to the testing set-up in which the CGT is placed at the center of the arch-section of the dam to excite the entire dam, and the responses to such *global* excitation are recorded at each of the different locations. Therefore, the important assumption underlying the following analysis is that the *global* CGT impulse can effectively "pluck" the anchors embedded in concrete dams and that the accelerometers are able to capture the anchor-specific responses to this *global* input from the CGT.

Another important assumption in this approach is that the dynamic responses to CGT impulse will indicate if the monolith and embedded anchors behave as one composite system or two separate systems. While the dynamic behavior of steel anchors embedded within concrete monoliths is not known, it can be assumed that if they behave as one system, anchored monoliths will exhibit higher frequency than do non-anchored monoliths, as anchors should make the monoliths "stiffer." On the other hand, if anchors and monoliths behave as separate systems, we can expect the anchored blocks to exhibit the separate responses of both the anchor and the monolith.

SPECTRAL ANALYSIS

The acceleration data reviewed herein was collected in both unanchored and anchored monoliths in the left side of the dam. Data were captured using a sampling rate of 10,000 Hz after being filtered using a 4th order Butterworth low pass filter with a cut-off frequency of 1,000 Hz. In order to characterize the dynamic responses of different monoliths, power spectrum density (PSD) plots were constructed from all acceleration data on the left side of the dam (i.e. Blocks 25, 26, 27, and 28). To ensure accurate result and minimized noise, Welch's method with a Hanning window was applied. In Figure 2, the PSDs of a non-anchored monolith and anchored monolith on the left side of the dam are compared. Note that both monoliths have the highest peak in the range of 170 -180Hz in stream direction, but the anchored monolith (right) has other high peaks of comparable amplitudes at 243Hz and 268Hz.

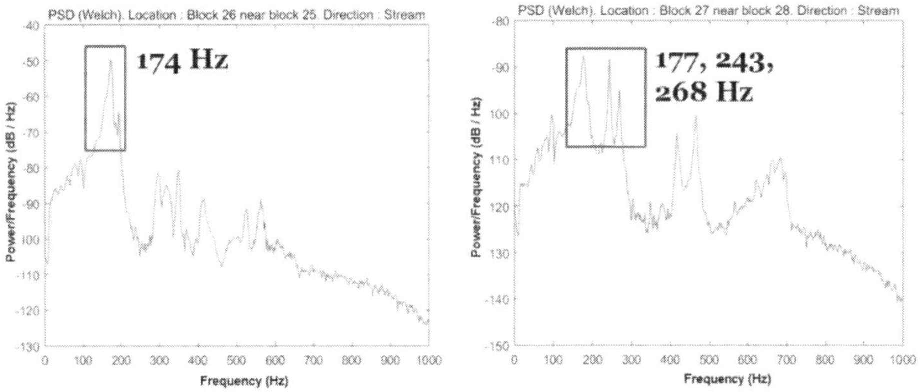

Figure 2: Power Spectral Density (PSD) of stream-direction acceleration data from Block 26 near Block 25 (non-anchored) and Block 27 near Block 28 (anchored)

At first glance, such high frequency peaks seem to contradict the study done by Southern California Edison Company Dam & Public Safety Group (2015), in which it is shown from the shaker tests in '95 and '09 that the fundamental resonance of the Shaver Lake Dam is in the range of 6 to 7 Hz. However, it is assumed that the wide peak around 175Hz is the response of the entire dam excited by the CGT impulse, and the two higher peaks in anchored monolith are specific to anchors in Block 27. This explanation will support the claim that the anchors are a separate system from the monoliths. Table 1 lists all such peaks from all accelerometers on the left side of the dam. It is clear that most locations share the same frequency near 175Hz, but the anchored monoliths exhibit two extra higher peaks, which could be anchor responses specific to different dimensions of anchors in each block. Notice that the response at Block 26 near Block 27, which is the response of the unanchored monolith at the joint adjacent the anchored monolith, show similar trend as anchored monoliths. This could be the response from anchors in Block 27 that has not attenuated much across the gap between Block 27 and Block 26.

Table 1: Frequency peaks of comparable amplitudes from PSD of all accelerometer locations in stream and cross-stream directions

Location	Orientation	PSD Peaks (Hz)
19 near 18	Stream	151, **173**
	Cross	**180**
25 near 26	Stream	163
	Cross	163
26 near 25	Stream	**174**
	Cross	**176**
26 near 27	Stream	**177**, 209, 245
	Cross	209, 245
27 near 26	Stream	**178**, 195, 217
	Cross	**178**, 196, 217
27 near 28	Stream	**177**, 243, 268
	Cross	243, 268
28 near 27	Stream	**178**, 238, 271
	Cross	**180**, 238, 271

TIME-FREQUENCY ANALYSIS

Time-frequency analysis is a technique that tracks how the dominant frequency content of the signal changes over time. This tool can be used to further strengthen the argument that on the anchored monoliths, separate monolith and anchor dynamic responses are observed. First, note that in Figure 3, the anchored monolith acceleration response (right) is not as smooth as the response of non-anchored monolith (left). The anchored monolith (Block 28) response is characterized by the delayed second and subsequent pulses with smaller amplitude. If the first transient in the response is induced by the CGT, the second transient may characterize the anchored monolith, but more investigation is needed.

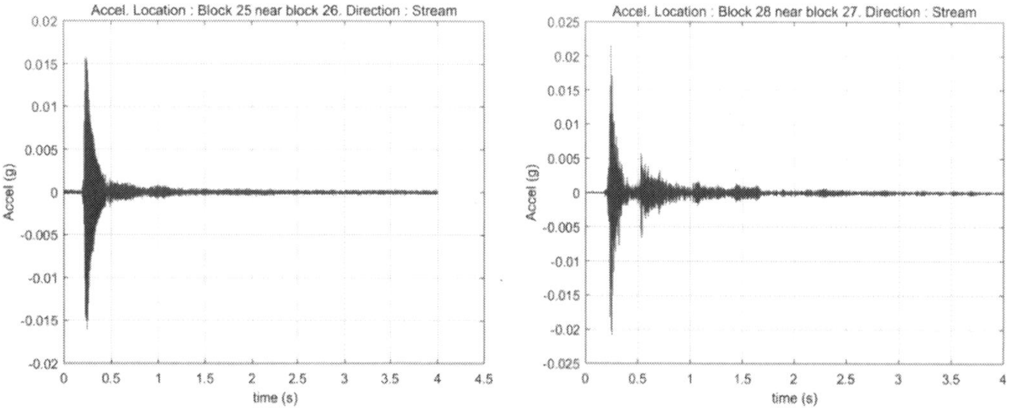

Figure 3: Acceleration time history on non-anchored and anchored blocks in stream direction

In fact, some of these transients have larger amplitudes than the amplitudes associated with the first transients, which indicates that they are probably not due to reflected waves since these reflections would be expected to have lower amplitudes. Moreover, the time lag between the first and second transients is too short for the pulse to be reflected from the end of the dam. Therefore, we consider these second transients to be associated with the anchored monoliths, and not with reflected wave propagation in the dam.

The plots of time-frequency analysis track the change in frequency content over time. Such plots are presented in Figure 4, in which the frequency spectra over the time history of acceleration in Figure 3 are shown. Notice that for a non-anchored monolith (left, Block 25), the response contains a similar frequency content throughout one pulse. However, in an anchored monolith (right, Block 28), the range of the frequency content shifts from near 175Hz to over 230Hz as the second pulse hits. From the analysis, it is evident that second pulse contains frequency information that is not the impact of CGT (first pulse – 175Hz). Specifically, the frequency shifted to higher range, which suggests that the frequencies higher than 175Hz in Table 1 are derived from the anchors in each anchored monolith.

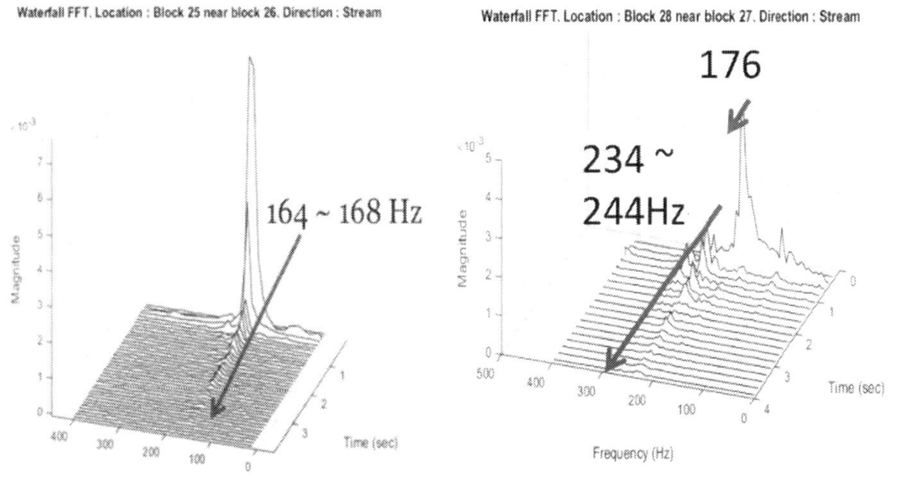

Figure 4: Time-Frequency plot of responses in Figure 3

MODEL VERIFICATION

In order to verify the assumption that high frequency peaks correspond to responses from the anchors placed inside the dam, a simple continuous model is built instead of a finite element model. This is because finite element models are complex and inefficient in simulating high frequency behavior beyond 200Hz. The model of an anchored monolith is shown in Figure 5, in which the steel anchors are modeled as an Euler-Bernoulli cantilever beam under tension (T), and monolith is represented as a lumped mass (M) connected at the other end of the beam.

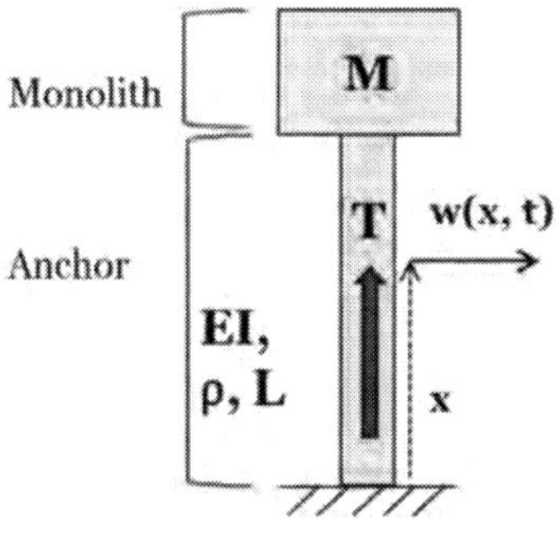

Figure 5: The continuous model with tip lumped mass

The parameters of a steel anchor are represented by the linear density (ρ) and the flexural rigidity, which is the multiplication of Young's Modulus (E) of the steel and area moment of inertia (I) of circular cross-section. In order to solve for the natural frequencies of the system, following governing equation is formulated for a uniform beam under tension:

$$EI\frac{\partial^4 w}{\partial x^4} - T\frac{\partial^2 w}{\partial x^2} + \rho\frac{\partial^2 w}{\partial t^2} = 0 \tag{1}$$

where $w(x, t)$ is the lateral displacement of the beam. In order to solve for Equation (1), separation of variable is applied to $w(x, t)$,

$$w(x,t) = u(x)\sin(\omega t + \theta) \tag{2}$$

where ω is the natural frequency of the system. The following form is assumed for the eigenfunction $u(x)$:

$$u(x) = Ae^{sx} \tag{3}$$

where A is a constant, and s represents the eigenvalue of the system. Equations (2) and (3) are substituted to Equation (1), and the following equations are obtained.

$$s_{1,2}^2 = \frac{T \pm \sqrt{T^2 + 4EI\rho\omega^2}}{2EI} \tag{4}$$

$$u(x) = A_1 \cosh(s_1 x) + A_2 \sinh(s_1 x) + A_3 \cos(s_2 x) + A_4 \sin(s_2 x) \tag{5}$$

Then, Equations (4) and (5) are used to solve the following four boundary conditions.

$$u(0) = 0 \tag{6}$$

$$\frac{du}{dx}(0) = 0 \tag{7}$$

$$EI\frac{d^2u}{dx^2}(L) = 0 \tag{8}$$

$$EI\frac{d^3u}{dx^3}(L) = -\omega^2 Mu(L) \tag{9}$$

Equations (6) to (9) represent zero-displacement, zero-slope at the fixed end, zero moment at free end, and discontinuity in shear due to a lumped mass at the top of the beam. Substituting Equations (4) and (5) into Equations (6) to (9) yields four equations with four unknown coefficients of $u(x)$, which can be put into the matrix form. In order to have a non-trivial set of solutions for the coefficients, the matrix should have zero determinant, which will result in a closed-form frequency equation in terms of the natural frequency, ω.

Table 2: Parameters of steel anchors in each anchored block

Block #	Number of Anchors	Number of Strands	Unbonded Length (ft)	Lock-Off Load (kips)
4	5	25	98	1026
5	5	25	107	1026
27	4	22	84	902
28	3	27	81	1108

In calculating the natural frequencies, the Young's Modulus of ASTM Grade 60 reinforcing steel (Fy = 60,000 psi) of 2.97E7 lbf/in^2 and density of 15.22 lbm/ft^3 are used. The circular cross-section with radius 0.3 inches per strand is assumed, and all anchors are modeled as one anchor at the center of monolith that takes into account the number of anchors and number of strands per anchors in the calculation of area moment of inertia (I) and linear density (ρ). For a lumped mass representing monolith, the volume of each monolith is approximated based on the base area of each monolith and the height shown in Figure 1, and the concrete density of 4.66 lbm/ft^3 is used to calculate the approximate mass of the monolith. The mass was then scaled by different amount to observe the effect of the ratio of mobilized mass to the resulting natural frequencies. It was shown that scaling the mass of monolith had little impact to the overall spacing of the frequencies, so the effective mass of 0.1M was assumed.

Table 3: Natural frequency comparison between continuous model and PSD peaks in Block 27 and Block 28.

Mode #	Natural Frequencies from Model (Hz)	Observed Peaks in PSD (Hz)	% Error	Natural Frequencies from Model (Hz)	Observed Peaks in PSD (Hz)	% Error
1	7.13	-		7.39		
…	…			…		
16	160.05	-		165.59		
17	175.35	**176**	0.37	181.39	**176**	-2.97
18	191.43	**198**	3.43	197.99		
19	208.29	**216**	3.7	215.4		
20	225.95	-		233.63	**234**	0.16
21	244.42	**243**	-0.58	252.7		
22	263.71	**268**	1.63	272.61	**273**	0.14

Notice the agreement between the frequencies observed in the PSD and the model. All the frequencies agree with less than 5% error in the range of 17th to 22nd mode. The model frequencies can be used as a reference point for evaluating future anchor condition. As the anchors loosen up or corrode over time, the natural frequencies are expected to drop, and change in difference of frequencies can be a measure of anchor performance.

SENSITIVITY TESTS

In order for the model frequencies to be used as a performance indicator, the sensitivity of anchor dynamics to tension change and corrosion must be investigated. Specifically, the amount of tension loss and corrosion needed to cause a 10% drop in the observed frequency is studied. In Figure 6, the 21st natural frequency of Block 27 is plotted with respect to varying tension and flexural rigidity (*EI*) separately. Originally in the model, a fully functional anchor has the 21st frequency at 244.4Hz when tension is set at the lock-off tension of 902 kips and EI is equal to 9.92E6 lbf · ft^2. In Figure 6, when tension and *EI* are varied separately while the other parameter is set at the original value, it can be observed that 10% drop in 21st natural frequency, which brings the frequency down to 220.0 Hz, is achieved when tension decreases to about 415 kips or when *EI* decreases to about 5.0E6 lbf · ft^2. These numbers correspond to half the original tension and flexural rigidity.

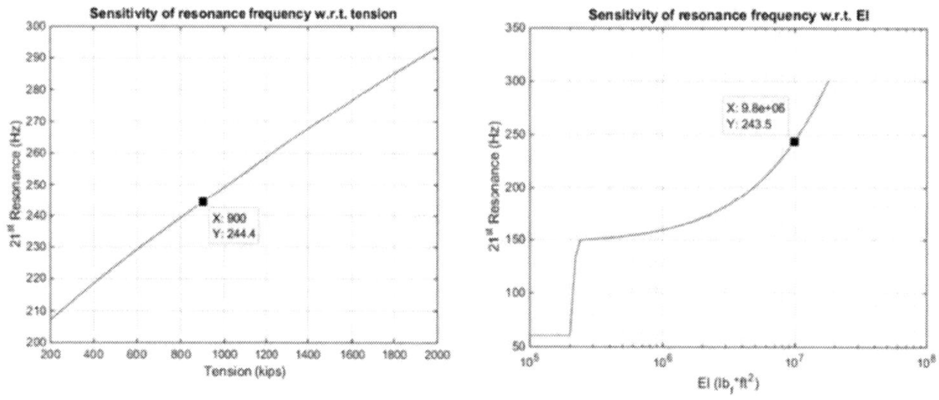

Figure 6: Sensitivity test of 21st natural frequency with respect to tension and EI variation.

Careful attention must be paid to the fact that this sensitivity test cannot indicate how much of tension loss and corrosion each affect the frequency of the anchors simultaneously. If the frequency from the PSD drops, it indicates degraded anchor functionality due to either tension loss or corrosion, but the model itself will not distinguish between the effects of lost tension and/or corrosion in the anchor strands or wires. Additionally, due to the simplicity of the model, in which multiple anchors within each monolith are modeled as one anchor with equivalent stiffness, the model will also not identify which of the anchors in a monolith has lost tension or suffers from corrosion. However, the model can be used to focus attention to a particular monolith in the dam where detailed inspections or testing may be considered.

PERFORMANCE INDICATOR FOR ANCHORS

The core of the analysis technique detailed above relates to the comparison between the anchor frequencies of a continuous model and those obtained from field-based testing using CGT. Theoretically, even though a continuous model cannot replicate the frequencies of a perfectly functioning set of anchors, it can be assumed that for a set of anchors with consistent parameters, constant reservoir level and other environmental factors, such as temperature, humidity, etc., the frequency peaks can be reproduced for different iterations of Performance-Based Testing. Therefore, based on the PBT analysis based on frequency studies and a simple model verification, a preliminary inspection guideline and performance indicator can be established to monitor the health of steel anchors embedded within the concrete dam on a non-contact, non-destructive manner.

The following is the proposed preliminary Anchor Condition Indicator (ACI) of an ith anchored monolith,

$$\boldsymbol{ACI_i = 100\left[1 - \frac{1}{NR}\sum_{k=1}^{NR}\left|\frac{f_n^{Model} - f_n^{PBT}}{f_n^{Model}}\right|\right], i = 1, \cdots, N} \ldots \qquad (10)$$

where NR is a number of responses (peaks) of comparable amplitude observed in PSD. ACI is basically an average of error between model frequencies and PBT frequencies, which must be calculated at the moment of installation. The error is manipulated such that a change in PSD frequencies from subsequent PBT results in decreased ACI.

Based on Equation (10), the ACI's for Block 27 and 28 are calculated in Table 4. Table 4 indicates that based on the analysis technique presented in this paper and the proposed performance indicator, the anchors in Block 27 and Block 28 are in good condition, assuming that the ACI at the time of installation is not drastically different from the values in Table 4.

Table 4: ACI values for Block 27 and Block 28

Block No.	ACI
27	98.06
28	98.91

In order to have consistent ACI values for regular inspection, it is important to perform PBT and identify the modes of model frequency to which frequency peaks in PSD correspond. Once the modes of frequencies are identified, the frequency peaks from subsequent tests must be compared to those specific modes in order to compute consistent and correct ACI values. For example, in the sensitivity test the 21st mode of Block 27 was studied. In the future tests, even if the frequency drops to the level of 20th mode, it must still be compared to the 21st mode of model frequency to accurately reflect the failed anchor status.

Moreover, ACI can be used to overcome the limitation of sensitivity test described previously. In the previous section, the sensitivity of 21st resonance of Block 27 with respect to tension loss and corrosion was studied. However, the collective drop or shift across the entire anchor resonant range must be considered in order to have more accurate and complete understanding of the anchor status due to corrosion or tension change, instead of investigating just one mode. Then, ACI can be used in exchange of conducting 5 sensitivity tests for each resonance. The sensitivity of ACI, which is an average error of all anchor resonances, will provide more reliable intuition on the tensile and corrosion effects on the anchors.

RECOMMENDATIONS

Improvements to the PBT that target the evaluation of anchor condition in a concrete dam include the application of the CGT load near each of the anchored blocks, or at least, on the unanchored blocks near the ends of the dam where the anchors are placed. Sensor placement can also be improved by monitoring both the center as well as the joints of each block, so as to improve the analysis and interpretation of the block responses.

The evaluation procedures described in this paper were limited to using the lateral responses acquired during PBT. Additional information may be obtained from the evaluation of the vertical responses on each block, and perhaps, a better ACI will result by including additional degrees of freedom into the formulation.

CONCLUSION

In this paper, an innovative non-destructive PBT-based method has been proposed to study and monitor the functionality of steel anchors embedded within the concrete dam. In the case of Shaver Lake Dam, the CGT was placed at the center of the arch section of the dam and the impulse was induced globally to the anchored blocks. From the responses recorded using the accelerometers placed on the monoliths, it has been shown that the anchored blocks show delayed responses that have a frequency content different from the initial response caused by the CGT. These delayed responses were considered as responses specific to the anchors, and a simple continuous model of anchors was constructed to show significant agreement in frequencies obtained via PBT and from the model. A simple sensitivity test was performed to show that a drop in frequency indicates either a loss of anchor tension or corrosion.

The work presented in this paper needs further validation. Specifically, a local approach is suggested to place CGT not on the center of the dam but on the monolith of interest. Doing so will help take more consistent and reliable data from all anchored monoliths present in the dam. Same analysis should be conducted on those data, and it must be validated that anchors do produce responses that can characterize its own status of functionality. Once it is established that the aforementioned analysis can be used to monitor the anchor functionality over time, a dam operator can periodically perform PBT on the anchored monoliths, and use the performance indicator (ACI) explained in this paper to track the change in anchor performance over time.

REFERENCES

Brown, E. T, "Rock Engineering Design of Post-Tensioned Anchors for Dams – A Review." *Journal of Rock Mechanics and Geotechnical Engineering.* 7 (2015) 1 – 13

Goldkamp, M.R., Savage, T.H., Edelman, J.B., Orme, P.J., Duron, Z., Knarr, M. (2014). "Performance-Based Testing of a Large Concrete Dam." San Francisco, California, 34th Annual USSD Conference.

Heslin, G., Bruce, D.A., Littlejohn, G.S., Westover, T. (2009). "Performance of Aging Post-Tensioned Rock Anchors in Dams." ASDSO West Regional Conference Proceedings

Poiroux, G. V. (2011). "Post-Tensioned Trunnion Anchor Rod Testing, West Point Dam and R.F. Henry Dam." San Diego, California, 31st Annual USSD Conference.

Southern California Edison Company Dam & Public Safety Group. (2015). "A Performance Based Evaluation of Shaver Lake Dam."

ON THE USE OF DISCRETE LUMPED ELEMENT MODELING FOR INSIGHT INTO CONCRETE DAM BEHAVIOR

Jonathan Lum[1]
Samantha Hoang[2]
Tae Ha Park[3]
Ziyad Duron, PhD[4]
Guy Lund, PE[5]

ABSTRACT

Discrete lumped element modeling is an old technique that has to a large extent been replaced by finite element modeling. Although the lumped element models (LEM) do not capture the detail included in finite element models (FEM), the LEM provides valuable insight into a dam's modal behavior and interactions between the components that make up the dam-foundation-reservoir system. The LEM is not intended to replace finite element models, but rather to provide insight into how to better construct them.

Using Monticello Dam as a test case, the dam, reservoir, foundation, and morning glory spillway are modeled as a coupled system of masses, springs, and dampers. To produce accurate system parameters, the masses in the system are normalized to the physical mass of the dam, and stiffnesses are chosen to match the resonant frequencies of each physical component. Frequency Response Functions (FRFs) and earthquake responses are developed by varying system parameters such as material properties, boundary and interface conditions, and the mass mobilized during the dynamic interactions. The FRFs and earthquake responses reveal dominant modal behavior as well as the influence of system parameters on dam response characteristics. This leads to a clearer understanding of the effects each component in the system has on the dam's response. The paper describes the development of the LEM and how it can be used to improve FEM.

INTRODUCTION

Finite element models (FEM) are frequently used to model structures such as large concrete dams. While an FEM is a powerful tool that can capture minute details of a structure and its behavior, it is important to ensure that the FEM accurately predicts a structure's true response to an input excitation. In the case of Monticello Dam, a blind challenge was issued to reproduce the measured linear-elastic dam response due to a 4.1 earthquake. None of the FEM produced were able to match the measured response. This suggests that improvements can be made to FEM to better characterize dam behavior. FEM accuracy can be increased if the interactions between the system and its surrounding structures are understood on a more basic level.

---

[1] De Pietro Engineering Fellow, Harvey Mudd College, Claremont, CA, 91711, jlum@hmc.edu
[2] De Pietro Engineering Fellow, Harvey Mudd College, Claremont, CA, 91711, shoang@hmc.edu
[3] De Pietro Engineering Fellow, Harvey Mudd College, Claremont, CA, 91711, tpark@hmc.edu
[4] Professor of Engineering, Harvey Mudd College, Claremont, CA, 91711, duron@hmc.edu
[5] Office Principal Engineer, Gannett Fleming, Englewood, CO 80112, glund@gfnet.com

Lumped element models (LEM) can be constructed prior to creating an FEM to deduce the dominant behavior characteristics of the system. Modelling the system as masses, dampers, and springs will not yield the finer details of an FEM, but can predict general modal behavior of the system as well as the strength of the interactions of components in the system that contribute to the overall system response. Thus, an LEM can be implemented to better understand interactions between a dam and its surroundings. Possessing knowledge of a dam's interactions before building an FEM can lead to a model that more accurately predicts the true behavior of a dam. Herein, Monticello Dam will be modelled using lumped elements as an example of how modal behavior can be understood prior to creating an FEM.

CONSTRUCTING A LUMPED ELEMENT MODEL OF MONTICELLO DAM

Monticello Dam is an arch dam located in Napa County, California. One of its most prominent features is its large morning glory spillway that has an inlet diameter of 72 ft. which reduces to 28 ft. as one moves along the spillway. The large spillway structure is fixed into the surrounding foundation. An overhead view of Monticello Dam and its spillway is depicted in Figure 1. To deduce modal behavior of the dam using LEM, the bodies which contribute to the dam's response must be identified. The dam and all of its surrounding components are treated as a single system composed of smaller components.

Figure 1: Overhead view Monticello Dam

Large bodies, namely the water, spillway, and the dam can be considered system components in the case of Monticello Dam. Other smaller bodies such as the debris rack and smaller spillway located at the center of the dam are assumed to have little contribution to the overall dam behavior and are subsequently ignored. Each component has an equivalent mass, M, stiffness, K, damping, C, and displacement, x in the LEM. Subscripts w, s, and d denote the water, spring, and dam respectively. The system components are attached together with springs and dampers, denoted by K_{ij} and C_{ij}, where i and j represent the two bodies that are coupled by the connection. These springs and dampers correspond to the interfaces and boundaries between components. In order to account for an input such as an impulse or earthquake into the system, a time varying signal $u(t)$, representing the forcing input to the system, is interfaced to each component through their respective stiffness and damping attachments. A visual depiction of the Monticello LEM is shown in Figure 2.

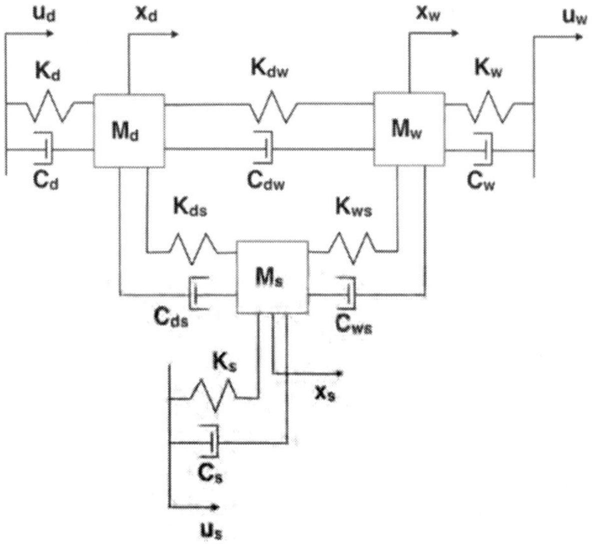

Figure 2: Monticello Dam represented as a coupled spring, mass, damper system

For simplicity, only the stream direction is considered in this particular model. However, this method can be extended to the cross stream and vertical directions by finding the proper equivalent mass, spring, and damper for each of the components of the system and reevaluating the model. It is important to note that the LEM can be tweaked and adjusted by adding or removing components and changing the values of the system parameters as necessary. A list of system parameters is found in Table 1.

Table 1: System parameters for Monticello Dam and their corresponding notation

System Parameter	Parameter Notation
Dam Mass	M_d
Spillway Mass	M_s
Water Mass	M_w
Dam Stiffness	K_d
Spillway Stiffness	K_s
Water Stiffness	K_w
Dam Damping	C_d
Spillway Damping	C_s
Water Damping	C_w
Dam-Spillway Interface Stiffness	K_{ds}
Dam-Water Interface Stiffness	K_{dw}
Water-Spillway Interface Stiffness	K_{ws}
Dam-Spillway Interface Damping	C_{ds}
Dam-Water Interface Damping	C_{dw}
Water-Spillway Interface Damping	C_{ws}
Dam Displacement	x_d
Spillway Displacement	x_s
Water Displacement	x_w
Dam Forcing Input	u_d
Spillway Forcing Input	u_s
Water Forcing Input	u_w

SELECTION OF SYSTEM PARAMETERS

To utilize the LEM, the system parameter values need to be judiciously approximated. An advantage of the LEM is that the dam behavior is driven by the relative parameter values, allowing for ratios of parameters to be used as inputs to the model, reducing the need for the absolute (correct) parameter values. Thus, only the ratios of the masses, springs, dampers need to be approximated. For convenience, all the parameters are normalized by the equivalent mass of the dam.

Four different spillway resonances were considered, along with conditions of water compressibility and incompressibility. The resonant values for the eight different test cases are listed in Table 4. The test cases described in detail in this paper were chosen to reveal the influence of system parameters on modal behavior of Monticello Dam.

In approximating the mass ratios of the system, the Westergaard equation is used to calculate the equivalent mass of reservoir water mobilized by the dam.[1] For the spillway, the mass of the spillway is normalized by the mass of the dam. The described method for determining mass is a rough estimation used as a starting point for the model.

When evaluating the system, the initially calculated parameter ratios are flexible and may be adjusted accordingly to better represent the actual system. In this case, the spillway ratio is decreased, starting from the initially calculated ratio. In the true physical system, the morning glory spillway is grouted into the foundation. If it were completely fixed, the spillway would not need to be included in the model as it would move perfectly in sync with the foundation and could simply be treated as rock foundation. Allowing the spillway to undergo small movements is a better assumption, as a completely fixed boundary condition to the rock foundation is unlikely. Because only small, restricted movement is allowed in the model, the stiffness ratio must be decreased, as only a small percentage of the spillway mass will be engaged if subjected to an excitation. Similarly, the spillway to dam mass ratio must be adjusted. Using the initial mass ratio of the spillway to dam assumes that the entire mass of the spillway is engaged, while in reality this is unlikely to occur.

The stiffness ratios are determined using the natural frequencies of the system. For a single degree of freedom discrete element system, the fundamental frequency is given by:

$$\omega_n = \sqrt{\frac{k}{m}} \qquad (1)$$

Using Equation 1, the normalized stiffness is calculated given the known resonance of the structure, and the normalized mass of the structure. The resonances of Monticello Dam and the water reservoir were experimentally measured in 1987 by Ray W. Clough, Yusof Ghanaat, and Xiong-Fei Qiu. The resonances for the both the dry and wet dam and the water reservoir are listed in Table 2.

Table 2: Measured and calculated frequencies of Monticello Dam and its reservoir, 1987

Mode Number	Wet Measured Freq. (Hz)	Wet Calculated Freq. (Hz)	Dry Calculated Freq. (Hz)	Infinite Reservoir Calc. Freq. (Hz)
1	3.12	3.08	3.85	6.02
2	3.55	3.51	4.24	11.42
3	4.70	4.62	5.46	14.19
4	6.00	5.77	6.69	-

[1] The Westergaard equation is one of the earliest formulations used to approximate the amount of reservoir water mobilized by a dam and is most applicable on gravity dams. While there exist newer equations which better approximate the body of added mass due to water for a variety of different cases, the Westergaard equation was chosen because of its simplistic calculation and is sufficient for this LEM. The estimate of the mobilized water mass from the Westergaard equation is used as a starting point for the dam to water reservoir mass ratio. The actual ratio used in the LEM can be varied from this starting point to better match experimental data.

The spillway resonances were developed through modelling in SolidWorks (Figure 3). Because the exact spillway resonances were not experimentally determined, a range of possible resonances were determined by varying the boundary conditions of the spillway. The upper end of the range was obtained by completely fixing the spillway into the rock foundation. To achieve the lower end of the frequency range, the fixed boundary conditions were relaxed, allowing for larger movements between the spillway and its rock interface.

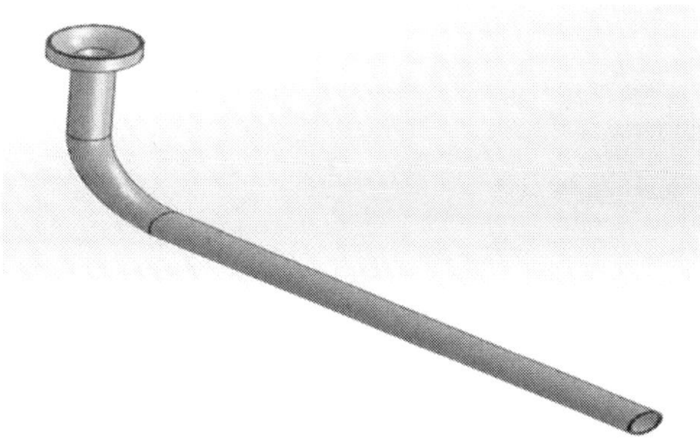

Figure 3: Solidworks model of morning glory spillway (boundary conditions not shown)

Frequencies on the higher side were initially selected when calculating the stiffness value for the spillway, as it is probable that the spillway more closely resembles the fixed interface condition. Table 3 contains a list of the frequency ranges for the first three modes of the spillway.

Table 3: Frequency ranges for the first four modes of the spillway

Mode Number	Lower Bound Freq. (Hz)	Upper Bound Freq. (Hz)
1	2.68	20.05
2	3.36	33.38
3	3.74	33.41
4	20.24	47.00

To calculate the proper (equivalent) damping, values between 0.1% and 5% of critical were assumed. The relationship between C and damping factor, ζ is given by:

$$C = \frac{2\zeta\omega_n}{M} \qquad (2)$$

Thus, C values for the dam, spillway, and water were selected and varied to set ζ to the desired damping percentage between 0.1% and 5% of critical damping. The selection of the interface parameters, C_{ij} and K_{ij} between elements in the model, is based on the following assumptions:

- The spillway-dam interface damping and stiffness are identical to the dam damping and dam stiffness[2]
- The water-spillway interface and water-dam interface have identical damping and stiffness
- Water stiffness is varied in the model to account for compressibility effects

The LEM can treat the water as compressible by leaving all the water coupled components intact, as in Figure 1. For the incompressible water case, the water coupling components are removed, and the equivalent water mass is added directly to the mass of the dam. In effect, the mobilized body of water moves with the dam. An illustration of the incompressible case is shown in Figure 4.

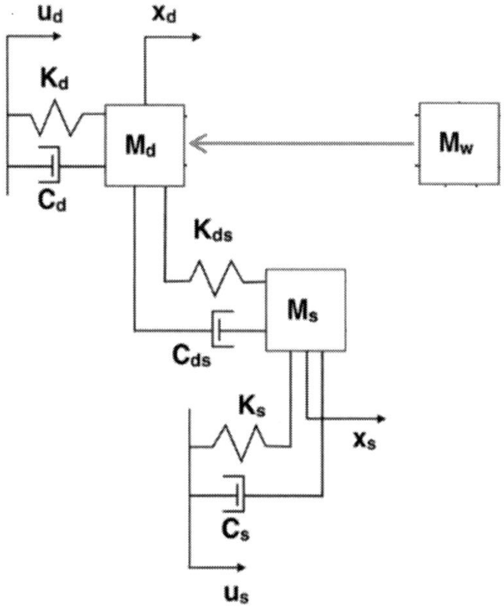

Figure 4: Reservoir water treated as incompressible in the LEM

[2] The dam and spillway are both connected to the foundation. Thus, the coupling between the two is approximated to be similar to the dam-foundation interface

Table 4: System parameter ratio values for various LEM test cases

Test Case	Dam Resonance (Hz)	Spillway Resonance (Hz)	Water Resonance (Hz)	Incompressible Water (Y/N)
1	3.85	2	--	Y
2	3.85	2	6.02	N
3	3.85	3.85	--	Y
4	3.85	3.85	6.02	N
5	3.85	8	--	Y
6	3.85	8	6.02	N
7	3.85	20	--	Y
8	3.85	20	6.02	N

LUMPED ELEMENT MODEL VERIFICATION AND FREQUENCY RESPONSE

The system can be analyzed by applying Newton's second law to each element in the LEM. Dividing by the dam mass yields the following normalized system of equations:

$$\ddot{x}_d + \frac{C_d+C_{dw}+C_{ds}}{M_d}\dot{x}_d - \frac{C_{dw}}{M_d}\dot{x}_w - \frac{C_{ds}}{M_d}\dot{x}_s + \frac{K_d+K_{dw}+K_{ds}}{M_d}x_d - \frac{K_{dw}}{M_d}x_w - \frac{K_{ds}}{M_d}x_s = \frac{C_d}{M_d}\dot{u}_d + \frac{K_d}{M_d}u_d \quad (3)$$

$$\ddot{x}_s + \frac{C_s+C_{ds}+C_{ws}}{M_s}\dot{x}_s - \frac{C_{ds}}{M_s}\dot{x}_d - \frac{C_{ws}}{M_s}\dot{x}_w + \frac{K_s+K_{ds}+K_{ws}}{s}x_s - \frac{K_{ds}}{M_s}x_d - \frac{K_{ws}}{M_s}x_w = \frac{C_s}{M_s}\dot{u}_s + \frac{K_s}{M_s}u_s \quad (4)$$

$$\ddot{x}_w + \frac{C_w+C_{dw}+C_{ws}}{M_w}\dot{x}_w - \frac{C_{dw}}{M_w}\dot{x}_d - \frac{C_{ws}}{M_w}\dot{x}_s + \frac{K_w+K_{dw}+K_{ws}}{M_w}x_w - \frac{K_{dw}}{M_w}x_d - \frac{K_{ws}}{M_w}x_s = \frac{C_w}{M_w}\dot{u}_w + \frac{K_w}{M_w}u_w \quad (5)$$

To solve the system, Equations 3-5 can be rewritten in matrix form, as follows:

$$[M]\underline{\ddot{X}} + [C]\underline{\dot{X}} + [K]\underline{X} = [F] \quad (6)$$

Where [M], [C], and [K] are the normalized mass, damping, and stiffness parameters respectively, [F] is the normalized forcing input matrix, and $\underline{X}$ is the displacement vector. Frequency response functions (FRFs) are produced by assuming $\underline{X}$ is a solution of exponential form to forcing input $\underline{F}$ also of exponential form:

$$\underline{X} = \overline{X}e^{j\omega t} \quad (7)$$

$$\underline{F} = \overline{F}e^{j\omega t} \quad (8)$$

$$\bar{X} = \begin{bmatrix} \bar{x}_d \\ \bar{x}_s \\ \bar{x}_w \end{bmatrix} \quad (9)$$

$$\bar{F} = \begin{bmatrix} \bar{F}_d \\ \bar{F}_s \\ \bar{F}_w \end{bmatrix} \quad (10)$$

where $\overline{X}$ is the magnitude of displacement vector, $\overline{F}$ is the magnitude of forcing input vector, and ω is the forcing frequency in radians per second. Substituting Equations 7-8 into Equation 6 and solving for $\overline{X}$ yields:

$$\overline{X} = (-\omega^2[M] + j\omega[C] + [K])^{-1}\overline{F} \quad (11)$$

The first row of the displacement vector, $\overline{X}_{1,1} = \bar{x}_d$, in Equation 11 corresponds to the dam's displacement. The dam's displacement is of interest because the focus of the model is on the dam's response, although the spillway and water response can also be examined by looking at the other rows in the displacement vector. An FRF is a plot of the magnitude of displacement versus forcing frequency. FRFs can be used to reveal the natural frequencies associated with the dam and verify the accuracy of the LEM. At the resonant frequencies of the dam the magnitude of displacement in the FRF should be large. Likewise, the magnitude of displacement should be suppressed at other forcing input frequencies. Therefore, the model is functioning properly if the FRF contains the resonant frequencies that correspond to the true natural frequencies of a system. To verify the model is working properly, two base cases are evaluated: a standalone dry dam case and a tuned absorber case. For the dry dam case, all the parameters except the dam mass and stiffness, M_d and K_d, are set to zero. This reduces Equation 6 to a single degree of freedom harmonic oscillator:

$$\ddot{x}_d + \frac{k_d}{M_d} x_d = F \quad (12)$$

Thus, $\sqrt{\frac{k_d}{M_d}}$ is the expected natural frequency of the system.

For the tuned absorber case, the process is repeated, except the spillway is coupled to the dam so that M_s and K_{ds}, are also included in the system. In this case, Equation 6 simplifies to:

$$\begin{bmatrix} 1 & 0 \\ 0 & 1 \end{bmatrix} \begin{bmatrix} \ddot{x}_d \\ \ddot{x}_s \end{bmatrix} + \begin{bmatrix} \frac{k_d + k_{ds}}{M_d} & -\frac{k_{ds}}{M_d} \\ -\frac{k_{ds}}{M_s} & \frac{k_{ds}}{M_s} \end{bmatrix} \begin{bmatrix} x_d \\ x_s \end{bmatrix} = \begin{bmatrix} F_d \\ 0 \end{bmatrix} \quad (13)$$

Setting the spillway mass to stiffness ratio equal to that of the dam, $\frac{k_d}{M_d} = \frac{k_{ds}}{M_s}$, results in both the dam and spillway having the same resonant frequency, leading to tuned vibration absorber behavior. In a tuned vibration absorber, the natural frequency of the original system is suppressed, and the system will respond at two frequencies near the original natural frequency. The FRFs for both verification cases are displayed in Figure 5.

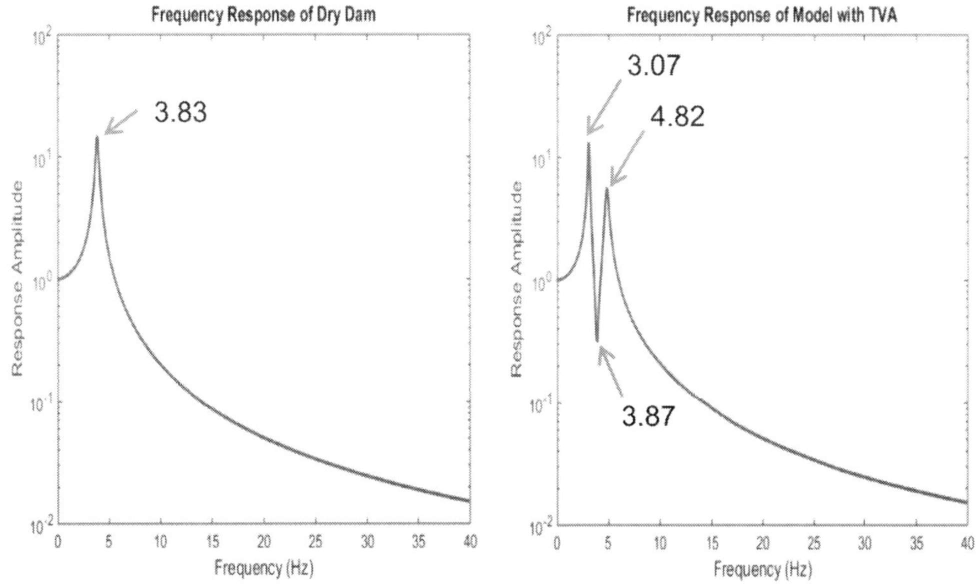

Figure 5: Frequency Response Function for the dry dam and dry dam-tuned spillway combination

Proper model behavior is verified by the single resonance that appears at 3.83 Hz in the case of the standalone dry dam. The true fundamental frequency of the dry dam is 3.85 Hz. The lone resonant peak in the FRF is within 0.5% of the true fundamental frequency.

In the test case where the spillway is attached to the dam and also set to vibrate at 3.85 Hz, a split peak in the FRF is observed, showing the characteristics of a tuned vibration absorber as expected. The suppressed resonance occurs at 3.87 Hz, also within 0.5% of the true fundamental frequency of the standalone dry dam. Two resonant frequencies are present at 3.07 and 4.82 Hz. These two frequencies lie just below and above the original natural frequency of 3.83 Hz. The FRFs produced by the model predict natural frequencies that almost exactly match the true frequency of the dry dam and also produced tuned vibration behavior when expected, suggesting the model is functioning properly.
FRFs were produced for the test cases listed in Table 4 to reveal potential natural frequencies of the dam. The FRFs for cases 3&4, 5&6, and 7&8 are shown in Figures 6-8, respectively.

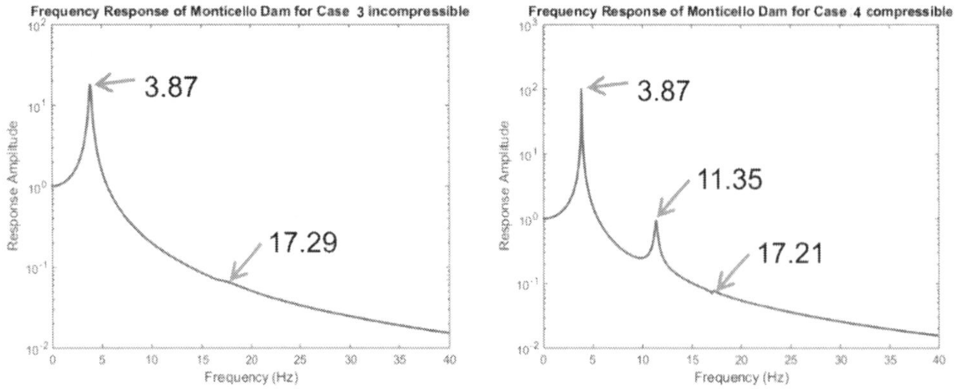

Figure 6: FRF of Monticello Dam, cases 3 and 4

Figure 6 depicts the case where the spillway resonance is tuned to that of the dam, similar to the second model verification case. In the model verification case, a tuned vibration absorber response was present (split peak in Figure 5) as there were no additional coupling terms. However, due to the coupling terms present in test cases 3&4, the FRF does not suppress the resonance of the dam and split it into two different distinct resonances. Instead, the original resonant frequency of the dam is present, along with two additional distinct peaks. The observed behavior is very different from that of a tuned vibration absorber and suggests that the interactions between components in the system can greatly alter the system's behavior. Thus, more scrutiny between a dam and its surrounding environment may be necessary when using an FEM.

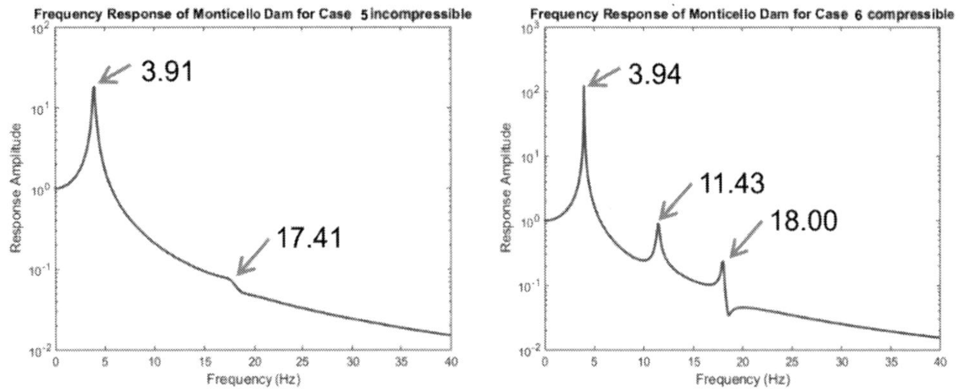

Figure 7: FRF of Monticello Dam, cases 5 and 6

In test cases 5&6, the spillway rigidity is slightly increased by tightening the boundary conditions between the spillway and rock foundation. Comparing the FRFs in Figure 7 to the FRFs in Figure 6, the resonant peaks do not vary significantly due to a small change in spillway frequency from 3.85 Hz to 6 Hz. The largest percent difference in resonant frequencies between cases 3&4 and 5&6 are less than 5%. However, a large change in the dam's resonant frequencies are observed when the spillway is made even more rigid by increasing its fundamental frequency to 20Hz, as in test cases 7&8. As seen in Figure

8, a rigid spillway results in the fundamental frequency of the dam shifting from between 3.8 and 3.9 Hz to 4.54 Hz. The third resonant peak in the compressible water case (Figure 8) for the rigid spillway is at 29.67 Hz. This value is large compared to the other two cases where the third peak is less than 20 Hz. The FRFs reveal that a relaxation of the fixed spillway condition will lower the resonant frequencies in the dam response. In the case of Monticello Dam, a rigid spillway is generally assumed, as the spillway should be strongly fixed into the foundation. This assumption may not be valid if low resonant frequencies are observed in the measured response of the dam. If low frequencies are observed, this may suggest the spillway fixture to the foundation has weakened.

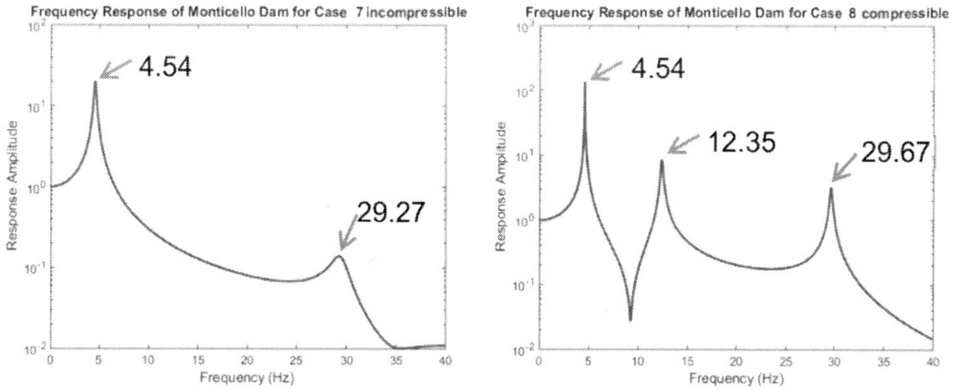

Figure 8: FRF of Monticello Dam, cases 7 and 8

Comparing compressible versus incompressible water across Figures 6-8, the LEM reveals the effect compressibility of water and spillway resonance has on the response of the dam. Three resonant frequencies are present in the FRF when water is treated as compressible in the LEM. In the incompressible case, only two resonant peaks are present. In the true response of Monticello Dam, three resonant frequencies were observed, suggesting water compressibility may be a factor. The effect of water can be utilized when constructing an FEM. Using the insight gained on water compressibility from an LEM could potentially benefit an FEM, as the water interaction in the system is already understood.

LUMPED ELEMENT MODEL EARTHQUAKE RESPONSE

In addition to producing FRFs, the LEM is also capable of producing predicted earthquake responses. Modal analysis takes advantage of the orthogonality of a system's modes to the mass matrix, $[M]$. Using modal analysis, a coupled system of equations, such as the system in Equation 6, can be decoupled by normalizing the modes. The displacement vector, $\underline{X}$, is converted into normalized coordinates by multiplying the normalized modal matrix. The impulse response of the decoupled system is solved for, which when convolved with and any arbitrary signal, such as an earthquake, produces the dam's response to the arbitrary signal. The convolution process is depicted in Figure 9.

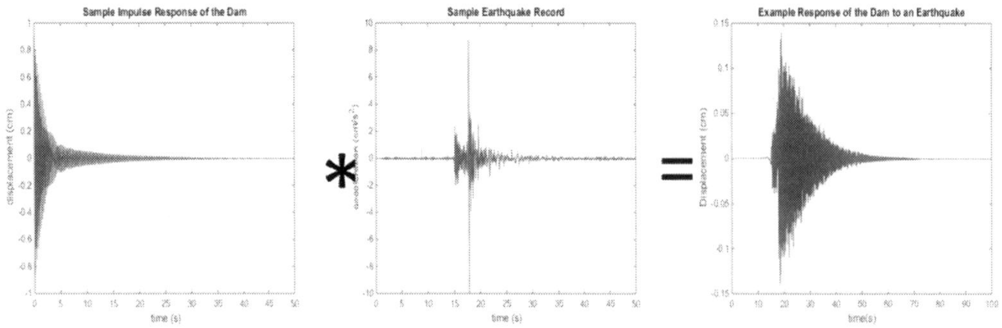

Figure 9: Sample convolution of earthquake in the LEM

For the case of Monticello Dam, the record of the 4.1 magnitude earthquake that shook the dam in the blind challenge was used as the arbitrary signal (excitation) in the LEM. The model was evaluated for each of the eight test cases to obtain the dam's response to the actual earthquake under the different system parameter cases. The earthquake response under test case 3 is displayed in Figure 10. The response under test case 4 is displayed in Figure 11. In both cases, the resonant peaks present are nearly identical to those in the FRFs. There is a downward trend in magnitude as frequency increases, which does not occur in the actual response of the dam to the earthquake, and is discussed later.

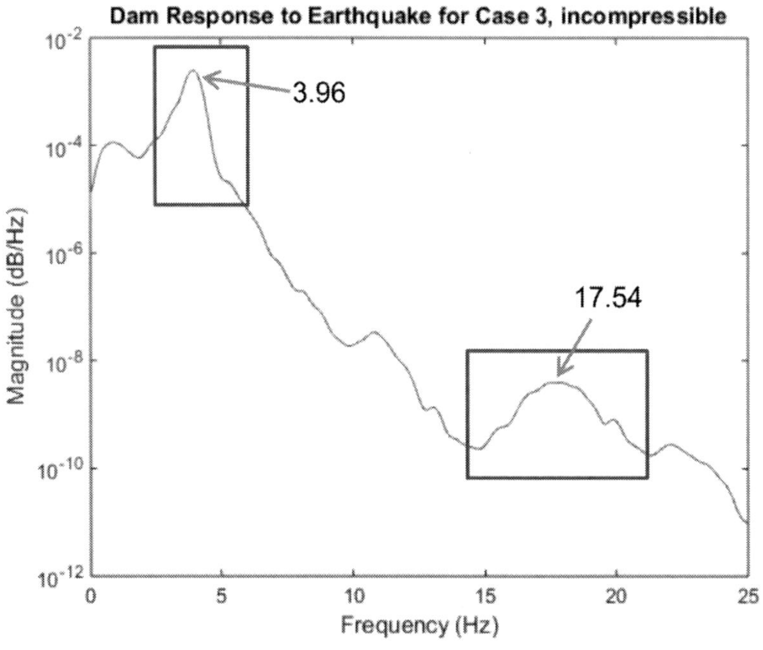

Figure 10: Earthquake response, case 3

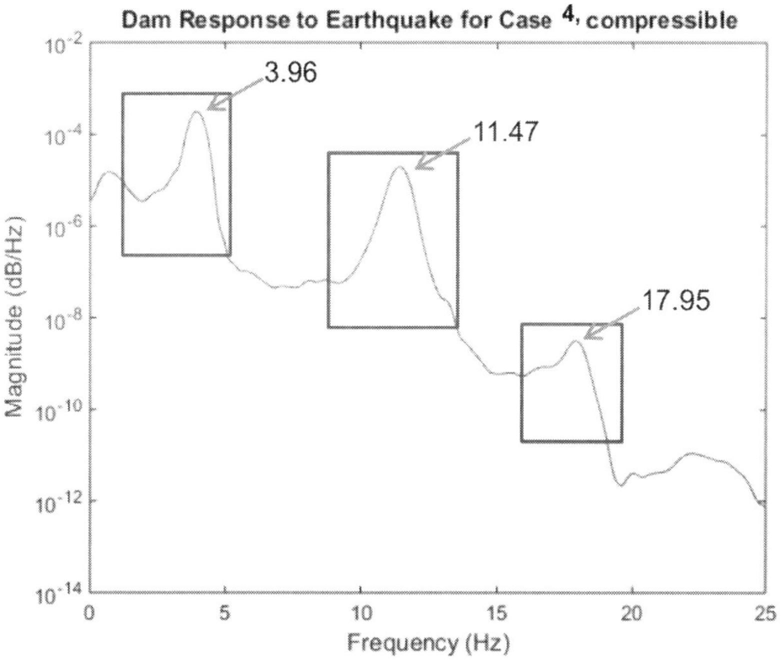

Figure 11: Earthquake response, case 4

LUMPED ELEMENT MODEL OBSERVATIONS

The practical use of an LEM comes from its short computational time and low computational cost combined with its ability to evaluate the dominant characteristics in the physical system. The LEM also allows for the evaluation of the components and couplings that drive the dominant behavior. Because many test cases can be evaluated in a short amount of time, the effect each individual parameter has on the overall behavior can be examined. In the example test cases of Monticello Dam, the effects of spillway rigidity and water compressibility were evaluated. Reevaluating the model to also include the smaller debris racks attached to Monticello could show the need to include or exclude the racks in an FEM. Comparing the modal behavior in each case to the measured response of the dam could reveal the need to incorporate various physical system components in an FEM. Thus, LEM has the potential to screen for components that should be included or excluded in an FEM.

LEM is also practical in the sense that the model does not depend on the true (exact) parameters of the real physical system. The parameters used in the LEM are ratios which are based on the estimations of the physical system parameters. The independence from using the exact parameters enables many test cases to be evaluated quickly in order to determine characteristic behavior.

The convolution approach employed in the LEM model only considers a single modal contribution to the physical system's earthquake response. This explains the declining amplitude of response at higher frequency ranges in the LEM (Figures 10-11). In the

Monticello Dam example, the system's first mode was evaluated. The higher modes that were not evaluated also contribute to the system response, but at the higher frequency ranges. The lack of higher modes in the model causes the declining amplitude response. To address this issue, the LEM can be extended using the principle of superposition. The system parameters can be reevaluated to align with the higher modes to produce the dam's response in higher frequency ranges. For example, the ratio of the dam mass and stiffness, M_d and K_d can be set to match the second natural frequency of the dam (4.24 Hz). The same can be done for the spillway and water. Evaluating the LEM at the second natural frequency should produce the dominant behavior of the system at its second mode. Combining the first and second modal responses could reduce the decline in amplitude seen in Figure 10 and Figure 11, producing a more accurate result. However, the proper weightings used to combine modal responses are difficult to determine. Still, the LEM allows for the analysis of each individual modal response of the system.

CONCLUSION

LEM is in no way intended to replace FEM. Its purpose is to aid in improving the understanding of how the entire dam system interacts with its surrounding environment. Utilizing an LEM may help guide and verify the more accurate FEM given that an LEM can reveal dominant behavior characteristics through FRF and earthquake response plots. Developing results using a wide range of system parameter cases prior to constructing an FEM can guide the FEM in the proper direction. In the case of Monticello Dam, LEM suggests coupling between the spillway and dam and compressibility effects in the reservoir. The LEM also suggests the spillway may not be completely fixed into the surrounding foundation. This information could possibly lead to results that better reflect experimental data in an FEM of Monticello Dam.

REFERENCES

Clough, Ray W., Yusof Ghanaat, and Xiong-Fei Qiu. *Dynamic Reservoir Interaction with Monticello Dam*. Rep. no. UCB/EERC-87/21. Berkeley: Earthquake Engineering Research Center, 1987. Print.

Orme, P. et al. *Analytical and Numerical Model of the Wanapum Pier-Ogee Spillway*. 35th Annual USSD Conference, 2015.

Westergaard, H. M. *Water Pressures on Dams during Earthquakes*. *Transactions of the American Society of Civil Engineers*, 1933, Vol. 98, Issue 2, pg. 418-433.

AUTHOR INDEX

Abbas, Syed 93
Abbaszadeh, Sam 1014
Abdoun, Tarek 405
Aboelata, Maged 1460
Adams, Gregg Batchelder 286
Alvarado, Gabriela 749
Ameen, Rozh 810
Anderlini, C. 1360
Antonaki, Nonika 405
Anyanwu, Emeruwa 765
Armenis, Petros 221
Armstrong, Richard J. 939
Arnold, Justin 74
Arnold, Michael 551
Atukorala, Upul 505
Baehr, A. C. 1762
Bagherzadeh, Harry 586
Baker, Bruce 90
Ballhausen, F. 1747
Barber, Todd 1724
Barfuss, Steven 1827
Barker, Bruce 924
Barker, Malcolm 221
Barry, Rob 1250
Bearlin, Andrew 74
Beaty, Michael 969
Beck, Kade 1827
Bector, Munit 1189
Beduhn, Bob 10
Benitez, Rolando 1334
Bergman, B. 1498
Bessette, Michael 1657
Bialek, Elizabeth 808
Bird, G. Richard 286
Bisnett, Rachael V. 1090
Black, Robert 1250
Blackett, Frank L. 174
Blair, Stephen 1119
Bocovich, Carolyne 1643
Boivin, Yannick 1498
Bonelli, Stéphane 1417
Booth, Paul 735
Borela, Rodrigo 749
Bouaanani, Najib 1658
Bouayad, Ahmed 539
Bouchie, Peter 304
Bourdeau, Philippe L. 749
Bowles, David S. 104
Boyer, Douglas D. 174
Bradner, Graham 1624
Bradner, Graham C. 1604
Brinker, Darren J. 1673
Brom, Kelly 73
Brown, Stephen 267
Brown, Will 765
Bukahir, Layne 378
Burke, Joseph 1250
Burns, Steve 1102
Bushey, Randall 421
Bynoe, Kareem 304
Caldwell, Jason 924
Carney, Shaun 90
Castillo-Rodríguez, Jessica 194
Cato, Kerry D. 139
Chan, Ccee 1102
Charlwood, Robin 1153
Chaudhry, M. Hanif 720
Chen, Wen Yan 1071
Chopra, Anil K. 1464
Chowdhury, Khaled 969, 1578
Christensen, Peter 221
Clarke, Bob 734
Clopper, Chris 1796
Cobb, Charles 404
Cochran, Brett 12
Collins, Kent L. 1
Conkle, Christopher 628
Corn, Aimee L. 1673
Cotie, Bruce 268
Crookston, Brian M. 1827
Crosby, Wesley 660
Cross, Jerry 800
Crouch, Michael M. 208
Curtis, Dan D. 1233, 1546
Curtis, Kedric 72
Cvijanovic, Vojislav 1709
Dall'Aglio, Milton 675
Daviero, Gregory J. 676
Davis, Peter 368
Dawson, Ethan 969
De Melo, Lucas 440, 954
deGroot, Dirk 550
Dolen, Timothy 469
Dolve, Blake 1709
Dominguez, Justin 696
Donaghy, Jud 800
Downing, Bruce 505
Doyal, Stephanie 661
Driller, Michael 1071, 1250
Duan, Wenbo 1658
Durkee, Dean B. 855
Duron, Ziyad 1206, 1218, 1425
Dutson, Greg 735
Ellenson, Sean 1189
Engel, John 10
Engemoen, William O. 422
England Jr., John F. 329, 345
Escuder-Bueno, Ignacio 194
Espinoza, Wilson 749
Fannin, R.J. 1360

AUTHOR INDEX

Faulhaber, Bob....551
Fernandez, Jorge....246
FitzPatrick, Brendan....613
Forrest, Michael....602
France, John W.....286
Friesen, Steven....1071
Friz, Peter....74
Galic, Dom....158
Garandet, Aurélie....1417
Garduno, A.....1747
Gassman, Sarah L.....1333
Gelo, Martina....1178
Gerlach, Mike....420
Gerressen, Franz-Werner....562
Ghaedi, Khaled....487
Ghanaat, Yusof....1528
Gheibi, Emad....1333
Gibson, Wayne....924
Gilbert, Daniel A.....1735
Ginther, Conrad....1334
Gleason, Erin....1673
Gloschat, J. A.....1762
Goulas, E.....470
Griffis, Wade....522
Grima, Nathan....1375
Groeneveld, Joe....74
Gross, Eric....765
Guangul, Seifu....93
Gunn, R.M.....470
Hageman, Eric....613
Hall, Brian M.....345, 839
Hamedian, Sherry....1685
Hanbali, Fauwaz....91, 674
Harder, Les....1657
Harder, Leslie F.....1250
Hare, John L.....867
Harper, Zachary....1528
Harrington, Graham....221
Harris, Jeff....907
Hasan, Husein....1233
Hasan, Husein A.....1735
Hasenbalg, Jaron....73
Hawson, Herb....505
Hayes, Chris....1153
Hill, Jason F.....613
Hoang, Samantha....1206, 1218, 1425
Hockett, Richard B.....1403
Hoffman, Dan....93
Hogan, Dennis....577
Horvath, Doug....1334
Hou, Susan....602
Hover, William H.....452
Hu, George....969, 1058, 1388
Huard, David....56
Huggins, Jennifer E.....1513
Hultstrand, Doug....886
Husted, Jack....377, 787
Hutton, Jesse....1143
Ibáñez-de-Aldecoa, Rafael....662
Ibrahim, Ibrahim A.....1359
Ibrahim, Zainah....487
Imran, Jasim....720
Iso-Ahola, Vik....1178, 1189
Ivanov, Dmitri....1334
Jaeger, Robert....1624
Jaeger, Robert A.....629, 1604
Jalayeri, Nikou....1102
Jamieson, Steve....268
Janes, Matthew....245
Johnson, Jeff....74
Johnson, Nicholas....1090
Jones, Rusty....1803
Kanning, Willem....1643
Kappel, Bill....886
Karlovits, Gregory S.....357
Kashiwada, Christina....1250
Keizer Jr., Richard A.....1604
Kelson, Keith I.....345
Kennedy, Eric....765
Killeen, Katharine....1775
Kimbrel, Sean....1, 1441
Kirksey, Brad....867
Kirschbaum, Daniel....74
Klein, Matthew....1424
Klippel, Sterling....377, 787
Klippel, Sterling E.....1178
Klipsch, Joan....1463
Kolakowski, Anna....1375
Koltuniuk, Roman....1513, 1685
Kost, Andrew D.....1028
Kovalishyn, Val....1102
Krivanec, Christopher....1657
Kropelnicki, Jesse....304
Krumm, Paul....1135
Kuo, Chun-Yi....1403
Kuttel, Jeanne....1250
Kynett, Michael....1578
Lamour, Vincent....1417
Laro, Katie....924
Leclerc, Martin....1498
Lee, Tedman....602
Lefkovitz, Nathan....1461
Léger, P.....1498
Lehrer, Gai....562
Leonard, Christina M.....345
Li, Chunling....440
Li, M.....1360
Li, Zhihua....697
Lilley, Keith....377
Lim, Yeo Howe....1442

AUTHOR INDEX

Lochu, Alexandre....578
Lokke, Arnkjell....1464
Lopez, Don....268
Lopez, Jose R....709
Lum, Jonathan....1206, 1218, 1425
Lund, Guy....1218
Lund, Guy S....1425, 1673, 1724
Lynch, Andrew J....855
Madera, Veronica....469
Maki, Ian P....629
Mallet, Mark....662
Manie, Jonna....1465
Manole, George....1178
Margo, David A....357
Marín, Jorge....662
Marshall, Kevin....1724
Mazighi, Hichem....487
McClung, Nicholas....613
McCoy, Andrew....1803
McGuire, Mathew....368
McHenry, Jim....765
McMahon, Michael....9
McPherson, Matthew....1462
McVicker, Jamie....1567
Mejia, Lelio....246
Mejia, Lelio H....1043
Melliger, Joshua....92
Menninger, John....93
Meshkat, M....825
Messio, Elisa....1658
Mihoubi, Mustapha Kamel....487
Miles, Patrick....867
Miller, Michael....313
Miquel, Benjamin....1658
Mishler, Andrew....867
Mitovski, Stevcho....645
Modini, George....674
Montero, M....1747
Montgomery, Jack....1014
Mooney, Michael....1643
Morales-Torres, Adrián....194
Morley, Derek S....139
Morrison, Michael A....954
Moses, Matt....378
Mostafa, Mohammadreza....1189
Moura, Danielli De Melo....749
Muehring, Deron....1803
Muhlestein, Geoff....886
Mulder, Henri V....1058
Musselman, Eric....627
Neamtu, Danielle....1119
Needham, Jason....104
Neff, Keil....90, 440
Nettle, Justin....765
Newman, Erik....602
Ng, Carman....602
Nickerson, James....696
Novembre, Nicole....924
O'Neal, Troy S....1403
Obermueller, Kristen....10
Oleksyn, Andrew....522, 1375
Olson, David....765
Orlovsky, Andrew J....1568
Ostadrahimi, Leila....1462
Ovalle, William....1333
Parekh, Minal....1643
Park, Tae Ha....1206, 1218, 1425
Partovi, M....470
Parzybok, Tye....924
Pathak, Chandra....906
Pattermann, Kenneth R....1388
Penner, O....1498
Perkins, J. Taylor....1735
Perlea, Mary....1578
Perlea, Vlad....969, 1058, 1578
Petkovski, Ljupcho....645
Pierce, Charles E....1333
Prashar, Yogesh....808
Prathivadi, Kishen....392
Preston, Al....440
Pupim, Pedro....675
Queen, D....1498
Rahman, Mostaqur....1333
Raitt, Douglas....12
Randle, Timothy....1
Rasekh, Ali....1189
Razavi-Darbar, S....1498
Redell, Chris J....709
Reed III, Gerard E....522
Reelie, Dane....245
Reynoldson, Thomas....522
Rice, John....1359
Richards, Adam....586
Rinehart, Robert V....587
Rittgers, Justin....658
Rittgers, Justin B....1344
Rix, Glenn J....954
Rizzardo, John....1775
Roach, Sharon Nicole....1828
Roadifer, John....602
Robbe, Emmanuel....1480
Rocha, Miguel....1513
Rodrigues, Larry....1375
Rogers, Michael....522
Rohrbach, Ben....907
Rondinel-Oviedo, Efrain....313
Rosen, Jamey....810
Royer, Joseph....1250
Ruark, Morgan D....208
Ruff, Jonathon....676

AUTHOR INDEX

Ruswick, Kevin J. 676
Rutherford, James 74
Rutherford, James H. 1102
Rutledge, John 867
S.Tarbox, Glenn 1178, 1189
Sabol, George 93
Salamon, Jerzy 1465
Samandi, Sam 867
Sanders, Cathi 1567
Santini, Mauricio 675
Sasaki, Reuben 839
Sasanakul, Inthuorn 405, 1333
Saucier, Christopher L. 954
Schaefer, Mel 90, 924
Schaeffer, Kelly R. 11
Scholz, Israel Christian 562
Schreck, David 304
Schreppers, G. 470
Schultz, Mark 1709
Scofield, David H. 1028
Scuero, Alberto 1317
Seed, Raymond B. 969
Seid-Karbasi, Mahmood 505
Serafini, David C. 1058
Serock, Adam 1440
Shannon, Alexander B. 1776
Sharp, Michael 1564, 1566
Shaw, Quentin 1167
Sherwood, Bill 734
Shoghli, Bahareh 1442
Siebenmann, Raphael 810
Silvertooth, D.L. 825
Simpson, Bryan K. 658, 1344
Smith, Haden 329, 839
Sooch, Gurinderbir S. 1233, 1546
Sookhu, Yogi 800
Sossenkina, Elena 1564
Srivastava, Anurag 208
Stanley, Mark 1604, 1624
Starcher, Ryan 1333
Stiady, James 522
Sutter, Amanda 764
Tabrizi, Ali Asghari 720
Tarbox, Glenn 522
Tarbox, Glenn S. 1153
Tatin, Maxime 1417
Taylor, George 924
Teal, Martin J. 1443
Tisdale, Rinda 1796
Toose, James 246
Torres, Raphael A. 1250
Treacy, John 268
Trudgeon, Ray 1102
Tullis, Blake 72
Unterberg, Jessie Drayton 286
Vaghti, Stewart S. 855
Valentyn, James 1090
Vaschetti, Gabriella 1317
Vazinkhoo, Saman 662
Vennapusa, Pavana 613
Venturini, Hillery 659
Vickers, John H. 452
Viducich, Jonathan M. G. 1443
Vroman, Noah 1564, 1565
Wade, Daniel 602
Wagner, Cassandra 658
Wagner, Cassandra M. 1344
Walker, Kent 73
Walker, William 809
Weber, Joseph 1604, 1624
Wehling, Timothy 1250
Weller, Ryan 1827
Werner, John 74
White, David J. 613
Whiteside, Stephen 1119
Whitten, Zachary 45
Wilkes, John 1317
Wong, Rachel 808
Wylde, Travis 787
Yamasaki, Kenji 1028
Yiadom, Atta 697
Yiadom, Atta B. 808
Young, Jeremy 627
Young, Matthew 1796
Zakrevskaya, Yulia 56
Zhou, Jim 522, 1375
Zimmerman, Peter 735
Zoeller, Ben 1461

United States Society on Dams (USSD)
9975 Wadsworth Pkwy.
Ste. K-2, 145
Westminster, CO 80021
USA

ISBN 978-1-5108-6027-8

CURRAN ASSOCIATES INC.
proceedings.com

37th Annual USSD Conference 2017

It's a Small World: Managing Our Water Resources

Anaheim, California, USA
3 - 7 April 2017

Volume 3 of 3

37th Annual USSD Conference 2017

It's a Small World: Managing Our Water Resources

Anaheim, California, USA
3 - 7 April 2017

Volume 3 of 3

ISBN: 978-1-5108-6027-8

Printed from e-media with permission by:

Curran Associates, Inc.
57 Morehouse Lane
Red Hook, NY 12571

Some format issues inherent in the e-media version may also appear in this print version.

Printed by Curran Associates, Inc. (2018)

For permission requests, please contact United States Society on Dams (USSD) at the address below.

United States Society on Dams (USSD)
9975 Wadsworth Pkwy.
Ste. K-2, 145
Westminster, CO 80021
USA

Phone: 303-792-8753
Fax: 303-792-8782

info@ussdams.org

Additional copies of this publication are available from:

Curran Associates, Inc.
57 Morehouse Lane
Red Hook, NY 12571 USA
Phone: 845-758-0400
Fax: 845-758-2633
Email: curran@proceedings.com
Web: www.proceedings.com

TABLE OF CONTENTS

VOLUME 1

SESSION 1A: CONFERENCE THEME

Reservoir Sedimentation and Sustainability 1
Timothy Randle, Sean Kimbrel, Kent L. Collins

Dam Safety and Resilient Water Management: A Case for Dynamic Reservoir Operations 9
Michael McMahon

The Nebraska Department of Natural Resources Coordinated Efforts in Flood Flows 10
Kristen Obermueller, Bob Beduhn, John Engel

SESSION 2A: ENVIRONMENT

Environmental Considerations in Dam Decommissioning 11
Kelly R. Schaeffer

Hydraulic Dredging of the Strontia Springs Reservoir — Lessons Learned 12
Douglas Raitt, Brett Cochran

Dam Decommissioning — Unique Approach by Conversion to a Levee 45
Zachary Whitten

The Prospects for Climate Scenarios in Hydropower Operations 56
Yulia Zakrevskaya, David Huard

SESSION 1B: PHYSICAL MODELING

Linear Weir Size Scale Effects: A Look at Flat Top Weirs 72
Blake Tullis, Kedric Curtis

Physical Model of Spillway and Reservoir Debris Interaction 73
Kent Walker, Jaron Hasenbalg, Kelly Brom

Boundary Dam Spillway No. 1 TDG Modifications 74
John Werner, James Rutherford, Andrew Bearlin, Peter Friz, Jeff Johnson, Daniel Kirschbaum, Joe Groeneveld, Justin Arnold

SESSION 2B: HYDROLOGIC MODELING

A Comprehensive Approach to Probabilistic Hydrologic Hazard Assessment and its Use to Support Risk-Informed Decisions 90
Shaun Carney, Keil Neff, Mel Schaefer, Bruce Baker

Meteorlogical Data Visualization and Analysis Using HEC-MetVue 91
Fauwaz Hanbali

Big Dam in a Small World: Reassessment of the Spillway Design Flood of Lake Sakakewea 92
Joshua Melliger

Application of Regional Hydrologic Criteria in the Rocky Mountains to Estimation of the PMF in Alberta 93
John Menninger, Seifu Guangul, Dan Hoffman, George Sabol, Syed Abbas

SESSION 1C: RISK

Approaches to Estimating Individual Risk for Dam Safety Tolerable Risk Evaluation 104
David S. Bowles, Jason Needham

Minimizing Bias in Geologic and Geotechnical Aspects of Risk Evaluation for Dams and Levees 139
Kerry D. Cato, Derek S. Morley

The Common Cause Adjustment in Dam Safety Risk Analysis: Scope, Purpose, and Applicability 158
Dom Galic

SESSION 2C: RISK

Potential Failure Mode Analysis – Insights and Lessons Learned for the Foundation of a Successful Dam Safety Management Program 174
Frank L. Blackett, Douglas D. Boyer

Epistemic vs Natural Uncertainty in Dam Safety Decision Making: Is it Fair Play? 194
Ignacio Escuder-Bueno, Adrián Morales-Torres, Jessica Castillo-Rodriguez

Consequence Estimation for a Large System of Dams: Our Experiences from TVA Projects 208
Michael M. Crouch, Anurag Srivastava, Morgan D. Ruark

Application of Risk Assessment to an Emergency Levee Design in an Urban Environment 221
Petros Armenis, Malcolm Barker, Peter Christensen, Graham Harrington

SESSION 1D: CONSTRUCTION I

Vibration Free Resonant Sheet Pile Driving in Dykes and Levees 245
Dane Reelie, Matthew Janes
Construction of Borinquen Dam 1E for the Panama Canal Expansion 246
Lelio Mejia, James Toose, Jorge Fernandez
Tsaile Dam — A Complete Approach Unearths a Safer Dam 267
Stephen Brown

SESSION 2D: CONSTRUCTION I

Replacement of Clear Lake Dam: Overcoming Challenges Constructing an RCC Dam on a Soil Foundation 268
Steve Jamieson, Bruce Cotie, Don Lopez, John Treacy
Dewatering — How to Address this Potential Achilles Heel for Construction 286
John W. France, G. Richard Bird, Gregg Batchelder Adams, Jessie Drayton Unterberg
Draft Tube Slab Analysis 304
Jesse Kropelnicki, Kareem Bynoe, David Schreck, Peter Bouchie
Core from Rock, Challenges of Core Construction for a Large High Hazard Dam 313
Michael Miller, Efrain Rondinel-Oviedo

SESSION 1E: RISK INFORMED H&H

Estimating the Reservoir Stage-Frequency Curve with Uncertainty Bounds for Cherry Creek Dam Using the Reservoir Frequency Analysis Software (RMC-RFA) 329
Haden Smith, John F. England Jr.
Recent Paleoflood Analyses for USACE Risk-Informed Dam Safety Evaluations 345
Keith I. Kelson, Christina M. Leonard, Brian M. Hall, John F. England Jr.
Risk Informed Freeboard Evaluation 357
Gregory S. Karlovits, David A. Margo

SESSION 2E: GATES AND VALVES

Fail Safe Design for Gated Structures Construction and Rehabilitation 368
Peter Davis, Mathew McGuire
Dams Inlet/Outlet Works Rehabilitation Program 377
Jack Husted, Sterling Klippel, Keith Lilley
Best Practices in a Tainter Gate Rehabilitation Program 378
Matt Moses, Layne Bukahir
Emergency Shutoff Valves 392
Kishen Prathivadi

SESSION 1F: EMBANKMENT DAMS I

A State Perspective on Tailings Dam Regulation and a Role for USSD and ASDSO 404
Charles Cobb
A Centrifuge Study of Co-Mixed Mine Tailings and Waste Rock Deposits 405
Nonika Antonaki, Tarek Abdoun, Inthuorn Sasanakul
Freeboard Assessment for Long Embankment Dams 420
Mike Gerlach

SESSION 2F: EMBANKMENT DAMS I

Offstream Reservoir Earthen Dam Seepage Systems — Design and Construction Challenges 421
Randall Bushey
Latest Compilation of Internal Erosion Incidents at Bureau of Reclamation Dams 422
William O. Engemoen
Modeling of Internal Erosion in Earthen Embankment Dam 440
Al Preston, Keil Neff, Chunling Li, Lucas de Melo
Turbid Discharge Incident at Cannonsville Dam 452
John H. Vickers, William H. Hover

POSTER SESSION

CONCRETE

Freezing and Thawing Durability of Laboratory and In-Situ Roller-Compacted Concrete Test Specimens 469
Timothy Dolen, Veronica Madera

Parametric Layout of Double-Curvature Arch Dams Using DIANA Finite Element Software 470
E. Goulas, G. Schreppers, M. Partovi, R.M. Gunn

Study of Uplift Pressure Effect in Roller Compacted Concrete Gravity Dam 487
Hichem Mazighi, Mustapha Kamel Mihoubi, Khaled Ghaedi, Zainah Ibrahim

Soil-Structure Interaction Dynamic Analysis for an Existing Old Earth-Rock Fill Dam with Concrete Core 505
Mahmood Seid-Karbasi, Upul Atukorala, Herb Hawson, Bruce Downing

RCC Post Construction Core Testing and Data Analysis For San Vicente Dam Raise Project 522
James Stiady, Andrew Oleksyn, Michael Rogers, Glenn Tarbox, Thomas Reynoldson, Gerard E. Reed III, Wade Griffis, Jim Zhou

CONSTRUCTION

Reconstruction of the Matala Spillway: Hydrotechnical and Safety Aspects 539
Ahmed Bouayad

Decommissioning Outdated Fish Bypass Systems and Preparation for Concrete Retrofit on Lower Granite Lock & Dam 550
Dirk deGroot

Sand Secant Pile Chimney Filter Installation at Pine Creek Dam 551
Bob Faulhaber, Michael Arnold

Execution of a Cut-Off Wall at the Dead Sea – Unique Conditions Require Unique Solution 562
Franz-WernerGerressen, Gai Lehrer, Israel Christian Scholz

Catawba Dam: The Long Road and Construction Update 577
Dennis Hogan

Water-Filled Bladders, an Innovative Bearing Interface for Arch Dam Bracing 578
Alexandre Lochu

Eductors — The Essential Alternative in Complex Dewatering Applications 586
Adam Richards, Harry Bagherzadeh

Application of Intelligent Compaction to Embankment Dam Construction 587
Robert V. Rinehart

Update No. 3 of the Calaveras Dam Replacement Project 602
John Roadifer, Michael Forrest, Erik Newman, Daniel Wade, Susan Hou, Tedman Lee, Carman Ng

VOLUME 2

Validated Integrated Compaction Monitoring at TVA's KIF Coal Combustion Product Stacking Facility 613
David J. White, Pavana Vennapusa, Brendan FitzPatrick, Jason F. Hill, Nicholas McClung, Eric Hageman

To Air is Human; To Succeed, Divine — Evolutionary Research on Grout Enriched RCC Air Entrainment 627
Jeremy Young, Eric Musselman

EMBANKMENT DAMS

Stability Analysis of a Concrete-Faced Rockfill Dam Using Parametric Evaluations 628
Christopher Conkle

Estimating the Peak Friction Angle of Sandy Soils In Situ with State-Based Overburden Normalized SPT Blow Counts 629
Robert A. Jaeger, Ian P. Maki

Comparison Analysis of the Behavior of Rock-Fill Dams with Clay Core at the Variation of Water Level in the Reservoir 645
Ljupcho Petkovski, Stevcho Mitovski

Using Non-Intrusive Geophysical Techniques in Embankment Dam and Levee Risk Assessment 658
Justin Rittgers, Cassandra Wagner, Bryan K. Simpson

Impact of Glacial Till Deposits — Structural Analysis of Green Mountain Dam 659
Hillery Venturini

HYDROLOGY AND HYDRAULICS

Using HEC-RAS 2 Dimensional Capability for Simplified Rapid Dam Break Analysis 660
Wesley Crosby

Hydraulic Modeling of a Complex Dam and Reservoir Project: A Case Study of Two Rivers Dam 661
Stephanie Doyal

Rainfall-Runoff Modeling for the Garrison Dam Drainage with HEC-HMS ..... N/A
Jeff Harris, Kevin Denn, Melissa Larsen

An Innovative Temporary Bulkhead for the Ruskin Dam and Powerhouse Upgrade Project ... 662
Jorge Marín, Saman Vazinkhoo, Rafael Ibáñez-de-Aldecoa, Mark Mallet
Risk and Uncertainty Features in HEC-ResSim ... 674
George Modini, Fauwaz Hanbali
Methodology Applied for Mapping Flooded Areas of the Lajeado HPP ... 675
Pedro Pupim, Mauricio Santini, Milton Dall'Aglio
Leveraging Multi-Sensor Hydrologic Data for Model Validation ... 676
Kevin J. Ruswick, Gregory J. Daviero, Jonathon Ruff

LEVEES

Design of Levee Modifications to Accommodate Long-Term River Scour ... 696
Justin Dominguez, James Nickerson
Detection of Levee Voids at Aqueduct Crossings Using Geophysical Imaging Techniques ... 697
Zhihua Li, Atta Yiadom
Risk Analysis in a Levee Safety Context: The Upper Wood River Levee Case History ... 709
Chris J. Redell, Jose R. Lopez
Effects of Compaction on Erodibility Characteristics of Levees Due to Overtopping ... 720
Ali Asghari Tabrizi, M. Hanif Chaudhry, Jasim Imran

MONITORING

Preventing Catastrophic Failure of Underwater Infrastructure: Condition Assessments by ROV ... 734
Bill Sherwood, Bob Clarke
Automated Data Acquisition System Basics for Dam Safety Instrumentation Monitoring ... 735
Peter Zimmerman, Paul Booth, Greg Dutson

RISK

Failure of the Acude da Nacao Earth Dam: A Probabilistic Perspective ... 749
Danielli de Melo Moura, Rodrigo Borela, Gabriela Alvarado, Wilson Espinoza, Philippe L. Bourdeau

SESSION 3A: MONITORING

Operation and Maintenance of an Instrumentation Program ... 764
Amanda Sutter
FERC Best Practices for Dam Safety Performance Monitoring ... 765
Will Brown, Eric Kennedy, Justin Nettle, Emeruwa Anyanwu, Eric Gross, Jim McHenry, David Olson
Calculation of Threshold Limits for Dam Safety Instrumentation ... 787
Jack Husted, Travis Wylde, Sterling Klippel
Somerset Dam – Innovative Remote Monitoring for Time-Sensitive Breach Detection ... 800
Jerry Cross, Yogi Sookhu, Jud Donaghy

SESSION 4A: MONITORING

Geospatial Tools for Routine Dam Safety Monitoring and Response Applications During Emergencies ... 808
Yogesh Prashar, Atta B. Yiadom, Elizabeth Bialek, Rachel Wong
Special Considerations When Installing Automated Instrumentation at a Site with Access Limitations ... 809
William Walker
Instrumentation and Information Management at Boone Dam ... 810
Rozh Ameen, Raphael Siebenmann, Jamey Rosen

SESSION 3B: HYDRAULIC MODELING

Modeling Dam Breaches and Flood Inundation Using HEC-RAS 5.0's New 2D Capabilities ... 825
D.L. Silvertooth, M. Meshkat
Estimating the Probable Maximum Flood for Whittier Narrows Dam Using a 2D Hydrodynamic Modeling Approach ... 839
Brian M. Hall, Haden Smith, Reuben Sasaki
When an Adjacent Watershed Inundates Your Offline Reservoir: Ukumehame Reservoir Dam Removal Challenges and Design ... 855
Andrew J. Lynch, Stewart S. Vaghti, Dean B. Durkee
Atoka Dam and Spillway Modifications ... 867
Patrick Miles, Brad Kirksey, John Rutledge, Sam Samandi, John L. Hare, Andrew Mishler

SESSION 4B: SITE SPECIFIC PMP

Recent Extreme Storms and Their Relation to PMP 886
Bill Kappel, Doug Hultstrand, Geoff Muhlestein
Needs to Update Probable Maximum Precipitation for Determining Hydrologic Loading on Dams 906
Chandra Pathak
Site Specific Probable Maximum Precipitation Development 907
Jeff Harris, Ben Rohrbach
Advances in Watershed Modeling Inputs and Methods for Dam Safety Risk Analyses 924
Jason Caldwell, Nicole Novembre, Katie Laro, Tye Parzybok, Bruce Barker, Mel Schaefer, George Taylor, Wayne Gibson

SESSION 3C: EARTHQUAKES I

Use of the Conditional Mean for Improved Prediction of Ground Motion Intensity Measures for Embankment Dams 939
Richard J. Armstrong
Seismic Deformation Analysis of Blue Ridge Dam 954
Glenn J. Rix, Lucas de Melo, Christopher L. Saucier, Michael A. Morrison
A Case History Evaluation of State of Practice for Seismic Deformation Modeling of Earthen Dams 969
Khaled Chowdhury, Vlad Perlea, Raymond B. Seed, Michael Beaty, Ethan Dawson, George Hu
Comparison of Two Constitutive Models for Simulating the Effects of Liquefaction on Embankment Dams 1014
Jack Montgomery, Sam Abbaszadeh

SESSION 4C: EARTHQUAKES I

Seismic Evaluation of Lookout Point Embankment Dam in Oregon – Cascadia Subduction Zone Implications 1028
Andrew D. Kost, Kenji Yamasaki, David H. Scofield
Performance of Fena Dam During the 1993 Guam Earthquake 1043
Lelio H. Mejia
Seismic Evaluation of the Remediated Isabella Auxiliary Dam 1058
George Hu, David C. Serafini, Henri V. Mulder, Vlad Perlea

SESSION 3D: CONSTRUCTION II

Seismic Remediation of Perris Dam Using CDSM 1071
Wen Yan Chen, Michael Driller, Steven Friesen
Controlled Blasting for the Red Rock Hydroelectric Project 1090
Rachael V. Bisnett, James Valentyn, Nicholas Johnson
Upriver Dam Spillway Rehabilitation 1102
Ray Trudgeon, James H. Rutherford, Steve Burns, Val Kovalishyn, Ccee Chan, Nikou Jalayeri
Lakeside Ranch Stormwater Treatment Area – Lessons Learned During Design and Construction 1119
Danielle Neamtu, Stephen Whiteside, Stephen Blair

SESSION 4D: CONSTRUCTION II

Design and Repair of a Power Intake Gate Sealing Surface Under Challenging Constraints .... N/A
Mike Carney, Don Thompson, Roland Lwiindi, Eli Lundberg, Joshua Phillips
Post-Tensioning in an Underwater Environment: Wanapum Dam + Monolith 4 Spillway Repair 1135
Paul Krumm
Balancing Water Resources and Power Generation for the Metlakatla Indian Community 1143
Jesse Hutton

SESSION 3E: CONCRETE I

Investigating the Structural Safety of Cracked Concrete Dams 1153
Glenn S. Tarbox, Robin Charlwood, Chris Hayes
A New Approach for the Design of the Köroglu RCC Arch/Gravity Dam 1167
Quentin Shaw
Upgrade and Spillway Modifications at Santa Anita Dam 1178
Sterling E. Klippel, Vik Iso-Ahola, Glenn S.Tarbox, Martina Gelo, George Manole
Understanding the Performance of Aging Infrastructure: A 99-Year-Old Multiple Arch Dam 1189
Vik Iso-Ahola, Glenn S.Tarbox, Mohammadreza Mostafa, Sean Ellenson, Munit Bector, Ali Rasekh

SESSION 4E: CONCRETE I

A Performance-Based Evaluation of Post-Tensioned Anchors Embedded Within a Concrete Gravity Dam 1206
Tae Ha Park, Samantha Hoang, Jonathan Lum, Ziyad Duron

On the Use of Discrete Lumped Element Modeling for Insight into Concrete Dam Behavior 1218
Jonathan Lum, Samantha Hoang, Tae Ha Park, Ziyad Duron, Guy Lund

VOLUME 3

Shear Strength of Concrete Lift Joints from Extensive Lab Testing Compared to Theoretical Results 1233
Dan D. Curtis, Husein Hasan, Gurinderbir S. Sooch

SESSION 3F: EMBANKMENT DAMS II

Seepage Distress and Canal Repairs on the California Aqueduct 1250
Leslie F. Harder, Jeanne Kuttel, Joseph Royer, Timothy Wehling, Joseph Burke, Robert Black, Michael Driller, Christina Kashiwada, Rob Barry, Raphael A. Torres

Geomembranes to Waterproof Embankment Dams 1317
John Wilkes, Alberto Scuero, Gabriella Vaschetti

Geotechnical Investigation of South Carolina Flood-Damaged Dams after the 2015 October Rainfall Event 1333
Inthuorn Sasanakul, Sarah L. Gassman, Charles E. Pierce, William Ovalle, Ryan Starcher, Emad Gheibi, Mostaqur Rahman

Rough River Dam: Phase 1B Exploratory Drilling and Grouting Program 1334
Conrad Ginther, Dmitri Ivanov, Doug Horvath, Rolando Benitez

SESSION 4F: EMBANKMENT DAMS II

Integrating a Multidisciplinary Field Investigation to Improve Assessment of Internal Erosion 1344
Cassandra M. Wagner, Justin B. Rittgers, Bryan K. Simpson

Laboratory Investigation of Backward Erosion Piping into Constricted Seepage Exits 1359
John Rice, Ibrahim A. Ibrahim

Evaluation of Internal Erosion Scenarios in an Embankment Dam 1360
R.J. Fannin, M. Li, C. Anderlini

SESSION 5A: MONITORING

Monitoring San Vicente Dam Raise During Initial Fill 1375
Andrew Oleksyn, Nathan Grima, Anna Kolakowski, Larry Rodrigues, Jim Zhou

First Filling Instrumentation Monitoring of the Folsom Dam Auxiliary Spillway Control Structure 1388
Kenneth R. Pattermann, George Hu

Effect of Antecedent Groundwater Level on Piezometric Response During Flood Events 1403
Chun-Yi Kuo, Richard B. Hockett, Troy S. O'Neal

SESSION 6A: MONITORING

Leak Detection In Earth Dams and Levees Using Optical Fiber Distributed Temperature Sensors 1417
Maxime Tatin, Vincent Lamour, Stéphane Bonelli, Aurélie Garandet

Elephant Butte Dam UAV Data Collection and Photogrammetry 1424
Matthew Klein

Proposed Guidelines for Conducting a Performance-Based Evaluation of a Concrete Dam 1425
Samantha Hoang, Tae Ha Park, Jonathan Lum, Ziyad Duron, Guy S. Lund

Reality Capture for Dam and Hydroelectric Facilities 1440
Adam Serock

SESSION 5B: RESERVOIR SEDIMENTATION

Numerical Sediment Transport Modeling of Operations at Paonia Reservoir, Colorado 1441
Sean Kimbrel

Impact of Climate Change on Sediment Accumulation in the Reservoirs of Embankment Dams: Case Study, Baldhill Dam, North Dakota 1442
Bahareh Shoghli, Yeo Howe Lim

Sediment Transport Through Lake Clarke and Lake Aldred 1443
Jonathan M. G. Viducich, Martin J. Teal

SESSION 6B: RESERVOIR OPERATIONS

To Release or Not to Release? Dam Operation Modeling Using Gridded Short-Term Forecast Rainfall 1460
Maged Aboelata
USACE CWMS: Case Studies of Water Management Decision Support Tools 1461
Ben Zoeller, Nathan Lefkovitz
Systems-Based Risk Analysis of Lake Mendocino Forecast-Informed Reservoir Operations with HEC-WAT 1462
Matthew McPherson, Leila Ostadrahimi
HEC-ResSim 3.3 — New Features to Support Complex Studies 1463
Joan Klipsch

SESSION 5C: EARTHQUAKES II

A Direct Finite Element Method for Nonlinear Analysis of Semi-Unbounded Dam-Water-Foundation Rock Systems 1464
Anil K. Chopra, Arnkjell Lokke
Numerical Assessment of Hydrodynamic Loads Induced During Seismic Interaction Between Reservoir and Concrete Dam 1465
Jerzy Salamon, Jonna Manie
Seismic Back Analysis of Monticello Arch Dam – Blind Prediction Workshop and Additional Analyses 1480
Emmanuel Robbe

SESSION 6C: EARTHQUAKES II

Non-Linear Time Domain Analysis of Ruskin Dam in LS-DYNA Including Reservoir-Foundation-Structure Interaction 1498
O. Penner, B. Bergman, S. Razavi-Darbar, D. Queen, P. Léger, Martin Leclerc, Yannick Boivin
Boca Dam Spillway Nonlinear Analysis and Validation Study 1513
Jennifer E. Huggins, Roman Koltuniuk, Miguel Rocha
Nonlinear Seismic Evaluation of Perris Dam Outlet Tower 1528
Yusof Ghanaat, Zachary Harper
Nonlinear Seismic Analysis of an Arch Dam Subjected to Very Intensive Earthquake Loading 1546
Gurinderbir S. Sooch, Dan D. Curtis

SESSION 5D: LEVEES

U.S. Levee Safety Coalition and International Community of Practice, Recent Activities and Updates 1564
Elena Sossenkina, Michael Sharp, Noah Vroman
USACE Levee Portfolio Report 1565
Noah Vroman
Levee Safety Guidance Comprehensive Document Review 1566
Michael Sharp
Inventory and Review of the Nation's Levees 1567
Jamie McVicker, Cathi Sanders
Levee Safety Practices in the Keystone State 1568
Andrew J. Orlovsky

SESSION 6D: LEVEES

Seismic Vulnerability Analysis of Slag Cement-Cement-Bentonite Seepage Cutoff Walls for Levees 1578
Vlad Perlea, Khaled Chowdhury, Mary Perlea, Michael Kynett
Advances in Design Procedures for Relief Well Lines in Variable Urban Settings 1604
Robert A. Jaeger, Richard A. Keizer Jr. Graham C. Bradner, Joseph Weber, Mark Stanley
Evaluation and Design of Irregularly-Positioned Relief Wells 1624
Joseph Weber, Mark Stanley, Robert Jaeger, Graham Bradner
Explaining Pore Pressure Changes Due to Piping Through Changes in Hydraulic Conductivity 1643
Carolyne Bocovich, Willem Kanning, Minal Parekh, Michael Mooney
Design and Construction of Feather River West Levee Improvements 1657
Christopher Krivanec, Les Harder, Michael Bessette

SESSION 5E: CONCRETE II

Simplified Nonlinear Seismic Analyses of Gravity Dams 1658
Elisa Messio, Wenbo Duan, Najib Bouaanani, Benjamin Miquel
Strontia Springs Dam: Methods to Better Understand Seismic Behavior 1673
Aimee L. Corn, Guy S. Lund, Darren J. Brinker, Erin Gleason

Cracking in Thin Arch Concrete Dam – Nonlinear Dynamic Structural Analysis 1685
Sherry Hamedian, Roman Koltuniuk

SESSION 6E: CONCRETE II

Comparison of Nonlinear Finite Element Analysis Methods Used to Simulate Post-Tensioned Anchorage in Concrete 1709
Vojislav Cvijanovic, Mark Schultz, Blake Dolve
Structural Inflection Points, and How They Enhance the Understanding of Structural Safety 1724
Guy S. Lund, Kevin Marshall, Todd Barber
Pseudo-Nonlinear Finite Element Analysis of Concrete Gravity Dams 1735
J. Taylor Perkins, Husein A. Hasan, Daniel A. Gilbert
Importance of RCC Mixture Lab Studies for the Design of Large Dams 1747
A. Garduno, M. Montero, F. Ballhausen

SESSION 5F: SAFETY & SECURITY

Managing Threats Through Effective Security Programs 1762
J. A. Gloschat, A. C. Baehr
Unmanned Aerial Vehicles – A Public Agency Approach to Deployment and Protection 1775
Katharine Killeen, John Rizzardo
The Economics of Dam Security Post 9/11 and the Threat to Consumer Confidence 1776
Alexander B. Shannon

SESSION 6F: SAFETY & SECURITY

Flood Risks from Spillways: Modeling and Communicating Non-Dam-Failure Impacts to Native American Tribes 1796
Rinda Tisdale, Matthew Young, Chris Clopper
City of Dubuque, Iowa: Defining Emergency Actions with a Levee Breach Evaluation 1803
Andrew McCoy, Deron Muehring, Rusty Jones
Public Safety and Unauthorized Extreme Activities at Dams 1827
Steven Barfuss, Brian M. Crookston, Kade Beck, Ryan Weller
Public Safety and Social Media 1828
Sharon Nicole Roach
Author Index

SHEAR STRENGTH OF CONCRETE LIFT JOINTS FROM EXTENSIVE LAB TESTING COMPARED TO THEORETICAL RESULTS

Dan D. Curtis (M.A.Sc., P.Eng.)[1]
Husein Hasan[2]
Gurinderbir S. Sooch (M.A.Sc., P.Eng.)[3]

ABSTRACT

During recent assessment of the Tennessee Valley Authority (TVA) dams, TVA has developed a systematic approach for determining shear strength parameters for use in the stability analysis. This systematic approach involved extensive field investigations which included drilling and sampling of concrete and rock cores. The core samples were then used in a laboratory strength testing program. As part of this work, concrete core strength testing involved testing for compressive strength, tensile strength (direct and splitting), and shear strength (lift joints and parent concrete). This paper will present the results of the strength testing (graphically) and provide discussion on the results. The results from several TVA dams will be presented. A comparison of direct tension and splitting tensile strength is provided and discussed. In addition the measured shear strength of lift joints and parent concrete is compared to results from previously published theoretical equations (Curtis, 2011). Finally, recommendations will be made for estimating shear strength of concrete (including lift joints) which can be used in the assessment of concrete gravity and arch dams.

INTRODUCTION

The shear strength of shear keys in concrete dams is important because during intense seismic loading a significant portion of the loads could be transferred to the abutments through arching and shear. Shear keys are important structural elements in both arch and gravity dams. A gravity dam in a narrow valley can transfer relatively large horizontal forces to the abutments during an earthquake provided the shear keys are sufficiently strong to transfer the additional loads. As a result of this lateral load transfer, the horizontal forces at the base of the dam blocks are reduced and the stability of the dam is enhanced.

In the evaluation of existing dams and spillways in high seismic zones, it is typically found that current maximum design earthquake peak ground acceleration is many times larger than the earthquake acceleration used in the original design, hence utilizing strength contributions from all available stabilizing resistance is important. Also, spillway piers are subjected to large shear stresses due to cross-valley motion.

[1] Senior Project Manager, HATCH Ltd, 4342 Queen St, Suite 500, Niagara Falls ON Canada L2E 7J7 +1 905 357 6998, dan.curtis@hatch.com
[2]Senior Program Manager, Tennessee Valley Authority, 400 West Summit Hill Dr. (WT10C-K), Knoxville, TN 37902; hahasan@tva.gov, 865.632.4194
[3] Civil/Structural Engineer, HATCH Ltd, 4342 Queen St, Suite 500, Niagara Falls ON Canada L2E 7J7 +1 905 374 0701 Ext 5252, gurinderbir.sooch@hatch.com

The objective of this paper to present a simplified shear strength equation and compare the shear strength estimates to that measured from extensive concrete core testing.

REVIEW OF FAILURE CRITERIA

It is informative to compare shear failure criteria for concrete. The following three criteria were compared:

- Mohr Coulomb Envelope
- Parabolic Mohr Envelope
- Griffith Envelope.

(a) Mohr Coulomb Envelope – The Mohr Coulomb strength criterion is given as:

$$\tau = c + \sigma_n \tan\phi \qquad (1)$$

Where:
c is cohesion (shear strength at zero normal stress)
σ_n is the normal stress applied perpendicular to the shear plane
ϕ is the friction angle.

Figure 1 shows the Mohr Coulomb strength envelope. In Figure 1, $\beta = \frac{\pi}{4} + \frac{\phi}{2}$. The cohesive strength can be expressed as a function of the uniaxial compressive strength and the friction angle as follows:

$$c = f_c' \frac{(1 - Sin\,\phi)}{2\ Cos\,\phi} \qquad (2)$$

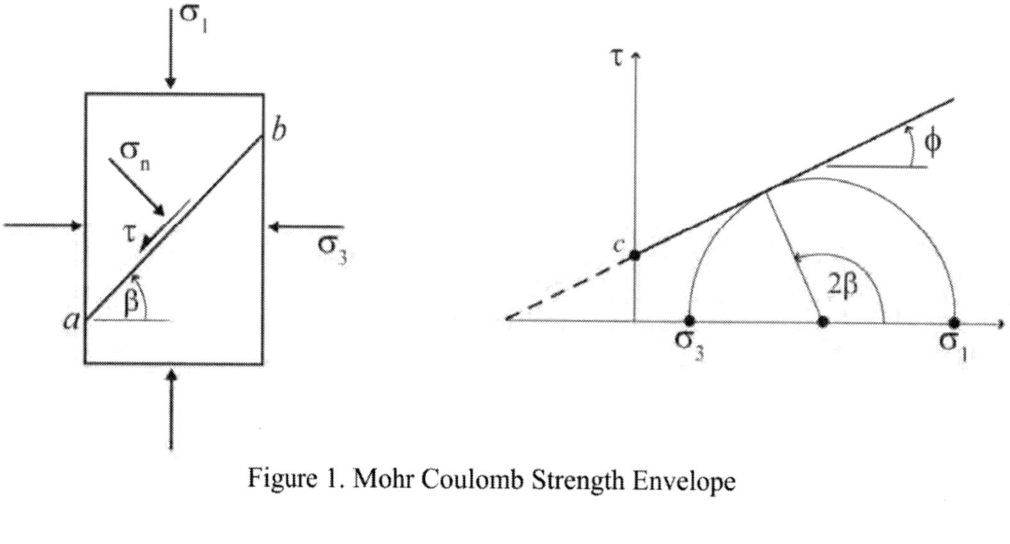

Figure 1. Mohr Coulomb Strength Envelope

(b) Parabolic Mohr Envelope – Typical shear strength test results for concrete show a curved failure surface. Figure 2 shows a parabolic fit to Mohr circles for uniaxial compression and tension test results. Mee (2001) developed a shear strength envelope using a parabolic fit to strengths from Mohr circles. The derivation is given in Appendix A. From Figure 2, the shear strength envelope is defined as follows:

(3)

$$\tau^2 = \left[f_c' - 2f_t \left(-1 + \sqrt{1 + \frac{f_c'}{f_t}} \right) \right] (\sigma_n + f_t)$$

at $\sigma_n = 0$

$$\tau = c = f_t \left(\frac{f_c'}{f_t} + 2 - 2\sqrt{1 + \frac{f_c'}{f_t}} \right)^{1/2}$$

(c) Griffith Criterion – The Mohr Coulomb criterion is empirical whereas the Griffith criterion is mechanistic. Griffith postulated that fracture of brittle materials is initiated at tensile stress concentrations at the tips of hypothetical minute elliptical cracks (Griffith cracks) in the material. The Griffith criterion (Griffith, 1921) is given as follows:

(4)

$$\tau = 2\sqrt{f_t(\sigma_n + f_t)}$$

Where:

τ is the peak shear strength

f_t is the tensile strength

σ_n is the normal stress

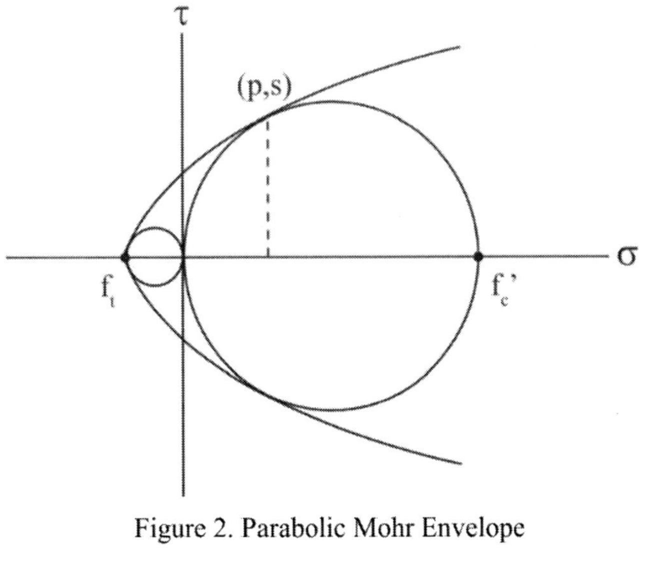

Figure 2. Parabolic Mohr Envelope

The Griffith Envelope is shown in Figure 3.

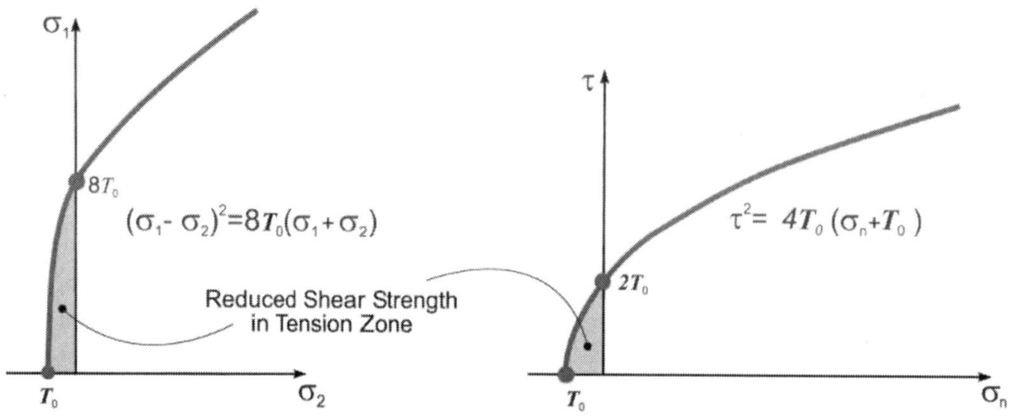

Figure 3. Griffith Envelope

SELECTED FAILURE CRITERION

The Griffith failure criterion was selected to estimate the shear strength of shear keys and lift joints, and to compare with results obtained from direct shear testing at low normal stresses. The Griffith criterion was selected for the following reasons:

- It is a mechanistic criterion based on thermodynamics.
- It assumes the presence of micro cracks which are now known to exist, i.e., real materials contain imperfections.
- It predicts failure at stresses lower than the empirical Mohr Coulomb criterion.
- It is directly applicable to shear failure of concrete which is a brittle failure mode, i.e., the Griffith criterion is not applicable to ductile failure.

From Figure 3, a tensile normal stress causes a significant reduction in shear strength. However, shear keys do not transfer tension across the joint, hence the lowest strength applicable to shear keys is the cohesive strength of concrete which is determined using $c = 2 \times f_t$ where f_t is the uniaxial tensile strength of concrete. The cohesive strength, c, is the shear strength at zero normal stress. The fact that shear keys do not transfer tension across the joint offers a significant benefit to limiting tensile forces on the shear keys themselves and preserving its ability to transmit shear forces across keyed vertical contraction joints in compression and shear. It is noted that Lo et al (1990) found the same relationship between tensile strength and cohesion as Griffith after testing well bonded dam/foundation contacts. In the work (Lo et al, 1990), the tensile strength was determined using direct tension tests.

FIELD INVESTIGATION FOR TENNESSEE VALLEY AUTHORITY DAMS

In recent years TVA has developed a systematic approach for determining material properties (i.e., of dam concrete and foundation rock) for hydro projects in TVA's portfolio. This systematic approach involved extensive field investigations which included drilling and sampling of concrete and rock cores. The core samples were then used in a laboratory strength testing program. As part of this work, concrete core strength testing involved testing for compressive strength, tensile strength (direct and splitting), and shear strength (lift joints and parent concrete).

COMPRESSIVE AND TENSILE STRENGTH OF MASS CONCRETE

Figure 4 shows the compressive strength gain over time for Guntersville, Cherokee, Douglas, Hiwassee, Pickwick, Kentucky, Fort Loudoun and Norris Dams. Figure 5 presents the ratio of present day compressive strength and 28 day compressive strength for tested TVA dams. Figure 5 also provides the present day age of various dams. From Figures 4 and 5, a considerable gain in compressive strength over time has been observed.

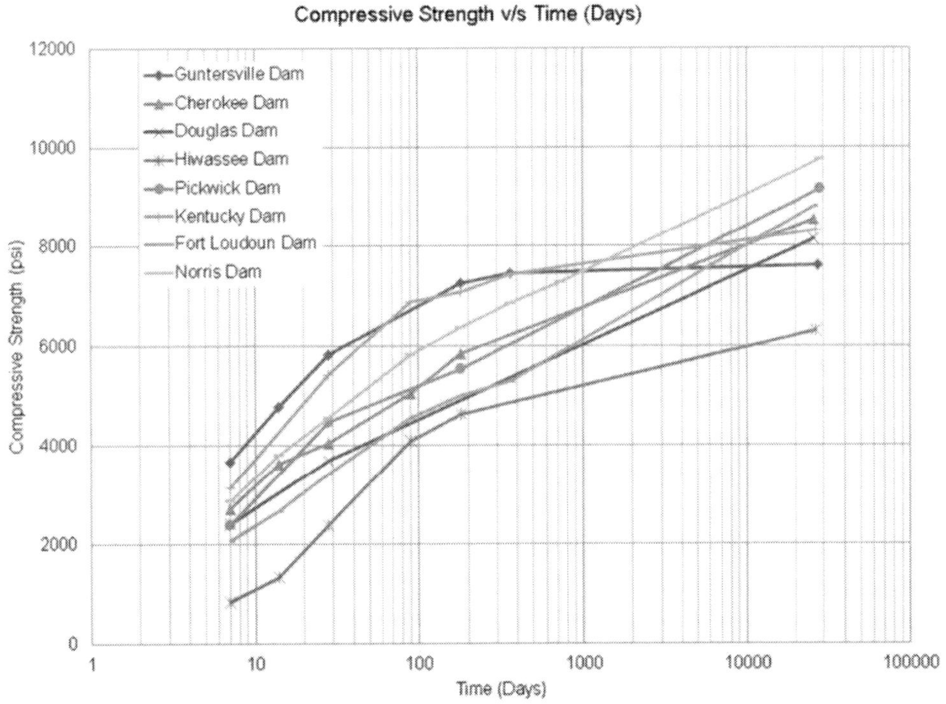

Figure 4. Compressive Strength for Tested TVA Dams

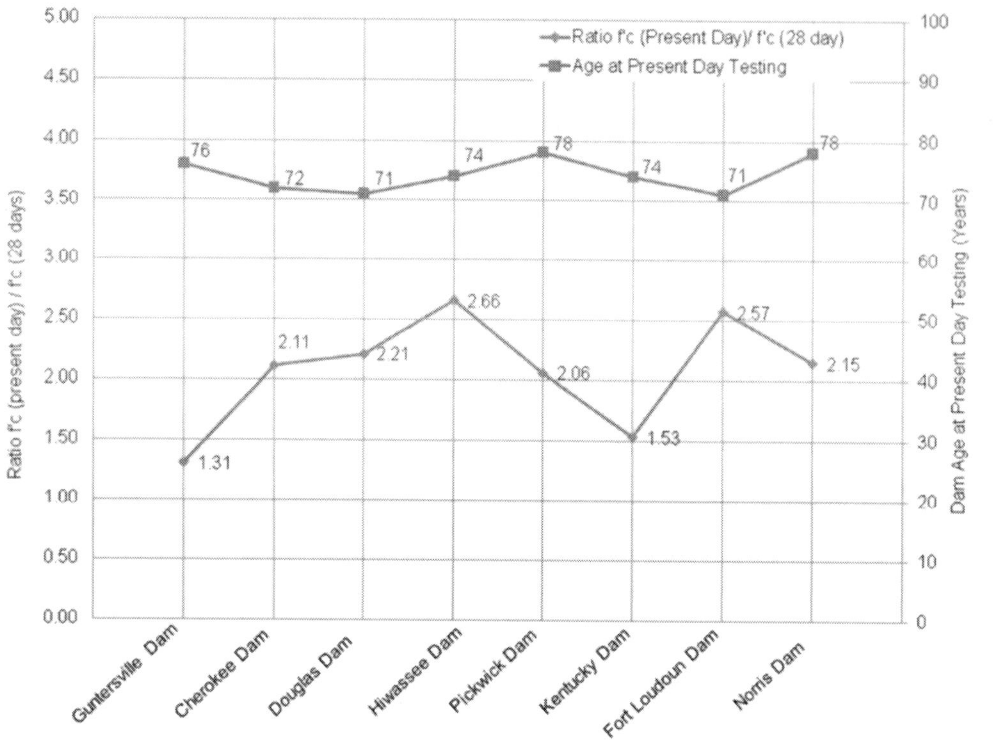

Figure 5. Ratio of Present Day Compressive Strength and 28 day Compressive Strength for Selected TVA Dams

Figure 6 presents the concrete strength data for Hiwassee Dam. The data shown in Figures 5 and 6, regarding the initial strengths, was taken from respective TVA project Brown Books (TVA, 1946). From Figure 6, the ratio of tensile strength and compressive strength varies from 0.17 to 0.12 over a period of 6 months (180 days). Thus, over a period of 6 months, a gain in tensile and compressive strength was observed.

From recent testing at Hiwassee Dam, the direct and splitting tensile strengths were 190 and 990 psi. Also, from recent testing the measured compressive strength is 6,300 psi. It should be noted the dam is affected by Alkali Aggregate Reaction (AAR) whereby the concrete is expanding over time. AAR causes micro cracking in the concrete, thus it tends to affect the direct tensile strength results to a much greater degree than the splitting tensile strength test results. It is further noted that the AAR-induced expansion is somewhat restrained in the dam and this causes an increase in concrete compressive stresses particularly in the longitudinal direction. The act of coring the concrete samples releases these compressive stresses and further increases the amount of micro cracking in the core samples. The measured early tensile strength test results used in Figure 6 were obtained from modulus of rupture test results (TVA, 1946) and then multiplied by 0.75 to obtain the concrete tensile strength as per Raphael (1984).

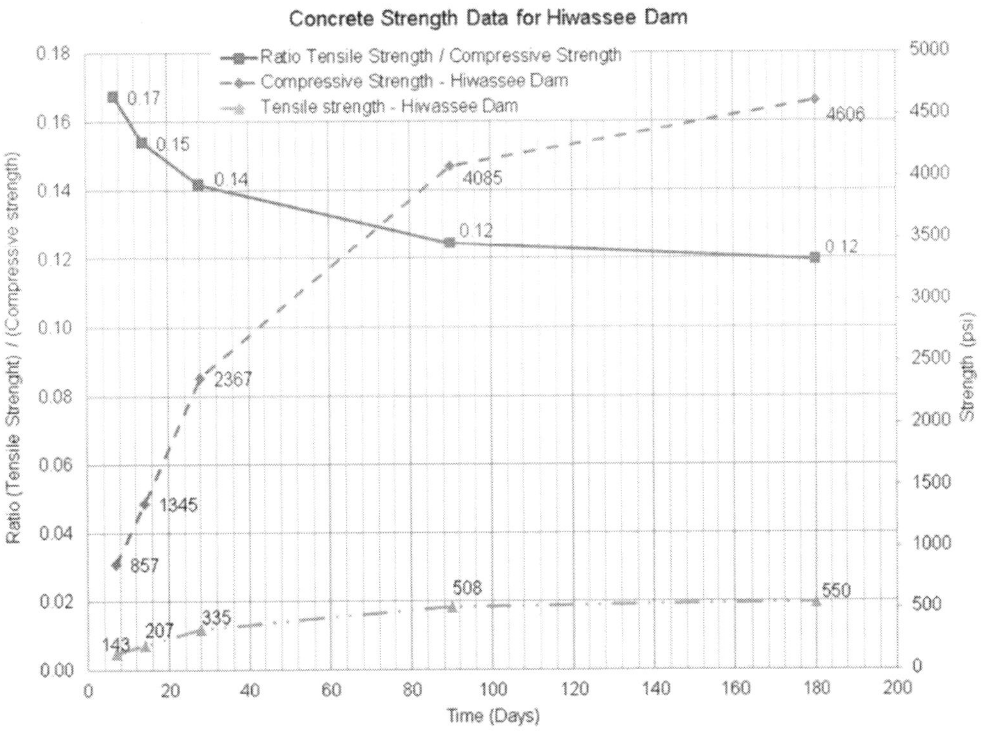

Figure 6. Concrete Strength Data for Hiwassee Dam

Figure 7 presents the compressive and tensile strength (includes both splitting and direct tensile strength) for tested TVA dams. The following equations representing the splitting and direct tensile strength were developed from Figure 7.

Splitting Tensile Strength: $$f_t = 0.1 f'_c \quad (5)$$

Direct Tensile Strength: $$f_t = 0.04 f'_c \quad (6)$$

It should be noted the direct tensile strength concrete samples include both intact concrete and concrete lift joint samples. Also, the concrete core sample diameter for various TVA dams varies from 3.3 to 6 inches. Also, the maximum aggregate size for various TVA dams varies from 3 to 6 inches. It was not considered practical to test larger diameter cores given the schedule restraints for the stability assessment projects. Figure 8 presents a comparison between the TVA Test Data Splitting; Direct Tensile Strength; and Raphael Tensile Strength (Raphael, 1984). From Figure 8, the tensile strength measurements using the direct tensile strength measurements from TVA were significantly lower than the splitting results from both Raphael and the splitting tests of TVA dams. Dolen et al (2014) present a good discussion on the splitting and direct tension testing of concrete cores taken from concrete dams.

It is noted that on many of the concrete cores taken from TVA dams it was difficult to locate lift joints due to the excellent condition of the lift joint. As a result, we believe Equation (6) provides a reasonable estimate of the concrete tensile strength of lift joints to be used in the shear strength equation provided in the next section of this paper. We do not recommend using Equation (6) for assessment of dams subjected to tensile stresses induced by bending under static or dynamic loading. In the case of dynamic tensile strength, it would be very useful to instrument a large number of concrete dams and record their responses to actual earthquakes and back-calculate the actual dynamic tensile strength.

It is of interest to note that Darbar et al (2016) presented average direct tensile results from a USBR database that ranged from $f_t = 0.03f_c'$ to $f_t = 0.04f_c'$ (excluding Ruskin Dam). The variation was dependant on the lift joint preparation. In general, the TVA dam lift joints were prepared using layer of grout that were wire broomed into the joint. From the database, the lift joints that were broomed and washed had a direct tensile strength corresponding to $f_t = 0.03f_c'$.

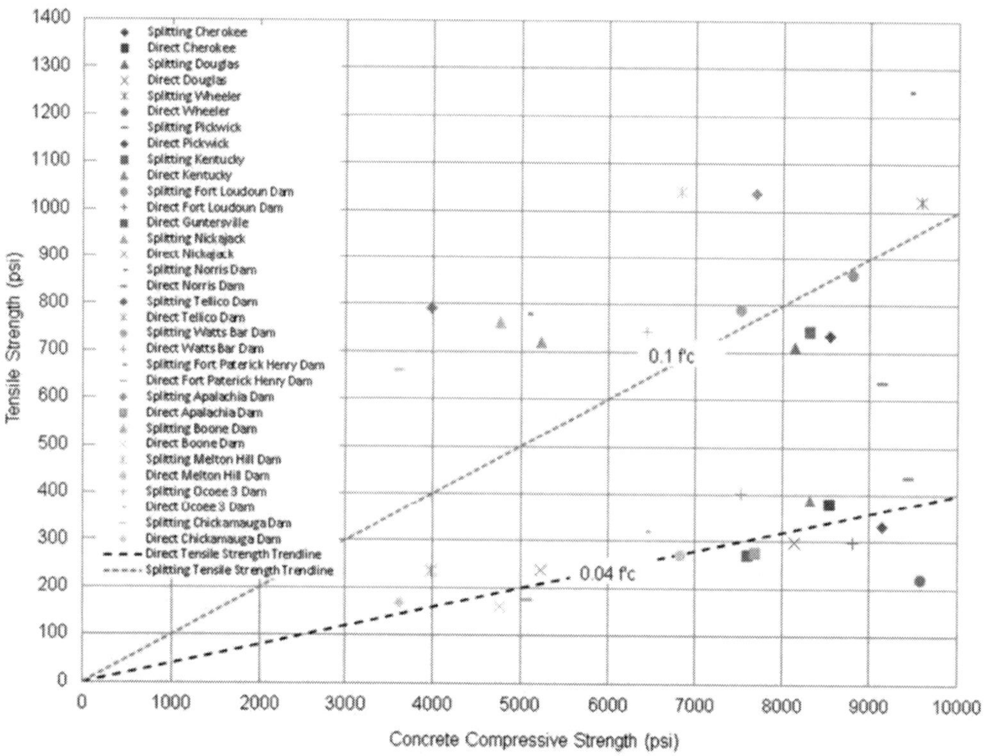

Figure 7. Compressive and Tensile Strength (includes both splitting and direct tensile strength) for Tested TVA Dams

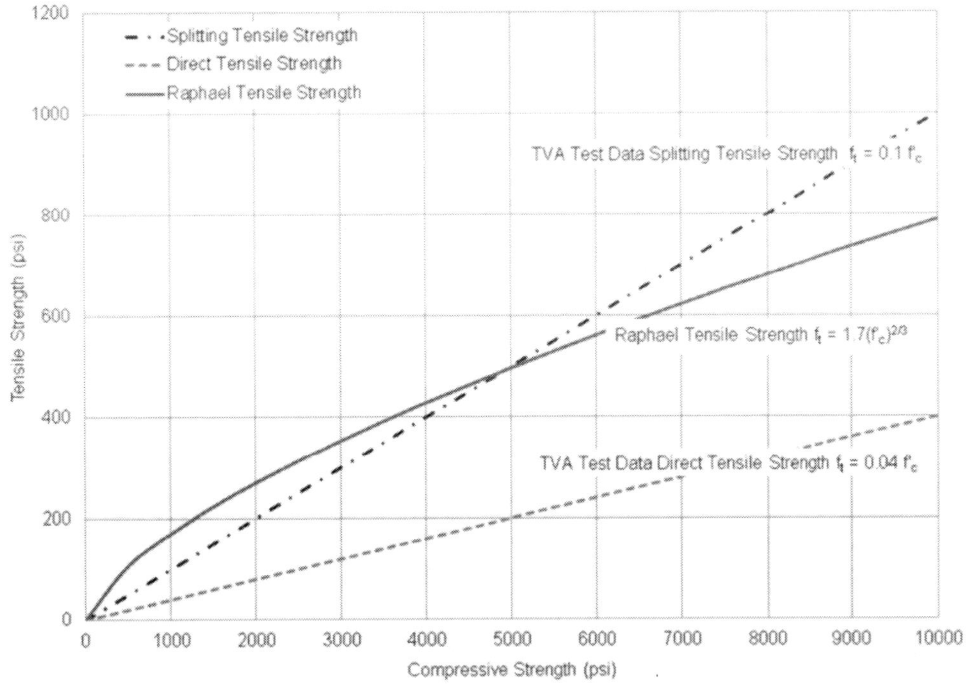

Figure 8. Comparison between the TVA Test Data Splitting; Direct Tensile Strength; and Raphael Tensile Strength

PEAK SHEAR STRENGTH OF BONDED CONCRETE LIFT JOINTS

A summary of the measured peak shear strength of bonded concrete lift joints for selected TVA dams is provided in Table 1.

Shear strength calculations were done using the Griffith criterion:

$$\tau = 2\sqrt{f_t(\sigma_n + f_t)} \tag{7}$$

Where:
τ is the peak shear strength
f_t is the tensile strength, $f_t = 0.04\ f'_c$ from Equation (6)
σ_n is the normal stress

It is important to note that the core size or aggregate size should not affect the test results for concrete lift joint as the aggregate do not cross the lift joints. Figures 9 to 19 present the measured and calculated shear strength of bonded concrete lift joints for TVA dams. From Figures 9 to 19, a reasonable match between the calculated and measured shear strength has been achieved. It should be noted that the results presented for Wheeler Dam in Figure 11 were an exception as the measured tensile strength does not fit the equation, i.e., $f_t = 0.04\ f'_c$ ($f_t = 0.04$ x 9,580 =380 psi). Therefore, we have used the measured tensile strength value, i.e., f_t = 220 psi, rather than the calculated tensile strength of 380 psi. However, it should be noted that with the measured tensile strength in the Griffith criterion, a reasonable comparison has been achieved between the measured and calculated shear strength of bonded concrete lift joints.

Table 1. Summary of Peak Shear Strength of Bonded Concrete Lift Joints for TVA Dams

S. No.	Description	Compressive Strength (f'_c) psi	Calculated Tensile Strength (f_t= 0.04 f'_c) psi	Calculated Peak Shear Strength at Normal Stress (σ_n) = 0 using Griffith criterion psi	Remarks
1	Cherokee Dam	8,540	340	680	-
2	Douglas Dam	8,130	325	650	
3	Wheeler Dam	9,580	220*	440	Wheeler Dam is an exception, the measured tensile strength, i.e., 220 psi was used, instead of calculated tensile strength, i.e., 380 psi from Equation (6)
4	Pickwick Dam	9,150	366	732	-
5	Kentucky Dam	8,316	332	664	
6	Nickajack Dam	5,230	210	420	
7	Norris Dam	9,750	390	780	
8	Apalachia Dam	7,688	307	614	
9	Boone Dam	4,670	186	372	
10	Melton Hill Dam	6,840	270	540	
11	Ocoee 3 Dam	6,020	240	480	

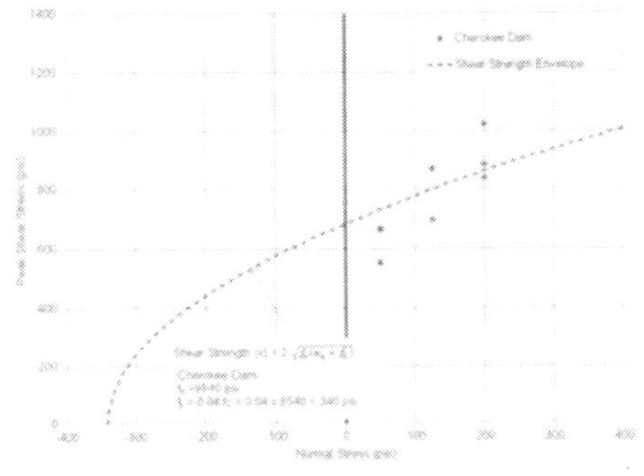

Figure 9 Measured and Calculated Peak Shear Strength for Bonded Concrete Lift Joints at Cherokee Dam

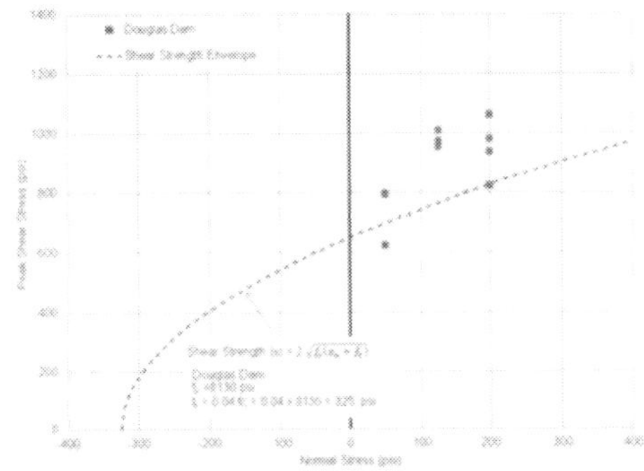

Figure 10 Measured and Calculated Peak Shear Strength for Bonded Concrete Lift Joints at Douglas Dam

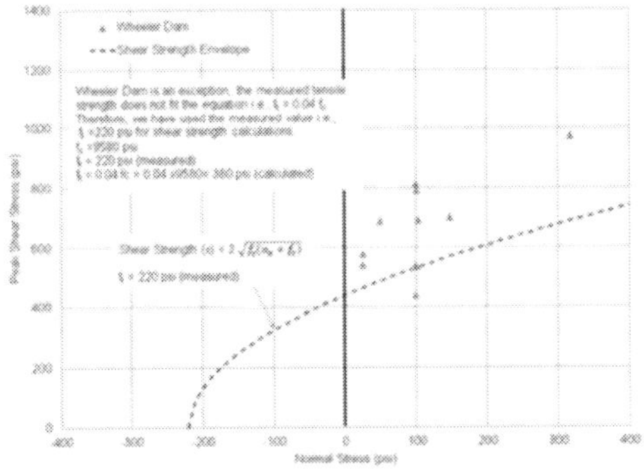

Figure 11 Measured and Calculated Peak Shear Strength for Bonded Concrete Lift Joints at Wheeler Dam

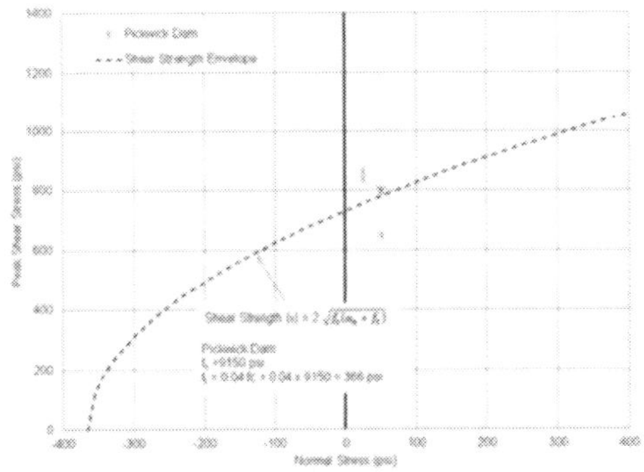

Figure 12 Measured and Calculated Peak Shear Strength for Bonded Concrete Lift Joints at Pickwick Dam

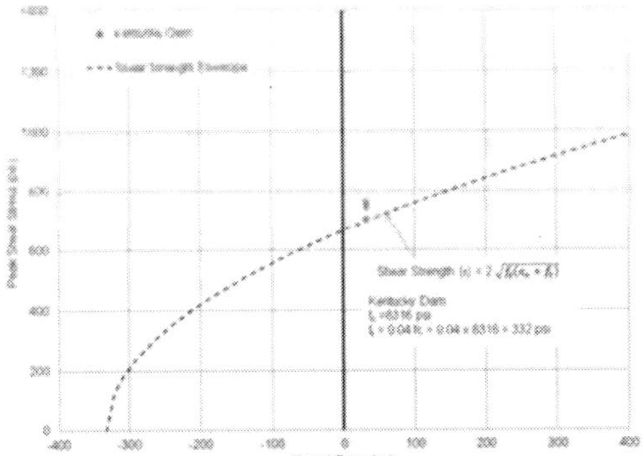

Figure 13 Measured and Calculated Peak Shear Strength for Bonded Concrete Lift Joints at Kentucky Dam

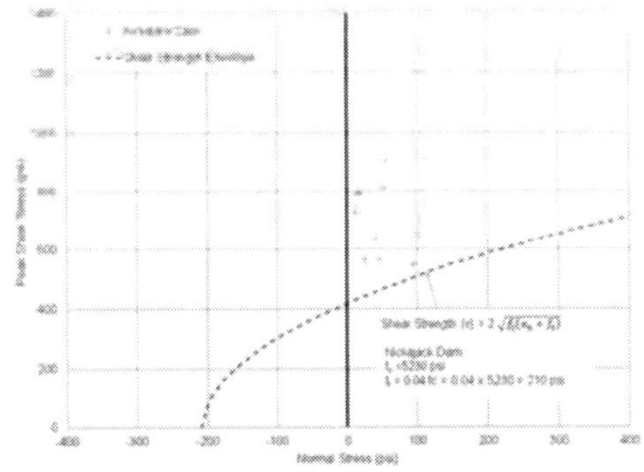

Figure 14 Measured and Calculated Peak Shear Strength for Bonded Concrete Lift Joints at Nickajack Dam

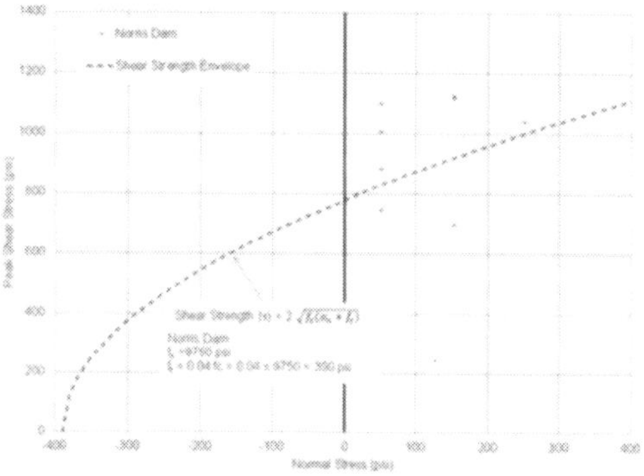

Figure 15 Measured and Calculated Peak Shear Strength for Bonded Concrete Lift Joints at Norris Dam

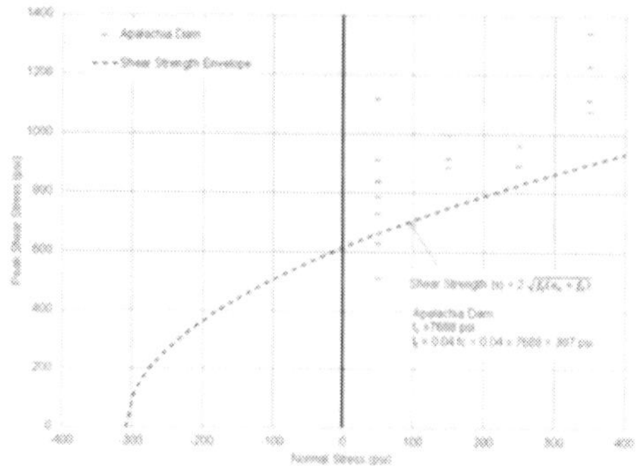

Figure 16 Measured and Calculated Peak Shear Strength for Bonded Concrete Lift Joints at Apalachia Dam

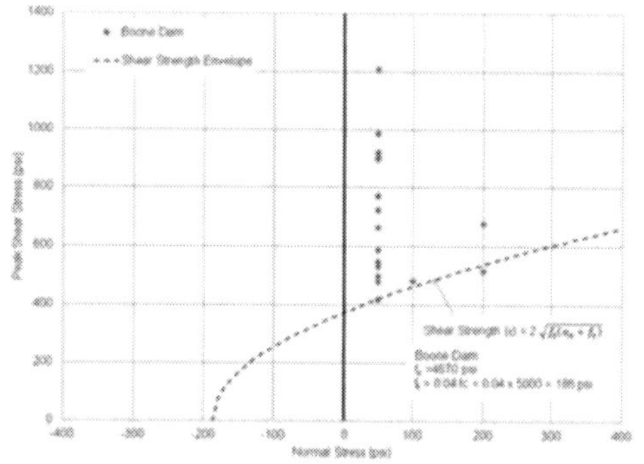

Figure 17 Measured and Calculated Peak Shear Strength for Bonded Concrete Lift Joints at Boone Dam

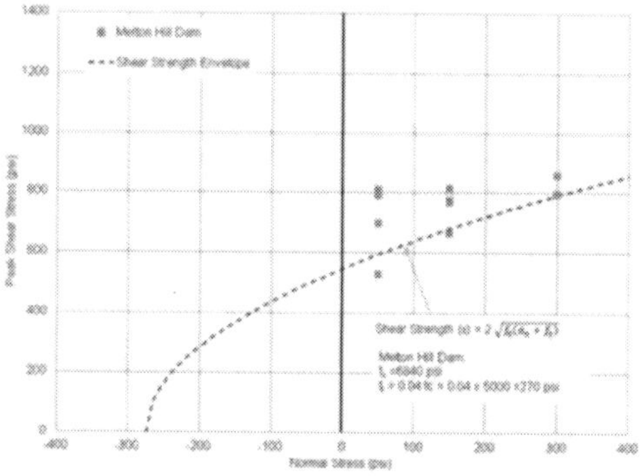

Figure 18 Measured and Calculated Peak Shear Strength for Bonded Concrete Lift Joints at Melton Hill Dam

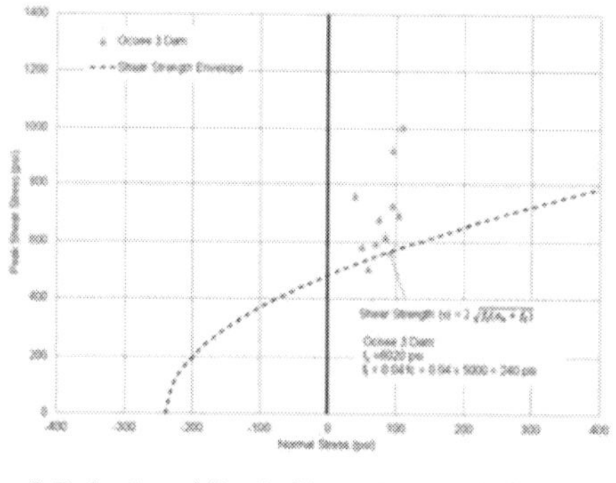

Figure 19 Measured and Calculated Peak Shear Strength for Bonded Concrete Lift Joints at Ocoee 3 Dam

BONDED LIFT LINE COHESIVE STRENGTH FROM OTHER DAMS

Dolen (2011) provided a summary of concrete material properties from the aging concrete information system (ACIS) from the Bureau of Reclamation. The database contains historical mass concrete core test results dating back to 1905. The Dolen paper also provides a summary of primary approaches used for lift line preparation methods used during construction. The focus of this current paper is on shear strength of bonded lift lines, therefore a comparison of cohesion results from the Dolen paper is provided in Table 2 along with the predicted values using the Griffith criterion, i.e., Equation (7). From Table 2, the cohesive strengths predicted by the Griffith criterion agree very well with the measurements from the ACIS database.

Table 2. Tensile and Cohesive Strength of Bonded Lift Lines from Dolen (2011) and Predicted Cohesive Strength Using Griffith Criterion

Years	Estimated % Bonded Lift Lines	Primary Surface Preparation Procedure	Average Tensile Strength (psi)	Average Compressive Strength (psi)	Tensile Strength % of Compressive Strength (%)	Measured Cohesion with 90% Tests Exceeding (psi)	Predicted Cohesion Using Griffith Criterion (psi)	Ratio: Predicted Cohesion / Measured Cohesion
1905 to 1933	50	Brooming and washing	100	3,680	2.7%	210	200	0.95
1934 to 1964	85	Green Cutting and Sandblasting	165	5,040	3.3%	350	330	0.94
1965 to 1993	74	Sandblasting or high pressure water blasting	195	5,620	3.5%	455	390	0.86

POST-PEAK SHEAR STRENGTH OF BONDED CONCRETE LIFT JOINTS

The post-peak shear strength parameters based on Patton (1966) for saw-tooth specimen are as follows:

$$\tau = \sigma_n \tan(\phi_b + i) \tag{8}$$

Where:

τ is the post-peak shear strength

σ_n is the normal stress

ϕ_b is the basic friction angle of the surface. The basic friction angle represents the frictional component of shear strength only when measured on naturally textured surfaces with no dilation during shearing. The basic friction angle of discontinuity or joint is independent of the size of the surface being tested.

i is the angle of the saw-tooth face or roughness angle.

The direct shear test samples for the post-peak shear strength have some roughness along the surface of the tested plane. The resistance provided by these discontinuities is represented by roughness angle (i) defined earlier in Patton (1966).

From Hencher and Richards (1989), the roughness angle (i) and basic friction angle (ϕ_b) are given in Equation (8), therefore phi basic (ϕ_b) can be expressed as follows:

$$\tau_c = \tau_u - \tan(i) \tag{9}$$

$$\phi_b = \tan^{-1}\left[\frac{\tau_c}{\sigma_n}\right] \tag{10}$$

Where:
τ_c is the corrected post-peak shear strength
τ_u is the un-corrected post-peak shear strength

Dilatancy can be defined as the change in the value of normal displacement as a result of applied shear force. As dilation occurs, work is done in lifting the sample over the discontinuities and this results in an increasing shear force being required to continue shearing. As mentioned earlier, Patton (1966) used Equation (8) to express the contribution due to the dilation.

Therefore, the dilation contribution to shear strength can be accounted for by the geometry of the sample. Therefore, the incremental roughness angle can be calculated throughout the test by considering the incremental vertical (*dv*) and horizontal (*dh*) displacements as follows:

$$dv/dh = \tan(i) \tag{11}$$

Therefore, once the roughness "i" is known, it is then substituted into Equation (9) to compute the corrected shear strength. Figure 20 presents a sample calculation for Kentucky Dam post-peak shear strength (psi) for combined parent concrete and bonded concrete lift joints with 6 inch diameter core samples.

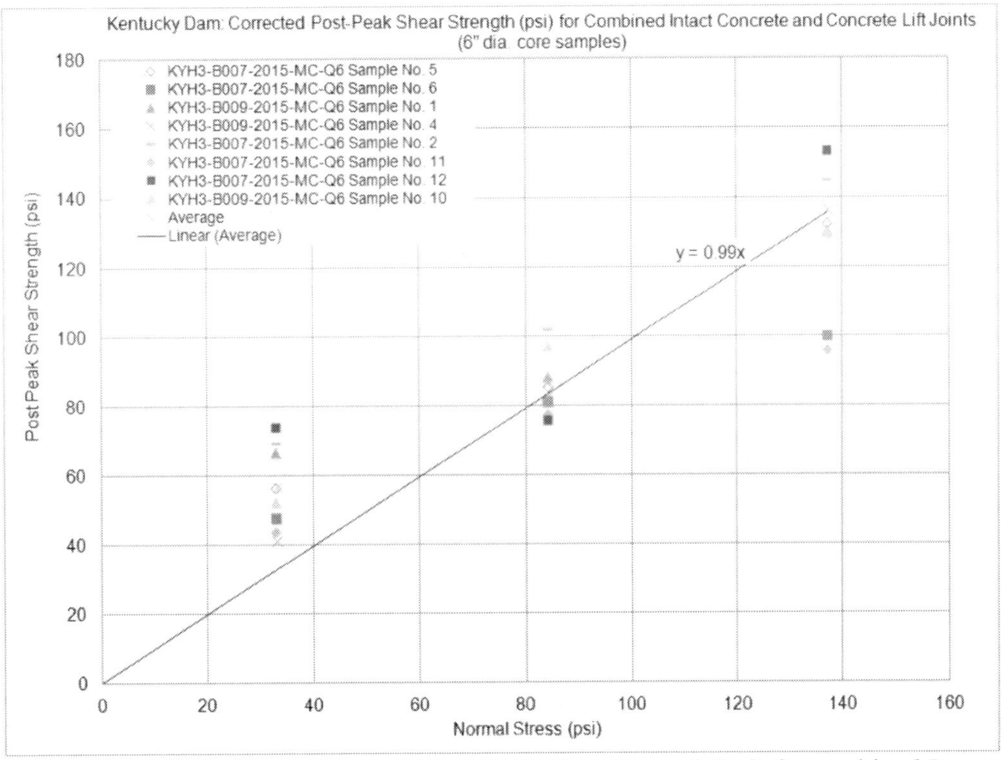

Figure 20. Kentucky Dam: Corrected Post-Peak Shear Strength (psi) for combined Parent Concrete and Bonded Lift Joints (6 inch diameter core samples)

CONCLUSIONS AND RECOMEMNDATIONS

The following conclusions are drawn from this paper:

- The tensile strength to be used in the Griffith shear strength criterion derived from direct tensile tests of concrete cores taken from TVA dams is estimated to be $0.04\,f_c'$. However, a lower bound estimate of $0.03\,f_c'$ can be used as a conservative estimate.
- The Griffith shear strength failure criterion using the above tensile strength estimates provide a very good envelope of shear strengths measured on concrete cores taken across lift joints of TVA dams.
- The Griffith failure criteria also provided a very good match to data from the Bureau of Reclamation ACIS database measurements of cohesive strength of bonded concrete lift lines which had wide range of joint preparation during construction.
- It is important that post-peak concrete lift line shear strength measurements be corrected for dilation measured during shear tests.

ACKNOWLEDGEMENTS

The authors would like to acknowledge the field and testing work done by other consultants i.e., Stantec, GEI, URS and Arcadis for TVA. We would also like recognize the tremendous diligent work done by TVA Project and Plant Staff.

REFERENCES

Griffith, A.A., "The Phenomena of Flow and Rupture in Solids", Phil. Trans. Royal Society: London, Vol. A221, 1921.

Curtis, D.D., "Estimated Strength of Shear Keys and Bonded Joints in Concrete Dams", 31st Annual USSD Conference, San Diego, California, April 11-15, 2011.

Darbar, S.R., Queen, D., Hatton, C., Dolen, T. and Bartojay, K. "Static and Dynamic Concrete Material Properties of a Concrete Gravity Dam", 36th Annual USSD Conference, Denver, Colorado, April 11-15, 2016.

Dolen, T.P., Selecting Strength Input Parameters for Structural Analysis of Aging Concrete Dams, 31st Annual USSD Conference, San Diego, CA, 2011.

Dolen, T.P., Harris, D.W. and Nuss, L.K., "Tensile Strength of Mass Concrete – Implication of Test Procedures and Size Effects on Structural Analysis of Concrete Dams", 34th Annual USSD Conference, San Francisco, California, April 7-11, 2014.

Hencher, S. R. and Richards, L.R.,"Laboratory Direct Shear Testing of Rock Discontinuities", Ground Engineering, March 1989.

Lo, K.Y., Lukajic, B., Wang, S., Ogawa, T. and Tsui, K.K., 1990. "Evaluation of Strength Parameters of Concrete-Rock Interface for Dam Safety Assessment: Session 2 – Dam Safety Assessments", Proceedings of Canadian Dam Safety Conference, September 17-20, 1990. Toronto, Ontario. pp 71-94.

Mee, A.L., 2001. "Personal Communication".

Patton, F. D., "Multiple modes of shear failure in rock", Proceedings of 1st congress international society of Rock Mechanics, Lisbon 1, 509-513, 1966.

Raphael, J. M., 1984. "Tensile Strength of Concrete", *ACI Journal*. Technical Paper, Mar-Apr 1984. Title No. 81-17.

Tennessee Valley Authority, "The Hiwassee Valley Projects", Technical Report No. 5, Volume 1, Washington, 1946.

APPENDIX A – PARABOLIC MOHR ENVELOPE – FROM A. L. MEE (2001)

Equation for a parabola

$\tau^2 = 4a(\sigma + \tau)$

differentiate $2\tau \dfrac{d\tau}{d\sigma} = 4a$

$\therefore \quad \dfrac{d\tau}{d\sigma} = \dfrac{2a}{\tau}$

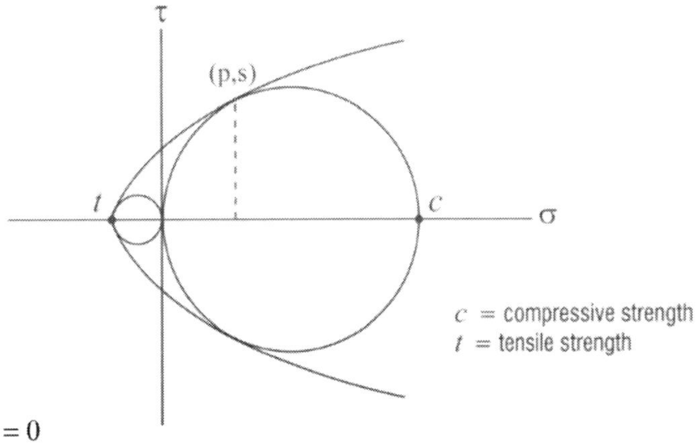

Compression circle

$\tau^2 + \sigma^2 - c\sigma = 0$

differentiate $2\tau \dfrac{d\tau}{d\sigma} + 2\sigma - c = 0$

$\therefore \quad \dfrac{d\tau}{d\sigma} = \dfrac{c - 2\sigma}{2\tau}$

At tangent contact, assume $\sigma = p, \ \tau = s$

$4a(p + t) = cp - p^2$

$\dfrac{2a}{s} = \dfrac{c - 2p}{2s}$

$4a = c - 2p$

$(c - 2p)(p + t) = cp - p^2$ or $p^2 + 2pt - ct + 0$ then $p = t\left(-1 + \sqrt{1 + \dfrac{c}{t}}\right)$

$\therefore \quad a = \dfrac{c}{4} - \dfrac{t}{2}\left(-1 + \sqrt{1 + \dfrac{c}{t}}\right)$

And $\tau^2 = \left[c - 2t\left(-1 + \sqrt{1 + \dfrac{c}{t}}\right)\right](\sigma + t)$

at $\sigma = 0, \ \tau = t\left(\dfrac{c}{t} + 2 - 2\sqrt{1 + \dfrac{c}{t}}\right)^{1/2}$

SEEPAGE DISTRESS AND CANAL REPAIRS ON THE CALIFORNIA AQUEDUCT

Leslie F. Harder, PhD, PE, GE[1]
Jeanne Kuttel, PE[2]
Joseph Royer, PE[3]
Timothy Wehling, PE[4]
Joseph Burke, PE[5]
Robert Black, PE, GE[6]
Michael Driller, PE, GE[7]
Christina Kashiwada, PE[8]
Rob Barry, CEG[9]
Raphael A. Torres, PE[10]

ABSTRACT

The California Aqueduct is the main conveyance facility for California's State Water Project. Numerous geotechnical issues were considered during the original design of the California Aqueduct in the 1960's and several design measures were implemented to address these issues. However, despite these measures, several reaches along this very long aqueduct have experienced various forms of seepage distress over time. The most common forms of seepage distress experienced have been associated with hydrocompaction, soluble soils, and rapid drawdown of the aqueduct pools. Many of these have required emergency repairs to prevent failures of the canal embankments over the last several decades of operation.

This paper describes the various mechanisms of seepage distress experienced along canal sections of the California Aqueduct over the last 45 years. It also presents the different types of emergency repairs implemented and describes how the approaches to the repairs have evolved over time. One of the major changes over time for many of the repairs has been to switch from a standard remove-and-replace approach to the use of layered water barriers composed of geosynthetic liners overlain by or sandwiched between thin layers

[1] Senior Professional Associate, HDR Engineering Inc., 2365 Iron Point Road, Suite 300, Folsom, CA 95630. Les.Harder@hdrinc.com
2 Chief, Division of Engineering, Department of Water Resources P.O. Box 942836 Sacramento, CA 95814. Jeanne.Kuttel@water.ca.gov
[3] Principal Engineer, Division of Engineering, Department of Water Resources P.O. Box 942836 Sacramento, CA 95814. Joseph.Royer@water.ca.gov
[4] Supervising Engineer, Division of Engineering, Department of Water Resources P.O. Box 942836 Sacramento, CA 95814. Tim.Wehling@water.ca.gov
[5] Senior Engineer, Division of Engineering, Department of Water Resources P.O. Box 942836 Sacramento, CA 95814. Joseph.Burke@water.ca.gov
[6] Senior Engineer, Division of Engineering, Department of Water Resources P.O. Box 942836 Sacramento, CA 95814. Robert.Black@water.ca.gov
[7] Senior Engineer, Division of Engineering, Department of Water Resources P.O. Box 942836 Sacramento, CA 95814. Mike.Driller@water.ca.gov
[8] Senior Engineer, Division of Engineering, Department of Water Resources P.O. Box 942836 Sacramento, CA 95814. Christina.Kashiwada@water.ca.gov
[9] Supervising Engineering Geologist, Division of Engineering, Department of Water Resources P.O. Box 942836 Sacramento, CA 95814. Rob.Barry@water.ca.gov
[10] Deputy Director, retired, Department of Water Resources P.O. Box 942836 Sacramento, CA 95814. torresraphael13@yahoo.com

of shotcrete. Other remedial measures included the use of various grouting measures, sheetpile cutoff walls, and seepage berms. The paper also summarizes the costs and the expedited construction times required to complete the emergency repairs together with a brief assessment of the performance of the repairs to date.

INTRODUCTION

The State Water Project (SWP) is the largest state-built, multipurpose water project in the United States. It was planned, designed, and constructed by the California Department of Water Resources (DWR) and currently includes 34 storage facilities and over 660 miles of canal and pipeline. Its main purpose is water supply, and it accomplishes this by storing surplus water during wet periods and conveying water to areas of need throughout most of the state.

The California Aqueduct is the main conveyance facility for the SWP. It begins at the Banks Delta Pumping Plant in the Sacramento-San Joaquin Delta, located to the east and upstream of San Francisco Bay, and runs south along the west side of California's Central Valley (see Figure 1). At the southern end of the valley, water is lifted by a series of pump stations. The largest, Edmonston Pumping Plant, lifts the water over 1,926 feet to cross the Tehachapi Mountains into Southern California. In Southern California, the California Aqueduct splits into two branches: the West Branch terminates at Castaic Lake, and the East Branch terminates at Lake Perris. Between the Delta and Lake Perris, the California Aqueduct runs over 444 miles and includes over 385 miles of concrete-lined canal. Additional branches of the California Aqueduct, consisting mainly of pipelines, include the South Bay Aqueduct, the Coastal Aqueduct, and the East Branch Extension.

CANAL DESIGN

General

The California Aqueduct is divided into a series of pools that can be isolated from each other by closing radial gates at check structures along its length. The main portion of the California Aqueduct has flow capacities ranging between 4,400 and 13,100 cfs, with capacities generally decreasing as the canal runs south to Edmonston Pumping Plant. The capacity of the East Branch in Southern California is currently only about 2,650 cfs. The canal prism is lined with unreinforced concrete panels, typically 2½ to 4½ inches thick with contraction joints generally spaced no more than 12½ to 15 feet in both directions. Elastomeric sealants were applied to grooved joints during the initial construction contracts, but in later contracts were replaced by cruciform-shaped waterstops in the lining joints. Interior canal slopes are typically sloped at 1.5H:1V, but flatten to 2:1 in areas of weaker soil and rock. In the larger canal sections in the northern portion of the California Aqueduct, the invert commonly has a width of approximately 40 feet and a wetted depth ranging up to 32.8 feet.

Figure 1. Location of the California Aqueduct and other facilities of the State Water Project (from DWR Bulletin 132-10, 2013)

The canal alignment along the western side of the Central Valley resulted in a series of balanced cuts and fills in order to provide construction material for the embankment sections. In fill sections, canal embankments were commonly composed of two zones: a "compacted embankment" zone adjacent to the canal prism and an outer "normal embankment" constructed under the embankment crown and landward. The compacted embankment was specified to have a minimum width of 7 feet at the top of the concrete lining and to have been placed in 6-inch lifts and compacted to 95 percent of DWR's compaction standard (20,000 ft-lb. per cubic foot), which is intermediate in compaction effort between ASTM D698 and D1557. The outer normal embankment was specified to be placed in 2-foot-thick lifts and compacted by construction equipment routed over the fill. The outer slopes of the normal embankment were generally constructed to 2:1 slopes and were designed to have a minimum factor of safety of 1.5 for normal operating conditions.

Geotechnical Challenges

Several geotechnical challenges were considered during the original design of the California Aqueduct. These included seismicity, slope stability, shallow subsidence or hydrocompaction, deep subsidence, swelling soils, soluble soils, and high groundwater. Design measures to address these issues included:

- Alignments were modified where practical to avoid several specific problem areas.
- Unstable canal slopes were either flattened or removed and replaced with compacted fill.
- Extra freeboard was provided to mitigate deep subsidence induced by future groundwater and petroleum extraction.
- Plastic soils with high swell potential were over-excavated a few feet and backfilled with imported compacted fill.
- In areas of high groundwater, finger drains of sand and gravel were placed at intervals along the canal slope immediately beneath the concrete lining. The finger drains led to longitudinal toe drains near the base of the slopes which were connected to sumps that could be pumped out when the canal was to be dewatered.
- Soluble soils and rock were generally not considered to be a major problem during the original design. However, in some areas, such soils or rock were over-excavated a few feet and backfilled with imported compacted fill.
- Shallow subsidence or hydrocompaction was mitigated in some areas by pre-saturating alluvial soils with ponded water.

Hydrocompaction

One of the most significant geotechnical challenges addressed during design was the potential for *hydrocompaction*, a form of shallow subsidence in collapsible soils induced by wetting. Galloway et al. (1999) defined hydrocompaction as:

> *"The process of volume decrease and density increase that occurs when certain moisture-deficient deposits compact as they are wetted for the first time since burial. The downward movement of the land surface that results from this process has also been termed 'shallow subsidence' and 'near-surface subsidence."*

Hydrocompaction had previously been recognized as a significant problem in the Central Valley for decades. Soils susceptible to hydrocompaction have a weak skeletal structure with a high void ratio. When wetted, the structure collapses causing significant surface settlements. Many of the soils susceptible to hydrocompaction were believed to have originated in debris flows flowing out of narrow canyons and spreading out onto alluvial fans. Such flows moved in the form of slurries of unsorted clays, silts, sands, gravels and boulders and dried rapidly after flow had ceased. During the design of the California Aqueduct, numerous laboratory and field tests were conducted along potential canal alignments to determine the potential for hydrocompaction. Many of the field tests consisted of test plots where an 8-foot-diameter corrugated metal pipe was embedded vertically into the soil. Within the CMP, gravel-packed wells ranging from 5 to 25 feet were installed. The CMP was then filled with water and settlement was allowed to occur. The resulting subsidence over the next few days was sometimes dramatic, commonly between 2 and 11 feet in susceptible areas, producing concentric settlement and cracking patterns (see Figure 2). The results of the geotechnical investigations and field tests led to the identification of areas along the California Aqueduct where the potential for hydrocompaction was believed to be significant (see Figure 3). In addition, DWR carried out a series of subsidence and pre-consolidation tests at its Mendota Test Site, located in the southern San Joaquin Valley. At this test site, different methods were tested to evaluate the most effective means of pre-compaction/consolidation of unstable soils susceptible to hydrocompaction. A prototype canal section was also constructed and evaluated. Figure 4 presents an aerial photograph of some of the test plots at the Mendota Test Site. Note that the settlement cracks on the ground surface induced by different pre-consolidation methods extended hundreds of feet beyond the footprint of the prototype canal section.

The mitigation measure that was finally adopted for canal reaches in the San Joaquin Valley considered susceptible to significant hydrocompaction was to pond water along the canal alignment in the early 1960's prior to Aqueduct construction. In this way, most of the shallow subsidence potential could be induced prior to constructing the Aqueduct. Hundreds of pre-consolidation ponds were built for this purpose, with many of the ponds containing gravel-packed infiltration wells to facilitate saturation. Pre-consolidation ponds were commonly 200 feet wide and 500 feet long. Ponding was continued for up to 2½ years and generally resulted in surface settlements ranging between 1 and 9 feet (see Figure 5).

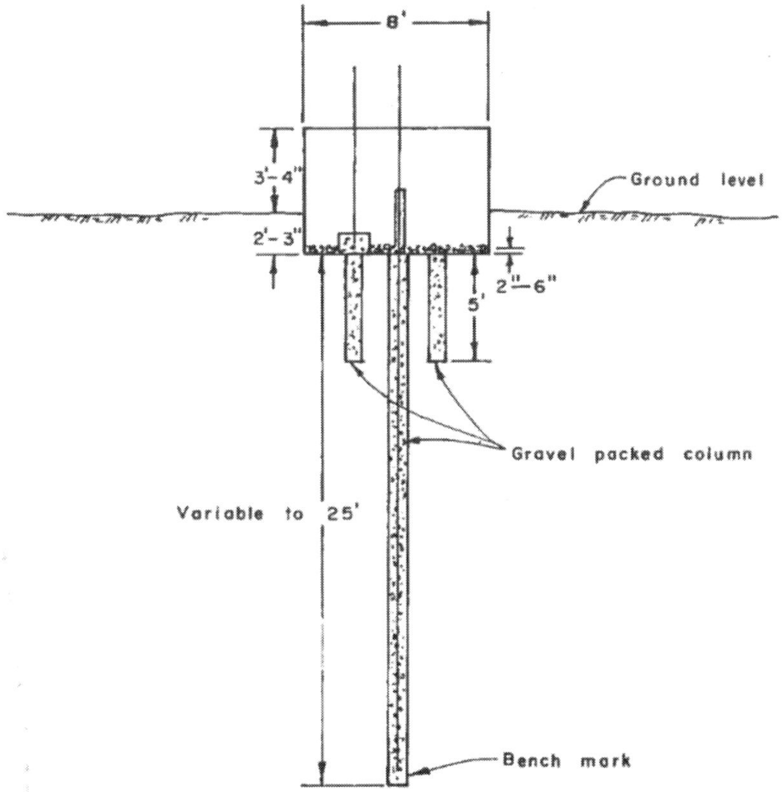

Figure 2. Subsidence Test Plot SP-7 located in the Southern San Joaquin Valley (from DWR, 1964)

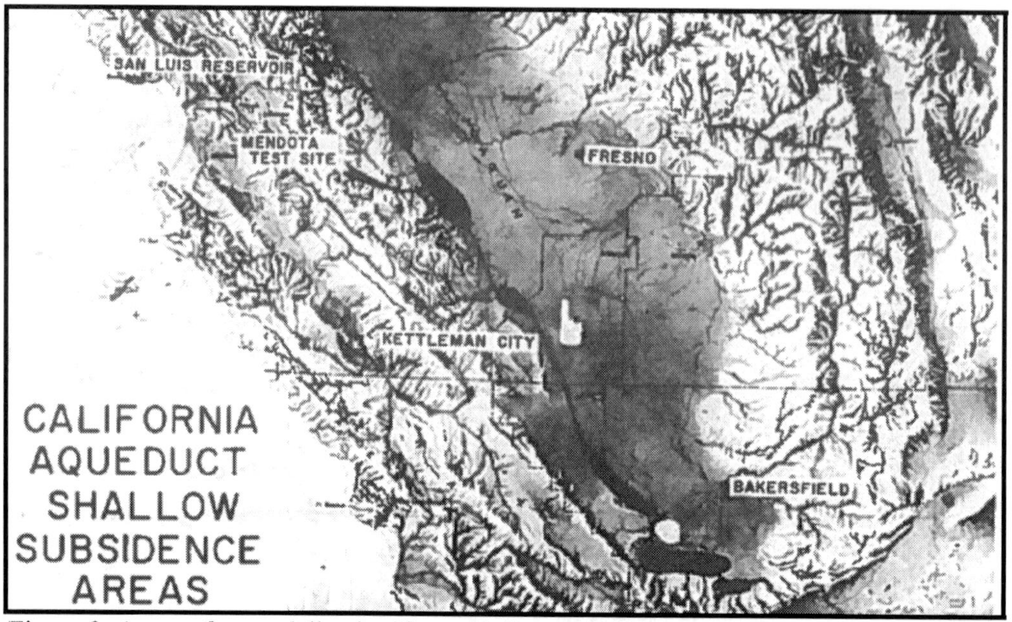

Figure 3. Areas of potentially significant shallow subsidence, or hydrocompaction, along the California Aqueduct (from DWR, Bulletin 200, 1974)

Figure 4. Pre-consolidation test plots at DWR Mendota Test Site (from DWR Photograph Library – circa 1964)

Figure 5. Pre-consolidation subsidence ponds constructed along the alignment of the California Aqueduct – note large cracking resulting from the induced settlements (from DWR Bulletin 200, 1974)

CANAL FAILURES INDUCED BY INITIAL FILLING

General

Canal failures along the California Aqueduct that have resulted in an uncontrolled release of aqueduct water have been relatively rare over the approximate 45-year life of the system. Two notable examples associated with initial filling are described in the following sections.

October 12, 1971 East Branch Canal Failure at Milepost 328

The first canal failure occurred in Pool 48 along the East Branch in Southern California on October 12, 1971 near Milepost 328 (see Figure 6). This location is approximately 32 miles upstream of the Pearblossom Pumping Plant. The failure occurred at the site of a box culvert running beneath a full embankment section. The embankment was founded on recent alluvium in a small drainage between two cut sections. Approximately 120 feet of canal was destroyed, and all but two of the 25-foot-long sections of the culvert were washed away. The pool where the failure occurred, Pool 48, had been filled for the first time only the previous day, less than 24 hours earlier. The failure and release flows had obliterated any evidence as to the cause of the failure, but engineers believed at the time that that a crack or cracks within the embankment allowed water to seep and pipe material out along the culvert. Excessive embankment settlement and/or foundation subsidence was not believed at the time to have contributed to the failure. Although there are soils along the East Branch that are susceptible to hydrocompaction, no pre-consolidation ponding had been done in this area. Subsequently, canal reaches in this area have exhibited significant distress which has been associated with hydrocompaction (e.g. Milepost 333 – see subsequent section). Therefore, the conclusion made at the time of the failure that hydrocompaction was not involved may not have been accurate.

Figure 6. Aerial views of October 12, 1971 Milepost 328 East Branch canal failure (from DWR Photograph Library)

The repair of this canal failure consisted of excavating the break area to solid foundation material and replacing the destroyed culvert with a 78-inch reinforced concrete pipe. Compacted embankment was then placed back to the original dimensions and the concrete canal lining was replaced. Water was flowing through the canal again on December 20, 1971 (see DWR Bulletin 200, 1974).

January 16, 1984 Lower Quail Canal Failure

Another early failure occurred on the West Branch along Lower Quail Canal on January 16, 1984. Lower Quail Canal is a large canal having a maximum design depth of 42.3 feet. It was constructed in three phases between 1967 and 1981 upstream of Pyramid Lake in southern California. The canal runs over a series of sandstone knobs and alluvial valleys and was constructed as a balanced cut and fill with a sidehill excavation on the left and a compacted embankment on the right. The canal has a concrete lining overlying a 2- to 3-foot-thick clay sub-lining. Up until December 26, 1983, the depth of water in the canal was no more than 27 feet. Between December 26, 1983 and January 11, 1984, the water level was raised in stages to have a depth of 37 feet and was held at this level until the canal break five days later on January 16, 1984.

The canal failure occurred as a result of a break in the right embankment at Station 283+25. The breach ran on a skewed alignment through the embankment to approximately Station 280+00 and had an average width of about 40 feet (see Figure 7). After the breach, a five-member investigative panel was convened to determine the cause of the failure. The panel concluded that pre-existing cracks in the embankment allowed water to seep through and erode relatively cohesionless embankment materials. These cracks were believed to have been caused by "*differential settlement of the embankment foundation where the foundation changes gradually from relatively non-compressive sandstone and thin overburden to deep, alluvial deposits of low-density silty soils.*" Surveys revealed that there had been relatively little settlement of the canal embankments where they were founded on the sandstone ridges, but up to 0.9 feet of settlement in areas where the embankments crossed alluvial valleys.

Figure 7. Aerial views of January 16, 1984 Lower Quail Canal failure
(from DWR Photograph Library)

Figure 8 illustrates the changing foundation conditions near the 1984 breach on Lower Quail Canal. In the upper cross section along Section A-A at Station 283+50, the canal embankment is founded on a relatively thin alluvial soil deposit overlying rock. However, in the lower cross section along Section B-B at Station 280+00, the alluvium has substantially thickened, thus allowing for significant potential differential settlement along a relatively short length along the canal. This is supported by the settlements measured between 1984 and 1992 along the upstream bench in the vicinity of the 1984 canal breach near Station 280 (up to 10 inches, see Figure 9). Note that the differential settlements that occurred prior to the 1984 failure are likely to have been much larger.

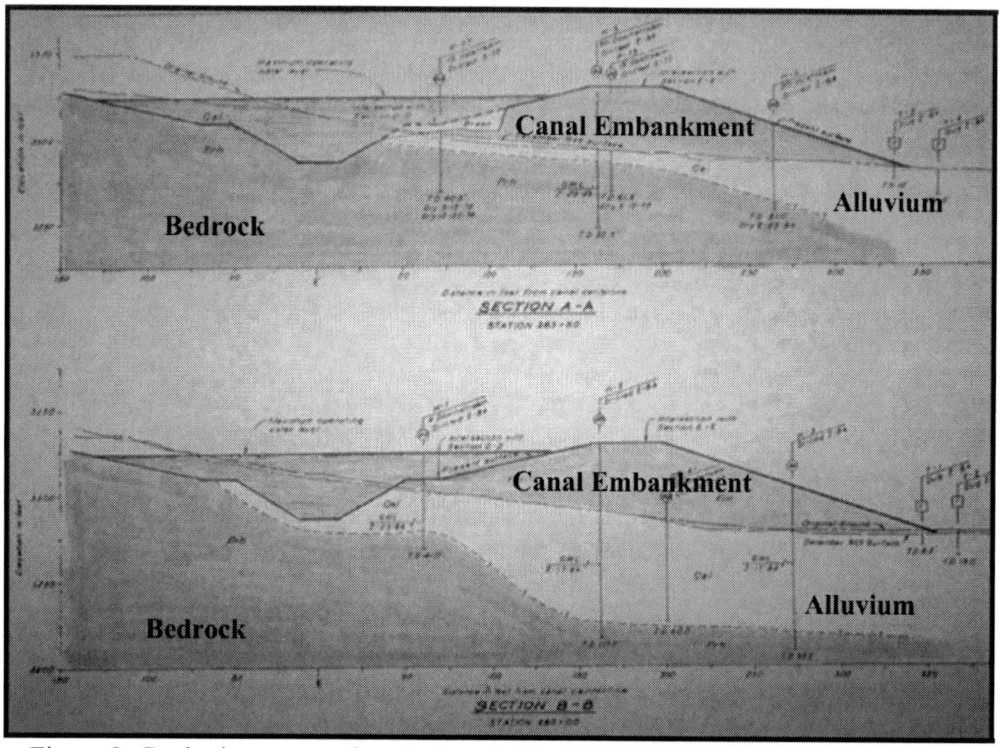

Figure 8. Geologic cross sections in the vicinity of the January 16, 1984 Lower Quail Canal Failure (from DWR Project Geology files)

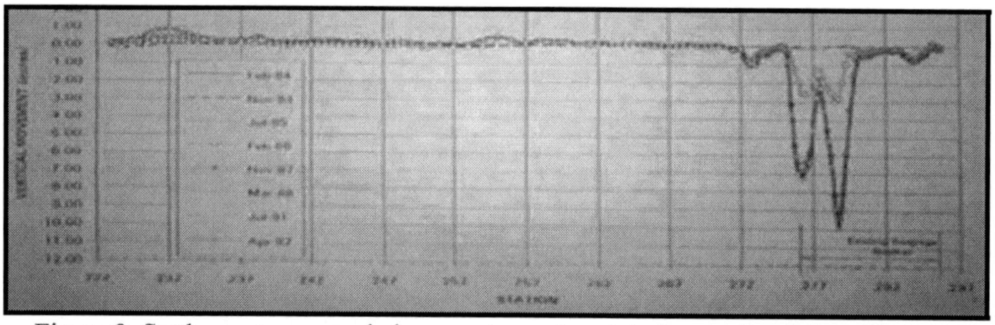

Figure 9. Settlement measured along upstream bench in Lower Quail Canal between 1984 and 1992 (from DWR Division of Engineering files)

The repair of the 1984 failure consisted of removing soft soil debris and replacing the compacted embankment and concrete lining back to the original dimensions. In addition, a seepage control blanket was placed along the outside slope of the right embankment along 5 selected reaches of the canal. This seepage control blanket was placed in areas of potential future differential settlement and was intended to prevent another erosion failure in the event of differential settlement and cracking of the canal clay lining and embankment. The blanket was composed of a woven filter fabric wrapped around crushed drain rock and a soil cover. The total cost of the repair and the seepage control blankets was $ 2.3 million.

Since 1984, many of the right canal embankments overlying alluvial deposits have continued to settle. There was also evidence that the foundation soils were susceptible to hydrocompaction. This was because survey monuments on the interior (wetted) midslope of the canal prism had settled as much as 11 inches over 8 years, whereas the monuments on the embankment crest at the same canal station had settled only about 2 to 4 inches under a much higher fill height. As a result, additional seepage control blankets have been added to selected sub-reaches over time (see 2006/2007 canal repair).

CANAL SEEPAGE, DISTRESS, AND REPAIRS OVER TIME

General

The California Aqueduct has generally performed excellently over time given its large size, long length, and the many difficult geologic conditions it traverses. However, the Aqueduct is now fifty years old in many places and has experienced several incidents of seepage distress in recent years. These incidents suggest that effects such as settlement, hydrocompaction, and foundation solutioning may be progressive over time and that they can induce seepage distress even decades after initial construction. Table 1 summarizes 46 Aqueduct repairs carried out over the last 45 years resulting from seepage and/or differential settlement distress. These 46 repairs have cost DWR over $77 million in construction contractor payments alone. The following sections present brief descriptions for twelve of these incidents and the repair techniques used to remediate them. The authors selected these particular repairs to illustrate the range of construction techniques and the evolution of repair approaches over time:

- 1976/1990 Urgent Canal Repairs at Milepost 56
- 1995 Emergency Canal Repair at Milepost 333.5
- 1997 Temporary Emergency Canal Repair at Milepost 55
- 1997 Temporary Emergency Canal Repair at Milepost 62
- 1999 Urgent Canal Repairs in Pools 48 and 49
- 1999 Emergency Canal Repair of rapid drawdown slides in Lower Quail Canal

- 2001 Emergency Canal Repair at Milepost 4.25
- 2006/2007 Canal Repair at Lower Quail Canal at Milepost 310
- 2009/2010 Canal Repairs at Milepost 88.3
- 2012/2016 Canal Repairs Repair at Milepost 230.88
- 2013 Canal Repair at Milepost 342.65
- 2011/2016 Canal Repairs at Milepost 248.9

1976/1990 Urgent Canal Repairs at Milepost 56

The California Aqueduct near Milepost 56 crosses a drainage known as Bennett Valley for a length of approximately 1,000 feet. This valley is located approximately 10 miles north of O'Neill Forebay in Northern California. In this reach, the canal is configured principally in fill with the invert elevation approximately 5 feet above original ground level. The original construction for this portion of the Aqueduct began in 1963 and was completed in 1968. In 1976, a seep near the toe of the left embankment abruptly increased from about 40 gpm to 1,500 gpm within a day. DWR declared an emergency and emptied the canal in this reach. Workers then completed a limited repair involving the removal and replacement of up to 10 feet of material beneath the invert for a length of about 250 feet (see Figure 10). This work was performed at a cost of about $171,000. During this repair, workers found two large "*gopher-type*" holes in the left side of the excavated slope and locally sealed them with grout. While referred to "*gopher-type*" holes, these were not considered to be animal burrows, but rather pipes created by internal erosion.

Table 1
Summary of Seepage Repairs on the California Aqueduct (1971 – 2014)

Spec.	Location	Type of Repair	Reason for Repair	Date of Construction	Construction Cost of Repair
71-29	Pool 48 – 32 miles upstream of Pearblossom Pumping Plant, SFD	Remove and replace embankment/ foundation, damaged lining & install new concrete lining	Seepage & piping failure of canal embankment	Oct – Dec 1971	$127,485
74-08	Upstream from Oso Pumping Plant, SFD	Remove and replace embankment and damaged lining and install new concrete lining	Emergency repair	Jan – Feb 1974	$20,720

Spec.	Location	Type of Repair	Reason for Repair	Date of Construction	Construction Cost of Repair
74-44	Milepost 234.43 SJFD	Remove and replace embankment, damaged lining & place shotcrete	Foundation failure (slide) after dewatering	Oct – Nov 1977	$374,907
*	Milepost 89, SLFD	Injection grouting	Seepage	1978	**
76-17	Milepost 56, DFD	Remove and replace embankment/ foundation, damaged lining & install new concrete lining	Seepage and piping, near failure	Jul – Aug 1976	$171,115
78-44	Oso Canal Mile 1.48, SFD	Remove and replace embankment, damaged lining & place shotcrete	Foundation failure (slide) after dewatering	Aug -Oct 1978	$1,092,206
79-28	Mile 260.42, 260.45, 273.48, SJFD	Remove damaged lining and install new pre-cast concrete lining with divers	Liner damage due to stormwater overtopping	Aug 1979 – Mar 1980	$ 267,767
81-21	Milepost 10.3, DFD	Remove and replace embankment/ foundation, damaged lining & install new concrete lining	Emergency repair	May– Jul 1981	$1,882,117
83-43	Various locations—Milepost 8.11 to 26.83 and SBA Milepost 7.3 to 27.7, DFD	Sediment material removal from inlets, outlets, channels, and culverts; Remove slide material and repair damaged slopes from slides	Repair slopes damaged from slides and remove slide material	1983	$36,388
84-30	Lower Quail Canal Sta. 283+25 SFD	Remove and replace embankment/ foundation, damaged lining & install new concrete lining	Seepage & piping failure of canal embankment due to diff. settle	Aug 1984 – Apr 1985	$2,256,513
87-42	Mile 287.15 SJFD	Install new concrete lining	Damaged liner	1987	$36,388
90-17	Mile 56 DFD	Remove embankment/ foundation, damaged lining, install geomembrane and concrete lining	Gypsum solutioning/ embankment & foundation piping	Apr – Aug 1990	$3,211,000
92-25	Coastal Mile 12.85 SJFD	Shotcrete	Seepage	Nov 1992	$29,648
93-29	Mile 329 and 336 SFD	Install geomembrane, shotcrete canal lining	Diff. settlement/ cracking due to hydrocompacti	Nov – Jan 1993	$374,000

Spec.	Location	Type of Repair	Reason for Repair	Date of Construction	Construction Cost of Repair
			on		
95-09	Mile 333.53 SFD	Remove concrete lining, replace embankment, spray on membrane and shotcrete	Seepage and near failure	Mar – May 1995	$1,538,000
95-19	Lower Quail Canal SFD	Install seepage control blanket composed of drain rock, filter fabric + embankment	To control potential seepage from future diff. settlement	Oct – Nov 1995	$623,000
95-25	Mile 331.89, 334.02, 334.64, 334.95 SFD	Repair existing lining, spray on membrane and shotcrete	Seepage	Sep – Dec 1995	$1,652,000
95-34	Mile 39.5 DFD	Remove damaged lining, install geomembrane and shotcrete lining	Seepage repair	Nov 1995 – Mar 1996	$2,294,000
96-13	Mile 333.8, 343.81, 344.14 SFD	Repair existing lining, install geomembrane and shotcrete	Seepage repair	Oct 1996 – Mar 1997	$2,257,608
96-19	Mile 206.10, 207.94 SJFD	New embankment, concrete lining	Subsidence	Oct 1996 – Feb 1997	$1,054,000
96-30	Mile 134.98 & 157.4 SLFD	Remove damaged lining, install fabric form concrete	Erosion due to flooding overtopping liner	Jan – Mar 1997	$512,000
97-13	Lower Quail Canal Slide Repair SFD	Remove slide debris, compact embankment, place concrete panels	Repair rapid drawdown slides	Dec 1998 – Mar 1999	$4,423,000
97-20	Mile 54.95, 62.29, 66.71 SLFD	Rebuild embankment and concrete lining	Seepage & piping/sinkhol e; Slope failure due to rapid drawdown in canal	Aug – Oct 1997	$1,915,000
98-01	Mile 326.77 to 330.82 SFD	Remove existing lining, install geomembrane and shotcrete	Urgent repair of damaged liner	Jan – Apr 1998	$5,917,000
98-01	Edmonston PP SJFD	Install articulating concrete mats over damaged/ removed concrete panels	Emergency repair of forebay slopes due to drawdown slides	Feb – Mar 1998	$1,967,000
98-06	Mile 55 SLFD	Compacted embankment, install geomembrane and shotcrete	Final repair of previous repair (97-20)	Mar – May 1998	$5,002,000

Spec.	Location	Type of Repair	Reason for Repair	Date of Construction	Construction Cost of Repair
98-25	Mile 305.1 to 305.44, 323.25 to 330.82, 365.73 to 395.71 SFD	Clean and repair canal lining, install geomembrane and shotcrete	Seepage repair	Dec 1998 – May 1999	$3,178,000
99-29	Mile 320 to 350 SFD	Clean and repair canal lining, install waterproof geomembrane and shotcrete	Seepage repair	Oct – Nov 1999	$3,977,000
01-19	Mile 4.25 DFD	Plug leak with concrete, remove damaged lining, place geomembrane and shotcrete	Seepage and piping, near failure	Jun – Sep 2001	$5,965,000
06-23	Lower Quail Canal Milepost 311 SFD	Seepage berm	Seepage	Jan – Mar 2007	$656,718
*	Mile 305 SFD	Repair uplifted panel joints with mortar	Emergency canal repair	Jan – Feb 2007	**
*	Mile 224.48 SJFD	Plug leaking panels with mortar	Seepage	Nov 2007	**
07-20	Mile 56.4 to 164.90 SLFD	Fabric form concrete; New concrete lining; Seepage berm	Damaged panels; Seepage	2007 – 2011	$8,385,960
09-07	Mile 88.3 SLFD	Sheet pile cut-off wall	Seepage	2010	$1,605,155
09-21 CO	Mile 248.97 SJFD	Plug leak with grout, compaction grouting of embankment	Seepage repair	Oct 2011 – Jan 2012	$630,000
09-21 CO	Mile 121.98 SJFD	Seepage berm	Seepage and boil	Dec 2012 – Jan 2013	$ 80,515
09-21 CO	Mile 88.9 SLFD	Sheet pile Cut-off, seepage and stability berm	Seepage and unstable embankment repair	June 2012 – Dec 2013	$1,571,602
*	Mile 230.88 SJFD	Grout embankment through liner with divers	Seepage, boils, embankment settlement	Sept 2012 (boil #1); Mar 1 – 2, 2016 (boil #2)	$ 321,534
13-10	Mile 342, Pool 52 SFD	Clean and repair canal lining, install geomembrane and shotcrete	Seepage, sinkhole due to water level above liner	Jul 2013 – Jan 2014	$1,333,405
*	Mile 90.28 SLFD	Grouted embankment through liner with divers and through drill holes in primary road	Seepage and boils	Jan 20 – 31, 2014	$67,000

Spec.	Location	Type of Repair	Reason for Repair	Date of Construction	Construction Cost of Repair
*	Mile 7.56 DFD	Grouted embankment through drill holes in primary and secondary road	Seepage and boils	Feb 3 – 14, 2014	$64,000
*	Mile 131.14 SLFD	Grout embankment through liner with divers	Seepage and boil	Feb – Mar 2014	$62,000
*	Mile 50.89 DFD	Grouted embankment through liner with divers	Seepage and boils	Nov 4 - 5, 2015	$ 184,000
15-16 CO	Mile 335.8 SFD	Remove and replace damaged lining, excavate sediment	Storm damage due to overtopping	Oct 2015 – Feb 2016	$592,348
	Mile 6.6 DFD	Grouted embankment through liner with divers	Seepage and boils	Mar 15 – 17, 2016	$27,000
15-16 CO	Mile 244.7, 245.09 & 248.9, Pool 30, SJFD	Remove damaged lining, excavate and replace unsuitable embankment, install waterproof geomembrane and shotcrete, repair roads	Seepage and boils, severely damaged lining	Jan – May 2016	~$10 million

* These repairs were not completed under a construction contract and thus a specification number is not associated with the project. Most of these repairs were completed in-house with DWR labor and materials.

** The construction costs of these repairs are unknown and have not been recorded within DWR files. For these in-house repairs, final costs are not always documented after the quick emergency repair is completed.

DFD, SLFD, SJFD, and SFD refer to DWR Delta, San Luis, San Joaquin, and Southern Field Divisions, respectively

Figure 10. Limited repair in 1976 at Milepost 56 on the California Aqueduct (from DWR Division of Engineering files)

The leakage was contained by the 1976 repair until 1987 when new seepage and distress in the replaced concrete lining were discovered. Divers reported that invert panels had relative displacements of as much as 12 inches and that the divers could extend their arms completely into the voids behind the panels. In addition, a boil formed at the base of the left canal embankment in 1989. Between March 1987 and November 1989, numerous grouting programs were carried out in an attempt to control seepage:

- April 1987 – 23 grout holes were drilled into the crest of the right canal embankment in an effort to seal a small seep at the base of the right embankment.
- October 1987 – DWR divers sealed open joints and cracks with grout.
- April 1988 – 23 grout holes were drilled through the right embankment crest and 1,146 lbs. of bentonite and 18,142 lbs. of cement were used to grout voids in the foundation.
- January 1989 – DWR divers pumped 75 lbs. of bentonite and 280 lbs. of cement beneath cracked concrete panels. Grout holes were also drilled into the left embankment crest and 300 lbs. of bentonite and 4,230 lbs. of cement were pumped into the lower embankment and foundation.
- May 1989 – Grouting was again carried out through the left embankment crest and 750 lbs. of bentonite and 19,740 lbs. of cement were pumped into the lower embankment and foundation to partially reduce seepage at the base of the left embankment (see Figure 11).
- June 1989 – DWR divers drilled 2-inch holes in the concrete panels and pumped bentonite slurry to reduce new seepage at the base of the left embankment. Over 6,500 lbs. of bentonite were used to reduce a seepage flow of 18 gpm down to about 12 gpm.
- September 1989 – DWR divers again pumped bentonite slurries through holes drilled in the concrete panels and used 6,055 lbs. of bentonite to temporarily stop a seepage flow of 61 gpm at the base of the left embankment.
- November 1989 – Seepage had risen to a peak flow of 172 gpm at the base of the left embankment (see Figure 12) and several attempts by divers to pump bentonite slurry beneath the concrete lining (over 5, 000 lbs. of bentonite used) failed to slow down the seepage. Divers then poured the contents of numerous packages of dry bentonite pellets into the open panel joints and cracks to again temporarily stop the seepage.

Figure 11. 1989 grouting of left canal embankment and foundation at Milepost 56 (from DWR Division of Engineering files)

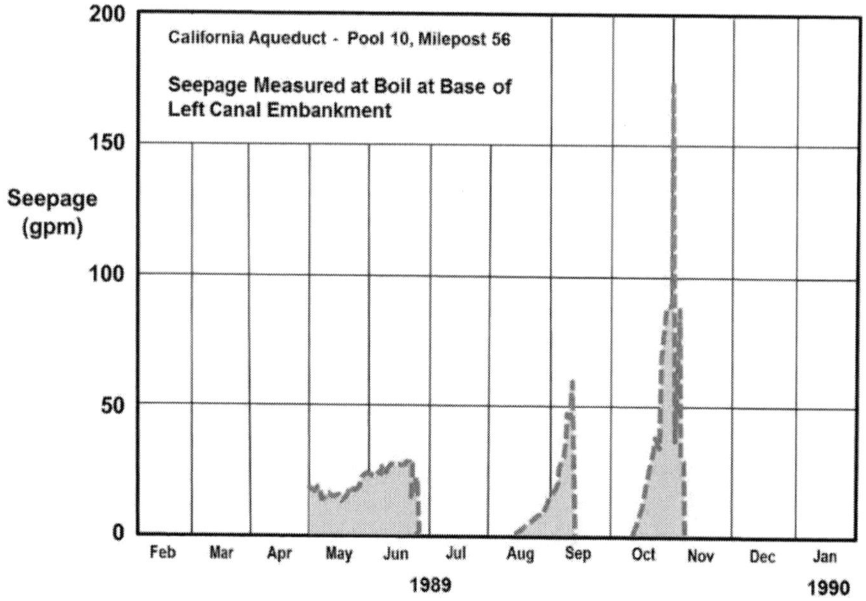

Figure 12. Boil seepage flows at base of left canal embankment at Milepost 56 between 1989 and 1990 (from DWR Division of Engineering Files)

The cause of the seepage distress was concluded to be associated with the solutioning of gypsiferous foundation soils beneath the canal embankment. Beneath the embankment, there was approximately 20 feet of soft clayey and sandy alluvial soil overlying sedimentary rock. The sedimentary rock, Kreyenhagen Formation, contains beds of gypsum up to 1 inch thick (see Figure 13). The alluvial soils originating from this formation and deposited in the valley contain high concentrations of fine-grained gypsum. In some areas, the soft alluvial soil contained almost pure sugar-like granules of gypsum (see Figure 14). It was theorized that the weight of the canal embankments caused the soft foundation soils to consolidate over time and that this led to differential settlements and cracking. This in tum led to concentrated seepage and eventual solutioning of gypsiferous soils. Over several years, the problem simply accelerated. Once preferred seepage paths were created, progressive solutioning occurred to the point that grouting approaches were able to only temporarily control the seepage.

Figure 13. Gypsum crystal found in sedimentary rock of the Kreyenhagen Formation underlying the alluvial foundation soils beneath the canal embankments at Milepost 56 (from DWR Division of Engineering Files)

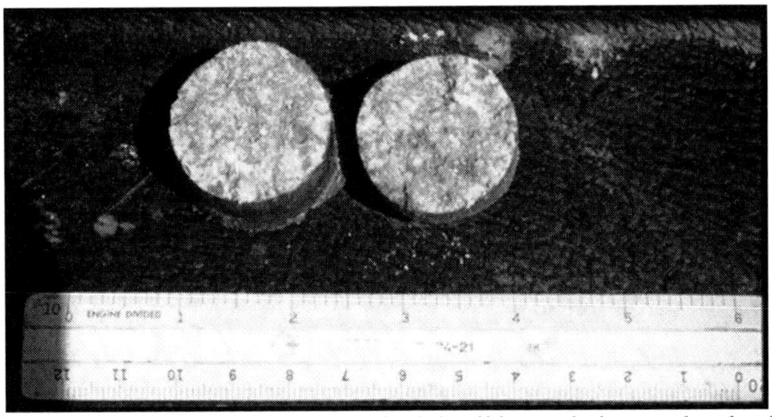

Figure 14. Fine gypsum crystals found in alluvial soil beneath the canal embankments at Milepost 56 (from DWR Division of Engineering Files)

By November 1989, DWR concluded that the grouting operations were not going to be a permanent solution, and that their effectiveness even as a temporary remedy was declining rapidly. Accordingly, DWR operators scheduled a canal outage for May-July 1990 for a permanent repair. Construction contractor workers placed dumped earthen cofferdams on either side of the repair area and dewatered the repair area at a rate of about 1 foot per day. This repair consisted of removing the interior 25 feet (normal distance) along both canal slopes within the canal prism and carrying the excavation down to the Kreyenhagen Formation about 25 feet below the canal invert (see Figures 15 and 16). The excavation was then backfilled with compacted clayey material back to the original canal dimensions (see Figure 17). This resulted in a firm foundation and essentially a 25-foot-thick clay liner to prevent future seepage. In addition, a composite 36-mil geomembrane bonded to a 16-oz. geotextile was placed beneath the invert concrete panels to provide additional seepage protection. Figure 18 shows the beginning placement of forms for the invert concrete panels on top of the composite geomembrane. This repair was carried out using double shifts and cost approximately $3.2 million in construction contractor payments.

Figure 15. Aerial view of 1990 Milepost 56 canal repair (from DWR Photography Library)

Figure 16. View of excavation of interior canal prism during 1990 Milepost 56 canal repair (from DWR Photography Library)

Figure 17. View of clayey interior fill placement during 1990 Milepost 56 canal repair (from DWR Photography Library)

Figure 18. View of concrete lining placement on rebuilt canal invert during 1990 Milepost 56 canal repair (from DWR Photography Library)

One of the more interesting observations made during the excavation work during the 1990 Milepost 56 repair was the discovery of grout that had been injected through boreholes drilled through the crown of the canal embankments the previous year. As shown previously in Figure 11, the grout was injected in boreholes drilled down about 35 to 45 feet below the embankment crown into the lower portion of the canal embankment and into the alluvial foundation. The boreholes were spaced only about 3 feet apart along the canal and grout pressures up to 60 or 80 psi were used. The grouting crews stated that it was sometimes necessary to use such high pressures in order to inject the cement grout into the foundation. In hindsight, it was recognized that the lack of proper grout refusal criteria and the use of such high injection pressures likely caused hydraulic fracturing within the embankment and foundation materials. Shown in Figure 19 is a photograph of some of the grout masses encountered during the remedial excavation. These masses likely resulted from the grouting program encountering internal voids in the foundation created by solutioning and internal erosion. They likely represent some of the more successful results of the grouting program and are indicative of the filling of open voids at lower grout pressures. On the other hand, Figure 20 shows a photograph of a grout-filled fissure leading away from its injection hole, which can be seen by the PVC grout pipe. These grout-filled fissures were generally parallel with the axis of the canal and were likely caused by hydraulic fracturing during the grouting operations on a vertical plane perpendicular to the minimum principal stress.

Figure 19. Grout masses encountered in excavation associated with the 1990 Milepost 56 canal repair (from DWR Photography Library)

Figure 20. PVC grout pipe and grout-filled fissure encountered during excavation associated with the 1990 Milepost 56 canal repair; fissure is parallel to canal alignment (from DWR Photography Library)

1995 Emergency Canal Repair at Milepost 333.5

In mid-March 1995, a boil appeared above a box culvert near the base of the left canal embankment at Milepost 333.5. This location is in Pool 49 of the East Branch, approximately 27 miles upstream of the Pearblossom Pumping Plant in southern California. The boil soon dramatically increased and was carrying soil away (see muddy boil in Figure 21). Emergency drawdown of Pool 49 was able to keep the canal from totally failing. However, piping along the culvert led to a near breach of the canal embankment as illustrated by the piping exit tunnel shown in Figure 22.

Figure 21. View of boil that developed over concrete culvert underchute at Milepost 333.5 in March 1995 (from DWR Photography Library)

Figure 22. View of erosion pipe exit created above box culvert in March 1995 during the near-failure of the East Branch of the California Aqueduct at Milepost 333.5 (from DWR Photography Library)

The geologic conditions at Milepost 333.5 are similar to those described for the 1971 canal failure at Milepost 328 in Pool 48, only about 5 miles away (see Figure 6). At Milepost 333.5, the left canal embankment is a fill section across a local drainage located between nearby cut slopes. The left embankment is approximately 25 feet high and founded on recent alluvium. ollowing the event and after the canal was dewatered and the interior inspected, it was discovered that the canal fill had settled, presumably by hydrocompaction, over and around the buried box culvert by about 10 inches. The two men shown in the upper photograph in Figure 23 are standing on concrete canal panels which had settled around the buried culvert to accommodate the 10-inch settlement (the buried culvert runs left to right in this photograph). Along with the panel settlement, panel joints had opened up by as much as 2 inches which had allowed water to enter the foundation and to flow out along the culvert. The joints within the culvert itself were also offset and some had opened up as much as 2 inches.

The 1995 emergency repair at Milepost 333.5 consisted of excavating disturbed soil down to the base of the culvert and backfilling the excavation back to original grade with imported fill (see Figure 24 for emergency design sketches). Much of the excavation work had to be done in snow and rain leading to a muddy construction site (see Figure 25). A geogrid and a ~80-mil spray-on elastomeric membrane, trade name ***Liquid Boot***, were placed approximately 1 foot above the buried culvert to distribute any future settlements and to restrict seepage. After the embankment was brought back to the original dimensions, replacement concrete panels were shotcreted into place. To prevent future hydrocompaction, an additional ~120-rnil thickness of Liquid *Boot* was applied on top of the concrete panels for a 500-foot length centered on the culvert crossing. The *Liquid Boot* geomembrane was applied by first tacking down a thin non-woven geotextile to the concrete panels (see Figure 26). The 120-rnil thickness of *Liquid Boot* was then sprayed on top of the thin geotextile and then covered with another thin non-woven geotextile. A 2-inch layer of shotcrete with polypropylene fibers was then placed on top of the geotextile to provide protection against abrasion and sunlight (see Figure 26). Both the geotextiles and the *Liquid Boot* geomembrane were anchored in trenches at the top of both slopes and at the upstream and downstream ends of the 500-foot application. The repair was carried out in double shifts for approximately 6 weeks at a cost of approximately $1.5 million.

Figure 23. Views of "hump" created by settlement of canal fill over box culvert at Milepost 333.5 (from DWR Photography Library)

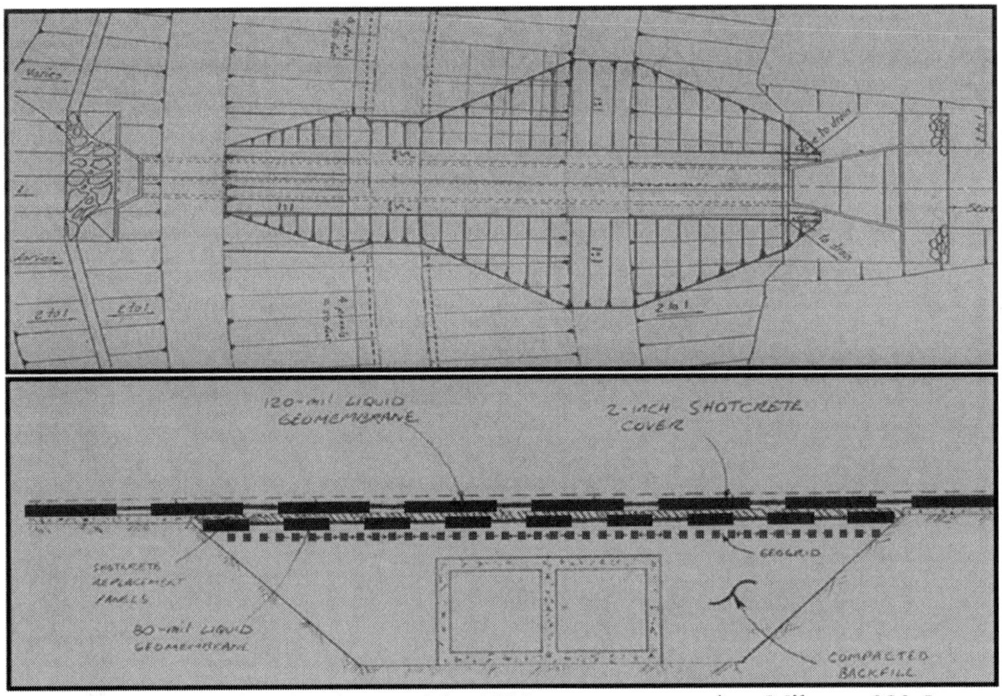

Figure 24. Design sketches used for 1995 emergency repair at Milepost 333.5 (from DWR Division of Engineering files)

Figure 25. Views of excavation work during 1995 emergency canal repair at Milepost 333.5 (from DWR Photography Library)

Figure 26. Views illustrating application of Liquid Boot geomembrane and shotcrete canal lining placement during 1995 emergency repair at Milepost 333.5 (from DWR Photography Library)

1997 Temporary Emergency Canal Repairs at Milepost 55

The Milepost 55 site lies approximately 1 mile upstream of the 1990 Milepost 56 repair and has several similar conditions. It is a site where the Aqueduct crosses a small valley and where the canal section is almost entirely in fill (see Figure 27). It has the same geologic conditions whereby the canal embankments are founded on soft, alluvial soils rich with gypsiferous deposits (see Figure 28). In 1996, a small seep began saturating an area beneath the left canal embankment. The seep kept the grasses green through the summer and had been flowing about 2 gpm (note green band in the photograph shown in Figure 27). This continued for about a year until the first week in August 1997 when the seepage had increased to about 8 gpm. On August 8, 1997, the seep abruptly increased to over 800 gpm (see Figure 29). Over the next few hours, the canal pool was lowered 12 feet, but this did not noticeably reduce the flow. Within the canal prism, depressed concrete panels were observed midway down the left embankment slope, and a small whirlpool was noted over a hole in the center of the depressed panels (see Figure 30). Over the next 24 hours, several failed attempts were made to slow down the seepage:

- Placing carpet over the depressed panels beneath the whirlpool
- Bulldozing soil into the water in the area of the whirlpool
- Throwing scores of 50-lb. bags of powdered bentonite into the center of the waterside depression
- Building a sand-bag ring chimney at the landside base of the embankment around the 800-gpm exit flow

While these measures did not slow down the flow, they may have contributed significantly in preventing the canal from failing during these initial hours. Finally, about 24 hours after the major flow had been discovered, the flow was stopped by pumping 9 yards of concrete into the hole at the center of the depressed panels early on the morning of August 9.

Figure 27. Aerial View of 1997 Milepost 55 seepage distress (from DWR Photography Library)

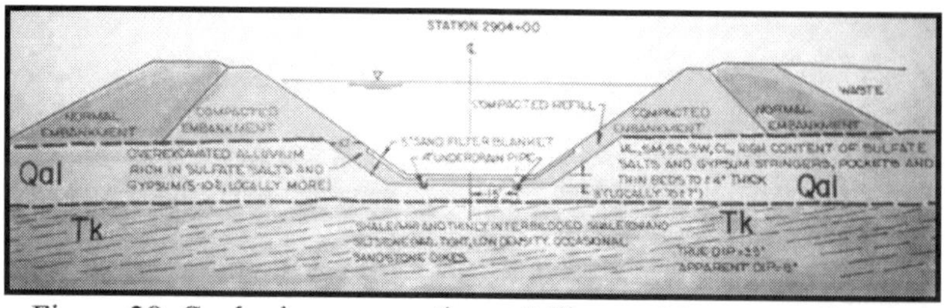

Figure 28. Geologic cross section at Milepost 55 seepage distress
(from DWR Division of Engineering files)

Figure 29. Large boil and flow (~800 gpm) into shotcrete-lined drainage ditch located at the base of the left canal embankment at Milepost 55 seepage distress
(photograph taken August 8, 1997 - from DWR Division of Engineering files)

Figure 30. Depressed concrete panels and whirlpool along interior slope of the left canal embankment opposite seepage area at Milepost 55
(photograph taken August 8, 1997 - from DWR Division of Engineering files)

To help hold this patch until a more permanent repair could be made during an outage the following spring, a PVC geomembrane was placed underwater by divers over the next few days against the left interior slope and invert (see Figure 31). In addition, approximately 300 compaction grout holes were completed in the next few weeks through the crest of the left embankment to also help hold the foundation in place until the permanent repair could be carried out. The permanent repair was carried out in March-May 1998 and used earthen cofferdams to help dewater the approximate 1,000-foot-long repair area. The permanent repair consisted of removing and replacing distressed concrete panels and applying a 120-mil elastomeric asphaltic membrane, trade name Teranap 331, over the entire canal prism for a length of approximately 800 feet. The geomembrane was covered with a 2-inch layer of fiber-reinforced shotcrete. The use of Teranap is discussed further in subsequent sections.

Figure 31. Installation of temporary PVC geomembrane and compaction grouting as part of the 1997 temporary emergency repair at Milepost 55
(from DWR Photography Library)

1997 Temporary Emergency Canal Repairs at Milepost 62

During the August 8-9, 1997 emergency associated with the Milepost 55 seepage distress, a 150-foot-long slide occurred on August 9th in the right interior canal slope near Milepost 62 (see Figure 32). The slide occurred at the location where a Tosco petroleum pipeline crossed the Aqueduct near Milepost 62 and was approximately 7 miles downstream of the Milepost 55 seepage distress. The slide was likely triggered by the relatively rapid 12-foot drawdown carried out to help save the canal at Milepost 55. At Milepost 62, the canal is founded mostly in cut within a sedimentary rock unit known as the Moreno Formation (Km). The Moreno Formation is commonly composed of weakly bedded shale. During the original construction of this canal reach in the 1960's, several slides occurred due to dip-slip movements on the right cut slope for the canal. Because of this, the right cut slope was over-excavated for thousands of feet during the original construction and replaced by compacted fill. However, this over-excavation was not done in the immediate vicinity of the Tosco pipeline, presumably because the pipeline was in the way (see Figure 33). Thus, the adverse geologic conditions on the right canal slope were left in place in the immediate vicinity of the pipeline only to be triggered by the rapid drawdown of the pool on August 8 and 9.

Figure 32. Aerial views of August 9, 1997 slide in right canal bank at Milepost 62 (from DWR Photography Library)

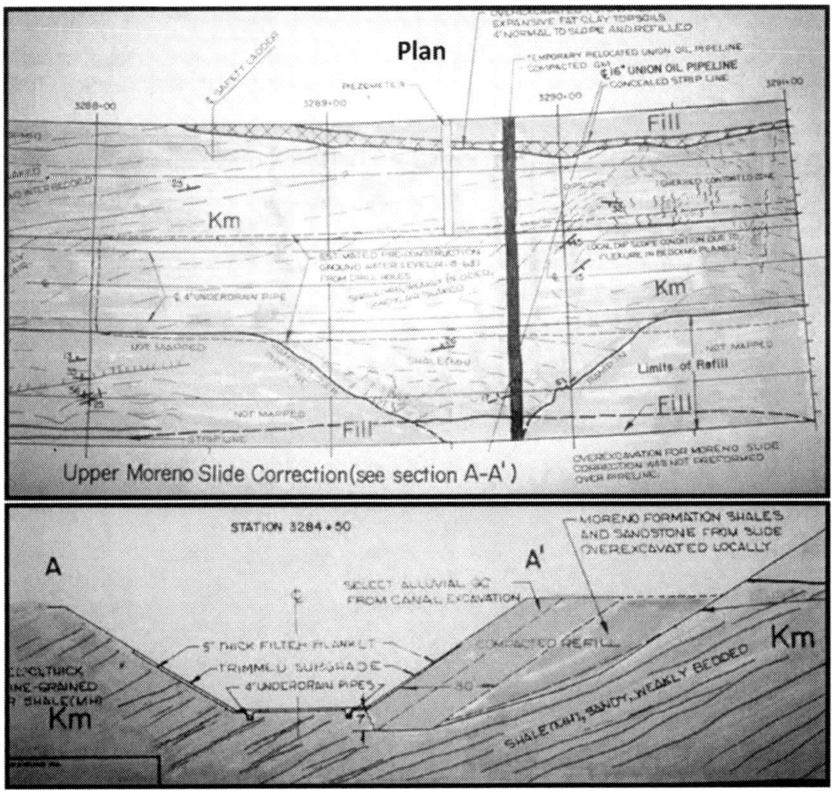

Figure 33. Geologic plan and section views of interior canal slopes at the location of the 1997 Milepost 62 slide (from DWR Division of Engineering files)

Complicating the situation and plans to repair the slide was the fact that petroleum was leaking out of the slope failure and into the Aqueduct water. At first, the oil was thought to be coming from a new break in the pipeline. However, it was later realized that the oil was actually coming from contaminated soil and groundwater which had resulted from a previous pipeline break in 1984 upslope of the Aqueduct. Because of the unstable ground and the need to keep the pipeline in operation, the temporary repair at Milepost 62 consisted of unloading the slide by removing the upper 8 feet of the slide mass and leaving the lower part of the slide below in place below the water surface. A woven geotextile and gravel were applied to the upper exposed slope for temporary wave and slope protection (see Figure 34). Tosco reinforced their pipeline by installing piles near the top of the slope to provide additional support to the pipeline. Tosco also placed containment booms within the Aqueduct and had hazmat teams mop up the released petroleum. Tosco then constructed a 25-foot-deep trench above the slide which was fitted with a vertically hanging geomembrane to act as a cutoff for migrating petroleum migration (see Figure 35). After the canal pool was raised back up to its normal pool elevation, petroleum leakage into the Aqueduct essentially ceased.

Cursory slope stability analyses performed during the emergency using Spencer's Method indicated a factor of safety of 1.19 for static dip-slip sliding of this slope prior to drawdown if a cohesionless friction angle of 24 degrees was assumed for the bedding shear strength in the shale. However, after rapid drawdown, the factor of safety dropped to only 0.89. By removing the upper portion of the slide mass and partially unloading the slide, the static factor of safety increased to about 1.4. This temporary repair has been in place for several years. The pipeline was relocated in 2015 and a permanent repair for the slide is scheduled for 2017 and will consist of excavation of the slide mass and reconstruction of the canal prism.

Figure 34. 1997 temporary repair of slide at Milepost 62 slide (from DWR Division of Engineering files)

Figure 35. 1997 geomembrane cutoff installed by Tosco to mitigate flow petroleum contamination into the canal (from DWR Division of Engineering files)

1998 Urgent Canal Repairs in Pools 48 and 49

Due to the increasing number of seepage distress incidents occurring along the East Branch in southern California, it was decided in the late 1990's to begin treating potential problem areas before they became emergencies. Consequently, a series of repair contracts were issued to repair problem and potential problem areas in several pools along the East Branch. In 1998, one of these repair contracts, Specification 98-01, was issued to repair 10 areas in Pools 47 and 48. A 60-day outage was given and the pools were dewatered. The repairs proceeded as follows:

- The interior canal prism was cleaned by high-pressure water application. Due to the winter weather and frequent rain/snow storms, the invert had to be cleaned several times.

- Cracks, joints, and other imperfections in the concrete panels were filled with epoxy mortar.

- Concrete panels with significant distress were removed and any underlying voids were backfilled with compacted fill. Replacement panels were shotcreted into place.

- Anchor trenches were constructed and a 120-mil elastomeric geomembrane, Teranap 331, was applied across the entire canal prism. Teranap is a distilled bitumen paving membrane previously used by DWR in the early 1990's in the construction of the hydraulic asphalt lining at Devil Canyon Second Afterbay to connect the hydraulic asphalt layers to the concrete inlet and outlet structures. For the 1998 canal repairs, the Teranap came in rolls to the site, was unwound from the rolls and placed perpendicularly across the canal prism and anchored at the top in the anchor trenches (see Figure 36). Joint connections were accomplished by overlapping the geomembrane and applying heat to the edges to weld the rolls together. Joint seams/welds were tested by injecting air to the seam with any leaks immediately patched with additional welds (see Figure 37).

- A 2-inch fiber-reinforced layer of shotcrete was applied to provide a wearing surface and protection for the geomembrane (see Figure 38). Contraction joints were tooled into the shotcrete cover to approximately match the joints in the original concrete panels underlying the geomembrane. The joints were then filled with an elastomeric joint sealer to provide additional seepage protection (see finished repair in Figure 39).

In general, the work progressed upstream to downstream for the 10 repair sites with a minimum 500-foot length of treatment made at each site. The construction costs for the repairs made at the 10 sites were approximately $5.9 million.

Figure 36. Placement of Teranap geomembrane at repair sites during the 1998 East Branch canal repairs in Pools 48 and 49 (from DWR Division of Engineering files)

Figure 37. Welding and weld testing of Teranap geomembrane during 1998 canal repairs in Pools 48 and 49 (from DWR Division of Engineering files)

Figure 38. Placement of shotcrete and use of roller screed for new canal lining above Teranap geomembrane during 1998 canal repairs in East Branch Pools 48 and 49 (from DWR Division of Engineering files)

Figure 39. Completed repair at one of the canal repair sites during the 1998 canal repairs in East Branch Pools 48 and 49 (from DWR Division of Engineering files)

1999 Emergency Canal Repair of Rapid Drawdown Slides in Lower Quail Canal

In late December 1998/early January 1999, the pool water in Lower Quail Canal was lowered to provide access for inspection and maintenance of the canal. Unfortunately, the pool was lowered relatively quickly and resulted in rapid drawdown slides of the lower interior canal slopes at two locations. At these locations, the lower slope was approximately 21 feet high with a 2:1 slope. At the top of this lower slope, there is a 12-foot-wide bench. The upper slope above the bench is much flatter at a 5:1 slope (see Figure 8). The slides involved only the steeper lower slope and portions of the bench as illustrated in Figure 40.

Figure 40. 1998/1999 rapid drawdown slides at two locations in Lower Quail Canal (from DWR Division of Engineering files)

The emergency repairs at these two sites consisted of excavating and removing the slide masses and broken concrete panels (see Figure 41), and then reconstructing the canal prism using imported fill. During the reconstruction, the lower slopes were flattened and the midslope benches were removed in the slide areas. This resulted in a much flatter ~3:1 slope instead of the previous 2:1 slope in these areas. New canal linings were added through the use of shotcrete. The finished repair for one of the repair sites is shown in Figure 42. The construction cost for this emergency repair was approximately $4.4 million.

Figure 41. Slide excavation during the 1999 repair of rapid drawdown slides in Lower Quail Canal (from DWR Division of Engineering files)

Figure 42. Completed slide repair following the 1999 repair of rapid drawdown slides in Lower Quail Canal (from DWR Division of Engineering files)

2001 Emergency Canal Repair at Milepost 4.25

On the morning of June 5, 2001, a boil was discovered in a ditch below an access road near the base of the left canal embankment of the California Aqueduct at Milepost 4.25 (see Figure 43). This distress site is located in northern California only a couple of miles downstream of the Delta Pumping Plant and only about a thousand feet upstream of Bethany Reservoir. The boil was flowing approximately 2 cfs with light brown, silty water (see Figure 43). At the time the boil was discovered, the pool elevation was at approximately Elevation 243.5 feet. With the boil located at about Elevation 185 feet, the maximum head on the boil was approximately 58.5 feet. DWR and contract divers reported that the source of the seepage was through cracks and opened joints in the concrete lining just above the canal invert. Vegetation was then cleared and sandbag berms were constructed around the boil to reduce the head differential. In addition, the pool began to be lowered at a rate of approximately 0.3 feet per hour. Despite these efforts, the seepage appeared to be increasing.

Figure 43. Location and nature of 2 cfs boil on June 5, 2001 at California Aqueduct Milepost 4.25 (from DWR Division of Engineering files)

On the following day, June 6th, construction crews and contract divers pumped concrete underwater into holes drilled by the divers in the concrete lining. DWR's grouting consultant, Mr. Jim Warner, was on site to help direct the work and to develop concrete mix designs (see Figure 44). A 10-bag mix with 30 percent minus 3/8-inch rounded pea gravel was batched with the objective of sealing small seepage channels that were believed to be feeding the main internal erosion pipe. The concrete was pumped using concrete boom trucks at a rate of approximately 2½ cubic yards per minute, or about 4 to 5 minutes to empty a 10-yard concrete truck. Seconds after initial pumping of the concrete began, cement was observed exiting the boil. Boil ejecta also included small gravel particles from the concrete mix. Two 10-yard truckloads of concrete were initially pumped, but only temporarily slowed the seepage out of the boil. A second batch included the use of ¾-inch fibrillated polypropylene fibers and the maximum amount of minus 1½-inch aggregate that could be pumped. Calcium chloride (3 percent by weight) was also added as an accelerator to rapidly set the concrete. After an additional 38 cubic yards of concrete was pumped, the drill holes in the canal lining stopped taking any more concrete and the boil decreased to only about 2 gpm. The next day, June 7th, contract divers then placed a 40-mil PVC membrane over the leak area within the canal to help further seal off the source of the seepage. Approximately 33 hours transpired between the discovery of the boil and the sealing of the leak. During this time, the canal had been lowered approximately 11 feet down to Elevation 232 feet.

Figure 44. Construction crews and contract divers pumping concrete into holes drilled in the concrete lining of the California Aqueduct at Milepost 4.25 to seal off concentrated seepage distress and internal erosion on June 6, 2001 (from DWR Division of Engineering files)

Immediately following the sealing of the canal leak by the concrete pumping, DWR staff and management concluded that a more permanent repair was necessary. The repair that was developed incorporated many of the techniques that DWR engineers had successfully used in previous emergency canal repairs, together with further refinements. The repair approach consisted of a multi-layered line of seepage defenses and the following actions:

- Two cofferdams were placed into the canal to dewater a 1,200-foot-long portion of the Aqueduct. The cofferdams were situated in cut sections within sedimentary rock beyond the embankment section in order to avoid loading soft alluvium present in the foundations beneath the embankment section. The cofferdams consisted mostly of minus 8-inch rockfill obtained from a rock quarry near San Luis reservoir that was placed mostly by bulldozing the material into the water by crews working in double shifts. After the rockfill section was placed, a finer gravel layer consisting of ¾-inch Class II aggregate was placed on the waterside rockfill slopes to reduce the permeability of the rockfill, and a PVC geomembrane was placed on top of the leveling course to further seal the cofferdams. Rockfill placement began on June 9th and was completed on June 13th.

- Dewatering of the 1,200-foot-long reach between cofferdams began on June 13th at a rate of approximately 2 to 4 feet per day, averaging about 2½ feet per day for 7 days between June 13th and June 20th. Dewatering was carried out through the use of submersible pumps. On June 20th, with the pool water at about Elevation 216.7 feet and about 2½ feet deep, fish seining operations were conducted to salvage trapped fish. A total of 193 fish were salvaged and released to the downstream side of the downstream cofferdam.

- Within the 1,200-foot-long dewatered canal length, the actual repair length was selected to be 800 feet long, approximately 520 feet downstream and 280 feet upstream of the leak location. After the fish salvage operation was completed, the remaining water between the cofferdams was emptied and the concrete panels within the 800-foot-long repair length were cleaned by power washing.

- To provide additional seepage protection and to investigate the potential presence of voids beneath the canal embankment, compaction grouting was performed through boreholes drilled through the left embankment and into the alluvial foundation. A second phase of permeation grouting investigated and filled voids in the fractured bedrock underlying the alluvium. Additional cone penetration testing and exploration drilling was also carried out.

- The concrete lining covering an area 10 panels wide and 7 panels high centered on the leak areas was removed from the left interior slope and the left edge of the canal invert. Local excavation in the area of the leak source disclosed three zones of material: a brown, stiff, silty clay (CL) compacted embankment; a 1.5-foot-thick black, strong-smelling zone of organic material marking the un-stripped horizon of original ground; and a higher plasticity, soft, red-brown silty clay (CL-CH) alluvium that extended below the excavation invert. The excavation also revealed a 4-foot-diameter funnel-shaped concrete bulb immediately behind the concrete panel (from the concrete pumping). This bulb tapered to an 18-inch plug of concrete at the bottom of the excavation, which was about 8 feet deep into the foundation. The large, symmetrical funnel shape indicated that the leak source was one large, erosion pipe rather than a system of small channels as earlier theorized. This internal erosion pipe entered the soft clay alluvium near the canal invert and exited at the boil in the drainage ditch at an angle that closely paralleled the strike of the underlying bedrock.

- The localized excavation at the source of the leak was backfilled with approximately 9½ cubic yards of concrete to fill the hole to approximately canal invert elevation (see Figure 45). The rest of this localized excavation was then backfilled with soil previously excavated and with imported borrow material. Grout was then used to fill any other voids created in the soil slope and invert areas by the removal of the concrete panels, and then a 24-foot-wide by 50-foot-long geogrid layer was placed over the leak area. On top of the geogrid, a layer of Teranap geomembrane was placed as a primary seepage barrier in the area where the concrete panels had been removed (see Figure 46). Over the Teranap geomembrane, a 4- to 6-inch thick layer of shotcrete was placed to conform to the surface of the concrete panels which had not been removed. On top of all this, another layer of Teranap was placed for the entire canal prism along the 800-foot-long repair reach as a second line of seepage protection. On the canal slopes, this second layer of Teranap (GTX) was bonded with a geotextile fabric on top to help facilitate the placement of a two-inch layer. On top of the upper Teranap layer, a two-inch-thick layer of shotcrete was placed (see Figure 47). As in previous canal repairs, a roller screed was used and contraction joints were tooled in to match the underlying panel joints. Sixty-eight rows of panels were completed in this fashion.

Figure 45. Local excavation and concrete backfill at source of leak within canal prism at 2001 emergency repair at Milepost 4.25 (from DWR Division of Engineering files)

Figure 46. Aerial views of 800-foot-long repair area and placement of lower layers of Teranap and shotcrete during at 2001 emergency repair at Milepost 4.25 (from DWR Photograph Library

Figure 47. Placement of upper2-inch-thick shotcrete panels overlying upper layer of Teranap GTX geomembrane during 2001 emergency canal repair at Milepost 4.25 (from DWR Division of Engineering files)

- The construction contractor finished the interior geomembrane and shotcrete work on July 1st, and the Aqueduct pool was re-watered on July 2nd. The two cofferdams were removed between July 2nd and July 10th. Partial operation of the canal resumed on July 2nd, less than 23 days after the start of the emergency repair on June 9th.

- After the Aqueduct pool was full, several other small seeps were observed in the drainage ditch. As an additional defensive measure, a filtered drain trench and berm were placed over the length of the embankment toe between June 28th and July 10th. The purpose of these features was to provide a filtered seepage exit and ballast over the critical toe area. The drainage ditch was also moved east of the embankment to increase the lengths of potential seepage paths.

The construction cost for the 2001 Milepost 4.25 emergency canal repair was approximately $4.7 million. An additional $1.8 million was expended to support operations and maintenance functions associated with the canal repair and outage. The cause of the seepage distress and internal erosion will never be fully known. However, two materials in the canal embankment foundation were implicated in the internal erosion: the soft clay alluvium overlying bedrock and the fractured bedrock itself (see Figure 48). Based on excavations at the inlet and the outlet of the erosion pipe, the path of seepage and internal erosion entered and exited the soft alluvium foundation. The original designers were likely aware of the weak condition of the alluvium, and their concern was demonstrated by the provision of a 4-foot-thick compacted clay subliner that was intended to cover the foundation of the canal prism. Unfortunately, the clay subliner was placed only below the canal invert, leaving the poor foundation soils at the base of the left canal slope exposed to full canal head. This where the source of seepage/piping was found. The fractured bedrock was also found to be capable of carrying significant seepage, with high permeation grout takes in the vicinity of the boil location. The potential failure modes include concentrated seepage and internal erosion through cracks in the soft alluvium and at the contact between the soft alluvium and the underlying bedrock. As noted above, the alignment of the seepage path closely parallels the strike of the bedrock.

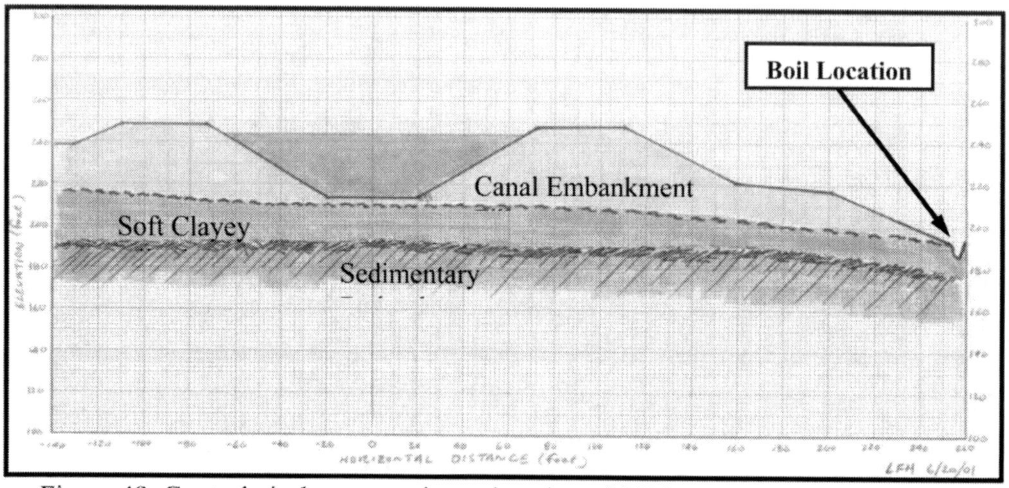

Figure 48. Geotechnical cross section at location of 2001 emergency canal repair at Milepost 4.25 (from DWR Division of Engineering files)

2006/2007 Canal Repair at Lower Quail Canal near Station 310

On February 10, 2006, a seepage problem was reported near Station 310 along the right canal embankment on Lower Quail Canal. A wet spot, approximately 10 feet by 15 feet, had been discovered at the downstream toe of the right embankment (see Figure 49). DWR divers located an area of cracking in the concrete liner panels along the interior bench with water seeping into the embankment. After the discovery of the wet spot, divers filled some of the longitudinal cracks immediately upstream of the wet spot. In addition, DWR staff began lowering the water level in the canal. Shortly after drawing down the water level, the wet spot dried up. A conservatively slow rate of drawdown was undertaken to avoid destabilizing the slope as occurred in December 1998/January 1999. The water level was lowered to expose the interior bench where divers had noticed distress in the concrete liner panels. A series of longitudinal cracks running parallel with the canal were observed along the interior bench. In addition to cracking, entire sections of the bench had dropped/collapsed up to 8 inches and joints within the concrete panels were observed to be pulled open as much as an inch along the bench. The pattern of cracking suggested that piping of soil led to voids under the bench which led to buckling of the panels. See earlier discussion of the January 16, 1984 Lower Quail Canal failure for a description of the geology, differential settlement, and potential for internal erosion of the foundation soils in this area.

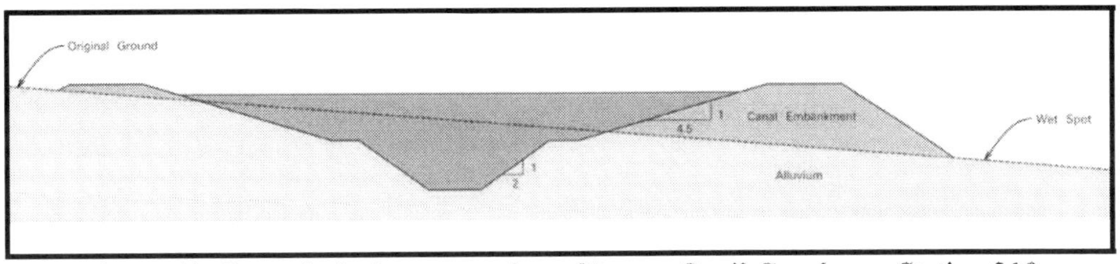

Figure 49. Geotechnical cross section of Lower Quail Canal near Station 310

A temporary grouting program was initiated to fill the voids, stabilize panels, and minimize potential seepage paths through the embankment until a permanent repair could be initiated. A water-cement-bentonite grout was gravity-fed into 1.5-inch diameter holes spaced about 3 feet apart. Approximately 80 cubic yards of grout were applied over a length of about 3,000 feet. The largest grout takes were at Stations 309-310, 320-325 and 334-335.

The permanent repair was performed between January and March 2007. The recommended repair included the placement of a downstream seepage control blanket between Station 310 and 316, where the embankment crosses an alluvial valley. Figure 50 shows the typical repair section. A filter fabric was used between the embankment and the drain material to reduce the potential for piping in the event of cracking and seepage. During such an event, the fabric would act to retain the foundation soil while allowing water to flow into the drain. The filter fabric was designed with a woven filter fabric, large apparent opening size (0.425 mm, 40 sieve) and high permittivity (2.1 sec-1) to reduce the potential for clogging. As shown in Figure 50, the filter and drain system connects to two risers located downstream with sensors that trigger a remote alarm which notifies DWR Maintenance staff when water is detected. To date, the seepage control blanket remains dry, which suggests that the grout injection alone may have been a successful permanent repair here. Figure 51 shows drain material being placed over filter fabric during construction.

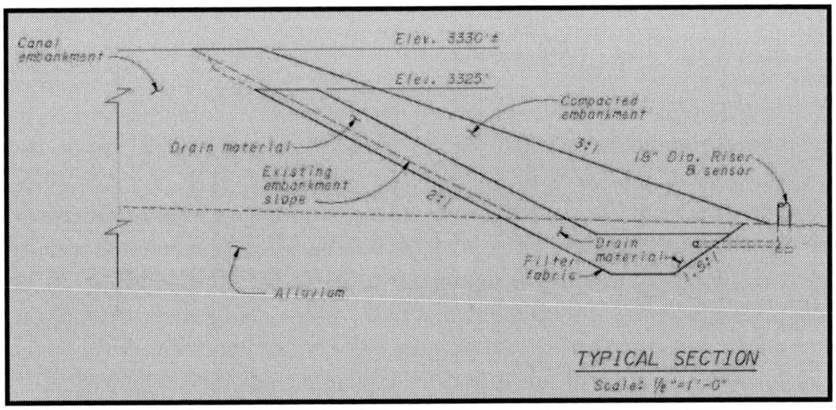

Figure 50. Seepage Control Blanket Design Drawing – Station 310
(from DWR Division of Engineering files)

Figure 51. Installation of filter fabric and drain rock near Station 310 during 2006 Lower Quail Canal repair

2009/2010 Canal Repair of California Aqueduct at Milepost 88.3

The site of the 2009/2010 canal repair at Milepost 88.3 is within Pool 14 of the San Luis Canal portion of the California Aqueduct approximately 12 miles south of the town of Los Banos. At this location, the left canal embankment is approximately 18 to 20 feet high. The interior canal prism has 2:1 slopes, an 85-foot-wide invert width, and a normal water depth of 24.9 feet. The canal's invert elevation is approximately 8 feet below original ground, and the operational elevation of Pool 14 is generally about 17 feet above the landside ground surface. Seepage has been observed at this location near the toe of the left embankment in the form of boils, wet spots, seeps, and ponded water since before 1975. Since that time, DWR had attempted numerous seepage repair techniques, including the installation of small seepage berms along the landside slope of the left embankment and over 9 repeated attempts to stop the seepage by pumping bentonite slurry into angled two-inch-diameter grout tubes installed in boreholes drilled in the crown of the left embankment. Much like the grout injections at Milepost 56 in the late 1980's, excessive injection pressures were used at Milepost 88.3. On November 1, 2006, DWR divers inspected the canal's concrete liner and found a 15-foot-wide row of panels on the slope immediately above the invert that had buckled. The buckled panels may have resulted from hydraulic fracturing and pressurizing of the panels during the grouting operations. In January 2008, the 15-foot wide row of buckled panels found during the 2006 dive was removed and the excavated area was backfilled with gravel. Over this area, a 36-foot-wide section of fabric-formed concrete was placed underwater.

Though many of the interim repairs made at this site over the last 30 years were temporarily successful in reducing the seepage, the water would eventually start flowing again or a new boil would develop. Over time the seepage increased to a rate requiring daily pumping to return the water back into the canal. As a result, the decision was made to make a more permanent repair. Geotechnical explorations were conducted in 2008 and remedial alternatives were developed in April 2009. The geotechnical investigations revealed that the left canal embankment at this location consists mostly of clay, silty clay and clayey silt. However, there are two zones of high permeability soils or open conduits located near the original ground elevation. These two zones corresponded with the two areas where boils and seeps had developed over time. The pore water pressures measured during cone penetration soundings through the crest indicated that parts of the embankment were saturated, but unsaturated below the embankment foundation. Finite element analyses using SEEP/W (GeoSlope International) were used to analyze seepage. A seepage zone of higher permeability material was incorporated in the model at the base of the embankment and material properties were varied until the model resulted in seepage pore pressures that matched observed values in a temporary observation well through the embankment crown (water level about 5 feet above the landside toe), and also values measured in the cone penetration soundings. The calibrated SEEP/W model was then used to evaluate the effectiveness of remedial alternatives. In the end, the remedial technique selected was to install a 1,000-foot-long row of steel sheetpiles through the embankment crown. The sheetpiles would extend 60 feet below the embankment crown, or about 40 feet into the foundation, to serve as a seepage barrier. The SEEP/W analysis for this alternative is shown in Figure 52 and the installation of the sheetpile row is shown in Figure 53.

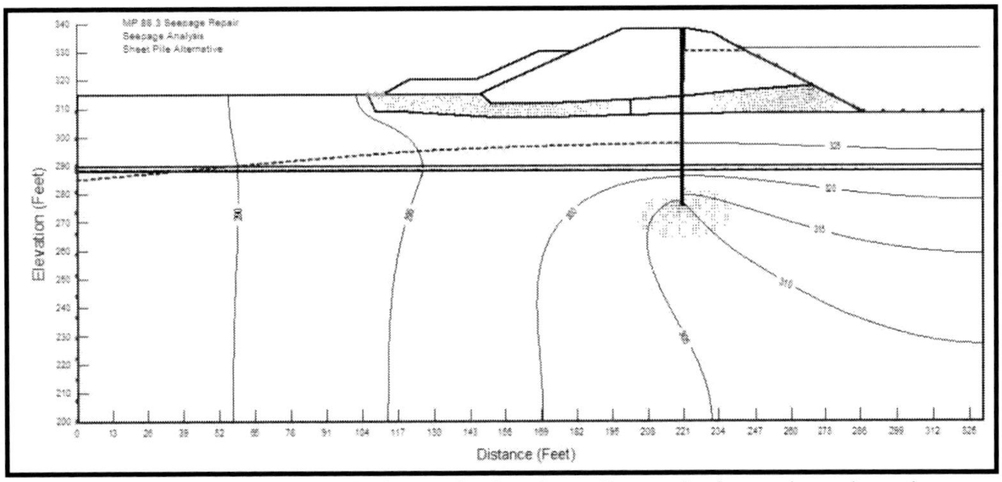

Figure 52. SEEP/W analysis results for sheetpile repair alternative selected for canal repair of the California Aqueduct at Milepost 88.3 (from DWR Division of Engineering, 2009)

Figure 53. Sheetpile installation for 2009/2010 canal repair at Milepost 88.3
(from DWR Division of Engineering, 2011)

The contract for the sheetpile construction was awarded on August 31, 2009, and the contractor mobilized to the site on November 30th. The excavation for the shallow trench for the 1,000-foot-long sheetpile wall and deliveries of the 60-foot-long sheetpiles began on December 2nd. Pile driving was done principally with a Giken 675 Pile Press between December 18, 2009 and March 19, 2010 (see Figure 53). The Contractor generally installed two sheetpiles simultaneously using previously installed sheets as reaction piles. Sheetpile interlock sealants were also applied to the sheetpiles prior to pressing them into the ground. There were 492 sheetpiles installed between Station 972+00 and 981+98 (998 feet), generally at installation rates ranging from 1 pair to 16 pairs of sheetpiles per day. The 1,000-foot-long sheetpile wall was intended to address two main seepage issues located between Station 973+20 and Station 975+00 (180 feet) and between Stations 977+00 and Station 978+20 (120 feet). Observations made during sheetpile installation indicated that seepage paths were not perfectly normal to the canal alignment as the seepage was sometimes intersected at stations further downstream than where the boils had developed. At the end of sheetpile construction, the seepage had been cut off. In fact, the vast majority of the seepage stopped when a single sheetpile pair was pressed into place, indicating that a single conduit or erosion pipe had carried the majoring of the seepage. The canal retained full water delivery operation during sheetpile installation.

The contract had been specified to be completed by November 27, 2009. However, several delays associated with pile installation difficulties, safety, and other issues led to the actual completion on July 19, 2010. The final construction cost was $2.65 million.

2012/2016 Canal Repairs at Milepost 230.88

A leak (boil) was reported at California Aqueduct Milepost 230.88 on September 17, 2012. The boil was located on undisturbed ground about 25 feet from the toe of the left canal embankment (see Figure 54). Initial examinations indicated no distress (displaced panels, cracking, large offsets at joints) in the concrete liner above the water surface in the area of the leak. However, a visible depression in the road and corresponding dip in the top of the concrete liner were observed in line with the boil. Several oblique cracks were also observed along the crown of the left canal embankment about 300 feet upstream of the boil location (see Figure 54). These cracks were reportedly first observed in May 2012 by Field Division staff. In this area, it appears that the embankment had pulled away from the liner over a 30- to 50-foot length. Several animal burrows were also observed along the top of the liner. Flow measured at the boil through a sandbag ring placed around it was approximately 180 to 200 gallons per minute (gpm) at its peak.

After the initial observations, it was decided that the leak should be grouted as soon as possible. Divers arrived on September 19, 2012 to begin the inspection and grouting. During the inspection, a depression in the concrete liner measuring approximately 6 feet in diameter was discovered about 60 feet upstream of the boil in about 12 feet of water in the canal (approximately near the original ground level under the adjacent embankment). Voids were also observed around the perimeter of the depression with significant suction. In order to assess the size of the voids behind the panel and provide injection points, 3 holes were drilled through the concrete liner around the depression. The voids were subsequently determined to be a maximum of 1 foot deep behind the liner at these locations. Grouting commenced and was completed by September 20, 2012.

Approximately 2,800 gallons of cement-bentonite grout were injected into the voids, stopping the boil flow.

Figure 54. Boil and cracking along the left canal embankment at the California Aqueduct at Milepost 230.88 (from DWR Division of Engineering files)

A review of geologic maps and pre-construction topography indicates that this section of the Aqueduct was constructed within alluvial fan deposits originating in the adjacent Elk Hills. Prior to construction, several drainages coalesced in the area where the boil was observed. Other areas of the Aqueduct constructed on similar deposits (e.g. MP 248 and Dos Amigos Pumping Plant) have experienced similar distress (e.g. boils/leaks, damaged lining, piping of materials). These geologic materials are more likely to contain collapsible soils and other problem soils, loose sands, gypsum, and dispersive soils, due to the depositional environment. Pre-consolidation ponds were employed in this reach due to the soil conditions encountered during the initial planning and design. A subsurface investigation consisting of CPT soundings and hollow-stem auger holes was completed between September 23 and September 25, 2013.

Materials encountered underlying the left embankment consisted of relatively dense clayey and silty sands (embankment fill) to depths of 8 to 12 feet, loose to medium dense silty to clayey sands and soft to medium stiff silts and clays with varying amounts of sand underlying the embankment fill to depths of 30 to 35 feet, and dense poorly graded sands and stiff sandy clays to the maximum depth explored of 60 feet. Materials encountered in the explorations located on the farm road east of the right-of-way fence were similar to those encountered below the primary road embankment fill. Groundwater was measured at depths ranging from 40 feet below the primary road to 30 feet below ground surface on the farm road during the drilling (all of the drill holes were backfilled soon after drilling, so a stabilized groundwater level could not be determined). However, pore pressure dissipation tests during most CPT soundings, as well as moisture contents from tube samples, indicated near-saturated conditions and elevated pore pressures at shallow depths, especially within the soft clay and loose sand layers below the left canal embankment. This is consistent with the leaks observed in the liner during the repair.

Possible causes for the embankment and roadway distress were investigated and it was concluded that the most likely cause is hydrocompaction of the sediments underlying the roadway embankment and in the areas beyond. The pre-consolidation ponds only extended under the normal embankment supporting the Aqueduct liner and not the roadway embankment which extends up to 50 feet beyond the limits of the liner (see Figure 55).

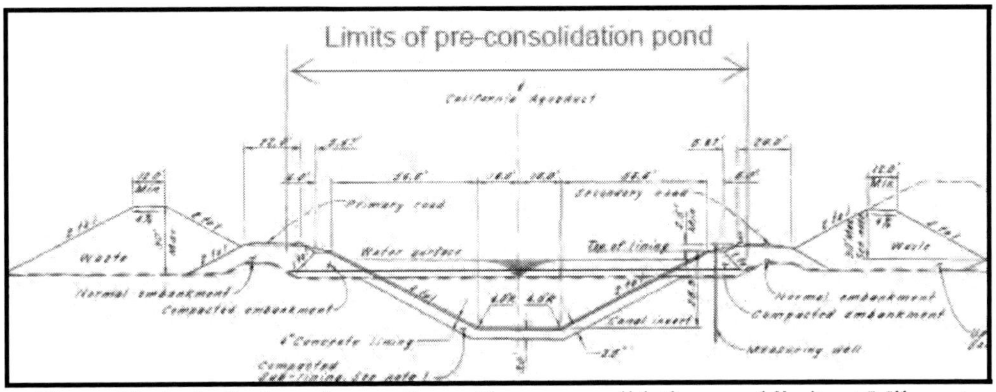

Figure 55. As-built Aqueduct section with pre-consolidation pond limits at Milepost 230.88 (from DWR Division of Engineering files)

Leakage through joints and subsequent saturation of the surrounding sediments likely led to hydrocompaction over time and the observed distress. The leak was situated near the transition from native (cut) to fill embankment, and the water likely followed the cut/fill transition through the lower density sediments.

Subsequent information provided by NASA/JPL during their work for DWR on subsidence in the region confirmed that the distress is ongoing and extends well beyond the Aqueduct right of way. The data was obtained between April 2014 and January 2015 by the airborne Synthetic Aperture Radar technique. Multiple flights over the same area and radar interferometry were used to measure changes in distance which equates to changes in elevation. The data shows subsidence of up to 4 inches in the primary road over the 8-month period. This measured subsidence matches up with the cracking observed during the initial work at the site in 2012. In addition, new roadway cracking reported upstream near an oil pipeline crossing aligns with an area of about 2 inches of subsidence in the primary road. The largest displacement (up to 6 inches) occurred about 200 to 250 feet east of the Aqueduct in what appears to be a storage area on adjacent farm property (see Figure 56).

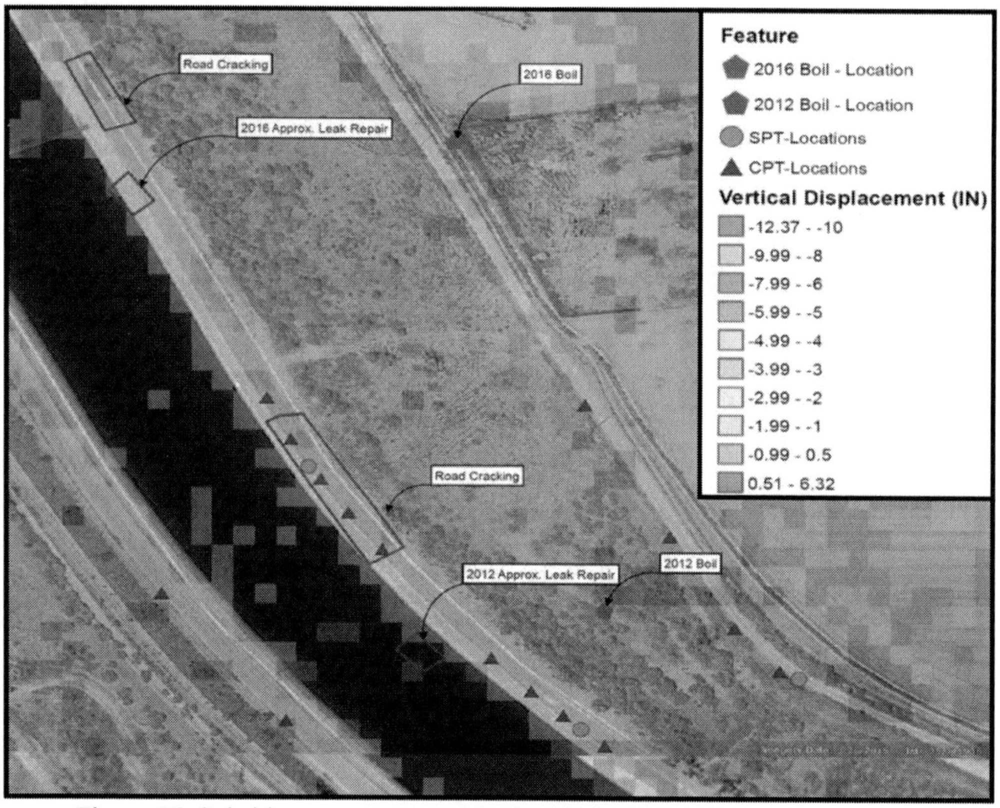

Figure 56. Subsidence survey provided by NASA/JPL indicates the amounts of settlement between April 2014 and January 2015 (from DWR Division of Engineering files)

Further evidence that the leakage and distress are continuing to adversely affect the Aqueduct was confirmed during a dive inspection carried out on March 1 and 2, 2016 by DWR's diving contractor and observed by DWR engineers and geologists. The divers found a new leak located upstream of the 2012 leak near a pipe crossing at approximately Milepost 230.70. The new leak was identified at a vertical joint in about 8 feet of water, and had a triangular-shaped void measuring approximately 10" x 9" x 9" by 15 inches deep. At this location, the Aqueduct is located almost entirely in cut, but there is some vertical relief between the top of the liner and the surrounding ground. Subsequent inspections revealed a new boil located 300 feet due east of the previous leak. Dye was injected at the leak and observed in the boil, confirming a connection between the two. The boil was estimated to be flowing at about 3 gallons per minute, and it was decided to attempt to grout the leak using divers.

Grouting commenced immediately using contract divers and the DWR field crew. Bentonite and other additives necessary for a stabile, zero-bleed grout mix were not immediately available, so a cement grout was used until the necessary supplies could be obtained. Approximately 715 gallons of cement grout were injected into the void until supplies ran out. At this point, the seepage flow exiting the boil was observed to be lower. On March 2nd, a sandbag ring was constructed at the boil, and the flow measured only about 1 gpm. Grouting commenced again, and an additional 230 gallons of cement-bentonite grout were injected into the void, stopping the flow at the boil. It should be noted that rags and oakum (oil soaked jute rope) were also inserted into the void by the divers during the grouting operation to facilitate grout set up by slowing the flow of water. After the grouting was complete, the divers patched the void with hydraulic cement. Additional grout was pumped into the roadway cracks as well.

DWR engineers and geologists are continuing to monitor the conditions at this seepage distress site and plan to carry out a comprehensive investigation plan to determine an appropriate long-term repair for this area.

2013 Canal Repair at Milepost 342.65

The Milepost 342.65 repair site is located on the California Aqueduct approximately 5 miles west of the city of Palmdale in northern Los Angeles County. This site sits directly on top of the San Andreas Fault and was showing considerable subsidence and fractures in the concrete liner. In December 2010, sinkholes were discovered on the left embankment. A full geologic investigation was performed in 2011 to evaluate the site. The decision was made to address this stretch of the Aqueduct with a large scale repair to address the poor condition of the embankment foundation and concrete panels.

The Aqueduct repair work was performed during a planned 3-week outage in December 2013 that was scheduled over two years in advance. Significant coordination was required to plan for the full outage and dewatering of the Aqueduct. In order to complete the repair during the short outage window, construction operations were performed 24 hours a day, 7 days a week. Additionally, due to the nature of the adverse winter weather conditions typically experienced in the high desert area of Palmdale, precautions were taken during the planning phase of the construction. It is not unusual for rain, snow, and high wind velocities to occur in the area. The elements of the repair included:

- Compaction grouting on the primary and secondary embankments to improve the foundation
- Removal of damaged concrete panels
- Removal of unsuitable embankment material behind the removed concrete liner
- Placement of new compacted embankment with 12-18" angular rock at the base of the excavation in most areas
- Placement of new concrete liner
- Placement of 600 linear feet of Teranap geomembrane across the limits of the repair site
- Placement of new shotcrete liner on top of the Teranap geomembrane

A multitude of challenges were initially experienced with the grouting subcontractor which delayed the start of the work. These challenges included inexperienced personnel, deficient submittals, unacceptable materials, and inadequate monitoring, testing, and grouting equipment. Finally, after the deficiencies were addressed, a total of 47 holes were drilled and compaction grouted to densify the soils in the embankment foundation, which were found to be soft and loose during the geologic investigation. The subcontractor completed this portion of the work in November 2013. A total of 52 holes were originally planned for the project, 13 holes in each of 4 reaches. However, in the southwest reach, 8 holes were deleted due to minimal grout take and observed cracking at the surface of the embankment during grouting. Meanwhile, 3 holes were added in the northwest reach due to high volumes of grout take. An approximate total of 2,000 cubic feet of low-mobility grout was injected into the 47 holes.

A week prior to the Aqueduct outage, DWR closed the Check 52 and Check 53 gates, placed stop logs, and lowered the top 5 feet of Pool 52 by gravity. On the first day of the outage, the Contractor started pumping water into the downstream Pool 53 to draw down Pool 52 while exercising the of 2 feet per 24-hour period maximum rate restriction per the operating procedures.

When the pool elevation was dewatered with approximately 2 feet remaining in the Aqueduct, the removal of the existing concrete liner began. Once the underlying embankment was exposed, DWR engineers determined that additional panels should be removed in 3 of the 4 reaches due to the very soft soils and voids discovered behind the adjacent liner panels. The excavation line of the unsuitable material in the embankment was also extended approximately 2 to 3 feet deeper after the existing conditions were evaluated. Additionally, a 1- to 2-foot-thick layer of angular 12- to 18-inch maximum rockfill was placed in 2 of the reaches in the benched portion of the excavated area due to the condition of the unsuitable material. The intent was to help provide a more robust base for placing the compacted backfill. After a competent base was created, compacted backfill was placed up to finished grade in all of the excavated areas.

A new 4-inch thick cast-in-place concrete liner was placed over all of the excavated areas after the new compacted backfill had been placed. Due to the very low ambient temperatures, the Contractor made adjustments to the concrete placement activities by increasing the air entrainment admixture and covering the freshly placed concrete with a thick blanket covering weighed down with sandbags to keep the temperature of the concrete from dropping too low overnight.

Early on in the design, Teranap was again selected for use over the 600-foot long repair site. Previous Aqueduct repairs utilized this material, and a stockpile of Teranap was available. The Contractor placed this stockpiled material along the length of the repair site and on the bottom 1.5-feet of the bridge piers. New Teranap rolls were purchased as part of this construction contract to replenish the stockpile. During placement, the overlapping seams of each Teranap roll were welded and air tested. The Teranap was then rolled into anchor trenches on the upstream and downstream ends of the repair act as a cutoff wall for any potential water traveling through the soils behind the concrete liner.

Lastly, a 2-inch-thick layer of shotcrete was applied to the entire surface of the Teranap repair area to provide a protective layer for the geomembrane. Along the side slopes and invert, roller screeds were used and construction joints were tooled to create the panels. Around the piers, an epoxy coated rebar was added to the design in the field prior to applying the shotcrete over the geomembrane to ensure that the shotcrete adhered to the vertical surface.

The outage work was completed on time, and water was released back into Pool 52 on December 19th, 2013.

2011/2016 Canal Repairs at Milepost 248.9

On October 18, 2011, a boil was observed in an irrigation ditch that runs parallel to the Aqueduct along its east side at Milepost 248.9 in the San Joaquin Valley (see Figure 57). At this time, the boil was estimated to be flowing at approximately 5 gpm. The irrigation ditch is approximately 110 feet to the east of the Aqueduct embankment toe. Initially, only one boil was observed. From October 25 to 27, 2011, DWR staff installed an earthen dike around the boil to reduce the net head on the boil after flow from the boil had increased to over 100 gpm. On October 28, 2011, the flow from the boil was measured to be 10 gpm from a pipe installed in the sand bag dike. However, after the water level within the dike began to rise, a second boil emerged in the irrigation ditch north of the first boil. It was reported that the second boil was flowing at a similar rate as the first boil. Dive inspections were conducted between October 25, 2011 and October 29, 2011 to determine the source of the boil. Divers found cracks and suction at several locations and attempted to repair and grout the liner from October 30, 2011 thru November 4, 2011. Divers also drilled 20 holes in the Aqueduct liner at the location of the cracks and injected stable cement-bentonite grout by gravity pressure alone. The flow in the boils stopped flowing even though grout never flowed out of the boil.

Figure 57. 2011 Boil in ditch east of California Aqueduct at Milepost 248.9 (from DWR Division of Engineering files)

A field investigation employing 30 cone penetration test soundings was carried out in early November 2011. The CPT investigation found several zones of sensitive fine-grained material. Based on construction geology reports, the Aqueduct in this reach (Pool 30) follows the northern and western margin of the former Buena Vista Lake. The lakebed sediments are composed of primarily interbedded lacustrine clays and micaceous sands. Accumulations of gypsum, calcium carbonate, and other minerals, concentrated by groundwater or evaporation, occur widely distributed in the soils. During the original construction, low density materials, expansive clays, and materials with high gypsum content were encountered and where necessary, over-excavation and replacement with suitable compacted material was used to mitigate them. Locally, shallow groundwater required the use of drag lines during excavation of the Aqueduct profile together with the installation of an underdrain system.

Following the field investigation, DWR then quickly designed and implemented a compaction grouting program to increase the density of the sensitive fine-grained material and to remediate any void spaces or piping conduits located in the foundation of the primary embankment. From November 29 to December 9, 2011, Geo Grout Inc. completed the compaction grouting program. Primary compaction grout holes were spaced 16 feet on center and secondary holes split-spaced the primary holes for a final spacing of 8 feet on center. A single line of 4-inch-diameter holes was completed in the primary embankment, extending upstream and downstream of the boil. This line of holes consisted of 39 compaction grout holes (20 primary and 19 secondary holes) to a depth of 50 to 60 feet below the embankment road surface. Zero-slump grout was injected in three-foot stages, using pressures less than 400 psi, a pumping rate of less than 1 cubic foot per minute, and each stage was limited to a take of 30 cubic feet.

Primary compaction grout holes took on average 10 cubic feet of grout per stage, while the secondary compaction grout holes took on average 5.5 cubic feet of grout per stage. A total of 4,013.6 cubic feet of compaction grout was injected. Higher grout takes were expected in the primary holes, as they were grouted first and the secondary holes were grouted about one day following the placement of the primary holes. An attempt was made to correlate grout take with soil type; however, a correlation was not evident. In December 15-16, 2011 drivers inspected and repaired cracks along the primary embankment and a total of 20 holes were drilled in an attempt to find voids to grout. No voids were found.

On January 2, 2016, more than 4 years after the compaction grouting program, a leak and boil on the order of 5,000 gpm erupted at the same boil location as the previous 2011 boil in the irrigation ditch (see Figures 58 and 59). However, a dive inspection and dye test confirmed that the leak originated on the opposite side (primary side) of the Aqueduct in a depressed panel area located in approximately 8 feet of water. On January 4 and 5, 2016, DWR staff stabilized the boil by plugging the entrance to the leak with low slump ready-mix concrete, similar to what was used to stop the leak at Milepost 4.25 in 2001. At this point in time in 2016, the demand for water was seasonably very low. As a result, DWR managers then authorized an outage of Pool 30 and began dewatering the pool on January 6, 2016.

The DWR design team decided early on to use the Teranap geomembrane liner as a primary repair element, principally because of past successes with this approach. Initial plans called for minimal removal of concrete panels and embankment around the boil entrance, and Teranap placement for a length of approximately 900 feet. After dewatering of the canal pool, several other areas of panel removal were added on the primary side, and one additional area on the secondary side that extended beyond the Teranap limits by 205 feet. In addition, an area of sunken panels extending upstream of the Teranap limits approximately 132 feet had epoxy sealant installed in the transverse joints. The Teranap limits were shifted 50 feet in the upstream direction to cover distressed panels observed after dewatering, but the overall length did not change (900 lineal feet along the centerline). After installation of the Teranap membrane, the typical 2 inches of shotcrete surfacing was placed on top of the Teranap for protection of the geomembrane. The construction cost for the 2016 repair, along with two other similar-sized repair areas within the same Aqueduct pool was approximately $10 million.

Figure 58. 2016 boil and internal erosion pipe in ditch east of California Aqueduct at Milepost 248.9 – note the large size of the erosion pipe associated with the boil, the blue dye, and that the boil produced so much sediment that it formed its own sand ring (from DWR Division of Engineering files)

Figure 59. Aerial view of 2016 boil and canal repair at Milepost 248.9 (from DWR Division of Engineering files)

CONSTRUCTION COSTS

Table 1, presented previously, summarized the construction costs for 46 canal repairs made between 1971 and 2016. The costs ranged from as low as $20,000 to as high as $10 million for any individual repair or set of repairs. These costs, as presented in Table 1, represent the construction costs expended at the time of the repairs. These costs have also been adjusted to represent expenditures in 2016 dollars and grouped to provide an adjusted yearly cost for Aqueduct repairs. As shown in Figure 60, the yearly expenditure by year, in 2016 dollars, ranged from zero to as high as $25 million. The data in Figure 60 also indicate a significant increase in expenditures over the last 20 years or so. Part of this increase is due to the aging of the California Aqueduct while another portion is associated with the goal of DWR to be more proactive in repairing seepage distress in advance of an emergency. The total amount of these expenditures is approximately $115 million in 2016 dollars. However, this sum represents only the construction costs paid to construction contractors. If we also consider the costs associated with design, construction administration, environmental documentation and mitigation, and monitoring, then the cost is likely to increase by at least 35 percent, leading to a total cost somewhere near $155 million. However, over the 50-year life of the California Aqueduct, this works out to only about $3 million per year – not bad for an aqueduct that spans hundreds of miles over difficult terrain and geologic hazards. Even if this rate is doubled to $6 million per year to account for the increased rate of expenditure over the last 20 years, it seems to be a reasonable cost of doing business.

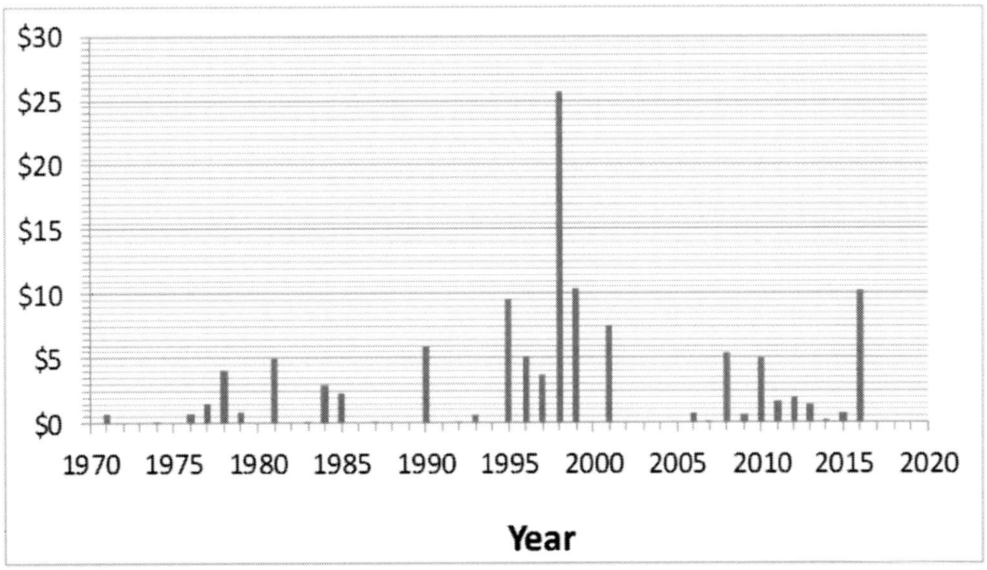

Figure 60. Construction costs expended by year for canal repairs on the California Aqueduct – shown in millions per year adjusted to 2016 dollar values (from DWR, Division of Engineering files)

PERFORMANCE OF CANAL REPAIRS AND LESSONS LEARNED

Over the last 45 years, a range of temporary, interim, and permanent canal repairs have been implemented to address various forms of seepage distress along the California Aqueduct. The following summarizes the general performance of these different types of repairs as well as lessons learned:

1. The performance of limited canal repairs has been mixed. As temporary measures, flood-fighting techniques such as temporarily lowering canal pool elevations or ringing boils with sand bags to reduce net hydraulic heads on the boils have been effective. However, they are usually only temporary at best.

2. When done improperly, bentonite or cement grouting through closely spaced boreholes (e.g. 3-foot spacing) drilled through the crowns of canal embankments can do just as much harm as it does good. While grout injected at excessively high pressures with a lack of refusal criteria can temporarily cut off preferred seepage paths and internal erosion, it can also cause hydraulic fracturing that may lead to additional seepage paths (e.g. see experience at Mileposts 56 and 88.3). Furthermore, when grouting in gypsiferous soils, the plugging of leakage has generally been temporary and commonly lasting only for weeks, or up to a short number of years, and often becomes completely ineffective over time.

3. When following proper grout mixtures and methods, grout injections in non-solutioning soils may cost-effectively last for decades. The grout, or sometimes concrete, needs to have the appropriate viscosity, anti-washout properties, and injection rate to satisfy the needs of the job. Thinner viscosities, as in high mobility grouts, are needed to fill finer interconnected piping voids, while thicker viscosities, or even low-slump concrete, are needed for single, large piping voids. Anti-washout admixtures, such as diutan gum, are needed when injecting in the presence of flowing water. Alternatively, higher dosages of bentonite or extra sacks of cement in the mix can mitigate the tendency of flowing water to dilute the mix. Finally, the flow rate of the grout or concrete needs to be comparable to the leakage rate in order to overwhelm the leakage and allow the mix to setup in place. If the grout mixer or concrete pump cannot produce mixtures fast enough, the flow of leakage can be reduced by injecting debris (e.g., oakum, burlap, or aggregate) along with the mix at the point of injection.

4. As an emergency measure when large preferred seepage and internal erosion is occurring, the most successful approaches have been to either rapidly lower the canal pool (e.g. see experience at Milepost 333.5), or to pump concrete underwater into the source of the leak/seepage to cutoff the seepage source (e.g. Milepost 4.25). This latter technique generally requires the use of special concrete mix designs, trained divers, and underwater drilling into the canal liner to be effective. Rapidly dewatering the canal pool to address seepage distress may also lead to rapid drawdown slides within the canal prism (e.g. see Milepost 62 and Lower Quail Canal rapid drawdown slides). However, the rapid drawdown slides may be the price that is paid to avoid a

failure of a canal embankment and uncontrolled release of Aqueduct water. Rapidly lowering the canal pool is also likely to be more successful when the canal capacity and pool depths are relatively small, such as on the East Branch in southern California, as opposed to major canal sections with large depths such as those in the northern portion of the San Joaquin Valley.

5. Approaches using thick clay liners such as was used at Milepost 56, or extensive multi-layered geomembrane/shotcrete layers such as was used at Milepost 333.5 and Milepost 4.25 have the best proven track record, with over 15 to 25 years of successful performance with no reoccurrence of the seepage distress that triggered them. The use of a geogrid over the seepage source beneath the geomembrane to distribute future differential movements, such as was used at the Milepost 333.5 and 4.25 repairs, provides an extra layer of resiliency.

6. Various geomembranes have been used over time for canal repairs on the California Aqueduct. However, DWR has settled on the use of Teranap, a bituminous paving membrane, for its standard canal repair. Teranap has demonstrated satisfactory performance over the last 20 years for the Department with only minimal maintenance required to date. The main component of the geomembrane is distilled bitumen which is inert, stable, and resistant to weathering and oxidation. Teranap has demonstrated excellent performance over time. Surface examinations of this geomembrane after the protective shotcrete layers were removed have shown no deterioration in these geomembranes, and they are relatively easy to install and seam, and standard construction contractors can install them with minimal training from the manufacturer. Application of two layers of the Teranap geomembranes and shotcrete protective cover layers is believed to be the most robust approach in making these repairs (e.g. see Milepost 4.25). The use of a fabric bonded to the Teranap geomembrane (Teranap GTX) is considered to be the most practical approach for applying shotcrete panels on top of this geomembrane on slopes.

7. Purchasing and maintaining a stockpile of Teranap for use in emergency canal repairs has been an important lesson that DWR has learned in order to avoid delays and extended canal outages and to minimize risks of canal failures.

8. DWR has used compaction grouting at several emergency canal repairs to strengthen soft soils and fill voids, provide a more erosion resistant foundation element, and as an exploratory tool to identify weak or soft foundation zones beneath the canal embankments. In terms of purely seepage improvements, the benefits of this method have been difficult to quantify. However, it is believed that compaction grouting is very worthwhile as a secondary or temporary seepage mitigation measure. Based on the performance at Milepost 248.9 where compaction grouting was used to strengthen soils and fill voids after high-mobility grouting sealed the leak, the compaction grout columns were likely the only things holding up the embankment when 5,000 gpm of water was rushing through the foundation 4 years later. However, if compaction grouting is used as a primary seepage remediation measure, it must be regarded as a temporary measure at best.

9. The use of steel sheetpiles to cutoff seepage as was done for the canal repair at Milepost 88.3 is very attractive as it does not require dewatering of the canal or interrupt project deliveries of water. However, this approach required a significant amount of time and its effectiveness over time has yet to be proven. In addition, the sheetpiles tend to raise the phreatic surface on the waterside portion of the canal embankment without improving any of the soils properties, thereby reducing the embankment's drawdown stability.

10. As the California Aqueduct continues to age, increasing surveillance and preventative repairs will be necessary to reduce the number of seepage-related emergencies on the canal system, and to better schedule complete repairs of seepage distress.

11. As part of an effort to minimize or defer extensive canal repairs and outages of the California Aqueduct, DWR recently entered into a 3-year construction contract for advance measures consisting of limited repairs that would have only limited impacts to canal operations. Approximately 20 locations where minor seepage distress has been occurring have been identified for such advance measures, most of which will consist of limited underwater grouting by divers and/or the replacement of concrete lining panels. It is hoped that such measures will be successful in preventing, or at least deferring, major repairs in the future.

12. In early 2016, DWR initiated a formal risk assessment program that identified current asset management deficiencies and will help develop a long-range policy for change by the end of 2018. Ultimately, it may take 10 or more years for DWR to integrate numerous elements of the ISO 55001 standard into its business practices to enable DWR to evolve into a mature asset management organization. Managing the aging California Aqueduct and canal seepage distress over time will be an important piece of this asset management program.

ACKNOWLEDGMENTS

The generally excellent performance of the California Aqueduct is a tribute to the people who planned, designed, constructed, and now maintain the system. Despite its large size, extraordinary length, and challenging geologic hazards, the Aqueduct has generally performed excellently over time. In recent years, seepage distress incidents required numerous quick and adaptive repair approaches. The authors are indebted to several past engineers, construction contract managers, consultants, and geologists who assisted them in developing and implementing these techniques, including Sonny Fong, Don George, Frank Glick, Mike Inamine, Ron Lee, Jim Peddy, Tom Tidyman, the late Ted Tsuruda, Jim Warner, and the late Bill Verigin.

REFERENCES

1. Department of Water Resources (1964), "*Design and Construction Studies of Shallow Land Subsidence for the California Aqueduct in the San Joaquin Valley*," Interim Report, Division of Design and Construction, Department of Water Resources, The

Resources Agency, State of California, December.

2. Department of Water Resources (1972), "*Concrete Lining Report, State Water Project, The California Aqueduct and South Bay Aqueduct*," Division of Design and Construction, department of Water Resources, The Resources Agency, State of California, July.

3. Department of Water Resources (1974), "*California State Water Project, Volume II, Conveyance Facilities*," Bulletin 200, Department of Water Resources, The Resources Agency, State of California, November.

4. Investigation Panel (1984), "*PANEL REPORT – Lower Quail Lake Canal Break*," Department of Water Resources, The Resources Agency, State of California, March 1, 1984.

5. Department of Water Resources (1994), "*Lower Quail Canal, Seepage and Erosion Control, Conceptual Design Report*," Memorandum Report, Division of Design and Construction, Department of Water Resources, The Resources Agency, State of California, September.

6. Department of Water Resources (2002), "*Final Construction Report, Emergency Canal Repair, California Aqueduct Mile Post 4.25, Specification Number 01-19,Contract Number C51247*," Draft Report, Subject to Revision, Division of Engineering, Department of Water Resources, The Resources Agency, State of California.

7. Department of Water Resources (1976-2016), Miscellaneous Investigation and Construction Data, Division of Engineering, Department of Water Resources, California Natural Resources Agency, State of California.

8. Department of Water Resources (2009), "*Seepage Repair, California Aqueduct, Mile Post 88.3, 89.4, and 90.2, Analysis and Recommendations*," DRAFT Memorandum Report, Division of Engineering, Department of Water Resources, California Natural Resources Agency, State of California, April.

9. Department of Water Resources (2011), "*Canal Seepage Repair, Mile Post 88.30, Final Construction Report, Specification No. 09-07, Contract No. C51392*," Division of Engineering, Department of Water Resources, California Natural Resources Agency, State of California, February.

10. Department of Water Resources (2013), "*Management of the California State Water Project*," Bulletin 132-10, Department of Water Resources, California Natural Resources Agency, State of California, June.

11. Galloway, Devin, Jones, David R., and Ingebritsen, S. E. (1999), "*Land Subsidence in the United States*," Circular 1182, United States Geological Survey, United States Department of the Interior.

12. Lucas, Clifford V. and James, Laurence B. (1976), "*Land Subsidence and the California State Water Project*," Publication No. 121 of the International Association of Hydrological Sciences, Proceedings of the Anaheim Symposium, December 1976.

GEOMEMBRANES TO WATERPROOF EMBANKMENT DAMS

John Wilkes, MS, PE[1]
Alberto Scuero, MS, PE[2]
Gabriella Vaschetti, MS, PE[3]

ABSTRACT

Polymeric geomembranes, with permeability some orders of magnitude lower than traditional water barriers, have been used in new embankment dams since 1959. The first applications were for new dams and with covered geomembrane. The use of upstream exposed geomembranes for new dams and to rehabilitate dams started at the beginning of the 1970s, due to greater confidence acquired through use of these relatively new synthetic materials. These dams had primarily hard subgrades (concrete, RCC, asphalt concrete), which allowed anchorage of stainless steel profile systems to tension the geomembrane against the wind. Since the 1970s, design and manufacturing of geomembrane materials, and design and installation techniques of waterproofing geomembrane systems, have been constantly developing. At the beginning of the new century, the technology became mature and its use has spread to embankment structures with granular subgrades. New anchorage systems employ extruded curbs, earth anchors, grouted anchors, and ballasting trenches. These techniques have now been installed on more than 10 dams worldwide, on 5 continents, including all 18 of the New Panama Canal Water Saving Basins. This paper will show the new anchoring techniques and the installation for both new and rehabilitated earth embankment structures. This includes the ability to install geomembrane systems underwater, even with flowing water. These techniques offer the ability to waterproof the upstream face of embankment structures for significantly less cost. The paper will present installations from Iran, Laos, Tahiti, Greece, Romania, Panama, and Egypt.

INTRODUCTION

Geomembranes are prefabricated, practically watertight flexible synthetic materials used in hydraulic projects for more than half a century. They are engineered to ensure adequate properties that allow them to sustain the loads imparted during service; since the membranes are manufactured in the controlled environment of a factory, the properties established by design are constantly maintained throughout the production by computer-controlled manufacturing processes, and constantly checked in the laboratory. This guarantees that the designed properties are constant for the whole production lot. At the job site, the properties, especially those crucial for good performance, such as watertightness and tensile properties, are checked with standard methods to verify that

[1] President, Carpi USA and Carpi Tech Canada, 4370 Starkey Rd Suite 4D, Roanoke, VA, USA 24018, john.wilkes@carpitech.com.
[2] President and Technical Director, Carpi Tech, Via Passeggiata 1, 6828 Balerna, Switzerland, alberto.scuero@carpitech.com.
[3] Vice President, Carpi Tech, Via Passeggiata 1, 6828 Balerna, Switzerland, gabriella.vaschetti@carpitech.com.

they are compliant to specifications and have remained constant during transport and storage. Since properties of geomembranes are independent of weather conditions during construction, the characteristics of the geomembrane water barrier when the dam is completed are consistent and verifiable. In CFRDs or in dams with impervious cores, the characteristics of the water barrier, on the contrary, depends on natural materials which are influenced by construction conditions, construction procedures, and workmanship.

Geomembranes are of particular interest when embankment dams are at stake. The major assets of geomembranes, in particular Polyvinylchloride (PVC) geomembranes, for embankment dams, are their outstanding elongation properties, which allow resisting settlements, differential movements, and seismic events that would destroy more rigid water barriers such as concrete facings (CFRDs); the possibility of engineering them to resist environmental aggression; and their permanent, consistent quality over large quantities, enables high quality installations.

The seaming of PVC geomembranes enables the face to be a waterstop, eliminating the need for the multiple defense lines of waterstops needed for a CFRD. Furthermore, geomembrane connections to the ancillary concrete structures can be designed to accept large differential movements ensuring watertightness at the perimeter. A geomembrane facing allows for building the dam on highly deformable foundations, allows for steeper upstream face, which means less volume of fill, and a shorter, smaller diversion tunnel.

From a construction standpoint, geomembranes have the advantage of reducing construction times, constraints and costs. In CFRDs, the installation/construction of the reinforced concrete face slabs, and the placement of copper and PVC waterstops, can have considerable impact on the overall construction schedule. In dams with clay or asphalt concrete cores, the construction of the dam body and the construction of the central core are strictly related. There are constraints imposed by weather conditions, or any disruption in placement of the filter material, or in placement/compaction of the impervious core. This affects the overall rate of construction of the clay or asphaltic core dam. On the contrary, installation of a geomembrane system is practically unaffected by weather. The geomembrane system does not need to be a serial task in the construction process. The geomembrane construction can be done in parallel with other activities, so that it is independent of most construction or operational constraints of the dam. The geomembrane can be installed when the dam is completed, or installed on the lower completed part of the dam while construction of the fill is ongoing in the upper part. Geomembrane installations offer the additional benefit that if there are floods during construction, the waterproofed lower part of the dam will be a barrier against the flood, increasing the safety of the project.

The complexity of the techniques needed to construct the waterproofing system must be taken into consideration when evaluating the time and costs of embankment dams. Inadequate placement of the waterstops and inadequate construction of the connections to the ancillary concrete structures, are critical, and can have highly negative effects on the future performance of the dam. Especially when the main contractor has little or no experience in construction of such dams, adopting a geomembrane system is "forgiving" and will avoid the problems related to inadequate construction, which could require future time consuming and expensive repairs.

Geomembranes were first used in construction of new embankment dams (1959 in Europe and in Canada). In these pioneering projects, when synthetic materials were relatively new to dam engineers, the geomembranes were covered. In the 1970s, once confidence in PVC geomembrane behavior and reliability had increased, a cover layer was no longer mandatory and exposed Polyvinylchloride (PVC) geomembranes began to be used. At present, PVC geomembranes are used in new construction of embankment dams, RCC dams, as external waterstops for peripheral and vertical joints in CFRDs, and as external waterstops for joints between monolith blocks in RCC dams. In rehabilitation, all kinds of dams have been waterproofed around the world, in both dry and underwater installations. This paper focuses on upstream exposed geomembrane systems in new construction and rehabilitation of embankment dams.

THE UPSTREAM EXPOSED GEOMEMBRANE SYSTEM: CONCEPTS AND COMPONENTS

Carpi upstream geomembrane systems are based on the concept of providing a flexible watertight barrier that can elongate well beyond the maximum expected deformations of the dam body. The flexible water barrier is anchored to the dam face with a site-specific anchorage system as described in this chapter and in the case histories that follow, and at concrete boundaries with a perimeter seal, designed to resist differential settlements while maintaining watertightness.

Upstream Water Barrier

In new construction as well as in rehabilitation, the most common exposed upstream water barrier is a flexible composite synthetic material (geocomposite) consisting of PVC geomembrane providing watertightness, with an anti-puncture geotextile backing, heat-bonded during manufacturing to the geomembrane. The PVC geocomposites used in the majority of these projects are known under the trademark SIBELON®, and are produced with a proprietary formulation under ISO 9001 certification.

One main reason for selecting a PVC geomembrane over other types of geomembranes is its tensile behavior. The tensile behavior of the geomembrane in a fill dam is extremely important, especially at locations where the geomembrane, supported by an embankment, is connected to a rigid structure (e.g. a concrete plinth). A study by Giroud & Soderman (1995) has shown that an appropriate combination of tensile strength and strain is essential. This optimal combination depends on the shape of the tension-strain curve of the geomembrane. Furthermore the same study shows it is quantified by the co-energy of the tension-strain curve. The co-energy is the area between the tension-strain curve and the tension axis, and quantifies the ability of a geomembrane to withstand differential settlements. Only the continuously increasing portion of the tension-strain curve can be used in the determination of the co-energy. Figure 1 which follows, shows at left the increasing portion of the tension-strain curve of two considered geomembranes: the PVC geocomposite (red line) has higher co-energy than the Low Linear Density Polyethylene (LLDPE) geomembrane (purple line). If a High Density Polyethylene (HDPE) geomembrane is considered, its co-energy in comparison to a PVC geocomposite is even less: since the tension-strain curve of an HDPE geomembrane has a yield peak at a strain of 12% to 15%, beyond the yield peak, an HDPE geomembrane ceases to function from a mechanical standpoint (Giroud 1984). Therefore, the co-energy of an HDPE geomembrane is calculated up to the yield peak, as shown in Figure 1 at right. The SIBELON® geocomposite, having a co-energy (i.e. area between the tension-strain curve and the tension vertical axis) much larger than the co-energy of the LLDPE geomembrane and even larger than the co-energy of the HDPE geomembrane, has higher ability to withstand differential settlements than LLDPE and HDPE geomembranes.

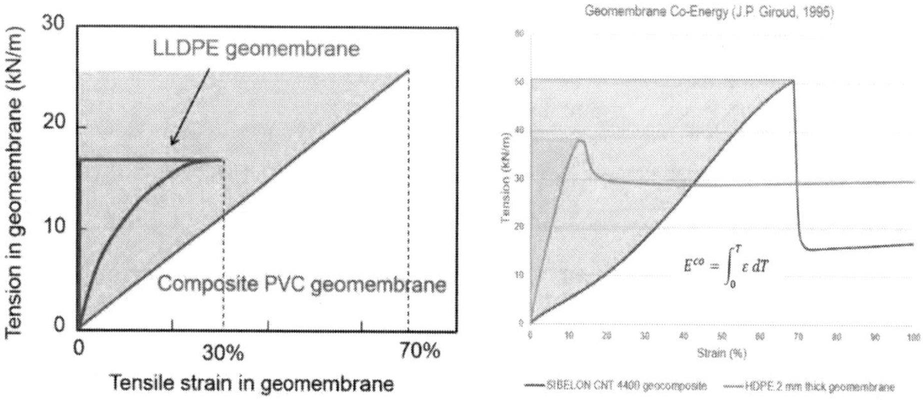

Figure 1. Left: co-energy of a 1.5 mm thick LLDPE geomembrane and a 2 mm thick composite PVC geomembrane. *Courtesy of JP Giroud*. Right: co-energy of a 3 mm thick composite PVC geomembrane (in blue) and of a 2 mm thick HDPE geomembrane (in yellow).

Anchorage to the Dam Face

The SIBELON® PVC geocomposite is anchored to the dam face to resist uplift by back-pressures, wind and waves. In the beginning of the 2000s, Carpi began extensive research to develop new techniques for anchoring geomembrane systems into the upstream fill of embankment structures without hard subgrades. Depending on the methodology of dam construction, different anchorage systems are available. All systems described below are Carpi patented:

1. In dams where the finishing layer is formed by extruded porous concrete curbs: the face anchorage system consists of PVC geocomposite anchor strips embedded in the extruded porous concrete curbs that form the stable finishing layer of the embankment. The PVC anchor strip fixed to one curb overlaps the PVC anchor strip fixed to the preceding curb, and the overlapping strips are heat-welded to form continuous anchor lines (Figure 2 at left). The distance between the anchor lines is designed in function of the design uplift loads. The PVC geocomposite liner is then deployed over the PVC

anchor lines and watertight heat-welded to them (Figure 2 at right). This system, presented in ICOLD Bulletin 135, applies to new dams.

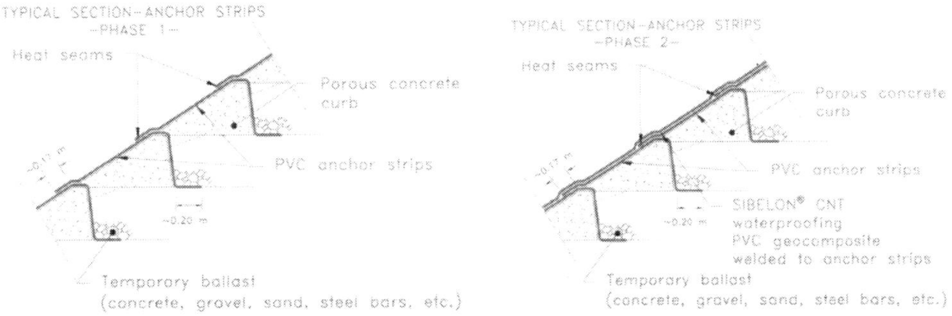

Figure 2. Face anchorage with PVC anchor strips embedded in curbs.

2. In dams with finishing layer formed by compacted granular materials: depending on site-specific construction method and type of uppermost layers, the anchorage system can be made either by deep anchors (grouted - Figure 3 at left - or Duckbill type) driven into the embankment at site-specific patterns, or by PVC geocomposite anchor strips embedded in trenches excavated in the base layer at site-specific spacing, and backfilled with draining material (Figure 3 at right). The PVC geocomposite liner is then deployed and punched over the deep anchors/welded to the anchor strips. These face anchorage systems can be used for new dams and for rehabilitation.

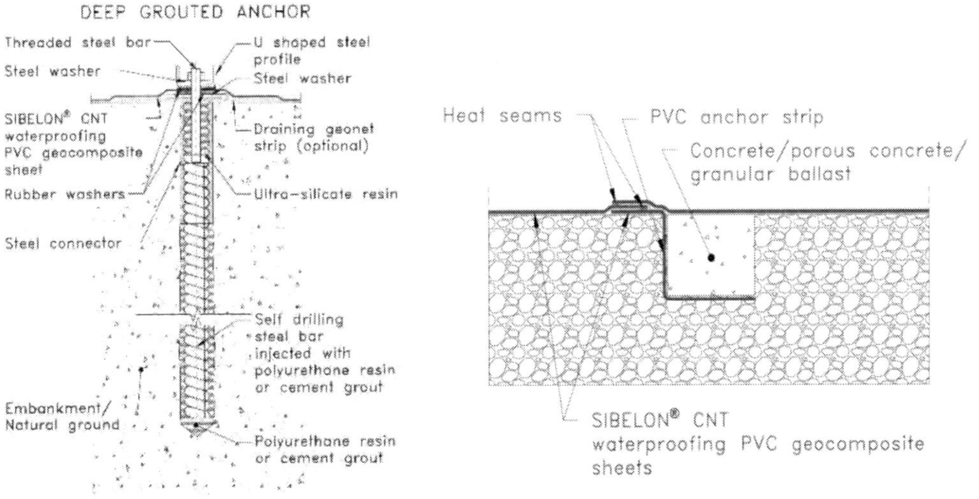

Figure 3. Face anchorage with deep anchors (left), and with PVC anchor strips embedded in trenches (right).

Perimeter Sealing

At all peripheries, the SIBELON® geocomposite liner is anchored by perimeter seals to avoid water infiltrating behind the liner. Perimeter seals have been tested to water pressures exceeding 800 meters. The case histories that follow will describe how the seal is designed in function of the underlying substrate.

CASE HISTORIES

Face Anchorage by PVC Strips Embedded in Curbs

This system applies to new construction only, when the dam body is raised placing the fill against extruded porous concrete curbs that form a draining finishing layer at the upstream face of the dam.

Sar Cheshmeh tailings dam raising (2008). This was the first embankment dam adopting this type of face anchorage system. Sar Cheshmeh's existing tailings storage in Iran, owned by National Copper Industries Co., included a 75 m high main embankment consisting of an inclined clay core and of outer colluvial gravel shells. The production escalation called for a set up comprising a 39.5 m high and 1000 m long downstream raise to the main embankment, in four separate stages, of which IIB and IIC have been completed. Future stages III and IV have yet to be executed. Stability analysis showed that the seismic stability of a raised clay core was not sufficient, due to the geometry of the raising. Furthermore, no suitable clay based materials were available at site. ATC Williams, designers of the dam raising, considered as alternative solutions--an asphaltic core, an upstream bituminous membrane, and polymeric geomembranes. An upstream exposed PVC geomembrane facing rockfill dam (GFRD) was selected because of superior safety in respect to earthquakes. ATC Williams deemed the GFRD system would be the most stable, efficient and buildable arrangement.

The finishing layer of the dam is made with extruded porous concrete curbs. The face anchorage for the PVC geocomposite is the patented method with PVC anchor strips discussed above. As the embankment & curbs were being raised, the PVC strips were nailed to the curbs and then permanently anchored by the fill compacted against the curbs. Overlapping PCV strips are joined by heat-seaming.

Figure 4. PVC anchor strips embedded in curbs.

The PVC strips heat-welded at the overlap, form continuous anchor lines. The SIBELON® geocomposite liner sheets are then deployed from the crest, after having been secured at top by a stainless steel batten strip on a conventional concrete curb. After cleaning the PVC anchor lines (Figure 5 at left), the PVC geocomposite sheets are temporarily anchored at the crest of the first stage, Stage IIB, and then unrolled down the slope (Figure 5 on right).

Figure 5. PVC anchor lines and deployment of geocomposite sheets.

The PVC geocomposite used for the anchor strips and for the liner is SIBELON® CNT 4400, consisting of a 3 mm thick PVC geomembrane, heat-bonded during manufacturing to a 500 g/m^2 non-woven polypropylene geotextile. The PVC geocomposite sheets are heat-welded to the PVC anchor strips. Adjoining PVC geocomposite sheets are watertight heat-welded at overlapping, shown in Figure 6.

Figure 6. The PVC geocomposite sheet heat-welded to the PVC anchor line (left). On right, adjoining PVC sheets are heat-welded to form a continuous watertight facing.

The bottom seal of Stage IIB was made by embedding the PVC geocomposite in a trench excavated in the existing clay core and then backfilled with clay. The bottom perimeter seal at the concrete plinth of the abutments is mechanical, referred to as the tie-down type. In tie-down seals, watertightness against water in pressure is attained by compressing the PVC geocomposite unto the concrete with flat stainless steel batten strips secured by stainless steel anchor rods embedded using chemical glass anchoring capsules at regular spacing. Smoothing epoxy resin, rubber gaskets, stainless steel bat ten strips and splice plates achieve even adequate compression necessary for watertightness, as extensively described in international literature, Figure 7.

Figure 7. Bottom perimeter seal - embedded type (left) and the tie-down type (right).

The intermediate seal between Stage IIB and Stage IIC is made by watertight joining of the geocomposite of the upper stage on the geocomposite lower stage, and covering the weld with a horizontal PVC geomembrane cover strip. The perimeter seal at top of Stage IIC is made by embedding the geocomposite in a trench then it is ballasted with a conventional concrete beam, as in Figure 8.

Figure 8. Intermediate seal between Stage IIB and Stage IIC (left) and top perimeter seal at Stage IIC (right).

Stage IIB and Staged IIC raisings reached 20 m of height. Total surface was 38,500 m^2. Installation of the waterproofing system, after completion of the curbs, took 14 weeks in total.

Nam Ou VI (2015/2016). Nam Ou VI dam is part of the Nam Ou VI Hydropower Project in Lao PDR, total installed capacity o180 MW. The scheme includes an 88 m high rockfill dam, designed as a Geomembrane Facing Rockfill Dam (GFRD), which will be the highest GFRD in Laos. The only element providing watertightness to the dam is an exposed composite PVC geomembrane. The dam body consists of slate rockfill for the whole cross-section, except for the bottom drainage facility. Upstream slope inclination is 1.6H/1V, the crest is 362 m long, and the total lined area is about 38,000 m^2.

The original design of a polyethylene geomembrane placed on a 3 m wide granular bedding layer, sandwiched between two anti-puncture geotextiles and covered by precast concrete elements, was modified following a study showing that the net expected settlement and horizontal displacement due to the embankment load and the reservoir hydrostatic head on the dam face, would be in the order of 100 mm vertical and 80 mm horizontal. Such expected deformations, well within the performance of a suitable geomembrane lining, could exceed the resistance of the foreseen concrete cover, possibly causing damage to the geomembrane. The initial solution was modified, eliminating the concrete cover and leaving the geomembrane exposed. A different type of geomembrane had to be selected to ensure long-term successful performance in an exposed position: a SIBELON® CNT 5250 geocomposite, consisting of a 3.5 mm PVC geomembrane, heat-bonded during fabrication to a nonwoven, needle-punched, continuous filament polypropylene geotextile, 700 g/m^2. The final design included an extruded porous concrete curbs facing, on which the PVC geocomposite is placed in exposed position, and anchored to the dam body by the above described system, using PVC anchor strips embedded in the curbs, forming parallel anchorage lines at 6 m spacing. The dam body was constructed in three separate stages: Stage 1 from the bottom of the excavation at El. 427 m, to a temporary work platform at El. 472.1 m, Stage 2 to from El. 472.1 m to a temporary platform at El. 512.1 m, and Stage 3, comprising a 3 m high parapet wall, to reach El. 515.0 m. Similarly the geomembrane system was installed in 3 stages (Stage 1 – 13,500 m^2, Stage 2 – 23,000 m^2, Stage 3 - parapet wall connection). Construction of the dam body of Stage 1, embedding the face anchorage system, started on July 23, 2014 and was completed on November 2, 2014. Installation of the geomembrane system of Stage 1 started on November 3, 2014 and was completed in 24 days for about 13,700 m^2 of upstream water barrier, a fraction of what a concrete facing would have required. See Figure 9.

Figure 9. Stage 1: PVC anchor lines, welding of geocomposite sheets to anchor line, Stage 1 waterproofing system completed in 24 days.

The perimeter seal at plinth was designed with extra geocomposite material, to further extend the elongation capability of the system at this critical connection. Construction of the dam body of Stage 2, embedding the face anchorage system, began December 12, 2014 and was completed on April 14, 2015. The geomembrane system of Stage 2, identical to that of Stage 1 and connected to it by watertight heat-welding the geocomposites of the two stages, was installed starting April 15, 2015 in 28 days, for about 23,000 m^2 of upstream water barrier, shown in Figure 10. Waterproofing of the wave wall was made in 2016.

Figure 10. Preparation of perimeter seal at plinth, starting on construction of dam body and face anchorage system of Stage 2, waterproofing system completed in 28 days. The geocomposite of Stage 1, already covered by dust, looks like concrete.

__Face Anchorage with Deep Anchors__

This system applies to rehabilitation and to new construction.

Vaité (2011). Vaité earthfill dam, in French Polynesia, is an example of rehabilitation using deep anchors of the Duckbill type, which consist of a stainless steel bar or wire, of a rotating Duckbill, and of a tendon, which are driven into the ground. When these anchors are fully inserted, an upward pull on the tendon, rotates the Duckbill into a perpendicular "anchor lock" position in the soil (Figure 11). In this project, the anchorage system has been designed to resist hurricane conditions, with wind speed u to 204 km/h in the top 1/5 of the section, and to 100 km/h winds in the remaining 4/5. The SIBELON® CNT 3750 geocomposite, consisting of a 2.5 mm PVC geomembrane heat-bonded during fabrication, to a nonwoven, needle-punched, continuous filament polypropylene geotextile, 500 g/m^2, extends on the bottom of the reservoir for about 50 m, and on the abutments, to provide some waterproofing of the foundations as well. The results of real scale field testing carried out prior to final design showed that the deep anchors could soundly anchor the geocomposite. On the bottom of the reservoir, anchorage is made by ballast.

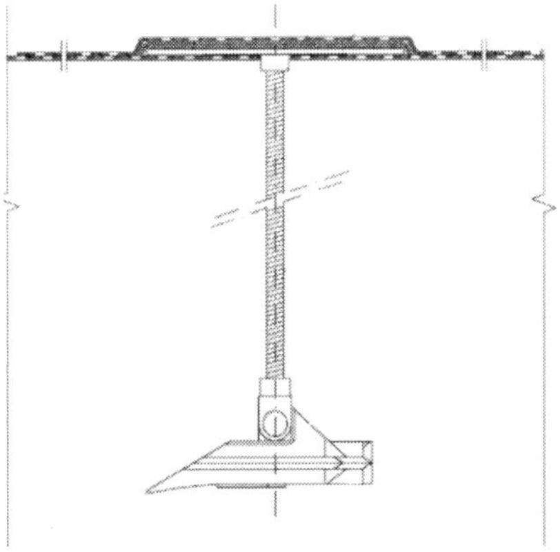

Figure 11. The Duckbill earth anchor, before and after rotation.

Figure 12. Deep earth anchors at Vaité earthfill dam.

Experience at Vaité dam and at canals in Austria and Germany, where deep grouted anchors were installed for rehabilitation, showed us that anchorage systems with deep anchors can be installed in any kind of subgrade, quickly and without any inconvenience. Application of the system to new embankment dams was then considered, to allow constructing very quickly and at low cost, impervious dams consisting of compacted fill providing stability, and with a flexible upstream liner providing imperviousness, and capable of accommodating settlements and differential movements (over 200% over large areas) exceeding any reasonable anticipated settlement.

Filiatrinos (2015). Filiatrinos 55 m high hardfill dam in Greece is an example of an application of deep anchors to new construction, and the first hardfill dam in the world to adopt a geomembrane facing. A hardfill dam, a sub-species of the cemented material dam (CMD), is a dam in which aggregates of even poor quality, together with low cement content (typically 50-80 kg/m^3), achieve a mix of low strength, sufficient for stability requirements. The avoidance of joints in the dam body, and the use of extruded or precast concrete elements at the external faces of the dam, simplify construction. A hardfill dam being semi-pervious, needs an upstream sealing element to fulfill watertightness requirements. An alternative to the concrete slabs traditionally used for watertightness, is a suitably designed geomembrane/geotextile sealing system, placed at the upstream face and drained underneath. The system can be part of the dam's design, or be retrofitted, to substitute a concrete facing, as it was at Filiatrinos. The geocomposite is secured to the hardfill by deep grouted anchors placed at 4 x 4 m spacing. Installation of the system was completed on August 8, 2015. Figure 13.

Details on this project can be found in recent literature (Scuero et al., 2016).

Figure 13. Deep grouted anchors keep the geocomposite stable against wind uplift at Filiatrinos hardfill dam.

Face Anchorage by PVC Strips Embedded in Trenches

This system has been developed for new construction and can be retrofitted when part of the dam body has already been constructed, as was the case at Runcu rockfill dam in Romania.

Runcu (2016). Runcu is a 91m high rockfill dam in Romania, which will be used for water supply and irrigation. After construction had already started, the original CFRD design was modified to incorporate an exposed geomembrane, instead of the concrete facing, to construct a GFRD. The reasons for changing the design were related to particularly long construction times required for the support layer of the slabs and for the slabs themselves, and because of the high costs involved with the CFRD design. The solution with concrete slabs would have cost about 130% more than the solution with the geomembrane facing, because it involved constructing three different zones of granular material, made on site, as support for the slabs.

The dam is a homogeneous rockfill placed in 1.5 m high compacted lifts, with quite a steep (1.4H / 1V) upstream face. In the original design a Zone 2 (5 to 250 mm) layer is placed on the fill, followed by a Zone 1 (5 to 90 mm) layer. Both layers, each about 5 m thick, were to be covered by a porous concrete layer before placement of the concrete slabs. The dam was to be constructed in phases, a first phase up to elevation +681 m, a second phase up to +710 m which would allow partial operation of the dam to a water level of +700 m, and then the third phase up to crest, at elevation 736 m. When the design was changed, the rockfill had already been constructed up to elevation +681 m, with a 0.2 m thick layer of sand and gravel placed on top of Zone 1. The design of the new waterproofing system had to adapt to the existing situation. Since the dam was constructed up (above Zone 1) to elevation +686.50 m, according to the original design, the PVC anchor strips had to be embedded in vertical trenches excavated in the existing surface at about 6 m spacing and filled with concrete. From elevation +686.50 m up to crest, the Zone 1 and Zone 2 layers were eliminated and the face is formed by porous concrete curbs extruded by a curb extruder, with PVC anchor strips embedded in the curbs. Today, Runcu is the highest embankment dam in the world having an exposed geomembrane facing as the only waterproofing element. See Figure 14.

Figure 14. Runcu dam. The PVC anchor strips are embedded in trenches excavated in the already existing dam body below elevation +686.50 m, while above elevation +686.50 m the dam is raised with extruded porous concrete curbs embedding the PVC anchor strips.

Panama Canal Expansion (completed 2016). The Panama Canal Expansion project, which opened to traffic in June 2016, involved the addition of 6 locks to the 77 km long navigation canal connecting the Atlantic with the Pacific Ocean. The new locks increased the canal width and depth necessary in order to support modern container ships. The New Panama Canal project was one of the largest construction sites in the world, over the past 5 years.

Gatun Lake is the natural lake in Panama that has been increased by a dam to allow ships to transit between the Atlantic and Pacific locks. In order to save the fresh water of Gatun Lake for the operation of the locks, 18 Water Saving Basins (WSBs) -- 9 on the Atlantic side locks and 9 on the Pacific side locks, were built and waterproofed with exposed PVC geomembranes and geocomposites. The contractors' consortium, with an international consultants' consortium, in close collaboration with the specialized subcontractor, developed the construction design adapting and optimizing the design baselines to the conditions encountered during the construction of the basins.

The waterproofing system is composed of a geocomposite, consisting of a PVC geomembrane (3.0 mm thick on the slopes and 2.5 mm thick on the bottom) heat-bonded to a 500 g/m^2 polypropylene geotextile, covering an 8 mm thick geodrain--composed of a geonet coupled with geotextile filters at both faces to avoid the intrusion of fines. The geodrain layer is double in the perimeter of the intakes under the maintenance road to increase the capacity of the drainage. The greater thickness chosen for the PVC geomembrane on the slopes was due to its exposure to UV radiation during normal operation of the basins (14 cycles per day). The waterproofing of the WSB at the Atlantic and Pacific sites covered the whole surface, bottom and slopes, joints at the dividing concrete walls of each basin, reaching the intake of the conduits, two at each basin. The whole surface lined by the waterproofing system, about 600,000.00 m^2, is in direct contact with the excavated soil or earthfill.

The forces considered in the design of the anchoring system were uplift by wind speed and hydrodynamic water pressure, erosion due to fluctuations on the operation level of the basins, sliding of slopes, and ground geotechnical conditions. This project is a good example of the use of different anchorage systems, designed to function on the large variety of subgrade encountered at different parts of the WSBs. The subgrade changed without a regular fixed pattern from gatun rock, to clay, to excavated basalt or rock/soil filling, to concrete, to shotcrete. Depending on the type of subgrade, face anchorage is made by deep soil anchors, or by ballast trenches filled with concrete, or by ballast concrete blocks, or by tensioning profiles, or by tensioning trenches, each anchorage type with the common purpose of holding the geomembrane against the terrain to reach a smooth lining in case of all the mentioned forces. The tensioning trenches concept, used mainly in the bottom of the basins and in the top anchorage, consist in a ballast of concrete placed on top of the impervious geocomposite, with the function of holding the system on place and tensioning all at once, in light of the 14 cycles/day of varying water levels. All concrete structures in contact with water, and not covered by geomembrane, have a perimeter watertight seal designed to guarantee the watertightness.

Figure 15. A general view of the lower chamber at Pacific side, and general view of the 9 WSBs at Atlantic side at the Panama Canal Expansion.

The installation of the waterproofing system of the Water Saving Basins was executed simultaneously with the civil works activities, in permanent coordination with the main contractor, to avoid major interferences and damages to the waterproofing liner. The design underwent some optimizations and changes adapting the baseline design to the real conditions of the ground site. This was possible only because of the flexibility and adaptability of the waterproofing system and the expertise of the specialized contractors and designers. The installation works were completed in less than a year, in difficult weather and site conditions, and they fulfilled all the contract requirements, passing all tests at completion.

A NEW TECHNOLOGY FOR UNDERWATER WATERPROOFING

Recently, a totally innovative waterproofing technology has been developed to restore watertightness when embankment dams or canals cannot be dewatered, without any impact on operation. The waterproofing system consists of two watertight geomembranes connected to form a mattress. The connection system between the two geomembranes allows free distribution of a ballasting filling material, such as inexpensive cement grout. The mattress is prefabricated in panels 10 m wide, with custom made length to minimize junctions and facilitate placement. Adjoining panels are connected by watertight heavy-duty zippers sealed to the panels during prefabrication. The lower geomembrane provides watertightness, the grout provides ballast anchoring the mattress, the upper geomembrane provides containment of the grout, protects the ballast during operation, and in canals grants hydraulic efficiency. Installation is performed without stopping operation or reducing water speed. The grout will crack allowing the geomembrane to accommodate settlement and adjust to settlements.

For the last 5 years, various testing has occurred on the components to develop a complete system. In the last 2 years full scale field tests have been installed on a test section in a 40 m wide canal in Italy, and on a 72 m wide test section in an irrigation canal in Egypt, without reducing the 1.2 m/s water speed.

Figure 16. Watertight PVC mattress installed underwater at Ismailia canal in Egypt.

The technology can be applied to embankment dams, to provide an impermeable upstream facing or an impermeable blanket, even on very aggressive irregular subgrade.

CONCLUSION

Over the last 15 years, various new anchoring techniques have been developed to allow geomembrane systems to be installed on embankment structures and left exposed. The elimination of the need for a cover layer can be a significant cost savings. Additionally, these new anchoring techniques can be applied to rehabilitate existing embankment structures, but require the removal or in-filling of rip-rap. These techniques offer the ability to waterproof the upstream face of embankment structures for significantly less cost. Finally, a recent technique involving a complete system of geomembrane mattresses will allow the waterproofing of hydraulic structures in-situ, with flowing water. This system promises the possibility to rehabilitate many embankment structures that currently cannot be dewatered.

REFERENCES

Giroud, J.P., "Analysis of Stresses and Elongations in Geomembranes", Proceedings of the International Conference on Geomembranes, Vol. 2, Denver, CO, USA, June 1984, pp. 481-486.

Giroud, J.P., and Soderman, K.L., "Comparison of Geomembranes Subjected to Differential Settlement", Geosynthetics International, Special Issue on Design of Geomembrane Applications, 1995 Vol. 2, No. 6, pp. 953-969.

ICOLD - International Commission on Large Dams, "Bulletin 135: Geomembrane Sealing Systems for Dams - Design Principles and Review of Experience", Paris, 2010.

A. Scuero, G. Vaschetti, Y. Thanopoulos, "Filiatrinos, the first Hardfill Dam with a Geomembrane Facing", 10th ICOLD European Club Symposium, Antalya, Turkey, October 25-30 2016.

GEOTECHNICAL INVESTIGATION OF SOUTH CAROLINA FLOOD-DAMAGED DAMS AFTER THE 2015 OCTOBER RAINFALL EVENT

Inthuorn Sasanakul[1]
Charles E. Pierce[2]
Ryan Starcher[2]
Mostaqur Rahman[2]
Sarah L. Gassman[2]
William Ovalle[2]
Emad Gheibi[2]

On October 2015, a historic rainfall event occurred resulting in extreme flooding conditions in South Carolina causing loss of life and more than one billion dollars in property damage. Dam failure was one of the major causes contributing to the damage. A total of 47 regulated dams breached during the flood event and over 200 dams were damaged. A majority of these dams are privately owned and had a lack of maintenance for several years before the flood event. Lack of funding is delaying the repair of these dams and many of them will possibly be removed.

The research team performed field investigations at several sites where dams ed and were damaged after the flood event. Visual observations, photographs, field measurements, and soil samples were collected during the investigation. Geotechnical laboratory testing was performed on collected soil samples to characterize engineering properties of the dam fill materials. This paper summarizes results of the field investigations conducted for selected dam sites. The paper provides the background of the historic flood event, condition of the sites before and after the flood, description of the dam fill materials, and key findings observed at the sites. The data collected from this project will serve as a set of case histories data to be used for future research study.

[1] University of South Carolina, sasanaku@cec.sc.edu
[2] University of South Carolina

ROUGH RIVER DAM:
PHASE 1B EXPLORATORY DRILLING AND GROUTING PROGRAM

Conrad Ginther, PE[1]
Dmitri Ivanov, PEng[2]
Doug Horvath[3]
Rolando Benitez, PG[4]

ABSTRACT

The Rough River Dam is an embankment dam with a history of high piezometric levels and problem seepage impounding the Rough River near Falls of Rough, Kentucky. The Dam is an approximately 130 foot tall homogenous earth fill dam with an internal filter and a crest length of 1,590 feet. The Dam impounds the Rough River to form a 5,100 acre reservoir which provides flood control, water supply, and recreational benefits to the surrounding communities. The Dam is largely constructed on alluvial materials which were not excavated or improved during original construction with the exception of along the alignment of the spillway conduit. The underlying rock units are a series of weakly cemented sandstones, limestones with variable amounts of solutioning, and soft clayey shales. A series of risk assessments and failure mode analyses have been performed for the Dam by the USACE, with the primary failure modes identified consisting of erosion of embankment materials into the untreated rock or scour along the contact between the embankment fill and foundation materials. The current preferred alternative for remediation consists of the installation of a grout curtain followed by the construction of a cutoff wall through the embankment, alluvium, and carbonate rock units.

This case history will discuss the methods employed to perform this state of the art exploratory program. Topics covered will include the use real time monitoring and integration of electronic records for dam safety instrumentation, drilling, water pressure testing, and grouting; the use of special designated downstages to treat and protect the foundation interface; and the use of proprietary instrumented packers capable of measuring injection pressure and column pressure simultaneously for all grout and water pressure test stages.

[1] Black & Veatch, 320C Boone Dam Road, Kingsport, TN 37663, 1-803-351-3657, GintherC@BV.com
[2] Advanced Construction Techniques, 3935 Lloydtown Aurora Rd, King, ON L&B-0E7, 1-877-373-7248, divanov@agtgroup.com
[3] Advanced Construction Techniques, 1000 N. West Street, Suite 1200, Wilmington, DE 19801, 1-877-373-7248, dhorvath@agtgroup.com
[4] RIZZO Associates, 500 Penn Center Blvd, Pittsburgh, PA 15235, 1-412-856-9700, Rolando.Benitez@rizzoassoc.com

INTRODUCTION

The Rough River Dam is an embankment dam with a history of high piezometric levels and problem seepage impounding the Rough River near Falls of Rough, Kentucky. To support the development of designs for remediation of the Dam, the Louisville District of the U.S. Army Corps of Engineers (USACE) awarded the Rough River Dam Safety Modification, Phase 1B – Exploratory Drilling and Grouting project to Advanced Construction Techniques (ACT) in May 2015. The program was initiated in July 2015 and completed in July 2016, and consisted of two grout lines of opposed holes drilled twenty degrees from vertical through the embankment and underlying rock. As shown in Figure 1, the upstream line extended from the left abutment at Station 9+16 to the approximate center of the river valley at Station 21+16, with an additional section of inclined holes installed to target the rock under the conduit from near the right abutment. The downstream line began at Station 19+82 and ended in the right abutment at Station 26+00. The upstream and downstream lines initially were specified a common termination depth of approximately 215ft, penetrating the Sample Sandstone. Subsequently the termination depth was reduced for most holes to within the Reelsville Limestone to limit the impacts caused by the hydrocarbons encountered in the Sample. A subsequent drilling and grouting phase is currently underway to complete the grout lines and further prepare the foundation for cutoff wall construction.

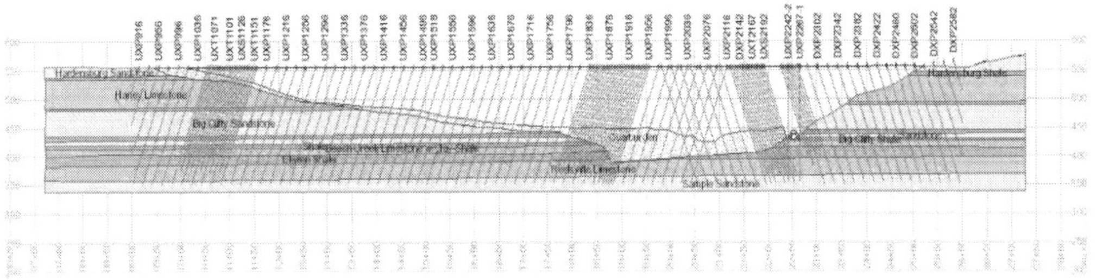

Figure 2. Rough River Dam Phase 1B Grout Lines Section

SITE GEOLOGIC SETTING

The Rough River Dam is in the northwestern portion of the Mississippian Plateau in south-central Kentucky in Falls of Rough, KY approximately 60 miles southwest of Louisville. The rock units at the site are generally flat lying and consist of a series of sandstone, shale, and limestone units. In general, the geologic features of interest with regard to hydraulic conductivity appear to be related to weathering mechanisms rather than specific sets of structural features or relative differences in rock type. High takes in pressure tests or grout stages were typically found to correspond inversely with depth below top of first encountered rock, regardless of rock unit; that is, the more cover the less measured permeability or grout take. Table 1 is arranged from highest unit to lowest encountered at the site, and summarizes general pertinent information for each rock unit.

Table 1. Summary of Rock Units

Rock Unit	Approximate Elevation Range (ft)	General Description	Station Range where Unit is first encountered Rock
Hardinsburg Sandstone & Shale	560 – 540	Light grey, fine grained, thinly bedded, with shale and siltstone beds.	0+00 to 10+00
Haney Limestone	540 – 490	Karstic, cherty, hard, light grey, fine to coarse grained Limestone	10+00 to 14+00 23+60 to 24+50
Big Clifty Sandstone	490 – 430	Light yellowish grey to olive, thinly bedded, interbedded with siltstone and shale beds, poorly cemented, soft to medium hard Sandstone	14+00 to 17+75
Big Clifty Shale	430 – 425	Dark brown, interbedded with sandstone, soft, clay rich Shale,	
Beech Creek Limestone	425 – 415	Karstic, hard, light grey to grey brown, fine to medium grained, fossiliferous Limestone	17+75 to 22+50 (depending on valley depth)
Elwren Shale	415 – 400	Soft, blue green to dark grey, clay Shale with limestone beds and hydrocarbon traces	17+75 to 22+50 (depending on valley depth)
Reelsville Limestone	400 – 370	Light to medium gray, fine grained, fossiliferous Limestone	N/A
Sample Sandstone	370 – 320	Oil bearing, medium hard to hard, white to light yellowish grey, fine to medium grained Sandstone with numerous olive green shale beds	N/A

The Hardinsburg sandstone and shale units were only encountered in the far right and left abutments, and typically experienced low grout takes. These units were typically not water tested because their extents were within the first 15 feet of top of rock and so were completely included in the initial interface and gravity treatments.

The Haney limestone generally exhibited low takes in water pressure testing (WPT) and grouting with some exceptions. Several stages in the Haney on the right abutment exhibited high Lugeon values and grout takes associated with intercepting high angle fractures. On the left abutment, a collapse feature or trough from approximately Station 10+60 to 11+00 dropped almost completely through the formation and high WPT and grout takes were encountered around the margins of the feature. In addition, a soft shale layer marking the boundary between the Haney and Big Clifty units around this feature required multiple downstages to stabilize and permit the deeper work to proceed.

The Big Clifty sandstone and shale units were generally tight, with moderate to high takes experienced where the sandstone was the uppermost unit and in the vicinity of the collapse feature in the left abutment. The weak cementation of the sandstone combined with interbedded zones with soft clayey shales sometimes made this unit difficult to advance through and required low take downstages on the left side of the dam.

The Beech Creek Limestone was found to have high hydraulic conductivity and correspondingly high grout takes where unprotected by the upper rock units, particularly in the vicinity of the outlet works conduit. Where protected by upper rock units, takes and open defects were minimal.

Where the Elwren Shale was the first encountered rock unit, the shale was clay rich and generally soft to very soft, but did not exhibit high takes. In the central portion of the dam, this unit was found to bear trace hydrocarbons.

Similar to the Beech Creek, the Reelsville Limestone was found to be tight where protected by upper rock units. In the center of the valley from approximate station 19+76 to 21+45 the Reelsville was either the first encountered rock unit or covered by a small amount of Elwren Shale and large takes in WPT and grouting were encountered.

The Sample Sandstone was explored in relatively few holes in this phase of the work because it was found to contain higher than expected volumes of oil when penetrated in the center of the valley. To close these holes an additional treatment process consisting of pumping the hydrocarbon out of the sandstone and drumming the waste for proper disposal followed by immediate grouting of the bottom stage of the holes was enacted. Several of these stages took significant grout volumes after being pumped clean. Where penetrated in the abutments, the Sample was found to contain free hydrocarbon but not at more than nuisance levels.

TECHNICAL APPROACH

Instrumentation Monitoring

Prior to the start of the exploratory drilling and grouting program, an extensive vibrating wire piezometer network was installed at the dam over the course of numerous geotechnical studies. Figure 2 shows a plan view of the Rough River Dam including locations of the borings and piezometers installed in previous studies. These instruments were automated and networked to provide real time monitoring of the embankment, foundation soils, and underlying rock during drilling, water pressure testing, and grouting activities during the exploratory program. During any production activities, several critical piezometers were monitored by the grouting engineers depending on the location of the work and the geologic unit being worked in. In addition, automated reports were developed detailing the instrumentation responses that occurred during each drill event, water pressure test stage, and grout stage. These reports provide valuable insight into the overall hydrogeologic regime in the foundation and served to further the project team's understanding of the overall site geologic setting.

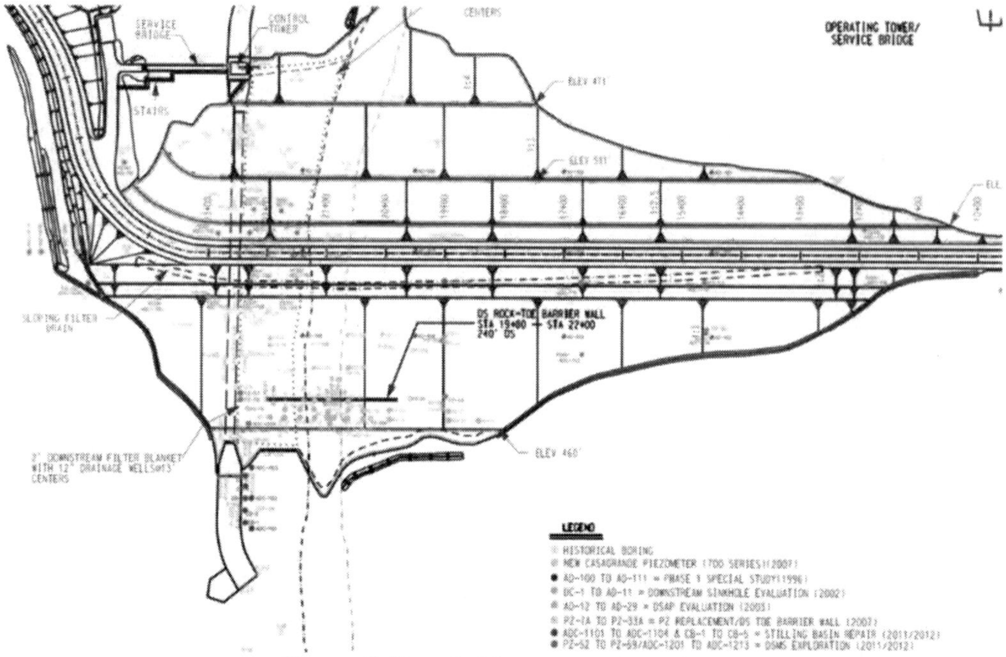
Figure 2. Rough River Dam Plan View

Progression of Treatment

Previous investigations indicated that the rock/soil interface (with the soil variably consisting of embankment fill, granular natural overburden, or cohesive natural overburden) and the upper portion of the rock were the likely conduits of problem seepage and potential erosion. Accordingly, the exploratory grout program was designed to effectively treat this interface zone while minimizing the risk of inducing damage to the zone. The following summarizes the typical treatment process for a hole:

1. Drill overburden and 2 foot rock socket using rotary sonic drilling without drilling fluids.
2. Install multiple port sleeve pipe (MPSP) in sonic drilled hole.
    a. Insert MPSP with grout bag provided 1 foot above top of rock and sufficient additional ports to support MPSP backfill and interface grouting. MPSPs were provided with a geotextile and plastic cap which allowed water to flow into the pipe during installation but prevented grout from leaving the MPSP except through the grout ports during installation.
    b. Inflate grout bag with a controlled volume at a suitable pressure to cause pressure filtration of the grout bag mix and ensure an effective barrier at the base of the MPSP.
    c. Backfill above barrier bag in stages according to allowable maximum pressure calculation in embankment and pull casing. Typically the MPSP annulus was backfilled in two stages, dependent on hole depth and material type at the base of the MPSP.

3. After MPSP backfill, grout interface zone through port or ports provided at the base of the MPSP below the barrier bag using gravity pressures.
4. Drill prescribed "gravity grout" downstage 15 feet from the bottom of the MPSP. Initially this was performed with rotary percussion drilling, however as the work progressed a change was made to drill the gravity grout stages with HQ3 coring in all holes. This change allowed for cleaner, more stable holes where the MPSP was tipped in the relatively soft sandstone and shale units. Subsequent rotary percussion drilling was performed with the benefit of the strength gain provided by the gravity grouting.
5. Grout "gravity grout" downstage using gravity pressures. Initially this was specified to be performed by filling the MPSP with grout and running a falling head test with grout. After successful performance of the gravity stages at gravity pressures using the instrumented packer, the remainder of the "gravity grout" downstages were performed using the instrumented packer, resulting in significant savings of batched grout quantities and hole washing time and effort. Gravity stages were completed for all orders (Primary, Secondary, etc) of prescribed holes in a given area prior to beginning the drilling and grouting work at depth.
6. Redrill through gravity stage and perform the work via typical upstage grouting methodology where possible, or via downstages if necessary. All WPT and grouting for the program was performed with the use of the instrumented packer, real time monitoring of grouting parameters, and real time monitoring of geotechnical instrumentation.

Instrumented Packers for WPT and Grouting

In the current general practice of grouting, the standard for measuring injection pressure has been to place a pressure gauge on the header at the top of the hole. The effective injection pressure is then calculated by adding line loss pressures and subtracting the difference between the head of the grout column and the head of the water table acting on the midpoint of the stage being grouted. When calculating pressures by using a gauge at the top of the hole, a key part of the calculation for determining the actual injection pressure is the dynamic line loss. The dynamic line losses depend heavily on the length of the grout line, the flow rate through the grout line and the viscosity of the mix used. The dynamic losses through the grout line become more severe with thicker grout mixes and greater flows. Dynamic line loss testing is time and resource intensive especially when one is faced with multiple grout line lengths (when there are multiple simultaneous points of injection on a project) and with multiple grout mixes. The process has to be repeated to account for any changes to any of the three variables.

Paul et al. (2013), presented the grouting community with an alternative to the above process, by introducing a sensor that could read the fluid pressure at the point of injection, thereby eliminating the need for the above calculations. Working from this foundational concept, ACT has designed and constructed a pressure sensing head that can be attached to a standard P sized packer, called an "IntelliPacker" that is suitable for use when injecting water or cement based grouts. A schematic of the IntelliPacker is shown in Figure 3. ACT has also built in the capability to measure pressure just above the inflated rubber bladder of the packer. This enables the packer to sense the column of liquid in the borehole and is very useful for determining if packer bypass is occurring. Identifying packer bypass is critical because it can lead to grouting in the packer, resulting in loss of expensive equipment and costly replacement of the boring. By using the column pressure sensor, any packer bypass was immediately noticed by the cart operator and/or the engineer monitoring the grouting operations. During the Exploratory Drilling and Grouting Program at Rough River Dam the IntelliPacker was used across all water testing and grouting activities, including: barrier bag inflation, casing annulus grouting, water pressure testing, and rock grouting.

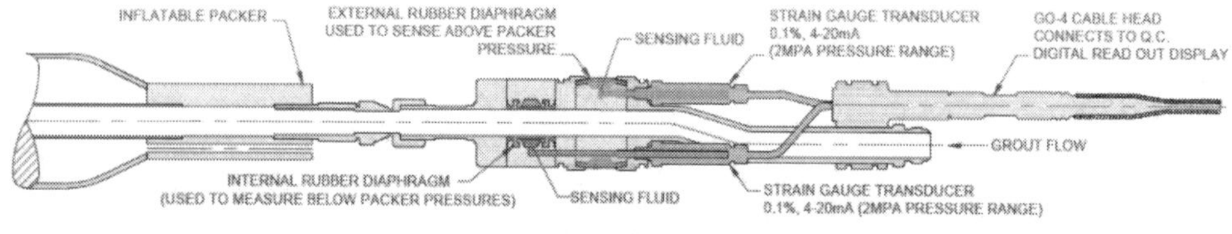

Figure 3. Schematic of IntelliPacker

Barrier Bag Inflation

The barrier bag is made of high strength geotextile fabric and is attached to the PVC casing using double punch lock clamps at each end of the bag. The barrier bag was inflated using an effective pressure of 35psi. This pressure was determined through pre-production testing as the optimal pressure for squeezing the water out of the barrier bag grout mix and leaving behind a solid grout cake. Use of the IntelliPacker was critical to ensuring the integrity of the barrier bag due to the fact that at depths below 61ft, the grout head pressure in the grout line would be greater than 35psi. For this reason, grouting barrier bags using the old method would lead to many burst bags due to over pressurizing the bag. During the grouting of the bag, the response was a linear increase in pressure at a high flowrate, followed by an almost sudden drop in flow and rise in pressure as the bag became full. In order to properly pressure filtrate the grout to leave a "grout cake" in the bag, the pressure was held at 35psi for 20 minutes.

Casing Annulus Grouting

To prevent fracturing the embankment soils during casing grouting, the USACE required calculations to be run on all MPSP annulus grouting to evaluate the pressures exerted on the soils. If the grouting pressures exceeded the embankment fracture pressure, the MPSP was required to be backfilled in stages. In general the MPSP's installed at Rough River required 2 stages of grout backfill, with the exception of those installed in the abutments where the depth to rock was shallow. The embankment fracture pressure was calculated as the horizontal total stress provided by the overlying materials with the addition of the shear strength of any cohesive soils at the MPSP tip depth. This pressure was calculated on a per hole basis based on the depth of the rock and the composition of the overlaying strata. A maximum grouting pressure equal to 90% of the embankment fracture pressure was considered allowable. Before beginning the casing annulus grouting the theoretical volume of grout that would be needed to achieve this maximum pressure was calculated. Since the annular space was open to the surface, the gradual injection of grout increased the measured injection pressure in a very linear fashion as the level of the grout head rose inside the annular space (Figure 4).

As evidenced by the actual volumes recorded, the theoretical and actual volume injected during casing installation were nearly identical. This was recorded throughout the majority of the casing installations at the Rough River site resulting in no over pressurization or hydrofracturing of the embankment or foundation soils. At the end of the grout injection, the pressure was kept at the maximum pressure for several minutes to evaluate the flow rate. If the flow rate dropped to zero with no loss of injection pressure, no grout losses were occurring and the hole was deemed "tight". In cases where grout loss was occurring, the pressure was held at the maximum until the flowrate dropped sufficiently to indicate no loss of grout.

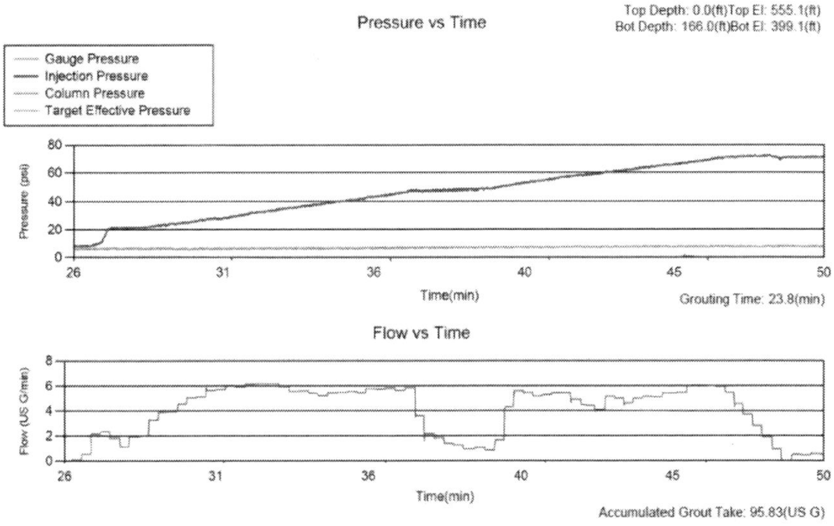

Figure 4. Typical first stage casing annulus grout trend plot

<u>Rock Grouting in Long Stages</u>

Figure 4 shows the process of thickening the grout mix due to a stage requiring a large amount of grout. At the end of the grouting process, at approximately the 250 minute mark, the effective pressure being recorded by IntelliPacker was relatively steady versus the effective pressure calculated at the grout cart. During normal operating procedures, as the cart operator would notice the collar gauge pressure rising, the operator would start to slowly close the valve supplying grout to the hole. Since the actual measured effective pressure was not responding in the same manner, this throttling of the grout supply would lead to false refusal.

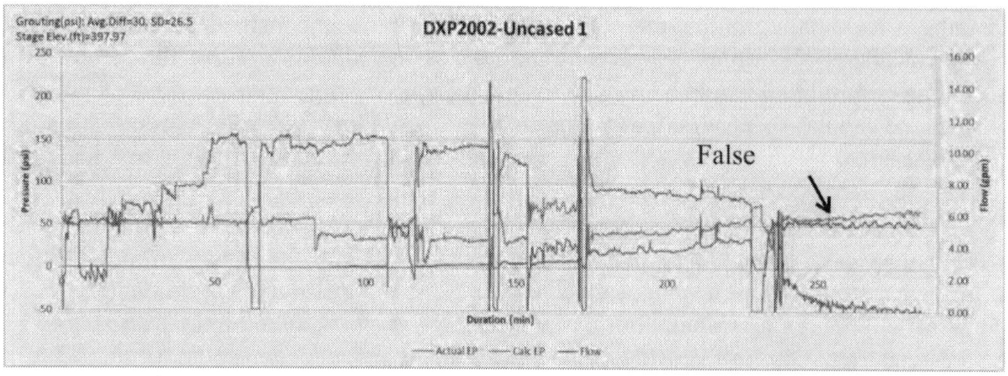

Figure 5. Calculated (Red) vs Measured (Blue) Effective Pressure

Integrated Electronic Records

During the performance of the exploratory grouting program, an integrated electronic records system was used to manage the flow of data coming as it was collected and interpreted and to manage and automate the submittal of completed records. Data from field drilling logs and real time monitoring data from water pressure testing and grouting activities were all collected and organized by ACT's comprehensive data management system called Intellisystem. Intellisystem streamlines output of the data to an FTP site for review by the client and provides a ready, searchable database to the grout operators on a day to day basis. A GIS based web portal was provided for easy records lookup and near real time presentation of grouting profiles and plan views, allowing for easy interpretation of ongoing results of the exploratory program. Figure 6 is a section view of the grout results from the GIS viewer showing the location of all takes greater than 50 gallons in a stage.

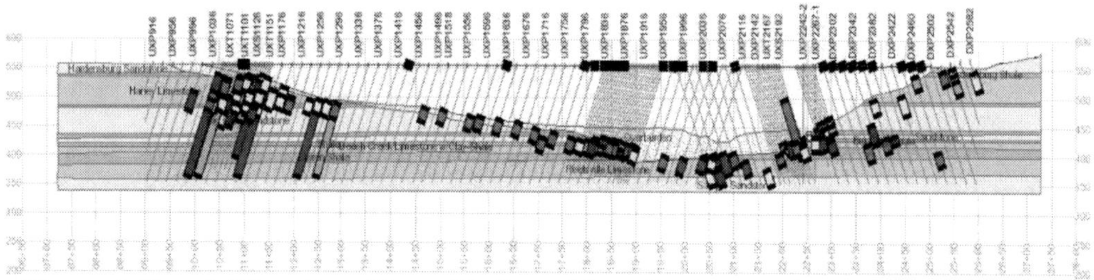
Figure 6. Rough River Dam Phase 1B Grout Lines Section – Takes Greater than 50 gal

CONCLUSION

The Louisville District of the U.S. Army Corps of Engineers (USACE) awarded the Rough River Dam Safety Modification, Phase 1B – Exploratory Drilling and Grouting project to Advanced Construction Techniques (ACT). The program was initiated in July 2015 and completed in July 2016. The exploratory program utilized real time dam safety instrumentation monitoring, real time computer monitoring of water pressure testing and grouting activities including the measurement and reporting of actual effective pressures at the point of injection, and a treatment process tailored to the unique conditions at the Rough River Dam to treat the foundation of the embankment and increase the USACE's understanding of the site geologic setting in support of the design of further improvements to the facility.

The work performed in this exploratory program demonstrated that the best way to effectively manage injection pressures during all phases of construction is to measure the pressure being exerted in the ground at the point of injection in real-time. The IntelliPacker assembly was used on 100% of the holes for all facets of the Exploratory Drilling and Grouting Program.

ACKNOWLEDGEMENTS

In addition, the authors would like to thank the Shop and Field team at Advanced Construction Techniques for their support and assistance in designing and implementing IntelliPacker's assembly head, as well as ACT's Technical Director, Haixue (Michael) Xu, Ph.D., P. Eng. for his early work and research into dynamic head losses.

REFERENCES

Paul, D., Shaeffer, J., Brosi, B., Conti, C., "Development and Field Testing of a Pressure Sensing Grout Packer", Proceedings of Annual USSD Conference, 2013.

INTEGRATING A MULTIDISCIPLINARY FIELD INVESTIGATION TO IMPROVE ASSESSMENT OF INTERNAL EROSION

Cassandra M. Wagner, P.G.[1]
Justin B. Rittgers, Ph.D.[2]
Bryan K. Simpson, P.G., P.E.[3]

ABSTRACT

El Vado Dam, built in the early 1930s by the Middle Rio Grande Conservancy District and transferred to the Bureau of Reclamation in 1954, is a distinctive structure due to its unique seepage protection of the embankment and foundation on a highly-fractured landslide. The dam was constructed with ¼-inch steel plating across the upstream face serving as the sole water barrier for the random earth-fill structure, and features a cutoff wall addressing seepage through the left abutment. Historical observations indicate that these design features were ineffective in mitigating the vulnerabilities of the site, leading to a multidisciplinary field investigation in support of overall risk reduction at the project, and in particular, assessment of conditions for internal erosion.

Investigations included drilling, strength testing, laboratory testing, geophysical imaging, hydrogeological testing, and assessment of instrumentation, aerial photography, and field mapping. All results were integrated into general internal erosion event tree nodes, including detecting and characterizing the existence and locations of flaws in the foundation which could represent geologic vulnerabilities to the embankment structure, identifying discontinuities in the rock foundation, identifying discontinuities located along the foundation contact with the embankment, providing evidence for or against the migration of embankment material into the foundation, characterizing flow characteristics of the discontinuity network, and assessing the continuity of any erodible features within the foundation. Collective interpretations of each node reduced uncertainty and provided a stronger understanding of the critical events supported by multiple scientific perspectives. The integration of a multidisciplinary program maximized the value of each individual component, and has helped to minimize the likelihood for subsequent redundant investigations.

INTRODUCTION

In 2015, the second of two field explorations was developed to detail exploration requirements to classify and characterize embankment and foundation materials at El Vado Dam, as requested in a 2013 Safety of Dams recommendation.

[1] U.S. Bureau of Reclamation, Denver Federal Center, PO Box 25007 (85-832000), Denver, CO 80225, 303-445-3147, cwagner@usbr.gov
[2] U.S. Bureau of Reclamation, Denver Federal Center, PO Box 25007 (85-833000), Denver, CO 80225, 303-445-3010, jrittgers@usbr.gov
[3] U.S. Bureau of Reclamation, Denver Federal Center, PO Box 25007 (85-832000), Denver, CO 80225, 303-445-3094, bksimpson@usbr.gov

El Vado Dam is distinctive in design and construction due to its unique seepage protection of the embankment and foundation on a highly-fractured landslide. El Vado Dam was originally designed and built by the Middle Rio Grande Conservancy District during the early 1930's. Ownership of the dam was later transferred to the Bureau of Reclamation in 1954. The project is one of only a few dams in the world where an upstream steel face operates as the sole barrier to seepage entering the embankment. This type of seepage element was chosen in part because of the lack of suitable earthfill material for a traditional core. Additionally, the left abutment of the dam was constructed on a large landslide, recognized prior to construction. Following the construction of the dam, reservoir induced movement of the landslide lobe contributed to buckling and deformation of the upstream steel face during first filling. High volumes of seepage through the left abutment were noted soon after, prompting the construction of an upstream cutoff wall in 1955.

The development of a landslide during the Quaternary at this location is understandable, given the regional setting and site geology. El Vado damsite lies within a north-south trending structural upwarp known as the Archuleta anticlinorium, which separates the San Juan Basin from the Chama Basin. The project, oriented with the downstream direction to the south, is located on the western limb of the anticline, proximate to the hinge line. This provides the setting for strata dipping up to 9 degrees toward the west under the site, and thus toward the floodplain in the left abutment. Refer to Figure 1 for the site location and regional geology.

The project site is underlain by a series of outcropping sandstones, sandy shales and shale beds. A prominent sandstone bed, which caps the left abutment above the dam, is recognized as the Tres Hermanos Sandstone Member of the Mancos Shale Formation. The sandstone of this unit was documented as being fractured and jointed prior to the incision of the Rio Chama River.

The downward cutting of the Rio Chama River, which incised a canyon approximately 200 feet near the dam site, has exposed rocks from younger to older age in a downstream direction. In the vicinity of the El Vado reservoir, the Rio Chama River is entrenched in the Mancos Shale of Cretaceous Period (138-63 million years ago [mya]). The down-cutting by the Rio Chama River caused unloading of the foundation down-dip along the limb of the Archuleta anticline, and caused the higher, more competent beds to slide down the dip on top of softer shales, toward the river channel. The distance of sliding at the site is believed to be several hundred feet. As a result of the landslide, the left abutment foundation of the dam is comprised of a sequence of the Tres Hermanos Sandstone Member that was presumed to be generally intact, but highly fractured. Refer to Figure 2 for the approximate extents of the landslide deposit.

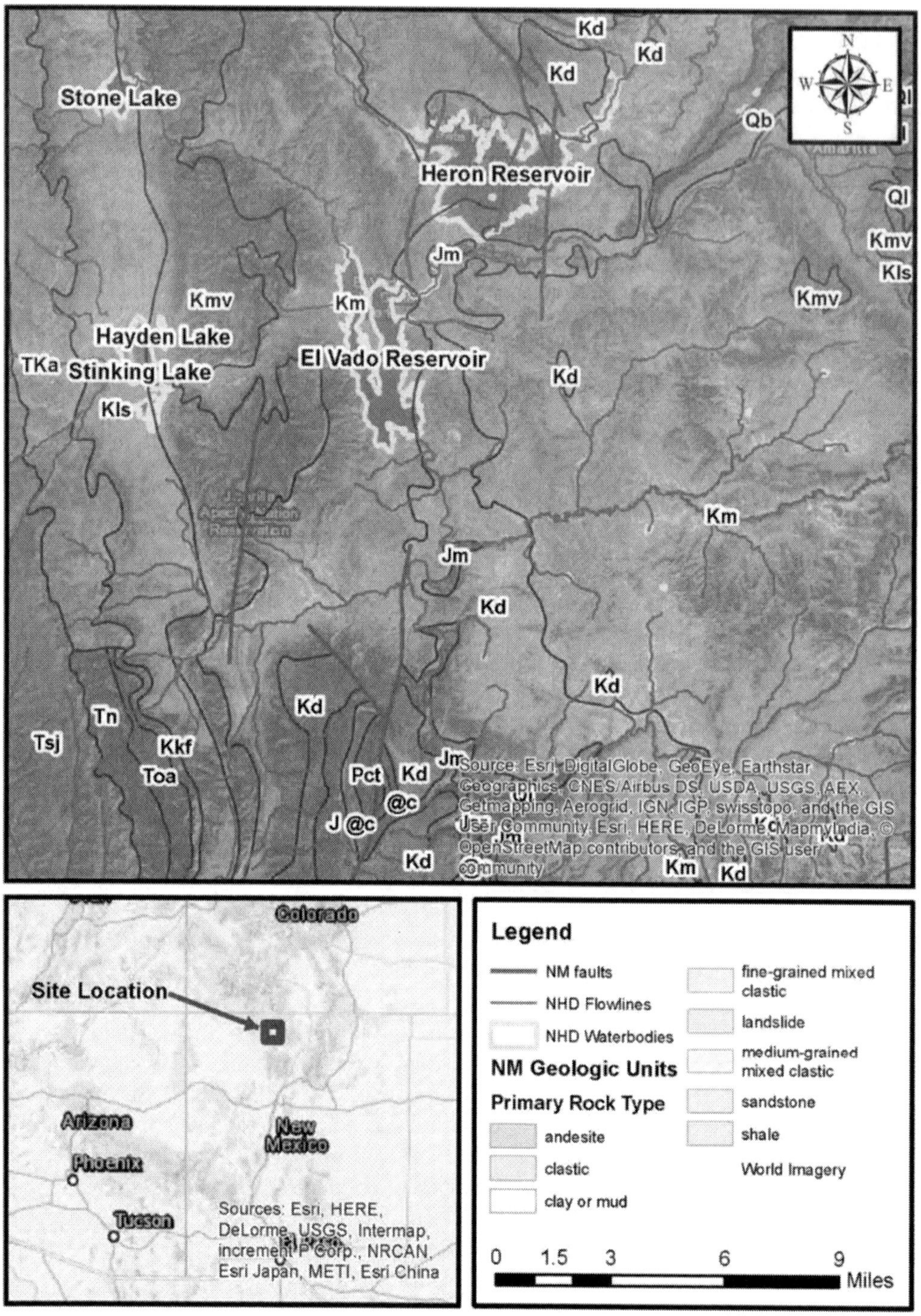

Figure 1. Project Location Map and Regional Geology.

Figure 2. Google Earth approximation of extents of left abutment landslide.

Performance History

The geologic conditions at the site invariably led to indications of distress. In summary, the notable events in the performance history of the project included observed left abutment seepage, evidence for embankment seepage, an upstream "whirlpool" in the reservoir, settlement and deflection along the embankment crest, evidence for material migration near the crest of the dam and beneath the upstream steel plating, discontinuous anomalies in the piezometric surface along the centerline of the dam, and deformation of the embankment due to movement of the landslide or imbricated landslide surfaces during first filling.

Seepage in the left abutment is observed directly in tall, vertically oriented joints and fractures in the rock. Additionally, a groin drain captures and carries seepage along the downstream left abutment, where flows are noted to fluctuate over time. Evidence for embankment seepage is given in the maximum section of the dam, where warm temperature anomalies and moist surface materials prevent the accumulation of light snow. During high pool events, the seepage in these sections is noted to increase. Refer to Figure 3 for approximate locations of some of these features.

An upstream whirlpool in the reservoir is observed at certain reservoir elevations, and is located proximate to the left abutment. Refer to Figure 3 for the projected location of this feature. Settlement and deflection is noted along the crest approaching the left abutment.

Upstream guardrails are noted to deflect toward the upstream direction, and the development cracking in the crest pavement indicate movement of material near the crest. Instrumentation installed in the vicinity of the left abutment does not conclusively identify directional creep or deflection of the left abutment materials, though small trends indicate subtle down-valley movement in recent years. Material is evidenced to be moving near the crest of the dam and underneath the upstream steel plating, where material losses near the crest allow for the formation of small sinkholes along the crest, and where previous investigations to assess the void distance beneath the upstream steel plating indicated gaps of approximately 5 inches between the steel and embankment.

The phreatic surface, plotted along the centerline of the dam, indicates discontinuous zones of relatively high readings. Due to the recent installation of the observation wells, there is an insufficient quantity of readings to be fully conclusive. Initial trends signify that these zones of relatively high readings approach the reservoir elevation and potentially indicate a location specific recharge. The phreatic surface of this recharge feature decreases in elevation before approaching the abutment, based on water level readings in flanking observation wells, reducing the potential for assigning abutment flow as a driver of this observation. Given that the upstream barrier to seepage inflow is the steel plating, a more probable supposition is incomplete cutoff in the steel plating. The location of the discontinuous high phreatic surface elevation roughly coincides with evidenced seepage exiting on the downstream face of the dam described above. Due to the spacing intervals of the observation wells, the observed "hump" in the centerline phreatic surface may be shifted left or right.

Deformation of the upstream steel plating was observed following first filling, and has been attributed to movement of the landslide or imbricated landslide surfaces during first filling. The observed pattern of deformation indicates a series of echelon, linear warps in the steel, determining a down-valley relative motion of the foundation. Analysis of the landslide following the deformation of the upstream steel facing concluded that the basal extent of the landslide appeared to be stable, though the stability of potentially imbricated surfaces was uncertain.

Identification of Potential Failure Modes

A number of static, hydrologic, and seismic failure modes were identified for this project. Of these, two overall failure mode categories are related to the geologic characterization of the site. The primary failure modes considered are related to internal erosion of the embankment into foundation or of the foundation itself, and hydrologic and/or seismic reactivation of the landslide. This paper will focus on the application of field exploration and laboratory methods for reduction of uncertainty pertaining to internal erosion. An assessment of risk associated with hydrologic and/or seismic reactivation of the landslide was completed using the same field exploration; however, the landslide risk alone was not highly significant to the total risk of the project.

METHODS

A field exploration was conducted in 2015 in support of overall risk reduction at the project. Investigations included drilling and strength testing, laboratory testing, geophysical imaging, hydrogeological testing, and assessments of instrumentation, aerial photography, and field mapping. These investigative approaches were used to characterize the embankment and foundation materials, and to detect and characterize the locations of flaws in the foundation which could represent geologic vulnerabilities to the embankment structure.

Figure 3. Site Investigation Map.

Prior to the investigations initiated in 2013, the landslide was of an unknown thickness. The initial objectives of the field exploration were to determine the geometry of the structure, evaluate the potential for imbrications in the landslide structure, and assess the safety of the basal and subsequent failure surfaces. Given the performance parameters, the development of the field exploration methods gave particular attention to constraining the characterization of risks associated with internal erosion in the embankment or landslide structure, and reactivation of the landslide surfaces(s) due to hydrologic and seismic loadings. The highly disturbed nature of the foundation brought particular attention to the potential for voids, soft and erodible materials, and preferential seepage paths. These facets of risk have been long understood at the project, but were accompanied by high uncertainty.

The field exploration program included advancement of nine (9) drill holes using sonic drilling, Hollow Stem Auger (HSA), and HQ3 Series 6 Wireline Drilling methods. A tenth drill hole was planned but not completed due to concerns of the drill hole creating a potential unfiltered seepage entrance from the reservoir at the location of the drill hole. Five drill holes were completed along the crest centerline, through the embankment materials, the underlying landslide materials where present, and terminating into the Tertiary bedrock once encountered. A set of paired drill holes were completed on the upstream left abutment of the dam, and two drill holes were completed downstream of the downstream toe of the dam. All holes, with the exception of one upstream drill hole, were completed as observation wells. The remaining upstream drill hole was completed as an inclinometer. A location map of the investigations is included in Figure 3. Site Investigation Map.

Geophysical analyses, including Electric Resistivity Tomography (ERT) and S-wave Seismic Refraction Tomography were conducted along the embankment and downstream toe as part of Phase 2 of FER #2. The primary purposes of these techniques were 1) to determine the sizes and continuity of soil-filled voids between rock blocks comprising the landslide (or other seismically slow/attenuative and anomalous electrical resistivity features/trends that may be associated with concentrated seepage/internal erosion), 2) further determine the contact between the landslide and in-place bedrock, and 3) to help understand potential seepage pathways believed to be extending from the upstream left abutment area to the downstream toe. Specifically, several seepage outlets are visible near the power house (left groin) and further downstream along a road-cut that exposes in-place Dakota Sandstone and Mancos Shale units, but the seepage pathway geometries and source/inflow areas are not well understood. Five (5) seismic lines were completed; four were completed along the downstream toe of the left abutment, and one was completed on the upstream toe of the left abutment. Seven (7) ERT lines were completed; six were completed along the downstream toe and outlet works channel, and one was completed along the length of the crest centerline. These locations are generally concurrent with the locations of drill holes for verification of results.

As part of the ongoing evaluation of the left abutment and cut off wall of El Vado Dam, a tracer test was performed. The tracer testing were conducted to evaluate the performance of the cutoff wall in the left abutment landslide, and characterize any seepage that is noted during the investigation.

Samples of the embankment materials were logged, collected, photographed and submitted for standard physical properties testing. Due to the employment of sonic drilling methods, large samples of embankment material were obtained from the sonic drilling boreholes were compiled to complete gradations including characteristics of oversized material. Additional specialized testing was assigned to characterize supplementary properties of the material. Samples of bedrock material were selected for petrographic analysis, in order to standardize descriptions of the bedrock material. Thirteen (13) samples were selected for both x-ray fluorescence (XRF) and x-ray powder diffraction (XRD) to determine the elemental and mineral composition of the samples, respectively. These thirteen samples represented three groups, and were used to evaluate the presence of embankment materials in the landslide deposit.

Specialized laboratory testing was also selected to test the embankment material for internal stability against erosion. A modified permeameter, termed internal erosion permeameter, was developed for this testing.

RESULTS

Quaternary Landslide Materials

The contact between the embankment materials and underlying landslide materials was sloped along the left abutment of the dam. The landslide materials were encountered in all boreholes except for one located in the max section of the canyon, and ranged from 60.3 to 135.6 feet in thickness. The landslide deposits were composed of a variety of materials, ranging from relatively intact foundation materials to very intensely fractured and decomposed foundation materials. Relatively intact foundation materials were largely comprised of boulder-sized materials of sandstone, often rotated approximately 5 to 35 degrees. Zones of intensely to very intensely fractured sandstone with interbedded shale were also frequently encountered. Some of these zones well-correlated between borings and geophysical investigations, and are interpreted to represent subsequent imbricated rupture surfaces within the compound landslide feature.

The landslide deposit discontinuously overlies the Mancos Shale (Km) and Dakota Sandstone (Kd) at various locations across the site. It has been assumed that the Km unit provided the rupture surface at the base of the landslide. Field observations found that the very intensely fractured shear zone at the base of the land slide consisted of shale with interbedded sandstone and shale.

Intact Bedrock

As indicated by petrographic analysis of bedrock materials, the intact bedrock consisted of alternating sequences of dark gray shale with interbedded sandstone, quartz sandstone, and dark gray massive shale. Intact Cretaceous Dakota Sandstone (Kd) was encountered in all drill holes except for one boring, which terminated in the Quaternary Landslide deposit (Qls). The Kd materials were encountered underlying intact remnants of the Cretaceous Mancos Shale (Km) in some of the boreholes.

Laboratory Testing

Specialized laboratory testing was selected to obtain evidence for migration of material from the embankment into the landslide, including X-ray fluorescence (XRF) and X-ray powder diffraction (XRD). Results of the XRF and XRD testing do not definitively support the migration of embankment material into the foundation. However, one of the five samples of soil material encountered in the landslide indicated a statistically relevant similarity to embankment material. The sample, located in DH-15-2 at a depth of 38.5

feet, is located 2.5 feet below the contact of embankment with Qls material. The sampled material was selected from within a jointed sandstone interval, due to its similarity of appearance to embankment material, though its appearance as a decomposed shale layer could not immediately be ruled out. There is ambiguity as to whether the sample was incorporated into the sandstone interval 2.5 feet below the Qls and embankment contact during construction of the dam, or if material has been transported from the embankment into the landslide. A void was not described at this location; however, vertical jointing and non-discernible bedding was noted. Given the depth of 2.5 feet from the embankment and landslide contact, and that the sample was taken from a vertical joint and not a void, it may be unlikely that this material was deliberately placed during construction.

Specialized laboratory testing was also selected to test the embankment material for internal stability against erosion. A modified permeameter, termed internal erosion permeameter, was developed for this testing. The testing device was constructed as a fully enclosed, rigid wall permeameter, with an 11.5" inside diameter. Porous end plates were sized to allow particle migration, and the device monitored application of normal (consolidation) stress up to 50 psi. Computer control of upward or downward gradient was used to take measurements of in- and out-flow rates. Additionally, the device recorded measurements of pore pressure at seven locations to compute average and local gradients, and normal stress at top and bottom of specimen. The results indicated that the embankment material, with a USCS classification of GC (borderline GP-GC), was internally unstable, and that the material was readily subject to suffusion. The methods used in the internal erosion permeameter may have prevented the suffosion of grains filtered by a 1-in gravel filter, and could not test for upward gradients less than approximately 1.4. The hydraulic conductivity of the embankment material was determined to be 1.4 x 10-3 cm/s.

Geophysical Investigation

Geophysical methods were selected to image and better characterize the location and distribution of potential void-like features intercepted during drilling activities at depth, and to help understand potential seepage pathways believed to be extending from the upstream left abutment area to the downstream toe, where several seepage outlets are visible near the power house (left groin) and further downstream along a road-cut that exposes in-place Dakota Sandstone and Mancos Shale units. Results of these geophysical surveys are highly correlated with as-built drawings, data collected during drilling activities, on-site visual observations, and visible/known seepage features.

Linear features/patterns have been identified in the recovered models that trend across multiple adjacent tomograms and reveal potential secondary shear zones within the landslide mass and likely interconnected joints/fractures/weathered zones within the Dakota Sandstone and Manco Shale units that are most likely providing active seepage pathways that generally extend from the left abutment/foundation contact with the landslide mass to the downstream toe/road-cut. Seven main seepage outlets observed near the downstream power house have been interpreted/identified/traced toward their respective upstream source areas using tomography results. Refer to Figure 4 for the alignment of possible seepage pathways.

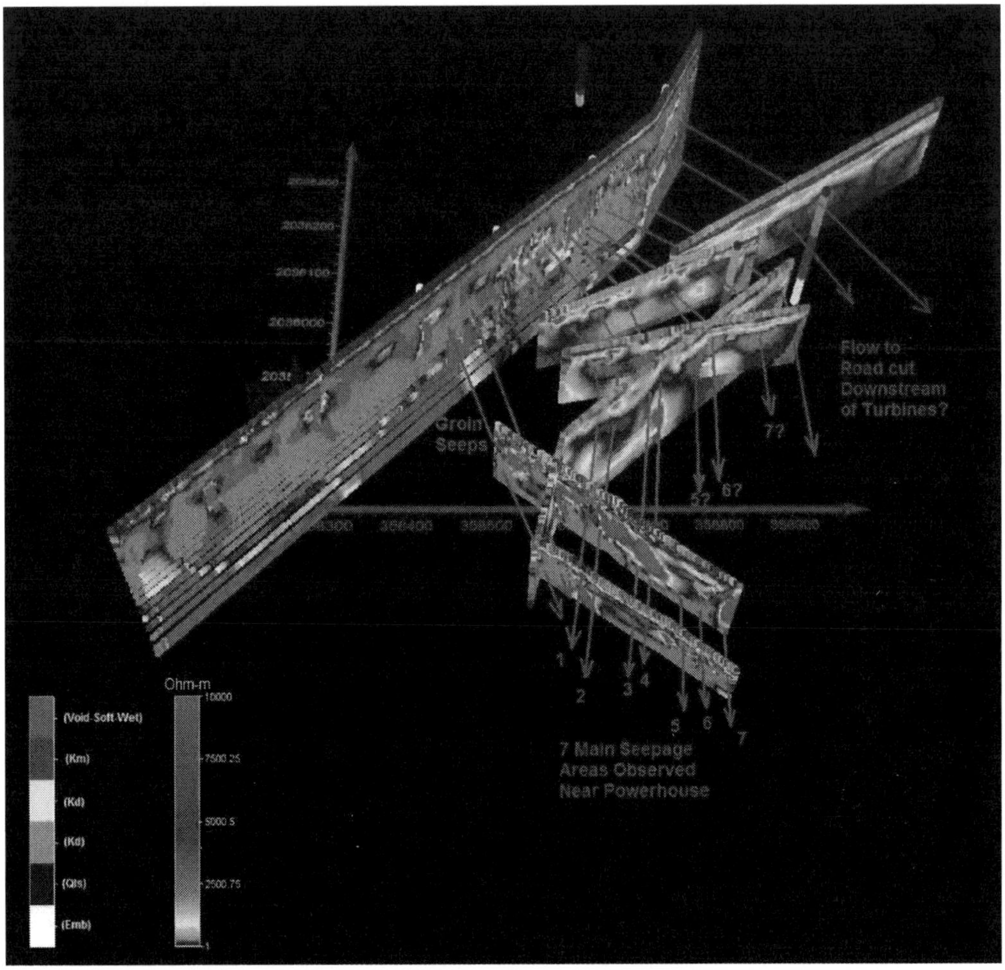

Figure 4. 3D perspective plots of all ERT tomograms with seven possible "main" seepage pathways interpreted across adjacent tomograms and extending upstream from observed seepage outlets along the road-cut. A separate possible seepage pathway has been interpreted along and just below the left groin. Viewer is looking downward and north.

Based on the results and interpretations presented here, in addition to historic photos and as-built drawings produced during construction in the 1930s, "void-like" features of concern that were encountered near the embankment contact with underlying landslide materials during drilling along the crest line, are believed to be open or poorly infilled jointing and fractured zones within the displaced Dakota Sandstone (Kds) unit. These features appear to generally align with potential seepage pathways/linear anomalies identified in other nearby tomograms, and are thus believed to be directly related to seepage observed along the road-cut near the downstream toe. Refer to Figure 5 for features identified along the crest of the dam.

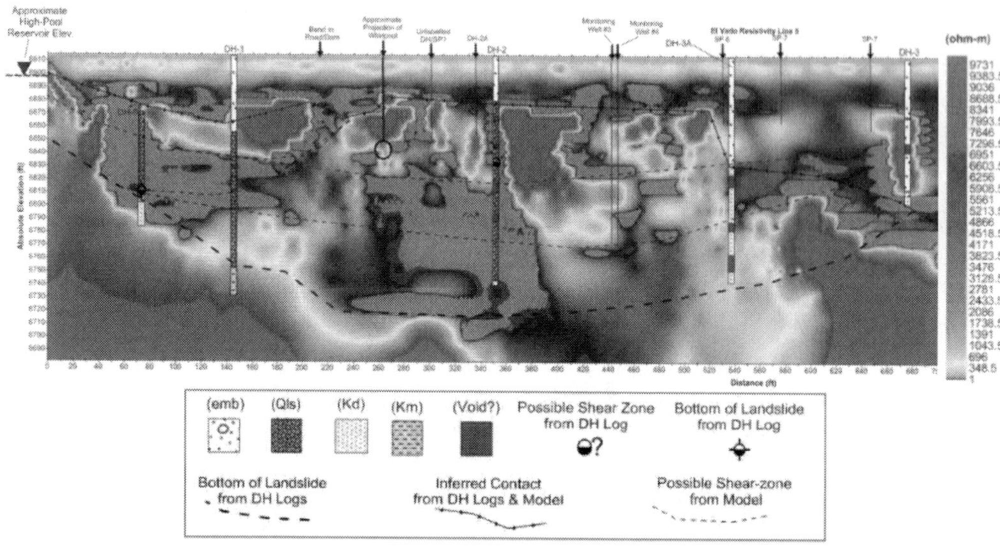

Figure 5. Electrical resistivity tomogram for ERT Line 5. This tomogram is approximately 2-3 times the dimensions of all other tomograms (utilized 112 channels). Viewer is looking downstream into the tomogram. Material types logged during drilling are projected onto the tomogram and demonstrate an extremely high level of correlation with ERT tomography results. Interpreted unit contacts (hashed lines), possible shear zones (dashed/dotted lines) and seepage zones in possible deeper open fracture sets and/or jointing (dashed circles).

Groundwater and Tracer Testing

Groundwater was encountered during sonic drilling within the landslide materials, indicating that some landslide materials are variably in a saturated state. Saturated materials within the landslide were generally underlain by non-saturated (moist) materials, indicating that there may exist preferable seepage paths through the landslide materials, and that this unit likely contains permeabilities that range several orders of magnitude. The general observation during the drilling program is that the embankment materials were relatively dry, and the underlying landslide materials were in a variable saturated state. Saturated materials within the landslide deposit were in some cases correlated to low strength zones, interpreted to represent subsequent imbricated rupture surfaces within the compound landslide feature.

A tracer test was performed on June 8, 2016 as part of the ongoing evaluation of the left abutment and cut off wall of El Vado Dam. Dye testing was completed using Rhodamine WT. Injection occurred upstream of the cutoff wall in the location of upstream drill hole DH-15-5, which is screened within the landslide deposit between 62 and 82 ft below ground surface. The downstream monitoring location was located approximately 880 feet from the injection well, DH-15-5, given the most direct route in plan view. The effective seepage path, however, is understood to be much more tortuous which would imply a relatively higher seepage velocity. Results of dye testing indicates a peak tracer concentration reaching the monitoring point 44.1 hours after injection. While peak concentration occurred at 44.1 hours after injection, an initial spike was observed after 23.5 hours. These results imply a continuity of flow-supporting features from upstream to the downstream monitoring location, and incomplete cutoff provided by the upstream cutoff wall.

DISCUSSION

Discontinuities were encountered in the landslide materials during drilling, sampling, and geophysical investigations, and consisted of voids, soft zones, highly fractured zones, and saturated materials. These terms are broadly applied, but it is important to differentiate between four main types of voids, or scenarios of void creation. Generally speaking, the findings of this investigation address the potential presence of three of the four general scenarios or types of voids within the subsurface. These four different scenarios include the following:

1. Interstitial voids within the embankment material caused either by differential settlement, internal erosion, cracking, or poorly placed fill materials due to poor construction practices or other unknown causes.

2. Pre-existing and naturally-occurring air-filled voids at the contact of the embankment with the foundation that have accommodated subsequent gravity-driven migration/repository of embankment materials from above, and now extend across the foundation contact.

3. Air-filled or fluid-filled voids within the embankment or extending across the foundation contact that were created during drilling and are initially confined to the immediate vicinity of the drill-hole.

4. Pre-existing/naturally-occurring air-filled, water-filled or poorly back-filled interstitial voids within the landslide mass.

Void-like features of concern were encountered near the embankment contact with underlying landslide materials during recent drilling along the crest line. These features are believed to be pre-existing and naturally-occurring open or poorly infilled jointing and fractured zones within the displaced Dakota Sandstone (Kd) unit. This interpretation is based on the results of the field exploration program, in addition to historic photos and as-built drawings produced during construction in the 1930s. It is believed that these void-like features encountered during drilling fall into categories 2 and 4 as listed above (pre-existing and naturally-occurring at or below the foundation contact).

Application of Results to the Assessment of Internal Erosion

During the planning, analysis, and application of these investigations, considerations were made to reducing the uncertainty of individual event tree nodes. During the course of identifying nodes related to internal erosion, interpretations of the integrated results were grouped at each node, and in general, included: detecting and characterizing the existence and locations of flaws in the foundation which could represent geologic vulnerabilities to the embankment structure; identifying discontinuities in the rock foundation; identifying discontinuities located along the foundation contact with the embankment; providing evidence for or against the migration of embankment material into the foundation; characterizing the flow characteristics of the discontinuity network; and assessing the continuity of any erodible features within the foundation.

Discontinuities within the landslide foundation and at the contact with the embankment were encountered in the landslide materials during drilling, sampling, and geophysical investigations. Discontinuities consist of voids, soft zones, highly fractured zones, and saturated materials. The locations of discontinuities were strongly correlated between investigations, providing evidence to support the existence of the discontinuities network.

During the drilling of DH-15-1, DH-15-2, DH-15-3, and DH-15-3A, voids or other discontinuities were encountered at or below the contact between the embankment and Quaternary landslide materials. Discontinuities were spatially consistent with resistivity anomalies. The resistivity anomalies were laterally continuous beyond the correlated materials encountered during drilling, and provide geometrical support for the potential for shear surfaces within the landslide, and void-like discontinuities extending to the embankment and Qls contact.

Observations from drilling and laboratory testing indicate the potential for embankment materials to have migrated since placement. During the drilling of DH-15-1, DH-15-2, DH-15-3, and DH-15-3A, voids or other discontinuities were encountered at or below the contact between the embankment and Quaternary landslide materials. In DH-15-3 and -3A, materials were observed to appear visually and texturally similar to embankment materials based on visual classification techniques. As such, materials were sampled above and below the contact of embankment materials overlying landslide materials. The samples were submitted for analysis, including standard physical properties, XRF and XRD. Material sampled from a vertical joint in DH-15-2, 2.5 feet below the embankment contact with the quaternary landslide contact was statistically related to embankment material, and not other sampled infilling material in joints and fractures of the landslide deposit. The presence of this material is interpreted as potential migration of material since placement.

Observations from drilling, tracer testing, and geophysical investigations indicate a potentially correlatable network of intensely fractured zones. Potential evidence of water induced weathering was noted in open joints and fractures of cored material as well. Drill holes DH-15-1, -2 and -3A intersected a possible failure surface within the landslide, in addition to the strongly correlated base of landslide deposit. Failure surfaces were visually logged as very intensely fractured shear zones, with some material saturation noted in the upper failure surface located within the landslide. There was no anomalous material saturation noted at the overall base of landslide materials. This would indicate increased shear strength of materials at the base of the landslide where material lacks saturation. Conversely, saturated materials that comprise the potential failure surface(s) within the landslide may be characterized with decreased shear strength and decreased resistance to movement, such as creep.

The results of geophysical investigations support the drilling observations of a potential secondary major rupture surface within the landslide, given that locations of discontinuities were strongly correlated between drilling and geophysical investigations. These locations may represent continuous features of low strength materials. The profile of potential shear zone features observed in the electrical resistivity tomogram correlate strongly to potential shear zone features in the drill hole logs. The geophysical investigation indicates a lateral continuity of discontinuity features between boreholes, based on the ground correlation achieved by layering logged features to geophysical features.

The continuity of any erodible features within the foundation was difficult to assess, and primary considerations were given to a path of erosion which initiates due to material migration into the landslide, and progresses due to the sustained erosion of embankment materials into the foundation discontinuity network. Unique testing of embankment materials, including use of in the internal erosion permeameter and standard physical properties, supported the quantification of internal instability of the embankment. Continuity in the foundation was assigned on the premise of the interconnectivity of discontinuous, void-like and open fracture spaces at or near the embankment and foundation interface.

CONCLUSION

By integrating the results of drilling and strength testing, diverse laboratory testing, geophysical imaging, hydrogeological testing, and assessments of instrumentation, aerial photography, and field mapping, collective interpretations were assigned to individual nodes related to the development of an internal erosion event tree. This resulted in reduced uncertainty at each node, greatly reducing the overall uncertainty in the risk estimate. While previous assessments of risk did not significantly differentiate between failure modes related to alternative routes of internal erosion or interpretations of factors related to the landslide, these investigations permitted the identification, with meaningful certainty, of not only the most critical path, but justified the reduction of risk to the other variations of internal erosion failure modes. The integration of a multidisciplinary program maximized the value of each individual component, and has helped to minimize the likelihood for subsequent redundant investigations. By winnowing out the variations of related failure modes to only the most critical event tree, and applying a stronger understanding of the critical events supported by multiple scientific perspectives, a clearer path forward is identified for pursuing engineering solutions to optimize the safety of the structure.

REFERENCES

Wells, E.H., "The Geology of the El Vado Damsite and Reservoir, Rio Arriba County, New Mexico", Middle Rio Grande Conservancy District, Albuquerque, New Mexico, September 1, 1927.

"Comprehensive Facility Review, El Vado Dam, Middle Rio Grande Project, New Mexico," U.S. Department of the Interior, Bureau of Reclamation, Denver, Colorado, August 2013.

"El Vado Dam Issue Evaluation - Risk Analysis", Middle Rio Grande Project, New Mexico, Upper Colorado Region, Technical Memorandum No. ELV-8130-RA-2007-1, U.S. Department of the Interior, Bureau of Reclamation, Technical Service Center, Denver, Colorado, August 2005.

"El Vado Dam Void Detection, 2016 Geophysical Surveys", TM-85-833000-2016-12, U.S. Department of the Interior, Bureau of Reclamation, Denver, Colorado, August 2016.

Field Trip Report, To: Chief Engineer, From: Simonds, A. W., Staats, B. H., and Thompson, W. A., "Inspection of El Vado Dam, Middle Rio Grande Project", Bureau of Reclamation, Denver, CO, April 28, 1949

LABORATORY INVESTIGATION OF BACKWARD EROSION PIPING INTO CONSTRICTED SEEPAGE EXITS

John Rice[1]
Ibrahim A. Ibrahim[2]

Internal erosion often occurs when seepage flow is concentrated into a small, unprotected opening. One such example is where sandy soil is eroded through a defect in an overlying clay layer, resulting in a sand boil in the process. The erosion initiates through the heave and backward erosion piping mechanisms and continues beneath the clay layer through the piping process, forming a pipe that progresses toward the source of the seepage. The initiation of erosion at the seepage flow concentration is a complex mechanism involving a number of hydraulic and soil mechanics principles, including: flow concentration, soil arching, heave, detachment of soil grains, and transportation of soil grains.

A laboratory testing program is performed to investigate the mechanisms of erosion into a concentrated, unprotected exit. The study builds upon previous research on the mechanisms of piping initiation performed at Utah State University and uses a similar apparatus. A number of different soils representing a range of grain size, grain shape, and gradations are forced to erode into constricted seepage exits having a range of aperture diameters. The exit is fixed with a riser pipe to model the upward transport of eroding soils. The results are compared with axisymmetric finite element analyses in order to develop a better understanding of the initiation process for backward erosion piping.

[1] Associate Professor, Utah State University, john.rice@usu.edu
[2] Graduate Student, Utah State University

EVALUATION OF INTERNAL EROSION SCENARIOS IN AN EMBANKMENT DAM

R.J. Fannin[1]
M. Li[2]
C. Anderlini[3]

ABSTRACT

Internal erosion represents a significant risk to be managed in dam safety, with respect to zoned embankment dams. Within a designated zone of soil, it may take the form of internal instability and manifest itself as either suffusion or suffosion, depending on whether or not the movement of finer soil grains leads to a localised collapse of the grain assembly. Between two adjacent zones of soil, it may take the form of filter incompatibility and manifest itself as movement of the base soil zone into the openings of the filter zone.

A general framework is presented to evaluate the susceptibility of a zoned embankment dam to internal erosion, with specific reference to the phenomena of internal instability (by suffusion-suffosion) and filter incompatibility. Consideration is given to (i) the utility of construction records, and more specifically the utility of grain size distribution curves obtained during placement of zoned fill material, (ii) the use of empirical "rules" to evaluate susceptibility to internal instability and filter compatibility, based on the shape of the grain size curve, (iii) the insights gained from laboratory permeameter testing to investigate internal instability and filter incompatibility, and (iv) the contribution of current research to inform a confident interpretation.

INTRODUCTION

Internal erosion occurs when fractions of the soil gradation within an embankment dam or its foundation are transported downstream by the action of seepage flow. The phenomenon varies both spatially and temporally over the service life of the dam, as reported, for example, at the site of the now decommissioned Coursier Dam in British Columbia (Garner et al., 2004). From a survey of internal erosion incidents across the portfolio of USBR embankment dams, Engemoen (2016) found them to occur in many structures and, on occasion, several times at the same dam.

Internal erosion initiates when the hydraulic forces exceed the ability of the materials in the dam and foundation to resist them, and develops in one of the following four mechanisms:

- concentrated leaks (most usually in plastic soil possessing cracks, fractures, or gaps);

[1] Professor, Department of Civil Engineering, University of British Columbia, Vancouver, BC V6T1Z4, jonathan.fannin@ubc.ca
[2] Senior Engineer, British Columbia Hydro, Burnaby, BC V3N4X8, maoxin.li@bchydro.com
[3] Senior Engineer, British Columbia Hydro, Burnaby, BC V3N4X8, charissa.anderlini@bchydro.com.

- backward erosion (most usually in non-plastic soil, if inadequately filtered for soil retention);
- contact erosion (for example, along the interface between a fine- and coarse-grained soil); and,
- suffusion-suffosion (most usually in an internally unstable non-plastic soil).

Internal erosion may be arrested if the adjoining material provides some kind of filter. In this paper, we present a framework that was adopted for evaluation of internal erosion scenarios at an embankment dam. More specifically, we address the mechanisms of (i) concentrated leaks, (ii) backward erosion, and (iii) suffusion-suffosion, which are caused by internal instability and/or filter incompatibility (USBR-USACE, 2015). These mechanisms may occur in non-plastic soils, and are of particular relevance to concerns for internal erosion in till-core dams. From analysis of the actual construction records at a dam-of-interest, we report on the use of empirical screening tools to evaluate the susceptibility of zoned Core and Transition materials at the dam. Laboratory testing was performed on select gradations of soil from the dam site with the objective of verifying the findings of the screening exercise. Discussion addresses the implications of the screening exercise, and laboratory testing, for vulnerability of the dam to internal erosion.

CONSTRUCTION RECORDS

Internal erosion assessment has been conducted for a dam "A" using the framework presented in the paper. Dam "A" is a zoned embankment dam, for which a representative cross-section is shown in Figure 1. It consists of a wide impervious central Core zone that is supported on both sides by random shell material. The Core is protected by an internal filter and drainage system that is comprised of a Transition, a Filter, a Chimney Drain, and a horizontal Blanket Drain. The seepage flows are directed by splitter dykes, constructed of the shell material, resting directly on the foundation bedrock. A portion of some of the splitter dykes is in direct contact with the Core.

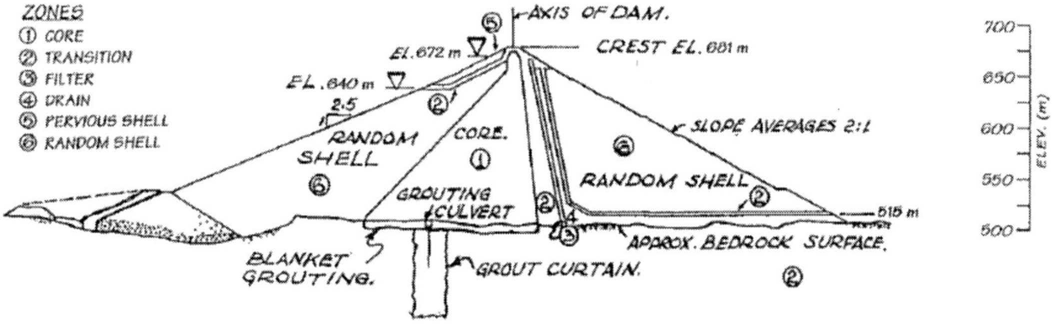

Figure 1 Representative Cross-Section of Dam "A"

The earthfill dam was constructed primarily of processed granular material from a nearby glacial deposit; the fines for the Core were supplemented from a local glacio-lacustrine silt deposit, and blended in at the material processing facility. The dam was constructed over a four-year period, and there was significant variability in the materials, which required modifications to the design and construction specifications. Due to the high placement rates and variability of the materials, methods of construction control and quality control were developed, and significant laboratory and field testing was performed including the following: particle size gradings, as-placed moisture content, optimum moisture content, density, and degree of compaction. The location from which each sample was taken was also recorded. In total, tThere are nearly 10,000 and 3,000 construction control tests of the core and downstream materials, respectively.

All of this information was digitized into a spreadsheet and incorporated into a GIS model to permit analysis of spatial and temporal distribution, statistics, internal stability, and filter compatibility. The analyses combined with review of construction photos and construction progress reports indicated that for internal stability and filter compatibility assessment, the dam fills could be characterized temporally based on the year of placement; in general, most of the changes in material handling and processing occurred early during each construction season.

ICOLD (2015) recognizes "the core and filter or transition in a dam have a range of particle size gradings, and it is important to consider the implications of this" (see Figures 2 and 3). Therefore, the finest base soil grading, the average base soil grading, and the coarsest base soil grading after regrading at the No.4 sieve were examined in analysis. It was found that these three material gradations do not necessarily represent the upper and lower filter requirements, especially for the Core materials which fall into two different base soil categories, as discussed in the following section. Li et al. (2009) used a statistical approach to account for the many gradation curves available in construction records. All of the analyses were performed using an Excel spreadsheet, and a VBA (Visual Basic for Applications) code was developed to analyze each gradation curve, from which statistical analysis was performed to obtain the range of filter requirements (and its average value).

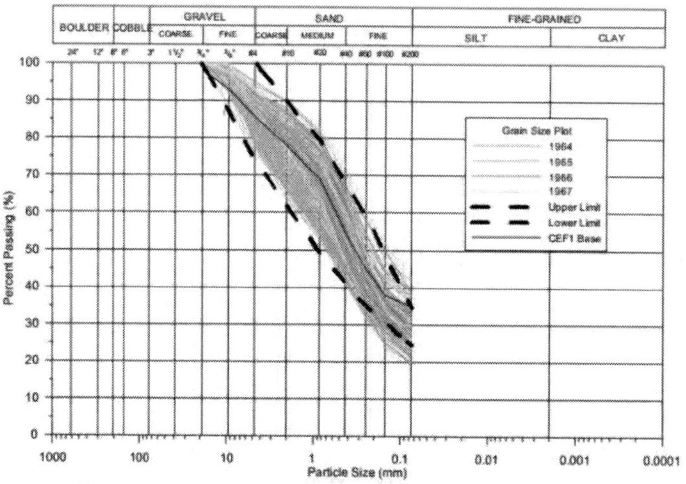

Figure 2 Range of As-Placed Gradations of the Core

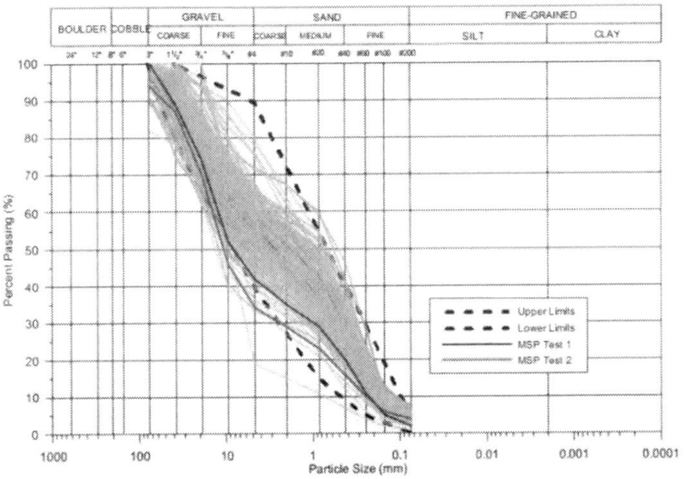

Figure 3 Range of As-Placed Gradations of the Transition

SCREENING EVALUATION BASED ON EMPIRICAL METHODS

Modern filter design guidance, such as USBR (2011), establishes control-points on the gradation envelope of a filter for both permeability and base soil particle retention. Base soil retention generally requires that the filter be capable of retaining the smallest particle that can be transported unaccompanied by movement of larger particles (no erosion). The design criteria give explicit consideration to both segregation and internal instability of the filter material, as each has potential to compromise the particle retention capability of the downstream zones.

Foster and Fell (1999, 2001) recognized that filters which are coarser than modern design guidance may still provide adequate protection - they developed a methodology for assessing the filtering performance of existing dams that is based on the results of Continuing Erosion Filter laboratory tests (CEF), and comparison with performance interpretations of existing dams. Foster and Fell observed that varying amounts of leakage and erosion can occur, but that continued erosion can be arrested if the larger base (core) particles can be mobilized within the base material and retained against the filter, thus preventing internal erosion of the medium particles, which in turn prevents loss of the finer particles. The critical particle size for the filter depends on the base soil type it is protecting, for which four different category-types were identified. All of the CEF tests on which the methodology is based were carried out on filters that were uniformly graded and internally stable. This section discusses and presents the screening evaluation that was performed for the Core and Transition materials from Dam "A" to consider the effects of internal instability, and also segregation, when applying the Foster and Fell method, per Foster (2007) and ICOLD (2015).

Internal Stability

Suffusion-suffosion is an internal erosion process by which the fine particles are removed through the constrictions between the coarser particles by seepage flow (Fannin and

Slangen, 2014). It initiates in internally unstable soils once the seepage forces exceed the resistance, typically defined as the critical hydraulic gradient. The critical gradient for an internally unstable soil is generally significantly less than the Terzaghi (1939) heave criterion (Skempton and Brogan, 1994). Given that critical hydraulic gradient is poorly understood and that local hydraulic gradients are difficult to monitor in a dam, it is assumed to occur if the materials are potentially internally unstable. The other parameter of importance in assessing the effects of internal instability on a grain size distribution curve, particularly that of a zoned filter material, is the maximum erodible particle size.

A number of researchers have developed empirical methods to assess the potential for internal instability of a soil. Most methods are based on quantification of the shape of the grain size distribution curve, and have been developed and/or verified against laboratory test data. Given the empirical nature of these methods, most methods are not applicable to all soil types, but are considered only suitable for the type of soils that were examined in the original development of the methodology. Accordingly, in this screening exercise, six methods that were developed for soils similar to the zoned earthfill dam 'A' were considered: Kezdi (1969), Sherard (1979), Kenney and Lau (1986), Burenkova (1993), Li and Fannin (2008), and Wan and Fell (2008).

The Kezdi (1979) and Sherard (1979) methods both involve splitting the gradation into a finer fraction and a coarser fraction at any size exceeding about 0.1 mm, and evaluating the coarser fraction as a filter to the finer fraction. Kezdi used the Terzaghi (1943) criterion to evaluate for soil retention, with reference to $D'_{15}/d'_{85} \leq 4$, where d'_{85} is 85th percentile of grain diameter of the finer fraction and D'_{15} is 15th percentile of grain diameter of the coarser fraction. The Sherard method is similar to that of Kezdi, except the base soil retention ratio ranges from 4 to 5.

The Kenney and Lau (K&L) method, and its Li and Fannin (L&F) adaptation, both involve quantifying the shape of grain size distribution curve using the stability number, H/F, where F = the percentage passing at grain size D, and H = the increment of percentage passing over a designated grain size interval of D to 4D. Kenney and Lau (1986) proposed a boundary between internally stable and unstable gradations at H=F over a designated finer portion of the gradation curve to F = 20 for a widely-graded primary fabric, and F = 30 for a narrowly-graded primary fabric. The Li and Fannin method is similar, except the boundary is amended to account for believed conservatism at F >15 on the gradation curve.

The Burenkova method, and its Wan and Fell adaptation, both invoke two parameters to assess the internal stability of a soil. The Burenkova method is based on d_{90}/d_{60} and d_{90}/d_{15} ratios where dx is xth percentile of grain size. The d_{90}/d_{60} ratio represents the slope of the coarse part of the grain size distribution curve. The d_{90}/d_{15} ratio can be regarded as a compound index involving the slope of the coarse fraction (d_{90}/d_{60}) and the slope of the middle fraction (d_{60}/d_{15}). The boundary for an internally stable soil is $0.76 \log(d_{90}/d_{15}) < d_{90}/d_{60} < 1.86 \log(d_{90}/d_{15}) + 1$. The Wan and Fell method is similar to that of Burenkova, but uses the d_{20}/d_5 ratio instead of d_{90}/d_{15}. The d_{90}/d_{60} ratio represents the slope of the coarse part of the grain size distribution curve, while the d_{20}/d_5 represents the slope of the finer fraction. The boundary for an internally stable soil is $15/\log(d_{20}/d_5) > 22$ or $30/\log(d_{90}/d_{60}) < 80$.

The Kenney and Lau and the Li and Fannin methods were developed from observations on sand and gravel soils. The Kezdi, Sherard and Burenkova methods are believed to be suitable for all soil types, regardless of the grain size and the shape of the gradation curve. The Wan and Fell method was proposed for broadly graded gravel-sand-silt soils, but not for gap-graded soils.

Where suitable, each of these methods was applied to all of the as-placed construction control laboratory grain size test result for each dam fill unit. An example of the results is shown on Table 1, for the Transition gradations, with reference to year-of-construction. Each of the methods examines the potential for internal instability in different proportions of this material. There is wide variation in the results, which is tentatively attributed to the suitability of the different methodologies for these materials. Thus, laboratory tests are necessary to verify the empirical results, as suggested by ICOLD (2015).

Table 1 Summary of Internal Stability Assessment of Transition Gradations

Year	Kezdi	Sherard	K&L	L&F	Burenkova	W&F
1964	47%	18%	21%	13%	44%	N/A
1965	66%	48%	13%	15%	38%	N/A
1966	46%	33%	5%	3%	51%	N/A
1967	35%	20%	4%	4%	65%	N/A
Total	51%	35%	8%	9%	49%	N/A

Note: NA = Not applicable to gap-graded materials

Maximum erodible particle size is an important consideration in assessing the effects of internal erosion, because it can affect the filtering capability of filter and drain materials. There are limited methods available for estimating the maximum erodible particle size of a soil. Wan and Fell (2004) recommended the method generally based on the Kenney and Lau (1985, 1986), in which the maximum erodible particle size is the particle size corresponding to the maximum F value less than 40 at which the stability number, H/F, is less than 1.3. This method is applicable to all soils. Fell et al. (2008) suggested that the maximum erodible particle size corresponds to the inflection point of the gradation. Li and Fannin (2008) noted that the most influential point on the gradation curve of a gap-graded soil is the maximum particle size of the finer fraction and, for a widely graded soil, the particle size corresponding to the $(H/F)_{min}$ location up to a maximum F = 41%. In this current study, the maximum erodible finer particle size is assumed to be that at $(H/F)_{min}$ up to a limiting particle size at 41% mass passing.

Segregation is a process that causes separation of granular materials into local finer and coarser sub-zones. It occurs during dumping and spreading operations in construction, where materials are "flowing". It cannot be completely eliminated, but its effects can be minimized by using narrowly-graded soils, so that zones of widely differing grain size do not develop. There are criteria to assess whether construction materials are prone to segregation, such as USDA SCS (1994), Kenney and Westland (1993), and Sherard and Dunnigan (1989). FEMA (2011), and USBR (2013), have adapted the USDA (1994) criteria in filter design. However, it is difficult to estimate the point of separation between the finer and coarser zones that is caused by segregation. ICOLD (2015) suggests using the point of inflection of broadly graded soils to separate the gradations, the same method recommended to identify the maximum erodible particle size for internally unstable soils. Kenney and Westland found that most materials segregate in a similar manner, and that the gradation of the coarser 25% of the segregated materials is bounded by the D_{50} and the D_{90} of the initial soil; the D_{50} of the Transition is about 4 mm. It was assumed that segregation is characterized by the absence of all minus No. 4 particles.

Filter Compatibility

Foster and Fell (2001) developed filter erosion criteria to assess the filter performance of existing dams based on statistical analysis of published filter tests as well as additional laboratory testing (CEF testing). They found that the performance of the base soils could be subdivided into four soil groups (Soil Groups 1, 2A, 3, and 4A) that are generally consistent with Sherard and Dunnigan (1989), but with minor modifications to reduce the subdivision boundary between Soil Groups 2 and 4 from 40% fines content to 35% fines content, termed the modified Soils Groups 2A and 4A.

Foster and Fell found that filter behaviour can be characterized into the following categories (ICOLD, 2015) and proposed three different maximum allowable, D_{15max}, filter boundaries (DF15 on Figure 4) to delineate four behaviours:

- *Seals with No Erosion (NE) – the filtering material stops erosion with no or very little erosion of the material it is protecting. The increase in leakage flows is so small that it is unlikely to be detectable.*
- *Seals with Some Erosion (SE) – the filtering material initially allow erosion from the soil it is protecting, but it eventually seals up and stops erosion. Leakage flows due to piping can be up to 100 l/s, but are self-healing*
- *Seals with Excessive Erosion (EE) – the filtering material allows erosion from the material it is protecting, and in the process permits large increases in leakage flow (up to 1000 l/s), but the flows are self-healing. The extent of erosion is sufficient to cause sinkholes on the crest and erosion tunnels through the core.*
- *Continuing Erosion (CE) – the filtering material is too coarse to stop erosion of the material it is protecting and continuing erosion is permitted. Unlimited erosion and leakage flows are likely.*

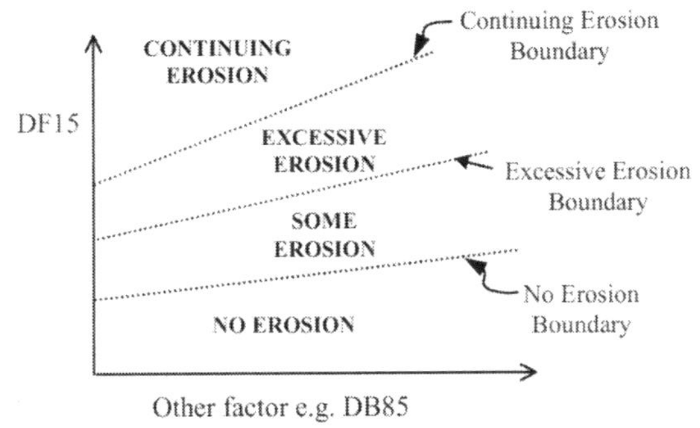

Figure 4 Filter Erosion Boundaries (after Foster and Fell, 2001)

The filter erosion criteria cannot be directly applied to dams in which the filtering material has internally unstable gradations. Thus, in this screening exercise, the filter gradation was adjusted by partially or completely removing some fractions of the original filter gradations to account for internal instability and segregation per Foster (2007) and ICOLD (2015). Note that although not all of the gradations of the filter were assessed to be potentially internally unstable, in this study the effects of internal instability were conservatively assumed for all gradations.

Thus, in this study of the zoned Transition material, four scenarios were considered for purposes of screening for internal erosion:
1) Design scenario – as-placed gradation
2) Likely wash-out scenario – assumed 50% of the minus maximum erodible finer particles (1mm) are removed by seepage flow
3) Complete wash-out scenario – assumed 100% of the minus maximum erodible finer particles are removed by seepage flow; and
4) Segregated scenario – characterized by the absence of all minus No. 4 particles.

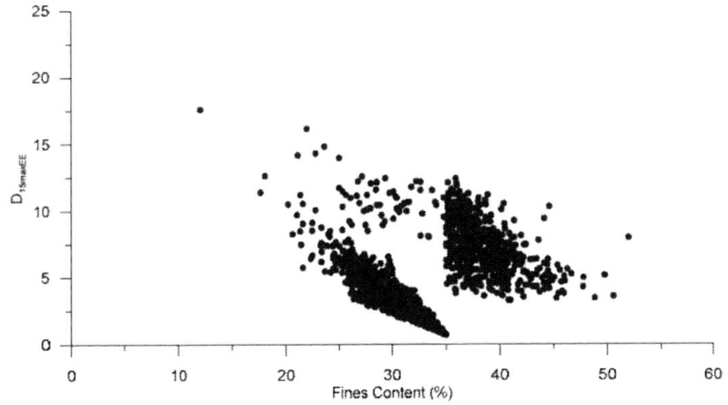

Figure 5 Variation in Maximum Allowable D_{15} of the Transition to Satisfy the Excessive Erosion Boundary with Fines Content

The Foster and Fell methodology requires that the base soil be re-graded to the No. 4 sieve for analysis. After re-grading at the No.4 sieve size, about one-third of the Core was categorized as the finer Soil Group 2A (fines content between 35 and 85%) while two-thirds were categorized as the coarser Soil Group 4A (fines content between 15 and 35%). The fines content of the core categorized as Soil Group 4A range from about 16% to 35%; the average fines content is about 30%.

The maximum allowable D_{15} value of the filter (DF_{15}), to satisfy each of the three erosion boundaries, was calculated for each construction record of the Core. As shown on Figure 5 and Table 2, the finest gradations of the Core do not yield the smallest and most conservative maximum allowable D_{15} boundary, particularly for the excessive erosion condition. Most of the Core gradations of the dam considered herein are categorized as Soil Group 4A, for which the database on which the boundaries were developed is smaller than for Soil Group 2A materials, particularly at a fines content similar to the core of Dam "A". In addition, the D_{15} values estimated for the 100% wash-out of the Transition fall close to the excessive erosion boundary of the Core.

Table 2 Estimated Maximum Allowable Filter Size for the Core based on Foster and Fell and Transition Filter Compatibility

Year	Core (2A) D_{15max} (mm)			Core (4A) D_{15max} (mm)			Transition D_{15} (mm)			
	NE	EE	CE	NE	EE	CE	As-placed	50% Wash Out	100% Wash Out	Segre-gated.
1964	0.7	5.2 (5.3)	31.7	3.7	5.8	32.5	0.3	0.6	4.5	7.8
1965	0.7	8.0 (9.5)	25.8	2.9	4.4	26.5	0.3	0.4	2.7	8.4
1966	0.7	9.8 (10)	22.4	2.6	4.5	30.2	0.25	0.4	2.4	9.0
1967	0.7	8.3 (6.0)	25.1	3.2	5.1	26.7	0.3	0.4	1.9	8.2

Note: The mean value of the D_{15} is given in the table based on statistical analysis. NE = No Erosion, EE = Excessive Erosion, CE = Continuing Erosion. Shown in parentheses under the EE boundary for Core materials characterized as Soil Group 2A are those calculated for the finest gradations of the Core.

LABORATORY TESTS

Laboratory testing was performed to verify the findings of the empirically-based evaluation of internal stability and filter compatibility. The most commonly used laboratory tests to assess internal stability and filter compatibility are Multi-Stage Permeameter (MSP) and Continuing Erosion Filter (CEF) tests, respectively. The method developed by Kenney and Lau (1985) was adopted for MSP tests, and the method by Foster and Fell (2001) was used for CEF tests.

Internal Instability Tests

Multi-Stage Permeameter tests were performed on the as-placed materials to assess the internal stability of select gradations, and to estimate the maximum erodible particle size.

The permeameter that was used in the study is a cast acrylic tube of inside diameter 279 mm and height of 998 mm, as shown in Figure 7 and described by Moffat and Fannin (2006). The specimen rests on a wire mesh screen of 2.76 mm square openings located within the acrylic tube. This device for allows application of varying surcharge loads on the top of the specimen to simulate a field stress condition, and measurement of the pore water pressure along the specimen to monitor the change in local gradients.

Figure 3 shows the two Transition gradations that were tested, together with representative portions of the construction records. The two selected gradations represent the as-placed gradation at the coarser end of the range of the material placed in construction. The maximum particle size tested was limited to 37.5 mm, which is close to 1/8 of the permeameter diameter; the oversize particles were replaced with an equal proportion of particles between 25.4 and 37.5 mm. The soil specimen was reconstituted in the permeameter cell by compaction, with control of compactive effort and moisture content used to reach a target dry density close to the field condition.

Multi-stage downward seepage flow was applied to the specimen over a range of hydraulic gradients to a maximum value of approximately 25, which is believed to be the maximum gradient in the dam. The seepage flow monitoring and observations indicate that both gradations exhibited particle migration, which is consistent with the assessment using the empirical methods (Li et al, 2014). The maximum erodible particle size was found to be about 1 mm, close to $(H/F)_{min}$ value for $F < 41\%$. However, not all of the soil particles smaller than 1 mm were eroded from the specimen; the amount may have been limited by the mesh opening size, as discussed in Moffatt and Fannin (2006).

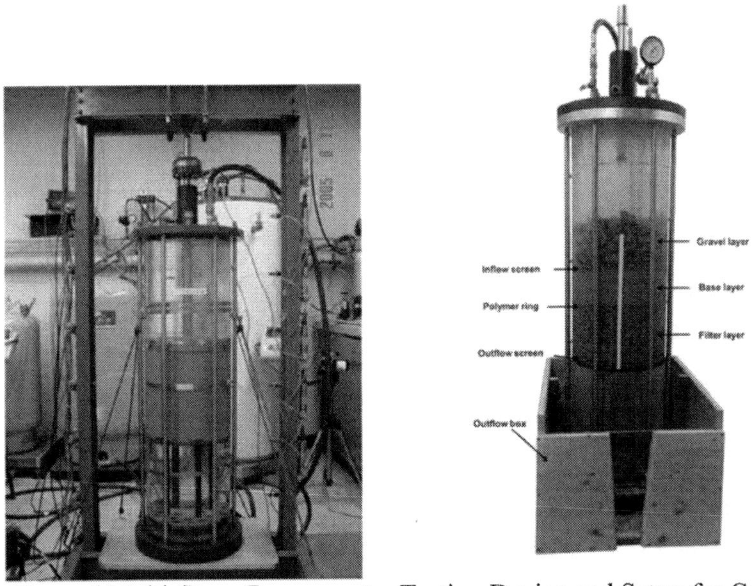

Figure 6 Multi-Stage Permeameter Testing Device and Setup for Continuing Erosion Testing

Continuing Erosion Filter Test

CEF tests were performed on the Core and Transition materials to assess the filter performance, and to verify the application of the Foster and Fell method to earthfill dam "A" materials. The Foster and Fell filter erosion criteria are empirical, and there is a limited amount of testing on Base Soil Type 4A materials in the database from which their criteria were developed, yet this soil category represents the majority of the Core of earthfill dam "A". Therefore the CEF tests on Soil Type 4A materials were conducted to verify Base Soil Type 4A criteria. The CEF tests were carried out using the same device used for the MSP tests, which was reconfigured for CEF testing as shown on Figure 6. The procedures described by Foster and Fell (2001) were used in the tests.

Foster (2007) modified the CEF test procedure to be used for an internally unstable filter by installing piezometers into the filter materials to allow the measurement of pore pressures. This approach is not applicable to segregated filters. Therefore, the Transition gradations that were tested were adjusted, such that all of the particles smaller than 1 mm and 4.75 mm were removed from its gradation to account for internal instability and segregation, respectively. The same gradations of the Transition that were tested in the MSP tests were used in the CEF tests.

Core material falling under both Base Soil Groups 2A and 4A were tested. The Core material was reconstituted by compaction to a moisture content and density representative of the as-placed field condition. A pre-formed axial hole of 5 mm diameter in the Core simulates the influence of a concentrated leak through the base soil. Test specimens of Transition material were lightly compacted to simulate wash-out. Upon flooding of the test specimen, water above the Core was pressurized to a value of 275 to 300 kPa, causing seepage flow in a downward direction and across the base soil and filter interface. Volumetric flow was measured during all tests, which provided useful insight to the process, as shown on Figure 7 for select tests. All of the tests sealed, and were characterized by initially turbid flow that increased in rate, and gradually became clean and diminished to a constant value. A thin eroded slot that formed over the bottom of some of the base specimens was observed during testing.

Foster and Fell (2001) proposed the following excessive erosion (EE) boundaries for interpretation of CEF laboratory tests:

1) Soil Group 2A: a mass loss greater than 1.0 g/cm^2
2) Soil Group 4A: a total mass loss greater than 100 g

Application of the above Foster and Fell threshold limits to interpret the results of the CEF tests yielded mixed results, as shown on Table 3. For Base Soil Group 2A, the Foster and Fell assessment criteria are in agreement with the CEF test results (Li et al, 2014). In contrast, for Base Soil Group 4A, in some tests the eroded mass exceeded the threshold of 100 g associated with excessive erosion response and is inconsistent with the empirical assessment.

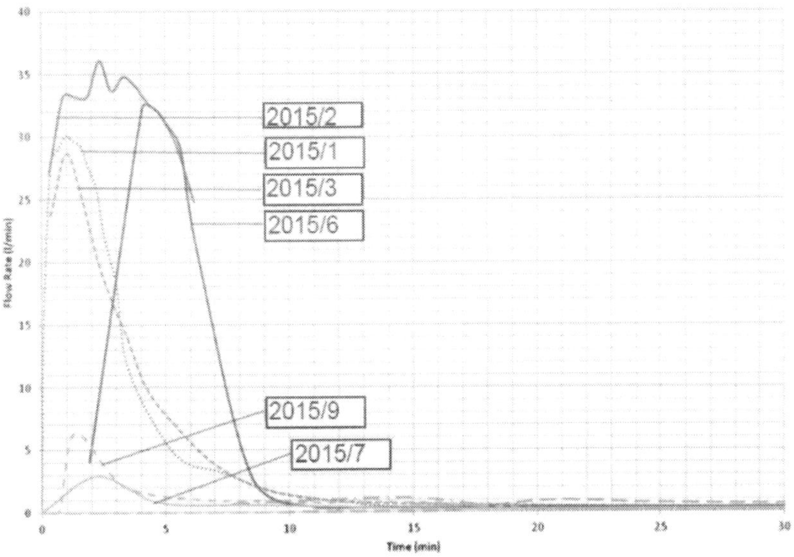

Figure 7 Measured Specimen Seepage Flows

This is not surprising given the larger data base of laboratory testing and historical dam filter performance information for Soil Group 2A versus Soil Group 4A, and given the size of the specimens tested herein is generally larger than those tested by Foster and Fell. However, if the eroded mass is expressed as unit mass loss rather than total mass loss, and the Soil Group 2A thresholds are applied to interpret the CEF test results for the Soil Group 4A materials, then good agreement is obtained between the empirical filter assessment and the CEF tests. Large seepage flows were also measured in the filter specimens for which the excessive erosion boundary was exceeded, indicating that volume flow measurement can also be used to interpret CEF laboratory testing results. Further research is required, particularly for Soil Group 4A materials.

Table 3 Summary of Select CEF Test Results

	Test Number						
	13/1	15/2	15/3	15/4	15/6	15/8	15/9
Base Category	2A	2A	2A	4A	4A	4A	4A
Fines Content (%)	35	35	35	23	22	28	27
$D_{15maxNE}$ (mm)	0.7	0.7	0.7	5.8	4.3	2.0	2.5
$D_{15maxEE}$ (mm)	7-11	7-11	7-11	9.1	6.7	3.2	3.9
Filter D_{15} (mm)	3.6	12.6	8.4	8.4	8.4	3.5	8.4
Erosion Category	SE	EE	EE	SE	EE	SE/EE	EE
Mass Eroded (g)	200	1935	926	353	1134	132	780
Mass Eroded (g/cm^2)	0.3	3.2	1.5	0.6	1.85	0.2	1.3
Max. Flow (L/min)	2	36	29	1	32	0.9	6.4
CEF Category	SE	EE	EE	EE(SE)*	EE	EE(SE)*	EE

* Assessments using the Soil Group 2A thresholds are shown in parentheses

CONCLUDING REMARKS

Construction records at a dam-of-interest were analyzed, using empirical screening tools, to evaluate (i) the susceptibility of one of the filter zone materials (Transition) to internal instability, and (ii) the susceptibility of the adjacent dam fill zones to filter incompatibility for four different scenarios. The four scenarios considered are the as-placed Transition, 50% wash-out Transition, 100% wash-out Transition, and segregated Transition. Laboratory testing was performed on select gradations of soil from the dam site with the objective of verifying the findings of the screening exercise. For the earthfill dam materials considered, the following observations are made:

- The combination of the empirical tools, and companion laboratory testing, provides a general framework to assess the potential for internal erosion in an embankment dam. Appropriate empirical methods should be selected for the assessment, and laboratory testing is recommended to verify the application of the empirical tools.

- Screening-level type analyses that consider the finest base soil and coarsest filter soil may not yield the most conservative filter compatibility assessment. Analyses should be carried out over the entire database of gradations available for the dam. The use of a statistical method to analyze all construction data is recommended.

- The use of scenario-based outcomes to assess the filter performance can account for internal instability and/or segregation of filters by adjusting their gradations, and contributes to a better understanding of the likely filter performance in the embankment dam.

- Laboratory testing of a number of combinations of Core and Transition material, in the Continuing Erosion Filter (CEF) test, were performed for Core material that is categorized as Soil Group 4A. The screening level assessment is inconsistent with interpretation of the CEF tests based on the Foster and Fell threshold limits for Soil Group 4A, but is consistent if the CEF test results are interpreted with the threshold limits for Soil Group 2A. There is significant potential in using seepage observations in interpretation of the CEF tests, particularly if they can also be used to interpret actual dam field measurements.

ACKNOWLEDGEMENTS

The authors wish to thank the BC Hydro and Power Authority for permission to publish this paper. We also acknowledge the work of Dr. Paul Slangen and Dr. Kaley Crawford-Flett for their analysis of the construction records, and performance of the laboratory tests, respectively. Dr. Mark Foster provided a review and input to the CEF test procedure used in the study.

REFERENCES

Burenkova, V. V. 1993. Assessment of suffosion in noncohesive and graded soils, in *Filters in Geotechnical and Hydraulic Engineering*, editors Brauns, Heibum & Schuler, Balkeman, 357–360

Engemoen, W. (2016). Reclamation experience with initiation rates of concentrated leak erosion. USSD Symp. on the mechanics of internal erosion for dams and levees, 8-10 Aug. 2016, Salt Lake City, USA.

Fannin, R.J. and Slangen, P. (2014). On the distinct phenomena of suffusion and suffusion. *Geotechnique Letters* 4:289-294.

Fell, R., Foster, M., Cyganiewicz, J., Sills, G., Vroman, N., and Davidson, R. Risk Analysis for Dam Safety – A Unified Method for Estimating Probabilities of Failure of Embankment Dams by Internal Erosion and Piping, UNICIV Report No. R-446, April 2008, The University of New South Wales.

Foster, M.A., 2007. Application of no, excessive and continuing erosion criteria to existing dams, In *Internal Erosion of Dams and their Foundations*, Editors R Fell and J-J Fry, Taylor Francis, 103-114.

Foster, M. and Fell, R., 1999. Assessing Embankment Dam Filters which do not satisfy Design Criteria, UNICIV Report No. R376, May 1999, The University of New South Wales.

Foster, M. and Fell, R., 2001. Assessing embankment dams filters which do not satisfy design criteria, *J. Geotechnical and Geoenvironmental Engineering*, ASCE, Vol.127, No.4, May 2001, 398-407.

Garner, S.J., Seyers, W.C. and Mathews, H.M. (2004). The Decommissioning of Coursier Dam – A Case for Dam Safety, Proc. CDA annual conference, Ottawa, Sept. 2004.

ICOLD (2015). Internal Erosion of Existing Dams, Levees and Dikes, and their Foundations, Volume 1: Internal Erosion Processing and Engineering Assessment, Bulletin 164. International Commission on Large Dams, Paris, France.

Kenney, T.C., and Lau, D. 1985. Internal stability of granular Filters. *Canadian Geotechnical Journal* 22: 215–225

Kenney, T.C., and Lau, D. 1986. Internal stability of granular Filters – Reply. *Canadian Geotechnical Journal* 23: 420–423

Kenney, T.C., and Westland, J. 1993. Laboratory study of segregation of granular filter materials. *in Filters in Geotechnical and Hydraulic Engineering*, editors Brauns, Heibum & Schuler, Balkeman, 313–319

Kezdi, A. 1979. Soil physics – selected topics. Elsevier Scientific Publishing Company, Amsterdam, 160p

Li, M., and Fannin, R.J. 2008. Comparison of two criteria for internal stability of granular soil. *Canadian Geotechnical Journal* 45: 1303–1309

Li, M., Fannin, R.J. and Garner, S.J. 2009. Application of a new criterion for assessing the susceptibility to internal erosion. CDA 2009 Annual Conference, Whistler, BC, Canada. October.

Li, M., Fannin, R.J. and Anderlini, C. 2014, Assessing performance of a potentially internally unstable filter, CDA 2014 Annual Conference, Banff, Alberta, Canada, October

Moffat, R. and Fannin, R.J. 2006 A large permeameter for study of internal stability in cohensionless soils, Geotechnical Testing Journal, Vol.29, N0.4, pp. 1-7.
Ripley, C.F. 1967. Portage Mountain Dam: 1. an outline of the project, *Canadian Geotechnical Journal*, 4 (2): 125-141.

Sherard, J.L. 1979. Sinkholes in dams of coarse, broadly graded soils. In *Thirteenth Internaltional Congres on Large Dams.* , New Delhi, Vol.2, pp 25-34

Sherard, J.L., and Dunnigan, L.P. 1989. Critical filters for impervious soils. *Journal of Geotechnical Engineering*, 115:927–947

Skempton, A.W. and Brogan, J.M. 1994 Experiments on piping in sandy gravels. Geotechnique, 44:449-460.

United States Bureau of Reclamation (USBR), 2011, Testing Standards No. 13 Embankment dams - Chapter 5 Protective Filters. Denver, CO

USBR-USACE (2015). Best practices in dam and levee safety risk analysis. A joint publication of the US Bureau of Reclamation and the US Army Corps of Engineers. Version 4.0 July 2015.

Wan, C.F. and Fell, R. 2004. Experimental investigation of internal erosion by the process of suffusion in embankment dams and their foundations, *ANCOLD Bulletin* No. 126, pp 69-78

Wan, C.F., and Fell, R. 2008. Assessing the potential of internal instability and suffusion in embankment dams and their foundations. *Journal of Geotechnical and Geoenvironmental Engineering* 134: 401–407

MONITORING SAN VICENTE DAM RAISE DURING INITIAL FILL

Andrew Oleksyn, PE[1]
Nathan Grima [2]
Anna Kolakowski PE[3]
Larry Rodrigues[4]
Jim Zhou, PE[5]

ABSTRACT

The San Diego County Water Authority (SDCWA) recently completed the San Vicente Dam Raise Project (SVDR) to increase local water storage capacity in San Diego County, California. The SVDR project raised the original 220-foot high concrete gravity dam (90,063 AF of storage) by 117 feet, increasing reservoir storage capacity by 152,000 AF. The dam raise utilized more than 600,000 cubic yards of roller-compacted concrete (RCC), making it the largest dam raise in the world using RCC.

The California Department of Water Resources Division of Safety of Dams (DSOD) certified the dam for initial filling above the height of the original dam in accordance with a controlled fill and monitoring plan. The plan stipulated fill increments of 30 feet of reservoir elevation behind the dam with a 3-week hold period between each fill. The progress of the filling and responses of the dam and foundation were closely monitored by a team comprising the city of San Diego (owner of the dam), the designer of record, and DSOD. Monitoring activities included daily inspection by the city, weekly review of the reports and instrumentation data by the designer and DSOD. Also during each hold period, the city completed emergency discharge valves testing, the designer completed his inspection and assessment, and DSOD witnessed the valve testing and accepted the monitoring results prior to proceeding with the next fill increment. To date, the expanded reservoir has been filled up two-thirds of the raised height of the dam.

This paper summarizes the filling and monitoring activities and results of the observations, surveys and instrumentation trends.

[1] San Diego County Water Authority (SDCWA), 4677 Overland Ave., San Diego, CA 92123, AOleksyn@sdcwa.org.
[2] City of San Diego Public Utilities Department (COSD) 2797 Caminito Chollas, San Diego, CA 92105, NGrima@sandiego.gov
[3] California Department of Water Resources Division of Safety of Dams (DSOD), 2200 X Street, Suite 200 Sacramento, CA 95818, anna.kolakowski@water.ca.gov
[4] City of San Diego Public Utilities Department (COSD), 12375 Moreno Ave., Lakeside, CA, 92040 LRodrigues@sandiego.gov
[5] San Diego County Water Authority (SDCWA), 4677 Overland Ave., San Diego, CA 92123, JZhou@sdcwa.org.

INTRODUCTION

The original San Vicente Dam had a spillway crest elevation of 650 feet above mean sea level. The reservoir and outlet works remained in operation during construction of the dam raise, but the water elevation was lowered to approximately 594 feet to accommodate the work. Once the RCC placement was completed, the State of California, Division of Safety of Dams (DSOD) allowed filling in increments starting with permission to raise the reservoir elevation to 650 near the spillway crest of the original dam in early 2014. Notice of Completion was issued in September 2014 and initial filling behind the raised portion of the dam was authorized in May 2015.

INITIAL FILLING AND MONITORING PLAN

The initial filling and monitoring plan was prepared by the San Diego County Water Authority in collaboration with the City of San Diego (owner). The plan was reviewed and approved by the Design Engineer of Record (Stantec, formerly MWH)[6] and DSOD. Following post-construction review, DSOD allowed the initial fill increment to a maximum water surface level of El. 680, approximately thirty feet above the previously authorized level.

The approved filling plan had the following conditions:

1. When the reservoir was at El. 643, the owner would make a baseline visual inspection, dam axis survey, and collected dam safety instrumentation data.

2. Reservoir filling would proceed at a rate no greater than 2 vertical feet per day (24 hours) as requested by the Designer.

3. Daily patrols would be conducted of the main dam and environs, its appurtenances, and areas of the reservoir including areas down-slope of the saddle dam. The results would be documented.

4. Once the reservoir level reached the midpoint of the planned fill – about El. 665, the owner would again make a dam axis survey and a visual inspection for comparison to the baseline data and observations.

5. Reservoir filling would continue without stopping to EL. 680, which was expected to be reached on or about September 1, 2015.

6. Once the reservoir was at the planned fill endpoint, El. 680, the owner would again make a dam axis survey and a visual inspection for comparison to the baseline and midpoint observations/data.

7. Existing dam safety instruments would be monitored throughout the filling

[6] Michael F. Rogers, PE, PMP, P. Eng., Stantec (formerly MWH) 9665 Granite Ridge Dr., San Diego, CA 92123, Michael.f.rogers@mwhglobal.com

cycle**.

8. The reservoir would be held at El. 680 for a minimum of 21 days, after which the owner would again make a dam axis survey and a visual inspection for comparison to the baseline and midpoint observations/data.

9. At each reservoir filling hold point, the Designer would conduct a visual inspection of the dam and facilities and review the survey and dam safety instrumentation data to make a recommendation on continued filling.

10. During the filling hold period, the owner would exercise the outlet system valves under the increased reservoir pressure, including the 108" butterfly valve, knife gate valves and emergency release plunger valves, with DSOD inspection.

11. Reservoir filling above the approved maximum elevation would not proceed until approved by DSOD.

The initial filling and inspection plan and notes are summarized in the Table 1.

Table 1. Summary of Inspection Plan and Notes.

Reservoir EL. (feet)	Estimated Date	Dam Axis Line Survey	Visual Inspection*	Instrumentation Data**	Submit Data to DSOD
643 (baseline)	April 15, 2015	Yes	Yes	Yes	Yes
665 (midpoint)	July 16, 2015	Yes	Yes	Yes	Yes
680*** (end of fill)	Sept. 1, 2015	Yes	Yes	Yes	Yes

Notes:

* Inspect the major structures including dam, outlet tower, DCF and LLO for abnormal settlements, heaving, deflections, movement, unexpected cracking/spalling, opening of joints, abnormal leakages and seepages through the dam, foundation or abutments. Also inspect reservoir area for early signs of slope movement, downstream channel and surrounding areas for erosion and abnormal subsidence, reservoir saddles for elevated groundwater and abnormal seepage.

** Monitoring frequency:

- *Daily* for automated instrumentation – reservoir level, wet well level, reservoir water temperature, foundation piezometers (7 locations) on the abutments, thermocouples, extensometers, and weirs.
- *Weekly* for manual instrumentation – manual piezometers (4 locations) in main adit, drains from Carpi liner (5 locations).
- At the *mid-point and end* of each 30-foot fill cycle – survey the crest of dam (axis).

*** End of Fill

- Hold for a minimum of 21 days.
- Designer to assess the conditions of dam and make recommendations.
- DSOD to approve or disapprove the next filling above EL. 680.

INSTALLED INSTRUMENTATION

New instrumentation was installed as part of the San Vicente Dam Raise project for surveillance of dam performance. This included an array of thermocouples that provided concrete temperature measurements during construction and for future long-term monitoring. A traditional survey monument system was installed on the dam crest and spillway to monitor movements of the dam structure. Multiple position borehole extensometers were provided beneath the dam to measure deformations in the foundation. Five strong motion accelerometers were installed at the dam crest, dam toe and on the left and right abutments to record motions during strong seismic events. A number of piezometers, both vibrating-wire and manual observation wells in the main gallery were installed. And finally, three flow measurement weirs were installed to measure seepage discharging from the galleries. Instrumentation locations are shown in Figure 1 and summarized in Table 2.

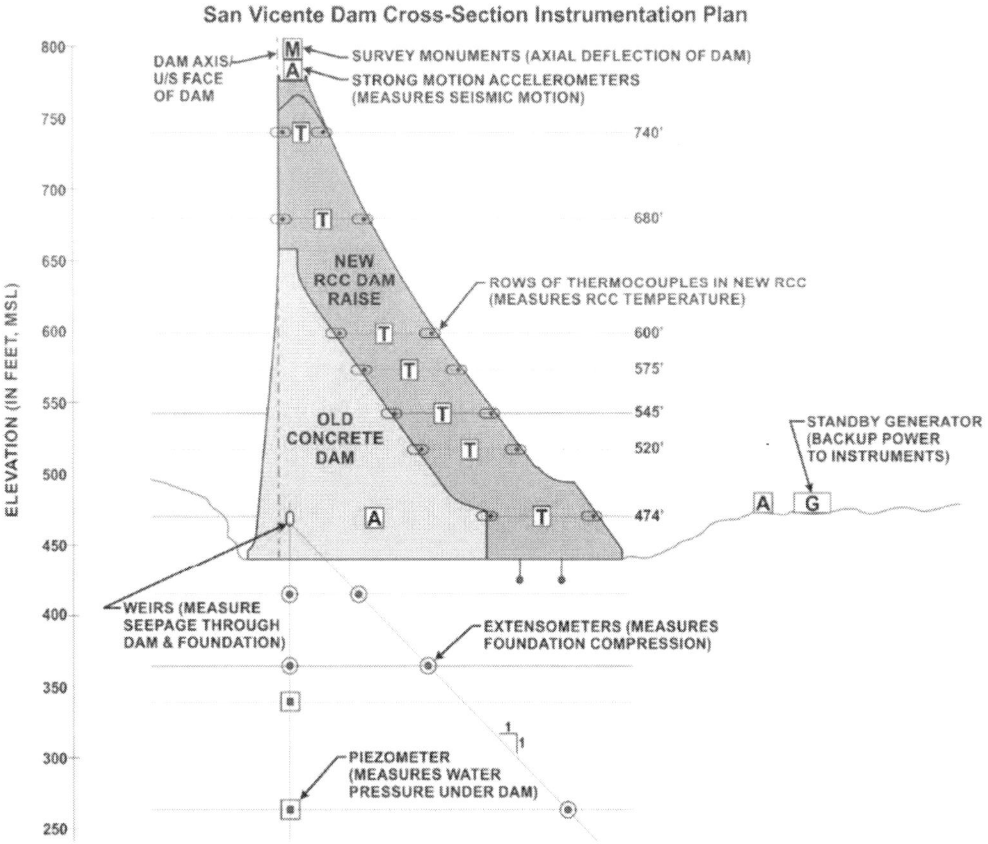

Figure 1. Instrumentation Locations Profile

Table 2. Instrumentation Summary

Instrument Type	Quantity	Notes
Foundation Piezometer	7	Vibrating Wire
Manual Piezometer	4	Main adit
Thermocouples	55	At various positions and Elevations
Extensometers	5	Foundation rocks
Accelerometer	5	Various locations
Weir	3	Inside gallery
Surface Monument	8	Crest of the dam

INSPECTIONS

City staff made site inspections each day (7 days a week) during the reservoir fill period between about May 15 and July 31, 2015 as the reservoir level was raised to elevation 680. During the hold period, the City conducted dam inspections once each week. The Designer of Record performed a site inspection of the dam with the City of San Diego Dam Keeper and Water Authority Staff in early September 2015. The Designer was given access to the previous City inspection forms, which were reviewed. The inspection was conducted along the same general inspection route followed by City inspectors, with the addition of an upstream face inspection by boat. The reservoir was at the hold point elevation of 680, thirty feet above the original spillway. Inspection areas included:

- General Mechanical Maintenance Issues
- Outlet Works
- Downstream Face and Toe of the Dam
- Left and Right Abutment Crests
- Spillway Crest (with binoculars)
- Upstream Face and Exposed Carpi Lining System (from boat)
- Drainage Gallery (Foundation Drains, Carpi Lining Drains,
- Instrumentation Review (Extensometers, Piezometers, Leakage and Rainfall, Thermocouples, Crack Gauge)

An interim fill inspection report[7] was submitted to DSOD in late September 2015 with the following summary observations.

General Maintenance and Outlet Works – At the time of the inspection, several warranty items were being addressed by the Contractor including binding of the outlet tower fish screens, and hydraulic power unit dirty filter alarms. However since construction of the facility was completed relatively recently, the mechanical works were observed to be in excellent condition.

Toe, Abutments and Dam Face – No seepage was noted. Several of the planned contraction joints (CJ) were evident, showing that the internal controlled cracks from the metal plates installed during construction every 50-feet were reflective to the downstream face.

Crests and Spillway – No noticeable changes were observed. Review of crest and monument surveys performed by the City showed little movement beyond survey tolerances.

Carpi Liner - The exposed geomembrane liner appeared in good condition throughout the face. Some small air sags in the liner were noted just above the old tower, requiring no action.

[7] Michael F. Rogers, *San Vicente Dam-Site Inspection at WSEL 680 Hold Point*, September 23, 2015

Galleries - There was continued minor seepage at the contact between the gallery access building and the lateral access gallery as was observed during construction. There were also a few minor areas of seepage along the precast concrete sections in different areas of the gallery. Very minor leakage was noted from the upstream facing system drainage that is connected into the gallery. Each of the five drains that connect to the five compartments of the upstream facing system was observed with only one that appeared to be dripping. The City measured the leakage through the upstream facing system drains at about 0.05 gallons per minute (gpm). The majority of the foundation drains had water levels either just at the top of the drain, or below – only a few had small flows. The overall leakage in the gallery was measured at about 2 gpm. Some red-colored mineral deposits were observed coming out of solution at the top of a couple foundation drains similar to the precipitate process observed at Water Authority's Olivenhain Dam. This precipitate appears to be iron and manganese due to iron bacteria common in foundation drains at dams (as reported by US Bureau of Reclamation).

Figure 3. Precipitate at Gallery Drain.

Piezometers - The vibrating-wire piezometers were found to have remained unchanged since the data recording started. The expectation was to see some response in the instrument located furthest upstream (FP-1) during the reservoir fill from 650 to 680. On the recommendation of the Designer, the method of calculating and plotting the piezometer data was reviewed and adjusted in the data acquisition software. Manual piezometer water levels in the main section of the dam showed little to no change, indicating the grout curtain is effective.

Thermocouples - All temperature data showed cooling of the RCC as expected. The seasonal trend of thermocouples subject to influence from ambient air temperature close to the upstream (above the water level) and downstream faces seemed reasonable.

Recorded temperature cycles were influenced primarily due to ambient air temperatures and solar radiation on the dam concrete. The thermocouples near the upstream side of the dam below the water level measured close to consistent values, which were expected as they were primarily influenced by the cooler water temperatures. There was also a thermocouple (#6) that is located at about the mid-point (upstream to downstream of the RCC section) at each instrumentation group elevation. At each elevation, this instrument was measuring near steady-state temperature of the concrete, which was expected given the insulating effect of the dam concrete. The RCC temperature in the center is currently at about 84 degrees F after approximately four and half years of the placement. The RCC at San Vicente has cooled down slightly faster than the Olivenhain Dam, as expected, mainly due to a relatively small volume. For the same reason, the upper-most mid-point instrument at El. 740 (#TC740-6) located in a narrower area of the RCC section close to the top of the dam responded more rapidly to the ambient air temperature influences.

Extensometers - The extensometers were measured throughout the duration of construction and recorded a noticeable but permissible change (less than 5 mm) as RCC was being placed. During the initial fill, however the instruments recorded little movement in the foundation, which was expected for the relatively small increase in load on the dam that had occurred thus far.

Crack #9 Gauge - The Crack #9 in the original dam had been monitored for a long time and did not change during the RCC construction and initial fill. This confirmed the visual observations during the inspection that there had been no discernable differential movement or significant cracking in the dam (other than the expected contraction joint reflections to the downstream face).

Water Temperature – Water temperature was not a dam safety concern but monitored to provide the City with the information to select an appropriate level for water withdrawal and help optimize performance of the water treatment plant owned by the city.

Leakage/Seepage - No changes in leakage were found at this point of reservoir increase or rainfall occurrence. It was recommended that this data be closely scrutinized after a significant rainfall event.

Carpi (Upstream Exposed Geomembrane Liner) Drains – The Carpi liner provides a water-tight barrier to slow down seepage through the RCC section. There was no significant leakage observed indicating excellent performance thus far.

The Designer concluded that there were no dam safety issues identified in any area of the dam and foundation inspected. The dam appeared to be functioning normally, meeting the design intent without any significant cracking or differential movement. The dam safety instrumentation confirmed that the dam and foundation were reacting normally with increasing reservoir level. Based on the results of this inspection and review of the dam safety instrumentation, the Designer recommended that the reservoir filling continue to the next planned hold point at reservoir elevation 710.

Coring program results (Rogers 2016 and Stiady 2017) and the Designers inspection results were presented to DSOD at the end of September 2015 showing that the in-situ RCC met the design requirements for strength. Based on this and an independent analysis by DSOD, the next fill increment was approved and certification to fill to the spillway crest elevation of the raised dam at 766 feet was issued in December 2015. The initial fill monitoring program requirements outlined above remained in place for the 30-feet fill increments and hold points at elevations 710, 740 and 766.

Delivery of imported water continued to San Vicente reservoir to a maximum elevation to date of 727 feet. Another interim hold point and inspection was conducted at elevation 710 with midpoint evaluations at Elevations 695 and 725. The following observations were made with the increased water elevation behind the newly raised dam at the 710 hold point.

General Maintenance and Outlet Works – A minor flange leak developed on the 108 inch discharge pipe. The flange bolts were subsequently tightened by the contractor under warranty and the flange leakage was stopped.

Toe, Abutments and Dam Face – Minor seepage on both left and right abutment groins was observed. This was not unexpected in the areas of the large shear zones in the foundation, although they had been extensively treated during construction. Also this part of the foundation was subject to hydrostatic pressures for the first time. Seepage is expected to decrease over time as remaining rock fracture seepage paths calcify.

Crests and Spillway - Changes in survey readings were small, with slight movement downstream as expected. One survey monument showed movement upstream from the previous measurement at the midpoint survey, but this was attributed to either survey instrument or human error. The subsequent survey during the hold point showed this monument at its previous location.

Carpi Liner – Drainage increased slightly from the membrane system drains from a trace at 680 to approximately 0.22 gpm at 710. This is still well within the 35 gpm leak rate guaranteed by Carpi and indicates continued good performance.

Galleries - Seepage increased in the gallery as was expected and is primarily attributed to the increased hydrostatic pressure on construction joints of the original dam below the Carpi liner. The seepage rate during the elevation 710 hold point inspection was approximately 10 gpm and gradually increased to approximately 18 gpm (excluding the Carpi drain) at elevation 725. Three locations of seepage were of note:

1. Steady seepage was coming into the Gallery Access Building near its connection with the gallery extension walkway to the main adit. This seepage was originally noted during construction during concrete curing with water spray on the downstream face. The seepage had increased with the reservoir filling but there were also rain showers prior to the inspection. The seepage water in this area was probably a combination of seepage from the reservoir and rainwater.

2. At Station 9+00 in the gallery, one of the 8-inch vertical formed drains had noticeably more seepage than during the previous inspection (Sept. 2015/WSEL 680). Water was also seeping from the gallery roof and downstream wall through this contraction joint (CJ). This appeared to be an active CJ as the City had previously set up crack monitoring pins on the upstream wall and a crack gauge on the downstream wall at this location. Given the pre-existing crack monitoring and the raised reservoir elevation, the seepage at this joint is not surprising but will be closely monitored.

3. Leakage was also observed from the upstream wall at the location of Drain #31, Sta. 4+21 (Figure 4). This is a construction joint in the original dam. Leakage was also slowly coming from the gallery crown and downstream wall at this location. Based on old stains and calcifications, this CJ appears to have previously experienced seepage water and possibly rain water infiltration. This seepage was not considered a dam safety issue but close monitoring of this joint is to be continued for the remainder of filling and beyond.

Figure 4. Construction Joint Leakage at Station 4+21.

Piezometers - – The vibrating-wire piezometers continued to show little or no change with increasing reservoir level. A review of the instrument outputs was conducted and adjustments to level calculations were made, but steady-state outputs continued. Review of the four manual piezometers showed modest increases in downhole water elevation as expected. The water elevations remained below the gallery floor elevation as shown in Figure 5 indicating uplift pressure on the bottom of the dam was still within the design intent. All piezometers continue to be monitored closely.

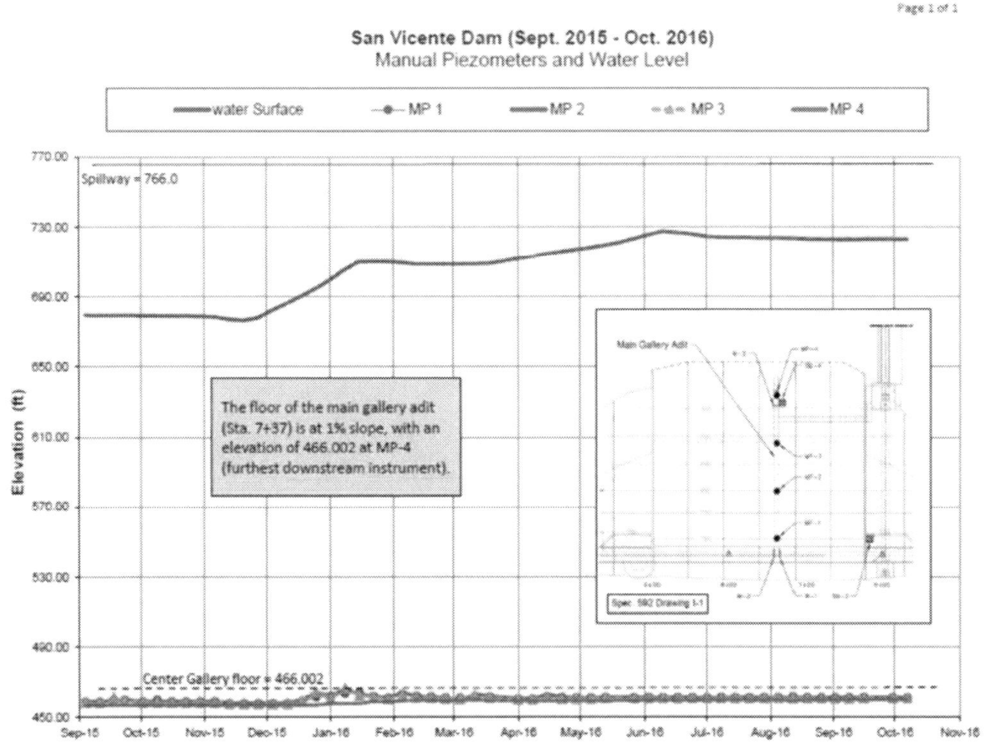

Figure 5. Main Gallery Manual Piezometer Readings.

Thermocouples - Temperature readings continued to show cooling of the concrete and seasonal trends from ambient air temperature as expected. Figure 6 shows that the data for the concrete deeper in the main body of the dam, with gradual cooling and fluctuations of the concrete temperature nearer to the surface as expected. This resembles the pattern observed at the Water Authority's Olivenhain RCC dam.

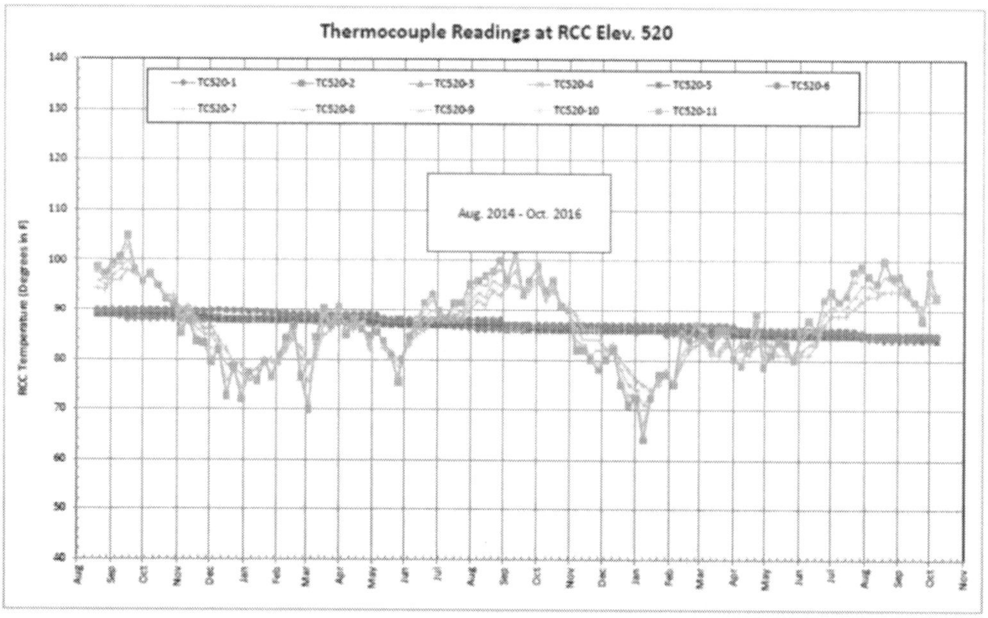

Figure 6. Temperature Readings at Lower Dam Elevation.

Higher in the dam profile, atmospheric conditions affect the concrete temperatures in the thinner cross section closer to the dam crest as shown in Figure 7.

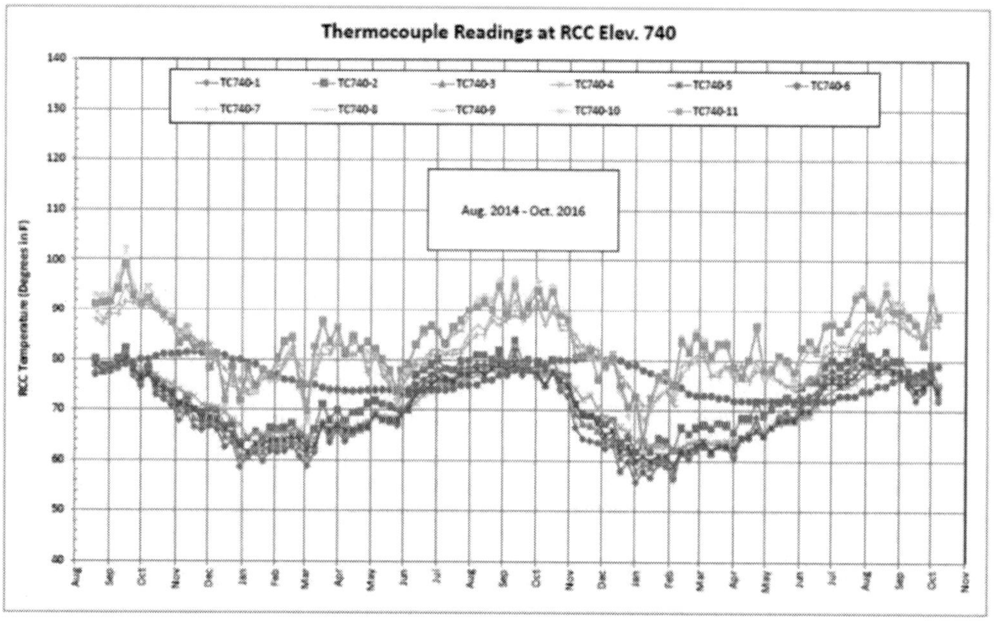

Figure 7. Temperature Readings at Higher Dam Elevation.

CONCLUSION

Following the inspections at the reservoir hold point elevation of 710 feet and interim instrument review at elevation 725, the Designer of Record again reported that there were no dam safety issues identified in any area of the dam and foundation inspected. No detrimental cracking or differential movements were observed. Weekly review of the dam safety instrumentation during filling confirmed that the performance of the dam and foundation were meeting the design intent as the reservoir level increased. DSOD reviewed the report and allowed the incremental fills to continue in accordance with the reservoir fill plan.

Gallery seepage increased modestly from historic flowrates from the original dam as the water level increased behind the raised dam. The increased seepage is predominantly from one construction joint near the right extreme of the original dam. This location is below the new upstream geomembrane liner and high in the elevation of the original dam, which was not subject to high hydrostatic pressures over the dam's history. The construction joint is the nearest to the new outlet tower where blasting occurred to excavate the tower foundation. Blast vibrations may have opened some seepage paths around the original copper water stop in the construction joint. It will continue to be monitored closely, but no corrective action is deemed necessary at this time.

REFERENCES

1. J. Stiady, A. Oleksyn, M. Rogers, G. Tarbox, T. Reynoldson, W. Griffis, J. Zhou, "RCC Post Construction Core Testing and Data Analysis for San Vicente Dam Raise Project", USSD Annual Conference, April 2017, Anaheim, California. [to be presented]

2. M. Rogers, A. Oleksyn, G. Tarbox, T. Reynoldson, J. Reed, W. Griffis, J. Zhou, J. Stiady, "Post-Construction Coring Evaluation of RCC at the San Vicente Dam Raise Project", Proceedings of ASDSO Annual Conference, September 2016, Philadelphia, Pennsylvania.

3. M. Rogers, "San Vicente Dam-site Inspection at WSEL 680 Hold Point", Memo, September 2015 (unpublished)

4. M. Rogers, "San Vicente Dam-site Inspection at WSEL 710 (Fill Hold Point)", Memo, February 2016 (unpublished)

5. MWH, "Technical Memorandum No.4, RCC Dam Design", Final Design, October 2010 (unpublished).

FIRST FILLING INSTRUMENTATION MONITORING OF THE FOLSOM DAM AUXILIARY SPILLWAY CONTROL STRUCTURE

Kenneth R. Pattermann, Geotechnical Engineer (PE, GE)[1]
George Hu, Geotechnical Engineer (PE)[2]

ABSTRACT

The Folsom Dam Joint Federal Project (JFP) consists of an auxiliary spillway concrete control structure that contains six (6) tainter gates that measure 23 x 36.5 feet each. The control structure consists of six monoliths with a maximum height of approximately 150 feet and a crest length of 350 feet. The control structure was placed on mostly slightly to unweathered lightly-jointed quartz diorite in the foundation with two shear zones, and varying degrees of weathered rock at the abutments. A single-line grout curtain was installed that also included drains downstream of the grout curtain. Instrumentation consisted of vibrating-wire piezometers installed in fully-grouted boreholes in the foundation and abutments. Inclinometers, joint meters, and accelerometers were also installed at the site. The first filling began on the morning of 26 January 2016, and the filling plan consisted of pumping water from the reservoir into the interspace between the control structure and the upstream cofferdam. The interspace was loaded relatively quickly at an average rate of 1.6 feet per hour over a 24-hour period. At the end of this 24 hour filling, the head across the structure was approximately equal to 38 feet and increased to 56 feet by 4 February 2016. The filling and monitoring team was a multi-disciplinary group of engineers, geologists, and safety personnel that monitored the structure for 24 hours per day for the first 6 days, and a reduced schedule was implemented thereafter depending on performance. Hourly readings were collected for the first filling using the automated data collection system but readings from the foundation drains were collected manually. Piezometer water elevations were estimated before the first filling using seepage and drain efficiency analysis. The predicted readings were fairly close to those recorded. Visual observations combined with instrumentation monitoring lead to a successful first filling of the project.

INTRODUCTION

Construction of the Folsom Auxiliary Spillway Control Structure, known as Phase 3 of the project, was completed in September 2016, and the primary purpose of the project is to provide flood risk reduction to the City of Sacramento, CA and surrounding communities along the American river. Phase 4 of the project was completed in early 2017, which included the approach channel and spur dike, upper and stepped chutes, and stilling basin. The control structure is a reinforced concrete structure that is comprised of six independent monoliths, two flow-through (Nos. 3 and 4), and four non-flow through monoliths. The flow-through monoliths contain 3 tainter gates each that are at a sill elevation of 370 feet. This is 50 feet lower than the tainter gates on the main dam (el. 420 feet), which was completed in 1956. The crest elevation of the control structure is

[1] U.S. Army Corps of Engineers, Sacramento District, California, kenneth.r.pattermann@usace.army.mil
[2] U.S. Army Corps of Engineers, Sacramento District, California, george.hu@usace.army.mil

equal to 482.84 feet. All elevations are given in the North American Vertical Datum of 1988 (NAVD 88). Major project features are shown in aerial Photo 1 which was taken the day after the first filling.

Photo 1. Folsom Auxiliary Spillway - Aerial View Looking Downstream from Reservoir

CONSTRUCTION

The foundation of the control structure was excavated into extremely strong, slightly to unweathered, fine to medium grained, unfractured to slightly fractured quartz diorite (granodiorite). There is a shear zone (S-1) that that strikes approximately east-west but was over-excavated by an additional 15 feet below the grade beam below monoliths 3 and 4. The S-1 contains intervals of intense fracturing and shearing, strongly developed fracture cleavage, strong brecciation, gouge seams, variable weathering and variable degrees of clay and other alteration. The left abutment consists of mostly slightly to unweathered quartz diorite, but the right abutment contained some weathered rock in isolated seams or smaller areas.

Piezometers and inclinometers were installed in the foundation of the control structure, and drilled before the monoliths were constructed or poured. Drilling was conducted with several different rigs, CME (Central Mine Equipment) 45 & 55 models with HQ #10, 12, and 14 size bits, and a Mozell MX 600 with HQ #10 and #13 size bits. The drill rigs operated on rubber mats to provide a more level surface than on the hard jagged granite foundation as shown in Photo 2.

Photo 2. Piezometer Boreholes Being Drilled Into Monolith # 2 Foundation – Looking from the Top of the Left Abutment.

As per design, the piezometers were installed in fully-grouted boreholes using a mix ratio of 2.5:1:0.35, water:cement:bentonite. Each borehole contained 2 piezometers typically 30 feet apart vertically, with a redundant piezometer 3 feet below the original, for a total of 4 piezometers per borehole. Although the piezometers were designed to be placed at a specified elevation, it was recommended during construction by the geotechnical engineer that they be moved to the nearest rock joint in order to improve response time. The piezometers in the foundation were heavy duty vibrating wire instruments with armored cable and were brought up with the construction of the monoliths and terminated in the grout gallery. The piezometer cables were secured to rebar hooks and covered with plastic wrap as the concrete lifts were added and the structure built-up as shown in Photo 3. Piezometers and an inclinometer in the right and left abutments were installed after the structure was completed. The piezometer series numbers along with the monolith Nos. and quantities are shown in Table. 1.

Photo 3. Piezometer Cables and Inclinometer Casing - Being Brought Up with Monolith # 2 Construction - Looking Upstream from the Downstream Toe of the Monolith

Table 1. Piezometers per Monolith

Monolith #	Piezometer Series (Quantity per Monolith*)						
	PZ1-X	PZ2-X	PZ3-X	PZ4-X	PZ5-X	PZ6-X	PZ7-X
1	**4**						
2		**8**					
3			**8**	**8**			
4					**8**	**8**	
5							**8**

*Not Including Redundant Piezometers

A cross section of the control structure can be seen in Figure 1 along with piezometers PZ4-# and PZ5-# series. Monolith Numbers (Nos.) can be seen in Photo 4.

The left abutment includes piezometer PZ9-# series and has 1 set (2 piezometers at different elevations) upstream of the grout curtain and another set downstream of the grout curtain, for a total of 4 piezometers (not including redundants). The right abutment includes the PZ8-# and PZ10-# series. The PZ-10 series has 2 piezometers upstream of the grout curtain and 6 piezometers in 3 boreholes downstream of the grout curtain (not including redundants). The PZ-8 series piezometers includes 4 piezometers in 2 boreholes along the right abutment and are all downstream of the grout curtain. See Photo 4 for locations.

Four (4) in-place inclinometers were installed in the control structure and one (1) in the right abutment (IN5). Two inclinometers are short (32 feet total and 10 feet into the foundation) and terminate in the grout gallery (IN3 and IN4), and two run the full-height of the control structure and into the foundation (148 feet total) and terminate on the crest of the dam (IN1 and IN2). IN1, IN2, and IN5 locations are shown in Photo 4.

Photo 4. Abutment Piezometer and Inclinometer Locations and Monolith Numbers – Aerial View Looking from Left Abutment

The joint meters are vibrating wire sensors that were installed in concrete block-outs that are located at monolith joints 1-2, 2-3, 3-4, and 4-5 in the mechanical gallery at elevation 444 feet, and two at monolith joints 2-3 and 4-5 along the tainter gate access walkway at elevation 406 feet. Each block-out contains 3 sensors that measure displacement in 3 directions including axial, transverse, and vertical (x, y, z) as shown in Photo 5. The joint meters are connected to a manual switch box (on the far right in Photo 5) and to a field data logger (in the stainless steel NEMA 4 enclosure in Photo 5).

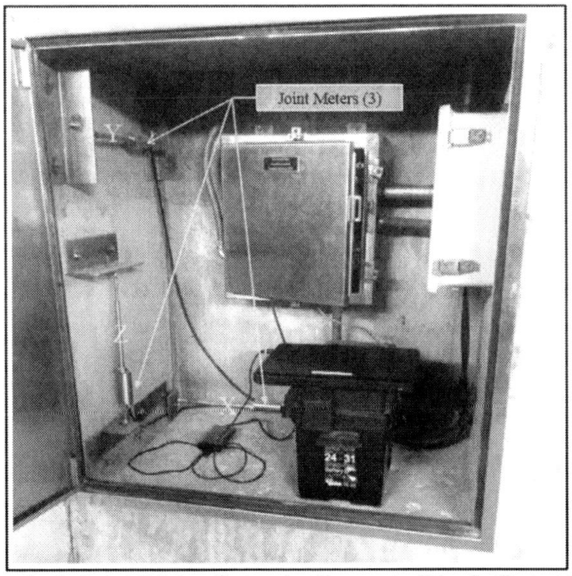

Photo 5. Three-Axis Joint Meter Setup in Concrete Block-Out at Monolith Joint

It's also interesting to compare these to the 1950s era joint meters installed for the old Folsom main concrete gravity dam shown in Photo 6.

Photo 6. Joint Meter Configuration used for Old Folsom Main Dam - 1950s era

Foundation and abutment drain boreholes were drilled from the grout gallery and stairwells and were angled 5 degrees in the downstream direction. The drain holes are spaced every 10 feet, are 60 feet deep and terminate in the grout gallery gutter where PVC riser pipes and v-notch weir boxes were installed (see Photo 7). Slotted PVC drain pipe was required in only 1 borehole in the right abutment that encountered some decomposed or weathered granite, but all the other drains are clean boreholes in rock. Existing small shrinkage cracks in the concrete floor were painted blue so that new ones could be easily detected (as shown in Photo 7). No new cracks were observed.

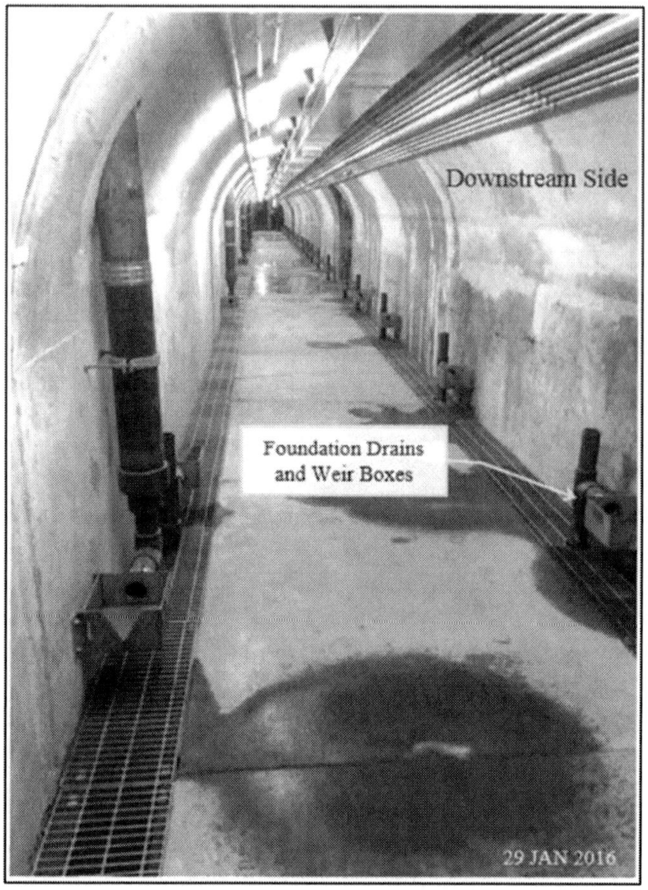

Photo 7. Grout Gallery Showing Foundation Drains and Gutters- Looking Towards the South End or Left Abutment

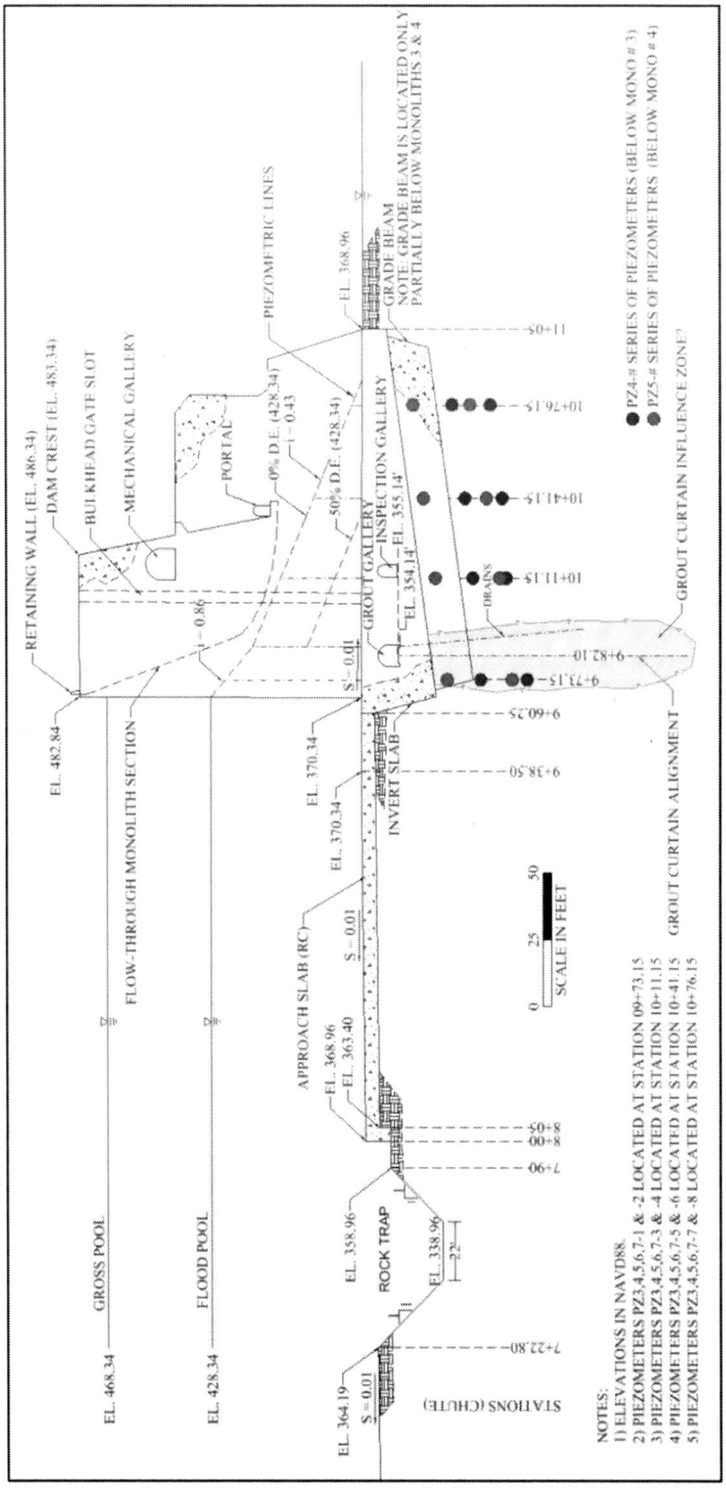

Figure 1. Flow-Through Monolith Cross Section with Piezometers

A three-dimensional (3D) view of the piezometric water elevation contours is shown in Figure 2. The contours represent the piezometric readings at the end of the first filling on 1 February 2016. It can be seen in Figure 2 how the piezometric water elevations drop to elevations 358 to 368 feet, which occur downstream of the grout curtain (red contours). There's a little higher water elevation in the left abutment piezometers (dark blue contours) due to the water table influence from the hills and valleys to the South of the control structure.

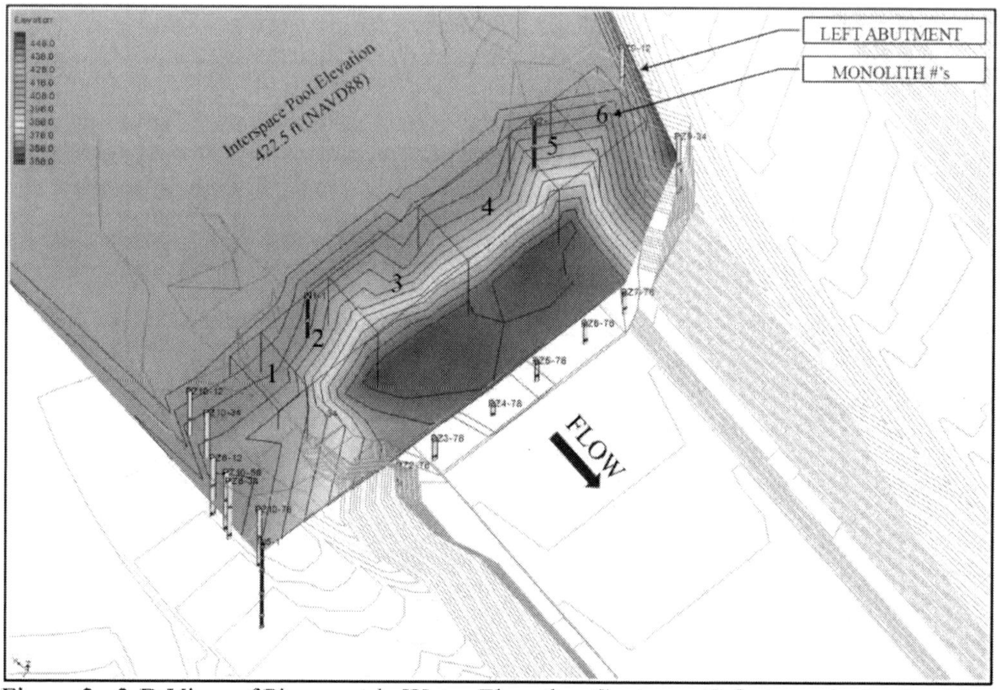

Figure 2. 3-D View of Piezometric Water Elevation Contours (5-feet spacing) – Looking From Above the Dam

FIRST FILLING

Team members developed a Filling and Monitoring Plan (Ref. 1) that was approved by the Sacramento District Dam Safety Officer and the JFP Chief of Engineering in December 2015. The first periodic inspection of the control structure was conducted in August of 2015.

The first filling began on 26 January 2016 by pumping water (from the reservoir) over the cofferdam and into the interspace between the cofferdam and the control structure. The pumping began about 7:15 am on the 26th and ended approximately 9:15 am on the 27th. At this time the water level in the interspace was about 1-foot higher than the reservoir elevation. The filling rate can be seen in Figure 3. The average rate the interspace pool rose over the first 24 hours was equal to 1.6 feet per hour. The interspace elevations were read from the upstream staff gage on the left approach wall.

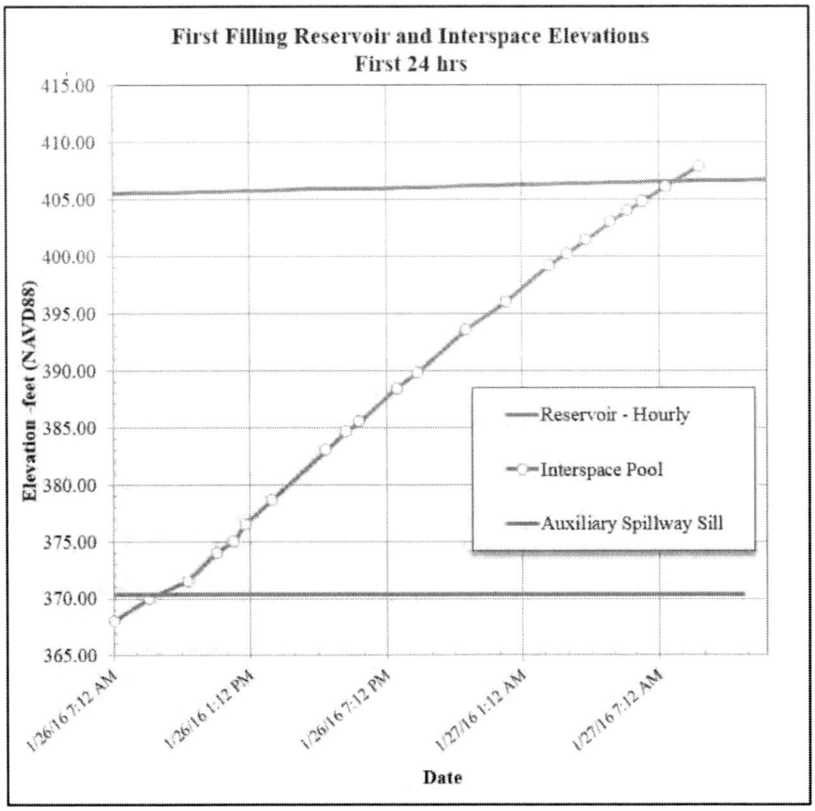

Figure 3. Interspace Filling Rate and Reservoir Elevation - First 24 Hours

The first filling team conducted 24-hour surveillance of the dam by using 3-teams per day each with 8-hour shifts. The team included geologists and mechanical, structural, and geotechnical engineers, instrumentation engineers and technicians, technical coordinators, and safety personnel.

The interspace between the control structure and the upstream cofferdam was scheduled to be filled within 24-hours. This was accomplished by using 8 portable pumps that moved water from the reservoir to the interspace through large HDPE pipes. The cofferdam was intentionally breached after the interspace was filled in order to keep the interspace at the same elevation as the reservoir. Reservoir water can be seen entering the interspace in Photo 8.

Photo 8. Reservoir Entering Interspace through Planned Cofferdam Breach – Looking from the Left Side of Approach Channel

To inspect the structure, the team used forms for taking inspection notes and manual readings, used the ADAS for automated instrumentation review, and also took photographs. The photographs were very helpful in examining any changes to the condition of the rock abutments and upstream concrete walls and riprap slopes. It is also very important to take baseline condition photographs before the first filling for comparison.

The control structure contains the following automated instruments:

i. Piezometers (vibrating wire) = 68 (not including redundants)
ii. Joint meters (vibrating wire) = 6 locations (3-axes each)
iii. In-Place Inclinometers (vibrating wire) = 5 Boreholes (23 sensors total)
iv. Force Balanced Accelerometers = 4

The piezometers, joint meters, and in-place inclinometers were set to read every 15 minutes during the first filling period, normally they would be set to read every hour. The ADAS computer is located in a stainless steel terminal enclosure. A staff gauge was installed on the approach channel wall and was very valuable during the first filling to determine the water elevation in the interspace. Reservoir elevation is available online.

The non-automated manually read instruments consists of the following:

i. Survey Monuments = 13
ii. Foundation and Abutment Drain Weir Boxes = 74
iii. Flow Meters (Sump Pumps) = 2
iv. Crack Monitors (Plastic) = 12

The piezometer response for Monolith No. 3 (PZ3-# and PZ4-# series) during the first 6 days of filling is shown in Figure 4. The piezometers upstream of the grout curtain (USGC) show a rapid response to changes in pool elevation. It can be seen that when the cofferdam was intentionally breached on 30 January, the USGC piezometers showed a rapid and paralleling response to the changes in the reservoir elevation. The difference in readings between the reservoir and the USGC piezometer water elevations is approximately 18 feet. The piezometers downstream of the grout curtain remained essentially the same.

The gradient across the grout curtain at the end of the 6th day, where the reservoir elevation was equal to 422.5 ft, as measured from PZ4-1 to PZ4-3, is shown in the following equation:

$$\mathbf{i} = \Delta H / \Delta L = (404 - 360) / (9 + 29) = 1.\mathbf{16} \quad \text{(Reservoir El. 422.5 ft.)}$$

where;
i = Hydraulic Gradient
ΔH = Piezometric Elevation difference between PZ4-1 & PZ4-3
ΔL = Horizontal Distance from PZ4-1 to PZ4-3

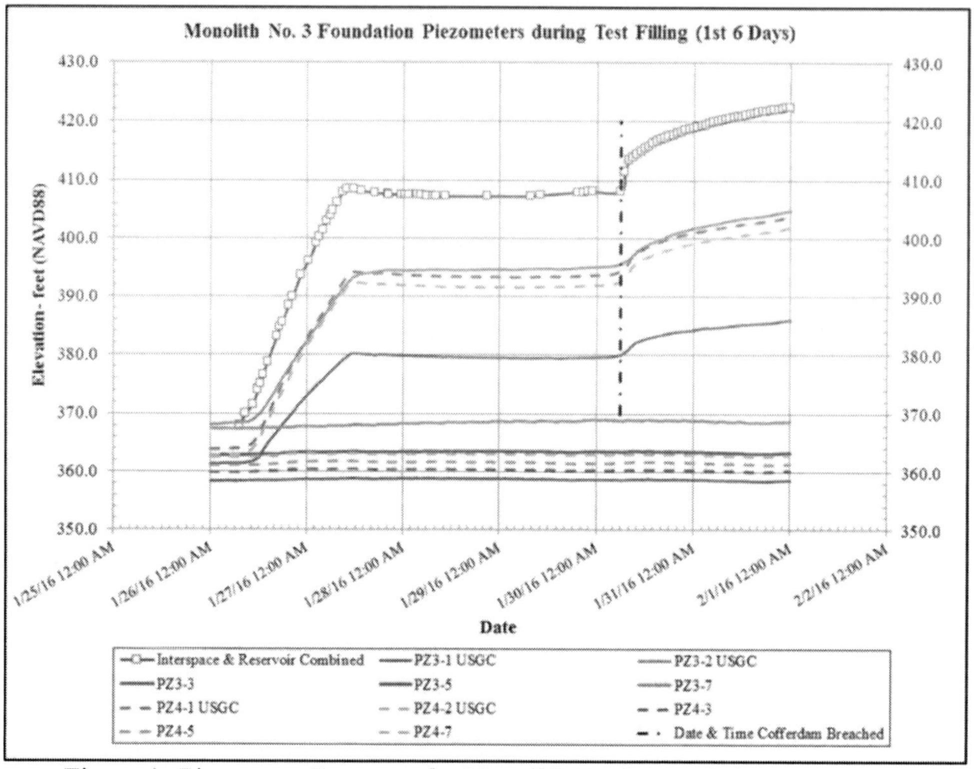

Figure 4. Piezometer Response for Monolith No. 3 Foundation (1st 6 days)

The piezometer response for Monolith No. 5 (PZ7-# series) during the first 6 days of filling is shown in Figure 5. The piezometers upstream of the grout curtain (USGC) show a rapid response to changes in pool elevation. Once the cofferdam was intentionally breached on 30 January, the USGC piezometers showed a rapid and paralleling response to the changes in the reservoir elevation. The difference in readings between the reservoir and the USGC piezometer water elevations is approximately 34 feet. The piezometers downstream of the grout curtain remained essentially the same.

The gradient across the grout curtain at the end of the 6th day, where the reservoir elevation was equal to 422.5 ft, as measured from PZ7-1 to PZ7-3, is shown in the following equation:

$$\mathbf{i} = \Delta H / \Delta L = (387 - 361.7) / (9 + 29) = \mathbf{0.67} \qquad \text{(Reservoir El. 422.5 ft.)}$$

where;
i = Hydraulic Gradient
ΔH = Piezometric Elevation difference between PZ7-1 & PZ7-3
ΔL = Horizontal Distance from PZ7-1 to PZ7-3

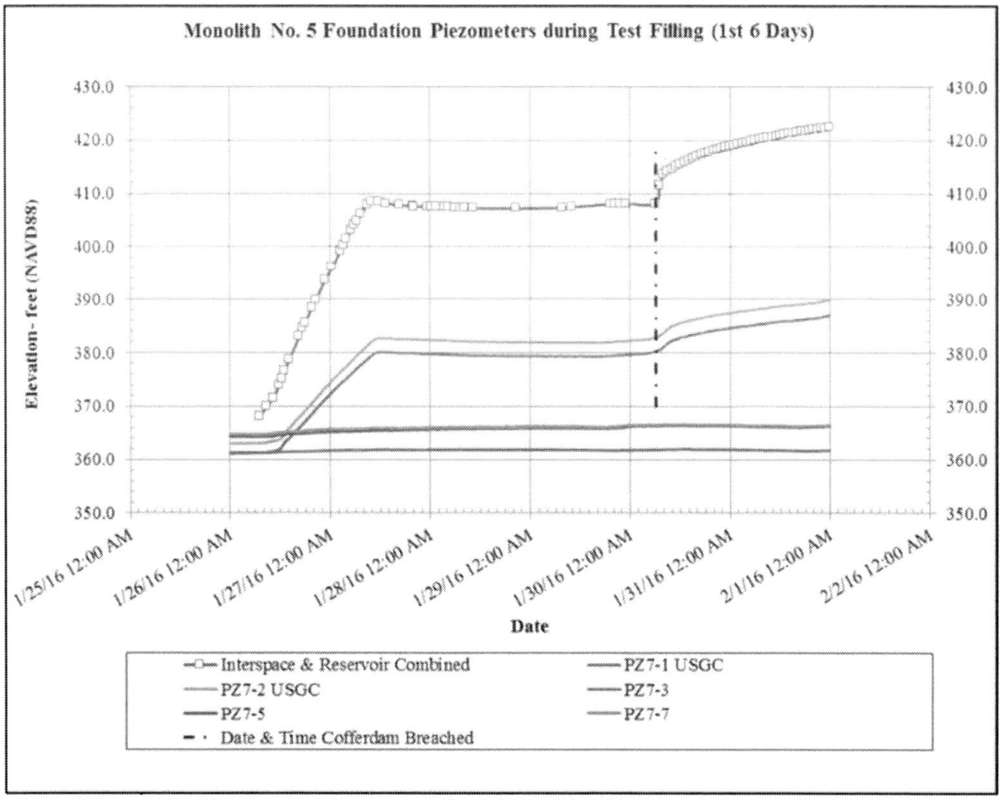

Figure 5. Piezometer Response for Monolith No. 5 Foundation (1st 6 days)

CONCLUSIONS

This report documents the conditions and readings of the instrumentation during the first six days of the first filling. All instruments have shown expected behavior and have been below maximum expected values. The piezometers upstream of the grout curtain showed a rapid response to reservoir elevation changes but piezometric water elevations were below maximum expected values. Piezometers downstream of the grout curtain showed no response to reservoir changes and were essentially flat.

Maximum expected piezometric water elevations for reservoir pools up to elevation 428.34 (flood pool elevation as shown in Figure 1), were determined to be 420 feet for the USGC piezometers (Ref. 1). The maximum reached for a reservoir elevation of 422.4 feet was equal to 405 feet for PZ3-2. Piezometers downstream of the grout curtain were predicted to have a maximum water elevation of 380 feet (Ref. 1), and all values were below this threshold.

Foundation drains and weir boxes showed very little flow (<0.1 gallons per minute) and were basically only dripping water during the first filling. This is certainly due to the tightness of the joints of the existing foundation rock and the effectiveness of the single-line grout curtain.

Inclinometers that run the full height of the control structure showed little movement. The maximum deflections were in the uppermost sensors that were 5 feet from the crest of the dam which had a maximum range of movement of 0.07 inches, but could be due to inherent instrument error. The difference in readings between the beginning and the end of the first filling was only 0.01 inches. The sensors in and just above the foundation of the dam showed no movement. The short inclinometers in the grout gallery, IN3 and IN4, showed no movement. The uppermost sensor (at elevation 432.9 feet) for inclinometer IN5 in the right abutment, showed a range of movement of 0.15 inches, but the difference in readings between the beginning and the end of the first filling was 0.0 inches. It is interesting to note that one of the inclinometer sensors near the crest of the control structure showed excessive movement before the first filling of up to 15 inches, which was not realistic based on visual inspection. This sensor was replaced and has since functioned normally.

Joint meters displacements for all sensors in the transverse direction showed a maximum displacement of 0.05 inches, essentially no movement in the axial direction, and a maximum of -0.07 inches in the vertical direction.

If there were any problems encountered with the new control structure or unstable conditions with the abutments or foundation during the first filling, the interspace could have always been emptied by opening the gates temporarily and then measures could be taken to correct the deficiencies. Fortunately, the first filling was a success and the project continues to operate well even at higher reservoir elevations for over a year now, thanks to the efforts of a great team.

REFERENCES

U. S. Army Corps of Engineers, "Filling and Monitoring Plan, Joint Federal Project (JFP), Folsom Auxiliary Spillway Control Structure," Folsom CA, Sacramento District, December 2015.

EFFECT OF ANTECEDENT GROUNDWATER LEVEL ON PIEZOMETRIC RESPONSE DURING FLOOD EVENTS

Chun-Yi Kuo, Ph.D., P.E.[1]
Richard B. Hockett, P.G.[2]
Troy S. O'Neal, Ph.D., P.E.[3]

ABSTRACT

Moose Creek Dam (MCD) experienced two back-to-back flood loadings in 2014. Although the floods were only moderate, higher than previously predicted piezometric responses and escalated boil activity were observed downstream of the stability berm. Assessment of piezometric responses indicates that the majority of the elevated uplift pressure was likely due to the back-to-back flooding conditions. The first flood raised the regional ground water table in the MCD project area, and that the second flood occurred before the first pool was fully drained. As such, the piezometric response reacted with greater magnitude during the second inundation. The 2014 floods induced even higher uplift pressure than the 1992 record pool, although the 2014 pool load was significantly lower than the 1992 record pool. In addition, the decommissioning of 31 non-functioning relief wells in permafrost areas appears to also contribute to the development of artesian pressure and boils downstream of the stability berm. The 2014 floods demonstrate that for dams founded on thick pervious material and with relatively shallow ground water level, antecedent ground water level could have significant effect on piezometric response during flood events and should be considered in risk assessment for more serious flooding.

INTRODUCTION

Piezometers are important in monitoring dam performance during routine operations and during flood events. Although over 100 working piezometers are present, the piezometric data at Moose Creek Dam (MCD) is very limited because the dam is rarely loaded. To understand and analyze piezometric response, "Piezometer Response Ratio" was utilized for meaningful analysis at MCD (Kuo et al. 2015). In 2014, MCD experienced two back-to-back floods which was the first operation since the 2003 flood season that the floodway was being inundated. Although the 2014 flood loadings were moderate relative to the 1992 pool of record, higher ground water level (and thus higher uplift pressure) and escalated boil activity were observed downstream of the stability berm. The elevated piezometric response and the escalated boil activity require assessment of the piezometric response during the 2014 floods so that the piezometric response to future pools can be accurately predicted and the dam can be adequately protected from seepage-induced distress.

[1] Geotechnical Engineer, Louisville District, US Army Corps of Engineers; chun-yi.kuo@usace.army.mil
[2] Geologist, Louisville District, US Army Corps of Engineers; Rick.B.Hockett@usace.army.mil
[3] Dam Safety Program Manager, LRD, US Army Corps of Engineers; troy.s.oneal@usace.army.mil

MOOSE CREEK DAM

The 7.5-mile long Moose Creek Dam is located upstream of Fairbanks, Alaska and reduces flooding on the Chena River by temporarily impounding and diverting floodwater to the much larger Tanana River (Figure 1) when the normally-open flood gates are closed during operations. The dam is constructed over a broad valley filled with mostly clean sand and gravel. Surficial material is predominantly 0 to 20 feet of non-plastic inorganic silt and sandy silt overlying the thick layers of sand and gravel. The silt and sandy silt were excavated within the embankment footprint during construction. There are localized areas of permafrost which have likely degraded (thawed) since construction. The depth to schist bedrock ranges from the ground surface at the north abutment to over 600 feet under most of the dam (O'Neal and Kuo, 2015).

Figure 1. Operational Schematic and Aerial View of Moose Creek Dam

Construction of MCD began in 1975 and the dam became operational in 1979. Initial test filling reached an elevation of 506.5 feet in July 1981 and the record pool occurred in June 1992, reaching an elevation of 512.7 feet. All elevations are referenced to North American Vertical Datum of 1988 (NAVD88) unless otherwise noted. During flood events, ground water levels in the piezometers are measured twice daily. MCD is typically a dry dam and only eight historical high pool events have loaded the dam longer than 3 days (Figure 2). As a result, even though there are over 100 piezometers, the data are limited. Figure 3 shows hydrologic loadings to be expected at Moose Creek Dam.

MCD can be subjected to rapid pool loadings. Sand boils have developed at the downstream toe of the dam during prior loading events. Since the 1981 test fill, remedial measures were constructed consisting of increasing the lateral extent of the upstream impermeable blanket and adding 188 relief wells. The original design of the embankment incorporated a constructed upstream silt blanket of varying width and thickness in areas where natural silt cover was less than three feet thick. The upstream silt blanket was extended to form a continuous section at least 600-foot wide with a minimum thickness of 3 feet along the length of the embankment in 1982-1983. In addition to six relief wells constructed before 1981, more than a hundred relief wells were added during 1982-1983, 1986, 1994, and 1997. The incremental installation of remedial measures complicates the interpretation and extrapolation of piezometric response due to their control on ground water levels during flood events.

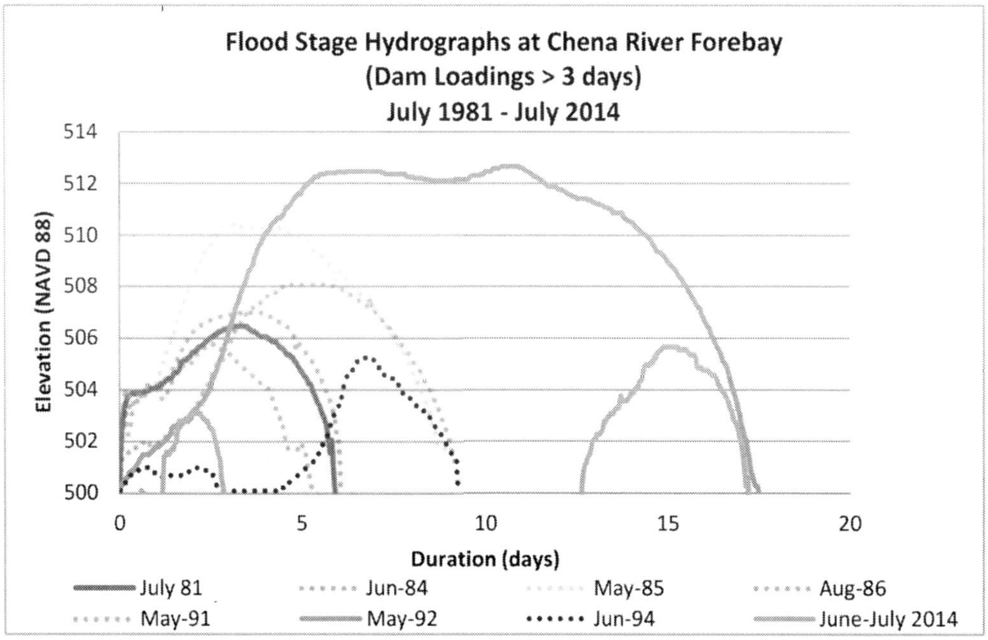

Figure 2. Pool Elevation and Duration of Major Reservoir Events at Moose Creek Dam

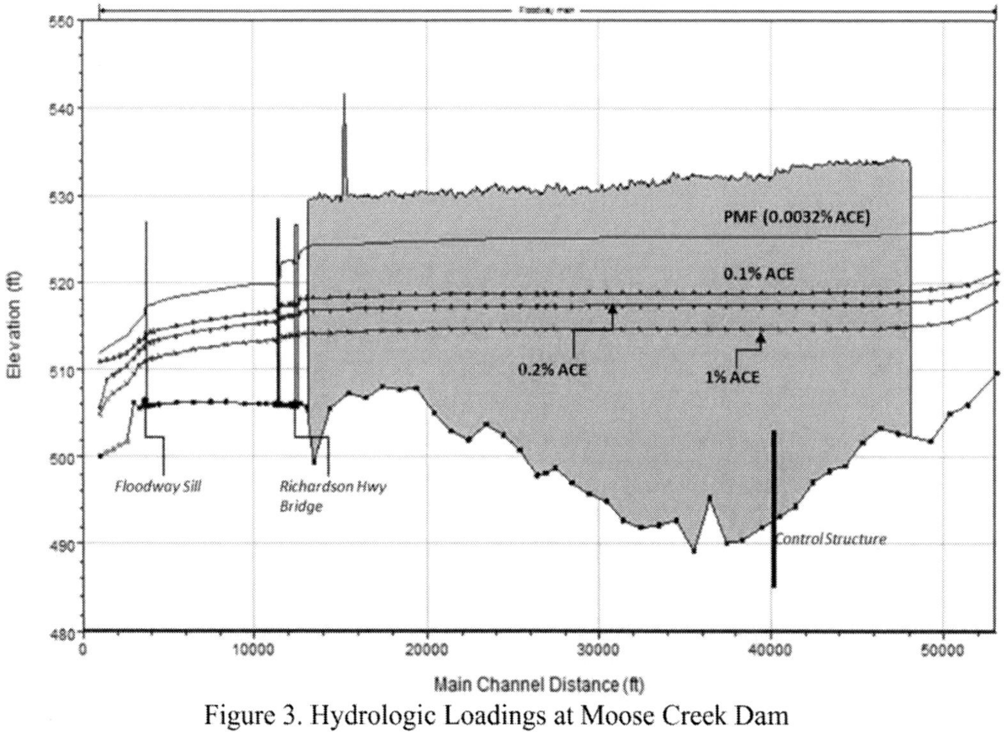

Figure 3. Hydrologic Loadings at Moose Creek Dam

2014 FLOODS

MCD was designed to restrict flows in the Chena River in downtown Fairbanks to 12,000 cubic feet per second (cfs) or less. Current operating procedures limit the discharge from the dam to 8,500 cfs to limit flooding in the Chena River between the dam and downtown Fairbanks. Two rainfall events in 2014 (late June, 24-hour maximum rain total of 1.10 inches; and early July, 24-hour maximum rain total of 2.83 inches) resulted in impoundment of floodwaters at MCD. Gates closures successfully restricted flow and prevented flooding in Fairbanks.

On June 21, the discharge through the dam with the gates in the normal, open configuration was 8,030 cfs. The monitoring team decided to partially close the gates until June 23, resulting in a maximum pool elevation of 503.1 feet, and a maximum discharge in downtown Fairbanks of 9,110 cfs. A second rainfall event resulted in a discharge through the dam of 7,930 cfs on July 2. The gates were partially closed again until July 7, resulting in a maximum pool elevation of 505.7 feet, and a maximum discharge in downtown Fairbanks of 11,000 cfs. Figure 4 shows the extent of the pools during the 2014 flooding. The pool elevation at 503.1 feet is shown in blue and the pool elevation at 505.7 feet is shown in purple. The 2014 pools did not load the entire length of the dam and no flow into the Tanana River occurred because the pool elevation did not exceed the highest floodway elevation between the outlet works and the Tanana River (Figure 3).

Figure 4. Extent of pools for the 2014 flood season. Light blue shows maximum June pool, purple shows maximum July pool.

Moose Creek Dam experienced surface erosion, boils, depressions, and seepage distresses as a result of the 2014 flooding. A total of 1,434 boils were found downstream of the stability berm, with the majority (57%) of the boils measuring 0.5 feet to 2.0 feet in diameter. small percentage of boils (2%) were 4 feet or greater in diameter. Figure 5 show typical small and medium cone boils. Piezometers in locations where a pool developed were measured and recorded at least once a day. This provided valuable data on how pore waters reacted within the dam and downstream. The data also showed that artesian pressures were typically found in permafrost areas, corresponding to the location of the majority of boils found downstream of the stability berm.

Figure 5. Typical small and medium cone boils with throats from 0.5 inches to three inches.

The 2014 flood season was the first time since the 2003 flood season that the floodway was inundated with a pool, although the inundation was partial. As can be seen from Figure 2, the 2014 flood loadings were less than the 1992 pool of record and other previous major events. The escalated boil activity from the two 2014 events was however unusual and it requires further assessment of piezometric response. The project team faced a number of new questions: What causes the relative high ground water level and escalated boil activity as compared to previous floods? Is this an indication of dam deterioration? Or are the elevated piezometric response and the escalated boil activity caused by different hydraulic loading characteristics that are different than experienced during previous floods?

PIEZOMETRIC RESPONSE AND ITS ASSESSMENT

Piezometers are the primary instrument used to monitor, measure and evaluate the performance of the embankment, foundation, horizontal and toe drains, upstream impervious blanket, and relief wells at Moose Creek Dam. Some of the key indicators to determine the performance of the dam are whether or not the pore water pressures react with the same delay (lag after pool loading) along the length of the embankment, and excessive embankment pressure. When properly used, piezometers are a very effective instrument in determining soil pore water pressure and phreatic surface. The observed sand boils discussed in the previous section are located in areas that exhibited artesian pressure. Sand boils and artesian pressures are not always an indication of a dangerous situation; however, it does indicate areas that would benefit from additional pressure relief or a larger resisting force such as thicker/wider downstream berm.

Throughout the history of Moose Creek Dam operations, many open tube piezometers have been installed and/or decommissioned. During the 2014 flood event there were 149 functioning open tube piezometers through the crest of the dam, along the embankment toe, and in selected downstream areas. Efforts were made to read the open piezometers twice a day, and data were collected throughout the event. Piezometer pipe top (lip) elevations change from time to time due to frost heave; therefore, techniques were employed to take reading from the ground surface to the water level for each data reading. All piezometer data for the 2014 flood season are presented in Appendix B of *Moose Creek Dam – After Action Report for 2014 Flooding* (USACE, 2015).

Table 1 lists the peak piezometric readings during 2014 flooding as recorded from piezometers located at the toe of the stability berm along Moose Creek Dam. For comparison, the peak piezometric readings during the 1992 pool of record (POR) are also listed. The maximum pool elevation during 2014 floods is 505.7 feet and the maximum pool elevation for the 1992 POR is 512.7 feet.

It can be seen from Table 1, P-380 and P-406 which are located north of control works where the dam was not inundated during the first 2014 flood (a maximum pool elevation of 503.1 feet) have peak readings less than those recorded during 1992 POR as expected. All the other piezometers, except P-452, where the dam was inundated during both 2014 floods (between Stations 262+00 to 400+00, approximately) have higher peak readings than those recorded during 1992 POR, even though the maximum pool elevation is only 505.7 feet as compared to the POR of 512.7 feet. It is very likely that the elevated piezometric response was due to the flooding conditions that occurred during the 2014 flood events. The first flood, with regulated flow between June 21 and 23, saturated the ground in the Moose Creek Dam project area. At the initiation of the second regulated operation on July 2, June's pool had not been fully drained. Therefore, the piezometric response associated with the project reacted with greater magnitude during the second inundation.

Table 1. Peak Piezometric Readings during 1992 & 2014 floods from Piezometers located at the Toe of Stability Berm

Piezometer	Station	2014 Peak Reading	1992 Peak Reading	2014-1992
P-425	277+05	499.1	497.8	+1.3
P-426	277+55	499.2	497.9	+1.3
P-428	278+20	500.0	497.9	+2.1
P-429	278+55	498.7	498.3	+0.4
P-430	279+05	502.1	498.5	+3.6
P-457	307+50	499.3	495.1	+4.2
P-415	321+00	497.7	494.4	+3.3
P-416	321+50	498.1	494.2	+3.9
P-417	322+00	497.3	494.3	+3.0
P-422	323+00	496.0	494.2	+1.8
P-424	324+00	495.0	494.3	+0.7
P-439	327+85	495.2	493.8	+1.4
P-438	328+85	495.3	493.8	+1.5
P-437	329+85	495.4	493.7	+1.7
P-436	329+90	496.4	494.1	+2.3
P-440	330+85	495.3	494.4	+0.9
P-275	345+10	497.7	494.5	+3.2
P-286	347+08	497.0	494.4	+2.6
P-278	347+12	497.4	494.3	+3.1
P-281	350+10	496.4	494.4	+2.0
P-284	352+10	498.3	495.0	+3.3
P-447	354+08	496.0	495.1	+0.9
P-450	359+48	496.9	495.8	+1.1
P-451	360+72	498.1	496.6	+1.5
P-452	361+98	494.6	496.1	-1.5
P-490	370+82	499.4	498.3	+1.1
P-492	382+40	500.2	499.9	+0.3
P-380	412+00	500.5	501.5	-1.0
P-406	492+22	499.9	500.6	-0.7

Similar conditions may have also occurred during the 1994 flood. As it has been discussed in Kuo et al. 2015: "The June 1994 flood was composed of two closely spaced floods (see Figure 2). The first flood, peak at 501 feet, stayed for about 3 days and the second flood, occurred two days later, peak at 505.3 feet. As a result, the Piezometer Response Ratios during the 1994 floods appear to be relatively high because the antecedent ground water for the second 505.3 feet flood was higher (due to the first 501 feet flood) ....".

The concept of "Piezometer Response Ratio" has been utilized for meaningful piezometer data analysis at Moose Creek Dam (Kuo, et al., 2015). "Response Ratio" of a piezometer during a flood event is defined as the ratio of the peak rise of the ground water level in the piezometer to the peak rise of the flood pool, both with respect to the antecedent groundwater level (Figure 6).

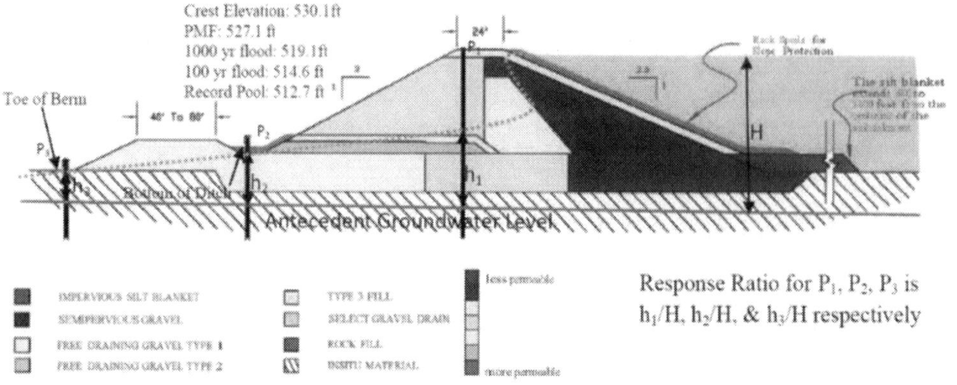

Figure 6. Illustration of Typical Dam Section & the Corresponding Piezometer Response Ratio

The projected piezometric responses based on "piezometer response ratio" if the first small but closely spaced flood was also taken into account are listed in Table 2. In this approach, the projected piezometric responses after first flood of 503.1 feet are the new antecedent ground water for the second flood (elevation of 505.7 feet) for projecting the peak response after second flood.

Recall the definition of response ratio which is the ratio of the peak rise of the piezometer to the peak rise of the flood, both relative to the antecedent groundwater.

$$RR = h_{flood1}/H_{flood1} = (PZ_{flood1} - GW_A)/(Flood1 - GW_A) \quad (1)$$

$$PZ_{flood1} = GW_A + (Flood1 - GW_A) \times RR \quad (2)$$

$$PZ_{flood2} = PZ_{flood1} + (Flood2 - PZ_{flood1}) \times RR \quad (3)$$

Where Flood1: Peak elevation of flood 1 (503.1 feet)
Flood2: Peak elevation of flood 2 (505.7 feet)
GW_A: Antecedent ground water level (reported by USACE, 1995, CRCP-PR-28)
RR: Response ratio (as determined in Kuo et al, 2015)
PZ_{flood1}: Peak piezometer reading during flood 1
PZ_{flood2}: Peak piezometer reading during flood 2
h_{flood1}: Peak rise of the piezometer during flood 1
H_{flood1}: Peak rise of the pool during flood 1

As shown on Table 2 above, the response ratio approach does predicts that the two smaller, back to back 2014 floods could result in higher piezometric response than the 1992 pool of record. The projection predicts a slightly higher (up to approximately 1 foot higher) piezometric response for the back to back 2014 floods if the antecedent ground waters and the response ratios are assumed to be the same for both the 1992 pool of record and the 2014 first flood event.

Table 2. Projected Peak Piezometric Readings during 2014 floods from Piezometers located at the Toe of Stability Berm

Piezometer	Station	Response Ratio	Antecedent Ground Water	Projected Peak After First Flood	Projected Peak After Second Flood	Peak Reading	1992 Peak Reading
P-425	277+05	0.33	491.2	495.1	498.6	499.1	497.8
P-426	277+55	0.34	491.2	495.2	498.8	499.2	497.9
P-428	278+20	0.34	491.2	495.2	498.8	500	497.9
P-429	278+55	0.34	491.2	495.2	498.8	498.7	498.3
P-430	279+05	0.37	491.2	495.6	499.3	502.1	498.5
P-457	307+51	0.22	490.5	493.3	496.0	499.3	495.1
P-415	321+00	0.2	490.1	492.7	495.3	497.7	494.4
P-416	321+50	0.2	490.1	492.7	495.3	498.1	494.2
P-417	322+00	0.2	490.1	492.7	495.3	497.3	494.3
P-422	323+00	0.19	490.1	492.6	495.1	496	494.2
P-424	324+00	0.2	490.1	492.7	495.3	495	494.3
P-439	327+85	0.19	489.6	492.2	494.7	495.2	493.8
P-438	328+85	0.19	489.6	492.2	494.7	495.3	493.8
P-437	329+85	0.18	489.6	492.0	494.5	495.4	493.7
P-436	329+90	0.19	489.6	492.2	494.7	496.4	494.1
P-440	330+85	0.2	489.6	492.3	495.0	495.3	494.4
P-275	345+10	0.25	488.9	492.5	495.8	497.7	494.5
P-286	347+08	0.25	488.7	492.3	495.7	497	494.4
P-278	347+12	0.25	488.7	492.3	495.7	497.4	494.3
P-281	350+10	0.25	488.6	492.2	495.6	496.4	494.4
P-284	352+10	0.27	488.5	492.4	496.0	498.3	495
P-447	354+08	0.28	488.3	492.4	496.2	496	495.1
P-450	359+48	0.32	488.3	493.0	497.1	496.9	495.8
P-451	360+72	0.34	488.3	493.3	497.5	498.1	496.6
P-452	361+98	0.32	488.3	493.0	497.1	494.6	496.1
P-490	370+82	0.43	488.2	494.6	499.4	499.4	498.3
P-492	382+40	0.5	488.1	495.6	500.7	500.2	499.9

Note: Piezometers had much higher than expected 2014 peak readings were highlighted

Comparing the 2014 projections with the 2014 readings, in general, the response ratio approach makes a fair prediction for the peak responses, with the majority differences within 1 foot. Piezometric response at some stations, however, is much higher than what was experienced in 1992 record pool and is also higher than to be expected by the response ratio approach. These higher than expected piezometric response, are highlighted in Table 2. The possible cause for the elevated piezometric response is discussed as follows:

P-457 & P-275: Decommissioned Relief Wells

P-457 located at Station 307+51, recorded peak reading of 499.3 feet while it recorded peak reading of 495.1 feet in 1992 POR. It is reported that during the first June 2014 flood (elevation of 503.1 feet), a small number of boils had initiated early at station 307+00 and P-457 showed artesian pressure on June 22 (groundwater reading of 498.7 feet vs. ground elevation of 496.7 feet).

Piezometer P-457 is located in the area that experienced the largest concentrations of boils during the 2014 flood (Figure 7). This area has extensive permafrost, as shown in blue dashed circles in Figure 7, and has been an area of concern. Letter Report 21(USACE, 1987) stated "Flow to the relief wells at the toe of the stability berm is cut off. The overall result is excessive uplift pressure between stations 285+00 and 310+00 beneath the downstream toe of the dam with no flow to the relief wells except those between stations 300+00 to 303+00." Letter Report 21 recommended installation of 31 additional relief wells in the ditch between the stability berm and the main embankment from Station 279+00 through 309+00 to alleviate this condition.

Frost action ("jacking") has sometimes prevented relief wells from functioning as designed. Many of the relief wells had jacked to the extent that they were judged to be marginally functional or non-functional. In the fall of 1998 a down-hole video camera survey of all the relief wells at the project was performed. It revealed that a significant number of the wells had sustained damage. 31 frost-jacked relief wells at the toe of the stability berm from stations 291+10 through 308+25 were decommissioned in 1999. The decommissioning consisted of excavating around each well to a minimum depth of 5 feet or to the water table, removing any concrete cast around the well, cutting the well casing off at a minimum of 5 feet below grade, using a tremie technique to completely fill the well casing with bentonite grout, placing a three-bag cap of bentonite chips around and over the top of the well casing, then backfilling and compacting the excavation to grade. It is believed that the decommissioned relief wells around P-457 are most likely the reason behind its much higher piezometric response as compared to the 1992 record pool. P-430 located at station 279+00 also exhibited higher than expected readings. Relief Well 279+00 installed in the ditch did not flow at all during the 2014 floods. The ditch relief wells installed in 1986 only had either 22.5 feet or 45 feet total depth which were very shallow with respect to the 600 plus feet thick aquifer. It appears that the relief wells installed in the ditch between the stability berm and the main embankment were ineffective during the 2014 floods in reducing the uplift pressure at the toe of the stability berm in the permafrost area. P-275 located at station 345+10 also showed much higher piezometric response. The nearby relief wells (343+75, 344+15 & 344+75) have also been decommissioned as shown in Figure 7.

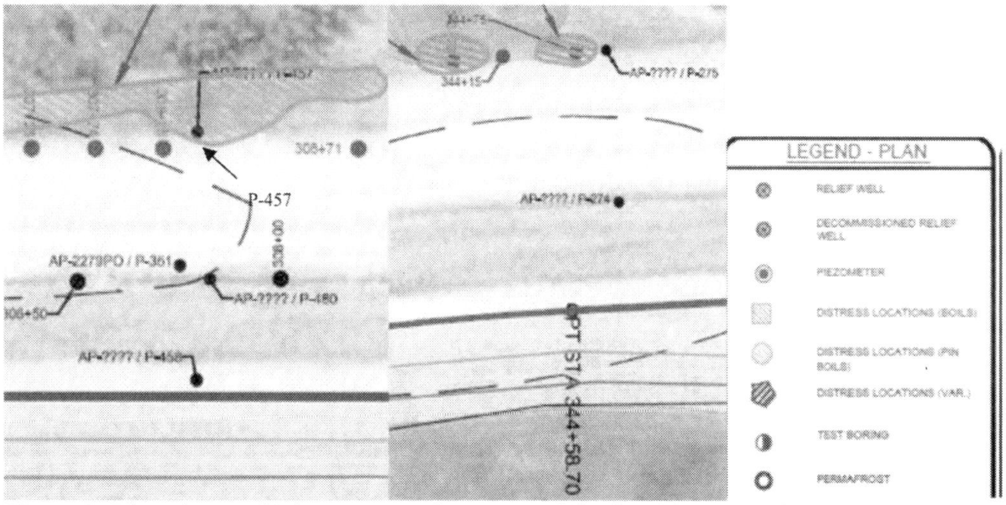

Figure 7. Plan view and Distress Locations around Station 307+00 and Station 344+50

<u>P-415, P-416, P-417 & P-436: Limited flow (ineffectiveness) of Relief Wells</u>

As shown in Table 2, P-415, P-416, and P-417 located near Station 321+50 and P-436 located near Station 329+50 also show much higher piezometric responses as compared to the readings during the 1992 pool of record and that predicted by response ratio approach. Figure 8 shows relief well flows during the 2014 flood events near stations 321+50 and 329+50. In the second event, flow was not measured only whether or not the wells were flowing was documented. It can be noted from Figure 7 that relief wells between 321+29 and 322+25 only have very limited flow during the events with much smaller flow rates as compared to the nearby Relief Well 320+85 as recorded in the first event. Similarly, relief wells 329+34 and 330+35 recorded much smaller flow rate as compared to the nearby relief wells 328+35 and 331+34. Though these relief wells are flowing, the suspected ineffectiveness of relief wells near stations 321+50 and 329+50 is likely the cause for the higher responses of the piezometers in these two areas.

<u>P-286, P-278, P-281, & P-284:</u>

Piezometers P-286, P-278, P-281, & P-284 located between stations 347+08 and 352+10 also exhibited water levels that are somewhat higher than expected, however, no boils were observed during the 2014 floods. These piezometers are located in the Low Point Drain area and currently no relief wells are present. Although no distress was observed, this area should receive special attention in future flood events due to higher than expected piezometric responses experienced in the 2014 floods.

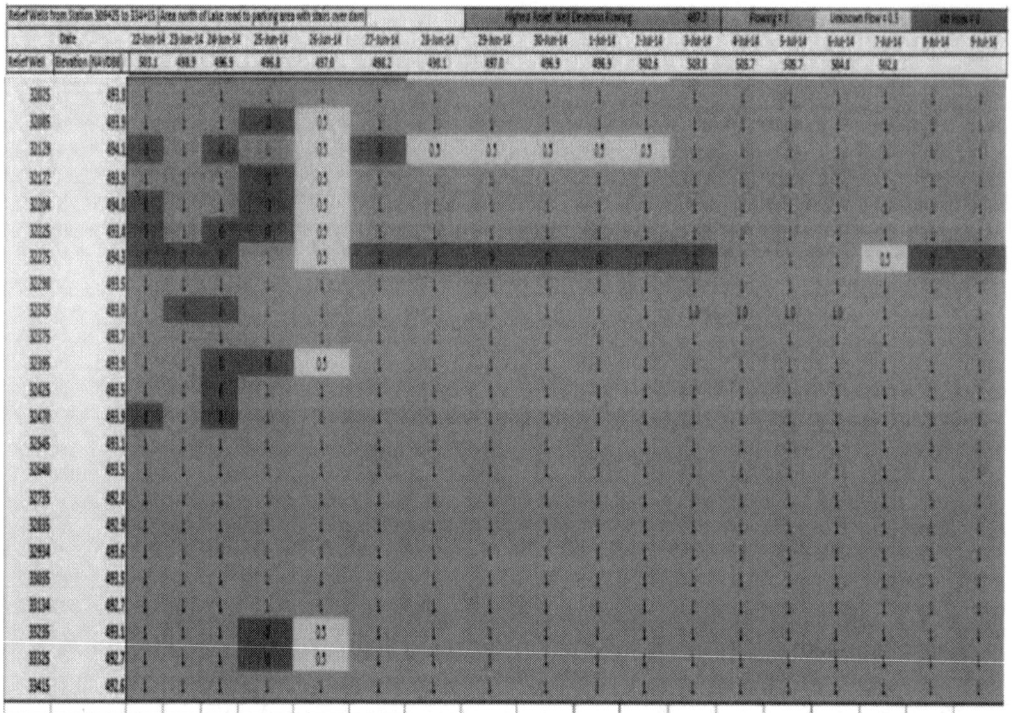

Relief Well	Elevation (NAVD88)	Flow Rate (gal/s)		
		22 June 2014	23 June 2014	24 June 2014
320+25	493.8	2.5	2.6	same
320+85	493.9	2.4	3.2	increase
321+29	494.1	0	0.04	0
321+72	493.9	0.2	0.7	increase
322+04	494.0	0	1.6	not observed
322+25	493.4	0	0.05	0
328+35	492.9	3.8	4.6	3.6
329+34	493.6	0.2	0.2	0.1
330+35	493.5	1.0	1.0	0.6
331+34	492.7	2.6	3.4	2.9

Figure 8. Record of Relief Wells near Station 321+50 and Station 329+50 during 2014 floods.

CONCLUSION

The 2014 floods at Moose Creek Dam resulted in numerous boils. The two back–to-back relatively moderate flood events caused escalated boil activity and higher than expected piezometric response at the toe of the stability berm as compared to previous floods when the single-event flood loadings are at about the same or higher magnitudes. Assessment of piezometric responses indicated that the elevated uplift pressure was likely due to the double-flood loading conditions. The first flood, with regulated flow between June 21 and 23, raised the regional ground water table in the Moose Creek Dam project area. At the initiation of the second regulated operation on July 2, June's pool had not been fully drained. Therefore, the consequent piezometric response reacted quickly and with greater magnitude during the second inundation.

It is demonstrated, through the utilization of piezometer response ratios, the 2014 floods could induced higher uplift pressure than the 1992 record pool due to the back-to-back flooding conditions. In addition, the 1999 decommissioning of 31 frost-jacked relief wells in permafrost areas from stations 291+10 through 308+25 also contributed to the development of artesian pressures and boils found downstream of the stability berm in 2014. Those relief wells, though functioning during 1992 record pool, were frost-jacked and were decommissioned in 1999. For piezometers located in the vicinity of decommissioned relief wells, the established response ratios based on pre-1999 flood events as demonstrated in Kuo et al. (2015) are not representative of current conditions and are likely underestimated. A few other piezometers that show higher than expected piezometric responses are located in areas with limited functioning relief wells, near stations 321+50 & 329+50, or areas with no nearby relief wells, between stations 347+08 and 352+10.

United State Army Corps of Engineers (USACE) requires consideration of an antecedent pool when routing the Probable Maximum Flood (PMF). However, the effect of antecedent ground water level is usually overlooked by the risk assessor in predicting the piezometric response for high pools such as a 1000-year flood or the PMF. For a site with thick pervious material, shallow antecedent ground water level and relatively thin surficial silt layer, back-to-back events could induce high uplift pressure as experienced at Moose Creek Dam. Extrapolation of piezometric response based only the single final pool magnitude may likely be underestimated.

REFERENCES

Kuo, Chun-Yi, O'Neal, T. S., and Hockett, R. B., 2015, Predicting Piezometric Response during High Loading Events Based on Limited Historical Data, United States Society on Dams Annual Conference, Louisville, KY.

O'Neal, T. S. and Kuo, Chun-Yi, 2015, Moose Creek Dam – Risk Assessment of an 8-Mile Long Dam Founded on Deep Sand and Gravel Deposits with Discontinuous Permafrost in Fairbanks, Alaska , United States Society on Dams Annual Conference, Louisville, KY.

US Army Corps of Engineers (USACE), 1987, Letter Report Number 21, Moose Creek Dam, Permafrost Effects on Seepage, Chena River Lake Project Test Filling, Alaska, Alaska District.

US Army Corps of Engineers (USACE), 1995, The Development and Application of Initial Hydrogeologic Flow Model for the Chena River Lakes Flood Control Project, CRCP-PR-28, Hydrologic Engineering Center, Davis, CA.

US Army Corps of Engineers (USACE), 2015, Moose Creek Dam – After Action Report for 2014 Flooding, Alaska District.

LEAK DETECTION IN EARTH DAMS AND LEVEES USING OPTICAL FIBER DISTRIBUTED TEMPERATURE SENSORS

Maxime Tatin[1]
Vincent Lamour[1]
Stéphane Bonelli[2]
Aurélie Garandet[3]

ABSTRACT

Internal erosion is a well-known menace for earth dams and levee safety. Internal erosion monitoring is achieved through identification and quantification of leakage flows inside the structure. Nowadays, the most promising method to detect leakage areas consists of thermal analysis of the dam embankment (water flow induces a change in the dam thermal field). Since the early 1990's, an innovative leakage monitoring system using optical fiber as a distributed temperature sensor has been emerging. As a distributed sensor, the optical fiber allows detection of early signs of internal erosion all along the structure.

This paper focuses on the deployment and exploitation of this new technology in a pathologic area of a levee located on the Rhône river in France. To measure temperature throughout a linear section of 400m, optical cables were installed in the levee downstream toe and in eighteen tubes distributed along the levee. The system also includes environmental sensors and has the benefit of being solar-powered.

The system is connected to a web interface displaying the data in real time, especially a two-dimensional diagram of the dam temperature field on which the leakage area is clearly identified. For advanced interpretation, a specific model based on the combination of a purely statistical approach (HSTP: Hydrostatic - Seasonal - Time - Pluviometry) and a physical-based approach (IRFA: Impulse Response Function Analysis) is employed.

INTRODUCTION

As a major menace for earth dam and levee safety, internal erosion has to be monitored through identification and quantification of leakage flows inside the structure.

Since the 1980's, thermometric methods are employed to detect leaks in earth dams (Armbruster et al, 1982), (Armbruster et al, 1989), (Johansson, 1991). These methods consist of thermal analysis of the dam body, based on the principle that a water flow induces a change in the dam thermal field. Without leakage, thermal transfers are induced by conduction in the dam body and the thermal field depends on the boundary conditions, mainly air and water temperature. In case of a leak, another thermal transfer mode is

[1] Cementys, 9 rue Léon Blum, 91120 Palaiseau, FRANCE, maximetatin@cementys.com
[2] Irstea, Unité de Recherche Ouvrages Hydrauliques et Hydrologie, 3275 Route de Cézanne, CS 40061, 13182 Aix-en-Provence Cedex 5, FRANCE
[3] Compagnie Nationale du Rhône, 2 rue André Bonin, 69316 Lyon cedex 4, FRANCE

appearing: advection. The flow induces a heat transport trough the porous material. If the flow is significant, the temperature near the flow path depends mainly of the water temperature. So, a thermal anomaly (temperature variation due to the water flow) corresponds with the leakage area and its near neighborhood.

Nowadays, thermometric methods based on fiber optic temperature sensors are considered as the most promising methods to detect leakage areas in earth dams. As a distributed sensor, the optical fiber allows detection of early signs of internal erosion all along the dam. It is in the 90's that the soundness and relevance of the method has been proved: in Sweden (Johansson, 1997), in Germany (Perzlmaier, 2007) and in Canada (Côté et al, 1991). In Sweden, the so-called "passive" method (Johansson, 1997) (Johansson, 2005) (Johansson et al, 2000) has been developed and consists in measuring the natural temperature along a horizontal optical fiber located in the saturated area. In Germany, it is mainly a spatial analysis of the temperature that is employed to detect an anomaly by comparison with a temperature profile in a known normal area (without leakage). Moreover, the so-called "active" method (Aufleger et al, 1998) has been developed in order to quantify locally the identified anomaly. It consists in artificially heating the soil near the optical fiber and to estimate the heat quantity carries away with the flow, which is directly correlated to the water flow rate. In Canada, temperature measurements are analyzed using a finite element model of the dam.

Within the scope of a R&D program aiming to produce tools and monitoring methods for leak detection inside earth dams, a pathologic section of a levee has been instrumented with optical fibers to complete the already in place monitoring system. The structure instrumented is a canal levee of a hydroelectric scheme owned by the CNR (Compagnie Nationale du Rhône) over the Rhône river in France (Figure 1). The section instrumented is 400 meters long and contains a 80 meter long area with potential leaks associated to internal erosion. Although, the levee length is several kilometers, this 400 meters section has been chosen because of the significant leakage flows observed at the downstream toe.

The already in place monitoring system includes:

- A gauging system equipped with a flowmeter
- Almost 30 vertical 5 to 10 meters deep tubes implanted in mesh over the 80 meter area for periodic manual piezometric measurements.

The new complementary and innovative monitoring system, enables leak detection using optical fiber as a temperature sensor and spatial analysis of the temperature with "active" and "passive" methods.

THE MONITORING SYSTEM

The new monitoring system installed on the levee, is composed of a THMLogger datalogger and a SensoLogger optical fiber interrogator. The system has the benefit of being solar-powered (Figure 1). Moreover the system is continuously connected to the web-based THMInsight software, providing a real time display of the data.

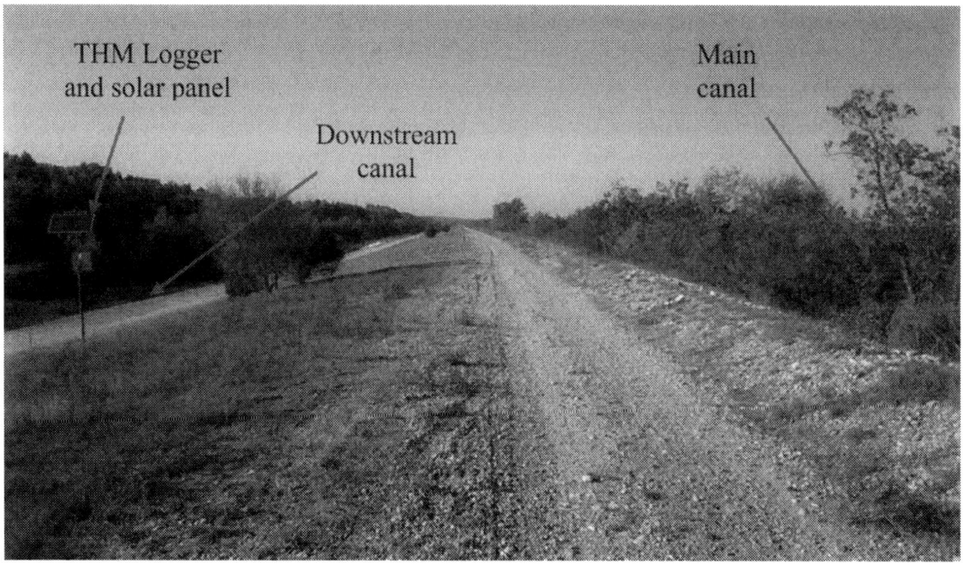

Figure 1: canal levee of a hydroelectric scheme owned by the CNR (Compagnie Nationale du Rhône) over the Rhône river. Mast with solar panel and THM Logger box.

The monitoring system includes several sensors (Figure 2):

- Distributed fiber optic sensors:

    - SensoLux Detect®: Distributed temperature sensors (Raman backscatter)
      Sixteen 12 meters depth thermometric tubes have been installed in the leakage area and two others have been installed in healthy area. All these tubes are instrumented with a SensoLuxDetect optical fiber cable and enable to produce a 2D diagram of the temperature in the leakage area.
      In the downstream toe, another optical fiber cable has been introduce in the soil at 50 centimeters depth in order to localize the leakage area with the longitudinal profile of the temperature.

    - SensoLux TM®: Distributed strain measurement (Brillouin backscatter) in the top the levee to observe the settlement of the ground.

- PT100 sensors: temperature measurement in the thermometric tubes in order to correct and calibrate the optical fiber measurements.

- TDR sensors: water content measurement in the soil around the leakage area to control the leakage evolution.

- Vibrating wire sensors: water level measurements in order to feed the statistical analysis.

- Meteorological sensors: environmental parameters measurements (atmospheric pressure, air temperature wind speed, humidity, rainfall) in order to the feed the statistical analysis.

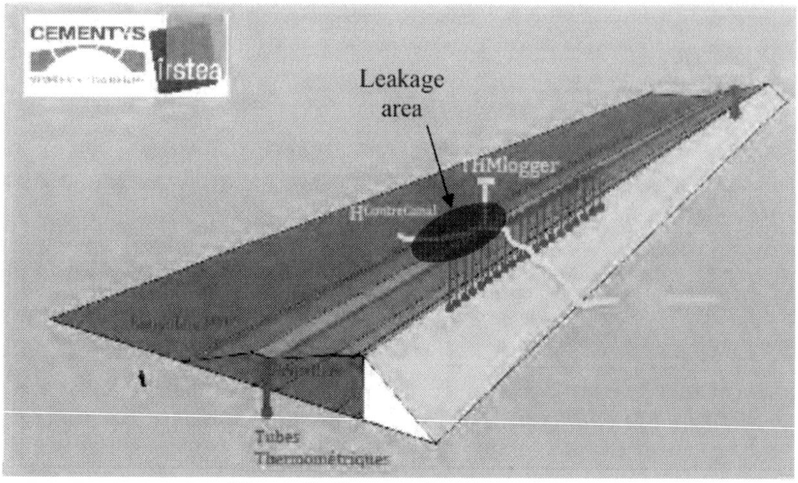

Figure 2: Sensors implantation

Temperature measurements are performed continuously without heating the soil ("passive method"). During punctual campaigns (twice a year), the soil is heated around the fiber ("active" method") in the downstream toe. These "active" measurements enable a better identification of the leakage area by increasing the thermal contrast between areas in which only the conduction phenomenon occurs and areas in which both conduction and advection phenomena occur. Moreover, the "active" method enables to highlight an evolution of the flow rate, without giving it a precise value.

MEASUREMENTS ANALYSIS

The first analysis performed on the measurements data is a 2D diagram of the temperature inside the levee. This diagram represents the difference between the temperature profiles in the tubes in the leakage area and the temperature profiles of the two healthy tubes. Doing so, the contrasts obtained enable to detect the leakage area (Figure 3), which is colder in depth than the healthy tubes. Indeed, the diagram presented here has been obtained in winter, when the water temperature is colder than the soil at this depth. This 2D diagram and its evolution is displayed in real time on the web application THMInsight.

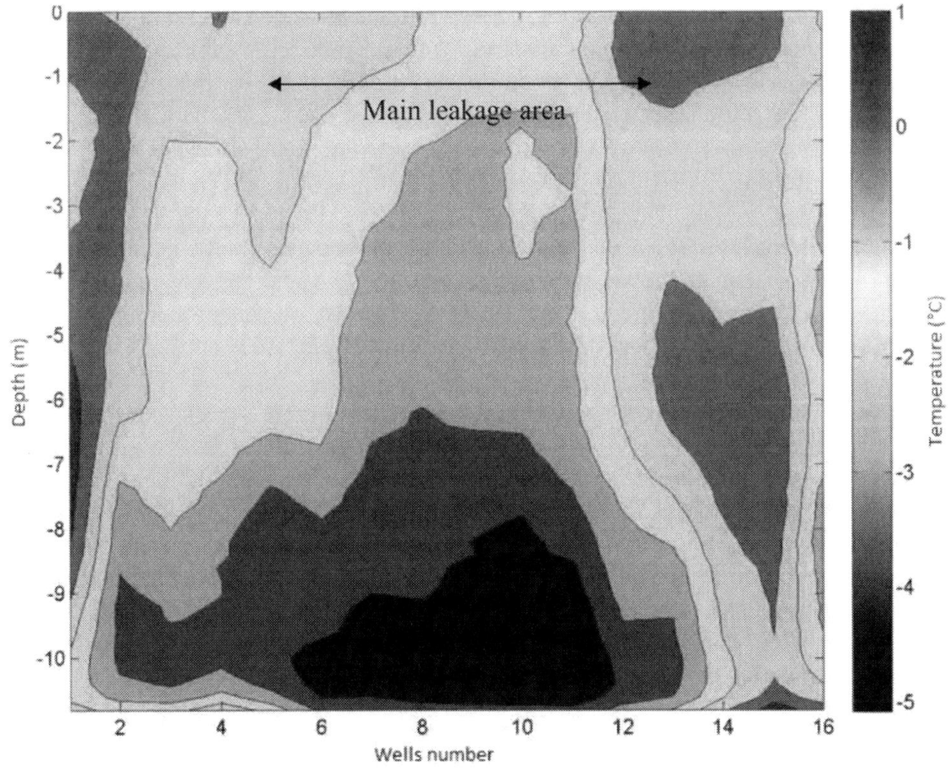

Figure 3: 2D diagram of the difference between the temperature profiles in the tubes and the temperature profiles of the two healthy tubes

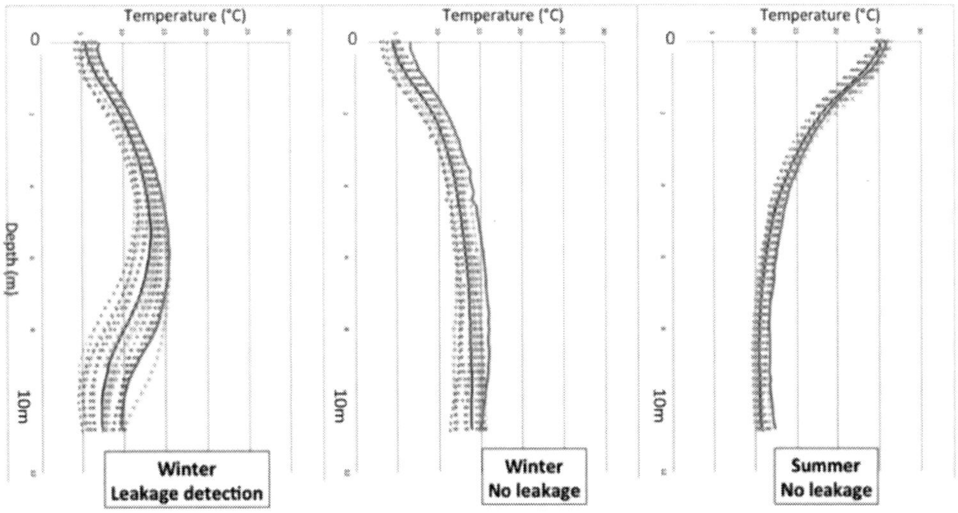

Figure 4: Vertical temperature profile for all standpipes.

The vertical temperature profile for standpipes are plotted in Figure 4. In winter time (January), the average air temperature is 5°C while the average soil temperature at depth in the foundation is 13 °C (middle curves). In summer time (July), the average air temperature is 25°C while the average soil temperature at depth in the foundation is 11 °C (right curves). Air temperature influences soil temperature only to a depth of about 2 m. Small seepage flows will not affect the seasonal temperature variation in the levee. This result in almost constant temperature beyond 2m depth (middle and right curves). In a zone subjected to a leakage, the temperature profile is affected by the heat transported by the concentrated seepage flow. In this case, the temperature profile is no longer constant in depth. This is shown in Figure 4 (left curves) where the seepage flows in the foundation are large (depth greater than 10 m). In this case, the heat transported corresponds to the water temperature of the reservoir several weeks previously.

CONCLUSIONS

This paper presents an innovative leakage monitoring system using optical fiber as a distributed temperature sensor, deployed in a pathologic area (internal erosion) of a levee located on the Rhône river in France. The specificity of the proposed system is that the optical cables were installed horizontally in the levee downstream toe, but also vertically in eighteen tubes distributed along the levee. This enables to plot a two-dimensional diagram of the dam temperature field which is displayed in real time on a web interface. The results show that the leakage area is clearly identified by the system.

REFERENCES

Armbruster H., Blinde A., Brauns J., Doscher H., Hotzl H., Merkler G., The application of geoelectrical and thermal measurements to locate dam leakage, *Detection of Subsurface Flow Phenomena*, volume 27, pages 31-47, 1989

Armbruster H., Merkler G., Measurement of subsoil flow phenomena by thermic and geoelectric methods, *Bulletin of International Association of Engineering Geology*, 26-27:135-142, 1982

Artières O., Bonelli S., Fabre J.-P., Guidoux C., Radzicki K., Royet P., Vedrenne C., Active and passive defences against internal erosion, in *Assessment of the Risk of Internal Erosion of Water Retaining Structures: Dams, Dykes and Levees, Intermediate report of the European working group of ICOLD*, N° 114, Essen: TUM, Technische Universität München, pp. 235-244, 2007

Aufleger M., Dornstadter J., Fabritius A., Strobl T., Fibre optic temperature measurements for leakage detection –applications in the reconstruction of dams, *66th ICOLD annual meeting, International Commission on Large Dams*, 1998

Côté A., Carrier, B., Leduc J. Gervais R., Water Leakage Detection Using Optical Fiber at the Peribonka Dam, *Seventh International Symposium on Field Measurements in Geomechanics,* Boston, 2007

Johansson S., Localization and quantification of water leakage in ageing embankment dams by regular temperature measurements, *17th International Congress on Large Dam*, pages 991-1005, 1991

Johansson S., Seepage monitoring in embankment dams, *PhD Thesis*, KTH, 1997

Johansson S., Detection of internal erosion in embankment dams-possible methods in theory and practice. Stability and Breaching of Embankment Dams, EBL, 2005

Johansson S., Farhadiroushan M., Parker T., Application of fibre-optic systems in embankment dams for temperature, strain and pressure measurements – some comparisons and experiences , *20th International Congress on Large Dam*, 2000

Perzlmaier S., Verteilte Filtergeschwindigkeitsmessung in Staudämmen. *PhD Thesis*, TU München, 2007

Radzicki K., Bonelli S. Physical and parametric monitoring of leakages in earth dams using analysis of fibre optic distributed temperature measurements with IRFTA model, *24th International Congress on Large Dam,* Kyoto, Japan, 6-8 June 2012

Radzicki K, Siudy A., Stolinski M., An innovative 3D system for thermal monitoring of seepage and erosion processes and an example of its use for upgrading the monitoring system at the Kozłowa Góra dam in Poland, *25th International Congress on Large Dam*, Stavanger, Norway, 13-20 June 2015

ELEPHANT BUTTE DAM UAV DATA COLLECTION AND PHOTOGRAMMETRY

Matthew Klein[1]

Recently the US Bureau of Reclamation completed an contractor-operated unmanned aerial system (UAS) data collection of Elephant Butte Dam near Truth or Consequences, New Mexico. The operation was performed in through support provided by Reclamation's Research and Development Office. The purpose of the project was to evaluate the usefulness of UAS for data collection.

The UAS collected HD video, high resolution digital images and thermographic infrared data. The aerial data collection platform was a commercially available single rotor heavy lift system with real-time video feed and GPS waypoint navigation.

The RGB and IR data collected was processed photogrammatrically to produce a 3D model of the structure. The model is used to provide spacial reference and evaluation for condition assessment. This allows for inspection of the facility regardless of weather conditions, negates the use of specialized access restrictions and equipment thereby increasing safety and reducing costs.

Combining the latest UAS technology with high quality optical sensors and software processing will offer increased analysis and inspection capabilities and is poised to become the inspection procedure of the future.

[1] Bureau of Reclamation, mjklein@usbr.gov

PROPOSED GUIDELINES FOR CONDUCTING A PERFORMANCE-BASED EVALUATION OF A CONCRETE DAM

Samantha Hoang[1]
Tae Ha Park[2]
Jonathan Lum[3]
Ziyad Duron, PhD[4]
Guy S. Lund, PE[5]

ABSTRACT

Performance-Based Evaluation (PBE) techniques are currently being developed to determine, monitor, and track the condition of concrete dams by interrogating the physical system of the dam itself. A PBE often includes Performance-Based Testing (PBT) of the structure in question, which involves collecting operational, maintenance, and forced-response data. Operational data are collected under normal operating conditions such as spilling, and forced-response data are obtained by using a Cold Gas Thruster (CGT) to induce large-amplitude, broadband, transient behavior in the dam by subjecting it to an impulse. The forced-response can be convolved to calculate the dam's response to earthquakes by taking advantage of broadband, linear analysis techniques. An important feature of a PBE is the development of Performance Indicators (PIs) that can be used to predict dam behavior. One example of this is fragility, which can be calculated directly from a dam's PBT responses. Overall, a PBE will provide dam owners with the ability to better monitor and track their dam's condition over time using PIs and also provide them with a better understanding of how the dam responds to hazard conditions such as earthquakes. The knowledge gained from PBE can improve confidence in the use and interpretation of complex, numerical models for predicting nonlinear behavior.

INTRODUCTION

In recent years, the evaluation of concrete dams has mainly focused on using complex numerical models to predict dam performance under extreme loading conditions. A recent failure of a large dam under its normal operating conditions highlights a need for a more complete understanding of how dams function and perform under both extreme and normal conditions. Since the early 1990s, advances in numerical techniques have greatly outpaced the development of field-testing techniques. This means that the results of these numerical techniques are not compared to observational data from the dam being modeled since there has been no data to compare with the model results. This paper will seek to contradict the notion that there is no field data available to validate numerical models and their results. Since 2012, Performance-Based Evaluation (PBE) techniques

1 De Pietro Engineering Fellow, Harvey Mudd College, Claremont, CA 91711, shoang@hmc.edu
2 De Pietro Engineering Fellow, Harvey Mudd College, Claremont, CA 91711, tpark@hmc.edu
3 De Pietro Engineering Fellow, Harvey Mudd College, Claremont, CA 91711, jlum@hmc.edu
4 Professor of Engineering, Harvey Mudd College, Claremont, CA 91711, duron@hmc.edu
5 Office Principal Engineer, Gannett Fleming, Englewood, CO 80112, glund@gfnet.com

have been developed and applied to large concrete dams in order to use the dam itself to produce information that will give insight into how a dam actually responds to extreme loading conditions and to validate numerical models and analysis techniques.

PBE consists of both Performance-Based Testing (PBT) and Performance-Based Analysis (PBA). As described in Goldkamp, et. al (2014), PBT is a technique that combines a field test and a numerical model to develop Performance Indicators (PIs) for a structure under certain loading conditions. PBA is used to predict how a structure will perform under various loading conditions in conjunction with the results from PBT. One example of the use of PBA is given in Edelman, et. al (2014), which describes the use of PBA on a multiple-arch concrete dam to predict performance under various loading conditions. Another important result from PBT is the ability to create fragility curves due to seismic loading conditions for a dam. These findings are described in Menda, et. al (2016). Together, these papers provide a basis upon which the guidelines for PBE can be developed.

This paper begins by reviewing the purpose of PBT as discussed by Goldkamp, et. al (2014) and then moves into a review of the different analysis techniques presented by Edelman, et. al (2014). Various PIs, such as the fragility curves described by Menda, et. al (2016), are then discussed in the context of how they can be developed from the results of PBT. Finally, the paper will include a sample proposal for the PBE of a concrete dam.

BACKGROUND

Alone, PBT can be used to derive performance indicators for a large structure, in this case a dam. PBT is based upon convolution which dictates that if the unit impulse response, $h(t)$, of a system is known, then the system's response, $y(t)$, to any input, $x(t)$, can be calculated using Equation 1 below.

$$y(t) = h(t) * x(t) = \int_{-\infty}^{t} x(\tau)h(\tau - t)\, d\tau \qquad (1)$$

The first step of performing a PBT is to measure the forced-response data of the dam. In addition to the forced-response measurements, it is also recommended that maintenance data and operational data, when possible, be recorded as well. Maintenance data refers to ambient data that are taken when there are no forcing inputs to the system. These data exhibit how the dam responds to the natural environment. It is important to examine maintenance data since they can give insight into how the maintenance staff can improve maintenance procedures for the dam. Operational data refers to the data collected when the spillways of the dam are allowed to spill. This data can reveal if any current operating configurations correspond to reduced safety margins for the dam. Based on the results of the PBE, analysts can then recommend a new operating plan for the dam that will not cause damage to the dam due to vibrations from the spillway gates.

Performance-Based Analysis Techniques

A large quantity of data is collected during PBT to ensure satisfactory data quality, and to allow flexibility in the analysis and interpretation of the PBT responses within the context of the dam's fundamental behavior and characteristics. In order to process the data, two approaches can be taken as described in Edelman, et. al (2014): the frequency-domain approach and the time-domain approach. These two approaches will yield different results since they each reveal different information about the dam.

The frequency domain approach can be used to determine the magnitude of the dam's response based on the frequency at which the response occurs. For instance, this approach can be used to examine how mechanisms or components in the dam behave. The response of features such as cracks in the concrete, connections between the monoliths of the dam, or anchors installed in the monoliths are examples of the mechanisms and components that can be evaluated using a frequency-based approach. If a monolith's frequency response appears significantly different from another, similar monolith's response, then the analyst can infer that there may be some damage in the monolith that is causing this difference and perform direct inspections based on the results of the analysis.

The time-domain approach can be used to examine the dam's overall behavior. In this approach the time history of the dam's response is examined for features such as the peak acceleration or displacement at the crest of the dam. The measured impulse response time histories can also be used to calculate the dam's approximate response to any earthquake. From the calculated earthquake responses, various properties of the dam's response can be calculated at each of the locations whose responses were measured. These properties of the dam's response can then be used to characterize which types of earthquakes are the most damaging to the dam. These two approaches allow the creation of various PIs for each dam dependent on what the dam owner is interested in measuring regarding the performance of in their dam.

Developing Performance Indicators

There are several common PIs that can be used to measure a dam's performance. Common PIs include fragility curves, base flexibility, and amplification from the bottom to the top of a monolith. Each PI is used for a different purpose. For example, fragility curves are used to determine what earthquake properties most strongly correlate to damage being inflicted on a dam. On the other hand, base flexibility examines how much the connection between the dam and foundation allows the dam to rock back and forth when an earthquake hits. This is important in determining how well connected the dam is to the foundation. Finally, amplification is used to determine how much the earthquake ground motion will magnify once it reaches the top of the dam. For dams that are taller a large amplification could indicate that the dam needs to be reinforced in order to withstand seismic loading conditions.

PROPOSED GUIDELINES FOR PERFORMANCE-BASED EVALUATIONS

There are many components that play a role in the success of a PBE. Before and during the PBT, several considerations must be taken into account, which include: the placement of the forcing input, the amount of force that should be used, the placement of the sensors that measure the dam's response, how to measure maintenance (ambient) data for this dam, and how to measure operational data for this dam, if it is applicable.

Force Location

The location of the forcing input is dictated by what the goals of the PBT are. When the goals include characterizing the behavior of the entire dam, the forcing input must be placed such that a global response is induced throughout the dam. Because eccentric mass vibrators only excite a single frequency at a time, their placement is especially important. To ensure that steady state responses are excited properly, the vibrator must be placed such that it does not coincide with any nodes of the modes of interest.

PBT offers the advantage of using a device that outputs a short-duration force pulse with a wide spectral band of energy at higher amplitudes than shakers or impact hammers. This load can be considered an ideal impulse for a dam due to its massive size even though it occurs in a time frame of one to ten milliseconds. The resulting impulse responses are transient, broadband decaying sinusoids that contain a wide range of modal behavior. In theory, the introduction of an impulse anywhere in the system should excite the dam's impulse response so the location of the excitation source should not matter. There are, however, practical limitations during any PBT that should be considered. These limitations include the placement of the excitation source since the placement will affect how much each mode of the dam is excited. The use of finite-element models (FEMs) can be helpful in ensuring effective PBT of the dam since they can provide insights into how the dam responds to impulses.

When the dam has an irregular geometry it is recommended that a FEM be constructed and then used to obtain the first several mode shapes of the dam. By examining the first several modes, the locations on the dam that correspond to the largest and smallest deformations of each mode can be determined. This allows the location of the input system to be chosen based on the predicted dam behavior so that it can be placed such that it will produce the largest displacement. Consider for example the first few modes of a generic concrete arch dam modeled in a finite element modeling software, such as Abaqus, shown in **Figure 1**.

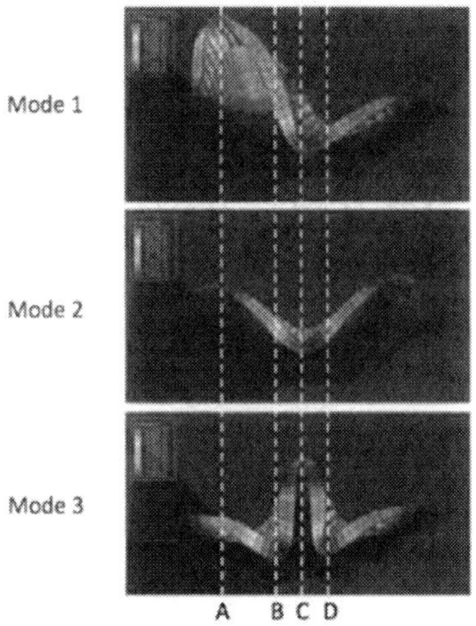

Figure 1: Locations of largest displacement for a generic arch dam. Blue indicates the locations with the least displacement and red indicates locations with the most displacement

Lines A, B, C, and D show the locations where each of the first three modes experience the most deformation. All of the lines pass through only one mode's location of maximum displacement, or antinodes, except for line C. Where line C intersects the third mode, the amount of movement is minimal. Because of this, the location indicated by line C should not be chosen as the location of the thruster since then the forcing input will not adequately excite the third mode of the dam. Instead the locations indicated by lines B or D should be chosen, as these are both closest to the antinodes of the first two modes and at the antinode of the third mode. Based on this preliminary FEM, the analyst can determine that the forcing input should be placed at either the location indicated by line B or line D since these two locations will most likely cause the largest displacement of the dam overall thus making it easier to measure the dam's response to the forcing input.

Input Force Magnitude

Another important consideration is the magnitude of the load delivered to the dam. This, like the previous consideration, is entirely dependent on the goals of the PBE. The location of the forcing input will also dictate the magnitude of the input force since the location will affect how the force propagates through the dam. For a PBE designed to evaluate the overall dam's condition, the magnitude of the forcing input should be large enough to mobilize the entire dam. For a PBE designed to evaluate local behavior, a smaller force may be used since only the area closest to the input needs to be mobilized. One example of when localized tests are preferred is when the goal of the PBE is to evaluate anchor behavior in certain monoliths of the dam. Take as a case study Shaver Lake Dam, where a PBE was conducted to determine the condition of its anchors as described in Park, et. al (2017). A previous test was performed where the forcing input was placed at the center of the central arch section. Signal decay across the unanchored section of the right gravity leg made the data on the right side of the dam unusable. While the forcing input was powerful enough to mobilize the majority of the dam, the properties of the pulse were too altered by the distance that it had to travel.

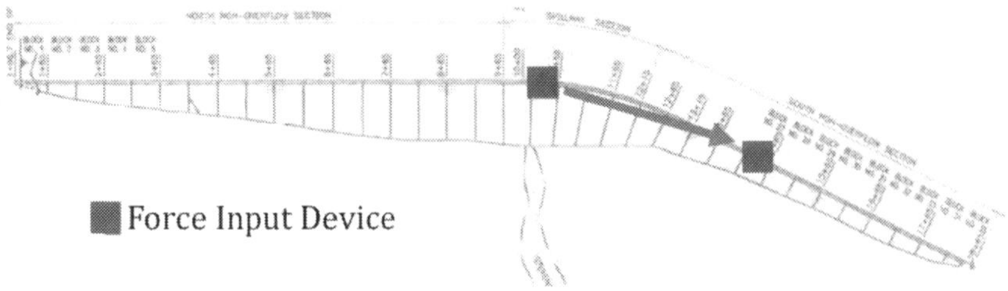

Figure 2: New forcing input position for localized testing

Figure 2 shows how the forcing input can be moved onto the anchored blocks in order to minimize the possibility of pulse decay occurring. This allows the forcing input to be placed directly on the anchored blocks that are being tested. By moving the forcing input closer, the forcing input output can also be decreased since the pulse only has to mobilize the monoliths around it rather than the entire dam.

Sensor Location

The locations of sensors to measure the dam's response are highly dependent on the location of the forcing input and the purpose of the PBE. For a PBE focused on evaluating the condition of the entire dam, the sensors must be evenly spaced across the crest and along the profile of the dam. To place the sensors along the dam's profile, the sensors must either be placed on the downstream face of the dam or in the operating gallery, depending on which is more easily accessible. For the Shaver Lake Dam example using the localized approach, the accelerometers should be placed on either side of the monoliths of interest since the anchors could cause torsional behavior as well as lateral and vertical behavior in the dam.

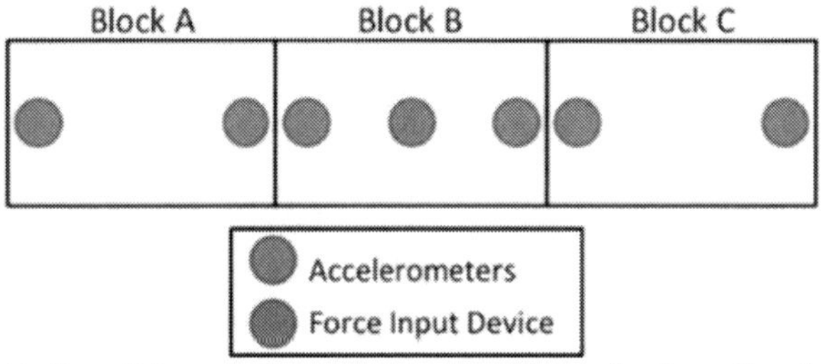

Figure 3: Possible accelerometer positions for localized testing

Figure 3 shows one possible arrangement of the accelerometers for localized testing. The accelerometers must be placed on both sides of the gaps between monoliths in order to characterize the behavior at the interface, since this behavior may affect the signal that is transmitted to the adjacent monoliths. In this configuration the integrity of the impulse is maintained and the resulting measurements should be representative of how individual monoliths react to an impulse load.

Maintenance Data Collection

Another important aspect of PBT is the collection of maintenance, or ambient, data. These are data that the operations and maintenance (O&M) staff of a dam will have the most access to since they only require the sensors be placed on the dam at specified locations. No forcing input is involved in collecting this data since the intent is to capture the dam's ambient behavior. The purpose is to acquire a baseline for the dam's performance under ambient conditions so that changes in dam performance under operating conditions can be evaluated with confidence. By establishing a baseline condition for the dam, O&M staff can routinely collect data from the dam and analyze it to determine how the condition of the dam changes overtime. The O&M staff can then adjust their maintenance procedures accordingly to properly maintain the dam's condition. As a result, dam owners would no longer have to run large, expensive tests, or even perform a PBE on the dams to evaluate their current condition.

Operational Data Collection

One final aspect of PBT that this paper will discuss is the collection of operational data. As stated previously, operational data consist of data collected while the dam is being operated. This is especially important for dams that have large spillways that could cause damage to the dam if operated incorrectly. During this part of the PBT, there is no forcing input, but the sensor locations must be adjusted to account for the locations of the spillways. For dams with spillways, it is of large interest whether there is torsional behavior occurring during the most common operating plans. The presence of torsional behavior near the spillways could cause more damage to the dam since it causes the spillway gate joints to exert a greater force on its connections to the concrete. It is especially important to place sensors near the features of interest since the induced behaviors by operating conditions could create localized dynamic behavior near these features.

Performance-Based Analysis Techniques

Before performing a PBE, the analysis techniques to be used in PBA must also be determined since the choice of analysis technique will dictate how some tests should bee run, what information should be recorded, and how much data should be collected for each of the forcing input response, ambient, and operational data collection efforts. Analysis techniques will also vary depending on what aspect of the dam will be characterized from the results. For PBEs with the goal of characterizing the dam's overall behavior, the time-domain approach is more useful since the forcing input response data can then be used to calculate the response of the dam to a variety of earthquakes to produce a fragility curve. Fragility curves are useful when the dam owner is interested in knowing what types of earthquakes have the largest effect on their specific dam.

If instead the analyst and dam owner are seeking information about localized features such as spillways or cracks in the concrete, then the frequency-domain approach may be more useful. Cracks act as a high pass filter for any forcing input so if frequency domain analysis techniques are used, then the frequency spectrum of the measured response can be used to determine the approximate location of the crack. The data taken from locations closer to the crack will exhibit more of the effect of the crack's high pass filter since low frequencies will be severely attenuated closest to the crack.

When the purpose of the PBE is to obtain PIs that will be used to monitor the dam in the future, both the time-domain and the frequency-domain approaches are recommended. Often a combination of the two approaches will give the most comprehensive set of PIs for the dam. PIs can consist of measures such as base flexibility and monolith amplification. The PIs must be determined before any part of the PBT has started since the PIs chosen will dictate where measurements must be taken.

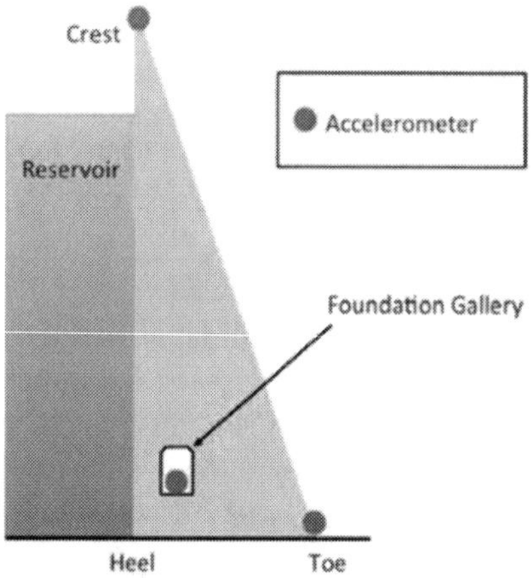

Figure 4: Dam cross-section showing sensor placement

With base flexibility, measurements must be taken at both the heel and the toe of the dam. For monolith amplification, measurements must be taken at the crest of the dam and at the heel of the dam. **Figure 4** shows where these measurements would have to be taken. The accelerometer shown in the foundation gallery could also be placed in a drain hole if the dam has ones that are easily accessible in order to have the sensor placed closer to the foundation.

To better understand the necessary steps to conduct a PBE, a proposal for the PBE of a concrete dam is included from the next page on. This proposal details the overarching plan for the PBE, with specific details for how the PBT will be completed. The PBA of the data is not heavily discussed in the proposal because the type of analysis will depend heavily on the content of the data collected during the PBT.

PROPOSAL FOR PERFORMANCE-BASED SEISMIC EVALUATION OF A CONCRETE DAM

A Performance-Based Seismic Evaluation (PBSE) procedure for a concrete dam is proposed in order to properly evaluate the effect of seismic hazard conditions and how these effects can be influenced by the effect of current operating conditions, and develop Performance Indicators (PIs) that indicate the current condition of the dam. The PBSE should return to the dam owner information about how current operating plans affect dam condition, how the dam will respond to seismic loading, PIs that can be used to evaluate the current condition of the dam, and recommendations for how to improve operating and maintenance plans.

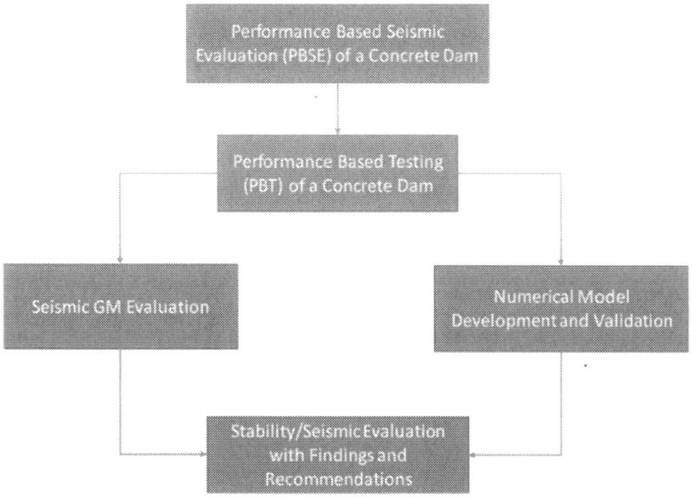

Figure 5: Overview of the PBSE for a Concrete Dam.

Figure 5 shows the overall plan for the PBSE to be conducted on a concrete dam, which includes both a ground motion (GM) analysis and the development and validation of a numerical model. Once it has been completed, the dam owner will be presented with information regarding how the dam responds to seismic loading conditions and PIs that can be used to evaluate the dam's condition in the future.

PBT on a concrete dam is the first step in PBSE. **Figure 6** shows the overview of the plan for the PBT and possible PIs that can be found for the dam. As listed in **Figure 6**, some possible PIs for a concrete dam include:

- Amplification from the base to crest of a monolith
- Relative motion between each monolith
- Energy dissipation within the seismic spectral frequency band (0.1-100Hz)
- Fragility estimates at multiple locations on the dam and spillways
- Effect of spilling operations on dam stability

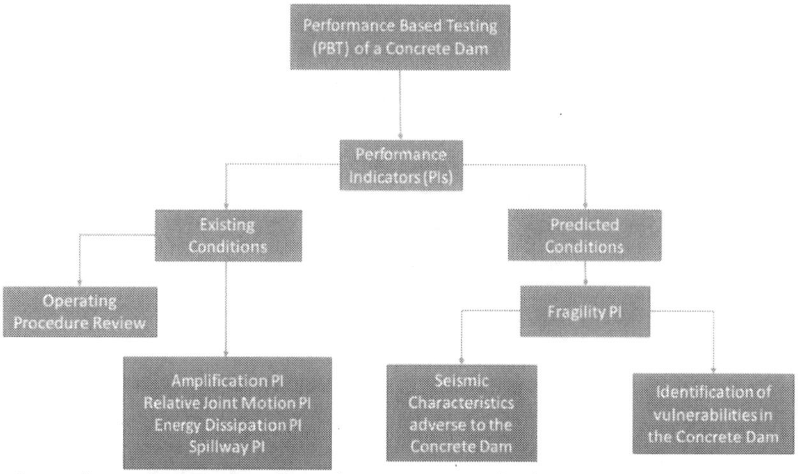

Figure 6: Overview of PBT plan for a Concrete Dam including the Deliverables of PBSE

The PBT can also give insight into which seismic characteristics are most harmful to the dam by providing analysts the ability to calculate the dam's response to an earthquake from the recorded forced-response. These fragility curves will give the dam owner insight into what types of earthquakes would cause the most damage to the dam. They also provide a guideline for the seismic evaluation of the dam since the earthquake engineers will have an idea of what characteristics to look for when performing the seismic hazard analysis.

Performance-Based Testing of a Concrete Dam

For the PBT of the concrete dam, the following responses will be collected for analysis to evaluate the dam's condition.

- *Ambient (Maintenance) Data* – The ambient data will be collected on the crest of the dam and over spillway gates, if possible. This broadband, low-level response will be used to set a baseline for the rest of the analysis. If possible, the data will also be analyzed to determine if it can be related to the condition of the dam.
- *Forced-Response Data* – The forced-response data are obtained by measuring the dam's response to an impulse load from a device that can produce broadband, high-level behavior in the dam.
- *Operational (Flow-Induced) Data* – These data will be collected for a number of spillway configurations that are commonly used by the operations and maintenance staff. They will provide insight into how current operating plans are affecting the dam's structural stability.

Measurements in the stream, cross-stream, and vertical directions will be obtained for all of these loading conditions. A minimum of two data sets for each loading condition is required in order to ensure data integrity.

The purpose of these tests is to

- Provide evidence of three-dimensional effects in the dam, foundation, and reservoir
- Develop PIs for the dam under operating conditions
- Develop PIs that can predict how the dam will respond to seismic loads
- Guide the development of numerical models for evaluating the dam's seismic performance capabilities
- Provide evidence for the validation of numerical models used in stability and seismic evaluations of the concrete dam

Preliminary Analyses

Before the testing begins, preliminary analysis must be completed on the dam. These preliminary analyses will give insight into the modal behavior of the dam and provide understanding on how to better conduct the PBT on the concrete dam. The preliminary analyses will include

- Finding free response of the concrete dam
- Finding forced-response of the concrete dam to a point load on the crest
- Determining resonant frequencies and mode shape estimates of a concrete dam

One additional objective for the preliminary analyses is to determine the equivalent static loading conditions created by different spillway configurations. Using these equivalent loading conditions, static analyses on the dam will be completed to determine the gate configurations that lead to or suggest adverse loading conditions.

The preliminary analysis results will be used to determine the best location to place the forcing device so that it will generate the largest response in the dam. The analysis on the loading conditions of the various spillway operating configurations will be used to determine which spillway configurations could be the most damaging to the dam. As a results, the collection of operational data on these damaging configurations can be prioritized in the subsequent testing.

Performance-Based Testing Procedures

The procedures used for the evaluation of the concrete dam are based on the procedures used in previous investigations of large concrete dams. These tests will collect data for the loading conditions described above in order to adequately characterize the dam's behavior. The dam configurations for each of these loading conditions are described here.

- *Ambient* – there will be no loading input on the dam and all the spillway and bascule gates will be closed.
- *Forced-response* – a device called a Cold Gas Thruster (CGT) will be placed on the dam and then fired to produce an impulse load on the system. All of the spillway and bascule gates will be closed for this configuration.
- *Flow-Induced* – the spillway and bascule gates will be operated based on the Spillway Operating Plan (SOP) with each configuration being held for several

minutes. The specific configurations from the SOP will be chosen based on the results of the preliminary analysis of the dam.

To obtain the forced-response of the dam, the dam must have an impulse load applied to it by the CGT. The CGT is a metal chamber that is sealed closed on one end and closed on the other end by a burst diaphragm and nozzle. **Figure 7** shows the CGT with a torn burst diaphragm on top.

Figure 7: CGT assembly showing various components that are necessary for it to function

The CGT must be mounted on the crest of a dam using the mounting bracket with the CGT bolted to the vertical back plate and the bottom of the mounting bracket bolted onto the crest of the dam. The bolts will ensure that the CGT and dam are securely fastened so that the measured responses are accurate. The chamber of the CGT is pressurized with a gas such as nitrogen until the burst diaphragm breaks, after which all the air in the chamber will escape through the nozzle. The force pulse produced by the CGT can effectively be treated as an impulse as shown by **Figure 8**, which indicates that the CGT burst lasts under 10ms.

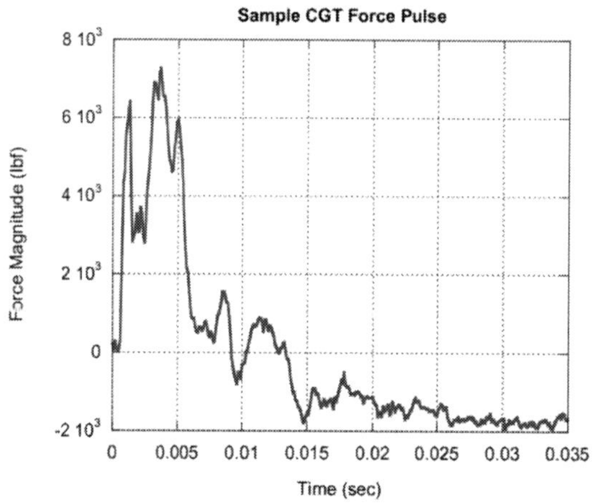

Figure 8: Sample plot of force input to the dam

The sensors for these tests will be three-axis accelerometers that output measurements in the stream, cross-stream, and vertical directions. As seen in **Figure 9**, these accelerometers are mounted on a mounting plate with a level included so that when bolted they can lay completely flat on the dam.

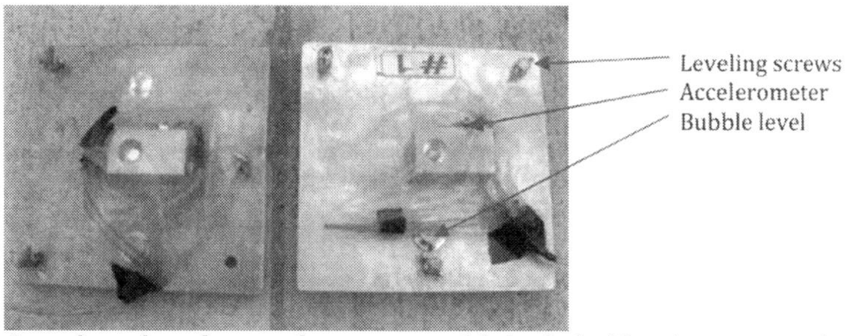

Figure 9: Mounting plate for accelerometers with included hardware to make installation easier

After installation of the accelerometers is complete, the ambient data can be collected immediately. The operational response data will be collected when the operations staff implements the SOP. The ambient and operational data must be collected over multiple periods of 5 minutes at a minimum.

For the forced-response measurements, the accelerometers must be moved to the appropriate locations and the CGT must be installed before forced-response testing can commence. Data acquisition for these tests will begin before the CGT chamber is filled with an inert gas and continue until the dam can return to pre-pulse conditions. Each test setup will have a minimum of 2 sets of data collected with the total number of times the CGT is fired being sufficient for all the necessary data to be collected. Due to the nature of this test, the duration of data collection and the number of times the CGT is fired will be determined on-site.

Measurement Locations

Measurements for the ambient and operational loading conditions will be acquired with sensors placed at the locations given in **Figure 10**. The most significant choice of placement in this arrangement is that there are two accelerometers placed on either side of the radial spillway and bascule gate. This is done intentionally so that any local torsional behavior around the gates can be easily detected.

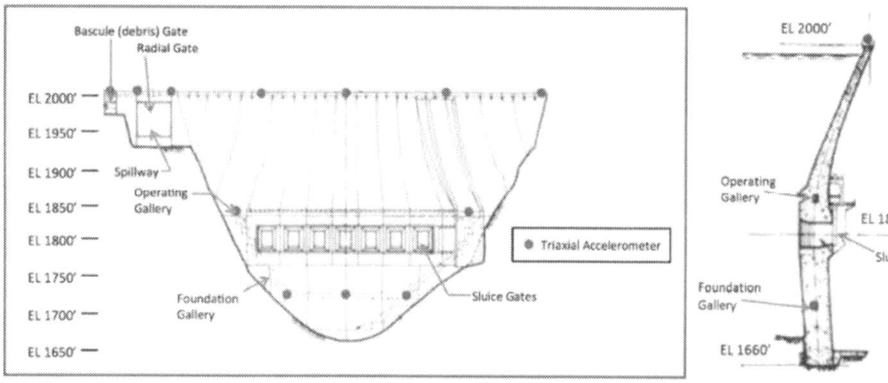

Figure 10: Typical accelerometer configuration for ambient and operational data collection

For the forced-response measurements, the accelerometer locations are adjusted to better capture the behavior of the dam at the crest by moving them so that they are evenly spaced across the top of the dam. **Figure 11** shows the suggested configuration for the accelerometers and CGT.

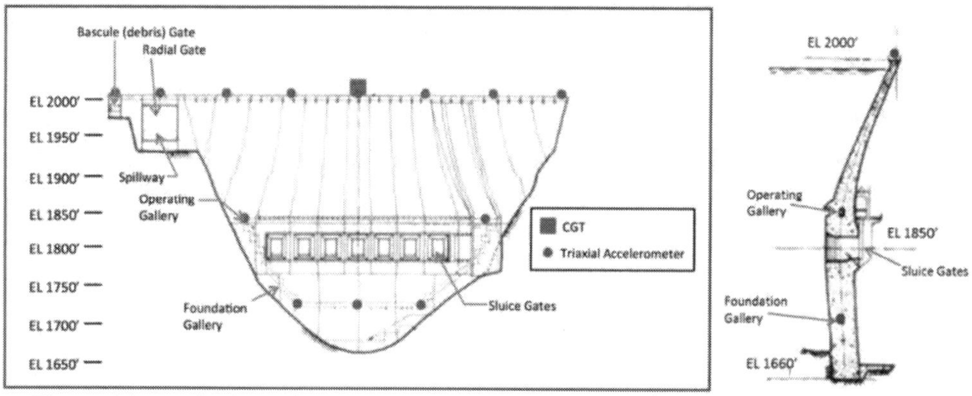

Figure 11: Typical accelerometer and CGT configuration for force response testing

The CGT is placed at the center of the arch in **Figure 11**, but the actual placement of the CGT will depend on the results of the preliminary analysis and on accessibility of the chosen site for CGT installation. A main measurement station will be installed inside a van or equivalent space from which all the sensors will be controlled. This station should provide protection from the weather so that it will not affect the main computer and data acquisition system's ability to record response data.

CONCLUSION

Performance-Based Evaluation is a powerful, new technique for evaluating the condition of a large, concrete dam. The Performance-Based Testing that occurs as part of PBE provides information about the dam's dynamic behavior from the dam itself. This removes the large amounts of uncertainty present in finite-element models where many assumptions have to be made about the system in order to construct them. From the ambient, operational, and forced-response data collected, a variety of performance indicators can be developed. The forced-response data can also be used to calculate fragility curves, which provide much insight into which earthquake characteristics are the most damaging to the dam. The flexibility of the data collected allows for a large variety of analysis techniques, two of which are described in this paper, to be used in analyzing the data to extract information about the system. This information about the system can then be used to validate numerical models of the dam with data from actual dam performance. These data and concepts can also be used in developing static equivalent loading functions from dynamic or flow-induced response measurements on a dam. The loading functions can then be used to develop estimates of the dam's ability to withstand potential flood conditions. Additionally, this proposed PBE is a significant improvement from traditional testing techniques. The costs are reduced, data quality is greatly improved, and the content of information and ability to extract it is much larger than what was possible with traditional techniques.

REFERENCES

Anderson, S., K. Menda, M. Zhao, and Z. Duron. (2016). "Selecting an Intensity Measure for Fragility Curves Derived from Performance-Based Testing." Denver, Colorado, 36th Annual USSD Conference.

Duron, Z. (2016). "A New Performance Based Seismic Evaluation Approach for Concrete Dams." Prepared for Southern California Edison.

Edelman, J., P. Orme, M. Goldkamp, T. Savage, Z. Duron, M. Knarr. (2014). "Performance-Based Analysis of a Large Concrete Dam." San Francisco, California, 34th Annual USSD Conference.

Goldkamp, M., T. Savage, J. Edelman, P. Orme, Z. Duron, and M. Knarr. (2014). "Performance-Based Testing of a Large Concrete Dam." San Francisco, California, 34th Annual USSD Conference.

Menda, K., M. Zhao, S. Anderson, and Z. Duron. (2016). "A Performance-Based Approach to Developing Fragility Curves for Structures Subject to Seismic Loading Conditions." Denver, Colorado, 36th Annual USSD Conference.

Zhao, M., S. Anderson, K. Menda, and Z. Duron. (2016). "Selecting a Performance measure for Fragility Curves Derived from Performance-Based Testing." Denver, Colorado, 36th Annual USSD Conference.

REALITY CAPTURE FOR DAM AND HYDROELECTRIC FACILITIES

Adam Serock[1]

Whether you're modifying, updating, expanding, or operating an existing facility you should have accurate information of the existing conditions. This process typically involves multiple site visits to gather photos, filed measurements, and notes. But could you imagine the benefits in terms of accuracy and money saved if you could walk in, push a button and simply capture reality, then pull that reality into a virtual world where the data could be leveraged time and time again?

This presentation will cover the process of using terrestrial Lidar (Light Detection and Ranging) to capture reality in the form of millions of data points. The process involves multiple setups around a site or facility to completely capture the existing conditions. The separate scans are stitched together by comparing overlapping geometry within each scan. The result is an accurate spatial representation of the existing conditions, commonly referred to as a "point cloud." Point cloud data can then be leveraged in many ways: It can be used to verify size and location of structures and project elements within the space. It can be used to capture accurate 3D information on hard to access and complex structures, such as spillway gates and movable machinery. It can be loaded into design modeling (Building Information Modeling, or BIM) software to aid in the routing of new systems or leveraged to create a model of the existing facility. It can be displayed in model review software for coordination with new work to be performed. It can be used to communicate with stakeholder not familiar with site. In addition, with multiple inexpensive periodic set ups, this information can be used to monitor any project for movement - enabling visual confirmation of deformation of a dam, retaining wall, spillway gate or any other structure. The presentation will discuss the technology including its benefits, limitations and typical cost and provide examples of potential applications for dams and hydropower projects.

From dams and spillway gates to pump station and turbines; points clouds generated through the Lidar process will allow you to capture reality and leverage the data for multiple uses.

[1] HDR, adam.serock@hdrinc.com

NUMERICAL SEDIMENT TRANSPORT MODELING OF OPERATIONS AT PAONIA RESERVOIR, COLORADO

Sean Kimbrel[1]

Paonia Dam and Reservoir are located on Muddy Creek, a tributary of the North Fork Gunnison River in western Colorado. Based on a recent bathymetric survey of the entire reservoir, conducted in June 2013, the estimated average annual rate of sedimentation has been approximately 100 acre-feet per year since completion of the dam in 1962. Nearly 25% of the reservoir's original capacity of 20,950 acre-feet has been lost to sediment deposition.

At the request of local stakeholders, staff at the Bureau of Reclamation's Sedimentation and River Hydraulics Group developed and calibrated a one-dimensional, unsteady flow, non-equilibrium sediment transport model called SRH-1D using observed sedimentation volumes in 2013 and 2014. It was then used to predict the distribution of reservoir sediment during the 2015 irrigation season and under potential dam operation scenarios in the fall 2015 to support decisions on short-term facility operations and sediment management activities.

This presentation will document the numerical sediment transport modeling efforts at Paonia Reservoir, which include:
1. Model development,
2. Model calibration to 2013 and 2014 sediment data collection efforts, and
3. Model simulation of potential summer/fall 2015 operational scenarios.

[1] Bureau of Reclamation, skimbrel@usbr.gov

IMPACT OF CLIMATE CHANGE ON SEDIMENT ACCUMULATION IN THE RESERVOIRS OF EMBANKMENT DAMS: CASE STUDY, BALDHILL DAM, NORTH DAKOTA

Bahareh Shoghli[1]
Yeo Howe Lim[2]

Almost all regions of the world have been encountered with negative impact of climate change on their water resources and fresh water ecosystems. However, the intensity and magnitude of these impacts vary considerably from one region to another. Some regions experience water shortages and others experience increasing precipitation magnitude. This increase, along with large storm events and early time of snow melting, causes river flooding and flash flooding.

The reservoirs formed behind dams, which essentially are large-capacity water storages, are vital to the world's economy in terms of electricity generation, flood control, water supply, and recreation. In most reaches of natural rivers, sediment movements are approximately balanced on the inflow and outflow components. Reservoir sedimentation is a serious problem in many regions, particularly in regions with high sediment yield. Small-capacity reservoirs in rapidly eroding regions are more vulnerable than other reservoirs.

The presentation involves a case study of the reservoir of Baldhill Dam, which is located on the Sheyenne River watershed in North Dakota, and the result will be vital in future dam design, current dam operation, and with considerations of dam safety and sustainability. Arc-SWAT is used for simulating the watershed area upstream of the reservoirs. With the help of these simulations, the inflow and outflow sediment regimes are determined. As a result, the trap efficiency of the sediment in the reservoir of the small dams will be studied and assessed.

[1] University of North Dakota, bahareh.shoghli@und.edu
[2] University of North Dakota

SEDIMENT TRANSPORT THROUGH LAKE CLARKE AND LAKE ALDRED

Jonathan M. G. Viducich, E.I.[1]
Martin J. Teal, P.E., P.H., D.WRE[2]

ABSTRACT

Recent interest in the impact of the Lower Susquehanna River reservoirs (Lake Clarke, Lake Aldred, and Conowingo Pond) on Chesapeake Bay water quality has prompted a variety of research and modeling efforts. WEST Consultants, Inc. was contracted by Exelon Generation Company, LLC to develop a one-dimensional, fully-unsteady HEC-RAS 5.0 sediment transport model of Lakes Clarke and Aldred as part of a multi-model initiative spanning these two reservoirs plus Conowingo Pond. The model was hydraulically calibrated using stage and discharge data from the United States Geological Survey (USGS) streamgage at Marietta, PA (USGS Gage No. 01576000), and calibrated for sediment transport using observed bed volume change for both reservoirs between 2008 and 2013. Modelers used a verification period from 2013 to 2015 to validate the model's performance during a time characterized by lower flows. The calibrated model successfully replicated bed changes for both reservoirs and time periods within target ranges defined by survey equipment accuracy, and simulated both deposition and scour processes at varying discharges and reservoir states. Sediment rating curves were developed for three particle size classes (sand, silt, and clay) at the upstream boundary at Marietta, at Safe Harbor Dam (impounding Lake Clarke), and at Holtwood Dam (impounding Lake Aldred). The curves will be used in studies by others to aid in parameterization of the Chesapeake Bay Watershed Model. In addition, a time series of discharge and sediment loading at Holtwood Dam will be used to develop inputs to a three-dimensional Conowingo Pond Mass Balance Model.

INTRODUCTION

As the largest tributary to the Chesapeake Bay (the Bay), the Susquehanna River plays a primary role in supplying fresh water, sediment and nutrients to the Bay system. Recent concern about storm-driven sediment and associated nutrient pulses, which can negatively affect Bay ecosystems and related industries, resulted in ongoing collaborative work between a wide consortium of public and private agencies and organizations. In addition to a focus on land use and wider watershed management, attention was given to the role of three Lower Susquehanna hydroelectric dams in sediment and nutrient flux dynamics. The dams—Safe Harbor, impounding Lake Clarke; Holtwood, impounding Lake Aldred; and Conowingo, impounding Conowingo Pond—which for many decades served as net sediment traps, are generally believed to have approached or reached states of dynamic equilibrium. In this state, the volume of stored sediment may oscillate in the short term, but is relatively stable when averaged over the long term. Based on the

[1] Staff Hydraulic Engineer, WEST Consultants, Inc., 11440 W. Bernardo Court, Suite 360, San Diego, California 92127-1644, (858) 487-9378, jviducich@westconsultants.com.
[2] Senior Vice President, WEST Consultants, Inc., 11440 W. Bernardo Court, Suite 360, San Diego, California 92127-1644, (858) 487-9378, mteal@westconsultants.com.

current understanding of the system, some large storms may, in addition to importing new sediment from the watershed, also scour previously-deposited sediment in the reservoirs and temporarily increase trapping capacity. Most recent research has considered the role of Conowingo Dam, which impounds the last of the three reservoirs with some remaining sediment trapping capacity. The present study focuses on Safe Harbor and Holtwood Dams, which together impound approximately 20 miles of the Lower Susquehanna River between Marietta, PA and Holtwood, PA.

Safe Harbor Dam was constructed in 1931, and its average height of 75 feet (ft.) impounds a reservoir covering about 7,360 acres. The dam is operated by the Safe Harbor Water Power Corporation (SHWPC) (Safe Harbor Water Power Corporation, n.d.). Construction of Holtwood Dam was completed in 1910, and with a height of 55 ft., the structure impounds a 2,400-acre reservoir (Porse, 2010). Ownership and operation of the dam transferred to Brookfield Renewable Energy Partners in April 2016 (Weissman, 2016).

Past studies of the Lower Susquehanna reservoir system included physical and chemical profiling of the reservoirs' bed sediments (Hainly, et al., 1995; Reed & Hoffman, 1996; Ott, et al., 1991; Takita & Edwards, 2001; Edwards, 2006), computations of changes in dam storage capacities and trap efficiencies (Hainly, et al., 1995; Langland, 2009), evaluation of the hydrodynamic impacts on deposition and scour rates (Hainly, et al., 1995; Hirsch, 2012; Langland & Koerkle, 2014), and other topics.

STUDY PURPOSE AND SCOPE

Exelon contracted WEST Consultants, Inc. (WEST) to develop an unsteady, one-dimensional sediment transport model of the upper two reservoirs, from near the town of Marietta, PA, to Holtwood Dam near Holtwood, PA. Originally, Gomez and Sullivan Engineers planned to collect gaged, inter-reservoir sediment data as part of Exelon's Lower Susquehanna River Integrated Sediment and Nutrient Monitoring Program, but insufficient data were collected due the absence of large storm events during the program's duration. The purpose of this study was to develop an improved understanding of sediment transport processes in Lakes Clarke and Aldred using a deterministic numerical model, the U.S. Army Corps of Engineers (USACE) Hydrologic Engineering Center-River Analysis System (HEC-RAS), in lieu of observed sediment transport data. The study's aim was ultimately to help parameterize the Chesapeake Bay Watershed Model and provide inputs for a three-dimensional sediment and nutrient flux model developed by HDR, Inc. (HDR) for Conowingo Pond.

While other sediment transport models, including HEC-RAS, were previously developed for the same reach, the current study aimed to build and improve upon earlier work using the new HEC-RAS 5.0, which was officially released in March 2016. Version 5.0 offers several key advantages over previous HEC-RAS versions. Instead of approximating the hydrodynamics as a series of steady flows (a quasi-unsteady simulation), the model's new, fully-unsteady sediment transport capabilities solve the unsteady flow equations, routing flow through the model and explicitly accounting for storage and travel time.

Volume is thus conserved, which is very important in reservoir systems. Other improvements include features designed to add more flexibility and precision to sediment inputs, such as the ability to apply sample-specific cohesive parameters for bed gradations.

MODEL DEVELOPMENT

HEC-RAS is an integrated software package designed to perform one-dimensional steady and unsteady hydraulics analyses, two-dimensional steady and unsteady hydraulics analyses, sediment transport computations, and water temperature and water quality modeling. The system is comprised of a graphical user interface (GUI), separate analysis components, data storage and management architecture, graphics, and reporting facilities (USACE, 2016b). HEC-RAS performs one-dimensional hydraulic calculations by solving the one-dimensional energy equation and/or momentum equation for flow through a channel comprised of two-dimensional cross sections separated by known reach lengths. Definition of the channel's shape and roughness allows the model to solve for energy losses due to friction and contraction/expansion.

Geometric Data

The model's geometry was based on a previous HEC-RAS model developed by the USGS using a combination of 1996 and 2008 surveyed bathymetric data, Light Detection and Ranging (LiDAR) data, a historic flood insurance study, and hand-interpolated cross sections (Langland & Koerkle, 2014; MDNR, 2014; Sullivan, 2016a). Several adjustments were made to reflect the current project's scope and new understanding of the available data. In all, the adjusted model covers a longitudinal distance of approximately 20.4 miles (mi.), represented by 43 cross sections including an inline structure at Safe Harbor. Figure 1 presents a schematic of the HEC-RAS model geometry; while the map shows the location of Holtwood Dam, the structure itself is not explicitly included in the model.

Manning's roughness coefficients (*n* values) from the USGS model (Langland & Koerkle, 2014) were initially assumed, and some horizontal variation in *n* values was added to account for vegetated islands and other channel features visible in satellite imagery. Flow roughness factors (1 to 1.2) were applied to compensate for decreasing roughness associated with increasing stages between the Marietta streamgage and cross sections approximately 1.5 miles downstream. Seasonal roughness change factors (0.8 to 1.7) were also applied at the same locations to account for changes in bed roughness.

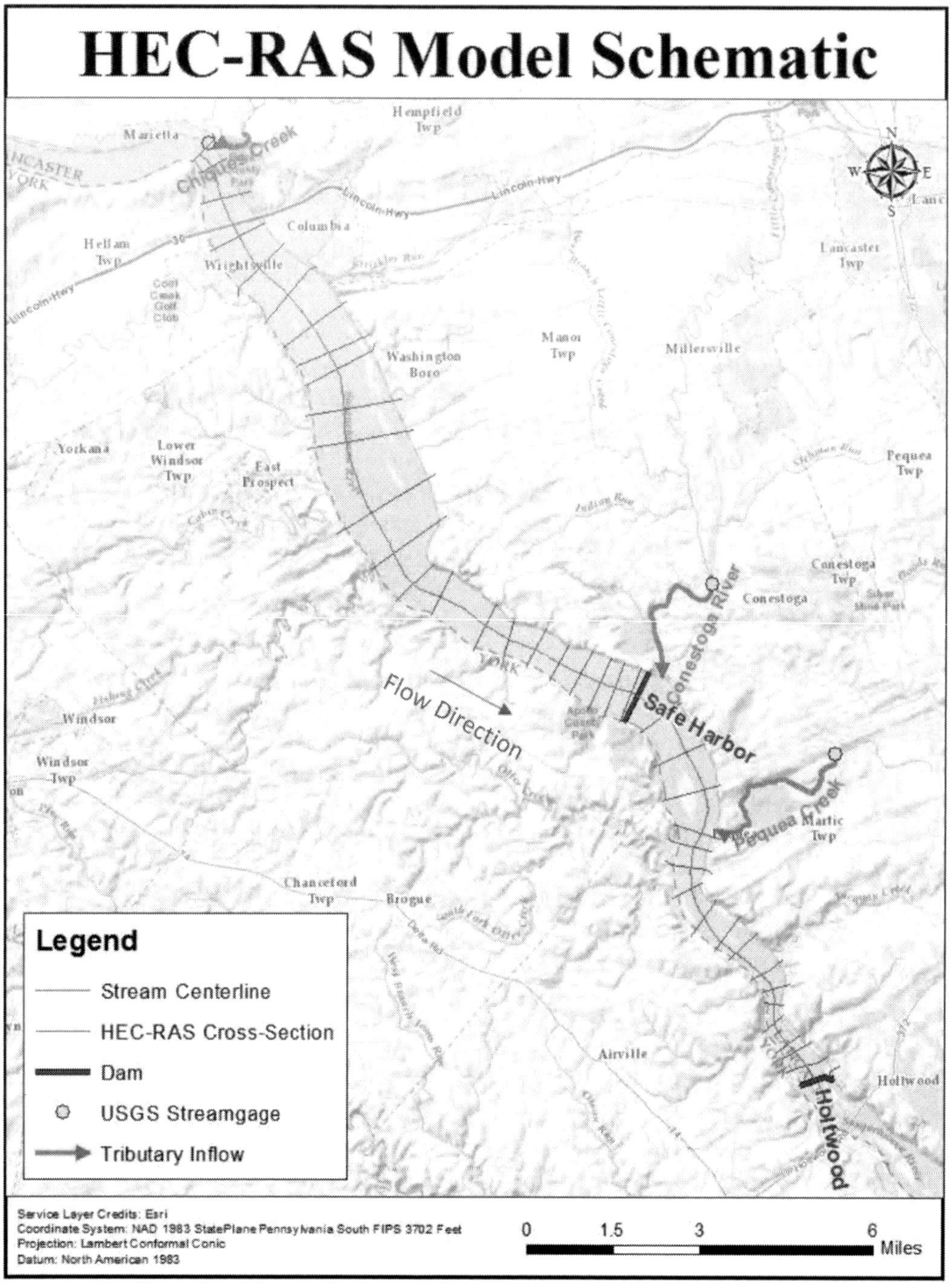

Figure 1. HEC-RAS Model Schematic.

Flow Data

HEC-RAS requires a boundary condition to determine flow at the most upstream cross section of a model, and additional inflows and other boundary conditions may be added at other locations. For the Lake Clarke and Lake Aldred HEC-RAS model, boundary condition discharges accounted for the flow contributions of the Susquehanna River and three tributaries: Chiques Creek, Conestoga River, and Pequea Creek. Together, the inflows account for 99.5 percent (%) of the drainage area at Holtwood Dam.

A USGS streamgage (USGS Gage No. 01576000) located on the Susquehanna River at Marietta, PA provided the upstream flow boundary condition. The gaged drainage area, which includes that of Chiques Creek, is approximately 25,990 square miles (mi^2). A 1-hour time series was developed for the full simulation period, with a maximum discharge of 665,000 cubic feet/second (cfs) on 9 September 2011 (during Tropical Storm Lee) and eight events with flows at Marietta, PA in excess of 250,000 cfs.

Lateral inflow data from the USGS streamgage (USGS Gage No. 01576754) on the Conestoga River at Conestoga, PA were applied at a model cross section just downstream of Safe Harbor Dam. At that location, the drainage area of the Conestoga River is approximately 470 mi^2. A USGS streamgage (USGS 01576787) at Martic Forge, PA provided the lateral inflow data for Pequea Creek, which has a drainage area of 148 mi^2 at the gage and enters the Susquehanna River about 5 miles upstream of Holtwood Dam.

Sediment Data

In order to simulate erosion, transport, and deposition processes, an HEC-RAS model requires information about both bed sediment and additional sediment entering the model at the upstream model boundary and other inflow locations.

Bed Sediment. Bed particle size gradations for Lake Clarke and Lake Aldred were developed using USGS core data sampled at 35 locations in 1990 and 1991 (Hainly, et al., 1995; Reed & Hoffman, 1996) and provided by the USGS. Langland and Koerkle (2014) used the same coring data to develop bed gradations in their HEC-RAS model, and reported good agreement between samples collected in 1990 and 1991 and samples collected in 2000 and described by Edwards (2006). The more recent sample gradations were not available at the time of the study. Ten grouped sediment gradations were initially developed to represent bed sediments in Lake Clarke and Lake Aldred, based on trends within sections of each reservoir. Twelve particle size classes ranging from clays (<0.004 millimeters (mm)) to fine gravel (4 – 8 mm) were modeled.

In their model, Langland and Koerkle (2014) defined maximum erodible bed depths in the upper reservoirs as estimates based on Hainly, et al. (1995), Reed and Hoffman (1996), and bathymetry data. Those depths varied from approximately 0 to 20 ft. in Lake Clarke, increasing in the downstream direction, and approximately 0 to 10 ft. in Lake Aldred. The same maximum erodible bed depths were assumed as a starting point in the current HEC-RAS model.

The beds in Lake Clarke and Lake Aldred contain significant percentages of silt and clay—cohesive materials—and erosion rates cannot be estimated accurately using standard sand and gravel transport equations. Cohesive forces resist applied shear stresses in beds containing high percentages of fine material, and those forces are highly sensitive to particle size distribution, particle coatings, fine sediment mineralogy, organic content, bulk density, gas content, pore-water chemistry, and biological activity (Scott & Sharp, 2014).

HEC-RAS 5.0 allows users to define cohesive parameters for individual bed gradations, in addition to the previously-available option of assigning global cohesive parameters. This is an important improvement over earlier versions of the model, especially given the high degree of variation in the cohesive properties of the bed sediments of the Lower Susquehanna reservoirs. Langland and Koerkle (2014) noted the inability to model variation in cohesion as an important limitation of their HEC-RAS model. Given the complexity of and variation in cohesive parameters, the HEC-RAS 5.0 manual recommends using direct measurements of particle and mass wasting erosion parameters.

The U.S. Army Corps of Engineers' Engineer Research and Development Center (ERDC) used the SEDflume apparatus to develop erosion rate data for Conowingo Pond cores in 2012, and published them in Attachment B-2 of the Lower Susquehanna River Watershed Assessment (LSRWA; Scott & Sharp, 2014). Unfortunately, no significant correlation was found between measured cohesive parameters and the cores' physical sediment parameters, and physical sediment properties reported by ERDC differed appreciably from those of samples collected by the USGS (analyzed by the USGS and Maryland Geological Survey (MGS)) and AECOM (Zeff, 2016) in Conowingo Pond.

Given uncertainties in the data, the percent clay and cohesive parameters were used as calibration tools, the latter bracketed by the combination of SEDflume-measured values most and least likely to produce scour (Table 1). Initial cohesive parameters for each area of the reservoirs were developed by grouping the ERDC cores by their relative locations in Conowingo Pond. Surface samples measured by ERDC were not included, given the thin, low-density, highly-erodible layer noted at the sediment-water interface.

Table 1. Bounding Cohesive Soil Parameters

		Most Scour Expected		**Least Scour Expected**	
Parameter	**Units**	**Particle Erosion**	**Mass Wasting**	**Particle Erosion**	**Mass Wasting**
Threshold	(lb/ft^2)	0.0021	0.0167	0.0334	0.0585
Erosion Rate	($lb/ft^2/hr$)	76.6792	238.1479	7.0781	64.1451

Inflowing Sediment Load. Instantaneous suspended sediment data collected at irregular intervals and corresponding to locations at or very near each USGS gage were downloaded from the USGS and Susquehanna River Basin Commission (SRBC) websites and combined (USGS, 2016; SRBC, n.d.). Given the infrequency of the sampling, it was not possible to create reliable sediment loading time series, and the measured values were instead used to generate loading-discharge rating curves for each inflow as an HEC-RAS input. Similar to the HEC-6 model developed by Hainly, et al. (1995), a positive bias of 2% was initially assumed and added to the load as an estimate of unmeasured bedload. Available suspended particle gradation data were also very limited, and gradation rating curves developed for the HEC-6 model using other historic data were adopted and applied to the corresponding discharges. Given the substantial uncertainty in the loading and particle gradation rating curves, the parameters were assigned initial values and later varied during calibration.

Transport Function and Sediment Parameters

Following sensitivity testing with several transport functions, the Toffaleti transport function was applied to the model. The Thomas (Exner 5) sorting method was used, along with the Report 12 fall velocity method.

HEC-RAS 5.0 features many options designed to improve predictions of spatially-variable sediment transport within a one-dimensional modeling framework. The reservoir deposition option, which deposits more material in deeper parts of the cross section rather than depositing a veneer of equal thickness across all wetted areas, was applied for the model.

While HEC-RAS 5.0 allows users to define sample-specific cohesive parameters for bed gradations, other sediment parameters such as specific gravity and dry unit weights are applied globally. Some bed and inflowing sediments may in reality be slightly lighter due to the presence of coal in the Lower Susquehanna bed sediments, but the default values for specific gravity and unit weight by size class were applied (Table 2).

Table 2. Global Values for Specific Gravity and Unit Weights.

Specific Gravity (dimensionless)	Unit Weight Sand (lb/ft^3)	Unit Weight Silt (lb/ft^3)	Unit Weight Clay (lb/ft^3)
2.65	93	65	30

Temperature Data

HEC-RAS sediment transport models require users to enter a time series of water temperature data, which is used to calculate fall velocities in sediment transport computations. A fifth-order polynomial was fit to 205 water temperature measurements taken between 2008 and 2014 at the Marietta, PA streamgage (USGS Gage No. 01576000) and used to relate temperature with the day of the year. A temperature time series was created for the simulation period, and observed data were used when available.

Dam Operations

While some information about the normal pool levels and physical dimensions of Safe Harbor and Holtwood Dams is available online, details of day-to-day dam operations are largely unavailable.

Rather than attempt to explicitly mimic hour-to-hour or day-to-day gate operations, rating curves were developed for both dams for the current model based on available information. According to the SHWPC (Safe Harbor Water Power Corporation, n.d.), the normal pool elevation for Safe Harbor is 227.2 ft., corresponding to the NGVD29 vertical datum. The powerhouse has a capacity of 113,000 cfs, while the flood gates have a capacity of 1,120,000 cfs. At no point in the simulation period does the river's discharge exceed Safe Harbor's gate capacity, and while water surface elevations in the reservoir vary at flows below the powerhouse capacity (based on peaking operations), the water surface elevation is maintained within a range of approximately 224 to 228 ft. Figure 2 presents the internal rating curve developed for Safe Harbor Dam.

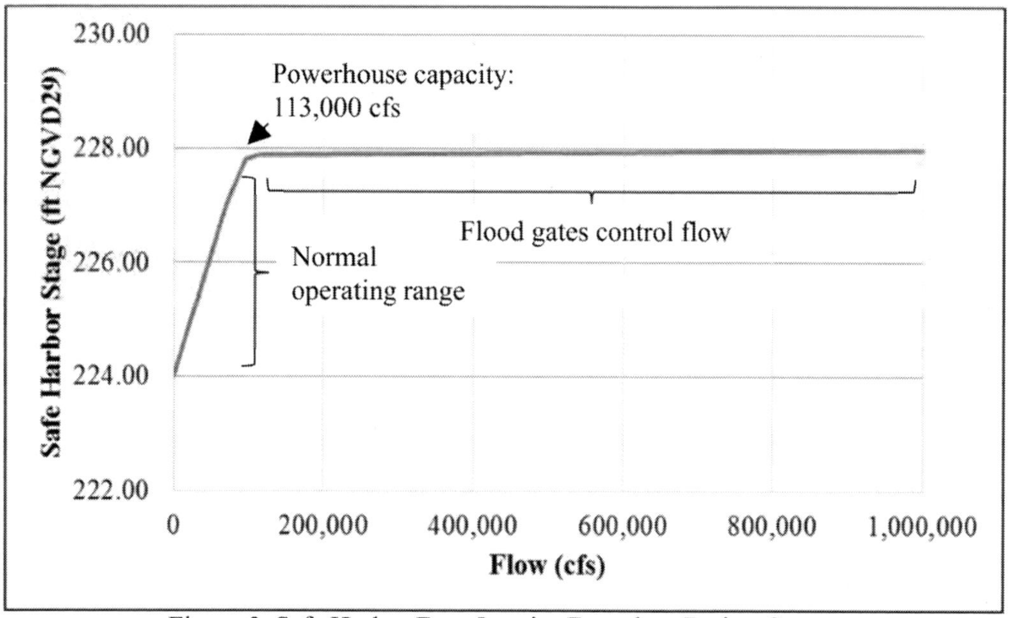

Figure 2. Safe Harbor Dam Interior Boundary Rating Curve.

Holtwood Dam has relatively limited control of the river, and aside from its power production units and a 4.75-foot inflatable rubber dam, the structure essentially functions as a weir. The published normal pool elevation—that is, the elevation at which the rubber dam is crested and the dam begins to spill—is 169.75 ft. (Safe Harbor Water Power Corporation, n.d.). The vertical datum is not listed on the website, and it was discovered that the value is actually referenced to a local datum at Holtwood (Sullivan, 2016b). When converted to the NGVD29 datum, the target pool elevation is 170.52 ft. A composite rating curve was developed to represent stage vs. flow at the downstream boundary. Stage heights ranged linearly from 165.77 ft. to 170.52 ft. for flows below

61,460 cfs, reflecting assumed operation of the flashboards/rubber dam within the operating capacity of the powerhouse. A weir coefficient of 3.7 was assumed for the flashboards/rubber dam, and those were assumed to fail gradually between 2 and 4 ft. above the flashboard height. The powerhouse was assumed to gradually shut down between a total flow of 350,000 and 375,000 cfs, at which point the dam functioned as a weir with an assumed coefficient of 4.06. Figure 3 presents the rating curve developed for Holtwood Dam.

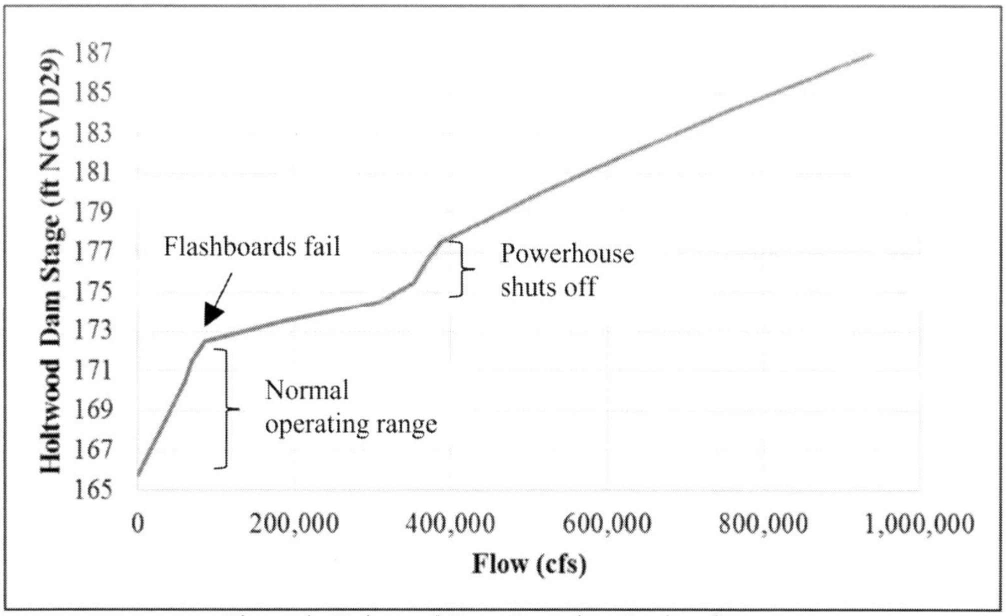

Figure 3. Holtwood Dam Boundary Rating Curve.

MODEL CALIBRATION AND VERIFICATION

The modeling effort was designed with a three-phase calibration-verification process. First, the hydraulics of the fixed-bed unsteady flow model were calibrated for the simulation period 1 January 2008 – 30 August 2013. Next, the unsteady sediment transport model was run and calibrated to bed volume change from the same period. Finally, the sediment transport model was verified using calculated bed volume change for 30 August 2013 – 15 October 2015.

Fixed-Bed Calibration

One challenge to the calibration of a fixed-bed hydraulic model for the Lower Susquehanna River system was the lack of historic, inter-system stage and flow data with which to compare the modeled results. The three hydraulic structures located between the USGS streamgage at Marietta, PA and the USGS streamgage immediately downstream of the powerhouse at Conowingo Dam control the flows to varying degrees, yet no gaged data are available to confirm modeled results between the dams. However, the gage at Marietta does provide a useful check on the upstream end of the unsteady, fixed-bed model, through a comparison of over 120 instantaneous stage measurements made during the calibration period with the water surface elevation predicted by the model.

Despite the challenges of limited input data, the fixed-bed model performed well with very little calibration. Modeled water surface elevations matched observed values to within an average of approximately ±0.34 ft. over the full range of discharges during the calibration period. Error was generally one of timing rather than magnitude, and was likely attributable to the combination of a relatively-coarse 12-hour hydrograph output interval and the steep hydrograph slopes. A visual comparison of the observed and modeled water surface elevations shows an offset of less than 12 hours between observed and modeled stages for almost all large events.

Sediment Transport Calibration and Verification

As with the hydraulic calibration, the lack of historic gaged sediment data between the dams presents a major challenge to calibrating sediment transport models within the Lower Susquehanna reservoir system. In order to perform the calibration and verification, a measure of change was developed and the model parameters were systematically altered within reasonable ranges to approach the desired results.

Sediment Volume Change. In lieu of inter-system gaged sediment transport data, changes in bed volume provided an alternate source of calibration data for this study. While the temporal resolution of the calibration was limited to the frequency of historic bathymetric surveys, the relatively high spatial resolution of the cross sections within the reservoirs provided some insight into the bed regions more prone to deposition and/or scour.

Several methods for computing volume change were evaluated, and Gomez and Sullivan Engineers ultimately applied a GIS-based method. The cross sections from Langland and Koerkle's HEC-RAS model (2014) were assumed to represent the 2008 bathymetry, and the 2013 and 2015 bathymetries were developed using data collected by Gomez and Sullivan Engineers. To minimize apparent bed volume changes due to differences in the number and location of cross sections, a set of common cross sections was selected based on those with good spatial agreement between the USGS survey and model datasets and the Gomez and Sullivan Engineers bathymetry data.

All volume change calculation methods indicated net deposition over both time periods for Lake Clarke and little change for Lake Aldred. Calculated changes were relatively small in terms of the system as a whole. The lakes were each split into a number of sub-areas, and the sub-areas with the greatest degrees of certainty in bathymetric data (five in Lake Clarke, four in Lake Aldred) were selected for the calibration. A range was calculated for each sub-area based on the assumed accuracy of the survey equipment.

Model Calibration. Calibration was performed by running the model from 1 January 2008 through 30 August 2013—a period characterized by several large events, including

Tropical Storm Lee—and systematically varying parameters to produce results approaching the observed bed volume changes for the same period.

Initially, far too much sediment deposited in the middle reaches of the Lake Clarke, and too little sediment was deposited in the two-reservoir system. The former was remedied by reducing the maximum erodible bed depth at the upstream end of Lake Clarke, near the gage, where the bed is primarily exposed bedrock and some boulders (Zarr, 2016), and removing sediment coarser than fine sand (0.25 mm) from the inflowing load at the upstream boundary. By reducing the amount of coarse sediment scoured or imported at the upstream end of the model, more material was transported downstream, better reflecting observed volume changes. Also, an erosion channel of the shape and size of the main channel was input in the deeper, island-free region which conveys the majority of flow during most events—this new feature in HEC-RAS 5.0 helps mimic the lateral variation in geomorphic changes within the confines of a one-dimensional hydraulic model. Finally, the Manning's *n* value was decreased slightly in the main channel to increase conveyance. Calibrated Manning's *n* values chosen for the main channel ranged from 0.023 to 0.029, and values in the overbanks and islands ranged from 0.03 to 0.1. These changes concentrated flow and focused the effects of shear stress in the main channel while permitting deposition near islands and areas of ineffective flow.

To increase deposition more generally throughout the system, the sediment loading at Marietta was increased by 20-30% at various flows—values still well within the range of scatter in the observed loading. Hybrid bed gradations were created for many cross sections, with the percent clay and cohesive parameters adjusted to promote or resist scour. The changes were relatively small: increases or decreases in clay composition of less than 4% of the total sample, and cohesive parameters limited to the range measured in the SEDflume analysis of Conowingo sediments. Overall, the calibration process was largely successful, and the modeled net volume change for each reservoir as a whole fell within the target range of the observed values.

<u>Model Verification</u>. Following the calibration process, the model was re-run using a longer simulation period extending from 1 January 2008 to 15 October 2015. Modeled bed volume change results were evaluated from 30 August 2013 through 15 October 2015 and compared with the observed volume changes for the same period to verify the model's performance. This period was characterized by unusually low flows with no peaks above 200,000 cfs, which provided a good contrast to the calibration period.

While the model performed fairly well from the outset, further calibration was required to balance differences in sediment dynamics between the two periods. The percentage of fines in the Conestoga River and Pequea Creek inflowing loads was increased slightly to shift the bulk of deposition further downstream within Lake Aldred, and the sediment loading for Pequea Creek was increased by 10-20% at various flows—again, using values still well within the observed data scatter.

The final model was somewhat limited in its ability to replicate volume changes on a sub-area basis, only matching the target ranges for half of the sub-areas, but successfully replicated whole-reservoir changes for both reservoirs and time periods. Figure 4 presents a comparison of modeled cumulative bed volume change and target ranges in the calibration sub-areas used for Lake Clarke, and Figure 5 presents the equivalent results for Lake Aldred.

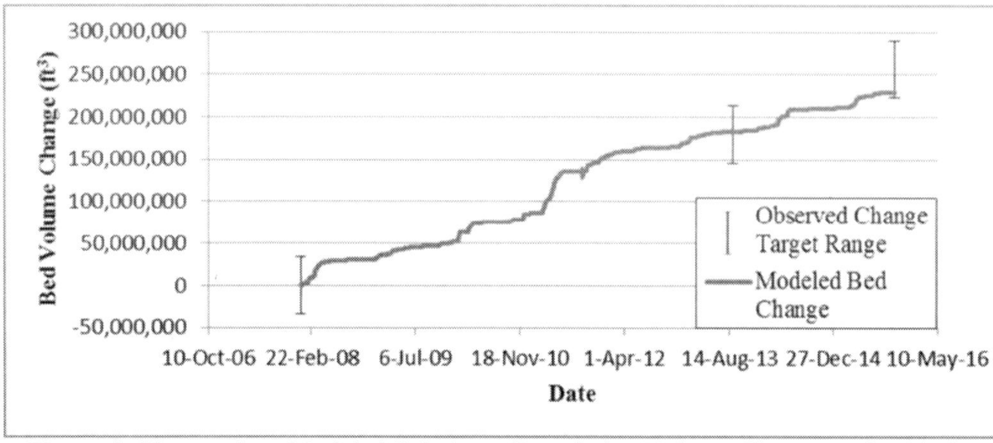

Figure 4. Observed vs. Modeled Sub-Area Cumulative Volume Change: Lake Clarke, January 2008 – October 2015.

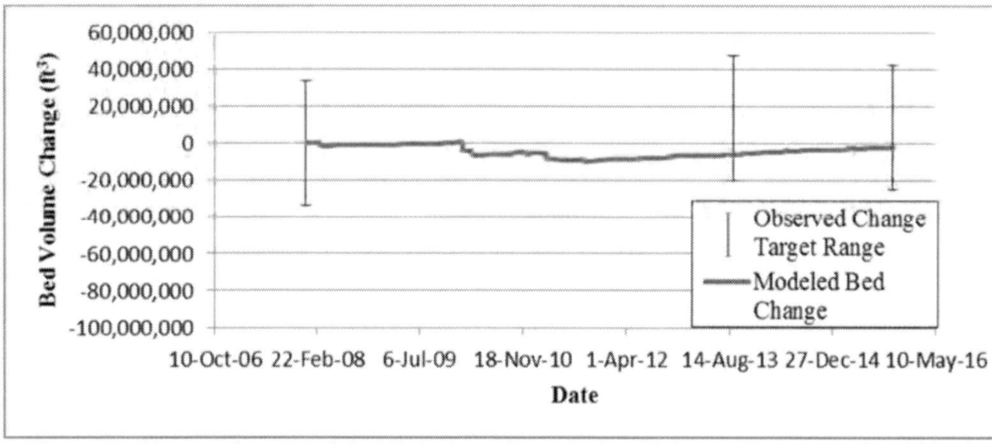

Figure 5. Observed vs. Modeled Sub-Area Cumulative Volume Change: Lake Aldred, January 2008 – October 2015.

SUMMARY OF RESULTS AND LIMITATIONS

After the model was calibrated and verified, four additional runs were executed to simulate historical flow events. In each case, one of four sets of flood hydrographs (peaks ranging 421,000 - 588,000 cfs) was added to the end of the flow time series for the model's upstream boundary and tributary inflows, to simulate the effects of a hypothetical large event given the reservoirs' states in October 2015. The purpose of the production runs was to augment the number of sediment output data at very large flows, for the purpose of strengthening the rating curve fits at higher flows.

The sediment mass transport results for the 12 particle size classes were aggregated into three classes—sand and gravel, silt, and clay—as requested for use in the Phase 6 Chesapeake Bay Watershed Model. Transport rates for each size class were then plotted versus discharge at the Marietta gage, Safe Harbor Dam, and Holtwood Dam.

Scatter in the results increased in the downstream direction, due to the effects of storage and other natural system factors. Variance was pronounced at large flows, especially for sands, and hysteresis was visible for individual storm events. Figure 6 shows the hourly mass transport results, aggregated into the three size classes, for the Holtwood Dam location. The green points furthest to the right of the plot track the modeled sand transport during Tropical Storm Lee in September 2011. Modeled transport was much greater on the rising limb of the storm than on the falling, likely due to both the asynchronous peaking of the tributaries and mainstem flows and the reduction in applied bed shear stress on the falling limb. Notably, modeled sand transport rates at the highest flows vary by more than 350,000 tons/day for a given discharge.

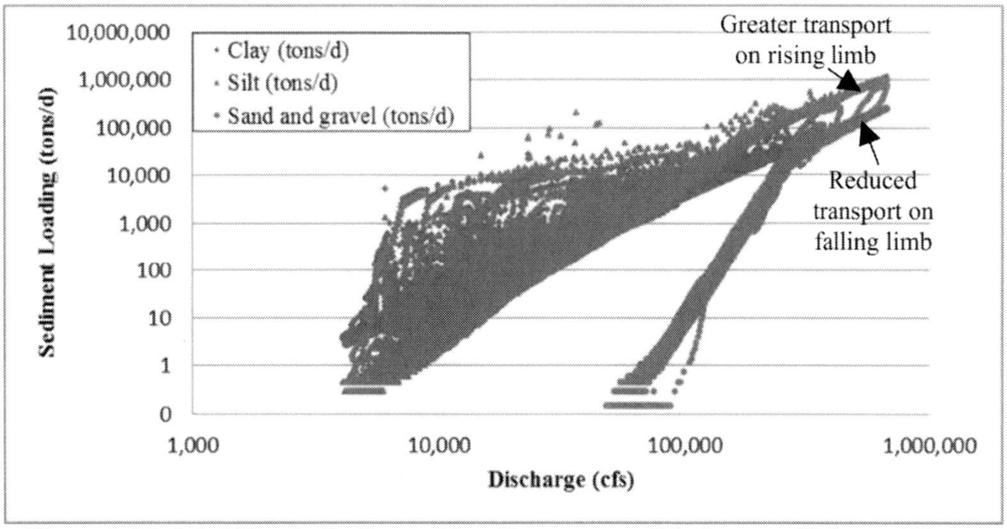

Figure 6. Modeled Hourly Sediment Mass Transport vs. Discharge at Holtwood Dam. Points Represent Hourly Flows and Associated Hourly Sediment Loads (in Tons/Day).

Piecewise rating curves were fit to mass transport data for each particle size class and location and extrapolated to 1,000,000 cfs. Comparison of the curves for each location gave additional insights into modeled results. However, a direct visual mass balance is not possible using the three curves alone, for two reasons. First, the three locations seldom experience the same discharge simultaneously, and the rating curves do not themselves represent temporal offsets. Second, a true mass balance for Lake Aldred would require explicit representation of all sediment inputs and outputs, and while the rating curves do represent the inputs and outputs at Safe Harbor and Holtwood Dams, respectively, they do not include the modeled sediment inflows from Conestoga River and Pequea Creek. Despite these caveats, the relative values of the overlain curves do reflect generally-expected behavior for the reservoirs. Figure 7 presents the rating curves.

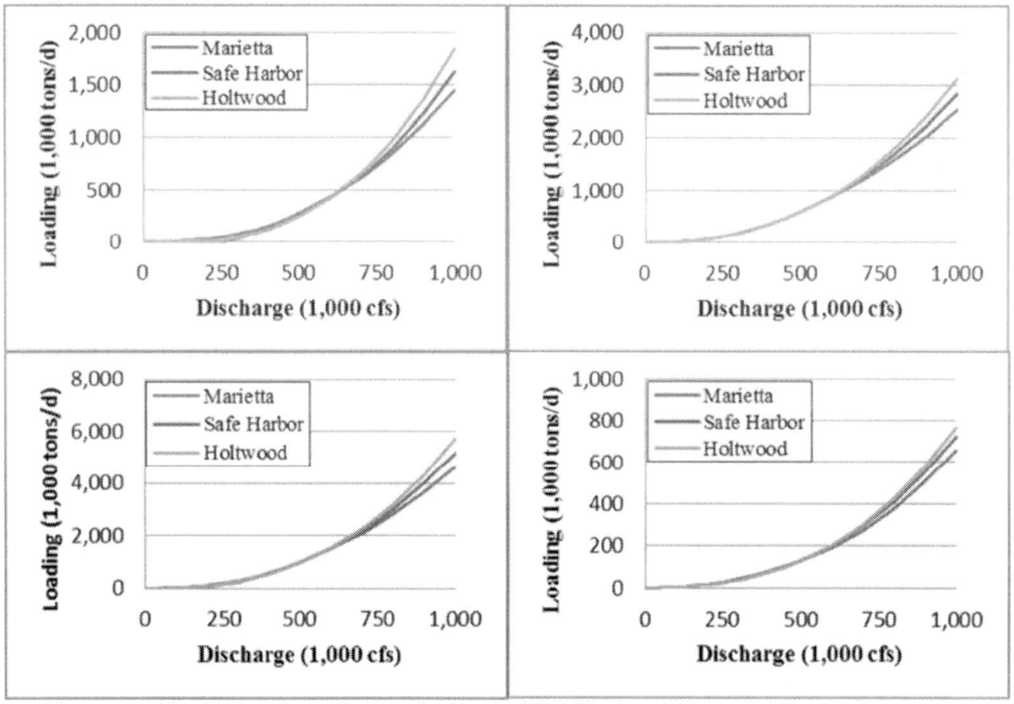

Figure 7. Clockwise from Top Left: Sand, Silt, Clay, and Total Sediment Transport Rating Curves Produced for Three Model Locations, Based on Modeled Results.

For all particle classes, loading was higher at the Marietta gage at low flows than at Holtwood Dam, indicating net deposition in the system. Conversely, transport was greater at the two reservoirs than at the Marietta Streamgage for higher flows, generally indicating net scour in the system.

Model Limitations

Several limitations and assumptions should be noted in order to understand the modeled results. First, while HEC-RAS version 5.0 offers many advantages over previous HEC models developed for the same reach, the one-dimensional representation of the reservoir system is limited in its ability to capture the effects of some two-dimensional flow dynamics. While the best available methods and data were used in the model's development, it was neither possible nor reasonable to fully capture the effects of all channel features, and the modeling effort's goal was to represent general trends rather than exact values for each sub-area.

The availability and quality of the input geometry also represent sources of uncertainty in the results. The method of calibrating to bed volume changes limits calibration to periods bracketed by available historical bathymetric data, and very few datasets were available for the two reservoirs. The 2008 channel geometry was based on Langland and Koerkle's HEC-RAS model (2014), which itself drew from several sources with varying levels of detail and accuracy. Also, the calculated volume changes were subject to uncertainties based on the methods used and the quality of the bathymetric data. The survey data collected by Gomez and Sullivan Engineers in 2013 and 2015 are of greater density and quality than the bathymetric data used for development of the 2008 geometry. Finally, the lack of dam operations data for Safe Harbor and Holtwood Dams represented a limitation to model development.

Other factors limited the reliability of the sediment output rating curves. The calibration and verification periods included few large storm events, and even with the addition of the production runs, the largest modeled output discharge was approximately 660,000 cfs. As a result, the sediment output values for discharges greater than 600,000 cfs were based on uncertain, projected curves fit to data at lower discharges.

The hysteresis and increased variance in modeled loading at high discharges also limited the rating curves' usefulness during extreme events. Sediment transport is a function of many variables, so it was not possible to fully capture the reservoirs' behavior using rating curves dependent solely on discharge. To evaluate the curves' ability to represent longer-term changes, a mass balance was performed on a monthly basis to compare modeled mass changes in each reservoir with expected mass change calculated using the modeled flow time series and the sediment rating curve at each location. (Tributary inflows were also included in the mass balance.) The rating curves appeared to be good representations for Lake Clarke, especially for silts and sands, but did not appear to perform quite as well for Lake Aldred. This is due to two factors. First, the overall magnitude of the bed changes was much smaller in Lake Aldred, which alternated between periods of relatively minor net scour and deposition. Also, the discharge-based rating curves fit the modeled sediment transport data better in Lake Clarke than in Lake Aldred, due the greater scatter in the latter. This is especially true at the upper end of discharges for sand. Due to the large variation in sand transport near the peak of storm hydrographs, along with the disproportionate importance of large flow events, sand transport during large events in Lake Aldred is not represented as well by the rating curves as other sediment size classes and periods.

CONCLUSION

WEST Consultants, Inc. was contracted by Exelon Generation Company, LLC to develop a one-dimensional HEC-RAS 5.0 sediment transport model for Lakes Clarke and Aldred on the Lower Susquehanna River, as part of a multi-model initiative. The model applied gaged sediment and flow records, sediment core data, and a variety of other inputs to replicate sediment volume change in the two reservoirs between 2008 and 2013, a period characterized by several large storm events. A verification period from 2013 to 2015 validated the model's performance during a time characterized by lower flows. The calibrated model successfully simulated bed changes for both reservoirs and time periods, within target ranges defined by survey equipment accuracy, and predicted both deposition and scour processes at varying discharges and reservoir states. The model results were used to develop sediment rating curves for three particle size classes (sand, silt, and clay) at three locations (Marietta, PA; Safe Harbor Dam; and Holtwood Dam), which will aid in parameterization of the Chesapeake Bay Watershed Model by other researchers.

REFERENCES

Edwards, R. E., 2006. Comprehensive Analysis of the Sediments Retained Behind Hydroelectric Dams of the Lower Susquehanna River, Harrisburg: Susquehanna River Basin Commission.

Hainly, R. A., Reed, L. A., Flippo, Jr., H. N. & Barton, G. J., 1995. Deposition and Simulation of Sediment Transport in the Lower Susquehanna River Reservoir System, Lemoyne: U.S. Department of the Interior, U.S. Geological Survey.

Hirsch, R. M., 2012. Flux of Nitrogen, Phosphorus, and Suspended Sediment from the Susquehanna River Basin to the Chesapeake Bay during Tropical Storm Lee, September 2011, as an Indicator of the Effects of Reservoir Sedimentation on Water Quality, Washington, D.C.: U.S. Geological Survey.

Langland, M. J., 2009. Bathymetry and Sediment-Storage Capacity Change in Three Reservoirs on the Lower Susquehanna River, 1996-2008, New Cumberland: U.S. Geological Survey.

Langland, M. J. & Koerkle, E. H., 2014. Calibration of a One-Dimensional Hydraulic Model (HEC-RAS) for Simulating Sediment Transport through Three Reservoirs, Lower Susquehanna River Basin, 2008-2011, Lower Susquehanna River Watershed Assessment, Phase 1, Appendix A with 3 Attachments, Washington, D.C.: U.S. Department of the Interior, U.S. Geological Survey.

MDNR, 2014. Quarterly Meeting Summary for April 30, 2012, Lower Susquehanna River Watershed Assessment, Phase 1, Appendix I-6, Baltimore, MD: Maryland Department of Natural Resources.

Ott, A. N., Takita, C. S., Edwards, R. E. & Bollinger, S. W., 1991. Loads and Yields of Nutrients and Suspended Sediment Transported in the Susquehanna River Basin, 1985–1989, Harrisburg: Susquehanna River Basin Commission, Resource Quality Management and Protection Division.

Porse, N. C., 2010. Taking Holtwood into the Next Century. Hydro Review, 29(4), p. 8.

Reed, L. A. & Hoffman, S. A., 1996. Sediment Deposition in Lake Clarke, Lake Aldred, and Conowingo Reservoir, Pennsylvania and Maryland, 1910-93 , Lemoyne: U.S. Geological Survey.

Safe Harbor Water Power Corporation, n.d.. Estimated Levels of Lake Aldred at Holtwood Dam. [Online]
Available at: http://www.shwpc.com/includes/hd_levels.htm
[Accessed May 2016].

Safe Harbor Water Power Corporation, n.d.. Facts & Figures. [Online]
Available at: http://www.shwpc.com/facts_figures.html
[Accessed May 2016].

Scott, S.H. & Sharp, J.A., 2014 (February 2014 draft). Sediment Transport Characteristics of Conowingo Reservoir, Attachment B-2: SEDflume Erosion Data and Analysis, ERDC: U.S. Army Corps of Engineers.

SRBC, n.d.. Sediment and Nutrient Assessment Program (SNAP). [Online]
Available at: http://www.srbc.net/programs/cbp/nutrientprogram.htm
[Accessed December 2015].

Sullivan, T., 2016a. Personal Communication. Gomez and Sullivan Engineers.

Sullivan, T., 2016b. Personal Communication. Gomez and Sullivan Engineers.

Takita, C. S. & Edwards, R. E., 2001. Nutrients and suspended sediment transported in the Susquehanna River Basin, 2000, and trends, January 1985 through December 2000, Harrisburg: Susquehanna River Basin Commission.

USACE, 2016a. HEC-RAS River Analysis System Hydraulic Reference Manual Version 5.0, Davis, CA: United States Army Corps of Engineers.

USACE, 2016b. HEC-RAS River Analysis System User's Manual, Version 5.0, Davis, CA: U.S. Army Corps of Engineers.

USGS, 2016. USGS Sediment Data Portal. [Online]
Available at: http://cida.usgs.gov/sediment/
[Accessed December 2015].

USGS, 2016. USGS Water Data for USA. [Online]
Available at: http://waterdata.usgs.gov/nwis
[Accessed December 2015].

Weissman, D., 2016. Talen Closes Popular Lancaster Properties Near Holtwood Dam. York Dispatch, 29 April.

Zarr, L., 2016. Personal Communication. U.S. Geological Survey, Pennsylvania Water Science Center.

Zeff, M. L., 2016. Personal Communication. AECOM.

TO RELEASE OR NOT TO RELEASE? DAM OPERATION MODELING USING GRIDDED SHORT-TERM FORECAST RAINFALL

Maged Aboelata[1]

A critical decision for dam operators in anticipation of a storm event is whether to reduce the pool level to create storage void for the incoming flow, or maintain the current storage level longer in case the storm changes course. The decision becomes more complex if the magnitude of dam releases could negatively affect downstream areas.

This presentation discusses an innovative modeling approach applied to two dams in series owned and operated by the Washington Suburban Sanitary Commission for water supply. The system consists of a detailed hydrologic model that receives 6-hour forecast rainfall data, a dam operation model to control releases from the dam under specific rules and constraints, and a hydraulic and mapping model to estimate flooding elevations at critical locations. The system is designed to work seamlessly between the three models without the need for manual data transfer. Model operators can test operation actions in advance and try to minimize loss of storage on one hand, and maintain the dam safety and the safety of the downstream community on the other hand.

[1] AECOM, maged.aboelata@aecom.com

USACE CWMS: CASE STUDIES OF WATER MANAGEMENT DECISION SUPPORT TOOLS

Ben Zoeller[1]
Nathan Lefkovitz[2]

The Corps Water Management System (CWMS) is the automated decision support tool used by water managers at the U.S. Army Corps of Engineers (USACE) to support its water control management mission, which includes the regulation of river flow through more than 700 reservoirs, locks, and other water control structures. Through their national implementation effort, the USACE has been systematically developing CWMS for watersheds throughout the U.S.

Managed by the USACE New England District, the Blackstone and Thames River Watersheds combined drain over 2,000 square miles of the States of Connecticut, Massachusetts, and Rhode Island. Beginning in late 2014, Amec Foster Wheeler, under contract with the USACE Mapping, Modeling, and Consequences (MMC) Center, developed and implemented CWMS modeling for the Blackstone and Thames River Watersheds. This presentation will highlight the unique model development challenges encountered in both watersheds. The HEC-HMS, HEC-ResSim, and HEC-RAS model development in the Thames River Watershed will be highlighted, including discussion on how those models interact during real-time forecasting and how they provide decision support for water managers. This presentation will also describe how the risk analysis model HEC-FIA was developed for estimating Population at Risk (PAR) and economic damages for modeled events in the Blackstone River Watershed and will underscore the importance of creating custom-built GIS tools to properly attribute highly accurate state supplied parcel-level building data for the purpose of HEC-FIA modeling.

[1] Amec Foster Wheeler, ben.zoeller@amecfw.com
[2] Amec Foster Wheeler

SYSTEMS-BASED RISK ANALYSIS OF LAKE MENDOCINO FORECAST-INFORMED RESERVOIR OPERATIONS WITH HEC-WAT

Matthew McPherson[1]
Leila Ostadrahimi[2]

Forecast Informed Reservoir Operations (FIRO) at the Corps' Coyote Valley Dam can potentially make more Lake Mendocino storage and Russian River flow available for water supply and environmental uses without increasing flood risk. Corps policy requires detailed consideration of flood risk impacts elsewhere in the Russian River watershed due to the different operations. The HEC Watershed Analysis Tool (HEC-WAT) model performs simulations of project operations for a baseline condition, and alternative operational plans that leverage forecast information about future precipitation, reservoir inflow volume, and available channel capacity at a key downstream location. The HEC-WAT model produced results for both a 60-year continuous simulation (i.e., Period of Record compute - PoR) and a probabilistic Monte Carlo approach (Flood Risk Analysis compute - FRA).

The HEC-WAT model takes precipitation data as the primary input, and uses HEC-HMS to compute the resulting stream flow. An HEC-ResSim model simulates operations at Coyote Valley Dam and Warm Springs Dam. An unsteady HEC-RAS model routes the reservoir releases and tributary flows through the Russian River to the ocean, determining peak water surface elevations and inundated extents. An HEC-FIA model uses a detailed parcel-based structure inventory and stage-damage information to compute flood damages. Differences in flood risk among alternatives are quantified by frequency analysis of flows and stages at key locations along the mainstem river, and annual average damage in the watershed. Other metrics include reservoir storage at the end of flood season, duration analysis of compliance with a representative in-stream flow requirement, incidence of downstream flows above a key threshold specified in the existing water control manual, and a hybrid metric that considers both reservoir storage for available and releases made for conservation purposes.

The results demonstrate that even very aggressive forecast-informed operations at Coyote Valley Dam do not necessarily increase flood risk on the Russian River.

[1] U.S. Army Corps of Engineers, Matthew.M.McPherson@usace.army.mil
[2] U.S. Army Corps of Engineers

HEC-RESSIM 3.3 - NEW FEATURES TO SUPPORT COMPLEX STUDIES

Joan Klipsch[1]

The Reservoir System Simulation (HEC-ResSim) software developed by the Hydrologic Engineering Center of the US Army Corps of Engineers simulates operations at one or more reservoirs to meet a variety of operational goals and constraints. The new version (v3.3) of HEC-ResSim, released in October 2016, introduces some new features to assist reservoir modelers in performing complex studies. These new features include Ensemble Alternatives, Firm Yield Analysis, and Storage Accounts.

The Ensemble feature applies a fixed set of operations to a predefined ensemble of inflows. The ensemble of inflows can be generated by an external "sampling" product or can be assembled from the results of a hydrology model that simulated a set of hypothetical or probabilistic events or from a period of record of observed flows. A separate run is performed for each member of the ensemble and standard ResSim output is produced. Special plotting features have been included for display of the ensemble output including the addition of several statistical curves for the min, max, mean, and 25 & 75 percentiles.

The Firm Yield feature can assist in determining the maximum yield that a reservoir or system of reservoirs can provide over period of record or critical period. This feature repeatedly executes the alternative, adjusting the limit of the identified rule until the limit is found that "just empties" the active storage of the reservoir operating for the identified rule.

The Storage Accounting feature tracks usage of storage to meet a specific demand and can restrict withdrawals when the account empties. The storage accounting is also tied in to the Firm Yield Analysis capability to assist in determining the yield of a specific storage account.

[1] U.S. Army Corps of Engineers, joan.klipsch@usace.army.mil

A DIRECT FINITE ELEMENT METHOD FOR NONLINEAR ANALYSIS OF SEMI-UNBOUNDED DAM-WATER-FOUNDATION ROCK SYSTEMS

Anil K. Chopra[1]
Arnkjell Lokke[2]

Earthquake analysis of concrete dams is greatly complicated by the fact that the structure interacts with the water impounded in the reservoir and with the deformable foundation rock that supports it and because the fluid and foundation domains extend to large distances. The standard finite element method (FEM) is unable to model such semi-unbounded domains.

This paper will present a direct FEM that overcomes the limitations of the standard FEM by introducing wave-absorbing boundaries at two locations: (1) upstream end of the fluid domain to model its essentially infinite length; and (2) the bottom and side boundaries of the foundation domain to model its semi-unbounded geometry. Included are water compressibility, reservoir bottom sediments; mass, stiffness, and material damping of the foundation rock. Thus, the untenable assumptions of massless rock and incompressible water in the standard FEM are eliminated.

The earthquake excitation defined at the bottom and side boundaries of the foundation domain is determined by deconvolution of the design ground motion, typically specified on level ground at the elevation of the abutments. The resulting spatially varying motions cannot be input directly at wave-absorbing boundaries; instead, tractions determined from the motions are specified.

The proposed FEM is applicable to nonlinear systems, thus permitting modeling of concrete cracking, as well as sliding and separation at construction joints, lift joints, and concrete-rock interfaces. This method will be validated against the well-established substructure method that rigorously models semi-unbounded domains. Results of example analyses will be presented to demonstrate accuracy and effectiveness of the method. Applications of the method to analysis of actual dams will be demonstrated and its usefulness in design of new dams and evaluation of existing dams will be discussed.

[1] University of California, Berkeley, chopra@berkeley.edu
[2] University of California, Berkeley

NUMERICAL ASSESSMENT OF HYDRODYNAMIC LOADS INDUCED DURING SEISMIC INTERACTION BETWEEN RESERVOIR AND CONCRETE DAM

Jerzy Salamon, P.E., Ph.D.[1]
Jonna Manie, M.Sc.[2]

ABSTRACT

During the 2016 Monticello Blind Prediction Analysis Workshop hosted by the USSD Earthquake Committee, the Monticello Dam case study was investigated. Participants in the study attempted to generate finite element (FE) results using various methods to determine seismically induced hydrodynamic loads on the concrete arch dam. The primary methods were based on the "added mass" approach, as well as the "acoustic fluid" and the "fluid-like material model" analysis techniques. Among the results presented in the workshop, significant differences were observed for all methods.

This paper presents a comparative analysis of these three methods. This comparison is illustrated using the results of FE analysis on two concrete gravity dams, one of small size and one of medium size. This paper also discusses the assumptions and limitations of the analysis techniques.

INTRODUCTION

The hydrodynamic loads induced during earthquakes are important for the design and structural evaluation of concrete dams. An in-depth understanding of the computation methods and their limitations are the key factor that determines the accuracy of the analysis results.

In this paper, seismic interactions between a reservoir and a concrete gravity type dam are investigated. Three approaches commonly used in the direct time integration analysis are presented and discussed: (1) the "added mass" approach, (2) the "acoustic fluid" analysis technique, and the (3) "fluid-like material model" method.

ADDED MASS APPROACH

General

The first systematic analysis of dam-reservoir interaction was performed by Westergaard [5] in connection with the Bureau of Reclamation's design of Hoover Dam. In a study of the earthquake response of a rigid dam with a vertical upstream face (Figure 1), Westergaard developed an analytical solution for the hydrodynamic pressure distribution

[1]Technical Specialist, Waterways & Concrete Dams Group, Bureau of Reclamation, PO Box 25007, Denver, Colorado, 80225, USA, e-mail: jsalamon@usbr.gov
[2]Consulting Engineer, Engineering Team, DIANA FEA BV, Delftechpark 19a, 2628 XJ, Delft, The Netherlands, e-mail: j.manie@dianafea.com

in the reservoir and along the upstream face of the dam, as a result of the dam harmonic horizontal motions. The approximate formula for the parabolic hydrodynamic pressure distribution over a vertical dam face, and the concept of added mass presented by Westergaard in a fundamental and in a simplistic form, have been used consistently in the engineering practice for the preliminary analysis and design of concrete dams.

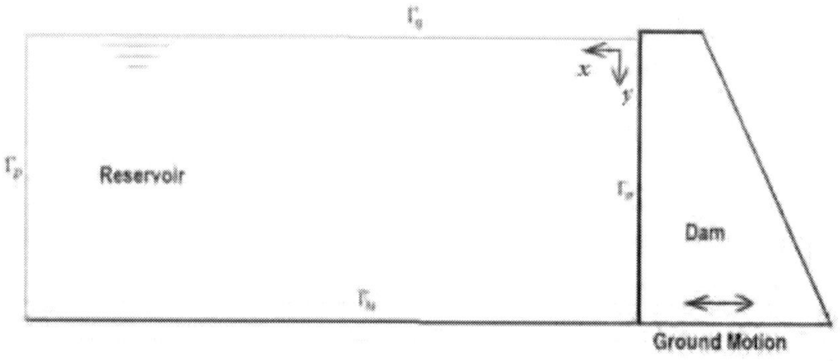

Figure 1. Westergaard's model.

<u>**Problem Formulation**</u>

The seismic motion of a rigid dam interacting with an infinite length reservoir of depth h was mathematically described by Westergaard [5] in terms of the theory of elasticity of solids, based on the formulation provided by Lamb [2]. The boundary value problem (BVP) was formulated by Westergaard in the following form:

Equations of Motion:

$$\frac{\partial \sigma}{\partial x} = \frac{w}{g}\frac{\partial^2 u}{\partial t^2} \qquad \frac{\partial \sigma}{\partial y} = \frac{w}{g}\frac{\partial^2 v}{\partial t^2} \qquad \text{(Eq.1)}$$

Geometric Equation:

$$\varepsilon_{xy} = \frac{1}{2}\left(\frac{\partial u}{\partial y} + \frac{\partial v}{\partial x}\right) \qquad \text{(Eq.2)}$$

Constitutive Law for Reservoir:

$$p = \frac{1}{3}\sigma_{kk} = K\,\varepsilon_{kk} \qquad \text{(Eq.3)}$$

where: u, v are horizontal and vertical displacements, respectively
x is the axis at the surface of the reservoir directed upstream (Figure 1)
y is the vertical downward axis (Figure 1)
p is the pressure in the reservoir
σ_{kk} *and* ε_{kk} are volumetric stress and volumetric strain, respectively
w is the weight of water per unit of volume
g is the acceleration due to gravity
K is the bulk modulus of water

The model in his original form was defined with the following assumptions:

- The dam upstream face is straight and vertical at Γ_σ.
- The dam does not deform and is considered to be a rigid block.
- The reservoir is infinite in length at Γ_p.
- Dam sinusoidal oscillations are horizontal.
- Small motions are assumed during earthquake.
- The problem is defined in two-dimensional (2D) space.
- The period of free vibration of the dam, T_0, is significantly smaller than the period of vibration, T, of the earthquake (resonance is not expected).
- Nondimensional horizontal acceleration of $\alpha = 0.1$.
- Horizontal harmonic ground/dam motion applied to the dam.

Exact and Approximate Solutions

Westergaard addressed the problem of horizontal and vertical motions of water (plane strain) in the form of a stress (pressure) distribution in the reservoir and along the upstream face of the dam. The maximum pressure distribution at the upstream face of the dam was expressed by the equations:

$$p = \frac{8\alpha wh}{\pi^2} \sum_{1,3,5\ldots}^{n} \frac{1}{n^2 c_n} \sin\frac{n\pi y}{2h} \quad \text{(Eq.4)}$$

with:

$$c_n = \sqrt{1 - \frac{16wh^2}{n^2 gkT^2}} \quad \text{(Eq.5)}$$

where:

h is the depth of the reservoir
α is the maximum horizontal acceleration of dam divided by g
T is the period of horizontal vibration at the dam

Westergaard obtained the approximate solution by approximation of the exact solution (Eq.4 and Eq.5) using a parabolic function expressed by Eq.6:

$$p = \frac{7}{8} w\alpha\sqrt{h(h-y)} \quad \text{(Eq.6)}$$

Concept of Added Mass

The parabolic hydrodynamic pressure distribution over the height of the dam was determined by Westergaard to be the same as a pressure developed by a certain body of water called "*added mass,*" which was forced to move with the dam during the ground motion. The volume of water of "*added mass*" per unit width was described by a parabola (Figure 2) with dimension b expressed by Eq.7:

$$b = \frac{7}{8} w \sqrt{h(h - y)} \qquad \text{(Eq.7)}$$

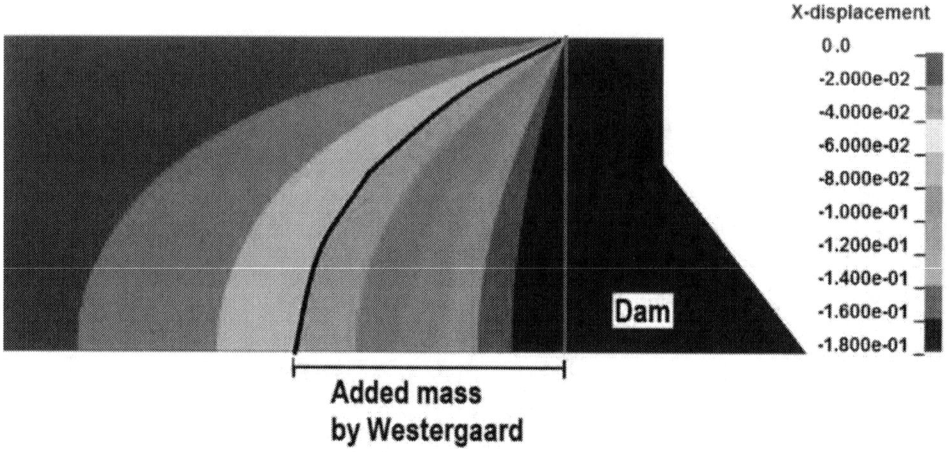

Figure 2. Horizontal displacement of a rigid dam obtained in FE analysis for a 100-feet-deep reservoir. Black line denotes the area of *added mass* volume determined by Westergaard (Eq.7).

DIRECT TIME INTEGRATION METHODS

General

This section presents a brief overview of the direct time integration methods used to calculate the seismically induced hydrodynamic pressures at the face of a concrete gravity dam using the FE analysis method. Two approaches are presented: (1) the classical "acoustic fluid" approach, and (2) the "fluid-like material model" approach.

Coupling of Acoustic Fluid and Structural Elements

In this method, the displacement field of the dam (structural domain) is coupled with the pressure field of the reservoir (fluid domain) via the interaction forces at the interface between the dam and the reservoir (Figure 3).

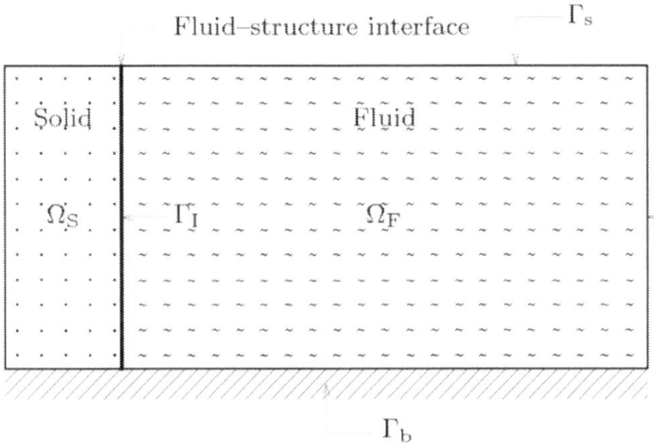

Figure 3. Acoustic fluid model.

In the structural domain Ωs, the discretization in the familiar form is given by:

$$M_S\ddot{u} + C_S\dot{u} + K_S u + f_I = f_S^{ext} \quad \text{(Eq.8)}$$

where: M_S, C_S, and K_S are mass, damping, and stiffness matrices, respectively
u is a set of unknowns describing the displacements of the structure
f_I stands for forces due to the interface interaction with the fluid
f_S^{ext} represents external forces

The dynamic pressure distribution in the fluid is described by a single variable p. Assuming the state of the fluid is linear, the governing equation is the wave (acoustic) equation:

$$\nabla^2 p = \frac{1}{c^2}\ddot{p} \quad \text{(Eq.9)}$$

where p is the dynamic pressure (compression positive)
c is the wave speed given by:

$$c^2 = \frac{\beta}{\rho_F} \quad \text{(Eq.10)}$$

where β is the bulk modulus
ρ_F is the fluid density

Coupling within the dam and the reservoir is achieved by considering the interface forces at the dam face. The conditions applying to the fluid-structure interface Γ_I can be written as:

$$\sigma\, n_S = p\, n_F \quad \text{(Eq.11)}$$

where n_F and n_s are the outward normal to the fluid domain and the outward normal to the structural domain, respectively.

Coupling between the fluid domain and the structural domain can be achieved by continuity between the normal displacements with the condition $u_F = u_S$ and is obtained by combining this condition with Eq.11:

$$\frac{\partial p}{\partial n} = -\rho_F \mathrm{n_F}^{\mathrm{T}} \ddot{\mathrm{u}}_{\mathrm{S}} \qquad \text{(Eq.12)}$$

At the free surface, Γ_s, and the fluid far field boundary (infinite extent), Γ_e, the prescribed dynamic pressures are assumed zero. For the bottom boundary, Γ_b, a full reflective bottom is assumed (i.e., no bottom absorption effects or prescribed dynamic pressures).

Implementing a standard FE discretization for approximating p, a system of algebraic equations for the fluid domain can be defined by:

$$M_F \ddot{p} + K_F p + r_I = 0 \qquad \text{(Eq.13)}$$

where M_F and K_F are the fluid mass and stiffness matrixes, respectively
r_I is the interface reaction force

The interface interaction forces within the fluid element can be written as:

$$f_I^e = -R^{eT} p^e \qquad \text{(Eq.14)}$$

where R^e is the interaction matrix for the dynamic pressures on the element level.

Likewise, the contribution r_I on element level as:

$$r_I^e = \rho_F R^e \ddot{u}^e \qquad \text{(Eq.15)}$$

After assembling contributions from each type of element, the following coupled system of equations for the fluid structure interaction problem is obtained:

$$\begin{bmatrix} M_S & 0 \\ \rho_F R & M_F \end{bmatrix} \begin{Bmatrix} \ddot{u} \\ \ddot{p} \end{Bmatrix} + \begin{bmatrix} C_S & 0 \\ 0 & 0 \end{bmatrix} \begin{Bmatrix} \dot{u} \\ \dot{p} \end{Bmatrix} + \begin{bmatrix} K_S & -R^T \\ 0 & K_F \end{bmatrix} \begin{Bmatrix} u \\ p \end{Bmatrix} = \begin{Bmatrix} f_S^{ext} \\ 0 \end{Bmatrix} \qquad \text{(Eq.16)}$$

If the compression effects are neglected (i.e., an incompressible fluid is assumed ($c = \infty$)), the fluid matrix M_F becomes zero. The dynamic pressure vector $\boldsymbol{p}$ can now be obtained directly in terms of $\ddot{u}$ as:

$$\boldsymbol{p} = -K_F^{-1} \rho_F R \ddot{u} \qquad \text{(Eq.17)}$$

Substituting Equation 17 in Equation 16, the structural matrix results in the following general equation:

$$(M_S + \widetilde{M}_F)\ddot{u} + C_S\dot{u} + K_S u = f_S^{ext} \qquad \text{(Eq.18)}$$

where the added mass is simply given by:

$$\widetilde{M}_F = \rho_F R^T K_F^{-1} R \qquad \text{(Eq.19)}$$

If an arbitrary transient loading is considered, the response of the model with incompressible fluid is obtained by a direct time integration method.

The "acoustic fluid" formulation is evaluated in this study using DIANA FE software developed by DIANA FEA BV [4].

Fluid-Like Behavior Model

Modeling fluid behavior with a "fluid-like" material model is an alternative approach for analyzing the fluid-structure interaction behavior. The fluid is defined in the BVP as a structural elastic material (Eq.20) with the shear modulus equal to zero.

Constitutive Law for Fluid:

$$\dot{p} = \frac{1}{3}\dot{\sigma}_{kk} = K\,\dot{\varepsilon}_{kk} \qquad \text{(Eq.20)}$$

where: $\dot{\varepsilon}, \dot{\sigma}, and\ \dot{p}$ are strain, stress, and pressure rates
K is the bulk modulus of fluid

The behavior of the fluid-like material model is implemented in LS-DYNA FE software developed by LSTC [3].

CASE STUDIES

General

To illustrate the analysis techniques discussed above, two models of the concrete dams are considered: (1) a small 100-foot-high dam, and (2) a medium 287-foot-high dam (Figure 4).

The foundation in the models has a length of 1200 feet and a depth of 300 feet. The foundation is modeled massless with linear elastic material properties. The water level is 100 feet for the small dam and 282 feet for the medium dam. The reservoir has a length of 600 feet for both dam models.

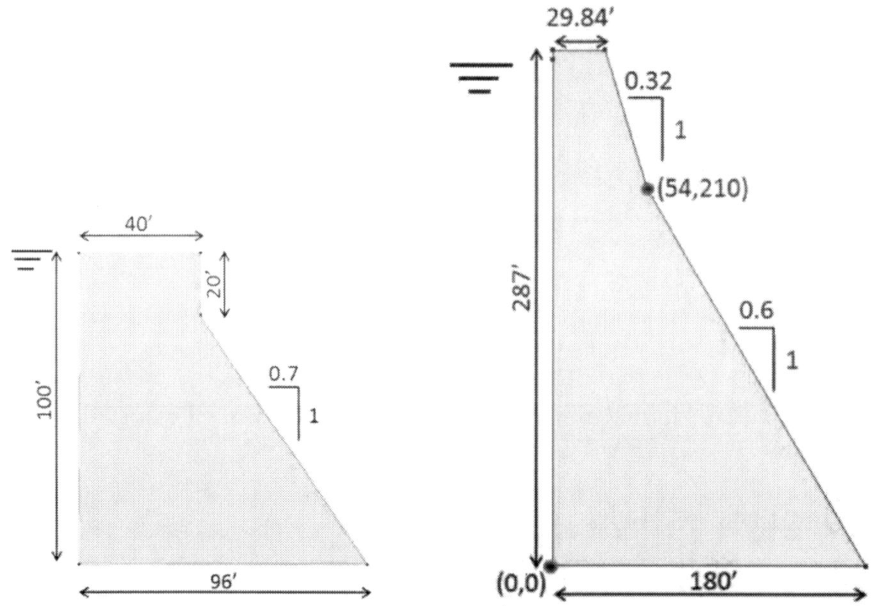

Figure 4. Model of small dam (left) and medium dam (right).

Description of the FE Model

The FE models are set up with plane strain elements. At the free surface and the fluid far field boundary (infinite extent), the prescribed pressures are assumed zero. For the reservoir bottom, a full reflective boundary is assumed (i.e., no bottom absorption effects or prescribed dynamic pressures). Table 1 lists the material properties used in the analysis for the dam, foundation, and reservoir.

Table 1. Material properties.

Material property	Dam	Foundation	Reservoir	Unit
Young's modulus E	4.0E6	3.0E6	3.0E5*	lb/in^2
Poisson's ratio ν	0.2	0.2	-	-
Density ρ	155.0	0.0	62.4	lb/ft^3
Sonic speed c_w	-	-	4869.0	ft/s

Note (*) – Bulk modulus

The amplitude of the excitation (i.e., prescribed harmonic acceleration in horizontal direction) was 0.1 g and was applied at the base of the dam for the model with a rigid foundation, and at the bottom and sides of the foundation block for the model with the elastic properties of the foundation. The harmonic excitations at 0.66 second and 1.33 seconds were considered in the analysis.

Modal Analysis

The ratio of the natural frequency of the reservoir to the natural frequency of the dam is the key parameter determining the effect of the reservoir compressibility in the reservoir-dam interaction during an earthquake [1]. The first natural frequency of the reservoir can be determined by the equation:

$$f_r = \frac{c_w}{4\,h} \qquad \text{(Eq.21)}$$

For reservoir depths of 100 feet and 282 feet, assumed for the small dam and medium dam models in Figure 4, respectively, the natural frequencies of the reservoir determined from Eq. 21 are 12.18 Hz and 4.32 Hz, respectively.

Four eigenfrequencies for the small size dam and medium size dam models are presented in Tables 2 and 3. The frequencies were determined by free-vibration eigenvalue analyses for the empty reservoir, added mass, and incompressible fluid case, and by a direct frequency response method for the compressible fluid case, considering elastic and rigid foundations.

Table 2. Eigenfrequencies [Hz] for small size dam.

Reservoir	Foundation	1st	2nd	3rd	4th
Empty reservoir	rigid	13.1	32.0	33.9	61.8
Empty reservoir	elastic	8.2	15.2	21.6	43.4
Added mass	rigid	11.5	28.7	32.2	50.9
Added mass	elastic	7.1	15.1	19.1	39.1
Incompressible fluid	rigid	11.9	30.2	32.4	58.9
Incompressible fluid	elastic	7.3	15.1	20.1	42.4
Compressible fluid	rigid	11.0	28.7	31.5	50.1
Compressible fluid	elastic	7.1	15.1	20.7	42.2

Table 3. Eigenfrequencies [Hz] for medium size dam.

Reservoir	Foundation	1st	2nd	3rd	4th
Empty reservoir	rigid	3.9	10.0	12.9	18.6
Empty reservoir	elastic	2.9	7.1	8.1	13.2
Added mass	rigid	3.0	7.8	12.8	14.3
Added mass	elastic	2.3	6.0	7.8	11.2
Incompressible fluid	rigid	3.2	8.6	12.8	17.0
Incompressible fluid	elastic	2.4	6.5	7.8	13.1
Compressible fluid	rigid	3.1	5.0	7.3	9.4
Compressible fluid	elastic	2.4	4.9	6.6	12.3

The results presented in Tables 2 and 3 show that the eigenfrequencies of the dam- reservoir-foundation system vary significantly with the type of the dam-reservoir-foundation model considered in the analysis. The highest natural frequency is obtained for the elastic dam model on rigid foundation without the presence of the reservoir. Note that elasticity of the foundation is a significant factor influencing the natural frequency of the dam-reservoir-foundation system.

Table 4 presents the ratio between natural frequencies of the reservoir and natural frequencies of the dam-reservoir-foundation model.

Table 4. Ratio between the first natural frequencies of the reservoir and the dam-reservoir-foundation model.

Reservoir	Foundation	Small Size Dam	Medium Size Dam
Empty reservoir	rigid	0.93	1.11
Empty reservoir	elastic	1.49	1.49
Added mass	rigid	1.06	1.44
Added mass	elastic	1.72	1.87
Incompressible fluid	rigid	1.02	1.35
Incompressible fluid	elastic	1.67	1.80
Compressible fluid	rigid	1.11	1.39
Compressible fluid	elastic	1.72	1.80

In general, the first natural frequency of the reservoir is higher than that of the dam-reservoir-foundation system of the two dam models (Figure 4) selected in the analysis (ratio larger than 1.0 in Table 4).

Analysis of a Rigid Dam on a Rigid Foundation

The classical model of a rigid dam placed on a ridged foundation is evaluated in this section. Water pressure distributions at the face of the rigid dam, obtained by both the approximate and the exact Westergaard's solutions [5], are compared in Figure 5 with the "fluid-like material model" FE analysis results for the small size dam (Figure 4).

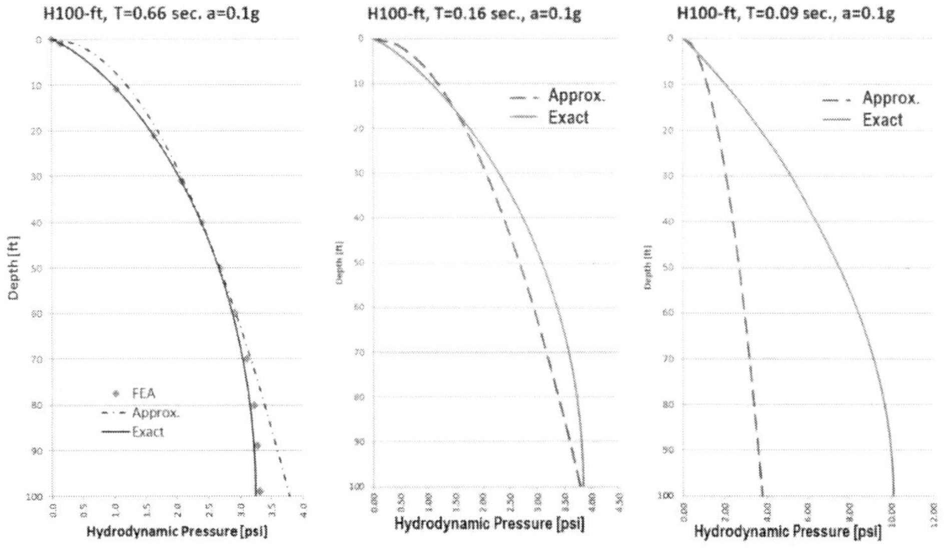

Figure 5. Hydrodynamic pressures at the face of a rigid dam obtained by exact and approximate Westergaard's solutions for the harmonic dam excitations at periods of 0.66 second (1.52 Hz), 0.16 second (6.25 Hz), and 0.09 second (11.1 Hz) and the acceleration amplitude of 0.1g.

In general, the approximate Westergaard's formula (Eq. 6) overestimates the hydrodynamic pressure at the upper part of a rigid dam when it is compared with the exact Westergaard's solution. The increase in the hydrodynamic pressure determined by the exact Westergaard's solution (Figure 5, right graph) could be explained by resonance between the reservoir (first natural frequency of 12.18 Hz) and the dam excitation frequency of 11.1 Hz. Note, that the Westergaard's solution is valid only if the frequency of excitations is significantly smaller than the first natural frequency of the reservoir. For comparison, FE analysis results are presented in Figure 5 (left), excluding higher frequency wave effects in the reservoir.

Time Analysis

The time analyses were performed using the "fluid-like material model" approach [3], assuming various arrangements for the dam-reservoir-foundation system. A rigid foundation was assumed for both analysis models and the harmonic excitation of the dam at periods of 0.66 second and 1.33 seconds. Figures 6 and 7 present time-history pressure distributions for both the small-size and medium-size dam models. These graphs illustrate a pressure at the dam face resulting from combining the dam excitation effect and the wave's generated by the reservoir compressibility. When higher frequency wave effects in the reservoir are excluded, the pressure as a function of time would be a sinusoidal function, with the frequency corresponding to the frequency of excitation. The local peaks in the pressure distribution (Figure 6) are separated by 0.81 second, which corresponds to the reservoir first natural frequency of 12.18 Hz (period 0.82 second) determined by Equation 21.

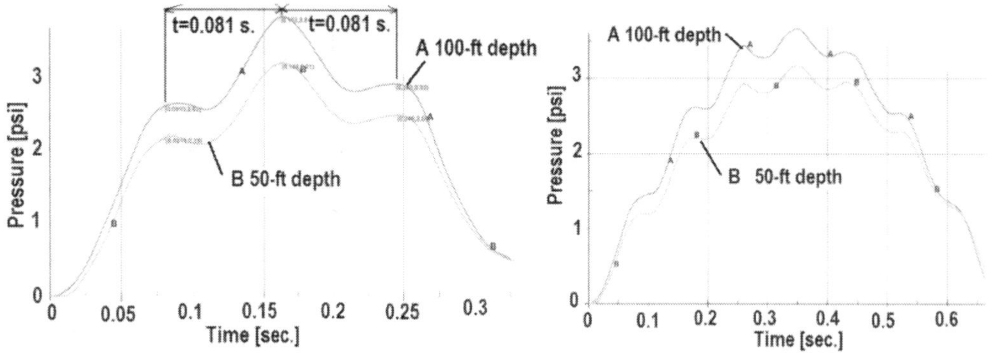

Figure 6. Pressure distribution at the dam face for small size dam with a rigid foundation at 100-foot and 50-foot reservoir depths and for the harmonic excitation at 0.66 second (left) and 1.33 seconds (right).

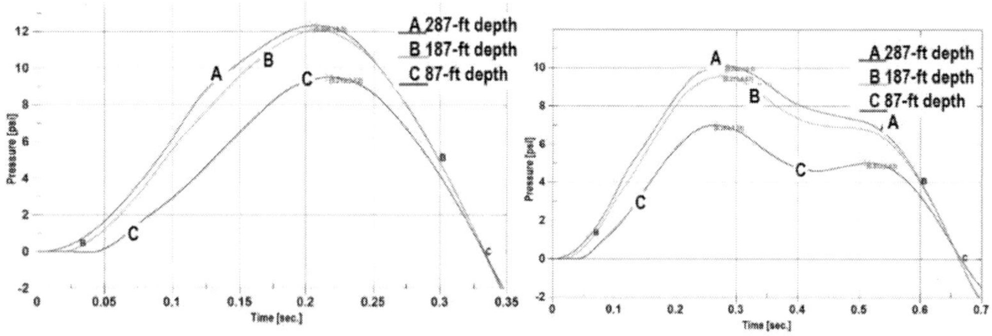

Figure 7. Pressure distribution at the middle size dam face with the rigid foundation at 287-, 187-, and 87-foot depths and for the harmonic excitation at 0.66 second (left) and 1.33 seconds (right).

The maximum pressures obtained from the FE analysis are compared with Westergaard's solutions in Table 5. The results show that higher pressures are obtained when the compressibility effect is included in the analysis.

Table 5. Comparison of maximum pressures at the bottom of the reservoir [lb/in^2]

Reservoir	Foundation and Period	Small Size Dam	Medium Size Dam
Incompressible fluid	rigid	3.43	9.79
Incompressible fluid	elastic	3.76	11.43
Compressible fluid	rigid & (T=0.66 sec.)	3.85	12.3
Compressible fluid	rigid & (T=1.33 sec.)	3.67	10.1
Westergaard approximate solution		3.79	7.58
Westergaard exact solution	T=0.66 sec.	3.25	6.68
	T=1.33 sec.	3.23	6.49

Figures 8 and 9 show the positive and the negative hydrodynamic pressures turbulence that develops in the reservoir during simulation of the dam-reservoir model at the various analysis times. Pressure waves that separate from the face of the dam spread along the length of the reservoir until they are absorbed by the "far field boundary conditions" (Perfect Matching Layers [3]). The turbulence in the pressure distribution is observed in the analysis for the linear material model with harmonic dam excitations and geometric non-linearities (large deformations).

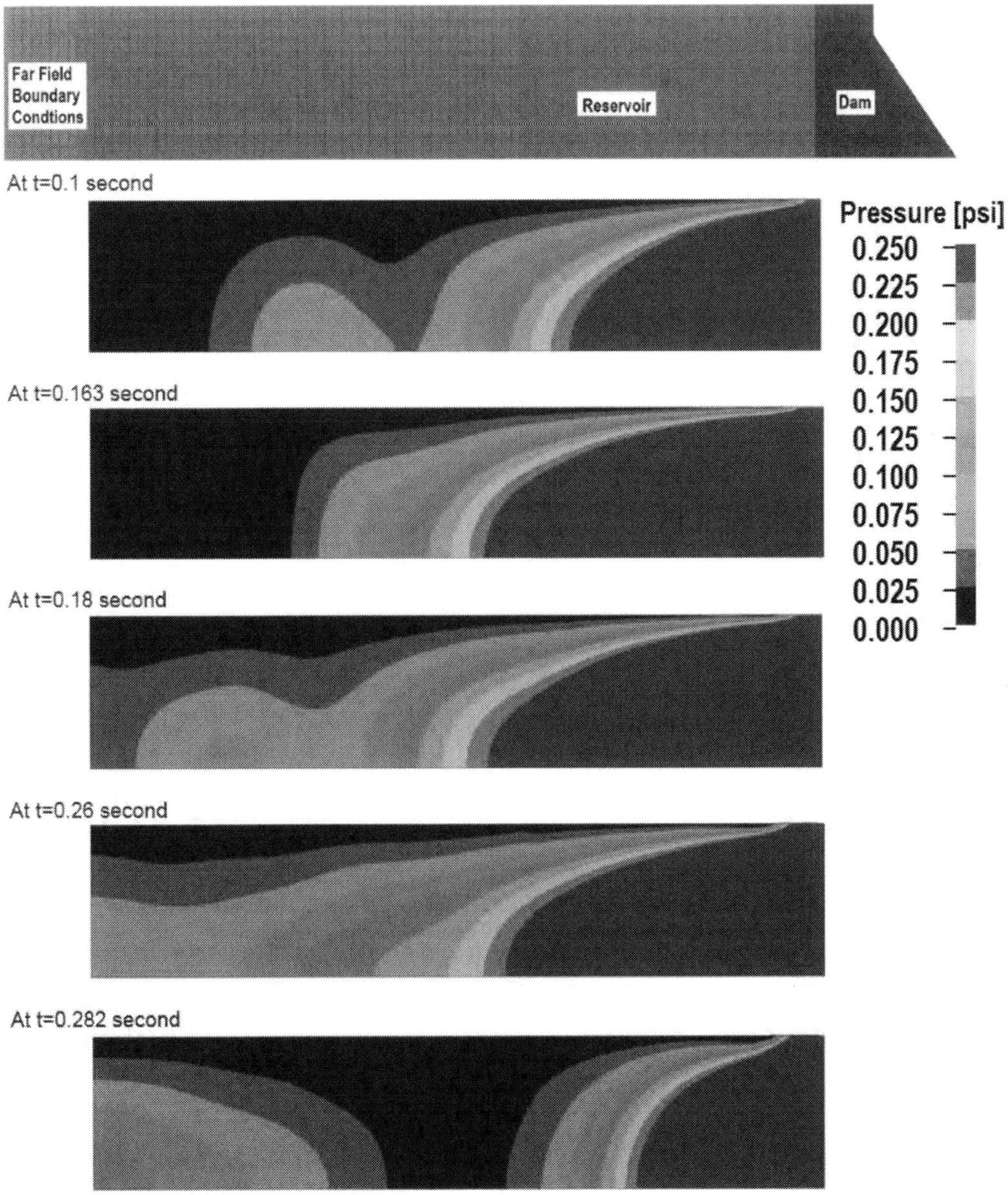

Figure 8. Positive pressure distribution in the reservoir corresponding to the time line presented in Figure 6 (left).

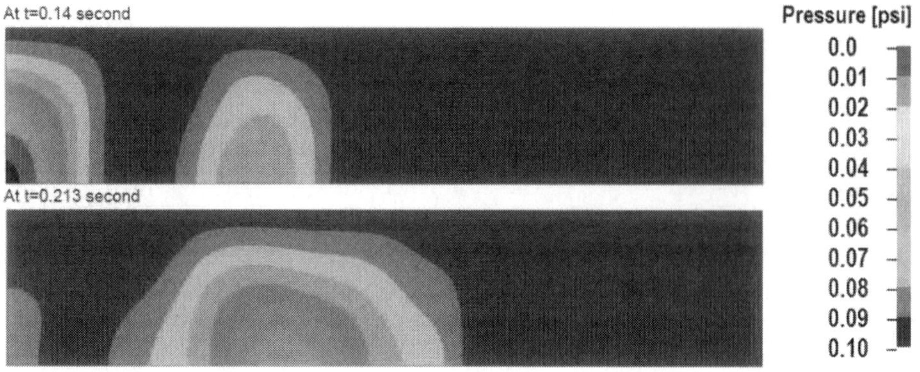

Figure 9. Negative pressure distribution in the reservoir corresponding to the time line presented in Figure 6 (left).

SUMMARY AND CONCLUSIONS

Summary

This paper evaluates three analysis methods that are commonly used in the time simulations of the dam-reservoir-foundation system during an earthquake. These methods include: (1) the "added mass" approach, (2) the "acoustic fluid" method, and (3) the "fluid-like material model" method. The theory for each of the methods is briefly presented and is illustrated by the results of two analyzed concrete dams: the small size dam and the medium size dam. In general, the results of the analysis could be summarized as follows:

- A good agreement existed between the exact Westergaard's solution and the FE results for a rigid dam with vertical upstream dam face when higher frequency wave effects are excluded from the analysis; however, the hydrodynamic pressure determined by the approximate Westergaard's solution significantly differs from the results obtained by the FE and the exact Westergaard's solutions.
- According to analysis results, eigenfrequencies of the dam-reservoir-foundation system vary significantly with the type of assumptions made for the physical model.
- The FE time-history simulation of a linear dam-reservoir model with harmonic dam excitations showed that compressibility of the fluid significantly influences the analysis results, resulting in a turbulent pressure distribution in the reservoir.

Conclusions

The following conclusions could be formulated based on the results of the investigations:

- The differing results of the blind predictions in the Monticello Dam analysis appear to be due primarily the various approaches used by participants. Of the three approaches, the "added mass" approach offers the least confidence for accuracy of the solution.

- The "added mass" approach, based on the approximate Westergaard's solution, only roughly estimates the mass of water interacting with the dam during an earthquake. The error is particularly large when the excitation frequency is similar to the first natural frequency of the reservoir. Even greater errors would exist in Westergaard's solution if the dam face was not vertical, or if the dam and the foundation is considered not rigid.
- It appears that the "added mass" approach should not be implemented in the advanced time-history analysis, and that it only be used in a preliminary estimation of seismically induced hydrodynamic loads on dams.
- The "added mass" may provide relatively good results only for "rigid" type dams (small size concrete gravity dams on stiff rock foundations) with vertical upstream face.
- Compressibility of water is the primary factor that influences the hydrodynamic pressure distribution in the reservoir. The time analysis with an "incompressible fluid" material model provides significantly different results when compressibility of the fluid is considered.
- It appears that a hybrid frequency-time domain analysis should be implemented in the seismic simulations of the dam-reservoir-foundation system.

REFERENCES

[1] Chopra A. K., *Hydrodynamic Pressures on Dams during Earthquakes*, Journal of the Engineering Mechanics Division, Proceedings of the ASCE, 1967.

[2] Lamb H., *Hydrodynamics*, Fifth Ed., Chapter X, 1924.

[3] LSTC, LS-DYNA Manual, Livermore 2013.

[4] Manie J. (Ed.), DIANA User's Manual, Delft 2016.

[5] Westergaard H. M., *Water Pressures on Dams during Earthquakes*, ASCE Transactions, November 1931.

SEISMIC BACK ANALYSIS OF MONTICELLO ARCH DAM – BLIND PREDICTION WORKSHOP AND ADDITIONAL ANALYSES

Emmanuel Robbe[1]

ABSTRACT

During the 2016 United States Society of Dams (USSD) conference in Denver, a blind prediction workshop was organized for purpose of comparing blind predictions of structural behavior of Monticello arch dam with the existing records of the dam as of May 22, 2015 when it experienced a M 4.1 earthquake. Interested analysts were given an information package containing drawings of the dam, material testing results and free field rock time history based on which they were then to propose a model for predicting the seismic response of the dam..

This paper presents the analyses performed by the author, on behalf of the Hydro Engineering Center of EDF (Electricité de France), and results thereof. The paper further provides an overview of how these results have been compared with existing records and gives a short descriptive summary of a methodology used to submit the prediction. In addition, this paper also compares predictions, including several other assumptions, with the existing records in order to improve the comparison.

INTRODUCTION

During the USSD conference in Denver in 2016, a blind prediction workshop was organized in order to compare predictions of the structural behavior of Monticello arch dam to what was recorded in 2015 during a small earthquake with epicenter in the vicinity of the dam.

This paper presents the analyses performed by the author on behalf of the Hydro Engineering Center of EDF (Electricité de France) by: 1) summarizing the methodology used to submit the prediction, 2) comparing the prediction with the existing records in order to understand the fault in the prediction, 3) performing additional analyses to improve the FE results compared to the records.

MONTICELLO DAM AND THE EARTHQUAKE RECORDED

Monticello is a concrete medium thick arch dam 304 ft (93 m) high from the foundation, 1,023 ft (312 m) long and 239 ft (73 m) above the riverbed. The dam is 100 ft (30 m) thick at the base, tapering to 12 ft (3.7 m) at the crest. The total volume of construction materials is 326,000 cubic yards (249,000 m^3). The dam was constructed between 1953 and 1957.

[1] Engineer, EDF Hydro Engineering Center, Savoie-Technolac, 73373 Le Bourget du lac, emmanuel.robbe@edf.fr.

Figure 1 Monticello arch dam

On May 22, 2015, Monticello arch dam experienced a M 4.1 earthquake with the epicenter approximately 16 km away from the dam. The instrumentation, located both at the bottom and at the crest of the dam, recorded the event. Figure 2 presents the acceleration time histories recorded in the foundation, at the foot of the dam. The us/ds peak ground acceleration is 0.01g.

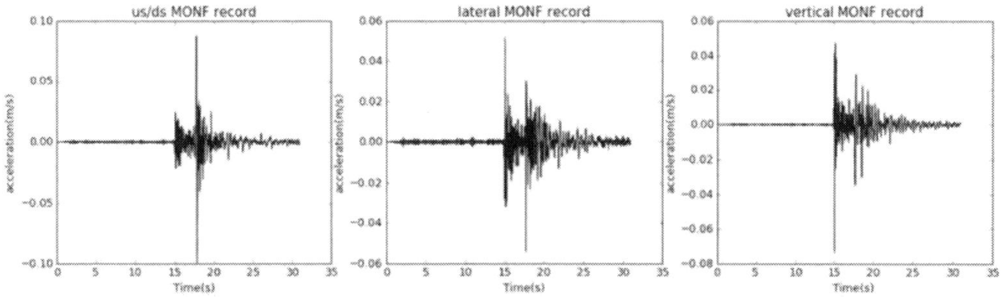

Figure 2 Acceleration Time histories at the foot of the dam

METHODOLOGY

Finite-Element Method

[Chopra 2008 and 2012] found that finite element analyses with massless foundation and Westergaard added masses assumption had strongly overestimated the response of concrete dams to earthquakes. In this paper, a method taking into account better soil-structure and fluid-structure interactions has been chosen:

- The mass of the foundation is taken into account with absorption of the waves radiating from the dam by a viscous-spring model on the boundaries of the foundation (see Figure 3 and more details in [Liu] and [Zhang])

- Potential-based fluid elements are used [Bouaanani] to take into account compressibility of the water. Wave absorption at the end of the modeled reservoir is also considered using absorbing boundaries. Absorption due to sediments at the bottom of the reservoir is not introduced at first.

These features are available in the open-sources finite element software Code_Aster (www.code-aster.org), developed by EDF. Back-analyses and comparisons between records and finite element analyses had been successfully performed on Japanese concrete dams with the use of one such model [Robbe 2017]. The Monticello workshop was a good opportunity to confirm these results.

With a view to the low intensity of the earthquake, the analyses in this paper observe a linear elastic behavior of the dam and its foundation, with no cantilever joints or dam-foundation contacts taken into account.

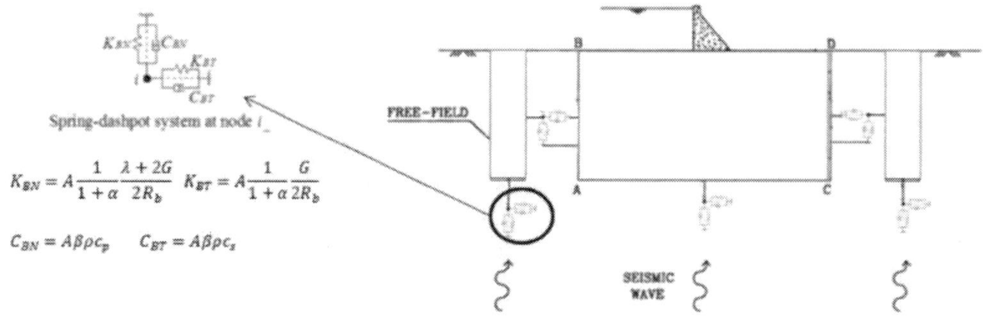

Figure 3 Viscous spring boundary model

Geometry and Mesh of the Dam

The dam, its foundation and reservoir are modeled and meshed based on the construction drawings. Considering the finite-element method chosen, the minimum element size has been set to 20 m so to allow for a propagation of the wave up to 15-20 Hz. Foundation and reservoir geometries do not precisely respect the topography around the dam but are rather defined as an extrusion in the upstream and downstream direction from the dam (Figure 4).

For the rest of the analyses described in this paper, X stands for us/ds, Y for transversal and Z for vertical, while MONF, MONC and MONQ describe the position of the accelerometers at the bottom and at the crest respectively (see Figure 4). An additional accelerometer, MONA, located on the left bank of the dam will also be mentioned later in the analyses.

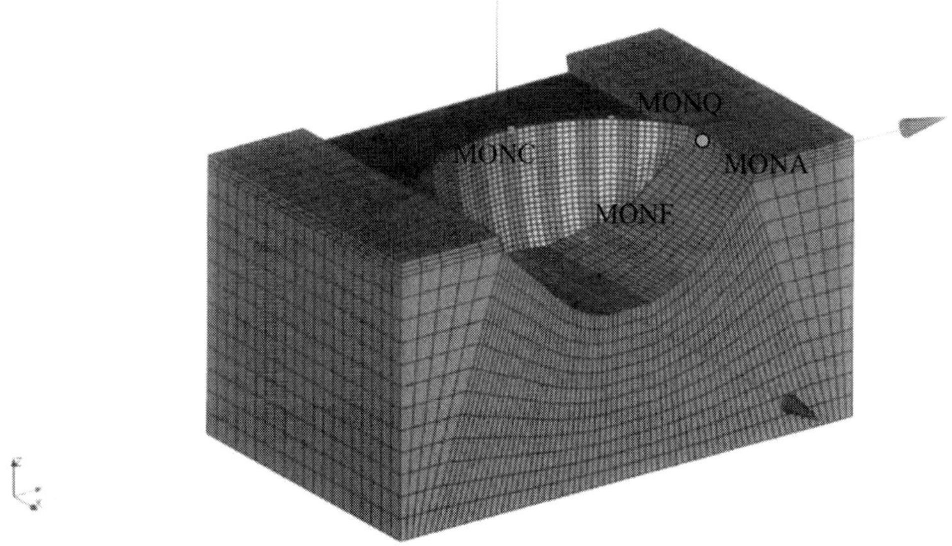

Figure 4 Mesh of the dam and location of the accelerometers

Material Properties and Calibration of the Model

The average density of concrete and foundation, each equaling approx. 2,450 kg/m^3 and Poisson coefficient of 0.2 for concrete and 0.25 for the rock have been chosen for purpose of the analyses.

A shaking test performed with the reservoir level at 435 feet (with a load of 83.8 m height of water) has identified the first modes of the structure, namely 3.13 Hz, 3.55 Hz, 4.68 Hz, 6.00 Hz, 7.60 Hz with the symmetrical modes 1, 3 and 5. The results of the test are provided in Figure 6.

To assess the calibration of the model, finite element analyses of the dam have been performed with a white noise input for different water level (from 0 to 85 m (i.e. 435 feet)) and the following Young Modulus for concrete and rock: $E_{concrete}$ = 35000 MPa, E_{rock} = 30000 MPa. Mechanical tests on concrete and rock cores show that concrete is slightly stiffer than rock. Acceleration transfer functions are computed between the crest (MONC) and the bottom (MONF) of the dam: this is a good way to evaluate the first eigenfrequencies of the structure by looking at the peaks. The results are presented in Figure 5.

With the first eigenfrequency at high water level (84m), the level measured during the shaking test, proving to be slightly too high (3.23 Hz) compared to the value estimated by the test (3.13 Hz), the value of the concrete modulus has been slightly decreased (30,000 MPa) for purpose of the analyses.

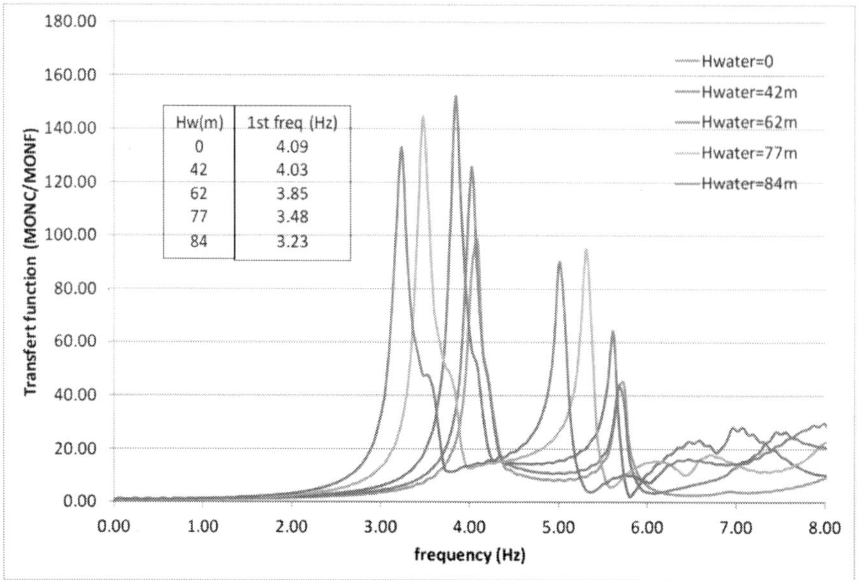

Figure 5 Transfer functions (MONC/MONF) for different level of water in the reservoir

In what concerns damping of the material, earlier simulations performed and compared to earthquake records on concrete dams [Robbe 2017] show that, with a method involving absorbing boundaries, concrete damping values between 1 % and 3 % can be used. Taking into account the low intensity of the recorded earthquake, the author of this paper has chosen 1% Rayleigh damping, however, with no damping considered for the foundation material.

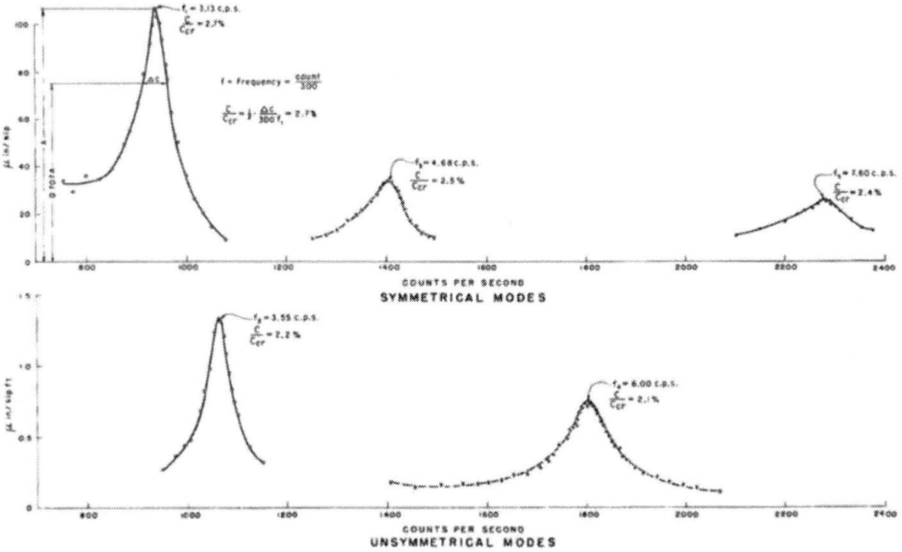

Figure 6 Frequency response curves of the shaking test

Seismic Event Used in the Model

While considering the previously described soil-structure interaction method, the free-field rock time histories recorded have been introduced as compression and shear waves travelling vertically from the bottom to the top of the foundation. Knowing that the MONF acceleration time history cannot be applied directly to the bottom of the model, deconvolution is performed prior to applying the time history to the base of the model without considering the cross terms:

- Finite element analysis is first performed with the initial input (i.e. accelerations recorded at MONF) introduced at the bottom of the foundation;
- Ratio between the Fast Fourier Transform (FFT) of the computed and recorded accelerations in the 3 directions at MONF is calculated:

$$R_i(f) = \frac{FFT(a_{i-computed}(t))}{FFT(a_{i-recorded}(t))} \; with \; i = X, Y, Z \quad (1)$$

- This ratio is then applied on the free-field records to define the new input to be used for the analyses:

$$input_i(t) = FFT^{-1}\left(\frac{FFT(a_{i-recorded}(t))}{R_i(f)}\right) \; with \; i = X, Y, Z \quad (2)$$

COMPARISON BETWEEN PREDICTION SUBMITED AND RECORDS

Following the 2016 USSD workshop, records at the crest of the dams were released. This section compares the prediction performed with the assumptions described in the previous section, compared to the real records on the dam during the earthquake. This prediction is slightly different from the prediction submitted for the workshop, following a correction applied to the fluid-structure interaction model. In order to compare the behavior recorded and predicted, the following results have been plotted:

- Acceleration time history gives a rough overview of the quality of the prediction, (this is quite useful for quick comparisons in maximum acceleration and for observing any change of behavior of the dam during the earthquake.
- FFT of the acceleration is useful to evaluate and compare the frequency response of the dam. FFT must be computed at same time intervals in order to be comparable.
- Displacement time history, which is directly related to the stress state of the structure, is probably the most important parameter to evaluate if the model overestimates or underestimates the response of the dam.

Acceleration Time History Comparison

The comparison of the time history accelerations recorded and computed at the bottom and at the crest of the dam, for each direction is proposed as shown in Figure 7. A focus between t = 14.5 s and t = 20 s is plotted.

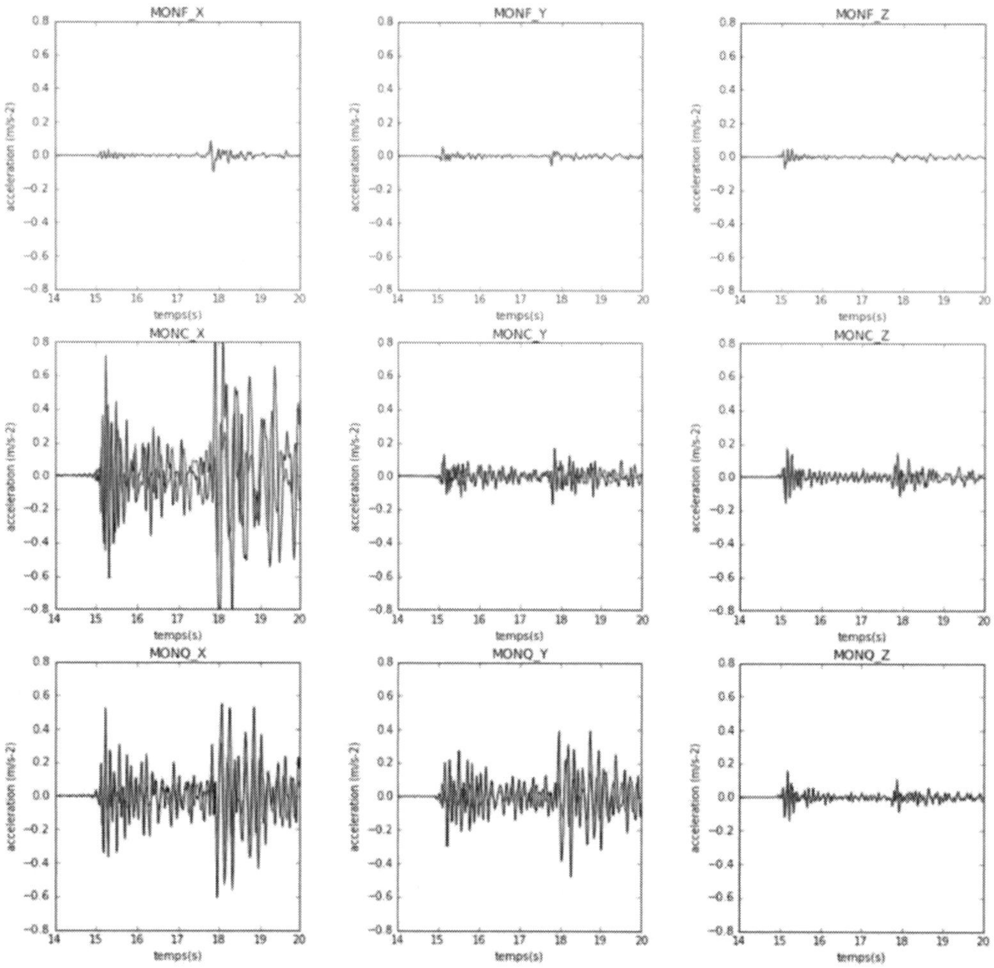

Figure 7 Comparison of time history accelerations at MONF, MONC and MONQ
Recorded (Red) vs. Calculated (Blue)

The comparison at the bottom of the dam confirms that the input is correctly introduced in the finite element model. At the crest, the finite element model greatly overestimates the response of the dam by a factor 2, particularly in horizontal directions.

Frequency Comparison

Figure 8 presents the FFT of the accelerations recorded and computed at the crest between t=0 and t=30 sec. The finite element model detects correctly the first mode of the structure, while strongly overestimating the amplitude.

Displacement Comparison

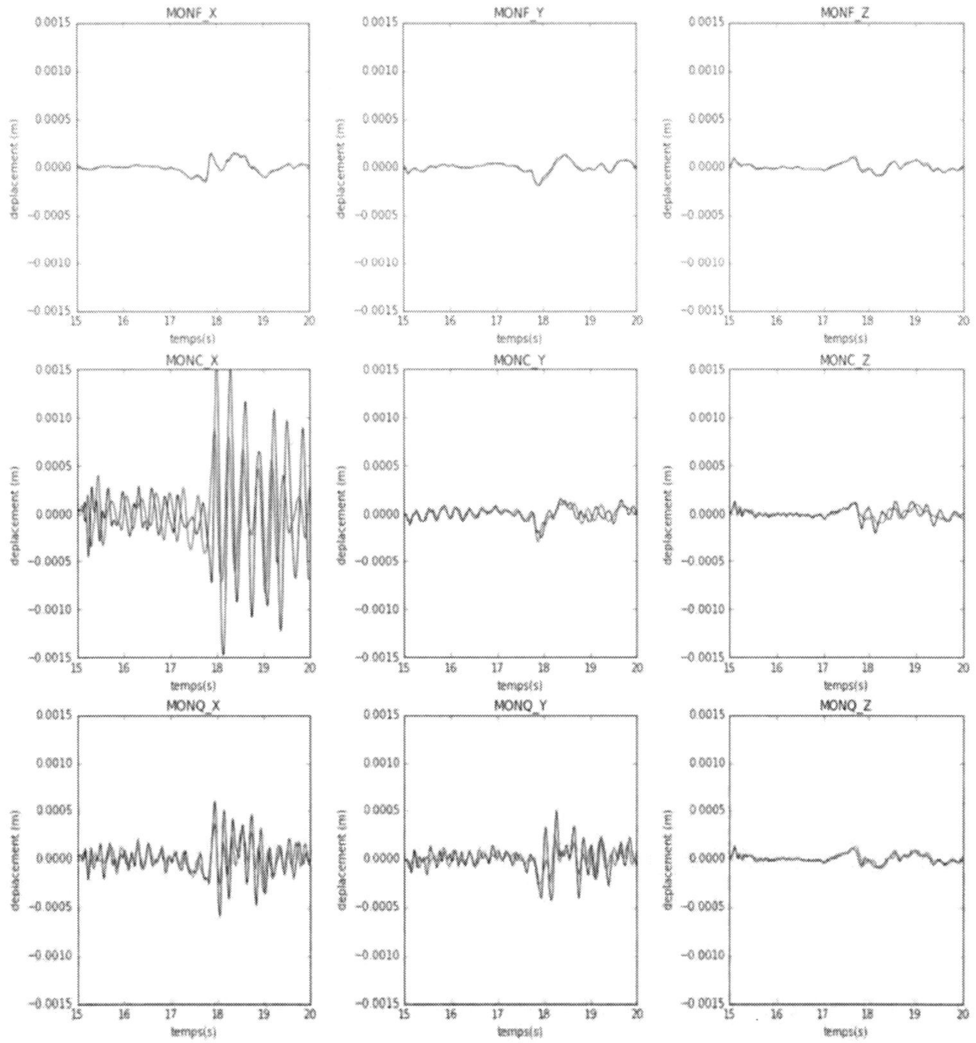

Figure 9 presents the comparison of the displacements at the foundation (MONF) and at the crest (MONC and MONQ) for each direction. Some displacements show a good mutual agreement (MONF in every direction, MONC_Y, MONC_Z and MONQ_Z), but displacements at the crest in the us/ds direction are clearly overestimated, particularly during the stronger phase of recording (starting at t=18 s).

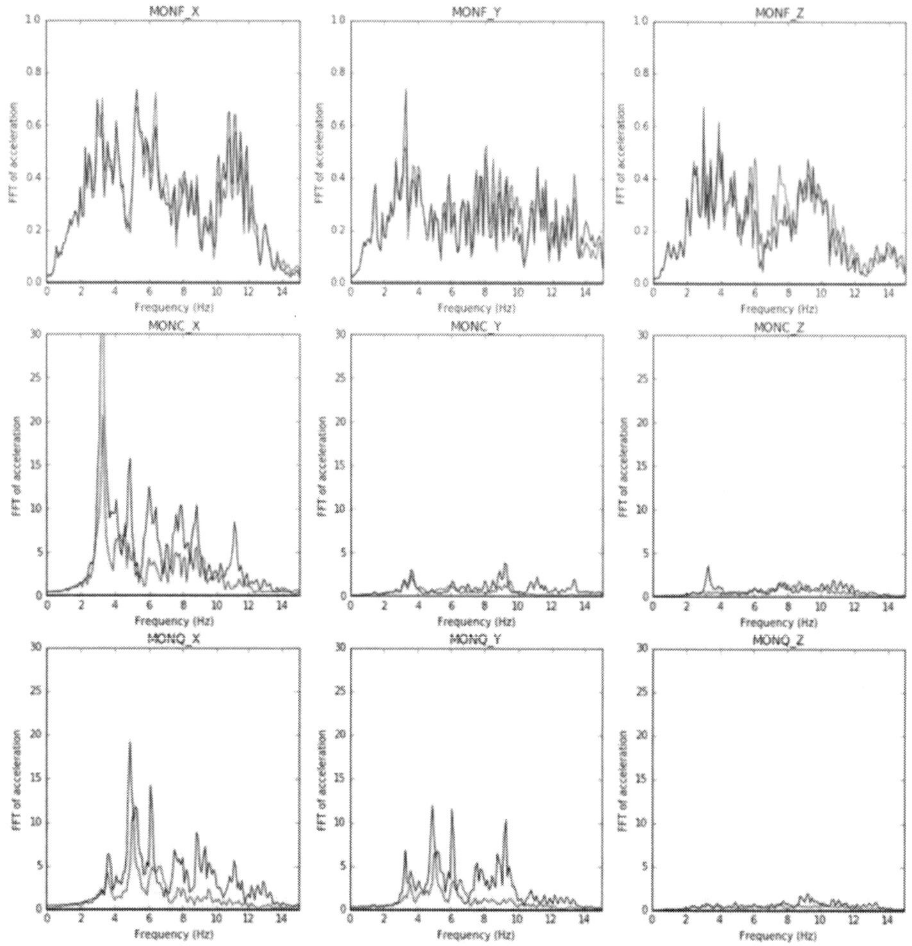

Figure 8 FFT of the accelerations at MONF, MONC and MONQ
Recorded (Red) vs. Calculated (Blue)

Conclusion on the Comparison

Despite the fact that the input has been correctly introduced in the finite model, the prediction cannot be considered completely satisfying as:

- Accelerations computed at the crest are much higher than the recorded ones.
- With a view to the accelerations' spectrum, eigenfrequencies are correctly computed, but amplitudes are not.
- Displacements at the crest, particularly in the us/ds direction, are overestimated.

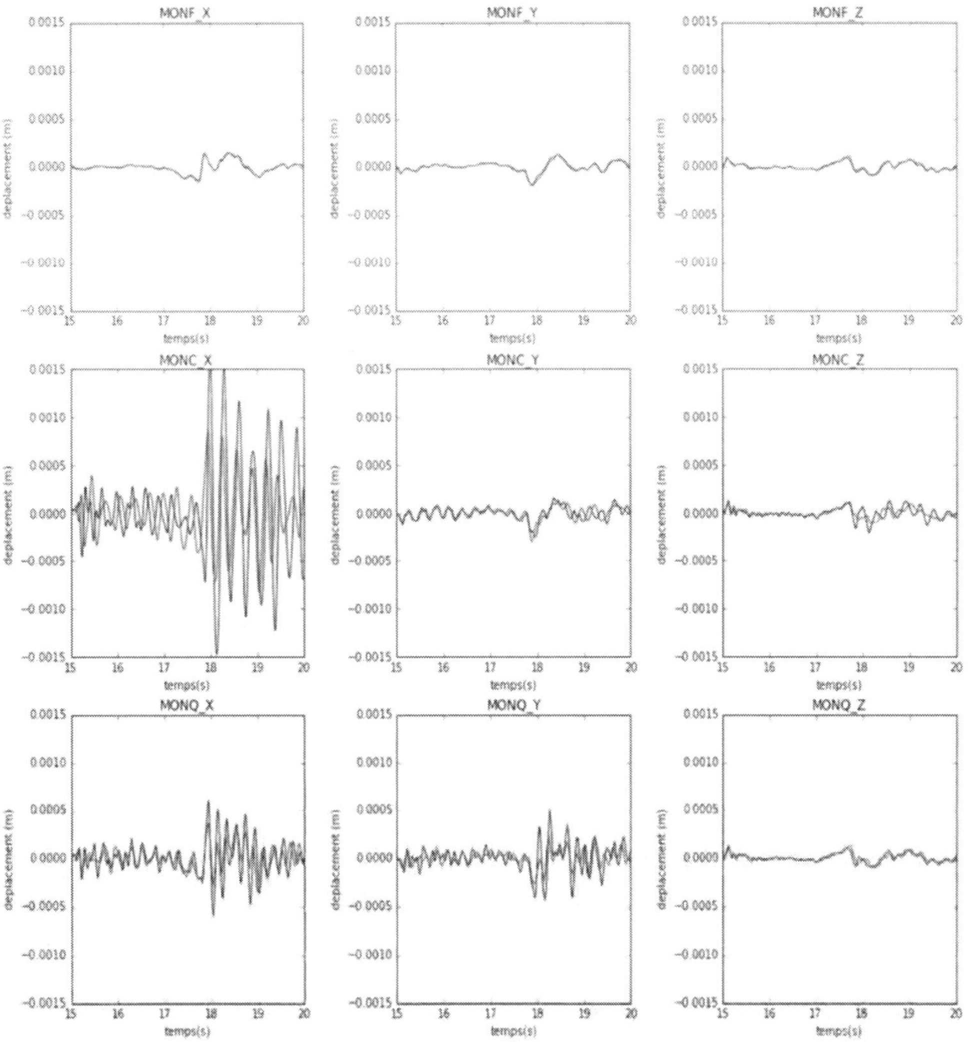

Figure 9 Comparison of displacements: Recorded (Red) vs. Calculated (Blue)

ADDITIONAL ANALYSES

Knowing the real records, the author has performed additional analyses in order to determine what would have been the best parameter to choose for the finite element analysis to better match the existing records.

Comparison Method

Comparison between finite element analyses and records can be a long process involving multiple parameters, as previously shown. This being the case, for purpose of the additional analyses only the following characteristics have been compared and a visual comparison tool provided (Figure 10):

- Peak acceleration (PGA)
- Peak displacement (PGD)
- Acceleration spectrum
- Energy: $I_{EI}(t) = \int_0^t v_i^2(\tau)d\tau$ where v_i is the velocity in one of the 3 directions.

For each characteristic, a ratio between the computed and recorded characteristic is evaluated. For the acceleration spectrum, a mean ratio is computed on 3 bands: 1-5 Hz, 5-10 Hz and 10-15 Hz. The ratios, computed for each direction, are then displayed in a circular chart as shown in Figure 10. Where the finite element results match the records perfectly, a regular circle should appear.

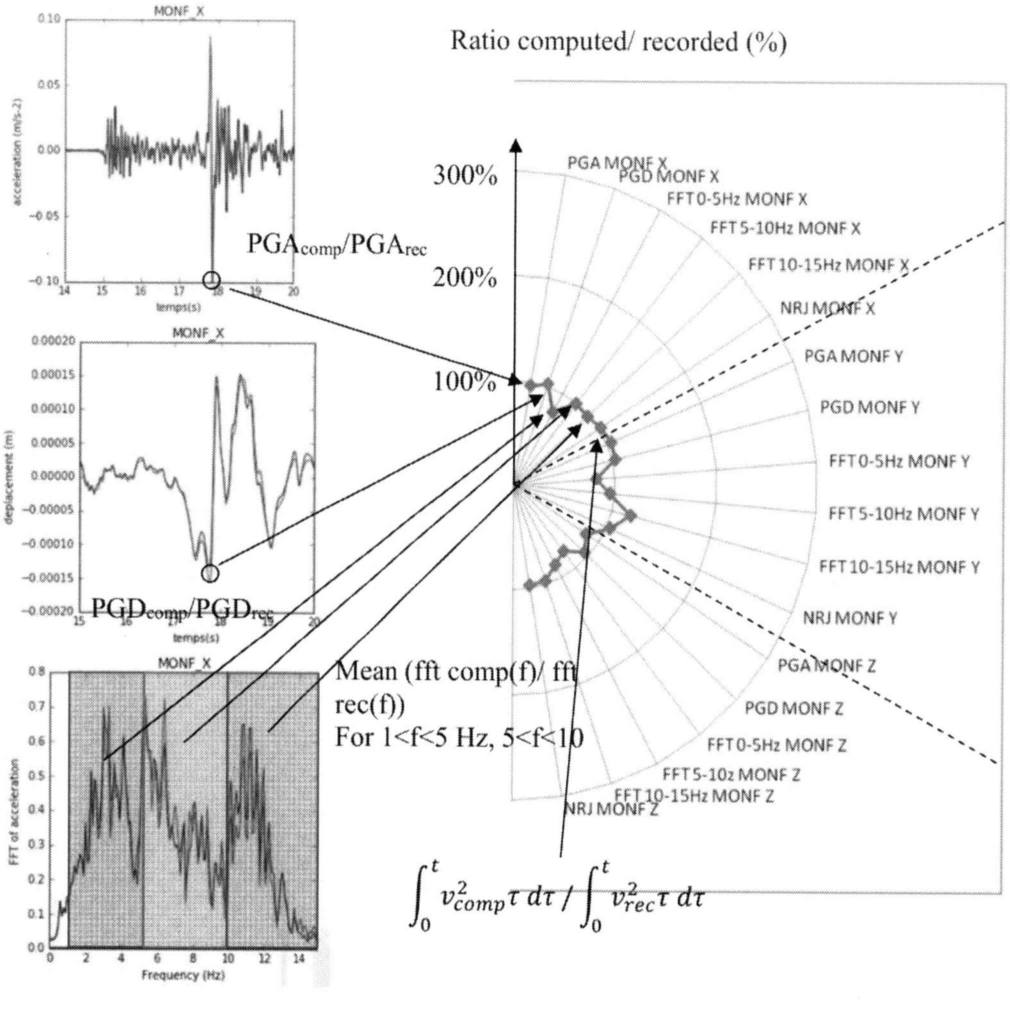

Figure 10 Example of construction of the visual comparison tool for the MONF record, Us/Ds direction (x)

This comparison method is applied with the results presented in the previous section. Figure 11 summarizes the results of the comparison. Taking into consideration the results at the crest (MONC and MONQ), it becomes obvious that the results from the finite-element model are overstimated, as previously stated.

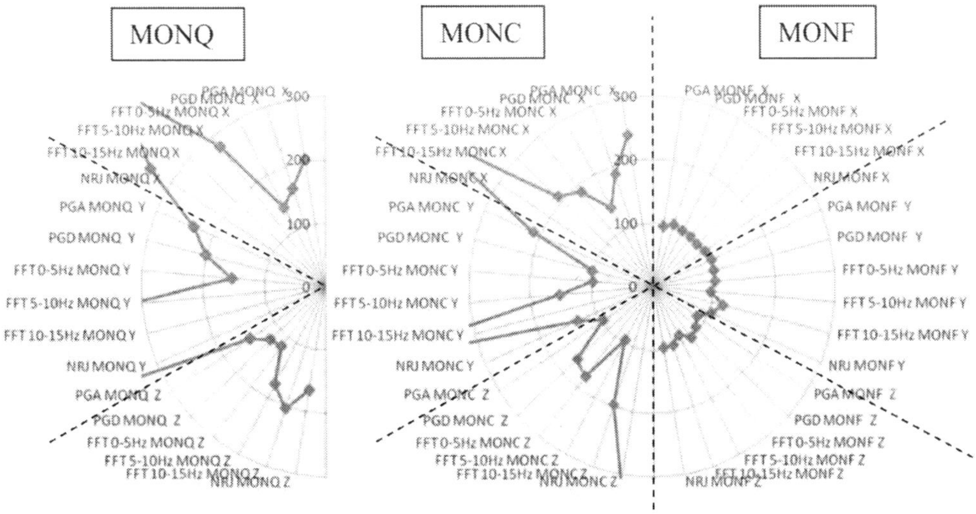

Figure 11 Comparison of a modified prediction with the records

Comparison with Massless Foundation & Westergaard Added Masses Model

This section compares the results provided by 2 finite-element methods with the following characteristics:

- The method presented in this paper introducing viscous-spring boundaries in the foundation and fluid element for the fluid-structure interaction. 1% Rayleigh damping is considered for the concrete.
- One of the most commonly used methods in engineering, considering a massless foundation and simplified Westergaard added masses for the fluid-structure interaction. 5% modal damping is considered (common practice).

Figure 12 shows the ratios computed, i.e. recorded for both methods. Despite the higher concrete damping of the model with massless foundation, results turn out to be rather similar and once again reveal an overestimation of the crest acceleration and displacement in the us/ds direction.

Overall, it can be concluded that the damping introduced by the absorbing boundaries is around 3 - 4 %. In this case, and despite what is usually thought [Chopra 2008 and 2012], the simplified method with massless foundation gives rather good results.

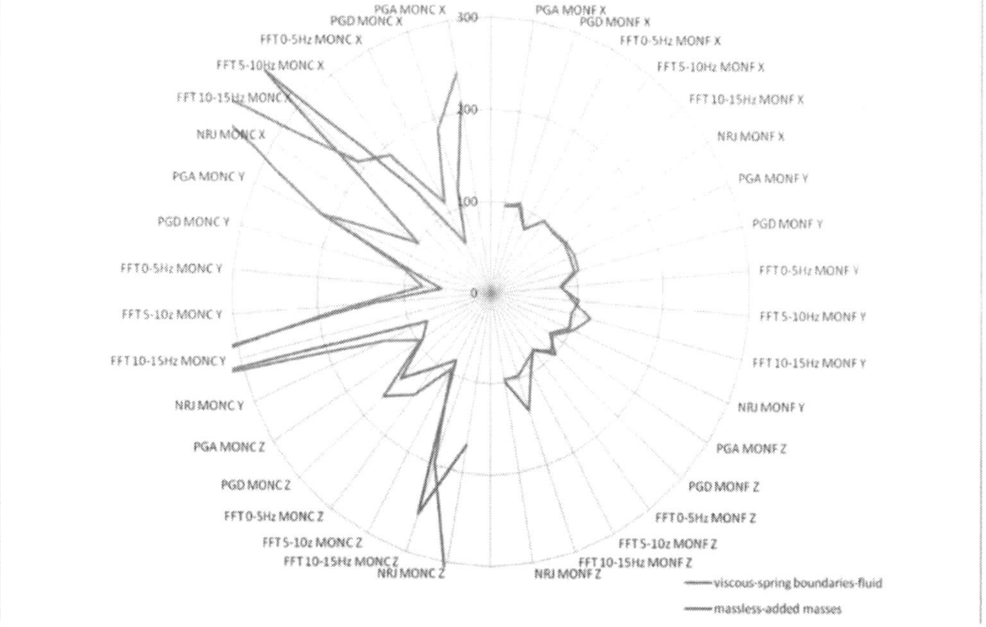

Figure 12 Comparison of FE methods: massless foundation & Westergaard added masses – viscous-spring boundaries & fluid

Correction of the Rock Modulus

Dynamic rock modulus of 30,000 MPa is here chosen knowing that mechanical tests on concrete and rock cores show that concrete is slightly stiffer than rock.

As presented in Figure 4, there are 2 accelerometers located in the foundation: at the toe of the dam (MONF) and near the top of the left abutment (MONA).

Displacement records at MONF and MONA are compared. With the records being rather similar, but with a slight shift in time, cross-correlation is used in order to determine this shift, direction by direction on (Figure 13).

The analysis shows that for the horizontal components, it takes 0.06 s for the wave to travel from MONF to MONA and 0.04 s for the vertical component. These values complement the assumption that waves propagate vertically with the velocity of compression waves that are higher than the velocity of shear waves. In view of a relationship between dynamic modulus and velocities in the rock, and knowing that the vertical distance between MONF and MONA devices measures around 84 m, a dynamic modulus of 10,000 MPa should be considered in the rock.

Figure 14 present the results of the earthquake back analyses of Monticello by using both the initial value 30,000 MPa and the new value 10,000 MPa for the rock. Taking into account this important change in the foundation, it becomes obvious that the deconvolution process needs updating. The comparison shows that with the new assumption, the frequency content of response at the crest is improved, however the peak accelerations and displacements remain overestimated by the FE analysis. The new rock modulus value is therefore considered for the rest of the study.

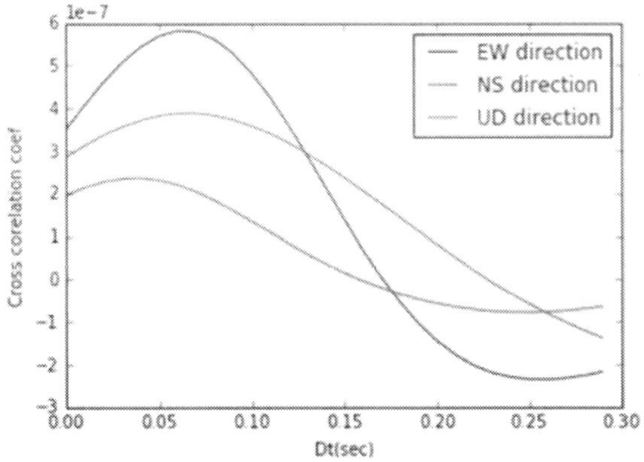

Figure 13 Cross-correlation between displacement records at MONF and MONA

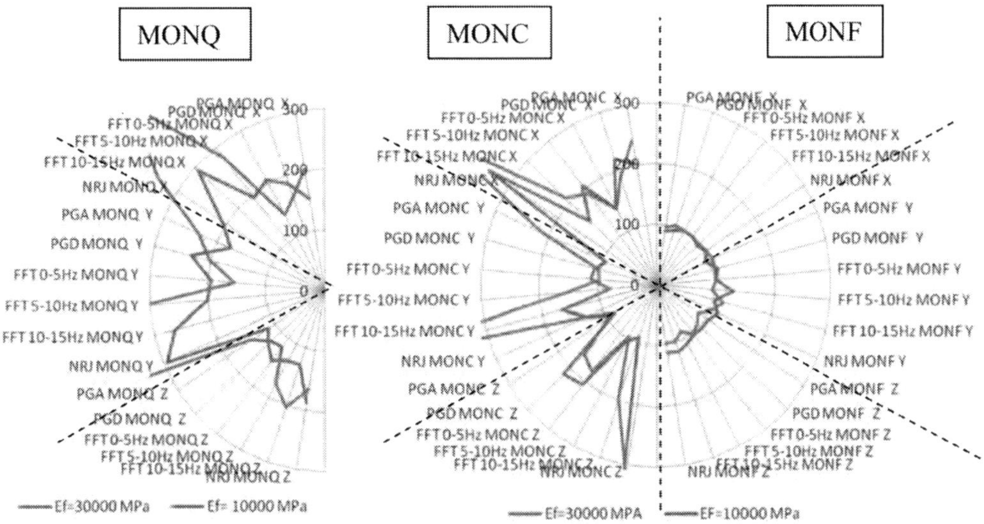

Figure 14 Comparison of the rock modulus

Concrete Damping

The analyses observe 1% concrete Rayleigh damping due to the low intensity of the earthquake recorded. Since this value might appear low, a sensibility analysis is performed while increasing values to 3% and 5%. The results of the analysis (Figure 15) show that despite a trend to lower the response of the dam at the crest (particularly the energy in the us/ds direction), concrete damping does not significantly improve the results.

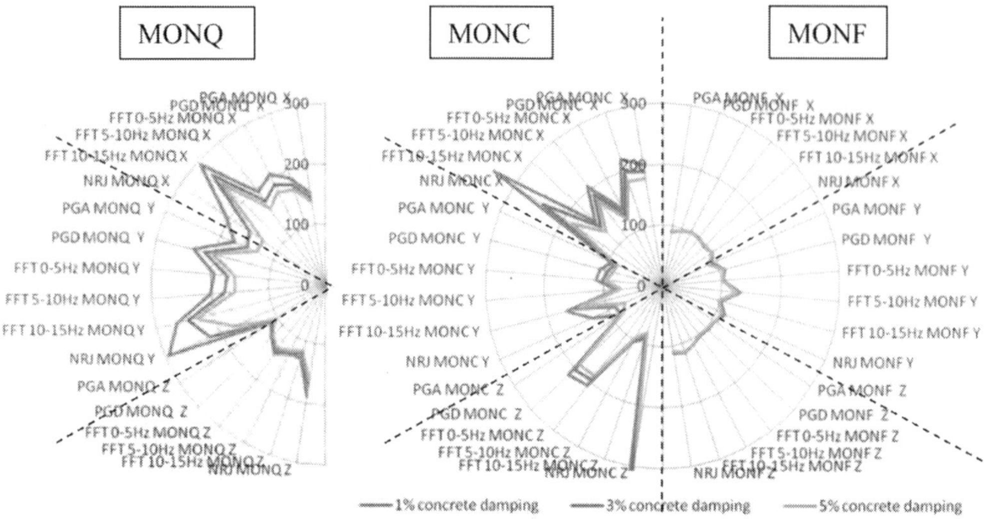

Figure 15 Comparison of the concrete damping

Partial Absorption of Pressure Wave at the Reservoir Bottom

As already established, the presence of alluvium and sediments at the bottom of the reservoirs affect the response of dams to earthquake loading. The absorption coefficient alpha (α=1 no absorption / α=0 full absorption) has been measured so far on several dams, including the Monticello Dam [Ghanaat], while refractions and reflections surveys have found alpha values to be around 0.4 and 0.6. Comparison between the finite element results (taking into account absorption at the bottom of the reservoir with α=1.0, 0.6, 0.4, 0.2) and the records is presented in Figure 16, with 1% concrete damping being considered. This additional damping slightly reduces the response of the dam at the crest but does not significantly reduce the overestimation by the finite element model.

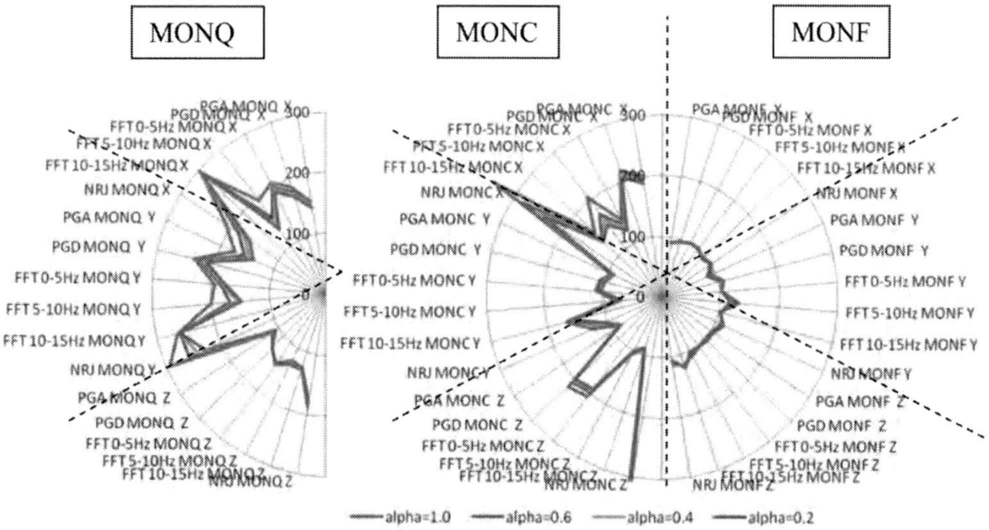

Figure 16 Comparison of the absorption due to sediments at the bottom of reservoir

Change of the Earthquake Input

As noted at the workshop, the earthquake input used for the existing records was based on the MONF recorded at the foot of the dam. The following section presents the results of the finite element analysis while using the MONA records as an input. The deconvolution process is performed to assess whether the computed response would match the record at the MONA device location. Figure 17 shows the comparison of the results, with the following remarks:

- Because of the deconvolution process, FE results match perfectly the records at MONA. Results at MONQ, located near the left abutment, have also improved.
- With the new input, results at MONF get slightly underestimated, which is probably why the results at MONC seem to be more precise.
- Taking into account the results at the rock accelerometers (MONF and MONA) and assuming that a wave propagates vertically, it can be noticed that the wave strength slightly decreases while reaching the top of the abutment.

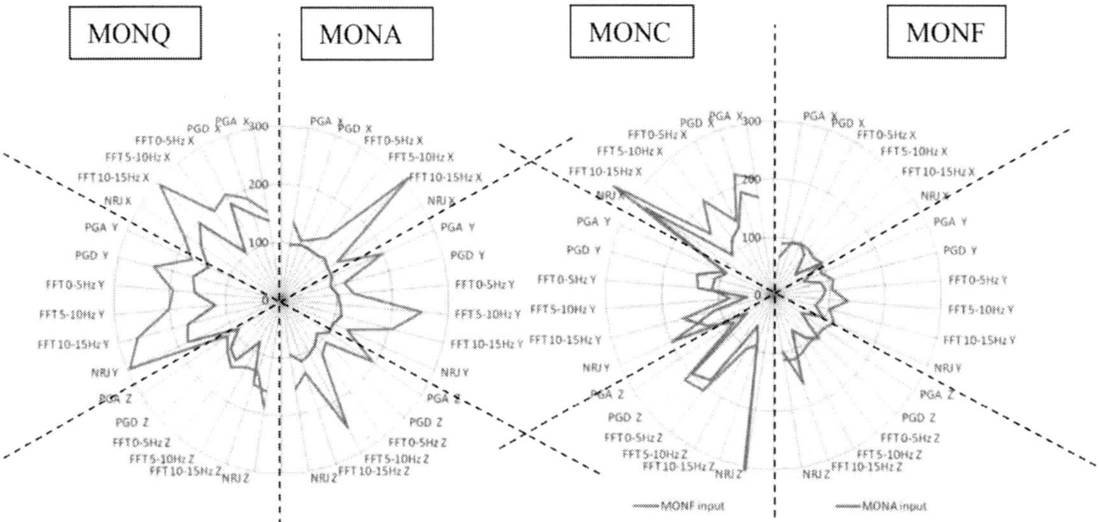

Figure 17 Comparison of the choice of input

Final Assumptions

Following up on the assumptions proposed in the previous chapters, Figure 18 presents the comparison between the FE analyses and the records, while considering the following characteristics:

- Earthquake input from MONA or MONF records,
- 3% concrete damping,
- Absorption of the wave at the bottom of the reservoir (alpha=0.6),
- 35,000 MPa concrete modulus,
- 10,000 MPa rock modulus,

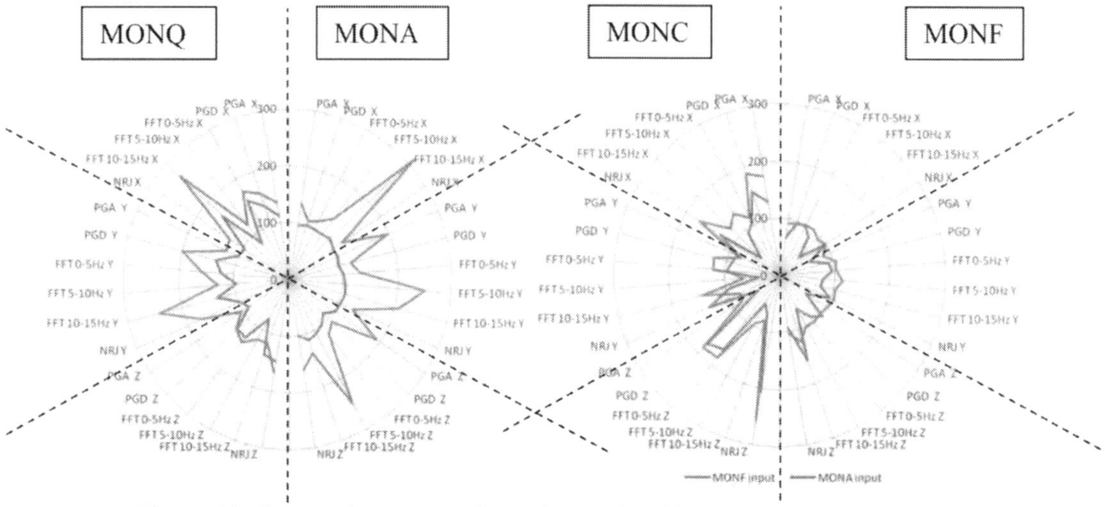

Figure 18 Comparison FE analyses / records with the final assumptions

By introducing different sources of damping: higher concrete damping, waves absorption by the sediment at the bottom of the reservoir and even by reducing the rock modulus (which increases the damping introduced by the viscous-spring boundaries in foundation), the results predicted by the finite element analysis get closer to the records. Nonetheless, having in mind the input at the toe of the dam, finite element results are still to be considered as overestimated by a factor 1.5 on average.

CONCLUSION

The initially submitted prediction of response of the Monticello arch dam to the M4.1 earthquake, using a finite element model with advanced fluid-structure and soil-structure interaction, and data available, was overestimated roughly by a factor 2.

Additional analyses, however, have shown that the finite element model could be somewhat improved, namely by: taking a lower rock modulus and introducing additional damping values in order to reduce the response of the dam, as well as by using the 3% concrete damping (despite the very low intensity of the earthquake). Wave absorption at the bottom of the reservoir slightly increases the overall damping of the system. Nevertheless, the dam's response was still overestimated by a factor 1.5.

A comparison of the results obtained while observing the earthquake input motion at the toe *vis a vis* that on the abutment of the dam shows that the earthquake signal was weaker than predicted on the abutment, which might explain the overestimation of the finite element analyses in this area. Spatial variation of the input should also be investigated.

ACKNOWLEDGEMENTS

The author would like to thank the U.S. Bureau of Reclamation for organizing the workshop which gave the opportunity to participating experts to work with, i.e. further analyze real earthquake records of the Monticello arch dam.

REFERENCES

Bouaanani N (2009): Assessment of potential-based fluid finite elements for seismic analysis of dam-reservoir systems. *Computer and structures,* **87** (3), 206-224

Chopra A.K., 2012, Earthquake response analysis procedure for evaluating and upgrading existing arch dams for seismic safety, ICOLD, Kyoto

Chopra A.K., 2008 Earthquake analysis of arch dams: factors to be considered, *14th World Conference on Earthquake Engineering,* Beijing

Ghanaat Y., Redpath B.B., 1995, measurement of reservoir-reflection coefficient at seven concrete dam sites

Liu YS, Chen DH (2013), Earthquake response analysis of a gravity dam considering the radiation damping of infinite foundation, *APCOM & ISCM*

Robbe E (2017), Seismic analyses of concrete dams: Comparison between finite-element analyses and seismic records, *16th World Conference on Earthquake Engineering,* Santiago de Chile

Zhang C, Pan J (2009), Influence of seismic input mechanisms and radiation damping on arch dam response, *Soil Dynamic and Earthquake Engineering*, **29** (9), 1282-1293

NON-LINEAR TIME DOMAIN ANALYSIS OF RUSKIN DAM IN LS-DYNA INCLUDING RESERVOIR-FOUNDATION-STRUCTURE INTERACTION

O. Penner, Ph.D., P. Eng.[1]
B. Bergman, P. Eng.[2]
S. Razavi-Darbar, Ph.D., P. Eng.[3]
D. Queen, P. Eng.[4]
P. Léger, Ph.D., P. Eng.[5]
Y. Boivin, Ph.D., P. Eng.[6]
M. Leclerc, P. Eng.[7]

ABSTRACT

Decision making regarding potentially costly seismic upgrades of dams is becoming increasingly reliant on the outcomes of large and complex finite element (FE) analysis models. As such, models should avoid introducing undue conservatism and reflect real conditions as accurately as practicable. Previous research has demonstrated the influence of reservoir-foundation-structure (RFS) interaction in particular on the dynamic response of concrete gravity dams, but incorporation of this interaction in time domain analysis has been difficult to achieve with the FE implementations in traditional codes.

This paper presents a two-dimensional non-linear time domain FE analysis of Ruskin Dam, a concrete gravity dam located in southwestern British Columbia, performed in LS-DYNA. The analysis of a gated overflow monolith, 53.4 m in height, includes RFS interaction, with earthquake excitation applied by the effective seismic input method (ESIM) and perfectly matched layers (PML) used to represent the semi-infinite nature of the foundation and reservoir domains. The first part of this paper presents validation of several aspects of the analysis using simplified models. The second part presents results for a complete two-dimensional model of the dam including the reservoir and foundation, and compares the response of the dam with that obtained using traditional methods such as a massless foundation and hydrodynamic added mass, as well as with that obtained using a deconvolved-input model in ANSYS. It is demonstrated that RFS interaction using the ESIM/PML approach in LS-DYNA captures the dynamic effects of the foundation and reservoir on the dam response with similar results as the deconvolved-input model. The seismic response can vary significantly compared to that obtained using traditional time domain methods based on added mass representation of hydrodynamic pressures and massless treatment of the foundation.

[1]Structural Engineer, BC Hydro, Burnaby, BC V3N 4X8, Canada, osmar.penner@bchydro.com
[2]Structural Engineer, BC Hydro, Burnaby, BC V3N 4X8, Canada, brent.bergman@bchydro.com
[3]Structural Team Lead, BC Hydro, Burnaby, BC V3N 4X8, Canada, soheil.razavidarbar@bchydro.com
[4]Structural Specialist Engineer, BC Hydro, Burnaby, BC V3N 4X8, Canada, david.queen@bchydro.com
[5]Professor, Department of Civil, Geological and Mining Engineering, École Polytechnique Montréal, QC H3T 1J4, Canada, pierre.leger@polymtl.ca
[6]Structural Engineer, Tetra Tech, Sherbrooke, QC J1J 2E8, Canada, yannick.boivin@tetratech.com
[7]Research Engineer, Department of Civil, Geological and Mining Engineering, École Polytechnique Montréal, QC H3T 1J4, Canada, martin.leclerc@polymtl.ca

INTRODUCTION

Ruskin Dam, completed in 1930, is the most downstream of three projects in the Stave River development in southwestern British Columbia. The impounded reservoir supplies water to a 3-unit 105 MW powerhouse[8] located adjacent to the dam on the left abutment. The dam consists of two minor and six major monolithic concrete gravity blocks, with a maximum total height of roughly 57 m. It is situated in a narrow valley and is predominantly founded on rock. The two minor blocks (B1, B2) are located on the right abutment while the six major blocks (B3–B8, from right to left) span the main valley section (Figure 1). The blocks are separated with contraction joints that are keyed and grouted. The dam is roughly 130 m long at the roadway deck, comprising a radial-gated spillway roughly 85 m long and non-overflow sections on either side totalling 45 m in length. In its original configuration, the spillway consisted of seven bays and eight piers. To improve the seismic withstand of the spillway, the piers and gates are being replaced with a five-bay, six-pier configuration as part of the ongoing Ruskin Dam and Powerhouse Upgrade Project (hereafter referred to as the Upgrade Project).

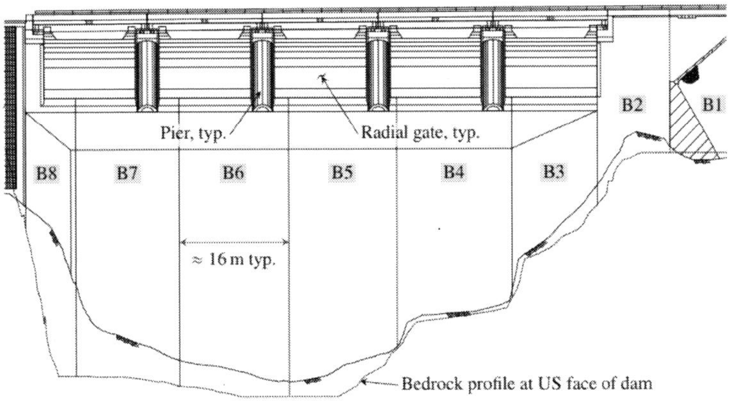

Figure 1: Upstream Elevation of Ruskin Dam, New Configuration

Previous studies on the stability of the dam body have identified a potential for sliding instability under seismic loading, particularly along lift joints near the crest of the dam. However, the analyses on which these studies were based have involved conservative simplifications. The main shortcomings identified in previous analyses were (a) the use of a massless representation of the foundation, which resulted in unrealistic amplification of input ground motions; and, (b) the use of elastic fluid elements tied to the dam, which resulted in unrealistic suction pressure occurring on the upstream dam face during large displacements. In addition, the seismic hazard at the site has been updated since the most recent stability assessment.

As part of the Upgrade Project, a new analysis and design effort (hereafter referred to as the Work) was undertaken with the intent to address the limitations of previous work and minimize conservatism. Analysis tasks were staged in order of progressively increasing

[8]The capacity will increase to 114 MW after completion of the Ruskin Dam and Powerhouse Upgrade Project.

complexity, with checking, verification, and validation carried out at each stage. The first stage consisted of simplified 2D and 3D analysis using a variety of traditional analysis tools. The second stage consisted of demonstrating proof-of-concept for 2D non-linear time domain analysis in LS-DYNA of the complete dam-reservoir-foundation system, including simulation of crack initiation and sliding along a discrete failure plane. The following stage will expand the methodology developed in the second stage into a 3D analysis of the complete system.

This paper presents the development, validation, and verification of key aspects of the second stage analysis using a 2D representation of Ruskin Dam. Achieving a reasonable simulation of fluid-structure interaction and foundation-structure interaction was of primary interest, and is the focus of this paper. To that end, only the first portion of the analysis, in which the dam was considered as monolithic (i.e. without the failure plane), is presented in this paper.

ANALYSIS CODE

The analysis code LS-DYNA is a multi-physics code with both implicit and explicit time integration capabilities (LSTC, 2015). The explicit method was adopted for the Work; this method is forward-stepping (non-iterative), making it well-suited to highly nonlinear problems. In this section, some of the main features and capabilities of LS-DYNA are discussed in the context of seismic analysis of a dam-reservoir-foundation system.

Foundation-Structure Interaction and Seismic Input

Recent versions of LS-DYNA include implementation of the effective seismic input method (ESIM) proposed by Bielak and Christiano (1984) and initially implemented by Cremonini et al. (1988). In this method, the seismic excitation is prescribed in the form of effective forces, and is applied on an interface *within* the discretized domain of the model. The problem is solved using a direct method, in which the total response of the generalized (nonlinear) structure and the scattered (or perturbed) response of the linear foundation domain are computed simultaneously. This eliminates the need to transmit the seismic excitation through the artificial foundation boundary, which therefore only needs to serve as an absorber of outgoing waves. In addition, no approximation is necessary in the specification of the input motion, so the absorbing boundary may be placed close to the structure. For a given discretized model, accuracy depends solely on the ability of the artificial boundaries to absorb scattered waves (Cremonini et al., 1988).

The method is illustrated conceptually in Figure 2. The free-field state of the system consists of an infinite half-space of linear foundation material subjected to an earthquake in the absence of any generalized structure (Figure 2a). When a structure is added to the free-field system, a response is induced in the structure, and the response in the foundation is changed due to the addition of scattered waves emanating from the structure (Figure 2b). Since the foundation is linear, the total state in the foundation can be considered as the superposition of the free-field and scattered states. By a series of substitutions and rearrangements of the equation of motion of the system, the free-field component of the

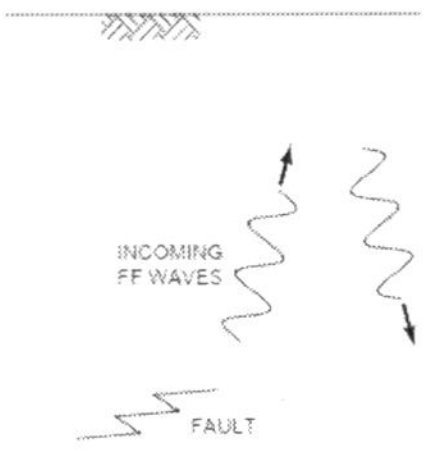

(a) Free-Field State—In Absence of Structure

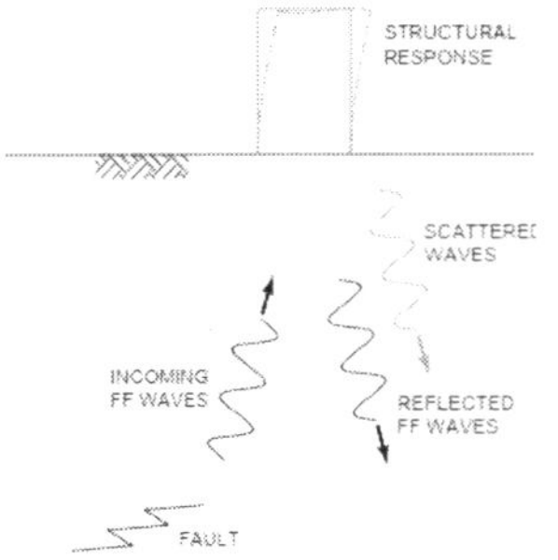

(b) Total State—In Presence of Structure

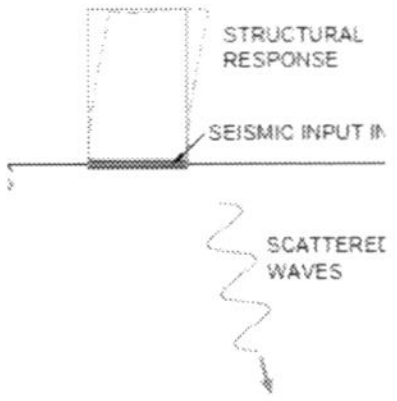

(c) Considering Scattered Response in Foundation, Total Response in Structure

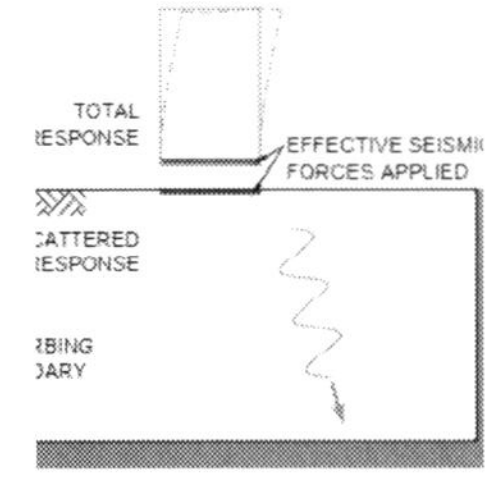

(d) Truncated Model with Scattered Waves in Foundation Absorbed by Boundary

Figure 2: Effective Seismic Input Method

response can be removed from the foundation when a particular set of forces—the effective seismic forces—is applied on the seismic input interface (SII), between the foundation and structure (Figure 2c). In this form, the equation of motion is formulated in terms of the total response in the structure, and the scattered response in the foundation. When truncating the foundation domain, as is necessary for a time-domain finite element (FE) implementation, the absorbing boundary is only required to absorb outgoing waves —nothing needs to be transmitted through the boundary into the model (Figure 2d).

In conjunction with the ESIM, LS-DYNA includes implementation of a novel absorbing boundary in the form of a perfectly-matched layer (PML), as presented by Basu (2009). The PML applies complex-valued coordinate stretching to the elastic wave equation such that waves of all frequencies and all angles of incidence are absorbed into the PML without any significant reflection from the interface. This makes the PML an appropriate model for the boundary of an unbounded domain (Basu, 2009).

The combination of the ESIM and the PML result in a highly accurate and efficient treatment of foundation-structure interaction and seismic input. The foundation domain can be much smaller than in the traditional approach, where the free-field input must be transmitted through the foundation domain to the structure, reducing computational demands. In addition, the effective seismic forces in the ESIM are formulated directly in terms of the free-field motion. This eliminates the logistical complications and inaccuracies associated with the traditional approach of computing and applying deconvolved motions at depth that are consistent with the target free-field motions. The ESIM/PML approach was adopted for the analysis performed as part of the Work.

Reservoir-Structure Interaction

LS-DYNA can simulate the response of compressible fluids in either Eulerian, Lagrangian, or arbitrary Lagrangian-Eulerian (ALE) frameworks. The Lagrangian approach was adopted for the Work for the purpose of simulating interaction between the dam and reservoir. This approach is computationally efficient, and is appropriate when the relative deformations of fluid elements are small. In addition, this approach integrates seamlessly with the ESIM/PML method; a PML is used to model the semi-infinite nature of the reservoir as in the foundation. The potential issue of unrealistic suction being produced on the upstream dam face is avoided by using a compression-only contact between the fluid and the dam, which allows the fluid to separate from the dam face.

VERIFICATION AND VALIDATION OF LS-DYNA

Verification and validation (V/V) was conducted for numerous aspects of dynamic analysis of dam-reservoir-foundation systems in LS-DYNA prior to and during the second stage analysis. This section of the report summarizes the methodology and the key findings for V/V of foundation-structure and reservoir-structure interaction.

Foundation-Structure Interaction

Treatment of foundation-structure interaction and seismic input using the ESIM and a PML is a novel approach that has not yet been widely adopted by the dam industry and had not previously been used at BC Hydro. Accordingly, thorough V/V of this approach was undertaken using a simple 2D model of a dam-foundation system subjected in turn to applied force pulses and to earthquake loading, both with and without a PML, and for varying sizes of the elastic foundation zone (between the SII and the PML). The main objectives included verifying the performance of the PML, the effect of the size of elastic foundation zone, the accuracy of the ESIM in producing the free-field motion, and the response trends of the foundation-structure system. The response of the system in all aspects was generally consistent with expectations. Key observations are summarized below.

- With the dam subjected to a short-duration applied force pulse, for a model with a 20-element thick elastic foundation zone ('medium'), the PML was observed to absorb outgoing shear and normal stress waves nearly perfectly relative to a model with an extended foundation mesh ($\approx$ 500 m deep) and no PML. When the elastic foundation zone was only 3 elements thick ('small'), the response was slightly less damped, but very similar (Figure 3).
- Under earthquake loading, when the weight and stiffness of the dam were dropped to 10 % of their actual values (i.e. approaching free-field conditions), the ground

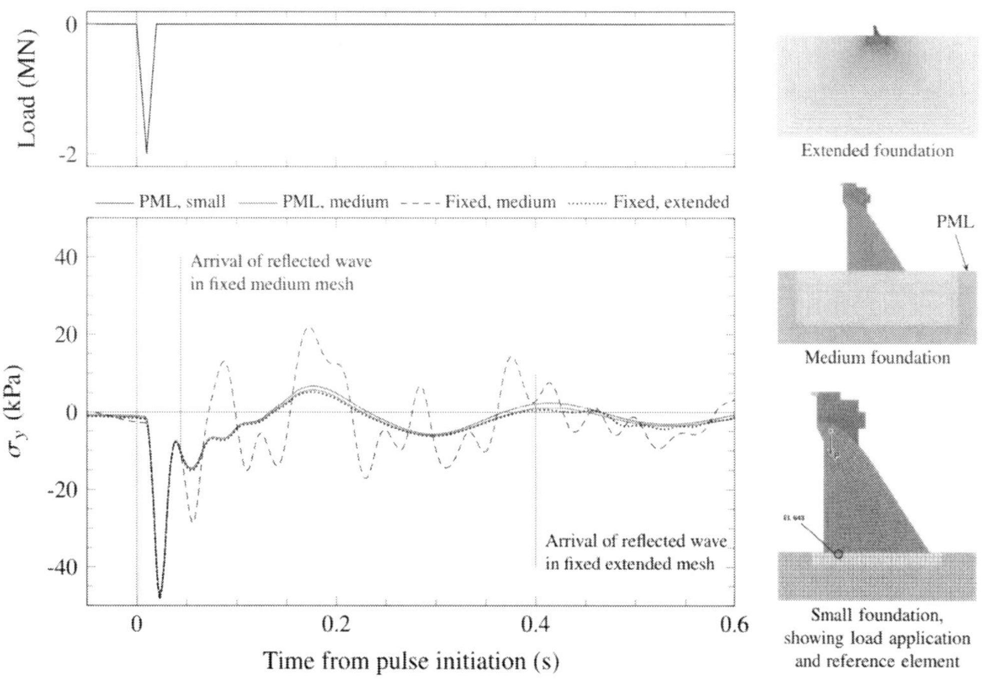

Figure 3: Vertical Normal Stress in Foundation Under Vertical Pulse Loading Applied in the Dam Crest—Models With and Without PML, Varying Foundation Size

motion observed at the SII was virtually identical to the free-field motion input to the model.

- The first mode response of the dam exhibited increasing total damping and decreasing natural frequency for decreasing foundation-to-concrete stiffness ratio, E_f/E_c. The observed trends were similar to those predicted for a generic dam section using the simplified frequency domain approach of Lokke and Chopra (2013). Similar results for the reference analysis are shown in Figure 8.
- The spectral response of the dam for $E_f/E_c = 8$ (very stiff foundation) was very similar to that of a model with no foundation and without the ESIM and PML, which was excited by directly applying the free-field motion to the nodes on the base of the dam.

Reservoir-Structure Interaction

A Lagrangian approach using elastic fluid elements was adopted for modelling of the reservoir. The main objectives of the V/V included verifying the performance of the fluid PML, the extent of fluid domain required in the model, the frequency response of the hydrodynamic pressures on the fluid-structure interface, and that the formation of suction (i.e. negative) pressure on the upstream face of the dam could be prevented. V/V was carried out using several simplified 2D models that simulated the excitation of a vertical wall impounding a reservoir, with and without a foundation. The response of the fluid-structure systems was generally consistent with expectations. Key observations are summarized below.

- The PML was observed to absorb outgoing pressure waves in the fluid nearly perfectly relative to a model with an extended upstream reservoir mesh.
- When including vertical excitation at the base of the reservoir in an ESIM/PML framework, the dynamic response of the system converged to a stable solution for upstream-downstream (U/D) lengths of the near-field (within the generalized structure) and buffer (between the near-field and PML) zones in the fluid at least 1.5 and 1.0 times the maximum depth of reservoir, respectively.
- The frequency response of hydrodynamic pressures on the fluid-structure interface with a pseudo-rigid wall closely matched the analytical solution given by Chopra (1967). The natural frequencies of the reservoir water column, the amplification of the hydrodynamic pressure around the natural frequencies, and the shapes of the pressure profile up the height of the wall at various frequencies were reproduced well by the LS-DYNA model (Figure 4).

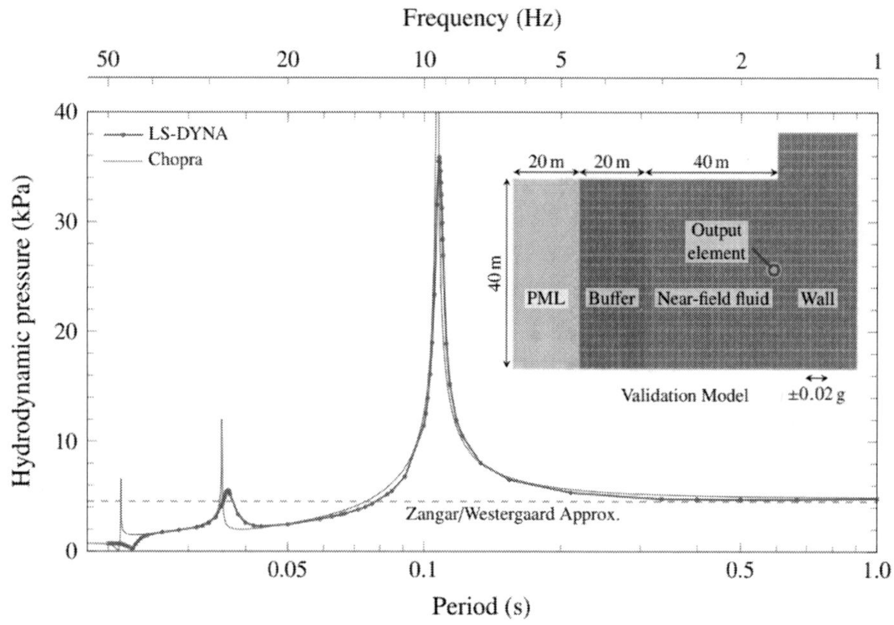

Figure 4: Frequency Response of Hydrodynamic Pressure Under Steady-State Harmonic Excitation—Near Mid-Height of Pseudo-Rigid Wall, Total Depth = 40 m

- Using a pressure cutoff in the fluid material formulation and a compression-only contact at the fluid-structure interface, contact pressures at the interface were effectively limited to positive values. When applying atmospheric pressure at the boundary of the fluid domain, this simulates separation of the fluid from the dam when vapour pressure is reached (near zero absolute pressure).
- The hydrodynamic pressures on the fluid-structure interface with a pseudo-rigid wall were sensitive to the contact penalty stiffness near the natural frequencies of

the reservoir, where large amplifications occur, but converged towards Chopra's analytical solution as the stiffness increased.

ANALYSIS METHODOLOGY

The 2D response of an U/D slice through Block 6 of Ruskin Dam to earthquake loading was examined using a variety of analysis models. A single spectrally matched ground motion based on a record from the 1994 Northridge earthquake was adopted for the proof-of-concept stage of the work, with the intent of developing and using a larger suite of records for later stages of the analysis. The horizontal and vertical components of this motion used as input for the various models are referred to as the free-field (FF) motions, and their response spectra are shown on plots in the results section. The main characteristics of the various analysis models are described in this section.

LS-DYNA Reference Model

A pseudo-2D analysis model representing an U/D slice through Block 6 of Ruskin Dam was developed. The model consists entirely of solid 8-noded hex elements, with one layer of elements in the cross-valley (C/V) direction. All nodes are restrained out-of-plane, thus producing conditions analogous to 2D plane strain. A schematic representation of the model is shown in Figure 5a, indicating the various parts, interfaces, and contacts. A detailed view of the mesh near the dam is shown in Figure 5b, in which nodes of interest are highlighted. As mentioned previously, this paper focuses on analysis not including a failure plane.

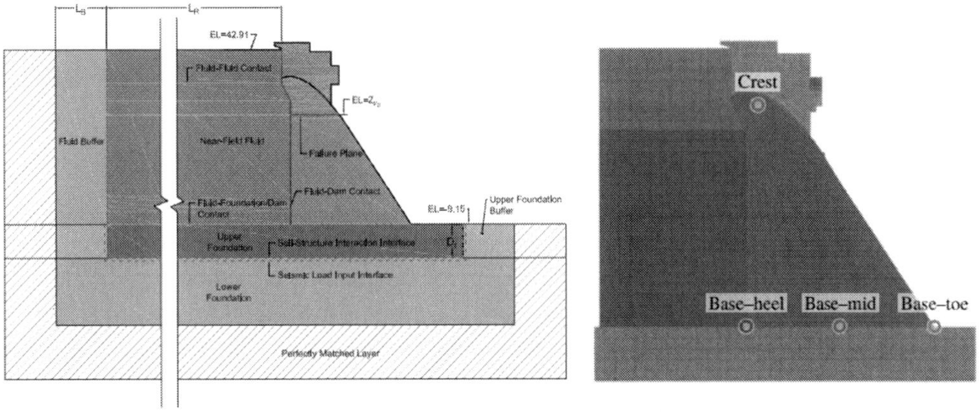

(a) Schematic Representation

(b) Mesh and Nodes of Interest

Figure 5: LS-DYNA Analysis Model

Material damping was applied using a combination of frequency-range damping and stiffness damping. Both of these methods function differently than traditional Rayleigh damping but discussion of these details is beyond the scope of this paper; it suffices to note that the applied material damping on the dam and foundation is roughly 1–2 % over frequencies around 4–60 Hz.

Traditional Models

Two additional models were developed to compare the LS-DYNA results with traditional approaches: (a) a modified LS-DYNA model, which represents the reservoir using added mass according to the method of Zangar (1952) instead of using fluid elements (referred to as the LS-DYNA added mass model); and, (b) an ANSYS model using the same added mass approach for the reservoir, and using a massless representation of the foundation (referred to as the ANSYS traditional model). The LS-DYNA added mass model uses the same ESIM/PML configuration as in the reference model to represent the foundation-structure interaction.

ANSYS was adopted to implement the massless foundation since massless elements can not be solved with explicit time integration. In the ANSYS traditional model, the foundation extends roughly 3 times the base length of the dam vertically and horizontally (Figure 6), and the free-field acceleration time history was applied at the outside boundaries of the foundation. Rayleigh damping in the ANSYS model was tuned to roughly match the total damping observed in the first mode of the LS-DYNA added mass model ($\approx 7\,\%$).

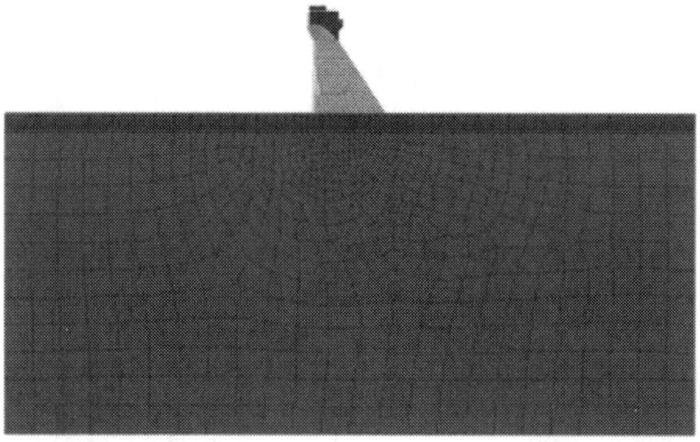

Figure 6: ANSYS Massless Foundation Model

ANSYS Reference Model

An additional ANSYS model using a deconvolved seismic input method was developed for validation of the complete dam-reservoir-foundation simulation (referred to as the ANSYS reference model). The foundation extends to three times the dam height in each direction. At the boundaries of the foundation and reservoir, 1D linear dampers simulate radiation into the semi-infinite domains. Dampers are applied to the upstream end of the reservoir in the normal (horizontal) direction, and to the foundation in the normal and tangential directions (Figure 7). The damping coefficients depend on the material densities (ρ) and the shear and compressional wave velocities (v_s, v_p) in the respective materials.

Seismic loads are input as 2D transient nodal stresses on the bottom surface of the foundation. The applied stresses are computed by using a 'flat box' calibration model, which consists of the same foundation model as shown in Figure 7, but excludes the dam and reservoir—this simulates the free-field state (Figure 2a). The goal is to determine the

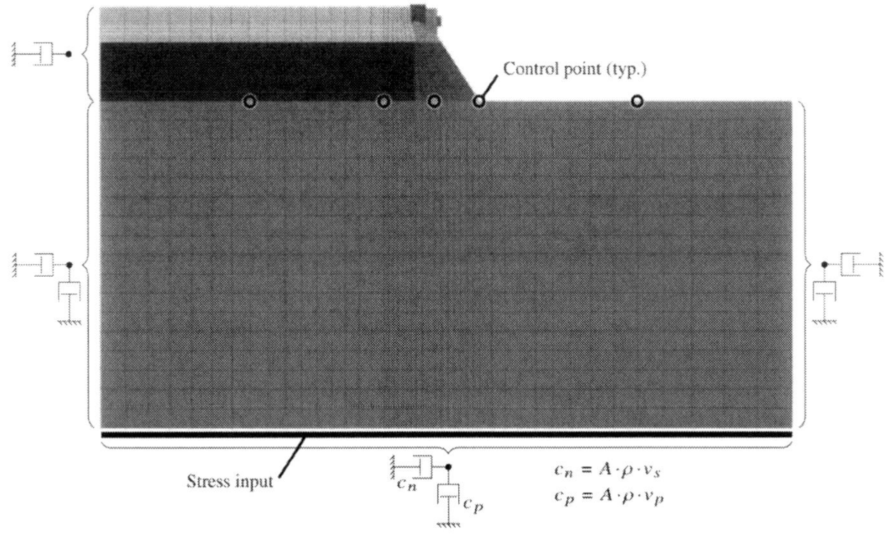

Figure 7: ANSYS Reference Model

required input stress time histories such that the output acceleration time histories at the surface of the box match the target free-field motion. The same stress time histories can then be applied to the complete model with the dam and reservoir to solve the complete dam-reservoir-foundation system. The flat box calibration involves the following steps, based on the method of USBR (2013):

1. Develop two layered 1D site models (e.g. in SHAKE)—one model for horizontal motion (shear wave propagation) and one for vertical motion (compression wave propagation). Compute the deconvolved velocity time histories, $V_x(t)$ and $V_y(t)$, at the bottom of the foundation.
2. Compute the transient nodal stresses to be applied to the nodes on the bottom of the foundation. Shear stresses are computed as $\tau(t) = 2 \cdot \rho_f \cdot v_s \cdot V_x(t)$, and normal stresses as $\sigma(t) = 2 \cdot \rho_f \cdot v_p \cdot V_y(t)$.
3. Run the flat box model with applied nodal stresses from step [2], and extract the acceleration time histories at the surface at selected control points (Figure 7).
4. Compare the extracted surface motions with the target free-field motion. Adjust the amplitude and phase of the input motions in the frequency domain and re-run the flat box model. Iterate until the extracted surface motions are sufficiently close to the target free-field motion.

Using frequency domain scaling, a close match was obtained between the surface motions near the center of the foundation (at the base of the dam) and the target free-field motions, while more deviation was observed further from the center. The uniformity of the flat box response could potentially be improved by applying stresses to the vertical sides of the foundation in addition to the base. However, the added complexity of this approach was beyond the scope of the Work.

RESULTS

The first-mode damping and natural frequency of the LS-DYNA reference model were assessed by applying a sine-sweep input ground motion to the model in the horizontal direction and computing the transfer function between the motion in the crest and the free-field motion. With reference material properties (concrete and foundation moduli of $E_c = 24\,\text{GPa}$ and $E_f = 14\,\text{GPa}$, respectively), the first mode occurs at roughly 3.65 Hz with 9.9 % damping. The first-mode frequency and damping are shown as a function of varying foundation stiffness in Figure 8. Consistent with the frequency-domain results of Lokke and Chopra (2013), the total damping in the first mode increases (at increasing rate) as E_f/E_c decreases. This is attributable to radiation damping in the foundation.

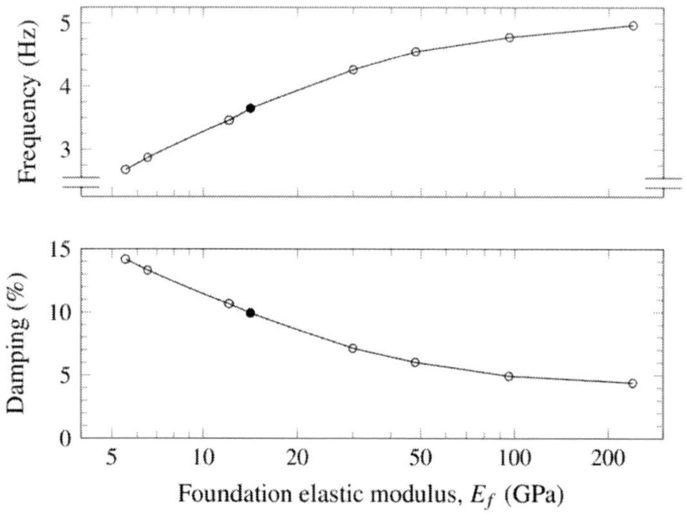

Figure 8: LS-DYNA Reference Model—First Mode Damping and Frequency, E_c = 24 GPa

The effect of fluid-structure interaction is examined by comparing the response of the LS-DYNA reference model with that of the LS-DYNA added mass model. The spectral response in the crest of the dam for each model is shown in Figure 9. Two results are shown for the LS-DYNA reference model: one with the input motion at full scale, and another with the input motion at 10 % scale, with the response of the latter scaled back up by a factor of 10. The difference between these two runs is evident in the high frequency range, where the full-scale run exhibits amplification due to impacts following separation of the fluid from the upstream face of the dam. In the scaled run, no separation occurs, eliminating this high frequency content from the response. At lower frequencies, the response of these two runs is very similar.

The spectral peak at the first mode is higher in the added mass model, for which the damping was measured at 7.1 % (vs. 9.9 % in the reference model). In contrast, the response is higher for the reference model at frequencies between about 5 and 15 Hz in the horizontal direction, even for the scaled run. This could be attributable to amplification of

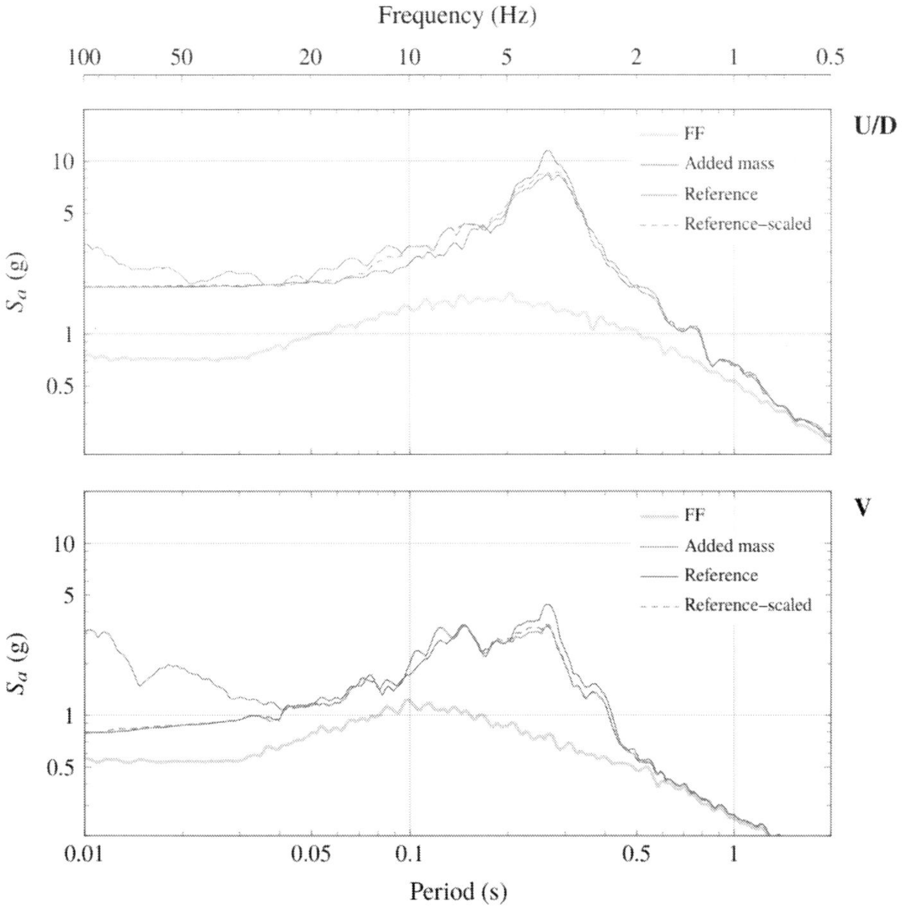

Figure 9: Fluid Elements vs. Added Mass—Dam Response at Crest

the hydrodynamic pressures at the first mode of the water column (as per Figure 4), which for the depth in question (52 m) occurs at roughly 7 Hz.

The effect of foundation-structure interaction is examined by comparing the response of the LS-DYNA added mass model with that of the ANSYS traditional model. The spectral response in the crest of the dam for each model is shown in Figure 10.

In the low-frequency range and around the first mode response (roughly 4 Hz), the two models exhibit very similar responses. The first-mode rocking response is reproduced nearly identically in both models. However, the responses diverge significantly at frequencies above roughly 5 Hz. While the response at the dam base in the LS-DYNA model is similar to the free-field signal, the response of the ANSYS model features a large spike with a peak around 9 Hz, most prominently in the vertical direction.

This frequency corresponds to the closely spaced 2nd and 3rd modes of this model, in which the dam exhibits rigid-body motion in the vertical direction with deformation

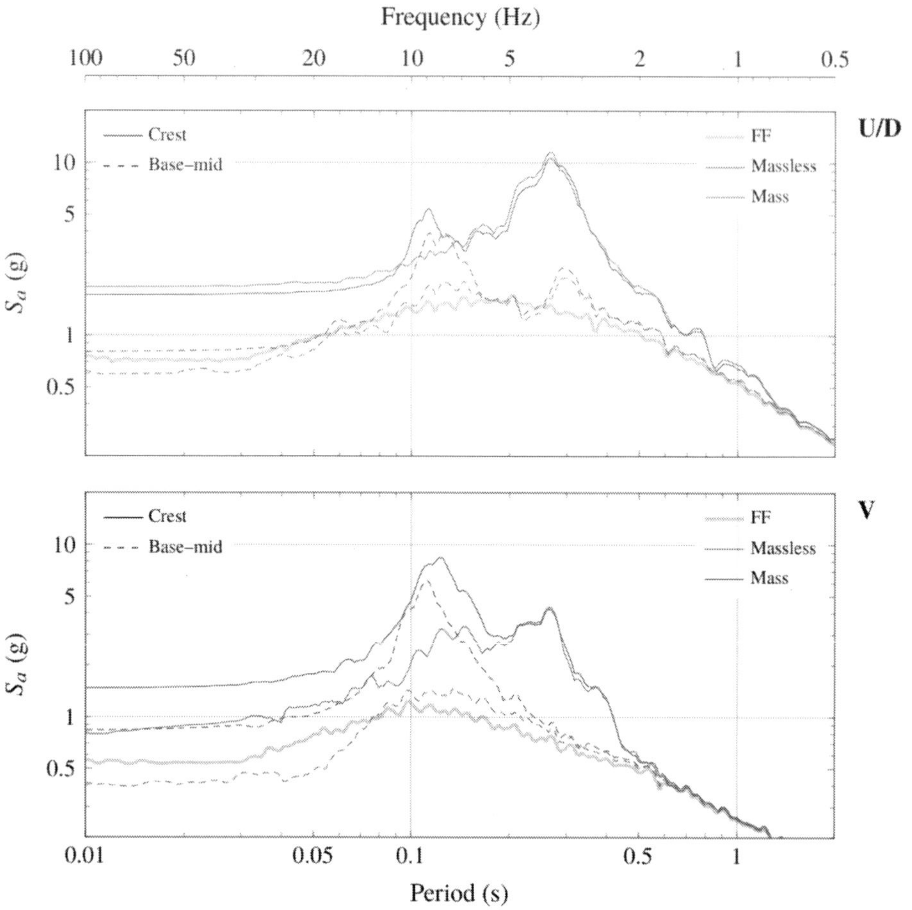

Figure 10: Massless vs. Mass Foundation—Dam Response—Base–mid and Crest

concentrated in the foundation, coupled with vertical elongation and bending in the dam body. The amplification is clearly transferred up to the crest of the dam, producing a peak spectral response in the vertical direction roughly 2.5 times higher than that from the LS-DYNA model.

As a final validation on the complete dam-reservoir-foundation simulation, the response of the LS-DYNA reference model was compared with that of the ANSYS reference model. The response in the crest and at the middle of the base are shown for both models in Figure 11. Two runs with different values of applied damping were conducted with the ANSYS model, for which the total first-mode damping was roughly 7.2 % and 10.7 % (vs. 9.9 % in the LS-DYNA reference model).

The response of the LS-DYNA model is similar to that of the ANSYS model over most of the frequency range. Around the first mode, the response of the LS-DYNA model is nearly the same as that of the ANSYS model with the higher damping in both directions. There is some deviation at higher frequencies, which is expected given that the response

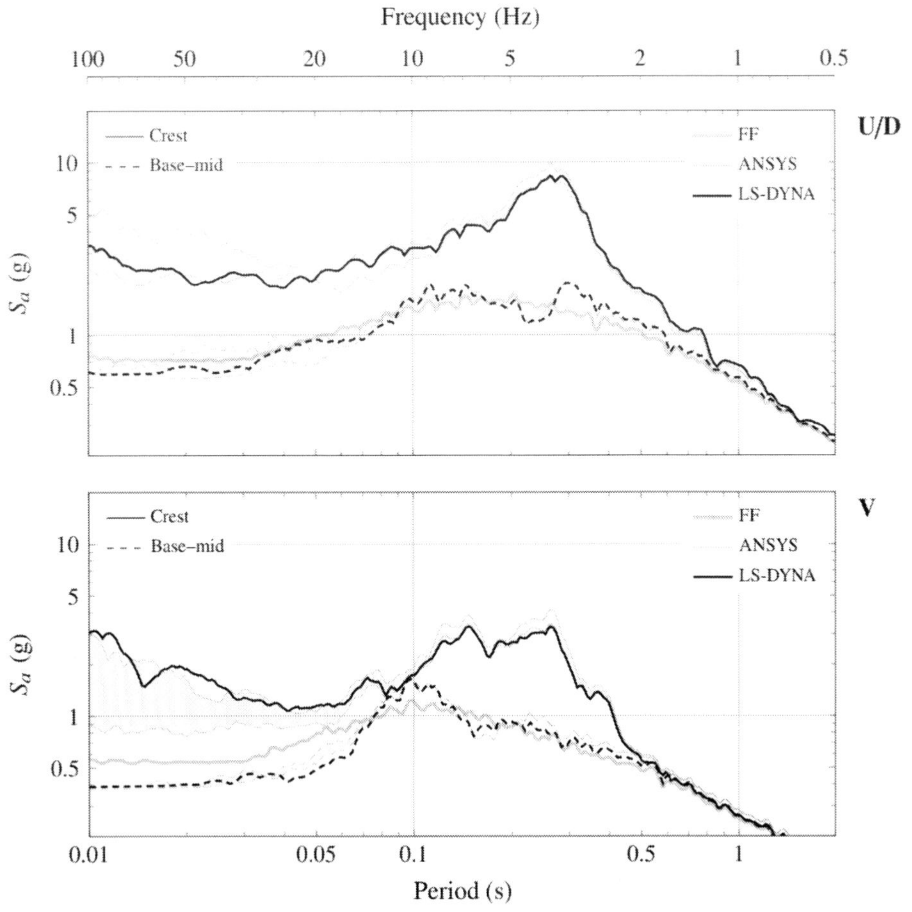

Figure 11: LS-DYNA vs. ANSYS Reference Models—Dam Response—Crest and Base–Mid

includes separation and impact between the dam and reservoir. Considering the substantial differences in damping, contact, and time integration methods between the two models, the results are remarkably consistent.

DISCUSSION AND CONCLUSIONS

The validation, verification, and analysis work documented in this paper demonstrates that time-domain dynamic analysis of a dam-reservoir-foundation system in LS-DYNA can be conducted efficiently, and that the results are consistent with the expected behaviour of such a system. Based on the results of the 2D analysis in this stage, it is expected that a 3D analysis with equivalent methodology can be successfully conducted in future stages.

Foundation-structure interaction can be appropriately simulated using the ESIM and a PML representing the infinite bottom and lateral extents of the foundation. Fluid-structure interaction can be appropriately simulated by Lagrangian fluid elements, a PML representing the infinite upstream extent of the reservoir, and a segment-based contact between the fluid and the upstream face of the dam. Results of an LS-DYNA model combining these methods were consistent with those of an ANSYS model with a deconvolved input motion and dashpot-type absorbing boundaries.

The ESIM/PML implementation in LS-DYNA is simple and efficient, producing accurate results with small foundation and fluid meshes and without the need for frequency-domain scaling of the input ground motion. Analysis results using this method can vary significantly compared to those based on traditional time domain methods based on added mass representation of hydrodynamic pressures and massless treatment of the foundation, with the potential to reduce the seismic demands on the dam considerably in some aspects. With decision making regarding potentially costly seismic upgrades of dams becoming increasingly reliant on the outcomes of such analyses, the proposed methodology will translate into greater confidence in decision inputs and more cost effective upgrades.

ACKNOWLEDGEMENTS

The authors of this paper acknowledge the BC Hydro Dam Safety Department for the opportunity to explore novel approaches to the seismic analysis of dams.

REFERENCES

Basu, U. (2009). Explicit Finite Element Perfectly Matched Layer for Transient Three-Dimensional Elastic Waves. *International Journal for Numerical Methods in Engineering*, 77(2):151–176.

Bielak, J. and Christiano, P. (1984). On the Effective Seismic Input for Nonlinear Soil-Structure Interaction Systems. *Earthquake Engineering & Structural Dynamics*, 12(1):107–119.

Chopra, A. K. (1967). Hydrodynamic Pressures on Dams During Earthquakes. *ASCE Journal of the Engineering Mechanics Division*, 93(6):205–224.

Cremonini, M. G., Christiano, P., and Bielak, J. (1988). Implementation of Effective Seismic Input for Soil-Structure Interaction Systems. *Earthquake Engineering & Structural Dynamics*, 16(4):615–625.

Livermore Software Technology Corporation (LSTC) (2015). *LS-DYNA - Keyword User's Manual, LS-DYNA R8.0, Revision 6319.* Livermore Software Technology Corporation.

Lokke, A. and Chopra, A. K. (2013). Response Spectrum Analysis of Concrete Gravity Dams Including Dam-Water-Foundation Interaction. *Pacific Earthquake Engineering Research Center, PEER Report 2013/17*, July 2013.

United States Bureau of Reclamation (USBR) (2013). State-of-Practice for the Nonlinear Analysis of Concrete Dams. *Technical Service Center.*

Zangar, C. N. (1952). Hydrodynamic Pressures on Dams Due to Horizontal Earthquake Effects. *U.S. Bureau of Reclamation Engineering Monograph No. 11.*

BOCA DAM SPILLWAY NONLINEAR ANALYSIS AND VALIDATION STUDY

Jennifer E. Huggins, PE[1]
Roman Koltuniuk, PE[2]
Miguel Rocha[3]

ABSTRACT

A suite of seismic simulations was done using LS-DYNA in order to make risk estimates and, subsequently, evaluate the need for dam safety modifications associated with seismic potential failure modes for the spillway gate structure within an embankment dam. A reduced model validation study was done to gain confidence in the simulation results, which capture the complexities of a soil-structure interaction problem. Both the large and reduced models contain all of the parts of interest to the analysis including the spillway gate structure, adjacent dam and dike embankments, reservoir, soil foundation layers, and bedrock; however, the extent of the model was reduced to facilitate faster run times. This paper examines the results obtained with LS-DYNA, in terms of wall behavior and seismic earth pressure, and compares them to results obtained from more traditional pseudo static methods.

INTRODUCTION

Boca Dam was completed in 1939 and is a zoned, rolled earthfill dam in California. The spillway crest structure is situated on a knoll between the main dam embankments and the dike, and the walls of the gated crest structure are directly adjacent to the embankments' impermeable core material. Because the spillway crest structure is in such a critical location, the Bureau of Reclamation (Reclamation) has been assessing the likelihood that seismic failure of the spillway crest structure walls could lead to a failure of the dam embankment.

Since 2009, Reclamation has been working to better understand the risks associated with the potential seismic failure of the spillway crest structure walls. This first evaluation used pseudo-static methods to compute seismic earth pressures and compared these loads to the spillway crest structure wall capacity in order to develop risk estimates (Bureau of Reclamation [Reclamation], 2009).

As Reclamation moved forward with efforts to better understand and reduce the risks associated with a possible seismic failure of the spillway crest structure walls, a finite element model was developed for use in risk analysis and evaluation of alternatives for dam safety modification of the spillway crest structure.

---

[1] Civil Engineer, Waterways and Concrete Dams Group 1, Bureau of Reclamation, P.O. Box 25007, 86-68110, Denver, CO 80225-0007, (303) 445-2491, jhuggins@usbr.gov
[2] Civil Engineer, Waterways and Concrete Dams Group 1, Bureau of Reclamation, P.O. Box 25007, 86-68110, Denver, CO 80225-0007, (303) 445-3225, rkoltuniuk@usbr.gov
[3] Program Manager, Mid Pacific Region, Dam Safety, Bureau of Reclamation, P.O. Box 25007, 84-44000, Denver, CO 80225-0007, (303) 445-2841, mrocha@usbr.gov

This paper will explore classical earth pressure methods and the use of finite element analyses (FEA). It will begin by describing the FEAs used to evaluate the Boca Dam spillway crest structure. Then it will compare soil pressure distributions and the demands placed on the counterforted walls as computed by for the various FEA and pseudo-static calculation methods.

CLASSICAL EARTH PRESSURE METHODS AND THE USE OF FEA

The determination of seismic soil loads on retaining walls has traditionally been done using two pseudo static methods: Mononobe-Okabe and Wood's.

Mononabe-Okabe extends Coulomb's theory of static active (and passive) earth pressures to include the effects of dynamic earth pressures on retaining walls. The Mononobe-Okabe theory incorporates the effect of earthquakes through the use of a constant horizontal acceleration in units of "g" acting on the soil mass comprising Coulomb's active wedge (or passive wedge) within the backfill (Mononabe & Matsuo, 1929). Reclamation's Best Practices in Risk Analysis manual (2009) notes an important limitation to the application of Mononobe-Okabe for very large seismic loads. As the horizontal seismic coefficient, k_h, increases, the angle of failure surface, α_a, decreases, until eventually it becomes equal to β, the slope angle of the backfill. The sliding wedge then becomes infinite, thus the calculated force on the wall also becomes infinite.

Wood's theory assumes that the retaining wall has a non-yielding backfill behind it. Sufficient wall movements do not occur and the shear strength of the backfill is not fully mobilized (Wood, 1973). Wood analyzed the response of a wall retaining non-yielding backfill to dynamic excitation assuming the soil backfill to be an elastic material. Wood's simplified solutions do not account for: 1) amplified accelerations from the base to the crest of the wall, 2) vertical or 2-compontent horizontal accelerations, 3) increased modulus with depth in the backfill, 4) the out-of-phase response along the height of the wall between the wall and the soil at any given time, and 5) the effect of the reduced soil stiffness with the level of shaking induced in both the soil backfill and soil foundation.

While Wood's and Mononabe-Okabe methods have sound bases in theory and can be applied fairly easily, they have limitations as discussed above and tend to yield results that may be overly conservative. That is, soil pressures computed with these techniques may give values for soil loads much higher than a structure would experience in the field. In a report prepared in cooperation with the State of California Department of Transportation, Geraili Mikola and Sitar (2013) say "Observations… of many different types of retaining structures… in recent earthquakes show that failures are rare even if the structures were not designed for the actual intensity of the earthquake loading." Geraili Mikola and Sitar used a combination of experimental and numerical methods to develop a recommendation for use in design of retaining structures.

A comprehensive finite element analysis was done for Boca Dam. This analysis takes into account the nonlinear behavior of soil during a seismic event, 3-D effects around the spillway area, seismic motions in three orthogonal directions, the site-specific ground motions at Boca Dam, and the wave propagation of the ground motions.

FINITE ELEMENT MODELS

The TrueGrid mesh generator was used to create the 3-D finite element model of the dam, reservoir, spillway crest structure, and the foundation with surrounding topography.

This suite of analyses focuses on the reinforced concrete spillway crest structure because this structure directly abuts the impermeable core material of the dam and its failure during a seismic event could result in an uncontrolled release of the reservoir. The spillway crest structure model consists of the ogee crest, gates, and wall panels supported by footings with three counterforts. The spillway crest structure concrete is modeled with brick elements and the reinforcement is represented by truss elements as seen in Figure 1 left and right, respectively.

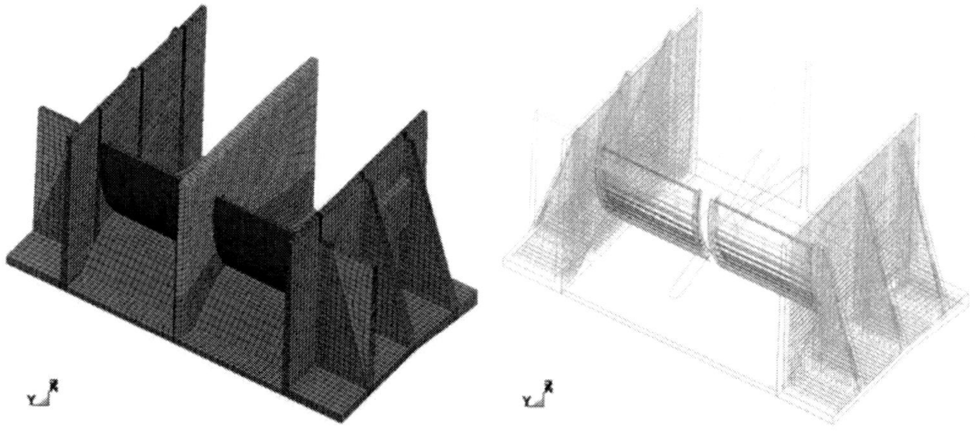

Figure 1. The LS-DYNA Model of the Spillway Crest Structure.

The 3-D FEM of the dam and dike consits of four zoned materials, using soild elements with nonlinear soil properties. Material properties were assed based on site investitgation and laboratory testing. See Figure 2 for a representation of the dam and dike as viewed from the right abutment foundation.

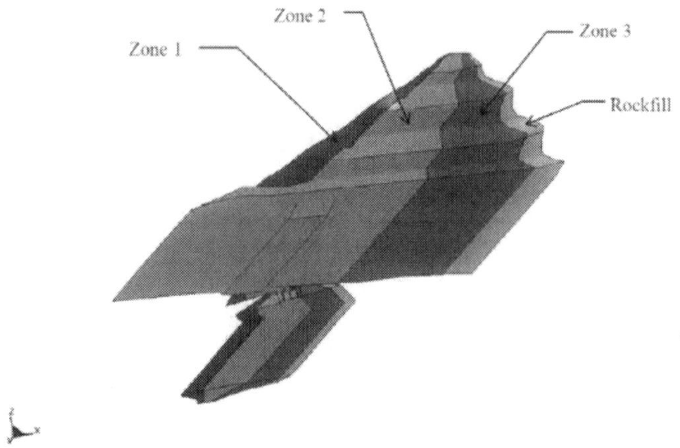

Figure 2. The LS-DYNA Model of the Zone Embankment Dam and Dike.

The foundation at Boca is composed of several materials. The tuff and tuff breccia bedrock (Tt/Tv), shown as the brown material in Figure 3, is modeled by linear material solid elements. In addition, there are quaternary units including channel alluvium (Qal), Donner Outwash (Qdo), and two Tahoe Outwash units (Qto1 and Qto2). All of these units are modeled with solid elements and a nonlinear soil material. Between the two layers of quaternary Tahoe Outwash is a narrow subunit of Basal Sand. This subunit, being a 1.5- to 3-foot-thick layer, is modeled by a sliding contact interface using a coefficient of friction computed from the estimated friction angle of the Basal Sand.

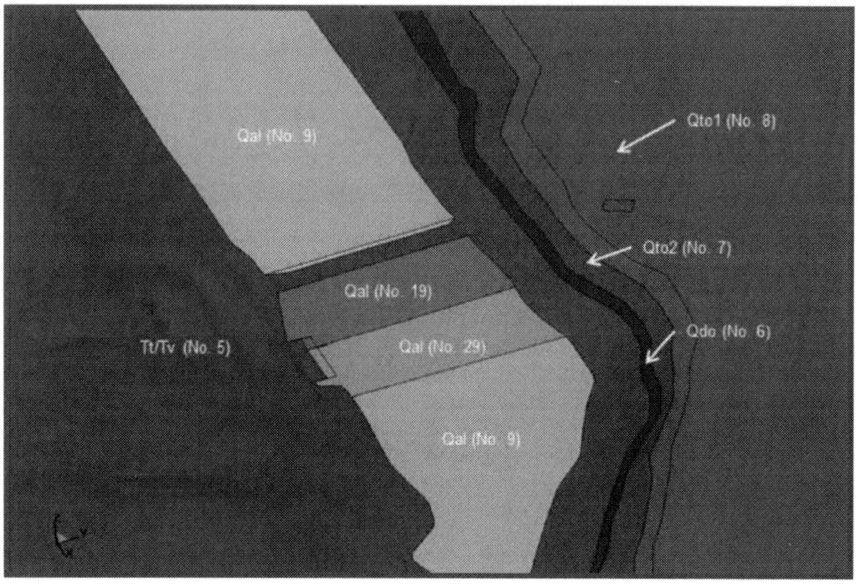

Figure 3. The LS-DYNA Foundation Model Parts.

The reservoir behind the dam and dike was modeled using solid elements with a fluid equation of state. The inlet channel reservoir adjoining the gates is also modeled with the same elements and fluid parameters. The reservoir is shown in blue in Figure 4.

Two sizes of models were used in these analyses; one very large and one much smaller. Both models contain all of the parts relevant to the analysis. The large model gives a much better spectral response to the ground motion and was used in risk assessment, while the smaller model offers much faster run times and was used for parametric studies. Figure 4 shows the large and small models in their entirety.

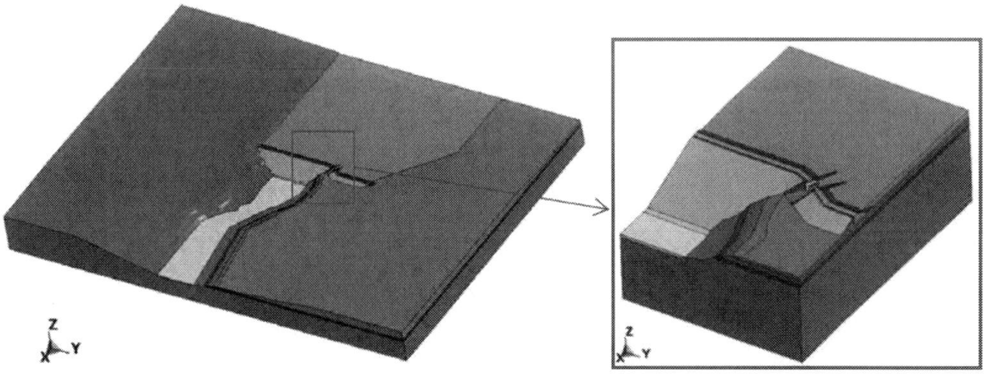

Figure 4. The LS-DYNA Large (left) and Small (right) Models.

SOIL PRESSURE DISTRIBUTIONS FROM THE VARIOUS METHODS

There are a few ways to assess the load, or demand, on a soil retaining structure under seismic loading including those proposed by Wood, and Geraili Mikola and Sitar. For counterforted retaining walls, such as those in the Boca Dam spillway crest structure, the passive earth pressure loading also represents a severe case.

This section of the paper will compare the demands placed on the structure as computed by Wood's Method, Geraili Mikola and Sitar, Rankine's passive earth pressure, and the Large and Small LS-DYNA FEAs. These calculations will assume that the peak horizontal accelerations correspond to the 10,000-year event.

Wood's Method

Figure 5 shows a schematic of the soil load to the spillway walls as computed by Wood's method. The at-rest component of the earth pressure is computed on a per-linear foot of wall basis by Rankin's theory as shown in equation (1), and the dynamic component can be found using Wood's method as shown in equation (2). In equation (2) k_h is taken to be the peak horizontal acceleration of the 10,000-year event, and F_p is estimated to be 1.0 from Kramer (1996) for $\nu = 0.3$ backfill. The dynamic component is assumed to act at a distance of 0.55.H from the base of the wall as shown in Figure 5.

$$P_o = \frac{1}{2}(1 - \sin\phi)\gamma H^2 \qquad (1)$$

$$\Delta P_{AE} = \gamma H^2 k_h F_p \qquad (2)$$

The moment acting at the base of the wall will be the maximum and computed as shown in equation (3).

$$M_o = {}^{1}/_{3} H P_o + 0.55H\ \Delta P_{AE} \qquad (3)$$

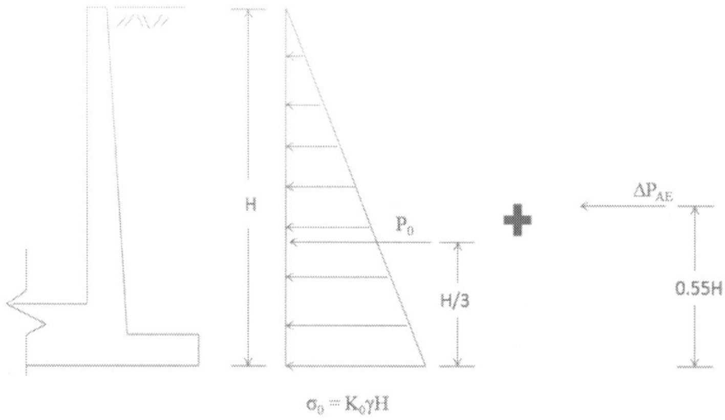

Figure 5. Spillway Wall Loading Schematic for Wood's Method.

Geraili Mikola and Sitar's Method

Figure 6 shows a schematic of the soil load to the spillway walls as computed by Geraili Mikola and Sitar's method. The at-rest component of the earth pressure is computed by Rankin's theory as shown in equation (1), and the dynamic component can be found using equation (4). In equation (4) K_{ae} is taken to be 0.4*(PGA_{FF}), and 0.4 times the peak free field ground acceleration is the upper limit recommended by Geraili Mikola and Sitar. The dynamic component is assumed to act at a distance of 0.3.H from the base of the wall as shown in Figure 6 and prescribed by the authors.

$$\Delta P_{ae} = \gamma H^2 K_{ae} \qquad (4)$$

The moment acting at the base of the wall will be the maximum and is computed as shown in equation (5).

$$M_o = {}^{1}/_{3} H P_o + 0.3H\ \Delta P_{ae} \qquad (5)$$

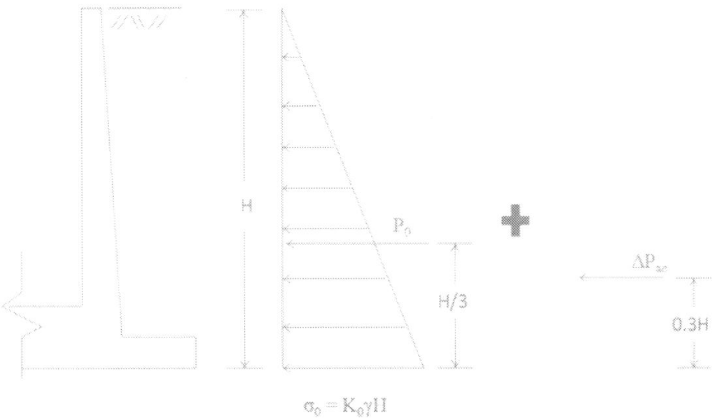

Figure 6. Spillway Wall Loading Schematic for Geraili Mikola and Sitar's Method.

Passive Earth Pressure

The passive earth pressure is computed by Rankin's theory as shown in equation (6), and acts as a triangular load distribution, like the at-rest component shown in Figures 5 and 6.

$$P_p = \tfrac{1}{2}.\tan^2(45 + \phi/2).\gamma.H^2 \qquad (6)$$

The moment acting at the base of the wall will be the maximum and computed as shown in equation (7).

$$M_o = {}^{1}/_{3}HP_p \qquad (7)$$

LS-DYNA Earth Pressures

Stress history data for element stresses was obtained from the LS-DYNA analyses for both the king size and fun size models. Elements in the backfill soil were selected on vertical intervals of approximately five feet from the top of the wall to the base of the wall, as shown in Figure 7.

Figure 7. Elements in Backfill Soil Selected for Stress Histories.

The stress histories are plotted in Figure 8 and Figure 9 for Small and Large model simulations, respectively. This seismic load represents the 10,000-year event. Both plots show that the stresses are highest at the base of the wall and approach zero near the top of the wall, indicating that we can assume a triangular load distribution increasing with wall depth. The Small model achieves a maximum stress of 26.5 lb/in^2 in element F at time t = 13.0s, while the Large model reaches a maximum stress of 24.6 lb/in^2 in element F at time t = 8.74s.

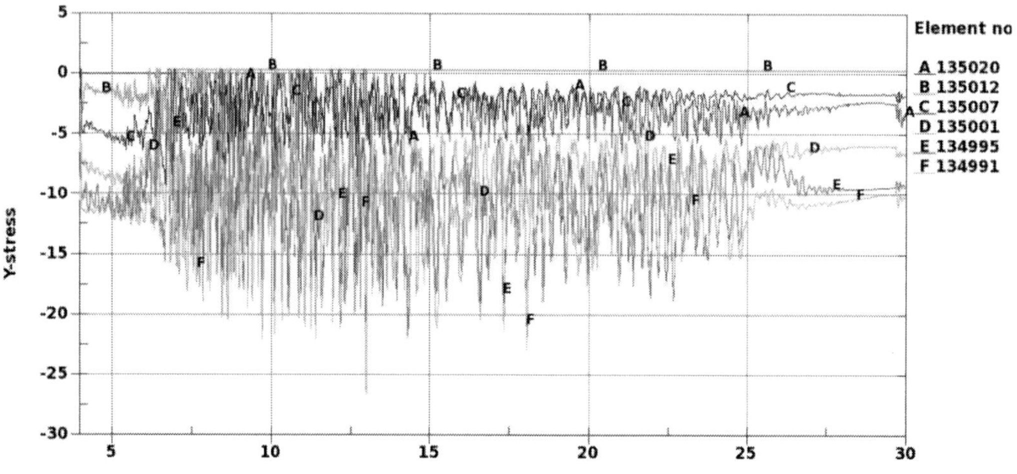

Figure 8. Stress Histories for Backfill Soil in the Small Model.

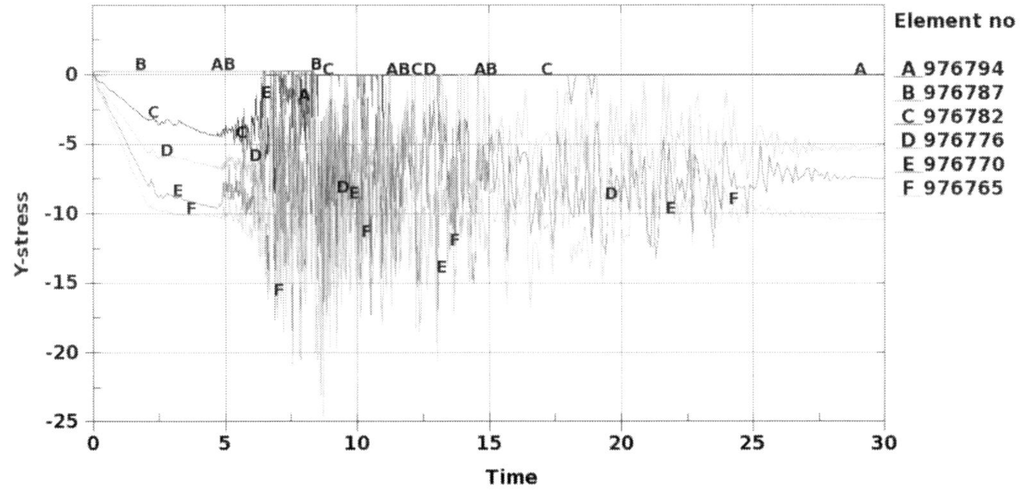

Figure 9. Stress Histories for Backfill Soil in the Large Model.

Summary of Computed Earth Pressures and Moments

Applying the equations enumerated in the preceding sections and assuming that the cohesionless soil has a friction angle of 308 and a moist unit weight of 130 lb/ft^3, we can compute the total force on the wall per linear foot. Peak free field ground acceleration will be assumed to correspond to the 10,000-year event for all seismic components. The wall height is taken as 26 feet. Table 1 shows a summary of computed loads to the spillway walls for the various methods.

Table 1

Method	Wall Load (kips/ft)	Moment at Wall Base (kip.ft/ft)
Wood's	104	1360
Geraili Mikola & Sitar	55	445
Passive	132	1140
LS-DYNA (large model)	46	400
LS-DYNA (small model)	50	430

Wood's method gives the highest moment due to the large dynamic component in conjunction with its lever arm being much longer than those used by Geraili Mikola and Sitar or Rankine. The LS-DYNA results are giving values close to those computed with the method prescribed by Geraili Mikola and Sitar. Passive earth pressures yield the highest total load to the wall; however, since it is applied so close to the base of the wall the moment is not as large as that obtained using Wood's method.

COMPUTED MOMENT DEMANDS FOR SPILLWAY WALLS

The preceding sections have assessed the various ways of computing the demands upon the spillway crest structure under seismic loads. This section will discuss the wall capacity and summarize moment demands upon the wall computed by the various methods. This paper will limit discussion to the potential failure of counterfort moment steel.

The tributary area will be based upon the uniform spacing of the counterforts. The effective flange width at the base of the wall is 78 inches was computed for moment capacity calculations using ACI 318-05, Section 8.10.2b. The area of the reinforcing steel is based upon the 5 pairs of rows of 1-inch-square bars found at the base of the wall. The resulting computed nominal design capacity is 4200 kip.ft.

Applying the tributary area to the previously computed moment demands on the spillway crest structure walls allows us to compute the total demand for the counterforted walls. See Table 2 for the results of these calculations.

Table 2

Method	Wall moment (kips.ft)
Wood's	15,753
Geraili Mikola & Sitar	5,155
Passive	13,205
LS-DYNA (king size)	4,633
LS-DYNA (fun size)	4,981

All of these calculations indicate that the moment capacity of the spillway wall will be exceeded under seismic loadings, regardless of the method used to evaluate the loading to the structure.

RESULTS FROM LS-DYNA

In the most recent risk analysis done for the Boca Dam spillway crest structure walls, the team did not evaluate the moment demand on the walls. That is done only in the paper for a complete comparison. The team assessed the likelihood that the concrete would crack under seismic loading and examined the state of the reinforcing steel.

Under seismic loading from the 10,000-year event one would expect a concrete retaining wall, such as the Boca Dam spillway crest structure, to crack. LS-DYNA models show concrete cracking for this seismic event. Damage contour plots for both the Large and the Small models are shown in Figure 11. In both cases, cracking through the counterforts occurs, although cracks in the footing are much more widespread in the large model.

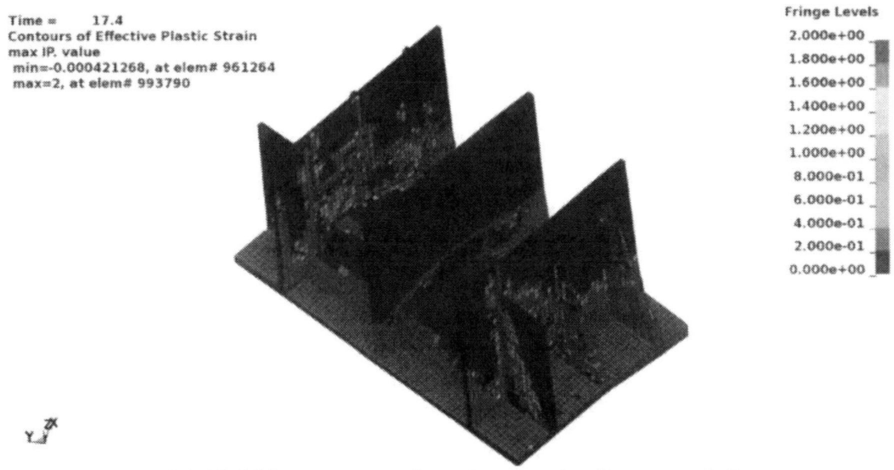

(a) 10,000-year ground motion results (large model)

(b) 10,000-year ground motion results (small model)

Figure 11. Concrete Damage Contour Plots (t = 17.4s).

The counterfort moment steel at the base of the wall is modeled so that it will yield at a force of 100,000 lb. The reinforcement at Boca was found to have a yield strength of 40,000 lb/in^2, and each of the beam elements shown in Figures 12 and 13 account for the area of two and one half bars. In addition, the steel elements were set to fail and be deleted upon reaching a failure strain of 0.05. The contour plots in Figures 12 and 13 do not shown any bars failing and dropping out of the solution during the earthquake for either the fun size or the king size model. However, this figure does indicate that several of the elements are reaching the yield force of 100,000 lb/in^2. The center counterfort on either side of the spillway has reinforcement being most heavily loaded.

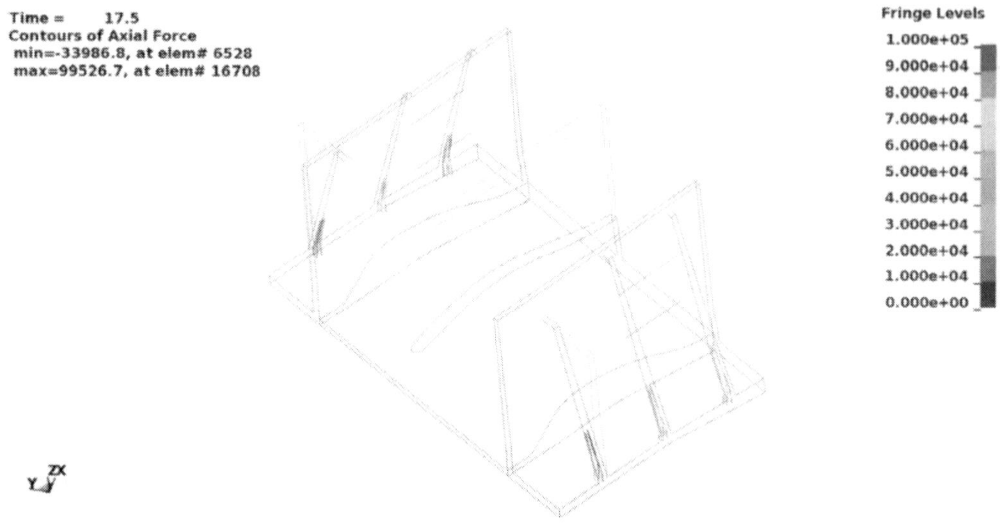

Figure 12. Large Model Counterfort Moment Steel Axial Force Contour Plots (t = 17.5s).

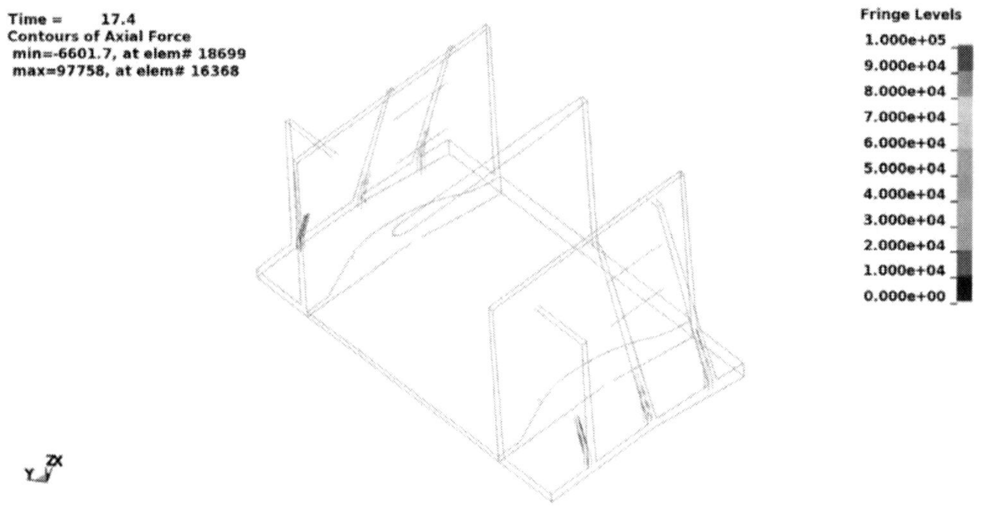

Figure 13. Small Model Counterfort Moment Steel Axial Force Contour Plots (t = 17.4s).

Force history plots for elements at the base of the center counterfort for the Large and Small models are shown in Figures 14 and 15, respectively. Both models show that some of the bars are approaching the yield strength, but that is the only trend they share. The Large model, in Figure 14, shows that after gravity has been applied (t = 4s), and in the early part of the earthquake (up until about 7s), the forces in the reinforcing steel are fairly flat and oscillate only a little. As the shaking becomes more intense, between time 7s and 12s, the forces begin to increase until they approach 100 kips. When the earthquake intensity begins to ebb after time 12s, the forces retreat to about 80 kips and remain fairly steady for the rest of the analysis. The Small model force histories in Figure 15 don't follow quite such a sensible progression with the ground motion, rather they steadily increase from time 4s, to time 15s, at which time they plateau.

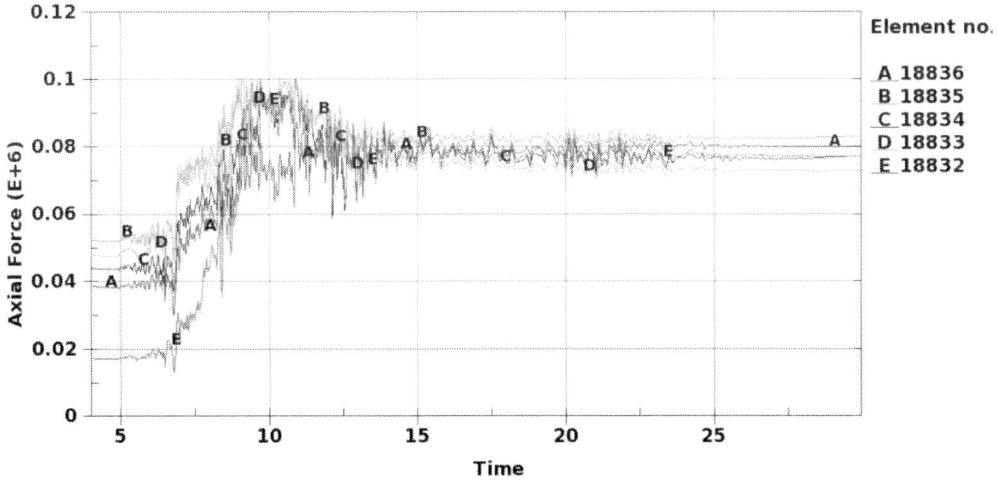

Figure 14. Large Model Right Center Counterfort Moment Steel Axial Force History.

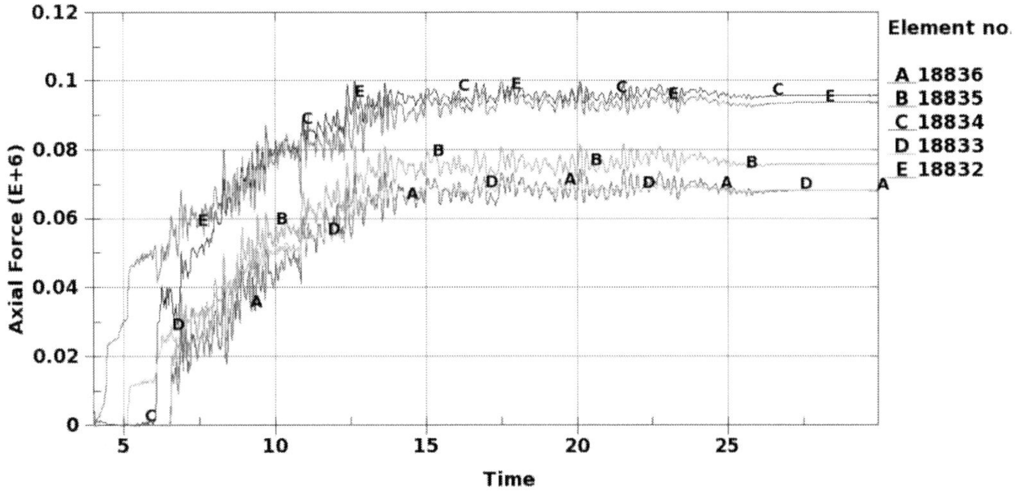

Figure 15. Small Model Right Center Counterfort Moment Steel Axial Force History.

The center counterfort is most highly stressed in the FEAs. This one is probably bearing the load most representative of the tributary area computed in the preceding sections. These results, at the center counterfort at least, may be considered indicative of a demand/capacity ration near 1.0. Using the D/C ratio to estimate risks has a few shortcomings, but can be a good tool for initial assessment of a structure. The tributary area and seismic earth pressure should be carefully considered, and it should be noted that ACI codes do include strength reduction factors for use in design.

CONCLUSION

The LS-DYNA results are able to take into account many factors that the pseudo static analyses cannot. More recently developed techniques for computing and applying seismic earth pressures, such as that prescribed by Geraili Mikola and Sitar, may yield results more in keeping with those obtained by FEAs. Using earth pressure distributions such as those computed by Wood's method or passive loading cases can serve to bracket a problem as "worst case" scenarios; however, their use in design and analysis may not give the optimal, economical engineering results.

Smaller FEMs can be very useful for parametric studies, but larger models should be used in risk assessments when decisions of public safety are at stake.

ACKNOWLEDGEMENTS

Thanks to Steve Dominic, and Hillery Venturini for their thoughtful review of this paper.

REFERENCES

American Concrete Institute. (2005). Building code requirements for structural concrete (ACI 318-05). , Farmington Hills, MI: American Concrete Institute.

U.S. Department of the Interior, Bureau of Reclamation. (2009). Boca Dam issue evaluation: seismic stability of the spillway (Technical Memorandum No. BOC-86-68110-IE-18-1). Denver, CO: Technical Service Center.

U.S. Department of the Interior, Bureau of Reclamation. (2009). Dam safety risk analysis best practices training manual (Version 1.1). Denver, CO: Technical Service Center.

U.S. Department of the Interior, Bureau of Reclamation. (2013). State-of-practice for nonlinear analysis of concrete dams. Denver, CO: Technical Service Center.

Geraili Mikola, R., & Sitar, N. (2013). Seismic earth pressures on retaining structures in cohesionless soils (Report No. UCB GT 13-01). University of California – Berkeley: Department of Civil and Environmental Engineering, Geotechnical Engineering.

Kramer, S. (1996). Geotechnical earthquake engineering. Upper Saddle River, NJ. : Prentice Hall.

LS-DYNA 971 Release 7.1.1 [Computer program]. (2014). Livermore, CA: Livermore Software Technology Corporation.

Mononobe, N., & Matsuo, H. (1929). On the determination of earth pressures during earthquakes. Proceedings, World Engineering Congress, 9, 179-187.

Wood, J. (1973). Earthquake induced soil pressures on structures (Report No. EERL 73-05). Pasadena, CA: California Institute of Technology.

TrueGrid Version 2.3.4 [Computer program]. (2009). Pleasant Hill, CA: XYZ Scientific Applications, Inc.

NONLINEAR SEISMIC EVALUATION OF PERRIS DAM OUTLET TOWER

Yusof Ghanaat, Ph.D., PE[1]
Zachary Harper[2]

ABSTRACT

Outlet Tower at Perris Dam in Southern California is a critical feature of an emergency outlet works to drain the reservoir to avert potential uncontrolled release of water should a dam emergency occur. The tower is a 105-foot-high tubular reinforced concrete structure with an inside diameter of 26 feet and outside diameter of 31 feet and is accessed by a bridge. Previous linear-elastic seismic evaluations of the tower with pre-NGA ground motions indicated significant damage, and results with NGA ground motions were inconclusive. This paper presents the most recent seismic assessments of the tower using two different three-dimensional (3D) nonlinear models for prediction and validation of the level of damage to determine whether the tower would remain stable, functional, and accessible to safeguard dam safety following a postulated maximum credible earthquake (MCE).

One model employed layered shells comprised of nonlinear concrete and nonlinear steel layers to allow concrete cracking and steel yielding. The second more advanced model used 3D nonlinear solid elements to predict concrete cracking in three orthogonal directions and steel yielding/rupturing in individual vertical and horizontal reinforcements. Both models included the access bridge and allowed the bridge to slide on rollers at the tower support. The nonlinear models were analyzed for three sets of three-component rock acceleration time histories applied along the principal axes of the tower. The results indicate that the tower would perform satisfactorily and remain operational when subjected to the postulated MCE. It would suffer only minor damage in the form of microcracks in the bottom third with no yielding of the vertical and horizontal steel. The access bridge would be accessible to foot traffic but may not be safe for vehicle crossing.

INTRODUCTION

A linear seismic analysis of Perris Outlet Tower performed in 2006 indicated that the tower shear and moment capacities would be exceeded during a pre-NGA estimated maximum credible earthquake (MCE) and the tower would suffer extensive damage. The Next Generation Attenuation (NGA) relationships published by Pacific Earthquake Engineering Research Center (PEER) in 2008 significantly reduced the ground motion at this site. The tower was reanalyzed in 2012 to examine several modeling assumptions and determine if the tower and bridge could resist the NGA-estimated ground motion.
The 2012 analysis concluded that the Perris Outlet Tower would perform adequately, but did not satisfy issues raised regarding shear cracking and degree of effectiveness of the

---

[1] Quest Structures, Inc., 3 Altarinda Road, Suite 203, Orinda, CA 94563, yghanaat@QuestStructures.com.
[2] Quest Structures, Inc., 3 Altarinda Road, Suite 203, Orinda, CA 94563.
Zachary.harper@QuestStructures.com

horizontal shear reinforcement. To address these issues, a detailed nonlinear analysis was performed to predict the extent of concrete cracking and the effects it might have on shear reinforcement and to provide a more accurate evaluation of the seismic performance of the tower and the bridge. This paper describes the nonlinear analyses and results obtained for Perris Outlet Tower.

Description of Tower

Perris Outlet Tower is owned and operated by the California Department of Water Resources (DWR). The tower is a 105-foot-high tubular reinforced concrete structure with an inside diameter of 26 ft, outside diameter of 31 ft, and a wall thickness of 2.5 ft (Figure 1). At the bottom, the tower connects to a 12.5-foot-diameter tunnel. The tower includes ten 72-inch butterfly valves along its height: two at each of five equally spaced elevations between 1,503 and 1,567 ft. Near the valves, the tower wall thickness increases to form "haunches" that create a flat vertical surface for embedment of the valves. A 52-foot-diameter concrete deck caps the tower at the top elevation of 1,600 ft. The tower includes two layers of longitudinal reinforcement near the outside and inside wall surfaces with transverse reinforcement hoops along the height. Each longitudinal layer consists of two-bar bundles spaced at 16 inches on-center with bar sizes varying from #18 bars at the bottom to #11 at the top. Transverse reinforcing hoops, each formed from multiple individual bars with lap-spliced connections, are laid at 11 inches on-center with differing bar sizes from #9 at the bottom to #8 at the top. The tower is founded on a 50-foot-diameter by 12-foot-thick concrete pad. The pad is anchored to the underlying rock with two rows of #18 bar anchors installed at 20- and 22-ft radii from the pad center. A 113-foot-long single-span deep-girder steel bridge provides access to the tower. The bridge is supported on a fixed bearing at the abutment and roller bearing at the tower. A 3-inch gap separates the bridge from the tower.

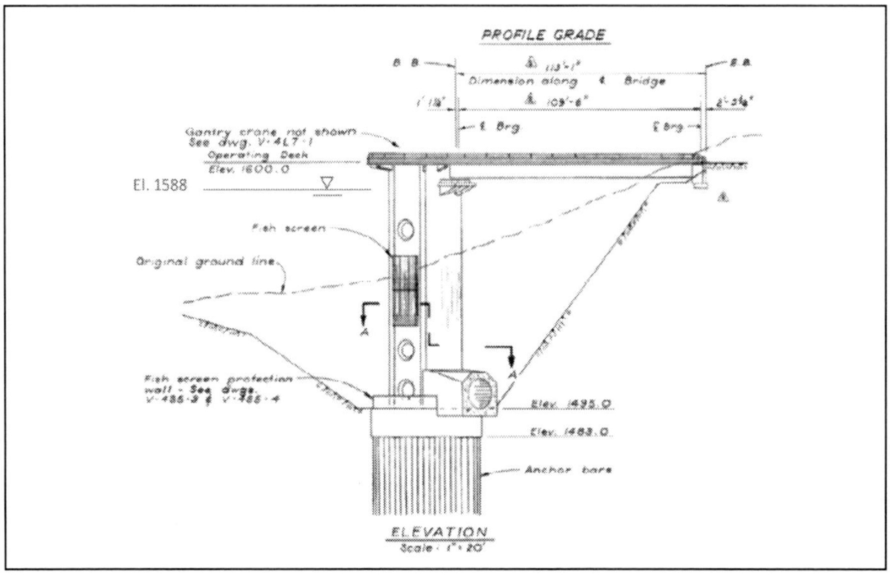

Figure 1. Elevation view of Outlet Tower, access bridge, and foundation.

Seismic Hazard

A site-specific probabilistic seismic hazard analysis (PSHA) was performed by URS (2009) for DWR to obtain ground motions for evaluation of Perris Dam and Outlet Tower. The probabilistic seismic hazard incorporated the most recent USGS characterization of the major faults in southern California and explicitly included the uncertainties in each ground motion parameter using logic trees. The site-specific probabilistic seismic hazard (PSHA) indicated that the San Jacinto fault was the dominant contributor to the probabilistic hazard at all return periods. The San Jacinto Fault Zone is a major right-lateral strike-slip fault zone. It is 210 km long including Coyote Creek fault and runs through eight communities in Southern California.

The deterministic seismic hazard assumed a M_w7.5 earthquake on the San Jacinto fault at the closest distance of 8 km from the site to represent the MCE. The earthquake peak ground acceleration and response spectra were developed for the hard rock at the 84th percentile confidence level using the 2008 Next Generation Attenuation (NGA) relationships. The estimated horizontal and vertical peak ground acceleration were 0.44g and 0.40g, respectively. Figure 2 compares the pre-NGA and the 2008 NGA target response spectra. Three sets of MCE response-spectrum compatible acceleration time histories were developed by DWR as the input for the seismic assessment of the tower. The seed acceleration time histories were selected from the 1999 M7.6 Chi-Chi, Taiwan; the 1992 M7.3 Landers, California; and the 1978 M7.4 Tabas, Iran.

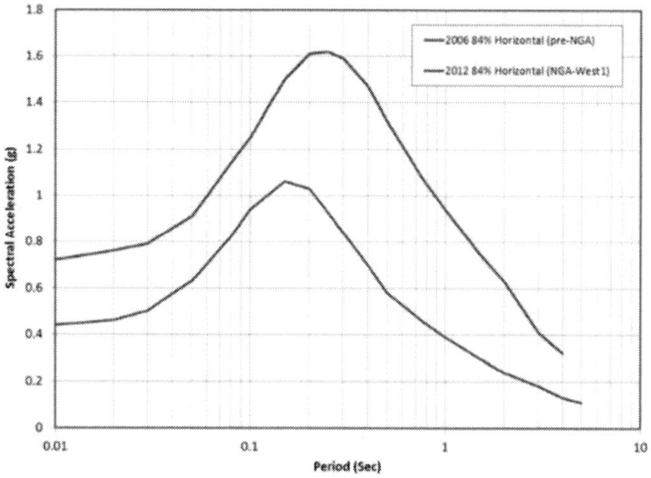

Figure 2. Pre-NGA and 2008 NGA response spectra for the site.

COMPUTER MODELS

Two different three-dimensional (3D) nonlinear modeling approaches were employed to validate the results. One approach used nonlinear layered shell elements to model concrete cracking and steel yielding. Another method used 3D solid elements with a concrete nonlinear constitutive model to allow concrete cracking in three orthogonal directions and nonlinear beam elements to allow steel yielding and rupturing (if any). The use of two different modeling approaches produced independent results that allowed for validation of the nonlinear results. In both cases, the access bridge was modeled using a combination of linear finite elements but allowed to slide at the roller support and potentially impact the tower if the gap between the tower and bridge was closed, or to become unseated from the tower if the bearing roller limit was exceeded.

Layered Shell Model

A layered shell model of the outlet tower and the access bridge was developed and analyzed using SAP2000 (Figure 2). The cylindrical tower (gray) and a portion of the portal integrated with the tower (blue) were represented with nonlinear layered shell elements. The shell elements were positioned at the centerline of the cylinder. The valve port openings were included in the model, and an adequate number of shell elements were used to capture the geometry of the openings and thicker sections surrounding the openings. Standard shell elements with linear-elastic material properties were used for the tower deck (green), brackets (yellow), bridge seat (red), and the access bridge.

An initial linear analysis of the tower indicated that potential concrete cracking and steel yielding would be confined to the bottom half of the tower. The nonlinear concrete and steel properties were, therefore, assigned to layered shells in the outlet portal and bottom half of the tower. Linear concrete and steel properties were assigned to the remainder of the tower in the upper half and to the access bridge to reduce computation time.

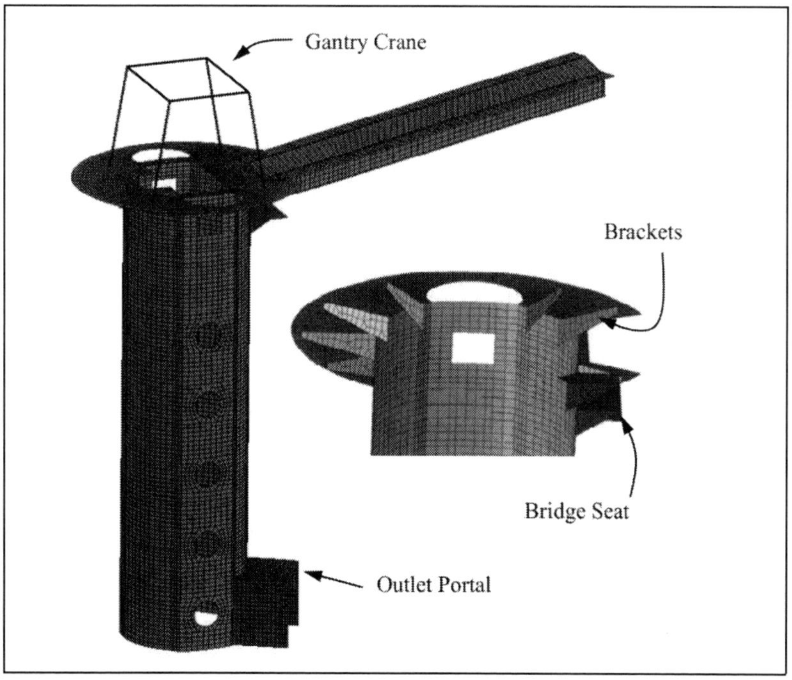

Figure 3. SAP2000 layered shell model of Perris Outlet Tower.

Layered Shell Elements. The layered shell element allows any number of composite layers to be defined through the thickness direction, each with an independent location, thickness, and material (CSI, 2014). Material behavior may be nonlinear. The layered shell usually represents full-shell behavior with translational and rotational degrees of freedom, capable of supporting forces and moments. The shell behavior can be controlled on a layer-by-layer basis to capture only the dominant nonlinear response behavior. For example, a steel reinforcing layer may be defined for unidirectional axial behavior only.

For Perris Outlet Tower, the mid-surface of the tower wall was chosen as the reference surface for the layered shells. Each layer was located by specifying the distance from the reference surface to the center of the layer. Each layer had a single thickness, measured in the local-3 direction of the element. For reinforcing bars, a very thin "smeared" layer with an equivalent cross-sectional area was specified. Over the majority of the tower wall, each shell consisted of 3 layers: a single layer to represent the full thickness of concrete, and two layers to represent the vertical rebar (one near the inside face and one near the outside face). Figure 4 depicts these as Layer 1 for concrete, and Layers 2 and 3 for the vertical rebars on the inside and outside face of the tower, respectively. The concrete layer was modeled to behave nonlinearly with respect to vertical stresses and linear with respect to horizontal stresses. This modeling assumption allowed capturing the most important nonlinear mechanism while reducing the computation time. The rebar layers were conservatively defined to act only in the vertical axial direction, providing zero stiffness in the horizontal (circumferential) direction. They did, however, provide shear stiffness through the thickness direction, representing rebar dowel action.

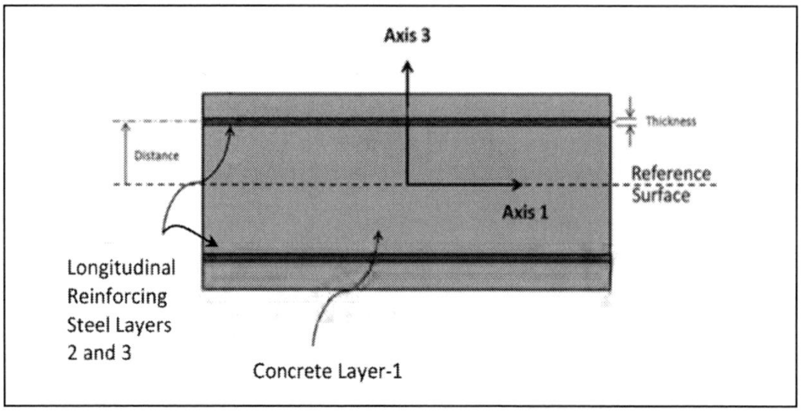

Figure 4. Layered shell model showing concrete as Layer 1 and vertical reinforcing bars as Layers 2 and 3.

3D Solid Model

A more advanced nonlinear analysis of the tower was conducted using LS-DYNA (LSTC, 2014), a multi-physics computer program that originated at the Lawrence Livermore National Laboratory. The LS-DYNA model (Figure 5) included both the tower and the access bridge. The concrete and steel rebars in the tower were modeled separately. The concrete was represented using 3D solid elements with Winfrith nonlinear concrete material model, which permits tensile cracking with shear transfer across the crack due to aggregate interlock. The rebars were modeled with truss elements using a kinematic hardening plasticity model for yielding in tension and compression. The vertical and horizontal rebars were modeled individually as a cage (Figure 5) at their exact locations. The rebar cage nodes did not need to coincide with the concrete nodes. Instead, they were constrained to move with solid elements in a compatible manner, by taking advantage of a feature available in LS-DYNA.

The dynamic interaction of the tower with the hard rock at this site was considered to be negligible and was not considered in the analysis.

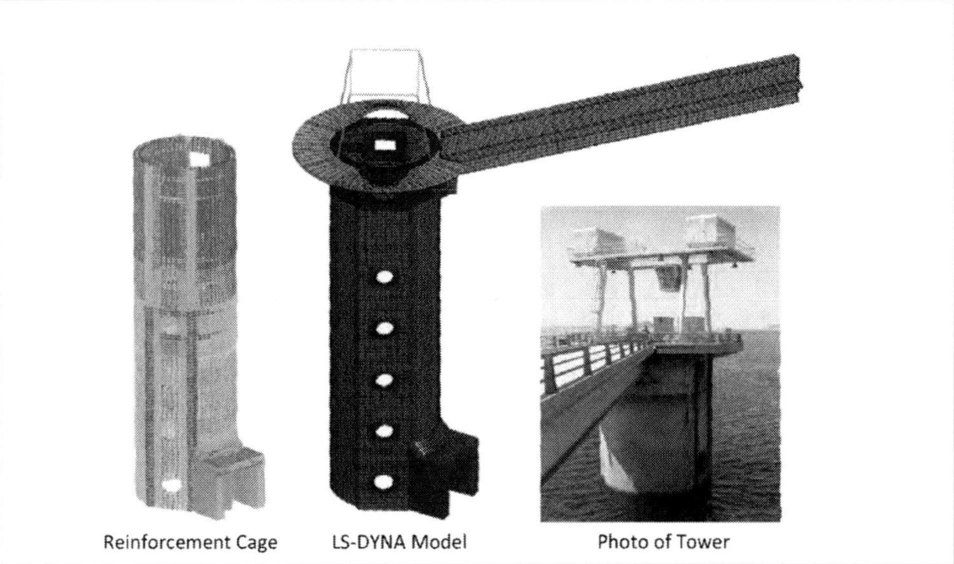

Figure 5. LS-DYNA models of reinforcement cage, outlet tower, and access bridge.

Winfrith Nonlinear Concrete Material Model. The concrete in the tower wall and outlet was modeled using the Winfrith nonlinear concrete model (without rate effects), which permits tensile cracking with shear transfer across the crack due to aggregate interlock. The basic underlying plasticity model used by the Winfrith model is based on a 4-parameter shear failure surface proposed by Ottosen (1977). The Winfrith model does not allow the user to adjust the fitting parameters directly. Instead, it uses proprietary correlations developed from empirical test data to compute them from the compressive and tensile strength of the concrete. A more thorough summary of the Winfrith plasticity model can be found in Schwer (2011).

When the state of stress in the element reaches the failure surface in the tensile mode, a crack forms. Each element can form up to three orthogonal crack planes, with the crack width (how far each crack is open) tracked over time. In a cracked element, the tensile stress normal to the crack plane goes to zero (linearly, over a short distance related to the fracture energy of the crack), but the element is still capable of carrying compressive stresses when the crack closes. Due to aggregate interlock, closed cracks continue to transfer shear across the crack plane, but as the crack opens the shear behavior softens and eventually goes to zero. The crack width at which this occurs varies, depending on the aggregate size. The Winfrith model requires the user to supply a minimum of six parameters to fully define the material. The Young's modulus and Poisson's ratio define its elastic properties, the compressive strength and tensile strength define the failure surface, and the facture energy and maximum aggregate size control post-crack behavior.

Access Bridge

The access bridge was explicitly modeled in three dimensions to assess its seismic performance and to account more accurately for its interaction with the tower. The steel girders and their stiffeners were modeled using linear shell elements. The structural tees in the intermediate diaphragms and structural tees with channels in the end diaphragms were modeled using linear beam elements with steel properties. The bridge deck was represented by shell elements having linear concrete properties in SAP2000 model, and 3D linear solid element in LS-DYNA model. Each girder was supported by a fixed boundary condition at the abutment and roller bearings at the tower.

The roller bearings were modeled using compression-only link elements in SAP2000 and contact surfaces in LS-DYNA that allowed relative movement between the bridge and tower along the bridge axis, but not in the direction perpendicular to the bridge axis. The possibility of the bridge deck colliding with the operating deck of the tower was also considered by link elements in SAP2000 and contact surfaces in LS-DYNA.

Hydrodynamic Added Mass

The mass associated with the seismic inertia forces of the water inside and outside the tower was computed using the procedure developed by Goyal and Chopra (1989). At a given elevation, this procedure computes the total added mass per unit height for the inside and outside water. This was divided by the tower circumference to calculate the added mass per unit area. Using this procedure, added mass was assigned to each submerged nodal point per its elevation and tributary area. Because hydrodynamic effects are not encountered in the vertical direction, the added masses were only applied to the horizontal degrees of freedom of the nodes.

NONLINEAR TIME HISTORY ANALYSIS

The nonlinear analysis of the SAP2000 layered shell model was performed using a direct step-by-step integration method, where the static and dynamic loads were applied in a cumulative sequence. The static loads consisted of the gravity, hydrostatic, and temperature changes. Dynamic load included seismic loads associated with the MCE acceleration time histories applied to the fixed base of the tower and bridge abutment support in three orthogonal directions. Mass and stiffness proportional Rayleigh damping equivalent to 5% critical damping in the ten lowest vibration modes of the tower was used.

The nonlinear analysis of the solid model was performed using the LS-DYNA explicit engine. However, the explicit engine is incapable of performing a true static analysis, and the instantaneous application of static loads, such as gravity, at the initial timestep, would result in dynamic excitation rather than static equilibrium. For situations in which a dynamic analysis requires a static preload, LS-DYNA offers a feature called *dynamic relaxation*. Dynamic relaxation is an initialization stage before the main analysis in which the model can be preloaded using a variety of different methods. The method selected in this case was to use the implicit engine to perform a static analysis. Once the dynamic relaxation is complete, LS-DYNA starts the explicit analysis, using the static displacements and internal stresses to initialize the model. When the static loads are applied instantaneously at time zero, the model is already in equilibrium.

For the LS-DYNA analysis, the gravity and temperature loads were applied during the implicit dynamic relaxation phase. (The hydrostatic loads were omitted, due to the negligible effect observed in the SAP2000 model.) Seismic loads were then applied during the explicit phase. A frequency-independent damping feature was applied, which provides an approximately constant damping ratio over a range of vibration frequencies. The underlying damping model adjusts automatically to changes in tangent stiffness, which makes it ideal for nonlinear structural applications. Additionally, it is applied at the level of element deformation, and thus does not damp rigid body motion. 5% of critical damping was applied for frequencies between 15 rad/sec and 263 rad/sec, encompassing the 80 most fundamental modes of vibration of the model.

ACCEPTANCE CRITERIA

There are no established criteria for assessment of the nonlinear analysis of outlet towers. As such, we formulated a set of criteria specific to this project by which the nonlinear response of the Perris Outlet Tower could be appraised to ensure that the tower would remain operational and function satisfactorily for emergency drawdown should it be required after an MCE event. The functionality requirement for an MCE event implies that the tower may experience some limited damage, but its load resistance capability should not diminish. This condition corresponds to a performance level referred to as "Damage Control Performance" in the USACE Engineer Manual EM 1110-2-6053 (USACE, 2007). The EM 1110-2-6053 provides three performance levels for the concrete hydraulic structures: Serviceability Performance, Damage Control Performance, and Collapse Prevention Performance, as illustrated in Figure 6. Serviceability performance applies to the linear-elastic response behavior for ground motions up to the operating basis earthquake (OBE). Damage control performance refers to limited nonlinear response behavior under the maximum design earthquake (MDE). The MDE is usually less than the MCE, except for dams and critical structures that must remain functional under the MCE. Collapse prevention performance is for MCE events where significant damage and loss of strength are acceptable, but the structure should not collapse.

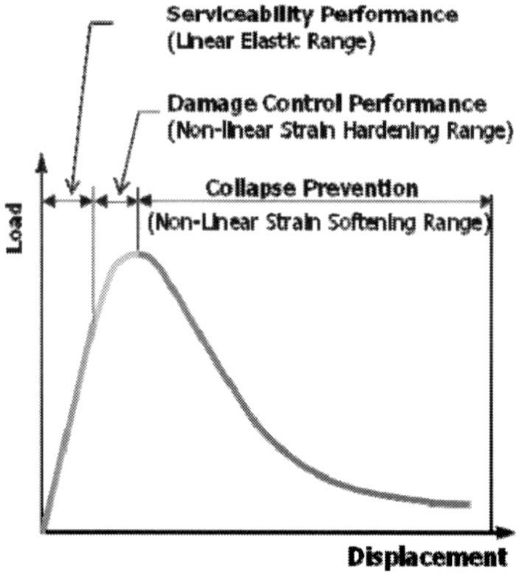

Figure 6. Performance levels for concrete hydraulic structures (USACE, 2007).

Considering that Perris Outlet Tower should remain functional in the event of an MCE, Damage Control Performance level offered a reasonable acceptance criterion for the assessment of the nonlinear response of the tower. To show that the tower response to the MCE meets this acceptance criterion, the nonlinear response of the tower should be modest and fall within the damage-control-performance range of behavior. This requirement was accomplished by examining several critical structural responses such as the tower and bridge displacements, concrete cracking, yielding and rupturing of reinforcing steel, and bridge reaction forces, as discussed below.

Displacements

Maximum displacement at the top of the tower should be less than 3 inches to avoid impact with the bridge or unseatment of the bridge by exceeding the roller bearing roller limit. Further, displacement histories of the tower should not exhibit any significant irreversible lateral movements or tilt indicative of significant steel yielding that could affect the operation and access to the tower.

Concrete Cracking

Concrete cracking is acceptable if it is not in the form of one or more concentrated open cracks that could lead to significant yielding and potential rupturing of the reinforcing steel and possibly the global instability of the tower. Only distributed cracks capable of engaging and transferring loads to reinforcing steel and inducing inconsequential or no steel yielding are acceptable. The distributed cracks are interpreted as a modest nonlinear response indicative of the damage-control-performance range of behavior.

Steel Yielding/Rupturing

To ensure that the nonlinear response of the tower is modest and falls within the damage control performance range of behavior, reinforcing steel should not yield or plastic deformation should be minimal. This condition is met if only distributed microcracking develops, ensuring limited steel deformations.

Access Bridge

The nonlinear behavior and potential failure modes of the access bridge are expected to occur at the roller or fixed supports, or both. At the tower support, the 3-inch gap between the bridge and tower should not close fully to avoid an impact. Also, tower separation from the bridge should remain below 5.8 inches, so as not to exceed the roller bearing limit. At the abutment support, reaction forces should not exceed the strength of the anchorage system in order to avoid failure modes such as tension and shear failure of anchor steel, tension and shear concrete breakout, anchor pullout, and shear concrete pryout.

RESULTS

Mode Shapes

Vibration periods and mode shapes were computed to check the SAP2000 and LS-DYNA models and gain insight into dynamic characteristics of the tower. Both models generated similar results with minor differences attributed to shell versus 3D modeling and eigensolution algorithms. Table 1 summarizes estimated vibration periods and corresponding frequencies for the lowest fifteen modes of vibration for both the SAP2000 and LS-DYNA models. The periods in the SAP2000 model vary from 0.378 seconds to 0.0767 seconds, while the LS-DYNA periods range from 0.437 seconds to 0.0736 seconds.

In both models, Mode 1 consists of swaying of the gantry crane, while the remaining modes (starting with Mode 2) include the three-dimensional participation of the tower, the access bridge, or both. Some examples of tower deflected shapes for the LS-DYNA model are shown in Figure 7. They include bending (Mode 2), twisting (Mode 9), ovalling (Mode 12), and higher-order bending (Mode 14) modes. Examples of bridge mode shapes are shown in Figure 8. They include bending (Mode 2), lateral-torsional twisting (Mode 5), higher order bending (Mode 10) and higher order twist (Mode 12) modes.

Table 1. Modal periods and frequencies.

Mode	SAP2000		LS-DYNA	
	Period (s)	Frequency (Hz)	Period (s)	Frequency (Hz)
1	0.378	2.646	0.437	2.288
2	0.300	3.336	0.298	3.356
3	0.287	3.489	0.290	3.449
4	0.282	3.543	0.283	3.529
5	0.245	4.090	0.233	4.288
6	0.209	4.795	0.226	4.430
7	0.164	6.110	0.163	6.151
8	0.161	6.222	0.149	6.697
9	0.106	9.397	0.105	9.562
10	0.100	10.046	0.0884	11.314
11	0.0916	10.921	0.0869	11.511
12	0.0908	11.012	0.0841	11.886
13	0.0887	11.268	0.0830	12.050
14	0.0786	12.728	0.0784	12.757

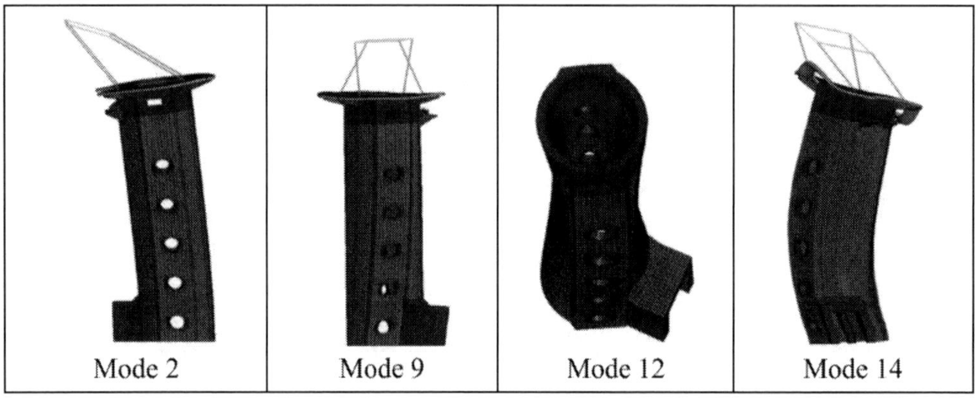

Figure 7. Examples of tower mode shapes (LS-DYNA).

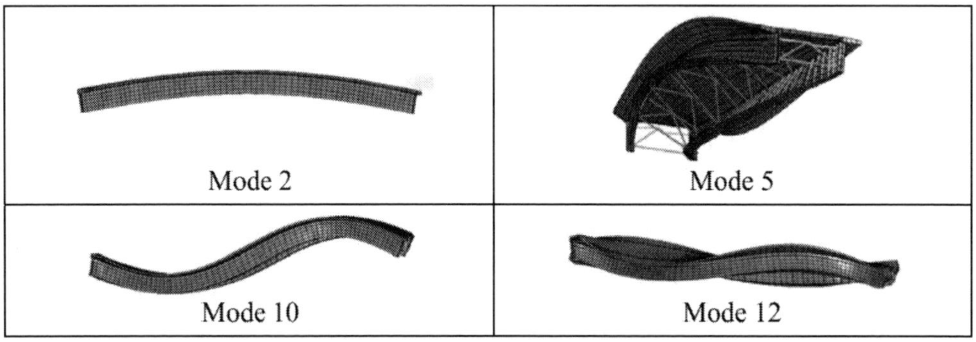

Figure 8. Examples of bridge mode shapes (LS-DYNA).

Stress Contours

The SAP2000 initial linear analysis indicated that the maximum vertical tensile stresses exceed the tensile strength of the concrete in the lower half of the tower. Thus, cracking is indicated throughout the lower half of the tower. The corresponding stress contours for the nonlinear SAP2000 analysis (Figure 9a) shows that all vertical tensile stresses larger than the tensile strength of the concrete (460 psi) have vanished. The yellow region in Figure 9a with vertical tensile stresses limited to 460 psi represents tensile cracking. Despite the assumption of linear concrete behavior with respect to horizontal stresses, the hoop stresses were generally small and approached the tensile strength of the concrete only in isolated small areas around the valve openings. Vertical stresses in the exterior vertical steel layer (Figure 9b) were negligible (less than 6 ksi) throughout most of the tower. The steel stresses were somewhat higher in the lower third of the tower as the loads redistributed from the cracked concrete to steel, but were limited to a maximum of 18 ksi, or 45% of the nominal yield strength of 40 ksi.

Figure 10 shows example stress envelope contours for the nonlinear LS-DYNA model. The envelopes of principal tensile stresses are consistent with the results of SAP2000, showing that high tensile stresses and cracking occured in the lower half of the tower. Cracking is indicated by "splotchy" red contours (Figure 10a) because the concrete failure criterion incorporates both tension and shear, so the concrete tends to crack before reaching the strength limit for pure tension. The corresponding rebar stress envelopes (Figure 10b) show moderate axial stresses at the bottom of the tower that are consistent with the location of concrete cracking, confirming the validity of the results. The rebar stress plots also corroborate the SAP2000 nonlinear results. The peak tensile stresses of vertical rebars were in the range of 20–24 ksi. Further, the reinforcing hoop stresses were also small, except in a few isolated locations between the valve openings where stresses in a short length of the horizontal steel hoops approached the yield strength (Figure 10b). However, an examination of stress histories showed that the isolated peak hoop rebar stresses were merely a couple of brief spikes; thus, we conclude that the hoop lap splices will hold. Another reason for this conclusion is because, as discussed below, horizontal cracks were small, at less than 0.01 inches wide, indicating that the concrete cover and bond between the concrete and reinforcing hoops would remain intact. Thus, hoop slippage would not occur.

Concrete Cracking

The Winfrith concrete material model keeps track of the location and size of cracks, and can plot the crack evolution over time. The resulting crack patterns show a mixture of flexural cracks (horizontal) and shear cracks (diagonal). Once an element has cracked in bending or shear, it can only develop new cracks in orthogonal directions to the initial crack. Thus, the orientation of cracks is determined by the direction in which the element was initially overstressed. The severity of the cracking is indicated by presenting multiple plots of crack width, each of which only shows cracks that have opened wider than a given threshold. As the threshold increases, only the wider cracks remain, indicating the regions where the crack width is the largest.

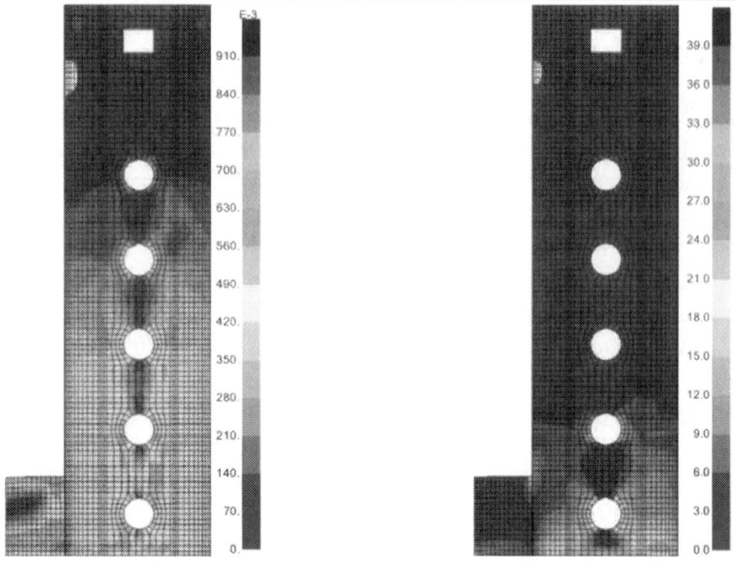

a. Concrete stress (nonlinear) b. Steel stress (nonlinear)

Figure 9. SAP2000 vertical stress envelope contours for Chi-Chi record (ksi).

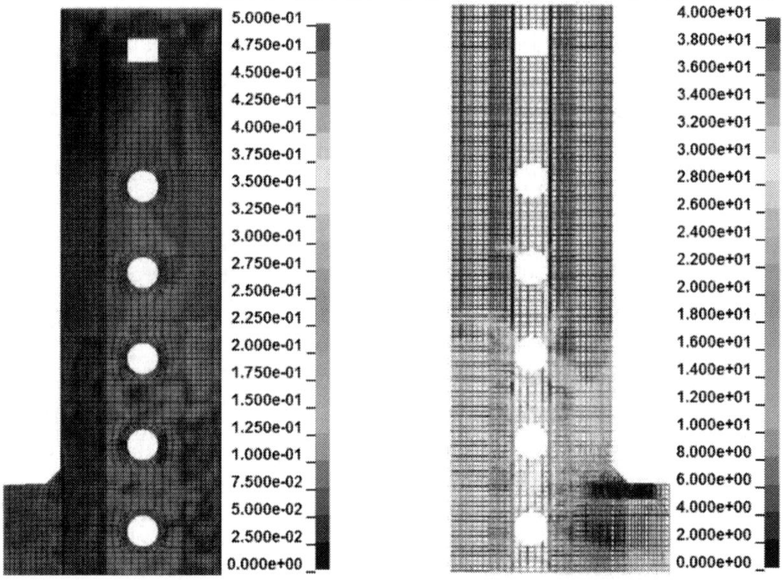

a. Concrete principal tensile stress b. Rebar axial stress

Figure 10. LS-DYNA stress envelope contours for Chi-Chi record (ksi).

Figure 11 is an example of a crack pattern at the time of peak seismic demand showing dominant horizontal bending cracks with some shear cracks crossing the valve openings. The crack pattern shows horizontal microcracks are distributed along the bottom third of the tower, which is expected behavior for adequately reinforced concrete and agrees with the stress contour plots. The final plot reveals the mild nature of the cracking, which shows that nearly all the cracks opened less than a hundredth of an inch. This observation proves that only microcracks developed; thus, the concrete cover remained intact, and load resistance capability of the horizontal reinforcing steel was not lost.

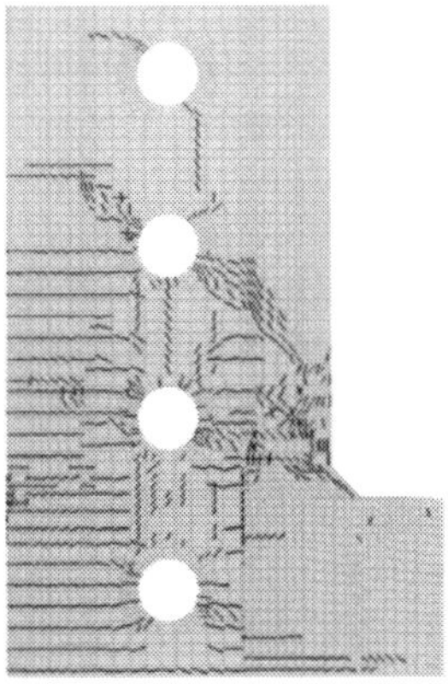

a) All cracks

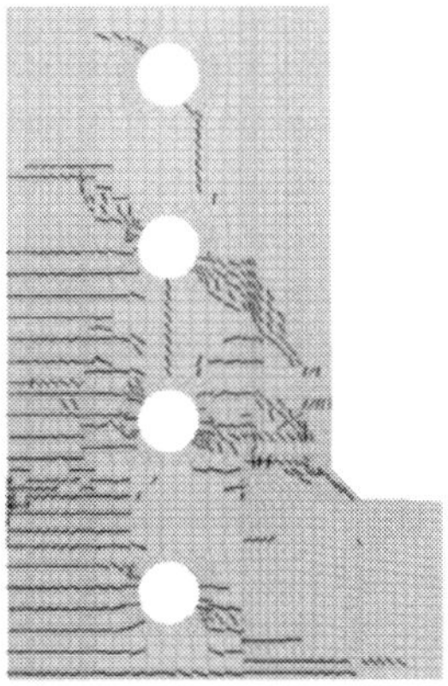

b) Cracks > 0.002 inches

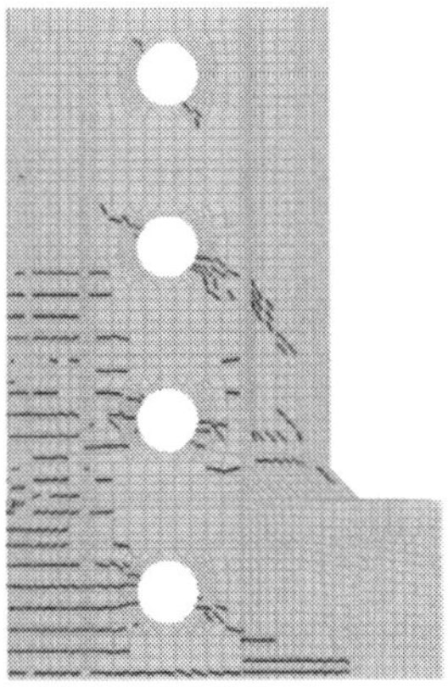

c) Cracks > 0.006 inches

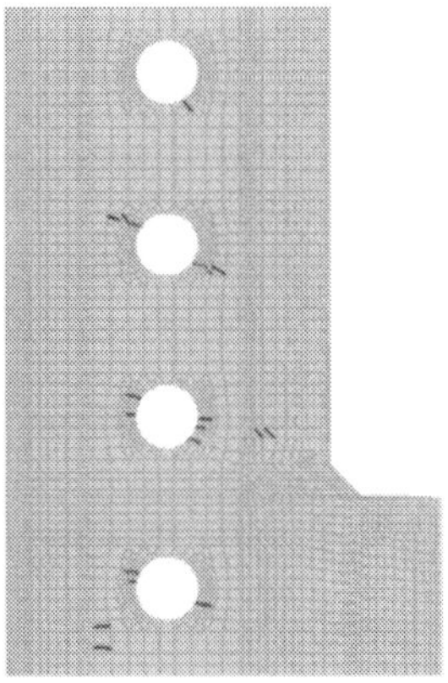

d) Cracks > 0.010 inches

Figure 11. Distribution of concrete bending cracks during Chi-Chi ground motion (t = 10.32 seconds).

The crack patterns examined at all other time steps indicated that at no point during the entire earthquake ground shaking do the crack widths exceed 0.01 inches. Such small cracks qualify as microcracks and, in combination with vertical rebars subjected to loads well below their yield strength, can maintain an intact concrete cover to protect and provide support for the vertical and horizontal hoop steel.

Compression Failure Check

Figure 12 illustrates bending damage at the base of the tower for six solid elements lined up radially through the tower wall. An examination of the vertical stresses in these elements shows that compressive stresses are well within the compressive strength of the concrete and that the peak value of 3.1 ksi at t = 13.3 seconds is less than half of the compressive strength of 6.374 ksi. The concrete, therefore, does not fail in compression, and the use of Winfrith concrete material model with no compression softening capability does not affect the results. All six elements are initially compressed equally under the weight of the tower (0 to 3.4 seconds). As the tower begins to sway back and forth, gradients of bending stress are created, with the outermost element (farthest from the center of the tower) subjected to the greatest amount of tension or compression. At around t = 6.8 seconds, the tension failure criterion is reached, and the five outermost elements crack in tension. In the next load cycle, at t = 7.1 seconds, the innermost element also cracks. Once the elements crack, they no longer carry any tensile stress but continue to carry compression, as demonstrated by one-sided stress histories (Figure 12). Once the concrete elements crack, flexural forces are transferred to the vertical rebars. A plot of the axial stress in a critical vertical rebar element at the tower base (Figure 13) shows that the majority of the time the stress is less than 15 ksi. Only a few spikes make it to 20 ksi and only 1 spike approaches 25 ksi. This observation indicates that stresses even for the most critical vertical rebar remain fully within the linear-elastic range and well below the nominal yield strength of 40 ksi.

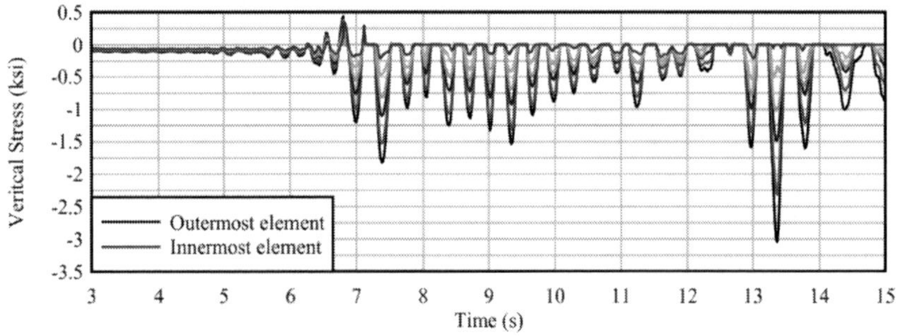

Figure 12. Vertical stress in concrete elements at base of tower (partial time history).

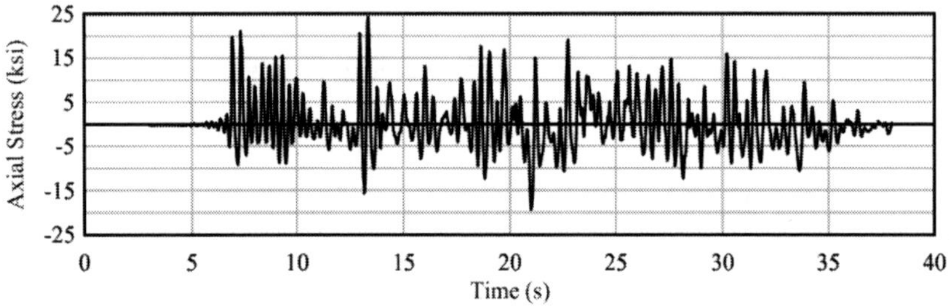

Figure 13. Axial stress in rebar element at base of tower.

Displacements

Figure 14 displays displacement histories at the top of the tower in the two lateral directions predicted by the LS-DYNA model. The maximum resultant displacement (not shown) is less than 1.5 inches and agrees with the SAP2000 predicted value. There is essentially no chance of structural collapse under such a small movement. It is also worth noting that the tower returns to its original position with no permanent offset or tilting.

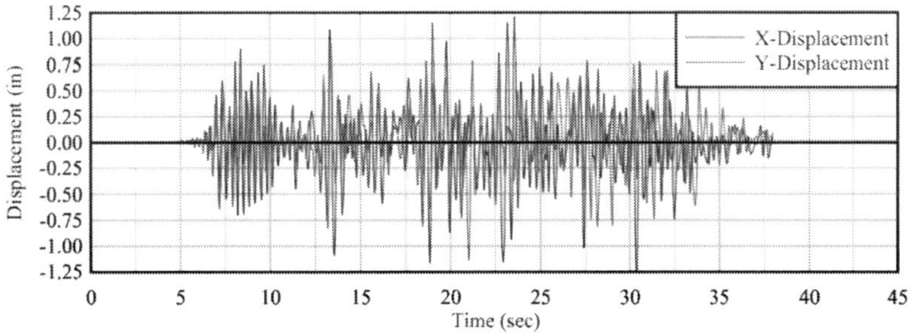

Figure 14. Lateral displacement at the top of tower for Chi-Chi record (LS-DYNA).

Figure 15 shows time histories of gap opening (red) and closing (blue) between the tower and bridge for Chi-Chi record as predicted by the LS-DYNA model. The zero point is defined as the even 3-inch gap at the reference design value. The gap closing begins at about −0.4 inches, due to bridge expansion caused by the initial temperature load. The predicted maximum gap opening for all three ground motion records varies from 0.60 to 1.30 inches, remaining well below the roller bearing limit value of 2.8 inches. The bridge roller bearing would, therefore, perform satisfactorily in providing support for the bridge. Maximum gap closing varies from −1.00 to −1.75 inches; this is also well within the 3 inches of expansion allowed by the design plans. The results indicate that there is no danger that the bridge could hit the tower and cause or suffer damage.

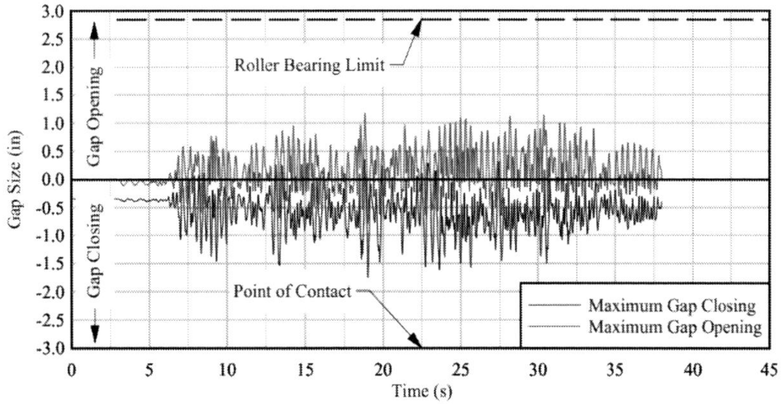

Figure 15. Gap between tower and bridge for Chi-Chi record (LS-DYNA).

Bridge Abutment Supports

The access bridge is connected to the abutment by fixed bearing assemblies, which are aligned with the bridge axis (and, therefore, skewed relative to the abutment and bearing pedestals). Each pedestal supports a steel masonry plate (affixed with a layer of cement paste), which is welded to a large steel mass that acts as a shear key (Figure 16). The bottom flange of each girder is welded to a sole plate with two keeper plates that fit around the shear key. To prevent uplift, the sole plate is anchored to the abutment with two 1¼″-diameter bolts ("tension bolts"), which pass through the masonry plate. Two additional "shear bolts" anchor only the masonry plate. Therefore, four bolts are available to transfer shear from the bridge, but only two bolts can transfer tension.

Strength calculations for the fixed bearing anchorage were based on ACI 318-11 with some modifications and conservative assumptions. Several failure modes were evaluated in tension and shear. In both tension and shear, concrete breakout was found to be the controlling failure mode. Comparison of shear demands with shear capacity showed that the shear strength of the bridge abutment bearings is not sufficient and that the bearings could fail primarily in shear breakout.

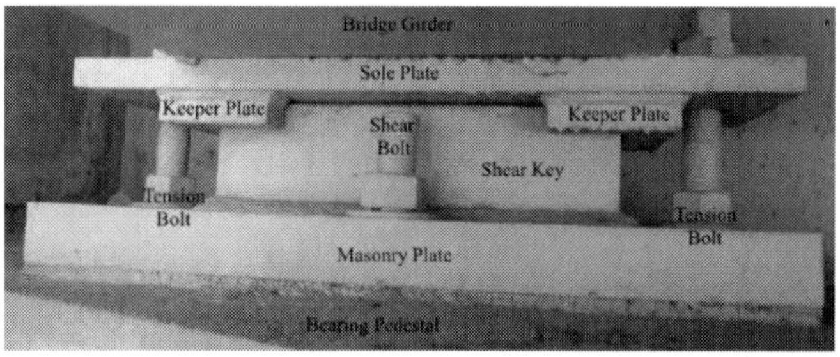

Figure 16. Photograph of fixed bearing assembly.

CONCLUSION

Nonlinear seismic analyses are a valuable tool for evaluating the performance of concrete structures under damaging earthquakes. A high-quality nonlinear seismic analysis study can provide more realistic assessments of damage and potential failure modes than are possible with linear analysis methods. However, the results of nonlinear analyses should be thoroughly examined and validated with another modeling approach and computer software, and in some cases with independent reviews and studies.

The nonlinear analysis results indicate that Perris Outlet Tower would perform satisfactorily and remain operational when subjected to the postulated MCE ground motion. The damage would be minor in the form of microcracks mostly less than 0.01 inches wide distributed over the bottom third of the tower. All vertical reinforcing steel bars remain elastic and stretch only 50-60% of the steel yield strength. The horizontal hoops show similar behavior, except that a few of them approach the yield strength momentarily for hundredths of a second at isolated locations between the valve openings.

The modest microcracking combined with the fully elastic behavior of the vertical rebars suggest that the concrete cover would remain intact and able to protect and provide support for the reinforcing steel. Further, lateral displacements are small, and exhibit no permanent lateral deflection or tilting. Relative motion between the bridge and tower is not large enough to unseat the bridge or cause an impact. The bridge abutment bearings may suffer damage and not be safe for vehicle crossing, but the bridge might still be accessible to foot traffic.

ACKNOWLEDGEMENTS

This study was performed for the California Department of Water Resources (DWR) under a subcontract to HDR Engineers, Inc. Any opinions, findings, conclusions or recommendations expressed in this paper are those of the authors and do not necessarily reflect the views of DWR. The authors wish to thank DWR for the opportunity to present this study.

REFERENCES

ACI, "318-11: Building Code and Commentary," American Concrete Institute, 2011.

CSI, *SAP2000 v17.1, Integrated Software for Structural Analysis and Design,* Walnut Creek, CA: Computer and Structures, Inc., 2014.

Goyal, A., and Chopra, A.K. (1989), "Simplified Evaluation of Added Hydrodynamic Mass for Intake Towers," ASCE Journal of Engineering Mechanics, 115(7), 1393-1433.

LSTC, *LS-DYNA, A general static and dynamic purpose finite-element code for large deformations static and dynamic analyses including soil-structure and fluid-structure interactions, Version LS971R7.1,* Livermore, CA: Livermore Software Technology Corp., 2014.

Ottosen, N.S. (1977), "A Failure Criterion for Concrete," *Journal of the Engineering Mechanics Division*, vol. 103, no. 4, pp. 527-535, July/August 1977.

Schwer, L. (2011), "The Winfrith Concrete Model: Beauty or Beast? Insights into the Winfrith Concrete Model," in *8th European LS-DYNA Users Conference*, Strasbourg, 2011.

USACE, "EM 1110-2-6053: Earthquake Design and Evaluation of Concrete Hydraulic Structures," US Army Corps of Engineers, 2007.

NONLINEAR SEISMIC ANALYSIS OF AN ARCH DAM SUBJECTED TO VERY INTENSIVE EARTHQUAKE LOADING

Gurinderbir S. Sooch (M.A.Sc., P.Eng.) [1]
Dan D. Curtis (M.A.Sc., P.Eng.) [2]

ABSTRACT

This paper presents the results of a state-of-the-art dynamic analysis where a nonlinear seismic analysis of an existing arch dam under an intense earthquake was recently performed. The 3D finite element model was developed and the analysis performed using the explicit finite element program LS-DYNA. This 3D model includes the effects of dam-foundation-reservoir interactions. Also, rock wedges within the abutments were modeled to simulate potential abutment block sliding. Non-reflecting absorbing boundaries were modeled along the foundation sides and the reservoir extents. The vertical contraction joints and the dam-foundation interface were modeled with contact elements capable of opening and closing along with sliding. Additionally, a nonlinear concrete model was used to simulate the concrete cracking during the earthquake. Finally, the analysis was performed under an intense earthquake with a peak ground acceleration (PGA) of about 0.9*g*. The results of this analysis are a reasonable match to the simulated failure mode created by a scaled shake table test of similar arch dams.

INTRODUCTION

This particular concrete arch dam was constructed in 1920s with a height and crest length of 275 feet and 412 feet, respectively. A general purpose finite element program LS-DYNA was used for the seismic analysis and the purpose was to evaluate the performance of the arch dam during a seismic event.

Two 3D finite element models (M1 and M2) were developed. The objective of the first model, M1, was to analyze the stresses in the concrete arch dam. Therefore, the rock wedges were not modeled in model M1. Similarly, the objective of the model M2 was to analyze the stability of the rock wedges underneath the right abutment of the arch so they were included in the second model.

The finite element model M1 consists of 651,886 nodes and 611,650 elements. The dam, reservoir and foundation were modeled with 8-noded elements with a reduced integration point for these elements (the element formulation is Type 1 in LS-DYNA). Figures 1 to 5 present the finite element models.

[1] Civil/Structural Engineer, HATCH Ltd, 4342 Queen St, Suite 500, Niagara Falls ON Canada L2E 7J7 gurinderbir.sooch@hatch.com, Phone: +1-905-374-0701 ext. 5252.
[2] Senior Project Manager, HATCH Ltd, 4342 Queen St, Suite 500, Niagara Falls ON Canada L2E 7J7 dan.curtis@hatch.com, Phone: +1-905-357-6998.

Figure 1 Finite Element Mesh of entire model

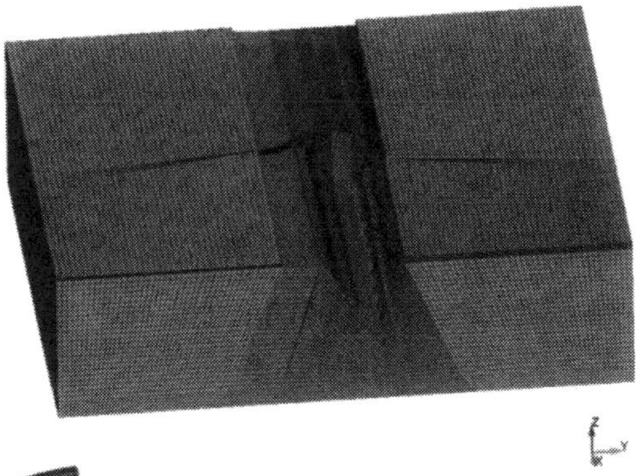

Figure 2 Finite Element Mesh of Dam: Upstream View

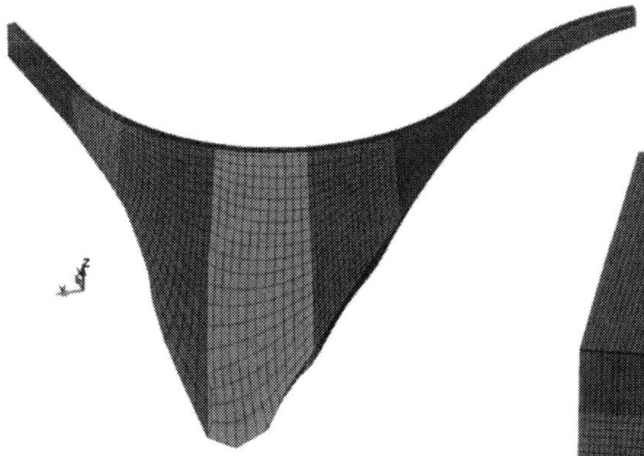

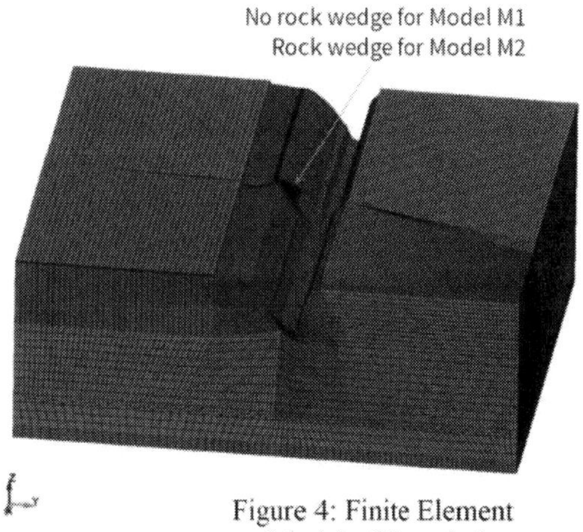

Figure 4: Finite Element Model of Foundation

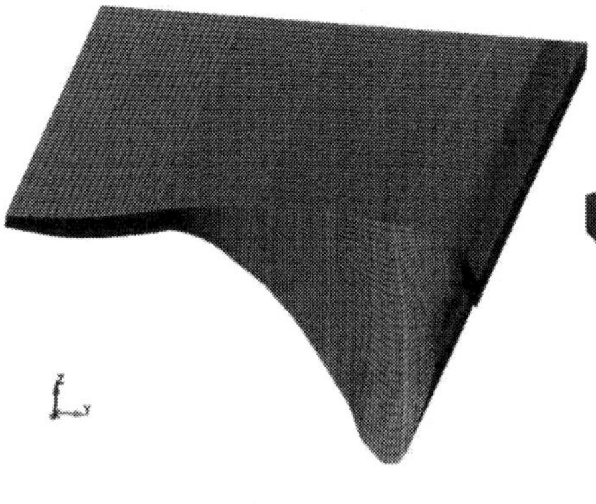

Figure 3 Finite Element Mesh of Reservoir

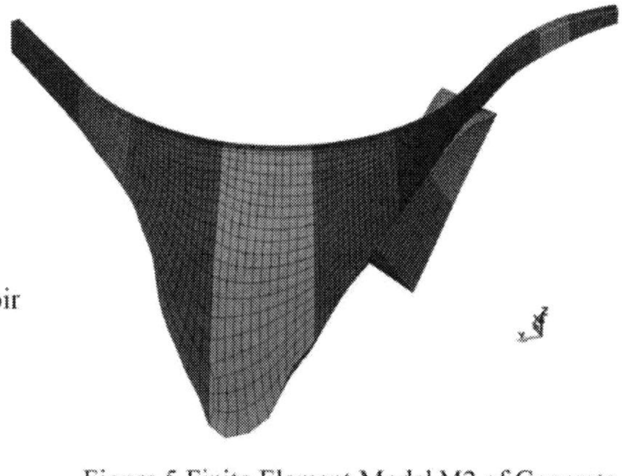

Figure 5 Finite Element Model M2 of Concrete Arch Dam and Foundation Rock Wedges

MATERIAL PROPERTIES

The material properties required as input to the finite element model are given in Table 1.

Table 1. Finite Element Model Input Parameters

Material	Parameter	Value	Units
Concrete	Elastic Modulus (Sustained) Elastic Modulus (Dynamic) Poisson Ratio Weight Density	3.0×10^6 3.75×10^6 0.2 150	psi psi - lb/ft^3
Foundation Rock	Elastic Modulus (Sustained) Elastic Modulus (Dynamic)	4.0×10^6 5.0×10^6	psi psi
General	Damping Compressive Strength Dam-Foundation-Rock Wedge Contact Friction Angle	5 4,500 55°	% of critical psi angle

The static tensile strength (ST) of concrete was assumed to be $0.05f'_c = 0.05 \times 4500 =$ 225 psi (Raphael, 1984). The dynamic tensile strength (DT) of concrete was assumed to be 1.5 x ST = 1.5 x 225 ~ 350 psi. From Table 1, 5% damping was assumed. It should be noted that only Rayleigh mass proportional damping (α) was used. Conservatively, the Rayleigh stiffness-proportional damping was ignored. The stiffness-proportional damping can greatly hinder sliding. It is noted that a sensitivity analysis on damping was envisioned and started with a higher value of 5%. However, even with 5% damping significant damage was computed, therefore additional sensitivity analysis for damping was not undertaken. Also, Hall (2005) highlights the problems associated with Rayleigh damping, i.e., development of unrealistically large damping forces during a seismic analysis. As mentioned earlier, even with 5% damping significant damage was computed. Therefore, it would be useful if more concrete dams were instrumented for seismic motion such that values of damping under intense seismic motion can be better estimated.

LOADS

The static and dynamic loads were applied step-by-step during the analysis. First static loads assuming normal reservoir levels were applied along a load curve over a period of zero to three (3) seconds. Then sufficient quiet time was given to achieve a steady state condition in the model. The dynamic analysis was started at 8.5 seconds. The duration of dynamic analysis varied according to the length of respective time histories.

During the initial static analysis, the sides and base of the foundation were constrained for movement perpendicular to their respective faces. The nodal reactions were then extracted from the nodes along these faces. In the dynamic analysis, the constrained nodes were replaced with the non-reflective boundaries and static nodal reactions. The non-reflective boundaries were then applied to restrict the artificial amplification of the seismic waves within the model. These nodal reactions were also ramped up at the same time as the gravity forces.

EARTHQUAKE GROUND MOTIONS

The dam was evaluated for a 1-in-5,000 year return period earthquake. The peak ground acceleration was about 0.9*g* and the earthquake ground motions used in this analysis were deconvolved. Deconvolution is a mathematical process which allows the adjustment of the amplitude and frequency contents of an earthquake ground motion applied at the base of the foundation to achieve the desired output at the dam-foundation interface. Initially, the ground acceleration applied at the base of the foundation is assumed to be the same as the specified one (i.e., the original free field ground acceleration time history). The acceleration time history at the foundation top surface is then estimated by performing dynamic analysis. The resulting ground acceleration at the foundation top is then compared to the original free field ground acceleration by transforming both signals into the frequency domain using Fourier analysis. An iterative adjustment is performed until the resulting ground acceleration at the top of the foundation matches with the target acceleration time history (Sooch and Bagchi, 2014).

SALIENT FEATURES OF MODEL

This nonlinear dynamic analysis was performed with inclusion of dam-foundation-reservoir interaction. The interaction between the dam-foundation was modeled with inclusion of foundation inertia and damping. Foundation rock wedges were modeled to simulate there movements during the earthquake. The reservoir was modeled with fluid elements to capture the dam-reservoir interaction. To model the semi-infinite domain of the foundation in a finite model, non-reflecting absorbing boundaries were modeled along the foundation sides and at reservoir extents.

Surface contacts along the vertical contraction joints in the dam were modeled with friction contact elements. The dam blocks along the vertical contraction joints were allowed to open and close during the dynamic analysis and the earthquake ground motions were deconvolved to effectively model the seismic loads. To simulate crack propagation during the earthquake Nonlinear Material Models LS-DYNA Mat 159 and Mat 84/85 were used.

SHEAR KEY AT DAM-FOUNDATION JOINT

Figures 6 and 7 show the presence of the open rock joint at the dam-foundation interface. The joint is most likely continuous from the left (north) to the right (south) abutments and was filled with concrete during construction. This concrete acts as a shear key for the arch dam and would certainly restrict the movement at the dam-foundation interface. Thus, the site condition for the dam-foundation interface would be somewhere in between the glued and frictional joints.

Figure 6:
Power Intake and North (Left) Abutment of Dam

Figure 7:
South (Right) Abutment of Dam

(A) Open rock joint was filled with concrete at dam-foundation interface.

(B) Concrete filled rock joint acts as a concrete shear key for the arch dam. North abutment joint continues to south abutment.

NON-LINEAR LS-DYNA MATERIAL MODEL

The modeling of crack development during a high intensity earthquake is very complex. The evaluation methodology by elastic material models is well established and multiple load cases can be evaluated quite easily. Also, with the linear elastic material model, the stress-state is evaluated with conventional evaluation methods (Area of high-tension region; observed tensile, compressive and shear stresses; demand/capacity ratio, etc). However, there are drawbacks when using linear elastic material models, such as, no information regarding the ultimate load capacity is attained from linear elastic material models.

As mentioned earlier, the modeling of crack development with nonlinear material models is very complex and requires considerable effort. LS-DYNA material 159 is a damage plasticity model and can effectively predict the behavior of dam mass concrete during a cyclic earthquake event. The model includes a smooth intersection between the shear yield surface and hardening cap. The model also includes isotropic constitutive equation (yield and hardening surfaces) and damage formulations to simulate softening and modulus reduction. Figure 8 shows the general shape of LS-DYNA material 159 yield surfaces.

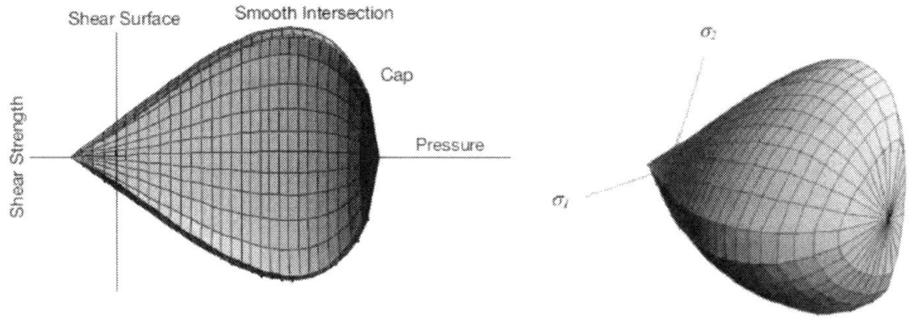

Figure 8: General Shape of the Concrete Model Yield Surface in Two Dimensions in Meridonal Plane and in Three Dimensions (FHA, 2007)

Also, by switching on the "ERODE=ON" option in LS-DYNA model 159 input parameters, the program enables the element to erode when the effective principal strain reaches a value of 1.0. It should be noted that by default the "ERODE" option is not active in LS-DYNA material 159. The eroded element loses all strength and stiffness effectively removing it from the model. It should also be noted that switching on the ERODE option is quite conservative as strain softening is ignored with the ERODE option.

It is important to note that LS-DYNA material 159 has a very complex input. Therefore, input parameters were adjusted to achieve the desired strength properties in the material model. For testing tensile and compressive strength, a finite element model of concrete cylinder with a diameter of 6 in. and a length of 12 in. was developed. In the case of the tensile strength test, uniform velocity was applied in the opposite direction to both ends of the cylinder and tensile stresses in the model were monitored. Similarly, in the case of the compressive strength test, the base of the cylinder was fixed and uniform velocity was applied in the downstream direction at the top of the cylinder and compressive stresses in the model were monitored. Figures 9 and 10 present the results of tensile and compressive strength testing for LS-DYNA material 159.

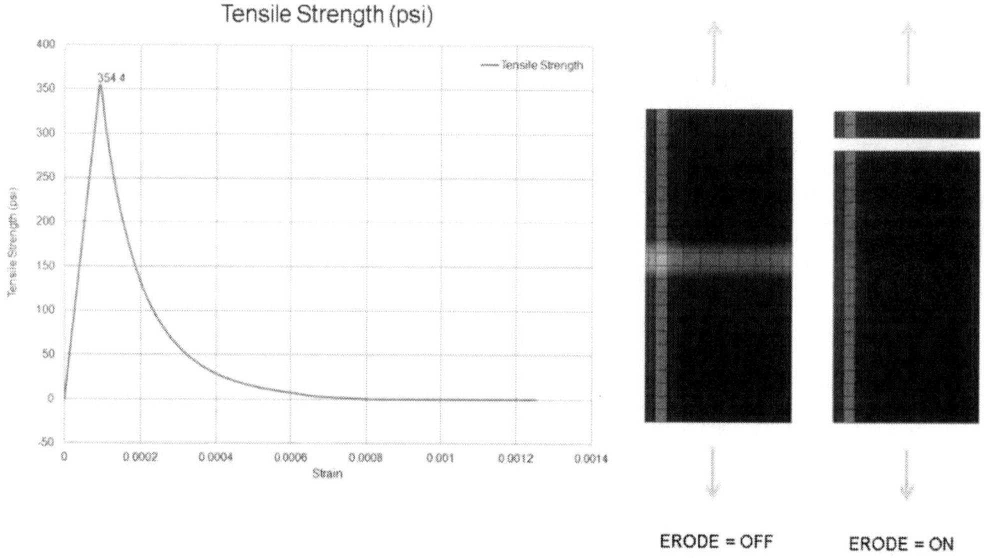

Figure 9: Tensile Strength Test of LS-DYNA Material 159 with 6 in. Diameter and 12 in. Long Concrete Cylinder

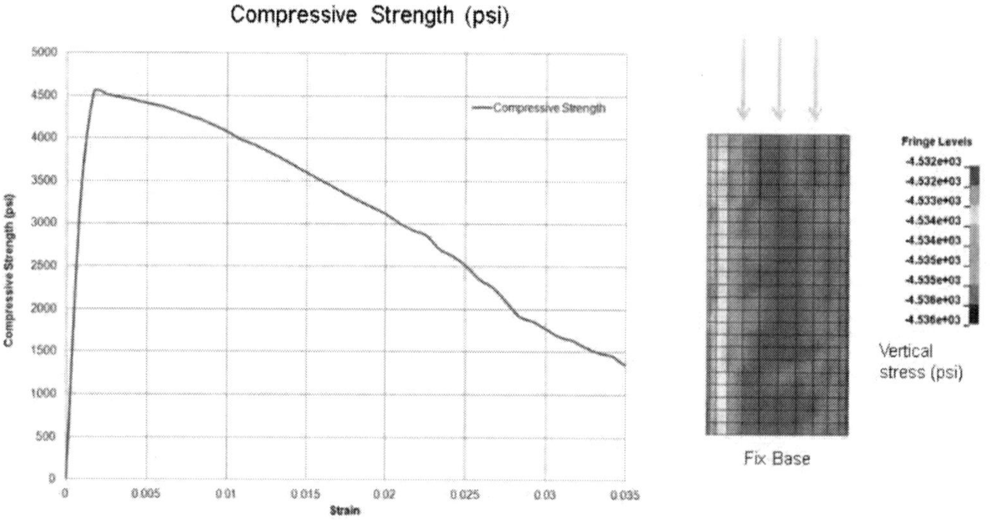

Figure 10: Compressive Strength Test of LS-DYNA Material 159 with 6 in. Diameter and 12 in. Long Concrete Cylinder

Also, it should be noted that during the seismic event the concrete elements transition between tension to compression depending on the earthquake load direction. The default input parameter for LS-DYNA material 159 assumes the undamaged modulus is recovered during the transition from tension to compression. Figure 11 presents the performance of LS-DYNA material 159 during the transitions from tension to compression or vice versa.

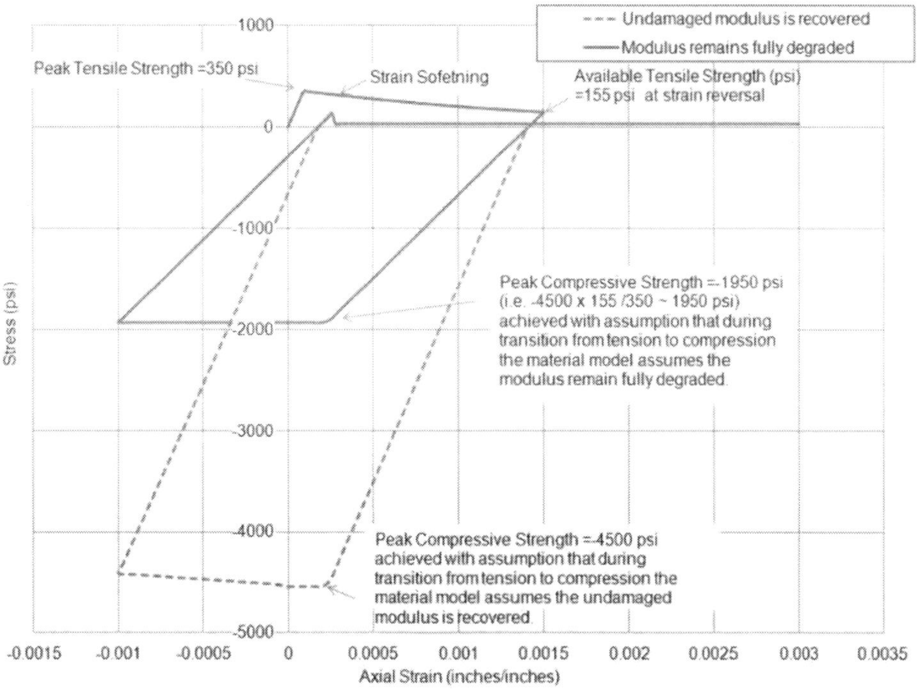

Figure 11: Performance of LS-DYNA Material 159 during Transition from Tension to Compression or Vice Versa

From Figure 11, the strain reversal at 155 psi tension, i.e., after yield at 350 psi, results in compressive strength of 4,500 psi and 1,950 psi (4,500 x 155 /350 ~ 1,950 psi) for both options, i.e., undamaged modulus is recovered and modulus remains fully degraded, respectively. It should be noted that in reality the response of the concrete lies in-between the two options, i.e., undamaged modulus is recovered and modulus remains fully degraded.

Also, it should be noted that in defining input of the LS-DYNA material 159, the rate effects (i.e., the strength of the material increases with the increasing strain rate) were ignored. As mentioned earlier, the assumed dynamic tensile strength (350 psi) is about 1.5 times the static tensile strength (225 psi). Therefore, a rate effect of 1.5 is already included in the material input values and the rate effects equations built into the LS-DYNA material 159 were ignored.

ANALYSIS RESULTS

A total of nine cases were analyzed using the M1 and M2 models. Table 1 presents the summary of nine cases analyzed using Models M1 and M2. From Table 1, the following failure mode was analyzed for Cases 1 to 6. In these cases, during the earthquake the vertical joints open up causing the central blocks to act as cantilevers. The upstream displacement of the cantilever central blocks causes overstressing of the arch dam on the downstream face. Also, from Table 1, the following failure mode was analyzed for Cases

7 to 9. In these cases, the potential for the rock wedges to pop out of the foundation during the seismic event was assessed. The sign convention for stresses in LS-DYNA is that tension is positive and compression in negative.

Table 1: Summary of Nine Cases Analyzed using Models M1 and M2

Case No.	Assumptions and Brief Description	Alteration to Model	Failure Mode Analyzed
1	a) Linear Elastic Concrete Model. b) Dam-Foundation contact was glued. c) Rock Wedges not included. d) FEA Model: M1.	-	Cases 1 to 6: During the earthquake the vertical joints open up; thus causing the central blocks to act as cantilevers. The upstream displacement of cantilever central blocks causes overstressing of arch dam on the downstream face
2	Same as Case 1	a) Non-Linear concrete model (Mat 159) was added. b) Assumed concrete tensile strength = 350 psi.	
3	Same as Case 1	a) Non-Linear Concrete Model (Mat 159) was added. b) Element erosion was switched "ON". c) Assumed concrete tensile strength = 350 psi.	
4	Same as Case 1	a) Non-Linear Concrete Model (Mat 159) was added. b) Assumed concrete tensile strength = 450 psi.	
5	Same as Case 1	a) Non-Linear Concrete Winfrith Model (Mat 84) was added. b) The assumed concrete tensile strength = 350 psi.	
6	Same as Case 1	a) Non-Linear Concrete Model (Mat 159) was added. b) The concrete tensile strength = 350 psi. c) Also, recovery of modulus during transition from tension to compression was switched "off", i.e., the modulus remains fully degraded during the transition from tension to compression.	

Case No.	Assumptions and Brief Description	Alteration to Model	Failure Mode Analyzed
7	a) Linear Elastic Model b) The rock wedge joints were modeled with a 55° friction angle c) Only neighboring dam blocks to rock wedges were modeled with friction angle (Φ = 55°) (see Figure 12). The rest of the dam blocks were glued to the foundation. d) The dam blocks and rock wedges were glued together e) FEA Model: M2	-	Cases 7 to 9: Assessed the potential of rock wedges to pop out of the foundation during the seismic event.
8	Same as Case 7	a) Non-Linear Concrete Model (Mat 159) added. b) The assumed concrete tensile strength = 350 psi.	
9	Same as Case 7	The dam blocks and rock wedges were not glued together	

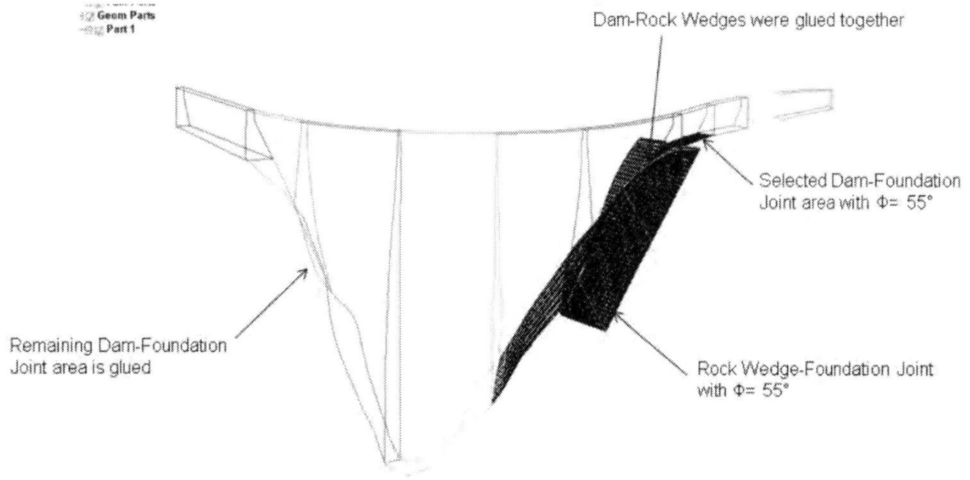

Figure 12: Selected Dam-Foundation Area with Frictional Joint (Φ =55°)

Case 1: Case 1 assumes a linear elastic model for concrete where the dam-foundation contact was glued due to the presence of a large shear key at the dam-foundation. Also, the objective of the Case 1 was to conservatively evaluate stresses in concrete arch dam.

Therefore, the rock wedges were not included and the analysis was performed with FEA Model M1. Figure 13 shows the time history of relative displacement at the top of Block 4 (Central Dam Block). The relative displacement time history shown in Figure 13 is calculated by deducting the displacements at the top of the dam with displacements at the base of the dam block. For this case, the maximum upstream displacement of 2.4 in. (shown in Figure 13) occurs at 17.85 seconds with the first 8.5 seconds taking place under the static load conditions. At 17.85 seconds, the corresponding maximum tensile stress is 1,000 psi (shown in Figure 14) on the downstream face.

However, at 16.4 seconds, the upstream displacement was 1.05 in. (shown in Figure 13) and considering the size of the dam, this displacement is significant with a maximum principal stress (i.e., maximum tensile stress) of 470 psi (shown in Figure 15). Figure 16 shows possible initial crack orientations (i.e., perpendicular to the maximum principal stress [Sigma1]) at t=16.4. The possible initial crack orientations match well with analysis results from Cases 2 and 3 and are discussed further in Case 3. It should also be noted that at t=16.4 seconds, the corresponding ground acceleration is only up to 0.63g and is well below the PGA of 0.9g. Consequently, the dam was found to suffer extensive damage during this seismic event.

Case 2: Case 2 is similar to Case 1. However, the dam concrete was modeled with nonlinear concrete model (LS-DYNA Mat 159). The objective of this case was to model cracking of the dam concrete during the same seismic event. The tensile strength of concrete was assumed to be 350 psi and Figure 17 shows the concrete damage at the downstream face of the concrete arch dam at 16.5 seconds which is 0.1 seconds after the 1.05 inches of displacement occurs. From Figure 17, extensive damage near the crown was predicted.

Case 3: For Case 3, the dam concrete was modeled with nonlinear concrete model (LS-DYNA Mat 159) which is similar to Case 2. However, for this case the ERODE option was switched “ON.” It should be noted that the ERODE option is quite conservative as strain softening is ignored. Also, the tensile strength of concrete was assumed to be 350 psi. Figure 18 shows the concrete damage at 16.5 second just after the 1.05 inches of upstream displacement and indicates loss of the upper portion of the dam. Figure 19 presents a photo of a shake table test performed at the ISMES facility in Italy (see reference 2). Therefore, the damage shown in Figures 17 and 18 is consistent with the initial crack orientations shown in Figure 16. Figures 17 and 18 are a reasonable match to the simulated failure by the scaled shake table test (shown in Figure 19) of a similar arch dam.

Case 4: In Case 4, the dam concrete was modeled with nonlinear concrete model (LS-DYNA Mat 159) which is similar to Case 2. However, the tensile strength of concrete was assumed to be 450 psi. Figure 20 presents the concrete damage at 16.5 second where some damage was still observed in the upper portion of the concrete arch dam. It should be noted that the results obtained from Case 4 and Case 2 are similar. Therefore, it can be concluded that the performance of the arch dam was not significantly affected with higher tensile strength in Case 4.

Case 5: For Case 5, the dam concrete was modeled with nonlinear concrete model (LS-DYNA Mat 84). The objective of Case 5 was to present a comparison between the Winfrith Concrete model (LS-DYNA Mat 84) and Case 2 with the LS-DYNA Mat 159. Figure 21 presents the concrete damage at 16.5 second for Case 5 which is comparable to Figure 17 (Case 2) so the effects of using the LS-DYNA Mat 159 model (Case 2) or the Winfrith Concrete Model LS-DYNA Mat 84 (Case 5) are insignificant.

Case 6: With Case 6, the dam concrete was modeled with nonlinear concrete model (LS-DYNA Mat 159) similar to Case 2 but the full recovery of modulus option during the transition from tension to compression in LS-DYNA Mat 159 was switched off. This allowed the modulus to remain fully degraded after transitioning from tension to compression. The performance of LS-DYNA Material 159 during the transition from tension to compression or vice versa is shown in Figure 11. Figure 22 shows the concrete damage at the downstream face of the dam at 16.5 seconds for Case 6 and the results are comparable to Figures 17 (Case 2). Therefore, recovery of modulus during transition between tension to compression or vice versa does not have significant effect on the overall results of this analysis.

Case 7: The objective of Case 7 was to assess the stability of rock wedges during the seismic event. Case 7 assumes linear elastic concrete model where the rock wedges and neighboring dam blocks to the rock wedges were modeled with a friction angle $\Phi = 55°$ (see Figure 12). The rest of the dam blocks were glued to the foundation and the dam blocks and rock wedges were glued together. This established finite element model M2 that was used for the analysis and in Case 7, small sliding displacements were observed at the rock wedges.

Case 8: Case 8 is similar to Case 7; however the dam concrete was modeled with a nonlinear concrete model (LS-DYNA Mat 159). The objective of this case was to model cracking of the dam concrete during the seismic event. The tensile strength of concrete was assumed to be 350 psi and in Case 8, considerable damage is predicted. It should be noted that damage propagation in Case 8 was different from Case 2 in that the damage starts at the base due to stress concentrations when the base joints open. For comparison, in Case 2, the damage started at the top portion and formed an arc at the upper central part.

Case 9: Case 9 is similar to Case 7 with the only exception being the dam and rock wedges were not glued together. Similar to Case 7, Case 9 provides an opportunity to model a potential failure mode where rock wedges try to pop out from the foundation after not being glued to the dam. In this case small sliding displacements were observed with the rock wedges and Figure 23 presents the time history of this displacement (inches) of Dam Block 5. From Figure 23, the maximum base sliding displacement at Dam Block 5 is -1.49 in downstream. Also, the arching action of dam will restrict the movement after the earthquake; therefore the downstream base sliding displacement of 1.49 in. was assumed acceptable.

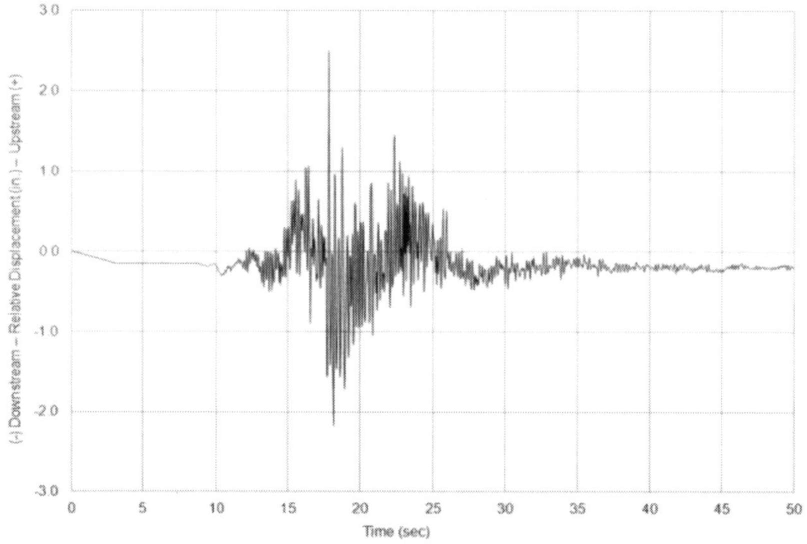

Figure 13: Relative Displacement Time History at Top of Block 4

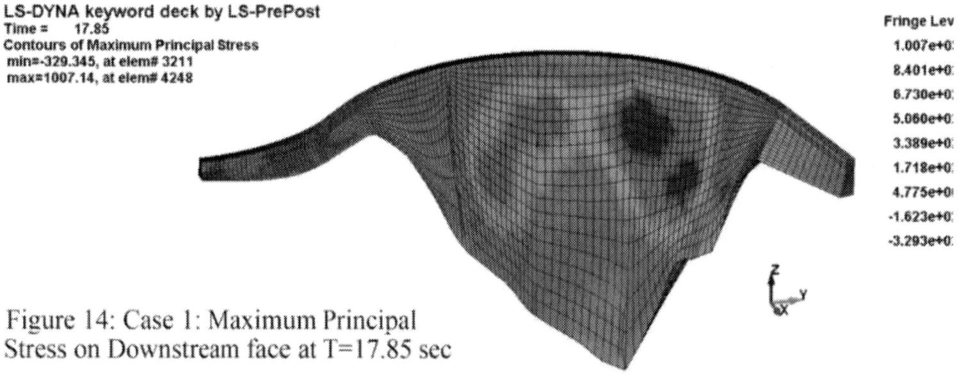

Figure 14: Case 1: Maximum Principal Stress on Downstream face at T=17.85 sec

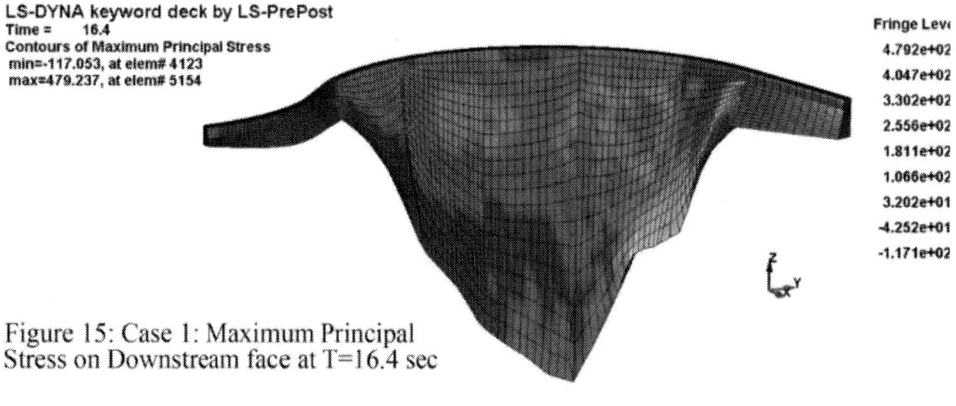

Figure 15: Case 1: Maximum Principal Stress on Downstream face at T=16.4 sec

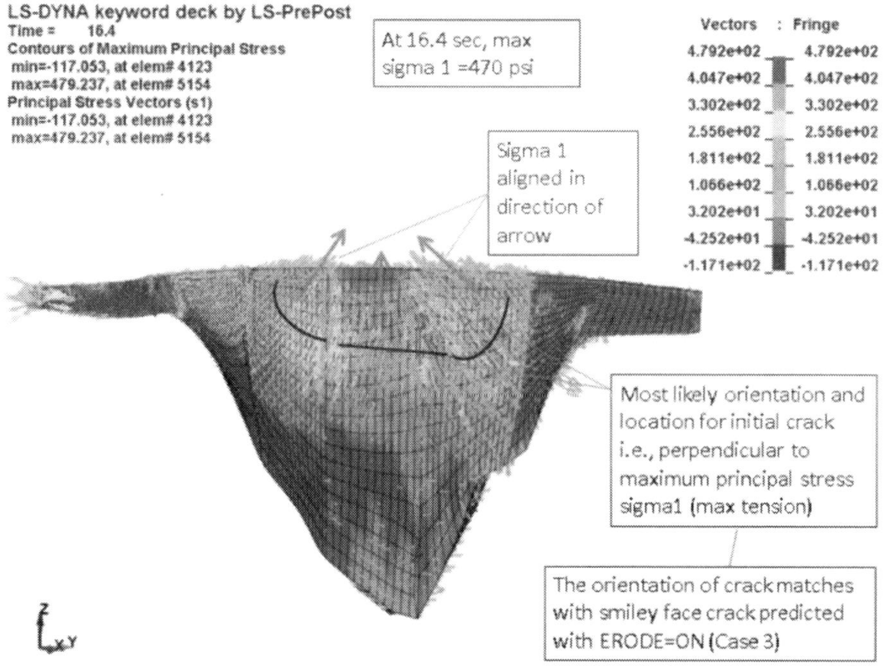

Figure 16: Case 1: Possible initial crack orientations (i.e. perpendicular to maximum principal stress (Sigma1)) at t=16.4

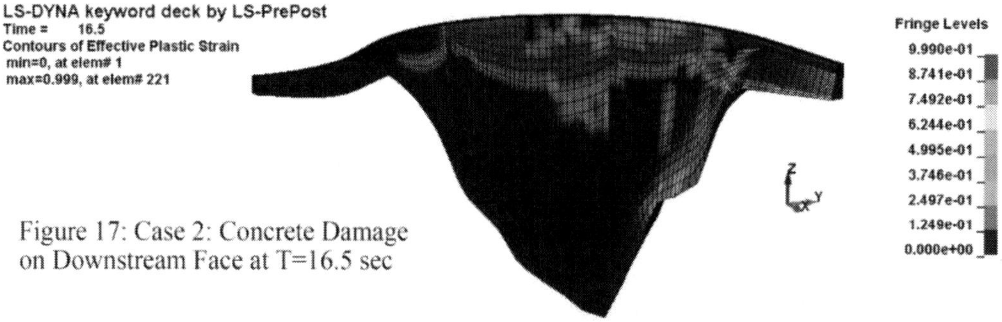

Figure 17: Case 2: Concrete Damage on Downstream Face at T=16.5 sec

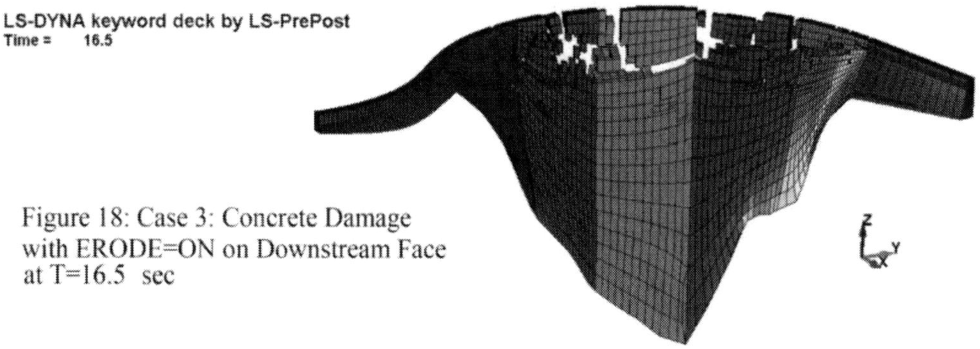

Figure 18: Case 3: Concrete Damage with ERODE=ON on Downstream Face at T=16.5 sec

Investigation of the failure modes of concrete dams physical model tests.

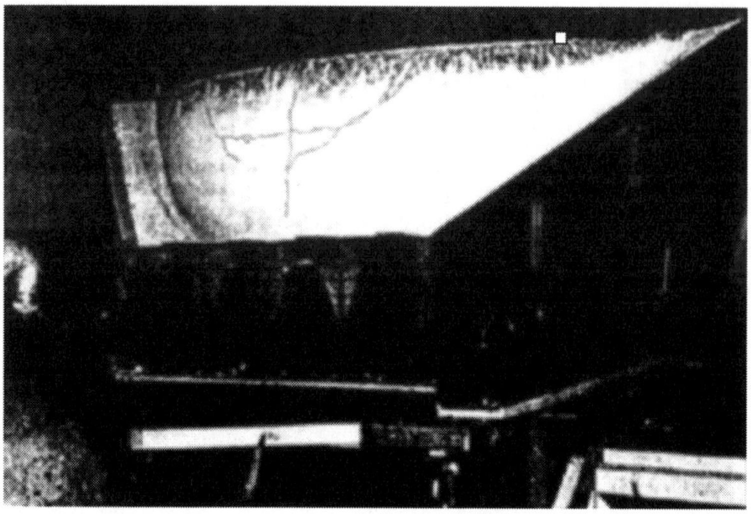

ISMES wide arch dam model failure. Homogeneous dam shaken to failure with earthquake simulation.

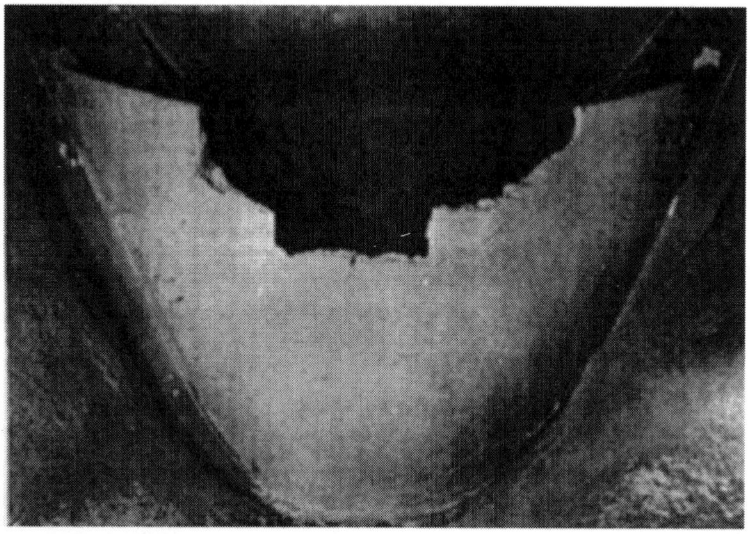

ISMES test of narrow canyon dam model.

Figure 19: Potential Failure Mode for Concrete Arch Dams (USBR, 2006), Reference 2

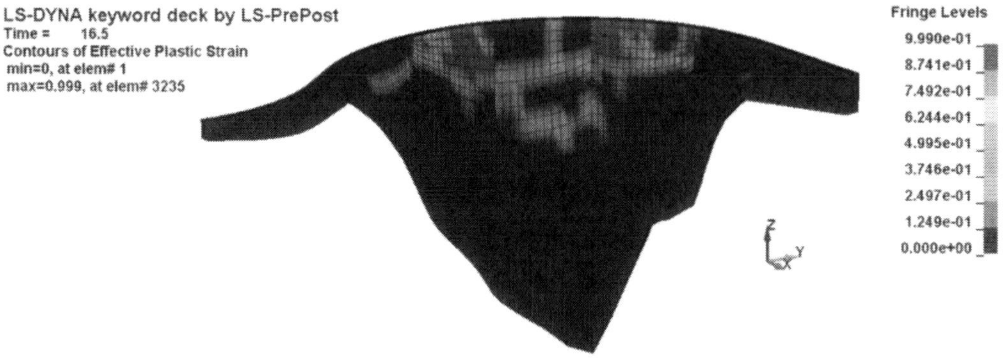

Figure 20: Case 4: Concrete Damage on Downstream Face at T=16.5 sec

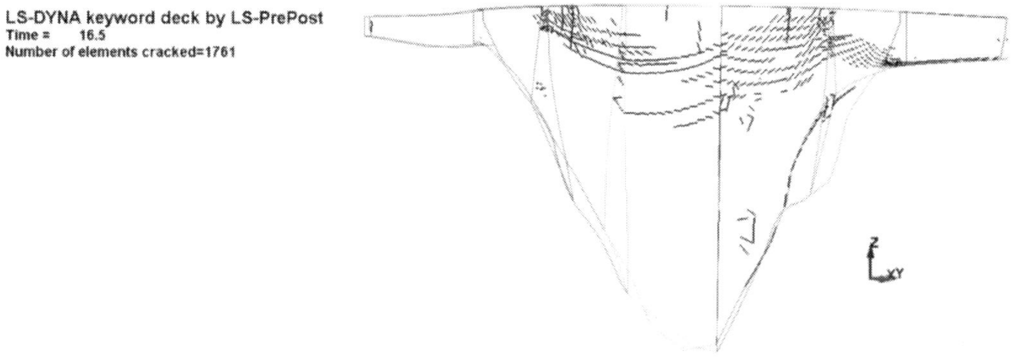

Figure 21: Case 5: Concrete Damage on Downstream Face at T=16.5 sec

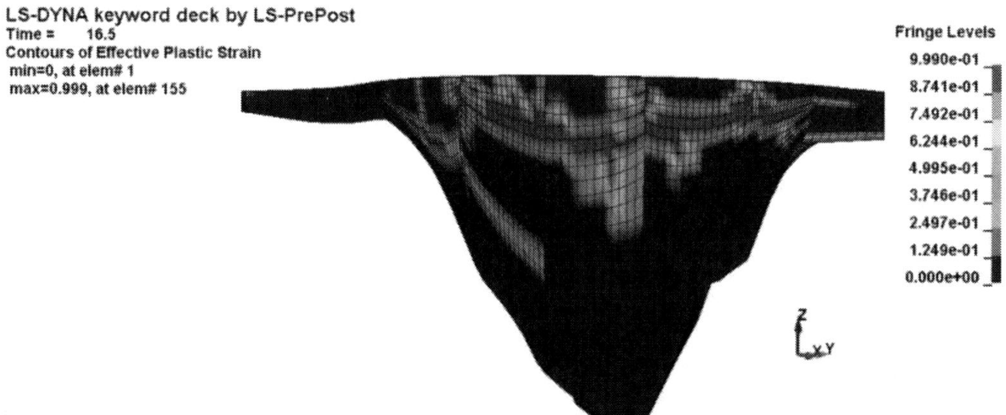

Figure 22: Case 6: Concrete Damage on Downstream Face at T=16.5 sec

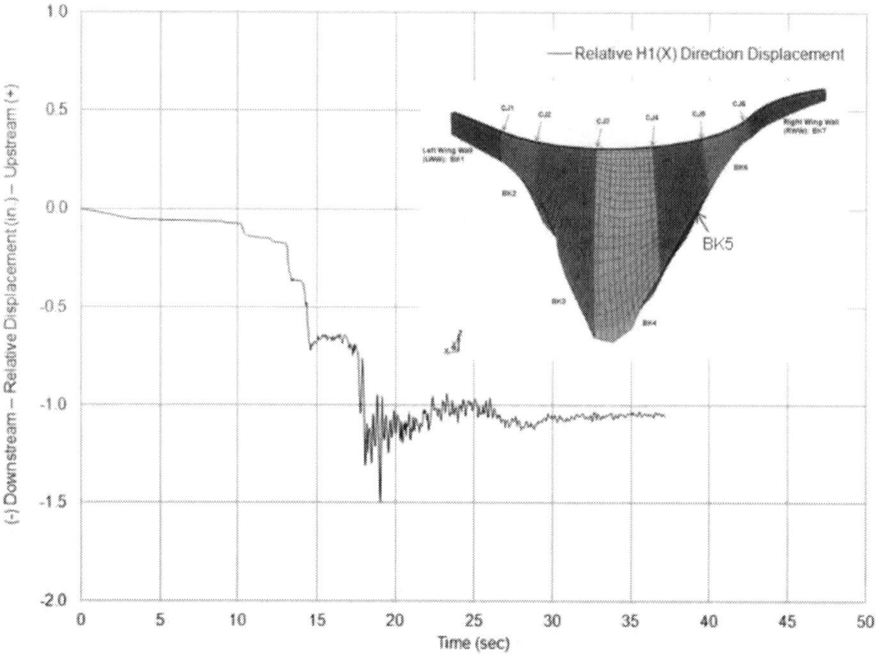

Figure 23: Case 9: Time History of Sliding Displacements (inch) at Block 5

CONCLUSIONS

The concrete arch dam was predicted to suffer extensive damage during a seismic event with the PGA of 0.9*g*. The non-linear concrete model shows the cracking would start at t=16.4 seconds with the first 8.5 seconds being the establishment of the static load. At t=16.4 seconds the corresponding PGA was only 0.63*g*. It is also noteworthy that the ERODE=ON model (used in Case 3) was conservative because strain softening was ignored. The model does provide a good sense of the location of the cracks and potential failure modes. Analysis of the movements on the foundation joints (Cases 7 to 9) show only small displacements in the right abutment. However, some movement of Dam Block 5 was observed in Case 9 and this movement was considered acceptable. The concentrated stresses observed at the dam-foundation joints are insignificant and can be ignored.

Further analysis is required to investigate the post-earthquake condition with the circular cracked surface at the upper central portion of the dam.

RECOMMENDATIONS

The following recommendations are presented:

- It would be very useful to have measurements from existing arch dams that have been subjected to intense earthquake ground motions, but this data does not exist.

- It is recommended that seismic monitoring of dams be strongly encouraged on dams in seismically active areas worldwide.
- Data obtained during actual earthquakes would provide valuable information on damping, dynamic tensile strength of concrete based on back calculations and model calibration efforts.
- There is considerable debate on the tensile strength of concrete. In a recent USSD paper i.e., Darbar et al. (2016), large concrete cores were tested dynamically in direct tension and relatively low tensile strengths were measured. The results appear to be debatable and the tensile strength of concrete needs to be resolved as it is directly contributing to the failure mode in this dam.
- In order to investigate the post-earthquake behavior consideration is being given to model a predefined circular crack identified by the analysis presented herein.

REFERENCES

1. Hall, J. F., "Problems encountered from the use (or misuse) of Rayleigh damping", Earthquake Engineering and Structural Dynamics, 35: 525–545, 2005.

2. USBR, "Investigation of the Failure Modes of Concrete Dams Physical Model Tests", Report DSO-06-03, December 2006.

3. FHA, " User's Manual for LS-DYNA Concrete Material Model 159", Publication No. FHWA-HRT-05-062, US Department of Transportation, May 2007.

4. Darbar, S.; Queen, D.; Hatton, C.; Dolen, T.; and Bartojay, K., "Static and Dynamic Mass Concrete Material Properties of A Concrete Gravity Dam", 36th Annual USSD Conference, Colorado, 2016.

5. Sooch, G. S. and Bagchi, A., "A New Iterative Procedure for Deconvolution of Seismic Ground Motion in Dam-Reservoir-Foundation Systems", Journal of Applied Mathematics, September 2014.

6. Raphael, J. M., 1984. "Tensile Strength of Concrete", ACI Journal. Technical Paper, Mar-Apr 1984. Title No. 81-17.

U.S. LEVEE SAFETY COALITION AND INTERNATIONAL COMMUNITY OF PRACTICE, RECENT ACTIVITIES AND UPDATES

Elena Sossenkina[1]
Michael Sharp[2]
Noah Vroman[3]

USSD is a charter member and a presiding Chair of the U.S. Levee Safety Coalition (Coalition), a group of professional organizations that have joined together to support and advance levee safety in the United States of America. The original members of the Coalition are the ASCE, ACEC, ASDSO, DFI, NAFSMA and USSD. Because of USSD's strong international connections and ties to ICOLD, our organization serves as a link between the Coalition and the international community of practice.

Recently, the Coalition has been asked to prepare a chapter on the state of levees in the United States for a report being developed by a working group on levees and flood defenses of the European Club of ICOLD. The report is titled "European Levees and Flood Defenses, Inventory of characteristics, risks and governance." A target completion date is May 2017. The report will consist of chapters prepared by individual countries. Each chapter will follow the same format and address topics such an overview of inventory of levee systems in the participating countries, legislative framework and governance, recent major floods and incidents, best practices and critical knowledge gaps. The U.S. chapter is being developed in close cooperation with U.S. Army Corps of Engineers and FEMA and builds upon the work completed by the National Committee on Levee Safety in 2009 and the ongoing efforts by USACE to characterize their portfolio of levees and flood risk infrastructure.

This paper will provide background on the U.S Levee Safety Coalition and its recent activities, and present key findings from the EURCOLD report. The report will serve as a stepping stone to establishing an international community of practice for levees and flood defense systems under the ICOLD umbrella and will provide useful information for dam and levee safety professionals around the world.

[1] HDR, elena.sossenkina@hdrinc.com
[2] U.S. Army Corps of Engineers
[3] U.S. Army Corps of Engineers

USACE LEVEE PORTFOLIO REPORT

Noah Vroman[1]

Levees within USACE authorities include over 2,500 levees systems totaling 14,580 miles in length. These levee systems provide benefits to over 10 million people and property valued at nearly $1.3 trillion dollars. USACE has performed engineering inspections and risk assessments to improve the understanding of the condition and flood risk related to these levee systems. USACE is currently reviewing the information collected through this effort to support and inform programmatic implementation of the USACE Levee Safety Program. This information aids in the evaluation of opportunities to manage flood risk and identification of key partners who have shared responsibility to manage these risks.

This includes an improved understanding of factors that contribute to flood risk related to these levees as well as magnitude of costs that may be necessary address levee system deficiencies.

[1] U.S. Army Corps of Engineers, noah.d.vroman@usace.army.mil

LEVEE SAFETY GUIDANCE COMPREHENSIVE DOCUMENT REVIEW

Michael Sharp[1]

The USACE is currently engaged in an effort to identify those documents with relevance to levee safety. The overall purpose of this effort is to enhance levee safety by providing a stronger foundation to adapt, learn and change. The immediate purpose is to encourage high safety standards in the practices and procedures of the USACE in dealing with the management of levees. The guidelines are intended to be a collection of all documents, USACE and non-USACE, that in some way support and promote levee safety. They are intended to outline national procedures that will continually stimulate technical methods in levee planning, design, construction, and operation for minimizing risk of failure. Those charged with administering these guidelines must recognize that the achievement of levee safety is through a continuous, dynamic process in which practices and procedures are examined periodically and updated. This effort has multiple objectives as detailed following: develop a methodology that provides a process to systematically and consistently evaluate any document for relevance to levee safety and then add perspective as to the importance of that document; develop a database of documents relevant to levee safety that could be easily maintained and that could also be searched and queried; provide an evaluation of the resulting documents to highlight critical features such as those that are dated, redundant, conflicting, have gaps, etc. This evaluation could also inform a better way to keep documents current and relevant.

[1] U.S. Army Corps of Engineers, Michael.K.Sharp@usace.army.mil

INVENTORY AND REVIEW OF THE NATION'S LEVEES

Jamie McVicker[1]
Cathi Sanders[2]

Congress has authorized the U.S. Army Corps of Engineers (USACE) to conduct a one-time inventory, inspection, and risk assessment of all levees in the National Levee Database (NLD). An inventory of the nation's levees is an important aspect of flood risk management. A comprehensive inventory and review will allow USACE and the nation to understand where public and private levees exist and the imposed flood risks. USACE will assess the reliability of the nation's levees, quantify the people and infrastructure that rely on them, and help levee operators and communities prioritize where limited public funds should be spent.

USACE plans to engage with State agencies and levee owners / operators about this voluntary opportunity to participate in this effort with the goal of gathering lessons learned and best practices; working with volunteer states, communities and levee sponsors to refine approaches, materials, and tools necessary to conduct inventory, inspection, and risk assessment activities. The project seeks to both inform the current USACE Levee Safety Program and to make these products more useful to interested parties outside of USACE.

[1] U.S. Army Corps of Engineers, Jamie.Mcvicker@usace.army.mil
[2] U.S. Army Corps of Engineers

LEVEE SAFETY PRACTICES IN THE KEYSTONE STATE

Andrew J. Orlovsky, P.E.[1]

ABSTRACT

This paper will document past, current, and future Levee Safety efforts in the Commonwealth of Pennsylvania. The focus of this paper will include the history of levee safety in Pennsylvania, the Commonwealth's current policies, and potential efforts, opportunities, and challenges with regards to the implementation of the National Levee Safety Program, in which Pennsylvania's program was chosen as a pilot.

Flood protection has been considered a high priority in the Commonwealth since the Pennsylvania Flood Control Law, signed August 7, 1936, authorized the design and construction of flood protection projects. Since then, as part of this program, the Pennsylvania Department of Environmental Protection (PA DEP) has constructed 52 distinct levee systems. While operation and maintenance of these levees is the responsibility of a local sponsor municipality, PA DEP has continuing involvement throughout the life of the levee. Their duties include annual inspections of each levee and serving as the liaison with the U.S. Army Corps of Engineers regarding federal assistance for the levees, notably with damage assessments and rehabilitation assistance under the PL84/99 Program, among others. As the National Levee Safety Program gets underway, Pennsylvania's current activities put it at a unique advantage as far as implementation of the program. This paper will discuss some of those advantages and the expectation for implementation.

The Commonwealth of Pennsylvania is one of the most flood-prone states in the nation. Due to Pennsylvania's mountainous terrain, it's more than 86,000 miles of rivers and streams (second to only Alaska) and an average annual rainfall of over 42 inches; flood protection has been considered a high priority over the past 80 years (Zomok, 2011). The Pennsylvania Flood Control Act was signed August 7, 1936 to complement the Federal Flood Control Act that was passed by Congress that same year. This act authorized the Pennsylvania Water and Power Resources Board and its successor agencies, currently the Pennsylvania Department of Environmental Protection (PA DEP), to design and construct structural flood protection projects. Since then, PA DEP has constructed more than 200 flood protection projects throughout the state at a cost of more than $800 million when translated to today's dollars. Of these 200 projects, 52 of them include distinct levee systems. Proper operation and maintenance of these projects is overseen by the Division of Project Inspection within PA DEP's Bureau of Waterways Engineering and Wetlands.

[1] Chief, Completed Projects Section, Bureau of Waterways Engineering and Wetlands, Pennsylvania Department of Environmental Protection, Harrisburg, PA 17105, aorlovsky@pa.gov

STATE OWNED LEVEE SYSTEMS

Pennsylvania is unique among other states as most of its non-federal levee systems were constructed by the Commonwealth with funding provided by the Public Improvement Project line-item in the Capital Budget. Generally, PA DEP constructs smaller levee systems on smaller streams to complement the larger levee systems along rivers constructed by the U.S. Army Corps of Engineers, though PA DEP has constructed projects on larger rivers. These facilities reduce flood damage to small communities as well as larger more urbanized areas.

The PA DEP-constructed levees vary between 600 and 12,000 feet in length and 5 to 24 feet in height. Several of the levee systems contain pump stations and many include mechanical closure structures such as roller gates, swing gates, and stop log structures. Every Commonwealth-built levee system is operated and maintained by a municipal sponsor.

As a condition of receiving a levee or any other flood protection project, the municipality signs a legal agreement with the Commonwealth prior to final design. This agreement, known as a Sponsorship Agreement, is intended to last for the lifetime of the project. In accordance with the agreement, the municipality agrees to secure all lands necessary to construct the project, to replace any necessary roads or bridges, to demolish or relocate any existing structures that will interfere with the construction of the project, and to perform any necessary relocation of utilities. After construction, the municipality also agrees to allocate necessary funding in their annual budget to properly maintain the project. PA DEP obligations under the Sponsorship Agreement include performing annual inspections of the levee and providing assistance to the local municipality.

The Commonwealth also strives to have all new levee systems entered into the Federal Rehabilitation and Inspection Program in accordance with Public Law 84-99 (PL84/99), which is administered by the United States Army Corps of Engineers (Corps). During the design phase, PA DEP will forward plans to the Corps for review and to ensure that the proposed project will be eligible. PA DEP will also attempt to enter into PL84/99 levee systems that were previously ineligible for PL84/99, but had undergone recent rehabilitations. The Commonwealth will provide the municipality with technical assistance and provide the sponsor also necessary documentation to submit to the Corps for an Initial Eligibility Inspection (IEI).

The Commonwealth of Pennsylvania contains three major watersheds, and therefore, PA DEP partners with three Army Corps Districts to conduct annual Continuing Eligibility Inspections of its levee projects. The Pittsburgh District handles the Ohio River Watershed, the Baltimore District handles the Chesapeake Bay Watershed (which includes the Susquehanna and Potomac River Basins), and the Philadelphia District handles the Delaware River Watershed. Small portions of Pennsylvania include the Lake Erie Watershed and the Genesee River Watershed, which are handled by the Buffalo District, but no levees are situated within those watersheds.

Services Provided

In this unique partnership, PA DEP and the Corps meet annually to review changes in inspection requirements, work out inspection assignments, and achieve consistency on inspections and reporting. As a condition of being in the PL84/99 program, the Corps performs biannual continuing eligibility inspections. On the off-years, DEP personnel will perform annual inspections. Both DEP and the Corps will review each other's inspection reports in order to consensually determine a rating for the project.

Weather Warnings. PA DEP will often contact project sponsors if the weather forecast calls for heavy rain so that they can prepare for possible flash flooding. The sponsors are encouraged to contract their respective Corps District if they are in need of flood fighting supplies. The sponsors are also directed to properly exercise their Emergency Action Plan (EAP) if necessary. Finally, PA DEP will direct the sponsor to inspect their project after the storm to determine if any damages resulted. If the project is active in the PL84/99, rehabilitation assistance will be provided by PA DEP and the Corps.

Repair Financing. While the Federal Government pays 80 percent of construction repairs to flood damaged projects, Pennsylvania is unique among states as it typically provides funding for the sponsor's 20 percent cost share. PA DEP levee safety engineers assist the Corps with the flood damage assessment and will also assist with the permitting and any necessary land acquisition. PA DEP will also spot check PL84-99 funded construction repairs.

Photo 1: Concrete Floodwall in Danville Borough, Montour County, PA rehabilitated though PL84-99 Program.

Technical Assistance. DEP also provides other services to its sponsor municipalities to ensure peak operation of their projects and to maintain their eligibility in PL84/99. These services tend to vary based on the staff of Division of Project Inspection and the funding available in any given year. For example, PA DEP's levee safety engineers are fully trained and equipped to enter and inspect confined spaces. In accordance with PL84/99 requirements, levee conduits are inspected on a 5-year cycle. Previously, levee conduit camera inspections were conducted by Division of Project Inspection personnel, but due to a change in policy, grants are now awarded to sponsor municipalities, with PA DEP providing the specification based on Corps guidance. PA DEP also developed a standard form for recording the results of each levee conduit video inspection, which will be forwarded to the Corps.

The video camera inspections are revealing many levee conduit deficiencies that are difficult and costly to address, and hence rehabilitation projects that incorporate more substantial improvements beyond routine maintenance are becoming more necessary. For these more complex projects, PA DEP will also provide the sponsors with design and construction inspection of rehabilitation projects or may offer grants for small rehabilitation projects or specialized equipment.

Emergency Action Plan Development. PA DEP will also provide technical assistance to project sponsors seeking larger state grants, seeking levee certification, or developing EAPs. In coordination with its sister agency responsible for emergency management, PA DEP has developed an EAP guidance document for levees. Since project sponsors are required to be in possession of an updated EAP during Continuing Eligibility Inspections, the need for EAP guidance has become a higher priority in comparison with previous years. In addition, proposed updates to the PL84/99 will place a much higher priority on risk and the sponsor's activities during high water events, which further stress the importance of a well-prepared and often-updated EAP.

O and M Plan Development. A module-based approach to generating the newly required Operation and Maintenance Manuals for the sponsors of levee projects has also been developed by PA DEP. Corps inspectors have reviewed and approved this approach to O & M Manual development. In the past, PA DEP has also provided funding for municipal sponsors to conduct Levee Certification Studies with the goal of having the levee accredited by the Federal Emergency Management Agency (FEMA), but no funding is currently available for this purpose.

Training and Workshops. Finally, PA DEP hosts a two-day bi-annual workshop which provides training to the municipal sponsors in areas of operation and maintenance, proper state and federal permitting, and also to educate the sponsors about seeking grant funding. The event is attended by nearly 150 participants and presenters including municipal sponsors of both state and federal flood protection projects, as well as officials from the three Corps districts, Corps headquarters, and many others state and federal agencies. These bi-annual flood protection workshops provide a great opportunity for project sponsors to receive training, share information, and learn about the latest levee safety issues. Vendors of flood protection-related materials and services, from flap gates to vegetation control, are invited to set up displays and equipment during the workshop. At an evening recognition dinner, awards are presented to project sponsors who have displayed exemplary effort in operating and maintaining their projects.

INACTIVE STATE LEVEES

There are a few levees that were constructed by the state and were previously inspected, but have fell into a state of disrepair. Under the policy that PA DEP had in place prior to 2008, the state would declare the municipal sponsors of these levees in violation of their Sponsorship Agreement and would remove them from their program, declare them inactive, and would stop inspecting them. Since 2008, PA DEP has made an effort to continue inspecting these inactive levees but finding documentation of levees that have been inactive for several decades is often difficult. The state has also constructed many spoil levees, but in these cases, the state did not enter into a maintenance agreement with the local municipality. These "push up" embankments, generally constructed of unconsolidated riverine gravel and sediments, provide some local flood protection benefits, but frequently fail and require repair. It is unknown exactly how many of these structures exist or where they are located. These structures do propose a risk, and many of the citizens protected by these structures are unaware of their presence. Therefore, PA DEP has attempted to reach out to the municipalities that own these levees, whether they are spoil levees or engineered structures, and is looking to restore them into the PA DEP inspection program.

Photo 2: Inactive State Levee in Rockhill Borough, Huntingdon County, PA. Note Presence of Large Trees on Embankment.

FEDERAL LEVEES

Pennsylvania has limited involvement with the federally-constructed levees within its borders. PA DEP generally keeps records of the Corps annual inspection reports for federal projects. When funds are available for the PA DEP Flood Protection Grant Program, federal sponsors are also eligible to apply (this grant program has not been active since 2010). Federal sponsors are also invited to the aforementioned bi-annual workshop which provides training to the municipal sponsors in areas of operation and maintenance, proper state and federal permitting, and also to educate the sponsors about seeking grant funding. Some federal levees are also situated on state lands and PA DEP is responsible for issuing license agreements for third parties that desire to occupy those lands. An example could be a utility company that needs to install a pipeline under the levee.

PERMITTING OF STATE LEVEES

PA DEP provides permitting assistance to both state and federal flood protection project sponsors. While dams and waterways encroachments permits are typically issued by regional engineers or those within the Division of Dam Safety, PA DEP Engineers in the Division of Project Inspection have authorized permits at the state level for new projects, project rehabilitation and improvements, and normal maintenance that may impact streams and wetlands. New and rehabilitation projects include projects designed in-house by DEP, or projects designed by consultants on behalf of either DEP or the project's sponsor. PA DEP has also worked with the USACE Regulatory Branch to streamline federal authorization for similar work at both state and federal levees.

Conversely, sponsors of both state and federal levees are encouraged to contact PA DEP when they become aware of proposed construction projects that may impact the levees they are tasked with maintaining. PA DEP engineers conduct a technical review of the proposed work from a levee safety perspective and provide recommendations on to the state level regulatory staff prior to issuance of any state level permits. In the case of federal levees, details on the proposed work are forwarded to the Corps for their review.

NON-STATE AND NON-FEDERAL LEVEES

As part of the National Levee Safety Pilot Program, Pennsylvania is currently undertaking an effort support the Corps' National Levee Database by locating private levees and levees constructed by local governments rather than the state or federal government. No non-state or non-federal levees in Pennsylvania are currently active in the PL84/99 Program. Most of these structures are spoil levees, but some are well-engineered structures. Under PA Environmental Code, levees are not specifically regulated as are dams, so DEP currently has little oversight over these structures and Pennsylvania does not require a special levee permit.

However, in many cases the state has jurisdiction over levees under other regulatory programs. Chapter 105 of the code, titled "Dam Safety and Waterways Management" gives DEP the authority to regulate the construction of all structures within the floodway, which is defined by state regulation as the channel of the watercourse and portions of the adjoining floodplains which are reasonably required to carry and discharge the 100-year flood. The boundary of the floodway is as indicated on maps and flood insurance studies provided by FEMA. In areas where no FEMA maps or studies have defined the boundary of the 100-year floodway, absent evidence to the contrary, it is assumed that the floodway extends from the stream to 50 feet behind the top of the bank of the stream (PA Code, 2011). Most levees are at least partially situated in the floodway. Other pertinent facets of the PA Environmental Code are Chapter 102, titled Erosion and Sediment Control, and Chapter 106 titled Floodplain Management.

These regulations allow PA DEP to conduct Levee safety reviews of proposed facilities that are not state or federally owned. Whenever a DEP permitting engineer receives an encroachment permit for a project that involves the construction of a new levee or a rehabilitation of an existing levee, the plans are forwarded to the completed projects section for a levee safety review. PA DEP currently relies upon the *USACE Manual for the Design and Construction of Levees* as its primary technical guidance document on the levee safety reviews. Concerning levees constructed by private owners, if a proposed levee that only impacts the owners personal property, DEP will provide suggestions to the owner , but not require the levee be constructed to the current standards. Private levees that protect multiple households will be evaluated on a case-by-case basis.

Photo 3: Private Levee in Bristol Borough, Bucks County, PA owned by Rohm and Haas Chemical Company

In 2008, the Commonwealth's Auditor General recommended that more comprehensive levee safety requirements be adopted (Wagner, 2008). Considering the implementation of the National Levee Safety Program, there is a good chance legislation specifically concerning the regulation of levees may be passed in the near future.

THE NATIONAL LEVEE SAFETY PROGRAM

In Section 3016 of the Water Resources Reform and Development Act of 2014 (amended Title IX of WRDA 2007) Congress provided responsibilities and authorities to the Corps (along with FEMA) needed to stand up a National Levee Safety Program. While the program has yet to receive a full appropriation from Congress, the Corps has begun to implement an initial set of component-based activities (Halpin, 2016). The Corps' stated purpose of the Pilot Program is to develop a nation-wide understanding of levee condition and risk by adding levees that are not currently in the National Levee Database (NLD) and also conducting a review and structural evaluation of levees that are not part of the PL84/99 Program. The stated plan is to share experiences, best practices, and lessons learned. Goals were stated that included the reduction of risk, the improvement of processes, the alignment levee safety activities, and providing a starting point for state levee safety programs where appropriate. Potential pilot states were identified through the Silver Jackets program (state interagency flood risk reduction programs) as well as Corps Divisions and Districts. Because Pennsylvania is already so involved with Levee Safety, the Commonwealth was selected by the Corps to participate in an Inventory & Review Pilot Program.

Representatives from the Corps met with a representative of PA DEP in June 2016 to discuss the next steps. The initial plan for the Pilot Program is twofold. The first goal is to identify any levees within Pennsylvania that are not on the NLD. Nearly all the PA DEP constructed levees are already listed on the NLD, but several levees that are either privately owned or owned by a local government were not. DEP Division of Project Inspection staff will continue to work with regional engineers and other state and local personnel to identify these levees, which could be challenging as many are quite small and have not been properly maintained in years.

The second aspect involves the goal to conduct inspections and risk assessments of levees of various ownership that are either not federally maintained or not part of the PL84/99 Program. The initial program will select and document 4 or 5 levees within one Corps District for inspection and screening-level risk assessment. These are anticipated to generally comprise levees that are planned to be entered or reinstated in the PL84/99 program in the near future. PA DEP identified a handful of potential candidate levees. These levees are large enough to pose a significant risk to residents should failure occur. The candidate levees include private levees, levees constructed by local governments, and inactive state levees. Generally, these levees have several deficiencies that would prevent them from passing an Initial Eligibility Inspection for entrance into the PL84/99 program unless a multi-million-dollar rehabilitation project was commence.

Currently, PA DEP is in the process of reaching out to the owners/sponsors of these levees for their consent to enter the Pilot Program. As of February 2017, the Corps plans to begin inspection and risk assessment of three levee systems; one was constructed by a city; one was constructed by the state, but has not been in good enough condition for entrance into the PL84-99 program; and the third system is of unknown construction (though thought to be federal by the locals). PA DEP and the Corps have partnered to gather information on these systems, which could be difficult as very limited data currently exists.

The implementation of the National Levee Safety program presents a new challenge for Pennsylvania's Flood Protection Program, as well as for state flood protection programs throughout the rest of the nation. As much of the United States' flood protection infrastructure continues to age and deteriorate, this program presents an excellent opportunity for the nation to improve on the D-minus grade Levees have received on the 2013 Infrastructure Report Card published by the American Society of Civil Engineers.

REFERENCES

Halpin, E., National Levee Safety Program, U.S. Army Corps of Engineers, Proceedings from the 36th Annual USSD Conference, 2016.

Pennsylvania Code, Chapter 102: Erosion and Sediment Control, Title 25 Environmental Protection, 2011.

Pennsylvania Code, Chapter 105: Dam Safety and Waterways Manager, Title 25 Environmental Protection, 2011.

Pennsylvania Code, Chapter 106: Floodplain Management, Title 25 Environmental Protection, 2011.

Wagner, J., A Special Performance Audit of the Department of Environmental Protection – Dam and Levee Safety, Pennsylvania Department of the Auditor General, 2008.

Zomok, R., Levee Safety Experiences from Pennsylvania, PA Department of Environmental Protection, Proceedings from Dam Safety 2011.

SEISMIC VULNERABILITY ANALYSIS OF SLAG CEMENT-CEMENT-BENTONITE SEEPAGE CUTOFF WALLS FOR LEVEES

Vlad Perlea, PhD, PE[1]
Khaled Chowdhury, PE, GE[2]
Mary Perlea, PE[3]
Michael Kynett, PE[4]

ABSTRACT

A slag cement-Portland cement-bentonite (SCCB) cutoff wall is one of the common seepage mitigation measures in the Central Valley of California. It is usually selected in levees where increased strength is needed, such as when the wall is located near the waterside toe or across features such as bridges, utilities, and roadways. As a SCCB wall is a cemented linear structure within a deformable soil mass, a concentration of stress resulting in cracks is considered a potential vulnerability during an earthquake. As SCCB wall is a rigid, cemented structure, cracks developed during an earthquake are expected to be irrecoverable. To evaluate the vulnerability of a SCCB wall during a seismic event, a dynamic numerical modeling program was implemented using software, FLAC 2D v7. The material properties for SCCB wall were developed based on a recently completed laboratory testing program for two Central Valley levee projects. The liquefaction potential of the foundation has been evaluated by modeling the liquefiable layers with the UBCSAND constitutive model and the non-liquefiable layers were modeled using Mohr-Coulomb constitutive model. Eleven different SCCB mix designs from the laboratory testing program have been considered, covering the range of UCS between 30 psi and 440 psi. These SCCB mixes have shown relatively low shear strain at failure (1 to 2 percent) in laboratory testing. The stress and strain in a SCCB wall were found to be within the elastic range for ground motions with 200-year return periods, when UCS strengths are greater than 100 psi was used and the phreatic surface was relatively low, as expected for coincidence with a strong seismic event. A sensitivity study on one cross section using a higher water table indicates that the strain potential becomes higher and may exhibit plastic failure. The results indicate that a higher strength requirement is needed for SCCB walls, if post-seismic effectiveness is considered during design.

[1] Senior Geotechnical Practitioner, AECOM Technical Services Inc., 2020 L St, Sacramento, CA 95811, Vlad.Perlea@aecom.com
[2] Senior Geotechnical Engineer, USACE, Dam Safety Production Center, 1325 J St, Sacramento, CA 95814, Khaled.Chowdhury@usace.army.mil
[3] Senior Geotechnical Practitioner, AECOM Technical Services Inc., 2020 L St, Sacramento, CA 95811, Mary.Perlea@aecom.com
[4] Levee Safety Program Manager, USACE, Sacramento District, 1325 J St, Sacramento, CA 95814, Michael.N.Kynett@usace.army.mil

INTRODUCTION

Seepage cutoff walls have been selected as the preferred alternative in the majority of Central Valley of California levees that require mitigation. The types of cutoff walls being considered include soil bentonite (SB) cutoff walls and slag cement-Portland cement-bentonite (SCCB) cutoff walls installed from the levee centerline. The purpose of the study described in this paper was to evaluate the potential of the SCCB wall to crack during a 200-year average return period seismic event. Although several distress types are possible, only one possible damage, believed the most potentially damaging, was evaluated: cracking due to shear stress exceeding the wall material shear strength. FLAC Version 7.00.424 (Itasca, 2009) and the C++/003.dll version of UBCSAND (Beaty and Byrne, 2011) were used in the analysis. A generalized cross section was developed in a relatively low seismically active fluvial region of the Central Valley of California, with shallow and moderately thick liquefiable layer (less than 20 feet).

SECTION ANALYZED

It is believed that the presence of liquefiable materials in foundation aggravates the potential for seismic induced strain, and therefore the probability of cutoff wall cracking. Figure 1 shows the most critical levee cross section of the evaluated levee system for seismic liquefaction potential. A relatively shallow coincident water surface elevation (WSE, Elevation 16.8) was considered in the base case. This coincident WSE would saturate the liquefiable sand near the ground surface (S_2 and S_3, Figure 1). Steady-state seepage and stability analyses showed that through seepage, underseepage, and slope stability criteria were met with the centerline cutoff wall improvement alternative.

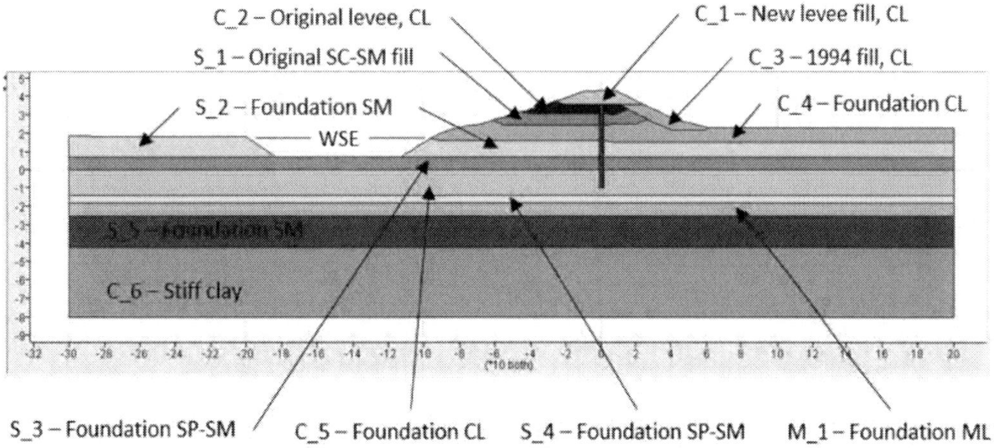

Figure 1. Materials in the analyzed cross section.

Properties of various materials, as defined in Figure 1, are summarized in Table 1.

Table 1. Geotechnical State and Strength Parameters for the Analyzed Cross Section.

Material	Thickness (feet)	Unit Weight Dry / Moist / Saturated (pcf)	V_s (fps)	Consolidated Drained		Consolidated Undrained	
				Φ' (°)	c' (psf)	$Φ_u$ (°)	c_u (psf)
C_1	7	99 / 114 / 125	400	31	100	17	200
C_2	5	91 / 105 / 120	475	27	50	18	100
C_3	n/a	99 / 114 / 125	515	31	100	22	150
S_1	7	100 / 110 / 125	540	32	10[1)]	-	-
C_4	8	99 / 114 / 125	605	31	100	17	200
S_2	9.1	108 / 119 / 130	670	32	10[1)]	-	-
S_3	6.5	108 / n/a / 130	700	33	0	-	-
C_5	14.5	99 / n/a / 125	730	31	100	17	200
S_4	4	108 / n/a / 130	750	36	0	-	-
M_1	7	100 / n/a / 125	770	32	75	18	150
S_5	12.9	108 / n/a / 130	800	36	0	-	-
C_6	46	107 / n/a / 130	1,200	n/a	n/a	-	-

Note: [1)] A small cohesion was assigned for preventing failure of triangular zones at slopes.

Shear wave velocity (V_s, fps) was estimated from an average Vs profile developed for West Sacramento Region based on seismic cone field testing performed for California Department of Water Resources, DWR (URS, 2008).

Two foundation layers, labeled S_2 and S_3 in Figure 1, were found potentially liquefiable based on evaluations by URS for DWR's Urban Levee Geotechnical Evaluations (ULE) project (URS, 2008). The liquefaction potential of the foundation has been evaluated by modeling the liquefiable layers with the UBCSAND model. The built-in FLAC version of this model (C++/003.dll) is calibrated based on Youd et al., 2001 criterion; liquefaction is considered triggered when the excess pore pressure of 70% of the initial effective stress has been reached. The residual strength, used in a post-earthquake stage, was evaluated as recommended by Seed and Harder, 1990; a curve within the recommended range, 1/3 from the lower bound and 2/3 from the upper bound, was considered.

With both liquefaction triggering and post-liquefaction residual strength, the main defining soil parameter is the standard penetration blowcount of an equivalent clean sand material. This parameter was considered $(N_1)_{60\text{-}cs} = 10$; the corresponding residual strength was $S_r = 235$ psf.

Water flow characteristics of various materials are presented in Table 2.

Table 2. Water Flow Characteristics of Various Natural Materials

Material	Effective porosity	Horizontal permeability		Vertical permeability	
		k_h (cm/s)	FLAC k11*	k_v (cm/s)	FLAC k22*
C_1	0.25	$1.0\ 10^{-6}$	$5.2\ 10^{-10}$	$2.5\ 10^{-7}$	$1.3\ 10^{-10}$
C_2	0.25	$1.0\ 10^{-5}$	$5.2\ 10^{-9}$	$2.5\ 10^{-6}$	$1.3\ 10^{-9}$
C_3	0.25	$1.0\ 10^{-5}$	$5.2\ 10^{-9}$	$2.5\ 10^{-6}$	$1.3\ 10^{-9}$
S_1	0.3	$4.0\ 10^{-4}$	$2.1\ 10^{-7}$	$1.0\ 10^{-4}$	$5.2\ 10^{-8}$
C_4	0.25	$1.0\ 10^{-5}$	$5.2\ 10^{-9}$	$2.5\ 10^{-6}$	$1.3\ 10^{-9}$
S_2	0.3	$1.2\ 10^{-3}$	$6.2\ 10^{-7}$	$3.0\ 10^{-4}$	$1.6\ 10^{-7}$
S_3	0.3	$1.0\ 10^{-2}$	$5.2\ 10^{-6}$	$5.0\ 10^{-3}$	$2.6\ 10^{-6}$
C_5	0.25	$1.0\ 10^{-6}$	$5.2\ 10^{-10}$	$2.5\ 10^{-7}$	$1.3\ 10^{-10}$
S_4	0.3	$1.0\ 10^{-3}$	$5.2\ 10^{-7}$	$5.0\ 10^{-4}$	$2.6\ 10^{-7}$
M_1	0.3	$4.0\ 10^{-6}$	$2.1\ 10^{-9}$	$1.0\ 10^{-6}$	$5.2\ 10^{-10}$
S_5	0.3	$1.2\ 10^{-4}$	$6.2\ 10^{-8}$	$3.0\ 10^{-5}$	$1.6\ 10^{-8}$
C_6	0.25	$1.0\ 10^{-6}$	$5.2\ 10^{-10}$	$2.5\ 10^{-7}$	$1.3\ 10^{-10}$

*Note: * FLAC permeability (ft^4/lb-s) is obtained by multiplying the values in cm/s by 5.2 10^{-4}.*

The stiffness parameters for various soil materials (C_1 through S_5 in Tables 1 and 2) were computed through routines incorporated in the FLAC project file; they were updated frequently, in accordance with the stress state at any given moment.

The Pleistocene clayey material C_6 was modeled elastic, with moduli evaluated as follows:

$$G_{static} = G_{max} = \rho\ V_s^2 = (130 / 32.185)\ 1200^2 = 5.816 \cdot 10^6 \text{ psf}$$

$$B = 2\ G_{static}\ (1 + \nu) / [3\ (1 - 2\ \nu)] = 2.17\ G_{static} = 1.262 \cdot 10^7 \text{ psf}$$

where $\nu = 0.3$ was the assumed Poisson Ratio.

The same moduli were considered with both static and dynamic analyses for the elastic material.

ARTIFICIAL MATERIAL PROPERTIES

Slag Cement-Portland Cement-Bentonite (SCCB) Materials

Table 3 summarizes the properties of various mix designs, with different slag cement/Portland cement ratios and cement/water ratios. The slag proportions in total cement contents were varied between 65% and 91%, and (slag cement + Portland cement)/water ratios were varied between 0.10 and 0.20. More details on the trial mix designs are available in Chowdhury et al., 2016.

Table 3. Trial Mix Design

Trial Mix	Slag / Total Cement (%)	Total Cement / Water Ratio	Bentonite Content (% of total weight)
A	65	0.12	5
B	65	0.15	5
C	70	0.15	5
D	70	0.20	5
E	75	0.15	5
F	75	0.20	5
G	80	0.12	5
H	83	0.15	5
I	91	0.10	5
J	91	0.15	5
K	91	0.20	5

Unconfined compressive (UC) strength tests (per ASTM D-2166) were performed at 1-day, 2-days, 3-days, 7-days, 14-days, and 28-days. Three mix designs (D, E, and G) samples with slag were additionally tested for unconsolidated-undrained (UU) triaxial compression strength (per ASTM D2850). The UU tests were performed with confining stresses of 3.5 psi (~500 psf), 7 psi (~1,000 psf), and 10.5 psi (~1,500 psf).

Significant difference in strength was not evident as a function of the confining stress; therefore, where both UC and UU tests were available, all parameters (ultimate strength, strain at ultimate strength, modulus of elasticity) were defined through the median of the four available results.

The unconfined compressive strength, q_u in psi, was found to vary significantly with both slag/total cement (S/C) ratio and total cement/water (C/W) ratio, as shown in Figure 2.

Based on the limited number of mix designs, approximate boundaries of mixtures with compressive strength greater than 200 psi and lower than 50 psi are drawn on this plot.

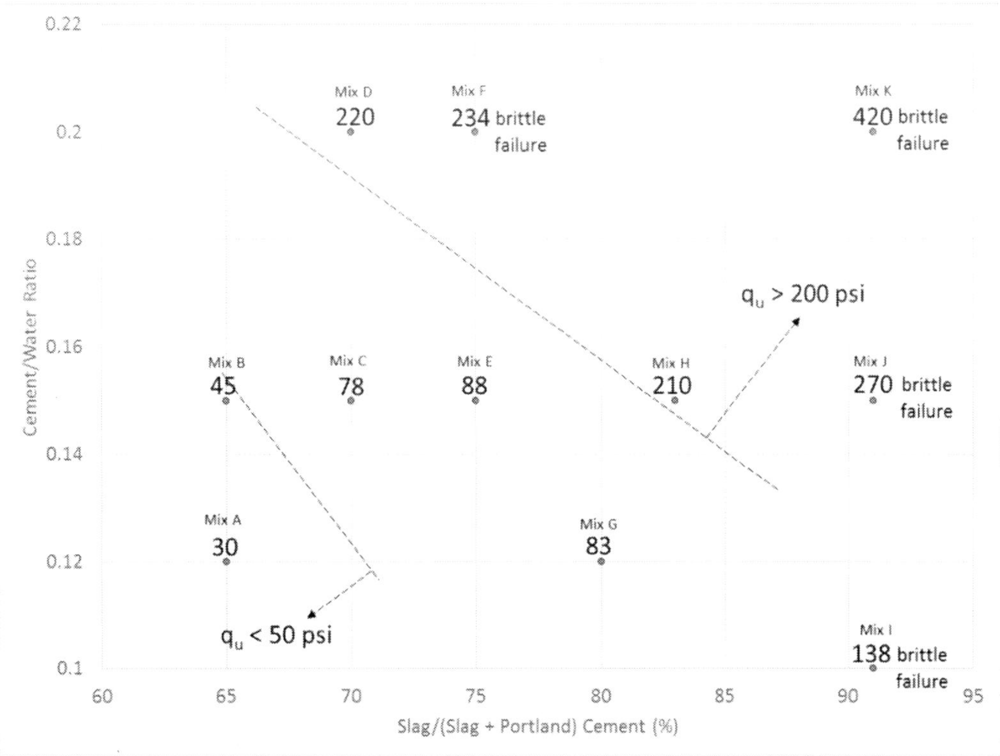

Figure 2. Unconfined compressive strength (psi) at 28-day curing of various mix designs.

Figure 3 presents the breakage strain of various mix design tested at 28-day curing. Where more than one UC test was available, the slightly conservative median of the four available results was considered. The contours suggest the probable trend of breakage strain (in percent axial strain) with the mix design parameters, S/C and C/W.

The 28-day secant Young's modulus was defined for the quasi-linear portion of the stress/strain relationship, either the secant modulus between $q_u = 0.2\ q_{u,max}$ and $q_u = 0.5\ q_{u,max}$, or the secant modulus between $q_u = 0.2\ q_{u,max}$ and $q_u = 0.8\ q_{u,max}$. The median values are shown in Figure 4.

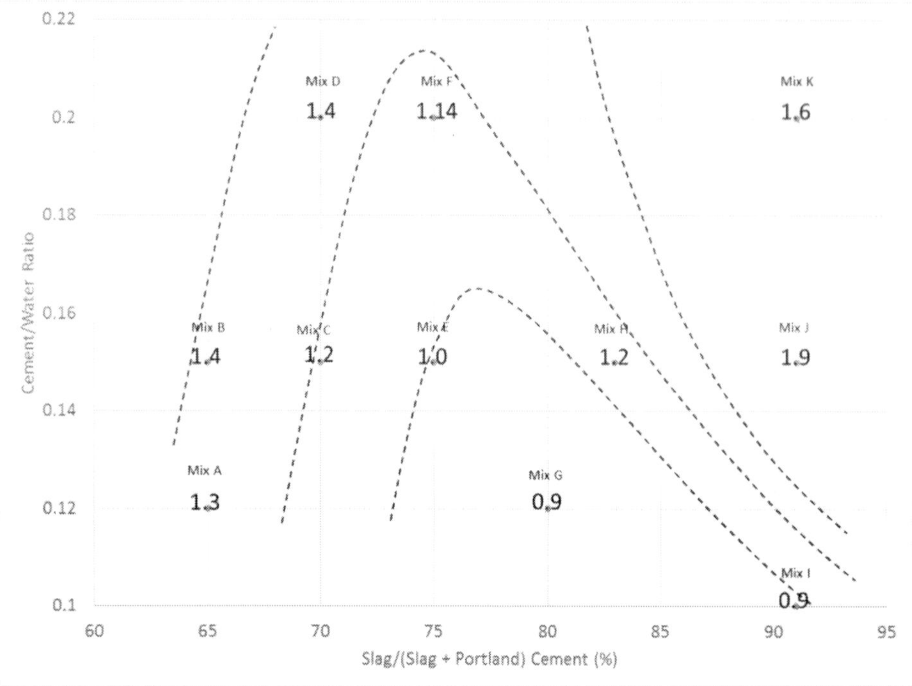

Figure 3. Approximate 28-day breakage strain (%) of various mix designs.

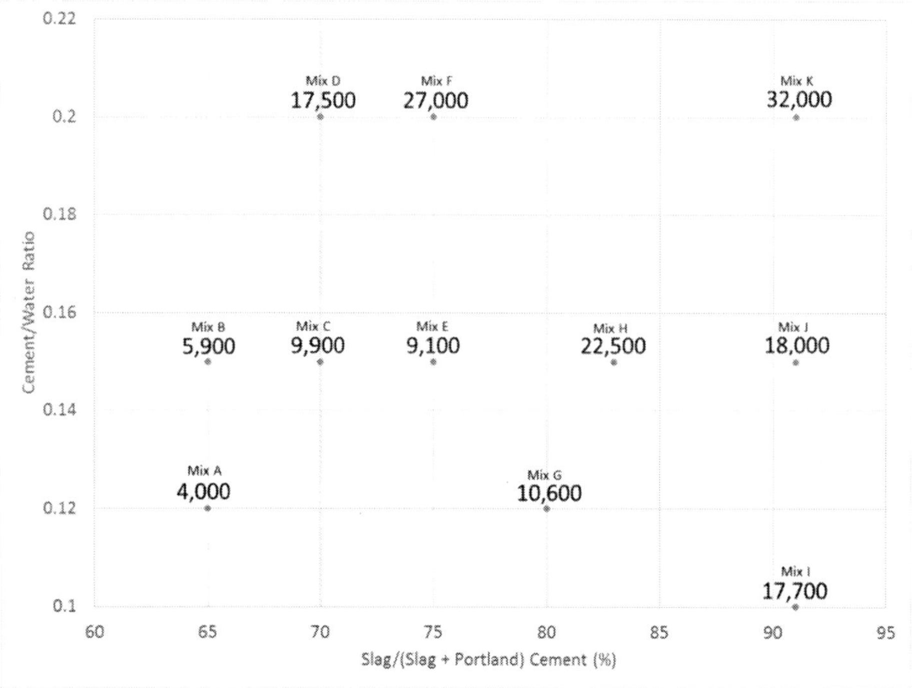

Figure 4. Median 28-day secant Young's modulus (E, psi) of various mix designs.

The bulk (B) and shear (G) static moduli were calculated with the following relationships:

$$B = E / [3 (1 - 2\nu)]$$

$$G = E / [2 (1 + \nu)]$$

where $\nu = 0.35$ was the assumed Poisson Ratio.

Although the dynamic shear modulus is expected to be several times greater than the static modulus, it was conservatively assumed that the secant dynamic modulus was equal to the static shear modulus. Anyway, it was assumed that the choice of wall deformation parameters has an insignificant effect on the global deformation pattern.

Soil-Bentonite (SB)

There are no tests available on the soil-bentonite material, but it is considered that, when saturated, behaves like a soft clay and can be modeled as Mohr-Coulomb material. There may be more appropriate constitutive models for modeling the soil-bentonite material than the Mohr-Coulomb model, but they would require input parameters that are not available. For FLAC evaluation purposes, a shear wave velocity should be associated with this material. Classical correlations with SPT blowcounts for clay and low N_{60} values were considered.

Table 4. Assumed Shear Wave Velocity, V_s

Reference	Equation	V_s (fps) for:	
		$N_{60} = 2$	$N_{60} = 5$
Ohta and Goto, 1978	$3.28 \cdot 82.4\ N_{60}^{0.34}$	342	467
Imai and Tonouchi, 1982	$3.28 \cdot 95.7\ N_{60}^{0.25}$	373	469
Average values:		358 (~ 360)	468 (~ 470)

It was found that the estimated $V_s = 470$ fps based on $N_{60} = 5$ is very similar to that for the SCCB Mix A, where the back-calculated Vs, from the moduli determined through testing, was 463 fps; therefore, SCCB Mix A results were considered an upper bound for SB material, and the evaluation performed under the present analyses, based on $N_{60} = 2$, a lower bound.

MODELING STATIC EQUILIBRIUM

Table 5 describes the detailed steps used in simulation of construction of the levee and cutoff wall, using the computer software FLAC 7.0.

Table 5. Construction Simulation for Modeling Static Equilibrium

Step	Description
1	Construct mesh
2	Bring original foundation to equilibrium. Ground Elevation 18 on the canal side, approximately 20 under the levee, and 23 on the protected side. Assume WSE = 6.9 for canal excavation in dry conditions.
3	Excavate the canal to Elevation 6.9, in stages. [1)]
4	Construct the existing levee up to Elevation 35.9, where the platform for cutoff wall construction will be built; it is considered that raising the levee to the actual current crest elevation and cutting down to the platform does not significantly affect the stress/strain condition. [1)]
5	Construct the cutoff wall in one single stage.
6	Build the landside addition and raise the levee up to Elevation 43, in stages. [1)]

Note:[1)] *Bring to approximate equilibrium at end of each stage. Stages for construction and excavation are typically sub-horizontal units that are one row high. Used flow=off.*

Step 1: Finite Difference Mesh

The mesh has essentially 5-foot by 5-foot zones and the following special features:

- Horizontal extent: The mesh was extended on each side of the levee to a distance from the levee toe of about twice the thickness of recent and older alluvium.
- About 40 feet of stiff clay was included at the base of the mesh and modeled as elastic.
- To keep the displacements within the mesh, three columns of zones at each lateral boundary of the mesh were modeled with elastic material.
- At the projected location of the cutoff wall the zones have the horizontal dimension of 2.5 feet, the minimum desired thickness of the wall.

The mesh is shown in Figure 5. Note that the automatically generated mesh was slightly altered for accommodating modeling the 2.5-foot wide cutoff wall between -1.25 and +1.25 from the levee axis.

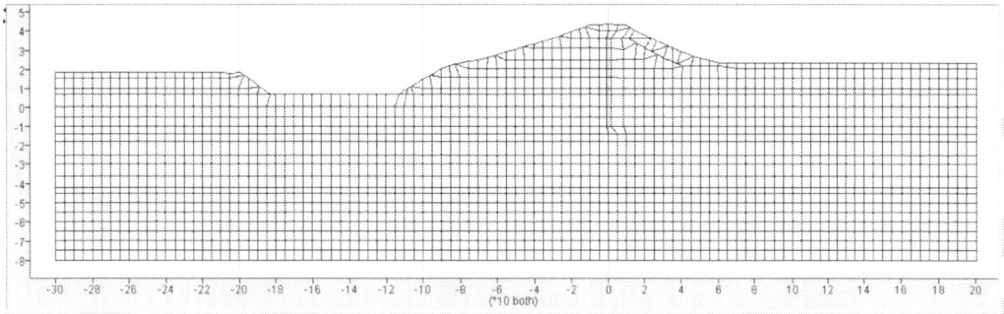

Figure 5. Finite difference mesh for FLAC analysis.

Step 2: Foundation Equilibrium

The foundation consists of 60 - 65 feet of recent and older alluvium (recent C_4, S_2, S_3, and older C_5, S_4, M_1, and S_5) over 38 feet of stiff clay (C_6). The base of the mesh is, therefore, at about 100 feet from the ground surface, as presented in Figure 6.

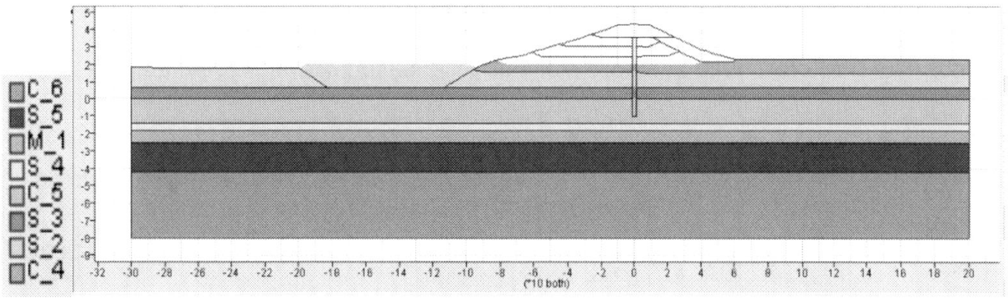

Figure 6. Foundation alluvial soil (Holocene) and portion of stiff clay (Pleistocene) included in the model. Abscissa: distance in feet from levee axis; ordinate: elevation in feet.

The phreatic water surface before the construction of the levee and the canal at the waterside toe was assumed at the Elevation 6.9, i.e. at the bottom of the future canal.

Step 3: Excavate Canal, One Row at a Time

Excavation simulation of the existing canal started from the ground surface and progressed to the assumed bottom of the canal, Elevation 6.9, one row of zones (elements) at a time.

Step 4: Construct Levee

Construct the existing levee up to Elevation 35.9, where the platform for cutoff wall construction will be built; it is considered that raising the levee to the actual current crest elevation and cutting down to the platform does not significantly affect the stress/strain condition. Construction simulation of the existing levee started from the ground surface and progressed to the top of the levee, building the embankment one row at a time. The section is presented in Figure 7.

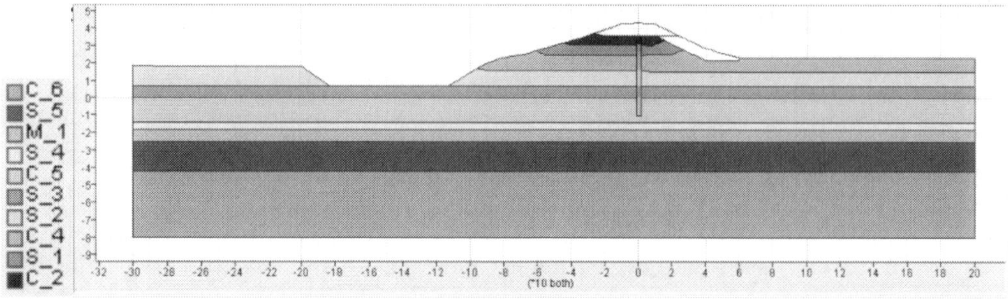

Figure 7. Existing levee after preparation for cutoff wall installation; crest Elevation: 35.9.

Steps 5 and 6: Construct Cutoff Wall and Raise Levee

Construction of the cutoff wall is simulated in one single stage. Building the landside addition and raising the levee up to Elevation 43, was performed in stages, one row at a time. The final cross section before dynamic loading simulation is presented in Figure 8.

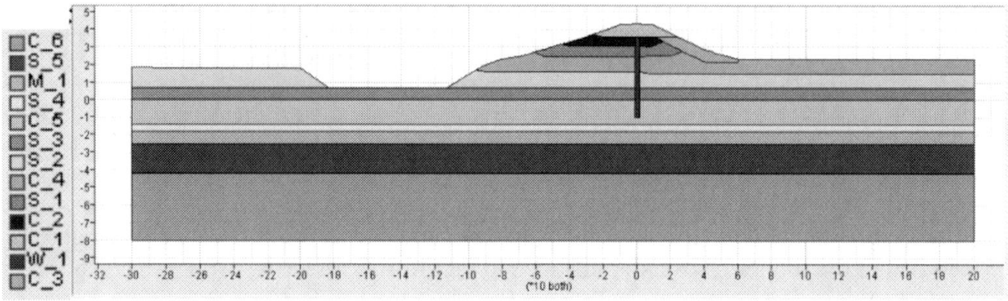

Figure 8. Analyzed cross section for seismic behavior. See also Figure 1 for description of the soil materials.

Static Equilibrium for Coincident WSE of 16.8

The Water Surface Elevation (WSE) = 16.8 ensures saturation of both potentially liquefiable layers, S_2 and S_3, with minimum stabilizing water pressure on the canal side toward the levee. This WSE is also approximately the highest of average summer or winter water table elevation, which is considered an appropriate coincident water level for a 200-year seismic event.

The corresponding pore water pressure and stress state distributions are shown on Figures 9 to 12.

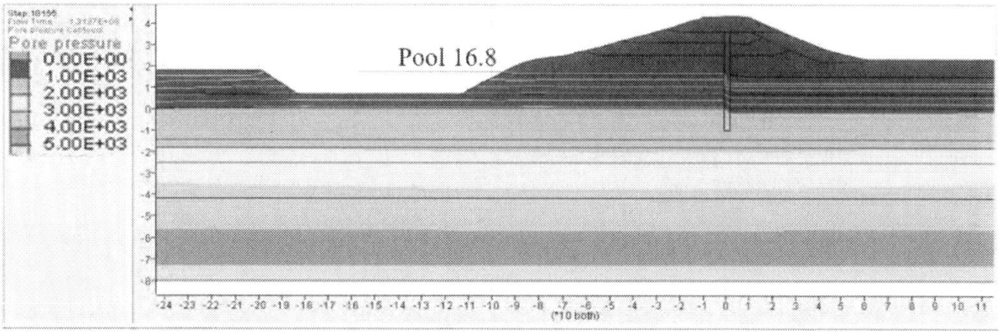

Figure 9. Pore pressure (psf) before simulation of the dynamic loading, with WSE=16.8.

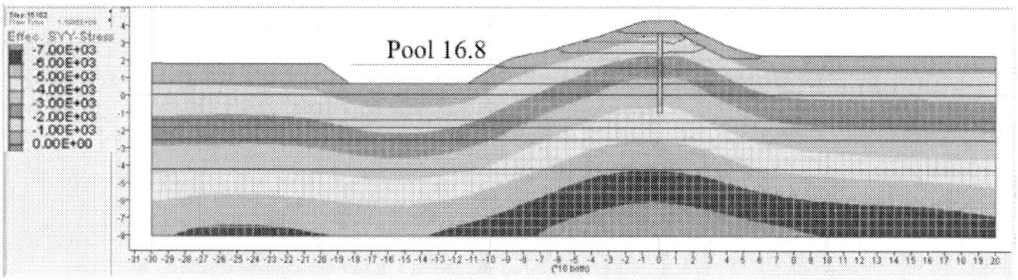

Figure 10. Effective vertical stress (psf) with WSE = 16.8.

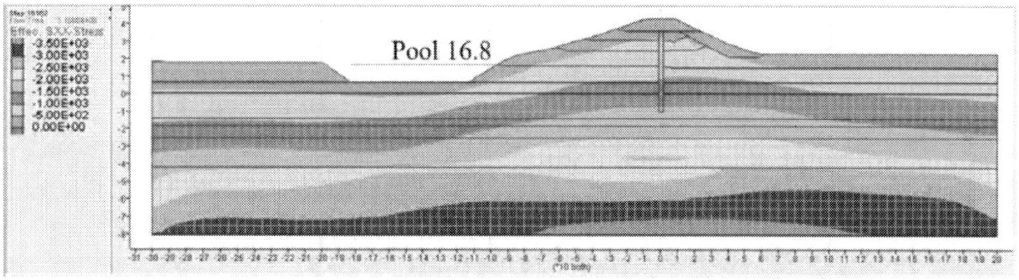

Figure 11. Effective horizontal stress (psf) with WSE = 16.8.

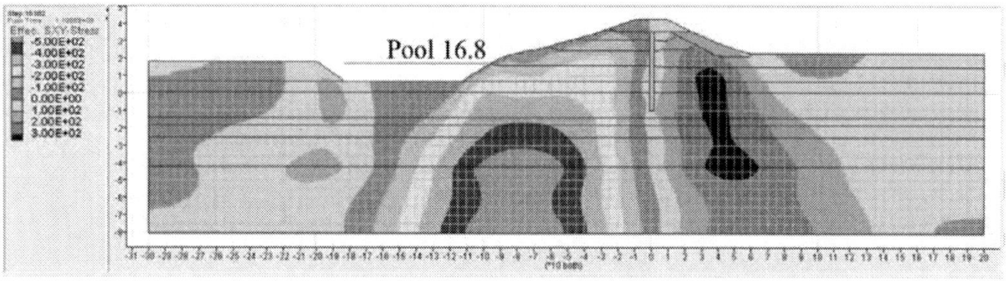

Figure 12. Shear stress (psf) on horizontal/vertical planes with WSE = 16.8.

Higher pools were also considered, as presented in Figures 13, 14, and 15.

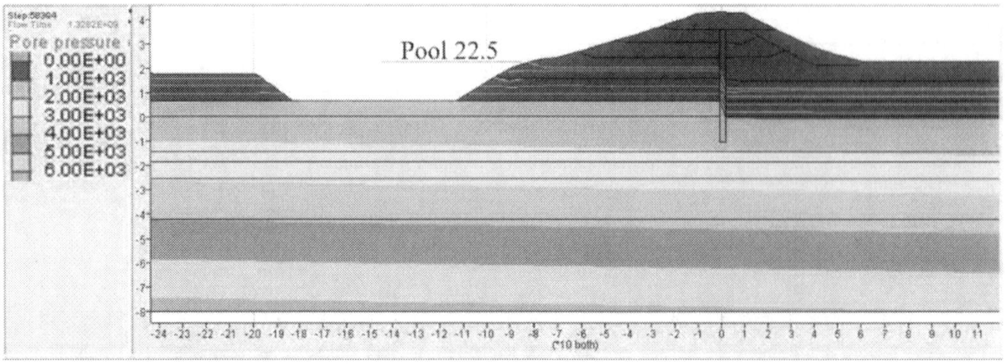

Figure 13. Pore pressure distribution (psf) with WSE= 22.5.

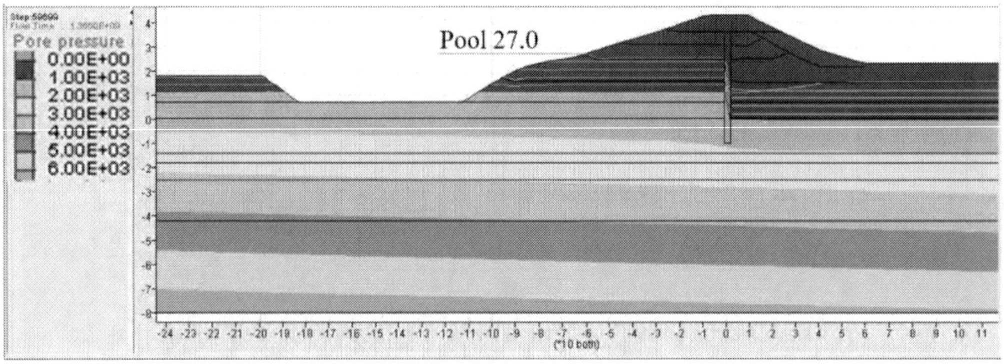

Figure 14. Pore pressure distribution (psf) with WSE= 27.0.

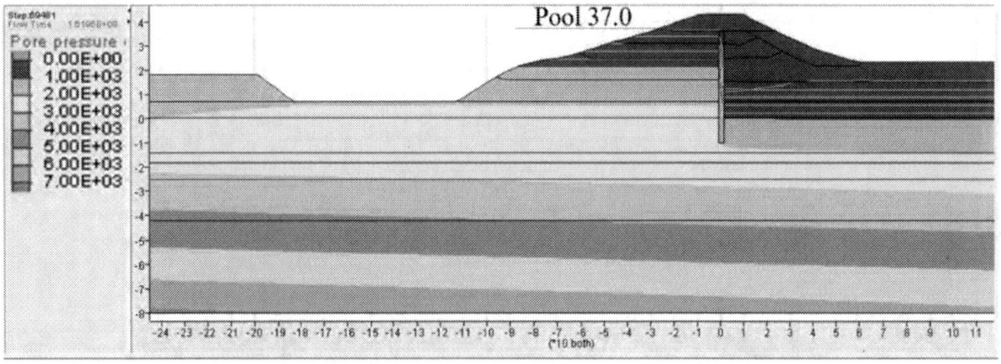

Figure 15. Pore pressure distribution (psf) with WSE= 37.0.

Pre-dynamic Condition

At the end of the static stage of the analysis, the model for liquefiable alluvium was switched from Mohr-Coulomb to UBCSAND (UBCSAND003.dll version). Figure 16 shows the zones where various constitutive models were used. With the assumption of WSE = 16.8 both S_2 and S_3 may become saturated. $(N_1)_{60,cs}$ = 10 was assigned to both these materials.

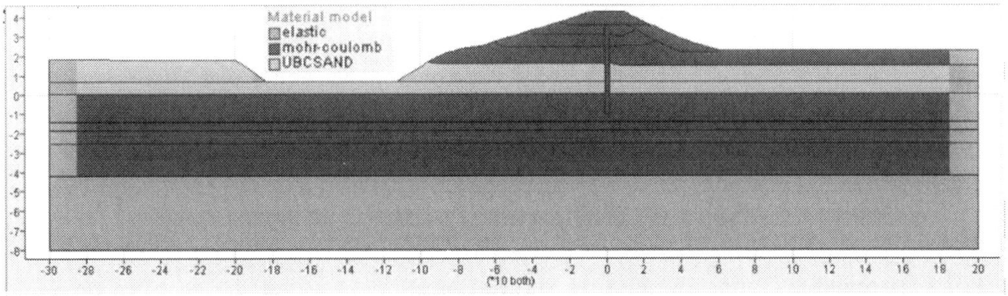

Figure 16. FLAC constitutive models with WSE = 16.8; for avoiding instability problems with triangular zones at the bottom of the canal, they were modeled as Mohr-Coulomb.

Layer S_2 is modeled with two rows of zones and layer S_3 with one row, for a total of three rows for the potentially liquefiable material. Beaty and Perlea (2011) noted that the number of liquefiable rows has a significant impact on the computed seismic deformation. At least four rows are needed for a correct evaluation of the excess pore pressure increase and, accordingly, the time of liquefaction occurrence and the extent of the liquefied zone. As shown in the referenced paper, this effect can be avoided by preventing the pore pressure dissipation toward adjacent zones, i.e. using flow = off instead flow = on. This option was selected.

UBCSAND is an effective stress constitutive model and using flow = on is generally recommended for consideration of partial dissipation of the pore pressure during shaking. However, in this case using flow = off is not excessively conservative for the following reasons: Dissipation of pore pressure from S_2 toward C_4 would be insignificant, as C_4 is clayey soil; dissipation of pore pressure from S_3 toward C_5 would also be insignificant, because C_5 is clayey soil; the only potentially significant path for pore pressure dissipation that is prevented by using flow = off is from S_3 toward the canal, but liquefaction of the material under the canal has insignificant impact on the levee and its foundation deformation. Therefore, it was considered that using the option of flow = off is not excessively conservative.

Undrained strength (consolidated undrained) parameters were set in saturated materials of relatively low permeability (CL materials); because FLAC assumes by default drained parameters, the Φ_u-angle was replaced with an equivalent cohesion by multiplying $\tan\Phi_u$ by the average total vertical stress (in the middle of each layer, in free field on the landside of the levee); it was checked that this equivalent cohesion plus C_u is less than the drained strength at each location ($C' + \sigma_v' \tan\Phi'$). Figure 17 presents the cohesion values in Mohr-Coulomb zones only, which for CL soils under water is the undrained strength.

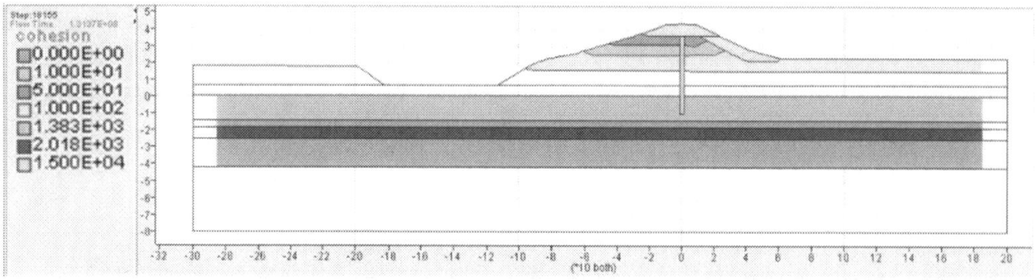

Figure 17. Undrained shear strength (psf) in CL saturated materials and effective cohesion in all other materials.

MODELING EARTHQUAKE ACTION

Time History for Dynamic Loading Simulation

For the project location, in Sacramento, California, the USGS grid data published in 2014 give the Peak Horizontal Acceleration, PHA = 0.1g for an approximate 200-year earthquake. The dominant magnitude for a 200-year event in that area is 6.5. It was, therefore, considered appropriate to develop the analysis time history from the 1987, Superstition Hills strong motion records at USGS 5052 Plaster City Station with the following characteristics:

Table 6. Key Characteristics of Superstition Hills, 1987, USGS 5052 Plaster City Station Seed Motion

Earthquake Motion Characteristics	Component 045 Degrees	Component 135 Degrees
Magnitude (M)	6.5	
Closest distance (km)	22.2	
Mechanism	Strike-slip	
Average shear wave velocity in top 30m, V_{s30} (m/s)	345	
Peak horizontal acceleration, PHA (g)	0.121	0.186
Peak ground velocity, PGV (cm/s)	9.48	20.64
Peak ground displacement, PGD (cm)	1.91	5.43
Arias Intensity (m/s)	0.30	0.63
Strong Shaking Duration (s)	~5	

The critical component and its most damaging polarity (045 component, negative polarity) were determined by comparing damages provoked to the analyzed cross section by time histories scaled to the same peak acceleration. The time history was scaled so that the desired PHA = 0.1g being induced at the top of the Pleistocene formation (noted C_6 – Stiff Clay in Figure 1), assimilated with bedrock. For achieving this goal, the time history applied at the base of the mesh should correspond to an accelerogram with the peak acceleration of 0.15g.

Dynamic Loading

The ground motion was applied to the base of the model using a compliant boundary. This was achieved by using a quiet boundary condition in both the X- and Y-directions and then applying the horizontal ground motion as an equivalent shear wave.

FLAC uses a viscous boundary scheme consisting of two sets of dashpots attached independently to the mesh in the normal and shear directions. In the shear direction the dashpots provide a viscous shear traction given by:

$$\sigma_S = \rho \; V_S \; v_S$$

where ρ and V_s are the density and shear wave velocity of the base material, and v_s is the shear-component of particle velocity at the boundary.

$$\rho = \gamma_{sat} \,/\, g = 130 \,/\, 32.185 = 4.039 \text{ slugs/cf}$$

$$V_s = 1200 \text{ fps}$$

Post-Earthquake Static Equilibrium

After the end of earthquake and 5 seconds of quiet time, UBCSAND model was switched back to Mohr-Coulomb everywhere the excess pore pressure ratio reached values greater than 0.7 (where liquefaction practically occurred) at any moment during the simulated earthquake. The residual strength of liquefied material, S_r = 235 psf, was assumed mobilized within the liquefied zones (per Seed and Harder, 1990, for $(N_1)_{60\text{-}cs} = 10$.)

The resulted value of the residual undrained shear strength was capped at the shear strength before the earthquake, which is defined by the effective stress parameters Φ' and C', and the effective stresses before the earthquake. This cap was applicable to the zones with low effective stress only, i.e. immediately below the canal bottom.

RESULTS OF THE SEISMIC LOADING ANALYSES

With the exception of stress and strain within the cutoff wall, the response of the cross section to the seismic loading was similar with all SCCB mix design and SB (soil-bentonite) option. Therefore, the general behavior of the levee and its foundation will be presented in what follows for one option only: design earthquake (PHA = 0.1g on top of the stiff clay, WSE = 16.8, SCCB mix design H).

Induced Liquefaction

Figure 18 presents the maximum ever occurred excess pore pressure ratio, defined as the ratio between the pore pressure in excess of static, divided by the initial (before earthquake) vertical effective stress: $r_u = (\sigma_{v0}' - \sigma_v') / \sigma_{v0}'$. This parameter was computed in the zones modeled with UBCSAND only. It is noted that per UBCSAND model the level $r_u = 0.7$ is considered 100% liquefaction.

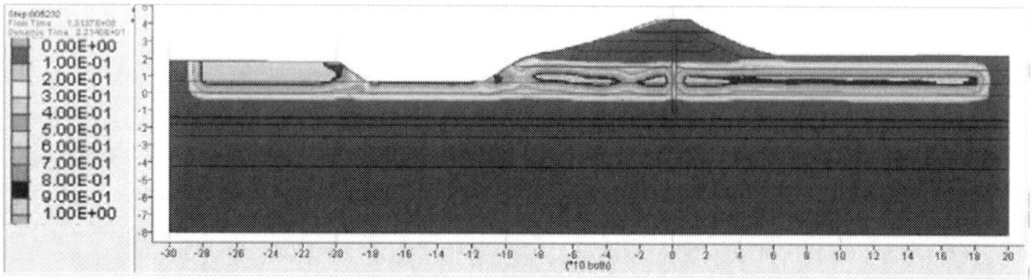

Figure 18. Pore pressure in excess ratio at the end of shaking (22.14 seconds).

It is evident that the potentially liquefiable zone nearly completely liquefied. It should be noted that in the plot of r_u the values plotted along the boundaries of the liquefiable layer are the average of r_u in the liquefiable layer and r_u = zero in the non-liquefiable material. For example, if r_u was 0.8 on the liquefiable side of the boundary, the average $r_u = (0.8 + 0) / 2 = 0.4$ was plotted.

Figure 19 shows the zones that were not considered liquefied in the post-earthquake stage, so that the constitutive model was not switched from UBCSAND to Mohr-Coulomb (zones in yellow).

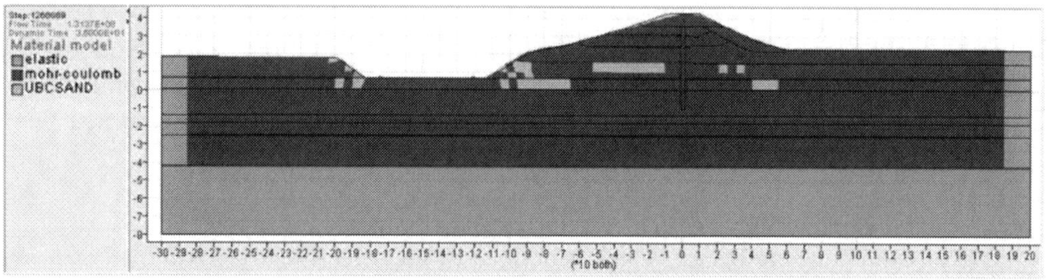

Figure 19. Constitutive models in the post-earthquake stage.

Displacements and Shear Strain

Final results (after post-earthquake stage) are presented in Figures 20, 21, and 22.

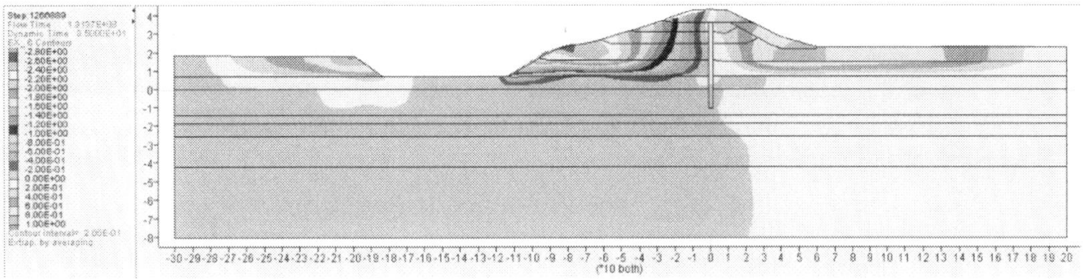

Figure 20. Horizontal displacement – 0.2-foot contour intervals.

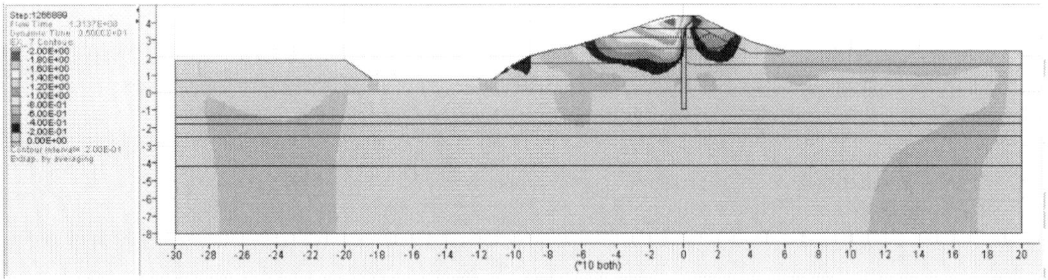

Figure 21. Vertical displacement – 0.2-foot contour intervals.

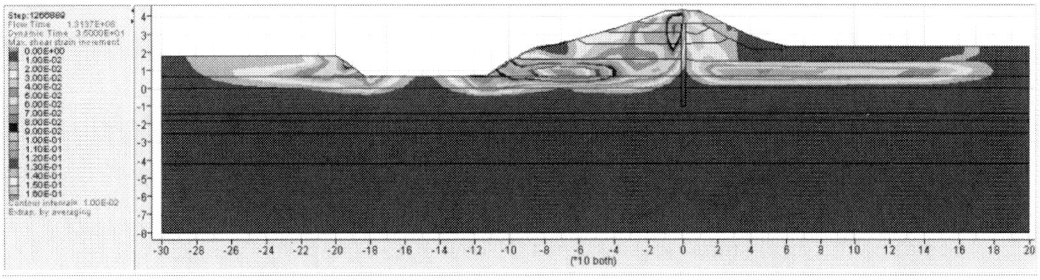

Figure 22. Maximum shear strain at the end of analysis, – each contour is equal to a shear strain γ of 2%.

Strain Affecting the Cutoff Wall

Locating the cutoff wall near the levee axis makes it less vulnerable to displacements in the adjacent soil materials than in any other location. However, the symmetry of a levee cross section is never perfect: the slopes are usually different, the ground surface and the thickness of the liquefiable layers may differ landside and waterside, etc. As a consequence of the unavoidable asymmetry, the cutoff wall would experience displacements during the earthquake shaking and the post-earthquake stage that may impact the wall integrity. This possible occurrence depends significantly on the wall material properties. The following graphs (Figure 23) present for the same variant as considered before (PHA = 0.1g on top of the stiff clay, WSE = 16.8, SCCB mix design

H) the time histories of the shear strain at various elevations within the cutoff wall. They are grouped within three portions of the wall: upper, middle and lower.

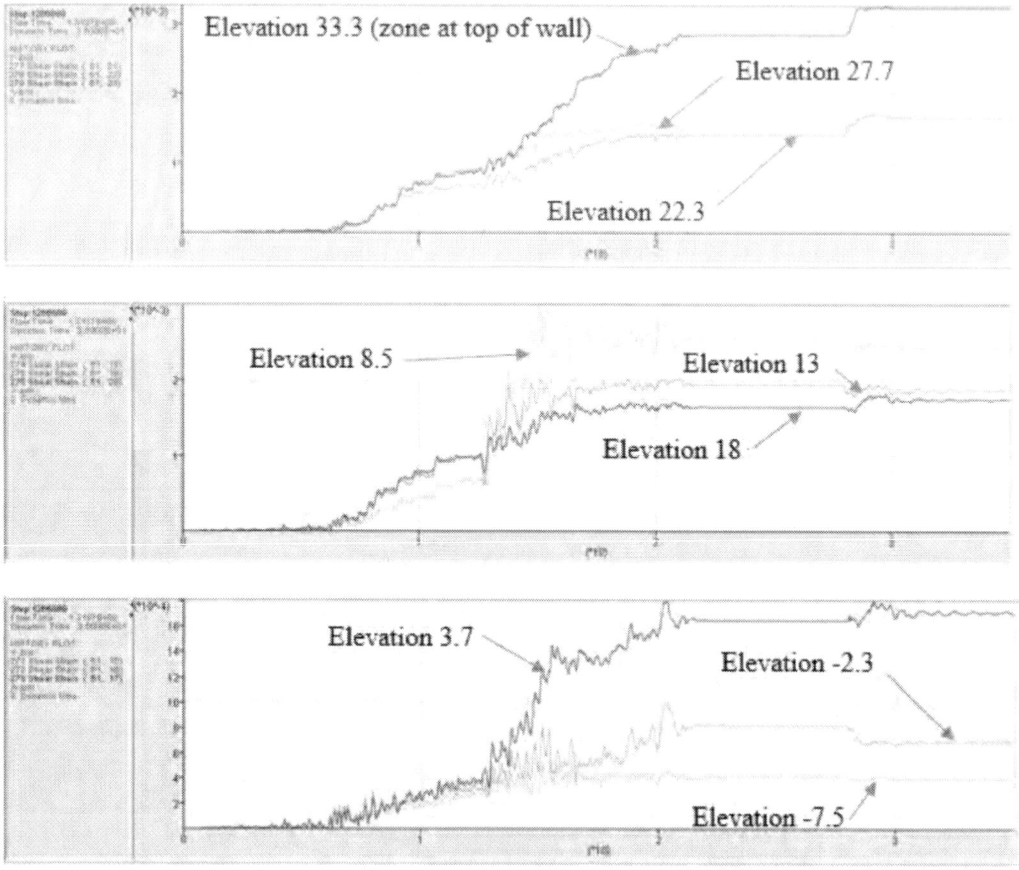

Figure 23. Shear strain time histories (PHA = 0.1g on top of bedrock, WSE = 16.8, SCCB mix design H). Abscissa: time in seconds; ordinate: shear strain (feet/feet).

Note that the dynamic loading occurred between zero and 22.14 seconds and the post-earthquake stage started at 28 seconds. Near the top of the wall (Elevations 33.3, 27.7, and 22.3) the maximum shear strain occurred during the post-earthquake stage, due probably to significant displacements once the residual strength had been mobilized. At lower elevations the maximum shear strain occurred generally during shaking.

Figure 24 shows the variation with depth of the wall of the maximum shear strain that occurred any time, during shaking or in post-earthquake stage. Plots for all mix design cases are shown. The graphs are grouped based on similarity in maximum shear strain value.

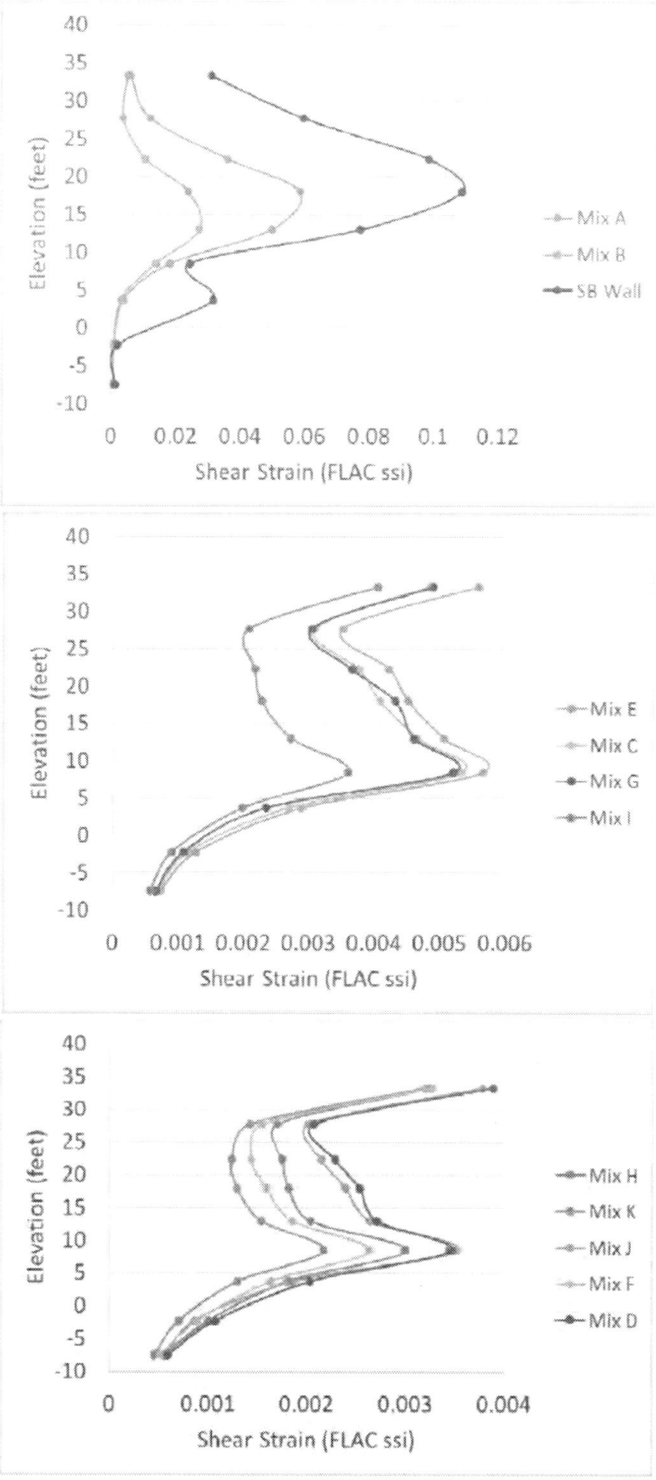

a

b

c

Figure 24. Variation with depth of shear strain within the wall for various mix designs.

From the graphs in Figure 24 it is evident that the peak shear strain occurred generally in the middle height of the wall, Elevation 8 to 9, with the exception of weak mixtures, A, B, and soil-bentonite (presented in Figure 24,a) where the peak strain occurred in the upper half of the wall. It is noted that the liquefiable layers were approximately located between Elevations 16 and 0; WSE was 16.8.

The evaluated maximum shear strains were plotted as a function of mix parameters in Figure 25.

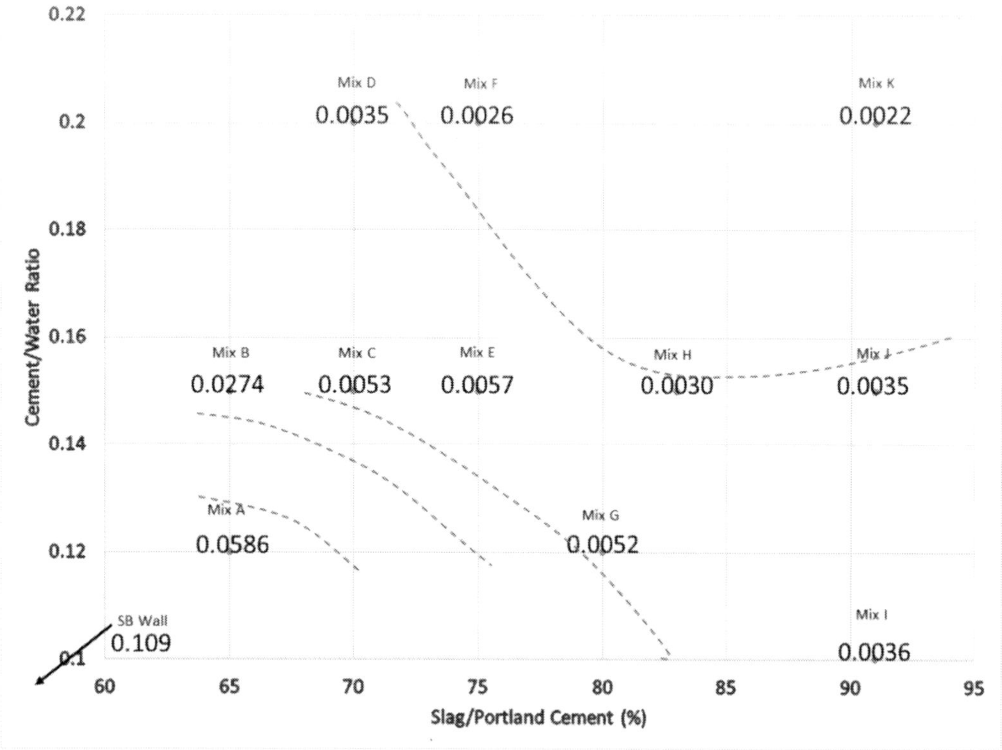

Figure 25. Maximum shear strain (FLAC ssi) for various wall materials and WSE = 16.8 case.

It is evident that the stronger the wall is (per Figure 2) the smaller are the maximum shear strains. The maximum shear strain in the wall built with Mix A (6%) is an order of magnitude greater than if Mix E were used (0.6%). Walls with Mix F and Mix K experienced the lowest values of maximum shear strain (less than 0.3%). The maximum computed shear strain, of the order of 11%, was obtained with the soil-bentonite wall.

CONCLUSIONS

Safety against Breakage

The FLAC "ssi" is defined as follows (per FLAC 7.0 Manual):

"For the purposes of printing and plotting, the term 'maximum shear strain' means the radius of the Mohr's circle in the *xy*-plane, as illustrated in Figure 26. Thus, for conditions of two-dimensional plane-strain analysis, the maximum shear strain, γ, is defined as

$$\gamma = \frac{1}{2}\left(\left(e_{xx} - e_{yy}\right)^2 + 4e_{xy}^2\right)^{1/2}$$

This is the equation used for calculating the maximum shear strain values, **ssi** (strains derived from displacements)."

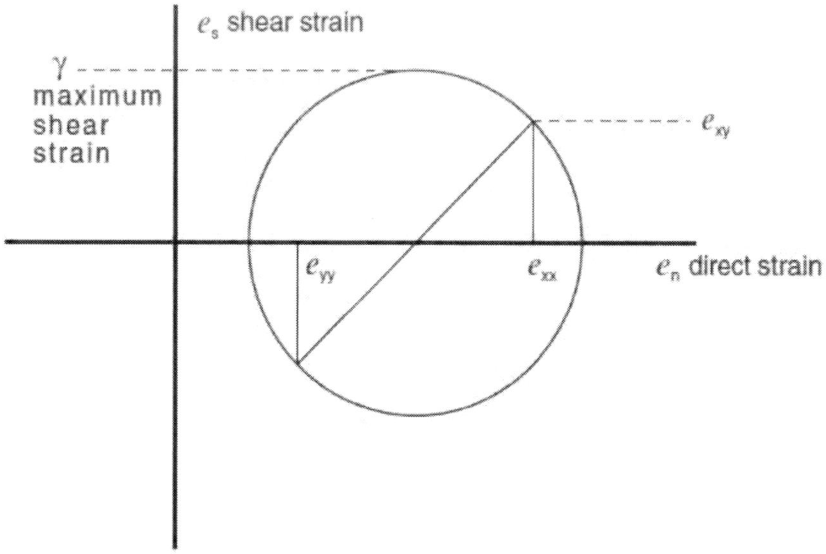

Figure 26. Mohr's circle of strain.

From Figure 3 it is evident that an axial strain, e_{xx}, of about 1.2% would break Mix H, for example; that value was derived from uniaxial compression, where $e_{yy} \leq 0$; it results that e_{xx} in the unconfined compression cannot be greater than 2γ, as defined by the FLAC ssi. A limit e_{xx} of 0.012 (1.2%) corresponds to a limit ssi of 0.006 (0.6%) or slightly more.

With the triaxial UU tests, as used with derivation of the breakage strain for mix designs D, E, and G, comparing the FLAC ssi with half of axial strain at failure should also be a good approximation. Comparing the maximum ssi obtained in various evaluated cases with half the slightly conservative limit strain presented in Figure 3 would indicate the likelihood of the shearing of the wall and can be viewed as a safety factor (SF).

Table 7 summarizes the results of the main results of the computations and presents the safety factor against failure of the cutoff wall following the action of the design earthquake.

Table 7. Summary of the Shear Strain Evaluations

Cutoff Wall Material	Maximum Allowable ssi	Induced ssi by the Design Earthquake	Factor of Safety against Breakage
A	1.3 x 0.005 = 0.0065	0.0586	0.1
B	1.4 x 0.005 = 0.007	0.02737	0.3
C	1.2 x 0.005 = 0.006	0.005333	1.1
D	1.4 x 0.005 = 0.007	0.003457	2.0
E	1.0 x 0.005 = 0.005	0.005665	0.9
F	1.14 x 0.005 = 0.0057	0.002635	2.2
G	0.9 x 0.005 = 0.0045	0.005215	0.9
H	1.2 x 0.005 = 0.006	0.002996	2.0
I	0.9 x 0.005 = 0.0045	0.003604	1.2
J	1.9 x 0.005 = 0.0095	0.003523	2.7
K	1.6 x 0.005 = 0.008	0.002175	3.7

Mix designs A and B appear not safe under the 200-year earthquake action if both potentially liquefiable layers (S_2 and S_3) may become saturated (i.e., with the evaluated WSE = Elevation 16.8).

Mix designs C, E, G, and I are marginally adequate, with a computed safety factor around 1.0 for the assumed coincident WSE of Elevation 16.8.

Mix D and Mix H, with the compressive strength in the range of 210 to 220 psi, appear to be safe under the ground motion of 0.15g that induced the target PHA = 0.10g, even if considering the conservative assumption of the dynamic shear modulus equal to the static shear modulus as determined from the unconfined compression test. However, they may become unsafe if the 200-year seismic event occurs during flood condition.

Mix designs F, J, and K, with compressive strength in excess of 230 psi, were determined safe with respect to seismic breakage, but have brittle behavior that should generally be avoided.

The soil-bentonite wall, modeled as made of soft clay, experienced under the design earthquake higher shear strain than any SCCB option, of the order of ssi = 0.11 (up to 22% axial strain); however, because the material is not brittle, it may remain functional, at least in zones where it was saturated during the earthquake.

The analysis for this study have shown that a peak compressive strength of 100 psi or greater of the wall material would have relatively low cracking potential during a 200-year design earthquake if it is not coincident with a flood event.

There is an evident correlation between the factor of safety against seismic breakage following the design earthquake action and the strength of the wall material, as shown in Figure 27.

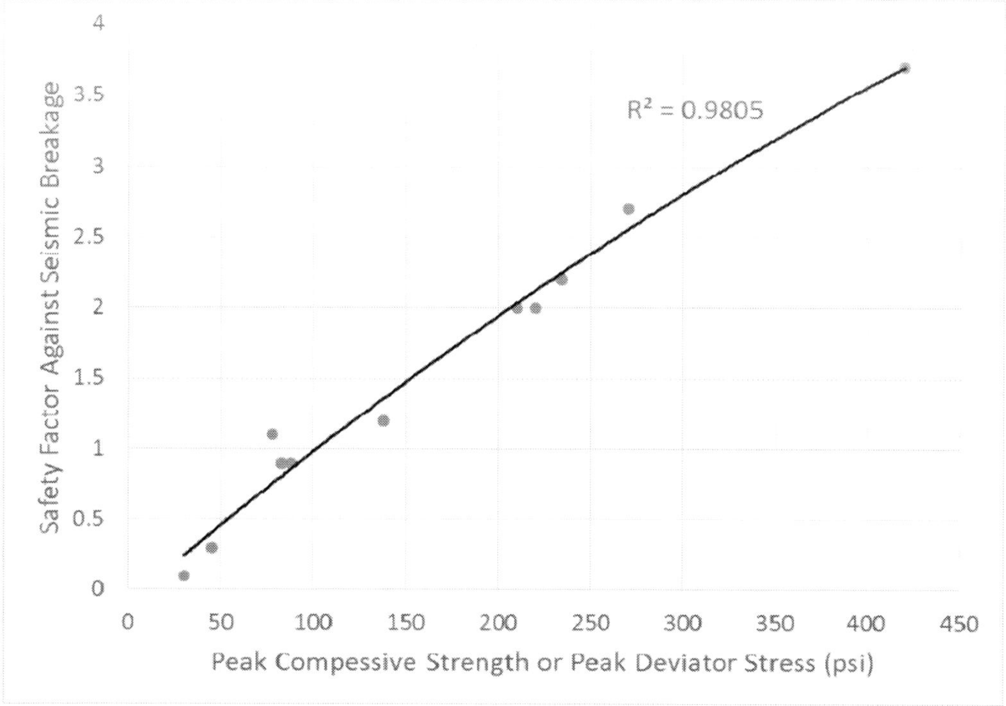

Figure 27. Factor of safety against breakage under the action of the design earthquake versus peak compressive strength of various mix designs.

Sensitivity to Assumed Coincident WSE

As previously stated, the water table (Elevation 16.8) coincident with the seismic event was assumed slightly below the ground surface (that exists at Elevation 18 on the water side and Elevation 23 landside), saturating the potentially liquefiable layers (their top elevation was between 15 and 16 under the levee). Because the 200-year earthquake may occur during higher water elevation stages, it was considered of interest to evaluate the effect of the assumed coincident WSE on the factor of safety against seismic breakage.

This evaluation was done for Mix H and the results presented in Figure 28 were obtained.

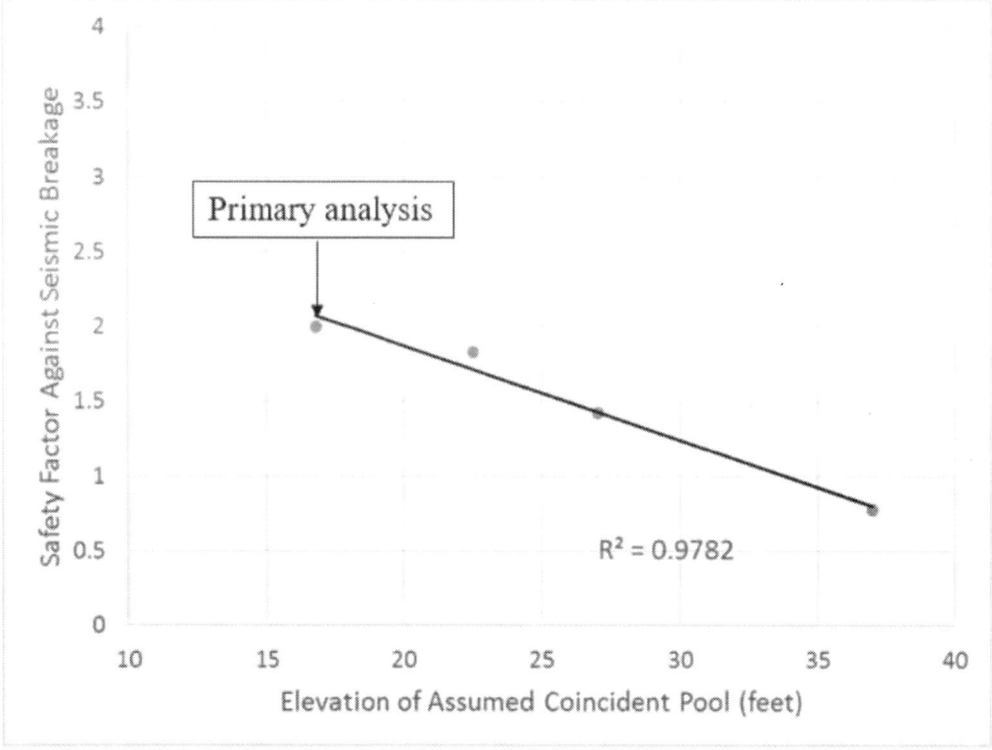

Figure 28. Factor of safety versus the assumed coincident WSE.

For the particular case of Mix H the factor of safety dropped with increased WSE reaching the threshold value of 1.0 at about WSE = 34 (the crest of levee is Elevation 43.1).

Additional Notes

The seismic deformation analyses in this study have been performed to evaluate possible effects of a 200-year earthquake event if the cutoff wall properties match with laboratory test derived properties. The actual cutoff wall properties depend on excavation methods, sediment mixed with slurry (amount and type), among other factors.

The authors are grateful to Darren Mack and John Rice for their careful evaluation of the draft paper and their comments and suggestions that improved the clarity of presentation.

REFERENCES

Beaty, M.H. and Byrne, P.M., 2011. UBCSAND Constitutive Model, Version 904aR. Documentation Report: UBCSAND Constitutive Model on Itasca UDM Web Site.

Beaty, M.H. and Perlea, V.G., 2011. Several Observations on Advanced Analyses with Liquefiable Materials, 31st USSD Annual Meeting and Conference, San Diego, 11-15 April 2011.

Chowdhury, K.H., Kynett, M., Millet, R., Nixon, R., Walberg, F., and Tollefson, N., 2016. The Use of Slag Cement in CB Seepage Cutoff Walls for Levees, in Proceedings, Dam Safety 2016 National Conference, ASDSO, Philadelphia, PA.

Imai, T. and Tonouchi, K., 1982. Correlation of N-value with S-wave velocity and shear modulus, in *Proceedings, Second European Symposium of Penetration Testing,* Amsterdam, pp 67-72.

Itasca, 2011. "FLAC, Fast Lagrangian Analysis of Continua, Version 7.0", Itasca Consulting Group, Minneapolis, Minnesota, USA.

Ohta, Y. and Goto, N., 1976. Estimation of S-wave velocity in terms of characteristic indices of soil, in *Butsuri-Tanko,* 29(4), pp 34-41.

Seed, R. B. and Harder, L. F., 1990. SPT-based analysis of cyclic pore pressure generation and undrained residual strength, in *Proceedings, Seed Memorial Symposium*, J. M. Duncan, ed., BiTech Publishers, Vancouver, British Columbia, pp 351–376.

URS, 2008. Levee Seismic Vulnerability Assessment Methodology. Urban Levee Geotechnical Evaluations Program Contract 4600007418, Task Order 8, June 2008.

USGS, 2014. http://earthquake.usgs.gov/hazards/products/conterminous/2014/data/

Youd, T.L., et al, 2001. Liquefaction resistance of soils: Summary report from the 1996 NCEER and 1998 NCEER/NSF workshops on evaluation of liquefaction resistance of soils, *Journal of Geotechnical and Geoenvironmental Engineering, ASCE*, Vol. 127, No. 10, October, pp 817-833.

ADVANCES IN DESIGN PROCEDURES FOR RELIEF WELL LINES IN VARIABLE URBAN SETTINGS

Robert A. Jaeger, PhD, PE[1]
Richard A. Keizer, Jr.[2]
Graham C. Bradner, PG, CEG, CHg, PMP[3]
Joseph P. Weber, PhD, PE[4]
Mark H. Stanley, PE, GE, D.GE[5]

ABSTRACT

Lines of equally-spaced relief wells can often be evaluated and designed using simplified design procedures. However, design of relief wells in an urban environment can pose significant technical challenges with respect to variable relief well spacing and offset from the landside levee toe, which are not readily accounted for in simplified relief well design procedures. Additionally, considerations such as variable aquifer conditions, topography, well screen lengths, and interfacing with cutoff walls are beyond the capabilities of simplified design procedures and thus require more advanced analyses.

This paper describes a staged analysis and design approach developed and implemented to design lines of relief wells along the Sacramento River East Levee (SREL) in northern California. The approach incorporates three analysis stages: (1) simplified analyses with the USACE EM 1110-2-1914 procedure, (2) SEEP/W plan view analyses assuming fully-penetrating wells, and (3) SEEP/W plan view analyses accounting for partially-penetrating wells, if applicable. For Stage 1, the USACE simplified procedure was implemented with corrections and modifications as presented in Guy et al. (2010) and Guy et al. (2014). Stage 2 analyses included SEEP/W plan view analyses assuming fully-penetrating relief wells with efficiency-related head losses calculated with the approach presented by Guy et al. (2010). Stage 3 analyses for partially-penetrating wells were analyzed by applying flow rate boundary conditions, where flow rates were estimated from Stage 2 analyses and partial-penetration flow factors per USACE EM 1110-2-1914.

An application of the proposed approach to the SREL is presented. Comparisons of aquifer heads and relief well flow rates from Stage 1 and Stage 3 analyses indicated similar results for wells near the middle of the relief well lines, but significant differences were calculated at the ends of the relief well lines due to end around effects and interactions with existing relief wells.

[1] Senior Engineer, GEI Consultants, Inc., 2868 Prospect Park Drive, Suite 400, Rancho Cordova, CA 95670, rjaeger@geiconsultants.com
[2] Project Engineer, GEI Consultants, Inc., 2868 Prospect Park Drive, Suite 400, Rancho Cordova, CA 95670, rkeizer@geiconsultants.com
[3] Senior Geologist, GEI Consultants, Inc., 2868 Prospect Park Drive, Suite 400, Rancho Cordova, CA 95670, gbradner@geiconsultants.com
[4] Staff Engineer, HDR, Inc., 2365 Iron Point Road, Suite 300, Folsom, CA 95630, Joseph.Weber@hdrinc.com
[5] Senior Technical Advisor, HDR, Inc., 2365 Iron Point Road, Suite 300, Folsom, CA 95630, Mark.Stanley@hdrinc.com

INTRODUCTION

The Sacramento River East Levee (SREL) system in Sacramento, California is composed of earthen embankments originally constructed in stages from the mid-1800s to early-1900s. At that time, levees were often constructed on the flood plain along the top of the existing natural levee on top of the riverbank, without much setback from the river channel. During the last three decades, the Sacramento Area Flood Control Agency (SAFCA) and the United States Army Corps of Engineers (USACE) have designed and constructed improvements along several portions of the subject levee segments. Improvements have included shallow and deep seepage cutoff walls, relief wells, berms, and erosion protection.

Recently, GEI Consultants, Inc., in association with HDR, Inc. (GEI/HDR), was contracted to evaluate 8.1 miles of the SREL previously found to not meet Federal Emergency Management Agency (FEMA) 100-year criteria (44 CFR 65.10) for geotechnical embankment and foundation stability, as shown in Figure 1. GEI/HDR collected additional information and reevaluated the 8.1 miles of the SREL relative to FEMA criteria and California's Department of Water Resources (DWR) Urban Levee Design Criteria (ULDC) dated May 2012 (DWR 2012). The evaluation considered the results of geotechnical seepage and stability analyses for existing conditions, understanding of foundation conditions and existing remediations, as well as a review of available records of past performance. Based on the geotechnical evaluations, GEI/HDR proposed improvements to 5.9 miles of the 8.1-mile study area, which include cutoff walls, relief wells, berms, waterside clay blankets, and waterside erosion and stability repairs. The selection of appropriate methods for repairs and improvements has been constrained by access limitations, lack of undeveloped land beyond the landside levee toe, and a limited amount of waterside berm along the waterside portion of the levee.

Relief wells have been proposed for approximately 1.1 miles of the SREL, including lines of relief wells and irregularly-positioned relief wells. Lines of relief wells were designed using a staged design approach combining relief well design procedures from USACE EM 1110-2-1914 (USACE 1992) with plan view seepage analyses. This paper presents the design approach for lines of relief wells, including an application of the approach to the design of relief wells in Sacramento's Pocket neighborhood; the approach for irregularly-positioned relief wells is presented in a companion paper (Weber et al. 2017).

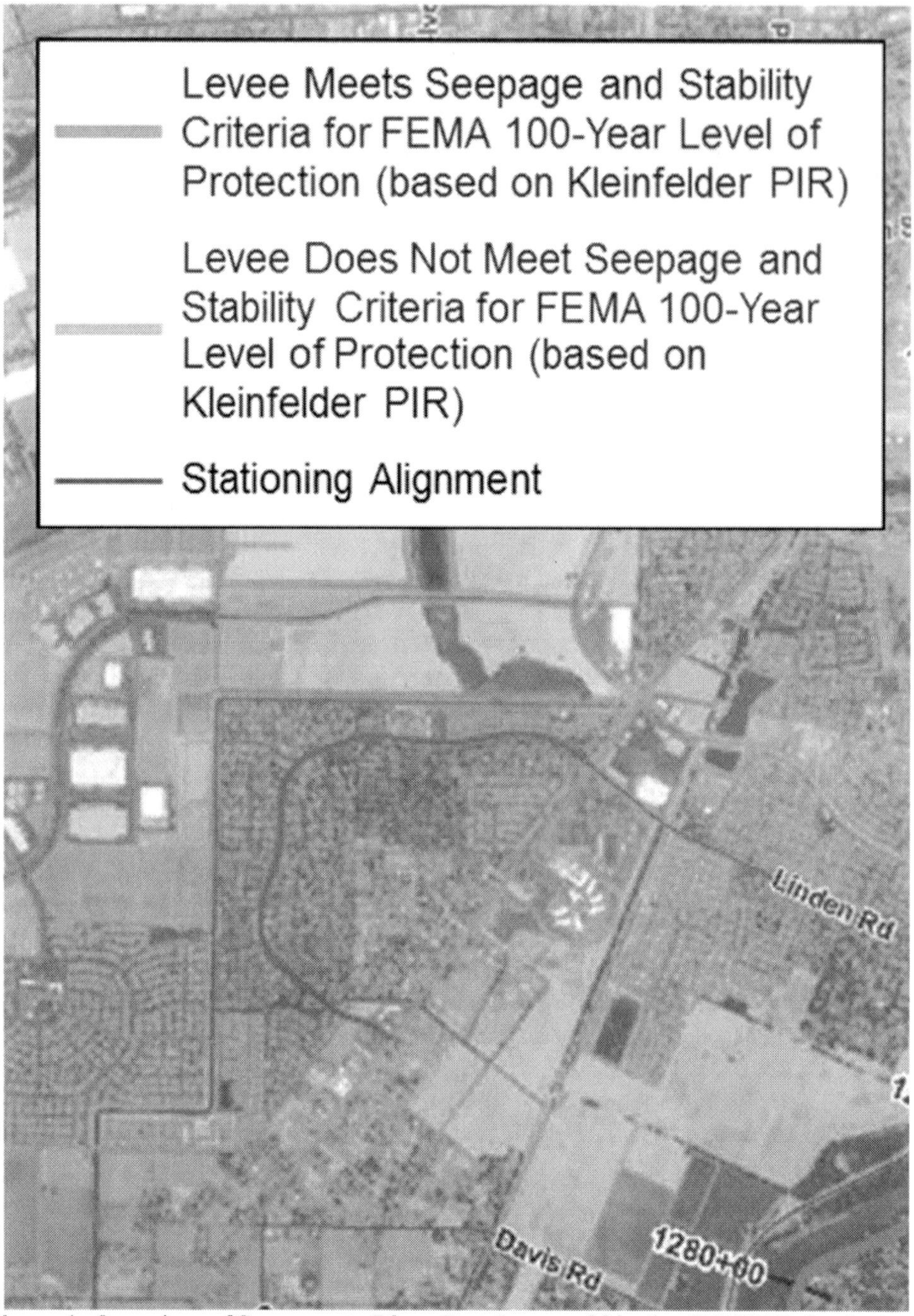

Figure 1. Overview of Sacramento River East Levee within the GEI/HDR Study Area

DESIGN METHODOLOGY

Overview

Relief well lines were designed using a staged analysis approach that combined the blanket theory approach in USACE EM 1110-2-1914 with two-dimensional (2D) plan view seepage analyses in the commercial finite element analysis program SEEP/W (GEO-SLOPE International, Ltd. 2015). A staged analysis approach was used to account for three dimensional effects associated with finite relief well lines, interactions with complimentary underseepage remediations (i.e., existing relief wells, fully-penetrating cutoff walls), and variations of relief well top of riser elevations, well offset from the landside toe, well spacing, relief well penetration (or screened) lengths, aquifer thickness, hydraulic conductivity, and boundary conditions, which are not accounted for in simplified blanket theory design procedures.

The design relief well calculation stages were:

- Stage 1 – Blanket theory analyses at cross sections within the relief well extents to identify combinations of relief well spacing and penetration (or screened) length meeting underseepage criteria.
- Stage 2 – SEEP/W plan view (aquifer) analyses of the design relief well alignment assuming all relief wells were fully-penetrating.
- Stage 3 – SEEP/W plan view reanalyzed by scaling Stage 2 flow rates to account for partially-penetrating relief wells. This stage is only performed if either existing or proposed partially-penetrating relief wells are present.

The design water surfaces, underseepage exit gradient criteria, and design calculation stages are described in detail below.

Design Water Surfaces and Underseepage Criteria

The water surfaces for design included geotechnical design 200-year and hydraulic top of levee (HTOL) water surface elevations (WSEs), which were specified based on levee station (MBK 2016). The geotechnical design 200-year water surface was set 1 ft above the 200-year water surface for added robustness and future sea level rise. The HTOL WSE is defined in the ULDC as the minimum of the 500-year WSE or 3 ft above the 200-year WSE (DWR 2012). For the remainder of the paper, the geotechnical design 200-year WSE will simply be referred to as the 200-year WSE.

For design of relief wells along the SREL, the ULDC maximum allowable exit gradient at the landside toe is 0.50 and 0.60 for the 200-year and HTOL WSEs, respectively, taken at the midpoint between two adjacent relief wells. For locations landward of the landside levee toe, the allowable gradients increase according to the following equations (DWR 2012):

$$i_{allow} = 0.50 + 0.3 * \left(\frac{d_{toe}}{150}\right) \leq 0.80 \text{ for the 200-year WSE} \qquad (1)$$

$$i_{allow} = 0.60 + 0.3 * \left(\frac{d_{toe}}{150}\right) \leq 0.90 \text{ for the HTOL WSE} \quad (2)$$

where d_{toe} is the distance landward from the landside levee toe.

Stage 1 Analyses

Stage 1 included analyses with the USACE EM 1110-2-1914 blanket theory approach for designing lines of relief wells. The purpose of the Stage 1 calculations was to identify combinations of relief well spacing and penetration (or screened) length meeting underseepage criteria. Additionally, the Stage 1 calculations were used to (1) calculate relief well head losses for lines of fully-penetrating wells for use in Stage 2 and Stage 3 analyses, and (2) calculate partially-penetrating flow rate reduction factors (G_p) for use in Stage 3 analyses, if applicable.

Blanket theory analyses of relief well lines were implemented as presented in USACE EM 1110-2-1914, but with modifications to the approach as presented in Guy et al. (2010) and typographical corrections to USACE EM 1110-2-1914 as presented in Guy et al. (2014). Guy et al. (2010) revised the equations for lines of relief wells in USACE EM 1110-2-1914 to include a well offset x_w from the landside levee toe, which required revisions to the equations for calculating net head corrected for well losses (Δh), average net head in the plane of the wells corrected for well losses (Δh_{av}), and the net head midway between the wells corrected for well losses (Δh_m). Additionally, Guy et al. (2010) developed a relationship between well-efficiency head losses and Δh_{av}, which was incorporated into the Stage 1 analyses using an efficiency of 80%. The revised equations are provided in Guy et al. (2010) and Guy et al. (2014).

As described in USACE EM 1110-2-1914, blanket theory calculations analyze a single impervious top stratum overlying a single pervious substratum. Analysis sections with multiple blanket layers or multiple aquifer layers were modeled through transformation of thicknesses and hydraulic conductivities, as presented in USACE EM 1110-2-1914. The transformed aquifer thickness ($\underline{D}$) is calculated as:

$$\underline{D} = \sqrt{\sum_{m=1}^{m=n} d_m k_{hm} \sum_{m=1}^{m=n} \frac{d_m}{k_{vm}}} \quad (3)$$

where for each layer m of the aquifer, d_m is the thickness, k_{hm} is the horizontal hydraulic conductivity, and k_{vm} is the vertical hydraulic conductivity. The transformed aquifer hydraulic conductivity ($\underline{k_e}$) is calculated as:

$$\underline{k_e} = \sqrt{\frac{\sum_{m=1}^{m=n} d_m k_{hm}}{\sum_{m=1}^{m=n} \frac{d_m}{k_{vm}}}} \quad (4)$$

Blanket theory analyses were performed at analysis cross sections within stretches of levee being remediated by lines of relief wells. Analysis section topography, boundary conditions, and representative subsurface stratigraphy were used in the blanket theory analyses.

Combinations of well spacing and penetration meeting underseepage criteria within a given reach of levee were identified for the 200-year and HTOL WSEs. A design chart showing combinations of well spacing and well penetration length (W) divided by aquifer thickness (D) meeting criteria at both WSEs was developed and used to optimize well spacing and penetration in the Stage 2 and Stage 3 analyses.

Stage 2 Analyses

Stage 2 included plan view seepage analyses using SEEP/W assuming all relief wells were fully-penetrating. The SEEP/W plan view model extended from the waterside seepage entry (i.e., where the top of the aquifer contacts the river channel) to the landside boundary, placed 2,000 ft landward of the levee alignment. Upstream and downstream boundaries were placed 1,000 ft beyond the levee reaches with relief wells to eliminate potential boundary effects. The boundary conditions along the waterside seepage entry were specified as a constant head boundary set to either the 200-year or HTOL WSE. The landside boundary was modeled as a constant head boundary set to the ground surface elevation 2,000 ft landward of the levee alignment. Where appropriate, the landside boundary head was varied to account for variations in topography or hydraulic features (e.g., reservoirs, canals). The upstream and downstream boundaries (side boundaries perpendicular to the levee alignment) were modeled as no flow boundaries.

Variations in aquifer thickness and hydraulic conductivity can occur along long lines of relief wells, which can significantly affect relief well flow rates. Although SEEP/W allows for variation of the aquifer thickness within plan view models, the implementation is limited to a single linear variation across the entire model and thus an alternative approach was used. The developed approach modeled variations in aquifer properties by specifying a constant aquifer thickness for the entire model and scaling the hydraulic conductivity of model regions such that the aquifer transmissivity in each region is equal to the aquifer transmissivity from its corresponding Stage 1 analyses. Using this approach, the hydraulic conductivity for use in the SEEP/W analyses was calculated as:

$$k_{model} = \underline{k_e}\left(\frac{\underline{D}}{D_{model}}\right) \tag{5}$$

where k_{model} is the scaled hydraulic conductivity for use in the SEEP/W plan view analyses, $\underline{k_e}$ is the aquifer's effective hydraulic conductivity calculated in Stage 1 (Eqn. 4), $\underline{D}$ is the transformed thickness of the aquifer calculated in Stage 1 (Eqn. 3), and D_{model} is the modeled constant aquifer thickness in the SEEP/W plan view analysis.

All relief wells were modeled as fully-penetrating wells by specifying a total head at the relief well locations. The total head boundary condition (H) assigned to each well was calculated as the sum of the well's top of riser elevation (El_{TOR}) and well efficiency head loss (H_h):

$$H = El_{TOR} + H_h \quad (6)$$

For lines of relief wells, the head loss (H_h) due to well efficiency was calculated in the Stage 1 analyses with the Guy et al. (2010) approach. Head losses were calculated separately for the 200-year and HTOL WSEs, giving slightly higher boundary conditions at the HTOL WSE compared to the 200-year WSE. For irregularly-positioned relief wells, H_h was iteratively calculated using the Guy et al. (2010) approach with the average net heads from the SEEP/W analysis. The average net head assigned to each well was calculated as the distance-weighted net head along lines extending halfway between the well of interest and the nearest adjacent wells. Similar to the lines of relief wells, H_h tended to be slightly higher for the design HTOL WSE compared to the 200-year WSE. Additional details regarding the calculation of head loss for irregularly-positioned wells is given in a companion paper (Weber et al. 2017).

Stage 3 Analyses

Stage 3 included plan view seepage analyses with SEEP/W with partially-penetrating wells modeled by specifying a discharge flow rate at the relief well locations, whereas fully-penetrating wells were modeled using total head boundary conditions. The flow rates assigned to partially-penetrating relief wells were estimated with the USACE EM 1110-2-1914 approach, which can be written as:

$$Q_{w,PP} = G_p * Q_{w,FP} \quad (7)$$

where $Q_{w,PP}$ is the flow rate for a partially-penetrating well, G_p is a flow rate reduction factor ranging from 0 to 1, and $Q_{w,FP}$ is the flow rate for a fully-penetrating well. The fully-penetrating well flow rates for each well were taken from the Stage 2 SEEP/W analyses at the 200-year and HTOL WSEs, which assumed fully-penetrating conditions. The Stage 3 SEEP/W plan view analyses were also run at the 200-year and HTOL WSEs.

The G_p factor was estimated using two approaches. For lines of relief wells, G_p was calculated as the ratio of flow rates for partially-penetrating to fully-penetrating calculated with the Stage 1 blanket theory analyses. For irregularly-positioned relief wells, G_p was calculated using the Muskat (1937) approach, as presented in USACE EM 1110-2-1914, assuming the radius of influence was equal to a representative spacing between the wells. The radius of influence was taken as the average spacing along irregularly-positioned relief wells. The sensitivity of the G_p factor to the radius of influence was evaluated for the range of well spacings in irregularly-positioned wells and G_p was found to be relatively insensitive to the assumed radius of influence.

Calculation of Exit Gradients

For uniform non-stratified blankets, exit gradients (i) were calculated directly with mid-well aquifer total heads from SEEP/W plan view analyses (Stage 2 for all fully-penetrating relief wells, Stage 3 if partially-penetrating relief wells were present). For stratified blankets, exit gradients were calculated from SEEP/W cross section analyses with mid-well aquifer total heads from either Stage 2 or Stage 3 plan view analyses, as appropriate, applied over the entire aquifer thickness at the location of the relief well line and mid-well heads applied to the bottom of the blanket at exit gradient calculation locations. Exit gradients across stratified blankets were then calculated at critical locations using total heads from the cross section analysis.

APPLICATION TO THE SACRAMENTO RIVER EAST LEVEE

An application of the proposed staged design approach for the design of proposed lines of relief wells is presented in this section. The proposed lines of relief wells are located in levee reaches Reach 22 (Sta. 1517+00 to Sta. 1530+30) and Reach 24 (Sta. 1534+40 to Sta. 1550+00), adjacent to the Sump 132 facility (Reach 23: Sta. 1530+30 to Sta. 1534+40) in the Pocket neighborhood of Sacramento, California (Figure 2). The design approach presented in the previous sections of this paper was used to design the proposed relief wells in Reaches 22 and 24 such that underseepage criteria were met in both reaches for the 200-year and HTOL WSEs.

Differences in the surface and foundation conditions in Reaches 22 and 24 required separate designs for each reach. Additionally, changes in aquifer conditions in Reach 24 required splitting the relief well designs in the reach into upstream (Sta. 1534+40 to Sta. 1543+30) and downstream (Sta. 1543+30 to Sta. 1550+00) portions. For brevity, this section presents the design for the downstream portion of Reach 24 at the 200-year WSE.

Figure 2. Plan View in the Vicinity of SREL Reaches 22, 23, and 24 (Google Earth, 2017)

Stage 1 Results

The geometry, blanket stratigraphy, and boundary conditions for the analysis section at Sta. 1544+30 were used in the Stage 1 blanket theory analyses (Figure 3). As shown in Figure 3, the analysis section at Sta. 1544+30 consisted of an approximately 20-ft-thick sandy silt (Sandy ML) blanket (Layer 2) overlying an approximately 48-ft-thick sand with silt (SP-SM) aquifer (Layer 3). The WSEs used in the design were 31.12 ft for the 200-year WSE and 33.12 ft for the HTOL WSE. The relief well top of riser elevation for Stage 1 analyses was taken as the highest El_{TOR} considering the nearest upstream and downstream wells adjacent to the analysis section location.

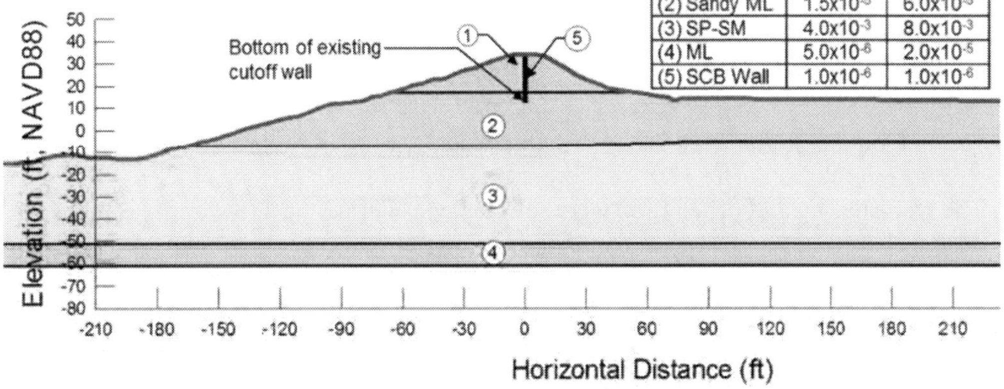

Layer Name	k_v (cm/sec)	k_h (cm/sec)
(1) SM	$5.0x10^{-4}$	$2.0x10^{-3}$
(2) Sandy ML	$1.5x10^{-5}$	$6.0x10^{-5}$
(3) SP-SM	$4.0x10^{-3}$	$8.0x10^{-3}$
(4) ML	$5.0x10^{-6}$	$2.0x10^{-5}$
(5) SCB Wall	$1.0x10^{-6}$	$1.0x10^{-6}$

Figure 3. Analysis Cross Section for Reach 24 at Station 1544+30 (Modified from GEI/HDR 2015)

Using the USACE EM 1110-2-1914 approach for analyzing an infinite line of relief wells, combinations of well spacing and penetration meeting underseepage criteria for the 200-year and HTOL WSEs were identified. The critical location with respect to underseepage for the analysis section at Sta. 1544+30 is a landside low area located 18 ft from the landside toe. The maximum allowable gradient at the critical location in the analysis section was 0.54 and 0.64 for the 200-year and HTOL WSEs, respectively.

The design chart for downstream Reach 24 is presented in terms of well spacing versus well penetration length (W) divided by aquifer thickness (D). Combinations meeting underseepage exit gradient criteria are shown in Figure 4 for the 200-year and HTOL WSEs. As shown in Figure 4, the design is controlled by the 200-year WSE, which indicates longer screened lengths would be required for the 200-year WSE compared to the HTOL WSE for a given well spacing. The maximum well spacing meeting both the 200-year and HTOL criteria is 390 ft and would require fully-penetrating wells (Figure 4).

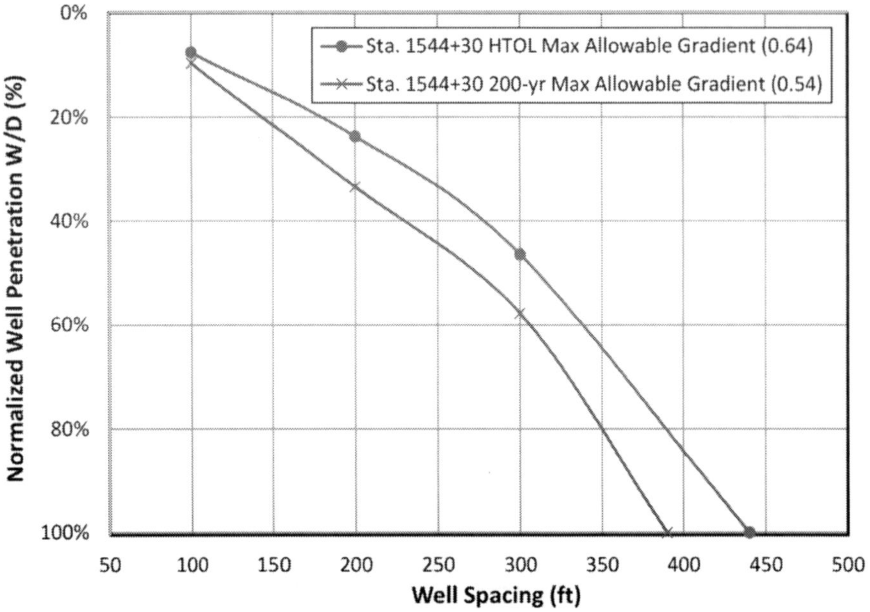

Figure 4. Relief Well Design Chart for the Downstream Portion of Reach 24

The well spacing and penetration were selected considering the site-specific relief well design chart (Figure 4), surface features (i.e., structures or other conflicts), construction costs, and robustness of the system. A well spacing of 150 ft was selected for the nine downstream wells and about 100 ft was selected for the well closest to Reach 23 (Sump 132). The selected design includes seven partially-penetrating wells in upstream Reach 24 and three fully-penetrating wells in downstream Reach 24. The seven relief wells located in upstream Reach 24 will be screened for about 50 ft in the aquifer (W/D of about 50-55%), whereas the three relief wells in downstream Reach 24 will be fully-penetrating wells screened for about 35-to-50-ft in the aquifer (W/D of 100%).

As shown in Figure 4, the selected combination of W/D of 100% and a well spacing of 150 ft for the downstream portion of Reach 24 was expected to provide sufficient underseepage relief such that underseepage criteria would be met. Although not presented herein, a separate design chart was prepared for upstream Reach 24, which confirmed a W/D of 50-55% with a well spacing of 100-150 ft would also exceed the requirements to meet underseepage criteria.

Stage 2 Results

A single SEEP/W plan view model was developed for Reaches 22, 23, and 24 (Figure 5). The model extended 1,000 ft upstream of Reach 22 and 1,000 ft downstream of Reach 24 (Sta. 1507+00 to Sta. 1560+00) to eliminate potential boundary effects. The model consisted of 13 proposed relief wells in Reach 22, 10 existing relief wells in Reach 23, and 10 proposed relief wells in Reach 24. The relief wells in Reach 22 were sequentially numbered from RW22-1 to RW22-13 in a downstream (southerly) direction. The relief wells in Reach 23 were numbered sequentially from waterside to landside, with RW23-1 through RW23-5 on the upstream side of the Sump 132 facility and RW23-6 through RW23-10 located on the downstream side of the facility. The relief wells in Reach 24 were numbered sequentially from RW24-1 to RW24-10 in an upstream (northerly) direction. An existing deep cutoff wall at Sump 132 (Reach 23) was conservatively not included in the analyses as it does not fully-penetrate the aquifer.

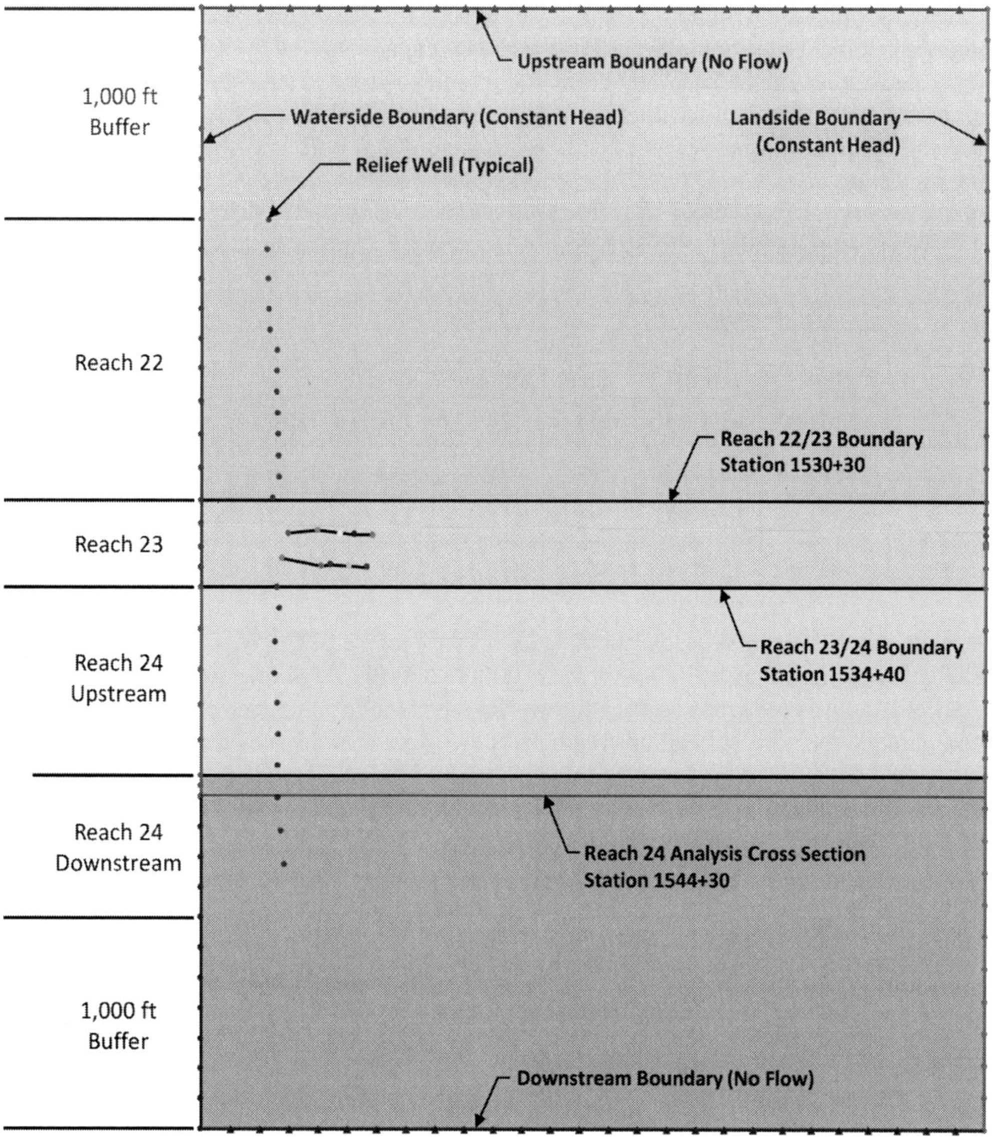

Figure 5. SEEP/W Plan View Model and Boundary Conditions

The presence of the Pocket canal in Reach 23 was considered in the development of the model. However, review of available subsurface explorations indicated the canal did not penetrate the confined aquifer and thus the canal was not expected to have a significant impact on flows or heads in the aquifer. The Pocket canal intersected the model landside boundary and thus the landside boundary condition at this location was specified as a head value consistent with an empty canal.

A single model thickness of 291 ft was assigned to the entire model to match the transformed thickness of the Reach 22 aquifer. Table 1 shows the actual aquifer thicknesses (D_{actual}), transformed hydraulic conductivities ($\underline{k_e}$), transformed thicknesses ($\underline{D}$), transmissivity (T), and modeled hydraulic conductivities (k_{model}) and aquifer thicknesses (D_{model}) assigned to each aquifer section in the plan view analyses. The hydraulic conductivities in the SEEP/W model were adjusted such that the aquifer transmissivity in the SEEP/W model is equal to the estimated aquifer transmissivity assuming a constant aquifer thickness. The upstream 1,000 ft buffer was modeled with Reach 22 aquifer properties and the downstream 1,000 ft buffer was modeled with Reach 24 aquifer properties.

Table 1. SEEP/W Plan View Model Aquifer Properties

Reach	Stationing		D_{actual} (ft)	Transformed Properties		T (ft^2/day)	Modeled Aquifer Properties	
	Start	End		$\underline{k_e}$ (ft/day)	$\underline{D}$ (ft)		$\underline{k_{model}}$ (ft/day)	D_{model} (ft)
22	1517+00	1530+30	94	24.5	291	7120	24.5	291
23	1530+30	1534+40	94	30.1	133	3984	13.7	291
24	1534+40	1543+30	94	16.1	133	2127	7.3	291
	1543+30	1550+00	48.5	16.1	69	1100	3.8	291

Results in terms of total heads calculated with the Stage 2 SEEP/W plan view seepage analyses are presented in Figure 6 for the 200-year WSE. As shown in Figure 6, the aquifer head was reduced significantly within the extent of proposed relief wells compared to the aquifer head upstream and downstream of the remediated length of levee. The high heads upstream and downstream of the relief wells caused seepage around the ends of the relief well lines (i.e., end around) and toward the relief wells at Sump 132 (Reach 23) and the middle of the model.

The relief wells in Reaches 22, 23, and 24 included both existing and proposed partially-penetrating wells. Therefore, Stage 3 analyses were required to calculate the aquifer head, relief well flow rates, and exit gradients. These analyses are described in the next section.

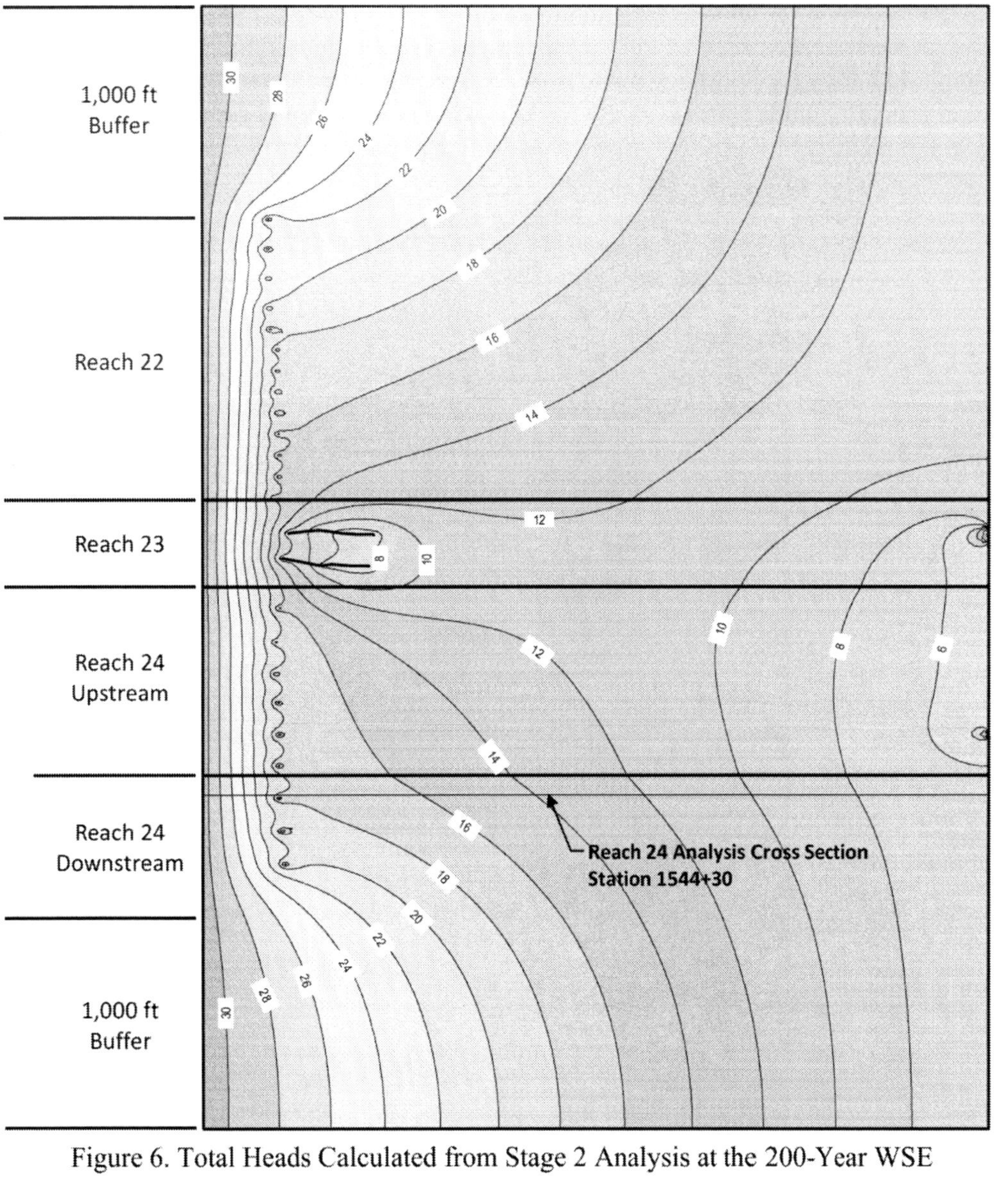

Figure 6. Total Heads Calculated from Stage 2 Analysis at the 200-Year WSE

Stage 3 Results

Total heads calculated with Stage 3 SEEP/W plan view seepage analyses are presented in Figure 7 for the 200-year WSE. As was observed in the Stage 2 analyses, the relief wells significantly reduced the aquifer heads within the extent of the relief wells. However, the reduction of aquifer heads in Stage 3 was less than for Stage 2 (i.e., higher heads were calculated in Stage 3). The higher heads calculated for Stage 3 analyses was expected due to the effects of partial penetration, which reduce the flow rate for partially-penetrating relief wells compared to fully-penetrating conditions and hence also reduced the amount of relief (or decrease of head) in the aquifer.

The high heads upstream and downstream of the relief wells caused end around seepage at the ends of the relief well lines, as was observed for the Stage 2 analyses. The end around seepage was directed toward the middle of the model, including the relief wells at Sump 132 (Reach 23). Although the pattern of seepage calculated in the Stage 3 analyses was generally landward, the flow toward the middle of the model was not insignificant and was different than the conditions modeled in the USACE EM 1110-2-1914 analyses (seepage landward of the wells is generally oriented landward).

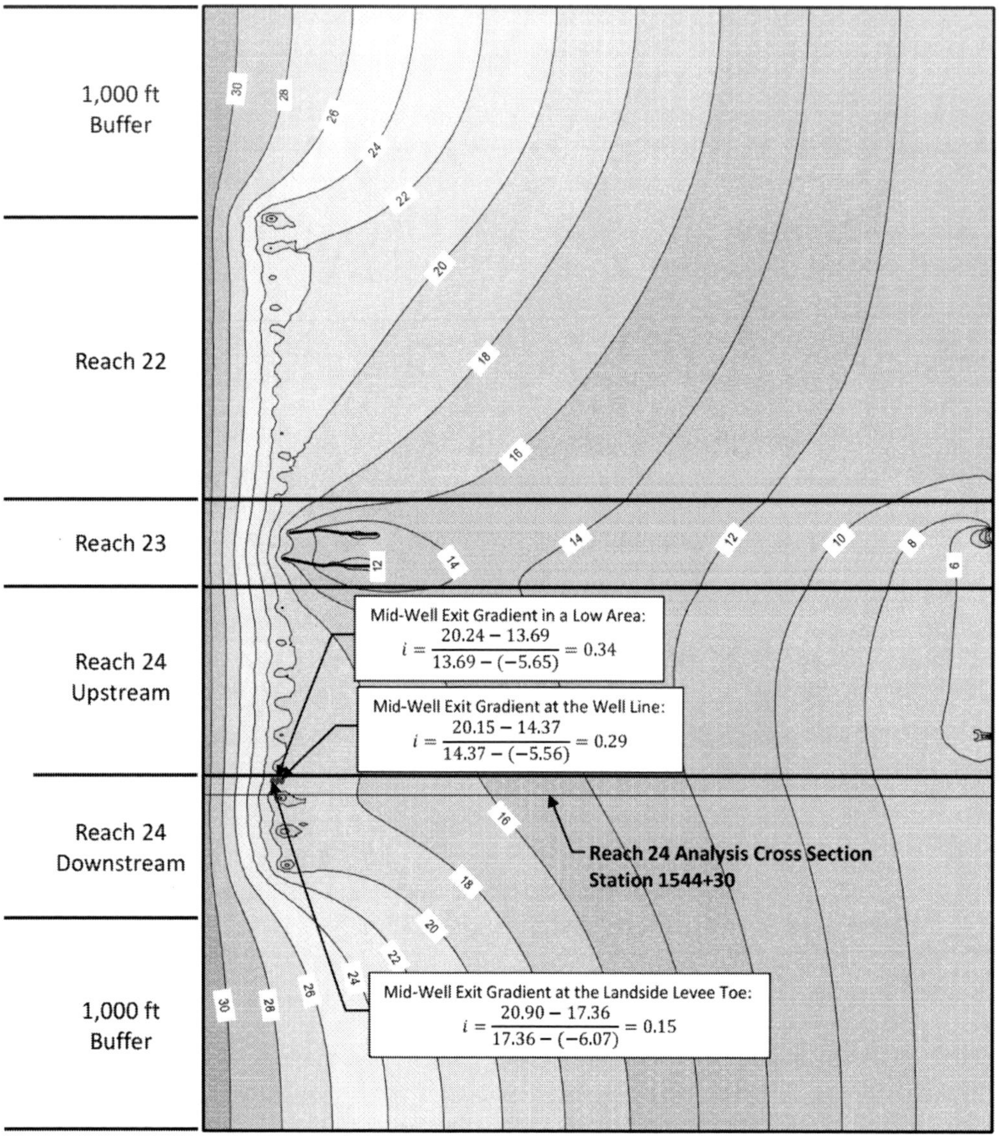

Figure 7. Total Heads Calculated from Stage 3 Analysis at the 200-Year WSE

Exit gradients at Sta. 1544+30 are presented in Table 2 for existing conditions (i.e., without proposed relief wells) and remediated conditions analyses. Existing conditions exit gradients were calculated using two-dimensional seepage analysis of the cross section presented in Figure 3. Remediated conditions exit gradients were calculated with Stage 3 analyses, which were then used to evaluate the relief well designs against underseepage exit gradient criteria. Exit gradients for remediated conditions using Stage 1 analyses are also included in Table 2 for the purposes of comparison between the proposed design approach and simplified USACE EM 1110-2-1914.

For existing conditions analyses at the 200-year WSE, Sta. 1544+30 was found to meet underseepage criteria at the landside toe, but did not meet criteria in a low area 18 ft from the landside toe where the exit gradient was 0.57 (allowable exit gradient of 0.54). As shown in Figure 7 and in Table 2, all exit gradients at Sta. 1544+30 were found to meet criteria at the 200-year WSE with the proposed relief wells based on the Stage 3 analysis results. Similar results were obtained for the HTOL WSE, but have been omitted for brevity. Therefore, the downstream portion of Reach 24 meets ULDC underseepage criteria for the 200-year WSE.

Comparison of exit gradients calculated with Stage 3 and Stage 1 analyses demonstrated the two methods were in agreement for Sta. 1544+30, with only small differences in the calculated exit gradients (0.05-0.07). Additionally, both analyses were in agreement that the proposed relief well spacing and penetration in downstream Reach 24 were sufficient to meet underseepage exit gradient criteria. Differences in the calculated exit gradients can be attributed primarily to several factors including variable relief well penetration into the aquifer (partially-penetrating wells in upstream Reach 24, end around seepage, and variable relief well top of riser elevations.

Table 2. Downstream Reach 24 Exit Gradients for Existing and Remediated Conditions for the 200-Year WSE

Location	Calculated Existing Conditions Exit Gradient	Exit Gradient with Proposed Relief Wells: Stage 3 SEEP/W Analysis	Exit Gradient with Proposed Relief Wells: Stage 1: Blanket Theory Analysis
Landside Toe, Mid-Well (i_{allow} = 0.50)	0.33	0.15	0.10
Low Area, Mid-Well (i_{allow} = 0.54)	0.57	0.34	0.27
Well Line, Mid-Well (i_{allow} = 0.54)	N/A	0.29	0.22

Stage 3 and Stage 1 analyses were also compared in terms of relief well flows and aquifer heads along the well lines (Figure 8). Stage 1 calculations from three analysis sections were included in the comparison: Sta. 1524+04 (Reach 22), Sta. 1538+31 (upstream Reach 24), and Sta. 1544+30 (downstream Reach 24). As shown in Figure 8a, the relief well flow rates from Stage 3 analyses for wells near the three analyses sections were approximately equal to the results from Stage 1 analyses (within about 5-15 gpm). On average the Stage 3 flow rates were 20% higher than the Stage 1 flow rates for all wells. However, at the ends of the relief well lines where end around seepage occurred, the Stage 3 flow rates were larger than Stage 1 flow rates by as much as 2.0-2.8 times for the upstream relief wells in Reach 22 and about 1.3-1.4 times for the downstream relief wells in Reach 24. For the proposed wells near the 10 existing relief wells in Reach 23 (Sump 132), the Stage 3 flow rates were smaller than the Stage 1 flow rates by as much as 80% of the Stage 1 flow. The lower flow rates in the Stage 3 analyses can be attributed to interactions with the existing Sump 132 relief wells, which have significantly lower top of riser elevations than the proposed wells (up to 11 ft lower). Overall, the Stage 3 flow rates tended to be about 20% higher than the Stage 1 flow rates, but at the extremes, could be about 0.1 to 2.8 times the Stage 1 flow rates where end around or interactions with existing wells was significant.

Aquifer heads along the relief well alignments were compared between Stages 1 and 3 by plotting the higher of either the average head or mid-well head in the plane of wells from Stage 1 (variables h_{av} and h_m from USACE EM 1110-2-1914), and head values sampled along the relief well alignments from the Stage 3 analyses (Figure 8b). As shown in Figure 8b, the Stage 3 Reach 22 mid-well aquifer heads were about 0.9 ft higher than for Stage 1 for the upstream 400 ft of the reach (Sta. 1517+00 to 1521+00), about 0.4 ft lower than for Stage 1 in the vicinity of the analysis section (Sta. 1521+00 to 1525+00), and about 1.8 ft lower for the remainder of the reach (Sta. 1525+00 to Sta. 1530+30). In Reach 24, the Stage 3 mid-well aquifer heads were about to 1.6 ft lower than for Stage 1 or upstream Reach 24, and about 1.9 ft higher than for Stage 1 for downstream Reach 24. Overall, the Stage 3 heads were generally higher near the ends of the relief well lines where end around effects were significant, but generally lower further from the ends of the relief well lines. The differences between the heads can be attributed to a number of factors, including end around seepage, interactions between adjacent fully-penetrating and partially-penetrating wells, variability in the relief well top of riser elevations and offset from the landside toe, and variations in boundary conditions.

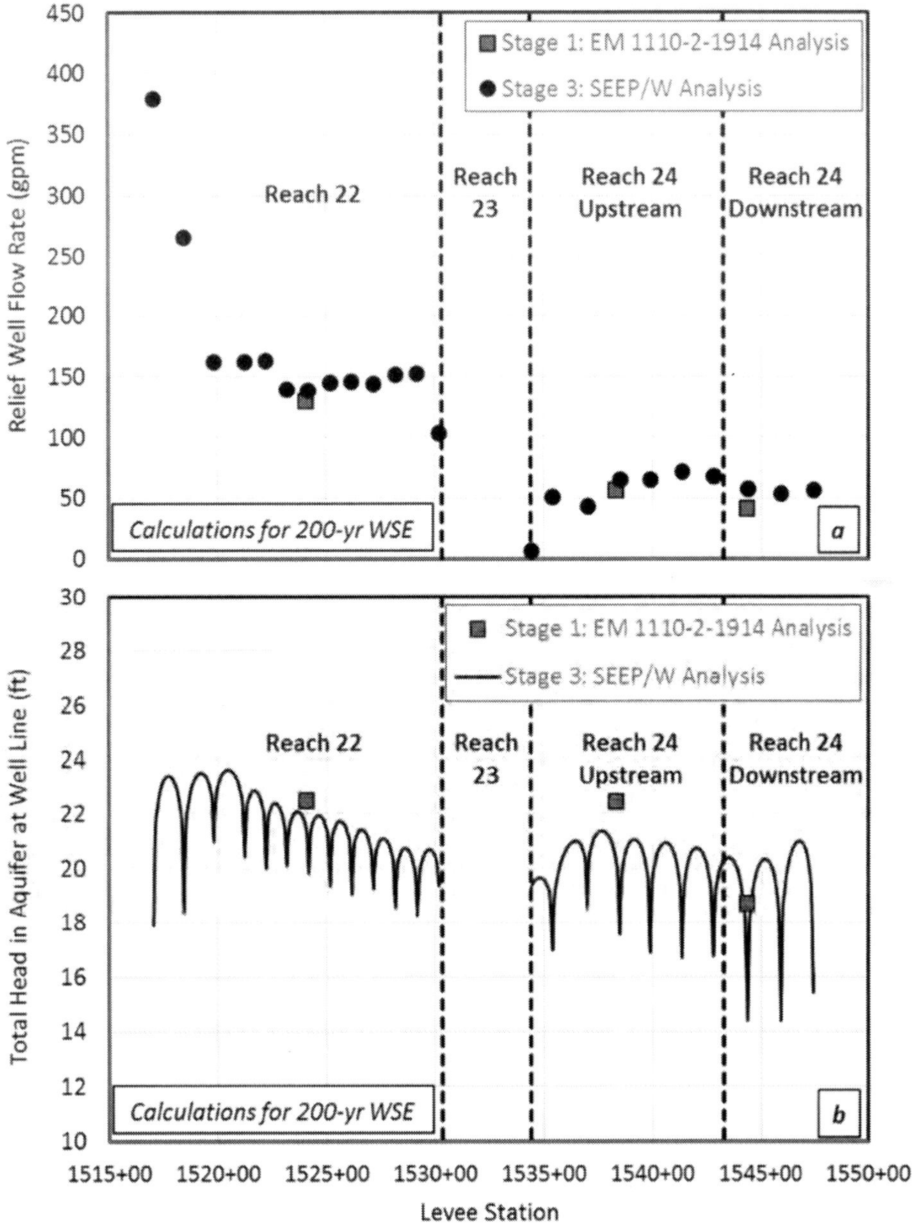

Figure 8. Comparison of Stage 1 and Stage 3 (a) Relief Well Flow Rates, and (b) Aquifer Head at the Well Line

SUMMARY

A staged analysis and design approach for lines of relief wells was presented. The approach incorporated three analysis stages: (1) simplified analyses with the USACE EM 1110-2-1914 procedure, (2) SEEP/W plan view analyses assuming fully-penetrating wells, and (3) SEEP/W plan view analyses with partially-penetrating wells. Application of the proposed method was demonstrated for a levee reach along the Sacramento River East Levee where lines of relief wells are currently in design. The staged analysis approach was used to account for three dimensional effects associated with finite relief well lines, interactions with complimentary underseepage remediations (i.e., relief wells, fully-penetrating cutoff walls), and variations of relief well top of riser elevations, well offset from the landside toe, relief well spacing, relief well penetration (or screened) lengths, aquifer thickness, hydraulic conductivity, and boundary conditions, which are often encountered, but are not accounted for in simplified blanket theory design procedures.

Results with the proposed stage analysis approach were compared with traditional blanket theory analyses with USACE EM 1110-2-1914. The comparison indicated flow rates estimated with the proposed approach were consistent with results from USACE EM 1110-2-1914 for wells near the middle of the relief well lines. The comparison also indicated aquifer heads estimated with the proposed method were of similar magnitude as USACE EM 1110-2-1914 for wells near the middle of the relief well lines. However, differences were encountered near the ends of the relief well lines due to end around effects, interactions with adjacent relief well lines, variations in boundary conditions, and variability of relief well top of riser elevations and offset from the landside toe. In contrast to USACE EM 1110-2-1914, the proposed method is capable of modeling these aspects. Thus, the proposed staged analysis approach is a rational and flexible method for analysis and design of partially-penetrating and fully-penetrating relief wells.

ACKNOWLEDGEMENTS

The authors would like to acknowledge Mr. George Sills, Mr. Stephen Verigin, and Dr. Leslie Harder, Jr. of the project's Value, Cost, and Constructability (VCC) panel for their detailed reviews of the relief well design approaches, analyses, and designs presented in this paper. In addition, the authors wish to thank the Sacramento Area Flood Control Agency (SAFCA) for their leadership and support during the project. Lastly, the authors would like to acknowledge Dr. Erich Guy for discussions regarding analyses of relief wells and procedures for estimating well losses.

REFERENCES

California Department of Water Resources (DWR) (2012). "Urban Levee Design Criteria". May 2012.

GEI/HDR (2015). "Embankment and Foundation Stability Memorandum – Sacramento River East Levee Improvement Project, Sacramento, California". October 2015.

GEO-SLOPE International, Ltd. (2015). Geostudio 2012, August 2015 update.

Google Earth (2017) Map image: Landsat/Copernicus. Accessed February 2017.

Guy, E. D., Nettles, R. L., Davis, J. R., Carter, S. C., and Newberry, L. A. (2010). "Relief Well System Design Approach: HHD Case Study". In Proc. Association of State Dam Safety Officials. Charleston, West Virginia. 19 pp.

Guy, E. D., Ider, H. M., and Darko-Kagya, K. (2014). "Several Relief Well Design Considerations for Dams and Levees". In. Proc. 45th Annual Ohio River Valley Soils Seminar. 26 pp.

MBK Engineers (2016). "SAFCA Levee Accreditation Project, Computed Water Surface Elevations". February 10, 2016.

Muskat, M. (1937). "The Flow of Fluids through Porous Media". Journal of Applied Physics, 8(4). pp. 274-282.

United States Army Corps of Engineers (1992). "Design, Construction, and Maintenance of Relief Wells". Engineer Manual 1110-2-1914. May 29, 1992.

U.S. Code of Federal Regulations (2002). Title 44 (44 CFR) Section 65.10. October 1, 2002.

Weber, J. P., Stanley, M. H., Jaeger, R. A., and Bradner, G. C. (2017). "Evaluation and Design of Irregularly-Positioned Relief Wells". USSD 2017 Annual Conference. In Press.

EVALUATION AND DESIGN OF IRREGULARLY-POSITIONED RELIEF WELLS

Joseph Weber, PhD, PE [1]
Mark Stanley, PE, GE, D.GE[2]
Robert Jaeger, PhD, PE[3]
Graham Bradner, PG, CEG, CHg, PMP[4]

ABSTRACT

Evaluation and remediation of existing levee segments with passive relief wells can present complexities that are not easily resolved using standard simplified approaches. This is especially true in urban areas where structures are often in proximity to the landside levee toe, as is the situation along the East Levee of the Sacramento River in Sacramento, California. Thus, traditional, simplified design procedures for lines of equally-spaced relief wells are not applicable.

The current studies along the Sacramento River East Levee (SREL) required the use of less common relief well evaluation and design techniques. The area around Sacramento's Pioneer Reservoir created such a scenario. The reservoir is in proximity to the levee with existing relief wells in localized areas of relatively low elevation near the landside toe. Based on the relief well configuration, it was judged that the most appropriate technique was a direct solution method using superposition and the Thiem equation as presented in USACE EM 1110-2-1914.

The selected methodology allowed the existing relief wells to be modeled in their relative positions. The technique also allowed LiDAR elevation data to be overlaid on a map of the estimated aquifer heads to efficiently identify critical locations and optimize placement of new relief wells.

Implementation of this approach included an evaluation of available procedures to evaluate the efficiency head losses that should be assumed for design. An approach for a line of relief wells, described in Guy et al. (2010), was generalized to the case of irregularly-positioned relief wells. This technique was compared to efficiency head loss estimates predicted using USACE EM 1110-2-1914 techniques to confirm the reasonableness of the results.

[1] Staff Engineer, HDR, Inc., 2365 Iron Point Road, Suite 300, Folsom, CA 95630, Joseph.Weber@hdrinc.com
[2] Senior Technical Advisor, HDR, Inc., 2365 Iron Point Road, Suite 300, Folsom, CA 95630, Mark.Stanley@hdrinc.com
[3] Senior Engineer, GEI Consultants, Inc., 2868 Prospect Park Drive, Suite 400, Rancho Cordova, CA 95670, rjaeger@geiconsultants.com
[4] Senior Geologist, GEI Consultants, Inc., 2868 Prospect Park Drive, Suite 400, Rancho Cordova, CA 95670, gbradner@geiconsultants.com

INTRODUCTION

The Sacramento River East Levee (SREL) system in Sacramento, California is composed of earthen embankments originally constructed in stages from the mid-1800s to early-1900s. At that time, levees were often constructed on the flood plain along the top of the existing natural levee on top of the riverbank, without much setback from the river channel. During the last three decades, the Sacramento Area Flood Control Agency (SAFCA) and the United States Army Corps of Engineers (USACE) have designed and constructed improvements along several portions of the subject levee segments. Improvements have included shallow and deep seepage cutoff walls, relief wells, berms, and erosion protection.

Recently, GEI Consultants, Inc., in association with HDR, Inc. (GEI/HDR), was contracted to evaluate 8.1 miles of the SREL previously found to not meet Federal Emergency Management Agency (FEMA) 100-year criteria (44 CFR 65.10) for geotechnical embankment and foundation stability, as shown in Figure 1. GEI/HDR collected additional information and reevaluated the 8.1 miles of the SREL relative to FEMA criteria and California's Department of Water Resources (DWR) Urban Levee Design Criteria (ULDC) dated May 2012 (DWR 2012). The evaluation considered the results of geotechnical seepage and stability analyses for existing conditions, understanding of foundation conditions and existing remediations, as well as a review of available records of past performance. Based on the geotechnical evaluations, GEI/HDR proposed improvements to 5.9 miles of the 8.1-mile study area, which include cutoff walls, relief wells, berms, waterside clay blankets, and waterside erosion and stability repairs. The selection of appropriate methods for repairs and improvements has been constrained by access limitations, lack of undeveloped land beyond the landside levee toe, and a limited amount of waterside berm along the waterside portion of the levee.

Relief wells have been proposed for approximately 1.1 miles of the SREL, including lines of relief wells and irregularly-positioned relief wells. The relief wells were designed using a staged design approach combining relief well design procedures from USACE EM 1110-2-1914 (USACE 1992) with plan view seepage analyses. The design approach for irregularly-positioned relief wells is presented in this paper and the approach for lines of relief wells is presented in a companion paper (Jaeger et al. 2017).

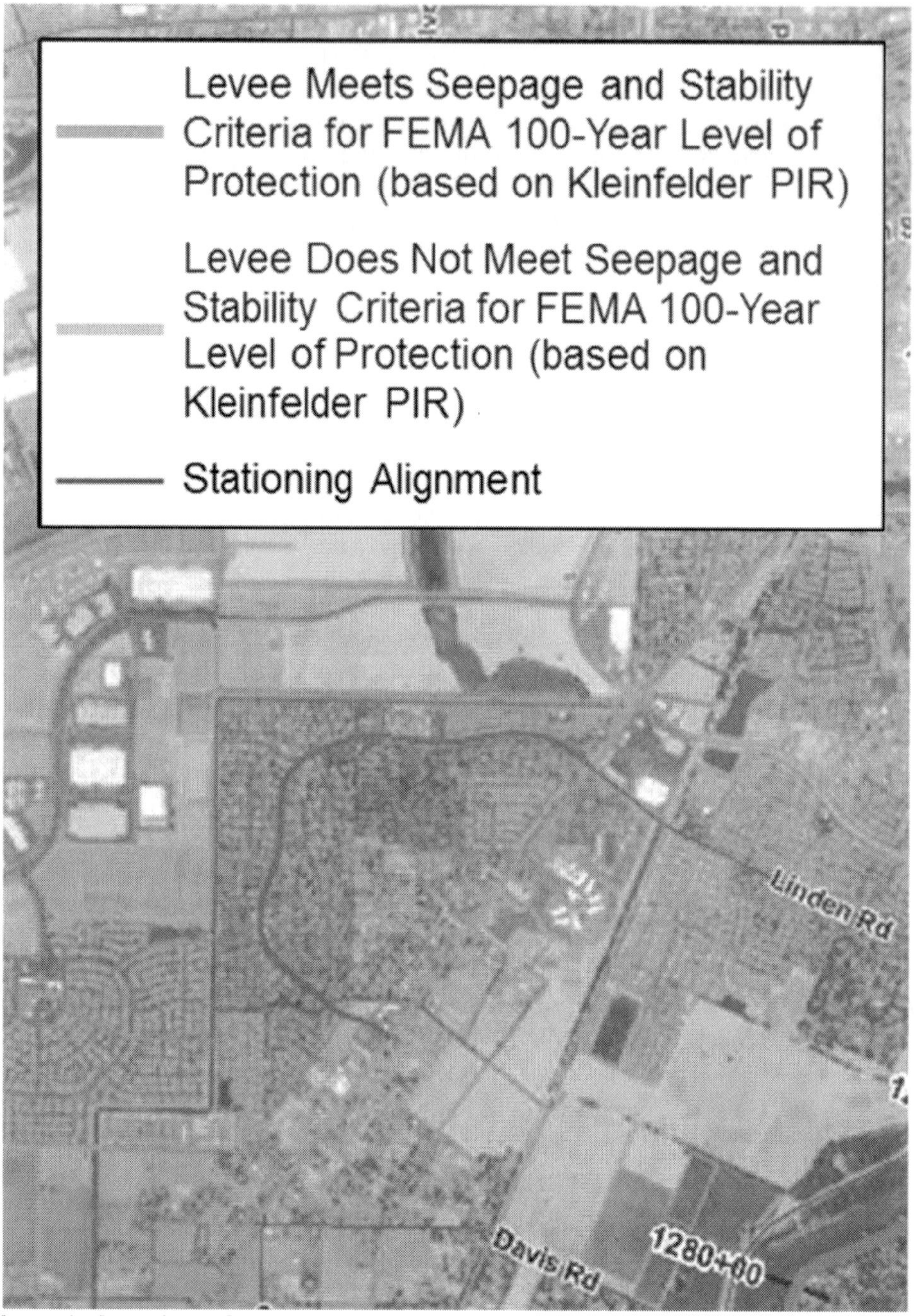

Figure 1. Overview of Sacramento River East Levee within the GEI/HDR Study Area

OVERVIEW

For irregularly-positioned relief wells at Pioneer Reservoir, relief wells were designed using well hydraulics with image wells. Analyses were performed to evaluate the flow and total heads in a confined aquifer considering a discrete number of irregularly-positioned relief wells screened in the aquifer. The analysis used the Thiem equation for confined aquifer well hydraulics (Thiem 1906) combined with the principle of superposition and image well theory to select the number and location of new relief wells required to meet underseepage exit gradient criteria.

The three step process used to evaluate irregularly-positioned relief wells can be summarized as:

- Step 1: Determine well flow rates not accounting for head losses in each of the wells
- Step 2: Estimate head loss and recalculate relief well flow rates for each well. Iterate until convergence is reached.
- Step 3: Calculate gradients at all points of interest.

The solution approach was implemented as presented in USACE EM 1110-2-1914 with adjustments for an assumed linear head distribution from waterside to landside of the model. Efficiency losses were estimated using a procedure based on the approach presented in Guy et al. (2010) for a line of relief wells. The procedure used to evaluate the irregularly-positioned relief wells at Pioneer Reservoir is described following the overview discussion below.

The water surfaces for design included geotechnical design 200-year and hydraulic top of levee (HTOL) water surface elevations (WSEs), which were specified based on levee station (MBK 2016). The geotechnical design 200-year water surface was set 1 ft above the 200-year water surface for added robustness and future sea level rise. The HTOL WSE is defined in the ULDC as the minimum of the 500-year WSE or 3 ft above the 200-year WSE (DWR 2012). For the remainder of the paper, the geotechnical design 200-year WSE will simply be referred to as the 200-year WSE.

RELIEF WELL DESIGN APPROACH FOR IRREGULARLY-POSITIONED RELIEF WELLS

The selected approach for irregularly-positioned relief wells is based on the Thiem equation based procedure presented in USACE EM 1110-2-1914, which utilizes the method of image wells with an infinite line source. Combinations of locations, number of wells, and well penetration that meet underseepage criteria were identified. Existing and proposed relief wells were included in the analyses. The location of the infinite line source was selected as the location of the waterside seepage entry, where the top of the aquifer contacts the river channel.

To model an infinite line source, an imaginary well is included for each existing and proposed relief well. Refer to Section 5-4 and Figure 5-3 of EM 1110-2-1914 for a description of the method of image wells for irregularly spaced relief wells of varying discharge elevations adjacent to an infinite line source. The location of the imaginary wells are mirror images of the existing and proposed relief wells in relation to the infinite line source, where the imaginary well has the same perpendicular offset from the infinite line source as the real well, although in the opposite direction from the infinite line source.

The following equation from EM 1110-2-1914, based on superposition of wells in a confined aquifer with an adjacent infinite line source, was utilized to determine the total head at any point in the zone of interest:

$$h_p = H_1 - \frac{1}{2\pi \underline{k_e}\underline{D}} \sum_{i=1}^{i=n} Q_{wi} \ln \frac{r_i'}{r_i} \tag{1}$$

where h_p is the head at a point p, H_l is the head at the infinite line source, $\underline{k_e}$ is the effective hydraulic conductivity of the aquifer, $\underline{D}$ is the transformed thickness of the aquifer, Q_{wi} is the flow for real well i, r_i' is the distance from point p to image well i, and r_i is the distance from point p to the real well i. The transformed parameters $\underline{k_e}$ and $\underline{D}$ are calculated using the EM 1110-2-1914 approach.

The principle of superposition was also utilized to incorporate a constant rate of head drop in the aquifer from the infinite line source to the landside boundary (infinite line sink). The above equation (Eqn. 1) from EM 1110-2-1914 was modified to the following equation to incorporate a constant head landside boundary condition and specify H_l as the WSE used in the analysis:

$$h_p = El_{flood} - \frac{d_{ILSo\_p}}{d_{ILSo\_LSi}} \left(El_{flood} - El_{lsgs}\right) - \frac{1}{2\pi \underline{k_e}\underline{D}} \sum_{i=1}^{i=n} Q_{wi} \ln \frac{r_i'}{r_i} \tag{2}$$

where El_{flood} is the WSE used in the analysis, El_{lsgs} is the total head at the landside boundary condition, d_{ILSo_p} is the distance from the infinite line source to a point p, and d_{ILSo_LSi} is the distance from the infinite line source to the infinite line sink (location of landside boundary).

Step 1: Solve the System of Equations to Determine the Flow at Each Well Ignoring Head Losses

In order to determine the head beneath the blanket any point of interest, the flows at each well must first be determined. The flows were calculated by forming and solving a system of equations. Applying Eqn. 2 to each well point creates a system of equations relating total heads to the well flow rates for each real well. Initially, the total head at each well, ignoring head losses, was set equal to each well's top of riser elevation (El_{TOR}). The radial distance between wells was then determined by calculating the distance between center points of the wells and subtracting the effective well radius of each well from that distance. The effective well radius, r_w, is determined per

EM 1110-2-1914:

$$r_w = \frac{1}{2}\left(d_w + T_{fp}\right) \tag{3}$$

where d_w is the diameter of the well screen and T_{fp} is the thickness of the filter pack.

Step 2: Determine Head Loss for Each Well

The head loss, H_h, due to well efficiency was calculated using a similar approach to that developed by Guy et al. (2010) for the EM 1110-2-1914 analysis for a line of relief wells. The Guy et al. (2010) approach utilizes the average net head in a plane of wells corrected for well losses, Δh_{av}. To generalize this approach to a group of irregularly-positioned wells, the average net head was assumed to be equal to the distance-weighted average net head along a line half the distance from the well of interest to adjacent wells. Examples of the implementation of the average net head calculation are presented later in this paper.

Once the Δh_{av} for each well was determined, the H_h values were determined assuming a well efficiency (E_w) of 80%. The total head assigned to each well was calculated as the sum of El_{TOR} and H_h:

$$H_h = \frac{\Delta h_{av}}{E_w} - \Delta h_{av} \tag{4}$$

$$H = El_{TOR} + H_h \tag{5}$$

The flows of each well were then reevaluated by reforming and solving the system of equations, as described in Step 1, with the updated total heads at the wells. The process was repeated until the difference between H_h for consecutive iterations was negligible (less than 0.01 ft).

Step 3: Determine Aquifer Head Distribution and Gradient at All Points of Interest

The aquifer head at all points of interest was determined using the flow rates determined from Steps 1 and 2 combined with Eqn. 2. Following the determination of the head beneath the blanket at all points of interest, the gradient was determined following the procedure described in EM 1110-2-1914. If the blanket was stratified, a transformed blanket calculation was performed in accordance with the procedure presented in Appendix B of EM 1110-2-1914.

The calculated exit gradients were compared to allowable exit gradients to determine whether the trial relief well configuration sufficiently reduced heads beneath the blanket in the area of interest. If the configuration did not meet underseepage criteria, the relief well configuration was modified and the process was repeated.

Allowable Exit Gradients

For design of relief wells along the SREL, the ULDC maximum allowable exit gradient at the landside toe is 0.50 and 0.60 for the 200-year and HTOL WSEs, respectively, taken at the midpoint between two adjacent relief wells. For locations landward of the landside levee toe, the allowable gradients increase according to the following equations (DWR 2012):

$$i_{allow} = 0.50 + 0.3 * \left(\frac{d_{toe}}{150}\right) \leq 0.80 \text{ for the 200-year WSE} \quad (6)$$

$$i_{allow} = 0.60 + 0.3 * \left(\frac{d_{toe}}{150}\right) \leq 0.90 \text{ for the HTOL WSE} \quad (7)$$

where d_{toe} is the distance landward from the landside levee toe.

EXISTING CONDITIONS EVALUATION

Existing Conditions Overview

Existing conditions analyses were performed considering only the existing irregularly-positioned relief wells at Pioneer Reservoir. The 200-year and HTOL WSEs were specified based on the water levels at Station 1096+00. While both the 200-year and HTOL WSEs were considered in the evaluation and subsequent design of relief wells at Pioneer Reservoir, only the result considering the 200-year WSE is presented in this paper for brevity.

The four existing relief wells are irregularly-positioned with varying El_{TOR} values, thus requiring the utilization of the previously described procedure. Figure 2 below presents an aerial view of the north side of Pioneer Reservoir, highlighting the location of the existing relief wells and the identified topographic low area, at approximately Station 1096+00.

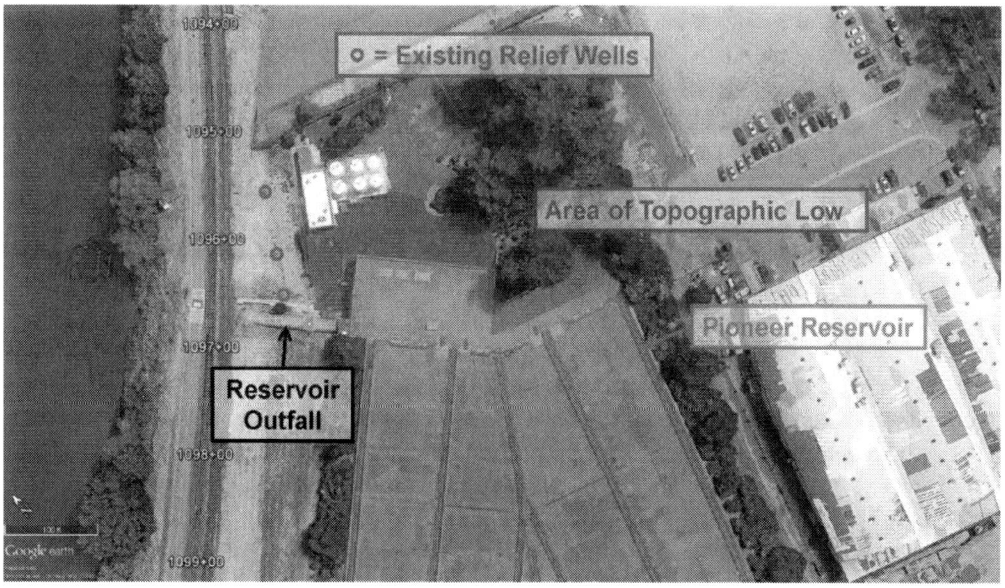

Figure 2. Aerial View of the North Side of Pioneer Reservoir (Google Earth 2017)

Step 1: Solve the System of Equations to Determine the Flow at Each Well Ignoring Head Losses

The geometry, blanket stratigraphy, and boundary conditions for the analysis section at Station 1096+00, presented in Figure 3, were used in the development of the irregularly-positioned relief well flow model. The 200-year WSE of 35.49 ft was applied 85 ft waterward of the levee centerline alignment, which corresponds to the location where the top of the aquifer is projected to daylight on the waterside riverbank. The landside boundary condition is set 2,000 ft landward of the levee alignment at El. 16 ft corresponding to the approximate ground surface elevation in that area.

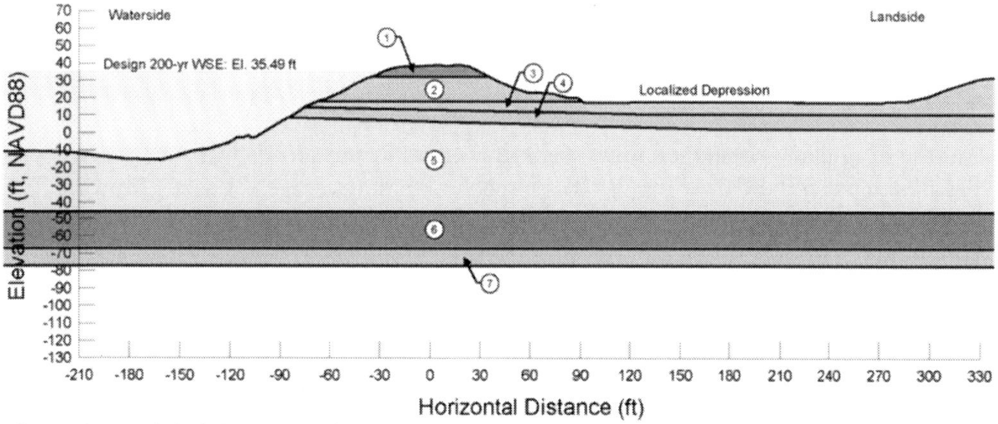

Figure 3. Modeled Cross Section at Station 1096+00 (Modified from GEI/HDR 2015)

The stratigraphy was modeled based on conditions in the vicinity of Station 1096+00 and is summarized in Table 1. The aquifer generally consists of a thick upper sand aquifer overlying gravel, identified in Figure 3 and Table 1 as Layers 5 and 6, respectively. The upper aquifer layer consists of SP with localized intervals of SM and ML. The lower aquifer layer consists of GP with lenses of GW, SP, and ML. Transforming the stratified aquifer, as described in EM 1110-2-1914, resulted with one layer of an equivalent hydraulic conductivity, $\underline{k_e}$, of $3.4\text{x}10^{-2}$ cm/sec and a transformed thickness, D, of 92 ft.

The ElTOR of each existing 8 in diameter relief wells (RW-3, RW-4, RW-5, and RW-6) were modeled with surveyed top of riser elevations and locations. Relief well RW-3 is located at the toe of the levee in the topographic low north of Pioneer Reservoir and the remaining three existing relief wells are located on an elevated bench at the toe of the levee. The relief well dimensions for the existing wells are based on the USACE (2013) Operations and Maintenance Manual for the Pioneer Reservoir Relief Wells. Based on the modeled stratigraphy and well dimensions, the existing wells are believed to be nearly fully-penetrating relief wells. The ElTOR and flow rates (Q_w) calculated ignoring head losses in Step 1 are summarized in Table 2.

Table 1. Modeled Layers and Seepage Properties at Station 1096+00

Layer Number	Material Type	Horizontal Hydraulic Conductivity, k_h	k_h/k_v
1	GW	$1.0\text{x}10^{-1}$ cm/sec	1
2	ML	$2.0\text{x}10^{-5}$ cm/sec	4
3	ML	$2.0\text{x}10^{-5}$ cm/sec	4
4	SM	$2.0\text{x}10^{-3}$ cm/sec	4
5	SP	$2.0\text{x}10^{-2}$ cm/sec	1
6	GP	$1.0\text{x}10^{-1}$ cm/sec	1
7	ML	$2.0\text{x}10^{-5}$ cm/sec	4

Step 2: Determine Head Loss for Each Well

For the irregularly-positioned relief wells at Pioneer Reservoir, H_h was iteratively calculated using the Guy et al. (2010) approach with the average net heads from the analysis. The average net head assigned to each well was calculated as the distance-weighted net head between the well of interest and midway to the nearest adjacent wells. The lines where the average net head was calculated and attributed to each relief well are shown in Figure 4. The Step 2 Q_w values calculated assuming 80% efficiency, H_h values, and resulting total head boundary conditions (H) assigned at each well location for the 200-year WSE are given in Table 2 below.

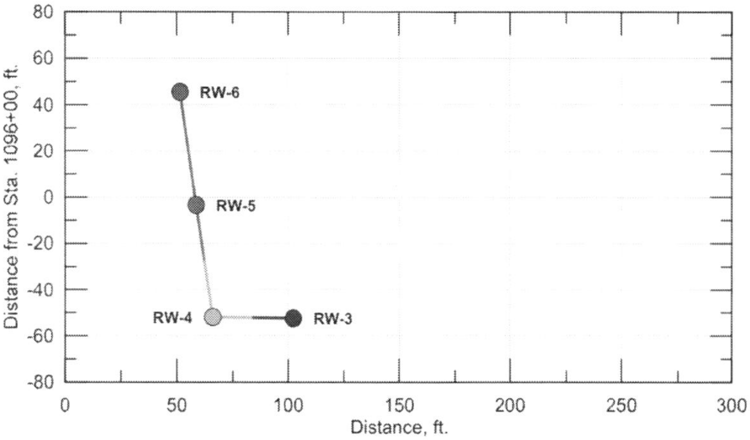

Figure 4. Average Net Head Calculation Lines for the Existing Conditions Evaluation

Table 2. Step 1 and Step 2 Results for the Existing Conditions Evaluation

		Step 1	Step 2 (Following Convergence)		
Well ID	El_{TOR} (ft)	Flow (Without Well Losses) Q_w (gpm)	Flow (80% Efficient) Q_w (gpm)	Head Loss, H_h (ft)	Total Head, H (ft)
RW-3	15.6	523	495	0.90	16.50
RW-4	18.2	368	357	0.63	18.83
RW-5	18.6	356	340	0.72	19.32
RW-6	19.0	459	434	0.78	19.78

Step 3: Determine Aquifer Head Distribution and Gradient at All Points of Interest

The distribution of heads in the aquifer from the 200-year WSE were determined in the vicinity of the localized depression on a 1 ft by 1 ft grid. The resulting head values in the localized depression for the 200-year WSE are depicted in Figure 5. Figure 5 and similar subsequent figures present results of a subset of the entire model defined as 80 ft upstream and downstream of Station 1096+00 on the project levee alignment and a distance of 300 ft landward of the levee alignment at Station 1096+00. Buildings and other identifiable feature footprints visible in Figure 2 are superimposed on Figure 5 and similar subsequent figures for reference.

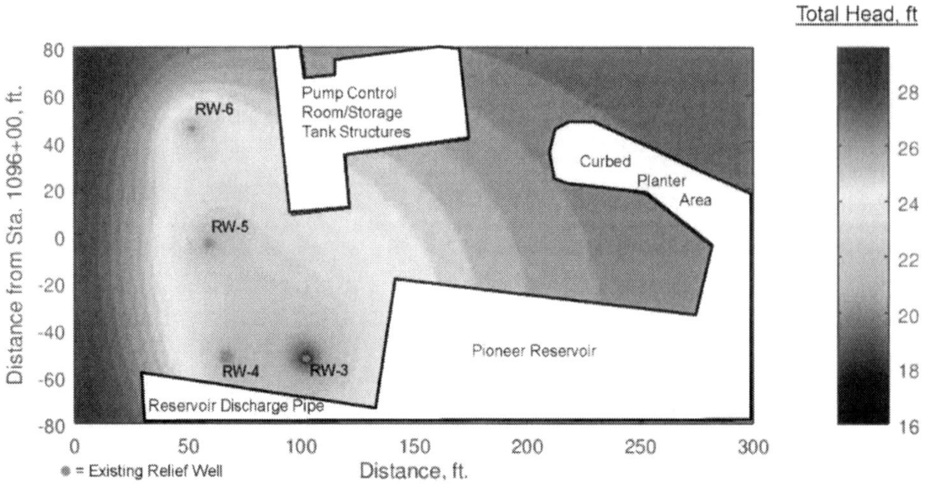

Figure 5. Existing Conditions 200-year Total Head Distribution

Layers 3 and 4, presented in Figure 3, form a stratified blanket in the vicinity of the localized depression. Figure 6(a) presents the ground surface elevations based on the LiDAR survey. Transformed blanket thicknesses were determined at each point of the grid based on the hydraulic conductivities reported in Table 1 and assuming the stratified blanked interface and lower boundary are consistent in the upstream and downstream directions. Figure 6(b) presents the transformed blanket thickness in the area of the depression.

Figure 6(c) depicts the gradient criteria in the area of concern for the 200-year WSE condition. Figure 7 presents the results of the gradient evaluation by plotting the ratio of allowable exit gradient to computed exit gradient:

$$FS = \frac{i_{allowable}}{i} \tag{7}$$

where $i_{allowable}$ is the allowable gradient and i is the calculated gradient.

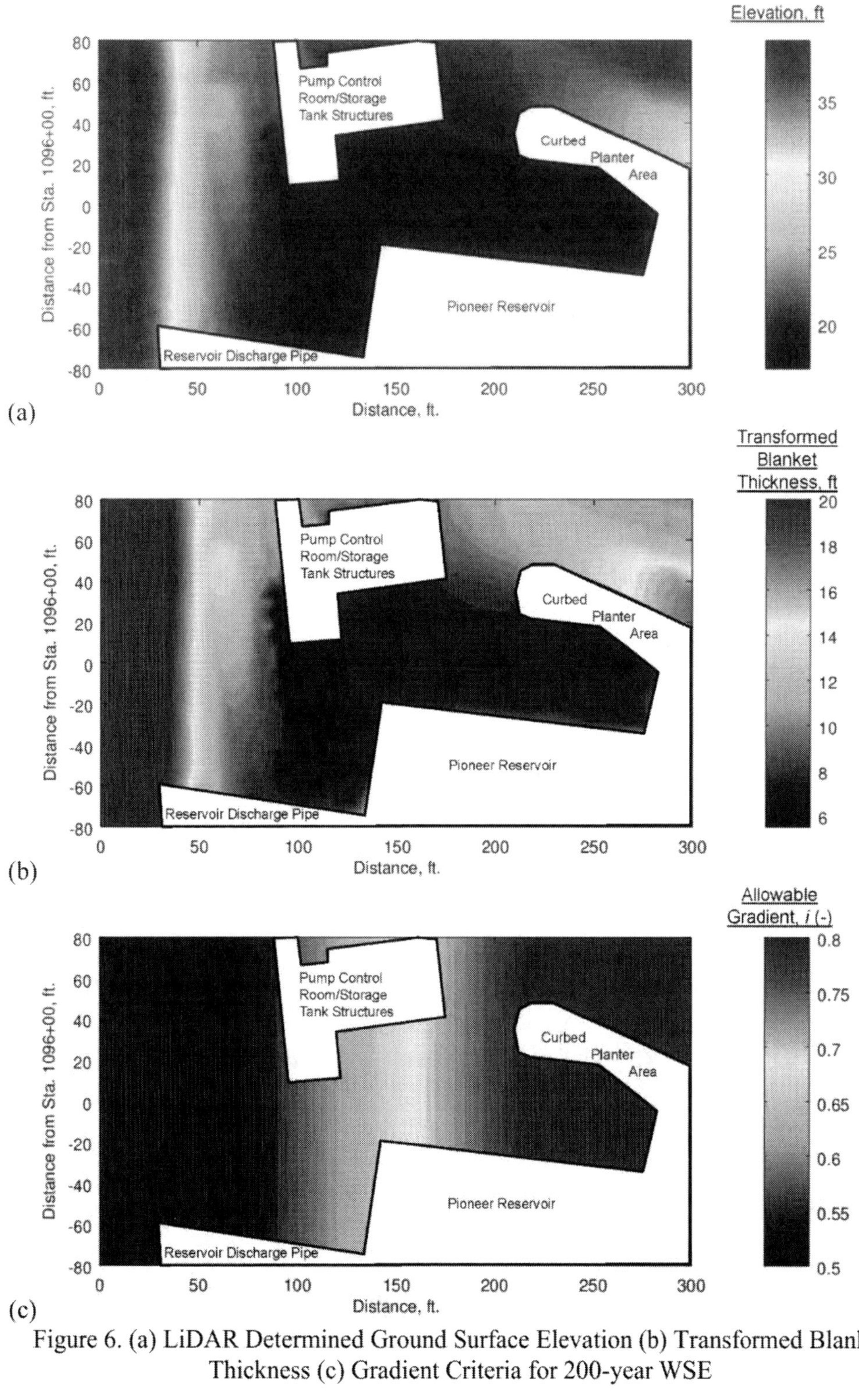

Figure 6. (a) LiDAR Determined Ground Surface Elevation (b) Transformed Blanket Thickness (c) Gradient Criteria for 200-year WSE

As shown in Figure 7, in the vicinity of the localized depression, the actual exit gradient is greater than the allowable gradient (Gradient FS < 1).

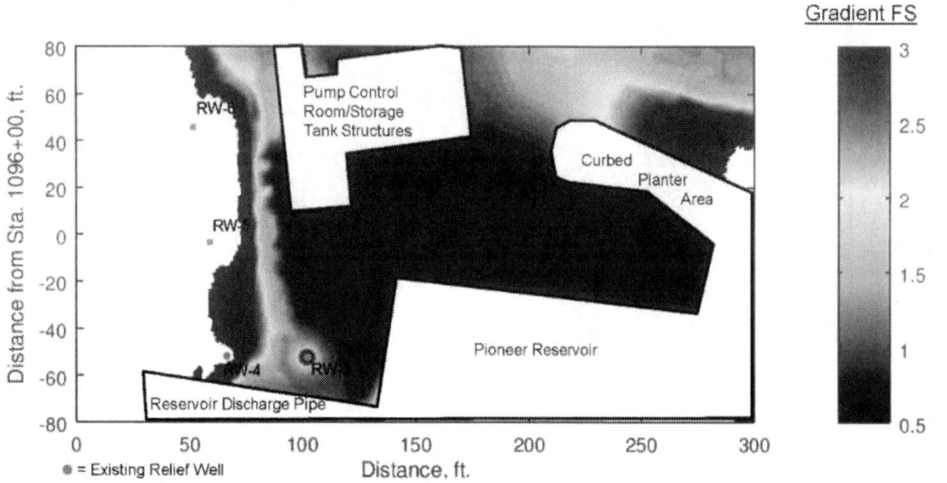

Figure 7. Existing Conditions Gradient Factor of Safety for the 200-year WSE

HEAD LOSS ESTIMATION PROCEDURE EVALUATION

To confirm the reasonableness of the head loss estimation procedure, the head loss estimates from Table 2 were compared to head loss estimates produced using the procedure presented in EM 1110-2-1914. As the present approach is an iterative procedure and the flows used should incorporate head losses, the 80% efficient flows from Step 2 were used in this comparison.

Relief well attributes from USACE (2013) and calculated values utilized to estimate the head loss using the EM procedure are presented in Table 3. As prescribed the EM procedure, the effective length of pipe for the purpose of friction loss determination is taken as half the screen length plus the length of the riser section. Also included in Table 3 are the tabulated values of flow per foot of screen and the average velocity of water in the riser pipe for each well, both of which are required when following the EM head loss estimation procedure. The flow per foot of screen is calculated dividing the flows for each well (80% efficient flows from Table 2) by the associated length of screened interval in Table 3. The average velocity of water in the riser pipe is calculated by dividing the flows in each well (80% efficient flows) by the cross sectional area of an 8 in diameter pipe, which is common to all of the existing wells.

Table 3. Well Attributes and Calculated Values for USACE EM 1110-2-1914 Approach

Well ID	Length of Screened Interval (ft)	Effective Length of Pipe (ft)	Flow per Foot of Screen (gpm/ft)	Average Velocity of Water in Riser Pipe (ft/ s)
RW-3	60	54.0	8	3.2
RW-4	60	56.0	6	2.3
RW-5	60	57.9	6	2.2
RW-6	60	55.3	7	2.8

Table 4 presents estimated values of entrance head loss, friction head loss, and velocity head loss for each well from Figure 6-2, Figure 6-3, and Equation 6-3 of EM 1110-2-1914, respectively. A range was also presented for the entrance head loss (and subsequent total head) estimate due to the fact that the entrance head loss data from 8 in diameter slotted wood well screens presented in Figure 6-2 is limited and potentially not applicable to the 8 in stainless steel well screens present at Pioneer Reservoir. Best estimate values of the entrance loss were also chosen, in-part, considering an analogous entrance loss estimation procedure presented in The Unified Facilities Criteria, Dewatering and Groundwater Control Manual, UFC 3-220-05. In the UFC 3-220-05 procedure, Figure 4-24(a) presents a recommended correlation for entrance loss based on well discharge from pumping test performed on a 16 in diameter well, 100 sq. in stainless steel screen with 5/32 in slots, and a 6 in thick filter.

Table 4. Head Loss Estimate from USACE EM 1110-2-1914 Approach

Well ID	Entrance Head Loss (ft)	Friction Head Loss (ft / 100 ft of pipe)	Friction Head Loss (ft)	Velocity Head Loss (ft)	Total Head Loss (ft)	Step 2 Head Loss, H_h (ft)
RW-3	0.50 (0.05 to 1.20)	0.83	0.45	0.16	1.11 (0.71 to 1.81)	0.90
RW-4	0.35 (0.05 to 0.75)	0.45	0.25	0.08	0.68 (0.38 to 1.08)	0.63
RW-5	0.35 (0.05 to 0.75)	0.42	0.24	0.08	0.65 (0.37 to 1.07)	0.72
RW-6	0.40 (0.05 to 1.00)	0.65	0.36	0.12	0.88 (0.53 to 1.48)	0.78

Table 4 compares the estimated total head loss from the USACE EM 1110-2-1914 procedure compared to the calculated head losses from Step 2 for 80% efficiency. The H_h values from Step 2 for RW-4 and RW-5 are similar to the best estimated values and fall in the middle of the estimated range from the EM 1914 procedure. Similarly, the H_h values for RW-3 and RW-6 are slightly lower than the best estimate (by no more than 0.21 ft) and is within the lower portion of the estimated range. Overall, the head loss estimates based on the EM 1110-2-1914 procedure compare well with the head losses calculated from the above described procedure.

PROPOSED RELIEF WELL OPTIMIZATION

Remediated Conditions Overview

Analysis of the existing relief wells indicated insufficient reduction of the aquifer heads beneath the blanket within the localized depression. Therefore, additional relief wells were considered as a remedial design option.

The proposed wells are intended to supplement the existing wells, and are also placed at irregular intervals within the localized depression to optimize reduction of head in the aquifer. An iterative procedure was used to determine the optimum location for additional wells. Additional wells were positioned in the most critical locations and reanalyzed until a configuration of relief wells meeting criteria was obtained. The procedure resulted in the proposal of four additional relief wells. The evaluation of the final proposed configuration of existing and proposed relief wells is described below.

Step 1: Solve the System of Equations to Determine the Flow at Each Well Ignoring Head Losses

Four proposed relief wells (RW4-7, RW4-8, RW4-9, and RW4-10) were placed within the depression. An El_{TOR} of 15.6 ft was selective for the new walls. Locations of the proposed relief wells are depicted in Figures 8-10. While both partially- and fully-penetrating wells were considered, fully penetrating wells were found to be required due to the deep GP aquifer unit and thin blanket in the vicinity of the depression. The resulting Q_w for each well, calculated ignoring head losses in Step 1, are summarized in Table 5.

Step 2: Determine Head Loss for Each Well

The head losses for each relief well were evaluated in a similar manner as the existing conditions evaluation, considering the distance to the nearest adjacent wells in the proposed configuration. However, unlike the existing conditions analysis, the addition of the proposed wells in the localized depression create two distinct groups of relief wells based on elevation. To prevent the artificial rise or lowering of average net heads by averaging between wells with significantly different El_{TOR} values, relief wells were grouped by El_{TOR} values for the average net head calculation, as they would be expected to have similar average net heads. As such, the relief wells in the localized depression with the same El_{TOR} value of 15.6 ft were grouped together, as were the existing relief wells on the elevated bench slope (RW-4, RW-5, and RW-6) with El_{TOR} values between 18.2 ft and 19.0 ft. This grouping for the average net head calculation is presented in Figure 8.

Table 5 summarizes the Step 2 Q_w values calculated assuming 80% efficiency, H_h values, and H values assigned at each well location for the 200-year WSE. As was performed and discussed above in Step 2 of the existing conditions evaluation, the calculated head losses compared well to head losses estimated from the USACE EM 1110-2-1914 procedure.

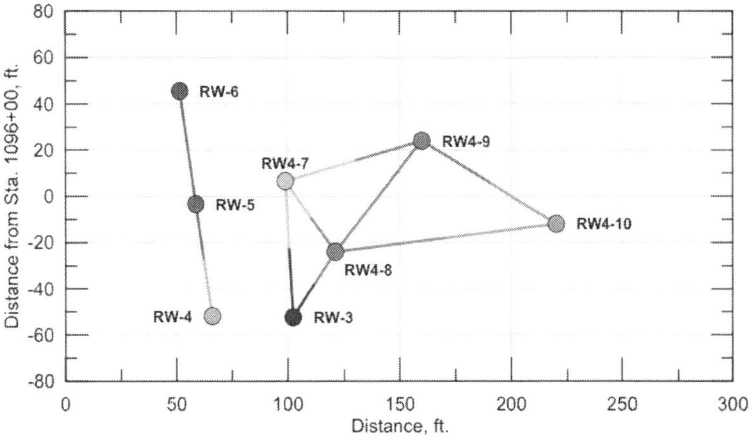

Figure 8. Average Net Head Calculation Lines for the Proposed Configuration

Table 5. Step 1 and Step 2 Results for the Proposed Configuration Evaluation

Well ID	El_{TOR} (ft)	Step 1: Flow (Without Well Losses) Q_w (gpm)	Step 2 (Following Convergence): Flow (80% Efficient) Q_w (gpm)	Step 2 (Following Convergence): Head Loss H_h (ft)	Step 2 (Following Convergence): Total Head H (ft)
RW-3	15.6	316	310	0.50	16.10
RW-4	18.2	261	254	0.46	18.66
RW-5	18.6	211	206	0.44	19.04
RW-6	19.0	330	317	0.51	19.51
RW4-7	15.6	266	262	0.50	16.10
RW4-8	15.6	216	210	0.54	16.14
RW4-9	15.6	290	278	0.62	16.22
RW4-10	15.6	340	321	0.72	16.32

Step 3: Determine Aquifer Head Distribution and Gradient at All Points of Interest

The distribution of heads in the aquifer for the existing and proposed wells was determined using the process previously presented in this paper. The resulting head values and gradient factor of safety for the 200-year WSE are depicted in Figure 9 and Figure 10, respectively.

As shown in Figure 10, the actual exit gradient is less than or equal to the allowable gradient (Gradient FS $\geq$ 1) in the vicinity of the localized depression for the 200-year WSE. Therefore, the proposed relief well configuration meets underseepage criteria.

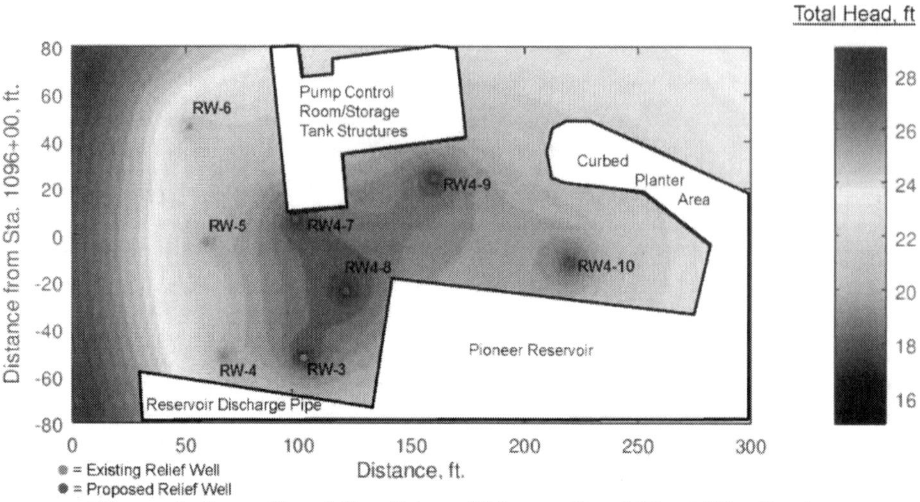

Figure 9. Remediated Conditions 200-year Total Head Distribution

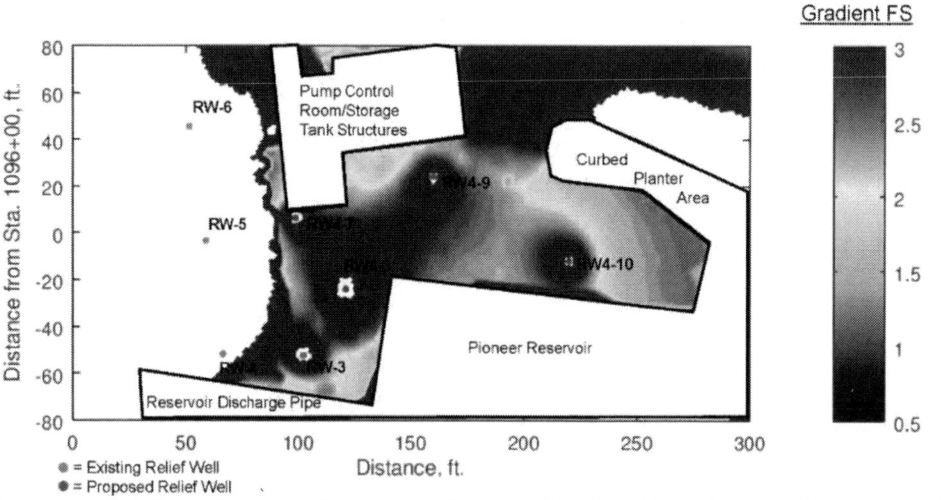

Figure 10. Gradient Factor of Safety with Proposed Relief Wells for the 200-year WSE

CONCLUSION

A design procedure based on the Thiem equation for analysis of irregularly-positioned relief wells adjacent to an infinite source incorporating an infinite line sink at the landside model boundary was presented. The procedure incorporated high-quality LiDAR data into blanket thickness calculations across the site to better identify areas of the site where ULDC exit gradient criteria was not met and support the subsequent optimization of proposed relief well configurations.

Efficiency losses determined using the Guy et al. (2010) approach, modified for use with irregularly-positioned clusters of wells, produced reasonable estimates of head loss at each well when compared to the EM 1110-2-1914 procedure.

The implementation of this procedure proved to be instrumental in both the identification of the extent of areas not meeting criteria and the optimization of the proposed remedial design. Overall, the procedure presented was shown to be effective for design, analysis, and optimization of irregularly-positioned relief wells.

ACKNOWLEDGMENTS

The authors would like to acknowledge Mr. George Sills, Mr. Stephen Verigin, and Dr. Leslie Harder, Jr. of the project's Value, Cost, and Constructability (VCC) panel for their detailed reviews of the relief well design approaches, analyses, and designs presented in this paper. In addition, the authors wish to thank the Sacramento Area Flood Control Agency (SAFCA) for their leadership and support during the project. Lastly, the authors would like to acknowledge Dr. Erich Guy for discussions regarding analyses of relief wells and procedures for estimating well losses.

REFERENCES

California Department of Water Resources (DWR) (2012). Urban Levee Design Criteria. May 2012.

GEI/HDR (2015). Embankment and Foundation Stability Memorandum – Sacramento River East Levee Improvement Project, Sacramento, California.

Google Earth (2017) Map image: Landsat, Map data: SIO, NOAA, U.S. Navy, NGA, Geboo. Accessed January 2017.

Guy, E. D., Nettles, R. L., Davis, J. R., Carter, S. C., and Newberry, L. A. (2010). Relief Well System Design Approach: HHD Case Study. In Proc. Association of State Dam Safety Officials. Charleston, West Virginia. 19 pp.

Jaeger, R. A., Keizer, R. A, Bradner, G. C., Weber, J. P., and Stanley, M. H. (2017). "Advances in Design Procedures for Relief Well Lines in Variable Urban Settings". USSD 2017 Annual Conference. In Press.

MBK Engineers (2016). "SAFCA Levee Accreditation Project, Computed Water Surface Elevations". February 10, 2016.

Thiem, G. (1906). Hydrologische Methoden, Gebhardt, Leipzig.

United Facility Criteria (2004) Dewatering and Groundwater Control, UFC 3-220-05. January 16, 2004.

USACE (1992). Design, Construction, and Maintenance of Relief Wells. Engineer Manual 1110-2-1914. May 29, 1992.

USACE (2013) Operation and Maintenance Manual, Pioneer Reservoir Relief Wells, Sacramento River Mile 58.5 – Left Bank, Sacramento, California, February 2013.

U.S. Code of Federal Regulations (2002). Title 44 (44 CFR) Section 65.10. October 1, 2002.

EXPLAINING PORE PRESSURE CHANGES DUE TO PIPING THROUGH CHANGES IN HYDRUALIC CONDUCTIVITY

Carolyne Bocovich[1]
Willem Kanning, PhD[2]
Minal Parekh, PE, PhD[3]
Michael Mooney, PE, PhD[4]

ABSTRACT

Pore water pressure measurements are used to investigate the piping failure mechanism in dams and levees through changes in hydraulic conductivity. The measured changes in pore pressure during piping development were collected during the second IJkdijk 2009 experiment. The geometry, pressure gauge locations, and soil properties were used to create an FEM model in COMSOL Multiphysics 4.3a. Using this model, various anomalies of different hydraulic conductivities were used to compare the modeled and measured pore pressure responses and trends. The modeled pore pressure have similar trends as observed in the measured data, indicating that this is a reasonable method to estimate piping behavior based on pore pressure data. However, the modeled pore pressure could not directly simulate the measured pore pressure. Further research will include more advanced parameter estimation tools used to estimate the spatial distribution of hydraulic conductivities based on measured pore pressure.

INTRODUCTION

Backwards piping erosion (referred to as piping in this paper) is a failure mechanism affecting earthen dams and levees (EDLs), occurring when sufficiently high pressure gradients erode material from underneath the embankment, forming sand boils downstream. It is one of the leading causes of EDL failure (Foster et. al. 2000) and the most dominant failure mechanism contributing to flood risk in the Netherlands (VNK1 2005). There are many models to predict whether piping will lead to failure including Bligh (Bligh 1910), Sellmeijer (Sellmeijer 1988, Sellmeijer et al. 2011), and the use of blanket equations with a limit equilibrium such as the ratio between the unit weight of buoyant soil to the unit weight of water (USACE 2000). Experiments and embankment failures have been studied to characterize and better understand the progression of piping (Fell and Fry 2007, van Beek 2015). Unfortunately, soil parameters affecting piping are highly uncertain and locally variable, resulting in uncertain failure probabilities. The observations of concentrated seepage and sandboils are largely used to detect the

---

[1] PhD Student, Department of Civil and Environmental Engineering, Colorado School of Mines, Golden, CO 80401, cbocovic@mymail.mines.edu
[2] Researcher/Consultant Dike Safety, Deltares, Delft, Netherlands, wim.kanning@deltares.nl
[3] Civil Engineer, United States Army Corps of Engineers, Risk Management Center, Lakewood, CO 80228, minal.l.parekh@usace.army.mil
[4] Professor, Department of Civil and Environmental Engineering, Colorado School of Mines, Golden, CO 80401, mmooney@mines.edu

occurrence of piping, however, they provide little information about the progression or rate of piping. Research has been conducted to provide insights into the progression of piping through pore pressure measurements as was shown by Parekh et al (2016). This paper aims to further explain pore pressure responses and trends caused by piping through spatial changes of hydraulic conductivity.

The spatial changes in hydraulic conductivity are modeled using the finite element method (FEM) to simulate pore pressure responses and trends from piping. This is done to gain a better understanding of the relationship between changes in hydraulic conductivity and measured pore pressures during piping and to address the use of FEMs to estimate changes in hydraulic conductivity. The results from the forward FEM model will be compared to measured pore pressures collected during the second IJkdijk 2009 experiment. The IJkdijk experiments were conducted to better understand the progression of piping and use of instrumentation and monitoring to detect piping. This paper will compare the trends in measured pore pressure (P) collected during the second IJkdijk 2009 experiment with modeled pore pressure ($\hat{P}$) calculated using FEM by varying local hydraulic conductivities. Measured and modeled pore water pressure are expressed in units of pressure head (meters of water) throughout this paper, and referred to as pore pressure. The goal of this paper is to investigate whether changes in hydraulic conductivity can explain pore pressure trends observed during piping.

IJKDIJK 2009

The IJkdijk experiments were conducted in the Netherlands to investigate levee behavior and study monitoring techniques (Flood Control IJkdijk 2015, Parekh et al. 2016). In 2009 a series of four experiments were designed and completed to study internal erosion, validate Sellmeijer's equations, and prove that internal erosion in levees can lead to failure (van Beek et al. 2010). Sellmeijer's equation can be used to predict whether, and at what upstream water levels, piping will lead to failure (USBR & USACE 2015, TAW 1999). The equation is based on physical relationships (Sellmeijer, 1988) and modified from small and medium scale experiments, as shown in Equation 1 (Sellmeijer et al. 2011).

$$\Delta H_c = L * F_R * F_s * F_G + 0.3 * d_{bl} \tag{1}$$

Such that:

$$F_R = \frac{\gamma'_f}{\gamma_w} * \{\eta * \tan(\theta)\} \; ; F_S = \frac{d_{70m}}{\sqrt[3]{\kappa * L}} * \left(\frac{d_{70m}}{d_{70}}\right)^{0.4} \; ; F_G = 0.91 * \left(\frac{D}{L}\right)^{\frac{0.28}{(D/L)^{2.8}-1}+0.04}$$

Here, ΔH_c is the critical head such that piping will lead to failure if the head difference across the embankment is not lowered, L is the horizontal seepage length of the embankment, γ_f' is the buoyant unit weight of the sandy aquifer (erodible material), γ_w is the unit weight of water, η is White's coefficient, θ is the rolling friction angle, d_{70} is the 70th percentile of sand diameter by mass, κ is the intrinsic permeability, and D is the thickness of the aquifer (Sellmeijer et al. 2011).

The IJkdijk experiments performed in 2009 were designed with a sandy aquifer underlying a clay embankment, as seen in Figure 1, to promote the progression of piping underneath the embankment. The aquifer was built with a depth of 3 meters and a horizontal seepage length of about 15 m (van Beek et al. 2010). This paper uses pore pressure measurements collected during the second IJkdijk 2009 test. The relevant soil properties are listed in Table 1.

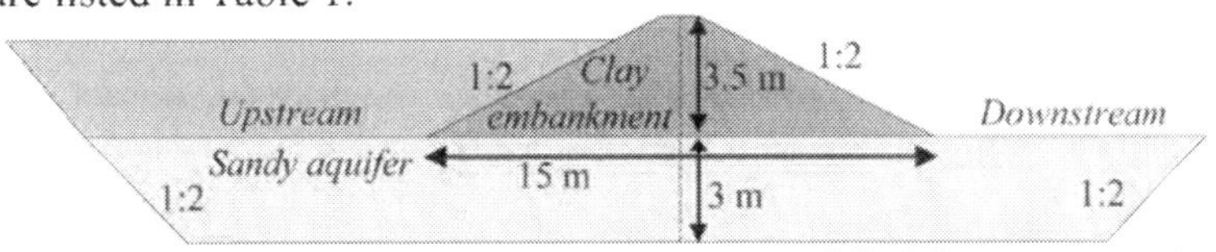

Figure 1. Cross Section of IJkdijk 2009 experiment (modified from van Beek et al. 2010)

Table 1. Soil properties of the aquifer in IJkdijk 2009, test 2 (Kramer 2014).

L	15 m
D	2.85 m
K (hydraulic conductivity)	$1.4*10^{-3}$ m/s
d_{70}	260 μm
γ_f'	24.7 kN/m^3
Θ	33

During the experiment, the downstream water level remained constant while the upstream water level (UWL) was increased incrementally by 0.1 m per hour until sand transport was observed at the downstream toe. Once sand transport was initiated, the UWL was not increased until sand transport ceased or 24 hours had passed and sand production remained stable (van Beek et al. 2010). To monitor the progression of internal erosion, densely spaced pore pressure gauges were installed just below the embankment, as seen in Figure 2.

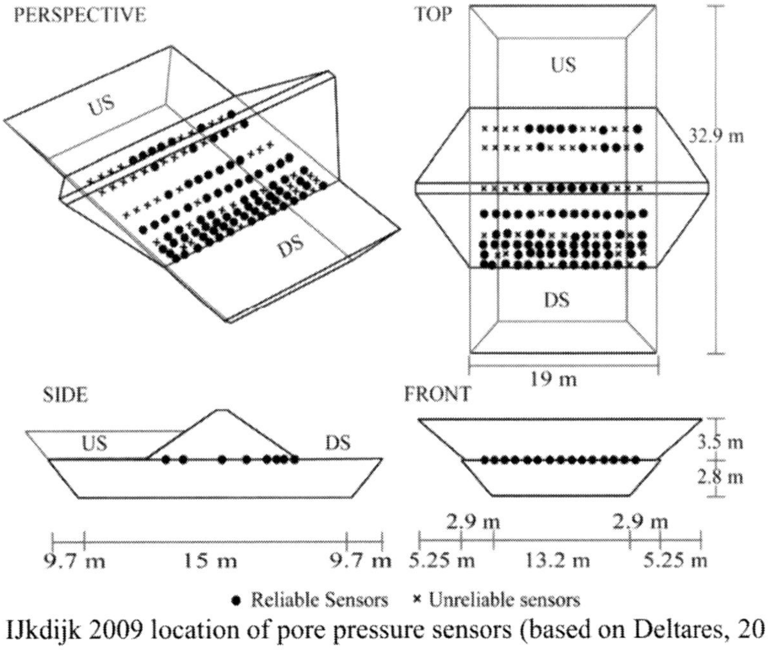

Figure 2. IJkdijk 2009 location of pore pressure sensors (based on Deltares, 2012 data)

COMPARING MEASURED AND MODELED PORE PRESSURE

FEM modeling

The FEM model was constructed using *COMSOL Multiphysics 4.3a* based on the IJkdijk 2009 geometry, pressure gauge locations, and soil properties, as shown in Figures 1 and 2, and Table 1. A simplified transient model using Darcy's law with a conservation of mass equation, as seen in Equation 2, is used to calculate modeled pore pressure ($\hat{P}$) at each sensor in order to compare with the measured pore pressure (P).

$$\frac{\partial \hat{P}}{\partial t} - \frac{K}{S_s}\left(\frac{\partial^2 \hat{P}}{\partial x^2} + \frac{\partial^2 \hat{P}}{\partial y^2} + \frac{\partial^2 \hat{P}}{\partial z^2}\right) = 0 \qquad (2)$$

where $\hat{P}$ is the modeled pore pressure, t is time, K is the model hydraulic conductivity, S_s is the specific storage, and x, y, and z are the spatial dimensions. It is assumed that the hydraulic conductivity and specific storage are isotropic through the sandy aquifer. It is also assumed that although the hydraulic conductivity might change within the aquifer, the specific storage remains a constant value of 10^{-8} 1/Pa, which has been shown not to significantly affect the relationship between the hydraulic conductivity and the modeled pore pressure (Cardiff and Barrash 2011).

The 3D FEM model is used to predict the pore pressure response to varied hydraulic conductivity anomalies. The mesh consists of triangular elements ranging in size from 0.02 m to 1.8 m, to best capture the small diameter of the pipe while maintaining reasonable convergence times. The smallest elements are located in the area of expected piping, directly underneath the embankment.

Most synthetic anomalies are positioned under the embankment, as was observed in small and medium scale experiments (van Beek 2015). The thickness of the anomaly is set to 0.25 m, which is large compared to expected piping diameters (van Beek 2015). The combined effect of hydraulic conductivity and anomaly size, or transmissivity, will be examined in this paper. Studies such as Olsen and Stephens (2016) show how pore pressure response is related to both the size of an anomaly and the hydraulic conductivity of an anomaly, indicating that this is a reasonable assumption.

Trends in measured pore pressure

The measured pore pressure has been examined by van Beek et al. (2010), Sellmeijer et al. (2011), and Parekh et al. (2016). Local decreases in pore pressure during a constant or increasing UWL were observed during sandboil production, and related to piping progression (van Beek et al. 2010). The pore pressures were studied to predict the progression of the piping channel in Sellmeijer et al. (2011). However, it is cautioned that only one transect is used for the analysis while piping occurs in 3 dimensions (Sellmeijer et al. 2011). In Parekh et al. (2016) it is noted that decreases in pore pressure are observed upstream and remote of localized erosion.

Two trends of interest are observed by studying patterns of measured pore pressure normalized to modeled pore pressure ($P/\hat{P}_h$) calculated from assuming homogenous hydraulic conductivity, referred to as the homogenous model ($\hat{P}_h$), and measured pore pressure normalized to the UWL (*P/UWL*). The first trend is denoted as first detection (*fd*), where the normalized pore pressure starts to decrease followed by the second trend of pressure stabilization (*ps*) where the slope of normalized pore pressure reduces toward zero (Parekh et al. 2016). The rise in UWL during the initial stages of the IJkdijk 2009 experiment creates noise in the measured pore pressure and results in difficulty detecting *fd* when normalizing the measured pore pressure to the UWL. These trends can be observed in pore pressure data shown from one transect in Figure 4.

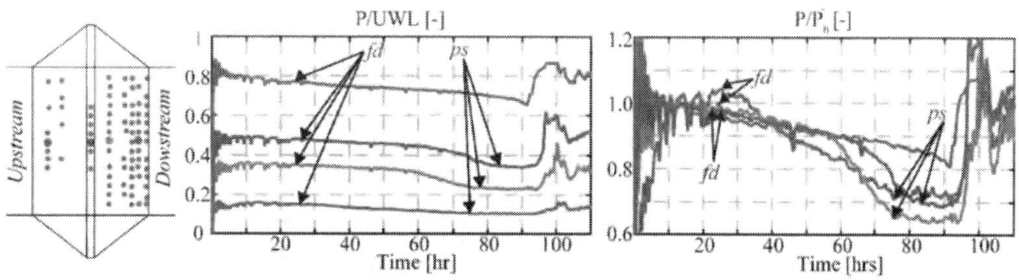

Figure 4. Trends in pore water pressure data

Pore pressure calculated from the homogenous model ($\hat{P}_h$) compared with measured pore pressure (P) shows that as piping progresses the measured pore pressure decreases from the modeled response, seen in Figure 5. This is discussed in Parekh, et al. (2016) and agrees with analysis from Olsen and Stephens (2016), as the hydraulic conductivity increases downstream of a sensor, the pore pressure will decrease from expected values.

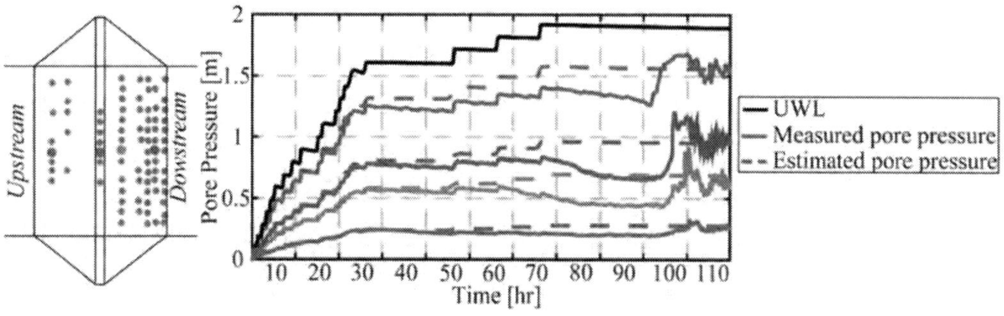

Figure 5. Comparison between measured P (solid lines), pore pressure calculated using a homogenous FEM model ($\hat{P}_h$, dashed lines), and the UWL (black line).

To further analyze the measured pore pressure (*P*) spatially, contour plots are produced using a linear interpolation and constant extrapolation of measured pore pressure. Two sets of contour plots are shown, the first row presents the measured pore pressure while the second row presents the deviation of measured pore pressure from modeled pore pressure (calculated using the homogenous model) ($\hat{P}_h - P$).

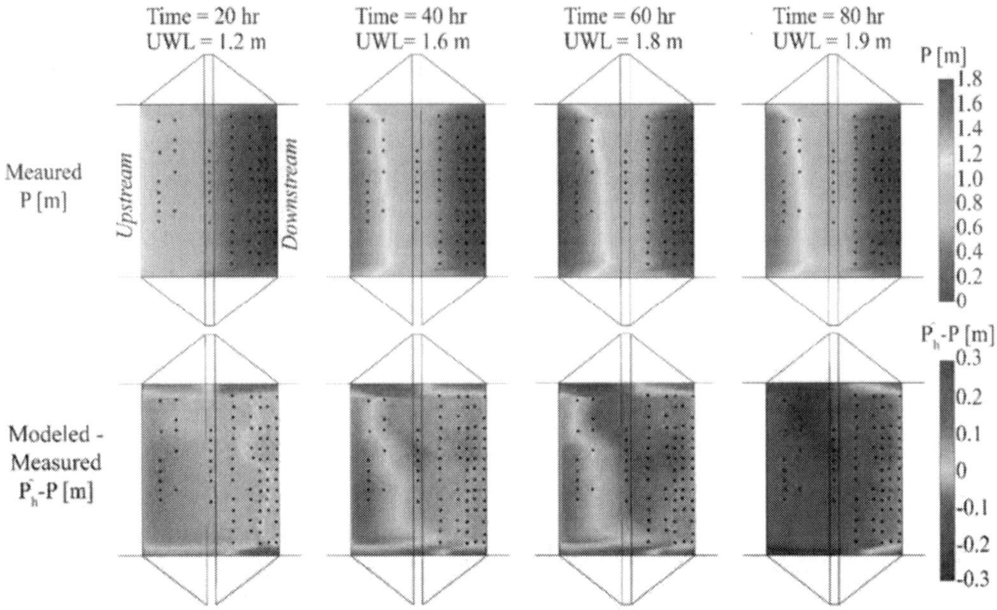

Figure 6. Contours of pore pressure, shown in plain view

The contour plots in Figure 6 show the deviation of measured pore pressure from the expected pore pressure. Due to simultaneous progression of piping and increases in UWL, these trends are difficult to interpret in the pore pressure contour plots. Contour plots showing the deviation of measured from modeled pore pressure (P-$\hat{P}_h$) show more clearly the locations and magnitudes of deviations. As time increases, the location of highest deviation moves upstream, capturing the influence of piping progression. The largest deviations of pore pressure occur on the lefts side of the embankment (looking downstream), corresponding with the location of sandboil production and embankment breach (Parekh et al. 2016).

Simulating pore pressure trends with FEM

In order to investigate the changes in hydraulic conductivity required to simulate the deviations in pore pressure, four cases with different synthetic anomalies are modeled in *COMSOL* using the FEM model. The anomaly cases are described below and Cases 1, 3, and 4 are shown in Figures 7-9.

- Case 1. Pipe anomaly with constant UWL: a thin (0.25 m X 0.25 m) anomaly with high hydraulic conductivity of 0.01 m/s compared to the background hydraulic conductivity of 0.00014 m/s with increasing lengths of 2.5 m, 7.5 m, 12.5 m, and 15 m starting at the downstream toe. The anomaly is located in the center of the embankment under transect 8, in order to study the pore pressure responses at and remote of the anomaly.
- Case 2. Pipe anomaly with increasing UWL: a thin anomaly (0.25 m x 0.25 m) with high hydraulic conductivity of 0.01 m/s compared to a background hydraulic conductivity of 0.00014 m/s with increasing lengths relative to increasing UWL based

on Sellmeijer's equations. The anomaly starts at the downstream toe, as in Case 1, increasing in length by 0.1 m, while the UWL is calculated using a linear relationship, $UWL = \frac{1}{5.4}(l + 0.35)$, where l is the pipe length. This relationship is based on Sellmeijer's equations as presented in Kramer (2014). These anomalies are not shown in Figures 7-9, but results are compared with the pipe anomaly with constant head (Case 1) in Figure 10. As in Case 1, the anomaly is located in the center of the embankment under transect 8, in order to study the pore pressure responses at the anomaly and remote of the anomaly.

- Case 3. Upstream clogging anomaly: a low hydraulic conductivity anomaly of 10^{-5} m/s compared with a background hydraulic of $1.4*10^{-4}$ m/s is applied just below the upstream reservoir with a thickness of 0.25 m, to study the effects of clogging. The UWL is constant at 1.8 m, and steady state seepage is assumed.
- Case 4. Transmissivity: high hydraulic conductivity anomalies starting at the toe are studied with increasing anomaly width relative to hydraulic conductivity to maintain a constant transmissivity. This is to verify that the hydraulic conductivity and the area of the anomaly determine the magnitude of pore pressure response. Hence a wider (or deeper), less permeable channel may result in the same pore pressure drop as a more narrow (less shallow) and more permeable channel. The anomaly width and hydraulic conductivities are shown in Table 2. The UWL is constant at 2 m.

Table 2. Pipe width to hydraulic conductivity in Case 4

Pipe width [m]	Hydraulic conductivity [m/s]
0.25	0.01
0.5	0.005
1	0.0025

To compare with similar studies, both the pore pressure and the deviation between the homogenous model ($\hat{P}_h$) and the anomaly models ($\hat{P}_a$) are studied. The first two cases relate increasing pipe length with normalized pore pressure, normalized to the UWL and to the homogenous model, which are compared with results from Parekh et al. (2016) shown in Figure 4.

RESULTS

The results from anomaly Cases 1-4 demonstrate the relationship between pore pressure and hydraulic conductivity, showing similar contours as interpolated from the IJkdijk 2009 measured pore pressure, shown in Figure 6. Figures 7-9 show anomalies from Cases 1, 3, and 4 and the corresponding pore pressure distribution resulting from these anomalies. The pore pressure distributions and the deviation of pore pressure from the homogenous model in Figure 7 show that the largest deviations in pore pressure surrounds the upstream tip of the anomaly where flow resistance to the downstream toe of the embankment is lowest. The pore pressure measurements downstream of this point also experience decreased resistance to flow to the upstream reservoir, allowing for the pore pressure to increase toward their homogenous values. Deviations in pore pressure occur remote of the anomaly but decrease with distance from the anomaly tip. This is similar to what is seen in Figure 6, where high deviations in pore pressure are surrounded by smaller deviations.

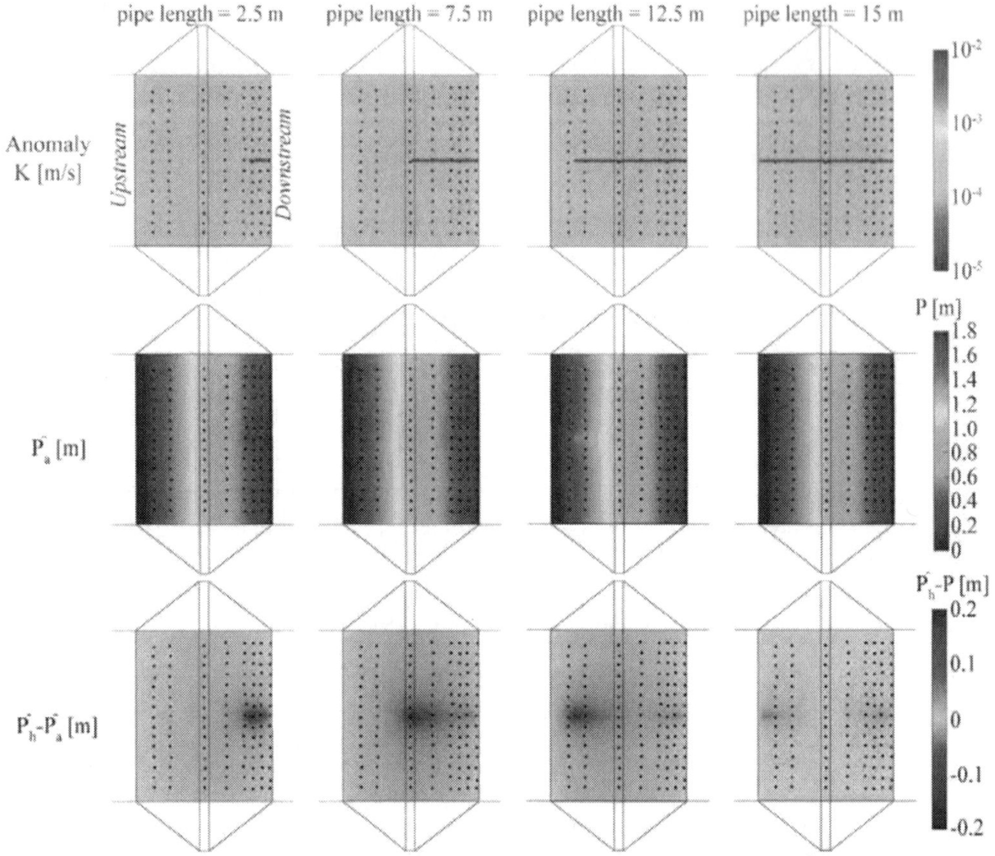

Figure 7. Test 1, showing anomalies (K [m/s]), resulting pore pressure ($\hat{P}_a$), and deviation of modeled anomaly pore pressure and homogenous pore pressure ($\hat{P}_a$-$\hat{P}_h$)

Results from Case 1 show that with increasing anomaly length, the deviations in pore pressure are highest at the tip of the anomaly, but do affect pore pressures away from this point.

The upstream clogging layer, from Case 3, causes a spatial decrease in pore pressure as compared to the homogenous model assuming steady state conditions. The deviation is highest at the upstream toe of the embankment and decreases toward the downstream toe. Unlike pore pressure trends seen at times 40 hrs, 60 hrs, and 80 hrs of the IJkdijk experiment (Figure 6), the spatial pore pressure deviations observed in Case 3 are consistent laterally across the embankment and decrease evenly from upstream to downstream, shown in Figure 8.

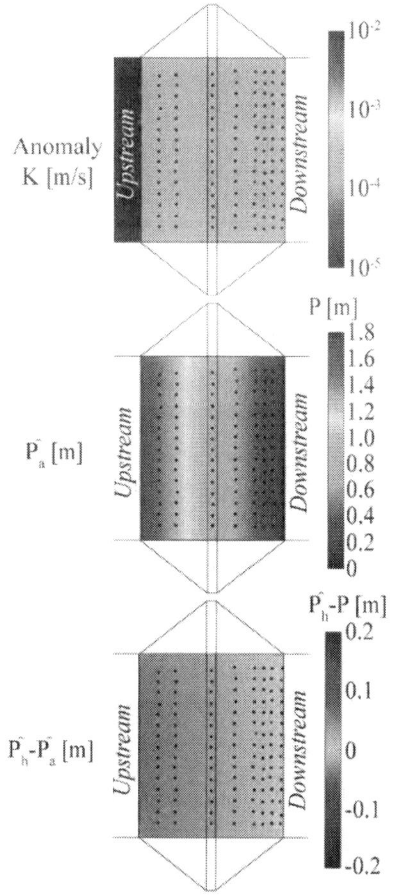

Figure 8. Test 3, anomalies (K [m/s]), resulting pore pressure ($\hat{P}_a$), and deviation of modeled anomaly pore pressure and homogenous pore pressure ($\hat{P}_h$- $\hat{P}_a$)

Results from Case 4 (Figure 9) show only slight differences in magnitude and spatial extent of the deviated pore pressure, indicating that the combination of area and hydraulic conductivity of the anomaly (or transmissivity) effects the patterns in pore pressure deviation from the homogenous model. This indicates it should be possible to reproduce pore pressure trends seen in the IJkdijk 2009 experiments by adjusting the hydraulic conductivity in the modeled piping channel, even when modeling a larger pipe cross sectional area than is expected to occur in the field.

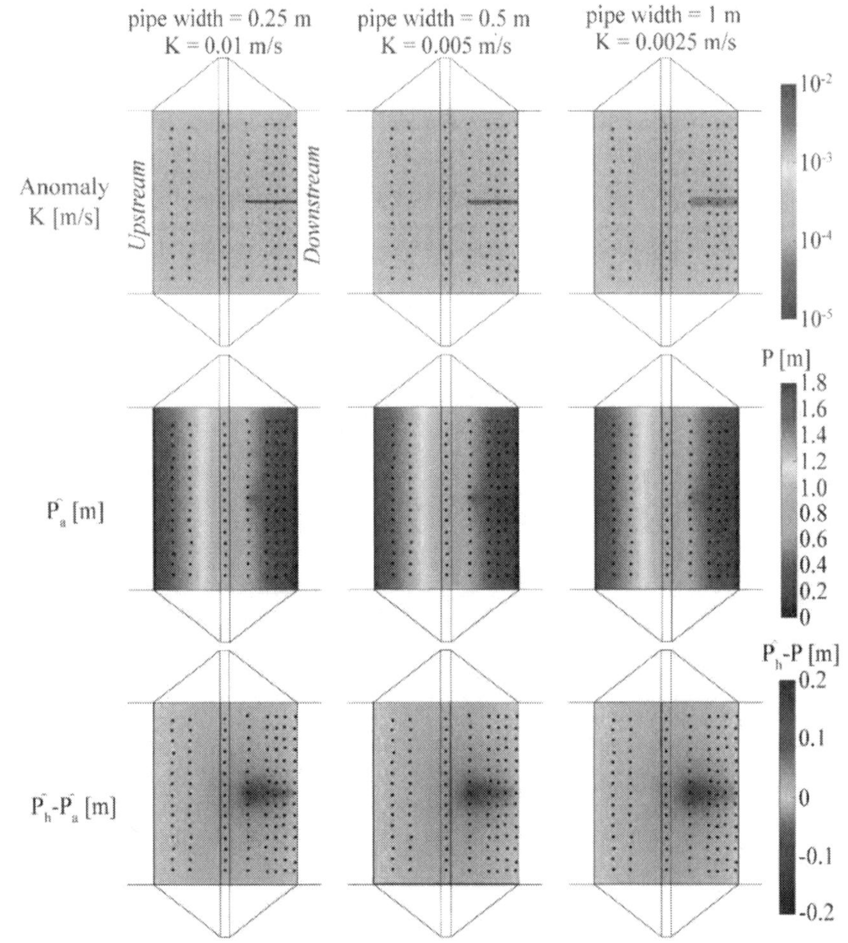

Figure 9. Test 4, contours of pore water pressure with corresponding anomalies

Further trends in the relationship between the length of modeled anomalies and pore pressure are presented. Figures 10 and 11 show the response to normalized pore pressures with increasing anomaly length, as investigated in Cases 1 and 2. Pore pressures are normalized to both UWL and pore pressure from a homogenous hydraulic conductivity model ($\hat{P}_h$) to compare modeled pore pressure trends to those investigated in Parekh et al. (2016), shown in Figure 4. The modeled anomaly is located along transect 8 for both tests, while transect 6 is included to compare trends seen remote from the anomaly.

The normalized pore pressures in transect 8 during constant UWL (Case 1) decrease as the anomaly remains downstream of the pressure gauge location but increases as the pipe length passes the gauge location, shown in Figure 10. This response clearly demonstrates the length of the anomaly. However, this only appears to be the case when the anomaly is directly over the pressure gauge sensors and the UWL is constant. Normalized pressure trends in transect 6 (removed from the anomaly) respond similarly, however to a lesser magnitude and with a time lag. This indicates that as the pore pressure starts increasing again, the anomaly length is already past the gauge location.

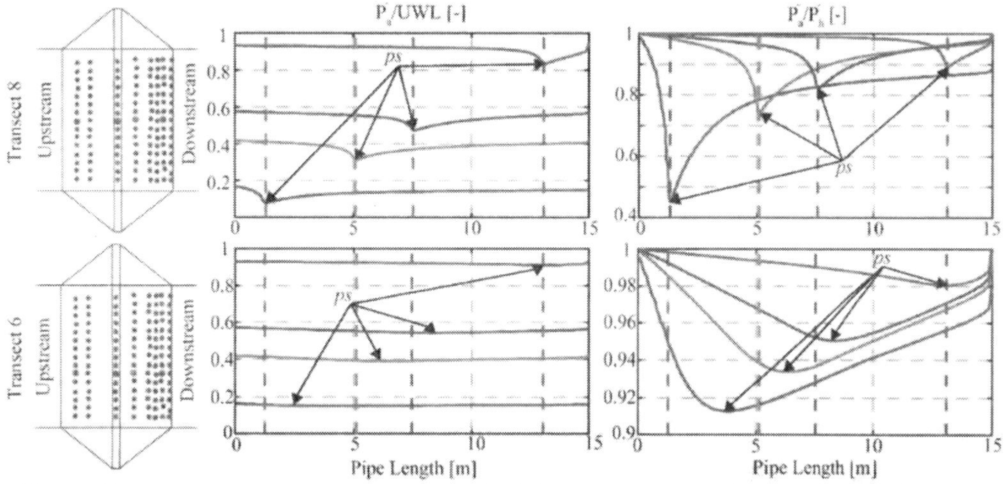

Figure 10. Case 1, pore pressure compared with seepage length for transects at (transect 8) and remote of (transect 6) the anomaly. Dashed lines indicates pipe length corresponding with sensor location.

The responses in both transect 6 and 8 during increasing UWL and increasing pipe length (Case 2) are less clear, shown in Figure 11. The response normalized to the UWL is controlled by the increasing UWL in the pressure gauges located towards the downstream of the embankment (rows 1-5), while slight trends similar to those during constant UWL are apparent in the upstream sensors (rows 6-8). The pore pressure normalized to a homogenous model increase after the anomaly length has passed the pressure gauge location, resulting in little information about pipe length.

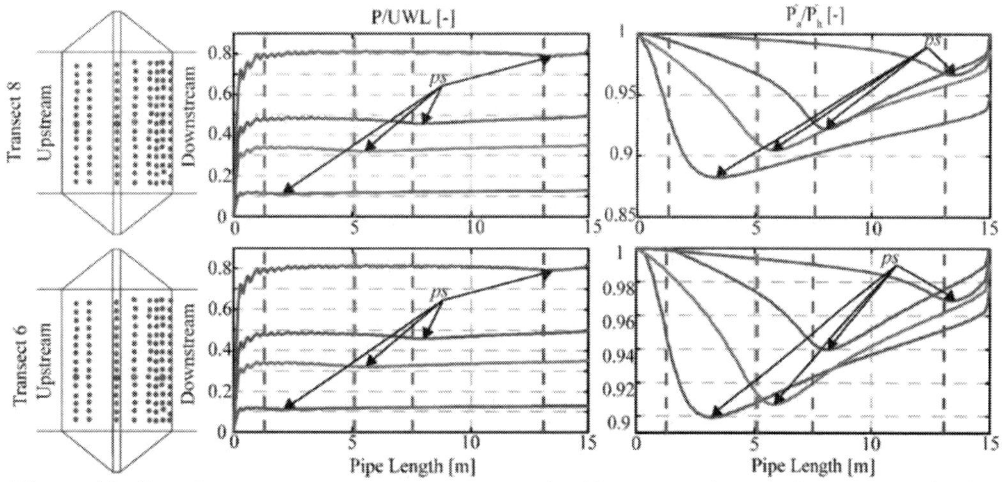

Figure 11. Case 2, pore water pressure compared with seepage length for transects in the anomaly (transect 8) and remote of the anomaly (transect 6). Dashed lines indicate the pipe length at the corresponding sensors.

The modeled pore pressure response with increasing anomaly length, as studied in Cases 1 and 2 and shown in Figure 10 and 11, result in similar trends as observed in the IJkdijk 2009 measured pore pressure. This comparison using one sensor is shown in Figure 12.

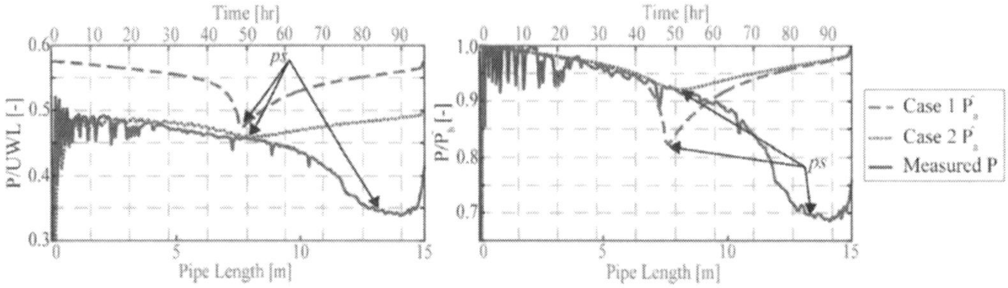

Figure 12. Comparison of normalized pore pressure trends from measured IJkdijk data and modeled data from Cases 1 and 2

A decrease in normalized pore pressure is followed by a slight increase or stabilizing of pore pressures, labeled as pressure stabilization (*ps*) due to piping progression is observed in the measured and modeled normalized pore pressure trends, seen in figure 12. However, the model pore pressures do not stabilize, but instead start increasing again. This indicates that one pipe with a consistent width is inadequate to model pore pressure responses to piping behavior, as described through Sellmeijer's equations (Sellmeijer, 1988). This is consistent with small and medium scale experiments that show that a zone of piping channels form rather than one channel and the channel(s) width increases at the downstream toe (van Beek et al. 2011).

CONCLUSION

This paper demonstrates that the relationship between spatial changes in hydraulic conductivity and pore pressure can be used to simulate the changes in pore pressure caused by piping. To do this, trends in measured pore pressure are compared with modeled pore pressure responses to spatial hydraulic conductivity anomalies. The FEM modeling was performed using COMSOL Multiphysics 4.3a. A transient model based on Darcy's law and the conservation of mass equation was used with the geometry, pressure gauge locations and soil parameters of the 2nd IJkdijk 2009 experiments to better understand piping.

Cases 1, 2, and 4 show that similar pore pressure responses are produced using the FEM model with changes in hydraulic conductivity as were observed in the measured pore pressure collected during the IJkdijk 2009 experiment. The pore pressure decreases relative to the expected homogenous pore pressure, with the highest decreases surrounding the tip of the anomaly. Normalized pore pressure trends experience a decrease before the high hydraulic conductivity anomaly reaches the pore pressure location. These trends are also experienced in the IJkdijk 2009 pore pressure measurements (Parekh et al. 2016). However, the modeled pore pressures do not directly simulate the changes observed in the measured pore pressure. Based on these results, further research will use parameter estimation techniques to estimate changes in hydraulic conductivity based on the measured pore pressure. This will be performed by comparing the modeled pressure using the described FEM forward model and the measured pore pressure from the IJkdijk experiments to estimate a spatial distribution of hydraulic conductivity.

REFERENCES

Bligh, W.G. 1910. "Dams, Barrages and Weirs on Porous Foundations." *Engineering News* 64: 708–10.

Cardiff, M., and W. Barrash. 2011. "3D Transient Hydraulic Tomography in Unconfined Aquifers with Fast Drainage Response." *Water Resources Research* 47: W12518. doi:10.1029/2010WR010367.

COMSOL Multiphysics version 4.3a. [Computer SeSoftware]. Comsol, Palo Alto. CA.

Fell, R., and J.J. Fry. 2007. *Internal Erosion of Dams and Their Foundations*. London UK: Taylor & Francis.

Flood Control IJkdijk. 2016. " Background." ⟨http://www.floodcontrolijkdijk.nl/en/ijkdijk/background ⟩ (Apr. 13, 2016).

Foster, M., R. Fell, and M. Spannagle. 2000. "The Statistics of Embankment Dam Failures and Accidents." *Canadian Geotechnical Journal* 37: 1000–1024.

Kramer, R. 2014. "Piping under Transient Conditions: Investigation of Time-Dependent Erosion under Dikes." Master, Netherlands: University of Twente.

Olson, R., and I. Stephens. 2016. "Referencing how to look at piezometer data for seepage evaluation". USSD2016.

Parekh, M.L., W. Kanning, C. Bocovich, M. Mooney, and A.R. Koelewijn. 2016. "Backward Erosion Monitored by Spatial-Temporal Pore Pressure Changes during Field Experiments." *Journal of Geotechnical and Geoenvironmental Engineering* 142 (10).

Sellmeijer, J.B. 1988. "On the Mechanisms of Piping under Impervious Structures." Dissertation, Netherlands: TU Delft.

Sellmeijer, J.B., J.L. de la Cruz, V.M. van Beek, and H. Knoeff. 2011. "Fine-Tuning of the Backward Erosion Piping Model through Small-Scale, Medium-Scale and IJkdijk Experiments." *European Journal of Environmental and Civil Engineering* 15 (8): 1139–54.

TAW. 1999. *Technical Report on Sand Boils (piping)*. Technical report, Technische Adviescommissie Waterkeringen.

U.S. Army Corps of Engineers – USACE. 2000. *Engineering and Design – Design and Construction of Levees*. EM 1110-2-1913. April 2000.

U.S. Department of the Interior, Bureau of Reclamation (USBR), and U.S. Army Corps of Engineers (USACE). 2015. "Chapter IV-4. Internal Erosion Risks for Embankments and Foundations." In *Best Practices in Dam and Levee Safety Risk Analysis*, 4th ed. https://www.usbr.gov/ssle/damsafety/risk/BestPractices/Chapters/IV-4-20150617.pdf.

van Beek, V.M., H. T. J. de Bruijn, J. G. Knoeff, A. Bezuijen, and U. Foster. 2010. "Levee Failure Due to Piping: A Full Scale Experiment." In, *Scour and Erosion GSP 2010.* 283–92. San Francisco, California, USA: American Society of Civil Engineers (ASCE).

van Beek, V. M., H. Knoeff, and H. Sellmeijer. 2011. "Observation on the Process of Backward Erosion Piping in Small-, Medium- and Full- Scale Experiments." *European Journal of Environmental and Civil Engineering*, no. 15: 1115–37.

van Beek, V. 2015. "Backward Erosion Piping: Initiation and Progression." Dissertation, Netherlands: TU Delft. http://repository.tudelft.nl/.

VNK1. 2015. "The National Flood Risk Analysis for the Netherlands." Rijkswaterstaat VNK Project Office.

DESIGN AND CONSTRUCTION OF FEATHER RIVER WEST LEVEE IMPROVEMENTS

Christopher Krivanec[1]
Les Harder[2]
Michael Bessette[3]

The Feather River West Levee, located east of Yuba City, California, runs over 44 miles along the Feather River. It was originally constructed using hydraulic and clam-shell dredging processes over 100 years ago. At the time it was constructed, it protected generally rural areas and was intended to promote agricultural development in the 165,000-acre Sutter-Butte basin. Today the basin is home to over 70,000 people, but is considered one of the most vulnerable regions in the Sacramento Valley in terms of risk of significant flooding. In 1955 the levee failed during a major flood, killing 38 people in Yuba City. Recent levee investigations have shown that the levees remain vulnerable to underseepage, through seepage, and erosion, and do not provide 100-year flood protection. The potential depths of flooding within the basin exceed 15 feet, and floodwaters would likely consist of cold snowmelt runoff from the mountains to the east.

To address this significant flood risk and threat, the communities in the Sutter-Butte basin formed the Sutter Butte Flood Control Agency (SBFCA). SBFCA has partnered with the State of California to rapidly complete levee improvements necessary to provide a 0.5 percent annual chance (200-year) level of flood protection to the northern urban portion of the basin by 2017. The overall Feather River West Levee improvements are expected to cost over $300 million.

This paper summarizes the challenges the levee design team faced while navigating the project through problem identification, alternatives analysis, design and construction. Key design and construction efforts have included the need to construct new seepage cutoffs to address through and underseepage issues over 85% of the project's length, the need to work within the existing levee geometry to minimize land use impacts, the reuse of existing materials through the use of a zoned embankment, addressing existing encroachments in or near the levee, meeting evolving and conflicting urban level design criteria, and working in areas of historic cultural significance. The first phase of the project was awarded the ASCE 2015 Project of the Year in the State of California.

[1] HDR, christopher.krivanec@hdrinc.com
[2] HDR
[3] Sutter Butte Flood Control Agency

SIMPLIFIED NONLINEAR SEISMIC ANALYSES OF GRAVITY DAMS

Elisa Messio, M.Sc.A.[1]
Wenbo Duan, Ph.D. Candidate[2]
Najib Bouaanani, Eng. Ph.D.[3]
Benjamin Miquel, Eng., Ph.D.[4]

ABSTRACT

World-wide experience shows that complete concrete dam failures under seismic loads are rather rare, but more or less extensive cracking has been observed. Such cracking might modify the dynamic properties of a concrete dam during the shaking, while indicating its capacity to dissipate seismic energy and to redistribute seismic forces accordingly. Limited cracking can then be beneficial to the seismic stability of concrete dams and might therefore avoid unnecessary rehabilitation interventions. Conventional cracking analyses based on the gravity method are unable to account for this potentially beneficial behavior while more adequate numerical simulations are rarely conducted in day-to-day engineering practice. Nonlinear dynamic analyses indeed require the use of finite element programs which are generally time-consuming and not adapted to account for phenomena specific to dams, such as uplift-pressures and interactions between the dam, impounded reservoir and underlying rock foundation. In this paper, we propose a new reliable simplified technique to effectively and rigorously assess the nonlinear seismic response of concrete gravity dams. The proposed procedure is based on an equivalent cantilever stepped beam system developed to account for the rupture of dam joints. The proposed methodology is applied to investigate the nonlinear sliding response of a typical concrete dam. The obtained results are compared to finite element simulations. Friction and cohesion at concrete joints and dam-rock interface are varied to investigate their impact on dam nonlinear seismic response.

INTRODUCTION

Many aging concrete gravity dams are located in moderate to high seismic activity zones, and it is therefore imperative to assess their dynamic response under the effect of seismic loads typical of these regions. Different approaches are available for this purpose, varying from: (i) sophisticated numerical methods accounting for several parameters such as dam deformability, water compressibility, wave absorption at reservoir bottom, or dam-rock and concrete joints nonlinearities, to (ii) more practical techniques where simplifying assumptions are introduced to facilitate the application of these methods in day-to-day engineering practice. In practice, a progressive approach where the degree of complexity of the seismic analysis is gradually increased from one level to the next is generally applied, i.e. from pseudo-static analysis, to pseudo-dynamic analysis, to linear dynamic analysis, to nonlinear dynamic analysis. In many cases, the pseudo-dynamic method

---

[1] Dept. of Civil, Geological and Mining Eng., Polytechnique Montréal, QC, Canada. Email: elisa.messio@polymtl.ca
[2] Dept. of Civil, Geological and Mining Eng., Polytechnique Montréal, QC, Canada. Email: wenbo.duan@polymtl.ca
[3] Dept. of Civil, Geological and Mining Eng., Polytechnique Montréal, QC, Canada. Email: najib.bouaanani@polymtl.ca
[4] Division of Expertise in Dams, Hydro-Québec, Montreal, QC, Canada. Email: miquel.benjamin@hydro.qc.ca

provides higher seismic forces than linear or nonlinear transient dynamic analyses, which might trigger the criteria for dam instability under earthquake effects (Ghrib et al. 1997). World-wide experience shows that concrete dam failures under seismic loads are rather rare (Nuss et al. 2012). The only case with significant damage reported is the Shih-Kang dam in Taiwan, which was affected by large fault displacements induced by the Chi-Chi 1999 earthquake. More or less extensive cracking of concrete dams has been observed during earthquakes (Nuss et al. 2012). Such cracking might modify the dynamic properties of a concrete dam during the shaking, while indicating its capacity to dissipate more seismic energy and to redistribute seismic forces accordingly. Limited cracking can then be beneficial to the seismic stability of concrete dams and might therefore avoid unnecessary rehabilitation interventions. However, conventional cracking analyses based on the gravity method are unable to account for this potentially beneficial behaviour while more adequate numerical simulation of cracking is rarely conducted in day-to-day engineering practice due to the complexity of relevant modeling approaches.

This paper presents a simplified method to assess the seismic stability response of concrete gravity dams including nonlinear sliding effects. First a lumped-mass multi-beam model is developed to represent the flexural and shear stiffness of gravity dam monolith. The proposed model can take into account of dam flexibility and nonlinearity at dam-rock and concrete joints. Time-history analysis results such as axial forces, bending moments and shear forces can be computed directly using beam theory. The proposed model is capable of predicting the amount of sliding at dam-rock interface and concrete joints. An example illustrating the application of the proposed technique to a typical 35-m high gravity dam is presented. The results obtained are verified against finite element simulations.

BASIC ASSUMPTIONS AND PROPOSED APPROACH

Basic assumptions

Figure 1 (a) shows a typical dam monolith and impounded reservoir. We assume that the dam-reservoir system is supported by a massless rigid rock foundation and that nonlinear responses of the dam are localized only at dam-rock interface. The dam concrete above the base is assumed to behave linearly and the base of the dam is assumed rigid. The gravity dam is subjected to its own weight, hydrostatic and hydrodynamic loads from the reservoir, uplift pressures and horizontal earthquake ground motion as illustrated in Figure 1 (b). The hydrodynamic loads are modeled using Westergaard added masses (Westergaard 1933). Uplift pressures are assumed constant during seismic excitation (USACE 1995).

Proposed Lumped-Mass Multi-Beam (LMMB) Model

The proposed approach to investigate nonlinear sliding response of a gravity dam under earthquakes is based on a Lumped-Mass Multi-Beam (LMMB) model shown in Figure 1 (c). The LMMB model is constructed using the dimensions of the actual dam monolith as illustrated in Figure 2. The dam monolith is divided into $(n + 1)$ layers delimiting n

volumes, numbered from bottom to top. Each volume corresponds to a two-node vertical beam element $i = 1 \dots n$, connecting lower node $n_i^{(\mathrm{inf})}$ to upper node $n_i^{(\mathrm{sup})}$. The degrees of freedom (DOFs) of each beam i are $u_i^{(\mathrm{inf})}$, $v_i^{(\mathrm{inf})}$, $\theta_i^{(\mathrm{inf})}$, $u_i^{(\mathrm{sup})}$, $v_i^{(\mathrm{sup})}$, $\theta_i^{(\mathrm{sup})}$, where u, v and θ denote the horizontal translation, vertical translation, and rotation, respectively.

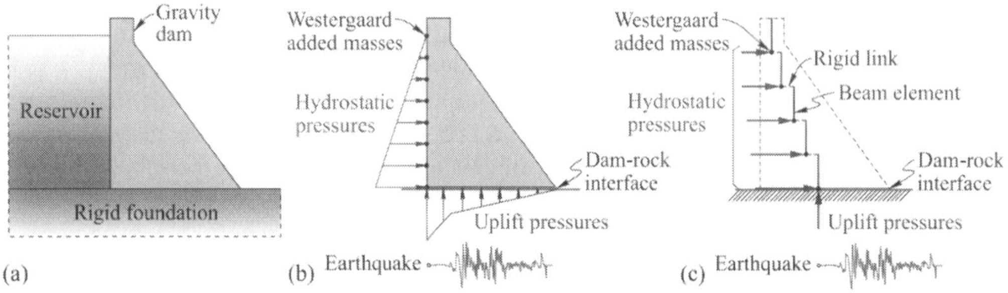

Figure 1. (a) Dam-reservoir-foundation system, (b) Applied loads on dam monolith, (c) Proposed LMMB model and applied loads.

The middle of each beam element is the same as the centroid of the corresponding volume. The area $A^{(i)}$, shear area $A_s^{(i)}$ and moment of inertia $I^{(i)}$ of the rectangular cross-section of the dam monolith located at the same height as the centroid of each volume i are determined. These properties are then assigned to the corresponding beam elements. The mass of the dam is lumped at the nodes of the beams, as well as Westergaard added masses. Rigid links are introduced to connect each node $n_i^{(\mathrm{sup})}$ to $n_{i+1}^{(\mathrm{inf})}$.

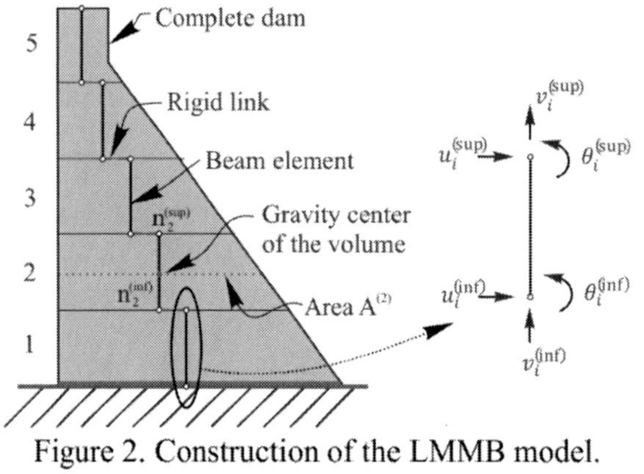

Figure 2. Construction of the LMMB model.

The beam elements are formulated according to the exact Timoshenko beam formulation (Reddy 1997). The global stiffness matrix **K** of the stick model is then assembled as well as the lumped mass matrix **M**. A mass proportional Rayleigh damping matrix **C**, equivalent to a modal damping ξ, is considered

$$\mathbf{C} = 2\pi\, f_1\, \xi\, \mathbf{M}^* \tag{1}$$

where f_1 is the fundamental frequency of the dam monolith, and $\mathbf{M}^*$ denotes the mass matrix excluding Westergaard added masses.

Evaluation of nonlinear sliding response at dam base

The dynamic response of the LMMB model is governed by

$$\begin{bmatrix} \mathbf{M_{aa}} & \mathbf{0} \\ \mathbf{0} & \mathbf{M_{bb}} \end{bmatrix} \begin{bmatrix} \ddot{\mathbf{U}}_\mathbf{a}(\mathrm{t}) \\ \ddot{\mathbf{U}}_\mathbf{b}(\mathrm{t}) \end{bmatrix} + \begin{bmatrix} \mathbf{C_{aa}} & \mathbf{0} \\ \mathbf{0} & \mathbf{C_{bb}} \end{bmatrix} \begin{bmatrix} \dot{\mathbf{U}}_\mathbf{a}(\mathrm{t}) \\ \dot{\mathbf{U}}_\mathbf{b}(\mathrm{t}) \end{bmatrix} + \begin{bmatrix} \mathbf{K_{aa}} & \mathbf{K_{ab}} \\ \mathbf{K_{ba}} & \mathbf{K_{bb}} \end{bmatrix} \begin{bmatrix} \mathbf{U_a}(\mathrm{t}) \\ \mathbf{U_b}(\mathrm{t}) \end{bmatrix} = \begin{bmatrix} \mathbf{F_a}(\mathrm{t}) \\ \mathbf{F_b}(\mathrm{t}) + \mathbf{R_b}(\mathrm{t}) \end{bmatrix} \tag{2}$$

where $\mathbf{U_b}$ is a vector containing the relative displacement of the node at the base of the LMMB model, i.e. $n_1^{(\mathrm{inf})}$, and $\mathbf{U_a}$ is a vector containing the relative displacements of the other nodes. The dot ($\dot{}$) and double dot ($\ddot{}$) denote first and second derivatives. $\mathbf{M_{aa}}$ and $\mathbf{M_{bb}}$ are mass sub-matrices, $\mathbf{C_{aa}}$ and $\mathbf{C_{bb}}$ are damping sub-matrices, $\mathbf{K_{aa}}$, $\mathbf{K_{ab}}$, $\mathbf{K_{ba}}$ and $\mathbf{K_{bb}}$ are stiffness sub-matrices. $\mathbf{F_a}$ and $\mathbf{F_b}$ are sub-vectors containing applied forces including dam self-weight, earthquake loads as well as hydrostatic and hydrodynamic loads. Uplift pressure is also included in $\mathbf{F_b}$. $\mathbf{R_b}$ is a sub-vector containing the reaction at the base of the LMMB model. For brevity of notation, the DOFs $u_1^{(\mathrm{inf})}$, $v_1^{(\mathrm{inf})}$ and $\theta_1^{(\mathrm{inf})}$ of the node at the base of the LMMB model are designated hereafter by u_b, v_b and θ_b, respectively.

During the non-sliding phase, the node at the base of the LMMB model is considered fixed, i.e. $\mathbf{U_b} = [u_b, v_b, \theta_b]^\mathrm{T} = \mathbf{0}$, where $^\mathrm{T}$ denotes transpose. The displacements $\mathbf{U_a}$ corresponding to the other nodes then satisfy

$$\mathbf{M_{aa}}\ddot{\mathbf{U}}_\mathbf{a}(\mathrm{t}) + \mathbf{C_{aa}}\dot{\mathbf{U}}_\mathbf{a}(\mathrm{t}) + \mathbf{K_{aa}}\mathbf{U_a}(\mathrm{t}) = \mathbf{F_a}(\mathrm{t}) \tag{3}$$

and can be obtained using a time integration algorithm in the present work, Newmark's method is used (Newmark 1959). The vector $\mathbf{R_b}$ of reactions is then obtained as

$$\mathbf{R_b}(\mathrm{t}) = \mathbf{K_{ba}}\mathbf{U_a}(\mathrm{t}) - \mathbf{F_b}(\mathrm{t}) \tag{4}$$

The reaction vector $\mathbf{R_b}$ contains shear force $\mathcal{V}_\mathrm{b}$, normal force $\mathcal{N}_\mathrm{b}$ and bending moment $\mathcal{M}_\mathrm{b}$ at the base node of the LMMB model. These forces will be used later to evaluate the nonlinear behavior at dam-rock interface and corresponding sliding displacement $\mathbf{U_b}$.

The Sliding Stability Factor $SSF(t)$ at time t can be expressed as

$$SSF(t) = \left| \frac{\mu\, \mathcal{N}_\mathrm{b}(t) + c\, \gamma(t)\, A^{(1)}}{\mathcal{V}_\mathrm{b}(t)} \right| \tag{5}$$

in which μ and c denote the friction coefficient and cohesion characterizing dam-rock interface, respectively. The coefficient $\gamma(t)$ is equal to 1 for $t < t_1$ and 0 otherwise, where t_1 denotes the time when the first sliding at dam-rock interface occurs. The friction coefficient is given by $\mu = \tan(\phi)$, where ϕ denotes friction angle at dam-rock interface. The expression of SSF obeys the Mohr-Coulomb failure criterion, considering both friction and cohesion to evaluate dam stability (USACE 2007, Alliard and Léger 2008, Renaud et al. 2016). The cohesion force $c\gamma(t)A^{(1)}$ acts against sliding when the interface is not completely cracked. In this paper, $\gamma = 0$ or 1, so partial cracking is not considered (Miquel et al. 2013). The value of SSF determines whether sliding occurs at dam-rock interface, i.e. sliding occurs as soon as $SSF < 1$.

When sliding is initiated at $t = t_1$, the acceleration $\ddot{u}_{\mathrm{b}}$ at the base of the dam can be determined as

$$\ddot{u}_{\mathrm{b}}(t) = \frac{\mathcal{V}_{\mathrm{b}}(t) - |\mu\, \mathcal{N}_{\mathrm{b}}(t)|\, \mathrm{sign}(\mathcal{V}_{\mathrm{b}}(t))}{M_{\mathrm{tot}}} \tag{6}$$

where M_{tot} denotes the total mass of the dam, $\mathcal{V}_{\mathrm{b}}(t)$ and $\mathcal{N}_{\mathrm{b}}(t)$ are the shear force and normal force obtained from linear analysis at time t. The sign function is defined by $\mathrm{sign}(x) = -1$ for $x < 0$, $\mathrm{sign}(x) = 0$ for $x = 0$, and $\mathrm{sign}(x) = 1$ for $x > 0$.

The sliding velocity and displacement at the base of the beam model are then calculated using the average acceleration method at each time interval Δt as follow

$$\dot{u}_{\mathrm{b}}(t) = \dot{u}_{\mathrm{b}}(t - \Delta t) + \frac{\Delta t}{2}\left(\ddot{u}_{\mathrm{b}}(t - \Delta t) + \ddot{u}_{\mathrm{b}}(t)\right) \tag{7}$$

$$u_{\mathrm{b}}(t) = u(t - \Delta t) + \dot{u}_{\mathrm{b}}(t - \Delta t)\,\Delta t + \frac{\Delta t^2}{4}\left(\ddot{u}_{\mathrm{b}}(t - \Delta t) + \ddot{u}_{\mathrm{b}}(t)\right) \tag{8}$$

Starting from $t = t_1$, the sliding of the dam also induces a shear force $\mathcal{V}_{\mathrm{b}}^{(\mathrm{d})}$ due to damping, which can be expressed as

$$\mathcal{V}_{\mathrm{b}}^{(\mathrm{d})}(t) = -2\pi\, f_1\, \xi\, M_{\mathrm{tot}}^{*}\, \dot{u}_{\mathrm{b}}(t) \tag{9}$$

where M_{tot}^{*} denotes the total mass of the dam without Westergaard added masses.

The sliding at the base of the dam influences the linear elastic response of the upper nodes of the structure, which results in an additional shear force $\mathcal{V}_{\mathrm{b}}^{(\mathrm{s})}$ at dam base. Sliding is approximated using a linear analysis of the LMMB model subjected to ground acceleration added to $\ddot{u}_{\mathrm{b}}$. The modified loading vector $\mathbf{F}^{*}$ is then used to obtain a new reaction vector $\mathbf{R}^{*}$ containing a modified shear force $\mathcal{V}_{\mathrm{b}}^{*}$. This force can be decomposed as follows

$$\mathcal{V}_{\mathrm{b}}^{*} = \mathcal{V}_{\mathrm{b}} + \mathcal{V}_{\mathrm{b}}^{(\mathrm{s})} \tag{10}$$

with

$$\mathcal{V}_{\mathrm{b}}^{(\mathrm{s})} = \mathcal{V}_{\mathrm{b}}^{(\mathrm{dyn})} + \ddot{u}_{\mathrm{b}}\, M_{\mathrm{tot}} \tag{11}$$

where $\mathcal{V}_{\mathrm{b}}^{(\mathrm{dyn})}$ denotes the force generated by the modification of the ground acceleration and $\ddot{u}_{\mathrm{b}} M_{\mathrm{tot}}$ is the static force induced by added loading.

The total shear force at the base during the sliding phase can be approximated as

$$\mathcal{V}_{\mathrm{b}}^{(\mathrm{tot})}(t) = \mathcal{V}_{\mathrm{b}}^{*}(t) - \ddot{u}_{\mathrm{b}}\, M_{\mathrm{tot}} - 2\pi\, f_1\, \xi\, M_{\mathrm{tot}}^{*}\, \dot{u}_{\mathrm{b}}(t - \Delta t) \tag{12}$$

and the acceleration at the base of the dam $\ddot{u}_{\mathrm{b}}$ can be estimated using

$$\ddot{u}_{\mathrm{b}}(t) = \frac{\mathcal{V}_{\mathrm{b}}^{(\mathrm{tot})}(t) - |\mu\, \mathcal{N}_{\mathrm{b}}(t)|\ \mathrm{sign}\left(\mathcal{V}_{\mathrm{b}}^{(\mathrm{tot})}(t)\right)}{M_{\mathrm{tot}}} \tag{13}$$

The sliding velocity and displacement at the base of the LMMB model are then calculated using equations (7) and (8) until $t = t_1'$, where time t_1' corresponds to $SSF(t_1') > 1$.

Once $SSF(t_1') > 1$, the sliding slows down and continues as long as $\dot{u}_{\mathrm{b}}$ keeps the same sign. To identify the exact time corresponding to the end of sliding phase, we proceed as previously by calculating $\ddot{u}_{\mathrm{b}}$, $\dot{u}_{\mathrm{b}}$ and u_{b} at each time step until $t = t_1''$, where t_1'' corresponds to $\mathrm{sign}(\dot{u}_{\mathrm{b}}(t_1'' + \Delta t)) = -\mathrm{sign}(\dot{u}_{\mathrm{b}}(t_1''))$. In this case, sliding stops within this time step, at instant $t = t_1'' + \tau$, yielding to

$$\dot{u}_{\mathrm{b}}(t_1'' + \tau) = \dot{u}_{\mathrm{b}}(t_1'') + \tau\, \ddot{u}_{\mathrm{b}}(t_1'') + \frac{\tau^2}{2\Delta t}\left(\ddot{u}_{\mathrm{b}}(t_1'' + \Delta t) - \ddot{u}_{\mathrm{b}}(t_1'')\right) = 0 \tag{14}$$

By solving equation (14), the exact time corresponding to the end of sliding can be obtained, and final dam sliding can be evaluated as

$$u_{\mathrm{b}}(t_1'' + \tau) = u_{\mathrm{b}}(t_1'') + \dot{u}_{\mathrm{b}}(t_1'')\, \tau + \frac{\ddot{u}_{\mathrm{b}}(t_1'')\tau^2}{2} + \frac{\tau^3}{6\Delta t}(\ddot{u}_{\mathrm{b}}(t_1'' + \Delta t) - \ddot{u}_{\mathrm{b}}(t_1'')) \tag{15}$$

The procedure detailed above can be repeated for other occurrences of sliding at instants t_i as illustrated through the flowchart in Figure 3. The main difference is that cohesion is no longer active in the subsequent sliding phases.

Evaluation of nonlinear sliding at multiple dam interfaces

Sliding may occur not only at a dam-rock interface but also at a weak upper concrete joint of a dam monolith. In this section, an approach is proposed to assess the seismic stability of a gravity dam with multiple potential sliding interfaces. For this purpose, we consider a dam monolith with N joints delimiting N blocks as illustrated in Figure 4.

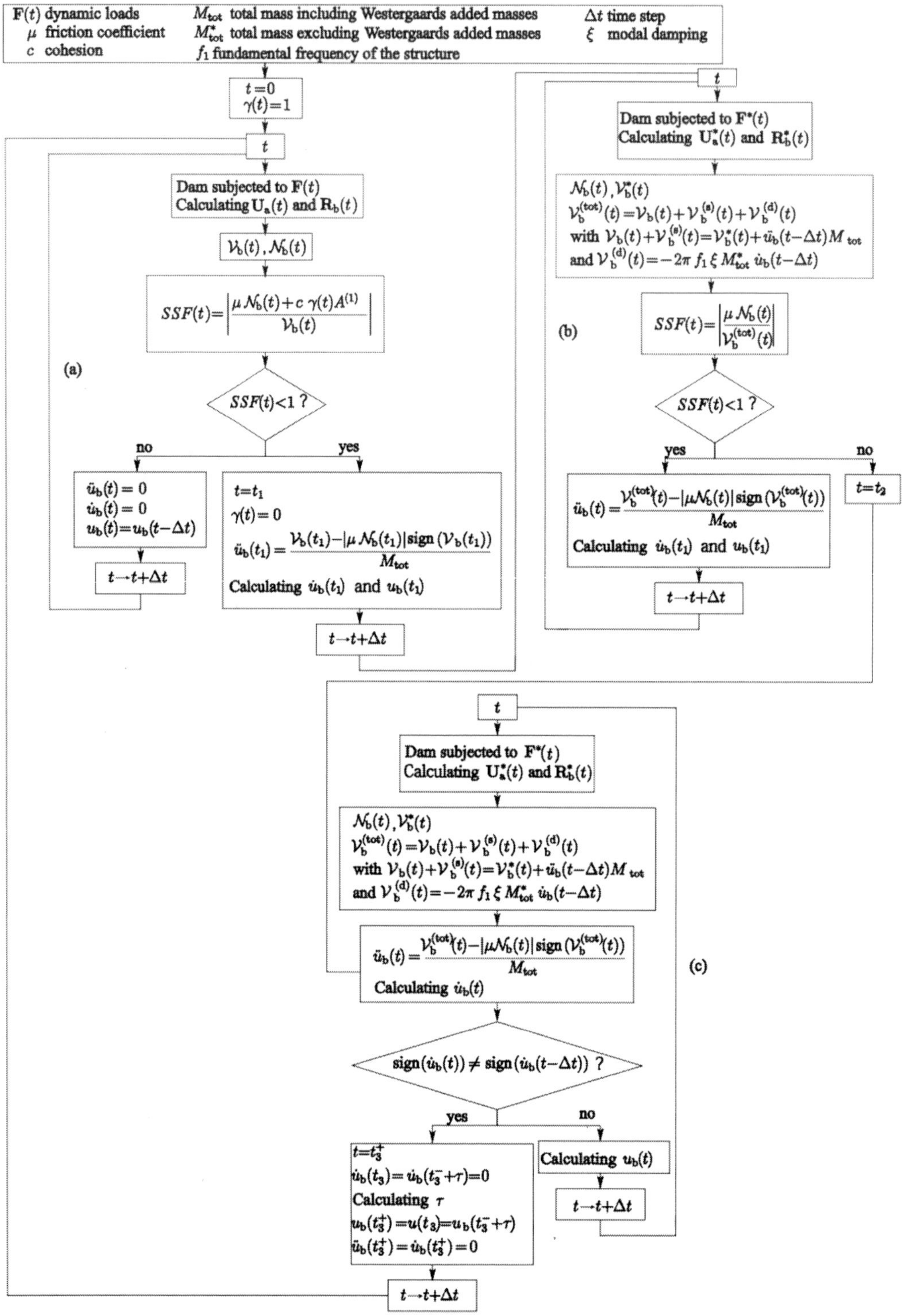

Figure 3. Flowchart illustrating the simplified non-linear analysis of a dam: (a) when the dam is stable, (b) when the dam is sliding and (c) at the end of the sliding phase.

First, all interfaces are considered fully bonded. The SSF at each joint, including dam-rock interface, is determined at the first time step of earthquake excitation through linear elastic dynamic analysis. The results of this analysis are used to determine the hierarchy of the sliding joints and to combine them into k groups G_i numbered from bottom to top as shown in Figure 4. Each group G_i is composed of n_i blocks $B_i^{(j)}$ for $j = 1 \dots n_i$ numbered from top to bottom of the group. The inter-block joints within each group are fully bonded during the first time step and the values of sliding at the other interfaces are determined. This process is repeated at each time step.

Knowing the sliding/no-sliding state of each interface, the linear seismic analysis of each block is performed considering the reaction applied on each block i which depends on the state of the upper blocks $i + 1$, as presented in Figure 5. If the upper interface is sliding, the loading at the top of the block is $-\text{sign}\left(\mathcal{V}_\text{b}^{(i+1)}\right)\left|\mu_{i+1}\mathcal{N}_\text{b}^{(i+1)}\right|$, $-\mathcal{N}_\text{b}^{(i+1)}$, $-\mathcal{M}_\text{b}^{(i+1)}$, where $\mathcal{V}_\text{b}^{(i+1)}$ and $\mathcal{N}_\text{b}^{(i+1)}$ represent the shear and normal force at the base of the block $i +$ 1, respectively. If the upper block is not sliding, the loading will be $-\mathcal{V}_\text{b}^{(i+1)}$, $-\mathcal{N}_\text{b}^{(i+1)}$, $-\mathcal{M}_\text{b}^{(i+1)}$.

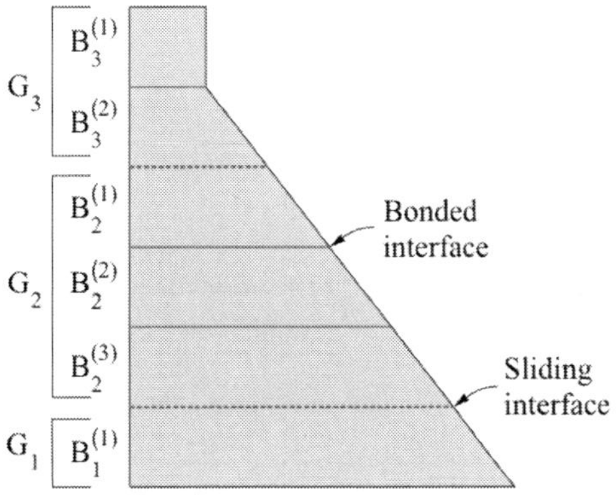

Figure 4. Sliding interfaces, bonded interfaces, and gathering of blocks into groups.

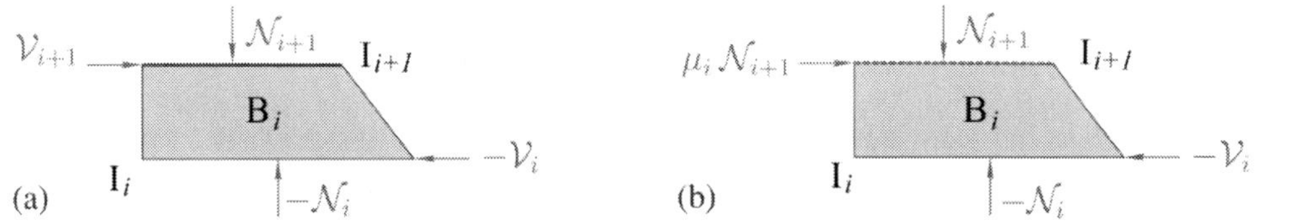

Figure 5. Analysis of blocks of the dam depending on the sliding/no-sliding state of their upper interface: (a) upper block is not sliding, (b) upper block is sliding.

The state of each interface is then determined through non-linear dynamic analysis, considering that the blocks within each group are bonded and slide monolithically together. The non-linear analysis of each block is then performed considering the total mass of this block and all the upper blocks of the same group. The flowchart in Figure 6 illustrates this process.

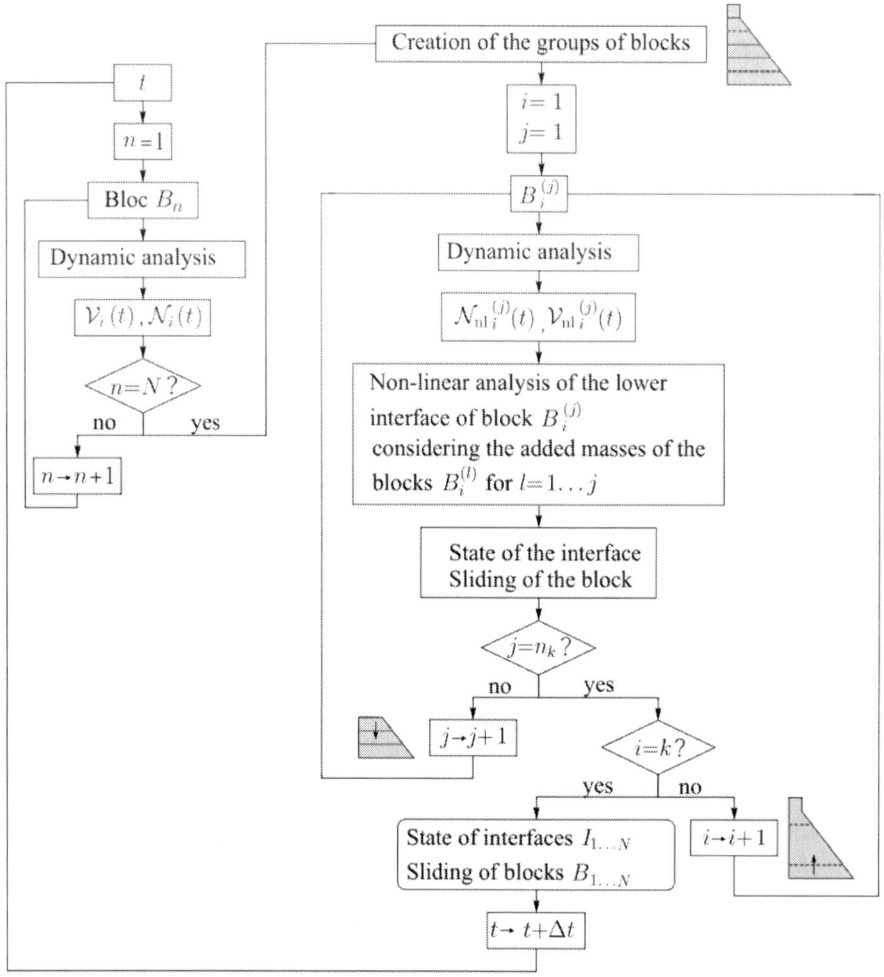

Figure 6. Flowchart illustrating non-linear analysis including multiple sliding interfaces.

NUMERICAL CASE STUDY

Description of the studied dam, numerical models and applied loads

To illustrate and verify the proposed modeling approach, it is applied in this section to a typical gravity dam. The dam has a height of 35 m (114.8 ft) and a base of 27.5 m (90.2 ft). The level of water in the reservoir is 32 m (105 ft) as illustrated in Figure 7(a). The modulus of elasticity of the concrete is $E = 25$ GPa (3,626 ksi). A very high elastic modulus of $E = 25000$ GPa (3,626,000 ksi) will also be considered to comply with a rigid body assumption for comparison purposes. The concrete density is taken as $\rho = 2400\ \text{kg/m}^3$ ($149.8\ \text{lb/ft}^3$). A drainage gallery at a height of 5 m (16.4 ft) and at 5 m (16.4 ft) from the upstream face of the dam is considered with an efficiency of 66%. The dam is subjected to two horizontal ground motions from Imperial Valley (1940) and Loma Prieta (1989) earthquakes. A mass-proportional Rayleigh damping, equivalent to a modal damping of 5% critical, is considered. A very small stiffness-proportional damping term is also included to enhance the stability of the numerical solutions.

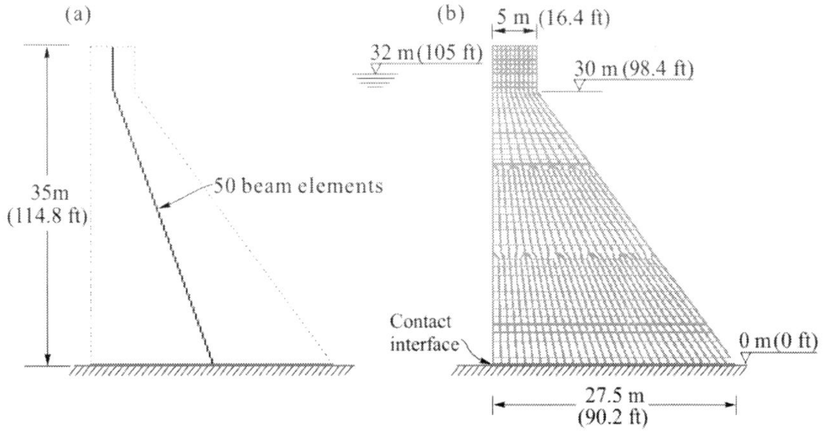

Figure 7. Gravity dam studied: (a) Proposed LMMB model, (b) Finite element model.

50 beam elements are used to build the LMMB model of the gravity dam as illustrated in Figure 7. For comparison purposes, the dam is also modeled in the commercial finite element software ADINA (2015) using 2D solid plane strain Finite Elements. Figure 7(b) shows the finite element mesh adopted. The sliding at the dam-rock interface is modeled using frictional contact elements programmed in ADINA according to the Mohr-Coulomb failure criterion. Cohesion in the finite element model is implemented using specific elements proposed by Renaud et al. (2016).

The dam is subjected to two ground motions with acceleration time-histories illustrated in Figure 8: (1) the horizontal components of the 1940 Imperial Valley earthquake at station El Centro, and (2) the horizontal component of the 1989 Loma Prieta earthquake.

Natural frequencies

First, the natural periods corresponding to the three first modes of vibration of the flexible dam monolith determined by the LMMB model and finite elements are compared in Table 1. A good agreement is found between the two models.

Table 1. First three natural periods of the studied gravity dam.

Mode number	LMMB model	Finite element model	Error
1	0.0909 (s)	0.0925 (s)	1.73 (%)
2	0.0376 (s)	0.0393 (s)	4.33 (%)
3	0.0333 (s)	0.0329 (s)	-1.22 (%)

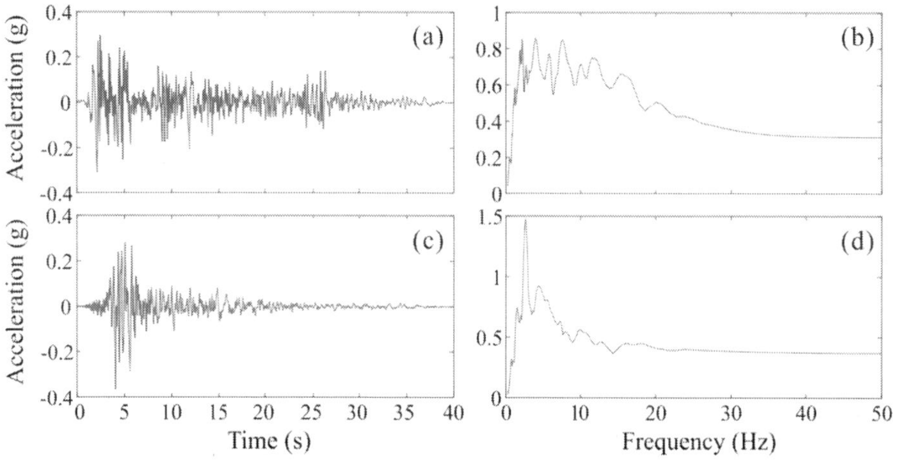

Figure 8. Acceleration time-histories and 5%-damped response spectrum of the ground motions considered: (a) and (b) Imperial Valley earthquake (1940) horizontal components at El Centro, (c) and (d) Loma Prieta earthquake (1989) horizontal components

Nonlinear analysis

The sliding displacement of the rigid dam with different friction angles at the dam-rock interface is evaluated using the proposed approach as well as 2D solid finite elements. The sliding displacements of the rigid dam under the effect of the two ground motions using differnt combinations of coeffcient of friction and cohesion are presented in Figure 9 and Figure 10. A very good agreement is found between the proposed technique and the finite element solution. As expected, the maximum sliding displacement of the dam decreases as the coefficient of friction μ at the dam-rock interface increases. The effect of cohesion can also be observed by comparing sliding displacements.

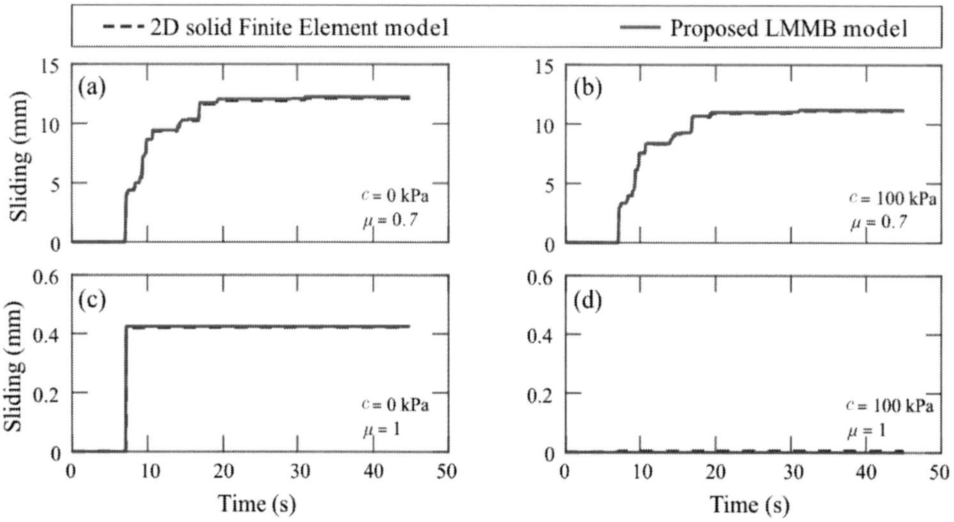

Figure 9. Contact sliding responses: (a) to (d) Rigid dam subjected to 1940 Imperial Valley ground motion.

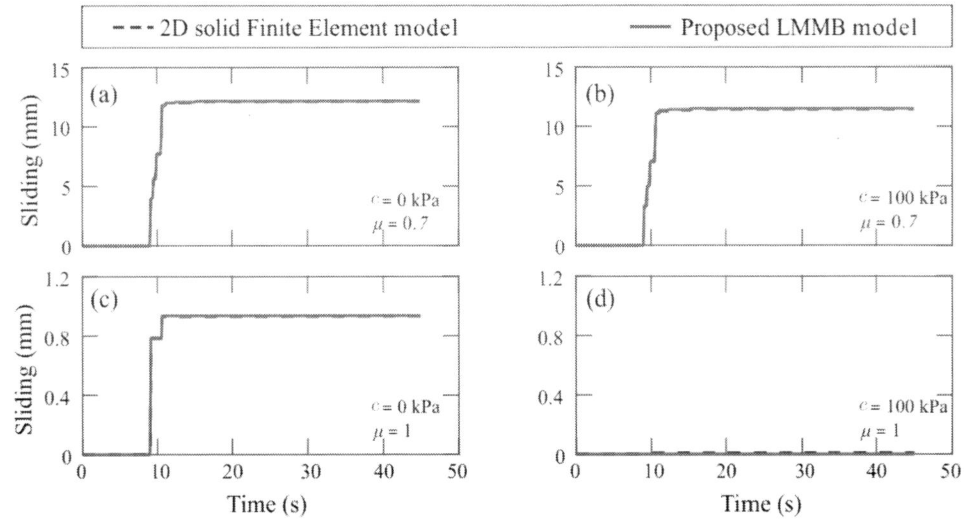

Figure 10. Contact sliding responses: (a) to (d) Rigid dam subjected to 1989 Loma Prieta ground motion.

Nonlinear analyses are also carried out for the flexible dam, i.e. modulus of elasticity of $E = 25$ GPa (3,626 ksi). The sliding of the dam with different friction angles and cohesion at dam-rock interface is presented in Figure 11 and Figure 12. The agreement between the proposed method and finite elements is very satisfactory irrespective of the friction or cohesion values. Comparing with the rigid dam case, the sliding displacements increase as the dam becomes more flexible, as expected. The influence of the friction and cohesion at dam-rock interface is observed as previously.

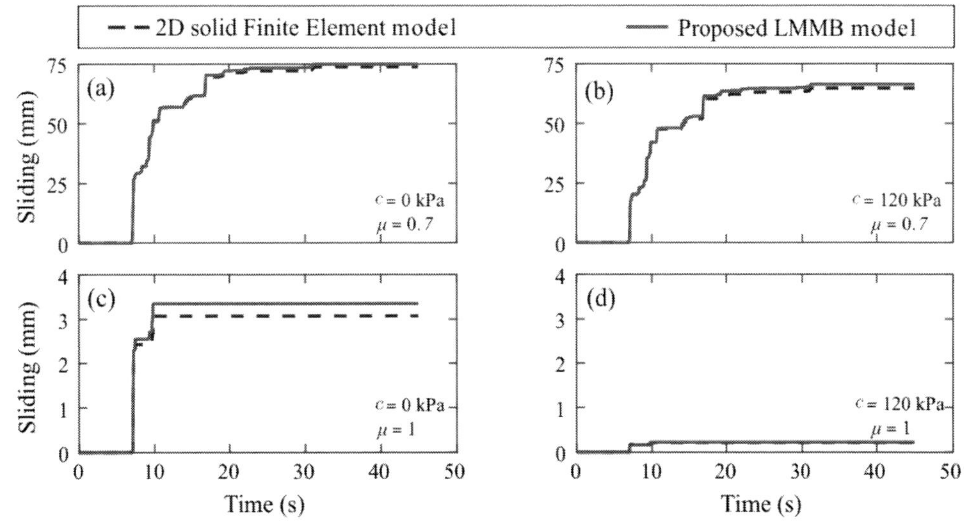

Figure 11. Contact sliding responses: (a) to (d) Flexible dam subjected to 1940 Imperial Valley ground motion.

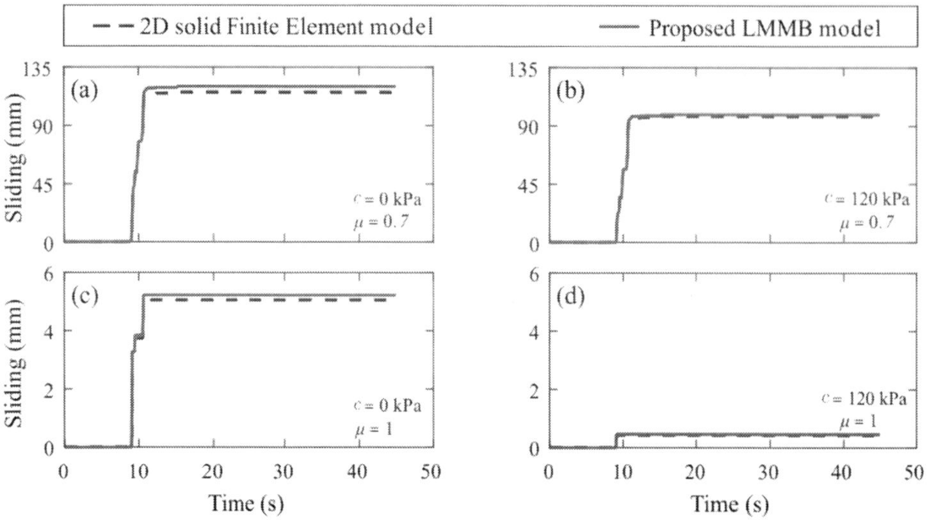

Figure 12. Contact sliding responses: (a) to (d) Flexible dam subjected to 1989 Loma Prieta ground motion.

The stability of the dam considering potential sliding joints other than dam-rock interface is investigated next. A total of three interfaces are considered for illustration purposes, as shown in Figure 13. Uplift pressure is considered constant for each interface and determined following the same criteria than previously.

Cohesion at all interfaces is set to zero and the friction coefficients are different for each interface, as follows: $\mu_b = \tan(55^o) = 1.42$ for the dam-foundation interface, and $\mu_1 = \tan(35^o) = 0.7$ and $\mu_2 = \tan(17^o) = 0.3$ for the joints located at 6.3 and 30 meters (20.7 and 98.4 ft) from the base, respectively.

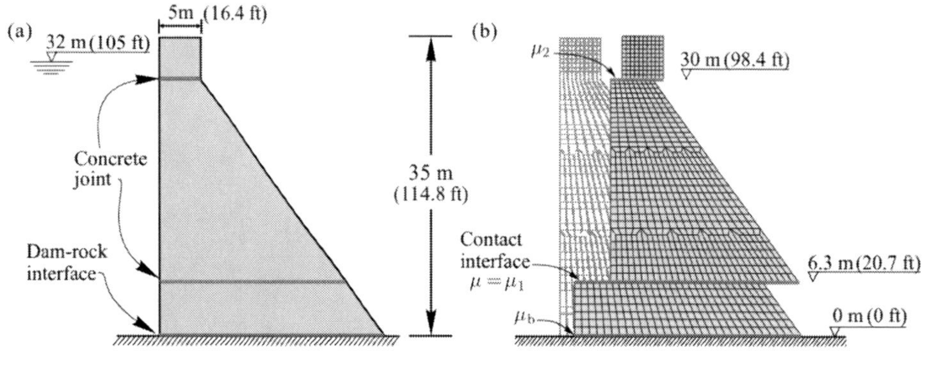

Figure 13. (a) Dam including three potential sliding joints, (b) Finite element model.

Figure 14 illustrate the sliding displacement relative to the ground at each nonlinear interface. Again, the results obtained using the proposed LMMB model are identical to those from finite elements.

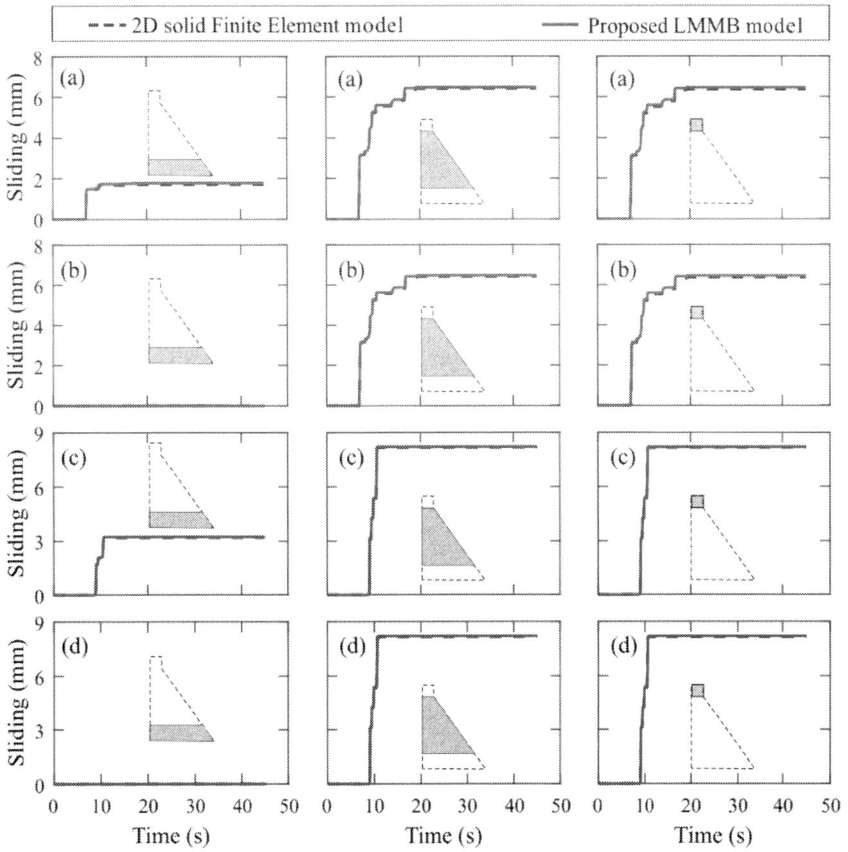

Figure 14. Contact sliding response of dam blocks: Dam with $\mu_b = 0.8$, $\mu_1 = 0.7$, $\mu_2 = 0.3$ subjected to (a) 1970 Imperial Valley ground motion and (b) 1989 Loma Prieta ground motion. Dam with $\mu_b = 1.42$, $\mu_1 = 0.7$, $\mu_2 = 0.3$ subjected to (c) 1940 Imperial Valley ground motion and (d) 1989 Loma Prieta ground motion.

CONCLUSIONS

A practical approach to investigate the nonlinear sliding seismic response of gravity dams was proposed. The developed technique is based on a Lumped-Mass Multi-Beam (LMMB) model, which can be built using conventional structural or finite element software. The effects of friction and cohesion are included, as well as potential sliding through dam-rock or multiple concrete joints. The technique was applied to a typical gravity dam and was shown to be accurate when compared to 2D finite element simulations. The proposed technique allows an efficient assessment of the sliding displacement at dam base as well as at concrete joints. It can be advantageously used to conduct extensive parametric analyses to reduce various uncertainties in seismic analyses such as those related to seismic hazard and mechanical properties.

ACKNOWLEDGEMENTS

The authors would like to acknowledge the financial support of the Natural Sciences and Engineering Research Council of Canada (NSERC) and Hydro-Québec.

REFERENCES

ADINA (2015). Theory and Modeling Guide. Report ARD 06-7. Watertown, MA, ADINA R&D, Inc.

Alliard, P. M. and P. Léger (2008). "Earthquake safety evaluation of gravity dams considering aftershocks and reduced drainage efficiency." Journal of Engineering Mechanics-Asce 134(1): 12-22.

Chávez, J. W. and G. L. Fenves (1995). "Earthquake analysis of concrete gravity dams including base sliding." Earthquake Engineering & Structural Dynamics 24(5): 673-686.

Ghrib, F., P. Léger, R. Tinawi, R. Lupien and M. Veilleux (1997). "Seismic safety evaluation of gravity dams." International Journal of Hydropower and Dams 4(2): 126-138.

Miquel, B., A. Lefrancois, N. Bouaanani and P. Léger (2013). Simplified seismic safety assessment of a concrete gravity dam using time history analysis. Canadian Dam Association Annual Conference. Montréal.

Newmark, N. M. (1959). "A Method of Computation for Structural Dynamics." Proc. ASCE 85(3): 67-94.

Nuss, L.K., Matsumoto, N., Hansen, K.D., 2012. Shaken, but not stirred-earthquake performance of concrete dams. 32nd Annual USSD Conference, New Orleans, LA, April 23-27, 1511-1530.

Reddy, J. N. (1997). "On locking-free shear deformable beam finite elements." Computer Methods in Applied Mechanics and Engineering 149(1-4): 113-132.

Renaud, S., N. Bouaanani and B. Miquel (2016). "Critical Appraisal of Common Simplified Assumptions in Seismic Stability Analyses of Gravity Dams." Journal of Performance of Constructed Facilities: 04016017.

USACE (1995). Gravity Dam Design. Washington, D.C., Dept. of the Army.

USACE (2007). Earthquake Design and Evaluation of Concrete Hydraulic Structures. Washington, D.C., Dept. of the Army.

Westergaard, H. M. (1933). "Water pressures on dams during earthquakes." Trans. ASCE 98: 418-432.

STRONTIA SPRINGS DAM: METHODS TO BETTER UNDERSTAND SEISMIC BEHAVIOR

Aimee L. Corn, EIT[1]
Guy S. Lund, PE[2]
Darren J. Brinker, PE[3]
Erin Gleason, PE[4]

ABSTRACT

Advancements in earthquake understanding over the last few decades have shown engineers that the seismic hazard in Colorado is significantly higher than originally thought. As a result, the seismic loads for Strontia Springs Dam have more than doubled since the project was designed and constructed in the early 1980s. Traditional stability evaluations have focused on the behavior of the structure for standards based seismic hazards, such as the maximum credible earthquake (MCE). Over the last few decades dam owners have seen hydrologic and seismic loads change (usually increased), primarily due to enhanced engineering understanding of the hazards.

The structural studies of Strontia Springs Dam used a three-dimensional linear and non-linear finite element model to evaluate the behavior of the dam. Because computation time can be costly, steps were taken to reduce the amount of data to sift through. This process involved using a linear transient analysis to first select the earthquake record. The response spectrum analysis of the MCE was run to determine "hot spots" in the dam where critical loads on the structure were expected to develop. A transient time history analysis was used to pull the stresses at the identified "hot spots," then the demand-capacity ratio was used to determine if cracking and non-linear behavior could develop. Finally, the non-linear analysis was performed to verify the conclusions of the linear analysis.

PROJECT DESCRIPTION

Strontia Springs Dam is located on the South Platte River, approximately 11 miles south-southwest of the Denver metropolitan area, and on the boundary between Jefferson and Douglas Counties. The drainage basin for the project encompasses 2,596 square miles of land on the eastern slope of the Rocky Mountains. The dam impounds a reservoir 1.7 miles long at normal reservoir water surface.

Strontia Springs Dam is a double-curvature, thin-arch, concrete dam with a maximum structural height of 292 feet, as shown in Figure 1. The crest of the dam is approximately 550 feet long (not including the thrust block and auxiliary spillway), and is at El. 6029 feet. The thickness of the dam varies from 30 feet at the base of the dam to 10 feet at the

[1] Gannett Fleming, 9605 Kingston Ct, Suite 130, Englewood, CO 80112, 720-439-4422, acorn@gfnet.com.
[2] Gannett Fleming, 9605 Kingston Ct, Suite 130, Englewood, CO 80112, 720-439-4544, glund@gfnet.com.
[3] Denver Water, 1600 West 12th Ave, Denver, CO 80204, 303-607-3130, darren.brinker@denverwater.org.
[4] Denver Water, 1600 West 12th Ave, Denver, CO 80204, 303-628-6862, erin.gleason@denverwater.org.

crest. The dam consists of 13 monoliths (blocks) that have a maximum length (along the axis of the dam) of 50 feet. The mass concrete for the monoliths was placed in 7.5-foot-high lifts. There are vertical contraction joints between the monoliths that contain shear keys. The contraction joints were grouted to assure monolithic behavior of the dam after the concrete was cooled by cooling pipes. The construction of the dam was completed in 1982, and a hydropower unit was put into service in 1986.

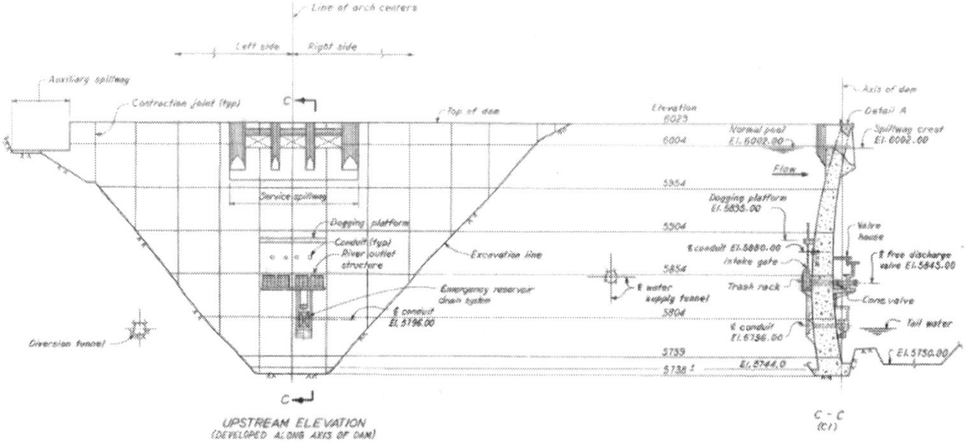

Figure 1. Profile and Section of Strontia Springs Dam

The Strontia Springs Dam site is in a steep V-shaped canyon approximately 1,500-feet-deep. The riverbed is approximately 50-feet-wide near the base of the dam and occupies the entire canyon bottom. The slope of the right abutment averages 50 degrees, and averages 45 degrees for the left abutment.

CONCRETE PROPERTIES

Strontia Springs Dam was designed assuming an unconfined compressive strength of 4,000 lb/in^2 at one year for the mass concrete. Laboratory tests on concrete cylinders that were formed during construction of the dam indicate the average unconfined compressive strength is 6,315 lb/in^2 for the 3 inch maximum size aggregate (MSA) mix, and 5,881 lb/in^2 for the 1.5 inch MSA mix (Harza, 1985). The concrete compressive strength used in these studies was conservatively assumed to be 4,000 lb/in^2, and is consistent with the original design.

The instantaneous modulus of elasticity and Poisson's ratio of the mass concrete used in these studies were based on laboratory tests on concrete cylinders that were formed during construction (Harza, 1985). The sustained modulus of elasticity, used for the evaluation of the static loads, was assumed approximately equal to 70 percent of the instantaneous value, as recommended by the USBR guidelines (USBR EM 19, 1977). The unit weight was estimated based on laboratory tests for concrete from other dams with similar aggregate rock type (USBR EM 34, 1977).

The ultimate strength of concrete is dependent on the rate of load and several studies have shown that under dynamic loads the strength of concrete will be greater than the static strength. The American Concrete Institute reports that the dynamic tensile strength will be approximately 50 percent greater than the static tensile strength (ACI, 1996), (FERC, 1999). Similarly, the dynamic compressive strength will be approximately 30 percent greater than the static compressive strength (ACI, 1996), (FERC, 1999).

The static tensile strength of the concrete is based on the modulus of rupture, and is computed using Equation 1. The seismic tensile strength was computed using Equation 2, which indicates the 50 percent increase in strength due to the rapid seismic load (ACI, 1996), (FERC, 1999). The material model in the finite element model is based on linear assumptions; therefore, for seismic loads the tensile strength of the concrete was more appropriately evaluated using the seismic apparent tensile strength, which was computed using Equation 3 (FERC, 1999).

Concrete Modulus of Rupture

$$f_t = 1.7 \times f_c^{2/3} \tag{1}$$

Concrete Seismic Tensile Strength

$$f_t = 2.6 \times f_c^{2/3} \tag{2}$$

Concrete Apparent Seismic Tensile Strength

$$f_t = 3.4 \times f_c^{2/3} \tag{3}$$

Where:

f_t = tensile strength
f_c = compressive strength

The concrete material properties are summarized in Table 1.

Table 1. Concrete Material Properties

Properties	Values	
Unconfined Compressive Strength	4,000	lb/in^2
Tensile Strength		
Modulus of Rupture	428	lb/in^2
Seismic Tensile Strength	655	lb/in^2
Apparent Seismic Tensile Strength	857	lb/in^2
Unit Weight	150	lb/in^3
Modulus of Elasticity		
Sustained	2,500,000	lb/in^2
Instantaneous	3,300,000	lb/in^2
Poisson's Ratio	0.2	

LOADS

The behavior of the dam was analyzed for static and dynamic loads. The usual static loads that were included in this model include gravity, normal reservoir elevation, sediment, and tailwater. The dynamic loads include the added mass due to the reservoir interaction with the dam and the ground acceleration associated with the Maximum Credible Earthquake (MCE).

The gravity load was based on an assumed unit weight of 150 pounds per cubic foot (lb/ft^3) for the concrete dam. The load was developed in the model using a staged construction analysis, which simulates the actual construction sequencing of the dam and more accurately represents how the weight of the dam will be transferred to the foundation during construction.

The hydrostatic loads corresponding to the normal reservoir level was simulated using a fluid density of 62.5 lb/ft^3. The normal water surface (NWS) was assumed to at the crest of the spillway, El. 6002 feet.

The sedimentation was assumed to be at El. 5840 feet for a normal reservoir level. The sediment load on the upstream face of the dam was simulated assuming a horizontal equivalent fluid density of 85 lb/ft^3 (FERC, 1999).

The tailwater surface (TWS) level was assumed to be at El. 5800 feet for a normal reservoir level. The tailwater was applied to the downstream face of the dam as a hydrostatic load with the same fluid density as the reservoir hydrostatic load.

Hydrodynamic loads are the results of the interaction between the reservoir and dam during an earthquake. Additional mass was included on the upstream face of the finite element model to simulate the added inertia due to the dam/reservoir interaction. The hydrodynamic masses were computed using the generalized theory for Westergaard's Added Mass (WAM) (Westergaard, 1932).

METHOD OF ANALYSIS

The finite element method of analysis (ANSYS) was used to evaluate the behavior of Strontia Springs Dam. Both linear and non-linear models were developed and used to estimate the stresses and deformations due to usual (static) and extreme (seismic) loading conditions. The models included a significant portion of the foundation in addition to the concrete dam. The foundation extends at least one dam height into each abutment and at least one dam height upstream and downstream from the extreme edges (toe and heel) of the dam.

Several different element types were used to simulate the behavior of the dam and the foundation rock. Eight-node brick elements simulated the solid bodies (i.e., mass concrete and foundation rock), single-node mass elements on the upstream face simulated the hydrodynamic dam/reservoir interaction, contact elements were used in the non-linear model at the dam/foundation interface and at contraction joints, and multi-point constraint (MPC) elements simulated the effects of the shear keys.

The geometry of the dam and foundation was based on data from record drawings. The global coordinate system for the finite element model is oriented such that the X-axis (i.e., cross-canyon) is positive toward the left abutment, Y-axis (i.e., vertical) is positive upward, and the Z-axis (i.e., upstream/downstream) is positive downstream, which is shown in Figure 2. The stress results were transformed into local cylindrical coordinate systems to evaluate the horizontal arch stresses and vertical cantilever stresses on the upstream and downstream face of the structure.

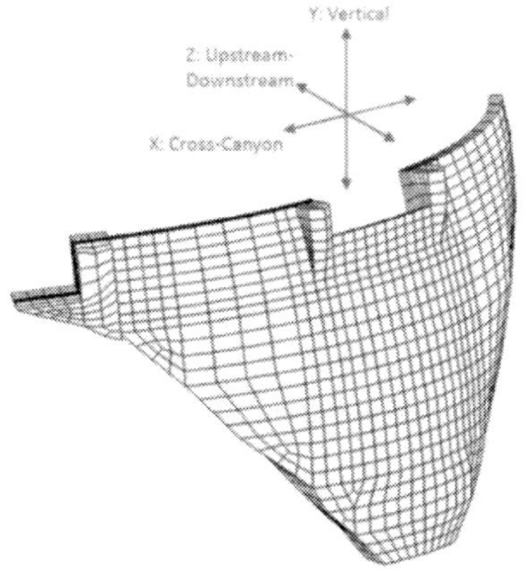

Figure 2. Looking Downstream at the Upstream Face of the Dam with the Global Coordinate System

LINEAR TRANSIENT ANALYSIS

Three time-history earthquake records were scaled to match the target seismic hazard spectrum, which were developed based on the MCE (magnitude 7.3 earthquake on the Ute Pass Fault located approximately 10 miles from the dam). The selected earthquakes used for these studies include the 1989 Loma Prieta Earthquake, the 1992 Landers Earthquake, and the 2010 El Mayor-Cucapah earthquake. Each earthquake record consists of two horizontal accelerations consisting of a strong and weak motion component and one vertical acceleration. These records were spectrally scaled to match the target spectrum (Lettis, 2016).

Transient analyses were performed with the finite element model using each of the three scaled earthquake time history records. The maximum horizontal component of acceleration was applied to the model in both the upstream-downstream and cross-canyon directions to determine the critical loading direction.

The deflection time-history results were used to identify the most severe seismic load on the dam. Selected nodes from various locations in the finite element model were used to evaluate the deformations. The most severe deformation is a good indicator of the most severe response of the dam, thus, the peak response tends to correspond to the peak deformation. The selected nodes for these studies are shown in Figure 3. The comparison of the deflection time-history results is shown in Figure 4.

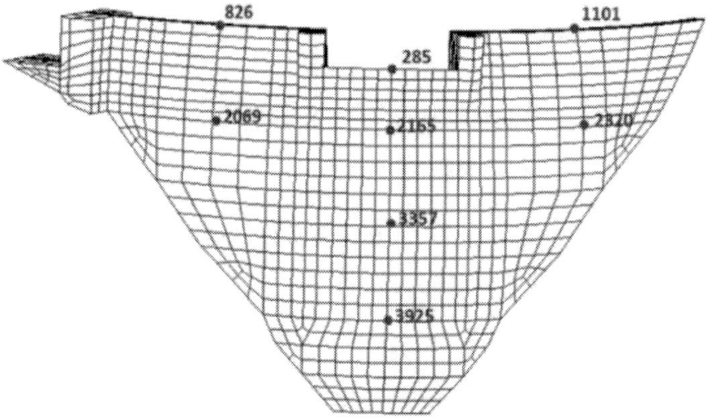

Figure 3. Upstream Face of Model Indicating Selected Nodes

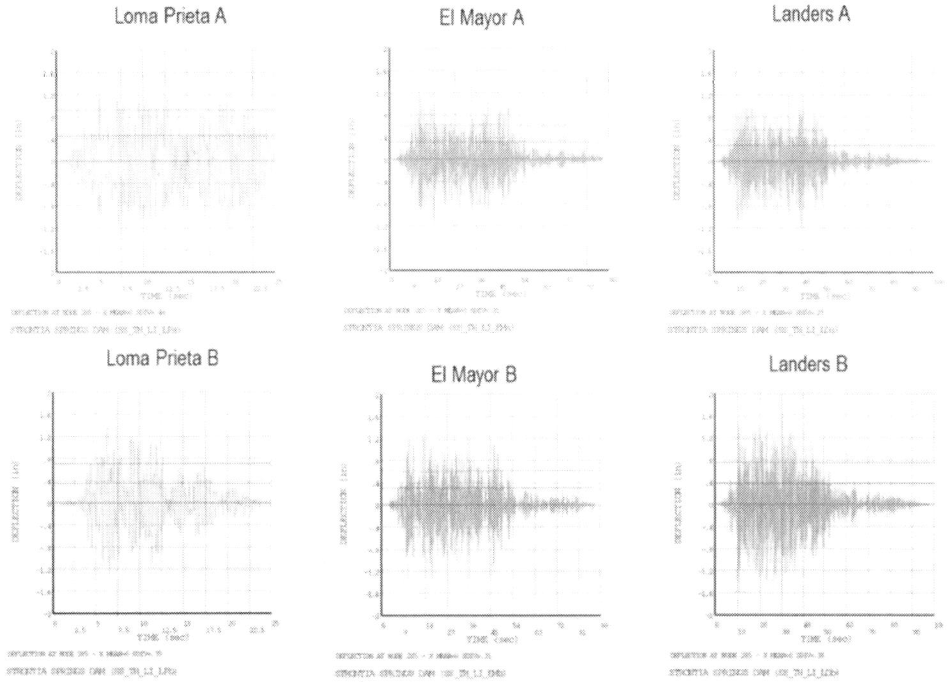

Figure 4. Deflection Time History plots for all Earthquakes at Node No. 285

The results from these studies indicated the Loma Prieta and Landers earthquakes would produce the most severe response in the dam. The period of strong shaking for the Loma Prieta event (25 seconds) is significantly less than the Landers event (100 seconds). Therefore, the Landers event was considered to be the controlling event because the period of motion is significantly longer.

RESPONSE SPECTRUM ANALYSIS

The excitation of a concrete arch dam due to the seismic loads is complex, and it can be difficult to determine the critical times of severe structural response. The response spectrum analysis (RSA) was used to better assess the maximum response of the dam. The RSA is a linear analysis, and provides the engineer a method to evaluate the modal contribution during the applied dynamic loads. The RSA combined each modal response using the square root sum of the squares (SRSS) method, and the results illustrate an envelope of behavior for the dynamic loading conditions. The acceleration response spectrum based on the MCE was used for this analysis, and the combined results are shown in Figure 5 and Figure 6.

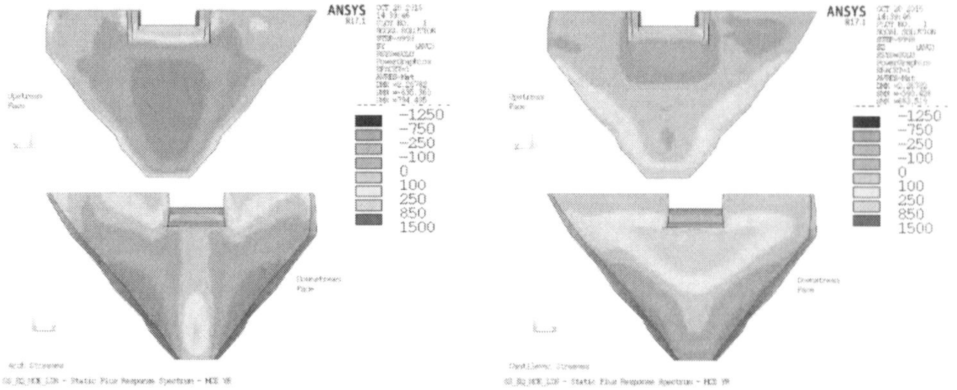

Figure 5. Contour Plots of Arch (left) and Cantilever (right) Stresses (psi) for Static Plus Response Spectrum Results

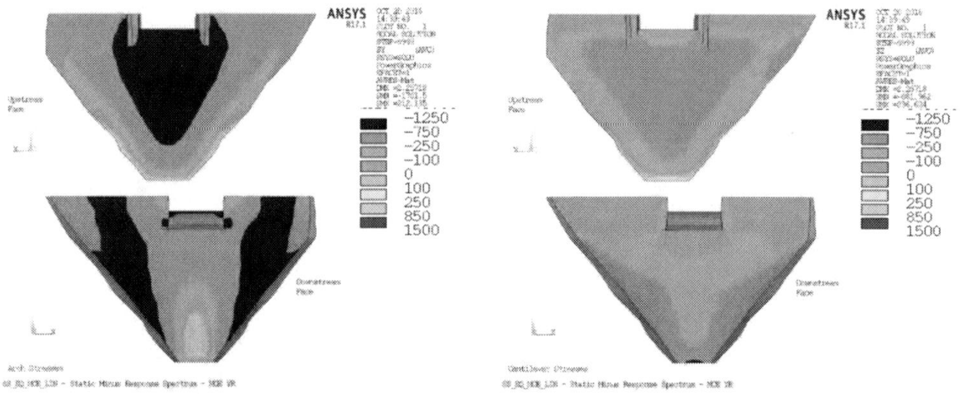

Figure 6. Contour Plots of Arch (left) and Cantilever (right) Stresses (psi) for the Static Minus Response Spectrum Results

The plots from the RSA show critical locations in the dam that can be used to evaluate the time-history response of the structure. For example, the critical location to evaluate the arch stresses is located on the intrados (downstream face) within the upper arch, and to the right and left of the central overflow spillway. Similarly, the critical location to evaluate the cantilever stresses is located on the downstream face near the crown, below the spillway. Selected nodes were based on the results from the RSA, for evaluation in the time-history analysis.

An interesting note, the conservative RSA results show all the calculated stresses from the finite element model are less than the allowable seismic tensile strength of the mass concrete. However, isolated areas of the dam develop stresses that are greater than the estimated tensile strength of the concrete. Therefore, the results from this conservative analysis indicate that the dam has adequate capacity to support the loads from the MCE event.

TRANSIENT TIME HISTORY ANALYSIS

The FEM was used to perform a linear transient analysis by modal superposition. The results were used to plot arch and cantilever time-history stresses for the selected nodes (critical locations based on results from the RSA analysis) in the dam. The stress time histories showed all compressive stresses to be less than the allowable compressive strength of the concrete (estimated to be 5,200 lb/in^2) and all tensile stresses are less than the static tensile strength of the concrete (428 lb/in^2) and significantly less than the estimated seismic tensile strength (655 lb/in^2).

DEMAND-CAPACITY RATIO

The U.S. Army Corps of Engineers (USACE) has published guidance on the structural assessment of mass concrete dams for seismic load using the demand-capacity ratio (DCR). The demand-capacity ratio (DCR) is defined as the ratio of the calculated stress (demand) to the tensile strength of the concrete (capacity). The DCR can exceed a value of 1.0 for short durations of time; however, if the DCR is greater than 2.0, then the structure would be considered to have experienced sufficient damage, such that the linear assumptions may not be valid (USACE, 2003). If the limits for the DCR are not exceeded, then it indicates that the structure will effectively behave in a linear manner, and the assumptions used for a linear elastic element analysis are considered valid.

The damage criterion employs the cumulative inelastic duration in conjunction with the DCR to account for the number of stress cycles that exceed the tensile strength of the concrete and access the probable level of damage. The cumulative inelastic duration of stress excursions refers to the total time the stress is greater than the tensile strength of the concrete (DCR > 1.0). For mass concrete arch dams, the cumulative inelastic duration of stress excursions above the concrete tensile strength is 0.4 seconds. Similarly for a DCR of 1.5 is 0.2 seconds, and greater than 2.0 is 0 seconds.

An evaluation of the DCR for stresses at the selected nodes was performed. The results are plotted in Figure 7 and show all of the selected nodes are within the criteria for the DCR ratio. As can be seen, there are no plotted values shown on the graphs, because all tensile stresses in the dam are less than the estimated static tensile strength of the concrete. This indicates the linear analysis is adequately modeling the behavior of the structure, and a more complex non-linear analysis is not necessary.

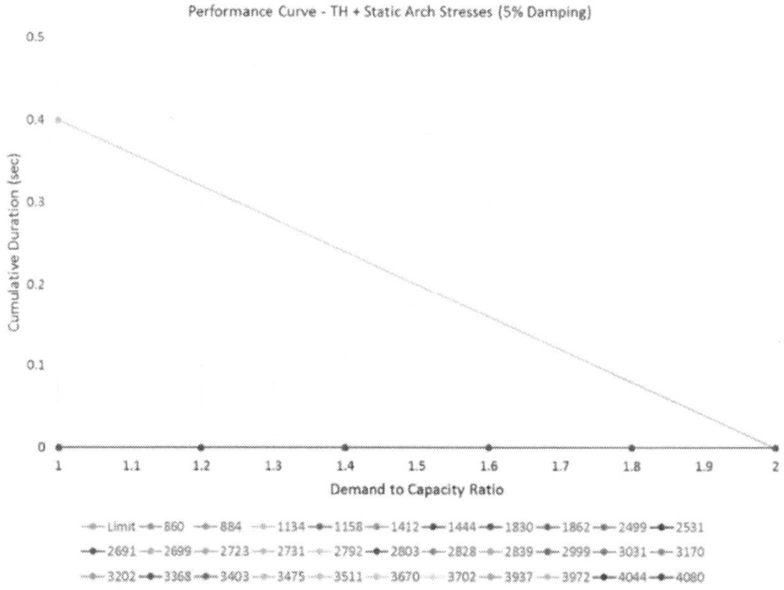

Figure 7. Performance Curve for the Time History plus Static Arch Stresses

NON-LINEAR TIME HISTORY ANALYSIS

A non-linear finite element model was used to evaluate the behavior of the dam for the seismic loads due to the MCE. The dynamic loads included the normal static loads, plus the effects from ground motions due to the MCE. The ground motions were simulated using the accelerations measured during the 1992 Landers event and spectrally scaled to match the MCE. The non-linear analysis was performed to verify the conclusions of the linear analysis, which indicated that the behavior of the dam remains in the effect linear realm and that the stresses do not exceed the allowable strength of the concrete.

A review of the results identified several conclusions, some of which include:

- Comparison of the linear and non-linear stress time-history results show a relatively good comparison. The amplitude of stress oscillation is similar for both the linear and non-linear studies.
- The stress time histories show that all the compressive stresses are less than the allowable compressive strength and all tensile stresses are less than the estimated static tensile stress. Thus, the concrete would not be expected to develop damage (i.e., cracking or crushing) due to assumed earthquake loads.

CONCLUSION

The conclusions from this study are summarized in the following bullets:

- Both a linear and non-linear FEM model was prepared to simulate the dynamic behavior of Strontia Springs Dam, and a process was developed to evaluate the potential for non-linear behavior.
- The dynamic loads due to the MCE were simulated using scaled records for the 1989 Loma Prieta, 1992 Landers and 2010 El Mayor Earthquakes.
- A transient analysis was performed and the deflection-time history results were used to evaluate which earthquake records resulted in the most severe response in the structure.
- A response spectrum analysis was performed and the results were used to identify critical locations in the structures that could be used to evaluate the transient response of during the earthquake.
- The results from a transient time history analysis were used to evaluate the demand-capacity ratio of the concrete in the dam. The results indicated that the response of the structure during the seismic event would be expected to remain in the linear range of behavior.
- The stress results from the transient analysis indicated that all stresses in the dam will remain less than the allowable compressive and tensile strength of the mass concrete.
- A non-linear analysis was performed to verify the conclusions of the linear analysis. The results indicated that the linear analysis adequately simulated the behavior of the dam, and thus the results from the linear analysis could be used to assess the adequacy of the dam during the seismic event.
- Based on the results, the capacity of the mass concrete is adequate to support the expected dynamic loads during the MCE event without damage (i.e., cracking or crushing).

REFERENCES

American Concrete Institute, "*Mass Concrete (ACI 207.1R-96)*", 1996.

FERC Guidelines. Federal Energy Regulatory Commission, Office of Hydropower Licensing, "*Engineering Guidelines for the Evaluation of Hydropower Project*", Washington, D.C., October 1999, Chapter 11-Arch Dams.

Harza Engineering Company, "*Final Report Volume III, Concrete*", May, 1985.

Lettis Consultants International, Inc. *Time Histories for Strontia Springs Dam, Rev. 1.* Walnut Creek, California, 2016.

U.S. Department of the Interior, Bureau of Reclamation, Engineering Monograph No. 19, "*Design Criteria for Concrete Arch and Gravity Dams*", Denver: GPO, 1977.

U.S. Department of the Interior, Bureau of Reclamation, Engineering Monograph No. 34, "*Control of Cracking in Mass Concrete Structures*", Denver, GPO, 1981.

U.S. Army Corp of Engineers, Engineer Manual 1110-2-6051, "*Time History Dynamic Analysis of Concrete Hydraulic Structures*", Washington, D.C., December 2003.

Westergaard, H.W., Paper, American Society of Civil Engineers, "*Water Pressures on Dams During Earthquakes*", 1932.

CRACKING IN THIN ARCH CONCRETE DAM – NONLINEAR DYNAMIC STRUCTURAL ANALYSIS

Sherry Hamedian, P.E., Civil Engineer[1]
Roman Koltuniuk, P.E., Structural Engineer[2]

ABSTRACT

An arch dam is a concrete structure which uses arch action to resist loads curving upstream in plan view. Arch dams obtain stability and resist movement from upstream reservoir pressure by transmitting loads into adjacent canyon walls. Successful arch action is dependent on a unified monolithic structure. Structural discontinuities such as open joints or cracks can affect the effectiveness of the inherent behavior of the arch. Classification of an arch dam depends on the thickness-to-height ratios, which can change the dynamic behavior of the structure.

An LS-DYNA [1] finite element (FE) dynamic structural analysis was used to model the behavior of a thin arch dam under seismic loading conditions coupled with the presence of existing and progressive cracks along the downstream face, which formed years after construction. The analysis also modeled foundation sliding wedges, defined by low angle discontinuities mapped in the right abutment.

The FE model used in the analysis consisted of a double-curvature, thin arch concrete dam with an uncontrolled overflow crest spillway, existing vertical contraction joints, the foundation surrounding the dam, the upstream reservoir and documented cracks on the downstream face using nonlinear material properties. Loadings included: a dynamic pre-cracking lateral load applied to the model with introduced low strength material at the existing crack locations to model the existing cracks; gravity; thermal; foundation uplift pressure; and three-component earthquake time histories.

A spectral analysis of surface waves (SASW) survey was conducted on the downstream face of the dam to estimate the extent, direction, depth, and orientation of the existing cracks. Results of this geophysical investigation were incorporated in the LS-DYNA FE analysis.

Results of the structural analysis illustrate the global effect of potential foundation block movement and cracking on the stability of a thin arch dam, including inherent behavior to redistribute stresses within the concrete and maintain compressive arch action during a seismic event.

[1] Sherry Hamedian, Waterways and Concrete Dams Group 1, Bureau of Reclamation, Denver Federal Center, Bldg. 67, 4th Floor, P.O. Box 25007 (86-68110), Denver, CO 80225-0007, shamedian@usbr.gov
[2] Roman Koltuniuk, Waterways and Concrete Dams Group 1, Bureau of Reclamation, Denver Federal Center, Bldg. 67, 4th Floor, P.O. Box 25007 (86-68110), Denver, CO 80225-0007, Rkoltuniuk@usbr.gov

INTRODUCTION

This paper presents the results of a FE analysis performed on a thin-arch concrete dam. The purpose of the analysis was to analyze the current condition of the arch dam with existing cracks and evaluate future condition of the structure under potential seismic loads.

Cracking of arch dams may be caused by the complex factors such as drying shrinkage, temperature variation, abnormal loads, deterioration of concrete, etc. Temperature variation is a major factor that can cause cracks in arch dams. Arch dams can expand upstream during the summer (especially if the reservoir is low) causing horizontal, vertical or diagonal cracks to form on the downstream face due to tensile stress greater than concrete tensile strength, as is the case for this dam. If the cracks develop and propagate completely through the structure's thickness and are adversely oriented (allowing a dam block to slide), the arch carrying capacity of the dam can be reduced resulting in instability of the arch. The downstream face of the subject dam and cracks are shown in Figure 1.

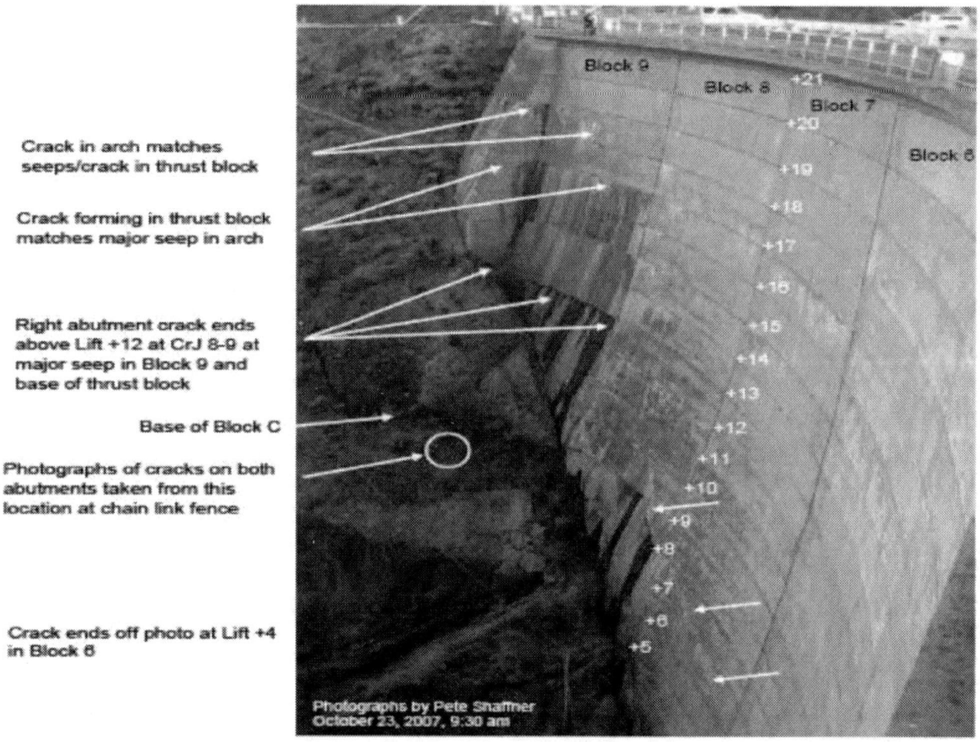

Figure 1. Cracks on the Downstream Face of the Dam.

LS-DYNA is a FE dynamic structural analysis tool used to model the behavior of the thin arch dam under seismic loading conditions. LS-DYNA can model the presence of existing and progressive cracks in the dam. LS-DYNA can also model removable foundation blocks, defined by low angle discontinuities mapped in the abutments.

MODELING DETAILS

Investigations

1. Independent Foundation Stability Analysis

The stability of a thin arch dam depends on the strength and integrity of the foundation. The most important factor affecting the abutment strength is the geologic structure relative to the foundation contact and the topographic surface. The geologic structure combined with the topographic surface can form potentially removable foundation blocks which could compromise the stability of the dam. Geologic investigations determined potentially removable foundation blocks as shown in Figure 2

Independent foundation stability analysis of the dam was performed to better understand the behavior of the removable foundation blocks in the right abutment and to calculate the uplift pressure acting on these blocks. The uplift water forces acting on foundation block planes were estimated and the static sliding factors of safety were computed for the foundation blocks. The uplift forces significantly impact the foundation stability. This conclusion was the impetus of including the removable foundation blocks in the FE analysis.

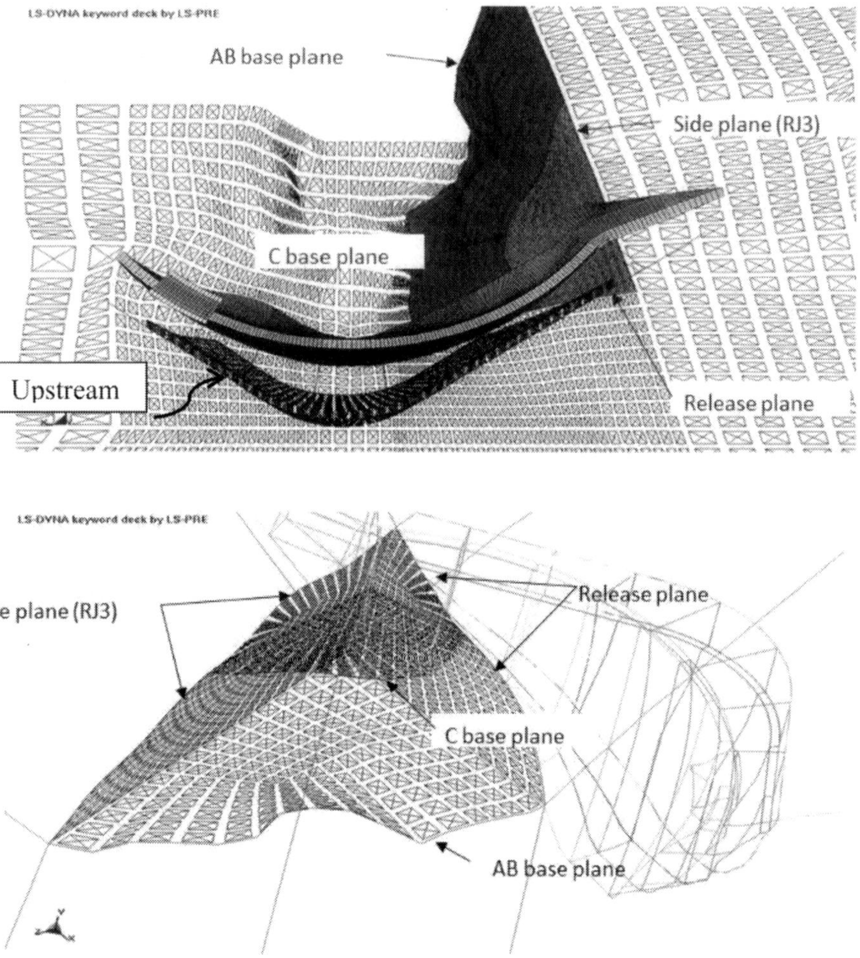

Figure 2. Removable Foundation Blocks in the Right Abutment.

2. *Geophysical Investigations Spectral Analysis of Surface Waves (SASW) Surveys*

This SASW survey was employed to better delineate the extent and orientation of the cracking into the dam structure at depths up to 10 feet into the downstream face of the dam. Figure 3 displays the field survey areas and crack locations on the structure.

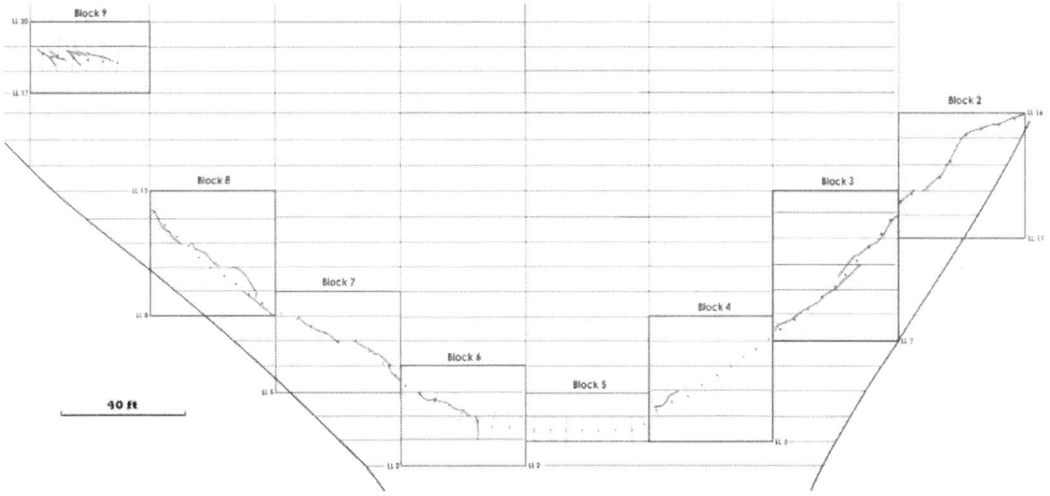

Figure 3. Location of the Survey Area on Downstream Face of the Dam.

Each survey area is associated with a block on the face of the dam. These areas are bound laterally by the vertical contraction joints between blocks and horizontally by the lift lines. The results of the SASW survey are presented in S-Wave Velocity Contours plots for each dam block (see Figure 4). Any recorded velocities between 5,500 feet/sec and 6,500 feet/sec indicate damaged concrete. Any zones with a velocity below 5,500 feet/sec are considered to be severely damaged concrete. This was confirmed empirically at various locations on the dam face by testing undamaged concrete away from known cracks. These areas had an S-wave average velocity of 8,000 feet/sec (with a range of 6,500 - 9,000 feet/sec).

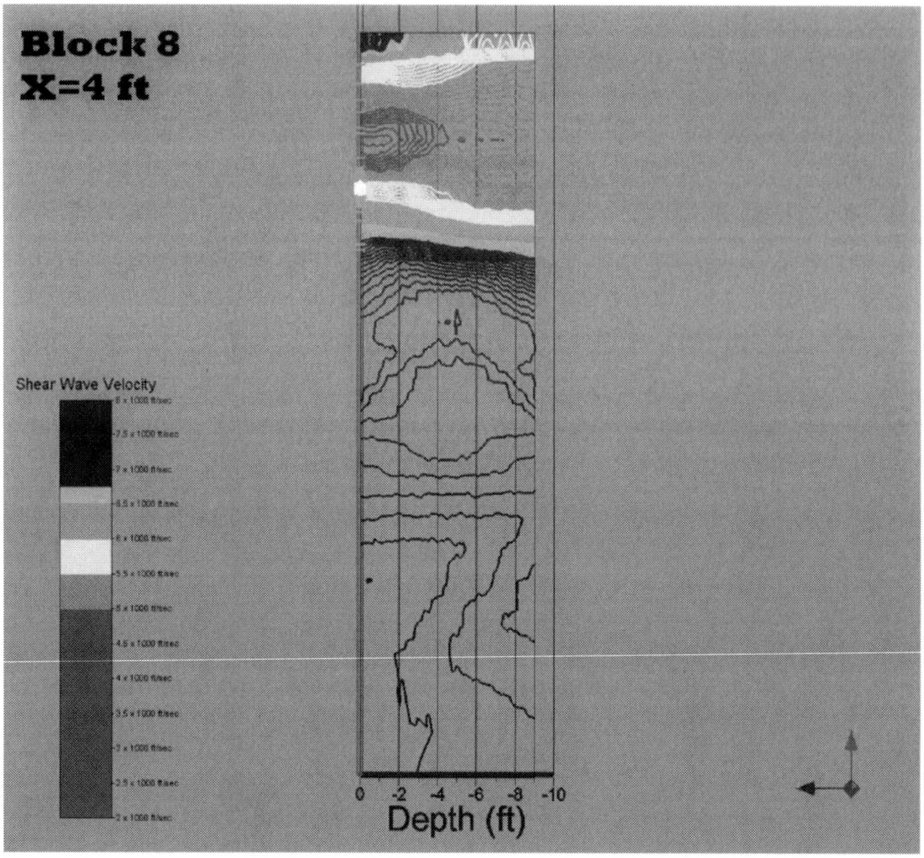

Figure 4. Example S-Wave Velocity Contour Plot for Dam Block 8 above Lift Line 8 and 4' From Left Vertical Contraction Joint (Left Side is Dam Face, Right Side is Upstream).

The results from these geophysical investigations were used to model the crack extents in the nonlinear 3-D dynamic FE analysis of the structure under static and seismic loading condition. The precise location, orientation and depth of the cracks were not determined from the SASW studies, but instead were used to approximately model the cracks.

Model Description

The FE model consisted of a double-curvature, thin arch concrete dam with existing cracks on the downstream face, an uncontrolled overflow crest spillway, existing vertical contraction joints without shear keys, a reservoir, foundation and potentially removable foundation rock blocks near the right abutment. Figures 5 shows the FE model of the dam with existing cracks and foundation blocks.

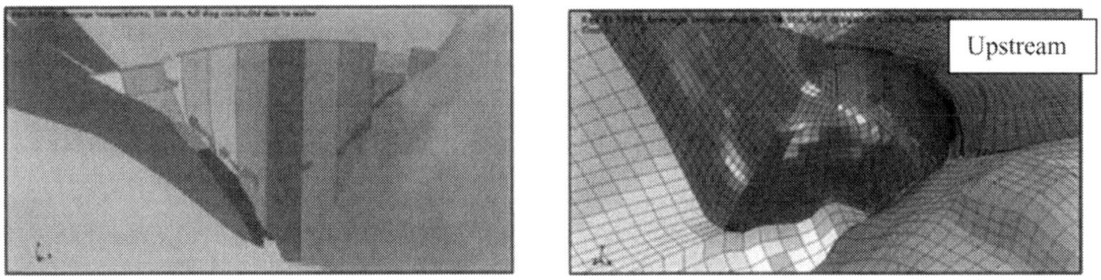

Figure 5. Finite Element Model of the Dam, Foundation and Reservoir

The arch dam was modeled using 10 monoliths separated by radial vertical contraction joints. Contact surfaces were used between dam monoliths, dam-to-foundation contact, dam-to-water contact and foundation-to-water contact. The dam geometry was created using the TrueGrid Mesh Generator [2]. This model also includes a thrust block on the right abutment, separated into two sections by a contraction joint. In addition, an unbonded lift-line was modeled to represent unbonded lift lines observed in the concrete core samples taken from this thrust block. The FE model also includes the existing cracks on the downstream face of the dam as shown in Figure 5 in purple.

The concrete dam and existing cracks were modeled using a nonlinear concrete material in order to understand the behavior of the dam and cracks under applied static and dynamic loading conditions. The foundation, removable foundation blocks and reservoir were modeled using an elastic material.

Two different concrete material models were used in the FE analysis; Continuous Surface Cap Model (Mat 159) [1] and Winfrith Model (Mat 84) [1]. This is done in order to compare results using two independent, nonlinear material models. Figures 6 and 7 are the graphical display of a typical cracking pattern for these two nonlinear materials.

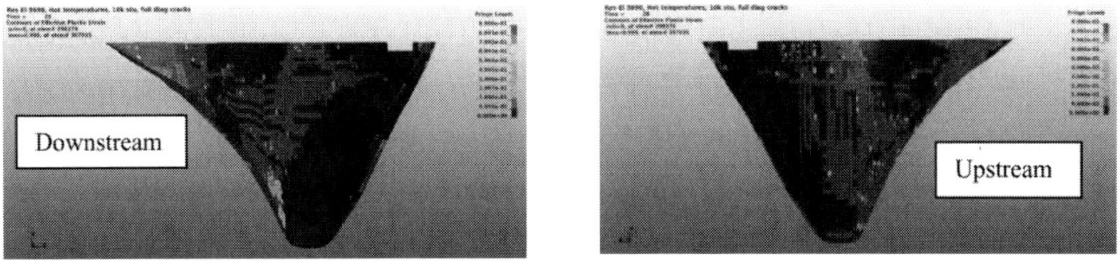

Figure 6. Typical Damage Pattern (Contour of Damage Parameter) – Mat159.

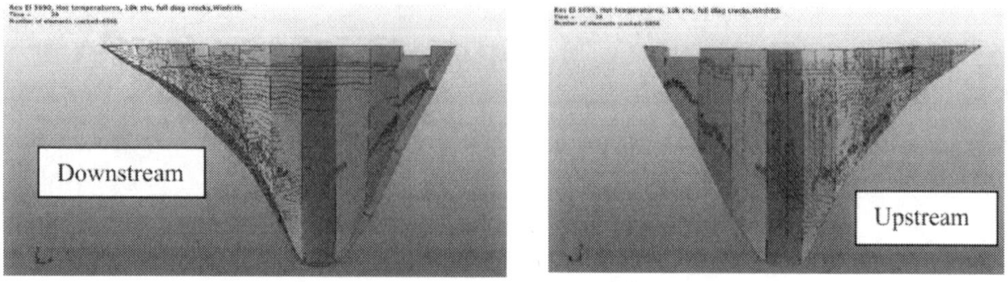

Figure 7. Typical Damage Pattern (Crack Location and Orientation) – Winfrith.

Material Properties

The material properties used in the nonlinear analyses were based on calibrations using laboratory test values and forced vibration testing performed by the Bureau of Reclamation's Material Engineering and Research Laboratory.

1. Foundation Properties

The foundation rock is hard, massive, and made up primarily of sub rounded to sub angular quartzite and limestone fragments with minor amounts of other rock types. The conglomerate is so well cemented that, when blasted, it breaks across the individual fragments rather than through the cementing material. The final foundation properties are shown in Table 1.

Table 1. Foundation Properties

Foundation Properties	Final Values
Average Density	*165 lb/ft*^{3}
Young's Modulus	8.3×10^6 *lb/in*^{2}
Poisson's Ratio	*0.19*

2. Concrete Properties

Based on laboratory test results, the following material properties were used for the dam in the FE analysis (see Table 2):

Table 2. Concrete Material Properties

Material Properties	Concrete Testing
Average Density	*147 lb/ft³*
Young's Modulus	*6.5 x 10⁶ lb/in²*
Poisson's Ratio	*0.17*
Compressive strength (f'c)	*5,500 lb/in²*
Tensile strength	*300 lb/in²*
Maximum aggregate size	*2.5 in*

The tensile strength of the concrete at the existing crack locations on the downstream face of the dam was set much lower than the undamaged concrete (to 40 lb/in²). This was done to initiate a cracked condition prior to applying the seismic motions.

Temperature Parameters

During construction, the dam was grouted at approximately 36 degrees, the "stress-free" temperature. A thermal analysis, which modeled cycles in temperature changes over 10 years, was performed to determine the temperature through each dam monolith for each of the temperature load cases. The following parameters were used as shown in Table 3:

Table 3. Thermal Properties

Thermal Properties	Concrete	Reservoir
Thermal conductivity	*0.76*	*1.966*
Heat capacity	*0.232*	*1.0*
Coefficient of thermal expansion	*6.7E-06*	

Reservoir

Solid brick elements using a null material with the following properties were used to approximate the reservoir water, as show in Table 4:

Table 4. Water Properties

Material Properties	Water
Modulus of Elasticity	*189.7 lb/in²*
Poisson's Ratio	*0.4999*
Density	*62.4 lb/ft³*
Bulk modulus *K*	*316,166.7 lb/in²*
Pressure Cutoff	*0.0 lb/in²*
Viscosity Coefficient	*0.0*

Loads

Static and dynamic loads were applied in LS-DYNA using load curves. In this model, gravity loads were applied at the beginning of the run followed by a period of time to reach steady state or equilibrium. Uplift forces on the removable foundation blocks, represented as pressure loads, were applied next followed by a quiet time. Ambient temperature loads reflecting winter, summer and average temperatures were applied to exposed surfaces of the dam and sustained throughout the analysis as shown in Figure 8.

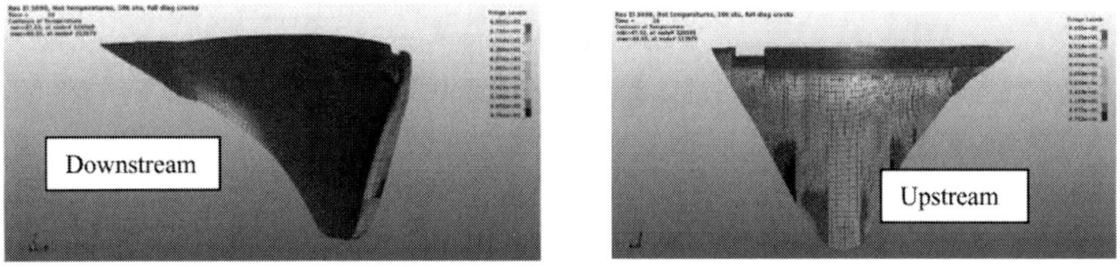

Figure 8. Example of Contour of Hot Temperature on the Downstream Face.

Deconvolved earthquake time histories were applied at the end of the static portion of the analysis followed by a quiet time. One set of ground motion with three return intervals of 5,000 year (0.27g), 10,000 year (0.33g) and 50,000 year (0.54g) were applied to the model at depth in three different directions:

- Upstream / downstream direction
- Cross canyon direction
- Vertical direction

A check must be done to insure that the propagation of the earthquake through the model is correct. This is done by evaluating the results of a flat box model (see Figure 9) with the same horizontal extent as the full dam/foundation/reservoir model and with a depth equal to the distance from the bottom of the canyon to the plane at depth where the deconvolved earthquake histories are applied. The deconvolved time histories were input at depth in this model and the response was captured at the free surface. The ground motion response calculated at the free surface of the box model should have been similar to the free-field target motions. However sometimes there were considerable differences between the calculated and target motions. In these cases the deconvolved input was modified by Reclamation's Geophysics and Seismology Group and was given back to be input again into the flat box model and compared with free-field target motions. This process was repeated until a suitable comparison was obtained. Figure 10 shows an example of a suitable comparison.

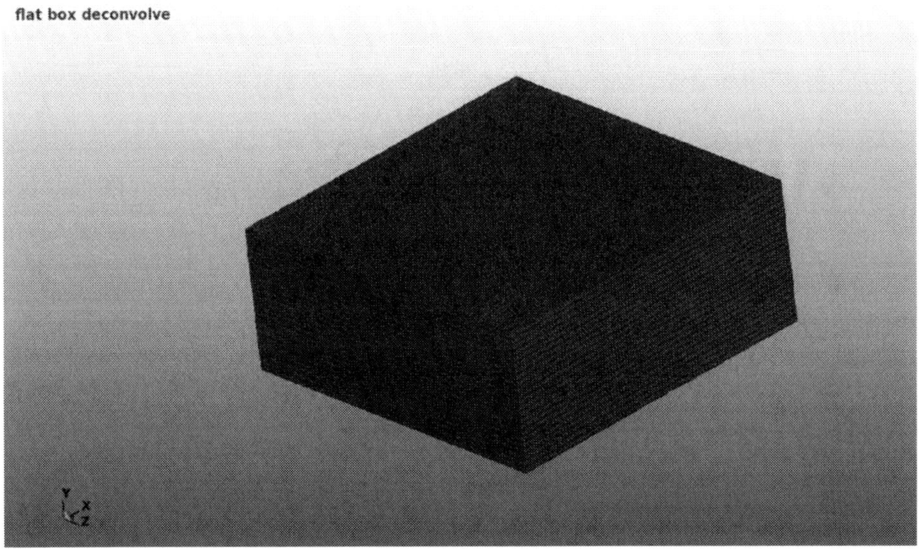

Figure 9. Flat Box Finite Element Model.

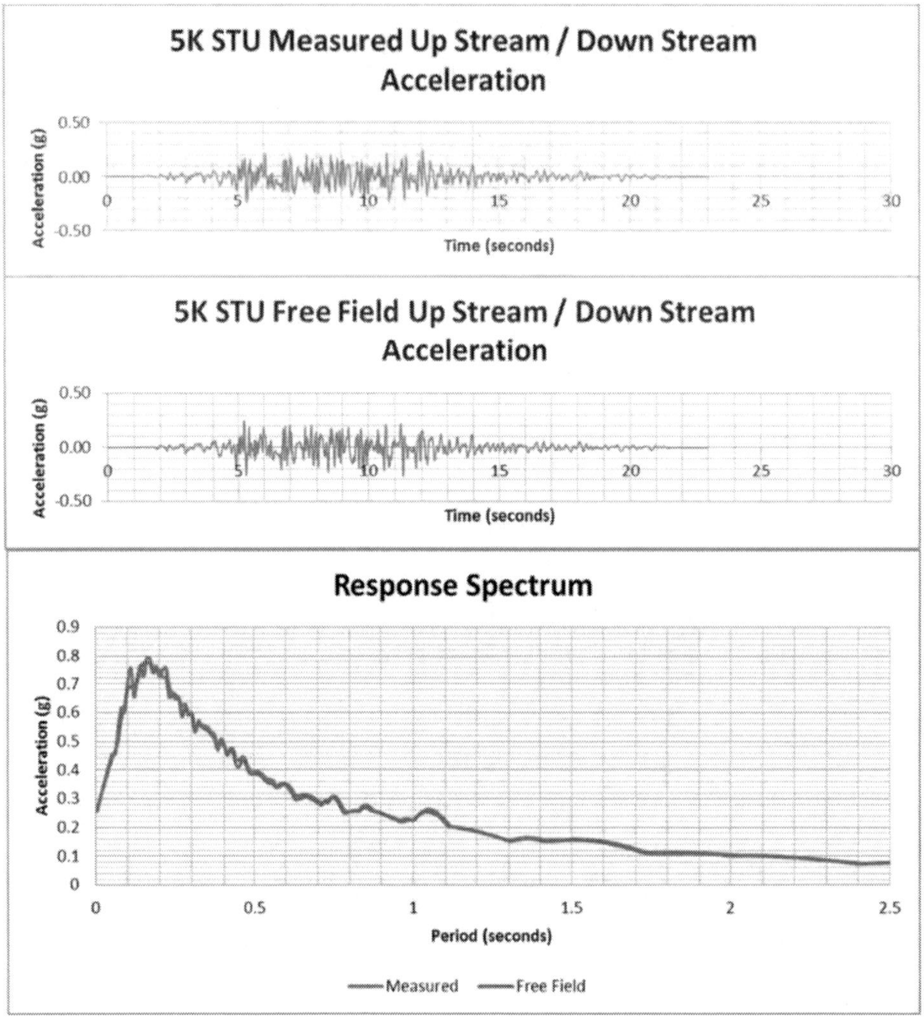

Figure 10. Free-Field versus Measured Motion-5k Sturno (STU)-Upstream/Downstream.

STRUCTURAL ANALYSIS RESULTS

Static Analysis Results

Results of the static portion of FE analysis were validated by comparing reservoir pressures estimated within the model to pressures calculated by hand assuming hydrostatic pressure on a vertical plane surface. Temperature loads were also checked to ensure the correct temperature was applied to both upstream and downstream faces of the dam.

Figure 11a shows the contoured reservoir pressures resulting from the application of gravity. Figure 11b shows a plot of the reservoir water pressure applied to the upstream face of the dam at various elevations and locations (see Figure 11a). The pressure time histories shows that the reservoir pressure increases as gravity goes on between 0.0 to 2.5 seconds and stabilizes during the quiet time before the application of the seismic load (5 seconds).

Hand calculations for water pressure at a selected element 370400 were performed as a check to the model results. A calculated water pressure of 78 lb/in^2 for a depth of 180 feet was very close to reported pressure history plots for this element (see Figure 11b). This indicates that the water elements were functioning correctly and that gravity was applied properly.

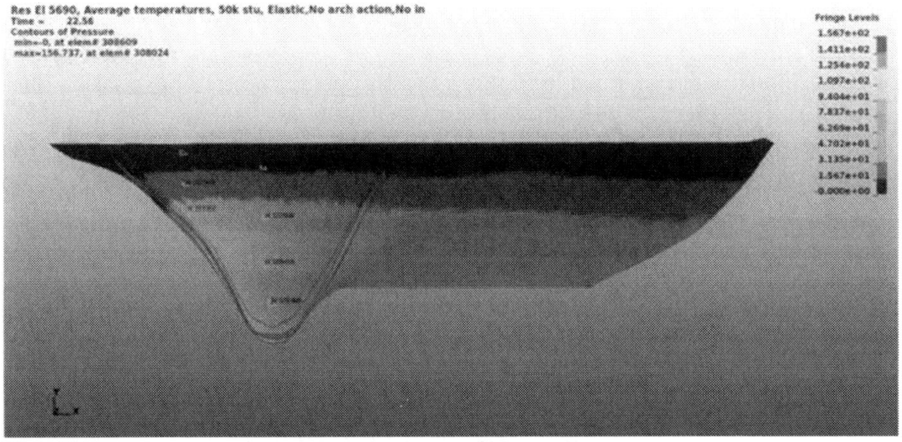

Figure 11a. Contour of Water Pressure.

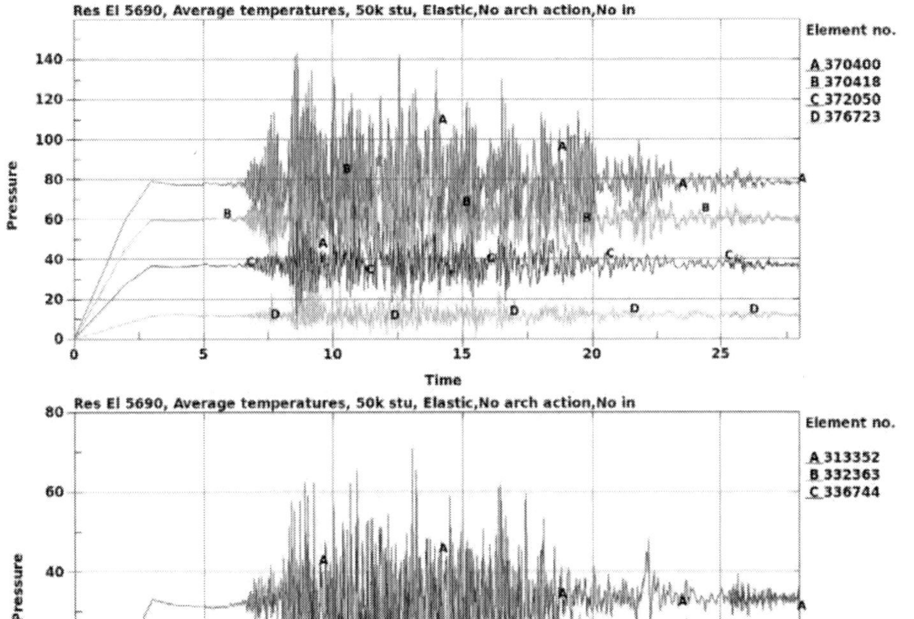

Figure 11b. Plot of Reservoir Water Pressure Applied to the Upstream Face of the Dam (Followed by the 50K STU Seismic Event).

Dynamic Analysis Results

1. Arch Dam Analysis Results

Dynamic analysis of arch dam was performed and post processed to determine the possibility of initiation/propagation of horizontal, vertical and diagonal cracks on downstream and upstream faces of the concrete dam and through the dam thickness and to evaluate the stability of the arch dam under the applied loads.

The pressure time histories shown in Figure 11b also record the water excitation during the seismic event. The reservoir pressure increases as gravity goes on and fluctuates during the application of the seismic load eventually stabilizing at the end of the earthquake. As shown in this figure, the water pressures stabilize at static values after the seismic event. No load is lost indicating contact is maintained throughout the analysis.

Below, results are presented in terms of comparing damage to the dam from different return periods (5K, 10K, and 50K) of the Sturno (STU) earthquake. The station Sturno recorded the November 23, 1980, M 6.9 Irpinia, Italy earthquake. Figures 12 and 13 show this concrete damage to the dam using Material 159 and Winfrith respectively.

As shown in Figure 12, the concrete damage for Material 159 is portrayed on a scale from zero (indicating no damage) to one (indicating a failed element). The concrete damage is portrayed by thin black lines indicating crack orientation for Winfrith as shown in Figure 13.

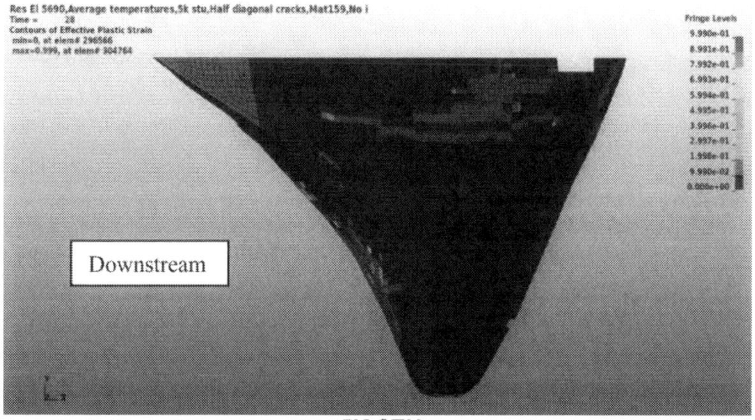

5K STU

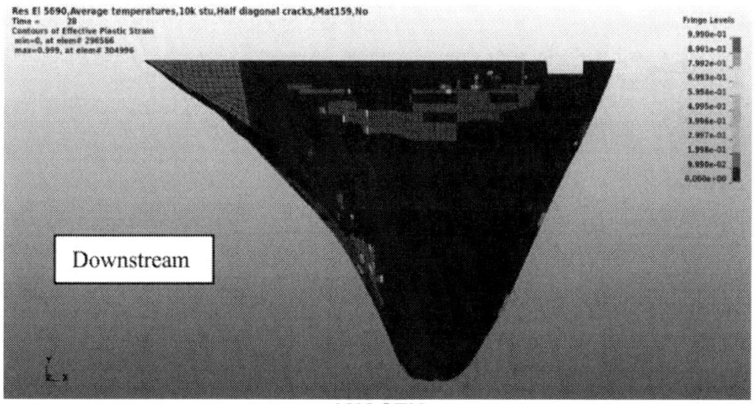

10K STU

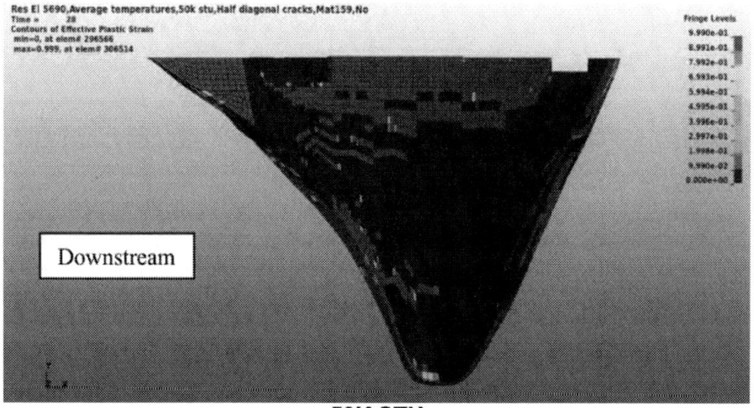

50K STU

Figure 12. Concrete Damage (Contour of Damage Parameter) -Material 159.

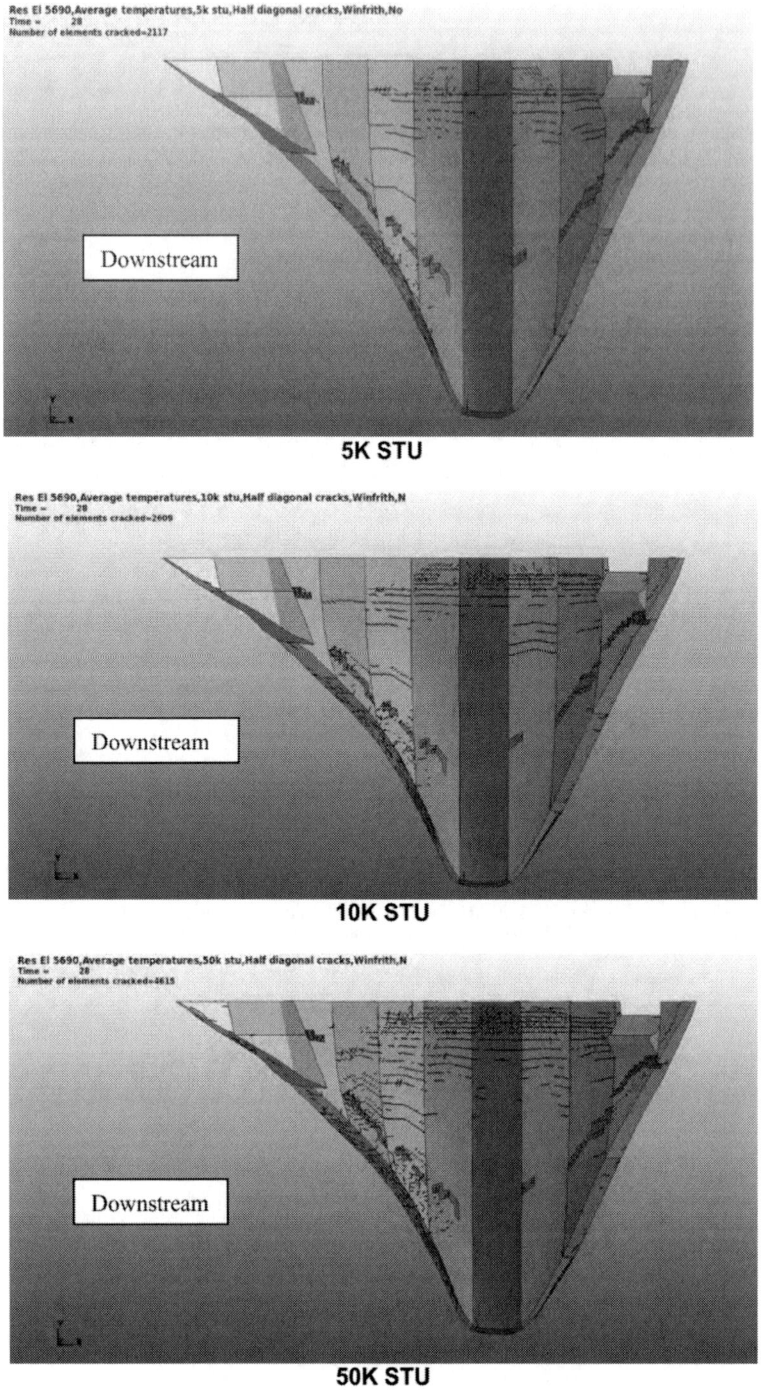

Figure 13. Concrete Damage (Crack Location and Orientation)-Winfrith.

As expected, damage increases as the level of shaking increases. Both materials exhibit similar cracking patterns with most of the damage occurring in the top half of the center monoliths. A time history of displacements of the dam at various locations show that the displacements stabilize after the seismic events, indicating stability of the arch.

Figures 14 and 15 compare concrete damage to the dam while varying the ambient temperature conditions using Material 159 and Winfrith respectively. This comparison is made using the 10K return period STU earthquake. There is not a significant difference in the concrete damage for the different temperature conditions.

COLD

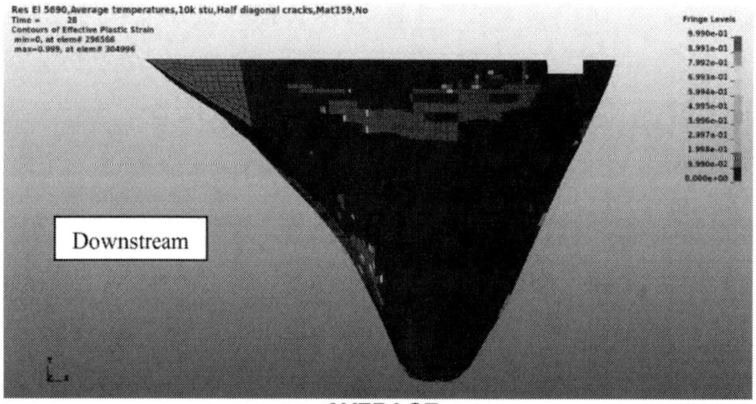

AVERAGE

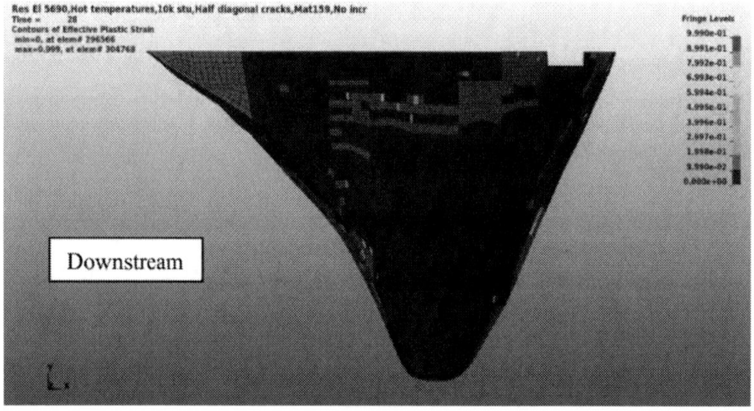

HOT

Figure 14. Concrete Damage for Various Ambient Temperature-Material 159.

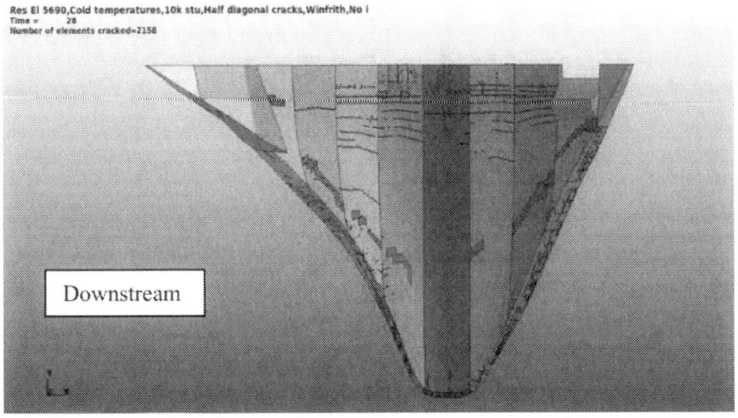

COLD

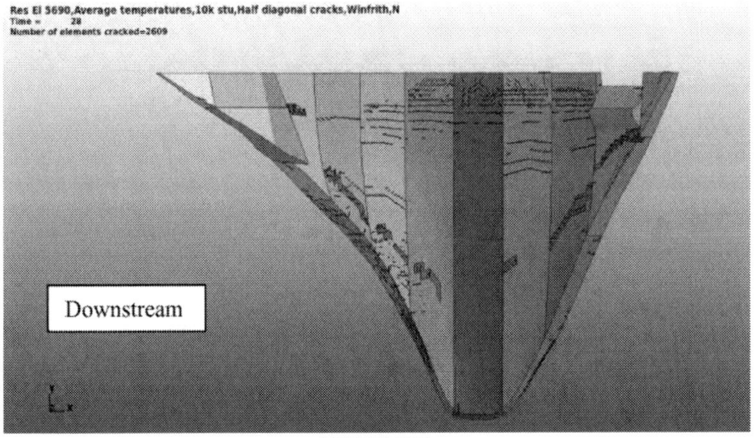

AVERAGE

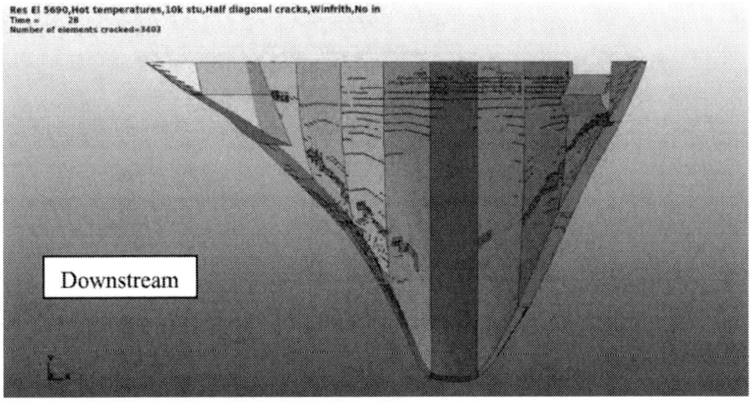

HOT

Figure 15. Concrete Damage for Various Ambient Temperature-Winfrith.

2. *Foundation Block Results*

One of the most important purposes of this dynamic analysis was to identify the potential displacement of the foundation rock blocks AB and C as shown in Figure 16 (see Figure 2 for description). Stability of the rock blocks was investigated under different loading conditions as well as stability of the arch dam due to rock block movement. Time histories of the relative displacement between nodes on the apex of each rock block and the corresponding node on the foundation were plotted to identify the displacement of the rock block relative to the rest of the foundation. Figure 16 shows the location of these nodes, while Figure 17 shows a typical rotation/displacement of the rock blocks after a seismic event (magnified x2). Figure 18 shows a time history plot of the resultant displacements (inch) versus time (second) of these nodes for the 50K STU earthquake. As can be seen, there is no unbound movement indicating stability of the rock blocks after the seismic event.

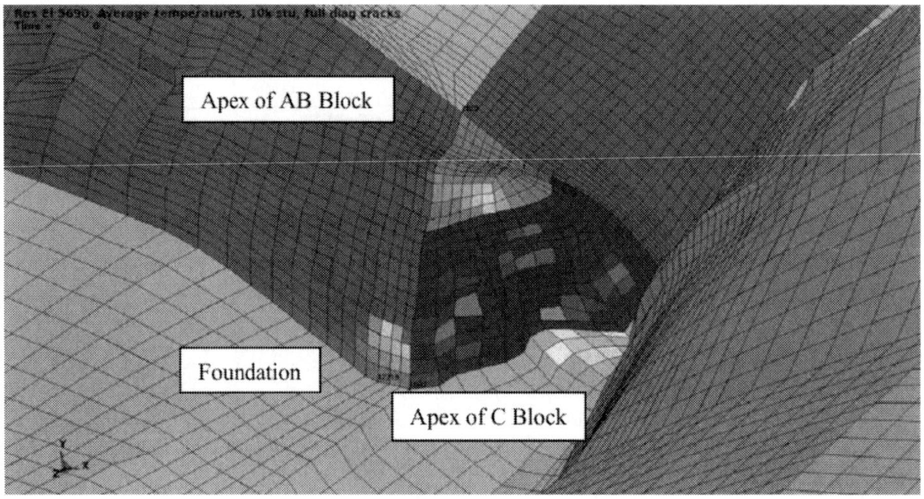

Figure 16. Node Numbers – Apex of AB Block (1007), Apex of C Block (2829), Foundation (12718).

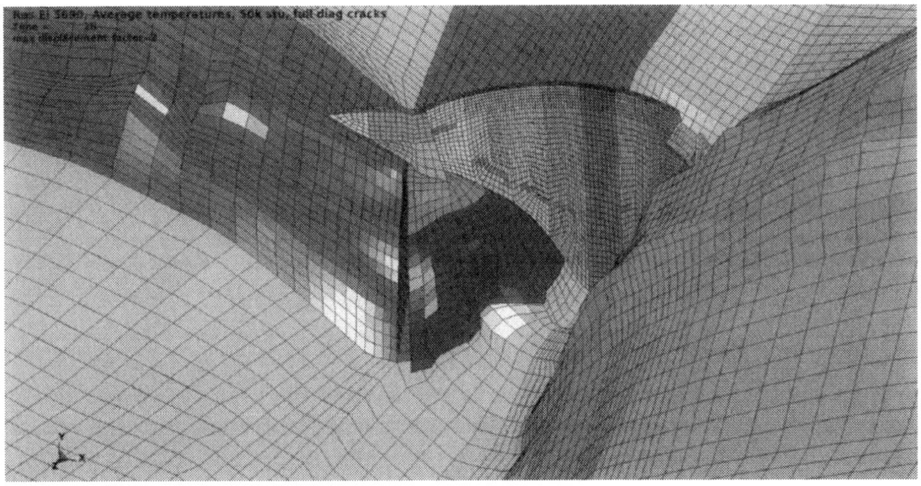

Figure 17. Typical Rotation of the Rock Blocks at the End of a Seismic Event.

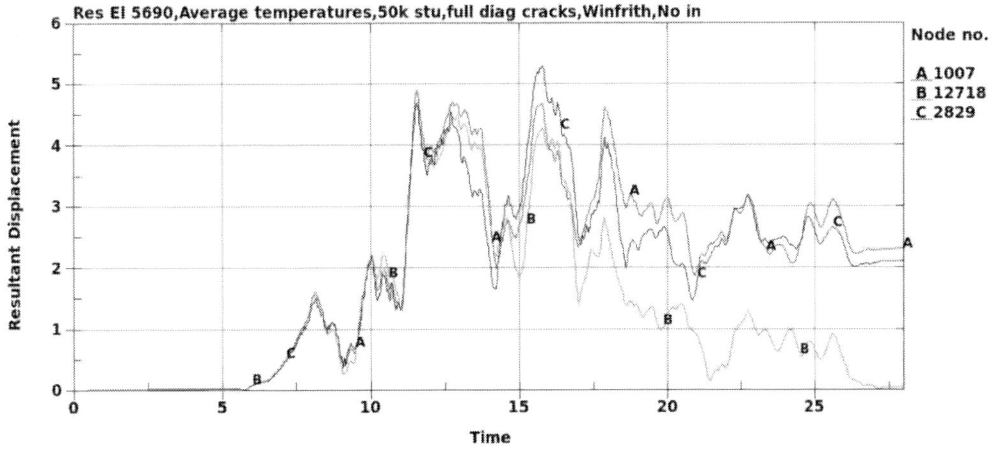

Figure 18: Resultant Displacement of AB Block (Node 1007), C Block (Node 2829), and Foundation (Node 12718).

As expected, the foundation blocks displace more under the larger seismic load. The uplift pressure has a significant impact on the movement of the blocks and stability of the blocks at the end of the earthquake. With increased movement of the blocks, comes a significant increase in cracking of the arch dam. This is because arch action is disturbed to a great extent with increased foundation block movement. Figure 19 and 20 compare this damage for two different displacements for Material 159 and Winfrith.

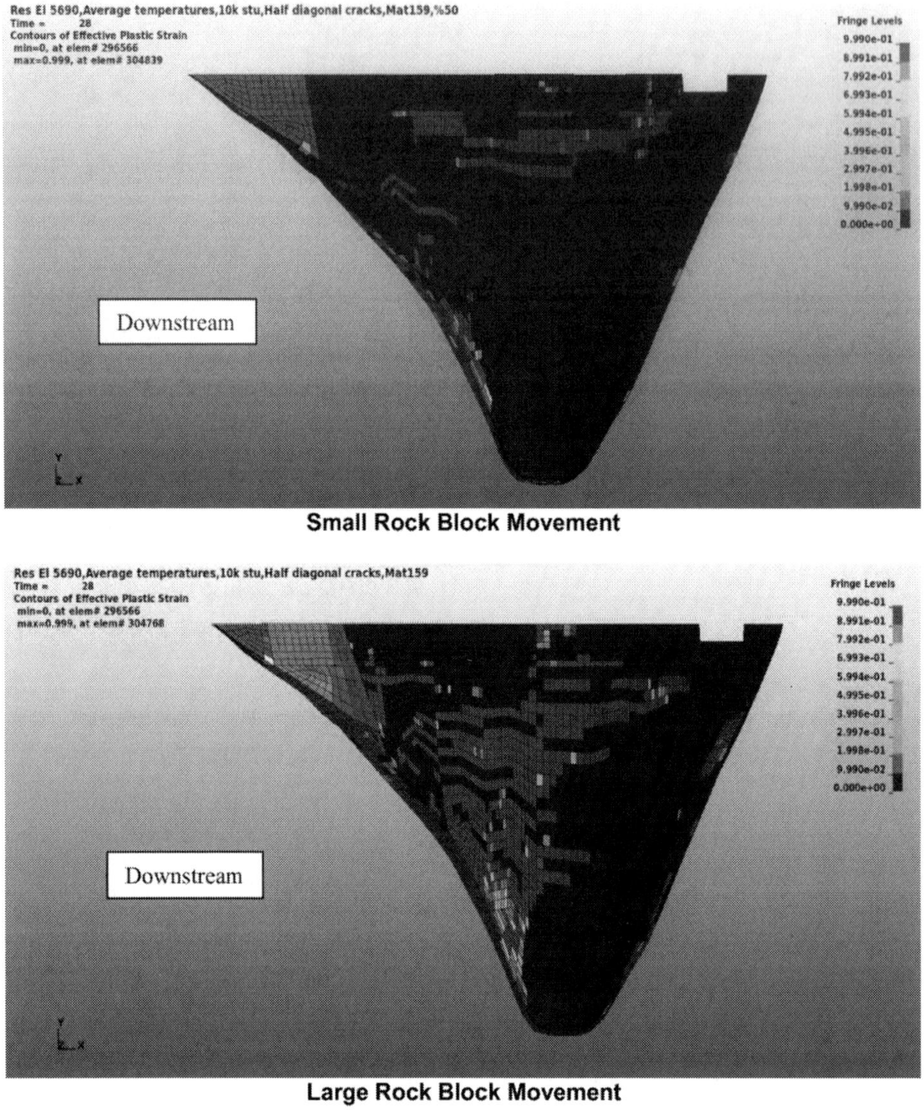

Figure 19. Concrete Damage Depending on Rock Block Movement, Material 159.

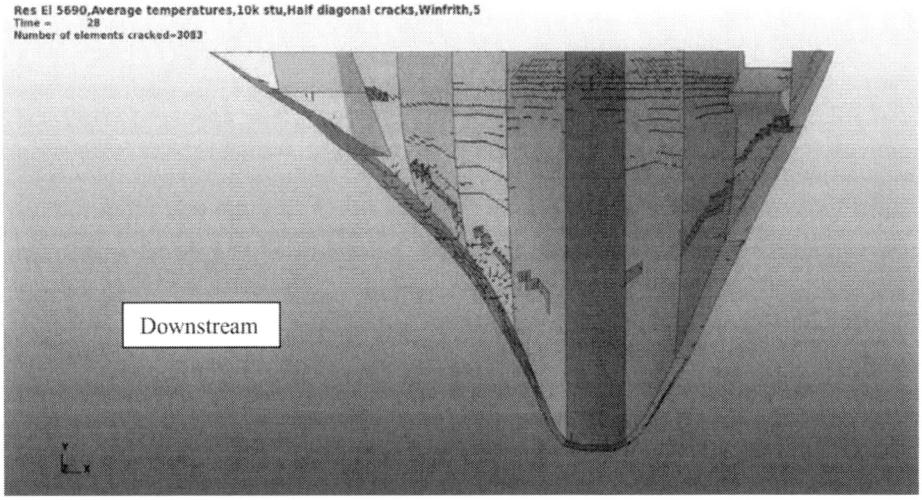

Small Rock Block Movement

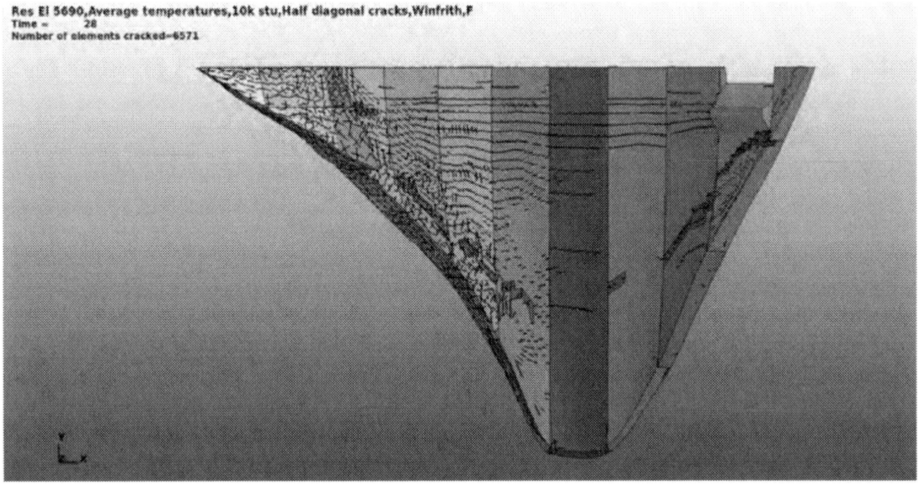

Large Rock Block Movement

Figure 20. Concrete Damage Depending on Rock Block Movement, Winfrith.

CONCLUSIONS

The nonlinear dynamic analysis of a thin arch dam with existing cracks on the downstream face and sliding foundation wedges indicate that the arch dam can remain stable under the combined application of seismic, thermal and uplift pressure loadings; though there would be significant cracking on the downstream face and extensive damage to the structure. Results of the structural analysis indicate the ability for an arch dam to redistribute stresses through arch action even in a damaged state resulting in stability of the dam. A time history of displacements of the dam at various locations show that the displacements stabilize after the seismic events, indicating stability of the arch.

The results of the structural analysis show that the earthquake magnitude has a significant impact on the stability of both concrete dam and foundation blocks. The arch dam could damage significantly under larger seismic loads. The movement of the foundation blocks downstream of the dam would also increase for larger seismic events.

Additionally, the analysis illustrated that uplift pressures are critical to understanding the movement of foundation blocks. Reduction of uplift force applied to the joints that define the foundation blocks resulted in reduction in displacement values. Therefore, the arch dam experienced less overall cracking on both upstream and downstream faces. In general, the stability of the concrete dam is related to the magnitude of the uplift pressure at the foundation block.

Finally, the analysis showed that the ambient temperature doesn't have a significant impact either on the arch dam cracking patterns or on the foundation block displacements. The FE analyses results for summer (hot), winter (cold) and spring/fall (average) temperature conditions showed very similar results for both the dam cracking patterns and the foundation blocks displacement.

REFERENCES

[1] "LS-DYNA – Version 971," Livermore Software Technology Corporation, 2876 Waverly Way, Livermore, California 94551.

[2] TrueGrid® Mesh Generator, XYZ Scientific Applications, Inc, Version 2.3.0, Livermore, California.

COMPARISON OF NONLINEAR FINITE ELEMENT ANALYSIS METHODS USED TO SIMULATE POST-TENSIONED ANCHORAGE IN CONCRETE

Vojislav Cvijanovic, PE[1]
Mark Schultz, SE, GE, PMP[2]
Blake Dolve, PE[3]

ABSTRACT

The numerical modeling of grouted post-tensioned anchors in concrete can present various challenges when analyzing load transfer through an anchor bonded to a grouted duct within a concrete pier. To capture this complex behavior, finite elements were used to model the steel anchor, duct, grout, and concrete. A non-linear plasticity model was used to simulate damage in the concrete and grout.

The fully inclusive approach incorporates all of the above components using solid elements and is considered to be a more accurate mathematical representation of the physical problem. This approach is computationally expensive and not practical in a complete dam model that considers numerous anchors and dynamic time-history loading. Therefore, a more efficient method that numerically constrains frame element nodes to solid element nodes was evaluated. The two approaches were compared to determine whether the simplified method can accurately transmit and propagate load through the bonded portion of an anchor.

The Post-Tensioning Institute (PTI) does not provide sufficient guidance for grout-bonded anchors in a concrete pier, which is a commonly used approach in providing trunnion anchorage for radial gates. Typical post-tensioning design is based on external bearing plate anchorage at each end of an anchor, which compresses the ends of a beam or slab. PTI provisions for ground anchors are more closely related to a grouted trunnion anchor, as they rely on grout-bonded load transfer. Unfortunately, PTI provisions assume bonding into ground and neglect edge and group effects associated with bonding into a pier of finite width. Edge effects, group action, and the potential for post tension induced damage to a pier are discussed in this paper.

INTRODUCTION

As finite element analysis tools and techniques continue to advance the state of practice, they provide valuable insight into the nonlinear behavior of complex physical problems. In the case of a post-tensioned anchor, characterizing load distribution can be difficult as it is heavily dependent on anchor configuration, grout strength, and various material and geometric nonlinearities.

---

[1] Senior Engineer, California Department of Water Resources, Vojislav.Cvijanovic@water.ca.gov.
[2] Supervising Engineer, California Department of Water Resources, Mark.Schultz@water.ca.gov.
[3] Engineer, California Department of Water Resources, Blake.Dolve@water.ca.gov.

To better understand the interaction, a series of nonlinear finite element analyses were considered to evaluate the behavior of a 3-inch grouted anchor inside of a 6.5-inch hole. All components of the anchor configuration were modeled explicitly using solid elements, and nonlinear material behavior was considered using the Winfrith constitutive material model available in LS-DYNA. The amount of detail included in the model was necessary to simulate interaction and capture how load gets distributed through the steel anchor via fracture of grout inside of and surrounding the duct.

The single anchor model was then extended to evaluate how group effects influence the performance of surrounding concrete. The final phase of the analyses considers a proposed trunnion anchor design within an 8-ft wide pier. The model considers 42 anchors and consists of 3.3 million solid elements. The bonded zones of the anchors are offset within the pier, which mitigates group effects to some extent.

MODELING OF A SINGLE ANCHOR

A single anchor model was used to characterize component interaction and the propagation of load through a grouted anchor (Figure 1). The 3-inch anchor, grouted duct, and concrete were all modeled explicitly using solid elements. The anchor and duct were modeled using an elastic material model with a modulus of 29,000 ksi. The concrete and grout consider a plasticity model (Winfrith) developed to capture non-linear concrete behavior. Compressive and tensile strengths of 8,000 psi and 474 psi were specified for grout, while compressive and tensile strengths of 5,500 psi and 195 psi were specified for the concrete.

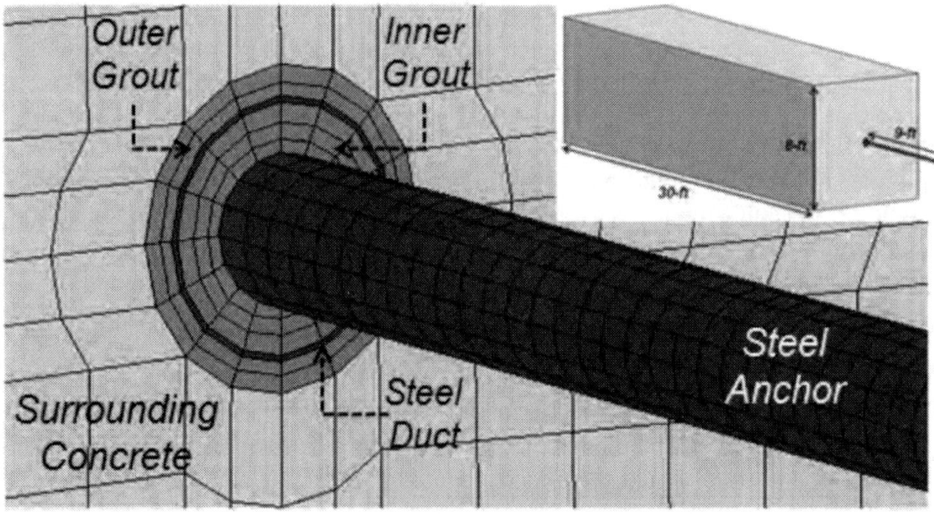

Figure 1. Fully Inclusive Single Anchor Model.

Tensile load was applied incrementally to the free end of the anchor as a prescribed displacement, while the propagation of cracks within the grout and the distribution of axial load throughout the anchor were tracked. Roller supports were used to restrain the anchor side face of the concrete block.

Figure 2a illustrates the distribution of grout cracks at various levels of applied load. Initially, load is heavily concentrated at the free end of the anchor. As additional load is applied, grout cracks propagate further into the duct, which allows for a deeper transmission of the anchor load. Figure 2c plots the distribution of anchor load at various levels of applied load.

In this analysis, anchor load propagation is primarily based on deformations and damage within grout elements. Bond between the grout and surrounding concrete is assumed to be adequate and differential displacements along interfaces (slip) are not considered, as the individual parts are modeled fully merged.

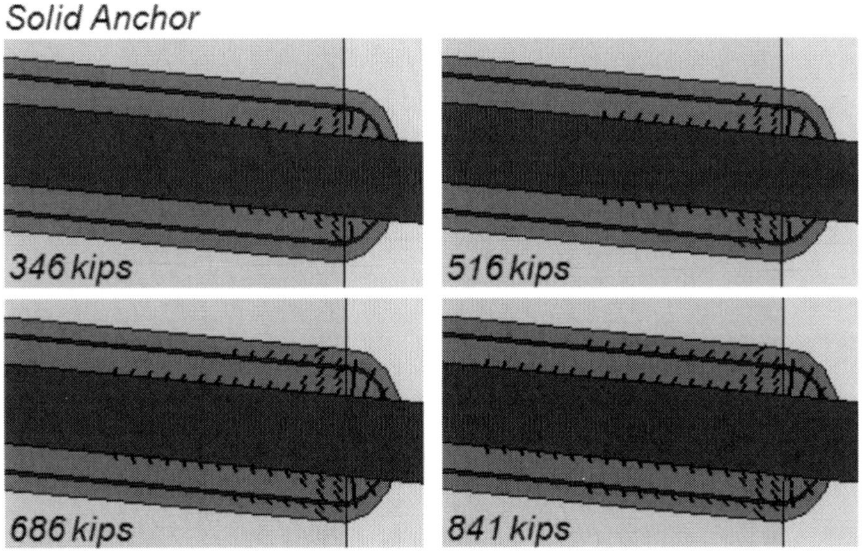

Figure 2a. Damage/Cracking in Fully Inclusive (Solid Anchor) Model.

The second approach simplifies the anchor and grouted duct by considering a frame element instead. The frame element is embedded into a solid block of 5,500 psi concrete, and the propagation of anchor load is transmitted via deformations and micro-cracking throughout the surrounding concrete (Figure 2b). The relatively higher-strength grout and confining effects associated with the sleeve are not considered and, once again, differential displacements along interfaces (slip) are not included.

Figure 2c compares the distribution of anchor load for the two approaches. In the second approach, load is transmitted slightly further into the bonded region of the tendon. Although the results seem comparable, this will not always be the case. The amount of discrepancy will typically be a function of anchor size and the ratio of grout to concrete strength.

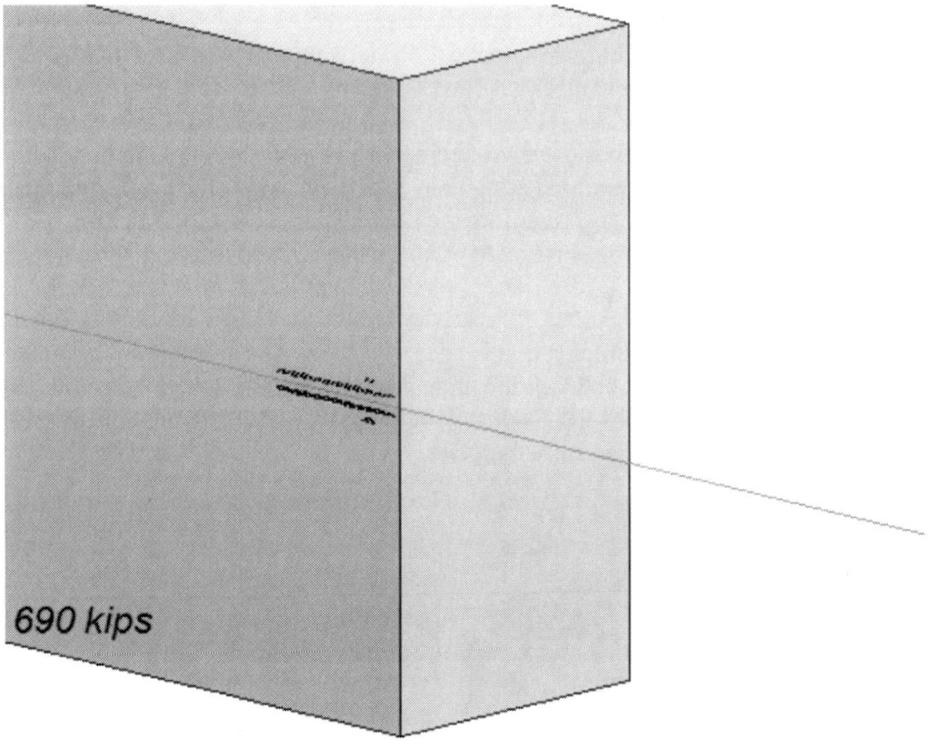

Figure 2b. Damage/Cracking in Simplified (Frame Anchor) Model.

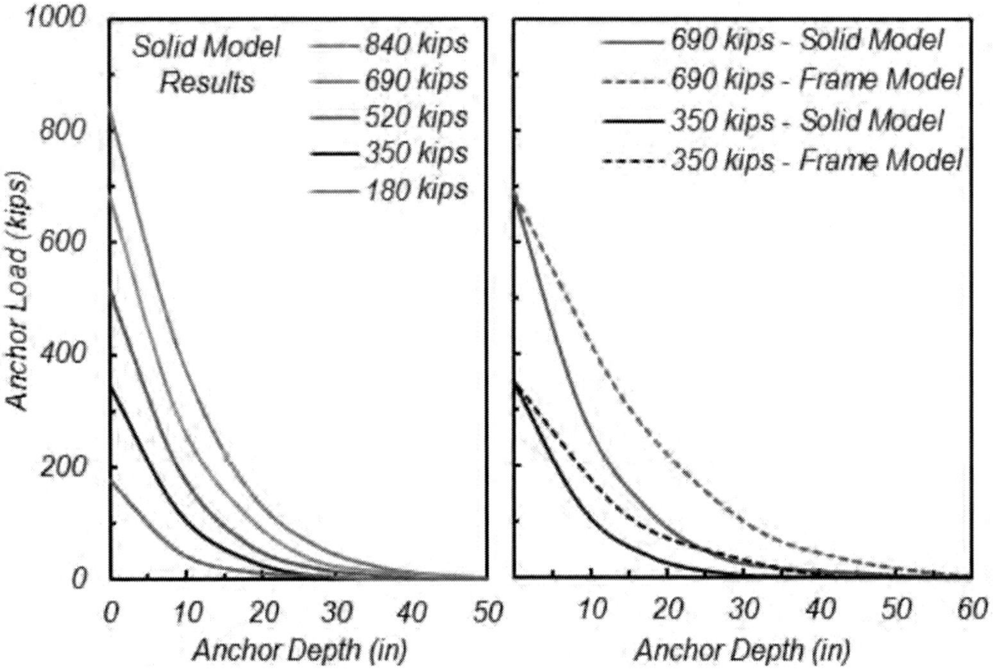

Figure 2c. Anchor Load Profiles.

The third approach is similar to the second in the sense that the anchor is represented with a frame element, and the grouted duct is not included. Therefore, propagation of anchor load is still transmitted via deformations throughout the surrounding concrete (as opposed to deformations throughout surrounding grout). The difference, however, is that the frame and concrete nodes are not merged and differential displacements are allowed. A one-dimensional sliding contact is used to model differential displacement (slip) along the grout-concrete interface. This LS-DYNA contact was developed specifically for the purpose of capturing bar slip within concrete.

A shear strength parameter specifies when interface slip occurs. The amount of slip then determines how load is transferred down the anchor. In this exercise, ultimate bond strengths of 200 psi and 600 psi were considered. This range covers typical PTI reference ranges used in determining required bond lengths, which PTI estimates by distributing the anchor load uniformly over the contact surface area. It should also be noted that in cases where interface slip is permitted, the anchor load propagation depths are consistent with PTI bond length estimates required to prevent pullout.

Figure 3 compares the distribution of anchor load for the various slip conditions. As the interface strength is decreased, load gets transmitted further down the anchor resulting in less incremental load being transferred into the concrete. When interface strength of 600 psi is specified, load is transmitted approximately 80 inches down the anchor (significantly greater than what was observed in the no slip case). Results indicate no damage to the surrounding concrete. With interface strength at 200 psi, load is transmitted 180 inches down the anchor resulting in even less incremental load transferred into the concrete. Extra care must therefore be taken when specifying interface strength, as conservative code based strength values will generate excessive slip and under predict damage to the surrounding concrete.

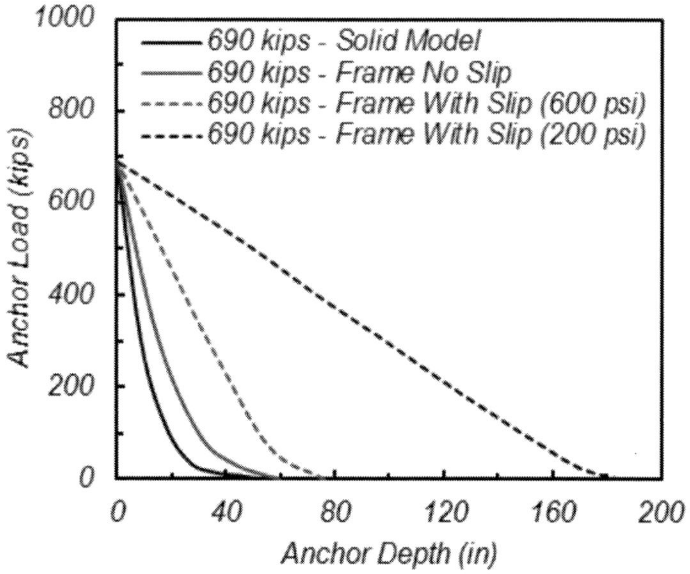

Figure 3. Comparison of Anchor Load Profiles.

SIMULATION OF GROUP ACTION

To demonstrate how group effects impact the propagation of anchor load and the performance of surrounding concrete, a group of nine identical anchors spaced at fifteen inches on center (typical spacing used in providing trunion anchorage) was considered. The anchor dimensions and material assumptions used in the single anchor exercise were adopted, and displacement-controlled load was applied to all nine anchors simultaneously.

The same three approaches considered in the single anchor exercise were used. Approach one explicitly models the grouted duct using solid elements. Analysis results demonstrate how when group action controls, significant cracking is not observed throughout the higher strength grout within the confined sleeve. Instead, the propagation of post tension is controlled by the strength of surrounding concrete, as cracks propagate through the concrete (Figure 4). This indicates that the explicit inclusion of a grouted duct is not as critical when group action governs.

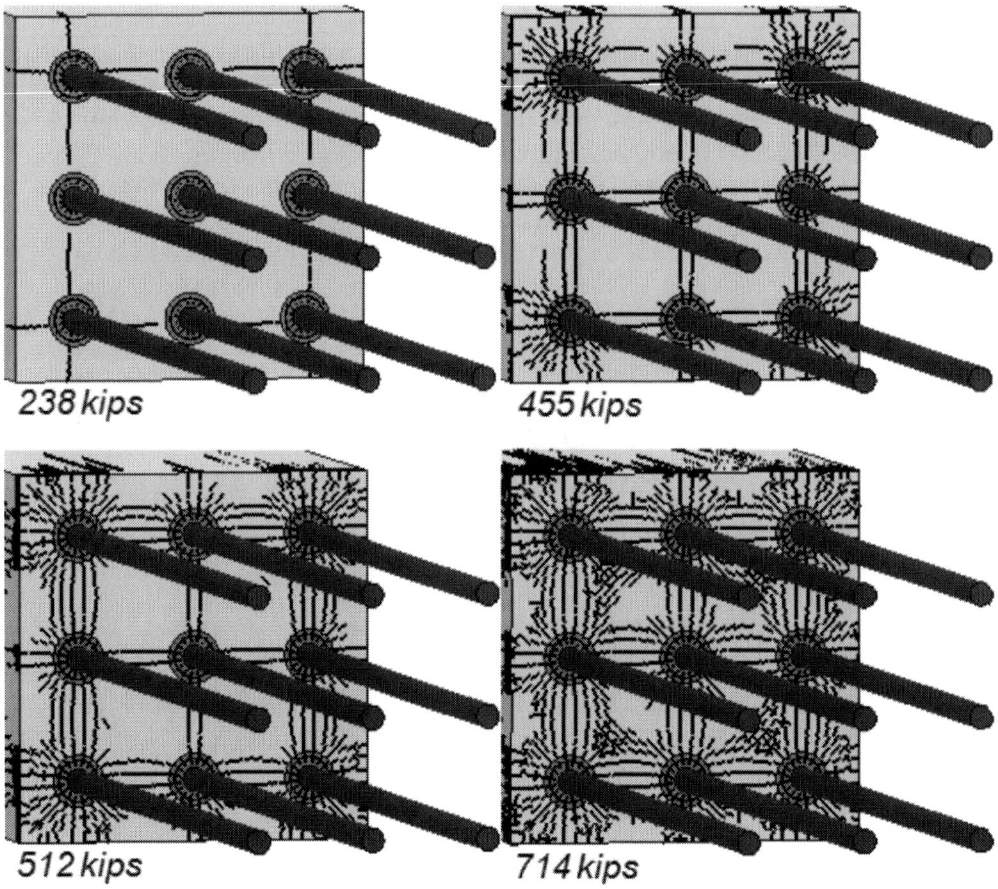

Figure 4. Damage/Cracking at Various Levels of Applied Load.

The second approach repeats the anchor group exercise using the simplified model, which does not include a grouted duct. Instead, frame elements embedded into a block of 5,500 psi concrete are utilized and differential displacements between the frame and concrete nodes are not allowed.

Approach three also considers frame anchor elements but allows for differential displacements between frame and concrete nodes by incorporating a one-dimensional sliding contact (based on 600 psi grout-concrete interface strength). Figure 5a compares the distribution of anchor load obtained using all three approaches and demonstrates why the explicit inclusion of a grouted duct may not be necessary to model cases where group action controls. Note that all three methods also produce similar concrete crack distributions at the various levels of applied load (Figure 5b).

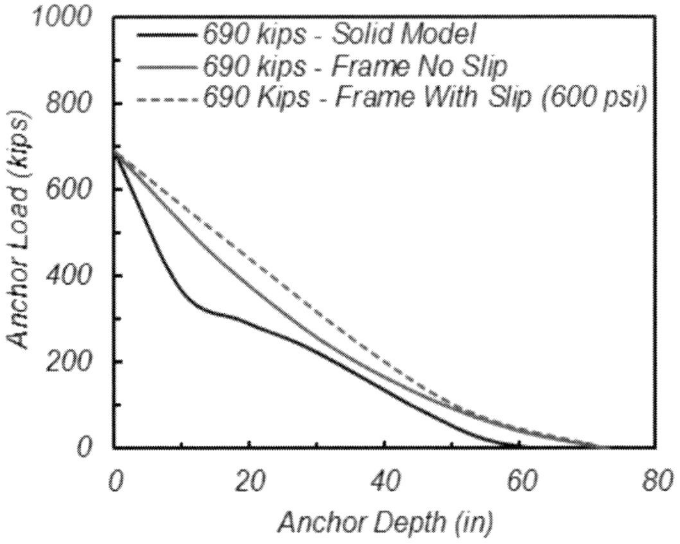

Figure 5a. Comparison of Anchor Load Profiles.

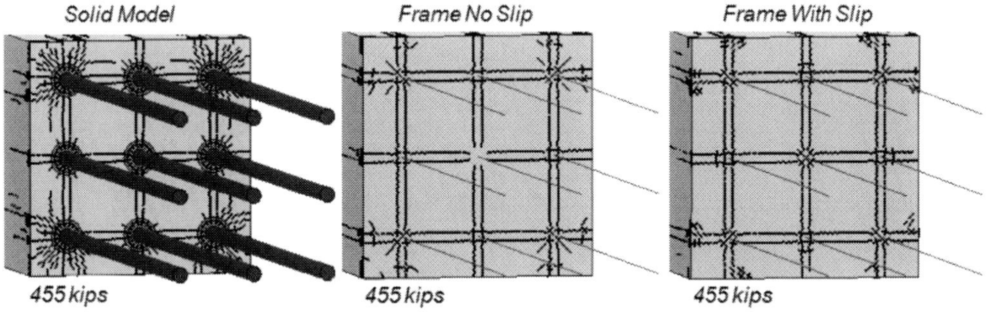

Figure 5b. Comparison of Concrete Damage/Cracking.

SIMULATION OF GROUP ACTION WITH BOND OFFSETS

The following exercise was completed to evaluate a proposed pier retrofit design, which considers forty two anchors spaced approximately 15 inches on center. This design includes bond offsets in an attempt to mitigate group action, however, it is not apparent whether the amount of offset will confine damage within the grouted duct and prevent cracks from forming throughout the pier.

This analysis is based on the fully inclusive approach, which explicitly models each grouted duct using solid elements. The model contains in excess of three million elements, some of which are very small and require that a reduced time step be used. Applying load over a fraction of a second therefore requires several hundred hours of run time. This approach is not practical for use in a time history analysis.

Incorporating bond slip through the application of surface contacts can introduce additional challenges as the interface strength is difficult to estimate. As previously discussed, conservative code based strength values will generate excessive slip and under predict damage to the surrounding concrete. Interface slip was therefore not allowed as the individual parts were modeled to be fully merged.

The proposed bond offsets are as follows. For interior anchors, grout is set back one to three feet from the concrete surface. Bonded zones for exterior anchors are set back ten to twelve feet. Bonded zones for the anchors between the exterior and interior anchors are set back five to seven feet (Figure 6). Once again, the anchor dimensions and material properties used in the analysis are consistent with those previously discussed, and 630 kips of displacement controlled load was applied to all 42-anchors simultaneously.

Following application of post-tension load, significant amounts of cracking were observed throughout the pier concrete. Analyses which apply load to anchors sequentially were also completed and found to produce similar results. Cracking patterns within the concrete pier originate at the beginning of the bonded region (Figure 7) and major cracking is not seen within the grouted duct. This implies that the propagation of anchor load into the bonded region of the anchor is limited by the strength of the surrounding concrete and that group action governs. In this case, the amount of offset is not sufficient in limiting group action, as it does not confine cracking to remain within the grouted duct.

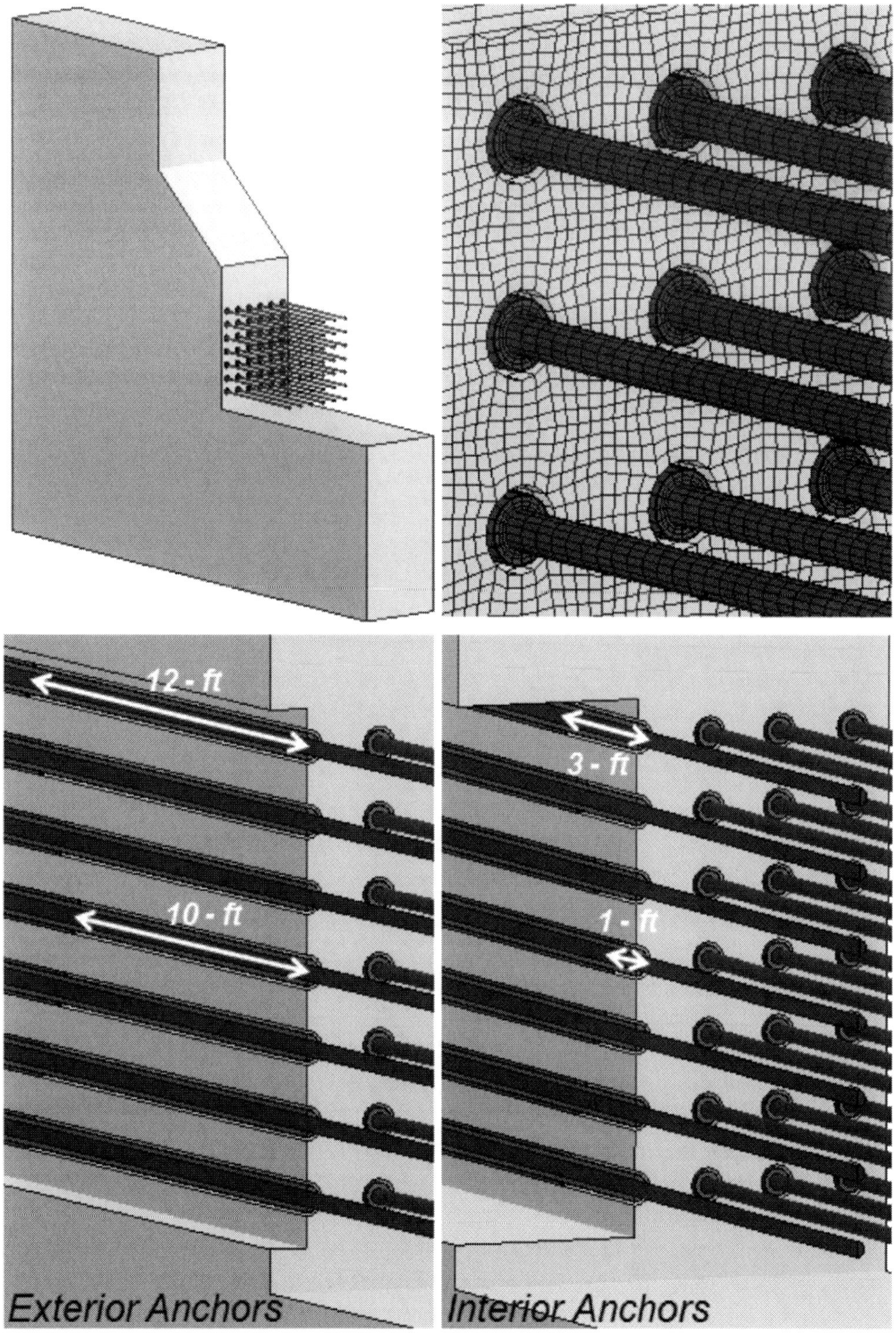

Figure 6. Fully Inclusive Pier Model.

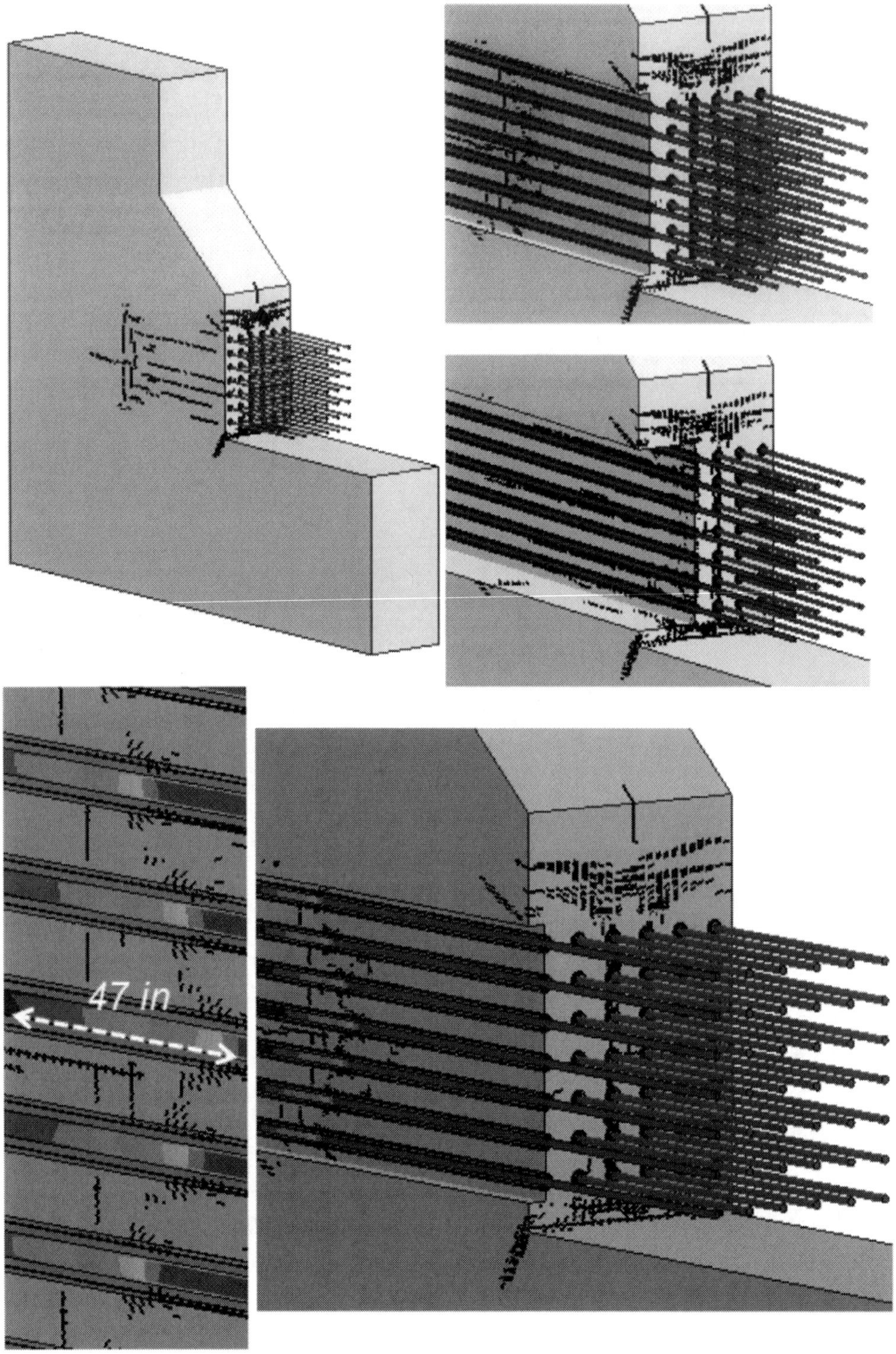

Figure 7. Damage/Cracking following Application of Post Tension.

The analysis described above was repeated using the simplified approach, which simulates anchorage using frames embedded into a mesh of eight node solid concrete elements (Figure 8). The model considers all forty two anchors (spaced approximately 15 inches on center) and includes the bond offsets, anchor dimensions, and material properties previously described. 630 kips of displacement controlled load was applied to all 42-anchors simultaneously. For consistency, differential displacements along interfaces (slip) were not permitted.

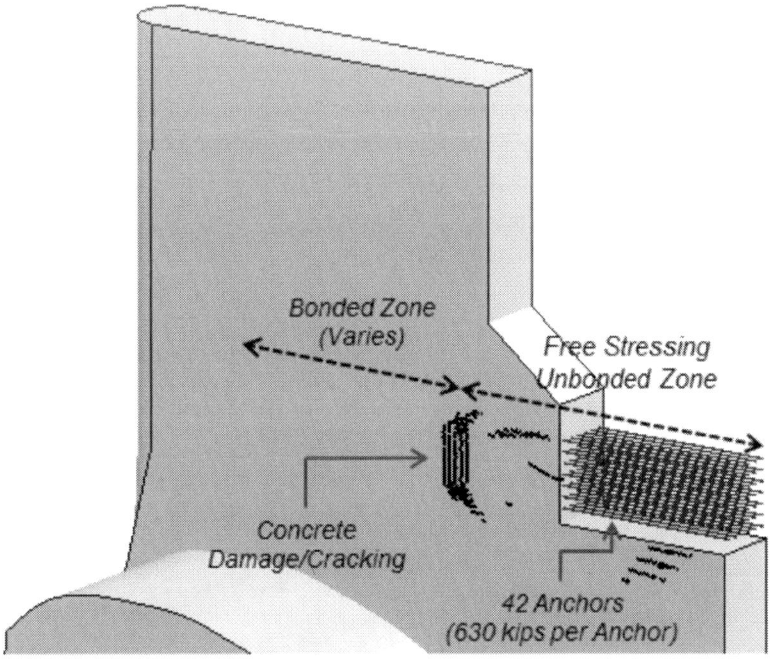

Figure 8. Simplified Pier Model.

Following application of post-tension load, significant amounts of cracking were observed throughout the pier concrete. Cracking patterns within the concrete pier originate at the beginning of the bonded region (Figure 8), which supports previous results indicating that the propagation of anchor load into the bonded region of the anchor is limited by the strength of the surrounding concrete. As was concluded by the fully inclusive model, the amount of offset is not sufficient in limiting group action.

To further mitigate group action, the anchorage configuration was modified. The number of anchors was increased to 44 (4 columns of 11), the spacing was increased to 24 inches in the vertical, and the amount of post tensioning was reduced to 545 kips per anchor (Figure 9b). Furthermore, the bonded zone offset stagger was adjusted to alternate between two and seventeen feet from one anchor to the next (Figure 9a). Following application of post-tension load to the modified pier, cracking/damage was not observed, as group action no longer controls.

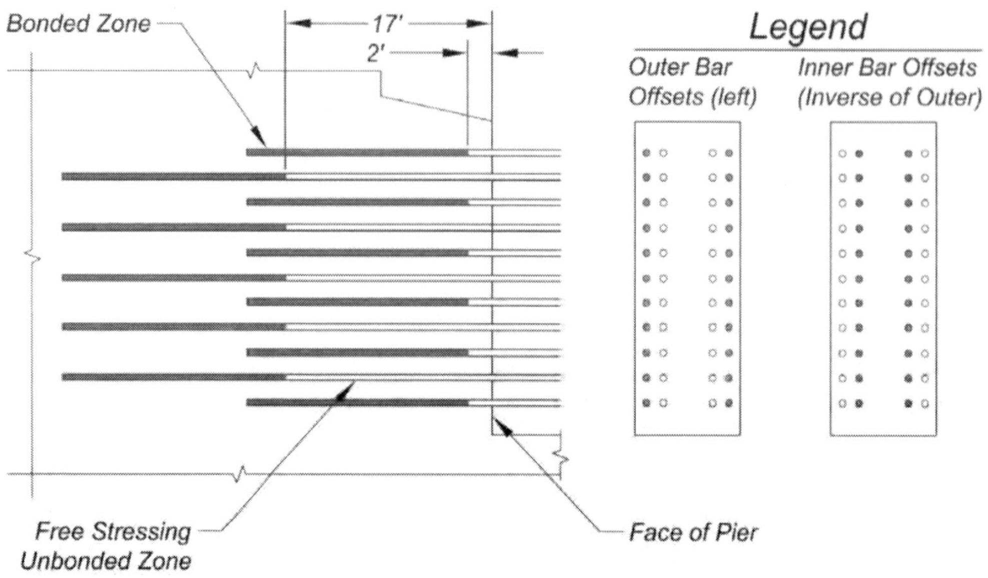

Figure 9a. Modified Bond Offsets.

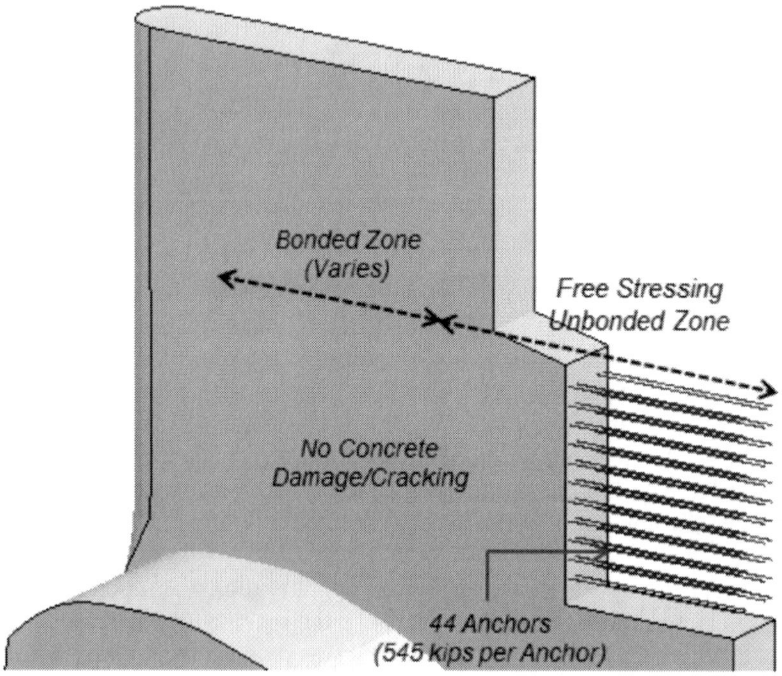

Figure 9b. Modified Pier Model.

COMPARISON TO TEST DATA

Pullout tests of large diameter bars were conducted by the Washington State Department of Transportation (WSDOT). WSDOT's research was intended to investigate bar anchorage for pre-cast bridge bent caps and produced results summarizing bar displacement as a function of load. These tests were modeled using LS-DYNA. The intent of the following exercise was to determine whether a numerical model could re-create the test data.

The test setup prepared by WSDOT emulated post-tensioned anchorage conditions where a single anchor is grouted into a corrugated steel duct previously cast into concrete. Results produced strain gage measurements along the surface of the bar and potentiometer measurements at the surface of the concrete and grout. Curves representing applied force and bar displacement near the grout face were then generated.

Stresses along the anchor-grout or grout-concrete interface are non-uniform and vary parabolically. Bond failure would first initiate near the surface of the grout or concrete face. As the bonds fail, higher stresses would travel down the anchor-grout or grout-concrete interface until completely delaminated, at which point friction becomes the leading residual strength parameter (Brown, 2014). Figure 10 illustrates this concept. The uniform bond stress approach used by PTI is a means of estimating the required bond length (to prevent pullout) and does not reflect the true distribution of anchor load.

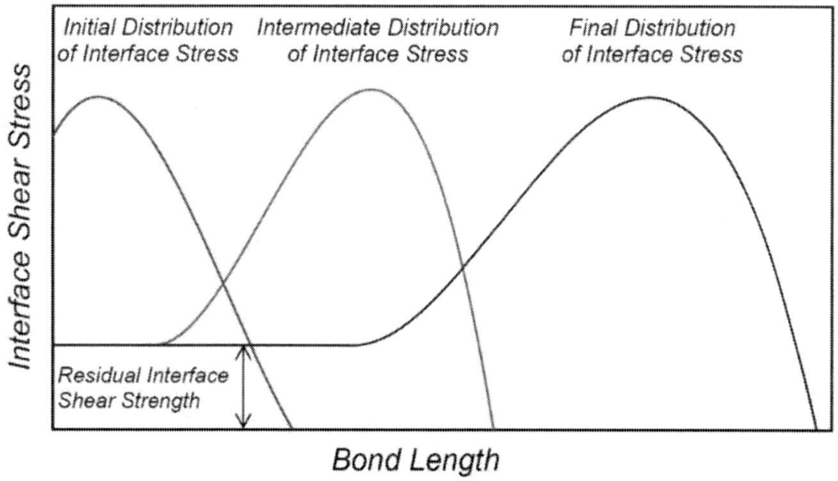

Figure 10. Interface Shear Stress at Various Levels of Applied Load.

The test setup was modeled in LS-DYNA and considers beam elements embedded within the grout (solid elements). Interface slip was permitted. The corrugated pipe was modeled using smooth shell elements tied to both the outer concrete and inner grout layers.

The #18 rebar was modeled as a beam element with diameter of 2.26 inches, and a grouted depth of 32 inches. The grout surrounding the bar was confined within an 8.5

inch diameter (0.068 inch thick) pipe running the full depth of the anchor. The outer limits of concrete had a diameter of 36-inches. One noteworthy aspect of this research is the relatively large diameter duct, which is typical in precast connections due to alignment tolerances. Larger diameter ducts tend to confine the cracking within the grout more than smaller diameter ducts.

Grout and concrete material properties reflect lab test results obtained by WSDOT. Unconfined compressive strengths for grout and concrete were 10,310 psi and 7,430 psi, respectively. The unconfined direct tensile strength for both concrete and grout is required by the Winfrith constitutive material model, but was not tested in the research. This parameter was therefore varied from two to four percent of the unconfined compressive strength, consistent with mass concrete test results used in typical dam piers. Ultimately, a value of three percent was used. Note that reducing the unconfined direct tensile strength from four percent to three percent of the unconfined compressive strength softened the model stiffness within the elastic and inelastic regions. No notable effect was witnessed below three percent.

The pipe sleeve elastic modulus was assumed to equal 29,000 ksi. The yield stress and elastic modulus of the anchor were 70.8 ksi and 27,000 ksi, respectively. Stress-strain test data was provided for the anchor and was input into a nonlinear steel constitutive material model. The results generated by LS-DYNA show how the model calculations correlate to the test data (Figure 11). In addition to re-creating the load-displacement curve, the LS-DYNA model was able to accurately predict the distribution of concrete cracking (Figure 11).

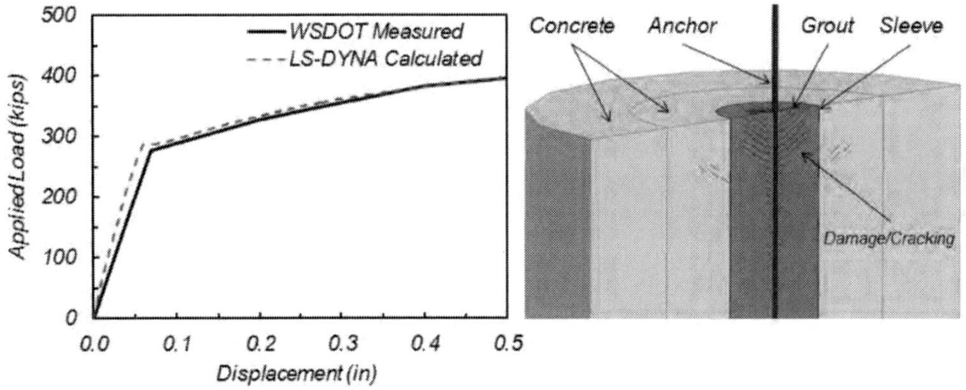

Figure 11. Comparison of Force-Displacement and Illustration of Damage/Cracking.

CONCLUSIONS

The explicit inclusion of a grouted duct is important when modeling load propagation that is primarily dependent on the formation of cracks within a grouted duct. In cases where group action governs, the transmittal of load is controlled by the strength of the surrounding concrete and therefore, explicitly including a grouted duct may not be necessary.

As a design evolves to mitigate group action effects through an increase in anchor spacing and bond offset, group action will begin to have less influence on behavior. As this occurs, care must be taken to ensure that anchor load is being propagated correctly. This is especially important in analyses which rely on a simplified frame element based anchor approach where the grouted duct is not modeled explicitly.

In cases where interface slip is expected, models can be developed to capture differential displacements between individual parts. For inclusive models, an interface contact between parts can be defined. If the simplified frame anchor approach is used, differential frame to solid node displacements can be permitted. In both cases, strength properties at an interface (which can be difficult to estimate in the absence of test data) have significant impact on how load gets transferred down an anchor and how the surrounding concrete performs. Specifying a conservative code based value will typically generate excessive slip and under predict damage to the surrounding concrete.

ACKNOWLEDGEMENTS

Although this research was supported by the California Department of Water Resources, the views expressed are those of the authors and do not necessarily represent the views of the Department.

REFERENCES

Brown, E. T. *Rock Engineering Design of Post-Tensioned Anchors for Dams - A Review*. Rockgeotech.org. Journal of Rock Mechanics and Geotechnical Engineering, August 18, 2014.

Bruce, D. A., et al. *Load Transfer Mechanisms in High Capacity Prestressed Rock Anchors for Dams*. (1991). Annual Conference, Association of State Dam Safety Officials, San Diego, CA.

Hedstrom, E. G., and Hogan, M. B. *The Properties of Masonry Grout in Concrete Masonry*. ASTM STP 1063, 47-62.

Littlejohn, G. S. *Rock Anchors - State of the Art Part 1: Design*. Geplus.com. Geotechnics Research Group, May 1975.

LSTC [2007], *LS-DYNA Keyword User's Manual*, Version 971, Livermore Software Technology Corporation, Inc., Livermore, California.

Steuck, Kyle P., et al. *Anchorage of Large Diameter Reinforcing Bars Grouted Into Ducts*. Washington State Transportation Center, 2008.

United States Department of Transportation - Federal Highway Administration [2007], *User Manual for LS-DYNA Concrete Material Model 159*, Publication No. FHWAHRT-05-062.

STRUCTURAL INFLECTION POINTS, AND HOW THEY ENHANCE THE UNDERSTANDING OF STRUCTURAL SAFETY

Guy S. Lund, PE[1]
Kevin Marshall, PE[2]
Todd Barber, PE[3]

ABSTRACT

For nearly all structures, the load versus response curve can be subdivided into two types of behavior, linear and non-linear. Between these two realms of behavior falls the structural inflection point (SIP), and understanding the loads that correlate to the inflection point helps engineers and operators better understand the margin of safety of the structures for a full range of loading conditions, as well as provide better tools to evaluate the measurements that are obtained from the project surveillance and monitoring program.

HISTORICAL EXAMPLE OF THE STRUCTURAL INFLECTION POINT

During a routine visual inspection of the Wanapum Dam in February 2014 an operator observed an unusual bend in the curb of the roadway bridge. Subsequent underwater inspections found a horizontal crack along the upstream face of monolith no. 4. The crack opening extended across the entire 65-foot-width of the monolith, and appeared to correspond to a concrete lift joint. Measurements indicated that the crack opening was more than 2-inches wide, and had a downstream displacement of approximately 1.5 inches.

Figure 1. Offset in roadway bridge curb at Wanapum Hydroelectric Project, Washington.

The Grant County Public Utility District (GCPUD) – owner and operator of the Wanapum Development – initiated lowering of the forebay pool (reservoir) to reduce the load on spillway pier. During the reservoir releases, GCPUD monitored the deflections of the spillway pier on Monolith No. 4. Initially, the data showed no response from the pier as the reservoir was lowered from the normal pool elevation (El.) 570. Then, when the reservoir level passed through

[1] Gannett Fleming, Inc., Denver, CO glund@gfnet.com
[2] Grant County Public Utility District, Beverly, WA
[3] Southern Co., Atlanta, GA

approximately El. 552.5 the spillway rebounded and the crack in the monolith closed suddenly. No further significant movement was recorded when the reservoir level was lowered below El. 552.5.

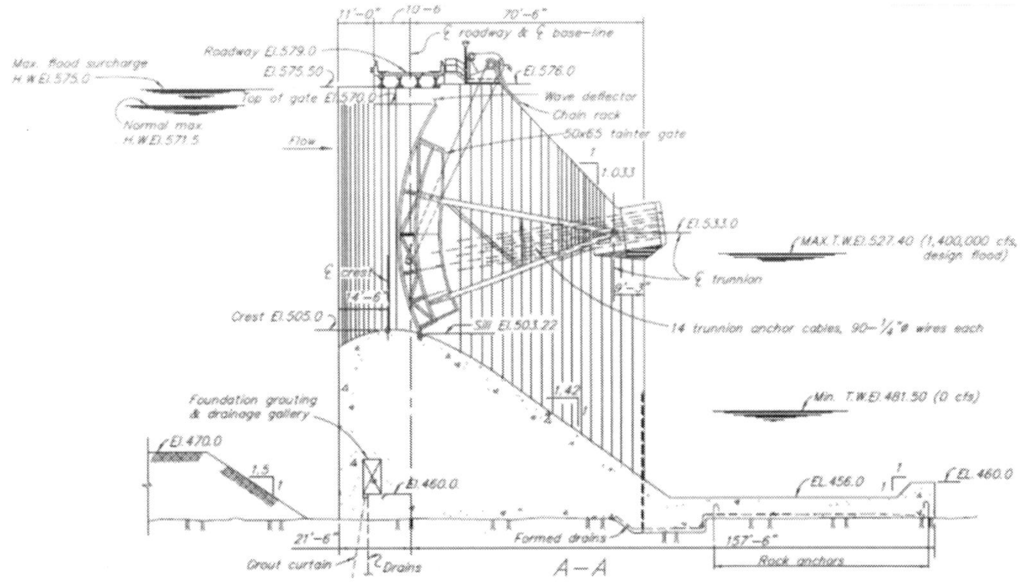

Figure 2. Section through typical monolith at Wanapum Hydroelectric Project, Washington.

An independent structural analysis of the spillway monolith was performed concurrently with the lowering of the reservoir level. The structural analysis used a non-linear finite element model (FEM), which included contact type elements to simulate the orientation of the crack as it was understood at the time, as shown in the Figure 3. It is interesting to note, that the results from initial structural studies over predicted the deformations of the spillway pier. Modifications had to be performed to calibrate the model to the measured deformations from the dam. Once calibrated, the model showed a similar deformation response as the measured data from the project, and indicated that the primary response of the spillway monolith would occur at approximately reservoir level El. 552.5.

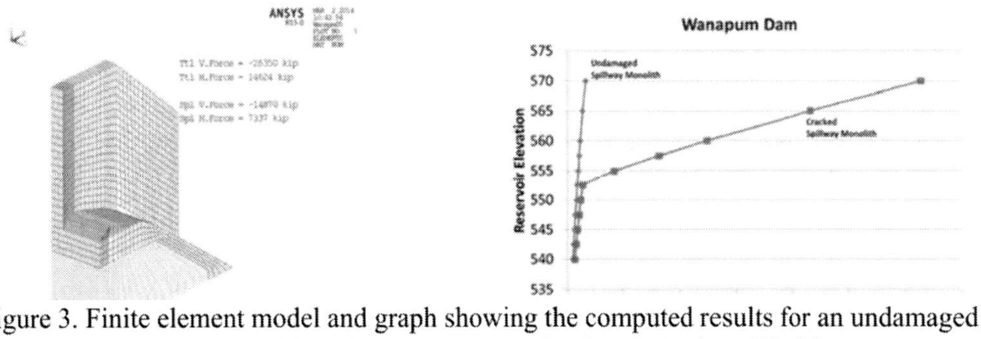

Figure 3. Finite element model and graph showing the computed results for an undamaged and cracked monolith at the Wanapum Hydroelectric Project, Washington.

By observation, the response of the structural model to the load hinged at the pool level El. 552.5. Below pool level El. 552.5, the behavior of the structure was effectively linear for both the cracked and uncracked monolith, as shown in Figure 3. Above pool level El. 552.5 the deformation response of the structure is significantly different for the cracked structure. Upon reflection, the behavior of the structural system was effectively linear below reservoir level El. 552.5, and non-linear above El. 552.5.

A SIMPLIFIED LOOK AT STRUCTURAL BEHAVIOR

Linear and Non-linear Behavior

Take most any structure and apply a load. The behavior up to some magnitude of the load will be linear; that is that when the load is removed, the structure will rebound to the original shape. As the load is increased, eventually the behavior of the structure will be non-linear, and when the load is removed the structure will not rebound to the original shape, but will have permanent deformation. For the purpose of this paper, we will define these two behaviors as linear and non-linear. Linear behavior means that the load has not caused any permanent deformation to the structure. Non-linear behavior means that the load has resulted in some permanent unrecoverable deformation.

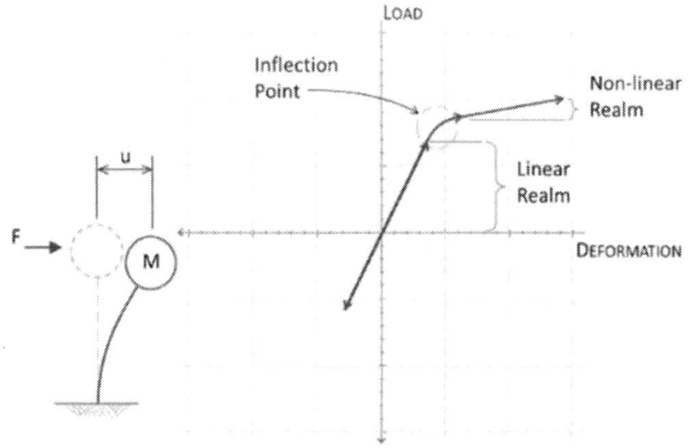

Figure 4. Illustration of the structural inflection point.

Ductile and Brittle Behavior

For the most part, if a structure's behavior is known to remain in the linear realm, then there is not much concern regarding safety. Failures do not typically occur in the linear realm of behavior. Since failures tend to occur in the non-linear realm it becomes important to further understand the types of non-linear behavior.

Understanding the difference between ductile and brittle behavior is an important characteristic of the non-linear realm. Steel, for example, tends to fail in a ductile manner. After passing the inflection point (typically called the yield point in steel design), there is significant deformation, with little increase in the load. Thus, a ductile failure can provide warning time in a monitoring program.

Mass concrete, on-the-other-hand, tends to fail in a brittle manner. A brittle failure occurs rapidly after the material has surpassed the inflection point. Brittle failures are difficult to predict because there is little, or no warning time.

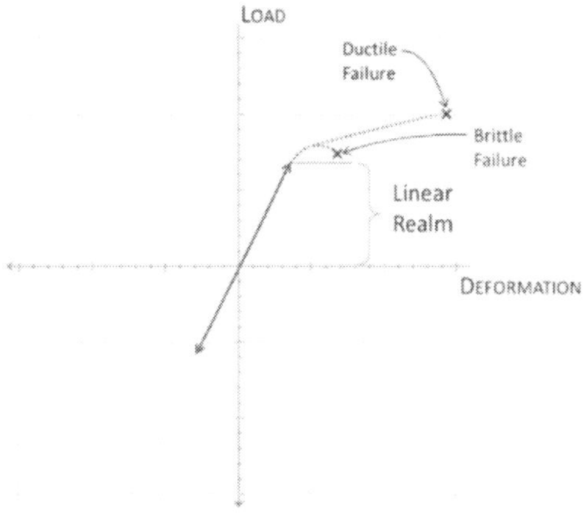

Figure 5. Illustration of brittle and ductile failure.

Concrete dams are primarily constructed of mass (unreinforced) concrete. Thus, engineers have assumed for years that the failure of concrete dams would be brittle in nature. Historical failures, such as the Malpasset Dam failure of 1959, support the brittle failure assumption. Malpasset Dam consisted of a concrete arch structure on the Reyran River, in southern France. It failed on 2 December 1959, killing 423 people in the resulting flood.

This presents several problems for owners and operators of concrete dams with regard to surveillance and monitoring programs (SMP). The safety programs in place for many dams use surveillance and monitoring to assess the development of potential failure modes, as the FERC states in the Engineering Guidelines, Chapter 14, Appendix J:

> *"The purpose/intent of the instrument or visual monitoring program should be documented relative to* ***potential failure modes****. . . .* ***Action Levels*** *should be developed to aid in immediate field-verification of instrumentation readings and/or to assist in determining if readings are approaching a level which would cause concern regarding the instability of a structure. It is highly recommended that Action Levels be established for instruments that are used to evaluate and monitor the development of a specific Potential Failure Mode. "*

If the non-linear behavior of a concrete dam is brittle in nature, then at what point does the owner/operator establish the Action level to assure adequate warning time before a potential problem? If the Action Level is conservative, then alarms may indicate problems when in fact the structure is well within the linear realm of behavior. This may be like crying wolf, and cause the owner/operator to become desensitized to the SMP. On-the-other-hand, if the Action Level is set at too high a level, it may not alarm the owner/operators to a problem, until it is too late. Either way, establishing the Action level is tricky. Understanding the load that corresponds to the structural inflection point can assist in better understanding the behavior and establish appropriate Action Levels.

INFLECTION POINTS

Identifying the load that the places the system at or near the inflection point can be accomplished through parametric or sensitivity analyses. A parametric analysis can be performed by varying the load and evaluating the behavior for the range of assumptions. The results can then be used to identify the load that pushes the system into non-linear behavior.

The following case history of Tallulah Falls Dam is presented to illustrate the benefit of the understanding the structural inflection point.

Tallulah Falls Dam

Tallulah Falls Dam consists of a cyclopean concrete curved gravity dam and reservoir near Clayton, Georgia. The project is owned and operated by the Southern Company Generation (also known as Georgia Power Company (GPC)), and regulated by the Federal Energy Regulatory Commission (FERC Project No. 2354).

Figure 6. Downstream view of Tallulah Falls Dam, Georgia

A structural analysis was performed for the usual (normal) and unusual (flood) loading conditions. The studies evaluated the adequacy of the dam to safely support the assumed unusual flood loads against concrete overstressing, overturning, and sliding stability. A non-linear, finite element model of the dam and foundation was used for these studies. The non-linear model consisted of contact type elements that simulate opening and closing of discontinuities, such as selected vertical contraction joints within the dam, and the interface along the dam/foundation contact.

The results showed internal stresses that develop due to the assumed flood loads were well within the allowable limits of the material. Thus, the assumed loads were not expected to place the material in a state that would cause cracking or crushing of the concrete. Therefore, based on the results, it was concluded that the dam has adequate safety against overstressing.

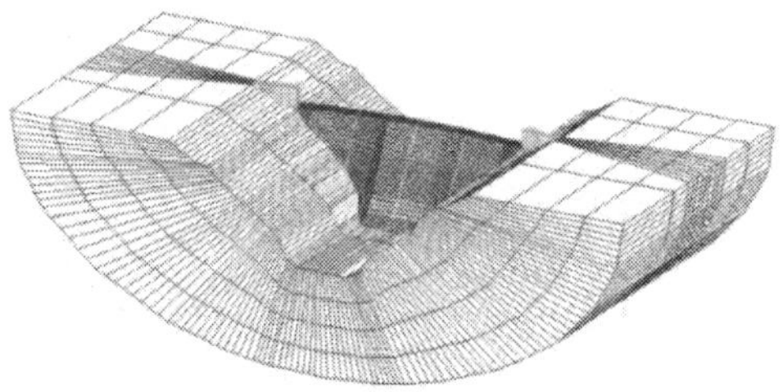

Figure 7. Finite Element Model used for analysis of Tallulah Falls Dam

The results from the analysis were also used to evaluate the overturning and sliding stability of the structure. The potential for overturning was evaluated using the calculated normal stresses along the dam/foundation interface. If the interface separated (i.e., cracked), then the uplift load was increased to full reservoir pressure, and simulates communication with the reservoir. The results from the analysis indicated that the dam/foundation interface would be expected to crack during the severe flooding events, and thus, subjected to increased uplift loads. However, the studies also showed that potential cracking along the base of the dam would stabilize prior to reaching the full thickness of the dam – even with increased uplift loads. Therefore, based on the results, it was reasonable to conclude that dam will satisfy moment equilibrium, and thus has adequate safety against overturning.

Finally, the results from the finite element analysis were used to estimate the factor of safety against sliding for the usual and unusual loads. The initial results showed that the computed sliding factor of safety (SFOS) for the usual (normal) loads is greater than the minimum required value (1.5) that satisfies the FERC guidelines. However, the sliding factor of safety for the unusual load due to the probable maximum flood (PMF) was less than the minimum required factor of safety. For these studies an effective friction angle of 55 degrees was assumed to simulate the shear strength of the interface and the effects of cohesion were neglected.

The effective friction angle of 55 degrees had been developed in previous studies and was considered to be a conservative assumption of the overall foundation shear strength. Based on the results from the finite element model, it was shown that using the *"conservative"* shear strength did not satisfy the requirements set forth in the FERC Guidelines. Therefore, it became necessary for additional study to evaluate the foundation shear strength parameters, and better understand the margin of safety for the dam.

Figure 8. Historical photograph of the irregular nature of right abutment foundation contact.

Historical Photographs

Historical documentation and photographs of the foundation excavation indicated that weathered and loose material had been removed, and the concrete material was placed against good, hard rock. Photographs show that the final excavated surface is very rough, with large undulations (several feet), as shown in Figure 8. These large undulations significantly increase the effective friction angle along the dam/foundation interface.

The photographs also showed horizontal trending discontinuities (foliation planes) in the rock mass that at first appeared continuous. However, upon further review and field inspections, the foliation planes were found to be dis-continuous, as shown by the large offsets in the excavation. Since the offsets were not continuous, the excavation had to break through intact rock between the foliation planes. Thus, without a continuous discontinuity, sliding would likely have to develop through the rock mass, or along the dam/foundation interface.

Field Data

Field mapping of the foundation rock had been performed in 1983. In addition, the following observations have been recorded from field reports:

- Field observations have not recorded any adversely oriented rock joints, bedding planes, or discontinuities that would impact sliding stability of the dam.
- There have been no identified continuous (through-going) upstream dipping joints that might daylight downstream of the dam and provide a shear plane through the foundation rock.
- The joints in the foundation rock mass are tight and do not contain soft seam infillings of lower strength than the host rock.
- There were quartz veins observed in the host rock but these do not present planes of weakness.

Evaluation of Effective Shear Strength Parameters

Additional study of the effective foundation shear strength parameters was performed considering the effect of asperities along the dam/foundation interface, and the geologic strength index parameters for the rock mass. Based on the evidence, the assumed effective friction angle of 55 degrees was considered overly conservative. Field evidence from similar rock types, as well and the evaluation of the field data, suggested that the effective shear strength for the foundation rock mass, neglecting the effect of apparent cohesion, would be in the low 70-degrees.

Estimated Inflection Point of the Structure

As previously stated, the results from the structural analysis indicated that the dam did not satisfy the minimum required factor of safety for the unusual PMF load combination. However, the study was based on a conservative estimate of the foundation rock effective friction angle. Overall, the study did not give the owner a comfortable understanding of the actual safety. It only identified that the structure had adequate safety for the usual load, and did not satisfy safety requirements for the unusual load.

Southern Company decided to invest in additional studies to better understand the overall behavior of the dam. This was done by identifying the inflection point of the dam, with regard to the sliding stability. The sensitivity studies were performed for different reservoir levels ranging from the normal level (El. 1500) to the peak PMF flood level (El.

1516.1). The total effective deformation of the crest of the dam at the right and left abutment were recorded from each analysis. An illustration of the crest deformation for the assumed range of friction angles is shown in Figure 9.

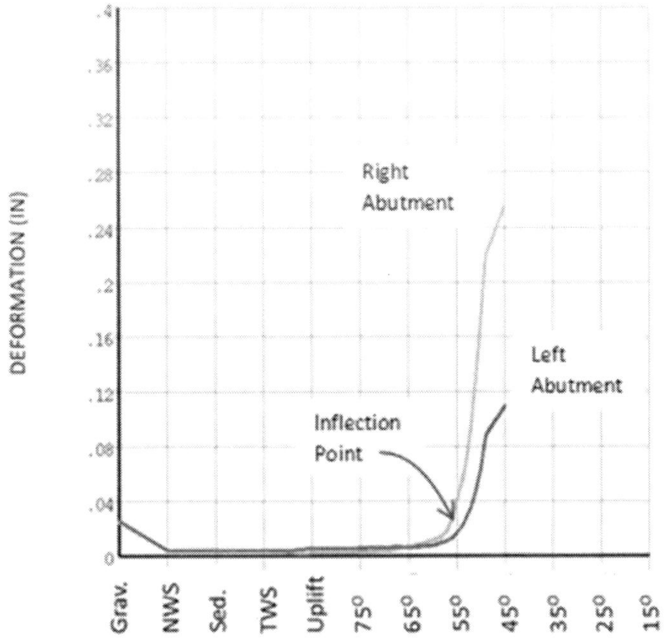

Figure 9 Plot of deformation for range of assumed foundation angles of friction (reservoir level El. 1510)

The evaluations first applied the gravity load, simulating the construction of the dam. Next, the hydrostatic pressures on the upstream face of the model were applies to simulate the reservoir load. Similarly, the reservoir sedimentation load and tailwater loads were applied to the model. Finally, the uplift load was applied, assuming full reservoir at the upstream heel of the dam, varying linearly to tailwater pressure at the downstream toe.

Initially, non-linear contact element simulates the shear friction along the dam/foundation interface as well as the potential for opening and closing of the base. The initial evaluation assumed a very large effective friction angle, equal to 75-degrees. The effective friction angle at the interface was then reduced to 15-degrees, and the deformation at the crest was measured.

As shown in Figure 9, the crest deformation remains effectively unchanged from the initial unusual load combination (with gravity, reservoir level El. 1510.0, sediment, tailwater, and uplift for effective friction angle of 75 degrees) until an effective friction angle of approximately 55-degrees. The inflection points for a reservoir level El. 1510.0

is approximately 55-degrees. This process was performed to several reservoir levels, simulating different flood magnitudes, and the estimated inflection points for the selected reservoir elevations are summarized in Table 1.

Table 1. Minimum Required friction Angle for selected Reservoir Elevations

Reservoir Level	SFOS = 1.0	SFOS = 1.5
El 1500.0	31 deg.	42 deg.
El. 1502.5	33 deg.	44 deg.
El. 1505.0	37 deg.	49 deg.
El. 1507.5	47 deg.	58 deg.
El. 1510.0	55 deg.	65 deg.
El. 1512.5	63 deg.	71 deg.
El. 1515.0	66 deg.	74 deg.

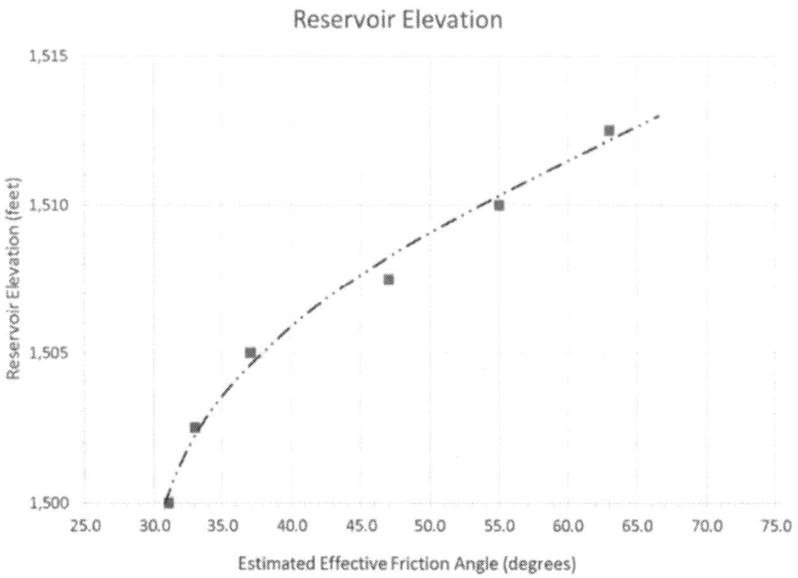

Figure 10. Plot of minimum required friction angle to satisfy sliding factor of safety equal to 1.0.

CONCLUSION

Several comments can be provided, based on the results from these studies.

1. The results show continuity in the behavior. The minimum required friction angle varies from approximately 31 degrees at a reservoir level El. 1500. The increase in require friction angle is continuous to the reservoir level El. 1515. Continuity of the results can provide additional confidence in the results of a numerical model study.

2. For a flood that results in a reservoir level El.1512.5 feet, the dam will have SFOS greater than 1.5 (adequate stability per the FERC Guidelines) if the foundation shear strength is equivalent to an effective friction angle of 71 degrees, or greater. The review of the geologic characteristics indicates that the effective friction angle may be as high as the 75 degrees. The high effective friction angle is due to the roughness (asperities) along the interface, and the lack of discontinuities that could provide a sliding plane in the foundation rock mass.

3. For the PMF (reservoir level El.1516.0 feet), sliding instability is considered unlikely for effective friction angle of 74 degrees, or greater.

4. For Tallulah Falls Dam, it is important to note the potential risk for a severe flooding event is considered very low. The population at risk (PAR) due to failure is located within the canyon downstream of the dam, and primarily consists of recreational users. The canyon is evacuated at lesser flood events for hiker safety as the trail becomes slick with any rain event and river crossings are required, which are more difficult with higher flows. Therefore, the PAR for Tallulah Falls Dam is considered to be very low to negligible for severe flooding events.

ACKNOWLEDGEMENTS

We wish to acknowledge the assistance of all who contributed to this paper. For Southern Company Generation, Mr. Todd L. Barber, P.E. (Senior Engineer - Hydro Services), who provided project data and project management for the owner. For Gannett Fleming, Inc., Mr. Guy S. Lund, P.E., Denver, Principal Structural Engineer, who was responsible for the structural analysis of the dam and overall review; Mr. Matthew Balven, PMP, P.E., Phoenix, and Mr. Brian H. Greene PhD, P.G., Pittsburgh, who reviewed geologic data and performed field inspection of the geologic rock mass.

REFERENCES

Historical photographs of construction.

Core Logs taken during field investigations May 1983.

Georgia Power Company Interoffice Communication, dated 15 November 1983, *"Determination of Rock Mass rating (RMR) and Modulus of Deformation (EM) at Tallulah Falls Dam by use of the Geomechanics Classification of Jointed Rock Mass."*

Georgia Power Company Report, dated September 1983, "Study to Determine the Benefits of the Arched Shape of Tallulah Falls Dam."

Law Engineering Testing Company, August 3, 1983, Laboratory Testing, Tallulah Falls Dam, GPCo Job Number NG-3201, LETCo Job Number G-8081.05

Law Engineering Testing Company, August 26, 1983, Laboratory Testing, Tallulah Falls Dam, GPCo Job Number NG-3201, LETCo Job Number G-8081.05

Law Engineering Testing Company, August 3, 1983, Report of Laboratory Testing of Concrete and Rock Cores, Tallulah Falls Dam, LETCo Job Number G-8081.05

URS Corporation, "Tallulah Falls Dam, Structural Stability Evaluation," July 2014

PSEUDO-NONLINEAR FINITE ELEMENT ANALYSIS OF CONCRETE GRAVITY DAMS

J. Taylor Perkins, SE, PE[1]
Husein A. Hasan[2]
Daniel A. Gilbert, PE[3]

ABSTRACT

Traditional concrete gravity dam stability analysis evaluates each failure plane independent of the other to determine the extent of cracking along a given plane. An alternative approach to traditional analyses consists of a pseudo-nonlinear approach to propagate cracking along the concrete-rock-interface to achieve static equilibrium. Failure planes are then evaluated based on the derived equilibrium condition versus evaluating failure planes independent of one another. This modified approach provides a more accurate representation of stress distributions in the structure and can reduce concerns related to over prediction of cracking along failure planes that can be common with the traditional approach. The pseudo-nonlinear approach can also provide many of the benefits of a complex non-linear finite element analysis at a fraction of the cost. This paper provides an overview of a pseudo-nonlinear approach to concrete gravity dam analysis, benefits of the modified approach and compares stress distribution, crack length predictions and calculated results to traditional concrete gravity dam analysis.

INTRODUCTION

Concrete gravity dam stability evaluation requires a cracked base analysis performed for each potential failure plane within the structure, along the concrete-rock interface, and within the foundation beneath the structure. For dams with multiple potential failure planes, the traditional approach is to perform the cracked base analysis on each failure plane independently regardless of cracking that may occur at other planes within the structure. This approach is well suited to hand calculation techniques where it is difficult to determine the stress redistribution that occurs when a crack initiates in a failure plane.

Often this traditional approach is extended to finite element based stability evaluations. This methodology, however, fails to leverage the full benefits of the finite element method; mainly, the capability to accurately predict stresses within the structure. Instead, it is possible to perform a pseudo-nonlinear analysis that considers all failure planes together within a single analysis. The pseudo-nonlinear analysis initiates cracking at the critical failure plane where cracking is expected to occur first in the physical world and then incrementally evaluates the other planes within the structure for cracking.

[1] Senior Structural Engineer, Stantec, 3052 Beaumont Centre Circle, Lexington, Kentucky, 40513, Taylor.Perkins@Stantec.com
[2] Senior Program Manager, Dam Safety, Tennessee Valley Authority, 400 West Summit Hill Dr. WT10C-K, Knoxville, Tennessee, 37902, hahasan@tva.gov
[3] Principal, Stantec, 3052 Beaumont Centre Circle, Lexington, Kentucky, 40513, Daniel.Gilbert@Stantec.com

Stantec has successfully applied the pseudo-nonlinear analysis approach to the stability evaluation of several Tennessee Valley Authority (TVA) owned concrete gravity dams. This paper outlines the methodology behind the pseudo-nonlinear analysis as well as provides the benefits over the traditional approach and the fully nonlinear approach. Results are compared between the pseudo-nonlinear and traditional approaches for a TVA's Ocoee No. 3 Dam.

STRUCTURE

In 2016, Stantec performed a stress and global stability evaluation of Ocoee No. 3 Dam; shown in Figure 1. The Ocoee No. 3 project is one of three hydroelectric projects owned and operated by the TVA on the Ocoee River in southeastern Tennessee. The dam is classified as a high hazard structure.

Figure 1. TVA Ocoee No. 3 dam

The Dam is a mass concrete gravity dam constructed in 1943. The structure can be roughly divided into three sections; the 230 ft. long non-overflow section, the 290 ft. long spillway section, and the 98 ft. long intake section (Figure 2). Each section consists of a number of monolithic blocks ranging from 37 ft. to 45 ft. wide. The blocks were constructed in 5 ft. lifts with total heights varying from 70 ft. to 110 ft.

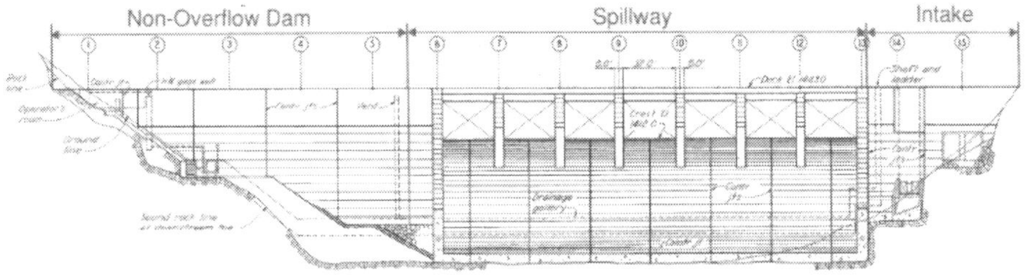

Figure 2. Elevation view showing the three sections of the Dam

ANALYSIS METHODOLOGY

A phased approach was used to evaluate the dam, where the analysis methodology starts simple in the early phases and incorporates more sophistication as necessary. Preliminary screening evaluations are first performed using simplified hand calculation techniques to identify potential trouble spots and help support field investigation. The next phase of analysis incorporates 2-dimensional finite element modeling and if warranted future phases incorporate more sophisticated 3-dimensional and nonlinear techniques on targeted portions of the structure. This phased approach is an economical way to filter the different portions of the structure and perform complex 3-dimensional modeling only on the sections of the structure that will benefit from complex 3-dimensional analysis.

The preliminary screening for the dam yielded four different cross-sections to be considered for 2-dimensional finite element modeling. The screening also helped to support an in-depth field investigation and lab testing program that determined critical input parameters for the stability evaluation such as material and strength properties. Two to five unique failure planes were strategically selected for stability evaluation in each analysis section; these failure planes coincided with seams in the foundation, concrete-rock interface (CRI), and critical lift joints within the structure. It total, 16 failure planes across the four analysis sections were evaluated. An analysis section for the non-overflow monolith including the potential failure planes is shown in Figure 3.

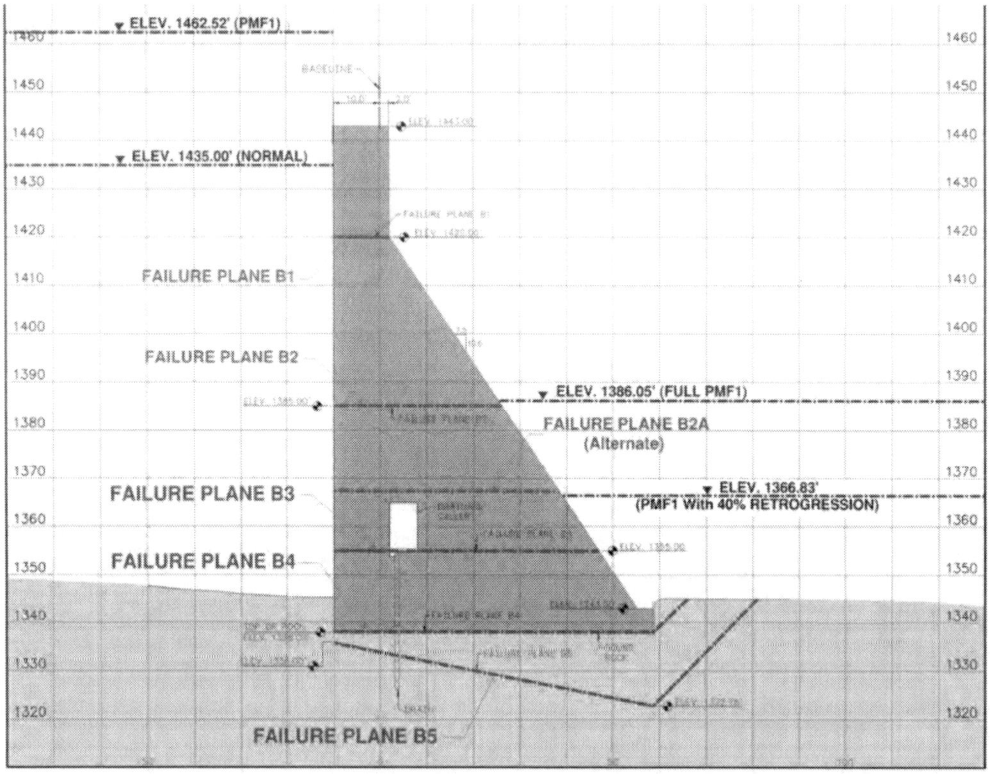

Figure 3. Analysis section for non-overflow monolith

The finite element analysis for the concrete monolith sections was performed using GT STRUDL 2015 (Intergraph Corporation, Version 34, 2015). GT STRUDL is a general purpose structural analysis software package with extensive finite element capabilities. GT STRUDL, originally from the Georgia Tech Research Corporation, is now distributed and maintained by Intergraph Corporation.

Each of the analysis sections were modeled along with a representative block of foundation material. The foundation block should represent a semi-infinite medium stretching laterally and below the structure. As such, the block should be large enough that the boundary conditions along the edges do not influence stress distributions in areas of interest, i.e. around potential failure planes, but not so large to significantly influence computational demands of the analysis. General guidance is to extend the foundation block upstream, downstream, and below the structure a distance equal to the height of the dam. The meshing of the structure was strategically planned such that the pre-determined failure planes coincide with element edges in the model. This facilitates extraction of failure plane forces for post-processing.

The models were constructed with 8-node isoparametric quadratic quadrilateral elements (IPQQ); this element is shown in Figure 4. These are 2-dimensional planar elements with two translational degrees-of-freedom at each node. The elements are formulated for either plane stress or plane strain; for this case the plane strain formulation was selected due to the long prismatic nature of the monoliths. The IPQQ elements yield excellent results for this type of finite element stress analysis. The additional mid-side nodes allow for quadric distribution of stresses through the element. This better represents stress distributions in regions of high stress gradients without requiring the smaller element sizes necessary for linear finite elements. The isoparametric formulation of the elements is also very useful in the geometric modeling of the structure due to the complicated geometry of the dam. The isoparametric formulation allows the geometry of the element boundary to match the order of the shape functions used for stress interpolation. This means that the edges of the isoparametric quadratic element can be quadratically distorted, which is especially helpful in modeling curved boundaries.

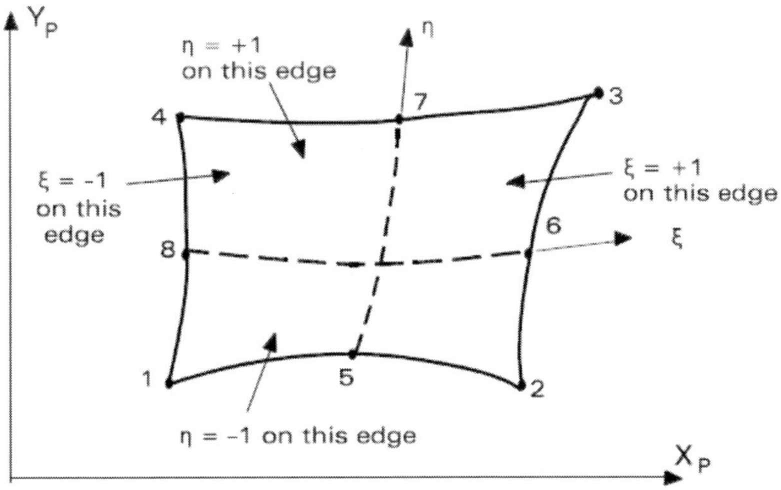

Figure 4. IPQQ Element (Intergraph Corporation, GT STRUDL User Reference Manual, Vol. 3, 2014)

Difficulties arise when modeling complex 3-dimensional structures as a 2-dimensional representation. In particular, the plane strain elements are prismatic in nature and cannot exactly model non-continuous voids through the width of the monolith section. This is a problem in particular for the intake section that has many voids and openings within the structure to facilitate the hydroelectric process. This is handled in the model by determining equivalent thicknesses for each element based on the non-voided dimension along the width of the monolith. By using this methodology, GTSTRUDL will internally scale the mass and stiffness of each element to appropriately smash the structure to a 2-dimensional section. The full finite element mesh for a non-overflow section is shown in Figure 5.

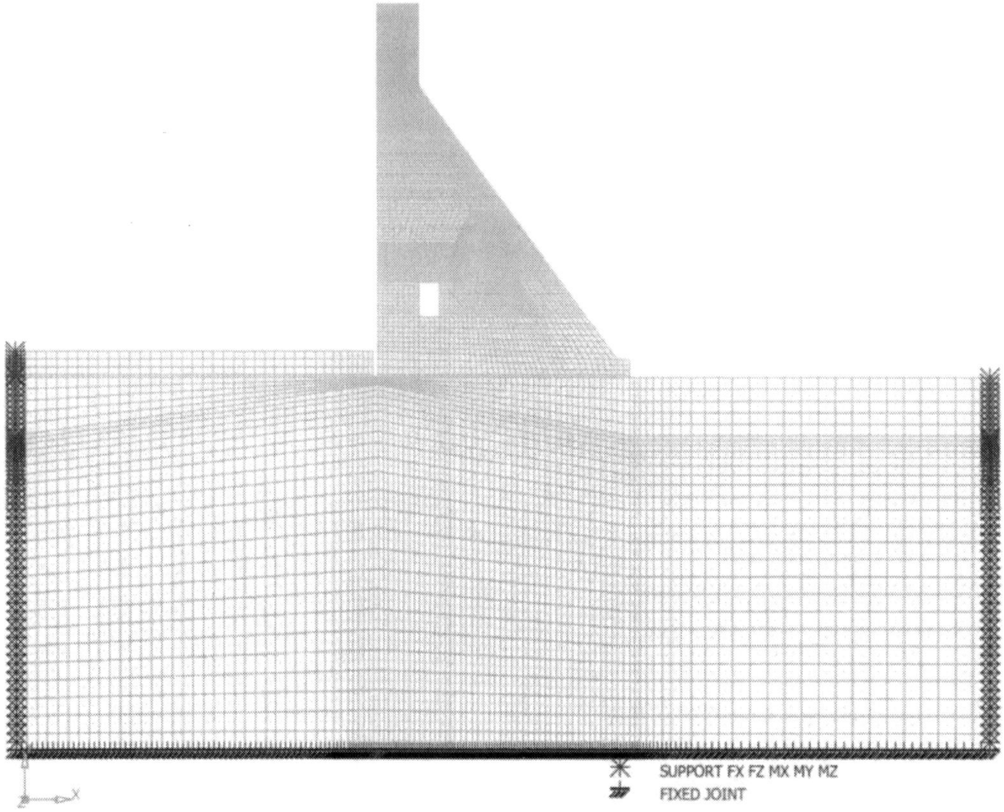

Figure 5. Finite element mesh for non-overflow section

The finite element analysis was performed using a pseudo-nonlinear approach for cracked base considerations. In order to understand the methodology and benefits of the pseudo-nonlinear analysis, first consider how a full nonlinear analysis is performed. A full nonlinear analysis would incorporate contact elements along the failure plane interfaces. For each failure plane, the tension limit for the contact element is set equal to the anticipated tensile strength of the lift joint. The contact element would then transfer forces between planar elements on each side of the failure plane under compression and up to the tension limit. After the tension threshold is exceeded the elements effectively disconnect and a crack is propagated in the model.

The analyst must specify how the loading is applied in the analysis; the loads may be incrementally increased until the total load is applied to the structure and analysis executed at each load step or the loads can be applied in a single step. One of the disadvantages of nonlinear analysis is that the solution is path-dependent, meaning that the analysis solution is dependent on the parameters of the analysis. A model that reaches a converged solution using small load steps may not converge when loads are instantaneously applied. Smaller load steps significantly increase computational demands of the analysis, as a convergent solution must be iteratively found at the end of each load step.

The analysis is performed at each load step. At the end of the analysis the tension within the contact elements is checked. If the tension limit is exceeded, the element connectivity is removed. The analysis is then rerun with the new structure configuration and the contact elements re-evaluated. This iterative process continues until the length of the crack converges. After convergence, the same process is repeated for the next load step.

The pseudo-nonlinear analysis works much like the full nonlinear analysis, however all analyses are linear elastic and the iterations are performed by the analyst. As an example, consider the non-overflow section. The concrete rock interface (CRI) is meshed as a row of thin elements. The initial analysis is run and the nodal stresses in the CRI are output from the program and assessed whether they exceed the tensile capacity of the interface. Elements that have nodes with tensile stresses exceeding the tensile limit are deleted from the model. The analysis is re-run and the process is continued until crack length converges.

Similar to full nonlinear analysis, the decision on the number of elements to delete in each step in the pseudo-nonlinear analysis could influence the results. In general, the best approach is to delete only a few elements per step in order to allow the interface to slowly unzip. This will produce the most reliable results. Removing too many elements at each step could erroneously detect overturning instability in the structure or over-predict the crack length.

One of the major issues in performing nonlinear analyses for these type structures and in particular modeling cracked interfaces is the inclusion of uplift forces. Uplift forces are an internal hydrostatic pressure that works to split an interface apart. These pressures have an assumed distribution, usually varying linearly from upstream to downstream end, in the intact configuration. As a crack propagates from the upstream face it is assumed to fill with water and the pressure remains constant across the length of the crack and then varies per the assumed distribution along the intact portion of the interface. This produces a situation where the uplift pressure distribution, and subsequently the loading in the model, is dependent upon the crack length. Most analysis software will not allow the application of loads to vary within the execution of a nonlinear analysis. The result is that the nonlinear analysis may have to be run iteratively, thus negating many of the automation associated with nonlinear analyses.

In the pseudo-nonlinear analysis the uplift pressure can be handled in an easy and effective fashion. It is possible that the nodal loads from uplift pressure be calculated based on tributary areas and then applied in equal and opposite direction to the nodes along the top and bottom edges of the interface elements. This will create a splitting force working to separate the interface. It is best, however, that the internal uplift pressures are not applied in the finite element model. Applying nodal loads to internal element nodes creates local singularities, or stress spikes, in the model. This can significantly complicate the evaluation of stresses along the interface output from the model. The better approach is to incorporate the uplift pressures externally after the stresses have been output from the program. The uplift pressures are simply added or subtracted from the program stresses before comparison to the tension limit. Once an interface element has been removed from the model, the uplift pressures are then explicitly applied as nodal forces to the remaining nodes in the model. Figure 5 shows the uplift forces and crack propagation at an iteration in the pseudo-nonlinear analysis.

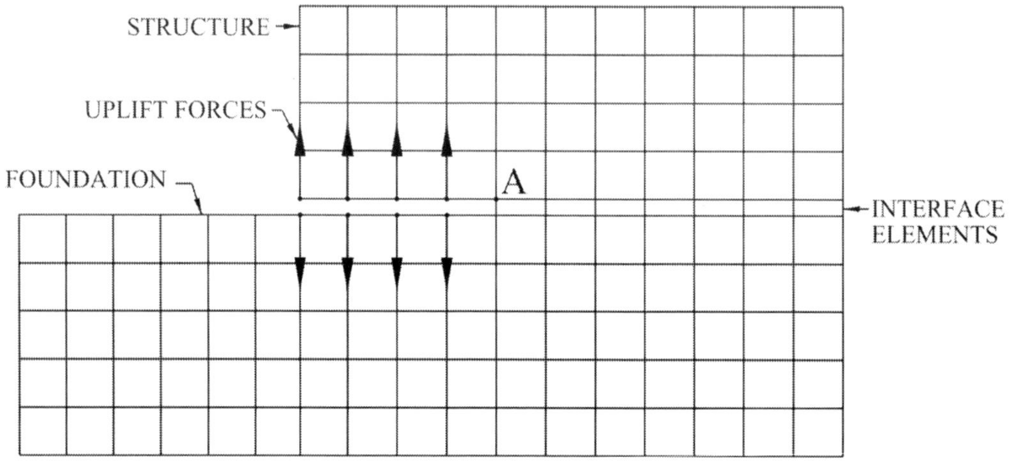

Figure 6. Crack propagation and uplift forces in finite element model

There is a potential pitfall in the pseudo-nonlinear analysis that can be avoided if it is understood. When elements along the interface are deleted, it leaves a reentrant corner in the model, denoted by "A" in Figure 6. The reentrant corner is a theoretical point of infinite stress according to the equations of elasticity. As such, it creates a singularity within the model, which is a localized zone of high stresses. These high stresses quickly reduce moving further away from the singularity. This can impact the analysis because the stress output may show cracking at the leading node, but the stresses there may be erroneously high due to the singularity. The analyst must use judgement to determine if the crack is truly propagating and whether additional elements should be deleted. Often it is necessary to look at stresses in other elements adjacent to the singularity in order to get a sense of the true stress at that location.

The pseudo-nonlinear analysis has another advantage over the full nonlinear analysis because it provides the analyst with much better understanding of the behavior of the structure and sensitivity to overturning. Nonlinear analysis can be a black-box type of approach where the analyst inputs the required information, executes the analysis, and processes the results without understanding what happens between the steps. With the pseudo-nonlinear approach, the analyst can physically see the interface unzip, track sliding factors-of-safety at each stage, and assess the impact of uplift pressure, all as part of the routine execution of the analysis. Because this process gives much better insight into the response of the structure and the driving cause for potential overturning and sliding issues, it is especially beneficial for cases that may require mitigation techniques for stabilization.

After the crack length has converged and analysis run to establish the final equilibrium state of the model, a section cut is defined across the remaining interface elements. The section cut sums resultant joint forces and outputs a normal force, shear force, and moment along the interface; when applicable the normal force from the analysis output is manually reduced by the uplift force. The normal force is used along with the Mohr-Coulomb strength criteria determined by the field investigation and lab testing program to determine the sliding resistance. The factor-of-safety against sliding is then simply taken as the ratio of the sliding resistance to the sliding force from the model. The resultant location is determined by the ratio of moment to the normal force from the results output.

The pseudo-nonlinear analysis approach is very similar to the classical cracked base analysis performed by hand calculations. The finite element method provides benefits over hand calculations, primarily due to the fact that the stress distribution along the interface is determined by the relative stiffnesses of the structure concrete and foundation material rather than being arbitrarily assumed linear. Often engineers are tempted to adapt new tools and methods to fit familiar old practices, and the pseudo-nonlinear analysis is no different. Traditional hand stability analysis requires that the cracked base analysis for each failure plane be performed independently. Failure planes in the foundation, along CRI, and in the structure are all evaluated separately assuming that the other failure planes remain intact. This same approach can be taken with the pseudo-nonlinear methodology, but fails to fully leverage the advantages of the finite element analysis.

The cracked base iterations in the finite element analysis should begin with the weakest failure plane. For the case where two failure planes have the same tensile capacity, the plane with the highest tensile stress should be considered first; and for cases where multiple planes have zero tensile strength, cracking should begin in the plane with the highest elevation. In most cases, this corresponds to the CRI. Once the CRI cracked base analysis is complete and the equilibrium condition is reached, the next weakest failure plane with the highest tensile stress is evaluated and the cracked base analysis is performed in the same model with the appropriate CRI elements deleted. This process continues until all failure planes have been assessed.

This methodology provides a much more accurate representation of the condition of the structure. Traditional stability evaluations that look at each failure plane separately may not accurately predict cracking along failure planes. When load is applied to the structure and the weakest failure plane begins to crack, stresses are redistributed. For cases where the upstream face is in compression in the un-cracked condition, the traditional method may under-predict cracking and for cases where the upstream face is in tension, the traditional method may over-predict cracking.

The pseudo-nonlinear approach can save significant amount of time and resources over the traditional stability evaluation. For the monolith with four failure planes, the finite element model must be duplicated four times and cracked base analysis performed for each failure plane. The pseudo-nonlinear approach can accommodate all failure planes in a single model and one iterative analysis as described above.

It is important to note that the pseudo-nonlinear analysis does not work for seismic analysis; this includes both linear modal superposition analysis and response spectrum analysis. The only way to get accurate crack lengths for a seismic event is to execute a full nonlinear dynamic time history analysis. When performing a response spectrum analysis the temptation is to iteratively perform to same iterative cracked base element deletion algorithm for the pseudo-dynamic response spectrum results. This is not appropriate since the response spectrum analysis yields a single set of enveloped results for the maximum response of the structure for the entire duration of the seismic event. As such, the analysis results do not yield a valid equilibrium state to perform the cracked base iteration. Further, the results of both the response spectrum and modal superposition analysis are dependent on the modal frequencies and stiffness of the structure. Modification of the structure in the form of interface element deletion significantly changes the dynamic characteristics of the structure. This means that there is no guaranteed convergence of the crack length.

COMPARISON OF RESULTS

In order to better understand the differences between the pseudo-nonlinear analysis and the traditional method of analyzing failure planes separately, the non-overflow section of the dam was evaluated using both approaches. This analysis section had five pre-defined failure planes for evaluation; three concrete lift joints, one CRI, and two foundation failure planes. The analyses only considered loading under the normal operating condition. In order to induce more cracking in the structure and yield a better comparison of the methods, the normally effective foundation drains were removed from the analysis and the uplift pressure was assumed to vary linearly from headwater pressure at the upstream face of the dam and tailwater pressure at the downstream face.

For the CRI interface the pseudo-nonlinear and traditional approaches are the same. The analysis predicted nearly 9.5 ft. of cracking from the upstream face of the dam. For the traditional approach, each of the concrete lift joints was evaluated with an independent GTSTRUDL finite element run with the CRI intact. The upstream face of the failure planes were in compression with the CRI fully intact. Table 1 presents the comparative stress results at the upstream face of the dam for the three structure lift joints shown in Figure 3. As expected, the net compressive stress at the upstream face of the lift joints reduced when the crack was induced at the CRI. This effect becomes more significant moving closer to the CRI. The failure plane near the top of the structure remained unchanged between the two analyses, the second failure plane located around the mid-height of the dam saw a 10% reduction in compressive stress and the third failure plane located just above the CRI saw a reduction of 140% in compressive stress. In fact the stresses at the upstream face of the third failure plane switched from compression in the traditional analysis to tension in the pseudo-nonlinear method. In this case, the resulting sliding factors-of-safety between the two methods were the same due to the relatively high tensile strength of the lift joints. The increased tensile stresses in the pseudo-nonlinear analysis were still below the cracking stress determined by the lab testing. If the lift joint were to have a lower tensile capacity, the traditional method could have significantly underestimated the sliding factors-of-safety and the resultant location of overturning.

Table 1. Comparison of stresses at upstream face of dam for structure lift joints

Failure Plane	Location	Stress at Upstream Face		Compressive Stress Reduction with Pseudo-Nonlinear Analysis
		Traditional	Pseudo-Nonlinear	
B1	Neck	Compression	Compression	0%
B2	Mid-Height	Compression	Compression	10%
B3	Near Base	Compression	Tension	140%

The foundation failure plane exhibited tension at the upstream end with an intact CRI and the traditional approach yielded a 10.5 ft. crack. As expected, the pseudo-nonlinear analysis not only reduced the tension, but eliminated it all together. This effect is not as significant in the sliding factors-of-safety, as the pseudo-nonlinear analysis only caused a reduction of 3%.

CONCLUSION

Pseudo-nonlinear finite element analysis is an effective method for stress and stability evaluation of concrete gravity dams. The method roughly mimics the process performed internally in most nonlinear finite element software packages. Because the pseudo-nonlinear analysis is performed with standard linear elastic methods it avoids many of the difficulties associated with full nonlinear analysis typically associated with solution convergence and results interpretation. The pseudo-nonlinear analysis overcomes the difficulties associated with incorporation of uplift pressures in the nonlinear method. In fact, uplift pressures can be incorporated into the pseudo-nonlinear analysis in a very efficient manner. Because the analyst manually performs the iterative steps in the pseudo-nonlinear analysis it provides insight into the behavior of the structure that is not possible with full nonlinear analysis.

The pseudo-nonlinear approach should be handled differently than the traditional approach to stability analysis. Unlike with traditional methods, all failure planes should be evaluated in the same analysis. This yields more accurate and dependable results since it more closely represents reality. The traditional approach may tend to either over-predict or under-predict cracking on a case-by-case basis. In addition to greater reliability of results, the pseudo-nonlinear analysis is much less labor intensive since a single analysis is executed for all failure planes.

Comparison of the traditional and pseudo-nonlinear approach revealed significant differences in crack lengths and localized stresses on the upstream face of the structure. For this particular case, the calculated sliding factors-of-safety were not sensitive to the different stress distributions between the two models. This, however, may not be the case for all situations. In particular, the traditional method has the potential to significantly over-predict sliding factors-of-safety for lift joint failure planes located near the CRI. This is especially a concern for cases that have low lift joint tensile strengths.

Due to its distinct advantages, the pseudo-nonlinear analysis approach should be considered in place of the traditional approach for cases that warrant 2-dimensional finite element stability evaluation. Further study needs to be performed to adapt the pseudo-nonlinear method to cases that require 3-dimensional analysis.

REFERENCES

Intergraph Corporation (2015). GTSTRUDL Software. Version 34, 2015

Intergraph Corporation (2015). "GTSTRUDL Users Reference Manual", Vol. 3, February 2015

IMPORTANCE OF RCC MIXTURE LAB STUDIES FOR THE DESIGN OF LARGE DAMS

A. Garduno[1]
M. Montero[2]
F. Ballhausen[3]

ABSTRACT

A literature review may be a good starting point as a source of information about roller compacted concrete (RCC) properties for a preliminary design of large dams. However, for optimization of the dam section and final design, reliable data about mechanical and thermal properties of the concrete, are essential.

Laboratory mixture studies should include different available sources for cement, aggregates, pozzolans and admixtures. Lab studies should be the base to establish the best materials for construction of a large dam.

RCC mixture study for a large dam, should evaluate mechanical properties of mixtures with different cementitious contents. This will enable the designer to improve zoning on the dam section, looking for safety in structural behavior, as well as economy and constructability. Accelerated curing samples may be useful to estimate long term properties within a couple of weeks.

For large hydro dams, such as Las Cruces gravity dam in Mexico and Santiago arch-gravity dam in Ecuador, there are properties that the designer needs to know, which require special tests, for example shear strength and direct tensile strength between layers (for different lift joint conditions), and creep. For the very important thermal analysis, coefficient of linear thermal expansion, thermal conductivity and adiabatic temperature rise, are necessary.

Some of these special tests are presented in this paper, explaining the difficulties for preparing samples at a laboratory scale and the test itself (equipment, duration, etc.), and representative results associated to Las Cruces and Santiago projects.

INTRODUCTION

It is a common practice to start a preliminary design of an RCC dam by consulting known literature sources of RCC properties, such as ACI 207.5R, Roller Compacted Mass Concrete, EM 1110-2-2006, USACE Roller Compacted Concrete, or the ICOLD Bulletin

Department of Concrete Studies, Comision Federal de Electricidad, Augusto Rodin No. 265, Col. Nochebuena, Cd. De Mexico 03720, MEXICO,
[1] amanda.garduno@cfe.gob.mx
[2] mario.montero@cfe.gob.mx
[3] fritzludwig.ballhausen@cfe.gob.mx

126, Roller Compacted Concrete Dams. If available, the designer could use data from other projects with "similar" aggregates, cement and pozzolans.

Initial laboratory mixture studies must include a full study of different available cement, aggregates, pozzolans and admixtures. Laboratory studies should be the base to establish the best materials for construction, the cost must be considered, but it shouldn´t be the main factor.

Studies for the selection of mixture proportions take into account common requirements, strength, density, impermeability, constructability, as well as durability and construction conditions (ICOLD, Bulletin 126, Chapter 4).

Once the designer is working on the final design of a large RCC gravity dam, he needs data from mixtures prepared with the materials that will be used in the construction, and for the joints treatments that will be used, to check the critical aspect, the strength of lift joints: tensile strength and shear properties. Tensile and shear strength of lift joints are the prevailing requirements for gravity dams under seismic loading, which is the case of Las Cruces.

On large RCC dams, because of the construction method, by layers, temperature distribution and thus thermal stresses, are a major design condition that requires information about the heat development on the mixtures.

In some cases, not only the strength of the mixture is decisive for choosing materials or selecting proportions, but thermal properties, particularly for large arch-gravity dams, where stress levels are usually higher than in gravity dams leading to high cementitious content mixtures. In these dams, the thicker section, compared to thin concrete arch dams, allows temperature inside of the body to become stable after many years, so thermal issues are greatly affected by the RCC mixture.

For thermal analysis, it is also required to gather information about long term deformation (creep) of RCC. The test itself is similar to the one on conventional concrete, with some extra details during sample preparation and during the test.

For final design of high RCC dams, it is recommended (ICOLD Bulletin 126, Chapter 2) to obtain tensile and shear strength parameters at lift joints from full scale trials (trial section), at site. For some projects, the parameters may be needed long before the dam construction starts, so a full scale trial is not always an option.

This paper addresses the Las Cruces project, in Mexico, and the Santiago project, in Ecuador. They are examples of large dams, gravity and arch-gravity, successively, which required full laboratory studies to provide the designer enough information for the design of the dams. In both cases, the seismicity was the main issue for the design. The paper will focus on some of the specials tests that helped to obtain parameters for the designer, mainly in the reproduction of samples in laboratory conditions.

OVERVIEW OF THE PROJECTS

Las Cruces, Mexico

The project is located 65 km North of Tepic, in the state of Nayarit, in the west coast of Mexico (Figure 1), on San Pedro River. Temperatures range from 10 °C to 40 °C, and the average relative humidity is 75 %. The wind speed is about 5 m/s. For seismicity, an Ordinary Base Earthquake, OBE, of 0.08 g, and a Maximum Credible Earthquake, MCE, of 0.16 g, are contemplated. The foundation rock is ignimbrite, which is fairly competent. The basin area is 24,175 km^2.

An RCC gravity dam of 185 m height, and about 1.8 million m^3 of RCC will be required The crown elevation is at 245 meters above sea level, masl, with a length of 815 m and a width of 8.0 m. The slope in the downstream face is 0.8:1.0 and in the upstream face is 0.1:1.0. There are two diversion tunnels in the left abutment, with 14 m diameter D-shaped section, and an average length of 586 m. The spillway is a channel in the left abutment. Cofferdams are proposed to be of rockfill. (Figure 1).

The power house, with 2 units, is located in the riverbed. The installed power is 240 MW, with 2 Francis type turbines, and the average annual generation is 783 GWh.

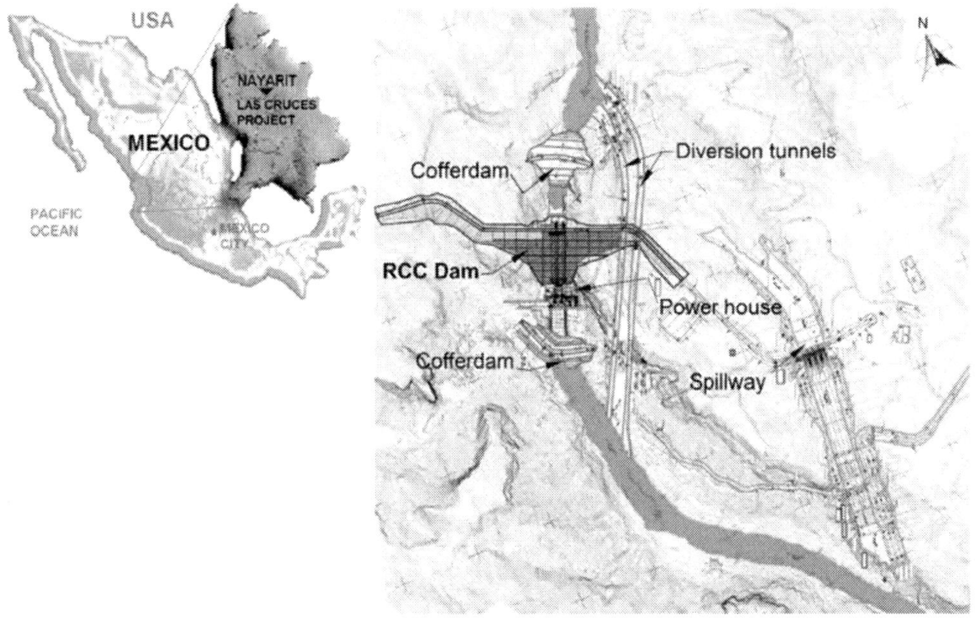

Figure 1. Location and general layout of Las Cruces project, in Mexico.

Santiago, Ecuador

The project is located in the southeast of Ecuador, downstream of the confluence of the rivers Zamora and Namangoza, on river Santiago, in the start of the Amazonian Hydrographic region (Figure 2). Temperatures are stable throughout the year, ranging from 14 °C to 32 °C; relative humidity is 94 %. Wind speed, in the north direction is about 4 m/s. Seismicity contemplates an OBE of 0.20 g and a MCE of 0.27 g. The foundation is a sequence of siltstone, sandstone breccia, sandstone tuff and volcanic breccia. The basin area is about 22,260 km^2.

It is considered an RCC arch-gravity dam that includes the spillway, which is controlled by seven gates, and six mid height outlets, in the dam body. It has about 3.2 million m^3 of RCC. The dam is 205 m in height, with the base at elevation 250 masl and a maximum width of 109.5 m. The crown elevation is at 455 masl, its length is 457.4 m, and the width is 12.5 m in the wings and 10.3 m in the spillway. The horizontal radius of the dam is 252 m in the center of the dam, and in the wings it is variable; the vertical radius increases with the height. The upstream slope is variable, the downstream slope is 0.6:1.0 in the spillway section and it varies in the wing sections. The dam is designed with 4 construction joints that include shear keys and a preventive injection system, as well as 2 m partial vertical joints to mitigate thermal cracking on the surface.

The project consists of three diversion tunnels, of 18.8 m horseshoe section, with an average of 830 m length, in the right margin; and rockfill cofferdams upstream and downstream. (Figure 2).

The powerhouse is underground in the left abutment. The installad power is 3,630 MW (6 Francis type turbines) and the average annual generation is about 14,800 GWh.

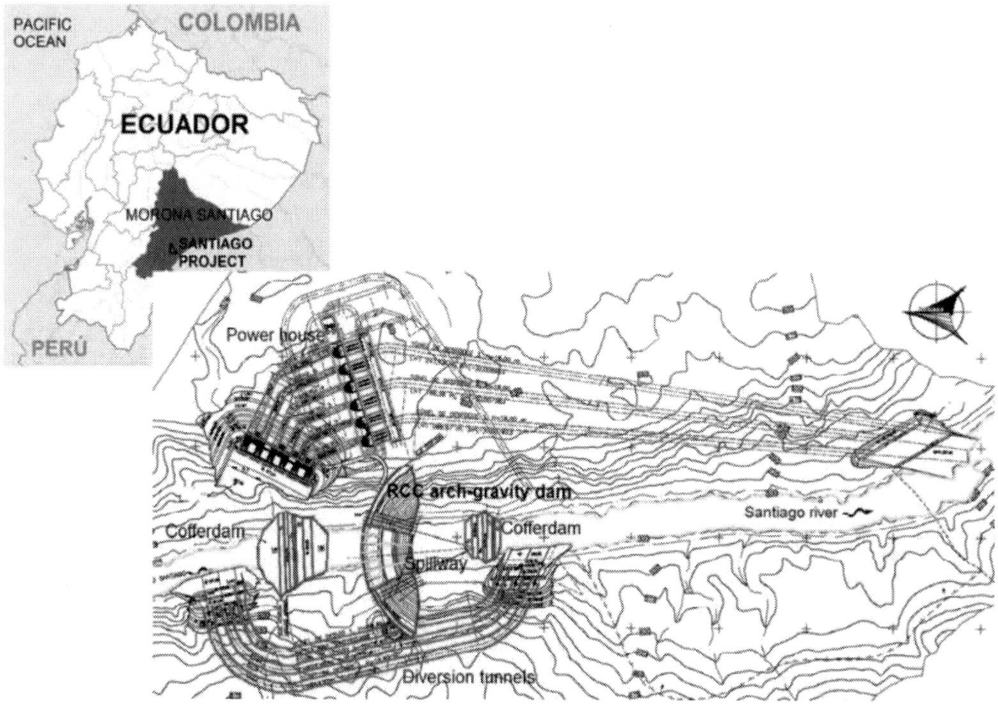

Figure 2. Location and general layout of Santiago Project, in Ecuador.

LABORATORY STUDIES

The reference projects are examples of large dams, which RCC laboratory studies, and others, such as seismic, geotechnical and hydraulics, were done by Comision Federal de Electricidad on site and in laboratories in Mexico City. Las Cruces RCC studies began in 2012, and were performed in different stages, and Santiago was studied mostly in 2015. Concrete studies included aggregates, cement and pozzolans, in order to choose the best option for the project. Some properties, for instance heat of hydration of cement, were evaluated for long periods (up to 1 year).

Each project has its particularities, for instance, in Mexico there are coal power plants, with the possibility to supply suitable fly ash (Garduno & Montero, 2010), but there aren´t any in Ecuador; on the other hand, Ecuador has sources of pozzolans of good quality.

The main limitation for the lab studies for Santiago project, was the amount of aggregates that was transported from Ecuador to Mexico. And, for both projects, time was also a limitation, because the dam designer needed information in parallel to the lab studies.

Some test procedures were improved from Las Cruces to Santiago, as will be mentioned later. For both projects the use of accelerated curing was intended to obtain in short term the compressive strength associated to 1 year, after an increase-decrease cycle (López & Schrader, 2012). This assumption was not always valid, and the age strength equivalence varied according to the type and amount of cement used on the mixture. Nevertheless, results were useful, and after 1 year compressive strength results, it is possible to establish the age strength equivalence.

RCC mixture studies have been widely documented (e.g. Garduno & Montero, 2016), this paper does not address the mixture studies. Just as reference it will be mentioned the materials studied for each project, and representative results of the RCC properties of fresh concrete and of strength, usually determined on these studies.

RCC MIXTURE STUDIES FOR LAS CRUCES

Cement and Pozzolan

The proposed cements, type II and type IS (with 45% of granulated slag) according to ASTM C150 and ASTM C595 cement designation, were studied. Composition and characteristics were determined, and these included heat of hydration (ASTM C186) and compressive strength on mortar cubes (ASTM C109) up to 360 days age. Figure 3 presents the results of heat of hydration, which were used afterwards for estimating the adiabatic temperature rise.

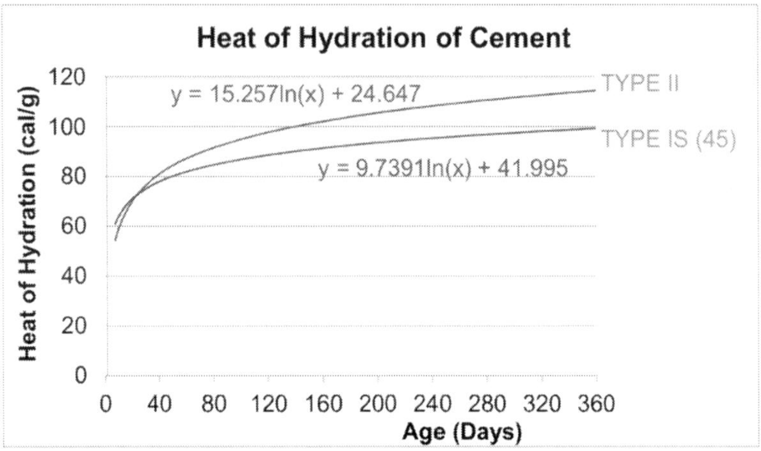

Figure 3. Heat of hydration of cement type II and type IS for Las Cruces.

Mixtures with cement type II were thoroughly studied for thermal properties and creep. Cement type IS is considered as an alternative to cement type II, but thermal properties and creep must be verified.

Several sources of pozzolan were studied: fly ash, natural pozzolans and fines (passing sieve No. 200, 75 μm) from the potential aggregates quarries. Their chemical composition (by X-ray fluorescence technique), and strength activity index with Portland cement (ASTM C311), up to 180 days, were evaluated. Although fly ash from Petacalco power plant proved excellent quality as cementitious material, it was not considered for the project because of the cost for its transportation. Natural silt and fines from quarries barely showed pozzolanic activity. Finally, ignimbrite dust was proposed as filler.

Admixture

A lignosulphanate retarding admixture (Plastiment RCC by SIKA) was used.

Aggregates

Alluvial deposits on riverbanks and potential quarries in both margins, were studied. Crushed aggregates from ignimbrite quarries presented low densities (<2400 kg/m^3) and high absorptions (> 6.2%), compared to the higher density (>2485 kg/m^3) and lower absorption (< 3.8 %) of crushed alluvial aggregates. It was proposed to combine aggregates in a: 50% crushed alluvial + 50 % crushed ignimbrite, proportion.

Fresh Properties

- VeBe time (ASTM C 1170): 8 s (lower than expected, water content can be reduced)
- Time of setting (ASTM C 403): from 12 to 24 hours (initial to final setting) curing on lab conditions
- Air content (ASTM C 231): 1.2 %
- Density (ASTM C138): 2260 kg/m^3

Mechanical Properties Results (180 Days Age):

Representative mixture: 90 kg/m3 of cement type II, combined aggregates,

- Compressive strength, fc = 7 MPa,
- Modulus of elasticity at 40% of the maximum load, E40 = 11 GPa,
- Indirect Tensile strength, T indirect = 0.7 MPa,
- Direct Tensile strength, T direct = 0.4 MPa

RCC MIXTURE STUDIES FOR SANTIAGO

Cement And Pozzolan

Production of cement type II and IP is feasible for the site. Studies included composition, heat of hydration and compressive strength on mortar cubes. Results of heat of hydration are presented in Figure 4.

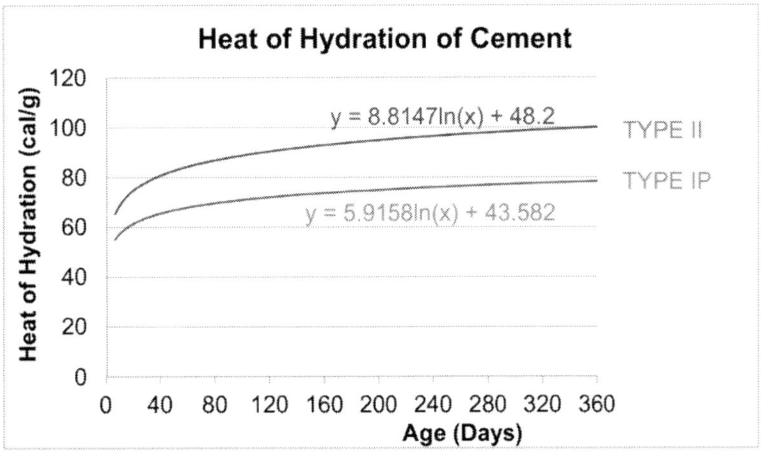

Figure 4. Heat of hydration of cement type II and type IP for Santiago.

A natural source of pozzolan was studied, which showed high strength activity index. This pozzolan was used in mixtures, as 30 % and as 60 % of the cementitious content.

Mixtures with cement type II and pozzolan (60%) were the preferred option because the high compressive strength obtained, and the desire for low temperature rise.

Admixture

The retarding admixture (Plastiment RCC by SIKA) was used in all the mixtures.

Aggregates

Crushed aggregates from two potential quarries were studied, calcareous and diorite. Both can be used for construction, they present high densities (≈2600 kg/m^3) and low absorptions (< 3.5%).

Fresh Properties

- VeBe time (ASTM C 1170): < 17 s
- Time of setting (ASTM C 403): from 12 to 18 hours (initial to final setting) curing on lab conditions
- Air content (ASTM C 231): 1.3 %
- Density (ASTM C138): 2390 kg/m^3

Mechanical Properties Results (180 Days Age):

Representative mixture, 80 kg/m^3 of cement type II + 120 kg/m^3 of pozzolan, and calcareous aggregates

- Compressive strength, fc = 24 MPa,
- Modulus of elasticity at 40% of the maximum load, E40 = 18 GPa,
- Indirect Tensile strength, T indirect = 2.9 MPa,

PROPERTIES FOR THE DESIGNER

Shear Strength of Lift Joints

Due to the construction process, by lifts, of RCC dams, shear strength and direct tensile strength of lift joints are critical values for the design of the dam, and the main issue is to reproduce representative samples at a laboratory scale. It is desirable to prepare trial sections on site, but frequently, the designer requires the mentioned properties long before the construction is started.

For Las Cruces is was possible to prepare several laboratory trial sections (about 1.5 x 2.0 m, with two lifts of 0.3 m each); each of them reproduced, the following lift joints conditions:

Hot lift joint: the second lift is compacted before the initial setting time of the first lift.
Warm lift joint: the second lift is compacted before final setting time of the first lift.
Cold lift joint: the second lift is compacted after final setting time of the first lift.

The use of bedding mix between lifts wasn´t evaluated, because this treatment would represent an expensive way of improving cohesion, and shouldn´t be used during construction. It is desirable that the rapid construction process would not require the use of bedding mix. But this is a condition that needs to be evaluated in the trial section before the construction of the dam starts.

Compaction of RCC on lab trial sections was achieved with a single drum walk-behind roller of 450 kg. Samples (about 30 x 30 x 60 cm blocks) for shear strength tests were sawed with a 120 cm diameter diamond blade, and banded to keep the two halves of the block together.

Sample blocks were tested to determine shear strength in joints (cohesion and friction angle), in a similar way than it is done on rock discontinuities (ASTM D 4554). The test was run with three levels of normal load until the maximum shear strength is obtained, and then, up to 20 mm displacement (sliding) of the joint, to obtain residual shear strength, for each lift joint condition. By plotting shear resistance vs. normal load, the friction angle and cohesion value can be determined. By continuing the test past the "break bond" level of load, the residual friction angle was determined after sliding.

After these tests, the original blocks were separated into two, and each of these smaller blocks was tested to determine shear strength on the parent RCC.

The next figure present results from Las Cruces.

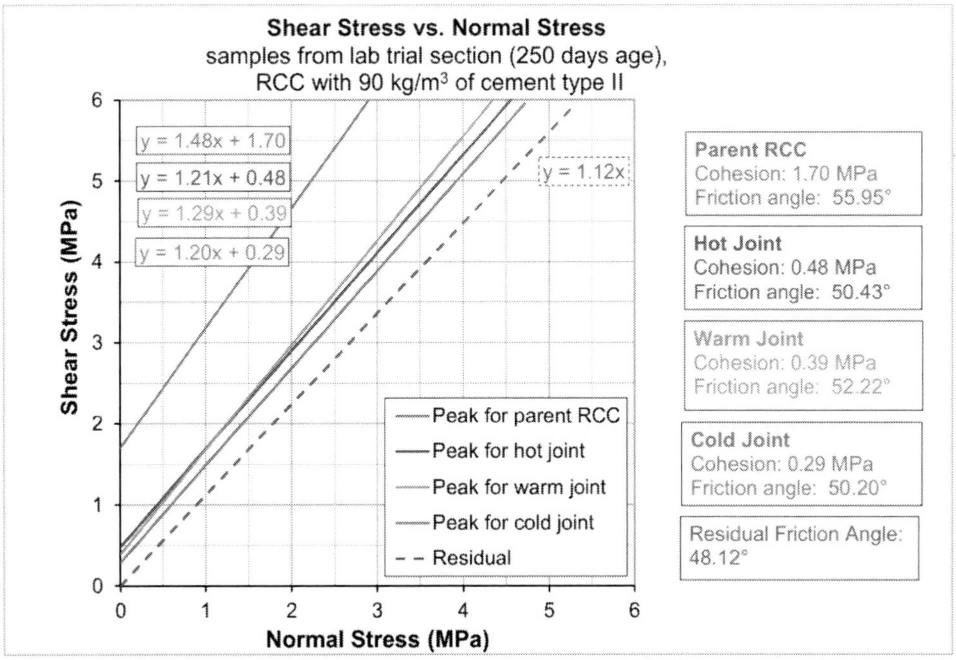

Figure 5. Results of shear strength test for Las Cruces.

For project Santiago, there was a limitation on the volume of aggregates, so it was not possible to prepare lab trial sections. The design took into account hot joints, and in case of cold or warm joints during construction, they must be treated with bedding mix. Samples were reproduced on cubic molds, one above the other, with the hot joint in between. Compaction was achieved with a vibrating hammer in a similar way than cylindrical samples are prepared (ASTM C 1435). The next figure presents results for the mixture with 80 kg/m^3 of cement type II + 120 kg/m^3 of pozzolan, at 90 days age.

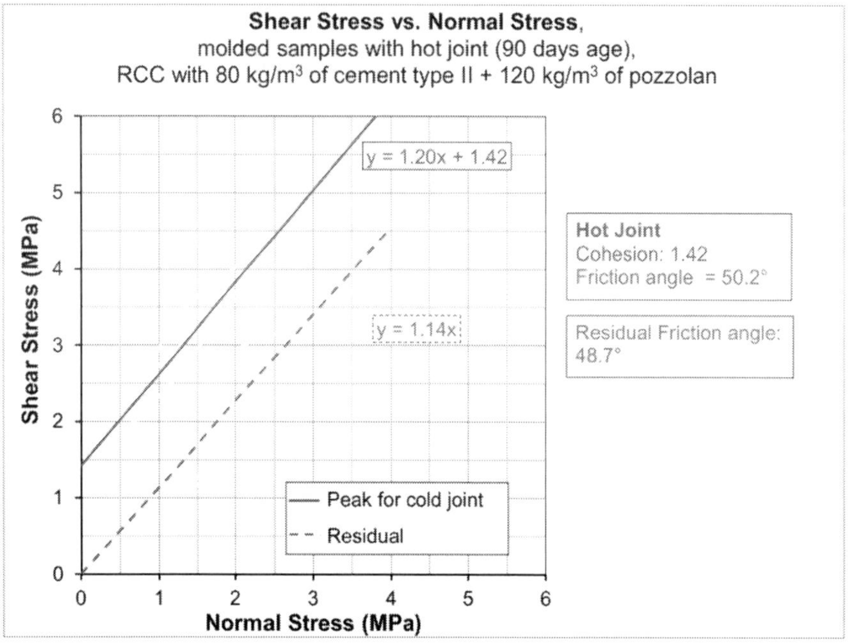

Figure 6. Results of shear strength test for Santiago

There has been the question if tests on molded samples are representative of a trial section, and thus of the dam. In Las Cruces study, results showed that the compaction of the lab trial section was deficient: hot joint results are closer to the cold joint than to the parent RCC results. And results from the molded samples from Santiago study present a feasible result, which is indicative of an adequate compaction of the lifts. In both studies, results will be verified when the trial section is built before the construction of the dam.

Figure 7 A) Preparation of samples from a laboratory trial test, for Las Cruces study, and B) compaction of molded samples for Santiago study.

Direct Tensile Strength of Lift Joints

For direct tensile strength on joints, coring from the laboratory trial sections prepared for Las Cruces was complicated, but the extraction of a couple of cores was a proof of good bonding (adequate compaction and enough paste) of layers. Obviously, it is not possible to extract intact cores through the cold joint. Direct tensile test (CRD C 164) is a difficult test to run (Dolen & Dunstan, 2012), but can be achieved in cores with hot/warm joint, as well as in parent RCC cores. The maximum result of direct tensile strength on lift joint, was 0.3 MPa, which is low but is expected to be improved when compacting during construction.

For Santiago project, direct tensile strength on joints was inferred from the indirect tensile results from the parent RCC, and then reduced for several factors proposed by Consultant Ernest Schrader due to the influence of the test (indirect to direct test), the quality of the joint, and its maturity (age and exposure conditions).

Long Term Deformation (Creep)

Creep frames were installed in a room with humidity and temperature controlled conditions. Preparation of samples required embedded accessories on the sample, used as references for measuring length changes, so compaction of samples was done carefully. Strain measurements were registered daily for initial 3 weeks, and afterward weekly for several months or even a year.

For Las Cruces tests, the strain measuring device, shown in figure 8B, didn't provide enough repeatability nor accuracy. So afterwards, for Santiago creep test, this problem was solved by attaching dial indicators micrometers (with an accuracy of 0.001 mm) for the whole duration of the test. Protection of the samples was improved for Santiago by covering them with latex, as shown in figure 8C, instead of wrapping them with self-adhesive PVC (used for food preservation).

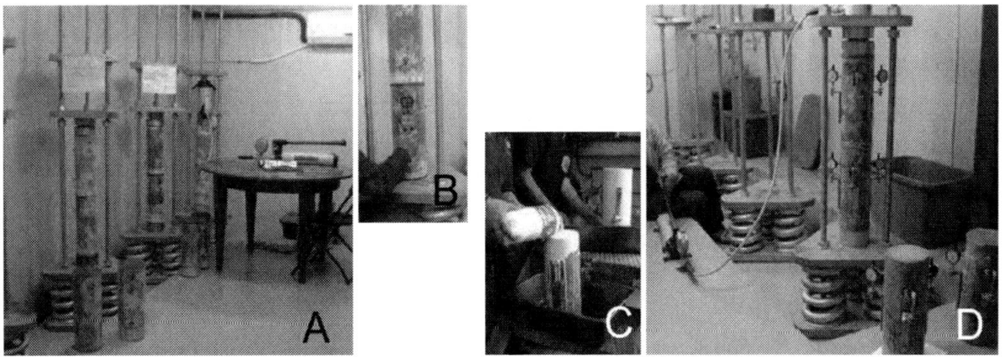

Figure 8. A) and B) Creep test for Las Cruces, C) and D) Preparation samples with latex and testing for Santiago; notice the fixed dial indicators micrometers.

RCC creep test has some differences from ASTM C512, used for creep in concrete samples, for instance, the number of specimens on each frame (2) and the size of specimens (40 cm height). The age for initial loading can vary, typically 28, 90 or 365 days, and then sustain the load for 1 year, or more. Recommended applied stress is 40% of the compressive strength at the age of loading. A pair of cylinders with the same characteristics are used for measuring autogenous change of volume, results which are used for "correcting" measurements of the loaded specimens.

Results obtained so far for Las Cruces are inconsistent and not reliable. For Santiago, results obtained are shown on figure 9, samples were loaded at 130 days age and results are up to 300 days from loading.

Creep values are useful to estimate if the RCC will slowly relieve stress and strain buildup due to foundation restraint, thermal and exterior loadings (EM 1110-2-2006).

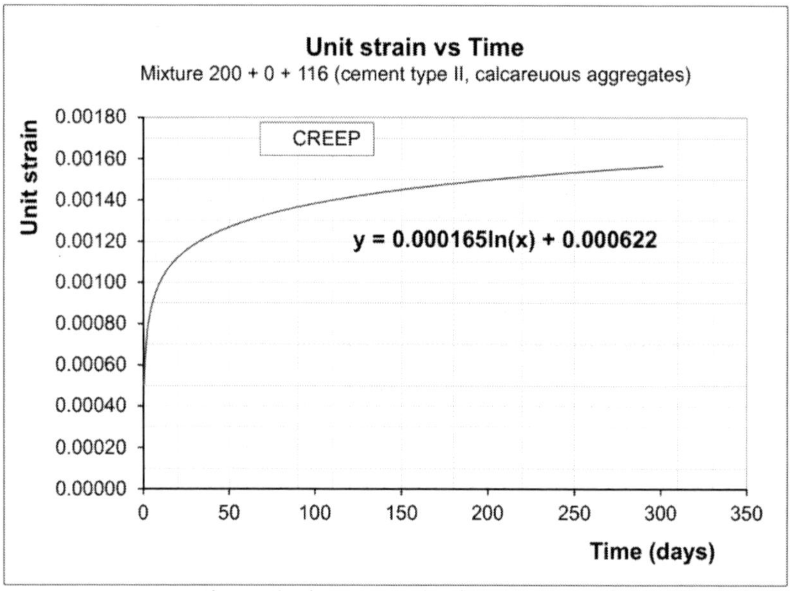

Figure 9. Creep results for Santiago.

Coefficient of Linear Thermal Expansion

This is measured according to CRD C 39, samples (bars of 10x10x30 cm) are prepared from RCC mixtures sieved by 38 mm, with a thermocouple embedded and submerged inside automated tanks that control water temperature. Temperature is gradually increased up to 90 °C (about 2.8 °C/hour), then decreased to 5 °C, increased again to 90 °C and finally decreased to 23 °C. Length changes are measured in a length comparator at 5, 30, 60 and 90 °C, both, during increase and decrease of temperature.

Results are presented in the next graph. For Las Cruces, measured samples were from the mixture with 130 kg/m^3 of cement type II and combined aggregates (crushed alluvial + ignimbrite), and for Santiago the samples were prepared with 80 kg/m^3 of cement type II

+ 120 kg/m^3 of pozzolan, and calcareous aggregates, as well as with 140 kg/m^3 of cement type II + 60 kg/m^3 of pozzolan and aggregates from crushed diorite.

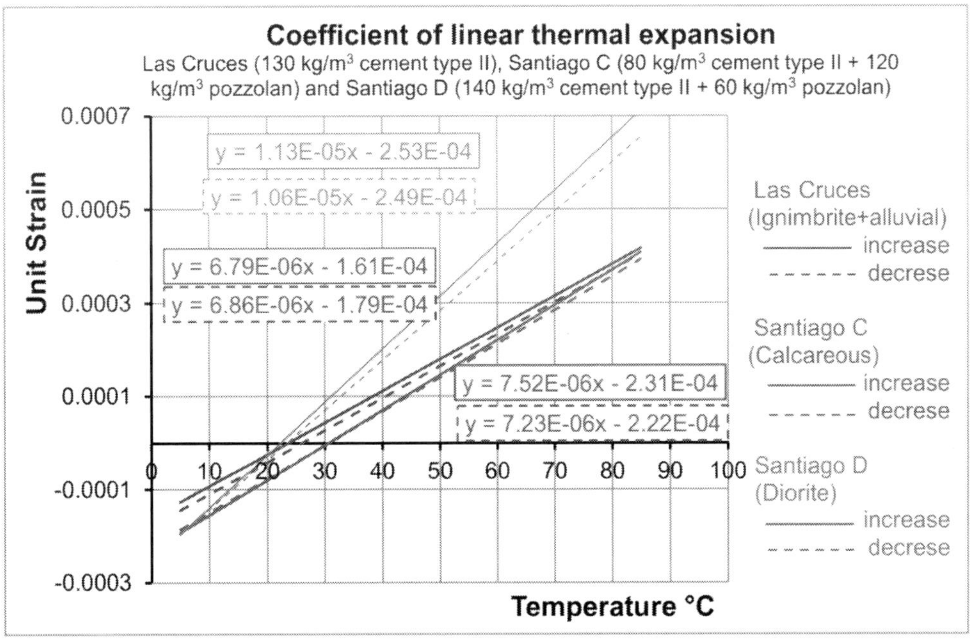

Figure 10. Coefficient of linear thermal expansion for Las Cruces and Santiago.

Adiabatic Temperature Rise

Temperature rise of RCC can be estimated from the heat of hydration obtained for the cement proposed for each project, and the pozzolanic materials included in the mixture, but results measured on adiabatic conditions (CRD C 38) are ideal. The test requires a huge RCC sample (cylinder of 65 cm diameter and 65 cm height) with Resistance Temperature Detectors inside (with an accuracy of 0.01°C). The sample is enclosed in a thermally insulated chamber which can adjust its internal temperature to the one in the sample. Information of the sensors is constantly registered.

Estimated temperature rise can be calculated up to 1 year. The test duration is usually 60 days, when most of the heat has developed and a tendency is reached, so results can be extrapolated for older ages.

For both projects the test was run, but as shown on the next figure, results do not match with the estimated adiabatic rise (based on the cement chemistry).

At this moment, it has been more reliable the estimated result, and next tests will reinforce the fast preparation of the sample and the preheating of the chamber before entering the sample.

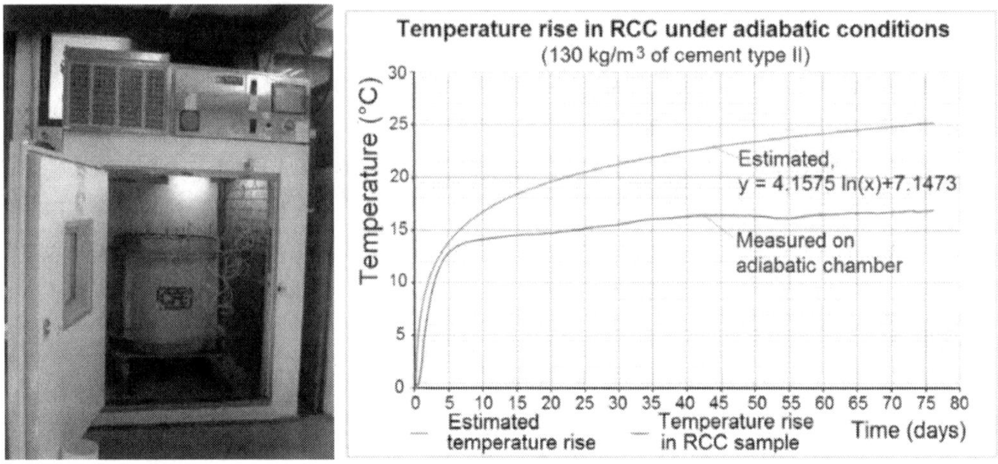

Figure 11. Adiabatic chamber and temperature rise for Las Cruces.

IMPLICATIONS OF NOT KNOWING RCC PROPERTIES

Some of the negative implications when the designer considers "conservative" properties of the RCC for the analysis are,

- If a low RCC strength is assumed, it can lead to a non-optimized (larger than required) section of the dam, or to ask for RCC mixtures with higher cement contents, or to demand improving lift joints with bedding mix where it is not needed.
- Concerning thermal properties, for instance, higher temperature rise or lower thermal conductivity than the real properties of the RCC, can orient to undesirable cooling treatments (postcooling), slow construction process, shortening joint spacing, etc. in order to mitigate thermal cracking.

On the other hand, if the designer overestimates strength properties, the design could be risky or even unsafe.

COMMENTS AND CONCLUSIONS

It is highly recommended that the designer relies on real values of the RCC properties for the design of large RCC dams. Conservative properties may not endanger the structure, but they may increase costs, slow construction, induce a larger dam section, with higher cement contents, and cooling methodologies, than the essential ones.

For large RCC dams, and especially if they will be in high seismicity zones, laboratory mixture studies should start with enough anticipation so the designer of the dam could have accurate data about usual strength and elastic properties (compressive strength, modulus of elasticity and tensile strength) at long ages; as well as thermal properties, creep, shear strength and direct tensile strength. These parameters require months, or even years, to obtain results.

A critical issue is to prepare representative samples under lab conditions, especially of lift joints. It is expected that cohesion and direct tensile strength of lift joints results, when sampling at site in a trial section and during construction, will be higher than results obtained on lab section, because the compaction process will be better than in laboratory scale specimens. Nevertheless, these results could be better than assuming properties.

A better understanding of the RCC properties will contribute to a better and more accurate design of the dam.

ACKNOWLEDGEMENTS

Authors are grateful to CFE for permitting the use of the information about Las Cruces project presented in this paper, as well as for Mr. Luis Alberto Ochoa, from CELEC, for approving the use of information about Santiago project.

REFERENCES

ACI 207.5 R-99, Roller Compacted Mass Concrete, American Concrete Institute, Manual of Concrete Practice, Detroit, Michigan.

Dolen & Dunstan, M (2012), The Tensile Strength of RCC for Dams. Published as part of proceedings of the Sixth International Symposium on Roller Compacted Concrete (RCC) dams in Zaragoza, Spain. 23-25 October 2012.

Garduno, A & Montero, M (2010). An option for the use of Fly Ash on Roller Compacted Concrete Dams in Mexico. Published as part of the proceedings of the Sixth International Conference on Concrete Under severe Conditions, Environment and Loading in Mérida, México 7-9 June 2010.

Garduno, A & Montero, M (2016). RCC Mixtures Study for the Design of a Large Dam in Mexico. Published as part of the proceedings of the International Symposium on Appropriate technology to ensure proper Development, Operation and Maintenance of Dams in Developing Countries, in Johannesburg, South Africa, 18 May 2016.

ICOLD Bulletin 126 (2003), Roller Compacted Concrete Dams, State of the art and case histories.

Lopez, J & Schrader, E (2012). Accelerate curing test: History, Development and applications. Published as part of the proceedings of the Sixth International Symposium on Roller Compacted Concrete (RCC) dams in Zaragoza, Spain, 23-25 October 2012.

USACE EM 1110-2-2006, Roller-Compacted Concrete, US Army Corps of Engineers, Engineering and Design, Washington D.C., 15 January 2000.

MANAGING THREATS THROUGH EFFECTIVE SECURITY PROGRAMS

J. A. Gloschat[1]
A. C. Baehr[2]

ABSTRACT

An effective security program is comprised of a myriad of components including people, paper, and procedures. Security programs must be considered dynamic, since the environment we live in is never static.

A security plan must be able to accurately provide prescriptive direction for both present and future events, capture potential scenarios and consequences, detail the organization's actions both during and following specific events, and educate the organization on their role in the plan itself. All of these actions are necessary so that when an incident, whether accidental, natural disaster, sabotage, or attack occurs, the impact to the organization is averted, mitigated, or minimized. This requires an organization to be aware of the type of adversary that is likely to negatively impact the facility and plan accordingly.

The ability to accurately track, document, and then model trends is a valuable tool for any organization as well as those that have a security department and a supporting law enforcement office.

This paper examines the steps and phases that are necessary to ensure the establishment of an effective security program. The following concepts are examined in detail:

- Identifying and understanding the threats, both internal and external, to the facility
- Accurate reporting, tracking, and analysis of threat-related incidents
- Design, Implementation, and Maintenance of the security program

In order to demand so much from a security program, it is critical that there is careful forethought and planning dedicated up front. The quality of a security program is directly proportional to the amount of time and effort an organization devotes to it.

HISTORICAL PERSPECTIVE

Why care about a security program? Security programs are an essential and integral part of any organization. It is imperative to understand their significance and the importance of their roles in defending against any threat to an organization, its facilities, and assets. Providing a summary of past historical events highlights the importance of understanding the present day threats. Historical data sheds light on the significance of events that have occurred. The examples below highlight previous threats faced by just one of the sixteen critical infrastructure sectors: dams.

[1] Senior Security and Intelligence Analyst, Gannett Fleming, jgloschat@gfnet.com
[2] Security and Intelligence Analyst, Gannett Fleming, abaehr@gfnet.com

In August 2007, Michael Keehn, a former system manager at the Tehama Colusa Canal Authority in Willows, CA, uploaded unauthorized software onto the computers housing the Authority's water management supervisory control and data acquisition (SCADA) systems. The software damaged the computer used to divert the water from the Sacramento River, but did not damage the canal management system or result in the loss of water. The canals could have been operated manually in the event of the loss of the SCADA system controlling them.[3]

On 23 February 2011, the FBI arrested Khalid M. al-Dawsari, a Saudi Arabian exchange student studying at Texas Tech University, on charges of plotting to destroy dams in Texas and California, including the San Luis Reservoir, which is a part of the California State Water Project. Al-Dawsari planned to stage a Vehicle-Borne Improvised Explosive Device (VDIED) attack against 12 reservoirs or dams in California and Colorado, specifically targeting hydroelectric dams.[4] Notably, he targeted the dams themselves rather than electricity-generating equipment co-located at the assets.

In 2013, Hamid Firoozi, employed by two Iranian companies sponsored by the Revolutionary Guards, attempted to hack into the cellular modem connection at the Bowman Avenue Dam in Rye Brook, New York. According to authorities, the dam was offline, under repair, and the hack was unsuccessful.[5] It is believed that the hacker's intended target was the much larger Arthur R. Bowman Dam in Oregon.

The above examples illustrate that malicious actors originate both internally and externally for organizations and have the capability, intent, and motivation to identify vulnerabilities. Sennewald (2003) writes, "Terrorists have tremendous advantages over defenses because they search for vulnerabilities and strike almost anywhere with surprise."[6] Organizations must be diligent in monitoring such events while continuously working to evolve their security program to thwart the intentions of a multitude of threats. However, they must become skilled at looking where to identify all these threats. While this paper will focus on the managing of threats for a successful security program within hydropower critical infrastructure facilities, the elements discussed within are applicable to any organization.

A SHIFT IN THE PARADIGM

What is a paradigm? Why do we need to shift ours?

Merriam Webster, in the most recent version of their web-based dictionary, defines paradigm as "a philosophical and theoretical framework of a scientific school or discipline within which theories, laws, and generalizations and the experiments

[3] Weiss, Joseph. Protecting Industrial Control Systems from Electronic Threats. Momentum Press. 2010.
[4] US Department of Justice. Texas Resident Arrested on Charge of Attempted Use of Weapons of Mass Destruction. The Federal Bureau of Investigation. 2011.
[5] Berger, Joseph. "A Dam, Small and Unsung, Is Caught Up in an Iranian Hacking Case." The New York Times. 2016.
[6] Sennewald, Charles A. Effective Security Management. Elsevier Science. 2003.

performed in support of them are formulated; *broadly*: a philosophical or theoretical framework of any kind."[7]

The following figure is a clever method of depicting the different ways we view things. It has been used innumerable times in both lectures and presentations to assist in explaining how the brain is pre-wired to view certain objects and events, despite evidence existing to the contrary.

Take a look at Figure 1and think of the first thing you see.

Figure 1. Paradigms

You may see a young woman looking over her right shoulder, exposing only the left profile of her face. You may, however, see an old woman looking to your left. You would see the profile of the left side of her face.

Why is this figure such a good example of paradigms? Everyone who sees the figure will immediately see the young female or old woman. Once they are made aware that there is another image depicted, they look longer and eventually see both of female images.

Our brains are programmed to see events and actions in a certain way. It is human nature to stick to the paradigms we are both familiar and comfortable with, and equally challenging to change those paradigms.

David Kahn (1999), in his book about the attack on Pearl Harbor, stated the following:

> American officials did not think Japan would attack their country. To start a war with so superior a power would be to commit national hara-kiri [suicide]. To

[7] "Paradigm." *Merriam-Webster.com*. Merriam-Webster, n.d. Web. 1 Dec. 2016.

Western modes of thought, it made no sense. This rationalism was paralleled by a racism that led Americans to underrate Japanese abilities and will. Such views were held not only by common bigots but by opinion-makers as well. These preconceptions blocked out of American minds the possibility that Japan would attack an American possession. . . . An attack on Pearl Harbor was seen as all but excluded. Though senior army and navy officers knew that Japan had often started wars with surprise attacks, and though the naval air defense plan for Hawaii warned of a dawn assault, officials also knew that the base was the nation's best defended and that the fleet had been stationed that far west not to attract, but to deter, Japan.[8]

On December 7, 1941, Japan attacked Pearl Harbor. It was a devastating loss to the United States Military and to the entire country, caused by an inability to clearly identify the threat, regardless of preconceptions.

In 2008, Dominic Johnson stated in his collaborative book, Paradigm Shifts in Security Strategy, stated the following:

> Prior to 9/11, U.S. counterterrorism policy and intelligence suffered from numerous problems. The striking feature about this is not the flaws themselves, but rather that these flaws were long appreciated and nothing was done to correct them. It took a massive disaster—3000 American deaths—to cough up the cash and motivation to address what was already by that time a longstanding threat of a major terrorist attack on the U.S. homeland. A second striking feature is that this failure to adapt is no novelty. Pearl Harbor, the Cuban Missile Crisis, and the Vietnam War were also belated wake up calls to adapt to what in each period had become major new challenges for the United States.[9]

Prior to the attack on the World Trade Centers in September 2001, it was the prevailing and strongly held opinion that flying an aircraft into a building as an act of terror was ridiculous. In fact, it was not even a documented credible course of action for which the Department of Defense prepared.

All of the above discussions illustrate the inability to comprehend a threat, based on preconceived assumptions about an enemy. Learning from the past allows security organizations to shift the paradigms in order to more accurately define, identify, and defend against present day threats.

UNDERSTANDING THREATS

The world is a dangerous place. Every day there are constant reminders in the news of nefarious characters that not only threaten individuals within a society, but also organizations. Threats come in all different shapes and sizes and they are not limited to

[8] Kahn, D. 1999. Pearl Harbor as an intelligence failure. In Pearl Harbor and the coming of the Pacific War: A brief history with documents and essays, ed. A. Iriye, 158–169. Boston: Bedford/St. Martin's.
[9] Johnson, DDP & Madin, EMP (2008) Paradigm shifts in security strategy. Why does it take disasters to trigger change? In: *Natural Security: A Darwinian Approach to a Dangerous World*, RD Sagarin and T Taylor (eds), Berkeley and Los Angeles, CA: University of California Press.

commonly thought of malicious actors. The most accurate way to identify the most likely threat to an organization is to conduct a threat assessment of the organization and its features.

It is not possible to develop an effective protective program without some knowledge of the likely adversaries and their tactics. However, threats are inherently uncertain and unpredictable. Even after the best efforts to identify likely adversaries and to determine their current capabilities (e.g., tactics, tools, weapons), there will always be some uncertainty about adversaries' future capabilities and their objectives. Potential adversaries range from trained terrorists to local criminals and vandals; they could also include present employees who pose an insider threat.[10]

A thorough risk assessment must include a comprehensive and well-documented threat assessment that covers as complete a spectrum of anticipated adversaries as is possible. Figure 2. depicts a roman arch. Common to all arch designs is the center arch stone called the keystone. It is the last stone put into place and the stone that holds all the other stones in place.[11]

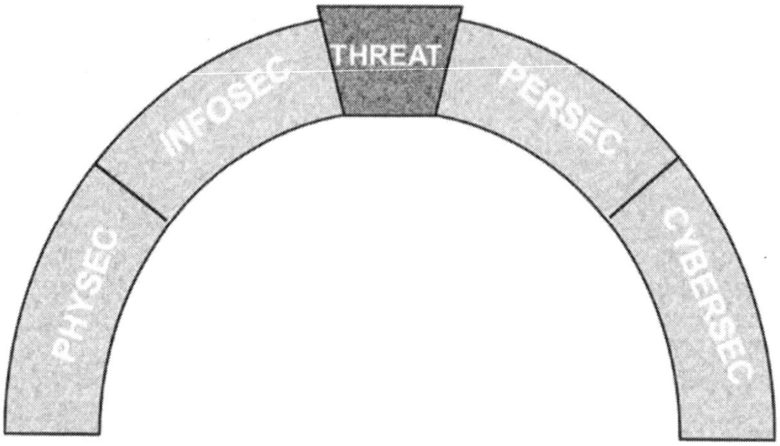

Figure 2. Roman Arch

This arch depicts the keystone as "Threat". It shows Threat, or for the purpose of this discussion, our understanding of the threat.

The other stones, Physical security (PHYSEC), Information Security (INFOSEC), Personnel Security (PERSEC) and Cyber Security (CYBERSEC), are supported by, and held in place by our understanding of the Threat. Of course, PHYSEC, INFOSEC, PERSEC and CYBERSEC are just fractional representations of the various facets of a comprehensive security program. Other areas would include Business Continuity and Crisis Management.

[10] Department of Homeland Security, Dam Sector Protective Measures Handbook, 2015.
[11] Ching, Francis D.K. (1995). *A Visual Dictionary of Architecture*. New York: John Wiley & Sons, Inc. p. 12.ISBN 0-471-28451-3.

If we do not have a clear understanding of the threat, both potential and ongoing, we will not be able to successfully maintain any other facet of security. This is because of three words, Design Basis Threat, or DBT, for short.

Design Basis Threat

In a presentation in August 2015, the Critical Infrastructure Protection Committee defined Design Basis Threat as what is used to "determine the level of appropriate and cost effective physical protection measures required to protect against malicious acts, i.e. theft / sabotage. [12]

Essentially, the DBT is the threat that an organization selects to base the standards of a security program against. The DBT assist the organization in establishing a baseline that it must defend against, while including additional measures to be implemented with any escalation of threat. Successful identification of the appropriate DBT come after the conclusions of a threat and vulnerability assessment, which allows the organization to objectively identify the most likely threats.

The identification of a DBT is so critical to a successful security program that the Nuclear Regulatory Commission has defined each adversarial course of action and the corresponding DBT. This data greatly assists the resident Security Manager in designing a more effective security system.[13]

Internal Threats

The example provided above in the Historical Perspectives section in regards to Michael Keen uploading malicious software highlights how important it is for an organization to look inside for the threats. An insider threat is defined as the following:

A security threat that originates from within the organization being attacked or targeted, often an employee or officer of an organization, or enterprise. An insider does not have to be a present employee or stakeholder, but can also be a former employee, board member, or anyone who at one time had access to proprietary or confidential information from within an organization or entity. [14]

Mary Lynn Garcia (2008) in her books distinguishes insiders with the following three characteristics:

1. System knowledge that can be used to their advantage;

2. Authorized access to the facility, assets or physical protection systems without raising suspicions of others; and

[12] Critical Infrastructure Protection Committee (CIPC), Atlanta, GA, December 15-16, 2015.
[13] Title 10, Section 73.1(a), of the *Code of Federal Regulations*.
[14] Lord, Nate. "What is an Insider Threat? An Insider Threat Definition." Digital Guardian. 2016.

3. Opportunity to choose the best time to commit an act.[15]

Insider threats are often discounted since organizations are likely to believe that the job screening process will reveal any potential disqualifying attributes. However, it is important to recognize that insiders might also have similar motivations as external threats and it cannot be assumed that all employees do not have negative intentions. [16] Insiders have access to sensitive areas and have the ability to by-pass access controls that would otherwise deter an external threat.

An organization's security program cannot discount the possibility of an insider threat. While insider threats present many challenges for security program designers, the possibility exists that one day an organization may face such a threat. When conducting any threat, vulnerability, and risk assessment, organizations must always consider an insider threat as a possible DBT on which to base the design of the security program. This includes identifying the vulnerabilities that insiders have access to.

External Threats

External threats to an organization are the most commonly thought of when naming examples of external threats. External threats include, but are not limited to, hackers, criminals, terrorists or extremists.[17] In order for an organization to determine the most likely threat, a thorough threat, vulnerability, and risk assessment must take place utilizing a close relationship with law enforcement to develop the most accurate threat picture, as previously mentioned.

Identifying the external threats also requires a thorough understanding of the motivations of the external threat. Adversarial motivation can originate from ideological motivations, economic motivations or personal motivations.[18] Garcia (2008) defines each of the three motivations below:

1. Ideological motivations are those that are linked to a political or philosophical system. They would include those of political terrorists, anti-nuclear extremists, and certain groups of philosophical or religious fanatics. Some examples include anti-abortion, animal rights, militia, and various hate groups.

2. Economic motivations involve a desire for financial gain. Criminals might view material or information as potentially attractive targets for schemes of theft for ransom, sale, or extortion.

3. Personal motivations pertain to the special situations of specific individuals. Personal reasons for committing a crime could range from those of the hostile

[15] Garcia, Mary Lynn. "The Design of Evaluation of Physical Protection Systems." Butterworth-Heinemann. 2008.
[16] Ibid.
[17] Ibid.
[18] ASIS International, Protection of Assets, Alexandria, Virginia, USA; Section 1.3.

employee with a grievance against an employer, to those of the psychotic individual.[19]

In addition to internal and external threat actors, the potential exists for a third type of threat: an insider working in collusion with an external threat. In this type of threat, an internal actor is collaborating with an external threat, such as an insider threat performing espionage for sabotage on behalf of the external threat. Personal motivations, while possible in an external threat to an organization, may also include the rationale behind an internal threat.

It is important to think outside of the box when considering both internal and external threats. For example, Purpura (2002) writes "…first consider the method used by terrorists, which go beyond bombings. The raw materials for chemical, biological, and radioactive weapons can be purchased on the open black markets." [20] Considering the tools an external threat might need to carry out an attack on an organization enables a more detailed threat picture.

REPORTING AND TRACKING THREATS

One of the most effective ways an organization can continually self-assess its security program is to develop a reporting system for the threats it faces. The documented attempts against the organization will reveal if the selected DBT on which the security program is designed to defend against is effective or not.

There are many ways in which an organization's security program can document threat incidents, both internal and external. Senneald (2003) writes, "…internal security incident reports are used to determine security weaknesses and problem areas, as well as to select crime countermeasures, calibrate countermeasure effectiveness, and consider future budget needs."[21] By documenting attempted incidents, the organization can continually monitor the overall threat picture, which over time is used to update future threat, vulnerability and risk assessments, and ultimately, the entire security program.

Internal documentation is not the only way to track threats. As mentioned above, one of the steps to conducting a successful threat assessment is to collect crime data from outside law enforcement agencies. Crime data/statistics are typically available from local law enforcement, and should be compared to internal security incident reports in addition to their use in developing periodic threat assessments for the organization.

---

[19] Garcia, Mary Lynn. "The Design of Evaluation of Physical Protection Systems." Butterworth-Heinemann. 20008.
[20] Purpura, Philip A. "Security and Loss Prevention An Introduction," Elservier Science, 2002.
[21] Sennewald, Charles A. "Effective Security Management," Elservier Science, 2003.

AN EFFECTIVE SECURITY PROGRAM

An effective security program is a living thing. Comprised of a myriad of components and procedures, security documents must be considered as "living" with their relevance and content changing from day-to-day. Maintenance of a security program takes constant shepherding to maintain its relevance. If not, it will lose value to its user.

The collection of documents forming the security program, must be, by design and intent, focused on three primary missions: remedial measures, preventative measures, and overlapping both of these, education. The security plan must be able to accurately describe situations both present and future, capture potential scenarios and consequences, detail the organization's actions both during and following specific events, and educate the organization on their role in the plan itself.

All of these actions are necessary so that when, not if, that accident, natural disaster, sabotage, or attack occurs, that the potential disruption, loss of life, impact to the facility or organization or region, is either averted, mitigated, or minimized.

In order to demand so much from a security program, it is critical that much forethought and planning be done up front. The quality of any security program is directly proportional to the amount of time and effort the organization has devoted to it.

There are three steps or phases necessary to ensure a quality Security Program. That they are not rushed through or omitted are equally important. They are: evaluate, establish and sustain.

Evaluate

Evaluation of an organization's security program includes examining the mission, assets, consequences, threats, and security system effectiveness. A security program must be tailored specifically to each organization. This can be accomplished by asking three questions:

1. What is produced?
2. How is it produced?
3. For whom is it produced?

By examining the mission through analysis, asset definitions are identified. Assets include employees, physical structures, knowledge, intellectual property, vendors and clients, and any activity that has a positive value to its owner. Performing asset analysis is essential to the identification of consequences which include revenue loss, impact to vendors, customers or operations, and the effects on the local community and region. Consequence analysis requires a threat picture in order to identify the types of adversaries capable of carrying out such devastating end results. All of these elements are essential in evaluating the types of security measures that best protect an organization. Figure 3.

provides an example of the importance of asset identification and the development of concentric rings of security to protect against an identified threat.

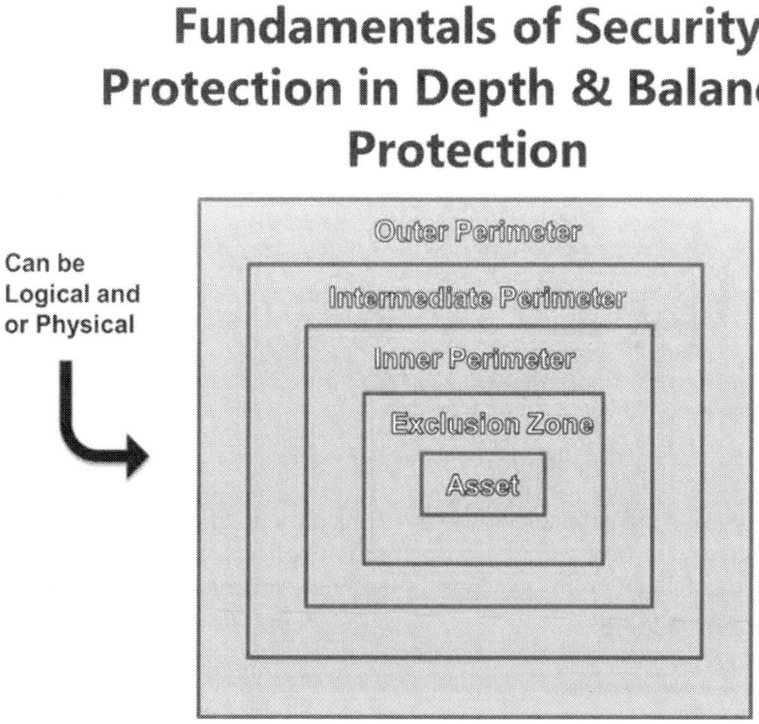

Figure 3. Layers of Security

Establish

Figure 4. represents the focal importance of people to the equation. Regardless of the quality of policies, procedures, access controls, manuals, etc., the linchpin of the whole program relies on the organization's personnel. They are either the strongest or the weakest link that holds the other three components together.

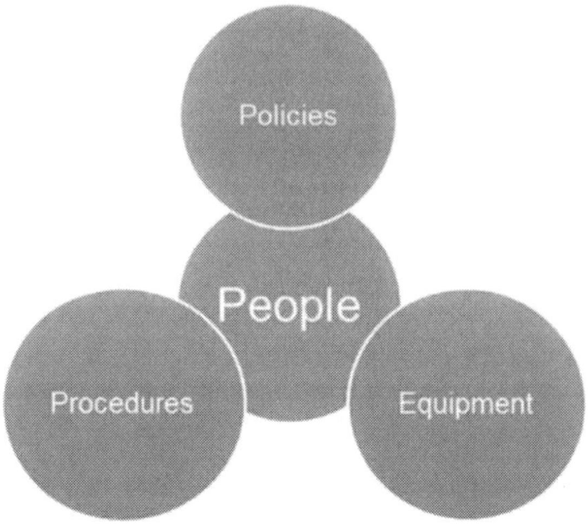

Figure 4. Effective Security Programs

Identifying the importance of personnel allows for the security program designers to identify gaps and create an in-depth program comprised of policies, procedures and equipment. Establishing this program also requires thorough training of personnel on policies, procedures and equipment.

Sustain

Much like the amount of detailed analysis that is required for a comprehensive Risk Assessment, that same amount of research is required to design a Security Program. In fact to that point, it is strongly recommended that a Risk Assessment be completed as a precursor to the program design phase. Only through such an assessment, can the organization know the Who, What, When, Where, and Why toward which the preventative and remedial measures are intended.

Sustaining an effective security program requires continuing education, exercises to test response measures, continuous building of relationships with external stakeholders, and constant reevaluation of the program design to identify holes.

Understanding the threats to an organization allows for the proper selection of the most likely threats to design a security system to meet the chosen adversary, and the testing of the system to determine its effectiveness against said threats is the cornerstone of an effective security program.[22]

---

[22] Garcia, Mary Lynn. "The Design of Evaluation of Physical Protection Systems." Butterworth-Heinemann. 2008.

CONCLUSION

A successful security program requires the comprehensive understanding of the various threats that an organization faces. Far too often security plans are generated in order to create a paper trail for compliance. But a security plan is more than that, it is a living entity that requires constant attention and adjustment in order to meet the demands of ever evolving threats. The essential first step to a successful and efficient security program is to identify the threats facing the organization. Threats are classified as internal, external, or internal working in collusion with an external threat.

Proper identification of threats is derived from a risk and vulnerability assessment, working in conjunction with local law enforcement to develop a comprehensive threat picture. Once the threat or threats have been identified an organization can develop processes and procedures to mitigate, reduce or accept the threat.

Tracking threat information is imperative in order for an organization to evaluate itself for future evolution. All of these actions are necessary so that when, not if, that accident, natural disaster, sabotage, or attack occurs, the potential disruption, loss of life, and impact to the facility, organization, or region, is either averted, mitigated, or minimized.

Mitigation, reduction, or acceptance measures are used in the design, implementation and maintenance of the security program. An integral part of the security program design includes processes and procedures for identifying, reporting, and tracking of threat related incidents. Equal attention must be paid to all sections in order for the comprehensive development of an organization's security program that can stand the test of time.

REFERENCES

ASIS International, Protection of Assets, Alexandria, Virginia, USA; Section 1.3.

Berger, Joseph. A Dam, Small and Unsung, Is Caught Up in an Iranian Hacking Case. The New York Times. 2016. http://www.nytimes.com/2016/03/26/nyregion/rye-brook-dam-caught-in-computer-hacking-case.html?_r=0.

Ching, Francis D.K. (1995). *A Visual Dictionary of Architecture*. New York: John Wiley & Sons, Inc. p. 12.ISBN 0-471-28451-3.

Critical Infrastructure Protection Committee (CIPC), Atlanta, GA, December 15-16, 2015.

Department of Homeland Security, Dam Sector Protective Measures Handbook, 2015.

Garcia, Mary Lynn. "The Design of Evaluation of Physical Protection Systems." Butterworth-Heinemann. 2008.

Johnson, DDP & Madin, EMP (2008) Paradigm shifts in security strategy. Why does it take disasters to trigger change? In: *Natural Security: A Darwinian Approach to a*

Dangerous World, RD Sagarin and T Taylor (eds), Berkeley and Los Angeles, CA: University of California Press.

Kahn, D. 1999. Pearl Harbor as an intelligence failure. In Pearl Harbor and the coming of the Pacific War: A brief history with documents and essays, ed. A. Iriye, 158–169. Boston: Bedford/St. Martin's.

Lord, Nate. "What is an Insider Threat? An Insider Threat Definition." Digital Guardian. 2016. https://digitalguardian.com/blog/what-insider-threat-insider-threat-definition.

"Paradigm." *Merriam-Webster.com*. Merriam-Webster, n.d. Web. 1 Dec. 2016.

Purpura, Philip A. "Security and Loss Prevention An Introduction," Elservier Science, 2002.

Sennewald, Charles A. Effective Security Management. Elsevier Science. 2003.

Title 10, Section 73.1(a), of the *Code of Federal Regulations.*

US Department of Justice. Texas Resident Arrested on Charge of Attempted Use of Weapons of Mass Destruction. The Federal Bureau of Investigation. 2011. https://archives.fbi.gov/archives/dallas/press-releases/2011/dl022411.htm.

Weiss, Joseph. Protecting Industrial Control Systems from Electronic Threats. Momentum Press. 2010.

UNMANNED AERIAL VEHICLES-A PUBLIC AGENCY APPROACH TO DEPLOYMENT AND PROTECTION

Katharine Killeen[1]
John Rizzardo[2]

Unmanned Aerial Vehicle (UAV) use and diversity of applications has increased and expanded exponentially over the last few years. It is estimated that over a million civilian UAVs have been sold through 2015. To date the laws and regulations allowing and governing the use of UAV's in the United States have struggled to keep pace with the evolving technology. From hobbyists to small and large businesses, first responders, government agencies, and the military, the diversity of interested users of the technology is substantial, but for every potential beneficial use of the technology, there remains many safety, privacy, and security concerns with the usage of UAVs.

The California Department of Water Resources (CDWR) has identified many useful applications of UAVs to assist with meeting its mission, including mapping, conducting environmental surveys, flood watch and warnings, inspections of levees and water conveyance facilities, and emergency response. DWR created a committee to consider these applications and determine the requirements, licenses, policies, and procedures that would be necessary for the Department's use of the technology. Taking advantage of the benefits offered by UAV while not sacrificing protection from hazards has been integral to our approach. Concurrently, DWR is sensitive to and considering worker and public safety and privacy while also looking at security considerations along State Water Project facilities. DWR will also describe legislative proposals that have been introduced in California as well as their compatibility with FAA regulations and no-fly zones. DWR will present its current plans for UAV use, its developing policies and procedures, the steps taken to get there, lessons learned and what issues remain to be worked out.

[1] California Department of Water Resources, kkilleen@water.ca.gov
[2] California Department of Water Resources

THE ECONOMICS OF DAM SECURITY POST 9/11 AND THE THREAT TO CONSUMER CONFIDENCE

Alexander B. Shannon[1]

ABSTRACT

In the post 9/11 era, the U.S. Department of Homeland Security has promoted securing vulnerable and valuable national resources and infrastructure from possible threats. Dams are unquestionably a critical national resource that must be protected. Terrorism poses a threat to the U.S.' water infrastructure, which is vulnerable to physical destruction, biological and chemical contamination, and cyber-attacks. This threat directly influences government and private security spending decisions, consumer costs, and confidence.

Consumer confidence is a driving influence behind economic decisions, and impacts consumption, spending habits, and investment. Actively reducing risk will ultimately influence an individual's economics choices, and have a positive effect on economic growth. The U.S. Government spent over $900m between 2001 and 2010 on dam security alone, providing assistance to state and privately owned facilities, while also increasing security regulations and identifying high risk facilities (Copeland, 2010).

Previous assessments of the economics of dam failures measured consequences based on the real costs of failure. While this approach is still crucial, the broader effects must also be considered. This paper examines the effects of risk mitigation on the nation's confidence in our water supply, and explores the economic impact of an attack on a U.S. dam. In the wake of a dam failure, goods that use water in production will experience higher prices and supply shortage. Complement and substitute goods will also experience shifts in demand, causing economic uncertainty. To further understand this relationship, this paper introduces the concept of water confidence, a measure of the level of confidence in having a readily available supply of water.

In an ever increasing financialized economy, water plays a major factor in its stability. It is essential to understand the impacts of securing infrastructure to bolster this confidence in water, and the risks of not doing so.

INTRODUCTION

Water is one of the most basic of human resources, and it follows that critical water infrastructure must be secured from potential threats. Outside of natural causes, the US has not experienced significant water supply issues compared to many developing nations. It is generally assumed that water will be accessible and treated to a high quality. Yet, as recent events such as lead contamination in Flint, Michigan have shown, failure to deliver on this assumption can catastrophically affect a community. There are numerous areas in which water supply is susceptible to issue, but this paper will focus on the threat of terrorism facing US dams, as we argue the effect of an attack would be contagious to multiple sectors of the economy, dealing a powerful fiscal blow beyond destruction. Prior

[1] Economic Consultant, CH2M Hill Engineers, Seattle, WA 98006, alex.shannon@ch2m.com.

to the September 11, 2001 terrorist attacks in New York, the concept of a massive attack on US soil was limited. While terrorism is not a new phenomenon, its recent global presence has revolutionized how the US government approaches security issues. Identified as one of 18 key components of US economic well-being, dams and other water resource facilities need protection from potential terrorist threats (Copeland, 2010). The Department of Homeland Security has recorded 25 attacks against dams worldwide between 2001 and 2011 (Homeland Security, 2012). Of these, 13 included explosive devices, and one incendiary device was used in an attempt to attack Black Rock Dam in the United States. Dams are a high profile target as they are significant national icons and provide essential services including drinking water, flood control, power generation, agricultural water, and industrial water, thus proving a significant target. The Patriot Act of 2001 changed the way that protection spending is distributed, and tasked many federal agencies with terrorism risk identification and mitigation (Prante, 2008). The federal government provides funds to states, which aggregately sums to 1.6 billion dollars over the last 12 years under the Homeland Security Grant Program (HSGP), which allows for states to secure their infrastructure (Prante, 2008). Congressional research focuses on informing the government of the risk of terrorism to water resources (Copeland, 2010). Providing this context has allowed for the appropriation of $923 million towards water security at facilities in the 10 years after 9/11, and funded research in determining effective ways to secure facilities and combat terrorism. Prior to this, the Bureau of Reclamation had only funded security measures at five dams, Hoover, Shasta, Grand Coulee, Glen Canyon, and Fulsom (Copeland, 2010). As a part of the effort to secure dams, there is need to identify high-risk facilities, with particular emphasis on water as a drinking resource and electrical generation. However, the US federal government owns and operates only 5% of the dams that could cause death under a failure scenario, but provides resources for dam owners and operators to secure facilities. Even though appropriations are provided, given this information, much of the private cost is still passed on to consumers. Copeland identifies gaps in security, including the lack of threat/vulnerability assessments, identification of potential biological/chemical threats, establishing community standards, and computer based monitoring systems. Under the management of the EPA, Homeland Security, FEMA, the Bureau of Reclamation, the US Army Corps of Engineers (USACE), and the Federal Energy Regulatory Commission (FERC) there is a push to establish more effective federal standards for municipal water districts and private water companies. As a part of these standards, agencies are providing training and physical security requirements. The Government Accountability Office (GAO) is responsible for oversight of these programs, and has analyzed security efforts, determining that physical and technological upgrades, training and law enforcement relationships will be the focus of future federal funding.

Legislation also adapted in response to this new analysis. The Drinking Water Security and Safety Amendments (SDWA) manifest Congress' new dedication to protecting drinking water as a valuable and scarce resource (Shermer, 2005). Water has no true substitutes, and the security regulations under SDWA provide impact over 265 million Americans which has been a crucial factor in motivating these organizations to change the way they approach water infrastructure security (Shermer, 2005). This paper will discuss the origination of this concept through consequence-based analysis, such that dams are seen as one of the key infrastructure components of the country.

This paper will further examine the costs that are generated by increased appropriations to dam security and the benefits that this spending brings. Both these costs and benefits are shared by industries that rely on water. In the event of an attack on a dam, the resulting costs would be similarly shared such that it is economically devastating to multiple sectors. Tracing this affect back, we can understand how consumers' view of dam and water stability influences their economic decisions. Ultimately, this paper will provide a theoretical analysis to assess the economic impacts of securing dams against terrorism threats. Current literature is examined to show the impact of a dam failure scenario, water contagion and disruption of distribution, and the lack of economic impact to consumer confidence analysis. This literature will follow a historical trend in which the approach to water facility safety has evolved from natural disaster based modeling to consequence based modeling. Finally, a cost-benefit analysis of security related spending, and an analysis of consumer confidence under different security scenarios is done to examine water resources under the threat of terrorism.

LITERATURE REVIEW

Fortunately the US has not experienced a damaging attack on water resources or dams, but the threat remains and proxies for understanding the impacts play a key role. Literature to date has used three separate techniques to evaluate water security. These include natural disaster modeling, consequence based modeling, and primitive economic analysis of costs and funding allocation. This paper will look at the historical trend of these models to show the evolution of consumer perception of preparedness. Homeland Security reports are the culmination of public analysis to show the impact of a terrorist attack, but this paper will look at the broader economic implications, and the risk of contagion to other markets. Finally, it will examine how water impacts consumer confidence, such that we can link dams and water supply to macroeconomic stability.

Natural Disaster Modeling

Prior to 2001, academic research focused on development and distribution models for water in the most efficient manner possible, as well as natural disaster preparedness (Haimes, 2002). At this point, natural disasters were the only recognized significant threat facing the water industry. Grigg (2003) suggests that by modeling the effects of a natural disaster to water distribution, we accurately predict the potential future consequences of a terrorist attack. These models used loss of life simulation to predict the costs of a disaster in order to quantify the potential losses under a dam failure scenario (Lehman & Needham, 2012). Despite general disagreement with the accuracy of natural disaster models, this approach has been the focus of research for decades and continues to influence how the US approaches water security. As such, natural disasters were previously viewed as a close proxy to what a terrorist attack would do to the economy. As outlined in studies done by Grigg (2003), Munger (2009) and Gleick (2006), in addition to the data presented to Congress by Copeland (2010), natural disaster models were the main mode used to distribute funding to secure dams from threats of terrorism.

Research has started to trend in the direction of disagreeing with Grigg's (2003) use of natural disasters as a complete proxy for risk analysis. Analyses use Sunny Day Failure (SDF) modeling in which dams fail under the most amenable conditions. Haimes (2002) argues that the Sunny Day Failure (SDF) model is inaccurate in analyzing terrorism failure scenarios, simply due to the diversity of facilities in their scale, structure and systematic configuration. SDF also assumes high warning time windows, allowing for minimal loss of life and potential mitigation of consequences, which would not be true in a terrorist attack. In addition, in an early study of this potential scenario, Haimes established the theoretical model for examining terrorist attacks on water facilities through showing that the natural disaster proxy is a lower bound of the economic damage potential. Research is trending towards considering a consequence-based analysis for terrorist attacks, in which risk analysis is based less on the type of disaster and more on the potential consequences in the wake of a failure scenario.

Consequence Based Modeling

After the popularity of natural disaster based modeling subsided, a new approach was formulated to evaluate preventative measures based on the respective potential consequences. From Homeland Security and Presidential directives originates a significant portion of this consequence-based research on water safety. Stemming from a desire to secure borders and prevent domestic terrorist activity, the specific focus of preventative measures is on infrastructure that could cripple the economy. In this way, the September 11th attacks were a significant turning point in water safety, prior to which this infrastructure had no special security measures in place (Haimes, 2002). Lack of security prompted internal review of which agencies are responsible for protecting the nation's assets, and how these facilities are evaluated. To provide resolution, the US government, under the Bureau of Reclamation, the Census Bureau, the Department of Homeland Security and the Congressional Research Service, collects most data regarding infrastructure spending. This data identifies preventative spending, but does not provide information on the mitigated consequences. Studies in the late 1900s identified water as at risk to human attack, however this was not thoroughly addressed in policy until 2001 through the Drinking Water Security Amendments (Haimes, 2002).

Scholars generally agree that there are three main modes of attack (Beering, 2002; Copeland, 2010; Gleick, 2006; Grigg, 2003; Shermer, 2006). These include physical, biological/chemical and cyber threats. For the purpose of this paper, we will focus on physical and biological/chemical threats, due to their extreme visibility and subsequent influence on consumer confidence. Cyber threats pose just as potentially devastating, but in the eyes of the consumer are less visible and therefore preventative measures may have less impact on behavior. These two modes have the potential to disrupt water distribution in a way that drastically affects how other industries operate (Qiao et al., 2005). Qiao et al. highlight the nature of water subsystems as necessary components of analysis when examining water infrastructure as a complete system. These systems are a series of independent water supply facilities that differ by geographic location and communities (Haimes, 2002). Systems serve to distribute water to different sectors, as water acts as a resource input for industries, as cooling and processing agents in power plants and manufacturing facilities (Folga et al., 2010). Folga et al.'s study shows the economic losses in each of these industries if water delivery was interrupted, with the top losses being in agriculture and manufacturing. Furthermore, the significant difference between localities presents a challenge in determining how spending is distributed. The interconnected nature of industry and agriculture becomes extremely important when analyzing the potential consequences. Water has always been considered important but never to the point that it could cripple production in one blow. Yet, with this research come problems in analyzing how different sectors will respond under an attack scenario. Compared to other researchers, Qiao et al. show the difficulty in creating a general model of response due to these differing systems.

With policy changes in the early 2000s, there has also been an evolution in the research conducted on water terrorism. While former studies used natural disaster dam failure scenarios to predict damage, the main current argument suggests there is a difference in how systems respond under terrorist conditions (Lehman & Needham, 2012). Furthermore, as evidenced in attacks on dams in other countries, workers and officials may be injured or killed in an attack, and unable to respond. Natural disaster modeling includes time for warning the general public. The warning time window, which is crucial to evacuation orders and preparation, would be significantly shorter in the terrorism scenario, meaning the breach would be far more catastrophic (Lehman & Needham, 2012). This model is still used in some consequence-based analysis, but other research has determined that this model would not hold true for terrorism, given that an attack would have little warning and response would be focused on saving lives over property and securing other industries (Munger, 2009). As the warning time window shrinks, the consequences grow. In a terrorist attack, the lack of warning time window would mean consequences would be at their peak. Almost all water facilities have Emergency Action Plans (EAPs) that dictate how a facility responds to an emergency or failure scenario. While these plans can account for the terrorism potential, facilities may unprepared or unable to respond to attacks that impact staff safety. Regardless of EAPs, the lack of warning time and the method in which the attack occurs could render an attack significantly more economically devastating.

As a major sector in the National Infrastructure Protection Plan (NIPP), Homeland Security has also taken the consequence based approach in determining priority dams that need to be secured (Homeland Security, 2015). Their consequence based model allows dam owners and administrators the ability to understand how their facility falls in the department's Consequence-Based Top Screen (CTS) methodology. This methodology incorporates human, economic, and critical function factors to determine the consequence of failure. The economic portion of their analysis looks generally at asset replacement value, remediation cost, and costs associated with business interruption. While critical function disruption would include interruption in water supply, irrigation, power generation and others, there is a broader macroeconomic impact associated with this that remain unquantified and unexamined in documented reports.

Current literature also revolves around identifying the means and feasibility of a terrorist attack. More specifically, Gleick (2006) analyzes the realistic nature of a chemical or biological attack by examining the means of dispersion of an agent into the water supply at a reservoir or in distribution control systems at a dam facility. Gleick determines contaminant stability in water, chlorine tolerance in parts per million, and whether the pathogen has been weaponized. Although speculative, this research coupled with historical attack data across the world shows the feasibility of a chemical or biological attack (Gleick, 2006). While this provides data on the potential of a threat, it does not provide an analysis of spending or operating changes in response to this type of an attack. Health consequences of water borne pathogens could be massive, hard to identify, and overwhelm response mechanisms (Meinhardt, 2005). Biological and chemical attacks have impacted the U.S. military abroad, where soldiers are at risk for water born bio-terrorism agents. Furthermore, the U.S. has experienced this domestically in an outbreak of Cryptosporidium in Wisconsin which caused over 100 people to become sick in a short period of time (Meinhardt, 2005; Gleick, 2006). On a larger scale, particularly when introduced to a community that relies on a dam for its water source, the magnitude of contagion could be exponentially higher. Bio-chemical terrorism could be difficult to detect given that some water systems exist or originate in a location that is not considered a high target, which is at odds with the general model of appropriations to regions where loss of life has a high potential (Meinhard, 2005). In examining biological and chemical terrorism, Gleick stipulates that confidence, communication, monitoring and response are key to addressing this threat.

Munger (2009) also highlights one of the major gaps in security analysis. Dams are often examined under what is considered to be a "Sunny Day Failure" (SDF) in which the failure occurs under normal or optimal conditions. However, it is important to note that this is not the case when terrorism plays a role in dam failure, as the conditions are likely less than optimal (Munger, 2009). In creating proxy variables for such an event, most research to date has used the SDF conditions model, which does not accurately depict the full expense of the economic costs associated with a dam failure in a terrorism model (Grigg, 2003). Haimes (2002) argues that this SDF model is inaccurate in analyzing terrorism failure scenarios, simply due to the diversity of facilities in their scale, structure and systematic configuration.

Economic Perspective of Costs and Funding Distribution

The September 11 attacks were a major turning point that started a conversation regarding how congressional appropriations should focus on protecting national infrastructure. Water's multitude of uses and lack of substitutes makes it extremely valuable to the US economy, and the potential for market contagion is high leading to potential high future national costs (Beering, 2002; Gleick, 2006; Meinhardt, 2005). Theoretical disagreements exist in how these costs impact the consumer, and this paper will supplement those gaps by providing a cost-benefit analysis related to protecting water facilities.

The price and value of water is a good indicator of the diffusion of costs to consumers. Water is an economic good and has value given that it is scarce resource with no true substitutes (Van der Zaag & Savenije, 2006). It is priced differently across municipalities based on distribution costs and availability. Still, to provide water to all levels of income, there is typically a block pricing system in which the price of water increases based on how much is used by the household per month. Van der Zaag and Savenije argue that the market for water is not homogenous in its consumable value, given that the preferences for quality differ between agricultural and industrial uses. The quantity demanded by agricultural sectors is much higher, but the quality needed is significantly lower, and it more restricted by budget. This makes determining the overall monetized value of water difficult, as the opportunity cost for each additional unit of water consumed by these markets different.

In the context of terrorism, Beering (2002) postulates that lack of warning systems in place, as well as a lack of institutional oversight and consistent regulation of infrastructure security make much of the costs that go into preparing for a potential attack more extreme. While security procedures and systems may be deterrents to an attack, many facilities lack these systems due to their high cost. However, this cost may be offset by the economic benefits enjoyed from stable confidence in the nation's dams and their services provided.

Quantitative analysis of the Homeland Security Grant Program, while offsetting the costs of security using public funding, begs the question: who is responsible for the security of water facilities? Given that spending is highly influenced by politics, it follows that this funding dispersion is controversial because many water facilities exist in rural areas, which some argue require less protection due to lower population density (Prante & Bohara, 2008). Placing water spending in an economic context, Prante and Bohara (2008) performed an econometric analysis of how funding is dispersed under the terrorism threat. In this model funding is a function of risk, politics, and power. This study highlights other factors that influence how funding is distributed, but also shows that one of the major difficulties in studying appropriations is that risk measures are often imprecise (Prante and Bohara, 2008). Similarly, Gleick (2006) shows that many of the issues in spending distribution originate from the very source of their necessity, which is the uncertainty of an attack. These recent studies focused on consequence based risk aversion, but have not focused on how uncertainty influences consumer confidence. Uncertainty provides great economic motivation that has not been empirically analyzed to a great extent in this issue. Gleick (2006) identifies that there was a significant, although not quantified, loss of productivity and wages during the Wisconsin cryptosporidium outbreak. We stipulate that this type of outbreak would cause higher damage under a terrorism scenario. What is lacking in this literature is the extent to which this threat generates new costs for consumers through the monitoring, prevention, and response measures recommended.

Analyzing the devastating market effect of an attack is crucial to show the reason behind spending. Munger (2009) shows the economic impact and devastating nature of a dam failure scenario. In this aspect, Munger highlights that potential future lost benefits can be analyzed as costs of a dam failure, dramatically increasing its impact. While not directly analyzing the data, Munger asserts that water loss to agricultural areas can be measured by looking at the opportunity cost for water in these areas, or by examining the value of water that is leased (or what would be considered the market value) by region. This highlights the importance of previous studies done in which appropriation divisions across states have been considerably debated. Typically, this debate revolves around population and potential physical damage, but it does not examine the differing costs of water in rural areas, and distribution costs associated with lower population density.

In essence, this is the difference between a short run impact and a long run impact analysis. As other research has shown, the long run impact will be far more costly to dependent industries, and this type of analysis has not been undertaken when examining the real cost of a threat. Furthermore, this long run impact also has a less visible impact to consumer confidence, as it will impact more sectors of output and consumption. Consumer confidence does play a large role in the overall economic effects that a water based terror attack would have. As roughly 70% of GDP, consumption plays a major role in economic growth and stability. An analysis by Adgrani and Macri (2010) found a strong correlational relationship between consumer perception of economic strength and stability, and confidence in the overall economy, and concluded that if consumer confidence were to decrease it would follow that economic stagnation, overall decreased production, decreased employment, and higher price levels would occur.

In the next section, this paper will address the gaps in the current research on dam security and the threat of terrorism in the form of examining the costs of terrorist threats to consumers and the role of consumer confidence in water resource terrorism. The literature presents a number of analyses but lacks models that quantify the loss of productivity, wages, and income, as well as broader macroeconomic impacts. Haimes (2002) identified the need for an economic evaluation of the consequences of water terrorism. Yet, literature does not analyze the economic cost of the threat alone, and the quantitative impact on consumer confidence that an attack would have.

ECONOMIC ANALYSIS

The questions raised by scholars establish the basis for the economic analysis of this paper. To fully understand the impact of terrorism to water resources, we must examine the costs and benefits that are created through securing water facilities, such that we can analyze the impact of terrorism on overall consumer confidence in both the macro economy and the water market. Literature emphasizes the potential consequence of a dam failure to other sectors that use water as an input. Whether this takes the form of agriculture or industry, if water supply and distribution are disrupted, the ripple effects would be enormous. While beneficial to modeling risk, no study has yet examined the impact to consumer confidence that would in turn affect global markets. This paper will focus on the abstract nature of consumer confidence in the water market, how dams play a stabilizing role ensuring consistent supply, and the psychological nature of a terrorist attack on a dam. The purpose in determining how funds are allocated is the perceived value to society associated with protecting water from a terrorist attack. It follows then that consumer confidence must become a necessary piece of this appropriations algorithm.

Before examining the effect of consumer confidence, we must first quantify the perceived value to society that security measures provide. The costs are measured in the form of dollars put towards security costs. Cash resources come from several sources, however the highest source of security funding comes from the federal government. The benefits come in the form of potential future costs that are avoided or mitigated through newly implemented security measures.

Consumer confidence cyclically fuels the macro economy in its ability to generate new levels of consumption and employment. Future expectation, which are heavily influenced by confidence, are one of the main drivers of demand. In this respect, it is important to add this future potential growth to the present benefits. In order to understand the confidence process, we must examine it under several different scenarios. As a constant, or control scenario, consumer confidence without terrorism will be the base for this analysis. Confidence will increase when consumers know that dams and water resources are protected, continuous and available. Furthermore, their confidence in other industries will increase knowing that prices are more likely to remain stable. To further fuel this argument for strong protective measures, confidence will drastically decrease if a terrorist attack occurs. In addition, confidence is drastically hindered under the constant threat of an attack. Even given the lack of true substitutes, consumers will likely change their levels of consumption of water if they perceive a threat. The threat alone thus has the potential to decrease growth, hurt jobs and decrease demand for water and other complementary goods and industrial production. To model consumer water confidence, security costs and benefits will be defined as inputs to show the effect of the protection on economic stability.

Quantifying costs

The costs of a terrorist attack on a dam can be broken in four categories: (1) damage, (2) the impact of disruption of water distribution or power generation, (3) increased security, and (4) the impact to consumer confidence and broader economic contagion. Dams are incredibly durable and able to withstand significant damage before failing, but even in a non-failure scenario, the damage would likely be in the order of magnitude of millions of dollars in repair and damaged equipment.

Homeland Security has published the direct impacts associated with the recorded attacks on worldwide dam facilities. Though these attacks did not result in failure leading to catastrophic loss of life, in many cases lives were lost as a result of the attack. This is the starting point for understanding how an attack on a dam will affect the broader economy. By understanding the risk and consequences of an attack, we can understand the benefits brought by mitigation and preventative security measures. This analysis also suggests that public spending to secure dams and other critical infrastructure is justified in comparison to the costs of an attack. Homeland Security quantifies the costs of lost benefits for a case study at Blue Dam in Colorado. Blue Dam was chosen due to its considerable size and its provision of recreation, power generation, irrigation, instream flow for fish and wildlife, and flood prevention. These expected lost benefits in power generation and other areas are estimated at $229m (Homeland Security, 2011). Furthermore, they estimate the repair or replacement cost to be $167m, and total damages from flooding to be $590m. Coupled with other remediation costs, the total cost of a Blue Dam failure is estimated at over $4.3 billion.

The roughly 100,000 dams in the US provide 10% irrigation to all US cropland, and 6.7% of all electricity (Homeland Security, 2015). In the Pacific Northwest, this number grows as dams provide 60% of electricity through hydropower dams. If we translate that to population impact, hydropower dams provide electricity to an estimated 2.6 million people nationally. As a major technology hub for companies such as Microsoft, Amazon, Facebook, and Google, interruption to power generation in the Pacific Northwest could have a globally catastrophic impact. An attack would costly to companies forcing decreased labor needs, lower wages, and goods supply shortages.

Table 1 below shows a summary of the attacks against dams worldwide between 2001 and 2011.

Table 1 Dam Attacks Summary by Incident Chronology[2]

Dam Attacks Summary					
Facility	Country	Date	Attack Type	Attacker Type	Result
Lhokseumawe Resevoir	Indonesia	August 17, 2001	Explosive Device	Separatist (Suspected)	Minor Damage
Panauti Plant	Nepal	November 24, 2001	Explosive Device	Communist Insurgent-Maoist (Suspected)	Damage - $500,000
Kidapawan Resevoir	Philippines	March 19, 2003	Standoff Weapons (Rockets)	Islamic Insurgent (Suspected)	Main pipline destroyed, disrupted water supply to 100,000 residents
Kajaki Dam	Afghanistan	May 2, 2003	Standoff Weapons (Rockets)	Islamic Insurgent	Failed Attack- None
Gomal Zam Dam	Pakistan	September 21, 2004	Assault Team	Islamic Insurgent	1 Civilian Fatality during hostage standoff
Zelenchuck	Russia	September 21, 2004	Assault Team	Islamic Separatist	Failed Attack- None
Dumarao	Philippines	December 15, 2004	Explosive Device	Communist Insurgent (Suspected)	Crane damage
Selaghat Dam Project	Nepal	December 19, 2004	Explosive Device	Communist Insurgent-Maoist (Suspected)	Electricity failure and damage to 200KW powerhouse
Mirani Dam	Pakistan	May 18, 2005	Explosive Device	Unknown	Failed Attack- None
Haditha Dam	Iraq	August 2, 2005	Explosive Device	Unknown	Damage to Dam (uknown cost)
Haditha Dam	Iraq	September, 2005	Standoff Weapons (Rockets)	Islamic Insurgent	No Damage, heavily guarded facility
Kajaki Dam	Iraq	September 17, 2005	Explosive Device	Islamic Insurgent	Attemped destruction of dam, failed. Work at facility haulted until NATO secured facility two years later
Hlaingbwe Dam	Burma	May, 2007	Explosive Device	Separatist (Suspected)	Employee Killed
Hlaingbwe Dam	Burma	May 2007 and September 2, 2007	Standoff Weapons (Mortar)	Separatist (Suspected)	Employee Killed
Waeng Station	Thailand	August 1, 2007	Explosive Device	Islamic Separatist (Suspected)	Control Station Damaged
Kajaki Dam	Afghanistan	March 30, 2008	Explosive Device	Islamic Insurgent	Two British Marines Killed
Tipaimukh Dam	India	April 26, 2008	Assault Team Explosive Device	Unknown	Machinery Destroyed, Security Increased
Mosul Resevoir Dam	Iraq	May 1, 2009	Explosive Device	Unknown	5-10 Shiite Killed
Balimela Power Station	India	December 19, 2009	Incendiary Device	Communist Insurgent-Maoist (Suspected)	Damage to power station valve house, water intake tunnels
Mytikyina Dam	Burma	April 17, 2010	Explosive Device	Ethnic Separatist	4 Chinese workers killed, 20 injured. Construction suspended indefinitely
Thawk Yin Kha Dam	Burma	April 27, 2010	Explosive Device	Ethnic Separatist	Failed Attack- 4 Injured
Black Rock Dam	United States	July 4, 2010	Incendiary Device	Unknown	Failed Attack
Baksan Power Plant	Russia	July 20, 2010	Assault Team Explosive Device	Islamic Separatist (Suspected)	Two guards killed, two generators destroyed
Machlagho Dam	Afghanistan	July 18, 2011	Assault Team	Unknown	Three police killed, several injured or captured
Thawk Yin Kha Dam	Burma	July 20, 2011	Standoff Weapons (Rockets)	Ethnic Separatist (Suspected)	Eight rocket grenades fired at construction site. No injuries.

Damage to a facility from an explosion, incendiary device, or mortar attack would have extremely high costs. If the facility was not damaged beyond repair, we would expect the costs of repairs to be proportional to the damage. It is difficult to quantify or monetize the widespread cost of lost water distribution to a community, agriculture, or industry, or loss of power generation provided by hydropower facilities. Given that 31% of dams have a high potential of hazard if they fail or even don't operate correctly, it follows that the economic consequence would be equally severe.

[2] U.S. Department of Homeland Security, 2012.

We must also consider the costs to the water industry through the amount of extra spending that has been used to provide new security measures to facilities. The following model attributes security costs as a function of the threat level, the marginal personnel costs under the threat, new training, recreation lost, and capital improvements, such that we can fully understand the impact to the industry that terrorism has.

$$\textbf{\textit{security costs}} = \{\boldsymbol{f}(\textbf{\textit{threat level, marginal personnel costs, training, recreation loss, captial improvements, facility size, location, consequence score}})\}$$

The threat level in this model represents the measure of how the public perceives the practicality of a terror threat, and any recent evidence that an attack is imminent. While some measure of a threat is constant, there is room for fluctuation of costs based on how direct the threat is. Marginal personnel costs exist in the increase cost for each additional unit of labor that is required to secure facilities under the terror threat, each which require training. Existing staff also requires additional terrorism specific training to be able to implement effective emergency action protocols. There is some measure of loss to the public that is derived from the inability to enjoy and use reservoirs and lakes with new security measures in place. Finally, capital improvements, including new security systems, access restriction measures such as fences and gates, structural improvements and warning systems add significant costs to facilities. There have been few attempts to quantify these important components that comprise security costs. Capital improvements, marginal personnel and training costs are easily measurable, although data is not readily available. The other variables are more abstract in nature.

Yet, we can summarily look at security costs through the funds provided by congress to address security issues at water facilities. Congressional appropriations make up a large portion of funds being allocated for security purposes and safety regulation that support federal, state, and private facilities. To secure the water industry, facilities must take several measures. This paper argues that the most effective measures include providing additional security personnel, training existing personnel, restricting recreational use of lakes and reservoirs, and monitoring water for chemical and biological contaminants. Thus far, most analyses focused on the real costs to the industry, as opposed to the abstract variables that can generate higher costs. As such, data represents capital, personnel, training, and evaluation funding to the industry. The following data (Figure 1) was compiled from a study prepared by the Congressional Research Service, which outlines the amount of funding allocated to water security since 2001.

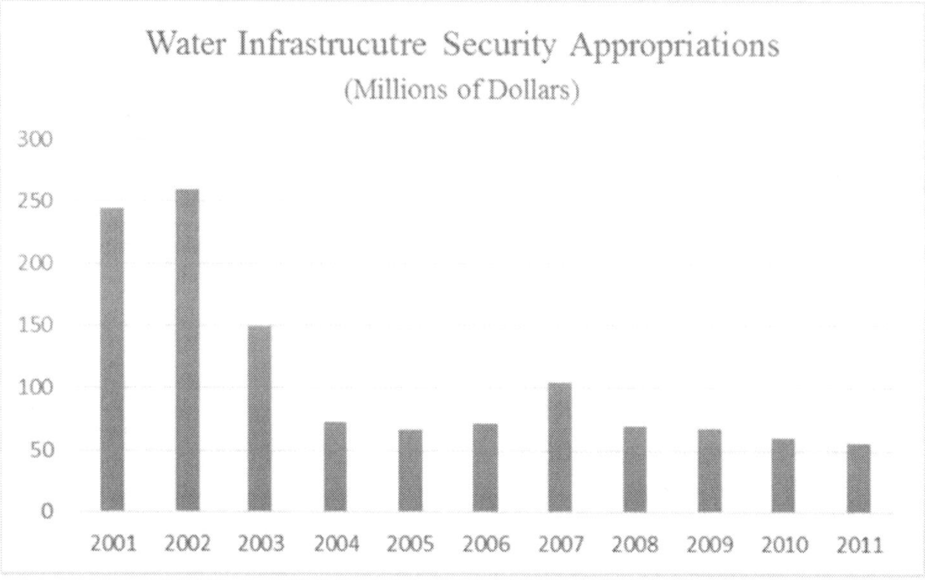

Figure 1. U.S. Water Infrastructure Security Appropriations (Millions of Dollars)[3]

Figure 1 shows a spike in spending after the September 11th attacks to 260 million dollars, with a decline in fiscal year 2003 and then relatively similar amounts of spending in fiscal years 2004 through 2011 at an average of 71.425 million dollars. Copeland (2010) divides government spending into three parent agencies, the Environmental Protection Agency, Bureau of Reclamation, and the Corps of Engineers. Appropriations to the EPA and the Bureau of Reclamation, while decreasing since 2002, have decreased at a slower rate. The Corps of Engineers saw the highest amount of spending in 2002 and 2003 with very few appropriations in subsequent years. This suggests that one of the primary goals was to provide analysis of structural integrity with respect to potential terrorism in the post 9/11 era. These agencies are then responsible for distributing funds to local water facilities, both to provide higher security and analyze vulnerability (Copeland, 2010).

Another form of cost is considerably difficult to measure, which is the change in water prices in response to increased costs. While actual prices vary across utilities, water prices have been generally rising in the last 12 years (McCoy, 2012). Given that increased water security spending is correlated with higher utility water prices, for the purposes of analyzing cost, consumption and confidence, this paper makes the assumption that security costs lead to higher water prices. Making this assumption leads to the conclusion that consumers directly experience the costs of securing the water industry, providing relevant significance to their level of confidence.

[3] Copeland, 2010.

Quantifying Benefits

Before we directly examine water confidence, we must establish the benefits consumers derive from water and dam security. These benefits are both in the form of avoiding potential future costs and enjoying stable economic progress. As such, there are four categories of benefits including: (1) mitigation of damage to facilities, (2) uninterrupted water and power distribution, (3) new security related jobs and investment, and (4) increased confidence in water and the government.

Benefits derived purely from the mitigation of consequence are more abstract, and are not as visible to consumers as their associated cost if recognized. Thus, the amount of investment and spending to secure the infrastructure is not purely a function of the benefits, but of the perceived level of risk, public visibility, and the amount of risk the owner is willing to assume.

To examine this, researchers have looked at dams and water facilities under different scenarios. The purpose of this methodology is to calculate the perceived value to society that is generated by avoiding a failure scenario. Consumers do not receive any direct benefits, and rather incur higher prices and taxes. Yet this spending is justified by policy makers through arguing that present costs are far less than potential future costs (present benefits). To model these costs, researches look at several different factors. These include loss of life simulations, potential capital repair costs to the industry, and property damage. To model this, we establish the following equation where benefits in time t_0 are a function of future costs, future loss of life, property damages, lost wages, current consumption, and the future price of water.

$$\begin{aligned}\boldsymbol{security\ benefits_t} &= \boldsymbol{\beta_0 + \beta_1(capital\ cost_{t+1}) + \beta_2(loss\ of\ life_{t+1})} \\ &\quad \boldsymbol{+ \beta_3(property\ damage_{t+1}) + \beta_4(lost\ wages_{t+1})} \\ &\quad \boldsymbol{+ \beta_5(\Delta consumption_t) + \beta_6(water\ price_{t+1})} \\ &\quad \boldsymbol{+ \beta_7(ripple\ effects_{t+1})}\end{aligned}$$

Time (t) represents a point in time before a terror attack, and t+1 represents a point in time directly post terror attack, thus articulating the present benefits derived from increased current security measures to avoid high future costs experiences in a terror attack. Loss of life simulations model the number of lives that would be lost under a failure scenario. As discussed, simulations have primarily focused on natural disaster response. Under natural conditions, we would argue that less people would be affected. Depending on the type of terrorism, crucial warning systems may be ineffective in preventing loss of life. An act of physical terrorism would likely have little warning, meaning that evacuations prior to the attack are unlikely. In the event of a biological or chemical attack, new detection systems would prevent the mass outbreak. This is a key component that has been upgraded to prevent successful terrorism. With early detection systems, water facilities will be able to control the flow of water and alert consumers to the safety risk. However, even with these systems in place there would still be a cost to many industries if this type of attack occurs.

Most importantly, and far more difficult to measure, are the potential costs and rippling effects to other industries that would occur under a dam failure scenario, which is considered to be a present benefit. The macro economy relies on the cyclical nature of income and demand to keep steady economic growth. If industries that use water as an input were interrupted the costs would be extremely high. Long-term water shortage will cause high production costs leading to fewer jobs, less disposable income and less consumption. Not only will many people find themselves in dire economic situations, water has the potential to cripple both regional and national economies. This type of situation is difficult to fully quantify given its abstract nature.

Measuring Confidence

Finally, though influenced by previously discussed costs and benefits, dams interact most significantly with the broader economy through consumer confidence. As an added aspect of consumption decisions, confidence affects all decisions made in the water sector. A terrorist has three potential inhibitions to confidence. First, an attack would be incredibly psychologically traumatic to the population, such that consumers would be less likely to participate in economic transactions. This phenomenon is largely attributed to fear. Second, an attack would cause consumers to become less confident in industries, meaning that they are likely to find substitutes, or decrease their demand for water. While we argue that water has no "true" substitutes, people will still likely decrease their normal consumption habits in the event of attack. Third, the threat of an attack alone has the potential to influence consumption choices. If the population has little confidence in the water industry's ability to secure itself, it will likely consume other more expensive water sources, or migrate. For a large population, this would likely cause extreme economic hardship. As demonstrated in Homeland Security's 2012 report, terrorism goes hand in hand with political goals. As critical infrastructure, forces seeking to destabilize a region, country, society, or government, would find them of considerable value. If consumers do not have confidence in the government to provide drinking water, power, industrial water, and agricultural water, fiscal policy would in turn take a direct hit.

Terrorism has the potential to impact consumer confidence such that all industries are hurt. While arguably water is such an essential resource that it plays a great role in all aspects of life, certain industries do not rely on it as production input, but would still feel cyclical decline. In particular, complementary goods will experience significant recession after a dam failure, and as there are no true supplementary goods, people will be forced to migrate to find safe water. Consumer confidence further changes how other markets respond after an attack. In this respect water is connected to industries that do not rely on water as an input. This represents a long run approach, in which the macro economy adjusts to changes in consumer confidence. As literature suggests, a decrease in consumer confidence will cause aggregate demand to decrease as consumption decreases. Investment spending will also see a decrease, as will an increase in savings. Government funds will be tied up repairing water facilities, redirecting water supply that is necessary to sustain life, and providing emergency assistance to areas affected. The 2012 Homeland Security report is a proxy, giving insight into the threats that face critical infrastructure across the world. Security in the U.S. arguably far exceeds that of other nations, and it is important to understand the direct correlative effect between increased security operations, intelligence gathering, and deterrence to promote confidence. By understanding what possibly threats face US dams, owners and operators are better prepared. Even in cases where terrorism was not a factor of water disruption, such as lead contamination in Flint, Michigan, the society and economies of such places have suffered greatly. In Flint, confidence in political leaders, infrastructure, and safety caused migration and drastic changes in consumption habits.

Previous studies have established consumer confidence beyond an economic indicator such that consumption is a function it (Dees & Brinca, 2011). We establish a new indicator in the short run as water confidence, or the confidence that consumers have directly related to the water industry. Water confidence will directly impact the consumption of water in the short run, thus affecting other industries and local economies. While water confidence is impacted by other infrastructure components, including treatment facilities, lift stations, and linear distribution assets, dams are the biggest and most significant component of water supply in the eye of the consumer. Applying the consumption model to water resources, ceteris paribus, the basic equation is given as:

$$\Delta C_{water} = \alpha + \sum_{i=1}^{n} \beta_1 \Delta waterconf_{t-i} + \epsilon_i$$

This model suggests that consumption is not only correlated with a change in consumer confidence, but is actually caused by this measure of perception. Alpha represents other controlled (in this equation constant) variables that influence confidence. In evaluating the change in confidence of water based on this model, we establish the following equation:

$$\boldsymbol{waterconf_t = percieved\ threat_t + \beta_1(security\ \ benefits - security\ costs) + government\ spending}$$

The perceived threat is assumed to be constant, leaving change in consumption to be derived from the amount of present value such that is a measure of perceived threat mitigation in the net amount of benefits derived from current security spending. Government response is also an incredibly important role in the wake of an attack. In discussing the impact of the threat alone, or the impact of current spending to confidence, government response will be equal to zero, and have no present effect. Only if an attack actually occurs will the government response help dictate how consumers view the water industry, as it ties into the psychological response of consumers. We are speaking in the abstract here, as data is unavailable for water and dam confidence, yet we must understand how these decisions are made as a function of relevant factors. As a supplier of water, power, flood prevention and more, dams represent the single biggest contributor to confidence in the U.S. water supply.

The confidence estimator incorporates all types of terrorism, but we stipulate that the resulting change would be different if an attack was physical or biological/chemical. This is because the failure effects are potentially far more catastrophic when the attack is visible and heavily damaging, and go far beyond disruption. Using the September 11,

2001 attacks as a base for this analysis gives us insight into how consumers will respond to a physical attack. 9/11 established that physical attacks on US soil are incredibly possible, drastically changing how consumers view infrastructure. Consumer response to a biological/chemical attack would be far more localized to drinking water in the short run, and agriculture in the long run. The impact to industry would be less severe, but the economy would still assume considerable costs through the health care industry and lost wages. Dams represent the most visible method of water supply and delivery, and as such have the biggest impact on confidence, and the biggest impact on economic stability in the water market. This research suggests that it is critical to protect dams to promote both economic growth, and prevent severe economic recession if a terror attack were to occur.

Future Spending on Security Measures

Because confidence is tied to so many facets of the economy, it must be the crux of the argument for investing in security at dams in the U.S. Providing stable, continuous supply of water is a top priority of ensuring a stable economy. Dam security is a designated sector of Homeland Security's infrastructure protection mandate. It is clear that as long as the threat exists there will be analysis and effort to secure dams from the types of attacks they are susceptible to. Compared to the relative consequences explored in this paper, we argue that the benefits of spending security far outweigh the costs. Based on the analysis of this paper, there are multiple implications for future water security measures. If we accept that confidence is key to consumption, then we must allow for continued future spending in the water industry to mitigate risk. This future spending should be proportional to the amount of perceived risk, such that future confidence is offset and consumption stays steady. If consumption is stable, then the aggregate economy will continue to be stable. This is the key necessity of water security, as the national economy relies on consumption and government spending for growth.

Based on the scenarios presented, future security measures may also evaluate protecting other industries from water disruption. This includes providing systems at farms that are meant to detect water based biological/chemical contamination, and the development of response plans at industries that rely on water distributions. Eliminating single points of failure related to water is critical to ensuring operations continue whenever possible. Having these types of mechanisms in place will allow these sectors to recover fast after an attack, and drastically check the decrease in confidence.

LIMITATIONS AND FUTURE ANALYSIS

One of the greatest limitations of evaluating the water industry is analyzing the change in prices relative to the change in spending. Data on water prices is not readily available, nor is it recorded in the national consumer price index. This is due in part to the vast array from prices, differing by municipality. In what may be a failure of economics, is our inability to accurately predict change in water prices due to perception of the water industry.

Given that this data is not readily available the models presented in this paper require the use of comparative statics to understand water consumption and confidence relationships. These models present the relationships between costs, benefits, consumption and confidence to explain the decisions behind government spending. While there are other factors that have a great influence on the water industry, this model holds all else constant in order to isolate the effect of terrorism. Understanding that other factors influence pricing, funding decisions, and confidence are still necessary when fully examining such an important infrastructure. Further examination of critical infrastructure is still necessary at the federal and state levels to understand how these decisions affect the nature of consumer confidence. As a critical part of individual and household decisions, the water industry cannot be secured in a vacuum.

CONCLUSION

This paper has highlighted much of the existing research around dam and water security and has shown the economic modeling necessary for a comprehensive analysis of water security. Previous research focused on providing national funds to facilities based on the level of perceived risk, but as our models show, there are other necessary factors that require attention to effectively determine the necessary expenditure.

Securing water infrastructure is an ongoing collaboration between many different federal agencies, coordinated by the Department of Homeland Security. Given that only 5% of dams are federally owned facilities, other facilities must make certain private decisions outside of what is mandated across the industry. While government programs address much of the risk to water, many facilities still go unprotected. The risk at these facilities may be lower, yet they are still vulnerable to terror attacks, which in some ways operates outside of traditional risk models simply due to the erratic nature of terrorism. While it is not the intent of this paper to showcase any means of disrupting water distribution or breaching physical security, it is important to note that the economic implications of a dam failure or water based attack are severe. Arguably these facilities have been secured in the last 10 years such that potential threats are deterred; yet there is still enough real possibility for a terrorist attack that we must explore the broader economic impact and on consumers.

The models presented here suggest that security costs are necessary to ensure the continued benefits that protection is offering. As discussed, confidence is the most volatile measure of how the market responds to a terrorist attack, given that consumer perception is one of the most powerful forces in the aggregate economy. Dam security, and for that matter overall infrastructure security, is becoming an important economic issue in national market. In a time when investment decisions are scrutinized like no other, understanding the impact that water resources have on consumer confidence presents a strong argument for ensuring that these assets remain functional, safe, and secure, is an unprecedented driver of economic stability.

ACKNOWLEDGEMENTS

The author would like to thank and recognize the following individuals for their support.

Del Shannon, PE, ASI Constructors
Jaason Englesmith, Vice President, CH2M Hill Engineers
Kate Stirling, PhD, Professor of Economics, University of Puget Sound
Beth Shannon, Mary and Bruce Neumann, Al and Marylin Shannon, William L. and Ellen Shannon

REFERENCES

Adrangi, B., & Macri, J. Consumer confidence and aggregate consumption expenditures in the United States.

Beering, P. S. (2002). Threats on tap: Understanding the terrorist threat to water. *Journal of Water Resources Planning & Management, 128*(3), 163.

Copeland, C., & Cody, B. (2010). Terrorism and security issues facing the water infrastructure sector.

Dees, S., & Brinca, P. S. (2011). Consumer confidence as a predictor of consumption spending: Evidence for the United States and the euro area. *European Central Bank,* (1394)

Folga, S., Allison, T., Peerenboom, J., Carr, J., Matheu, E., Seda-Sanabria, & Y. (2010). Incorporating critical infrastructure interdependencies into dam failure consequence analysis. *US Society on Dams,* 505-517.

Gleick, P. H. (2006). Water and terrorism. *Water Policy, 8*(6), 481-503.

Grigg, N. (2003). Water utility security: multiple hazards and multiple barriers. *Journal of Infrastructure Systems, 9*, 81-88.

Haimes, Y. Y. (2002). *Strategic responses to risks of terrorism to water resources* American Society of Civil Engineers.

Lehman, W., & Needham, J. (2012). Consequence estimation dam failures. *US Society on Dams,*

McCoy, K. (2012,). Water costs gush higher. *USA Today*

Meinhardt, P. L. (2005). WATER AND BIOTERRORISM: Preparing for the potential threat to U.S. water supplies and public health. *Annual Review of Public Health, 26*(1), 213-239. doi:10.1146/annurev.publhealth.24.100901.140910

Munger, D. (2009). *Methodology for estimating economic consequences for dam failure scenarios.* (2009). Denver: Bureau of Reclamation.

Prante, T., & Bohara, A. K. (2008). What determines homeland security spending? An econometric analysis of the homeland security grant program. *Policy Studies Journal, 36*(2), 243-256.

Qiao, J., Jeong, D., Lawley, M., Richard, J. P., Abraham, D. M., & Yih, Y. (2007). Allocating security resources to a water supply network. *IIE Transactions, 39*(1), 95-109.

Shermer, S. D. (2006). The drinking water security and safety amendments of 2002: Is America's drinking water infrastructure safer four years later? *UCLA Journal of Environmental Law & Policy, 24*(2), 355-457.

U.S. Department of Homeland Security. "Estimating Economic Consequences for Dam Failure Scenarios". 2011. https://www.damsafety.org/media/documents/owner%20documents/Dam%20Security/Dams%20Sector%20-%20Consequence%20Estimation%20-%20Economic%20Consequences.pdf

U.S. Department of Homeland Security. "Worldwide Attacks against Dams: A Historical Threat Assessment for Owners and Operators". 2012. http://www.cowarn.org/uploads/news/Worldwide%20Attacks%20against%20Dams%20-%202012.pdf

U.S. Department of Homeland Security. "Dams Sector-Specific Plan: An Annex to NIPP 2013". 2015. https://www.dhs.gov/sites/default/files/publications/nipp-ssp-dams-2015-508.pdf

Van der Zaag, P., & Savenije, H. (2006). Water as an economic good: The value of pricing and the failure of markets. Unesco-IHE.

FLOOD RISKS FROM SPILLWAYS: MODELING AND COMMUNICATING NON-DAM-FAILURE IMPACTS TO NATIVE AMERICAN TRIBES

Rinda Tisdale, M.E.P., C.E.M.[1]
Matthew Young, P.E., C.F.M.[2]
Chris Clopper, P.E.[3]

ABSTRACT

The Bureau of Indian Affairs (BIA) Branch of Dam Safety, Security, and Emergency Management Safety of Dams (SOD) Program endeavors to reduce the potential loss of human life and property damage caused by dam failure by making BIA dams as safe as practicably possible. The SOD Program currently administers over 900 dams on Indian reservations throughout several states. Of those dams, 137 are classified as high- or significant-hazard potential dams based on the prospective downstream consequences associated with uncontrolled releases from each dam (refer to Figure 1). These dams comprise a significant part of the water resources infrastructure and trust assets for numerous Indian reservations and tribes. Benefits from the dams include irrigation water storage, recreation, flood control, conservation, and power generation (refer to Figure 2).

BIA EAP PROGRAM BACKGROUND

The Indian Dam Safety Act of 1994 (Public Law 103-302) stipulates that the BIA has a responsibility to protect and alert communities with respect to dam safety incidents on recognized Indian reservations. The BIA Dam Safety Program strives to accomplish this initiative through its Emergency Management Program. Emergency management helps reduce the likelihood of life loss and damage to property from a dam failure. The goal of the Emergency Management Program is to prepare key personnel and emergency responders for dam safety emergencies, increasing the likelihood that lives can be saved and damage to property minimized in the event of a dam failure or incident. The Emergency Management Program conducts tabletop and functional exercises and updates Emergency Action Plans (EAPs) for each SOD Program dam on a regular basis. The Emergency Management Program focuses on preparedness and response to incidents at BIA dams through the preparation and exercising of EAPs. These plans outline procedures for notifying the downstream jurisdictions in the event of a potential failure of the dam, such that warning and evacuation can be effectively performed to minimize loss of life and property. Emergency Action Plans are required for all high- and significant-hazard potential dams in the SOD Program.

[1] Emergency Management Coordinator, Bureau of Indian Affairs, Branch of Dam Safety, Security, and Emergency Management, Lakewood, CO 80401, 303-231-5235, rinda.tisdale@bia.gov.
[2] Dam Safety Engineer, Bureau of Indian Affairs, Branch of Dam Safety, Security, and Emergency Management, Lakewood, CO 80401, 303-231-5115, matthew.young@bia.gov.
[3] Hydraulic Engineer, Managed Business Solutions, LLC, Lakewood, CO 80401, 303-231-5221, christopher.clopper@bia.gov.

Figure 1. BIA Safety of Dams Program National Map.

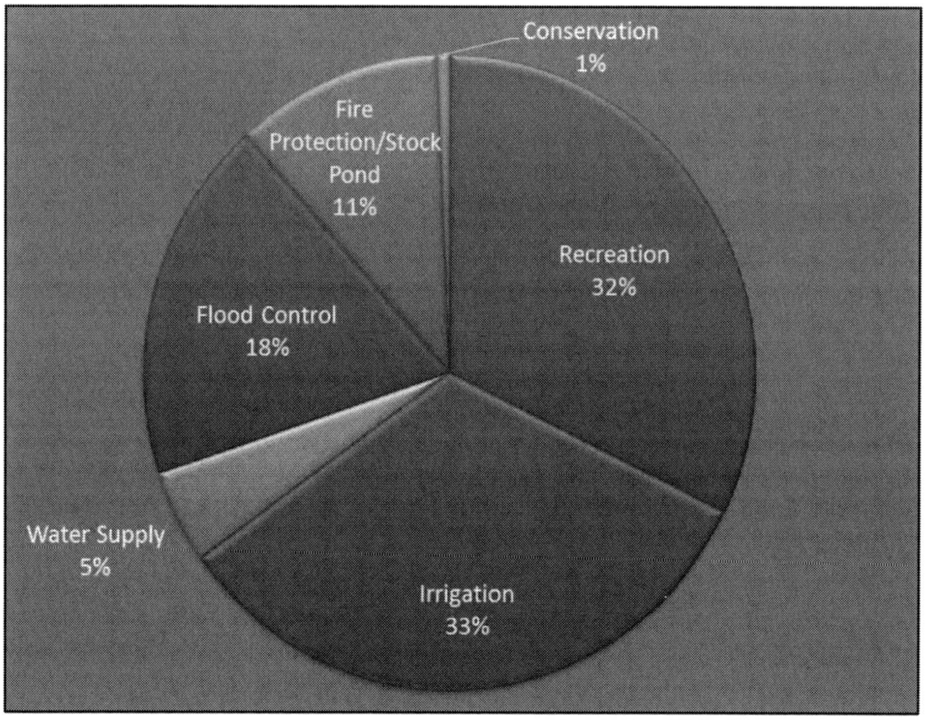

Figure 2. Primary Uses of BIA Dams.

CURRENT BIA EAP PROGRAM

In accordance with the inclusions of the Indian Dam Safety Act, BIA responsibilities also include floodplain management matters downstream of dams. Flooding is one of the most common and destructive natural hazards confronting tribal communities. Despite this, natural flood risks throughout most tribal lands are poorly understood. In 2013 the Government Accountability Office published a report to Congressional Committees titled, "Flood Insurance: Participation of Indian Tribes in Federal and Private Programs." This report stated that "as of August 2012, just 37 of 566 federally recognized tribes (7 percent) were participating in the National Flood Insurance Program." It was noted that the Federal Emergency Management Agency (FEMA) "has not placed a high priority on mapping rural areas, including many Indian lands, for flood risk, and most tribal lands remain unmapped." The report also stated that "without flood-hazard maps, tribal communities may be unaware of their flood risk, even in high-risk areas."

Rainfall Maps

Conventionally, dam failure inundation maps contained within SOD Program EAPs have depicted worst-case probable maximum flood (PMF) boundaries associated with the uncontrolled release of maximum water storage following a hydrologic dam failure. While this information is valuable to first responders in preparing for statistically worst-case dam failure flood events, these scenarios are not representative of the range of flood risks that can affect communities located downstream of dams. To better communicate these more common flood risks, the SOD Program has developed a number of supplementary planning and preparedness tools that include rainfall maps, Rate of Rise Graphs, and Spillway Discharge Curves (refer to Figure 3). Rainfall maps depict watershed-specific precipitation-frequency data in a geospatial format over multiple storm durations. These maps are included in SOD Program EAPs with the intent of giving responders an accessible means of relating rainfall intensity to mapped flooding events in near real-time.

Rate of Rise Graphs

Rate of Rise Graphs are used by dam personnel to predict the approximate timing of dam overtopping during a worst-case hydrologic scenario with steady rainfall, and accordingly, provide additional response time for less severe, more frequent flood events. The primary purpose of the Rate of Rise Graph is to provide a visual representation of the EAP response levels associated with hydrologic events while also incorporating an element of advance warning time into emergency response at the dam and downstream communities. EAP response levels specified on the Rate of Rise Graphs are defined accordingly: Response Level 1 is a slowly developing, unusual situation at the dam where the potential for adverse impacts is not yet serious, but could progress into a potentially threatening event if it continues or intensifies; Response Level 2 is a rapidly developing situation, more serious that Response Level 1, where the dam may be becoming unstable and populations at risk (PAR) should be notified to standby and prepare to evacuate potential inundation areas; Response Level 3 is identified as either dam failure is imminent, is occurring, or has occurred and immediate evacuation of the PAR is recommended.

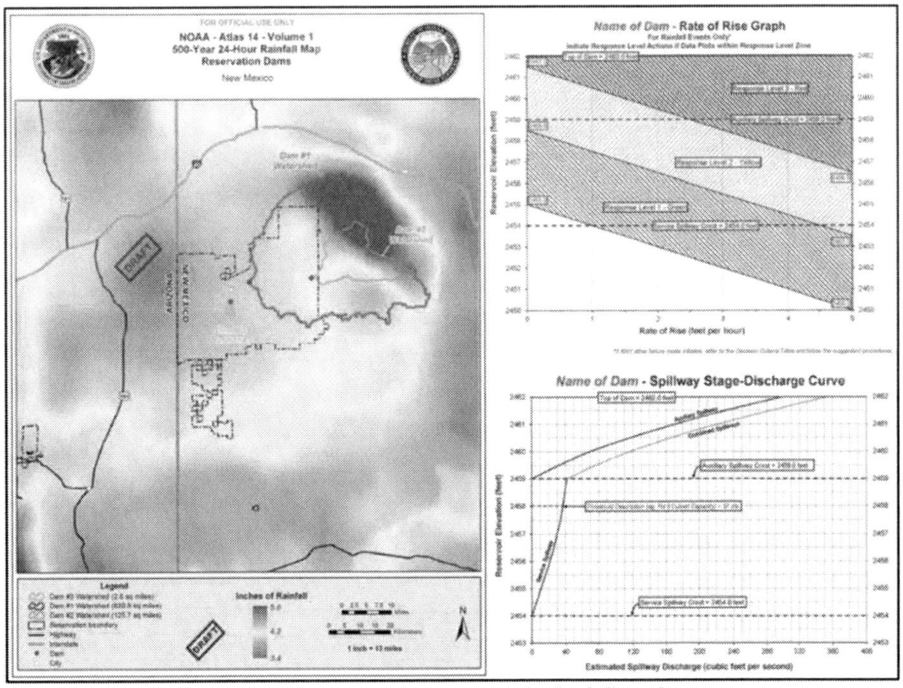

Figure 3. SOD Program EAP Technical Supplements.

Spillway Discharge Curves

Spillway Discharge Curves are included in the EAP in an effort to provide relevant information associated with major storm events that may result in significant releases from the dam. The primary purpose of the Spillway Discharge Curve is to provide an estimate of the discharge released from the dam spillway(s) in correlation with increasing reservoir water surface elevations. More importantly, by working with tribal members familiar with the facilities, specific reservoir elevations are indicated on the Spillway Discharge Curve to identify elevations that correspond to known downstream threshold discharges. Such thresholds may include, but are not limited to, downstream culvert and road crossing capacities, estimated safe channel capacities, in addition to dam infrastructure elements, such as bridge decks and dike elevations, which act to provide visual indicators that can improve understanding of the flood risks and impacts due to natural spillway releases. These curves are included in SOD Program EAPs to better communicate the magnitude of floodwater releases to downstream communities.

Advisory Flood Maps (Non-Dam-Failure)

Additionally, tribes have repeatedly requested improved inundation maps that are more useful for their emergency and flood preparedness needs. The SOD Program has attempted to further supplement tribal understanding of flood risks through the development of Non-Dam-Failure Advisory Flood (Advisory Flood) maps (refer to Figure 4). These maps have been created with tribal community input to depict flooding scenarios downstream of BIA dams that can be used by community leaders for land-use planning and increased public awareness of natural flood risks. To date, the SOD Program efforts to support tribal floodplain management have been confined to waterways directly downstream of BIA dams.

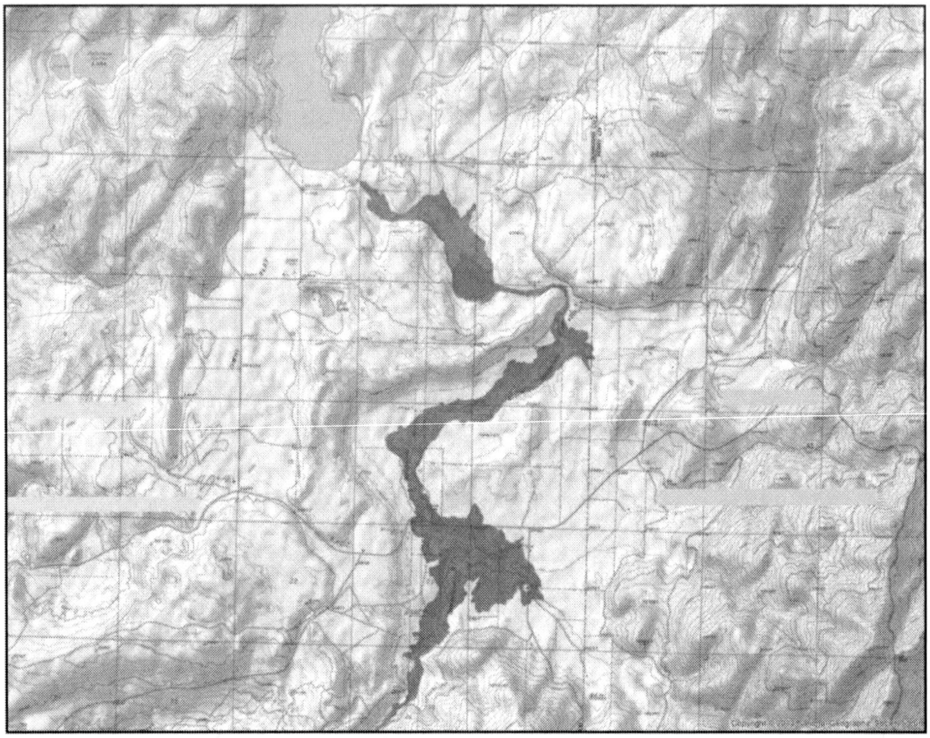

Figure 4. Example of PMF scenario (in red) and Non-Dam-Failure Advisory Flood scenario (in blue).

As part of BIA's EAP development and exercise process, the BIA SOD Program has developed an Advisory Flood map standard that is included as an appendix to the EAP, and presented in detail during Tabletop Exercises. This Advisory Flood scenario depicts flood risks associated with select hydrologic loadings, whereby the dam operates as designed to allow floods up to the Inflow Design Flood (IDF) to bypass the facility through the spillway (refer to Figure 5). The hydrologic scenario selected as the Advisory Flood will typically vary within a range of flood magnitudes from the estimated safe channel capacity as a lower limit to a full spillway capacity flood as an upper limit. In most cases, these floods will be correlated to a statistical return period to further enhance usability to the tribal community.

SUMMARY

The inclusion of rainfall maps, Rate of Rise Graphs, Spillway Discharge Curves and Advisory Flood maps as flood risk communication tools has proven to be both useful and highly desirable among tribal communities. While not specifically dam-failure preparedness tools, the incorporation of these products within SOD Program EAPs has shown great potential to enhance community understanding of risks associated with natural flooding downstream of dams, while acting to greatly improve coordination and collaboration between dam owners and affected downstream communities.

Figure 5. Typical Advisory Flood Scenario.

REFERENCES

US Government Accountability Office. "Flood Insurance: Participation of Indian Tribes in Federal and Private Programs." GAO-13-226. Jan 4 2013. Available: http://www.gao.gov/products/GAO-13-226

CITY OF DUBUQUE, IOWA: DEFINING EMERGENCY ACTIONS WITH A LEVEE BREACH EVALUATION

Andrew McCoy PhD, PE[1]
Deron Muehring PE[2]
Rusty Jones, PE[3]

ABSTRACT

The City of Dubuque, Iowa, has evaluated the effects of the resultant flooding from a hypothetical Mississippi River levee breach along the City's levee system, the consequences of its failure, and has identified opportunities to mitigate the impacts from a levee breach. The United States Army Corps of Engineers (USACE) had previously performed 2D hydraulic modeling (HEC RAS 5.0) to simulate the resultant flooding from five levee breach locations along the City's levee system and was an excellent starting point for additional more detailed analysis of evacuation routes, potential flood hardening, and overall resiliency planning and design. The additional modeling further defined the depth of inundation, time to inundation, and inundation paths at critical infrastructure and transportation corridors as well as how sensitive the hydraulic characteristics of the resultant flooding are to different breach locations (total of seven), characteristics (breach width and time to development), and river conditions (4 different events). As a result of this study, the City further developed their emergency action plan, identify locations for flood fighting, and determine the best strategies within the City of Dubuque for protection from a potential levee breach event. This has helped the City further understand their risk and helped drive the identification of risk mitigation measures.

INTRODUCTION

The City of Dubuque (City) is located in Northeastern Iowa, along the Mississippi River. It had a population of 58,800 in 2015. It is situated between the river and by steep bluffs. The City is protected by a levee and floodwall system and is approximately 30,000 ft long. Most of the levee embankment has a sand core and a clay covering, but nearly 7,000 ft is comprised of I-wall, T-wall, and gates where earthen embankments are not possible. Construction began in 1968 and was completed in 1973. The levee protects significant residential, industrial, and recreational property, much of which has had recent investment (Figure 1).

The Dubuque levee system was designed for a minimum of 3-ft of freeboard at the water surface associated with a river discharge of 362,000 cfs (USACE 1966) located at Mississippi River Mile 579.9. The discharges associated with the 1-percent and 0.2 percent annual chance exceedance events are 274,000 cfs and 326,000, respectively. This means that a discharge and water surface that would result in overtopping the levee

[1]HDR, 300 East Locust, Ste. 210, Des Moines, Iowa, 50312 (515) 280-4950 andrew.mccoy@hdrinc.com.
[2]City of Dubuque, 50 W 13th St, Dubuque, IA 52001 (563) 589-4276 dmuehrin@cityofdubuque.org.
[3]HDR, 300 East Locust, Ste. 210, Des Moines, Iowa, 50312 (515) 280-4961 russell.jones@hdrinc.com.

system is associated with a probability much lower than the 0.2 percent annual chance, or the so-called 500-year flood.

Even though the probability of levee overtopping is lower than the 0.2 percent annual chance flood event and there is no known imminent threat of levee breach in the City of Dubuque, the consequences of a levee failure would be catastrophic to the residents and businesses present in the community. As the levee system ages and is flood-tested, it continue to be inspected, repaired, and strengthened. In addition to overtopping, levee failure could occur due to piping failure (undermining) during a flood event smaller than the design event, gate failure, or other man made impacts such as a barge accident. Risk of damage due to flooding can be reduced, but is always present for communities located behind levees.

In response to this risk, the City of Dubuque has begun evaluating the effects of a Mississippi River levee breach along the City's levee system, the consequences of its failure, and identifying opportunities to mitigate the impacts from a breach.

In August 2014, the United States Army Corps of Engineers (USACE) received funding to prepare a levee breach analysis for the city of Dubuque to assist in planning for flooding of the Mississippi River into the City of Dubuque if the levee was breached in some way. Collaboration between city of Dubuque stakeholders and USACE began to address managing residual risk associated with the leveed area. USACE performed hydraulic modeling to simulate the resultant flooding from five levee breach locations along the City's levee system.

The City of Dubuque contracted with HDR in 2015 to take the USACE analysis a step further. The USACE levee breach analysis was an excellent starting point for an additional more detailed analysis of evacuation routes, potential flood hardening, flood fighting, and overall resiliency planning and design.

The additional modeling provided an evaluation of how sensitive the hydraulic characteristics of the resultant flooding are to different breach locations, characteristics, and river conditions.

As a result of these studies, the City further developed their emergency action plan, identified locations for flood fighting, and determined the best strategies within the City of Dubuque for protection from a potential breach event. This has helped the City further understand their risk and helped drive the identification of risk mitigation measures.

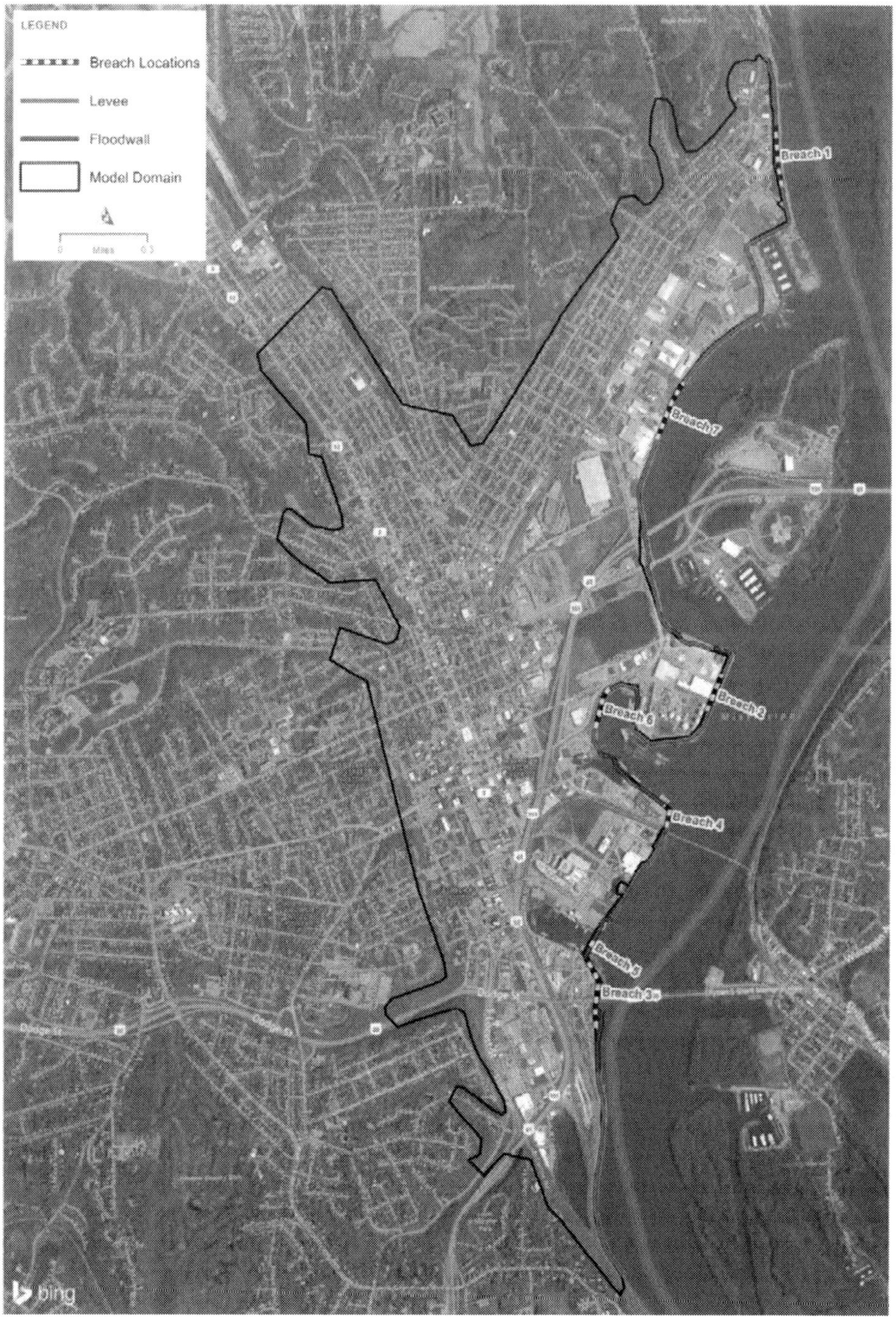

Figure 1. Levee System in the City of Dubuque.

HYDRAULIC MODEL DEVELOPMENT

The study methodology included further development of a hydraulic modeling tool that could simulate resultant flooding due to hypothetical levee breaches, explore the sensitivity of results to modeling parameters and Mississippi River conditions, and possess fine enough detail that the results could support detailed emergency planning; including evacuation route planning and establishing flood fighting locations.

Terrain Development

For the City to further develop an emergency action plan, identify locations for flood fighting, and determine the best strategies within the City of Dubuque for protection from a potential levee breach, a 2D hydraulic model was developed and tested. Figure 2 shows the model domain extents as well as elevations within the City.

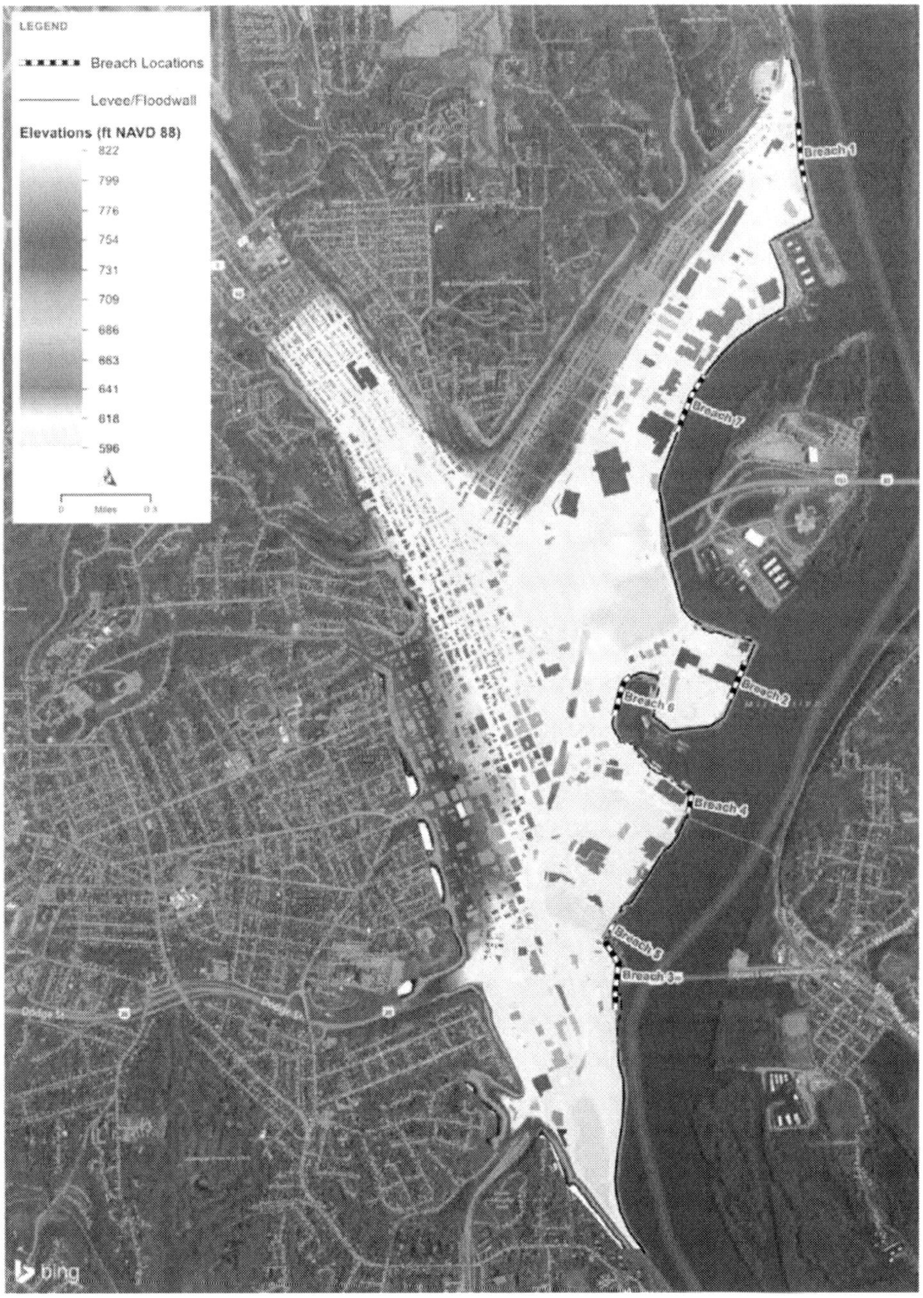

Figure 2. Model Domain and Elevations in the City of Dubuque.

The starting model files were provided by USACE along with a technical memorandum (USACE 2015). The USACE model was built using a beta version of HEC-RAS version 5.0 (October 2014). The current model results use HEC RAS version 5.0.3. The USACE model was based on the state of Iowa LiDAR database. The state of Iowa LiDAR database is available for public use and is generally of good quality.

However, the City has more recently obtained photogrammetry, which better represents the terrain in the City than the State of Iowa LiDAR (especially features such as road embankments that function as hydraulic controls in flooding). In addition, building outlines and heights provided by the City were incorporated into the terrain. Figure 3 illustrates the differences between the terrain data developed with State of Iowa LiDAR versus City photogrammetry.

Buildings were explicitly added to the terrain as well. By decreasing the volume available for floodwaters, representing buildings explicitly shortens travel times. This leads to more conservative travel times (faster) during a breach event and was deemed appropriate for evacuation route planning.

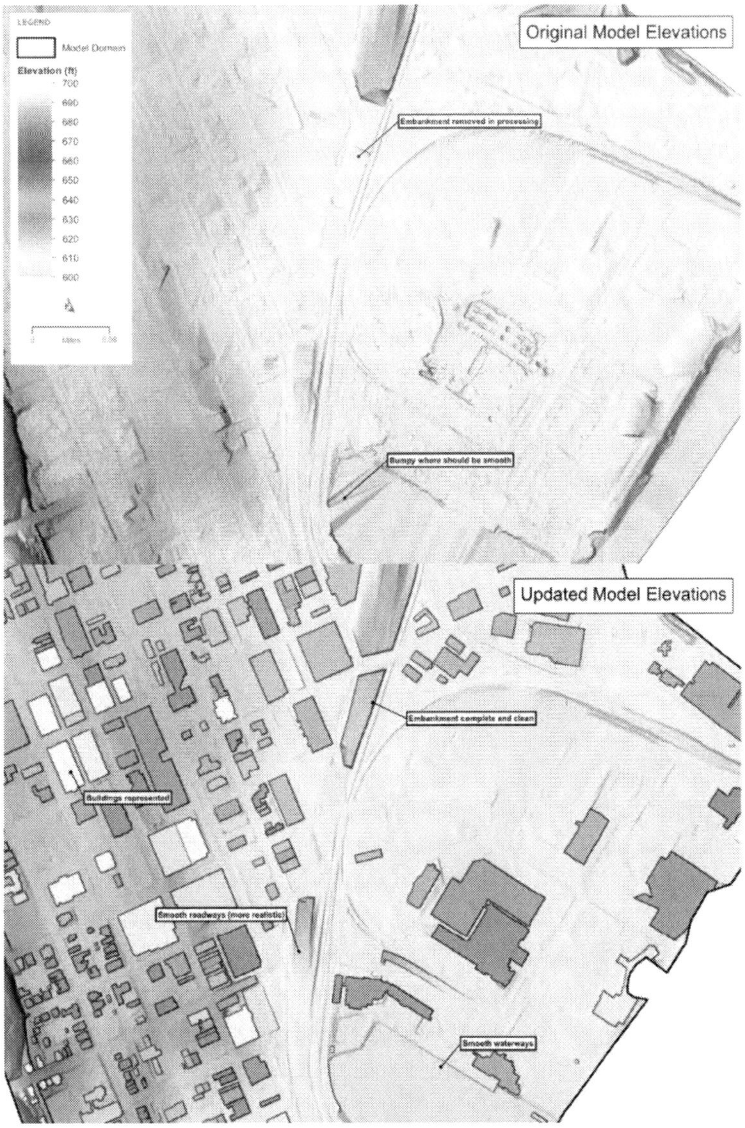

Figure 3. Comparison Between State of Iowa LiDAR Derived Terrain (Original) and City of Dubuque Photogrammetry(Current).

Hydraulic Controls

Breaklines were created to enforce the hydraulic controls (road and railroad embankments or other linear features) in the model. If hydraulic model cell edges don't follow embankments or other hydraulic controls, flow can move freely from one side of the embankment to the other negating the impact of these controls. The breakline feature allowed accurate representation of hydraulic controls. The original model effort was performed with a beta version of HEC-RAS 5.0, and the incorporation of breaklines into the model was not possible. However, in the current version of HEC-RAS 5.0.3, which the current modeling is based upon, breaklines were readily added using the breakline editor. Figure 4 shows a comparison between the two model configurations.

Figure 4. Comparison Showing Use of Breaklines to Define Hydraulic Controls.

Spatially Varied Roughness and Manning's Roughness Coefficient

The Manning's roughness coefficient was varied within the domain based upon land use types within the City. The option for more than one Manning's roughness coefficient was not available in previous versions of HEC-RAS. That was a limitation with respect to capturing travel times accurately. Land use data was provided by the City and the Manning's roughness coefficients were assigned (Figure 5). The coefficients used are shown in Table 1.

Table 1. Land use types and associated Manning Roughness Coefficients.

Land Use Type	Manning Roughness Coefficient
Commercial	0.060
Heavy Industrial	0.080
Light Industrial	0.060
Open Space	0.040
Residential	0.060
Default (pavement)	0.025

Figure 5. Land Use Types Correlated to Manning's Roughness Coefficients

Computational Parameters

The computational parameter choices made during model development were not modified during the simulations. The simulations were run with the full momentum solver rather than the diffusive wave solver. The eddy viscosity transverse mixing coefficient was set to 0.7 to help capture horizontal velocity gradient present in the urban setting. The lateral structure flow stability factor and the weir flow submergence decay exponent were both set to 3.0 to help with model stability at breach locations as the breaches become submerged.

Boundary Conditions – Mississippi River

Flood events on the Mississippi River are a precursor for resultant breach flooding. Several Mississippi River conditions were incorporated into the hydraulic modeling and they are provided in Table 2. Considering a range of Mississippi River conditions provides a larger database of resultant flooding conditions inside the City. Previous modeling considered a breach near the design event only, which has a recurrence interval of greater than the 500 year event. This study broadened the conditions leading to levee breach and resultant flooding. An extremely rare event like the design event puts more pressure on the levee system, but other smaller Mississippi River flood events have two to five times greater chance of occurring. In total, four Mississippi River flood events were used as boundary conditions for this levee breach study. The river stage, description, maximum discharge (rounded to the nearest thousand), and recurrence interval for each boundary condition analyzed can be found in Table 2. The recurrence interval is based upon the Upper Mississippi River System Flow Frequency Study (USACE 2004).

Table 2. Boundary Conditions – Mississippi River.

Max River Stage (ft)	Description	Max Discharge (cfs)	Recurrence Interval
28.3	Design	361,000	> 500 year
25.5	200-Year	296,000	200-Year
24.3	2001	267,000	77-Year
22.1	2011	233,000	28-Year

Boundary Conditions - Levee Breach Details

Travel times associated with a breach event are dependent on breach progression speed and final breach size. Not knowing the travel times associated with flooding will impact the ability for the City to flood fight. Given the uncertainty, multiple simulations were performed to inform the range of travel times, depths, and velocities anticipated from a breach at one of the seven breach locations. The simulations were intended to simulate reasonable breach sizes and growth rates as well as more conservative estimates (Conservative in this context refers to selecting parameters that lead to faster flood inundation and higher flood depths). The simulations were performed at seven locations along (Figure 1) the City riverfont and during four Mississippi River floods (Table 2). In all, 44 simulations were performed.

For the breach locations along earthen levee segments two parameters were varied, the maximum breach width and the breach growth rate. The maximum breach widths were estimated from a study identifying the median (likely) and the 84% confidence level (not as likely, and larger) in similar levee reaches (URS 2013). The breach growth rates were informed by research at Mississippi State University for the Department of Homeland Security where researchers concluded that lateral growth rates for similar earthen levee breaches generally vary between 30 ft/hour in erosion-resistant soils to 200 ft/hour in erodible soils (Saucier, Howard, and Tom 2009). The median breach events simulated in this study use a horizontal growth rate of 100 ft/hour. The 84% confidence level breach events use a horizontal growth rate of 200 ft/hour.

The exceptions were breach 5, which represents a failure of the Ice Harbor gates and breach 3, which is a floodwall failure. Both are assumed to fail very quickly and independent of the neighboring floodwall.

There is less guidance for estimating floodwall and gate failure times than for levees although failures are often classified as catastrophic (happening over a very short amount of time – minutes instead of hours). To represent the floodwall failure at breach 3, a set of shorter and longer breach widths were chosen to provide a range of resultant flood inundation information. The shorter floodwall breaches were failed over a 15 minute interval while longer sections were failed between 0.5 hour and 1.0 hour. Breach 5 represents failure of the Ice Harbor gates. A larger failure was not simulated since the breach width is limited by gage geometry. The gate failure is assumed to occur over a 15 minute interval.

Because the levee is not overtopped in any of the events, all of the failures simulated are piping failures. The river elevation where piping is initiated and the river elevation when the levee failure initiates are input parameters to the model and captured in Table 3. These elevations were chosen based upon the ground elevation and peak river elevation at the breach location. Some breach and event combinations were not evaluated because the peak river elevation was too low compared to the ground elevation to generate significant pressure required for a piping failure. Levee dimensions were based on accreditation reports (GEI 2013).

Table 3. Levee breach parameters and Mississippi River conditions for all simulations

Levee Breach Parameters and Mississippi River Conditions										
Breach	*River Stage (ft)*	*Event*	*Peak WSE (ft)*	*Ground Elevation (ft)*	*Piping Elevation (ft)*	*Trigger Elevation (ft)*	*Median*		*84th Percentile*	
							Width (ft)	*Time (hours)*	*Width (ft)*	*Time (hours)*
1	28.3	Design	615.3	606	612	615.2	460	4.0	1000	2.0
1	25.5	200-Year	612.4	606	610.5	612.0	390	3.0	880	2.0
1	24.3	2001	610.6	606	608	609.0	390	3.0	880	2.0
1	22.1	2011	608.9	606	607.5	608.0	360	3.0	800	2.0
2	28.3	Design	614.7	610	613.2	614.5	260	2.5	600	1.25
2	25.5	200-Year	611.9	610	610.5	611.6	260	2.5	600	1.25
3	28.3	Design	614.4	602	608	614.15	400	0.25	1200	0.5
3	25.5	200-Year	612.3	602	608	611.3	300	0.25	N/A	N/A
3	24.3	2001	609.8	602	604	606.0	300	0.25	1000	0.5
3	22.1	2011	608.2	602	604	606.0	300	0.25	1000	0.5
4	28.3	Design	614.5	608.5	612	614.2	360	3.5	720	1.75
4	25.5	200-Year	611.7	608.5	611	611.5	360	3.5	720	1.75
5	28.3	Design	614.4	597	608	614.15	85	0.25	N/A	N/A
5	25.5	200-Year	611.6	597	608	611.5	85	0.25	N/A	N/A
5	24.3	2001	609.9	597	604	604.5	85	0.25	N/A	N/A
5	22.1	2011	608.2	597	604	604.5	85	0.25	N/A	N/A
6	28.3	Design	614.7	608	612	614.2	260	2.5	600	1.25
6	25.5	200-Year	611.8	608	611	611.6	180	1.0	400	0.5
6	24.3	2001	610.1	608	609	609.5	180	1.0	400	0.5
7	28.3	Design	614.7	607	612	614.2	460	4.0	1000	2.0
7	25.5	200-Year	612.2	607	610	612	390	3.0	880	2.0
7	24.3	2001	610.1	607	608	609.0	390	3.0	880	2.0

Note: Breach location 3 is a floodwall failure. Breach location 5 is a gate failure. Breach locations 1,2,4,6, and 7 are earthen embankments that experience failure due to piping, not overtopping.

MODEL RESULTS

A total of 44 simulations were run with varying breach locations, boundary conditions, and breach parameters. The range of model results were used to gain insight into the range of resultant modeled travel times and ultimately inform breach results to use in emergency planning efforts.

Hydraulic Results

The hydraulic results included maps of maximum depth and time to 1-ft and 2-ft of inundation depth for each breach location. Maps were produced that focused on time of inundation related to evacuation routes and critical structures. Figure 6 is an image showing the maximum depth resulting at critical structures in the City during a breach at location 6 during the approximate 200-year event. Figure 7 shows the inundation time to 1-ft of depth resulting at critical structures in the City during a breach at location 6 during the approximate 200-year event. Figure 8 shows the inundation time to 1-ft of depth at evacuation routes, which renders them impassable, in the City during a breach at location 6 during the approximate 200-year event.

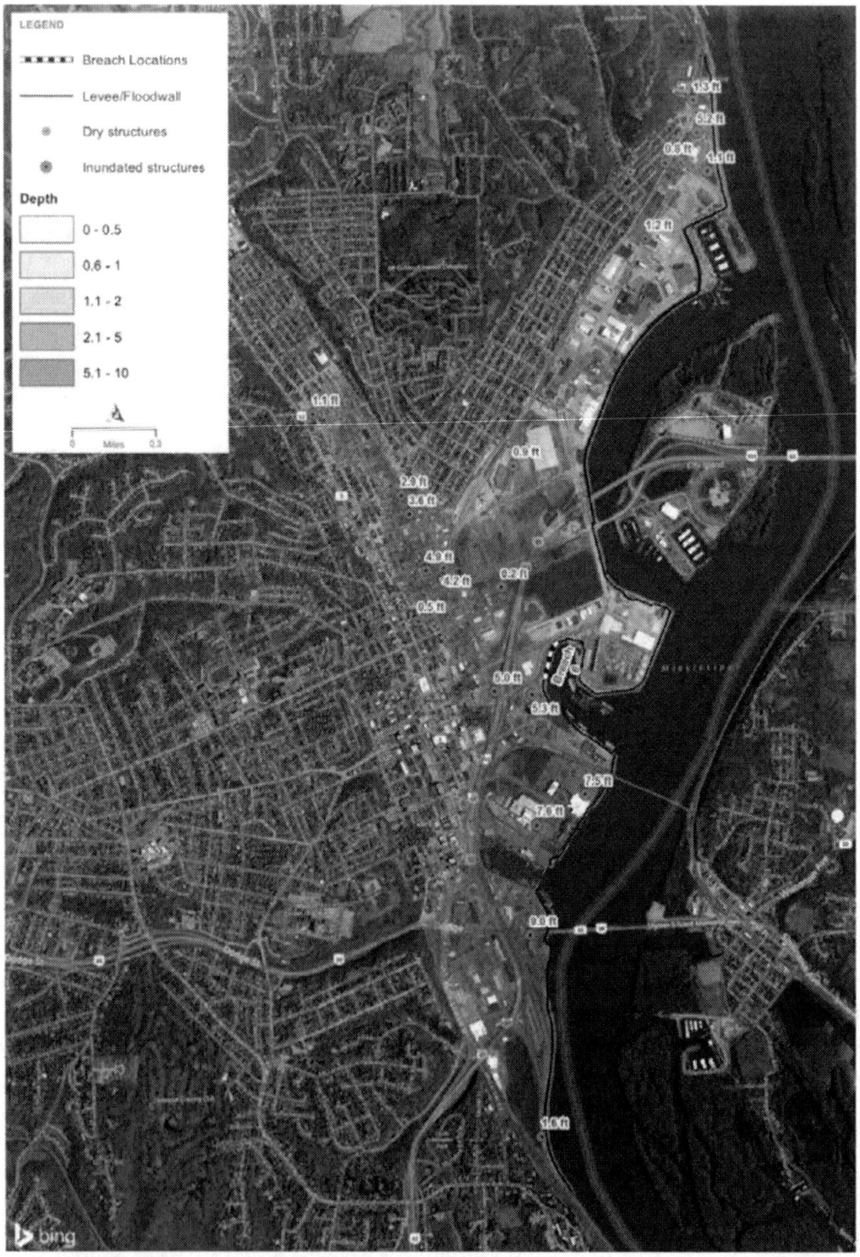

Figure 6. Maximum Depth at Critical Structures at Location 6 Breach - 200-year Mississippi River Event

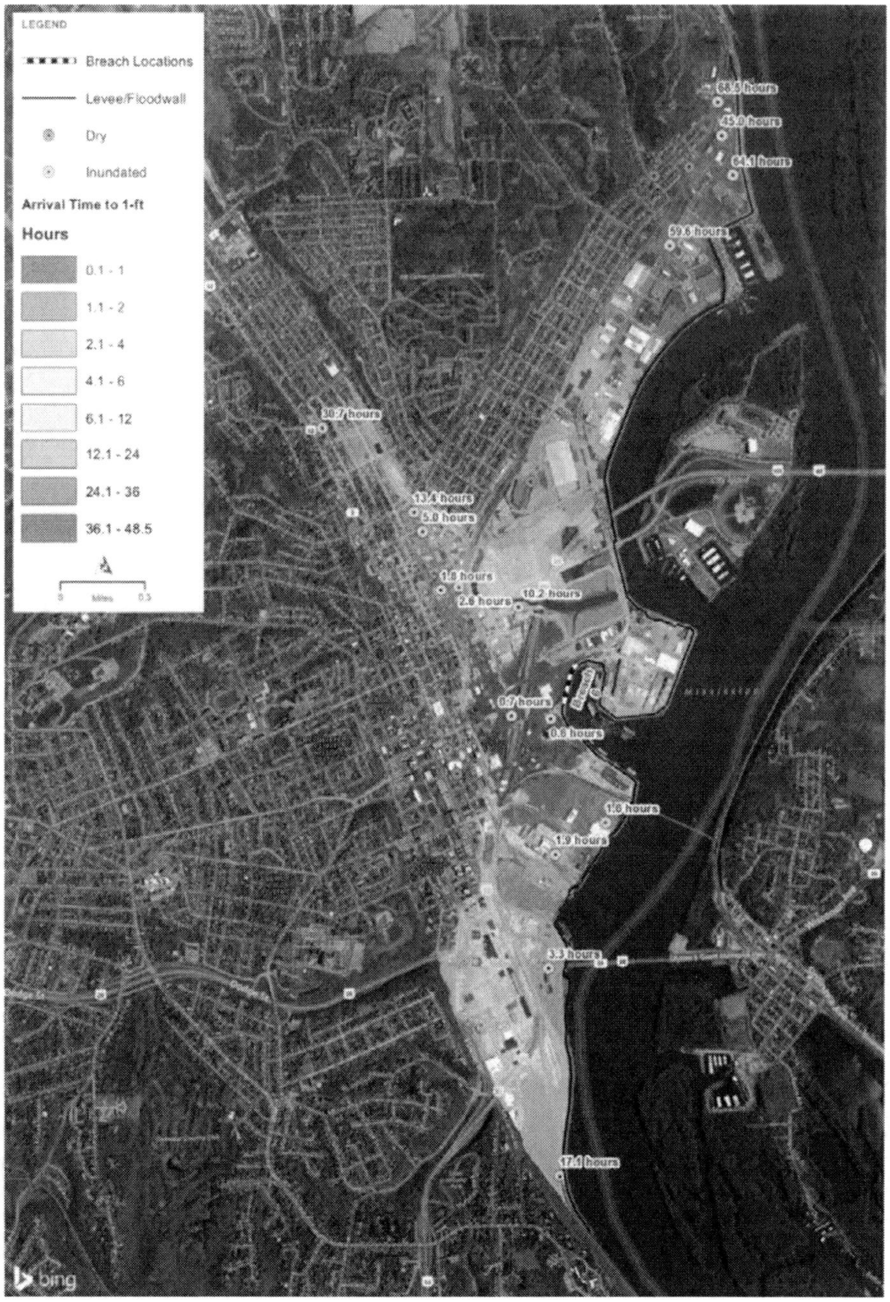

Figure 7. Time to 1-ft Depth at Critical Structures at Location 6 Breach - 200-year Mississippi River Event

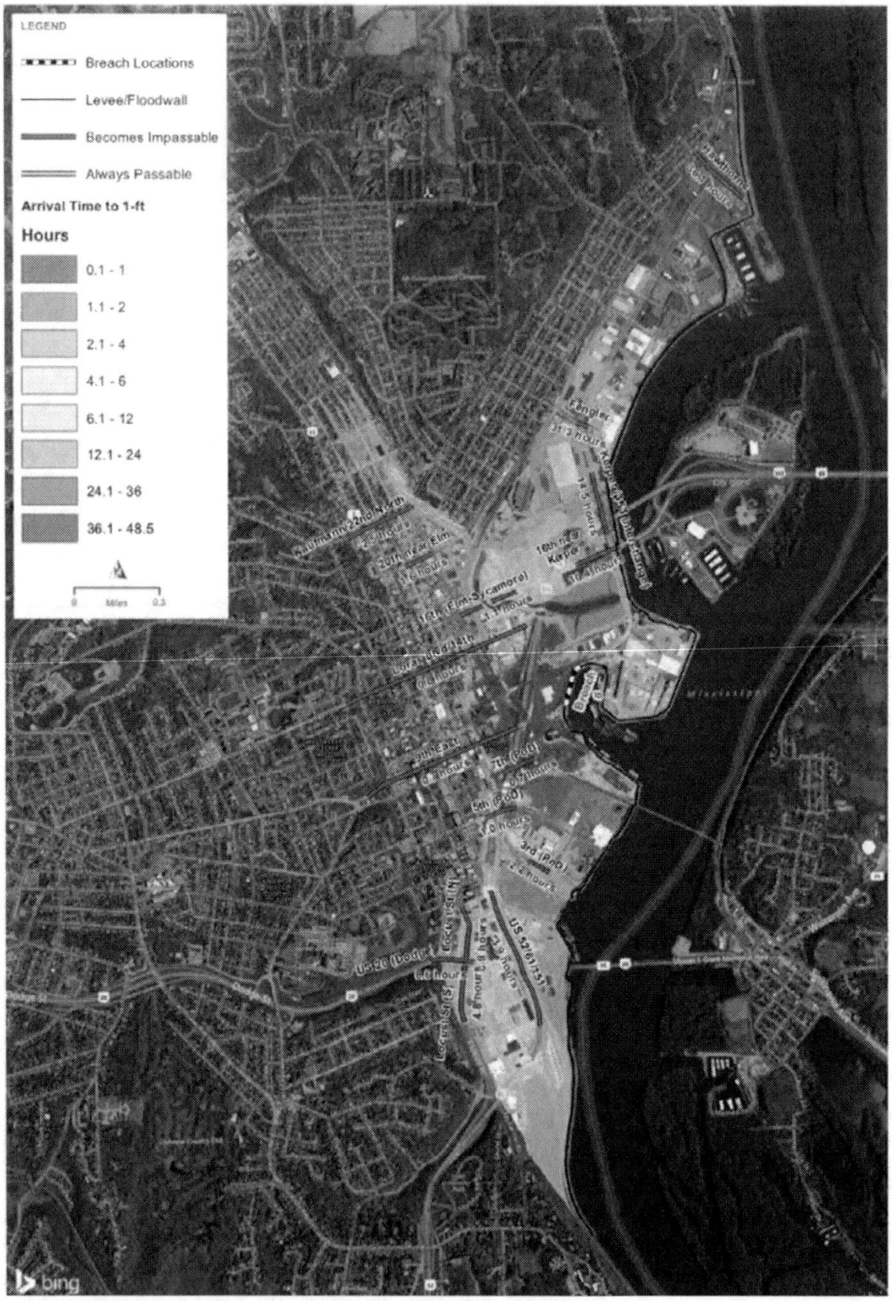

Figure 8. Time to 1-ft of Depth Along Evacuation Routes During Breach at Location 6 Breach -200-year Mississippi River Event

Sensitivity Evaulation

A rigorous evaluation of model result sensitivity to all relevant parameters is beyond the scope of this paper. Rather, the range of results at a particular location gives both insight into the importance of model parameters and what opportunities for flood fighting may exist.

The strategy behind running the simulations was to gain insight on how to best use the model results for emergency planning opportunities. Tables 4-6 are examples at key evacuation routes in the City. The evacuation route locations are shown on Figure 8. Once the corridor receives 1-ft of water, it is impassable. The breach locations are on the left (shown in Figure 1), three Mississippi River levels are reported (Design, 200-year, and 2001), Levee Breach Parameter A (Median) and Levee Breach Parameter C (84th Percentile – Larger breach, quicker breach) are also reported. Breach Locations 3 and 5 are left out since they are not earthen embankments.

The tables show that at the location of the evacuation routes it is seen that if a breach happens close by the evacuation routes are rendered impassable within 1-2 hours. The different breach parameters also impact the travel times between 10 and 50 percent. For this reason, the results associated with Levee Breach Parameter C are deemed more conservative and thus adopted for planning purposes.

The higher the Mississippi River elevation through the breach, the faster the inundation. Comparing breach locations across the three events shows a general decrease in inundation time with Mississippi River water surface elevation.

Evaluating the arrival time information for a particular area of interest across events and breach locations also allows for an assessment of feasible evacuation actions and potential flood fighting strategies. For the points of interest, there is very little response time available during a Design or 200 year event for a levee breach at location 4 or 6. This underscores the importance of using this information to prioritize evacuation procedures by updating the emergency action plan for the City. However, for breaches at locations 1, 2, and 7 for all events, the corridor stays open for a minimum of 4-19 hours during the design event. This underscores the variable nature of inundation within the City as well as helps identify opportunities for flood fighting through construction of temporary cross levees at strategic locations.

Table 4. Time to 1-ft of Water Depth Resulting from Various Breach Scenarios at US52/61/151

US 52/61/151	Design			200-Year			2001		
	A	C	Diff (%)	A	C	Diff (%)	A	C	Diff (%)
Breach Location									
1	14.1	13.2	-5.9%						
2	11.3	7.0	-38.2%		44.7				
4	3.8	2.4	-35.6%	9.2	5.3	-41.8%			
6	2.3	1.8	-21.4%	3.9	2.8	-29.8%	11.2	7.8	-30.4%
7	13.0	11.8	-9.0%	60.8	53.5	-11.9%			

Table 5. Time to 1-ft of Water Depth Resulting from Various Breach Scenarios at Kaufmann/22nd N

Kaufmann/22nd N	Design			200-Year			2001		
	A	C	Diff (%)	A	C	Diff (%)	A	C	Diff (%)
Breach Location									
1	8.3	7.7	-8.0%						
2	8.4	4.7	-44.6%						
4	5.1	3.3	-36.1%	24.9	11.4	-54.2%			
6	1.8	1.5	-18.2%	2.5	1.8	-30.0%	9.2	5.2	-42.7%
7	7.6	6.6	-13.2%	41.8	37.3	-10.8%			

Table 6. Time to 1-ft of Water Depth Resulting from Various Breach Scenarios at 16th Near Kerper

16th Near Kerper	Design			200-Year			2001		
	A	C	Diff (%)	A	C	Diff (%)	A	C	Diff (%)
Breach Location									
1	4.0	3.4	-14.6%	80.0	69.0	-13.7%			
2	18.8	8.2	-56.6%						
4	13.3	8.4	-36.9%	40.7	22.3	-45.4%			
6	3.3	2.5	-25.0%	10.4	6.6	-36.8%	33.2	24.2	-27.3%
7	4.9	2.4	-50.8%	34.1	30.0	-12.0%			

DISCUSSION OF FINDINGS

As a result of this study, the City further developed an emergency action plan, identified locations for flood fighting, and determined the best strategies within the City of Dubuque for protection from a potential levee breach event.

Evacuation Routes

One major outcome of the study has been the development of flood inundation information as it relates to transportation corridors. In the future, as the City enters into flood fighting mode, they will also be able to plan for potential levee breach scenarios based on a) the level of the Mississippi River and b) potential weak spots that become apparent during the flood fight. As it relates to evacuation routes and population centers, they have enough information to understand, as Figure 9 indicates, how long evacuation routes will be open, and which population centers should receive priority evacuation. This information has been developed for each levee breach location, for each applicable Mississippi River level, and for the 84th percentile levee breach parameters as applicable.

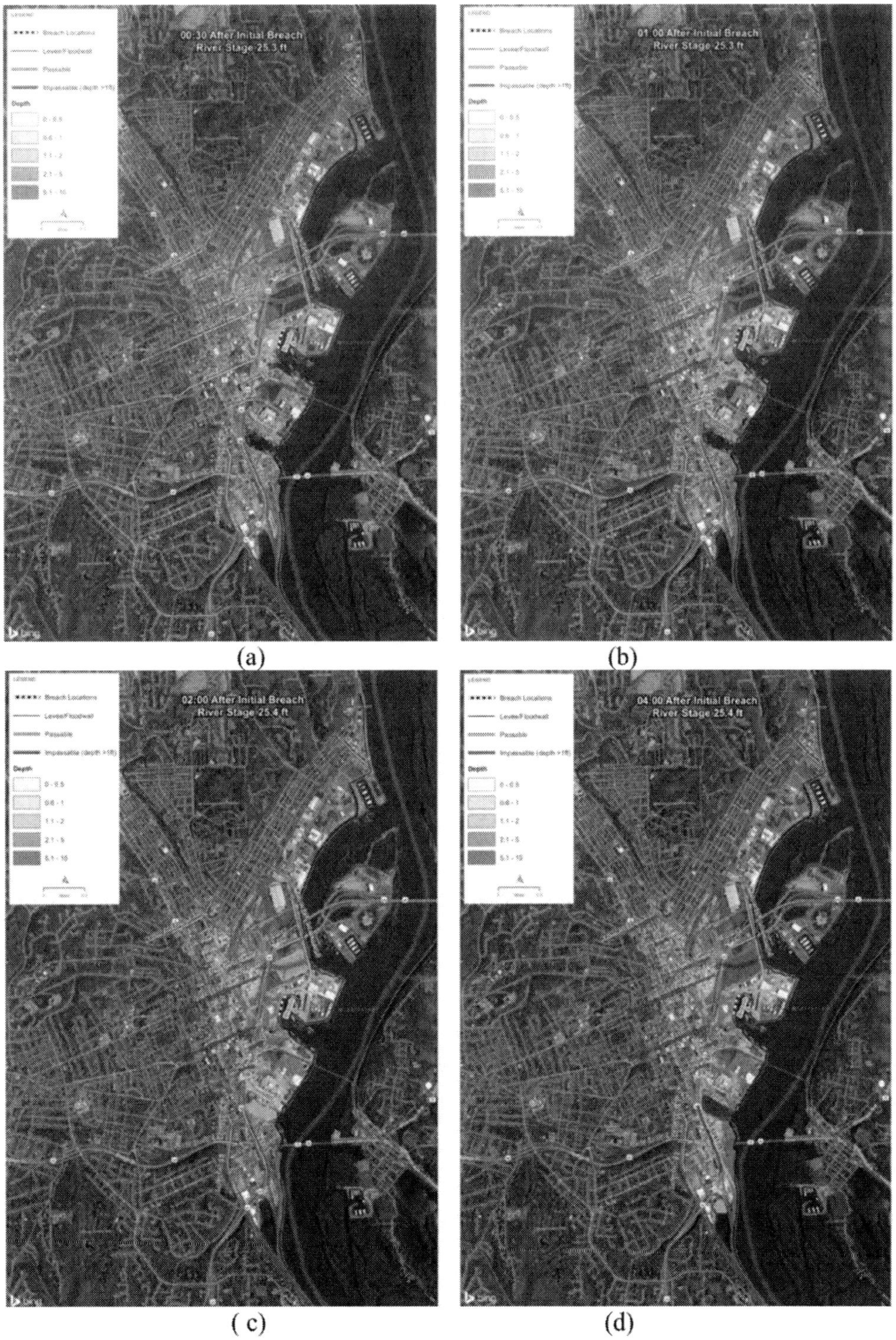

Figure 9. Inundation Depths and Transportation Corridor Status (Green = Open, Red=Closed) During Location 6 Breach - 200-year Mississippi River Event a) after 30 min; b) after 60 min; c) after 2 hours; and d) after 4 hours.

Strategic Flood Fighting Locations

The other major outcome of the study has been developing an understanding of the timing and severity associated with each breach location and Mississippi River level. As the City enters flood fighting mode during high Mississippi River levels, there may be weak spots that develop in the embankment or floodwall, this information can be used to prepare for construction of emergency cross levees to cut-off any flooding associated with a levee breach. It is possible to have a catastrophic failure without warning, such as gate failure or runaway barge impacting a floodwall. In this case, the study information can be used to make strategic decisions on construction of emergency cross levees.

Figure 10 illustrates twelve locations that make ideal strategic locations for construction of emergency cross levees in the event of an actual levee breach, or an imminent levee breach during a Mississippi River flood event. Depending on the location of the breach event, the City can choose to isolate flood waters by building cross levees.

Figure 10. Potential Flood Fighting Locations

Temporary Flood Fighting Strategies

The City calculated quantities and estimated construction time associated with temporary flood fighting barriers using conventional sand bags, trap bags, or Hesco barriers. Once the quantities and construction times were calculated, the information was added to the emergency action plan for flood fighting.

The breach database was mined to find the lengths and highest water surface elevations along each alignment, regardless of breach location. Figure 11 is an example of maximum depths reached during a breach event along strategic flood fighting alignments within the City during four Mississippi River flood events regardless of breach location. This information guided the flood fighting preparedness plan.

For material estimates it was assumed that conventional sand bags can be used up to 3-ft high and barriers (such as Hesco) can be used in four foot tiers. An eight foot and twelve foot barrier requires a pyramid, which contain 3 barriers and 9 barriers respectively. It was assumed that material cost $14,000 for 100 ft of 4 ft high wall barrier, $42,000 for 100 ft of 8 ft high wall barrier, and $84,000 for 100 ft of 12 ft high wall barrier. It was assumed that each100 ft of 4 ft high wall barrier took 6 workers 2 hours to complete.

Quantities for each temporary berm were calculated and used to plan for future flood preparedness. The City can purchase clay, barriers, and sand and stockpile them in strategic quantities locations.

The modeling results indicate that breach flood travel times are very short, within a few hours, evacuation routes and buildings can become submerged. By taking the time to plan the best locations to build temporary flood fighting barriers as well as to what height, given the Mississippi River Conditions, the City is in a much better position to protect it's people and property.

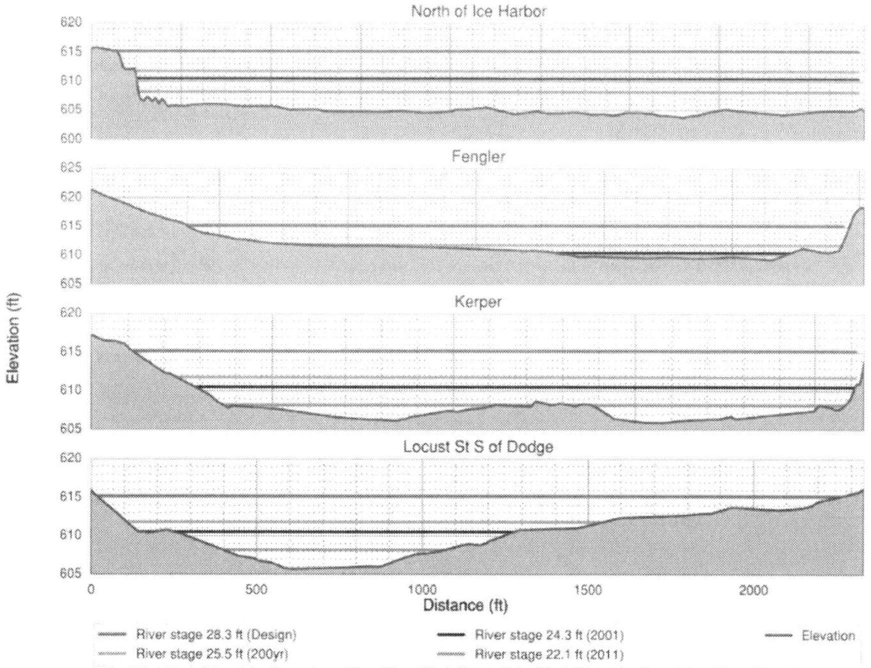

Figure 11. Maximum Depths at Strategic Flood Fighting Locations during Various Mississippi River Elevations, Regardless of Breach Location

CONCLUSION

This paper described a levee breach modeling study which defined the depth of inundation, time to inundation, and inundation paths at critical infrastructure and transportation corridors for levee breach scenarios as well as how sensitive the hydraulic characteristics of the resultant flooding are to different breach locations (total of seven), characteristics (breach width and time to development), and river conditions (4 different events). As a result of this study, the City further developed emergency action plans, identified locations for flood fighting, and determined the best strategies within the City of Dubuque for protection from a potential levee breach event. This has helped the City further understand their risk and helped drive the identification of risk mitigation measures.

ACKNOWLEDGEMENTS

The authors wish to acknowledge the City of Dubuque public works, engineering, emergency management, police, fire, and planning departments as well as the City Manager for their input in this study.

REFERENCES

GEI (2011). FEMA Accreditation Seepage and Stability Evaluation. Dubuque Local Protection Project. Prepared for City of Dubuque. May 2011.

Saucier, Howard, and Tom (2009). Levee Breach Geometries and Algorithms to Simulate Breach Closure" October 2009.

URS (2013). Memo: DWR Levee Breach Database. September 2013.

USACE (1966). Dubuque Urban Levee System. General Design Memorandum. April 1966

USACE (2003). Upper Mississippi River System Flow Frequency Study. Rock Island District, August 2003.

USACE (2015). Dubuque Urban Levee System Study Breach Analysis. Dubuque, Iowa. Flood Plain Management Services Special Study. Section 206 (PL 86-645) of the 1960 Flood Control Act as Amended. June 2015.

PUBLIC SAFETY AND UNAUTHORIZED EXTREME ACTIVITIES AT DAMS

Steven Barfuss[1]
Brian M. Crookston[2]
Kade Beck[3]
Ryan Weller[4]

Because of an increasing need for reliable water supplies, flood control, navigation, and hydropower, dams continue to play a critical role in our communities. The water reservoirs created by dams are also recreational destinations for many, and it is not uncommon for the public to place a high level of importance on the recreational opportunities provided at dams, while perhaps giving little thought to the more critical benefits dams provide or the associated risk of their recreational activities. Well-known recreational activities in and around reservoirs and rivers include swimming, boating, camping, and fishing. However, a wider range of activities also occurs at dams, many of which are illegal and can potentially result in damage or death. Indeed, there is a surprisingly broad range of poorly documented illegal recreational activities occurring at spillways throughout the US, such as bowling, surfing, wakeboarding, sledding, snowmobiling, skating, biking, and diving, to name a few.

Currently, dam safety programs focus on the structural integrity and operation of dams, including potential hazards due to flooding if a failure were to occur. However, when compared to fatalities due to dam performance, incidents and failures; far more deaths reported at dams in the United States are related to personal accidents and injuries related to both appropriate as well as unregulated and inappropriate public use. Furthermore, illegal and unintended use of spillways, chutes, and stilling basins is poorly documented. This paper focuses on the hazardous and illegal use of spillways by the public. A database of activities occurring in the USA, prepared by the authors, is summarized. The experiences of a number of dam owners is also documented, including unique challenges they have encountered and corresponding approaches for mitigation. The authors believe that increased awareness of these recreational pursuits will assist in efforts to improve public safety at dams.

[1] Utah State University, steve.barfuss@usu.edu
[2] Schnabel Engineering
[3] Utah State University
[4] Utah State University

PUBLIC SAFETY AND SOCIAL MEDIA

Sharon Nicole Roach, PE[1]

ABSTRACT

Hydroelectric industrial facilities have the illusion of safe recreation. That illusion, combined with ill-advised legislation and an increasingly thrill seeking yet risk adverse and litigious public, has created a perfect storm for facility owners and operators. Ignorance of unauthorized recreational use of a facility may be a less plausible defense of public safety hazards. At the least, a proactive approach to mitigating the dangers of the hydroelectric facilities will help to keep the public safe and ensure the owners and operators are maintaining a standard of care.

Keeping the public safe around hydroelectric industrial facilities can be an onerous task. Looking for typical signs of unauthorized public recreation is a reactive approach to public safety. This includes making note of unauthorized, non-malignant access to the facility, left over debris and sometimes evidence of heavy trafficked areas that were not intended for recreational use.

Monitoring social media can better educate facility owners, operators and regulators as to how the public is using the facilities. This is a reactive approach that allows the observer to see the trends, at both their own and other facilities, and proactively develop a comprehensive public safety plan.

INTRODUCTION

Social media has changed how we get our news, how we interact with our neighbors and even how we keep in touch with high school friends. It can be an asset to hydroelectric industrial facility owners, operators and regulators by giving them actual video and pictures of how the public uses their facilities. It can also be a tool used to drive up copy-cat trends of reckless public use of the facilities. Monitoring and reacting to social media has become a standard of care for hydroelectric facilities in respect to promoting public safety. It also can alert owners of trending issues and is a more proactive approach than simply observing the clues of use that were left in-situ.

GETTING TO KNOW SOCIAL MEDIA

Becoming more familiar with the types of social media that are available can be an invaluable tool when assessing the effectiveness of a facilities public safety plan. These types of social media include online reviews, e-commerce, social news, microblogging, interest based networks, discussion forums, social publishing, personal networks, social bookmarking and media sharing networks. Many social media platforms cross over into several categories of social media. For instance, Google is known for online reviews, e-

[1] Civil Engineer, Division of Dam Safety and Inspections, Office of Energy Projects, Federal Energy Regulatory Commission, San Francisco, CA 94105, sharon.roach@ferc.gov

commerce, and social (trending) news. It has also launched its own personal networking platform.

Each social platform can cross over into various social media types. Being aware of the types and what is available will help an owner, operator or regulator become more familiar with possible popular recreation happening in their facilities. With that information, changes to the Public Safety Plan can be proactively directed to keeping the public out of harm's way, or making concession to allow the popular recreation to continue, with authorization.

USING SOCIAL MEDIA TO FIND PUBLIC SAFETY ISSUES

Google & Google Earth

Google and Google Earth are operated by the same parent company. Google is a search engine which crosses several social media together, as mentioned. It aggregates data and brings it to the user based on their search parameters. Google Earth is a geobrowser that has satellite and aerial imagery, ocean bathymetry and other geographic data. It creates a virtual globe. Google Earth also is a media sharing network.

Figure 1 displays the Google homepage. Googling, or using a similar search engine, may be a shot in the dark to find information about your facility. However, the amount of information that will be returned will be impressive and broad.

Figure 1. Google Home Screen.

Queries can be made for information on your facility, area or just general searches. The query will bring back all of your results including articles, pictures, videos, maps or news as seen in Figure 2 and 3. From that page, the user can select and research from the results. As mentioned, this may be time consuming, but an all-encompassing method to obtain information about a facility without logging into a social media platform.

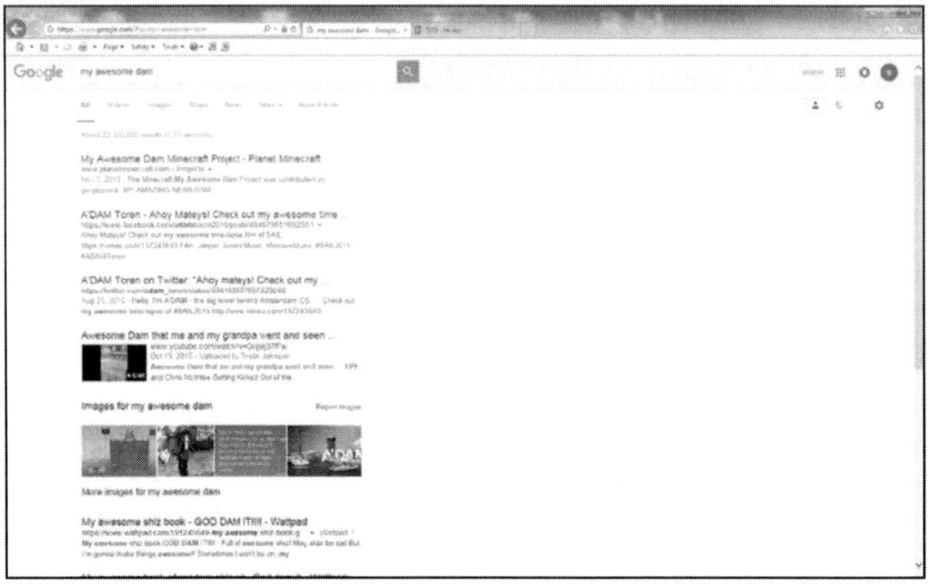

Figure 2. Google Query Return.

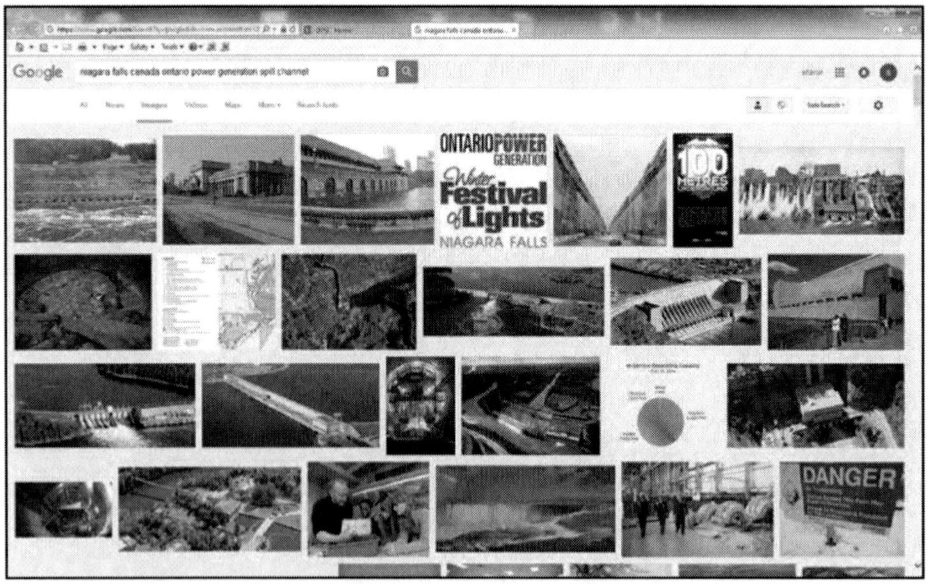

Figure 3. Google Image Query Return.

From this type of search, a user can see pictures from various social media, news articles and websites that meet the search criteria as seen in Figures 2 & 3. The owner may see video links, or their company's own news brief from 1962.

Google Earth, as seen in Figure 4, is a platform that allows users to post pictures, pinned to the location that they were taken. This is a program that must be downloaded to your computer. A user can customize some portions of the information that is displayed- like road names, borders, national landmarks or buildings. Once running, the user can zoom into their location or any other location around the world. An owner or operator can see what pictures have been uploaded near their facility, as seen in Figure 5 and 6.

Figure 4. Google Earth Home Page.

Figure 5. Google Earth Bird-Eye View of Area.

Figure 6. Google Earth User-Uploaded Content.

This type of search differs from a Google search because a user can see pictures from just this program. The pictures are also georeferenced to the location that they were taken. An owner can see the pictures of their facilities that are being uploaded. The uploaded images are attached to a user profile. Other photos uploaded by the same user profile may be displayed as well. This is a beneficial tool in identifying where members of the public are accessing a hydroelectric facility and what other portions of the facility they may be accessing. It also identifies what may be of interest to that user. However, there is no information regarding what the user has done and their intent.

Facebook, Instagram, Pinterest & Twitter

Similar to Google Earth, Facebook, Instagram, Pinterest and Twitter are social media platforms that require user profiles and allow users to upload media to share. Unlike Google Earth they are not downloaded programs. They have subtle differences, like Twitter allows limited messages to be sent within a user's network, Pinterest is more of a social bookmarking media, Instagram comprises of uploaded media and Facebook seems to encompass all plus e-commerce and microblogging.

Facebook houses the most features. Owners and operators can connect with users globally to form their network. Depending on security settings, people using the application can view user uploaded photos, video or personal thoughts. Users can search within the program for locations that other users have tagged in their posts as seen in Figure 7. This will allow owners and operators to see photos or articles that users have tagged to their facility. Some of these photos are not georeferenced and a user may get some results that are not quite what they were expecting, as seen in Figure 8. This is a dam that was tagged as "Union Valley Dam" however; it is clearly not the dam or reservoir that exists at the tagged location on the search. This does allow users to see photos of other users recreating at their facility and read some articles. If something is noted in the photos, the facility owners can take measures to correct it. Also, if articles or posts are made, tagging a facility, the owner can become aware of a user's thoughts on the facility.

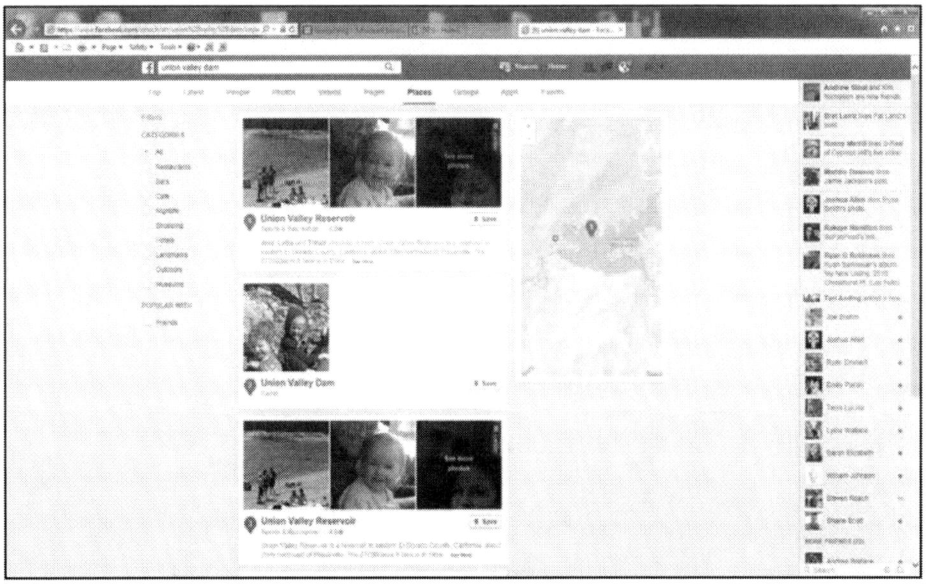

Figure 7. Facebook Query Return.

Figure 8. Facebook Query Return.

Owners and operators should be particularly aware and interested in increasing the size of their social network. Their network could be either driven from a personal page, or the facilities public social media page. A focus on included local citizens into a facilities network will bring invaluable information.

Figure 9. Concrete Double Arch Dam.

Figure 9 is a public image of a concrete double arch dam. It stands around 215 feet tall. Three areas of interest are circled in red: an area on the dam crest which has a make-shift tripod secured to the top, the intake building to the right and in the background, towards the rear of the reservoir, a boat launch. The boat launch is a public area, accessible to all. The dam and intake are behind a vehicle gate, which restricts public access to foot traffic only. The dam has a second gate to restrict all access which features a chain link fence and gate and razor ribbon. Occasionally signs of public access to the dam have been identified, like empty beverage containers and a worn foot path where the public is known to sneak, precariously, around the security features.

Figures 10 through 13 show user uploaded images which were shared within the user's network of a day recreating at a hydroelectric industrial facility. An owner, operator or regulator within the user's network would be allowed to view the photos. They would also be able to see how their facilities are being used by the public and make any necessary changes to their Public Safety Plan. If warranted, the user can be contacted and requests made to take down any photos that may inspire other users to do similarly unsafe actions.

Figure 10. Public Use of Boat Launch.

Figure 10 shows a member of the public diving head-first into the reservoir just near the boat launch (circled in red). While this area is meant to be accessible to the public, diving head first into shallow water may not be safe.

Figure 11. Public Use of Intake Building.

Figure 11 shows members of the public standing on both the water and concrete drive side of the vehicle safety railing of the intake building (circled in red). This area is not restricted from foot access by the public. Current standards encourage recreational access to some areas of a hydroelectric industrial site. This site is included in that. However, the ledge is approximately 30 feet above the water surface elevation. A fall from that height could be dangerous. Subsequent photos did include the men jumping into the water below from the intake building. It should be noted that the depth of water below decreases dramatically due to the steep canyon. Some of the men would have dove into water that was very shallow compared to the height that they were jumping from. Also, the intake building while operating creates its own drowning hazard particularly if a swimmer were to dive deep enough to be overcome by the suction.

Figures 12 and 13 shows a member of the public hanging from the long line strong arm over the side of the dam (circled in red). This area is not meant to be accessible to the public and has fencing with razor ribbon and barbed wire to discourage public access. The long line tri pod is meant to transport materials and tools down to the base of the dam for maintenance work. A subsequent photo pictured the man holding the strong arm with a single hand and a beverage in his free hand. This is an obvious public safety issue that can be addressed in the Public Safety Plan.

Figure 12. Public Use of Dam, Angle 1.

Figure 13. Public Use of Dam, Angle 2.

Figure 14. Safety Modification to Strong-Arm of Tripod.

Figure 14 shows the change that the owner of this facility did to the strong arm by positioning and partially disassembling the tripod when not in use. The owner also requested that the users' photos be taken off of the social media platform as to not encourage copy-cat public safety issues.

Meetup & Geocaching

Meet up and geochaching are similar in that they encourage users to collaborate online, but most of the activity that users experience is offline. Meetups bring together users with similar interests, much like clubs or activity groups. Geochaching is somewhat similar to

Google earth, as they also allow users to upload photos that are geolocated but they differ in that geochaching is encourages users to go on a world-wide scavenger hunt for user hidden caches given clues to the location of the cache by other users.

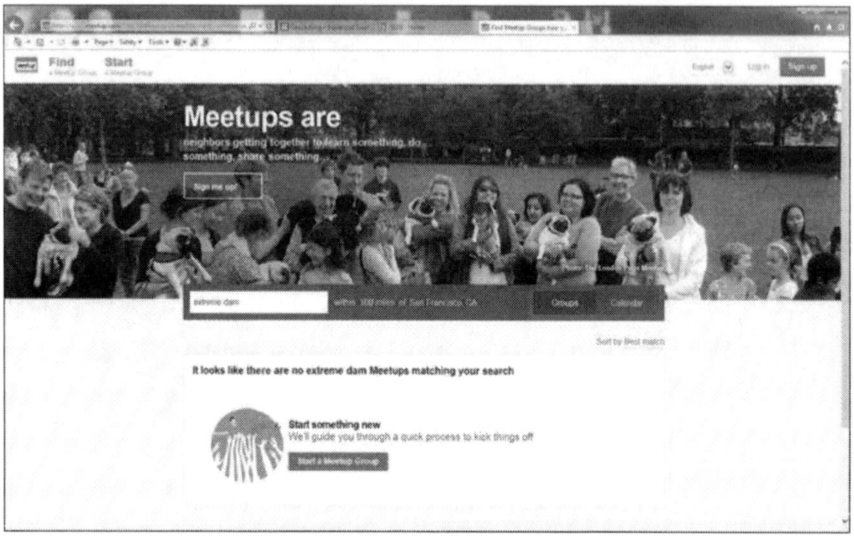

Figure 15. Meetup Homepage

Figure 15 shows the meetup homepage where user can search for groups based on their search parameters. Some meetups are hiking groups, others are extreme sports. Meetup groups can post photos online so that interested parties can see what they are getting themselves into. An owner can see if anyone is meeting up near their facility, or has taken pictures in areas around their facility.

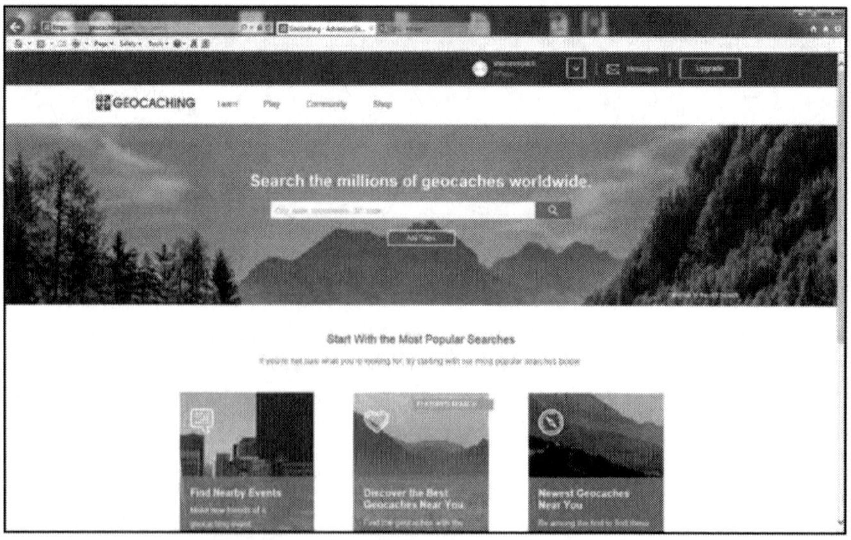

Figure 16. Geocaching Homepage

Figure 16 shows the Geocaching homepage. From that page users can search within a given area for caches that are in their area. Owners can search the location of their facilities to see if there are any caches near their industrial sites. Once they hone in on a site, the user can see who created the cache, any photos linked to the cache and the clues to find the cache. Figure 17 shows a cache that was discovered by an owner. The location of the cache is circled in red, behind the trash rack entrance to a tunnel between two reservoirs.

Figure 17. Geocaching Query Return and Cache location

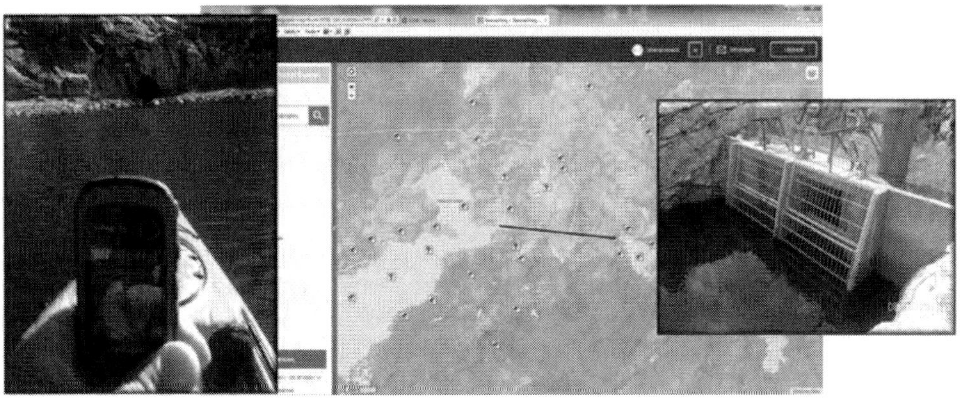

Figure 18. Geocaching Cache Clues

Note that Figure 17 is actually inside the tunnel. To access the cache, a user would have to boat or walk in from the downstream reservoir. The underground tunnel is mapped out in Figure 18 as the red line. Users have uploaded photo clues of the downstream entrance to help other geocachers find the entrance as seen on the left insert of Figure 18. The right insert of Figure 18 is trash rack on the upstream reservoir leading to the tunnel. Users have even uploaded photos, as seen in Figure 19, of the inside of the tunnel near where the cache is located.

Figure 19. Geocaching Cache Clues

Currently, the owner saw no threat to public safety with this recreation given the location and nature of current standard operating procedures. However, this may not be the case given other circumstances.

You Tube

You Tube is a social media platform that allows discussion forums and media sharing. Users can search through the uploaded content and see videos based on location or keywords as seen in Figure 20. User uploaded content can be discussed and rated by fellow users. Users can select to follow other users. This platform gives owners a sometimes startling and intimate view of what users are doing in their industrial facilities. Owners can watch what is happening, how users are entering the facilities, what they are doing and if the use is an isolated event or somewhat normal occurrence. Links to several videos can be viewed in the reference section. Owners can see informal tours of their facilities, professional thrill-seekers jumping into the plunge pool or riding their bikes down the downstream face of a concrete dam. They can view partiers using a flume as a lazy river or the spillway as a slip and slide. They can see other users literally jump off of a dam, landing on the upstream face, injuring themselves and begging onlookers to not call 911, because they do not want to get into trouble.

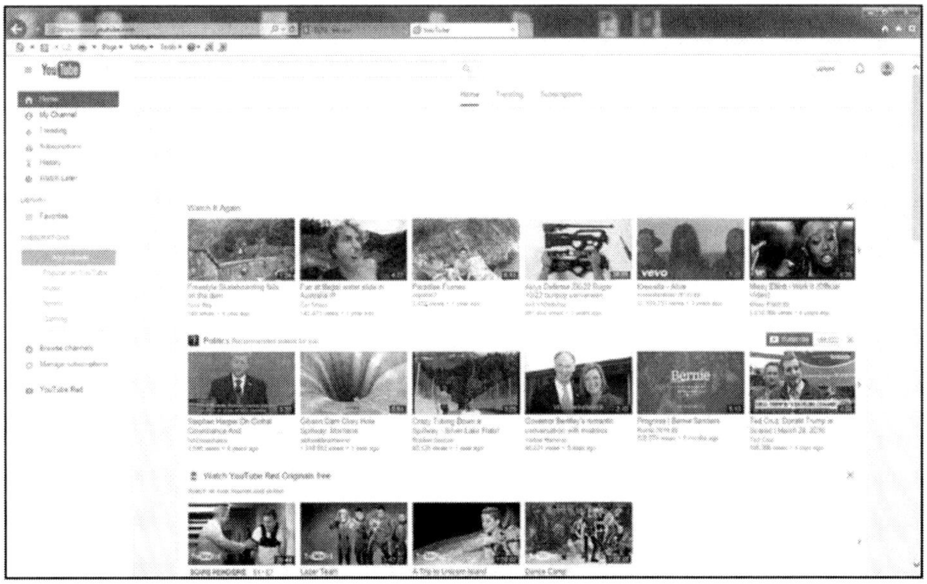

Figure 20. You Tube Home Page

CONCLUSION

Becoming familiar with social media, and using it to the advantage of owners, operators and regulators of hydroelectric industrial sites can influence how Public Safety Plans are used and adapted to real life circumstances. Using social media can illuminate owners understanding of the risks that they are passively or unknowingly taking when not including social media searches as part of the standard of care for public safety. It is imperative that owners or operators regularly monitor social media for changing trends, address possible copy-cat behavior, change the safety features in the field to address the current trends that are observed in social media and educate the public of the dangers of these industrial sites.

REFERENCES

www.facebook.com

https://www.geocaching.com/play

www.google.com

https://www.google.com/earth/

www.instagram.com

www.meetup.com

www.pinterest.com

www.twitter.com

www.youtube.com

http://swf.tubechop.com/tubechop.swf?vurl=2mA1vgnpsi4&start=10&end=40&cid=7830870

http://swf.tubechop.com/tubechop.swf?vurl=EH6t0pKtlqw&start=18&end=80&cid=7830889
http://swf.tubechop.com/tubechop.swf?vurl=uYJxxO4II2Y&start=170&end=210&cid=7830902
http://swf.tubechop.com/tubechop.swf?vurl=Cjl_BJgK3bY&start=23&end=41&cid=7830790
http://swf.tubechop.com/tubechop.swf?vurl=N2l803qEkow&start=40&end=110&cid=7830847
http://swf.tubechop.com/tubechop.swf?vurl=ktv3mF_D0Bg&start=13.21&end=59.88&cid=7830862
http://swf.tubechop.com/tubechop.swf?vurl=eyJO7uQwq2g&start=0&end=55&cid=7830921
http://swf.tubechop.com/tubechop.swf?vurl=Xpzd8kJbVi8&start=138&end=215&cid=7830928
http://swf.tubechop.com/tubechop.swf?vurl=Hm1YJM_N-wg&start=0&end=102&cid=7830930
http://swf.tubechop.com/tubechop.swf?vurl=NHbRx4mWuqo&start=0&end=69&cid=7830933
http://swf.tubechop.com/tubechop.swf?vurl=1UScyMoMTVg&start=55&end=110&cid=7830938

AUTHOR INDEX

Abbas, Syed 93
Abbaszadeh, Sam 1014
Abdoun, Tarek 405
Aboelata, Maged 1460
Adams, Gregg Batchelder 286
Alvarado, Gabriela 749
Ameen, Rozh 810
Anderlini, C. 1360
Antonaki, Nonika 405
Anyanwu, Emeruwa 765
Armenis, Petros 221
Armstrong, Richard J. 939
Arnold, Justin 74
Arnold, Michael 551
Atukorala, Upul 505
Baehr, A. C. 1762
Bagherzadeh, Harry 586
Baker, Bruce 90
Ballhausen, F. 1747
Barber, Todd 1724
Barfuss, Steven 1827
Barker, Bruce 924
Barker, Malcolm 221
Barry, Rob 1250
Bearlin, Andrew 74
Beaty, Michael 969
Beck, Kade 1827
Bector, Munit 1189
Beduhn, Bob 10
Benitez, Rolando 1334
Bergman, B. 1498
Bessette, Michael 1657
Bialek, Elizabeth 808
Bird, G. Richard 286
Bisnett, Rachael V. 1090
Black, Robert 1250
Blackett, Frank L. 174
Blair, Stephen 1119
Bocovich, Carolyne 1643
Boivin, Yannick 1498
Bonelli, Stéphane 1417
Booth, Paul 735
Borela, Rodrigo 749
Bouaanani, Najib 1658
Bouayad, Ahmed 539
Bouchie, Peter 304
Bourdeau, Philippe L. 749
Bowles, David S. 104
Boyer, Douglas D. 174
Bradner, Graham 1624
Bradner, Graham C. 1604
Brinker, Darren J. 1673
Brom, Kelly 73
Brown, Stephen 267
Brown, Will 765
Bukahir, Layne 378
Burke, Joseph 1250
Burns, Steve 1102
Bushey, Randall 421
Bynoe, Kareem 304
Caldwell, Jason 924
Carney, Shaun 90
Castillo-Rodríguez, Jessica 194
Cato, Kerry D. 139
Chan, Ccee 1102
Charlwood, Robin 1153
Chaudhry, M. Hanif 720
Chen, Wen Yan 1071
Chopra, Anil K. 1464
Chowdhury, Khaled 969, 1578
Christensen, Peter 221
Clarke, Bob 734
Clopper, Chris 1796
Cobb, Charles 404
Cochran, Brett 12
Collins, Kent L. 1
Conkle, Christopher 628
Corn, Aimee L. 1673
Cotie, Bruce 268
Crookston, Brian M. 1827
Crosby, Wesley 660
Cross, Jerry 800
Crouch, Michael M. 208
Curtis, Dan D. 1233, 1546
Curtis, Kedric 72
Cvijanovic, Vojislav 1709
Dall'Aglio, Milton 675
Daviero, Gregory J. 676
Davis, Peter 368
Dawson, Ethan 969
De Melo, Lucas 440, 954
deGroot, Dirk 550
Dolen, Timothy 469
Dolve, Blake 1709
Dominguez, Justin 696
Donaghy, Jud 800
Downing, Bruce 505
Doyal, Stephanie 661
Driller, Michael 1071, 1250
Duan, Wenbo 1658
Durkee, Dean B. 855
Duron, Ziyad 1206, 1218, 1425
Dutson, Greg 735
Ellenson, Sean 1189
Engel, John 10
Engemoen, William O. 422
England Jr., John F. 329, 345
Escuder-Bueno, Ignacio 194
Espinoza, Wilson 749
Fannin, R.J. 1360

AUTHOR INDEX

Faulhaber, Bob 551
Fernandez, Jorge 246
FitzPatrick, Brendan 613
Forrest, Michael 602
France, John W. 286
Friesen, Steven 1071
Friz, Peter 74
Galic, Dom 158
Garandet, Aurélie 1417
Garduno, A. 1747
Gassman, Sarah L. 1333
Gelo, Martina 1178
Gerlach, Mike 420
Gerressen, Franz-Werner 562
Ghaedi, Khaled 487
Ghanaat, Yusof 1528
Gheibi, Emad 1333
Gibson, Wayne 924
Gilbert, Daniel A. 1735
Ginther, Conrad 1334
Gleason, Erin 1673
Gloschat, J. A. 1762
Goulas, E. 470
Griffis, Wade 522
Grima, Nathan 1375
Groeneveld, Joe 74
Gross, Eric 765
Guangul, Seifu 93
Gunn, R.M. 470
Hageman, Eric 613
Hall, Brian M. 345, 839
Hamedian, Sherry 1685
Hanbali, Fauwaz 91, 674
Harder, Les 1657
Harder, Leslie F. 1250
Hare, John L. 867
Harper, Zachary 1528
Harrington, Graham 221
Harris, Jeff 907
Hasan, Husein 1233
Hasan, Husein A. 1735
Hasenbalg, Jaron 73
Hawson, Herb 505
Hayes, Chris 1153
Hill, Jason F. 613
Hoang, Samantha 1206, 1218, 1425
Hockett, Richard B. 1403
Hoffman, Dan 93
Hogan, Dennis 577
Horvath, Doug 1334
Hou, Susan 602
Hover, William H. 452
Hu, George 969, 1058, 1388
Huard, David 56
Huggins, Jennifer E. 1513
Hultstrand, Doug 886
Husted, Jack 377, 787
Hutton, Jesse 1143
Ibáñez-de-Aldecoa, Rafael 662
Ibrahim, Ibrahim A. 1359
Ibrahim, Zainah 487
Imran, Jasim 720
Iso-Ahola, Vik 1178, 1189
Ivanov, Dmitri 1334
Jaeger, Robert 1624
Jaeger, Robert A. 629, 1604
Jalayeri, Nikou 1102
Jamieson, Steve 268
Janes, Matthew 245
Johnson, Jeff 74
Johnson, Nicholas 1090
Jones, Rusty 1803
Kanning, Willem 1643
Kappel, Bill 886
Karlovits, Gregory S. 357
Kashiwada, Christina 1250
Keizer Jr., Richard A. 1604
Kelson, Keith I. 345
Kennedy, Eric 765
Killeen, Katharine 1775
Kimbrel, Sean 1, 1441
Kirksey, Brad 867
Kirschbaum, Daniel 74
Klein, Matthew 1424
Klippel, Sterling 377, 787
Klippel, Sterling E. 1178
Klipsch, Joan 1463
Kolakowski, Anna 1375
Koltuniuk, Roman 1513, 1685
Kost, Andrew D. 1028
Kovalishyn, Val 1102
Krivanec, Christopher 1657
Kropelnicki, Jesse 304
Krumm, Paul 1135
Kuo, Chun-Yi 1403
Kuttel, Jeanne 1250
Kynett, Michael 1578
Lamour, Vincent 1417
Laro, Katie 924
Leclerc, Martin 1498
Lee, Tedman 602
Lefkovitz, Nathan 1461
Léger, P. 1498
Lehrer, Gai 562
Leonard, Christina M. 345
Li, Chunling 440
Li, M. 1360
Li, Zhihua 697
Lilley, Keith 377
Lim, Yeo Howe 1442

AUTHOR INDEX

Lochu, Alexandre 578
Lokke, Arnkjell 1464
Lopez, Don 268
Lopez, Jose R. 709
Lum, Jonathan 1206, 1218, 1425
Lund, Guy 1218
Lund, Guy S. 1425, 1673, 1724
Lynch, Andrew J. 855
Madera, Veronica 469
Maki, Ian P. 629
Mallet, Mark 662
Manie, Jonna 1465
Manole, George 1178
Margo, David A. 357
Marín, Jorge 662
Marshall, Kevin 1724
Mazighi, Hichem 487
McClung, Nicholas 613
McCoy, Andrew 1803
McGuire, Mathew 368
McHenry, Jim 765
McMahon, Michael 9
McPherson, Matthew 1462
McVicker, Jamie 1567
Mejia, Lelio 246
Mejia, Lelio H. 1043
Melliger, Joshua 92
Menninger, John 93
Meshkat, M. 825
Messio, Elisa 1658
Mihoubi, Mustapha Kamel 487
Miles, Patrick 867
Miller, Michael 313
Miquel, Benjamin 1658
Mishler, Andrew 867
Mitovski, Stevcho 645
Modini, George 674
Montero, M. 1747
Montgomery, Jack 1014
Mooney, Michael 1643
Morales-Torres, Adrián 194
Morley, Derek S. 139
Morrison, Michael A. 954
Moses, Matt 378
Mostafa, Mohammadreza 1189
Moura, Danielli De Melo 749
Muehring, Deron 1803
Muhlestein, Geoff 886
Mulder, Henri V. 1058
Musselman, Eric 627
Neamtu, Danielle 1119
Needham, Jason 104
Neff, Keil 90, 440
Nettle, Justin 765
Newman, Erik 602
Ng, Carman 602
Nickerson, James 696
Novembre, Nicole 924
O'Neal, Troy S. 1403
Obermueller, Kristen 10
Oleksyn, Andrew 522, 1375
Olson, David 765
Orlovsky, Andrew J. 1568
Ostadrahimi, Leila 1462
Ovalle, William 1333
Parekh, Minal 1643
Park, Tae Ha 1206, 1218, 1425
Partovi, M. 470
Parzybok, Tye 924
Pathak, Chandra 906
Pattermann, Kenneth R. 1388
Penner, O. 1498
Perkins, J. Taylor 1735
Perlea, Mary 1578
Perlea, Vlad 969, 1058, 1578
Petkovski, Ljupcho 645
Pierce, Charles E. 1333
Prashar, Yogesh 808
Prathivadi, Kishen 392
Preston, Al 440
Pupim, Pedro 675
Queen, D. 1498
Rahman, Mostaqur 1333
Raitt, Douglas 12
Randle, Timothy 1
Rasekh, Ali 1189
Razavi-Darbar, S. 1498
Redell, Chris J. 709
Reed III, Gerard E. 522
Reelie, Dane 245
Reynoldson, Thomas 522
Rice, John 1359
Richards, Adam 586
Rinehart, Robert V. 587
Rittgers, Justin 658
Rittgers, Justin B. 1344
Rix, Glenn J. 954
Rizzardo, John 1775
Roach, Sharon Nicole 1828
Roadifer, John 602
Robbe, Emmanuel 1480
Rocha, Miguel 1513
Rodrigues, Larry 1375
Rogers, Michael 522
Rohrbach, Ben 907
Rondinel-Oviedo, Efrain 313
Rosen, Jamey 810
Royer, Joseph 1250
Ruark, Morgan D. 208
Ruff, Jonathon 676

AUTHOR INDEX

Ruswick, Kevin J. 676
Rutherford, James 74
Rutherford, James H. 1102
Rutledge, John....867
S.Tarbox, Glenn 1178, 1189
Sabol, George 93
Salamon, Jerzy 1465
Samandi, Sam....867
Sanders, Cathi 1567
Santini, Mauricio 675
Sasaki, Reuben 839
Sasanakul, Inthuorn 405, 1333
Saucier, Christopher L. 954
Schaefer, Mel....90, 924
Schaeffer, Kelly R. 11
Scholz, Israel Christian 562
Schreck, David 304
Schreppers, G. 470
Schultz, Mark....1709
Scofield, David H....1028
Scuero, Alberto....1317
Seed, Raymond B. 969
Seid-Karbasi, Mahmood....505
Serafini, David C. 1058
Serock, Adam 1440
Shannon, Alexander B. 1776
Sharp, Michael 1564, 1566
Shaw, Quentin 1167
Sherwood, Bill 734
Shoghli, Bahareh....1442
Siebenmann, Raphael 810
Silvertooth, D.L. 825
Simpson, Bryan K. 658, 1344
Smith, Haden....329, 839
Sooch, Gurinderbir S. 1233, 1546
Sookhu, Yogi 800
Sossenkina, Elena....1564
Srivastava, Anurag 208
Stanley, Mark....1604, 1624
Starcher, Ryan 1333
Stiady, James....522
Sutter, Amanda 764
Tabrizi, Ali Asghari 720
Tarbox, Glenn 522
Tarbox, Glenn S. 1153
Tatin, Maxime....1417
Taylor, George 924
Teal, Martin J....1443
Tisdale, Rinda 1796
Toose, James....246
Torres, Raphael A. 1250
Treacy, John....268
Trudgeon, Ray....1102
Tullis, Blake....72
Unterberg, Jessie Drayton....286
Vaghti, Stewart S. 855
Valentyn, James....1090
Vaschetti, Gabriella....1317
Vazinkhoo, Saman....662
Vennapusa, Pavana....613
Venturini, Hillery 659
Vickers, John H. 452
Viducich, Jonathan M. G. 1443
Vroman, Noah 1564, 1565
Wade, Daniel....602
Wagner, Cassandra....658
Wagner, Cassandra M. 1344
Walker, Kent 73
Walker, William 809
Weber, Joseph 1604, 1624
Wehling, Timothy....1250
Weller, Ryan 1827
Werner, John....74
White, David J. 613
Whiteside, Stephen....1119
Whitten, Zachary 45
Wilkes, John 1317
Wong, Rachel....808
Wylde, Travis....787
Yamasaki, Kenji 1028
Yiadom, Atta....697
Yiadom, Atta B. 808
Young, Jeremy....627
Young, Matthew 1796
Zakrevskaya, Yulia 56
Zhou, Jim 522, 1375
Zimmerman, Peter 735
Zoeller, Ben....1461

United States Society on Dams (USSD)
9975 Wadsworth Pkwy.
Ste. K-2, 145
Westminster, CO 80021
USA

ISBN 978-1-5108-6027-8

CURRAN ASSOCIATES INC.
proceedings.com